J. R. M Culloch

A geographical, statistical, and historical dictionary

Of the various countries, places, and principal natural objects in the world

J. R. M Culloch

A geographical, statistical, and historical dictionary
Of the various countries, places, and principal natural objects in the world

ISBN/EAN: 9783741198359

Manufactured in Europe, USA, Canada, Australia, Japa

Cover: Foto ©Klaus-Uwe Gerhardt /pixelio.de

Manufactured and distributed by brebook publishing software
(www.brebook.com)

J. R. M Culloch

A geographical, statistical, and historical dictionary

M'CULLOCH'S DICTIONARY

GEOGRAPHICAL, STATISTICAL, AND HISTORICAL.

VOLUME II.

LONDON
PRINTED BY SPOTTISWOODE AND CO.
NEW-STREET SQUARE

A DICTIONARY

GEOGRAPHICAL, STATISTICAL, AND HISTORICAL

OF THE VARIOUS

COUNTRIES, PLACES, AND PRINCIPAL NATURAL

OBJECTS IN THE WORLD.

BY

J. R. M'CULLOCH.

NEW EDITION, CAREFULLY REVISED,

WITH THE STATISTICAL INFORMATION BROUGHT UP TO THE LATEST RETURNS

BY

FREDERICK MARTIN

AUTHOR OF 'THE STATESMAN'S YEAR-BOOK.'

IN FOUR VOLUMES.

VOL. II.

LONDON:

LONGMANS, GREEN, AND CO.

1866.

LIST OF MAPS.

A DICTIONARY

GEOGRAPHICAL, STATISTICAL, AND HISTORICAL.

CASPIAN SEA

CASPIAN SEA (the *Mare Hyrcanum* of the ancients), a great salt lake of W. Asia, between 86° 35' and 47° 25' N. lat., and 46° 15' and 55° 10' E. long. It is wholly inclosed, having no outlet whatever to the ocean, and is surrounded by Tartary, Persia, the Caucasian countries, and the Russian governments of Astrakhan and Orenburg. Its direction is from N. by W., ¼ W. to S. by E., ¼ E., but at its N. end it turns due E., terminating in a considerable gulf called Merveli Kuitoh, or the dead sea. It is here almost 400 m. from E. to W., but in general it is not much more than half that width, and at its narrowest part (about 40° 20' N.) it does not exceed 120 m. across: its greatest length from N. to S. is 760 m., and its area may be estimated at 118,000 or 120,000 sq. m. (Hanway's Travels, i. 344, &c.; Great Russian Map, 1800; Rennell's Gen. View of W. Asia, pl. 6, 10, 11; Arrowsmith's Atlas.)

The coast of the Caspian is considerably broken, but its gulfs and bays are more remarkable for their number than their size; the most important after Merveli, is the Balkhan Gulf, or lake, as it is sometimes, though improperly, called, which projects from the main body of the sea, near its N.E. corner, and stretches E. over nearly 2° of long. The others are mostly little more than very large harbours, nearly surrounded by the land; such as Alexander Bay, Karabugas Lake, Astrabad Gulf, and others on the E. coast; the gulfs of Kuzilgutch, Agrakhan, Kolpichi, and others on the W. The N. coast has an almost unbroken line, but the N. is frittered in pieces, especially towards the W., by a countless number of sandy marshy islands, the shores and positions of which are continually changing. The depth of the Caspian is very variable; on the N. shore there is nowhere more than 12 ft., and usually not more than 5 ft., water; and this extraordinary shallowness continues for more than 20 m. from the land; on the E., W., and S. shores, on the other hand, the depth is sometimes 160 ft.; though here, also, shoal water is far from uncommon. In the middle the bottom has not been reached at a depth of 2,900 ft. From the general result of the soundings it would appear that, in some parts at least, the bed of the sea descends by terraces; for, on the S.E. coast, the depth lies very regularly between 12 and 15 ft. for some distance from the land, when it suddenly increases to 40 or 50 ft., at which depth the soundings run in a line, equal in extent and parallel to the former one. A similar phenomenon is observed on the N. shore, and in several other parts. (Georgi, Geog. Phys. and Stat. des Russ., i. 257-260; Gmelin's Reise durch Russland, iii. 231, &c.; Hanway's Travels, i. 135, 155, 397, &c.)

The basin of this sea is extremely limited on the S., as well as on the E. side. On the N., the Elburz mountains press so closely on the water that the fact of their allowing a passage for the road at one point on the SW. corner is remarked as a singularity,—the roads and passes being generally so impracticable that many lives are annually lost in travelling them, without reckoning those who fall victims to the robber population. (Hanway, i. 271-277, &c.) It seems that there is good reason to believe that the Caspian was formerly much more extended towards the E., but it is now shut in, in that direction, by high cliffs and sand hills close to its shore, beyond which a flat desert, full 80 ft. higher than its present surface, stretches to the shores of Lake Aral. (Hanway, i. 138, et seq.; Pallas's Trav. in S. Russia, i. 90, &c.) On these sides, therefore, the drainage is insignificant; the Elburz, indeed, gives forth a great many streams, but they are all of the nature of mountain torrents; and in the dreary desert to the E. scarcely a single rivulet is found between the Attruck, at the SE. corner, and the Yemba, at the extreme NE. (Hanway, i. 130-134.)

The W. shore presents a singular appearance. As high as 43° of lat. the whole space between this sea and the Euxine is filled by the immense masses of the Caucasus; yet from this region the Caspian receives rivers which have their sources at nearly 800 m. distant from its coasts; they flow, however, over high plateaux, and through narrow ravines, apparently cut by their own action, and which are sometimes scarcely wide enough to afford them passage. (Col. Monteith, Geog. Journ., iii. 89, et pass.) Col. Monteith believes the narrow bed of the Terek to be the Pike Caspiæ of the ancients; and it answers exactly, in both description and situation, to the pass which Pliny says (vi. 11-13) was erroneously so called; but the true Caspian gates were an artificial opening cut through the Elburz mountains on the S. coast. (See CAUCASUS.) N. of the Caucasus, the country W. of the Caspian spreads into a wide flat; but, remarkably enough, between the Terek and the Wolga, there is only one river mouth, that of the Kuma (an. *Cambyses* or *Udon*); for the land

B

though flat and sandy, is elevated suddenly at a little distance from the sea, so that the edge of the latter consists of extremely swampy ground, and all the running water that is not absorbed in the soil flows N. and W. to the Don or the Black Sea. (Pallas, i. 78, &c.; Gmelin, iii. 236, &c.) On the NW. and N. the Caspian opens on the great European plain; its mighty rivers run courses varying from 500 to upwards of 2,000 m. (see URAL, WOLGA, &c.), and its basin becomes so mingled with those of the Euxine, Baltic, and Arctic oceans, that it is impossible to assign, with any accuracy, the limits of each. (See BALTIC SEA.) So closely, indeed, do the several branches of these waters approach each other, that a short canal near Tver, by uniting the little rivers Tvertza and Tschlina, has connected the Caspian with the Baltic for upwards of a century; and much of the timber used in the imperial yard at Petersburg is cut in the woods of Kasan, being conveyed up the Wolga to this point of artificial communication. This canal was the work of Peter the Great; and the same prince projected the union of the Caspian and Euxine, by another, between two small streams, affluents respectively of the Wolga and Don, which in the neighbourhood of Tzaritzen approach each other within 2 m.; the whole distance between the larger rivers being here less than 15 m. (Algarotti's Letters, 67; Hanway, i. 96; Tooke's Russia, ii. 144; Pallas, i. 91.) With respect to its basin and drainage, therefore, the Caspian is much more of a European than an Asiatic lake; the former is extensive only on the side of Europe, and the latter carries off at least 1-6th of all the running water belonging to that division of the world. The rivers which descend from the Caucasian mountains, the only ones of consequence which the Caspian receives from Asia, are quite insignificant when compared with such streams as the Wolga and Ural; the former of which alone drains 140,000 sq. m. (Lichenstein's Cosmog., i. 328.)

There are, of course, no tides in this close lake, nor do there seem to be any regular currents, in the usual acceptation of the word; but, from the freedom with which the wind blows over so large a surface, many considerable and very irregular changes are effected in its motions and character. A strong breeze from the N. drives the waters over the low lands of the N. coast, sometimes to the distance of several miles; vessels, at such times, are said to have been carried on for inland, that, on the retiring of the sea, it was found necessary to break them up where they lay, from the impossibility of transporting them back to the shore. It must be remembered, however, that these vessels are of peculiar construction, the numerous and extensive shoals preventing the general use of any (on the N. coast) that would require much depth of water. Such a wind, too, by driving the sea into the mouths of the great rivers, causes these to rise in their beds, and, consequently, when the wind subsides, a very violent N. current is produced by the water returning to its usual level. A N. wind produces the same effect on the S. shore; only, from the nature of the coast, the water cannot extend so far over the neighbouring land; but it is frequently raised from 3 to 4 ft. above its natural level; the return to which, therefore, causes a rushing and confused motion of the waters to all points of the compass. Vessels drawing 9 or 10 ft. are, during these changes, exposed to great hazard, and, as the winds are extremely uncertain, the navigation of the Caspian, like that of most confined sheets of water, is one of very considerable danger. (Hanway, i. 142, 393, &c.; Georgi, i. 358; Monteith, G. J., iii. 23.)

There is another motion of the sea much more remarkable, however, than the preceding. It appears to increase and decrease in actual bulk, in periods, according to native report, of about 30 years each. When navigated by Hanway, its surface was incontestably rising. If the united testimony of the inhabitants upon the coasts be credited; and this testimony received confirmation from the appearance of the coasts themselves. Tops of houses were seen in water several feet in depth; the sea had visibly risen on the walls of fortified towns; and these encroachments were going on equally on all parts of the coast at the same time; so that the natives round the whole circuit were living in a state of great alarm. (i. 155-157, 871, &c.; see also Algarotti, 78, et seq.) Now Hanway makes his remarks in 1743, when the sea had certainly been rising more than 20 years—that is, from before the expedition of Peter the Great, in 1722 (i. 155); and, therefore, if the native tradition were founded on fact, it had nearly reached its greatest height. It is, at least, a remarkable circumstance that, in 1764, the sea was again (or still) rising, having, by its action, levelled the outer wall of Baku, which was standing in the time of Hanway. (Forster's Travels, 227); while, between 1811 and 1828, it had very sensibly decreased (Col. Monteith, Geog. Journ., iii. 23), and, in 1832, is had received from the N. shore full 300 yards. (Burnes's Travels to Bokhara, ii. 121.) It is clear that, in the 41 years between the observations of Hanway and Forster, there had been time, upon the native hypothesis, for the sea to reach its greatest depression, and begin again to rise. At all events, the facts, meagre as they are, seem to warrant the conclusion of periodic variations; though what law these follow, the data are at present far too limited to determine. If a conjecture may be hazarded, they probably depend upon meteorological causes, and the general state of the atmosphere. Hanway (though he disbelieves the periodic variation) appears to hold an opinion similar to this; for he remarks, that the summers, from the time of Peter the Great to that of his own observations, had been less hot than formerly; that consequently evaporation had been less, while the supply of water had continued the same. (i. 156.) It would be a corroboration of this theory, could it be established that, from 1811 to 1832, when the Caspian was unquestionably and rapidly sinking, the summer heat had been peculiarly great; but on this point nothing certain is known. In the meantime it is worthy of remark, that, between the observations of Hanway and those of Monteith and Burnes, 90 years (a multiple of the asserted period) had elapsed; that the same during which the sea was known to be constantly rising in the one case, and sinking in the other, was the same, namely, 21 years; and that, on the supposition of the tricentennial alternation of the phenomena, it should have been found sinking, as it was, by the last named travellers. That there is something very peculiar in the atmosphere of this region is evident. Monteith found its extreme pressure to be equivalent to a column of 890 ft. in height (Geog. Journ., iii. 23); Burnes, some 4 or 5 years later, to one of 800 ft. (Travels, ii. 172.) These results were obtained, not by the barometer, but by the boiling point of water; the difference of pressure would, however, cause a rise of nearly ½ an inch in the former (Nettleton, Phil. Trans., xxxiii, 808), and consequently a depression of almost 7 inches in the surface of the Caspian. This co-existence of phenomena is similar to that observed in the Baltic, only much more powerful and longer continued; it is, therefore, at least probable, that in both cases the varying level depends upon the varying pressure, and that, with extended knowledge, the explanations of the Swedish mathematicians may be brought to bear, generally, upon this peculiar feature of physical geography. (See BALTIC SEA).

But whatever may be the variations in the present surface of this lake, there can be little doubt but that it was formerly much more extensive on three sides—the N., NW., and E.; and it is still, most likely, diminishing. The fact that it never increases, in any thing approaching to the ratio of the water poured into it, has been, combined with its want of outlet to discharge that water, a standing wonder for centuries; and the most extravagant hypotheses have been adopted to account for a phenomenon apparently so paradoxical; among others, a filtration through a shelly sand into an imaginary abyss; and a subterranean communication with the ocean. The latter is the prevalent opinion among the natives; the former, to the discredit of philosophy, has found abettors among Europeans, who should have known better. (Hanway, i. 156; Algarotti, 67; Tooke, i. 138; Burnes, ii. 189, &c.) But evaporation is in these regions great, beyond belief; not from the temperature, which is lower than might be expected, but from the extreme dryness of the air. In an experiment made by Dr. Gerard in the E. desert, a bowl full of water disappeared altogether in two days. (Burnes, ii. 189.) Without pretending to decide the proportion between this exhaustion and the supply afforded by the rivers, which could not be done without much more extensive data, it is clear that the first, unchecked by the latter, would be sufficient to dry up the Caspian, or a much larger body of water, in the course of a few years; and that the waste is, or at all events *has been*, greater than the supply, is shown by the appearance of the plain country in the neighbourhood of this sea. It has been observed that the present bed appears to descend in terraces, and on the E. and NW. shores the land rises in the same manner. This land presents, also, incontestable proofs of having been formerly covered with sea water; it is uniformly flat, except where it rises in sandy ridges, to form the terraces before mentioned; it is uniform in soil, consisting of sand combined with marine slime, without a trace of terrestrial vegetation except the common desert plants, or the slightest indication of minerals; the substratum is clay, at a considerable depth from the surface; and the surface itself abounds in sea salt, sea-weed marshes, salt pits and lakes, together with innumerable shells exactly resembling those of the Caspian Sea, and which *are not found in any of the rivers*. This uniform and dreary country terminates suddenly towards the N., at a comparatively high tract running from the Wolga to the Ural, near the 51st parallel; and on the NW. at a similar tract between the Wolga and Don, a little to the E. of the 46th meridian. The change of soil is here striking and instantaneous; salt, sea-weed, shells, and sand disappear, and are replaced by black mould, solid turf, and all the usual appearances of vegetable soil upon reasonably old land, though still belonging to a period *geologically* recent, as is demonstrated by its horizontal strata and the continued absence of mineral productions. These high grounds formed, therefore, in all probability, the ancient shores of the Caspian; but that to the NW. terminates abruptly on the little river Manitysh, near the 46th parallel, between which and the Caucasian mts., a low and narrow tract, exactly resembling that on the immediate borders of the Caspian, stretches without interruption to those of the Sea of Azoph, having every appearance of the deserted bed of a strait formerly uniting the two waters. Towards the E. the whole country has the same appearance of a deserted sea-bed; and the conclusion, therefore, appears inevitable, that, at comparatively no distant period, the Sea of Aral, the Caspian, and the Black Sea formed one body of water, uniting the present anomalous salt lakes of Asia with the ocean. This conclusion is further strengthened by the presence of the same species of fish, seals, &c., in the three seas; a fact which it is impossible to account for on the supposition that they were always separated. (Pallas, i. 78–87, 278–304, &c.; Gmelin, iii. 231–248; Georgi, i. 259, &c.)

When it is considered that Russia is extremely flat; that its slope from the Arctic Ocean to the Caspian is uninterrupted; and that this slope is so considerable, that the Wolga, though rising in a flat country, has *rather* a rapid current (Pallas, i. 25),—it will be evident that the position of the Caspian must be very low. A suspicion having long existed that it was lower than the level of the ocean, Messrs. Englehardt and Parrot, in the beginning of the present century, performed a series of barometric levelling between its shores and those of the Black Sea; the result of which gave a depression of 333 ft. (51 *toises*) for the surface of the Caspian. (Reise in die Krym und der Kaukasus, ii. 55.) It is to be remarked, that at this time the sea was *sinking*; and, therefore, if the opinion hazarded on the cause of its variations be correct (see *ante*), the atmospheric pressure on its coasts was greater than it would be in the ordinary state of the air. This seems, also, to be borne out by the observations of Monteith and Burnes, upon the boiling point of water. Results depending upon the height of the barometric column would, under such circumstances, be inevitably excessive; and this was, upon other grounds, suspected by Humboldt, from the very first publication of Messrs. Englehardt's and Parrot's memoir. (Parrot, Voy. à l'Ararat, ii. 192.) To determine the question, the Russian government, in 1836, despatched an expedition, which, after two years' labour, completed, in 1838, a splendid series of trigonometrical levelling; from which it appears that the Caspian is 101·2 Prussian ft. (about 116 ft. English) below the Black Sea. (Geog. Journ., viii. 135.) The known elevation of the desert steppe E. of the Caspian will, according to this survey, place the Sea of Aral very nearly on a level with, or even something higher than, the Euxine.

Considering its lat., that of S. France and Italy, the temperature of this sea and its neighbourhood is extremely low; the N. part is very frequently frozen, and the ice in the mouth of the Wolga (lat. 46°) does not usually break up till April. (Hanway, i. 140; Pallas, i. 89, &c.) Even the Aral, as low as 45°, is *sometimes* frozen; and the inhabitants have a tradition that one of its islands was peopled by a colony which crossed the ice, with all their flocks and herds. (Burnes, ii. 189.) This fact is the more remarkable, from the low, level, and S. aspect of the region round the Caspian; but the want of mountains towards the N. exposes it to the influence of chilling winds from the Arctic Sea, while the intervention of snow-capped ranges on the S. prevents the counterbalancing effects of the hot breeze from the equator. The summer heat is, however, generally great, and is towards the S. attended with a humidity, which renders it very unhealthy; though, from this very cause, the S. and SW. districts present a luxuriance of vegetation strongly contrasted with the bare salt deserts on the NW., N., and E. Rice, maize, cotton, fruits of all kinds, and a countless variety of forest trees, are among the productions of these districts; which, with the exception of the Russian colony in the steppe of Astrakhan, are the only parts of the coast possessing a settled population; but such is the deadly nature of the climate, that all who are able leave the towns in the beginning of summer, and retire to the mountains, where the atmosphere is of course more salubrious. The deserts are occupied by the wandering Kalmucks, Kirghis, and Turkomans, who preserve unaltered the roving and predatory habits of their earliest ancestors. (Pallas, i. 92, 115, &c.; Fraser's Trav. on the S. Bank of Casp., 11, 15, &c.; Conolly's Narrative, i. 35–49, 116, &c.; Burnes, ii. 101–127, &c.)

The waters of this sea are less salt than those of the ocean, and considerably less so near the mouths of rivers than at a distance from the shore. The waters of Lake Aral are even drinkable (Burnes, ii. 109); but all have a bitter taste, ascribed by some to the great quantities of naphtha with which the soil abounds, but by others to the presence of glauber salts, among the substances held in solution. The fish are principally salmon, sturgeons, and sterlets; a kind of herring is also found, and there are likewise porpoises and seals. It has been already said, that the same inhabitants are found in the waters of the Caspian, Aral, and Black Seas. The fisheries employ many vessels annually, and the shores abound in aquatic fowl, storks, herons, bitterns, spoonbills, red geese, red ducks, &c. (Gmelin, iii. 233–357; Pallas, i. pass.; Tooke, i. 234, &c.)

It is somewhat remarkable that, though situated on the confines of Europe, this sea should have remained nearly unknown, except by name, till the beginning of the last century. It is scarcely less remarkable that the oldest observer, Herodotus, described it truly as an ocean by itself, communicating with no other and of such size that a swift-oared boat would traverse its length in fifteen days, its greatest breadth in eight days. (Clio, 203.) These proportions are accurate according to the best modern observations, and at 60 m. per day for the swift boat's progress, would give the actual measurement. After this clear account, it is startling to find the Caspian transformed by Strabo into a gulf of the Northern Ocean, and otherwise distorted, according to a theory which must be regarded as purely fanciful. (Geog., xi. 507.) Ptolemy restored the Caspian to its lake-like form: he had some knowledge of the Wolga, which he calls Rha; but he gives the greatest length of the sea from E. to W., and makes it a vast deal too large. (v. 2, vi. 9, 13, &c.) It is to be remarked, that Herodotus does not state in what direction lay the greatest length; but it may be very readily deduced from his descriptions of the surrounding countries, that he meant it to be understood as stretching N. and S. The authority of Ptolemy remained paramount and unquestioned for many centuries; and the first modern account of the Caspian, at all consistent with the truth, is due to Anthony Jenkinson, an Englishman, who, in 1558, traversed its waters, and gave an account of its dimensions and bearings, agreeing in all its main points with the more brief description of Herodotus. (Hakluyt's Voy., i. 326–348.) Jenkinson's voyage did not, however, gain much attention; and in 1719 a regular survey was commenced, by command of Peter the Great. Vanverden's map, the result of that survey, and which was partly constructed by the emperor himself, is still and justly, held in high estimation. The voyages of Hanway had for their object the establishment of a trade (in English hands) between Russia and Persia. The failure of that object was owing to the ambition of a Mr. Elton, who, attaching himself to the Persian court, gave such offence to that of Russia, that the latter eventually prohibited the English commerce on the Caspian. (Hanway, ii. 279, et pass.) A mass of valuable information was, however, collected during these transactions, by Hanway himself. Elton, Woodroffe, and others. The more modern travellers, Gmelin, Georgi, Pallas, Englehardt, Parrot, Fraser, Fraser, Conolly, Burnes, Monteith, Fuss, Kohler, and Sawitch, have added immeasurably to that information; but much still remains to be done; and as the Russian government seems fully alive to the importance of accurate knowledge on geographical subjects, and as their power or influence is nearly established on all parts of this sea, it may be reasonably hoped that every year will make W. Europe better acquainted with this very remarkable region.

The largest class of vessels that navigate the Caspian, are called by the Russians schuyts, and belong wholly to Astrakhan and Baku; their burthen varies from 50 to 100, and sometimes 150 tons. They are not built on any scientific principle, and are constructed of the worst materials—that is, of the timber of the barks that bring corn down the Wolga to Astrakhan. There are supposed to be in all about 100 sail of these vessels. A second class of vessels, called rascheivas, employed on the Caspian, carry from 70 to 140 tons, and sail better than the schuyts, and there are great numbers of small craft employed in the rivers, in the fisheries, and as lighters to the schuyts. But steamboats will, no doubt, in the end supersede most of these vessels; they have already, indeed, been introduced, not only upon the rivers, but upon the Caspian itself. The trade of the sea is entirely in the hands of Russia; and, whatever objections may, on other grounds, be made to her conquests in this quarter, it is certain that, by introducing European arts and sciences, and comparative good order and security, into countries formerly immersed in barbarism, she has materially improved their condition, and accelerated their progress to a more advanced state.

The Caspian Sea, Κασπιη θαλασσα (Herod. Clio, 203), is the oldest name of this water. It was derived from the Caspii, a people who inhabited its banks; as the more modern term Hyrcanian Sea, Ὑρκανον θαλασσα (Strabo, xi. 507), was similarly derived from the more important Hyrcanii, a principal branch of the great Persian family. In the present day it is called More Gwalenshoi, by the Russians; Kulzem, by the Persians; Bahr Kurzem, by the Arabs; Kulzum Dengibi, by the Turks; and Abskaphis, by the Tartars. (Tooke, i. 232.)

CASSANO, a town of Southern Italy, prov. Cosenza, cap. cant., in the concave recess of a steep mountain, round an insulated rock, on which are the ruins of an ancient castle, 7 m. ENE. Castrovillari, and 10 m. from the Gulf of Tarentum. Pop. 8,125 in 1862. The town is well built; is the residence of a bishop; has a cathedral, four convents, a seminary, and a workhouse. The inhabitants are industrious, and manufacture maccaroni, stamped leathers, and table-linen. Cotton and silk are also grown, spun, and woven; and the environs are productive of excellent timber, fruits, and corn.

CASSAY, KATHEE, or MUNNEEPOOR, a country of India beyond the Ganges, between lat. 24° and 26° N., and long. 93° and 95° E.; having N. Assam and the Birman empire; S. a hill country, inhabited by independent Khyens (see Birman), Koukies (see Cachar), and W.

Cachar. Area about 7,000 sq. m. Cassay consists of a central fertile valley, of comparatively small extent, surrounded on every side by a wild and mountainous country. The Naga mountains bound it N., averaging in height 5,000 or 6,000 ft. above the sea; although in some parts they are as much as 8,000 or 9,000 ft. high. Two branches, passing S. from the Naga mountains, inclose the Cassay valley E. and W., and the S. boundary, from the confluence of the Chikoo nullah, or rivulet, with the Barak, is formed by the same ranges, which run E. and W., bounding Cachar S., and Tipperah N.E. The W. mountain range is more elevated and extensive than any other, and runs from the banks of the Barak S.S.W. for 80 m., steep and precipitous, towards Cachar; but in some parts almost cleared of forest, and annually cultivated with rice and cotton. This range has nine principal peaks, varying in height from 5,790 to 8,300 ft. above the sea, which, from superstitious motives, are left covered with wood by the inhabitants of the hills, and are often capped with a dense stratum of clouds. The E. hills vary from 4,900 to 6,730 ft. above the sea. The valley thus inclosed is about 36 m. long and 18 m. broad, having an area of 650 sq. m. of rich alluvial soil, 2,500 feet above the level of the sea.

The chief rivers are the Khoogta, or Munnereepoor river, Eerll, and Thobal. The first rises in the Naga mountains, in lat. 25° 12' N., long. 94° E.; it completely traverses the central valley N. to S. and falls into the Ningthee or Kyen-dwen river. It is the only outlet for the waters of the Cassay valley; and, as the latter is 2,000 feet above the Ningthee, it is probable there are several considerable falls in its course through the mountains. Almost all the centre of the Cassay valley is a series of jheels and marshes; there is a small lake (Logta) at its S.W. corner; compact sandstone, slate, and limestone are the prevailing geological features of this region.

Iron is the only metal found in Cassay; it is met with under the form of titaniferous oxide, and is detected by thrusting spears into the ground, and, where iron is present, small particles soon adhere to them. (Pemberton.) The Cassay valley is rich in salt springs, especially on its E. side; and more than enough salt for home consumption is made. The climate of the valley is lower by many degrees than in Calcutta, but not so low as might have been expected from the elevation. There are more rainy days in the year, but less rain falls than at Calcutta: from March the showers become continual; the permanent rise of the streams begins in May, and continues till the middle of October, from which time they rapidly decrease. From Nov. to Jan. fogs settle during the whole night in the valley, and hoar frosts prevail on the hills; yet the climate of the former region is decidedly salubrious, and peculiarly healthy to European constitutions. The surrounding mountains are, in most instances, covered with the noblest varieties of forest trees, common both to tropical and colder climates; and, according to Capt. Pemberton, there is no part of India where the forests are more varied and magnificent; but, from the small number of streams, and the want of good roads, their utility is entirely local; there being at present no means of conveying the timber to any distance. The valley is perfectly free from forest, though every village is surrounded by a grove of fruit-trees; the soil of the detached hills, and their N. faces especially, are highly adapted to the culture of fruit. Herds of wild elephants are constantly seen in the glens and defiles of the N.; wild hogs and deer of the largest size abound everywhere; and the chase is a favourite sport with the Cassayers. Tigers are not common, and have retired to the mountain fastnesses; there are no jackals; but wild dogs, greatly resembling that animal, abound on the hills, where they hunt in packs. With the exception of woollen cloth, this country furnishes every article essential to the comfort and prosperity of its inhabitants. All the tribes N., W., and E. of the central valley partake strongly of the Tartar countenance, and are probably the descendants of a Tartar colony who passed hither from the N.W. borders of China, during the sanguinary struggles for supremacy between the Chinese and Tartar dynasties, in the 13th and 14th centuries. They have much more affinity, both in person and manners, with the Hindoos, than with the Burmese, to which latter race they bear little similarity. They differ from the Koolies of the S. hills in their superior height, finer complexions, higher foreheads, unharmonious voices, and harsh language. They are highly ingenious, and are final horsemen, on which account they were formerly exclusively employed in the Burmese cavalry service. The upper classes are worshippers of Vishnu, and this country may be regarded as the extreme E. limit of Brahminism; the Cassay tongue is, however, widely different from Sanscrit. There are many other distinct tribes in different parts of Cassay and its neighbourhood. All cultivate tobacco, cotton, ginger, and pepper, and manufacture cloths; which articles they barter for others with the inhabitants of the neighbouring plains of Bengal, Assam, and Kirnah. In the central valley rice is the chief object of agriculture, and the land there is well irrigated, and highly suited to it; but scarcely ½ part of the land available for it is under culture, owing to a paucity of inhabitants. The whole pop. of the valley in 1835 was barely 20,000. Tobacco, sugar-cane, indigo, mustard, dhal, and opium are also grown, and each house is surrounded by a little garden, in which culinary vegetables are raised in large quantity. Almost all the garden produce of Europe is found here, having been introduced by the British since the Burmese war; and the pea and potato are found so acceptable, that their culture is nearly universal, and they are constantly exposed for sale in the bazaars. The pine-apple attains an excellence in Cassay not surpassed in any part of the world. Buffaloes are used for ploughing; there are about 3,000 in the central valley, and perhaps an equal number of bullocks, which are superior, both in size and symmetry, to those of Bengal.

The ponies of Munnereepoor are much and deservedly esteemed, by both the Cassayers and Burmese, who use them for the élite of their cavalry. They average from 12 to 13½ hands, and are rarely more than 13 hands in height: they are hardy and vigorous, and have a peculiar blunt appearance, but are now nearly extinct; and scarcely more than 200 could be found fit for active service. Formerly, every inhabt. had two or three; and the Cassayers affirm that, in a military sense, they have lost one of their arms by the decrease of the breed. Sheep were unknown till introduced by the British; they thrive on the slopes of the central valley; goats are bred by the Naga tribes on the hills, but invariably deteriorate if brought into the lowlands; poultry are plentiful in the latter districts, and the mountaineers purchase fowls thence at a very high price. The chief manufactures are coarse white cottons; a very soft and light muslin; a coarser kind, used for turbans and jackets; silks, remarkable for the brilliancy of their colours, and which are much prized at Ava; iron articles; and salt. The chief iron articles made are axes, hoes, ploughshares, spear and arrow heads, for horse use; and blades, 1 or 2 ft. in length, which, fixed into wooden or other handles, form the axe, the inseparable companion of the Cassayer, Shan, and Singpho. Salt is got from wells, sunk in the valley to about 40 or 60 ft.; all of which are the property of the rajah, who levies a tax of 1-5th upon the water drawn. The quantity of salt obtained by evaporation is about 1-28th the weight of the water, or nearly double the quantity obtained by evaporation from sea water at Newcastle; the labourers engaged are paid in salt to the value of 3 or 4 rupees a month each, which they barter for other commodities. Wax, cotton, and elephants' teeth, form part of the tribute of the hill tribes; the same articles, with ponies, &c., are bought by the Chinese merchants of Yun-nan; and smaller products, with silks, iron, dammer, wood, oil, sandal-wood, camphor, thread, &c., were taken in lieu of money payments by the British, for assistance to the rajah about the middle of the last century.

The records of Cassay bear some character for truth, and, it is said, reach back to a remote epoch. In 1475, the Kubo valley was annexed to Cassay by conquest; and in 1754, the Cassayers conquered Birmah, and took its then capital, Sakaing. Subsequently, Cassay was frequently invaded and devastated by the Birmese; and from 1774 to 1824 was subject to Ava. By the treaty of Yandabu, in 1826, it became independent. In 1833, the valley of Kubo was ceded to the Birmese by British authority.

CASSEL (anc. *Castellum Cattorum*), a town of W. Germany, prov. Lower Hesse, of which, and of the electorate of Hesse-Cassel, it is the cap., and residence of the elector. It is finely situated on both sides the Fulda, 72 m. N. by W. Hanover, and 89 m. NNE. Frankfurt-on-the-Mayne, on the main line of railway from Frankfurt to Berlin. Pop. 38,920 in 1861. The town is divided into three separate parts, and has three suburbs. The Old Town and Upper New Town, with the Wilhelmshöhe and Frankfort suburbs, are built on the left or W. bank; while the Lower New Town, and the Leipzig suburb, are on the E. bank of the river. The two divisions are connected by a stone bridge across the Fulda, 273 German feet in length. Cassel is walled, and has numerous gates; it was formerly well fortified, but its ramparts were demolished in 1764. The Old Town, by the river, consists of narrow dirty streets; but the Upper or French New Town, so called because originally built by French refugees, on a height above the former, is one of the best laid out and handsomest towns in Germany. It contains, among others of less dimensions, the largest square in any German city (the Friedrichs Platz), and one street, nearly a mile in length, and proportionally broad. Houses in the New Town and the Wilhelmshöhe suburb, generally well and tastefully built. In this quarter of Cassel are the elector's palace, a structure nowise remarkable; the museum, the handsomest building in the city, containing a library with 70,000 volumes; an observatory; and cabinets of natural history, mineralogy, coins, artificial curiosities, statuary, and antiquities; the latter comprising several interesting Roman relics found in Hesse Cassel; a picture gallery, containing some valuable paintings by Rembrandt, Rubens, and Vandyke; the Bellevue palace, with others belonging to the electoral family: the electoral stables, and riding-school, mint, town-hall, arsenal, old and new barracks, and an opera-house. In the Old Town are the Kattenburg, a large unfinished structure, begun upon the site of the old electoral palace destroyed by fire in 1811: the old townhall; government offices; and St. Martin's, the principal church in the city, and the burial-place of the sovereigns of Cassel. The Lower New Town contains the castle, an ancient fortress, now used as a state prison; and several other prisons.

Cassel has 9 churches, 7 of which belong to the Lutheran or Reformed faith; and 1 synagogue. It has altogether 20 edifices devoted to military purposes, and 51 other public buildings. Amongst the institutions for public education are, a lyceum, academies of painting and design, a teachers' seminary, a military school, and a school of mechanical employments, called the *Bau-und-Handwerksschule*. There are societies for the promotion of agriculture, trade, and manufactures, and numerous charitable establishments; the latter include the *Waldau Institut*, at which many poor are provided for, and taught different trades. Notwithstanding the Fulda is navigable, and that Cassel is on all sides surrounded by large commercial towns and districts, with which it has abundant railway communication, its own trade is not very considerable. It possesses manufactures of cottons, silk and woollen fabrics, leather, hats, carpets, snuff, gold and silver lace, porcelain, earthen and lacquered ware, playing-cards, wax-lights, chemical products, dyes (Cassel yellow and black), soap, starch, hardware, musical instruments, linen, damask, chicory, and some machinery. It has two fairs annually. S. of the Upper New Town is the *Aue*, or *Auegarten*, a fine park containing an orangery, a pheasantry, and a marble bath; but the land is overloaded with ornament, and in bad taste. A straight and handsome road, shaded by an avenue of limes, 5 m. in length, conducts from the Wilhelmshöhe gate to Wilhelmshöhe, the summer palace of the elector, a magnificent residence, with costly fountains and waterworks, sometimes called the German Versailles.

During the short period that Jerome Bonaparte was on the throne of Westphalia, Cassel was the cap. of his king, and the place of his residence.

CASSEL, a town of France, dép. Nord, cap. cant., on an isolated mountain in the middle of an extensive plain, 24 m. NW. Lille. Pop. 4,200 in 1861. The town is well built, and notwithstanding its situation, is well supplied with spring water. It has fabrics of lawn, thread, hats, oil, and earthenware. It is very ancient, having been the capital of the *Morini* when Cæsar invaded the country. It was united to France in 1678, by the treaty of Nimeguen. Several battles have been fought in its vicinity.

CASSIS, a sea-port town of France, dép. Bouches-du-Rhône, in a narrow valley on the Mediterranean, 10 m. SE. Marseilles, on the railway from Marseilles to Toulon. Pop. 2,035 in 1861. The town has a tribunal of *prud'hommes*, an office of health, a workhouse, and yards for the building of small vessels. Its port is confined, and admits only vessels of small burden. The figs and greengages of Cassis are held in much estimation; and it has a considerable trade in excellent muscatel wine, produced in the environs. This is the native country of the learned and excellent Abbé Barthélemy, author of the "Voyage d' Anacharsis," who was born here on the 20th of January, 1716.

CASTEL-A-MARE, a city and sea-port of Southern Italy, prov. Naples, on the Gulf of Naples, 15 m. W. Salerno, on a branch line of the railway from Naples to Salerno. Pop. 25,543 in 1862. It is the seat of a bishopric, and the residence of a sotto' intendente; and is well built, partly along the shore, but principally on the side of the mountain, rising immediately from it. It has a royal palace, a cathedral, 5 churches, several convents, a military hospital, fine barracks, a royal dockyard, and hot baths. There are manufactures of linen, silk, and cotton, with tanneries. The port, which is small, is defended by two forts. Being exposed to the N., and elevated, Castel-a-mare has acquired great celebrity as a summer residence, in consequence of its coolness, the salubrity of its air, and the beauty of its environs. But in autumn it becomes damp, chill, and disagreeable.

Castel-a-mare is built on the site of the ancient *Stabiæ*, which, having been destroyed by Sylla during the civil wars, was afterwards principally occupied by villas and pleasure-grounds. It was here, A.C. 79, that the elder Pliny, wishing to approach as near as possible to Vesuvius during the dreadful eruption that overwhelmed Herculaneum and Pompeii, fell a victim to his curiosity and thirst for knowledge.

CASTEL-A-MARE, a sea-port town of Sicily, prov. Trapani, cap. cant., on a gulf of its own name. 6 m. NW. Alcamo; lat. 38° 1' 51" N., long. 12° 52' 43", E. Pop. 11,959 in 1862. It is a mean, dirty town, with a castle falling fast to decay. The bay is spacious, but it is not safe with northerly winds which throw in a heavy sea. The neighbouring country is well cultivated; and considerable quantities of wine, fruit, grain, manna, and opium are exported.

CASTELLON, or CASTELLON-DE-LA-PLANA (an. *Castulo*), a town of Spain, Valencia, cap. dep. 4 m. from the coast, and 41 m. NNE. city of Valencia, on the railway from Valencia to Barcelona. Pop. 19,340 in 1857. The town is finely situated in a well-watered, extensive, and fertile plain. This fertility is entirely the result of industry, the water which gives life and verdure to the plain being brought by an aqueduct, cut in great part through the solid limestone rock, from the Mijares, which flows about 5 m. N. from the town. This great work has been ascribed to the Romans and Moors; but others assert that it was constructed, about 1240, by James the Conqueror, king of Aragon. The town, which is well built, has 8 churches, 6 convents, 1 hospital, 3 houses of charity, and a public granary. The beauty of the situation, the mildness of the climate, and the abundance and excellence of the fruits, make this one of the favourite residences in the prov.

CASTELNAUDARY, a town of France, dép. Aude, cap. arrond., in an elevated fine situation, contiguous to the Canal du Midi, 71 m. WNW. Carcassonne, on the railway from Toulouse to Narbonne. Pop. 9,584 in 1861. The town is very indifferently built, and there are few edifices worth notice, except the church of St. Michael, said to be the finest in the dep. It has a tribunal of primary jurisdiction, a departmental college, and a philharmonic society. The canal has a superb basin contiguous to the town, surrounded by fine quays and warehouses, which, with the vessels by which it is sometimes crowded, give it the appearance of a sea-port. The public promenade commands this basin and a fine view extending as far as the Pyrenees. There are here manufactures of cloth and silk, with establishments for the spinning of cotton, print-fields, and tanneries; and a considerable trade is carried on in the manufactures of the town, and the produce of the adjoining country.

In 1632, in an encounter under the walls of the town, the Duc de Montmorenci, commanding the troops of Gaston, duc d'Orleans, was wounded and taken prisoner; and being conveyed to Toulouse, was convicted of treason, and executed in the same year.

CASTELO BRANCO, a city of Portugal, prov. Beira, on a hill on the Liria, 51 m. NE. Alcantara. Pop. 5,893 in 1858. The town is the see of a bishop, and the residence of the captain-general of

Lower Beira. Streets narrow and steep, and the houses mean, except some modern ones without the walls; the latter are double, and flanked with seven towers. The cathedral also is without the city; and there is an old ruined castle on the summit of the hill on which the town stands. It has a college and two collegiate churches.

CASTEL-SARRASIN, a town of France, dép. Tarn-et-Garonne, cap. arrond., pleasantly situated in a fertile plain on the Séoguière, 1 m. from its confluence with the Garonne, 13 m. W. Montauban. Pop. 6,856 in 1861. The town is well built, and the walls and ditches by which it was surrounded have been converted into promenades. It is the seat of a court of primary jurisdiction, and of a departmental college; and has manufactures of serges and other woollen stuffs, hats, and tanneries.

CASTELVETRANO, a town of Sicily, prov. Trapani, cap. cant., on a hill 8 m. from the sea, and 12 m. E. Mazzara. Pop. 14,540 in 1862. The town is well built with stone, the streets being spacious, and disposed with some attention to regularity; and there are several churches and convents. It has a good trade in wine and olives, the former grown in the neighbourhood, and much renowned.

CASTIGLIONE-DELLE-STIVIERE, a town of Northern Italy, prov. Brescia, on a hill 17 m. NW. Mantua. Pop. 5,227 in 1861. The town is surrounded by a low wall, and contains several churches, the ruins of a castle, and a conventual seminary; but is chiefly noted for a decisive victory gained here by the French over the Austrians, 5th August, 1796; from which Marshal Augereau derived his title of Duc de Castiglione.

CASTILE, the central and largest division of Spain, lying between lat. 38° 25' and 42° 50' N., and long. 1° 7' and 5° 37' W.; it has N. and NE. the territory of Reinosa, Alava, and Navarre; E., Aragon and Valencia; SE., Murcia; S., Andalusia; W., Estremadura and Leon; length about 300 m. from N. to S.; mean breadth about 160 m. Area about 48,600 sq. m. It is divided into two parts by a range of high mountains, called in different parts Urbiana, Carpetano, Sierra de Guadarama, Gata, Somosierra, and de Estrella. The country to the N. of the ridge, having been the first recovered from the Saracens, is called Old, whilst that to the S. is named New Castile. Old Castile comprises the modern provinces of Burgos, Soria, Segovia, and Avila, so named after their chief towns. New Castile comprises the provinces of Madrid, Guadalajara, Cuenca, Toledo, and La Mancha, each also so called after the names of their chief towns, except La Mancha, whose cap. is Ciudad Real. Principal towns, exclusive of the capitals, are (besus, Calahorra, Logroño, Calzada, Haro, Alfaro, Miranda, Briviesca, Almazar, Toledo, Aranjuez, Alcala de Henares, Talavera de la Reina, Illescas, Zurita, Trembleque, Villanueva, &c. The Ebro, Douro, Tagus, and Guadiana have their sources in this province. The first flows SE., along the NE. boundary, to the Mediterranean; the Douro and Tagus, to the Atlantic; and the Guadiana, WSW. to the same. There are many other rivers, affluents of the above. The Xucar, flowing E. to the Mediterranean, also rises in this province. Besides the chain of mountains that separate Old and New Castile, there are three other important chains that traverse these provinces. First, the Sierra de Toledo, which winds semicircularly past Daroca, from the Castilian chain, and then runs SW. nearly parallel to it, to the hills of Santa Cruz, near Merida. Next, the Sierra Morena, or Black Mountains, beginning above Alcaraz, near the source of the Guadalquivir, and running like the two former, nearly SW., to the narrow pass of

Montegil. Lastly, the Sierra Nevada, or Snowy Mountains, that commence between the sources of the Xucar, Guadiana, and Guadalquivir, and extend into Andalusia. These last are here extremely steep and bare, mostly schistose, and often coated with limestone. They have white quartz in considerable veins; and valuable dark green, and a profusion of other marbles. The N.E. part of the Sierra Morena is of considerable height, and rather resembles table-land than a ridge of hills. The seasons are very different on the two sides of this range. In Andalusia, the vines are all in leaf, and the fruit is set, when, on the N. side, hardly a leaf is to be seen, or a bud to be found in the vineyards. There are here a few remains of former forests, which might have existed when Cervantes made these parts the scene of the exploits of his hero; and a variety of flowering shrubs, particularly the rock-rose, or gum cistus, from which manna is procured, and suchlike. In this chain are vertical beds of argillaceous schist, and beds of grained quartz, with entire hills of pudding stone, and some porphyry, and the finest jasper. It is the richest in minerals of any in the kingdom; and has veins of gold and silver. The quicksilver mines at Almaden have been worked for nearly 8,000 years, and furnished the vermillion sent to ancient Rome. They produce annually 2,000,000 lbs. of quicksilver. (Bowles, Historia Natural de España, p. 12; A Year in Spain by a Young American, i. 199.) The Castilian mountains are composed of gneiss granite, which often terminates in peaks of great height; schist, limestone, sandstone, breccia, quartz, marble, gypsum, &c. The Guadarama mountains, about 20 m. N.W. Madrid, are bleak, dreary and barren near their summits, which in many places, are covered with nearly perpetual snow, indicating that they must be 8,000 or 9,000 ft. above the level of the sea; the limit of perpetual snow in these latitudes being about 9,500 ft. The height of Moncayo, the highest mountain in Castile, is estimated at 9,000 ft. The rock, being partly decomposed, forms a light soil that produces the juniper europæus, Daphne mezereon, matricaria maria, genista, thyme, and a great many other aromatic herbs. The cistus tribes abound at every level on the granite mountains, not covered with snow; pines appear on the summits; the noble oak and the elm near their bases. (Townsend, ii. 106.) The scenery is often of the wildest description; the mountains full of deep cuts and ravines, mostly the beds of winter torrents; aged and stunted pines hang upon their edges, and are strewn upon the brown acclivities around; and bare rocks frequently project over the passes, and force them to the very edge of undefended precipices. (Inglis, i. 355.) The quality of the soil is various; in some parts a blackish or brown nitrous clay, which is extremely fertile; in others, light and stony, and little productive. New Castile is in great part clayey, and covered with molehill. Besides the minerals mentioned above, the Castiles produce calamine, ochre, bole armeniac, fine emery, rock crystal, salt, many curious stones and fossil shells, hot and cold saline springs; and in the mountains are many remarkable caverns, that contain beautiful stalactites, in a variety of fantastic forms. Near Molina is the hill of La Platilla, which has a remarkable mass of copper, in masses of white quartz. Though the ore is near the surface, the hill is covered with plants. Townsend had no doubt that there is tin near Huesca. (i. 214, 219, 343; ii. 106; Miñano, Diccionario Geográfico, ii. 467, et seq.; Dillon's Travels through Spain, p. 110, 112, 115, 196, 202, 205—207, 227, 229; Antillon, Geographie d'Espagne, p. 8—14.)

The climate of the Castiles is in general healthy; that of Old Castile is rather cold and moist. In new Castile it is excessively dry; but rendered healthy by the purity of the prevailing winds, and the great elevation of the country; but this altitude sometimes exposes it to strong dry winds, which, not meeting with the thick woods by which they were formerly tempered, are found very unpleasant, and at times even dangerous, as Madrid, in winter, by producing pulmonary complaints. The height of the plateau of Castile reduces the mean temperature to 59 Fahr., while on the coasts of Spain it is from 62° to 75°. The ordinary extremes of temperature, in Madrid, are 90° Fahr. in summer, and 32° in winter; but the thermometer often rises to above 100°, and falls below 14°.

Products.—The principal product of the Castiles is corn, some of which they export to Valencia, Andalusia, and Estremadura. No other province of Spain has wines so strong, and yet so sweet, though but little exported, or known abroad. The most celebrated is that of the Val de Peñas, or 'Valley of Stones,' in La Mancha. It is a dry, strong, red wine of the Burgundy species, and is said to be so plentiful and cheap that a bottle may be had in the country for 1½d. It is drunk by the better classes all over the Castiles; but in the greatest perfection in its native district, on account of the taint given it by the skins in which it is carried to a distance. The Castiles produce also pulse, and some fruit and oil. Hemp, flax, madder and saffron are partially cultivated. Garden stuffs are not abundant. On the mountains and in the pastures considerable numbers of black cattle, sheep, and mules are raised; but the increase of the latter has almost annihilated the race of good horses in the Castiles. There are fallow deer, wild boars, wolves, hares, peacocks, and all kinds of poultry and small game in abundance. The larger game has decreased through the breaking up of the land near the royal seats during the absence of Ferdinand VII. Bears are seen in some parts, and lynxes are not uncommon in the high mountains. Not only the fallow land, but the cultivated fields in New Castile, are full of two species of broom (genista sphærocarpa and monosperma), and the Daphne gnidium. They grow to nearly six feet in height, and have a great effect on the prospect. These plants, with the asphodelus ramosus, and several other bulbous plants that abound in the pasture fields, give a peculiar character to the landscape of Spain. There is a want of trees, which is partly attributable to the flat and unsheltered nature of the plains, and the dryness of the climate, but chiefly to a prejudice against them, entertained from time immemorial; the peasantry thinking that they are good for nothing, unless it be to attract and shelter vermin. They dislike them so much that they destroy those planted by government along the high roads. It is believed that the want of trees to attract humidity has prevented that drought which, next to bad government, is the curse of the Castiles. From the Duero to the Tagus there is not a stream ankle deep, except when swollen by floods. Agriculture is in the most backward state; the consequence of a comparatively thin population, having little interest in the soil, which is monopolised by the clergy and nobility. Irrigation, which in such a country is indispensable, is but very little practised, and even manuring is all but neglected; and thus, while three-fourths of the country remain fallow, the rest produces only poor crops of grain or potatoes. The great distance between the towns, the badness of the roads, and still more the insecurity of life and property, which prevents the farmer from living insulated on his farm, are additional checks to agriculture. Eight or ten miles frequently intervene without a single habitation, and the country looks poor and miserable in the extreme. Nothing can be more gloomy than the appearance of the towns, with old-fashioned towers projecting out of a dismal group of houses plastered over with clay. At the entrance of each is a gate for receiving the duties on all articles that pass; and in the centre a square, round which are the buildings occupied by the ayuntamiento, or municipality, the posada, or inn, and the butcher, baker, tailor, cobbler, and village surgeon, or barber. Most of the towns exhibit every symptom of decline, (Slidell, i. 136; Inglis, i. 56.) Before the construction of railways, there was nearly a total want of free communication, all but the main road to France being neglected. The old road between Madrid and Toledo was mostly carried over ploughed fields, sometimes with hardly a visible track. The new iron roads, established chiefly by English capital, and built by English 'navvies,' have greatly improved this state of things, and bid fair to raise even Castile from its state of poverty and misery. (See SPAIN.)

Manufactures, though formerly considerable, are now at a very low ebb. The cloths of Segovia were once the best in Europe; and there are still some woollen fabrics, among which is the famous vicugna cloth and coarse camlets, serges, and flannels, and some of wrought silks, silk stockings and gloves, galloons, blond lace, coarse linens, hats, caps, soap, saltpetre, gunpowder, the celebrated plate-glass of St. Ildefonso, white earthenware, tanned leather, and paper, but they are all inconsiderable. Castile has little commerce: wool is the staple commodity. The exportation of sheep was always strictly forbidden, till by the treaty of Basle the French were allowed to purchase 5,000 Merino rams and as many ewes; and from this stock, and subsequent exportations from Spain, the quality of the wools of France, England, Germany, and other parts of the world, has been greatly improved.

The following table shows the area and population of the two Castiles according to the census of 1857. Valladolid and Valencia are sometimes included in the Castiles; but they did not formerly belong to them, and are excluded in this table.

Provinces		Area in Eng. Sq. Miles	Population in May, 1857
New Castile	Madrid	1,015	475,785
	Guadalajara	1,912	199,088
	Toledo	5,176	329,756
	Cuenca	11,304	279,309
	Ciudad Real	7,515	244,328
	Total	**30,892**	**1,477,918**
Old Castile	Logroño	7,474	435,556
			173,812
	Santander		814,141
	Oviedo	3,820	524,529
	Soria	4,776	147,604
	Burgos	8,164	146,878
	Avila	7,562	164,143
	Leon	5,691	319,756
	Palencia	1,733	185,970
	Valladolid	2,379	744,022
	Salamanca	5,892	262,611
	Zamora	5,562	219,164
	Total	**71,441**	**6,473,936**

The Castilians have the character of probity, sobriety, and moderation; they are serious and contemplative, which makes them, at first, seem gloomy and haughty; but, after a time, they are found not deficient in the agreeable qualities.

They have to boast of many illustrious men: at the head of whom stand Cervantes, the inimitable author of Don Quixote, and Lope de Vega. They are not what would be called hospitable, but they are, notwithstanding, generous. The middle and upper classes are fond of display and ostentation to an extraordinary degree, while inconsiderate-ness and carelessness are conspicuous in the characters both of the lower and middle classes. Almost every one lives up to his income; even the employees, whose tenure of office is so uncertain, seldom lay by anything, and generally die penniless. But the love of ease and pleasure, and proneness to indolence, is less marked, perhaps, in Castile, than in the southern provinces. Their want of industry is the result of the circumstances under which they have been placed, and of their various institutions. No man will be industrious, where industry does not bring along with it a corresponding reward; and this is very rarely done in Spain. Had the Castilians the means of improving their condition by labour, their apathy and listlessness would speedily give place to activity and enterprise. In Madrid, and generally in Castile, there is something more of luxury at the table than in the N. provinces, though the Spaniards in general are abstemious, and little addicted to its pleasures. The dining-room is generally the meanest apartment; but the houses of respectable persons are scrupulously clean, particularly the kitchens and bed-rooms. Female education begins to improve; besides embroidery and music, a little history and geography is taught in the schools, though not in the convents, where the higher orders are educated. In the time of the constitution of the Cortes, there were two Lancastrian schools for boys and one for girls at Madrid; but three for the boys were suppressed on the king's return. The influence of the regular clergy is diminished much more than that of the monks, who are still, through the austerities they practise, and the alms they distribute at the convent doors, held in considerable veneration, except in Madrid, where less attention is paid to religious ceremonies and processions than in any other city of Spain. The large towns have a sombre aspect, the women being nearly all in black, without a bonnet or a ribband. Every one has a mantilla or scarf thrown over the shoulders, which varies in quality with the station of the wearer. Besides a waistcoat and jacket of cloth, covered with abundance of silver buttons, the men usually wear a sheepskin jacket with the woolly side outwards; or, instead of this, an ample brown cloak, the right fold of which is thrown over the left shoulder with a Roman air. The head is covered with a pointed cap of black velvet, the ends of which bring drawn down over the ears, leave exposed a high forehead and manly features. They have tight breeches, sustained above the hips by a red sash, and fastened the whole way down the outside of the thigh by brill buttons, woollen stockings, stout shoes, and leather gaiters, curiously embroidered, and fastened at top with a gay-coloured string. The love of dancing is universal among them; the ladies equally dance well, but in a style quite different from the French; they laugh and talk while they dance, and are strangers to that burlesque silence and gravity that prevail among the quadrillers of France and England. Music is much cultivated; and it is rare to find a female even in the middle ranks who is not a good pianist. Among their amusements, the bull-fights, to which all classes are passionately addicted, must not be forgotten. These have been prohibited several times; and the cruelties practised at them may seem sufficient to stamp them with the character of brutality and barbarism. Yet there is nothing of deliberate cruelty in the character of the Spaniards, and they have as little, perhaps, of hard-heartedness as other people. The use of the toledo, or bravo, to revenge private wrongs, is now unknown. Horse-racing was attempted to be introduced by the Duke of San Carlos, at Madrid, in 1830, with an English horse against a Spanish one; but the English horse was beaten by foul play, and the duke insulted as he left the ground.

The Castilian is the standard dialect of the Spanish language. During the struggles with the Moors, many dialects of the Romanzo, or mixture of the Latin with the Germanic tongues, grew up in Spain, which finally melted into three—the Gallician, Castilian, and Catalonian. On the marriage of Isabella, queen of Castile, with Ferdinand of Aragon, the Castilian Romanzo became the language of the court, and has maintained its pre-eminence ever since.

History.—The Castiles anciently formed parts of *Cantabria*, and the country of the *Celtiberi*, *Oretani*, and *Oxyresani*; and, like the rest of Spain, were successively overrun by Romans, Goths, and Saracens. After the expulsion of the Saracens, and various vicissitudes, the sovereignty of Castile came by marriage to Sancho III., king of Navarra, whose son Ferdinand was made king of Castile in 1034. He married the sister of Verimund III., king of Leon, but afterwards killed his father-in-law in battle, and was himself crowned king of Leon, in 1037. The crowns of Castile and Leon were afterwards separated and again united several times, till, by the marriage of Isabella, who held both crowns, with Ferdinand, king of Aragon, in 1479, the three kingdoms were, as at present, consolidated into one. Castile, as well as the rest of Spain, has for a lengthened period been exposed to the scourge of a civil war carried on without zeal on either side, but with the most detestable perfidy and cruelty.

CASTILLON, a town of France, dép. Gironde, cap. cant., on the Dordogne, 11 m. ESE. Libourne. Pop. 8,516 in 1861. In 1451, an obstinate engagement was fought under the walls of this town between the English and French, when the latter were victorious. In the commune of Castillon are the remains of the *Château de Montaigne*, to which the illustrious essayist of that name retired in 1578, and where he breathed his last on the 13th of September, 1512.

CASTLEBAR, an inl. town of Ireland, prov. Connaught, co. Mayo, at the N. extremity of the lake of the same name, 126 m. W. by N. Dublin, on the Midland-Great-Western railway. Pop. 6,373 in 1831, and 8,022 in 1861. The town was taken by a French force under General Humbert, which landed at Killala in 1798, but was shortly after evacuated on the approach of the main army of the British under Lord Cornwallis. It is the assize town of the co., and consists of a square, and a long street with some branches. The par. church and a R. Cath. chapel are new, large and elegant buildings; there are also a meeting-house for Methodists, a large parochial school, a national school, an infirmary, and two dispensaries. There are barracks for artillery and infantry, fit to accommodate 650 men. The constabulary and the revenue police have stations here. By a charter of James I. in 1613, the corporation consists of a portreeve, 15 burgesses, and a commonalty, which returned 2 mems. to the Irish H. of C. till the Union, when it was disfranchised. The assizes for the co. are held here; also general sessions in Jan. and Oct., and petty sessions every Saturday. The court-house is a well-arranged building. The county prison, erected on the radiating principle, has 129 cells, and 83 other sleeping rooms. Linen

and linen yarn are manufactured to some extent, and sold in the linen-hall; there are also tanneries and soap manufactories, a tannery, and a brewery. There is an extensive trade in grain, and other agricultural produce. Markets on Saturdays; fairs, 11th of May, 9th of July, 16th of Sept., and 18th of November.

CASTLECOMER, an inland town of Ireland, prov. Leinster, co. Kilkenny, on the Deen, an affluent of the Nore, 52 m. SE. Dublin. Pop. 2,436 in 1831, and 1,435 in 1861. The town, which suffered much in an unsuccessful attack by the insurgents in 1798, consists of a main street planted on each side, and of some others branching from it, and is remarkable for neatness and good order. The par. church on a neighbouring hill, a large R. Cath. chapel, a convent, a Methodist meeting-house, a court-house, a dispensary, and a barrack, are the principal buildings. Little trade is carried on, the place deriving its support chiefly from the neighbouring collieries, which furnish a copious supply of fuel to the adjoining counties. The mineral is of the carbonaceous or stone coal species, which burns without flame, being the slaty, glanty coal of Werner. Fairs are held on Mar. 27, May 8, June 21, Aug. 10, Sept. 14, Oct. 28, and Dec. 14. General sessions in June, and petty sessions every Friday; also a manorial court for small debts.

CASTLEDOUGLAS (formerly *Carlinwark*, from the name of a lake in its immediate vicinity), an inland burgh or barony of Scotland, co. of stewartry of Kirkcudbright, par. Kelton, on the railway from Dumfries to Portpatrick, 18 m. from the former, and 68 from the latter. Pop. 2,261 in 1861. The town is neat and well built, and consists of a main street along the road, with several lesser streets running at right angles or parallel to it. It is quite a modern town, and is wholly indebted for its existence and prosperity to the advancing wealth of the thriving agricultural district by which it is surrounded. Its consequence has been of late years materially increased by the transfer to it of the weekly corn and cattle markets, the most important in the co., originally held at Rhonehouse, a small village, distant 1½ m. The famous horse-fair of Kelton Hill is still held at Rhonehouse; but it has lost much of its original importance, as horses from Ireland, which formed its staple, are now generally sent direct to the fairs in England by steam, instead of taking a circuitous land route by Kelton Hill. It has an extensive retail trade, but no manufactures.

CASTLETON, a par. of England, co. Derby, hund. High Peak. Area, 10,100 acres. Pop. 1,157 in 1861. The village is 143 m. N. by W., London. The vale of Castleton is in the heart of the Peak district, about 1,000 ft. below the level of the surrounding hill ranges, and is 6 m. in length, and from 1 to 2 m. in width, with several smaller dales opening to it on the N. and S. It is a fertile tract watered by several rivulets, and approached from the Chapel-le-Frith side, through a long and deep chasm, crossing the mountain range, and called the 'Winnats,' or windgates, from the strong gusts and currents of air that usually prevail: the road winds down a considerable declivity, between precipices rising upwards of 1,000 ft. on each side, and opens, by a sudden turn, on the vale, in which there are three villages, Hope, Brough (both in the parish of Hope), and Castleton. The latter is at the base of a steep rock, whose summit is crowned by the ruins of the Castle of the Peak, considered a genuine specimen of the Saxon period; though the traditions of the neighbourhood ascribe it to Wm. Peveril, a natural son of the Norman Conqueror. The keep is still nearly entire, and some portions of the outer walls, in many places 20 ft. high and 9 ft. thick. The church is small, but considered a very interesting relic of the early painted style; here are also a Wesleyan chapel, and an endowed charity school, in which 23 scholars are educated. The inhabitants are chiefly employed in the mines of the surrounding district, which produce lead, calamine, and the coloured flour spar called 'blue John,' much in request for vases and other ornaments. The whole of the calcareous strata in the vicinity are remarkably deranged, and are also characterised by numerous cavernous fissures and the frequent disappearance of streams (through what are termed swallow-holes), which, after subterranean courses of various lengths, again emerge to the light. The outer chamber of the Great Peak, or Devil's Cavern, has a natural arch of about 120 ft. span; several small cottages have been built in it. The rest of the chambers are only to be explored by torches; they extend about 2,200 ft. from the entrance to the innermost end, where, though there are probably others beyond, the rocks close down so near a subterranean stream as to prevent further access; this stream has to be crossed two or three times in proceeding, and at one part a small boat is kept for the purpose. The average depth from the floors to the upper surface of the mountain is about 650 ft. The strata abound in marine fossil remains. The Elden hole, 3 m. W. of Castleton, is of a similar character, and also that approached by the level of the Shrodwell mine, near the Winnats. This mine has been given up; but the Odin mine, in the vicinity, which was worked in the Saxon period, is still productive. Mam Torr, or the Shivering Mountain, rises 1,300 ft. above the vale, and is composed of alternating strata of shale and micaceous grit. There is an ancient encampment on its summit, and British and other ancient remains are frequent in the district, which is one of the most remarkable in the kingdom for its picturesque character, and the abundance of natural objects and phenomena interesting to science. On the attainder of the grandson of William Peverell of the Peak (for poisoning the Earl of Chester), the castle was granted by Henry II. to his son, afterwards King John; subsequently Edward III. gave it to John of Gaunt; since which it has formed part of the duchy of Lancaster, and is at present leased by the Duke of Devonshire.

CASTRES, a town of France, dép. Tarn, cap. arrond., in an agreeable and fertile valley, on the Agout, 23 m. SSE. Alby, on the railway from Alby to Narbonne. Pop. 21,538 in 1861. This, though not the capital, is the principal town of the dép., and is thriving and industrious. It is divided into two parts by the river, over which it has two bridges. It is but indifferently built, and the streets are narrow and winding. The principal building is the old episcopal palace, now the sous-préfecture; it has also barracks, workhouses, an exchange, a theatre, and a fine promenade. It is the seat of a court of primary jurisdiction; and has a model school, a diocesan seminary, with 113 pupils, a Protestant consistorial church, a class of linear design, and a public library with 6,000 volumes. There are here extensive manufactures of cloth and woollen stuffs, with establishments for the spinning of cotton, linen fabrics, paper fabrics, dye-works, bleach-fields, and tanneries. It has also copper forges and foundries.

Castres espoused, in the 16th century, the Protestant party, and Henry IV. resided in it for a lengthened period. Its ramparts were demolished by Louis XIII., and the bishopric was suppressed at the Revolution. It is the birthplace of Dacier,

the critic, of Rapin the historian of England, and of the Abbé Sabatier.

CASTRO, a seaport town of Southern Italy, prov. Lecce, on the Adriatic, 28 m. SSE. Lecce, with which it is connected by railway. Pop. 5,300 in 1862. The town has an old castle and a cathedral, and is the seat of a bishopric. It was sacked by the Turks in the 16th century; and since then has suffered much from the inroads of Barbary cruisers. Its harbour admits only small vessels. The environs are productive of corn, wine, cotton, and fruits.

CASTRO DEL RIO EL LEAL (an. Castro Julia), a town of Spain, prov. Cordova, on the Guadajus, 16 m. SE. Cordova. Pop. 8,945 in 1857. The town has two churches, two hospitals, a foundling hospital, two seminaries for the education of boys and girls, and a castle; with manufactures of wool and hemp.

CASTROGIOVANNI (an. Enna), a town of Sicily, prov. Catania, cap. cant., almost in the centre of the island, 55 m. ESE. Palermo, in a plain about 5 m. in circ., being the summit of a lofty and almost inaccessible mountain, more than 4,000 ft. above the level of the sea. Pop. 13,747 in 1862. This city, celebrated in antiquity as the birthplace of Ceres, and the site of her most sacred temple, is now one of the poorest towns in the island. It still, however, commands an extensive and delightful prospect, is well supplied with excellent water, and has a clear salubrious atmosphere. The surrounding country, which is very fertile, was, in antiquity, ornamented with innumerable groves and temples, appropriated to the worship of Ceres and Proserpine. Livy has correctly described the city as built in arcchis loco ac praerupto; and Cicero has given an eloquent description of the town, temple, and statue of Ceres, carried off by the wholesale plunderer, Verres: 'Simulacrum Cereris Ennæ et aliud ac domo sustulit, quod erat tale, ut homines, quum viderent, aut ipsam videre se Cererem, aut effigiem Cereris, non humana manu factam, sed cælo delapsam, arbitrarentur.' But all traces of the temple, as well as of the worship of the goddess, have disappeared. The castle in the modern town, which is going fast to ruin, is evidently of Saracen or Norman origin.

About 5 m. from the town, at the foot of the mountain, is the famous lake, on the borders of which

    'Proserpine gathering flowers,
Herself a fairer flow'r, by gloomy Dis
Was gathered.'

The orators and poets of antiquity have exhausted their powers in describing the beauty and sublimity of this famous lake. (See, among others, Cicero in Verrem, iv. § 48; Ovid, Met. lib. v., lin. 385.) But it no longer wears the livery of perpetual spring; its groves have been cut down, and its temples levelled with the dust! All is desolate and deserted:—

   'Pro molli viola, pro purpureo narcisso,
Carduus, et spinis surgit paliurus acutis.'

Its naked borders are fœtid and loathsome, and in the summer months exhale a pestilential air.

   'Tantum avi longinqua valet mutare vetustas.'

Enna was the head-quarters of the revolted slaves under Eunus, during the first servile war in Sicily. Here they defied for several years the power of Rome, and defeated three Prætorian armies. At last they were entirely defeated under the walls of Messina, by the consul Piso; and Enna was subsequently taken by the consul Rupilius, and the slaves put to the sword or crucified.

CATALONIA (Span. Cataluña), an old prov. of

Spain, occupying the N.E. portion of the kingdom, between lat. 40° 30′ and 42° 51′ N., and long. 0° 15′ and 3° 21′ E. It is of a triangular shape, and has the E. Pyrenees, which separate it from France on the N.; the Mediterranean on the E.; and Aragon, and a small part of Valencia on the W. Greatest length and breadth, 180 and 130 m.; area about 12,150 sq. m., including Andorra. Offsets from the Pyrenees spread themselves through the whole prov. from N. to S., forming valleys of larger or smaller extent, like those of Ampurdam, Urgel, Aran, and Lerida. Towards the middle of the prov., 29 m. NW. from Barcelona, is the celebrated Montserrat, 4,500 ft. in height; and farther N. on the Ebro, is the Sierra de la Llena. The Pyrenees are not so rugged on this as on the French side, and descend gradually towards the Mediterranean. They are mostly granitic. The other mountains of Catalonia are in many respects similar. The mountain of Cardona, 17 m. NW. Montserrat, almost in the centre of the prov., is a solid mass of pure rock-salt, without the least crevice or fissure, between 400 and 500 ft. high, and 3 m. in circ. This prodigious mass of salt is unparalleled in Europe, and perhaps in the world. In almost any other country it would be turned to great account, and be made the means of an extensive trade; but here, owing to the badness of the roads and the difficulty of access, this inexhaustible source of wealth is but little known, and comparatively neglected. (Dillon's Travels in Spain, p. 590.) Near Olot, in this prov., about 55 m. N. Barcelona, is a remarkable district of extinct volcanos, that has been visited and described by Mr. Lyell. It contains about 14 distinct cones, with craters. The greatest number of perfect cones are close to Olot; and the level plain on which the town stands has clearly, according to Mr. Lyell, been produced by the flowing down of lava from the adjoining hills. Most of these volcanos are as entire as those near Naples, or on the flanks of Etna. Some of them contain caverns called bufadors, from which a current of cold air blows during summer. There is no record of any eruption here; but the town of Olot was nearly destroyed by an earthquake in 1421. (Principles of Geology, ii. 39, 3d. ed.) The mountains in the S. of the prov., near the coast, are limestone. On the E. of Cervera gypsum only is met with; but more to the W. it gives place to chalk. The coast is mostly bold and rugged. In the N. is Cape Creus, the most E. point of Spain, being the extremity of a rocky peninsula stretching out into the sea, and separating the Gulf of Lyons from that of Rosas, lat. 42° 19′ 53″ N., long. 3° 20′ 10″ E. The prov. is well watered. One of the affluents of the Ebro, the Noguera, forms for nearly 60 m. the line of demarcation between it and Aragon. The Ebro itself enters the prov. at Mequinenza, and flowing through its most S. portion by Tortosa and Amposta, falls into the Mediterranean 15 m. E. from the latter. The Segre, with its affluents, unites with the Ebro at Mequinenza. The principal rivers, unconnected with the Ebro, are the Llobregat and Ter, the one flowing SE., and the other E. to the Mediterranean.

The Pyrenees furnish iron, copper, zinc, and manganese. There are lead mines in various districts. Coal is abundant, but much difficulty has always been encountered in working it, from the want of capital and of improved means of communication. Townsend says, that copper and silver abound in the valley of Aran, and that coal, silver, and gold, have all been found in the vicinity of Lerida. There is abundance of alum in the valley of Aran; nitre is produced spontaneously in the plains of Urgel, and cathartic

salts at Cervera. The mountain of rock-salt at Cardona has been already noticed. There are marbles, jasper, and other stones useful in architecture and sculpture; alabaster, amethysts, topazes, and coloured rock crystal; quartz, barytic spa, fluor spa, limestone, chalk, and gypsum, in all varieties; amianthus, talc, serpentine, and chalcedony. There are many mineral waters and hot springs.

The air is dry and unusually bright and clear in the interior; but on the coast it is variable and moist; and in summer pestilential diseases not unfrequently prevail. The mountains are everywhere covered with snow during the winter, and in the Pyrenees frequently even in June.

Soil and Produce.—About half the surface is susceptible of cultivation, the rest consisting of rocks, naked barren hills, and woodland. The mountain land is stony, and full of fragments of granite; but the valleys are mostly fertile. All sorts of grain are grown, viz. wheat, rye, maize, barley, oats, and millet. The plains of Ampurdan are suitable for rice; but its cultivation is prohibited, as prejudicial to health. (Miñano.) Pulse is produced in all parts. Hemp, flax, saffron, madder, woad, anise, liquorice, and barilla are also produced. The E. districts yield good strong wines, which are frequently employed to give body to the wines of other provs., and are sometimes exported for that purpose to Cette, and thence to Bordeaux. Oranges, lemons, and citrons, are found on the coast; figs and almonds are grown in the plain of Tarragona; and apples, pears, cherries, quinces, medlars, apricots, peaches, walnuts, chesnuts, and filberts, in all the plains. Oil, though not of the best quality, is produced in all the warmer parts of the coast district. Silk, honey, and wax are also produced in considerable quantities. Timber is plentiful, especially the noble-oak, beech, fir, elm, evergreen poplar, cork-tree, &c. Nuts and cork constitute important articles of export from the prov., being in this respect second only to linen and cotton goods and brandy. Bears and wolves are sometimes seen in the Pyrenees. Laborde estimated the produce of wool at 30,000 quintals.

Catalonia is the best cultivated, and the people the most industrious, of any of the Spanish provs. This is owing to a variety of causes, but principally, perhaps, to its exemption from the alcavala and other oppressive imposts (See SPAIN), and to the mode in which lands are occupied. Generally, throughout Spain, the land is divided into vast estates, held under a system of strict entail, and administered by stewards on account of the proprietors. The disastrous influence of this system is apparent in the low state of agriculture, and the wretchedness of the peasantry, in most parts of the monarchy. But in Catalonia its influence is materially modified by the landlords having power, by what is called the emphyteutic contract, to lease a portion of their estates. This they may do for a term of years, either absolute or conditional, for lives or in perpetuity; always reserving a quit-rent, as in the English copyhold, with a relief on every succession, a fine on the alienation of the land, and other seigniorial rights dependent on the custom of the district. The reserved rent is commonly paid in money; but the agreement is often for wine, oil, corn, or poultry. If the tenant quits before the end of his term (which he may do), he loses all claim for improvements, for which he must otherwise be paid. Persons occupying land under this tenure have an obvious interest in its profitable cultivation; and wherever it prevails the country is in a comparatively flourishing state.

Irrigation is the leading feature in the husbandry of the prov., and is carried to a great extent by means of canals and trenches cut from every available source; the maintenance of which, together with the distribution of the water, is committed to the care of a particular junta. Great numbers of farms are also watered by means of the noria, a machine introduced by the Saracens for raising water from wells. The soil is in parts so very light that it is ploughed with a couple of oxen, and sometimes with one horse, or even mule; but with the help of the water it is rendered fertile, and produces on the same spot corn, wine, oranges, and olives.

The silk and woollen manufactures of Catalonia were formerly carried on to a great extent, and are still of considerable value and importance. In the latter part of last century the cotton manufacture was introduced; but it has not succeeded. Exclusive of silks, cottons, and woollens, a good deal of linen is made, with paper, hats, and cordage. All kinds of weaving are carried on upon the slopes of the Pyrenees, where wages are lowest, the webs being brought to Barcelona to be bleached and printed. Leather is largely manufactured, and shoe-making used to be one of the principal employments. In 1786, the export of shoes from Barcelona only was estimated at 700,000 pairs, mostly for the colonies. Since the emancipation of the latter, this trade has greatly declined. Distillation is extensively carried on; the exports of brandy amounting, on the average, to 35,000 pipes a year. Cannon and small arms, soap, glass, sheet-iron, and copper utensils, are also produced. Women, in the agricultural districts, are employed in the making of blond and other laces. The shipbuilding, formerly carried on at Barcelona, Mataro, and other places on the coast, where timber was cheap, has nearly ceased. Tarragona is the chief place in the prov. for the export of nuts, almonds, wines, brandy, cork wood, and cork bark. (See TARRAGONA.)

The pop. of Catalonia was estimated in 1788 at 814,412. According to the census of 1857, it contained 1,652,291 inhabitants. Catalonia is now divided into the four provinces of Barcelona, Tarragona, Lerida, and Gerona. The principal towns are Barcelona, Tarragona, Gerona, Lerida, Reus, Mataro, and Tortosa.

The language of the Catalans is a dialect of the Romance or Provençal, at one time the common language in the S. of France, and in some other parts. But it is now a good deal intermixed with Castilian and other words. Letters were successfully cultivated at the court of Barcelona; and some of the counts attained to distinction as troubadours.

Catalonia had for a lengthened period its states, composed of the clergy, nobility, and commons, who shared the legislative power with the sovereign. It had, also, particular and very extensive privileges, and a peculiar form of jurisdiction in the hands of magistrates, called regidores, whose districts are named reguerias. The highest court of appeal was the royal council established in Catalonia. Their contributions to the king were not considered as imposts, but as voluntary gifts; the Catalans were to be tried by the laws of Catalonia only, and by native judges; and their estates were never to be confiscated, unless for treason. But these privileges were suppressed by Philip V. when he subdued the province; and the laws of Catalonia were then assimilated to those of Castile. They have always been exempted from the alcavala, rionas, and millones, in lieu of which they paid 10 per cent. on all fruits, whether belonging to individuals or communities,

and on the supposed gains of merchants and mechanics.

The Catalans are hardy, active, and industrious; and used to be distinguished by their attachment to their privileges, and their opposition to arbitrary power. But in this respect they seem to have undergone a material change, being now distinguished by their veneration for the apostolical party in church and state—a consequence probably of their ignorance and subserviency to the priesthood. There seems, indeed, to be little or no provision made for education. Philip V. suppressed the universities of Barcelona, Lerida, and Gerona, and established in their stead only that of Cervera. There are academies in the principal towns; but the great bulk of the people appear to be without the means of instruction. Their improved condition is not therefore in any degree owing to their superior intelligence, but to the comparatively favourable circumstances under which they have, in other respects, been placed.

The difference between the cottages of Catalonia and those of the other provinces of Spain is very visible. The houses and cottages here have an air of convenience and comfort; there is glass in the windows, and the insides display the articles of furniture in common use. No beggars, and few ragged people, are seen; industry is every where active; stones are removed from the ground and collected in heaps; fences are more general and more neatly constructed; nobody is seen basking in the sun; even the women and girls who attend the cattle do not sit idle, wrapped up in their plaids, but every one has her spindle in her hand.

Catalonia anciently made a part of the Hispania Tarraconensis of the Romans. The Goths were its next masters, who reared themselves from it over the rest of Spain. On the fall of the Gothic empire, the Catalans submitted to the Moors, but the dominion of the latter was not of long duration. In the 8th and 9th centuries, Catalonia, with the adjoining country of Roussillon, became an independent state, subject to the counts or earls of Barcelona. Under their government, liberal institutions were established in the prov.; it was distinguished by its naval power, commerce, and proficiency in the arts; and its fleets and armies frequently interfered with decisive effect in the contests of the time. In 1137, Catalonia was united with Aragon by the marriage of one of its counts with the heiress of the latter; but the Catalonians retained their separate legislature, and distinct privileges. In 1640 the prov. revolted against Philip IV., and was not recovered till 1659. In the war of the succession, the Catalonians were the most zealous adherents of the Archduke Charles; and even after England and Austria had withdrawn from the contest, they refused to submit, and defended Barcelona with an obstinacy of which there are but few examples. On its capture, their ancient cortes, and most of their peculiar privileges, were suppressed.

CATANIA, an ancient and celebrated city and sea-port of Sicily, cap. prov. same name, on the E. coast of the island, at the foot of Mount Etna, at the extremity of a vast plain, 31 m. NNW. Syracuse, on the railway from Messina to Syracuse. Pop. 64,296 in 1862. The city, though it has suffered much from earthquakes, by one of which, in 1693, it was all but totally destroyed, has always risen from its ruins finer and more magnificent than ever. Catania has a noble appearance from the sea; and what is rare in an Italian town, the effect is not diminished on landing; for the streets are regular, spacious, and handsome; and the numerous churches, convents, palaces, and public establishments, principally constructed of lava, faced with magnesian limestone from Malta and Syracuse, and enriched with marbles from the ruins, are magnificent. The city is nobly situated, on the route of Ætna, its prosperity and its benefactor. Overwhelmed, as it has often been, by torrents of liquid fire, it has risen, like the phœnix, more splendid from its ashes. The very substance which once ravaged its plains has, by its own decomposition, covered them with soil fertile as the fabled gardens of the Hesperides; and on all sides the material of destruction is turned to the purpose of ornament and utility. The streets are paved with lava; houses, palaces, and churches, are built of lava; of lava they form ornamental chimney-pieces, tables, and a variety of toys; whilst a natural mole of lava defends the shipping from the fury of the tempest. The cathedral, founded in 1094, was rebuilt on a simple and grand scale, after the earthquake of 1693; the senate-house, mont di pietà, theatre, and most of the municipal establishments, are also fine, appropriate buildings. Near the cathedral is a fine square, ornamented with an antique statue of an elephant bearing on its back an obelisk. It has 49 churches, of which that of St. Maria dell' Ajuto, and several others, are magnificent structures; it has also 19 convents for men, and 11 for women. The Benedictine convent of San Nicolò d'Arena has long been justly celebrated for its vast extent, superb church, excellent organ, large museum, ancient masons, and great riches. Among the charitable establishments, exclusive of the monte di pietà, are several hospitals, a workhouse, a foundling hospital, a lying-in hospital, and a Magdalen asylum. The university, founded in 1445 by Alphonso of Aragon, is an extensive foundation with an annual revenue of above 2,000l. It has able professors, and is well attended; its library and museums are open on holydays to the public. The heirs of Prince Biscari and others have also fine museums. Catania is the seat of a bishopric, of a court of appeal, a criminal court, a civil court, and of the provincial authorities; and enjoys extensive privileges. The humanity, hospitality, and good-breeding of the inhabitants have been eulogized by all travellers. On many occasions they have shown a singular unanimity in public affairs; they had the courage to practise inoculation so early as 1742, and to introduce the potato while an ignorant prejudice existed against it among their neighbours. The principal manufacture is that of silk, which is largely carried on. The working of the yellow amber found on the S. coast of the island affords employment to some thousands of the population. The snow of Mount Ætna is also a great source of wealth. The harbour is not equal to the importance of the city; but it is generally full of small craft that resort thither for corn, macaroni, potatoes, olives, figs, silk, wine, almonds, cheese, oil, salt, manna, cantharides, amber, snow, and lava. The environs are fruitful, and well cultivated.

Catania is very ancient. It is believed to have been founded by the Chalcidians, and had Charondas for its early legislator. Under the Romans, it was the residence of a pretor, and was adorned with many noble buildings. Owing, however, to the repeated occurrence of earthquakes, and the irruption of lava from Ætna, its ancient monuments have been mostly destroyed; but the remains of its amphitheatre, the circumference of which exceeds even that of the colosseum, as well as of its theatre, odeum, hippodrome, temples, aqueducts, baths, &c., attest its former extent and magnificence.

CATANZARO, a town of Southern Italy, prov.

Cosenza, in a healthy and agreeable situation, on a mountain near the Gulf of Squillace, 29 m. ESE. Cosenza. Pop. 31,464 in 1862. The town suffered very severely from the dreadful earthquake of 1783, which overthrew several of its principal buildings; it still, however, has a cathedral, several churches and convents, a seminary, a royal academy of science, a lyceum, a foundling hospital, a monte di pietà, and two hospitals; and is defended by a castle. It is the seat of a bishopric, of one of the four great civil courts of the kingdom, of a criminal court, and of an ordinary civil tribunal. There are considerable manufactures of silk, velvet, and cloth, and a good deal of trade is carried on in silk, corn, cattle, wine, and oil. The inhabitants are affable and industrious, and the women are reckoned the handsomest in the three Calabrias.

CATEAU-CAMBRESIS, a town of France, dép. du Nord, cap. cant., on the Selle, 16 m. ESE. Cambray, on the Northern railway. Pop. 9,212 in 1861. The town was formerly fortified; and has manufactures of starch, soap, and tobacco, with tanneries, and some trade in lace. It is celebrated in diplomatic history for the treaty concluded in it, in 1559, between France and Spain.

CATHERINA (SANTA), or NOSSA SEN-HORA DO DESTERRO, a marit. city of Brazil, cap. prov. St. Catherine, on the W. side of the island of same name, on the narrow strait separating it from the mainland, 520 m. SW. Rio Janeiro; lat. 27° 36' S., long. 48° 40' W. Pop. probably from 5,000 to 6,000. From the landing place in the harbour, which is at the bottom of a verdant slope of about 500 yards, the town has a most beautiful appearance, and the perspective is nobly crowned by its fine cathedral. The green is interspersed with orange trees, and forms an agreeable parade. The houses are well built, have two or three stories with boarded floors, and are provided with neat gardens well stocked with excellent vegetables and flowers. Besides the church of Nossa Senhora do Desterro, which gives name to the place, there were some years ago two chapels, a convent, an hospicio, and good barracks. Notwithstanding its excellent port and convenient situation, the trade of the town is not very considerable; but it is frequently visited by ships passing to and from the Pacific, and by those in the S. Sea whale-fishery. Sperm-whales used to be frequent on this coast, and even in the bay of St. Catherine, but they are now comparatively rare. There are some manufactures of coarse cotton and linen stuffs, and earthenware.

The island of St. Catherine may be entirely circumnavigated, and many good anchorages are found between its W. coast and the continent; but the N. part of the channel is the only one suitable for large vessels. Here they anchor in 5 fathoms on a mud bottom which holds well, and are protected from all winds, except from the NE., which are rarely dangerous. Opposite to the town the channel narrows, and the depth of water decreases to 2 fathoms. The roadstead is defended by two forts. This is one of the very best places at which to refit; excellent water may be had in any quantity for nothing, and provisions of all kinds are cheap and abundant.

The island of St. Catherine is about 25 m. in length, N. to S., and from 4 to 8 m. in width. Its shores rise abruptly from the sea to such a height, that in fair weather it is visible 45 m. off. Its most N. extremity, Point Rapa, is in lat. 27° 22' 31" S., long. 48° 37' 7" W. The surface of the island is singularly varied, presenting granite mountains, fertile plains, swamps fit for the growth of rice, lakes stocked with fish, and several small streams. Mandioc and flax are the chief articles of culture; but wheat, maize, pulse, cotton, rice, sugar, cotton, indigo, and an abundance of fruit are also grown. The climate is rather humid, but temperate and salubrious.

CATMANDOO, or KHATMANDU, an inland city of N. Hindostan, cap. of the Nepaul dom., built in a mountainous region, 154 m. NNW. Patna, and 4,784 ft. above the level of the plains of Bengal. Estimated pop. 20,000. It extends for about 1 m. along the bank of a river; and contains many wooden and brick temples, with the palace of the Nepaul rajah. The houses are mostly mean brick or tile buildings, often three or four stories high; streets narrow and dirty.

CATHINE, a manufacturing town of Scotland, co. Ayr, parish Kern, on the S. bank of the Ayr, 3⅝ m. S. Glasgow. Pop. 2,484 in 1861. Cotton-works were erected here by a company as early as 1786, and a bleaching-work in 1821. Both works are carried on by means of water-power, but in case of a deficient supply of water, steam-engines make good the deficiency. The bleaching establishment, in addition to what is manufactured at Catrine, bleaches all the cotton produced at the other mills belonging to the same company, the quantity varying from 15,000 to 25,000 yards per day. Every part of the process is carried on within doors, and without interruption, at all seasons of the year. There are seven schools, six of which are supported by the school fees, and one maintained by a fixed salary paid by the company; four libraries, one of which is attached to a Sunday-school; and several places of worship connected with the established church, or belonging to Presbyterian dissenters.

CATTARO, a town of the Austrian states, cap. circ. of same name, at the SE. extremity of the Gulf or Borsa di Cattaro, 210 m. SE. Zara; lat. 42° 25′ 16″ N., long. 18° 46′ 16″ E. Pop. 3,970 in 1857. The town is walled, and is farther defended by a fort built on an adjoining eminence. Streets narrow, dark, and gloomy. Notwithstanding its small size, it has a cathedral, a collegiate church, seventeen other Roman Catholic churches and chapels, a Greek church, six convents, and a hospital. It is the seat of the administration of the circle and of a bishop, and has a government high-school. The harbour is one of the best in the Adriatic. At its mouth there are two rocks dividing the entrance into three separate channels, two of which admit the largest ships. Internally the gulf is spacious and secure, though little frequented by shipping. The trade of Cattaro is chiefly with the Turkish district of Montenegro. The vicinity is very picturesque; but from being surrounded on three sides by mountains, Cattaro has this disadvantage, that the sun rises an hour later and is lost an hour earlier than in other places under the same latitude. The district of Cattaro was the seat of a Roman colony; but the town itself only dates from the 6th century. It has suffered much from earthquakes, especially in 1563 and 1667. It was long the cap. of a small republic, which, falling into debt, placed itself under the government of Venice on the single condition of having its debts paid. Previously to the treaty of Tilsit this town was for some time in the occupation of the Russians.

CATTEGAT, or KATTEGAT, a portion of the N. Sea, or of the Baltic, between Jutland and Sweden. (See BALTIC.)

CAUDUL, or CABUL (an. Aria and Arachosia), an extensive region of Central Asia, formerly the centre of a powerful kingdom reaching from Meshed to Cashmere, and from the Oxus to the ocean, but now comprising only the country be-

tween lat. 29° and 37° N., and long. 60° 30′ and 72° E.; and divided into four chiefships, independent of each other, viz. those of its principal cities, Caubul, Peshawur, Candahar, and Herat. Caubul, in its extended sense, includes the greater portion of Afghanistan, Seistan (an. Drangiana), and Beweestan, with parts of Khorassan, Cofristian (the Kobistan), and Lahore; length and breadth each about 600 m. The pop. was estimated by Mr. Elphinstone, in 1808, at about 14,000,000, but this estimate is believed to have been too high when it was framed; and since that period civil wars and foreign conquests have deprived Caubul of the prov. of Iribochistan, Sinde, Mooltan, Damaun, Cashmere, Bakh, &c., and have diminished the pop. to little more than 4,000,000. At present, besides the cities already named, the chief towns are, Ghizmee, Doushak, and Furrah.

The N. and E. portion of Caubul is a lofty table-land, its mountains belonging to the Hindoo Kush (or Indian Caucasus), and two of its offsets, viz. the Sollmaun and Paropamisan ranges. The Koosh mountain, about long. 69° E., gives its name to the range which extends from it both W. and E., and beyond the Indus is continuous with the Himalaya, running generally SW. to NE., and in the Kobistan forming the N. boundary of Caubul. Between long. 70° and 72° it makes a remarkable curve to the S., opposite to which the Hoho-Tugh (or cloudy mountains) unites with or approaches it, from Buddukhshan on the N. The highest, as well as the most S. point of this curve, is apparently a mountain, called Coond, or Koaner, near long. 71°, where the Afghauns believe the ark to have rested after the deluge; a tradition current, however, regarding the Tukhte Sollmaun also. The Koosh is covered with perpetual snow; its peaks are visible from Bactria, India, and even Tartary, and one of them, measured by Sir A. Burnes, was found to be 20,493 ft. high. Mr. Elphinstone observed at Peshawur three inferior mountain ranges, progressively decreasing in height beneath the former; the description of which will serve, he says, to give an idea of the rest of the Koosh chain; the lowest range was destitute of snow, and its sides were clothed with forests of pine, oak, and wild olive, European fruits and flowers, fern, and elegant shrubs. The tops of the second range are covered with snow, and the third are so to half their height. On the high central range Mr. Elphinstone observed that 'no diminution in the snow could be perceived in any part in the month of June, when the thermometer in the plain of Peshawur was at 112° Fahr.' The Koh-i-Baba range, between Caubul and Baumian, is the continuation W. of the Koosh; but its peaks are not so lofty, probably not more than 18,000 ft. (Burnes, iii. 203), although 'covered with eternal snow for a considerable distance beneath their summits.' The passes of Hajeeguk and Kaloo on this range are respectively 12,400 and 13,000 ft. above the sea; the other passes are none more than 9,000 ft. in height, and all, without exception, are free from snow by the end of June. In the defiles the road often winds at the base of a mural precipice, rising to 2,000 or 3,000 ft. perpendicularly, and in one part, called Dura-i-zendan, or the 'Valley of the Dungeon,' the height is such as to exclude the sun at noon-day; at the height of 10,000 ft., however, the ground in some parts is ploughed when the snow disappears, the grain sown in May being reaped in October. The ranges N. of the Koh-i-Baba are much inferior in height, and often free from snow; but rise from the plains of Balkh in a bold and precipitous line, 2,500 ft. high. The valley of the Caubul river separates the Koosh from the Teera mountains, which run in a parallel direction, decreasing in size to the E.; but in their higher parts are covered with perpetual snow, and are certainly as much as 15,000 ft. high. (Burnes, ii. 105.) The Solimaun range commences with the Sufaed-Koh, S. of the Caubul valley; across which it may be considered as connecting itself with the Koosh, by means of cross ranges, causing many cascades and acclivities in the bed of the river. This range stretches from nearly 34° to 29° N. lat., where it becomes connected with the high table-land of Kelat (Beloochistan). It is not so high as the Koosh; its principal points are the Sufaed-Koh, or 'White Mountain,' and the Takhte Solimaun, or 'Throne of Solomon,' the last near lat., 31° 30' N.; the former is always covered with snow, and the latter so for three months in the year. Between these two points this range decreases considerably in height, especially where it is intersected by the Gomal river. The Solimaun chain has several parallel ridges, and gives off many lateral and other ranges, especially a remarkable one to the SW., including the Khojeh Amraun hills; a broad range, though of no great altitude, which appears to join the table-land of Kelat. On the E. a high and broad range, abounding in salt, passes off near the Teera mountains, across the Indus, into the Punjab, with a NE. direction. The Paropamisan mountains (for which as a whole there is no modern name) occupy a large space of country, extending 350 m. E. to W., and 200 m. N. to S.; W. of the Koosh, and between the Helmund river and Tourkistan. They are a mass of mountains, difficult of access, and little frequented; their E. portion is cold, rugged, and barren, although some where covered with perpetual snow: in the W. they contain rather wider valleys, and are somewhat better cultivated. Their greatest declivity is on the N. side, from which they send off several ranges towards Balkh; the slope of the whole tract is towards the W.

The Koosh, collectively called the Caubul Kohistan, or 'Land of Mountains,' contains, in its higher ranges, a number of narrow valleys; in its lower portions the valleys are of same size; Mr. Elphinstone calling them 'plains.' Many open laterally into the valley of Caubul, which occupies the space between the Indian Caucasus and the Solimaun and Teera mountains, and which in some places is 25 m. wide. The narrow plain, or valley of the Swaut river, is well watered; yields two harvests of most sorts of grain; and abounds in orchards, mulberry-gardens, and plane-trees; others are by no means so wide or productive, and are often bounded by a number of narrow glens. There are many fertile and well-watered valleys on both sides the Solimaun range.

Besides those of the desert, which extend over the S. and W. parts of Caubul, there are many extensive and productive plains: that of Peshawur, about 35 m. in diam., is well watered; its streams fringed with willows and tamarisks; and has numerous gardens and orchards scattered over it; the latter contain a profusion of apple, plum, peach, pear, quince, and pomegranate trees. The greater part of this plain is highly cultivated and irrigated by canals, and the uncultivated parts covered with a thick elastic sod, scarcely equalled except in England; its villages are generally large, very clean and neat, and surrounded with groves of slate, peepul, and tamarisk. The valley of Caubul encloses some small plains, of which that of Jellalabad is the principal. Most of the cities and large towns are in fertile plains; one of,

great luxuriance surrounds Herat; and the site of Furrah, and other places in the W., as well as the banks of the Helmund, seem 'rich oases in the midst of a waste.' The desert in Seistan, Gurmseer, and Shorawuk, has an ill-defined boundary, and often encroaches on the habitable country.

The Indus forms, for a short distance, the E. boundary, and excepting it, there is no river which is not fordable throughout its course for the greater part of the year. The principal of the minor rivers are the Caubul, Helmund, Furrah-Rood and Lora. The only lake of any importance is that of Seistan, or Zurrah (Aria Palus), which receives the waters of the Helmund (Etymander).

The climate varies with the elevation; the temperature is much higher at Peshawur and Candahar than at Caubul and Ghizner; but, generally speaking, the average heat of the year does not equal that of India, nor the cold that of England. At Caubul the snow lies on the ground for five months, and Burnes found the thermometer stand no higher than 64° Fahr. during the hottest period of the day in the month of May. The prevailing winds throughout Caubul are westerly. The rains brought by the SW. monsoons are much diminished in power by the time they reach the NE. part of the country, where the rainy season is limited to a month of cloudy weather, and occasional showers. At Candahar the influence of this monsoon is not felt in the least degree; at Caubul there is no regular wet season; but showers are frequent at all times of the year, as in England. At Peshawur, by the first week in March, peach and plum trees begin to blossom, and by the end of that month are in full foliage: from July to Nept., the weather is cloudy; the winter lasts from the latter month till Feb. Caubul generally is healthy; the most prevalent diseases are fevers, small-pox, and ophthalmia. Sir A. Burnes found the inhabitants of the Koosh, at 10,000 ft. above the sea, quite free from goitre, so common in the lower ranges of the Himalaya.

Geology and Minerals.—A core of granite, and resting on it a deep bed of slate, are the prominent geological features of the Koosh: the slate formation includes gneiss, mica, and clay-slate, chlorite, carbonate of lime, and quartz; gneiss generally occupying the lower portion. The Solimaun chain is composed of a hard black stone; its accompanying ranges on the E. of an equally hard red stone, and a friable grey sandstone; the hills between Herat and Joonbah consist partly of a mixed reddish and black rock, streaked with ore, and partly of greywacke slate. Iron, lead, copper, antimony, tin, and zinc are found in various parts of the mountain region, and 10 or 12 lead mines near Baurnian, and elsewhere, are worked; gold is washed down by the rivers that come from the Hindoo Koosh; there are extensive deposits of sulphur in Seistan, at Cobat, &c.; coal, naphtha, and petroleum are met with in the latter district; salt in the E. part of the country, both in springs and beds; and saltpetre is procured from the soil in many places.

Many of the forest trees, and most of the finer fruits of Europe grow wild. The timber in the mountain region consists chiefly of pine, oak, cedar, gigantic cypress, and wild olive: the Hindoo Koosh is destitute of wood, and in many places of verdure. Some of the hills produce the birch, holly, hazel, and mastic, the wild vine, barberry, blackberry, and many other bushes bearing edible berries; the valleys abound with extensive orchards, particularly of apricot-trees: the other trees most common on the plains are the mulberry, tamarisk, plane, willow, and poplar. The assafœtida plant grows luxuriantly at an elevation of

7,000 ft.; hemlock, fennel, peppermint, nettles, and other such plants common in Europe, are equally common in the higher parts of Caubul, with a profusion of roses, poppies, hyacinths, and jessamines. The vegetation of the lowlands approximates more to that of India; and, on descending into them, the contrast with the country just passed is so striking that it is thus adverted to by the Emperor Baber in his commentaries:—' I saw another world. The grass, the birds, the trees, the animals, and the tribes of men: all was new! I was astonished.'

Lions of a small species are said to have been found in the hilly country about Caubul; tigers are met with in most of the wooded tracts; wolves, hyenas, jackals, wild dogs, the elk, and various other kinds of deer, wild sheep, and goats, on the E. hills; the wild ass in the desert: foxes, hares, porcupines, ichneumons, and ferrets are also found. Birds are very numerous, and include several kinds of eagles, hawks, and other birds of prey; herons, cranes, wild fowl, and game, in plenty; doves, magpies, thrushes, and nightingales: parrots and birds of rich plumage are found only in the E. Turtles and tortoises are numerous; there are no crocodiles in the rivers; the snakes are mostly harmless. Large scorpions infest Peshawar; mosquitoes, except in Kabistan, are less troublesome than in India; large flights of locusts are rare, but occasionally cause a famine in Khorassan.

Races of Men.—The Afghans, who call themselves Pooshtoon, bear a considerable resemblance to the Jews; and, though they consider it a reproach to be called Jews, they claim descent from a son of Saul. Sir W. Jones and Sir A. Burnes contend for their Jewish origin; Mr. Elphinstone discredits it. They are divided into a number of tribes, often at war with each other, especially those in the E. of Caubul, and each under the authority of a chief, who, however, is usually assisted by a council (jeerga), consisting of the heads of the tribe. Mr. Elphinstone conceives their political condition to bear a strong analogy to that of the Scottish clans, in former times; but the genius of the Afghans is more decidedly republican; they resist every encroachment of their rulers, and have a boldness and elevation of character unknown to most other Asiatic nations. They are Mohammedans of the Soonite sect, but use the Persian alphabet; their literature bears a similarity to that of the Persians; but it has a superior dignity and refinement, and in many respects is not unlike that of Europe. The Afghans are hospitable, and tolerant in religion; but extremely superstitious and addicted to astrology, divination and alchemy. They are plunderers by profession; in the W. they live in tents, in the E. in fixed habitations; only a few of them reside in the large towns. Their chief amusements are the chase, feasting, songs and recitations: they have slaves, but traffic very little in them. (See AFFGHANISTAN.)

The Eimauks and Hazzarehs, two races of Tartar origin, although using dialects of the Persian tongue, inhabit the Paropamisan mountains. The Eimauks, who are divided into four principal tribes, subdivided into numerous clans, each governed by its chief, occupy the lower parts of the country, between Caubul city and Herat; Mr. Elphinstone estimated their number at about 450,000. In war they are ferocious and cruel; they retain many Mogul customs, mixed with others of Persian origin; they live almost entirely in camps, and use the same kind of tent as the Afghans, with the addition of brushwood and bread of an oily kind of nut. They cultivate wheat, barley, and millet;

keep many sheep, and rear a small but active breed of horses: they are Mohammedans of the Soonite sect. The Hazzarehs have been estimated at about 350,000; they inhabit a higher region than the Eimauks, a cold and sterile country, where little corn can be grown: their sheep, oxen, horses, and the produce of the chase, furnish them with their principal articles of food; sugar and salt are the foreign commodities most in demand amongst them. They live in villages of thatched houses, and are divided into different clans, constantly at war with each other, and each governed by an absolute chief. The Hazzarehs have strong Tartar features, and many similarities in costume and dress with the Uzbeks; the women, who are frequently good-looking, possess an unexampled licence and ascendancy over their husbands. These people are passionate, fickle, and capricious; but convertible, hospitable, and very fond of music, recitation, visiting, and other sociable kinds of amusement. Many of them are performers on a guitar, poets, and improvisatori. They belong to the sect of Ali. The Taujiks, or Taujika (see Ilmtiana), are probably descendants of the original Persian inhabitants of the country, and of the Arabs who conquered it in the first century after the Hegira. They live mostly in and round the larger towns, and everywhere reside in fixed habitations, having settled employments. They are traders, farmers, mikti, sober, peaceable, and industrious; and assimilate much more with the Afghans than their brethren of Hindustan do with the Uzbeks. The Taujiks are most numerous towards the W. of Caubul; as the Hindkees (Hindoos, Jats, Sindians) are towards the E. The Hindoos are, however, to be met with all over the country, chiefly as money-changers and tradesmen; they are mostly of the Kshatriya or military caste. The Kuzzilbashes, or Persian Turks, inhabit the towns; the Hindkees are generally almost confined to the N.: there are about 2,000 Arab families, besides Armenians, Abyssinians, European Turks, Jews, and Caulirs amongst the population.

Agriculture.—There are five classes of cultivators—1st, proprietors, who cultivate their own land; 2nd, tenants, who pay a fixed rent in money, or a proportion of the produce; 3rd, buzgurs, or croppers; 4th, hired labourers; 5th, riflkins, who cultivate their lords' lands without wages. The lands are more equally divided in Caubul than in most countries, and the first class, or that of small proprietors, is very large, as by the Mohammedan law every man's estate is at his death divided equally amongst his sons. The class of tenants is not numerous. Leases are generally from 1 to 5 years, and the rent varies from 1-10th part to half the produce; the landlord generally providing the seed, cattle, and farm implements. Labourers are principally employed by the buzgurs; they are fed and clothed by their employers, and paid for 9 months' work about 80 rupees. The riflkins are many of them of foreign descent, and always attached to the service of some master; they are subject to taxation, and even death-punishment from their lord, but have the privilege of removing from the service of one master to another: they are most numerous amongst the Eusofzyes and other Afghan tribes in the N.E. There are two harvests in the year; one crop, consisting of rice, millet, jowarri, and maize, is sown in the spring, and reaped in autumn; the other, which consists of wheat, barley, and legumes, is sown at the end of autumn and reaped in summer. Rice is grown in most parts of the country, but wheat is the common food of the people: barley is usually given to horses. The vegetables and pot-herbs of Europe and India are cultivated largely, especially turnips and carrots; melons and cucumbers are abundantly grown in the neighbourhood of the towns; and ginger, turmeric, and the sugar-cane in the E.; but the latter plant is confined to rich plains, and most of the sugar, as well as the cotton, used in Caubul is brought from India. The palma Christi, sesamum, and mustard, are grown for the sake of their oil; tobacco is cultivated in most parts; madder abounds in the W.; and Caubul furnishes to India its chief supply of that article; lucerne and other artificial grasses are sown for the cattle. Much of the land fit for culture has been brought into that state by irrigation undertaken by individuals singly, or associated for the purpose. Cultivable land in Caubul is generally valued at from nine to twelve years' purchase. Irrigation is effected by means of canals and subterranean conduits, beneath the slopes of hills, terraced courses, which are common in Persia. The plough is heavier and makes deeper furrows than that of India, but still only employs one pair of oxen. All grain is sown broadcast; and drill husbandry is unknown. The place of a harrow is supplied by a plank dragged over the field, on which a man stands. The sickle is the only instrument used for reaping. The flail is unknown; and the corn is trodden out by oxen, or forced out by a frame of wood filled with branches, on which a man sits, and is dragged over the straw by cattle. It is winnowed by being thrown against the wind, and, when cleaned, is kept in hampers plastered with mud, unbaked earthen pots, and coarse hair-cloth bags.

For grinding the corn, windmills are used in the W., but these are very different from ours, for the sails are inside, and there is an opening in the erection to admit the wind. Water-mills are not unknown, but handmills are most generally used. The manure employed is composed of dung, straw, and ashes, but the dung of camels is carefully avoided. Horses are employed in ploughing only by the Eimauks; in Seistan camels perform this work. There are no carts. The horses of Herat are very fine, and somewhat similar to the Arabian breed; and there is a strong and useful breed of ponies, especially about Bameean. Mules preferable to those of India; but asses, camels, and dromedaries mostly are used for carriage. The ox resembles that of India; sheep chiefly of the broad-tailed kind; and the goats, which are numerous, have often long and tortuous horns. The greyhounds and pointers are excellent. A great number of horses are annually sold in the N. and W. of India, under the name of Caubul and Candahar breeds; but no horses are bred in large numbers in Caubul, nor are those of Candahar exported in any quantity.

Trade.—Exports.— The principal foreign trade is with India, Persia, and Toorkistan; the exports to the first-named country are principally horses and ponies; furs, shawls, chintz, madder, assafetida, tobacco and fruits; those to Toorkistan are shawls, turbans, chintz, white cloth, indigo, and other Indian produce; to Persia the same articles, with the carpets of Herat. The latter-named article, with woollens, furs, madder, cheese, and sugar piece-goods, are sent from the W. to the E. provs.; and Lhawrpoor and Mooltan cloths, silk, cotton, and indigo, are sent back in return. Iron, salt, alum, sulphur, and the other natural produce, are also exported.

Imports.— From India are coarse cotton cloths, worn by the mass of the people; muslins, silks, and brocade; indigo, in great quantities; ivory, chalk, bamboos, tea, tin, sandal-wood, sugar, and spices; from Toorkistan, horses, gold, and silver; rock-mineral, broad cloth, and tinsel; cast-iron pots, cutlery, hardware, and other European articles,

from Russia, via Bokhara. Silks, cottons, embroidery, and Indian chintz come from Persia; slaves from Arabia and Abyssinia; silks, satins, tea, porcelain, dyes, and the precious metals, from the Chinese dominions; and dates and cocoa-nuts from Beloochistan. The merchants are chiefly Tadjiks, Persians, or Afghans, and Hindkers in the E.; but no Afghan ever keeps a shop, or exercises any handicraft trade. Caubul is the great mart for the trade with Toorkistan; Peshawur for that with the Punjab; and Candahar and Herat for that with Persia. The demand for British manufactures has increased so much latterly, that Russia, which before 1816 supplied a great many articles, now only sends nankeen and broad chintz, of a description not manufactured in Britain, into the market. The greater part of the trade between India, Caubul, and Bokhara is conducted by the Lohanees, a pastoral tribe of Afghans, often of considerable wealth. About 1,000 camel-loads of Indian goods are annually consumed in Caubul. The Caubul merchants have latterly begun to frequent the annual fairs on the borders of the Russian dominions, and most of the Russian trade with Bokhara has fallen into their hands. Sir A. Burnes remarks, that were such fairs to be established on our NW. frontier, and encouragement given to the Lohanee merchants, who are every way deserving of it, a large export of British manufactures would take place.

Roads.— In an inland country, without navigable rivers, and not suited to wheeled carriages, traffic must be carried on by means of beasts of burden; camels are the principal of these in Caubul, and constitute great part of the wealth of many individuals, as they are let out to merchants by those who cannot afford to trade themselves. The merchants commonly travel in bodies, called caravans, and place themselves generally under the conduct of some chief whom they elect as a reniffe-bashee, or an officer with absolute command over all the arrangements of the journey. There are but two great routes through the country; one from Balkh across the mountains at Bameean, through Caubul to Peshawur, and thence into the Punjab; and the other from Herat to Candahar: on this line there are few obstacles to oppose a European army, and the latter city could furnish abundant supplies. From Candahar there are two routes; the former through Ghizner to Caubul, not difficult for nine months in the year, but next to impassable in the winter, from the snow and intense cold; the second through the valley of Pisheen and Quetta to Shikarpore in Sinde; a country furnishing supplies of food, but deficient in wood and water. There is another road across the Soliman range from Candahar to Dera Ghazee Khan, in Damaun; but it is said to be hardly practicable for a European army, and is not travelled by merchants. The Khyber Pass from Peshawur to Caubul has, in consequence of enormous exactions on merchandise at the former place, been deserted by traders, and is unsafe. Camels, horses, mules, &c., are cheap enough throughout Caubul; but fuel is very scarce and dear, and water is not generally to be had in abundance: two great drawbacks in travelling. (Conolly, ii. 373.)

The Public Revenue, in settled times, amounts, according to Mr. Elphinstone, to nearly 3,000,000l.; but, before the revolution which dethroned Shah Shoojah, 1-3rd part was remitted to different tributary princes, who consented to hold their dominions as grants from the khan of Caubul; of the rest, half was assigned for military services to the chiefs, and the remainder for the maintenance of moollahs and dervises. The chief sources of the revenue under the present khan are, the land, the tribute of certain tribes, the town duties and customs, certain fines and forfeitures, and the profits of the mint. The land revenue is collected by the head man of each village, and paid either through the head of the tribe, or the *hakim* or governor of the province: great speculation is often practised by the *hakim*, as the current expenses are paid before the balance is sent to the treasury.

Government.—Under the monarchy, the crown was hereditary in the family of the Suddozyes, who belonged to the tribe of the Douranees, said by Mr. Elphinstone to be the greatest, bravest, and most civilised of all the Afghan tribes. The right of succession was not always vested in the eldest son; but the future heir was determined either by the reigning sovereign or a council of the great officers of state.

Justice is administered in the cities by the *cauzy* (or *cazi*), assisted by muftis and other officers; but where the khan happens to reside, criminal complaints are made to him. The *cauzies* have deputies over the whole country. The police of towns is managed under one head, in three departments, viz. watchmen, inspectors of public morals, and superintendents of weights and measures. In the country the people to whom the land belongs are answerable for the police. In cases of robbery and theft, if the chief of the village or of the division of a tribe in whose lands a crime was committed, fail to produce the thief, he pays the value of the property stolen, and levies it on the people under him. The police is very bad, and does not interfere in murders for retaliation, except in towns and their vicinity.

Religious Establishment.—Moollahs or priests always fill the duties of inspectors of public morals: under the police established in the country, they have grants of land from the head-man of the tribe, and a tax similar to tithes, but by no means equivalent to them in amount: in the towns they are maintained by fees on marriages and burials, and the gifts of their congregations. A superintendent priest and a registrar are established in each city; several are connected with the royal household; and at the visit of Mr. Elphinstone, there was a professor and a body of students in theology at the king's palace, each of whom received a daily allowance for his support.

Armed force is chiefly cavalry, 3-4ths of whom are Kuzzilbashes. They are collected in bodies, varying in number from 6 to 300, under their several chiefs, and tolerably mounted. Their dress is a *kummur* or turban, one end of which is tied under the throat in the field; a *kummurbund* or garment, which serves for a coverlid at night; a *koorta* or shirt, *ollenbeg* (low trowsers), and boots to the knees, and over all a *ruffian* or cloak: their arms are a sabre, a gun, with a good flint lock, and long bayonet; a powder and ball pouch round the waist, and always a shield: their saddles are high both behind and before, and they all carry a rope with a twisted chain attached, by which they can secure their horses at any place or time. There are about 12,000 infantry, all Afghans, armed with a sword, shield, and match-lock, which carries twice as far as a musket; but being too heavy to be brought up to the shoulder, is furnished with a prong or rest, which is fixed in the ground. These troops are but skirmishers, and fight generally in ambush: there are besides two regiments raised seven years ago in Bombay, one of 800, and the other of 300 men, dressed in European uniform, but ill paid and disciplined: and wretched artillery of about fifty field-pieces of different sizes, only half of which are used.

History.—Caubul was amongst the countries invaded by Alexander, and several spots may be almost confidently identified with those mentioned by the historians of that conqueror. A remarkable rock near Bajour is probably the celebrated Aornus; Jellalabad is supposed to be in the neighbourhood of the spot where Alexander revelled in imitation of Bacchus: many topes or artificial mounds are situated along the skirts of the mountain ridges, and on the banks of the Caubul river, some of which having been opened, have been found to contain Grecian coins, gems, boxes, cups, lamps, &c. A. D. 997 Caubul was conquered by the Tartars under Sebuctaghi, whose successors extended their empire over great part of India, Khorasan, Balkh, and Hodukshan. In 1737 Nadir Shah possessed himself of the country; and in 1747 Ahmed Shah Abdalli, the founder of the Dooranee dynasty, was crowned at Candahar. His successor Timour Shah died in 1793 without naming an heir, and, in consequence of the uncertainty of the succession, a protracted civil war broke out among his three sons. One of them, Shah Shoojah-ul-Moolk, having succeeded in placing himself on the throne, was defeated and deposed, in 1809, by Futteh Khan, chief of the Baurikzye family, who espoused the cause of Mahmoud, brother of Shah Shoojah. But notwithstanding his great services, Futteh Khan was treacherously murdered, in 1818, by Mahmoud. On this event taking place, the brothers of Futteh Khan, who had been made governors of provinces, revolted; and one of them, Dost Mohamed Khan, established himself on the throne of Caubul. Runjeet Singh seized about the same time on Cashmere and Peshawur; and Herat and its dependencies were the only part of the old monarchy that continued in the possession of the Dooranee dynasty. Dost Mohamed having assisted the Persians in their attempts on Herat; and having, it is alleged, on various occasions evinced his hostility to British interests, the Indian government determined upon dethroning him, and on placing Shah Shoojah on the vacant. For this purpose a powerful army crossed the Indus, and advanced as far as Ghizney without meeting any opposition, other than that arising from the nature of the country, and the deficiency of supplies. The latter having been taken by storm, after a short but sharp contest, on the 23rd June 1839, a panic seized the troops of Dost Mohamed, who immediately disbanded themselves; and Shah Shoojah was shortly after enthroned at Caubul, whence he had been driven thirty years before. But he was unable to maintain himself on his slippery elevation even with the assistance of the strong British force left in Caubul. We have elsewhere noticed the singularly disastrous retreat of that force from that city early in 1842; with the subsequent invasion of the country by the British, and their final withdrawal from it. (See AFFGHANISTAN in this Dict.; Elphinstone's *Caubul, passim*; Conolly's *Journey to India*; Burnes' *Trav. into Bokhara*, 1836.)

CAUBUL, the ancient cap. of the above country, under the Dooranee dynasty, situated in the plain, and on both banks of the river of same name, 6,600 ft. above the level of the sea; 56 m. NNE. Ghizney, 140 m. WNW. Peshawur; lat. 34° 27' N., long. 69° 15' E. Pop. about 60,000. The city is completely built: on three sides it is enclosed by a semicircle of low hills, along the top of which runs a weak wall, with an opening surrounded by a rampart towards the E., by which the principal road enters through a gate, after passing a bridge over the river. The Balla Hissaur, or 'palace of the kings,' which stands on the part of the hill N. of this eminence, is a kind of citadel, and contains several halls, distinguished with the royal ornament of a gilded cupola; there is an upper citadel, formerly used as a state prison for princes of the blood; but as fortresses both are contemptible. In the centre of the city is an open square, whence issue four bazaars, with shops about two stories high; the houses are constructed of sun-dried bricks and wood, but few of them have any pretensions to elegance. Cabul is, however, a bustling place; the chief mart of trade is the country; and its bazaars are superior to most in the E.; the great bazaar is a handsome roofed arcade 600 ft. long by 30 ft. broad. Each different trade has its separate quarter. Provisions in summer are moderate, but both wood and grain are dear in winter. Its climate, and the scenery around it, are both very fine; the banks of its river are beautifully adorned with poplar, willow, and mulberry; but the most pleasing spot in its vicinity is the tomb of the Emperor Baber, who made Cabul his capital. His grave is marked by two erect slabs of white marble, situated in a small garden at the summit of a hill overlooking the city; outside Caubul also stands the tomb of Timour Shah, an unfinished octagonal brick building 50 ft. high.

In the 7th century of our era, the Arabian writers mention Cabul as the residence of a Hindoo prince; it was, as already stated, the capital of the empire of Baber, and taken by Nadir Shah in 1739. At his death it was taken by Ahmed Shah Abdalli, and remained the capital of Affghanistan till the destruction of the monarchy.

The chiefship of Cabul extends N. to the Hindoo Koosh and Bamean; E. to Nowsla half way to Peshawer; N. to Ghiznee, which city it includes; and W. to the country of the Hazarehs. Much of the country is mountainous, and of great natural strength but small resources; there is plenty of fruit, and forage for cattle, but grain grows scantily. The revenues of Cabul amount to 18 lacs rupees a year; these derived from the city customs are 2 lacs annually, which amount they have reached in consequence of the encouragement given to trade by the lately deposed khan.

CAUCASUS, a great mountain-range, extending in a NW. and SE. direction, between the Black and Caspian Seas. Its extreme points are those of the main ridge or back bone of the system, which, commencing at Anape, on the Black Sea, in lat. 44° 50' N., runs first SE. as far as the parallel of 43° 55', and meridian of 40° 45'; then almost due E. to the king, of 46°, and finally, again NE., to Baku, on the Caspian Sea, in lat. 40° 20', where it terminates. The direct distance between Anape and Baku is 650 m., but, following its windings, the ridge of the Caucasus measures 900 m. The extent of the mountains towards the N. is very well marked by the courses of the rivers Kuban and Terek; the one flowing W., along the back to the Black Sea, the other E. to the Caspian. The natural N. limit is the Araxes; so that the breadth of this range, in its widest part, is about 5°, or nearly 250 m.; and in its narrowest, along the shores of the Caspian, not much short of 250 m. The area enclosed by these two seas and three rivers, taken as the boundaries of the Caucasian system, is not less than 100,000 sq. m., but it must be remarked, that within these limits there is, though so much, some level land; and that the least elevation is found, not in the bed of the Araxes, but in that of the Kur. (Klaproth's Trav. in Cauc. and Geor. p. 154, et seq.; Bigram's Winter

Journey through Russia, I. 27, et seq.; Col. Monteith's Geog. Journ., lll. 21-57.)

The highest peak of the Caucasus attains an altitude of 17,785 ft., which is more than 2,000 ft. higher than Mont Blanc. (Hostskoff-hal's Lettre sur le Caucase, p. 72.) This peak, or rather mountain-knot, is found nearly at the intersection of the 43rd parallel with the 42nd meridian. Among European geographers it has been called, improperly enough, Elbrous, Elbourz or Elbruz; a name which, in the spread of information concerning E. countries, is likely to be productive of no little confusion. It is already applied to a peak of the Caucasus, and a range on the N. of the Caspian Sea, and may, unless care be taken, be multiplied indefinitely, since it is not a proper name, but a common designation for any mountain which reaches the snow line. (Klaproth, p. 170.) From this point, as from a centre, the mountains descend in all directions, but much more rapidly towards the N. and W. than towards the E. and S. (Klaproth, p. 276.) The Mqsinvari peak, to which the Russians have improperly given the name of Kazbek, is said by Klaproth to attain an elevation of 4,419 metres, or of 14,500 ft. (Lettres sur la Caucase, p. 40.) Farther E., the ridge declines towards the Caspian; and where it approaches that sea, as in the Cape of Absharon, or at the town of Derbend, the eminences do not probably exceed 1,500 or 2,000 ft. The ridge W. from Elbours is very considerably lower, and presents fewer peaks; it appears to descend gradually, till at Anape, on the Black Sea, its elevation is only about 100 feet above the water; but this height rises perpendicularly, and the face of the rock is continued downwards for several hundred fathoms; such being the depth of the sea at this point. The N. ranges run nearly parallel to the main ridge, and extend about 100 m., when they suddenly and abruptly terminate in the low steppe of the Don and Wolga. This frontier, as it may be termed, of the Caucasus, is called the Black Mountains (Schernye Gory). The Bechtau, the highest point, is probably not less than 6,000 ft. in height, and there are several summits which appear to have a nearly equal elevation; extreme ruggedness is, however, a stronger characteristic of these hills than altitude. The Elbours (Usha Makhua) appeared to Pallas to rise in the horizon to more than double the height of the Bechtag, when viewed from a station very near the base of the latter. S. of the main chain, the country spreads into table-lands, terraces, and slopes, broken and intersected by transverse ranges and peaks, of which the highest is Ali Gou, in 40° N., 44° E., its elevation being about 15,000 ft. Almost 50 m. N. of this, but on the other side of the Araxes, is Mount Ararat; but it cannot with any propriety be reckoned as part of the Caucasus. Towards the SE., between the Kur and Araxes, the mountains spread into a level but considerably elevated plain, 24 m. in width, and terminated by a strong defile towards Erivan. The various plains, valleys, and defiles of this part of the mountains seem to vary between 4,000 and 6,000 ft. in height. On the N. the Caucasus is absolutely unconnected with any other mountain-range, unless the chain of the Crimea may be regarded as an exception; but on the S. it mingles with the high land of Azerbijan; on the SW. it combines with the mountains of Armenia, and through them with the Taurus; and on the SE. its offshoots appear to be continued by the mountains of Ghilan and Mazanderan, to the Elbours (Persian), Parvpamisus, Hindoo Koosh, and Himalaya. (Guldenstadt, Reise durch Russland,

i. 133, et seq., ii. 22, et seq.; Gmelin, Reise durch Russland, iii. 34, et seq.; Annales des Voy., xii. 3, et seq., 167, et seq.; Pallas, i. 339, et seq.; Klaproth, 158, et seq.; Monteith's Geog. Journ., iii. 51, et seq.)

The above results as to the extent and elevation of the Caucasus are derived from a very full comparison of the authorities cited, and of others not named. It is right, however, to state that they cannot be wholly depended upon. There is the most extraordinary discrepancy among authorities as to the extent of the mountain-systems, its elevation, &c.; but the above results seem to be those on which most reliance may be placed.

The ancients mention two principal passes of the Caucasus, the Caucasian Gates and the Albanian Gates; of which the former is at present the great, indeed almost the only frequented pass. It runs close by the base of the Kazbek mountain, in lat. 42½° N., long. 44½ E., and is, in fact, a deep ravine, through which the Terek seems to have cut its way in a channel, sometimes scarcely wide enough to allow of its passage. The commencement of this cleft on the N. is 4,000 ft., and it continues to rise, till, at the neck of the pass, it is full 8,000 ft. above the sea. The precipitous walls of porphyry and schist, 8,000 ft. in height, press upon its sides; and awful abysses open beneath it, sometimes, it is said, to the depth of 10,000 ft. Avalanches are frequent in the pass, carrying with them not only any unfortunate travellers who may be in the defile, but very often the road itself, and even when the snow does not descend in masses, its meltings in the spring and summer cause occasional floods, which carry every thing before them. The direct length of this defile may be about 120 m., from Mosdok to Gory; and some idea may be formed of its difficulty from the fact that Strabo (xi. 500) describes it as occupying four days in the passage. This must be understood also of summer travelling, since in winter the pass was wholly impracticable. The Russians have, however, made it passable even for carriages; and in January, 1834, it was crossed by the Persian embassy, but this winter transit employed six days. (Mignan, i. 46.) About midway stands the old castle of Dariel, in the narrowest and highest part of the gorge, where the statement of Pliny (vi. 12), that an iron gate would be sufficient to close the opening, seems to be any thing but an exaggeration. This castle is therefore, in all probability, the fortress which, according to the Roman naturalist, was called, though improperly, the Pyla Caspiæ. (Klaproth, p. 311; Monteith, G. J., iii. 39.) The Albanian Gates appear to answer to a pass between Georgia and Daghestan, in lat. 42° N., long. 47° E. (Ptolemy, v. 9; Lapie's Map; An. Voy., xii. 1.) This is, however, very little known; it is almost wholly in the possession of the native tribes, and probably is not passable except for hunters, and in the summer. Ptolemy's E. Sarmatian Gates (Geog. v. 9) appear to be the pass of Derbend, on the Caspian Sea; this is always available; its narrowness makes it a strong military position, and the swampy nature of the shore renders travelling along it often difficult. A similar remark applies still more forcibly to the pass between the W. termination of the Caucasus and the Black Sea. Along this road Prince Gortschakoff, in the last war, succeeded in marching an army, with incredible difficulty, from Anapa to Sukhumkale (about 150 m.); but here he found it equally impossible to advance or retreat, and was compelled to return by sea. (Monteith, G. J., iii. 37.) The impediments to the coast roads appear, however, to consist only in the number

and power of the mountain torrents, which, without bridge or boat, are quite impassable; and as a very wide bank of hard sand stretches along the whole shore, it may be practicable to throw bridges over all the streams; but very considerable height and strength will be necessary to secure them from the effects of sudden floods. These are all the passes over the main ridge, and the transverse ranges do not seem to be better provided; one only appears to exist between Imeritia and Georgia, and that has been rendered available only within modern times, and is still encumbered with great difficulties.

Geology.—The bases of the Caucasus on the N. seem to be covered with sand or a sandy marl, from which the first eminences rise in low but abrupt hills of sandstone, tufa, and iron-stone. These are rapidly succeeded by higher and more mountainous elevations of white calcareous limestone, many of which exhibit unquestionable evidence of decay, the rivers that flow through and round them depositing thick layers of a yellow and grey sandy consistence. Occasionally the limestone rises into great rocky peaks and ridges, between which marshy plains of sandy mud are not unfrequent, apparently formed by the debris of the mountains themselves. This limestone, of which the Herihing, the Metahaka, and nearly all the frontier line of the Caucasus is formed, is very ancient, and exhibits scarcely any petrifactions; behind this rises a ridge of slate, from the appearance of which the term *Black Mountain* is given to the range. The higher ranges, which rise to the snow line, consist of basalt, schistus, porphyry, granite, and other old formations, so that whether its actual material, or the absence of organic remains, be considered, it is probable that the Caucasus is one of the oldest mountain systems in the world. The S. slope exhibits the same succession of formations, as far as regards the three principal strata, but much less rapidly. Sandstone is far less abundant in the S. than in the N., but, on the other hand, calcareous spar, milk-quartz, and other fossils, are frequently met with, indicating a much greater degree of wealth in mineral ores. Lava and other volcanic matter is common enough among the formations; but, though mud volcanoes exist in various parts of the Caucasus, igneous eruptions are unknown; and neither Klaproth nor Pallas could come to any satisfactory conclusion as to their former existence. Monteith is of opinion that the volcanic rocks are rather to be ascribed to the sudden rise of a great extent of country, than to emissions from particular mountains. (Güldenstädt, i. 434–441, ii. 23–29; Pallas, i. 337, 347, 358, 345, &c.; Klaproth, pp. 896–890; Monteith, G. J., iii. 49.)

Hydrography.—The Caucasus, like the Alps, does not form the dividing line between rivers flowing in opposite directions; other ranges rise immediately on its S., which shut it out from communication with the Persian Gulf and the Mediterranean; while, on the N., the great plain of the Wolga and Don, after rising from the beds of those rivers for some distance S., subsides again, leaving a positive, though scarcely perceptible, ridge between the sources of Manytch and Sarpa and the base of the Caucasus, with a positive though very gentle slope towards the latter. In consequence of this formation, every drop of water from the Caucasus falls into the Black or Caspian Sea. The principal stream, besides the Kuban, Terek, and Araxes, already mentioned, are the Kur (anc. Cyrus), and the Phasis, rising on opposite sides of the transverse range which divides Imeritia from Georgia, and running, the first SE. to the Araxes, the other W. to the Black Sea. The

Rhwusk or Jorsk (an. *Aparrus*) is another tolerably large river, running to the Black Sea, and the Koran (an. *Cambus*), a still larger, falling into the Caspian. The torrents that run short courses to these are from the flanks of the mountains in their neighbourhood are quite innumerable, as are the affluents of the principal streams which pass from the mountain sides in every direction, sometimes with respectable length of course, and always in immense volume. It may, indeed, be reasonably concluded that the store of moisture in the Caucasus cannot possibly be exceeded by that of any other country of like extent; and since, from the causes before named, it is prevented from spreading beyond the bounds of the mountains. It follows necessarily that no land can be more abundantly watered. Most of the streams are flooded by the melting of the winter snows; and their action on the substance of the mountains is at all times very violent, especially on the slate and limestone. The former is brought down in the form of a black glistening sand, the latter, in that of a soft white substance, so fine and so abundant, that it is used by the natives, in its natural state, for whitening their houses. (Klaproth, p. 586; Pallas, i. 365.)

There is, perhaps, no other mountain region in the world so destitute of lakes as Caucasus. The lake of Sevan or Goukcha, between the Kur and Araxes, is the only one of any size in the whole region, and it can hardly be regarded as belonging to the Caucasus. It is a salt lake, of the kind so common in Central Asia, without any outlet, and occupying nearly the whole extent of a small elevated plain about 48 m. long, by 12 m. in width, 5,700 ft. above the sea. (Pallas, i. 337, &c.; Klaproth, pp. 169, 341–407, &c.; Monteith, G. J., iii. 43, *et seq.*; Spencer, *pass.*)

Climate.—This, of course, varies with the elevation; but perhaps still more with the degree of shelter afforded by the neighbouring ranges from the different winds. Some of the N. valleys, notwithstanding their exposure to the bleak gusts from the Snowy Mountains, are so perfectly protected from the N. wind, that their winter is as mild as in the S. parts of the Crimea. (Pallas, i. 338.) They are subject, however, to sudden and fierce, though brief, vicissitudes; and the very shelter which they possess, by confining the air, makes them unhealthy. With the sharp ascent of the land, the temperature rapidly decreases, and a few hours serves to convey the traveller from the climate of the temperate zone to that of everlasting winter. The cold in the upper ranges is intense; but observations are wanting on which to found any conclusion as to its average; Mignan, at a comparatively low part of the range, found it, in Jan. 1830, a very cold winter, at 1° Fahr., or 2° below the freezing point. (L. 34.) A better idea may, perhaps, be formed on this point, from the quantity of snow deposited in the defiles: 1,400 men were employed a fortnight in cutting a road for the Persian embassy, which, after all, was scarcely passable. (Mignan, i. 40.) Notwithstanding this intensity of cold, the plague is very common on the mountains.

On the S., the countries on the Black Sea and Caspian may be described as warm; those of Imeritia and Georgia as rather cold; but this, again, must be taken with considerable limitation, the N. parts of the two seas being subject to winter frosts. The melting of the ice in them frequently causes chilly summers on their shores, while, on the other hand, some of the higher valleys are among the coldest spots in the Caucasus. The abundance of running water, and the neighbourhood of the two bounding seas, cause a great accumulation of vapour; indeed, so extensive is the exhalation

constantly going on, that it may be said every wind, if long continued, brings with it a mist, which nothing can disperse except a storm. These last are, consequently, frequent and terrible. Luckily, however, the cause that produces them gives warning of their approach; the vapours, when grown too heavy for the atmosphere, collect themselves in dense masses round the sides and tops of the mountains; and the mountains, warned by this clothing of their Alps, prepare for the explosion, which they know, by long experience, will speedily follow. They wrap themselves in their *tcherukas* (large cloaks made of wool and goat's hair, and perfectly waterproof), and under a low tent made of felt, expressly for such emergencies, or under the lee of a rock or tree, await, generally in safety, the passing of the tempest. (Spencer's W. Cauc., p. 129.) Sometimes, however, the falling of the cliff or tree destroys those who have sought its shelter; but these accidents are of rare occurrence, as it is not often that the natives are compelled, for want of their felt tents, to run such risks; but to strangers unprovided with the means of combating these storms, the effect is sure to be ultimately fatal. One or, at most, two years' exposure to the varying influence of a Caucasian climate, sends the Russian soldier either to his grave or to the hospital, with a constitution irrecoverably broken. The uncertain temperature and the humidity of the atmosphere appear indeed to make it very unhealthy to strangers, especially on the slopes and flats towards the sea. Intermittent and bilious fevers of a very grave kind are endemic, and exceedingly obstinate; and the plague, as before observed, is also very common. (For a singular statement connected with the climate of the Caucasus, see Herodotus, Clio, p. 103; Klaproth, p. 160.) The varying humidity, or some other cause, seems likewise to impair the air with very peculiar qualities; observations of altitude by the barometer, or the boiling point of water, give very inconsistent results at different times (see CASPIAN SEA), and the extent of horizontal vision is frequently quite startling. The Caspian Sea is sometimes seen from the summit of the Berhing, 164 m. distant; and the Snowy Mountains from Sarepta on the Wolga, a length of 322 m. (Pallas, i. 570; Klaproth, p. 158.) The distance of the visible horizon, exclusive of refraction, would be in the first case about 94½ m., in the second about 163½ m.; the amount of refraction is, therefore, equal to more than 1°, and nearly 2½° respectively; but, in ordinary states of the atmosphere, the maximum being only 33', the excess of 27' and 1° 57' indicates a variable density in the medium which is truly surprising. Some of the larger clefts are said, in the traditions of the natives, to have been caused by earthquakes; but there are no authenticated records of these phenomena. (Güldenstädt, i. 217–432; Pallas, i. 340, 358, 447, *et pass.*; Gmelin, iii. *pass.*; Klaproth, pp. 163, 165, 309, 833, &c.; Chardin, p. 165; Spencer, W. Cauc., p. 125, *et seq.* 320, &c.; Circass., i. 286, &c.; Monteith, G. J., iii. 31, &c.)

Productions.—I. *Minerals.*—Except in its deficiency of lakes, the Caucasus has many points of resemblance to the Alps; among others, an apparent poverty of mineral treasures. It is true that this, in the case of the Caucasus, may be apparent only. The ancients unquestionably believed these mountains to be rich in the precious metals, but they also believed of most other districts that were but slightly known to them; and the limited observations of scientific men in modern times tend to the opposite conclusion. A yellow mineral, called cat gold, is indeed found, which may, perhaps, have occasioned the stories as to the gold mines of the Caucasus; but it is perfectly worthless. Iron, copper, saltpetre, sulphur, and lead, are found, the last in tolerably large quantities. Salt is almost wholly wanting, and of gems there does not appear to be any vestige. Indications of coal have lately been discovered; and, from the enormous quantity of lime deposits, it is likely that marbles may be found. (Pallas, i. 479; Guldenstadt, i. 441, 456; Klaproth, p. 391; Spencer, W. Cauc., i. 331.)

2. *Vegetables.*—In amount and variety of vegetation, the Caucasian regions seem to be unrivalled. Chardin, writing in 1692, says, 'Mount Caucasus, till ye come to the very top of it, is extremely fruitful;' and Spencer, in 1838, says, 'However high the ascent, we see luxurious vegetation, mingling even with the snow of centuries.' Nearly every tree, shrub, fruit, grain, and flower, found from the limit of the temperate zone to the pole, is native to or may be raised in the Caucasus. The N. basin consist of arable land of an excellent quality, meadows of the finest grass, and dwarf wood in great abundance. At a very little distance the increase of wood indicates a higher and colder country, but the plants which delight in a warm situation still continue to be very numerous. From the more rapid rise of the ground, bare rocks are more numerous on the N. than on the S., but every shelf, however limited, is marked by a rich vegetation to a bright almost inconceivable. The N. slopes and table-lands are still more abundant and varied in their productions than those on the N.; to say nothing of the swampy shores of the Euxine and Caspian, which are, in most cases, nearly impenetrable jungles of the rankest and most varied vegetation. The rising country consists of a succession of small flats, each covered with a most productive earth. The mountain sides and higher plains are clothed with dense forests, and the rivers are frequently unapproachable for a great distance. The forest trees consist of oaks of every species, cedars, cypresses, beeches, savins, junipers, hazels, firs, larees, pines, alders, and a host of others. Among the standard fruits are found the date palm, the jujube, quince, cherry, olive, wild apricot, and willow-leaved pear. Pomegranates, figs, and mulberries grow wild in all the warmer valleys; and vines twine round the standard trees to a very great elevation up the mountains. A hard-wood tree, called by the natives astchelin, is apparently peculiar: it is of a deep rose colour, very closely grained, and susceptible of an extremely high polish. In addition to the vine, the other climbing plants are innumerable, which mixing with the standards, the bramble fruits, such as raspberries and blackberries, and other dwarf woods, form a density of vegetation which it is impossible to penetrate, unless a passage be hewn with the hatchet. Rye, barley, oats, wheat, and millet are abundantly raised, even as high as 7,500 ft. above the sea; and besides these grains, the warmer plains and valleys produce flowers of every scent and dye, cotton, rice, flax, hemp, tobacco, and indigo, with every variety of cucumber and melon. This list is of necessity very imperfect, as will be evident when it is stated that Guldenstadt has filled eighteen quarto pages with the mere names of the various plants seen by him on the banks of the Terek and in Georgia. (i. 168-197, 418-450.) It may serve, however, to exhibit the vegetable riches of a region which seems to produce everything necessary for the existence and, with the exception of salt, even for the luxurious accumulation of man. (Guldenstadt, as above, et pass.; Gmelin, iii. 32-58, et pass.; Pallas, i. 310, 357,

864, 368, 879, &c.; Klaproth, pp. 167, 309, 391; &c.; Spencer, Circassia, i. 317, 430, ii. 283, 316, 357, &c.; W. Cauc., i. 29, 188-195, 216, &c.; Monteith, G. J., iii. 51-33.)

3. *Animals.*—Animal life in the Caucasus is on a scale of magnitude and variety equal to its vegetation. Wolves, bears, lynxes, jackals, foxes, wild cats, a peculiar beast of prey called chrou, together with many varieties of deer, wander in the forests and on the sides of the mountains. The smaller fur-bearing tribes are also common, as weasels, polecats, ermines, and moles of many varieties. Hares and every other species of game abound, with chamois and goats, of which the Caucasian goat (*Capra Caucasica*) seems peculiar. Sheep with peculiarly long wool are numerous; and it is even doubtful if, among the mountains, this creature be not yet living in a state of nature. This also is one of the homes of wild cattle; the largest species (the aurochs) being found in its forests; while of the domesticated kinds the varieties are numerous and serviceable. The horses of the Caucasus have been famous from a very high antiquity, the Becting Mountain having formerly been called Hippicon (*Ίππικον*), from the number of these animals which were grazed upon its sides. (Ptolemy, v. 9.) They are not less numerous in the present day, and are among the very finest varieties of the species. Of birds there are pheasants, partridges, grouse, and the whole tribe of mountain game, a great variety of the crow kind, nearly every species of birds of prey and passage, and some of the best specimens of the domestic varieties. Among insects, the bee and silkworm claim pre-eminence: they are both numerous, and their productions, particularly the honey, formed a considerable branch of trade with Turkey, till the power of Russia sealed the ports of the Black Sea. Other insects are equally numerous, as are also the reptile tribes, among which are some fine species of tortoises and snakes, both harmless and venomous. (Guldenstadt, i. 418, et passim; Gmelin, iii. 58, et passim; Pallas, i. 341, 410, &c.; Klaproth, p. 344, &c.; Spencer, passim.)

Inhabitants.—There is probably no other part of the world, except Africa, S. of the Sahara, where so many nations and languages are collected within so small a space as in the Caucasus. Guldenstadt gives a list of seven different nations, besides Tartars, who speak languages radically different, and who are again subdivided into almost innumerable tribes, among whom the varieties of dialects are nearly infinite. The principal nations in them enumerate:—1. Georgians; 2. Reslans; 3. Abchasians; 4. Tcherkessians; 5. Ossetians; 6. Kistians; 7. Leughians; 8. Tartars. (Reise, i. 438-485.) Of these the most numerous and important are the Georgians and Circassians or Tcherkessians; but the Abchasians and Ossetians, called by Pallas and Klaproth Abassians and Ossetians, are also powerful tribes. In habits and manners a strong resemblance is observed among them all; they are usually wandering hunters and warriors, for which occupations their country is peculiarly fitted, and only in an inferior degree shepherds or agriculturists. A partial exception must, however, be made to this general character in favour of the Georgians, who reside in towns, and have long possessed a fixed form of government and internal polity; but, for the rest, they appear to possess the erratic disposition, reckless courage, boundless hospitality, and much of the predatory habits which mark the Arab and other half barbarous people. (See Circassia, Georgia, &c.) It is well known that Blumenbach looked here for the origin of his first and most intellectual race of men (the Caucasian); but for this, as has been proved, there is not a particle of evidence, historical or philological. The Caucasians, though surrounded by the means of improvement, and occupying a country more favourably situated than that of Switzerland, have made no progress either in arts or arms; and continue to this day the same unlettered barbarians as in the days of Herodotus. (Clio, 203.) They have fine physical forms, but their mental endowments are of the most inferior description.

Name.—This has in all ages been the same among neighbouring nations, though, according to Strabo (xi. 500), the range was called by the natives Kaspios ipos (Caspian Mountains). The names Caspian and Caucasus have, in the opinion of Klaproth (p. 188), a similar etymology, namely, *Kuh-Casp* or *Casp*, the mountain of Casp, so called from the Caspii, a powerful people on its sides. (See CASPIAN SEA.) Pliny (vi. 9) derives the name, but with no great appearance of probability, from *Graucasus*, which, he says, in the Scythian tongue, means *nive candidus*. At present the term Caucasus is but little used by the Asiatics, the name for the mountains among the Tartars being *Jal-bus*; among the Turks, *Chafdaghi* (Mount Chaf); and among the Armenians, *Jalban-ser*, a modification of the Tartar term; but Caucasus is still in use among them.

CAUDEBEC, a sea-port town of France, dép. Seine Inférieure, cap. cant., on the Seine, at the mouth of the Candelare, 6 m. S. Yvetot. Pop. 2,164 in 1861. The parish church, built in the 16th century, is remarkable for the boldness and delicacy of its architecture. It has some manufactures of cotton goods. Previously to the revocation of the edict of Nantes, it was comparatively flourishing; but that disastrous measure gave a blow to its manufactures and commerce, from which it has not recovered. Its port, though safe, commodious, and advantageously situated between Havre and Rouen, is but little frequented.

CAUDETE (an. *Bigerra*), a town of Spain, prov. Murcia, 8 m. NNW. Villena, 12 m. NE. Yecla. Pop. 6,573 in 1857. The town was formerly fortified; and has a church, 2 convents, a hospital, several distilleries, and a palace of the bishop of Orihuela. On the heights in the vicinity a battle was fought in 1705, the day after the great victory gained by the Duke of Berwick at Almanza, between a detachment of the combined French and Spanish forces and those of the Archduke Charles, which ended in the defeat of the latter.

CAUFIRISTAN, or CAFFRISTAN, a region of Central Asia, occupying a great part of the Hindoo Koosh and a portion of the Indoo Tagh mountains, chiefly between lat. 35° and 36° N., and long. 70° E. and the W. limits of Cashmere; having N. Badakshan, E. Little Thibet, S. the dom. of the Punjab and Caubul, and W. those of Caubul and Kunduz. The hills N. of Bajaur and Kaneer form its S. limit; its other boundaries have been very imperfectly defined. The whole of this country is a lofty Alpine tract of snow-capped mountains, deep pine forests, interspersed with small but fertile and often populous valleys, and table-lands sometimes 10 or 15 m. across. Torrents and rivers are numerous, and are crossed by stationary wooden bridges or hanging bridges of rope and osiers. The cold of the winter is severe, but the valleys afford an abundance of grapes and other fruits, and the hills good pasture for sheep and goats. The Caufirs (infidels) who inhabit this region are an independent nation, said by Baber and Abul Fazel, and believed by themselves, to be descended from the troops of Alexander the Great.

They are supposed by some to have been driven thither from the valley of the Oxus, on its being overrun by the Mohammedans; but Sir A. Burnes and Mr. Elphinstone suppose they had emigrated, through a similar cause, from the neighbourhood of Candahar. They are remarkable for the fairness and beauty of their complexions; are liberal, social, and extremely hospitable: they never combine in war against their neighbours, but retaliate invasions fiercely, and fight with great bravery and determination. They indulge an excessive hatred against Mohammedans, and a Caufir adds an additional ornament to his dress, or another trophy to a high pole before his door, for each Mussulman he has slain. All wear tight clothes; those of some tribes made of black goat skins, and of others of white cotton: all suffer their hair to hang over their shoulders, and each looks upon every one else as a brother who wears ringlets and drinks wine: to the latter they are much addicted, and grape juice is given to children at the breast. They eat the flesh of all kinds of animals, except the dog and jackal, and use both tables and chairs of a rude construction; the women perform the business of tillage, as well as all laborious domestic occupations. Fine rice, wheat, and barley are the principal grains cultivated; honey, vinegar, cheese, butter, milk, bread and fruit, constitute the rest of their food. Both sexes drink wine to excess. Their dwelling-houses are usually built of wood upon hill-slopes, the roof of one row of houses forming the street to those above it: the only roads in the country are footpaths. Their weapons are spears, scimitars, and bows and arrows. After battle the victors are crowned with chaplets of mulberry-leaves. Both sexes wear ornaments of gold, silver, and other metals; and drinking-cups of the precious metals are often used, and much prized by them. Their language is unintelligible to Hindoos, Usbecks, or Affghans: it contains a mixture of words from the Hindoo, Affghan, and Persian tongues; but the major part of its roots are different from either: they have no books, and neither understand reading nor writing. They adore a supreme being, whom they call Imgun, and to whom they sacrifice both cows and goats; but address themselves to subordinate deities, represented by idols of wood or stone, who, they say, intercede with the chief deity in their behalf: fire is a requisite in every religious ceremony, although no veneration is paid to that element itself. They neither burn nor bury their dead, but expose the corpse in an open coffin, in a forest jungle or on a mountain, and after a certain time collect as many of the bones as possible, and deposit them in a cave: these ceremonies are solemnized with triumph, dances, and sacrifices. Music, dancing, which is eagerly practised by all classes, conversation, and carousals, form their chief amusements. They have priests, but they do not possess an extensive influence: they live under different chiefs, but little farther is known respecting their government. The slavery of such as have lost their relations is universal; some of the Caufirs possess many slaves and cattle, and much land. By old writers this region is often named Kuttaor: it was invaded by Timour, and in 1780, unsuccessfully, by a confederacy of the surrounding Mohammedan nations. (Elphinstone's Caubul, ii. 373–377; Burnes's Trav., iii. 183–185.)

CAUNES (LES), a town of France, dép. Aude, on the Argent-Double, 11 m. NE. Carcassonne. Pop. 2,347 in 1861. The town has a fine parish church, formerly belonging to the Benedictine abbey suppressed at the revolution: with distilleries, tanneries, dye-works, marble-works for working the marble found in the neighbouring mountains.

CAUSSADE, a town of France, dép. Tarn-et-Garonne, cap. cant., in a fertile country, near the Candé, 12 m. NE. Montauban, on the railway from Montauban to Vivien. Pop. 4,033 in 1861. The town is handsome, well-built, and has broad and straight streets; has numerous flour-mills, with manufactures of woollen and linen stuffs, and carries on some trade in corn, saffron, and truffles.

CAVA, a town of Southern Italy, prov. Salerno, cap. cant., in the middle of the agreeable valley of Frascatro, 25 m. ESE. Naples. Pop. 24,378 in 1862. The town has a cathedral, three other churches, a convent for noble ladies, a charity workhouse, a hospital, and a seminary. Silk, cotton, and woollen stuffs are manufactured in the town and the adjacent villages. The territory is not very fruitful, but the inhabitants have become rich by their industry and commerce. About a mile from the town is the magnificent Benedictine convent of La Trinità, with a fine library.

CAVAILLON (anc. Cabellio), a town of France, dép. Vaucluse, cap. cant., on the Durance, near where it is joined by the Coulon, at the foot of a mountain, 13 m. SE. Avignon. Pop. 7,797 in 1861. The town is mostly ill-built, with narrow and dirty streets. The fortifications by which it was formerly surrounded were destroyed during the revolution; the bishopric of which it was the seat has been also abolished. It has a considerable trade in dried fruits and preserves, shoes and nuts.

Cavaillon is a very ancient town. The Romans are believed to have planted a colony in it, and, at all events, they embellished it with several magnificent edifices. But having been since repeatedly overrun and pillaged by barbarians, and having suffered much from an earthquake in 1731, comparatively few remains of antiquity are to be found either in the town or its vicinity. The best preserved, though even that is much dilapidated, is a fragment of a triumphal arch supposed to belong to the age of Augustus.

CAVAN, an inl. co. of Ireland, prov. Ulster, having N. Fermanagh, E. Monaghan, S. Longford, Meath, and Westmeath, and W. Leitrim and Longford. Area, 473,749 imperial acres, of which 80,000 are unimproved mountain and bog, and 21,967 water, remaining principally of lougho Sheilin, Ramor, and Oughter. The Shannon has its principal source in the NW. part of this co., and it is traversed by the Erne, Annalee, &c. Surface hilly, and soil generally poor. There are some large estates, but the greater number are of moderate size. About 4-5ths of the land under tillage. Agriculture in the most depressed state; holdings generally small, and the competition for them excessive. Spade cultivation is very general, so much so that in some parishes there is hardly a plough. Oats and potatoes principal crops, but some wheat is raised, and flax. Cottiers have generally pigs and goats; the former being sold to pay the rent, and the latter kept for their milk. Linen manufacture widely diffused, having not a little contributed to the subdivision of the co. It is affirmed that the condition of the peasantry has been materially deteriorated during the last 70 years. Minerals little known. Cavan is divided into 7 baronies and 80 parishes, and sends 2 members to the H. of C. for the co. Registered electors 5,989 in 1863. Principal town Cavan. The co. had a population of 243,824 in 1841; of 174,260 in 1851; and of 153,906 in 1861. These statistics of population tell, more than words can do, a sad history of decline.

CAVAN, an inland town of Ireland, co. Cavan, prov. Ulster, 60 m. NW. Dublin, on the railway from Dublin to Enniskillen. Pop. 3,209 in 1861. Cavan, though the assize town, is with few exceptions meanly built, long lines of suburbs being formed of thatched mud cabins. The public buildings are a large parish church and R.C. Cath. chapel; an endowed school of royal foundation, having accommodation for 180 resident students; a fine court-house, a co. prison on the radiating plan, and an infirmary. A garden of Lord Farnham's, near the town, has been thrown open as a promenade for the inhabitants. The corporation, under a charter of James I., in 1610, consisted of a sovereign, 2 portreeves, 2 burgesses, and an unlimited commonalty; but having been deprived at the Union of the right of sending mem. to the H. of C., it has fallen into desuetude. The assizes for the co., general sessions at Hilary and Midsummer, and petty sessions every week, are held here. Trade inconsiderable, and chiefly in oats and butter. Markets are held on Tuesdays; fairs on Feb. 1, April 4, May 14, June 30, Aug. 14, Sept. 25, and Nov. 12.

CAVERY, a river of S. Hindostan, the most considerable and useful S. of the Krishna; bath Mysore and the Carnatic owing much of their agricultural wealth to the water it distributes. It rises in Coorg, bounds Coimbatoor NE., and after a winding course of 450 m., chiefly in a E. direction, falls into the sea by various mouths in the district of Tanjore; where it is industriously made use of for irrigation. It is filled by both monsoons, but is not navigable for large vessels.

CAVERYPAUK, a town of Hindostan, prov. Carnatic, 57 m. WSW. Madras, in the neighbourhood of which is an immense tank 5 m. long by 3 m. broad, faced with large stones, and supported by a mound of earth 30 ft. high. This is one of the finest works constructed for the purpose of irrigation throughout the S. of India.

CAVITE, a town of Luzon, one of the Philippine Islands, in the Bay of Manilla, 8 m. SW. that city, of which it is the port; lat. 14° 31' N., long. 120° 48' E. Estimated pop. 8,000. It is the naval depôt of all the Spanish possessions in the East, and is built on the E. extremity of a low bifurcated peninsula, stretching into the sea for about 3 m., having between its two extremities the outer harbour, while the inner harbour is situated to the N. of the town: neither has more than four fathoms water, though very large ships moor in the inner harbour. The houses of Cavite, which are two stories high, are built chiefly of wood, their windows being furnished with a semi-transparent shell instead of glass. It has an arsenal, a marine hospital, some well-built churches, and several convents; but has of late years greatly decreased in size and importance.

CAWNPORE, or CAUNPOOR (Khanpura), a district or collectorate of Hindostan, prov. Allahabad, presid. Bengal, composed of crosshans from the nabob of Oude, between lat. 26° and 27° N., and long. 79° 30' and 80° 30' E., having NW. the distr. of Etawah, Belah, and Furruckabad, NE. the Oude reserved territories, SE. the Futtehpoor and Kalpee distrs., and SW. Bundlecund. Area 2,650 sq. m. Pop. probably nearly a million. This distr. is bounded NE. by the Ganges, and intersected in its entire length by the Jumna; it is therefore almost wholly comprised within the Doab. Surface flat; soil highly productive, and upon the whole tolerably well cultivated, though in some parts there are extensive wastes. Maize, barley, and wheat, turnips, cabbages, and other European vegetables; grapes, peaches, &c., are grown, and the sugar-cane flourishes in great luxuriance. Agriculture prospers in the neighbourhood of the cap., owing to the presence of a Euro-

... market, and consequent high prices. The assessment on the land is high, and the prov. was on its first coming into British possession very much over-assessed, and suffered greatly in consequence. There are about 2,000 villages in this distr., which possess lands; but the perpetual settlement is also established. Nearly all the pop. are Hindoos, the heads of the villages being mostly of the Rajpoot caste. Offences are frequent, but yearly diminishing as the efficiency of the police increases; dacoity, or gang-robbery, was formerly frequent, but was committed only by gangs out of the Oude reserved territory. Thuggee, or murder by professional murderers, also prevailed greatly in this distr.; and from 1830 to 1840 the average was about 10 thuggees yearly. The principal towns are Cawnpore, the cap., Rawalabad, Jaugrosur, and Arbutpoor.

CAWNPORE, the cap. town of the above distr., and chief British military station in the ceded provinces, on the W. bank of the Ganges, 85 m. SW. Lucknow, and 100 m. NW. Allahabad; lat. 26° 30′ N., long. 80° 15′ E. The town extends irregularly for 6 m. along the bank of the river, which is here a mile broad, and lined by the bungalows of European officers. It is built in a very straggling manner, with the exception of a tolerable main street nearly parallel with the military lines, composed of well-built brick houses two or three stories high, with wooden balconies in front. Excepting its size, few circumstances about Cawnpore attract much notice; the European public buildings are of simple architecture, and confined to works of absolute necessity; the chief are the military hospital, gaol, assembly-room, and custom-house. A Protestant church has been erected by public subscription within the last few years; most of the other religious edifices are mosques, some of which are handsome. Shops large and tolerably well supplied, provisions being about half the price they bring in Calcutta. The European private houses are roomy, one story high, with sloping roofs, first thatched and then tiled. The officers' bungalows along the banks of the Ganges are enriched by gardens surrounded by mud walls. At the NW. extremity of the town are the public magazines protected by a slight entrenchment; and farther on, in the same direction, is the old town of Cawnpore, a place of no consequence, and containing no interesting relics of antiquity. A free-school was established here in 1822, which is attended by Europeans, Mohammedans, and Hindoos, who receive instruction together, and the progress of which is most satisfactory. It is supported partly by a government grant of 4,800 rupees a year. Cawnpore is not a pleasant place of residence for Europeans. Its great heat and the clouds of dust to which it is subject are represented as most distressing.

Cawnpore derived a sad notoriety during the Indian mutiny of 1857. The small British force stationed in this town having surrendered, by capitulation, to Nana Sahib, they were allowed to leave; but had no sooner embarked in their boats, on the 17th of June, when they were fired upon, and nearly all cruelly murdered. A number of women and children escaped the slaughter only to be killed, soon after, with unexampled brutality. The tale of these horrors is perpetuated by a monument erected at Cawnpore.

CAXAMARCA, a city of Peru, cap. prov. of same name, in a fertile and well-cultivated valley in the Andes, 370 m. NNW. Lima; lat. 7° 8′ 36″ S., long. 78° 36′ 15″ W. Pop. about 7,000, chiefly Indians and Mestizors. Its name is equivalent to ' place of frost,' and has been probably derived from its being sometimes visited by frosty winds from

the E.; but, in general, the climate is excellent. Most of the houses are tiled and whitewashed. The churches, which are numerous and handsome, are built of stone richly cut, and are ornamented with spires and domes. They were formerly celebrated for the quantity of gold and silver decorations they contained. There are also some convents and nunneries. The inhab. are industrious, and considered the best silver and iron workers in Peru. ' I have,' says Mr. Stephenson, ' seen many very handsome sword-blades and daggers made here; pocket-steels and bridle-bits most curiously wrought, besides several well-finished pistol and gun locks. Literature would prosper here, were it properly cultivated; the natives are fond of instruction, and scholars are not rare; many of the richer inhab. send their children to Truxillo and Lima to be educated.' (Stevenson's Peru, ii. 162.) The inhab. of the interior resort thither to sell their own produce and manufactures, and to purchase such other as they may require. Hence a considerable trade is carried on with Lambayeque, and other places on the coast, to which Caxamarca furnishes manufactured goods, such as baizes, coarse cloth, blankets, and flannels; and receives in return European manufactures, soap, sugar, cocoa, brandy, wine, indigo, Paraguay tea, salt-fish, iron, and steel. Some of the shops are well stored with European goods. The markets are well supplied with fresh meat, poultry, bread, vegetables, fruit, butter, and cheese, at very low prices. About a league E. from this city are some hot and cold springs, which were used by the Incas for baths, and are still employed for the same purpose.

Caxamarca is a place of considerable celebrity in the history of Peru, and of Spanish atrocity. The Incas had a palace here; and it was here that Friar Vincente Valverde delivered his famous harangue to the Inca Atahualpa, which was immediately followed by the butchery of the Peruvians, and by the imprisonment, accusation, and murder of the Inca.

CAYENNE, a sea-port town of French Guyana, cap. of that colony, at the NW. extremity of the isl. of same name, at the mouth of the Oyaque; lat. 4° 56′ 15″ N., long. 52° 14′ 45″ W. Pop. 6,230 in 1861. The town covers a surface of about 70 hectares, and contains about 600 houses, mostly of wood. It is divided into the old and new towns: the former, which is ill-built, contains the government house and the ancient Jesuits' college; it is separated from the new town by the Place d'Armes, a large open space planted with orange-trees. The new town is larger than the old, and was laid out at the end of the last century; its streets are wide, straight, mostly paved, and clean; it has a handsome church, with some large warehouses and good private residences. The old town is commanded by a fort, which, with some low batteries, protects the entrance of the harbour. The latter is shallow, but otherwise good, and well adapted for merchant-vessels of moderate size. There are two quays for lading and unlading. The roadstead at the mouth of the Oyaque, though small, is the best on the coast. Its holding-ground is good, and it has everywhere from 12 to 18 ft. water; trading vessels lie in it within 1 m. of the land, and 7 m. of the town. Ships drawing more than 15 ft. water anchor about 6 m. from Cayenne, near a rocky islet called ' L'Enfant Perdu.' Cayenne is in the centre of the whole trade of the colony. (See GUYANA, FRENCH.) It is the seat of a royal court, a court of assizes and of tribunals of the peace and original jurisdiction. The town was founded about 1635. The Emperor Napoleon III., on establishing himself on the throne of France, sent a number of political prisoners here, many of whom perished on account of the unhealthiness of the climate.

CAYENNE. See GUYANA (FRENCH).

CAYLUS, a town of France, dép. Tarn-et-Garonne, near the right bank of the Bonnette river, and the high road between Montauban and Rhodez, 24 m. NE. the former city, Pop. 4,973 in 1861. It has a considerable trade in corn, and eleven fairs annually.

CAZALLA, a town of Spain, prov. Seville, on the crest of the Sierra Morena, 13 m. NE. Guadalcanal, Pop. 6,852 in 1857. The town has a church, five monasteries, and two hospitals. Its environs have many Roman and Arabic antiquities, and ruins of country residences of more modern date; with mines of silver, iron, sulphur, antianthus, and copper; and quarries of beautifully variegated marble. The mountains are the resort of wild boars and wolves, which make much havoc among the cattle.

CAZÈRES, a town of France, dép. Haute Garonne, cap. cant., on the Garonne, 31 m. SW. Toulouse. Pop. 2,633 in 1861. A handsome promenade separates the town from the suburbs. There are fabrics of hats, with dye-works and tanneries.

CEFALÙ, a sea-port town of Sicily, prov. Palermo, on the Tyrrhenean Sea, at the foot of a rock, 40 m. ENE. Palermo; lat. 38° N., long. 14° 13' 67" E. Pop. 11,183 in 1861. The town is surrounded by a battlemented line wall, but the works are old and weak. The streets are tolerably regular, and there is a good cathedral and some other churches, with a school of navigation. The port is small, and the trade of the place but inconsiderable. On the summits of the hill above the town are the ruins of a Saracenic castle.

CEHEJIN (Segura), a town of Spain, prov. Murcia, on the river Caravaca, 8 m. E. Caravaca town, and 40 m. WNW. Murcia. Pop. 8,710 in 1857. The town is situated in a well cultivated and fertile district. The principal streets are well paved, and the houses good—some of them magnificent, marble being abundant in the neighbourhood. It has a church, a convent, and an ancient castle, with several distilleries, and manufactures of coarse paper, linen, and sandals.

CELANO, a town of Southern Italy, prov. Aquila, cap. cant., near the lake Fucino or Celano, 30 m. SSE. Aquila. Pop. 8,525 in 1861. The town has one collegiate and some other churches, and a manufactory of paper. For an account of the Lake of Celano see FUCINO (LAKE OF).

CELEBES, a large island of the E. Archipelago, forming the centre of its 2nd division; stretching from lat. 2° N. to nearly 6° S., and from long. 119° to 125° E.; having N. the Sea of Celebes, W. the Straits of Macassar, E. the Moluccas and Pitt's Passages, and S. the Flores Sea. Area estimated at 75,000 sq. m. Pop. supposed to be between 1,500,000 and 2,000,000. Its shape is singularly irregular; it is deeply indented by three great bays, separated by four peninsulas, diverging N., E., and S.

Celebes, unlike most of the other great islands of this archipelago, abounds in extensive grassy plains, free from forests, which are looked upon as the common property of the tribes who live upon them, by whom they are carefully guarded from the intrusion of aliens. There are only three rivers of any consequence; the Chianura, which rises near the centre of the island, and running N. through the state of Boni, falls by several mouths into the bay of the same name; a second stream, having a S. direction; and a third, which discharges itself on the W. coast, S. of Macassar.

The Chianura is navigable for ships to some distance; and native boats pass up it considerably farther into a fresh-water lake. Volcanos are said to exist in the N. division of the island. Gold is found in Celebes; but in a less quantity than in Borneo, and chiefly in the sands of the streams. Timber is not very plentiful; teak-trees are generally few; but a large forest of them exists in one part of the island, which the natives report to have been raised from imported seed. The vast plains afford abundant pasture and cover for a variety of the best game, deer, wild hogs, &c. The tiger and leopard, though common in the W. parts of the archipelago, are here unknown. The horses of Celebes, though seldom exceeding 16 hands high, are larger built, and unite a greater share of blood and strength than any other breed of the E. islands; they are regularly trained for hunting, and are noted for fleetness and perseverance. Rice, maize, and cassava, with cotton and tobacco, are the chief articles grown. The N. peninsula being the most healthy, is by far the most extensively peopled, and contains the two principal states of the island, those of Boni and Macassar. The centre of the island is said to be inhabited by Horafuras (see E. ARCHIPELAGO), supposed to be aborigines: the brown race consists of a number of tribes agreeing remarkably in person, but divided into four or five different nations, of which that of the Bugis is by far the most considerable. They are usually squat, robust, and somewhat heavily formed, though not ill built; their medium height is a little above 5 ft.; faces round; cheekbones high; nose small, and neither very prominent nor flattened; mouth wide, and teeth fine, when not discoloured by art. They are more distinguished for a revengeful disposition than any of the other natives of this archipelago. Notwithstanding most of the tribes have long passed that stage of society in which the chase is pursued for subsistence, they follow it with great ardour; and no sooner is the rice seed cast into the ground, than the chiefs and their retainers turn with enthusiasm to the sports of the field, in parties of frequently not less than 200 horsemen.

The Wadju, or Tawadju tribe, inhabiting the body of the island, are distinguished as a commercial and enterprising people. The natives of Celebes and Bali are the most celebrated in the archipelago for their manufactures of cloth, their fabrics ranking before all others for fineness and durability; they are, however, ignorant of the art of printing cloths, or of giving them the brilliant colours of the fabrics of the Asiatic continent. The inhabitants import cotton, birds' nests, tripang, sharks' fins, tortoise shell, agar-agar, &c.; and, together with gold in small quantities, and birds, re-export these articles to China, by the junks which annually trade to Celebes. The several chiefs have often a monopoly of some article of produce, as brass, betel-nut, opium, and salt.

The various independent nations of Celebes have each their peculiar form of government; but these are for the most part limited monarchies, the sovereigns being controlled by the subordinate chieftains, and these again frequently by the mass of the people. The federal state of Boni consists of eight petty states, each governed by its own hereditary despot; while the general government is vested in one of the number elected from among the rest, but who can do nothing without the assent of the others.

In the state of the Goa Macassar, the king is chosen by ten electors, who also choose an officer invested with powers similar to those of the mayors of the palace of France, or the ancient justizas of Aragon, and who can, of his own authority, remove the king himself or any one of the council, and direct the electors to proceed to a new election.

In the Bugis state of Wadje, forty chiefs constitute the great council of the nation, which is divided into three chambers, from each of which two members are nominated, who, in their turn, elect the chief of the confederacy. The 'Council of Forty' decide on all questions of peace and war. Women or infants of the privileged families in Celebes are commonly eligible to the throne; and women very frequently actually exercise the powers of sovereignty; they are throughout the island associated on terms of equality with the men, taking active concern in all the business of life. They appear in public without scandal, and are often consulted on public affairs. Though the husband invariably pays a price for his wife, she is never treated with contempt or disdain.

Notwithstanding the symptoms of a considerable advance in civilisation now enumerated, a great deal of rudeness and barbarity exhibit themselves among the inhabitants. Crimes are frequent; thefts and robberies extremely so; a total disregard of human life seems to prevail, and murder and assassination for hire are by no means rare. Mohammedanism is the predominant religion, especially in the S. part of the island; it was introduced by the Malays; but the inhab. generally are by no means strict as to its injunctions. The languages spoken belong to the great Polynesian family, but differ from those common in the W. of the archipelago, in being more soft and vocalic, and having less internal trace of Sanscrit: the two dialects of the Bugis and Macassars are the principal, and amongst the most improved tongues of the archipelago: the Bugis have a literature by no means contemptible. In their costume, the people of Celebes avoid showing the knee; they wear a long coloured cloth, the end of which they throw over the shoulders. They blacken the teeth, and use unctuous cosmetics; their ornaments are flowers, gold trinkets, and diamonds, &c. They appear to have no scientific treatises; but are not wholly ignorant of some of the constellations, by the observation of which they navigate their proas.

Celebes was first visited by the Portuguese in 1512, who were expelled by the Dutch in 1660. In 1811 the territories belonging to that nation fell under the British dominion; but in 1816 were restored. The principal Dutch settlement is Macassar, which contains Fort Rotterdam, the residence of the governor. The Dutch have other settlements on the bays of Tolo and Tominie; and most of the native states are subordinate to them. (Crawfurd, Hist. of the Indian Archipelago, 3 vols.)

CEPHALONIA (an. *Cephallenia*), an isl. in the Mediterranean, and the largest of those composing the former Ionian republic, now forming part of the kingdom of Greece, near the W. coast of Greece, opposite the Gulf of Patras; between lat. 38° 5' and 38° 29' N., and long. 20° 21' and 20° 49' E.; 5 m. N. Zante, 5 m. S. Santa Maura, and 64 m. SSE. Corfu. Length, NSW. to SSE., 32 m.; breadth, very unequal. Area 348 sq. m. Pop. 70,120 in 1860. Its aspect is generally mountainous and barren, and though some spots are rich and fertile, the soil is, for the most part, only scantily spread over the limestone rock, of which the country consists. The shores are indented by numerous bays, of which that of Argostoli in the NW. is the principal. It extends for 7 or 8 m. inland, and has, in most parts, deep water and good anchorage. In the interior of the island an elevated range, called the Black Mountain, runs NW. to SE., the highest point of which (an. M.

(Enos), is 5,000 ft. above the level of the sea. Surface generally uneven; the only plain is in the SW. near Argostoli, which is also the most densely inhabited part of the island. Climate mild; but storms and heavy rains, sudden changes of temperature, and earthquakes are frequent. The island contains about 40,000 acres of cultivated, and 180,000 acres of uncultivated land. Wheat, Indian and other corn, pulse, currants, olive oil, wine, cotton, flax, and salt, constitute the chief products. The principal article of export is currants; and next to it, wine and oil. The annual produce of currants is estimated at from 5,000,000 to 6,000,000 lbs. The Valonea oak abounds. Tenures of land are mostly annual, on the metayer system. Property is much divided, few proprietors having a revenue of 1,000£ a year. Cephalonia is represented by ten deputies in the parliament of the kingdom of Greece. Argostoli and Lixuri are the chief towns; they are situated on either side the Bay of Argostoli. At the mouth of this inlet there is a lighthouse; and at Lixuri, a mole for the security of trading vessels has been constructed. Near Argostoli, a curious undershot water-mill was built by an English merchant in 1835. The roads were formerly very bad, but have been greatly improved during the time that the island was under British protection. Most of the pop. belong to the Greek church; the remainder are chiefly Roman Cath. Lixuri is the seat of a Roman Cath. bishop. The inhabitants of this island are active, enterprising, and noted for their industry and commercial spirit. A great number of them are physicians; and, like many other of their countrymen, emigrate and settle elsewhere. The island was anciently known by several names: Thucydides calls it Tetrapolis, from its four principal cities, Samos, Pali, Krani, and Pronos, remains of which still exist. The site of Samos exhibits very extensive ruins, amongst which many medals, vases, statues, &c. have been found, and Dr. Holland traced the Cyclopean walls of Krani, at the head of the Gulf of Argostoli, in almost their entire extent. Cephalonia belonged successively to the Byzantine empire, Norman, Venetian, Turks, and Venetians again; from whom it was taken by the French in 1799. In 1815 it was, with the rest of the Ionian Islands, placed under the protectorate of Great Britain, but ceded to the kingdom of Greece in 1864.

CERAM, a considerable island of the E. Archipelago (third division), chiefly between lat. 8° and 40 S., and long. 128° and 131° E.; length, E. and W., about 185 m. by 30 m. average breadth; area 5,500 sq. m. A mountain chain runs E. and W., through the centre of the island, the highest peak of which is apparently about 7,000 ft. above the level of the sea. Ceram is chiefly distinguished for its large forests of sago-palm and its fine woods for cabinet-work; in one portion of it great quantities of nutmegs and cloves were formerly produced; but the trees were extirpated by the Dutch about 1657. The shores of Ceram abound with rare and beautiful shells; its interior is peopled by tribes of Horaforas. (See ARCHIPELAGO, EASTERN.) A cluster of small islands, called Ceram Laut, lies off the E. end of Ceram.

CERET, a town of France, dep. Pyrenées Orientales, cap. arrond., near the Tech, 15 m. SSW. Perpignan, and 5 m. from the frontier of Spain. Pop. 3,583 in 1861. The town is the seat of a departmental college and of a court of primary jurisdiction. It was here that the plenipotentiaries met to fix the limits between France and Spain, in 1660.

CERIGNOLA, a town of Southern Italy, prov. Foggia, cap. cant., 23 m. SE. Foggia. Pop. 18,317 in 1881. It is a well-built town, with a college, several convents, and a hospital. In the neighbourhood of this town, in 1503, Gonsalvo de Cordova gained a decisive victory over the French forces commanded by the Duc de Nemours, who was killed in the action.

CERIGO (an. *Cythera*), the most southerly of the seven principal Ionian Islands, which formerly constituted the Ionian republic, situated at a considerable distance from the others, near the S. extremity of the Morea, between lat. 36° 7' and 36° 23' N., and long. 22° 55' 30'' and 23° 7' 30'' E. Length, N. to S., 20 m.; greatest breadth, 12 m. Area 115 sq. m. Pop. 11,100 in 1860. The surface is mountainous, rocky, and mostly uncultivated; but some parts of it produce wheat, maize, pulse, cotton, flax, wine, and olive oil; the latter of which is highly esteemed. The honey of Cerigo is also of very good quality. It has a greater number of horned cattle than any of the other islands. The shores are abrupt; the sea round Cerigo is much disturbed by currents; and gales dangerous to shipping are frequent. The best anchorage is at St. Nicolo, on the E. coast. The principal town is Kapsali, at the S. extremity, with a pop. of about 5,000; houses mostly of wood and ill-built. Though now comparatively insignificant, Cythera was formerly a place of considerable importance, and probably of wealth, if we may judge from the ruins still extant in various parts of the island. It was the birthplace of Helen, and sacred to Venus, in honour of whom a temple, said to have been founded by Æneas, was erected. (Larcher, *Mémoire sur Venus*, 144.) Cythera was originally called *Porphyris*, from the nature of its rocks. It was long a naval station of the Lacedæmonians; and belonged successively to Macedon, Egypt, Rome, and Venice. The little island of Cerigotto, (an. *Ægilia*), 4 m. long, and inhabited by about thirty families, lies midway between Cerigo and Crete, about 20 m. from either.

CERRETO, a town of Southern Italy, prov. Benevento, cap. cant., on the declivity of Monte Matara, near the Cusano, 10 m. ENE. Piedimonte. Pop. 6,981 in 1862. It is well built, and is one of the most agreeable towns in the province: it has a fine cathedral ornamented with superb pictures, a collegiate church, three convents, a seminary, and considerable manufacture of coarse cloth. In 1686 it was wasted by the plague, and in 1688 an earthquake destroyed great part of the town.

CERVERA, a city of Spain, prov. Catalonia, on an eminence, 57 m. NW. Barcelona, 102 m. E. Saragossa. Pop. 4,499 in 1857. The town stands on a considerable eminence, is surrounded by walls, and has an ancient decayed castle. It has a church, five convents, a hospital, and five colleges. Some of its streets are well paved. The church is a Gothic building, with three naves; and the university, established in this city by Philip V., is a large, magnificent structure. The vicinity produces wine, oil, almonds, grain, pulse, cattle, and plenty of game.

CERVIA, a town of Central Italy, prov. Ravenna, near the Adriatic, with which it communicates by a canal, 11½ m. SE. Ravenna. Pop. 5,733 in 1862. The town is a seat of a bishopric; is regularly built; has a cathedral and several churches and convents. To the W. of the town is a vast morsel, called the *Valle di Cervia*.

CESENA, a town of Central Italy, prov. Ferrara, on the Savio, at the foot of a mountain, 10 m. SE. Forli. Pop. 38,758 in 1862. The town is the seat of a bishopric; is well built; has a cathedral, a handsome town-house, fourteen convents for men, and seven for women, a seminary, a society of agriculture and of arts, with silk factories, and a con-

siderable trade in wine and hemp, produced in its territory.

CETTE, a fortified sea-port town of France, dép Hérault, cap. cant., on the narrow tongue of land separating the lagoon of Thau from the sea, and on the declivity and at the foot of a calcareous hill, which advances into the Mediterranean in the form of a peninsula, 15 m. SW. Montpellier, on the railway from Montpellier to Narbonne. Pop. 22,438 in 1861. The town is well built, but it derives its chief importance from its harbour, and from its being the port, on the Mediterranean side, of the Canal du Midi. The harbour is formed by two lateral moles, with a breakwater across the entrance. There are forts on both these moles, and on the principal is a lighthouse, the lantern being elevated 84 ft. above the level of the sea. The harbour is perfectly safe in all weathers; has from 18 to 19 ft. water; and can accommodate about 400 sail of large and small ships. A broad and deep canal, bordered with quays, establishes a communication between the port and the lagoon of Thau; and, consequently, with the Canal du Midi on the one hand, and with the canals leading to the lihouse on the other. Cette is the centre of a great deal of traffic, particularly of the coasting description; and from about the middle of November to the end of March freights are generally to be met with. There is regular steamboat communication with Algiers and the chief ports on the eastern coast of Spain; but the principal articles of export and import are those conveyed by the canal. About 80,000 tons of wine, and 4,000 tons of brandy, are annually exported. A good deal of Benicarlo wine from Spain, for mixing with claret, is imported. It has a court of summary jurisdiction, a school of navigation, an exchange, barracks, and a theatre. Ships are built here, and there are glass, soap, and tobacco-works, with distilleries, and a manufactory of highly esteemed liqueurs. The fishery of sardines is successfully carried on along the coast; and the salt-works on the adjoining lagoon are extensive, and furnish employment to many individuals. Cette is of modern date, having been founded in 1666, to serve as a port for the great canal.

CEUTA (an. *Septum* or *Septa*), a sea-port town of N. Africa, in the possession of Spain, coast of Morocco, directly opposite Gibraltar, and at the SE. extremity of the straits, on a narrow peninsula stretching about 5 m. ENE. into the Mediterranean, and having a capacious bay on its S., and a smaller one on its N. side. Pop. 7,144 in 1857. The E. part of the peninsula is occupied by the mountain of Almina, on the highest point of which is the castle of Ceuta, 14 m. S. by E. from Europa Point; lat. 35° 54' 4'' N., long. 5° 17' W. This mountain, which, towards the sea, is formed round by inaccessible rocks, is the *Abyla* Propre of the ancients, and is famous as one of the pillars of Hercules; the rock of Gibraltar (*Mons Calpe*) being the other. The citadel, a very strong fort, is built across the narrowest and lowest part of the peninsula, at its junction with the mainland. The town, immediately to the E. of the citadel, is situated at the foot and on the declivity of the mountain. Ceuta has many points of resemblance with Gibraltar, and, like it, if properly garrisoned, would be all but impregnable. It is well supplied with water; is the seat of a bishopric; has a cathedral, two convents, a hospital, and a bagne or prison for criminals employed on the public works. It is also used as a place for the confinement of state prisoners. It is the most important of all the Spanish presidios or settlements in Africa, and is the seat of a military governor, a royal tribunal, and a financial intendant. Most of the provisions and other necessaries required for the supply of the town and garrison are brought from Spain. Ceuta was taken from the Moors by John, king of Portugal, in 1415. Since 1640 it has belonged to Spain. It has been several times besieged by the Africans, especially in 1697.

CEVA (an. *Ceba*), an inl. town of Northern Italy, prov. Cuneo, cap. mand., at the confluence of the Cevetta with the Tanaro, 10 m. E. by N. Mondovì. Pop. 4,590 in 1862. It is built, at the foot of a rock, formerly surmounted by a castle, which was used as a state prison previously to its destruction by the French revolutionary forces. The town was formerly surrounded with walls; but these were in great part destroyed by an inundation of the Tanaro, in 1564. It contains a church, and several convents; some forges, and silk factories; and, in both ancient and modern times, has been celebrated for its cheese.

CEYLON (an. *Taprobane*), a large island belonging to Great Britain, near the S. extremity of Hindostan, bearing the like relation to the Indian that Sicily does to the Italian peninsula. It lies between lat. 5° 56' and 9° 50' N., and almost entirely between long. 80° and 82° E.; having NW. the Gulf of Manaar and Palk's Straits, which separate it from Hindostan; S. and SW. the Indian Ocean, and E. the Bay of Bengal. It tapers to a point towards the N., and is shaped like the section of a pear cut lengthwise through the middle. Length N. to S. 270 m.; average breadth nearly 100 m.; area 24,500 sq. m. Pop. 2,075,231 in 1862, of whom only 7,102 whites.

The *Coasts*, on the N. and NW., are low and flat; those on the N. and E. bold and rocky, and in some places fenced with reefs; in many parts they are deeply indented by the sea, and present some large and many small harbours. Trincomalee harbour, on the NE. coast, is one of the finest anywhere met with. Point de Galle, in the S., is the next in importance; the inferior harbours are Batticaloa, Matura, and Calture, on the S. and E., and Negumbo, Chilaw, Calpentyn, Manaar, and Point Pedro, on the W. coasts. The deep water along the E. shores admits the safe approach of large vessels, but the harbours on the N. and NW. are full of sands and shallows, whose position varies with the monsoon. Columbo, the marit. cap., has merely a roadstead, which is practicable for large ships only from the beginning of Dec. to the latter end of March. So large a number of inlets causes a corresponding proportion of small islands, promontories, and peninsulas; of the latter the principal are the peninsulas of Jaffnapatam, on the NW., and that of Calpentyn, on the W. coast. At its N. extremity especially, the shores of Ceylon are studded with numerous small rocky and verdant islets. The ridge of sandbanks called Adam's Bridge, which crosses the Gulf of Manaar from Ceylon to the island of Ramisseram, near the opposite coast of India, is connected by the natives with a variety of curious traditions, and forms a great obstacle to the more speedy communication with the continent, by its hinderance to navigation. It consists of loose sand, resting on firm foundations, but constantly varying in form from the action of the monsoons. There are three principal openings or channels through this ridge; one near the island of Manaar, another 8 m. farther to the W., and a third about 11 m. from the island of Ramisseram; but all of them are impracticable except for small native boats in fine weather, and even then the navigation is attended with some danger. Near these openings the bank rises above the water for some miles, broken occasionally by smaller channels,

but towards the centre it is mostly covered by water, the depth of which does not in any part exceed a few feet. By the late accounts (see *Asiat. Journ.*, April, 1859), attempts at enlarging the passage between Ramisseram and the continent are now in progress.

Interior — Mountains. — The belt of country along the shore surrounding the interior, or old kingdom of Candy, is, for the most part, flat, varying in width from 8 to 30 m., and, in the N. to nearly 80 m.; its extensive green plains giving to the shores of Ceylon an advantageous appearance when contrasted with the barren and sandy shores of the opposite continent. The interior consists of three distinct natural divisions — the low country, the hills, and the mountains. The centre of the island N. of lat. 80° N. is occupied by an extensive tableland, 87 m. in length, by about 30 m. in width, and estimated at from 2,000 to 3,000 ft. above the sea. The interior of the N. and central divisions consists of ranges of mountains running mostly NE. and NW., and varying from 1,000 to 4,000 ft. above the sea, clothed to the summits with magnificent forests, and intersected by numerous ravines, cataracts, and cascades. From these regions various conical-shaped hills rise up at intervals to an additional height of from 2,000 to 3,000 ft. The most conspicuous summit is that which is known by the name of Adam's Peak (the Samanella of the Singalese), in lat. 7° N., and long. 80° 40' E., 46 m. ESE. Columbo, rising to 6,152 ft. above the sea. Nuwara-Ellia, Kandy, the next in elevation, is about 5,548 ft. above the sea.

The mountains are generally in continuous ranges, and are seldom or never found isolated. This region is skirted by a hilly country, from 10 to 20 m. wide, and varying in elevation from 100 to 500 ft., with occasional summits of more than twice that height. This tract is destitute of the ravines and other bold features of the mountainous country.

Rivers and Lakes. — Ceylon has numerous small rivers and perennial streams; but few of them are navigable, even by a canoe, to many miles from their mouths. The principal is the Mahavilly Ganga; it rises near the highest part of the central tableland, about 30 m. S. Candy; and having received many tributaries, falls into the sea, a little S. of Trincomalee, after a course of about 200 m. It is the only river navigable for any considerable distance. The next most important river is the Kalani-Ganga, which has its source in the country at the foot of Adam's Peak, and empties itself into the ocean by several mouths in the neighbourhood of Columbo; it is made considerable use of for internal traffic.

There are no lakes of any consequence in the interior, the largest being no more than 4 m. across; but along the E. coast, from Batticaloa northward, there are several extensive lagoons, which, by means of artificial channels, are made serviceable to traffic; other lagoons exist in the neighbourhood of Negumbo and Columbo. (Davy's *Account of the Interior of Ceylon*, pp. 1–5; Percival's *Account*, pp. 55–60.)

Geology and Minerals. — The rocks met with in Ceylon are mostly primitive, and consist, with little exception, of granite or gneiss, with large veins of quartz, hornblende, and a snow-white dolomite; limestone occurs only in Jaffnapatam, and the N. districts. A belt of grey or black sandstone, together with coral formations, nearly encompasses the whole island. The upper soil is in general sandy, with but a small mixture of clay, and chiefly derived from the disintegration of primitive rocks; the cinnamon soil near Columbo is perfectly white, and consists of pure quartz. Ceylon is rich in valuable minerals; its metallic products are, however, comparatively unknown: ores of iron, lead, tin, and manganese are found in the interior, but are made little use of: plumbago is the only article amongst them which has become of any commercial importance. Mines of quicksilver were formerly worked by the Dutch. It has numerous gems: and common salt-beds are found in various places. No volcanos exist in Ceylon, nor are mineral waters very abundant; but they are met with near Trincomalee.

Climate.—The mountain ranges which separate Ceylon almost completely into two parts, by arresting the course of the monsoons, occasion a radical difference at the same moment in the climate of the E. and W. parts, whole floods of rain deluging the island on one side, while on the other the natives are carefully husbanding all the water left from previous inundations. In the N. and SW. the climate is moist, temperate, and similar to that of Malabar; in the E. and SE. it is hot and dry, and more like that prevalent on the Coromandel coast. The SW. monsoon lasts from April to Sept.; the NE. from Nov. to Feb.: in the intervening months the winds are variable. The SW. monsoons are usually accompanied by violent storms of thunder and lightning, and torrents of rain, which sometimes extend themselves to the central table-land, especially in March and April: but this high region is generally out of the influence of either monsoon, and both its winds and temperature are greatly modified by its own physical character, and the directions of its principal ridges. The quantity of rain which falls during the year is about three times as great as in England; the rains being, though not more frequent, far heavier, so much so that a fall of two or even three inches in twenty-four hours is not uncommon; 84 inches is the annual estimate in the alpine region, and 100 inches at Columbo. The seasons depend more on the monsoons than on the course of the sun; and the coolest season is during the summer solstice, while the SW. monsoon prevails. The heat is, however, nearly the same throughout the year, and much less oppressive than on the continent of India. Along the coast, the annual mean temperature is about 84° Fahr.; at Candy, 1,467 ft. above the sea, it is 78°; at Columbo the annual variation is from 76° to 84°; at Galle, 70° to 90°; at Trincomalee, 74° to 91°. For a tropical country, Ceylon has a comparatively salubrious climate; but some of the less inhabited parts, and the low wooded hilly country between the mountains and the sea, are highly insalubrious. Near Columbo and Trincomalee, where the jungle has been cleared away, and the land drained, the country has been rendered perfectly healthy. The prevalent diseases are those affecting the liver and intestines, often accompanied by fever: diseases of the lungs, urinary organs, and nervous system, are very rare; gout is unknown. Elephantiasis, *Lichen tropicus*, and other cutaneous complaints, are common. The small-pox was formerly very destructive, but is now guarded against by vaccination, to which the natives raise no objection: measles and hooping-cough both occur in a mild form. The beri-beri (*Hydrops asthmaticus*) is a disease nearly peculiar to Ceylon.

Vegetable products are numerous and valuable. The most important, next to rice and other grain, is the cinnamon (*Laurus Cinnamomum*), called by the Singalese *corunda*, which here arrives at its greatest perfection, and has always been a chief article of export. It delights in a poor sandy soil, with a moist atmosphere, and is almost exclusively confined to the SE. part of the island, between Negombo and Matura. In the N., where the climate is dry and sultry, it is totally unknown, and the endeavours to propagate it at Batavia, in the W. Indies, and on the opposite coast of Tinnevelly, have not been so successful as was anticipated. In its wild state it grows to the height of 20 or 30 ft., and bears a white blossom in January: while in bloom, the cinnamon forests have a very beautiful appearance; but the aroma of the plant revives wholly in the bark, and the fragrance of the groves is not nearly so great as strangers have been led to believe. The soil is peculiarly suitable for the growth of coffee; and its culture has of late years been so much extended that it is now the principal article of export. The cocoa-nut tree flourishes with singular vigour, and is of great importance to the native population, almost every part of the tree being converted into articles of food or domestic use: the best trees produce from 50 to 100 nuts annually, and grow so close to the sea, that the roots are even washed by its surge. The Palmyra palm grows principally in the N. part of the island, and is scarcely of less importance than the cocoa-nut tree. The tallipot palm, the leaves of which are large enough to shelter many individuals, grows luxuriantly here, though rare on the continent of India. The bread-fruit-tree attains an immense size; cotton is not equal to that of India; indigo is found wild, but its culture is neglected; the areca and betel nut, as well as tobacco, all of which are of excellent quality, grow abundantly: the cardamom seeds are inferior to those of Malabar. Gum-lac and gamboge are also produced in this island. The Flora of Ceylon is not so extensive as beautiful and various: the rose, pink, mignonette, &c. are as fragrant as in England, and the jessamine much more so; the glorious superba and amaryllis grow in profusion, and the jamba, or rose-apple, strews the ground with its scarlet blossoms. (Heber's Narrative, iii. 143-145, &c.; Percival, pp. 319-337.)

Animals.—Ceylon has been from an early period celebrated for its breed of elephants, which, though inferior in size to those of other countries, are more valued for their greater strength and docility. The chase of these animals has always been with the Singalese an object of great importance; but the avidity with which they have been pursued has greatly diminished their numbers, and they are now chiefly confined to the N. and NE. districts. The royal tiger is not met with, but bears, leopards, the cheta (a small species of leopard), hyenas, jackals, and tiger-cats are numerous: besides this, deer, gazelles, buffaloes, wild hogs, and monkeys. Near Jaffna a large baboon is very abundant, and fearless: a large variety of the monkey tribe, porcupines, racoons, armadilloes, squirrels, and mungooses, are met with. There are no foxes; but the flying fox and rats are very common and troublesome. Pheasants, snipes, red-legged partridges, pigeons, peacocks, and a great variety of birds; with serpents, alligators, and reptiles of all sorts, are abundantly plentiful. The fishing of the pearl oyster is an important branch of industry.

People.—The pop. of Ceylon, exclusive of the various colonists who have at different times possessed themselves of the coasts, may be divided into four classes:—1st, the native Singalese of Ceylonese, who may be again subdivided into those occupying the Candian territories, and those of the coasts; 2nd, the Moors, who are found in all parts of the island, and form the chief population of the district of Putlam; 3rd, the Veddahs, a savage race, who are supposed to be the aborigines, and inhabit the mountainous regions and uncxplored fastnesses, almost in a state of nature; 4th, the Malabar and other Hindoos, who are chiefly confined to the N. and E. coasts. The Singalese of the coasts, whose complexion, features, language, and manners closely resemble those of the Maldivians, are about 5 ft. 6 in. in height, of a slim figure and fair complexion, especially the women; they are represented as remarkably mild, bashful, and timid, and rather deficient in intellect. The Candian Ceylonese are in all respects superior to those of the coasts, and differ from Europeans less in feature than in colour; they are taller, better made, and more robust, than the Singalese; and for Indians are stout, with large chests and broad shoulders. They have small bones, rather short but muscular legs and thighs, and small hands and feet; heads well formed, and, like those of other Asiatics, larger than those of Europeans; features often handsome. The colour of their skin, eyes, and hair varies from brown to black; they have a profusion of hair, which is allowed to grow to a considerable length. The Candian character differs essentially from that of the Singalese, having none of the effeminacy and timidity which distinguish the latter, and there is a certain haughtiness and independence in their whole bearing and demeanour. They will not generally, however, attack an enemy in the open field; but resort to ambush, in the same manner as the Singalese. Indolence, hypocrisy, and revenge may be regarded as national vices. Some traits may be recognised as common to the natives of Ceylon with the Bengalese, but they are still more closely allied, both in physical and moral characteristics, as well as language, religion, and traditions, with the Indo-Chinese nations, and especially the Birmese. The Malabars of Ceylon differ but little in any respect from those of the continent, though varying somewhat in their manners and customs. They retain, in great measure, the religion and manners of their congeners of S. India, and are much less numerous than formerly. The Moors have a tradition that they are the descendants of a tribe of the posterity of Hashem, expelled by Mohammed from Arabia. They retain many customs similar to those of the ancient Jews.

Of the Veddahs little more is known than that they chiefly inhabit the great forests which extend from the S. to the E. and N., and also the most inaccessible parts of the central table-land, having neither clothing nor habitations, subsisting upon wild fruits and animals, and having the branches of large trees for their resting-places. They are conjectured by some to be a portion of the original inhabitants, who, upon the invasion of the island, retreated to the inaccessible haunts in which they are now found. They are divided into two tribes, —the Village and the Forest Veddahs; the former, who are the more civilized, occasionally go down into the lower districts to exchange their game and cattle for rice, cloth, and iron. They live in huts and cultivate the ground; though, in common with their more savage brethren, they seek their chief subsistence in the forests. They are peaceable and inoffensive, never commencing, although easily persuaded to join in any insurrection; and in times of disturbance they have occasionally been employed as mercenaries.

The other inhabitants of the coast consist of Dutch, Portuguese, and English colonists; some Caffres and Javanese; a few Chinese and Parsee traders; and a various pop., springing from the intermixture of these with each other and with the native races. The burghers, many of whom fill public offices and subordinate situations under

government, are the descendants of Europeans and half-castes. The distinctions of caste are recognised, and in some instances scrupulously preserved, by the Ceylonese; but they respect them only in their civil, rejecting their religious, influence.

Till lately, the pop. had been diminishing for four or five centuries. But a considerable increase has taken place in the pop. of the maritime provinces during the last thirty or forty years. Several parts of the interior are, however, very thinly peopled, there being, in some districts, not more than four, five, or six persons to a square mile. In the central prov. the pop. is denser in certain parts; but with the exception of the country round Candy, and the districts of Ouva and Mattele, seven-eighths of the ground is covered with wood and jungle, and nearly unpeopled.

Ceylon is now divided into six provinces, the area and population of which, according to a census taken in the year 1852, is shown in the following table:—

Provinces	Area in Square Miles	Population
Western	8,850	725,812
North-Western	8,367	216,994
Southern	7,147	834,350
Eastern	4,733	93,733
Northern	5,427	419,162
Central	5,191	885,882
Total	**34,700**	**3,073,234**

Not included in these population returns are 4,647 military persons, which added make the total population 2,979,881.

It will be seen, from the preceding table, that the pop. of Ceylon is very unequally distributed, the western province being the densest populated part—180 inhabitants per square mile—and the eastern province the least dense—only seventeen inhabitants on the square mile. This inequality is only partly explained by differences of soil and climate.

Agriculture.—The tract of country near the Coromandel coast is only in some parts fit for tillage, the ground for many miles exposing only a barren and naked surface. The soil of the central parts is capable of producing luxuriant crops were it properly cultivated. All products requiring a moist soil and climate flourish most in the SW., and rice is grown chiefly in the level lands there, or in the valleys of the hill region, but often also on the slopes, on account of the facilities they present for irrigation. Around the fields, on the level lands intended for its reception, small embankments, about three feet in height, are raised, and water let in upon them; they are afterwards trodden over by buffaloes or turned up with a sort of light plough. On the hill slopes the rice-fields are dammed up, and form a succession of terraces, for irrigating which the water is conveyed sometimes for a mile or two along the mountain sides, and let off from one terrace to another, as the state of the grain requires it. There are two rice harvests during the year; the first crop is sown from July to October, and reaped from January to March; the second is sown from March to May, and reaped from August to October. What is called a plough consists of a piece of crooked timber shod with iron, which tears rather than ploughs up the ground. After the first ploughing, the fields are flooded; then ploughed again, and carefully weeded. Rice is industriously cultivated by the Malabars of the N. and NE. districts; but the produce is insufficient for the consumption of the island, and large quantities are annually imported from both the Malabar and Coromandel coasts. Hemp is raised in abundance, the sandy soil of the maritime districts being well adapted for it. Cotton of different sorts grows with the greatest facility, the buds ripening within four months after being sown. Each village or hut has its sugar and tobacco plantation; coffee is raised of a very superior quality.

As cinnamon forms a chief article of export and revenue in Ceylon, its cultivation is one of great interest, and is conducted with much care. The neighbourhood of Colombo is particularly favourable for its growth, being well sheltered, and having a high and equable temperature. About 2,000 acres of land, chiefly near that town, are laid out in cinnamon plantations, furnishing employment to 80,000 individuals, and yielding annually about 500,000 lbs. of bark, worth 150,000l. sterling. In its wild state the plant grows to the size of a large apple-tree; but when cultivated, is not allowed to attain to more than 10 or 12 ft. in height, after seven or eight years' growth. May and June are the months for stripping the bark from the plant, which is done by two methods. In the first, the rough bark is removed with knives, and the inner rinds stripped off by a peculiarly shaped instrument; by the other method, the outer bark is not artificially removed, but the process of fermentation which the strips undergo when tied together in large quantities spontaneously removes it. The bark, in drying, gradually contracts, and rolls itself into a quill-like form; and, after being subsequently dried in the sun, the smaller are inserted in the larger pieces, and the whole are made up into bundles of about 30 lbs. weight. Layers, shoots, and transplanted stumps are the best means of extending the growth of the cinnamon plant.

Wages are considerably higher, and provisions proportionally dearer, in Ceylon than in Bengal. Those of the poorer classes, who possess small portions of land, rarely derive their support from it exclusively, but employ themselves in fisheries, trades, manufacture, and the petty traffic of the country; the wages of mechanics and artisans are proportionally higher than those of the labouring population, but still very moderate. A very minute subdivision of property often exists, and the inheritance of one person will sometimes consist of 9-10ths of a seer of rice land, 6-12ths of the produce of a cocoa-nut-tree, or 2-3rds of that of a jack-tree. Notwithstanding this, the peasantry of Ceylon are generally in better circumstances than those of the adjoining continent. They are not under either a zemindary or ryotwarry settlement, and the demands of the government on the land rarely exceed 1-10th part of the produce, and are sometimes less. Under the Candyan government, the tenures of land were of three kinds. Some lands belonged wholly to the sovereign; others were cultivated by individuals at a government rent, of some fixed proportion of the produce; and others, again, were granted as payment for the performance of specific services to the headmen of different districts, chiefs, and reverted again to the crown on the death of such individuals. The latter could neither be mortgaged nor alienated; the second class of lands might be transferred in any way as long as the permanent rent continued to be paid. The lands belonging to the sovereign himself were cultivated on his account, or let out to the highest bidder, and sometimes brought a rent of 1-3rd or half the produce. The plan of redeeming the whole rent, above 1-10th part of the produce, has been adopted by the British government with much success, and in those districts where the practice

has prevailed the revenue has increased rather than diminished; for more lands having been brought into cultivation, 1-10th part of the crops now yields as much as 1-5th or 1-4th part formerly did. Domestic animals are not numerous. The horse is a degenerate breed, and not aboriginal; oxen, though small, are well tamed, and the chief food of the British troops, though eaten by none else; poultry of all kinds are abundant.

Pearl Fishery.—The pearl fishery in the Bay of Condatchy, which was formerly a government monopoly, is now free; but, whether from the banks having been over-fished, or otherwise, the produce is now of comparatively little importance. The pearl banks are formed by coral ridges from 6 to 10 m. off shore, and of a variable depth, but commonly from five to seven fathoms below the surface. The oysters are attached by threadlands to these ridges from within a short time of their hatching from the egg, to about 6½ years old, when they loose their hold, and drop to the sandy bottom, where they lie in heaps. Soon after attaining the age of seven years, the animals are said to perish. As many as sixty pearls have been found in one oyster; but such instances are rare, as it, indeed, the presence of pearls generally. The season commences in Feb. and finishes in April; six weeks or two months, at the utmost, is the time allowed for its continuance. Each of the boats carries a tindal, or master, and twenty-three men, ten of whom are divers, and relieve each other, five divers being constantly at work during the hours of fishing. After they are taken out of the boats, the oysters are left to open spontaneously, die, and rot. the stench of their putrefaction filling the air for many miles round Condatchy, till it is swept off by the SW. monsoon. The Ceylon pearls are whiter than those of Ormuz, or the Arabian coast; and the natives are extremely expert in cutting and drilling them. The usual Ceylonese boats are like the catamarans of Madras and other parts of the peninsula. A great number of chank shells are found, and exported to India from the N. shores of Ceylon. (Haerlenberger, Dr., in Martin's Statistics, p. 400; Percival, pp. 86-100; Sturt, in Phil. Transac., iii. 8.)

Salt is a government monopoly, and its manufacture, in lewarys and pits on the sea-shore, is carried on to a great extent in the N. and E., where it is of fine quality, and may be procured in greater abundance than the government requires, or has been able to collect. Before the Dutch monopoly existed, this coast supplied Bengal with salt; and, indeed, the Ceylon salt may be imported at Calcutta for two-thirds the price of the salt produced in India. There are no other manufactures of any extent or importance, except that of arrack, which is distilled from the blossoms of the cocoa-nut-tree, as toddy and jaghery are from the juice; while ropes, brushes, baskets, brooms, matting, rafters, and thatch for cottages are obtained from the various parts of the tree, in addition to the valuable oil now in extensive use in England. Saltpetre is made from the chipping of rocks, in which nitrate of lime is prevalent, mixed with wood ashes; the mixture washed, and the liquor evaporated to a concentrated solution, and suffered to crystallise. Lime of excellent quality, and possessing a power of adhesion much greater than that procured from shells, is made by burning the coral found upon the shore. Gunpowder is made by a rude process; the native petry is coarse and unglazed. Little progress has been made in weaving, the loom is somewhat similar to the primitive loom of Ireland; all the cloths used are of domestic manufacture; no muslins are woven, nor indeed anything but coarse cottons, and some silks. Rude images and implements of husbandry are made of the native metals, and the Singalese can work with dexterity and taste in gold and silver. They are generally more capable of setting gems than cutting them; and excel in the manufacture of lacquered ware.

Trade.—Since the Dutch monopoly system has been abandoned, both the internal traffic and foreign trade have greatly increased. Subjoined is a table of the imports and exports of Ceylon, in the two years 1862 and 1863:—

Ceylon—Imports		1862	1863
PRINCIPAL ARTICLES			
Coals and Coke .	Tons	23,779	41,164
	£	89,490	177,729
Cotton Manu-factures .	Pieces	1,089,311	1,094,653
	Packgs.	12,938	11,648
	£	563,344	190,444
Cotton Twist .	Cwts.	1,087	901
	£	977	89
	£	46,893	60,694
Cutlery and Hardware .	Packgs.	5,415	3,774
	Cwts.	740	149
	Pieces	950	11,665
	£	17,343	65,417
Curry Stuffs .	Packgs.	—	7
	Cwts.	68,763	57,607
	Baskets		
	£	87,974	49,617
Fish, Salted and Dried .	Cwts.	51,643	64,945
	£	53,643	63,345
Grain : Paddy .	Bushels	597,477	794,790
		50,453	117,741
— Rice .	Bushels	1,390,991	1,415,620
	£	1,263,541	1,731,714
Haberdashery and Millinery	Packgs.	1,301	1,659
	Pieces	40	4,513
	£	48,853	84,383
Specie and Bullion	£	1,549,114	1,642,974
Total Value of principal and other Articles		£4,643,140	£5,422,667

Ceylon—Exports		1862	1863
PRINCIPAL ARTICLES			
Areca Nuts .	Cwts.	55,773	69,108
	£	41,429	41,704
Cinnamon .	Lbs.	575,474	751,999
	£	43,778	36,349
Coffee, Planta-tion .	Cwts.	418,535	519,944
	£	1,272,317	1,947,104
Coffee, Native .	Cwts.	177,935	156,517
	£	344,48	317,054
Cotton Manu-factures .	Packgs.	4,404	5,961
	Pieces	773,415	341,604
	£	140,591	349,763
Cotton Twist .	Packgs.	714	944
	£	87,574	69,105
— Wool .	Bags	3,273	4,417
	Cwts.	494	1,892
	£	36,193	83,361
Oil, Cocoanut .	Cwts.	115,344	152,074
	£	143,316	180,878
Specie and Bullion .	£	706,453	406,064
Spirits, Arrack .	Gallons	164,642	109,250
	£	5,341	7,489
Tobacco, un-manufactured	Cwts.	19,730	27,113
	£	19,369	27,144
	Bales	—	—
Total Value of principal and other Articles		£7,894,120	£12,467,234

Both the imports and exports of Ceylon have enormously increased since the year 1850. In this year, the imports were 1,406,679*l.*, and the exports 1,246,956*l.* The rise took place very gradually, but in imports was chiefly visible in grain, and in exports in coffee.

There is a canal between Calpentyn and Columbo, by which cargoes are conveyed during the SW. monsoon. A fine road has been constructed from Columbo to Candy, on which a mail-coach runs; carriage-roads also extend from Columbo N. to Chilaw, and S. to Matura. Many rapid and unfordable streams have had iron and wooden bridges thrown across them, amongst which is that of Parabdenia, across the Mahavilly Ganga, which consists of a single arch, with a span of 205 ft., principally composed of satin-wood.

English weights, measures, and moneys are becoming universal in Ceylon.

The public revenue of Ceylon consists chiefly of import duties on merchandise and indirect taxes. It amounted to 767,401*l.* in 1860; to 751,597*l.* in 1861; and 759,136*l.* in 1862. The public expenditure is principally for costs of administration, and was 705,445*l.* in 1860; 635,250*l.* in 1861; and 670,654*l.* in 1862. The cost of governor and principal officers amounted to 67,865*l.* in 1862; while there were expended in the same year for works and buildings 23,896*l.*, and for roads, streets, and bridges 94,167*l.*

The administration of the colony is vested in the hands of a British governor, assisted by a council of European civil servants, selected either by the governor himself or the secretary of state for the colonies; but the power of the council is limited, and subservient to the authority of the governor. The governor has complete control over the financial department in the interior, while in the maritime provinces he is restricted to a certain sum for contingent expenditure, unless authorised in exceeding it by his council, to whom, except on this point, he refers, or not, at pleasure, being empowered to carry into effect any law without their concurrence. All laws, before being acted upon, are published in the official gazette, for the purpose of their general diffusion, with translations into the Singalese and Malabar languages.

The native business of the government is conducted by individuals of three different classes. Officers of the first and second classes are usually filled by Europeans; the subordinate situations by natives; but, by recent regulations, any person judged to possess sufficient qualifications may fill the most important offices without reference to nation or faith; a knowledge of the English language being, however, considered indispensable. Each village and caste has its elected headman, who is recognised by the government, which commonly selects native servants from amongst this class of people; the *modeliars* of castes, or lieutenants of districts, are appointed from this body.

Armed Force.—Exclusive of native troops there are in Ceylon, on the average, some 3,000 British troops. The cost of these is chiefly borne by the home government, and amounted to 110,206*l.* in 1862. The contribution of the colony towards this military expenditure was only 24,000*l.* in this year 1862.

Justice.—A supreme court of justice is established at Columbo, with powers equivalent to those of the Court of Queen's Bench and Court of Chancery. It is presided over by three English judges, aided by two other functionaries, all of whom are appointed from England. Trial by jury was introduced into Ceylon by Sir A. Johnston, and is now established in every district. Exclusive of Columbo, the whole island is divided into three circuits, viz. the N., S., and E.; the last of which comprises the old kingdom of Candy, with all the country to the E. of it. The circuits are subdivided into many districts, each of which has its own court, with a judge and three assessors, and with jurisdiction in all cases not punishable with more than a fine of 10£, one year's imprisonment, or 100 lashes. The supreme court in Columbo is the sole court of appeal. Excepting in the maritime provinces, where attack drinking is prevalent, atrocious crimes are in general rare; so that the courts are more occupied with petty litigations than serious offences.

The Religion of the Singalese is Buddhism; but the upper classes profess Christianity, and many of the others have been converted to Mohammedanism. There are 16 Protestant churches in the island, subordinate to the archdeacon of Colombo, and 32 dissenting places of worship. Roman Catholic chapels are very numerous, and 10 years ago it was believed that half the Ceylonese population were Christians, following the ritual mostly of the Romish and Dutch churches. There is a tradition amongst the natives that Buddh himself visited this island, which, before his advent, had been inhabited by demons. There are numerous temples to that deity in the island, especially in the central parts, where the Buddhic sect is most prevalent; and the British government, having succeeded to the temple patronage and other privileges belonging to the old kingdom of Candy, has the appointment of the Buddhic priests. When the palace of Candy was taken by the British, a celebrated relic, believed by the natives to be a genuine tooth of Buddh, was captured; the possession of which is considered to insure its possessor the sovereignty of the whole island. This relic is annually exposed with great state and ceremony, and is worshipped by multitudes flocking from all parts of the country, and bringing offerings of various kinds to the priests, who thereby realise considerable sums. (Journal of the Asiat. Soc., iii. 101.)

Public Education.—Education is making great progress in Ceylon. There were, in 1832, above 600 schools in the colony, attended by 25,000 pupils. Of these, 5,518 were in 'public;' 49 in 'orphan;' 822 in 'regimental;' 13,511 in 'free;' and 5,500 in 'private' schools. Unfortunately, the benefit of this education did not include the female sex. In 1832 there were but 876 females in the 'public;' 21 in the 'orphan;' 35 in the 'regimental;' and 116 in the 'private' schools. The free schools, which had 13,511 male pupils, were not attended by a single female. The public schools are supported by government. The others have been established by the Church Missionary and Dissenters' Missionary societies. Free elementary education in the English language, arithmetic, and geography is given in these schools. The government schools are chiefly in the Singalese maritime districts. At Columbo there is a superior academy, where the usual branches of a classical and mathematical education are taught.

Civilization and Arts.—In civilization the Singalese appear to be nearly, if not quite, on a par with the Hindus; in courtesy and polish of manners they are inferior to none, but in intellectual acquirements, and proficiency in the arts and sciences, they have made little advancement. Many of the male Singalese read and write in their own tongue, but this is no part of female education. They write with a sharp iron style, on talipot leaves, and colour the traces afterwards with lamp-black. They excel more in lacquered painting than in any other art. Their statuary is better than their pictures, though the figures of Buddh have been subject to no innovation of style, and are always in the same posture, of whatever material they may be formed. The Singalese colour the statues of their gods, and give a pupil to the eye; which last ceremony is supposed to confer all the holiness belonging to the figure, and is done with much mystery and solemnity. There seems to be no peculiar national style of architecture; the Buddhic temples are like Tartar structures. The Ceylonese rise at dawn, and retire at nine or ten o'clock at night; they sleep either on mats on the floor, or on couches. Their meals are short and unsocial, the men and women not often eating together; there are two principal meals, one taken at noon, and the other at seven or eight o'clock in the evening. The standing dish consists of rice with curry; some eat eggs and poultry; but beef is never eaten excepting by a very low class, who are in consequence held in great abhorrence: milk, ghee, oil, and fruits are the other important articles of diet. The best of their houses are commonly of mud, with tiled roofs, and a single story in height; built on a low terrace, presenting outwardly dead walls, and having in the interior an open space, into which the rooms open by doors, which, as well as the windows, are very narrow. The floors are composed of clay, plastered with manure, to keep off the insects, and the walls are covered with the same material, or a coat of white clay; lime is used for the walls of temples only. The furniture of the houses consists of two or three stools, a few mats, and porcelain dishes, a stone hand mill, a pestle and mortar for rice, a rattan bag for compressing seeds to procure their oil, and a few other indispensable articles. The dress of the men is a handkerchief wrapped like a turban round the head, leaving the top exposed, and a long cloth, called *toprity*, reaching from the loins to the ankles. That of the women is very similar; they leave the head uncovered, but the end of their dress is thrown across the left shoulder. On occasions of ceremony, both sexes wear a small jacket. Rings, and silver and crystal bangles and other ornaments, are commonly worn, and certain privileged persons are permitted to wear gold and silver chains and trinkets; but the Ceylonese look with extreme jealousy on every assumption of these which is not strictly in conformity with the caste of its wearer. Like the Hindus, they admit of the four chief subdivisions of castes, viz. the religious and military orders; *Teesera*, cultivators, merchants, &c.; and *Raband-ras*, artisans: the first two ranks have, however, scarcely any actual existence in Ceylon, and all the honours and hereditary rank in the island are monopolised by the cultivators, at the head of the third class, with whom all Europeans are ranked, while the Moors are classed with the fishermen at the head of the fourth order. The male Singalese marry generally at the age of eighteen or twenty, the females earlier. Matches are determined on and concluded by the parents of the parties to be affianced; the dowry of the women generally consists of household goods, or rather, seldom of land; the husband always pays a price for his wife. The women seldom have more than four or five children; but sometimes suckle them for as many years; the latter are in consequence very backward, and often neither speak nor walk till upwards of two years old. Infidelity is little regarded, provided it is not an intrigue with a person of inferior caste: concubinage and polygamy are indulged in by the men, but plurality of husbands is more common than that of wives, one woman belonging equally to several brothers of the same family. This, as well as other usages, is, however, fast disappearing before new habits, acquired by the extending intercourse with Europeans. The Ceylonese appear to be sincere and warm in their attachments. Dr. Davy disbelieves the report of the practice of exposing female infants, 'excepting in the wildest parts of the country, and then nerve from choice, but necessity, and when the parents are on the brink of starving.' The sick and dying, though not openly exposed, are certainly removed to temporary buildings. Every respectable family burns its dead; low castes are not allowed to do so, but bury them with the head towards the west. Immediately after a decease, the relations, with their hair dishevelled, and beating their breasts, cry and embrace each other, giving utterance to lamentations of a highly poetical nature. (See Journal of Asiatic Society, ii. 63, 64.) A common exhortation is, 'When I die, pay me due honours.' The common language of the Singalese is a dialect of the Sanscrit; the sacred language, like that of the Birmans, is the Pali. (For further details of Ceylon, see the works of Sir J. Emerson Tennent:—'Christianity in Ceylon,' 1850; 'Sketches of the Natural History of Ceylon,' 1861; and the admirable and most exhaustive 'Ceylon, an Account of the Island, Physical, Historical, and Topographical, 6th edit. 1861.)

Antiquities and History.—The proper name of this island is Singhala; but there is considerable uncertainty whence the people originated who gave it that name, and who are called Singalese. They have a tradition that their ancestors came hither from the eastward nearly 2,400 years ago: some modern authors think, on the other hand, that they were a colony of Singhs, or Rajpoots, who arrived here about 500 years B.C. Tsijeya (perhaps of the royal house of Satya Singh of Magadha, the native country of Budh, but evidently the same as the Sanscrit Vijaya) is the first king of Ceylon mentioned in history. The numerous ruins of cities, tanks, aqueducts, extensive canals, bridges, temples, &c., show that Ceylon had been, at a remote period, a rich, populous, and comparatively civilised country. In 1505 the Portuguese formed settlements on the W. and S. coasts, and received a tribute of cinnamon from the king of Candy, on condition of defending Ceylon against the Arabian pirates. They, as well as the Dutch who expelled them, after a long and sanguinary struggle in the next century, and the English, who superseded the latter, became, soon after the conquest of their first enemies, involved in hostilities with their native allies. In 1815 the Candyans entrusted the interference of the British, to drive a tyrannical sovereign from the throne. This was soon effected, and Candy has since become a part of the British dominions.

CHABLIS, a town of France, dép. Yonne, cap. cant., on the Serny, 10 m. E. Auxerre. Pop. 2,385 in 1861. The town is principally distinguished by its excellent white wines, which the French epicures take with oysters.

CHAIBAR, or KHEIBAR, a town of Arabia, in El-Hedjaz. Lat. 25° N., long. 39° 30′ E., 150 m. N.E. Medina. Pop. said to be 50,000. It is the cap. of, and given name to, an independent sovereignty of Jews, the descendants, according to their own assertion, of the Trans-Jordanic tribes, Reuben, Gad, and Manasseh. They have a character for bravery and learning; but the term Beni-Chaibar is so odious among Mohammedans that its application is regarded as an insult. In manners and appearance the Jews of Chaibar do not differ from other Arabs; their state has existed upwards of 1,100 years; and though the town was captured by Mohammed in the 7th Hejira, A.D. 628, it is still said to be flourishing and powerful. It was here that Mohammed received from a Jewess a poisoned cup, professedly to test his prophetic powers, which laid the seeds of the disorder under which he finally sank, about four years afterwards.

CHALONS-SUR-MARNE, or CHALONS, a city of France, cap. dép. Marne, on the Marne, in the middle of extensive meadows, 27 m. SE. Rheims, on the railway from Paris to Strasbourg. Pop. 16,576 in 1861. The Marne formerly traversed the town, but since 1788 it has skirted it in a new channel dug for the purpose, and crossed by a magnificent stone bridge. Two small affluents of the Marne run through the town. It is surrounded by old walls in pretty good preservation. With the exception of that which leads from the bridge to the Hôtel de Ville the streets are narrow and crooked; houses generally mean, and a few being of wood. The cathedral, consecrated in 1117, and rebuilt in 1672, is a large fabric, partly of Greek and partly of Gothic architecture. The Hôtel de Ville and the Hôtel de Préfecture are both fine buildings; the Porte St. Croix has a good effect, and there is a splendid promenade, called the Jard. It is the seat of a bishopric, and has a court of primary jurisdiction, a commercial tribunal, a departmental college, a primary normal school, a diocesan seminary, a school of practical geometry, a botanical garden, a society of agriculture, commerce, and a public library, with 20,000 vols. But the most important establishment belonging to the town is the public school of arts and trades, at which 450 pupils are maintained, at the expense of government, exclusive of those who pay. It has also a theatre. Different branches of the woollen, linen, and cotton manufactures are carried on in the town; there are also extensive tanneries, and a good deal of trade is carried on with Paris in wine, corn, wool, hemp, and rape-oil. La Calle, the astronomer, and D'Aldancourt, the translator, were natives of Chalons.

This is a very ancient town: it has been repeatedly taken and pillaged, and was once much more considerable than at present. Attila was defeated under its walls in 451. In 1591 and 1592 it burned the bulls of Pope Gregory XIV. and Clement VIII. against Henry IV. In 1814 it was for a while the central point of the operations of Napoleon.

CHALONS-SUR-SAONE, or CHALOS, a town of France, dép. Saône-et-Loire, cap. arrond., in a fertile plain, on the right bank of the Saône, which here forms an island, in which is situated the suburb St. Laurent, 34 m. N. Macon on the railway from Paris to Lyon. Pop. 19,709 in 1861. The town is pretty well built, but the streets are narrow and ill paved: it has a fine quay on the Saône, and is connected with its suburb by a stone bridge of five arches. There is a cathedral, and a hôtel de ville; but the objects most worthy of attention are the Hospice St. Laurent, in the suburb of that name, and the Hôpital St. Louis, both large establishments, and exceedingly well managed. The latter is an asylum for indigent old persons and orphans. There are some fine promenades, one of which at the head of the Canal du Centre, is ornamented with an obelisk in honour of Napoleon. The bishopric has been suppressed; but it has a court of primary jurisdiction, a tribunal of commerce, a dep. college, a school of design, a public library with 10,000 volumes, and a theatre.

Chalons is very favourably situated for a commercial entrepôt, communicating with the Mediterranean by the great line of railway from Paris to Marseilles, which has a station here, as well as by the Rhone and Saône, and the canals connected with them, and with the North Sea by the canal of the centre, constructed in 1792.

The town is very ancient, and was for some time the capital of the kingdom of Burgundy. It suffered severely during the civil wars of the 16th century, and not a little from the invasion of the allies in 1811. It was formerly very unhealthy; but in this respect it has been materially impovered, by the better drainage of the surrounding country, and the greater attention paid to cleanliness in the town, though in both these respects it might still be very considerably improved. The famous Alelard died here in 1142.

CHAMAS (ST.), a town of France, dép. Bouches-du-Rhône, on the N. bank of the lagoon de Berre, 23 m. NW. Marseilles. Pop. 2,692 in 1861. The town is well built, has a handsome church, and is celebrated for its oils and olives, which it ships from its port on the lagoon. It is divided into two portions by a hill, through which a large tunnel has been cut for a channel of communication. It has an important powder magazine, which supplies Toulon and the fortresses dependent upon it. In the vicinity is a Roman bridge, of a single arch, having a triumphal arch at each extremity.

CHAMBERTIN, a famous vineyard of France, dép. Côte d'Or, a few miles NE. Beaune. It occupies about twenty-five hectares, and produces at an average from 130 to 150 pipes of Burgundy. Chambertin was the favourite wine of Louis XIV. and of Napoleon.

CHAMBERY, a city of France, dép. Savoie, cap. of dép., on the left bank of the Ayrve, in an elevated and fertile valley, 110 m. WNW. Turin, and 42 m. SSW. Geneva, on the railway from Paris to Mont Cenis, which is to be prolonged by means of a gigantic tunnel under the Alps, to Turin. Pop. 19,950 in 1861. The city presents little worthy of notice; it has one good street, but most of the others are crooked, dark, and somber. There are several squares adorned with fountains; and most of the houses are three stories in height. Chief public buildings, the cathedral, the Hôtel Dieu or principal hospital, the barracks constructed by the French, and the manufactories of silk-gauzes, for which Chambery has long been celebrated. The palace is an old castle, in no way remarkable. The churches exhibit gaudy decorations; in one, however, there is some good painted glass. The city was formerly fortified; but the walls have been removed, and the space they occupied is laid out as public walks.

Chambery is the seat of the superior judicial tribunal, and of an archbishop. It has societies of agriculture and commerce, a public library, theatre, public baths, and many charitable institutions. Besides gauze, other silk fabrics, lace, hats, leather, and soap are manufactured; and there is some trade in liqueurs, wines, lead, copper, and various other articles. The environs abound in vineyards, woods, and picturesque scenery. Near Chambery is the country house of Les Charmettes, once the residence of Mad. de Warens and Rousseau. The city is supposed to stand near, though not upon, the site of the ancient Lemincum. It was taken by the French in 1792, who made it the cap. of the dép. of Mont Blanc, and retained it till the second treaty of Paris, in November, 1815, when it was made over to the king of Sardinia, who, however, gave it up, together with the whole province of Savoy, to France, in 1861.

CHAMBORD, a village and famous castle of France, dép. Loire-et-Cher, on the Cosson, 10 m. E. Blois. The village—pop. 527 in 1861—is inconsiderable, and the place derives its entire importance from its castle, one of the most magnificent and best preserved in France. This noble edifice was commenced by Francis I., after his return from Spain. He is said to have employed 1,800 workmen for twelve years upon it; and here, in 1540, he entertained his illustrious rival Charles V. The building was still further enlarged by Henry II., and finished by Louis XIV., who frequently inhabited it during the early part

of his reign. The Bourgeois Gentilhomme of Molière was acted, for the first time, at a fête given here by Louis, in October, 1670. Stanislaus Lezinsky, king of Poland, occupied this castle for nine years previously to his being put in possession of the duchy of Lorraine. In 1748 it was assigned by Louis XV. to Marshal Saxe, who spent in it the evening of his days in almost regal splendour. After many vicissitudes, it was given by Napoleon to Marshal Berthier; and having been sold by his widow, in 1820, it was bought by subscription for the Duc de Bordeaux, to whom its possession has since been confirmed by a decision of the courts. Since the expulsion of the elder line of the Bourbons from France, the head of the family has taken his name from this property.

The castle is buried in deep woods, and its situation is rather low and damp. It is of vast extent, in the Gothic style, and has a profusion of towers, turrets, and minarets. Being built of black stone, it has a heavy appearance. The interior is very magnificent. The grand staircase is so contrived that persons ascending and descending do not see each other; it has two fine chapels, and many spacious apartments and splendid ceilings. Its gorgeous furniture was sold by auction during the Revolution; and the beautiful tapestry that adorned the apartments of Francis I., Louis XIV., and Marshal Saxe, was burned, as the surest way of getting at the gold and silver with which it was embroidered; but the castle itself was not injured. The park is of great extent, comprising above 12,000 acres.

CHAMOND (ST.), a town of France, dép. Loire, cap. cant., in a fine valley at the confluence of the Gier and the Ban, 8 m. NE. St. Etienne. Pop. 11,620 in 1861. It is a thriving, industrious town, is well built, has a handsome promenade, a departmental college, a fine parish church, and public baths. On a hill above the town are the ruins of the ancient castle, destroyed during the revolution. The manufacture of ribbons and lacets (laces) is very extensively carried on. It has, also, considerable cast-iron and nail-works.

CHAMOUNY, or CHAMOUNIX, a celebrated valley of the Alps, dép. Haute-Savoie, France, immediately NW. of Mont Blanc, by which, and others of the Pennine Alps, it is bounded on its S. and E. sides, and on the W. and N. by Mont Breven and the Aiguilles Rouges. Its length, NE. to SW., from the base of the mountains, is about 12 m., and its breadth at the bottom in most parts exceeds a mile; but including the mountain slopes and sides, it is as much as 9 m. in breadth, and may be reckoned 22 m. in length from its head at the Col-de-Balme to its outlet at the torrent of the Diosa, near Servoz. The average height of this valley above the sea is about 3,400 ft.; the Arve rises at its upper end, and intersects it in its entire length, escaping into the valley of Servoz through a ridge of granitic rock. The pines and larches which clothe the lower parts of the mountains give a somber appearance to the W. end of the valley; and this effect is increased by the unvaried masses of Mont Blanc, which hang over it. But after passing the jetory of Chamounix, the scene changes, and to this dreary magnificence succeeds a series of majestic pyramids, called Aiguilles, or needles, of astonishing height, and too steep to admit of the snows resting on them at any season. The valley, which becomes narrower, is richly ornamented with trees; and the Arve, rushing between finely-clothed rocks and precipices, adds life and beauty to the scene. The little village of Argentière, with its church and glittering spire, and the two *Aiguilles* above it, together with the cheerful appearance of cultivation, from a landscape sublimely picturesque. The average height of the mountain-range on the R. side of Chamounix is about 6000 ft.; but the principal *Aiguilles* on this side, viz. those of Charmos; the A. Verte, de Dru, d'Argentière, and de la Tour, rise from 11,000 to 13,100 ft. above the level of the sea. Between these *Aiguilles* are situated the numerous glaciers which constitute the chief interest of the valley, to the very bottom of which they descend. Nowhere else in the Alps are the glaciers of equal magnitude.

These mountains of ice are formed by the consolidation of the snow lodged in the high Alpine valleys. As the surface of the snow thaws and percolates through the mass, it is again frozen, and acts as a cement; and by a repetition of this process, the whole mass is converted into solid ice; not so compact, however, as that of rivers or lakes; for it is full of air-bubbles, owing to the mode of its formation. Entering the valley from the NW., the first glaciers met with are those of Tacunay and de Boisson, succeeding which are the more considerable ones of Montanvert, de Boix, d'Argentière, and de la Tour. The glacier de Boix, at the foot of the Aiguille de Dru, and about a league E. of the village of Chamounix, is the largest of all: it is upwards of 7 m. in length, and in some places more than a mile broad; it is, in fact, the terminus of the Mer de Glace, (See MONT BLANC.) Near its foot, the Arveiron, a tributary of the Arve, has its source in an ice-cavern, which varies in size at different periods of the year; but is sometimes as much as 100 ft. in height. On the W. side of the valley, Mont Breven, and the *Aiguilles Rouges* (so called from their reddish colour) form an unbroken ridge, but of a much less elevation than that on the opposite side of Chamouny. The Col-de-Balme, at the NE. end of the valley, and 8,000 ft. above it, affords a full and magnificent view of the gigantic group. Across this mountain one of the roads from Chamouny into the Valais passes. The climate is rigorous; the winter in the valley of Chamounix lasts from October to May, during which season the snow usually lies to the depth of 3 ft., while at the village of Tour, the highest in the valley, it often attains the depth of 12 or 13 ft. In summer, the thermometer at noon commonly stands no higher than from 57° to 68°; it rarely reaches 68° Fahr. Barley and other kinds of corn, pulse, hemp, and some fruits, are grown, and a great many cattle are reared. The honey of Chamounix is of a very fine quality. The total pop. of the valley was about 4,000 in 1861. There are several small villages; that of Prieuré, or Chamounix, *par excellence*, on the right bank of the Arve, towards the centre of the valley, has a pop. of about 1,700, and several good inns. It originated in a Benedictine convent, founded here at the end of the 11th century by Count Aymon of Geneva. The other chief villages are Ouches, Argentière, Le Bossons, and Tour.

CHAMPAGNE, the name of an old prov. of France, in the E. part of the k. adjacent to Franche Comté and Lorraine, now distributed among the depts. of the Ardennes, Marne, Haute Marne, Aube, Gonne, and Seine-et Marne. Champagne is also the name of several small towns in different parts of France.

CHAMPLAIN (LAKE OF), a long and narrow lake, principally in the U. States of N. America, between New York and Vermont, and having its N. extremity in Lower Canada. This lake occupies a considerable part of what has been called the Great Glen of N. America; that is, the remarkable hollow or chasm, stretching N. from New York to the St. Lawrence, a distance of about 500 m. The glen is occupied from New York to Glen's Falls, 130 m., by the Hudson; thence for 21 or 22 m. to Lake Champlain, by a table land which, in its highest part, is only 140 ft. above the level of the tides in the Hudson. The lake extends N. and S. 110 m., with a breadth varying from 1 to 14 m.; but it is, in general, very narrow; the distance, 67 or 70 m. from the lake to the St. Lawrence, is traversed by the river Richelieu, or Chambly, the outlet of the lake, which is partly navigable by vessels of 150 tons, and throughout by river barges. A canal has been constructed uniting Lake Champlain and the navigable portion of the Hudson; so that there is now a direct inland navigation, which, by a little outlay on the Richelieu, might be made suitable for steamers, from New York to the St. Lawrence, between Montreal and Quebec. (Darby; Gordon's Gaz. of New York.)

CHAMPON, or CHOOMPHOON, an inl. town of Lower Siam, on the road between Ligor and Bankok, on the E. bank of a river about 7 m. W. the Gulf of Siam; lat. 10° 51' N., long. 99° 27' E. Estimated pop. 8,000. In 1826 it was stockaded, and considered by the Siamese an important military post. Tin, good timber for ship-building, and excellent rattans, are found in its vicinity.

CHANDA, an inl. town of Hindostan, prov. Gundwanah, cap. distr. of same name, between two small rivers, 62 m. S. Nagpoor; lat. 20° 4' N., long. 79° 22' E. Its walls are 6 m. in circuit, and from 16 to 20 ft. in height, built of freestone, well cemented, and flanked by round towers. Its interior consists of straggling streets, detached houses, gardens, and plantations. In 1803 it contained 5,000 houses; in 1822 only 2,000. In its centre there is a fort called Bala Killa. Chanda was taken by the British in 1818, when it was found to contain a good deal of treasure and valuable property, brought thither for security.

CHANDERNAGORE, a marit. town of Hindostan, prov. Bengal, belonging to the French, built on the W. bank of the Hooghly river, 16 m. NNW. Calcutta, and in point of situation, in every respect superior to that city; lat. 22° 49' N., long. 88° 26' E. In 1814 it had a pop. of 41,000, but which has been reduced now to less than half that number. The streets are straight and well-paved, but present a scene of solitude and desertion; and the trade, formerly so flourishing, is almost annihilated. There are some manufactures of cotton cloths; the commerce is chiefly in opium. The territory originally attached to this town extended to 2 m. along the river, and 1 m. inland; about 2 m. below Chandernagore are the ruins of a superb house, the country residence of its former governors. The French, in 1676, obtained permission to establish this settlement, which they subsequently appropriated and fortified. In 1757 it was taken by the British, who destroyed the fortifications.

CHANDORE, a considerable inl. town of Hindostan, prov. Candeish, pres. Bombay, 60 m. WNW. Aurangabad, lat. 20° 19' N., long. 74° 19' E. It has a most formidable position on a rock, commanding one of the best passes on the range of hills on which it is situated, and is quite inaccessible everywhere but at the gateway, where it is strongly fortified. It however surrendered without much resistance to the British arms, both in 1804 and 1818.

CHANTIBUN, a large inl. town of Siam, cap. of the rich distr. of the same name, at the foot of the mountain chain separating it from Cambaja, on the S. bank of a river 18 m. E. the Gulf of Siam, and 150 m. SE. Bankok; lat. 12° 45' N., long.

10° 19′ E. It is a place of considerable trade; its chief export is pepper, to the amount of 30,000 or 40,000 pécule yearly. Cardamoms, rose-wood, dye-woods, ship timber, hides, horns, ivory, lac, and beeswax, are products of the Chantibon distr. Near the town are mines of precious stones.

CHANTILLY, a neat town of France, dép. Oise, on the Nonette, and on the road from Paris to Amiens, 24 m. N. of the former, on the Northern railway. Pop. 2,930 in 1861. It has a fine hospital, endowed by the last prince of Condé. This town is distinguished by its industry and manufactures of cotton and porcelain; but it owes its celebrity to its having been, since 1632, the seat of the family of Condé, and to the vast sums they expended on the formation and embellishment of its castle, park, and gardens. The castle was one of the largest and finest structures of the kind in France; the 'grand Condé' lived here in regal magnificence; and the entertainments given by him to Louis XIV. were so splendid as to excite the jealousy of the monarch. But the glories of Chantilly have disappeared, and cotton-mills occupy the sites where Racine, Molière, and Boileau used to recite their chefs d'œuvres amid the applauses of all that was beautiful and chivalrous in France.

The Grand Château, rebuilt in 1779, was destroyed during the revolution, and all that now remains is the Petit Château, the Château d'Enghien, and the stables; the latter, constructed between 1719 and 1735, are unequalled in Europe. The remains of the Admiral de Coligni, butchered at the massacre of St. Bartholomew, are interred in the parish church of Chantilly. The forest of Chantilly occupies a space of about 3,905 hectares.

CHAPEL-EN-LE-FRITH, a market town and par., England, co. Derby, hund. High Peak, on the declivity of a hill rising from an extensive and fertile vale, surrounded by lofty eminences, 11 m. NW. by W. Derby, 167 m. NW. by N. London. Pop. of par. 4,261 in 1861. The town is only partially paved. There is one cotton mill, employing about 120 hands, and many of the lower classes are employed in weaving for the Manchester houses. At White Hall Mill is a considerable manufactory of paper. There is a brewery in the town, and nails are also made. Here is an establishment for warehousing goods, the place being a medium of communication between Manchester and Sheffield, and having in consequence a large carrying trade. The town is one of the polling places for the election of members, for the N. div. of the co. Besides the par. church, a neat edifice with a square tower, there is a chapel for Wesleyan Methodists. There is also an endowed school at Chapel-en-le-Frith, and another at Bowden's Edge. Lead and coal mines and quarries are worked in the vicinity. The Peak Forest lime-works lie 3 m. E. of this town, and communicate by railway with the Peak Forest canal. The par. includes the townships of Bowden's Edge, Bradshaw's Edge, and Combs's Edge.

CHARD, a town and bor. of England, co. Somerset, hund. Kingsbury East, in an elevated situation, near the S. border of the co., 11 m. N. Lyme Regis, and 170 m. WSW. London, by London and South-Western railway. Pop. of bor. 2,276, and of par. 5,316 in 1861. The old municipal bor., which is a parish of itself, comprises an area of fifty-two acres; but the area of the new municipal bor. has been increased. It has an old town-hall, an extensive market-place, a church with a tower and bells, a well-endowed hospital for the maintenance of old and infirm persons belonging to the parish, and is well supplied with water. Fairs, 1st Wednesday in May, August, and November.

Market-day, Monday. Chard was made a bor. by Edward I. and elected mems. to nine parliaments, when it lost the privilege.

CHARENTE, an inland dép. of France, distr. of the W., formed principally out of the ancient prov. of Angoumois; it takes its name from the Charente, by which it is traversed; and has N. the Deux Sèvres and Vienne, E. Haute Vienne, S. Dordogne, and W. the Charente Inférieure. Area 561,234 hectares, or 8,270 Eng. sq. m.; pop. 379,081 in 1861. Surface diversified by a great number of little hills. Soil various, being mostly thin or clayey, and encumbered with moisture; the latter prevails in the arrond. of Confolens, where there are no fewer than sixty-two shallow lakes, or étangs, none of them of considerable extent; there is also in the latter arrond., and in that of Barbezieux, a large extent of heath and waste land. Principal corn crops, wheat, maslin (a mixture of wheat and rye), maize and millet, rye, barley, and oats; but, owing to the inferiority of the soil, the returns are among the poorest in France, and the produce is insufficient for the consumption. The principal wealth of the dép. consists in its vineyards, which cover about 100,000 hectares. Their produce is mostly converted into eaux-de-vie, the superiority of that made at Cognac being universally acknowledged. Hemp, flax, and potatoes are extensively cultivated. The woods cover about 74,000 hectares; and the produce of chestnuts averages 200,000 hectolitres. Truffles are abundant, the value of those sold being estimated at about 500,000 fr. a year. There are, comparatively, few horses; but cattle, sheep, and hogs are abundant; wolves, foxes, and otters are pretty common, but wild boars have become rare. The minerals are antimony, lead, iron, and gypsum; the last two being wrought to a considerable extent. Besides the iron-works, there are very extensive distilleries, with paper-works (see Angoulême), tanneries, and manufactures of linen, canvas, cordage, cloth, hats, and earthenware. The dép. is divided into five arrondissements. The principal towns are Angoulême, Cognac, Ruffec, and Confolens.

CHARENTE INFÉRIEURE, a maritime dép. of France, on the W. coast, deriving, like the foregoing, its name from the Charente, by which it is intersected; having N. Vendée, NE. Deux Sèvres, E. Charente, S. the Gironde, and W. the Atlantic Ocean. Area, including that of the islands of Oleron, Ré, and Aix, 682,569 hectares, or 2,762 sq. m.; pop. 481,060 in 1861. Surface flat, and in part marshy; soil partly light, calcareous, and gravelly, and partly heavy and clayey. Principal crops, wheat, maslin, rye, barley, maize, and oats. The rotation is, 1st year, wheat; 2nd rye, or some other grain; during the 3rd year the ground remains untilled, serving as a kind of pasture for sheep; in the 4th year the old routine recommences. Rent of arable and pasture land varies from 8s. to 36s. an acre. About half the dép. is cultivated by proprietors, who possess from 50 to 100 and 150 acres; the other half is occupied by farmers, whose farms may vary from 300 to 700 acres, and who are said to be prosperous. About 112,000 hectares are occupied by vineyards, whose product, like those of the Charente, is mostly converted into eau-de-vie or brandy. The forests cover above 70,000 hectares. Pastures extensive and excellent, furnishing food for a great number of cattle, excellent horses, and sheep. Minerals not of much importance; but there are in the dép. very extensive salt marshes, particularly in the neighbourhood of Marennes, which furnish large quantities of salt. In summer, the marshes are unhealthy, but otherwise the climate is mild and salubrious. This dép. has great facilities for commerce. It has several deep bays and excellent ports, and, exclusive of the Charente, which has Rochefort near its mouth, it is watered by the navigable rivers Seudre and Sèvre, from the latter of which there is a canal to La Rochelle, and is skirted on the N. by the Gironde. The fishery of sardines and oysters is extensively carried on, and vessels are also fitted out for the cod fishery. La Rochelle, Rochefort, and the other ports have also a considerable share of the colonial and coasting trade of France. With the exception of the salt manufacture and distillation, manufacturing industry is not prosecuted on a large scale; but coarse woollen stuffs, soap, fine earthenware, and glass are produced; and there are also tanneries and sugar refineries. The dép. is divided into six arrond. Principal towns, La Rochelle, Rochefort, Saintes, and St. Jean d'Angely.

CHARENTON-LE-PONT, a town of France, dép. Seine, cap. cant., agreeably situated on the Marne, near its confluence with the Seine, 4 m. SE. Paris, on the railway from Paris to Troyes. Pop. 5,551 in 1861. The town has several country houses, among which is the one occupied by the famous Gabrielle d'Estrées. The Marne is here crossed by a bridge, the possession of which has always been regarded as of material importance to the defence or attack of Paris; and it has frequently been the scene of obstinate conflicts, the last of which took place in 1814, when it was forced by the allies. The bridge unites the town with the village of Charenton St. Maurice. There is here an excellent lunatic asylum, founded in 1741, and capable of accommodating 400 patients. The Protestants had formerly a large church in this village, in which synods were held in 1623, 1631, and 1644; but it was demolished in 1685, after the revocation of the edict of Nantes.

CHARITÉ (LA), a town of France, dép. Nièvre, cap. cant., at the foot of a hill planted with vines. The Southern railway has a station here. Pop. 5,297 in 1861. The town is situated on the right bank of the Loire, over which there are two bridges. It was formerly fortified, and much more considerable than at present. It is celebrated for its manufactures of coarse jewellery, buttons, glass, earthenware and woollen stuffs.

CHARKOFF, See Kharkoff.

CHARLEROY, or CHARLEROI, a fortified and important manufacturing town of the prov. of Hainault, in Belgium, on the navigable river Sambre, 33 m. S. of Brussels, on the railway from Brussels to Paris. Pop. 10,800 in 1856. The town is built on the side of a steep hill, and the inhab. are occupied chiefly in working the extensive coal mines of the district, and in numerous iron foundries and glass works. The town is in the centre of the great coal-basin of Charleroy. Adjacent quarries of slate and marble are also important sources of industry and wealth; and the neighbourhood contains numerous mills for sawing marbles. The manufactures of glass comprise all kinds of vessels and sheet glass, of various qualities; and the iron works include the manufacture of fire-arms, cutlery, tools, and utensils. There are, besides these principal establishments, several factories for spinning wool and weaving woollen cloths; dye-houses, tanneries, snuff mills, rope walks, soap-houses, salt and sugar refineries, breweries, distilleries, and brickyards. The communication with Brussels by means of the railway, as well as the Charleroy canal, affords great facilities for commerce. Between 200 and 300 capacious barges are constantly employed in exporting from Charleroy to Brussels coal, iron, slate, glass, and soap. A large fair for cattle and merchandise is held during 10 days, commencing on the 5th of Aug.

The fortress of Charleroy was built in 1666, by Rodrigo, Spanish gov. of the Netherlands, and named after Charles II. king of Spain. The lower and middle town were added by Louis XIV. In 1676, Charleroy has sustained several memorable sieges; and by various treaties has been transferred from Spain to France, from France to Spain, from Spain to Austria, and from Austria to France. The fortifications were materially improved under the direction of the Duke of Wellington, after the campaign of 1815. Near Charleroy are the ruins of the magnificent abbey of Alne, in a beautifully romantic solitude, about 9 m. from the town. The cloisters of this superb establishment were supported by 800 columns of coloured marble, and its revenue amounted to 250,000l.

CHARLESTON, a city and sea-port of the U. States, one of the principal in the S. part of the Union, and the largest town of S. Carolina, on a low point of land at the confluence of the Cooper and Ashley rivers, 6 m. W. by N. the nearest point of the Atlantic, 118 m. NE. Savannah, and 500 m. NNW. Baltimore; lat. 32° 46' N., long. 79° 49' W. Pop. 51,200 in 1860. Charleston was, till 1787, the seat of the state government. This city was visited, in 1838, by a most destructive fire, which raged with great fury in its most populous part, destroying several streets and an immense amount of property. Previously to this disaster, the streets, which were rather narrow, crossed each other at right angles, and were often planted with pride-of-India trees (Melia azederach); the houses were mostly of brick, and generally furnished with verandahs. Charleston was partly destroyed a second time in 1861, when it was taken possession of by the troops of the United States, after having been for four years in the hands of the Confederate government, serving as the chief port of entry for foreign vessels into the Southern States, and the principal refuge of 'blockade runners.' The town has a college, townhall, exchange, custom-house, guard-house, theatre, citrus, orphan asylum, hospital, two markets, two arsenals, and numerous churches. The college, established in 1785, was reorganised in 1824; it possesses a commodious edifice, with a library and philosophical apparatus. There are two medical schools, and various learned and charitable societies. The harbour is large and convenient, but rather difficult of access, in consequence of its entrance being obstructed by a range of sandbanks. Through these there are but two channels suitable for ships of large burden. In the principal or S. channel the depth of water in the shallowest part, 8 m. SE. from the town, at ebb tide, is only about 12 ft., and at flood tide from 17 to 18 ft. A lighthouse, 80 ft. high, with a revolving light, has been erected on a small island bearing 2½ m. NW. from the bar, at the entrance to the S. channel. After crossing the bar, there is deep water up to the city, where vessels lie moored alongside wharfs or quays. Charleston is a place of very extensive trade, it being the part whence more than three-fourths of the whole foreign trade of S. Carolina is carried on. Its exports consist chiefly of cotton and rice. Most of the imports are from the N. and middle states, and consist of wheat and flour, fish, shoes, and all kinds of manufactured goods. The foreign imports are mostly brought at second hand from New York, and consist of cottons, woollens, linens, hardware, iron and steel, coffee, sugar, tea, wine, and spices. Like most other cities in the S. part of the United States, Charleston formerly had a large slave pop., and the slaves were treated with a severity revolting to those who lived in countries free from this moral contamination. Happily all this has ceased by the emancipation of the slaves in 1863, in consequence of the great civil war in the United States. The yellow fever occasionally commits great ravages here; but it is more fatal to foreigners than to the native pop. The fever is supposed to be owing, in a considerable degree, to the marshy nature of the soil on which a part of the town has been built; but the swampy ravines by which it was formerly intersected have been gradually filled up and drained, and the city has, in consequence, become much more healthy. The town is badly supplied with water, having mostly to depend on the rain water collected in cisterns. Charleston was founded in 1680, and was the seat of government till the building of Columbia, in 1787.

CHARLEVILLE, an inland town of Ireland, prov. Munster, N. extremity co. Cork, 24 m. S. Limerick, on the railway from Limerick to Cork. Pop. 4,766 in 1851, and 2,468 in 1861. The town consists of four main streets crossing each other at right angles. In it are the par. church, a large R. Cath. chapel, a building for public meetings, a national school, and an endowed grammar school. The corporation, under a charter of Charles II., in 1671, consists of a sovereign, two bailiffs, twelve burgesses, and an indefinite commonalty. It returned two members to the Irish H. of C. till the Union, when it was disfranchised. A manor court has jurisdiction in pleas to the amount of 20£, and is a civil bill court. Petty sessions are held on alternate Mondays. The court and market house are in the same building. Tanning and blanket making are carried on to some extent, and there are two large flour mills. Markets on Saturdays; fairs on 10th Jan., 10th March, 12th May, 15th Aug., 10th Oct., and 10th Nov. The town is a constabulary station.

CHARLEVILLE, a town of France, dép. Ardennes, on the Meuse, at a short distance from Mezières, on the railway from Chalons to Sedan. Pop. 9,307 in 1861. The town is extremely well built; streets straight and broad, intersecting each other at right angles; houses nearly all of the same bright, and slated, having a comfortable, gay appearance. In the centre of the town is a fine square, surrounded by arcades, and ornamented with a superb fountain. The river is crossed by a suspension bridge. It is the seat of a court of primary jurisdiction, and of a commercial tribunal; and has a departmental college, a primary normal school, a secondary ecclesiastical school, a course of geometry and mechanics applied to the arts, a public library, with 21,000 vols., a cabinet of natural history and antiquities, and a theatre. The royal manufactory of arms, formerly established here, has been transferred to Tulle and Châtellerault; but arms are still largely manufactured on account of individuals. The nail-works produce about 3,500,000 kilog. of nails a year; and there are, besides, copper foundries, where large quantities of copper-wire, and plates, are produced, with map-works and tanneries. It has a considerable part on the Meuse, and a considerable trade in wine, spirits, coal, iron, slates, marble, and manufactured goods. Through the canal of Ardennes, as well as the railway, it has also an easy communication with Paris.

The foundations of Charleville were laid in 1606, by Charles of Gonzaga, duke of Mantua Nevers, who gave it his name. Having passed from his heirs to the house of Bourbon, the fortifications were razed in 1686, by order of Louis XIV.

CHARLOTTENBURG, a town of Prussia, prov. Brandenburg, on the left bank of the Spree,

5 m. W. Berlin, with which it is connected by railway. Pop. 12,431 in 1861. The town consists chiefly of villas and taverns, the summer residence of the rich, and the resort of the humbler classes from Berlin; is well built, and has handsome straight streets, ornamented with rows of trees. There is a magnificent palace, built by Frederick the Great, and furnished with a collection of antiquities. The gardens, which are finely laid out, are always open to the public, and are much visited by Sunday parties and strollers from the capital. Within the gardens is the mausoleum, erected by King Frederick William III., over the remains of his beautiful and unfortunate queen, Louisa of Mecklenburg. It contains the celebrated recumbent marble statue of Louisa, by Rauch, admitted to be not only the masterpiece of that eminent sculptor, but one of the finest modern works of art.

CHAROLLES, a town of France, dép. Saône-et-Loire, cap. arrond., at the confluence of the Semence and the Reconce, 28 m. W.N.W. Mâcon. Pop. 3,264 in 1861. The town is agreeably situated, neat, and well built; has a communal college, tribunals of primary jurisdiction and commerce, an agricultural society, iron forges, and fabrics of earthenware, and crucibles. A hill above the town is crowned with the picturesque ruins of the old castle of the counts of Charolais. One of these, a prince of the blood royal, who lived during the reign of Louis XV., achieved an infamous notoriety.

CHARTRES, a city of France, dép. Eure-et-Loir, of which it is the capital, on the Eure, 48 m. S.W. Paris, on the railway from Paris to Nantes. Pop. 19,531 in 1861. The town is surrounded by walls and ditches, and is situated partly on a hill, and partly on low ground. The Eure, which here divides into two branches, runs through and encircles the lower town. Streets narrow and crooked; those forming the communication between the upper and lower towns being so very steep as to be inaccessible to carriages. The cathedral is reckoned one of the finest Gothic edifices in France. Here are, also, two fine steeples, a monument to General Marceau, barracks, a theatre, and some fine promenades. It is the seat of a bishopric; has a court of assizes, tribunals of primary jurisdiction and commerce, a departmental college, a public library, with 30,000 vols., a school of design, and a botanical garden. The manufactures consist principally of hosiery and hats, and there are also tanneries and dye-works. Chartres is the centre of the corn trade of the dép., its corn-markets being among the most important in France, and providing in a great measure for the supply of Paris. It is the native country of Regnier the poet, of Brissot, and Pétion, members of the convention, and of General Marceau.

This is a very ancient city, being reckoned before the Roman conquest, as the capital of Celtic Gaul. It was for a considerable time in the possession of the English. Henry IV. was crowned here in 1594.

CHARTREUSE (LA GRANDE), a famous monastery of France, dép. Isère, 14 m. N. Grenoble, among rugged mountains, at an elevation of 3,781 ft. (1,000 metres) above the level of the sea. The access to it is very difficult. This monastery was founded in 1084; but having been several times pillaged and burnt down, the present building has been erected since 1676. It is of vast extent, and has cost an immense sum. During the revolution, the monks were driven out, and their property, including their valuable library, confiscated and sold. But, in 1816, the building, which had escaped the revolutionary tempest, was restored to its original destitution. Some of the old monks, accompanied by several neophytes, returned to the building; and the Chartreuse existed once more, but shorn of its old lustre, importance, and wealth.

CHARYBDIS. See Scylla and Charybdis.

CHATEAUBRIANT, a town of France, dep. Loire Inférieure, cap. cant., on the Chere, near the pond or lake of Grand Lieu, 26 m. W.N.W. Ancenis. Pop. 4,636 in 1861. The town is old and meanly built, round the ruins of the old castle, founded in 1015, whence it derives its name. Françoise de Foix, celebrated for her beauty and gallantries with Francis I., died here in 1537, and was buried in the church of the Trinity, with an epitaph on her tomb written by Clement Marot. The town has a court of primary jurisdiction, an agricultural society, and manufactures of coarse woollen stuffs, and its pastry and confitures are held in high estimation. It has some trade in iron, coal, and wood, and a considerable corn-market.

CHATEAU-CHINON, or CHINONVILLE, a town of France, dép. Nièvre, cap. arrond., near the Yonne, in the middle of mountains, at an elevation of 1,968 ft. (600 metres) above the level of the sea, 50 m. W.N.W. Autun. Pop. 2,561 in 1861. The town was formerly surrounded by fortifications, and was defended by a vast castle, of which there exist considerable ruins. It has a court of primary jurisdiction, an agricultural society, and some fabrics of coarse woollens and linens. Having been taken by the royalists in 1591, after an obstinate resistance, the garrison and the greater part of the inhabitants were put to the sword.

CHATEAUDUN, a town of France, dép. Eure-et-Loire, cap. arrond., near the left bank of the Loire, 25 m. S.S.W. Chartres. Pop. 6,719 in 1861. Having been almost wholly burnt down in 1723, it has been rebuilt on a regular plan, with broad straight streets, and uniform houses. The principal square, the Hôtel de Ville, and the buildings of the communal college, are worthy of notice. Besides the college, it has a court of primary jurisdiction, a public library, with 6,000 vols., and some manufactures of woollens and tanneries. On a rock, commanding the town, are the remains of the old castle of the Counts of Dunois, the chapel attached to which has the tomb of the famous general of Charles VII., and some other tombs of less distinguished members of the family.

CHATEAU-GONTIER, a town of France, dép. Mayenne, cap. arrond., on the Mayenne, 18 m. S. Laval. Pop. 7,311 in 1861. The town is badly laid out, but well built; has a stone bridge over the river, by which it is united to its principal suburb, a fine Gothic church, a communal college, 8 hospitals, public baths, an agricultural society, &c.; and is the seat of a court of original jurisdiction. It has considerable manufactures of fine linen and linen thread, with extensive bleachfields; is the entrepôt of a great proportion of the wines, slate, coal, and tufa of the dep.; and the centre of the trade in fine thread. The town was formerly surrounded by walls, and had a castle, whence it took its name. It suffered a good deal during the wars of Vendée.

CHATEAULIN, a town of France, dép. Finistère, cap. arrond., in an agreeable valley, on the Aulne, which there takes the name of Châteaulin, 22 m. S.E. Brest, on the railway from Brest to Quimper. Pop. 2,832 in 1861. The town is ill-built; has a court of primary jurisdiction, and an agricultural society. Vessels of from 60 to 80 tons come up to the town, which has a good deal of trade in slates, procured from quarries in the neighbourhood, cattle, and butter.

CHATEAUNEUF-DE-RANDON, an inconsiderable town of France, dép. Lozère, cap. cant., on

a mountain, 12 m. N.E. Mende. Pop. 1,463 in 1861. The town was formerly fortified; and an English garrison was besieged in it, in 1380, by a French force under the famous constable Duguesclin; the constable having died during the course of the siege, the English governor laid on his coffin the keys of the town, which he had engaged to deliver up to him if not relieved within fifteen days. A monument was erected here in 1820 to the memory of Duguesclin.

CHATEAUNEUF-SUR-CHARENTE, a town of France, dép. Charente, cap. cant., on the Charente, 12 m. W.N.W. Angoulême. Pop. 3,565 in 1861. The town has a considerable trade in wine, brandy, and salt. It was anciently called Bordeville, and was defended by a castle burnt down in 1081. A new castle having been built to replace the former, the town took from it the name of Châteauneuf.

CHATEAUROUX, a town of France, dép. Indre, of which it is the cap., in an extensive plain on the left bank of the Indre and on the railway from Paris to Bordeaux. Pop. 16,176 in 1861. Châteauroux continues to be one of the worst built towns in France. Streets narrow, crooked, and ill-paved; houses small, irregularly built, and gloomy. It has, however, some finely shaded agreeable promenades, and some good buildings. It is the seat of a court of assizes, of tribunals of primary jurisdiction and commerce; and has a theatre, a public library, a public garden, a society of agriculture, science and arts, and an annual exhibition of the products of the industry of the dep. The cloth manufacture is very extensively carried on; cotton hosiery and hats are also produced, and there are establishments for the spinning of wool, with tan-works, and tile-works.

Châteauroux was founded in 950; was burned down in 1868, and rebuilt shortly after. Louis XIII. erected it into a duchy; and it was given by Louis XV. to one of his mistresses, Madame de Mailly, better known by the name of the Duchess de Châteauroux.

CHATEAU-THIERRY, a town of France, dép. Aisne, cap. arrond., on the Marne, 25 m. S. Soissons, on the railway from Paris to Soissons. Pop. 5,925 in 1861. The town is built on the declivity of a hill, the summit of which is surmounted by its ancient castle, a vast mass of thick walls, towers, and turrets. It has a considerable suburb on the left bank of the Marne, the communication between them being kept up by a handsome stone bridge of three arches. It has a court of primary jurisdiction, a communal college, an establishment for the spinning of cotton, and tanneries. The famous poet La Fontaine, not less original by his character and conduct than by his talent and genius, was born here on the 8th of July, 1621. The house which he inhabited is still preserved; and a marble statue was erected to his memory on the end of the bridge in 1824. Château-Thierry suffered considerably during the campaign of 1814.

CHATELLERAULT, a town of France, dép. Vienne, cap. arrond., on the Vienne, 20 m. N.N.E. Poitiers on the railway from Paris to Poitiers and Bordeaux. Pop. 11,210 in 1861. The town is situated in a fertile, agreeable country, but is ill built. It is joined to its suburb on the opposite side of the river by a stone bridge, built by the Duc de Sully. Besides several churches, it has a communal college, a theatre, an exchange, a hospital, and a royal manufacture of arms, the buildings of which are among the finest in the town; and some fine promenades. This town has been long famous for its cutlery, and has manufactures of clocks and watches, and lace. It serves as a kind of entrepôt for the towns of the S. and the N. of France; particularly for wines, spirits, salt, slates, iron, corn, hemp, and timber. The French Earl of Arran, ancestor of the Dukes of Hamilton, was created Duc de Châtelherault in 1548.

CHATHAM, a par., town, parl. bor., naval arsenal, and sea-port of England, co. Kent, lat ho Aylesford, on the Medway, 29 m. E. by N. London, by road, and 31 m. by London, Chatham and Dover railway. Pop. of parl. bor. 36,177 in 1861. Chatham is separated from the city of Rochester by a merely artificial line; and the latter being connected with Strood by a bridge, the three towns form a continuous street of upwards of 3 m. along the old Dover road from London. For about ½ m. below Rochester, the town extends along the bank of the river, which there bends NNE. and E. by N. till it falls into the estuary of the Thames at Sheerness. Notwithstanding the shortness of its course, the Medway has very deep water. At Chatham the tide rises 18 ft. at springs, and 12 ft. at neaps; and from Sheerness to Chatham there is water to float the largest ships; and the ground being soft, and the reaches short, it forms an admirable harbour for men-of-war; and it is to its facilities in this respect that Chatham and the contiguous towns are mainly indebted for their rise.

The principal church, a plain brick structure, was rebuilt in 1788; and a more modern one was erected in 1821, by the parliamentary commissioners. Several more churches have since been built, as also a number of dissenting chapels, a national school, a proprietary classical school, a philosophical and literary institution, to which a museum is attached, and two public subscription libraries. Here is also a chapel, on the site of one attached to a monastery, founded in 1078, the endowment of which supports four brothers, two of them in orders. Sir J. Hawkins' hospital for decayed seamen and shipwrights, chartered in 1594, supports 10 individuals. There are three or four minor charities. 'Chatham chest,' which originated with Sir F. Drake and Sir J. Hawkins, after the Spanish Armada, and at first consisted of voluntary contributions from seamen, soon became compulsory, and was ultimately removed, in 1803, to Greenwich. Down to the 11th Wm. IV. it was supported by deductions from the monthly wages of seamen, but an act of that session made it chargeable on the consolidated fund. The town was considerably improved under an act passed in 1772, but many parts of it still remain narrow and irregular. It is in the jurisdiction of the co. magistrates, with the exception of a small part, comprised within the municipal limits of Rochester. The Reform Act, which made it a borough, conferred on it the privilege of returning one member to the H. of C. The limits of the parliamentary borough include a considerable area N. and E. of the town. Registered electors 1,741 in 1865. Market-day, Saturday; annual fairs, May 15, September 19, each lasting three days; annual races in August.

The town is almost wholly dependent on the great naval and military establishments at the emption, in its immediate neighbourhood, but separated from it by a line of fortifications. The dockyard, which lies along the E. side of the river, is including the arsenal, above 1 m. in length; and is defended by Gillingham Fort, Upnor Castle, and several bastions. Fort Pitt, on the S. or land side of the town, was erected in 1805. The dockyard contains between 500 and 600 houses for the artificers employed in the different works, and is abundantly supplied with every means and accommodation required for the building and fitting out of the largest ships. It has five large tide docks, capable of receiving first rate men-of-war, and six building-slips for vessels of the largest dimensions; a mast-house, attached to which are saw-mills worked by steam, and two large floating basins for the reception of the timber for the masts; a smithery, where anchors of the largest size are forged; a rope-house, where cables above 100 fathoms in length, and 25 inches diameter, are twisted by powerful machinery; a mill-house; and numerous warehouses, containing every article required for the building and equipment of ships of war. Here also is a spare set of Brunel's block machinery, in the event of that at Portsmouth getting out of order; dwellings for the civil officers of the establishment, and a handsome chapel. Near the entrance (which is a spacious gateway flanked by two towers) is a general marine hospital, built in 1828, and capable of receiving 300 patients. Four hulks moored off the dockyard, one for juvenile, two for adult offenders, and one as a hospital, form the convict establishment, usually containing from 500 to 1,000 individuals, employed in the common drudgery of the arsenal. The ordnance wharf, to the W. of the dockyard (on the slip of land between the church and river, quarter of a m. from the high street of Chatham), contains the guns belonging to each vessel respectively, in separate tiers, piles of shot and shells, a well-arranged armoury, and a large building in which lead is rolled and print ground by steam machinery. The military establishments, comprised within the lines, consist of large infantry, marine, engineer, and artillery barracks, with a park of artillery. There is also a school, established in 1812, where young engineering officers and recruits are trained to a practical acquaintance with their duties. The naval arsenal was first formed a short time previously to the Spanish Armada, on the site of the present ordnance wharf: Upnor Castle was also built about the same time. The dockyard was removed to its present site by James I., and was subsequently enlarged and improved, by the formation of floating docks, by Charles I., at which period Gillingham Fort was built; but the present establishments were principally formed subsequently to 1758, when an act was passed for their construction. Previously to this, the security of the arsenal depended mainly on the river forts, especially that of Sheerness, and on the guard ships stationed in the river. These, however, were not adequate for its protection. A memorable instance of their insufficiency occurred in 1667, when a powerful Dutch fleet, under De Ruyter, having suddenly appeared in the Thames, took Sheerness, broke a strong chain that had been drawn across the Medway, and, sailing up the river as far as Chatham, destroyed several sail of the line and a great quantity of stores. The Dutch accomplished this brilliant and daring achievement without incurring any material loss; but the fortifications were soon after very materially strengthened, and are now such as to render any coup de main of this sort quite out of the question.

To shorten the distance by water, and facilitate the communication between London and Chatham, an open canal and tunnel was made, at the beginning of the present century, from the Thames, opposite Tilbury Fort, to Chatham, a distance of about 9 m. of which about 2 m. are tunnelled. But notwithstanding the obvious importance of this channel of communication as a means of saving distance, the too great height of the rates prevented it from being much used, and it was ultimately sold to the North Kent railway company, who used the tunnel for the railroad. Ceccham, or the Village of Cottages, is the name of Chatham in Domesday, and many British and Roman remains have been found in its vicinity; but the greater part of the modern town has been built since the reign of Elizabeth. Chatham has given the title of earl to the Pitt family.

CHATILLON-SUR-LOING, a town of France, dép. Loiret, cap. cant., on the Loing, 14 m. SSE. Montargis. Pop. 2,501 in 1861. This town belonged to the family of Coligny; and in its old castle, on the 16th of February, 1517, was born the famous Admiral de Coligni, the most illustrious victim of the massacre of St. Bartholomew. The mangled remains of the admiral having been deposited, by the care of some of his servants, in the chapel of the castle of Chatillon, were transferred, in 1786, to Maupertuis, where a monument was erected to his memory.

CHATILLON-SUR-SEINE, a town of France, dép. Côte d'Or, cap. arrond., on the Seine, 28 m. NNE. Semur-en-Auxois, on the railway from Paris to Mulhouse. Pop. 4,836 in 1861. The town is neat, well built, and well laid out; it has a fine castle, a communal college, a small public library, a hospital, and a school of design, a society of agriculture. It has also fabrics of coarse cloth, hats, jewellery, iron-plates, glass, beet-root sugar, and casks. There was formerly, within the park belonging to Marshal Marmont, a very perfect agricultural establishment, and an establishment for the preparation of iron and hardware articles; but since the death of the marshal, the establishments in question have been dismantled and the articles sold. Chatillon was, in 1814, the seat of the unsuccessful negotiations between Napoleon and the Allies.

CHATRE (LA), a town of France, dép. Indre, cap. arrond., on the left bank of the Indre, 22 m. SE. Châteauroux. Pop. 5,038 in 1861. The town is agreeably situated on the side of a hill, and was formerly defended by an immense castle, now in ruins, and of which one of the towers serves for a prison. It has a handsome church, and a fine promenade; with a court of primary jurisdiction, a communal college, very extensive tanneries and leather manufactories, and fabrics of serge and other coarse woollen stuffs. Chestnuts are very plentiful in its vicinity; and it has a considerable trade in them, and in cattle, wool, and hides.

CHATSK, a town of Russia in Europe, gov. Tambof, cap. distr., on the Chatcha, 55 m. N. Tambof. Estim. pop. 8,000. The town was founded in 1553, and peopled with Streltsi, Pushgars, and Cossacks, and was formerly fortified; and has a good deal of trade in corn, cattle, tallow, honey, hemp, and iron.

CHATSWORTH, a famous seat belonging to the Duke of Devonshire. See HARDWICK.

CHATTERPOOR, a town of Hindostan, province Allahabad, about 140 miles WSW. that city, formerly a flourishing place, and still possessing extensive manufactures of coarse cotton and paper.

CHAUDES-AIGUES, a town of France, dép. Cantal, cap. cant., in a narrow, deep gorge, on one of the affluents of the Truyère, 11 m. SSW. St. Flour. Pop. 1,950 in 1861. This town is indebted for whatever importance it may possess to its hot springs, which were known to the Romans, by whom they were called Aquae Calentes, of which its modern name is a translation. Their temperature varies from 77° to 80° Reaumur. In winter, the houses are warmed with the hot water conveyed through the streets and into the houses in wooden pipes. It is also successfully employed in the incubation of various species of eggs. It

has some trade in hinglass, and carries on some branches of the woollen manufacture.

CHAUMONT (formerly Chaumont-en-Bassigny), a town of France, dép. Haute Marne, of which it is the cap., on a height between the Marne and the Suize, about 1½ m. from the confluence of these rivers, 18 m. NNW. Langres. Pop. 7,140 in 1861. The town is indifferently built; streets straight and clean, but some of them steep and of difficult access. It formerly laboured under a deficiency of water; but now it possesses several fine fountains, supplied by means of a hydraulic machine. It has several good public buildings; and in the upper part of the town are some fine promenades. Louis XII, Francis I, and Henry II, surrounded it with walls and ditches; but these are in a state of disrepair, and in most places, indeed, are thrown down and filled up. It has tribunals of primary jurisdiction and of commerce; a departmental college, a society of agriculture, commerce, and arts; a public library, with 35,000 volumes; a theatre, a hospital, and a house of correction; manufactures of coarse woollens and druggets, with important fabrics of hosiery and gloves; and a considerable trade in linen and cutlery. The congresses of Austria and Russia, and the king of Prussia, signed here, in 1814, a treaty against Napoleon.

CHAUNY, a town of France, dép. Aisne, cap. cant., at the point where the Oise is joined by the canal of St. Quentin, half the town being built on an island in the river, 18 m. W. Laon, on the railway from Paris to Mons. Pop. 8,163 in 1861. A good deal of cider is made in the town, which has also a considerable amount of trade, being favourably situated for commerce.

CHAVES, a fortified frontier town of Portugal, prov. Tras os Montes, on the right bank of the Tamega, over which is has a Roman bridge of eighteen arches, 40 m. W. Braganza. Pop. 6,720 in 1858. The town has mineral baths, which were formerly much frequented. It was taken by the French, under Marshal Soult, on his entry into Portugal in 1809, but was recaptured by the Spaniards in the following year.

CHAYENPOOR, a town and distr. of Nepaul, N. Hindostan; the former is fortified, and is 170 m. E. by S. Catmandoo. The distr. is altogether mountainous; it exports to Thibet rice, wheat, oil, butter, iron, copper, cotton and woollen cloth, planks, spices, indigo, tobacco, sugar, furs, and pearls; and imports thence, salt, gold, silver, musk, musk deer skins, chowries, blankets, Chinese silks, borax, and medicinal herbs.

CHEADLE, a market town and par. of England, co. Stafford, S. div., hund. Totmonslow, 15 m. NW. London, by London and North-Western railway. Pop. of town 8,191, and of par. 1,903 in 1861. The town is pleasantly seated in the most fertile part of the Moorland, in a vale surrounded by hills, planted with forest trees, and in a district abounding with coal. It consists of one principal and four small streets, and is intersected by the roads from Newcastle to Ashbourn, and from Leek to Uttoxeter. The church is an ancient structure, in the decorated style of English church architecture. The chapel of ease, a neat building, was erected by subscription in 1832. The town is governed by a constable and headborough, nominated annually at the court-leet, held by the lord of the manor. It is also a station for receiving votes at the election of mems. of the H. of C. for the N. div. of the co. The living is a rectory, in the archdeaconry of Stafford, and diocese of Lichfield and Coventry. Patron, master and fellows of Trinity College, Cambridge. There are various chapels for dissenters and R. Catholics. It has a free school, endowed in 1680; a national school, and sundry bequests for the poor of the par. In the vicinage are very extensive copper, tin, and brass works, and a considerable tape manufactory. There are also in the town numerous blacksmiths, braziers, and tin-plate workers; iron merchants, nail-makers, curriers, and tanners; rope-makers, flax-dressers, saddlers, and maltsters. Copper ore has been found in the neighbourhood, but not in sufficient abundance to make its working advantageous. The Caldon branch of the Trent and Mersey canal passes within 4 m. of Cheadle. Market-day, Friday; and fairs are held in January, March, Holy Thursday, 16th August, and 4th October, for cattle and horses.

CHEDDER, a par. and village of England, co. Somerset; hund. Winterstoke. Area 6,090 acres. Pop. 2,632 in 1861. The village, 15 m. S. by W. Bristol, at the base of the Cheddar cliff, a part of the Mendip hills, has three irregular streets branching from a centre. The church is a spacious structure, with a lofty pinnacled tower; there is a charity school for 35 boys and 12 girls, supported by a portion of a bequest left in 1751, the remainder being appropriated to the apprenticing of poor children, and the relief of the poor generally. There are fairs for sheep and cattle, May 4 and Oct. 29. The inhabitants are chiefly employed in agriculture; but a paper-mill in the immediate vicinity employs several hands, and many females are engaged in knitting stockings. The Cheddar rocks, close to the town, form a huge chasm of gorge, apparently torn apart by some convulsion of nature, presenting irregular precipices and extensive caverns, characteristic of calcareous strata. The extensive downs comprised within the par. are clothed with fine pasture; and the dairies of the district have long been famous for the production of an excellent species of cheese, known by the name of Cheddar.

CHEDUBA, an island in the Bay of Bengal, about 10 m. SW. Ramree, Aracan, to which prov. it belongs, constituting one of its four chief divisions. It lies between lat. 18° 36' and 18° 46' N., and long. 93° 26' and 93° 41' E.; shape, nearly round; length and breadth, about 20 m. each; area, 400 sq. m. Pop. between 5,000 and 6,000. Nearly the whole of its surface consists of a rich productive soil; the interior is much more free from jungle than that of any other island upon this coast. The sugar cane, tobacco, hemp, cotton, and rice grow most luxuriantly, and the cattle are the finest in the whole prov.

CHELMSFORD, a town and par. of England, co. Essex, hund. Chelmsford, at the confluence of the Wid, or Cann, and Chelmer, 26 m. NE. by E. London by road, and 29½ m. by Great Eastern railway. Pop. of town 8,513, and of parish 8,407 in 1861. The town, which is almost in the centre of the co., consists of one principal street and three others branching from it: houses mostly well built, many of them having gardens extending to the rivers. It is lighted, and well supplied with water from a spring distant ½ m., conveyed to a handsome reservoir in the town. The church, a stately fabric of the early part of the 15th century, has been repaired within the last few years, but the original pointed style has been carefully preserved. It has a chapel of ease, several dissenting chapels, four sets of almshouses (the oldest founded in 1625); a public dispensary, and many minor charities and benevolent societies; a grammar school, founded by Edward VI., which participate alternately with those of Malton and Brentwood in an exhibition to Caius College, Cambridge; 2 charity schools (one founded in 1713, one in 1714), which respectively clothe and educate 50 boys and 20 girls; a national, a Lancastrian, and an infant school; a neat theatre; public baths, with a reading-room attached; and a handsome hall, in which the courts of assize and of quarter sessions for the county are held, and which also contains a spacious assembly-room. The present co. gaol, on a hill about 1 m. from Chelmsford, in the par. of Springfield, where it occupies an area of 8 or 9 acres, was built in 1828 on the radiating plan. The former gaol, in the same par., is now only used for prisoners previously to conviction, and debtors; attached to it is a house of correction for females. During the last war, two sets of barracks, capable of containing 4,000 men, were erected near the town; but they have since been taken down. A line of entrenchments defended by star batteries may still be traced, erected during the threatened invasion in 1805, to protect the approaches to the metropolis from the E. coast. The Chelmer is crossed by a handsome iron bridge. Below the town, the river has been formed into a navigable canal, 12 m. in length, for barges, by means of locks and artificial cuts, to Maldon, at the head of the estuary of the Blackwater. A handsome stone bridge of one arch has been thrown over the Cann, to replace an older bridge of three arches built in the reign of Henry I. Chelmsford is on the line of the Great Eastern railway, and has long been the main thoroughfare to the E. parts of Essex, and to those of Norfolk and Suffolk. This and the general co. business of assizes are the chief support of the place, for there is no manufacture, and the principal part of the labouring pop. are employed in agriculture, or as carriers and drovers to the metropolis. There are well-frequented annual races in July, held on Galley Common, 2 m. from the town. Chelmsford is near the Commeroys of the Roman period.

CHELSEA, a town and par. forming part of the W. suburbs of the metropolis of England, co. Middlesex, hund. Ossulstone, Kensington div. Pop. of par. 32,371 in 1831; 40,179 in 1841; and 63,439 in 1861. Chelsea is situated on the N. bank of the Thames, along the widest of its reaches above London Bridge, and is connected with Battersea by a modern suspension, and an old wooden bridge. The lower, or old town, is irregularly built, and on the whole of mean appearance: its best houses are those of Cheyne Walk, along the side of the river above the hospital, anciently a fashionable resort, where many distinguished individuals resided. The upper and more modern town, which extends towards Hyde Park, and comprises Sloane Street and Square, Cadogan Place, and part of Knightsbridge, consists of handsome and regularly built houses. The original parish church near the river (the oldest part of which is of the 14th century) contains many interesting monuments; amongst others, one to Sir Thomas More, and in its churchyard is one to Sir Hans Sloane, who resided here, and was lord of the manor. This original church has now become a chapel of ease to a splendid church, built in 1824, in the decorated and later Gothic style, of which it is a very fine specimen; it has 927 free sittings, in consequence of the parliamentary commissioners having contributed several thousand pounds towards its erection. There are numerous other religious edifices, among them an episcopal chapel in Park Street, built in 1718; another in Sloane Street, in the later pointed style, in which there are 630 free sittings; several dissenting chapels, a charity school founded in 1694, in which 40 scholars are educated, and 30 of the number clothed; a national school behind the church, and others connected with the Park and Sloane Street chapels; besides several minor charities. The most important public estab., however, is that of the military hospital, finished in

1690, on a plan of Sir C. Wren, at an expense of 150,000£.; it is of brick, with stone quoins, columns, and cornices, and forms three quadrangles in the centre of extensive grounds; those at the back of the structure being planted with avenues, those in front, occupied by gardens which extend to the river, to which the central quadrangle opens, fronting the S. front, with wings on either side, ornamented with porticos and piazzas. The estab. has a governor and lieutenant-governor, and usually about 550 in-pensioners, consisting of veteran soldiers, who, besides food and clothing, receive weekly pay, varying according to rank and service. The affairs of the hospital are managed by a board of commissioners. Sir Stephen Fox, the chief promoter of this noble institution, contributed 13,000£. towards its formation. York Hospital is connected with the Royal Hospital, having been built for the reception of wounded soldiers from foreign stations, who are taken into the other as vacancies occur. A military asylum was established by the Duke of York in 1801, for soldiers' orphans, and the children of those on foreign stations. It is a handsome building, not far from the Royal Hospital; 700 boys and 300 girls being maintained, clothed, and educated in it on Bell's plan; the boys, on leaving, enter the army; the girls are apprenticed. Between Chelsea Hospital and Cheyne's Walk are the botanical gardens of the Apothecaries' Company, occupying four acres on the bank of the river, granted by Sir H. Sloane, whose statue by Rysbrach is placed there; a hothouse green-houses and library are connected with them, and annual lectures are given. There are similar gardens near Sloane Street, estab. in 1807, comprising six acres, in which lectures are also given in May and June; the plants are arranged in compartments on the Linnæan system.

Chelsea continued, through the 17th and 18th centuries, a favourite and fashionable resort, and was noted for its taverns and public gardens; the Ranelagh Gardens, adjoining those of the Royal Hospital, were closed in 1805.

CHELTENHAM, a town, parl. bor. and fashionable watering-place of England, co. Gloucester, hund. Cheltenham, in a fertile vale opening to the S. and W., at the base of the Cotswold Hills, on the Chelt, a small stream, whence it derives its name; 9 m. NE. Gloucester, 97 m. WNW. London by road, and 121½ m. by Great Western railway. Pop. of parl. bor., which is identical with the parish, 39,693 in 1851. The increase of the town since the commencement of the present century, occasioned by the great influx of wealthy invalids and others, attracted by the celebrity of its spas, the mild and equable temperature of the site, and the beauty of the surrounding neighbourhood, has been quite extraordinary. In 1801 the pop. amounted to only 3,076; in 1811 it had increased to 8,325; in 1821, to 13,396; in 1831, to 22,492; and in 1841, to 31,411. The High Street, running NW. and SE., is upwards of 1½ m. in length; several others branch from it at right angles on each side, leading to the various squares, terraces, detached villas, and spas; each of the latter being surrounded by extensive pleasure-grounds. On the N. side of the town, amongst other fine ranges, are Columbia Place, St. Margaret's Terrace, and Pittville Lawn; on the S. the Upper and Lower Promenades (on the plan of the Louvre), and the Crescent; and up the ascent in that direction, Lansdown Place, Crescent, and Terrace, commanding fine views of the Malvern Hills. The spas, to which the town is indebted for its rapid growth and celebrity in the fashionable world, originate in a considerable number of saline springs, rising in different parts of the vale,

and having their source in the new red sandstone formation, which appears at the surface at the base of Cleeve Hill, NW. of the town, whence it dips gradually, and is about 700 ft. beneath the surface of the chief streets and squares. In all the springs, chloride of sodium is the predominating ingredient, and prevails the most where the red sandstone is approached the nearest. The other mineral components consist chiefly of the sulphates of soda, magnesia, and lime, oxide of iron, and chloride of manganese—the three last in smaller proportions. Iodine and bromine have also been detected in several of the springs. Though the ground has been bored to the depth of 300 ft., none of the present wells exceed 130 ft. in depth. The waters, not only of different springs, but those of the same spring, at different times, probably vary much in their analysis, as several eminent chemists have arrived at different results. The various ingredients, except chloride of sodium, are supposed to be derived from the lias incumbent on the red sand, the waters becoming impregnated in their ascent through the different marls and clays of that formation. They are chiefly efficacious in bilious and dyspeptic cases; and are taken as aperients, usually to the extent of 2 or 3 half-pint glasses before breakfast, at intervals of a quarter of an hour between each. The alkaline form the most numerous class; the magnesian occurs in 2 or 3 wells of recent origin; and at the old wells and Montpellier are sulphureous springs used in cases of scrofula. The earliest of these saline springs first attracted attention in 1716, and was subsequently enclosed and resorted to by a few invalids. It was not, however, till the visit of George III. in 1778, that the waters obtained any extensive repute; since which period, or a little later, Cheltenham has increased, with singular rapidity, and with every prospect of its still continuing a favourite resort of the fashionable world, and of wealthy invalids from the E. Indies, and other hot climates. The Original Establishment, or Old Well, has been greatly extended and improved; it is approached by a fine avenue, and has the crest of a pigeon on various parts of the structure, in allusion to the discovery of the first spring, from its being resorted to by flocks of those birds. The Montpellier Spa (about ½ m. S. of the town) was first opened in 1809; this has also been greatly augmented, and is at present the most fashionable resort during the season, which, at all the spas, begins May 1 and ends Oct. 31. During this period they are opened at 8 in the morning; and at Montpellier there is a numerous band in attendance from 8 to 10 o'clock, the usual time for drinking the waters and promenading. The evening musical promenades at the same spa are also amongst the principal attractions of Cheltenham; and, during the season, the weekly assemblies take place in the rotunda of this spa. In winter they are held at a splendid suite of rooms in the High Street. The Montpellier baths comprise every variety of warm, cold, vapour, air, and shampooing, and adjoining them is an extensive laboratory for manufacturing the various kinds of 'Cheltenham Salts.' They form altogether an extensive range of buildings, and are supplied with the mineral water of 80 different wells, conducted by one main pipe to the establishment. The monthly exhibitions of the Horticultural and Floral Society are held at the Montpellier and Pittville Spas. The latter is in the Grecian style, and is a splendid structure, on an eminence N. of the town, commanding fine prospects, with extensive walks and drives round it. A few public breakfasts are given at this spa during the season, but hitherto the southern quarter of Cheltenham has always been the most frequented and fashionable. The Pittville establishment was opened in 1830, having cost in all about £60,000. The Cambray Spa is a small Gothic structure, built over a chalybeate spring. The whole of these spas are more or less frequented throughout the year: there are also good public baths in the High Street. The parish church is an ancient Gothic building, with a lofty spire, in the midst of an extensive churchyard, planted with noble avenues. There are also 6 modern churches; built partly by private subscription, and partly by grant from the commissioners; a Catholic, and various dissenting chapels. A free grammar school was founded in 1586, for at least 60 boys; but grammar being held to mean Latin, though the scholars are instructed gratuitously in that language, they have to pay for instruction in English. Various efforts have been made to obviate this anomaly and get the school placed on an improved footing. It has 2 exhibitions at Pembroke College, Oxford, worth 60£ each; and 41 church livings are exclusively open to the scholars of this school who have obtained exhibitions. A charity school was founded in 1682, for boys of this and several other pars., who, on leaving, have an apprentice fee allowed them; a national school, established in 1817, has between 500 and 600 children daily, and 200 on Sundays; a female orphan asylum, founded in 1806 by Queen Charlotte, maintains and educates about 27 children.

But the principal educational establishment connected with the town is the Proprietary College. It was set on foot by a large body of subscribers, with the view of furnishing a complete course of classical instruction to the sons of the upper classes. The building, a magnificent fabric in the Tudor style, opened in 1843, has a front 210 ft. in length, with a tower rising to the height of about 80 ft. Inside it has a school-room 90 ft. by 45 do., a gymnasium of the same dimensions, and lecture-rooms.

The principal charitable institutions are, the general hospital, accommodating 100 patients from all parts; the dispensary and casualty hospital, established 1813; the benevolent and anti-mendicity society, established in 1827, and affording relief in kind, by means of tickets; the Columly (for women in childbirth), Dorcas, and numerous others; alms-houses, founded 1574, for six old people; and several minor charities. There are public libraries and reading-rooms at each of the spas, and five or six others in the town; a literary and philosophical institution, established 1833, at which lectures are frequently given, with a good library and museum; and zoological gardens. The General Association for Scientific and Literary Instruction has weekly meetings and courses of lectures—it is on the plan of a Mechanics' Institute. There is a neat theatre, usually open in summer, but enjoying no great share of patronage. The assembly rooms in Regent Street, opened in 1810, cost 60,000£. The ball-room is 87 ft. by 40, and 40 ft. high. The market-place is an extensive structure, built in 1823, with an entrance, through an avenue, from the High Street. Market, Thurs. and Sat.; there is usually an abundant supply, at moderate prices. Annual fairs for cattle and cheese are held the 2nd Thursday in April, August 5, 2nd Tuesday in Sept., and 2nd Thursday in Dec.; there are also two statute fairs, on Thursday before and after Old Michaelmas Day. Malting is carried on to some extent, but the chief trade of the place is caused by the great influx of visitors to the spas, and by its being a considerable thoroughfare. Cheltenham is connected by railways with all parts of the country, and has profited much by the consequent facility of communication. Coals and other articles of general consumption are brought,

by a railway, from the Gloucester and Berkeley Ship Canal to the W. side of the town (9 m.), where there are convenient wharfs and warehouses. Water, for domestic use, is conducted from springs in the Cotswold Hills to a large reservoir, and thence, by pipes, to the upper stories of most of the houses; this and the gas (with which the whole of the town and suburbs are well lighted) are supplied by private companies. The Reform Act conferred on Cheltenham, for the first time, the privilege of returning 1 mem. to the H. of C. The limits of the parl. bor. coincide, as already mentioned, with those of the par. Registered electors, 2,664 in 1862. Gross annual value of real property, 201,082£ in 1857, and 218,162£ in 1862. The government of the town is vested in commissioners. The scenery in every direction is very beautiful and nightingales abound in the vicinity. Stennal's Wood, about 1 m. from the town, has been named, from the numbers that frequent it, Nightingale Grove. From some of the neighbouring summits extensive prospects are commanded, especially from Cleeve Cloud, Birdlip, Charlton Deer Park, and 'the Castles,' so named from the remains of some ancient encampments. Sudeley Castle, a splendid old ruin; Southam, a curious specimen of domestic architecture of the Tudor period; Witcombe, where the remains of a Roman villa were discovered in 1818 (Archæolog. vol. II.); and Toddington, a splendid modern seat, are in the vicinity.

CHELVA, a town of Spain, prov. Valencia, on a river of the same name, 39 m. NW. Valencia. Pop. 4,489 in 1857. There are vestiges of an ancient Roman aqueduct, on the NE. of this town, that served to convey water to Liria. The neighbourhood is planted with mulberries and vines, and produces wheat, barley, rye, oats, maize, wine, and oil.

CHEMNITZ, a town of the k. of Saxony, circ. Zwickau, cap. distr. of same name, on the Chemnitz river, 29 m. ENE. Zwickau, and 37 m. WSW. Dresden, on the railway from Dresden to Nuremberg. Pop. 45,132 in 1861. The town was formerly walled, but its fortifications have been levelled, and their site is now laid out in public walks. It has some good streets and squares, a castle, five churches, four hospitals, a town hall, cloth hall, lyceum, and school of design, and has handsome and thriving suburbs. Chemnitz is the principal manufacturing town of the kingdom. It has extensive cotton manufactures, and that of cotton hosiery, mitts, &c., to which it is mainly indebted for its rapid growth, is said to employ from 15,000 to 20,000 looms in Chemnitz and the neighbouring villages. The stockings and mitts manufactured here are now very widely diffused over the states comprised within the German Customs League; and considerable quantities are also shipped for the U. States. In 1862, there were in the town 51 factories of woollen stuffs; 18 factories for stockings and mitts; and 16 for cotton. There were also, at the same date, 4 iron foundries, and 20 establishments for the manufacture of spinning machinery, with which it supplies a considerable part of the Continent. The town has besides manufactures of linens, and dyeing and bleaching establishments. The district of Chemnitz contains fourteen villages, and had, in 1861, a pop. of above 80,000 inhab., most of whom are employed in the above branches of industry. Chemnitz was for 400 years a free imperial city. It was the birthplace of Puffendorf.

CHENONCEAUX (CASTLE OF). See BLERÉ.

CHEPSTOW, a sea-port town and par. of England, co. Monmouth, hund. Caldecot; on the Wye, 2½ m. from its embouchure in the Severn, 110 m.

W. Lond. by rail, and 141½ m. by Great Western railway. Pop. 3,364 in 1861. The town stands on a gradual slope betwixt bold cliffs rising from the W. bank of the river, and is surrounded by some of the finest scenery in England. Streets broad, well paved, and lighted with gas, but badly supplied with water. There are many good houses, and the town looks neat and cheerful. The church has a fine Norman entrance, and many curious specimens of the early pointed style. It has also a Cuth, and several dim. chapels; an endowed charity school for thirteen children; a national school, two ancient hospitals, in which twenty-five aged persons are supported; and several minor charities. Market, Wed. and Sat. Fairs, Frid. and Sat. in Whitsun-week; Sat. before June 20, Aug. 1, and Frid. before Oct. 29. It has no manufactures; but a considerable trade, being the chief port of most of the places on the Wye and Lug, including Herefordshire and the E. part of Monmouth. Ship-building is carried on to some extent; and about 70 vessels of the aggregate tonnage of 4,600 tons, belong to the port. The tide runs with great rapidity in the river, making its navigation a little dangerous; and it rises at ordinary springs between 40 and 50 ft., and at high springs it sometimes reaches between 60 and 60 ft.; hence very large ships may come up to the town, and barges of 50 tons burden ascend the river to Hereford. A handsome iron bridge was thrown over the river in 1816 at the joint expense of the two cos. separated by the Wye. The castle, on a steep cliff overhanging the Wye, dates from the 11th century, though most of the existing remains, which occupy a considerable space, appear to be of more recent origin; it was alternately in the hands of both parties during the last civil war; and after the restoration, Henry Martyn, the regicide, was imprisoned for life in one of its towers, where he died after thirty years' confinement. The co. magistrates hold petty sessions in the town, and a small theatre is occasionally opened.

CHER, an inl. dep. of France, reg. Centre, formed of part of Berri and Bourbonnais, having N. the dép. Loiret, E. Nièvre, S. Allier and Cher, and W. Indre and Loire-et-Cher. Area. 710,000 hectares, or 2,833 Eng. sq. m. Pop. 823,398 in 1861. It derives its name from the Cher, by which it is intersected, and is included in the basin of the Loire, which, with the Allier, forms its E. boundary. Surface generally flat, soil various: in the E. and along the Loire, it is very fertile; N. it is of a medium quality, while in the N. it is sandy, and covered in great part with heath. Agriculture backward. Principal crops, wheat, maslin, rye, barley, and oats. Hemp is largely cultivated, the crop being estimated at about 750,000 kilog. a year. The natural meadows, which are extensive and valuable, are principally depastured by sheep and cattle. The stock of sheep is estimated at about 500,000 head, producing annually 570,000 kilog. of wool. The stock of black cattle is estimated at 83,000 head. In the reign of Henry IV., the horses of Berri enjoyed a high reputation; but the breed is now greatly deteriorated. Hogs and goats numerous. The forests occupy about 120,000 hectares; and furnish timber for the navy. The vineyards cover nearly 13,000 hectares; those in the arrond. of Sancerre furnish the best wines. Iron is abundant, and is pretty extensively wrought. The cloth manufacture, once the staple of the dep., has greatly fallen off; and the glass works that were formerly to be met with have ceased to exist. The cutlery of Bourges is much esteemed; and there are fabrics of coarse cloth and linen, with earthenware manufactures, breweries, and tanneries.

The dep. is divided into three arrondissements. Principal towns, Bourges, St. Amand, Vierzon, and Sancerre.

CHERASCO, an inl. town of N. Italy, prov. Cuneo, advantageously situated on a point of land between the Stura and Tanaro, near their confluence, 31 m. SNE. Turin. Pop. 8,052 in 1861. The town was formerly an important military post, and is still surrounded with walls; but its citadel was dismantled in 1796. It is well built and laid out, and supplied with water by a canal cut from the Stura, which also turns several silk mills. Trade chiefly in wine and silk.

CHERBOURG, a principal sea-port and fortified town of France, dép. Manche, on its N. shore, nearly opposite the W. extremity of the Isle of Wight, at the bottom of a bay formed by Cape Levi on the E. and Cape La Hogue on its extreme W., at the mouth of the Divette, 41 m. NW. St. Lo, and 185 m. WNW. Paris, at the terminus of the Paris-Cherbourg railway. Pop. 41,512 in 1861. The streets are narrow and dirty, notwithstanding there are many public fountains. Houses mostly of stone and slated. Chief public buildings: the military and marine arsenals; a spacious marine and several other hospitals; the parish church, a singular edifice; the town hall and prison, both new and handsome buildings; a theatre; public baths and barracks. From its advanced position in the English Channel, it has long been a favourite object with the French government to render Cherbourg a great naval arsenal, and a secure asylum for ships of war; and, to accomplish this, vast sums have been expended upon it. The harbours for merchantmen and ships of war are quite distinct from each other. The last, which was constructed by the Emperor Napoleon I., is a magnificent work. It is mostly excavated out of the solid rock, is 828 yards long by 280 wide, and is capable of accommodating 50 ships of the line, which may enter it at all times, there being 25 ft. water at low ebb. It has four fine covered granite docks, 85 ft. deep, for the building of ships, and a basin for those undergoing repair. Near the naval port is the dockyard of Chantereyne for the building of frigates, containing a large timber yard, and a rope walk 546 yds. in length. The commercial port, formed by the mouth of the Divette, and easy of access, consists of an outer harbour and a basin, the former 262 yds. long, by 219 wide; the latter 446 yds. long, by 138 wide. Between the two divisions is a sluice; the outer harbour communicates with the sea by a canal 550 yds. long, and 54 wide, bordered in its whole length by a granite jetty, within which a depth of 19 ft. water is always retained. The roadstead of Cherbourg is one of the best in the Channel, and capable of containing 400 sail. It is defended on all sides by batteries, and is protected from the northerly winds, which would otherwise throw in a heavy sea, and by a great measure also from the Channel currents, by a vast artificial digue, or breakwater, similar to that in Plymouth Sound, constructed in the centre of the bay, opposite to, and about 2½ m. from, the mouth of the river. This great work, formed for the most part of granite and sandstone, was commenced under Louis XVI. in 1784, and continued till 1791; it was re-commenced by Napoleon I. in 1808, again discontinued in 1813, and finally completed by the Emperor Napoleon III. in 1853. Its foundation was laid by sinking many massive wooden frames, which were afterwards filled with blocks of stone. The length of the digue is 8,768 metres (4,120 yards); breadth at its base, 262 ft., at its summit, 101 ft.; its central part, which is 9½ ft. above the water at the highest spring thier, a battery has been erected. The E. channel between it and the shore is 1,060 yards in width, that on the W. side 2,550 yards.

Cherbourg is the seat of a tribunal of original jurisdiction, of a maril. tribunal and prefecture, and is the cap. of the 1st naval arrondissement. It has a departmental college, a royal academical society, a public library with 3,500 vols., a naval library, and several museums. Cherbourg, which is very ancient, was in the 10th century called Carusbur. It was long in the possession of the English, and was the last place they retained in Normandy.

CHERIBON, a sea-port town of Java, cap. div. and prov., at the head of a wide bay on the N. coast of the isl., 120 m. SE. by E. Batavia; lat. 6° 48' N., long. 108° 37' E. In the early part of the present century it suffered from a pestilence, which destroyed more than a third of its inhab.; and, from this and other causes, it is said to have declined of late; but it still continues to be the residence of a Dutch governor, and enjoys considerable trade. The town and harbour are protected by a fort. The district of Cheribon is remarkable for its fertility, and the excellence of its coffee, indigo, teak and timber.

CHERSO and OSERO (an. *Crepsa* and *Absorus*, together called *Absyrtides*), two contiguous, long and narrow isls. of the Adriatic, belonging to Illyria, gov. Trieste; between lat. 44° 30' and 45° 20' N., and long. 14° 15' and 14° 30' E., separated from Istria by the Gulf of Quarnero; united length nearly 50 m., breadth varying from 1 to 8 m. Area 85 sq. m. Pop. of Cherso, 17,500; of Osero, 3,535 in 1837. Surface generally mountainous, stony, and barren; but in some parts the olive, vine, fig, and various other fruits, and a little corn, are grown, and in several parts there are good pasture lands for sheep. Oil is the most valuable product of Cherso, wine of Osero. In the N. part of the former island there are some fine woods; and shrubs and plants for dyeing are very abundant. The breed of sheep is very indifferent, and the wool bad. Other domestic animals are few. Many of the pop. subsist by the tunny and anchovy fisheries. There are a few manufactures, chiefly of coarse woollen cloth and liqueurs; and vessels are built at the principal towns:—three are Cherso, Osero, Lessin Grande, and Lessin Piccolo. Cherso, the cap., on the W. side of the island of same name, has a good though small harbour, and 8,000 inhab. It contains a cathedral and numerous other churches; its streets are narrow and dirty; but its inhab. clean and industrious. Osero, also on the W. side of the island of Cherso, in an unhealthy situation, has only 1,500 inhab.; but it has a cathedral with a fine steeple, and was formerly the seat of a bishopric. It was sacked by the Saracens in 840. Its inhab. have some trade in timber. Lessin Grande and Piccolo are two insignificant towns on the island of Osero. The two islands are connected by a bridge.

CHERSON. See Kherson.

CHERTSEY, a town and par. of England, co. Surrey, hund. Godley, 21¼ m. SW. London, by the London and South Western railway. Pop. of town 2,910, and of par. 6,589 in 1861. The town, situated on the S. bank of the Thames, is neatly built of brick, partially paved, and well supplied with spring water, but not lighted. It is connected with the Middlesex side of the river by a stone bridge of 7 arches, built in 1785, at the joint expense of the two counties. The church, a handsome structure, erected in 1808, in the later pointed style, contains a tablet to the memory of the celebrated statesman C. J. Fox, who resided

for a lengthened period at St. Ann's Hill, near the town. There are also several dissenting chapels, and a school, founded in 1725, for 20 children of this and three adjoining parishes; its present revenues above 400l. a year, and it has been arranged on Bell's plan, and now educates 230 boys and 150 girls, of whom 30 of either sex belonging to Chertsey are clothed. Market day, Wednesday. Fairs, First Monday and Tuesday in Lent, for cattle; May 14, for sheep; Aug. 6 and Sept. 25, for pleasure and pedlery. The chief business of Chertsey consists in the manufacture of malt, flour, iron hoops, and brooms; great quantities of bricks are also made in the neighbourhood; and vegetables are largely cultivated for the London markets. Cæsar is supposed to have crossed the Thames near this place to attack Cassivelaunus; the stakes then driven into the bed of the river by the Britons to obstruct the passage of the Romans are noticed by Bede as remaining in the 8th century; and vestiges of them are still traceable ½ m. below the bridge. During the Heptarchy, Chertsey was the residence of the S. Saxon kings: at Hardwick Court, in the par. (now a farm), Henry VI. resided when a child; and in an ancient monastery (founded by Edgar, and existing till Henry VIII.) he was privately interred, though his remains were subsequently removed to Windsor. Cowley the poet died in this town, where his study is still preserved.

CHESAPEAKE BAY, a noble bay on the Atlantic side of the U. S. of N. America, having its embouchure on the coast of Virginia, between Cape Charles, lat. 37° 7' N., long. 76° 2' W., and Cape Henry, lat. 36° 56' N., long. 76° 4' W., about 13 m. apart. It stretches nearly due N. from Cape Henry to the mouth of the Susquehannah river, in 39° 33' N., a distance in a direct line of above 180 m. Its average breadth N. of the Potomac river, in lat. 38°, is about 10 m.; but S. of that point it is about 25 m. Its coast line is very irregular, inasmuch as it terminates not on both sides into an immense number of bays; but including these, and its numerous islands, its area is estimated at 3,660 sq. m. (Darby.) It is wholly within the states of Virginia and Maryland. Chesapeake Bay differs from the other sounds on the Atlantic slope of the U. States in having only one outlet, as well as in its greater depth of water, which is generally about nine fathoms, affording many commodious harbours, and a safe and easy navigation for ships of the largest burden. At its head it receives the Susquehannah; and on its W. side the Potomac, Rappahannock, York, and James rivers. On the same side are Baltimore, Annapolis, Norfolk, Hampton, &c.; and on its E. shore, Chester and Cambridge. Dismal Swamp canal connects Chesapeake Bay with Albemarle Sound; the Chesapeake and Ohio canal, from the tide water of the Potomac to Pittsburg, was commenced in 1828.

CHESHAM, a town and par. of England, co. Bucks, hund. Burnham, in a fertile vale, through which a small brook flows to join the Colne. Area of par. 11,840 acres. Pop. of town 2,298, and of par. 5,986 in 1861. The town, 26 m. NW. London, consists of three streets. The church, an ancient cruciform structure, has an embattled tower and spire; there are also four dissenting chapels, an almshouse for four old people, and a national school. Market on Wednesday for corn, Saturday for general provisions. Fairs, April 21 and July 22, for cattle; a statute fair Sept. 28. Chesham was formerly noted for the manufacture of wooden turnery-ware, which, though still carried on, has greatly declined. The lace manufacture is wholly discontinued. Shoemaking, for the supply of the

E

metropolis, is the chief business; but the making of straw plait employs many females; there are also several paper-mills, and a small silk-mill in the vicinity.

CHESHIRE, a marit. co. of England, having N. the Irish Sea, the estuary of the Mersey, Lancashire, and a small part of Yorkshire; E. the cos. Derby and Stafford; S. Salop, and a portion of Flint; and W. Denbigh, Flint, and the estuary of the Dee. Area, 673,200 acres, of which about 600,000 are supposed to be arable, meadow, and pasture. Pop. 470,174 in 1841. The surface is generally low and flat, with some considerable hills along its E. border, and a broken ridge on its W. side extending from Malpas to Frodsham; in this ridge, near Tarporley, is the insulated rock of Beeston. It is watered by the Dee, Weaver, and other streams, and the Mersey forms the line of demarcation between it and Lancashire; it is also intersected by several canals. It has mines of coal, copper, lead, and cobalt; but its most valuable mineral consists of an inexhaustible supply of rock-salt, vast quantities of which are annually dug up, and used partly for home consumption and partly for exportation; a great quantity of salt is also procured from the brine springs contiguous to Northwich, Middlewich, &c. The soil consists, for the most part, of a red, rich, sandy or clayey loam, much improved by marling, and generally very fertile. The climate is mild and humid; and the country being low and well sheltered, and divided by hedges and hedge-row trees, is remarkable for its verdure and the luxuriance of its pastures. Hence Cheshire is one of the finest grazing districts in England, and has been long celebrated for its dairies. Cheese is the principal product; and is not only highly esteemed throughout England, where it is consumed in immense quantities, but also in many parts of the Continent and of America. Arable husbandry is a secondary object, and is less suited to the climate; but potatoes are grown in large quantities. Estates for the most part large; this is one of the cos. in which the least change has taken place, for a lengthened period, in the ownership of land; farms mostly small, a great many under 10 acres; but, excluding these, the average is probably about 70 acres. Though there are but few extensive woods, Cheshire has, owing to the prevalence of hedge-row trees, a very woody appearance, and a large supply of available timber. Manufactures of cotton and silk are carried on with great spirit and success at Macclesfield, Congleton, Stockport, and other places. Cheshire has 7 hund. and 20 par., exclusive of the city of Chester. It sends ten members to the H. of C., viz. four for the co., and two each for the city of Chester, and the bors. of Macclesfield and Stockport. Regist. electors for the co. 19,184 in 1843, of whom 6,985 for the northern and 6,981 for the southern division. Gross annual value of real property—in the northern division, 896,416l. in 1857, and 862,777l. in 1862; in the southern division 1,144,024l. in 1857; and 1,067,528l. in 1862.

Cheshire is called a co. palatine, from the sovereign power in it being formerly exercised by the Earl of Chester as fully as by the king. But it has been long held by the crown. It had, however, separate courts and law officers till the passing of the Welsh Jurisdiction Act of Geo. IV., when they were abolished, and its courts assimilated to those of the rest of the kingdom.

CHESTER, a city, co., parl. bor., and seaport of England, locally in the co. of Chester, hund. Broxton, on a rocky elevation on the N. bank of the Dee, by which it is half encircled, on the S. border of the co., about 6 m. above the confluence of the Dee with its estuary, 37 m. S. by E. Liverpool, 164 m. NW. London, by road, and 179 m. by London and North Western railway. Pop. 31,110 in 1861. The city is enclosed within an oblong quadrangle by walls of great antiquity, and which are most probably built on the site of those constructed by the Romans. They make in all a circuit of 2,620 yards, and are of great thickness, and kept in a complete state of repair. The ancient gateways having been removed and replaced by modern arches, a continuous walk on the top of the walls, 6 ft. wide, defended on one side by a parapet, and on the other by a railing, extends all round the city, and affords a great variety of fine prospects. 'The form of the city,' says Mr. Pennant, 'evinces its Roman origin, being in the figure of their camps; with four gates, four principal streets, and a variety of lesser, crossing the other at right angles, so as to divide the whole into lesser squares. The structure of the four principal streets is without parallel; they run direct from E. to W. and N. to N.; and have been excavated out of the earth, and sink several feet below the surface. The carriage drive far below the level of the kitchens, on a line with ranges of shops, over which, on each side of the streets, passengers walk from end to end, secure from wet or heat, in galleries (or rows, as they are called) parfinned from the floor of each house, open in front, and balustraded. The back courts of all these houses are level with the rows; but to go into any one of these four streets it is necessary to descend a flight of several steps.' (Tour in Wales, i. 147, 8vo. ed.) The city has of late years been much modernised and improved, and a handsome new street has been formed from near the centre of the town to Grosvenor Bridge,—a noble stone structure of a single arch, 200 ft. in span, with a roadway 33 ft. in width. Previously to the erection of this bridge, the communication across the river was by an old, narrow, and inconvenient bridge of seven arches. The suburbs have also been considerably extended. The whole is paved, lighted by gas, and supplied with water, raised by a steam-engine, from the Dee, and conducted by pipes to a large reservoir. The cathedral is a large Gothic pile, with a low massive tower; the interior is fine, with several lateral chapels in the earlier, and a clerestory in the later pointed style: the bishop's throne, and several ancient monuments, are highly interesting. Contiguous to the cathedral are the remains of St. Werburgh's Abbey, which for nearly seven centuries was one of the wealthiest in the kingdom. The bishop's palace (rebuilt 1752), the prebendal, and other good modern houses (forming the Abbey Square), occupy the rest of the precinct. There are nine parish churches, and two others not parochial. St. John's church is a magnificent specimen of Saxon architecture; in Trinity Church are monuments to Parnell, the poet, and Matthew Henry, the celebrated commentator, interred within its walls. It has also a Catholic and several dissenting chapels; a grammar-school, founded in 36 Hen. VIII. for 24 boys, from whom the cathedral choristers are selected; its annual revenue is 106l., and it has one exhibition to either university; two charity schools founded in 1717, on the site of the ancient hospital of St. John, one for 30 boys, of whom 20 are also maintained; the other for a like number of girls; the Marquis of Westminster's school, established in 1811, and wholly supported by him, educating between 400 and 500 children; a diocesan school, on Bell's plan, for 150 boys; three infant schools, and several large dissenting and Sunday schools,

The co. infirmary, and the co. lunatic asylum, each have accommodation for 100 patients; and it has a lying-in hospital, a house of industry, several sets of alms-houses, and various charitable bequests,—the chief of which (called Jones's) produces about 400l. a year, which is shared by the members of the ancient city guilds. The old Norman castle (with the exception of one tower) was removed in 1790, and a magnificent co. hall and gaol, together with government barracks, and an armoury, subsequently built on the site. These structures are in the Grecian style, and have great architectural merit; they form three sides of a large quadrangle, the entrance to the area being by a splendid Doric portico. The city courts of justice are held, and corporation business transacted in the Exchange, a plain brick edifice on pillars. There are three commercial halls; one built by the Irish Linen Company, in 1780, for their trade, but at present used for the cheese fairs,—that of linen, once so considerable, having wholly ceased; a second hall, built in 1809 by the Manchester manufacturers for their business; and a third, in 1815, for general purposes, as a private speculation: they are all on the same plan, forming a quadrangle, round which are pillared arcades and shops. There are also commercial rooms, comprising a good public library, news-room, a small theatre, and a good modern market-place. Market, Wednesday and Saturday. Fairs, last Thursday in Feb. for horses and cattle; July 10 and Oct. 10 for general merchandise; these last are of great antiquity, and continue several days; there are also eight annual cheese fairs of recent origin; and the city being situated in the principal cheese-making district of the empire, these fairs have become of considerable importance. Annual races are held in the first clear week of May on 'the Rood-hee,' a level pasture tract of about 80 acres at the base of the city walls. Manufactures inconsiderable; skins and gloves once formed the staples; but these have greatly diminished; there are a few small fabrics of tobacco-pipes, large flour-mills by the old bridge, and a shot-tower beside the canal, on the N. side of the city, where also are several wharfs and warehouses, chiefly for the convenience of the traffic between the city and Liverpool; articles of general consumption being now chiefly supplied from the latter.

At the era of the Conquest, and for long after, Chester was a place of very considerable importance as a commercial and shipping port; but the gradual filling up of the channel of the river, and latterly the superior facilities enjoyed by Liverpool, have proved destructive to its trade. In 1737, in order to obviate the difficulties of the river navigation, an artificial channel was excavated, on a plan suggested long previously by the celebrated Andrew Yarranton, from Chester to the sea. It has since been improved, and vessels of 300 tons may now be admitted to the city; but it has not recovered any portion of its former importance as a maritime town.

Chester is a boro. by prescription; its three earliest charters are without date, but were probably granted in the early part of the 13th century. There are many others, the latest of which dates in 41 Geo. III.; the governing charter (previously to the Municipal Reform Act), in 21 Hen. VII., considerably extended the former privileges, and made Chester a distinct co.; under it were a mayor, deputy mayor, 24 aldermen, and 40 common councillors. The governing body were self-elective, despite the provisions of the charter, and of much litigation, which in the twenty years preceding 1832, cost upwards of 70,000l. Chester has returned two mem. to the H. of C. since 1541. Previously to the Reform Act, the elective franchise rested in the governing body and in the resident freemen. The limits of the parl. bor. include the greater part of the township of Broughton and some other patches, the registered constituency numbering 2,535 in 1862, of whom 1,896 old freemen. The limits of the municipal have since been made to coincide with those of the parl. bor.; and it is now divided into five wards, and governed by a mayor, 10 aldermen, and 30 councillors. The gross annual value of real property assessed to income tax, in the city, amounted to 253,168l. in 1857, and to 174,664l. in 1862. There are 24 ancient guilds or trades still subsisting, though at present possessing scarcely any property of importance, except that of the goldsmiths, who have an assay master and office, and claim the examination of all plate manufactured for sale in Cheshire, Chester, Lancashire, and N. Wales. The crown mote is the criminal court, with jurisdiction over the highest officers; the pret-mote is the chief civil court where actions to any amount are tried; the pentice and passage courts are subordinate to the latter, the sheriff presiding in them. There are three general sessions a year, held in the superior courts, attended by barristers, and presided over by the recorder and mayor; petty sessions for the city are held twice a week.

The city is most probably of Roman origin. Originally it had the name of Deva, from its situation on the Dee, and subsequently of Cestria, from its being a castrum, or camp. It was the head-quarters of the 20th legion, which came into Britain previously to A.D. 61; and not only does the figure and construction of the town attest its Roman origin, but fragments of Roman arches and other buildings existed down to a recent period, and probably some still remain; and pavements, many coins, and an altar dedicated to Jupiter Tanarus by the principino (principal centurion) of the 20th legion, have been dug up. William the Conqueror bestowed the title of Earl of Chester, with sovereign power over the whole of Cheshire, on his nephew Hugh d'Avranches, or Lupus; and his successors to the reign of Henry III. continued in the exercise of like authority. In the last civil war Chester sustained a memorable siege under Lord Byron, by whom it was ultimately surrendered on honourable terms. In 1745 it was garrisoned against the Pretender, which is the last event of any importance in its history.

Eaton Hall, the magnificent seat of the Marquis of Westminster, is about 3 m. S. of Chester; its chief approach being by a triple avenue of limes extending from the end of the new Grosvenor Bridge (where there is a Gothic lodge) to the principal front, through a park abounding in fine forest trees. The structure is an adaptation of the pointed ecclesiastical style to modern domestic purposes; that of Edward III., as seen in York Minster, is chiefly followed, and emblazoned shields are profusely dispersed; in the compartments of some of the windows are several fine portraits executed from cartoons by Singleton; among others those of the six first earls of Chester, who held sovereign power previously to the title being bestowed by Hen. III. on his eldest son; since which period it has uniformly been conferred on the eldest sons of his successors.

CHESTER-LE-STREET, a township of England, co. Durham, near the Wear, 5 m. N. Durham. Pop. of tn. 3,013, and of par. 25,076 in 1861. It stands in a valley, on the line of the Roman way called Ermine Street, leading to Newcastle. The Saxons called it Cunecastre, or Cunceceastre, and while that name it was the seat of the episcopal see of Durham for 113 years, till its removal to Durham in 995. The town is nearly 2 m. in length, and has a bridge over the Wear, opened in 1821. The church, formerly collegiate, and dedicated to St. Mary and St. Cuthbert, has a tower surmounted by a very fine spire 160 ft. high, and contains monuments with effigies of members of the Lumley family from the Conquest to the time of Elizabeth. The Independents and Primitive and Wesleyan Methodists have places of worship. An endowed school educates twelve children. A mechanics' institute is held in a handsome building erected for the purpose. Copyhold courts are held in April and Nov., in which debts under 40s. are recoverable; and petty sessions are held on alternate Thursdays. The place is a station for receiving votes at elections for the S. div. of the co. The manufacture of nails, ropes, and tiles is carried on here; but the inhab. are mostly employed in the surrounding collieries and other works.

CHESTERFIELD, a bor. and market town of England, co. Derby, hund. Scarsdale, 20 m. N. Derby, 150 m. NW. by W. London by road, and 151½ m. by the Midland railway. Pop. of bor. 9,836, and of par. 10,976 in 1861. The town, which is irregularly built, covers a considerable extent of ground, and is pleasantly situated between the rivers Rother and Hyper, in the vale of Scarsdale. The church, a beautiful and spacious edifice of the 13th century, is remarkable for its crooked spire, 230 ft. high. There is also an elegant assembly-room, and near the town is a racecourse, on which races are annually run in the autumn. There are two or three manufactories of silk and cotton, but they are not considerable. Just out of the bor. there are some large iron-works. The chief source of support for the town is the weekly market for agricultural produce, which is well attended. It is governed by four aldermen and twelve councillors, but is not divided into wards. The lord of the manor holds a court leet in Oct., when a constable is chosen; and a court of record for the recovery of debts not exceeding 20l. The petty sessions for the division are held here in the town-hall, on the ground-floor of which there is a prison for debtors. Chesterfield is one of the polling places at the election of M.P. for the N. division of the co. The town is lighted under an act passed in 1825. There are various places of worship for dissenters, a free grammar-school, founded 2 Eliz., and formerly well attended, was closed in 1832. It has still, however, infant, Sunday, and national schools; several well-endowed alms-houses, a dispensary, a savings' bank, a mechanics' institute, and a literary and philosophical institution. The N. Midland railway between Derby and Leeds passes by Chesterfield. It gives the title of earl to a branch of the Stanhope family. Market-day, Saturday. Fairs, Jan. 27, Feb. 28, first Sat. in April, May 4, July 4, Sept. 25, Nov. 25. The par. of Chesterfield includes an area of 13,160 acres.

CHEVIOT HILLS, a range of hills in Great Britain, on the confines of England and Scotland, partly in Northumberland and partly in Roxburghshire. They extend from Kirknewton N. to Carter Fell on the S., where they unite with the hills that stretch across Dumfriesshire and Galloway. The hill to which the name Cheviot is especially given, is in Northumberland, on the borders of Roxburghshire, 8 m. NSW. Wooler, and is 2,658 ft. in height. The Cheviot hills are mostly pointed, the sides smooth and rapidly sloping, and their bases separated by deep narrow glens. They are mostly covered with a close green sward; but in a few instances, as in that of the Cheviot itself,

there are considerable tracts of heath. These hills are depastured by the valuable and peculiar breed of sheep called the Cheviots, now widely diffused over England and Scotland.

CHIAPA DOS INDIOS, a considerable inl. town of Mexico, state of Chiapas, advantageously placed in a valley near the Tabasco, 30 m. WNW. Ciudad de Las Casas. It is chiefly inhabited by Indians, whence its name, of whom there are said to be as many as 4,500 families. It is the largest town in the state, the chief trade of which it engrosses. Its principal export is logwood, which is sent down the river to Tabasco, on the Gulf of Mexico; but a great deal of sugar is also grown in its neighbourhood. Its inhab. are said to be rich. Chiapa enjoys many privileges; it was founded in 1527.

CHIARAMONTE, a town of Sicily, prov. Syracuse, cap. cant., on a hill, 11½ m. NNW. Modica. Pop. 8,995 in 1861. The town is regularly built, with broad and straight streets. From the Capuchin convent there is one of the finest and most extensive views in Sicily. The environs produce good wine, and the town is thriving.

CHIARI, a town of Northern Italy, prov. Brescia, cap. distr., near the left bank of the Oglio, 15 m. W. by S. Brescia. Pop. 9,420 in 1861. The inhabitants are chiefly occupied in spinning silk and tanning leather. The town preserves some remains of its ancient fortifications, and has a handsome collegiate and many other churches, a hospital, and a public library.

CHIAROMONTE, a town of Southern Italy, prov. Potenza, cap. cant., on a high mountain. Pop. 2,921 in 1861. It has two parish churches, a convent, and a seminary. Its environs produce wine and silk, and there is a fine chartreuse about 3 m. off.

CHIAVARI, a marit. town of N. Italy, prov. Genoa, cap. prov., at the head of the Bay of Rapallo, 22 m. ESE. Genoa. Pop. 10,501 in 1861. It is a handsome and flourishing place, surrounded by hills, the rich produce of which supplies a profitable commerce. The Genoese, from the earliest times, appreciating its natural advantages, surrounded it with a strong wall, and gave it many privileges to encourage the resort of merchants. It has a hospital and many fine edifices, and several lace and silk twist factories. Marble and slate are quarried in its neighbourhood, and it has a productive anchovy fishery.

CHICAGO, a town of the U. States, Illinois, at the embouchure of the Chicago river, in the SW. corner of Lake Michigan; lat. 42° N., long. 87° 37' W. Pop. 4,853 in 1840; 29,963 in 1850; and 109,260 in 1860. The river, which is formed of two branches that unite about ⅓ m. from the lake, divides the town into three portions, the principal seat of business being on the S. side of the main stream. The growth of Chicago has been quite extraordinary, as will be seen from the preceding statistics of population, and there is every probability that it will continue rapidly to increase for many years to come. It is indebted for this wonderful development to its situation and the enterprise of its inhabitants. It is the natural entrepôt for the trade between the flourishing state of Illinois and the vast regions watered by the great lakes; its importance in this respect having been very greatly increased by its having been united by a canal, of the largest class, with the navigable waters of the Illinois river, an affluent of the Mississippi; so that it communicates, on the one hand, with New Orleans and the Mexican Gulf, and, on the other, with Quebec and the St. Lawrence. Hence the value of its exports and imports, which, in 1840, were respectively 228,636 and 562,105 dollars, had risen, in 1860, to 3,576,150 and 4,123,761 dollars. The harbour, which is partly artificial, is formed by means of piers, at the extremity of one of which is a lighthouse, projecting from the river into the lake. The trade of the port employs a great number of steamers and sailing vessels, many of which belong to the town. The situation, though low, is above the level of the inundations, and is said to be healthy. The streets cross each other at right angles, and the wooden buildings of the first settlers have given way to substantial brick edifices. It has some handsome churches, a medical college, various elementary and superior schools, a merchants' academy, banks and insurance offices. Five different lines of railway centre at Chicago. The most important of them are the Chicago and Alton line, 220 m. long; the Chicago and Rock-Island line, connecting Lake Michigan with the Mississippi river; and the Chicago and North-Western, 213 m. long, extending from Chicago to Appleton, Wisconsin. Fort Dearborn, which acquired some celebrity in the last war between this country and the United States, is in the immediate vicinity of Chicago.

CHICHESTER, a city, co., and parl. bor. of England, co. Sussex, 55 m. SW. by S. London by road, and 70 m. by London, Brighton, and South-Coast railway. Pop. 8,059 in 1861. The city is about 1½ m. E. from the extreme NE. angle of the bay or arm of the sea called Chichester Harbour. It is situated on a gentle eminence, sloping in every direction, amidst the widest part of the plain named from it. The Lavant (a small rivulet usually dry in summer) bounds it on the E. and S. Its walls, forming a circuit of about 1½ m., are still in tolerable preservation, within which a mound extends all round in the Roman fashion, planted in parts with fine elms. Chichester is well built, lighted, watered, and drained. It consists chiefly of four principal streets, diverging at right angles from a common centre, occupied by an octagonal cross, erected towards the close of the 15th century, and said to be the most beautiful of this class of structures in the kingdom. The present cathedral was built in the 13th century, on the site of an older one founded in 1108. It is an inferior building of its class, partly in the Norman, and partly in the earlier pointed style; the old tower and spire of the 14th century fell in Feb. 1861, and a new spire was completed in 1865. The cathedral contains many ancient and several well-executed modern monuments; among the latter is one to the memory of the poet Collins, a native of the town. The collegiate establishment was, from the first, for secular canons, and so left unaltered at the Reformation; it consists of a dean, thirty prebendaries, and other ecclesiastical officers. The see comprises the entire co. of Sussex, with the exception of twenty-two parishes, which are peculiars; the episcopal palace is within the city walls, and has fine gardens attached to it. Except that of St. Paul, which is a handsome modern structure in the pointed style, the other churches are small, mean buildings. There is a grammar school, founded in 1497, and a blue-coat school, founded in 1702. There are several charitable institutions, the most ancient of which is that of St. Mary's Hospital, with a chapel attached to it. The other public buildings are the guildhall, town-hall, market-house, and corn exchange; the buildings of the mechanics' institute, of the Literary and Philosophical Society, and a small theatre.

There are no manufactures, the town principally depending on the surrounding agricultural district. Market-days, Wednesday and Saturday; the former for corn, the latter for general provisions: an important cattle-market is held every second Wednesday; and four large cattle and horse fairs, May 4, Whit-Monday, Oct. 10 and 20. The transit of corn through the town to the metropolis and to the W. of England is also considerable. The harbour is rather difficult of access; but at spring-tides vessels of 170 or 180 tons reach the quay, about ½ m. below the town; but its communication with the sea is kept up by the Arundel and Portsmouth Canal, a branch from which is carried to the city. It is divided into two wards, and governed by a mayor, six aldermen, and eighteen councillors. Chichester has returned two mem. to the H. of C. from the 23rd of Edw. I. Previously to the Reform Act the franchise was vested in the corporation and freemen and scot-and-lot payers within the bor. The Boundary Act extended the limits of the parl. bor., which is identical with the municipal bor., so as to embrace the suburbs. Registered electors, 585 in 1865. Annual value of real property, 42,731l. in 1857, and 37,408l. in 1862. Chichester is supposed to occupy the site of the Regnum of the Roman period. It was destroyed by Ella in the 5th century, and restored by his son Cissa, whence the name. Some additional importance was given to it by the removal of the see from Selsea thither, after the Conquest. It gives the title of earl to the Pelham family.

CHICACOLE, or CICACOLE, an inl. town of Hindostan, formerly the cap. of the N. Circar of same name, on the high N. bank of the river Chicacole, 4 m. NW. the bay of Bengal, and 50 m. NE. Vizagapatam. It is of considerable size, but irregularly built, being a collection of all sorts of houses and huts. It contains some neat European barracks, several large bazars, and numerous mosques and other Mohammedan buildings.

CHICLANA, a town of Spain, Andalusia, prov. Cadiz, 12 m. SE. Cadiz. Pop. 9,097 in 1857. The town is situated between two hills, on one of which are the ruins of an ancient Moorish castle; has two churches, two convents, a hospital, a workhouse belonging to Cadiz, a theatre, and some good private houses. It is much resorted to by the wealthy classes of Cadiz, who have here country residences and pleasure grounds. The adjoining heights command a fine view of Cadiz and its bay, the isle of Leon, &c. on one side; and, on the other, the ancient city of Medina Sidonia, and plains of Andalusia, towards Algeciras and Gibraltar. The battle of Barosa, in which, after an obstinate engagement, the Anglo-Spanish army under Sir Thomas Graham (Lord Lynedoch) defeated a French force under Marshal Victor, was fought, a few m. S. from Chiclana, on the 5th of March, 1811.

CHIETI (an. Carreus Palvatia), an inl. town of N. Italy, prov. Turin, cap. mand., on the declivity of a vine-clad hill, 8 m. SE. Turin, on the railway from Turin to Alessandria. Pop. 15,032 in 1861. The town is well built, has four handsome squares, and a collegiate church, said to have been originally a temple of Minerva. Its fortress, La Rochetta, was destroyed in the 16th century. It has some cotton and linen thread and woollen-cloth factories.

CHIETI, a city of Southern Italy, prov. Chieti, of which it is the cap., on the narrow crest of a range of hills, on the right bank of the Pescara, about 10 m. from the Adriatic, on the railway from Ancona to Naples. Pop. 20,192 in 1861. The streets of the town are generally narrow and crooked, and in many parts dark and dirty; but the houses and shops are good, and approach nearer to the standard of the metropolis than those of most provincial towns. It has a large cathedral, and four other churches, a lyceum, or college, a large seminary;

numerous convents; a society of agriculture, arts, and commerce; a hospital; a workhouse; and a handsome theatre. It is the seat of an archbishopric, of the civil and criminal tribunals of the place, and has manufactures of woollens and silks. The surrounding country is well cultivated and fertile, and the population have an appearance of ease, cheerfulness, and activity. The Abbé Galiani was a native of Chieti, having been born here in 1728.

Chieti is very ancient, being built on the site of *Teate*, the capital of the small but not unimportant tribe of the *Marrucini*. Silius Italicus calls it *Magnum et Clorum*. The remains of a theatre of considerable dimensions, a large public edifice, two temples, a gateway and Mosaic pavement, with numerous coins and inscriptions, evince its ancient magnitude and importance.

CHIHUAHUA, a city of Mexico, state of Chihuahua, of which it is the capital, 740 m. NNW. Mexico, 190 m. E. Guaymas, and 540 m. from the mouth of the Rio Grande del Norte; lat. 28° 47′ N., long. 107° 30′ W. It is situated in an arid plain, on a rivulet which falls into an affluent of the Rio Grande. Pop. at one period said to have been 70,000; in 1803, 11,600; at present estimated at 10,000. Streets regular; houses well built and well supplied with water, conveyed to it by an aqueduct 3 m. long. The cathedral, a very large and highly ornamented structure, was erected at an expense of 1,500,000 doll. raised by a duty on the produce of the adjoining mines. The town is chiefly maintained by supplying necessaries to the surrounding mining districts; and from being a depôt for goods to and from Guaymas. Charcoal is conveyed thither for the mines and domestic purposes from a distance of 30 leagues. There are several large monasteries in the town; but they are much diminished in their income and in the splendour of their buildings and establishments since the revolution. The country surrounding the city is occupied by extensive haciendas, or farms, in which large herds of mules, horned cattle, and sheep, are pastured. But, notwithstanding the great capabilities of the soil, agriculture is in a very depressed state, the mines being the great objects of attention. Of these the most celebrated for the quantity of the precious metals drawn from it is El Parral in the SE. part of the state; but it is now in so dilapidated a condition, that the amount of capital required to re-establish it is too great to justify a well-grounded expectation of its returns being sufficient to repay the outlay. Batopilas, 80 leagues W. of Parras, once one of the most productive of the Mexican mines—a single mass of pure silver weighing 425 lbs. having been found in it—is but feebly worked. One of its veins was discovered by an Indian, who, on swimming across a branch of the Rio del Fuerte after a flood, perceived the crest of a rich lode laid bare by the force of the current, the greatest part pure silver, sparkling in the sun. Santa Eulalia in the E. has long been abandoned. The pop. of the plain country is almost wholly of European descent, the natives having retired before them into the mountainous recesses of the Sierra de Mapimi. Their principal tribes are the Apaches, Comanches, and Chichimeques.

CHILI, or CHILE, a republic of S. America. In the SW. part of that continent, consisting of a long and narrow strip of country between the Andes and the ocean, extending from lat. 25° 20′ to 42° S., and between long. 70° and 74° W.; having N. the southern extremity of Bolivia, E. the territ. of La Plata, SE. and S. Patagonia and the Gulf of Ancud, and Strait of Chacao (which separates it from the Archipelago of Chiloé), and W. the Pacific. Length, N. to S., 1,150

m.; average breadth between 110 and 120 m. Estimated area 219,352 sq. m. Pop. 1,439,120, according to the census of April 19, 1854; and 1,618,894 according to official returns of the year 1861. The country is divided into fifteen provinces, the pop. of which, by the census of 1854, was as follows:—

Province	Population
Atacama	58,680
Coquimbo	110,340
Aconcagua	111,594
Valparaiso	116,024
Santiago	272,499
Colchagua	192,704
Talca	79,139
Maule	156,243
Nuble	100,791
Concepcion	110,891
Arauco	43,486
Valdivia	39,253
Chiloe	61,240
Llanquihue	3,475
Magallanes	153
Total	1,439,170

Besides these territories, the islands of Juan-Fernandez, Mocha, and some others in the Pacific belong to Chili.

Topography.—The country rises successively from the coast to the Great Cordillera of the Andes; but not by a number of successive terraces running parallel to each other and to the sea, except in the N. Elsewhere, the surface, as Mr. Miers says, 'is not formed by a series of table heights, reaching from the sea to the foot of the Cordillera; but it is a broad expansion of the mountainous Andes, which spreads forth its ramifications from the central longitudinal ridge towards the sea, diminishing continually, but irregularly, till they reach the ocean. . . . These mountain branches are of considerable height, being seldom less than 1,000 ft., and more generally 2,000 ft. above the bottom of the valleys which intersect them: it may, therefore, be readily conceived that there is but little level country between the smaller branches of these chains; the more valuable portions were formed by the beds of the rivers now comparatively small, although there is evidence of their having been once the courses of greater streams. Some of these valleys present broad expansions of surface, such, by way of illustration, as that portion of the country called the Valley of Aconcagua. These are the patches which constitute the finest and loveliest portions of the middle portion of Chile.' (Miers' Trav. in Chile, i. 378, 379.)

The Great Cordillera of the Andes has in S. Chile a mean elevation of 13,000 or 14,000 ft. above the level of the ocean; but it presents many peaks which rise to a considerably greater height. These peaks, most of which are volcanic, begin to be numerous beyond lat. 30°, and increase in number as we proceed farther S. The principal one is that of Aconcagua, about lat. 8° 10′, which has been proved to be at least 23,200 ft. in height (Capt. Fitzroy's Paper in Geog. Journ., vii. 143), and therefore ranks first among the mountains of S. America. At intervals it is an active volcano. N. of 33° 30′ the Cordillera is divided into two separate ranges, enclosing the immense valley of Uspallata, so celebrated for its mineral riches, and other valleys. The principal road across the Andes—from Santiago and the Vale of Aconcagua to Mendoza—crosses Uspallata; several other passes from Chili into the La Plata territories exist farther S. (See ANDES.) Between the ramifications of the mountain chains and the sea some small plains line the coast. The shores are mostly high, steep, and rocky, as in general along the whole of the W. coast of S. America. They have almost everywhere, however, deep water near them, and there are many tolerable harbours, the best being those of Valdivia, Conception, Valparaiso, and Coquimbo, though some are safe only during certain seasons of the year. The rivers of the middle and S. parts are sufficiently numerous, but they are all small. The N. part of the country is scarcely watered by any; and from Mayro to Atacama, a distance of 1,000 geog. m., all the rivers and streams together would not form so considerable a body of water as that with which the Rhone enters the Lake of Geneva, or as that of the Thames at Staines. (Schmidtmeyer's Trav., p. 28.) The rivers retain pretty much the same quantity of water throughout the year: they are not augmented much at any particular season by the melting of the snows, since, while in the summer the snow on the upper mountain ranges melts, that on the lower heights liquefies even in the winter. They are generally unfit for the purposes of trade. In the N., there is no stream navigable for laden boats for more than 6 m. inland: in the middle provs., the Maule is the only one which brigs of 160 tons burden can enter at high tide, and these cannot ascend far; and in the N. the Callacalla, or river of Valdivia, is the only one capable of being entered with safety by ships carrying 60 guns. Some lakes, or rather lagunes, are scattered over the country; they are most numerous in the S., and in the prov. of Valdivia and in Araucania are of some size. A few are 60 or 70 m. in circumference.

Climate is equable and healthy: epidemic diseases are rare. The interior is hotter than the coast: in the former, during Jan. and Feb., the thermometer often rises to 90° and 95° Fahr. in the shade; on the latter, at the same season, it rises to about 85° in the day, and sinks to 70° or 75° in the night. At Santiago the mean summer heat from December to March at midday is about 81°, and at night 58°. A cool and pleasant breeze arises at sunset. Winter begins in June. No snow falls on the coast, and frost is rare; on the Andes the snows remain from June to November. About April the rains set in, and fall at intervals till Aug.; but this is only in the N. provs. N. of Santiago the rainy season is limited to a few occasional showers, and in the arid prov. of Coquimbo no rain whatever falls; the want of it being occasionally supplied by heavy night dews. The N. provs., being at a distance from the volcanoes of the Cordillera, which apparently act as safety-valves, are especially subject to earthquakes. Shocks are felt in some parts almost daily, and the country is continually desolated by them. In 1819 the town of Copiapo was totally destroyed; and in 1835 Conception, and other towns on the coast in the middle provs., were nearly ruined by an earthquake. (Miers, i. 378-399; Schmidtmeyer, p. 25, &c.; Campbell's Geog. Journ., vol. vi.; Molina; Voyage of the Adventure and Beagle, &c.)

Geology.—According to Schmidtmeyer, the high chain of the Andes is chiefly composed of argillaceous schist, while the lower chains and mountain groups are principally granite. Sienitic, basaltic, and felspar porphyries, serpentines of various colours, quartz, hornblende and other slates, pudding-stone, gypsum, abound in the Cordillera, and fine statuary marble is said to abound in the department of Copiapo. Chili is extremely rich in metals: silver is found there at a greater elevation than any other metal; it is also met with in the valleys or bowls in the lower ranges, but, generally speaking, its quantity decreases in proportion to its distance from the Andes. Gold is most frequently situated at a much less elevation than

silver; it is found chiefly in the bowls, and perhaps few of the lower mountain ranges throughout Chili are without it. Most, or perhaps all the rivers, wash down gold. The copper mines are one of the chief sources of national wealth. Lead and iron are found in abundance, but neither is much sought after. Zinc, antimony, manganese, arsenic, tin, sulphur so pure as not to require refining, alum, salt, and nitre, are plentiful. Coal mines have been opened near Conception; the coal improves with the depth of the mine, and has already become a considerable article of trade and consumption at Valparaiso. The soil of the N. provs. is sandy and saline; and in the opinion of Mr. Miers, not 1-50th part of the N. half of Chili can ever be cultivated. Some of the valleys in the central provs., as that of Aconcagua, present broad and fertile expansions of surface, and others, being considerably inclined, admit of irrigation wherever water can be procured; but the hilly parts being dried and parched during the greater part of the year, are incapable of culture. S. of the river Maule, however, the proportion of cultivable land is larger, the soil becoming progressively more stiff and loamy. (Miers; Schmidtmeyer.)

Vegetable Products.—Fertility increases in proportion as we proceed S. Capt. Basil Hall observes: 'At Conception, in the S. of Chili, the eye is delighted with the richness and most luxuriant foliage: at Valparaiso, which lies between 100 and 200 m. farther N., the hills are purely clad with a stunted brushwood, and a faint attempt at grass, the ground looking everywhere starved and naked: at Coquimbo even this brushwood is gone, and nothing is left to supply its place but a wretched sort of prickly pear bush, and a scanty sprinkling of wiry grasses. At Guasco, there is not a trace of vegetation to be seen, all the hills and plains being covered with bare sand, excepting where the little solitary stream of water, caused by the melting of the snow amongst the Andes, gives animation to the channel which conducts it to the sea. The respective latitudes of these places are 37°, 33°, 30°, and 24°.' (Hall's Extracts from a Journal, in Constable's Misc., iii. 9, 10.) Extensive forests cover Araucania and the S. provs. The banks of the Andes also exhibit a profuse vegetation. The Mimosa farnesiana flourishes over most of the country, and the algaroba is nearly as common. The quillai, the bark of which produces a natural soap, is brought to the towns as an article of trade; laurels, myrtles, cypresses, and other evergreens, grow to such a size as to be highly useful for their timber. Most European fruits flourish, but tropical plants are few. Schmidtmeyer observes, that the numerous groves of palm and cinnamon trees, spoken of by Molina, have disappeared since his time. Chili produces many hard woods, which, in a great measure, supersede the use of iron in the country; and Mr. Miers says that 'the herbaceous plants and flowers are so rich, various, beautiful, and novel, that to a botanist no treat can be greater than a journey through the Cordilleras.'

Animals.—The cougar or puma, the jaguar, llama, guanaco, numerous monkeys, and other wild animals common to this continent, inhabit Chili. A kind of beaver (Castor huidobrius) frequents the rivers, and the chinchilla abounds in the desert country of the N.; both are hunted for their fur, which is much prized. The great condor, several vultures, pelicans, and many other water fowl, flocks of parrots, partrequets, &c., are among the birds; whales, dolphins, cod, pilchards, &c., are caught around the coasts. The skunk, which, when pursued, emits an intolerable odour, is a native of Chili; but in other respects this country enjoys a singular freedom from annoying or ve-

trenous quadrupeds, noxious insects, and reptiles.
(Miers, vol. i.; Schmidtmeyer.)

Agriculture and Cattle Breeding.—The climate
and soil of the N. and central parts of Chili are
highly suitable for the culture of European grains,
N. of lat. 30°, the limit at which they cease to
attain perfection varies from 3,700 to 5,200 ft.
above the ocean; but at the height of 3,000 ft. the
harvests are extremely good. Only the middle
provs., however, produce sufficient corn for exportation, after supplying the wants of their inhab.
Aconcagua is by far the best cultivated prov., and
that which exports most corn. Its produce goes
chiefly to the market of Valparaiso. Wheat is the
staple, and in the N. almost the only grain cultivated. Barley is grown in the N.; maize, buckwheat, and oats are but little raised, and rye is
unknown. Kidney beans are exported to Peru,
and occasionally to Brazil; all kinds of pulse are
common; and potatoes are extensively cultivated,
though they fail in flavour. Culinary vegetables
are raised, especially near the towns. Water melons
are very fine, and gourds of a good flavour are produced in great abundance; the latter are appendages to every Chilian dish of boiled meat. Hemp
of good quality is grown chiefly in Aconcagua.
The sugar-cane has been tried, but does not succeed. Rice and cacao are imported. At Quillota
there are some good gardens: in Aconcagua prov.
the vineyards and olive grounds yield an abundance of good fruit: and in that of Concepcion,
which was once celebrated for its wine, the vineyards are still extensive, and the grapes thin-fla-
voured. Elsewhere, according to Poeppig (Reise
in Chili, i. 125–127), both orchard and garden cultivation is in the back-ground. The olive crops
are good, but the oil is ruined by a bad mode of
treatment, and rendered unfit for European markets. Little care is taken in the culture of corn.
The art of agriculture is greatly in arrear. The
plough, which is everywhere alike throughout the
country, consists of only a part of the trunk of a
tree, with a crooked branch which serves as a
handle, the forepart of the trunk being wedge-
shaped and having nailed to it 'a somewhat
pointed flat plate of iron, which performs the necessary operation of coulter and share, neither of
which were ever heard of by the natives.' (Miers.)
The yoke is fastened not to the shoulders, but to
the horns, of the oxen, according to the approved
ancient Spanish method. The substitute for a
harrow is a heap of bushes weighed down with
stones; the turning up of the soil by spade digging and the use of the English hoe are unknown;
and what little weeding is practised is performed
by the hand or the blade-bone of a sheep. Lands
are cultivated until worn out, with the interval of
a fallow every four or five years; no manure is
used. The productiveness of the soil in Chili
appears to have been formerly much overrated.
Mr. Miers observes, that a piece of ground recently
cleared 'may produce to the extent of 100 or even
300 fold during the first year; but such lands are
now scarce in the cultivated parts of Chili; and
the average of the wheat fields may be from 8 to
12, or of the best crops, from 12 to 20 fold. (Miers,
i. 371.) Reaping is performed by means of a rough
sickle; and the corn, in quantities of about 100 or
150 quarters at a time, threshed out in a hard dry
spot of ground, by being galloped over by horses.
It is then generally left in the open air for some
months, not being housed till the rainy season
begins.

Few farms are wholly arable, and such as are so
are small and situated in narrow valleys. Cattle
breeding is the most important branch of rural industry. In the middle provs., the *haciendas*, or

farms, feed often from 10,000 to 15,000 head of
cattle, in some cases as many as 20,000; and on
the smallest grazing farms from 4,000 to 5,000 head
are reared. The black cattle in some parts are
strong and heavy, but in the N. small; they are
dull and neither the beef nor milk they yield is
very good. The horses of Santiago are said to be
excellent, well broken, and more docile than those
of Buenos Ayres. Those of the country generally
are well made, and gallop, though they do not
trot, well. Schmidtmeyer says (Trav., p. 93) that
they are 'so strong and hardy as to be able to
carry their riders above 80 m. a day at a gallop,
with very little rest, and no other food than lucerne grass.' The mules and asses are of a good
size, hardy and strong; the former are the general
beasts of burden, and are especially used in travelling across the Cordillera. Goats are plentiful,
being more fitted than sheep for the pastures of
Chili. The sheep are said to be very inferior, and
both the mutton and wool bad. Hogs are not
very good, and very little of their flesh is consumed. In the dry season the cattle are often
reduced to great straits for want of food. (Poeppig,
i. 121–129.)

After its conquest by the Spaniards, Chili was
divided into 360 portions, which were given to as
many individuals; and though by the Spanish
law of succession these portions have been, and
continue to be, subdivided frequently, most estates
still remain very large. The proprietors of these
large grazing estates usually reside with their
families in the towns, and keep on their farms a
major-domo or steward, under whom are a head
and a few subordinate herdsmen, and these are
assisted sometimes by a few tenants who hold
their dwellings under the proprietor by a kind of
feudal tenure, being obliged to give their services
in any kind of labour that is required of them,
without pay, or for a very small remuneration.
Land is never leased out to the agricultural
tenants, but from year to year; the latter have
neither oxen for ploughing, means for thrashing,
nor capital to get in their crops; and all these, and
all other kinds of assistance, come from the proprietor, who is repaid out of the produce of the
land, which he besides generally buys up at two-
thirds or half what the former might sell it for,
could he command the necessary funds to harvest
it. The cultivator, in short, is rather worse off
than the day-labourer, and is even in the habit of
hiring himself out as such at times to recruit his
means. He is destitute of most comforts, can
seldom read or write, nor has any means within
his reach of educating his children. The moment
his harvest or the produce of his garden is reaped,
the landlord enforces his right to the stubble and
pasture for the benefit of his cattle, and large
droves are even frequently turned in before the
produce is cut, either utterly destroying the crops,
or obliging them to be gathered half ripe. The
tenant is scarcely ever allowed to build his hut on
cultivated grounds, to enclose his rented land with
fences, or to possess any cattle; and a multitude
of other arbitrary practices tend to keep the man
in that state of servitude in which it is the object
of the proprietor to retain him. (See especially
Miers, i. 311–376.)

Fisheries.—The coasts present good fishing
ground, and with good boats, good nets, and good
government regulations, the Chilians might be
made tolerable fishermen; but, owing in part to
some ill-advised measures adopted by the government, Mr. Miers affirms that in his time the fishers
were the most abandoned, lazy, and worthless
class in the country. They seldom fish more than
a mile from shore, using only canoes of the rudest

possible construction, or rafts supported on large seal-skin air-bags, both urged onward by means of a double-bladed paddle, used first on one side and then on the other.

The country has abundance of minerals of the richest quality, from which, however, little profit is drawn, owing to the constant civil strife and political disturbances under which the republic is suffering. Nevertheless, several mines of silver, gold, and copper are being worked in the province of Coquimbo, and, since the year 1859, some valuable coal mines are worked at Lota and Coronel, in the prov. of Talca. At Lota the whole of the mines are the property of Messrs. Cousino and Kar, natives of Chili, and are wrought by English and native coal miners on the English system, with the assistance of railways, steam engines, and wharves, and are now formed into a very complete establishment, at an expense to the owners of upwards of 1,000,000 dollars. The establishment was commenced in 1852, but only got into proper working order during the year 1859. The produce of the mines is at present from 4,000 to 5,000 tons of clean coal per month, and can be greatly increased when more labourers can be obtained. The coal of these mines is brought mined, screened, and embarked by contract, and at current prices leaves a clear profit to the owners of more than three dollars and a-half per English ton. Messrs. Cousino and Son have also furnaces for the purpose of smelting copper ore with the refuse or small coal. The mines of Coronel are being wrought by several individuals to a much greater extent than even the mines of Lota, and the produce is greater. Samples of very rich silver ore, gold quartz, and copper ore have been found in the Araucanian territory; but nothing can be done with either, until the Indians come under the dominion of some civilised government. (Report by Mr. Cunningham, British Vice-Consul at Talcahuano, Chili, in 'Consular Reports,' 1863.)

It is a common saying in Chili, that 'a diligent man who works a copper mine is sure to gain; that he who opens one of silver may either gain or lose; but that if the mine be of gold, he will certainly be ruined.' This is owing in great part to the circumstance of many mines having been opened or wrought by persons without capital, who are very soon obliged to suspend their operations; land carriage being difficult and laborious, and fuel, water, and fodder very scarce in those districts which are the richest in ore. The mines are mostly wrought by two parties, one the proprietor of the mine, who supplies the labour, the other the habilitador, who advances the capital. The proprietor, who usually resides on the spot and superintends the works, is seldom wealthy enough to conduct them on his own resources, and it is generally the habilitador, or moneyed individual, who resides at the port where the metal is shipped, who alone derives any ultimate benefit from the mine. (Meyen; Hall; Schmidtmeyer.)

Manufactures and Trade.—The Chilians are good potters, and make light and strong earthenware jars, which ring like metal. Hempen cloths, indifferent hemp, cordage, soap, copper wares made in a very rough manner, leather, brandy, tallow, and charcoal, are amongst the chief articles manufactured. The rest are mostly domestic, and conducted by women.

Chili is supposed to be the only American state, formerly subject to Spain, whose commerce has increased since the separation from the mother country. Most of the foreign trade is with Great Britain, the imports from which, consisting chiefly of cotton and woollen goods, hardware, iron, &c., amounted, in 1853, to 935,176£; in 1858, to 413,617£;

in 1859, to 1,510,176£; and, in 1860, to 1,474,010£. A portion of the merchandise imported from Great Britain is subsequently sent to other parts of America. Linens, &c. are imported from Germany; silks, paper, perfumery, jewelry, wines, and brandy, from France; silks, nankeens, tea and sugar from China and the E. Indies; tobacco, spermaceti, candles, oil, sugar, and manufactured goods, from the U. States; dyes, coffee, pearls, sugar, cacao, tobacco, cotton, rice, salt, and spirits, from Peru and Central America; and cotton, Paraguay tea, and European goods, from La Plata and Brazil. The exports are chiefly bullion, copper, hides, tallow, pulse, wheat, fruits, drugs, and European goods re-exported to Peru, Bolivia, and Central America. The exports to the United Kingdom have rapidly increased of late years. They were of the value of 1,869,547£. in 1859; of 2,416,935£. in 1861; and of 2,296,603£. in 1863. Copper was the principal article of these exports, furnishing about three-fourths of the value. Valparaiso is the chief port, and centre of the foreign trade.

But little accommodation exists for internal commerce. The only towns of any importance, except the cap. Santiago, viz. Valparaiso, Coquimbo, Conception, and Valdivia, are near the sea, and at a great distance from each other; and, except between Valparaiso and Santiago, the latter city and Talca, there are no good roads. Latterly, however, the want of ordinary roads has to some extent been mitigated by the construction of railways. In the year 1863 there existed nearly 460 miles of railway, among them lines from Valparaiso to Santiago, from Santiago to San Fernando, from Caldera to Pabellon, and from Coquimbo to Las Canas.

Government.—The public revenue, which, in 1851, amounted to 1,517,537 dollars, has since been progressively increasing in amount, and in 1860 had risen to 7,494,750 dollars, or 1,498,950£. The expenditure, in the latter year, amounted to 7,307,025 dollars, or 1,461,405£. There was a public debt, at the end of the year 1861, of 15,251,500 dollars, or 3,050,300£.

Chili is a republic under a president, elected for a term of years. It has a congress of 50 members elected by the different provs. The executive power is in the hands of the president and a council of ministers.

The national religion is the Roman Catholic. The clergy are not numerous; they are subordinate to the bishop of Santiago. Other religions are tolerated; but the exercise of their public worship is not allowed.

People—are mostly of Spanish and Indian descent, but there are some negroes and mulattoes. 'The Chilians,' says Mr. Miers, 'though they may be said to possess in no degree a single virtue, have the credit of possessing fewer vices than other creoles; there is a passiveness, an evenness about them approaching to the Chinese, whom they strongly resemble in many respects; even in their physiognomy, they have the broad low forehead and contracted eyes; they have the same cunning, the same egotism, and the same disposition to petty theft.' (Travels, ii. 321, 324.) They are moderate in their food, but frequently very dissipated and prodigate in their habits, and in the towns very fond of dress and display. Highway robbery is rare, and so are murders in the country, but not in the towns. Education, or any taste for the fine arts, have hitherto made but little progress.

History.—Previously to the Spanish conquest, Chili belonged to the incas of Peru. In 1535 Pizarro sent Almagro to invade the country, and in 1540, Valdivia; the latter of whom conquered most of the country excepting Araueania. The revolution, which separated the colony from Spain, broke out in 1810; from 1814 to 1817 it was kept under by the royalist forces; but in the latter year the victory of Maypu gained by San Martin, permanently secured the independence of Chili, and opened for it a career, which promises a high state of national prosperity, unless prevented by internal dissensions, which, unfortunately, have been very frequent of late years.

CHILKEAH, an inl. town of Hindostan, prov. Delhi, on the borders of the Kumaon distr., 110 m. NE. Delhi; lat. 29° 24′ N., long. 79° 5′ E. It is a chief mart of trade for the W. provinces, with Kumaon, Thibet, and Tartary, but abandoned on the approach of the unhealthy season, when dangerous malaria prevails.

CHILLAMBARAM, a marit. town of S. Hindostan, prov. Carnatic, 34 m. S. Pondicherry, and a short distance N. the mouth of the Coleroon river; lat. 11° 26′ N., long. 79° 47′ E. In its vicinity there are some celebrated Hindoo temples, of considerable antiquity.

CHILMARRY (Chalmari), a town of Hindostan, prov. Bengal, distr. Rungpore, on the Brahmaputra, 65 m. SE. Rungpore. A festival is annually held here, which is usually attended by 60,000, and sometimes by 100,000 Hindoo pilgrims and others.

CHILOE (ISLAND AND ARCHIPELAGO), a province of Chili, consisting of a large island in the S. Pacific, near the S. coast of Chili and the NW. coast of Patagonia, between lat. 41° 48′ and 43° 50′ S., and having on its E. side 63 small islands, 36 of which are inhabited. The group, including the town of Maulin on the main land of the continent, forms the most S. prov. of Chili. Shape of the island of Chiloe, oblong; length, N. to S. 120 m.; average breadth, 40 m. Area, 4,800 sq. m. Pop. 61,606 in 1854. The island is mountainous, and covered with wood, chiefly a lowland cedar, very durable, and exported in great quantities to Peru and Chili. There are several good harbours, in all of which vessels of any size may anchor with the greatest safety; and in those of St. Carlos (the cap. in the NE. part of the island), and Castro, ships ride quite land-locked close to the shore in good holding ground. Climate healthy, but damp; at an average, ten months of the year may be called rainy. Cold, however, is not severe; water seldom freezes, and a fall of snow is unknown. Little ground is cleared; the soil is rich, though never manured; it consists of dark mould and fine loam upon chalk, and produces good crops of wheat, potatoes, fruit trees, especially apples, which yield a large quantity of cider. Wine is prohibited, and spirits are rarely seen. Tobacco, being a government monopoly, is very dear. Domestic animals are largely reared. The sheep are bred solely for their wool, and are never eaten. The island swarms with hogs, and the hams of Chiloe are celebrated in S. America. Poultry and fish are very abundant. Principal export—planks about 200,000, and hams 7,000 annually; besoms, hides and woollen cloths, to the value of about 25,000 dollars a year. The archipelago possesses about 1,500 coasting vessels. Money is here nearly unknown, and traffic is conducted by barter, or payment in indigo, tea, salt, or Cayenne pepper. All three articles are much valued, especially the first for dyeing woollens, for the weaving of which there is a loom in every house.

The archipelago sends one mem. to the Chilian congress. The public revenue is chiefly derived from a tithe on all produce, paid in kind. There are numerous churches and chapels, but few priests.

The chief towns are San Carlos, which is fortified, and has about 7,000 inhab., Castro, and Maulin. A good road, 54 m. long, runs between the two former towns. According to Captain Blanckley, the golden age would seem to be revived in this part of the world. 'Murders,' says he, 'robbery, or persons being in debt, are never heard of; drunkenness is only known or seen when European vessels are in port; not a private dwelling in the towns or country has a lock on the doors; and the prison is in disuse. (Blanckley, in Geog. Journal, iv. 344–361.) The inhab. are passionately fond of music and dancing. Chiloe was the last possession held by Spain in the Pacific.

CHILTERN HILLS, a ridge of chalk hills in England, traversing the co. of Bucks, and reaching from Tring, in the co. of Hereford, to Goring on the Thames in Oxford. Wendover Hill, in Bucks, the highest part of the range, is 905 ft. above the level of the sea. Camden says that these hills were once thickly covered with trees, which were a receptacle for thieves till they were cleared by the abbot of St. Alban's. (Gibson's Camden, i. 327.) An office, called the stewardship of the Chiltern hundreds, was established at a remote period. Whatever were formerly its duties, they have long since ceased; and it is now nominal only, being kept up to afford means of vacating their seats.

CHIMBORAZO, one of the highest summits of the Andes, which see.

CHINA, a vast country of SE. Asia, between lat. 21° and 50° N., and long. 70° and 144° E.; in form nearly square, being bounded on the E. and SE. by those arms of the Pacific Ocean known as the Gulf of Tartary, the Sea of Japan, the Yellow Sea, the Strait of Formosa, the Chinese Sea, and the Gulf of Tonquin; on the land sides by Tonquin, Laos, and Birmah; SW. and W. by Independent Tartary; and N. for the immense extent of 3,500 m. by Asiatic Russia. Its extent from the borders of Kokhan and Badukshan to the Sea of Ochotsk is 8,350 m., and its greatest width from the frontiers of Dzungaria N. to Tonquin S., is 2,100 m.; inclosing altogether a space of about 5,300,000 sq. m. Thus the Chinese empire includes all the table land of Eastern Asia—about a third part of the whole continent—or a little less than a tenth part of the habitable globe; and contains, within its enormous area, the largest amount of population and of wealth united under one government in the world. The coast line has an extent of above 3,530 m., and the total circumference of the empire is about 12,550 m. (More detailed particulars of the surrounding possessions of China must be sought in the articles ASIA, THIBET, MONGOLIA, MANCHOURIA, Islands of HAINAN, FORMOSA, and TCHUSAN.)

The area of China Proper does not exceed a fourth part of the whole empire. It is true that its dimensions have not been satisfactorily determined, and the following estimate of the extent of the empire, as well as of China Proper, differs from the calculations of many geographers, which, in their turn, widely disagree with each other, except where the mistakes of one writer have been copied by another. To determine the extent of the empire, seventeen linear measurements have been made; two upon native maps, which have been carefully compared with European maps, and the result in reference to China Proper stands thus:—for its length, from N. to S., 1,474 m.; breadth, from W. to E., 1,355 m. But these are not the longest straight lines that may be made to intersect its surface; since, from the NE. corner to the frontiers of Birmah the distance is 1,669 m.

and from the NW. extending to the Isle of Amoy it is 1,557 m. The entire area contains 1,359,870 sq. m. The coast is upwards of 2,500 m. in length, while the land frontier occupies a space of 4,400 m. Thus China Proper is about eight times the size of France, and eleven times that of Great Britain. (Staunton; Tab. Geog. Chin. Native; Ogilby, i. 7, and Map; Du Halde's General and Particular Maps; Lord Macartney's do.; Arrowsmith's Atlas, pl. 77, 32, 32, 33; Gutzlaff's China Opened, i. 21–57.)

General Aspect.—The first object that invites attention in the general aspect of China is its Great Plain, which, occupying the NE. part of the country, is above 700 m. in length, and varies in width from 150 to near 500 m. The entire area belongs to no less than six provinces, and a space of 210,000 sq. m. being seven times greater than the plain of Lombardy. It is extremely populous; and if we might depend upon the census of 1813, no fewer than 170,000,000 months'—the Chinese expression for souls—are fed upon its surface. The N. portion, bounded by the great wall, is dry and sandy, and its E. portion, bordering on the sea, and between the two great rivers the Hoang-ho and the Yang-tse-Kiang, by which it is intersected, is low, swampy, and studded with lakes. But, notwithstanding these defections, it may be said to be, on the whole, extremely fertile. It has few trees, but is everywhere well watered; is cultivated with the utmost care, and produces vast quantities of rice, with cotton, wheat, &c.

Mountains and Hills.—The mountainous and hilly districts of China comprise about half its area. A portion of the great mountain system of E. Asia entering this country at its NW. and SW. frontiers, subsides previously to its terminating near the sea-coast into low hills; or that, tracing their course westwards from E. to W., they gradually ascend in terraces or slopes, and give to the N. and W. districts a mountainous, and to the E. divisions a hilly character. NW., at about 34° N. lat. and 100° E. long., the great Pe-ling range, which has already traversed a portion of Thibet from W. to E., is joined by the Yun-ling chain, which, entering China at about 31° N. lat. and 101° E. long., descends southward nearly to the prov. of Yun-nan. These mountains form the easternmost edge of the high table-lands of E. Asia, are snow-capped, and inaccessible to the natives, being actually left blank in the Chinese maps. (Davis, i. 131.) Another ridge, joining the Pe-ling at the same point, takes an opposite or NNE. direction, and entering the empire in the prov. of Shen-se, reaches nearly to 110° of E. long. Another arm of the Pe-ling—the Ta-pa-ling chain —intersects the country from W. to E. to about 115° E. long.; the Pe-ling itself continuing in its former course, gives out various branches, which traverse the central provinces. The other mountain chains join the stupendous Himalaya ridges, and enter the country at its SW. extremity in the province of Yun-nan, from whose high table-lands the most extensive Chinese ranges rise. The Yun-ling, the most southerly of these chains, runs nearly E. into the prov. of Quang-tung. But by far the most important mountain range is the Nan-ling, which, branching off from the northern edge of the Yun-nan highlands, runs eastward to within 150 m. of Canton; it then inclines to the NE. to its termination near the harbour of Ningpo; having given out many branches, some of the mountains belonging to which rise above the snow-line. (Macartney's Embassy, pp. 207, 246, 259; Barrow, ii. 241, iii. 29, 121; Malte Brun, ii. 551, 552; Davis, pp. 130, 131.) Most of the mountains here encountered end in low hills in the

eastern provinces, which consequently comprise the hilly districts. These are the most picturesque portions of China; and being covered with noble forests, crowned with pagodas, and with cities along their sides, give to the country a magnificent aspect, without interrupting its culture.

Rivers and Lakes.—It is to her mighty rivers that China is chiefly indebted for that fertility which is at once the source of her riches, and of her vast population. The Hoang-ho, or yellow river, and the Yang-tse-Kiang, or 'son of the ocean,' rank in the first class of rivers. These two great streams, similar both in rise and destination, descend with rapidity from the great table lands of central Asia, and each of them meets a branch of mountains which forces it to describe an immense circuit, the Hoang-ho to the S., and the Yang-tse-Kiang to the N. Separated by an interval of 1,100 m., the one seems inclined to direct itself to the tropical seas, while the other wanders off among the icy deserts of Mongolia. Suddenly recalled, as if by a recollection of their early brotherhood, they approach one another like the Euphrates and Tigris in ancient Mesopotamia; where, being almost conjoined by lakes and canals, they terminate, within a mutual distance of 110 m., their majestic and immense course,' (Malte-Brun, ii. 556.) The waters of the Hoang-ho bring down from its sources large quantities of yellow clay, which not only tinge them with that colour, but supply the banks with alluvial soil. Large deposits of this clay are constantly being made at the mouth of the Hoang-ho; so that the depth of the Yellow Sea has sensibly diminished. The Yang-tse-Kiang is, however, the pride of China. It is the chief artery of the country, and undoubtedly one of the largest rivers of Asia. This stream is also heavily charged with alluvium, for at its exit into the sea—near which it is from 15 to 20 m. broad—continued deposits have formed the I. of Tsung-ming, besides numerous banks. The tributaries received into this river during its course, which is about 2,300 m., are innumerable; and, with the canals, connect it with the whole empire. Both the rivers, especially the Hoang-ho, which has a very rapid course, occasionally overflow their banks, and, in spite of many strong artificial mounds, cause the most destructive inundations. The river next in importance is the Eu-ho or Yun-liang river, which flows NE. till it joins the Pei-ho at Pekin river; the latter rises in the mountains NW. of Pekin, near which city it becomes navigable for boats; and is, during the rest of its course, the most populous stream of a country where a large proportion of natives live upon the water in junks; their united waters flow into the sea in the most W. angle of the Pe-che-lee Gulf. The Ta-si-Kiang, Choo Kiang, or Canton river, rising in the prov. of Yun-nan, takes an E. course to the plains of Canton, and having received the Pe-ki-ang, the Ta-ho, and other smaller streams, forms an estuary known as the Bocca Tigris, by which it is finally discharged into the China Sea, after a course of 600 m. There are a vast number of other rivers, some of which fall into the sea, and others into the great lakes. The Brahmaputra, Irawaddy, Thaluen, Meinam, &c., have their sources in the SW. parts of China. (Journal Royal Geog. Soc., iii. 305; Lindsay's Voyage in the Lord Amherst, passim; Gutzlaff's Voyage, passim; China Opened, i. 79 and 61–164; Malte-Brun, ii. 555–557.)

The principal lake in China is the Tonting-hoo, 220 m. in circ. It receives the waters of many considerable rivers, and furnishes an important affluent to the Yang-tse-Kiang, which passes near its N. extremity. After a further course of be-

(tween 200 and 300 m., this great river receives the surplus waters of the Po-Yang-hoo lake, which also is of great dimensions, and is the recipient of many considerable streams. This lake is surrounded by picturesque and finely-wooded hills. Indeed, its scenery is so much admired, that its shores are the favourite spot where Chinese poets muse and write their versified prose. It is, however, subject to sudden tempests, which render its navigation dangerous. The environs of the Tai-hoo lake, near the E. coast, lat. 31° N., long. 120° E., are even more picturesque than those of the Po-yang, having gained the name of the 'Chinese Arcadia.' The Hong-tse-hoo, being situated near the junction of the Grand Canal with the Yellow River, is much frequented on account of its advantageous position. All the lakes, in fact, furnish intermedia of communication, and are abundantly stocked with fish. China contains several smaller lakes, but the whole do not occupy any great proportion of her vast surface. (China Opened, i. 31; Barrow, ii. 587, 291, &c. &c.)

Coast.—The coast of China has yet to be described. If our statement be correct, that the sea-coast extends for 2,500 m., there is only one mile of coast to every 559 m. of territory; but internal navigation is carried on so extensively that this deficiency has no ill effect upon Chinese commerce. Commencing at the NE., the coast opposite Corea is bold and rocky, has, on approaching the Gulf of Pe-che-lee, presents a low and sandy shore, scarcely perceptible from the sea. The bar formed in this bay, at the mouth of the Pei-ho, makes its bed inconveniently narrow, and, when the N. winds blow, the whole adjacent country is overflowed to a great extent. The coast of the Shan-tung peninsula is bold and rocky, so indented as to afford excellent harbours; but, once reached, the low swampy character of coast is again presented as far as the Tchusan islands. Meantime, the two great rivers have brought down their immense deposits from the interior, which give its name to the Yellow Sea. The mud is so thick as to retard the headway, and affect the steering of ships; and this great gulf will, in process of time, become a vast alluvial district, like Bengal and Egypt. 'The present inclination of the bottom is about a foot in a geographical mile, or somewhat less than 1 in 5,000; and it is probable that the bottom of the Yellow Sea, as it rises, will likewise gradually approximate to a horizontal plain.' (Hall's Voyages, i. 27.) This sea is nearly surrounded with islands. The coast down to the strait of Formosa continues low, and, except where it faces the Tchusan islands, and in the prov. of Fokien, is but little indented. The strait itself abounds with headlands, and is also so thickly studded with islands, which are but imperfectly notified even in the best charts, that navigation is, by Captain Hall's account, 'exceedingly trying to the nerves.' The Quang-tong shore is bold and high, except in the recesses of the numerous bays and harbours. A narrow peninsula is thrust out far into the sea at the W. extremity of Quang-tong, and forms with the island of Hainan, a narrow channel, which is shoal, full of sand banks and rocks, so that even the native flat-bottomed junks are exposed to great danger. The rest of the shore is washed by the Tonquin Gulf, which is studded with small islands. (Hall's Voyages, Ttno, edit. i. 29-46; Gutzlaff's Voyage, passim; Lindsay's Voyage; Journal Geog. Soc., iii. 291-310.)

Public Works.—Aspect of Cities and Towns.—An amount of human labour, probably unmatched by any other nation in the world, except ancient Egypt, has been expended on the public works of China, by which the natural aspect of the country has been materially varied. The first and most stupendous of these is the great wall, built several hundred years before the Christian era, to protect China from Tartar incursions. It extends along the whole N. frontier, from the Gulf of Leatong, in 120°, to the NW. extremity of the empire, in about 99° E. long., and 40° N. lat., being including its windings, about 1,250 m. in length; it is carried over the tops of the highest mountains, through the deepest valleys, and continued by bridges over rivers. Its height varies from 15 to 30 ft. It is 15 ft. across at the top; and, at short intervals, square towers are erected, some of them 37 ft. high. The wall is composed of earth faced with masonry, the top or platform being paved with square tiles. It is now in a state of decay, being no longer required, since the union of the Tartar with the Chinese territory, for its original purpose. (Davis, i. 136; Bell's Travels, ii. &c.)

The Great Canal commences at Hang-tchou, near the mouth of the Tching-tang-chiang river, in about 30° 22' N. lat., and 119° 45' E. long., and, extending N., unites first with the Yang-tse-Kiang, and then with the Hoang-ho, terminating at Lin-tcing, on the Eu-ho river, in about 37° N. lat., and 116° E. long. The direct distance between the extreme limits of the canal is about 512 m., but, including the bends, it is above 650 m. in length; and as the Eu-ho, which is a navigable river, unites with the Pei-ho, also navigable, an internal water communication is thus established between Hang-tchou and Pekin, across 10° of lat. And by the junction of smaller canals and numerous rivers, the Great Canal not only assists in the irrigation of immense tracts of land, but affords a ready means for conveying its produce to all parts of the empire. But, apart from its utility, the Great Canal does not rank high as a work of art. A vast amount of labour has, however, been expended upon it: for though it mostly passes through a flat country, and winds about to preserve its level, its bed is in parts cut down to a considerable depth, while in other parts it is carried over extensive hollows, lakes, &c., on vast mounds of earth and stone. (Barrow, 511.) The sluices, which keep its waters at the necessary level, are all of very simple construction. In the public roads, and where rugged steeps are only accessible by means of laboriously formed passes, Chinese industry is fully apparent. Three mountain paths (traverse the Nan-ling; one, N. of Canton, is estimated by Sir G. Staunton to rise 8,000 ft. above the sea; yet vast quantities of goods are conveyed over this pass from Canton to the interior by coolies or porters. The obstacles to communication presented by the Pe-ling and Ta-pa-ling ranges are greatly diminished by an artificial road sometimes conducted over yawning clefts by arches, in other places deeply cut through high mountains, and extending altogether for 150 m. In short, wherever intercourse is expedient between any two parts of China, no natural impediments are too gigantic, no labour or expense too great, to overcome them.

The following summary of the general appearance of the cities and towns of China is supplied by Gutzlaff:—'The districts on the sea-coast are generally the least inhabited and the richest; the tracts along the Yang-tse-Kiang the most fertile. Large and flourishing cities are found only where a ready water communication with other parts of the empire can be carried on. The greatest sameness exists in all the cities. In the larger ones are a few well paved streets, lined with shops; but the greater part of the streets are very narrow, extremely filthy, and planted with mere hovels.

The suburbs of many cities are much larger than the cities themselves; and it is by no means extraordinary to see an immense walled space without any houses, where formerly a city stood. Villages and hamlets have a beautiful appearance at a distance; but on entering them one sees nothing but a heap of houses irregularly thrown together, the outside fair to behold, but the inside without furniture or comforts, and more filthy even than a stable. This does not apply to one district only, but it is common to most. Although the fields and gardens are beautifully laid out, there yet appears in them little attention either to elegance or pleasure. The gardens are very few; and a Chinese grandee delights more in artificial landscapes laid out in a small compass, than in an extensive park or a flower-garden. Utility is studied in preference to pleasure. The grandeur of natural scenery is in many parts of China so striking as in many parts of the world. Mountains, crags, rivulets, and valleys, both picturesque and romantic, are found in most provinces. Commanding situations are chosen for temples, the haunts of superstition and idolatry. These serve likewise for taverns, stages, public halls, and gambling-houses. The building of houses is regulated by law; none are allowed to exceed a certain dimension. Public halls have little to recommend them; the Chinese were never great architects; they understand the building of dwelling houses, but not of palaces. (China Opened, i, 57, &c.)

Climate.—Connected with this subject there are some singular circumstances. Situated between the 20th and 42nd degrees of N. lat., and the most E. long. of any part of the Old World, the temperature of China is very low for its geographical position. Its climate may also be said to be one of extremes; and while at Pekin, which is nearly 1° farther N. than Naples, the mean temperature is that of Brittany, the scorching heats of summer are greater than at Cairo, and the winters as rigorous as in the northern provinces of Sweden. But in so extensive a territory there are necessarily many variations. The W. districts are much influenced by the colds diffused by the mountains, while the climate of the maritime provinces is modified by the sea. At Canton, which is under the tropic, the heat during July, August, and September, is excessive: then occur those frightful tornadoes, called typhoons, spreading devastation in their course, which, however, do not extend far beyond Canton. At the breaking up of the hurricanes, the transitions from the heat of day to cold and foggy nights are more violent and sudden than in any other part of the globe. The N. winds set in about November, and bring with them cold as intense as the preceding heats. The mean temperature of Canton is 70° Fahr. The climate of the interior is not however, with few exceptions, so extreme, particularly towards the N. frontier, where the summers are genial; and though the winter be cold, it is dry, and does not check the growth of fruit; but the N. winds bring clouds of white sand, which afflict the natives with ophthalmia. The W. frontier districts of Yun-nan and Sze-chuen are said to be unhealthy, and are selected as places of banishment for Chinese convicts. The central provinces present a striking contrast to those already named. There the climate exhibits a happy medium between the rigour of the N. regions and the enervating heats and sudden colds of the S. The Kiang-tse is the most favoured in this respect. The fall of rain in China varies considerably in different years. Humboldt states—without naming on what authority—that the average quantity per an. is 70 in.; though it has been known to exceed 90. Many violent earthquakes have been felt in China. (Malte-Brun, art. 'China'; China Opened, i, 31, 60, 90, 102, 164, 185; The Fan-qui in China, by C. T. Downing, Esq., i, 191, 192; Lyell's Geology, ii, 50.)

Population.—China has long been very generally believed to be the most densely peopled country of any considerable extent in the world. The Jesuit Semedo, writing in 1615, remarks that, after living in the country twenty-two years, he was no less surprised on leaving them on his first arrival, at the immense number of persons he met with, not only in the towns and cities, but on the highways, 'where,' says he, 'there is at all times as large a crowd as is usually to be met with on some great festival or public occasion.' The Jesuit Amiot, founding on official documents, estimated the pop. in 1743 at about 148,000,000, which, adding for some classes that he had omitted, may be carried to about 150,000,000; and in 1792, Lord Macartney was informed, by a mandarin, 'a plain, unaffected, honest man, whose statement is said to have been made on the authority of official documents, that the pop. was 333,000,000, and later accounts carry it up to above 360,000,000.

It must be confessed, however, that, with the exception of that of Amiot, these statements appear altogether incredible, and that, in point of fact, there is no certain information as to the pop. of China. According to the statements in Chinese official works, the pop. of the empire amounted, in 1393, to 60,515,000; and in 1578 to 60,692,000. It is supposed to have continued at or about this amount till the Tartar conquest in 1644, a year before the publication of Semedo's work. But it appears from an imperial proclamation quoted in the 'Chinese Repository,' issued in 1792, and said to be founded on official data, that the pop. had been reduced in 1711 to 28,605,716 (vol. i, p. 356, Canton, 1833). This extraordinary diminution is attempted to be explained in the work referred to, by the mortality occasioned by the long and bloody wars that accompanied the establishment of the Manchoo dynasty, by the fact of some of the provinces, in the S. not having been fully subdued when this census was taken; and by the circumstance of a poll-tax being then imposed, which made it for the interest of individuals to escape being enrolled in the census. But even admitting the force of some of these statements, and allowing that but for the wars occasioned by the Tartar conquest, and the imperfectly subdued state of parts of the country, a correct census taken in 1711 would have given a pop. of sixty or seventy millions, still it can scarcely be credited that the pop. should have increased from even that amount, in 1711, to above 300,000,000 in 1792. Had China been a new country, or had the Tartars, by whom she was overrun in the 17th century, been distinguished by their superior intelligence and industry, an increase of this sort might have been possible. But the reverse of all this is the fact. China has been settled and civilised for many centuries; the great works undertaken and completed by her inhabit, at a very remote period, show that she had then been pretty thickly peopled; and it is admitted, on all hands, that in China the arts have been for ages in a nearly stationary state. The Tartars imparted to her little that was new. They were, in truth, mere roving hordsmen; and though they might have given the Chinese some instruction in predatory warfare, they could communicate to them no useful art, science, or invention. Under these circumstances it must be admitted either that the former official accounts of the pop. were greatly underrated, or that the later ones were greatly ex-aggravated. (For a further discussion of this subject, see De Guignes, Voyages à Pékin, iii. 55-56.)

Subjoined is an account of the area of the different provs. as given by Lord Macartney, and their pop. as given by Amiot in 1743, by Lord Macartney in 1793, and by the official returns in 1813.

Provinces	Area in sq. m.	Pop. 1743 (Amiot)	Pop. 1793 (Macartney)	Pop. 1813 (Official)
Northern				
Pe-che-lee	58,949	18,703,745	38,000,000	27,990,871
Shan-se (W. of Hoai.)	55,268	9,768,190	27,000,000	14,004,210
Shen-se (W. of Pass.)	154,008	14,804,625	18,000,000	10,207,256
Kan-suh			12,000,000	15,193,125
Central				
Ho-nan	65,104	19,637,980	25,000,000	23,037,171
Kiang-se	72,176	6,651,250	19,000,000	23,046,999
Hoo-pih	144,770	4,904,650	14,000,000	27,370,098
Hoo-nan			13,000,000	18,652,507
Kwei-choo	64,554	8,692,777	9,000,000	5,288,219
Southern & Eastern				
Shan-tung	65,104	17,169,000	24,000,000	28,958,764
Kiang-su	92,961	28,705,285	32,000,000	37,011,300
Gan-hwuy	89,150	16,673,950	21,000,000	28,956,784
Che-Kiang	53,460	7,513,025	16,000,000	14,777,410
Fo-Kien	73,456	6,806,000	21,000,000	10,174,000
Quan-tong	78,250	1,142,150	10,000,000	7,313,895
Kwang-se	107,969	3,190,925	8,000,000	8,261,230
Yun-nan				
Western				
Sze-chuen	166,800	15,181,710	27,000,000	21,435,678
Leau-tung		926,620		
Total	1,397,989	156,205,478	333,000,000	360,279,897

The census for 1812 adds an additional 1,413,902 souls as the pop. of Ning-hing, Kirin, Turfan, Ilchmat, and Formosa; and 188,526 families as engaged in the service of the emperor. Supposing the latter to consist of four members each, the total pop., according to the census of that year, will be 362,447,183.

A glance at the above table will show that the account of the pop. furnished to Lord Macartney in 1792, and the census of 1813, cannot both be accurate. The last shows an excess over the former of 29½ millions in the aggregate; but it would appear that in the majority of the provinces there has been no increase; but, on the contrary, a diminution. In the evidence adduced before the British parliamentary committee, in 1830, 1831, and 1832, the area of China was computed at 1,372,452 English statute square miles, and the number of inhabitants at 141,470,000, or 103 to the square mile; to which was added 1,182,000 for the standing army, and 12,000,000 for Tartary. But the information was very obscure with regard to the population. Thibet, Kurra, the Manchus, and other Tartar and Mongolian states, were computed to have a population of more than 50,000,000, which would increase the whole population of China and its assumed dependencies to nearly 400,000,000 inhabitants.

Local Divisions.—Though the geography of the world be not much studied in the 'Celestial Empire,' the more minute details of local topography are no where better understood. The survey of the Jesuits, made by order of the emperor Kang-he, is said to be very correct; and every district of any importance has since found a geographer, who describes it, if not so scientifically as the Catholic missionaries, with the utmost minuteness, so that, with little difficulty, a library of 3,000 vols. might be collected treating exclu-

sively of Chinese geography. Nothing can be more systematic than the manner in which the whole empire is divided. Each prov. is partioned off into provincial districts; while the towns and cities are divided into the 1st class (*foo*), 2nd class (*chow*), and 3rd class (*heén*). Formerly China Proper consisted of fifteen provs.; but in Kefa-Lang's time the largest were bisected, and there are now eighteen.

Northern Provinces.—1. Pe-che-lee (the independent) is subdivided into sixteen districts, the most W. of which are very flat; the central ones somewhat hilly; while those on the sea-coast along the Pe-che-lee Gulf are low and marshy. Pekin, the metropolis of Northern China and residence of the court, is situated in this prov., about 60 m. from the great wall, and 100 m. from the sea. The Pei-ho flows through Pe-che-lee, disembarking at the small sea-port of Takoo. The chief ports are Tong-choo and Tein-sing. It is a curious fact, and one which does not square well with the popular notions of absoluteism, that, despite the residence of the court, the bulk of the population are probably more depressed in this than in any other prov. (Barrow, 463.) 2. Shan-se, or Chan-se (west of the mountains), is divided from Mongolia by the great wall, a branch of which (the inner great wall) separates its E. limit from Pe-che-lee. It is said to have been the most early occupied part of China. Its mountainous portions are not, however, habitable, and many other localities afford but a scanty subsistence. Hence it has no large or remarkable cities. 3. Shen-se, or Chan-se (west of the pass), is also separated from the Mongolian borders by the great wall, which in this place is kept in good repair. The mountains in this prov., which are more rugged than high, contain gold mines, but these are not allowed to be worked, lest the attention of the people should be withdrawn from agriculture. The valleys through which the Hei-ho and the Han-Kiang run are fertile in millet, wheat, and pulse, but are too dry to produce much rice. Swarms of locusts frequently appear in Shen-se, destroying the harvest, and converting smiling valleys into wastes. The chief town is Se-gan-foo, one of the largest in the empire. 4. Kan-suh (voluntary awe) and Shan-se, formerly called, made one large prov., extending over a space of 154,000 sq. m. Kan-suh consists principally of a narrow neck of land thrust out upon the edge of the great Gobi desert; hence the soil is cold and barren. Kan-suh forms the NW. limit of China, the great wall ending at Kwang-lan.

Central Provinces.—5. Ho-nan (south of the river) is one of the most fertile provinces of the great plain, and is called the garden of China. Shen-se, Pe-che-lee, and a part of Shan-tung join its N. boundary, while branches of the Pe-ling enclose it to the W. The Hoang-ho, or Yellow River, runs nearly parallel with the N. boundary, and intersects the finest parts of the prov. 6. Kiang-se (west of the river) has its boundaries well defined by the Nan-king range and its branches, which surround it on three sides, the W., S., and E. Its N. part contains the great Poo-Yang lake, and its contiguous marshes, said by Mr. Barrow to be the sink of China. It has, however, many well cultivated valleys, in which rice, cotton, indigo, and sugar, are produced. It has also extensive manufactures, amongst which must not be forgotten the China-ware, so highly esteemed all over the world, till European imitators exceeded the original manufacture in beauty and cheapness. Still, however, no fewer than a million persons are said to be exclusively employed in this manufacture, which is chiefly

carried on at the capital King-te-chin. Here
(40) furnaces are constantly burning. 7. Hoo-pih
(north of the river), and, 8. Hoo-nan (south of
the river), form the ancient prov. of Hoo-Kwang,
divided into two parts by the Yang-tse-Kiang.
The former is divided into eleven and the latter
into thirteen districts; the whole covering an area
of 144,770 sq. m. Both prows. are extremely fertile, and the capital of Hoo-pih yields to few
cities of the empire in extent and prosperity.
The tea grown in its neighbourhood is of superior
quality, and the bamboo-paper manufactured
within its walls is extensively exported. This
city is called Woo-chang-foo. Hoo-nan bears a
great resemblance to the Hoo-nan prov., but is
richer in minerals. A very active trade is carried
on, on both banks of the Yang-tse-Kiang. Hoo-
pih and Hoo-nan are both within the great plain.
9. Kwei-chow has been designated the Switzerland of China, being traversed by the highest
portion of the Nan-ling range. To the S. it is
peopled by wild and intractable highlanders
(Meaou-tse), who, though in the centre of the
empire, preserve their independence, and frequently make predatory descents on the adjoining provinces. Kwei-chow has no large towns,
but several fortresses.

Maritime and Southern Provinces.—10. Shan-
tung (east of the mountains) is partly in the
great plain and partly consists of a promontory
jutting into the Yellow Sea, S. of Pe-che-lee,
and N.E. of Ho-nan. Its W. part is traversed
by the Great Canal; but the country is poor,
and the climate, though bracing, bleak. There
are, however, some valuable coal mines, which
supply the whole empire with that article. The
coast is bold, and affords good shelter. The principal port is Tong-chow-foo. 11 & 12. The Kiang-
nan (river Sea) and Gan-hway (fixed excellence)
prov. were once united under the name of Kiang-
nang. The two great rivers, the Hoang-ho and
Yang-tse-Kiang, cross both districts, and fall into
the sea 70 apart, forming the Chinese delta. Kiang-
hway has 13 districts, and the Kiang-soo 11;
their united extent being 92,961 sq. m. 'If we
consider,' remarks Gutzlaff, 'their agricultural resources, their great manufactures, their various
productions, their excellent situation on the banks
of the two largest rivers in China, their many
canals, and amongst them the Great Canal and
tributary rivers, they are doubtless the best territory of China.' Enjoying these blessings, chiefly
conferred by their two great rivers, these provinces
are also the most liable to the evils they produce,
namely, frequent and destructive inundations.
The staple products are grain, cotton, green teas,
and silk. Hien suits admirably with the black
marshy loam of which most of the soil consists.
Nanking (capital of the S.) is situated on the S.
bank of the Yang-tse-Kiang, but at the distance
of a league from the stream (Nankin). The
Kiang-soo prov. only faces the ocean. The scene
which appeared at the junction of the Yang-tse-
Kiang and Great Canal is thus described by
Barrow :—' The multitude of ships of war, of
burden, and of pleasure ; some gliding down the
stream, others sailing against it ; some moving
by oars, and others lying at anchor ; the banks
on either side covered with towns and houses as
far as the eye could reach ; presented a prospect
more varied and cheerful than any that had
hitherto occurred. Nor was the canal on the
opposite side less lively. For two whole days we
were continually passing among fleets of ships
of different construction and dimensions. Cities,
towns, and villages were continued along the
banks without intermission. The face of the

country was beautifully diversified with hill and
dale, and every part in a high state of cultivation,' (p. 516.) 13. The Che-Kiang (river Che), or
Tche Kiang, is the smallest Chinese prov. It
occupies the N.E. corner of the great plain. The
Yun-ling chain ends here in innumerable low
hills, the most barren of which produce abundance of tea. In fact the whole district is most
assiduously laid under contribution by the inhabitants, every inch of ground being tenanted.
At the port of Cha-poo, a large trade is carried
on with Japan. 14. Fo-Kien (happy establishment), which forms the W. shore of the Formosa
channel, is mountainous. Barren hills and sandy
plains are, in truth, the natural characteristics of
Fo-Kien, but Chinese industry has made the land
fruitful. The tea-plant thrives in perfection, and
the 'China orange' is chiefly derived from this
prov. The maritime commerce of Fo-Kien is
extensive, its merchants monopolising most of
the Chinese shipping trade. Emigration though
discouraged by the government, is here very prevalent. 15. Quan-tung (eastern breadth) joins
Fo-Kien to the E.; its shores stretch along the
whole S. coast of China, to the borders of Cochin
China, the S. boundary being formed by the
Nan-ling mountains. Quan-tung has 18 districts,
and an equal number of trading emporiums, and
to this prov. alone are Europeans allowed to
trade. It has many wide valleys, particularly
the plain around Canton, which is of great extent, and many valuable products; but, though it
be the great entrepôt for tea, that article is not of
the number. The capital, Kwang-chow-foo (Canton), is the greatest emporium of the E. 16.
Kwang-se (western breadth) joins the W. limits
of Quang-tung, the Nan-ling range divides it
from Hoo-nan on the N., while its S. border
unites it with the Cochin Chinese prov. of Tonkin.
The mountainous portions of the prov.—by far
the greatest part of it—are said to contain gold
and other metals; the lowlands and valleys produce rice, silk, and timber. Both the language
and manners of the inhabitants differ from those
of their countrymen. 17. Yun-nan (south of the
clouds), the most W. of the S. prov.; is conterminous on the S. with Cochin China and the
Birman empire; and towards the W. with Thibet.
Its mountains, which are remarkably high and
bold, furnish the copper that supplies the currency
of China. It is in Yun-nan that the Yang-tse-
Kiang enters China; and by the aid of a high
road, which has been made parallel to its banks
for a great distance, communication between it
and the rest of the empire is rendered constant
and easy. The same road branching off to the S.,
extends into the heart of the Birman empire.

The western province, Sze-chuen (five Rivers), is
the largest in China. Plains, mountains (the
Yun-ling), and extensive deserts are its principal
components. The Yang-tse-Kiang having taken
a S. bend at the Yun-nan frontier, traverses its
whole extent; and, during this part of its course,
receives several tributaries. The capital, Ching-
too, was once the metropolis of an independent
state, which then surrounded it; and its inhab.
still boast of greater independence of character
than their neighbours; which they evince by frequent rebellions. (China Opened, i. 155-168.)

Natural Productions of China.—The climate of
China, exhibiting occasionally such severe cold,
forbids the presence of some members of the
animal kingdom met with in the similar latitudes
of India. The universal cultivation of China
Proper, and the thickness of its population, have
long expelled most of the wild animals which
still abound in the surrounding regions. There

are also fewer domestic ones than inhabit most European countries. Beasts of burden are in a great degree superseded by the means of transit so copiously afforded by canals and water-courses, and by that fine race of men the coolies or porters; while the canal boats are dragged along by trackers. Add to this, that animal food is considerably less in use among the Chinese than vegetable diet. There are no members for feeding cattle; and even if there were, the natives have a singular aversion to butter and milk. Tigers, though they have been seen in the forests of Yun-nan, are scarcely known; and the lion is almost deemed fabulous in China. There are wild cats, which are caught, confined, and fed in cages, and considered a dainty for the table. Monkeys are found in the southern districts. The Chinese horse and ass are small and spiritless, and so is the buffalo, which is sometimes employed in ploughing. Dromedaries are much used between Pekin and Tartary. Pigs are reared with great care; sheep are smaller than those of England, and goats, of various colours, have uniformly straight horns. The dog of China is almost the size of a spaniel, and is uniformly met with of the same variety. Rats emigrate occasionally from one place to another in large troops, when they devour crops and harvests; they are very large, and are used by the common people as an article of food. There is a genus of rat peculiar to China, which bears some resemblance to the bamboo rat of Sumatra. The ornithology of China presents, in the first place, the eagle, which frequents the mountainous districts; the kestrel, a kind of falcon, abounding in the province of Che-keang, is considered imperial property, while the magpie, which is so numerous as to be the farmer's worst nuisance, is considered sacred by the reigning family. Crows and sparrows are also abundant in China. Among others of their manifold stratagems for catching fish, the Chinese have trained the fishing cormorant; but that the bird may not help itself too bountifully, the owner puts an iron ring round its neck, which obliges it to deliver up a portion of its prey. Curlews and quails are found in great quantities in the N.; the latter are esteemed chiefly for their fighting qualities, as cocks used to be in England; and, when tamed, good fighting quails sell at enormous prices. Larks are numerous, and sing admirably. But the greatest boast of Chinese ornithology is its splendid varieties of pheasants. One, the medallion pheasant, takes its name from a membrane of brilliantly coloured feathers, which are displayed or contracted at the will of the bird. The gold and silver pheasants have also a most brilliant appearance, and are so plentiful as, in some districts, to furnish the tables of the poor with an excellent dish. Pigeons of different sorts are not rare, but the natives seldom domesticate them. Aquatic birds are naturally inclined to a country which has so many lakes and rivers. The most celebrated of these is the mandarin duck, a species of teal, so celebrated for the strong mutual affection between the male and female that it is used by the Chinese as an emblem of conjugal fidelity; their plumage is beautiful. The snow-white rice-bird of Siam is of great use in China in extirpating vermin from the marshy rice-fields; which it is enabled to accomplish by means of its long legs and long beak.

From the fishes peculiar to China we derive the gold and silver fish, which are kept there, as in Europe, for ornament in glass globes. The edible fish peculiar to China are, first, one of a yellowish colour, caught in the Yang-tse-Kiang, which,

while fresh, is insipid; but is considered a great delicacy after having been kept for a time in ice. The sheng-tong, sea-eel, and a sort of rock cod, called tsang-yu, are also much esteemed, and so are sturgeon, mullet, carp, perch, sea-bream, &c. Crab-fish of various kinds are plentiful. On parts of the rocky coast, oysters are successfully preserved and fattened in oyster-beds.

Though the larger species of reptiles are unknown in China, the smaller lizard tribes are numerous in the hot months; several fresh water tortoises have been discovered, and also two new species of frogs. Venomous serpents are but little known. The insect tribes of China furnish its greatest plague and its greatest blessing. The plague of locust-swarms is terribly inflicted upon the N. and W. prov. Nothing can exceed their voracity; and it is not uncommon for them to occasion so much destruction, as to reduce thousands of human beings to starvation; while another insect, the silk-worm, furnishes employment and riches to an immense part of the pop. In rearing these profitable worms, the Chinese excel all other nations. Scorpions and centipedes are plentiful. A spider, peculiar to China, which inhabits trees, devours small birds, after entangling them in its enormous web. Butterflies of gigantic size, and brilliant colours, abound E. of Canton. Multitudes of white ants are very destructive in the N.; and the mosquito is found in some parts of the country during the summer months. There is a singular sort of bee, called the white-wax insect, which furnishes the whole nation with that article, which it deposits upon a particular sort of tree, furnished by the natives with nests to attract the insects.

The vegetable kingdom of China is remarkable for not containing any very large trees, and timber is comparatively scarce. The oak is seldom seen, fir trees chiefly supplying its place, every ridge of mountain where it is likely to grow being planted with the fir. Palms, laurel, cassia, and caper trees are often met with, especially in the S. provinces, and the cultivator grows together the banana, guava, orange, papaw, cocoa, litchi, peach, apricot, vine, pomegranate, and chestnut. There is also a singular production called the tallow-tree, which resembles the birch, but the bark is white and the branches slender; the fruit, growing in bunches, is enclosed in a brown capsule, which encloses three kernels, all coated with tallow, themselves containing an oil much used for the lamp, while the tallow is converted into candles. There is also the tse, or varnish tree, resembling the ash, which exudes a valuable essential oil, but produces a cutaneous disease if dropped upon the skin. It is the white blossoms of the tsi-pih which attract the wax-fly. The camphor-laurel is extremely productive of that drug in China. The dorm-tsu contains a pith which, when ground to powder, answers all the purposes of flour. A species of sycamore, the koo-shu, supplies paper to the Chinese from the rind; thin, riband-like strips are peeled and made into paper. Mulberry trees, as food for silk-worms, have much pains bestowed on their culture.

We come now to the shrub which has brought China into nearer contact with foreigners than her sages ever desired, or her government seem willing to render closer. The tea-plant, called by the natives cha, rises from three to five feet in height, and bears a strong resemblance to the myrtle, but the flower is not unlike small white hedge roses. Although European botanists have only discovered two varieties, black tea and green tea, native writers enumerate as many hundreds; an obvious exaggeration. Though this plant will

grow in the most sterile ground, the quality of
the leaves depends upon the soil which nourishes
them, and the age of the tree. The best are taken
from three year old shrubs. There are three in-
gatherings of the leaves; the first in early spring,
the second at the commencement, and the third
at the end of the summer. They are carefully
manipulated, dried in various ways, and then
packed. The coarsest leaves are beaten into
cakes and exported, principally into Tartary,
under the name of *brick-tea*, or brick tea. But
the finer descriptions of tea require a vast deal of
labour in their preparation, and could only be
produced in a country where the inhabitants are
universally industrious, and wages low. That
giant of the grass tribe, the bamboo, is most extensively used; besides being an important instrument for enforcing the laws, the Chinese
build cottages and fashion all sorts of furniture
with it. The tender shoots make an excellent
food, and supply the material for a coarse sort of
paper. Tobacco, the cotton plant, and sugar-
canes, are also profitably cultivated. The growth
of garden flowers is not much encouraged, every
available inch of ground being used for the production of edible plants. Even the more opulent
natives are content with a few flower-pots, with
some pretty flower for the sake of ornament. The
water-lily not only produces a beautiful flower,
but its fruit provides an excellent meal, and unlike gruel, is much prized among the Chinese.
They have almost unlimited varieties of the
camellia. A plant, the name of which has not
yet reached this country, furnishes that delicate
material for strawing upon, and making into
artificial flowers, falsely called rice-paper.

The great crop of China, and the foodcrop of
the people for vegetables, came a great number of
table-plants to be reared. Turnips, carrots, sweet
potatoes, and pot-herbs of every kind, are produced in abundance. A white cabbage, called
pâtsoe, and not unlike the Roman lettuce, constitutes the principal food of every class, and is
really delicious. Of grain, the plenitude of water
in China causes rice to be so successfully cultivated, that it is brought to greater perfection there
than in any other part of the globe. Indeed,
there is scarcely any sort of grain but may be
found in some part of the country or other. No
medical root is in such high favour as the *ginseng*, which is administered as a sort of universal
panacea, and is a good tonic. It was formerly
found only in Shau-tung, Leao-tung, and Tartary; and brought a very high price. But it has
been discovered in different parts of America,
and is now extensively imported into Canton by
the American traders. The *ti-tsang*, a plant very
similar to liquorice, is also much used as a re-
storative. The other roots are *Radix Chins* (a
sort of truffle), galangal, rhubarb, ginger (often
exported as a sweetmeat), and poppy, whose juice
is made a substitute for opium, and is extensively
cultivated in spite of the strictest government
regulations to the contrary.

But scanty information is to be obtained of the
mineral kingdom of China; but the portion of the
mountain districts that has been explored is found
to possess great mineral riches. The gold mines
are worked exclusively by governments, but their
situation is kept a secret, though that metal is
supposed to be derived from the Kerri-ebou and
Yun-nan mountains. Gold-dust is found in the
Yang-tse-Kiang during its course through Sze-
chuen. Iron is produced throughout the empire.
Several sorts of copper are found in abundance,
the most famous of which is the *pe-king*, or white
copper, dug up in Yun-nan. Mercury is also very

common, as are arsenic, cobalt, and orpiment.
There are coal mines in various parts of China.
The beautiful *lapis lazuli* is met with in the W.
provinces. Salt, produced from the earth, and by
the evaporation of sea-water, is an article of great
traffic: it is collected in immense amounts, chiefly
on the banks of the Pei-ho. China also furnishes
the crystal, ruby, amethyst, sapphire, topaz;
but diamonds are little valued. There are small
resembling basalt, which, when struck, give out
a sound. Marble, porphyry, and jasper are produced from the quarries of N. China, beside excellent granite and quartz. (Dr. Abel's Narrative
of a Journey into the Interior of China, *passim*;
Downing's Fan-Qui in China, ii. 146-157; China
Opened, i. 35-54; Malte Brun, art. "China.")

Trade and Commerce.—The Chinese are famous
for their industry. Of the immense territory they
inhabit, there is scarcely a rood of arable ground
that is not assiduously cultivated; and such importance do they attach to agriculture, that once
a year the sovereign of the Celestial Empire—so
seldom seen in public—exhibits himself holding
a plough. But it is the misfortune of the Chinese
that their patient enduring industry is allowed
to usurp the place of ingenuity and science.
Their farming instruments are of the most primitive kind, their ploughs being inferior to the very
worst of ours. Owing to the smallness of the
farms, there is no room for the subdivision of
employments; and agriculture, as a science, is
but little advanced in China. But they accomplish all that can be effected by the most persevering industry. They spare no pains in the
collection and preparation of manure; and they
are superior to every other people in the irrigating
of land. By the aid of chain-pumps, they draw
water from the numerous canals and rivers, while
the highest mountains are cut into terraces so
constructed as to retain the requisite quantity of
water, and to allow what is superfluous to pass
off; by these means, and a good system of manuring, they are able, in many parts, to produce two
crops a year, without intermission.

But notwithstanding their remarkable industry
and economy, the bulk of the population have
usually so little to spare, and are so completely
without the ability to retrench in periods of distress, or to resort to a less expensive species of
food, that the failure of a crop never fails to involve them in the extremity of want; and, despite
the supplies brought from other parts of the
country, is frequently occasions the death of vast
numbers, and the committal of all sorts of out-
rages. There can, in fact, be no real security for
a country at all approaching to the condition
of China, unless the food of the people in ordinary circumstances be such as to permit of their
retrenching in adverse seasons, and thus counter-
vailing the deficiency of the crops by increased
economy.

As a *manufacturing* people, the Chinese are
highly distinguished: the fabric of porcelain originated entirely with them; and though the forms
of their articles will not bear a comparison with
those of the classic ages of antiquity again brought
into use in modern Europe, the fabric is excellent,
and the colours unmistakable. The art of spinning
silk was also given to the W. world by the Chinese;
and that light cotton stuff we call nankeen derives
its name from the ancient capital of China. The
lacquered ware, though eclipsed by that of Japan,
is very beautiful; but it is in the minute arts of
carving and inlaying that the Chinese excel. The
articles brought here in mother-of-pearl and ivory
are too well known to need description. Gunpowder, though a Chinese invention, is manufac-

tured only on a small scale, and is exceedingly bad; which, indeed, could hardly be otherwise, as it is a part of the soldier's employment to make his own gunpowder. (Barrow, p. 301.) Paper is also a Chinese invention, and seems to have been first manufactured A.D. 95. The materials used in making it are very various. It is thin, silky, and very absorbent of ink. Chinese books are printed only on one side the leaf. The government is jealous of everything new; but the people discover no lack of genius to conceive, or of dexterity to execute. Their talent for imitation is well known. During the course of the present century, a Chinese sailor, who came to England in an Indiaman, frequented a manufactory in Southwark where Prussian blue was prepared; and having made himself master of the process, without exciting the suspicion, or attracting the notice of anyone, he established, on his return home, a similar work; and so well has it succeeded, that the whole empire is now supplied with native Prussian blue, whereas it was formerly wholly imported.

Money in China consists of the cash, about the size of an English farthing, made of copper; from 720 to 1,000 of them being, according to their quality, equal to a dollar. Silver is employed rather as an article of traffic than as a circulating medium; that used as money is cast into the shape of a horse's hoof, and called fuel, being equal to a little over 6s. of English money. Gold is also seldom used as currency; but when it is, comes into the market beaten into thin leaves. Credit is little known, except at Canton; consequently paper money has not a very extensive circulation. There are, however, banks in the large commercial towns, which issue paper. The Chinese trade has the peculiarity of being for the most part internal, the country supplying most articles necessary for the subsistence or luxury of its inhabitants, and is carried on by means of canal and river boats. The primitive expedient of barter is still resorted to on account, perhaps, of the inconvenience of the circulating medium. Salt may be almost designated the standard commodity, as being an article of the most extensive commerce.

The foreign trade of China is chiefly in the hands of the English and Americans. The first attempt on the part of Great Britain to open a trade with China was made in 1637, when four merchant vessels arrived at Macao; but through the intrigues of the Portuguese there established, the enterprise failed. Afterwards the East India Company carried on a small traffic at the different maritime ports, and chiefly at Canton. In 1792, Lord Macartney's embassy attempted to put the trade on a more liberal basis, but with little success. In 1816, Lord Amherst's mission for a similar purpose also failed, though the English trade continued for the next twenty years. In 1834 the exclusive trade of the East India Company with China terminated, and the country was thrown open to general traders. However, the government placed many obstacles in the way of trade, and, in 1839, went so far as to confiscate 20,000 chests of opium belonging to English merchants at Canton. This led to war with Great Britain, ending in the Treaty of Nankin—concluded August 29, 1842—which virtually unlocked, for the first time, the gates of the CELESTIAL EMPIRE.

The following is the official return of the declared annual value of British produce and manufactures exported to China and Hongkong, from 1831—the year when the distinction was first made in the Custom-house returns between the exports to China and to India—to 1863:—

Year	To China	To Hongkong	To China and Hongkong conjointly
	£	£	£
1831	—	—	845,191
1832	—	—	3,024,709
1833	—	—	1,378,4xx
1837	—	—	614,375
1838	—	—	2,994,854
1839	—	—	851,869
1840	—	—	521,1xx
1841	—	—	bo2,57x
1842	—	—	969,5x1
1844	719,42x	736,4x7	1,456,1xx
1845	384,256	1,912,741	2,515,617
1846	855,1x4	1,589,63x	2,395,9x7
1846	563,212	1,229,x77	1,791,4xx
1847	745,04x	766,xx1	1,541,8x9
1848	793,163	851,4x4	1,435,55x
1849	845,110	851,9x9	1,547,109
1850	975,963	599,13x	1,574,145
1851	1,575,96x	631,799	7,161,76x
1852	1,918,211	645,2xx	2,580,5xx
1853	2,10,5x0	57x,xxx	1,719,597
1854	852,6x9	4xx,x11	1,001,7x1
1855	814,x25	78x,263	1,577,841
1856	1,417,476	895,611	2,316,678
1857	1,70,xx5	721,097	2,459,942
1858	1,789,724	1,16x,x49	2,956,667
1859	7,56x,x9f	1,941,576	4,152,875
1860	7,852,915	2,445,591	5,314,054
1861	8,311,x01	1,755,561	4,814,657
1862	7,911,111	1,119,774	3,127,341
1863	3,155,705	1,474,222	3,649,927

There is no separate record of the exports to Hongkong prior to 1843.

By the terms of the commercial treaty signed on August 29, 1842, by the plenipotentiaries of the Queen of Great Britain and the Emperor of China, five ports of the empire were opened to European trade. The five ports are those of Canton, Amoy, Foochowfoo, Ningpo, and Shanghai. Some minor ports were added to these by the treaty of peace of June 26, 1858. The exports from China—including Hongkong—to the United Kingdom are of great value, and consist of two principal articles, namely, tea and silk, to which lately there has been added a third in cotton. The total value of the exports amounted to 9,054,310l. in 1859; 9,523,761l. in 1860; 9,970,445l. in 1861; 12,137,055l. in 1862; and 11,185,510l. in 1863. The sole article tea figures to the amount of two-thirds in the sum total of these exports. The computed real value of tea exported from China to the United Kingdom amounted to 5,529,662l. in 1859; to 6,603,991l. in 1860; to 6,419,540l. in 1861; to 8,759,764l. in 1862; and to 10,031,662l. in 1863. Compared with this article, the other exports of China to Great Britain seem insignificant. Of raw silk, the exports amounted to 3,031,280l. in 1862, but only to 528,336l. in 1863. On the other hand, the export of raw cotton was but of the value of 106,995l. in 1862, and rose to 2,164,095l. in 1863. In return for the vast quantities of tea, silk, and cotton which China sends to the United Kingdom, she accepts little else but a few manufactured cotton goods of about one-third the value. Thus the principal article of British imports into China, was of but the value of 1,162,546l. in 1863, while the tea exports amounted to 10,031,662l.

History, Government, and Laws.—It may be almost said that China has no history, for she has so few revolutions or political changes to record, that her annals rise but in a small degree above the limits of chronology. The antiquity which the Chinese have claimed for their origin, is most, even by the enlightened among themselves, considered fabulous. Almost the first names mentioned in their annals are Shing-noong, "the divine husbandman,' who taught their ancestors the arts of agriculture; and Hoang-ty, who partitioned their lands and contrived a cycle of 60 years, to enable them to register events, and to mark the progress of the seasons. Then comes the period of the 'five kings,' the last two of whom, Yaou and Shan, are held up as patterns for future sovereigns, being the exemplars of royalty down to the present reign. Yu, the successor of Shan, made himself conspicuous by his transcendent merit in draining the country that had suffered from a great deluge. The Chinese have no existing records older than the compilations of Confucius (born 550 B.C.), which must have been made from tradition. From that period the annals of the empire have been carefully noted and preserved, and descend in an unbroken line down to the present day. These, 'the successive labours of twenty-one historians,' consist of 500 vols. Formed into a prosperous and comparatively civilised community, under the Tsin dynasty, the Chinese became objects of envy to their neighbours, of whom the Tartars were the most troublesome; and, to guard against their incursions, the great wall was built. A.D. 184 was the era of the 'three states,' into which the empire was divided; but in 585 it was again united under one ruler. The 8th and 10th centuries were much occupied in civil wars, caused by the contending claims of several aspirants to the throne; but these were finally adjusted A.D. 950, by the consolidation of the Soong dynasty, under Tae-tsou. This was the first great literary age of Chinese history; and printing having been invented 500 years before it was known to Europeans, authors and books were much multiplied. Under this dynasty the Chinese, unable to resist the Tartars, called in the aid of the Moguls; and they, by a policy of which history affords numerous examples, soon exchanged the character of allies for that of conquerors; and, under the famous Kublai-Khan, founded the Mogul dynasty. This able sovereign established the seat of his government at Pekin, or Kambalu, as it is called by Marco Polo, and constructed the great canal. But his successors rapidly degenerated; and the ninth Mogul monarch surrendered the throne to a Chinese, A.D. 1368. Twelve emperors of this native dynasty of Ming reigned in comparative peace till, in 1618, during the sway of Wan-lie, the 13th in succession, the Manchuns, a race sprung from the expelled Moguls and the Kin or E. Tartars, after a war of twenty-seven years, established themselves firmly in the empire. The seventh in descent from Shunchy, the first of the Ta-tsing dynasty of Tartars, occupies the throne of China at this day. (Davis, i. 157, 100.)

The most conflicting statements have been made with respect to the government of China; while some writers have represented the whole empire as trembling under the yoke of a capricious despot, others have represented the government as administered according to the inflexible rules of justice, and with the greatest moderation and humanity. Both these representations seem to be alike inconsistent with the facts. According to the theory of the constitution, the emperor is absolute: his will is law; and he is not responsible to any earthly tribunal for any of his actions. In China, as in ancient Rome, fathers have full power over their families, and, on the same principle, the emperor is held to be the father of the entire Chinese people; and to have the same unlimited power over them that each individual has over his own children. Practically, however, his power is comparatively circumscribed. In China everything is determined by custom, or by immemorial practice, from which it would be highly dangerous

for even the emperor to depart. The Chinese is emphatically a government of precedent; and his celestial majesty is, in reality, the creature of custom and etiquette. All employments are bestowed, according to fixed rules, on those who have obtained certificates of proficiency after passing their examinations. The penal laws of the empire are printed in a cheap form, and widely diffused; and one of the sixteen discourses annually read to the public, inculcates the propriety of every man making himself acquainted with them, and with the penalties consequent on their infraction. Although, therefore, the government of China be despotical in its form, and every device is employed to give to the emperor not merely a paternal, but a sacred character, he in fact governs according to long-established rules; and with probably as little admixture of despotism as is to be found in most governments.

The great defect of the Chinese, as of all similarly constituted governments, is the want of any effectual control over the inferior agents. The emperor is not omniscient; and notwithstanding the various devices put in motion to learn the real conduct of the subordinate authorities, and their liability to punishment if they abuse their power, it would seem that these checks are, in many instances, of comparatively little avail; and that rough injustice and oppression on the part of persons in power, escape detection and punishment.

The emperor is called ' the son of heaven' (Tien-tsze), and the mandarins and other natives not only prostrate themselves when in his presence, but also before a tablet with the inscription 'the lord of a myriad years' (Wansuy-yey). In his character of patriarch, his imperial majesty is not only looked upon as the father of that multitudinous family, the jus, of his empire, but is also considered the sole dispenser of the blessings of heaven; for the prime canon of belief is, that 'the duty of affording to the people sustenance and instruction is imposed on The One Man;' while, on occasions of national calamity, he publicly confesses his errors, and acknowledges his misconduct to be the cause of the divine displeasure. (Quarterly Review, xxv. 416.) The parallel between the relations in which every person stands to his own parents and to the emperor is carried out from the most important functions of the legislature, down to the minutest observances of ceremony, all of which are regularly prescribed by law. (Davis, i. 201.) The union of the avenger with the father, in the emperor, is well illustrated by Davis. A man and his wife had severely ill used the mother of the former, which circumstance was reported to the emperor. The very place where the crime was committed was made accursed. The principal offenders were put to death; the mother of the wife was beheaded, branded, and exiled, for the daughter's crime; the scholars of the district were not permitted to attend the public examinations for three years; and their promotion was thereby stopped. The magistrates were deprived of their office, and banished. ' For,' says the edict published on the occasion, ' I intend to render the empire filial.' Every device is employed to create the impression of awe. Dressed in a robe of yellow, the colour worn, say the Chinese, by the sun, the emperor is surrounded by all the pageantry of the highest dignity in the world. All ranks must bow the head to a yellow screen of silk; in the great man's presence no one dares speak but in a whisper, though his person is too sacred to be often exhibited in public, and an imperial despatch is received by the kneeling of incense and prostration. But with all this he is not allowed to lean back in public; to smoke, to change his dress, or, in fact to indulge in the least relaxation from the fatiguing support of his dignity. (Chinese Hist. ; Davis ; Quarterly Review, lvi. 499; Ellis's Account of Lord Amherst's Embassy, p. 307.)

Next, after the emperor, the court is composed of four principal ministers, two Tartars and two Chinese, who form the great council of state, assisted by certain assessors from the Han-lin or Great College, who have studied the sacred books of Confucius, which form the basis of Chinese law. These may be considered as the cabinet ; but the real business of the empire is executed by the Le-poos, or Six Boards. No. 1. Le-poo is the board of official appointments, which has cognizance of the conduct of all civil officers ; 2. Hoo-poo, the board of revenue, which regulates all fiscal matters ; 3. Le-poo, board of rites and ceremonies, which enforces the customs to be observed by the people ; 4. Ping-poo, military board ; 5. Hing-poo, or supreme court of criminal jurisdiction ; 6. Kung-poo, board of public works. There is also a colonial-office, composed of Manchoos and Mongols, so that the respective tributary princes may have confidence in referring whatever concerns their interests to their own countrymen. To each of the poos, a viceroy is appointed by the chief, or Le-poo board ; and every town is presided over by a magistrate, who takes rank according as he is at the head of a foo, tchoo, or heen. Subordinate officers superintend the lesser divisions. All these functionaries are removed every three years ; and that no ties of kindred may interfere with the strict discharge of their duties, the viceroys and magistrates are forbidden to form any matrimonial connection with a family within the limits of their rule. It is honourable to the Chinese that, for these and other state offices, merit alone is the qualification ; the son of the poorest peasant or artificer may offer himself as a candidate, and, by talent and application, rise to the highest employments. A singular expedient is adopted to ascertain with what fidelity the viceroys and magistrates perform their duties. There is a board, headed by a Tartar and a Chinese, on whom it formerly devolved to watch over the words and actions of the emperor, and freely censure him for any misdemeanour ! The duties for which this office was originally established have, for reasons easily understood, long fallen into disuse ; and the members are now employed as censors for the emperor, being sent as inspectors into the poos, to see how the viceroys and magistrates do their duty, and to report their delinquencies. But these functionaries are less formidable than might be supposed. If they did their duty honestly, they would, no doubt, be of singular advantage ; but in China, as elsewhere, it is usually found that inspectors look with an indulgent eye on the faults of those in authority ; and it has been doubted whether their visits be not as often the means of stifling the complaints of the public, and of preventing and delaying justice, as of facilitating its course. Nothing can be more lucid and methodical than the code of laws promulgated for the guidance of the boards and their subordinate officers. Each district has a separate code, adapted to the habits and disposition of those for whom it is framed ; and offences, with their punishments, are classed under six different heads, corresponding with the six boards, so that each case is referred to the tribunal against whose authority the offence may have been committed, unless it be one admitting of summary punishment.

The Ta-tsing Leu Lee, being the fundamental laws, and a selection from the supplemental statutes of the penal code of China, has been ably translated by Sir George Staunton. The most

remarkable thing in this code is its great reasonableness, clearness, and consistency ; the business-like brevity and directness of the various provisions, and the plainness and moderation of the language in which they are expressed. There is nothing here of the monstrous verbiage of most other Asiatic productions ; none of the superstitious delirium, the miserable incoherence, the tremendous non-sequiturs, and eternal repetitions of those oracular performances ; nothing even of the turgid adulation, the accumulated epithets, and fatiguing self-praise of other eastern despotisms ; but a clear, concise, and distinct series of enactments, savouring throughout of practical judgment and European good sense ; and if not always conformable to our improved notions of expediency in this country, in general approaching to them more nearly than the codes of most other nations. (Edin. Rev., xvi.)

This is high, but not undeserved praise. At the same time, however, the Chinese code is not without very serious defects. There is an elaborate attention to trifles ; and a perpetual interference on the part of the legislator to enforce duties and observances of no importance, or that had better be left to the discretion of individuals. But its greatest defect is the vagueness of some of its clauses ; so that a person may be punished if his conduct be "contrary to the spirit of the law ! " The frequency of corporal punishment seems extraordinary to Europeans. It is, in fact, the universal penalty ; offences the most trivial and the gravest, whether committed by persons in the highest or the lowest walks of life, being visited by so many strokes of the bamboo ! These, however, are not always inflicted. Persons under fifteen or above seventy, or maimed, may redeem themselves from all but capital punishments, by a small fine ; in other instances the punishment may be commuted by paying a sum of money proportioned to the number of blows. But there are crimes for which even those who are rich enough to escape whipping for ordinary offences are not suffered to make a pecuniary compromise. Indeed the bamboo seems in universal requisition, from the emperor down to the meanest of his subjects ; and not only the number of blows, but the length and thickness of the instrument to be used for each offence, are minutely prescribed. The prerogative of mercy is not unfrequently extended, with, however, one exception. In a country which has preserved its institutions unchanged, and its laws unaltered, for 2,000 years, it is not surprising that seditious offences should be severely dealt with. The crime of treason is visited with remorseless severity. In 1803, Mr. Davis states, a single assassin attempted the life of the emperor. He was condemned to a lingering death ; and the criminal's sons, being of tender age, were ' mercifully ' strangled ; for it seems to be the peculiar barbarity of the Chinese criminal code, that it involves the innocent family of an offender in the retribution for his crime. There is much in use a sort of pillory, called the cangue ; and torture is employed to extort confession. The police of China is said to be vigilant and efficient ; but, as a safeguard against oppression, the name of every person in any way connected with the government is published in a sort of Red Book, of which a corrected edition appears four times a year.

Another trait of the patriarchal form of the Chinese government is to be found in the mode in which the state revenue is produced ; it consists principally of tithes ; not paid in the nature of taxation, but as rent, the emperor uniting the character of universal landlord with that of king and father ; but though the whole people be tenants at-will, ejectment is seldom resorted to; and it is his own fault if a Chinese be ever deprived of his lands. There are here no great estates; but if any one happen to hold more land than he can conveniently cultivate, he lets it to another, on the *metayer* principle, or on condition of his receiving half the produce, out of which he pays the whole of taxes. A great part of the poorer peasantry hold land-in this way. (Barrow, p. 398; De Guignes, iii. 341.) The revenue is paid partly in money and partly in kind. The greatest possible discrepancy exists amongst the estimates that have been given of its amount. It is believed, however, that the entire revenue remitted to the imperial treasury may amount to about 12,000,000*l.* sterling, that is 16,000,000*l.* in money, and 2,000,000*l.* in produce. But it is essential to bear in mind that this is not the whole amount of Chinese taxation, inasmuch as the expenses of a collection, and many local and provincial charges, are deducted before any remittance be made to the imperial treasury.

The *Military* service of China is nominally composed of 1,000,000 soldiers, besides the militia and numerous standards of Mongol cavalry; but from this vast number many names must be deducted which are merely entered in the books, and perhaps the whole force does not exceed 700,000. The soldiers are enrolled in the corps quartered in the provinces in which they are born, and which are never quartered any where else; the Chinese government being impressed with the opinion, that soldiers living with their families and bring, in fact, more than half citizens, will exhibit greater bravery in the defence of their country, should any occasion arise for their services, than if they were cooped up in barracks or fortresses, and subjected at all times to strict discipline and martial law. The troops are only embodied at certain seasons, being at other periods their own masters. The Tartar troops, inasmuch as they belong to a standing army at a distance from home, receive higher pay, and are more efficient soldiers than the native Chinese; though they also seem to be enervated by their long residence in this tranquil region. The whole army is divided into standards, distinguished by their different borders and colours. These corps—not unlike our brigades—are subdivided into camps and wings; the right, left, and middle. The officers are all raised from the ranks, and are looked upon by the civilians as little better than police agents; but, like the latter, are obliged to take their regular degrees to obtain promotion, which is rapid. Their grades are precisely similar to ours, from the 1st tah, commander-in-chief of the forces, down to the Waou-wei, or serjeant. The principal weapons are bows and arrows; but they also use clumsy match-locks and iron guns, without carriages, and more recently, have imported tolerably good rifles and superior ordnance from Europe. The theory of tactics is well understood; but the practice is very deficient. In so peaceful a country there is but little occasion for military skill; and without intelligent officers, or improved weapons, it is not to be supposed that they should make any effectual opposition to European troops. A standing army, in the European sense of the word, is not in existence. The soldiers do not live in barracks, but in their own houses, pursuing as chief business some civil occupation, frequently that of day-labourers, and meeting only on certain occasions, pursuant to orders from the military chieftains. (Meyer, Marquis de, Recollections of Baron Gros's Embassy to China, Lond. 1860.)

The Chinese *Navy* is extensive, but inefficient; it includes, perhaps, 1,000 sail; but the men-of-war are mere junks, which mount a few guns; and there are few large vessels. This imperial navy is commanded by three high admirals and their inferior officers, all of whom are so profoundly ignorant of their business, that the merchant junks are better managed than the imperial cruisers. Gutzlaff draws a deplorable picture of the condition and discipline of the mercantile navy. Few sailors are regularly bred to the service, but are chiefly wretches who have been obliged to flee from their homes. Though there be a nominal commander in every junk, his authority is uniformly disregarded. Every one having the liberty of putting a certain quantity of goods on board, is a sort of shareholder, and does nearly what he pleases. The Chinese make use of a compass, invented by themselves, divided into 24 parts, beginning at the S., the needle moving freely in a box placed upon a bed of sand. Their pilots having been accustomed to the sea from their youth, and always performing the same voyage, have a perfect knowledge of the various localities. In the construction of river craft, the Chinese are more skilful; many of these vessels are indeed floating habitations, and thousands of families live in them during their whole lives. (Sketch of Chinese Hist. by Gutzlaff, i. Introd. 1–40; Sir G. Staunton's Trans. of the Leu-lee, or Criminal Code; Davis's Chinese, i. 294 *et seq.*; Quarterly Review, No. vi.)

Character and Social Condition.—The Chinese are said by Mr. Davis to be a nation of 'incurable conservatives.' Their rule is to adhere to all that is established, and to reject all that is new. They are the very transcript of the ancient world living in the present day; they wear the same costumes, are subject to the same laws, which are administered precisely in the same way, and they exist to all intents and purposes in the same social and intellectual condition as their forefathers did 2,000 years ago. This uniformity may be almost said to have been ordained by nature, for it is a remarkable fact that the Chinese are so much like each other in personal appearance, that it is difficult for a European to distinguish between them. We find no diversity in the colour of their hair, no variety of eye, no prominent and striking feature which indicates the place of their birth. (China Opened, i. 296.) They have black, stiff and strong hair, shaved so as to leave a much cherished tail depending from the crown; a depressed face, wherein the distinguishing features are not strongly marked, a flat nose, small angular eyes, round and prominent cheeks, a pointed chin, thin eyelids, small beards, middle stature, and strong bones, long ears and plumpness form their beau ideal of beauty; consequently, to attain the latter, they exercise but little agility. (Ib. p. 293.) The aristocracy of rank and wealth are unknown in China. *Distinction is solely to be obtained by learning*; and dignity is only conferred by office. Even the sons of the emperor and their families merge into the common mass, should they not study, so as to become qualified for some official employment. The mandarins, or literary aristocrats, do not obtain their rank except by passing repeated examinations, as to the fairness of which no doubt has ever been surmised, and establishing their superiority over their competitors to the satisfaction of the Board of Examination. There are nine degrees of mandarins, the highest being viceroys or governors, and the lowest, collectors of the revenue, &c.; promotion can only be obtained by superior proficiency in the study of the law. The different functionaries are distinguished by the number of buttons in their caps, and other variations of costume. As the pay of all persons in office is unreasonably small, they often resort to extortion to make up this deficiency, and there is scarcely a number of the 'Pekin Gazette,' that does not record some instance of a public officer being degraded for that crime. The natural characteristics of the Chinese are summed up by Davis in these words:— 'The advantageous features of their characters, as mildness, docility, industry, peaceableness, subordination, and respect for the aged, are accompanied by the vices of specious insincerity, falsehood, mutual distrust, and jealousy.' The lower orders are passionately addicted to gambling, for which they have their peculiar cards and dice. That honesty is more valued than practised has been inferred from the notification to be frequently seen in shop windows, that 'there is no cheating here,' and from a caution placarded in most public conveyances for travellers, to 'take care of their purses;' but we doubt whether such motives really go for much. The insincerity and falsehood laid to their charge, in so far as they really exist, are the natural consequences of the restraints, under which they are laid from infancy, of the interference of the law with all their actions, and of their being obliged to suppress and conceal those feelings and emotions to which, in other countries, full vent would be given. Their attention to etiquette is a consequence of the same principle. Even when presents visit each other, complimentary cards—the size of which determines the rank of the sender—and polite answers are exchanged. 'On the arrival of the guest, considerable difficulty is found in arranging who shall make the lowest bow, or first enter the door, or take the highest seat, or assume precedency at table, though the host contrives to place his guest in the most elevated position. When conversation commences, the mutual assent to every proposition, the scrupulous avoidance of all contradiction, and the entire absence of every offensive expression or melancholy allusion, show what a sense these people entertain of politeness.' (Medhurst's China: its State, Prospects, &c., 1838.) The condition of the poor is wretched in the extreme: they are frequently destitute of food, and many are said to perish in the winter season from cold, for want of fuel. (Gutzlaff's Voyages, p. 67.) Begging is common in the large cities, but not more so than in Europe. It is a curious fact, that though the Chinese be remarkable for assisting each other, particularly their own relatives, with money or food, they will on no account step out of their way, in case of accident, to save a fellow-creature's life; but this arises from their laws making the person last seen near a corpse answerable for the death. Robbery is not uncommon, but is very seldom accompanied with murder. The people, generally so quiet and submissive, when once roused by the oppression of an intolerant magistrate, will rise en masse against him, and subject him to lynch law: in such cases the government of Pekin generally concludes that the magistrate has been in fault; and the outrage is allowed quietly to fall into oblivion. The drowning of infants, particularly of females, has been said to be customary in China; but this is a most unfounded statement. That an enormity of this sort is sometimes committed is certainly true; but we believe that it is of exceedingly rare occurrence. Mr. Davis says, that 'the Chinese in general are exceedingly fond of their children, and the attachment seems to be mutual.' (i. 216.)

The whole of the Chinese nation is divided into families, each of which bear the same surname, and consider each other cousins. These clans are bound to assist each other in any way that may be required; and the most powerful of them act as a salutary check upon local despotism. The women

of China occupy a lower scale in the estimation of their countrymen than those of other nations. A broad face, diminutive waist, pale features, and feet small to deformity, constitute female beauty in the eyes of a Chinese. To insure this last, their feet are confined from tender age in shoes calculated to stop their growth, so that the feet of some ladies only measure 3 in. from toe to heel. Females are universally objects of traffic. When young they are purchased by dealers for the harems of the great, where they remain in splendid seclusion. Marriages depend entirely upon the will of the parents, who sell their daughters at from 5,000 to 6,000 dollars a piece, according to the beauty or rank of the female. Early marriages are universal; no man who can afford the expense of the ceremony deferring it after the age of 20, and parents get rid of their daughters as soon as they can, even at the early age of 14. The Chinese may be said to be an omnivorous people. The principal part of their food consists of rice, which is generally eaten dry; but in the S. provinces it is mixed with the sweet potato in a sort of soup. Vegetables are the chief provision of all ranks, who do not consume a fifth part of the animal food that Europeans do. Pork is the favourite dish, and the head of the sow is esteemed a great delicacy. To eat every thing which can possibly give nourishment is the comprehensive principle upon which Chinese diet is regulated; so that dogs, cats, and even rats and mice, are not rejected by them. They are the most expert fishermen in the world; no aquatic creature escapes their vigilance, whether it inhabit the sea, lake, canal, or river; even pools and the ridges of fields are searched for fish. Every kind of meat is minced into small pieces, and is eaten with chop-sticks. The Chinese epicure delights in soups made of edible birds' nests of the swallow species (Hirundo esculenta), and imported in great quantities from the E. islands. It appears that the birds make use of great quantities of a peculiar sea-weed (Sphæro-coccus cartilagineus), and when it is sufficiently softened in their stomachs, it is returned and used as a plaister to cement the dirt and feathers of the nest. These nests, after having been purified in immense manufactories, are eaten with great gust by the Chinese. The favourite beverage is tea, drunk out of small cups, which are seldom washed, for that process is thought to diminish the flavour. In this article the Chinese are as great connoisseurs as Europeans are in wines. Distilled liquors are chiefly made from rice; rum is much used, but grape wine has not been met with. Drunkenness prevails, especially in the N. provinces; but the usual species of debauchery is opium smoking, which, when carried to excess, deprives the victim of strength; he becomes a walking shadow; his eyes are vacant and staring; his whole frame is deranged, and he soon sinks into a premature grave. But it should be observed that there are the consequences of the abuse of this practice: when used in moderation, it is said to be comparatively innoxious. The fumes of the drug are inhaled through a peculiar pipe, in a recumbent position, and the smoker soon sleeps. When he awakes, he drinks a cup of tea, and smokes again. The Chinese delight in the drama: they will attend a play for a whole night without being wearied, and recount with ecstacy what they have seen; in their pastimes the women are never associated.

The accounts of Chinese architecture are not very satisfactory, a consequence of its being necessary to employ terms in its description that convey to foreigners impressions very different from the reality. According to Mr. Barrow, it is 'as unsightly as unsolid; without elegance or convenience of design, and without any settled proportion; mean in its appearance, and clumsy in the workmanship.' (p. 830.) Perhaps, however, this opinion is founded too much on preconceived notions of the absolute superiority of the European standard. But without entering on this, it is sufficient to observe that the walls of the houses are of brick, stone, or wood, but principally of the first. The roofs are always supported on columns, that is, on upright pieces of timber, without either capital or base. In the country they are rarely more than one story in height, but in the great towns they are frequently two. Their roofs, which are curved, are usually covered with tiles. Their pagodas are polygonal buildings, of 5, 7, or 9 stories or roofs. Mr. Barrow says, that the pagoda erected by George III. in Kew Gardens is 'not inferior to the very best' he met with in China—a statement which certainly does not tend to exalt our opinions of this species of buildings.

Religion.—There is no religion in China actually supported by the state, and Fo, the doctrine of Confucius, is the only one countenanced by it, but there are two other sects; Fo, or Buddhism, and Taou, or that of the 'rationalists.' The first acknowledges a Supreme Being, and believes the emperor his sole vicegerent on earth. Heaven, earth, the elements, Confucius, gods of various attributes, saints, the emperor, &c., are objects of worship; the rites in performing which are watched over with the most jealous care by the Le-pau, or Board of Rites. The doctrine of Confucius fills the world with genii, demons, and the spirits of deceased worthies, who are supposed to have each their separate duties and influences assigned to them. No worship is so strictly observed as that of ancestry, so that filial piety is carried to an excess, even beyond the grave. The religious edifices of the Taou sect are said to be very splendid. They chiefly consist of one large hall approached by steps, with the idol placed upon an altar, or table; the walls are adorned with pictures, and the ceiling with gilded griffins and dragons. An apparatus for sacrificing various animals is also provided. There is no congregational worship. Buddhism is a despised creed in China, and is entirely supported by the mendicancy of its priests. The latter practise celibacy, dress in a similar manner to monks, and the devotees use holy water, and a rosary to keep account of their prayers. Mr. Malcom, the missionary, has given a very favourable view of Buddhism. 'It has no mythology of obscene and ferocious deities; no sanguinary or impure observances; no self-inflicted tortures; no tyrannising priesthood; no confounding of right and wrong, by making certain iniquities laudable in worship. In its moral code, its descriptions of the purity and peace of the first ages, of the shortness of man's life because of his sins, &c., it seems to have followed genuine traditions. In almost every respect it seems to be the best religion man ever invented.' (Travels, i. 324.) The professors of Taouism pretend to magic, alchemy, and to be possessed of the elixir of long life; practise glaring impositions, and inculcate the most puerile superstitions. They encourage a belief in ghosts and evil spirits; make use of spells and talismans, lucky and unlucky birds, and a system of tricks called *fung-shuey*, by which they pretend to choose lucky situations for building houses and tombs, and a hundred other fallacies, by which these impostors contrive to fill their purses. Religion, of whatever kind, has always, we believe, been reckoned a matter of secondary importance in China. But this is a subject as to which our information is comparatively little to be relied on. The ancient and modern missionaries,

how much soever they may have admired many parts of the Chinese character and institutions, have generally represented their morals and religion in the most unfavourable point of view. That there is much about them that is objectionable is certainly true; but it is so much the interest of the missionaries, by depreciating the moral and religious character of those they are labouring amongst, to exalt their own utility and importance, and to justify their claims to the patronage and support of the Christian public, that their statements can hardly be supposed to be free from bias. Many endeavours have been made to introduce Christianity into China, but with less success than has attended similar efforts in other nations. It was first introduced by the Nestorians in the 13th century. These were followed by the Jesuits, whose missionaries were more successful than those of any other sect; for at the Tartar invasion there were no fewer than thirty Catholic churches in the province of Keang-nan alone; the first of the Tartar princes openly espoused the cause of the missionaries, by taking a German Jesuit, Adam Schaal, for his instructor. The abolition of that order, and the continual wars in Europe, reducing their funds, the Catholic missions declined, and but few native converts at present remain. The late Dr. Morrison was the first Protestant missionary who landed in China; he compiled a dictionary (having been provided in that arduous task by Dr. Galignes) and grammar; translated the Scriptures into the Chinese language, and established printing-presses at Canton, from which a judicious selection of tracts has issued. These pious efforts have been ably seconded by Mr. Milne and the Rev. Charles Gutzlaff, the latter of whom has published several valuable works on China, of which we have made considerable use. The Mohammedan, Jewish, and many other religions are to be found in China, but in a very languishing condition. A semi-political, semireligious movement, which broke out in China about the year 1850, and, according to make reports, threatened for a time the destruction of the actual government, was long believed to be owing to the teaching of Christian missionaries, but this belief was scarcely founded on fact. The insurgents, commonly called Taepings, whatever their religious faith, were certainly not Christians, for the many atrocious acts committed by them—acts completely inexcusable even by the direst necessities of warfare, and warfare in its bitterest form, civil strife—showed them entirely unacquainted with the fundamental precepts of the divine Gospel of Christ. Thus, too, was the conviction of the leading statesmen of Europe, with whose help, and the aid of British and American officers, the Taepings were finally crushed in 1864. The valuable help thus afforded went far to reconcile the Chinese government to European progress, and to enter upon a liberal fulfilment of the treaty of peace concluded with Great Britain, June 26, 1858, by the terms of which Christianity will be tolerated throughout the whole of the Chinese empire.

Language, Education, and Literature.—Distinct as the Chinese are from the rest of mankind in habits, manners, and religion, their total dissimilarity is rendered complete by their languages, which, arrested between the hieroglyphic and alphabetic systems, present a singular phenomenon. The most obvious expedient for expressing substantive ideas otherwise than by speech, would be to figure a representation of the object intended to be expressed; and this was unquestionably the plan first adopted by man to communicate and record what he thought through the medium of the eye instead of the ear. As civilization and knowledge advanced, and the necessity for communicating it increased, more concise forms or conventional letters were substituted; but in the case of the Chinese, the primitive mode is still the principle upon which their characters are constructed; so that their system may be called the perfection of the hieroglyphic method of written language. Having pictorial representations of natural objects for their basis, the elementary signs of the Chinese language are few and simple. A horizontal, a perpendicular, two oblique lines drawn in different directions, and an acute angle and dot, are the elements of which the Chinese characters consist. These marks are so combined in the first instance as to form 214 keys or generic characters. Thus, the symbol for 'man' is always present in a word which has direct or indirect reference to him; this character, for example, combined with the symbol for field, signifies a farmer. The Chinese notion of government is well expressed in another example: the verb 'to govern' is represented by the two characters that stand for 'bamboo' and 'stroke.' The keys are divided into 17 classes, and the number of words thus formed, upon a system more complete than that of any of the W. languages, to be found in the most copious Chinese dictionaries, amounts to 40,000, each of which stands so arbitrarily for the thing or idea intended to be conveyed as a figure does in a painting for the object it is meant to represent. Thus the character presents an object to the eye which enters the mind with a striking and vivid certainty; it forms a feature which really is, or by early associations is considered, beautiful and impressive. Chinese writing is also more permanent than the alphabetic system, which is ever varying its spelling with the continually changing pronunciation of the living voice. Perhaps the Chinese written language has contributed in some degree to the unity of the Chinese nation. (Dr. Marsham's Clavis Sinica; Elements of Chinese Grammar, Introduction, p. xi.; De Guignes, Dictionnaire Chinois, Introduction; Quarterly Review, lvi. 540; China Opened, i. 391.) The causes, however, which operate to make the written language in China the most complete and beautiful in the world, render oral communication the most difficult and confused. That systematic regularity which so continually requires the presence of the keys, as parts of words bearing different meanings, and thus precludes a necessary variety of sounds, leaves the spoken language as meagre and defective as, when written, it is rich and complete. The sound corresponding with our r has at least 2,000 signification, and 'one might write a perfectly intelligible treatise in which only the sound of s was employed.' (China Opened, i. 383.) Thus, in conversation between even two of the best educated Chinese, constant misapprehensions occur. 'They understand each other,' says Mr. Davis, 'perfectly on paper, but are mutually unintelligible in speech.' And in the most common-place colloquy it is not unfrequent for the speakers to resort to pen, or rather brush, ink, and paper, to make themselves understood; in the absence of these materials, they draw the figure of the root or key in the air with their fingers. So that oratory is entirely unknown in China; and all affairs of importance, such as lawsuits, civil or criminal, are carried on in writing. The deficiencies of the oral language are in a small degree supplied by the different tones in which the same words and their various significations are uttered. But these inflections are so nice as to be only distinguishable by a native ear. The difficulty of free intellectual intercourse must have had a very considerable effect in preventing the Chinese

from advancing a step further in civilization than they had attained so many hundred years ago.

Education in China is more encouraged and favoured even than in Prussia; and such is the estimation in which it is held, that all state employments are given by competition, as school and college prizes to the best scholars. Schools for youth are abundant in every part of the empire; and education is so general, and its cost so reasonable, that reading and writing may be almost said to be universal. Language is taught to very young pupils by means of rude pictures which represent the names of the chief objects in nature and art. Then follows the San-tse-king, or summary of infant erudition, conveyed in chiming lines of three words or feet. They soon after proceed to the 'Four Books,' which contain the doctrines of Confucius, and which, with the 'Five Classics,' subsequently added, are, in fact, the Chinese Scriptures. Writing is taught by tracing the characters with a hair-pencil, on transparent paper placed over the copy. This is a most important article in Chinese education, for no man who does not write a good hand can lay claim to literary distinction. The emperor himself, when bestowing a great reward, writes a few characters on a piece of paper, and sends it to his favourite, and this is more valuable than conferring an order. (Davis, i. 290; China Opened, i. 524.) Females of the higher class are allowed to acquire a little reading and writing, and have been known to write poetry; but the great object of their education is to inculcate obedience. The schools established all over the empire are superintended by various officers appointed by government. In every district there is a sort of literary chancellor; but early aspirants are examined by superintendents, who make the circuit of their district twice a year for that purpose. The pupils they approve of repair to the chief, and should they pass that ordeal, and thus obtain the approbation of the officers of their native district, they are eligible for the lowest literary honour of the state. This is called Tse-tsaе (flowery talent). For this degree the examinations take place twice in every three years in face of every province; the scholars having each a theme given them from the 'Five Classics,' in a large hall, are confined to separate boxes to prevent their receiving assistance from others; and every avenue is strictly guarded by soldiers. The Tse-tsaе degree having been obtained, the aspirant has to acquire two other honours in the metropolis of his province, and he is placed on the books as eligible for employment corresponding with his advancement. To procure the highest state offices, an examination before the national college, or Han-lin, is necessary; but the very pinnacle of fame is only arrived at by being examined by the emperor himself. Every literary honour confers the title of mandarin, and each degree is distinguished by a difference of the dress, which is, in some instances, very splendid. Genius and originality amongst a people so blindly enthusiastic in their admiration of the ancients, are considered rather a blot upon, than as an ornament in, the character of a student. Memory is the chief object of admiration—memory to repeat the greatest number of the wise sayings of the ancient sages.

From what has been already stated, it will be readily conceived that the literature of the Chinese is most extensive. 'Books,' says Mr. Medhurst, 'are multiplied at a cheap rate, and to almost an indefinite extent, and every present and pedlar has the common depositories of knowledge within his reach. It would not be hazarding too much to say, that in China there are more books and more people to read them than in any other country in the world. Amongst the 360,000,000 of Chinamen, it had 7,000,000 are literati.' (China Opened, i. 417.) Yet it may appear strange that there is hardly one original writer among them: it is generally believed in China, that whatever is to be known has already been discovered and communicated by the ancient sages; and should an author be bold enough to start any thing new, if that should happen to vary in the smallest particular from the orthodox writers, he would be severely punished. It is this which keeps the knowledge and civilization of China at a stand-still. The historical writings are nothing more than elaborate chronologies; and, where real dates have been wanting, the writers are suspected of having supplied them from their own imaginations. The scientific and philosophical works of the Chinese are by the 'ten philosophers,' or Confucius and his disciples and commentators. Chinese literature has, however, been in several respects unjustly depreciated. It has been said, for example, that they are so ignorant and ostentatious as to suppose that China occupies the centre of the world, and that it is surrounded with a few insignificant and petty territories, all its tributaries. But the accounts that have been translated from Chinese writers of several foreign countries, how defective soever in many respects, are sufficient to show that this is a most unfounded statement. It seems highly probable that the vast empire of China will, notwithstanding the extreme conservative character of its inhabitants, be gradually led within the pale of Western civilization and European modes of thought. Recent events, among them a war with the two greatest nations of Europe, have powerfully contributed to this effect. The handling, by Chinese soldiers, of a small vessel, the 'Arrow,' Oct. 8, 1856, and other trifling matters, having led to a war between Great Britain and China, in which France was made to join, the government of Pekin was in a short time reduced to such straits as to sue for peace on the most humiliating terms. According to the stipulations of the treaty of peace concluded between Great Britain and China, June 26, 1858, the empire is open to European travellers, especially British subjects, while British men-of-war may visit any Chinese port. More than this, it is stipulated that missionaries shall be allowed freely to preach the Gospel of Christ, and that Christianity shall be tolerated throughout the Chinese empire.

CHINACHIN, a large town of Nepaul, N. Hindostan, 250 m. WNW. Chitamunioo. Its houses are of brick and stone, with flat roofs: it has two Hindoo temples, and an export trade in horses, cow tails, sheep, salt, musk, drugs, and woollen cloth; and imports metals, spices, cloth, &c., from other parts of Hindostan.

CHINAUB (an. Accsines), the largest river of the Punjab, rising in the Himalaya, in lat. about 32° 10' N., long. 77° 50' E.; running at first with a NW. but afterwards with a SW. course between the Ravee (Hydraotes), and Jhylum (Hydaspes). It unites with the latter river below Jhung with considerable noise and violence, as remarked by the historians both of Alexander and Timour, and with the Sutlege (Hyphasis) near Ouch; after which it joins the Indus, in lat. 29°, long. 70° 30'. About 50 m. S. Lahore, it has been found to measure 1½ m. across in the month of July; but, in the dry season, is there only 300 yards wide. It is no where fordable S. of the hills, though in many places easily crossed. Kishtawar, Vuzirabad, and Jhung are on its banks.

CHINCHILLA, a city of Spain, prov. Albacete, cap. dist., in an elevated situation, on the high road from Valencia to Madrid, 116 m. SE. Madrid,

72 m. NNW. Murcia. Pop. 6,044 in 1857. The city has a church, convent, a hospital, barracks, and an ancient ruined castle, which was partly restored during the war of independence. There are mines of silver in the neighbourhood; and it produces earthenware and some coarse linen and woollen cloths.

CHINCHOOR, an inl. town of Hindostan; prov. Aurungabad, pres. Bombay, on the road between that city and Poonah, 10 m. NNW. the latter. Estimated pop. 5,000, including 300 Brahmin families. It is chiefly remarkable as the residence of the Chintamun or Surrain Deo, an individual whose honours are hereditary, and who is believed by a large proportion of the Mahratta nation to be an incarnation of their favourite deity theumpotty.

CHINGLEPUT, or 'the Jaghire,' a distr. of Hindostan; prov. Carnatic, pres. Madras; between 12° and 14° N., and intersected by long. 80° E.; having N. the distr. Nellore; WS. Arcot; and E. the Bay of Bengal. Area, 2,733 sq. m. Pop. estim. at about 350,000. Surface generally low, but with hills interspersed; there are several rivers, the principal of which is the Palaur, which rises among the Nundydroog hills in Mysore, and after a winding course of 310 m., chiefly E., past Vellore, Arcot, Conjeveram, and Chingleput, falls into the sea, near Sadras. There are some lakes and lagoons, or inlets of the sea, the chief of which is that of Pulicat. Granite is the most abundant of the primitive formations, and often projects in detached masses from the surface. Soil sandy and indifferent, and the country often barren, or overrun with low prickly bushes. Owing partly to the scarcity of water, but quite as much to the oppressiveness of the government, a large portion of the land does not enjoy the cost of cultivation; but the rest supplies the Madras market with grain, betel, fruit, oil, vegetables, &c.; the palmyra (borassus flabelliformis) thrives without trouble, and is both cheap and abundant. There are no manufactures, excepting some of cloth. The great mass of the people are Hindoos. Chief towns, Chingleput and Conjeveram. This distr. was obtained by the E. I. Comp. in 1763, from the nabob of the Carnatic, who rented it till 1780, when the Madras pres. assumed the entire control over it. It was twice invaded by Hyder Ali, and was afterwards nearly depopulated by famine and emigration. During the present century it has been gradually recovering.

CHINGLEPUT (Singhapatta), an inl. town of Hindostan; presid. Madras, cap. of the above distr.; in a small valley, in great part covered by a beautiful artificial lake; 40 m. W. the Bay of Bengal, and 36 m. SSW. Madras; lat. 12° 46' N., long. 80° E. Though much reduced in extent, it has a fort of great strength, and in a good state of defence: the latter incloses an inner fort, in which the public functionaries hold their several courts and offices.

CHINON, a town of France, dép. Indre-et-Loire, cap. arrond., on the Vienne, 28 m. SW. Tours. Pop. 6,005 in 1861. The town was formerly fortified; and the ruins of its walls and those of its castle are its most important and interesting objects. It has a court of primary jurisdiction, a commercial college, and some manufactures of linen and woollen stuffs. The celebrated Rabelais was born within a short distance of Chinon, in 1483.

CHINSURAH, an inl. town of Hindostan, prov. Bengal, formerly a Dutch settlement, but latterly transferred to the British government, on the W. side of the Hooghly river, 18 m. N. Calcutta, and about 2 m. NNE. Chandernagore; lat. 22° 53' N., bing, 88° 30' E. In appearance it has quite a Dutch character. There are many small neat houses, with green doors and windows. A pretty little square, with grass-plot and promenades, shaded by trees; a fortified factory; and a gloomy old-fashioned government-house, are the more remarkable features.

CHIO. See Scio.

CHIOGGIA, or CHIOZZA (perhaps the Portus Edro of the ancients), a sea-port town of Austrian Italy, prov. Venice, cap. distr., on an island of the same name, at the S. extremity of the lagoon of Venice, 14 m. S. that city. Pop. 26,000 in 1858. The town is about 2 m. in circuit; well built; contains a wide and handsome street lined with porticos, a cathedral, hospital, orphan asylum, and theatre; and is connected with the mainland by a stone bridge of forty-three arches. It has a harbour with 17 ft. water, protected by two forts: there are other batteries, and Chioggia is deemed one of the most strongly defended points of the Venetian lagoon. It is a bishopric, and has an episcopal palace, a gymnasium, a high seminary, conventual female school, and an evening rudimental school attended by nearly 300 poor children. In its vicinity are some important saltworks, which, together with the manufacture of cordage, the building of vessels, for which there are thirty-six slips, navigation, and fishing, occupy many of the inhab. Trade active in Italian and German produce, and facilitated by canals communicating with the Brenta, Adige, and Po.

CHIPPENHAM, a parl. bor., town, and par. of England, co. Wilts, hund. Chippenham, 87 m. W. London, 20 m. E. Bristol, on the Great Western railway. Pop. of municipal bor., 1,603, and of parl. bor. 7,075 in 1861. The town is situated on the Avon, which is here crossed by a bridge of twenty-two arches. It is well built, paved, lighted with gas, and amply supplied with water. From its situation at the intersection of two great roads, the Malmesbury and the London and Bath lines, many daily coaches formerly used to pass it, and it had a bustling appearance. It is now on the line of the Great Western railway, and a branch of the Berks and Wilts canal terminates in the town. The church is a spacious structure of various dates, some portion being as old as the 13th century; there are also several dissenting chapels, a free school for twelve children, and other charitable and benevolent institutions. The market, which was formerly very extensive, is held on Friday. There are large cattle-fairs, May 17, June 22, Oct. 29, and Dec. 11.

Though one of the oldest towns in the kingdom, Chippenham received no charter till 1554. Under the Municipal Reform Act, it is governed by four aldermen and twelve councillors; and the limits of the bor. have been extended for municipal purposes, so as to include the whole town and a pop. of about 4,000. The corporation revenue amounts to about 350l. a year, derived principally from an estate left for the maintenance of the bridge and of a road to Derryhill in the vicinity. A court of requests for debts under 40s. sits successively here and at Calne and Corsham.

Chippenham has sent two members to the H. of C. from the reign of Edward I. Previously to the Reform Act, the right of voting was restricted to the occupiers of 129 burgage tenements within the ancient bor. The extension of the limits of the parl. bor. by the Boundary Act has been noticed above. Registered electors, 375 in 1865.

CHIPPING NORTON, a town and par. of England, co. Oxford, hund. Chadlington, 85 m. NW. London by London and North Western railway. Pop. 3,137 in 1861. The town is built partly on low and partly on high ground. It has a large Gothic church, with a low tower; a free school, founded by Edward VI.; a subscription school, for educating and clothing forty girls; and almshouses founded in 1640. It returned two members to the H. of C. in the 30th of Edward I., and the 22nd and 33rd of Edward III. Its bailiffs were empowered by a charter of James I. to decide actions under 40s.

About 8 m. from Chipping Norton is the Rowldrich monument, formed of upright stones, arranged in a nearly circular form. This monument is ascribed by Dr. Stukeley, though probably without any good foundation, to the Druids. (See Avebury.)

CHISWICK, a par. and village of England, co. Middlesex, Kensington div. of Osulston hund., on the N. bank of the Thames, 4½ m. from Hyde Park corner by road, and 6½ m. from Waterloo Bridge by London and South Western railway. Pop. of par. 6,505 in 1861. The church, which has been frequently repaired and altered, has several interesting monuments; and in the churchyard is the tomb of Hogarth. There are many fine villas; but the great ornament of the place is Chiswick House, belonging to the Duke of Devonshire. It was built after the model of a villa by Palladio, by the famous Earl of Burlington, and has a choice collection of paintings. The illustrious statesmen, C. J. Fox and George Canning, breathed their last in this villa.

CHITORE, a city and strong fortress of Hindostan, prov. Rajpootana, and formerly the cap. of the rajahship of Oodeypore, 64 m. ENE. that city. The fortress, situated upon a rock scarped by nature and art to the height of from 80 to 120 ft., is surrounded by a rude wall with semicircular bastions, the circuit of which is said to be 12 m., but which incloses only a narrow, irregular, and disproportionately small area. Its outworks are massive and striking, and its appearance picturesque; its interior contains numerous temples, several palaces, some minarets, one of which is a square tower of white marble, nine stories high, and surmounted by a cupola; and many wells, fountains, and cisterns. All the public buildings are of Hindoo origin, excepting one erected by a son of Aurungzebe. The town, seated below the fortress, is chiefly inhabited by weavers and dealers in grain.

CHITTAGONG (Chatergram), a dist. of India beyond the Ganges and Brahmapootra, but included in the prov. of Bengal, of which it forms the SE. extremity, lying chiefly between lat. 21° and 23° N., and long. 91° 30' and 93° E., having N. Tipperah, E. the country of the indep. Khyens, S. Arracan, and W. the Bay of Bengal. Length, N. to S., about 165 m.; breadth uncertain. Pop. estimated at 750,000. The islands of Hattia, Sundeep, and Ramgony, with Manral and others contiguous to its shores, are under its jurisdiction. Its coast, S. of the mouth of the Kurnaphuli or Chittagong river, abounds with openings and harbours; but unfortunately none of them are available for ships of any size, their mouths being choked up with sandbanks and shoals. Surface along the coast low and flat; the interior is hilly; and the E. frontier is formed by the same extensive mountain chain which bounds Sylhet, Tipperah, and Arracan, to the E., and which in this portion of its extent varies from 2,000 to 5,000 ft. in height. In this region many streams arise which discharge on the Chittagong coast. Climate in many respects similar to that of Bengal; but the rains set in earlier, and last longer: in the hill region the crops often suffer from the inundations of the mountain torrents, as they do on the coast from invasions of the sea. Chittagong is in many parts particularly healthy, and is, therefore, often frequented by Europeans from Bengal. Many of the valleys and plains possess so fertile a soil that very little labour insures redundant crops. Much of the country is overgrown with jungle, and the whole of the mountain chain is covered with lofty forests. The hilly region, when cleared, is believed to be well adapted for the culture of coffee, pepper, and spices. The low hills are interspersed with many hamlets inhabited by Mughs, who emigrated thither after the conquest of Arracan by the Burmese in 1783, in the neighbourhood of which, on small plots of cleared land, they raise plantains, ginger, betel-leaf, the sugar-cane, cotton, indigo, tobacco, and capsicum. The hills in the N. are inhabited by Tiparrah, Joomea, and other tribes, apparently without any dependence on particular chiefs, who cultivate cotton and rice, and rear hogs, goats, and poultry, which they exchange with the Bengalese for salt, iron, earthenware, and fish.

Notwithstanding the fertility of its soil, Chittagong is, upon the whole, but thinly inhabited; towards the end of last century it was estimated that there was twice as much unproductive hilly country as cultivated arable land. Landed property is mostly divided into very small portions, among numerous proprietors. The waste lands, when cleared, become liable to assessment under the decennial land settlement. Except on the sea-coast, towns and villages are very scarce. The Mughs or Arracanese inhabit either temporary hamlets, which they change together with the spots they cultivate, or else permanent dwellings about 10 ft. long by 20 broad, elevated on posts several feet from the ground, after the fashion of some Ultra-Gangetic nations, ascended by a ladder or notched stick, and much more comfortable in their interior than the huts of the Bengalese peasantry. The male Mugh pop. have adopted the dress and habits of Bengal, while the females retain those of Arracan and Ava; all are Buddhists. The Mohammedans in this dist. are to the Hindoos as 2 to 7; but are extremely tolerant, and have adopted many Hindoo habits and customs. The chief exports of Chittagong are timber, planks, canvas, coarse cloths, stockings, umbrellas, &c.; on the sea coast salt, which is a government monopoly, is extensively manufactured. Coal is believed to exist, but no mines have yet been worked. The elephants of Chittagong have been celebrated both for size and excellence. They are admirably adapted for the camp and the chase, and hunting them still forms a chief occupation of some of the forest inhabitants. Many were formerly caught and exported, yielding a considerable profit to the sovereign; the trade in them is now farmed by the government to a contractor.

Chittagong probably once formed part of the extensive kingdom of Tiparrah. In the sixteenth century it was successively possessed by the Afghan kings of Bengal and the Arracan rajah; in 1760 it was finally ceded by its nabob to the British.

CHITTELDROOG (*Sittala durga*, the spotted castle), an inland town and fortress of Hindostan, prov. and dom. of Mysore, but occupied by a British garrison; cap. of a dist. on a cluster of rocks at the extremity of a ridge of hills, 110 m. NNE. Seringapatam, 200 m. WNW. Madras; lat. 14° 4′ N., long. 76° 30′ E. The town, which stretches along the base of the ridge or fortress at the NE., is surrounded by dilapidated ramparts of granite with round towers at intervals, a glorious ditch excavated from the rock, and a wide moated glacis; it is neither very large nor populous, but

its principal street is remarkably spacious. The fort, enclosed by the town, is probably the most elaborate specimen of a defensive rock to be found in S. India; an endless labyrinth of walls of solid masonry winds irregularly up to the summit, guarding every accessible point, and forming enclosure within enclosure; the more exposed points are crowned with batteries, and the ascent is partly by steps, and partly by superficial notches cut in the rock, and scaled with great difficulty. Such is the intricacy of the works, that an enemy might be master of the outer walls and yet not materially advanced towards the reduction of the fort. The lower enclosure contains the former pollgar's palace, now occupied by the British commandant, other ancient structures, the officers' bungalows, and a reservoir of good water which supplies all the town; in the other enclosures there are two other tanks, various Hindoo temples, a deep magazine sunk in the rock, and a depôt for glue. At a short distance W. of Chitteldroog is a curious suite of subterranean chambers, apparently the former habitations of devotee worshippers of Siva. This station is noted above all others in India for the great variety and excellence of its fruits.

CHIUSA, an inl. town of N. Italy, prov. Coni, cap. mand., on the Proiа, 7 m. SE. Coni. Pop. 6,514 in 1861. The inhabitants are chiefly occupied in the manufacture of silk goods and mirrors, and vine cultivation. The town is well built. A continuation of the ancient Emilian way passes through its vicinity.

CHIVASSO, an inl. town of N. Italy, prov. Turin, cap. mand., on the Po, in a fertile plain, 13 m. NE. Turin. Pop. 8,734 in 1861. The town was formerly one of the strongest places in Piedmont, but is now surrounded by only a simple wall with two gates leading to two suburbs. It has a square, a church, and several convents, and some trade in corn and cattle.

CHOLET, or CHOLLET, a town of France, dép. Maine-et-Loire, cap. cant., on the Maine, 17 m. SSE. Beaupreau, on the railway from Paris to Nantes. Pop. 12,733 in 1861. The town is finely situated, and had formerly several religious houses and a superb castle, destroyed during the revolution. Extensive manufactures of cottons were established here and in the neighbouring communes during the last century; but the town having been the theatre of a battle, in 1793, between the Vendéans and the republicans, the manufactures were all but destroyed, and the workmen either put to death or dispersed. In 1795, however, after the first pacification of Vendée, the expatriated manufacturers returned to Cholet; and, instead of being dispirited by their disasters, entered with fresh vigour on a new career of industry, and have succeeded in carrying the manufactures of the town and its vicinity to a higher pitch of prosperity than ever. At present there are established here for the spinning of cotton and wool, with extensive bleach-fields and dye-works. A great variety of cotton, linen, and other goods are produced in the town.

CHOLULA, an inl. town of Mexico, state of La Puebla, in a fertile plain S, of the Cordillera of the Malinche, 8 m. WNW. Puebla, and 64 m. SE. Mexico; lat. 19° 2′ 6″ N., long. 9° 15′ 15″ W. Pop. when visited by Humboldt, 16,000; but it has fallen off in the interval. It was conquered by Cortez, in the early part of the 16th century, with the most populous cities of Spain; but it declined with the rise of Puebla. It still, however, covers a large space of ground, and the size of its great square indicates its past importance. It contains many churches, and regular and broad streets; the houses are mostly of one story, and flat roofed.

There are some manufactures of cotton cloth. The principal extant relic of its ancient grandeur is a huge pyramid, or *teocalli*, to the E. of the town, now covered with prickly-pear, cypress, and other evergreen shrubs, and looking at a distance like a natural conical-shaped hill. As it is approached, however, it is seen to consist of four distinct pyramidical stories, the whole built with alternate layers of clay and sun-dried bricks, and crowned with a small church. According to Humboldt, each side of its base measures 439 metres (1,440 ft.), being almost double the base of the great pyramid of Cheops (which stands on an area equal to that of Lincoln's Inn Fields); its height, however, is only 50 metres (164 ft.). It appears to have been constructed exactly in the direction of the four cardinal points. The ascent to the platform on the summit is by a flight of 120 steps. This elevated area comprises 4,200 sq. metres (5,023 sq. yds.). The chapel erected on it is in the shape of a cross, about 90 ft. in length, with two towers and a dome. It was dedicated to the Virgin by the Spaniards, and has succeeded to a temple of Quetzalcoatl, the god of the air. This pyramidal pile is, however, conjectured to have served for a cemetery, as well as for the purposes of religion; and Humboldt and other authorities regard it as bearing a remarkable analogy to the temple of Belus, and other ancient structures of the Oriental world. The Indians believe it to be hollow, and have a tradition that during the abode of Cortes at Cholula a number of armed warriors were concealed within it, who were to have fallen suddenly upon the Spanish army. At all events, it is certain that Cortes, having some suspicion or information of such a plot, unexpectedly assaulted the citizens of Cholula, 6,000 of whom were killed. In making the present road from Puebla to Mexico, the first story of this pyramid was cut through, and a square stone chamber discovered, destitute of an outlet, supported by beams of cypress, and built in a remarkable way, every succeeding course of bricks passing beyond the lower, in a manner similar to some rude substitutes for the arch met with in certain Egyptian edifices. In this chamber, two skeletons, some idols in basalt, and some curiously varnished and painted vases, were found. There are some other detached masses of clay and unburnt brick in the immediate vicinity, in one of which, apparently an ancient fortress, many human bones, earthenware, and weapons of the ancient Mexicans, have been found. The view from the great pyramid, embracing the Cordillera, the volcanoes of La Puebla, and the cultivated plain beneath, is both extensive and magnificent. Cholula is surrounded by corn fields, aloe plantations, and neatly cultivated gardens. (Humboldt, Researches, i. 88, Eng. Trans.; Bullock, Six Months in Mexico, pp. 111-116; Ward, Antiq. of Mexico.)

CHOOROO, an inl. town of Hindostan, prov. Rajpootana, in a naked tract of sand hills, 100 m. ENE. Bicanere; lat. 28° 12' N., long. 74° 55' E. It is 1½ m. in circ., exclusive of its suburbs, and has a very handsome external appearance. The houses are all terraced, and, as well as the walls of the town, are built of a kind of limestone found in vast quantities in this part of the prov., of a very pure white, but soft, and apt to crumble. In 1817 Chooroo was plundered by one of Meer Khan's sirdars; in 1818 it was visited by a British detachment, and afterwards transferred to the rajah of Biramere; its chief, however, is rather a dependent than a subject of that prince.

CHORLEY, a par. and market tn. of England, co. Lancashire, hund. Leyland, on the Chor, 70 m. NW. Manchester, 8 m. N. Wigan, and 175 m.

NWX., London, by London and North Western railway. Pop. 15,013 in 1861. This thriving town, which takes its name from the stream near the source of which it is situated, stands on a rising ground about a mile above the confluence of the Chor and Yarrow. It is well built; streets broad, lighted with gas, and abundantly supplied with water from a reservoir, into which the stream is thrown up by steam machinery. The par. church of St. Laurence is an ancient structure in the Norman style; that of St. George, a handsome edifice, was built by the parl. commissioners in 1835, at an expense of 13,750*l*. The Independents, Unitarians, Methodists, and R. Catholics have places of worship, to some of which Sunday-schools are attached. A free grammar school was founded in 1634, and a national school in 1824. The town is governed by a constable chosen annually at a court leet. The increase of population—from 4,516 in 1801—is a consequence of the still more rapid increase of the cotton trade. As early as 1790, spinning-mills began to be erected in the town. Exclusive of yarn, the fabrics principally produced are muslins, jaconets, and fancy goods. Bleach-greens and print-works are established on the banks of the neighbouring streams. The coal mines in the neighbourhood have contributed greatly to the improvement of the town; there are also valuable quarries of slate, and gritstone for mills, with lead and iron mines. The Liverpool and Leeds canal, which passes within half a mile of the town, and is joined by that from Lancaster and Preston at a short distance from it, affords great facilities for conveying the produce of the factories and mines throughout all the N. counties. Markets are held on Tuesday; fairs on 26th March and 5th May for horned cattle; 21st October for horses; and 4th, 5th, and 6th Sept. for woollens and general purposes.

CHOWBENT, or ATHERTON, a township of England, co. Lancashire, hund. W. Derby, par. Leigh, 10 m. WNW. Manchester, and 6 m. ESE. Wigan. Pop. 6,907 in 1861. This is a thriving place. Previously to the American war, the making of nails was extensively carried on here; and, though the manufacture has declined, considerable quantities are still made for exportation. It is also remarkable for several inventions and improvements in cotton machinery; and it is said that the value of the application of heat in the production of some kinds of cotton fabrics was discovered here. The Bolton and Leigh railway passes within a short distance of the town. Fairs, at which premiums for the best cattle are given, take place on the first Saturday in May, and the last Saturday in October.

CHRISTCHURCH, a parl. bor. and par. of England, co. Hants, New Forest, W. div., hund. Christchurch, 99 m. SW. London by London and South Western railway. Pop. of parl. bor. 9,368, and of par. 7,042, in 1861. It is situated at the confluence of the Avon and Stour, about 1 m. from where their united streams fall into Christchurch Bay, 90 m. SW. London. The town presents no symptoms of activity or industry. No trade nor manufacture is carried on. The church was the collegiate one of the ancient priory, and is a large, fine structure; the older part in the Norman, the rest in the earlier and later pointed styles; the fine tower is of the fifteenth century. It has a very ancient and curiously carved altar, and many beautiful chapels. There are also two episcopal chapels (one of them built by parliamentary grant in 1823, with 462 free sittings), a Roman Catholic chapel, a dissenting ditto, a free school of uncertain foundation, educating ten boys, a national and a Lancastrian school, and several small charrities. Market on Mondays; fairs, Trinity Thursday, and Oct. 17, for horses and cattle. It returned two members to the H. of C. in 35th Edw. I. and in the 2nd Edw. II. No other return appears till the 13th of Eliz.; since which period it regularly returned two members, till the Reform Act deprived it of one of them. The franchise, previously to this act, was vested in the corporation, which consisted of a mayor and an unlimited number of burgesses. The Boundary Act very materially extended the limits of the port, but, Registered electors, 351 in 1861. The harbour has a shifting bar, with not more than 5 or 6 ft. water over it, so that it is accessible only at spring tides for the smaller class of coasters. There are several breweries in the town; and the manufacture of watch springs employs a few hands. The name is derived from its ancient priory, of very remote origin. There are traces of many ancient camps and barrows in its vicinity.

CHRISTIANIA, a sea-port town of Norway, of which it is the cap., on the Agger, at the bottom of a very deep gulf or fiord, to which it gives name; 162 m. ENE. Bergen, 232 m. S. by E. Drontheim, and 255 m. W. by N. Stockholm. Pop. 44,212 in 1860. The town is surrounded by an amphitheatre of hills, and its situation is extremely picturesque. It is well laid out; streets spacious and regular, and some of them even handsome. Houses in the town all brick or stone; those of wood having been prohibited, on account of the former frequency of fires. They are airy and well built, though seldom more than two stories high. In the best quarters they are built round an open square court, and are generally occupied by several families. It is the residence of the viceroy, and the seat of the diet; has a cathedral, and three other churches; a military and a lunatic hospital, two orphan asylums, a house of correction, a new town hall and exchange, and two theatres; but none of the public buildings in any wise remarkable. Four suburbs part from the town as a centre, one of which is the old town of Opslo, from which Christiania originated. In these, wooden houses are not prohibited; and, as the suburbs are mostly inhabited by the lower classes, the dwellings are chiefly of wood. A short distance beyond the walls is the royal palace, a plain brick building of modern construction. The whole vicinity of the town is sprinkled with the country houses of citizens. The gulf of Christiania unites with the farthest N. point of the Skagerrec; though in parts narrow, and difficult of navigation, it has deep water throughout, there being 6 or 7 fathoms close to the quay. Christiania is the seat of the higher courts of law, and a university. The latter is attended by about 600 students, and has attached to it a public library, with 115,000 volumes, collections of natural history and mineralogy, a museum of northern antiquities, an observatory, and a botanic garden. There is a military school, with schools of commerce and design, elementary schools, and several learned and philanthropic societies. Manufactures not very extensive; the chief are those of woollens, tobacco, glass, hardware, soap, leather, and cordage. Principal exports, timber, deals, glass, iron and nails, smalts, bones, oak-bark, and salted and pickled fish, a staple mostly sent to Bergen. The deals of Christiania have always been held in the highest estimation, in consequence of the sap being carefully cut away. Christiania was built by Christian IV., king of Denmark, in 1624.

CHRISTIANSAND, a sea-port and fortified town of Norway, near its S. extremity, cap. diocese of same name, dist. Mandahl, on the Skagerrec, at the head of a deep fiord, 160 m. SW.

Christiania. Pop. 10,336 in 1860. The town is regularly laid out; streets long and wide, houses generally built of wood, and separated by gardens. Chief public building the cathedral, a Gothic structure, and, next to that of Drontheim, the finest ecclesiastical edifice in Norway. Here is an asylum for the poor, a sail-cloth manufactory, and docks for the construction of vessels, ship building being the principal branch of industry. The harbour is very secure, and sheltered on newly every side by lofty and rocky heights. It is well supplied with fish; and lobsters are taken in great numbers, and exported to the London markets. Timber is another principal article of export. Christiansand ranks as the fourth town in Norway; it is a bishopric, and the residence of a governor. It was founded in 1641 by Christian IV., king of Denmark, who intended to make it the principal naval port of his dominions.

CHRISTOPHERS (ST.), or ST. KITTS, one of the W. India islands belonging to Great Britain, lying about lat. 17° 20' N., and long. 62° 46' W., and about 50 m. W. by N. Antigua, of the government of which island it constitutes a part. Length, NW. to SE., about 15 m.; breadth in general about 4 m., but no more than 3 m. towards its SE. extremity, where it is divided by only a narrow channel from the island of Nevis. Total area 108 sq. m.; pop. 23,177 in 1851, and 24,440 in 1861. The island contains many rugged precipices and barren mountains, the principal of which, Mount Misery, an extinct volcano, rises to 3,711 ft. above the sea. The climate is healthy, but violent hurricanes sometimes occur. Of 43,726 acres of land, the extent of the surface of the island, it is estimated that nearly half is unfit for culture. The soil of the plains, however, which is of a volcanic origin, intermixed with a fine loam, makes amends by its fertility for the barrenness of the mountains. Sugar is the great article of cultivation, the only articles raised in addition to it being a little cotton, coffee, and arrow-root. The value of the principal articles of produce imported into the U. Kingdom from St. Christopher's amounted to 68,226l. in 1850; to 131,329l. in 1851; to 164,519l. in 1860; and to 118,929 in 1861. The exports from the U. Kingdom to St. Christopher's were of the value of 92,143l. in 1850; of 96,054l. in 1851; of 130,934l. in 1860; and of 151,552l. in 1861. The island is divided into nine parishes, and contains four towns, Basseterre, Sandy Point, Old Road, and Deep Bay. The first two are ports of entry established by law. Basseterre, in the SW., is the cap. It contains about 800 houses, and, as well as Sandy Point and some other parts of the island, is defended by several batteries. St. Christopher's was discovered, in 1493, by Columbus, who gave it the name it bears; but it was not settled till 1623, when a party of English took possession of it. After many disputes for its occupation with the French and Spaniards, it was finally ceded to Great Britain at the peace of Utrecht in 1713.

CHUDLEIGH, a town and par. of England, co. Devon, hund. Exminster. Area of par., 6,230 acres. Pop. of ditto, 2,100 in 1861. The town, on an acclivity near the Teign, 8 m. S. by W. Exeter, consists chiefly of one wide street of well-built houses, being part of the main line of road from Exeter to Plymouth. The church is an old structure amidst fine trees; the vicarage in the patronage of such of the parishioners as have freeholds to the amount of 5l. a year and upwards. There are two dissenting chapels, a grammar school, founded 1668, with a residence for the master, and three exhibitions to the university of Cambridge, a national school, and several charities

Market on Saturdays. Fairs, Easter Tuesday, third Tuesday and Wednesday in June, and Oct. 2, for cattle and sheep. The serge manufacture was formerly carried on to some extent, but at present there is no manufacture of any kind, and the labouring part of the pop. are chiefly engaged in agriculture. Ugbrook Park, in the immediate neighbourhood (the seat of Lord de Clifford), is considered one of the finest in the kingdom.

CHUMBUL, (supposed to be the *Nambus* of Arrian), a river of Hindostan, which rises in Malwah prov., and falls into the Jumna river, about 25 m. below Etawah, after a course of about 600 m., generally in a NE. direction.

CHUMPANEER, a town and large district of Hindostan, prov. Gujerat; the former, called also Powanghur, stands on a scarped rock 25 m. NE. Baroda, and is supposed to have been the cap. of a Hindoo principality, before the Mohammedan rule in India. The remains of an ancient city stretch for several miles on either side of it. This town was taken by Humayoon in 1534, and by the British in 1803.

CHUPRAH, a town of Hindostan, prov. Bahar, distr. Sarun, of which it is the cap., on the N. side of the Ganges, along which it extends for nearly a mile; 33 m. W. by N. Patna. Pop about 30,000. It has some trade in cotton and sugar.

CHUQUISACA (formerly *La Plata* or *Churcas*), an inland city of S. America, cap. Bolivia, in a small plain surrounded by heights, on the N. bank of the Cachimayo, and on the high road between Potosi and Santa Cruz de la Sierra, 55 m. ENE. the former, and 220 m. SW. the latter; lat. 19° 20' S. long. 65° 40' W. Estimated pop. 10,000, partly equally divided amongst Spaniards, Indians, and mixed races. The city contains a large and handsome cathedral, with some good paintings and decorations, several monastic establishments with splendid churches, a conventual hospital, three nunneries, and a university. The best houses are but one story in height, but roomy, and have pleasant gardens; it is supplied with water from several public fountains. The climate is mild; but the rains are of long continuance, and during the winter violent tempests are not unfrequent. Chuquisaca was founded in 1538, made a bishopric in 1551, the seat of a royal *audiencia* in 1559, and an archbishopric in 1600.

CIEZA, or ZIEZA (an. *Cattioa*, or *Carteia*), a town of Spain, prov. Murcia, on the Segura, in a rich well-cultivated plain. 24 m. NW. cap. Pop. 8,316 in 1857. The town has convents for both sexes, a workhouse, public granary, &c.; with manufactures of coarse linens. On the opposite side of the river are ruins supposed by some to be those of the ancient Carteja.

CINCINNATI, a city of the U. S. of America, Ohio, cap. co. Hamilton, and, next to New Orleans, the largest and most flourishing commercial town in the W. part of the Union, on the N. bank of the Ohio, about 410 m. W. by N. Washington. Pop. 24,831 in 1830; 46,338 in 1840; 115,436 in 1850; and 161,044 in 1860. The town is built on two inclined plateaux rising from the river, one about 50 ft. higher than the other, and both running parallel to the Ohio. It is regularly laid out; streets wide and clean, and intersecting each other mostly at right angles. They are generally lined with trees on either side, and most of the houses have a small enclosure in front filled with flowering shrubs. Houses mostly of red and particoloured brick; but many are stuccoed, and a few are of stone. A square in the centre of the city is appropriated to public buildings. Here, and in other parts of the town, are numerous

churches; the city has also the Cincinnati college, 2 theatres, 4 market-houses, one 600 ft. in length, a court-house, medical college, mechanics' institute, Catholic orphan-asylum, 2 museums, a lunatic asylum, with hospitals, and numerous schools. Manufactures extensive and increasing; the principal are those of iron; next in importance are cabinet work, steam-boat building, and hat-making; the manufacture of cotton and woollen stuffs, and extensive distilleries and flour-mills. Cincinnati is the largest pork-market in the Union. Two-thirds of all the hogs fed in the forests of Ohio, Kentucky, and W. Virginia, are driven here for slaughter and exportation. The buildings for this branch of trade are very extensive, and occupy many acres.

The Ohio is 600 yards wide at Cincinnati, and navigable for small steam-vessels as far as Pittsburgh, 454 m. higher. Lane Seminary, founded in 1829, chiefly for theology, and situated about 2 m. from the city, has 2 commodious edifices, and contains 100 rooms for students. It possesses a library of 10,000 volumes. One of the museums contains a number of enormous organic remains and antique vases, excavated from some of the ancient mounds in Ohio. There are a great many religious and benevolent associations, several academies, a public library, and some excellent hotels. The pop. is composed of emigrants from all the states of America and most of the countries in Europe. There are said to be no less than 30,000 German settlers.

The advance made by Cincinnati has been wonderfully rapid. It was founded in 1789, and in 1800 the population was only 800; in 1810, it was 2,500; in 1815, about 6,500; in 1820, 9,600; and in 1850 it amounted, as already seen, to 161,044. Its picturesque situation, and the beauty of its environs and of the surrounding scenery, have gained for it the title of 'Queen of the West;' while its central position, the abundance of its railway communication, and its rapid increase in population and commerce, make it probable that it will speedily rival in wealth and importance the principal cities of the United States.

CINTRA (*Mons Cynthia*), a town of Portugal, 12 m. WNW. Lisbon. Pop. 4,460 in 1858. This Richmond of the Portuguese capital is situated at the head of the rich and beautiful valley of the Collaris, and at the foot of a rugged rock or mountain. The latter is in part covered with scanty herbage; in parts it rises into conical hills, formed of immense stones, and piled so strangely that all the machinery of deluges and volcanos must fail to satisfy the inquirer for their origin. On one of the mountain eminences stands the Penha convent, visible from the hills near Lisbon; on another are the ruins of a Moorish castle. From this elevation the eye stretches over a bare and melancholy country, to Lisbon on the one side, and on the other to the distant convent of Mafra, the Atlantic bounding the greater part of the prospect.' (Southey's Letters, ii. 202.) In summer, the citizens of Lisbon resort on the Saturday nights to Cintra, where they spend the Sundays, returning home on Monday. Many of the nobility and the wealthier merchants, especially the English, have villas in the vicinity of the town, which is so much celebrated for its fine air as for the beauty of its situation. It has also a palace, occasionally occupied by the court; in one of its apartments are painted the armorial bearings of all the noble families of Portugal.

The convention agreed to in 1808, after the battle of Vimiera, a memorable incident in the war with Napoleon, by which the French forces under Junot, with their arms and artillery, were conveyed to France, is usually described as the Convention of Cintra.

CIOTAT (LA), a seaport town of France, dep. Bouches du Rhône, cap. cant., on the W. side of the Bay of Leques, 15 m. SE. Marseilles, on the railway from Marseilles to Toulon. Pop. 8,111 in 1861. The town is surrounded by an ancient rampart of considerable extent, and is a tolerably perfect condition. Streets regular, and well paved; houses well built. It possesses some good quays, a large pier, church built in the 16th century, and a fine public promenade, but is ill-supplied with water. Its port, sheltered by a mole and defended by a fort, is commodious, secure, and accessible to vessels of 300 tons burden. A lighthouse, in the fort, has the lantern elevated 82 ft. above the level of the sea. Ships are built, and oil is manufactured here; and it has a considerable trade in wines and dried fruits, the vicinity being interspersed with vineyards, olive grounds, and plantations of oranges and figs. La Ciotat is said to occupy the site of the ancient Citharista; the modern town was, however, founded in the 13th century, and did not acquire municipal rights till 1429.

CICARS (NORTHERN), a large marit. prov. of Hindostan, extending along the E. coast for 470 m., between lat. 16° and 20° N., and long. 79° and 85° E.; having N. and W. Orissa, Gundwanah, and Hyderabad, and S. and E. the Carnatic and the Bay of Bengal. It comprises portions of the ancient territories of Orissa and Telingana, and, previously to the British rule, consisted of five divisions or 'circars' viz. Ganjam, Condapilly, Ellore, Rajamundry, and Cicacole. At present it is wholly included within the territories of the Madras presidency. Area 25,760 sq. m.; pop. estimated at 3,000,000. The territory (islands) W. by a chain of mountains continuous with the E. Ghauts, but no where of any great height. Vizagapatam, between lat. 17° and 19° N., is the most mountainous district, and contains a considerable range of hills, running parallel to the former and to the coast, often closely approaching the latter, and enclosing an extensive and fertile valley, together with the principal range. From Ganjam to Coringa, the coast generally appears mountainous, but thence is low, flat, and sandy, with numerous small streams. Chief rivers, the Godavery and Krishna; the first has an extensive and fertile delta at its mouth below Rajamundry. The Chilka lake constitutes the N. limit of the prov.; the only other lake of note is that of Colair in the Masulipatam distr.; but several lagunes of some size are met with on the shores. A black soil prevails in the N. parts of the prov., highly suitable to the cultivation of cotton. N. of the Godavery the climate is extremely hot, and for a month preceding the rains, the thermometer in the country round the mouth of the Krishna sometimes stands for a whole week at 110° Fahr.; in other parts it has been known to stand at 112° at 8 o'clock in the evening, and at midnight as high as 100°. At such times, wood of all kinds readily warps, and glass cracks and flies in pieces; in all the hilly regions and round Masulipatam, a very noxious state of the air prevails throughout the different seasons of vegetation.

The circars are extremely productive of grain, and have long been the granary of Madras during the NE. monsoon, though at present the distr. of Masulipatam annually imports large quantities of rice from Calcutta and Arracan for home consumption. Large crops of paddy and dry grains, cotton, and tobacco of excellent quality, the sugar-cane, and esculent vegetables, are produced in the S.;

the same articles, with ginger, yams, turmeric, chilies, &c., in the central parts; a great deal of sugar in the delta of the Godavery; and wheat, maize, the sugar-cane, and an abundance of rice and other grains in the N.

Agriculture is least advanced in Vizagapatam, owing chiefly to an oppressive revenue assessment: many of its hills are wild, and destitute of vegetation. In Masulipatam distr. there are extensive tracts of grass. The total number of black cattle in the circars is about 1,300,000, of sheep 500,000. The Ganjam distr. is interspersed with numerous bamboo jungles. The forests of Rajamundry abound with teak, which tree is found no where else on the E. side of Hindostan. The chief manufactures are chintzes, carpets, and cotton stuffs, in the central; and indigo, punjam cloths, muslins, and silks, in the N. distr.; the piece goods of the circars, which were formerly their staple, are now rather objects of curiosity than made in any considerable quantity. Rum was formerly distilled in the N.; the sugar of Ganjam is in much request, and exported in large quantities; the other exports are wax, salt, pepper, betel, ivory, indigo, tobacco, and other agricultural produce. The external trade is chiefly with Madras, Calcutta, Hyderabad, and the central Deccan. The exports to Europe are chiefly the cotton goods; all the raw silk used is imported. The natives are mostly Hindoos; Mohammedans are few. The Oriya and Telinga races have become much intermixed, though they still retain distinct dialects, and have distinct traits and customs. The villages consist of mud huts and houses; but the peasantry are not on the whole incommodiously lodged. The roads are amongst the worst in India, and unfit for wheeled carriages; there are but few tanks, bridges, or ferry-boats. The lands appear for a long period past to have belonged either to the government or to zemindars; for no instance has occurred since the British have come into possession of the prov., of any great claimant being cultivated by him. The chief towns of the circars are Chicacole, Ellore, Coringa, &c., besides those which bear the names of the several districts. Religious temples are not numerous; but in Ganjam, where Juggernaut is the favourite object of worship, their architecture is peculiar; they consist of groups of low buildings, each with a groined and pyramidical roof, terminating in an ornamented conical cupola. In 1571, the rajah of Hyderabad conquered this prov., which, together with Hyderabad, fell under the dom. of Aurungzebe, in 1687; it however became again independent of the Mogul empire in 1724. The English obtained the four most N. circars in 1765; the French had become possessed of Guntoor in 1753; but it also came into British possession in 1788.

CIRCASSIA, more properly TCHERKESSIA, or TCHERKESKAIA, the largest and most important country in the Caucasus, of which mountain-range it occupies nearly the whole N. slope; extending from 42°30' to 45° W N. lat., and from 37° to 45° 48' E. long. At its NW. corner it reaches the Black Sea, but, with this exception, it is bounded on the N. and W. by the main ridge of the mountains which divide it from Georgia, Mingrelia, Imeritia, and Great Abchasia. The N. limit is formed by the rivers Kuban and Terek, which separate it from the lowlands of the Cossacks, Turkomans, Nogay Tartars, and the Russian colonies in the Caucasian steppes; towards the E. it terminates at the junction of the little river Sunsha with the Terek, at which point a host of small streams divide it from the country of the Tchetchentzes. In extreme length, from NW. to SE., Circassia is about 470 m.; in its greatest width, about 100 m.; in its least, about 40 m., and, at an average, about 70 m. Its area may therefore be calculated at about 33,000 sq. m. Estimated pop. between 3,000,000 and 4,000,000. (Guldenstädt, Reise durch Russland, i. 466–469; Pallas's Trav. in S. Russia, i. 390, 391, 395, 422, &c.; Klaproth's Trav. in Caucasus and Georgia, pp. 252, 311, &c.; Lapie, Annales des Voy., xii. 36.)

Political Divisions.—The Circassians are divided into a great number of tribes, who lead a partially wandering life, so that no very precise arrangement can be made with regard to the districts of their country. The E. portion, or that enclosed by the Terek, is divided by Russian geographers into two provinces—*Great Kabardah*, to the SW., and *Little Kabardah*, to the NE. These divisions are not, however, recognised by the Circassians, who know but of one Kabardah, and that in the SW. portion, called by the Russians *Great*, (Klaproth, p. 354.) Between the sources of the Kuban and Terek, and along the courses of those rivers, as far as they run N., the land is wholly occupied by a tribe called the *Abkhasians* or *Abaze*; and forms the *Little Abaza* of Pallas, the *Altikesek Abchasia* of Guldenstädt. The *Great Abaza* of Pallas, *Basiana* of Guldenstädt, occupies likewise a very considerable part of the Kubanian Circassia; among the rest, the Natukhaitsi district, mentioned by Spencer. It appears indeed, that the Abaze are the lawful proprietors of all Kubanian Circassia, and that the Circassians have only the right of conquest to the W. portion of their country; that right is, however, very fully established, not only on the N. slopes of the mountains, but even to a very great degree on the W. side, along the shores of the Black Sea (the *Great Abchasia* of Guldenstädt). Spencer makes but little distinction between the Abaze and Circassians, and frequently speaks of them as one people: this must, however, be an error, since the former display a very peculiar physical conformation, and their language, with the exception of a few Circassian words, is totally unlike that of their conquerors, and of every other known people, European or Asiatic. The Circassian princes are cruel and oppressive tyrants to their Abasekan subjects, so much so, that the latter have in many instances sought the protection of the Russian government; but it does not appear that they are in any moral attribute superior to their taskmasters, since in every age they have been infamous for their robberies by land, their piracies by sea, and their reckless cruelties everywhere. (Guldenstädt, i. 460, 463, 466, 469; Pallas, i. 383–391; Klaproth, pp. 247–263, 283, 311; Spencer's Circassia, ii. 412, &c.; W. Caucasus, i. 29, 203, 312, 247, &c.)

Physical Features.—These have been generally described in the article CAUCASUS (which see), and what is peculiar to Circassia is only the consequence of that country's occupying the N. slope of the mountains. With the exception of the lowlands on the banks of the Kuban and Terek, the whole territory is broken into precipitous mountains, small table-lands, and valleys of the most picturesque and romantic description. Its hydrography belongs to two systems, the waters of Kabardah being all conveyed by the Terek to the Caspian, and those of W. Circassia by the Kuban to the Black Sea. The former river rises near the Kazibeck, and, forcing its way through the pass of Dariel (an Caucasian Gate), receives, directly or indirectly, thirty-five streams before it quits the Circassian country. Of these, the Malk, which joins it at its E. bend, is scarcely inferior in size to the principal river. It rises near the E. base of the Elburz (Osha Makhua), and is itself the recipient of a considerable number of tributaries.

The Kuban rises on the N. base of the Elburz, not far from the sources of the Malk, and receives the water of more than fifty rivers, thirty of which fall directly into its bed. It has every reason to be considered, exclusively, a Circassian river; for though no part of its N. bank be inhabited by Circassians, it does not receive a single drop of water, in its whole course, that does not rise within their territory. A similar remark will apply, in a modified sense, to the Terek, which, like the Kuban, does not receive a single stream from the N., and only one of consequence after entering the Tartar country E. of Little Kabardah. The country between the sources of the Malk and Kuban is watered by various streams; and when it is recollected that, in addition to these, innumerable torrents pour from the upper ranges of the mountains, it will be evident that no land can be better irrigated. The water is in general clear and good, but occasionally impregnated with mineral and other extraneous matters. The tributary streams become flooded in winter, and extremely shallow during the heats of summer; the currents of all are extremely rapid, as are those also of the Terek and Kuban, except where the latter forms morasses, which it does in some parts of the flat country, when its course becomes sluggish, and its water thick and muddy. (Guldenstädt, i. 469, and map; Klaproth, pp. 212–247, 255, 259, 261, 351, &c.; Pallas, i. 385–389, 418–417; Spencer's W. Caucasus, i. 106; Circassia, ii. 412, et passim.)

Climate, Soil, and Natural Productions.—These are also the same with those of the Caucasus generally (see CAUCASUS), but the temperature is rather lower than on the S. slopes, except on the banks of the Kuban, where the greater depression more than compensates for the difference of aspect, and where the extensive marshes and the exuberant vegetation create miasma, which render it more pestilential than any other district in the whole region. (Spencer's W. Cauc. i. 106; Circassia, ii. 304.) There is a greater proportion of bare rock in Circassia than in Georgia and the other countries S. of the main ridge, but on every shelf and in every rift, trees, grain, vegetables, and fruit of almost every kind, are produced from most fertile soil. The animals, also, are on the same scale of abundance and variety, whether the wild or domesticated tribes be considered; the quadrupeds, birds, fishes, insects, or reptiles. (See CAUCASUS.) The Circassian horses are nearly as famous and quite as good as those of Arabia. Cattle of all kinds are abundant in the extreme, and in addition to the herds forming the numerous stocks of the pastoral population, the aurochs and argali (wild ox and sheep) still wander among the mountains, with the ibex, and another beautiful variety of the goat. Game of all kinds, winged, hoofed, or clawed, are found in equal abundance, but differing in kind, in the mountains and plains; nor are beasts of prey, as jackals, wolves, bears, lynxes, and tiger cats. &c., much less numerous, though they do not seem to be much regarded by the natives. Wild boars are found, especially among the swamps of the Kuban, and it is affirmed that the tiger is not wholly unknown. The reptile and insect tribes are equally numerous. In one of the late campaigns of the Russians, besides the thousands who fell victims to the bad air, numbers died from mortified bites of mosquitoes. (Spencer's Circassia, ii. 317.) Both natives and Russians believe that the mountains abound in gold and silver, but apparently on no good grounds. (See CAUCASUS.) Iron, however, lead, and copper are found; and saltpetre is very abundant. Salt is nowhere found within the limits of Circassia; and since Russia has excluded the natives from the brine pits in the Caucasian steppe, and sealed their ports against the trade of Turkey and Persia. They have been almost totally deprived of this necessary. (Guldenstadt, i. 104, 411, &c.; Pallas, i. 339-380, &c.; Klaproth, pp. 309, 324, 356, et passim; Spencer's Circassia, ii. 72b, 233, 342, 220, 305, 317; W. Carr., i. 334-341, &c.)

Inhabitants.—The Circassians have long been proverbial for their beauty of form and figure, especially the women, and though it seems they have in this respect been confounded with the Georgians, who are a totally distinct nation, yet all the statements of the modern and most accurate travellers, concur in describing them as an extremely handsome people, tall, finely formed, slender in the loins, small in the foot and hand, elegantly featured, with fresh complexions, and extremely intelligent countenances. (Pallas, i. 398; Spencer, passim, &c.) It would be well did their moral and intellectual attainments correspond with their physical appearance; but it is obvious, even from the statements of their eulogists, that they are mere semi-barbarians, whose darling occupation is robbery and plunder, and who seem to be radically deficient in most of the requisites necessary to form a civilised and flourishing community. They have many points in common with the Arabs; and, like the sons of Ishmael, are quite as barbarous at the present day as in antiquity.

The Circassians are divided into five classes. 1. *Pschi*, or *pschech* (princes); 2. *Uork* (ancient nobles); 3. the freedmen of these princes and ancient nobles, who, by their manumission, become themselves noble, and are called *uork* of *uork*; 4. the freedmen of these new nobles, called *beguolis*; and 5. the vassals or *tchokotl*. Between the ancient and recent nobility there is no real distinction, except that, in military service, the latter are still under the command of their former masters; nor is there any great practical difference between the beguolis and the tchokotl, or vassals. The latter are, of course, the labourers; and are subdivided into such as are engaged in agriculture, and such as serve the superior classes in the capacity of menial servants. Of the former, many are wealthy, nor is the state of any, one of great degradation, since there are very few, if any, officers of labour which prince or noble would consider as derogatory to himself. To every princely house belongs a certain number of uork, or nobles, as they are called by the Russians; and the latter are the direct proprietors of the vassals. Of these last, though all are unquestionably slaves, those engaged in agriculture cannot be sold singly; and the sale of any is so rare as almost to be prohibited by custom. On the other hand, it appears the vassal may transfer his duty to another uoden; which is, of course, a great protection from ill usage. The vassals pay no money tax, and though they are compelled to supply their lord with all he wants, yet this, from the check upon the noble's power just alluded to, extends no farther, usually, than to bare necessaries; since, should the latter carry his demands too far, he runs the risk of losing his vassal altogether. The relation between prince and uoden is precisely the same as that between uoden and vassal; the noble must supply the necessities of his sovereign; but should the exactions of the latter become excessive, the former may transfer his allegiance to another prince. The uoden must pay the debts of their prince, and the vassals those of their uoden; and, in each case, the inferior must make good all losses sustained by his superior, whether from robbery or accident; by which arrangement it is evident that all losses or expenses are defrayed, ultimately, by the vassal. The head of the princely house is the leader in war; and his

uoden are bound to attend him with all their retainers, or as many as may be required. There is no people, not even the Arabs, among whom pride of birth is carried to a greater height than among the Circassians, especially those of Kabardah. In this district, if an uoden were to marry or seduce a princess, he would forfeit his life without mercy; and the same result would attend the attempt of a beguolis or vassal to ally himself to a noble house; an Alanelan prince is, in this respect, considered equal only to a Circassian uoden, and can obtain a Circassian wife only from that class. The rigorous enforcement of this custom has preserved the different ranks very distinct, though Pallas has observed, even in the Kabardah, some traces which indicate a descent from Tartar mothers. (i. 398.) It must be observed, however, that there does not appear to be any restriction upon a man's taking a wife or concubine from an inferior class; and the issue of such connexions take rank from the father, but are not accounted equal to the descendants of a pure stock from both parents. Thus there are princes of the 1st, 2d, and 3d class, &c., according to the greater or less degree of inferior blood which they inherit from their maternal ancestors. This state of society, closely resembling the feudal institutions of the Gothic ages, seems to imply the division of the Circassians into two distinct people, a conquering and a conquered race; but when or how the present relations were established is involved in obscurity. (Klaproth, p. 314, et seq.; Pallas, i. 395, 401; Spencer, passim.)

Customs, Habits, and Manners of the Circassians.—The whole of the Circassian and Abchasian tribes live in small villages scattered here and there, without the slightest approach to anything resembling a city or walled town; indeed, the peace or noble has an unconquerable aversion to any castle or place of artificial strength, which he regards as only fitted to restrain his state of wild freedom. He lives, therefore, in the centre of his village, which usually consists of 40 or 50 houses, or rather huts, formed of plaited osiers, plastered within and without, covered with straw or grass, and arranged in a circle, within the area of which the cattle are secured at night. These primitive dwellings, which strongly resemble, in form and appearance, the humbler residences in Arabian towns, have, however, the peculiar recommendation of being unexceptionably clean, which is also the case with the persons, dress, and cookery of the inmates. From the slender nature of the buildings, they are evidently not formed for long continuance, and a Circassian village is, in fact, by no means a fixture. The accumulation of dirt in their neighbourhood, the insecurity of the position, and frequently even the caprice of the inhabitants, cause them to be from time to time abandoned. On such occasions the dwellings are destroyed, the household utensils packed up, and the whole colony migrate in search of a new abode. While stationary, however, there is much comfort in a Circassian's hovel, for those who can dispense with superfluities; but, as may be supposed, their domestic arrangements are of the most simple kind. The usual occupations of the higher classes are the chase and war, on which expeditions, or on those of a predatory kind, they depart with no other provision than a little millet, or wheat, and that without the slightest fear of suffering from want, since every man who possesses and can use a rifle is sure of finding provisions on every hedge. In these expeditions the Circassians carry with them tent covers of felt, but chiefly for the purpose of protecting themselves from sudden storms (see CAUCASUS), as, in fine weather, the hardy mountainer throws himself on the ground, and sleeps with no other covering than the heavens. While in his hut, the Circassian of whatever rank is his own carpenter, weaver, carver, and shepherd. It does not appear, however, that the higher classes often take part in agricultural pursuits, not so much because it is considered derogatory, as from that species of indolence (quite consistent with great occasional exertion) which recoils from regular and continuous labour. The occupations of the women consist in spinning and needle-work. They make the clothes of their household down to the very shoes, and also saddle-cushions, housings, and horse trappings, and sheaths for the warriors' swords and poniards. They frequently excel in embroidery, are skilful dairy-women, and sometimes even noblewomen may be seen taking a part in field labour. As in other half-barbarous societies, the greater portion of labour falls upon the females; but their condition is far superior in Circassia to what it is in most other Eastern countries. As Mohammedanism is little more than a profession among these people, their habits, with the exception of some formal observances with regard to food, have undergone but little change by its introduction. The sexes mix freely together while unmarried, and, under the restriction of caste, love matches are probably as numerous here as in other parts of the world. The husband has, however, to purchase his bride of her father, and neither husband nor wife, from the moment of their union, is permitted to appear in the presence of the parents for a year, or till the birth of the first child. It is a still more remarkable custom, that the husband must never be seen in company with his wife; and though the latter is permitted to receive without restraint the visits of strangers, yet the former is never present on such occasions, and the matrimonial correspondence is always carried on by stealth, and in the utmost secrecy.

The greatest insult that can be offered to a prince, or usden, is to inquire after the health of his wife or family. The son of a prince is committed, at the age of three days, to the care of an usden, by whom he is brought up, and never again seen by his father till he is married; the son of an usden remains in the paternal household till he is three or four years old, when he, in like manner, is consigned to the care of a stranger. The foster father stands in every respect in the place of the natural parent. He receives no payment for his trouble, but claims all the duty and service of his ward. The cause of this very remarkable custom is said to be the wish to prevent the effect of indulgence consequent on a home education, in enervating the character; but though it destroys the usual affection subsisting between father and son, it establishes another not less strong between the guardian and his ward, which is usually as intense as any exhibited in the social connections of other countries. The daughters are brought up at home, and at the age of ten or twelve years have their waists enclosed by tight-fitting stays, or a broad band of untanned leather, *which is never removed nor loosened* till they are married. On the wedding night the bridegroom cuts this bodice open with his dagger, an operation which is frequently attended with danger. As a fine waist is considered the great beauty of a Circassian, men are also subjected to a very heavy compression on that part, but nothing to that which the females endure. The girdle remains on the latter for a period varying from two to six years, (a girl unmarried at seventeen rarely obtains a husband,) during which the victim is growing, and, in addition to this, they are, still farther to improve the form, so sparingly fed, that the young unmarried females have generally a look of ill health. The finest looking women are the young wives.

The dress of both sexes is rather long, that of the men consisting of shirt, tunic, and cloak, much resembling those of the Kalmuck Tartars, but formed of better materials, and in general richer; the female costume is not very different, except in being longer. According to the plate (18, p. 898) in Pallas's first vol., the outer robe reaches to the instep, and is furnished with hanging sleeves. The men shave or crop the head, leaving only a single lock of hair hanging from the crown; they wear thick moustachios, and the *learned* classes (priests and physicians) suffer the whole beard to grow. The women's heads have luxuriant tresses, but both sexes eradicate every appearance of hair on all other parts of their bodies, by means of a caustic ointment of unslaked lime and orpiment. The princes and usden rarely go out unarmed, and in his coat of mail, helmet, musket, pistols, bow, quiver, and shield, the Circassian chief forms a most imposing and picturesque object. In this dress they pay their visits of state, and in this also they ride out on their warlike or predatory expeditions. The Circassian, like the Arab, is a strange mixture of ferocity and hospitality; the unfortunate traveller who approaches his country without securing the protection of some chief, is seized as a slave by the first native who meets him; but, on the other hand, should this protection be extended, the whole power of the host, or *konak*, as he is called, is strained to procure not only the safety, but accommodation, of the guest. The form of granting protection is remarkable. The wife of the konak gives the stranger her breast to suck, after which ceremony he is regarded as her son, and the whole tribe as his adopted brethren. Robbery and plunder are honourable occupations; but the charge of thieving is accounted an insult, because it implies detection. The custom of blood revenge is precisely similar, in all its details, to the same custom in Arabia (see ARABIA), and is known by a name *thlil-wase*, which is said to be similar in etymology to the *thar* of that country. The ransom by fine is, according to Pallas (i. 405), never taken; but Spencer (Circass., ii. 387), on the contrary, affirms, that it is almost always preferred. The exclusive nature of Circassian marriages has been already noticed. It is, however, a little inconsistent, that while a Circassian prince would unhesitatingly slaughter an usden of his own tribe, or Abchasian, who should presume to wed his daughter, he *will unhesitatingly sell her* to Turk, Persian, Turcoman, Nogay Tartar, or Kalmuck! Spencer, who professes to admire every institution of these people, has ingeniously discovered (Circass., ii. 375) that this practice has tended to *refine* and *civilize* the inhabitants of the Caucasus! He admits, indeed, that it has occasioned wars and feuds innumerable among the petty tribes, from the rapacity with which they have overrun each other's territory in search of beauty for the foreign market. The greater portion of the females thus sold have, however, always been Imeritians, Georgians, and Mingrelians; the Circassian slave trade having been chiefly confined to the male sex, from which they supplied the Mamelukes and other slave troops of Egypt and Turkey. The fact is, that the suppression of this infamous traffic by the Russians has been one of the main causes of the resistance made to their government by the Circassian chiefs. (Pallas, i. 396–409; Interiano, la Vita de Zichi, &c., pp. 1–10; Klaproth, pp. 315–320, 322–

825, &c.; Spencer's Circass., ii. 222, 253, 242, 246, 325, 375, 3rd, &c.)

Laws.—These might have been included in the last article, since they rest only on long-established custom. They are administered in a council of elders, but not always by the reigning prince of the tribe, if any other of his rank possess the requisite qualities in a higher degree. The council consists not of princes and nobles only, but also of the wealthiest and more aged vassals, who, in the judgment-seat, are regarded as on an equality with the higher classes. The laws themselves are based upon the principle of retaliation, and the business of the court seems to consist of little else than the assessment of damages. Robbery of a prince is punished by the restitution of nine times the property stolen; of an uden by simple restitution, and a fine of thirty oxen. The prince or uden can scarcely commit a robbery on a vassal, since his abstract right to all the property of the latter is tacitly acknowledged, and the punishment of robbery by one vassal of another appears to vary with the circumstances of the case. Fine, as among the Arabs, seems almost the universal punishment, except in cases of murder and adultery; in both which cases the punishment is left in the hands of the injured party. The offending wife has her head shaved, her ears slit, the sleeves of her garment cut off, and in this trim is sent back, on horseback, to her father, who, if he cannot sell, generally kills her. The paramour is certain of death, being a marked man by all the husband's tribe. Polygamy is allowed, but very rarely practised. The Circassians are very attentive to their breeds of horses, and have distinct marks to show the noble races from which they have descended. The stamping a false mark upon a filly is a forgery, for which nothing but life can atone. (Klaproth, p. 319; Pallas, L 411; Spencer's Circassia, ii. 342, &c.)

Learning is a complete blank. The people, from whom Rumenckoff took it into his head to suppose that the Europeans are mostly all descended, have not even an alphabet, and consequently neither book nor manuscript in their own language. The few who read, and they are very few, use the Tartar or Arabic tongues, both of which the former especially, are very generally understood. The Circassian language is itself totally different from any other at present known, and what is singular, considering the total absence of letters, there is a secret dialect, apparently an old barbarous gibberish, peculiar to the princes and udens, and used by them chiefly on their predatory excursions. (Klaproth, p. 321; Pallas, &c.)

Arts, Manufactures, Commerce.—These also are at the lowest ebb; the doctors are simply conjurors or saints, who profess to cure diseases by charms and the roughest applications of actual cautery. Their success may be surmised from the fact, that notwithstanding the length and inveteracy of the war with the Russians, scarcely a single instance of a maimed Circassian warrior is to be met with; to be wounded among these people is to die. Of artificers and skilled mechanics there are only cutlers, armourers, and goldsmiths, who, however, exhibit great ingenuity in the construction and decoration of the warriors' arms. The art of preparing gunpowder has been known for ages in the Caucasus, and the abundance of saltpetre renders the inhabitants independent of other countries for this important element of warfare; their mode of manufacture is, however, very primitive. It has already been stated that the women are the great manufacturers of clothes, which may be said to be the only manufacture

which these people possess. They formerly traded with Persia and Turkey for their chain and other armour, and with Tartar tribes northward for salt, the equivalents on their parts being their children and cattle. The Russians have annihilated both trades; and this, as already stated, is one great cause of the hatred entertained against them by the Circassians. (Klaproth, p. 324; Pallas, i. 400, &c.; Spencer, ii. 246, &c.)

Name, History.—The word *Tcherkessia* is Tartar, and literally means *cut the road*; that is, highwayman or robber, one who makes communication unsafe. The general name for these people, in the Caucasus, is *Kossack*, whence it has been inferred that they are of the same race with the Cossacks of the Don and the Wolga; but etymology has indeed run mad upon this point; for this term, like the former, has a general, not a national, signification, and means a man who leads a wandering and martial life. The Circassians themselves recognise neither term; they style themselves *Adige*, which has been derived by some authorities from the Turco-Tartar *adah* (island), whence it has been inferred that these people came originally from the Crimea. This may be the case, but it acquires no strength from the etymological proof, since the Circassians have no word for island (how should they, being necessarily ignorant of the thing?) and their language, as before observed, has no connection with either Turkish or Tartar. From a resemblance in sound between the Tartar name (*Tcherkess*), they have been pretty generally supposed to be identical with the Zygen (Zoyoi) of Strabo (ii. 129, xi. 492). (Stephen of Byzantium, art. Zoyoi, and Procopius, De Bel. Got., iv. 4.) This, again, is not improbable, but the premises are far too weak and uncertain to found a conclusion upon. The Kabardians have a tradition that they are Arab (Pallas, i. 392); but in the W. mountains they say that before their ancestors arrived here, the land was inhabited by men so small, that they rode hares instead of horses (Spencer's Circass., ii.); and, as to the time when this settlement took place, they are profoundly ignorant. Among all this confusion, naturally to be expected in speaking of a barbarous and but little known people, all that can be inferred with certainty is, that the Circassians have inhabited their mountains for many centuries, and that they have always been the same hardy, reckless, daring robber warriors, that we find them at this hour. Christianity is supposed to have found its way among them in the very early part of the Christian era; but, in the palmy days of Turkish power, they nominally embraced Mohammedanism, preserving, however, many Christian ceremonies, and acknowledged a kind of doubtful dependence on the Porte. Their first connection with Russia took place in 1555, when the princes of the Bessch Tag submitted to the Czar Iwan Vassilievitch. From that time the Russian power has been constantly increasing in the Caucasus; and, by the treaty of Adrianople (1829), Turkey made over to it the whole Circassian country. By the end of 1864, Russia had become master of the whole of Circassia, and introduced her own form of government.

CIRCENSTER (usually called Cicester), a parish, town, and par, of England, co. Gloucester, hund. Crowthorne, on the Churn, 89 m. W. by N. London by road, and 95 m. by Great Western railway. Pop. 6,336 in 1861. The limits of the parl. bor. are identical with those of the par. The town is on the line of road from Oxford to Bath, and consists of four principal and several smaller streets, paved and lighted; houses mostly of stone, and well built; many of the more respectable are detached, and have shrubberies round them. Portions of its ancient walls (2 m. in circuit) are still traceable, showing that the modern town occupies only a portion of the ancient site, a large part of the enclosed area, on the NE., being occupied by gardens and meadows. The church is in the decorated style of the 14th century, with a lofty tower, and several lateral chapels and ancient monuments of great interest; both within and without, it is elaborately ornamented, and is one of the finest par. churches in England. There are four dissenting chapels; a free grammar-school, founded in the reign of Hen. VII., which had Dr. Jenner for a pupil; blue-coat and yellow-coat schools with small endowments, clothing and educating about forty children; three ancient hospitals, or almshouses, and several other charitable institutions. In the vicinity is an agricultural college of considerable repute. The building, about 1½ m. from the town, is in the Elizabethan style, the principal front being 190 ft. in length, and commanding an extensive view. It includes a private chapel, dining hall, library, museum, and lecture rooms. An experimental farm of about 450 acres is attached to the college, and it has extensive kitchen and botanical gardens. The course of instruction comprises the science and practice of agriculture, chemistry, natural history, veterinary practice, surveying and practical engineering. The business of education is carried on by a principal and professors, under the superintendence of a council of noblemen and gentlemen. Cirencester is a polling place for the E. div. of the co.; and has itself returned two mem. to the H. of C. from the 13th of Eliz., the franchise previously to the Reform Act having been vested in the inhab. householders being parishioners. Registered electors, 459 in 1864.

Cirencester was the Corinium of the Romans; and was a place of considerable importance from its being situated at the intersection of three military roads. Numerous Roman remains have been discovered; and near it is an amphitheatre (now called the Bull-ring), being an ellipse of 63 by 45 yards, enclosed by a mound 20 ft. high, on the inner slope of which were turf seats, which are still partially traceable. A magnificent abbey of Black Canons was founded here by Hen. I., whose abbot was mitred, and had a seat in parl. Its revenue at the general dissolution was 1,051l.; some slight remains of it still exist. Oakley Park, the seat of Earl Bathurst, is in the immediate vicinity. The ancient annalist, Richard of Cirencester, was a native of the town.

CIUDAD DE LAS CASAS (formerly Ciudad Real), an inl. city of Mexico, cap. of the state of Chiapas, in a fertile plain near the border of Guatemala. 150 m. SE. Mexico. Pop. about 4,000, one-eighth of whom are Indians. It has a cathedral, another church, and several chapels. It was the see of the celebrated bishop Las Casas, the protector of the Indians, to whose memory a monument is here erected.

CIUDAD REAL, a city of Spain, prov. Ciudad Real, of which it is the cap., in a plain about 5 m. S. and E. from the Guadiana, 102 m. S. Madrid, 162 m. NE. Seville, on a branch of the railway from Madrid to Seville. Pop. 10,159 in 1857. The city was built after the expulsion of the Moors from La Mancha, to serve as a check upon those who still maintained themselves in the Sierra Morena. Extensive remains of its ancient walls and towers still exist. Streets long and straight, but narrow. The grand square is surrounded by two rows of houses for viewing the bull-fights and public festivals. It has five churches, eight convents, three hospitals, barracks for troops, a magnificent workhouse, including a school for the

instruction of poor children in useful occupations. It was the head-quarters of the famous Santa Hermandad, or Holy Brotherhood, an order founded in 1249, for the extirpation of highway robbers.

CIUDAD RODRIGO, a city of Spain, prov. Salamanca, on an eminence on the right bank of the Aqueda, which is here crossed by a bridge of seven arches; 55 m. SW. Salamanca, 146 m. W. Madrid, and 16 m. from the frontiers of Portugal. Pop. 5,730 in 1857. The city has a castle, and is strongly fortified. It is tolerably well built, and has some good public buildings, including a cathedral, founded in 1770, with numerous churches and convents, an episcopal seminary, and a hospital. In the great square are three Roman columns, with inscriptions. The city has two suburbs, and its environs are fertile. Ciudad Rodrigo was taken by the French under Marshal Massena, in 1810. The Duke of (then Lord) Wellington, having come upon it by surprise, with the allied English and Portuguese forces, on Jan. 8, 1812, after a vigorous siege, took it by assault on the 20th of the same month. A large battering train and immense quantities of ammunition were found in the town. The allies lost about 1,200 men, and 90 officers, in the siege and assault. This important achievement procured for the general the title of Duke of Ciudad Rodrigo from the Spanish gov., and of Marquis Torres Vedras from the Portuguese.

CIUDADELA, a city of the Spanish island of Minorca, of which it was formerly the cap., at the head of a deep and narrow bay on the W. coast of the island; lat. 39° 59' N., long. 3° 51' E. Pop. 5,726 in 1857. The city has walls, partly of Moorish construction, and partly modern, with stone bastions. In the centre of the town is a large fine Gothic church. The streets are narrow, but it has a considerable number of good houses, inhabited by many of the nobles of the adjacent country.

CIVITA VECCHIA (an. Centum Cellæ), a fortified sea-port town of Central Italy, cap. deleg. of same name, on the Mediterranean, 36 m. WNW. Rome, of which it is the port, and with which it is connected by railway. Pop. 24,985 in 1858. Though the streets are narrow, the town is tolerably well-built and laid out; it contains several convents, a lazaretto, a theatre, an arsenal, building-docks, and warehouses, and has a very considerable import and export trade. Its harbour, which was constructed by the Emperor Trajan, is formed of three large moles—two projecting from the mainland, and inclined the one a little to the N., and the other to the S.; and a third constructed opposite to the gap between the others, and serving to protect the shipping from the heavy sea that would otherwise be thrown in during W. gales. The latter mole clearly appears from a passage in Pliny's letters (lib. 6, epist. 31) to have been formed in a precisely similar manner to the breakwater at Plymouth, by sinking immense blocks of stone, which became fixed and consolidated by their own weight, till the structure was raised above the waves. Its extremities are about 90 fathoms distant from those of the lateral moles, and at its S. end there is a lighthouse, with a lantern elevated 74 ft. above the level of the sea. The S. entrance to the harbour is the deepest, having from 8 to 4 fathoms water. Ships may anchor within the port, in from 16 to 18 ft. water, or between it and the outer mole, where the depth is greater. Civita Vecchia is a free port,—that is, a port into which produce may be imported, and either made use of or re-exported free of duty; but quarantine regulations are very strictly enforced. Its imports consist chiefly of cotton, woollen, silk, and linen stuffs; coffee, sugar, cocoa, and other colonial products; salt and salted fish, wines, jewellery, glass, and earthenware. The exports are principally staves and timber, corn, wool, cheese, potash, pumice-stone, alum, and other Italian produce. The shipping, in 1862, consisted of 1,9nt vessels, of 300,659 tons, which entered, and 1,995 vessels, of 302,751 tons, which cleared the port. Among them were 65 British vessels of 10,512 tons entering, and 65 vessels of 12,701 clearing. For more important than the British commerce with Civita Vecchia is the French, which includes almost one half of the whole shipping.

This city was originally called *Trajanus Portus*, and it is to be regretted that it did not continue to bear the name of its illustrious founder.

CLACKMANNAN, the smallest co. of Scotland, on the N. side of the Forth, being, except for a short distance on the E., where it adjoins Fife, every where surrounded by the cos. of Perth and Stirling. Area 29,711 acres, or 46 sq. m. The co. is traversed by the Devon, an affluent of the Forth. The range of the Ochill hills crosses and mostly occupies the part of the co. to the N. of the Devon; but the other and far largest portion consists, for the most part, of clay and carse land, and is remarkably fertile and well cultivated, producing excellent crops of wheat and beans. Estates middle-sized; farms large; farm buildings excellent. There are valuable mines of coal, large quantities of which are shipped at Alloa; ironstone is also abundant. There are some large distilleries and breweries, but little other manufacture is carried on. Alloa is the largest, but Clackmannan is the co. town. Clackmannan is divided into five parishes, and had 2,936 inhab. houses, with a pop. of 21,450 in 1861. It is united with Kinross in returning 1 mem. to the H. of C. Registered electors 680 in 1865. The old valued rent was 2,807.; the new valuation for 1864-5 amounted to 74,000l.

CLACKMANNAN, a town of Scotland, cap. of the above co., on an eminence 190 ft. high, on the left bank of the Frith of Forth. Pop. 1,150 in 1861. The town consists principally of one long unpaved street, and is a very unimportant place. On the W. of the town is Clackmannan Tower, the palace of King Robert Bruce, long the residence of a branch of the Bruce family, and now the property of the Earl of Zetland. The par. church is a modern Gothic building. Debtors and criminals are sent to Stirling, the jail of which is partly maintained by the co. of Clackmannan.

CLAGENFURTH (Germ. *Klagenfurt*), a town of Illyria, gov. Laybach, cap. duchy of Carinthia, on the Glan, an affluent of the Drave, in an extensive plain, 21 m. E. Villach, and 40 m. NNW. Laybach, on a branch of the railway from Vienna to Trieste. Pop. 15,478 in 1857. The town was formerly fortified, but its works were destroyed by the French in 1809. It has four suburbs, is well built, with broad and regular streets. There are five squares, one of which has a leaden statue of the Empress Maria Theresa, and a group (indifferently executed) representing Hercules destroying the hydra. Another square contains the residence of the Prince-Bishop of Gurk, with its galleries of paintings, statuary, a rich cabinet of minerals, and an obelisk erected in honour of Francis I. There are seven churches, two hospitals, several infirmaries, a lying-in hospital, workhouse, house of correction, lyceum with a public library, college, normal high school, Ursuline school for girls, an agricultural society, and a theatre. This town is the seat of the court of appeal for the gov. of Laybach, and of municipal, provincial, and other courts of justice. It has a few manufactures of fine woollen and silk fabrics, and white lead. Clagenfurth is supposed by some to derive its name from the Emperor Claudius, and in its vicinity

there are some ruins believed to be those of the ancient *Tiburnia*. It has several times been partially destroyed by fire.

CLAMECY, a town of France, dép. Nièvre, in which it holds the second rank, at the foot and on the declivity of a hill on the left bank of the Yonne, where it is joined by the Beuvron, by both of which it is intersected; 38 m. NE. Nevers. Pop. 5,672 in 1861. Little remains of its ancient castle, and the massive walls by which it was formerly surrounded. It, however, contains several old Gothic churches, and a handsome modern castle surrounded by fine gardens, which stands in the *Place de l'ouvert*. Clamecy has manufactures of common woollen cloths, fulling mills, dyeing houses, tanneries, and a considerable trade in wood and charcoal, most of which are sent down the Yonne to Paris. There are good coal mines in the neighbourhood.

CLARE, a marit. co. of Ireland, prov. Munster. It is in a great measure insulated, having Galway Bay on the N.; the Atlantic on the W.; the Shannon, by which it is separated from Kerry, Limerick, and Tipperary, on the S. and SE.; and Galway on the NE. Area, 832,352 acres, of which 258,581 are unimproved mountain and bog, and 18,665 water. Surface in parts almost mountainous; but it has a large extent of low level land. The low grounds, known by the name of the Corcasses, on the banks of the Shannon and Fergus, are almost equal to the very best grazing lands in Lincolnshire. The arable lands are mostly light, but fertile. Estates large; tillage farms very small, many lying below 5, and very few above 60 or 70 acres. Agriculture bad, but improving; it is still common in many parts to take a succession of corn crops till the land be completely exhausted. Principal crops, oats and potatoes; but wheat and barley are now rather extensively cultivated. Sea-weed and sea-sand are a good deal used as manure; and in the hilly parts the bog, or spade, is much employed in cultivation. Cottagers mostly of stone, but without lime or other cement. Condition of the occupiers of small tillage farms and cottiers quite as bad as in most other parts of Ireland. Lime is the most important mineral. Manufactures have hardly any footing. Exclusive of the Shannon, the Fergus is the principal river. Clare has 9 baronies and 79 parishes, and sends 3 mem. to the H. of C., viz. 2 for the co. and 1 for the town of Ennis, the principal town in the co. Registered electors for the co. 5,300 in 1865. The pop. was 286,323 in 1811; 212,784 in 1851; and 166,305 in 1861. Consequently, the decrease of population, in the year 1841-61, amounted to 42 per cent. In 1841 Clare had 44,870 inhab. houses, and 286,394 individuals, of whom 144,109 were males, and 142,285 females.

CLAUSTHAL, or KLAUSTHAL, a town of the k. of Hanover, cap. of the mining captaincy (*Berghauptmannschaft*) of the same name, and the principal mining town of the Hartz; in a bare and bleak region on the top and slopes of a hill 1,740 ft. above the sea, 26 m. NE. Göttingen, and 56 m. SW. by S. Hanover. Pop. 8,918 in 1861. The inhab. are mostly miners or persons connected with the mines and smelting-houses. The town has a desolate appearance; its houses are chiefly of wood, and even its principal church is of the same material. It contains a mining-school, supported by the king, and possessing an extensive collection of models of mines, mining buildings, machinery, and a cabinet of the Hartz minerals. The chief lead and silver mines in the Hartz are in the neighbourhood, next to which are the Silberzeya. The shaft of one of these mines reaches to 2,000 ft. below the level of the Baltic. The mines are drained by a tunnel, cut through the mountain to the small town of Girard, a distance of 6 m. The total length of this tunnel, however, with its branches, is nearly double this distance; it was commenced in 1777, and finished in 1799. Nearly all the machinery used in the mines being set in motion by water-power, every little stream around Clausthal is carefully made use of to form a reservoir; and the canals conducting the water thence to the different mills, machines, &c., are said to have an aggregate length of 125 m. There are numerous forges; besides which, camlets, and a few other articles are manufactured.

CLERMONT-DE-LODEVE (see LODÈVE). There are many other small towns in France named Clermont; but none of any importance.

CLERMONT-FERRAND (an. *Augustonemetum*), a city of France, dép. Puy-de-Dôme, of which it is the cap., on an affluent of the Allier; 82 m. W. Lyons, and 208 m. S. by E. Paris, on the railway from Bourges to Le Puy. Pop. 32,275 in 1861. The city is finely situated on an eminence, surrounded on the S. and W., by an amphitheatre of mountains, of which the Puy-de-Dôme is the culminating point, and overlooking on the N. and E. the picturesque and rich plain of the Limagne. The city itself is about 1½ m. in circuit, being separated by a boulevard partially planted with trees, from several considerable suburbs. Though it has some fine structures, it is in general badly laid out; streets crooked, narrow, and dirty; houses lofty, mostly old, and gloomy looking from being built of the lava found in the neighbourhood, with which also the streets are paved. The more modern buildings, however, which are rapidly increasing in Clermont and its suburbs, have a more cheerful and agreeable aspect. It has several squares ornamented with handsome fountains, and is exceedingly well supplied with good water, conveyed to it by subterranean conduits from Royat, a league distant. The principal edifice is the cathedral, a work of the 13th century, and the third, according to Hugo, which has been constructed in this city. Externally it has nothing to recommend it, being unfinished, and crowded amongst a number of mean buildings; but its interior is considered one of the finest existing specimens of Gothic architecture. It is built of Volvic lava, a material well in keeping with its style, and has a choir, and chapels of great beauty, a number of handsome columns supporting a lofty nave and aisles, and much elegant carving and stained glass. Of the five towers it possessed before the revolution, only one remains. Of the other churches, that of Notre Dame du Port, built in 863, is the most ancient, and is elaborately ornamented externally with mosaic work, bas-reliefs, &c. The corn and linen halls, the ancient college, town-hall, cavalry barracks, *Hôtel Dieu*, and another hospital, the prefecture, a public library with 16,000 vols., founded by Massillon, and the theatre, are the other principal public buildings. It has also a botanic garden, museums of natural history and antiquities, and a cabinet of mineralogy, particularly rich in specimens of the volcanic products of the neighbourhood. It is the seat of a bishopric which has to boast of Massillon for one of its incumbents, and of tribunals of original jurisdiction, and commerce; and has a royal college of the third class with about 300 pupils, a primary school, and an academy of sciences and belles lettres. Trade considerable, it being the entrepôt for the produce of the surrounding dép., consisting of hemp, flax, corn, wines, cheese, leather, and linen fabrics, and for a part of the merchandise of Provence and Languedoc intended for Paris, besides being on the great line of communication between

Bordeaux and Lyons. Four large fairs are held annually. Manufactures not very important; the chief are those of silk stockings, druggets, tinsed paper, coarse woollens, linen, cutlery, porcelain, cotton yarn, twine, sweetmeats, preserved fruits, and chemical products. There is also a saltpetre refinery. In and round Clermont there are numerous warm chalybeate springs, holding in solution carbonates of lime, and which, on cooling, deposit very extensive sediments. The most remarkable of these is in the suburb of St. Allyre, where a streamlet having rolled its bed to a considerable height by means of successive deposits, and subsequently formed a cascade over another streamlet which it had previously run, has effected the formation of a natural bridge over the latter, 25 ft. in length by 16 ft. high. The little town of Mont-ferrand, formerly containing the stronghold of the Counts of Auvergne, is now one of the suburbs of Clermont, with which it is connected by a fine avenue of willow and walnut trees.

Anterior to the Roman conquest, this city was named Nemossus, and was the cap. of the Arverni; Augustus embellished it, and gave it his name. In the 3rd century it was erected into a bishopric. It was several times demolished in the succeeding ages, and especially by Pepin-le-Bref. The counts of Clermont and Auvergne afterwards possessed it. It was here that the celebrated council, which bears its name, was held in 1095, when the first crusade was resolved on. Philip Augustus united this city to his dominions in 1212. Clermont has been the birth-place of many illustrious men, amongst whom may be specified Gregory of Tours; Pascal, born here on the 19th June, 1623; Thomas, Chamfort, Delille the poet, and General Desaix, in honour of whom an obelisk has been erected in one of the squares.

CLEVELAND, a town of the U. States, Ohio, on the S. shore of Lake Erie, at the mouth of the Cuyahoga river. Pop. 36,125 in 1860. The town —which had only a pop. of 6,071 in 1840—has grown-up very rapidly, owing to its advantageous commercial position, and at the point where the Grand Canal, connecting the Ohio river (and consequently the Mississippi) with Lake Erie unites with the latter. The opening of this canal has made Cleveland, which was previously quite unknown, a place of much importance, heightened by the subsequent construction of several lines of railway, among them the Cleveland, Columbus and Cincinnati, 141 m. long, and the Cleveland and Toledo, 87 m. long, which place the town in direct communication with the whole railway system of the United States.

CLEVES, an ancient town of the Prussian states, Rhine prov., formerly the cap. duchy of Cleves and now of a circ., on the railway from Cologne to Utrecht and Amsterdam. Pop. 8,055 in 1861. The town stands on the declivity of some hills, nearly at the NW. extremity of the prov., about 2½ m. from the Rhine, with which it is united by a canal. It is neatly built in the Dutch style, and surrounded by walls, but is not a place of any strength. It has a gymnasium or college, a handsome town-house, with iron foundries, and manufacture of flannel and cotton.

CLITHEROE, a town and parl. bor. of England, co. Lancaster, hund. Blackburn, on the Ribble, 188 m. NW. by N. London, and 29 m. SE. by E. Lancaster, on the Lancashire and Yorkshire railway. Pop. of town 6,990, and of parl. bor. 10,864 in 1861. The town stands at the foot of Pendle-hill, which rises 1,800 ft. above the level of the sea. The houses are of stone; the streets paved, well kept, and plentifully supplied with water from springs. The parish church, rebuilt in 1828, is a plain building; there is another church, and chapels belonging to the Methodists, Independents, and R. Catholics. In the churchyard is the free grammar-school endowed by Queen Mary in 1554. Clitheroe is a fee, by prescription, and has returned 2 mem. to the H. of Com. since 1 Elizabeth. Under an order of the H. of Com. in 1694, the right of election was vested in the burgesses and freemen, who held in right of freehold in houses or land within the bor.; out-burgesses, holding free burgage tenures in the bor., had also the right of voting. Previously to the Reform Act, the number of burgage tenures was 192, of which not more than a half were occupied by burgesses, and in fact it was a mere nomination bor. The Reform Act deprived it of one of its members; and the electoral limits were at the same time extended so as to comprise various adjoining chapelries and townships. Registered electors 436 in 1865. Under the new municipal corporation act, the bor. consists of one ward, and is governed by 4 aldermen and 12 councillors. Several branches of the cotton manufacture are extensively carried on; they consist principally of the weaving of calicoes by hand and power looms, cotton spinning, and calico printing. In the neighbourhood are extensive beds of limestone, of which large quantities are burnt for manure and building. A mineral spa near the town is much resorted to. In 1669, the town and neighbourhood suffered severely from an extraordinary outbreak of water from the higher part of Pendle-hill. Markets are held on Tuesday; cattle shows on alternate Tuesdays; fairs on 24th and 25th March; 1st and 2nd Aug.; Thursday and Friday before the fourth Saturday after 29th Sept.; and 7th and 8th Dec.

CLONAKILTY, a marit. town of Ireland, co. Cork, prov. Munster, at the bottom of the bay of the same name; 19 m. SW. Cork. Pop. 3,667 in 1851, and 3,108 in 1861. The town is formed of four streets, that meet in the centre, and of a square. It has a par. church, a R. Catholic chapel, a Methodist meeting-house, an endowed grammar-school, a dispensary, a public library, three reading-rooms, a court-house, bridewell, linen-hall, and market-house. The corporation, under the charter of James I, in 1613, consists of a sovereign, twenty-four burgesses, and a commonalty. It returned two mems. to the Irish H. of C. till the Union, when it was disfranchised. A manor court, held every third Wednesday, has cognisance of pleas to the amount of 2l. Markets, Fridays; fairs on 5th April, 1st June, 1st Aug., 10th Oct. and 12th Nov. A party of the constabulary is stationed here. The trade is much limited by the badness of the harbour, which is nearly impracticable for vessels of any size, in consequence of its shallow and shifting bar. Sea-sand is raised here in large quantities, and carried to the adjoining country for manure. Corn is exported to Cork, and coal received in return, chiefly viâ Kinsale, from which there is a railway to Cork.

CLONES, an inl. town of Ireland, co. Monaghan, prov. Ulster; 66 m. NW. by N. Dublin. Pop. 2,591 in 1851, and 2,390 in 1861. The town consists of a triangular market-place, in which is an ancient stone cross, and a few streets with mean thatched houses. It has a par. church, a R. Cath. chapel, two Presbyterians and two Methodist meeting-houses, and two dispensaries; and is a constabulary station. A manorial court is held monthly, and petty sessions on alternate Fridays. Fairs are held on the last Thursday of every month. The Ulster canal passes near the town.

CLONMEL, an inl. to. and parl. bor. of Ireland, prov. Munster, partly in Tipperary and partly

in Waterford on the Suir, 90 m. SW. by W. Dublin, and 26 m. NW. by W. Waterford, on the railway from Waterford to Limerick. Pop. 13,012 in 1821; 13,505, in 1841; and 11,771 in 1861. The town chiefly lies on the N. side of the river in Tipperary; the communication with the other portion in Waterford being maintained by three stone bridges. The streets, which consist of a main thoroughfare upwards of a mile in length, intersected by several smaller, are well paved and lighted with gas. The co. club house is at the E. end of the town, and near it are extensive barracks for cavalry, infantry, and artillery. It has a parish church, a modern building, with some good monuments, two Rom. Catholic par. chapels, a Franciscan and a Presentation chapel, and meeting houses for Presbyterian Calvinists, Unitarians, Baptists, Primitive and Wesleyan Methodists, and Quakers. An endowed school has been rebuilt at an expense of 3,000l.; besides which there are parochial schools for boys and girls, and others are maintained by voluntary contributions. The co. infirmary and dispensary, the fever hospital and the house of industry for the reception of well-conducted paupers and the confinement of vagrants, are in the town, as are two orphan establishments, a mendicity association, and a savings' bank. Here also is the district lunatic asylum for the co., built to accommodate 170 patients.

The bor. was incorporated at a very early period, but its ruling charter was granted by James I, in 1608. The governing body consists of a mayor, two bailiffs, twenty other burgesses, and an unlimited number of freemen: the right of freedom is enjoyed by the eldest son, by apprenticeship or by marriage with a freeman's daughter. Previously to the Union, the bor. sent two mem. to the Irish H. of C., and it now sends one mem. to the Imperial H. of C. The elective franchise is vested in the burgesses and freemen resident within 7 m., and in the 10l. householders. No. of registered electors 566 in 1862. The electoral boundary comprises 331 acres, but the municipal jurisdiction for other purposes extends over 4,040 Irish acres, of which 1,000 are on the Tipperary side, and the remainder on the Waterford side of the river.

The woollen manufacture was introduced into the town in 1667, when a number of German manufacturers were induced to remove thither; it declined at the Revolution, and has never revived. The cotton manufacture has been introduced, and there are extensive flour mills in the town and its vicinity. The town is well situated for inland trade, being on the main lines of road from Dublin to Cork, and a chief station on the railway from Waterford to Limerick, and having the advantage of river navigation for barges of 50 tons burden to Waterford, a distance of 23 m. There is an extensive salmon fishery on the Suir, and the influence of the tide is perceptible beyond Clonmel. The principal trade is in grain, provisions, cattle, and butter, with all which it supplies the Liverpool, London, and Bristol markets. A considerable portion of the produce goes to Waterford, and numerous carriers conduct the inland trade with all the surrounding country. The butter market is a spacious building, with suitable offices for inspecting and marking the article before it is exposed for sale. Market-days, Tuesdays and Saturdays; fairs are held on 5th May and 5th Nov., and on the first Wednesday of every other month; they are chiefly for cattle.

CLOSTER-SEVEN, a small village of Hanover, duchy of Bremen, on the Aue, 26 m. NE. Bremen. It deserves notice only from its being the place where the famous convention, which bears its

name, was agreed to on Sept. 10, 1757, by which
an army of 38,000 Hanoverians and Hessians,
commanded by William, Duke of Cumberland,
was dispersed and sent into cantonments. This
convention was alike unpopular in England and
in France: in the first it was looked upon as the
result of imbecility and misconduct; and in the
latter it was believed, and probably on good
grounds, that had Marshal Richelieu not assented
to the convention, the Duke of Cumberland must
have surrendered at discretion. (The convention
is given in Smollett's Hist. of Eng. iii. 413.)

CLOYNE, an inl. town, or rather city, of Ireland, co. Cork, prov. Munster, in a fertile valley,
3 m. E. Cork harb., with which it is connected by
railway. Pop. 1,217 in 1831, and 1,434 in 1861.
The town, which is small, irregularly built, and
far from prosperous, has a large old cruciform
cathedral, in which are some good monuments;
among others one to Dr. Woodward, bishop of
Cloyne, who died in 1794, and was one of the
earliest advocates for the introduction of poor laws
into Ireland. A little distance from the cathedral
is one of those extraordinary round towers, the
origin and object of which have given rise to
much learned conjecture. It is 102 ft. in height.
The old episcopal palace at the E. end of the
town is now a private residence; the bishopric of
Cloyne having, on the death of Dr. Brinkley, the
last bishop, in 1835, been merged in that of Cork.
The famous Dr. Berkeley, one of the subtlest of me-
taphysicians, and most amiable of men, was bishop
of Cloyne from 1732 to 1753. The R. Cath. cathedral is a plain building, without any pretensions to architectural beauty. Crowe's charity-
school, founded in 1719, gives instruction to 35
pupils in reading, writing, and arithmetic. There
is here a constabulary station. Market-day
Thursday. A court leet is held annually; a
minor court every week; and petty sessions on
alternate Wednesdays. At Carrigacrump, near
the town, is a quarry of close-marble, of which
from 2,000 to 6,000 tons are raised annually.

CLYDE, a river of Scotland, and the only important one on the W. coast of that part of the
U. Kingdom. It has its source near the S. extremity of Lanarkshire, on the borders of Dumfries-shire and Peebles-shire, in the highest part
of the S. mountain-land of Scotland, contiguous to
the sources of the Tweed and Annan. Its course
is at first N., with a little inclination to the E.,
till near Biggar it turns NW.; it then makes a
sweep round by the SE., till, being joined at Har-
peround by the Douglas water, it re-assumes its
NW. course, and, passing by Lanark, Hamilton,
and Glasgow, unites with the Frith of Clyde, a
little below Dumbarton. The distance in a direct
line, from its source to Dumbarton, is only about
62 m., but including its windings, the course of
the river is near 75 m. Soon after its junction
with the Douglas, it is precipitated over a series of
falls celebrated for their picturesque beauty; of
these the principal are the falls of Bonington,
Corehouse, Dundaff, and Stonebyres. The distance from the highest to the lowest fall is
about 6 m.; during the whole of which the river
dashes along with great impetuosity. Corehouse
Fall is about 70 ft. in height. The Clyde has
been rendered navigable at high water as far as
Glasgow for vessels of 350 and 400 tons. (See
GLASGOW.)

COAST CASTLE (CAPE), or CABO CORSO,
the cap. of the British settlements on the Gold
Coast of Africa, empire of Ashantee; lat. 5° 6' N.,
long. 1° 61' W. The first colonial establishment
formed here was by the Portuguese in 1610, but
the Dutch dislodged them after a short period.

Finally the British obtained possession of the settlements, in whose hands it has remained since 1661.

The castle is built upon a rock about 50 ft. high,
projecting into the sea, its walls being washed by
the surf that rolls impetuously along the coast.
It is of a quadrangular shape, with bastions at
each angle; has barracks, with accommodations
for 16 officers and 200 men; but is of little
strength, the walls being out of repair, and commanded in every direction by the adjacent heights
(but on some of these forts have been erected).
The water for the garrison is obtained from tanks,
in which the rain from the buildings is collected.
(Captain Tulloch's Report on W. Africa.)

The town is situated behind the castle, and presents a dirty and irregular appearance. The
native houses have a few small rooms scantily
furnished with mats and stools; the fires are made
in a corner, with no other escape for smoke than a
hole in the roof. There are, however, some superior residences belonging to Europeans, and the
merchants have built themselves a neat club-
house. The scenery of the neighbourhood has
been described by a late distinguished female
poet, Mrs. Maclean, better known as L. E. L.,
whose melancholy death at this place, in 1837,
has given an interest to it which it did not previously possess. 'The land view, with its cocoa
and palm trees, is very striking—it is like a scene
in the Arabian Nights. The native huts I first
took for ricks of hay, but those of the better sort
are pretty white houses with green blinds. The
English gentlemen resident here have very large
houses, quite mansions, with galleries running
round. Generally speaking, the vegetation is so
thick that the growth of the shrubs rather resembles a wall. The hills are covered to the top
with what we should call calf-weed, but here it is
called bush.'

The climate of this settlement is characterised
by excessive humidity. The heat is, however, not
so great as might be supposed. In the hottest
weather, owing to the tempering influence of the
sea breeze, the thermometer seldom rises above
82° Fahr., and rarely, in the coldest, falls below
76°. It has generally been described as exceedingly unhealthy, and the official statements show
that such is the fact. During the four years
ending with 1826, two-thirds of the white troops
in garrison died annually; and in 1871 the mortality was in the enormous ratio of 9072 in 1,000.
It is true that these were singularly unhealthy
seasons, and that the vice and intemperance prevalent among the troops added considerably to
their sickness and mortality. But still, to use
Captain Tulloch's words, 'there is unquestionable
evidence that in every year, and to all classes of
Europeans, the climate proves extremely fatal.'

The imports consist of cottons, hardware, and
gunpowder, from Great Britain; sugar, rum, and
tobacco from the colonies; and of foreign produce,
beads, silks, and tobacco. The exports are gold dust,
ivory, palm-oil, pepper, cam or dye-wood, tortoiseshell and malsize. But the value of the trade is inconsiderable. The total exports from the Gold Coast to
the United Kingdom amounted to 42,763l. in 1859;
to 74,460l. in 1861; and to 89,209l. in 1863. The
imports were of the value of 63,905l. in 1859;
144,194l. in 1861; and 80,849l. in 1863.

COBLENTZ (the Confluentes of the Romans),
a town and fortress of the Prussian states, prov.
Rhine, cap. reg. and circ. on the railway from
Cologne to Mayence. Pop. 26,525 in 1861, exclu.
of 3,810 military persons. The town stands in a
beautiful situation on the point of land at the confluence of the Rhine and Moselle. It has a freestone bridge across the latter, and one of boats

across the Rhine. The streets are mostly regular, and many of the public buildings are handsome; but, being a fortress, Coblentz has derived but little advantage from its fine situation for commerce. The principal public building is the magnificent castle, erected in 1779 for the elector of Treves. It was converted into barracks by the French; but has since been repaired, and is now used for the holding of the civil and criminal courts. Coblentz has a court of appeal for the regency, a theatre, a gymnasium or college for Catholics, and some other literary establishments. Commerce pretty extensive. Prince Metternich, the late prime minister of Austria, was a native of Coblentz.

Coblentz has been rendered one of the strongest places in the Prussian monarchy, and is deemed one of the principal bulwarks of Germany on the side of France. The fortifications by which it is surrounded are constructed partly on the system of Vauban, and partly on that of Montalembert. They enclose a large extent of ground, and are capable of accommodating 100,000 men. Ehrenbreitstein, 'the Gibraltar of the Rhine,' on the right bank of the river, the fortifications of which had been blown up by the French, has been rendered stronger than ever, and is one of the principal outworks of Coblentz.

COBURG, or more properly SAXE-COBURG-GOTHA, a duchy of Central Germany, and the most N. of the Indep. Saxon principalities, consisting of several small detached portions of territory; between lat. 50° 7′ 30″ and 51° 17′ N., and long. 10° 15′ and 12° 40′ E., surrounded mostly by the territories of Bavaria, Prussia, Saxony, Meiningen, Hildburghausen, and Weimar. The area and pop. of its two great divisions are, according to the census of Dec. 1861:—

	Area in sq. m.	Pop. 1861
Saxe-Coburg	230	47,015
Gotha	500	112,017
Total	818	159,432

Coburg Proper is on the S. side of the Thüringer Wald (Thuringian Forest), and is included within the basin of the Rhine, having a general slope to the N. Gotha and Altenburg are situated wholly on the N. side of the Thüringer Wald, and belong to the basins of the Elbe and Weser. The most mountainous parts of the country are the N. of Coburg and the N. of Gotha; through these the Thüringian forest-range passes, the highest summits of which,—the Beerberg, 3,205 ft., and the Schneekopf (snow-cap), 3,443 ft. in elevation, —are in the latter principality. Both divisions are, however, interspersed with fine valleys and fertile plains: Gotha is watered by the Unstrut, Gera, Hörsel, and Saale; and Coburg by the Itz, a tributary of the Mayn, and other rivers. Climate healthy and mild, especially N. of the mountains. The principal occupations of the people are tillage and cattle breeding; but the mountains, which are covered with pine forests, contain little cultivable land, and the forest economy there forms the chief branch of industry. In the valley of the Itz, the vine is cultivated, and hops, flax, and hemp, are also grown in the N.; the other agricultural products are corn, pulse, culinary vegetables, fruits, aniseed, coriander, cummin, safflower, and other medicinal plants: potatoes are a principal article of nourishment. Many hogs are fattened in the woods and sent down the Mayn to Frankfurt and elsewhere: considerable quantities of timber, pitch, tar, charcoal, and potash are obtained from the forests. Iron, coal, excellent millstones, marble, alabaster, gypsum, potter's clay, and salt are mined or quarried. Agriculture flourishes most in Coburg, manufacturing industry in Gotha. The principal manufactures are those of linen cloth, tick, linen, thread, woollen and cotton fabrics, leather, steel, iron, and copper wares, glass, earthenware, buttons, and paper. There are also numerous sawing-mills, linen-bleaching factories, breweries, and distilleries; and great numbers of toys are made at Neustadt in Coburg. A good deal of advantage accrues from the transit trade, the duchy being on the road between Leipzig and Frankfurt. Gotha is the principal trading town, and has several considerable mercantile establishments. The government is a constitutional monarchy; each of the principalities has its own elective assembly, and the two unite into one chamber, composed of 80 members. Every man above the age of 25, who pays taxes, has a vote, and any citizen above 30 may be elected a deputy. New elections take place every four years, for which period also the budget is voted. The annual public revenue for the period July 1, 1861, to June 30, 1863, amounted to 83,925l. and the annual expenditure to 57,851l. The greater part of the surplus thus produced went into the private purse of the reigning duke.

Education is well attended to in the duchy. There are 3 gymnasiums and classical schools, 1 academical gymnasium, 2 seminaries for schoolmasters, 35 town schools, and about 850 village schools in the duchy. The ducal house, and nearly all the pop., profess the Lutheran religion, there being only about 1,000 Roman Catholics and 1,000 Jews. Difference of religion, however, does not affect the equal enjoyment of political rights. The Duke of Saxe-Coburg-Gotha holds, together with the Duke of Saxe-Altenburg and the Duke of Saxe-Meiningen, and the Grand Duke of Saxe-Weimar, the twelfth place in the German diet; and the duchy is bound to furnish a contingent of 1,866 men for the service of the confederation. Coburg belonged successively to the counts of Henneberg, the house of Saxony, and that of Saalfeld. In 1816, its territories were enlarged by the cession of the principality of Lichtenberg, on the left bank of the Rhine; but the reigning duke disposed of that possession to Prussia. In consequence of the extinction of the line of Gotha in 1826, the Duke of Saxe-Coburg became possessed of the territories of Gotha and Altenburg, for which, by a family compact, Saalfeld was exchanged. The house of Saxe-Coburg is famous as one of the most fortunate of all the existing great families of Europe in respect to marriages. The late king Leopold married, first, the heiress to the British throne, next a daughter of the King of the French, and was then seated on the throne of Belgium, after having refused that of Greece. Prince Ferdinand married one of the richest heiresses of the Austrian empire, and his son became king-consort of Portugal. One princess married the Grand Duke Constantine, heir presumptive to all the Russias; another became the Duchess of Kent; and, finally, the late Prince Albert, in wedding the sovereign of the British realms, became progenitor of a new race of kings—'father of our kings to be.' (Tennyson).

COTTBUS, a town of Central Germany, cap. of the above duchy, on the left bank of the Itz, 106 m. E. by N. Frankfurt-on-the-Mayn, and 130 m. SW. Dresden, on a branch line of the railway from Frankfurt to Dresden. Pop. 11,110 in 1861. The streets of the town are mostly narrow and uneven; but it is surrounded by some agreeable public walks, which separate it from its suburbs and has several handsome public buildings. The Ehrenberg palace, built in 1549, contains a collection of pictures, a library of 26,000 vols., and some apartments adorned with figures in alto-relievo, the finest of which is a state banqueting-room, called the *Saāle de Gram*, from some colossal caryatides which surround it. On an eminence commanding the town stands an ancient castle of the dukes of Coburg, now in part converted into a prison and house of correction, but containing also a collection of armour, and some rooms once occupied by Luther, with the bedstead on which he slept. This castle was unsuccessfully besieged during the 30 years' war by Wallenstein, who had for some time his head-quarters here. Coburg contains five churches, a government house, a gymnasium, with an observatory, and two libraries, a superior ladies' school, a teachers' seminary, a large workhouse, and other charitable institutions, and a riding-school. The principal places of amusement are the theatre, casino, redoute, and musical club. The town is the seat of gov., and of the high board of taxation for the duchy, and of the superior judicial courts and church consistory for the principal of Coburg. It has manufactures of woollen, linen, and cotton fabrics, porcelain, earthenware, and gold and silver articles, with bleaching and dye-works. The fine seat of the duke, Rosenau, is in the immediate neighbourhood.

COCENTAYNA, a town of Spain, prov. Alicante, 30 m. N. Alicante. Pop. 7,469 in 1857. The town has 2 churches, 2 convents, a hospital, and a house of charity for poor travellers. Neither the streets nor the houses correspond with the number and wealth of the inhabitants, who are more intent upon increasing their substance by agricultural and manufacturing industry, than on beautifying the town. They manufacture cloths, taffeties, handkerchiefs, and other articles. Their fields, which are well irrigated, produce wheat, maize, pulse, wine, oil, and silk.

COCHIN, a small rajaship of Hindostan, near its S. extremity, extending along the Malabar coast, chiefly between lat. 9° 30′ and 10° 30′ N., and long. 76° and 77° E.; having N. and E. the territory of the Madras presidency, S. Travancore, and W. the ocean: average length and breadth about 45 m. each; area, 1,900 sq. m. Its E. boundary is formed by the W. Ghauts, which are here covered with forests of teak and rūf (a black wood), of large dimensions, which obliges both to be cut into short logs, in order to reach the coast; with pʳᵐᵘ, jack, and iron woods, &c. Towards Caʳᵃᵘᵘ the hills are covered with grass instead of trees; but though their soil appears good, they are but little cultivated; in the N. there are narrow and well-watered valleys, in which rice is raised, and sometimes two crops a year are reaped. The houses of the cultivators are often embosomed in groves of palms, mangoes, jacks, and plantains. A considerable portion of the rajah's revenue is derived from the teak forests, the timber of Cochin being in great demand in Bengal, and, since 1814, having been sent to the dockyards of Bombay, from which, previously to that period, it was excluded. There are many villages inhabited by Christians and Jews; the latter are settled mostly in the interior, but have a synagogue at Cochin town. This country was for a long period badly governed, and its inhab. much oppressed. The rajah for a time was tributary to Tippoo Sahib, and subsequently became subject to the British.

COCHIN (*Cochhi*, a morass), a marit. town of Hindostan, prov. Malabar, on a small island near the S. extremity of India; formerly cap. of the above rajahship, but since 1796 it has belonged to the British. Next to Bombay, it is the most eligible port on the Malabar coast; it is 150 m. NW. Cape Comorin, 80 m. SSE. Calicut; lat. 9° 51′ N., long. 76° 17′ E.; and is built on the N. extremity of the island, along the entrance from the sea to the 'Blackwater,' an inl. harbour or lagoon, which extends nearly 120 m., being separated from the sea by a narrow peninsulated tract. Under the Portuguese and Dutch, by whom it was successively possessed, Cochin was a flourishing town; but since it has belonged to the English, who in 1806 demolished the fortifications and many of the buildings, it has progressively declined, and the inhab. are now very much impoverished; it still, however, trades with the rest of the Malabar coast. China, the E. Archipelago, and the Arabian and Persian Gulfs. Large supplies of teak floated by the rivers from the forests into the Blackwater, are shipped for the ports of the two last-named countries; the other exports are sandal wood, pepper, cardamoms, cocoa-nuts, coir, cordage, cassia, and fish-maws. It is the only place on the coast S. of Bombay where ships of any size can be built. Under the walls of the old fort there is always from 25 to 30 ft. water, and ships obtain supplies of fresh water without difficulty. Provisions are extremely cheap, and as a port, as well as a place of trade, it is said to be much superior to Calicut. Jews of both the black and white castes are numerous, and have a synagogue in Cochin, almost the only one in India. Cochin is also the see of a Roman Catholic bishop, whose diocese includes Ceylon, and comprises more than 100 churches. Here in 1503 Albuquerque erected the first fortress possessed by the Portuguese in India.

COCHIN-CHINA, a prov. of the empire of Anam, which see.

COCKERMOUTH, a market-town and parl. bor. of England, co. Cumberland, at the confluence of the Cocker and Derwent; 24 m. SW. Carlisle, 12 m. NE. Whitehaven, 306 m. NW. London by road, and 319 m. by London and North Western railway. Pop. of town 6,908, and of parl. bor. 7,057 in 1861. Cockermouth has but few houses of a better sort, and little seems to have been done towards its improvement. The streets are narrow in many places, with a want of foot-pavement everywhere; and though the lower classes seem to be better off than in many other towns in the same co., yet there appears to be little about the place tending to improvement. There are bridges over both rivers, that over the Derwent being 270 ft. long. Though unpaved, the streets are clean, and well supplied with water. A castle on a hill over the town, built shortly after the Conquest, was taken and razed by the parl. forces in the war of 1641. The church of All Saints, erected in the time of Edward III., was rebuilt in 1711, and enlarged in 1825. St. Mary's church, rebuilt in 1850, has a memorial window to the poet Wordsworth, who was a native of the town. The Independents, Methodists, and Society of Friends have places of worship. There are also a free grammar school and some almshouses. The borough returned two mem. to the H. of C. in 23 Edward I., after which the privilege was not exercised till 16 Charles I., since which it has been uninterruptedly enjoyed. Previously to the Reform Act, the franchise was exclusively vested in the holders of burgage tenures in the town of Cockermouth. The boundaries of the parl. bor. were then extended. Registered electors 413 in 1865. The bor. is also a polling-place at elections for mem. for the W. div. of the co. There are collieries at Greysouthern and Broughton, about 8 m. distant.

CODOGNO, a town of Northern Italy, prov. Milan, cap. distr., in a fertile territory, between the Po and Adda, 15 m. SE. Loui. Pop. 9,620 in 1841. The town has broad streets and good private buildings, some handsome churches, several colleges and schools, with a hospital and theatre. It is a place of considerable trade, especially in Parmesan cheese, and has some silk manufactures. Near this town the Austrian troops were defeated, in 1746, by the Spaniards, and in 1796 by the French.

COGGESHALL, a town and par. of England, co. Essex, hund. Lexden, the town being on a hill on the NE. bank of the Blackwater, 10 m. W. Colchester. Pop. 3,116 in 1841. The town is ill-built; and the clothing trade, particularly the manufacture of baize, formerly carried on, has almost wholly disappeared; but some branches of the silk manufacture have been introduced; and a few of the inhab. are engaged in the making of toys. The church, a spacious structure, in the perpendicular style, has a large square tower. The river is here crossed by an ancient bridge of three arches. It has an endowed school, three unendowed almshouses; and an annuity of 150l. a year, payable by Pembroke Hall, Cambridge, goes to the support and education of the poor. The Cistercian monks had an abbey here, a portion of the ruins of which still remains.

COGNAC, a town and river port of France, dép. Charente, cap. arrond., on the navigable river Charente, 22 m. W. by N. Angoulême. Pop. 8,167 in 1841. The town is ill-built, and contains no edifice worthy of notice, except an ancient castle, now converted into warehouses. The brandy, for the shipment of which the town is celebrated, and which is everywhere known by its name, is made from white wine: that made from red wine is very inferior. In good years wine yields about 1-5th part of its volume of eau-de-vie, whereas in bad years it does not yield more than from 1-8th to 1-11th part. All the brandy of Charente is sold under the name of Cognac; but the best qualities are produced in the canton of that name, and in those of Hliansac, Jarnac, Rouillac, Aigre, and Haller. The park belonging to the castle is an agreeable public promenade, and in it is a bronze statue of Francis I., erected on the spot where he was born, in 1494. Three councils have been held in Cognac.

COIMBATORE, a British prov. of S. Hindostan, presid. Madras, between lat. 10° 8' and 12° 10' N., and long. 76° 50' and 78° 10' E., having N. the Mysore dom., E. the provs. Salem and Carnatic, S. the latter, and W. Cochin and Malabar; area, 8,282 sq. m. Pop. estimated at near 1,000,000. Generally it is a flat open country, with a medium height of 900 ft. above the sea; its surface gradually ascending from the Cavery on the E. to the Ghauts and Neilgherry hills on its W. borders. The W. Ghauts rise from 1,500 to 2,000 ft. above the Coimbatore plain, and have in one place a remarkable opening, about 31 m. in length, called the Palighautcherry Pass, presenting a clear level way from the Malabar to the Coromandel coast. Next to the Cavery the principal rivers are the Bowany, Noyel, and Amberawatty, all which run more or less E., and join the Cavery before it leaves the distr. Climate on the whole healthy and pleasant; and except in that part facing the Palighautcherry Pass, this prov. is protected by the Ghauts from the violence of the SW. monsoon. There are some marshes in the N. and in the vicinity of the hills; but the soil in general is dry, and well adapted for the dry grain culture, to which nearly ten times as much land is appropriated as is occupied by wet, and twenty times as much as is occupied by wet cultivation. In the N. rice is the chief crop; cotton of several kinds is grown in considerable quantities both above and below the Ghauts, and almost all the tobacco that supplies Malabar comes from this distr. There are altogether about 578,700 acres of pasture land; cattle and sheep numerous. Chief mineral products, salt and nitre, which are occasionally obtained from certain earths impregnated with muriates and nitrates abundantly scattered throughout the distr. In 1818, an aquamarine mine was opened and worked. Weaving is the only art that has attained any perfection. Some of the towns are large and well built; but, excepting in these, most cottages with red tiled or thatched roofs are almost the only houses. The peasantry, however, are contented, and enjoy comparative comfort. Pagodas or temples are not numerous; and excepting that of Peowra, a little W. of the cap., which contains some well-carved granite figures, they have little notoriety. The areas in front of most of them are ornamented with gigantic groups in pottery covered with chunam of cauris coured horses, elephants, and grotesque figures. Near the Ghauts the ox is adored, and every village possesses one or two bulls, to which weekly or monthly worship is paid. The prov. became subject to the Mysore rajahs nearly 200 years ago, and to the British in 1799. It was greatly depopulated by an epidemic fever, which prevailed from 1809 to 1811.

COIMBATORE, an inl. town of S. Hindostan, cap. of the above distr. and seat of a collector of revenue under the Madras presid., in an elevated situation on the N. bank of one of the affluents of the Cavery, 90 m. SSE. Mysore, and 270 m. SW. Madras; lat. 10° 52' N., long. 77° 6' E. It is tolerably well built, and has a mosque erected by Tippoo, who sometimes resided here. The water is brackish, and 2 m. off both salt and nitre are procured by lixiviating the soil. Five m. to the N. iron is smelted from black sand. Peowra, not far distant, has a temple dedicated to Siva, highly ornamented with Hindoo figures, but destitute of elegance, which was spared by Tippoo when he demolished most other idolatrous buildings. In 1783 and 1790 Coimbatore was taken by the British, to whom it has permanently belonged since 1799.

COIMBRA, a city of Portugal, prov. Beyra, cap. distr., and see of a bishop, partly on a steep rocky precipice, and partly on a plain contiguous to the Mondego, 115 m. NNE. Lisbon, on the railway from Lisbon to Oporto. Pop. 15,710 in 1838. The town was fortified at a very early period, and has undergone many sieges. The ancient walls and towers still remain, and form its only defence. It has an imposing appearance when seen at a distance, the summits of the adjoining heights being crowned with convents and public buildings; but the interior of the town by no means corresponds with the exterior view, the streets being narrow, steep, crooked, and dirty. The principal public building is the university, the only one in Portugal, transferred thither from Lisbon in 1308. It consists of eighteen colleges, and is divided into six faculties, viz. those of theology, the canon law, civil law, medicine, natural philosophy, and mathematics. It has also attached to it grammar schools with schools of philosophy and rhetoric, ecclesiastical and civil colleges or seminaries, and a royal college of arts, at which those who intend entering at the university complete their preliminary studies. Different degrees are taken in the respective faculties, the student applying himself principally to the particular branch most connected with his intended profession, which, as Lord Carnarvon ass-

nisms, is probably an improvement upon the English system of college education, where the same degree is taken by all, without reference to the nature of their future occupations. (Caernarvon's Portugal and Galicia, I. 42.) The collection of subjects of natural history is tolerably good, the observatory complete, and the instruments in perfect order, the greater part having been made in London and Paris. The present system of education was introduced by the Marquis Pombal, in 1773; it is, however, indebted, for various improvements in the course of study, to Englishmen, who have been instructors; but, with all this, it is still very far behind; and many important branches of knowledge are either not taught at all, or are taught in the worst possible manner. The university is extremely well endowed; and the inferior class of nobles are sometimes competitors for the vacant chairs. The annual expenses of the students do not exceed 50l. each, any excess being defrayed from the revenues of the institution. The library consists of three large saloons, containing about 30,000 vols, but they are nearly all of ancient date. The College of Arts, which formerly belonged to the Jesuits, is a remarkably handsome building. The monastery of Santa Cruz, an immense Gothic building in the worst taste, belongs to the order of Augustines, who, in addition to numerous important privileges, enjoy the right of appointing their prior to the office of chancellor of the university. The monks are, for the most part, of noble descent and polished manners, and are often seen mounted on fine horses splendidly caparisoned, being forbidden by the regulations of the monastery to appear on foot beyond its walls. (Lord Carnarvon, I. 43.) On a hill opposite to the town is the superb convent and church of the nuns of St. Clara. Besides these public buildings, there are the cathedral and eight churches, five of which are collegiate, with several other convents, hospitals, &c. There is a fine stone bridge over the Mondego, whose bed, which is progressively rising, is nearly dry in the summer, while in the winter it becomes an impetuous torrent, and overflows the surrounding country. The town is well supplied with water, conveyed to it by an aqueduct. Near Coimbra, on the S. bank of the river, is the Quinta das Lagrimas, or Villa of Tears, the residence of the beautiful Ines de Castro, whose murder forms the subject of the fine episode in the third book of the 'Lusiad.' Earthenware of good quality is produced here, with woollen and linen cloths.

Coimbra is said to occupy the site of Conimbrica, founded by the Romans 300 years B.C. It suffered severely by the earthquake of 1755, and was a scene of great distress in 1810, when the Duke of Wellington retreated on the lines of Torres Vedras.

COLABBA, an island on the Malabar or W. coast of Hindostan, immediately S. the Island of Bombay, with which it is connected by a causeway, and on which a fine lighthouse and cantonments for the British troops have been erected. (See Bombay.)

COLAPOOR, a small rajahship of Hindostan, in the presidency of Bombay, partly above and partly below the W. Ghauts, including the towns of Colapore, Parwellah, Mulenpore, and Cukerog. The rajah of Colapore is descended from the eldest branch of the family of Sevajee, the founder of the Mahratta empire. He formerly possessed Malwan, and some other parts on the Malabar coast; but his subjects being notorious for piracy, the British compelled him to cede these places in 1812; and in 1829 assumed the government of the country.

COLARSON, an inl. town of Hindostan, cap. of

the preceding dist., in a valley surrounded on three sides by hills: 125 m. SSE. Punaah; lat. 16° 19' N., long. 74° 25' E. It has a citadel; but its chief protection is in two hill forts in the vicinity. The town is neatly built, and contains some lofty trees, gardens, and good tanks.

COLBERG, a fortified sea-port town of Prussia, reg. Coslin in Pomerania, on the Persante, near where it falls into the Baltic, and on the terminus of the railway from Berlin to the Baltic Sea. Pop. 11,760 in 1861, exclus. of a garrison of 1,678. The principal public buildings are the cathedral, townhouse, and the aqueduct for supplying the town with water. There is in the ancient ducal castle a foundation for the daughters of nobles and landgraves. It has a gymnasium, a house of correction, and some manufactures; but its salmon and lamprey fisheries, and its shipping, are the principal sources of wealth. There are salt springs in the vicinity; but, owing to the want of coal and timber, they are of comparatively little use.

COLCHESTER, a parl. bor. and river port of England, co. Essex, dir. Colchester, hund. Lexden; 50 m. NE. London by road, and 51½ m. by Great Eastern railway. Pop. 23,815 in 1861. The town stands on the declivity of a hill rising from the Colne, which cuts off a small suburb. It is well built, has several good streets, is paved, lighted with gas, and adequately supplied with water. Great improvements in its interior have been effected, and are still going on. There are three bridges over the river. A part of the remains of the ancient castle, said to have been founded by Edward the Elder, is occasionally used as a prison. There are eight parish churches: St. Peter's, built previously to the Conquest, has been modernised and enlarged; St. James's dates previously to Edward II., and is a handsome structure; St. Leonard's is also large and convenient; besides these, there are a French and a Dutch Protestant church, and nine dissenting chapels. The remains of the church of St. Botolph's priory, founded in the early part of the 12th century, are said to afford some of the finest specimens of Norman architecture in the kingdom.

Colchester has a free grammar school, founded in the 26th of Elizabeth, with one scholarship in St. John's college, Cambridge, annexed to it; two others, in the same college, revert to this school on failure of applicants of the surname of Gilbert (that of founder) or Torbington; and four founded in Pembroke college, Cambridge, on failure of any boys being sent from the Ipswich grammar school. It educates from thirty to forty scholars; two charity schools, founded in 1708, have been joined to the national school, in which about 400 boys are educated, of whom 148 are clothed by the charity; a Lancastrian school, and an endowed school founded in 1816, for children of Quakers, with a library attached to it. The principal charitable institutions are, a hospital, founded by James I.; several almshouses; and the Essex and Colchester Hospital, built in 1820. A commodious theatre was erected in 1812; and there are literary and philosophical, medical, botanical, and musical societies, all in a flourishing state. Market-days, Wednesdays and Saturdays: the latter a large corn market; but general provisions are on sale daily in the large and commodious market-place. There are large annual cattle fairs on the 5th and 6th of July, 23rd and 21th of the same month, and 20th Oct, and three following days.

Colchester is a bonding port, but the foreign imports are comparatively insignificant: they consist chiefly of wine, oil-cake from Holland, and timber from the Baltic. The trade coastwise is more extensive, the imports being chiefly colonial produce, and home manufactures, from London; with coals, &c. from the northern counties; the exports, corn and malt. The river is navigable for vessels of 150 tons to 'The Hythe,' a little below the town, where there is a custom-house and commodious quay, warehouses, and bonding, coal and timber yards; larger vessels (chiefly colliers) discharge at Wivenhoe, still lower down, into lighters. On Jan. 1, 1864, there belonged to the port, or rather river, exactly 300 vessels; but of these no fewer than 202 were under 50 tons burden; and their aggregate tonnage and that of the 98 vessels of above 50 tons, amounted to only 16,168 tons. The oyster fishery of the river has been long celebrated, and was granted to the burgesses by Richard I.; it employs a considerable number of the inhabs, and a large proportion of the small craft belonging to the town. There is a large distillery at Hythe. A silk manufactory in the town employs between 300 and 400 hands, chiefly females. The weaving of baize (introduced by the Flemings in the reign of Elizabeth) used formerly to be carried on to some extent, but has wholly ceased. At present, the prosperity of the town mainly depends on its retail trade, by which an extensive agricultural district is supplied. During the last war a large military establishment was stationed here, the withdrawal of which caused some deterioration to the borough. Under the Municipal Act its boundaries are contracted to an area of about 2,000 acres immediately round the town; and it is divided into two wards, and governed by a mayor, six aldermen, and eighteen councillors. Borough revenue 3,670l. in 1862, of which about one-fourth from mines.

Colchester has (with some interruption) returned two mem. to the H. of C. from the 23rd of Edward I. Previously to the Reform Act the right of election was vested in the free burgesses not receiving alms. The parl. bor. (co-extensive with the ancient liberties) extends over a space of 11,770 acres, divided by the Colne into two nearly equal parts. Number of registered electors, 1,314 in 1862, of which 413 are freemen. Ann. val. of real prop. assessed to income-tax 66,820l. in 1857, and 65,727l. in 1862.

Colchester has claims to high antiquity, and is supposed by some to have been the Camulodunum of the Roman period, though this has been disputed. There is, however, no place in the kingdom where more numerous Roman remains have been discovered. It had many monastic institutions previously to the Reformation; of these, St. John's Abbey, of which the noble gateway is the sole relic, was the chief.

Colchester was made the seat of a suffragan bishop in the 29th Henry VIII. There were two consecrations only, the first in 1536, the other in 1592; on the death of the last diocesan, in 1607, no successor was nominated. In 1648 the town was held by insurrectionary royalists, and endured a siege, by Fairfax, of eleven weeks, when it was starved into surrender, and the leaders hung; half the fine subsequently levied appears to have been paid by Dutch refugees, who had escaped from the Duke of Alva's persecution. It gives the title of baron to the Abbot family.

COLDSTREAM, one of the border towns of Scotland, co. Berwick, on the Tweed, 11 m. SW. Berwick-upon-Tweed. Pop. 1,834 in 1861. Formerly the communication between England and Scotland was here effected by a ford, by which Edw. I. entered the latter with a powerful army in 1296; and it continued to be the chief passage for the Scottish and English armies till the union of the crowns in 1603. It was by this ford, also, that the Coronators entered England in 1640. A bridge

of five arches spans the river, which formed one of the greatest thoroughfares between the two kingdoms previously to the construction of railways. At present, the iron roads have completely thrown the old highway into the shade, and Coldstream bridge lies silent and deserted. The town is irregularly built, and quite Scotch in appearance. It has a weekly corn-market, and a monthly sheep and cattle market, both of considerable importance. There is a par. church and two Presbyterian dissenting chapels, three subscription libraries, and four friendly societies. The means of education are good. General Monk resided at Coldstream in 1659–60, previously to his going to England and effecting the Restoration. During his stay here, he raised a horse regiment, to which he gave the name of the 'Coldstream Guards,' which name the regiment still retains.

COLERAINE, a marit. town and parl. bor. of Ireland, prov. Ulster, co. Londonderry, on the Lower Bann, 4 m. from its mouth, and 47 m. NNW. Belfast, on the railway from Belfast to Portrush. Pop. 4,851 in 1821; 8,143 in 1841; and 5,631 in 1861. The town was built and fortified by the Irish Society of London, to whom the district was granted by James I. in 1613. The town consists of a square, called the Diamond, a main street, and several others, in which are many well-built houses. A wooden bridge, constructed in 1716, and renovated in 1743, connects it with the suburb of Killowen or Waterside, on the W. bank of the Bann. The par. church is a large plain building. The Rom. Cath. chapel, an elegant structure, is in Killowen. The other places of worship are, two for Presbyterians, and one each for Methodists, Independents, and Seceders. The manufactures in the town and immediate neighbourhood are trifling; a few paper-mills and some small tanneries. It has an endowed school, built by the Irish Society; a town-hall, with a dispensary, loan fund, and a mendicity association. The corporation, consisting of a mayor, 12 aldermen, 24 burgesses, and an unlimited number of freemen, is become extinct; and its property is now vested in commissioners. Its jurisdiction extended over the town and liberties, the limits of which were fixed by the charter at 8 m. in every direction from the centre of the town. The town returned two members to the Irish H. of C. until the Union, since which it has sent one member to the Imperial H. of C. Registered electors, 274 in 1865.

There are numerous bleach-greens in the neighbourhood. The salmon and eel fisheries on the Bann, in the vicinity of the town, are valuable. The principal trade is in the export of corn and meal, provisions, including pork, and linens of a fine kind, called 'Coleraines.' The entries at the port, in the year 1863, comprised six British vessels, of 1,850 tons, and three foreign vessels, of 709 tons. The customs duties revived amounted to 7,561l. in 1839; to 7,911l. in 1861; and to 6,168l. in 1865. Formerly, the trade of the town was much impeded by the bar at the mouth of the river, which had but 9 ft. water over it at springs, and 5 at neaps; but this defect has been in a great degree obviated by the formation of a harbour at Portrush, 4 m. NE. from the mouth of the Bann, in which vessels drawing 17 ft. water may anchor, being sheltered by a projecting rock from the swell of the ocean. The outlay on this harbour amounted to about 13,000l.; and it affords great facilities to the trade of Coleraine, there being also a railway from the town to Portrush.

COLESHILL, a town and par. of England, co. Warwick, Birmingham div., hund. Hemlingham; 116 m. NW. London by London and North Western railway. Pop. of par. 2,053 th 1861. The town derives its name from its being situated on a hill, near the Cole. It has a handsome Gothic church with a lofty spire, several good houses, and a school supported out of lands purchased by the inhab. after the dissolution of the monasteries.

COLLUMPTON, or CULLOMPTON, a town and par. of England, co. Devon, hund. Hayridge, 12 m. NE. Exeter, and 1814 m. W. London by Great Western railway. Pop. of town 2,935, and of par. 8,185 in 1861. The town is situated in an extensive vale beside the Culm, a tributary of the Exe, and consists of one large street, along the road from Exeter to Bath, and of several smaller streets diverging from it on either side; many of the houses are ancient, and some of them favourable specimens of their day. The church, originally collegiate, is a spacious structure, in the later pointed style, with a lofty and highly ornamental tower, and a beautiful chapel attached. There are seven dissenting chapels; a national school, in which above 200 boys and girls are educated; with other schools, and several extensive charities. Market, Saturdays; fairs, first Wednesdays in May and Nov., for cattle and cloth. There is a woollen mill; and the manufacture of narrow woollen cloths and serges employs a considerable portion of the pop. though the business has much declined. There is also, in the immediate vicinity, a paper-mill, two large flour-mills, and four tanyards. A monthly session for the district is held in the town.

COLMAR (an. Columbaria, or Columaria, a city of France, dép. Haute Rhin, of which it is the cap., in a fertile plain, on the banks of two tributaries of the Ill; 36 m. NNE. Strasbourg, and 254 m. ESE. Paris, on the railway from Strasbourg to Mulhouse and Basel. Pop. 22,629 in 1861. The city was fortified previously to 1673, when Louis XIV, having taken it from Germany, destroyed its defences, and united it to the dominions of the French crown. The city is now surrounded only by boulevards, planted with trees, and serving for public walks. It is tolerably well built, but contains few public edifices deserving of notice. The principal are the cathedral, built in 1363, the theatre, and prison. The other public buildings and establishments are the hall of justice, city hall, prefecture, college, with a public library containing 64,000 vols. and several paintings by Albert Durer and others; the deaf and dumb asylum, civil and military hospitals, church of the Dominican convent, now a corn-hall, Protestant church, and museum, containing, amongst other curiosities, a remarkable stedite, which descended near Ensisheim in 1492, and originally weighed 260 pounds.

Colmar is environed by pleasant walks, gardens, and country houses; and possesses an orangery and departmental nursery grounds. It is the seat of a royal court, and of tribunals of primary jurisdiction and commerce. It has numerous manufactures of cotton stuffs and printed goods, a large cotton and silk ribbon factory, besides makers of cutlery, paper, brushes, combs, and leather; and an extensive trade in iron, spices, drugs, and wine, which, with its manufactured goods, it exports largely to Switzerland. The Columbaria of the Romans is believed to have replaced the more ancient Argentovaria. This town was several times destroyed by the barbarians, and in after times suffered greatly during the wars between the houses of Hapsburg and Nassau. The Swedes took it in 1632.

COLMENAR DE OREJA, a town of Spain, prov. Madrid, 13 m. ESE. Aranjuez. Pop. 4,833 in 1857. The town contains a fine church, two

convents, and two hospitals; and is finely situated in a plain productive of wine, oil, and fruit. It has manufactures of woollens, pottery, and Spanish rush; and mill-stones, and fine white stone for building, are found in the vicinity.

COLNE, a market town and chapelry of England, co. Lancaster, hund. Blackburn, par. Whalley, on the Colne, an affluent of the Calder; 26 m. N. Manchester, 15 m. NE. Blackburn, and 2284 m. N. London, by London and North Western and Midland railway. Pop. 7,905 in 1861. This is a place of great antiquity; but antiquaries are undecided whether it be the Colunio of the Romans, or the Colne of the Saxons. Many Roman coins have been found here; and Castor Cliff, about 1 m. distant, retains evident traces of a military station, having a regular quadrangular rampart, surrounded by a fosse. The town is situated on an eminence, on a tongue of land formed by the river and the Leeds and Liverpool canal, which passes through a tunnel about 1 m. from the place, and is surrounded by the fine grazing district of Craven. It is a brisk second-rate town, and has of late years been greatly improved. It is well supplied with water by pipes from Flass spring, 2 m. E. The parochial chapel of St. Bartholomew, supposed to be coeval with the reign of Henry I. but repaired in that of Henry VIII., and more recently in 1815, is said to be a "spacious and decent building." The Methodists, Baptists, Independents, and Inghamites, have places of worship. A gallery in the first named of these gave way in 1777, from the pressure of the crowd assembled to hear John Wesley, the founder of the society, preach on its opening; but though many were injured by the accident, no lives were lost. A free grammar-school, rebuilt in 1812 by subscription, on the site of one more ancient, educates six boys; Archbishop Tillotson was a pupil in it. The co. magistrates hold sessions here, and a constable for the gov. of the place is chosen annually by the rate-payers. The lord of the manor holds a court baron, and courts leet or halimote are held in May and Oct. This is one of the most ancient seats of the woollen manufactures; a fulling-mill existed in 1311, and about the same period a coal-mine was worked in the vicinity. In addition to the woollen fabrics, shalloons, calamancoes, and tammies, were made in considerable quantities; and a piece-hall, on the principle of those at Bradford and Halifax, was erected in 1775. It is a substantial stone building, containing two rooms, each 162 ft. by 42 ft. The upper room has been used for the sale of woollens during the fairs, and, owing to the decline of the worsted trade, the whole building is now thrown open for the sale of general merchandise on the same occasions. The cotton trade having been introduced towards the close of last century, has nearly superseded the woollen trade, and the pop. is now principally employed in manufacturing cotton goods for the Manchester market. The spinning power is chiefly water supplied in abundance from the streams, steam-engines being used to obviate their occasional failure. The first power-loom was introduced into the district in 1832. The canal already noticed affords a ready mode of conveyance for the coal, slate, lime, and stone raised here. Markets on Wednesday; fairs, March 7, May 13 and 15, Oct. 11, Dec. 21; also a fair on the last Wednesday of the month for cattle and cloth.

COLOGNE, or COLN (Germ. Köln), an ancient and celebrated city of Prussia, formerly the cap. of the electorate of the same name, and now of the Rhine prov., and of a reg. and circ. of the same, on the left bank of the Rhine, and at the junction of the great lines of railway from Berlin to Paris, and from Amsterdam to Frankfort-on-the-Mayn. Pop. 120,568 in 1861, exclusive of a garrison of 7,485. The city, one of the most flourishing in the Prussian dominions, is connected by a fine bridge built of stone, as well as by a bridge of boats, with the town of Deutz, on the opposite side of the river. It is built in the form of a crescent, close to the water; and is strongly fortified. The walls have a number of towers à la Montalembert, and form a circuit of nearly 7 m.; but a part of the included space is laid out in promenades and gardens. Though finely situated on the banks of a noble river, on a slightly elevated ground, Cologne has many wood houses, and is ill-built, having been laid out in the Middle Ages, when the object of architects was more directed towards defence against external enemies than interior comfort and beauty. The city has a great many interesting buildings, chief among them the cathedral or minster of St. Peter, a vast and imposing but incomplete Gothic edifice, begun about the year 1248. It is about 400 ft. in length, and the choir rises to the height of 160 ft. To complete the vast structure and add to it a suitable tower, has been the object of all Germany for the last forty or fifty years, and large sums have been collected for the purpose. The church of St. Mary is remarkable for its antiquity, and that of St. Peter for the famous altar-piece painted by Rubens. Several of the other churches are also interesting, particularly that of St. Gereon. The town-house is a fine old building. The hall for the courts of justice was erected in 1824. In the arsenal are preserved many curious specimens of ancient armour. Cologne is the seat of an archbishopric, of the provincial authorities, and of the courts of appeal for the province. Its university, established in 1388, was suppressed during the occupation of the country by the French. The city has two gymnasiums or colleges—one for Catholics, to which is attached a very valuable library, and one for Protestants; there is besides an archiepiscopal seminary for the education of clergymen, a normal school, a commercial school, a public library, with numerous literary institutions, and a theatre. Manufactures important; they consist principally of cotton yarn and stuffs, woollen stockings, hosiery, silks, velvets, tobacco, soap, hats, lace, thread, and clocks. There are tan-works and several distilleries, the most esteemed product of the latter being the well known eau de Cologne. The city has a very good port on the Rhine, and is the principal entrepôt of the extensive and increasing commerce between the Netherlands and the countries included within the German customs' union. Rubens was born in Cologne in 1577, and several of its churches are ornamented with his chef-d'œuvres.

Cologne was anciently called Oppidum Ubiorum, from its being the chief town of the Ubii, a German tribe. A Roman colony was planted in it by Agrippina, the daughter of Germanicus, who was born in it; hence it obtained the name of Agrippina Colonia, and latterly of Colonia and Cologne. (Tacit. Annal. lib. xii. § 27; Cellarii Notit. Orbis Antiqui, i. p. 387.) In the middle ages, Cologne was much more populous and wealthy than at present. It was for a lengthened period one of the most important cities belonging to the Hanseatic League. It suffered much at different periods from the intolerance of its magistrates, by whom all Protestants were expelled from the city in 1618.

COLOMBIA, a vast territory of S. America, formerly one country, but, since the year 1831, divided into the states of Ecuador, New Granada, and Venezuela. The territory occupies the N. part of South America, between lat. 12° 25' N. and 5° S., and long. 60° and 83° W.; having N. the Caribbean Sea, E. British Guiana and Brazil, S. Brazil and Peru, and W. the Pacific Ocean and the republic of Central America; length E. to W. 1,320 m.; breadth N. to S. 1,080 m.; area 1,155,000 sq. m.

Colombia is naturally divided into 3 distinct zones, or tracts of country. The first comprises the country between the Pacific Ocean and the Caribbean Sea and the Andes; the second, the mountainous region; the third, the immense savannahs which stretch S. and E. from the Andes to the neighbourhood of the river Amazon, and the mountains which border on the Orinoco. Colombia has as much as 2,000 m. of coast on the Caribbean Sea and the Atlantic, and 1,300 m. on the Pacific. The former is a great deal more indented with bays and inlets than the latter; the principal are the Gulfs of Paria, Maracaybo, and Darien, on the Caribbean Sea; with Panama, Choco, and the Gulf of Guayaquil, on the Pacific. Several islands belonging to Colombia surround its coast; as those of Margarita, Tortuga, &c. (Venezuela); I. Roy, Quito, &c. (N. Granada); and Puna (Ecuador). (Hall's Colombia, &c., pp. 26-28; Med. Trav., xxvii. 7, &c.)

Mountains.—The great Cordillera of the Andes enters the prov. of Loxa from the S., between lat. 4° and 5° N.; in 3° 23' S., where it is nearly 13,000 ft. in height, it divides into two parallel ridges, in the elevated valley between which, 9,000 ft. above the level of the sea, Quito and other towns are situated. E. of this valley rise the summits of Cayambe, 16,380, Tunguragua, 16,720, Cotopaxi, 17,200, and Gaya mbe, 18,190 ft.; and on its W. side, those of Chimborazo, 20,100, Ileniza, 16,372, and Pichincha, 14,280 ft. high; all covered with perpetual snows, from amidst which torrents of flame and lava have frequently burst, and devastated the surrounding country. These two ranges afterwards unite, but near 1° N. again separate, enclosing the lofty valley of Pasto, bounded by the still active volcanoes of Arubal and Cumbal, and the extinct one of Chiles. Beyond Pasto, the Cordilleras consist of three ranges, the most W., the elevation of which is generally less than 5,000 ft., follows the coast of the Pacific, and terminates in the isthmus of Panama; the central range is interrupted between the valleys of the Cauca and Magdalena rivers, and terminates near Mompox, between lat. 9° and 10° N.; and the third, bring the most E. and highest range, extends to the extremity of the Parian promontory, in long. 62° E. This last-named range divides the waters which flow into the Orinoco on its E., from the Magdalena, Zulia, Tocuyo, &c., and their affluents, on its W. side. Many of its summits reach above the limit of perpetual snow; and it has numerous lower summits, called paramos, which rise to 10,000 or 12,000 ft. above the level of the sea, and are constantly enveloped in damp and thick fogs. The city of Bogota, 8,100 ft. above the sea, is built on a table-land formed by this mountain range; as are the towns of Niragua, San Felipe el Fuerte, Barquisimeto, and Tocuyo; but these are at a much lower elevation than Bogota, the mountains decreasing in height very considerably N. of Merida. The mean elevation of the Andes in Colombia is about 11,100 ft.; their altitude is greatest near the equator. In Venezuela, between the parallels of 8° and 7° N. lat., there is another mountain system, unconnected with the Andes, from which it is separated by the Orinoco and the plains of Caraccas, Varinas, and those in the E. parts of New Granada. This system has been called the Cordillera, or Sierra of Parima. It is less a chain than a collection of granitic mountains, separated by small plains, and not uniformly disposed in lines; its mean height is not above 3,500 ft., although some summits rise to upwards of 8,000 ft. above the level of the sea. (Humboldt's Pers. Narr. and Researches; Hall's Colombia, pp. 2–8; Mod. Trav., vol xxvii.)

Plains.—Colombia includes the most northerly of the three great basins of the S. American continent, the Llanos of Varinas and Caraccas; which, like the Pampas of Buenos Ayres, consists of savannahs or steppes devoid of large trees. These, in the rainy season, appear from the high lands as a boundless extent of verdure, but in time of drought they are a complete desert. Humboldt remarks, that ' there is something awful, but sad and gloomy, in the uniform aspect of these steppes.' ' I know not,' he says, ' whether the first sight of the Llanos excites less astonishment than that of the Andes. The plains of the W. and N. of Europe present but a feeble image of these. All around us the plains seemed to ascend towards the sky; and that vast and profound solitude appeared like an ocean covered with sea-weeds.' The chief characteristic of these steppes, like those of N. Asia, is the absolute want of hills and inequalities. An uninterrupted flat of 180 leagues extends from the mouths of the Orinoco to Araure and Oxyloco; and from San Carlos to the savannahs of the Caqueta for 200 leagues. This resemblance to the surface of the sea strikes the imagination most powerfully where the plains are altogether destitute of palm-trees, and where the mountains of the shore and of the Orinoco are so distant that they cannot be seen. Occasionally, however, fractured strata of sand-stone, or compact limestone, stand 4 or 5 ft. higher than the plain, and extend for three or four leagues along it; and convex eminences, of a very trifling height, separate the streams which flow to the coast from those that join the Orinoco. The phenomena of the mirage, and the apparitions of large lakes, with an undulating surface, may frequently be observed. These savannahs are watered by the numerous streams which form the Meta, the Apure, and finally the Orinoco; and the periodical overflowings of which convert the whole country, during four months of the year, into an inland sea. The equally well-watered plains of Ecuador are intersected by numerous large branches of the Amazon, and form a part of the great central basin of the continent. (Humboldt's Pers. Narr.; Hall, p. 8; Mod. Trav., pp. 19–21, 226–230.)

Rivers.—The chief are the Amazon, which, in the earlier part of its course, runs almost entirely through Ecuador, near its S. border; and the Orinoco, which, together with all its branches, is wholly included within the territories of Venezuela and New Granada. Besides these, there are the Magdalena, Cauca, Atrato, Zulia, Tocuyo, and Guarapiche, whose waters go to the Caribbean Sea; the Patia, Mira, Esmeralda, and Guayaquil rivers falling into the Pacific; the Yapura, Putumayo, Napo, Piçaena, Pastaça, Maranon, Santiago, Huallaga, &c., affluents of the Amazon; the Guaviare, Meta, Arauca, Apure, with its numerous branches, Ventuari, Caura, and Carony, which discharge themselves into the Orinoco; and the Cayuni, which passes into the territory of British Guiana.

Lakes.—The most considerable is that of Maracaybo, which is rather a kind of inland fresh water sea, and communicates with the gulf of the same name by a channel about 2 leagues broad and 8 long. (See MARACAYBO.) The lake of Valencia,

which is the next in importance, is larger than that of Neufchatel in Switzerland; there are others, both in the plains and in the mountainous regions; the most celebrated of them is that of Guatavita, not far from Bogota, into which, it is affirmed, large sums were thrown by the natives during the period of the Spanish conquests. Some extensive salt marshes are to be met with in different parts of the NW. coast. (Mod. Trav., vol. xxvii.; Account of Colombia, pp. 19–25.)

Minerals.—The Cordilleras teem with metallic wealth; and, though imperfectly explored, have already produced large quantities of gold, silver, platina, mercury, copper, lead, and iron : the gold is mostly obtained by washing the auriferous soil, and comes chiefly from the prov. of Choco, Antioquia, and Popayan; silver is found in the prov. of Pamplona and the valley of the Cauca; platina, on the coast of the Pacific; mercury and cinnabar, in several parts, as well as lead; and iron and pit-coal in abundance near Bogota: copper, in great plenty, is found, especially at Ania, in New Granada. There are mines of rock salt in the mountains NE. of Bogota, and caves producing nitre near the lake Guavita. Hot sulphureous springs abound in several parts; those of Las Trincheras, about 10 m. from Valencia, are believed to be the hottest hitherto discovered, excepting those of Offimo in Japan. Colombia abounds in stupendous natural wonders; amongst the rest are the natural bridges of Icononzo, not far from Bogota; the fall of Tequendama, the loftiest cataract, and the Silla de Caxorras, the loftiest cliff yet discovered. (Humboldt's Pers. Narr. and Researches; Delahecha's Geolog. Manual, pp. 410, 411; Present State of Colombia, pp. 297–314.)

The climate of the country between the Cordillera and the Caribbean Sea is extremely hot, and generally unhealthy. In the valley of the Orinoco the heat is also intense; but this tract is not so insalubrious as the sea coast, and is often refreshed by strong breezes. The middle region possesses every gradation of temperature, according to elevation : when at the level of the sea, the thermometer has been found to stand at 115° Fah.; at the height of 1,900 ft. it has descended to 77°; at 8,800 ft. to 50°; at 9,000 ft. high, it becomes extremely cold; and at 15,700 ft. all vegetation ceases. At Caraccas, most rain falls in April, May, and June : Dec., Jan., Feb. are the months of greatest drought. Violent storms, accompanied with thunder and lightning, are frequent at Maracaybo. Earthquakes are very common; many took place at the end of the last century, and one in 1812 overthrew most of the principal towns on the N. coast, with great destruction of human life. Intermittent, putrid, and bilious fevers and dysenteries are the most prevalent diseases on the coast; goitre is nearly universal in the mountainous regions. (Hall's Colombia, pp. 6–10; Account of Colombia, pp. 15–18; Mod. Trav., vol. xxvii.)

Vegetable Products.—The vast forests that line the shores of the rivers, and cover the mountains, abound with fine timber, which would yield a large revenue, if the means of transit to the coast were better. Mahogany, cedars, and an infinite number of woods of great beauty and durability, a very hard species of oak (Quercus cerus, Linn.), iron-wood, ebony of various kinds; Nicaragua, Brazil, and numerous other dye-woods; the cocoa and other palms; bananas, plantains, and gigantic mimosas, are found in profusion. Humboldt observes, ' it might be said that the earth, overloaded with plants, does not allow them space enough to unfold themselves. The trunks of the trees are every where concealed under a thick carpet of verdure; and if we carefully transplanted the

Orchids, the palms, and the pothos, which a single courtyard or American fig-tree nourishes, ought to cover a vast extent of ground.' Venezuela is, generally speaking, more fertile and richly wooded than New Granada. Mangroves and Cacti grow thick upon the coast; the tamarind, date, and various other tropical fruits, are nearly every where plentiful, and the Ficus gigantea sometimes reaches the height of 100 feet. The cocoa-nut, indigo, cotton, tobacco, yam, and potato, are indigenous to Colombia, as are vanilla, cassia-fistula, cochineal, &c.; the prov. of Loxa and Mariquito are famous for their cinchona bark; camphor, sarsaparilla, sassafras, squills, storax, and a multitude of other medicinal plants, gums, resins, and balsams, are natives of this country. Arborescent ferns of an enormous size are met with; and the earth in some parts is covered with graminaceous plants occasionally 80 ft. high. (Humboldt's Pers. Narr. and Researches; Mrs. Trav.; Hall's Colombia, pp. 30, 31, &c.; Account of Colombia, pp. 144–151.)

Animals.—Nature has been equally prodigal of animal as of vegetable life. Jaguars, tapirs, wild horses, hogs, deer in immense numbers, wild dogs, and monkeys of different kinds, are amongst the most common quadrupeds; so vultures, parrots, and parroquets, in large flocks, macaws, scarlet cardinals, flamingoes, pelicans, and an abundance of water-fowl, are plentiful among birds. Immense alligators inhabit the larger rivers and llanos, where, together with large serpents of various kinds, they lie buried in the mud during the dry season, and revive at the first appearance of the rains. The rivers and lakes are well stocked with fish; and the stagnant pools in the llanos abound with the gymnotus, or electrical eel. (For a description of this remarkable animal, see Humboldt's Pers. Narr., 345–377; or Mod. Trav. xxvii. 223–237.) Scorpions, millipedes, scolopendras, termites, mosquitos, and myriads of other insects abound; the pearl oyster inhabits several parts of the coast.

Agriculture.—Cocoa, coffee, cotton, indigo, sugar, tobacco, hides, cattle, and Brazil-wood, are the principal articles of culture and commerce; the grain, and the nutritious roots known in the West Indies by the name of ground provisions, are produced only in sufficient quantities for home consumption. Maize is grown every where, and, when ripe, is pounded in wooden mortars into a coarse meal, then bring no more perfect machinery for grinding it. Wheat is grown on the higher lands, especially in New Granada, where it succeeds as well as in England, and often yields 40 bushels an acre; two crops may be produced in a year. A substitute for bread is found in cassava, which is procured, by a process similar to that for making starch, from the yuca; and the plantain is to the mass of the natives what the potato has become to the poor of Ireland; the rice of Colombia is indifferent. Cocoa (properly the cacao nut) is principally grown in Venezuela, on the low rich soil of the coast, in Varinas, and near Guayaquil. It does not come into full bearing till after eight or nine years' growth; but, after that, continues to produce from 20 to 50 years, bearing two crops a year, with little trouble or expense. The cultivation of cocoa has however diminished, that of coffee having been in part substituted for it. Coffee has been introduced into almost all the temperate valleys of Venezuela, and the prov. of Santa Martha and Mariquita in New Granada; but its culture is conducted with less care than in the W. Indian Islands. Its produce and the trade in it have, however, increased rapidly since the revolutionary war, and it now forms by far the

greatest article of export. Cotton is grown in all parts of the country; but principally in the valleys of Aragua, and the provs. Cartagena and Maracaybo. The produce is said to be inferior in quality to that from the uplands of N. America, which is in great measure owing to the defective mode generally followed of cleaning and depriving it of the seed. In the prov. Cartagena, the plant is grown upon newly cleared land, between successive crops of maize. Indigo is cultivated principally in the valleys of Aragua and the prov. Varinas, and formerly was exported in large quantities; but the competition in this article, which British skill and capital has produced in Hindostan, materially affects this branch of agriculture. The tobacco of Cumana is greatly superior to that of Virginia, yielding only to that of Cuba and the Rio Negro; in some places, as at Cumanacoa, it is even superior to the latter. Under the Spanish regime, the culture and sale of tobacco were monopolised by the government. All individuals authorised to raise it were registered, and the entire produce was brought to the government depôts (estancos), and sold to its agents at a certain fixed price, who again sold it to the consumer at a large advance. The Colombian congress originally abolished this among other monopolies; but finding that they could not spare the revenue, of which it was productive, it was again revived. The cultivation of the plant had, however, from some cause or other, so much declined, that the revenue derived from the monopoly ceased to be of any material importance; and a law passed the congress for its abolition, on the 1st of June, 1834.

Previously to the arrival of Columbus the horse and ox were unknown in the New World; but the llanos are now covered with herds of both. M. Depons, in the early part of the present century, estimated that there were, from the mouths of the Orinoco to the lake Maracaybo, 1,200,000 oxen, 180,000 horses, and 90,000 mules; an estimate which Humboldt thought too low. Sheep and goats are plentiful in the table-lands of Bogota; animal food is cheap and much consumed; and hides, wool, and cheese form a principal portion of rural produce. Agriculture generally is in a very low state, and the government have been lately desirous to promote its improvement by encouraging foreign settlers, and disposing of the waste lands to them at a low rate, and exempting them for a period from taxes. Few people possess estates of 5,000l. a year; 5,000 dollars are reckoned a good income. Near Pamplona the grounds are surrounded with stone wall hedges, which give an air of proprietorship too often seen; and in the valley of Serinza (New Granada), a similar plan is adopted, and cultivation is in a tolerably advanced stage. Commonly, however, the natural indolence of the natives precludes this, and 'the Colombian who can eat beef and plantains, and smoke cigars as he swings in his hammock, is possessed of almost every thing his habits qualify him to enjoy, or which his ambition prompts him to attain—the poor have little less, the rich scarcely covet more.' In the llanos the indolence of the inhabitants is such that, after having suffered for half the year from inundations, they patiently expose themselves during the other half to the most distressing want of water, though they know that almost every where they may obtain a good supply at 10 ft. below the surface of the earth. The fertility of the soil and the warmth of the climate have, in fact, indisposed and unfitted the people for any vigorous exertion. (Humboldt; Mod. Trav.; Hall.)

Pearl Fisheries.—Along the coast many of the inhabitants subsist as fishermen, bartering the fish they catch for maize and other inland produce. There are three pearl fisheries; two on the shores of the Atlantic, and one on those of the Pacific. The first are situated on the coast of the islands Margarita, Cubagua, and Coche, and at the mouth of the Rio Hacha; in the 16th century they were much celebrated, and yielded pearls to the value of half a million dollars annually. The pearls of this coast are remarkable for their beautiful play of light, in which they are much superior to those of the East. The other fishery is at Panama; all of them are now much neglected, and do not yield more than 180,000 dollars a year. The Indians of Cariaco have a singular method of catching wildfowl, which may here be noticed: they leave calabashes continually floating on the water, that the birds may be accustomed to the sight of them, 'When they wish to catch any of these wild fowl, they go into the water with their heads covered each with a calabash, in which they make two holes for seeing through. They thus swim towards the birds, throwing a handful of maize on the water from time to time, the grains of which scatter on the surface. The birds approach to feed on the maize, and at that moment the swimmer seizes them by the feet, pulls them under water, and wrings their necks before they can make the least movement, or, by their noise, spread an alarm among the flock.... Many have no other trade in the neighbourhood of large towns, and daily take multitudes of these birds, which they sell at a low rate.' (Humboldt's Pers. Narr., ii. 271, 276; Present State of Colombia, pp. 372, 373; Hall's Colombia, pp. 28, 29; Mod. Trav., xxvii. 30, 51, 53, &c.)

Manufactures.—Such of these as are not merely domestic are chiefly leather, hammocks, baizes, blankets, coarse cloths of various kinds, hats, and salt; but none of them is of any importance. The principal salt works are at Araya and Santa Martha. The whole process is left to nature, and consists simply in the washing of the muriatiferous soil by the rains, into shallow basins, where the salt is found incrusted, after evaporation, in a state of great purity. The common pottery is rude, and made by Indian women only. At Carupe, oil is manufactured by the Indians, from the fat of young guacheroi birds; and on the Magdalena, the negroes stuff their pillows with the wool obtained from the fruit of the mahagua (*bombax*). Such expedients often supply the place of better manufactures, all of which must be procured from abroad, and are comparatively scarce and dear. (Mod. Trav.; Humboldt; Hall's Colombia, &c.)

Trade.—The ports of La Guayra, Rio del Hacha, Santa Martha, Cartagena, Chagres, Puerto-Cabello, Panama, and Guayaquil are those most frequented by foreign traders. The value of the imports and exports of the three states of Colombia from and to the United Kingdom, in the years 1861 and 1862, is shown in the subjoined statement:—

		1861	1862
		£	£
New Grenada	Imports from	811,394	774,411
	Exports to	838,781	1,615,638
Venezuela	Imports from	...	23,167
	Exports to	375,301	411,919
Ecuador	Imports from	85,132	88,680
	Exports to	1,026	16,064

The internal trade of the Colombian states is of no great importance. The want of internal communication is a considerable disadvantage: throughout the whole country there is scarcely a road passable for wheel carriages; and every species of commodity is conveyed on mules. The ways generally are mere tracks, formed by the tread of successive travellers, and even in what were formerly termed royal roads, all that has been done was to cut down the trees. Bridges are few, and except those of Valencia and Capitanejo, consist of only a few rough planks, with branches laid across; or of ropes, upon which a suspended basket is made to run from one end to the other. In the more precipitous and dangerous passes, where mules can scarcely be used, it is customary for travellers to be carried in chairs fastened to the backs of men, who obtain a miserable livelihood by continually exposing themselves to risks, such as those which level the chamois-hunter. (See Andes.)

Government is vested, in each of the states, in a senate and a house of representatives, both consisting of members elected by the cantonal deputies of the provinces, in a provisional assembly, held once in four years. In Colombia, previous to its partition, the right of suffrage in the election of deputies required the parochial voter to be a Colombian, above the age of 21, the owner of property worth 100 dollars, or exercising some trade or profession, and able to read and write (this last qualification to be peremptory after 1840). To be a cantonal elector, it was requisite to be a native of the canton, possessed of property worth 500 dols., or an income of 300 dollars; to be a senator, it was necessary to have an income of 500 dollars, or to be of a learned profession. The executive power was vested in a president and vice-president, the former of whom could not continue in office longer than eight years successively; and neither he, nor any of the ministers, could be members of the congress. With some variations, this government has been adopted by the existing states. The political government of each department is, by law, vested in the hands of an *intendente*, appointed by the president, with the sanction of the congress, with authority over the administration of justice, police, finance, and defence; but without the command of an armed military force. The provinces are under the administration of governors, with powers similar to those of the intendente; the cantons and parishes have each their own officers.

The civil and criminal codes are an ill-digested collection of the laws of Castile and of the Indies, royal ordinances and other Spanish decrees, and colonial regulations; and their administration is very unfavourably spoken of. The judges were elected by the congress, from lists given by the president. Trial by jury, and the liberty of the press, were amongst the first enactments of the Colombian congress.

Religion, the Roman Catholic, the ceremonies and festivals of which are celebrated with great splendour. The Inquisition was abolished in 1821; but the clergy still possess considerable power, and though general toleration is afforded to persons of other creeds, they are not at liberty to perform their rites in public. The clergy are paid by the state: convents are still numerous, but diminishing, and dissent from Catholicism is spreading. Many Indians have embraced Christianity.

The ranks of the different armies are filled with Indians and mixed races, in a tolerable state of discipline. In addition to these, there is a militia, consisting of the whole male population between 16 and 40 years of age. Considerable pains have been taken by the states of Colombia to raise a navy; but their maritime force is inconsiderable. A marine school has, however, been established at Carthagena.

During the Spanish regime elementary education was sadly neglected, and all the more important branches of useful knowledge professed at the universities of the Caracas, Bogota, and Quito, were so taught as to be really more than useless; and instead of expanding and enlightening the mind, served rather to imbue it with the grossest prejudices. But considerable progress has since been made towards the establishment of a better order of things. Primary schools were ordered to be established in every parish, by the congress of 1821; Lancastrian schools exist in the principal towns, and the universities have been remodelled. The Columbian congress applied certain property formerly belonging to the clergy to the aid of public education; and the legislatures of the present republics have been anxious to carry into effect the system adopted by it. Several public journals are established in different parts of the country.

Architecture has made but little progress, and almost the only specimens worth notice are confined to Bogota. Painting is successfully cultivated in that city and Quito, and music in Caracas; but, generally speaking, the fine arts are in a very backward state. The besetting vice of the Colombians is indolence, which retards all their social progress: they are courteous, hospitable, and, when intimately known, friendly and cordial; temperate in their habits, and grave in their department; but suspicious, reserved, slow, and imbued with much national pride. The manners, dress, habits, and amusements of those of European descent resemble those of their Spanish ancestors.

History.—Ecuador, and especially the valley of Quito, contains many monuments of the sway of the Incas. Venezuela was the first part of the new continent discovered by Columbus in 1498. The Spaniards found more difficulty in conquering this than any other part of their American territories; but, before the middle of the 16th century, both Venezuela and New Granada had been erected into captaincies, governed by viceroys from Spain. In 1808, after the invasion of Spain by Napoleon, a spirit of insubordination broke out in these colonies; in 1811, their independence was declared; and, in 1819, Venezuela and New Granada united into one republic, under the name of Colombia. In 1822, the royalists in Ecuador were defeated by Gen. Sucre; Bolivar headed the revolutionists elsewhere; and in 1823 the struggle ended with their complete independence. In 1829, Venezuela separated from the other states; rejoined them for a short period in 1830; but in Nov. 1831 separated anew; since which period Colombia has remained divided into the above three republics. But such is the state of insecurity in which all these governments exist, that it is highly probable that, for the sake of gaining the necessary power to resist foreign aggression, they will, before long, be again consolidated into one united state of Colombia.

COLUMB (ST. MAJOR), a town and par. of England, co. Cornwall, hund. Pyder. Area of par. 11,640 acres. Pop. of ditto 2,879 in 1841. The town is situated on an eminence, at the foot of which is a small river, 4 m. from the sea, and 14 m. NE. Truro. It had formerly a communication with the sea by means of a canal, now fallen into disuse. It has a large old church, and two methodist chapels. Market-day, Thursday.

COLUMBIA, a distr. of the U. S. of America, lying between the states of Virginia and Maryland, on both sides the Potomac, about 120 m. from its mouth; length and breadth, 10 m. each; area, 100 sq. m. Pop. 75,000 in 1860; and 39,834 in 1830. Surface gently undulating; soil naturally thin, sandy, and sterile. Climate healthy; mean temp. of the cap. about 55° Fahr. The Po-

tomac traverses the distr. chiefly in a SE. direction, receiving in its way through it a tributary from the E. by its junction with which a peninsula is formed, on which the city of Washington is built. At the confluence of the two rivers there is an excellent harbour and a navy-yard, to which ships of the largest tonnage may ascend. The yard covers a space of 37 acres, and in it are made all the anchors, cables, and blocks required for the service of the U. S. navy.

Washington is the cap. of the U. States, the seat of the general government, and the residence of the president and other principal officers of state. (See WASHINGTON.) The other chief towns are Georgetown and Alexandria; the former is separated from Washington by Rock Creek, another affluent of the Potomac. Alexandria is on the right bank of the river, 7 m. below Washington.

Considerable quantities of flour and other domestic produce are brought down the Potomac, but neither the commerce nor shipping of the distr. are of great importance. Alexandria and Georgetown have together about 19,000 tons shipping. There are three colleges in the district, all in active operation:—the Columbia Institute at Washington; the Roman Catholic university at Georgetown; and the theological seminary at Alexandria; connected with which is a medical department, and a preparatory school. The district is under the immediate government of congress. It was ceded to the U. States by Maryland and Virginia in 1790; and in 1801 it was enacted that the laws of these states should continue in force in the portions ceded by each. Congress first met here in 1800.

COLUMBIA, a town of the U. S. of America, cap. S. Carolina, and seat of the state government, in an elevated plain near the centre of the state, near the Congaree river; 100 m. SNW. Charleston, and 68 m. NE. Augusta. Pop. 7,058 in 1842. The streets, which are 100 ft. wide, intersect each other mostly at right angles, and it has many good houses. It has a state-house, court-house, gaol, and several places of worship. The S. Carolina college, founded in this town in 1804, has two large brick edifices, and possesses a philosophical apparatus, cabinet of minerals, and library of 10,000 vols. Here is also a theological seminary, established in 1829. Columbia was founded in 1787.

COLUMBIA RIVER, a large river of N. America, the principal in the Oregon territory, with an extremely tortuous course. It rises in the Rocky Mountains, in about the 51st deg. of N. lat., and the 116th deg. of W. long. Its course is first NW. till about the 53rd deg. lat.; and then nearly S. for about 215 m. till its junction with the Flathead or Clarke's river. It then pursues a WSW. course, being precipitated over some very high falls, till it reaches Fort Okanegan, in about 48° N. lat. and 119° W. long., when it flows S. to Fort Nezperces, a distance of 155 m., where it is joined by the Great Snake river from the SE. After receiving the latter it turns to the W.; and pursuing that direction during the remainder of its course, it falls into the Pacific Ocean, between Cape Disappointment on the N. and Point Adams on the S., in 46° 18′ N. lat. and 124° W. long. Its embouchure is 6 or 6 m. in width. It has not, where deepest, more than from 4 to 5 fathoms over its bar, on which the sea breaks with considerable violence, making its ingress and egress, to sailing vessels, a work always of considerable difficulty, and practicable only, it is said, at certain seasons. Vessels of 400 tons may ascend the river to Fort Vancouver,

H 2

about 100 m. (dir. dist.) from its mouth; and along may as end it for about 80 miles farther. At the Long Narrows, by which the navigation is first interrupted, the river is precipitated over an upper and a lower fall respectively 20 and 8 ft. in height.

COLUMBO, a sea-port town of Ceylon, the modern cap. of the island, and seat of government, on the W. coast, towards its S. extremity; lat. 6° 55' N., long. 79° 45' E.; pop. estimated at 60,000. The town has a fort, defended by walls flanked with several bastions, and is built upon a peninsula projecting into the sea, having on the land side a fresh water lake of same size. It contains the residence of the governor and most of the British inhabitants. The pettah, or inner town, a few hundred yards E. from the fort, has a mixed pop. of Dutch, Portuguese, and their descendants. The native Ceylonese reside chiefly in the suburbs. The town within the walls is regularly laid out, and built in the European style; houses, chiefly of stone, clay, and lime, are seldom more than a story in height, but each has in front a large wooden verandah. The English have substituted Venetian blinds in their houses for the glass windows used by the Dutch. The fort contains the government house, a handsome building of two stories, the English church, court-house, library, museum, several hotels, and a lighthouse 97 ft. high. There are also in Columbo a Dutch and a Portuguese church, several Protestant dissenting chapels, with missionary and other schools. To the N. of the fort is a small semicircular bay, on which a wooden quay has been built, but the depth of water is not sufficient to admit of vessels above 100 tons burden coming alongside. The bay is sheltered and defended by a projecting rock on which two batteries are erected; but from this rock a bar of shifting sand stretches across the mouth of the bay, within which the larger class of ships can venture only during the fine weather of the safe season. Besides its small bay, Columbo has an open roadstead, which, however, is safe only during the NE. monsoon: were the town more favoured in this respect, it would be the most eligible port in the island, since it is placed in the centre of the cinnamon country, is the depot for nearly all the foreign trade of the island, and has a somewhat extensive traffic by means of internal navigation. Columbo is ill supplied with water. Its climate is healthy, though damp and destructive of books, clothing, &c. The Portuguese erected a fort here in the early part of the 16th century, of which the Dutch dispossessed them in 1656; and the town was taken from the latter by the English in 1796, which change of masters was afterwards ratified by the peace of Amiens.

COLUMBUS, a city of the U. States, cap. Ohio, of which it is nearly in the centre, on the banks of the Scioto, immediately above the point where it is joined by the Whetstone river, 100 m. NE. Cincinnati; lat. 39° 57' N., long. 82° 3' W. Pop. 14,350 in 1860. The town was founded so late as 1812, the land on which it stands having previously been a wilderness. It is well situated on land rising gradually from the river; the streets, which are broad and straight, cross each other at right angles, being for the most part lined with substantial houses. It has a square which comprises 10 acres; and a convenient wharf extends along the margin of the river. But the navigation of the latter (an affluent of the Ohio) being liable to interruption, the city is united by a canal to the Ohio canal, which opens an easy communication with the lakes on the one hand, and the Mississippi on the other; and its trade is farther

promoted by its being on the line of railway from Indianapolis to Zanesville. A bridge across the river unites the city with the suburb of Franklinton. The public buildings comprise a state house, commanding a fine view of the surrounding country; an edifice for the accommodation of the officers of the state; a state penitentiary; a lunatic asylum, and asylums for the blind, and for deaf and dumb persons; a Lutheran theological seminary, and numerous churches. Here, as in the other towns of the U. States, there is ample provision for the education of the young in elementary and superior schools. The town has factories of various sorts, with tanneries, breweries, distilleries, and printing-offices.

COMBOCONUM, an inl. town of Hindostan, prov. Carnatic, distr. Tanjore, 20 m. NE. that city, Pop. estimated at 40,000. It was anciently the cap. of the Cholas, one of the most ancient Hindoo dynasties in the S. of India; of which any traces have been discovered, and who gave their name to the whole coast of Cholamandal or Coromandel. Its ancient splendour is evinced by its pagodas and tanks. It is chiefly inhabited by Brahmins.

COMILLAH, an inl. town of Hindostan, prov. Bengal, distr. Tipperah, of which it is the cap. on the S. bank of an affluent of the Brahmaputra river, 50 m. SE. Dacca. The roads round it have been much improved by the labour of convicts. Six m. W. of Comillah are the remains of many brick buildings, and of a fort 700 ft. square, the residence of the former rajahs of Tipperah.

COMO (CITY OF) (an. Comum), a city of Northern Italy, cap. of the province of same name, at the SW. extremity of the Lake of Como, 23 m. NNW. Milan, with which it is connected by railway. Pop. 21,614 in 1861. The city is encircled by an amphitheatre of hills, one of which to the N. is surmounted by the old fort of Baradello. It is defended by double walls, flanked with massive towers, and has four gates. Its interior is crowded with dark streets, numerous old churches, and dismantled dwellings of the citadini. The suburbs, however, in which more than half the pop. resides, contain many good streets and buildings; Borgo de Vico, the chief, stretches along the shore of the lake for a considerable distance, and is adorned with the Olivelobi and Loriani palaces, besides numerous other handsome edifices. Como has 12 churches, the principal of which, the cathedral, commenced in 1396 and finished in 1513, is an imposing building, notwithstanding its incongruous character. It is of white marble, the front is of light and not inelegant Gothic, the nave is supported by Gothic arches, the choir and transepts are adorned with composite pillars, and a dome rises over the centre. In front of the cathedral is a statue of Pliny the younger, a native of Como, with a bas-relief alluding to his writings, and an inscription to his honour on each side the grand entrance. In one of the squares a monument is erected in honour of Volta, also a native of this city. Como possesses a lyceum erected by the French, with some fine philosophical apparatus, and a library of 15,000 vols., an ecclesiastical college, 3 gymnasia, 2 female seminaries, a hospital, workhouse, orphan asylum, and many other charitable institutions, a cabinet of nat. history, and botanic garden, a new theatre, and an amphitheatre. Como is a bishopric, and the seat of the provincial council, and of civil, criminal, and commercial tribunals. At one period it was the principal seat of the Inquisition. It has manufactures of woollen cloths, silks, cotton yarn, and soap, for which latter article it is celebrated. Its trade, which is facilitated by a port on the lake, is chiefly with the Swiss canton of Tirino, and with Germany, to which it sends rice, and raw and manufactured silks. The artisans of Como have, in all ages, been noted for their disposition to emigrate as hawkers of goods, or in search of employment, and they may be met with all over Europe, as venders of telescopes, spectacles, and barometers. The fine climate and situation of Como attract many visitors. Como is said to have been founded by the Orobii, the earliest inhabit. of this district. It was taken by the Romans 196 B.C.; and owed its principal importance under them to a colony of Greeks planted in it by Julius Cæsar, when it took the name of *Novocomum*. Near it is the Villa d'Este, once the property and residence of Queen Caroline of England. In the middle ages it belonged to the Ghibelline party, and was the rival of Milan. Under the French it was the cap. of the dép. of the Lario.

COMO (LAKE OF), (It. *Lago di Como*, an. *Larius Lacus*), a famous lake of N. Italy, which, in modern times, has derived its name from the above city. This fine sheet of water is very irregularly shaped, being divided by the triangular district which has Bellagio at its apex, into three great arms, one of which stretches from Bellagio SW. to Como, another N. to Riva and Novate, near the mouth of the Maria river, and a third SE. to Lecco, and the outlet of the Adda. These divisions of the lake are sometimes called from the chief towns on their banks, the lakes of Como, Bellano, and Lecco. Its greatest length, following its windings, may be about 45 m.; but it is no where above 4 m. in width. The depth is said to vary from 40 to 600 ft. It receives the waters of the Upper Adda, and several other rivers, but its only outlet is by the Lower Adda. Owing to the great height of the surrounding mountains, which expose it to sudden squalls, and the influence of currents, its navigation is rather dangerous to sailing vessels; but steamers traverse it in all directions with ease and expedition. The climate round the lake is mild and delightful; and, except in its more N. part, near the mouth of the Upper Adda, its banks are remarkably healthy. Throughout its whole extent its banks are formed of precipitous mountains, from 2,000 to 3,000 ft. high; in some places overhanging the water, and in others partially clothed with wood, and studded with hamlets, cottages, villas, chapels, and convents. The most beautiful point of view is at Bellagio. The upper waters are there seen winding up to the very foot of the higher chain of the Alps, and terminating within a short distance of the territic pass of the Splugen; the loftier hills that border the lake of Lecco rise on one side, and on the other the wider expanse of the lower lake retires behind the beautiful foreground, rocks and hanging woods that form the point of Bellagio.

The younger Pliny had several seats on the border of this lake. The principal of these stood, one upon a height commanding a view of the lake, and the other so close to its edge as to admit of fishing lines being thrown into the water from the bed-rooms. (Epist. lib. ix. § 7.) Many attempts, but very unsuccessful ones, have been made to identify the site of these villas. The *Villa Pliniana*, 5 m. NE. from Como, is, from its having near it an intermittent fountain, usually supposed to occupy the site of one of these villas. But Pliny does not say that the intermitting fountain which he describes was on his estate, or near his seat (iv. c. 30); and there is no real ground for supposing that the *Villa Pliniana*, which was built near the middle of the 16th century, has anything in common with either of the villas described by Pliny.

COMORIN (CAPE), a promontory forming the

S. extremity of Hindostan, in Travancore, lat. n. NW. Colombo, in Ceylon; lat. 8° 4' N., long. 77° 14' 30" E. Its approaches are lined with rocks. Notwithstanding its remarkable position, it never attracted the least attention from the Hindoo geographers; and, what is more singular, modern authorities differ considerably as to its lat. The above is that given by Heywood.

COMORN (Hungar. *Komarom*), a fortified town of Hungary, in the NW. part of that king. cap. co. of the same name, on a point of land formed by the confluence of the Waag with the Danube; 46 m. WNW. Buda, on the railway from Buda-Pesth to Vienna. Pop. 12,175 in 1858, excl. of garrison. The citadel, built by Mathias Corvinus, in the 15th century, is held to be impregnable, and its works have been so much strengthened during the present century, that it is now one of the strongest fortresses in Europe. The town is irregularly built, and the streets are narrow and dark. It contains 4 Catholic and 2 Protestant churches, a Greek church, and a synagogue, a county hall, town council house, many large magazines and barracks, a hospital, bath, and Prot. high schools, and an assurance-office for vessels navigating the Danube, which river is here crossed by both a flying bridge and a bridge of boats. It has manufactures of woollen cloths, and considerable trade in corn, wine, honey, fish, and timber, by the Danube. There are numerous vineyards in its neighbourhood.

COMPIEGNE, a town of France, dép. Oise, cap. arrond. on the Oise, which is here crossed by a handsome bridge of three arches; 33 m. E. by S. Beauvais, on the railway from Paris to St. Quentin. Pop. 12,137 in 1861. The town is ill laid out and ill built, but contains many public and private edifices worthy of notice; amongst them the town-hall, a curious Gothic building, and several churches. But the glory of Compiègne is its royal palace, one of the most remarkable in France for extent and magnificence. A palace was originally built here by the Merovingian kings; but the present edifice was commenced under Louis XV., finished by his successor, and renovated by Napoleon. It has a noble front towards the forest of Compiègne, 623 ft. in length; all the apartments are on a single floor, communicating with each other. The *peristyle*, *salles des gardes*, ball-room, theatre, and a superb gallery, are especially deserving of admiration. The gardens surrounding this palace are much more extensive than those of the Tuileries, which they rival in beauty. Compiègne contains a public library with 28,000 vols., and a theatre. It was formerly fortified by walls flanked with towers, and entered by seven gates. Charles the Bald established an abbey here, and gave the town the name of *Carlopolis*, after which it rose considerably in importance, and became the seat of many national councils and assemblies, as well as the burial-place of several of the French kings. But in proportion as the consequence of St. Denis increased under the kings of the third race, that of Compiègne declined. It was at the siege of this place, in 1430, that the famous heroine Joan of Arc fell, through the mean jealousy of the governor, into the power of the English.

COMPOSTELLA, an inl. town of Mexico, state Guadalaxara, 86 m. from the Pacific Ocean, and 100 m. W. by S. Guadalaxara. In its vicinity there are some silver mines; and to the NW. of it, tobacco of a superior quality was formerly grown.

CONCAN, a narrow tract of country, prov. Bombay, comprising a portion of the ancient Hindoo subdiv. of *Konkana*, whence its name. It extends both N. and S. of that city, along the

Malabar coast, between lat. 15° 50' and 20° 15' N., and long. 72° 40' and 73° 54' E., having N. the collectorate of Surat, and a detached portion of the Guicowar's dom.; E. the distr. Ahmednuggur and Poonah and the Sattarah dom.; from which it is separated by the W. Ghauts; S. a portion of the Sattarah territory, and W. the ocean. Length N. to S. 810 m.; breadth varying to nearly 60 m. The territory is commonly divided into Southern and Northern Concan, the former with an area of 6,770 and the latter of 5,400 sq. m. The general aspect, though there are many fertile tracts, is that of a congeries of steep and rocky mountains, intermixed with a multitude of ravines and chasms, and interspersed with jungle. It formerly abounded in fortified heights, difficult of access, most of which have been dismantled by the British since their conquest of the country in 1819. The coast has a very straight general outline, but is broken by a great number of shallow harbours, which, previously to the British rule, were the resort of numerous pirates. The W. Ghauts, which bound the Concan to the E., rise to the elevation of from 2,000 to 4,000 ft., with an abrupt face towards the W. The passes over them are impracticable for wheeled carriages. They are mostly composed of primitive trap-rocks; but their summits are covered with a thick crust of laterite or ferruginous clay-stone, of which material much of the surface of the Concan is composed. In the S. shelly sand-stone is met with. There are many mountain streams, but none deserving the name of a river. Concan produces all the grains of Malabar, but is chiefly celebrated for its hemp and cocoa-nuts. Oil grains, the sugar-cane, turmeric, ginger, &c., are grown in the S. The land in S. Concan is assessed on the ryotwar, and in the N. on the village system. In some instances ill-cultivated tracts of land are allotted for a term of years at a low rent to a speculator for the purpose of improvement. A large proportion of the inhab. are Hindoos, and Suttees (burnings of widows) are said to have been more frequent here than in any other part of India, Bengal excepted. Many Bheels and Coolies inhabit the Ghauts and N. Concan. A large portion of the Bombay native army was formerly, and to some extent is still, recruited from these districts. The Angria family once possessed nearly the whole of Concan; it subsequently belonged to the Peishwa, on whose fall it came into the possession of the British.

CONCEPTION, a city of Chili, in the S. part of the Republic, cap. prov. of same name, on the right bank of the Biobio, 8 m. E. from its mouth, and about 270 m. SSW. Santiago; lat. 36° 43' 25" S., long. 73° 5' 25" W. Estimated pop. 12,000. It stands upon a low neck of land between the Biobio and the SE. angle of the Bay of Conception, and occupies a surface of about a sq. mile. Streets intersect each other at right angles; houses mostly only one story in height in consequence of the great frequency of earthquakes, and many are built entirely of unbaked bricks. Conception was formerly a flourishing town, containing several good buildings, and 20,000 inhab.; and, previously to 1835, it possessed a massive cathedral, but this and the greater part of the city were in that year totally destroyed by an earthquake. It is the residence of a bishop and the military governor of the prov. Manufactures and trade are said to be at present of little importance.

The Bay of Conception is a large square inlet, open on the N., while the S. and W. sides are flanked by a high promontory jutting out from the main land, and bending into the shape of an elbow, each side being 3 or 4 leagues long. The diameter of the space thus enclosed is about 5 m. The

mouth is divided by the island Quiriquina, which lies across it, into 2 channels; the N. entrance has 30 fathoms water, diminishing gradually to 12 fathoms in the middle of the bay; the S. entrance has 30 fathoms at its commencement, and 11 fathoms at its entrance into the Talcahuano anchorage. There are 8 harbours; that of Talcahuano, close to the small fortified town of the same name, under the promontory in the SW. angle, is the most secure from winds, and that in which ships generally lie. Full 12 fathoms water are found in all parts of the bay within ¼ m. of the beach; the holding ground is excellent, and the bottom free from rocks.

Conception was founded in 1763, after the destruction of the old city of Penco by inundation, during an earthquake.

CONCORD, a town of the U. S. of America, cap. New Hampshire, and seat of the state government, co. Rockingham, on the Merrimac, 65 m. NNW. Boston. Pop. 10,890 in 1860. The town consists chiefly of two streets, extending for above 2 m. along the W. side of the river, which is here crossed by two bridges. It contains the state-house, a handsome stone building, and the state prison. The courts were removed to Concord from Portsmouth in 1823. It is a town of considerable trade, and has a water communication with Boston by means of the Merrimac and Middlesex canal.

CONDE, a town of France, dép. du Nord, cap. cant., at the confluence of the Hagne with the Escaut (Scheldt), 25 m. SE. Lille, on the railway from Lille to Valenciennes. Pop. 5,804 in 1861. The town is strongly fortified by works constructed by Vauban; is well built, and contains a handsome town-hall and a fine arsenal. A canal, 15 m. in length, connects Condé with Mons, in the Netherlands. It was taken by Louis XI. in 1476.

CONDE SUR NOIREAU, a town of France, dép. Calvados, cap. cant., on the road between Caen and Domfront, 23 m. SSW. the former. Pop. 7,254 in 1861. The buildings are generally heavy; the town contains, however, two old churches worthy of notice. It formerly possessed a castle with a large tower, but little now remains of that edifice. It has some commercial activity, and fabrics of woollen, cotton, and linen articles, and cutlery.

CONDOM, a town of France, dép. Gers, cap. arrond., on a height the foot of which is washed by the Baïse, which is here crossed by two bridges, 23 m. NW. by N. Auch. Pop. 8,070 in 1861. The town is ill-built, but improving; is surrounded by boulevards planted with trees, and has numerous villas in its environs. In its centre is a large open space, in which is the parish church, formerly the cathedral, which, despite the mutilations it has undergone, is still a magnificent Gothic edifice. Pens, corks, earthenware, brandy, woollen yarn, and leather are produced here; and there is a brisk trade in corn, flour, and wines. It has a tribunal of original jurisdiction and a communal college. It owes its origin to a monastery, which existed in the 5th century, but was of a much earlier date. It was formerly the seat of a bishopric, once filled by Bossuet.

CONDRIEU, a town of France, dép. Rhône, at the S. extremity of which it is situated, cap. cant., on the Rhône, 21 m. S. Lyons. Pop. 2,305 in 1861. The town has acquired some celebrity for excellent white wines, the original plants producing which were, it is said, brought thither from Dalmatia by order of the emperor Probus.

CONGLETON, a market town and bor. of England, co. Chester, hund. Northwich; 27 m. S. Manchester, and 161½ m. NW. London by London and North Western and North Staffordshire railway. Pop. 12,344 in 1861. The town stands in a remarkably healthy situation, on the Dane, in a deep valley bordering on Staffordsh. The principal street is upwards of a mile in length, paved, and lighted with gas: it contains many ancient houses of timber framing and plaster; at the W. end are many detached mansions, surrounded by gardens and shrubberies, and chiefly occupied by the more opulent manufacturers. It has an episcopal chapel, in the patronage of the corporation; a Catholic and several large dissenting chapels; a grammar-school, nominally free for the sons of burgesses; an infant school, established in 1835; several large Sunday-schools; and many charitable institutions and bequests, the latter chiefly held in trust by the corporation; a town-hall; and public assembly-rooms, built in 1622, contiguous to which is a modern market-place. Silk is the staple manufacture of the town; the silk-mills being mostly erected along the banks of the river. The trade consists chiefly in the throwing of raw silk, the spinning of waste ditto, the manufacture of thrown silk into plain ribands by power looms, of which there are about 251 in the town, and the weaving of ribands and broad cloths by hand-looms. There are also cotton spinning factories, and a few tanneries and leather manufactories. Certain lands reserved under an enclosure act are held in trust for the benefit of the poor. The tow. is divided into 3 wards, and governed by 6 aldermen and 18 councillors.

CONGO, otherwise LOWER or S. GUINEA, a country in SW. Africa. to which various boundaries have been assigned by the old and more recent travellers. The Portuguese, who discovered it in 1487, included in Congo all the coast of W. Africa from Cape Lopez Gonsalvo (Loango), in lat. 0° 37′ S. long. 8° 45′ E., to Cape Negro, in lat. 15° 50′ S. long. 11° 65′ E.; for they found the whole of that tract inhabited by negro tribes. resembling each other in every respect, and subject to one paramount chief, called Mani-Congo (Sovereign of Congo); but in process of time this empire became dismembered; inferior chiefs threw off their allegiance and erected separate kingdoms, which are at present known as Angola (a name now most frequently applied to the district over which all these kingdoms extend), Loango, Benguela, and lastly Congo Proper.

The boundaries of Congo Proper are at present marked N. by the river Congo or Zaire, which at about lat. 6° 5′ separates it from Loango; S. by the river Dande, in lat. 8° 20′ S., dividing it from Angola; W. the Congoese coast is washed by the S. Atlantic ocean, while to the E. it has the unknown countries of Fungeno and Matamba, the Mountains of the Sun. &c. According to the investigations of Ritter, Congo consists of two distinct regions; that next to the sea, or the littoral, is low and flat, is traversed by many streams, and abounds in sandy deserts, but is elsewhere very fertile. The climate in this region is exceedingly unfavourable; and pestilential emanations, and swarms of noxious animals, expose the lives of the inhab. to perpetual danger. The other region consists of the terraces, or acclivities, ascending from the plain to the high table-land in the interior. This is by far the finest part of the country, and the richest and most populous. The river Zaire, which descends from the interior to the coast, has its great cataracts in passing through this region.

This river is a most conspicuous object in the topography of Congo: it is a magnificent stream, particularly towards its embouchure: it overflows during the rainy season, and fertilises the surrounding country; but these risings take place also in the dry season, elevating the current 7 ft., —increased to 12 ft. by the rains. It is exceedingly deep: Maxwy's sounding-machine having indicated 118 fathoms, and yet the lead had not touched the bottom. In the upper parts, the current varies in strength from 2½ to 5 m. an hour, but is sufficiently strong in the channel to prevent a transport entering the river without the aid of a powerful sea-breeze. At about 110 m. from its mouth, the Zaire narrows to from 800 to 800 yds. for about 40 m.; its banks bristling with precipitous masses of slate, which sometimes interrupt the stream, and form rapids and cataracts, called by the natives yellala. Beyond these craggy regions, the Zaire expands in breadth to 2, 3, and even to 4 m.; and near the place where Captain Tuckey was compelled to abandon his journey, the width and majestic appearance of the river, the verdure of the land, which was here well peopled, combined to render the scene agreeable in the highest degree. (Tuckey's Expedition, pp. 337-346; Journ. Royal Geog. Soc., iii, 220.)

The banks of the Zaire, from its mouth to Embomma (about 60 m.), are clothed with a most exuberant vegetation, presenting to the eye a continued forest of tall and majestic trees, clothed with foliage of never-fading verdure.

The supposed identity of the Congo with the Niger was long a question agitated among geographers; and its decision was one of the objects of Tuckey's expedition. This question has been, as every one knows, set at rest by the Messrs. Lander. But it is sufficiently clear from the information collected by Tuckey, that the Zaire, at no great distance from the point to which he had ascended, divides into two great arms, the most N. of which has its source in a lake or marsh.

The natural productions of Congo have been admirably arranged by Professor Smith, a member of Tuckey's expedition (who unhappily lost his life in the course of it), and Mr. Brown. Large trees are only found in the valleys, or thinly sprinkled over the sides and summits of the hills, and consist for the most part of the Adansonia, Bombax pentandrum, Anthocleista, Musanga (native term, but allied to Cecropia), Elaïs guineensis, Raphia vinifera, and Pandanus candelabrum. Intermixed with these, on the alluvial banks of the Quorra, large patches of the Egyptian papyrus form a grand feature in the vegetation. The edible productions are maize, cassava, sweet and bitter, two kinds of pulse, the Cytisus cajan, a species of Phaseolus, and ground nuts (Arachis hypogæa). The common yam, besides another species of Dioscorea, so bitter as to require four days' boiling before it be eatable, with the sugar-cane, capsicum, and tobacco, are alimentary plants of secondary importance. The most valuable fruits are plantains, papaws, limes, oranges, pine-apples, pumpkins, tamarinds, and a fruit about the size of a small plum, called safu. The plant, however, of most importance to the natives is the oil palm (Elaïs guineensis), from which is extracted the best palm wine; this and two other species of palm (Raphia vinifera and a Hyphæne) are to the Congoese what the cocoa-tree is to many of the Asiatic islanders. The indigenous fruits are the Anona senegalensis, Sarcocephalus, a species of cream-fruit, Chrysobalanus, Icaco, a species of Ximenia, and another of Antidesma. (Professor Smith's Journal in Tuckey's work, with remarks thereon by Mr. Brown, passim; Quarterly Review, xviii, 350, 351.)

The animals appear to be those chiefly which are found in every part of this great continent; lions, leopards, elephants, buffaloes, antelopes, wild hogs, porcupines, hares, and monkeys. The river abounds with good fish, and also with those huge monsters the hippopotamus and crocodile. Domestic animals are few and scarce; those mostly met with are hogs, goats, fowls. Muscovy ducks, and pigeons, and a few sheep, generally quoted with hair instead of wool. The natives eat these animals in a manner quite characteristic of their rooted laziness. They remove neither skin, feathers, nor hair; and scarcely warming them by the fire, tear the meat in pieces with their teeth. (Dr. Leach and Mr. Cranch, in Appendix to Turkey's work; Quarterly Review, xviii. 351.)

Government and Population.—If we may depend on the traditions of the people, who have neither annals nor history, Congo was formerly a powerful empire under a single sovereign, or rather absolute despot. But it is evident from the accounts of the early travellers, little as they are, in many respects, to be depended on, that, when first visited by Europeans, the government of Congo did not differ materially in its form from what we find it at the present day; and that it consisted of a sort of confederacy of small states under a principal sovereign. (Prevost, Histoire Générale des Voyages, v. 1-7.) It would appear, however, to be pretty certain that the power of the superior monarch has materially declined during the last 300 years. At all events, Congo is now split into an infinite number of petty states or chieftainships, each governed by a chenin or chief. These chieftainships would, in Europe, be said to be fiefs, held under a principal sovereign, called *king* or *Mwly N'Congo*, residing at Banza Congo. But it would seem that most of these chiefs affect a nearly total independence; and being all despots in their own limited spheres, and frequently at war with each other, and with the principal sovereign, the country is uniformly almost in a state of the most frightful anarchy. At the death of a chenin, it is not his son, but his brother or maternal uncle that succeeds him.

The inhabitants are said to be a mixed race; but the Portuguese never visited the country in such numbers as to produce any impression on the physical character of the people; and the Congoese are certainly one of the least favoured negro varieties. Speaking generally, they seem to be sunk in the lowest state of degradation. They are incorrigibly indolent; have little or no clothing; and though they raise Indian corn, agriculture is in the lowest state, and they frequently suffer the extremity of famine. Their religion is the grossest species of fetichism. The Portuguese having established missions in different parts of the country, the natives sometimes exhibit in their religion an odious mixture of Christianity and idolatry. They are prone to all sorts of excesses and debauchery. The women are degraded to the condition of beasts of burden; and prostitution to strangers is considered as a necessary part of hospitality. Still, however, they are said wholly destitute of good qualities; and are said to be sincere, hospitable, and companionate. Having been long a principal seat of the slave trade, a considerable part of the disorders that prevail in the country are with much probability ascribed to the enormities growing out of that detestable traffic. This is said to isolate one petty state from another, and to occasion perpetual wars; the slaves being mostly prisoners taken in battle, or kidnapped on the public roads. But, admitting the influence of these causes, still we apprehend that the intellectual inferiority of the negro race is at bottom the real cause of the degraded condition of Congo, and of all the other negro states. The Congoese are said frequently to decapitate their prisoners, and burn their bodies; and if such barbarity be practised when the prisoners may be sold, the presumption would seem

to be that it would become much more prevalent were the traffic put an end to. (See Turkey, *passim*; and Ritter's Geography of Africa, French translation, i. 879-397.)

The country has been represented as very populous, and as studded with towns and villages swarming with inhabitants. Carli, one of the early missionaries, gravely reports that a king of Congo marched against the Portuguese at the head of an army of 900,000 men. (Prevost, *ubi supra.*) But it is evident that a country in the state we have described cannot be thickly peopled; and, in point of fact, Turkey states that the most considerable towns, or caps, of a petty state that he visited did not contain more than 100 huts and 600 persons. In Embomma he found 60 huts, with 300 inhabitants; and at Inga 70 houses, in which not more than 300 persons resided. It is true that his observations in the interior were not very extended; and he admits that the upper banks of the Zaire (where his operations unhappily ended) were considerably more populous than those towards the coast; but still it is abundantly certain that the accounts of the extraordinary population of the country have no better foundation than the imagination of the writers. According to the statements of the missionaries, the cap. of the country, which they divided into six provinces, was built on a mountain about 150 m. from the sea, and was called by them St. Salvador. They speak in the most extravagant terms of the beauty and salubrity of its situation.

CONGOON, a sea-port town of Persia, prov. Fars, on the Persian Gulf, 130 m. S. by E. Schiraz. Pop. from 6,000 to 7,000. It has an excellent roadstead, where a frigate may ride in safety in the most tempestuous weather, and good water and firewood may be procured in abundance. (Kinneir's Persian Empire, p. 81.)

CONI, or CUNEO, a town of N. Italy, cap. div. and prov., on a hill at the confluence of the Stura and Gesso, 43 m. S. by W. Turin, with which it is connected by railway. Pop. 22,510 in 1861. This was formerly a strong fortress, and sustained without capture various sieges, till being delivered up to the French they dismantled it in 1800. It is still, however, surrounded by a wall, with two gates; it has a cathedral, three other churches, a royal college, hospital, workhouse, and some public baths. Its principal street is wide and handsome, and is lined throughout with porticoes: the other streets are, in fact, mere lanes. Coni is the seat of a court of primary jurisdiction and a bishopric, and the residence of the Intendente and military commandant of the div. It has some silk fabrics, and carries on a considerable trade, being a sort of *entrepôt* to Turin and Nice.

CONGEVERAM (Canchipuram, the golden city), a considerable town of Hindostan, prov. Carnatic, distr. Chingleput, in which it is the chief military station under the Madras presidency. It stands in a valley 36 m. WNW. Madras, and 25 m. E. Arcot; lat. 12° 49' N., long. 79° 41' E. It is tolerably populous, and covers a large space of ground, which is in great part occupied by extensive gardens and cocoa plantations. It has two remarkable pagodas; one, dedicated to Siva, contains many pillars handsomely sculptured, and some well-carved figures of elephants, &c.; the other, which is smaller, has a great deal of curious workmanship and sculpture, which, for truth of proportion and delicacy of execution, is scarcely surpassed by any other Hindoo edifice. There are numerous weavers amongst the pop.; who manufacture red handkerchiefs, turbans, and cloths for native dresses. Small pagodas and choultries, or travellers' houses, abound both in the town and its vicinity; the valley of Congeveram is fertile, contains many substantial tanks, and appears in a prosperous state.

CONNAUGHT, one of the four provs. into which Ireland is divided, on its W. coast, containing the cos. of Galway, Leitrim, Mayo, Roscommon, and Sligo. (See IRELAND.)

CONNECTICUT, one of the smallest of the U. States, in the N. part of the Union, between lat. 40° 58′ and 42° 2′ N., and long. 71° 55′ and 73° 50′ W., having N. Massachusetts, E. Rhode Island, W. New York, and S. Long Island Sound; length, E. to W., 90 m.; average breadth, about 54 m.; area, 4,674 sq. m. Pop. 460,147 in 1860. The state ranks third in the Union as to density of pop., having 96 individuals to the sq. m. Surface generally undulating. A chain of mountains of inconsiderable height runs N. and S. through the W. part of the state. The principal river is the Connecticut: it rises in New Hampshire, and having passed through Massachusetts, intersects this state nearly in its centre; and then bending to the E., falls into Long Island Sound, a little below Newhaven, after a course of 410 m., 250 of which have been made navigable by means of locks and canals. Along the coast are several excellent harbours; the best are those of New London and Newhaven. Climate very variable; an extreme degree of heat and cold are experienced at different seasons; but the sky is usually serene, and the country healthy. There are some sterile districts; but the soil is for the most part fertile, and (for America) well cultivated. European grains, Indian corn, flax, hemp, and culinary vegetables, are raised in abundance; orchards are numerous, and apples so plentiful that cider is a considerable product. The pasture-lands are good; large herds of cattle are reared, and butter and cheese are made in large quantities. Farms vary in size from 50 to 200 acres. There are mines of iron ore, lead, and copper; but, excepting the first, none of them are wrought. Marble, black-lead, porcelain clay, and freestone, are found in many parts. The chalybeate waters of Stafford are celebrated. Manufactures occupy more attention than rural industry, and are more considerable, in proportion to the population, than in any other state of the Union. Rhode Island excepted. The principal are those of cotton and woollen stuffs, iron and tin ware, leather, fire-arms, carriages, powder, clocks, gin, and snuff. There were 49 savings banks on April 1, 1863, with an invested capital of 23,446,936 dollars. A considerable coasting trade and traffic with the W. Indies are maintained. The principal articles of export are cattle, horses, mules, grain, fish, candles, soap, butter, and cheese. The state is divided into eight counties. Hartford is the chief city, and is, in conjunction with Newhaven, the seat of government; the other principal towns are Middletown, New London, and Norwich. There contain several colleges, learned societies, and public schools. The state school-fund, founded in 1831, is the most considerable of any in the Union; the capital amounted, Feb. 28, 1863, to 2,019,426 dollars, while the revenue was 132,589 dollars. Yale College, founded at Saybrook in 1700, and removed in 1716 to Newhaven, contains the finest cabinet of minerals in the Union, and an extensive library. The legislature of the state consisted, in 1863, of a senate of 21 mems., and a H. of Representatives of 237 members. The senators, representatives, governor, and lieut.-governor are all elected every year, on the first Monday in April, by the vote of all male citizens who have resided one year in the state, and have attained the age of 21. The judges of the supreme courts are appointed by the assembly, and hold their offices during good behaviour, or until they are

70 years of age, when they must retire. Connecticut sends four members to the national H. of Representatives, and two senators to the national senate. This portion of the Union was first colonised in 1635 and 1638, by two colonies united in 1665. Its subsequent progress has been one of almost uninterrupted prosperity.

CONSTANCE (au. Constantia, Germ. Kostnitz or Costnitz), a city of the grand duchy of Baden, cap. circ, same name, or Seekreis (Lake Circle), finely situated on the Rhine, at the point where it emerges from the Lake of Constance, 100 m. SSE. Carlsruhe, 26 m. E. Schaffhausen, on the terminus of the Basel-Constance railway. Pop. 7,818 in 1861. Constance is a highly interesting city, from its historical associations. In the 15th century it is said to have contained from 50,000 to 40,000 inhab.; and its streets and many of its buildings remain unaltered since that period, though several of them are wholly, or almost wholly, deserted. It is fortified by a wall flanked with towers, and surrounded by a ditch; has three suburbs, one of which, Petershausen, is on the opposite bank of the Rhine, but communicates with the city by a long covered wooden bridge built upon stone piers. The cathedral or minster, begun in 1052, is a handsome Gothic structure with a lofty steeple, commanding an extensive view of the lake and country, as far as the mountains of Vorarlberg and the Grisons. The doors of the main portal are curiously carved; and the choir is supported by sixteen pillars, each of a single block. A fine high altar, and several interesting tombs and relics, attest the ancient wealth and grandeur of the see, which was formerly the most considerable in Germany, and had large possessions in, and jurisdiction over, Switzerland. A plate of metal let into the floor of this cathedral, near the entrance, marks the spot where John Huss stood when he was condemned in 1415. The Franciscan convent, the first prison of Huss, is now a ruin; and the Dominican convent, to which he was afterwards removed, has been converted into a cotton factory. The kaufhaus (market-hall), erected in 1388, is interesting, as being the place of meeting of the famous Council of Constance, held from 1414 to 1418. The concourse of ecclesiastics and others, from all parts of Christendom, at this council was such that not only the houses in the town were crowded, but booths were erected in the streets, while thousands of pilgrims were encamped in the adjacent fields. Religious processions, dramatic representations, and entertainments of every description, hourly succeeded each other; and thousands of individuals were employed solely in transporting thither the choicest delicacies of Europe. The great object of this council was to vindicate the authority of general councils, to which the Roman pontiff was declared to be amenable. And having done this, the council proceeded to depose three popes or antipopes, John XXIII., Gregory XII., and Benedict XIII.; they next elected Martin V., and thus put an end to a schism which had lasted forty years. But, notwithstanding its merit in these respects, the Council of Constance is justly infamous, for the treacherous seizure and execution of John Huss and Jerome of Prague, notwithstanding the safe-conduct granted to the former by the Emperor Sigismund, the president of the assembly, who wanted power or inclination effectually to vindicate his pledge. Huss suffered at the stake, on the 6th of July, 1415; and Jerome, who had attracted him to the council, was burnt on the 30th of May, 1416. The opinions of Wycliffe were also condemned; and an order was issued to commit his works and bones to the flames. Various relics of this period, and a collection of

Roman and German antiquities found in the neighbourhood, are preserved in the *bonifacus*.

Constance contains an ancient palace, a lyceum, a hospital, a conventual school for females, several collections of art and science, and a theatre. The suburb of Peterhausen contains a grand ducal residence, formerly a Benedictine abbey; that of Kreuzlingen is fortified, and possesses a convent, in the church of which there is some elaborate carving. The suburb of Brald is the scene of the martyrdom of Huss and Jerome. On the bridge across the Rhine there are mills for various purposes.

Constance is the seat of the circle and district government. It was a place of considerable commercial importance till the period of the Reformation, since which it has, until very recently, progressively declined. The chief resources of its inhabitants are derived from the culture of fruit and vegetables, some trade, the navigation of the lake, and a few manufactures, chiefly of cotton cloth and yarn, and silk fabrics, which have latterly been a great deal extended. This is one of the oldest towns in Germany. It was founded or enlarged by the Romans in the 4th century. It was a free imperial city till 1548, when Charles V. placed it under the ban of the empire; next year it was attached to the Austrian dominions, and in 1805 to Baden.

CONSTANCE (LAKE OF), (an. *Lacus Brigantinus* or *Suevicus*, Germ. *Bodensee*), a lake of Central Europe, the largest belonging to Germany, between lat. 47° 29' and 47° 49' N., and long. 9° 2' 30" and 9° 45' E., surrounded by the territories of Baden, Würtemberg, Bavaria, Austria (Vorarlberg), and Switzerland. Length, N.W. to S.E., about 34 m., greatest breadth about 8 m.; area, about 200 sq. m.; elevation above the level of the sea, 1,255 ft.; greatest depth, 964 ft. Its most N. portion consists of a narrow prolongation, called the Ueberling Lake. The Rhine enters the Lake of Constance on the S.E., and issues from its N.W. extremity at the city of Constance, connecting it with the lake called the Unter or Zeller-see, which contains the fertile isl. of Reichenau, and is sometimes considered part of the Lake of Constance. The banks of the latter are mostly flat or gently undulating, and distinguished for their fertility. They abound with corn-fields and orchards, and some tolerable wine is grown on them. The S. shore especially is studded with a picturesque line of ruined castles and other remains of the middle ages; and both sides are crowded with numerous towns and villages, the principal of which are Landau, in Bavaria; Friedrichshafen, a summer resort of the king of Würtemberg, Meersburg, and Ueberling, in Baden; Arbon, in Switzerland; and Bregenz, in the Austrian dominions. The waters of this lake are green, clear, and subject to sudden risings, the cause of which has not been satisfactorily explained. Numerous aquatic birds and crustacea inhabit this lake; and it is abundantly stocked with fish. Its navigation is somewhat dangerous, owing to sudden squalls; considerable traffic, however, takes place upon it, and a number of steamboats run almost hourly from Constance to the different ports situated around it.

CONSTANTINA (vulg. *Kossantinah*), an inland city of N. Africa, Algeria, cap. of its E. prov., beyond the Lesser Atlas, on a peninsulated height, surrounded on three sides by the Rummel, or Wad-el-Kebir (*Ampsaga* of the ancients), which runs in part through a deep ravine, crossed by an ancient bridge. 111 yards above the water, and 113 yards in length; 190 m. E.S.E. Algiers; lat. 36° 21' N., long. 6° 8' E. The hill, on which the city stands, appears to have been separated from the opposite heights of Setub-el-Mansurah by an earthquake, or some other natural convulsion. On the S.W. side it gradually declines downwards to the plain, and on that side only the city is accessible. The present city is about 1½ m. in circ. Pop. 34,500 in 1861, of whom 6,500 Europeans, about a half Kabyles, a fourth Moors, and the rest Turks and Jews. The ancient city was much larger, extending on the other side of the ravine, and down into the plain.

Constantina is strong, as well by art as by nature; the walls on the land side are 5 ft. thick, and have, in many parts, encountre behind them. There are 4 gates, all of Arabic construction, built however, in great part, of the materials of Roman edifices; the superb gates, with columns of red marble, mentioned by former travellers, no longer exist. On its N. side, on the most elevated part of the plateau, is the Kasba, or citadel, occupying the site where was formerly the Numidian citadel, and more recently the Roman capital, parts of both which edifices still exist. The palace, built within these few years, is a large edifice, handsomely fitted up. There are said to be 18 mosques, exclusive of chapels, but none of them deserve any especial notice. Streets narrow and dirty; houses generally two stories high, covered with tiled roofs, à dos d'âne; they are constructed of brick, raised on a foundation of stones, the remains of the ancient buildings. Many of them are large and well furnished, and there are no indications of extreme poverty in any class of the inhabitants. There are many remains of antiquity; but these have suffered much of late years, having been taken down, and employed as materials for the fortifications. The bridge over the ravine, already alluded to, was originally constructed by the Romans. There are also several Roman cisterns, and a church, probably of the era of Constantine, with arches. The inhabitants are industrious; the principal manufactures are those of saddles, bridles, boots, slippers, and garters; a few coarse blankets are also made; and the late bey employed 25 men in the manufacture of gunpowder. A considerable trade is carried on with the S., the inhab. receiving gold-dust, ostrich feathers, slaves, and the finer sort of haiks, both silk and wool, in return for corn, saddlery, and articles of European manufacture. From 1,500 to 1,800 mule-loads of corn used to be annually sent to Tunis. The land round the town is fertile, and mostly belongs to the community. The actual cultivators pay four-fifths of the produce as rent.

A French force of 8,000 were foiled in an attempt to take this city in 1836, and suffered much on their retreat. In the following year another French army, proceeding from Bona, sat down before it on the 6th of October, and took it by storm, after a desperate resistance, on the 13th of the same month.

CONSTANTINOPLE, so called from its founder, or rather restorer, Constantine the Great (Turk. *Stamboul*), a famous city of Turkey in Europe, cap. of the Turkish dominions, and the first city of the Mohammedan world; a distinction which it has held since 1453, when it ceased to be the cap. of the Eastern empire. Its situation, whether considered in a commercial or political point of view, is the finest imaginable; and it seems naturally fitted to be the metropolis of an extensive empire. It occupies a triangular promontory near the E. extremity of the prov. of Roumelia (an. *Thrace*), at the junction of the sea of Marmara with the Thracian Bosphorus, or Channel of Constantinople, being separated from its suburbs of Galata, Pera, and Casim-Pacha by the noble harbour called the Golden Horn; lat. 41° 0' 17" N., long. 29° 50' 2" E. Pop. uncertain, but supposed to amount, including the suburbs, to above a million.

Constantinople is shaped somewhat like a harp; the longest side of the triangle being towards the sea of Marmora, and the shortest towards the 'Golden Horn.' Its length, E. to W., is about 3½ m.; breadth varies from 1 to 4 m. Its circ. has been variously estimated at from 10 to 23 m.; but measured upon the maps of Kauffer and Le Chevalier, it appears to be about 12½ m. in circuit, and contains, according to Dallaway and Gibbon, an area of about 2,080 acres. Like Rome, Constantinople has been built on seven hills, six of which may be observed, distinctly enough, from the port, to rise progressively above each other from the level of the sea to 200 ft. above it; the seventh hill, to the S.W. of the others, occupies more than one-third of the entire area of the city. Each of these hills affords a site to some conspicuous edifice. The first is occupied by the Seraglio; the second crowned with the Burnt Pillar, erected by Constantine, and the mosque of Othman; the mosques of the sultans Solyman, Mohammed, and Selim stand on the summits of the third, fourth, and fifth; the W. walls of the city run along the top of the sixth; and the Pillar of Arcadius was erected upon the seventh.

This amphitheatre of peopled hills, with its innumerable cupolas and minarets interspersed with tall dark cypresses, and its almost unrivalled port, crowded with the vessels of all nations, has, externally, a most imposing aspect, to which its interior forms a lamentable contrast. The expectations of the stranger are, perhaps, nowhere more deceived. The streets are narrow, crooked, steep, dark, ill-paved or not paved at all, and dirty; though, by reason of the slope of the ground on either side towards the sea and harbour, and the great number of public fountains, much of the filth is conveniently cleared away. Adrianople Street, running from the gate of the same name to the Seraglio, is the only one deserving the name of street; the rest are mere lanes. The houses are mostly small and low, being built of wood, earth, or, at the best, of rough or unhewn stone. It is the palaces, mosques, fountains, bazars, khans, &c. that make so splendid a show at a distance. Dallaway (Constantinople, p. 70) and Mr J. Hobhouse believe that its streets were anciently not more regular than at present; and that from the frequent and sudden devastations by fire, mentioned by the Byzantine historians, its houses were formerly, as now, built mostly of wood or other fragile materials. About a century after its restoration, Constantinople is reported (Gibbon, ch. xvii.) to have contained 'a capitol, or school of learning, a circus, 2 theatres, 8 public and 153 private baths, 52 porticos, 5 granaries, 8 aqueducts, or reservoirs of water, 4 spacious halls for the meetings of the senate or courts of justice, 14 churches, fourteen squares, 344 streets, and 4,388 houses, which for their size or beauty deserved to be distinguished from the multitude of plebeian habitations.' It contains, at present, 14 royal and 332 other mosques, or houses of Mohammedan worship, 40 colleges of Mohammedan priests, 183 hospitals, 36 Christian churches, several synagogues, 130 public baths, nearly 300 khans, and numerous coffee-houses, caravanserais, and public fountains; besides some extensive subterranean cisterns, the aqueduct of Valens, several remarkable pillars and obelisks erected by the Greek emperors, and other monuments which, together with the walls, the castle of 'Seven Towers,' &c. are interesting remains of antiquity, and for the most part in a tolerable state of preservation. (André-ossy, p. 124; Cours Méthodique de Géographie, p. 625; Hallhouse.)

Constantine surrounded the city with walls, chiefly of freestone, flanked at variable distances by towers. These have been, in many parts, demolished at different periods by the violence of the sea, and by frequent earthquakes, and on the side facing the port are especially in a very ruinous state. The city was increased towards the W. by Theodosius II., who built the walls on the land side which still bear his name. These consist of a triple range, rising one above another, about 16 ft. apart, and defended on the outside by a ditch 25 to 30 ft. broad, and 12 to 16 ft. deep. The outer wall is now very much dilapidated, and in many places is only a little above the level of the edge of the ditch; it seems never to have had any towers. The second wall is about 12 ft. in height, and furnished with towers of various shapes, from 50 to 100 yards apart. The third wall is above 20 ft. high, and its towers, which answer to those of the second, are well proportioned. These walls are constructed of alternate courses of brick and stone; and the inner ones, notwithstanding the ravages of time, earthquakes, and numerous sieges, are still tolerably perfect. On both the other sides of the city the walls are only double, and, generally speaking, not so lofty. They are frequently adorned with crosses and other ornaments, which have not been removed by the Turks; and in many parts there are bas-reliefs, and inscriptions by the Greek emperors who have built or repaired the several portions. When Dr. Clarke visited the place, he says there were in all 478 mural towers, and probably about the same number still exist.

Constantinople originally possessed 48 gates, 16 of which opened on the land side, 19 towards the Golden Horn, and 13 towards the Propontis. Only 7 gates now exist, or are at present used, on the land side, the centre one of which, the Top-Kapousi, or Cannon Gate, is the Porta Sancti Romani, through which Mohammed II. made his triumphal entry into the city. Near the S.W. angle of the city is the Heptapyrgium, or castle of 'Seven Towers' (though it has now but four towers), an irregular fortress, supposed to have been built about the year 1081. It was enlarged in succeeding ages, and in great part rebuilt by Mohammed II., who made it a state prison, it being useless as a fortress. The Golden Gate, erected by Theodosius to commemorate his victory over Maximus, was originally profusely ornamented with beaten gold, and surmounted by a gilded bronze statue of Victory. Mohammed II. walled it up. When Wheeler saw it, it was still adorned with bas-reliefs, in white marble, representing several scenes of classic mythology; but these must have disappeared, since more recent travellers speak of it as only an ordinary arch between two large marble pillars, and ornamented with Corinthian pilasters, 'd'un style assez médiocre.'

The ancient Byzantium, founded by Byzas the Megarean, B.C. 656, and ultimately destroyed by Severus, not long before the building of Constantinople, occupied the first hill or apex of the triangle, at present the site of the Seraglio. Its walls, according to Herodian, were Cyclopean, and so skilfully adjusted that they seemed like one entire mass. Most authors say that there are no vestiges of Byzantium; but Dr. Webb affirms that 'part of the walls of this very ancient city are actually standing, and cut off the gardens from the adjoining streets.' The Seraglio, which is believed to be of about the same extent as the ancient Byzantium, is nearly triangular, about 3 us, in circuit, and entirely surrounded by walls; those of the city forming its boundary towards the port and sea of Marmora, while on the W. it is shut in by a lofty wall with gates and towers, built by Mohammed II., soon after the capture of Constantinople. Its whole surface is 'irregularly covered with detached suites of apartments, baths, mosques, kiosks, gardens, and groves of cypress.' The apartments are chiefly on the top of the hill, and the gardens below, stretching to the sea. Though externally picturesque, from the contrast of its light and elegant minarets with its dark, solemn, and stately trees, the Seraglio is unmarked by anything to characterise it as the habitation of royalty. The greater part of its interior is not open to the public; but those acquainted with it say that it contains little worthy of admiration, and that that little has been imported from Europe. The palace consists of various parts built at different times, and according to the taste of successive sultans, without any regard to uniformity or architectural rule; and it is, therefore, a heap of houses clustered together without any kind of order. Outside are two courts, the first of which is free to all persons, and is entered by the Bab-a-hoomajön or Sublime Porte, the principal of the gates on the city side,—a ponderous, unsightly structure, covered with Arabic inscriptions, guarded by fifty porters, and having a niche on either side in front, in which the heads of state offenders are publicly exposed. The irregular but spacious area into which this gate leads, formerly the Forum Augusti, contains the mint, the vizier's divan, and other state offices, the infirmaries for the sick belonging to the Seraglio, and the church of St. Irene, believed to have been built by Constantine, and in which the second general council was held by Theodosius, (Androssy, 16.) This church resembles St. Sophia on a small scale, and contains much marble and mosaic work; the Turks have converted it into an arsenal. The second quadrangle is smaller, being about 300 paces only in diameter; but is more regular and handsomer than the former. It is laid out in turf, intersected by paved walks, and supplied with fountains. On the left hand are the treasury, the divan, or hall of justice, and the smaller stables (the larger stables, containing, according to Tournefort, 1200 horses, are in another place, facing the sea of Marmora). On the right are the offices of the attendants, nine kitchens, and the entrance to the private apartments. All round the court runs a low gallery, covered with lead, and supported by columns of marble. At its farther end is the tall Corinthian column erected by Theodosius the Great to commemorate his victory over the Goths; and near it are the Bala-Saadi, 'Gates of health and happiness,' which lead to the throne-hall, the royal library, the apartments of the sultan, the harem, and other suites of rooms, embellished with a costly but tasteless magnificence. The throne-hall is isolated, lofty, built in great part of marble, and adorned with handsome marble columns and stained glass windows. The throne itself is a canopy of velvet fringed with jewels, supported by four columns enriched with gold, pearls, and precious stones; but its effect is destroyed by horse-tails, and other paltry ornaments, suspended from the roof. The state apartments closely resemble each other; their chief furniture consists of sofas, carpets, and mirrors. The walls are wainscotted with jasper, mother-of-pearl, and ornamented ivory inlaid with mosaic flowers, landscapes, and sentences in Arabic. The pavilions of the harem are built upon arches, and roofed by domes covered with lead or spires with gilded crescents. They have many balconies, galleries, cabinets, &c. Baths of marble and porcelain, rich pavilions overlooking the sea, marble basins, and spouting fountains, are sprinkled over the rest of the surface within the Seraglio. The number of inmates, and others connected with the Seraglio, have been estimated at upwards of 10,000; but this is probably much beyond the mark. All are provided for by the sultan. And Tournefort (Lett. v. vol. ii. p. 108) states that, when he visited the place, besides 40,000 oxen yearly, the purveyors furnished for the use of the Seraglio daily 200 sheep, 100 lambs or goats, 10 calves, 200 hens, 200 pairs of pullets, 100 pairs of pigeons, and 50 green geese. But, notwithstanding this general accuracy of Tournefort, we have no doubt that in this instance he was misled, and that Mr. Elliott (i. 395) has done right in rejecting this statement.

On the third hill is the Eski Serai, or Old Palace, said to have been the residence of the later Greek emperors; a building surrounded by a lofty rectangular wall about 1 m. in circuit, and to which, when a sultan dies, his harem is removed. It presents nothing remarkable.

The mosques of Constantinople have all an open space around them, generally planted with trees, and refreshed by fountains. The principal mosque, the celebrated St. Sophia, stands on the W. declivity of the first hill, near the Sublime Porte of the Seraglio. It was begun and finished under the Emperor Justinian, between the years 531 and 537. It is in the form of a Greek cross, 269 ft. in length, by 243 ft. wide, or about 3-5ths the length of St. Paul's, London, by nearly the same width; and surmounted in its centre by a dome, the middle of which is 180 ft. above the floor. The dome is of an elliptical form, and much too flat to be externally beautiful, its height not exceeding 1-6th part of the diameter; which is 115 ft., or 15 ft. more than that of the dome of St. Paul's, and 10 ft. less than that of St. Peter's at Rome. It is lighted by twenty-four windows ranged round its circumference, and rests upon four strong arches, the weight of which is firmly supported by four massive piles, strengthened on the N. and S. sides by four columns of Egyptian granite. The present dome is not coeval with the building; the original one, which was less lofty and more circular, having been thrown down by an earthquake twenty-one years after its erection. There are, besides, two large and six smaller semi-domes, the whole of which blending internally with the principal one, form altogether a magnificent expanse of roof. Four minarets, but each of a different shape, have been added to this mosque by the Mohammedans. The building has been outwardly so much patched and propped up in different ages, that it has lost whatever beauty it may have originally possessed, and is now a heavy, unwieldy, and confused-looking mass. It is entered on the W. side by a double vestibule, about 30 ft. in breadth, which communicates with the interior by nine bronze doors, ornamented with bas-reliefs in marble. The interior is spacious and imposing, not being broken by aisles or choirs; but the variegated marble floor is covered with mats and carpets; the mosaics of the dome, &c., have been whitewashed over by the Turks; the coloured seraphim and other sculptures have been in great part destroyed, and the general coup d'œil is spoiled by 'a thousand little cords depending from the summit to within 4 ft. of the pavement, and having at the end of them lamps of coloured glass, large ostrich-eggs, artificial flowers, tails, vases and globes of crystal, and other mean ornaments.' (Hobhouse.) The building is said to contain 170 columns of marble, granite, porphyry, verd-antique, &c., many of which were brought from the temple of Diana at Ephesus, and other ancient structures. The cost of the building, owing to the ambiguity of the Byzantine historians, cannot be accurately determined; but Gibbon observes (Decline and Fall, ch. xl), that 'the sum of our million sterling is the result of the lowest computation.' Yet with all this, Justinian seems to have failed in making St. Sophia a really fine edifice. Mr J. Hobhouse says of it.— 'My impression was, that the skill of the one hundred architects, and the labour of the ten thousand workmen, the wealth of an empire, and the ingenuity of presiding angels, had raised a stupendous monument of the heavy mediocrity which distinguished the productions of the sixth century from the perfect specimens of a happier age.'

Most travellers agree in preferring the mosques of Sulyman the Magnificent and Achmet to St. Sophia. The former of these, called the Sulymania, was built in 1556, of the ruins of the church of St. Euphemia at Chalcedon. It is 216 ft. in length by 210 ft. broad, and has a handsome dome, supported on four columns of Theban granite, 60 ft. high, pavements, galleries, &c., of marble, several minor cupolas, four fine minarets at the angles, a spacious court-yard leading to it, with galleries of green marble on either side, and twenty-eight leaded cupolas, and a very handsome gate of entrance ascended to by a flight of at least twenty marble steps. The whole of this mosque is in very good taste. Behind it, in an enclosed court shaded with trees, is the mausoleum of Sulyman, an octagonal building, and the handsomest of all the royal sepulchral monuments, which are very numerous in the city. The mosque of Achmet I., between St. Sophia and the Propontis, was constructed in 1610, and has a very beautiful marble pavement. It is the only mosque which possesses six minarets. These are of extraordinary height and beauty, and each has three Saracenic galleries surrounding it. The Osmanie, or mosque of Osman, completed in 1755, has a light and elegant dome, and is tastefully ornamented. The other principal mosques are those of Mohammed II., Bajazet, Selim II., Mustapha III., the Valida, &c. The last named, founded by the mother of Mohammed IV., contains a double row of fine marble pillars, chiefly brought from the ruins of Troy. Another mosque has become an object of curiosity, from its containing a sarcophagus, supposed to have been that of Constantine the Great. Many of the mosques have, like St. Sophia, been formerly Greek churches; the remainder have been erected mostly by the Turkish sovereigns, the viziers, or wealthy individuals. The royal foundations comprise a college, with a public library, a hospital, and an almshouse; and the mosques in general have attached to them some charitable institutions. They derive their revenues from villages and lands belonging to them, and held by a tenure not dissimilar to that of our churchlands. The incomes of some of the mosques are very large; that of St. Sophia has been said to amount to 800,000 livres annually (Hobhouse); Dallaway says 3,000l. (p. 58.)

The largest space in Constantinople is the At-Meidan, or Horse-course, the ancient Hippodrome. It is at present 300 yards long by 150 wide. (Elliott.) In it formerly stood the celebrated group of four horses, originally transported thither from Rome, and afterwards removed to the cathedral of St. Mark, at Venice. It still contains the granite obelisk from Thebes, set up by Theodosius the Great; the broken pyramid of Constantine Porphyrogenitus, shorn of its bronze plates; and between the two, the hollow spiral brass column, which originally supported the golden tripod in

the temple of Delphi. The last consists of three serpents, twisted together. Mr. Elliott describes it as being at present about 12 ft. high, mutilated at the top, and much injured in the centre. Close to the Hippodrome formerly stood the imperial palace, the senate-house, and the forum. No remains of these exist. The Hippodrome continues to be used by the Turks for feats of activity, both on horseback and on foot.

In the Adrianople Street is the 'Burnt Pillar,' so called from its having been blackened by repeated conflagrations. It was erected by Constantine the Great, and was originally 120 ft. in height, and composed of ten blocks of porphyry, each upwards of 9 ft. high, and 35 ft. in circumference, resting on a marble pedestal 20 ft. in height. The joints of the column were concealed by embossed brass or iron hoops, and the whole supported a colossal bronze statue of Apollo, said to have been the work of Phidias. (Gibbon, ch. xvii.) The statue and three of the blocks were thrown down by lightning in 1150, and the whole height is now only 90 ft. In the centre of the city the pillar of Marcian may be seen, enclosed in a private garden. It is of granite, with a Corinthian capital of white marble, surmounted by an urn of the same material. The finest of all, the Arcadian or Historical column, erected early in the 5th century, and covered with a series of bas-reliefs, representing the victories of Theodosius the Great, was taken down at the end of the 17th century, and only 14 ft. of it are now above ground. (Dallaway, pp. 113, 114.) Dallaway readily traced the vestiges of the Bucoleon palace, built by Theodosius II., opposite the Sea of Marmora.

The means for the supply of Constantinople with water are worthy of remark. The aqueduct of Valens, which communicates with another and more extensive, though similarly constructed aqueduct, beyond the walls, continues, as anciently, to convey water into the city. It was originally built by the Emperor Hadrian; and rebuilt first by Valens, and again by Sulyman the Magnificent. It runs from the summit of the third to that of the fourth hill, consisting of a double tier of forty Gothic arches in alternate layers of stone and brick. It is in some parts considerably dilapidated, and its E. extremity especially is much injured. Andreossy estimates that it was originally nearly 1,280 yards in length; it is now, he says, 660 yards long, and about 74 ft. in height. (Andreossy, p. 412.) There are several other aqueducts on both sides the port, which, as well as the bends, or reservoirs, without the walls, were chiefly the work of the Greek emperors, though they have been augmented and kept in repair by the Turkish sultans. All the water that supplies Constantinople comes from Belgrade, a village a little to the S.E. of the city. An American traveller (Sketches in Turkey in 1831–32) has estimated the quantity brought into the city at 15,000,000 gall. every twenty-four hours, and states that the various water-courses about Constantinople must exceed 60 m. in length. The whole of these important works are under the superintendence of an officer with great powers, and are annually inspected by the sultan.

The Greek emperors constructed many large cisterns within the walls, both open and subterranean; the former have been gradually filled with earth, and converted into gardens; but several of the subterranean ones still remain entire. The principal are contiguous to the Hippodrome. The largest, or Cisterna Basilica, is a vault of brickwork, covered with terrace composition, 336 ft. in length by 182 ft. broad, and supported by 336 marble pillars, each 40 ft. 9 in height. (Clarke, pp. 170, 171.) It still affords water to the inhabitants, being supplied by the city aqueduct, and many wells are sunk into it. Another vault, the *Cisterna Maxima*, called by the Turks 'the thousand and one columns', is, according to Mr. Elliott, 240 ft. long by 201 wide, 5 fathoms deep, and sustained by 14 rows of 16 double columns of white marble, the capital of one pillar forming the base for another. This cistern is now dry, and half filled with earth: it is at present used as a rope-walk, or place for spinning silk. Not far from it is another cistern, also dry, but capable of holding 1,500,000 gallons of water. (Elliott.)

The fountains are amongst the chief ornaments of the city. There are almost as many as there are streets; one is to be found in every piazza, market-place, and mosque. They are uniformly square, with a spout at each side and a leaden roof; and are generally gilded, painted, inscribed with sentences from the koran, or otherwise decorated. The public baths are built mostly of marble, on a uniform plan, and covered with little flat domes: their interior is generally handsome and spacious; and the price of a bath, the first of oriental luxuries, is so low that a poor man can enjoy a hot bath for a penny. In the better sort, coffee, sherbet, and pipes are furnished to the bathers. Few houses of consequence are unprovided with a commodious bath.

The greater number of the *khans* (bazaars) and *bezestins* (or changes) are built of stone or brick. The *khans* and *serais*, or inns, are for the most part royal or charitable endowments, each capable of accommodating from 100 to 1,000 persons. They consist of open squares, surrounded by rooms, in several stories, and possess recommendations for outweighing their want of architectural elegance. Most of them are intended for travelling merchants. Excepting a small present to the servant at departing, strangers are gratuitously lodged in them, and during their residence in the city are masters of their rooms, of which they keep the keys. 'They are for all men, of whatever quality, condition, country, or religion soever, and the construction of them has contributed to attract the merchandise of the furthest boundaries of Africa and Asia to the capital of Turkey. During fires or insurrections, their iron gates are closed, and they afford complete security to the persons as well as goods of the merchants.' (Hobhouse.)

The covered bazaars have more the appearance of a row of booths in a fair, than a street of shops. Each is appropriated to a separate article of merchandise. The shops are all open in front, and under cover of a common roof; the stalls of the windows, as in ancient Pompeii, forming the counters. (Elliott.)

The better sort of coffee-houses are open on one side, and have a fountain playing in the midst of a range of marble seats, and recesses furnished with pillows, mats, and stuffed carpets. A row of them, near the Solymania, is frequented by opium eaters; but there are not nearly so many of these individuals in the Turkish capital as is generally imagined. All the public buildings of Constantinople are crowned by cupolas, in consequence of which, their number, at a distance, seems to be as great as that of the private houses. The domes, as well as the minarets of all the sacred structures, are terminated by a crescent.

The houses of opulent Turks are built, like the khans and most other large houses in the E., round a court, which has always a fountain playing in its centre. Occasionally these residences are not ill-constructed: but the common dwellings are mere comfortless wooden boxes, with unglazed

windows, and without fire-places. (Dallaway.) House-rent is said to be higher in Constantinople than in any other city in the world; this is ascribed to the frequency of fires, a house not being reckoned worth more than five years' purchase, if so much. The fact is, that these fires are very often intentional; and that they are resorted to for the same purpose that public meetings and petitions are got up in England—to make the sultan aware of the public discontent, and of the necessity of appeasing it. A striking instance of this sort is given by Porter (Observations on the Turks, p. 92), and similar instances may be found in other travellers. We do not know that anything could better evince the atrocious nature of the despotism under which Turkey has so long groaned, than the circumstance of its making oppression a sort of constitutional resource!

The Golden Horn (an. *Sinus Byzantinus*) occupied the ancient name of the promontory on which Byzantium was built, and which was first called *Keras Sphaireos*, *Chrysorrhoas*, or Golden Horn, (Clarke's Trav. viii. 170, 182.) It is one of the finest and most secure harbours in the world, capable of containing upwards of 1,000 sail of the line, and of a depth sufficient to admit of goods being landed on the quays from the largest ships, in many places without the assistance of boats. It extends from the Seraglio Point inland, for about 4½ m. NW., with a breadth varying from a furlong to half a mile. At its entrance it has a light-house on either side, and is defended by some batteries on the Seraglio Point. At its upper end the ancient *Lycus*, now called the Sweet Waters, falls into it, and it is continually cleared by the stream of that river, in conjunction with a current setting into it from the Bosphorus. It exhibits a most picturesque and animated scene, covered as it always is, with merchant vessels, steamers, ships of war, and craft of all descriptions. Along the SW. side of this harbour, the *Fanas*, or Greek quarter, extends nearly the whole way from the seraglio to the western walls of the city. Beyond the walls, on the same side, is the suburb of Aivali or Eyoup, in the mosque of which the new sultan is always installed in his office. The upper extremity of the harbour, anciently called the *Mariciderm Mare*, is now, as formerly, a low, marshy, unwholesome tract; but about 1½ m. beyond, in the Valley of the Sweet Waters, the Sultan Achmet III, laid some grounds laid out in the French style, with the addition of gaudy kiosques, coffee-houses, &c. to which the inhabitants of the city and suburbs frequently resort.

On the NE. side of the harbour are the suburbs of Galata, Tophanah, Pera, and Cassim Pasha. The first two stand side by side on the shore opposite to the Seraglio, and E. end of the city. Pera is on a hill to the NE. behind both; and Cassim Pasha to the NW. of all, opposite the Fanar. Galata was built by the Genoese in the 13th century, and walled in the 16th. It is about 4 m. in circuit, divided into three quarters, and inhabited chiefly by Europeans and other merchants. It has twelve gates and contains a citadel or tower, 140 ft. high, built by the Emperor Anastasius, a very fine fish-market, several mosques, a handsome fountain, and a great number of shops. Tournefort remarks that 'one tastes in Galata a smatch of liberty not to be found elsewhere in the Ottoman empire. Galata is, as it were, Christendom in Turkey; taverns are tolerated, and the Turks themselves freely resort thither to take a cheerful glass.' Tophanah (or arsenal) contains an arsenal, artillery-barracks, and magazines, and a cannon foundry. Pera is beautifully situated, but irregularly built and ill-paved. It is about 2 m. in length; its pop. is almost wholly Frank, and it contains the residences of most of the European ambassadors, besides four Catholic and one Greek church, a monastery of dervishes, and a Mohammedan college. In 1831 it suffered severely from a fire, which destroyed 10,000 houses, amongst which were the palaces of nearly all the ambassadors, and property estimated to be worth 8,000,000 dollars. Casim Pasha contains the great naval arsenal, dock-yards, barracks, quarters for slaves and workmen, the palace of the capitan-pasha, &c. There are no suburbs on the W. side of Constantinople, only a few cemeteries and scattered cottages beyond the walls. The immediate vicinity towards Thrace consists generally of an expanse of open downs; the solitude and desolation which prevail on this side are remarkable. On the Asiatic continent, about a mile across the Bosphorus from the Seraglio Point, stands the town of Scutari (an. *Chrysopolis*); and about 2 m. S. of it, the ancient Chalcedon.

Manufactures few; the principal are those of silk and cotton fabrics, arms, morocco leather, saddlery, horse-trappings, shoes, and other articles of ordinary use and consumption, together with those of tobacco-bowls, tubes, and mouth-pieces. The latter branches of industry employ many hands, and one bazaar is devoted solely to those articles. The *keff-kil* earth is dug in several parts of Asia, rudely fashioned into pipe-bowls in Constantinople, and exported in large quantities to Hungary, Germany, and France, where the bowls are re-manufactured, and receive the name of *merschaums*. The best tubes are formed of the stems of the cherry or jessamine tree, both of which are largely cultivated in the neighbourhood for the purpose. The rank of a person in this city being determined by his pipe, it is often adorned in a very costly manner, and the price of a *tchibouque* may vary from 20 paras to 20,000 piastres. The fisheries of Constantinople are by no means unimportant: the sea and harbour abound with shoals of tunny, swordfish, &c., and the "sweet waters" with a profusion of fresh-water fish.

The foreign trade is considerable. Imports, chiefly corn, iron, timber, tallow, and furs, from the Black Sea; cotton stuffs and yarn, tin, tin-plates, woollens, silks, cutlery, watches, jewellery, paper, glass, furniture, indigo, cochineal, orpiment, &c., from England and other parts of Europe; corn and coffee from Alexandria; a good deal of coffee from Brazil and the W. Indies, in American bottoms, which traffic has latterly much increased; sugar, partly from the E., but chiefly from the W. Indies; wax, copper, drugs, gums, porcelain, overland from China (a trade which existed in the time of the Romans); and slaves, chiefly from Georgia, Circassia, and Africa. Exports comparatively trifling; chiefly silk, carpets, hides, wool, goats' hair, potash, wax, galls, bullion, and diamonds. The trade, which, as a whole, is less than might have been expected in a city of such size, is for the most part in the hands of English, French, Armenian, and Greek merchants, and Jew-brokers. The more wealthy Armenians (a nation constituting a considerable proportion of the pop.) are money-changers, bankers, jewellers, physicians, and apothecaries; the lower classes are employed in the most laborious occupations. As china-printers and muslin-painters, the Armenians here surpass most European artisans. The Greeks are much less numerous than before the Greek revolution.

Constantinople is the residence of a Greek, an Armenian, and a Catholic-Armenian patriarch. The first has now in authority in the newly erected kingdom of Greece. Elementary schools are to be met with in every street; and in every quarter there are Turkish free-schools for the poor, the expenses of which, as well as the board and lodging of many of the pupils, are defrayed out of the revenues of the mosques. The number of these elementary schools amounted to above 1,300 in the year 1844, according to an official return, while of upper schools, or colleges, there were 322. Some of the *medresses*, or colleges attached to the mosques, have between 400 and 500 students, who are lodged and educated on the foundation, and have each several professors, the salaries of the principal among which are equivalent to about 100l. a year. In these seminaries all the members of the *ulemah* are educated, and no one can be admitted into the hierarchy or the law without having first graduated in one of them. The Mohammedan law had prohibited the Turks from learning European tongues; but the late sultan established a school for the instruction of native youths in French, outside the Seraglio. The French and Austrian embassies have schools for the acquisition of Turkish by their members. There are 10 public libraries, 9 or 10 of which are attached to the royal mosques, and contain about 2,000 manuscripts each, mostly copies of the koran and commentaries on it. The private library in the Seraglio is richer than any of the rest, and contains some valuable Greek and Latin MSS.

'Amid the novelties that strike the European on his arrival, nothing surprises him more than the silence that pervades so large a capital. He hears no noise of carts or carriages rattling through the streets; for there are no wheeled vehicles in the city, except a very few painted carts, called *arabahs*, drawn by buffaloes, in which women occasionally take the air in the suburbs, and which go only at a foot-pace. The contrast is still more strongly marked at night. By ten o'clock every human voice is hushed.' Constantinople is not a healthy place of residence for strangers; it is subject to sudden changes of temperature; and the strong *cursins* or N. winds, which prevail in the summer, act to injury to trade, by preventing the access of ships from the Ægean and Mediterranean, are also detrimental to public health and comfort. Earthquakes, the plague, and devastating fires, often consuming 2,000 or 3,000 houses, cause great destruction of life and property. In other respects, too, it is a most unpleasant place of residence to a European or other stranger. In many cases property is not secure, justice is notoriously corrupt, the police is bad, the place is infested with cats, rats, and, as most travellers say, with herds of wild dogs, and birds of prey, which act as scavengers. Sir J. Hobhouse states, that 'Constantinople is distinguished from every other capital in Europe, by having no names to its streets, no lamps, and no post-office.'

The history of this renowned city for a lengthened period is given by Gibbon. It was originally founded by Byzas, from whom it derived the name of *Byzantium*, some 656 B.C.; and having been destroyed by Severus, was rebuilt, A.D. 328, by Constantine, who made it the cap. of the Roman empire. On the subjugation of the Western empire by the barbarians, Constantinople continued to be the cap. of the Eastern empire. Its wealth and magnificence were celebrated during the middle ages. It has sustained numerous sieges, but has only been twice taken: first, in 1204, by the Crusaders, who retained it till 1261; and, lastly, by the Turks, under Mohammed II., May 29, 1453, when the last remnant of the Roman empire was finally suppressed.

CONSUEGRA, a town of Spain, prov. Toledo, on the Amarguillo, 30 m. SE. Toledo. Pop. 6,879 in 1857. The town has 2 churches, 3 convents, a palace, and a variety of Roman inscriptions and antiquities. On a neighbouring hill are the remains of its ancient castle. Streets tolerably regular, but narrow and steep. The vicinity produces grain, wine, oil, barilla, and soda, and has quarries of azure-coloured marble, jasper, and other stones. It has fabrics of coarse stuffs, baize, and serge.

CONWAY, a town and parl. bor. of N. Wales, co. Caernarvon, hund. Isaf, on the estuary of the Conway river: 15 m. NW. by W. from the Menai Bridge, and 224½ m. NW. London by the London and North Western railway. Pop. of par. 1,355, and of parl. bor. 2,523 in 1861. The town, which is of a triangular shape, stands on a steep slope, and is surrounded by lofty walls, fenced with 24 round towers. The lower face of the triangle borders on the river; and at its farthest angle, on the verge of a slate rock, its magnificent castle

'Frowns o'er old Conway's foaming flood.'

This noble structure was built by Edward I. in 1284. 'A more beautiful fortress never arose. Its form is oblong, placed in all parts on the verge of the precipitous rock. One side is bounded by the river; another by a creek full of water at every tide, and most beautifully shaded by hanging woods. The other two sides face the town. Within are two courts; and on the outside project eight vast towers, each with a slender one of amazing elegance issuing from its top, within which had been a winding staircase. In one of the great towers is a fine window, in form of an arched recess, or bow, ornamented with pillars. The great hall suited the magnificence of the founder. It extended 130 ft. in length, was 32 broad and of a fine height. The roof was supported by eight noble arches, six of which still remain. There were two entrances into the fortress, one from the river, and one from the town. (Pennant's Tour in Wales, iii. 133, &c., ed.) The town is poor and inconsiderable, without trade or manufacture of any sort. Much of the ground within the walls is used for gardens. The bor. is one of the contributory bors. to Caernarvon in returning a mem. to the H. of C. The limits of the bor. extend to a considerable distance beyond the walls of the town. The port dries at low water.

The old and dangerous ferry over the river has been superseded by a magnificent suspension bridge completed in 1826. The length of the bridge between the centre of the supporting towers is 327 ft.; and it is elevated 18 ft. above high-water mark. The construction of this and the Menai Bridge, and the excavations and improvements that have been made at Penmaenmawr and other places, have made the road, formerly so dangerous, from St. Asaph and Conway to Bangor and Anglesea, one of the best and safest in the empire.

COOCH-BAHAR, or VIHAR, a rajahship of Hindostan, prov. Bengal, between lat. 26° and 27° N., long. 89° and 90° E.; having N. Bootan, and on all other sides the distr. of Rungpore, with which it is incorporated; length about 90 m.; greatest breadth, 50 m. Its rajah also possesses some tracts beyond the Mogul limits of Bengal, not subject to tribute, and on which opium is extensively cultivated. The S. part of this country is rice and fertile, but N. of the cap. it is low, marshy, and interspersed with jungle and coarse rank vegetation. The Cooch or Rajbansgi tribes eat various kinds of flesh, and are considered by the Bengalese and other Hindus as very low and impure. Notwithstanding provisions are cheap as compared with other districts, and rents low, many of the natives, especially in the N., are so indigent

as to be frequently obliged to sell their children for slaves. The cultivation is common. In 1582, Abul Fazel relates that the chief was a powerful sovereign, having Assam and Camroop under his government, and able to bring into the field 1,000 horse and 100,000 foot; in 1661 this territory was conquered by the Moguls, and devolved, with the rest of Bengal, to the British in 1765.

COOKSTOWN, an inl. town of Ireland, prov. Ulster, co. Tyrone, on the Ballinderry river, 9 m. W. from Lough Neagh. Pop. 2,093 in 1831, and 3,257 in 1861. The town consists of one long street, planted on each side, with a transverse street crossing it. The par. church is a large Gothic structure; there are three meeting-houses for Presbyterians, two for Methodists, and one for the town is a Rom. Cath. chapel. There is also a dispensary and a constabulary station. Linens are manufactured here, and bleached in the vicinity. Markets for grain are held on Tuesdays, and for general sales on Saturdays; fairs on the first Saturday of every month.

COORG (Kodagoo), an anc. rajahship of Hindostan, prov. Mysore, formerly independent, but now under the prov. of Madras. It lies for the most part between lat. 12° and 13° N. and intersected by the 76th parallel of E. long.; having N. and E. the Mysore territories, and on all other sides those of the Madras presidency; length N. to S. about 70 m., breadth very irregular: area, 2,340 sq. m. To the W. it is hemmed by the W. Ghauts, parallel to which there is a succession of lofty narrow ridges, enclosing valleys of various extent. The chief elevations are, Tadiamalmole 5,781 ft., and Soobramany 5,662 ft. above the sea; the principal valley is that between Markara and Naknaad, 18 m. long, by 15 m. broad, with an extremely uneven surface, in the lowest part of which runs the Cavery. The geology of Coorg strongly resembles that of the Neilgherries; the principal rocks being sienite, granite, and greenstone, and the subordinate ranges uniformly capped with the detritus of these, cemented by argillaceous earth, and coloured by oxide of iron; porcelain clay frequently occurs. The whole country, with few exceptions, is covered with forests, but not over-limited with jungle, excepting in the vicinity of the Mysore dominion; where elephants, game, and other wild animals are found. Saadal and other valuable woods abound. Both the botany and zoology of this region offer a rich field to observers, but have hitherto been but little studied. From the greater elevation, the temperature is much below that of either Malabar or Mysore, and remarkable for its equality. The climate is, in general, highly suitable to European constitutions, though the monsoon rains, from June to Sept., often fall with great violence. The Coorgas are a Nair tribe of martial habits; they have few towns, or even villages, of any size, preferring to live in jungles and wilds. They cultivate rice in the valleys, which are very productive, though the quantity of land under culture be very trifling. The pastures are excellent, and cattle are abundant. Manufactures limited to the blankets worn by the pop. Cotton cloths are imported. Contrary to the custom in Malabar on the other side of the Ghauts, hereditary rights and possessions in Coorg descend in the male line, and some family disputes arose in 1808, in consequence of Beer Rajendra (who had expelled the troops of Tippoo from Coorg) having left, at his death, the government of his dom, to his daughter, to the prejudice of his brother, who was ultimately established in possession by the British government. The country was annexed to the British dominions in 1832.

COOTEHILL, an inl. town of Ireland, prov.

Ulster, co. Cavan, on a small river of the same name, 26 m. W. by N. Dundalk. Pop. 2,176 in 1831, and 1,924 in 1861. The town consists of four broad streets, neatly laid out and well kept. It has a par. church, a Rom. Cath. chapel, two Presbyterian meeting-houses, places of worship for Moravians, Quakers, and Methodists: a market-house, a court-house, and a bridewell. There is an extensive trade in grain and coarse linens. The corn markets are held on Saturdays, the general markets on Fridays; and fairs on the second Friday in every month.

COPENHAGEN (Kiöbenhavn, merchants' haven), the cap. of Denmark, a well-built city, principally on the E. coast of the island of Zealand, but partly also on the contiguous small island of Amak, the channel between them forming the port. Pop. 155,143 in 1860. The town is well fortified. The ramparts, which extend for about 8 m., are flanked with bastions, and surrounded by a deep ditch filled with water. It is also defended by a very strong citadel, and by the Three Crowns battery, constructed at the entrance of the port on a bank of sand, about 1,500 fathoms from shore. The city is usually divided into the old town, the new town, and Christianshavn. The first is the largest and most populous, and having at different periods suffered much from fire, most part of it has been rebuilt on an improved plan, though some of the streets are still narrow, crooked, and inconvenient. In the new town the streets are straight and broad, though generally ill-paved, the squares regular and spacious, and the private houses and public buildings the finest in the city. The part called Christianshavn, from its being built by Christian IV., stands on the island of Amak. It is intersected by various canals, and communicates with the other parts of the town by bridges. Public buildings numerous, and many of them superb. Among others may be specified the castle of Christiansborg, destroyed by fire in 1795, and since rebuilt. It has a picture-gallery, comprising a complete collection of Danish pictures, with a fine collection of the Dutch school; a chapel ornamented by bas-reliefs from the chisel of Thorwaldsen; and the royal library, one of the best in Europe, containing, exclusive of manuscripts, above 450,000 volumes. The part of the new town called Amalienborg was entirely rebuilt by Frederick V., between 1743 and 1765. It consists chiefly of an octagon, divided by four broad rectangular streets, in which is the palace of the king. In the centre is a bronze equestrian statue of Frederick V., erected by the East India Company. There are also the royal palaces of Rosenborg and Charlottenborg, appropriated to public purposes; the university, the town-house, the theatre, the exchange, and the barracks. The cathedral church of Notre Dame, nearly destroyed during the bombardment in 1807, has been rebuilt; and is enriched by statues of Christ and the Apostles, by Thorwaldsen. The tower of the church of the Trinity, 115 ft. in height, is used as an observatory; it also contains the library of the university, and the great globe of Tycho Brahe. The church of Our Saviour is reckoned the finest in the town: its spire, nearly 300 ft. in height, is a masterpiece of art. The educational, literary, and scientific establishments of Copenhagen, rank with the first of their class, and reflect infinite credit on the government and the people. Besides the university, to which we have elsewhere alluded, there is a polytechnic school, a metropolitan school, a royal school of marine, a royal school for the higher military sciences, and a normal school. There is, also, a royal society similar to that of London, a Scandinavian society, and a society of
VOL. II.

northern antiquaries. The academy of arts is and has long been in a flourishing condition. Besides the royal library in the palace of Christiansborg, the university library has above 100,000 volumes, and a large collection of manuscripts. The Classen library, bequeathed to the public by the general of that name, is mainly devoted to science and natural history; and, exclusive of these, there are several other minor but still valuable collections.

The hospitals are numerous and well conducted. The most splendid is that of Frederick V. The lying-in hospital has attached to it a school of midwifery and a foundling hospital. The royal institution for deaf and dumb admits patients from all parts of the kingdom.

If distillation be excepted, the manufactures of Copenhagen are neither very extensive nor important. There are about 700 distilleries, mostly on a small scale, and about 50 breweries, with sugar refineries, tobacco manufactories, and soap-works. Cotton and woollen goods, linens, silks, gloves, and hats are also produced, but in limited quantities.

The trade of the port is considerable. There arrived, in 1860, 4,013 vessels, of 115,392 lasts (of 2 tons each); in 1861, 3,252 vessels, of 127,224 lasts, and, in 1862, 3,243 vessels of 126,363 lasts. Of these, there were British vessels 120 in 1860; 160 in 1861; and 219 in 1862. The principal articles of import are—anchors, pitch, and tar, from Sweden and Norway; flax and hemp, masts, sail cloth, and cordage from Russia; tobacco and rice, from the United States; wines and brandy from France; and coal, earthenware, cotton, and colonial produce from England.

The harbour is formed, as already stated, by the channel or arm of the sea running between Zealand and the opposite island of Amak. The entrance to it is narrow; but the water is sufficiently deep to admit the largest men-of-war. There are dry docks, and every facility for the building and repairing of ships. Copenhagen is the station of the Danish navy. The bank of Copenhagen, founded in 1736, was remodelled in 1818: it is now a private institution. The charge of the public health is entrusted to a commission. The police is under a special establishment; and besides the garrison, the citizens are formed into a national guard.

Copenhagen is not a very ancient city, having been founded in 1168. It has at different periods suffered severely from fire, particularly in 1728, 1794 and 1795; but how disastrous soever of the time, these visitations were in the end advantageous, the narrow streets and wooden houses of which the town formerly consisted having been replaced by broad streets and handsome stone buildings. Besides the loss of her fleet Copenhagen suffered severely from the bombardment by the English in 1807, and by an inundation in 1824. But she has fortunately recovered from both these disasters, and by her literary and other establishments has placed herself at the head of civilisation in the north of Europe.

The environs of Copenhagen are celebrated for their beauty. Fredericksberg, a magnificent castle, the summer residence of the king, stands on a rising ground within a moderate distance of the city. Its gardens are open to the public, and are a favourite resort. Frederiksborg, another royal residence, is situated about 21 m. N. Copenhagen. It is a vast, but incongruous pile, partly brick and partly stone, and partly of Greek and partly of Gothic architecture. It has some fine pictures and a series of portraits (partly imaginary) of the sovereigns of Denmark.

COPIAPO, the most N. town of Chili, formerly the cap. of the prov. of same name, now incorperated with that of Coquimbo. It stands on the right bank of the rivulet of Copiapo, 80 m. from the Pacific, and 170 m. NNE. Coquimbo; lat. 27° 10′ S., long. 71° 5′ 15″ W. Pop. estimated at 3,000. The town is connected by railway with Caldera. Most of the houses are built of sundried bricks whitewashed; and, the better to resist earthquakes, used to be constructed with great solidity; but in 1819 it was destroyed by the earthquake that caused such devastation throughout a great part of Chili. In 1822 it suffered severely from another earthquake. The harbour of Copiapo on the Pacific is good; and at a small village on the shore most of the ore from the mines of the prov. is smelted, and the metal is exported.

COQUIMBO, or LA SERENA, a sea-port town of Chili, in the N. part of the republic, cap. of the prov. of same name, on the Chuapa, near its mouth; 270 m. NNW. Santiago; lat. 29° 54′ 43″ S., long. 71° 18′ 40″ W. Estimated pop. 7,000. The town is clean, and tolerably well laid out; streets intersect each other at right angles; houses mostly of sun-dried bricks, and only one story in height, but interspersed with numerous gardens of fruit-trees and evergreens. It has several churches and convents, a public school, and a hospital. It is the seat of the intendent of the prov., and is the residence of many families, and in some sort the cap. of N. Chili, as well as the chief mercantile port. The exports amounted to 8,201,268 dollars in 1863, and to 4,888,870 dollars in 1864; the imports to 818,356 dollars in 1863, and to 678,041 dollars in 1864. (Report by Mr. Consul Tait, dated March 18, 1865.) The harbour or bay of Coquimbo is large, well-sheltered, and secure at all seasons. There is sufficient depth of water for ships of large burden, 9 fathoms being found 300 yards off shore, and nearly 3 fathoms close in shore. A railway connecting Coquimbo and Serena with Los Cardos and the mines in the interior was opened on 26th April, 1862. The line was entirely constructed and is worked by Englishmen. Coquimbo was founded by Valdivia in 1544. About 25 m. up the valley of Coquimbo are some singular parallel roads, of which Captain Hall has given an account.

CORDOVA (an. Corduba and Colonia Patricia), a famous city of Spain, cap. prov. and kingdom of the same name in Andalusia, on the Guadalquivir, 73 m. NE. Seville, and 145 m. SSW. Madrid, on the railway from Madrid to Seville and Cadiz. Pop. 42,909 in 1857. The city occupies a large oblong space of sloping ground, enclosed by walls flanked with towers originally erected by the Romans, and afterwards, repaired, strengthened, and extended by the Moors. But a great part of this space is now covered with gardens and ruined buildings, and but little remains of its ancient grandeur. Streets narrow, crooked, and dirty; and a few either of the public or private buildings are conspicuous for their architecture; the latter seldom exceed two stories in height. The great square, Plaza Real, or de la Constitucion, is, however, large and regular; the houses surrounding it are lofty, and furnished with porticoes and balconies. There is a suburb of some extent on the S. bank of the river, with which the city communicates by means of a stone bridge of 16 irregular arches, 860 ft. in length, and 23 ft. in width, constructed by the Moors towards the close of the 8th century, and the approach to which is guarded by an old Saracenic castle, still maintained in a state of defence. The city contains a cathedral, 13 parish churches, about 40 convents, 7 hospitals, a foundling and another asylum, city-hall, bishop's palace, 3 colleges, be-

sides other schools. By far the most remarkable public edifice is the cathedral or mosquita, formerly a mosque, built by the Moors at the latter end of the 8th century upon the ruins of a Gothic church, which is itself believed to have replaced a Roman temple. Both of these edifices have apparently furnished many pillars and other materials for the present building. The mosquita externally is unprepossessing, and little calculated to attract notice; but the singularity of its interior strikes every one with astonishment. It is a gloomy labyrinth of pillars, 856 ft. in length N. to S., by 394 ft. broad E. to W., and lighted only by the few doors that remain open, and some small cupolas in different parts of the roof, which latter is flat, and only 35 ft. above the pavement; being supported in most places by a kind of double arcade of horse-shoe arches. The columns supporting these arches, and which amount to several hundreds, are of jasper, marble, porphyry, granite, verd-antique, and various other materials, and differ as much in their architectural as in their geological character. They are all, however, of the same height; 'for the Arabs, having taken them from Roman buildings, served them in the same manner that Procrustes did his guests: to the short ones they clapped on monstrous capitals and thick bases; those that were too long for their purpose had their bases chopped off and a diminutive shallow basset placed on their head.' (Swinburne's Travels, ii. 89.) The number of aisles or naves is lengthwise 19, and transversely from 32 to 35. A considerable space at the S. end was parted off for the use of the Jesuits, and now serves for the chapter-house, sacristy, and treasury of the cathedral. In the front of this space is what is called the sancarium, an octagon Moorish sanctuary, 15 ft. in diameter, richly ornamented without and within, and domed over by a single block of white marble, carved into the form of a scallop-shell. Adjoining this, in 1815, another small apartment was brought to light, preserving, in a remarkable degree, its pristine decorations. The gorgeousness of this little chamber will perhaps give an idea of that of the building generally in the time of the Moors; for the splendour of almost all the rest of the mosquita has entirely disappeared; the gilding and ornaments of the roof, the arabesques and inscriptions on the walls, and the mosaics of the pavement, have nearly all vanished; and of the 24 gates, formerly plated with brass, and curiously embossed, only 6 remain open. The sacristy contains some tolerable paintings, and the church is very rich in jewels, plate, and silks. The mosquita stands within a court planted with orange-trees, palms, and cypresses, and surrounded with a cloister, on the N. side of which a square belfry has been built.

The bishop's palace is a large and rather handsome building, containing a suite of state apartments, in one of which there is a large collection of portraits of the bishops of Cordova. Previously to the late civil war, 2,000 poor persons were daily supplied with food from the bishop's kitchen, which mistaken bounty seems sufficiently for the swarms of beggars with which the town is infested. The famous palace of the Moorish sovereigns is now unoccupied; it had been converted into a royal stud-house, where the best horses in Spain were reared; the stables are now empty. The manufactures have participated in the general decay of the place; there are at present only some trifling fabrics of ribands, lace, hats, baize, and leather after the Moorish fashion: the latter article was formerly very extensively manufactured; and was known in commerce by the name of cordovan, and from it the term roadwaiter has been derived. In 1833, a handsome quay was erected above the bridge, but as there is but little trade, and the river is for 9 months in the year navigable only for boats, the quay would seem, like many other public works in Spain, to be more for show than for use.

Cordova is said by Strabo to have been founded by the Romans under Marcellus; but as there were several distinguished persons of that name, this leaves the epoch of its foundation uncertain. No mention is made of it before the age of Cæsar and Pompey, but it soon after attained to great distinction as a rich and populous city, and a seat of learning. (Cellarii, Not. Orbis Antiqui, i. 96.) In 572 it was taken by the Goths, and in 692 by the Moors, under whom it became the splendid cap. of the 'Caliphate of the West,' and subsequently of the kingdom of Cordova. In 1236, however, it was taken and almost wholly destroyed by the impolitic zeal of Ferdinand III. of Castile, and has never since recovered its previous prosperity. Cordova has given birth to some illustrious men, among whom may be specified the two Senecas, Lucan the poet, and the famous Arabic physicians, Avicenna and Averroes.

Cordova, an inl. town of Mexico, state Vera Cruz, at the E. foot of the volcano of Orizaba, and on one of the roads between Vera Cruz and La Puebla; 50 m. SW. the former, and 72 m. ENE. the latter city. Estimated pop. 6,000. Streets wide, regular, and well paved; houses built mostly of stone. In the centre of the town there is a large square, three sides of which are ornamented with Gothic arcades; the fourth is occupied by the principal church, an elegant structure, richly decorated within. Cordova contains two convents, each with a hospital attached; many of its edifices have domes, towers, or steeples. Cotton and woollen fabrics and leather are made here; and there are besides numerous distilleries, sugarmills, and bee-hive farms; but the principal employment of the inhab. is the culture of tobacco and coffee. The vicinity is extremely fertile, and abounds in fruits, timber, game, and fish.

COREA (called by the natives, Chaou-Sien, by the Chinese, Kaou-le, and by the Manchoo Tartars Sol-ho), a marit. country of NE. Asia. tributary to China, consisting of a large oblong-shaped peninsula, with an adjoining portion of the continent, and a vast number of islands, which are especially numerous on the W. coast. The whole of the dominions lie between lat. 33° and 43° N., and long. 123° 50' and 129° 30' E.; having E. the Sea of Japan; N. the Straits of Corea; W. the Yellow Sea and Gulf of Leao-tong; NW. the prov. Leao-tong; and N. Manchoo Tartary. From the latter it is separated by a mountain chain, and the Tha-men-Kiang river, and from Leao-tong mostly by a wooden wall or palisade. Length, NW. to SE., 550 m.; average breadth of the peninsula, about 130 m. Total area, inclusive of islands, probably about 80,000 sq. m. Corea is generally mountainous. A mountain range runs through it longitudinally, much nearer its E. than its W. coast. The E. declivity of this range is steep and rugged; to W. one declines gradually into a fertile and well-watered country. All the principal rivers run W. and discharge themselves into the Yellow Sea; the chief is the Ya-lu-kiang in the NW., which is navigable for large ships to about 22 m., and for small vessels for a distance of nearly 170 m. above its mouth. The coasts, as well of the islands as of the continent, are generally rocky and difficult of access; though there are some spacious and secure harbours. The climate

of the N. is very rigorous; the Tha-men-kiang, for six months in the year, is thickly frozen over, and barley is the only kind of corn capable of being cultivated in that region; even the N., though in the same lat. with Sicily and Malta, is said to experience sometimes very heavy falls of snow. The climate of this part of Corea, however, must be on the whole mild, since cotton, rice, and hemp are staple products; and Gutzlaff conjectures (Voyages, &c., p. 319), that many other plants, common to the S. of Europe, flourish. Gutzlaff observes, 'In point of vegetation, the coast of Corea is far superior to that of China, where barren rocks often preclude any attempt at cultivation; but here, where the land is fertile, the inhab. do not plough the ground.' (p. 337.) Agriculture may be better farther inland, but on the coast it is much neglected; wheat, millet, and ginseng are amongst the chief articles cultivated. Tobacco was introduced by the Japanese about the beginning of the 17th century, and potatoes, by Gutzlaff and Lindsay, in 1832. The orange, citron, hazle-nut, pear, chesnut, peach, mulberry, Morus papyrifera, Ficus saccharinus, and the wild grape, are common; but the art of making wine from the latter seems to be unknown. An ardent liquor is, however, made from rice. The mountainous parts of the N. are covered with extensive forests; pines are very abundant on the coasts; and in the interior there is a species of palm producing a valuable gum, from which a varnish, giving an appearance little inferior to gilding, is made. Oxen, hogs, and other domestic animals common to Europe are reared; there is a spirited breed of dwarf horses not exceeding 3 ft. in height; panthers, bears, wild boars, cats, and dogs, sables (whose skins form an important article of tribute), deer, and an abundance of game. Storks, and water-fowl of many sorts, are found; raywans of 30 or 40 ft. in length are said to be met with in the rivers, and venomous serpents are not rare. In the winter, whales, seals, &c. visit the shores. The mineral kingdom produces gold, silver, iron, rock salt, and coal.

People.—The pop. has been estimated at 15,000,000, but there are no real grounds for this estimate, which, we have little doubt, is greatly beyond the mark. Gutzlaff represents the coasts as thinly inhabited. We have elsewhere stated that the Coreans are superior in strength and stature to the Chinese and Japanese, but that they are inferior to either in mental energy and capacity, (See Asia.) They are gross in their habits, eat voraciously, and drink to excess. The dress of both men and women is very similar to that of the Chinese, though the Coreans do not, like that people, cut off their hair. Their houses are also like those of China, being built of bricks in the towns, and in the country are mere mud hovels; each house is surrounded by a wooden stockade. Their language or languages are peculiar, differing from those of their immediate neighbours. In writing they use alphabetic characters, though the symbolic characters of the Chinese are also understood and sometimes resorted to. They have a copious literature, and are very fond of reading, as well as of music, dancing, and festivities. Polygamy is permitted, but the women do not appear to be under such restraint as in China. (M'Leod.) The religion of the upper orders is that of Confucius, while the mass of the people are attached to Buddhism; but neither appears to have much influence. Christianity, which was introduced by the Japanese, appeared to be extinct when Gutzlaff visited Corea in 1832.

Manufactures and Trade.—The manufactures are few; the principal are a kind of grass-cloth, straw-plait, horse-hair caps, and other articles for domestic use; a very fine and transparent fabric woven from filaments of the *Urtara japonica*, cotton cloth, and a very strong kind of paper made of cotton, and rice-paper; which articles, together with ginseng, skins, some metals, horses, and silk, constitute the chief exports. What trade there is, is principally with Japan, from which they import pepper, aromatic woods, alum, buffaloes', goats', and bucks' horns, and Dutch and Japanese manufactured goods. There is, however, some trade with China carried on at Fungwang-ching (the *Phœnix-town*), beyond the Leao-tong border; but this trade is conducted with great secrecy, in consequence of the jealousy of the government of any intercourse with foreigners. This jealousy is so great, that no Chinese is allowed to settle in Corea, nor any Corean to leave his own country; Europeans are scarcely ever suffered to land, or remain any length of time on the coast; and the N. frontier is abandoned for many miles, in order that no communication should take place with the Manchoo Tartars. Little skill in ship-building is displayed by the Coreans; their junks do not carry more than 200 tons, and are quite unmanageable in a heavy sea. In the construction of their fishing-boats not a nail is used. Metallic articles and money are rare. The only coin in circulation is of copper, but payment is often made in silver ingots.

Corea is divided into 8 prora, King-hi-tao, the cap., is placed on the Kiang river, in about 37° 40' N. lat., and 127° 30' E. long., or about the centre of the kingd. The gov. is said to be despotical; most of the landed property in the country belongs to the king, of whom it is held in different preteus as fiefs, which revert to the sovereign at the decease of the occupier. Besides the revenues from these domains, a tenth part of all kind of produce belongs to the king. Justice is in many respects very rigid. Rebellion, as in China, is punished by the destruction of the rebel with his entire family, and the confiscation of their property. None but the king may order the death of an official person; the master has always power over the life of his slave. For minor crimes the general punishment is the bastinado, which is pretty constantly at work. The Chinese interfere but little with the internal administration of Corea; but the king can neither assume the government, nor choose his successor or colleague, without the authority of the court of Pekin, to which he sends tribute four times a year; the tribute consists of ginseng-root, sable-skins, white cotton paper, silk, horses, and silver ingots. The Corean ambassador is treated at Pekin with but little consideration. There seems reason to believe, that, like some other states in Asia, Corea is tributary to the more powerful nations on either side, and that it also sends a yearly tribute to Japan, consisting of ginseng, leopards, skins, silks, white cotton fabrics, and horses; but for which an acknowledgment is made in gold articles, fans, tea, and presents of silver to the ambassadors.

History.—Corea was known to the Chinese from a very early period, and is reputed to have been civilised by the Chinese sovereign Khilau, about 1,120 years before our era. After experiencing several revolutions, it was invaded and conquered by the Japanese in 1692, who, however, abandoned their conquest in 1694. The Coreans having called in the aid of China during that struggle, Corea has since formed a subordinate part of the Chinese empire.

CORELLA, a city of Spain, prov. Navarre, in a fertile plain on the Alama, 13 m. W. Tudela, 12 m. SE. Calahorra. Pop. 5,023 in 1857. The town has two churches, 4 convents, a hospital, and some remains of an ancient castle. The inhabitants are employed in the extraction of liquorice and madder juice, and in the manufacture of brandy, oil, and flour.

CORFE-CASTLE, a market town and bor. of England, co. Dorset, Blandford div., hund. Hasilor, in the Isle of Purbeck, 32 m. SSW. Salisbury. Pop. of par., 1,901 in 1861. The town is most probably indebted for its origin to its castle, on a steep rocky hill, a little to the N., formerly a place of considerable strength. But its importance, in more modern times, was owing to its having enjoyed the privilege of returning two mems. to the H. of C. from the 14th of Elizabeth down to the passing of the Reform Act, by which it was disfranchised. The inhab. are mostly employed in the neighbouring clay-works and quarries.

CORFU (an. *Corcyra*), an island in the Mediterranean, forming (since 1864) part of the kingdom of Greece, and the most important, though not the largest, of the Ionian Islands. It lies between lat. 39° 20' and 39° 50' N., and long. 19° 35' and 20° 6' E.; off the N. part of the coast of Albania, from which it is separated by a channel only 3-5ths of a m. wide at its N. extremity, 6 m. at its S. extremity, and 15 m. in the centre. The shape of Corfu is elongated; the island describes a curve, the convexity of which is towards the W.; length NW. to SE. 41 m.; breadth greatest in the N., where it is 20 m.; but it gradually tapers towards its S. extremity. Area, 227 sq. m. Pop. 69,414 in 1860, including 3,765 aliens and strangers. The native pop., in 1860, was composed of 33,520 males and 30,129 females—a rather remarkable preponderance of the male sex, particularly in a seafaring population. Surface hilly, particularly in the NW., where the peak of St. Salvador rises 2,979 ft. above the level of the sea. The streams watering it are few and small, and mostly dried up in summer. Climate mild; the mean maximum temp. in the open air for the five years ending December, 1858, was about 80° Fahr.; and the average minimum 31° Fahr.; but Corfu is subject to sudden transitions from heat to cold, owing, amongst other causes, to the proximity of the snowy mountains of Epirus. Earthquakes also are frequent. The more elevated lands are rugged and barren, but the plains and valleys are fertile, and productive of wheat, maize, oats, olive-oil, wine, cotton, flax, and pulse. Corfu yields no currants. Oil is the great staple of this isl., which has, in fact, the appearance of a continuous olive wood, a consequence partly of the extraordinary encouragement formerly given to the culture of the plant by the Venetians. There is an oil harvest every year, but the great crop is properly biennial, the trees being suffered to repose for a year. Next to oil, salt, obtained from saltpans along the shores, oranges, citrons, and other fruits, besides honey and wax, are the other chief articles produced. Corfu is divided into 6 cantons; it sends 12 mems. to the legislative assembly of Greece. Corfu, the cap., is the only town worthy of mention; the rest are mere villages.

The city and port of Corfu lie on the E. side of the island, on the channel between it and the opposite coast, which is here about 5 m. wide; lat. 39° 37' 39" N., long. 19° 56' 34" E. It consists of the town and citadel, both fortified; and has several suburbs, one of which is supposed to occupy the site of the ancient city of Corcyra, founded by the Corinthians about the same time with Syracuse. The citadel, separated from the town by wet ditches and outworks, and an esplanade, is built upon a rocky cape projecting into the sea, and contains the barracks, several military hospital, the former residence of the British lord high commissioner, now the seat of the Greek government, and a lighthouse erected upon a point 233 ft. above the level of the sea. The town has three gates towards the sea, and one on the land side. It is not well built; streets narrow and irregular, and houses mostly small and ill-contrived. Corfu is strengthened by two other fortresses besides its citadel—Fort Neuf and Vido. The latter is built on a small island of the same name (an. *Ptychia*), nearly 1 m. N. from the city, and has had much pains and (British) expense bestowed on its improvement. Corfu contains a cathedral, and several Greek and Roman Catholic churches and chapels, a university, gymnasium, ecclesiastical seminary, and several primary schools. Around it there are some pleasant walks, interesting from classical associations; the esplanade is well planted with trees, and forms an agreeable promenade. The town is well supplied with water, which is conveyed by means of iron pipes from Benizza, a distance of 7 m. Roads have been made from Corfu to most of the principal towns and villages in the island. The harbour between the island of Vido and the city is safe and commodious, and vessels anchor in from 12 to 17 fathoms water. The canal, or channel of Corfu, is a little difficult of navigation, but has deep water throughout; there is a lighthouse on the rock of Tignoso at its N. entrance, and a floating light is moored off the point of Lecchimo near its S. extremity. The city of Corfu is the seat of the supreme court of justice, the chief special courts for the island, and of a Greek archbishop. In 1716 it was unsuccessfully besieged by the Turks, and did not fall into their hands until the end of last century.

Corfu is the chief seat of the external trade of the Ionian Islands. The roads in it are good, having been greatly improved since it has been under British protection. Most of the inhabitants belong to the Greek church. It is believed to be the country of Phæacia, or Scheria, mentioned by Homer, on which Ulysses was wrecked, and afterwards hospitably entertained by King Alcinous. It became afterwards a celebrated colony and naval station of the Corinthians, and a quarrel between it and the mother country led to the Peloponnesian war. It was also an important naval station under the Romans. It belonged successively to the Eastern Empire, the Normans, and Venetians, and shared the fate of the Venetian republic in 1799. The island, with the rest of the Ionian republic, was placed under the protectorate of Great Britain by the congress of Vienna, but ceded to Greece in 1864. A Greek garrison arrived at Corfu, and took possession on the 28th of May, 1864.

CORINGA (*Coringa*), a considerable sea-port town of Hindustan, prov. N. Circars, distr. Rajahmundry, and 33 m. NE. that town; lat. 16° 40' N., long. 82° 44' E. Excepting Blackwood's Harbour, Coringa Bay contains the only smooth water to be found on the W. side of the Bay of Bengal, during the SW. monsoon. A wet dock has been formed, and many small vessels are annually built here. In 1764, a remarkable inundation of the sea took place, destroying much property and many inhabitants.

CORINTH (*Kórinthos*), a famous city of Greece within the Morea (an. *Peloponnesus*), near the isthmus of the same name, between the gulfs of Lepanto (*Corinthiacus Sinus*) on the W., and of Egina (*Saronicus Sinus*) on the E., 7 m. from the canal point of the latter, and 2 m. from the nearest point of the former; lat. 37° 55' 37" N., long. 22° 52' 6" E. Pop. 2,350 in 1860. The town is situated at the N. foot of a steep rock, 1,886 ft. in height, the *Acrocorinthus* or *Acropolis* of Corinth, the summit of which is now, as in antiquity, occu-

pied by a fortress. The present town, though thinly peopled, is of considerable extent, the houses being placed wide apart, and much space occupied with gardens. The only Grecian ruin at present to be found in Corinth is a Doric temple, with but a few columns standing. There are some shapeless and uninteresting Roman remains, supposed to have been baths; but there is nothing approaching to a well-defined building, and we may exclaim with the poet,—

'Where is thy grandeur, Corinth? shrunk from sight,
Thy ancient treasures, and thy rampart's height;
Thy god-like forms and palaces!—Oh, where
Thy mighty myriads and majestic fair!
Relentless war has pour'd around thy wall,
And hardly spared the traces of thy fall!'

The situation of Corinth is extremely advantageous, being placed on a narrow isthmus between the seas that wash the E. and W. shores of Greece, she could hardly fail to become an important emporium; while the Acrocorinthus, if properly fortified, would be all but impregnable, and the possession of the isthmus would enable her to command all access by land between the two great divisions of Greece. No wonder, therefore, that Corinth was early distinguished by the wealth, commerce, luxury, and refinement of her citizens. In the earlier ages of antiquity, the attempt to sail round the Peloponnesus, or to double Cape Malea, was regarded as an undertaking of the greatest hazard; and to obviate this danger, the usual practice was to land goods, coming from the W. shores of Greece, Italy, and Sicily, destined for the E., at the harbour of Lechaeum (the nearest point to Corinth), on the Corinthian Gulf, and to convey them across the isthmus to Cenchreae, on the Saronic Gulf, where they were again shipped for their final destination. The products of the E. coasts of Greece, Asia Minor, and the Black Sea, destined for the W. parts of Greece, Italy, &c., were conveyed through the Corinthian territory in an opposite direction; so that the city early became the seat of perhaps the most important transit trade carried on in antiquity. In addition to this, Corinth at an early period founded Corcyra, Syracuse, and other important colonies; established within her walls various manufactures, particularly of brass and earthenware; had numerous fleets, both of ships of war and merchantmen; and was the centre of an active commerce that extended to the Black Sea, Asia Minor, Phoenicia, Egypt, Sicily, and Italy. In the magnificence of her public buildings, and the splendour of the chefs-d'œuvre of statuary and painting by which they were adorned, she was second only to Athens. The opulence, of which she was the centre, made her a favourite seat of pleasure and dissipation, as well as of trade and industry. Venus was her principal deity, and the temple and statue of the goddess were prominent objects in the Acropolis. Lais, the most famous of the priestesses of Venus, though of Sicilian origin, selected Corinth as her favourite residence; and so highly was she esteemed, that a magnificent tomb (described by Pausanias) was erected over her remains, and medals struck in commemoration of her beauty! In consequence, Corinth became not only one of the most luxurious, but also one of the most expensive places of antiquity, which gave rise to the proverb—

'Non cuivis homini contingit adire Corinthum.'
Hor., Epist. I. 17 36.

The Acropolis is one of the most striking objects in Greece. It has some famous springs, and is in most parts precipitous. Livy calls it, '*Arx inter omnia in immensum altitudinem edita, praeter fonti-*

bus' (lib. 15, § 79) ; and Nonnus says, that it throws its shadow over both seas—

     —' quæ summas caput Acrocorinthus in altum
Tollit, et alterna geminans mare protegit umbra.'
                Theb., lib. 7, lin. 106.

If properly fortified, it would render all access to the Morea by land impracticable ; and as a fortress, it might be rendered not less secure than Gibraltar, (Clarke, vi. 568, 8vo. ed.) It is, in fact, one of the keys of Greece ; and was, therefore, aptly said by the oracle to be one of the horns which a conqueror should lay hold of to secure that valuable heifer the Peloponnesus. The view from its summit is one of the most extensive, and at the same time richest in classical associations, of any in Greece. Athens is seen in the distance ; and the eye wanders over six of the most celebrated of the Grecian states—Attica, Achaia, Bœotia, Locris, Phocis, and Argolis.

The government of Corinth, like that of the other Grecian states, was originally monarchical. It then became subject to the oligarchy of the Bacchidæ, and was again, after a period of ninety years, subjected to kings or tyrants. Periander, the early part of whose reign was that of a Titus, and the latter of a Tiberius, was the last of its sovereigns. At his death the Corinthians established a republican form of government, including, however, more to aristocracy or oligarchy than democracy. It seems to have been judiciously devised ; and the public tranquillity was less disturbed in Corinth than in most Grecian states.

When the Achæans became involved in a war with Rome, Corinth was one of their principal strongholds. Though the Roman senate had resolved upon the destruction of the city, Metellus was anxious to avert the catastrophe ; but his offers to bring about a reconciliation, which might have saved Corinth, were contemptuously rejected, and his deputies thrown into prison. The Corinthians suffered severely for this inconsiderate conduct. The consul Mummius, having superseded Metellus, appeared before Corinth with a powerful army ; and after defeating the Achæans, entered the city, which had been left without any garrison, and was deserted by the greater number of its inhabitants. It was first sacked, and then set on fire ; and it is said that the accidental mixture of the gold, silver, and copper, melted on this occasion, furnished the first specimens of the Corinthian brass, so much esteemed in subsequent ages! Not satisfied with the total destruction of the city, the natives of Corinth who had escaped were carefully hunted out and sold as slaves, their lands being at the same time disposed of to strangers, mostly to the Sicyonians. The destruction of Corinth took place anno 146 B.C. ; and it is worthy of remark that this also was the epoch of the destruction of Carthage, both these great cities having been sacrificed nearly at the same moment to the insatiable rapacity and ambition of Rome. According to Strabo, the finest works of art which adorned Rome in his time had been brought from Corinth ; but it seems pretty clear that many, if not the greater number, of these masterpieces had been destroyed. Polybius, who was present at the destruction of the city, had the mortification to see the Roman soldiers playing at dice on a picture of Aristides, a contemporary of Apelles, for which Attalus king of Pergamus subsequently offered 600,000 sesterces, or about 5,000l. of our money. (Strabo, lib. viii. ; Plin. Hist. Nat., lib. 35, cap. 4, &c.) We need not, indeed, be much surprised that the soldiers should have made use of such a fire-board, when we find the consul himself assuring the masters of the vessels se-

lected to convey the pictures and statues to Rome, that if any of them were lost or injured, he should compel them to supply others in their stead at their own cost! (Velleius Paterculus, lib. I. cap. 13.)

Corinth remained in the ruinous state to which it had been reduced by Mummius, till a colony was sent thither by Julius Cæsar. Under its new masters it once more became a considerable city, as is evident from the account given of it by Pausanias (lib. II.), and is much distinguished in the gospel history. After being sacked by Alaric, it came, on the fall of the Eastern empire, into the possession of the Venetians. The Turks took it from the Venetians, however, retook it in 1687, but lost it again to the Turks in 1715. It is now a principal place in the monarchy of Argolis and Corinth, kingdom of Greece. For some time after the establishment of Greek independence, the city prospered, but it was almost entirely destroyed by an earthquake in 1858.

CORINTH (ISTHMUS OF). Where narrowest, about 6 m. E. from Corinth, this celebrated isthmus is about 5 m. across. The advantages that would result to Corinth, and to the commerce of Greece, by cutting a canal or navigable channel through this isthmus, were perceived at a very early period ; and attempts to accomplish so beneficial a work were made by Periander, Demetrius Poliorcetes, Julius Cæsar, and other Roman emperors ; all of them, however, proved abortive, though parts of the excavations are still visible. This want of success has been variously accounted for ; but we incline to think that it was wholly owing to the difficulty of the ground. The isthmus is high and rocky ; and at a period when the construction of locks was unknown, the canal must either have been excavated to the required level, or been partly excavated and partly tunnelled, either of which operations would have been all but impracticable. As the next best resource, ships were drawn by means of machinery from one sea to another ; but it is clear that none but the smaller class of vessels could be so conveyed.

The isthmus has been repeatedly fortified. The first instance of this of which we have any certain accounts took place on the invasion of Greece by Xerxes. It was afterwards fortified by the Spartans and Athenians in the time of Epaminondas. During the decline of the Eastern empire, the defence of the Peloponnesus principally depended on this bulwark, which was strengthened and renovated under Justinian. It was restored for the last time by the Venetians in 1694, (see Dodwell's Greece, ii. 185, and the authorities there quoted.)

The Isthmus of Corinth was also famous in antiquity for the games celebrated there, every fifth year, in honour of Neptune and of Palæmon or Melicertes, with the utmost splendour and magnificence. They continued in vogue after the Olympian and other public games had fallen into disuse. After the destruction of Corinth the Romans committed the superintendence of the Isthmian games to the Sicyonians ; but on its restoration by Julius Cæsar, Corinth recovered its ancient proficiency. Dr. Clarke discovered at the part of Schœnus, on the E. side of the isthmus, the remains of the temple of Neptune, the theatre, stadium, and other public buildings, described by Pausanias as connected with the Isthmian solemnities.

CORK, a marit. co. of Ireland, prov. Munster, in the SW. part of the island, having N. Si. George's Channel, E. Waterford and Tipperary, N. Limerick, and W. Kerry and the Atlantic Ocean. It is the most extensive of all the Irish cvm., containing 1,769,383 imp. acres, of which about one-third are unimported mountain and bog. It has every variety of surface and soil; in the W. it is rugged and mountainous, but the N. and E. districts are distinguished by their richness and fertility. There is a great deficiency of timber, otherwise the country would be eminently beautiful. Climate extremely mild, but moist. Property principally in very large estates. Tillage farms for the most part small; those of larger size are frequently held in partnership, or have been divided amongst the family of the occupant. Where such practices prevail, agriculture cannot be otherwise than in a very backward state. Potatoes engross a great part of the attention and labour of the smaller class of occupiers; and after them the ground used to be subjected to a series of corn crops, as long as it was capable of bearing any thing. But an improved system has been introduced of late years on several large estates; and better implements and breeds of cattle are now generally met with. Oats is the principal corn crop, but wheat is also extensively produced. There are extensive dairies in the vicinity of Cork and in other districts; and the exports of corn, flour, provisions, and other articles of agricultural produce from Cork, are very extensive. The average value of land, per 100 acres, was £10½ in 1841; £8¾ in 1851; and £6¾ in 1841. (Census of Ireland, part v. 1864.) Different branches of the linen manufacture have been established at Cork and other towns, and there are some large distilleries. The coast of Cork is deeply indented by the sea, and has some of the finest bays and harbours in the world, among which Bantry Bay and Cork Harbour are pre-eminent. Principal rivers, Lee, Bandon, Blackwater, Ilen, Funcheon, Bride, and Awing. Principal towns, Cork city, Youghal, Bandon, Kinsale, Mallow, Fermoy. Cork contains, exclusive of the city of Cork, 22 baronies and 269 parishes, and returns eight members to the H. of C., viz. two for the co., two for the city of Cork, and one each for the boro. of Youghal, Bandon, Mallow, and Kinsale. Registered electors for co. 15,716 in 1861. In 1841, the co. of Cork had a population of 773,360; in 1851, of 565,751; and in 1861 of 464,697. The pop. per square mile was 296 in 1841; 215 in 1851; and 189 in 1861. Consequently the decrease of pop. from 1841 to 1861 amounted to 107 per square mile.

CORK, a city and river-port of Ireland, prov. Munster, on the Lee, 11 m. above where it discharges itself into Cork harbour; 136 m. SW. Dublin by road, and 164¾ by Great Southern and Western railway. Pop. 85,745 in 1851, and 80,121 in 1861. Cork is the third city of Ireland in respect of pop. and commercial importance, and forms a co. in itself, having a local jurisdiction separate from that of the co. of Cork, by which it is surrounded. The co. of the city extends over 48,000 acres, of which 2,683 are comprised within its municipal boundaries. The city lies in the vale of the river Lee, and is surrounded by hills of considerable elevation, which render the climate moist, though not unhealthy. It owes its origin to a religious establishment founded at a remote period. Previously to the arrival of the English, it was inhabited by a colony of Danes, and then, and for a long time after, consisted of a single street in an island formed by the river. Even so lately as the reign of Elizabeth, it is described as 'a little trading town of much resort,' but consisting of a single street. After the revolution it began to improve, and at length, chiefly in consequence of its vicinity to Cork harbour, a principal place of rendezvous for the Channel fleet

during wars with France, and its being a great mart for the supply of the fleets and colonies with provisions, it rose rapidly to wealth and importance, until it became the second city of Ireland. The pop. in 1821 amounted to 100,658 souls, and in 1831 to 107,016; after this period, a decline set in, and continued steadily to the present time, as shown in the statistics of pop. above given.

The city, situate on the river Lee, which here diverges into several branches, and forms an island, is 11 miles inland from the entrance of the river into Cork harbour. The public buildings are, the cathedral, 6 parish churches, and 2 chapels of ease, 4 Roman Catholic parochial chapels, 4 monasteries, and 7 nunneries, with a chapel attached to each; 3 Presbyterian, 4 Methodist, 1 Baptist, 1 Independent, and 1 Friends' meeting-house; the episcopal palace of the bishop of the consolidated dioceses of Cork, Cloyne, and Ross; the diocesan library; the county court-house; the military barrack; the queen's college; the county and city prisons; the house of correction; the bank of Ireland, provincial, national, and savings banks; the north and south infirmaries; the lunatic asylum; the custom house; the commercial buildings; the chamber of commerce; and the Royal Cork Institution. The head-quarters and staff of the Cork or southern military district of Ireland are stationed here. Near the city is a cemetery, after the plan of Père La Chaise, on the site of the old botanic garden. The New Wall is a picturesque public walk, 1½ miles in length along the S. bank of the river, from Albert Quay to the pier opposite the convent at Blackrock; and the Mardyke, a public walk, a mile in length, on the W. of the city. A park has been enclosed, containing about 240 acres, extending from the Victoria-road along the south bank of the river to Blackrock. There are 9 bridges over the river and its branches; and in Patrick Street there is a handsome bronze statue to the memory of Father Mathew, the apostle of temperance.

The corporation consists of the mayor, 16 aldermen, and 48 town councillors. The number of burgesses on the roll in the year 1864 was 1,850; and the revenue of the city in 1863, 11,793£. The borough returns 2 members to parliament; constituency 8,143 in 1865. The assizes for the county and city are held here.

The principal manufactures are tanning, distilling, brewing, iron foundries, gloves, ginghams, and friezes. The trade is extensive, chiefly in grain, provisions, and butter; and there are 19 markets in different districts.

The harbour, pre-eminent for its capacity and safety, is situate 11 miles below the city; it is 3 miles long, 2 broad, completely land-locked, and capable of sheltering the whole British navy. Its entrance is by a channel, 2 miles long and 1 broad, defended by batteries on each side, and by others in the interior. The upper portion extends for about 5 miles below the city to Passage, and this part since 1820 has been considerably deepened by steam dredging, so that vessels of 600 tons can unload at the quays, where at low water there is a depth of 7 feet. The tide flows up 1½ miles above the city. Within the harbour are Great Island, Little Island, Foaty Island; Spike Island, on which is a bomb-proof artillery barrack, and where a convict depot has been lately established for the reception of persons sentenced to transportation; Hawlbowline Island, containing an ordnance depot, and Rocky Island, in which there are 7 powder magazines, excavated from the rock. The number of vessels entered inwards in 1863 was 375 — tonnage, 115,634; and the number cleared outwards, 36 — tonnage, 28,691. The Great

Southern and Western railway connects the city with Dublin. The Cork, Blackrock, and Passage railway runs along the river through the city park, from the road near the Monerea marsh; the Cork, Bandon, and Kinsale railway terminates at Albert Quay; and the Cork, Queenstown, and Youghal, at Summer Hill.

The net annual value of property under the Tenement Valuation Act is 122,114£; and the property and income tax for the year ended 5th April, 1853, amounted to 77,053£. (Thom's Directory, 1853.)

The corporation derived its privileges from a series of charters, commencing with one from King John, when Earl of Morton and viceroy of Ireland. The mayor, recorder, and aldermen are justices for the city. The corporate business is transacted by the court of common council, composed of the mayor, recorder, sheriffs, and aldermen; and by the court d'oyer hundred, formed of the freemen at large. The mayor resides in the mansion-house, a large and elegant building on the Mardyke. The courts are those of the mayor and sheriffs, which have jurisdiction in pleas to any amount above 40s.; those of a lower rate are adjudicated in the court of conscience. The former of these courts sits weekly, as does the city sessions court, for criminal cases. The mayor, sheriffs, recorder, and aldermen are the recognised judges of these courts; but virtually the recorder presides. A post-office, or magistrates' court, is also held. The city court-house is a fine building erected at an expense of 40,000£. The prison is divided into 82 wards, besides day and work-rooms.

There is also a bridewell for the temporary confinement of persons under examination. The assizes for the co., and one of the general sessions for its E. riding, are held here. The county gaol and house of correction are situated a short distance from the city. A female convict depôt, for the reception of prisoners from all parts of the country, till the arrival of the transport ship to convey them to their destination, is in the S. suburb.

The foreign trade is carried on with Portugal, whence wines and salt are brought; with the Mediterranean, for wine and fruit; and with the Baltic, for timber and articles for naval equipment; timber is also imported from Halifax and Canada. The West India trade has declined, in consequence of the great facilities for supply from those colonies through the English ports. During war, Cork harbour is a great naval station, and the place of rendezvous for most of the outward-bound convoys. Naval arsenals and stores, which have now become nearly useless, having been abandoned by the government, though in the best state of preservation, were fitted up on the smaller islands.

The appearance and habits of the citizens of Cork are exclusively mercantile. The attempts that have been made to elevate the city in the scale of literature and science have not had that success which their more sanguine promoters anticipated; though they have probably succeeded better than a careless observer might suppose. Some rather distinguished persons have been natives of Cork, among whom may be specified Arthur O'Leary, O'Keefe, Barry the artist, Maclise the artist, and Sheridan Knowles. The newer part of the city indicates an increasing state of prosperity; in it are the town residences of the wealthy merchants; while the adjoining country, for several miles round, is studded with their villas and country seats. But, on the other hand, several extensive districts of the suburbs evince the existence of comparative destitution; lines of cabins being built and peopled like those in the surrounding rural villages. But improvement is notwithstanding, said to be advancing, even in those quarters in which there is the greatest poverty, and where old habits and prejudices are sure to linger longest. The food of the working classes consists chiefly of potatoes, which is all but equivalent to saying that their wages are low, and their condition alike degrading and precarious. Several remains of antiquities, chiefly monastic, are to be traced, as are considerable remains of the ancient walls, some parts of which are in a perfect state. Coins struck at a royal mint in the time of Edward I. have been occasionally found.

CORLEONE, an inland town of Sicily, prov. Palermo, cap. dist., near the source of the Belice, on the declivity of a hill rising from a fruitful, well-cultivated plain; 22 m. S. by W. Palermo, near the railway from Palermo to Girgenti. Pop. 13,123 in 1861. The town is well built, and has several churches and convents, a royal college, a prison, and some other public buildings.

CORNWALL, a marit. co. of England, forming the extremity of the SW. peninsula, being everywhere surrounded by the sea, except on the E., where it adjoins Devonshire, from which it is separated nearly in its whole length by the Tamar. Area, 851,289 acres, of which about 631,000 are arable, meadow, and pasture. In many parts Cornwall is rugged and mountish; but though its general aspect be bleak and dreary, it has numerous valleys of great beauty and fertility. The temperature is particularly equal, being so far embosomed in the Atlantic that it is neither so cold in winter, nor so warm in summer, as the co., more to the E. The winds, however, are very variable, and often violent; and the air being surcharged with moisture, harvests are late, and fruit is inferior in flavour to that raised in the E. and midland cos. The raising of corn and potatoes are the principal objects of Cornish agriculture, which has been much improved of late years. Property much divided and "vexatiously intermixed." Farms for the most part small, and held under lease for 14 or 21 years. The principal wealth of Cornwall is derived from its mines of tin and copper. It is believed that the Phœnicians traded thither for tin, and that the mines have been wrought ever since. The total quantity of tin produced in Cornwall amounts to about 5,000 tons a year. The Cornish copper mines, though they were not wrought, with spirit or success, till the beginning of last century, are now become of great value and importance. Their produce, which a century ago did not exceed 700 tons pure metal, amounts at present to about 12,000 tons. The copper and tin mines number about 240, giving employment to 60,000 persons. Ores of lead, antimony, manganese, &c., are also met with. Gold is sometimes found in the stream-works, or places where the alluvial deposits are washed in order to procure grain tin. Silver is also found intermixed with the lead ores, and is extracted to a considerable extent. About 5,000 tons of soapstone, and about 7,000 tons of China clay, are annually shipped for the Potteries and other seats of the porcelain manufacture. The miners and others engaged in the Cornish mines are under the especial jurisdiction of the stannary courts; these were much improved by a late act, and are said to transact the business brought before them expeditiously, cheaply, and well. The oppressive duties formerly imposed on the coinage of tin were repealed in 1837. The pilchard fishery is extensively carried on along the Cornish coasts, particularly at St. Ives, Mount's Bay, and Mogavissey; and is a considerable source of employment and of wealth to the co. Principal towns, Truro, Helston, Penzance, St. Ives,

Falmouth. Previously to the Reform Act, Cornwall sent forty-two members to the H. of C., but now it sends only fourteen, viz. four for the co., two each for the bors. of Bodmin, Falmouth, and Truro, and one each for the bors. of Launceston, Helston, St. Ives, and Liskeard. Registered electors for the co., 10,643 in 1845, of which number 5,908 for the east division, and 4,785 for the west division. The pop. of the co. was 355,558 in 1831, and 369,390 in 1861. Gross annual value of real property assessed to income tax—in eastern division £67,179/, in 1857, and 655,614/. in 1862; in western division, 548,263/. in 1857, and 611,277/. in 1862. Cornwall is divided into 9 hundreds; 203 whole parishes, with parts of 3 others; 14 registry districts; 13 poor-law unions; and 11 county courts.

CORO, a marit. city of Venezuela, cap. prov. of the same name, in a sandy arid arid plain, near the head of El Golete, an arm of the Gulf of Maracayba, 3 m. SW. the Caribbean Sea, and £10 m. WNW. Caracas; lat. 11° 23′ N., long. 69° 44′ W. Estimated pop. 10,000. It is well situated for commerce, and has had a considerable trade with the West India Islands, especially Curaçoa, in mules, goats, hides, skins, cheese, and pottery; but this has now very much dwindled, and the inhabitants are poor. The streets of Coro are regular, but unpaved, and the houses mean: the only public buildings are, two churches, a convent, several chapels, and a hospital. The climate is dry and hot, but not unhealthy; so great, however, is the scarcity of water, that it has to be brought thither daily, on the backs of mules, a distance of 2 m. Coro was the second European settlement formed on this coast, and was considered the capital of Venezuela till the transference of the seat of government to Caracas, in 1576.

COROMANDEL (Cholamandala), COAST OF, forming the E. shore of Hindostan, from Point Calymere, lat. 10° 20′, to the mouth of the Krishna river, 15° 56′ N., probably deriving its name from the Chola dynasty, who formerly ruled in Tanjore. It is destitute of any good harbours, and, from the great surf, it is usually difficult anywhere to effect a landing. The monsoons on this coast are always in a contrary direction to those on that of Malabar. From the middle of October to the middle of April, winds from the NE. prevail, during which period the storms are so violent and dangerous that all British ships of war are ordered to quit the coast by the 15th of October. In the middle of April the SW. winds set in, and a period of great drought commences.

CORREZE, a dep. of France, reg. South, formerly part of the Limousin, having N. the deps. Haute Vienne and Creuse, E. Puy-de-Dôme and Cantal, S. Lot, and W. Dordogne. Area, 586,049 hectares. Pop. 310,118 in 1861. Surface hilly and mountainous. Its N. part is intersected by a mountain chain, dividing the basin of the Loire from that of the Dordogne. The latter, which runs through the SE. part of the dep., is the only navigable stream, the Corrèze, from which the dep. derives its name, being available only for rafts and boats. Climate comparatively cold; soil stony and inferior, except in some of the larger valleys. Heaths and wastes occupy more of the surface than the arable lands; sufficient corn, however, chiefly rye and buckwheat, is grown for home consumption. Agriculture is in a backward state, partly owing to the obstinate attachment of the cultivators to ancient routine practices, and partly to want of capital, and to the minute division of the land. Chestnuts, buckwheat, and potatoes constitute the principal dependence of a large pro-

portion of the pop.; and when these fail, the inhabitants suffer severely. Vineyards occupy about 15,200 hectares. Some of the wines are tolerably good, and though no great quantity of wine be produced, still, as few of the labouring classes can afford to drink it, some is exported. The meadows are extensive, and considerable numbers of oxen are reared for the Paris market and the plough. There are upwards of 460,000 sheep, chiefly an indigenous breed, yielding annually about 450,000 kilogs. of wool. Property much subdivided, there not being in the whole dep. above a dozen properties which pay a government tax of 1,000 fr. Corrèze has mines of copper, iron, argentiferous lead, antimony, and coal; but, with the exception perhaps of coal at Lapleau, none of them are wrought to any considerable extent. Manufacturing industry is even in a less prosperous state than agriculture. There is, however, a large gun manufactory at Tulle, and a cotton mill at Brives. Tulle is generally supposed to be the grand seat of the manufacture of the species of point lace called point de Tulle; in point of fact, however, there is not a single lace-worker in the dep. nor has there been, time immemorial, a lace-frame in Tulle. Trade chiefly in cattle, wine, poultry, agricultural produce, and truffles. The dep. is divided into three arronds. Chief towns, Tulle, the cap., Brives, and Ussel. There exists a general usage (for it is inconsistent with the law of France) in this dep., whereby the eldest son becomes entitled to a clear fourth of the paternal property, over and above an equal share with each of the other children. The peasantry exhibit a remarkable dislike to enter the military service, but prove afterwards very good soldiers. Marmontel, Calmeis, and Latreille were natives of this dep.

CORSHAM, a par. and village of England, co. Wilts, hund. Chippenham; 19½ m. W. London by Great Western railway. Pop. of par. 3,196 in 1861. The village, in an open pleasant district, 8 m. NE. Bath, consists chiefly of one long street of neatly-built houses, with a market-house near the centre, erected in 1784. The church is a cruciform Gothic structure, with a tower. There are also two dissenting chapels; and an almshouse, founded in 1688, at present supporting six old women. A free school for boys and girls was built by the Methuen family, to which the manor belongs; and who have a fine mansion, with a good collection of pictures, near the village. The manufacture of woollens, formerly carried on to a considerable extent, has long been discontinued, agriculture being now the chief employment of the inhabitants. Sir E. Blackmore, the author of various epic poems, now known only by the satirical allusions made to them by Pope and other wits of the time, was a native of Corsham.

CORSICA (Fr. Corse), a large island of the Mediterranean, belonging to France, of which it forms a dep.; between lat. 41° 27′ and 4° 1′ N., and long. 8° 37′ and 9° 30′ E. Its S. extremity is 10 m. N. Sardinia, from which it is separated by the strait of Bonifacio. Piombino, about 55 m. distant, is the nearest town in Italy, and Antibes, 120 m. NW., the nearest point in France. Shape somewhat oval, with a projecting appendage at the NE. extremity; length, N. to S., 108 m.; greatest breadth, 48 m.; area, 874,741 hectares. Pop. 252,889 in 1861.

The E. shores of Corsica are generally low and sandy, and in many parts marshy; the W. shores are more lofty, and indented with several extensive harbours or bays, the principal of which are those of Valinco, Ajaccio, Sagone, Porto, Calvi, and St. Florent. Corsica has several small islets,

especially at its S. extremity. It is, generally
speaking, hilly. A chain of mountains traverses
it from its N. to its S. extremity, for the most part
nearer to its W. than to its E. coast; the highest
summits of this chain are Monte Rotondo, 8,766
ft., and Monte d'Oro (the *Mons Aureus* of
Ptolemy), 8,700 ft. above the level of the sea.
The declivities of the central chain are steep; it
abounds in clefts and gorges; valleys are few, excepting in the lower hill ranges, and even there
they are narrow. The plains along the E. coast
amounting to about 1-24th part of the whole surface, though rich and densely peopled in the time
of the Romans, are now mostly abandoned. Were
they drained and cultivated, they would be again,
as of old, the best part of the island. The majority of the rivers run W., but the two largest,
the Golo and Savignano, have an E. course; most
of them are mere torrents, and none of them are
navigable or adapted even for rafts, by reason of
their rapidity. There are a few insignificant lakes
in the centre of the island; but the largest collections of waters are some lagunes on the E.
coast, a topographical feature which this part of
Corsica shares with the opposite coast of the Tuscan Maremme and the Campagna di Roma.
These stagnant waters render the adjacent parts
unhealthy, giving rise to intermittent fevers, &c.,
similar to those of the corresponding Italian
shores; but elsewhere the climate is sufficiently
salubrious. The temperature of course varies with
the elevation; in the low lands the maximum is
94½° Fahr., in the mountains the minimum is
25½° Fahr. The most prevalent winds are—the
sirocco, or SE., which brings rain; the N., which
often brings snow; and the SW., which is commonly very violent. The aspect of the country
is, in the words of Hugo, 'a vast elevated region,
the culminating points of which are covered with
snow, surrounded by lower ranges of mountains,
their summits bare, but their sides covered with
thick forests of fir and oak; narrow and dark
glens, through which roll impetuous torrents; and
here and there an isolated human habitation,
perched on some military crag, like the inaccessible eyrie of an eagle. As we approach nearer the
sea the valleys enlarge, and show traces of culture, and villages begin to enliven the banks of
the rivulets; the hill-sides are covered with olive,
orange, and laurel trees; while their tops are
crowned with woods of chestnut, whose time-ho-
noured trunks, notwithstanding the little depth
of soil they grow in, have attained an enormous
size. On the sea-shore, obscured by an unhealthy fog, ruined habitations, corn-lands, maquis
(close copses), and marshes alternate with each
other, and the traveller hastens to quit this pestiferous tract for a brighter sky and a purer air
upon the uplands.' Granite, mica, porphyry, alabaster, and marble of various colours, serpentine,
jasper, and asbestos of remarkably long fibre, are
plentiful in Corsica. The island probably contains neither gold, silver, nor copper; but there is
a vein of lead at Barbaggio, and iron mines are
worked in several places; the produce of the last
occupies ten forges at Catalane. Quarries of statuary marble are worked; pipe-clay, emeralds,
and globular masses of granite and porphyry are
found; the last, which are prized as gems, have
been hitherto met with no where but in the bed
of one of the torrents. There are an abundance
of warm, mineral, and saline springs. The upper
soils consist chiefly of decomposed granite and
silex, with a small proportion of chalk and other
calcareous matters, and the remains of animal and
vegetable substances. In many parts the land is
very fertile; agriculture is, however, in a very

backward state, and artificial irrigation almost
unknown.

Landed property in Corsica is extremely subdivided, and is almost all occupied by owners.
'For centuries the laws have promoted an equal
succession among children; the Genoese, when
rulers, abetted this system, and the French law
of succession, which found Corsica in an extra-
vagantly parcelled state, has continued and aggravated it. These ancient and modern agrarian-
isms, unaccompanied by the remedies of capital
and of various funds to industry, have made a
proprietor of almost every Corsican, and have, it
is true, averted here mendicity, but also generally
created a narrow situation, without resources,
pregnant of family intrigues, and not unlikely
dissensions, litigious propensities, and various
checks on population; and, combining with these
incidents, they have fostered maxims which again
serve to the same end of disconnecting all landed
property. It is a distinctive trait, that *the Corsican rather starves than sells land*; that inheritances
which lose in value by division still must submit
to it; and advantageous offers are the more
readily refused the more such land would aggran-
dise and connect the purchaser's estate.' (Consular Report.) The inhab. do not live in cottages
dispersed over the country, but in villages, many
of which are built on the summits and declivities
of the mountains.

The forests are remarkably fine, and abound
with timber of the best quality, and which supplies the best masts for the dockyards at Toulon;
but such is the indolence of the inhab. that this
source of wealth is comparatively neglected. The
maquis, previously mentioned, are dense thickets of
cystus, bay, myrtle, and thorn, which rapidly
grow up on rich untilled lands, into inextricable
masses of 8 to 12 ft. in height, and which, when
burnt—the usual mode of getting rid of them—
form admirable manure. The orange, citron, and
pomegranate grow in the open air, and yield
excellent fruit. The olive is badly managed;
but much more oil is produced than is required in
the island, and is therefore exported. The vine is
tolerably well cultivated in most of the cantons;
and, notwithstanding that but little art is displayed in the manufacture of wines, the red wines
of Nari, and the white of Cape Corsica, are very
good, and exported to the Continent. The corn
grown is not adequate to the demand, but its deficiency is made up by the abundant supply of
chestnuts. Vast quantities of honey are produced
in the island. The honey has a bitterish taste,
supposed to be imparted by the abundance of boxwood and yew. A great portion of the immense
quantity of honey consumed in France is supplied
from Corsica. The island produced so much wax
in ancient times that the Romans imposed on it
an annual tribute of 100,000 lb. weight. Subsequently the inhabitants revolted, and they were
punished by the tribute being raised to 200,000 lb.
weight annually, which they were able to supply.
Wax is to honey in Corsica as one to fifteen, so
that the inhabitants must have gathered 3,000,000
kilogrammes of honey. When Corsica became a
dependency of the papal court it paid its taxes in
wax, and the quantity was sufficient to supply the
consumption not only of the churches in the city
of Rome, but those in the Papal States. Brittany
likewise supplies a great quantity of honey, but
of inferior quality to that of Corsica. The annual
value of the honey and wax produced in Corsica
is estimated at 5,000,000f., or 200,000l. Tobacco,
though little cultivated, is said to be preferable to
that of France; and the mulberry and flax are
grown with advantage. Cattle constitute the

principal wealth of the farmers and peasantry. Most kinds are small, but the ox, horse, mule and ass are all strong and active; the cows afford good milk, from which much cheese is made. The sheep are black, with four or even six horns; there are about 300,000 in the island; hogs very plentiful. Goats are large and strong; the mouflon, considered by Buffon to have been the original of the sheep, is found in the island. Game is extremely abundant, as are wild boars and foxes; turtles are obtained in great number, and are important articles of trade. There is a great profusion of the most excellent fish in the surrounding seas, and the Corsican mullet was among the delicacies supplied to the Roman tables. (Juv., Sat. v. 1, 92.) Red coral of a fine deep colour is found in many places round the coast. But, owing to the indolence and apathy occasioned by the dependence of the people on small patches of land, and the want of capital and manufactures, everything is conducted according to a system of routine, and very few improvements are either attempted or even so much as thought of. Agricultural implements are all of the most wretched description, and they hardly know anything even of the advantages of manure. All the more laborious employments are devolved upon the females, who are the slaves rather than the companions of their husbands, or upon emigrants from Lucca, Tuscany, and other parts of Italy, by whom the island is annually visited. The fisheries are wholly abandoned to the Genoese and Neapolitans. Their manufactures are limited to the fabrication of some coarse woollens used by themselves, a few forges and tanneries, a glass factory, a pottery (in which asbestos is used), a manufactory of tobacco-pipes, and one of soap. The exports are nearly confined to timber, firewood, wines, dried fruits, oil, silk, leather, and fish, in comparatively trifling quantities. The roads are wretched; those called royal being in parts almost impracticable even for mules.

In 1793 Corsica was divided into two departments—those of Golo and Liamone; but since 1811 these have been again united; the seat of the prefecture is Ajaccio. A royal court is established in the capital; there are five courts of original jurisdiction, one in each arrond, and three tribunals of commerce, viz. at Ajaccio, Bastia, and Ile Rousse. There are no churches but those of the Catholic establishment in Corsica; the dep. is a bishopric suffragan to Aix. Corsica forms the 17th military division of France; it contains ten fortresses.

In person, habits, and disposition, the Corsicans bear a considerable resemblance to the natives of Italy. They are brave, sober, and hospitable; but subject to violent gusts of passion, and in the last degree revengeful and implacable. This, in fact, is the distinguishing trait of their character, and has been supposed to indicate a peculiar ferocity of disposition. It appears, however, rather to have originated in the long-continued misgovernment of the Genoese, when the grossest corruption prevailed, and money or interest could procure impunity for the most nefarious crimes. Under such circumstances, the avenging of injuries became, as it were, a private duty; and the Corsican would have considered himself degraded who had not obtained that redress for himself that was denied by law. It is needless to point out the sanguinary practices, crimes, and enormities to which such a state of things must necessarily lead. The impoverished and more vigorous government introduced by the French has, however, done a great deal to lessen the temptations to vengeance; though it will be long before the passion be wholly subdued

among a people in the situation of the Corsicans. They use an Italian dialect, with a large number of Arabic words and Spanish idioms intermixed. The dress of both sexes bears a similarity to that of the Italians: the men wear a kind of Phrygian bonnet, and commonly go armed with a long knife, pistol, musket, and bayonet. At Cargese, on the W. coast, there is a Greek colony of Mainot origin, consisting of about 700 individuals, the descendants of some Greeks who settled in Corsica in 1676, who preserve their dress and religion, but have adopted Catholic rites of worship. The tract they inhabit is the best cultivated in the isl. The Phocæans, who afterwards founded Marseilles, and the Phœnicians, have both been considered the first inhab. of Corsica; and by them the island was called Cyrnus. It was afterwards conquered by the Carthaginians, from whom it was taken by the Romans about B. C. 231. In the middle ages, the Goths, the emperors of the East, Saracens, Franks, House of Colonna, Pisans, and Genoese, successively possessed it. Insurrections against the latter continued at intervals for several centuries, till the Genoese finally ceded it to France in 1768. The pop. under the gallant Paoli made a determined resistance; but ultimately they were forced to submit, and the island has since belonged to France, with the exception of two short periods, in 1794 and 1811, when it was occupied by British troops. The names of Pascal Paoli and of Napoleon, both natives of Corsica, are sufficient to confer on it an enduring celebrity.

CORTONA, or COTRONE (an. Crotona), a city and sea-port of Southern Italy, prov. Catanzaro, cap. district and cant., near the mouth of the Esaro (an. Æsarus), on the Ionian Sea. Pop. 3,910 in 1861. The town is surrounded by walls and defended by a strong citadel. The latter fronts the sea, and is separated from the town by a ditch and drawbridge. It has a cathedral and several other churches, 2 convents, a seminary, and 2 hospitals. The harbour is protected on the N. by the projecting tongue of land on the side of which the town is built, and on the N. by a mole; but it is too shallow to admit of vessels of considerable size, and is not very safe.

Cortona was once one of the richest, most populous and powerful cities of Magna Græcia. Various accounts have been given of its origin, but it is sufficient to say that it was founded by emigrants from Greece at a very remote period. It speedily rose to eminence. Pythagoras resided here for a considerable period after leaving Samos; founded a very extensive school; and is said, by his example and his precepts, to have effected a very considerable change in the manners and conduct of the inhab. It had also a celebrated school of medicine. Ancient writers have praised its invigorating air, which was said to give superior strength to the men, and beauty to the women. Milo, famous alike for his success as a wrestler at the Olympian and Pythian games, and for his tragical end, was a native of Crotona. It produced many other celebrated wrestlers, so that it became a proverbial saying, that the last wrestler of Crotona was the first of the other Greeks. (Strabo, ii. 262.) The mode which Zeuxis took to paint his famous picture of Helen is a sufficient compliment to the beauty of the fair Crotonæans. (The curious reader will find this subject thoroughly discussed in Bayle, art. 'Zeuxis.') In the third year of the 67th Olympiad, some exiles from Sybaris, having taken refuge in Crotona, the latter, on refusing to give them up, was attacked by 30,000 Sybarites; and though the Crotoniats are said to have been able only to bring 10,000 men into the field, they gained a complete victory over the Sybarites, and took and sacked their city. (Ancient Universal History, vi. 421, 8vo. edit.) But their success in this conflict is said to have been followed by a renewal of that corruption of morals which Pythagoras had done so much to correct, and by a decline of the martial virtues. At all events, the Crotoniats were not long after signally defeated by the Locrians, and do not appear to have again recovered their former power or influence. Still, however, Crotona was a large city at the epoch of the invasion of Italy by Pyrrhus, though it appears to have suffered severely in the contests in which it led. Livy says, ' *Urbs Croto murum in circuitu patentem 12,000 passuum habuit, ante Pyrrhi in Italiam adventum. Post vastitatem eo bello factam, vix pars dimidia habitabatur; flumen (Æsarus) quod medio oppido fluerent, extra frequentia tecta loca præterfluebat.*' (Liv. 24, § 3.) It was afterwards taken by the Carthaginians, and the inhabitants removed to Locri. Subsequently, however, it received a colony from Rome. In the war between Charles of Anjou and Frederick of Arragon, it was taken by surprise, and sacked; and it has since continued in the depressed state in which we now find it.

About 6 m. SE. from Crotona, at the extremity of the narrow projecting tongue of land, now called Capo Nao or Delle Colonne (the *Lacinium Promontorium* of the ancients), stood a famous temple of Juno, hence frequently called *Juno Lacinia*. It is said by Livy to be *nobile templum, ipsá urbe nobilius*. It was of great antiquity, was surrounded by magnificent groves, and was held in such veneration that it was annually resorted to by crowds of pilgrims from all parts of Italy and Greece. The Helen of Zeuxis was placed, with many other articles of great rarity and value, in this sacred edifice, whose sanctity was respected both by Pyrrhus and Hannibal. But succeeding conquerors have had less forbearance; and a solitary Doric column is now all that remains of this once venerated and splendid edifice.

CORTONA, a town of Central Italy, prov. Firenze, on the declivity of a steep hill, which commands a magnificent prospect of the Thrasimene lake, the mountains of Radicofani, and the wide and variegated vale of Chiana, 52 m. SE. Florence, and 22 m. NW. Perugia, on the railway from Florence to Perugia. Pop. 27,960 in 1861. This, which was one of the 12 principal cities of Etruria, is supposed to have been founded by the Pelasgi, and is probably among the most ancient towns in Italy. 'Its original walls still appear round the city, as foundations to the modern, which were built in the 13th century. These Etruscan works are most entire towards the N. Their huge, uncemented blocks have resisted, on that side, the storms of near 3,000 winters; while on the S. they have yielded to the silent erosion of the air, &c. None of the stones run parallel; most of them are faced in the form of *trapezia;* some are indented and inserted in each other like dove-tail. This construction is peculiar to the ruins of Tuscany; it is far more irregular, and therefore, I presume, more ancient than the Etruscan work of Rome. No part of these walls is fortified.' (Forsyth's Italy, p. 99.) The town is commanded by a castle built by the Medici, on the summit of the hill on which it stands. It has a cathedral, which possesses some fine works of art, several other churches, and a theatre. There is a temple of Bacchus, and the remains of some baths ornamented with mosaic work. Next to the city walls, however, the most interesting relic of antiquity is a small sepulchral chamber a little below the town, formed of large blocks of sandstone, the construction of which proves that the architects of the

Etruscan period were acquainted with the principle of the arch. Cortona is the residence of a bishop; it has an ecclesiastical and some other seminaries, and was the seat of the Etruscan academy, founded in 1726, which had been a library, a cabinet of natural history, a museum of antiquities, engravings, and gems; but these collections have been dispersed. In the middle ages, Cortona was attached to the Ghibelline party; since the early part of the 15th century it has always been subject to Florence, except during the short interval it belonged to the French under Napoleon.

CORUNNA (Span. *Coruña*), a city and sea-port of Spain, prov. Galicia, NW. extremity of the kingdom, on the E. side of a small peninsula, forming the S. extremity of the Betanzos Bay; 13 m. SW. Ferrol, 315 m. NW. Madrid, on the terminus of a railway from Madrid. Pop. 27,354 in 1857. Corunna is divided into the Upper and Lower Towns, the former, situated on more elevated ground, is surrounded by walls and bastions, and defended by a citadel; the other is situated lower down, on the isthmus joining the peninsula to the mainland, from which it is separated by ramparts and a ditch. The streets in the Upper Town are comparatively steep and narrow. Among the public buildings are 4 churches, 5 convents, a palace for the captain-general, and the supreme court of justice of the prov.; 3 barracks, an arsenal, 2 hospitals, and a school of design, mathematics and navigation, supported by the commercial consulate.

There is a fine commodious quay, and a good building yard. The harbour, which is safe and well-sheltered, is commanded by Fort St. Anthony, on an insulated rock at its mouth, and by Fort St. Diego on the mainland. It is the station for steamers between Spain and the Havannah, and between Spain and Falmouth. At the bottom of the harbour is the island of St. Lucia. On the S. shore of the peninsula is the famous lighthouse, called the Tower of Hercules, or the Iron Tower, 92 ft. in height, and which, being built on high land, is visible at sea in clear weather 60 m. off. The tower is said by Humboldt to be of Roman construction, and is believed to be of the æra of Trajan. It was repaired in 1791. The principal manufacture carried on in the town is that of fine table and other linen, with which the royal palaces used to be supplied, and of coarse linen. It has also fabrics of hats, canvass, and cordage, and a royal manufactory of cigars, in which about 500 women are employed. Corunna is famous, in the history of the struggle between Spain and Napoleon, for being the point to which Sir John Moore directed his disastrous retreat in 1808; and for his death in the engagement which took place under its walls, on the 16th of January, 1809, previously to the reembarkation of the British, when a superior French force under Marshal Soult was repulsed with great loss.

CUSALA, a town of Mexico, state of Sonora, in a mountainous district, 300 m. SE. El Fuerte, and 160 m. from the Pacific Ocean. Pop. estimated at 7,000. The town is the third in the state in point of size. It derives importance partly from being a depot for goods passing to and from the port of Guaymas, on the Gulf of California, but chiefly on account of its mines, one of which, called Guadalupe, contains an extremely rich vein of gold; and, being at a considerable elevation, is free from water.

COSLIN, or KOSLIN, a Prussian town, prov. Pomerania, cap. reg. and circ. of same name, on the Nieswbecke, about 4 m. from where it falls into the lagoon Jamund, which communicates with the Baltic, and on a branch line of the railway from Berlin to Dantzic. Pop. 12,110 in 1861. Having been nearly destroyed by fire in 1716, it was rebuilt on a regular plan by Frederick William I., whose statue has been erected in the market-place by the citizens to commemorate the beneficence of the monarch and their gratitude. It is the residence of the governor of the regency, and has a court of appeal, and a society for the promotion of agriculture, and various schools. Mount Gollen, a little to the E. of the town, is one of the highest elevations on the Pomeranian coast.

COSSEIR, KOSSAIR, or KOSIR, a sea-port town of Upper Egypt, on the W. shore of the Red Sea, 93 m. E. by S. Ghenneh, or Kenné, and 102 m. ENE. Thebes; lat. 26° 6' 59", long. 34° 23' E. Pop. estimated at from 1,500 to 2,000. It is situated near the centre of a semicircular bay, about 5 m. across, sheltered on the N. by a sandy point of land, where vessels may lie in 5 fathoms water within 60 yards of the shore. The town is meanly built; the houses being low, and built of sun-dried bricks made of a white calcareous earth; only a few have two stories. Immediately on the NW. is a small citadel defended by round towers, on which a few small guns are mounted. This fortress is the residence of the governor and garrison. A caravan road leads from Ghenneh to Cosseir, which is the centre for all the traffic between the upper valley of the Nile and the Arabian ports; and to this circumstance it owes its existence, as it has neither trade nor manufactures of its own, and the surrounding country is perfectly bare of all vegetation. Old Cosseir is about 10 m. NW. of the modern town, on the N. bank of a small inlet, from which the sea has now mostly retired. Of the latter town only a few ruins exist. Berenice, the great port for the eastern traffic of Egypt under the Ptolemies, was situated a good deal further S.

COSSENZA, or COSENZA (an. Consentia), a city of Southern Italy, cap. of prov. of same name, on the margin of a valley surrounded by hills, at the confluence of the Crati and Busento, 12 m. E. from the Mediterranean. Pop. 8,250 in 1861. The city is intersected by the Busento, which is here crossed by two bridges, and the lower parts of the town are said to be unhealthy. It has only one good street, the others being narrow, crooked, and dirty. The tribunale, or palace of justice, is a fine edifice; an old castle, now converted into barracks, crowns the summit of an eminence on the opposite side of the river. It has also a cathedral, several churches and convents, a grand seminary, a royal college, a hospital, a foundling hospital, 2 academies of science and belles-lettres, and a theatre. It is the seat of the provincial courts and authorities, and of an arch-bishop. Earthenware and cutlery are made here; and it has a considerable trade in silk, rice, wine, fruits, manna, and flax. In the 16th century there was here a famous academy, founded or improved by Bernardino Telesio.

In antiquity Cosentia was the cap. of the Brettii. Alaric, lay's born It was besieged anno 410, died before its walls, and was buried in the bed of the Busento. It was taken and sacked by the Saracens, who were expelled from it by the Normans, and has suffered much from earth-quakes, particularly from those of 1638 and 1783. The extensive forest of Sila lies a little to the W. of Cosenza.

COSSIMBAZAR, an inl. town of Hindostan, prov. Bengal, distr. Moorshedabad and about 1 m. S. of that city, of which it is the port; on the left bank of the Bhajirathi or Hooghly river; lat. 24° 10' N., long. 88° 15' E. It is one of the most considerable trading towns in Bengal, and during the rainy season has an unequalled variety and extent of water carriage. A vast quantity of raw silk is thence exported to Europe, and to almost every part of India; and a great deal consumed annually by the natives in the manufacture of carpets, sattins, and other stuffs. Cossimbazar is also noted for its stockings, which are wire-knitted, and esteemed the best in Bengal. Its vicinity is flat and sandy, and abounds with a great variety of wild animals.

COSTAMBOUL, or COSTAMANI, a town of Asiatic Turkey, Natolia, cap. pachalic, 236 m. E. Constantinople, and 50 m. S. from the nearest point of the Black Sea, in a dreary and unfertile country, intersected by deep ravines and numerous water-courses. Estimated pop. 12,500. It stands in a hollow, in the centre of which rises a lofty and perpendicular rock crowned with a ruined fortress, formerly possessed by the Comneni. The houses are built of wood and stone; and the palace of the pacha, a poor edifice, opens into the meydan or square. There are 30 mosques, with minarets, 25 public baths, 6 khans, and a Greek church. The trade of the town is but inconsiderable, and there are no manufactures. In the later ages of the Greek empire, Costamboul was the cap. of an independent prince, who was first expelled by Bajazet, reinstated in his possessions by Timour, and finally subdued by Mahomet I.

COTE-D'OR, a dép. of France, in the E. part of the king., between lat. 46° 55' and 48° 2' N., and long. 4° 7' and 5° 3' W., formerly part of the prov. of Burgundy, having N. the déps. Aube and Haute Marne, E. Haute Saône and Jura, S. Saône-et-Loire, and W. Yonne and Nièvre. Area, 876,116 hectares; pop. 384,140 in 1861. Surface mostly hilly and mountainous. The principal chain connecting the Faucilles with the Cévennes runs nearly through its centre, separating the affluents of the Saône. A part of this range gives its name to the dep., having been termed the Côte-d'Or, from the number and excellence of the vineyards on its declivities. Both the Seine and Armançon have their sources in this dep.; and the Saône winds along its SE. border. Climate temperate; but said to have become colder within the last 30 years, from the woods having been extensively cut down. Soil for the most part gravelly or calcareous; and in the E. and S. very fertile. The arable land is estimated at 457,000 hect., forests 196,000, meadows 63,000, and vine-yards 26,450 do. The vine culture is by far the most important branch of industry carried on in this dep. It has been said that the wines of the Côte-d'Or have degenerated within the last forty or fifty years; but this is not really the case, though, from the extension of vineyards in less favourable situations, the quantity of secondary and inferior growths bears a larger proportion to the superior growths, the supply of which is limited, and apparently unsusceptible of increase. The best wines are produced in two contiguous tracts to the SE. of the Côte-d'Or range. One tract, called the Côte-de-Nuits, extends between Dijon and Nuits; the other, the Côte Beaunoise, is comprised between Nuits and the Dheune. To the Côte-de-Nuits belong the first class wines of the Clos Vougeot, Romanée, Chambertin, Cortou, and Richebourg; to the Côte Beaunoise the celebrated but secondary growths of Volnay, Pomard, Beaune, and others, and some fine white wines, as Mon-trachet, and Mrursault. The total annual produce of wine is estimated at 700,000 hectolitres, or 18,500,000 gallons. Agriculture is in a medium state of advancement. More than sufficient corn is grown for home consumption, principally wheat, oats, barley, and rye. Hemp, flax, and some leguminous and oleaginous plants are also cultivated. Dijon is famous for its mustard. Cattle abundant; both the ox and horse are used for the plough, except in the mountainous districts, where the mule is employed. The first attempts to improve the breeds of sheep in France were made in this dep., and here they have been eminently successful. The annual produce of wool is estimated at 245,600 kilogs. There are some fine natural pastures on the banks of the Rhône, but the system of irrigation pursued in the Vosges and elsewhere is not adopted. Hogs are numerous, and bees are extensively reared. Property in this is less subdivided than in most other deps. in France. Mineral products numerous and valuable, especially iron and coal. There are above 100 furnaces for smelting iron, and its production and manufacture into different articles constitute a very considerable branch of industry. There are also numerous breweries and distilleries, with establishments for the manufacture of beet-root sugar, mustard, and vinegar; tanneries, potteries, and cloth fabrics. Wine, however, forms the principal article of export. The trade of the dep. is much promoted by the canal of Burgundy, by which it is intersected. It is divided into 4 arronds. 36 cantons, and 727 communes. Chief towns, Dijon, Beaune, and Chatillon-sur-Seine. There are several Roman antiquities in this dep., especially a sculptured column near Cussy, supposed to have been erected in the time of Diocletian.

COTES-DU-NORD, a marit. dép. of France, region of the NW., formerly part of the prov. of Brittany, having E. Ille-et-Vilaine, S. Morbihan, W. Finistère, and N. the British Channel. Area 604,562 hectares. Pop. 628,676 in 1861. Coast generally steep, rocky, much indented with the mouths of small rivers, the chief of which is the Rance, and surrounded, particularly towards its W. end, by many small islands. A chain of heights, called the ' Black Mountains,' runs through the centre of the dep. E. and W., sending off numerous branches on either side : the highest point of these is the Mènes-Hant, about 1,115 ft. above the level of the sea. Soil mostly stony, primitive formations being everywhere found near the surface; the plain on both sides the mountain-chain are often sandy and sterile. Arable lands occupy 411,000 hectares, meadows 54,500 do., heathy wastes and forests about 170,000 do. Agriculture is in a very backward state; in some cantons oxen only are employed in farm labour: more corn is however grown than is required for home consumption; it is mostly oats, wheat, and rye. This dep. is beyond the limits of the vine culture, but the annual produce of cider is estimated at 500,000 hectolitres. The sheep are generally small and weak, but the rearing of black-cattle and horses engrosses a considerable share of attention; and the latter especially are strong and much esteemed. The fisheries of cod, mackerel, and pilchards yield an annual sum of about 600,000 fr., and while they constitute one of the most important resources of the dep. are useful as preparatory schools for seamen. The forests are extensive, and abound with wild animals. Iron and lead mines are wrought; but the dep. is not rich in other minerals. The culture of flax, and its manufacture into linen, are pursued to a great extent. The linens of Brittany are mostly exported to S. America. Sailcloth, woollens, parchments, leather, shoes, and beet-root sugar are amongst the other principal articles of manufacture. Two canals, that of the Ille and Rance, and that between Nantes and Brest, pass through different parts of this dep. It is divided into 5 arronds, 48 cantons, and 375 communes.

Chief towns St. Brieuc, the cap., Dinan, Guingamp, Lannion, and Loudéac. The Bas-Breton is the language commonly spoken, but most of the upper classes understand French. Many Celtic and Roman antiquities are scattered over this dep., of which the temple of Lanleff is the principal.

COTHEN (Germ. Köthen), a town of Central Germany in the duchy of Anhalt, on the Ziethe, 76 m. SW. Berlin, and 33 m. NW. Leipsic, on the railway from Leipsic to Magdeburg. Pop. 11,312 in 1861. Cothen is divided into the old and new town, and is well built. Among the public buildings are the old ducal palace, with a gallery of paintings, cabinet of natural curiosities, and a good library; the new ducal Schloss—former residence of the reigning family of Anhalt-Cöthen, which became extinct in 1847—three churches, a synagogue, orphan and female asylums, a teachers' seminary, and a school for the indigent. Gold and silver lace, woollen cloth, linens, tobacco, and leather are manufactured here; and there is some trade in corn, butter, cheese, and wool.

COTOPAXI, a celebrated volcano of S. America, in the republic of Ecuador (Colombia), belonging to the E. or more inland chain of the great Cordillera of the Andes ; in lat. 0° 40' S., and long. 78° 39' W., 84 m. SSE. Quito. Its shape is a perfect cone ; it consists chiefly of mica, but in part of obsidian ; its absolute height is 18,878 ft. above the level of the ocean, the upper 4,400 of which are covered with perpetual snow. Its summit is not more than about 9,000 ft. above the great longitudinal valley between the two chains of the Cordillera; but such is its steepness that Humboldt was unable to ascend it above the point at which the perpetual snow commences. The crater appears to be surrounded by a kind of circular wall, which, especially on the S. side, has the aspect of a parapet ; and, probably owing for the most part to the heat, this summit of the cone is never covered with snow, and looks at a distance like a dark stripe. On the SE. side of the mountain, near the snow-limit, there is a comparatively small projecting mass of rock, studded with points, and called the ' Head of the Inca ' by the Indians, who have a popular tradition that it formed originally a part of the summit of Cotopaxi. Humboldt himself inclines to the belief that the cone supporting the present crater, like the summits on Vesuvius, is composed of a great number of strata of lava heaped upon each other, ' Cotopaxi is the most dreadful volcano of the kingdom of Quito, and its explosions are the most frequent and disastrous. The mass of scoriæ, and the huge pieces of rock thrown out of this volcano which are spread over the neighbouring valleys, covering a surface of several square leagues, would form, were they heaped together, a colossal mountain. In 1738, the flames of Cotopaxi rose nine hundred metres (14 furlongs) above the brink of the crater. In 1744, the roarings of the volcano were heard as far as Honda, a town on the borders of the Magdalena, and at the distance of 200 common leagues. On the 4th of April, 1768, the quantity of ashes ejected was so great that in the towns of Hambato and Tacunga day broke only at three in the afternoon. The explosion that took place in the month of January, 1803, was preceded by a dreadful phenomenon, the sudden melting of the snows that covered the mountain. At the port of Guayaquil, 52 leagues distant in a straight line from the crater, we heard day and night the noises of the volcano, like continued discharges of a battery ; we distinguished these tremendous sounds even on the Pacific Ocean, to the SW. of the island of Puna.' (Humboldt's Researches, English trans., i. 115-125.)

COTTBUS (Germ. *Kottbus*), a town of Prussia, prov. Brandenburg, cap. circ. same name, on the Spree, 12 m. S. by W. Frankfurt-on-the-Oder, and 67 m. SE. Berlin, on a branch line of the railway from Berlin to Breslau. Pop. 11,112 in 1861. The town is walled, and has four churches, two hospitals, a gymnasium, and library, an orphan asylum, and a girls' school. It has three suburbs, and is commanded by a castle built on a height to the E. Cottbus is the seat of the council for the circ., and of a municipal court. There are considerable fabrics of woollen and linen stuffs and stockings, with breweries and distilleries. This town was made over to Prussia by the congress of Vienna, previously to which it belonged to Saxony.

COVE OF CORK. (See QUEENSTOWN.)

COVENTRY, a co. and city of England, within the co. of Warwick, 10 m. NNE. Warwick, 18 m. ESE. Birmingham, 85 m. NNW. London, and 94 m. by London and North Western railway. Pop. of mun. city 40,936 and of parl. city 41,647 in 1861. Coventry stands on a gentle declivity on the N. Western railway, and is watered by the Radford and Sherbourne brooks. Streets of the old town (with the exception of Cross Cheaping, where the splendid cross formerly stood, and which is now used as a corn market), generally narrow and ill-paved, and the upper parts of a few of the houses, which are high, project and present a sombre appearance. Within the last forty years, however, the suburbs have been considerably extended, several new lines of streets having been laid out, and many new houses erected. The principal buildings are, St. Michael's church, one of the finest specimens of the lighter Gothic in England, with a beautiful steeple, 303 ft. in height; St. John's and Trinity churches, Christ Church, attached to the old and beautiful spire of the Greyfriars' monastery; a Catholic chapel; several dissenters' meeting-houses; the county hall, erected in 1785; St. Mary's hall, erected (Henry VI.) for the Trinity guild, now used for meetings of the town council, and public concerts; a neat and commodious theatre; the drapers'-hall; the canal office; the free school; the gaol, and the barracks. Coventry was, conjointly with Lichfield, the see of a bishop, but on the recommendation of the ecclesiastical commissioners, it has been joined to the diocese of Worcester.

Under the Municipal Corporation Act the city is divided into six wards; and is governed by a mayor, 10 aldermen, and 30 councillors. The jurisdiction of the corporate authorities extends over the city and the co. of the city, including, in all, an area of 15,970 acres. The recorder holds a court of quarter sessions, and a court of record for the recovery of debts to any amount. The sheriff holds a county court monthly. Coventry has regularly sent 2 mems. to the H. of C. since 1453. Previously to the Reform Act the right of voting was exclusively in the freemen of the city who had served a seven years' apprenticeship in the city or suburbs. Registered electors 5,578 in 1862. The limits of the parl. bor. correspond with the ancient limits of the pars. of St. Michael and the Holy Trinity, except that it does not include the hamlet of Keresley. It embraces an area of 4,920 acres. The municipal boundary is co-extensive with the co.

Coventry has been the seat of 12 parliaments: one (Henry IV.) in 1404, called, from lawyers being excluded, *parliamentum indoctum*; the other (Henry VI.) in 1459, called *parliamentum diabolorum*, from its numerous acts of attainder. The city was incorporated by Edward III., and the first mayor chosen in 1345. It was erected into a county by Henry VI., with the hamlets

belonging thereto, and lying within the vill, or townships.

This city has many extensive and well-endowed charities; of these, one of the most celebrated is the free school, founded by John Hales in the reign of Henry VIII., in which the celebrated antiquary, Dugdale, received the early part of his education; it has a revenue of 600l. a year, and exhibitions to both universities. Here are also various charity, national, and infant schools, as Bonds' hospital, at Hablake, for 45 old men, with a revenue of 1,050l. a year; and Wheatley's school and hospital at the same place, for 40 poor boys, with nearly 600l. a year; Ford's hospital, in Grey-friars-lane, for 35 old women; Fairfax's school, in St. John's par., for 40 boys; Mrs. Catherine Bailey's school, in St. Michael's par., for 35 boys; the Blue Coat school in Trinity par., for 50 girls; White's charity, amounting to about 2,500l. per annum; and the House of Industry, formerly the White-friars' monastery. A library was established here in 1751; it is regulated by a committee. A mechanics' institute was founded in 1828. Here is also a society for the diffusion of religious and useful knowledge; general and self-supporting dispensaries, and a public hospital.

Previously to 1436, woollen cloth caps and bonnets were an important article of manufacture. In the early part of the 16th century, Coventry became famous for the production of a blue thread, called 'Coventry true blue.' But this was given up before 1581, after which woollen and linsed cloths continued the staple until the destruction of the Turkey trade in 1694. The manufacture of striped and mixed tammies, camlets, shalloons, and calimancoes, flourished during a part of the last century, but is now almost discontinued. This was succeeded by silk throwing and riband weaving, now the staple business of the place, and watch making. When first introduced, about a century and a half ago, the riband trade was for some time confined to a few hands, but it afterwards increased so as to exceed that of every other town in England. The alteration of the law as to the silk trade in 1826, and the commercial treaty with France of 1860, though productive of considerable loss and injury at the time, have, by introducing a spirit of competition, and stimulating the manufacturers to call all the resources of science and ingenuity to their aid, been the causes of great improvement. Lute-strings may now be purchased more cheaply in Coventry than in France. Plain goods of English manufacture are fully equal to those of the French; but the latter have the advantage in style and fashion, and in the brilliancy, though not in the permanency, of their colours. It is the general practice for the work to be given out to be executed in the houses of the workmen. The manufacturers employ girls and young women, who work together on the premises of the manufacturers, in winding and warping the silk for the out-door weavers. In 1839 it appeared, from the report of Mr. Fletcher to the commissioners of inquiry into the condition of the hand-loom weavers, that the operative hand owners in the city and suburban villages held 3,867 looms, of which 5,115 were worked by members of their own families, and the remaining 821 by journeymen and half-pay apprentices. It further appears from the same report, that 27 master manufacturers employed in loom shops or factories 1,962 looms. No official report of the state of manufactures in Coventry has been made since that time, and it is probable that no great changes have taken place. Large quantities of ribands are exported but the principal demand is for the London and country markets. There are several large dychouses, for dyeing the silk, employing from 300 to 500 hands.

The manufacture of watches was introduced about a century ago, and has continued progressively to increase. Large quantities are prepared for the home and foreign markets; some manufacturers employing, when the trade is in a state of activity, great numbers of hands. The wages of the workmen vary from 15s. to 70s. per week, the larger amounts being paid to those only who are proficients in making of the patent lever and other superior watches, which are now produced here equal in quality to those made in London. Coventry is advantageously situated for commercial operations, lying nearly in the centre between the four greatest parts of the country—London, Bristol, Liverpool, and Hull, and having direct communication by railroads and canals with the metropolis and principal towns in the kingdom. Corp. revenue, 14,808 in 1802. Gross annual value of real property assessed to income tax £29,931, in 1857, and 157,2421. in 1862.

During the monastic ages, Coventry had a splendid monastery, and a large and beautiful cathedral, similar to that at Lichfield. The latter was destroyed by a barbarous order of Henry VIII., and only a few fragments of it now remain. The city was formerly surrounded with walls of great strength and grandeur, with 32 towers and 12 gates. It has been always renowned for its pageants and processions, and particularly, in the monastic ages, for the performance of Mysteries. The legend of Peeping Tom, and the Lady Godiva, is too well known, through the exquisite poem of Tennyson, 'I waited for the train at Coventry,' to require any special notice. An effigy of the over-inquisitive tailor may be seen in the upper part of a house at the corner of Hertford Street. The trades-men of Coventry were formerly famed for their affluence. In 1448, they equipped 600 men armed for the public service. Many eminent persons have either been born or lived at Coventry, among whom were, Nehemiah Grew, curator, in 1672, to the Royal Society for the anatomy of plants, and in 1677, sec. to the Royal Society. Coventry gives the title of earl to the descendants of John Coventry, mayor of London in 1425. Market-days, Wednesdays and Fridays. The principal fair, held first Friday after Trinity Sunday, is called Show Fair, and continues eight days, on the last of which the representation of the Countess Godiva's procession is sometimes enacted.

COVILHA, a town of Portugal, prov. Beyra, on the E. slope of the Sierra de la Estrella; 20 m. SW. Guarda. Pop. 6,156 in 1858. The town rises amphitheatrewise between two streams. In the upper part there is an antique castle and tower, and in the lower part, on the margin of one of the streams, is a manufactory of fine cloths, druggets, and baizes, carried on by a company in Lisbon, containing above 120 looms. There are nine churches, with a hospital and a workhouse.

COURLAND, a government of Russia in Europe, on its W. frontier, having N. the Gulf of Riga and Livonia; E. the gov. of Witepsk; S. that of Wilna, and a small portion of Prussia; and W. the Baltic. Area about 10,000 sq. m. Pop. 533,300 in 1816, and 567,956 in 1838. Near Mittau, and along the shores, the surface is flat, and is overspread with marshes and sandy heaths; but the interior is mostly undulating, there being a chain of hills along the bank of the Duna, which sends numerous tions over the whole country. The Duna forms the E. and a part of the N. boundary; the other principal rivers are the Aa and Windau. There are

many lakes. Speaking generally, the atmosphere is damp, the sky cloudy, and the temperature low and variable. Soil generally light and sandy, requiring much manure; it is most fertile towards the E.; two-fifths of the surface is occupied by forests, chiefly of pine, fir, birch, alder, with a considerable intermixture of oaks. Agriculture is the principal occupation of the people, and notwithstanding the badness of the soil, has advanced more than in any of the neighbouring provs. More corn is grown than is necessary for home consumption; it is chiefly rye, barley, and oats. Flax and hemp, and a few fruits and pulse, besides a little tobacco, are also cultivated. Pasturage is scarce, and but few cattle are reared; the oxen and horses are both of a bad quality, and the sheep yield only a coarse species of wool. Bees are kept only to a trifling extent. Iron, lime, and turf, and occasionally amber, are found. Manufactures quite insignificant, and mostly domestic; in respect to these, Courland ranks nearly last amongst the Russian govs. There are a few of paper, copper articles, and earthenware, and some brandy distilleries and tile factories. Mittau, the cap. is the only town of any size; the principal sea-ports are Libau and Windau, both on the W. coast. The exports, which are principally corn, flax, hemp, and hemp-seed, skins, and salted meat, are said to amount to about 7,500,000 roubles a year, and the imports about 600,000. The inland trade is almost entirely in the hands of the Jews, of whom there are about 20,000 in the gov. Most of the pop. are Lutherans, and of Letton origin. Courland was anciently a part of Livonia, and was conquered in the 13th century by the Teutonic Knights; in 1561 it became a fief of Poland. After the fall of that power, it remained for a short time independent under its own dukes; but in 1795 it was united to Russia.

COURTRAY, or COURTRAI (Flem. Kortryk, Lat. Cortoriacum), a fortified and manufacturing town of W. Flanders, 17 m. E. Ypres, 25 m. S. Bruges, on the railway from Ghent to Tournay. Pop. 24,652 in 1856. The town is situated on the navigable river Lys, by which it communicates with the principal towns of Flanders. Houses well built; streets spacious and remarkably clean. The principal public buildings are the town house and the cathedral of Notre Dame, which are fine old Gothic edifices beautifully ornamented. The church of St. Martin is also a handsome structure. There is a nunnery, a collegiate school, an excellent academy of design, two orphan asylums, a savings' bank, and an exchange and a chamber of commerce. The spinning of linen thread, and the weaving of plain and damask linens, employ a large portion of the inhabitants. The fine linens of Courtray are known throughout Europe. Nearly all the weaving is performed on the handloom at home, and much of it by cottage farmers. The annual quantity of unbleached linen brought to the Courtray market is about 80,000 pieces, two-thirds of which are bought by the merchants of the town, and the rest by those of Belgium, France, and England. The spinning of cotton yarn, and the manufacture and dyeing of various cotton fabrics, constitute an important branch of industry. Courtray has also establishments for the manufacture of soap, candles, salt, tobacco, chicory, chocolate, oil, wax, paper, and pottery; besides numerous breweries and tanneries.

The surrounding plain is abundantly productive of all kinds of field and garden crops, especially flax, of which immense quantities are grown of the finest description, and the vicinities of the town are picturesquely varied by numerous bleaching-fields. Courtray was first built in the 6th century.

It was anciently known under the name of Cortoriacum, and in the 7th century it was a municipal city. Like the other towns of Flanders, it has been subject to many vicissitudes, has sustained several memorable sieges, and been burnt and plundered in war. Under its walls was fought, in 1372, the famous battle of the Spurs, between 20,000 Flemings, consisting chiefly of weavers of Ghent and Bruges, and a French army composed of 7,000 knights and noblemen, and 40,000 infantry. In this conflict the flower of the French chivalry was slain, and the victorious Flemings collected from the battle-field about 6,000 pairs of gold spurs worn by their proud and defeated foes. Among the antiquities that have been found, are numerous medals of the Cæsars. Fairs for all kinds of merchandise are numerously attended on Easter Monday and Aug. 24.

COUTANCES (an. *Constantia*), a town of France, dép. La Manche, cap. arrond., on a hill on the S. bank of the Soulle, 6 m. E. from the sea, and 16 m. WSW. St. Lô. Pop. 8,062 in 1861. Streets narrow, steep, and ill-paved; houses mostly of stone, roofed with slate. It contains several old churches worthy of notice, especially a Gothic cathedral, having two spires in front, and a large square tower surmounting the centre of the cross; it is a conspicuous object, and a landmark for ships in the Channel. The town has a bishop's palace, a communal college, a public library with 5,000 volumes, and a small theatre. Druggets, cutlery, and parchments are produced here; it has also marble-works, and a brisk trade in corn, butter, poultry, flax, hemp, and horses. In its immediate vicinity are the remains of an ancient aqueduct, with many of the arches still very perfect. Coutances was the birthplace of the Abbé de St. Pierre.

COWES (WEST), a town and sea-port of England, co. Hants, Isle of Wight, liberty West Medina, par. Northwood, 75 m. SW. London, 10 m. W. Portsmouth, on the acclivity and summit of a hill rising immediately from the W. bank of the Medina, at its embouchure in the channel between the Isle of Wight and the opposite coast of Hampshire. Area of par., 4,270 acres: pop. 4,691 in 1861. Streets narrow and very irregular; but, as the houses rise above each other from the water's edge to the summit, they have a striking effect, many of the upper and more modern ones being handsome structures commanding splendid and extensive views. In the immediate neighbourhood are numerous elegant villas. The town, which is much resorted to as a fashionable sea-bathing place, possesses ample accommodations for visitors, in hotels, lodging-houses, assembly-rooms, and reading-rooms. A crescent-shaped battery, defending the entrance to the harbour, has some heavy pieces of ordnance and accommodation for a company of artillery. E. Cowes, on the opposite side of the river, (¼ m. from W. Cowes,) is a small irregular built hamlet, of the par. of Whippingham, at the foot of a hill. Pop. 1,954 in 1861. Here is the custom-house of the port. The harbour and roadstead of Cowes are amongst the best and most convenient in the English Channel, and form the rendezvous of the Royal Yacht Club, and the station where their annual regatta is held. Many merchant vessels and yachts are built in the harbour. Many large ships, outward or homeward bound from or to London, are accustomed to touch at Cowes before proceeding on their voyage. It has also a considerable coasting trade. The exports consist chiefly of agricultural produce and malt; the imports of coals, manufactured goods, colonial produce, and other articles of general consumption. There are hourly steamers to Portsmouth and Southampton, and passage boats to Newport, up to which the tide flows.

CRACOW, a small and formerly—until Nov. 16, 1846—a nominally independent state of Central Europe, once part of the k. of Poland, at the present time a circle of Galicia; between lat. 50° and 50° 15' N., and long. 19° 8' and 20° 12' E. Length, E. to W., 46 m.; breadth varying from 3 to 15 m. Area, 480 sq. m. Surface generally undulating, consisting of the last ramifications of the Carpathian mountains. The Vistula, which bounds it on the N. in its whole extent, receives several small streams from the N. in this part of its course, one of which, the Brinkra, forms the W. boundary of the Cracow territory. Climate healthy and temperate; mean annual temp. 47½° Fahr. Soil very fertile, producing sufficient corn for home consumption, and an abundance of pulse, culinary vegetables, and fruit. The territory contains rich mines of coal, zinc, and alum; some iron is also found; and there are quarries of marble, building stone, and freestone. By the third partition of Poland, in 1795, Cracow passed under the dominion of Austria; but it was reconquered by the Poles in 1809, and incorporated with the grand duchy of Warsaw. At the Congress of Vienna, in 1815, the territory was erected into an independent neutral republic, under the protection of Russia, Austria, and Prussia. Agreeably to the amended constitution of 1833, the government was vested in a senate composed of a president and eight senators, two of whom were elected for life, and the other six, as well as the president, for six years. One of the latter was elected by the clergy (*chapter*) of Cracow. There was a legislative chamber composed of two senators (one of whom, chosen by the chamber, presided at its deliberations), 4 justices of the peace, 2 delegates of the clergy, 3 of the university, and 20 representatives, chosen by the electoral colleges of the city and territory. This assembly was convoked every three years to vote the budget, to inquire into the administration of the public funds, to elect the members of the senate and the different tribunals, and to discuss the laws presented for its sanction by the senate. But from 1836 to 1846 the territory was garrisoned by Austrian troops; and at the latter date, as above stated, it was incorporated into the Austrian empire.

CRACOW (an. *Cracovia*), a city of Central Europe, previously in the 17th century, the metropolis of the k. of Poland; on the N. bank of the Vistula, where it is joined by the Rudawa, 160 m. SSW. Warsaw, and 800 m. NE. Vienna, on the railway from Vienna to Lemberg. Pop. 41,086 in 1857, excl. of garrison. The city is divided into three portions, one of which, the Jews' quarter, is built on an isl. in the Vistula. The city has, besides, several suburbs. Cracow has near it Mount Wawel, a rock of moderate elevation, but considerable extent, on which are the castle and cathedral; and two barrows, said to be the burial-places of the founder of the city and of his daughter Venda. The city itself is old, and irregularly built; but its streets are broad, and its churches and other public buildings, having many of them interesting monuments, and being associated with some of the most important events in Polish history, invest it with much interest. It was formerly fortified, but the ramparts have been converted into public walks. The royal castle of Cracow, built in the 14th century, and formerly the residence of the kings of Poland, though now in ruins, is greatly decayed. It has been partly destroyed by fire at different times, and imperfectly restored; but it has suffered more from the effects of war, having been in great part demolished by Charles XII. in

1702, and still more from its change of masters; at one time it was used by the Austrians for barracks, and now serves for a workhouse. Of the 76 churches formerly in Cracow, about 40 are in ruins; the cathedral alone has retained its splendour and costly decorations, for which, and for its monuments, it is celebrated. Around its interior are 20 small chapels, crowned with domes in the Byzantine style. Most of the Polish kings and many illustrious men are buried in it; among others it contains the tombs of Casimir the Great, of John Sobieski, the deliverer of Vienna, and of the 'last of the Poles,' Kosciusko and Poniatowski. The other churches and palaces have fine paintings, statues, and ancient monuments. The episcopal palace is the most striking of the modern edifices, its walls being adorned with paintings in fresco, representing the most remarkable events of Polish history.

The university, founded and endowed by Casimir the Great, and improved by Ladislaus Jaghellon, has lost most of its ancient importance. Cracow contains a college, a school of arts, an academy of painting, a public library with 30,000 vols., and 4,000 MSS., an observatory, and a botanical garden. The articles of export and import consist principally of skins, linen, wax, corn, wood, Hungarian wines, and manufactured articles from England and Germany.

About a league W. of the city is an artificial tumulus erected to the memory of Kosciusko. On the 16th of Oct. 1820, the senate of Cracow, accompanied by vast numbers of the nobles and the people from all the different provinces, proceeded to deposit the first load of earth upon an eminence not far from the walls of the city, which had been selected to bear a mountain tumulus in honour of the patriotic general. For four years this great work was eagerly pursued; citizens of every rank toiled at the wheelbarrow; parcels of the sacred soil were sent to join the mass from all the great battle-fields which had been sprinkled with Polish blood; and the mound gradually rose to an altitude of about 150 ft. This monument of clay, planted on the soil which has been most frequently and grievously convulsed by political revolutions, will probably maintain its place as long as the world is habitable by men. Of all the structures of our age, if structure it can be called, this alone seems raised for all time—a thing lasting in itself, lasting for the name it bears, and lasting by the spirit which made it, when those who raised it shall all be scattered in uncollected dust.' (Reeve's Sketches of Bohemia.)

The city is said to have been built about the year 700, by Krak, a Polish duke, from whom it derived its name. It successively belonged to the Moravians and Bohemians, and was taken from the latter at the end of the 10th century by Boleslaus the Great, who made it the cap. of Poland. In the 16th century it contained three times its present number of inhab.

CRAIL, a royal and parl. bor., of Scotland, co. Fife, 9 m. from the East Neuk of Fife, or Fife Ness. Pop. 865 in 1861. It is a decayed place, destitute of trade or manufactures. Many of the houses, however, are of that massive description that indicates former greatness. David I. had a palace here, which is now entirely demolished. The par. church was once collegiate, with a provost, sacrist, and ten prebendaries. The famous James Sharp, afterwards archbishop of St. Andrew's, murdered by the Covenanters on Magus Muir in 1679, was once minister of Crail. Coal is abundant in the neighbourhood.

CRANBOURNE, a town and par. of England, co. Dorset, div. Blandford. Area of par., 15,700

acres. Pop. of ditto, 2,656 in 1861. The town is situated in an open pleasant district, 12 m. SSW. Salisbury. The church is a fine old structure, partly in the Norman, and partly in the earliest Gothic, with a noble tower in the later Gothic style. There is an almshouse for three old people, and a few smaller charities. The ribbon manufacture, formerly carried on here, has declined, and the inhabitants are now chiefly employed in agriculture. This par. is the supposed arena of the battle between the British, under Boadicea, and the Romans. Numerous barrows are dispersed over it, in which bones and urns have been found. On the Castle-hill, N. of the town, are the remains of a circular fortification, enclosing an area of six acres, Cranbourne Chase, a tract extending nearly to Salisbury, was celebrated during both the Saxon and the Norman periods. An old embattled manor house, called the Castle, still exists, which was occasionally the royal residence; in its hall courts were held; and there is a dungeon for the confinement of those who infringed on the game laws. Bishop Stillingfleet was a native of Cranbourne.

CRANBROOKE, a town and par. of England, co. Kent, lathe of Scray, hund. Cranbrooke. Area of par., 10,000 acres. Pop. of ditto, 4,128 in 1861. The town, on the Crane (a small stream traversing the Weald district), 50 m. SSE. London, consists of a main street, nearly 1 m. in length, and a smaller one diverging from it. Many of the houses are well built, and it is partially paved and lighted, and amply supplied with water. The church, re-built about 1730, in the later Gothic style, has a lofty embattled tower. There are also six dissenting chapels; a grammar-school, endowed by Queen Elizabeth; a writing-school, founded in the same reign, with a small endowment; and a national subscription school. The woollen trade, introduced here by Edward III., and long considerable, has disappeared; and the trade in hops is now the staple business of the place. Sir R. Baker, the antiquary, and Huntington, the founder of a religious sect, were natives of this place.

CRAYFORD, a town and par. of England, co. Kent, lathe Sutton-at-Hone; 11 m. E. by S. London by road, and 14½ m. by London, Chatham, and Dover railway. Area of par., 2,380 acres. Pop. of ditto, 3,013 in 1861. The town, situated on the Cray, about 1 m. above its confluence with the Darent, and on the great road from London to Dartford, consists of a long irregular street. The church is in good modern structure, on an acclivity at the higher end of the town. Its market has been long discontinued, but an annual fair is held Sept. 8. Until a recent period, extensive printworks were carried on a little below the town; and a mill for flattening iron and splitting iron into loops, one of the first of its sort constructed in England, was, until recently, in operation. In the parish are numerous artificial caves, upwards of 100 ft. in depth, increasing in magnitude as they recede from the earth's surface. Some of them contain several distinct apartments, excavated in the chalk, supported by pillars left at intervals for the purpose. Their origin is a matter of dispute; some having supposed them to be mere chalk quarries, while by others they are supposed to be places of security excavated by the ancient Britons or Saxons as receptacles for their families and goods during periods of danger. The Roman station Noviomagus is supposed to have been near Crayford, contiguous to which, A.D. 457, was fought the great battle between Hengist and Vortigern, which ended in the total defeat of the Britons.

CRECY, an inconsiderable village of France, dép. Somme, 11 m. N. Abbeville, famous in history for the victory gained here on the 25th of August, 1340, by the English forces under Edward III. over the French under their king Philip of Valois. The French army is believed to have amounted to about 120,000 men, while that of the English was under 40,000; but the superior discipline and good order of the latter more than counterbalanced their inferiority in point of numbers, and enabled them to achieve one of the greatest victories of which we have any account. The loss of the French, in the battle and pursuit, has been estimated at 1,200 knights, 1,400 gentlemen, 4,000 men at arms, and about 30,000 inferior troops. Besides the king of France, there were in the defeated army the kings of Bohemia and Majorca, both of whom were killed. The crest of the former, consisting of three ostrich feathers, with the motto *Ich Dien (I serve)*, was adopted by the Black Prince, the eldest son of Edward, whose bravery was most conspicuous on this occasion; and has been continued as the crest and motto of all subsequent princes of Wales down to the present times. The loss on the part of the English was comparatively trifling. It has been said that cannon were first employed by the English in this battle, and that they contributed not a little to their success. (Rapin's Hist. of England, iii. 458, 8vo. edit.; Hume's ditto, cap. 15.)

CREDITON, a town and par. of England, co. Devon, hund. Crediton, 7 m. NW. Exeter, on the London and South Western railway. Area of par., 11,410 acres. Pop. of town, 4,046, and of par., 5,731 in 1851. The town is situated in a narrow vale between two steep ridges, through which the Creedy flows and joins the Exe a little lower down. It is divided into two distinct parts, the E. or ancient town, and the W. more modern and larger part, consisting chiefly of a broad street along the principal line of road from Exeter to N. Devon. The church, a noble building in the later pointed style, with a fine tower springing from the centre, was rebuilt in 2 Henry VII. There are four dissenting chapels; a free grammar school, founded by Edward VI., for boys of Crediton and Sandford par.,—it has three exhibitions to either university; a blue-coat school, founded 1760, and incorporated with a national school established 1814, in which 150 boys are instructed, 80 of whom are clothed; a mathematical school, founded 1794, for 12 boys; two sets of ancient almshouses; and several minor charities. The majority of the labouring pop. are now employed in agriculture. Formerly there were several large woollen and serge manufactories: at present, however, there are no resident manufacturers, though many females weave long ells at their own dwellings, for manufacturers resident in N. Tawton. This town sent members to the parl. at Carlisle, in Edward I. (Willis's Not. Parl.) It was several times the head-quarters of each party during the last civil war. In 1743 it was nearly destroyed by fire, and was also seriously injured by fire in 1769.

CREETOWN, a neat marit. village of Scotland, co. of stewartry of Kirkcudbright, par. Kirkmabreck, at the head of Wigtown Bay, where it receives the Cree, and on the road between Dumfries and Portpatrick. Pop. 909 in 1861. The hills in the neighbourhood of Creetown seem to be almost entirely composed of granite; and an extensive granite quarry, within 2 m. of the village, has furnished materials for the Liverpool docks. They used formerly to be large beds of sea-shells in the vicinity, the shipment of which for manure to other places was a considerable source of employment; but these are now nearly exhausted. The late Dr. Thomas Brown, the celebrated ethical philosopher, was born here in 1780, his father being minister of the parish.

CREFELD, a thriving town of Rhenish Prussia, cap. circ. same name, in a fertile plain, 6 m. W. from the Rhine, and 13 m. NW. Dusseldorf, on the railway from Cologne to Utrecht and Amsterdam. Pop. 50,584 in 1861. It is the principal town in the Prussian dom. for the manufacture of silks, silk velvets, and silk thread. It has also fabrics of woollen, cotton, and linen stuffs, lace, oil-cloth, camlets, and earthenware; with tanneries and distilleries. The town is well built, with wide streets and neat houses. It has four churches, an orphan and a deaf and dumb asylum, a hospital, a high school, police and commercial courts, and is the seat of a court of justice. In its vicinity is an old castle, now used for a silk-dyeing establishment. In the latter half of the seventeenth and beginning of the eighteenth centuries, its pop. was greatly augmented by many reformists and Mennonites, expelled from the neighbouring duchy of Juliers, and who, in return for their hospitable reception, introduced those manufactures to which the town owes all its prosperity.

CREMA, a town of Northern Italy, prov. Cremona, on the Serio, 25 m. ESE. Milan. Pop. 8,240 in 1861. The town is surrounded by a brick wall, a ditch, and some other old fortifications, and has a castle, which, before the use of artillery, was considered one of the four strongest fortresses in Italy. It is well built; streets spacious; palaces and public edifices numerous, including a cathedral and many other churches, a hospital, three separate charitable asylums, and two theatres. It has manufactures of lace, hats, linen thread, and silks, and is celebrated for the excellence of its flax. Very good wine, fruit, and fish are obtained in its vicinity. Crema was founded about 570 A.D., during the reign of Alboin, the first Lombard king of Italy. In 1159 it was sacked by Fred. Barbarossa; it was taken by the French in 1797, the day after the rupture of Lodi.

CREMONA, a city of Northern Italy, cap. deleg. same name, on the left bank of the Po, 44 m. SE. by E. Milan, and 26 m. NW. Parma, on a branch of the railway from Milan to Venice. Pop. 28,591 in 1861. The town is of an oval shape, about 6 m. in circ.; is surrounded by walls, bastions, and wet ditches, and defended by a citadel. It is well laid out, but has a melancholy appearance, from the evident signs of decay, and large tracts of grass being seen in many of the broad and regular streets. Among its 44 churches, the *Duomo* alone has any particular attractions. This is an ancient edifice in the style of architecture approaching to Saxon, mixed with a sort of mongrel Italian. If not beautiful, it is at least picturesque; and its lofty tower, 372 ft. in height, is singularly so, being adorned with a sort of rich open work; it is one of the highest in Italy. The interior is composed of a nave with two aisles, divided by eight immense pillars, above which are a series of paintings by Bordonone. Near the cathedral is an octagon baptistery, said to have been once a temple of Minerva. In the town-hall, among others, there is a fine picture by Paul Veronese.

Cremona is the residence of the delegate of the prov. and seat of a bishopric; it has civil, criminal, and commercial tribunals, a lyceum, gymnasium, superior and female schools, several well-attended infant schools, which were the first institutions of the kind opened in Italy, a public library, numerous collections of works of art, two theatres, barracks, a *monte-di-pietà*, and several hospitals, asylums, and other charitable institutions. The manufactures of silk and cotton fabrics are considerable, and there are others of porcelain and earthenware, dyes, and chemical products. During

R 2

the 17th, and the earlier part of last century, Cremona was highly celebrated for its musical instruments, especially its violins made by the Amati and Straduarius. Instruments by these makers are now very scarce, and fetch an extraordinary price; and the manufacture of violins and strings has greatly declined. Cremona has a brisk trade in corn, flax, cheese, silk, oil, honey, wax, &c.; the flax grown in its vicinity is much esteemed. This city is very ancient; it was probably founded originally by the Gauls, and, together with Placentia, was the seat of the first colony established by the Romans in Cisalpine Gaul; but its antiquities have been swept away by the successive revolutions it has undergone. Having espoused the cause of Brutus, Augustus divided its territory among his veterans; and this being insufficient for the purpose, he added to it the territory of Mantua, as is well known from the line of Virgil:—

'Mantua væ miseræ nimium vicina Cremonæ!'
                                        Eclog. ix. 28.

But it speedily recovered from this disaster, and rose to great wealth and eminence. Certainly, however, it was, as Tacitus says, 'bellis civilibus infelix.' In the struggle between Vitellius and Vespasian it was occupied by the troops of the former, and, being taken by those of the latter, it was sacked and burnt by the infuriated soldiery. (Tacit. Hist., lib. III. §§ 26–33.) It was again, in as far as practicable, restored by Vespasian. From the 12th century, downwards, its history is identified with that of Milan. In 1796 it opened its gates to the French; and from 1800 to 1814 was the cap. of the dép. Alto-Po. Vida, bishop of Alba, one of the best modern Latin poets, was born at Cremona in 1490.

CRETE (vulg. CANDIA), a large and celebrated isl. of the Mediterranean, belonging to the Grecian Archipelago, of which it forms the S. boundary. It lies between 34° 57' and 35° 41' N. lat., and 23° 29' and 26° 20' E. long., its NW. extremity being 80 m. SE. Cape Matapan, in Greece, and its NW. termination 110 m. SW. the nearest point of Asia Minor. It is long and narrow, its length from E. to W., being about 160 m., with a breadth varying from 6 to nearly 50 m., but averaging about 20 m. Area, 3,281 sq. m. Pop. estimated at 150,000; of whom 100,000 are native Greeks, 44,000 Turks, and the remainder Hellenes, Jews, and other foreigners. Previously to the breaking out of the Greek Revolution, the pop. was estimated at about 270,000. At the period when it was acquired by the Venetians, Crete had probably a pop. of 500,000 or 600,000, but it fell off greatly under their oppressive sway. Its fertility, and the number and magnitude of its ancient cities, warrant the supposition that the pop. in antiquity may have amounted to 1,000,000 or 1,200,000. (Pashley, ii. 326.) The isl. at present belongs to Turkey, and is divided into the three prov. of Candia, Retimo, and Canea, so named from their respective capitals. These prov. are subdivided into 20 eparchies, or districts, of which Candia comprises 11, Retimo 4, and Canea 5.

Topography.—Crete is almost wholly covered with mountains. A serrated range stretches through its whole extent E. to W.; in the E. although rugged and barren, it attains no great elevation; but as it proceeds westward, its peaks increase in height, and are covered with snow even in June. At the W. extremity of the island, the range of the White, or Sphakian mountains, rises to perhaps 5,000 ft., and Ida (now Psiloriti) the loftiest as well as the most famous of the Cretan mountains, nearly in the centre of the island, is, according to Sieber 7,674 ft. high. Ida, however,

has little besides its height and classical celebrity to recommend it. The different mountain ranges abound with grottos and caverns, some of which are alike extensive and celebrated. Every classical reader must be acquainted with the history of the famous labyrinth in which Minos kept the Minotaur killed by Theseus. A cavern of great extent and intricacy, and which answers in all the most essential particulars to the accounts given of the labyrinth, in a hill at the N. foot of Mount Ida, about 3 m. from the ruins of Gortyna, has been visited and described by Tournefort (I. 65), Cockerell (Walpole's Memoirs, I. 405), and others. It has been supposed by some that this cavern, which consists principally of many long, winding, and narrow passages, which can only be safely explored by means of a clue, was a quarry whence the stones used in the building of Cnossus and Gortyna had been derived; but any such supposition seems wholly out of the question; it is not possible to imagine, had it been a quarry, that it should have been excavated in narrow winding passages, as that would have added immeasurably to the difficulty and cost of procuring the stones. Tournefort has supposed it to have been originally a natural cavern, and that it had been improved and perfected by art, to make it a place of concealment, or refuge, in periods of distress.

On every side of the island, but especially on the N., the mountain region extends quite to the coast, which is generally lofty and inaccessible. The N. shores present several remarkable headlands, as capes Busa (Corycum), Spada (Psacon), Melek (Cyamon Pr.), St. John, Salmone, &c., and are indented by many extensive bays, the chief of which are those of Kissamos, Khania, Sudha, Armyro (Amphimalla) and Mirabel. There are some tolerable harbours on this shore; but of these the S. coast is entirely destitute, and presents only one point worthy of notice, Cape Matala, the most southerly of all, belonging to Europe. Several small islands surround Crete, as Graham, Dhia, Gozo, &c., and in the bay of Sudha are the Leuca, supposed to be the isles of the Syrens celebrated by Homer. The plains are few; the chief are those in the N. of Crete, surrounding the towns of Canea, Candia, &c., and the larger one of Gortyna or Messara in the S., through which the Messara, the largest stream, flows. There are no rivers of any importance, but every little ravine in the furrowed sides of the mountains bears its tribute of melted snow to the rich alluvial valleys lying at their feet, rendering them abundantly fertile. At the E. and W. extremities of Crete there are a few unimportant lakes.

Climate and Natural Products.—In the lower parts of the country it is never freeze, and in summer the heat would be intolerable if not tempered by N. winds, which are then prevalent. Rains occur mostly in the spring and autumn. The country is generally healthy, and subject to few endemic diseases. Granite, schist, slate, &c. are amongst the primary rocks of the mountains, but calcareous formations, as in Greece, are the most common. Crete is not rich in metals; there are no mines, though Diodorus Siculus and other ancient writers preserve the tradition that iron was first discovered here. The mountains are clothed with woods of oak, chesnut, walnut, and pine trees, and the plane, cypress, myrtle, wild olive, vine, carob, aloe, arbutus, ficus indicus, and a multitude of fine fruits and vegetables grow spontaneously, while the ground is fragrant with aromatic herbs. For luxuriant vegetation it presents a wide and favourable contrast with some of the arid regions of continental Greece. The wild boar, wild goat, wolf, &c. are met with in the forests, and game of various kinds is plentiful. Birds of prey are numerous, but reptiles are few. (Pashley, Scott, &c.)

Agriculture.—From 1821 to 1830, Crete suffered the worst evils of a sanguinary and devastating war, and though its agriculture be now somewhat revived, it is still in a deplorable state. Its trade has been more than decimated, its olive plantations and vineyards uprooted, its villages burned down, and much of its most productive land been overgrown with rank vegetation. The soil is for the most part light, and but little adapted for the culture of grain. Wheat, barley, and oats are, however, grown, and, previously to the Greek revolution, wheat was annually exported; but sufficient corn is not produced for home consumption, and Crete is obliged to depend for supplies on Egypt and Barca. The chief products are oil, silk, wine, raisins, carobs, valonea, wool, oranges, lemons, wax, honey, linseed, and almonds. Cotton and flax are also cultivated, and in the mountains many of the fruits and vegetables of colder climates. The oil is good. Cretan wine is frequently eulogized by ancient authors. In the middle ages it held the first place amongst the exports, and under the names of Malmsey and Muscadine, considerable quantities were sent to England. The pastures are fine, and cattle of all kinds are reared, but their exportation is prohibited. Poultry are everywhere plentiful. Almost every peasant has his own farm; those who have not, cultivate the lands of the aga, or district governor, on a kind of metayer system, the lessor furnishing the seed and all the necessaries of husbandry, and dividing the crops in equal proportions with the cultivator, after deducting the seventh, to be paid to the government, and the seed previously advanced. The Mussulman rural population has been diminishing ever since the island fell under the Egyptian rule. Finding they are no longer able to obtain the forced labour of the Greeks, they are eagerly selling their lands, which are as eagerly purchased by the Greeks, who often borrow money for the purpose at an interest of 20 to 30 per cent. per ann. Landed property gives at an average a nett profit of 8 to 10 per cent. per ann.

Commerce and Trade.—The subjoined table exhibits the exports of the island (Report by Mr. Consul Frank Hay on the Trade of the Island of Crete, dated April 26, 1865) for the year 1864:—

Articles	1864				
	Quantity	Rate		Value	
		£ s. d.		£ s.	
Olive Oil . tons	7,426	60 0 0		118,000 0	
Soap . . . ,	5,079	34 0 0		170,300 0	
Silk . . lbs.	18,150	1 2 0		19,965 0	
Wax . . cwt.	12	9 10 0		114 0	
Honey . . lbs.	46,920	0 5 0		1,048 15	
Almonds . cwt.	950	8 0 0		7,877 0	
Carobs . . ,	78,628	0 5 0		19,754 0	
Valonea . tons	535	10 0 0		5,350 0	
Cheese . . cwt.	554	3 5 0		1,850 10	
Oranges } per	6,603	0 16 0		5,453 17	
&Lemons } 1000					
Wool . . lbs.		—		—	
Wine . gallons	18,200	0 1 0		910 0	
Linseed bushels	7,277	0 5 0		848 11	
Chestnuts value		—		3,570 0	
Raisins . cwt.	5,042	0 7 0		1,800 15	
Lamb Skins No.	7,040	0 0 0		176 0	
Cotton . cwt.	503	9 0 0		4,577 0	
Total . £		. .		344,407 14	

The manufacturers of the island are inconsiderable. The chief are those of soap, leather, and spirits; the rest consist only of domestic manufactures, as overticks, stocking, and coarse cloths, woven by women and children. There are twenty-four soap manufactories at work, capable of producing 6,000 tons a year, though little more than half that quantity is made. The article is of good quality, highly esteemed in the Levant, and fetches the highest price in the market at Trieste.

Government.—Crete is governed by a pasha, and each province by a president with a large salary, who is either a European or Asiatic Turk. In each province there is a council consisting of the cadi, treasurer, and other functionaries, and of a Turkish and a Greek representative from each of its districts, chosen however not by the districts they represent, but by the pasha himself, from whom they receive a salary. These councils decide on all judicial questions within their respective provinces, and proceeding according to the code Napoleon. The will of the president determines the council.

The *armed force* amounts to about 4,500 men, chiefly Arabs and Albanians. There are eight fortresses, mounting altogether 464 pieces of cannon. The fortifications of the principal towns are kept in good order; but those of the others are in the most neglected state.

Before the Greek revolution, the Christians and Mohammedans were nearly equal as to numbers; the balance is now greatly in favour of the former. The island is divided into eight bishoprics, the metropolitan bishop residing at the town of Candia. There are thirty large monasteries and many small ones in the island; and, like the mosques, they are all endowed, and possess extensive lands. The patriarch of Constantinople receives annually from Crete about 250,000 piastres (2,500l.). The priesthood are generally very ignorant.

People.—The Cretans are stronger built than the Ionians, of the other Greek islands; but it is said that generally they have not the same intelligence or vivacity. They are frugal, inoffensive, and superstitious in the extreme. Both ancients and moderns have accused them of being excessively addicted to lying and thieving; but Pashley (i. 38) thinks that in the interior, at least, they hardly deserve this character. They are polite and circumspicious, and dress like other Greeks, except that the men all wear high boots, and the women, when abroad, cover the face. Their dwellings are mean and comfortless; the food of the peasantry consists mostly of barley bread, cheese, olives, pulse, and vegetables, cooked with an abundance of oil. The language is modern Greek.

Antiquities and History.—Crete is highly interesting from its classical associations. Its history leads us back to the earliest mythological ages. It was the birthplace of Jupiter, 'king of gods and men.' Adventurers from Phœnicia and Egypt introduced arts and sciences into Crete, while Greece and the rest of Europe were involved in the darkest barbarism. The laws of Minos served as a model to those of Lycurgus; so that Crete became, as it were, a channel by which the civilisation of the East was transferred to Europe. Its wealth, and the number (100) and flourishing condition of its cities, particularly those of Cnossus, Gortyna, Cydonia, &c., are repeatedly referred to by Homer. Unluckily, however, the most violent animosities usually subsisted among the principal cities of the island, which formed so many independent republics; and Crete was thus prevented from playing any conspicuous part in the affairs of Greece, or from making that figure in history it could hardly have failed to make had it been a single state. It was conquered by the Romans, after an obstinate resistance, anno 67 B.C. After being possessed for a while by the Byzantine emperors, the Saracens took it in the 9th century; but being expelled in 952, it was again restored to the Eastern empire. The Genoese, and the Marquis of Montferrat, afterwards successively possessed it. The Venetians bought it of the latter in 1204; and in 1669, after a 24 years' war, it was conquered by the Turks. The revolution in Greece was followed by one in Crete, which deserved, and would doubtless have obtained, a happier issue had not the allies confirmed the gift of the island, in 1830, by the sultan, to Mehemet Ali, for his services during the war. Before the outbreak of the Greek revolution, Crete was the worst governed and most oppressed province of the Turkish empire. Since it has belonged to Egypt, notwithstanding the tyrannical rule of the viceroy, some amelioration has been experienced; but the Cretans still skill to be united to Greece, or to be taken under the protection of some European power, a protection to which their ancient fame, and their sacrifices in the cause of freedom, give them a well-founded claim.

CREUSE, a dép. of France, reg. centre, having N. the dép. Indre and Cher, E. Allier and Puy-de-Dôme, S. Corrèze, and W. Haute-Vienne. Area, 556,830 hectares; pop. 270,055 in 1861. Surface mostly mountainous, with a general slope towards the N. Some of its mountains are so environed with volcanic products as to leave little doubt that they were formerly active volcanoes. Plains of any extent few. Rivers numerous, including the Creuse (whence the dép. has its name), Cher, Tardes, &c., but none navigable. Climate rather severe; the summer being comparatively short, and the winter long and rigorous. Soil, except in the valleys, sandy and little productive. Arable lands occupy about 240,000 hect., pastures, 132,000 do., and heaths, wastes, &c. 122,000 do. Agriculture is in general very backward, and is so where pursued on a large scale. Corn, the chief part of which is rye, is not grown in sufficient quantity for home consumption. Fruits of various kinds are cultivated, but wine is furnished from the neighbouring dép. Cattle-breeding is rather an important branch of industry. The oxen, which are of a middle size, fatten readily, and form a portion of the supply for the Paris market. The sheep supply annually about 350,000 kilog. of wool, but it is mostly of inferior quality. Hogs are reared both for home consumption and for exportation. The management of bees is well understood, and the honey and wax are excellent. Property is here very much subdivided; more than three-fourths of the estates in the dép. being assessed below 20 fr. a year. Some coal mines, and quarries of granite, building-stone, and plastic clay, are worked. Manufactures very few: the chief are those of carpets, at Aubusson and Felletin; a porcelain factory at Bourganeuf, and some fabrics of paper, coarse woollen and linen cloths, glass, earthenware, and leather. The exports are limited to some thousand head of cattle, timber, coarse woollens, carpets, and pottery, with a very curious article, namely, hair, which the females of this dép. supply in exchange for articles of dress, to the extent of many cwt. a year, sent to the coiffeurs of Paris. The imports include most articles of prime necessity, including all the wine and nearly all the wheat consumed, with iron, salt, colonial produce, hemp, silks, and drugs. The depressed state of agriculture and manufactures, and the consequent want of employment, occasion the annual emigration of from 22,000 to 28,000 labourers, who resort to other parts of the kingdom in search of work and wages. They leave home in small parties of from 4 to 12, which sometimes augment on the road to 300. Each of these parties travels under

the conduct of a master, who undertakes work, and engages and pays those who travel with him. The period of emigration is from March to December. Creuse is divided into 4 arrond., 25 cantons, and 269 communes. Chief towns, Guéret, the cap., Aubusson, Bourganeuf, and Felletin. Generally speaking, this dép. is remarkably free from crime. The whole are poor and economical, but excessively litigious. The women share in the most laborious occupations.

CHEWKERNE, a town and par. of England, near the S. border of the co. of Somerset; in a vale watered by the Parret and Axe, 16 m. SE. Taunton, on the London and South Western railway. Area of par., 6,810 acres. Pop. of town, 3,555, and of par., 4,705 in 1861. The town consists chiefly of five streets, diverging from a central market-place, and is paved, lighted with gas, and amply supplied with water. The church, a cruciform structure in the later Gothic style, has a fine elaborately-ornamented tower, and the windows and interior also present rich specimens of tracery. A free grammar-school, founded in 1419, has an annual revenue of 300l., and there are four exhib. from it to any college in Oxford. There is also a national subscription school, and two sets of almshouses, founded in 1707; the one for six old men, the other for six old women. There are manufactures of sail-cloth, dowlas, and stockings, each of which employs a considerable number of hands.

CRICKLADE, a parl. bor. of England, co. Wilts, hunds. Highworth, Cricklade, and Staple, in an open level tract, at the junction of the Churn and Key with the Isis; 76 m. WNW. London by road, and 81¼ m. by Great Western railway, and Purton station. Pop. 36,593 in 1861. The borough consists chiefly of one long street of meanly built houses, paved, but not lighted, and very inadequately supplied with water. It comprises two par., St. Mary and St. Sampson, and a township, including in all an area of 2,810 acres. The church of the former par. is small and antique, while that of St. Sampson is a spacious cruciform building, with a lofty and highly ornamented tower. It has numerous escutcheons, bearing the cognizances of the earl of Warwick, and other eminent individuals, and is a fine specimen of the Gothic. In the churchyard is a well-preserved cross, with canopied niches, which was removed from the High Street, and placed here when the old town-hall was demolished. The remains of a priory, founded in the 1st of Henry III., are now used as tenements for paupers. There are two national schools, supported by subscription; formerly an ancient free school existed, but the endowment has been lost; a charity, producing 122l. a year from land, is appropriated to the apprenticing of poor children. The Thames and Severn canal passes through the N. end of the town; and a branch, joining the Wilts and Berks canal at Swindon, crosses within 1 m. of it. The inhab. are chiefly engaged in agriculture.

Cricklade returned 2 mems. to the H. of C. from the 21st of Edward I. to the 1st of Henry VI., with some interruptions; and from the latter reign, continuously to 1780, the right being exclusively vested in freeholders and copyholders of the bor. lands, and leaseholders of the same for not less than 3 years. In 1780 (after a contested election) the bor., in consequence of its notorious corruption, was thrown open, and the freeholders of the 5 adjoining divisions of Highworth, Cricklade, Staple, Kingsbridge, and Malmesbury, admitted to a participation in the elective franchise. Registered electors, 1,719 in 1861. The bailiff of Cricklade is returning officer. This town has considerable claims to antiquity; but the story of the

University of Oxford being founded by the professors and students of an ancient school established here, appears to be wholly destitute of foundation.

CRIEFF, a burgh of barony of Scotland, co. Perth, on a gentle acclivity on the N. bank of the Earn (a tributary of the Tay), 17 m. W. Perth. Pop. 2,363 in 1801. The place lies near the foot of the Grampian Hills, at the mouth of one of the important passes to the Highlands, and is the second town in the co. It formed, more than once, the head-quarters of the Duke of Montrose, during the civil wars in the reign of Charles I., and was burnt by the Highlanders in 1715. It was formerly the greatest cattle market in Scotland, but that was transferred to Falkirk in 1770. Its chief distinction now consists in its manufacturing industry. There are in Crieff about 500 hand-loom weavers, chiefly employed in the cotton trade. There is, also, a considerable trade in tambouring and flowering webs for the Glasgow manufacturers, carried on by females. About 300 acres of land in the immediate vicinity of the town are let to the inhabs in small patches, technically called acres; or in still smaller portions, called perds. There are three places of worship connected with the established church, and several chapels belonging to Presbyterian dissenters, and an episcopal chapel.

CRIMEA, the *Chersonesus Taurica* of the ancients, a peninsula of Russia in Europe, government of Taurida; between 44° 26′ and 46° N. lat., and 32° 33′ and 36° 22′ E. long. It is united on the N. to the mainland by the isth. of Perekop, 5 m. in width, and has on its E. the *Siracke*, or Putrid Sea (which see), the Sea of Azof, and the Straits of Yenikale, by which it is separated from the Isle of Taman, being everywhere else surrounded by the Black Sea. It is estimated to contain about 15,000 sq. m. Pop. estimated, in 1838, at 500,000. The Crimea is divided into two distinct parts, one lying N. and the other N. of the river heights, which flows from W. to E.; and is the only stream of any importance in the peninsula. The former consists almost entirely of vast plains, or steppes, destitute of trees, but covered with luxuriant pasture, except where they are interspersed with heaths, salt-lakes, and marshes. The climate of this region is far from good; being cold and damp in winter, and oppressively hot, and very unhealthy in summer, particularly along the Putrid Sea. The aspect and climate of the other, or S. portion of the peninsula, are entirely different. It presents a succession of lofty mountains, picturesque ravines, chasms, and the most beautiful slopes and valleys. The mountains, formed of strata of calcareous rocks, stretch along the S. coast from Caffa, on the E., to Balaclava on the W. The Tchatyrdagi, or Tent mountain, the highest in the chain, rises to the height of about 5,110 ft. above the level of the sea, and several of the other summits attain to a considerable elevation. The climate of the valleys, and of the slopes between the mountains and the sea, is said to be the most delicious that can be imagined; and, besides the common products, such as corn, flax, hemp, and tobacco, vines, olives, fig-trees, mulberry-trees, pomegranates, and oranges, flourish in the greatest profusion. Pallas, Dr. Clarke, and others, have given the most glowing descriptions of this interesting region. According to Clarke, 'If there exist a terrestrial paradise, it is to be found in the district intervening between Kutchukoy and Sudak, on the S. coast of the Crimea. Protected by encircling alps from every cold and blighting wind, and only open to those breezes which are wafted from the

S., the inhabitants enjoy every advantage of climate and of situation. Continual streams of crystal water pour down from the mountains upon their gardens, where every species of fruit known in the rest of Europe, and many that are not, attain the highest perfection. Neither unwholesome exhalations, nor chilling winds, nor venomous insects, nor poisonous reptiles, nor hostile neighbours, infest their blessed territory. The life of its inhabitants resembles that of the golden age. The soil, like a hot-bed, rapidly puts forth such variety of spontaneous produce, that labour becomes merely an amusing exercise. Peace and plenty crown their board; while the repose they so much admire is only interrupted by harmless thunder, reverberating on rocks above them, or by the murmur of the waves on the beach below.' (Clarke, ii. p. 252, 8vo. ed.) But if this description be as faithful as it is eloquent, it will not certainly apply to any other portion of the Crimea, not even to the famous valley of Baidar. At certain seasons of the year the finest parts of the peninsula are infested with swarms of locusts, which frequently commit the most dreadful devastations, nothing escaping them, from the leaves of the forest to the herbs of the plain. Tarantulas, centipedes, scorpions, and other venomous insects, are also met with in most parts; and even to the S. of the mountains the air in autumn is not everywhere salubrious, and malignant fevers are not uncommon.

Owing to the thinness of the population, and their want of industry, the Crimea, which in antiquity was the granary of Athens, and whose natural fertility is nowise diminished, does not produce a tenth part of what it might do. The steppe, or N. portion, is in general more suitable for grazing than for tillage, and is depastured by immense numbers of sheep, horses, and black cattle. Some of the rich Nogai Tartars are said to have as many as 50,000 sheep, and 1,000 horses; and the poor classes have 100 of the former and 10 of the latter. Thousands of cattle often belong to a single individual; camels also are abundant. Breed of horses improved by crossing with Arabs. Sheep mostly of the large-tailed species peculiar to the Kirghises. The buffalo is domesticated, and yields a rich milk; and the culture of bees is a good deal attended to. Though they have renounced their migratory habits, the Tartars, who constitute the bulk of the population, have little liking to, or skill in, husbandry. Exclusive of milk and other animal food, they subsist chiefly on millet, producing, however, in some years, as much as 150,000 chetverts of wheat for exportation. The mountainous, or S. portion of the peninsula, furnishes large quantities of indifferent wine, with flax, fruits, timber, honey, and wax; but the cultivation of corn is so little attended to, that even in the best years its inhabitants have to import a large proportion of their supplies. The most important and valuable product of the Crimea is the salt derived from the salt-lakes in the vicinity of Perekop, Kaffa, Kozlov, and Kertsch. It is monopolised by the gov., and yields a considerable revenue. The quantity exported from the lakes near Kertsch amounts to from 1,500,000 to 2,000,000 poods a year; the lakes of Perekop are even more productive. At Kozlov there is only a single lake. About 13,000 men are employed in the works; each pood costs the treasury 4 copecks, or thereabouts, the expense of production being seldom greater than from 6 to 10 copecks. Government sells this salt at 80 copecks per pood, except the portion destined for the consumption of the peninsula, which only pays 15 copecks.

Exclusive of salt and corn, the other principal articles of export are wine, honey (of an excellent quality), wax, morocco leather, hides, a considerable quantity of inferior wool, with lamb-skins, which are highly esteemed. Silks and cottons, in the style of the Asiatics, form the basis of the import trade; and there are also imported woollen stuffs, wine, oil, dried fruits, tobacco, jewellery, drugs, and spices. The only manufacture worth notice is that of morocco leather. Principal towns—Kertsch, Caffa, Balaclava, and Kuslow, or Eupatoria. Sevastopol, the finest harbour in the peninsula, is one of the chief stations of the Russian fleet. Baktchiserai was the capital under the Khans; Simpheropol is, however, the modern capital, not of the Crimea only, but of the entire gov. of Taurida.

The population consists of Tartars, Russians, Greeks, Germans, Jews, Armenians, and gipsies. The variety of different nations found in the Crimea, and the fact that each lives as in its own country, practising its peculiar customs, and preserving its religious rites, is one of the remarkable circumstances that render the peninsula so curious to a stranger. The number of Tartars has declined considerably, by emigration and otherwise, since the occupation of the country by the Russians; but they still form the nucleus and main body of the population. They consist, 1st, of Nogai Tartars, living in villages, who pique themselves on their pure Mongolian blood; 2d, of Tartars of the steppe, of less pure descent; and 3d, of those inhabiting the S. coast, a mixed breed, largely alloyed with Greek and Turkish blood, and despised by the others, who bestow on them the contemptuous designation of Tat, or renegade. They are all attached to the Mohammedan faith, and Simpheropol is the seat of one of the two muftis of the Russian empire. The Tartars are divided into the classes of nobles (mourzas), of whom there are about 250, priests (mollahs), and peasants. A mullah is at the head of every parish, and nothing is undertaken without his consent. The peasants plough his land, sow and reap his corn, and carry it home; and it is seldom that the proprietor takes tithe of the priest. In summer the feet and legs of the peasantry are bare, but in winter they are clothed after the Russian fashion. They are simple in their manners and dress; and their sobriety, chastity, cleanliness, and hospitality have been highly eulogised, and probably exaggerated; they live principally on the produce of their flocks and herds; are wedded to routine practices; and if they be not, as Pallas seems to have supposed, decidedly averse from labour, they at all events are but little disposed to be industrious. The emigration that took place after the occupation of the country by the Russians was owing quite as much to the efforts of the latter to convert the Tartars into husbandmen, as to the excesses they committed. (Brailly, p. 176.) In their diet they make great use of honey, and are much addicted to smoking. Every family has two or more copies of the Koran, which the children are taught to read; but in despite of this, and of the schools established in their villages, they are, for the most part, exceedingly ignorant.

The Greeks established themselves in the Crimea, and founded several colonies upon its coasts, nearly six centuries before the Christian era. The country fell successively into the possession of Mithridates, and of the Romans, Goths, and Huns. In 1237 it was taken possession of by the Tartars. About the same time its ports were much resorted to by the Venetians and Genoese: the latter of whom rebuilt Caffa, the ancient Theodosia, and made it the centre of their power and of the ex-

tensive commerce they carried on in the Euxine. In 1475 the Turkish sovereign Mahomet II. expelled the Genoese, and reduced the peninsula to a sort of colonial dependency of the Ottoman empire, leaving it to be governed by a khan or native prince. This state of things continued for about three centuries, or till Catherine II. stipulated for the independence of the Crimea. In 1783, the khan having abdicated, the affairs of Russia took forcible possession of the country, which was secured to her by the peace of 1791. The Crimea became the theatre of one of the most sanguinary wars of modern times in 1854. Great Britain and France having taken part in a dispute between Turkey and Russia, and not finding themselves able to attack the latter power with sufficient energy at the mouth of the Danube, resolved to invade the Crimea in the summer of 1854. Having effected a landing, there followed, Sept. 20, the battle of Alma, the capture of Balaclava, and the siege of Sebastopol, extending from Oct. 17, 1854, to Sept. 8, 1855. The treaty of Paris, of March 30, 1856, net result of the war, nominally crippled the power of Russia in the Black Sea, by reducing the fleet of war and the aggressive strength of the maritime forces in Sebastopol. Succeeding years, however, proved the entire ineffectiveness of these treaty stipulations.

CROATIA (AUSTRIAN), called by the inhab. Horváth Orszag, a prov. of the Austrian empire, regarded as forming the marit. portion of Hungary; between lat. 44° 7' and 46° 23' N., and long. 14° 23' and 17° 51' E.; having NW. Carniola and Styria, NE. Hungary Proper, E. and SE. Slavonia, Turkish Croatia, and Dalmatia, and SW. the Adriatic. Shape very irregular; length NE. to SW., 150 m., breadth varying from 30 to 126 m. Area, 9,500 sq. m. Pop. 876,009 in 1857. The S. portion of Croatia is mountainous, being intersected by the Julian Alps and their ramifications. N. of the Save the surface is rather hilly than mountainous, but a continuation of the Carnic Alps traverses the N. portion of the country, dividing the waters which flow into the Drave from those which flow into the Save and Unna. The valleys are numerous, and there are some considerable plains. The principal rivers are the Drave, separating Croatia from Hungary; the Unna, which for the most part forms its boundary on the side of Turkey; and the Save and Kulpa by which it is intersected. Climate varies very much in different parts. Along the Adriatic, it is similar to that of the opposite coast of Italy; and the olive and other fruits of S. climates grow in perfection; in the N. also it is warmer than in Hungary; but in the elevated mountain region of the S., snow frequently falls in Aug. or Sept. and lies till the following April or May. The mountain ranges are composed chiefly of limestone; they however afford not only fine marble, alabaster, and gypsum, but porphyry, gneiss, clayslate, and quartz. The upper soil is frequently gravelly or sandy; it is less fertile in the S. than in the N., where maize, barley, buckwheat, millet, and oats are grown in considerable quantities. But little wheat and rye are cultivated, and the flax and hemp produced are sufficient only for home consumption. The most abundant fruit is the Damascene plum, of which the favourite beverage of the Croats and Illyrians is made. The vine is, however, cultivated to some extent in the N., and a strong and full-flavoured wine is made, most part of which is consumed in the prov. There are large forests, and timber is an important product. The pastures are limited, and but little fodder is grown, so that the rearing of cattle is but little attended to. Hogs, which feed in the woods, are the most plentiful domestic animals. Iron, copper, lead, and a little silver are found in various parts; and small quantities of gold are obtained by washing the sands of the Drave. Coal, sulphur, and salt are the other chief mineral products. Manufactures very few, and of the rudest kind. Croatia is divided into six cos.; its principal cities, Agram, the cap., Warasdin, Carlstadt, Bellovar, Kreutz, and Fiume, the principal sea port. It has its own provincial diet, the same as all the other provinces of the empire (see AUSTRIA), and is likewise represented in the reichsrath, or central parliament. The inhab. are either Roman Catholics, or of the united Greek Church; the former are under the bishop of Agram; the latter have their own bishop, who resides at Kreutz. The Croats are of a Slavonian stock, speaking a dialect which has a greater affinity with the Polish than any other language; they are the descendants of the Chrobaks, who settled here in 610, and established several extensive empires, or duchies. Towards the end of the 11th century, Croatia was erected into a kingdom, which acquired dominion over parts of Dalmatia and Bosnia; about 1100, it was incorporated with Hungary. Its present constitution, which made it an integral part of the Austrian empire, was proclaimed Feb. 26, 1861.

CROATIA (TURKISH). See BOSNIA.

CROMARTY, a small co. of Scotland, consisting of various detached portions, about 14 in number, almost wholly included in Ross-shire, with which it is connected in the return of a member to the H. of C. Its area, incl. Ross, is 8,157 sq. m., or 2,016,573 acres; pop. 81,406 in 1861. The old valued rent was 1,074l.; the new valuation, for 1863-4, was 8,178l. Registered electors 48 in 1864.

CROMARTY, a sea-port town and parl. bor. of Scotland, cap. of the above co., on a low alluvial promontory, at the N. entrance to the Cromarty Frith. Pop. 1,491 in 1861. Though irregularly built, it is neat and clean. Owing to its situation, its communication with different parts of the country is interrupted by friths and arms of the sea. The Cromarty Frith, the mouth of which is formed by two richly wooded hills, nearly alike, and about 3 m. apart, extends about 10 m. inland, forming a most spacious bay, with deep water, and sufficient to afford safe anchorage for every navy in the world. Cromarty, though in former times a royal burgh, was disfranchised by the Scottish parliament in the 17th century, and is now only a burgh of barony. It has an excellent pier and harbour, vessels of 400 tons coming close up to the quay. The inhabitants have long engaged extensively in the herring fishery. In some instances, not fewer than 20,000 barrels are stated as having been cured in the town in a single year. Cromarty has long carried on a considerable trade in the hempen manufacture, including sacking and sailcloth. It also enjoys an extensive trade in pork for the English market, the value of the quantity exported varying from 15,000l. to 20,000l. annually. Ship-building is carried on to a trifling extent. A steamboat plies between Cromarty and Leith; and there is also regular steam communication with London. Cromarty unites with Dingwall, Dornoch, Kirkwall, and Tain in sending a member to the H. of C. Registered electors 33 in 1864. Sir Thomas Urquhart, the eccentric but learned author of the 'Jewel,' 'Logopandecteision,' and numerous other works, was proprietor of the whole co. of Cromarty.

CROMER, a sea-port town and par. of England, co. Norfolk, hund. N. Erpingham, on a high cliff on the N.E. coast, 21 m. N. Norwich. Area of par., 800 acres; pop. of do. 1,232 in 1831, and 1,367 in 1861. Cromer was formerly but a small fishing station; but of late years it has been much resorted to by sea-bathers, attracted by the fine beach and picturesque scenery of the vicinity. The older part consists of mean, badly arranged tenements; but the more modern houses, near the sea, are much superior, and pleasantly situated. The church, in the later Gothic style, has a pinnacled tower, 160 feet in height. There is also a dissenting chapel, and a national subscription school. Some remains of an ancient abbey, and of the old walls which surrounded the town, are still traceable. A fort and two half-moon batteries were erected during the late war on an adjoining eminence. About 1 m. E. of the town is Foulness lighthouse, furnished with a revolving light, and having the lantern elevated 274 feet above the level of the sea. In consequence of the dangerous character of the coast, there are three other lighthouses betwixt this place and Yarmouth. The parish was formerly of much greater extent, and at the period of Doomsday Book included the town of Shipden; which subsequently with its church, and also a considerable number of houses in an adjoining parish, were swept off by an inroad of the ocean. The sea is here, in fact, constantly gaining on the land. In the winter of 1825, some cliffs contiguous to the lighthouse, 250 ft. in height, were precipitated into the sea, their fragments covering 12 acres. (Lyell's Geology, i. 336.) The inhabitants are mostly engaged in the fishery; the coasting trade is also carried on, though under considerable difficulties, from the want of a proper landing-place, which makes it necessary to employ carts to load and unload the vessels lying on the beach at low water. Cromer Bay is exceedingly dangerous, and has thence obtained from the sailors the expressive name of the 'Devil's Throat.' Exports chiefly corn; imports coals, tiles, and oil-cake. Many attempts have been made to construct a pier, but it has always been swept off. Life-boats are kept in constant readiness on the beach, and have been the means of rescuing many from destruction.

CROMFORD, a chapelry and town of England, co. Derby, hund. and par. of Wirksworth, on the Derwent, near the S. end of Matlock Dale; 13 m. N. by W. Derby, and 145 m. NW. London by Midland railway. Pop. 1,291 in 1831, and 1,140 in 1861. The town is mostly on the N. side of the stream, and is surrounded on the N., S., and W. by lofty calcareous rocks; the houses are mostly small neat buildings, occupied by work-people employed in the adjoining cotton factories. There is a neat episcopal chapel, founded by Sir R. Arkwright; a Wesleyan chapel; two good schoolrooms, built in 1832; and almshouses for six poor widows. The town owes its rise to Sir R. Arkwright, the great founder of the British cotton-manufacture, who built here two large cotton mills—(the first in 1771, the other a few years subsequently)—where his great improvements were brought into successful operation; these and another factory are still in the possession of his family. Lead and lime mines are worked in the immediate vicinity. The S. terminus of the Cromford and Peak Forest railway is at this town; and from it a canal extends to the Erewash canal near Langley bridge.

CRONSTADT (Ger. Kronstadt; Hung. Brasso), a town of Transylvania, near its SE. extremity, being the largest and most populous, as well as the principal manufacturing and commercial town in that country; cap. co. of the same name in the 'Saxon-land,' in a narrow valley, 120 m. SE.

Klausenburg, Pop. 26,826 in 1857. 'If the reader will understand the situation of Kronstadt, let him imagine an opening in the long line of mountains which separate Transylvania from Wallachia, in the form of a triangle, between the legs of which stands an isolated hill. Within this triangle lies the town of Kronstadt, and on the top of the isolated hill there is a modern fortress of some strength. The mountains come so close down on the little valley, that the walls are in many places built part of the way up their sides.' (Paget, Hungary, ii. 434.) Cronstadt Proper, or the 'Inner Town,' is small, rectangular, surrounded by walls, towers, and ditches, and entered by five gates. It is regularly and well built, with paved streets. The inhabitants are mostly of Saxon descent. Blumenau, the E. suburb, is chiefly inhabited by Szeklers, as Bolgarey, the S. suburb, is by Wallachs; the latter is built on a height interspersed with gardens, and separated from the inner town by a large open esplanade, ornamented with avenues of trees and a Turkish kiosk. Altstadt, the other suburb, is on the N. side. The chief public edifices in Cronstadt are the great Lutheran church, a venerable Gothic building of the 14th century; the Lutheran college, Wallach and Roman Catholic churches, the former rebuilt by Elizabeth, empress of Russia, in 1751, town-hall, barracks, two hospitals, the workhouse, several different schools, and the great market-house. In the latter, Saxons, Greeks, Armenians, Jews, Moldavians, Szeklers, Hungarians, Turks, Wallachs, and gipsies meet to make up the bustling and motley crew. Its proximity to Turkey has introduced a good deal of Turkish habits and manners. But Cronstadt is principally distinguished by its industry.

'A rapid stream rushes in various channels through the streets, and makes itself useful to a host of dyers, fellmongers, tanners, and millers, with which this little Manchester abounds. Kronstadt and its neighbourhood are, in fact, the only parts of Transylvania in which any manufactured produce is prepared for exportation, and here it is carried on to a considerable extent. The chief articles produced are woollen cloths of a coarse description, such as are used for the dresses of the peasants, linen and cotton goods, stockings, skins, leather, wooden bottles of a peculiar form and very much esteemed, and light waggons on wooden springs. The principal part of its exports are to Wallachia and Moldavia. A considerable transit commerce between Vienna and the principalities is likewise carried on through Kronstadt, which is chiefly in the hands of a privileged company of Greek merchants.' (Paget, ii. 435, 436.) The first paper-mill and printing press in Transylvania were established at Cronstadt.

CRONSTADT, or KRONSTADT, a strongly fortified marit. town of Russia in Europe, gov. Petersburg, of which city it is the port, besides being the principal station of the Russian navy. Pop. 29,416 in 1838. The town stands on the SE. extremity of the sandy island of Kotline in the Gulf of Finland, about 20 m. W. Petersburg, with which it is connected by regular steamers. Its shape is triangular, its base being towards the S. Being, as it were, the outwork of Petersburg, it is very strongly fortified. The narrow channel which bounds the island of Kotline S. and is the only practicable passage from the Gulf of Finland to the cap. is protected on the side of Cronstadt by a fortress erected on a detached islet; and on the opposite side by the batteries of the Riesbank and the castle of Cronslot. The streets of Cronstadt are regular and generally paved; but the houses are mostly of wood, and only one story in height.

There are about 160 stone buildings, most of which belong to the government. The town is divided into two grand sections, those of the commandant and the admiralty, and into four subdivisions; it is traversed by two navigable canals, those of Peter the Great and of Catherine. The former, commenced in 1721 and finished in 1752, is 3½ fathoms in length, by about 30 yds. wide, and bordered with stonework. It is in the form of a cross, one of its arms communicating with a dock paved with granite, in which ten ships of the line may be repaired at once. The Catherine canal, begun in 1782, is much more extensive, and bordered with granite; it communicates with the mercantile part, and is used chiefly for commercial purposes. Between these two canals is the Italian palace, built and formerly inhabited by Prince Menschikoff, now a school for Baltic pilots with 300 pupils. The other principal public buildings and establishments are, the naval hospital with 2,500 beds, the civil hospital, arsenal, cannon and ball foundry, admiralty, barracks, custom-house, Protestant college, several schools, nobility's club, three churches, and two chapels appropriated to the Greek faith, and Lutheran, English, and Roman Catholic chapels. Peter the Great had a residence and a garden here; the latter continues to be a public promenade; but of the trees planted by the creative hand of Peter, only a few remain: here is, however, a bust of the great emperor on a column, which bears an inscription stating that he founded Cronstadt in 1703. On the S. side of the town are the three ports; the E. or imperial port will accommodate 35 ships of the line, besides small vessels; the second or middle part, used chiefly for the equipment or repair of ships, has been already noticed, and has attached to it some building docks and pitch-houses, and a powder-magazine; the W. or mercantile part is capable of accommodating 600 vessels of any size. All these ports are very strongly fortified, of a convenient depth, and safe; but the freshness of the water injures ships which remain long in them; and the bay of Cronstadt is liable to be blocked up with ice for several months of the year.

Two-thirds of the whole external commerce of Russia is carried on through Cronstadt. Most ships load and unload here, and goods are conveyed to and from Petersburg by means of lighters, the channel higher up being generally available only for vessels drawing not more than 7 or 8 ft. of water. (For further particulars respecting the trade of Cronstadt see PETERSBURG.)

CROWLAND, a town and par. of England, co. Lincoln, parts of Holland, wapent. Elloe. Area of par., inc. Deeping Fen, 20,070 acres. Pop. of town 2,413 and of par., 3,148 in 1861. The town is situated in a low flat district, 8 m. NE. Peterborough, on the rivers Welland and Nene, and the Catwater drain. The communication between its different parts was formerly kept up by a bridge of singular construction, impassable for carriages, built in the reign of Edw. II.; but as the two streams have both been covered in, sewer-like, in recent years, the 'triangular bridge,' as it is called, stands now in the middle of a rather broad thoroughfare; a singular object for the curious traveller. At Crowland was formerly one of the most celebrated of English abbeys. The present church forms but a small portion of that originally attached to the abbey, but it is, notwithstanding, a very fine specimen of the later Gothic style; its W. front is elaborately ornamented, and has statues of several kings and abbots. The windows and interior tracery are also very splendid. The remains of the abbey are highly interesting. It was built on piers, of which many remain. The ruins are partly in the Norman and partly in the different periods of the Gothic style. It was originally founded by Ethelbald, in 716; though several times destroyed, it was as often rebuilt with augmented splendour: its endowments were most ample; and its revenue at the dissolution in the reign of Henry VIII. amounted to 1,217l. 5s. 11d. a year. From this period it fell into decay; and during the civil war (after being for some time occupied as a garrison) was almost wholly demolished. A market formerly held in the town has long been removed to Thorney; but there is still an annual fair, on Sept. 5, for cattle and flax. The inhab. are chiefly employed in agriculture. The par., formerly for the most part an unprofitable morass, has, by dint of draining, been converted into rich arable and pasture land; to assist in this are several powerful windmills, which pump up the superfluous water into channels, which conduct it off. An extensive fishery (formerly belonging to the abbey, and now to the crown) includes many decoys for wildfowl of which this parish furnishes a large supply to various markets. Geese are also largely reared. (Stukeley's Itinerarium Curiosum, p. 33.)

CROYDON, a town and par. of England, co. Surrey, hund. Wallington, 9 m. S. London by road, and 10 m. by South Eastern and by London, Brighton, and South Coast railway. Pop. of town 20,325, and of par. 30,240 in 1861. The town is situated on the borders of Banstead Downs, near the source of the Wandle. The parish church, the finest in the co., in the later pointed style, has a lofty tower with pinnacles, and contains many fine old monuments, chiefly of archbishops of Canterbury. This originated in the circumstance of the archbishops of the metropolitan see having formerly resided in a palace here, the remains and grounds of which were sold in 1780. There are several other churches, built by parliamentary grant, within a recent period; one near Croydon Common, with 400 free sittings, the other at Norwood, with 632; also a number of dissenting chapels; a free school, founded 1710, for 20 children; a school of industry for girls; a school for educating 150 children of Quakers, removed from Islington, 1825; a Lancastrian school; and a national school. The last occupies the schoolroom of the Trinity Hospital, founded by Archbishop Whitgift in 1596, for a warden, schoolmaster, chaplain, and not less than 30 or more than 40 poor brothers and sisters: the income, which originally amounted to about 200l. a year, is now nearly 2,000l.; the Archbishop of Canterbury is visitor. The building (with a chapel annexed) forms three sides of a quadrangle, in the domestic style of that period: there are also two sets of almshouses, and several minor charities: a small theatre, seldom opened. A handsome town-hall, surmounted by a dome, was built in 1807, in which the summer assizes of the co. are held, alternately with Guildford; when not thus used, it is occupied as a corn-market. On the site of the old town-hall is a structure used as a prison during the assizes, and at other times as a poultry and butter market. The co. magistrates hold petty sessions weekly for the district; and there is a court of requests for debts under 5l., whose jurisdiction extends over the hundred. Market, Saturday. Fairs, July 6, for cattle; Oct. 2, horses, cattle, sheep, pigs; the latter is also a crowded pleasure fair, and noted for the large quantity of walnuts brought to it. The principal line of road from London to Brighton passes through Croydon, and it also communicates with the metropolis by two lines of railways. In consequence of this facility of intercourse a great

many persons engaged in London during the day have taken up their residence at Croydon, which has led to the erection of a vast number of 'villas' and other houses of a similar description. The members for the E. division co. Surrey are elected here. Croydon is the centre of a poor union of 10 parishes. At Addiscombe, 1½ m. distant, a military college was established in 1809, for cadets in the E. I. Company's service; it had, till its extinction, which took place with that of the company, about 14 professors and masters in the various departments, and usually from 120 to 150 students.

Croydon is the supposed site of the Noviomagus of Antonine's Itinerary. On Broad Green, near it, are traces of the Roman road from London to Arundel, and many Roman coins have been found; there are also many remains of an older period; amongst others, a cluster of twenty-five tumuli, on a hill, between the town and Addington Park (the Archh. of Canterbury's seat), and a circular encampment with a double moat.

CRUZ (SANTA), the most S. of the Virgin Islands in the W. Indies, belonging to Denmark, and situated in the Caribbean Sea, about lat. 17° 45' N., and long. 61° 40' W.; 60 m. ENE. Porto Rico. Length, E. to W., 20 m.; average breadth, 5 m. Area about 100 sq. m. Estim. pop. 32,000. There is a chain of hills in the N.; but the island is generally level. The coasts are much indented, and present numerous harbours, the best of which are those of Christiansstadt and Friedrichsstadt. The rivulets are dried up during a part of the year, and water is then scarce and bad. The climate is unhealthy at certain seasons. Soil fertile, producing the sugar-cane, cotton, coffee, and indigo. The average value of the produce of sugar amounts to about 1,200,000 rix dollars, and that of rum to 500,000 rixdollars a year. Timber is scarce. The principal town, Christiansstadt, the cap. of all the Danish possessions in the W. Indies, is situated on the declivity of a hill on the NE. shore of the island; it is well built, and has 5,000 inhab. Its port is secure, and defended by a battery. Friedrichsstadt, on the W. coast, has 1,200 inhab. This island was discovered by Columbus in his second voyage. The Dutch, English, French, Spaniards, and Danes alternately possessed it till 1814, when it was finally ceded to Denmark.

CSABA, a large market town of Hungary, in the Great Hungarian plain beyond the Theiss, 63 m. SSW. Debreczin. Pop. 27,865 in 1857. The inhab. are mostly Protestants. Previously to 1840, Csaba was but a village, 'the largest village in Hungary.' It has an extensive trade in corn, wine, cattle, fruit, hemp, and flax.

CSANAD, a town of Hungary beyond the Theiss, cap. co. of same name, on the Maros, 7 m. SE. Mako. Pop. 2,903 in 1857. It was formerly a populous and flourishing place; but its castle is now in ruins, its bishop non-resident, and the county meetings have been transferred to Mako.

CSONGRAD, a market town of Hungary, between the Danube and Theiss, on the right bank of the latter, immediately after the influx of the Koros; 31 m. N. Szegedin, on the railway from Szegedin to Pesth. Pop. 16,200 in 1857. The town is well built, and contains the ruins of an ancient castle. It was the original cap. of the co. of same name; but the county meetings are now held at Szegedin.

CUBA, an isl. belonging to Spain, being the largest, most flourishing, and important of the Antilles, or W. Indian isls. It was discovered by Columbus, Oct. 28, 1492; and was first called Juana, in honour of Prince John, son of Ferdinand

and Isabella; afterwards Ferdinand, in memory of the Catholic king; then successively Santiago and Ave Maria, in deference to the patron saint of Spain and the Virgin; and by Spanish geographers *La lengua de pájaros*, as being descriptive of its form. The name Cuba was that in use among the aborigines at the time of its discovery.

Form, Position, and Extent.—Its figure is long and narrow, approaching to that of a crescent, with its convex side looking towards the Arctic Pole; its W. portion, lying between Florida and the peninsula of Yucatan in Mexico, leaves two entrances into the Gulf of Mexico; the distance from Cape St. Antonio, the most W. point of the island, in lat. 21° 54′ N., long. 84° 57′ 15″ W., to the nearest point in Yucatan, is 126 m. across; and that from Point Icacos, the most N. point in the island, in lat. 23° 10′ N., long. 81° 11′ 45″ W., to Cape Tancha, the S. extremity of Florida, being 130 m. across. Point Maysi, the E. extremity of Cuba, lat. 20° 16′ 40″ N., long. 74° 7′ 53″ W., is 49 m. NE. by E. from Cape San Nicholas Mole, in Hayti; and Cape Cruz, in Cuba, is about 95 m. N. from the nearest point of Jamaica. The greatest length of the island, following its curve, is about 800 m.; its breadth, which is very irregular, varies from 130 to 25 m. The total area of the island is stated, in the census of 1861, to embrace 48,480 sq. miles, or about as large as Belgium, Holland, Denmark, and Switzerland taken together. Its coasts are very much indented, and it is surrounded by many islands, islets, and reefs. Notwithstanding the general difficulty of approaching its shores, it has several excellent harbours, that of the Havannah being one of the best in the world. The land along the sea-shore, almost all round the island, is so low and flat as to be scarcely raised above the level of the sea, which greatly increases the difficulty, especially in the rainy season, of communicating with the interior. In the lagunas, near the shore, especially on the N. side of the island, where the tides are filled with sea water during spring tides, sufficient salt is collected for the use of the inhabs. A cordillera stretches from the one end of the isl. to the other, dividing it into two unequal sections, that on the N. side being for the most part the narrower of the two. Of the geology little is known beyond what may be found in Humboldt. The cordillera is one great calcareous mass, which is found to rest on a schistose formation. Its summit presents a naked ridge of barren rocks, occasionally interrupted by more gentle undulations. It attains, in some parts, to an elevation of about 7,000 ft.

Climate.—In the W. half of the isl. the climate is such as to be expected along the N. limit of the torrid zone, presenting many inequalities of temp. from the near neighbourhood of the American continent. The seasons are spoken of as the rainy and the dry, but the line of demarcation is not very clearly defined. The warmest months are July and August, when the mean temp. is from 2° to 3° of the centigrade, or from 82° to 84° Fahr. The coldest months are Dec. and Jan., when the mean temp. is nearly 10° Fahr. less than under the equator. During the rainy season the heat would be insupportable but for the regular alternation of the land and sea breezes. The weather of the dry season is comparatively cool and agreeable. It never snows, but hail and hoar frost are not uncommon; and at an elevation of 800 or 400 ft. above the level of the sea, ice has been found several times in thickness, when the N. wind has happened to prevail for several weeks in succession. Hurricanes are not so frequent as in Hayti and the other W. Indian isls., and seldom do much damage on shore. In the E. part of the isl., par-

ticularly in the neighbourhood of Santiago, earthquakes are not unfrequent. The most severe on record are those which took place in 1675, 1682, 1766, and 1826.

Animal Kingdom.—The only indigenous quadruped known in the island is the *jutía* or *hutia*, shaped like a rat, but from 12 to 18 in. in length, exclusive of the tail; of a clear black colour, feeding on leaves and fruits, and inhabiting the hollows and clefts of trees. Its flesh, though insipid, is sometimes eaten. Amphibious oviparous animals, the crocodile, cayman, manati, tortoise, and fishes; the first on the coast, and the others in the rivers and lagunas. The *perro jíbaro* is the domestic dog restored to a state of nature. It becomes fierce and carnivorous, though not so much so as the wolf of Europe; never attacking man until pressed in the chase. Whatever be their original colour, they uniformly degenerate into a dirty black, with a very rough coat. In spite of the efforts made to extirpate them, they increase in numbers, and do great damage among the cattle. The domestic cat, called the *gato fibaru*, when it becomes wild, commits similar depredations on the poultry yard. The most valuable of the domestic animals are the cow and pig. The sheep, goat, and ass are not in such general use, although within three few years the great jackass of the peninsula has been introduced with some success, for the purpose of breeding mules. The feathered race are remarkable for the beauty of their plumage; but are far too numerous for separate notice. The rivers, though not large, are well supplied with excellent fish, as are the bays and inlets with the natives of the deep. Oysters and other shell-fish are also numerous, but of inferior quality, and adhere to the branches of the mangrove trees which surround the coast. Snakes of a large size are of rare occurrence, though some have been seen 10 or 12 ft. long, and 7 or 8 in. in diameter. Of insects the bee is turned to valuable account by the exportation of its wax, and the use made of its honey. The mosquito tribe are troublesome, and the phosphorescent family are remarkable for the brilliancy of the coloured lights they exhibit.

Vegetable Kingdom.—The forests are of vast extent. Mahogany and other hard woods are indigenous, and several sorts are well suited for ship-building. The palm tribe are as remarkable for beauty as utility, and of vines there is great variety, some of such strength as to destroy the largest of the forest trees in their parasitical embrace. The tropical fruits are plentiful and various; of these the pine-apple, orange, and its varieties, are the most highly valued. Of the alimentary plants, the *plátano*, or plantain, is by far the most important. Next in order come the sweet and bitter yam, the sweet root being eaten as a vegetable, and the bitter converted into bread after its poisonous juice has been extracted. The sweet potato, the yam, and other farinaceous roots are also known, although not in such general use as in the British West Indies. The maize or Indian corn is indigenous, and in extensive use; the green leaves for fodder, under the name of *millaja*, and the grain in various forms for man and beast. Rice is cultivated in considerable quantity; and a variety of beans, especially the *garbanzo*, as well known in the peninsula. Garden stuffs are scarcely known, except in the Havannah and other large towns, and there only in the dry season. The culture of flowers is still less attended to.

Mineral Kingdom.—The pursuit of the precious metals was the great object of the first discoverers, but if gold was found at all, it was probably in washing the sands of some of the rivers, as no traces of the supposed mining operations are now to be found. The gold and silver sent to Spain from Cuba, Hayti, and Jamaica, soon after the discovery and conquest of these islands, consisted, most likely, of the accumulations of the aborigines. In the course of the 17th century, the copper mines near Santiago, in the E. part of the island, were wrought with some success, but were abandoned upwards of 100 years ago, from the imperfect knowledge which then existed, of the art of extracting the metal from the ore. When the mines were abandoned, a large quantity of the mineral, amounting to several hundred tons, was left on the spot as worthless, but having been subjected to analysis by one of the present English proprietors, it was found to be so rich in metal as amply to repay the expense of sending it to Swansea for smelting. In consequence of this discovery, the old workings were explored, and companies formed for the purpose of renewing the mining operations on a scale of considerable magnitude. One of these, called the English Company, has been highly successful, employing many miners and labourers, some of them slaves, some emigrants from the Canaries, and some articled servants from Cornwall. Powerful steam engines have been erected by this company to assist in preparing the ore for shipment at Santiago. In the neighbourhood of Santa Clara, another copper-mine has been opened by an American company. At first, the mineral thence obtained was sent to be smelted at New York; but latterly, like that from Cobre, it has been shipped to the great smelting-houses in Wales. Of 51,207 tons of copper-ore imported into the U. Kingdom in 1848, no fewer than 30,579 tons were brought from Cuba. Coal of tolerable quality has been found in the neighbourhood of the Havannah, but though several pits have been opened, the English coal, carried out in the sugar ships as ballast, may still be sold at a cheaper rate. The coal of Cuba is highly bituminous, and in some places degenerates into a form resembling the asphaltum which is found in the pitch lake of Trinidad, and in various parts of Europe. The ships of the discoverers were careened with this bitumen, which is often found near the coast in a semi-liquid state, like petroleum or naphtha. Marbles and jaspers, of various colours, and susceptible of a high polish, are found in many parts of Cuba, and in its chief dependency, the Isle of Pines. The mineral waters of San Diego, Madruga, and Guanabacoa have obtained some celebrity, but with the exception of the last, which is within a few miles of the Havannah, they are difficult of access, and therefore not much resorted to.

Population and Industry.—There have been various censuses of the population: the first in 1775, when it amounted to 170,370; a second in 1791, when it was 272,140; a third in 1817, when it was 651,998, and with transient persons, 630,980; and a fourth in 1827, when the permanent population was 704,487, and with transient persons, 730,562. According to the last census, of the year 1861, the pop. numbered 1,390,530; among them 793,484 whites. The coloured population, numbering 603,046, was divided into 225,813 free persons, 6,650 called 'emancipated,' and 370,553 slaves.

The increase of the slave population has been very rapid, being due chiefly to the continued importation of slaves from Africa. In some years, since the peace of 1815, as many as 30,000 blacks are believed to have been imported into Cuba in a single year. Spain had indeed agreed by treaty in 1820 to abolish the trade; but this treaty was

little better than a dead letter, and it is only since 1855, when a more efficient treaty with Spain was entered into, that the trade sustained any considerable diminution. It is highly probable that slavery will soon entirely cease in Cuba, having lost its chief support in the United States, by the downfall of the slave-holding Southern states in 1865.

The raising of sugar constitutes by far the most important branch of industry carried on in Cuba. Its culture has advanced with extraordinary rapidity, especially since 1809, when the ports of the island were freely opened to foreigners. It is principally shipped from the Havannah; and its export from that city, which in 1760 amounted to about 5,000,000 lbs., had increased in 1800 to above 40,000,000 lbs., in 1820 to above 100,000,000 lbs., and in 1840 to 240,000,000 lbs. The exports from the whole island in the year 1861, amounted to 1,127,351,750 libras, or 10,655,610 cwts., valued at 67,641,105 pesos, or 14,373,735l.

The culture of coffee advanced for a while with equal or even greater rapidity than that of sugar. In 1800 there were but 80 plantations in the island: in 1817 there were 779; and in 1827 there were no fewer than 2,067, of at least 40,000 trees each. But the low prices of coffee which subsequently prevailed, not merely checked this astonishing progress, but occasioned the abandonment of a great many coffee plantations. While, in 1837, the exports of coffee exceeded 50,000,000 lbs., they only amounted to 17,553,425 lbs. in 1848. In 1861, the total exports of coffee amounted to 150,277 cwts., valued at 2,523,500 pesos, or 536,262l. More important than that of coffee is the cultivation of tobacco, celebrated for its excellence in all parts of the world. The exports of tobacco, in 1861, were no less than 6,163,396 cwts., valued at 16,912,500 pesos, or 3,593,906l. Since the outbreak of the civil war in the American Union, cotton, once an important article of culture, but subsequently neglected, has been again raised in small quantities. Indian corn, rice, beans, plantains, and even wheat, are also raised, but not in anything like sufficient quantities for the demand, flour and rice being, in particular, very largely imported. Cattle have become extremely numerous, being estimated at about 1,300,000 head; but while hides form a large article of export, fresh and salted meat, and jerked beef, nevertheless, occupy a prominent place among the imports. Horticulture is very little attended to. Of manufactures, the most important are the making of sugar, molasses, and rum, the preparation of coffee, the making of cigars, the bleaching of wax, and the manipulation of the minor staples of the island.

Internal Communication.—Down to a recent period the means of communication between the different parts of the island were very deficient. The common roads were in general badly constructed; and during the rainy season were, for the most part, impracticable for wheel carriages. The long narrow shape of the island, by lessening the distance from the interior to the sea coast, obviated in some degree three difficulties. But down to a very late period it was customary in most parts for the negroes to be employed in the severe drudgery of carrying produce in baskets on their heads to and from the estates, to the seaports, or to the public roads. Within these few years, however, this system has been wholly abandoned by the introduction of a very well-planned system of railways. At the commencement of 1863, there were 27 different lines, of a length of 818 miles, either finished, or in course of construction. The principal line, as well as the first constructed, runs from Havannah to Güines and La Union; it was commenced as early as November, 1835. Another important line, from Cardenas to Macagua, was started in 1838; and a branch, from Cardenas to Jucaro, in 1839. The other lines were constructed since 1840, and the whole of them afford the most rapid and perfect means of communication to the inhabitants of every important place in Cuba. The carriages on some of the railways are drawn by horses, so that they have been constructed at a comparatively small cost. They have, however, been of the greatest service to the island; and may, perhaps, be regarded as the principal cause of the late extraordinary extension of cultivation and general prosperity of Cuba.

Currency.—Paper money is unknown. The coins in use are Spanish doubloons of ounces, which are a legal tender for 17 hard dollars, and at the exchange of 5 per cent. are worth 3l. 10s. 10d.; also the subdivisions of these doubloons, the half being 16d dols.; the quarter, 4¼ dols.; the eighth, 2¼ doll.; and the sixteenth, 1¼ doll. Mexican and Columbian doubloons, or ounces, are also in circulation, and are legal tender for 16 hard dollars, equal to 3l. 6s. 8d.; they are sometimes in demand for exportation, at a premium. Their aliquot parts are worth eight, four, two, and one dollar respectively. Of silver coins, the Spanish pillar dollar is worth 4s. 2d.; and is only legal tender at its nominal worth; but it is generally in demand for export, at a premium of from 2 to 5 per cent. Mexican, U. States, and S. American dollars are also legal tender at their numerical value, and are occasionally in demand, at a trifling premium. For small payments, the coins in circulation are the four, two, one and half real pieces, which are equal to the half, quarter, eighth, and sixteenth of a dollar respectively.

Trade.—The total exports of Cuba, in the year 1861, amounted to 101,887,041 pesos, or 22,208,187l. As already stated, sugar is the principal article exported, engrossing about two-thirds of the value of the whole. Next follows tobacco; then coffee, wax, and honey. The total value of the miscellaneous articles exported in 1861, was 14,748,746 pesos, or 77,288,187l. The exports of Cuba to the United Kingdom are not separately given in the Board of Trade returns, but figure together with those of Puerto Rico. For both they amounted to 4,271,783l. in 1861. There is no obstacle whatever to the establishment of foreigners as merchants in the island. The law says that those who are naturalised in Spain may freely carry on trade with the same rights and obligations as the natives of the kingdom, and that those who have not been naturalised, or have a legal domicile, may still carry on trade under the regulations stipulated in the treaties in force between the respective governments; and in default of such conventional relations, the same privileges are to be conceded as those enjoyed by Spaniards carrying on trade in the country of which such foreigners are natives. In practice this last condition is not much attended to, as foreigners are allowed to establish themselves as merchants without any inquiry as to the right or privileges enjoyed by Spaniards in the country they come from.

Government and Social State.—Public Education is not much attended to; but in this respect there has recently been a great improvement. Elementary schools have been extended; and an institution has recently been established for the instruction of engineers. There are two colleges in the Havannah, with numerous and eminent professors, and literary societies. Several daily newspapers, some of them conducted with considerable ability,

are published in the capital and other large towns.

Morals and Religion are both at a low ebb, a consequence partly and principally of the Inquisition and of the degrading superstition so long established in the island, and partly of the institution of slavery. But improvement is not less perceptible in the character and comfort of the people than in their industry and physical comforts.

As respects its civil jurisdiction, Cuba is divided into three provs. of which the Havannah, Santiago, and Trinidad are the caps. The captain-general, governor, or supreme military chief of the island, is, at the same time, civil governor of the W. prov.; but, except in military matters, the governors of the other prov. are perfectly independent of the captain-general, and are responsible only to the court of Madrid. The island is also divided into three military divisions—a western, central, and eastern; the chiefs of which are, of course, subordinate to the captain-general. The royal court (Real Audiencia) of Puerto Principe, of which the captain-general is the ex officio president, has the supreme jurisdiction in all civil and criminal affairs. In the principalities there are Ayuntamientos, and in the rural districts Jueces Pedaneos, who combine the exercise of judicial functions with those of police commissioners. Spain ordinarily keeps a marine force of from 10 to 50 vessels, most of them small vessels, stationed at the island. In 1864, the navy thus employed consisted of 4 frigates, 15 steamers, and 52 small craft.

CUCKFIELD, a market-town and par. of England, co. Sussex, rape Lewes, the town being in a commanding situation on the high road from London to Brighton, 34 m. S. from the former, and 13 m. N. from the latter. Area of par. 10,840 acres; pop. of do., 3,539 in 1861. It is a neat little town. The church is a spacious structure, has a lofty spire, covered with wooden shingles, that have assumed the colour and appearance of blue slate. It has a free grammar school, founded in the reign of Queen Elizabeth.

CUCUTA (formerly Rosario, or San Jose de Cucuta), an inl. town of New Granada, prov. Pamplona, near the border of Venezuela, 28 m. NNE. Pamplona; lat. 7° 37′ N., long. 72° 14′ W. Its situation is extremely pleasant; it is well-built, neat, and clean; streets paved, with currents of water running through them. The par. church is celebrated as the place in which the first congress was held, and the constitution of Colombia formed, in 1820. It is of Moorish architecture, and contains a respectable copy of one of Raphael's Madonnas, by a Mexican artist.

CUDDALORE, a marit. town of Hindostan, Carnatic, S. dir., Arcot, and one of the most extensive and populous towns in the S. of India, 80 m. SSE. Madras; lat. 11° 43′ 24″ N., long. 79° 49′ E. It is naturally strong, being enclosed between two arms of the Pannaur. Streets broad, and it contains many houses of the better class. N. the Pannaur is a suburb called the New Town, with a large Portuguese church, and some handsome European dwelling-houses and other buildings; and beyond this is a large and beautifully situated edifice, formerly the residence of the chief-governor of the British settlements on this coast. Some English looms have been established in this town, and a paper manufactory. Cuddalore was taken by the British in 1760, but obliged to surrender to the French in 1782. It was restored to the British in 1785.

CUDDAPAH (Cripa, mercy), an inl. town of Hindostan, presid. Madras, on the banks of the Cuddapah river, 507 ft. above the sea, 120 m. NW.

Madras. It has a small fort, containing the palace of the former nabobs, now converted into a court of justice, and a prison for both debtors and felons. Cuddapah is not a place of much trade; it was the cap. of an indep. Patan state, which survived the destruction of the other Deccany kingdoms; a great deal of sugar and jaggery is made in its vicinity.

CUENCA, a city of Spain, cap. prov. same name, on a high mountain, between two others higher still, and separated from them by the deep beds of the Jucar and the Huecar rivers, near their confluence; 96 m. ESE. Madrid, 146 m. SW. Saragossa. Pop. 7,610 in 1857. The town is surrounded by high walls, and its streets are extremely steep, crooked, and narrow. It has seven gates; six bridges over the Huecar, and two over the Jucar, one of the latter being of very superior construction. Cuenca is the see of a bishop, and the residence of the principal authorities of the prov., and contains a vast cathedral built by Alphonso IX. in the 12th century; a fine episcopal palace; 14 parish churches; 13 convents, some of them built on precipices overhanging the river, and containing paintings of great merit; 3 colleges, and an ecclesiastical seminary; 2 hospitals for the sick, and 1 for foundlings; a public granary, and several public fountains. It has some fabrics of paper and wool. The latter were formerly much more considerable than at present; and the town was also much more populous and important. It is the native country of the painter Salmeron, and of the famous Jesuit Mollna. Cuenca was given in dowry by the Moorish king of Seville, Ben Abad, with his daughter Zaida, to Alphonso VI., king of Castile, when he left the cloisters to succeed his brother in 1072. The Moors again retook it, but it was finally wrested from them in 1176.

CUENCA, an inl. town of Ecuador, cap. prov. same name, in a spacious plain, nearly 9,000 ft. above the level of the sea, 186 m. S. Quito; lat. 2° 56' S., long. 79° 12' W. Pop. estimated at 30,000, of whom about 3,000 are Indians. Its streets are broad and straight; but the houses are low, and built of unburnt brick. It contains a cathedral, two par. churches, several monasteries, a college, and a hospital; has manufactories of confectionery, cheese, and hats; and some trade in these, together with grain, cinchona, bark, and other productions of its vicinity. Its climate is temperate as to heat, but it is subject to violent storms. A little to the N. is the Mountain of Farqui, chosen by the French astronomers for their meridian in 1742. In its neighbourhood there are several remains of the works of the Peruvian Incas.

CUEVAS, a town of Spain, prov. Castellon, 61 m. NE. Almeria. Pop. 3,026 in 1857. The town is almost surrounded by the river Almanzor, and there are between the town and the Mediterranean, about 8 m. distant, a number of very deep caverns in the mountains, supposed to have been opened by the Moors, in search of minerals or water; from these the town takes its name. It contains a church, a convent, and a public granary. There is a castle on the coast, and a small island belonging to the town.

CULIACAN (an, Hevivahomos, Mex.), an inl. town of Mexico, state of Sonora, on the right bank of the river of the same name; 105 m. ENE. Cinaloa, and 170 NE. El Fuerte. Pop. 10,825 in 1858. It is a depot for goods passing to and from the port of Guaymas, on the Gulf of California. During the Spanish rule it was the cap. of a prov. The country around is well watered and highly productive.

CULLEN, a marit. royal, and parl. bor. of Scotland, co. Banff, on an eminence at the mouth of a little rivulet, 12 m. W. Banff. Pop. 3,543 in 1861. Though an ancient burgh, the present town is comparatively new, the old town having been superseded, and the site on which it stood enclosed within the park of Cullen House, the mansion of the Earl of Seafield. The linen manufacture, so common on all the E. coast of Scotland, N. of Dundee, has found its way to Cullen, but is carried on to an inconsiderable extent. The inhab. engage in the herring fishery, and in that of cod, skate, ling, and haddock, which abound on their shores; so that dried or cured fish form their chief export. The harbour is bad, and the town, on the whole, not flourishing.

Cullen unites with Banff, Inverury, Kintore, and Peterhead, in returning a mem. to the H. of C. Registered electors, 44 in 1861.

CULLERA, a sea-port town of Spain, Valencia, 11 km on the Jucar, near its mouth, and to the S. of the mountain and cape of the same name, on the Mediterranean coast, 25 m. S. Valencia. Pop. 9,014 in 1857. The town has a church, a convent, a hospital, a public granary, and barracks for troops on their march, bring on the shortest and most frequented road from the coast to the capital. It carries on a considerable coasting trade, as many as forty or fifty vessels being sometimes seen at a time, principally about 30 tons burden, taking in fruit for France, rice for the Balearic Islands, and the coasts of the Peninsula. The neighbourhood produces rice, wheat, maize, muscatel raisins, wine, oil, and garden stuff.

CULPEE, an inland town of Hindostan, prov. Bengal, in a jungly and unhealthy situation, on the left bank of the Hooghly river, about 30 m. SSW. Calcutta; lat. 22° 6' N., long. 88° 25' E.

CULROSS, a royal and parl. bor. and marit. town of Scotland, in a detached corner. co. Perth, on a steep acclivity on the N. shore of the Frith of Forth, about 10 m. NE. Edinburgh. Pop. 517 in 1861. Culross was made a royal burgh by James VI. in 1588; and though it had once a considerable trade in salt and coal, the latter of which was wrought at a very remote period, trade of every kind has now entirely left it, except, perhaps, a little traffic in fish caught in the Firth, and a little damask weaving for manufacturers in Dunfermline. There are vestiges of an old harbour; but the smallest yawls can now approach the town only at high water. But though of no modern importance, Culross can boast of many remains of antiquity, which throw an air of interest over a place otherwise mean and decayed. At the E. end of the town once stood a chapel dedicated to St. Mango or Kentigern, said to have been born here. A monastery, dedicated to the Virgin and St. Serf, was founded here in 1217, by Malcolm earl of Fife, for Cistercian monks; of which considerable remains are extant, a part of it serving as the parish church. Culross Abbey, occupying a magnificent terrace overlooking the sea, and successively the seat of the Bruces and the noble family of Dundonald, is now the property of the heirs of the late Sir Robert Preston, Bart. The present parish church is collegiate, having two clergymen.

Culross unites with Queensferry, Inverkeithing, Dunfermline, and Stirling, in sending a mem. to the H. of C. Registered electors, 23 in 1861.

CUMANA, a city of Venezuela, cap. of the dep. and prov. Cumana, in an arid and sandy plain on the E. bank of the Manzanares, and near the mouth of the Gulf of Cariaco, about 1 m. from the sea-shore, and 180 m. E. Caracas; lat. 10° 27' N., long. 64° 16' W. Pop. 8,580 in 1858. The city is commanded by Fort St. Antonio, built on the extremity of a hill immediately to the E.; the Manzanares encompasses the town on the S. and

W., dividing it from its principal suburbs. It has two parish churches, two convents, and a theatre. Having suffered greatly at different times from earthquakes, its buildings are generally low; but in the early part of the present century great improvements were introduced into the buildings, and its prosperity was much augmented by the judicious conduct of its governor. It has a roadstead capable of receiving all the navies of Europe, with excellent anchorage for large ships. It is protected by a shoal and the battery of lines at its entrance. Export—mules, cattle, smoked meat, salted fish, cacao, and other provisions; fish, wild fowl, and other necessaries, are obtained here in great plenty, and very cheap. Climate intensely hot, from June to October the temperature being usually 90° or 88° F. during the day, and seldom so low even as 80° in the night. The inhab. are distinguished for their assiduity in business, and their polished manners. This is the oldest European city in the New Continent, having been built by Diego Castellon in 1521. It was totally destroyed by the earthquake of 1760.

CUMANACOA, an inl. town of Venezuela, prov. Cumana, in a valley surrounded by lofty heights, 21 m. SE. Cumana, and noted for the prodigious difference between its climate and that of the latter city; lat. 10° 15' N., long. 64° 5' W. Pop. 2,470 in 1858. Cumanacoa has seven months of wintry weather, though only 730 ft. above the level of the sea. It is small, ill-built, with houses mostly of wood.

CUMBERLAND, a marit. co. of England, having N. Scotland and the Solway Frith, E. Northumberland and Durham, S. Westmoreland and Lancashire, and W. the Irish Sea. Area 1,565 sq. m., or 1,001,273 acres, of which about 800,000 acres are mountain and lake. The co. has some of the highest mountains in the kingdom; on its E. border, adjoining Northumberland and Durham, these consist of a portion of the Pennine or great central chain; while the W. group has received the name of the Cambrian range, from their being principally in this co.; the two ranges are divided by the plain of the Eden (see ENGLAND for an account of these mountains, and of the lakes interspersed among them). Principal rivers, Eden, Esk, Irthing, Derwent, and Caldew. Soil in the lower districts, and in parts of the W. mountains, light, and well adapted to the turnip husbandry; but there is also a good deal of wet loam on a clay bottom. The soil of the E. or central moors and mountains is mostly peat earth, and they are bleak, heathy, and extremely barren. Climate rather humid. Principal crops, wheat and oats. Agriculture is much improved; a judicious rotation is observed; and turnips are extensively cultivated according to the most approved principles of the drill-husbandry. Property is much divided. There are a few large estates, but by far the greatest portion of the co. is divided into small properties, worth from 10l. or 20l. to 200l. a year, belonging to 'statesmen,' or 'lairds,' formerly distinguished by their attachment to routine practices, their supplying themselves with all sorts of domestic manufactures, and their economy and independence. But their habits have materially changed during the present century; domestic manufactures have been wholly abandoned, and their habits approach much more nearly than before to the common level of cultivators. There are valuable coal mines near Whitehaven, and in other places; plumbago, or black lead, is found in the greatest perfection in Borrowdale in this co.; and limestone and slate are abundant. The cotton manufacture is extensively carried on at Carlisle and Penrith; and cordage and canvas are made, and ships built, at Whitehaven and other places. Principal towns, Carlisle, Whitehaven, Workington, and Cockermouth.

Cumberland is divided into 5 wards and 106 pars. It returns 9 mem. to the H. of C., viz. 4 for the co., 2 each for Carlisle and Cockermouth, and 1 for Whitehaven. Registered electors for the co. 10,164 in 1865; of which number the E. division had 5,441, and the W. division 4,723. Pop. 206,276 in 1861, inhabiting 40,552 houses. Annual value of real property assessed to income tax: in E. division 462,574l. in 1857, and 479,513l. in 1842; and in W. division 424,296l. in 1857, and 511,372l. in 1842.

CUMBERNAULD, a manufacturing village of Scotland, co. Dumbarton, 13 m. E. Glasgow, on the highway leading from that city to Falkirk and Stirling. Pop. 1,561 in 1861. The chief employment of the people is cotton weaving. The Forth and Clyde canal runs within a ¼ m. of the town; and the Edinburgh and Glasgow railroad has a station here. Cumbernauld was erected into a burgh of barony in 1649; and has for five centuries been the property of the family of Fleming, whose seat is in its immediate vicinity.

CUMNOCK, or OLD CUMNOCK, a village of Scotland, co. Ayr, on the Lugar water, 12 m. E. Ayr. Pop. 2,510 in 1861. This place has been famous for above 30 years for the manufacture of what are known by the name of Cumnock, or Lawrencekirk, snuff-boxes. An artist of the name of Crawford caught the first idea of them from a box made at Lawrencekirk, which had been sent him to repair. The excellence of the Cumnock snuff-boxes lies in the hinge, which is both ingenious in point of contrivance and delicate in point of execution; so that it is styled the 'invisible wooden hinge.' The wood used in the manufacture is plane, by reason of its peculiarly close texture. One set of artists make the boxes; another set paint those designs that embellish the lids; while women and children are employed in varnishing and polishing. The principle on which the hinge is formed, as well as the instruments employed in making it, were for many years kept secret. The manufacture exists also in the neighbouring village of Mauchline, as also, to a less degree, in Lawrencekirk, Montrose, and one of two other places.

CUPAR-ANGUS, a burgh of barony of Scotland, partly in co. Perth, and partly in Angus, on the Isla, a tributary of the Tay, on the high road between Perth and Aberdeen, about 12½ m. from the former. Pop. 3,624 in 1861. The place is neatly built, well paved, and lighted; has a townhouse and jail, an elegant parish church, two chapels belonging to Presbyterian dissenters, and an episcopal chapel; a weekly cattle-market, and five annual fairs. The town enjoys its share of the weaving of the coarser kinds of linen fabrics, for the manufacture of which the various towns and villages of Angus are distinguished. The webs are generally absent from Dundee. It has also extensive bleach-fields and tan-pits; but weaving is the staple employment of the place.

CUPAR-FIFE (so called to distinguish it from Cupar-Angus), a royal and parl. bor. of Scotland, co. Fife, of which it is the cap. 25 ft. above the level of the sea, in the centre of the Howe of Fife, and on the L. bank of the Eden, 10 m. W. St. Andrew's. Pop. 5,029 in 1861. Though ancient, Cupar has all the characteristic appearances of a modern town. The streets seem as if they had been recently built; and are wide, well built, lighted with gas, and partially paved. The county-hall is a handsome modern structure. The manufacture of the coarser fabrics of linen form the maple trade of the town. There are also corn, barley, and flour mills, reckoned the best in the co., a snuff-mill which manufactures 60,000 lbs. of snuff a year, a washing or fulling mill, a glue manufactory, three breweries, two tan-works, a tile and brick work, at which coarse earthenware is made, and a rope-work. Cupar has long had a flourishing joint-stock academy, with numerous other schools; there is a bequest by Dr. Bell of 10,000l. for educational purposes according to the Madras system. Besides the par. church, there are Presbyterian dissenting chapels, one Episcopal and one Glassite chapel. Cupar is associated with St. Andrew's, the two Anstruthers, Crail, Kilrenny, and Pittenweem, in returning a memb. to the H. of C. Registered electors 224 in 1865. Corporation revenue, 1304l. The borough is governed by a provost, 3 bailies, and 23 councellors. Cupar was a royal bor. so far back, at least, as the reign of David II. On a mound at the E. end of the town, called the Castle-hill, formerly stood a castellated fortress, the chief residence of the family of Macduff, the feudal thanes or earls of Fife. At the foot of this mound was a convent of Dominican or Black Friars, founded by the Macduffs, and afterwards annexed to St. Monance in the same co. (Keith's Scot. Bishops, ed. 1824, p. 445); but of these two buildings no traces are now extant. The patrimonial estate of the famous Scottish poet, Sir David Lindsay of the Mount, was within a short distance of Cupar; and on a verdant esplanade, still called the Play Field, in front of Macduff castle, was acted, in 1535, his witty drama of the 'Three Estates,' a popular satire on the priesthood, and which is thought to have had no mean effect in hastening the Reformation.

CUBACOA, or CURANSAO, an isl. in the Caribbean Sea, belonging to the Netherlands, off the N. coast of Venezuela; between lat. 12° and 12° 15' N., and long. 68° 44' and 69° 13' W. Length, NW. to SE., about 43 m.; average breadth about 14 m.; area, 600 sq. m. Pop. 19,546 in 1861, of whom about one-third slaves. The shores of the island are bold, and its interior is in parts hilly. It has several harbours, the chief of which is that of Santa Anna, in the SW., where its principal town is built. The soil is in general poor and rocky, and there is a great deficiency of water; but by the industry of the inhab. some tobacco, sugar in considerable quantities, and indigo are grown; and a good deal of salt is obtained from the margins. Maize, cassava, figs, oranges, citrons, and most European culinary vegetables, are cultivated; but provisions are not produced on the island in sufficient quantity for its inhab. The government is conducted by a stadtholder, assisted by a civil and military council. Wilhelmstadt, the cap. and seat of government, is one of the neatest cities in the W. Indies; its public buildings are magnificent, the private houses commodious, and the clean streets remind the traveller of those in the Dutch towns. The port of Curaçoa, St. Barbara, has a narrow entrance, but is large and safe. It is protected by the fort of Amsterdam and other batteries; but was taken by a squadron of four English frigates in 1807. Two smaller islands, one on either side, Buen Ayre and Oruba, also belong to the Dutch. Their inhab. are chiefly cattle-breeders. Curaçoa was discovered by the Spaniards, but taken from them by the Dutch in 1632. Great Britain took possession of it in 1798, but returned it to Holland in 1814.

CUSTRIN, or KUSTRIN, a strongly fortified town of Prussia, prov. Brandenburg, on the Oder, where it is joined by the Warta, 52 m. E. Berlin, on the railway from Berlin to Königsberg. Pop. 9,037 in 1861, excl. of garrison of 1,061. The

Oder is here crossed by a bridge nearly 900 ft. in length, uniting the citadel with the town; being surrounded by marshes, it is strong as well by nature as by art. The Russians burnt the town without, however, taking the fort in 1758. It was soon after rebuilt on a greatly improved plan. The fortifications have been much improved since the peace of 1815.

CUTCH-GUNDAVA, an inl. prov. of Beloochistan, differing in some important respects from all the others, being by far the most valuable portion of that country, and its only prov. E. the Brahooick mountains. It lies between lat. 27° 40' and 29° 15' N., and long. 67° 20' and 69° 30' E. Length N. to S. about 120 m.; breadth of its habitable and fertile part a little more than 60 m.; having N. Sewestan (Cutchel), E. and S. Sinde, and W. the prov. Thalawan. It is for the most part a plain, bounded by deserts on the N., S., and E.; and watered by several rivulets communicating by numerous aqueducts. Soil rich and loamy, and so exceedingly productive that it is said, were it all cultivated, the crops would be more than sufficient to supply all Beloochistan; as it is, considerable quantities of grain, besides cotton, indigo, and oil, are exported. It is alleged, but probably without foundation, that rice will not grow in Cutch-Gundava, notwithstanding the luxuriance of all other crops, and the plentiful supply of water. Climate oppressively hot throughout the summer, when the almanac is frequently experienced; during winter it is so mild that the chiefs and principal inhabitants of the adjoining W. provinces resort thither. The bulk of the pop. are Jats; there are a few Hindoos in the towns and villages, who live by barter, and transport grain. Villages extremely numerous. The chief towns are Gundava, the cap., Dadur, Bhag, and Lhere.

CUTTACK, a large marit. dist. of Hindostan, prov. Orissa, presid. Bengal, between lat. 19° 30' and 21° 40' N., and long. 84° 30' and 87° E.; having N. the dist. Midnapore and the Berar ceded districts, W. the latter, S. Ganjam, and E. the Bay of Bengal. Area 9,000 sq. m. Pop. 1,341,610. It consists of three different tracts of country,— the marshy coast, the dry central region, and the hilly country to the W. The latter abounds with trees, valuable either for cabinet-work, dyeing, or varnish-making. Rivers numerous; the chief are the Mahanuddy, Brahminy, Coyle, and Salonoreeka; all these are of considerable size, and even the minor streams swell, during the rains, to an enormous magnitude, rendering the construction of extensive and solid embankments necessary in many parts of the dist. The periodical rains are not so early here as in Bengal; the summer heats are very oppressive, and the forests of Cuttack are generally highly insalubrious. They are also much infested with ferocious wild animals, especially leopards; and reptiles, many of which are venomous. Rice of different qualities, wheat and maize, in the hilly tracts, the sugar-cane, pulse, arsenals nuts, spices, and dyeing drugs are the chief articles of culture. Several kinds of granite, slate, and iron ore are found, and gold dust in the beds of the mountain torrents. The land is not assessed under the permanent settlement, as in the rest of the adjoining prov. of Bengal; but an agreement is usually made between the government and the land-holders for a certain term, the amount of the land-tax being by no means fixed. A considerable proportion of the territory in the W. or mountainous region is in the possession of a number of nearly independent zemindars, each of whom maintains a kind of sovereign state, and pays but a light tribute. A more valuable source of revenue to the government than the land-tax has been the monopoly of salt, much of which remarkable for whiteness and purity, is made on the coast of this district. The chief towns are Cuttack the cap., Balasore, and Juggernaut, the seat of the celebrated temple of that name. (See JUGGERNAUT.) Cuttack was acquired by the British, on the expulsion of the Mahrattas, and the reduction of the Juggernaut rajah in 1803-4. In 1817, the too rapid introduction of the revenue and judicial systems established in Bengal amongst the rude and barbarous inhabitants of Cuttack, together with the evils of over-government and mismanagement, excited a rebellion in this dist., which was subdued in the ensuing year, but at the expense of much treasure, and the loss of many lives.

CUTTACK (Cutak, a royal residence), a town of Hindostan, cap. of the above dist., seat of its principal judicial court, &c., on the Mahanuddy, and in the rainy season insulated by two of its branches, 220 m. SW. Calcutta; lat. 20° 27' N., long. 86° 5' E. Pop. estimated at 40,000. Its principal street is well built, and it has many houses two and three stories high, a spacious market-place, some handsome Mohammedan structures, and some military cantonments. The dwellings of the civil establishment are dispersed over the environs. This town is secured from inundation by large and solid embankments along the river; the value of these was sufficiently proved in 1817, when during the heavy rains the waters of the river rose in one night 18 ft. or 8 ft. above the general level of the town, which was only preserved by their means. Cuttack is believed to have been a capital as early as the 10th century.

CUXHAVEN (Germ. Cuxhaven), a sea-port town of N. Germany, immediately within the estuary of the Elbe, on its SW. side, in a detached portion of territory belonging to Hamburgh, from which it is distant 55 m. WNW.; lat. 53° 52' 21" N., long. 8° 47' E. Pop. 1,110 in 1861. The town has a good harbour, with deep water, a lighthouse, and is a quarantine station. It was formerly the rendezvous of most passengers to and from England and the Elbe; but since the establishment of steam-packets they are conveyed direct to and from Hamburgh. Vessels entering the Elbe generally leave to opposite Cuxhaven for pilots, by whom it is mostly inhabited. In summer it is resorted to by sea-bathers.

CUZCO, an inland city of Peru, formerly the cap. of the empire of the Incas, at the foot of some hills, having an extensive valley opening to the SE. 11,380 ft. above the level of the sea, about 400 m. ESE. Lima; lat. 13° 30' 55" S., long. 72° 4' 10" W. Pop. 45,251 in 1858, mostly Indians. The cathedral and convent of St. Augustine are said to rank amongst the finest religious edifices in the New World; and it has besides six churches, eight convents, four well-endowed hospitals, three monasteries, a university, and three collegiate schools. But Cuzco derives most part of its interest from the historical associations connected with it, and from the remains of the architecture of the Incas. Even a great number of the private houses belong to that era; and by the size of the stones, and the immense and peculiarity of the buildings, give to the city an imposing air. The Dominican convent, a magnificent structure, is raised on walls that formed part of the famous temple of the sun, destroyed by the fanatical zeal of the Spaniards. Ulloa (Voyage d'Amerique, i. 505) says that the high altar stands on the very spot formerly occupied by the golden image of the sun. Upon a hill to the N. of the city are the ruins of a very extensive fortress, the work of the Incas, the walls of which are of the species named Cyclopean, and have a striking analogy to the so-called structures found in various parts of Greece and Italy. Some of the stones, which are all of angular shapes, are of such an enormous size that their weight is said to exceed 150 tons, and, though no cement be used in the building, they are so admirably jointed and fitted together, that the interstices are hardly perceptible. It is very difficult to imagine how such vast blocks could have been conveyed from the quarries and placed on the walls without the aid of powerful machinery. In the plain to the S. of the city are extensive remains of ancient edifices in the same style; and it is said by Ulloa that a subterranean passage led from the palace of the Incas to the fortress; and that a road was constructed from the city to Lima.

The inhabitants are industrious, and excelling in embroidery, painting, and sculpture. There are manufactures of cotton, linen, and woollen stuffs, and of leather and parchment. A considerable trade is carried on in these and in the products of the adjacent district.

Cuzco is the most ancient of the Peruvian cities. Its origin dating from the era of Manco Capac, the founder of the empire of the Incas, probably in the 12th century. Pizarro took possession of it in 1534, and was shortly after besieged in it by the whole Peruvian force. During this siege a great part of the town was destroyed. The city, as well as the province, of Cuzco, after being torn from the Spanish dominion, formed part of Peru from 1821 to 1836; it then fell to Bolivia, but was subsequently again united to Peru.

CYPRUS, or KIBRIS, a famous and considerable island, in the NE. angle of the Mediterranean, between Asia Minor and Syria, at present belonging to Turkey, 44 m. S. Cape Anamour in the former, 60 m. W. Latakia in the latter, and 250 m. E. Crete; between lat. 34° 34' and 35° 42' N., and long. 32° 18' and 34° 37' E. Shape somewhat oval, with a considerable promontory projecting ENE. from the main body of the island; greatest length 132 m.; average breadth from 40 to 45 m. Pop. estimated at 110,000, of whom about one-half are Greeks. The island is intersected lengthways, or from E. to W., by a range of mountains, the highest point of which, St. Croce (an. M. Olympus), is about 15 m. S. Nicosia. The principal river, Pedia (an. Pediaeus), consists of two main branches; it flows E. through the centre of the island, having its embouchure near the ruins of Constantia, on the E. coast; but this, like most of the other rivers, is but of limited dimensions, and is nearly dried up in summer. Cyprus is also otherwise ill supplied with water, that obtained from most of the wells being brackish. The principal plains lie along the banks of the Pedia, and the S. coast of the island. The climate differs in different parts: along the N. shore it is comparatively temperate; the winds coming from the cold mountainous districts of Asia Minor temper the heat in summer, and in winter produce piercing colds on the mountains, which are covered with snow for several months. But it is otherwise in the plains along the S. and E. coasts: these consist, for the most part, of a whitish soil which has an offensive glare, and being defended from the N. and NW. winds by the mountains, at the same time that they are exposed to the full sweep of the E., SE., and S. winds from the Syrian, Arabian, and Lybian deserts, they have a higher temperature than any other place in the Levant. During the summer heats malaria is frequently generated; and long droughts, combined with the want of industry and the neglect of irrigation, not unfrequently destroy the crops. The soil is naturally fruitful, and, in antiquity, Cyprus was famous for its fertility, and the variety and excellence of its products. Even now, though only a very small portion of the land is cultivated, and that in the most wretched manner, the merchants of Larnica annually export several cargoes of excellent wheat to Spain and Portugal. The best as well as the most agreeable parts of the island are in the vicinity of Cerina and Baffa, the ancient Paphos. (See BAFFA.)

Cotton of a superior quality is produced in the island. The cultivation was much extended after the outbreak of the American civil war, and in 1863 the total produce amounted to 8,000 bales of 2½ cwt., or 2,016,000 pounds. (Report of Mr. Vice-Consul White on the Trade of Cyprus, dated May 10, 1864.) Cyprus was formerly famous for its cotton, and, under the Venetians, the island annually exported about 30,000 bales. It then also exported considerable quantities of sugar, produced from plantations of cane in the vicinity of Limasol and Baffa. There are extensive forests of oak, beech, and pines; groves of olives and plantations of mulberries. It is remarkable for the fineness of its fruits, and its rich sweet wine, oil, and silk. The latter is of two kinds, yellow and white, but the former is preferred. The wheat is of a superior quality, affording excellent bread; and rice, madder, and an endless variety of other valuable products, might be cultivated in several parts of the island.

The wines of Cyprus, particularly those produced from the vineyard called the Commandery, from its having belonged to the knights of Malta, were formerly more highly prized for desserts than even those of Crete. In the earlier part of last century, the total produce of the vintage was supposed to amount to above 2,000,000 gallons, of which nearly half was exported; but now, the wine grown and exported does not amount to a tenth part of these quantities. 'Perhaps,' says Dr. Clarke, 'there is no part of the world where the vine yields such redundant and luscious fruit: the juice of the Cyprian grape resembles a concentrated essence. The wine of the island is famous all over the Levant. Englishmen, however, do not consider it as a favourite beverage; it requires nearly a century of age to deprive it of that sickly sweetness which renders it repugnant to their palates. Its powerful aperient quality is also not likely to recommend it. When it has remained in bottles for 10 or 12 years, it acquires a slight degree of fermentation, upon exposure to the air; and this, added to the sweetness and high colour, causes it to resemble Tokay more than any other wine. It will keep in casks, to which the air has access, for any number of years. If the inhabitants were industrious, and capable of turning their vintage to the best account, the red wine of the island might be rendered as famous as the white, and, perhaps, better suited for exportation. (Travels, iv. 19.)

Cyprus was formerly far more densely populated than it is at present. In antiquity, the population, probably fell little short of 1,000,000; and in 1571, when it was conquered by the Turks, it had a population of about 300,000, or nearly four times its present amount. 'Nowhere,' as Mr. Kinneir states (Journey through Asia Minor, pp. 176, &c.), 'is the baleful influence of the Ottoman dominion more conspicuous than in Cyprus, where it has literally turned cities into miserable villages, and cultivated fields into arid deserts.' In describing his journey from Larnica to Nicosia, Dr. Clarke (Travels, iv. 55) observes, 'The soil everywhere exhibited a white marly clay, said to be exceed-

ingly rich in its nature, although neglected. The Greeks are so oppressed by their Turkish masters, that they dare not cultivate the land; the harvest would instantly be taken from them if they did. Their whole aim seems to be to scrape together sufficient, in the course of the year, to pay their tax to the governor. The omission of this is punished by torture or by death; and, in cases of their inability to supply the impost, the inhabitants from the island. So many emigrations of this sort happen during the year, that the population of all Cyprus scarcely exceeds 60,000 persons, a number formerly insufficient to have peopled one of its many cities. The governor resides at Nicosia. His appointment is annual, and as it is obtained by purchase, the highest bidder succeeds; each striving, after his arrival, to surpass his predecessors in the enormity of his exactions. From this terrible oppression, the consuls and a few other families are free, in consequence of a protection granted by their respective masters.'

Mr. Kinneir (Journey, pp. 182-3) states, that 'the governor and the archbishop deal more largely in corn than all the other people of the island put together; they frequently seize upon the whole yearly produce, at their own valuation, and either export or retail it at an advanced price; nay, it happened more than once, during the war in Spain, that the whole of the corn was purchased in this manner by the merchants of Malta, and exported without leaving the lower orders a morsel of bread.' More recently, the conditions of the people seem to have somewhat improved, to judge from consular and other reports. The exports of produce are also steadily increasing. The total amount of exports for 1863 was 276,580l., being an increase of 80,556l. upon the preceding year, chiefly due to augmented culture of cotton. The total amount of imports in 1863 was 129,034l., exceeding that of the year 1862 by 20,000l. Greece is the chief importing country, next Austria and then France. (Report of Mr. Vice-Consul White on the Trade of Cyprus, dated May 10, 1864.)

Sheep and cattle are bred in considerable numbers. There is abundance of game, such as partridges, quails, woodcocks, and snipes; there are no wild quadrupeds, excepting foxes and hares, but many kinds of serpents, and the tarantula. Clouds of locusts sometimes devastate the country. The ancient mines of Cyprus now wholly neglected, afforded large quantities of the finest copper (Æs Cyprium), whence, though that is very doubtful, the name of the island has been supposed to be derived. It is also said to contain ores of gold, silver, and other metals, and has a species of rock-crystal called l'opus diamond. Amianthus, or asbestos, of a very superior quality, is found near Baffa; it is flexible as silk, white, and more delicately fibrous than that of any other country. Maritti states that a village, called Amianthus, existed in Cyprus in his time; and it was most probably the spot where the Amianthus or incombustible cloth, used by the ancients to wrap up the bodies of distinguished persons when laid on the funeral pile, was principally produced. (Travels, I. 177.) Salt is obtained by evaporation at various places on the S. coast. The inhabitants manufacture small carpets, some silk and cotton fabrics, and excellent Turkey leather. Under the Turks this island was divided into three sanjacks—those of Baffa, Cerina, and Nicosia. Nicosia, in the centre of the island, is the capital. The other principal towns are Larnica, on the site of the ancient Citium, Limasol, Famagosta on the E., Cerina (an. Cerinia) on the N., and Baffa (Paphos) on the W. coast. Even the ruins of most of the

L 2

ancient cities mentioned by Strabo have disappeared; but at Constantia, near Famagusta, Kinneir traced the site, of the ancient walls, and the foundations of some buildings; and at Larnica medals and other antiquities are frequently dug up. The remains of a monastery, built by a princess of the house of Lusignan, stand about 4 m. NE. l'erina. Cyprus was originally peopled by the Phœnicians. It was colonised by the Greeks, and successively possessed by the Egyptians, Persians, Greeks, and Romans. In antiquity, it was so famous for the worship of Venus as Dione for that of Apollo and Diana. This, in fact, was the favourite seat of the goddess, 'dira potens Cypri.' Divine honours are supposed to have been first paid to her at Paphos (See IIAPFA,) where she had a magnificent temple—

> 'ubi templum illi, centumque Sabæo
> Thure calent aræ, sertisque recentibus halant.'
> —*Ænid, i. 416.*

But the whole Island was sacred to Venus; and, besides Paphos, other three cities were celebrated for her worship.

> 'Est Amathus, est celsa mihi Paphos, atque Cythera,
> Idaliæque domus.' —*Ænid, x. 51.*

Hence the epithets Cyprian, Paphian, and Idalian, applied to Venus. It is alleged that the ladies of the Island are still devotedly attached to the worship of the goddess.

After the fall of the Western empire, Cyprus formed part of the Byzantine empire, from which it was taken by the Saracens. Isaac, a prince of the Comneni family, having assumed the sovereignty, was dethroned, in 1191, by Richard I., king of England. The latter having conferred the Island on Guy de Lusignan, to indemnify him for the loss of Jerusalem, it continued in possession of his family for three centuries, or till 1489, when, on default of heirs, it fell to the Venetians. The Turks took it from them in 1571. Bragadino, the gallant defender of Famagusta, after exhausting every resource, at last capitulated on honourable terms. No sooner, however, had the place been delivered up than the capitulation was disregarded, and Bragadino himself was skinned alive and impaled—a dreadful augury of what the population was to suffer under the dominion of the warlike followers of Mahomet. However, it seems probable that the better government of Turkey, inaugurated in recent years by the influence of the Western powers, will also make itself felt before long in this magnificent island.

CZEGLED, a large market town of Hungary, between the Danube and Theiss, on the high road between that city and Debreczin, 39 m. SE. the former, and 84 m. WSW. the latter. Pop. 18,130 in 1857. The inhabitants are chiefly Protestants. A great deal of ordinary red wine is made here, as well as beer.

D

DACCA, or DIIAKA, an inl. city of Hindostan, prov. Bengal, formerly very extensive, populous, and rich, and still one of the principal cities of the Bengal presidency, and the seat of a court of circuit and appeal for the seven E. distr. of Bengal. It extends, with its suburbs, for 8 m. along a river which, uniting with the Ganges on the one hand and the Brahmaputra on the other, affords the greatest facilities to commerce; lat. 28° 42' N., long. 90° 17' E. 127 m. NE. Calcutta, with which it is connected by the East Bengal railway. Pop. estimated at 70,000. Like other native towns, it is a mixture of brick, thatch, and mud houses, with narrow and crooked streets. The huts of the houses are so very combustible, that they are usually burned down once a year. According to Heber, Dacca is like the worst part of Calcutta, near Chitpore, but with some really fine ruins intermingled with the huts, which cover three-fourths of its area. There are few European houses, and these mostly small and mean, compared with those of Calcutta. Some Greek buildings, which were the favourite residence of the late nabobs, were ruined by the encroachments of the river, in the 17th century Islam Khan built a palace and fort here, the ruins of which form an imposing object; and toward the end of the same century a grandson of Aurungzebe constructed and finished a magnificent palace, now also in ruins. The pagodas are few and small, owing to the ascendancy of Mohammedanism, and almost every brick building has its Persian or Arabic inscription. There is a small but pretty Gothic English church; and a burial ground about a mile from the city, containing some handsome tombs, both Christian and Mussulman. There are several obelisks in and around the city; and about 1 m. off is a beautiful Gothic bridge, said to have been constructed by a Frenchman, but, like most of the other public edifices, in a state of ruin. All the buildings beyond the inhabited portion of the city are surrounded by ruins and rank vegetation; and the castle, factories, and churches of the Dutch, French, and Portuguese, have all fallen into decay. English goods and manufactures, or imitations of them, are to be met with in the bazaars; but no vessels larger than small country-built brigs come up the river. The trade of the city, however, has greatly improved in recent years by the establishment of the Eastern Bengal railway, a line running from Calcutta to Dacca, via Paksa, with a branch to Jessore. The total length of this railway is 110 m. and it was opened throughout on the 15th November, 1862. The striped and flowered muslins of Dacca were formerly regarded as inimitable, and were in great request at the Mogul court, and other native Indian courts, as well as at the old court of France. The manufacture was hereditary in several families, but has been annihilated by the destruction of the native courts and the wealthy native nobles. Its ruin has been very generally ascribed to the importation of the cheaper muslins of England, but this is a mistake: it was wholly suppressed before a yard of British muslin or calico found its way to India. The manufacture, in fact, was never carried on upon a large scale; and being one of luxury only, it fell with the fall of the wealthy class, who alone purchased its products. The cotton grown in the district is now mostly exported to England. There are some respectable Greek, Portuguese, and Armenian merchants. The country round Dacca being always covered with verdure during the dry months, it is comparatively free from violent heats, and is reckoned one of the healthiest stations in Bengal.

Dacca is comparatively modern: it is not mentioned by Abul Fazel. From 1608 to 1639 it was the metropolis of Bengal, and again attained to that dignity in 1655. the commencement of the era of its greatest splendour, when, judging from its ruins, it must have vied in extent and wealth with the largest cities of India. Its decline began with the disorders consequent to the invasion of Nadir Shah.

DACCA, and DACCA JELALPORE, two districts of Hindustan, prov. Bengal, chiefly between lat. 23° and 24° N., and long. 89° 30′ and 91° E.; having N. the distr. Mymunsing, E. Tipperah, S. Backergunge, and W. Jessore and Rajshahye. The area of Dacca is 1,870, and of Dacca Jelalpore 2,565 sq. m.; pop. of both districts 1,237,000 in 1801. The country is almost a dead flat, studded with lakes, and intersected by the two great rivers, Brahmaputra and Ganges. During the rainy season it exhibits the appearance of an inland sea, over which the villages, raised on artificial embankments, are scattered like so many islands. The land fertilised by such extensive inundations is extremely productive; but a large proportion of it is covered with jungle, and infested with elephants, tigers, and other wild animals, which do considerable damage to cultivation. These, however, are much less numerous now than formerly; and a great deal of the land that had been overspread with jungle has latterly been cleared, and brought into cultivation. The banks of the Commerally river, one of the arms of the Ganges, are populous and well cultivated, producing rice, sugar, cotton, and indigo; a species of cotton called banga, though not of a superior quality, very well adapted for the fine striped muslins, for which this prov. was long famous, used to be grown in large quantities. The land is subdivided into extremely small estates, and the constant shifting of the river-courses alters their extent and boundaries so much that the assessment and collection of the revenue have always been matters of much difficulty. Dimities, cloths resembling diaper, and damask linen, are now the chief manufactures. About half the pop. are Hindus, and half Mohammedans. Slavery is pretty prevalent. These districts had formerly an unenviable notoriety, from the number and enormity of the crimes committed in them, but in this respect they have lately very much improved. There are numerous Hindu schools, for instruction in the Bengalee language, religion, and laws. Chief towns, Dacca, Narraingunge, Sunnergunge, and Rajanagur.

DAHOMEY, a country of Africa, on the Guinea coast, of which the boundaries are far from being well defined, but which is supposed to extend between about 6° and 8° or 9° N. lat., and from 1° to perhaps 3° E., long.; having W. Ashantee, E. Yarriba and Benin, and S. the Atlantic Ocean. As far as has been hitherto discovered, this country is destitute of any considerable hills, and consists of an immense plain rising gradually from the sea to the Kong Mountains, which are here from 150 to 200 m. inland. The Volta and Loha rivers bound it on the W., but excepting these, there seem to be no stream of any considerable importance. The country is, however, well watered, and interspersed with small marshes. The soil is wholly alluvial, not a stone is to be met with; the surface is covered with a vegetation of unbounded luxuriance; and the beauty and excellence of the country are spoken of in terms of the highest admiration. (See the statements of Bosman and Phillips, in the Histoire Générale des Voyages, iv. 374, &c.) Oranges, limes, guavas, and other tropical fruits, melons, pine-apples, and yams, grow wild; and maize, millet, and other grains, potatoes, indigo, cotton, sugar, tobacco, and spices are successfully cultivated. In some parts the country is covered with dense forests, the retreat of lions, hyenas, leopards, elephants, and overgrown serpents. Deer and domestic animals are plentiful. Previously to the early part of last century this country was divided into a number of petty states, and is represented as having been populous and well cultivated. The Dahomans, by whom it was overrun and laid waste, came from the interior of the Continent. They are said to be hospitable to strangers, brave and resolute; and these, if they exist, would appear to make up the whole amount of their good qualities. Their disposition seems, from their conduct, to be a compound of that of the tiger and the spaniel, exhibiting the utmost ferocity and thirst for blood with the most abject servility. All the most arbitrary forms of eastern despotism seem to be mild and free, when compared with that established in this wretched country. It is singular, too, that this despotism is not founded upon force and terror, nor is it connected with anything timid or effeminate in the character of the people. It rests on a blind and idolatrous veneration for the person of the sovereign, as for that of a superior being. He is the absolute master of the lives and properties of his subjects, and disposes of them at pleasure. It is a crime in the latter to suppose that the king eats, drinks, sleeps, or performs any of the functions of an ordinary mortal. A sovereign of the name of Bossa having succeeded to the throne, caused all the persons of the same name in his dominions to be put to death, conceiving it to be an unpardonable presumption that any subject should bear the same name with his master. The greatest lords can only approach the king lying flat on their faces, and rolling their heads in the dust. The attempts thus made to inspire the people with reverence for their monarch, seem to have been completely successful. The Dahoman rushes to battle in obedience to the orders of his king with a blind, unthinking, brute confidence. Norris having asked a Dahoman before battle if he did not think the enemy too numerous; the latter replied, 'I think of my king, and then I dare engage five of the enemy myself.' He declared his indifference whether he survived or not; adding, 'It is not material; my head belongs to the king, not to myself; if he pleases to send for it I am ready to resign it; or if it is shot through in battle, it is no difference to me. I am satisfied.' It is not surprising, therefore, to learn that human skulls form the favourite ornament in the construction of the palaces and temples. The king's sleeping chamber has the floor paved with the skulls, and the roof ornamented with the jaw-bones of chiefs whom he has overcome in battle. Every year a grand festival is held, which lasts for several weeks, and during which the king waters the graves of his ancestors with the blood of hosts of human victims. The bodies of these unhappy men are not even interred, but are suspended by the feet to the walls, and left hanging till they putrefy. The ceremony is known as the 'grand customs.'

Perhaps the most extraordinary fact connected with this barbarous horde is, that all the women are monopolised by the sovereign; and that no individual can possess himself of either a wife or a concubine except by gift of, or purchase from, the king; and whether the lady be young or old, handsome or the reverse, she must be equally acceptable to the slave to whom she is given or sold. The king keeps a vast seraglio for himself; and at his death his wives and concubines fall to murdering each other, till the carnage is stopped by the interference of the new king. After these statements, it will only appear consistent and natural that the tiger should be the principal fetiche, or object of worship, among the Dahomans,

late reports state that, despite their ferocity, this most detestable of barbarian hordes has been checked in its devastating course. A number of the petty states it had subdued have emancipated themselves; and it appears probable that the sovereign of Dahomey is now tributary to the sovereign of Yarriba. Next to Abomey, the cap. and residence of the king, about 80 m. inland, Whydah, Ardrah, Ausna, and Calmina, are the chief towns or villages. (For farther accounts of Dahomey, many of them greatly contradictory, see F. E. Forbes, 'Dahomey and the Dahomans, being the Journals of Two Missions to the King of Dahomey in the Years 1849-50,' 2 vols. Lond. 1851; T. B. Freeman, 'Journal of various Visits to the Kingdoms of Ashanti, Aku, and Dahomey,' Lond. 1844; Mollien, G., 'Voyage dans l'Intérieur de l'Afrique,' Paris 1820; Lesal, J. M., 'Voyage to Africa, with some Account of the Manners and Customs of the Dahomian People,' Lond. 1820; and Dalzel, A., 'History of Dahomey,' Lond. 1795. Some of the more recent books about Dahomey are chiefly compilations from these older works, spiced with a good deal of romance.)

DALECARLIA, a prov. of Sweden, which see.

DALKEITH, a bor. of barony and market-town of Scotland, co. Mid-Lothian, on the road from Edinburgh to Coldstream, 5½ m. SE. Edinburgh, on the Edinburgh and Hawick railway. Pop. 5,450 in 1861. The town is situated on a peninsular neck of land between the N. and S. Esks, which unite about a mile E., and fall into the Frith of Forth at Musselburgh. It is a clean, well-built town; the principal street, which is wide and handsome, runs from E. to W., and there are several subordinate streets. Its public buildings are,—a parish church (an old Gothic edifice, used as a collegiate church before the Reformation), three chapels belonging to Presbyterian dissenters, and one belonging to the Independents. A new parish church was erected in 1820. Dalkeith has long been eminent for its educational institutions, particularly its classical school. This town, like other burghs of barony, was originally under the exclusive management of the baron or superior and his bailie; but, in 1759, an act of parliament was obtained, appointing certain trustees to superintend the paving, cleaning, and lighting the streets, to supply the burgh with water, and to provide a revenue for these purposes by imposing a small tax on the ale, porter, and beer consumed in the parish. Dalkeith is chiefly celebrated for its grain market, which is held every Thursday, and is reckoned the largest market of the kind in Scotland. The Dalkeith and Edinburgh railway, which connects these towns, was commenced as early as 1827, and opened for goods and passengers in 1831. The Duke of Buccleuch at his own expense, brought the Dalkeith line into the centre of the burgh, prolonging it, by a viaduct over the N. Esk, so as to communicate with coal mines in that quarter. Coal abounds throughout the whole neighbourhood of Dalkeith. Dalkeith Palace, the principal residence of the Duke of Buccleuch in Scotland, is within 300 yards of the E. termination of the town. This palace, which formerly belonged to the Douglases, earls of Morton, was acquired, in 1642, by the family of Buccleuch, who still retain it, and are superiors of the burgh. Anne, heiress of Buccleuch, was married to the Duke of Monmouth, a natural son of Charles II., beheaded for rebellion in 1685. George IV., on his visit to Scotland in 1822, resided in Dalkeith House. The parliamentary electors of the burgh unite with the county constituency in returning a member to the H. of C.

DALMATIA (an. part of *Illyricum*), a marit.

country of Europe, being the most S. prov. of the Austrian empire, comprising a long and narrow territory lying along the NE. shore of the Adriatic, and numerous islands in that sea, between lat. 42° 8' and 44° 55' N., and long. 14° 30' and 19° E., having N. Hungarian Croatia; E. Turkish Croatia, Herzegovina, and Montenegro; and S. and W. the Adriatic: length of the continental portion, NW. to SE., 240 m.: breadth greatest towards the N., where it averages nearly 40 m.; but it tapers thence gradually to its S. extremity, and in its lower half is never more than 15 m. in width. Area 220 Austrian, or about 5,900 Engl. sq. miles. Pop. 404,499 in 1857. Dalmatia is generally mountainous. The Dinaric Alps bound it on the E., and the whole country is intersected in a direction parallel to the coast by some of their subordinate ranges, the highest point of which, Mount Biocova, near lat. 43° 30', is 4,850 ft. in elevation. Here, as elsewhere, the Dinaric Alps are chiefly of calcareous formation, and full of clefts and ravines; they are rugged, and often destitute of soil, in consequence of which the country has in most parts a sterile and desolate aspect. Narrow valleys are abundant, but plains of any extent few. There are numerous small lakes, and one of a tolerable size, near Zara; but, generally speaking, Dalmatia is ill watered. The principal river, the Narenta, in the S., has not a course of more than 15 m. in the Austrian territory; the other chief rivers are, the Zermagna, Kerka, and Cettina, but none is of any great size. The Cettina is remarkable for a fine cascade, 170 ft. in altitude. The coast is indented with numerous harbours, of which those of Cattaro, Sebenico, and Ragusa are the best; it has also numerous brand-lands, and is fenced by a great number of elongated islands, lying in a direction parallel to the shore. The principal are, Arbe, Pago, Isola Grossa, Brazza, Lesina, Curzola, Lissa, Meleda, &c.; they are mountainous, and present the same general aspect as Continental Dalmatia. The climate is warmer than in any other part of the Austrian dominions. In the S. the date-palm flourishes in the open air, and the olive grows in the lowlands everywhere throughout the country. Frost and snow are almost unknown in the plains and valleys, and are of very short duration in the mountains; the mean temp. of the year at Ragusa is 57° 3' Fahrenheit. The winter is limited to six weeks of pretty constant rain; yet, on the whole, less rain falls in Dalmatia than in any other prov. of the empire, and the country often suffers from excess of drought. Except in the marshy tracts along the shore, the air is pure and salubrious. The arable land of Dalmatia is not more than 214 Austr. sq. miles in extent, or 11 per cent. of the whole area. (Arrenstein, Dr. Jos., Oesterreich in der Weltausstellung, Vienna, 1863.) Agriculture is in every respect extremely backward. Maize and barley are the principal kinds of grain cultivated; but not two-thirds of the corn necessary for home consumption is grown; the rest of the quantity required comes mostly from Turkey and Hungary. The Dalmatian wines are strong and deep-coloured, but are apt to acquire a taste from the leathern flasks in which they are kept. They, however, bear transport well, and considerable quantities are sent to Fiume, Trieste, and Venice. The total quantity produced annually is officially estimated at 8,828,000 gallons. Fruits are abundant and excellent. Figs may be considered the chief staple of Dalmatia; they grow without culture all along the coast, but the best are those of Lesina. During their period of maturity, they make a large part of the food of the village pop., and about 845,000 libbre are annually expected. The climate is highly suitable for the olive, and the oil is better than that produced in most parts of Italy. Nearly 17,000 cwt. are annually obtained. Cattle breeding is pursued to a great extent, but the breeds are mostly inferior. According to an official return published in 1865, there were in Dalmatia, at that period, 22,000 horses; 114,775 cattle; 815,632 sheep; 121,987 goats; and 42,216 swine. The wolf, wild dog, fox, and lynx are amongst the wild animals; game (excepting deer) abounds, as do waterfowl and birds of prey. The anchovy and tunny fisheries are important, though not so much so as during the last century; they furnish employment to about 8000 inhab. Dried and salted fish form an important article of commerce. There are some coral fisheries, of which that near Sebenico is the chief. The fish caught in the lakes, &c. form a chief part of the subsistence of many of the inhab. Excellent timber for ship-building and other purposes abounds in the interior; but is next to useless from the absolute want of roads, canals, or navigable rivers, to convey it to the sea. The large forests which formerly existed on the coast have been cut down, and that part of the country is now almost bare of wood. The attention of the Austrian government is now, however, directed to the forest economy of the prov., at the view of supplying the dockyards at Fiume and Venice with Dalmatian timber. Coal is found in several parts, and considerable quantities are exported to Trieste. Ship-building, and the distillation of maraschino and rosoglio, are the chief branches of manufacturing industry. Maraschino is extensively consumed at Vienna, and it is well known in this and most other countries. Besides these, a few articles of primary necessity only are manufactured; for all others, the inhab. are obliged to have recourse to the neighbouring countries. This prov. enjoys the important advantage of being placed without the Austrian customs line, the duty on foreign goods imported being only 3½ per cent. ad valorem. But the strictness with which quarantine regulations are enforced have gone far to nullify the important benefits that would otherwise have resulted from this valuable privilege. The Dalmatians are amongst the best sailors of the Adriatic. There entered the port of Zara, in 1843, 1,836 vessels, of a burthen of 89,152 tons, while at the next important port, Spalatro, there entered 2,804 vessels, of 70,347 tons burthen. The province is divided into four circles, named after their respective capitals, Zara, Spalatro, Ragusa, and Cattaro: the last two circles are separated from the rest of Dalmatia, and from each other, by two narrow slips of land belonging to Turkey, which stretch down to the sea coast. The other chief towns are Sebenico, Trau, and Macarsca. Zara is the cap. and seat of the government and council of the prov.

The inhab. of Dalmatia are Slavonians of the same race with the Croatians, Servians, and Illyrians. The names of the rivers and mountains are all Slavonic. The vicinity of, and constant intercourse with, the Italian harbours, has however introduced the use of the Italian language amongst the commercial part of the inhab., as German is the principal tongue heard amongst the civil and military official circles. Some descendants of Hungarian families are found amongst the nobility of the N. circles, and the Jews, who are not very numerous, are said to descend from the exiles of that nation driven from Spain in 1492. Near Verliko and in other parts, Ciazari, or gipsies, are found. Even amongst the Slavonic inhabitants different tribes are distinguishable. The most backward, in point of civilisation, are the Mor-

lacchi, the mountaineers of the circles of Zara and Spalatro. They are addicted to a nomadic life, and wander about as shepherds, sleeping in summer in the open air. The comforts of the agriculturist and labourers are few, as is usually the case in warm climates; their houses are small and badly built, and furniture is nearly dispensed with. Fish and vegetables are the chief articles of nourishment, and both are abundant. The dress of the inhab. of the coast consists in blue tight pantaloons, a blue waistcoat, and in winter a capmet, with a coarse brown cloak shaped like that of the Italian boatmen. The mountaineers wear a linen dress in summer, and in winter throw their sheep-skins about their shoulders, which are proof against all the vicissitudes of the weather. The inhab. are generally active, courageous, and of quick perception; but, until they came under the Austrian sceptre, were not only neglected, but living on terms of constant warfare with their Mussulman neighbours, from which state of things the recent border feuds are an inheritance. The large knife and pistols which the Morlacchi still wear in their girdles, and the gun which the sharp-eyed sling over his shoulder from custom, remind the stranger no less strongly than the shaven heads of some of the mountaineers, of the affinity, in descent and in manners, existing between the Slavonic tribes that inhabit both sides of the mountains. The inhab. are Rom. Cath., except about one-fifth part who belong to the Greek church, and a few Jews and gipsies.

Dalmatia, like the other provs. of the Austrian empire, has a provincial diet or representative assembly, instituted by imperial diploma of Oct. 20, 1860, followed by the 'Patent' of Feb. 26, 1861. (See AUSTRIA.) Besides, certain of its towns and some districts, especially that of Pagliaza near Spalatro, retain their own jurisdiction, and the same privileges they possessed before their union with Austria. The highest authority in Dalmatia is the governor, who resides at Zara, the seat of the Gubernium. In this city the court of appeals and the highest criminal court are established, with dependent courts in the four circle towns, Zara, Spalatro, Ragusa, and Cattaro. Each circle has several districts, the chief magistrate in which is named pretor, and takes cognisance of judicial and police affairs, besides directing the rural economy of the district. The districts divide into greater and lesser parishes or communes under headboroughs (Capi rode and Podesta), who receive no salary, but are exempted from taxation, as are also the Serdari, a description of gens-d'armes, formed by the government out of the peasantry. The guarding of the frontiers towards Turkey is an important charge in Dalmatia, and a strict watch is also kept along the coast. For purposes of trade, 8 bazaars or markets are held on the frontier, and 7 casella, or parlatoria, at intervening stations. Lazaretos are established at Zara, Spalatro, Ragusa, and Castelnuovo.

Dalmatia formed, from the commencement of the 12th century down to 1419, a portion of the kingdom of Hungary; at the last-named epoch it passed under the sway of the Venetians, who had made themselves masters of Ragusa nearly 100 years previously. During the 16th and 17th centuries this country was the constant seat of wars between the Venetians and Turks, until it was finally conquered by the former, who held it till 1797, when it was ceded to Austria. In 1805, Austria gave up Dalmatia to the French, who incorporated it into the kingdom of Italy. Napoleon I. made it a duchy, and conferred the title of duke of Dalmatia on Marshal Soult. On the downfall of Napoleon it reverted to Austria.

DAMASCUS (called by the natives Es-Sham, an. Dimishk, Heb. Dammesek, (Greek Damaskos), a city of Syria, cap. of an important pachalic of the same name, and the virtual metropolis of Syria, in a plain at the E. foot of the Anti-Libanus, about 200 m. N. by W. Aleppo; lat. 33° 27' N., long. 36° 25' E. Pop. from 120,000 to 150,000, of whom 12,000 are Christians, and as many Jews. A splendid mosque of great antiquity, the construction of which is disputed by Christians and Mussulmans, is the chief architectural ornament. The form of the building (a cross), with a similarity in arrangement to the sacred edifices of Italy, seems to evince its Christian origin, while the abundance of Saracenic ornaments prove that the Arabs, if not its founders, have contributed extensively to its decoration. It is 630 ft. in length, by 150 in width; a fountain plays in the midst of a magnificent court, and the pillars and other ornaments are superb. A skull, said to be that of the Baptist, and his sepulchre, give such sanctity to this mosque, that it is death for even a Muhammedan to enter the room where the relics are kept. A Christian was formerly liable to the bastinado for merely looking into the court; and the western world is indebted for its knowledge of the interior of the building to the works of Ali Bey and Buckingham, who, in their character of Mussulmans, were allowed to inspect what no known Christian is permitted to approach. There are many other mosques. According to Ali Bey (ii. 266) and Addison (ii. 151), they are unworthy of notice; but Robinson (ii. 224) says they are only less splendid than those of Constantinople. The bazaars are extremely numerous, and well supplied with merchandise: but the private residences of the gentry are, after all, the most striking objects to a stranger, and for their exterior appearance, which presents nothing but a gloomy wall of mud, or sun-dried bricks, but for the combination of convenience, magnificence, and taste, which mark the interior arrangements, and realise all that can be imagined of eastern splendour. 200,000 piastres (2,000l.) is sometimes expended on the fittings up of a single apartment. There are 31 khans, or establishments for the reception of merchandise, and that of Hussein Pacha, built of alternate layers of black and white marble, with its fountain, arcades, and corridors, is a very beautiful and imposing object. A mosque of dancing dervishes deserves notice, less as one of the principal edifices of the town, than from the singular contrast in the occupations of its inmates, who, every Friday (the Mohammedan Sunday), pirouette and twirl themselves about from morning till night, while, during the other six days, they are industrious silk weavers. There are also Greek, Maronite, Syrian, and Armenian churches, 3 convents of Franciscan monks, and 8 Jewish synagogues. Hospitals numerous; the principal, in which great numbers of sick and lame poor are lodged and fed gratuitously, is a fine building, with a mosque belonging to it. There are about 70 large schools for children, a great number of smaller ones, beside which public lectures are given daily in the great mosque, and in some others, but education is confined to the religion and laws of Mohammed. The serai, or palace of the pasha, is a large fortified building in the centre of the city. The latter is surrounded by walls and towers, but they are in a half ruinous state, and pressed upon by extensive suburbs on every side. Damascus is essentially a commercial town; some hundred merchants are permanently settled in it; and there are great numbers of tanners, painters, potters, dyers of various stuffs, silk-winders, dealers in damask cloth, grocers, saddlers,

tent-makers, coppersmiths, ironmongers, farriers, furriers, bakers, millers, and other artisans and traders. There are also a certain number of armourers, and though the ancient celebrity of Damascus sabres has very much declined, they still bear a good name. Saddlery, cabinet work, jewellery, and silk, are now the staple manufactures. Foreign trade is carried on, by the great Mecca caravan, which, in favourable times, departs once a year; the Bagdad caravan, which usually performs two or three journeys a year; the Aleppo caravan, two or three times a month; and by several small caravans to Beirout, Tripoli, Acre, &c., which arrive and depart daily. Beirout is reckoned the port of Damascus. This city is watered by two rivers, the Barrada and Fichre, which, after uniting, divide again into seven branches, again reunite, and finally deposit their waters in a lake (Lake of the Meadow), which has no outlet. This abundant supply and natural diffusion of water has rendered the neighbourhood of Damascus very fertile. The inhabitants do not remember a year of scarcity; wheat, barley, hemp, with every kind and variety of fruit, are produced in almost unlimited abundance, and the gardens, or enclosures, form a forest of trees, and a labyrinth of hedges, walls, and ditches, of more than 21 m. in circ. The natives speak with delight of the beauty of their home, especially as seen from the hills behind Salahieh, a large village on the N.; but, according to Dr. Richardson (ii. 481), the scenery is inferior to that seen from the summits of Highgate, Hampstead, and Richmond hills. The climate of Damascus is mild; the summits of the Anti-Libanus are covered with perpetual snow, which sometimes falls in the city. The people are said to enjoy good health, but blindness is frightfully prevalent, and leprosy, fever, and dropsy, are common. The plague, however, is almost unknown, and the ordinary duration of life is said to be from seventy to eighty years, but that, no doubt, is exaggerated. Damascus is very ancient: it is mentioned in Gen. xiv. 15, as existing 1918 years B.C., and was then, as subsequently, probably the capital of an independent Syrian kingdom. It was subdued by David (2 Sam. viii. 6), but recovered its independence, if not earlier, at least during the reign of Solomon. (1 Kings xi. 24.) It then became the capital of the kingdom of Ben-hadad and his successors (1 Kings xv. 18), and remained so till its subjugation by Tiglath-Pileser, about 740 B.C., a little before the downfall of its rival Samaria. (2 Kings xvi. 9.) From this time it followed the fortunes of the rest of Syria, falling successively under the power of the Persians, Greeks, and Romans. As a Roman city it attained great eminence, and figures very conspicuously in the history of the apostle Paul. (Acts ix.) Damascus was taken by the Saracens in 632, after a siege of seven months, and was for many years the cap. of the khalifate. It was unsuccessfully besieged by the Crusaders in 1148, captured by Timour Bec or Tamerlane in 1400, and destroyed by an accidental fire in the following year. In 1516 it fell into the hands of the Turks, who retained it till 1832, when it was captured by Ibrahim Pacha of Egypt. Damascus is remarkable as being the only city of the East which has not dwindled from its former greatness. Its pop. seems to be as great now as ever; while Babylon, Nineveh, and Palmyra have wholly vanished, and Antioch and Aleppo are but the shadows of their ancient glory. Damascus is one of the sacred cities of the Mahometans, and its inhab. had formerly the character of being the most intolerant and fanatical of all the prophet's followers. Till within the last thirty years, the appearance of a Frank costume was the signal for a riot. Christians and Jews were alike prohibited from riding any beast but an ass (in 1807 even this was forbidden); and the appointment of an English consul in 1831 caused an insurrection, which lasted several months. The conquests of Ibrahim Pacha, however, produced a great change, if not in the feelings of the people, at least in their mode of exhibiting them. Christians of all sects and Jews now walk in procession, openly rejoicing in the avowed protection of the present government, exposed only to the impotent threats of those who, retaining the will, have lost the power to annoy them. In spite, however, of their general intolerance, most travellers bear honourable testimony to the hospitality of the Damascenes. (For further accounts of Damascus, see William of Tyre; Addrisonius, Tav. Nouv.; Abul-Feda, Tab. Syr.; Maundrell, and Volney.)

DAMAUN, a marit. town of Hindostan, prov. Gujerat, belonging to the Portuguese, 82 m. N. Bombay, and 45 m. SSW. Surat; lat. 20° 25′ N., long. 72° 56′ E. Pop. estimated at 7,000. The town stands on the banks of a small river, which in spring tides, during the SW. monsoon, has from 18 to 20 ft. water. The buildings are mostly whitened, and give it a handsome appearance from the sea; its walls are incapable of defence, and its streets narrow and dirty. It contains several churches and convents, and a Parsee temple, in which it is affirmed a sacred flame brought from Persia has been kept up for 1,200 years. It has a roadstead, where vessels lie 3 m. off shore in 3 fathoms water. Damaun is most celebrated for its docks and ship-building: its ships wear well, and sail well before the wind, but some time since they were too short for their breadth, so that they laboured in a head sea. Damaun was taken by the Portuguese in 1531, and has belonged to them ever since.

DAMAUN, a large distr. of Affghanistan, now subordinate to the Maharajah of the Punjab, but formerly belonging to Caubul; between lat. 31° and 34° N., and long. 69° 30′ and 71° E., bounded N. by Sungur, in Sinde, W. by the Sulimano Mountains, N. by the salt range diverging from the latter, and E. by the Indus. Along the banks of the latter the country is a plain tract of grass, the soil apparently composed of the slime deposited by the river, by which it is regularly inundated; in the N. parts, especially, a good deal of this flat ground is overspread with low, thick tamarisk jungles, abounding in wild bears, hog, deer, and game of all sorts. Round the villages large woods of date trees are often seen, but no other trees of any size; where there is cultivation the country is rich, but by far the greater part of it is waste. The central parts are composed of arid sandy plains, divided by hill-ranges, and depending entirely upon rain for cultivation: the more uneven country skirting the W. mountains is more fertile, and produces wheat, bajree, jowaree, and other Indian grains. The winter in Damaun is cooler than in Hindostan, but the heat of summer is extreme. This distr. is inhabited by various turbulent clans, principally Jats and Beluches, living in perpetual contention with each other, and who, having been at a distance from the seat of government, had never rendered much more than a nominal obedience to the Caubul sovereign. Some of the Damaun tribes are nomadic, others fixed agriculturists, and many are shepherds, the country in many parts yielding good pasture land.

DAMIETTA, a town of Lower Egypt, the third in rank, pop., and importance in the country, on the E. bank of the branch of the Nile bearing its name, 6 m. N. from its mouth (the anc. *Phatnitium*

Ostium), 80 m. E. Rosetta, and 97 m. NNE. Cairo. Lat. 31° 25′ 45″ N., long. 31° 49′ 30″ E. Pop. estimated at 30,000; but this is probably much overrated. The inhab. are principally natives of Egypt, with a few Syrians and Levant Greeks. A bend in the river gives to the town a somewhat crescent shape. It is irregularly and ill built; though there are some good mosques, several bazaars, and some marble baths. Some of the latter sort of houses, which are of brick, have terraces and pavilions; and such as are near the Nile, have little ports, whence to embark on the water; but there are no open spaces, nor buildings, worthy of much notice, and, generally speaking, it is but a collection of miserable mud hovels. There is a school for industry officers, with 100 pupils; as well as an extensive collection of buildings for drying, husking, and cleaning rice, some mills, and a cotton factory. The latter supplies a great deal of coarse cotton cloth, which forms the wear of the labouring classes. The bar at the mouth of this branch of the Nile prevents the access of any large vessels to the town; so that merchant ships have to lie outside the bar, and load and unload by means of small Greek craft, Egyptian *djerms*, and other vessels of from 30 to 60 tons burthen. But, despite these difficulties, Damietta has a considerable trade. Its chief article of export is the rice grown in its neighbourhood, which is the best in Egypt. Dried fish of the Lake Menzaleh, dates from the numerous plantations round the town, with coffee, beans, and linen, are the other principal articles of export. Most European nations have vice-consuls here. It has a governor, and a municipal administration similar to that of Cairo and Alexandria.

DANTZIC (Germ. *Danzig*; Pol. *Gdansk*), an important commercial city, sea-port, and stronghold of the Prussian states, prov. Prussia Proper, cap. reg. and circ. of same name, on the left bank of the Vistula, about 5 m. from its mouth on a branch of the railway from Berlin to Königsberg. Pop. 82,765 in 1861, excl. garrison of 10,843. The city is traversed by the small rivers Motlau and Radaune, and is very strongly fortified. It is ill built, and the streets are narrow, irregular, and gloomy. The cathedral church of St. Mary is the principal public building; it was finished in 1503, and has a fine brass font and a magnificent picture of the last Judgment. The town-house, arsenal, and the *Artushof* or exchange, also deserve notice. There are 14 Lutheran churches and chapels, 4 Catholic churches, and a chapel, 2 synagogues, and an English church, with several monasteries and convents. The town has also a gymnasium, two grammar-schools, and many inferior schools, with schools of navigation, midwifery, and commerce; a school of arts and trades, a good public library, an observatory, a museum, a society of natural philosophy, an orphan and foundling hospital, a large workhouse, and various hospitals. Dantzic is the seat of the provincial authorities, of a court of appeal for the circle, a council of admiralty, and a tribunal of commerce. It has a vast number of distilleries and breweries, the latter of which produce the black-beer in such general demand; it has also large establishments for grinding flour, with dye-works, sugar-refineries, and manufactures of fire-arms, tobacco, silks, vitriol, &c., and some jewellery business. The harbour, called *Neufahrwasser*, is at the mouth of the river; but vessels drawing 8 or 9 ft. come up to the city, being the emporium of the extensive and fruitful countries traversed by the Vistula and its affluents. Dantzic has a very extensive commerce; and is, after Odessa, at the head of all the corn-shipping ports, not of Europe only, but of the world. Wheat forms the principal article of export; it is of the last quality, and very large quantities are exported, as many as 500,000 quarters having been shipped in a single year. There is also a large exportation of flour, rye, barley, pease, and oats, with timber inferior only to that of Memel, linseed and rapeseed, staves, pearl ashes, bones, zinc, flax and hemp, linens, leathers, beer and spirits. The subjoined table—compiled from the official report of Mr. Lowther, H. M.'s secretary of embassy, dated Berlin, July 28, 1864—shows the exports of Dantzig during the year 1863, the first column giving the total exports, and the second the exports to the United Kingdom:—

Exports in 1863		Total Quantities	To the United Kingdom
Refuse Bones, &c.	quantity	14,979	14,979
Raw Iron	,,	22,113	—
Wrought Iron	,,	35,379	86
Rails for Railways	,,	67,363	—
Iron and Steel Goods	,,	31,463	15
Corn—Wheat	scheffel	3,645,407	2,199,974
Rye	,,	7,747,329	89,150
Barley	,,	376,852	270,107
Oats	,,	8,509	1,525
Beans and Peas	,,	874,452	641,848
Linseed	centner	35,350	30,853
Wood, Masts and Bowsprits	,,	6,467	663
Beams and Blocks of Hard Wood	,,	63,891	41,573
Do. of Soft Wood	,,	767,963	716,779
Sleepers	,,	1,271,968	957,394
Laths	,,	19,741	10,324
Staves	,,	16,476	7,154
Boards, Lath Wood	,,	6,192	4,491
Matting	,,	10,271	6,451
Mill Utensils	,,	17,763	—
Soda	,,	81,564	—
Coal	,,	188,517	14,012

The principal articles of import consist of woollens, cottons, and other manufactured goods, colonial produce, dye-stuffs, wine, oil, spice, fruit, salt, and coals. The importation of the last-named article from Great Britain is increasing from year to year. In 1863, the imports of coal from this country amounted to 2,184,818 centner, or about 110,000 tons.

The harbour accommodation of Dantzig is very good. The usual depth of water at the river's mouth is from 13 to 14 ft.; but in the roads, which are protected by the long, low, narrow tongue of land called the Heel, there is good anchorage for ships of any burden. The greater part of the trade of Dantzic is in the hands of foreigners, particularly English. The granaries for storing the corn brought down the Vistula are generally seven stories high; and these, with the warehouses for linens, ashes, hemp, &c., are all situated on a small island surrounded by the Motlau.

Dantzic was founded in the 10th century. It was occupied by the Knights of the Teutonic order in 1310, and was held by them till 1454, when it emancipated itself from their yoke, and became a free independent state, under the protection of Poland. For a lengthened period Dantzic was a principal member of the Hanseatic Confederacy, and had under it several other cities. During its independence, the citizens were engaged in frequent contests with the Poles, Swedes, and Russians; and notwithstanding the protection of England, Holland, and Prussia, Peter the Great exacted from them considerable contributions. The pretension of Dantzic to the exclusive navigation of the Vistula, or to demand a toll from such ships as passed in and out of the river, was at all times submitted to with reluctance. After the first partition of Poland in 1771, Frederick the Great, having acquired a large accession of territory on the

Vistula, approaching almost to the gates of Dantzic, claimed for his subjects the right of free navigation on the river. This having been refused by the citizens, gave rise to some acts of hostility, and to lengthened negotiations. These, however, were cut short in 1793 by the second partition of Poland, when Dantzic was assigned to Prussia. During the invasion of France, the city was occupied for several years by a French garrison and suffered much from the hostilities and exactions to which she was exposed; but since the peace of 1815 she has recovered much of her ancient prosperity. The fortifications have been also greatly strengthened and improved, and magnificent works have been constructed, by which the whole adjacent territory may be laid under water.

During the independence of Dantzic, there were attached to it the Werder, an alluvial island formed by the Vistula and the Motlau, and the Frische Nehrung, a long narrow tongue of land between the Frische Haff and the sea. The former is very fertile, but the latter consists principally of sand.

DANUBE (an. Danubius, and in the lower part of its course Ister, Germ. Donau, Hung. Duna), a celebrated river of Central and S.E. Europe, being, though inferior in point of size to the Wolga, in every other respect the first among European rivers. Its general course is from W. to E.; it extends between long. 8° 10' and 29° 10' E., its extreme N. point of lat. being 49° 2', and its extreme S. point 43° 38' N. Its total course from its source to its mouths, on the W. shore of the Black Sea, is from 1,750 to 1,800 m.; during which it passes through the territories of Baden, Wurtemberg, Bavaria, and the Austrian empire, and divides Turkey from Wallachia, Moldavia, and Russia. It receives above 30 navigable and a vast number of inferior tributaries, the principal being the Iser, Inn, Drave, Save, Theiss, Morava, Sereth, and Pruth. The cities of Ulm, Ratisbon, Passau, Linz, Vienna, Presburg, Comorn, Gran, Waitzen, Buda, Pesth, Peterwardein, Neusatz, Semlin, Belgrade, Semendria, Widin, Nicopoli, Rustuk, Rasschuk, Silistria, Ibrahilov, and Galacz, are situated upon its banks.

The basin of the Danube and its tributaries has been estimated to comprise about 1-13th part of the entire surface of Europe. It is bounded N. by the Alps and the Balkans; and on the S. at first by the Black Forest and some minor Alpine ranges, and afterwards by the Bohemian Forest and Carpathian Mountains. It includes the plains of Bavaria, Hungary, and Turkey in Europe; and the course of the Danube has been generally considered under three grand divisions, each embracing one of these plains. As this division is not only natural but convenient, we shall adhere to it in the following statements.

The Danube originates in two streams, the Bregosch and the Brege, which have their source on the E. declivity of the Black Forest, in the grand duchy of Baden, in about 48° 10' N. lat. and 8° 15' E. long. These streams having united at Donaueschingen, where they are augmented by a spring sometimes regarded as the head of the river, the united stream takes the name of the Danube. It thence proceeds at first SE., but afterwards in a NE. direction as far as Ratisbon, near which city it attains its extreme N. lat. It then runs again in a SE. direction to about long. 13°, and from that point mostly E. to Vienna, where the first division of its course may be said to terminate. Within this division it receives on the right hand the streams of the Iller, Genz, Mindel, Lech, Isar, Inn, Traun, Ens, &c.; many of which are navigable for a considerable distance. Its affluents on the opposite side are, on the contrary, generally small; and indeed, throughout the whole upper half of its course, the principal tributaries of the Danube (excepting the Theiss) are from the S. or right side, while, in the lower division, those from the N. or left side are by far the most considerable. It receives, however, from the N. in the first division of its course, the Sala, Altmühl, Naab, and Regen, all of which are navigable streams. At its source the Danube is 2,175 ft. above the level of the sea, and runs through an alpine country to Ulm, where its elevation is 1,552 ft. From Donauwörth to Passau it traverses the Bavarian plain; its height above the sea being at the former 1,115 ft., and at the latter 816 ft. At Passau it leaves the Bavarian dom., and thence to Vienna, intersects a second mountainous region. At Linz its elevation is 755 ft., and at Vienna 512 ft. At Ulm, the Danube first becomes navigable for flat-bottomed vessels of from 60 to 100 tons burden, though its depth there measures little more than 7 ft., and its breadth little more than 100 ft. Through the Bavarian plain its average depth is 10 ft. This increases considerably when it becomes again enclosed between the mountains at Passau; but above Vienna its navigation is rendered difficult, not only by its general shallowness, but by its rapidity, and the frequent rocks, shoals, and whirlpools in its channel.

In the second division of its course, the Danube at first runs generally E. to Presburg, next through the lower Hungarian plain SE. to its confluence with the Raab, and then E. to Waitzen. At this point it turns S. through the great Hungarian plain, and runs parallel with the Theiss for nearly 2½° of lat. to its junction with the Drave, about lat. 45° 30′. Here it turns SE., in which general direction it continues to Orsova, where it leaves the Austrian dom.; the second division of its course terminating at the cataract or pass called the 'Iron Gate,' about 4 m. lower down. It is within this division that the Danube receives its largest and most important tributaries, including the Raab, Drave, Save, and Morava on its right, and the March, Waag, and Theiss on its left side. At Presburg, its waters are 331 ft. at Buda, 230 ft., and at Belgrade, 203 ft. above the level of the sea. From Vienna to the mouth of the Drave, the Danube runs through an expanse of plain country broken only in a few places, as at Presburg, Buda, and Waitzen. Near the latter it passes through a ravine formed in a chain of mountains, separating the two Hungarian plains. From its union with the Drave, its S. banks in Slavonia and Servia are usually mountainous, while its N. continue low and marshy as far as Moldova. Previously to its reaching Buda, it is about 700 yards wide; soon after passing that city it attains a width of upwards of 1,000 yards; and by the time it has arrived at Belgrade it is considerably more than ¾ of a mile across. (Dict. Geog.) From Vienna to Pesth, its bed is sprinkled with rocks, but they are not such impediments to navigation as in the upper portion of its course. Shifting sand banks, which prevail all down the river as far as Moldova, are greater obstacles; but when the water is tolerably high, they may generally be avoided by good pilotage. (Austria and the Austrians, i. 315.) At Gönyö, 70 m. above Pesth, the Danube first becomes navigable for vessels drawing more than 2 to 2½ ft. water. Near Moldova, a mountain range from the Balkhan, and another from the Carpathians, begin to confine the river on either side as far as Gladova in Servia. Throughout this distance, about 80 m., it is greatly contracted in width, abounds with rapids, and is beset with rocks. Near the termination of this defile, a short distance below

Orsova, is the famous pass of the 'Iron Gate' (Turk. Demir-Kapu), already alluded to. This is a gorge about 2,000 yards in length, enclosed on either side by a mountain of calcareous slate, a material very difficult to break or blast, through which the river rushes with great velocity, over an inclined plane, with a fall of about 15 ft. a mile. The rocks here divide it into three channels. The centre one is of considerable width, and vessels of 400 tons may pass down it, when the river is very full; the two others are but shallow; and that on the Wallachian or E. side is never used. According to Strabo (vii. 317), it was here that the Danubius ended, and the Ister commenced; but there is a great discrepancy as to this point among the ancient authorities.

In the third division of its course, the Danube runs at first generally S. by E. to Widin; thence its direction is mostly E. by S. to near Sistova, where it attains its most N. lat.; and from this point ENE. to Rassova. It then turns N. to Galacz, and finally runs from this town generally E. to its efflux in the Black Sea, about lat. 45°. As far as Galacz it forms the boundary between Turkey and Wallachia and Moldavia; and between Galacz and the sea it is the boundary between Russia and Turkey, its principal N. and central mouths being included within the Russian territory. While the Danube is running S. by E., its right bank is mountainous, but the elevated lands soon afterwards recede from its banks, and throughout the rest of its course the river flows through a low plain, which E. of Silistria becomes marshy. In this division it receives on its left side the Schyl, Aluta, Vede, Argis, Jalomnitza, Sereth, and Pruth. Its affluents on the opposite side are much less considerable; the principal are the Isker, Osma, Tatam, &c. In its progress through Turkey, the Danube varies in breadth from 1,400 to 2,100 yards; and its average depth is upwards of 20 ft. Ships of large size ascend as far as Silistria. About 50 m. from the Black Sea, it divides into three principal arms, besides giving origin to a considerable lake (Rassein) on its S. side, from which several minor arms proceed. The delta of the Danube is a vast swampy flat, interspersed with lagoons covered with bulrushes, the resort of vast flocks of water fowl. The N. principal arm of the river (Kilia) and the S. (Ederilis), which form the boundary between the Russian and Turkish dominions, are shallow and of little value; but the middle one (Sulineh) has from 10 to 12 ft. water over the bar at its mouth. This is said, however, to be gradually filling up from the deposit of mud brought down by the river, which the current has not sufficient strength to clear away, its fall and rapidity being very much diminished during the last 200 m. of its course.

Were it not for the rapids between Moldova and Gladova, the Danube would be at all times navigable from Ulm to its mouth. Great efforts have been made at various periods to overcome this interruption. The Roman emperor Trajan constructed, with great labour and sagacity, a road along the edge of the Servian side of the river, to facilitate the towing of ships against the current. Some remains of this extraordinary work still exist, with part of an inscription in honour of Trajan. In more recent times, attempts have been made to deepen the channel of the river, and to cut lateral canals in the most dangerous places; but these, owing to the almost insuperable obstacles to be overcome, have had but little success. Looking at the map, the best way would appear to be to cut a navigable canal from opposite Moldova to Herra Palanka, below the 'Iron Gate,' which would not only avoid the rapids, but shorten the distance, by avoiding the great bend of the river by Orsova. But the nature of the ground is said to oppose insurmountable obstacles to such a project, though prudently it would admit of the construction of a road, or, better still, a railway. The Hungarian government has constructed an excellent and very expensive road from Moldova to Orsova, along the left bank of the river. Unfortunately it terminates above the 'Iron Gate;' and passengers going down the river, unless when it is sufficiently high to admit of flat-bottomed boats going through the 'gate,' have to be ferried over to the Servian side of the river, where, after a land journey of about 8 m., they re-embark. Those ascending the river have also to cross at Orsova.

The Danube abounds with islands. They are especially numerous and large in the middle part of its course. The Great Schütt island extends between two arms of the river, from Presburg to Comorn, a distance of 64 m. The Crepel and Margitta islands, below Buda, formed in a similar way, are also of considerable size. The Danube has been said to wind more than other European rivers; this is peculiarly the case in its progress N. through the great Hungarian plain. It is also one of the swiftest rivers in Europe; its rapidity is such as in some places to render any navigation against its current impossible, except by the agency of steam. According to Mr. Quin (Steam Navigation, i. 216) it rushes through the 'Iron Gate,' at the rate of not less than 8 m. an hour; but it is clear that the velocity must vary materially with the volume of water. This rapidity for a long time prevented the erection of any stone bridges up the Danube below Ratisbon; nor was there a permanent bridge of any other kind below Linz previously to the commencement of that constructed at Buda. There are flying bridges at Presburg and Comorn, and bridges of boats at Pesth and Peterwardein; beyond the latter place no direct communication between the opposite banks exists. In antiquity, however, it was very different. About 5 m. below Gladova, Trajan constructed his famous bridge, the remains of which are still visible, and form one of the most interesting and remarkable monuments of the most brilliant era of Imperial Rome. This great structure consisted of 20 or 22 stone piers, with wooden arches. The greatest depth of the river is here 18 ft., and the length of the bridge between the pillars or buttresses that still remain on either bank was about 3,100 English feet. But the breadth of the river is less than this; and at present does not exceed 2,800 feet. This neighbourhood of Gladova is one of the widest parts of the river; and was no doubt selected for the site of the bridge partly on account of the ample channel that was thus afforded to carry off the sudden floods to which the river is subject; its bed is here also sound, and its depth less than in most other parts. When lowest, the heads of some of the piers are seen above the surface of the water. The noble work was destroyed by Adrian, the successor of Trajan, lest the barbarians should overpower the Roman troops in Dacia, and make use of the bridge to invade the empire. (Eutrop. in Adrian.) But it was not Adrian, but Aurelius, who abandoned Dacia.

The steam navigation of the Danube is of paramount importance. This undertaking was first actively commenced by Count Szechenyi, who, in 1830, established a joint stock company for the purpose, of which he was the managing director. The Austrian government soon afterwards took up the scheme, greatly enlarged the plans of the company, granted it a charter for the exclusive navigation of the river for a number of years, and accorded it the privilege of drawing, gratuitously, the necessary supplies of coal from the imperial mines of Moldova, on the banks of the river. The first steamboat was launched on the Danube, at Vienna, in 1830. The enterprise proved most successful, and led to the formation of several other establishments of the same nature after the monopoly of the first company had ceased. The barges and ordinary packet-boats on the Danube are unwieldy flat-bottomed boats, covered with sheds of rough planks; the rafts in use are large and clumsy fabrics of the rudest kind; sails are unknown on the Upper Danube; and the boats are steered only by paddles.

So far back as the 8th century Charlemagne contemplated uniting the Danube and the Rhine by means of a canal; and the remains of a work commenced with that view are still visible at Weissenberg. After the lapse of more than 1,000 years, an undertaking of a similar kind was commenced under the auspices of the Bavarian government, and completed in the reign of King Ludwig I., after whom it was named the Ludwig's Canal. The canal commences at Bamberg, on the Maine, and runs in a slight curve, by way of Forchheim and Erlangen to Nuremberg, and from thence, in a larger curve, to Dietfurt, on the river Altmühl, where it ends, the Altmühl being a tributary of the Danube. The canal is from 34 to 54 ft. broad, and 5 ft. deep throughout. It has 69 locks, and on its highest point it is 630 ft. above the river Maine at Bamberg, and 770 ft. above the junction of the Altmühl and the Danube.

DARABJEND, a town of Persia, prov. Fars, 155 m. SE. by E. Shiraz. It is finely situated on the banks of a river, and in an extensive plain, surrounded with groves of orange and lemon trees, which yield much an abundance of fruit that the juice is exported to all parts of Persia. Though much fallen off from its former splendour, and partially in ruins, it has still a pop. of from 15,000 to 20,000. The culture of tobacco is here carried to a great extent.

DARDANELLES (an. *Hellespontus*), the narrow strait

'Longus in angustum qua cluditur Hellespontus.'

connecting the Sea of Marmora with the Ægean, and separating part of the SE. coast of Europe from the most W. part of Asia. Its modern name is derived from the castles, called the Dardanelles, built on its banks. In general direction is NE. and SW. Length about 40 m.; breadth unequal, but where least, not more than 1 m. across. Being, as it were, the key to Constantinople and the Black Sea from the W., this strait is pretty strongly fortified. The entrance is 2 m. wide, and defended by a fort on either side: that of the Asiatic coast (*Koum Kaleasi*) mounting 80 guns and 4 mortars, and that on the European side (*Sertil Bahr Kaleasi*) mounting 70 large guns and 4 mortars. The adjacent heights are also crowned with batteries, and about 3 m. above the New Castle of Europe there is one mounting 12 guns. Proceeding onward 12 m. above the New Castles, are the Dardanelles, or Old Castles of Europe and Asia; these defend the narrowest part of the strait, which is here only 1 m. wide. The Sultanieh Kaleasi, or Asiatic castle, is the strongest, and is the residence of the aga of a pasha, whose authority extends over the forts on both sides. It has 2 connected forts, and 192 guns, 18 of which are of the largest calibre. The European castle is built in the form of a crescent, and in 1832 was furnished with 64 guns; it has 2 collateral batteries recently built; the most N. of which mounts 48, and the N. 30 guns. 1½ m further on the Asiatic side is a battery of 46 guns; and 5 m. above the European castle is a battery called Kémuké Bourses, with 80 guns, near the small town of Maito, supposed to occupy the site of the ancient *Madytos*. The last forts on both sides are Borelli Kaleasi, on the site of the ancient *Sestos*, and *Nagara*, near *Abydos*, which see. The direct distance between them is about 1½ m. A strong current runs always from the Sea of Marmora, through the Dardanelles, at the rate of from 2 to 4 m. an hour, according to circumstances. The wind also generally sets in the same direction. There are shoals in some places; but deep water is everywhere to be found in some part of the channel. The Asiatic shore presents the most beautiful scenery; that of Europe is, on the contrary, generally steep and rugged. To each of the Dardanelles a town is attached: the Asiatic is the larger, and contains 2,000 houses; but the streets are narrow, ill paved, and dirty, and almost all the buildings are of wood. It has manufactures of pottery. Gallipoli is the principal town on this strait, which see. This strait has been famous from the remotest period. It derives its name from Helle, daughter of Athamas, king of Thebes, drowned in it, (Hygin, Poet. Astron., lib. ii. § 20.) It is also memorable as the scene of the death of Leander, and of the important rage of Xerxes, whose [illfated host crossed over it on a bridge of boats between Sestos and Abydos.

DARFUR, a country of Central Africa, between 11° and 16° N. lat., and 25° and 30° E. long. It lies between Bornou and Abyssinia; almost due N. from Egypt, and W. of Sennaar, whence it is separated by Kordofan. Standing, however, like an oasis in the midst of the Great Sahara desert, Darfûr is situated at a great distance from all the above-named territories. The country is of the most dreary character, without rivers, lakes, or much cultivable land, with a few mountains rising from its sandy plains.

Of the topography and real extent of Darfûr we possess but limited information. The principal town appears to be Cobbé, in lat. 14° 11′, and long. 28° N., which is 2 m. in length, from N. to S., but very narrow, each house being separated from the others by a cultivated enclosure. The inhab. are supplied with water from shallow wells dug, in most instances, beside their houses, but so unskilfully that the soil often collapses, and the same well is seldom of use longer than four months at a time. This place is chiefly inhabited by merchants, and from it a caravan starts at irregular intervals to Cairo. 5,000 persons are said to reside at Cobbé. A neighbouring village, called El Fasher, is the residence of the sultan and his court. Sweini, another Fûrian town, lies almost N. of Cobbé, at the distance of about two days' diligent travelling, and in the direct road to Egypt; hence it is principally resorted to by merchants. Its environs are more fertile than those of Cobbé, and when the *jelabs* (traders) remain there, it boasts of a daily market. Cubcabia, due W. from Cobbé, at a distance of 2½ days, is a more considerable place, being the depôt of merchandise brought from the W. It has also a manufactory for leather and of *tobras*, a coarse cotton cloth from 5 to 9 yards long, and about 24 in. wide, which form the covering of all the lower class of both sexes. The other towns are Hil, Conr, Shoba, Gidid, and Gellé. (Browne's Travels, pp. 268–276.)

The inhab. of Darfûr, which have been generally estimated not to exceed 200,000 in number, Dr. Henry Barth says (Journal of Royal Geographical Society, 1856, xiii. p. 123), 'and more than 1,000,000 inhabitants, and perhaps much less' —

are a mixture of Arabs and Negroes. They are governed by a sultan, whose power is not altogether absolute, he being, in some degree, amenable to the *ulema*, or ecclesiastics, and frequently standing in some awe of his own troops. His power is delegated to the *meeks*, to governors, called *meleks*. Though the Fûrians are bigoted Mohammedans, they do not abstain from intoxicating liquors; the crime of drunkenness, committed by means of a decoction of hemp, is frequent among them. Snuff and tobacco appear to be almost necessaries of their existence; but for the endurance of hunger and thirst they are unequalled even by the inhab. of surrounding arid regions, among whom such a qualification is an essential. They are not remarkably cleanly in their persons; and, having no baths, rub their bodies with a kind of farinaceous paste as a substitute. The Fûrians are, unlike other Moslems, jovial, and even licentious, in their manners, and are particularly fond of dancing, each tribe having a dance peculiar to itself. At Cobbé education is in some degree provided for by four or five *meetebs* (schools), where reading and writing are taught. A *kadum* also lectures occasionally on the Koran, and what they call *elm*, philosophy. The language is a dialect of the Arabic peculiar to the Fûrians.

Agriculture in Darfûr is at a very low ebb; indeed, the soil which was presented to Mr. Browne's observation, consisting of bare rocks, sand, a small portion of clay, and a still smaller part of vegetable mould, seemed to offer no encouragement to that respect. Entirely devoid of rivers or lakes, the country solely derives irrigation from heavy periodical rains, which are preserved in numerous water-courses. At the commencement of the farmer digs innumerable holes in his fields, into which he throws the seed, and covering it over with his foot, leaves it without further care until the grain becomes ripe. (Ibid. p. 291.) The harvest is gathered by women and slaves, who break off the ears with their hands; so that the farming implements of the Fûrians are few and rude. The grains chiefly raised are wheat, *dokn* (Holcus dochna, Forskkal), *kreseb*, and *simsim* (*simsim*, Arabic term); the pulse consists of kidney-beans, a bean called *fûl*, and another denominated *sheû*, together with other leguminous plants peculiar to that part of Africa. The occasional drought is not favourable to water-melons, though many are grown. Tamarinds, dates of an inferior quality, the *Rhamnus mictorus* of Forskkal, and tobacco, which is said to be indigenous, are all cultivated in Darfûr. (Browne, pp. 306–313.)

Commerce.—Although the Fûrians have but a limited variety of articles to exchange for those necessaries of life which their own country does not produce, yet commerce, from their central situation, affords the chief means of support to the nation. Many of their towns are entirely peopled by merchants. The caravans from Egypt, Sennaar, &c., are laden with jewellery, swords, firearms, coffee, raw and manufactured silks, shoes, writing paper, Syrian soap, French and Egyptian cloths, with Indian muslins and cottons, wire, brass, silver, &c. For these the Fûrians give in exchange slaves, camels, ivory, ostrich feathers, gum, pimento, tamarinds, leather sacks for water (*rag*), others for dry articles (*gerams*), parroquets, monkeys, and guinea fowls. (Browne, pp. 346, 349.)

The *climate* of Darfûr is chiefly influenced by the perennial rains, which fall from the middle of June till September with frequency and violence, and suddenly invest the face of the country, till then dry and sterile, with a delightful verdure. July appears to be the hottest month, &c. according to Browne's meteorological journal, kept during the years 1794–5, the thermometer never sank below 80° at 3 P.M., but more frequently rose to 90°. In the April of 1794, however, it ranged from 84° to 101°, while the same month of the succeeding year exhibits an average far below that of either of the July months. The thermometer seldom sank, according to Browne's register, lower at 9 P.M. than 70°, or at 7 A.M. below 56°, which happened most frequently in February; December and January, also, exhibit low degrees. N. and N.W. winds are those which blow with the greatest frequency over Darfur. (Appendix to Browne's Travels, pp. 501–508.)

Among the animals to be found in Darfur are horses, of which there are but many; sheep, which also are scarce, yield meat of a poor quality; goats are more numerous; but horned cattle form the chief wealth of the Furians, as in the more S. African nations. The milk of the cows is not very palatable; but the beef is good. Camels of every variety of breed are exceedingly numerous; but the *Gerab* camel is much subject to the mange; the males are sometimes castrated. Dogs are employed both in hunting the antelope and for guarding sheep; the household cat is also met with. The wild animals are the lion, leopard, wolf, jackal, wild buffalo, &c. Elephants assemble in large herds of four or five hundred; though they are much smaller than the Asiatic elephant, the animal is a source of great profit to the Furians, who make a lucrative sale of his tusks, hold his flesh in great esteem as food, and manufacture the fat into a much-used unguent. Several sorts of monkeys, and the civet-cat, are also mentioned by Browne. Ostriches, vultures, parroquets, partridges, pigeons, and quails were also seen by him. Locusts, horned serpents, mosquitos, and white ants, infest the country in large numbers. (Travels, pp. 293–304.) Of the minerals found in Darfur, the best is copper; but iron is produced in the greatest abundance, and is formed into domestic utensils and arms. All the silver, lead, and tin is brought from Egypt. The other geological features of Darfur are scarcely known.

DARIEN. See PANAMA (ISTHMUS OF).

DARLINGTON, a market-town and bor. of England, co. Durham, Darlington Ward, S. div., on the Skerne, an affluent of the Tees, 241 m. N. by W. London by road, and 236 m. by Great Northern railway, *viâ* York. Pop. 15,789 in 1861. The town consists of several well-built and well-lighted streets, which branch out from a spacious market square. The river is crossed by a bridge of three arches. The church, formerly collegiate and dedicated to St. Cuthbert, was built about 1160; it has a fine tower and spire 180 ft. high. The Prim. and Wesl. Methodists, Independents, R. Catholics, and Soc. of Friends, have places of worship. A grammar-school was founded by Q. Eliz. in 1567, and a blue coat school by Lady Calverley in 1716. There are also Lancastrian, national, and Sunday schools, a dispensary, lying-in charity, and two alms-houses. It is a bor. by prescription, governed by a bailiff, who holds a court twice a year for the manor of Bondgate, and a bor. court also twice a year, in both of which debts under 40s. are recoverable. Petty sessions are held on alternate Mondays in the town-hall, a neat building; having a house of correction connected with it. The election for members for the S. division of the county is held here. The manufacture of linen, which was formerly carried on to such an extent as to give employment to 500 looms, has declined, but it is still pretty considerable. A great many persons are also employed in wool-combing; and there are several tan-yards, rope-walks, breweries, and iron and brass works. The Stockton and Darlington

railway, one of the first constructed in the kingdom, commences at Witton Park Colliery, near W. Auckland, and proceeds by Darlington and Yarm to Stockton, a distance of 24½ m. (For an interesting account of this railway, see Smiles, Samuel, 'Lives of the Engineers.') Darlington has cattle markets, on alternate Mondays. Fairs on the 1st Monday in March, Easter and Whit-Monday, and 10th Oct.; statute fairs on 19th May and 23rd Nov.

DARMSTADT, a town of W. Germany, cap. of the grand duchy of Hesse-Darmstadt, seat of the gov. and residence of the sovereign, prov. Starkenberg, in the great Rhenish plain near the N.W. extremity of the Odenwald, and on the *Bergstrasse*, or high road between Frankfort-on-the-Maine and Heidelberg (see HESSE-DARMSTADT), 17 m. S. the former city, 54 m. N. by E. Carlsruhe, and 8 m. E. by N. the Rhine, on the railway from Frankfort to Heidelberg. Pop. 29,526 in 1861. The town is rather dull, has little trade, nor, for a capital, does it present much deserving of notice. It consists of an old and a new town; both encircled by walls: the former is ill built, and its streets are narrow and dark; while the latter has broad, straight, and handsome streets, and good houses, many of which stand singly. The town is well lighted at night. It has four suburbs, six entrance-gates, three of which are handsome structures, and about sixty public edifices. Amongst the latter are the opera-house, built in the Italian style, and 270 (Rhenish) ft. in length, by 150 ft. broad. The riding-school, converted into a depôt for artillery. 319 ft. in length, by 157 ft. in breadth, is another conspicuous object. The grand-duke resides in a new palace of no great architectural pretensions. The old ducal palace, surrounded by a dry ditch, which has been changed into a shrubbery and garden, is a structure of the various ages from the 16th to the 18th century, and contains a picture-gallery with about 600 paintings, mostly second-rate, a museum of natural history with some valuable fossils, a museum of ancient and modern sculpture, a hall of antiquities, a collection of cork models, armoury, and a library of 120,000 vols. open to the public. The remaining principal public buildings are—the palaces of the hereditary prince and the Landgrave Christian; the Catholic church, a brick edifice, the interior of which is an elegant and imposing rotunda, 178 ft. in diameter, 121 ft. in height, and surmounted by pillars 50 ft. high; the Casino, in which the chambers of the duchy meet; the military hospital, royal stables, and orphan asylum.

Darmstadt is the seat of the high court of appeal for the grand duchy, and various other judicial tribunals and government offices. It has a gymnasium, a teachers' academy, a practical school of arts and sciences (*Realschule*), schools of artillery and military duty, and of sculpture and drawing. It has manufactures of tobacco, wax-candles, carpets, silver articles, coloured paper, cards, and starch. The majority of the inhab. depend, however, for subsistence on the presence and expenditure of the court. Scarcely any but military garments are seen in the streets, even the teachers of the public schools being obliged to dress in uniform, or court livery.

DARTFORD, a town and par. of England, co. Kent, lathe Sutton-at-Hone, hund. Axton, Dartford, and Wilmington; on the Darent, about 1 m. from its embouchure in the Thames, 15 m. ESE. London by road, and 17 m. by North Kent, or South Eastern railway. Pop. of town, 5,314, and of par., 6,597 in 1861. The town, situated in a narrow valley, consists chiefly of one main street, along the ancient high road from London to Dover,

and of two smaller ones branching from it. The
river is crossed, at the E. end of the town, by a
bridge of the era of Edw. III., widened and repaired in the last century. The church is a large
structure, with two burial-yards, one surrounding
it, the other on the summit overlooking its tower.
There are several dissenting chapels; a free grammar-school, founded in 1676, for eight boys; a
national school, and two sets of almshouses. There
is a co. bridewell near the town, and sessions for
the upper div. of the lathe are held in it. During
the reign of Elizabeth, the co. assizes were frequently held here; and at present a court of requests for debts under 5l., whose jurisdiction extends
over the town of Gravesend and four adjoining
hundreds. Market, Saturday; fair, August 2, for
horses and cattle. The chief business of the town
is caused by the numerous large gunpowder, paper,
oil, and flour mills on the Darent: there is also a
large steam-engine manufactory, and a foundry
connected with it. The river is navigable for boats
to the town, where there is a small wharf, used
chiefly by the colliers which supply the neighbouring factories. The Roman Watling Street is traceable near the town. In one of the chalk hills
between which it stands are several ancient excavations, supposed to have been scooped out for granaries during the Saxon period. There are some remains of an Augustine nunnery, subsequently made
a royal residence by Henry VIII. and by Elizabeth.
Dartford was the scene of the insurrection headed
by Wat Tyler, who, being a blacksmith in the
town, killed the poll-tax collector by a blow of his
hammer, for an insult offered to his daughter.

DARTMOOR. (See ENGLAND.)

DARTMOUTH, a parl. bor., town, and sea-port
of England, co. Devon, hund. Coleridge; 170 m.
WSW. London by road, and 228 m. by Great
Western railway, rid Brixham Road station. Pop.
4,114 in 1861. The town is situated on the W.
bank of the estuary of the Dart, near its embouchure in the English Channel, where it forms
a spacious harbour, capable of containing several
hundred sail of vessels of the largest size. The
entrance to the harbour is narrow, and protected
by a battery on its W. side, on the site of an
ancient castle, from which to a castle on the opposite bank (now in ruins) a chain used to be extended for the purpose of defence. The streets,
which are narrow and irregular, rise from the
margin of the river, and parallel with it, one over
another, along a steep acclivity, being mostly connected by flights of steps; houses mostly antique,
with projecting upper stories; the whole is paved,
well supplied with water, and partially lighted
with gas. There are three principal churches:—
St. Saviour's, built 1372, a curious old structure,
usually called the Mayor's Chapel; Town-hall
Chapel, on the summit beyond the town, with a
tower forming a sea-mark; and St. Petrock's, adjoining the battery at the entrance to the harbour.
There are also several dissenting chapels; two sets
of almshouses, one of which, founded 1671, is for
decayed mariners; and several minor charities.
There are large ship-docks, adapted for the repair
and building of vessels, and some activity in ship-
building. There are also establishments for sail
and rope-making, a spacious quay, and several
private wharfs. The exports consist chiefly of
woollen goods and cider, sent thither from the
interior, and shipped coast-wise; and of various
articles of general supply for the Labrador fisheries, in which several vessels belonging to the
port are directly engaged, though this trade has
greatly declined from its ancient importance.
There are regular steamers up the river to Totness.
There belonged to the port on the 1st of Jan. 1851,

166 sailing vessels of under 50, and 267 sailing
vessels of above 50 tons. There were also belonging to the port, at the same period, six small
steamers, of a total burthen of 154 tons. In ancient times, however, its mercantile marine was
comparatively much more considerable, as is evident from the fact of its having furnished 31 vessels and 757 seamen to the fleet of Edward III.
against Calais. The port is a bonding one, its
jurisdiction extending about 40 m. along the coast
(from the Teign to the Erme), and up the Dart to
Totness bridge (10 m.). The Dart is navigable
thus far for vessels of 150 tons, the channel having
been deepened and improved.

Dartmouth claims to be a bor. by prescription,
under the name of Clifton-Dartmouth. It regularly sent two members to the H. of C. from the
14th Edw. III. down to the Reform Act, which
deprived it of one member. The elective franchise
had been previously vested in the corporation and
in the freemen made by them, the inhab. of the
loc. not being entitled to their freedom in right of
birth, servitude, or residence. But the Reform
Act, besides giving the franchise to the 10l. householders, extended the limits of the bor. to the dimensions already stated. Registered electors, 255
in 1865. The municipal bor. is governed by a
mayor, four aldermen, and twelve councillors. The
income of the corporation, chiefly derived from
lands and houses, is about 1,100l. a year. The
scenery around Dartmouth is extremely picturesque. Flavel, an eminent Calvinistic writer, and
Newcomen, the inventor of the atmospheric engine,
were natives of this town; which also gives the
title of earl to the Legge family.

DAVENTRY, a loc. and par. of England, co.
Northampton, hund. Fawsley, 68 m. NW. London
by road, and 72 m. by London and North Western
railway, rid Weedon station, from which it is distant 4 m. Pop. 4,121 in 1861. The town is situated on the high road from London to Birmingham,
near the source of the Nen. It has a good modern
church, a free school, founded in 1576; five boys
are also educated by means of a legacy of Lord
Crew, bishop of Durham, and twelve at the expense of the corporation. The remains of a priory,
founded in 1090, are now occupied as dwellings by
the poor. Though incorporated at an early date,
the bor. does not appear ever to have been represented in the H. of C. On a neighbouring lofty
eminence, called Borough Hill, is an encampment
occupying the whole of the summit. A spring
rises in the outer ditch of the encampment, which,
according to Dr. Stukeley, is one of the highest
in England. (Stukeley's Itinerarium Curiosum,
II. 18.)

DAVID'S (ST.), a small decayed city of Wales,
co. Pembroke, hund. Dewisland, near the extreme
W. point of the principality, on a small stream
called the Allan, about 1 m. from the sea, and 16
m. NW. Milford Haven. The par., an extensive
one, had in 1861 a pop. of 2,350, of which the
'cathedral close' had 57. A bishopric was established here at a very early period; and to that
circumstance the place is most probably indebted
for its origin. The cathedral, the bishop's palace,
St. Mary's college, and other buildings appropriated to purposes connected with the establishment
and the residence of the clergy, are enclosed within
a lofty wall above 1,200 yards in circ. The cathedral, which occupies the site of one more ancient
destroyed by the Danes, was completed in the
reign of King John. It is a cruciform structure,
347 ft. in length within the walls, with a square
tower at the W. end; it has many interesting
monuments, but is, in great part, in ruins. The
bishop's palace, reckoned one of the most magni-

DAVIS'S STRAITS, the sea stretching NNW. and SSE., and uniting Baffin's Bay with the N. Atlantic ocean, having Greenland on its E. and Cumberland Island on its W. side. Where narrowest, under the Arctic circle, it is from 170 to 180 m. across; but its length is not accurately determined. It derives its name from Davis, by whom it was discovered between 1585 and 1587. Strong currents set towards the S. from this strait, which is also much encumbered with ice and icebergs. It has been for many years past the principal resort of the ships engaged in the N. whale fishery; the whales having been nearly exterminated in the seas round Spitzbergen, the original seat of the fishery. (See art. BAFFIN'S BAY.)

DAUPHINÉ, one of the provs. into which France was divided previously to the revolution. It is now distributed among the deps. of Isère, Drome, and Hautes A'pes.

DAX, AX, or **ACS**, a town of France, dép. Landes, cap. arrond., in a fertile plain on the Adour, 23 m. SW. Mont-de-Marsan, on the railway from Bordeaux to Bayonne. Pop. 5,846 in 1841. The town is well built, is surrounded by walls of Roman construction, and has an ancient episcopal palace, cathedral, hall of justice, and prison. Dax is, however, chiefly celebrated for its numerous hot saline springs, accounted efficacious in rheumatism and paralysis; and which being known to the Romans, they gave it the name of Aquae Augustae. The principal of three springs pours its waters into a large basin in the centre of the place, and the evaporation from it is so great, that in cool mornings the whole town is sometimes involved in a fog. There are several bathing establishments contiguous to the town. Dax communicates by a bridge across the Adour, with a suburb on the opposite side of the river. It has a tribunal of primary jurisdiction, a chamber of commerce, a communal college, and a theatre. Manufactures of earthenware, pitch, oil, thread, vinegar, leather, and some trade in corn, wine, brandy, and wool.

DEAD SEA (Lat. *Lacus Asphaltites*, Arab. *Bahr-el-Lout*), a lake of Palestine, celebrated in scripture history, in about 31½° N. lat., and 35° 40' E. long. Its dimensions have been variously stated, but it is probably about 40 m. in length, and 13 in extreme width. On the E. and W. it is bounded by lofty mountains; on the N. it opens to the plain of Jericho and the valley of the Jordan; on the S. the valley of El-Ghor extends, as if it were a continuation of its bed, though with a gradual rise, to the Gulf of Akabah. (See JORDAN.)

Nothing can be more dreary than the scenery around this famous lake: the temperature is very high; the soil, impregnated with salt, is without vegetation, the air is loaded with saline particles, and the bare crags of the surrounding mountains furnish no food for either beast or bird. Hence its neighbourhood is generally deserted by animated beings, and the dreary stillness of the place is increased by the nature of the lake itself. Intensely salt, its waters are not moved by a gentle breeze, and, owing to the bolderness of its basin, being seldom affected by a strong one. Its usual appearance is that of stagnation, agreeing well with the death-like stillness and desolation around.

This absence of life has given to the lake its popular designation of Dead Sea, and is the source of the common tradition that its waters are fatal to fish, and its exhalations to birds and other animals. This is, however, incorrect; birds fly over its surface uninjured; and Maundrell found upon its shores some shells, which seemed to imply that it was not altogether tenantless. The water is very limpid, but extremely bitter and numerous, the substances held in solution amounting to a fourth part of its whole weight:—

In 100 parts, as follows:—

Muriate of Lime	.	0·920
Magnesia	.	10·246
Soda	.	13·500
Sulphate of Lime	.	0·054
		24·880

It has also a strong petrifying quality, which accounts for the want of any great variety of fish; and it is peculiarly buoyant, though the assertion that nothing sinks within its bosom is wholly fabulous. Asphaltum (whence its classical name) floats in great quantities on its surface; and a bituminous stone, very inflammable, and capable of receiving a high polish, is found upon its shores.

The valley of the Jordan has been long known to be considerably depressed below the level of the ocean. This depression is, however, much greater than was formerly supposed. The Dead Sea is the lowest part of the valley; and its surface has recently been ascertained to be sunk above 1,300 ft. under the surface of the Mediterranean, being by far the greatest depression below the sea-level of which we have any authentic account. It consequently belongs to that class of lakes that have no visible outlets; it receives six streams besides the Jordan, but gives forth none; the surplus water being carried off by evaporation. Its depth, which varies in the dry and rainy seasons, exceeds, in some places, 360 fathoms; but towards its N. extremity it is so shallow as to be in parts fordable.

Its Arabic name, *Bahr-el-Lout* (Sea of Lot), refers to the connection between the history of this lake and that of the nephew of Abraham, in whose days its bed, or a portion thereof, the fertile valley of Siddim, contained, according to the sacred writer, 5 cities (Gen. xiv. 2); and according to Stephen of Byzantium (art. *Sodoma*) 10, and Strabo (xvi. cap. 2, 764), 13. In the visitation by which they were all destroyed, with the exception of Zoar, the neighbouring country underwent an extraordinary change; and is said by Moses (Deut. xxix. 23) to have become "a land of brimstone, and salt, and burning," characteristics by which it still continues to be marked. In Scripture this collection of water is called the Salt Sea (Gen. xiv. 3; Deut. iii. 17; Josh. xv. 5); the Sea of the Plain (Deut. iii. 17); and the East Sea. The best, as well as the most recent account of the Dead Sea, its geological formation and other features, is in a work by the Rev. H. B. Tristram, 'The Land of Israel; a Journal of Travels in Palestine, undertaken with special reference to its Physical Character,' p. 672. Lond. 1865.

DEAL, a parl. bor. and sea-port town of England, co. Kent, in the St. Augustine, hund. Bewsborough, 66 m. ESE. London by road, and 102 m. by South Eastern railway. Pop. 7,531 in 1861. The town is situated on the E. coast of Kent, opposite the Goodwin Sands, and about half way between Ramsgate and the S. Foreland. It consists of Upper, Middle, and Lower Deal. The latter, containing the great bulk of the pop., is built, principally in three parallel streets, close to the shingly beach, extending along the roadstead called the Downs. Streets mostly narrow and irregular, but paved and lighted. A row of houses connecting the lower with the upper village, constitutes Middle Deal; in these last the houses are detached, and are mostly occupied by the wealthier class. The par. church is in Upper Deal: there is a chapel of ease in the lower town, several dissenting chapels, and a national school. Walmer forms a continuation of Lower Deal, and owes its rise to the naval arsenal, hospital, and barracks, formed there during the last war. Since the Municipal Reform Act, it has been included in the bor. of Deal (of which it forms a ward); and the Reform Act conferred on both parishes, in conjunction with Sandwich, the privilege of returning two members, to the H. of C. Registered electors, 1,011 in 1862. Deal was annexed to the Cinque Ports soon after the Conquest; a decree exempting it from co. taxation shows it to have been so in 1229; a charter of 11th Wm. III. made it a bor. independent of Sandwich, Walmer included. It is now divided into three wards, governed by six aldermen and eighteen common-councilmen. There are no manufactures, the inhabitants being mostly shopkeepers, pilots, fishermen, and boatmen, mainly dependent on the resort of shipping to its famous roadstead, the Downs. The latter is a spacious and convenient anchorage, bounded seaward by the Goodwin Sands, and tolerably safe, except in heavy gales from the N. and E. Most outward and homeward-bound vessels touch here to take or land pilots, letters, and passengers. This business, however, has greatly fallen off since the last war with France, when the Downs was much resorted to by men-of-war and merchantmen waiting for convoy. The shipping belonging to the port of Deal consisted, on Jan. 1, 1864, of eight sailing vessels under 20, and three vessels over 20 tons; the total tonnage of the former being 132, and of the latter 302 tons. There were no steamers. Coals form almost the only article of import. Of late years, Walmer has been resorted to as a sea-bathing place, and there are several good lodging-houses for the reception of visitors during the season. Deal Castle, on the W. side of the town, is a round tower, built by Hen. VIII., with a moat and drawbridge. Deal is supposed to be the spot where Cæsar effected a landing on invading Great Britain.

DEBRECZIN, a town of Hungary, and, next to Pesth, the largest in the kingdom, cap. co. Bihar, in a flat, sandy, and arid plain, 114 m. E. Pesth, and 110 m. NW. Clausenburg, on the railway from Pesth to Kaschau. Pop. 37,850 in 1857. Debreczin is one of the most singular places in Europe. Notwithstanding its size, its general appearance is rather that of a large village than a town; and notwithstanding its manufactures and trade, both of which are considerable, none of the advantages ordinarily met with in large commercial cities are here to be found. Its

streets are broad, unpaved, and in rainy weather a mass of liquid mud. Scarcely any of the houses are above one story in height, and few are built on any regular plan. The greater part are thatched, which has rendered Debreczin subject at various times to severe ravages from fire. In the spring of 1811, not fewer than 2,000 habitations were reduced to ashes in the course of six hours. There are, however, five churches, three hospitals, two infirmaries, an orphan asylum, and a town-hall. The principal college of the Calvinists in Hungary, with a library of 30,000 vols., and upwards of 1,000 students, is at Debreczin. It has also a Piarist college, a Catholic high school, and a monastery. Shoes are manufactured in large quantities, as also tobacco-pipes, prepared sheep-skins, coarse woollen cloth, a spongy kind of soap greatly esteemed throughout the Austrian empire, with leather, furs, combs, coopers' and turnery wares. There is an extensive market for all three articles, as well as for oxen, sheep, horses, hogs, wheat, millet, wine, tobacco, water-melons, lead, wax, honey, and various other kinds of produce, especially at the fairs held at Debreczin every three months. On these occasions the country round the town is covered to an extent to which the eye can scarcely reach, with flocks and waggons, bales and cases, tents and huts, round which thousands of people are constantly gathered; presenting, in fact, all the appearance of an immense herd of nomades. A great deal of business is transacted at these fairs. Debreczin is, indeed, the great mart for the produce of the N. and E. parts of Hungary. By far the greater part of the pop. are Magyars; and it is here that the true Magyar character may be most advantageously studied. During the revolution of 1848-9, Debreczin became the last seat of the Hungarian parliament, but being an entirely open place, it was taken without resistance by the Austrian troops.

DECCAN (Daka-hina, the South), a term of Sanscrit origin, and formerly applied to the country comprising all that part of India to the S. of the Nerbudda river; but since the Mohammedan invasion, the term has been restricted so as to apply only to the countries between the Nerbudda and Krishna, that is, between the parallels of lat. 16° and 23° N., extending from the Arabian Sea to the Bay of Bengal, and including the provs. Candeish, Aurungabad, Beeder, Hyderabad, Bejapoor, Berar, Gundwanah, Orissa, and the N. Circars. British Deccan comprises the collectorates of Candeish, Ahmednuggur, Poonah, and Darwar, under the presidency of Bombay; and the ceded districts on the Nerbudda under the presidency of Bengal. The remainder of this region is mostly comprised within the dominions of the rajah of Berar, the nizam, the rajah of Nattarah, the Guicowar, and Scindia. (For further particulars, see the various provs., districts, and states referred to under their respective heads.)

DEE, a river of England, which has its source in Bala Lake, co. Merionethsh, N. Wales. At first it pursues an easterly course through the beautiful vale of Llangollen, till it passes Wynnstay. It then takes a northerly direction, and forms the line of demarcation between the cos. of Denbigh and Flint in Wales, and Cheshire in England. It nearly encompasses the ancient city of Chester, and is thence conveyed by an artificial channel, about 8 m. in length, to its spacious estuary on the Irish Sea. Its principal tributary is the Alwyn, which unites with it at Holt. Its estuary is much encumbered with sand banks.

The Dee is also the name of two considerable Scotch rivers, one of which falls into the N. Sea at

Aberdeen, and the other into the Irish Sea at the Little Ross, about 6 m. below Kirkcudbright. The latter is navigable as far as Tongland-bridge, 2 m. above Kirkcudbright, for vessels of large burden.

DELAWARE, one of the U. S. of America, and excepting Rhode Island, the smallest of the Union. It occupies a part of the peninsula, lying between the bays of Chesapeake and Delaware; extending from lat. 38° 30' to 39° 50' N., and long. 74° 55' to 75° 47' W.; having N. Pennsylvania, W. and S. Maryland, and E. Delaware bay and river. Length, N. to S., 95 m.; average breadth about 23 m. Area, 2,120 sq. m. Pop. 112,216 in 1860. Surface hilly in the N., more level in the S., and low alluvial, and marshy along the coast. One of the most elevated ridges in the peninsula passes through this state, dividing the waters that flow into either bay. The chief river, the Delaware, rises in New York, runs mostly S., and, after dividing that state and New Jersey from Pennsylvania, falls into the Bay of Delaware, near the N. extremity of the state, after a course of about 310 m. It receives several tributaries, and is navigable for ships of the greatest burden to Philadelphia, 55 m. from its mouth; and for small steam-vessels and boats, to nearly 135 m. higher. The other rivers are inconsiderable. There are no harbours on the seacoast; the only one in the state is that of Newcastle, 5 m. above the mouth of the Delaware river. The climate is healthy; but the degree of cold experienced in the N. is much greater, compared with that of the S., than could be expected from a difference in lat. of only 1° 20'. The soil in the N. is a rich clay; in other parts, and especially along the shore, it is sandy, and of inferior fertility; but it is everywhere well cultivated, at least for America. Principal crops, wheat, Indian corn, rye, barley, oats, flax, and buckwheat. The flour is of superior quality, and much esteemed for its softness and whiteness. The Cypress Swamp, a tract 12 m. in length and 6 in breadth, in the S. part of the state, has supplied a great deal of fine timber. Few minerals are met with, excepting large masses of bog iron along the banks of the smaller streams. Manufactures have made considerable progress. The mills situated on Brandywine Creek are considered the finest in the U. States; vessels are built, and there are ironfoundries and other extensive works at Wilmington. Wheat and flour are the principal articles of export.

The state is divided into three coun., and eight judicial circuits. Dover is the cap., but yields to Wilmington and Newcastle in size, trade, and pop. There is no college in the state; one planned in 1803 at Wilmington has not come into operation; but there are good academies in this and in several of the other towns. The state has a fund for the support of free schools, begun in 1864 had a capital of 411,392 dollars; and the objects of which are assisted by voluntary contributions from the different districts. The total taxation in 1863 amounted to 116,104 dollars, the largest item, the county tax, producing 85,853 dollars. A canal 14 m. in length, and navigable for small sea-vessels, unites the Delaware river near its mouth with the head of Chesapeake Bay.

The legislature consists of a senate and house of representatives, each co. sending three senators and seven representatives; the former are elected for four, and the latter for two years, by all the male citizens above twenty-one years of age who have resided in the state for a year, and paid taxes for six months preceding the election. The executive power is exercised by a governor chosen by the citizens, who retains office for four years, but is not re-eligible. Judges retain office during 'ap-

pointed conduct.' Most of the pop. are Presbyterians and Methodists.

Delaware was colonized by the Swedes in 1627. In 1655 it was acquired by the Dutch; and in 1664 came into the possession of the British. In 1704, when under the proprietorship of the celebrated W. Penn, it became a separate colonial establishment, and so remained until the independence of the states. Its constitution, formed in 1776, was amended in 1831. It sends 1 rep. to the Congress of the United States.

DELAWARE BAY is an arm of the sea between the states of Delaware and New Jersey, 65 m. in length, and about 30 m. wide in its centre, and 18 at its mouth, between Cape Henlopen, lat. 38° 47' N., long. 75° 6' W., and Cape May, lat. 38° 57' N., long. 74° 52' W. It has deep water throughout, and a line-of-battle ship may ascend the river Delaware to Philadelphia, 55 m. above the head of the bay, and 120 m. from the ocean. There is a magnificent breakwater at the entrance of Delaware Bay, near Cape Henlopen, forming an artificial harbour for the protection of vessels from the winds from the E. to the NW., round by the N., and from the floating ice descending the bay from the NW. The breakwater consists of two parts, one 1,200, and the other 500 yards in length. It was formed like the Admiralty pier at Dover, and the breakwaters of Plymouth and Cherbourg, by sinking blocks of granite in the sea.

DELFT, a town of S. Holland, on the Schie and on the canal between Rotterdam and the Hague, 4 m. SSE. the former, and 8 m. NW. the latter town, on the line of railway between Rotterdam and Amsterdam. Pop. 19,846 in 1861. Delft is an old-fashioned brick town, as Dutch as possible in its appearance, with old gateways, and lines of trees and havens in the middle of the streets. The chief building, not ecclesiastical, is the palace, in which William I., the most illustrious of all the princes of the house of Orange, and the founder of the independence of his country, was assassinated, July 10, 1584; it is a plain brick building within a court-yard, and is now used as a barrack. The new church, at the E. end of the market-place, is a fine old Gothic edifice, with a conspicuous lofty tower, and one of the best peals of bells in Europe. This church contains the tomb of William I., considered one of the most magnificent objects of art in Holland. It consists of a highly ornamented canopy, supported by a number of black and white marble pillars. In the centre, on a sarcophagus, lies the figure of the prince, in his robes, sculptured in white marble; and at his feet is his faithful dog, celebrated for having on one occasion saved his master's life in a midnight attack. There are several good figures in bronze round the tomb; that which is most admired is a figure of Fame blowing a trumpet, and resting lightly on one toe, as if about to take its flight. Beneath is the burial vault of the present royal family of Holland. Adjacent to this monument is that of the most illustrious individual Delft ever produced, Hugo Grotius, born here on the 10th of April, 1583. The Oude Kirke, or old church of Delft, is a structure remarkable for its extreme antiquity and huge size. It is situated in a mean street, and on approaching it the stranger is amazed at the enormous mass of brick, grey with age, which meets his eye. It is some 500 or 600 years old, and seems indebted for its protracted existence to the clusters of parasitical houses and shops built within the recesses of its buttressed walls. It contains the tombs of the famous Admiral Van Tromp; of Hein, another admiral who fell in battle at Tromp's side; and of the naturalist Leeuwenhoek, a native of this town.

Delhi was in former times the great seat of the manufacture of the common kind of earthenware, known by its name. England, however, has long since acquired a decided ascendancy in this branch of industry, and but very few persons are at present engaged in it in Delhi. In fact, nearly all the 'Delft ware' in use in Holland, and over the greater part of the Continent, is exported from England. Delhi, however, has a large woollen cloth factory, and others of carpets, coverlets, and soap, besides several distilleries and breweries. Its trade, however, is languishing, and it has little intercourse, except with Amsterdam and Delfshaven, a little town—with a pop. of 4,189 in 1851 —on the Maese, at the mouth of the canal which connects it with the Hague.

Delft was founded in 1074; it suffered severely from fire in 1536. Besides the eminent natives who have been already mentioned, it has produced many painters of celebrity, amongst whom was Berk, a pupil of Vandyke.

DELHI, a prov. of Hindostan, presid. Bengal; chiefly between lat. 28° and 31° N., and long. 75° and 80° E.; having N. the prov. of Lahore, and Gurwal, E. Gurwal and Inde, S. Agra, and W. Rajpootana. Like the other Mohammedan statelets of India, this prov. is not a modern subdivision under the British rule: the collectorates which have been formed out of it are subordinate to the court of Bareilly, the judicial capital in the Upper or W. provinces. The jurisdiction of Delhi at present extends only over the country W. the Jumna. Most of this prov. is flat; but at Wuzeerabad, near Delhi city, begins the long range of hills that extends through the Macberry chain towards Jyepore. The chief rivers are the Ganges, Jumna, Caggur, Chittang, and, in the NW, the almost extinct Serswatti, formerly a distinguished stream: the principal of these run through the prov. in a SE. direction. The land is sandy arid and sandy, and in the W. suffers greatly from drought in the hot season, when the water, which is of a brackish quality, from the natron and other salts with which the ground is impregnated, can be procured only at from 120 to 500 ft. below the surface. Still, however, no part of Hindostan is susceptible of greater improvement by irrigation. The British government has latterly directed much attention to the restoration of ancient canals and the construction of new ones in this prov. The canal of Ali Merdan Khan, which had been previously choked up for 100 m., was reopened in 1820, at an expense of 22,500l.; and, as its waters gradually advanced, the country for 5 or 6 m. on either side became fertilised in a most astonishing manner, and numerous wells, previously thought useless, became again serviceable. The canal of Sultan Feroze Shah, the bed of which passes from the former W. through Hurriana to the frontiers of Bicaneer, has been also surveyed preparatory to its restoration. A considerable tract between the Jumna and the Ganges, though now sterile and waste, was formerly highly cultivated and populous, having been fertilised by the great Doab canal. Between the Jumna and Sutlege mango trees are numerous, and the soil produces wheat, barley, and other dry grains; but the periodical rains are not sufficient to insure a crop. During the rainy season the temporary streams overflow, after which the pasture is good, and the climate tolerably healthy and temperate; but in the hot season the heat becomes so oppressive that the natives are often obliged to seek refuge from it in underground habitations. The land is assessed under a modification of the village system; but the mooddim, or head man, is not responsible for the payment, but is merely the agent for the rest of the village, removable at their pleasure, and not holding his office by any kind of hereditary tenure. Neither does he derive apparently any emolument from his office, nor is he analogous to the zemindar in the lower provinces, or the putmail in other parts; there being here no middle man to enjoy any portion of the land-tax, standing between the people and the government, which last receives from one-fourth part to a half perhaps of the produce of the land, according to circumstances, after the shares of the village functionaries, and certain other village expenses, have been deducted. The panchayet system of arbitration is in common use. In the zillah courts the European judges are assisted by both Mohammedan and Hindoo law officers, but the people in this prov. do not seem so disposed to litigation as in some others; they are, on the contrary, contented, orderly, and prosperous. At the commencement of the British rule, in 1803, there were about 600 villages deserted, the inhabitants of most of which had, before 1821, returned, and claimed and cultivated the lands they formerly possessed; and both the pop. and revenue had at that period very considerably increased. Mohammedans are most numerous in Delhi city, but Hindoos everywhere else, except in the NW., where the Seik religion is predominant, and the country is almost entirely occupied by petty Seik states. The chief towns are Delhi, Bareilly, Pillibheet, Shahjehanpoor, Ramjaur, Moradabad, Anoopshehr, and Meerut.

DELHI (Sanscrit, Indraprast'ha), a celebrated city of Hindostan, presid. Bengal, lieutenancy of Agra, cap. of the above prov., and anciently the metropolis of the Patan and Mogul empires, on the Jumna; 112 m. NNW. Agra, 426 m. NW. Benares, and 850 m. in the same direction from Calcutta, with which it is connected by the East Indian railway. Estimated pop. 160,000. That Delhi, in its period of splendour, was a city of vast extent and magnificence is sufficiently evinced by its ruins, which are supposed to cover nearly as large a surface as London, Westminster, and Southwark. The present inhabited city, E. and N. the ruins, built by the emperor Shah Jehan, and called by him Shahjehanabad, about 7 m. in circuit, is situated on a rocky range of hills, and is surrounded by an embattled wall, with many bastions and intervening martello towers, faced along its whole extent with substantial masonry, and recently strengthened with a moat and glacis by the British government. It has many great houses, chiefly of brick; the streets are in general narrow, but the principal are wide, handsome, and, for an Asiatic city, remarkably clean; the bazaars have a good appearance. There were formerly two very noble streets; but houses have been built down their centre and across, so as to spoil them: along one of these, running from the palace N. to the Agra gate, is the aqueduct of Ali Merdan Khan, reopened by Captain Blane in 1820. The principal public buildings are the palace, the Jumma Musjeed, or chief mosque, many other mosques, the tombs of the emperor Humayoon and of Seffdar Jung, and the Cuttub Minar; and, within the new city, the remains of many splendid palaces belonging formerly to the great dignitaries of the Mogul empire. Almost all these structures are of red granite, inlaid in some of the ornamental parts with white marble: the general style of building is simple, yet elegant; those of Patan architecture are never overdone with ornaments so as to interfere with their generally severe and sublime character. The palace, as seen from a distance, is a very high and extensive cluster of Gothic towers and battlements, towering above the other buildings. It was built by Shah Jehan,

m 2

is surrounded by a moat and an embattled wall, which, toward the street in which it stands, is 60 ft. high, and has several small round towers and two noble gateways. Some of the apartments are magnificent, even in the ruinous state in which they now are. There are rooms lined with white marble, inlaid with flowers and leaves of green serpentine, lapis lazuli, and porphyry, and also pavilions of marble, with many mosaic paintings, of birds, animals, and flowers. The Shalimar gardens (so highly extolled in 'Lalla Rookh') were also formed by Shah Jehan, and are said to have cost a million sterling; but 'laughing Ceres has reassumed her reign,' the gardens having been reconverted to agricultural purposes. The Jumna Masjeed, the largest and handsomest place of Mussulman worship in India, was built in six years by Shah Jehan, at an expense of ten lacs of rupees. It stands on a small rocky eminence, scarped for the purpose; the ascent to it is by a flight of 35 stone steps, through a handsome gateway of red stone, the doors of which are covered with wrought brass. The terrace on which it is built is about 1,100 yards square, and surrounded by an arched colonnade, with octagon pavilions at convenient distances. In the centre is a large marble reservoir, supplied by machinery from the canal. On the W. side is the mosque itself, of an oblong form, 261 ft. in length; its whole front coated with large slabs of white marble, and two apartments in the cornice inlaid with Arabic inscriptions in black. It is approached by another flight of steps, and entered by three Gothic arches, each surmounted by a marble dome. At the flanks are two minarets, 130 ft. high, of black marble and red stone alternately, each having three projecting galleries, and their summits crowned with light pavilions of white marble, the ascent to which is by a winding staircase of 130 steps of red stone. This noble structure is in tolerably good repair, being maintained by a grant from the British government. Not far from the palace is a mosque of red stone, surmounted with three gilt domes, in which Nadir Shah sat and witnessed the massacre of the unfortunate inhabitants. There are above forty other mosques; one, erected by the daughter of Aurungzebe, contains the tomb in which she was interred in 1710; some bear the marks of great antiquity, especially the Kala Masjeed, or black mosque, built of dark-coloured granite by the first Patan conquerors. It is exactly on the plan of the original Arabian mosques. The prospect S. the Shalimar gardens, as far as the eye can reach, is covered with the remains of extensive gardens, pavilions, mosques, and sepulchres, connecting the village of Cuttub with the new city of Delhi, from which it is nearly 10 m. distant SW., and exhibiting one of the most striking scenes of desolation to be anywhere met with. The celebrated Cuttub Minar is a very handsome round tower rising from a polygon of 27 sides, in 5 stages, gradually diminishing in circumference, to the height of 242 ft.; its summit, which is crowned by a majestic cupola rising from 4 arcades of red granite, is ascended by a spiral staircase of 344 steps, and between each stage a balcony runs round the pillar. The old Patan palace, a mass of ruin larger than the others, has been a solid fortress in a plain and unornamented style of architecture: it contains a high black pillar of cast metal of Hindoo construction, and originally covered with Hindoo characters, but which Feroze Shah afterwards enclosed within the court of his palace, covering it with Arabic and Persian inscriptions. The tomb of Humayoon is of Gothic architecture, surrounded by a large garden with terraces and fountains, nearly all of which are

now gone to decay. The garden is surrounded by an embattled wall and chubter, and in its centre, on a platform ascended by four flights of granite steps, is the tomb itself, a square building, with a circular apartment within about as large as the Radcliffe library at Oxford, surmounted by a dome of white marble. From the top of this building, the desolation is seen to extend to the W., in which direction Indraput stand, apparently to a range of barren hills, 7 or 8 m. off.

The soil in the neighbourhood of Delhi is singularly destitute of vegetation; the Jumna annually overflows its banks during the rains; but its waters in this part of its course are so much impregnated with nitre, that the ground is thereby rendered barren rather than fertile. In order to supply water to the royal gardens, the aqueduct of Ali Merdan Khan was constructed, by which the waters of the Jumna, while pure and wholesome, are conducted for 120 m. to Delhi, immediately after the river leaves the mountains. During the troubles that followed the decline of the Mogul power, the channel was neglected; and when the English took possession of the city, it was found choked up in most parts with rubbish. It is the sole source of vegetation to the gardens of Delhi, and of drinkable water to its inhab.; and, when re-opened in 1820, the whole pop. went out in jubilee to meet the stream as it flowed slowly onwards, throwing flowers, ghee, sweetmeats, and other offerings into the water, and calling down all manner of blessings on the British government. The deficiency of water is the greatest drawback upon the city and its prov., since Delhi is otherwise well fitted to become a great inland mart for the interchange of commodities between India and the countries to the N. and W. Cotton cloths and indigo are manufactured, and a shawl factory, with weavers from Cashmere, has of late been established here. Shawls, fruits, and horses, are brought from Cashmere and Caubul; precious stones and jewellery are good and plentiful; and there are perhaps few, if any, of the ancient cities of Hindostan which at the present time will be found to rival modern Delhi in the wealth of its bazaars or the activity of its pop. At the SW. extremity of the city stands the famous observatory, built, like that of Benares, by Jye Singh, rajah of Jyepoor, and formerly containing similar astronomical instruments; but which, together with the building itself, have been since partially destroyed. Near the Ajmeer gate is the Mudressa, or college of Ghazee-ud-Deen-Khan, an edifice of great beauty, for the repair of which, and the revival of its functions, the government has very liberally contributed. The Delhi college is now divided into the Oriental and the English departments; astronomy and mathematics are taught on European principles; and, in 1830, there were 287 students. According to Abul Fazel, no less than seven successive cities have stood on the ground occupied by Delhi and its ruins. Indraprestha or Indraput was the first, and the residence of the Hindoo rajahs before 1198, when the Afghans or Patans conquered it; it was the seat also of the first eight sovereigns of that dynasty. Sultan Hueen built another fortified palace; Moazuddeen another, on the banks of the Jumna; and others were built in different parts by succeeding sovereigns, one of which was near Cuttub; and lastly, Shah Jehan, towards the middle of the 17th century, chose the present spot for its site, which is certainly more advantageous than that of any of the preceding cities. In 1011 Delhi was taken and plundered by Mahmoud of Ghizni; in 1398 by Timour; in 1525 by Baber, who overturned the Patan dynasty, and commenced that

of the Moguls; In 1738 the Maharattas burned
the suburbs; and in 1739 Delhi was entered and
pillaged by Nadir Shah, who did not retain possession of it. Since 1803, together with its territory, it has virtually belonged to the British, and
from that time until 1857, was the seat of a 'Resident,' who took charge of the emperor and royal
family, to whom the liberal stipend of 150,000£ a
year was allowed. The Indian mutiny, however,
in which the people of Delhi took a leading part,
changed this state of things. On May 12, 1857,
the king of Delhi was proclaimed emperor by the
insurgents, and retained his nominal power till the
21st of September of the same year, when the
British forces, under General Wilson, stormed the
city, and made him a prisoner. On the 22nd of
September, the hiding places of the king's son and
grandson were discovered, and they were slain on
the spot. They had both been participators in
the massacre of Englishmen, as well as instiga-
tors of the mutiny; their bodies were therefore
brought into the city, and exposed to the view of
the public. Thus ended all hope of the restoration
of the Mogul dynasty, the once all-powerful rulers
of Delhi.

DELOS, a small, and now barren and deserted
but once famous island of Greece, in the strait between Mycone and Rhenea, or the greater Delos,
almost in the centre of the Cyclades; lat. 37° 25'
N., long. 25° 16' E. This island was regarded in
antiquity with peculiar veneration, from its being
supposed to be the birth-place of Apollo and
Diana, to whom it was sacred. Magnificent temples were erected in honour of these deities. The
temple of Apollo, of which the ruins still remain,
raised at the joint expense of the Grecian states,
is celebrated as having been one of the most
splendid in the ancient world; and his oracle here
was second only to that of Delphi. Pursuant to
a practice begun by Theseus, a vessel sailed annually from Athens to Delos with offerings, conveying at the same time deputations appointed to
perform sacrifices in honour of Apollo and Diana,
and choruses of youths and virgins, who danced
and sang hymns in their praise. Quinquennial
games were also celebrated with great pomp, and
were attended by deputations from all the Grecian
states and islands. Delos was repeatedly purified;
and to keep it from all pollution, neither births
nor deaths were allowed to take place within its
sacred precincts; but all women about to be confined, and all sick persons, were conveyed to the
greater Delos. Such was its character for sanctity, that it commanded the respect even of barbarians; and the Persian admirals, who laid waste
the other islands, would not touch at Delos. After
the Persian war, the Athenians made it the treasury of the Greeks, and all meetings relative to
the affairs of the confederacy were held in it.

Its sacred character, the security which it consequently enjoyed, its good harbour, and central
position, made Delos a favourite seat of commerce
as well as of religion and pleasure. Its festivals
were attended by the merchants of Greece, Asia
Minor, Phœnicia, Egypt, and Italy, who brought
thither the products of their respective countries.
On the destruction of Corinth, many of its principal merchants sought an asylum in Delos, which
acquired a large portion of the traffic that had
been driven from the former. It was a principal
seat of the ancient slave trade; and Strabo states
that thousands of slaves were brought thither
from Cilicia, and sold in its markets. Cicero says
of it, *Insula Delos, tam procul a nobis in Ægeo mari
posita, quo omnes undique cum mercibus atque mercibus commeabant, referta divitiis, parva, sine aqua,
nihil timebat.* (Pro Lege Manil., c. 18.) A bill in

the centre of the island was called *Mons Cynthus*,
and hence the epithets *Cynthius* and *Cynthia* so
frequently applied to Apollo and Diana. The
heaps of marble, and the fragments of columns,
architraves, &c., which are everywhere met with,
attest the ancient grandeur of this famous island.
But it has long since deserted; and Tournefort states that, in the early part of last century,
the inhabs. of Mycone were in the habit of holding
the greater Delos for the purposes of pasturage,
paying for it to the grand seigneur a rent of 20
crowns a year. (Tournefort's Voyage du Levant,
i. 290-325.)

DELPHI, DELPHOS, or PYTHO (at present
Castri), a famous city of ancient Greece, the cap. of
Phocis, and the seat of by far the most celebrated
oracle of the ancient world ('*commune humani ge-
neris oraculum*,' Liv. lib. 38, § 48), at the S. foot of
Mount Parnassus, 45 m. NW. Corinth, and 84 m.
N.E. from the nearest point of the Crissæan Sea
(Gulf of Lepanto). Delphi had every attribute
that could invest it with interest and inspire awe.
It was supposed to be situated in the centre of
the world, was built on the declivity of the mountain on successive terraces formed of Cyclopean
masonry, and rising above each other like the
seats in a theatre. Overhanging the city on the
N. rose the two famous peaks of Parnassus, the
chasm between them affording an outlet for the
waters of the *Castalian* spring, the source of
poetical inspiration. If we add to these natural
advantages, the fact that Delphi was the chosen
abode and principal oracle of Apollo; that she
was the seat of the council of the Amphictyons,
and the place where the Pythian games were
celebrated, we need not wonder at the extraordinary respect and veneration in which she was
held. She was not fortified by walls, but by precipices, and the especial protection of Apollo; so
that the ancients reckoned it doubtful '*utrum
munimentum loci, an majestas dei plus hic admi-
rationis habeat.*' (Justin, lib. 24, § 4.)

The origin of this famous city, and of the oracle
to which it owed all its glory, are buried in impenetrable obscurity. The most probable account
seems to be, that a mephitic vapour, similar in
some degree, perhaps, to that of the *Grotto del
Cane* at Naples, having issued from one of the
clefts of the rock, violently affected those by
whom it was inhaled, making them utter strange
incoherent sayings. On this narrow foundation
was built one of the most extraordinary fabrics
ever raised by superstition, fraud, and imposture.
The ravings of those affected by the vapour were
believed to be indications of future events; they
were said to be inspired; and the ejaculations
which they uttered were affirmed to have been
owing to their being filled with the breath or
spirit (*divinus afflatus*) of Apollo, the guardian god
of the place; the fame of the oracle rapidly increased, and it was soon seen how rich a harvest
might be derived from it. The sacred cavern was
forthwith enclosed; a tripod was placed over the
chasm whence the vapour issued; priests and
priestesses were appointed for the service of the
god; and a series of temples, each more magnificent than its predecessor, were erected in his
honour. States and princes were anxious to
learn their fate, or the success of any contemplated
enterprise, from the responses of the oracle; and
private individuals crowded to the city for the
same purpose. The answers of the god were not
gratuitous; and it would seem that an opinion
had early gained ground, that the nature of the
responses was to a considerable extent dependent
upon the value of the offerings! Hence there
arose a kind of competition among these consult-

ing the oracle who should be most liberal; and the wealth accumulated at Delphi came, in the course of time, to be prodigiously great. The responses were, apparently at least, delivered by a priestess. After being purified by bathing in the Castalian spring, she mounted the tripod, and having inhaled the intoxicating or stupifying vapour, she became violently convulsed—

> 'Subito non vultus, non color unus,
> Non comtæ mansère comæ; sed pectus anhelum,
> Et rabie fera corda tument; majorque videri,
> Nec mortale sonans; afflata est numine quando
> Jam propiore dei.'   *Æneid*, vi. line 47, &c.

The incoherent wraps of sentences which the Pythia uttered during this paroxysm having been collected and arranged in verses by the priests, formed the desired responses.

The responses of the Pythia were said to be comparatively precise; and she was sometimes resorted to in order to clear away the mystery in which three of other oracles were involved. It may, indeed, be reasonably enough supposed, that superior address and information on the part of the Delphic priests might enable them in many instances to give pretty distinct responses, that could not fail frequently to square with the event. But, even if no evidence of the thing had come down to us, we might have been assured that, speaking generally, their responses would be ambiguous, and so contrived that, however the event might turn out, the credit of the oracle would be preserved: and this, in point of fact, was the case. The answer of the oracle to Crœsus, that in making war upon the Persians he should destroy a great empire (Herod. i. § 53), is an instance of this; as it is plain the credit of the oracle would be equally secured whether Crœsus conquered or was himself conquered by the Persians. The answer of the oracle to Pyrrhus is another instance of this sort of ambiguity—

> 'Aio te, Æacida, Romanos vincere posse.'

as it might either be interpreted in favour of or against Pyrrhus. This equivocation was not, however, the worst feature of the imposture carried on at Delphi. The oracle was at once ambiguous and venal. A rich or a powerful individual seldom found much difficulty in obtaining a response favourable to his projects, how unjust or objectionable soever. Herodotus states distinctly that the Alcmæonidæ, who rebuilt the temple at Delphi, bribed the Pythia to recommend the Spartans to assist in delivering Athens from the tyranny of the Pisistratidæ (v. § 60, 90); and such were the base motives that made the oracle falsely pronounce Demaratus, king of Sparta, to be illegitimate, and obtained responses favourable to Lysander when he endeavoured to change the succession to the Spartan throne. This also was, no doubt, the sort of inspiration that dictated the responses favourable to Philip, which made Demosthenes declare that the Pythia *philippised*! But such and so powerful is the influence of superstition, that this threadbare system of fraud and quackery maintained a lengthened ascendancy; and that the responses of frantic girls, interpreted by venal priests, frequently sufficed to excite bloody wars, and to spread desolation through extensive states.

The credit of the oracle had been materially impaired before Christianity obtained an ascendancy in the ancient world; and the triumph of the latter was destructive of this as well as other oracles. Constantine carried off some of the finest and most costly ornaments of the Delphian temple to decorate his new capital. And there is still to be seen in Constantinople the brazen pillar, formed of three serpents twisted together, that supported the golden tripod which, after the defeat of Xerxes, was consecrated in the temple of Delphi by the victorious Greeks. (See CONSTANTINOPLE; Gibbon, cap. 17, &c.)

The vast wealth of the temple of Delphi exposed it to many attacks. A party sent by Xerxes to plunder the sacred edifice are said to have been defeated by the manifest interposition of Apollo himself, (Herod. viii. § 37.) But, on other occasions, the god was less vigilant or less successful. The fane was successively plundered by the Phocians under Philomelus, by the Gauls under Brennus, by Sylla, &c.; and Nero is reported to have deprived it of no fewer than 500 bronze statues! and yet, despite all these deductions from its ancient stores, it had, when visited and described by Pausanias, a vast number of statues and ornaments of all sorts. But its treasures had disappeared long previously; and the rich offerings of Gyges, Alyattes, Crœsus, and Midas were no longer to be seen.

Except its grand natural features, every thing at Delphi has undergone a total change. Not a vestige remains of the great temple, by which to form even a satisfactory conjecture as to its position. The prophetic cavern is searched for in vain; 'antraque mæsta silent, incunabula remota.' The village of Castri, that occupies a part at least of the site of the ancient city, is poor and miserable, and does not contain above 400 or 500 inhab.:—

> 'Tantas evil longinquum ruier omnes ruinas?'

DELVINO, a town of Turkey in Europe, prov. Albania, cap. of a sanjiack or distr., 43 m. WNW. Yanina. Estimated pop. 10,000. The vicinity contains some orange plantations; but is chiefly noted for its olive cultivation. The trade of the town is chiefly in oil, and other agricultural produce.

DEMERARA. See GUIANA (BRITISH).

DEMONTE, an inl. town of N. Italy, prov. Cuneo, cap. mand., on the Stura, 18 m. SW. Cuni. Pop. 6,166 in 1861. The town is commanded by a fortress placed on an isolated height, and contains three churches and a hospital. It was formerly fortified, but its works were demolished by the French in 1801.

DEMOTICA, or DIMOTIKA, a town of Turkey in Europe, prov. Roumelia, on the Maritza, at the foot of a conical hill, crowned by a citadel, containing a palace, occasionally occupied by the Turkish emperors during the period that Adrianople was the cap. of the empire, from which city Demotica is distant 24 m. S. by W. Pop. about 8000. The town is tolerably well built; it contains a mosque, and several Greek churches, schools, and public baths. The citadel is supplied with water by an aqueduct. It is the residence of a Greek archbishop, and has manufactures of silk and woollen stuffs, and earthenware. Charles XII. of Sweden resided in this town for more than a year subsequently to the battle of Pultowa.

DENAIN, a village of France, dép. du Nord, in the cant. of Bouchain, 6 m. SW. Valenciennes, on the Northern railway. Pop. 10,254 in 1861. The place has numerous forges, and beet-root factories, and there are extensive coal mines in the neighbourhood. Denain is famous in modern history as the scene of the decisive victory gained in 1712 by the French under Marshal Villars over the allies under Prince Eugene. This victory, which is partly to be ascribed to the improvidence of the allies, and partly to the skilful combinations of Villars, saved Louis XIV. from the disgrace of having the terms of peace dictated to him in his own capital. It changed, in fact, the whole aspect of public affairs; and brought the negotiations at Utrecht to a speedy conclusion.

DENBIGH, a marit. co. of N. Wales, having N. the Irish Sea, E. the cos. of Flint and Cheshire, S. Salop, Montgomery, and Merioneth, and W. Caernarvon. Shape very irregular. Area 648 sq. m., or 388,052 acres. Surface and soil much diversified; for the most part, however, it is rugged, wild, and mountainous; but it has some very fertile tracts, particularly in the far-famed vale of Clwyd, on both sides the river of that name, lying mostly in this co., and which is eminently beautiful and fertile, producing the necessaries of life not only in abundance for the inhab., but in ample sufficiency to spare to supply the wants of their neighbours. The vale of Llangollen, in the E. part of the co., though inferior in point of richness to that of Clwyd, is notwithstanding pretty fertile, and there is a considerable extent of good land in the vicinity of Wrexham. The climate in the valleys is remarkably mild, but rain is very prevalent, and considerable damage is sometimes done by the overflowing of the rivers. Agriculture, though a good deal improved, is still very backward. Barley, oats, and potatoes are the principal crops; wheat, beans, and pease being also raised in some of the more fertile districts. There is no regular rotation of crops; whichever grain happens to be most in demand is sown. It is also a frequent practice to burn the surface both of fresh enclosed lands and old clover leys; but this, though at the time it yields good crops of oats and turnips, impoverishes and ultimately exhausts the land. Farms are usually very small; and being let only by the year, and without any conditions as to management, we need not wonder at the low state of agriculture. Average rent of land, in 1810, 8s. an acre. The hills are depastured by large flocks of sheep, and large herds of cattle are found in the valleys. The dairy husbandry is carried on to a considerable extent, particularly in the E. parts of the co., adjacent to Cheshire. The minerals are valuable, coal, lead, and iron-mines being wrought in different parts of the co.; it also furnishes slate and mill-stones. The woollen manufacture is carried on to some extent, and gloves and shoes are produced in considerable quantities in Denbigh. It is bounded E. by the Dee, and W. by the Conway, and is traversed by the Clwyd and Elwy. Denbigh is divided into six cantrefs or hundreds, and sixty-four parishes. It returns three mem. to the H. of C., viz. two for the co., and one for Denbigh and its contributory bor.; county constituency, 4,346 in 1865. According to the census of 1861, the co. had a pop. of 100,778, inhabiting 21,310 houses. Gross annual value of real property assessed to income tax, 473,355l. in 1857, and 486,776l. in 1862.

DENBIGH, a town and parl. bor. of N. Wales, cap. of the above co., hund. Yale, near the middle of the vale of Clwyd, at the base and on the side of a steep hill, crowned with the magnificent ruins of its old castle; 22 m. W. Chester, and 180 m. N.W. London, on a branch line of the London and North Western railway. Pop. of parl. bor. 5,946, and of par. 4,054 in 1861. The town consists of three principal and some smaller streets and lanes; it is well paved and lighted, but many of the houses have a dilapidated appearance. The par. church, 1 m. E. from the town, has many interesting monuments; there are two other churches, besides a Catholic and four dissenting chapels. Other public buildings are, a town-hall, free grammar-school, with a small endowment, a blue-coat school, a national school, and several Sunday schools supported

by the various sects; a reading-room, and a literary society. There is also a dispensary for the poor of the town and neighbourhood. The old staple trades of the town are shoe-making, glove-making, and tanning, particularly the first. The limits of the parl. bor. were the subject of much dispute till they were definitively fixed in 1829. Denbigh, with Ruthin and Holt, has returned one mem. to the H. of C. since the 27th of Henry VIII., the right of voting being in the resident burgesses. The Reform Act added Wrexham to the contributory bors. Registered electors in the Denbigh parl. district, 845 in 1862. The present municipal bor. is restricted to the space immediately contiguous to the town; and the governing body consists of four aldermen and twelve councillors. The waste lands of the par. belong to the corporation, whose annual average revenue from these and other sources was 442l. in 1862. The Easter and Michaelmas quarter sessions of the co. are held in the town, which is a polling place for the co. The castle, both from its situation and structure, was anciently of great importance. It was founded in the 2nd of Edward I. by Henry Lacy, earl of Lincoln. A magnificent pointed archway, with a statue of the founder, is still in tolerable preservation; but the rest is entirely ruinous. There is a bowling-green and several cottages within the enclosure. The prospect from the castle is extensive and magnificent. In the last civil war the castle withstood a siege by the parliamentary forces in 1645; and thither the king retreated from Chester. It was taken in the following year, and soon after dismantled.

DENDERAH (the Tentyra of the Greeks), a ruined town of Upper Egypt, celebrated for its temple, the best-preserved of all the remains of antiquity with which Egypt, particularly the Said, abounds; near the W. bank of the Nile, 31 m. N. Thebes, lat. 26° 10' 20" N., long. 83° 40' 27" E. The town, 1½ m. E. from the temple, stands in an extensive and well-cultivated plain, which expands on both sides the Nile, and is surrounded by mountains, so as to give it the appearance of a beautiful circular basin, shaded by thick groves of palm trees. The temple stands on the very verge of the Lybian desert, the encroachments of which have buried a large portion of the buildings under heaps of sand; but enough is still visible to indicate its magnitude and magnificence, and to impress the spectator with the deepest sense of the wealth, power, and civilisation of the illustrious but long extinct people by whom so noble a fabric was raised. The temple and the buildings appertaining to it, with the exception of one propylon, are enclosed within a square wall of sun-dried bricks, each side measuring 1,000 ft. and in some parts 35 ft. high, and 15 ft. thick. After passing a small stone-building, and a gateway or propylon entirely covered with well-executed sculptures and hieroglyphics, the spectator, proceeding through the dromos (avenue lined on each side with sphynxes) arrives at the temple. It is nearly in the form of the letter T; and its simplicity, vastness, the durability of its structure, and its ornaments and sculptures, in perfect preservation, though no longer intelligible, excite the strongest feelings of awe and astonishment. The front of the pronaos, or portico, is adorned with a beautiful cornice, supported by six square columns, with capitals formed of colossal heads of Isis. Within, twenty-four cylindrical columns, ranged in six rows of four deep, support the roof; the capitals of these columns are quadrangular, and exhibit on each face the representation of a temple with a divinity under the portico of the sanctuary; between the capital and the shaft, heads of Isis

again appear; including their base and capitals.
the height of the columns is about 16 ft.; the shafts
are sculptured with hieroglyphics and figures in
basso-relievo, as are the front and ceiling; the
designs on which last have been supposed to be
intended to represent a zodiac. Indeed there is
no where in the whole apartment a space of 2 ft.
that is not covered with sculptures, in low relief,
of human beings, animals, plants, emblems of
agriculture or of religious ceremony. The temple, which is equally enriched with sculptures,
consists of several apartments, partially lighted
by circular holes cut in the ceiling. The sanctuary is, however, quite dark. Access is provided
to the roof by means of a staircase, with steps so
low that priests might convey up and down the
weighty paraphernalia of sacrifice. But the most
remarkable object, in the estimation of Europeans,
belonging to the temple, was the ceiling of an
upper chamber, exhibiting in twelve compartments, like that of the pronaos, a variety of mythological figures, which correspond very closely
with the Greek signs of the zodiac; it was enclosed within three concentric circles, and supported by eight male figures kneeling, and four
females standing, most harmoniously grouped.
The remains of a smaller temple stand to the
right of the propylon, supposed to have been dedicated to the malignant deity, Typhon.

A great deal of curious and learned discussion
has taken place with respect to the antiquity of
the zodiac of Denderah. (Notice sur le Zodiaque
de Denderah, par, M. M. Martin, Paris, 1824.)
Dupuis, Fourier, and other writers, concluded
from the places of the figures of the constellations
on it, compared with their present places, and the
precession of the equinoxes, that it had been constructed about 5,000 years ago. But Littrow,
Playfair, and some other learned astronomers, inferred from the same data, and with infinitely
more of probability, that the age of the zodiac did
not exceed 2,226 years. Subsequent researches
by Visconti, Letronne, St. Martin, and others,
have, however, gone far to show that the calculations referred to had no real foundation, and that
the figures on the so-called zodiac are probably
astrological or mythological representations, and
have nothing of an astronomical or scientific character. At the same time, however, it must be
admitted, that the purpose of the supposed zodiac,
and its antiquity, are still involved in the greatest
uncertainty. With respect to the temple itself, it
would seem, from its being one of the most perfect
and beautiful in the country, to belong to the
period of the later Egyptian kings, when the arts
had attained to their highest perfection; but there
are not, perhaps, any really good grounds for the
notion that it is of so late a date as the era of the
Ptolemies, though alterations may then have been
effected in it. The zodiac, or planisphere, that
gave rise to these discussions, is now in Paris.
The pacha having consented to the desecration
of the temple, this extraordinary monument was
skilfully cut out, and conveyed to France, in
1822, by a M. Lelorrain. It was subsequently
purchased by the French government for 15,000
fr., and placed in the Louvre.

DENDERMONDE (Belg. *Termonde*), a fortified town of Belgium, prov. E. Flanders, cap.
arrond. on the Scheldt, at the point where it is
joined by the Dendre, 15] m. E. Ghent, on the
railway from Ghent to Mechlin. Pop. 9,530 in
1856. The town has 4 churches, 5 chapels, a
town-hall, a hospital, lunatic and orphan asylums,
2 convents, a college, 14 schools, and a prison. It
is defended by a citadel constructed under the
Duke of Parma in 1584; is the seat of a court of

original jurisdiction; and has manufactures of
woollen stuffs, cotton yarn, hats, lace, tobacco,
soap, oil, and earthenware; bleaching and dye-
houses, breweries, distilleries, and flour and other
mills, with a considerable trade in corn, hemp,
flax, and oil. It is believed to have been founded
no earlier than the 8th century, though many
Roman antiquities have been dug up in it at different periods. It was unsuccessfully besieged by
Louis XIV. in 1667, but fell into the hands of the
French in 1745. It suffered severely from an
inundation in 1825.

DENHOLM, a manufacturing village of Scotland, co. Roxburgh, 4 m. NE. Hawick, on a rising
ground 1 m. S. from the river Teviot. Pop. 755
in 1861. The inhab. are almost entirely engaged
in the weaving of woollen stockings, on account
of the Hawick manufacturers. There is a flour-
mill here; as also a dissenting chapel, and a subscription library. Dr. John Leyden, the celebrated poet and linguist, was a native of this place.

DENIS (ST.), a town of France, dep. Seine,
cap. arrond. in a fertile plain near the Seine, and
on the canal which unites that river with the
canal of Ourcq, 5 m. N. Paris on the Northern
railway. Pop. 22,052 in 1861. The town is
chiefly remarkable for its abbey-church, built in
the 7th century by Dagobert I., who was buried
within its walls; since which time it has been
the customary burial-place of the kings of France.
It was materially improved by Suger, abbot of
St. Denis, in 1150, and has been further enlarged
by different sovereigns in succeeding ages, so that
it has a great variety of architectural style. It is
an imposing Gothic edifice, in the form of a cross,
4154 ft. in length, by 104] ft. broad, and 80 ft.
high. Its front has two towers, one of which is
surmounted by a spire. Most of the tombs of the
kings of the first, second, and third races are in a
subterranean vault. In 1793, during the revolution, many of these tombs were destroyed, and the
remains they enclosed, not excepting even those
of Henry IV., were thrown together and buried
under a heap of earth in the environs of Paris.
The demolition of the building itself was subsequently ordered, but this was not effected. The
town has some good infantry barracks, an establishment for the education of 500 girls, orphans of
members of the Legion of Honour, founded by
Napoleon, which occupies the celebrated abbey of
St. Denis, founded by Dagobert I. in 613; a public
library, and theatre. St. Denis is well built; it
is the seat of a sub-prefect; has manufactures of
woollens, cottons, and leather; and a brisk trade
in flour, wine, vinegar, wool, and timber.

DENMARK, one of the secondary European
kingdoms, on the south side of the entrance to
the Baltic, between 53° and 58° N. lat., and 8°
and 13° E. long. It consists partly of the peninsula, stretching from the river Königsaue, or
Kongeaae, the northern frontier of Schleswig, to
the Skaw or Skagen, and comprising the prov. of
Jutland; and partly of the Danish Archipelago,
or of the islands of Zealand, Funen, Laland, Falster, &c., between the Baltic and the Cattegat,
and the island of Bornholm, in the Baltic. Except
on the S., where it is bounded by the duchy of
Schleswig, continental Denmark is everywhere
surrounded by the sea, having E. the Baltic, the
Little Belt, and the Cattegat; N. the Skagerrac;
and W. the North Sea. Iceland, the Feroe Isles,
part of Greenland, and some possessions in the E.
and W. Indies, belong to Denmark. Exclusive
of these, the kingdom contains an area of 11,393
sq. m., with a pop. of 1,609,551, according to the
census of 1860.

Surface and Soil.—There are no mountains in

Denmark, and the few hills by which it is marked, are little more than undulations. It is generally low and level, the coasts being seldom elevated much above the sea. In parts of the W. coast of Jutland, the country, which has partly been wrested from the sea, is defended, as in Holland, against its irruptions by immense mounds or dikes, managed by a government board. Soil various, but in some districts, particularly in the SW. part of Jutland, it is exceeding fertile, being very rich marsh-land, producing the finest pasture and excellent crops. In other parts, more especially in central and NW. Jutland, the soil is arid, sandy, and barren, large tracts being heath. The soil of the islands consists of clay mixed with sand and lime.

Rivers and Lakes.—Denmark having no mountains, and every part of it being within a short distance of the sea, has no rivers of any magnitude. Fresh water lakes numerous, but not large. The most remarkable feature in the physical geography of Denmark is the number and extent of the inlets of the sea, or rather lagoons, by which the continental part of the country is intersected. The principal of these lagoons, the Lymfiord, formerly communicated only by a narrow channel with the Cattegat, stretching thence in a W. direction, with long windings, and expanding in various places into immense sheets of water, encompassing large islands, across the peninsula of Jutland almost to the North Sea. In 1825, however, during a violent storm, the isthmus between the North Sea and the Lymfiord was broken down in two places, so that it now isolates the N. portion of Jutland; but the newly opened channel is too shallow to be of much use for the purposes of navigation, and the depth of the opening to the Cattegat has also decreased, so as only to admit vessels of comparatively small burden. There are other fiords, but none so extensive as this. They, as well as the bays and rivers, are well stocked with fish, the fishery being a principal business and dependence of the inhab.

Animal and Vegetable Products.—These are almost the same in Denmark as in Great Britain. The horses and cattle of W. Jutland are amongst the best that are anywhere to be met with; those that belong to the islands and N. Jutland are of a smaller breed, but strong and active. The wool of the sheep is short and coarse, but latterly it has been a good deal improved by crossing with merinos. The feeding of pigs is prosecuted to a great extent, and quantities of bacon are yearly exported. Poultry is so abundant that their feathers alone make an article of export. All the common grasses, with potatoes, flax and hemp, madder, and tobacco, are raised in Denmark. The forests are not very extensive. They lie principally along the eastern shores of Jutland, and in Zealand and Funen; consisting principally of birch, but also of ash, alder, and oak. Pine and fir are rare.

Mineral Products, in Denmark, are but of little value. The subsoil chiefly consists of sand and clay, and no metals have been discovered that would repay the expense of working. There is a brine spring near Oldesloe; but it does not furnish salt sufficient for the consumption of the kingdom. The want of coal is in part compensated by the abundance of turf.

Climate.—Being almost everywhere surrounded by the sea, the climate is humid, and in its principal features approaches pretty closely to that of Scotland. The transition from winter to summer, and from summer to winter, is, however, a good deal more abrupt, so much so, indeed, that spring and autumn, particularly the first, are but faintly marked; the heat of the summer is, at the same time, greater than in Scotland, and the cold of the winter more severe. These differences arise from the greater proximity of Denmark to the continent. The winds not being broken by any mountains, often sweep along with great violence. The NW. wind, called *Skai*, which is especially felt in May and June, is so severe on the W. coast of Jutland, as to wither the tops of the trees. The Sound is sometimes frozen over; but this is said to arise more frequently from the drifting of ice formed in higher latitudes than from the intensity of the cold at the place. In 1659, the Swedes marched an army on the ice across the Sound to besiege Copenhagen. Fogs are very prevalent.

Agriculture.—In Denmark, as in most other European countries, the peasantry or occupiers of the soil were at no very distant period in the most depressed state imaginable. 'In Zealand,' says Lord Molesworth, and the same observations then applied to the rest of the kingdom, 'they are all as absolute slaves as the negroes are in Barbadoes; but with this difference, that their fare is not so good. Neither they, nor their posterity to all generations, can leave the land to which they belong; the gentlemen counting riches by their stocks of boors, as here with us by our stocks of cattle, and the more they have of them the richer they are. In case of purchase, they are sold as belonging to the freehold, just as timber trees are with us. There is no computing there by numbers of acres, but by numbers of boors; who, with all that belongs to them, appertain to the proprietor of the land. Yeomanry, which is the strength of England, is a state not known nor heard of in Denmark; but these poor drudges, after they have laboured with all their might to raise the king's taxes, must pay the overplus of the profits of the lands and their own toil to the landlords, who are almost as poor as themselves. If any of these poor wretches prove to be of a diligent and improving temper, who endeavours to do a little better than his fellows, and to that end has repaired his farm-house, making it convenient, neat, and pleasant, it is forty to one but he is presently transplanted from thence to a naked and uncomfortable habitation, to the end that his griping landlord may get more rent by placing another on the land that is thus improved: so that in some years 'tis likely there will be few or no farm-houses, when those already built are fallen through age or neglect.' (Account of Denmark in 1692, 4th ed. p. 54.) In 1761, the queen Sophia Magdalen had the honour of being the first to set a better example to the Danish proprietors, by publicly enfranchising the peasantry on her estates; and the example was soon after followed by Count Bernstorff and others. At this period about a sixth part of the land was supposed to belong to the crown; but the crown estates were soon after divided into farms of a moderate size, and a large portion of them disposed of to any one who chose to become a purchaser. Previously to this period very few peasants were proprietors; but their number now began speedily to increase, partly in consequence of the sale and division of the crown estates, and partly of their purchasing up their leases from their lords. In 1788, the peasantry of Denmark, that is of Jutland and the islands, were finally emancipated from all political bondage; and a commission was at the same time appointed to regulate the rents and services to be paid by those tenants holding hereditary leases, or leases for lives, where the parties could not come to an agreement. In 1791 and 1799 fresh ordinances were issued on the same subject, having for their object to reduce the number of such tenants, by converting them, under equitable condithena, into proprietors, and for restraining the right of free way.

In consequence of these measures a very great change has taken place in the distribution of property in Denmark. Large estates have been so much broken down, that at this moment at least one-half of the soil of the kingdom belongs to petty proprietors. The division of property has been going on at an increasing rate since the year 1850, when the last privileges attaching to the possession of landed estates were taken away. From an official return of the year 1861, it appears that at that time there were in the whole kingdom little more than a thousand *hovedgaards*, that is, estates taxed at above 12 tons of hartkorn, against nearly 80,000 *bondergaards*, or cottage-farms, taxed under 12 tons of hartkorn. The ton of hartkorn represents an area of no definite size, varying according to the nature of the soil and its fertility, the ton, or barrel, of grain being equal to 5/8 imperial lasterle. On the average, the ton of hartkorn may be said to be equivalent to 5½ acres. It will thus be seen that the parcelling out of the land has gone very far, and, indeed, it is generally admitted that the principal drawbacks upon agriculture in Denmark are the great division of property and the consequent want of capital. (As regards the taxation of land, *see d'Ismere below*.)

The average earnings of ordinary agricultural labourers in Denmark may vary from 15l. to 20l. a year. Their situation is decidedly comfortable. Mr. Macgregor, British consul at Elsinore, an intelligent and careful observer, gives the following details illustrative of their command over necessaries and comforts:—' The Danes are great eaters, and they eat at all times of the day. The following quantities of food are usually allowed to male farm servants per month: bread, 68 lbs.; potatoes, half a barrel; groats, half a bushel; butter, 4 lb.; bacon, 10 lb.; meat, 4 lb.; salted herrings, 30 lb.; salt fish, 8 lb.; beer, 60 quarts; milk, *ad libitum*. The Danish peasants make 5 meals a day in summer. Early in the morning they have, 1st, breakfast, consisting sometimes of coffee, but generally of warm milk and bread; 2d, at 9 o'clock, follows bread and butter and a dram; 3d, at 12 o'clock, dinner, the introduction to which consists of spoon-meat, such as milk porridge, beer soup, curds with warm milk or beer, or of fish, boiled groats, cheese, greens or dried peas, after which follows fresh or dried fish, bacon or meat, with potatoes or other vegetables, or boiled or poached eggs, or pancakes; 4th, at 5 o'clock, bread and butter and a dram or beer, especially in harvest time; 5th, supper after sunset, sour groats, curds, with milk or buttermilk. In winter, when they get up later, they have one breakfast, and consequently, they only make 4 meals a day. The poorer families seldom boil their kale upon meat, but upon a piece of hog's lard or bacon. In most of the cottages a sheep or a lamb is killed before the winter. The more substantial peasants kill a pig, a cow, or an ox, and they dispose of what they do not require themselves to their neighbours. They also kill a certain number of geese and ducks, salting them down for the winter, and using the feathers for their beds. This mode of living applies chiefly to peasants in districts of a middling soil, but where it is richer, they have more of bacon, meat, and fish, in lieu of other dishes; also is the beer they drink of greater strength. Fish is almost their diurnal food in villages adjacent to the sea, and they often use dried fish instead of bread, especially where the rye crops have failed, when their rye bread is often found mixed with barley. Amongst the poorer cottagers who have no land, it

would sometimes happen that they must content themselves with a crust of dry bread, and milk and water in lieu of beer; but such cases are not of frequent occurrence; at least, all the reports on the agricultural state of the country which have been published these later years, concur in stating that the generality of peasants are well off, and that there is plenty of employment in the country for all labourers that choose to work.

'I shall conclude these observations by stating the annual expenditure of a labourer with a wife and three children in this neighbourhood (Elsinore), the several items reduced into sterling:—

	£	s.	d.
House Rent and Taxes	0	10	0
Turf for Fuel	0	11	0
Rye for Bread, 2½ quarters, at 13s. 6d.	7	4	6
Barley for Bread and Groats, 6½ qrs., at 10s. 6d.	2	7	8
Meat and Bacon, 80 stone, at 1s. 8d.	3	12	6
Potatoes, 15 quarters, at 2s. 2d.	1	7	0
Coffee, 1 lb.; sugar, 1 lb. per week	1	5	0
Butter, 1½ Pskls.; cheese, 170 lb.	1	5	0
Milk, 8 quarts per week, at ½d.	0	17	4
Soap, Candles, and Groceries	0	14	0
Clothing, Brandy, and Lottery Tickets	3	0	0
School Rate, 2d.; Books, 2s. 2d.	0	7	2
Religious Teaching	0	2	0
	£19	8	4

'This is nearly what the amount of their joint labour would produce, provided they be employed during an average number of days in the year. At a certain distance from the large towns, the items of coffee, sugar, and brandy must in a great measure be omitted, by which the whole expenditure would be reduced to about 15l. sterling per annum.'

Barley, oats, and wheat are largely cultivated in Denmark. Wheat, though plump, is coarse and damp; the barley is heavy; oats of a medium quality; rye, being the principal bread corn of the country, especially of Jutland, is grown in large quantities; this also is the case with rape, beans, tares, buck-wheat, and potatoes, particularly the first, which is a leading article of export to Holland and England. But the principal attention of all the more extensive and intelligent Danish farmers is directed to grazing, salting, and the dairy. The pastures in many parts are little, if at all, inferior to those of Lincolnshire. Horses, cattle, salted pork and beef, butter, wool (which has been much improved), and other animal products are, in fact, in ordinary years, the principal article of export from the country. It appears from an official return ('Statistical Tables relating to Foreign Countries,' part ix.') that there were at the end of 1861, in the whole of the kingdom, 324,550 horses, 1,118,774 cattle; 1,761,950 sheep; and 300,924 swine. The dwelling-houses of the farmers and their office houses are generally contiguous in the same building, but they are notwithstanding sufficiently distinct; and the houses of the better class of farmers are neatly and comfortably furnished.

Manufactures in Denmark are not prosecuted on a considerable scale, nor is their condition at all prosperous. The peasantry in most parts of the kingdom spin and weave linens and woollens, and knit stockings for their own use. Woollens, silks, cottons, and linens are manufactured at Copenhagen and other towns; but the business is languishing and unprofitable. Distillation and brewing are prosecuted to a great extent, and with more success than any other branch of industry, in the capital. Coarse earthenware is made in various places, and a porcelain manufacture is carried on upon account of the crown, and, as might be expected, to its loss. There are also sugar refineries, paper mills, soap works, tanneries, and hat manufactories. With the exception of the manufacture of cannon and arms at Frederickswerk and Hellebæk, the iron and hardware works are quite unimportant. Within recent years numbers of these mills have been constructed, and large quantities of flour are now exported from Copenhagen and Flensburg.

The low state of manufacturing industry is ascribable partly and principally to natural, and partly to political causes. Denmark is essentially an agricultural country, being mostly destitute of coal, of water power, and of the useful minerals, she has no natural facilities for the successful prosecution of manufactures; and, in addition to this, she has little capital, and is deprived of the indispensable stimulus of domestic competition. All, or nearly all, the branches of industry carried on in the kingdom are subjected to the government of guilds or corporations. No person can engage in any business until he has been authorised by its particular guild; and as this is rarely obtained without a considerable sacrifice, the real effect of the system is to fetter competition and improvement. However, the education of mechanics is beginning to be improved by the formation of mechanics' institutions and similar establishments.

Trade and Commerce.—Subjoined is an account of the quantity of the principal articles, the produce of Denmark, exported from that kingdom in each of the years 1861 and 1862 :—

Exports		1861	1862
Animals:			
Horned Cattle	No.	446	463
Calves	„	19	8
Horses	„	198	902
Swine	„	1,801	11,133
Corn, Meal, and Flour	Tönder	2,621,668	2,738,627
Horse Beans	„		
Corn & Potatoe Brandy	Viertels	918,365	109,008
Rape Seed	Tönder	20,191	23,718
Oil Cakes	Lbs.	9,756,926	8,037,651
Potatoes	Tönder	4,562	1,045
Oil	Lbs.	199,624	199,347
Oil, Train	Tönder	10,291	3,525
Meal	Lbs.	1,176,744	1,433,774
Fish	„	1,381,427	649,108
Butter	Tönder	99,646	79,645
Horns and Lard	Lbs.	2,182,786	2,282,846
Milk	Pott		
Hides and Skins	Lbs.	1,848,862	1,885,660
Bones and Glues	„	813	1,330
Turf	Fader		
Wool	Lbs.	2,865,978	2,406,920
Wax	„	7,901	10,194
Tallow	„	294,844	147,400
Sugar and Molasses	„	644,640	669,871

The commerce of Denmark is less than what might be expected from its insular position, surrounded by excellent harbours on every side. But from the earliest times the people, notwithstanding their advantageous situation, could hardly, having but little native produce to export, engage extensively in any branch of foreign trade, except as carriers for others, and in this department they were far surpassed, first by the Hanse Towns, and afterwards by the Dutch. However, since the peace of Stockholm, in 1720, the commerce and navigation of Denmark have gradually improved. During the war between France and Great Britain, down to 1807, the neutrality enjoyed by the Danes gave them great advantages, and occasioned a considerable increase of their mercantile navy. But the attack on Copenhagen by the English in the last-mentioned year, and the hostilities in which the

Danes were consequently involved, deprived them of these advantages, and materially depressed their trade. The loss of Norway, at the general peace of 1814, though it detracted little, if any thing, from the real strength of the monarchy, greatly diminished the importance of Denmark as a naval power, which was again much depressed by the loss of the duchies in 1864. In fact, since the latter period, the commerce of the kingdom has been anything but progressive. At the end of 1862, the mercantile navy comprised 2,763 sailing vessels, of a total burthen of 68,003 lasts (of two tons each), and 44 steamers, of 8,101 lasts, and 2,518 horse power.

Colonies.—In the West Indies, the Danes possess the small but well-cultivated island of St. Croix, producing annually about 23,000,000 lbs. of sugar, and 1,400,000 galls. of rum. Previously to 1803, when the Danes, much to their honour, suppressed the slave trade, they had a considerable intercourse with Africa. But this has since nearly ceased. The trade with the East was formerly in the hands of a company, which was dissolved in 1838. The actual colonial possessions of Denmark consist of the islands of Faroe, Iceland, and Greenland in Europe ; the first-named—17 in number—having a population of 8,651 ; Iceland of 64,603; and Greenland of 9,892 souls. The West India possessions, St. Croix, St. Thomas, and St. John, with a number of smaller islands, have a population of 37,137, according to the census of 1860. The establishments on the coast of Guinea, forts Christiansborg, Fredensborg, and various other places, were ceded to Great Britain, by purchase, in 1850. The town of Tranquebar with the surrounding district, on the Coromandel coast, ceded to Denmark by the rajah of Tanjore, in 1620, and the small territory of Serampore—Danish Frederiksnagor—in Bengal, founded by the Danish East India Company in 1755, were transferred to Great Britain in 1846. The Nicobar Islands, in the Bay of Bengal, were taken possession of by the Danish government in 1756, and for some time were in a flourishing state, the population amounting to above 5,000 in the year 1840. Eight years later, however, in 1848, they were abandoned as useless, nominally on account of their insalubrity.

Races.—Population.—The prov. of Jutland received in antiquity the name of Cimbrica Chersonesus, from the earliest inhabitants being Cimbri or Celts, the ancestors of the Welsh. The Goths, in their progress from the N. and E., took possession of the country of the Cimbri ; and the expatriated inhabitants having been joined by some other displaced tribes, were wandering in quest of settlements, when they were met and entirely defeated by Marius in two great engagements, about 100 years before the Christian æra. After the expulsion of the Cimbri, the peninsula was parcelled among several Gothic tribes, who also took possession of the islands, now forming the principal part of the kingdom of Denmark.

The increase of population has been very considerable for a long time past, and particularly since the beginning of the present century. During the latter period, it was owing, no doubt, to the emancipation of the peasantry ; the breaking down of large estates, and the consequent increase of small properties and farms; the enclosure of commons and the progress made in agriculture ; the introduction of vaccination ; and the improved condition of the bulk of the people. The increase of pop. in the townships has exceeded that in the merely rural districts: but the town pop. is not very considerable ; in fact, if we except Copenhagen and Odense, no town in the kingdom has 12,000 inhabitants, and but very few approach nearly to that amount.

The proportionate increase in the population of Denmark for the last ten years has been larger in the towns than in the country districts. In Copenhagen it has been 8½ per cent, in the other commercial towns together 10·29 per cent, whilst in the country districts it has only been 6·99 per cent.

The titles of nobility in the kingdom of Denmark are only two, count, or earl, and baron; but there is a large untitled noblesse, consisting of the most ancient families in the country, which rank higher in public estimation than many of the modern houses ennobled by the crown.

The occupations of the people are stated as follows in the census of 1860. Out of an average of 1,000 people, 395 live exclusively by agriculture; 220 by manufactures and trades; 187 are day labourers; 53 are commercial men; 29 mariners; 20 paupers; 16 ministers and schoolmasters, or connected with education; 15 pensioners, or people living on 'aftagt' (an allowance to those who cede their farms from old age, &c.); 13 servants; between 11 and 12 hold appointments in the civil offices; 9 are commissioned and non-commissioned officers in the army and navy; 9 capitalists; 7 follow scientific and literary pursuits (including students at the Universities); about 5 have no fixed means of living; and a little over 1 are in prison for crimes or misdemeanours. The increase in the population by births has, on an average, been at the rate of 165 children to every 1,000 women between 20 and 50 years of age. Out of the above number of children, 1 in every 10 or 11 has been illegitimate, and between 4 and 5 per cent. still-born.

Government.—Previously to 1660, the crown of Denmark was elective. The supreme legislative authority was vested in a diet, or assembly, composed of deputies chosen by the nobility, clergy, and commons. But the influence of the nobles predominated very much in this assembly; and they also shared the executive power with the king and enjoyed many immunities. The dissatisfaction of the people with this distribution of power, and still more with the oppressions they too frequently suffered at the hands of the nobles, was greatly inflamed, at the period referred to, by the humiliating treaty concluded in the course of the year with Sweden, and by the refusal of the nobles to submit to bear an equal share of the burdens required by the state of public affairs. In this crisis the partisans of the crown prevailed on the deputies of the clergy and the commons to make a voluntary surrender of their rights, and as the only way of putting an end to the existing dissensions, and of rescuing themselves from the tyranny of the nobles, to confer absolute hereditary power on the sovereign. The nobility, taken by surprise, and unable to make any effectual opposition, were reluctantly compelled to concur with the clergy and the commons.

It is due to the sovereigns of Denmark to state that they exercised these great powers with singular moderation, and there can be no question that the mass of the people were gainers by the revolution of 1660, the results of which were embodied in all subsequent charters and forms of constitutions. The changes in these forms were many; the last of them, sanctioned June 5, 1849, with modifications adopted in January, 1855, containing the constitution now in force. According to this charter, the executive power is in the king and his responsible ministers, and the right of making and amending laws in the *Rigsdag*, or

diet, acting in conjunction with the sovereign. The king must be a member of the evangelical Lutheran church, which is declared to be the religion of the state. The *Rigsdag* consists of the *Landsthing* and the *Folksthing*, the former being a senate or upper house, and the latter a house of commons. The *Landsthing* consists of 50 members. Of these, 12 are nominated by the crown for the term of 12 years, and the rest are elected. To the *Landsthing* any male subject is eligible who is forty-one years of age, who does not labour under mental incapacity, and who either pays 200 rixdollars, or 22l. 14s. 2d. direct taxes, or has a yearly income of 1,200 rixdollars, or 136l. 5s. To the *Folksthing*, consisting of 101 members, any householder twenty-five years of age is eligible, provided he does not labour under any incapacity which would deprive him of the right of voting. This right belongs to every citizen who has reached his thirtieth year, who is not in the actual receipt of public charity, or who, if he has at any former time been in receipt of it, has repaid the same so received, and who does not labour under mental incapacity. The elected members of the *Landsthing* hold their seats for eight, and those of the *Folksthing* for three years.

At the side of these two houses of parliament—the lords and commons of Denmark—is placed a third body, called the *Rigsraad*, or supreme council of the nation. It consists of 47 members, appointed as follows:—

Nominated by the crown	12
Elected by the Landsthing	6
Elected by the Folksthing	12
Elected by the qualified voters in different districts	17
Total	**47**

The *Rigsraad* sits every second year for two months. It may be prorogued once in two years for a period not exceeding four months; and the king can dissolve it at his pleasure. If dissolved it must be re-assembled within four months, and more than two dissolutions cannot take place within a period of two years. The qualifications for a seat in this council are—complete citizenship (that is, the possession of all rights and privileges to which a native-born subject is, as such, entitled), an unblemished personal character, and the absence of any legal claim upon such property as the candidate may possess. The qualification for the direct electoral franchise is, in addition to the first above-named condition—thirty years of age, and the annual payment of 200 rixdollars, or about 22l. in direct taxes; or, an annual income of 1,200 rixdollars, equal to 136l. Private members of the Rigsraad cannot introduce bills, but can petition the crown for their introduction. The ministers take part in the debates of this body in virtue of their office, but cannot vote unless they are members.

The executive government is conducted under the king, by a privy council, and by departments or colleges, each having a minister at its head, in which the public business is transacted. The provinces are all divided into *stifters* or dioceses, and these again into *amter* or bailiwicks; but in the first the government and the administration of justice is committed to different parties, whereas the bailie, *amtman* (prefect), or chief of the administration in the duchies, is also chief judge in their civil and criminal courts. The lowest courts consist of a judge and a secretary, chosen by the proprietors of the district, and confirmed by the king. From these an appeal may be made to the provincial courts, and thence to the supreme court of appeal at Copenhagen. But in order to diminish the expense of justice, all civil cases must in the first instance be carried before a commission of conciliation, composed of the most intelligent and respectable men of the vicinage. Its sittings are private. If both parties agree to abide by the decision of this commission, it is registered, and has the effect of law; if not, either is at full liberty to proceed in a court of justice. The proceedings of the commission are upon unstamped paper, and must be concluded within fifteen days.

The Lutheran is the established religion, and though the most perfect toleration be practised, the numbers attached to other sects is quite inconsiderable. The bishops are nominated by the crown.

Education in Denmark is very widely diffused, there being very few persons, even among the lowest classes, unable to read and write. Besides the university of Copenhagen, there are grammar schools and academies in all the considerable towns. Parochial schools are almost everywhere established; and here, as in Prussia, attendance at school is not optional; for, by a law, all children from the age of seven to fourteen years must attend some public school. Children whose parents are unable to pay the usual school fees are educated at the public expense. The instruction in these schools includes reading, writing, and arithmetic, history, geography, and natural history. The grammar and parish schools are under the superintendence of a royal college of commission, consisting of three assessors and a president. This commission regulates the course of study, and appoints all professors in the university of Copenhagen as well as the masters in the grammar schools. The university of Copenhagen was founded in 1479, by Christian I., and has been augmented and amply endowed by his successors. It is divided into theological, medical, juridical, and philosophical faculties. The professors are either ordinary or extraordinary, their total number being generally about thirty. The examinations are strict, and the proficiency of the pupils very considerable. It is attended by about 600 students.

There is also an asylum for the education of the deaf and dumb in Copenhagen, with two seminaries for the education of schoolmasters, and two for cadets.

Army and Navy.—The army consists partly of regular troops, and partly of a militia or landwehr that is only occasionally called out to be exercised. The peasantry are all, with few exceptions, liable to compulsory service in the army for six years, during two of which they are constantly on duty; while during the other four they are only on duty for a month each year. At the end of the six years they may be enrolled in the militia. A certain number of soldiers are annually chosen by lot, in each district, according to its population, and the exigencies of the state. At present the regular army nominally amounts to 25,000 men; but latterly it has been much more considerable. During the war with Austria and Prussia, 1863–4, there were in the field 49,500 infantry, 10,500 cavalry, and 9,000 artillery, with 144 guns.

The navy consisted in September, 1864, of 19 sailing vessels, carrying 704 guns, and of 24 steamers, with 540 guns. It was manned by very nearly 2,000 men. Since then, however, great reductions have been made in the naval establishment.

Finance.—Previously to the late war, the revenue of Denmark, inc. that derived from the duchies, amounted to about 1,850,000l. a year. It consisted of excise and customs duties, a land-tax which produced nearly 600,000l. a year, a house-tax, the Sound dues amounting of late years to about 217,000l., and other items. But the serious falling off in the amount of the public revenue caused by the loss of the duchies, and the increased charge the crown has had to sustain on account of the war, have occasioned considerable financial embarrassment, and the addition of large sums to the national debt. The financial estimates for the year commencing April 1, 1865, and ending March 31, 1866, which were laid before the Folkething October 8, 1864, give the calculated revenue at 8,772,301 rixdollars, or 969,144l., and the expenditure at 6,161,291 rixdollars, or 681,267l. The loss of Schleswig-Holstein and Lauenburg, it appears from these estimates, brought down the revenue of Denmark to one-half of the former amount. The annual budgets of Denmark show large deficits, amounting, in 1849, to 10,235,911 rixdollars; in 1850, to 9,868,817 rixdollars, and little less in the following years. The kingdom was saved from financial disorder by the payment, in 1858, of the sum of 30,476,325 rixdollars, or 3,524,632l., given in purchase of the Sound dues. To this sum Great Britain contributed the principal share, amounting to exactly one-third. The capital was chiefly employed to pay off a part of the national debt. There still remained, on March 31, 1862, a debt of 96,261,793 rixdollars, or 10,726,179l., to which was added, in January, 1864, a new loan of 1,200,000l., to cover the cost of the war.

As already stated, from 1-5th to 1-4th part of the public revenue is derived from a land-tax, which is charged according to the quantity and quality of the land which each cultivator possesses, and which is measured in *tons of hardcorn.* The Danish acre, or ton of land, is equivalent to 56,000 sq. Danish ft., and 4 such acres are equal to a standard ton of hardcorn, one of the latter being consequently equal to 5½ English acres. But as the same amount of tax is laid on each ton of hardcorn, the size of the latter varies according to the fertility of the land, from 274,000 ft. to 2,240,000 ft. The ton of hardcorn is therefore, in fact, an imaginary measure, which contracts as the quality of the land to which it is applied improves, and expands as it deteriorates.

Provision for the Poor.—A compulsory provision for the support of the destitute poor was introduced into Denmark early in the present century. Each market-town, of which there are 65, and each parish, forms a separate poor district, the affairs relating to the poor of which are managed by a particular board. Every man residing for three years in a parish acquires a settlement in it, and a right to be supported in the event of his becoming unable to support himself; but the principle of the law is, that the pauper shall be supplied only with those things that are absolutely necessary for his support. All begging is strictly prohibited. Opinions differ as to the influence of this law. It took effect in 1803, and the rate is said to have since progressively augmented. The too great multiplication of cottages has been specified as one of the principal causes of the multiplication of the poor. But the probability seems to be, now that the feudal system has been subverted, and that a large portion of the country has got into the hands of small proprietors, that the increase of cottages would have been greater had there not been an advancement for the support of the poor. Savings' banks were introduced into Denmark in 1816; and since then upwards of ten millions of dollars, or above one million sterling, has been lodged in them. It very rarely happens that any one of the petty proprietors either solicits or obtains parish relief.

History.—The early history of Denmark is oberate and uninteresting. In 1385, Margaret, daughter of Waldemar king of Denmark, and wife of Haquin king of Norway, ascended the throne of these kingdoms; in 1380 she was chosen by the Swedes their sovereign; the three crowns being united, it was supposed, for ever, in 1897, by the treaty of Calmar. This great princess, who has been styled the Semiramis of the North, and whose reign is the most glorious in the annals of Denmark, died in 1412. After her death the Swedes began to evince their discontent with the union with Denmark; and, after a lengthened struggle, finally emancipated themselves from the Danish yoke in 1523. In 1448 the race of the ancient kings of Denmark having become extinct, Christian I., of the house of Oldenburg, was raised to the throne, which his posterity still possess; and by this means the valuable provinces of Sleswick and Holstein have been united to the crown, the first immediately, and the latter in 1761 and 1773. The reformed faith was established in Denmark with little difficulty. Lutheranism having been introduced in 1521, Catholicism was suppressed in 1537, the church lands being at the same time annexed to the crown. We have already noticed the memorable revolution of 1660, which had been preceded by a disastrous war, and the loss of the provinces previously held by the Dane in the south of Sweden. From thence down to a late period, there is little of interest in Danish history, other than the introduction of the reforms already alluded to, and the events of the last war. The attack on Copenhagen by the British in 1807, which ended in the capture of the Danish fleet, was an act of very questionable policy on our part, and of which no sufficient justification either has been or perhaps can be made. From this period down to the general pacification in 1815, the Danes were amongst our bitterest enemies. At the conclusion of the war Norway, which had been so long united with Denmark, was assigned to Sweden; the former obtaining in exchange the duchy of Lauenburg and a sum of money. The Danes felt this sacrifice very acutely; but it was one of apparent rather than of real power. A loss certainly greater was that of the duchies of Schleswig-Holstein and Lauenburg, attached to the crown of Denmark for centuries. Unwise legislation on the part of the government, and particularly interference with the language of the majority of the population—a matter on which all subject nationalities are more or less sensitive—brought about a rising in these German provinces in the revolutionary year 1848. But though Prussia took the part of the duchies, the matter was satisfactorily settled in 1852. Eleven years after, however, on the 15th of September, 1863, occurred the death of king Frederick VII., last of the direct line of the house of Oldenburg. Then a new rising took place in the duchies, and Prussia and Austria sending large armies to aid the insurgents, both Schleswig-Holstein and Lauenburg were wrested, after a sanguinary struggle, from the crown of Denmark. The cession of these provinces was legally confirmed by the treaty of Vienna of Oct. 30, 1864, followed by a proclamation of the king of Denmark, of Nov. 17, releasing the inhabitants of the duchies from their allegiance.

DENNY, a market town of Scotland, co. Stirling, 7 m. S. Stirling, and 14 m. NE. Glasgow, on the S. bank of the river Carron, which falls into the Frith of Forth at Grangemouth. Pop. 1,191 in 1861. It is irregularly built. The only public buildings are, a parish church and a dissenting chapel; but it is eminent for the various manufactures carried on either within its bounds or in its vicinity, viz. paper-mills, print-fields, mills for

spinning wool, one for preparing dye-stuffs, and collieries. Handloom weaving, in connection with the Glasgow market, is also carried on to a considerable extent.

DEPTFORD, a town and naval arsenal of England, mostly in co. Kent, laths Sutton-at-Hone, hund. Blackheath, a part being in co. Surrey, hund. Brixton, on the Ravensbourne, at its confluence with the Thames, 4 m. ESE. London, on the railway to Greenwich. Pop. of the par. of Deptford, St. Nicholas, in Kent, 8,139, and of Deptford, St. Paul, in Surrey, 37,634 in 1861. Deptford is contiguous to Greenwich, the two appearing to make only one large town. The lower town, next the river, has narrow irregular streets, and is meanly built, but the upper town is much superior in these respects, and has many handsome modern houses; the whole is lighted by gas, is paved, and supplied with water from the Kent water-works. The old church of St. Nicholas was rebuilt in 1697; that of St. Paul, a handsome structure in the Grecian style, was built in 1730, at which period Deptford was divided into 2 pars. There are also several dissenting chapels, two charity schools, a dispensary for the poor of the town and neighbourhood; a savings' bank; a mechanics' institute; and two sets of almshouses for decayed pilots and masters, or their widows; one founded in the reign of Henry VIII., with 28 dwellings; the other at the end of the 17th century, with 56. This society was incorporated by charter in 4th Henry VIII., when the ancient rights and privileges of the company of the mariners of England was confirmed to them, and they were styled the master, wardens, and assistants of the guild of the Holy Trinity, in St. Clement's, in Deptford Strond. Other charters were granted them by Eliz. and Chas. II. They are now governed by a master, 4 wardens, and 18 other elder brethren; the master and 2 wardens being chosen annually from among the elder brethren, who are elected for life. The number of younger brethren is unlimited, any master or mate sufficiently skilled in navigation being admissible; but they take no part in the business of the corporation, though, like the elders, they enjoy certain immunities, such as exemption from serving on juries, &c. The principal chartered functions of the society are—the examination of the mathematical students of Christ-church, and of masters in the royal navy; the appointment of pilots for king's ships, as well as for piloting merchant vessels on the several coasts and ports of England, except such as are specially placed under other jurisdiction (such as those of the Cinque Ports and the Bristol Channel), and of fixing the rates of pilotage; the erection and maintenance of lighthouses, beacons, buoys, and other sea-marks (with the exceptions previously stated); and the bearing and determining complaints between merchant officers and seamen, the appeal from them being to the Admiralty Court. They have also the power, under certain circumstances, of licensing seamen to ply on the Thames. Their revenue is derived from ancient endowments, contingent benefactions, and lighthouse and other dues, and the surplus, after defraying the expenses of maintaining these and other sea-marks, and other necessary expenses, is, by their charters, to be appropriated exclusively to the relief of decayed seamen and their widows. Between 7,000 and 8,000 is the usual number annually receiving periodical or casual relief, to various amounts. Their affairs were conducted at Deptford till 1787, when the Old Trinity House was pulled down, and they removed to the present structure on Tower-hill. The government dockyard is an enclosed area of 31 acres, with a double and single little dock, 3 building slips, 2 mast-ponds, a mast-house, smithy for forging anchors, several ranges of storehouses, and dwellings for the officers. The victualling office is close to the Thames, and has extensive buildings annexed for baking, brewing, slaughtering cattle, curing meat, and cooperage. During the war, 1,500 artificers and other workmen were employed in the dockyard, but this and the other establishments have since been somewhat reduced. There are a number of private docks, in the largest of which several line-of-battle ships were built during the war. The Ravensbourne forms a small estuary at its entrance, called Deptford Creek, over which is a bridge connecting the lower town with Greenwich. The Surrey Canal locks into the Thames at the north end of Deptford, whence a branch extends from it to Croydon. A railway, raised on brick arches, and extending from Charing Cross to Greenwich, crosses the upper town. There is an earthenware manufactory, a foundry for gun-barrels, and a large establishment, with a laboratory, and several furnaces, for refining the precious metals, and making sulphuric and other acids. The Reform Act included Deptford in a parliamentary borough, comprising also Greenwich, Woolwich, and part of Charlton, which returns 2 members to the House of Commons. It had 5,662 registered electors in 1865. Deptford was anciently called West Greenwich, and after Deep-ford Strand, and was a small fishing village previously to the re-establishment of the dockyard in the 4th Henry VIII. At Saye's Court (the site of the present workhouse), Evelyn, the author of the 'Sylva,' &c., resided, who lent it to Peter the Great in 1698, when that monarch passed some time in Deptford dockyard.

DERA ISMAEL KHAN, an inland town of Afghanistan, cap. dist. Damaun, now belonging to the Maharajah of the Punjab, about 100 yards from the W. bank of the Indus, and 200 m. W. Lahore; lat. 31° 50' N., long. 70° 53' E. It stands in a large wood of date trees, and, when Mr. Elphinstone visited it, was surrounded by a ruined wall of unburnt bricks about 1½ m. in circ. Its inhab. are mostly Beloochees, but some are Afghans and Hindoos.

DERBY, one of the central cos. of England, having N. Yorkshire and a part of Cheshire; E. the cos. of Nottingham and Leicester; S. the latter, Stafford, and a small part of Warwick; and W. Chester and Stafford. Length, from N. to S. about 55 m.; breadth very various. Area, 1,029 sq. m., or 658,848 acres, of which 560,000 are arable, meadow, or pasture. The Pennine mountain chain (see ENGLAND) terminates in this co., and occupies great part of its N. and E. districts. The hund. of High Peak, comprising the NW. angle of the co., is one of the most celebrated mountain districts in England; for though its hills do not soar to the height of those of Cumberland, Westmoreland, and Wales, nor afford the romantic beauties of lakes, cascades, and hanging woods, yet its situation in a more central part of the island, and its extraordinary caverns, perforations, and other curiosities, have made it an object of the greatest interest and attraction. The S. parts of the co. are comparatively flat, and consist generally of strong, heavy land. The climate varies with the elevation of the land and the nature of the soil; but, speaking generally, it is rather cold and bleak. Agriculture is in rather a backward state; farms generally small, and mostly held at will; there are no restrictions on the mode or frequency of cropping. In the N. and some of the W. parts of the co., the dairy is the principal dependence of the farmer. Oats is the principal crop in the High Peak, and wheat and beans in

the S. The drill is but rarely used, and there is a great waste of horse power in ploughing. No particular breed of cattle is preferred. Derby is famous for its minerals and manufactures. The coal-field is of great extent and value; and both lead and iron mines are wrought to a considerable extent. Zinc and copper are also obtained, though in no great quantity; and the spars, which are very elegant, are wrought into a variety of ornamental articles. Silk and cotton manufactures are extensively carried on at Derby, Belper, Chesterfield, Hope, Glossop, and other places. The flax and woollen manufactures are inconsiderable. Porcelain, of a superior quality, is made at Derby; and nails, hats, &c. in various parts of the co. Principal river the Derwent, which traverses nearly the whole extent of the co. from N. to S., dividing it into two pretty equal parts. The Trent crosses the S. angle of the co., and the Dove forms, for a lengthened distance, the line of demarcation between it and Stafford. Derbyshire is divided into 6 hund. and 189 parishes; it returns six members to the H. of C., viz. four for the co., and two for the bor. of Derby. Registered electors for the co. 12,057 in 1865, of whom 5,213 for the northern and 6,851 for the southern division. Pop. 850,527 in 1861, living in 69,262 houses. Gross annual value of real property assessed to income tax—in northern division 624,661l. in 1857, and 679,932l. in 1862; in southern division 1,541,014l. in 1857, and 1,991,785l. in 1862. Principal towns, Derby, Belper, and Chesterfield.

DERBY, a town and bor. of England, in the above co., of which it is the cap., being locally in the hund. of Morleston, but possessing separate jurisdiction, on the Derwent, in a fine valley; 48 m. SxE. Manchester, 119 m. NNW. London by road, and 127 m. by Midland railway. Pop. of bor. 43,091 in 1861. Besides the Derwent, the town is traversed by the Markeaton brook, both of them being crossed by several handsome bridges. It is a very thriving place, and has of late years been much improved. The streets in the older parts are narrow and crooked; but all of them are clean, well paved, and well lighted with gas. There is here a county-hall, a town-hall, assembly rooms, a co. gaol, which is one of the best in the kingdom, an infirmary, and theatre. The most conspicuous, and one of the finest, buildings in Derby is the Athenæum, also containing the Post Office. It stands in the centre of the town, is of Grecian architecture, has two fronts, one of 90, and one of 185 ft. The market place is a large open space in the centre of the town, and there is also a good covered market. There are fourteen churches. All Saints' Church is a Roman Doric edifice; the tower, erected about the time of Henry VIII, is in the perpendicular English style, 178 ft. high. The original church of St. Werburgh is supposed to have been built prior to the Conquest. The church of St. John's is a fine Gothic building; and notable also are St. Alkmund's, rebuilt 1844, with a spire 205 ft. high; and St. Michael's, opened 1858. There are chapels for most classes of Protestant dissenters, and the Catholics have a chapel with a fine Gothic tower. Derby has received many charters; the first from John; its last and (prior to the passing of the Municipal Reform Act) governing charter, from Charles II. It claims to be a bor. by prescription. Under the new municipal act, it is divided into 6 wards, and has 12 aldermen and 36 councilmen. Corp. revenue, 9,125l. in 1862. Annual value of real property assessed to income tax, 145,340l. in 1857, and 168,123l. in 1862. Derby has sent two members to the H. of C. since the 23rd Edward I. Previously to the Reform Act, the right of voting was vested in the freemen and sworn burgesses. Registered electors, 2,564 in 1865. Derby is also the place appointed for the election of the members for the S. div. of the co.

The town has many excellent charitable, educational, literary, scientific, and other institutions. Amongst them are the infirmary, a fine structure, erected by subscription at a cost of about 18,000l., and is replete with every convenience; the self-supporting charitable and parochial dispensary; a lady's charity, for assisting poor women during their confinement; several friendly societies and benefit clubs; almshouses, some of which were founded by the Countess of Shrewsbury, in 1599, for eight men and four women; others by R. Wilmot, in 1636, for six men and four women, now four of each; Large's Hospital, founded 1709, by Edward Large, for five clergymen's widows, subsequently enriched by sundry donations; and 13 neat and substantial almshouses, erected from the fund of a charity bequeathed 300 years ago by Robert Liversage to the par. of St. Peter; with various benefactions for different purposes. There are, also, national, infant, Sunday, and other schools, furnishing instruction to great numbers of children. A free school, founded in the reign of Henry II., is supposed to be one of the most ancient endowments of the kind in England. It was formerly in a very flourishing state; subsequently, however, it fell off very much; but it has latterly been getting into somewhat better repute. The literary and scientific institutions are the Philosophical Society (originally held in the house of Dr. Darwin), with a good library, a collection of fossils, and mathematical and philosophical apparatus; the Town and Country Library, which has been much enlarged, and has a public news-room and museum attached to it; and the Mechanics' Institute.

The town is remarkably well situated for manufactures, having an extensive command both of water power and coal; and mills for the manufacture of silk and cotton have been established either in it or its immediate vicinity. Early in the beginning of last century, Mr. John Lombe, who had, at considerable risk, and by dint of great ingenuity and application, made himself acquainted with the machinery in Italy, erected at Derby a mill for throwing silk on a very large scale; and the town has ever since continued to be a principal seat of the silk-throwing business. The other manufactures comprise stocking, lace, tape, pottery, nails, needles, paper, and railway carriages. The cotton manufacture is not carried on to nearly the same extent as that of silk; but of late years it has been increasing very rapidly in the co.

The town was formerly a great wool mart, and the art of dyeing woollen cloth was supposed to be practised here with peculiar advantage, in consequence of the water of the Derwent being specially adapted for that purpose. Hosiery has long been an important business in Derby. There are also large manufactures of bobbin net; and the weaving of silks and velvets has been introduced of late years. The manufacture of porcelain was originally established here about the year 1750. The ware is not, perhaps, of equal fineness with the French and Saxon, but its workmanship and ornaments are at least equal. The manufacture of figures and ornaments in what is termed biscuit, is extensively carried on. The Blue spar, or 'blue John,' of the vicinity, is wrought into vases and other ornaments; and the black marble of Ashford into vases, columns, and chimney-pieces. Various other factories, besides those specified, are conducted here on a large scale, such as for patent shot, for the construction of steam-engines, for

slitting and rolling iron, for smelting copper ore, for making tin plates. There are also red lead, colour, and varnish works; bleaching-grounds, in which the processes are performed by chemistry; tanneries; soaperies; extensive malting concerns, and corn mills.

Derby communicates by railways and canals with all parts of England. The river was several years since, made navigable from the town to its junction with the Trent; but, since the opening of the Derby canal, the navigation has been disused. This canal branches from the Trent and Mersey canal at Swarkestone, a few miles S. of Derby, runs N., and intersects the Derwent at Derby, a towing bridge being thrown across that river. The Derby canal supplies the town with coals, building stone, gypsum, and other things. Three railways meet at Derby:—1. The Derby and Birmingham; 2. The Midland Counties railway, which connects Derby and Nottingham with each other, and both with the London and North-Western railway at Rugby; 3. The North Midland railway, which connects Derby with Leeds, York, &c. Derby, in fact, is one of the centres of railway communication in England.

Many learned persons have either been natives or inhabitants of this town; among whom may be specified Dr. Thos. Linacre, a learned physician in the reign of Henry VIII.; Joseph Wright, an eminent painter; William Hutton, an industrious antiquary and topographer; Flamsteed, the celebrated astronomer, said to have been educated in the free school; the first Earl of Macclesfield, who, after practising here as an attorney, rose to the highest rank in his profession, having been lord chancellor; John Whitehurst, a scientific mechanist; and the celebrated Dr. Darwin.

DEREHAM (EAST), or MARKET DEREHAM, a town and par. of England, co. Norfolk, hund. Mitford, 15 m. NW. from Norwich, and 1254 NEN. London by Great Eastern railway. Area of par., 6,280 acres. Pop. of do., 4,568 in 1861. The town having suffered much from fires at different periods, has been rebuilt on an improved plan, and is neat and clean. The church, a very ancient structure, with a tower in the centre, has some interesting relics; and in it were deposited, in 1800, the remains of Cowper the poet. There are two annual fairs.

DERG (LOUGH), a lake of Ireland in the SE. angle of the co. Donegal, about 9 m. in circ. This lake, or rather a small island in it, is famous in the history of Irish superstition. In this island there was formerly a cave, called St. Patrick's Purgatory, a pilgrimage to which was long held to be of the greatest efficacy. The cave was, however, shut up in 1630, by order of government, the chapel on the island demolished, and the monks dispersed. It was supposed that this rough treatment had put an end to the delusion; and Bonte, writing soon after, says that 'the pilgrimage to purgatory has quite come to nothing, and never hath been since undertaken.' (Boate's Nat. Hist. of Ireland, p. 73, ed. 1652.) But if so, the practice revived at no distant period, and the island continues, down even to the present day, to be annually visited by crowds of pilgrims. Pope Benedict XIV. wrote a sermon recommending the pilgrimage; and, in 1830, the Catholic bishop of the diocese publicly notified that he would hold a 'station' here. The 'station,' or period for the resort of pilgrims, begins on the 1st of June, and terminates on the 15th of August. The average annual number of pilgrims are estimated at from 13,000 to 20,000. At present the rites are not performed in a cave, but in a chapel. A river called Derg falls into this lake.

DESSAU, a town of N. Germany, cap. of the duchy of Anhalt-Dessau, residence of the duke and seat of government, on the left bank of the Mulde, near its confluence with the Elbe, 67 m. NW. Berlin, and 29 m. N. by W. Leipzig, on a branch of the railway from Berlin to Leipzig. Pop. 15,608 in 1861. The town is walled round, except on the side next the river, which is here crossed by a fine bridge. Dessau is divided into the Old and New Town, the Sand, and three other suburbs, one of which is on the opposite side of the Mulde. It is one of the best built cities of Germany, and contains five public squares, and upwards of thirty good streets, which are well lighted at night. The ducal palace, a part of which was built in 1846, contains a theatre, a good collection of paintings, and other works of art; the palace of the dowager-duchess, the high school, Amelia asylum, riding-school, Catholic church, three Lutheran churches, and the synagogue, are the other public edifices most worthy of notice. Dessau is the seat of the high court of appeal for the duchy and other judicial courts. Its public schools are numerous, and include a gymnasium, teachers' seminary, citizens' primary and female schools, academies of music and singing, the Louisa school of industry, a high female school, and a celebrated Jewish commercial school with which a Jewish classical seminary is united. There are many public charities. The inhab., among whom there are a large number of Jews, long settled here, mostly derive their subsistence from employments connected with the court; but they also manufacture woollens and hats, and have tanneries, distilleries, and an extensive trade in corn. The public cemetery of Dessau is very handsomely laid out, and in the vicinity of the town are the ducal country residences, Luisium and Georgium, surrounded by extensive gardens. Dessau was the birthplace of the philosopher Moses Mendelssohn, born in 1729, deceased in 1786.

DETMOLD. See LIPPE-DETMOLD.

DETROIT, a city of the U. S. of America, cap. Michigan; on the W. side of the strait or channel uniting lakes Erie, St. Clair, and Huron, and consequently in one of the best positions for commanding a large share of the internal navigation of America. It is above the W. extremity of lake Erie; lat. 42° 19′ 53″ N., long. 82° 58′ W. Pop. 45,670 in 1860. The town is irregularly built; but there are some fine edifices. Three of the streets are each 200 ft. wide, the others vary from 60 to 120 ft. in width, and cross each other generally at right angles. There are several squares, and some good private mansions. The R. Catholic cathedral, finished in 1848, is an imposing edifice. Among the other public buildings are the statehouse, city hall, banks, markets, a theatre, museum, state penitentiary, co. gaol, mechanics' hall and various public offices. There are several extensive manufactories, including iron-foundries, a brass-foundry, and breweries. Ship building and the sawing of timber are the most important branches of industry. The city is the great commercial mart and emporium for the state, and the centre of a vast network of railways. The markets are usually well supplied; the fish-market, especially, is one of the best in the W. states. Among numerous charitable institutions there are two orphan asylums, several free schools, a hospital and a poorhouse. There are scientific and literary societies, and good male and female academies. The first steamboat visited Detroit in 1818; in 1852, the city had above a thousand steamers. Among the inhab. are many French, by whom the city was founded in 1670. The Detroit river, or strait, between lakes Erie and St. Clair, is 25

m. long, and upwards of a mile broad. The French settlements extend for a considerable distance along its banks, which are fertile and well cultivated.

DETTINGEN, a small village of Bavaria, on the Mayne, 8 m. NW. Aschaffenburg. Pop. 620 in 1861. Here, on the 26th June, 1743, the allied British and Hanoverian army, under George II. and the Earl of Stair, defeated a very superior French force under Marshal Noailles. The latter lost above 5,000 men killed and wounded; the allies about 2,000. It was the last time a king of England drew his sword in battle.

DEUX-PONTS (Germ. Zwei-brücken), a town of Rhenish Bavaria, formerly the cap. of the duchy of the same name, and at present of the Bavarian circle of the palatinate, on the Erbach, near its confluence with the Serre, 42 m. W. by N. Landau, and 47 m. SSE. Strasburg, on a branch line of the railway from Metz to Mayence. Pop. 8,285 in 1861. The town is pleasantly situated and well built. Here are the ruins of the ancient palace of the dukes of Zweibrücken, formerly one of the most magnificent residences in Germany, but which was for the most part destroyed by the French. What remains of it has been converted into a Catholic church. The cathedral and Lutheran church are amongst the other chief edifices. There are here two bridges across the Erbach, whence the town derives its name. Deux-Ponts is the seat of the high court of appeal for the circle, and contains a lyceum and a gymnasium. It has manufactories of woollen cloth, leather, cotton twist, and tobacco. Here, in 1779, was commenced the publication of the series of editions of the classics, known by the name of the Bipont edition. The undertaking was not, however, completed here, but at Strasburg.

Deux-Ponts and its duchy successively belonged to its own counts, of a branch line of the house of Bavaria, and then to Sweden and Bavaria, previously to the French revolution. It was afterwards taken by the French, and formed a portion of the dép. of Mont Tonnerre; but since 1814 it has again belonged to Bavaria.

DEVENTER, a fortified town of Holland, prov. Overyssel, cap. arrond., on the Yssel, 18 m. N. Zwoll. Pop. 16,342 in 1861. A cathedral 6 other churches, and a town-hall, are amongst its chief public buildings. It is the seat of a court of assize, a tribunal of primary jurisdiction, and several associations of public utility, and the residence of a military commandant. It has manufactures of stockings, carpets and linen fabrics, an iron foundry, and considerable trade in cattle, corn, butter, and other goods; and sends 7 members to the states of the prov. Deventer has sustained numerous sieges, and been several times taken.

DEVIZES, a parl. bor. and town of England, co. Wilts, hund. Potterne and Cannings, on an eminence near the N. limit of Salisbury Plain, on the principal road from London to Bath, and nearly in the centre of the co.; 85 m. W. by N. London by road, and 86 m. by Great Western railway. Pop. 6,555 in 1861. The town consists of several wide streets, branching from a large market-place. The houses are mostly well built. There are two ancient churches, affording specimens of the Norman and pointed styles, with some curious monuments. There are also 5 dissenting chapels; a charity school, educating and apprenticing 40 boys; Lancastrian, national and infant schools; a town-hall, with a circular front and Ionic columns; and a handsome cross, erected in 1815, by Lord Sidmouth. Among the other notable public buildings are the corn exchange, completed in 1857, in the Grecian style, 142 ft. long, and the county lunatic asylum, built 1851, on a site of 80 acres. The markets are Monday and Thurs., the latter for corn, and one of the largest in the W. of England. Malting is extensively carried on. The woollen business, formerly important, has wholly declined. The Kennet and Avon canal passes the town, giving it a water communication with Bristol and London. Devizes claims to be a bor. by prescription, but has several charters. It has returned 2 mems. to the H. of C. since the 4th of Edward III. Previously to the Reform Act, the elective franchise was vested in 36 burgesses and an unlimited number of free burgesses; but of these few were made. The Boundary Act extended the limits of the parl. bor., which had 322 registered electors in 1862. The present municipal coincides with the parl. bor. It is divided into 2 wards, and has 3 aldermen and 18 common-councilmen. Revenue of the corporation, 544l. in 1862. Annual value of real property, 23,639l. in 1857, and 24,732l. in 1862. Petty sessions for the division are held in the town, and quarter sessions for the co., alternately with Salisbury, Warminster, and Marlborough. The origin of the name in old records, Divisa and Divisio, is supposed to be from the division of the place between the king and the bishop of Salisbury. The town owes its rise to an important castle or fortress built here in the reign of Stephen, of which nothing but the mound remains.

DEVON, a marit. co. of England, forming part of its SW. peninsula, and having E. the cos. of Dorset and Somerset, N. the Bristol Channel, W. Cornwall, and S. the English Channel. It is of a rhomboidal shape; area 2,589 sq. m. or 1,657,199 acres, of which about 1,200,000 are arable, meadow, and pasture. Surface and soil various. A great portion of the W. district of the co., from Okehampton on the N. to Oldsworth on the S., and from Ilsington on the E. to near Tavistock on the W., is occupied by Dartmoor, one of the most barren tracts in the kingdom. It includes a space of above 250,000 acres, and is said to have a mean elevation of more than 1,700 ft. above the level of the sea, but we suspect this to be an exaggeration. Its surface is, in most places, extremely rugged; the soil, where it is not encumbered with broken fragments of rock, is thin and poor; and in the most elevated part of the moor there is an immense morass, covering about 80,000 acres, and which is, in parts, incapable of supporting even the lightest animals. That part of the moor, called the Forest, is parcel of the duchy of Cornwall; and on this, and some other of the less barren portions, some improvements have been effected, particularly in the way of planting. But, with the exception of this and a few other districts of very inferior dimensions, the country is alike beautiful and fertile. The vale of Exeter, comprising from 120,000 to 130,000 acres, is one of the richest in the kingdom; and the district called the South Hams, extending from Torbay round to Plymouth Sound, is frequently called the garden of Devonshire, and is finely diversified, and very productive. Climate mild, but moist, though not so much so as in Cornwall. Agriculture, though much improved, is still backward; there is throughout the co., a great want of any regular system of cultivation, and the crops are inferior. Potatoes are extensively cultivated; cyder is largely produced, especially on the W. parts of the co., and is a common beverage; but it is harsh and acid; and these qualities, and the freedom with which it is drunk, are said to occasion the colic prevalent among the natives. Devon is principally a grass-

ing and dairy co. The breed of cattle is excellent; they are of a high red colour, fatten easily, and yield capital beef; are well adapted for field labour, bring, though rather light, docile, and ready to exert themselves to the utmost. The dairy farmers not unfrequently let their cows to dairymen at so much a head. Stock of sheep estimated at between 600,000 and 700,000 head. Property much divided. Farms of all sizes from 10l. to 500l. a year; but the great majority small. Minerals important and valuable; copper and tin mines are wrought to a considerable extent; and lead, iron ore, and manganese are met with. About 28,000 tons of fine clay, raised near Kingsbridge, Bovey, and other places in that part of the co., are annually shipped for Staffordshire and other seats of the china-ware manufacture. The woollen manufacture, though a good deal fallen off, is still carried on to a considerable extent. Principal rivers Exe, Dart, Tamar, Taw, and Torridge. Principal towns, Plymouth, Devonport, Exeter, Tiverton, and Tavistock. Devonshire is divided into 33 hund. and 465 par.; it returns 22 mems. to the H. of C., viz. 4 for the co.; 2 for the city of Exeter; 2 each for the bors. of Plymouth, Devonport, Tiverton, Barnstaple, Honiton, Tavistock, and Totness; and 1 each for the bors. of Ashburton and Dartmouth. Registered electors for the co. 18,215 in 1862, of whom 8,774 for the northern, and 9,441 for the southern division. Pop. 584,373 in 1861, inhabiting 101,355 houses. Annual value of real property assessed to income tax, in northern division 768,649l. in 1857, and 873,053l. in 1862; in southern division, 1,129,729l. in 1857, and 1,262,180l. in 1862.

DEVONPORT, a sea-port town and parl. bor. of England, co. Devon, par. of Stoke Damerell, formerly called Plymouth Dock. Pop. of munic. bor. 50,440, and of parl. bor. 61,783 in 1861. The borough adjoins that portion of Plymouth called Stonehouse on the W.; but though it received its present distinctive appellation in 1824, has a separate municipal government, and returns 2 mems. to the H. of C., it is quite so much a part of Plymouth as the bor. of Marylebone is of London. It will, therefore, be described with the dockyard, breakwater, &c., under the head Plymouth.

DEWSBURY, a manufacturing town, par. and township of England, W. riding co. York, 216½ m. N. London, by London and North Western railway, via Stockport. The par., which contains 9,520 acres, is situated principally in the wapentake of Agbrigg, but partly also in that of Morley; the town of Dewsbury, however, which contains 1,830 acres, is wholly in the former. Pop. of township 18,148 in 1861. The town, situated at the foot of a hill, on the Calder, is 8 m. SsW., Leeds, 9 m. NE. Bradford, and 9 m. NE. Huddersfield; in the very centre, in fact, of the clothing district. The approach to the town by the London road, cut through a deep chasm, has a fine effect; the town lies low, and the smoke of the factories in the distance give it an enlarged appearance. It has a good market-place, with some good streets, and is well lighted with gas and supplied with water. All Saints, the principal church, is of great antiquity; it was rebuilt in 1766, but a good deal of the interior was preserved. Churches have been erected at Dewsbury Moor, Earlsheaton, and Hanging Heaton, and there are several Dissenting chapels. A charity school was founded here in 1769; it has an endowment of about 100l. a year, and about 80 boys are educated as free scholars. Wheelwright's free school, conducted on the national system, was founded in 1727, and is attended by 100 boys, and as many girls. There are here 2 almshouses; but, excepting these, the other charities are of little importance.

Dewsbury is at the head of what is called the *Shoddy* trade. Here refuse woollen rags are collected in vast quantities from all parts of the kingdom; and, after undergoing certain preparations, are torn to pieces, and reduced to their original state of wool, by the aid of powerful machinery; and this wool, being re-spun, is again made into cloth. Formerly, shoddy cloth was used only for padding, and such like purposes; but now blankets, flushings, druggets, carpets and table covers, cloth for pilot and Petersham great coats, &c., are either wholly or partly made of shoddy. The clothing of the army, and the greater part of that of the navy, consists principally of the same material, which, in fact, is occasionally worn by everybody. Large quantities of shoddy cloth are exported. Great improvements have recently been effected, not only in the fabric of the cloth, but also in the dyes; this is especially seen in the cloth for soldiers' uniforms, which is no longer of a brick dust colour, but makes a much nearer approach to scarlet. The beautiful woollen table cloths are made wholly of shoddy, being printed by *manufacta* from designs drawn in London and Manchester, and cut on holly and other blocks on the spot. The trade is of comparatively recent origin, and is rapidly extending itself. About 1,500 men, and as many women and boys, are employed in the mills, and in the manufactures of shoddy in Dewsbury.

DEZPHOUL, a town of Persia, in Kuzistan, in a fine plain, on the Abzal, 25 m. W. by N. Shuster. Pop. estimated at 14,000 or 16,000. Its only ornament is a noble bridge of 72 arches, constructed by command of Sapor. The piers are of stone, and the arches and upper parts brick. It is 450 yards in length, 70 in breadth, and about 40 in height.

DHAN, or DHARANUGGUR, an ancient inl. town of Hindostan, prov. Malwah, cap. of a small Maharatta state under British protection; 24 m. WNW. Indore, and 1,500 ft. above the level of the sea; lat. 22° 35' N., long. 75° 51' E. At one period it is said to have contained 20,000 houses; in 1820 there were less than 5,000, but the pop. was then increasing. It is surrounded by a mud wall, and contains some good buildings and several tanks. The fort, detached from the town, is surrounded by walls about 36 ft. high, with round and square towers. This town is of great antiquity; it belongs are of a most distinguished Maharatta family, and formerly had precedence of both Scindia and Holcar.

DHOLPOOR, an inl. town of Hindostan, prov. Agra, cap. of a small Hindoo principality under British protection; 34 m. SSW. Agra, 25 m. NNE. Gwalior, and 1 m. N. the Chumbul river; lat. 26° 42' N., long. 77° 44' E. It is frequently mentioned by the Emperor Baber in his memoirs, and is still of considerable size; its environs are rich and productive.

DIAMOND HARBOUR, a harbour in the river Hooghly, Hindostan, 34 m. below Calcutta, where the Company's ships usually unload their outward, and receive on board the greater part of their homeward cargoes. Here are government warehouses for ships' stores and rigging, protected by an embankment from inundation; and about thirty years ago an excellent brick road was constructed from hence to Calcutta. The place is very unhealthy; but the adjacent rice lands are in a high state of cultivation.

DIARBEKR, a city of Turkey in Asia (Armenia), cap. pachalik of same name, on the Tigris, by which it is nearly encircled, in a noble plain or table-land, 160 m. SSW. Erzeroum; lat. 37° 55'

37' N., long. 39° 52' E. Pop. estimated 50,000 in 1851. The city is surrounded by a prodigious wall of black stone, which, for height and solidity, is far superior to anything of the kind I have seen, either in Europe or Asia: it has, however, been much neglected, and is now in a ruinous condition. The houses are of stone, and have a good appearance, but the streets, though paved, are narrow and filthy. The castle is on the N. side of the town; it is also surrounded by a strong wall, and divided into many courts and handsome buildings. The bazaar is well supplied with corn and provisions, and the adjoining country is fruitful and well cultivated: cotton, silk, copper, and iron are manufactured, and sent to Bagdad and Constantinople. When viewed from a distance, it has a fine appearance. The elevation of the surrounding mountains, the windings of the Tigris, and height of the walls and towers, with the cupolas of the mosques, give it an air of grandeur far above that of any city which I have visited in this quarter of the world. The river is generally crossed on a bridge of twelve arches, about ½ m. below the town. (Kinneir, Memoir of Persian Empire, p. 854.) The city, formerly very important, fell subsequently into complete decay, but is now arising from its ruins. According to the estimate of former writers, the ratio of decline took place on the following scale:—

Arezini	in 1757, gives	Diarbekr 400,000 souls
Ives	1758	400,000
Niebuhr	1766	50,000
Kinneir	1792	50,000
Gardanne	1808	50,000
Dupré	1808	75,000
Kinneir	1810	38,000
Buckingham	1816	50,000
Heude	1817	45,000
Brant	1841	45,000
Southgate	1847	15,000
Mohle & Mühlbach	1856	20,000
Holmes	1857	70,000
Taylor (Consul)	1864	50,000

According to a report of Mr. Consul Taylor, addressed to the Foreign Office, in 1864, Diarbekr has entered upon a new prosperous career. The city was founded, or, more probably, restored, by the emperor Constans, anno 349. It is sometimes called Amid or Emid, and is described under this name by Abul-Feda.

DIE (an. *Dea Vocontiorum*), a town of France, dép. Drôme, cap. arrond., on the Drôme, 26 m. SE. Valence. Pop. 3,365 in 1861. It is surrounded by a wall flanked with numerous towers; is clean and well built; has a cathedral, an ancient episcopal palace, many Roman remains, and silk fabrics, tanneries, rope-walks, and paper-mills. In the 16th century the Calvinists were very numerous, and had a university here.

DIE (SAINT), a town of France, dép. Vosges, cap. arrond., on the Meurthe, 24 m. ENE. Epinal. Pop. 6,786 in 1861. It is well situated and well built; is surrounded by an ancient wall; and has a communal college, and a public library with 8,500 vols. It is the seat of a sub-prefecture, court of original jurisdiction, and a bishopric, of which the dep. Vosges forms the diocese. There are some fabrics of cottons, handkerchiefs, stockings, and potash.

DIEPPE, a marit. town of France, dép. Seine-Inférieure, cap. arrond., at the mouth of the Arques, on the British Channel, nearly opposite Beachy Head, from which it is distant 67 m. SSE., 31 m. N. Rouen, and 92 m. NW. Paris, on a branch line of the railway from Paris to Le Havre. Pop. 20,187 in 1861. The town is well built; streets broad, regular, and one of them 3-4ths m. in length; houses mostly of brick, and ornamented with bal-

N 2

contins. It consists of two parts,—the town properly so called, and its suburb of *Le Pollet*, separated from it by the port, but communicating with the town by a bridge. Dieppe is well supplied with water, which is conveyed by an aqueduct excavated in solid rock for 3 m., and distributed to 60 public, and above 300 private, fountains. Its port, enclosed by two jetties, and surrounded by quays, is capable of accommodating a great number of vessels of from 60 to 600 tons; but it dries at low water, is with difficulty kept from filling up, and is rather of dangerous access from its narrowness and the rapidity of the current both inwards and outwards. It is protected by an old castle on a cliff to the W. of the town, and by some batteries. The town has two churches, from the steeple of one of which, St. Jacques, the English coast may be seen. Since 1822, when a handsome establishment for sea-bathing was formed, Dieppe has become a favourite watering-place, and the number of visitors has continued to increase. It is the seat of a court of original jurisdiction; has a communal college with a public library containing 4,000 vols., a school of navigation, a hospital, and a theatre. Ivory articles are made here in greater perfection perhaps than in any other part of Europe; and there are some sugar-refineries, tanneries, rope-walks, and building docks for trading vessels. The manufacture of lace, for which this town was once distinguished, has now much diminished. Dieppe is an entrepôt for salt and colonial produce, and has a considerable trade; but by far the greatest portion of the inhabs. depend for support on the fisheries, especially those of mackerel and herrings, for the supply of the capital. Vessels are also fitted out for the cod and whale fisheries. There are two oyster-tanks, whence about 12,000,000 oysters are annually sent to Paris. A regular intercourse is kept up by steam-boats between Newhaven, near Brighton, and Dieppe; and as the journey to Paris by land is much shorter by this than by Calais or Boulogne, it is preferred by many travellers. Dieppe was bombarded and all but destroyed by the English and Dutch in 1694. The inhab. have been distinguished by their enterprise. They discovered Canada, founded Quebec, and explored the coasts of Africa to some distance a century previously to Vasco de Gama. Bruzen de la Martinière, the author of the Grand Dictionnaire Géographique, was a native of Dieppe.

DIEST, a town of Belgium, prov. S. Brabant, cap. cant., on the Demer, and on the railway from Antwerp to Liège, 32 m. ENE. Brussels. Pop. 8,521 in 1856. The town is about a league in circ., but this space is in great part occupied with fields and gardens. It has a college, manufactures of stockings, woollens, &c., and some excellent breweries. It was taken by the Duke of Marlborough in 1705, but retaken and dismantled by the French in the same year.

DIGNE (an. *Dinia*), a town of France, dep. Alpes-Basses, of which it is the cap.; at the foot and on the declivity of a hill, on the Bléone, 55 m. NE. Aix, 78 m. SSW. Grenoble, and 573 m. SE. Paris. Pop. 5,341 in 1861. The town is encircled by ancient walls flanked with square towers. Streets generally narrow, and the houses mean. In its vicinity are some saline baths, serviceable in rheumatic, paralytic, and cutaneous affections and gunshot wounds. One of the avenues to the town is planted with trees and bordered with handsome houses. Digne is the seat of a court of primary jurisdiction, a court of assize, a tribunal of commerce, a communal college, and a bishopric; but neither its principal church nor episcopal palace deserve notice. It has a public library with 3,000 vols., a society of agriculture, departmental nursery grounds, tanneries, and some trade in printed, almonds, corn, hemp, cloth, cattle, and leather.

DIJON (an. *Divio*), a celebrated town of France, dep. Côte d'Or, of which it is the cap.; as it was formerly of the duchy and prov. of Burgundy, in a fertile plain at the foot of the Côte d'Or Mountains, on the Ouche, at the confluence of the Suzon; 195 m. N. Lyons, and 160 m. SE. Paris, on the railway from Paris to Lyons and the Mediterranean. Pop. 37,054 in 1861. It is surrounded by ramparts planted with trees, and is for the most part well built. Its streets are broad, well paved, and clean; and it contains several large and fine squares. Its environs are extremely beautiful, and few towns in France possess such fine public walks. The Suzon, running in various subterranean channels through different quarters, contributes to the cleanliness for which Dijon is conspicuous. In the *Place Royale*, constructed in the form of a horse-shoe, is the palace which has succeeded to the ancient castle of the dukes of Burgundy, the greater part of which was destroyed by fire in 1502. A large square tower formerly belonging to this castle, and called *La Terrasse*, now serves as an observatory; the palace, which was finished in 1784, was destined for the reception of the states of the province, and for the residence of the Princes of Condé, who, under the old régime, were its hereditary governors; its magnificent suite of apartments is now occupied by the museums of painting and sculpture. The castle of Dijon, commenced by Louis XI., and terminated under Louis XII. in 1513, became in the 16th century a state-prison, in which the Duchess of Maine, Mirabeau, and other distinguished persons were confined; it now serves for the quarters of the gens-d'armes. Several of the churches are well worthy of notice. That of St. Bénigne, in which the installation of the dukes of Burgundy took place, was founded in the 5th century, and rebuilt in 1108. It suffered materially during the revolution; but its spire is still standing, and reaches to the height of 96 mètres, or 324 ft. above ground. The churches of Notre Dame and St. Michael are remarkable alike for their antiquity, the beauty of their architecture, and the magnificence of their ornaments. The church of St. Anne is an elegant modern structure, with a fine dome. The hall of justice is in a large ancient edifice, and the theatre is next to that of Bordeaux, the handsomest in France out of Paris. There are two public libraries, one of which has 40,000 printed vols., and 500 or 600 MSS.; a cabinet of natural history, and a botanic garden, 2 hospitals, an orphan asylum, 2 prisons, a town-hall, hotels of the prefecture and academy, and many private residences built during the independence of the duchy, which give to the city a venerable and interesting appearance. Dijon is the seat of an imperial court for the deps. Côte d'Or, Haute Marne, and Saône et Loire, a court of assize, tribunals of primary jurisdiction and of commerce, and of a bishopric; and is the head-quarters of a military division. It has an academic university, with faculties of law, science, and literature, a royal and 7 other colleges, a superior school of the fine arts, a secondary school of medicine, a primary normal school, and numerous learned societies. It has some fabrics of linen, cottons, and woollen stuffs, vinegar, mustard, for which it is famous, wax-candles, hats, earthenware, soap, &c., leather, sugar and wax refineries, tanneries, and breweries; but its principal dependence is on its wine trade, it being the principal depôt and market for the sale of Burgundy. This town existed previously to the period of Roman domination; it was fortified, and, according to some, rebuilt by Marcus

Aurelian; and enlarged and embellished by Aurelian, anno 274. In 1117 it was burnt down, but soon rose from its ashes: it was annexed to France, with the rest of Burgundy, in 1447. It has produced many very distinguished men; among others may be specified Bossuet, the glory of the Gallican church, born here on the 77th Sept., 1627; Crebillon, Piron, Longepierre, Daubenton, and Guyton de Morveau.

DINAGEPOOR, an inl. distr. of Hindostan, prov. Bengal; between lat. 24° 48' and 26° 16' N., and long. 88° 1' and 89° 11' E.; having W. and N. the distr. Purneah, E. Rungpoor, and S. Rajshaye; length N. to S. 105 m., breadth 60 m.; area, 5,374 sq. m.; pop. estimated at 2,500,000, nearly two-thirds of whom are Mohammedans. The district contains no mountains, nor even hills; but its surface is undulating; it is everywhere intersected by rivers, the principal of which are the Mahananda, Atreyi, and other tributaries of the Ganges. There are no lakes, but in the rainy season some of the rivers swell out into extensive marshes; and as they are constantly changing their courses, their deserted channels often contain a considerable expanse of stagnant water. The winds are more variable here than in any other part of India, but for the most part E. The rainy season commonly lasts from the middle of June to the middle of Oct. Towards the end of this season the nights are hot and oppressive, but the maximum heat is not so great as at Calcutta. From Nov. to Feb. the climate suffers much from cold, and then are agreeable to Europeans. The E. winds are accounted very unhealthy; and intermittent and other fevers annually destroy a great many of the pop. The soil is in some parts a red and stiff clay, unusual in Bengal; but by far the larger portion is light and sub-enhanced. Nitre was formerly made in this district, but the soil is not peculiarly adapted for it, and its manufacture has been removed to more favourable situations. The banyan, mango, areca, &c. flourish; palms, generally do not thrive. There are some small seed forests; but, generally speaking, timber is inferior, and useless for boat-building. There are few tigers or leopards, no wolves or hyenas, and the wild elephant or rhinoceros is very rarely seen. Deer, hares, porcupines, ichneumons, otters, &c., are very plentiful; and wild boys and buffaloes do much damage to the crops. Birds are abundant, and so are fish; the last form by far the greater part of the animal food consumed. Crocodiles are not uncommon, but are little dreaded; tortoises, and some lizards, are eagerly sought for as articles of diet; insects are not very troublesome. About two-thirds of the land is fully occupied and cultivated; rice is the principal article of culture, but is inferior to that of Patna; wheat, barley, millet, legumes, and oily seeds are successively the produce next in importance. The cotton raised in the N. is very bad, but that of the S. is finer than that imported from the W. of India; the sugarcane is largely cultivated, and is of a good quality; indigo and tobacco are also raised, but the latter not in sufficient quantity for home consumption. The husbandry of the district is deplorable; the plough is without a coulter or mould-board, and in some parts wants even the share; all the other farm implements are nearly as bad; and 5s. will buy all that are deemed necessary for the culture of five acres of land. Both the oxen and horses are wretched, except a breed of ponies from Bootan; all cattle are, however, ill fed, and on some but natural pastures. There are very few carts or conveyances of any kind in the district. The farms are generally small; about one farmer only in sixteen may rent from 30 to 100 acres; estates are

also generally small; most of the land belongs to Hindoos. There are, however, very few individuals in a state of beggary, and such as are so are readily relieved, the dispositions of the people generally being charitable. Except those of Europeans, and some Mohammedans, no houses have any other than a thatched roof: mud walls are most common; but, in some instances, the huts are wholly constructed of straw and reeds. The furniture of both Mohammedans and Hindoos is nearly alike, and the whole, amongst the labouring classes, not worth more than a rupee. Most of these classes sleep on mats-cloth or mats on the ground. They are generally very ill-clothed, but both Hindoos and Mohammedans wear many ornaments; the women of both ranks colour their eyelids with lamp-black. For food, the people are generally better off than for lodging, furniture, or clothing; and few are distressed by hunger, although their food is seldom very nourishing; the lower classes are obliged to use the ashes of the plantain root, &c. for salt, and often want for tobacco, their favourite and almost only stimulus. Slaves are few, and servants, especially female ones, scarce; for early marriages are so universal, that nearly every woman is married by the period of puberty, or is else subject to a stigma. The inhab. as might be expected from their poverty, are feeble, sickly, and subject to various diseases; and are also ignorant, mendacious, and occasionally rapacious. Thievity was formerly a very prevalent crime. Education has proceeded to very little more than rudimental instruction among about 1 in 16 of the male sex. Christianity has made but little progress.

DINAGEPOOR (Dinajpur, the abode of beggars), an inl. town of Hindostan; cap. of the above distr.; seat of the British judicial and revenue courts; between two tributaries to the Ganges; the m. ESE. Purneah, and 86 m. NNE. Moorshedabad; lat. 25° 37' N., long. 88° 43' E. Pop. estimat. at 35,000, chiefly Mohammedan. It is, as its name implies, a very poor place: its houses are chiefly thatched huts, there being, according to the latest accounts, but eight brick dwellings incl. of 5,000, exclusive of the European residences and public offices, which are built in the worst Anglo-Indian style. Its most densely peopled portion has for its centre a square surrounded with shops; in the English quarter, and other portions, the houses are detached from each other, and intermixed with gardens and pasture lands. What may be considered the port of the town, on the bank of the Punabhadra, is occupied by merchants and warehouses. It is clean and well watched, but not lighted; the roads round it are kept in good repair by convicts, but bridges are wanted. It contains no public building of any importance, excepting the house of the late rajah, built in 1780, a strange mixture of European, Moorish, and Hindoo style, surrounded by a ditch and rampart; but now in great measure gone to decay. The vicinity of Dinagepoor has a sandy soil, is ill supplied with water, and chiefly occupied by pastures.

DINAN, a town of France, dép. Côtes-du-Nord, cap. arrond., beautifully situated on a height on the left bank of the Rance, 18 m. E. St. Brieuc, on the Northern railway. Pop. 8,089 in 1861. The town is surrounded by walls of extraordinary height and thickness, the works outside of which are now converted into gardens, and laid out as public walks. Streets mostly ill built, narrow, and dirty; though of late years some parts of the town have been much improved. The principal public buildings are,—a castle, built about 1300, now used as a prison, two churches of Gothic architecture, the town-hall, clock-tower, hospital, and convent-hall. Vessels of from 70 to 80 tons come up to Dinan at high water: it communicates with Rennes by the canal of Ille and Rance, the river being navigable only as far as this town. Dinan is the seat of a tribunal of primary jurisdiction, and of a communal college: it has a school of design, a public library, and a society of agriculture, with manufactures of sail-cloth, cotton stuffs, flannels, shoes, and hats for the troops and colonies, leather, tanning, sugar-factories; and has some trade in butter, hemp, and thread.

DINANT, a town of Belgium, prov. Namur, cap. arrond., on the Meuse, 14 m. S. by E. Namur, on the railway from Namur to Luxembourg. Pop. 7,210 in 1856. The town is built on the declivity of a rocky hill, on the summit of which is its castle. It is divided into the ' Town-proper' and the 'Island,' and has a suburb, a Gothic cathedral, several other churches, two hospitals, and a Latin school. Its manufactures are chiefly woollen, hats, cutlery, cards, verjuice, mead, gingerbread, paper, and glass; but it has several oil, flour, and hemp mills, with mills for cutting and polishing marble, and numerous salt-refineries, tanneries, and breweries. It is the seat of a court of primary jurisdiction, and the residence of a military commandant. Dinant is very ancient. In the 14th century it was a prosperous commercial town; in 1466 it was sacked and burnt by Duke Philip of Burgundy; and again sacked in 1554 by the Duke of Nevers.

DINAPOOR, an inl. town, and British military station in Hindostan, prov. Babar, on the S. side of the Ganges, 14 m. W. Patna; lat. 25° 37' N., long. 85° 5' E. The cantonments are large and handsome, with a fine quay, three extensive squares of barracks for the European troops, uniformly built, of one lofty ground story, well raised, stuccoed, and furnished with verandas; there are also large barracks for the native troops. The garrison consists of about 6,000 men, one-fourth of them Europeans. The town is well supplied with European goods; and in its neighbourhood potatoes are largely cultivated by both Europeans and natives.

DINGLE, a marit. town of Ireland, co. Kerry, prov. Munster, on a slope at the bottom of Dingle Harbour, on the N. side of Dingle Bay, 26 m. W. by S. Tralee. Pop. 4,327 in 1831, and 2,360 in 1861. Many of the houses are built in the Spanish fashion, it having formerly maintained an intimate communication with Spain. The par. church and Rom. Cath. chapel are modern buildings: a second chapel is attached to a nunnery; and a large national school-house has been erected. It has a dispensary, and is a constabulary and coastguard station; and is frequented during summer as a bathing-place. The harbour is fit only for small vessels, which lie aground on mud at low water. The corporation, under a charter of Jas. I., in 1607, consists of a sovereign, 12 burgesses, and a commonalty. It has jurisdiction over a district of land extending ¼ Irish m. in every direction from the par. church; and that of the sovereign, as admiral of the harbour, is determined by the flight of an arrow discharged from the harbours of Dingle, Ventry, Smerwick, and Fernter's Creek. It returned 2 mems. to the Irish H. of C. till the Union, when it was disfranchised. General sessions are held twice a year, and petty sessions on alternate Fridays. The linen manufacture, which formerly flourished here to a considerable extent, is now confined to that of coarse cloth in small quantities: grain and butter are the chief articles of trade. The fishery, also, which had been very productive, large numbers of herring, flat and round fish, having been taken, is in a very depressed state, so much so, that the town is supplied with cured fish from foreign markets.

DINGWALL, a royal and parl. bor. and seaport town of Scotland, co. Ross, on the W. extremity of the Cromarty Frith, 19 m. NNW. Inverness, on the railway from Inverness to Inverness. Pop. 1,789 in 1841, and 2,081 in 1861. The town is built in the Dutch fashion, and is rather neat, consisting of one leading street, with several inferior ones branching from it. The harbour was originally at an inconvenient distance; but in 1815–17, a canal was formed (at an expense of 4,364l.), by which vessels of considerable burden are now brought to the immediate vicinity of the burgh. The annual revenue derived from the canal is about 130l., which is not more than sufficient to keep it in repair. The beautiful valley of Strathpeffer, at the head of which is a famous mineral spring, stretches W. 5 m. from Dingwall. The town-house is a venerable edifice, with a spire and clock; a plain parish church and a gaol are the only other buildings worth notice. Dingwall is a place of little or no trade. There are but few vessels belonging to it; and its exports consist exclusively of wheat and other country produce, and its imports of lime and coals. The charter of its erection into a royal burgh was granted in 1277, by Alexander II. On the E. of the town may still be seen the remains of the Castle of Dingwall, a fortified place, long the chief residence of the noble family of Ross. Near the church, on an artificial mound, stands an obelisk 57 ft. high, erected as a family burial-place by George, first earl of Cromarty, secretary of state for Scotland, in the reign of Queen Anne. Dingwall unites with Wick, Cromarty, Dornoch, Kirkwall, and Tain, in sending a member to the H. of C., and had 112 registered voters in 1864.

DION, NOMBRE DE, a town of Mexico, state Durango, on the road between Durango and Sombrerete, 60 m. SE. the former city. Pop. estim. at 8,000. Its chief source of wealth is an extensive trade in Vino Mescal, a spirit obtained from the American aloe.

DIZIER (ST.), a town of France, dép. Haute Marne, cap. cant., on the Marne, at the point where it becomes navigable, 47 m. NNW. Chaumont, on the railway from Vitry to Chaumont. Pop. 8,077 in 1861. The town was formerly well fortified, and in 1544 sustained a memorable siege by the emperor Charles V.; but its ramparts have been converted into agreeable promenades. It is a handsome town, with broad streets, and houses mostly of stone. The town-hall, of recent construction, is much admired. It has a hospital, the ruins of an ancient castle, with cotton fabrics, iron-foundries, and a considerable trade in wood. Many vessels are built here for the navigation of the Marne, the town being environed by a forest, whence the materials are easily procured. A part of the allied army which invaded France in 1814, was defeated with great loss at St. Dizier on the 27th Jan. and the 27th March by Napoleon. It was here also, after the breaking up of the congress of Chatillon, that the Duc de Vicenza (Caulaincourt) announced to Napoleon that he must abandon all hope of treating with the allied sovereigns.

DJEBAIL, or GIBYLE (an. Byblos, Byblus), a coast town of Syria, N. by W. Tripoli; lat. 34° 7' N., long. 35° 37' E. Pop., according to Volney, 6,000, but this is probably exaggerated; the inhab., according to the older Maundrell, and more recent Robinson, being few. They are chiefly Maronite Christians. An old castle on the N., built with stones of an enormous size, and the wreck of a very handsome church of great antiquity, are the principal remains; but shafts, columns, and other ruins are scattered about in great profusion. The walls are 1½ m. in circ., with square towers at interrals; an artificial harbour formerly existed, but has been long destroyed; and the town is evidently in a state of gradual, if not rapid decay. At a few m. distance on the S. flows the Nahr Ibrahim (an. Adonis), a short, but deep and rapid stream, over which is a well-built stone bridge of one arch. The surrounding soil is fertile, and peculiarly favourable to the growth of tobacco. The land of the Gibhites (Bollites) is mentioned in Josh. xiii. 5, and this town was evidently a place of considerable importance in the mercantile and maritime kingdom of Tyre. (Ezek. xxvii. 9.) Byblus occupies a distinguished place in Syrian mythology, from its being the scene of the death of Adonis or Thammuz, and a principal seat of the religious rites connected therewith. It may be mentioned, in reference to this subject, that wild boars are still very common in the surrounding mountains, and that the phenomenon mentioned by Lucian, of the river acquiring a reddish colour at certain seasons of the year, has been observed by Maundrell and other travellers, and is occasioned by the washing down of particles of red earth during heavy rains. Milton has beautifully alluded to this legend :—

    ' Thammuz came next behind,
Whose annual wound in Lebanon allur'd
The Syrian damsels to lament his fate
In amorous ditties all a summer's day,
While smooth Adonis, from his native rock,
Ran purple to the sea, supposed with blood
Of Thammuz yearly wounded.'

Byblus was a considerable sea-port under the Greek kings of Syria, but the existing remains are mostly of the Roman period. It was a favourite with the emperor Adrian, who appears to have been peculiarly attached to the worship of Adonis (see BETHLEHEM), and to whom an inscription near the land-gate still exists in good preservation. At an early period of the Crusades, it was captured by the Christians, who built its present walls; but in the furious wars of those fanatical age, the port and trade of Djebail shared the ruin of the other cities of the coast. Still, in its decay, it is the cap. of the Kesrouans (the coast between Tripoli and Beirout) and the see of a Maronite bishop. (Strabo, xvi. 755; Lucian, De Deâ Syriâ, 7; William of Tyre, xi. caps. 9 and 14; Maundrell, 44–46; Volney, ii. 148; Burckhardt's Trav. Syr., 179; Robinson, ii. 49–52.)

DJIDDA, or JIDDA, a marit. city of Arabia, in El-Hedjaz, being the port of Mecca, and one of the chief entrepôts for foreign commerce in the peninsula. Lat. 21° 13' 47" N., long. 30° 6' E. Resident pop., according to Ali Bey, 5,000, but this number is often much increased by the influx of strangers. The inhab. are nearly all foreigners, or settlers from other parts of Arabia; the only natives being a few sheriff families attached exclusively to the offices of religion and law. Five mosques, poor and mean, the governor's house, and a small castle, mounting nine or ten guns, are the only public buildings, except the khans, which are numerous and handsome. The houses in the town, built of stone and madrepore, are, from the perishable nature of the material, not very lasting; but in the suburbs they are mere huts, constructed of reeds and brushwood, inhabited principally by Bedouins. The streets are unpaved; but Djidda is, notwithstanding, cleaner, and in other respects superior to most Eastern cities of equal size. It is one of the holy places of Mohammedanism, and its sanctity is increased by the neighbourhood of the reputed tomb of Eve, a rude stone structure, about 2 m. to the N. The surrounding country is a bare desert, destitute of running streams; and though well water is easily

procurable, it is generally bad. The inhab. collect the rain in cisterns, and the commonest necessaries are brought from a distance. Corn, rice, butter, sugar, tobacco, oil, clothing, &c., are imported in very large quantities from Egypt, the Abyssinian coast, and (excepting butter) even from Persia and India. Djidda depends, therefore, for its existence upon its trade, which is very extensive, and wholly of the transit kind. From the interior dates, and the celebrated balm of Mecca, are brought for shipment westward; musk, civet, and incense are procured from Abyssinia; muslim, cloths, cambrics, teak timber, cocoa-nuts, cocoa-nut oil, pepper, ginger, turmeric, shawls, and tissues, are brought from India; the Malay Islands send spices and female slaves for sale at the Mecca market. The coffee trade, which, next to that of grain, was formerly the most important, has much declined of late, partly owing to the free admission of American produce to the Mediterranean, but principally to the impolitic exactions of the pachas of Egypt upon this branch of commerce. A trade in slaves is carried on with the Mozambique coast; and, altogether, it is calculated that the port of Djidda employs 250 vessels, great and small. The imported articles are conveyed by ships to Suez, whence they find their way to the Mediterranean ports, or by caravans to Mecca and Medina, from which cities they are again dispersed to Syria, Asia Minor, and Turkey. The caravans to Mecca start daily, those to Medina every forty or fifty days; but, besides these, Djidda carries on no land trade, except occasionally with N. Yemen for corn. The duties upon coffee were formerly 7½ per cent., they are now double that amount; those upon Indian goods are from 6 to 10 per cent, according to quality; the trade in grain is monopolised by the Egyptian government. Twice at least in every year Djidda is crowded with strangers, viz. on the arrival of the Indian fleet (about May), when merchants from all quarters pour in to purchase at the first hand; and during the hadj, when pilgrims come from all the African ports in vast numbers. In some years above 20,000 pilgrims land either at Djidda or Yembo, but mostly at the former. There is no manufacture in the town; everything, for use as well as for consumption, is imported, and the occupation of the poorer as of the richer inhab. consists almost exclusively of barter.

Abul-Feda (At. Dev., 60) supposes Djidda and its neighbourhood to be the Hades Hepium (Badia Basilevae) of Ptolemy (vi. 7, viii. 6); but Niebuhr with more reason believes the ground on which the city stands, to have been recovered from the sea within a short period. At some distance from the shore, he describes high sand hills, full of shells and corals; and the general appearance of the coast makes it impossible, in his mind, that the modern town can occupy the same site with its namesake, in the days of Mohammed. ' Djidda,' he says, ' s'avancera depuis en plus vers l'ouest;' and in fact, although a city of this name has been, for ages, the port of Mecca, yet the town now existing is evidently of modern origin. The sultan sheriff of Mecca, as sovereign of the Beled-el-Harem (Holy Land), has possessed Djidda since the first days of Islamism; a pacha, first appointed by the caliphs, and then by the grand signior, as head of the Mohammedan faith, was indeed the nominal governor; and, professedly, the customs were to be divided equally between him and the sultan sheriff. The latter, however, in the declining days of Turkish power, paid little regard to this arrangement, and in the end expelled the Turks entirely from El-Hedjaz. Scarcely was this effected when the growing power of the Wahhabees

became more formidable than that of the party.
Mecca and Medina were taken, and the sheriff,
shut up in Djidda, made a public but doubtful
profession of the Wahabee faith. In 1811, Mo-
hammet Ali established his power in El-Hedjaz, the
reigning sheriff was carried to Cairo, and his successor, appointed by the Egyptian pacha, retained
only a shadow of authority, with a monthly stipend in lieu of the port dues. (Abul-Feda, Ar.
Des. 59, 60 ; Niebuhr, Des. de l'Ar. 303–309 ;
Voy. Ar., i. 217–228 ; Lord Valentia, iii. 301–352 ;
Ali Bey, ii. 40–46 ; Burckhardt, i. 1–109 ; Well-
sted, ii. 208–289.)

DNIEPR (the *Borysthenis* of the ancients), a
large river of European Russia. It has its source
near the village of Dnieprousk, in the government
of Smolensk, and, pursuing a S. course past Smolensk, where it becomes navigable, Mogheleff, Kieff,
Ekaterinoslaff, and Kherson, unites with the Black
Sea about 60 m. below the latter, after a course of
above 1,200 m. Its principal affluents are the
Pripet, Beredna, and Desna. It is broad and
deep, and may be navigated with ease and safety,
from Smolensk as far as Ekaterinoslaff; but from
the latter to Alexandrofsk it is interrupted by
cataracts, which cannot be passed by any sort of
craft, except in spring after the *debacle*, and in the
latter part of autumn. Works were begun in 1833
for obviating these obstructions, an object of vast
importance to N. Russia; but we have not learned
what has been their success. What is called the
bar of the Dnieper lies about 15 m. below Kherson,
and between it and the town the water is shallow,
and the channel encumbered with shifting sands.
There are valuable fisheries below Kherson, and
in other parts of the river. (See KHERSON; see,
also, Hagemeister's Report on the Black Sea, p. 63,
English trans.)

DNIESTER (the *Tyras*, or *Danaster*, of the ancients), a large river of SE. Europe. It has its
source in the Carpathian mountains in Galicia,
and flowing in a SSE. direction along the E. frontier of Bessarabia, falls into the Black Sea between
Ovidiopol and Akerman, after a course of about
600 m. It has no very considerable affluents,
and being in most parts shallow and rapid, it is
of little service to internal navigation, except
during spring and autumn.

DODONA, a town of Epirus, famous in antiquity for its being the seat of an oracle of Jupiter,
the most ancient in Greece, and second only to
that of Delphi in celebrity and importance. It
appears to have been instituted by emigrants
from Egypt; at least this is the opinion of Herodotus, and seems to carry with it the greatest
probability, (Lib. ii. §§ 52–58.) The temple was
enriched by vast numbers of costly statues and
other offerings, presented by the states and individuals who had consulted the oracle. Adjoining
the temple was a grove sacred to Jupiter; and in
it was a divine or prophetic oak, by which the
responses of the god were sometimes manifested !
The imposture carried on here was, in fact, even
more gross and glaring than at Delphi. There
the priest framed a response from the ravings of
the Pythia; but at Dodona the priestess went
into the sacred forest, and listening to the cooing
of the doves, or the rustling of the leaves or branches of the sacred tree, drew thence her auguries !
Sometimes she deduced them from the sounds
emitted by the clashing of copper basins hung
round the temple, and from those emitted by a
brazen vessel placed on the top of a column, and
struck by the figure of a child put in motion by
the wind ! The responses, in ordinary cases, were,
of course, characterized by the usual ambiguity,
so that, let the event be what it might, the

credit of the oracle should be preserved; but here,
as at Delphi and elsewhere, a rich or powerful individual had little difficulty in getting such an
answer as he wished for. (See Ancient Universal
History, x. 67, 8vo. ed.; Voyage d'Anacharsis,
cap. 36, &c.) The site of this famous oracle is
now matter of dispute among the learned. It is
fixed by some at Proctagaso, near the lake Lab-
chista, 12 m. NNW. Yannina; but others place
it a good deal nearer the coast.

DOHUD (two frontiers), an inl. town of Hin-
dostan, on the boundary of Malwah and Gujerat;
Holcar's dom.; lat. 22° 55' N., long. 74° 20' E.
It is of some size, well built and well supplied
with grain and water; is much frequented by
traders, being on the high road between Upper
Hindostan and the Gulf of Cambay; and com-
mands the principal pass into Gujerat from the
NE. It has a fort said to have been built by
Aurungzebe.

DOL, a town of France, dép. Ille-et-Vilaine,
cap. cant., on an eminence among marshes which
have been dried, and are very fertile; 18 m. SE.
St. Malo, and 30 N. Rennes, on the railway from
St. Malo to Rennes. Pop. 4,191 in 1861. The
town is surrounded by walls and ditches, the
remains of its old fortifications; it having formerly been a bulwark of Brittany against the
invasions of the Normans. The glacis of the
ramparts has been converted into a fine promenade. It is ill built, and has but one tolerable
street; but its cathedral is one of the largest and
finest in the prov. Dol was a bishopric as early
as the 6th century.

DOLE, a town of France, dép. Jura, cap. arrond.,
finely situated at the foot of a hill planted with
vines, on the Doubs, and on the canal between
the Rhone and Rhine; 28 m. N. Lons-le-Saulnier,
on the railway from Paris to Besançon. Pop.
10,603 in 1861. The town was formerly fortified,
but its defences have been long since destroyed.
Its chief public buildings are the cathedral, with
a large square tower and three lofty naves, supported by enormous columns; the new prison,
Hôtel Dieu, general hospital, tower of Vergy, hall
of justice, barracks, the old college of the Jesuits,
and theatre. The bridge over the Doubs, and the
port on the canal, are also worthy of notice. Dole
has several Roman remains, including those of an
amphitheatre, some aqueducts, and part of the
superb Roman road leading from Lyraea to the
banks of the Rhine. It is the seat of tribunals of
primary jurisdiction and of commerce; has a
depôt de mendicité, an orphan asylum, a communal
college, a gratuitous school of design, painting,
schools of geometry and music, a public library
with 6,000 vols., and a society of agriculture; it
has, also, manufactures of straw-hats, leather,
chemical products, and agricultural implements;
and a considerable trade in agricultural produce.
Dole is very ancient; in the 14th century it became the occasional residence of the emperor Frederick Barbarossa, and in 1422 had a parl. and
university of its own. In 1479 it was taken by
the troops of Louis XI., when most of its buildings were destroyed or damaged, and many of the
inhab. put to the sword. It subsequently came
into the possession of the Spaniards, and being rebuilt by Charles V., many of its houses preserve
the Spanish style of architecture. Ultimately it
was united to France, in the reign of Louis XIV.

DOLGELLY, or DOLGELLEU, a town of
North Wales, co. Merioneth, on the Mynach, at
the foot of Cader-Idris, 46 m. W. Shrewsbury.
The par. of Dolgelly comprises 870 acres, and had,
in 1861, a pop. of 3,455, of which the town had
2,217. It is very irregularly built, but has some

grand houses; a bridge over the river, built in 1859; a co. hall, erected in 1825, and a church with a handsome tower and large nave. The co. gaol, situated outside the town, was built in 1811, at an expense of 5,000£. The town has long been noted for the manufacture of a coarse woollen fabric, called webs, principally shipped for America. Webs were formerly made in different parts of Montgomeryshire, but the manufacture is now entirely confined to this town and neighbourhood. The name of the town is derived from its situation in a dale, abundant in hazels.

DOLLAR, a village of Scotland, co. Clackmannan, 12 m. E. by N. Stirling, on the Scottish Central railway. Pop. 1,540 in 1861. The village is noteworthy as the seat of an academy, established by Mr. M'Nab a native of the place, who appropriated nearly 100,000£ for its foundation. The academy, a beautiful Grecian building, was erected in 1819. The branches taught in it, in addition to English, writing, arithmetic, and geography, are drawing, mathematics, and natural philosophy, French, Italian, and German, Latin, Greek, and the Oriental languages. An infant and a female school are attached to the institution, and a library. The session commences on Oct. 1, and terminates on the third Wednesday of August. The academy is governed—under 10 & 11 Vict. c. 16—by a body of trustees, comprising the lord-lieutenant of the county and other eminent men.

DOMINGO (ST.). See HAYTI.

DOMINICA, one of the Windward Islands in the W. Indies, belonging to Great Britain, situated between the islands of Guadaloupe and Martinique, 2½ m. from either; in lat. 15° 13' to 15° 36' N., and long. 61° 17' to 61° 31' W. Length N. to S., about 29 m.; greatest breadth 16 m.; area 186,436 acres. Pop. 25,065 in 1861. The island is the most elevated of the lesser Antilles, and contains many high and rugged hills, interspersed with fertile and well-watered valleys. The soil is, however, generally very light, and more fitted for the growth of coffee than of sugar. Maize, cotton, cocoa, and tobacco, are amongst the other staples. The higher parts produce abundance of rice and other woods used in cabinet-making. Hogs, poultry, and game are plentiful; the fisheries on the coast are very productive; and turtle, supposed to have been introduced from Europe, abound in a wild state. The island bears unequivocal marks of volcanic action, and sulphur is found in great plenty. The principal exports are sugar, rum, molasses, coffee and cocoa. The total value of exports, in the year 1853, amounted to 78,726£, and of imports to 47,765£. The public revenue, in 1853, amounted to 11,787£, and the expenditure to 12,035£. The government is under a lieutenant-governor—subordinate to the governor-in-chief at Antigua—an executive council of seven members, appointed by the crown, and a representative assembly of nineteen members. The island is deficient in good harbours; that of Roseau on the W., and Prince Rupert's Bay on the N. coast, are the only tolerable ones. Roseau and St. Joseph are the principal towns. Dominica was discovered by Columbus in 1493, ceded to England by France in 1763, retaken by the French in 1778, but restored at the peace of 1783.

DOMREMY LA PUCELLE, a small village of France, dép. Vosges, 7 m. N. Neufchâteau. Pop. 359 in 1861. Domremy is celebrated as the birthplace of the famous Joan of Arc, born here in 1412. The house once inhabited by the heroine is still extant. It has been purchased by government, and is preserved with a kind of religious care and veneration. Opposite to it, in 1820, a

handsome monument, surmounted by a colossal bust of Joan, and bearing an appropriate inscription, was erected to her memory by the dép.; and, at the same time, a school of mutual instruction for young girls was founded in the village. This village also gave birth to a female of a very different character from Joan, Madame Dubarry, the mistress of Louis XV.

DON (the anc. Tanais), a large and celebrated river of Russia in Europe. It rises in the distr. of Epifan, in the government of Tula; and passing by the town of Letrillou, flows N. to Voronezg and Kalitva; it then turns to the E., till, at Kachalinsk, it approaches within about 36 m. of the Wolga; here it takes a WSW. direction, it pursues till it falls, by various mouths, into the NE. corner of the Sea of Azoff, a little below the town of the same name. Altogether, its course, which is very circuitous, may be about 1,000 m. Principal affluents, Donetz, Sosna, Voruna, Medveditsa, &c. Its turbid and unwholesome waters are well stocked with fish. Its mouths are so encumbered with sand banks that they only admit of being entered by flat-bottomed vessels drawing from 5 to 8 ft. water; and in summer it is in most parts so very shallow that it is of little consequence as a channel of internal navigation, except during spring and autumn, when the products of the various provinces it traverses are brought down to Rostof, Nakhitchevan, and Taganrog. (Hagemeister on the Commerce of the Black Sea, p. 30. English trans.) Peter the Great projected a canal between the Don and the Wolga, where they approach nearest to each other; but, owing to the difficulty of the ground, it has not yet been accomplished. The former is, however, connected near its source by a canal with the Oka, an affluent of the Wolga, and, consequently, by a very circuitous course with the latter. Europe is now generally and properly extended, on the S., to the ridge of the Caucasus; but in antiquity the Don (Tanais) was held, during the latter part of its course, to be the line of demarcation between Europe and Asia. Lucan notices this circumstance, as follows:—

'　　　　—quâ vertice lapsus
Rhipæo Tanais diversi nomina mundi
Imposuit ripis, Asiæque et terminus idem
Europæ, mediæ dirimens confinia terræ,
Nunc hunc, nunc illum, quâ flectitur amplial ortum.'
Lib. iii. line 272.

DONAGHADEE, a sea-port town of Ireland, co. Down, prov. Ulster, on the nearest point of the coast to Portpatrick, in Scotland, from which it bears NW., distant 22 m. Pop. 2,986 in 1831, and 2,671 in 1861. The town, which is 19 m. E. Belfast by railway, is built like a crescent, on one side of the harbour, which has been much improved by a new pier carried out so as to have a depth of 16 ft. at low water, and having a lighthouse at its extremity. The par. church is an ancient cruciform structure; and there are two meeting-houses for Presbyterians, and one for Methodists, an infirmary, and a dispensary. A manor court, with jurisdiction to the amount of 5£, is held in the court-house, as are a court leet annually, and petty sessions every Wednesday. The constabulary and coast-guard have stations here. The embroidering of muslin is carried on to a considerable extent, and there are numerous flax-mills in the neighbourhood. The port is a creek to Belfast, and a station for the regular steamers to Portpatrick, the voyage being usually made in less than three hours.

DONCASTER, a handsome corporate and market town of England, W. riding co. York, on the

tion, which, including a branch called the Cher-wold is crossed by two fine stone bridges, 162 m. NNW, London by road, and 164 m. by Great Northern railway. Pop. 16,406 in 1861. The town, which is approached from the S. by a magnificent range of elm trees, is extremely well built, and the High Street, extending about a mile on the Great N. Road, has a remarkably fine appearance. It is a place of much importance both in its civil and ecclesiastical characters, through the whole period of British history. At the point where the town now stands, one of the great Roman highways crossed the river. This road connected two great stations, Lincoln and York; and was an improved British track-way, used for a communication between Lincolnshire and the interior of the Brigantian territory. It is the station *Danum* in the 'Itinerary' of Antoninus. In the middle ages it had a convent of Carmelites and White Friars, and received the grant of a charter from Richard I. The property belonging to the corporation of Doncaster amounts to nearly 9,000*l*. a year. During the old corporation it was greater. But that body having incurred a debt of above 100,000*l*., the new municipal body sold the Rossington estate to James Brown, esq., of Leeds, for the sum of 92,500*l*., to pay off the debt; the purchase was completed in 1839. The income is principally expended on objects of public utility, as the paving, lighting, cleaning, and watching of the town, and supplying it with water; the support of educational and charitable institutions; with the erection of buildings for the purposes of public utility and amusement, and the attraction of visitors. The par. church, dedicated to St. George, is a large imposing structure, in part very ancient, with a beautiful square tower 140 ft. high. There is another church, built by bequest from John Jarratt, esq., a native of the town, at an expense of 18,000*l*. The Wesleyans, Primitive Methodists, Unitarians, Independents, Quakers, and Catholics have also places of worship. Exclusive of the churches and chapels, the principal public buildings are—the mansion-house, a handsome structure, erected in 1744, but improved in 1800; the town-hall, the theatre, gaol, public library, news-room, and lyceum, with the splendid betting-room, 90 ft. in length, which, except during the races, is used for concerts, lectures, and exhibitions. Among the educational institutions are a grammar-school for the sons of freemen, supported by the corporation; a national school, a British school, and Sunday schools, supported by subscription. Of the charitable institutions the principal are—St. Thomas's Hospital, founded in 1555, by Thos. Ellis, for decayed housekeepers, with a revenue of about 3300*l*. a year; Kay's and Jarratt's charities; a dispensary, and sundry minor charities. The Yorkshire institution for deaf and dumb, a flourishing charity, is situated adjoining the race-ground; and the workhouse for the Doncaster union is near the town. Under the Municipal Act the town is divided into three wards, and has two aldermen and six councillors for each.

Doncaster is not a manufacturing town, but it has some small iron-foundries; a large water corn-mill on the Don bridge, and a steam corn-mill on the opposite bank. It is in the centre of a rich and highly cultivated district, and has an extensive retail trade. The Don is navigable as far as Sheffield by vessels of 50 tons burden. Doncaster used to derive considerable advantage from its situation on the Great N. Road, and the number of travellers, by coaching and posting, passing through it. But since the opening of the Great Northern railway the influx of travellers has

greatly diminished. The iron roads, however, with which Doncaster is now connected, have given an additional impetus to trade, and more than compensate for the loss of the Great N. Road.

Doncaster is principally indebted for its celebrity to its races, and the high station which they hold in the sporting world. The races were established in 1703, and from a small beginning have become almost unrivalled; they are held in September, and have been sedulously patronised by the corporation, the surrounding nobility and gentry, and the first names in turf annals. In 1776, the famous St. Leger stakes were established by Colonel St. Leger, who resided at Park Hill, near the town; hence their name. The first race was won by the Marquis of Rockingham; and the list of winners includes the finest horses that have been bred in England. The race-course, about 1 m. SE. from the town, adjoining the Great N. Road, is, in every respect, one of the finest in the kingdom. The course, nearly 2 m., is railed round; it is ornamented with a magnificent grand stand, for the accommodation of the principal company; the nobleman's stand, the stewards' or judge's stand, commodious booths, minor stands, and rubbing-houses. The interest excited by these races is quite extraordinary; they attract visitors from all parts of Great Britain and Ireland, and even from foreign countries.

DONEGAL, a co. of Ireland, prov. Ulster, of which it forms the NW. portion; having N. and W. the Atlantic, E. the counties of Tyrone and Londonderry, and S. Fermanagh and Donegal Bay. Area, 1,165,107 acres, of which 644,371 are mountainous and bog. It is deeply indented by bays and arms of the sea: and its surface is, in most parts, rugged, mountainous, and dreary. It has, however, some extensive tracts of good level land, which, under good management, would be exceedingly productive. Climate very wet, and unfavourable for the ripening of grain. Property in very large estates, but some of them are let on interminable leases; farms of various sizes, in the low grounds from 5 to 30 acres; in the mountainous districts from 30 to 500 do. Partnership leases common, but on the decline. Agriculture in the worst possible state. Potatoes, oats, and flax the principal crops, the first being the main dependence of the farmer. More work is done with the hoe or spade than with the plough. Average rent of land 5s. an acre, being the lowest of any in Ireland. Bulk of the people very badly off; English little known in some districts. The linen manufacture was widely diffused, but is on the decline. Fishing carried on to some extent in some of the bays along the coast. The barony of Innishowen, famous for its smuggled whisky, occupies the NE. portion of the co. between Lough Foyle and Lough Swilly. Donegal has five baronies, and forty-two parishes, and returns two members to the House of Commons, both for the county. Registered electors 1,807 in 1865. Pop. 296,040 in 1841; 255,257 in 1851; and 237,395 in 1861. Gross annual value of real property assessed to income-tax, 268,031*l*. in 1857, and 250,596*l*. in 1862.

DONERAILE, an inland town of Ireland, co. Cork, prov. Munster, on the Awbeg, an affluent of the Blackwater, 64 m. NNE. Mallow, and 23 m. N. by W. Cork. Pop. 2,253 in 1831, and 1,475 in 1861. The town consists of a long street, in which are the par. church, a mansion-house, Cath. chapel, a nunnery, market-house, and dispensary. Kilcolman Castle, in the vicinity, was some time the residence of Spenser, the poet. The town, though not incorporated, sent two members to the Irish H. of C., but was disfranchised at the Union.

Markets on Saturdays, and fairs on Aug. 12 and Nov. 12. It is a constabulary station.

DONGOLA, a prov. of Upper Nubia, consisting of that portion of the valley of the Nile which lies between 18° and 19° 50′ N. lat., bounded on the N. by Mahass, and on the S. by the country of the Sheygya negroes; but, like all the fertile districts reserved from the surrounding deserts by the inundations of the Nile, Dongola is extremely narrow, only in one instance exceeding 8 m. in breadth.

The Nile, which enters this prov. at about 18° 15′, near Korti, flows at first in a S. direction, but, immediately taking a circular bend to the W. and N., traverses the rest of Dongola parallel to its former course, and with but trifling deviations continues to follow the same line down to the Egyptian delta. The widest portion is that nearest its first entrance into Dongola; and at 'high Nile,' the flat, or low lands of the prov., are subject to inundations similar to those of Lower Egypt. The river makes its exit into the Nubian prov. of Mahass, at the island of Tombos, where rocky and rugged surface forms the third cataract, in lat. 19° 30′. (Waddington's Visit to some parts of Ethiopia, p. 10; Burckhardt's Nubia, p. 66; Map of Nubia in Arrowsmith's Atlas.) Navigation is exceedingly difficult in this part of the Nile, for, besides a strong current which the upward voyager has to contend against, the bed is shallow and bristles with rocks. (Narrative of Ismael Pacha's Expedition to Dongola and Sennar, by an American in the service of the Pacha, passim.)

The mountains of Dongola are a continuation of the same chains which, with slight interruptions, accompany both sides of the Nile during its whole course. Perhaps the most extensive of these intervals occurs here at the immense and fertile Dongolese plain, which forms the exception to the otherwise narrow breadth of the prov. A large solitary hill, about 4 m. E. of the river, called Mount Arambo (many-coloured, or chameleon), has from time immemorial marked the boundary between Mahass and Dongola. The great plain then intervenes, and the mountains recommence near New Dongola, and stretch beside the river without farther interruption to the S. frontier; those on the E. bank being by far the most considerable. Here the mountains are two hours' journey in breadth, reach close to the river, and form a natural boundary to Sheygya. Granite and sandstone are the chief components of these hills. (Waddington and Hanbury, p. 61; Burckhardt's Journey in Nubia, p. 68.)

The Valley of the Nile lies for the most part in this district, on the W. bank; for the sands of the desert, encroaching close upon the water's opposite edge, render the E. side barren and unproductive, while the more favoured district has generally a harder surface. S. of the town of Hannek commences the great plain of Dongola, called Wady Jarjar, which can hardly be exceeded in richness and fertility. At the period of the inundation it presents a watery surface of from 12 to 15 m. in breadth (Burckhardt, p. 66); while at low Nile, the river, bursting from its banks through small channels, seems as if it had divided itself into natural canals to irrigate as much ground as possible, and save man the trouble of cultivation. (Waddington, p. 61.) This plain is covered with acacia trees as far as the eye can reach. Further N. the mountains contract the valley, which to Wady Hemmowah is fertile and separated into well-cultivated patches by rows of acacias. Ruins of towns and tombs of Moslem saints are frequently met with in this portion of the valley, which is much infested with hordes of the Nubian wolf (Canis

Anthus, Rüppell), Wady Jebriah, situated towards the N. limit of Dongola, is overgrown with trees, amongst which cottages are thickly and irregularly strewed for some distance along the banks of the stream. Near Amhakol, about 8 m. W. of it, is a waste called Haagharlak. The superficial stratum here is a coarse sandstone, curious and interesting from its containing many siliceous fossil trees. 'I observed,' says Mr. Holroyd (Journal of the Royal Geog. Soc., ix. 1841, 'five or six, the largest of which, situated twenty minutes' walk from the river, is 51 ft. in length, and 20 in. in diameter at its largest extremity. It is partially buried in the sand. The peasantry splinter off fragments, and use them for gun flints and to strike a light.'

None of the islands with which the river is studded in its course through Dongola is so celebrated as Argo, situated above the island of Tombos, and a large granite rock called Hadjar-el-Dahab (the golden stone). Argo is upwards of 80 m. long, and is one of the most beautiful islands that spring up from the bed of the Nile. The scenery is highly picturesque, principally composed of small plains enclosed by rows of sycamore trees. Several remains of antiquity are strewed over the island, the most remarkable of which are two colossal statues cut in grey granite, the headless form of a female sculptured out of black granite, and the figures of four hippopotami standing side by side. The colossi are broken into fragments, lying close together, and 'really look as white and clear, and as free from the injuries of time, as if they were now fresh from the hand of the sculptor.' (Waddington, p. 46.) A peculiar breed of mosquitoes, not so large nor so many as others, annoy the inhabitants of Argo. Several other islands occur at short intervals, among which may be enumerated Sadghs, Tángér, and Gerfi, as the most important and fertile. At the island of Gertassi, near Amhakol, the Dongolese country centres.

The towns and villages are thickly scattered along the margins of the Nile, most frequently on the E. bank. The first of any consequence is the town of Hannek, opposite the isle of Tombos, where the cotton plant is said to be productively cultivated. But by far the most important is Maráhah, or New Dongola (situated, according to Linnant, in 19° 7′ 30″ N. lat., and 29° 54′ 35″ E. long.; but placed by Rüppell in lat. 19° 10′ 19″, and long. 30° 27′ 18″ E.), the present pop. of which has been estimated at 6,000, including 100 Copts. The bazaar is daily increasing, and is supplied from Cairo with shoes, printed cottons, calicoes, sugar, rice, cloth, hardware, &c.; but, on account of a heavy duty levied upon all articles of consumption, they are four times the price that they are in Cairo. Dongola boasts of a coffee-house and a manufactory for indigo; the government is also building baths. The thermometer on Christmas-day, 1838, stood, in the shade at 2 P.M., at 86°, and at 8 P.M. at 80°. (Journal Geog. Soc., ix. 164.) Property is valued according to the number of water-wheels an individual possesses, and he is taxed accordingly. (Burckhardt's Nubia, p. 66.) The chief places from New to Old Dongola are the dilapidated town of Hannek, Badeyn, and Radobol; between which numerous villages intervene, many of them in ruins. Tonga, or Old Dongola, the cap. of what was once a powerful Christian kingdom, is now a miserable ruin, situated on a rock which slopes down to the water's edge; it is covered with sand, a large mass of which has evidently buried the centre of the town, and divided the remains into two sections; the N. part only is inhabited by about 300 persons. The sand is of a bright yellow colour, and has accumulated in such quantities that its surface is level with the roofs of many of the houses, the only entrance to which is through the ceilings of the rooms. (Geog. Journal, ix. 104.) There is a mosque, on rather an elevated site, which commands a fine view of the surrounding country. This consists principally of drifted sand, with, at rare intervals, a few feet of cultivable soil. Amtuhot, the last Dongolese town, is one of little importance.

Dongola is now an appendage to Egypt, together with Lower Nubia, which territories were conquered by the late celebrated ruler of Egypt, Mehemet Ali. It was formerly one of the numerous kingdoms divided between the Sheygya Arabs, amongst whom, at their expulsion from Egypt, the Mamelukes sought refuge. The fugitives however, had scarcely been a month at Argo, when, upon some slight pretext, they murdered their benefactor, the Sheygya king, and spread themselves over the country, establishing a government of their own at New Dongola. The pacha of Egypt, upon pretence of punishing this breach of justice and hospitality, sent an expedition into the country, and, meeting with little resistance, took possession of it, which he has quietly retained ever since 1820. (Burckhardt's Nubia, p. 85; Quarterly Review, xxvii. 217.) The people possess the same characteristics as the rest of their countrymen (see Nubia), except that they are unusually 'dirty, idle, and ferocious' (Narrative of Ismael Pacha's Expedition, p. 189); but they are also, in common with their neighbours, extremely hospitable. Mr. Waddington describes the women as ugly in person, and unfeminine in conversation and manners; they wear scarcely any clothing.

The Dongolese horse must not be passed over without particular notice, though the natural history of this region must be sought for in the art. NUBIA. This animal, so celebrated all over the East, possesses the beauty of the finest Arabian breeds, with greater size and more bone. The mares are seldom ridden, and the stallions fetch a high price; from five to ten slaves being the value usually given for them. Most of them are fed for ten months in the year on little else than straw, and in spring upon green crops of barley. (Burckhardt's Nubia, p. 67.)

DONOBEW, an inl. town of the Birmese empire, Pegu, on the E. arm of the Irrawadi, 50 m. NW. Rangoon; lat. 17° 8' N., long. 95° 55' E. In 1825, its stockade extended for nearly a mile along the bank of the river; in 1827, the British embassy found this place considerably enlarged and strengthened. It is noted for the action, in the first named year, in which Bandoola, the Birman leader, was killed by a stray bomb.

DOONGURPOOR, an inl. town of Hindostan, prov. Gujerat, 82 m. NE. Ahmedabad; lat. 23° 54' N., long. 73° 50' E. Little is recorded respecting this town or its territory; the mounds encircling the Doongurpoor lake are said to be built of solid blocks of marble. The rajahs are acknowledged to be the senior branch of the reigning sovereigns of Odeypoor; the majority of their subjects are Bheels. Bauds of Arabs and Sindies, previously in the service of the rajah, harassed and laid waste this district, till a stop was put to their ravages by the British troops.

DOOSHAK, a town of Persia, prov. Seistan, of which it is the cap., near the Hirmund, and about 51 m. E. from Zurrah; lat. 31° 5' N., long. 62° 10' E. The modern city is small and compact, but the ruins cover a vast extent of ground. It is populous, has a good bazaar, and the inhab., who dress

in the Persian manner, have a more civilised appearance than the other natives of Seistan. The country in the vicinity is open, well-cultivated, and produces wheat and barley in sufficient quantities to be exported to Herat: the pasturage is also good and abundant. Its ruins show that it was formerly of much greater extent than at present; and is supposed to be identical with the Zarunga of Ptolemy.

DORCHESTER, a parl. bor. and town of England, cap. co. Dorset, div. Dorchester, hund. Ugrountombe, on a gentle elevation adjoining the Frome, 125 m. SW. by W. London by road, and 140½ m. by London and South Western railway. Pop. 6,823 in 1861. The town consists chiefly of three wide streets, diverging from a central area in the direction of the lines of road to London, Exeter, and Weymouth. It is well built, partially paved, and lighted with gas; and is very clean. It is more than two-thirds surrounded by a fine avenue, commanding extensive and diversified views. Fordington Field, an unenclosed tract of fertile land, 7 m. in circ., adjoins the town on the S.; it is partly arable, partly pasture, and belongs lives from the duchy of Cornwall. There are 3 churches—2 modern, on ancient sites, and 1 old, with many curious monuments, and a lofty pinnacled tower; 4 dissenting chapels; a free grammar school, founded in 1579, with 2 exhibitions to St. John's Coll., Cambridge, and one to either university; 8 sets of almshouses; a small theatre; a town-hall built in 1791, with a market-place under it; a shire hall, in which the county assizes and quarter sessions are held, and a county gaol and house of correction, built on Howard's plan, at an expense of above 16,000l., and occupying the site of the ancient castle. There are large barracks in the vicinity. Market, Sat. and Wed. Fairs, Candlemas day, Trinity Monday, St. John's day, St. James's day. These are large sheep and lamb fairs; large flocks of a valuable breed, named from the place, being kept on the extensive sheep walks of the vicinity. Formerly the town was a considerable seat of the woollen manufacture; but at present its chief dependence is on commercial business. It has breweries noted for the superiority of their ale; and there are annual races in September. Dorchester has returned 2 mems. to the H. of C. from the 21st Edw. I. Previously to the Reform Act, the franchise was confined to inhabitants of the bor. paying to church and poor in respect of their personal estates, and to such persons as paid to church and poor in respect of their real estates within the bor. Registered electors 439 in 1865. Under the Municipal Act it is governed by 4 aldermen and 12 councillors: its municipal limits coincide with the parl. ones. Annual value of real property assessed to income tax, 23,914l. in 1857, and 25,515l. in 1862.

Dorchester was one of the principal stations of the Romans in England. It was called by them Durnovaria and Dunium, and has still to boast of many interesting relics of its Roman masters. They had surrounded it with a wall and a fosse; part of the former having been standing as late as 1802, and 'great part' of it was standing in 1773, when visited by Stukeley. (Itinerarium Curiosum, p. 163.) Maiden Castle, about 1 mile SW. of the town, is also supposed to have been constructed by the Romans as a summer camp, remains entire. It is an irregular ellipse, surrounded by double ditches and ramparts; the former of great depth and the latter high and steep. The inner area comprises about 44 acres. Poundbury Castle, nearer the town, on its NW. side, is also supposed to be a Roman work; but, though of the same character, it is of very inferior dimensions to Maiden Castle,

But the most interesting Roman remain near Dorchester is the amphitheatre, about ⅓ m. SW. from the town, the most perfect structure of its kind in England. The arena, or inner floor of the amphitheatre, is level with the surrounding plain; while the sloping sides, on which were seats for the spectators, and which are formed of masses of chalk, rise 30 ft. above it. Its dimensions are very large; the length of the longest external diameter being 343 ft., and that of the shortest external diameter 339½ do.; its longest internal diameter is 218, and its shortest 163 ft. When complete, it is supposed to have been capable of accommodating about 13,000 spectators. In modern times, it has been occasionally used as a place of punishment; and on one occasion, on a woman being burnt in the arena, 10,000 persons are reported to have been congregated within the amphitheatre, to witness the horrible spectacle. It is to be regretted that this classical remain has not been preserved with due care, and that its arena has been repeatedly profaned by the plough. The assizes held at Dorchester in September, 1685, are famous, or rather infamous, for the judicial murders of Judge Jeffries.

DORDOGNE, one of the largest déps. of France, comprising the ancient prov. of Perigord, and part of Guienne; between lat. 44° 35' and 45° 42' N., and long. 0° and 1° 27' E.; having N. Charente and Haute Vienne, E. Corrèze and Lot, S. Lot-et-Garonne, and W. Gironde, Charente, and Charente-Inférieure. Greatest length and breadth, about 70 m. each. Area, 918,256 hectares; pop. 501,687 in 1861. Several hill-ranges intersect Dordogne, those in the N. belonging to the Limousin, and those in the S. to the Auvergne mountain chains. The principal summits are in the SE., but none is more than about 640 ft. high. Chief rivers, the Dordogne, Vézère, Isle, Dronne, &c., all of which have a SW. course. The Dordogne, resulting from the union of the rivulets Dor and Dogne, rises in the Mont d'Or, Puy-de-Dôme, flows at first SW., and afterwards due W. through Corrèze, Lot, Dordogne, and Gironde, and ultimately joins the Garonne, about 13 m. below Bordeaux, after a course of nearly 220 m., 167 of which are navigable. Climate rather damp, but upon the whole healthy; the winter and spring are rainy seasons; the summer is very dry; violent storms frequently occur. There are but 46,400 hectares of rich land, principally in the valleys of the Dordogne and the other larger streams; the smaller valleys are for the most part narrow and unproductive, and a large portion of the dép. consists of arid heaths and wastes, over which the traveller may journey for leagues without seeing a single hamlet. Sufficient corn is, however, grown for home consumption; principally rye, maize, and millet. The chesnut crops are important, and a good deal of walnut oil is made. The culture of the vine is pursued to a considerable extent, the average annual produce of wine being about 650,000 hectolitres. The white wine of Bergerac is greatly esteemed, though it is mostly on the left bank of the Dordogne that the red white wines of the dép. are grown; the right bank is more famous for its red wines. There are few meadows. Game is very plentiful. Iron, copper, lead, cadmium, manganese, coal, and lignite are mined; and marble, alabaster, granite, lithographic stone, &c., quarried. Working in metals, especially in iron and steel, and the manufacture of paper, are the chief branches of manufacturing industry. Coarse woollens, serges, leather, kid gloves, earthenware, good beer, liqueurs, brandy, and blue vitriol are, however, also made in the dép. The pâtés of Perigueux, and its truffled turkeys and other poultry, are held in the highest estimation both in France and other countries, and support a considerable trade. Dordogne is divided into 5 arrond., 47 cantons, and 583 communes. Chief towns, Perigueux, the cap., Bergerac and Sarlat. Perigord was from the 9th to the 15th century under the jurisdiction of its own counts; Henry IV., a part of whose patrimony it was, united it to the French crown.

DORKING, a market-town, and par. of England, co. Surrey, hund. Wotton, near the Mole, and on the high road from London to Brighton; 23 m. SSW. the former by road, and 29 m. by South Eastern railway. Pop. of town 4,061, and of par. 6,597 in 1861. Area of par. 10,150 acres. Dorking is finely situated on the side of a sandstone hill, many of the houses having cellars excavated in the rock; it has wide streets, and is a well-built, well-paved, neat country town. The country round is remarkably beautiful; it is well wooded, and presents a succession of fine bold hills and rich valleys, with a great number of fine seats. The church is a large ancient structure, and there is a good town-hall and some almshouses in the vicinity. Dorking has the finest breed of fowls in England; they have six claws, and the capons fatten to an immense size. The custom of Borough English, by which the youngest son succeeds to copyhold property, prevails in this manor.

DORNOCH, a market-town, and the only royal burgh in Sutherland, Scotland, on a low sandy beach, NE. coast of the Dornoch Frith, 53 m. N. Inverness. Pop. 483 in 1861. The sea approaches to about 150 yards of the town, yet does not confer on it the advantages of a sea-port, there being no harbour. It is a mean-looking town, with many marks of poverty and decay. It has no source of municipal revenue, except the customs levied at six annual fairs; but as these are on the decline, the income of the town is suffering accordingly. It was made a royal burgh by Charles I. in 1628. Dornoch is chiefly remarkable for its cathedral, and as having once been the seat of the bishop of Caithness. The cathedral is supposed to have been built by Richard Murray, bishop of the see, who died in 1245, and who was afterwards canonized. (Keith's Scottish Bishops, 1824, p. 209.) The remains of the buildings are extensive and magnificent. The present parish church consists of three ailes of the old cathedral; and underneath it is the burying-place of the noble family of Sutherland. A portion of the bishop's palace serves as the county court-room and gaol. A monastery of Red Friars was founded here by Sir Patrick Murray in 1271, of which the ruins have entirely disappeared. (Ib. 397.) Dornoch unites with Wick, Cromarty, Dingwall, Tain and Kirkwall, in sending a mem. to the H. of C.

DORPAT, or DERPT (Russ. Juriev), a town of Russia in Europe, gov. Riga, cap. distr., on the Embach, and on the high road between Riga and Petersburg, 150 m. NE. the former, and 170 m. SE. the latter city. Pop. 14,650 in 1858. The town, which is well built, is divided into three separate portions—Dorpat Proper, and the suburbs of Riga and Petersburg. It has a fine market-place, a stone bridge over the Embach, and a cathedral, now partly in ruins, but which formerly had a nave supported by 24 arches and surmounted by two towers. The old fortifications, with some of the ditches, have been converted into ornamental gardens, shrubberies, and public walks. It is surrounded by hills, which, as well as the banks of the river, offer many fine points of view. Dorpat is the seat of a university, which in 1862 had 65 professors and above 500 students. This institution was originally founded by Gustaine Adolphus of Sweden, in 1632. After suffering numerous vicissitudes during the wars between Sweden and Russia, and having been removed to Pernau, it was re-established in Dorpat in 1802 by the emperor Alexander I. The university possesses a library of 70,000 vols., a museum of arts, an observatory with some excellent instruments, cabinets of physical, chemical, mineralogical, zoological, and pathological subjects, an anatomical museum, a collection of agricultural models, and a botanical garden containing many rare plants; it has attached to it a hospital, theological and philological seminaries, and an institute for the education of professors. Though considered as especially belonging to this and the adjacent governments, it is much resorted to from many other parts of Russia. Dorpat also contains a gymnasium and a normal primary school.

The town is believed to have been founded in 1030. It was subsequently taken by the Teutonic knights, who erected it into a bishopric in 1224. Its commerce now began to flourish, and at one period it ranked as one of the Hanse Towns. It was afterwards alternately in the power of the Poles, Swedes, and Russians; the latter have retained possession of it since 1704.

DORSET, a marit. co. on the S. coast of England, having N. the British Channel, E. Hants, N. Wilts and Somerset, and W. Devonshire. Area, 987 square miles, or 632,025 acres, of which about 200,000 are arable, 440,000 are meadow and pasture, and the rest heath. Surface beautifully diversified; climate mild and salubrious, not being so rainy as in some districts more to the W. Soil principally chalk, sand, gravel, and loam. The vale of Blackmore, traversed by the Stour, containing 170,000 acres, and some other tracts in the W. part of the co. and along the coast, are eminently fertile and beautiful; but the distinguishing feature of the co. is the extent of its chalky downs, depastured by large flocks of sheep, and round Poole Harbour there are large tracts of heath. Agriculture in a medium state of advancement; but more improved in the E. than in the W. districts. Hemp and flax are a good deal grown, but less now than formerly. Water meadows extensive, and their management well understood. The greater part of the co. is in grass. There are some very large dairies; they are not generally looked after by the farmers, but let, at so much per cow, to dairymen, many of whom have made large fortunes. Stock of sheep estimated at between 600,000 and 700,000. Property in large estates. Farms of various sizes, but mostly large; they are let for 14 or 21 years, the rents, in most places, being paid once a year. St. Paul's Cathedral, Somerset House, and others of the principal buildings in London, as well as in the greater number of the towns in the S. of England, have been constructed of stone brought from the freestone quarries in the Isle of Portland in this co.; and the Isle of Purbeck supplies the potteries of Staffordshire with the clay used in the manufacture of the finer sorts of earthenware. There are considerable manufactures of flax and hemp at Beaminster, Netherbury, and Bridport. Shirt buttons are made at Shaftesbury and Blandford; silk is spun at Sherborne and Gillingham, and wool at Fordington and Lyme Regis. Principal rivers, Stour and Frome. Principal towns, Poole, Shaftesbury, Weymouth, and Melcombe Regis. Dorset has 34 hundreds and 271 parishes, and returns 13 mems. to the H. of C., viz. 3 for the co., 2 each for the boro. of Bridport, Dorchester, Poole, and Weymouth, and 1 each for Shaftesbury and Wareham. Registered electors for co. 6,221 in 1865. Pop. 188,748 in 1861, inhabiting 37,709 houses. Annual value of real

property assessed to income tax, £653,715L in 1857, and £956,571L in 1862.

DORT, or DORDRECHT, a partially fortified town of S. Holland, on an island formed by the great inundation of 1421, on the S. side of the Waal, a branch of the Maas, 10 m. SE. Rotterdam, on the railway from Antwerp to Rotterdam. Pop. 23,260 in 1861. Dort is a dull, though a tolerably well-built town; its streets are lined with houses of an antique fashion, the gables of which are turned outwards. They rise with many grotesquely ornamented windows and crow-steps to a considerable altitude; while the practice of painting the bricks a bright red, and the ornamental stones and cornices a light colour, adds to their fantastic appearance. A number of the houses, as appears from the dates carved on their exterior, were erected during the period of Spanish occupation, previously to 1572. The principal public buildings are the town-hall, a fine edifice, and the church, an old Gothic structure, 300 ft. long by 150 broad, with a heavy square tower conspicuous from a great distance. The latter building is paved entirely with flat monumental stones, some of which are of great antiquity; and its walls are surrounded with monuments, which the Dutch ingeniously preserved during the occupation of the country by the French, by concealing them with a screen of plaster. The church also contains a marble pulpit, highly ornamented with elaborate and elegant carving. The hall in which the famous synod of Dort held its sittings is still in excellent preservation, but is now a theatre. Dort is surrounded on the land side with fortifications; on the side of the Waal it has several quays, and a good harbour, from which two canals lead into the middle of the town. It is the centre of a considerable trade in flax, which is grown in great quantities in its vicinity, and a good deal of which is exported for England and Ireland. It has also a large trade in corn, salt-fish, train-oil, and timber; the latter article is floated down from the Upper Rhine in immense rafts, which, when sold, often realise from 25,000l. to 30,000l. There are many windmills for sawing deals in and near Dort, some sugar and salt refineries, linen-bleaching, tobacco, and white lead manufactories, and building docks. Dort is one of the oldest cities in the country; was the original residence of the counts of Holland, and, in 1572, the seat of the first meeting of the states at which the independence of the Seven United Provinces was declared; but the most memorable era in its history is that of the Synod of Dort, to which reference has been already made, held in consequence of a schism in the reformed church. James Arminius, professor of divinity in the university of Leyden, having rejected the doctrine of Calvin with respect to predestination and grace, obtained the support of Grotius, Barneveldt, and other learned and eminent persons, as well as of a considerable number of the middle and lower classes. His tenets were, however, opposed with extreme vehemence, and were represented as of the most dangerous description. The disputes that grew out of this controversy being not unfrequently attended with tumult and bloodshed, the States General at last agreed to refer the subject to dispute to a council or synod for its decision. This synod, which excited the greatest interest throughout Protestant Europe, assembled on the 13th of November, 1618, and continued its sittings till the 25th of May, 1619; it was attended not merely by all the most eminent divines of the United Provinces, but also by deputies from the reformed churches of England, Scotland, and Switzerland. The Calvinists having a decided majority in the assembly, all its decisions were in conformity to their views. The distinctive doctrines of Arminianism were pronounced to be predilectial errors and corruptions of the true faith; and this was followed up by the excommunication of the Arminians, the suppression of their religious assemblies, and the deprivation of his ministers.

These unjust and violent proceedings, being aggravated by political animosities, led to the most deplorable results. In the persecution to which they gave rise, the eminent statesman Barneveldt, though at the age of seventy-two, lost his life on the scaffold; many distinguished Arminians were driven into exile; and even Grotius was condemned to a perpetual imprisonment, from which he was only extricated by the sagacity, courage, and devotion of his wife. But after the death of Prince Maurice, the great enemy of the Arminians, in 1625, this persecution relaxed; and most of the exiles were soon after allowed to return to Holland. The Arminian doctrine is now very widely diffused, even among those who profess to differ from it. (See Mosheim, iv. 439–466, &c. edit.)

DOUAI, a strongly fortified town of France, dép. du Nord, cap. arrond., situated very advantageously for commerce, on the Scarpe, 18 m. N. Lille, on the railway from Paris to Lille. Pop. 21,695 in 1861. The town is well built, and the principal square is large and handsome: it is surrounded with old irregular walls, flanked with towers, and is farther defended by a fort on the right bank of the river, about 2 m. N. Douai. The town contains large establishments of artillery, a superb arsenal, and one of the three royal cannon foundries in the kingdom. It is the seat of a sub-préfecture, of an imperial court for the dépts. du Nord and the Pas-de-Calais, a tribunal of primary jurisdiction, a royal college with 262 pupils, a royal school of artillery, an Académie Universitaire, which has replaced its celebrated university, founded in 1562; with schools of design and music, a primary normal school, a public library with 28,000 printed vols, and 600 MSS., museums of painting and antiquities, cabinets of natural history and medical science, a botanical garden, two hospitals, an orphan asylum, and a theatre. Industry and the arts are alike thriving in Douai. It has fabrics of lace, tulles, gauze, cotton stuffs, linen and earthenware, glass and soap works, and salt and sugar refineries; with a considerable trade in flax, which is extensively cultivated in its neighbourhood. Douai is very ancient, having existed previously to the invasion of Julius Cæsar. Its possession was guaranteed to France by the treaty of Utrecht.

DOUBS, a frontier dép. of France, in the E. part of the kingdom, formerly comprised in Franche-Comté, having N. and NW. the dép. Haut-Rhin and Haute Saône, SW. that of Jura, and E. Switzerland. Length, NE. to NW., about 60 m.; breadth varying from 20 m. in the N. to 50 m. in the S.; area, 522,765 hectares. Pop. 298,200 in 1861. Four collateral mountain chains belonging to the Jura system intersect the dep. in nearly its entire length, decreasing in height from E. to W., and naturally dividing the surface into a mountain, hill, and plain region. The highest summit of the E. range, Mount Sechel, is 5,768 ft. above the level of the sea; the principal elevation of the W. range rises to only 955 ft. The plain country to the W. of the latter range is the most fertile, and well fitted for the growth of all kinds of corn, and of the vine; the rest of the country is not generally productive. The mountains are all of calcareous formation, and abound with narrow gorges, grottoes, and caverns: the more elevated ranges are covered with pine forests, and in many

parts with ice and snow for six months of the year. Chief rivers Doubs, Loue, and Ognon. The former rises at the foot of Mount Risoux, and, after a very tortuous course through the dep., it proceeds SW. through that of Jura, and a part of Saône-et-Loire, and ultimately joins the Saône at Verdun. From Besançon to near Montbéliard, the Doubs forms a part of the navigable canal between the Rhine and the Rhone. There are many small rivers and some large marshes. Climate variable and rather cold, but generally healthy. Wheat, rye, maize, hemp, pulse, fruits, wines, &c., are grown in the valleys and low country, which the inhabs. exchange with those of the mountainous districts for barley, flax, cheese, drugs, and timber. Agriculture very backward: fallows are so common as usually to occupy nearly a third part of the cultivable land, — a waste that ought to wholly, or almost wholly, avoided by the substitution of green crops, at the same time that a great additional supply of food for cattle and of manure would be obtained. According to official tables, 120,646 hectares of land are occupied with forests; and this is one of the few French deps. in which the planting of trees is actively going on. Meadow lands are extensive: in the arrond. of Montbéliard they are well irrigated. The rearing of cattle is pursued to a considerable extent, as well as the manufacture of cheese similar to that of Gruyère. This branch of industry is usually conducted either by the proprietors of from 40 to 60 cows, or by associations of small proprietors, whose share of the cheese is in proportion to the quantity of milk they respectively furnish. The total annual product of cheese is estimated at 2,300,000 kilogrammes, worth 1,650,000 fr.; of butter, 260,000 kilogr., value 260,000 fr. Iron, coal, and lignite are mined, and gypsum, marble, and building stone, quarried. There are about 20 iron-works in the dep. which supply yearly 1,700,000 kilogr. of bar iron, 7,450,000 kilogr. of cast do., 2,400,000 kilogr. of iron wire, 150,000 kilogr. of pointrs, 540,000 kilogr. of iron plates, and 30,000 sheets of tinned ware. The establishment at Audincourt alone yields 4,000,000 kilogr. of cast and forged iron. Watchmaking employs about 2,000 artisans, and about 40,000 watches are made annually in Besançon. Cutlery, copper wares, paper, leather, liqueurs, buttons, and a few fabrics of different kinds, are amongst the other principal manufactures. The exports of the dep. are chiefly cattle, cheese, butter, timber, iron, hardware, watches, and agricultural implements; its imports corn, wines, brandy, cotton, woollen, and other fabrics. Doubs is divided into 4 arrond., 27 cantons, and 640 communes. Chief towns, Besançon, the cap., Pontarlier, and Montbéliard. About 25,000 of the pop. are Protestants. This dep. formed a part of the circle of Burgundy under Charles V.; it was annexed to the French crown by Louis XIV, in 1660.

DOUGLASS, the principal town of the Isle of Man, on the E. coast of which it is situated, at the mouth of the Blackwater, on a circular bay, 30 m. NW. Liverpool; lat. 54° 12′ N., long 4° 23′ 47″ W. Pop. 9,894 in 1861. The town has some good streets and buildings; but, speaking generally, the former are narrow and dirty. It has, however, been a good deal improved of late years, in consequence of the influx of visitors from Liverpool and other places, in summer, attracted by the facilities for sea-bathing, and by the partial exemption from taxation enjoyed by residents in the island. (See Man, Isle of.) The steam-packets to and from Liverpool, Belfast, and Glasgow frequently touch at Douglas. Castle Mona, near the beach, a little NE. from the town, formerly the property and residence of the dukes of Athol, has been sold, and is now converted into a hotel. There is here a pier 520 ft. in length, with a light-house at its head. The harbour dries at low water; but vessels drawing 10 ft. water may enter it at high-water neaps, and those drawing 14 ft. at high water springs. The anchorage in stormy weather is but indifferent. The parish church is 2 m. from the town; but it has three other churches, one of which is a handsome structure, with chapels for Catholics, Methodists, and Independents. It has also assembly-rooms, a public library, a Lancastrian school, and several charitable foundations. The custom-house is one of its best buildings.

DOULLENS, or DOU'LLENS, a town of France, dép. Somme, cap. arrond., on the Authie, 16 m. N. Amiens. Pop. 4,372 in 1861. Its citadel, formerly considered one of the bulwarks of Picardy, was repaired by Vauban, and is very strong. The church of St. Martin is remarkable for beauty and lightness of style: the town has two hospitals, a theatre, and a large cotton-spinning factory.

DOUNE, a market town of Scotland, co. Perth, on the N. bank of the Teith, a tributary of the Forth, 7 m. NW. Stirling. Pop. 1,256 in 1861. The town consists of three streets, radiating from a centre where the market-cross stands. Its only public building is the parish church, a Gothic edifice with a handsome tower. It is famous for its annual cattle, sheep, and horse fairs, six in number, one of them lasting three days. The cattle and sheep are from the highlands, and are lean, and purchased to be fattened either in the Lowlands of Scotland or in England. The cotton manufactory of Deanston is within less than a mile of the town, on the bank of the Teith, and is driven by water. It belongs to a Glasgow company, and gives employment to 700 individuals in spinning, weaving, and bleaching. Doune Castle, which is within a few hundred yards of the town, on an elevated peninsula formed by the junction of the Ardoch with the Teith, was one of the strongest Scottish fortresses. It was originally the seat of the earls of Menteith. It was anciently the residence of Mary Queen of Scots. It was, for a while, in the hands of the rebels in 1745. It gives the second title to the noble family of Moray, whose property it has long been. It has a square tower 80 ft. high; the walls are 10 ft. thick. The bridge of Teith, in the immediate vicinity of the town, was built in 1535 by Robert Spittal, tailor to Margaret, wife of James IV. and daughter of Henry VII.

DOURO (Span. Duero, an. Durius), one of the principal rivers of Spain and Portugal, through the N. part of both which it flows. It rises in the Sierra de Oliban, prov. Soria, Old Castile, about lat. 42° N. and long. 2° 50' W. At first it runs SE. and then S. to near Soria, but thence onward its direction is generally W., through the kingdoms of Leon and Portugal to its mouth in the Atlantic: in lat. 41° 8' N., long. 8° 34' W., 2 m. W. Oporto. From near Miranda to beyond Torre de Moncorvo, however, it flows almost due SW., forming the boundary between the Spanish prov. of Salamanca and the Portuguese prov. of Tras-os-Montes. It afterwards separates the latter prov. and Minho from Beira. The length of its entire course is estimated at 500 m.; it receives the Pisuerga, Seguilla, Esla (its principal tributary), Sabor, Tua, and Tamega on the right, and the Grado, Eresma, Tormes, Agueda, Coa, Tavora, Paiva, &c., on the left side; its basin may be considered the most extensive in the whole peninsula. It runs for the most part through deep and narrow valleys; its bed is generally narrow, and its current very rapid. It is, how-

ever navigable as far as San João de Pesqueira, about 70 m. E. by N. Oporto; and since the Wine Company of the Upper Douro have partially removed some obstacles that existed at that point, it has been rendered available for flat-bottomed boats as high as Torre de Moncorvo, 100 m. from the ocean. It has a bar at its mouth, and its navigation is liable to be seriously affected by freshes, or sudden swellings, occasioned by rains, &c., to which it is very subject. (See OPORTO.) Soria, Aranda-de-Duero, Toro, and Zamora in Spain; and in Portugal Miranda, San João de Pesqueira, and Oporto, are situated on its banks. Sixteen stone bridges cross it at various points, besides which it presents numerous fords.

DOVER (vulgarly DOVOR), a Cinque Port, parl. borough, and town of England, co. Kent, hunds. St. Augustine, hund. Bewsborough, 69 m. SE. London by road, and 81 m. by South Eastern railway. Pop. 25,325 in 1861. The town stands on the SE. shore of the co., on the straits of Dover, in a valley formed by the depression of the chalk strata, 27 m. NW. by W. Calais, and 21 m. from the nearest part of the French coast. It is traversed by a small stream, which empties itself into the harbour. The town consists of one principal street, extending upwards of a mile in the direction of the valley, shorter ones branching from it on each side, and ranges of houses on the shore. What may be called the New Town of Dover, built chiefly for the reception of occasional visitors during the bathing season, is under the castle cliffs on the E.: the old part of the town is irregular, and the streets narrow; but the whole is improving. In consequence of the increase of building, the villages of Charlton and Buckland have become continuous portions of the town. It has two ancient par. churches, St. Mary's and St. James's, the former rebuilt in 1844; another St. James erected in 1862, at a cost of 10,000l., and twelve dissenting chapels. There is also a school, founded in 1789, for forty-five boys and thirty-four girls, now incorporated with a national school, which educates 400 children; a girl's school of industry, established 1819; an infant school; a savings' bank; a dispensary, and many minor charities; a town-hall and gaol; theatre and assembly rooms, built in 1790; public libraries, reading-rooms, and baths, on the Marine Parade. The harbour, formed by the mouth of the small stream which runs through the town, called the Dour or Pent, consists of an inner and outer harbour, of 6¼ and 7½ acres respectively. Vessels of 300 tons can come up to the quays, and those of 400 tons can enter the port. There is a great Harbour of Refuge outside the port, formed by a granite pier a mile in length, known as the Admiralty pier. The mail steamers to and from France land and discharge passengers at this pier, and the railway trains run along it, close to the boats. By a charter of James I., the lord warden of the Cinque Ports, and ten other commissioners, were appointed conservators of the harbour. On an eminence bounding the NE. side of the valley stands the castle, an immense collection of ancient and modern works, occupying an area of about thirty acres: it is approached by a bold ascent, but is itself commanded by the higher ground on the W. and SW. There are remains of ramparts, and of a temple, bath, and Pharos, supposed to be of Roman construction. Previously to the last French war, the works were much dilapidated, but they were then repaired, and greatly augmented. There are upper and lower courts, surrounded (except towards the sea) by curtains and large dry ditches. In the centre of the former is a spacious keep, built by Henry III., and now forming a bomb-proof magazine; the curtain of the lower court is flanked, at irregular intervals, by ten towers of various construction—the oldest built by Earl Godwin, the others at different times during the Norman dynasty; with these, subterranean passages communicate from the ditch; there are also four or five ancient wells, excavated to the depth of 370 ft. The modern works consist of batteries with heavy artillery, casemates, covered ways, a large vault, excavated in the chalk, and barracks capable of lodging 2,000 troops. The lord warden of the Cinque Ports is always constable of the castle. The heights on the E. side the valley were also strongly fortified during the last war, and the fortifications have been greatly strengthened in recent years, annual grants of parliament being allowed for the purpose. The grant for the financial year 1864–5 amounted to £81,332. These fortifications are garrisoned by 7,500 troops, under the command of a brigadier-general. There is a military hospital on the N. side of the town. Dover has a busy, thriving appearance. Its chief traffic being derived from the influx of passengers to and from the Continent: of late years, also, its popularity, as a fashionable sea-bathing place, has considerably increased. There are large paper mills in the vicinity, and in the town a brewery and private docks, where ship-building is carried on to some extent, and rope, sail, and other establishments connected with the supply of shipping. The intercourse with Calais and other French ports, and also with London, is almost wholly carried on by steamers. The coasting trade consists chiefly of corn exported to London, and coals imported from the northern counties. The port comprises the creek of Folkestone, and the stations of Hythe and Romney. About ninety-five vessels, of the aggregate burden of 5,000 tons, belong to the port.

Dover, under the Municipal Reform Act, is divided into three wards, with six aldermen and eighteen councillors, and the port and municipal limits coincide. Previously to the act, the governing body consisted of a mayor, twelve jurats, and thirty-six common councilmen, who, like the magistrates of the other Cinque Ports, enjoyed several peculiar privileges in the trial of crimes, &c.; but these are now either wholly done away with, or greatly abridged. The constable of the castle has still, however, the jurisdiction of a sheriff within the Cinque Port limits; writs from the superior courts are directed to him, and his warrant is executed by an officer called Bodar; the debtors' prison being in the castle; a court of Lodemanage is also still held for licensing and regulating pilots.

Dover has returned two members to the H. of C. from the 18th Edw. I. Previously to the Re-form Act, the right of voting was in the freemen; the right of freedom being acquired by birth, by marriage (during the wife's life), by the possession of a freehold within the town and port, by gift and purchase. Registered electors, 2,207 in 1865. Gross annual value of real property assessed to income-tax 121,015l. in 1857, and 117,502l. in 1862.

Dover was a station of the Romans, by whom it was called Dubris; and being situated nearer to the Continent than any other town in England, it was long regarded as of the highest importance, and as being, in fact, the key of the kingdom. At Swingfield, near the town, are the remains of a preceptory of the Knights Templars, where King John surrendered his crown, and received it back from the Pope's legate, in acknowledgment of superiority. In 1216, the castle was successfully de-

fended against the Dauphin of France, by Hugh de Burgh, earl of Kent. In the civil war it was taken by stratagem, in 1642, by the Republicans.

Dover cliffs lie both on the E. and W. sides of the town. The noble description in Shakespeare is applicable to the latter; but the cliff to which the poet alluded having been undermined and thrown down, those that remain do not quite come up to the description.

DOWLETABAD (The Fortunate City; Hind. Deoghir), an inland town and fortress of Hindostan, prov. Aurungabad, and its original cap., dom. of the Nizam, 7 m. NW. Aurungabad; lat. 19° 57′ N., long. 75° 45′ E. The fortress stands upon an isolated conical granite rock, the summit of which is about 500 ft. above the plain below, and which has been scarped for one-third nearly of its height, so as to present all round the appearance of a perpendicular cliff. An outer wall of no strength surrounds the fort; but three other lines of walls and gates must be passed before arriving at the ditch, the causeway across which will admit of only two persons abreast, and which is defended by a building with battlements on the opposite side. The mode of access to this singular hill fortress is thus described by the Earl of Munster:—'The governor led the way through an excavation into the heart of the rock, so low that I was obliged to stoop nearly double. But after a few paces, a number of torches showed me I was in a high vault, and we began to ascend on a winding passage, cut through the interior of the body of the hill, ... This passage was about 12 ft. high, and the same broad, and the rise regular. At certain distances from this dismal gallery are trap-doors with flights of small steep steps leading to the ditch below, only wide enough to admit a man to pass, also cut through the solid rock, to the water's edge, and unexposed to the fire of the assailants, unless they were on the very crest of the glacis. We might have been in all ten minutes mounting by torchlight, and came out in a sort of hollow in the rock about 70 ft. square. On one side, leaning against the cliff, was a large iron plate, nearly of the same size as the bottom of the hollow, with an immense iron poker. On the besiegers having gained the subterranean passage, this iron is intended to be laid down over the outlet, and a fire placed upon it.' Near it is a perforated hole in the rock, intended to act as a bellows to the fire. The road hence to the summit is very steep; in some places it is covered with brushwood, in others with small houses, towers, and gates: it passes through the governor's residence, a good building, surrounded by a verandah with 12 arches. On the peak the Nizam's flag flies, and a large brass 24-pounder is mounted; but, excepting this, in the whole fortress there are but a few 2 and 3 pounders. The pettah presents the remains of many buildings of a rough dark-coloured stone, but is now in great measure deserted: the interior of the lower fort is a similar collection of ruins, and contains a column of great diameter and perhaps 160 ft. high, deformed, however, by a huge gallery, which encompasses it at about a fourth part of its elevation from the ground. From its natural strength, and the labour that has been bestowed upon it, this fortress is looked upon as impregnable; and as there is plenty of water (one tank cut out of the rock is only about 100 yards from the summit), if properly defended, it could only be reduced by famine. Notwithstanding these advantages, it was one of the first fortresses that fell into the hands of the Mohammedans, who took it by surprise, and plundered it of immense riches, a.d.

1293. Early in the 14th century, Mohammed III., who made it his residence, nearly ruined Delhi by the absurd project of making its inhabitants remove to his new capital. It was afterwards successively possessed by the dynasties of Ahmed Nizam Shah, Malik Amber, Shah Jehan, and the French: since 1759 it has belonged to the Nizam's dom. The pagodas of Ellora (which see) are in the vicinity of Dowletabad.

DOWN, a marit. co. of Ireland, prov. Ulster, on its W. coast, having N. and E. the Irish Sea, and the N. Channel, N. Belfast, Lough, and Antrim, and W. Armagh and Louth. Area, 611,404 imp. acres, of which 108,569 are unimproved mountain and bog. The extent of arable land, in square miles, was 803 in 1811; 818 in 1851; and 821 in 1861 (census of Ireland for 1861). The mountains of Mourne, in the S. part of the co., are amongst the highest in Ireland; but, with this and a few other exceptions, the surface is abundantly level. Soil of a medium degree of fertility. There are some large estates; but there is also a fair proportion of those of medium size. Farms very small: those occupied by the better class of farmers run from 20 to 50, and a few to 100, acres; but the inferior holdings, which are the great mass, do not, perhaps, average 5 acres. The occupiers of the latter formerly depended, in a great degree, on the linen trade; but since its decline, or rather since the manufacture began to be principally carried on in factories, they have had nothing but the land to depend on, and the competition for the smallest patches is extreme. In this, as in most other parts of Ireland, a new tenant must not only pay the stipulated rent to the landlord, but he must also pay a sum to the previous occupier, whatever may have been the cause of his leaving the farm, to ensure his quiet possession. This latter sum is called the tenant's right; and in Down it frequently amounts to 10l. an acre! (Hiun's Miseries and Beauties of Ireland, Lett. &c.) Still, however, a good many improvements have been introduced of late years, though, where the holdings are so small, it would be absurd to suppose their agriculture can be far advanced. Potatoes, oats, and flax are the principal crops; turnips rare; potatoes mostly planted in 'lazy beds,' though drilling is now pretty common. Average rent of land, 10s. an acre. Cottages very generally whitewashed and neat. The condition of the cottiers or peasantry is much superior to what it is in most other Irish cos.; and would have been much more so had for that custom, the bane and curse of Ireland, of dividing and subdividing farms, which is nowhere more prevalent than here. Principal rivers, Bann, Lagan, and Newry. Principal towns, Newry, Ballynacarret, and Downpatrick. Down is divided into eight baronies and sixty parishes, and sends four members to the H. of C., two for the co., and one each for Newry and Downpatrick. Registered electors, 11,367 in 1865. Pop. 361,487 in 1841; 320,924 in 1851; and 299,302 in 1861. Gross annual value of real property assessed to income tax, 611,811l. in 1857, and 601,871l. in 1862.

DOWNHAM (MARKET), a town and par. of England, co. Norfolk, hund. Clackclose, 78 m. N. by E. London by road, and 87½ m. by Great Eastern railway. Pop. of town 2,488, and of par. 3,133 in 1861. Area of par. 2,890 acres. The town, on an acclivity near the E. bank of the Ouse, which is here crossed by a bridge, has three streets of well-built houses, and is paved and amply supplied with water. The church, on the summit of the acclivity, is an antique Gothic structure, with a low tower and spire, approached on the S. by a noble avenue, and on the N. by a flight of steps. There are also several dissenting chapels, a Lancastrian school for 65 boys, and a national school for 70 girls. Market, Sat., noted for the supply of fish and wild fowl from the fens. Fairs, March 8 for horses (one of the largest in the kingdom), May 8, cattle, and Nov. 13. There is an extensive bell foundry in the town, and in the immediate vicinity is a large mustard manufactory. It is chiefly a dairy parish, and has been long celebrated for its supply of butter; but its famous butter market, held on Monday, has been removed to Swaffham. Petty sessions are held weekly, and a court baron and leet quarterly, by the lord of the manor.

DOWNPATRICK, a marit. town and parl. bor. of Ireland, co. Down, of which it is the cap., prov. Ulster, near the Quoyle, a short distance from its embouchure, in the SW. angle of Lough Strangford, 21 m. S. by E. Belfast, with which is is connected by railway. Pop. 4,886 in 1841, and 3,849 in 1861. The town consists of four main streets, meeting in a confined valley, and extending up the declivities of the surrounding steep hills. Like other southern towns, it is divided into the English, Scotch, and Irish quarters. There is a quay about 1 m. from the town, on the river, accessible to vessels of 100 tons, and a new quay, about 1 m. nearer the Lough, is accessible to vessels of much larger burden. The town was formerly the seat of the bishopric of Down, but since the union of the see with that of Connor, the ecclesiastical business is transacted at Lisburn. The ruins of the ancient cathedral, and those of a neighbouring pillar tower, still remain. The new cathedral is built in the ancient style; besides which, there is a par. church, Rom. Cath. chapel, and meeting-houses for Presbyterians and for Methodists. The diocesan school for Down and Dromore diocese is held here, as also a subscription school, the co. infirmary, fever hospital, dispensary, an almshouse with schools annexed, endowed by the Southwell family, an asylum for clergymen's widows, a mendicity institution, and large barracks. A constabulary force is stationed here. In the immediate vicinity is a remarkable rath, or artificial mound, 60 ft. high, and surrounded by three ramparts, the outermost of which is nearly 1 m. in circ. About 2 m. distant, at the foot of the hill of Sleibh-na-griddle, are the Struel wells, much frequented at midsummer by Rom. Cath. pilgrims for devotional purposes, and for the supposed miraculous efficacy of their waters. The corporation, which consisted of a mayor, bailiffs, and commonalty, no longer exists, its powers being vested in commissioners. The bor. returned two members to the Irish H. of C. till the Union, since which it has sent one member to the imperial H. of C. The parl. bor. extends over a space of 1,486 stat. acres. Registered electors, 200 in 1862. Manor courts, with jurisdiction to the amount of 10l., are held every third Tuesday; courts leet in spring and at Michaelmas. The co. assizes are held here in the court-house, a modern building; as are also general sessions in March and October, and petty sessions on Thursdays. The co. gaol, a spacious building, contains 200 cells, and 15 other rooms for prisoners. The linen manufacture is carried on in the neighbourhood. Markets on Saturdays; fairs on the second Thursday in Jan., March 17, May 19, June 22, Oct. 29, and Nov. 19. This is a very old town, being formerly the residence of the kings of Ullagh or Ulster.

DOWNTON, a bor. town and par. of England, co. Wilts, near its S. border, hund. Downton, on the Upper Avon, which here divides into 3 branches, each crossed by a bridge; 78 m. SW. by W. London. Area of par. 11,120 acres. Pop. of do. 3,768 in 1861. The town has one principal street, with a few respectable houses. Exclusive of the church—a large cruciform structure with a tower—there is a chapel of ease in the parish, and three dissenting chapels. A free school, founded in 1679, educates 12 boys; and another, founded in 1757, 6 girls. Market discontinued. Fairs April 23, for cattle, Oct. 3 for horses and sheep. The bar. returned two mems. to the H. of C. from the reign of Edw. I. down to the passing of the Reform Act, when it was disfranchised. This is a place of considerable antiquity. At its NE. end is a conical mount, on which stood an ancient castle, whose entrenchments are still visible. Stand'inch of Trafalgar House, a national gift to the heirs of Lord Nelson, is within 2 m. of Downton.

DRAGUIGNAN, an inl. town of France, dép. Var, of which it is the cap., in a fertile valley, on an affluent of the Artenby, 40 m. NE. Toulon, and 410 m. SE. Paris, on a branch line of the railway from Toulon to Nice. Pop. 10,069 in 1861. Its climate is temperate and salubrious, and being situated in a basin, surrounded by vine and olive clad hills, it offers a delightful place of residence. Though without any particular beauty, the town is sufficiently well built, and has numerous public fountains. Chief public buildings—the hall of justice, prison, clock-tower, and hospital. Draguignan has a public library with 14,000 vols., an excellent botanic garden, cabinets of natural history, and a society of agriculture and commerce; with tribunals of primary jurisdiction and commerce, a chamber of manufactures, and a communal college. There are fabrics of broad-cloth, thrown silks, stockings, and soap, and distilleries.

DRAMMEN, a sea-port town of Norway, distr. Buskerud, on both sides of the river of the same name, near its mouth in the Christiana-fiord, and 20 m. NW. Christiana. Pop. 10,172 in 1860. 'It is a long straggling place. Though to us it seemed to have little of the bustle of trade, it is said to export more timber, chiefly in logs, than any town in Norway. Its women are reckoned among the greatest beauties of the North; and we can some who fully support its reputation in this respect. Most travellers, however, will recollect it better as the place in which is carried on the principal manufacture of the delightful little carriole.' (Bremner's Excursions, p. 96.)

DRAVE (Germ. Drave), a river of Europe, and one of the principal tributaries of the Danube. It lies wholly within the Austrian empire, extending between lat. 40° 34' and 46° 30' N., and long. 12° 20' and 19° E. It rises on the Toblach-heath, near the E. extremity of the Tyrol, in what is called the Puster-thal, about 17 m. ESE. Brunecken, and runs at first ENE. to Lienz, where it is augmented by the Isl. From this point its course generally is ESE. to its mouth in the Danube, near the castle of Erdödy, 17½ m. E. Esseeg. It traverses Carinthia and Styria, and afterwards forms the boundary between Croatia and Slavonia on the S., and Hungary Proper on the N. It receives the Möhl, Gurk, Lavant, and Mur (its chief affluent) on the left; and the Gail, Dran, Bednya and some other rivers of minor importance on its right side. Lienz, Greifenburg, Spital, Villach, Völkermarkt, Marburg, Pettau, Warasdin, and Esseeg, are the chief towns situated on its banks. It runs through a mountainous country and narrow valleys, as far as Warasdin, but thence onward its course is through a plain country. Its entire length is estimated at 370 m. In its upper part the Drave is extremely rapid; its navigation in many parts is greatly impeded by the number of trees torn down by its violence, which afterwards block up the current. At present this river is

made but little use of for commercial purposes; but in case of an extensive steam-navigation of the Danube, its value as a means of transit would be greatly enhanced. It is said that the Austrian government has in contemplation to form a communication between the Adriatic and one of the great tributaries of the Danube; and if so, this would probably be the one chosen, the country between the Upper Drave and the sea apparently presenting the fewest obstacles to such an undertaking. (Turnbull's Austria, ii. 876, 877.) The author of 'Germany and the Germans,' vol. ii., gives a spirited sketch and description of Hungarian peasants descending the Drave on rafts of empty barrels, after having disposed of their wine in the mountains of Carinthia.

DRESDEN, a city of Germany, cap. of the kingdom of Saxony, on both sides the Elbe; 61 m. ESE. Leipzig, 283 m. ENE. Frankfurt on the Mayne, 270 m. NNE. Munich, 100 m. N. by E. Berlin, and 290 m. NW. Vienna, on the main line of railway from Berlin to Prague and Vienna. Pop. 61,227 in 1811, and 128,152 in 1861. The city is more than 400 feet above the level of the sea, and is delightfully situated in the midst of the Saxon wine district, occupying the most beautiful and richly-cultivated portion of the valley of the Elbe. The banks of the river have, however, a very different appearance. The right is abrupt, rocky, and woody, and, having a S. aspect, is in great part covered with vineyards. The left is more flat, presenting a succession of meadows, groves, gardens, and orchards, studded with numerous villages: the whole landscape gradually rising till it becomes united with the distant Erzgebirge mountains. The city itself has been termed the 'German Florence,' and is certainly, on the whole, very handsome.

Dresden is divided into the Old and New Towns, —the first on the right or N. bank of the river, and the latter on the N. bank; and has seven suburbs, extending all round the Old Town, of which that called Friedrichstadt, lying to the W. of the small river Weiseritz, near its confluence with the Elbe, is the best built and most important. Immediately adjoining the town, are the New Anlagen, consisting of public walks and gardens. The Old and New Towns are connected by two bridges. The first, a noble stone bridge of sixteen arches, 1,420 ft. in length, and 36 ft. in width, is considered the longest and finest structure of the kind in Germany. It has a foot pavement and an iron balustrade on each side, with a bronze crucifix on its centre pier, and an inscription commemorative of the destruction of part of the bridge by Marshal Davoust, to facilitate his retreat in 1813, and its restoration in the same year by the Emperor Alexander of Russia. The other bridge, forming a portion of the railway leading from Leipzig, through Dresden to Prague, was opened in 1850, and is also a fine structure. The Old Town was formerly provided with fortifications; but these were demolished by the French in 1810, and the place they occupied is now laid out in public walks. That portion of these walks facing the Elbe, is called the Brühl Terrace, and is approached from the foot of the bridge by a grand flight of broad steps. From its own beauty, and the grandeur and variety of the scenery it commands, it is at all times a favourite resort of the inhabitants. As in most other fortified towns, the streets in the Old Town are narrow, the houses lofty and gloomy looking, and the squares irregular. In the construction of the buildings, generally, which are chiefly of sandstone, strength has been more studied than elegance: the principal of the public edifices are, however, in the part of

Dresden. The Schloss (castle), or royal palace, opposite the bridge, is a large antique and ungainly looking building, having the appearance of a fortress rather than of a royal residence; but, internally, it is in every respect worthy of its destination. It has halls of audience, ceremony, and various other state rooms, a royal library, the hall in which the Saxon legislature is opened, and a Catholic chapel with a tower 378 ft. high. It contains the celebrated state treasury, or Green Vault (Grüne Gewölbe), which occupies a suite of vaulted apartments on the ground floor. They contain an immense collection of precious stones, curiosities, and objects of vertu, and are reputed to be worth at least a million sterling. Adjoining the royal palace is the chamber of archives, and near it the palace of princes, containing a handsome chapel, gallery of portraits, and library. On the opposite side of the royal palace, and also communicating with it, is the far-famed gallery of paintings, the grand attraction of Dresden, being not only the finest collection in Germany, but the finest, taking it as a whole, to be found N. of the Alps. Amongst its valuable specimens of art, not one of which can be pronounced bad, few mediocre, numbers good, and several incomparable, are the celebrated Madonna di San Sisto of Raphael; the Notte, and five other works, by Correggio, in his best style; the St. Cecilia of Carlo Dolci; the Christo della Moneta, and a Venus, by Titian; other paintings, by Paul Veronese, Annibal Carracci, Guido, &c.; altogether 856, by Italian artists. In the works of the later German and Flemish masters, this gallery is also extremely rich: it contains magnificent specimens of Rembrandt, Rubens, Vandyke, Teniers, Hans Holbein the younger, Ruysdael, Wouwerman, &c. Of the French school, there are several paintings by Claude, Nic. Poussin, &c.; and beneath the gallery there is a fine collection of plaster casts of the most famous statues, made under the superintendence of Raphael Mengs. This gallery, founded by the Elector Augustus II., has remained untouched and unharmed amid the innumerable revolutions that have, in the interval, convulsed Germany. When Frederick the Great bombarded Dresden, battered down its churches, and laid its streets in ruins, he ordered the artillery to keep clear of the picture gallery; and Napoleon treated Saxony with so much consideration, that not one of her pictures made the journey to Paris.

The Zwinger, erected in 1711, and originally designed as merely the vestibule to a new palace, intended to be built by Augustus II., is a fine group of buildings, surrounding an enclosure planted with orange trees, and forming a favourite promenade. It contains the armoury (second only to the Ambras collection at Vienna), cabinets of natural history, mineralogy, and mathematical and philosophical apparatus, and a gallery of engravings, which possesses at least 500,000 specimens of that art. Immediately contiguous to one of the wings of the Zwinger, is the grand opera-house, a building capable of accommodating 3,000 spectators. It communicates, by a covered way, with the palace of the princes, but is now only used for court festivities; theatrical performances take place in a smaller theatre, near the Catholic church; the latter, occupying a very prominent situation between the royal palace and the bridge, is a large structure in the Italian style. Externally it is profusely decorated, and generally considered deficient in taste; but internally it is chaste, elegant, and imposing. It contains an altarpiece by Raphael Mengs, and a fine organ by Silbermann: the music in this church is celebrated throughout Germany. As a whole,

however, it is inferior in elegance to the Frauenkirche (church of Our Lady, or St. Mary) in the new market, a beautiful stone building, adorned with a cupola, constructed on the model of that of St. Peter's at Rome, 300 German ft. high. The other churches do not demand particular notice. The remaining principal edifices in the Old Town are, the Brühl palace, with a collection of 50 landscapes by Canaletto; the mint, arsenal, medico-chirurgical school, house of assembly, royal guard-house—a beautiful specimen of Grecian architecture, new post-office, trades' hall, and hall for the annual exhibition and sale of the works of Saxon artists. The town hall is the chief ornament of the old market, and the only regular square in the Old Town. The New Town is altogether much better laid out, and contains fine squares, spacious streets, and elegant faubourgs. In this quarter stands the Japanese palace, now called the Augusteum, in honour of its founder, Augustus II. This magnificent palace, appropriated wholly to public purposes, is beautifully situated on the banks of the Elbe, amid pleasure grounds, which form a most agreeable promenade for the citizens. It contains the museum of antiquities and modern statuary, which occupies 10 saloons, and is enriched by some of the finest antique statues in Germany; a cabinet of coins; a public library with 250,000 volumes, 4,000 MSS., 100,000 pamphlets, and 20,000 maps; and the celebrated porcelain cabinet. The last is a collection of more than 60,000 pieces of China, including the finest Meissen, Chinese, Japanese, Italian, and Sèvres ware, and specimens of the manufacture of every country, altogether filling 18 apartments. Here are to be seen the three splendid China vases that Augustus II. purchased of the Elector of Brandenburg, at the price of a regiment of dragoons fully equipped!

Through the centre of the New Town runs a broad handsome street, planted with linden trees, near the upper end of which are some extensive infantry and cavalry barracks. The other chief public buildings are, the commandant's residence, several military academies, the town hall, and the church of the Trinity. The market place is embellished with an equestrian statue of Augustus II., in ancient Roman costume. The Frederickstadt contains the Marcolini palace and the Roman Catholic cemetery, but this quarter is mostly inhabited by the working classes. The Pirna suburb boasts of Prince Anton's handsome villa and extensive gardens; and the Wildruf suburb has the palace, gardens, and observatory of Prince Maximilian. Dresden has a great number of literary and scientific institutions, and establishments devoted to education. Among these are an academy of arts, two colleges, a botanic garden; schools of medicine, surgery, and veterinary medicine; a high school, 2 normal schools, numerous free elementary schools, with schools for the reformation of depraved children, and the deaf and dumb, and blind; it has also many charitable institutions, including orphan asylums of various kinds, a foundling hospital, and 5 other hospitals. Amongst other conveniences, the city possesses excellent public baths of all kinds, the prices of admission to which being low, the poorest person is able to indulge in the use of what is found to contribute materially to the public health.

Dresden has no very considerable external trade. It has numerous painters, designers, sculptors, engravers, and other workers in the fine arts; and some manufactures of woollen and silk, leather, gold and silver articles, carpets, sealing wax, maccaroni, white lead, straw hats, artificial flowers, musical, mathematical, and philosophical instruments, with a brass and cannon foundry, and a large sugar refinery. What is called Dresden china is not made in this city, but at Meißen, 14 m. distant. The greater proportion of its external commerce has hitherto consisted in its transit trade by railway and by the river Elbe; its general trade is, however, increasing. Since 1828, a wool market has been established.

Few European capitals have such pleasant environs as Dresden. Nearly all the roads leading out of it, and especially from the New Town and Friedrickstadt, are planted with rows of trees. The Elbe to the NW. of the city is lined on either side with fine avenues for a considerable distance. SE. of the Pirna suburb is the *Grosse Garten*, a large park filled with fine trees, near which is the small village of Räcknitz, and the monument erected to Moreau on the spot where he received his death wound, 27th Aug. 1813. On the right bank of the Elbe is the *Linkbad*, a hotel surrounded by some beautiful gardens, containing a theatre, &c., about 1 m. from the New Town; and 2 m. beyond this is *Findlater's Vineyard*, a villa and grounds laid out with much taste by a deceased Scotch nobleman. To these different places people of all ranks delight to resort, which they do especially on Sunday afternoons, to take refreshments and dance, or listen to the excellent bands of music with which all the public places are provided.

Dresden and its environs have been the scene of some of the most important conflicts in modern warfare, particularly on the 26th and 27th August, 1813, when Napoleon defeated the allies under its walls. This city has been the favourite residence of many distinguished literary men; in its immediate neighbourhood, Körner lived. Schiller wrote great part of his 'Don Carlos,' and Weber composed his highly celebrated opera 'Der Freischütz.' Its inhabitants generally are great lovers of the fine arts, and devoted to music.

DREUX, a town of France, dép. Eure-et-Loire, cap. arrond., on the Blaise, a tributary of the Eure, which partly encircles it, 20 m. NNW. Chartres, on a branch line of the railway from Paris to Chartres. Pop. 6,940 in 1861. The town stands at the foot of a hill, on which are the ruins of an ancient castle, which belonged to the counts of Dreux: it is well built, and has a fine promenade along the river's bank, a hospital, public baths, a theatre, town-hall, and church. Louis Philippe, when Duke of Orleans, built in the castle a chapel, which he intended for his family burial-place. It is the seat of tribunals of primary jurisdiction and commerce, and of a communal college. Near it, in 1562, was fought the celebrated battle in which the Prince of Condé, then at the head of the Protestants, was taken prisoner. Dreux was the native place of Jean de Rotrou, the tragic poet, and of Philidor, the famous chess-player.

DRIFFIELD (GREAT), a market-town and township of England, E. Riding, co. York, near one of the sources of the Hull; 27 m. E. by N. York, on the Great Northern railway. Area of township, 4,910 acres. Pop. of do. 4,734 in 1861. The town, at the foot of the Wolds, consists chiefly of one long street, parallel to which flows the brook above noticed, which, at the S. extremity of the town, is enlarged into a navigable canal that joins the Hull below Frodingham Bridge. All Saints' church is an ancient structure in the Gothic style. The Independents, Wesleyan, and Primitive Methodists, and Baptists, have places of worship. There is a national school for 100 children, and a dispensary. The chief officer is a constable appointed annually; a court for the recovery of small debts is held here. The town is a station for receiving votes in elections of members for the E. Riding. Market-day, Thurs., and well attended cattle markets every fortnight.

DROGHEDA, a parl. bor. and sea-port town of Ireland, being a co. in itself, but locally in the cos. of Meath and Louth, prov. Leinster, on the Boyne, 4 m. above its embouchure in the Irish Sea, and 25 m. N. Dublin, on the railway from Dublin to Dundalk and Belfast. Pop. 17,365 in 1831, and 14,740 in 1861. From the time the English settled in Ireland, this town, formerly called Tredagh, was considered of great importance. Parliaments have been frequently held in it, and it was made the site of a university, but the privilege was not acted upon. In 1649 it was stormed by Cromwell, who put its inhabitants to the sword, with the exception of a few that were transported to the American settlements.

The Boyne divides the town into two unequal portions, the larger of which, on the N. bank of the river, is connected with the lesser by a bridge of three arches; part of the ancient walls, and the gate of St. Lawrence, still remain, but the buildings now extend considerably beyond them. The churches within the town are St. Peter's in the N. div., St. Mary's in the S., and a chapel of ease. The R. Cath. chapel of St. Peter, considered the cathedral of the archdiocese of Armagh, is a large and elegant building, as is also that of St. Mary. There are friaries of the Augustine, Dominican, and Franciscan orders, and convents of the Dominicans and the Presentation. The Presbyterians and Wesleyan Methodists have places of worship. There are here a classical school on the foundation of Erasmus Smith, and other public schools which give instruction to nearly a thousand pupils. It has also an infirmary, a workhouse for the accommodation of 940 inmates, a linen hall, a building for the widows of Protestant clergymen, and an almshouse. There is an infantry barrack in the town, and another in the vicinity of Richmond Fort. It is in general pretty well built; the streets are paved, lighted, and cleaned, by a committee of the corporation; but its appearance is unfavourable, and the streets swarm with beggars.

Drogheda originally consisted of two distinct corporations, one on the side of Meath, the other on that of Louth. These were united under Henry IV., who granted the newly formed bor. a charter, under which it is still regulated. Its jurisdiction extends over 5,780 acres. The corporation consists of 6 aldermen and 18 common-councilmen, elected by the three wards into which the town is divided. The assizes are held twice a year, and general sessions of the peace by the mayor and recorder in January, April, June, and October. Petty sessions are held every fortnight. The gaol is a well arranged building. It has 6 wards and 18 cells, for an average number of 26 prisoners. The bor. sent two mems. to the Irish H. of C.; and since the Union it has sent one mem. to the Imperial H. of C. Registered electors 689 in 1865. Gross annual value of real property assessed to income tax, £5,000l. in 1857, and £2,748l. in 1862.

An extensive manufacture of coarse linens was formerly carried on here, which gave way to that of cottons; but the latter is nearly extinct, while the former has revival. Flax spinning is at present the principal branch of industry carried on in the town. It has, also, an extensive foundry, where steam engines and other articles are made; with numerous corn-mills, salt-works, breweries, tanneries, and soap-works. Drogheda ale is in much demand both in England and in the foreign market.

The chief trade, which consists in the export of agricultural produce and of linens, is carried on with Great Britain by steamers, which ply regularly between the port and Liverpool. The coasting trade and coasting trade employ also many sailing-vessels. The greater part of the foreign trade is with the British colonies in N. America; timber is the principal article of import. The harbour and river have undergone several improvements, by means of which vessels of 500 tons may now discharge at the bridge, and barges of 70 tons may proceed inland as far as Navan by means of the Boyne navigation. The customs duties received at the port amounted to 12,804l. in 1859; to 9,782l. in 1861; and to 4,404l. in 1862. The railway from Drogheda to Dublin was opened in 1844. Fairs are held on March 10; April 11, May 10, June 22, Aug. 26, Oct. 29, Nov. 21, and Dec. 19. Hence and wool are the chief articles for sale. The shipping belonging to the port on the 1st of January, 1864, consisted of 3 sailing vessels under 50, 26 sailing vessels above 50 tons. There were, besides, 5 steamers, of a total burthen of 1,579 tons.

DROITWICH, a parl. and munic. bor. of England, famous for its salt springs, co. Worcester, 7 m. NE. by N. Worcester, 118 m. NW. London by road, and 125½ m. by Great Western and West Midland railway. Pop. of munic. bor. 3,124, and of parl. bor. 7,946 in 1861. Though locally in the upper division of the hund. of Halfshire, it has exclusive jurisdiction, and is pleasantly situated on the side of a narrow valley, at the bottom of which runs the Salwarp, on the road from Birmingham to Worcester. It has three parishes and three churches, of which St. Andrew, rebuilt after being destroyed by fire, in 1293, is the most ancient and interesting. The town was originally incorporated by charter from John, confirmed by Henry III. and some of his successors, previously to the charter of Imperimus, granted by James I. It is governed by a mayor, four aldermen, and twelve councillors; hor. income, 567l. in 1862. Gross annual value of real property assessed to income tax, 56,416l. in 1857, and 64,230l. in 1862. The bor. returned two mems. to the H. of C. under Edward I., and to the parliaments held in the 2nd and 4th Edward II., from which period the privilege ceased until 1554, since which time it regularly returned two mems. until the passing of the Reform Act, which deprived it of one of its members. Its boundaries were at the same time considerably extended. Registered electors, 380 in 1863; the bailiffs are the returning officers. The election of members for the E. division of the co. is held here. There are three chapels; a chapel of ease, one for Independents, and one for Wesleyans; a hospital for thirty-eight aged men and women, founded by Henry Coventry, in 1696; and a charity school for forty boys and forty girls, who are educated and clothed, and on leaving school apprenticed. The salt trade is the main support of the place; malting and tanning are also carried on, and there are some mills for grinding corn.

Droitwich has been celebrated from a very remote period for its brine springs, or wiches, a name of Saxon origin, though its meaning be not well known. (Campbell's Political Survey, i. 78.) Reference is made to these springs in Domesday book, and it is certain that they were known, and that salt was obtained from them, long before its compilation, as is evinced by the grants by different Saxon kings to the grants of Worcester, in all which the wiches are specially mentioned. (Camden's Britannica, Gibson's ed. i. 100.) Most probably indeed they had been known to, and

wrought by, the Romans. The springs are in the middle of the town, and the salt is obtained by boiling and evaporating the brine. About a century ago the usual depth of the brine-pits was about 30 ft., but now they are generally sunk to a much greater depth, and a far more copious supply of brine is obtained. An ounce of brine is said to contain 140½ grains muriate of soda, 2¼ grains sulphate of lime, 2¼ grains sulphate of soda, and a trace of muriate of magnesia. A canal from the Severn to Droitwich is used in the conveyance of the salt for shipment, and of the coals made use of in the works.

DROME, a dep. of France, in the SE. part of the kingdom, formerly a part of the prov. of Dauphiny, having N. and E. Isère, E. the Hautes and Basses Alpes, S. Vaucluse, and W. Ardèche, from which last it is separated by the Rhone. Length, N. to S., about 85 m.; greatest breadth, 50 m. Area, 652,155 hectares. Pop. 326,684 in 1861. This dep. is naturally divided into two portions, an easterly or mountainous, and a westerly or plain region. The former includes about 400,000 hectares, or nearly two-thirds of the total surface, and is intersected by ramifications of the Alps, with a mean elevation varying from 4,000 to 5,000 ft. The loftiest summits attain to about 5,750 ft. The chief rivers, after the Rhone, are the Isère and Drome, but the latter is not navigable. There are a number of streams, which, though usually small, become during the melting of the mountain snows devastating torrents. In the elevated parts it is almost always cold, while along the banks of the Rhone the summer heats are very overpowering: the climate is, however, generally healthy. In the lower parts of the dep. there are about 100,000 hectares of rich land, the rest being generally of inferior fertility. The cultivable lands comprise about 259,100 hectares; vineyards, 29,996 do.; and forests, heaths, and wastes, 800,550 do. Wheat, maize, and oats are the chief kinds of grain cultivated; but the corn grown is insufficient for home consumption. The other articles of culture are very various, including pulse of different kinds, hemp, walnut, olives, chestnuts, almonds, madder, and other dyeing plants and fruits. The vine culture is the most important branch of rural industry, and about 150,000 hectolitres of wine of the best quality are exported annually. The finest growths are the red wines of Hermitage, Crozes, Mercurol, and Gervans, and the white wines of Mercurol and Chevarroux, and the Clairette de Die. The genuine hermitage bears a comparison with the finest growths of the Bordelais and Upper Burgundy. The hills, called Mas, which produce it, have a S. aspect, and are usually covered with a thin calcareous soil; they are so steep, that the mould has to be sustained by rows of low walls. The wine of the Mas of Beau, which differs in several respects from the others, is principally bought up by the Bordeaux merchants to give body and flavour to the secondary clarets. The rearing of silkworms is carried on to a great extent, and there is a greater number of mulberry trees in Drome than in any other dep. of France. Gard alone excepted. A great many bees are kept, and the honey is of very good quality. The middle mountain region is covered with woods of oak, beech, fir, &c., supplying excellent timber; above these there are extensive pasture-lands, feeding in summer numerous flocks of sheep and goats, many of which come from Provence. Mines of iron, lead, and coal, and quarries of marble, granite, rock-crystal, and limestone, are wrought. Manufactures not very important; the chief are those of woollen cloths, serges, silks and silk-twist, coloured linens, stockings and gloves at Valence, hate, paper, leather, brandy, oils, steel articles, chemical products, and earthenware. The trade is principally in the products of the soil, which include excellent truffles. Drome is divided into 4 arrondissements, 29 cantons, and 350 communes. Chief towns, Valence, the cap., Montelimart, and Crest. Drome was annexed to France in 1343.

DRONTHEIM. See TRONDHEIM.

DUDHOY, or DUDHOI, an int. town of Rhotan, prov. Gujerat, dem. of the Guicowar, cap. of a pergunnah containing 84 villages, 38 m. NE. Barroach; lat. 22° 9′ N., long. 73° 25′ E. Toward the end of the last century it contained 10,000 inhab., a few of whom were Mohammedans, and none Parsees. It is nearly an exact square, and has been elaborately fortified, though only a portion of its works remains in any degree of preservation. The ancient walls have been built entirely of large square stones; the city gates are all strong and beautiful, especially the E. portal, called the 'Gate of Diamonds,' which, together with the temple connected with it, present a most complete and elegant specimen of Hindoo taste. 'In proportion of architecture, and elegance of sculpture,' says Mr. Forbes (Mod. Trav., ii. 162), 'it far exceeds any of their ancient structures I have met with, and the groups of warriors on horseback, on foot, and on fighting elephants, approach nearer to the classical bas-reliefs of Greece than any performances in the excavations of Elephanta.' Within the walls there was a magnificent tank, ¼ m. in circuit, lined with hewn stone, and with a flight of steps all round, and partly supplied with water by means of a stone aqueduct from receptacles without the walls. In the district around Dudhoy the soil is generally rich and loamy, producing fine crops of rice, jowaree, bajree, &c.; various legumes, cotton, sesamum, palma Christi, sugar-cane, hemp, flax, ginger, and plants for dyeing.

DUBLIN, the metropolitan co. of Ireland, on the E. coast of the island, having E. the Irish Sea, or St. George's Channel; S. Wicklow; W. Meath and Kildare; and N. Meath. Area, 218,931 acres, of which about 9,000 are unimproved, mountain, and bog. The extent of arable land, in square miles, was 308 in 1811; 304 in 1831; and 305 in 1841. (Census of Ireland of 1841.) Principal river, the Liffey, by which Dublin is intersected. Surface mostly flat or undulating; soil shallow, and naturally poor, the subsoil being a retentive clay. Agriculture is by no means in an improved state; there is a want of a proper rotation and drainage, and white crops still not unfrequently follow each other. A good deal of land in the vicinity of Dublin is appropriated to garden culture. Average rent of land, exclusive of that portion called the co. of the city of Dublin, and of country houses, 22s. an acre, being as high an average rent as is paid by any co. in Ireland. Property a good deal subdivided. Farms near the city small, but larger at a distance. In 1841 the co. of Dublin—excl. of the city—had a pop. of 142,665; in 1851, of 149,219; and in 1861, of 155,144. The return at the latter period showed 73,152 males and 82,992 females. The increase of pop. amounted to 4·57 between 1841 and 1851; and to 4·17 between 1851 and 1861.

DUBLIN, a city, the seat of a University, and sea port of Ireland, of which it is the cap., co. Dublin, on the E. coast of the island, at the mouth of the Liffey, by which it is intersected; 292 m. WNW. London; 138 m. W. Liverpool; 65 m. W. Holyhead. The movement of the population of Dublin, unlike that of other towns of Ireland, has grown on increasing for nearly two centuries. An enumeration of the year 1682 showed 64,483 inhabitants, while in 1755 there were 129,756, and

in 1798 there were 182,370 inhabitants. The pop. in 1821 had risen to 185,881; in 1831, to 205,650; in 1841, to 233,651; and in 1851, to 291,700. In the next ten years there was a decline, and the census of 1861 only showed 254,808 inhabitants. Of this number there were 118,283 males, and 136,525 females. The increase of pop. between 1841 and 1851 amounted to 10·98 per cent.; but the decrease between 1851 and 1861 was 7·63 per cent., leaving a net increase in the twenty years of 3·5 per cent. The city is supposed to be the Eblana of Ptolemy, and was called by the native Irish Bailyath-cliath, 'the town on the ford of hurdles;' and by the Danes Divelin or Dubhlin, 'the black pool,' from its vicinity to the muddy estuary at the mouth of the river. At the period of the English invasion under Strongbow, A.D. 1169, the city was of very limited extent; its buildings being confined to the summit and declivities of a hill on the S. side of the Liffey, and enclosed by a wall little more than 1 m. in circ. For many years afterwards its increase in extent and population was extremely slow. At the commencement of the 17th century its suburbs extended but a very short distance beyond its ancient walls. In the wars of 1641, the additional works thrown up for the defence of the place lay between the castle and the college, which was then considered as outside the city. After the Revolution, the progress of improvement was comparatively rapid; new lines of streets were opened, particularly to the N. and E.; many of the confined old avenues were enlarged; several squares were laid out, and the buildings, both public and private, were constructed with greater regard to architectural elegance as well as internal convenience. An avenue, called the Circular Road, which nearly surrounds the city, encloses an area of 1,264 acres; of which, 785 are on the S., and 479 on the N. side of the Liffey. The river is bordered on each side by broad and well-constructed quays.

The figure of the city is elliptical, its longer axis extending along the line of the river, from W. to E., 2¼ m.; its shorter, from N. to S., nearly 2 m. Sackville Street, on the N. side, is remarkable for its great width and far its buildings; St. Stephen's Green, the largest of the squares, has in its centre an equestrian statue of George II.; College Green, an irregular and confined area near the centre of the city, where most of the main avenues meet, contains some of the finest public buildings, and has in its centre the equestrian statue of William III., so famous in Irish party history. The other public monuments of note are, Nelson's Pillar, in Sackville Street; the Wellington Memorial, a lofty obelisk in the Phoenix Park; an equestrian statue of George I., and pedestrian statues of George III. and IV., Dr. Lucas, and Messrs. Grattan and Drummond.

To a traveller frequenting only the principal streets, Dublin appears to be one of the handsomest cities in Europe. The public buildings are all on a grand scale, and the principal streets and squares are capacious, handsome, and well laid out. But there is notwithstanding, especially in the older parts of the town, a vast number of crowded, dirty thoroughfares, with mean, wretched houses, destitute of all the elements of comfort and cleanliness. Wealth and poverty, comfort and misery, are brought into immediate and painful contrast; and Dublin may, in this respect, be taken as a fair representation of the island of which it is the capital.

Dublin Castle stands on the E. verge of the hill upon which the city was primarily built. It was originally a square fortress, with towers at the angles; it now consists of a quadrangle, 280 ft.

by 139, surrounded with buildings containing the state apartments of the lord-lieutenant, and accommodations for the meetings of the privy council and other public functionaries. Attached to it is the vice-regal chapel, a small but elegant structure of Florid Gothic architecture. Offices for the ordnance and quartermaster-general's departments, and for the commissariat, are also attached to it. A guard of honour, of cavalry and infantry, is mounted here daily. The lord-lieutenant's usual place of residence is in the Phœnix Park, an enclosed tract of about 1,750 acres, of which about 1,500 acres are open to the public, and serving also as a place of exercise for the troops of the garrison. In it is a powder magazine, a barrack, the offices of the trigonometrical survey of Ireland, an institution for soldiers' orphans, a military infirmary, and residences for some of the inferior officers of the government. Near its centre is a pillar, surmounted by a phœnix rising out of the flames.

The head-quarters of the military establishment for Ireland are at the Royal Hospital, Kilmainham, originally a priory of the Knights Templars, which, after the suppression of that order, was granted to the Knights of St. John of Jerusalem; and, having become the property of the crown on the dissolution of the monasteries, was converted by Charles II. into an hospital for superannuated and disabled soldiers. The building is a large square, three sides of which contain the lodgings of the veterans, and the fourth a chapel, a dining hall, and a suite of apartments for the commander of the forces.

The principal barracks are on the N. side of the city, near the Phœnix Park. They consist of several large quadrangles, containing accommodations for a general officer and his staff, and for 7,000 men, cavalry and infantry. There are also barracks at Portobello, for cavalry; at Richmond bridge, the recruiting depôt, and Gt. George's Street, for infantry; and at the Pigeon-house Fort and Island Bridge for artillery; having in all accommodation for 5,500 men. The military infirmary in the Phœnix Park, near its W. entrance, can receive 250 patients. The superior courts of Justice are held in a magnificent edifice on the N. Quay, consisting of a central circular hall, opening into the courts of Chancery, Rolls, Queen's Bench, Exchequer, Common Pleas, Nisi Prius, and Admiralty; and wings, in which are several repositories, and offices for the despatch of legal business. The King's Inns, or inns of court, which are at the N. extremity of the city, contain halls for meetings and dining; the courts, offices, and record repositories of the Prerogative and the Consistorial courts of the see of Dublin, and the Registry of Deeds; near the main building is the library, containing a large collection of books. The privilege granted it under the Copyright Act of receiving a copy of every work published in the United Kingdom has been commuted for an annual grant, applicable to the purchase of books, at the discretion of the benchers. The number of barristers on the rolls of the courts is about 800, and of solicitors and attorneys, 1,600; but many of these whose names are entered never practised, and many others have withdrawn from the active duties of their respective professions.

The municipal boundary of the city differs considerably from that of the police and electoral franchise. On the E. side it extends to the village of Harbrook, 5 m. from the centre of the city; while on the N., W., and S., several parts of parishes, in close contiguity with the rest of the city, are beyond it. The extent of the franchise, which was accurately laid down at a very remote period, is still ascertained by means of a triennial perambulation by the civic authorities. The limit on the sea side is determined by the place where a javelin, thrown by the lord mayor standing at low-water mark, falls into the water.

Under the new Municipal Act the city is divided into 15 wards, and the corporation consists of 15 aldermen, one of whom is elected lord mayor, and 45 councillors. The lord mayor is the civil and military governor of the city, in which he ranks next after the lord-lieutenant; he is admiral of the ports of Dublin and Harbour, and a justice of the peace; he presides at the court of city quarter sessions; sits on the bench at the commission of Oyer and Terminer; holds a separate court for trial of petty offences; is chief judge of the lord mayor and sheriffs' civil court; and has the regulation of the public markets, and the inspection of weights and measures. He is personally distinguished by wearing a gold chain, called 'the collar of S S,' and has a cap of dignity, and a sword and mace, borne before him on public occasions. He resides in a plain old-fashioned brick building, attached to which is a large circular hall, erected for the purpose of entertaining George IV. in 1821, but without any pretensions to exterior architectural beauty. The recorder, when elected by the aldermen and approved by the common council, retains his office during good behaviour. He is the legal adviser of the corporation, and presides in the city criminal court. The corporate meetings are held in the Assembly House, a plain building, originally erected for the exhibition of pictures.

The corporation holds a criminal court four times a year for minor offences, capital cases being referred to the superior judges. The court must be opened by the lord mayor and two aldermen; but, virtually, the recorder is the ruling judge. The lord mayor's court holds pleas of personal actions above 2l.; those under that amount are decided in the court of conscience, over which the lord mayor of the preceding year presides; its meetings take place in an apartment of the Assembly House. The recorder presides in the civil bill court, which is held four times a year, with power to decide by summary process in all cases of debt above 2l. arising within the city or liberties. The judicial business is transacted chiefly at the sessions-house; where also elections for the city representatives in parliament take place.

The city returns 2 and the university 2 members, to the H. of C. City court, 10,371 in 1835. University const., consisting of Masters of Arts whose names are on the books, 1,700.

The prisons for criminal offences are—1. Newgate, or the city gaol, a massive square building, for untried prisoners, felons condemned to death, who are executed from a balcony in its front, and convicts sentenced to transportation; there is also a ward for debtors under curator's process: 2. Richmond Bridewell, to the S. of the city, for adult males sentenced to imprisonment and hard labour: 3. Smithfield Penitentiary, for juvenile male offenders; and, 4. Grangegorman Penitentiary, N. of the city, for females under sentence of imprisonment by the civic courts, and for female convicts for transportation, from all parts, previously to their embarkation. The debtors' prisons are—1. the Sheriffs' Prison, near Newgate, for debtors not arrested under civic writs: 2. the Four-courts Marshalsea, for debtors under process of the superior courts: and, 3. the City Marshalsea, for those under process of the civic courts.

The supply of water was originally drawn from the Dodder; but in consequence of its insufficiency, arising from the enlarged demands of an increasing population, additional supplies have been procured from the Grand and Royal Canal companies, at the rate of 12½ per cent. from the former, and of 15 per cent. from the latter, on the gross amount of the pipe-water revenue. The inhabitants have since received a copious supply of excellent water from three reservoirs, two N. and one N. of the river. In 1809 the corporation was empowered by act of parliament to levy an additional rate, in order to substitute cast-iron service-pipes in lieu of those of wood. The levy of the rate became the subject of legal dispute with the rate-payers, which was finally decided in favour of the latter, on an appeal to the House of Lords, the decree of which declared the corporation to be indebted to the inhabitants in the sum of 74,500l. on this account, and that the pipe-water rents are received and held by the corporation in trust for the benefit of the city.

The expenditure required for the erection and repair of public buildings; the formation and repair of roads, the salaries of civic officers, and public charities, are defrayed by assessments made by the city grand jury, selected by the sheriff, who is appointed by the crown. The amount of taxation thus levied was 228,112l. in 1845, distributed as follows:—police rate, 24,500l.; North Dublin poor rate, 24,515l.; South Dublin poor rate, 40,750l.; improvement rate, 54,756l.; district sewer rate, 8,900l.; Grand Jury cess, 51,470l.; vestry cess abolition rate, 2,215l.; domestic water rate, 29,947l.; and public water rate, 9,502l. The expenditure for public buildings, roads, salaries of officers, and public charities, formerly under the absolute control of the grand jury, is now vested in the corporation, to whom the functions of the paving and lighting commissioners have also been transferred.

Within or adjoining the civic bounds are five local jurisdictions mostly independent of the authority of the corporation. They are, 1. the manor of St. Sepulchre; 2. the liberty or manor of Thomas Court and Donore; 3. the liberty of the deanery of St. Patrick; 4. the manor of Grangegorman, which includes the liberty of Christ Church; and, 5. the manor of Kilmainham. The three first are popularly called the Liberties. The manor of St. Sepulchre lies to the NE. of the city, and enjoys extensive powers, granted and confirmed to it by a succession of charters from the reign of John. It holds courts-leet and baron, an I a court of record. Its criminal jurisdiction extends to capital cases, but the right, as far as respects them, has fallen into desuetude. A small court-house and debtors' prison is attached to it. The archbishop of Dublin is lord of the manor. The liberty of Thomas Court and Donore lies SW. of the city; Thomas Court being within the county of the city, and Donore in the county at large, of which it forms one of the baronies. Its separate rights are secured by a series of charters, and it holds a court-leet, a court of civil bill, and a court of record for personal pleas to any amount. It has a court-house and small prison: the Earl of Meath is lord of the manor. The liberty of St. Patrick is a small district of about 54 acres surrounding the cathedral of the same name. It holds its privileges by prescription, and had courts-leet, and a court for the recovery of small debts, both of which have fallen into desuetude; hence it has become a kind of sanctuary for debtors of small sums from the adjacent parishes. Attempts to abolish an exclusive jurisdiction, which interferes with the claims of the just creditor, have been successfully resisted by the corporation of the dean and chapter, which is lord of the manor. A seneschal appointed by it receives a trifling salary, but has

no duties to perform. The pop. is small, and very poor; there are not more than 24 good houses in the deanery. The manor of Grangegorman or Glasnevin comprises the greater and wealthier portion of the bounds in the N. city parishes, and extends in some directions 7 m. N. and 10 m. S. It claims under an ancient charter, confirmed by another of 1 Jas. 1. The corporation of the dean and chapter is lord of the manor. The right of holding courts-leet and criminal courts has fallen into disuse. The seneschal holds a civil bill court on Friday morning for the N. part of the manor, and on every alternate Friday evening for the S.: its sittings are held in each case in an apartment in a tavern. There is no prison, debtors being sent to the county prison at Kilmainham. The liberty of Christ Church comprises the area in the centre of the city on which the cathedral is built. The manor of Kilmainham, in which the royal hospital is built, lies W. of the city, and extends 9 m. W.: Lord Cloncurry is lord of the manor. The seneschal holds a civil bill court six days in every quarter, with unlimited jurisdiction, but practically confined to actions under 5l.: the court sits in the county court-house at Kilmainham.

The police is vested, by an act passed in 1835, in 2 commissioners, under whom are 7 superintendents, 75 inspectors, 100 sergeants, 1,000 constables, and 20 supernumeraries. The city, with the liberty, is divided into the Castle, College, Rotunda, Barrack, Donnybrook, and Kingstown districts, in each of which there is an office, where an alderman and a barrister, both appointed by the lord lieutenant, sit daily. The police jurisdiction extends over a district of 8 m. round Dublin, in every direction. The expenses of the establishment are defrayed by a parliamentary grant, by a tax on the inhabitants, amounting, as before enumerated, to 25,500l. in 1845; by fines, and by carriage licenses.

The linen, woollen, silk, and cotton trades, which had been carried on to some extent in the city and its vicinity, have all declined. The sales of linen were chiefly effected in a large suite of buildings erected in 1728 by government, in the N. division, and rented to the factors: attached to it is a yarn-hall. The number of factors has decreased so much, in consequence of the decline of the trade, that most of the offices and stores are appropriated to other purposes. A pedestrian statue of George IV. was erected in one of the halls, in commemoration of his visit to the establishment in 1821. The woollen trade was long carried on to a considerable extent in the SW. liberties; a large building was erected there in 1814 by the late Thomas Pleasants, esq., for tentering the cloth, a process previously carried on in the open air, and therefore subject to interruption from changes of weather; but since the repeal of the protecting duties, the manufacture has been nearly extinguished. The silk trade was introduced by emigrants from France, who settled in Dublin in the beginning of last century. The favourite manufacture was a fabric of silken warp and woollen weft, called tabinet or Irish poplin, which is still in demand. The other branches of the silk trade have been for several years in a very depressed or extinct state. The same may be said of the cotton trade. Beer is extensively produced; and large quantities of porter and stout are exported to Great Britain and foreign countries; there are also several distilleries. A few ironfoundries are employed chiefly in executing orders demanding immediate attention. Cabinet-making is largely carried on, as are the various trades required to meet the demands of a large and concentrated population.

Smithfield market, which is within the civic jurisdiction, is held on Mondays and Thursdays for cattle, and on Tuesdays and Saturdays for hay and straw. A new cattle market, opened in November, 1863, has been erected by the corporation on the North Circular Road, where ample accommodation is provided, at a cost of about 15,000l. Spitalfields and Kevin Street markets are in the manor of St. Sepulchre; the principal commodities sold in both are bacon, butter, and potatoes; and in the latter hay and straw. A wholesale fish-market is held in Boot Lane; one for potatoes, fowls, and eggs, and another for fruit in the neighbourhood. The corn-market, formerly held in Thomas Street, is now carried on by a joint stock company, in a building erected for the purpose on Burgh Quay, where the grain is sold by sample. The retail markets are all private property, but their management is under the control of the officers of the jurisdiction in which they are held; those in the city being under the lord mayor.

The inland trade of Dublin has been greatly promoted by the Grand and Royal canals, both of which terminate in the city, and communicate with the sea through the Liffey. Still more conducive to the increase of trade has been the establishment of a network of railways centering in Dublin and spreading all over Ireland. There are five railway termini in the city, which it is intended to connect by a girdle railroad.

Banking business is transacted by the Bank of Ireland, established in 1783; the Hibernian Joint Stock Company, 1824; and by the Provincial, the National, the Royal, and London and Dublin Joint Stock Banks; and branches of the Ulster Bank, the Union Bank of Ireland, the latter opened in 1855. There are, besides, five private banking-houses and 2 savings' banks. The affairs of the Bank of Ireland are managed by a governor, who must hold 4,000l. stock; a deputy governor, with 3,000l.; and 15 directors with 2,000l. each. It is the place of deposit for all government monies. The buildings, formerly the Irish parliament house, form a quadrangle, standing on an area of 1½ acre, presenting three fronts; that to the E. of the Corinthian order, and those to the S. and W. of the Ionic. There is a very ingenious system of steam machinery for printing the bank notes, so as to render frauds extremely difficult. A statue of George III. occupies the spot on which the throne stood in the former House of Lords, now the directors' board-room.

An exchange was erected in 1787, in the centre of the city, partly by a parliamentary grant, and partly by subscription. The merchants hold their meetings in it until 1796, when the greater facilities afforded by the Commercial Buildings in College Green induced them to transfer their dealings thither; and the exchange has been since nearly useless. The building presents a fine specimen of Grecian architecture, and contains pedestrian statues of George III., Grattan, and Dr. Lucas.

The mercantile society of the Ouzel Galley, for deciding disputes relative to shipping and mercantile dealings by arbitration, was formed in 1705, and took its name from that of the vessel on which the first decision was pronounced. A chamber of commerce was established in 1820.

The river and port were vested in the corporation in 1270, by a charter of Henry III. Admiralty jurisdiction between Arklow, S., and the Nanny Water, N., was granted by Elizabeth. In 1787, it was empowered to erect a ballast-office, the annual expenses of which were 4,400l. at an average of thirteen years, from 1753 to 1786. In 1786, the management of the office was committed to a new board, with control over the ballastage,

tonnage, wherries, quayage, and pilotage of the port, including the harbours of Dunleary (now Kingstown) and Dalkey.

The commerce of the port of Dublin had increased so much towards the close of the last century, that the accommodation afforded in the river by shipping was found insufficient, and Parliament, consequently, granted 45,000l. for forming docks on both sides of it. The docks communicating with the Grand Canal, on the south side, were opened in 1796, and St. George's, the latest of the Custom House Docks, in 1821. These latter cover an area of 8 acres, have 16 feet depth of water, and 1,200 yards of quayage, and are capable of accommodating 40,000 tons of shipping. The docks on the south side afford commodious wharfage for merchantmen and colliers, exclusive of that supplied by the river-quays. The receipts of the Dublin Ballast Corporation for tonnage and quay-wall dues levied on vessels entering the port in 1863, was 35,871l.

The principal lighthouse of the port is at Poolbeg, on the extremity of the South Wall, and opposite to the great Northern Wall or breakwater, between which is the entrance crossing the bar to the harbour and quays; it is a bright light of 26 burners, 63 ft. in height. The other harbour lights are a floating light on the Kish Bank off Dalkey Island, the Bailey of Howth lighthouse, and a light on the extremity of the North Quay Wall. At the entrance to Kingstown Harbour there are lighthouses on each pier; that on the E. pier is a revolving light, every half minute, that can be seen 9 miles in clear weather. The mail packets to Holyhead start from Kingstown Harbour, which is 6½ miles from the city.

There were in 1863 registered at the port of Dublin 518 sailing vessels, with a tonnage of 38,167 tons, and 81 steamers, burden 11,900 tons. Most of these vessels were employed in the coasting or cross-channel trade, there having been but 6 or 8 in that of the West Indies, the same number in that of France and the Spanish peninsula, and 20 or 30 in the North American timber trade.

The brewing of porter is carried on extensively, and the number of barrels exported in 1861 was 170,384; 1862, 158,077; 1863, 174,941, nearly one-half was shipped by the eminent firm of Guinness.

The shipments of grain, &c., from Dublin, in 1863, were as follows:—

Wheat	81,779 quarters
Indian Corn	7,745 —
Oats	40,573 sacks
Barley	8,931 —
Flour	80,846 —
Oatmeal	36,674 —

The exports of provisions for the same period were:—Butter, 182,443 firkins; beef, 392 hogsheads, 1,901 tierces and casks; bacon, 6,672 bales, 501 boxes; hams, 340 hogsheads, 403 tierces and casks; pork, 4,503 barrels; lard, 6,231 barrels, 127 firkins and kegs; and of live stock, 162,712 head of cattle, 145,825 sheep, 1,124 calves, 90,904 pigs. Of wool, 16,204 bags were shipped.

The cross-channel trade is now carried on chiefly by steamers, which sail to Liverpool, Holyhead, and Bristol, London, Glasgow, Cork, and Belfast. Coals pay a duty of 4d. per ton, imposed to compensate the coal-meters, whose services have been rendered nearly unnecessary by the regulation allowing coal to be sold either by weight or measure. The amount of the customs' duties received at the port was 1,055,511l. in 1859; 1,004,276l. in 1861; 1,025,092l. in 1862; and 974,091l. in 1863.

The subjoined table shows the comparative amount of duties received at the port of Dublin in each of the years 1862 and 1863:—

Articles	1862	1863
	£	£
Tea	355,776	285,388
Muscovado Sugar	61,377	61,679
Refined Sugar	84,723	83,544
Coffee	5,208	4,643
Wine	67,725	71,663
Spirits	47,725	49,683
Tobacco	384,142	431,547
Timber	5,369	4,381
Other Articles	44,629	40,677
Total	**1,725,071**	**956,093**

The fiscal business of the port is carried on at the custom-house on the N. side of the river, near its mouth; a very extensive and magnificent structure, capable of serving as a custom-house for the empire. The transfer of part of the business to London, in consequence of the union of the British and Irish boards of customs and excise, having rendered great part of the building useless, many of its apartments have been appropriated to the use of the stamp office, the vice-treasurer's revenue department, the board of public works, the poor law commissioners, &c. Adjoining the main building are a floating dock and extensive stores, which were materially injured by a fire in 1834, but have since been in a great measure restored. The business of the post-office is transacted in a large and stately building in Sackville Street. The exports of home produce from Dublin to foreign countries are altogether not very considerable. The declared real value of the total exports of such produce to foreign ports amounted to 64,270l. in 1859; to 72,192l. in 1860; to 29,338l. in 1861; to 64,777l. in 1862; and to 38,196l. in 1863.

Dublin is the seat of an archbishop's see, and of the second of the archiepiscopal provinces into which Ireland is now divided. The provincial jurisdiction is nearly coextensive with the two civil provinces of Leinster and Munster. The see, including the bishopric of Glendalough, with which it was incorporated with it in 1214, includes the counties of Dublin, and Wicklow, and Kildare. The landed property contains 80,040 acres, of which 22,926 are profitable. There are two cathedrals: Christ Church, built near the summit of the hill on which the city stands, is the more ancient and superior. The building is plain, with no exterior architectural embellishments; it contains several remarkable monuments; among which is that of Strongbow, earl of Pembroke. St. Patrick's Cathedral, in the valley, S. of Christ Church, also contains some remarkable monuments. The chapters and installations of the Knights of St. Patrick are held in it. The city contains 20 parishes or parts of parishes.

According to the Roman Catholic ecclesiastical arrangements, these 20 parishes are consolidated into 9 unions, each having a place of worship; besides which, there are ministry chapels attached to friaries or nunneries. There are nearly 100 places of worship. St. George's Church in the N.E. part of the city is a splendid structure, in the Grecian style. It is the only place of worship, except the cathedrals, which has a peal of bells. The Roman Catholic church of the Conception, in Marlborough Street, considered the archbishop's cathedral, is of very large dimensions, and highly embellished internally, but not yet complete as to its exterior. The Roman Catholic chapel of St. Francis Xavier is also an elegant building of the Ionic order.

Dublin had, by the census of 1861, a pop. consisting of more than two-thirds of Roman Catholics. There were 23,807 males, and 25,644

females belonging to the established church; 89,337 males and 107,212 females who were Roman Catholics; 3,499 males and 1,976 females who were Presbyterians; 946 males and 851 females who were Methodists; and, finally, a few hundred persons entered as belonging to other sects. There were but few Jews, the total number, in 1861, being 154 males and 170 females.

Dublin was the seat of a university as early as 1320, but the institution gradually declined in consequence of the unsettled state of the country and the deficiency of funds. The existing university of Trinity College was founded in 1593, in the buildings of the dissolved monastery of All-hallows, applied to this purpose by the corporation, to which it had been granted at the dissolution of the monasteries. It consisted originally of a provost, 3 fellows, and 3 scholars; but at present it consists of a provost, 7 senior, an undefined number of junior fellows (at present 17), and 70 scholars. It has, also, 27 professors, with lecturers and assistants, all endowed. A school of engineering, founded in 1842, has 7 professors, and is said to furnish a very complete course of theoretical and practical instruction. The university is presided over by a chancellor and vice-chancellor, one of whom holds occasional visitations, and by a board, consisting of the provost and senior fellows, which sits weekly. The period of undergraduate instruction is four years; the number of students above 1,500. The course of studies for candidates for a fellowship is logic, mathematics, natural philosophy, ethics, history, Latin, Greek, and Hebrew. The examinations, which are public, are carried on in Latin. Exclusive of the fees of students, the university derives a large income, said to exceed 15,000l. a year, from lands; and it has, also, the patronage of 32 benefices. It enjoys the right of returning two members to the H. of C., who are elected by the fellows, scholars, and all those who at any time have been fellows or scholars, and have kept their names on the books. The buildings, which present an extended front to College Green, are large and elegant: the principal are a library, containing upwards of 120,000 volumes, and entitled to a copy of every work published in the empire; a chapel, an examination hall, a museum, a dining-hall, a theatre of anatomy, and a printing-office; it also maintains a small but well kept botanical garden in the SE. suburb. The College of Physicians is connected with the university; some of the courses of lectures are given in that institution, others in Sir Patrick Dun's Hospital. The College of Surgeons, St. Stephen's Green, was founded in 1784. The Incorporated Company of Apothecaries has established courses of lectures in pharmacy and other branches of medical science, at their hall in Henry Street. There are also several private medical and surgical schools, much frequented by students.

The chartered scientific and literary societies are —the Royal Dublin Society, for the promotion of the useful arts, having professorships in botany, chemistry, and experimental philosophy; drawing schools, a library, a museum, and a large botanic garden; the Royal Irish Academy, founded in 1786, for the encouragement of abstract science, polite literature, and antiquities, with a small but increasing library, containing a good collection of Irish MSS., and a museum; it has published nearly 20 vols. of Transactions. The Royal Hibernian Academy, founded in 1823, for the encouragement of the polite arts, meets in a building erected for its use, at an expense of 15,000l., and presented to it by the late Francis Johnston, architect. An exhibition of the works

of native artists takes place annually. These institutions usually receive grants of public money.
The principal libraries, besides those already
noticed, are Marsh's or St. Patrick's Library, near
the cathedral of that name; and the Dublin
Library, confined exclusively to subscribers. There
are smaller collections of books, some of which are
open to the public, as Sir Patrick Dun's Hospital,
Steevens's Hospital, the Royal Hospital, Christ
Church, and the Presbyterian meeting-house at
Strand Street. The unchartered societies for science,
literature, and the fine arts, supported wholly by
voluntary contributions, are—the Royal Irish Institution for Painting; the Zoological Society,
which has a handsome garden in the Phoenix
Park; the Horticultural, which maintains an
annual show of flowers and fruit; the Agricultural, with an annual show of cattle; the
Historical, for historical and political discussion;
the Civil Engineers' societies; the Natural History Society; and the Mechanics' Institute,
formed in 1837.

The model schools of the Board of National
Education are held at their respective establishments. Schools, on the foundation of Erasmus
Smith, are founded on the Coombe and in St.
Mark's parish. Most of the parishes and congregations have free schools attached to them.
The total number of schools maintained by grants
of public money and voluntary contributions is
about 700; the total number of pupils is about
15,000.

The principal charitable institutions which
maintain as well as educate orphans and destitute
children are—the Foundling Hospital, now very
much circumscribed; King Charles's, or the Blue-
Coat Hospital, a large and handsome range of
buildings, maintains about 100 boys, the sons of
reduced citizens. The Hibernian Society, in
the Phoenix Park, was founded for soldiers' children; the Marine school, on the NE. quay for
sailors' children; the Protestants' Orphan Society; and the Female Orphan House, N. Circular
Road, for female orphans. The principal institutions for the relief of disease and accidents are—
Steevens's Hospital, near Kilmainham; Sir Patrick
Dun's; the Meath Hospital, which is also the
county Infirmary; the City Hospital; Jervis
Street Infirmary; St. Mark's and Ann's; the
Westmoreland Lock and Netterville Hospitals;
the Hospital for Incurables; two fever hospitals,
one in Cork Street, the other on the N. Circular
Road; and 10 lying-in hospitals, of which that in
Rutland Square is the principal. Attached to
this last-named is a fine suite of apartments and
an enclosed garden or pleasure ground for public
amusements, the profits of which contribute to
the maintenance of the institution. There are 10
dispensaries, supported partly by parliamentary
grants and partly by private contributions. The
institutions for cases of mental derangement are
the District Richmond Lunatic Asylum for the
city and county, and for Louth, Meath, and Wicklow etc., supported by grand jury presentments;
Swift's Hospital, supported chiefly by the founder's
bequest; an asylum near Donnybrook; and 5
private institutions. Since the introduction of
the compulsory provision for the support of the
poor, Dublin has been divided into 2 Poor Law
Unions, the N. and S. each of which has a separate
workhouse, on a large scale, and board of guardians. The chief asylum for the aged and impotent is the House of Industry, established in
1773, and supported wholly by grants of public
money. Its buildings, yards, and gardens extend
over an area of 11 acres.

The minor asylums for age, debility, and want,
are numerous. There are 2 for the blind, the
Richmond, in Sackville Street, for males; the
Molyneux, in Peter Street, for females; the inmates in each contribute to their maintenance by
their labour. There are 2 houses of refuge for
females of good character, and 10 for penitent
prostitutes.

The places of public amusement are few, and
not much encouraged. They comprise the Theatre
Royal, the Queen's Theatre, the Rotunda Gardens,
and the Portobello Gardens. Clubs for social and
convivial purposes are numerous. The principal
are the Dublin University, Leinster, Kildare
Street, Sackville Street, United Service, Beef-steak,
and the Friendly Brothers. The Royal St. George's
and the Royal Irish Yacht clubs hold annual
regattas at Kingstown.

The environs of the city in every direction are
very beautiful; the view of the valley of the
Liffey from the rising grounds on the S. boundary
of the county commanding the highly cultivated
lands inclining to the sea-side, well planted, and
studded with numerous seats and villages, the
bay with the hills of Killiney on the one side,
and the city spread out on the other; the Hill of
Howth, Lambay, and Ireland's Eye, in the back-
ground; and, in clear weather, the Mourne mountains in the remote distance, present a landscape
of superior tranquil beauty. The external appearance of the city itself is equally striking. The
main avenues to it, particularly on the E. side,
are spacious, airy, and bordered with large dwelling-houses; the public buildings, both civil and
ecclesiastical, numerous, as compared with the
size of the city, mostly of elegant architecture,
and placed in imposing points of view. But, as
already stated, this description is by no means
applicable to a large portion of the city. A line
drawn N. and S. through Dublin Castle would
divide it into 2 parts, extremely different in appearance. The E., in which are the residences of
the more wealthy class, contains most of the
public buildings, all the squares, and streets of
fashionable resort, both for amusement and trade;
the W., once the principal seat of the trade of
the town, is now in a state of dilapidation and
extreme destitution.

A love of convivial enjoyments pervades all
ranks. The habits of the higher and middle
classes are social to a degree often bordering on
profusion. Letters of introduction from strangers
are the never failing harbingers of rounds of dinner parties, evening entertainments, assemblies,
balls, and suppers. The dinner hour varies from
five to seven, and scarcely any business is transacted afterwards. Neither do the daily occupations commence at an early hour in the morning.
The courts of justice seldom meet before eleven,
and generally close before four. Dancing is a
favourite amusement; cards are every year getting
less fashionable. Jaunting-cars, both open and
covered, carrying four persons, supply the place of
the London cabriolets, and have wholly supplanted hackney-coaches. The appearance of the
lower classes, however, exhibits, particularly in
the W. division of the city, every indication of
wretchedness. The habitations are mean and
neglected, their clothes tattered, and they seem
as if they maintained a constant struggle with
poverty; but, despite all this, there is a light-
heartedness about them that not only enables
them to bear up under the pressure of want, but
which, by rendering them comparatively insen-
sible to its existence, paralyses their efforts to improve their condition, and makes them contented
with the abject poverty in which they live. They
are equally fond of amusement as their superiors;

equally ready to indulge in dance and song. Intoxication is less frequent than formerly, and the spirit of riot and turbulence, which not many years since was the all but invariable consequence of festive meetings, is rapidly subsiding—a change partly owing to a stricter and better system of police, and partly to the moral influence of an improved state of society. Several meetings are held annually in the neighbourhood, ostensibly for the transacting of business, but in reality almost solely for festive purposes. The most celebrated is Donnybrook fair, in August, which formerly continued for a fortnight, but is now restricted to a few days, and has, in fact, quite lost its old character. The fairs of Rathfarnham, Palmerston, and Finglas are of the same character, but in a minor degree.

The principal events in the history of Dublin are identified with that of the island in general, and are therefore to be found in the article IRELAND. But a few facts may be stated with respect to it. In 1169 it was taken by storm by the English, under Richard de Clare, better known by the name of Strongbow; and the Danes, who two years after laid siege to it with a numerous naval and land armament, were defeated with the loss of their leader, and forced to raise the siege. This was their last attempt to recover the dominions they once held in Ireland. In 1172, Henry II. landed, and held his court here in a temporary building erected outside the town, which was too small to afford suitable accommodations for the monarch and his retinue. In 1209, the castle was erected, and four years after the citizens were treacherously attacked while amusing themselves in Cullen's Wood, now a suburb, by a party of Irish from the Wicklow mountains, and forced to seek the protection of the fortifications, after the loss of many lives. In 1210, King John held his court in Dublin, and about the same time the first bridge was built across the Liffey. In 1316, Edward Bruce was repulsed in an attempt to take Dublin. It was twice visited by Richard II., who took his final departure from it in 1399, the year of his dethronement and death. In 1486, the citizens declared for Lambert Simnel, and crowned him in Christ Church. About the same time the mayor was compelled to walk barefooted through the city, as a penance for a violent outrage committed by the citizens in St. Patrick's church. In 1534, Lord Thomas Fitzgerald, having rebelled against Henry VIII., laid siege to the city, on which occasion his batteries were mounted at Preston's Inn, now almost in its centre; but the obstinate resistance of the citizens, who burnt great part of the NW. suburb to check his approach, compelled him to raise the siege. In 1583, a litigate between two of the Irish family of O'Connor was decided by wager of battle in the castle, before the lord-justices and council. About the same time, the king's exchequer, which was kept between College Green and the castle, was plundered by a party of Irish from the mountains. During the civil wars of 1641, the battle of Rathmines, in which the Duke of Ormond was totally defeated by the garrison of Dublin, was fought in the neighbourhood. The Grand Canal was commenced in 1765. A penny post-office was opened in 1773. In 1778, the first regiment of Dublin volunteers, arrayed for the defence of the kingdom against the threatened invasion by the French, appeared under arms. The Royal Canal, to the N. of Dublin, was commenced in 1789. The first steam-engine was set up in 1791; next year the buildings of the House of Commons took fire, while the members were assembled, and were completely burnt down: the cause of the fire was

never clearly ascertained. The insurrections of 1798 and 1803 form part of the general history of the island. A jubilee was celebrated in 1809, in commemoration of George III. having entered on the 50th year of his reign. In 1816, the first steam packet sailed from the harbour. In 1821, Dublin was visited by George IV. who landed on his birth-day at Howth. In 1834, the railway between Dublin and Kingstown was opened. In 1849, on the 6th of August, her Majesty Queen Victoria and Consort landed at Kingstown. A great International exhibition of works of art and industry took place in Dublin in the summer of 1865, and was visited by above a million of people.

DUBNO, a town of European Russia, govern. Volhynia, on the Irwa, 86 m. NE. Brody. Pop. 7,590 in 1858. The town belongs to the prince Lubomirski, and is ill built, with narrow, crooked, and unpaved streets.

DUDLEY, a town and parl. bor. of England, in a detached part or enclave of the co. of Worcester, surrounded on all sides by Staffordshire; 8½ m. W. by N. Birmingham, 119 m. NW. London by road, and 141 m. by Great Western and West Midland railway. Pop. of parl. bor. 44,975 in 1861. The town consists principally of a long street, with a church at each end; the houses are generally good, and the streets paved, macadamised, and lighted with gas. St. Thomas's church, rebuilt in 1819 at an expense of £23,000, is a fine Gothic structure, with a lofty conspicuous spire. There are altogether five churches, and twelve chapels for Catholics, Methodists, Baptists, Independents, Unitarians, and Quakers. It has a grammar-school, founded in the reign of Eliz., and endowed with land worth about 300£. a year. There is a charity school for clothing and educating 40 girls, and a charity for clothing 7 poor men, established in 1819. A school was also founded in 1732, for clothing and educating 50 boys, exclusive of about 200 not on the foundation. There is likewise a blue-coat school, where many boys are educated, and a school of industry. The Unitarians have a school for girls, and there are Sunday schools attached to the several places of worship. There are several book societies, and a well-supported subscription library.

Dudley is a principal seat of the iron trade; its vicinity furnishing the inexhaustible supplies of coal and iron ore, while the canals with which it is connected afford the means of readily conveying its products to all the great markets of the empire. The inhabitants are principally engaged in nail-making, which is the staple trade of the town, mining, the smelting of iron ore, and the manufacture of flint glass. Exclusive of nails, a great variety of iron implements are made here. In 1862 above 6,000 hands were employed in the coal and metal works.

The workmen comprise engineers, able mechanics of almost every description, such as pattern makers, carpenters, first-rate masons, founders, men of great science for working the lime-stone, coal, and many others. Boys are employed in the pits and mines to attend to the fires, and various light work about the furnaces, to fill the boxes, barrows, &c. for the men. The price of coal in this district varies from 6s. to 8s. and 10s. a ton; the men engaged in most of the works are supplied with the coal at prime cost. The custom mostly is, to pay the men by the ton: some masters, however, pay them by the day. It takes a long time to make a man a collier. He is first apprenticed to a person, himself a collier, either his father or fellow workman. The labour is severe, and the workmen generally live upon good food,

Earl Dudley, the chief landowner, is the largest ironmaster in the kingdom.

A mayor and other officers are annually appointed by the lord of the manor, but the town is within the jurisdiction of the county magistrates, who hold petty sessions every Monday. A county court is established here. In the 23 Edward I, Dudley sent 2 members to the H. of C.; but the privilege was afterwards withdrawn, and the town remained unrepresented till the passing of the Reform Act, when the right to send 1 member to the H. of C. was conferred on it. Registered electors 1,127 in 1865, all 10l. householders. The returning officer is appointed by the sheriff of the county.

To a stranger, for the first time approaching the town at night, the appearance presented by the numerous fires rising from the furnaces, forges, and collieries, is particularly imposing, their lurid glare illuminating the country for a considerable distance round. There are, in many places, subterranean fires, which generally continue until the fuel which supplies them is nearly exhausted. This phenomenon has been observed, more or less, in the neighbourhood, for upwards of a century. At Russell Hall a stratum of from 25 to 30 ft. of argillaceous substances, lying between the upper stratum of coal and the surface of the earth, has been transmuted into a species of stone by the heat arising from these subterranean fires.

In the Saxon times a strong castle was built here, which has since undergone many vicissitudes. In 1644, it withstood a siege; and the occurrence of a fire, in 1750, completed its demarcation. Its ruins, which are very extensive, stand on an elevated situation, and command very fine and extensive views.

One of the most striking objects at Dudley consists of the remarkable development of the mountain limestone in the hills under and immediately adjoining the castle. The peculiar stratification incident to a force acting powerfully from beneath, which has elevated a portion of the previously deposited beds of limestone, leaving them to dip on both sides from a central ridge, is developed with great distinctness. The stratification of this locality is still farther exhibited by the very extensive excavations in the limestone itself; some of these are open, and consequently very readily inspected, but the more extensive consist of long, horizontal galleries, whose extent and brilliancy can only be observed with the aid of torches, but which well repay the labour required in gaining a view of their dark and secret recesses. These workings extend 1½ m. under the hill, and a canal, for the conveyance of the produce of the mine, extends the greater part of this distance. The organic remains of former races of animals are very numerous. Several species of trilobites (Dudley locust) and crinoïdes are met with not unfrequently, and corals and madrepores are in great profusion. Perhaps it may be said, that few localities in the kingdom present so many curious and interesting subjects of observation to the geologist as this. Seldom have the operations of nature and of art united in bringing so much of the secret economy of the interior of the globe under the observation of the inhab. of its surface. Many noble seats and spacious residences, lie within a circuit of a few miles of the town. At Ladywood, within the par. and about 2 m. from the town, is a valuable spa, possessing similar qualities to the Cheltenham and Leamington waters, and equally efficacious for cutaneous diseases: here are also commodious hot and cold baths open to the public. There are also several chalybeate springs in the neighbourhood. The

celebrated nonconformist divine, Richard Baxter, was for some time master of one of the schools in the par. Dudley conferred the title of earl upon Lord Ward in 1860.

DULCIGNO (Turk. *Olgun*), a maritime town of Turkey in Europe (the ancient *Olcinium*), prov. Albania, on the Adriatic; 10 m. SW. Scutari, and 40 m. SSE. Cattaro: lat. 41° 55′ 50″ N., long. 19° 11′ 40″ E. Pop. estimat. at 7,000. The town possesses a citadel and a harbour, has some little trade, and is the residence of a R. Cath. bishop. Its inhabitants are the only natives of Albania who have a taste for a sea-faring life, or rather, perhaps, for piratical excursions by sea. When Mr J. Hobhouse visited this town in 1809, they were accustomed to enter into the naval service of the Barbary powers, or to issue, 'as the Illyrians did of old, from the same port of Olcinium, to plunder the merchant ships of all nations.' (Journey through Albania, p. 168.)

DULWICH, a hamlet of England, co. Surrey, par. of Camberwell, hund. Brixton, 5 m. S. London by road, and 6½ m. by the London, Chatham, and Dover railway. Pop. of ham. 1,733 in 1861. It is a quiet rural place, mostly consisting of groups of respectable mansions scattered round a large open area planted with avenues of trees. Here is the celebrated Dulwich College, established in 1619, by Edward Allen or Alleyne, a contemporary of Jonson and Shakspeare, and the most celebrated tragic actor of his day. He endowed it with the manor of Dulwich, and certain lands and tenements in the parishes of Dulwich, Lambeth, and St. Botolph, Bishopsgate; the ann. rev. being at the time 800l., but at present it is very much larger. The college was originally built by the founder in the Elizabethan style, from a design of Inigo Jones; it has of late years been renovated and augmented, and forms three sides of a quadrangle, with offices, a picture-gallery, and a large garden. It was founded for a master, warden, 4 fellows, 6 poor brethren, 6 sisters, 12 scholars, 6 assistants, and 80 out-members. According to the statutes, the master and warden must each be of the blood and surname of Alleyne, or, in default of relatives, of the same surname; they must be 21 years of age, and unmarried. The 2 senior fellows are required to be of the degree of M.A. and unmarried; and the 2 junior fellows graduates in holy orders. The brethren and sisters must be 60 years old, and single, when admitted. On the death of the master, the warden succeeds, and a new warden is chosen by lot from amongst candidates qualified as above. The fellows are also chosen by lot, when vacancies occur. The poor brethren and sisters are chosen in the same mode, from the 80 out-members, who must be parishioners of St. Saviour's, Southwark, St. Botolph, Bishopsgate, or St. Giles's, Cripplegate (10 from each par.), and are lodged in almshouses appropriated to the purpose: the churchwardens of the above pars. are ex officio assistants in the government of the college. The Archbishop of Canterbury is visitor. A library was bequeathed to it by Edward Cartwright, a comic actor, who died about the end of the 17th century, which contained a large, curious, and unique collection of old plays, subsequently (and with very questionable propriety) assigned to Garrick in exchange for some modern works. A respectable collection of pictures was also left to the institution by the founder, and by Cartwright; and to this a most valuable and splendid addition was made in 1810 by a bequest of Sir Francis Bourgeois, R.A., who also left 2,000l. to build a gallery for their reception, and to defray the expense of their preservation, &c. This fine collection of the old masters is open (except on Fridays and Sundays) to the public, admission tickets being obtainable by any respectable person, on application in London. It consists of about 300 pictures, mostly of the cabinet size, and was formed by M. Dennistoun, an eminent collector, who bequeathed them, on his decease, to Sir Francis; and he, in turn, to the widow of his friend, for life, with reversion to the college; a mausoleum in the college chapel contains the remains of Sir Francis and Dennistoun. Public service is regularly performed there, and it serves as a chapel of ease to the hamlet. There is a free school in Dulwich, founded in 1741, by James Alleyne, then master of the college, for 60 boys and 60 girls: the present revenue amounts to £90 a year. There are many elegant villas in the vicinity; and in summer the villa is much resorted to by temporary visiters.

DUMBARTON, or DUNBARTON, a marit. co. of Scotland, consisting of two detached portions, of which the principal, or most westerly, lies between Loch Lomond on the NE., Loch Long on the W. and NW., the Clyde on the S., and the Milngavie burn on the E.: the other and much smaller portion lies on both sides the Forth and Clyde canal, from Cumbernauld to Kirkintilloch. Total area, 297 sq. m., or 189,814 acres, of which nearly 70,000 are water, being principally part of Loch Lomond. It consists mostly of lofty, rugged mountains incapable of cultivation; the arable lands being principally in the S. part of the co., between Loch Lomond and the Clyde, and along the Forth and Clyde canal. The low ground is very fertile, and is pretty well cultivated. Estates mostly large; but arable farms are rather small, and even stock farms are not so large as in most highland cos. Oats and potatoes principal crops, but very good wheat is also raised. Cattle in the upper parts chiefly of the Highland breed; but in the low grounds, where dairying is extensively carried on, Ayrshire cows are almost exclusively met with. Sheep partly black-faced, and partly Cheviots. There are mines of coal and iron, and freestone and limestone quarries. There are large cotton mills at Duntocher in this co.; and paper-making, &c., are carried on to a considerable extent; there are extensive print-fields on the Leven, the only river of any importance. The co. returns 1 mem. to the H. of C. Registered electors, 1,597 in 1865. The town of Dumbarton unites with Renfrew, Rutherglen, Kilmarnock, and Port Glasgow, in sending a mem. to the H. of C. Dumbarton is divided into 12 parishes; and had, in 1861, a pop. of 57,024, living in 8,593 houses. The old valued rent was 2,777l.; the new valuation for 1843–4 was 242,598l.

DUMBARTON, or DUNBARTON, a royal and parl. bor. and sea port of Scotland, cap. of the above co., on the W. bank of the Leven, within 100 yards of its junction with the Clyde, 18 m. NW. Glasgow, on the railway from Glasgow to Helensburgh. Pop. 8,733 in 1861. The town consists of one well-built, crescent-shaped street, and of some smaller ones. The houses are crowded closely together, so that many of them are ill-aired. There is a suburb E. of the Leven, connected with the burgh by a bridge of five arches. Chief public building par. church, a modern structure, with a spire and clock. There are also two chapels, belonging respectively to the United Associate Synod and the Rom. Cath. At high water, the Leven is navigable for large vessels to the quay at Dumbarton; but not so at low tides. This is owing partly to a bar across the mouth of the river, and partly to sandbanks between the entrance and the quay. The burgh has long been celebrated for its excellent schools. Among the distinguished individuals to whom they have fur-

nished instruction may be specified Sir John Smollet, of Bonhill, one of the commissioners for framing the articles of union between England and Scotland; his grandson, Smollet, the celebrated novelist; Dr. Colquhoun, author of a Treatise on the Police of London, and other works. Shipbuilding and rope-making are carried on to a considerable extent. Bleaching, the printing of cottons, and other branches of industry are carried on, along the line of the Leven from Loch Lomond, whence it flows, to the Clyde, a distance of 7 m. Dumbarton was made a royal burgh by Alex. II., in 1222; but on or near its site there had been a still more ancient town, called Alcluid, the cap. of the Strathclyde Britons. The most important object connected with the town is the castle, on a steep, isolated, basaltic rock, at the mouth of the river, once surrounded by water. It has two summits, the highest being 290 ft. above the sea, and is a conspicuous and interesting object from the Frith of Clyde and the opposite coast. The date of the erection of the castle is not known, but it has existed from a very remote period. It is intimately connected with the history of Scotland; and was successively in the possession of Edward I., Bruce, Queen Mary, Charles I., and Cromwell. It is one of the forts which it was stipulated in the treaty of Union should be kept in repair. The Dumbartonshire railway, from Balloch on the N. of Loch Lomond to Bowling on the Clyde, passes the town; it is intended to extend the line to Glasgow. Dumbarton unites with Port Glasgow, Renfrew, Rutherglen, and Kilmarnock, in sending a mem. to the H. of C.; and had 288 regist. voters in 1865.

DUMBLANE, or DUNBLANE, a market town and formerly a bishop's see, Scotland, co. Perth, on the Allan, a tributary of the Forth, 4 m. N. Stirling, and 21 m. SW. Perth on the railway from Perth to Stirling. Pop. 1,709 in 1861. Though once a city, having been the seat of a bishop, it is now only a small village, destitute of importance, and consisting of a single street, with a few lanes. But little business is carried on, except what results from a weekly market and four annual cattle fairs. A few strangers are attracted to it in summer, owing to an excellent mineral well in its vicinity. It is chiefly celebrated for the remains of its cathedral and other episcopal edifices. The former is pretty entire, but no portion of it is converted to use except the choir, which serves for the parish church. The dean's house is now used as the minister's manse or parsonage-house. Robert Leighton, afterwards archbishop of Glasgow, held the see of Dumblane from 1662 to 1670. This celebrated scholar bequeathed his library, consisting of 1,400 volumes, to the cathedral and diocese of Dumblane. It is still extant, and has received great accessions by subsequent bequests. It is open not only to the clergymen of the presbytery, but, on easy terms, to the public. The battle of Sheriffmuir, on the 13th of November, 1715, between the constitutional forces, under the Duke of Argyle, and those of the Pretender, under the Earl of Mar, was fought near this town. Though indecisive, the result of this conflict was eminently favourable to the revolutionary establishment.

DUMDUM, a military village and extensive cantonment in Hindostan, prov. Bengal, 6 m. ENE. Calcutta. It is the head-quarters of the Bengal artillery, and consists chiefly of several long low ranges of buildings of one story, ornamented with verandahs, the lodgings of the troops, and some small but convenient officers' quarters; the whole adjoining a large plain, used as a practice ground. A battalion of European artillery is usually stationed here: it has a church and a free school.

DUMFRIES, a marit. co. in the S. of Scotland, having S. the Solway Frith, E. Cumberland, N. Roxburgh, Selkirk, Peebles, and Lanark, and W. Ayrshire and Kirkcudbright. Area, 1,129 sq. m., or 722,813 acres, of which only about 1-4th or 1-5th part is supposed to be arable. With the exception of Annandale and Nithsdale, that is of the low grounds traversed by the rivers Annan and Nith, the principal in the co., it is for the most part mountainous; the mountains, however, are not generally rugged or heathy, but are mostly of an easy ascent, and afford good sheep pasture. This, like most other Scotch cos., has been wonderfully improved in recent years, principally through the facilities afforded by steam navigation for the conveyance of fat sheep, cattle, and other farm produce to Liverpool, and the consequent extension of the turnip culture, the introduction of bone manure, and furrow draining. Roads, fences, and farm buildings have been astonishingly improved, and are now, speaking generally, as good as any in the kingdom. Cattle are mostly of the Galloway breed; and Cheviots have been, for some years past, a common breed of sheep among the hills, where at no remote period the principal flocks were black faced. But such has been the progress of improvement, that it is now found not only practicable but more profitable to introduce extensively half-bred sheep, or a cross between the Leicester ram and Cheviot ewe. This has, latterly, been the favourite stock, and the numbers are increasing rapidly. There are other crosses between the same rams and blackfaced ewes, a hardy breed, which thrive well on the coarser grasses, and are in great request in certain districts of England and Wales for their feeding qualities. The formation of the Caledonian railway has been of very great advantage to the store-masters of Annandale. English lime, formerly brought 25 m. and upwards by cart, is now conveyed by railway in trucks, and deposited at stations within trifling distances of the homesteads along the line; a saving and convenience the good effects of which are obvious in the increased use of the mineral, and the improved appearance of a great extent of hill pasture land. Hogs extensively raised (see next article). Property, mostly in very large estates; that of the Duke of Buccleugh, in this co., is one of the finest in Scotland. Farms in the lower districts vary from 100 to 500 acres; in the hill district they vary from 540 to 10,000 acres. There are valuable coal and lead mines in the par. of Sanquhar; and freestone is abundant, particularly in the vicinity of Dumfries. Manufactures unimportant. The co. is divided into forty-three parishes, and sends 1 mem. to the H. of C. Registered electors, 2,097 in 1865. The tons. of Dumfries, Annan, Sanquhar, and Lochmaben (which are the principal towns), unite with Kirkcudbright in sending a mem. to the H. of C. In 1861 Dumfries-shire had a pop. of 75,878. living in 13,182 houses. The old valued rent was 13,218l.: the new valuation, for 1864-5, was 375,141l., exclusive of railways.

DUMFRIES, a sea-port and parl. bor. of Scotland, co. Dumfries, of which it is the cap., on the E. bank of the Nith, about 9 m. from its influx into the Solway Frith, 64 m. S. by W. Edinburgh, and 72 m. W. by N. Carlisle, on the Glasgow, Dumfries, and Carlisle railway. Pop. 11,025 in 1861. The town is well and handsomely, though irregularly, built; the High Street, which stretches nearly 1 m. in length, does not run in a straight line, and is greatly obstructed at one point by a steeple, or building, in which the town council holds its meetings, placed in its very centre. The other streets lie either at right angles to the High Street, or parallel to it. The houses are generally built of red freestone, which the neighbouring country produces in unlimited abundance; and such of the buildings as are of old date are generally whitewashed; while many in the modern part of the town are painted in imitation of Portland stone. Altogether, the town is clean, neat, and substantially built, with comparatively few marks of poverty or destitution; and is regarded as the provincial capital of the S. of Scotland. There has been no material increase in the streets or buildings of the town for years past, but there has been a considerable increase in the number of villas in the vicinity. The suburb of Maxwelltown is connected with it by two bridges, one built in the 18th century, and consisting originally of thirteen arches, of which only seven are now visible; the other, a very elegant structure, erected in 1795. The public buildings are numerous. There are two parish churches; St. Michael's, rebuilt in 1745, and the New Church, erected in 1727. The former is chiefly remarkable for its extensive and crowded burial-ground, and the vast number and variety of its monuments. It has been estimated that, exclusive of ruinous and dilapidated monuments, the cost of erecting those in preservation could not have been less than 100,000l. In this cemetery was erected, by public subscription, in 1815, at an expense of 1,500l. a mausoleum in memory of Burns, who spent the last years of his life in Dumfries, and whose remains are deposited in a vault below. An emblematic piece of marble sculpture, executed by Turnerelli, in the interior of the structure, represents the genius of Scotland finding the poet at the plough, and throwing her mantle over him. The house in which the poet lived and died, with some adjoining properties, was purchased on the 3rd July, 1850, for Lieut.-Colonel Burns, the second son of the bard. Near the churchyard gate are deposited the remains of Andrew Crosbie, esq., microcate, once the ornament of the Scotch bar, who exemplified in real life the character of Counsellor Pleydell, as portrayed by Sir Walter Scott. A third church was erected in 1840; and there are sundry dissenting chapels, some of them favourable specimens of architecture. The steeple in the High Street, already mentioned, is a handsome structure, the work of Inigo Jones. In Queensbury Square, off this street, is a handsome Doric column, erected in 1780, in honour of Charles, duke of Queensberry. The other public buildings are the trades' hall, court-house, county gaol containing a bride well, infirmary, dispensary, academy, assembly-rooms, theatre, and lunatic asylum. The latter, called 'The Crichton Royal Institution,' was founded in 1838, by Mrs. Crichton. An additional building was erected in 1849, capable of accommodating 200 pauper patients. The total expense exceeds considerably 100,000l. The infirmary, which was opened in 1776, is the only institution of the kind in the S. of Scotland. There is a workhouse, founded and endowed by two brothers of the name of Muirhead, in 1753, which accommodates, at an average, thirty old and twenty young paupers, besides dispensing charity to about forty widows, who live out of the building. There are three parochial schools in the parish, and four endowed seminaries under the patronage of the town-council, united under the name of the Dumfries Academy.

The chief manufactures carried on are those of hats and hosiery. Formerly checked cottons were produced here; but this branch has disappeared, and the cotton weavers who remain are employed through the medium of agents, by Carlisle or Glasgow manufacturers. The trade of tanning has declined; but the quantity of leather prepared by pit and bark processes is still considerable, and is esteemed for its durable qualities. There are several breweries, and the largest basket-making establishment in Scotland. The manufacture of clogs, or strong shoes, with thick wooden soles, the use of which is almost entirely confined to the inhab. of the S. of Scotland, is with one or two slight exceptions peculiar to Dumfries; but it does not employ many hands, the use of the article being on the decline. Shoemaking is here a flourishing branch of industry.

Dumfries has long been celebrated for its weekly cattle-markets, and its four great annual fairs, for the sale of cattle and horses, which, with the markets, are held on the Sands, an open space between the town and the river. Most part of the cattle raised in the co. of Dumfries, and a considerable part of the peculiarly fine breed of cattle raised in Galloway (viz. Kirkcudbright and Wigtown), are disposed of in the Dumfries markets. At an average, 25,000 head of cattle are annually sent up from Dumfries and Galloway to England, principally to Norfolk, where they are fattened for the London markets. From 400 to 600 horses are annually sold at each of the two great horse-fairs. Dumfries is also the principal pork-market in Scotland. The pigs come principally from Galloway, but they are produced to a greater or less extent in all parts of the district. The principal foreign trade is with America and the Baltic for timber, of which the annual value imported varies from 8,000l. to 10,000l.; the remainder is coasting trade. The imports are coal, slate, iron, tallow, hemp, bones, timber, wine, and colonial produce; the exports, wool, freestone, hosiery, shoes, pork, fat cattle and sheep, grain, wool, nursery plants, and grass seeds. There belonged to the port on the 1st January, 1864, 52 sailing vessels under and 63 above 50 tons burden; there were no steamers. The customs' revenue amounted to 8,464l. in 1859; to 7,502l. in 1861; and to 6,451l. in 1863. Vessels of above 60 tons burden can approach the town, the river having been much deepened; there is also a quay about 700 yds. distant; one for vessels of greater burden about a mile and a half farther down; and a fourth near the mouth of the river for foreign vessels, and such as draw too much water to approach nearer to the town. Dumfries is governed by a provost, three bailies, and twenty-five councillors. Corporation revenue, 1,515l. in 1863-4. Annual value of real property, 33,043l. in 1863-4. Dumfries unites with Annan, Kirkcudbright, Sanquhar, and Lochmaben, in sending a mem. to the H. of C. Registered electors, 677 in 1865.

Dumfries is a place of great antiquity, though it was not made a royal bgh. till the 12th century, in less than a century afterwards. Devorgilla, daughter of Alan, last lord of Galloway, and mother of John Baliol, erected a monastery here for Franciscan friars; and, for the sake of this religious house, she built the old bridge, the toll on which formed part of the endowment of the institution. It was here that John Comyn, the heir and representative of Lady Devorgilla, and one of the competitors for the throne, was assassinated, under circumstances of great provocation, in 1306, by his rival, the illustrious Robert Bruce. The castle belonging to the Comyns was situated on a spot in the immediate vicinity of the town, which still bears the name of Castledykes. A strong castle once stood on the site now occupied by the new church. Being in some respects a border town, Dumfries frequently fell into the hands of the

English. It was for some time in the possession of Edward I. It was burnt by the English previously to 1440, and again in 1536. In 1570, the castle was taken and sacked, together with the town, by the Earl of Essex and Lord Scrope. Queen Mary and her privy council, in 1565, ratified at Dumfries a peace with England. James VI. in passing through the town, in 1617, on his return to England, presented the trades with a small silver gun, to be awarded, from time to time, to the best marksman; but this dangerous pastime has been discontinued. The inhab. in 1715, displayed their opposition to the union of the two kingdoms, by burning the articles and the names of the commissioners at the market-cross. They evinced great loyalty towards the reigning family in 1715, and so fortified their town, that a large body of insurgents, who had determined to attack it, found it expedient to change their resolution. But, in 1745, it suffered severely from the rebel army, which was stationed here a few days on its return from England.

DUNBAR, a royal and parl. bor. and sea-port of Scotland, co. Haddington, on a slight eminence on the German Ocean, 27 m. E. by N. Edinburgh, and 29 m. NW. Berwick, on the Edinburgh and Berwick railway. Pop. 3,796 in 1861. The borough consists of a long and well-built street running E. and W., with inferior streets, towards the sea, and one on the S. introducing the road from Edinburgh. Its public buildings are a new parish church of Gothic architecture, with a tower 107½ ft. high; Dunbar House, the ordinary residence of the family of Lauderdale; the town-hall; and burgh schools. It has a subscription and mechanics' library; an English and Latin school under one master; a mathematical school; and several private seminaries; a sailors' society for the benefit of superannuated seamen and their widows; three dissenting chapels, two belonging to the United Associate Synod, and one to the Wesleyan Methodists. Dunbar is governed by a provost and 12 councillors; corporation revenue 1,226l. in 1863-4. The harbour has 9 ft. water at neap, and 14 at spring tides, but owing to rugged rocks the entrance is dangerous. Coal is imported to the extent of about 20,000 tons a year; foreign grain to a considerable extent. Corn of various kinds, including beans and pease, exported to the amount of about 25,000 qrs. White fish of all kinds are caught off the coast. The cod is pickled, and sent to London; the haddocks are smoked, and sent chiefly to Edinburgh and Glasgow; the lobsters are preserved in pits, cut in the rock within sea-mark, and sent to London. The herring fishery is, also, considerably productive. Dunbar was created a royal bgh. by David II., but existed as a burgh long before that date. It evidently grew up under the protection of the castle of Dunbar, a fortress which stood on a lofty rock within sea-mark; but the date of its building is unknown. The castle and lands of Dunbar were conferred, in 1072, by Malcolm Canmore, on the Earl of Northumberland, whose descendants, created earls of Dunbar and March, retained possession of them till their forfeiture in 1434. This fortress rendered Dunbar the theatre of many warlike exploits. It was taken by Edward I. in 1296. Edward II. took refuge in it after his defeat at Bannockburn. It was often besieged, and seems alternately to have belonged, for longer or shorter periods, to the English and Scotch. Four times it received within its walls Queen Mary. In 1567 parliament ordered it to be demolished, and scarcely a vestige of it now remains. The 6th earl of Dunbar, in 1218, founded in the neighbourhood a monastery of Red Friars, of which some traces yet remain; and the

P

7th earl founded a monastery of White Friars, but of it no vestige can now be seen. The title of earl of Dunbar was revived, in 1605, by James VI., in the person of George Home, of Manderston, lord high treasurer of Scotland, at whose death, as he left no heirs male, it became extinct. A splendid marble monument was erected to his memory in the old, and is now preserved in the new, church of Dunbar. Dunbar unites with N. Berwick, Haddington, Lauder, and Jedburgh, in sending a mem. to the H. of C., and in 1865 had 148 registered voters.

DUNDALK, a sea-port town and parl. bor. of Ireland, co. Louth, prov. Leinster, 45 m. N. Dublin, at the extreme E. point of Dundalk Bay, near the mouth of Castletown river, on the railway from Dublin to Belfast. Pop. 10,782 in 1841, and 10,428 in 1861. The town consists of two main streets, each 1 m. in length, intersecting each other near the centre, with several transverse thoroughfares. They are paved, lighted, and kept in order by commissioners under the watching and lighting act. A bridge crosses the Castletown river on the N. There is an assembly-room, a literary society, and two news-rooms. A hunting club holds its meetings here, and races take place occasionally in the neighbourhood. Near the seaside is a large cavalry barrack. The parish church is a spacious building: there is also a large R. Cath. chapel, and meeting-houses for Presbyterians, Independents, and Methodists. It has an endowed classical school, to which the sons of freemen are admissible at a low quarterly fee, a school called the Dundalk Institution, under the Incorporated Society; one on the foundation of Erasmus Smith; and some others supported by the contributions of individuals: these educate in all about 600 pupils. It has also a co. infirmary, a mendicity association, a savings' bank, and several minor charitable institutions.

Though incorporated by charter of Richard II., the bor. is governed under a charter of Charles II. The ruling body consists of a bailiff, 16 burgesses, and an unlimited number of freemen chosen by the burgesses. Dundalk returned 2 mem. to the Irish H. of C., and now returns 1 to the imperial H. of C. The parl. bor. comprises 415 acres. Registered electors, 304 in 1862. The assizes and general sessions of the peace for the co. are held here twice a year, and petty sessions every Thursday. A guildhall contains apartments for municipal purposes, an assembly-room, and offices for several branches of public business. The co. court-house, an elegant modern structure, is built on the model of the temple of Theseus at Athens. There are several distilleries, tanneries, salt-houses, a malthouse, and a foundry. The trade consists principally in the export of a large portion of the agricultural produce of Louth, Cavan, and Monaghan; comprising wheat and wheat-flour, oats and oatmeal, barley and malt, with cattle, sheep, and pigs. The introduction of steam navigation has occasioned a great increase in the export of eggs and poultry. The harbour, which is safe though shallow, has been much improved. A lighthouse has been erected on the bar at the mouth of the river. The anchorage ground has from 4 to 6 fathoms water. There belonged to the port, on the 1st of January, 1861, five sailing vessels under, and 23 over, 50 tons burthen; besides 4 steamers of a total burthen of 1,703 tons. The customs revenue amounted to 30,575*l.* in 1859; to 51,797*l.* in 1861; and to 88,313*l.* in 1863. Markets on Mondays; fairs on the Monday next but one before Ash-Wednesday, May 17, first Monday in July, last Monday in August, second Monday in October, and second Monday in November.

Dundalk was one of the fortresses erected by the English shortly after their settlement, for the defence of the northern pale; but its defences have since been suffered to fall into decay, and few remains of them are now in existence.

DUNDEE, a flourishing royal and parl. bor. and sea-port of Scotland, co. Forfar or Angus, on an acclivity on the N. bank of the Frith of Tay, on the railway from Perth to Arbroath; 37½ m. N. by E. Edinburgh, 57 m. SW. Aberdeen, and 9 m. W. of the lighthouse on Buttonness Point at the mouth of the Frith. Pop. 90,417 in 1861. The pop. has increased considerably since 1841, when it amounted to 62,794. The town stretches upwards of a mile along the Tay, and inland about half a mile up the acclivity which terminates in Dundee Law, an insulated conical hill, 525 ft. above the level of the river. In the centre of the town is a spacious parallelogram, 360 ft. long by 100 broad, called the High Street, and seven of the principal streets diverge from it, the Nethergate and Overgate to the W., the Murraygate and Seagate to the E., Castle Street and Crichton Street to the S., and Reform Street to the N. There is generally great irregularity in the streets, except in the modern portions of the town; and there are many narrow and mean lanes which contrast strikingly with the new streets. The suburbs along the Tay are marked by many elegant villas. Of public buildings the most imposing is St. Mary's Church with its tower 156 ft. high, a splendid edifice built in the 12th century. The only part of the original building, however, which remains is the tower; the other parts having been rebuilt in the 14th and present centuries. It is in the form of a cathedral, the tower at the west end, next to it the nave, then the transept, and at the east end the choir. After the Reformation it was divided by partition walls, into different Presbyterian churches, of which at one time there were four. In 1841, three of these were burned down by an accidental fire. They have been rebuilt conformably to the original style of the structure, but now it has only three places of worship, the transept containing only one instead of two, into which it was previously divided. There being a large open space in front of the building, it has a fine appearance, and is well seen by strangers passing through the town on their way to or from Perth. There are two other churches belonging to the Establishment, and numerous chapels belonging to Dissenters, but some of them are remarkable for their architecture. The town house is a fine building, on the S. side of the High Street, erected about the year 1734. The custom house was erected in 1841 after a design by Mr. James Leslie, the engineer of the harbour, the cost having been defrayed by government. A splendid arch, in commemoration of her majesty's landing at the harbour in 1844, was constructed at a cost of about 3,000*l.* The public school, better known by the name of 'The Public Seminaries,' is a handsome building in the Grecian style. The exchange buildings contain a splendid reading-room and other conveniences for the mercantile and professional classes. The 'Watt Institution,' so called in honour of the illustrious James Watt, is a neat building, containing a lecture-hall, library, &c., for the use of the working classes, for whose benefit it was erected by public subscription.

The chief of the public works connected with the town is the harbour. Previously to 1815 it was of very limited extent, and quite unsuited to the trade of the place. In that year an act of parliament was obtained for enlarging it, and placing it under the management of commissioners elected by the various incorporated public bodies of the town and county. The first plan of enlargement contained only one wet dock, but now there are three, measuring together about 26 acres, and capable of accommodating vessels of the largest size. Besides the wet or floating docks, there is a large graving dock capable of containing three vessels at a time, and a graving slip on Morton's plan, on which sailing-vessels or steam-vessels of almost any size can be hauled up for repair. The harbour cost nearly 1,000,000£ sterling. The cost has been defrayed from the dues collected on goods and vessels entering and departing, except about 200,000£ of borrowed money. Dundee has ample railway accommodation. W. there is the 'Dundee and Perth,' E. the 'Dundee and Arbroath,' N. the 'Dundee and Newtyle,' & the 'Edinburgh and Dundee.' There are two gas companies in the town. Until the year 1845 Dundee was ill-supplied with water; but a joint-stock company was established at that time, and they brought a plentiful supply from Monikie parish, distant between 8 and 9 m. A large reservoir is formed there to collect the rain-water of the district, and a covered conduit conducts the water to the town. There is a smaller reservoir within a mile of the town on high ground, from which, by means of pipes, there is a constant pressure of water, so that it rises to the attics of the highest houses, and is very convenient for extinguishing fires, there being fire-cocks in all the streets, and leather hose or flexible pipes to lead the water to the houses in which the fire breaks out. The whole cost of the water-works has been about 130,000£.

The staple trade of Dundee is the manufacture of linen, chiefly of the coarser descriptions, such as are used for the clothing of the lower orders, both at home and abroad. Large quantities of sacking, bagging, and sail-canvas are also manufactured both for the home and foreign markets; and some of the manufacturers have of late turned their attention to finer fabrics, such as drills, striped and checked linen, and fine shirting. The linen trade of Dundee is the most extensive in the town, the greater part of what is made in Forfar, Kirriemuir, and the other inland villages of Forfarshire, is sent here to be dressed, packed, and sent off; so that the quantity exported from Dundee exceeds considerably that from any other part in the kingdom. The declared value of the exports of home produce from the port of Dundee amounted to 137,473£ in 1859; to 71,424£ in 1860; to 97,081£ in 1861; to 65,549£ in 1862; and to 55,941£ in 1863. The raw materials for the linen manufacture of Dundee are received chiefly from Russia and Prussia. The gross amount of customs duties received was 66,257£ in 1859; 69,747£ in 1860; 58,804£ in 1861; 60,470£ in 1862; and 65,049£ in 1863.

The linen trade seems to have been introduced into Dundee early last century; but for a length-ened time it was quite inconsiderable. In 1745 only 74 tons of flax were imported. From that period to 1791 the progress of the manufacture was more rapid; in the latter year, 8,444 tons flax, and 799 tons hemp being imported, and about 8,000,000 yards of linen, sail-cloth, &c., exported. Previously to this period all the yarn used in the manufacture was spun upon the common hand-wheel, partly in the town and partly in the adjacent country. But the spinning of yarn by machinery soon after to be introduced, and the increased facility of production consequent to the erection of flax spinning mills has been such, that the cost of the yarn, including of course the raw material, is now less than the mere expense of spinning amounted to forty years ago. The re-

sult has been the total extinction of hand-spinning in all parts of Scotland, and a wonderful increase of the quantity of yarn produced, and of the manu-facture. In 1862, from 20,000 to 24,000 persons were engaged in the manufacture. The weaving of the linen, which was formerly done on hand-looms, is in course of being superseded by ma-chines, or 'power-looms.' There are many calendering and packing establishments, all of them of considerable extent. At these works the linens are passed through machines named 'calen-ders,' for the purpose of making them smooth and close in the texture. They are then cut down into short pieces, folded into such shapes as are saleable for the markets to which they are to be sent, and packed into bales by means of hydraulic presses, which squeeze them into very small bulk, in order that they may require less room in the vessels in which they are shipped to foreign countries.

More than half the linens sent from Dundee are exported to foreign countries, the remainder being sent to London, Glasgow, Manchester, and other large towns for home consumption. The linens which are exported to foreign countries are for the most part sent to Liverpool, London, and Glasgow, to be shipped there, it being found more advan-tageous to send them abroad as parts of general cargoes of goods, than to send whole cargoes of them direct from Dundee. Some of the manu-facturers are the exporters of their own linens; others sell to exporting merchants in Dundee, who employ their capital in that branch of trade. The linens are in general sent on consignment to agents in foreign countries for sale, and are sold by them to retailers on credit of six to twelve months. The foreign countries to which Dundee linens are ex-ported are the United States, Canada, Mexico, the West India Islands, Brazil, Peru, Chili; and some are also sent to Spain, Portugal, and the countries on the Mediterranean. A large amount of capital is required to carry on the linen trade of Dundee, as the raw material is mostly paid for in ready money, and the manufactured article is sold on long credit. For that portion of the latter which is sent abroad it is in general about two years from the time when the manufacture of it is commenced till the time when returns are received. It may, therefore, be estimated that nearly 3,000,000£ ster-ling are constantly employed in the Dundee trade, and in that of the small towns of which it is the sea-port. The gross annual value of real pro-perty, including railways, assessed to income tax, amounted to 214,900£ in 1867, and to 231,853£ in 1862.

The other important branches of trade carried on in Dundee, besides the retail trade common to all large towns, are the shipping trade, ship build-ing, and machine making. The number of ships or vessels belonging to the port on the 1st of Jan., 1864, was as follows:—8 sailing vessels under 50 tons, of a total burden of 195 tons; and 199 ves-sels above 50, and of a total burden of 12,874 tons. There were, besides, 2 small steamers, of a total burden of 69 tons, and 15 larger steamers, of 4,631 tons burden. Many of these vessels are employed in foreign trade not connected with Dundee; a considerable number is required for the importation of flax from the Baltic, some are engaged in whale fishing, and the rest in the coal and coasting trades. There are seven ship-building yards in Dundee, and there are commonly 10 or 12 vessels on the stocks.

Dundee is well supplied with schools for the children of the middle and lower classes, and a plain education can be got on very moderate terms. At the public seminaries, which are under the management of directors appointed by the town

r 2

council and those who contribute to the cost of the building, classical education, to a certain extent, as well as the elementary branches, can be obtained at a very moderate cost. There are teachers of Latin, Greek, French, German, Mathematics, and Natural Philosophy. The number of pupils of all ages and ranks at the public seminaries is generally between 600 and 700.

The principal charitable institutions are the infirmary or hospital, the industrial school, and the orphan house. The infirmary has accommodation for about 150 patients, and is supported entirely by donations and yearly contributions from benevolent individuals. The industrial, or ragged-school is for poor boys and girls who are deserted or not cared for by their parents, and is also supported by voluntary contributions. The number of children in it averages about 120. The orphan house is likewise maintained by contributions and bequests.

The municipal government of the town is vested in a provost, 4 bailies, a dean of guild, treasurer, and 21 councillors. The revenue of the corporation, which is derived from rent of property, and petty customs on provisions, amounted to 8,513l. in the financial year 1863-4. Previously to the passing of the Reform Act in 1832, Dundee was joined with Forfar, Perth, Cupar, and St. Andrew's in returning a member to the H. of C. But the act referred to conferred that privilege on Dundee singly. The pop. const. was 2,896 in 1868.

The origin of Dundee is involved in obscurity. In the early centuries of the Christian era it had probably been a village inhabited by a few fishermen, who gained a livelihood by supplying the neighbouring country with fish. In the eleventh century King Malcolm erected a residence here, and lived in it occasionally with Margaret his queen, a daughter of one of the Saxon kings of England. The next remarkable occurrence concerning Dundee was the building of a church in honour of the Virgin Mary, of which the old steeple or tower is now the only remaining original part, by David, earl of Huntingdon. Sir William Wallace was educated at the grammar school of the town, and was often in its neighbourhood with his gallant companions during the struggles which they made for the independence of Scotland. After he was overpowered by the forces of Edward I., Dundee suffered much for its adherence to the cause of Scottish independence. In the civil wars, during the reign of the Stuarts, it was also frequently the object of contention between the two parties; and ultimately it was almost totally destroyed by the army of General Monk, in 1651. After 1745 it gradually recovered, and towards the end of last century it had attained to a considerable degree of prosperity. It was formerly a walled town, but of its walls and gates no traces remain, except the 'Cowgate Port.' Dundee was made a royal burgh in 1210, when it received a charter from William the Lion. The charter was renewed at different times by the sovereigns of Scotland, and the existing charter was given by Charles I. in 1641.

DUNFERMLINE, a royal and parl. bor. and eminent manufacturing town of Scotland, co. Fife, 3 m. N. Frith of Forth, 16 m. NW. Edinburgh, and 6 m. NW. North Queensferry, on the railway from Stirling to Dundee. Pop. 13,506 in 1861. The town is about 300 ft. above the level of the sea, and occupies an agreeable but rather inconvenient situation, being placed on the face of an extensive eminence, difficult of ascent from the S. The town stretches fully a mile in length from E. to W., and its average breadth is about ¼ m. The main street, which is handsome and substantially

built, is pretty regular. Almost all the other streets are more or less irregular; and while some are handsome, not a few are of an opposite description. A large suburb having risen up on the W., and being separated from the town by a deep ravine, formed by the Tower-burn, a bridge was thrown over the rivulet in 1770; and the ravine having been so far filled up, buildings have been erected on both sides. The only remarkable public building is the parish or Abbey church, being part of a monastery founded here by Malcolm III., surnamed Canmore, and which served as the parish church till 1821, when a new church was erected to the E. of the old building, and in immediate connection with it. When digging in what was called the Psalter-churchyard (on which spot the choir formerly stood), for a proper site for the new edifice, the tomb of the most illustrious of the Scotch sovereigns, Robert Bruce, was discovered in 1818. His skeleton, which was pretty entire, and 6 ft. in length, was disinterred, and a cast of his skull taken. It was re-interred amidst much state, by the barons of exchequer, the bones being placed in a new coffin, filled up with bituminous matter, calculated to preserve them. The spot is below the pulpit of the new church. This building, which is of Gothic architecture, harmonises well with the old structure, of which it is a continuation; and is surmounted by a high square tower, round the sides of which, in open hewn work, are the words ' King Robert the Bruce,' in capital letters 4 ft. in height. The Abbey church of Dunfermline is altogether one of the most imposing and magnificent structures of the kind in Scotland. It has 2,051 seats; but is only available, from the obstruction of pillars and otherwise, for about 1,400 hearers. There are numerous other churches in the town and parish, some recently built, and neat in their construction. The largest church of the United Presbyterian Synod is a huge barn-looking building, which raises 'its enormous rectilinear ridge' over all the other buildings in the town. In front of it was placed, in 1849, a statue of Ralph Erskine. The other public buildings are the town-house, county-court buildings, grammar and commercial schools, gaol, poor's-house, and fever hospital. The town-house consists of three stories, and is surmounted by a steeple 100 ft. in height. The 3rd story was formerly used as the town gaol, but being extremely ill-suited for that purpose, a new gaol, erected on the town green, was opened in 1844. The county court buildings were originally named the guildhall, afterwards the Spire Inn, on account of the lofty spire (132 ft.) that distinguishes the edifice.

The means of instruction are ample; there is a grammar-school, established prior to the Reformation, of which Robert Henryson, an ingenious poet of the times of James II. or III., is believed to have been master; a commercial school under the patronage of the guildry; the M'Lean and various other schools. There is a mechanics' institute, and a scientific association for popular lectures on science and literature; the fees of admission to these lectures being low, they have been well attended. There are several subscription libraries and a local museum.

In addition to the Abbey church, which is collegiate, there are several churches of the establishment; besides Free churches; United Presbyterian churches, Episcopalian, Baptist, and Independent chapels. The Secession from the established church in 1733 originated here. Of the Messrs. Erskine, regarded as the fathers of the Secession, one, Mr. Ralph Erskine, was minister of the Abbey church of Dunfermline. The Relief church, also, originated here in 1752, by the deposition of Mr. Thomas Gillespie of Carnock, in honour of whom an elegant church was erected in 1810. More than three-fourths the inhab. of the par. are Presbyterian dissenters.

Dunfermline is distinguished by its proficiency in the manufacture of fine table linen and coloured table-covers. The business is of considerable antiquity, having been introduced towards the beginning of the 16th century; but the original fabrics were of a coarse description, namely, ticks and checks. Damask and diaper looms were introduced early in the last century, by an ingenious mechanic of the town, of the name of Blake; but for a lengthened period the trade increased very slowly. In 1778 a new epoch commenced in the manufacture, by the introduction of the fly-shuttle; and many improvements have since been effected in the construction and working of the loom, and in other particulars. Among the most important of these was the Jacquard loom, introduced in 1825, and now universally employed.

Previously to the end of last century, all the yarn was spun by the hand-wheel; but at that time machinery was introduced, and has now entirely superseded the former clumsy and expensive system. The manufacturers are supplied with the finer sorts of yarn chiefly from Yorkshire and Ireland, and the other sorts from the neighbourhood, Dundee, and elsewhere. They do not spin exclusively for the local market, but prepare such articles as linen thread, shoe thread, and twist, for the general market.

This branch of manufacture has found its way to a small extent to other parts of Fife, and to the north of Ireland.

Among other manufactories Dunfermline has breweries, candle-works, rope-works, tan-works, iron foundries, and flour-mills.

Dunfermline unites with Stirling, Culross, Inverkeithing, and Queensferry, in returning a mem. to the H. of C. Parl. constit., 502 in 1865. Annual value of real property assessed to income-tax, 25,168l. in 1857, and 28,945l. in 1862. The borough is governed by a provost, 2 bailies, and 22 councillors. There are several endowments in the bor. for the support of decayed widows and other poor persons.

The town is connected by railways with Edinburgh, Dundee, Perth, Alloa, and all the principal towns of Scotland. There are three harbours in the parish, each about 3 m. from the town, viz. Charleston, Bruce haven, and Limekilns. The last does not admit vessels of more than 200 tons burden. Charleston admits vessels of 400 tons. Its basin is capacious, and perfectly sheltered from every wind.

The parish of Dunfermline abounds in coal, lime, and ironstone. The coal has been wrought for upwards of 500 years. The quantity worked is nearly 150,000 tons a year. Bruce haven and Charleston, two of the three harbours referred to, were originally constructed by Lord Elgin, whose collieries and limeworks are on a very extensive scale.

Dunfermline can boast of great antiquity. A tower or fort, built here by Malcolm Canmore in the 11th century, gave origin to the burgh. The same king also founded a spacious Benedictine monastery, which ultimately became one of the most wealthy and important institutions of the kind in Scotland; and ordained that its precincts should form the burying-place of the Scottish kings. His own remains and those of his consort, Queen Margaret, were interred there, as also those of eight others of the royal line, including Robert

Bruce. Dunfermline continued to be a favourite royal residence as long as the Scottish dynasty existed. Charles I. was born here; as also his sister Elizabeth, afterwards queen of Bohemia; and Charles II. paid a visit to this ancient seat of royalty in 1650. The Scottish parliament was often held in it. The date of the erection of the palace is unknown; but it is believed to have been much extended and adorned by James IV. and James V. There now remains only the S. wall, and a vaulted apartment, which was the king's cellar, having the kitchen above. Of the tower, erected by King Malcolm, only a mouldering fragment is seen. Of the monastery, which was once of great extent, nothing remains entire except the S. and W. walls of the fratery, or refectory, in the latter of which is a fine Gothic window; and the nave of the old abbey church, which, as above stated, forms the vestibule to the new church. But ancient as the place is, it was not made a royal burgh till 1588.

DUNGANNON, an inland town and parl. bor. of Ireland, co. Tyrone, prov. Ulster, 12 m. N by W. Armagh, and 7 m. W. Lough Neagh, on the railway from Dundalk to Londonderry. Pop. 3,911 in 1841, and 3,584 in 1861. The town consists of a square, with several good streets branching from it along the sides of a hill. The par. church is a large ancient building, and it has also a Rom. Cath. chapel, and meeting-houses for Presbyterians, Seceders, and Methodists; a classical school, founded in the reign of Charles I., well endowed, and capable of accommodating 100 resident pupils; a dispensary, and a mendicity institution. The corporation, which consisted of a portreeve, burgesses, and commons, is now extinct. The town returned 2 mems. to the Irish H. of C. till the Union, since which it has returned 1 mem. to the imperial H. of C. Previously to the Reform Act the franchise was vested in the portreeve and burgesses. The ancient liberties of the bor. comprised 856 acres, but the parl. bor. has been restricted to 280 acres. Registered electors, 221 in 1865. A manor-court, with jurisdiction to the amount of 20l., is held every three weeks; as also general sessions twice in the year, and petty sessions every fortnight. The court-house, with a bridewell attached, is a handsome modern building; a party of the constabulary is stationed here. The linen manufacture, though much fallen off, is still carried on pretty extensively, and there are several bleach-greens in the neighbourhood; earthenware and pottery are also manufactured, and there are iron-works, a brewery, and a large distillery. Markets on Tuesdays and Thursdays, in a spacious and convenient market-house; fairs on the first Thursday of every month. It is the seat of a poor law union. Dungannon is famous in Irish history from its being the place where the delegates of the Ulster volunteers met in 1782; and whence they issued their resolutions declaratory of the independence of Ireland.

DUNGARVAN, a marit. town and parl. bor. of Ireland, co. Waterford, prov. Munster, principally on a peninsula in the estuary of the river Conigar, 25 m. W. by S. Waterford. Pop. 8,625 in 1841, and 5,888 in 1861. As vessels of above 150 tons cannot come up to the town, it is not a place of much trade, though some corn and other produce is shipped for England. Recently it has been much improved, principally through the exertions of the Duke of Devonshire, who has built, at his own expense, a handsome bridge, connecting the main body of the town with the suburb of Abbeyside, on the opposite bank of the river. It has a neat appearance, and is a good deal resorted to for sea-bathing; but is not rich in proportion to its population. The public buildings are the par.

church, a new Rom. Cath. chapel, with three others belonging to convents, a school-house for 300 pupils, a court-house and bridewell, a barrack, and a fever hospital and dispensary. It returned 2 mems. to the Irish H. of C. till the Union, since which it has returned 1 mem. to the imperial H. of C. Previously to the Reform Act, the franchise was vested in the occupiers of 5l. houses in the town, and the resident 40s. freeholders of the manor. But the extent of the existing parl. boundary, as fixed by the Boundary Act, is only 392 stat. acres. Reg. electors, 295 in 1865. A manor court is held every three weeks; also general sessions in Jan., April, and Oct., and petty sessions on Thursdays. Markets on Wednesdays and Saturdays; fairs. Feb. 7, June 28, Aug. 27, and Nov. 8. The deep sea fishery was formerly carried on here pretty extensively, but has latterly much declined.

DUNKELD, a brt. of barony and market town of Scotland, co. Perth, on the N. bank of the Tay, 15 m. N. by W. Perth, and 49 m. N. by W. Edinburgh, on the railway from Perth to Inverness. Pop. 1,096 in 1841, and 929 in 1861. Little Dunkeld is a suburb, though in a different parish, being divided from the bor. by the Tay, which is here crossed by an elegant bridge of seven arches, built in 1809. Except a handsome new street leading from the bridge into the town, the houses are generally old and of mean appearance. But the situation of Dunkeld and the surrounding scenery are most beautiful, and have long been objects of admiration to every stranger. The town is situated in the centre of a valley surrounded by mountains of considerable elevation, presenting a great variety of picturesque forms, and covered to their summits with trees of every species. It is, besides, regarded as the great pass to the Highlands on the E.: the bulk of its inhab. are of Highland origin, and speak the Gaelic language. The banks of the mountain stream Braan, which joins the Tay nearly opposite to Dunkeld, present some of the most striking scenery connected with the place. Dunkeld House, the residence of the ducal family of Atholl, is on the verge of the town, and the style, extent, and natural and artificial beauties of the pleasure grounds are not equalled by any in Scotland. The most imposing object in Dunkeld is its cathedral, situated on the banks of the Tay; an edifice partly Saxon and Gothic, and the remains of which, owing to the care of the family of Atholl, are both extensive and in good preservation. The choir of the building is used as the parish church. Different portions of the cathedral were erected at different times, but the oldest portion, the choir, was built in 1318. Gavin Douglas, who translated Virgil's 'Æneid,' and Henry Guthrie, author of 'Memoirs of Scottish Affairs from 1637 to the Death of Charles I,' were both bishops of this see. The Culdees had a monastery here as early as 729. When Iona, the original and chief seat of that order, was ravaged by the Danes in the 9th century, the primacy resided for some time in Dunkeld, but was afterwards transferred to St. Andrews. 'But the rank of the abbots of Dunkeld,' says Pinkerton, ' one of whom was the father of a royal race in Scotland, and another, Ethelred, the son of Malcolm III., sufficiently marks the estimation in which that dignity was long held.' (Early Hist. of Scotland, ii. 271, 272.) The monastery, however, was changed by David I. into a cathedral in 1127, at or about which period the system of the Culdees was superseded throughout Scotland by that of the Roman Catholics.

DUNKIRK (Fr. Dunquerke, the Church of the Dunes, or Sand Banks), a sea-port town of France, and the most northerly in that kingdom, dep. du

Nord, cap. arrond., on the Straits of Dover, 40 m. NW. Lille, and 47 m. E. Dover, on a branch of the railway from Paris to Calais. Pop. 32,215 in 1861. The town is well built, and has broad and well-paved streets. The Champ-de-Mars and the Place Jean Bart are large and fine squares: the latter, which is planted with trees, has a bust of the brave sailor whose name it bears, and who was a native of Dunkirk. The greatest drawback upon the town is its want of good water, it being indebted for this necessary wholly to the rain-water collected in cisterns. Its defences consist of a rampart and ditch, a citadel, and Fort Louis, about 5-8ths of a mile distant: the fortifications were formerly more formidable, but having been demolished, according to the stipulations in the treaty of Utrecht, they have not been completely re-established. Principal public buildings are the church of St. Eloi, with its fine parties, the naval storehouses, barracks, town-hall, and college. The Tour des Pilotes serves for a landmark, and was one of the positions whence Cassini, and more recently Biot and Arago, conducted their observations relating to the map of France, and the measurement of the earth: it has a very fine chime of bells. Dunkirk has also a communal college, a public library containing 18,000 vols., a school of hydrography, a theatre, and concert-hall.

The harbour of Dunkirk, though in a great degree artificial, is large and commodious; but a mud bank, which drive at low water, being interposed between the town and the road-tead, it is rather difficult of access, and is apt to fill up; but these inconveniences have been to a considerable extent obviated by works constructed in 1826. Dunkirk has both an inner and an outer roadstead, defended from the violence of the sea by sand-banks parallel to the shore, and having deep water and good holding ground. Being connected, by means of numerous canals, as well as a line of railway, with a very fertile district, Dunkirk is a considerable emporium. The inhabitants have always been distinguished for enterprise. During the late and former wars between England and France, great numbers of privateers were fitted out here. At present several vessels belonging to the port are engaged in the herring-fishery, and in the cod-fishery on the Dogger Bank, and the banks of Newfoundland. Dunkirk was made a free port in 1828, since which its commerce has materially increased, particularly its trade in French wines destined for the supply of Belgium, of which it is a depôt. It has extensive soap-works, with starch-works, rope-works, tanneries, and iron-foundries. It has also considerable Geneva distilleries, breweries, and sugar-refineries. It has a general and a foundling hospital, a military and civil prison; and is the seat of a sub-prefect and of tribunals of primary jurisdiction and commerce. There is regular communication by steamers between the port and London. Dunkirk is said to have been founded by Baldwin, count of Flanders, in 960; in 1388 it was burnt by the English; and in the 16th and 17th centuries alternately belonged to them and to the Spaniards and French. Charles II. sold it to Louis XIV. for 300,000l., who, aware of its importance, had it strongly fortified at a vast expense. But, as already stated, Louis was compelled, by the treaty of Utrecht, to consent to the demolition of its fortifications, and even to the shutting up of its port. It was unsuccessfully besieged by the Duke of York in 1793.

DUNLOP, a par. of Scotland, celebrated for its manufacture of cheese, partly in the co. of Ayr, and partly in that of Renfrew, 8 m. N. Kilmarnock. The village of Dunlop in the par., had also

Inhab. in 1861. Dunlop cheese has for nearly a century and a half held a high character. Previously to this date, or between 1684 and 1700, cheese here, as well as throughout Scotland, was made of skimmed milk, as is still the case in various districts. A female of the name of Barbara Gilmour, who had fled to Ireland during the persecuting times of Charles II., returned at the Revolution, and, having married a farmer, was the first to introduce the practice of using the unskimmed milk in the making of cheese. This practice, which succeeded admirably, was for a time confined to the par., but it gradually extended to almost every part of the W. and S. of Scotland, all the cheese made in these districts with unskimmed milk being called *Dunlop*. The fact, however, is, that cheese made in the par. of Dunlop is not superior but inferior to that made in other districts. Besides the cheese produced in the par., a great proportion of what is manufactured in other parts of Ayrshire passes through it on its way to the consumer. Being a convenient *entrepôt* between the producing country to the S. and W., and Glasgow, Paisley, &c., a considerable number of persons resident in Dunlop follow the business of cheese dealers, purchasing it from the farmers, and supplying the victuallers in the manufacturing towns and districts.

DUNMANWAY, an inl. town of Ireland, co. Cork, prov. Munster, near the junction of three streams, which form the Bandon, 24 m. W. by S. Cork, on the railway from Cork to Skibbereen. Pop. 2,734 in 1831, and 2,008 in 1861. The town has a par. church, a Rom. Cath. chapel, a market-house, and a bridewell. The linen trade, after being for some years rather flourishing, has declined; but tanning and brewing, and the corn trade, are largely carried on. A manor court is held every third Saturday, and petty sessions on alternate Mondays.

DUNSE, a bor. of barony and market town of Scotland, co. Berwick, in a plain at the S. foot of Dunse Law, an eminence 630 ft. above the level of the sea, 13 m. W. Berwick-upon-Tweed, and 36 m. SE. Edinburgh, on a branch of the Edinburgh and Berwick railway. Pop. 3,556 in 1861. Dunse is neat and regularly built, but devoid of public buildings, except the town-hall and Dunse Castle, in its vicinity, the residence of the feudal superior of the bor., of Gothic architecture, the greater part modern, but added to an ancient tower said to have been built by Randolph earl of Murray, in the time of Robert Bruce. The par. church is a plain building; as are the three dissenting chapels belonging to the Associate Synod and the Relief. The means of education are ample; a par. school, an eminent unendowed academy, six other unendowed schools, besides private seminaries for females, and several Sabbath schools. A subscription library was commenced so far back as 1768. There are two circulating libraries, and a reading-room. The assessment for the poor of the bor. and par. is 710l. There are two friendly societies, a savings' bank, and two branch banks. There is a weekly market, three fairs for black cattle and horses annually, and a quarterly fair for sheep.

Dunse was erected into a burgh of barony by James IV. in 1489; it was then situated on the NW. side of Dunse Law; but having been afterwards burnt by the English, it was rebuilt in 1588, and its present site adopted, in order that it might be more immediately under the protection of Dunse Castle. After Berwick-upon-Tweed was ceded to the English (1482), and ceased to be the co. town, Dunse enjoyed that distinction in common with Lauder. It was afterwards (1600) trans-ferred by act of parliament to Greenlaw; but Dunse was not altogether deprived of the privilege till 1696. It is, however, by far the largest and most important town in the co., and more country business is done in it than in both the towns referred to. In 1639, when Charles I. lay on the S. side of the Tweed with the intention of reducing the Scotch Presbyterians to submission, General Leslie took up his station on Dunse Law, with a body of 20,000 Covenanters, to defend the country from invasion. After the two armies had continued in this position for three weeks, a treaty of peace was concluded, and both were dissolved. Dunse has given birth to many distinguished men, among whom may be specified, John Duns Scotus, the Subtile Doctor, descended of the ancient family (not long extinct) of Duns of Duns, or of that Ilk; Boston, author of the Fourfold State and other works; and Dr. M'Crie, the historian of Knox.

DUNSTABLE, a town and par. of England, co. Bedford, hund. Manshead; 32 m. NW. London by road, and 34 by Great Northern railway. Pop. 4,470 in 1861. The town, situated on the N. acclivity of the Chiltern Hills, near the source of the Lea, has four streets, and is pretty well built. A celebrated priory was founded here by Henry I., in 1131, of which the par. church contains the nave. The Baptists and Methodists have also places of worship. Here is a charity school, founded in 1727, for 40 boys and 15 girls; with 12 almshouses for poor widows, and 4 do. for decayed maiden ladies. Dunstable is the principal seat of the British straw plait manufacture, which employs many females in the town and vicinity. Ladies' straw hats were, and still are, not unfrequently called Dunstables.

DUNWICH, a sea-port bor. and par. of England, co. Suffolk, hund. Blything, on the E. coast of the co.; 20 m. NE. London, and 26 m. NE. Ipswich. Pop. 232 in 1831, and 227 in 1861. Though now a poor fishing station, this was once an important sea-port, having an extensive trade, a large population, 2 abbeys, and several churches. It has been reduced to its present state of insignificance by repeated inroads of the sea; and would probably have been totally abandoned, but for its having had the privilege of returning two mem. to the H. of C. The encroachment of the sea began previously to the Conquest. In the reign of Edward III., an inundation swallowed up more than 400 substantial houses. The last great encroachment was in 1740; but the sea has continued progressively to encroach on the land; and at present there remains only the ruins of one of its many churches. It was disfranchised by the Reform Act; and no longer attracts any attention, except from those who visit the coast to study the great natural revolutions of which it has been the theatre. (Campbell's Survey, i. 277; Lyell's Geology, i. 403, 3d ed.)

DURANGO, a town of Mexico, cap. of the state of the same name, in the Sierra Madre, 6,815 ft. above the level of the sea; 450 m. NW. Mexico, and 160 m. NW. by W. Zacatecas; lat. 24° 25' N., long. 104° 15' W. Pop. estimat. at 20,000. The town is regularly built, and contains a cathedral and other churches, several convents, a mint, and a theatre. It is the seat of a bishopric. Its inhab. are industrious; they manufacture many woollen articles, woollen goods and leather, and have a considerable trade in cattle. Iron mines are worked in the vicinity.

DURAZZO (an. *Epidamnus* and *Dyrrachium*), a sea-port town of Turkey in Europe, Albania, on the E. shore of the Adriatic, and on the S. side of a projecting tongue of land, 7 m. S. Cape Pali;

lat. 41° 17′ 35″ N., long. 19° 26′ 41″ E. Estim.
pop. 6,000. This town, which has greatly declined from its ancient importance, is surrounded
by walls, and is indifferently fortified. It has
some trade in the export of corn. The bay, on
the N. side of which it stands, is 5 m. broad from
N. to N., with from 7 to 8 fathoms water, the best
anchorage being about 1½ m. S. by E. from the
town.

According to Plautus, the inhab. of Dyrrachium
were immersed in every sort of debauchery and
vice; wherefore, says he,—

'—— huic urbi nomen Epidamno inditum est,
Quia emo fermo huc sine damno divortitur !'
Menæchmi, Act II. Sc. 1.

According to the statements of a modern traveller,
M. Pouqueville, the descendants of these con-
temporaries of Plautus, if they be less luxurious,
exhibit few other symptoms of improvement.
He calls their town ' une anarchie, un repaire de
pirates, un séjour d'assassins, et le réceptacle impur
des misérables qui peuvent s'échapper des côtes de
l'Italie !' (Voyage dans la Grèce, i. 326.)

Dyrrachium was founded by a colony from Corcyra, anno 625 B.C. After it fell into the hands
of the Romans, it became a place of great importance, from its being the port which vessels
from Brundusium, bound for the opposite coast,
endeavoured to make, and from its being the
usual place of departure for ships crossing the
Adriatic with despatches or passengers from Greece
or Italy. It became the seat of some important
strategical operations during the struggle between
Cæsar and Pompey, which terminated advantageously for the latter. (Cæsar, de Bello Civili,
iii. § 41). It was made a Roman colony by Augustus; and, after various vicissitudes, was subjected to the Turks, under whom destructive sway
it still continues, by Bajazet II.

DURHAM, a marit. co. in the N. of England,
having E. the German Ocean, N. Northumberland,
W. Cumberland and Westmoreland, and S. Yorkshire. Area, 978 sq. m., or 625,476 acres, of which
about 200,000 are waste above ground, but rich in
mines below. In its W. parts is is occupied by
offsets from the Pennine range of mountains, and
by black heathy moors. Soil in parts good; but
generally it rests on a sub-soil of stiff clay, and is
cold and infertile. It is a curious fact, however,
that the W. parts of the co., though naturally the
least productive, are the best cultivated. Principal crops, wheat, oats, barley, beans, and peas.
A mixture of rye and wheat, provincially called
maslin, is also rather extensively cultivated. Turnips are generally introduced, particularly in the
W. districts. Lime, of which there is an abundant
supply, is principally used as manure, the quantity applied being from 70 to 80 bushels an acre.
Drainage is much neglected in the E. parts of the
co. which, in consequence, are in a comparatively
backward state. The Teeswater breed of short-
horned cattle, so called from the river Tees, which
bounds the co. on the S., is admitted to be one
of the very best, both for feeding and milking,
and is now very widely diffused. Sheep mostly
Cheviots; stock estimated at between 200,000
and 250,000 head. A great deal of property belongs to the church, and there are besides some
large estates: but property is, notwithstanding, a
good deal subdivided. Farms of all sizes, but the
greater number rather small; and the condition
of the occupiers of the small farms is said to be
very unfavourable. Durham has some of the
most extensive and valuable coal-fields in the
kingdom; and also valuable lead and iron mines.
Vast quantities of grind-stones are produced from

the quarries at Gateshead Fell. Manufactures
various, but not very extensive or important.
Principal rivers, Tees, Wear, and Derwent. Durham has 4 wards and 75 parishes, and returns 10
mem. to the H. of C., viz. 4 for the co., 2 each for
the city of Durham and Sunderland, and 1 each
for Gateshead and S. Shields. Registered electors
for the co. 12,717, namely 5,722 for the Northern
and 6,995 for the Southern division. Pop. 508,666
in 1861, inhabiting 84,817 houses. Gross annual
value of real property assessed to income tax—
Northern division 619,266l. in 1857, and 943,756l.
in 1862; Southern division 802,730l. in 1857, and
1,189,436l. in 1862. Principal towns, Durham
city, Sunderland, Gateshead, S. Shields, and Darlington.

DURHAM (originally Dunholme, from dun, a
hill, and holme, a river), an ancient and celebrated
city of England, cap. co. same name, and nearly
in its centre, on a bend of the river Wear, 251 m.
N. by W. London, and 65 NNW. York, on the
York and Newcastle railway. Pop. of city 14,088
in 1861, and of distr. 70,374. The chief objects of
interest in the city are the cathedral and castle:
their appearance from the surrounding country is
striking, being situated on a rocky peninsula, elevated about 80 ft. above the Wear, by which it is
nearly encircled. The first of these structures,
begun in the reign of William Rufus, but much
enlarged and improved in subsequent ages, is a
large and majestic pile of Norman architecture:
it is 461 ft. in length, by about 200 in extreme
breadth, from the N. to the S. transept; it has a
central tower, 214 ft. in height; and at the W.
end are two low towers, once topped with spires.
The inside has much of the clumsy though venerable magnificence of the early Norman style.
The pillars are vast cylinders, 23 ft. in circumference, and variously adorned. In the Galilee,
or lady's chapel, at the W. end of the cathedral,
is the tomb of the venerable Bede, his remains
having been transferred thither from Jarrow in
1870; and in the Nine Altars, at the E. end of
the cathedral, is the shrine of St. Cuthbert, the
patron saint of the sacred edifice. Dr. Johnson
says of this noble structure, that 'it strikes with a
kind of gigantic dignity, and aspires to no other
praise than that of rocky solidity and indeter-
minate duration.'

The bishop of Durham was, till deprived of it
by the act 6 and 7 William IV. cap. 19, custos
rotulorum and chief civil governor of the co.,
which has distinct courts and law officers; he
presided at the assizes, and all writs were return-
able to him, and not to the king. The practice in
the palatinate courts is now, however, assimilated
in a great measure to that of the superior courts
at Westminster; and as actions may be commenced in them for any sum, however large, the
change has been productive of great public benefit.

Cromwell founded a university in Durham in
1657, assigning to it the houses and part of the
lands belonging to the dean and chapter. This
institution, which, had it survived, must have been
of great service to the N. counties, fell to pieces
on the Restoration, when the church recovered her
old possessions. No new attempt, or at least no
successful one, was made to establish another university at Durham till 1831. In that year, however, a university, endowed by the dean and chapter, the bishop, and other wealthy individuals, was
founded, to afford instruction, and grant degrees
in the different faculties. It was incorporated by
royal charter in 1837, and consists of a warden,
professors, tutors &c.; but, however creditable to
the liberality of the founders, it is far from adequately meeting the existing wants of society, its

grand object being to furnish instruction for candidates for holy orders in connection with the church of England. There are 25 fellowships and 20 scholarships attached to the university. It has two colleges—University and Bishop Hatfield's Hall.

The castle, founded by William the Conqueror, and intended partly to bridle that part of his own dominions, and partly as a defence against the irruptions of the Scotch, stands a little to the N. of the cathedral, and on the same elevated peninsula. It was long used as the bishop's palace, and has some magnificent apartments. It is now (with the exception of a suite of rooms reserved for the accommodation of the bishop, on his visits to the city) appropriated to the purposes of the university, being occupied by students. Adjoining the castle, on an artificial mound of considerable elevation, is the ancient tower, or keep, which had fallen into decay, but has recently been restored, and presents a most imposing appearance. On the palace green is the exchequer, a large square building, in which the bishop's courts of exchequer and chancery were held, and near it is a magnificent library, erected by Bishop Cosins; and a little further distant is the office for the registry of wills. The cloisters adjacent to the cathedral are 147 ft. square, and very neat. The chapter-house, a plain building, opens into them.

The shelving sides of the peninsula, from the cathedral, castle, and other buildings, down to the river, are laid out in hanging gardens, intersected with walks, and planted with trees, that have a most imposing effect.

Besides the cathedral, there are here six parish churches. Those of St. Nicholas and St. Oswald, the latter remarkable for its vaulted roof of wood, are of considerable antiquity. The Primitive and Wesleyan Methodists, Rom. Catholics, Independents, and Quakers have also places of worship.

The grammar school connected with the cathedral has four exhibitions, of 25l. each, for the sons of clergymen at the school, and of 50l. each at either university; in addition to this, it has five scholarships, of 10l. each, at Peterhouse, Cambridge. There is a blue-coat school, an infant school, and an endowed charity school, which together furnish gratuitous instruction to about 1,000 children. It has an infirmary, founded in 1791, dependent on voluntary contributions, a lying-in hospital, and several almshouses. There are here also a public reading-room, a library, and assembly rooms, and two weekly newspapers. Races take place annually in May.

The city is divided into several distinct portions: that which contains the cathedral and castle has some excellent and finely situated houses, principally occupied by the dignitaries of the church. What is called the old town adjoins the castle on the N.; in it are the market-place, the theatre, and principal shops, but it has a great many inferior houses. In the market-place is a fountain for the supply of the city with water, conveyed thither from springs about 1½ m. distant, consisting of an octagon building, surmounted by a statue of Neptune. The township of Elvet lies on the E. side of the cathedral peninsula, being separated from it by the Wear. It is comparatively well built. In this quarter of the town is the magnificent new co. gaol and co. court house, erected in 1809, at an expense of nearly 140,000l. On the W. side of the Wear and of the cathedral are the townships of Crossgate and Framwellgate; they are occupied by a very inferior class of houses, and have the appearance of great poverty. The communication between the different parts of the town is kept up by two old and one comparatively modern bridge. The latter, called Prebend's Bridge, was erected by, and is maintained at the expense of the dean and chapter. It is placed in a very romantic situation, and connects the public walks called 'the Banks,' already alluded to. The city is well paved and lighted.

Durham, for a lengthened period, made little or no progress. It used to be one of the dullest and most stagnant of cathedral cities. But from the flow thitherward of the tide of pop., through the opening of the collieries, previously mentioned, and the construction of several railways in the district, connecting it with Shields, Sunderland, Hartlepool, and Stockton, none of which are more than 20 m. distant, a powerful impetus has lately been given to its trade. This is visible in the crowded state of the weekly markets, the improvement of the shops, and other signs of the increasing importance and trade of the town. An extensive carpet manufactory and a worsted spinning mill are situated on the banks of the Wear.

The Great N. of England railway connects the city directly with Newcastle-on-Tyne, 14 m. distant; and the Newcastle and Carlisle railway opens an easy communication with the W. coast. The neighbourhood of Durham abounds with situations well adapted for the establishment of various manufactures, having numerous rivulets and inexhaustible stores of stone, lime, coal, and iron. The want of means of outlet have hitherto rendered these resources unavailable; but as the railways now constructed supply facilities of transit, this part of the kingdom may, in the course of a few years, be expected to afford an excellent field for the advantageous employment of capital and industry.

Durham has sent two members to the H. of C. since 1675. Previously to the Reform Act, the right of voting was in the freemen who had acquired their freedom by patrimony or servitude within the city of Durham, the bor. of Framwellgate, 'or the streets and suburbs adjoining thereto.' Hence the limits of the parl. bor. were not definitely fixed, but varied with the varying size of the suburbs: a definite limit was, however, given to it by the Boundary Act. Registered electors, 1,161 in 1865, including 560 freemen. Gross annual value of real property assessed to income tax, 60,977l. in 1857, and 85,975l. in 1862. Under the Municipal Reform Act, the city is divided into three wards, and is governed by a mayor, 6 aldermen, and 18 councillors.

DURLACH, a town of the grand duchy of Baden, circ. Central Rhine, 3 m. ESE. Carlsruhe, with which it communicates by a road lined throughout with poplars. Pop. 5,017 in 1861. Before Carlsruhe was built this town was the residence of the margraves of Baden-Durlach, an old castle belonging to whom, now in ruins, stands upon a contiguous height; and has attached to it a fine garden, in which several Roman antiquities have been discovered. A palace of a later date has been in part converted into a cavalry barrack. The inhabitants manufacture porcelain, tobacco, and sealing-wax; and have some trade in corn, wine, fruit, and other agricultural produce.

DUSSELDORF, a town of Prussia, prov. Rhine, cap. reg. and circ. same name, at the confluence of the Dussel with the Rhine, on the main line of railway from Cologne to Berlin. Pop. 41,292 in 1861, exclusive of a garrison of 3,276. Dusseldorf has quadrupled its pop. in the course of the present century, at the commencement of which, as the name implies, it was but a village. It is a well-built, handsome, thriving town. It stands so close upon the Rhine, that vessels sail up to its quays, and there take on board the merchandise of which the town is the depôt. Near the river the streets are narrow, and full of symptoms of industry; but beyond these the town consists of handsome white stone houses, disposed in rows as streets, or as open squares and places with trees in the centre, all which are remarkably clean and quiet. The castle and other fortifications were destroyed by the French in 1791. The town is the seat of the provincial states or part of the Rhine prov., has a court of appeal for the regency, a gymnasium or college, an academy of sciences, an observatory, a fine public library, a theatre, and some remains of the noble collection of paintings transferred to Munich. Recently the school of painting at Dusseldorf, under Schadow, has attained to very considerable celebrity. There are considerable manufactures at Dusseldorf; but it derives its principal importance from its position on the Rhine, nearly opposite to where it is joined by the canal leading to Venlo on the Maese, and from its being the entrepôt and principal port of the contiguous flourishing manufacturing district, of which Elberfeldt is the capital. Cottons, cloths, &c. are imported from the latter; hardware, iron, and steel, from Solingen and Remscheid; and linen from Hattingen. Large quantities of coal brought from the mines on the Ruer, are shipped here for the Netherlands; and there is also an extensive trade in corn, oil, and wine.

DWARACA, or JUGGUTH, a marit. town of Hindostan, prov. Gujerat, the most W. point of which it occupies, dom. of the Guicowar, on a sandy shore 95 m. NW. Joonaghur; lat. 22° 16′ N. long. 69° 7′ E. It is the most sacred place in this part of India, and is annually frequented by about 15,000 pilgrims from all parts of that extensive country. Its principal pagoda is a magnificent carved stone building of high antiquity, dedicated to Runchon, an incarnation of Krishna, with an entrance towards the sea by a very long and noble flight of stone steps, succeeded by a massive gate, where the whole front breaks upon the view with a striking effect; its great pyramid is 140 ft. high, and much ornamented. There are numerous subordinate temples, having flags with representations of the sun and moon. In front of the large temple is the sacred place of ablution, formed by a creek of the sea, which is lined for some distance by small temples with stone steps down to the margin of the water, on which prayers are made, and idols, rings, and amulets sold by the Brahmins; the town itself is small, but surrounded with walls and towers washed by the tide. The devotees here are usually stamped by means of a hot iron, with the insignia of the god, and this rite is often practised upon young infants. The chalk with which the Brahmins mark their foreheads comes from Dwaraca, whence it is carried by merchants all over India. The revenue of the temples, derived from pilgrims, is estimated at about one lac of rupees, and was formerly swelled by the plunder of many piratical vessels, fitted out in the name of the idol. Dwaraca submitted to the British forces in 1816; but in the following year was transferred to the Guicowar, to whom its sanctity rendered it a highly acceptable acquisition.

DWINA, the name of two Russian rivers, one of which falls into the White Sea by several mouths, 35 m. below Archangel, and the other into the Gulf of Riga in the Baltic, 9 m. below Riga. The first, or Northern Dwina, is a large and important river. It is formed by the junction of the Soukhona, which rises in the farthest W. part of the government of Vologda, with the Joug rising in the central S. part of the same government. From the point of confluence, near Oustie-

oug-Veliki, the united river flows in a deep and broad stream, NW. to its embouchure below Archangel, a distance of about 350 m. Its principal affluent is the Vitchegda, flowing W. from the confines of Perm. The extent of natural navigation for boats and barges on this river and its affluents is very great, extending W. to the city of Vologda, S. to Nikolesk, and E. to the frontier of Perm. At Vologda an artificial navigation begins, which, by means of the Lobinski canal and the lake Beelo, connects the Soukhona with the Neva; while, on the E., the Severnoi canal connects the Vitchegda with the Kama, one of the principal affluents of the Wolga. Hence, goods imported at Archangel may be sent by water to either Petersburg or Astrakhan, and conversely. The ebb and flow of the sea is perceivable in the Dwina many m. above Archangel. Opposite to the latter it is above 4 m. in width; it is also very deep, though, owing to the sand-banks at its mouth, it does not admit vessels drawing more than from 12 to 14 ft. water. It is frozen over for about half the year. (See Archangel.)

The second, or Southern Dwina or Duna, though of inferior dimensions to the preceding, is also a large and important river. It rises in the Valdai hills, not far from the source of the Wolga; and following a NW. course to Vitebsk, it thence pursues a WNW. course to its embouchure below Riga. It is navigable from near its source, or for about 625 m. Near Dunaburg, however, it is a good deal interrupted by cataracts, and in other places it is encumbered with shoals, so that it can only be navigated with safety after the breaking up of the ice in the spring, and after the setting in of the autumnal rains. It has few affluents of any considerable magnitude. At Riga it is about 2,600 ft. broad. Its mouth is encumbered with banks, which render it inaccessible to vessels drawing more than from 12 to 15ft. water. It begins to freeze over about the end of Nov., and the breaking up of the ice, or débacle, usually takes place in the beginning of April, when there are inundations that frequently occasion great injury to Riga and the adjacent country.

This river has always been the principal channel by which the masts and other timber exported from Riga were conveyed to it. But owing to the gradual exhaustion of the forests, it is necessary to go much farther S. than formerly, to the provs. of Tchernigoff and Kieff, the timber from which is conveyed by water, against the stream, up a part of the Dnieper, and then carried across the country separating that river from the Dwina, to be embarked on the latter. This, however, is a very expensive and tedious process, requiring about two years for its completion; and hence the mast trade, that formerly centered wholly at Riga, is now beginning to be transferred, in part, at least, to Kherson, to which place the trees are, at the proper season, easily and rapidly floated down the Dnieper. (Hagemeister on the Black Sea, p. 172, English trans.)

DYSART, a royal bor. and sea-port of Scotland, co. Fife, on the N. coast of the Frith of Forth, 11½ m. N. by E. Edinburgh, and 1 m. E. Kirkaldy, on the Edinburgh-Perth railway. Pop. 7,857 in 1841, and 8,166 in 1861. Dysart consists chiefly of three narrow streets, with a square in the centre. The central or High Street is full of antique substantial buildings, the fronts of which are generally decorated with inscriptions and dates, and, in one part, with piazzas, the latter being the places in which, in former times, merchants exposed their goods to sale; but the greater part have been built up. In the middle of the town stands the town-house, erected in 1617, but rebuilt, after having been accidentally burnt by Cromwell's soldiers. Under its roofs are, the council-chamber, the prison, the public weigh-house, the guard-room, the black hole. Dysart House, the residence of the Earl of Rosslyn, stands on the W. of the town, being separated from it only by a wall. The par. church is a plain building; also the two dissenting chapels, which belong respectively to the Relief and Associate Synod. Dysart is a collegiate charge, or has the services of two parochial clergymen.

There are 14 schools in the par., all unendowed except three. There are four subscription libraries in the par., two reading rooms, several friendly societies, and a savings' bank.

Before the union between England and Scotland, Dysart was a place of such commercial eminence as to have been called 'Little Holland.' But its importance in this respect is now greatly reduced. Salt-making flourished in the bar, and neighbourhood, particularly at Gallaton, for a hundred years previously to the end of last century. But that trade has now entirely disappeared. It was in reference to Gallaton that Adam Smith remarked, in his 'Wealth of Nations,' published in 1776, 'There is at this day a village in Scotland where it is not uncommon, I am told, for a workman to carry nails, instead of money, to the baker's shop or the ale-house.' The manufacture of linen cloth, once extensively carried on here, has also disappeared. Salt was made here from sea-water at so early a period as 1450; and the trade continued to flourish till 1823, when the duty being repealed, it was relinquished. The principal trade at present is the manufacture of checks and ticks, a branch of the Dundee staple trade. This business was introduced into Dysart between 1710 and 1720. The number of looms employed in the manufacture of this fabric is not less than 2,000;

and the value of the cloth annually produced is estimated at about 150,000l. The number of hands employed by the manufacturers out of the par. is above 1,000.

Dysart coal was among the first wrought in Scotland, operations having been begun upwards of 350 years ago. Upwards of 100,000 tons are dug yearly. Sandstone, limestone, and ironstone, also abound, and are in considerable demand, particularly the two latter. The harbour is one of the safest on the Frith of Forth, except with easterly winds. It has a wet dock. There are only, however, a few brigs and sloops belonging to the port, and no foreign vessel approaches it, except occasionally from Holland or the Baltic, laden with flax, or when coals are wanted. The bor. is governed by a provost and 6 councillors; corp. revenue, 1,882l. in 1863-4. Annual value of real property, 13,113l. in 1863-4. Dysart unites with Kirkaldy, Burntisland, and Kinghorn in returning 1 member to the H. of C., and in 1865 had 187 registered voters.

Dysart is a place of great antiquity. It is mentioned in history so early as 874, when the Danes invaded Fife. But it was not made a royal burgh till the time of James V. The town was taken by Cromwell. There is a place at the harbour called the Fort, said to have been fortified by the Protector, but no remains of any work on it can now be seen. To the W. of the burgh is the castle of Ravenscraig, standing on a steep crag fronting the sea, but now a ruin. It has been the property of the Sinclairs, now earls of Rosslyn, for 500 years. On the N. or lower part of the town, there are the remains of a chapel, said to have been dedicated to St. Dennis. The ruins of the old church of Dysart are nearly at the same spot. One of the windows bears the date of 1570.

E

EAGLESHAM, a market-town and burgh of barony, Scotland, co. Renfrew, on a tributary of the White Cart, 9 m. S. Glasgow. Pop. 1,769 in 1861. The town, which is modern, though on the site of an ancient village, consists of two rows of well-built houses, all of free-stone, with a space between varying from 100 to 250 yards, laid out in fine green fields interspersed with trees, with a beautiful streamlet running down the middle. Length of the town nearly 3 furlongs. The cotton manufacture was introduced here about the year 1826. Besides cotton-spinning, there are about 400 persons in the town engaged in weaving. The noble family of Eglinton are the feudal superiors of the place, and appoint the baron bailie.

EARLSTON (formerly Ercildoun), a village in the par. of the same name, Scotland, co. Berwick, 30½ m. SE. Edinburgh, and situated in the middle of a pastoral district, within ½ m. of the Leader, a tributary of the Tweed. Pop. 700 in 1861. The village is straggling and irregularly built; but is well known in manufactures. 'Earlston ginghams' being familiar to most persons in the N. of Scotland. 'Thomas the Rhymer,' whose proper name was Thomas Learmont, who flourished in the 13th century, and is famous both as a poet and an alleged prophet, belonged to this place. An account of this celebrated person is given in Sir W. Scott's edition of 'Sir Tristram,' a poem ascribed to the Rhymer. The walls of the castle,

called 'Rhymer's Tower,' in which he lived, are still standing within ½ m. of Earlston.

EBORA, or EVORA, a city of Portugal, cap. prov. Alentejo, 85 m. E. Lisbon, 42 m. SW. Elvas, on the railway from Lisbon to Badajoz and Madrid. Pop. 17,121 in 1858. The city is built on an eminence, in the centre of a fertile plain, and is remarkable from the appearance of its ancient towers, as well as striking from its elevation. It is surrounded by ramparts, and has two forts in ruins. Streets narrow, crooked, and filthy; but it has some good houses. It is the see of an archbishop, and has a magnificent Gothic cathedral, with an altar in the Italian style, decorated with various marbles. Exclusive of the cathedral, there are four churches, several convents and hospitals, a house of charity, and fine barracks. There is a good collection of books in the bishop's library, and the museum is one of the finest in Portugal. It was formerly the seat of a university, suppressed on the expulsion of the Jesuits.

The city was for a lengthened period the headquarters of the famous Roman general Quintus Sertorius, by whom it was fortified, and adorned with several fine public buildings. An ancient temple, supposed to have been dedicated to Diana, (though much dilapidated), has still in front of some noble columns, evidently raised during the best period of Roman architecture, but this fine ruin has been greatly neglected. There is also a magnificent aqueduct, said to have been built by Sertorius, in fine preservation, and still applied to its original purpose. The city has manufactures of hardware, tanneries, and a fair for cattle on St. John's day, which is much frequented. Julius Cæsar made it a municipal town, and gave it the name of *Libreolitan Julia*. The Moors took it in 715. It has been the residence of many of the Portuguese sovereigns.

EBRO (an. *Iberus*), one of the principal rivers of Spain, through the NE. part of which is flows, uniformly almost in a SE. direction, being the only great Peninsular river that has its embouchure in the Mediterranean. It rises at Fontibre, prov. Santander, on the S. declivity of the Sierra Sejos, about lat. 43° N., and long. 4° W., near the sources of the Pisuerga, an affluent of the Douro. It afterwards separates the provs. Santander, Biscay, and Navarre from Old Castile, intersects Aragon in its centre, and disembogues near the S. extremity of Catalonia, about lat. 40° 40' N., and long. 0° 55' E. Its entire length is estimated at somewhat above 400 m.; its principal tributaries are, the Nela, Aragon, Gallego, and Segre, with the Cinca on the N., and the Oca, Tiron, Nagrillo, Xilon, Guadaloupe, &c., on the S. side. Reynosa, Miranda, Logrono, Tudela, Saragossa, Mequinenza, and Tortosa, are the chief cities and towns upon its banks. It runs mostly through a succession of narrow valleys till it reaches Mequinenza; after which it enters Catalonia, and flows through a more level country. At Amposta, 12 m. W. from its mouth, it is about 1800 yards wide. It immediately afterwards forms a kind of delta; a navigable canal having been cut from the port of Alfaquez, or San Carlos, at its N. mouth, to Amposta. The Ebro is navigable for boats as high as Tudela, but its current is very rapid, and its bed in many parts encumbered with rocks and shoals. To avoid these obstacles, and the numerous windings of the river, the Aragon canal has been cut along its right bank from near Tudela to Sastago. An ancient Moorish canal, now dry, formerly connected the town of Alcanez on the Guadaloupe, with the Ebro. The principal commercial utility of the Ebro is the transport of grain from Saragossa to Tortosa, together with the floating down of timber from the Pyrenees. This river, before the second Punic war, formed the boundary of the Roman and Carthaginian territories, and, in the time of Charlemagne, between the Moorish and Christian dominions.

ECBATANA. See Hamadan.

ECIJA (an. *Astigi*), a city of Spain, prov. Seville, finely situated on the banks of the Xenil, which is here crossed by a fine ancient bridge. 47 m. ENE. Seville, 31 m. NW. Cordova, on the railway from Seville to Cordova. Pop. 28,769 in 1857. The city is surrounded by walls, and has narrow crooked streets. Its churches, of which there are six, are built of brick, fitted up in the old taste, and crowded with pillars, loaded with impertinate ornaments, and covered with gold. The most notable is the church of *Nostra Senora del Rosario*, in the convent of the Dominicans. Exclusive of churches there are twenty convents, four hospitals, a foundling hospital, and a public granary. The Plaza Mayor, a fine spacious square, has a double row of balconies the whole way round. Along the river's side is a handsome alameda, or public promenade, planted with elms and other ornamental trees, provided with seats, and decorated with statues. It has manufactures of coarse cloth, serges, camlets, friezes, and linens; and the vicinity produces wheat, wine, and oil. Ecija is a very ancient city, having

been called by the Romans *Astigi* and *Augusta Firma*, (Plin. Hist. Nat., iii. s. 1.) It was for a lengthened period a border town between the Moors and Christians, and is famed in many a romance; but it is no longer of any importance as a fortress, and its walls are covered with brambles.

ECKMUHL, an inconsiderable village of Bavaria, circ. Regen, on the great Laber, 13 m. S. by E. Ratisbon. Pop. 110 in 1861. Here, on the 22nd April, 1809, the great French army, under Napoleon, gained a decisive victory over the Austrians, under the Archduke Charles. Marshal Davoust having particularly distinguished himself on this occasion, was raised by Napoleon to the dignity of Prince of Eckmuhl. The battle of the 22nd was preceded by partial actions on the 19th, 20th, and 21st, all of which terminated favourably for the French.

ECLOO, a town of Belgium, prov. E. Flanders, cap. arrond. on the road between Ghent and Bruges, 16 m. E. the former city. Pop. 8,790 in 1846. It is generally well built, and has several squares and well paved streets. It has 2 churches, a town-hall, an ancient convent, 8 schools, and a prison. Its manufactures are chiefly of coatings and other woollen stuffs, cottons, starch, soap, and chocolate; it has also breweries, distilleries, salt refineries, and various mills. Its trade, which is very active, especially at its weekly markets, which are the largest in the prov., is mostly in corn, linens, timber, and cattle.

EDDYSTONE LIGHTHOUSE. This, which is one of the most remarkable structures of its kind, is built on one of the points of a reef or ridge of rocks, from 600 to 700 ft. in length, in the English Channel, about 9 m. S. by W. from the Rambead, and 14 m. from Plymouth; lat. 50° 10' 56" N., long. 4° 15' 8" W. The Eddystone rocks are covered at high water; and being much exposed to heavy swells from the Bay of Biscay and the Atlantic, the waves frequently break over them with tremendous fury. In consequence of the many fatal accidents occasioned by ships running against these rocks, a lighthouse was erected on one of them in 1696; after standing many storms, it was overthrown in the dreadful tempest of the 27th Nov., 1703. A second lighthouse, erected in 1708, was burnt down in 1755. The present edifice, built by the celebrated engineer Smeaton, and finished in 1759, is universally admired for its solidity and the skill displayed in its construction, and bids fair to last for ages. The total height of the lighthouse is 100 ft.: the lantern being elevated 72 ft. above the sea at high water. The light is fixed, and is of the first magnitude. This lighthouse has served as a model for that on the Bell Rock, and others of the same kind.

EDEN, a river in the NW. of England, which has its sources on the borders of Westmoreland and Yorkshire, near Pendragon Castle, close to the sources of the Swale, in one of the highest parts of the Pennine or central range of mountains. It pursues a NW. course through the valley between the Pennine and Cumbrian mountains past Carlisle, 7 m. below which it falls into the Solway Frith. It is navigable to Carlisle; but the navigation being tedious and difficult, a canal has been cut from Carlisle to Bowness, lower down the Frith, a distance of 11½ m., which admits vessels of from 60 to 80 tons burden.

EDER, a town of Hindostan, prov. Gujerat, cap. of a principality of the same name; 17 m. N. by W. Ahmednuggur, and 117 m. SW. Odeypoor; lat. 23° 58' N., long. 73° 8' E. Estimated pop. 12,000. It is but a poor town, though built within the walls of a magnificent fortress constructed by the Mohammedan kings of Gujerat.

EDFOU (the *Apollinis urbis* of Strabo, and *Apollinopolis Magna* of the Romans), a town, or more correctly a large assemblage, of mud huts, congregated around and amidst the superb ruins of an ancient temple on the W. bank of the Nile, in Upper Egypt, about 2 m. from the river, and 62 m. S. by E. Thebes. Lat. 24° 58' 43" N., long. 32° 51' E. Pop. from 1,500 to 2,000, consisting principally of Ababdeh Arabs, with a few Coptic families, who manufacture blue cotton, cloth and pottery, and boast of inheriting from their ancestors the art of making earthen vessels; and it must be admitted that their kilns and the forms of their vases exactly resemble those of ancient Egypt, as represented on the monuments. Dr. Richardson says that the inhabitants are 'civil and dirty,' and the place would be unworthy notice were it not for its antiquities; but two noble temples, placed opposite to each other, though half buried in the sand, and an ancient quay, still remain to evince the former grandeur of *Apollinopolis Magna*. The great temple, on a small eminence, commands a view of all the surrounding country, and is therefore called, in Arabic, *Qala*, or 'the citadel.' Its propylon, or entrance, consists of a doorway, 17 ft. 4 in. wide, between two vast truncated rectangular pyramids or moles. The base of each of these pyramids is 104 ft. by 87 ft.; their height is 114 ft.; and the horizontal section of each at the top 64 ft. by 20. The door is surmounted by the often repeated sculpture of the globe with the serpent and wings, and three rows of immense figures are sculptured on the sides of the pyramids. These gigantic structures are not solid, but have chambers, to which and to the top access is practised by means of staircases. Within the doorway is an open rectangular court, now filled with mud and rubbish, 161 ft. by 140 ft., enclosed by high walls, which also renders the temple itself, and are 414 ft. on each of the longer sides, and 154½ ft. on the shorter. Notwithstanding these vast dimensions, the walls are elaborately covered with hieroglyphics. On each side of the longer walls in the court there is a row of pillars, so disposed that a space intervenes between them and the walls, which being roofed form two covered ways, leading from the propylon to the portico or pronaos of the temple. The columns, of which there are 52, present a most magnificent perspective. There is a gradual ascent in the court to the portico, the outside of which is adorned with six columns, having various capitals; and within are several apartments and corridors, supported by columns, and ornamented with sculptures. The adytum, or sanctuary, is an oblong apartment, about 33 ft. by 17 ft. The terraced roofs of the temple, from the pronaos to the extremity, are covered with mud huts, and the sanctuary and adjoining chambers are now either used as repositories for grain or other produce, or are half filled with sand, and with filth and rubbish, shut down by the Arabs through the apertures that formerly lighted the chambers. (Egyptian Antiquities, 'Library of Entertaining Knowledge;' Modern Traveller, vi. 176, &c.)

The plan and arrangement of this temple is simple and symmetrical. Its largest columns are 6 ft. 4 in. in diameter, 33 ft. in circ., and 42 ditto in height; the capitals are 87 ft. in circ. The palmiform capital, peculiar to Egyptian architecture, is here seen to great advantage. It represents the trunk of a palm, of which the spreading foliage forms a graceful stem. Art has here copied Nature with great fidelity; it has preserved the same number of leaves, the exact form of

the fruit, and the scales of the trunk, and the capitals gradually augment in size till they balance the leaves. The 82 capitals of the peristyle, and the 6 of the pronaos, exhibit in alternate columns the dactyliform and lotiform figure, which last is as faithfully borrowed from Nature as the palmiform. This is the account given by Jomard in the 'Description of Edfou' (p. 20), and copied by Hitter, who praises the simplicity and pure antique style of the temple. But a French authority, M. Champollion, is of a wholly different opinion:—' Ce monument,' says he, 'imposant par sa masse, porte cependant l'empreinte de la decadence de l'art égyptien sous les Ptolemées, au regne desquelles il appartient tout entier. Ce n'est plus la simplicité antique; on y remarque une recherche et une profusion d'ornements hieroglyphiques, et qui marquent la transition entre la noble gravité des monuments pharaoniques et le papillotage fatiguant, et de si mauvais goût, du Temple d'Esneh, construit au temps des empereurs.' (Lettre, p. 191.) This, however, is probably too unfavourable an opinion.

Notwithstanding its truly colossal character this temple is not, as was long supposed, of the Pharaonic era, but is comparatively modern, being, as now stated, the work of the Ptolemies. This is proved by the date of its decorations, the most ancient of which, according to Champollion, belong to the age of Ptolemy Philopater. It would appear from the same authority, that the meaning of the sculptures, and the object of the temple, had been completely misunderstood by Mr. Hamilton, M. Jomard, &c. who supposed that it was sacred to Osiris, the beneficent deity. M. Champollion affirms that this magnificent edifice was consecrated to a triad consisting of Isit, the god Harbat, the personification of heavenly science and light; 3dly, the goddess Hathor, the Egyptian Venus; and 4ndly, their son Harsunt-Tho, the Eros of the Greeks and Romans. (Lettres d'Egypte, p. 192.)

The other and much smaller temple at Edfou is peripteral, and was supposed to be devoted to the worship of the malignant deity, Typhon, whose image was believed to be represented above the capitals of the columns, and elsewhere on the walls. But Champollion has shown that this temple is really one of those *mammisi* that were always erected near the grand temples devoted to the worship of a triad, and that it represents the birth-place of the third person of the triad, or of Harsunt-Tho, son of Harbat and Hathor. The bas-reliefs on this temple are of the age of Ptolemy Energetes II. and Soter II. (Lettres, 192.)

Between Edfou and El Cab, one of those transverse valleys which frequently divide the mountain ranges of the E. desert, opens to the E., and is called the valley of Edfou. It extends from the Nile to the Red Sea, near Berenice; and upon it have been traced the tracks of a great commercial road, over which the traffic of the ancient sea-port of Berenice, and the produce of the celebrated emerald mountains were conveyed. It was also, formerly, a much frequented caravan route.

EDINBURGH, or MID-LOTHIAN. See LOTHIAN.

EDINBURGH, a celebrated city, the metropolis of Scotland, co. Mid-Lothian, 2 m. S. from the Frith of Forth, built principally on three parallel ridges, running E. and W., and separated by deep depressions; 337 m. NNW. London by road, and 399 m. by Great Northern railway. The central ridge of the ground on which the city stands, is terminated on the W. by a rock, nearly 400 ft. above the level of the sea, surmounted by the castle, and on the E. by the palace of Holyrood.

100 ft. above the same level. The circumference of the city, exclusive of Leith, its sea-port, lying between it and the Forth, is rather less than 8 m. When compared within its ancient limits, the pop. of Edinburgh was extremely dense. It is said by Maitland (Hist. of Edin., p. 7), referring to the year 1600, to have been 'so full of inhabitants that probably there is no town elsewhere of its dimensions so populous.' At the Union, in 1707, the pop. was estimated at 30,000; in 1755, before the New Town was commenced, and when the southern districts did not exceed a fifth part of their present extent, the pop. was estimated at 50,000; in 1775, soon after the commencement of the New Town, the pop. was 60,000; in 1791, about 71,000; but these enumerations exclude Leith, the pop. of which in 1755 (ibid. p. 600) was 7,290. The pop. of Edinburgh, city and suburbs, exclusive of Leith, according to the censuses since 1801, has been as follows: viz. 1801, 66,544; 1811, 81,787; 1821, 112,235; 1831, 136,301, and 1841, 138,182. By the census of 1865, Edinburgh had 168,121 inhabitants living in 9,760 houses. In Edinburgh a house often accommodates several families, each story (provincially flat) constituting, in such cases, a separate dwelling, to which access is obtained by means of a common stair. Nay, a story is sometimes subdivided into two or more separate residences, each being accessible by its own door opening to the same common stair. In the Old Town common stairs are all but universal. They are general also in the southern districts; but more rare in the New Town, separate or 'self-contained houses,' as they are termed, generally prevailing in this fashionable and wealthy quarter of the city. The loftiest houses are in Mound Place, in the Old Town; they extend to 11 stories, including the attics; and as each story is generally divided into two lodgings, each house is supposed to contain, as an average, about 30 families, or 100 individuals. With the exception of the older buildings, which range from five to six stories in different districts, the usual height is three stories, exclusive of the attics and the basement floor, which latter is generally half sunk under the level of the street. This is the case, with very unimportant exceptions, throughout the New Town. The word land is used in Edinburgh to signify a house or tenement from top to bottom, whether it be occupied by one family or several. Previously to the houses being numbered, they were distinguished by such names as Todrig's land, Menzie's land, Gavenlock's land, &c. Similar remarks apply to Leith.

The situation of this city is eminently romantic. It stands, as previously stated, on three separate ridges, of which that in the middle, having the castle at its W. extremity, is at once the most striking and the best defined. The castle is peculiarly picturesque. The rock on which it is built is on three sides, N., W., and S. high, steep, and in parts almost perpendicular. On the E. side the ground declines in a sloping ridge to Holyrood Palace; and on it,

'Piled deep and massy, close and high,'

stands the greater part of the Old Town. The neighbourhood is also marked by lofty hills, except towards the N., where the ground gently declines to the Frith of Forth. The Calton Hill, 347 ft. above the sea, on the E. side of the city, now surrounded with fine terraces of houses, affords the remarkable spectacle of a verdant hill, except where covered with monuments, within the precincts of a large town. Arthur's Seat (822 ft. above the level of the sea) and Salisbury Crags (547 ft.), the latter divided from the former by a deep and gloomy ravine, lie on the SE. of the city. Each of these hills rises abruptly from its base, and commands varied and very extensive views. Blackford Hill, the Braid Hills, the Pentland Hills, and Corstorphine Hill, rise at different distances on the S. and W. These eminences form a magnificent amphitheatre, within which, on elevated but lower ground, the Scottish capital is situated.

The ridge on which the Old Town is built was not inaptly compared by Arnot to a turtle, of which, says he, 'the castle is the head, the High Street the ridge of the back, the wynds or closes the shelving sides, and the palace of Holyroodhouse the tail.' (Hist. of Edin., 4th ed. p. 179.) It is separated from the New Town on the N. by a deep valley, which for centuries formed a lake, called (as it is still) the North Loch; but having been drained in 1763. It is now laid out in gardens, and is traversed by the Edinburgh and Glasgow railway. On the S. the Old Town is divided from the southern districts by a similar valley, the site of the Cowgate, now a narrow and mean, though once a fashionable street. From the High Street, on the summits of the ridge, descend, on both sides, in regular rows, numerous narrow lanes, which are mostly steep and difficult of passage, being rarely more than 6 ft. in width, and in general very dirty. Those of the greatest width, or which admit of a cart or carriage, are termed wynds, as Blackfriars' Wynd, St. Mary's Wynd, &c., while those which admit foot passengers only are called closes. A few have no thoroughfare, being in the form of culs de sac.

The High Street, which (including the Castle Hill, Lawn Market, and Canongate) stretches in nearly a straight line from the castle to the palace, a distance, as already stated, of more than 1 m., is a magnificent street; the houses, which vary from five to six or seven stories in height, have been mostly rebuilt; but a few, especially on the Castle Hill, are of great antiquity. One of these lofty buildings fell in 1861 from sheer age. This street, with its sheltering lanes and appendages, constitutes the whole of what is properly the 'old town.' It is connected with the southern districts by the Cowgate, and by two bridges which stretch over the valley in which that street is built, viz. the South Bridge, opened in 1788, and George the Fourth's Bridge, opened in 1836. On the other hand, the Old and New Towns are connected by the North Bridge, which spans the North Loch, and forms a continuation of the line of the South Bridge, and by the 'Earthen Mound.' The North Bridge, which consists of three central arches, with several smaller ones at each extremity, was opened in 1768; while the Mound, which was begun in 1784 from the accumulation of the rubbish from the excavations of the New Town, was formed into a thoroughfare about the beginning of the present century, but it has since received great additions. It is supposed to contain 800,500 cubic yards, or about 1,500,000 cartloads of earth. W. of the Cowgate lies the Grass Market, a wide, open street, used as a market-place for the sale of horses, sheep, and cows.

The New Town, which, as well as the more modern parts of the southern districts, is built of light coloured freestone, procured in abundance in the immediate vicinity of the city, stands on an eminence, which slopes to the water of Leith, the small river at the mouth of which Leith is built. The leading streets run in straight lines from E. to W., and are crossed at the distance of about every 250 yards, by streets running in an opposite direction; so that great regularity, elegance, and beauty characterise this quarter of the city. George's Street, which stretches along the top of the ridge, is terminated on the E. by St. Andrew's Square, and on the W. by Charlotte Square. Great King Street, which lies considerably down the declivity, and nearer the Water of Leith, has, in like manner, the Royal Circus on the W., and Drummond Place on the E. There are, also, James's Square (the oldest in the New Town), and Rutland Square. Another portion, built between 1822 and 1825, has Moray Place in the centre, and Randolph Crescent on the W. This is the most elegant and fashionable part of the city. The feus, or building leases, in this quarter fetch from 4s. to 6s. annually per foot of frontage. The New Town is terminated by the steep banks of the Water of Leith, and is connected with the grounds N. of that stream by the Dean Bridge, an elegant structure, consisting of 4 arches, each 96 ft. span, the height of the road-way above the bed of the river being 106 ft. One of the most celebrated streets in the New Town is Princes Street, forming a species of terrace, and facing the Old Town, of which it commands a fine view, which, especially by moonlight, is probably unequalled. Waterloo Bridge connects this street with the Calton Hill, being thrown over a deep ravine occupied with ancient but shabby buildings, called the Low Calton. The line of road, to which this bridge leads along the E. side of the Calton Hill, forms a grand approach to the city in this direction. The Queen Street Gardens, a piece of ground which extends from E. to W., about ¼ of a mile, by about 200 yards in width, may be regarded as bisecting the New Town. Elegant streets have, at different periods, been built W. of Princes Street and Charlotte Square, of which the most important are Atholl and Coates's Crescents.

The situation of the southern districts is considerably more elevated than that of the New Town; but the buildings are of an inferior order, nor has much regularity been observed in the laying out of the streets. The houses are high, mostly four stories, and common stairs prevail, with partial exceptions, particularly in George's Square: this, which is the handsomest place in this quarter of the town, was built in the last century, and is of large dimensions. It has on the W. the public walk leading to the Meadows; and on the N. it is separated from them by Buccleuch Place. The principal line of buildings is Nicolson Street, which stretches from the South Bridge, already mentioned, to the country on the S., and now forms the main approach to the city in this direction. The former approaches on this side were parallel to Nicolson Street, being an old street, called the Pleasance, on the E., and the Causeway Side on the W. While the Meadows bound the southern districts on the W., a valley or ravine, fronting Salisbury Crags, forms their termination on the E. Not a few of the public buildings, including the university, are in this district.

The original royalty, or 'borough roods,' embraced only the Old Town, excluding even the Canongate, which intervenes between it and the Palace. But the 'extended royalty,' as it is called, obtained from Parliament in 1767, while it excludes the Canongate, embraces the whole of the New Town, with the exception of a few streets which have stretched beyond its limits. The suburbs of Edinburgh may be briefly enumerated: the Canongate, including the Calton, a contiguous hamlet at the base of the hill of that name, the superiority of which is vested in the city of Edinburgh. The town council of the city has a veto on the election of two resident bailies for the Canongate: that body, besides, appoints one of its own members as baron-bailie. Wester and Easter Portsburgh, the former lying W. of the Grass Market, and the latter, now called the Potter Row, NE. These two places, which are of considerable antiquity, and which took their names from ports or gateways in the Old Town Wall, are also subject to the city of Edinburgh, being governed in a way similar to the Canongate. Leith was formerly in the same predicament; but it has of late years been rendered entirely free and independent. Broughton, a burgh of regality under the same jurisdiction, and lying on the site of the streets in the New Town, which now bears its name, has been nearly obliterated, and will soon entirely disappear. Its separate jurisdiction was destroyed when the act for extending the royalty was obtained.

Edinburgh was first walled in 1450. But the wall was confined to the town as it then existed; that is, it did not embrace the Canongate, nor did it extend so far N. as the site now occupied by the Cowgate. But after the battle of Flodden, in 1513, a new wall was built, comprising not merely the Cowgate, but the acclivity S. of that street, and running parallel to it throughout its whole length. Some remains of this wall, which enclosed the ground now occupied by the workhouse, the university, infirmary, Old High School, &c. are yet standing. A number of ports, or gates, gave access to the city in different directions, the last of which was removed in 1765. The Netherbow port, between the High Street and the Canongate, removed in 1764, was ornamented with a spire.

Public Buildings.—If these the castle deserves the first notice. The date of its foundation is unknown. It was originally called *Castrum Puellarum*, because the daughters of the Pictish kings were educated and kept in it till their marriage—a necessary precaution in those barbarous times. Queen Margaret, widow of Malcolm Canmore, died in this fortress in 1093. James VI. of Scotland, and afterwards I. of England, was born here in 1566. The fortress, which corresponds with some of the rules of art, being built according to the irregular form of the precipice on which it stands, is anything but impregnable. It has been repeatedly taken and retaken by contending parties, and was often in the hands of the English. It is, in short, of little or no strength, and is interesting only from its romantic situation on the top of a rugged basaltic rock, perpendicular on all sides except on that next the Old Town, the splendid view which it commands, and the many historical associations connected with it. It was originally used as a royal residence. In an apartment called the crown room were deposited the Scottish regalia at the Union in 1707; these relics, which consisted of the crown, sceptre, sword of state, and the lord treasurer's rod of office, were long supposed to have been removed or lost, but they were discovered, in 1818, in a large oaken chest in the crown room, by royal commissioners appointed to conduct the search. They are now open to the gratuitous inspection of the public. Queen Margaret's chapel, in the Norman style of the 11th century, was restored in 1859.

The Palace of Holyrood, which stands at the E. extremity of the city, next claims attention. It is a fine castellated edifice, of a quadrangular form, with an open area in the centre, 94 ft. square. The most ancient parts of the present palace were built by James V. in 1528. It was partially burnt by the English during the minority of Queen Mary, and again by the soldiers of Oliver Cromwell; but after the Restoration it was repaired and altered by Charles II., and underwent again considerable repairs in 1850. The mean and unsightly buildings by which it is hemmed in on the S. and on the side next the city, should be removed. The Pretender took up his residence here in 1745.

George IV., on his visit to Scotland in 1822, though he resided at Dalkeith Palace, held levees and drawing-rooms in this ancient abode of his ancestors. Meetings of privy council were also held here. The Count d'Artois, afterwards Charles X., of France, and other royal and noble French refugees, obtained a refuge here in 1793; and in 1831 the same apartments served a second time as an asylum for nearly the same individuals. It has a peculiar interest, from the circumstance of the apartments occupied by Queen Mary having been carefully preserved in the state in which she left them. Her bed is an object of interest to strangers; and many relics of her Majesty's needlework exist in the rooms. The spot where Darnley and his accomplices murdered her favourite, David Rizzio, and other interesting localities, are carefully marked. The closet in which Mary was at supper with the Countess of Argyle, Rizzio, and others, when this tragical scene was acted, is only 12 ft. square. In what is called the picture gallery, a hall 150 ft. in length, and 27 ft. in breadth, are hung the portraits (most of them fanciful) of 111 Scottish monarchs, painted towards the end of the 17th century by De Witt, an artist of the Flemish school, by order of James II. of England, when Duke of York. In this hall the election of the sixteen Scottish representative peers took place.

In immediate connection with the palace on the N., are the ruins of the Abbey of Holyrood, founded by David I. in 1128. The king conferred a large endowment and other privileges on the monks (of the order of St. Augustine) whom he established here; among these, the privilege of erecting a burgh between the abbey and the town of Edinburgh. Hence the origin of the Canongate, the superiority of which at the Reformation passed from the hands of the monks to the Earl of Roxburgh, from whom it was purchased in 1636 by the city of Edinburgh, which still retains it. At the Reformation, the buildings connected with this abbey suffered much; and it is now in a state of ruin, the roof having fallen in so long ago as 1775. The area of the royal chapel, which formed the nave of the Abbey church, has long been used as a burial-place by several of the Scotch nobility. In the NE. corner of the chapel is the royal vault, in which are deposited the remains of several of the Scotch sovereigns, and branches of their families. The precincts of the Abbey of Holyrood, including Arthur's Seat and Salisbury Crags, constitute a sanctuary for insolvent debtors.

The buildings of the Royal Institution, an edifice in a pure classical style, situated at the N. termination of the Earthen Mound, and fronting Princes Street, have a range of Doric pillars on each side, and another range surmounted with a pediment in front. The Royal Society of Edinburgh, the Royal Society of Arts, and the Board of Trustees, which last was instituted in 1727, for encouragement of trade and manufactures in Scotland, have also apartments under the roof of this institution. The Board of Trustees, besides the primary object for which it was founded, pays 500l. a year to the Royal Academy for the encouragement of the fine arts. The Royal Scottish Academy of Painting, Sculpture, and Architecture has hitherto had annual exhibitions in the Royal Institution. On the Mound, a ridge 800 ft. long, stands the National Gallery, a modern building, in the Grecian style. The Assembly, or Victoria Hall, is a handsome Gothic building, 141 ft. long, with a spire 212 ft. high; it was built in 1842, and is used for the meetings of the General Assembly.

The Calton Hill is the site of several interesting monuments: that of Nelson, though by no means in the best taste, is the most prominent; it stands

on the edge of a precipice, and consists of a lofty circular hollow tower, having a stair inside, and battlements at the top. Here, also, is the National Monument, in commemoration of the naval and military glories of the late French war. The foundation stone of the latter was laid in 1822, when George IV. was in Scotland. It is meant to be a facsimile of the Parthenon in the acropolis of Athens, except that it is of sandstone, whereas its great prototype is of marble. On the same hill are monuments to Dugald Stewart, the celebrated metaphysician, and Professor Playfair; the former singularly chaste and beautiful, being a reproduction, with some variations, of the choragic monument of Lysicrates at Athens. On the S. of the hill, on a detached eminence overlooking the Canongate, is a monument to Robert Burns, belonging to the Corinthian order. The Calton Hill is also the site of the Observatory and of the High School. On the SW. corner of the hill, along the right of the road leading from Princes Street to the country on the E., stand bridewell and the gaol, two heavy and plain but well arranged buildings.

There are various other monuments in different parts of the town; that to the late Lord Melville, in St. Andrew's Square, consists of a column, surmounted by a statue, total height 163 ft., after the model of Trajan's pillar at Rome; but the shaft, instead of being ornamented with sculpture, as in the case with its archetype, is fluted. Bronze statues of George IV. and William Pitt, by Chantrey, are placed on granite pedestals in George Street, at the crossings, respectively, of Hanover Street and Frederick Street; and a bronze statue of the late Earl of Hopetoun, by the same artist, is placed within a vacant space, opposite to the office of the Royal Bank, in St. Andrew's Square. Another bronze statue, by Campbell, of the late Duke of York, has been erected on the Castle Hill, between the High Street and the castle. A statue of the Duke of Wellington, erected in 1852, stands near the register house. The monument to Sir Walter Scott, on the vacant ground south of Princes Street, at the foot of St. David's Street, though not, perhaps, in the best situation that might have been selected, is one of the most striking and magnificent of this class of buildings. It is open Gothic, 200 ft. in height, and, including the statue, cost 15,650l. An equestrian statue of the Duke of Wellington is placed in front of the register office. The monument of David Hume, the historian, within the old Calton Hill burying-ground, is a conspicuous and interesting object.

The Register Office, a building erected to preserve the public records of Scotland, was constructed after a plan designed by Mr. Rob. Adam, and though begun in 1774 was not completed till 1822. It is situated at the E. end of Princes Street, and fronts the North Bridge. The building, which is of two stories, exclusive of the basement floor, consists of a square of 200 ft., with a quadrangular court in the centre, covered by a dome of 50 ft. diameter. It has great architectural beauty. Its front is ornamented with Corinthian pilasters, supporting a pediment, within which are the royal arms of Great Britain, with a fine entablature of the same order. It is disposed in nearly 100 small arched apartments entering from long corridors on both stories; and, though heated by flues, is, from the total absence of timber, proof against fire.

Churches and Chapels.—Edinburgh originally consisted of one parish, and John Knox was, for a time, the only minister of the city, that is, of the ancient royalty, independent of the suburbs. The single place of worship at that time was St. Giles's,

of the High Church. In 1635, the royalty was
divided into four parts.; in 1641, into 6; in 1641,
into 9; and subsequently into fifteen parishes.
There is exclusive of the Canongate, whose church
is collegiate, of the par. of St. Cuthbert, of South
Leith, and three others.

The most important ecclesiastical edifice is St.
Giles, so called after the tutelary saint of Edinburgh. It stands in the High Street, and forms
the N. side of the Parliament Square. It is an
ancient Gothic building, the date of its erection
being unknown; and is built in the form of a
cross. Its length is 206 ft., its greatest breadth
129. It is adorned with a lofty square tower, the
top of which is encircled with open figured stonework, while from each corner of the tower springs
an arch, which, meeting together in the centre,
form a magnificent imperial crown. A pointed
spire, elevated 161 ft. from the ground, terminates
this stately tower. Shortly after the Reformation,
St. Giles was divided into separate places of worship. In 1822-29 it was thoroughly repaired, with
the exception of the tower, renovated, and greatly
improved in appearance by an entire casing of
new freestone walls. Its ancient character being at
the same time carefully preserved. It now contains only three churches. The High Church, or
Easter St. Giles, has an ornamented seat for the
sovereign, with a canopy supported by four handsome columns. It has, also, the official seats of
the magistrates of the city, and of the Judges of
the court of session.

The next church, in respect of antiquity, was
Trinity College church, founded in 1462, by Mary
of Gueldres, widow of James II. The building,
which was Gothic, and in the cathedral form, appears never to have consisted of more than the
choir or E. part, and the transept or cross, the W.
part having been begun but not finished. But this
interesting relic of a bygone age has been removed
to make room for the terminus of the North British
railway.

The Tron church, which stands at the point of
intersection of the South Bridge and High Street,
is of Gothic architecture, blended with Roman
ornaments and details. The present spire of this
church, 160 ft. in height, replaces a former spire of
wood, burnt down in 1824. Among the other and
more modern churches are St. Andrew's, erected
in the extended royalty, in 1781, with a spire 168
ft. high; St. George's, opened in 1814; St. Mary's,
in 1824; St. Stephen's, in 1828; Greyfriars, built
1612, and rebuilt 1846; and Greenside, in 1839.
St. George's, on the W. side of Charlotte Square,
is a large, heavy, tasteless square fabric. From
the centre rises a tower surmounted with a dome
150 ft. in height, in imitation of St. Paul's; the
building cost 33,000l. The church for the Tol-
booth parish is situated on the castle hill, has
commodious apartments, inc. Victoria hall, already
mentioned, for the use of the General Assembly,
and is the most conspicuous object in the city.
The town council of Edinburgh are the patrons of
the fifteen city parishes.

The churches and chapels, nine in number, belonging to the Scotch episcopal church, are generally handsome structures. Of these, the principal
are, St. John's, the west of the dean, at the W.
end of Princes Street, in the florid Gothic style,
with a square tower, 120 ft. high; St. Paul's, the
seat of the bishop of Edinburgh, in York Place,
of Gothic architecture; and Trinity chapel, at the
N. extremity of Dean Bridge, also in the Gothic
style.

The Rom. Cath. have three places of worship,
besides a convent of nuns, called St. Margaret's,
at the head of Bruntsfield Links, attached to

which is an establishment at Milton House, in
the Canongate.

The chapels of the various dissenting denominations (including the Free Kirk) are all respectable,
and many of them spacious, elegant, and costly.
The following is the number of places of worship
in the city and suburbs (exclusive of Leith),
with the denominations to which they severally
belong:—

Established Church	.	.	.	17
United Presbyterian Church	.	.	.	19
Associate Synod of Original Seceders	.	.	2
Free Church	.	.	.	41
Independents	.	.	.	3
Episcopalians	.	.	.	9
Roman Catholics	.	.	.	3
Baptists	.	.	.	7
Methodists	.	.	.	4
Unitarians, Quakers, Unitarians, Jews, New Jerusalemites, Bereans, 1 each	.	.	4

Total number of Churches and Chapels 111

The city parochial clergy are supported chiefly by
an assessment (called annuity tax) of 6 per cent.
levied on all houses and shops within the ancient
and extended royalty, with the exception of the
dwelling houses of the members of the College of
Justice, that is, of the legal practitioners before
the court of session. The annuity bring a very
unpopular impost, its payment is often evaded,
even at the risk of imprisonment or distraining of
goods, so that great defalcations are experienced
in its collection. The clergy drew, till 1838, certain share dues at Leith, and other trifling imposts; but, by an act of parliament passed in that
year (Edinburgh and Leith Agreement) 1838, cap.
55), the sum of 2,000l. was secured to them, in
lieu of all such claims. Their average income of
late years has exceeded 500l.

Education.—University.—Edinburgh is not more
celebrated for anything than for her literary and
educational institutions: of these, the university
deserves the first notice. The building of this
seminary, the only foundation of the kind established in Scotland since the Reformation, began
in 1581, after many unsuccessful efforts had been
made by the citizens of Edinburgh to obtain for
their city the advantages of such an institution.
It received a charter from James VI. in 1582; and
in 1583 the college was opened for the reception
of students, the number of whom was forty-eight.
(Crawfurd's History of the University of Edinburgh, p. 51.) On the first institution of the
college there was but one professor or regent; a
second was soon afterwards added, then a third,
and so on, till there were six; a principal, who
was also professor of divinity; four regents of
philosophy; and a regent of humanity. Each of
the regents of philosophy conducted his class for
four successive years, including, in his course of
study, almost every department of science and
literature—the classics, logic, metaphysics, ethics,
mathematics, and physics. A division of labour
in teaching was gradually introduced, as new professorships were founded; but it was not till 1708
that the old system was entirely superseded. In
the year just mentioned, the number of professors,
including the principal (from whose duties the
office of regent of theology had been withdrawn
in 1620), was fifteen; but such has since been
the increase, that, in the year 1864, there were
thirty-four.

The medical school of Edinburgh, of late years
so famous, had its origin so recently as the end
of the 17th century, there being no professor of medicine previously to the year 1685. The magistrates, whose predecessors may be regarded as the
founders of the university, and who have been at

all times its munificent guardians, are its general patrons, and have power to institute new professorships, and to alter or modify the academical discipline. Out of the 32 appointments, they possess the exclusive right of presentation to the chairs of principal and of 14 professors; they unite with other parties in the right of election to 7 other chairs; the crown enjoys the patronage of 8; while the principal and professors are invested with the patronage of 1, viz. music, instituted in 1839. The chair of clinical medicine is taught in rotation by certain of the medical professors, according to an arrangement among themselves. The crown is the patron of those chairs only instituted by itself. No party except the crown (and even that was at one time disputed) has a right to found a professorship without the sanction of the magistrates. The income of the professors depend chiefly (some of them entirely) on the fees paid by the students. The crown endowed most of the chairs which it has founded; while such of the others as have salaries attached derive them either from the patrons of the university, their respective founders, or the bequests of private individuals. The chair of music, founded and endowed by General Reid, has attached to it the comparatively large salary of 300l.

The above sums include, in the case of the older chairs, allowances for house rent, as the professors and also the students originally lived within the walls of the college; but such is no longer the case. Both parties now live wherever they choose; and no discipline is exercised over a student, except when within the walls of the college. The professorships are divided into the four faculties of philosophy, law, medicine, and divinity. The students wear no particular academical dress. There is no such officer as a chancellor or rector except that the functions of the latter are said to be officially vested in the lord provost of Edinburgh. A standing body, called the college committee, appointed by the town council out of their own number, has charge of the seminary. There is but one session annually, from the first of November till the end of April.

The exhibitions, or bursaries, attached to the university are 34, their benefits being extended to 80 students; their aggregate amount is 1,172l. a year. There are of the annual value of 100l., six of 30l., ten of 20l., four between 20l. and 15l., one of 15l., five between 15l. and 10l., forty-two between 10l. and 5l., and three under 5l. The fees paid by the students are—for each class in the faculty of divinity, 2l. 2s.; in that of arts, 3l. 3s.; in those of law and medicine, 4l. 4s. There is, also, 1l. paid annually on matriculation.

The number of students increased pretty regularly from the institution of the university till 1825, when it was at its maximum. There were in that year 2531 students on the books. In 1840, the number had declined to 2023; and in 1850 to 1,561. The average number of students in recent years has been 1,700.

The great diminution of students is generally allowed to be owing, not to any inefficiency that attaches to the university of Edinburgh, but to a combination of circumstances, particularly to the institution of the Free Church College, and of several colleges in England, to an increased emigration to the British colonies, and to the country having become more commercial, and supplying more advantageous channels of employment than those afforded by the learned professions.

The university library consists of about 100,000 vols. It is open on payment of the matriculation fee, referred to above, to all students, who may borrow from it and carry to their lodgings as many books as they please, on depositing a sum equal to their value, which is returned to them when the books are replaced. The library is supported by the matriculation fee, by 6d. paid by each professor on his election, and by a portion of the fees of graduates both in medicine and in arts. It was formerly one of the institutions that were entitled to a copy of every book entered in Stationers' Hall; a right commuted for a certain fixed sum paid by government. The library hall is 190 ft. in length by 50 in width, and is certainly one of the largest and finest halls in the kingdom. There are various other subsidiary apartments. The theological faculty has a library, consisting of about 6,000 vols. appropriated to the use of its own students. The college museum, which occupies two large and elegant rooms, besides minor apartments, is particularly rich in objects of natural history.

The present university buildings, which are on a very magnificent scale, were begun in 1789, the expense being defrayed partly by public subscriptions, but chiefly by repeated grants from government. The structure is quadrangular, 358 ft. by 255, enclosing a court. A handsome portico, supported by massive Doric columns, forms the chief entrance. This is to be surmounted by a dome, the only thing that is now wanted to complete the building.

Free Church College.—In addition to the old, Edinburgh has now a new college in connection with the Free Church. The latter, situated at the N. end of the Mound, is a handsome and commodious building. Though complete in itself, the present structure forms only one of three quadrangles, embraced in the original plan. The N. front has a church at the E. end, and in the centre two large towers rise on each side of the portico of entrance. The buildings around the area of the quadrangle consist of the hall of the senatus academicus, the library, museum, divinity hall, and several class rooms. The S. part of the ground, which is unoccupied, extends to the High Street. It was founded in 1843, and though principally intended for the education of students belonging to the Free Church, it is not confined to any denomination; and the classes of moral philosophy, logic, and natural science, as well as those of theology, are open to all who choose to avail themselves of them. Besides the principal and two professors of theology, there are professors of divinity and ecclesiastical history, Hebrew, exegetical theology, moral philosophy, logic, natural science, and a classical tutor. The number of pupils at the college amounts to nearly 500 on the average. The professors have 100l. a year of salary, and there are several scholarships. The necessary funds for the maintenance of the college are derived from contributions and collections throughout the church and the fees of students. The library exceeds 10,000 vols, and the museum possesses several valuable specimens in the department of natural history.

The celebrity of Edinburgh as a medical school has depended materially (but formerly more than during the last fifty years) on the schools of a number of private lecturers of eminence in their separate departments, particularly in medicine. They are generally members of the Royal College of Surgeons, and attendance on their courses of lectures is allowed by that body to qualify for examination. This college grants diplomas in surgery, but not in medicine; so that a person may obtain the rank of surgeon in Edinburgh without attending a single class in the university. The lectures delivered under the auspices of the Royal College of Surgeons are recognised by the Uni-

versity of London, and qualify for examination
before that body. The Royal College of Surgeons,
incorporated by charter in 1778 has recently built
a hall in Nicolson Street, which ranks amongst
the handsomest buildings in the city.

The Royal College of Physicians was established so early as 1681 by a charter from Charles
II. The number of its fellows, resident and non-
resident, is 196.

The *High School* is at once the oldest and most
celebrated of all the Edinburgh schools; and is
surpassed by few classical seminaries in the empire.
It was instituted in 1519, but having fallen into
decay, was re-erected in 1577. It consists of a
rector, and four other Greek and Latin masters,
each of whom begins an elementary class yearly,
and at the end of four years hands it over to the
rector, under whom, generally during two additional years, the curriculum of study is completed.
The present building, one of the ornaments of the
city, is situated on the S. slope of the Calton Hill:
it was opened in 1829, is composed of a central
body and two wings, and cost 31,000l. The number of scholars has been (1820) as high as 960;
but, for some years past, the number has been
between 400 and 500. This decline is not, however, aceribable to any falling off in the reputation
of the school, but to the institution, in 1824, of a
more aristocratical establishment of the same
kind, called the Edinburgh Academy, conducted
by a committee of subscribers. A Naval and
Military Academy, instituted in 1825, embraces
all the classes necessary for the two professions
from which its title is derived, as well as all the
branches implied in a liberal education. The
other more eminent schools are the Normal
Schools in connection with the Church of Scotland
and the Free Church, the latter being held in the
house that once belonged to the Regent Murray;
the Southern Academy, situated in George Square,
embracing not merely classical literature, but all
the branches requisite in a commercial or general
education; the Hill Street Institution in the New
Town, of which a similar character may be given;
the Circus Place School, a seminary for English
literature; the Ladies' Institution for the southern
Districts; the Scottish Institution for the education
of young ladies; Dr. Bell's Schools; Lancastrian
School; the Sessional School, supported by the Kirk
Sessions of Edinburgh; and School of Arts, or Mechanics' Institute. There are, also, ragged and industrial schools. Literary and scientific associations
are common in Edinburgh, such as the Royal Society, the Astronomical Institution, the observatory
attached to which on the Calton Hill is in the
purest classical taste, the Society of Antiquaries,
Wernerian Society, Royal Physical, Royal Medical, Cuvierian, Phrenological, Speculative, &c. There are
also various subscription libraries, some of them of
great extent and value.

Charitable institutions are very numerous in
Edinburgh. The most important is George
Heriot's Hospital, from the name of its founder
the goldsmith and jeweller of James VI. This
noble structure, which is of a quadrangular form,
with a court in the centre, and of Gothic architecture, from a plan of the celebrated Inigo Jones,
is devoted to 'the maintenance and education of
poor fatherless boys, freemen's sons of the town of
Edinburgh.' It was opened for the reception of
boys in 1659, when thirty were admitted. It now
contains 180; but, by a recent act of parliament,
the governors of the hospital are empowered to
erect schools from the surpluses of income,
throughout the town, for the gratuitous education
primarily of freemen's sons; but if circumstances
admit, to be open to the children of poor pa-

rents generally. There are ten such schools, including three infant schools; aggregate attendance
about 3,500. The management of the charity is
vested in the eighteen city clergymen, and in
the members of the town council. The revenue
of the hospital is upwards of 17,000l. a year. The
other charitable institutions are George Watson's
Hospital, founded in 1741, containing eighty boys;
John Watson's Hospital, founded in 1828, and
containing 120 children, male and female; the
Merchant Maiden and the Trades' Maiden Hospitals; the Orphan Hospital; Gillespie's Hospital,
for the reception of old decayed men and women,
having attached to it a free school, attended by
about 160 poor children; Trinity Hospital, founded
by the widow of James II. in 1461, for the benefit of 'burgesses, their wives, or children not married, nor under the age of fifty years;' Cauvin's
Hospital, for the maintenance and education of the
sons of poor teachers, and of poor but honest
farmers; the Institution for the Deaf and Dumb;
Asylum for the Blind; Magdalene Asylum; Lunatic Asylum; House of Refuge; Royal Infirmary, founded in 1736; Society for the Relief
of the Destitute Sick; Lying-in Hospitals; Dispensaries. In addition to these, and other less
important charities, some large bequests have recently been made for benevolent purposes. James
Donaldson, printer, Edinburgh, who died in 1830,
bequeathed 210,000l. for the endowment and
erection of a hospital for the maintenance of poor
boys and girls, of whom a certain number are to
be deaf and dumb. The building for this hospital,
opened in 1850, is quadrangular, in the Elisa-
bethan style, and is one of the finest of all the
structures belonging to Edinburgh. Sir William
Fettes, who died in 1836, left the greater part of
his large fortune to form an endowment for the
maintenance, education, and outfit of young people
whose parents have fallen into adverse circumstances. George Chalmers, plumber, who died in
1836, bequeathed 30,000l. for the erection and
support of a hospital 'for the sick and hurt.'
There is, finally, a hospital for the maintenance
and education of poor boys, from a fund which
amounts to 90,000l., bequeathed by Mr. Daniel
Stewart.

Courts of Law.—Edinburgh is the seat of the
supreme courts of Scotland, or College of Justice,
founded by James V. in 1532. Of these, the principal is the Court of Session, or supreme civil
court, which possesses in itself all those peculiar
powers exercised in England by the Courts of
Chancery, Queen's Bench, Common Pleas, Admiralty, and others, being a court both of law and
equity. The constitution of the court has undergone various modifications in its different departments during the last 300 years. At present it
consists of thirteen judges, called lords, and separated into the first and second divisions. In the
former there are six lords, in the latter seven.
The two divisions form distinct courts, but they
may, and on important questions do, sit in judgment together. From the first division are detached two judges, called Lords Ordinary, and
from the second there are taken three. Before one
or other of these ordinaries, all cases must be
brought in the first instance; but an appeal lies
from their judgment to that division before whose
ordinary the case was primarily tried. Cases may
be appealed from the Court of Session to the
House of Lords, the decision of the latter being
final. The court has a winter term of four months,
and a summer term of two months. Trial by jury
in civil cases was introduced into Scotland, under
a separate court, in 1816; but in 1830 this tribunal merged in the court of session. In the same

q 2

supreme court has been vested the jurisdiction of the Teind or Tithe Court (the peculiar duty of which was to regulate the stipends of the clergy of the established church of Scotland), of the Commissary or Consistorial Court, and the Court of Exchequer. The High Court of Justiciary, or supreme criminal court, was instituted in 1672. It is composed of a president called the Lord Justice Clerk, and of other five judges, who must, at the same time, be lords of session, but the crown may appoint any of the other lords to act should such a step be thought expedient. (See SCOTLAND.)

The edifice which, since the Union, has been the place of meeting of the College of Justice, was the parliament house of Scotland, from 1640, the date of its erection, down to 1707, when the Union extinguished the separate legislature of Scotland. The building is situated in the centre of the Old Town, being separated from the High Street by the cathedral of St. Giles. A small space called the Parliament Square intervenes between it and that church. Nearly half the buildings which formed this square were burnt down in 1824; but both St. Giles and the Parliament House escaped. A new front, though but little in harmony with the surrounding buildings, has been given to the latter, and great changes have been effected in its interior in the course of the present century. There is in the court occupied by the second division an admirable statue by Roubilliac, of Duncan Forbes, of Culloden, president of the court of session; and in the court occupied by the first division is a statue of President Blair; and in the hall, where the lords ordinary sit, is a statue of Henry Dundas, Lord Melville; the last two are by Chantrey, but they are poor and spiritless, compared with the masterly production of Roubilliac.

The faculty of advocates is an association of barristers (but not incorporated), entitled to plead before the supreme or any other courts of record. The society of writers to the signet is an incorporated body, qualified to conduct causes, as agents, before the same courts, and enjoying the exclusive right of preparing such papers or warrants as are to receive the royal seal or signet, whence their designation. The solicitors before the supreme courts form a body of attorneys incorporated in 1797, but of inferior grade and dignity to the writers to the signet. Advocates' first clerks may practise before the supreme courts on undergoing the usual examination, and paying certain fees.

The legal practitioners, all ranks included, may be regarded as the most important class in Edinburgh. Public opinion is, to a considerable extent, affected by their influence; they form a very numerous body; but while they have greatly increased in numbers during the last 40 years, the business of the court of session, before which almost all of them exclusively practise, has undergone a remarkable diminution. It appears from official returns, that while the number of cases annually enrolled in the court of session is at present only about 2-3rds of what it was in 1708, the number of advocates has almost doubled, and that of agents of all kinds has nearly trebled. As, however, the capital and pop. of the country have more than doubled within the time specified, it is probable that conveyancing and such departments of business have greatly increased, though not nearly to the same extent as the number of lawyers.

In immediate connection with the parliament house are numerous apartments, some of them spacious and highly ornamented, fitted up for the libraries belonging to the faculty of advocates,

and the writers to the signet. The library of the former body was established in 1682. This collection, which exceeds 150,000 volumes, is by far the most extensive and valuable in Scotland, and is, in fact, a very noble national library. It receives a copy gratis of all works entered in Stationers' Hall. The library of the writers to the signet is also large and very valuable.

Places of Amusement.—Among these may be specified the theatre, which is tolerably well attended, and the assembly rooms. The former, situated at the N. end of North Bridge Street, is a plain building externally, but is handsomely and conveniently fitted up. The assembly rooms in George Street are large and elegant. Golf is a favourite game; and curling and skating are very favourite amusements in winter, when the lochs of Duddingstone and Lochend happen to be frozen over. Cricket is now also beginning to be practised, and various cricket clubs have recently been formed.

Manufactures.—Edinburgh can scarcely be regarded as a manufacturing town. The brewing of ale has for upwards of two centuries been established in Edinburgh, and there are many coach-making establishments. Figured shawls, in imitation of those of Cashmere, were first successfully made in Edinburgh. This took place about 1805, and the honour of it belongs to a Miss Thorie, who, with her father, had been for a number of years engaged in the gold lace manufacture. The invention of the Jacquard loom gave for a time the superiority in shawl-making to our French neighbours. But a knowledge of the invention having reached this country, produced a reaction in favour of the Scotch manufacture; and while this business was being cultivated with greater or less success in France, it established itself in Norwich, and in Paisley and Glasgow. Edinburgh, from the commencement of this manufacture, has taken the lead in most of the improvements connected with it, always producing the best goods of the kind; but from the circumstance of labour of various kinds being lower in Paisley and Glasgow, the manufacture has mostly been transferred to those places.

Literature has long been not only the principal glory of Edinburgh, but has also afforded a principal source of employment to the population. The publication of the Edinburgh Review, which commenced in 1802, made the celebrity of Edinburgh as a literary mart, which was not long after still farther extended by the appearance of the earlier productions of Sir Walter Scott. Since then a vast number of works of the highest eminence, in almost every department of literature, philosophy, and science, have appeared in Edinburgh. There are in Edinburgh about 60 printing offices, employing from 1,000 to 1,200 workmen, exclusive of masters. The business of bookbinding gives employment to about 500 persons, exclusive of masters.

The linen manufacture, both as respects the coarser and finer fabrics, long flourished in Edinburgh. 'The number of looms,' says Amot, 'employed in Edinburgh in the linen trade is extremely fluctuating; the largest number that has been known is about 1,500; at present (1779) it is supposed there are upwards of 800. This city has long been famous for making the finest damask table linen, and linen in the Dutch manner, equal to any that comes from Holland.' (Hist., p. 561.) But so thoroughly has the linen trade disappeared, that there are not at this moment 50 looms employed in the city. Dunfermline and Dundee have become the chief seats of the manufacture, the former devoting itself chiefly to damask and diaper, the latter to Osnaburghs and the coarser fabrics.

Canals and Railways.—The Union Canal, which commences at Port Hopetown, on the W. of Edinburgh, and joins the Forth and Clyde Canal, near Falkirk, forms a continuous line of water communication between the Scottish capital and Glasgow and the W. of Scotland. The course of the Union Canal is 31½ m., its depth 5 ft., its width at the surface 40 ft., and at the bottom 20 ft.

Edinburgh is, also, extremely well supplied with railway accommodation, and it has in this respect the peculiar advantage that, with one exception, the railways have their termini in the very centre of the city, in the hollow contiguous to N. Bridge. They consist of the railway to Glasgow; the N. British railway to Berwick, with a branch to Hawick; the railway to Perth and Dundee: the latter is carried under the New Town by a tunnel, whence it extends to Granton, and begins again at Burntisland on the N. side of the Frith of Forth. The Caledonian railway, which has its terminus at the Lothian Road, in the W. of the city, extends to Carlisle. The express trains from Edinburgh reach London in 11 hours.

For a lengthened period, Edinburgh was very indifferently supplied with water. There are no springs of any importance within the city, the water required for its consumption being conveyed in pipes from a considerable distance. The first of these pipes was laid in 1681; and additions were made to it in 1722, 1787, and 1790. Still, however, the supply, owing to the increase of population, was very defective, and it became necessary to take more efficient measures for increasing its quantity. With this view a joint-stock company was established by act of parliament in 1819, which conveyed into the town the water of the Crawley and Glencorse springs, about 7 m. SW. from the city, and afterwards, in 1849, the Ravelaw and Harlaw springs, 7 m. directly W. of the city. The works constructed to effect this object are on a scale of great magnificence, and the cost amounted to upwards of 300,000l. Edinburgh is well lighted with gas; and the pavement of the streets and lanes has long been celebrated for its excellence. The best material for paving is found in the neighbourhood.

The Scotch metropolis had long the enviable reputation of being one of the dirtiest towns in Europe; and though vast improvements have been effected in this respect, the reproach is not yet completely obviated. The dirtiness of the Old Town seems to have been mainly attributable to the crowded state and height of the buildings, and to the want of water. These circumstances hindered the formation of water-closets, and of common sewers; and down to the commencement of the American war, there was probably not a drain of the former, and certainly not one of the latter, in the city. Both are now universal in the New Town, but they are still wanting in very many parts of the Old Town; and notwithstanding the regulations laid down and enforced as to the emptying of filth on the streets, they can never, under the circumstances, be perfectly clean. In very many, too, of the stories (*flats*) or houses, especially those in the narrow closes or wynds on each side the High Street, there is no supply of water, save what is obtained from the public pumps in the vicinity; and this circumstance, combined with the want of ventilation, and with the poverty and usually crowded state of the inmates, render them the abode of misery and disease, to an extent that would not easily be believed.

None but burgesses were till lately entitled to carry on any trade or manufacture within the

royalty. But there does not now exist any such prohibition or exclusion. None, however, but burgesses or their children have a claim on the charity of the Trinity Hospital, and none but the sons of burgesses are entitled to admission to Heriot's Hospital.

Representation.—Before the passing of the Reform Bill, in 1832, the town council of Edinburgh, which consisted of thirty-three members, may be said to have been self-elected. With the exception of six, who were retained by certain incorporated trades, the council for the time being had the exclusive right of nominating their successors, the public having no voice or right to interfere in the matter. The town council thus elected possessed the exclusive right of choosing a representative in parliament for the city. Owing to the unpopularity that necessarily attached to this self-elected and irresponsible body, the passing of the Reform Bill was nowhere more strenuously insisted upon, or received, when framed into a law, with more sincere rejoicing, than in Edinburgh. By this law two representatives were given to the city. In 1864 the registered voters were 9,752. Under the Municipal Reform Act, Edinburgh is divided into five wards, and is governed by a lord provost, four bailies or aldermen, and forty-one councillors. The corporation revenue amounted in 1864-5 to 48,942l., exclusive of police revenue. The annual value of real property was 971,200l. in the financial year 1864-5.

History.—The origin of Edinburgh is involved in obscurity. So early as the beginning of the seventh century it had obtained the name of Edwinesburgh, derived, it is supposed, from Edwin, a prince of Northumberland, who overran a great part of the S. of Scotland. In the year 1128, it is called by David I. *his burgh of Edinburgh*; whence we infer that it was then a royal burgh. It was not a walled town, as previously stated, till the middle of the fifteenth century. James IV. encouraged the erection of its first printing press, in the beginning of the sixteenth century; but it was not till the succeeding reign that it was recognised as the undoubted capital of Scotland. From this time its history merges in that of the kingdom. It was converted to the Protestant faith at an early period of the Reformation; and the great bulk of its inhabitants, in successive ages, and under various forms of persecution, adopted the Calvinistic creed, and adhered rigidly to the Presbyterian form of worship. John Knox was, for some time, minister of Edinburgh; and the house which he inhabited (at the Netherbow, near the E. extremity of the High Street) is still standing, and is regarded with an ordinary degree of reverence. The union of the kingdoms excited great tumults in Edinburgh with the view of intimidating those members of the Scotch parliament who were favourable to the obnoxious measure. The act, however, was eventually passed (1st May, 1707) without bloodshed. In the rebellion of 1715, an unsuccessful attempt was made by the Jacobites to surprise the castle. In the subsequent rising of 1745, the rebels got possession of the city, a party of the Highlanders having secured the Netherbow Port; and they remained masters of the town from the 15th Sept. to the 31st Oct. But finding it impossible to reduce the castle, they abandoned the city, and proceeded on their march to England.

In 1736, a remarkable occurrence took place in Edinburgh, known by the name of the Porteous mob. On the 14th of April, at the execution of a smuggler of the name of Wilson, a disturbance arose, and the executioner and city guard were assailed by the populace. Porteous, the captain of the guard, having ordered his men to fire on

the crowd, six people were killed and eleven
wounded. Having been tried for the offence before the high court of justiciary, Porteous was
condemned to death, but was reprieved by the
crown. Hundred, however, that he should not
thus escape the fate which they thought he me-
rited, the mob, on the evening of the day previously to that on which he was to have been
executed, broke into the gaol in which he was
confined, and having dragged him out, led him to
the usual place of execution, and there hanged
him by torch-light on a dyer's pole. It being
supposed that the municipal authorities had neglected their duty on this occasion, the city was
ordered to pay a fine of 2,000 sterling to the
widow of Porteous; and, what is remarkable,
though a reward was offered for the discovery of
the perpetrators, they never were discovered, and
their names continue to be unknown.

Few events worth notice have since occurred in
the annals of Edinburgh. On the 2nd of Feb.
1779, during the parliamentary discussions on the
subject of the Catholic claims, an infuriated mob
burnt one Catholic chapel, plundered another, and
threatened to demolish the house of Principal
Robertson. Soon after the breaking out of the
French Revolution, a number of the inhabitants
of Edinburgh, sympathising with the principles
which then prevailed in France, formed themselves into societies for obtaining parliamentary
reform, and similar political objects. The proceedings of these associations, the members of
which styled themselves 'the friends of the people,'
were, on the whole, neither wise nor constitutional.
After doing much mischief, they at length attracted the notice of government; and the servility of the judges, and the wretched state of
jury trial in Scotland at the time, afforded a
ready means of inflicting on them the utmost
penalty of the law. One of the prosecuted men,
named Watt, was beheaded for sedition; and
Muir, Skirving, and others were transported.
Among the other events connected with Edinburgh which may, perhaps, be worth notice, may
be specified the visits of George IV., in 1822,
being the first sovereign who had entered the city
since the year 1650, and of Queen Victoria and
her consort, in 1842.

EGER (Hub. Cheb), a town of Bohemia, ranking
third in that kingdom, near its W. frontier, circ.
Ellnogen, on a rock on the Eger, 84 m. W. Prague,
on the railway from Pilsen to Hof. Pop. 11,172
in 1857. The place was formerly an important
fortress; but its walls are now almost destroyed,
and its ditches gradually filling up. It contains
some handsome buildings, inclusive of a fine par.
church and town-hall. In the centre of the town
is a large market-place, at the E. end of which is
the Burgomaster's house; in a bed-room of which,
Wallenstein was assassinated in 1634. In an angle
of the fortifications overhanging the river, stand
the ruins of the imperial castle, containing an ancient square tower built of black lava, supposed
by some to have been constructed in the time of
the Romans, a singular double chapel, and the
hall in which the principal friends of Wallenstein
were treacherously put to death at the same time
with their master. Eger has a gymnasium, 2 convents, a high school, a school for the children of
soldiers, 2 hospitals, an orphan asylum, 3 workhouses, a foundation for 12 old men, and manufactures of chintz and cotton fabrics, wool, hats,
and soap.

EGHAM, a par. and village of England, in the
N. part of the co. of Surrey, hund. Godley, 18 m.
W. London by road, and 21 m. by South Western
railway. Pop. of par. 4,864 in 1861. The village,

situated near the Thames, is connected with
Staines on the other side of the river by an iron
bridge, erected in 1807. The church, though of
mean appearance, is ancient, and contains some
curious monuments. There are two alms-houses,
one for 5 poor women, and one for 6 poor men and
as many women, N. from Egham, between the
village and the Thames, is Runnymede, famous in
English history from its being the scene of the
conferences between King John and the Barons,
that led to the signing of Magna Charta by the
king, in 1215. In this parish is Cooper's Hill,
which commands a fine prospect, and is the subject of the well-known descriptive poem of the
same name, by Sir John Denham.

EGINA, or ENGIA (an. Ægina), an island of
Greece, in the centre of the gulf to which it gives
name (Saronicus Sinus), 16 m. S. by W. Athens,
34 m. E. by N. Corinth, and 6 m. from the nearest
point of the promontory of Methana. It is about
8 m. from E. to W., and 8 from N. to S.: surface
diversified with hills and valleys; in the N. part
of the island there are rocks of lava. Soil rocky
and of a light colour. The low and cultivated
grounds are however fertile, and produce good
crops of corn, with wine, cotton, olives, figs, almonds, and other fruits. The hilly and uncultivated portions are deficient in water, and are
covered with pines, small cypresses, and junipers.
The red-legged partridge is very abundant. The
pop. was estimated to amount to about 6,000 in
1861; during the revolution it was much greater,
Egina having been then resorted to by crowds of
emigrants from the adjoining continent and islands,
but since the peace these have mostly returned
home. The inhab. who are industrious, carry on
a considerable trade. The port and principal town,
called Egina, or Engia, is on the W. side of the
island, near the extensive ruins of the ancient
city of the same name. There are from 15 to 16
fathoms water in the roadstead, on a tough clay-
ground. There is another and smaller town in the
N. part of the island.

Though unimportant in modern times, in antiquity Egina was early celebrated for its wealth
and population. Its position is very favourable
for commercial pursuits; and it was indebted for
its greatness to the zeal and success with which it
carried them on. At one period its naval power
was superior even to that of Athens; and it sent
30 ships to the battle of Salamis, to whom the
prize of valour was accorded by the suffrages of
the Greeks. But the proximity of Egina to the
Piræus awakened the jealousy, and provoked the
vindictive hostility of the Athenians, who, having
defeated the Eginetans and taken their city,
treated them with the utmost severity—"Itaris
etiam Athenienses, qui nervrunt ut Ægineta, qui
classe valebant, pollere providerentur: hoc enim
ut utile; simul enim immortui, propter propin-
quitatem, Eginam Piræus." (Cic. de Offic., lib. iii.
§ 11.) After various vicissitudes, Egina was restored to a nominal independence by Augustus;
since which period it has usually followed the
fortunes of the adjacent country of Greece.

The temple of Jupiter Panhellenius in the N.E.
part of the island, is among the most interesting
of the Grecian ruins. The hill on which it stands,
though of no great height, commands the greater
part of the island, the whole coast of Attica, with
the city of Athens, part of Peloponnesus, and
several of the islands in the gulf. It is built on a
platform, supported on all sides by terrace walls.
The temple, said to have been erected by Æacus,
grandson of Jupiter, is certainly one of the most
ancient in Greece. It is of the Doric order, being
90 ft. in length, measured at the base of the

columns, by 45 in breadth. Originally it had 36 columns, exclusive of those in the cella, of which 25 were standing when it was examined by Mr. Dodwell. The greater number of the statues that occupied the tympanum of the pediment, were dug up in 1811; and having been carried off, were purchased by king Ludwig I. of Bavaria for 10,000 sequins, and are now in the Museum at Munich. They are in the peculiar style of sculpture called Eginetan, and are amongst the most interesting relics that have ever been conveyed from Greece. (Chandler's Greece, caps. 3 and 4; Dodwell's Greece, i. 556–574.)

EGYPT (the Mizraim of the Hebrews, and Aiyuptos of the Greeks), a country on both banks of the Nile, occupying the NE. angle of the African continent; one of the earliest seats of art, science, and literature, and famous alike for the historical events of which it has been the theatre, its magnificent monuments, and physical character.

Boundaries and Extent.—There have been very different statements as to the boundaries of this famous country. There cannot of course be any doubt as to its N. limit, which is formed by the Mediterranean; and it seems to have been generally agreed from a very remote period, that its S. limit should be fixed at Syene, or rather at Philæ, in lat. 24° 3′ 45″ N. But the difficult point is to determine its breadth. From Philæ to near Cairo, the Nile in most parts flows through a narrow valley, bounded on either side by a ridge of hills, or inferior mountains; at Cairo these ridges diverge, that on the E. to Suez, and that on the W. in a NW. direction to the Mediterranean. Some authors identify Egypt with the tract lying between the mountain chains now referred to; while others, regarding the Nile as the source of life and vegetation in Egypt, restrict its territory within the limits covered by the inundation of the river. (Strabo, lib. xvii. p. 544.) But from the age of the Ptolemies down to the present day, the desert country lying between the valley of the Nile and the Red Sea has been uniformly included in Egypt. On the W. side the mountain ridge already noticed seems to be its only natural boundary. Still however, it has been usual to reckon the oases that lie within 100, or even 200 m. of this limit, as belonging to Egypt.

From Cape Bourlos, on the coast, lat. 31° 36′ N., to Philæ, the distance N. and S. is 7° 32′ 15″, about 522 geographical, or 520 English m. But the distance by water and the extent of the alluvial territory are considerably greater than would appear from this, because of the many and considerable bends of the river. The breadth of the Egyptian coast is 160 m.; but in ascending to Chibo (104 m. from Cape Bourlos), the cultivated tract tapers off to a point, and the rest of the country is chiefly comprised in the narrow valley of the Nile; which, however, at Beni-souef, 83 (by water) m. higher, spreads to the W. to form the vale of Fayoum, a circular valley of great fertility and beauty, measuring about 40 m. from E. to W. and 30 m. from N. to S. Thence to Syene, the valley of the Nile is mostly confined within very narrow limits. The whole cultivable territory of Egypt, including its lateral valleys, has been estimated at about 16,000 sq. m., or about half the area of Ireland. (Malte-Brun, iv. 21, 29; Modern Trav. art. 'Egypt,' i. 6; Hoven's Researches, ii. 210, Engl. trans.)

The Nile, so important among the great rivers of the world, is also the most striking object in the general aspect of a country which not only is wholly comprised within the sphere of its influence, but is entirely indebted to it for existence. As already stated, the Nile enters Egypt at the island of Philæ; and from it to Assouan (Syene), a distance of about 6 m., it has cut a passage for itself, through a ridge of granite rocks with which its stream is much encumbered. At Assouan is the last of the cataracts of the Nile, so celebrated by ancient authors. (Senec. Nat. Quæst, lib. iv. § 2; Plin. Hist. Nat. lib. v. § 9; Lucan. lib. x. line 320, &c.) These statements with respect to it seem to be not a little exaggerated, though there can be no doubt that the cataract must have been much more magnificent 2,000 years ago than at present, as the attrition of the water for so long a period could not fail materially to deepen and smooth its bed; at all events, however, it is now rather a rapid than a cataract. According to Sir F. Henniker, it is in not really more formidable than the fall in the Thames at low water at Old London Bridge, previously to its demolition. (p. 147.) But it is clear that its height and rapidity must depend materially on the state of the river. When the inundation is at its height, the fall is hardly perceptible, but at low water it varies from 8 to 10 f. After leaving Assouan, the river runs on in a placid quiet stream, till, a little below Cairo, at Batn-el-Bakara, it divides into two great arms, the most E. of which falls into the sea at Damietta, and the most W. at Rosetta; but it has other, though very subordinate, outlets. For the immense distance of 1,230 m.,—that is, from lat. 17° 45′, and about 34° 5′ of E. long., where it is joined by the Atbara, or Tacazzé,—the Nile rolls on to its mouths in the Mediterranean in solitary grandeur, without receiving a single affluent: an unexampled instance in the hydrographic history of the globe. The periodical inundations, which water the country and cover it with mud, have given occasion, in all ages, for much discussion; and modern discovery has confirmed the conjecture of the ancients (Herodotus, Euterpe, §§ 20–28; Strabo, xvii. 543), that these overflowings result from rains falling near the mountains amongst which the Nile has its source, or early course. Bruce has explained this phenomenon as follows: "The air is so much rarified by the sun during the time he remains almost stationary over the tropic of Capricorn, that the winds, loaded with vapours, rush in upon the land (to restore the equilibrium) from the Atlantic Ocean on the W., the Indian Ocean on the E., and the cold S. Ocean beyond the Cape. Thus a great quantity of vapour is gathered, as it were, into a focus; and as the same causes continue to operate during the progress of the sun N., a vast train of clouds proceed from S. to N. In April all the rivers in the S. of Abyssinia begin to swell; in the beginning of June they are all full, and continue so while the sun remains stationary in the tropic of Cancer. When the sun approaches the tropic of Cancer, the Etesian winds along the coast of Egypt begin to blow from the N., and convey vast quantities of aqueous vapours to the mountains, which are there precipitated in torrents along with the vapours derived from the oceans already specified. The Etesian winds also contribute to increase the inundation, by determining the waters of the Mediterranean to the coast of Egypt, and obstructing the exit of those of the river. On the sun again turning to the S. the rains begin to abate, and on his passing the equator they cease in the N., and commence in the S. hemisphere. The torrents, descending in their rapid course the soil from the upper country, bring down supplies of alluvium, so that the valley of the Nile is constantly gaining in elevation. Nor is the delta of Egypt exempted from this peculiarity; though, from there being a wider space for the deposits to spread over, the inerease of soil is not nearly so great; indeed, the accumulation decreases, even in Upper Egypt, in proportion as the river approaches the sea. 'According to an approximate calculation,' says Wilkinson (Journal Geog. Soc., ix. 432), 'the land about Elephantine, at the first cataract, in lat. 21° 5', has been raised 9 ft. in 1,700 years; at Thebes, in lat. 25° 43', about 7 ft.; and at Heliopolis and Cairo, in lat. 30°, about 5 ft. 10 in. At Rosetta and the mouths of the Nile, in lat. 31° 30', the diminution in the perpendicular thickness of the deposit has lessened in a much greater decreasing ratio than in the straightened valley of Central and Upper Egypt, owing to the great extent E. and W., over which the inundation spreads.'

Were it not that the bed of the river rises in the same proportion as its banks, the country would cease to be inundated,—an apprehension which till lately was strongly entertained. It is impossible to find anywhere among terrestrial objects a more striking instance of the stability of the laws of Nature than the periodical rise and fall of this mighty river. We know by the testimony of antiquity that the inundations of the Nile have been the same, with respect to their season and duration, for 3,000 years. They are so regular that the value and annual certainty of this gift regulates the public revenue; for when, by means of Nilometers, it is ascertained that the waters promise an unusually prosperous season, the taxes are proportionally increased. (Russell's Egypt, p. 46.) Sometimes, however, when the river exceeds its ordinary height, it becomes a calamity; occasioning the loss of life and property. In September, 1818, Belzoni witnessed a scene of this sort; the river having risen 3½ ft. above the highest mark left by the former inundations, it ascended with uncommon rapidity, and carried off several villages, and some hundreds of inhabitants. The swellings of the Nile in Upper Egypt are from 30 to 35 ft.; at Cairo, 23 ft.; in the N. part of the Delta, owing to the breadth of the inundation and artificial channels, only 4 ft. Pliny says of the inundation:— 'Justum incrementum est cubitorum 16. Minores aquæ non omnia rigant; ampliores detinent tardius recedendo. Hæ serendi tempora absumunt aut madidata; illæ non dant sitientia. Utramque reputat provincia. In duodecim cubitis famem sentiret, in tredecim etiamnum esurit: quatuordecim cubita hilaritatem afferunt, quindecim securitatem, sexdecim delicias.' (Hist. Nat. lib. v. § 9.) The depth and rapidity of the river vary at different times in different places. It is seldom that any vessel exceeding 80 tons burden can ascend as high as the cataracts. The mouth of Damietta is between 7 and 8 ft. deep when the waters are low, that of Rosetta does not exceed 4 or 6 ft.; but when the waters are high, canavels of 24 guns may sail up to Cairo. (Mod. Trav. i. 52.) As a beverage the water of the Nile is considered delicious; Maillet declares that it is among waters what champaign is among wines. The mud of the river gives on analysis one-half of argillaceous earth, one-fourth carbonate of lime, the remainder being water, oxide of iron, and carbonate of magnesia. (See Nile.)

The Mountain system of Egypt is very peculiar. Two ranges, already noticed, pressing closely on each bank of the river, extend from Syene to Cairo, and form the valley of the Nile, protecting it from the ravages of the deserts on either side. That to the E. gives out an arm at Kenneh (lat. 26° 12'), and inserts the desert to the Red Sea at Cosseir in nearly the same latitude; while the Libyan or W. range branches off from Assouan to the Great Oasis. (Ritter, ii. 307.) Near Cairo the mountains diverge on both sides; one ridge running in a NW. direction to the Mediterranean, the other due E. to Suez. (Malte-Brun, iv. 22.) The geological components of the hills, from Philæ through the cataract region to Syene, are chiefly granite, and a peculiar highly crystallised red formation called syenite marble. This primitive rock is remarkable for durability and the fine polish it is capable of receiving. From quarries of this stone the Pharaohs, Ptolemies, and Antonines drew materials not only for the stupendous monuments which still make Egypt a land of wonders, but also for many of the public buildings of Italy, the remains of which attest the genius of the Roman artists. Some days' journey S. of Thebes extends the limestone region, dug out into innumerable catacombs, their entrances artfully contrived to conceal the abode of the ancient dead, a precaution suggested by a prominent superstition of the Egyptians. Between this district and the most S. one, the mountains are composed of sandstone, evidently a recent deposit; for it is so very soft that the buildings constructed of it would not have long resisted the weather, had they not been covered with a coloured varnish. Towards the valley of Suez the mountains contain limestone. On the W. side of the Delta not the least remarkable object presented by this wonderful country is the Neets, or valley of Natron Lakes, bounded on one side by a lofty ridge of secondary rocks, which, perhaps, proves the means of concentrating the saline deposit which gives its name to the place. The banks and waters of these lakes, six in number, are covered with crystallisations, consisting of sea-salt and natron, or carbonate of soda, sometimes united; at others, found separately in different parts of the same lake. (Russell's Egypt, p. 48.)

The most considerable of the Egyptian lakes are those of Menzaleh, Bourlos, Edko, and Mareotis, lying along the shore of the Delta. But though called lakes, they are more properly lagoons, and bear a striking resemblance to the haffs that skirt the shores of Prussia. Some of the lagoons, especially that of Menzaleh, E. of Damietta, are of large dimensions. They are all shallow; are separated from the sea, with which they communicate, by a narrow bank or ridge of sand; and are in the course of being gradually, though slowly, filled up. In antiquity, the Nile is said to have disembogued itself by seven channels—Neptune-grani cutia Nili; but of these some were certainly artificial; and then, as now, there were two principal mouths—the Pelusiac, or Eastern, and the Canopic, or Western. The Sebennytic mouth, in the centre of the Delta, was also of considerable importance. But considering the nature of the soil, and the efforts that have been made from the remotest times to divert a portion of the river by canals and otherwise into new courses, we need not be surprised that very great changes should have taken place in the channels by which it pours its waters into the Mediterranean.

Exclusive of the lagoons in the Delta, there is a considerable lake occupying the NW. part of the valley of Fayoom. The principal canal of Egypt, the Bahr Jousef, communicates with this lake. It branches out from the Nile at Deirout-el-Sherif, S. of Minieh, traversing the valley of the Nile at the foot of the Libyan chain, till it reaches the waters of Fayoom at Ilahun, and thence continues still parallel to the Nile, the Rosetta branch of which it finally joins at Alkam. Under the name of Bouhaid the same canal is continued to Farhout in Upper Egypt. The whole of the Delta is intersected with canals in every direction, in which the overflowings of the Nile are preserved after the inundations, to afford communication between the various towns, and to keep a constant supply for the irrigation of the cultivated lands. (Browne's Travels, pp. 177-187, &c.)

Egypt is naturally divided into—1. The Delta, or Lower Egypt. 2. The Valley of the Nile, comprising Central and Upper Egypt. 3. The E. Desert. 4. The W. Desert and Oases.

1. *The Egyptian Delta*, which derived its name from the similarity of its figure to the Greek Δ, is a triangular tract, formed by the bifurcation of the Nile. The soil consists of the mud of the river, resting upon desert sand. Near the banks of the two branches this alluvium has collected to a thickness in some places of more than 30 ft., while at the extremity of the inundation it does not exceed 6 in. This constant accumulation and spreading of the deposit E. and W., has gradually extended the limits of the Delta further into the adjoining deserts than they reached in antiquity (Wilkinson on the Levels of Egypt, in 'Geographical Journal,' ix. 457), so that the arable land of the country is constantly increasing; and though the sand in its turn frequently encroaches in various places, yet the injury it inflicts is only partial and temporary, while the alluvial deposit goes on steadily increasing in extent. The greatest length of the Delta is at present about 85 m. from E. to W., and from the fork of the Nile to the sea about 90 m. intervene; but the inundations extend very considerably beyond these limits. The Delta is covered with meadows, plantations, and orchards, and presents a more fertile aspect than any other part of the country; but various causes have combined to prevent the spread of husbandry and cultivation, proportionally to the increase of territory rescued from the deserts by the annual overflows, (Ibid. p. 457.) This district, from its comparatively low situation, and from the absence of those mountains which enclose the Valley of the Nile and confine its waters, aptly designated by Browne 'the walls of Egypt,' is more influenced by the inundations than the upper lands: and when the river is at its greatest height, it presents the aspect of an extensive marsh. The river begins to swell in June, and continues to increase till Sept.; at which period the fields of the Delta are completely submerged, its villages, towns (which are built on natural or artificial mounds), and trees, only appearing above the water. After remaining stationary for a few days, the waters begin to subside, and by the end of Nov. leave the land altogether, having deposited a rich alluvium. An Egyptian spring, corresponding to our winter, gives to the Delta its most smiling and verdant appearance. The rice fields, having been sown before the water has entirely receded, are covered with a vivid green, trees put forth their blossoms, and the whole country bears at this season the aspect of a fruitful garden.

The question as to the origin of the Egyptian Delta, has engaged the attention of the ablest inquirers from the remotest period. The most probable as well as most ancient theory is, that which represents it as wholly formed of the deposits brought down by the Nile, and as constantly, though slowly, gaining on the sea. (Herodotus, ii. § 5.) Originally the sea is said to have flowed as far S. as the Pyramids; but in the course of ages, through the gradual accumulation of the mud of the river, assisted in some degree by the construction of canals and dykes, the land rose above the level of the sea, and ceased to be submerged, except during the period of the inundation. (Savary's Letters on Egypt, Letter I.) This opinion has, however, been stoutly denied; and though it be admitted on all hands that the land of Egypt and the bed of the river are both slowly rising, it is contended that the limits of the Delta to the N. are the same now as in the remotest antiquity. This opinion is supported by the high authority of Sir J. G. Wilkinson; and it is also supported by the learned author of the very able and elaborate article on Egypt, in the new edition of the 'Encyclopædia Britannica.' But though it were admitted that the limits of the Delta on the N. had continued nearly stationary from the age of Herodotus, that would not invalidate his statement that the cultivated portion of Egypt is the *gift of the river*. The chain of sand-banks skirting the Delta on the N. probably existed long before the Delta attained its present form; and the lakes, or lagoons, already reduced, lying to the S. of this chain, are apparently the last remains of the sea by which it was anciently covered. That the Delta should owe its existence to the Nile, is perfectly agreeable to what is observed in all similar situations; and no positive evidence has been brought forward to controvert, or even materially weaken, the strong and all but conclusive presumptions in its favour. (Shaw's Travels, 385, &c. 4to. ed.; Rennel's Geog. of Herodotus.) But few traces are now to be found of the many famous cities with which this part of Egypt was formerly studded; and that, except Alexandria, the only places of consequence in the Delta, at the present day, are Rosetta and Damietta, situated at the two mouths of the Nile. At the former the river is 1,941 ft. wide, but at Damietta only 800. The villages are numerous, and generally large; but the houses seldom exceed from 10 to 12 ft. square. They are built of sun-dried bricks, and are covered with flat roofs of straw and Nile mud. (Dr. Richardson's Travels, i. 40; Clarke's Travels, iii. 13; Modern Traveller, i. 190-232, &c.)

2. *The Valley of the Nile of Central and Upper Egypt.*—Ascending the river from its fork, the cultivable land at the apex of the Delta and for some distance is found to decrease; for here the banks are much more elevated, and are seldom quite covered with water, even during the highest inundations. (Geog. Journal, ix. 454.) Hence the alluvium does not reach the interior at this point. The E. or Arabian mountain chain terminates abruptly at Mount Mokattem, near Cairo, and diverges towards Suez; while the opposite or Libyan range ends at Faioum, having turned off to the W. to enclose that valley. Throughout the entire district the E. chain has generally more transverse breaks and ravines, is more lofty and rugged, and comes closer to the river, than the hills on the opposite side. Between Faioum and the Nile the Libyan ridge has nearly a level summit, overlooking the country below; and this table-land was chosen for the site of the Pyramids. The space left between both ridges seldom exceeds 10 m. in Central Egypt, while in the upper country they press even more closely upon the sides of the river; thus that part of the Valley of the Nile which belongs to Egypt has but a contracted breadth, and even that is not all available for the labours of the husbandman, a great portion of it being, from the height of the banks, out of the reach of the overflowings and their beneficent deposits; hence a strip of desert usually runs along at the foot of the hills. Where, however, the land is laid under water at high Nile, communication is kept up between one village and another by means of elevated roads or dykes, which commence on a level with the banks of the river; and, as they extend to the interior, rise to so great a height above the fields as to leave room for the construction of arches for the passage of the water. As the river

enters the Egyptian territory from Nubia, the granitic hills bear the appearance of having been rent by the stream. Hence, between the Isle of Philae and Assouan the current is interrupted by innumerable islands. Others, of a less rocky character—some of them extensive, considering the breadth of the Nile—spring up out of its bed at various intervals during its progress to the Mediterranean. The Isle of Elephantine, opposite to Assouan, wears so beautiful an aspect that it is called by the natives the 'Isle of Flowers' (*Djeziret-el-Nabiit*); and most European travellers describe it as a sort of terrestrial paradise. The Egyptian valley is strewed with those stupendous monuments of human labour, those beautiful remains of ancient art, which have excited the wonder and admiration of ages; and which seem the more marvellous, the more closely they are examined.

3. *The desert E. of the Nile* is broken by rugged mountains, and intersected by numerous wadys or ravines, sometimes thickly, but more frequently scantily, clothed with verdure. It has, however, the advantage of numerous springs; beside which are traced ancient caravan tracks that are still traversed in exactly the same manner as when the company of merchants found Joseph in the pit. The leading characteristic of this desert, particularly in its N. part, is its gradual ascent from the Nile to a certain distance E., where commences a plain nearly level, and of some extent, from which all the valleys or torrents running in a W. direction empty themselves into the Nile, and those to the E. into the Red Sea. Of such a character are the Ataka hills, mentioned before as branching E. from the Mokattem mountains, near Cairo. These are joined at a right angle by a series of eminences which skirt the shores of the Red Sea into the Nubian country, under the names of the Zarafana, Doffa, and Jaffatine ranges; and form the E. edges of the plateaux raised by the transverse hills, a chain of which appears again in lat. 29° between Beni-suef and that part of the Suez gulf called Birket Faraa. These are entirely of limestone, and present a gradual ascent from the Nile to a distance E. of 30 m.: the high plain which ascends is about 16 m. broad, and the descent down to the Red Sea occupies a space of about 50 m. At the S. declension of the N. Keilila mountains is a copper-mine, which appears from the ruined huts, furnaces, scoriæ, &c. found by Wilkinson to have been extensively worked. (Geog. Journ. ii. 52.) The Wady Arabah intervenes its desert of sand to the N. Keilila or Kolzim mountains, at the foot of which are situated the two celebrated convents of St. Anthony (17 m. from the sea) and St. Paul, placed about 11 m. apart. Between these convents and the gulf at Wady Gârâ are the remains of houses and watercourses, which appear to belong to the Greek period. In lat. 28° 36', the limestone formation, which continues with little interruption throughout the N. hills of this desert, is joined by primitive rocks, which present more irregular surfaces, but rise from the banks of the Nile with a gentler declivity than the series already described, and slant with proportionate abruptness upon the shores of the Red Sea. Mount Gebrib (28° 15'), one of these rugged eminences, is the highest of the hills in this desert, being 6,000 ft. above the sea. Four hours S. of Gârib are two copper-mines, with the same appearances of having been worked as those before mentioned. In lat. 26° the character of the levels again changes, being higher and more uniform from the Nile to where they make a descent to the sea, which is gradual

till they reach Mount Azzrit, which gives them an abrupt termination. Near Mount Dukhan (lat. 27° 15') are the ruins of a town, and vast quarries of red porphyry, strewed with the materials of a small temple, which seems to have never been completed. At Cossir, whose bay indents the Red Sea, at about lat. 26° 6', real the primitive hills that intersect the desert in a direction parallel to the Nile and the Red Sea, and join a transverse range, upon which extends the caravan route from Kenneh on the Nile to Cossir, where pilgrims embark to pay their devotions at the shrine of Mecca. (See Cossir.) The valley of Cossir extends down to about 25°, where another transverse range occurs, which contains, near the sea, some lead mines. Mount Zaburah, celebrated by ancient writers for its emeralds, rises a little further inland. Attempts have been made to re-open the sources of wealth which these mines are said to have afforded, but without success. (Caillaud's Travels, fol. Paris, 1822, p. 60.) Nearly on a line with Assouan (lat. 23° 56') are the ruins of Berenice. The whole of the desert of Egypt is the resort of distinct tribes of Arabs, who confine themselves to particular localities: they consist of the Maazy, occupying the country to the E. of Beni-suef, Atouni, and Beniha-el, S. of the Maazy and the Atabilie Arabs, who are scattered over the N. part of the desert, and breed camels for the market of Eneeh.

4. *The desert W. of Egypt* presents a scene so formidable to travellers, that few have visited the oases by which it is here and there interspersed. The most N. of these is Siwah, or Ammon; SE. from which, and nearer to the Nile, is the Little Oasis, or Wah-el-Bahtyeh; the chief village of which lies in lat. 7[?]° (8' N., and long. 28° 55' E. S. and W. are the small oases of El Haya, Farafreh, and Zerzoura; and still further N. is the Dakhleh oasis, whose first European visitant was Sir A. Edmonstone, in 1819. Its chief village stands in about lat. 25° 35' N., and long. 28° 55' E. Three days' journey to the E. brings the traveller to the Great Oasis of Wah-el-Khargeh, extending in length from 24° 20' to near 26° N. lat. Instead of islands of the Libet (Mare aux rooms) springing up amidst the surrounding and desolate ocean of sand, as the ancients describe them, the oases are valleys or depressions of the lofty plain which forms the extensive table-land of E. Africa. On descending to them, they are found to bear, in many respects, a similarity to a portion of the Valley of Egypt, being surrounded by steep cliffs of limestone, at some distance from the cultivated land, which vary in height in the different oases, those rising from the S. oases being the highest. Neither do they present a continuation of cultivable soil, all of them being intersected by patches of desert. They, no doubt, owe their origin to the springs with which they abound, the decay of the vegetation thence arising having produced the soil by which they are now covered. Their fertility has been deservedly celebrated; but the glowing eulogisms of travellers on their surpassing beauty are probably, in a great measure, to be ascribed to the striking contrast they present to the surrounding deserts of arid, burning sand. It may appear contradictory, considering the high opinion the ancients entertained of the fertility and beauty of the oases, that they should have selected them for places of banishment; but that such was the case, at least under the Romans, is certain. A law of the Digest (lib. 48, tit. 22) refers to this practice; and it has been supposed that the poet Juvenal was one of those who suffered a temporary banishment (*relegatio*) to the oases, though the evidence of this is by no means clear. (Biographie Univer-

selle, art. 'Juvenal.') But the fact of their being selected as places of banishment is not in anywise inconsistent with the received opinions as to their salubrity and fertility. They were selected, not because of their being naturally noxious or disagreeable, but because of their being, as it were, out of the world, and from the extreme difficulty of escaping from them. The larger oases have some fine remnants of antiquity, the most celebrated of which is the temple of Jupiter Ammon, at Siwah. (Edmonstone's Visit to the Oases, passim; Geog. Journal, ix, 440, 441, &c.)

The climate of Egypt is extremely hot: this is a consequence, no doubt, of the horizon of its elevation, of its being surrounded on all sides except the N. by vast tracts of burning sand, and of the scantiness of the rain. According to Volney, two seasons only are distinguishable, spring and summer; or, rather, the cool and the hot season. The latter continues from February or March to October; and during the greater part of this period the air is inflamed, the sky sparkling, and the heat oppressive to those unaccustomed to it: during this season the average height of the thermometer is about 90° Fahr. But the heat of the atmosphere is so much tempered by the inundations of the Nile, by the vapours brought by the Etesian winds from the N., and by the dews in the nights, that the natives and even Europeans occasionally complain of cold. During the remainder of the year, the average height of the thermometer is about 60° Fahr. It is necessary at all times to avoid exposure to the night air.

It might be imagined that Egypt, being for about three months of the year either wholly or partially inundated, and being subjected, at the same time, to the action of a powerful sun, producing an excessive evaporation, would be extremely unhealthy. But such is by no means the case. The exhalations from stagnant waters, so fatal in Cyprus and at Iskenderoon, and most other parts of the Levant, are here comparatively innoxious. They are not, however, entirely divested of their bad qualities. On the retiring of the waters, in November, which is the Egyptian seed-time, W. winds and fogs are prevalent, which produce ophthalmia, fever, diarrhoea, and catarrh. From December to March the winds blow mostly from the E.; the nights are cold, but the temperature during the day is that of June in France; the various productions of the earth are then vigorously on the increase; its surface is covered with the finest verdure; and all nature, reanimated by the fertilising influence of the river, and the moderate temperature, seems to grow young again. In Upper Egypt, the exhalations being comparatively few, the climate is proportionably healthy.

This general salubrity of the climate, notwithstanding the powerful deleterious influences to which it is exposed, is inferable to the natural dryness of the air; the proximity of the African and Arabian deserts, which incessantly absorb the humidity; and the currents of wind that sweep over the country without meeting with any interruption. The aridity is such that butcher's meat exposed, even in summer, to the N. wind does not putrefy, but dries up, and becomes hard as wood. In the desert dead carcases are found dried in this manner, so light that a man may easily lift the entire body of a camel. But near the sea the air is much less dry than farther up the country, and that at Alexandria and Rosetta iron exposed to the air speedily rusts.

It has been mentioned already, that on the approach of the sun to the tropic of Cancer the winds invariably blow from the N. or NW.; but as the sun recedes to the tropic of Capricorn the winds become variable, blowing from the E. and W., passing to the N. about the vernal equinox, and blowing from this quarter till about the end of May or the beginning of June. During this season Egypt is at intervals visited by the pestilential hot winds of the desert, here called khamsin, but identical with the simoom of the Arabs, and the samiel of the Turks. They have the same effects as in Arabia and other contiguous countries. (See Arabia.) Their heat is sometimes excessive; the soil is parched, and broken by chasms; the trees are stripped of their foliage, and the herbs of their verdure. The fine impalpable sand with which they are loaded obscures the sun, and gives to everything a dusty appearance. During this simoom the streets are deserted, and are as silent during day as during night. The rising of the Nile terminates these seasons of heat and drought, and again diffuses life and gladness over the land. The beneficent river

'From his broad bosom life and verdure flings, And broods o'er Egypt with his watery wings.'

The saline properties of the earth, or, as Volney supposes, of the air, in conjunction with the heat of the climate, give to vegetation an activity in Egypt unknown in cold climates. Wherever plants have water the rapidity of their growth is prodigious. But it is a curious fact, that the soil is exceedingly unfavourable to exotics, and that the seeds of those raised in the country require to be annually renewed. (Volney, Voyage en Syrie et en Egypte, i. 61-65, ed. 1787.)

In consequence of the extreme dryness of the air, comparatively little rain falls in Egypt; and some seasons have passed away without the occurrence of a single shower. But this is not usually the case, and occasionally the rains are pretty heavy. In this respect there is a great variety in the seasons; and, according to Marshal Marmont, falls of rain would appear latterly to have become comparatively frequent. He says that in Lower Egypt they have now pretty generally from thirty to forty rainy days in the year; and that the pacha has constructed immense warehouses for the securing of products in harvest, which were formerly exposed without inconvenience to the open air. (Voyage, &c. iii. 177.) No doubt, however, the rains have been quite as frequent and heavy in Egypt in past times as at present. In proof of this we may mention, that the learned and accurate Mr. Greaves, who visited Egypt in 1638 and 1639, states that the rains were heavier at Alexandria in December and January, than he had known in London; and that there were also, at the same time, very heavy falls in Cairo. (Pyramidographia, 'Works,' i. 303.) Hail showers occasionally occur in winter at Alexandria, and sometimes, though rarely, in Cairo. Snow is totally, and thunder and lightning nearly, unknown in Egypt. Earthquakes occur but seldom, but they are not unknown.

Diseases.—The inhabitants of Egypt are subject to a variety of diseases, some of which seem to be, at least in their extent, to a considerable degree peculiar. Of these ophthalmia is one of the most prevalent: nothing appears more extraordinary to a stranger in Cairo, than the number of persons whose sight is either lost or impaired. It is more common in Lower than in Upper Egypt. 'It generally arises from checked perspiration, but is aggravated by the dust and many other causes. Where remedies are promptly employed, this disease is seldom alarming in its progress; but vast numbers of the natives of Egypt, not knowing how to treat it, or obstinately resigning themselves to fate, lose one or both their eyes.' (Lane, i. 1.) Small-pox and leprosy are also very frequent. Elephantiasis is met with among labourers in the rice fields; and, in the marshy districts of the interior, the legs often swell to an enormous size. Syphilis is exceedingly prevalent; and malignant fevers prevail in April and May. The plague occasionally breaks out with great violence in Egypt; and in 1835 it destroyed 80,000 persons in Cairo only, and in 1835 its effects were still more fatal. Scarcely any year passes without this formidable disease making its appearance. It generally, though not always, breaks out during the prevalence of the Khamsin, or hot wind from the desert. But notwithstanding this formidable list of diseases, it is still true, as already stated, that no part of Egypt can be justly characterised as insalubrious. The diseases to which the people are subject are mostly to be ascribed to their depraved circumstances—their filth, miserable accommodations, and the bad quality and deficiency of their food. Much also is owing to their apathy, their belief in the doctrine of predestination, and, consequently, in the inutility of remedies and precautions, and the inefficiency of the police.

Plants and Animals.—The vegetable productions of Egypt are of a nature peculiarly fitted to its exigencies. The absence of rain forbids the existence of forests; and there being no high mountains, alpine productions are nowhere found. The native plants of Egypt are of a loose, plethoric texture; so that their proper aliment is prepared in continually distended veins, whose widely-opened mouths receive and retain the copious dews, and cause the leaves to perform the functions of so many roots. Hence, great transpiration is excited, and the continually moist roots enable the plants to pass from the extreme drought of summer to the humidity of a three months' flood. These characteristics will be found in the celebrated papyrus, the lotus, and its three varieties. Egyptian arum and safflower, Bulbs find a congenial soil in Egypt, and the gourd and cucumber tribe are everywhere planted. The acacia of the Nile, and date palm, and sycamore, are scattered rather than grouped over the country. The constant use to which the soil is put in rearing valuable plants prevents the accumulation of such as are noxious and weeds; so that the country is remarkably free from them. The number of fruit trees in Egypt by no means answers to the culture and fertility of the soil.

The peculiar hydrography and vegetation of Egypt exercise a great influence over its zoology. The larger species of wild animals find no forests in which to prowl—no recesses for their dens; and except those monsters of the Nile—hippopotami and crocodiles—are banished from the land. Birds also, that inhabit mountains and groves, avoid the exposed deserts and scorching fields around the Nile. The country is also unfriendly to some insects; their eggs and chrysalides being either washed away by the overflowings of the river, or smothered in the stagnant pools formed by its overflow.

The only primeval animals now left in Egypt are the hippopotamus and crocodile. The former, so poetically yet accurately described in the book of Job (xl. 15–24), has been known to measure 16 ft. long, 15 ft. in circumference, and to stand 7 ft. high. The skin is sufficiently thick and tough to withstand the effect of a musket-ball. Though amphibious, the animal is not nearly so powerful on land as in the water. Its appetite is enormous. The Nile crocodile is a lizard of enormous size, covered with a complete armour of bull-proof scales; its feet are provided with strong sharp claws; an immense mouth, opening as far as the ears, exhibits two rows of teeth like saws,

fitting into each other when closed. This is also an amphibious animal; but more than one-fourth part of its existence is passed in water, and, like the hippopotamus, it is a most voracious eater. The Ichneumon is a persevering destroyer of the eggs of crocodiles and serpents. The jerboa, or jumping mouse, Nilotic fox. Egyptian and Alexandrian rat and arvicola, complete the list of wild animals. The domestic and tame animals are chiefly oxen and buffaloes, which are employed in agriculture: the Egyptian goat; dogs, of which there is a peculiar breed at Alexandria; and the true cat, a native, it is supposed, of Egypt. Horses are much esteemed, and the Egyptian grooms are reckoned among the best in the world. Asses are in requisition all over the country. Lastly, the camel and dromedary yield their important services to the inhabitants of this desert-bounded land. (See ARABIA.)

Of the feathered tribe peculiar to Egypt, the first to claim attention is the Ibis, so often mentioned by ancient writers, and identified by Bruce with the abuhannes,—a species of curlew, placed by Cuvier amongst the grallæ, or wading birds. Its size is equal to that of a hen, with white plumage, except the tips of the quill feathers, which are black, the largest of them having violet reflections. Part of the head and neck are naked; black in the adult, but clothed with short black feathers in the young. (Cuvier's Animal Kingdom, by Blyth and others, 243.)

The Egyptian vulture and stork perform the office of scavengers in towns, by feeding upon the animal substances that would be otherwise left to corrupt the air. Pelicans are numerous along the banks of the Nile, and have a beautiful plumage. Pigeons are kept by almost every farmer in the country for the sake of their dung, and are provided with curious conical huts. Poultry abounds in Egypt; and the artificial mode of hatching eggs forms an important branch of Egyptian industry. Plovers, bustards, and partridges are often met with; quails visit the land in immense flocks, from the interior of Africa; and sea swallows abound along the base of the Delta, and on the shores of the Red Sea. History, sacred and profane, attests the predilection of the Egyptians for *fish* as an article of food; and the Nile abounds with it. Nile salmon is highly esteemed. The fishermen of the coast form an important and turbulent community. Besides the crocodile, the reptiles of Egypt are numerous. Serpent-charming is a regular profession; and some of the Arabs really perform extraordinary feats with the most venomous snakes. The horned and hooded viper (*Coluber cerastes* and *C. Haje*) are the most dangerous. Insects abound in Egypt during a great part of the year, particularly flies and mosquitoes, (Lane, i. 3.) Locusts also occasionally avenge the land, visiting it in such immense flights as to obscure the sun's rays, and destroying when they alight every vestige of herbage. The breeding and keeping of bees forms an extensive branch in the rural economy of the country. The beetle peculiar to Egypt (*Scarabæus sacer*), so often represented on the sacred monuments, is rather larger than the common beetle, and is entirely black. The Egyptian bat is also much larger than that of other countries. Zoophytes abound in the Red Sea, and it is the red coral which supplies its name. Sponges, various corallines, polypes, and madrepores, are also found on its shores, (Hasselquist's Appendix to Voyages and Travels in the Levant, &c.; Richardson's Travels, *passim*; Russell's Egypt, 464, &c.; Coude's Egypt, *passim*.)

Population, Manners, and Customs.—The political revolutions to which Egypt has been sub-lect from the earliest historical era have—as the Persians, Greeks, Romans, Arabs, Turks, and other nations, gained in their turn the ascendency—introduced into the country people of all their races. These, added to the Copts, descendants from the ancient Egyptians, slaves from the Upper Nile countries, a small number of Jews, and a few Europeans, make up the motley congregation at present assembled in the land of the Pharaohs. But of all its conquerors, Mohammed has left the most permanent traces in Egypt. The descendants of the Saracens who fought under his banner form by far the greatest portion of the present population. In the absence of more precise data, the total number of inhabitants may be estimated at about 5,000,000; of whom Arab-Egyptians 3,000,000; Copts or Christian Egyptians, 500,000; Turks, 200,000; Greeks and Armenians, 200,000; and the rest belonging to various tribes scattered over the country.

The Arab-Egyptians are divided, by Volney, into three classes. The first are the fellahs or husbandmen, the posterity, he says, of the Arabs, who emigrated from the peninsula after the conquest of Egypt by Amrou in 640. They still retain the features of their ancestors, but are taller and stronger. In general they reach 5 ft. 4 in., and many 5 ft. 6 or 7 in. Their skin, tinged by the sun, is almost black. They have oval heads, prominent foreheads, large but not aquiline noses, and well-shaped mouths. They constitute the bulk of the Egyptian peasantry. The second class of Arabs are Moghrebins, or settlers from Mauritania. They are very numerous in the Said, where they live in villages by themselves; they likewise are fellahs. The third class are Bedouins of the desert, or wandering tribes.

The Arabs, particularly the Bedouins, wherever they are found, have a remarkable identity of appearance and character. (See ARABIA.) But the fellahs or husbandmen of Egypt, having been subjected for centuries to a despotical government, and deprived of that wild freedom that is now, as of old, enjoyed by their brothers of the desert, have lost several of the distinguishing traits of the Arab character. They are rigid Mussulmen, and strictly observant of the religious rites and ceremonies laid down by their sheiks, or priests. Mr. Lane says:—' Very few large or handsome houses are to be seen in Egypt, excepting in the metropolis and some other towns. The dwellings of the lower orders, particularly those of the peasants, are of a very mean description: they are mostly built of unbaked bricks, cemented together with mud. Some of them are mere hovels. The greater number, however, comprise two or more apartments; though very few are two stories high. In one of these apartments, in the houses of the peasants in Lower Egypt, there is generally an oven (fooru), at the end farthest from the entrance, and occupying the whole width of the chamber. It resembles a wide bench or seat, and is about breast high; it is constructed of brick and mud; the roof arched within, and flat on the top. The inhabitants of the house, who seldom have any night-covering during the winter, sleep upon the top of the oven, having previously lighted a fire within it; or the husband and wife only enjoy this luxury, and the children sleep upon the floor. The chambers have small apertures high up in the walls, for the admission of light and air—sometimes furnished with a grating of wood. The roofs are formed of palm branches and palm leaves, or of millet stalks, &c., laid upon rafters of the trunk of the palm, and covered with a plaster of mud and chopped straw. The furniture consists of a mat or two to sleep upon, a few earthen vessels, and a hand-mill to grind the corn. In many villages large pigeon-houses, of a square form, but with the walls slightly inclining inwards (like many of the ancient Egyptian buildings), or of the form of a sugar loaf, are constructed upon the roofs of the huts, with crude brick, pottery, and mud. Most of the villages of Egypt are situated upon eminences of rubbish, which rise a few feet above the reach of the inundation, and are surrounded by palm trees, or have a few of these trees in their vicinity. The rubbish which they occupy chiefly consists of the materials of former huts, and seems to increase in about the same degree as the level of the alluvial plains and the bed of the river.' (Lane, Modern Egyptians, 30, 31.) The dress of the peasantry consists of coarse woollen cloths; and, like all Orientals, they are fond of attending coffee-houses, and listening to the tales of professed magicians, or the rude music of strolling musicians. They submit, without murmuring, to every species of ill-treatment; principally, it seems, from a deep-rooted conviction of its inutility, which has degenerated into an apathy that now forms the main feature of their character. They are—in spite of diet both poor in quality and scanty in quantity—robust, healthy, and capable of undergoing great severity of labour and fatigue, being muscular without deobesus or corpulency. Like Bedouins, they have a habit of half-shutting their eyes, from constant exposure to the sun. The women are in a most degraded condition, and perform all the laborious and menial offices. The Bedouins, or wandering Arabs, have a great contempt for the established peasantry of Egypt, and apply to them the name of fellahs, as one of contempt, signifying boors; distinguishing themselves as true Arabs (bedaweea). The latter, whenever they please, take the daughters of the former in marriage, but will not give their own daughters in return. Should a Bedouin be slain by a fellah, blood revenge is often perpetrated upon the offending tribe three or four fold.

The Egyptian Christians, or Copts, are usually regarded as the descendants of the ancient Egyptians; and it is believed that their written language is identical with that spoken by their ancestors. Some learned men have supposed, from certain resemblances between the Hebrew and Coptic, that the latter was a dialect of the former, or that it belonged to the Semitic languages. But this opinion is now all but abandoned. Michaelis says, that ' every person competent to form an opinion knows that the Coptic and the Hebrew have not the slightest original affinity; and that although many words occur in the former that resemble Semitic vocables, they are to be attributed to the influence which the proximity and intercourse of Semitic nations have exercised over the idiom of the native Egyptians.' (Quoted by Prichard, ii. 211.) The characteristics of the Coptic language are shortness of the words, and the simplicity of its grammatical construction; its genders and cases are expressed by prefixes and infixes, and not, as is usual with Asiatic and European languages, by terminations. (See Quatremère, Recherches sur la Littérature Egyptienne.) The modern Copts, however, speak Arabic, their original tongue being understood but by few persons; and though their liturgy be written in Coptic, it is expounded in Arabic. (Lane, ii. 312; Quarterly Review, lix. 170.) They are sober and steady; are much employed as secretaries in public offices; and are the head accountants in the country, few respectable traders being without a ' Coptic clerk.' They are held in great esteem by the government, and possess certain immunities, being accumulated in their religion, and exempted from military conscription; for which privilege, however, they compound by payment of a tribute. Their patriarch, though called the patriarch of Alexandria, resides in Fostादt, or Old Cairo. Many conflicting opinions have been entertained as to the physical characteristics of the ancient Egyptians. Their early and high civilization, and their great works, show conclusively that they were of a very different race from most other African nations. Cuvier, who states that he had examined the heads of more than fifty mummies, declares that not one of them had any of the distinguishing characters of the Negro or Hottentot races; and he concludes that they belonged to the same race of men as the Europeans. Even at this day the appearance of the Copts contrasts most advantageously with that of the Arabs. M. Pugnet, an intelligent and discriminating physician, observes, 'A l'extérieur déif et misérable des Arabes, les Coptes opposent un air de majesté et de puissance; à la maissse de leurs traits une affabilité soutenue; à leur abord inquiet et sournois une figure très-épanouie.'

The Turks settled in Egypt, though comparatively few in number, occupy important social positions, being masters of the country. They fill all the high offices of state, which are, however, often enough made the objects of bargain and sale, and administered with little impartiality. The Greeks and Armenians are chiefly devoted to mercantile business, and many of them persons of great influence; but the Jews, about 5,000 in number, are a despised, and therefore a distinctive class in Egypt, and have a particular quarter of every large town set apart for their residence—generally the most confined and dirty portion of the place. They are usually bankers, money changers, gold and silver-smiths, &c., and enjoy a fair share of religious toleration. Slaves, chiefly from Nubia, Abyssinia, and Darfur, are introduced in large numbers, and are sold in public markets belonging to every moderately sized town. (Lane, ii. 311–353; Burckhardt's Arabic Proverbs, passim; Niebuhr's Travels.)

The following statements, as to the condition of the labouring classes in Egypt, apply principally to the fellahs; they were supplied by an English gentleman long resident in the country:—

'With the labouring classes of Egypt bread is the great article of food, and may be said to be there more properly the staff of life than in any other country. Beans and lentils are next in importance. With bread, as a sort of seasoning, they use the yam, radish, cucumber, date, onion, and at certain seasons the melon, of which there is great abundance, and occasionally also cheese, and a sort of butter or ghee, in common use for cooking. Fish, too, and particularly the dried fish of Lake Menzaleh, is a favourite article of food. Rice is less used here than in most eastern countries, being dearer than bread; but still their favourite dish of pillau, or rice and butter mixed, with the addition of a fowl or meat, if the party happen to be of the better sort, is sometimes to be seen. Butcher's meat is beyond the reach of the labouring classes; and unless at their great festival of the Bairam, when the duties are taken off, they rarely taste it. But though thus living in a great measure on vegetable food, they are a robust and healthy people, capable of undergoing great fatigue; and in despite of the general unprepossessing appearance of both sexes, there are often to be seen specimens of the human form of matchless symmetry and beauty, particularly among the boatmen on the Nile. Their only luxuries are coffee and tobacco; the latter a coarse description produced in the country, yet still such

a solace to the poor man, that while he has it he seldom complains, though all else were wanting. The dress of both sexes consists of a coarse blue cotton shirt manufactured in the country, without anything else, except the red and often less showy cap, which covers the head, the shawl and broad cloth so ambitiously worn by the upper classes being far beyond the reach of the humble labourer. The richer classes of natives, including those in offices of trust under the government (which is the major part), or engaged as retailers or handi-craftsmen in such pursuits as yield a decent livelihood, live as well and as fully, and are as well clothed, as the same classes in any other country.

'In Egypt, where there is no personal liberty—where the government claims and enforces its right to the labour of every man, willing or not willing, on his own terms,—where among the native traders there is no property, or if it exist is not seen,—where no enterprise can be undertaken but with consent of the government, or at the risk of clashing with some of its private interests, there can be no proper rate of wages as applicable to any particular trade, nor any chance of the remuneration for labour being bottomed otherwise than upon favour or caprice. The native artisans, as cutlers, silk weavers, shoe-makers, saddlers, coppersmiths, &c., confine their operations to their own little booths and shops, and usually find in themselves and their families sufficient hands for all their work; and the same applies to all retailers of silk and cotton goods, coffee, tobacco, sugar, and every other article of consumption. Were a capitalist, supposing him possessed of the authority of the government, to embark in any enterprise, he would be almost sure to come in competition with the pacha, and to be driven out of the field, commanding, as the latter does, all the labour of the country at his own price, besides having monopolies of nearly every thing consumed in the country. Hence it will be seen that it is upon the pacha the whole labouring classes must mainly depend for support; and it has been generally stated, that whatever be the nature of the work, the average rate of wages paid by him does not exceed a piastre per day for a full grown man; one half usually in bread, upon which he has his profit, for he is a large baker also, and the other half in money. To women and children he pays from 10 to 20 paras per day. A Frank cannot command the labour of the same people for less than double the money. In the manufactories men who have made themselves remarkable for their skill are occasionally to be found drawing from 3 to 6 piastres per day, but these are rare exceptions. The common rate of one piastre per day may be said just to preserve the parties in existence, and that is all.'

Government and Laws.—Egypt, whose history commences with the history of civilised man and organised government, which gave laws to the old world and art to the Greeks, after being, for many centuries, subjected to foreign masters, became, at length, a province of the Ottoman empire. Under the Turkish sway it was long her fate to suffer that worst kind of despotism resulting from the delegation of arbitrary power by a careless tyrannical master to a scarcely responsible servant. The bold, innovating spirit of the first independent ruler of Egypt, the celebrated Mehemet Ali, has, however, introduced several reforms into the administrative constitution of the government, which have been upon the whole beneficial. The government of Egypt, under the successors of Mehemet Ali, and as at present organised, consists of, 1st, the viceroy, or pacha, whose power is unlimited and despotic. 2d, His deputy, called Kilsay'a. 3d, Seven councils of state, who have each a distinct department of the government to preside over. 4th, Governors (*Nazir*) appointed to each prov. By the imperial *Hatti-Scheriff*—lit. the illustrious writing—of Jan. 13, 1841, and a firman of the sultan of June 1, of the same year, the government of Egypt was made hereditary in the family of Mehemet Ali, subject to an annual tribute of 80,000 purses, or 400,000l. to the Turkish government. Under the new form of government, the country is divided into three great provinces, viz. *Saïd*, or Upper Egypt; *Vostani*, or Middle Egypt; and *Bahari*, or Lower Egypt. These provinces again are divided into 7 intendencies, and subdivided, after the French system, into departments and arrondissements. In other respects, the successors of Mehemet Ali have tried to imitate the forms of government of Imperial France. The civil and criminal laws are administered by a *cadi*, or chief judge, and his deputy, or *naïb*. But most of these officers being filled by Turks, who speak their own language, an official interpreter is necessary. The court of the cadi has also its *bash raoul* (chief sergeant of arrests, which are executed by his inferior officers); its *bash kâtib*, or chief secretary; and *shâhids*, or recorders, who prepare the business of the court, and relieve it of such details as would unnecessarily take up the time of the cadi. Petty cases are at once decided by a *nâib* or magistrate. The police is numerous and effective, and consists of the military and the magistrates, or zabtï police. Though still very defective, the administration of justice in Egypt has been vastly improved under the government of the successors of Mehemet Ali. Except in rare cases convicts are usually punished by being compelled to labour at the public works.

The pacha is, with some few exceptions, proprietor of all the land of Egypt; and he is, in fact, the only considerable agriculturist, manufacturer, and trader in the country. The pacha is the sole manufacturer, printer, and bleacher of cotton goods; the sole maker of sail-cloth and Fez caps; the sole glass-blower, paper-manufacturer, iron-founder, gun-maker, gunpowder-manufacturer, &c.; he has the monopoly of opium, indigo, saltpetre, and flaxseed oils; he is the only tanner in his dominions; he is the owner of all the mills and manufactories, and of more than half the camels, horses, buffaloes, and cattle in the country; and of half the lands. He specifies the employments in which the bulk of the pop. shall engage; the crops of produce they are to raise or furnish, and the prices at which, when produced, they are to deliver them to his agents. This system of administrative interference, described in detail by Marshal Marmont, is carried out in the following manner. The head cultivator (*chef de culture*), in conjunction with the head civil authority (*cheyk-el-beled*) of each village, makes every year a division of the lands to be cultivated by the inhabitants; this division having been made, the kind of culture to which each portion is to be applied is determined,—so much being devoted to dhourrah; so much to wheat, barley, pulse, and trefoil; so much to sugar, rice, cotton, and indigo. The quantity of dhourrah to be cultivated is regulated according to the quantity presumed to be necessary for the support of the cultivator's family; and the produce is given up wholly to him for their support. The other products are divided into two classes. The different kinds of wheat, barley, pulse, and trefoil belong to the cultivator, after the quantity of each demanded by the pacha has been deducted; this quantity varies every year, but is most commonly one half of the produce. The remainder, including rice, cotton, sugar, indigo, opium, and wood,

are reserved exclusively for the pacha. The cultivator is prohibited, under the heaviest penalties, from retaining the smallest portion of any one of these articles; they are deposited in the public magazines established throughout the country, and placed to the account of the fellahs at a price fixed by the pacha, which never exceeds two-thirds of the market price.

The fellah has to pay to the pacha the *miry*, which may be regarded either as a land-tax, or the rent of the land. This impost is regulated according to the quality of the land; the maximum is 28 patakis (15s.), the minimum 17 patakis (9s. 1d.) the faddan. The average may be about 9s. or 10s. an acre. The fellah pays, moreover, a personal tax, which varies, according to the presumed circumstances of the individual, from 15 piastres to 5 cents. His cattle is also taxed; oxen and cows at 30 piastres, and at 70 piastres when they are sold to the butcher; on the animal being killed, the skin belongs to government. An account is opened by the village tax-gatherer with each inhabitant: the fellah is credited with the value of the produce which he has deposited; and debited with the miry, and the other imposts, as well as the prices of the articles with which he has been furnished, which always exceed their value. The accounts are balanced every four years.

A cultivator, included in that portion of a district on which the corn required by the pacha is ordered to be grown, if he wish to commute for the delivery of that article by a money payment, is charged at the rate of thirty-six piastres the ardeb, and he will generally rather pay this sum than double the sum at his credit with government; such credit being of no service to him, since it is never paid; while, by selling his corn, even with the duties and the thirty-six piastres which he pays, he receives at least, in money, a fourth or fifth part of its value.

It must be admitted, notwithstanding the grinding oppressiveness of this system, that it has materially improved the agriculture of the country; and that some new and important branches of culture have been introduced, as that of cotton, now a staple product. Marshal Marmont states, and the fact can scarcely be doubted, that these improvements never could, under any system, have been effected by the fellahs, who are ignorant, attached to old habits, and easily satisfied. But the vice of the present system is, that the fellahs reap no advantage whatever from this increased production. On the contrary, it has stripped them of not a few of their limited enjoyments, and rendered them more impoverished and depressed than they ever were at any former period of their history; their increased labour, instead of bringing with it an increase of comfort, brings only an increase of privations. Hence, were anything to occur that should overthrow the government of the successors of Mehemet Ali, the whole fabric would fall to pieces. It is forced, factitious, and unnatural; and is certainly not based on or associated with the interests or affections of the people.

The gross yearly revenue of Egypt and its dependencies is unknown, and various estimates have been framed of its amount. It probably amounts to 790,000 purses, or about 3,950,000l. The expenditure, for a number of years, has been larger than the revenue. The debit has created a floating debt which, in the beginning of 1860, amounted to 5,000,000l. In August of this year the government contracted a loan of 28 millions of francs, or 1,120,000l., in Paris; and a second loan of 40 millions of francs, or 1,600,000l., was contracted in March, 1862. The accession of the

Egyptian government still increasing, a third loan of £3,000,000 sterling was effected in Paris and London in October, 1861.

Army and Navy.—The regeneration of the army was one of Mehemet Ali's first projects on attaining to the pachalic of Egypt. To accomplish this, to consolidate his government, and to pave the way for his other reforms, the reconstruction, or, if that was impossible, the destruction of the Mameluke force that had so long ruled in Egypt, was indispensable; and this Mehemet accomplished, partly by force and partly by treachery. This superb cavalry being destroyed, with the exception of a small party who enrolled themselves under the banners of the pacha, the latter commenced his work of military reform with equal vigour and success. He had long been sensible of the vast superiority of European tactics and discipline over the brave but tumultuary onsets of Asiatic troops, and he was determined at all hazards to introduce the European system into his dominions. With this view he had his troops drilled and disciplined in the European fashion, chiefly through the instrumentality of some Italian officers. But the natives were naturally disinclined to the change; and the injudicious severity with which it was attempted to be introduced and carried into effect, gave rise to a dangerous mutiny, that threatened to put an end to the projects and power of the pacha. Mehemet having succeeded in suppressing this formidable insurrection, saw his error, and resolved to proceed with greater caution. With this view he formed a depot of fellahs in Upper Egypt, and had them trained in the European manner by a French officer, Colonel Selves, who changed his name and title into Solyman Pacha. This officer, who had served with distinction under Napoleon, undertook the arduous task of new-modelling the army of the pacha, and of organising and disciplining it according to the most approved models; and by a rare combination of firmness, bravery, and good sense, he succeeded in gaining the confidence both of the pacha and the troops. The army organised in this manner is raised by conscription, which, in consequence of the limited pop. of the country, is very severe. The number of troops, in 1838, including veterans and invalids, amounted to 127,286, besides from 10,000 to 12,000 irregular Turkish troops, and the Bedouin Arabs, who can furnish 30,000 more. But the successors of Mehemet Ali, especially his grandson, Ismail Pacha, who succeeded to the government in 1863, greatly reduced this large army. From a semi-official statement of Sept. 1864, it appears that at that time the regular army numbered but 14,000 men, namely, 8,000 infantry, 3,000 cavalry, artillery, and engineers, and 8,000 black troops.

The Egyptian navy, also a creation of the founder of the present dynasty, comprised, in 1863, seven ships of the line, six frigates, nine corvettes, seven brigs, and eighteen gunboats and smaller vessels, besides twenty-seven transports. Many of these ships, constructed by native builders, are beautifully modelled; and though the crews have not attained to the proficiency of English or American sailors, they have, regard being had to the circumstances under which they have been placed, made the most extraordinary advances.

Literature and Education.—The literature of the Arabs is very comprehensive. The works on religion and jurisprudence comprehend about one-fourth of the entire number of Arabic books. Others on grammar, rhetoric, philology, history, and geography, are also numerous; as are also their poetical compositions. There are many large libraries in Cairo, most of which are attached to the mosques. A system of public instruction has been organised by the late Mehemet Ali, which deserves high praise. The pupils are first sent to the 'primary' schools, of which there are fifty throughout the country. The youth having acquired the rudiments of education at these, they are advanced to the 'preparatory' schools. The next step is to the 'special' schools, which are ten in number, each devoted to particular studies; namely, medicine, midwifery, veterinary surgery, languages, music, and agriculture, the other three being military schools, to fit the scholars for the cavalry, artillery, or infantry service. At Abouzabel in Cairo, the pacha established a military hospital and a medical college. The success that attended this establishment has been quite extraordinary; and notwithstanding their old prejudices, many of the Arab pupils have become expert anatomists and clever surgeons.

It is impossible to appreciate too highly the beneficial influence of these establishments; they have already effected, and will, no doubt, continue to effect, a very great revolution in the public mind in the East; and will pave the way for reforms and changes of which, at present, no one can form any distinct idea. Almost every mosque or public fountain has a school attached to it, mostly endowed by benevolent persons. At these schools, getting the Koran by heart forms the chief employment; but reading and writing are also taught; those who aspire to the higher branches of learning become students of the University of El-Azhar at Cairo, the principal seat of learning in the East. In this building are certain riwaks, or colleges, set apart for the natives of particular provinces. The regular subjects of study are grammar, rhetoric, Mohammedan theology, and the traditions of the Prophet; law, religious, civil, and criminal; algebra, and arithmetic. The sciences are but imperfectly understood in Egypt, though great improvements have been made in medical science, in consequence of the introduction of European practitioners, and natives being sent to Europe to study. Egyptian geography describes the earth as a flat surface; and astronomy, beyond merely computing the calendar, is studied for the purposes of astrology. Music affords a favourite study for pastime, but the theoretical system is complicated, as each note has three intervals or gradations of sound instead of two. Their melodies are mostly of a plaintive kind; but a kind of recitative, in which they chant their romances, has some bold measures. (Lane, l. 2?c, et seq.; Wagbuen's Egypt in 1838, Appendix; Egypt, a Popular Description, 182–190.)

Productive Industry.—No soil can be better adapted for agriculture than that brought down by the Nile, and deposited on its banks. The earliest authentic records of the human race represent Egypt as the granary of the old world, to which less fortunate nations resorted in times of scarcity; while she received from them, in exchange for the necessaries of life, all those luxuries and riches which enabled her people to make such early progress in the arts, and to leave behind them monuments surpassing even the remains of the classic world in continuous, extent, and grandeur. The supplies of slime annually brought down by the river considerably abridge the labours of the husbandman, and have enabled the country, with but little of his assistance, to bear for the last 3,000 years three, and sometimes four annual crops, without the least impoverishment. The husbandry of Egypt is divided into two great classes:—the upper, or sherdee lands, where the banks are too high for the country beyond them to benefit by the inundation; and the *sri*, or low lands, which are watered by the natural overflowings of the river.

About four millions of *feddáns* are now under cultivation in Egypt, of which from 700,000 to 800,000 are occupied with cotton; 1,000,000 with flax, indigo, sugar, dates, hemp, &c.; and the other 2,000,000 or 2,700,000 feddáns with grain, principally millet (*dhoorra*), maize, wheat, and rice. In Lower Egypt sowing commences immediately after the waters subside; the seed only requiring to be strewed over the land, and it either sinks into the soft earth by its own weight, or is trodden down by cattle driven over it. This is generally done in November; in February the fields are verdant, and in May the harvest takes place. In July rice and maize are again planted, and yield a second harvest in September. In Upper Egypt the constant artificial supply of irrigation required by the land gives to the farmer unceasing employment. Deprived of rain, and exposed almost always to a burning sun, the land would be arid and barren if not constantly refreshed with moisture. After the water has been preserved in canals and wells, it is raised by Persian water-wheels, worked by oxen, or by means of a hand-machine of a more simple construction. Sowing begins here about November, and the corn begins to spring up before the end of the month, and by December given to the country the appearance of a verdant spring. In January lupines, dolichos, and rumina are sown; and towards the end of the month the first barley harvest commences. In February sugar-canes are cut for the press. By April flax has ripened, and the plants are pulled up; tobacco leaves are gathered, and the wheat harvest is got in. In July there is a third crop of trefoil, and a second of rice. October is the month for all sorts of leguminous seeds to be sown. (Malte-Brun, iv. 45-46; Burckhardt's Arabic Proverbs, 134; Wilkinson's Topography of Thebes.)

The efforts of the present government have been principally directed to the culture of cotton, so that the crops of wheat have greatly fallen off, and Alexandria has ceased to be a port for the shipment of this species of grain. The cotton of Egypt is long-stapled, of good quality, and the soil is well suited to its growth. By far the largest amount of this produce is exported to the United Kingdom. These exports more than doubled in quantity and more than quintupled in value in the five years 1859 to 1863. The exports of raw cotton to the United Kingdom, which were 836,313 cwts., valued 1,241,577£ in 1859, had risen to 835,283 cwts., of the value of 9,941,857£ in 1863. The shipments, however, decreased in 1864, and still more in 1865, owing to the restoration of peace in the cotton growing districts of the great American republic.

As the productiveness of Egypt depends wholly on the extent of the inundation and the command of water, it would be of vast importance to the country if means could be found of regulating the inundation, and preserving the waste of water, which is here the one thing needful. These important considerations, which engrossed a large share of the care and attention of the ancient rulers of Egypt, who excavated the lake Mœris with this view, have not been overlooked by the present rulers, who have projected and, to some extent, carried out great works for the regulation of the inundation. It is believed that it is by no means impracticable so to regulate the flow of the river that it might always be equally diffused over a much larger extent of country than at present, and that an inexhaustible supply might be secured for irrigation in the dry season. Fully realized, such works could not fail to double or treble the productive capacities of the country; and with these and private enterprise and industry, the wealth and population of Egypt in modern times might be as great as under the Pharaohs.

Commerce.—No country can be better situated for commerce than Egypt. She forms the link that connects the Eastern and Western worlds; and it is to her admirable situation in this respect, and to the commerce of which she in consequence early became the centre, that her ancient wealth and civilisation are mainly to be ascribed. It has been customary to trace the ruin of commerce in Egypt, in modern times, to the discovery of the route to India by the Cape of Good Hope; but more stress has been laid on this event than it really seems to deserve. No doubt it must probably would, under any circumstances, have diverted a portion of the trade with the extreme western states of Europe, and in the bulkier articles, into a new channel; but had the same facilities for conducting the commerce with the East existed in Egypt in the 15th and 16th centuries that existed in antiquity, the trade between India and the countries on the Mediterranean, and in the lighter and more valuable products, would, there is every reason to think, have continued to a great extent in the old channel. The truth seems to be, that the extinction of the trade through Egypt, at the epoch referred to, was mainly owing to its having become subjected to the lawless and arbitrary dominion of the Mamelukes, who loaded all articles passing through the country with oppressive exactions, and treated all foreigners, especially Christians, with insolence and contempt. But a new æra has begun; and the intercourse with the East has already in part reverted to its old channels. The establishment of a steam communication between Europe and India by way of Alexandria and Suez, with a railway through Egypt, is one of the most striking and important events in recent times. It has shortened the journey to India, from six months to a month, and has thus immensely contributed to strengthen the hold of Great Britain over her vast possessions in the East. At the beginning of 1845, Egypt had no less than 860 miles of railway, the most important of the lines being that of the 'Overland Route' from Alexandria to Suez, 223 m. in length. This line, constructed chiefly by English engineers and with English capital, has proved of the greatest benefit to the commerce and trade of Egypt.

It was one of the grand projects of the late Mehemet Ali to reconstruct the famous canal that formerly connected the Red Sea and the Nile. According to Herodotus, this canal was commenced by Necho, king of Egypt, and finished by Darius. (Lib. ii. § 158, iv. 39.) Under the Ptolemies, by whom, according to some authorities it was completed, this canal became an important channel of communication. It joined the E. or Pelusiac branch of the Nile at Bubastis, the ruins of which still remain; it thence proceeded E. to the bitter or natron lakes of Temseh and Chrih-Amded, whence it followed a nearly N. direction to its junction with the Red Sea at Arsinoe, either at or near where Suez now stands. It is said by Strabo (lib. xvii. p. 805), to have been 1,000 stadia (172 m.) in length; but if we measure it on the best modern maps it could hardly have exceeded from 85 to 95 m. Herodotus says that it was wide enough to admit two triremes sailing abreast. This great work having fallen into decay after the downfall of the Ptolemaic dynasty, was renovated either by Trajan or Adrian; and it was finally renewed by Amrou, the general of the caliph Omar, the conqueror of Egypt, anno 639. (Herodotus, par Larcher, iii. 450.) The French engineers traced the remains of this great work for a considerable distance; and during Mehemet Ali's lifetime great efforts were made to form a new 'Isthmus of Suez Canal.' In the end, an enterprising engineer, M. F. de Lesseps, succeeded in forming a company to carry out this object, and in the spring of 1865 the works were so far advanced that a party of delegates, representing the chief states of Europe and America, actually passed from the Mediterranean to the Red Sea by the new canal. The delegates performed the journey from the Nile to Ismailia by the freshwater canal, and from Ismailia to Port Said by the maritime (saltwater) canal. The boats were towed by camels and horses, and for a distance also on the maritime canal by steamers. After inspecting the works in progress at Port Said the delegates returned again south, sailing back to Ismailia with a fresh northerly wind. The maritime canal, in the spring of 1865, was not navigable for boats beyond Ismailia; the remainder of the journey therefore to the Red Sea had to be performed by the freshwater canal. This latter canal takes its source at the Nile, close to the town of Zagazig, and runs nearly due east through the once fertile land of Goshen until it reaches Ismailia; at about 2 m. from the latter town it branches off to the southward, leaving Lake Timsah and the Bitter Lakes on the east, and joins the Red Sea at Suez.

Ismailia is a flourishing and picturesque little town of 3,000 inhabitants, situated in the centre of the isthmus, 75 kilometres from Port Said, midway between the two seas at the north end of Lake Timsah, and owes its existence to the works of the canal. The width of the maritime canal varies at present, according to localities, from about 16 to 58 metres, except in the immediate vicinity of Port Said, where it attains 81 metres. Its average depth at present is not more than from 2 to 3 ft. The depth of the freshwater canal is much the same at present as that of the maritime canal, but, of course, it is deeper during high Nile; its width varies from about 16 to about 25 metres. The company of M. F. de Lesseps, it is stated, have contracted with various French firms for the completion of the whole of the works from the Mediterranean to the Red Sea by the 1st of July, 1868, at which date the company expect that the canal will be opened to navigation, at a cost to the shareholders not exceeding the subscribed capital of 8,000,000l. sterling. (Malta Times, May 4, 1865.)

The whole foreign trade of Egypt centres in Alexandria; and we beg to refer the reader to the article on that city for an account of the imports and exports of the country, and of the mode in which the trade is at present carried on.

Money.—Accounts are kept in Egypt in current piastres, each equal to something under 3d., there being 100 of them to the pound sterling. There are, besides, coins to represent the ½ piastre (sum edirah) and 1-40th (paddah or parah) and 5 and 10 faddah pieces. The most-used is a small gold coin, of the value of four piastres; and the khrayregh is equal to nine piastres, or 21d. and 5-8ths. These are the only Egyptian coins. There is, however, the nominal sum of purse, which stands for 500 piastres, and the kuruck, or treasury of 1,000 purses, or 5,000l. sterling. The coins of Constantinople are current in Egypt, but scarce. European and American dollars are pretty generally exchanged for 20 piastres. The English

sovereign is called gin gah, for guinea, and is freely taken.

Antiquities of Egypt.—A contemplation of the remains of antiquity scattered throughout Egypt, carries us back to a period of which history furnishes no other records than those derived from the monuments themselves. The temples, the palaces, and pyramids of the country, mark the spot where idolatry began—where civilisation commenced its career; while the annals of other nations prove that this land of gigantic fabrics had attained to a high degree of civil and social order and architectural proficiency, when the rest of the world was involved in barbarism. The range of objects presented to the archæologist may be classified thus :—1. Pyramids ; 2. Temples ; 3. Colossi and Sphinxes ; 4. Sculptures and hieroglyphics ; 5. Tombs and paintings.

1. The Pyramids, which, for vastness and duration, stand at the head, not only of all the monuments of Egypt, but of the ancient world, are placed at irregular intervals along the E. foot of the Libyan hills, at some distance from the W. bank of the Nile. They commence at Gizeh, nearly opposite to Cairo, in about 30° lat., and extend S. to about 29°. The pyramids of Gizeh, three in number, are the best known, the largest, and most celebrated. They stand on a plateau of rock, elevated about 150 ft. above the desert, about 7 m. W. by S. from Cairo. The pyramidal form seems to have been adopted in order to ensure stability. Their plan is that of a perfect square, and their sides contract by regular gradations till they terminate in a point, but so that the width of the base always exceeds the perpendicular height. They are not solid; at least chambers and galleries have been explored in some of the principal pyramids. The greatest of the pyramids of Gizeh, and indeed of Egypt—that of Cheops, the building of which is described by Herodotus, is a gigantic structure. The sides of its base, which are in the line of the four cardinal points, measure, at the foundation, 764 ft., so that it occupies a space of more than 13 acres. Its perpendicular height is about 460 ft., being about 100 ft. higher than the summit of St. Paul's. This huge fabric consists of successive tiers of vast blocks of calcareous stone, rising above each other in the form of steps. The thickness of the stones, which is identical with the height of the steps, decreases as the altitude of the pyramid increases, the greatest height being 4 ft. and the least 1⅚ ft. The mean breadth of the steps is about 1 ft. 9 in. The best authorities agree in estimating the number of steps or tiers of stone at 203. According to the information communicated to Herodotus by the priests, 100,000 men were employed for twenty years in the construction of this prodigious edifice; and ten years were employed in constructing a causeway by which to convey the stones to the place, and in their construction. (Lib. ii. § 124.)

The other pyramids are of inferior dimensions; but they are mostly all, notwithstanding, of vast magnitude—some smaller: they are not all of stone, some of them being of brick.

Many learned dissertations have been written, and many fanciful and a few ingenious conjectures have been framed to account for the original use and object of these imperishable structures. But the difficulty of the subject is such, that hitherto no satisfactory conclusion has been arrived at. Even in the remotest antiquity their origin was matter of doubt, and nothing certain was known with respect to them or their founders. (Plin. Hist. Nat., lib. 36, § 12.) On the whole, however, it would seem to be most probable that they were intimately connected with the religion of the ancient Egyptians; and that they were at once a receive of tombs and temples, but participating more of the latter than of the former character. (For some remarks on this part of the subject, see Shaw's Travels, p. 170, &c. 4to edit.; and Greaves's Pyramidographia, in his works, vol. i.)

It has long been customary to regard the pyramids as monuments merely of the power and folly of the monarchs by whom they were raised, and of the bondage of their subjects. This, however, seems to be a very superficial prejudiced view of the matter. The varying magnitude of the pyramids, the fact of their being scattered over a space extending lengthwise about 70 m., and their extraordinary number, appear to show pretty conclusively that they must have been constructed from a sense of utility or duty; and not out of caprice, or from a vain desire to perpetuate the names of the celebrity of the founders. If we had sufficient knowledge of antiquity, it would probably be found that the motives which led to the construction of the pyramids were, at last tom, nearly identical with those which led to the construction of St. Peter's and St. Paul's; and that they are monuments of the religion and piety, as well as of the power, of the Pharaohs.

It is impossible for any one to look at these stupendous piles without a deep sense of their sublimity. Their prodigious magnitude, the impenetrable mystery that hangs over their origin, and the purpose to which they were applied, and the conviction that they will endure long after the proudest existing monuments of human greatness have been levelled with the dust, awaken feelings that cannot be excited by any other display of the power and industry of man. The pyramids, too, are associated with some of the most interesting events in the history of the human race. They were probably gazed upon by Moses, and certainly were regarded with wonder and admiration by Homer and Herodotus, Pythagoras and Plato: Alexander the Great and Napoleon marshalled their hosts under their shadow; and they are no doubt destined to survive the homage of poets, historians, and philosophers, and to witness the exploits of warriors, through the all but endless series of future ages. (For further details as to the Pyramids, besides the authorities already referred to, see the Description de l'Egypte, tom. ix.; Modern Traveller, 'Egypt,' vol. i.; Clarke's Travels, vol. v. 8vo ed.; Greaves's Works, i. 1–164, ed. 1737; Ancient Universal History, i. 425–445; Herodote, par Larcher, lib. ii., with the notes; and a host of other works.)

2. *Temples.*—The remains of buildings devoted to religious worship form, next to the pyramids, the most considerable reliques of antiquity in Egypt. Reared after one uniform design, gigantic in size, massive in detail, and calculated to strike awe to the heart of the worshipper, they show how large a share religion occupied in the policy of the rulers, and in the moral condition of the people. Egyptian architecture has—unlike that of Greece—found few imitators; for the vastness and solidity it demands, the enormous proportions it exhibits, require an amount of labour and material only to have been furnished in the land of the pyramids. Hence the unvarying uniformity which all the specimens of it present, unmixed as they are with the additions of modern taste, untouched by the hand of improvement, renders the architecture of Egypt, above that of all other nations, the most characteristic and unique. The plan and appurtenances of an Egyptian temple consist, first, of the approach to it, or dromos; a sacred avenue, lined on each side with sphinxes,

and in some instances a mile long. This conducts to the entrance, or propylon, a principal feature in the building, consisting of pyramidal moles, with a rectangular face and sides, inclining one to one another than in the perfect pyramid, upon which the most elaborate sculptures were cut. Between them is the door; but before the door sometimes two obelisks rise beside two colossi, as in the temple of Luxor (Thebes). The number of these propyla and dromi is indefinite; occasionally three must be passed before arriving at the pronaos, or portico of the temple itself, which has a massive façade, supported by pillars. A doorway leads to the sekos, or cell, which is always divided into several apartments. A second door generally leads to an hypæthral hall, having a flat roof, supported by huge pillars. (Denderah.) Some of these halls are of immense size. Other chambers succeed, until the holy recess presents itself; an oblong room, with an altar and several holes sculptured in stone. To almost every apartment there are staircases leading to the terraced roofs, many of which are of such dimensions that at present Arab villages are built upon them. Although many of the temples are more than a mile in length, their interiors are uniformly covered in every part with the most elaborate sculptures. The structures will be found more minutely described under Denderah, Edfou, and Thebes. (Strabo, Ed. de Casaubon, 805; Egyptian Antiquities, i. 69–77.)

3. *Colossi, Sphinxes, &c.*—Although these have been invariably found as appendages to the temples, yet the important place they occupy in the antiquities of Egypt demands a separate notice. Immensity of size, no main an element in producing grandeur of effect, was the chief end of the Egyptian artist; and that this might take a stronger hold upon the imagination of the spectator, the largest colossi have usually placed near them a small figure for contrast and measure of magnitude. These representing men are always the figures of some deity, and were placed in pairs opposite the propylon. They are naked, except a head-dress and cloth bound round the waist. Some are sculptured of one entire stone (hence called monoliths), and were cut out of the quarries and transported to the temples at an enormous expense of time and labour. On the plain of Thebes, about half way between the W. desert and the Nile, are two colossal figures, about 50 ft. in height, seated each on a pedestal 14 ft. long, 14 ft. broad, and 6 ft. high. One of these, supposed to be the 'Memnon,' the most celebrated by far of the Egyptian statues, is said to have emitted sounds at sunrise or soon after, and when the sun's rays fell on its lips. Strabo saw the statue, and heard the mysterious sound; and Tacitus tells us that Germanicus visited the 'Memnonis Sacræ effigies, ubi radiis solis icta est, vocalem sonum reddens.' (Annal. lib. ii. § 61.) A portion of a similar statue, but of smaller dimensions, may be seen in the British Museum (No. 4, Egyptian Saloon), which was brought by Belzoni from the Memnonium. Besides these gigantic representations of deified human beings, those of other gods are met with throughout the country. The strangest are those ideal figures called sphinxes, some having a man's head, and lion's limbs and body (andro-sphinxes); others, the most numerous, with a female head; others again displaying a man's head.

Sphinxes were usually placed in those double rows which formed the avenues, or dromi, of the temples, and vary very much in size. The largest is that placed E. of the second pyramid of Ghizeh. It is an andro-sphinx, much of it buried in sand, but the head and a portion of the body are visible; the first measuring, from the chin to the top of the forehead, 28 ft., the body being above 100 ft. long; the face has been much mutilated. The excavations of M. Caviglia disclosed some curious approaches to this gigantic monster. On a stone platform, between the fore-paws, is a block of granite 14 ft. by 7 ft., and 2 ft. thick, highly embellished with sculptures in bas-relief; and on the second digit of the southern paw, a Greek inscription is deeply cut (given with others in the Quarterly Review, xix. 411), with a translation by Dr. Young. Between the legs of the sphinx, and on the ground in front of it, is a small temple, a plan of which may be seen in the Quarterly Review. (xix. 416.) Appearances around the sphinx indicate that it was originally enclosed within a wall. Besides the human colossi and sphinxes, other figures belonging to the Egyptian mythology are of frequent occurrence. All the colossi, of whatever denomination, were, it is supposed, coloured over in every part, many of them still exhibiting traces of paint. (Heeren's Researches, ii. 214, Engl. trans.; Quarterly Review.)

4. *Sculptures and Hieroglyphics.*—The preceding chapters only give an account of those specimens of Egyptian architecture and sculpture whose immortality, and, when compared with the classic elegance of Grecian models, whose uncouth forms might be deemed the first rude, though gigantic, efforts of the Egyptian artists; but a close examination of the ornaments with which the ancient buildings are profusely enriched, shows the great proficiency to which they had attained in the more refined branches of art. The obelisks, the walls, and all the apartments of the edifices described above, are covered in almost every part with sculptures executed with the most minute finish and exquisite skill. The ruined temples and obelisks of Egypt are, in fact, so many historical records. The wars and triumphs of the Egyptian sovereigns were, for the most part, the theme of the sculptor. The immense propylæa and walls of Luxor and Karnac, for example, give a vivid picture of the forms of pursuit, the attitudes of the victors, the wounded, and the dying,—the sea fights, the religious sacrifices and processions.

The *hieroglyphics* or figures, symbolical devices, and characters with which the Egyptian obelisks and other monuments are covered, are highly interesting, from the insight which they afford into the steps by which men were led to the use of a written language. The most obvious expedient for communicating substantive ideas would be by drawing figures of the objects: thus, a battle might be represented by the figures of armed men contending with each other. But this is a very clumsy and inconvenient mode of conveying information, and cannot be applied to represent mental feelings or abstract ideas. Hence pictorial are very soon superseded by or mixed up with symbolical or allegorical representations, which depict facts, qualities, or circumstances, by conventional or arbitrary marks; and these sorts of characters being, in the course of time, still further simplified, lose a great portion of their original pictorial character, and degenerate into what may be called a common, demotic, or cursorial writing. The Chinese is the most perfect example of this sort of conventional writing; and Dubable has given an interesting account of the steps by which it was derived from pictorial writing. (Dubable, Description Géographique, &c., i. 272, ed. 1756.) The present Chinese characters are, in truth, nothing but a refined and improved species of hieroglyphics, each character presenting to the eye a distinct object or quality. At this point the Chinese have stopped; and it seems never to have occurred to them to attempt

to mark the different sounds of the voice by characters or letters, and by combining these to form a written language. Now, it was long supposed that, like the Chinese, the characters on the Egyptian monuments were wholly hieroglyphical, and much learning and ingenuity have been expended in efforts to decipher them. It was lately, however, conjectured by Zoega (De Origine et Usu Obeliscorum, p. 454), that some of the characters on the monuments might be neither pictorial nor symbolical, but phonetic (from *φωνή*, vox); that is, that they might represent sounds, and not things, and be either alphabetic or syllabic, or both. Warburton had already shown how the refined symbolic writing might pass into the phonetic, but he erroneously concluded that the monuments afforded no specimens of the latter. (Divine Leg. iii. 161.) The surmise, for it was little better, of Zoega has since, however, been established by Dr. Young, Champollion, and others. But in doing this they had facilities unknown to Warburton, Zoega, and previous inquirers. The French, when in Egypt, discovered at Rosetta a stone, now in the British Museum, on which three inscriptions are sculptured; and it appears from the last and most perfect of them, which is in Greek, that the inscriptions are either entirely or substantially identical with each other, being the same royal decree which, it says, was ordered to be cut in sacred characters or hieroglyphics, in enchorial characters (that is, in modified or conventional hieroglyphics), and in Greek. The inscriptions are a good deal mutilated, particularly the hieroglyphical; but they are still sufficiently distinct to allow the hieroglyphical and enchorial to be compared with each other and with the Greek. The study of this trilingual stone enabled Dr. Young to determine, or rather perhaps conjecture with considerable probability, which of the enchorial and hieroglyphical signs were phonetic, and to fix their value. M. Champollion and others have since zealously followed up the path thus opened, but with no great or marked success. If, indeed, the Egyptian writing were either wholly figurative or wholly phonetic, a key to its mysteries might be discovered, and its long hidden treasures be again brought to light. But the most probable conclusion seems to be, that it is partly the one and partly the other; or that the characters are in a state of transition from the former to the latter. This, also, is the matured opinion of Champollion, who lays it down distinctly, in the second edition of his Précis du Système Hieroglyphique, that 'the hieroglyphic mode of writing is a complex system —a system *figurative, symbolical, and phonetic*, in the same text, in the same phrase, I would almost say in the same word.' An examination of the hieroglyphic writings must go far to satisfy every one that this is a tolerably correct statement. Many of the characters are purely pictorial; while others are mere arbitrary symbols, and may be, and most probably in some instances are, phonetic; or, which is the same thing, alphabetic or syllabic. In fact, no certain conclusions can be, or, at all events, have been drawn with respect to it. No doubt it was sufficiently intelligible to those who were instructed in its mysteries, but to those destitute of such instruction its interpretation must be a work of all but insuperable difficulty; so that there seems but little probability that the veil which covered this in antiquity should ever be wholly removed. (Besides the authorities already referred to, the reader may consult the art. 'Hieroglyphics' in the Encyc. Britannica, one of the most able and elaborate treatises on the subject that has ever appeared.)

5. *Tombs and Paintings.*—Every relic of the ancient Egyptians appears to have been originally

designed for an almost perpetual endurance. Their architecture,—the forms of which are mostly pyramidal, with bases that have withstood the most needless and continued destruction; their colossal sculptures,—many of them monoliths cut out of the solid rock; and even the bodies of their dead,—all seem to have been intended for eternity. Thousands of years have passed since many of the mummies recently unrolled were embalmed, yet every feature, every fibre, still remains. Even the colours of the paintings with which their sepulchres were adorned are still as vivid as if they had been laid on yesterday. So deep were their religious sentiments concerning dissolution, that they bestowed more labour and ornament upon the dwellings of the dead than upon the habitations of the living. 'They call,' says Diodorus Siculus, ' the houses of the living inns, because for a short space we inhabit them; but the sepulchres of the dead they call eternal mansions, because they continue with the gods for an infinite space. Wherefore, in the structure of their houses, they are little solicitous; but in exquisitely adorning their sepulchres, they think no cost sufficient.' (Diod. Sic. lib. L.) It was not enough that the bodies of individuals should be preserved by the laborious and expensive process of embalming, but their actions and employments during life were elaborately recorded, and, as it were, perpetuated by the hand of the painter on the walls of the tombs in which they were laid. In every instance the entrances of the tombs were artfully concealed, presenting an exact resemblance to the rest of the rock in which they were cut; for all the tombs of Egypt are excavations, those of the people being dug in the side of the mountains, and those of the kings within the enclosures of the temples, the most remarkable of which is Biban-el-Meluk at Thebes. The expedients employed to secure the dead from desecration are elaborate in the extreme: not only were their entrances a secret, but devout to the chambers where the bodies were laid is only to be made by deep shafts and endless winding recesses. The mummy was enclosed in a sarcophagus profusely ornamented, and standing in the midst of a chamber. Besides human bodies, those of animals held to be sacred were also often embalmed. (Belzoni's Operations and Discoveries; Wilkinson's Topography of Thebes.)

As the monuments unravel, in some degree, the mystery of Egypt's ancient history, so an examination of the paintings that cover the tombs gives us some insight into the domestic condition and usages of its people. Every employment and amusement is vividly pourtrayed around these sepulchral walls, each according to the station of life of the person to which it refers. The forms of every article of furniture, of ships, of carriages, of every thing, in short, pertaining to civilised life, are there accurately figured. As pictures, however, these efforts of the primeval artists are far from pleasing. The colours, though still bright and vivid, are all positive, seldom being blended or softened; and perspective, or any approach to it, is no where to be detected. But the details of private life that they present are wonderfully minute and copious; and by a long and careful study of them, assisted in parts by an active imagination, and by a large infusion of what Dugald Stewart has called conjectural history, Sir J. G. Wilkinson has produced a singularly interesting and instructive work. In fact, if we might trust to his ingenious suggestions and deductions, we should have a clearer insight into the habits, manners, and every-day life of the ancient Egyptians, than we have into those of most European nations. (Manners and Customs of the Ancient Egyptians, 3 vols. Lond. 1837. See also Rosellini, Monumenti dell' Egitto, Pisa, 1834.)

History.—The origin of the Egyptian nation, and the history of their native princes, are involved in the greatest obscurity and uncertainty. This much, however, is established beyond the possibility of doubt, that the Egyptians had attained to great wealth and civilisation, and had established a regular, well-organised, and (if we may estimate it by its results) wisely-contrived system of government, while the greater number of the surrounding nations were involved in the grossest barbarism. At length, however, Cambyses, emperor of Persia, added Egypt to his other provinces. It continued attached to Persia for 193 years, though often in open rebellion against its conquerors. Alexander the Great had little difficulty in effecting its conquest; and it has been inferred from his foundation of Alexandria, which soon became the centre of an extensive commerce, that he intended to establish in it the seat of the government of his vast empire. On the death of Alexander, Ptolemy, the son of Lagus, became master of the country. Under this able prince and his immediate successors, Egypt recovered the greater portion of its ancient prosperity, and was for three centuries the favoured seat of commerce, art, and science. The feebleness and indolence of the last sovereigns of the Macedonian dynasty facilitated the conquest of Egypt by the Romans: Augustus possessed himself of it after a struggle of some duration, and for the next 666 years it belonged to the Roman and Greek empires, constituted their most valuable prov., and was for a lengthened period the granary, as it were, of Rome. In 640 Egypt submitted to the victorious Amrou, general of the caliph Omar; under whose successors it continued till about 1171, when the Turkmans expelled the caliphs: these again were in their turn expelled in 1250, by the Mamelukes. The latter raised to the throne one of their own chiefs with the title of sultan; and this new dynasty reigned over Egypt till 1517, when the Mamelukes were totally defeated, and the last of their sultans put to death by the Turkish sultan Selim. The conqueror did not, however, entirely suppress the Mameluke government, but merely reconstructed it on a new basis, placing at its head a pacha appointed by himself, who presided over a council of twenty-four Mameluke beys or chiefs. So long as the Ottoman sultans preserved their original power and authority, this form of government, though about the worst that could have been devised had the interests of the country been ever so little attended to, answered their purpose of preserving Egypt in dependence, and of drawing from it supplies of men and money: but the power of the pachas declined with that of their masters; and latterly the whole executive authority centered in the beys, who, except upon rare occasions, paid little more than a nominal deference to the orders of the sultan.

This state of things continued till 1798, when a French army, commanded by Napoleon, landed in Egypt. The Mameluke force having been annihilated or dispersed in a series of engagements with the French, the latter succeeded in subjugating the country. Napoleon having returned to France, the French in Egypt were attacked in 1801 by a British army, by which they were defeated, and obliged to enter into a convention for the evacuation of the country. The British having not long after evacuated Egypt, it relapsed into its former state of anarchy and barbarism, from which it was at last rescued by the good fortune and ability of Mehemet Ali. This extraordinary man,

a native of an obscure village of Albania, having
entered the military service, attained, partly by
his bravery, and partly by his talent for intrigue,
to the dignity of pacha in 1801. His subsequent
history is well known. The massacre of the
Mamelukes, in 1811, raised him to almost absolute power; and his victorious arms subsequently
wrested Syria from the Grand Seignior. But he
was compelled, in 1840, by the interference of the
European powers, to relinquish all his Asiatic
possessions. The treaty of London, in 1841, ratified by the imperial edict of June 1, made the
government of Egypt hereditary in the family of
Mehemet Ali. The fifth viceroy—more truly
king—of the new dynasty, Ismail Pasha, who assumed the government in January, 1863, was the
eldest surviving son of Ibrahim Pasha, eldest son
of Mehemet Ali.

EHRENDREITSTEIN, a town and strong fortress of Rhenish Prussia, on a steep and picturesque rock, 773 ft. in height, on the E. bank of
the Rhine, opposite to Coblentz, with which it is
connected by a bridge of boats. Pop. 8,837 in
1861, excl. of garrison of 1,356. A tower or fortress is said to have been constructed on the summit of this rock by the Romans; and in modern
times it was regularly fortified, a well was cut in
the rock to the depth of 584 ft., and it was justly
regarded as one of the principal bulwarks of Germany. It was unsuccessfully besieged by the
French in 1795, 1796, and 1797; but it fell into
their hands on the 27th of January, 1799, the
garrison having been previously reduced to a state
of famine. The French blew up the fortifications
subsequently to the treaty of Luneville. They
have, however, been reconstructed by the Prussian
government since 1815, and rendered more extensive and formidable than ever. Ehrenbreitstein, with the new fortresses on the hill of the
Chartreuse and the Petersberg, forms a portion of
the grand military position of which Coblentz
(which see) is the centre. The town of Ehrenbreitstein is situated at the foot of the castle rock.

EICHSTADT, a town of Bavaria, circ. Regensburg (Ratisbon), on the Altmühl, 41 m. WSW.
Ratisbon. Pop. 7,835 in 1861. The town is well
built, and contains the summer residence of the
ducal family of Leuchtenberg, with a Brazilian
cabinet, and other collections of art and science;
a cathedral, in the Gothic style, commenced in
1259; with several other churches, a Capuchin
convent, bishop's palace, Latin school, ecclesiastical seminary, public library, and museums of
painting, antiquities, and natural history. It has
four suburbs. About 1 m. distant is the Will-
baldsburg, a castle on a height, believed to have
replaced a Roman fortress. It has a well of great
depth, and its trenches have been cut in the solid
rock; but it is now in a state of decay. Eichstadt
has manufactures of hardware, earthenware, and
woollens; besides breweries and stone quarries.
The town originally belonged to the prince-bishops,
successors of St. Willibald, and was given by
Napoleon I. to Prince Eugene Beauharnais, to
whose memory the citizens have erected a handsome monument in the vicinity.

EIMBECK, or EINBECK, a town of Hanover,
cap. principality Grubenhagen, distr. Hildesheim,
on the Ilme, by which it is surrounded, 37 m. S.
by E. Hanover, on the railway from Hanover
to Hildesheim. Pop. 5,600 in 1861. The town
is enclosed by walls and broad ditches, and is
ill built and dirty. It has two hospitals, and a
superior school. Eimbeck was formerly celebrated
for its beer, which, like London porter, was sent
all over the empire. At present, Eimbeck is less
celebrated for its beer; but it has some breweries,

with fabrics of woollen and linen cloth, linen yarn,
stockings, shoes, leather, and chemical products,
and a brisk trade in flax and other agricultural
produce. In 1856 it suffered severely from a fire.

EISENACH, a market-town of Central Germany, duchy of Saxe-Weimar-Eisenach, cap. of the
principality and prov. of same name, on a gentle
declivity at the confluence of the Nesse and
Hörsel, encircled by wooded hills; 44 m. W. by S.
Weimar, and 91 m. NE. Frankfort-on-the-Mayne,
on the railway from Frankfort to Leipzig. Pop.
11,517 in 1861. It is the principal town in the
Thuringian Forest, is well built, and laid out,
paved and well lighted. It has five suburbs, with
four churches, a handsome market-place, in which
is the ducal residence, and the new citizens'
academy, estab. 1825; a mint, four hospitals, a
workhouse, house of correction, town-hall, gymnasium, teachers' seminary, school for females,
schools for the indigent, and various other public
and benevolent institutions. Formerly, this was
the most flourishing of all the manufacturing
towns between Leipzig and Frankfort. It was
formerly noted for its manufactures of serge, plush,
and other woollen stuffs; but during the period
of the 'Continental System,' the capitalists of
Eisenach forsook the manufacture of wool for that
of cotton, which, on the re-opening of the continental ports to British goods, was all but annihilated. About 1½ m. S. of the town, is the celebrated castle of Wartburg, on a hill, 1,243 ft. above
the level of the sea, in which Luther passed his
10 months' durance, under the friendly arrest of
the Elector of Saxony. Travellers are still shown
the room he occupied, though the castle is, in
great part, in a state of decay.

EISLEBEN, a town of the Prussian states,
prov. Saxony, distr. Merseburg; 19 m. W. by N.
Halle, and 25 m. SW. by S. Magdeburg, on the
railway from Halle to Nordhausen. Pop. 11,120
in 1861. The town is situated on elevated ground,
near the Bee, and is divided into an old and a new
town, the former of which is encircled with walls
and ditches. It has several suburbs; an ancient
castle, formerly the residence of the counts Mansfeld; four churches; a Protestant gymnasium,
and two hospitals; and is the seat of a council
for the circle, a judicial tribunal for the circle
and town, and a board of mines. Eisleben is celebrated as the native place of the great reformer,
Martin Luther, born here on the 10th of Nov.,
1483; and who also died here on the 18th of Feb.,
1546. The house in which he was born, and where
he breathed his last, was almost wholly destroyed
by fire in 1689. Being afterwards rebuilt it was
converted into a gratuitous school for poor children, and a teacher's seminary; the cap, cloak,
and other relics of Luther, are preserved in it, and
shown to visitors; and his bust is placed over the
door. In one of the churches of the town is a
pulpit, from which he occasionally preached; and
here, also, are busts of himself and Melancthon.
Luther was the son of a miner at Eisleben, and
the greater part of its inhab. continue to work in
the copper and silver mines in its vicinity; but it
has also some potash and saltpetre factories, and
one of tobacco, besides several breweries.

EKATERINEBURG, a town of the Russian
empire, gov. of Perm, near the bottom of the E.
declivity of the Oural chain, on the Iset, and in
the line of the great road leading from Perm to
Tobolsk. Pop. 17,380 in 1858. The town was
founded by Peter the Great in 1723, and is regularly built and fortified. Besides being the key
of Siberia, it is the cap. of the richest mining district of the empire; has a board for the general
direction of the mines, a mint for the coinage of

copper, and extensive iron and copper foundries in its immediate vicinity. Its inhab. who consist mostly of emancipated serfs, formerly belonging to the crown, are almost wholly employed in the mines and working metals.

EKATERINOSLAF, a gov. of European Russia, having the sea of Azoff, and the gov. of Taurida on its S. frontier. Area, 25,630 sq. m. Pop. 870,100 in 1846, and 1,012,601 in 1858. Nearly two-thirds of the surface consist of a vast steppe or plain, without trees, and with a thin arid soil. The portions on this side the Dnieper, by which it is traversed, are the most fertile. Grazing is the principal occupation of the inhab. who possess immense numbers of horses, cattle, sheep, hogs, and goats. The breed of sheep has been materially improved. Bees are abundant; and the silk-worm is raised in the vicinity of Mariupol. The pop. consists principally of Russians and Cossacks; but there are several other races, among whom 10,881 German colonists. Principal towns, Ekaterinoslaf, Bakhmout, and Mariupol.

EKATERINOSLAF, the cap. of the above gov., on the Dnieper, immediately below the cataracts; lat. 48° 27' 70" N., long. 31° 55' E. Pop. 11,650 in 1858. Catherine II. laid the first stone of this town, in presence of the emperor Joseph II., in 1787. It is designed on a large scale, and its broad rectangular streets are still very far from being completely filled up. Exclusive of the gov. offices, it has a gymnasium, and some other literary as well as charitable institutions.

ELBA (the *Æthalia* of the Greeks, and the *Ilua* or *Ilva* of the Etruscans and Romans), an island of the Mediterranean, or rather of the Tyrrhene sea, belonging to Italy, from which it is separated by the strait of Piombino, 7 or 8 m. across, between lat. 42° 43' and 42° 50' N., and long. 10° 5' and 11° 25' E. Shape irregular, but not very unlike that of the letter T, having the upper end towards the E. Length, E. to W., 16 m.; breadth, varying from 2 to 12 m.; circumference, about 64 m.; area, 150 sq. m. Pop. 18,450 in 1861. The island is covered with mountains; a central chain runs through its whole extent, the principal summit of which, towards its W. extremity, is 2,624 ft. in height. Granite abounds, especially in the E. part of the island, and it in a great measure constitutes the numerous rocky shelves with which the coasts are bristled. Geologically the island affords no traces of the action of fire. Secondary and tertiary formations, calcareous, aluminous, or magnesian, are plentiful in the W.; on the E. shore the surface is covered with a reddish vegetable earth, many feet in thickness, and furrowed with ferruginous veins. Iron is everywhere abundant:

'Insula inexhaustis Chalybum generosa metallis.'
Aeneid, x. 174.

besides which, copper, calamine, antimony, alum, selenium, opal, tourmaline, and various kinds of marble are found. There is no navigable river, but there are many small rivulets used to turn mills; the largest are on the N. side of the island, where there are also some salt marshes. Climate excellent, the heats being neither excessive, nor of long duration; nor the cold severe. Except in a few particular localities, Elba is decidedly healthy. The appearance of the island is far from prepossessing; and the cultivable land is but of very limited extent. 'Ruins scattered over the face of the country, wretched hamlets, two mean villages and one fortress—these, generally speaking, are all that meet the sight on the side of the island which extends along the channel of Piombino. The traveller, however, finds the scene changed on visiting Monte-Grosso (in the N.E.), covered with myrtles, rosemary, the mastick tree, laurel,

thyme, &c.; and Monte-Giove, where the green balm oak, cork tree, laurel, yew, and a small number of wild olives, afford an agreeable repose to the eye. The branches of the hills, which stretch towards Longone (S.E.) present only naked rocks, almost destitute of verdure. In the centre of the island the hillocks are overspread with olives, mulberries, and vines. On the W., the summits and declivities of the mountains consist of granitic rocks. Industry and toil render fertile the small quantity of earth which is collected at their base.' (Bernoud's *Voyage to Elba*, pp. 94, 95.)

Though the soil is throughout hilly, and the vegetable earth generally shallow, little labour suffices to render it productive. Agriculture, however, is nearly confined to the lowest hill ranges, and the sheltered valleys between them. The corn crop is trifling; at the beginning of the present century De Hernaud says it would have hardly supplied the wants of the inhab. during 1 part of the year. Maize and pulse are grown. The produce of flax is very small, and hemp is not cultivated; the thread that is used is manufactured from the leaves of the numerous aloes with which the fields of Longone are covered. All kinds of fruit trees common to Europe grow, excepting the apple; but they are generally ill cultivated, and their fruit inferior. The vintage takes place in September. Both white and red wines are produced; the former are chiefly for home consumption; the latter in small quantity, and good; constitute a chief article of export. The most esteemed is the *Aleatico*, obtained from a superior red Muscadine grape. The oak, beech, chestnut, poplar, alder, and buckthorn, are amongst the forest trees; but timber fit for carpenter's work is rare, the island affording little more than mere underwood. Pasturage is scarce, and cattle few; they consist of mares, some mules, and a few stunted horses, oxen, and cows. The number of pigs, sheep, and goats is more considerable; but the breeds are very inferior. The sea around Elba swarms with fish, including tunnies, anchovies, soles, the doradillæ (*Labrus rufa*, Linn.) and mullet (*Mullus barbatus*). Of these the tunny and mullet are taken in large quantities, and from 5,000 to 6,000 tons of the former are annually exported, besides a considerable supply of the latter.

The chief wealth of Elba is in its mines of iron and salt, which have been wrought from a very remote epoch. The principal mine near the little town of Rio, on the E. side of the island, consists of an entire mountain about 530 ft. in height, which, to use the words of Pliny, *in totum est ed materiâ*. It supplies iron ores in every known variety; some yielding from 0.75 to 0.85 of excellent iron, from which a very good steel is obtained. The ancients made many deep excavations and winding galleries in this mine; and pickaxes, nails, lamps, and various other antique articles have been from time to time discovered in it. The average produce of iron ore from Elba has of late years been nearly 18,000 tons a year, worth about 21s. a ton; the whole of which is taken to the opposite coast of Italy to be smelted. The miners work eight or nine hours a day, and are paid 40 lire (about 25s.) a month, 5 per cent. of which is deposited for a pension from the government, for themselves or their widows. Marine salt is manufactured by evaporation in four basins, near Porto Ferrajo. About 4,000,000 lbs. are produced annually, and nearly 100 persons employed in the manufacture. The other branches of industry are principally domestic.

Commerce is chiefly limited to the importation from Leghorn and Marseilles of grain, cheese, cattle, and other articles of prime necessity; and the exportation of tunny, salt, iron ore. Vermouth and Aleatico wines, vinegar, and granite. There are two towns—Porto Ferrajo on the N., and Porto Longone on the E. coast. The former, which is the cap., is built on a peninsula, between which and the main land is a spacious and good harbour, 1'rp, about 3,000. It is fortified; its streets, which are wide, clean, and well paved, are mostly terraces cut out in the rock; houses small, badly divided, built of brick, and generally two stories high. It is the residence of the governor of the island and of a military commandant, the seat of a civil and criminal court, and contains two churches, with a prison, lazaretto, hospital, and some subterranean corn magazines. Porto Longone, with 1,500 inhab., has a tolerable harbour, and is well fortified and difficult of access. The ordinary food of the pop. consists of dried palm, cheese, bacon, smoked provisions, coarse bread, fresh fish, and a few vegetables; fresh meat and white wine are used only on holydays. Their houses and furniture are equally simple and solid. Bowls, nine-pins, quoits, tennis, and firing at a mark, are the chief sports of the men; there is not much gaiety exhibited in the amusements of the island generally. Robbery is rare, murder still more so; the number of paupers inconsiderable.

The Etruscans, Phocians, Carthaginians, and Romans successively possessed Elba; in the middle ages it was subject to the Saracens, Pisans, Genoese, Lucchese, the counts of Piombino and Orsini. In the 16th century it was ravaged by Barbarossa, and soon afterwards fell to the crown of Naples. Under the French empire it formed part of the kingdom of Etruria. Its chief historical interest is derived from its having been the residence and empire of Napoleon from the 3d of May, 1814, to the 26th of Feb. 1815. During this short period a road was opened between the two principal towns, trade revived, and a new era seemed to have opened for Elba.

ELBE (an, Albis, flumen inclytum et maxime olim, Tacit. Germ., § 41.), a large and important river of Europe, through the central part of which it flows, generally in a NW. direction from Bohemia to the German Ocean. Its total length is about 730 m. during which course it passes through Austria, Saxony, Prussia, Anhalt-Dessau, Hanover, Mecklenburg, Denmark, and Hamburg. Its principal affluents are—on the left, the Moldau, Eger, Mulde, Saale, Ohre, Jetze, Pinnau, and Oste; and on the right, the Iser, Schwarz Elster, and Havel, with the Spree. Dresden, Meissen, Torgau, Magdeburg, Lenzen, Lauenburg, Harburg, and Hamburg, are situated upon its banks. It originates in several streams on the S. side of the Schneekoppe (Snow-cap), one of the Riesengebirge chain in the cir. of Hirschow in Bohemia, about 4,400 ft. above the level of the sea. At first its direction is E., next S.; at Pardubitz it turns W., and at Köllin NW., from which direction it does not afterward greatly vary. After leaving Tangau it runs for the most part through a flat country. Near Königgratz, about 40 m. from its source, its elevation above the sea is only 654 ft., at Melnik 431 ft., at Schandau 341 ft., at Dresden 279 ft., at Magdeburg 236 ft., and at Areuburg (Brandenburg) 176 ft. only. Above Melnik it is navigable for only small craft, but vessels of 1,500 centners burden may come up to that town. Its volume receives a considerable augmentation by the union of the Moldau; and when it enters Saxony the Elbe is upwards of 350 ft. in width. Between Hamburg and Harburg it is divided into several arms, enclosing some large islands; but these soon afterwards reunite, and the river proceeds in an undivided stream to its mouth. Its estuary, op-

posite Cuxhaven, 12 m. wide, is encumbered with sand banks, which render its navigation difficult; but ships drawing 14 ft. water come up to Hamburg at all times, and those drawing 18 ft. come up safely at spring tides.

The bridges across the Elbe are numerous above Hamburg; but below that town communication between the opposite banks takes place by means of ferries only. It is connected by the Finow and Frederick William canals, within the Prussian dom., with the Oder and the Vistula, and by that of Steknitz with the Trave near Lubeck; while the short railway from Boitzenbrück to Lietz connects its affluent, the Moldau, with the Danube.

In a commercial point of view, the Elbe is a river of much importance, being the channel by which the countries of NW. and Central Germany, from Hamburg to the E. parts of Bohemia, export more of their heavy products. By the treaty of 1815 it was provided that its navigation should be free throughout its whole course, but the governments through whose dominions the river flows have contrived to evade this provision, and a series of vexatious tolls and heavy duties are imposed on foreign merchandise. Prussia obliges the transfer at Magdeburg of many goods passing downward to her own vessels, and the government of Mecklenburg-Schwerin levies heavy taxes at Boitzenburg. Above Hamburg, the river has lost much of its former importance by the establishment of railways.

ELBERFELD, a town of Rhenish Prussia, circ. Elberfeld, distr. Düsseldorf; 15 m. E. by N. Düsseldorf, and 2d m. SNE. Cologne, on the railway from Berlin to Düsseldorf. Pop. 56,307 in 1851. The town stands on both sides of the river Wupper, and is irregularly built, but contains many good houses, most of which have gardens attached to them. It is the seat of the council for the circle, of the judicial and police courts, a commercial tribunal, and a board of taxation, and has two Protestant churches, a R. Cath. church, gymnasium, citizens' and commercial schools, a school of industry, numerous elementary schools, a town-hall, exchange, theatre, general hospital, two orphan asylums, two workhouses, and a savings bank. There are several casinos, or clubhouses, and a promenade. In the winter there are frequent balls and concerts. Its principal manufactures are silk, which employ about 8,000 looms; with cotton and linen fabrics, linen and cotton thread, velvet, lace, ribands, with establishments for calico printing. In the cotton factories many steam engines are employed, and there are numerous water-mills and establishments for the bleaching of linen. But the most celebrated of the Elberfeld factories are those appropriated to the dyeing of Turkey red. In this art, whether it be owing to the air or the water, or to some peculiar process or mystery, the dyers of Elberfeld have attained to unrivalled excellence. Considerable quantities of yarn were formerly exported from Glasgow and other places in the United Kingdom to be dyed at Elberfeld, and again imported to be wrought up. Elberfeld is the seat of the Rhenish Foreign Trade Company, the German-American Mining Union, the Rhenish Prison Society, a Bible and a scientific society, and many benevolent institutions.

Adjoining Elberfeld, and forming, in fact, a kind of suburb of it, is Barmen, a long straggling place, made up by the union of several villages. It has four churches, one of which, erected in 1830 for the use of the R. Cath. pop., was liberally contributed to by the Protestants; a high school, a deaf and dumb asylum, exchange, two discount banks, a police court, and a commercial tribunal. Its manufactures are the same as those

of Elberfeld, with the addition of steel and plated
articles, hardware, chemical products, and earthenware.
Along the banks of the river are some
extensive meadow grounds, used for bleaching
linen, which branch of industry contributed greatly
to the rise of both towns. Numerous kitchen gardens
surround Barmen, the cultivation of which
occupies many individuals. The road through
the valley of the Wupper, for a distance of perhaps
6 m., adjacent to Elberfeld, is lined on either
side with mills, factories, and habitations; this
being the most populous as well as the most industrious
district of the Prussian monarchy. It
is estimated that altogether nearly 16,000 hands
are employed in manufactures in and near Elberfeld
and Barmen, and that the value of the manufactured
goods annually amounts to 12,000,000 or
31,000,000 thalers, or from 1,000,000, to 2,000,000.
Wages, owing to the increasing demand for labour,
are high at Elberfeld, and the working classes are
comparatively well off.

ELBEUF, a town of France, dép. Seine Inférieure,
cap. cant., on the Seine, a tributary of which
intersects it, 11 m. S. by W. Rouen, on the railway
from Rouen to Paris. Pop. 21,692 in 1861.
The town is generally ill built, but possesses a
tolerably good square, and some handsome buildings.
It has no public edifices worthy of notice
except two churches, one of which has some stained
glass, presented by the cloth manufacturers of the
town in 1865, exhibiting a curious emblematical
device indicative of their profession. Elbeuf has
been long celebrated for its woollen manufactures,
and is at the present moment the principal seat of
that branch of industry in France. In 1787, Elbeuf
produced about 18,000 pieces of cloth yearly:
in 1815, the quantity had increased to from 20,000
to 25,000 pieces: and at present the produce is
estimated at about 100,000 pieces, valued at
75,000,000 francs, or 3,000,000. It is stated that
about 20,000 men, women, and children are employed
in the different departments of the business;
but of these many belong to the surrounding
districts, and return from town at night to their
lodgings in the country. 'The working classes of
Elbeuf,' says an official report, by the maire of the
town, 'enjoy, in general, easy circumstances: they
have always lived happily, for two very powerful
reasons: the first, because the manufacturers are
constantly in their workshops, work themselves
with their workmen, know their wants, and identify
themselves with all that happens to them for
good or evil; the second, because the price of
weaving varies little, the proportion between times
of prosperity and times of distress being 90 per
cent. at most on the amount of wages; and that
only in certain departments. The work-people
are divided into three classes; the adults, the day
labourers, and the weavers.' There are in Elbeuf
a gratuitous school of mutual instruction for boys,
a gratuitous institution for girls, an infant school,
and a gratuitous Sunday school for the adult
workmen; and, independently of these public institutions,
there are a number of private schools.
M. Villermé (État Physique et Moral des Ouvriers)
states that, compared with the work-people of
Rouen, those of Elbeuf are much the more correct
in their morals and habits. They are, he says,
for the most part industrious and economical; and
many of them are supposed to have saved a certain
portion of their earnings, especially those who live
out of town.

Elbeuf is said to have existed in the 9th century,
but its origin is uncertain. During the administration
of Colbert, its manufactures were in a
comparatively flourishing state; but they suffered
severely by the revocation of the edict of Nantes.

At the beginning of the 18th century, its manufactures
had begun to establish commercial relations
with Spain and Italy; and it now has a
direct trade not only with those countries, but
with America, Germany, and the Levant.

ELBING, a town of Prussia, prov. Prussia, cap.
circ. on the Elbing, about 5 m. from where it flows
into the SW. angle of the Frische Haff, 31 m.
SE. by E. Dantzig, and 50 m. SW. Königsberg,
on the railway from Dantzig to Königsberg.
Pop. 25,540 in 1861. The town is divided into
the old town, new town, and suburbs, part of
which are enclosed, together with the old and
new town, within a line of fortifications. The
ramparts and walls are lofty, flanked with towers,
and surrounded with ditches, but they have not
been in a state of efficient defence since 1772. The
town is entered by 7 gates. The new town is
well built, but it is quite otherwise with the old
town. Elbing is well lighted; it has a Catholic
and 9 Protestant churches, a synagogue, a gymnasium
with a library, 6 hospitals, an orphan and
other asylums, a convent for old women, a house
of industry, established by an Englishman named
Cowle, in which 100 children are educated, and
numerous schools for both sexes and all classes,
education among the poor having made great
progress in this town. It is also the seat of a
council, a judicial court for the circle, and a municipal
tribunal. It has a garrison, a bank, exchange,
fire assurance office, numerous warehouses,
principally in one of its suburbs, and many
sugar refineries, with pearl-ash, vitriol, tobacco,
linen, sail-cloth, oil, starch, soap, chicory, and
other factories, in some of which large steam
engines are employed. The trade of Elbing is
extensive; its exports consist chiefly of corn,
timber and staves, hemp and flax, the produce of
its own manufactures, feathers, horse-hair, wool,
fruit, butter, and pearl-bread. The Krafohl canal
connects Elbing with the Nogat. The Frische
Haff is too shallow to be navigated by vessels of
any considerable burden, so that the trade of the
town by sea has to be carried on, by means of
small vessels or lighters, through Pillau at the
mouth of the Frische Haff. About 25 ships, besides
river craft, belong to merchants of the town.
Elbing was founded about 1237, and became
afterwards one of the Hanse Towns. It was united
to the Prussian dom. in 1772.

ELCHE (an. *Ilici*), a town of Spain, prov.
Valencia, near the left bank of the Elda, in a
plain almost entirely covered with palm trees, 16 m.
WSW. Alicante, and 8 m. W. from the Mediterranean.
Pop. 10,353 in 1857. The town is surrounded
by walls, has some good streets and squares, and 6
public fountains, but of these one only has potable
water; 8 mc, churches, the principal of which is
a fine building, with a majestic dome; 3 convents;
a magnificent old castle, belonging to the Duke
of Arcos, on whose estate the town is built; a
barrack for cavalry; 2 primary schools, and a
grammar-school. It has manufactures of coarse
linens and cottons; 10 flour-mills; with distilleries
and tanneries.

Elche might, with propriety, be called the 'city
of dates,' being everywhere surrounded by plantations
of palms. Besides its large produce of
dates, the country round abounds in barilla, that
exported from Alicante being chiefly raised in the
vicinity of Elche. A great proportion of the
dates imported into England as the produce of
Barbary, are from this city. The wages of field
labour here are 3 or 4 reals, and every thing is
proportionably cheap. Elche is the native country
of Don George Juan, a distinguished mathematician
and natural philosopher, the companion

of Ulloa, in the commission sent to Peru, towards the middle of last century, by the French and Spanish governments, for the measurement of a degree of the earth's surface. Elche was recovered from the Moors in 1363.

ELCHINGEN, a small village of Bavaria, on the N. bank of the Danube, about 7 m. NW. Ulm. Pop. 550 in 1861. This village was the scene of an obstinate engagement between the French, under Marshal Ney, and the Austrians, on the 14th Oct., 1806; the former at length succeeded in carrying the bridge and position of Elchingen, and by this success contributed materially to the capture of Ulm, which, three days after, surrendered to Napoleon. Ney was rewarded for his gallantry on this occasion with the title of Duke of Elchingen.

ELEPHANTA, a small island on the W. coast of Hindostan, prov. Bombay, prov. Aurangabad, on the E. side of the harbour of Bombay. It is about 6 m. in circumference, and consists of two long hills and a narrow valley between them. It is named Garapuri by the Hindoos; the Portuguese gave it the name of Elephanta, from a colossal elephant, about three times the natural size, hewn out of the solid rock, and standing about ½ m. from the landing-place, but which has now almost entirely fallen to decay. A gentleman who visited the island in 1830 reports, that only three legs and a part of the fourth were remaining. This island is celebrated for some remarkable cave-temples, as many of which exist on the W. side of India. In the face of a hill, about ½ m. from the landing-place, is the first cave; little of which, however, appears to have been completed. About ¾ m. further is the great cave, an excavation 130½ ft. from N. to E., by 133 ft. from E. to W.; its ceiling flat, varying from 15 to 17½ ft. in height, and supported by 26 pillars and 16 pilasters. It has three entrances—on the N., E., and W.; the front of each consisting of 2 pillars and 2 pilasters; but the N. front is the principal, and directly faces the remarkable triad or three-headed figure—the principal object within the temple. This is a gigantic bust, 15 ft. high, composed of three colossal heads; the front face having a placid and agreeable physiognomy; that on the left being to all appearance a female, and also mild looking; but that on the right, according to most travellers, having a repulsive aspect. The latter, as well as the front face, has the third eye in the forehead, so characteristic of Siva. Indeed, in the opinion of the best authorities (see Erskine, in Trans. of the Bombay Lit. Soc., i.; Sykes, in Journ. of the Asiat. Soc., v. 81–88, &c), the whole three-headed figure relates to Siva only, and not to a trinity of Brahma, Vishnu, and Siva, as has sometimes been imagined. Similar busts abound in the Brahminical caves at Ellora (which see) appropriated to the worship of Siva. This figure has originally had 8 arms, each of the hands of which held some objects; but all are now greatly mutilated. The niches on either side the triad are of considerable size, and crowded with figures, among which, as well as in the other compartments around the temple, Brahma, Vishnu, Parvati, Kartik, Ganesa, and other Hindoo divinities, may be recognised, but always in a condition inferior to Siva. On either side of the principal figure is a small dark chamber, probably originally devoted to the cult of the lingam; and there are three separate sanctuaries within the temple, each containing a figure of the lingam. The columns and other portions of this cave are ornamented in a most elaborate manner, and, altogether, the temple within presents an imposing

appearance. From some cause, however, it is not much frequented by pilgrims; several of its pillars have been thrown down; it is in part mouldering away with damp, and becoming choked with earth; and, unless some effectual means be speedily taken for its preservation, it will in a few years be in a state of irreparable decay. (Erskine, in Bombay Trans.; Sykes, Asiat. Soc.; Grindlay's Views.)

ELEPHANTINÉ, the last of the larger islands, at the extremity of the cataracts of the Nile, immediately opposite to Assouan, near the N. boundary of Egypt; lat. 30° 5' 3'' N., long. 32° 54' 49'' E. Placed at the threshold of the kingdom, Elephantiné has been justly called the key of Egypt, and claims some importance as a military post. Under Psammetichus it contained an Egyptian garrison, to protect the country from the inroads of the Ethiopians. Herodotus (Euterp. § 30) found it occupied with Persian troops; and, in Strabo's time, the Romans had three cohorts there, to guard, what Tacitus has expressively called the Claustra Romani imperii. (Strabo, lib. 17; Tacit. Annal. lib. 2, § 61.)

The base or kernel of the island is a granite rock, covered with the rich alluvial soil brought down by the river; and to prevent this from being again washed away, it has been protected by quays, which have been repaired from time to time, so that it is impossible to fix the epoch of their first construction. The richness of its soil admits of the island being cultivated in every part; and though it be less than 1 m. in length, and not ½ m. broad, it presents a verdure and fertility equal to the finest spots of Egypt, and forms a refreshing contrast to the sterility to which, for many miles round, beyond the banks of the Nile, the country is doomed. Hence, the Arab name for Elephantiné is Djeziret el-Chof, 'the islet of flowers.' The N. extremity of the island only is rocky and elevated, and the bare rock comes down to the edge of the river; but the rest of it is covered with shrubs, groups of palms, mulberry gardens, acacias, dates, and sycamores interspersed amongst human habitations, mills, canals, and the ruins of temples. (Ritter's Africa, 3rd division, § 26; Richardson's Travels.)

The wreck of the ancient town forms a sort of plateau, and gives to the island its greatest elevation. Here, till recently, were the remains of two temples, one dedicated to Cnuphis by the Pharaoh, Amenophis III., and one dedicated to a triad consisting of Cnuphis, Sate, and Anuke, the latter being of the age of Alexander, son of Alexander the Great. But we regret to say that these interesting ruins no longer exist, having been barbarously demolished in order to employ the stones in building barracks and warehouses at Assouan. In the quay Champollion found fragments of edifices that had been constructed by the Pharaohs Moris, Mandouei, and Rhamses the Great, at Nemertes. (Lettres de l'Égypte, p. 174.)

The most interesting part of Elephantiné is its quarries. These furnished, in the reign of Amasis, one of the greatest marvels Herodotus (Euterpe, § 175) saw at Sais—a single block of granite, out of which was cut an entire temple. No fewer than 2,000 men are said to have been occupied during three years in transporting this huge monolithic edifice down the Nile to its destination. The quarry affords ample proofs of the mechanical skill and patient labour of the ancients. Immense columns have been evidently cut out of the solid rock in one mass. The marks of the workman's chisel and wedge are as fresh as if they had been imprinted yesterday, and the tracks of carriage wheels are equally distinct. Some sculptures are merely blocked out, while others appear in a more advanced stage, and a large sarcophagus is two-thirds cut out of the rock.

Besides the remains of Egyptian architecture, others have been found which would appear to belong to the Romans, particularly a large wall in the N. Another, from 40 to 45 ft. high, and 609 ft. long, of a convex construction, had a Nilometer fixed in it, which, there can be little doubt, was the one mentioned by Strabo (lib. xvii.). Champollion, however, says nothing of the Nilometer; and it may, perhaps, have been destroyed as well as the remains of the temples. Over the ruins of the ancient town are strewed many fragments of pottery, among which other memorials of the Romans have been found, consisting of tokens or coins of red earthenware, having the name *Antoninus* inscribed on them in a Greek running hand. (Ritter; Jowett's Christian Researches, p. 40; Condèr's Egypt, ii. 191–194.)

Elephantine is inhabited by Nubians, who are said to be kind and hospitable to strangers. The women are described by Dr. Richardson as possessing much personal beauty, somewhat too freely displayed. (Light's Travels, pp. 51–53; Richardson's Travels.)

ELGIN, a royal bor. and market town of Scotland, co. Elgin or Moray, on the Lossie, 5 m. from its bottom into the sea at Lossiemouth, 120 m. N. Edinburgh, and 38 NW. Aberdeen, on the railway from Aberdeen to Inverness. Pop. 7,543 in 1861. The situation of the town is very agreeable, having the Lady Hill, a beautiful verdant mount on the W., and the Quarrywood Hill on the E., clothed with wood to the summit. The town consists of one street, about a mile in length, with a few small streets intersecting it at various distances. The principal street is handsome, well paved, and so wide that a new church stands in the middle of it, on the site of an old church, called St. Giles. This new church, which has a richly ornamented cupola 112 feet high, and a spacious Doric portico, is one of the best of the numerous public buildings which Elgin contains. Grey's hospital (founded in 1819 for the reception of the sick poor of the town and county of Elgin), a building of two stories, of Grecian architecture, with a projecting portico of four Doric columns, and the centre crowned with a dome, stands on a rising ground at the W. end of the town, and forms a beautiful termination of the High Street. At the opposite end of the town stands the Elgin Institution, a quadrangular building of Grecian architecture, founded by the late General Anderson, for the education of youth, and the support of old age. This institution, which cost 17,000*l.* (its founder having bequeathed 70,000*l.* altogether for the charity), is calculated to contain 10 aged and indigent persons, and 60 children, and to afford gratuitous education for about 250 children belonging to the town and parish. The other public buildings are the academy, assembly rooms, Trinity Lodge rooms, jail and court-house, and chapels belonging respectively to the Episcopalians, the United Associate Synod (two), the Independents, and the Roman Catholics. But Elgin, which was the seat of the bishops of Moray, is principally celebrated for the ruins of its cathedral, one of the most magnificent in the kingdom. It was built in 1224, the cathedral establishment having been transplanted at that time from Spynie to Elgin. The original structure (with other sacred buildings, and no small portion of the town) was burned in 1390 by the Earl of Buchan, youngest son of Robert II., known by the name of the 'Wolf of Badenoch.' It was rebuilt by the bishops of Moray, in the form of a Passion or Jerusalem cross, having

3 towers, one at each end, and one in the centre. The length of the building was 264 feet; the breadth of the transept 114; while the height of the centre tower was 198. The cathedral was unroofed in 1668, by order of the Regents Morton, for the sake of its lead; and this venerable specimen of architecture and sculpture has since been allowed to fall into decay. The great centre tower fell in 1711. But the chapter-house, the turrets and walls of the east choir, and the towers on the west, are still remaining. Of the walls of the nave and transept only a few fragments remain. Steps have been taken by the barons of exchequer in Scotland to prevent any further dilapidation. A college was attached to the cathedral, and contained not only the church and grave-yard, but also the bishop's house and those of 22 canons. The eastern gateway and part of the wall are still standing. The ruins of a convent of Greyfriars, settled here by Alexander II. in 1284, are still to be traced N. of the town. Of the convent of the Observatines, established here in 1479, no remains can now be seen. A Maison Dieu, or religious hospital, once stood on the site now occupied by the Elgin Institution. (Keith's Scot. Bishops, by Russell, Edin. 1824, pp. 138, 141, 142, 441, 453.)

In addition to the two charitable institutions already mentioned (Grey's and Anderson's), there are eight other charitable endowments of a subordinate order, most of them old. One of them is Grey's charity (the founder of the hospital), for the support of reputed old maids of the town of Elgin, with funds amounting to 8,000*l.* The seven incorporated trades, and the guildry, are each, in one respect, of the nature of provident institutions. The academy, which is partly endowed, and partly supported from the town's funds, contains three separate schools, and has long been a distinguished seminary. There are no fewer than ten schools in the town. There is a subscription and other libraries, as also a reading-room, with numerous benevolent and religious societies. There are no manufactures, except a tannery and a brewery. The town has ten fairs yearly for live stock, and a weekly market for grain and other agricultural produce.

Elgin can boast of great antiquity. In the 12th century it was a considerable town with a royal castle situated on the Lady Hill. The earliest charter of guildry was granted in 1234. It unites with Cullen, Banff, Peterhead, Kintore, and Inverary, in sending a member to the H. of C., and, in 1861, had 314 registered voters.

ELJEM. See Tysdrus.

ELORA, or ELLORA (*i.bera*), a village of Hindostan, dom. of the Nizam, prov. Aurungabad, in about lat. 19° 58' N., and long. 75° 43' E.; celebrated for some remarkable cave temples, excavated in the solid rock, about 1 m. to the E. which in magnitude and perfection of execution, surpass all other structures of the kind in India. The site of these curious monuments of art is a crescent-shaped hill, of moderate elevation, the concavity of which faces W. or NW. Its constituent rocks are chiefly basalt, a hard vesicular rock, and a rock of a loose, gritty, absorbent, and crumbling nature, interspersed with veins of quartz, siliceous stone, and blood-stone. The caves are cut in the W. slope or concavity of the hill above mentioned, extending, with intervals of various length between them, for about 1 m. from one extremity to the other. They may be divided into three groups: the N., which appear to have belonged to the Jain sect, since the purely Buddhic sculpture and emblems in them are intermixed with many Brahminical ones; the central, which are by far the most numerous, and are solely Brahminical; and the S., which are as decidedly Buddhic. Beginning at the N. extremity, a few hundred yards up the hill, cut in a mural rock of black basalt, is what is called the *Parismesh*, a colossal figure of Buddh, 10 ft. high, apparently in a triumphal car, and seated on the folds of a large snake, whose seven heads form his canopy. Six attendant figures surround this statue, over which a handsome stone porch was erected about a century since. This idol is still held in much reverence by the Jains, many of whom make an annual pilgrimage thither. About 200 yards below this idol is what is called the *Indra Subbah*, or 'Court of Indra,' a temple consisting of three caves, opening one into another, and situated behind an area cut out of the rock, in which stand an elaborately sculptured pagoda, a handsome obelisk, and the figure of an elephant. The front of this temple is in many parts covered with sculptures in relief; and at the extremities of the verandah before it are two figures, a male and a female, the former seated on a couchant elephant, and the latter on a lion. These figures have been generally called Indra and Indranee; but Col. Sykes contends that they represent the prince and his consort who founded this temple. (*Journ. of the Asiat. Soc. of Bengal*, vi. (1837), 1034.) The caves consist of two stories each; but the lower stories are greatly injured by damp, and partially choked up with earth. The three chambers on the story above vary from about 60 to 70 ft. in length, by nearly as much in breadth, and from 18 to 15 ft. in height, and their ceilings are supported by numerous pillars and pilasters. Each contains a colossal figure of Buddh, similar to that already described; and in the first and second chambers there are figures of other personages. The compartments round the walls of each of these rooms contain figures of Buddh, in various attitudes, 'some standing and some sitting; the attendants are riding on elephants, tigers, and bulls.' (Sykes.) None of these caves have any cells opening from it, which appendages are almost universally found in temples strictly Buddhic. About 40 or 50 paces farther to the E. there is a fourth cave, and still farther on, another; but both are much choked up with earth.

The first of the series of Brahminical temples, proceeding from the W., is about 200 yards distant from the latter, and entitled *Ramasar Lyena*, 'the Nuptial Palace.' This is the most extensive chamber of all; under one roof it is 185 ft. in length, by 150 ft. broad; its ceiling averages 19 ft. in height, and is supported by 20 pillars and 70 pilasters. The entrance to this excavation is through a passage cut in the solid rock, 100 ft. long by 9 ft. broad. On the left-hand side of the W. entrance is an eight-armed figure of a revengeful character, representing Siva in one of his forms; on the right are Siva and Parvati together in a heaven, which Kaernn (the Hindoo *Ameras*), a figure with numerous heads and arms, is endeavouring to shake. At the end of the central colonnade is a square sanctuary, entered by four doors, each guarded by two gigantic figures, 14 ft. 8 in. in height, and containing the *lingam*, which emblem is found in nearly all the second group of caves at Elora. There are numerous small caves, all of which are considered to have been devoted to the worship of Siva: in the front of each there is a bust of the celebrated triad, a mutilated specimen of which exists at Elephanta. (See ELEPHANTA.) Over the door of one cave is the image of Luximee, attended by elephants; and another, a noble hall, 90 ft. long, 26½ ft. wide, 15 ft. in height, and adorned by highly-finished pillars, has numerous compartments full of figures, amongst which is a group

supposed to represent the marriage of Siva and Parvati.

But the most splendid temple at Ellora is that called *Kylas*, or 'Paradise,' a pagoda of a sugar-loaf form, 100 ft. in height, surrounded by five chapels, nearly similar in form; the whole, together with the area in which they are situated, being excavated in the solid rock, and covered with sculptures from top to bottom, both within and without. The extreme depth of the excavation is 401 ft.; the area itself is 823 ft. in depth, by 185 ft. in its greatest breadth (on the E. side). On the N. S., and E. it is surrounded by colonnades, varying in length from 185 to 115 ft., and having from 15 to 18 square pillars each; the walls which these colonnades surround are covered with sculptures, and in the front of the wall by which the area is enclosed on the W. side are niches filled by gigantic figures. *Kylas* contains the representations of nearly all the Hindoo Pantheon; but, as Col. Sykes observes, notices of its figures alone would fill a volume, and the temple must be seen to be duly appreciated. (Those who wish for farther information may resort to Captain Seely's work, and to the accounts of the Ellora Caves, by Col. Sykes, in the Trans. of the Lit. Soc. of Bombay, iii. 281, &c.; Sir C. Malet, in the Asiatic Researches, vi. 382-424.)

The southern group of caves is very interesting. There are four principal ones: the first has three stories; the second, 2; the third, 80 ft. long by 42½ broad, and 85¼ in height, is in beauty inferior to none, and has an arched roof, supported by ribs of wood similar to that of Cartee, or the great cave at Kennery; the fourth is accompanied by several smaller ones, and all are very highly finished. Each temple of this group contains a large figure of Buddh, and other characteristics of Buddhic temples. (For some speculations as to the era of these caves, see Journ. of the Asiat. Soc. of Bengal, vi. (1837), 1034-41.)

ELSINEUR, or ELSINORE (Dan. *Helsinger*), a marit. town of Denmark, on the E. shore of the isl. of Zealand, at the narrowest part of the Sound, or principal channel leading from the N. Sea to the Baltic, 7 m. W. Helsingborg in Sweden, and 23½ m. N. by E. Copenhagen; lat. 56° 2' 17" N., long. 12° 36' 49" E. Pop. 8,442 in 1861. The town stretches irregularly over sloping ground towards the shore. It is well built, and has some good edifices. There are two churches, one of which, though externally very plain, contains many interesting objects of antiquity, and a lofty altar gorgeously ornamented. The public cemetery of Elsineur is a large and handsome enclosure, immediately adjacent to the town, on the N.E., is the castle of Cronborg. This edifice, built by Frederick II. in the boldest style of Gothic architecture, is said to be one of the finest structures of its kind in Europe. 'Though of great extent, yet so elegant are its proportions, that it seems as light and graceful as a building raised more for ornament than for use. So far, however, from being a mere thing of show, it is a strong and substantial fortress, strengthened by all the advantages that military science can give to a position which, though very low, is still extremely important, from its sweeping the Sound most completely, both up and down. The approach, therefore, is garnished with lines and demi-lunes, scarps, ditches, stockades—in short, all the imposing externals of a fortress kept in the highest order.' (Bremner, i. 253.) From the summit of the lighthouse of this fortress the scene is one of surpassing beauty. Cronborg is now chiefly used as a prison; it was the place of confinement for some years of the unfortunate Queen Matilda, sister of George III., of

England. Before the abolition of the Sound dues —bought off, in 1856, by the payment of 3,374,632l., one-third of which was contributed by Great Britain—all merchant ships passing to and from the Baltic were obliged, under certain reservations depending on the weather, to salute Cronberg Castle by lowering their topsails when abreast of the same; and no ship, unless belonging to Sweden, was allowed to pass the Sound without clearing out at Elsineur and paying toll. The Sound duties had their origin in an agreement between the King of Denmark on the one part, and the Hanse Towns on the other, by which the former undertook to construct light-houses, land-marks, &c., along the Cattegat, and the latter to pay duty for the same. The duties varied at different periods; and the greater part of the inhab. of Elsineur were, in some way or other, connected with their management or collection. Now that the Sound dues are abolished, the place is very quiet, though not in a state of decay. The principal communication between Denmark and Sweden takes place here, and regular boats sail three times a day to and from Helsinburg.

Elsineur is well known from its being the scene of Shakspeare's noble tragedy of 'Hamlet.' 'The principal incidents of the play are founded on fact, but so deeply buried in remote antiquity, as to make it difficult to discriminate truth from fable. Saxo-Grammaticus, who flourished in the 12th century, is the earliest historian of Denmark who relates the adventures of Hamlet. His account is extracted, and much altered, by Belleforest, a French author; an English translation of whose romance was published under the title of 'The Historye of Hamlet;' and from this translation Shakspeare formed the groundwork of his play, though with many alterations and additions.' (Laue's Travels in the N. of Europe, v. 90.)

ELVAS, a fortified city of Portugal, on the frontiers of Spain, prov. Alemtejo, 120 m. E. Lisbon, 12 m. W. Badajoz, on the railway from Lisbon to Badajos and Madrid. Pop. 18,516 in 1858. The town is picturesquely situated, on a hill covered with olive trees and orchards, between two other hills which command it, and on which are the fortresses of Santa Lucia and La Lippe. These and the other defences of the town, reckoned the chef-d'œuvre of the Count de La Lippe Schomberg, and a model of their kind, render it so strong, that no impression could be made upon it, except by a large army and a regular siege. The principal street, Rua de Cadea, has an antique, venerable appearance, from the remains of Moorish houses and towers. The castle, or prison, stands at one end, and opposite to it is the hospital for the townspeople, which is well conducted, and divided into wards, as in England, with separate apartments for infectious diseases. On the whole, however, the town is ill built, and the streets mostly narrow and dirty. The principal edifices are—the cathedral, arsenal, bomb-proof barracks for 6,000 or 7,000 men, and theatre. It has several churches and convents, with a college and a seminary. There are manufactures of arms and jewellery; but the principal dependence of the inhabitants is on the contraband trade carried on across the Spanish frontier. The Plaça, or great square, is remarkable for a singularly formed tower in front of the cathedral, and the houses exhibit specimens of domestic architecture from the days of Moorish splendour and elegance down to modern times. Several of the grotesque carvings are executed with great richness and delicacy. The rooms in the modern houses are large, lofty, and paved with bricks

arranged in various figures, the windows and bring glazed, but merely closed with latticed blinds. The decorations of some of the chapels in the cathedral are extremely elegant, the walls and ceilings being covered with a profusion of gilded carving, but the pictures are execrable. The grand altar is supported by Corinthian pillars of grey marble, surmounted by a canopy of crimson and gold silk, beneath which is a large picture of the birth of Christ: the altar itself is covered with crimson and gold silk, and is crowded with silver candlesticks. There is no room in the town for public gardens, but the covered way from the Porta d'Esquina to the Ollivença gate is planted with trees, and each place of arms has a fountain, and is tastefully laid out. The walk round the ramparts is extremely fine, commanding a view of the country for many miles in all directions. The town is furnished with water, brought from an eminence about 3 m. W. from it by an aqueduct constructed by the Moors, which supplies numerous fountains, one of which is of very large dimensions. In crossing the valley 1½ m. in width, this aqueduct has four tiers of arches, each above the other, making together 250 ft. in height. It is supported by strong buttresses; and, to add to its strength, it is built in a zig-zag direction. The environs are fertile in grain, wine, oil, and fruit. Manufactures, arms and hardware.

Elvas was a post of great importance during the Peninsular war. Marshal Jmnot took possession of it in March, 1808, and held it till it was given up, under the convention of Cintra, in August following. It has bomb-proof barracks for 6,000 or 8,000 men, and furnished the artillery and stores for the siege of Badajos. The Duke of Wellington had a powerful telescope placed in the tower of La Lippe during the operations, by which the interior of the castle of Badajos could be plainly looked into, and all the operations discovered. (Napier's Peninsular War, i. 144, 160, 202; ii. 126; iii. 310; iv. 185, 401.)

ELY, a city of England, co. Cambridge, in the district called the Isle of Ely, on an eminence near the Ouse, 16 m. NNE. Cambridge and 71 N. London by Great Eastern railway. Pop. 7,424 in 1861. The city includes the parishes of Ely, Trinity, and St. Mary's; the extra-parochial district of Ely college, and the chapelry of Chettisham, comprising, in all, an area of 17,480 acres, of which about 5,000 may belong to the city properly so called. The latter consists principally of one long street, with a market-place in the centre; several of the houses are built of stone, and have an antique venerable appearance; and the place seems to have been but little affected by those changes that have so materially modified the appearance of most other towns. It owes its entire distinction to its being a seat of a bishopric, established here in 1107. Its cathedral is one of the most celebrated in England. Being partly of the reigns of William Rufus and Henry I., and partly of subsequent periods, it displays a singular admixture of the Saxon, Norman, and English styles of architecture; but notwithstanding the dissimilarity of its parts, it must, when considered as a whole, be regarded as a truly magnificent edifice. Its extreme length from E. to W., is 535 ft.; the length of the transept is 190 ft.; the height of the lantern on the summit of the dome over the celebrated octagon tower, is 170 ft.; the extreme height of the W. tower, one of the finest in the kingdom, is 270 ft.; the height of the E. front to the top of the cross is 112 ft. It has many interesting monuments. St. Mary's chapel, contiguous to the cathedral, now

Trinity church, was commenced in the reign of Edward II., and is one of the most perfect structures of the age: it is 200 ft. in length inside, by 46 ft. in breadth; the height of the vaulted roof being 60 ft.: it has neither pillars nor side aisles, but is supported by strong buttresses. The cloisters and other buildings, which belonged to a monastery founded here at a very early date, have been long since demolished, with the exception of the refectory, that has been converted into a deanery. The episcopal palace, near the W. end of the cathedral, retains few traces of its ancient architecture. The bishops of Ely formerly possessed powers within the isle similar to those enjoyed by the bishop of Durham, appointing their own chief justice and magistrates; but these were taken away by the act 6 and 7 William IV. cap. 87. The assizes are held here in the new shire hall, erected in 1821. Ely has a grammar-school, founded by Henry VIII.; a free school endowed by a lady of the name of Needham; and a national school supported by voluntary contributions. A considerable landed property left for the benefit of the city poor is vested in a body of incorporated trustees. There is an earthenware and tobacco-pipe manufactory within the city; but the inhabitants are principally employed in gardening, which is extensively carried on in the vicinity. Ely sent two members to the H. of C. in the 23rd of Edward I., but has not subsequently been represented. The Isle of Ely is included within the great level of the Fens, and is extremely fertile.

EMDEN, or EMBDEN, a sea-port town of Hanover, being the second in that kingdom in respect of size and importance; prov. Aurich, cap. cant., on the N. bank of the estuary of the Ems, or mouth of the bay called the Dollart, 15 m. SW. Aurich, and 46 m. WNW. Oldenburg, on the terminus of the railway from Hanover to the bay of Dollart. Pop. 13,170 in 1861. The town is surrounded by walls and wet ditches, and divided into the old town and the Faldern; the latter being the best built. Emden has 6 churches, one of which is a fine edifice, a council-house, judicial tribunal, custom-house, exchange, commercial weighing-house, naval assurance office, school of navigation, house of correction, orphan asylum, lying-in-charity, gymnasium, and society of natural history. A navigable canal connects it with Aurich, and various others intersect the adjacent country and the town, communicating with the port. The latter, which consists of two inner harbours opening into an outer harbour, is large, but shallow; so that vessels drawing more than 11 ft. can enter it only at high water, unless lightened of a portion of their cargo. But the roadstead, which is well protected, has water sufficient to float vessels of any size, and the holding ground is good. Emden has manufactures of linen and linen yarn, stockings, tobacco, brandy, leather, hats, soap, and starch; its herring fishery was formerly of considerable importance, and employed 1,500 hands, who took about 13,000 tuns of fish annually; but this branch of industry has greatly declined, and from 60 ships formerly engaged in it, the number is now reduced to 15. The general trade of the town has also declined. In the 16th century it had 600 sea-going vessels; and, in 1769, 273 of the aggregate burden of 19,290 lasts. In 1863, the shipping had declined to 105 sea and river vessels, chiefly coasters, of a total burden of 4,790 lasts. By far the greater number of the vessels that now frequent the port are inland craft, but there are also regular steamers to Hamburg, Hull, and London.

Though Emden is a free port, the advantage it thence derives is very insignificant. It has little communication with the interior of Germany, except with E. Friesland and the co. of Munster, of which it continues to be the emporium. The import trade is formerly carried on in colonial produce has been almost entirely transferred to Amsterdam, Hamburg, and Bremen, whence it is supplied at second hand. Its chief imports are hemp, potash, and timber, from the Baltic and Norway. The imports of timber are very considerable, the vicinity of Emden being singularly deficient in wood. It also imports considerable quantities of French wine. Its chief exports are oats, wheat, beans, rapeseed, rye, barley, herrings, butter, cheese, gin, tallow, honey, wax, wool, and hides.

Emden belonged, in the Middle Ages, to the counts of East Friesland. It subsequently became a Hanse town; but fell, in 1806, to Holland; in 1809, to France; and in 1814 to Prussia. The latter power ceded it in 1815, to Hanover.

ENGLAND AND WALES. This populous, wealthy, and important portion of the U. Kingdom of Great Britain and Ireland, comprises the most southerly, largest, and most fertile part of the island of Great Britain. It lies W. from and opposite to France. Belgium, Holland, and the S. parts of Denmark, between 49° 57' 30", and 55° 47' N. lat., and 1° 46' E. and 5° 41' W. long.; being bounded by the German Ocean on the NE. and E.; by the British Channel on the S.; by St. George's Channel and the Irish Sea on the W.; and on the NW. and N. by Scotland, from which it is separated by a waving line extending in a NE. direction from the mouth of the Sark, in the NE. corner of the Solway Frith, by Peel and Carter Fell, and the Cheviots, to Carham, and thence along the Tweed to Berwick. Its SE. extremity, at Dover, approaches to within 21 m. of the opposite coast of France. (See British Empire.) Its shape approaches nearest to that of a triangle, of which Berwick may be considered the apex, and a line from the Land's End to the N. Foreland (342 m.) the base; a line from the former along the W. side (476 m.), and from the latter along the E. side (334 m.) complete the figure. The sea-coast, if measured from one headland to another, is about 1,900 m. in extent; but if its principal indentations are followed, it will be found to be fully 2,000 m. The bays and harbours on the S. and W. shores are numerous, and some of them rank among the finest in the world; but on the E. side there are few that can be called safe, or easily accessible; the ports of London and Harwich being the only really good ones between the S. Foreland and the Tweed. The area amounts to 58,320 sq. m., or to 37,324,883 statute acres. The area of England alone is 32,590,397, and that of Wales 4,734,486 statute acres. (Census of England and Wales, 1861, vol iii., General Report, 1863.)

Aspect of the Country.—England combines within itself all that is most desirable in scenery with all that is most necessary for the subsistence and comfort of man. Although its features are moulded on a comparatively minute scale, they are marked with all the agreeable interchange which constitutes picturesque beauty. In some parts plains clothed in the richest verdure, watered by copious streams, and pasturing innumerable cattle, extend as far as the eye can reach; in others, gently rising hills and bending vales, fertile in corn, waving with woods, and interspersed with flowery meadows, offer the most delightful landscapes of rural opulence and beauty. Some tracts furnish prospects of the more romantic and impressive kind; lofty mountains, craggy rocks,

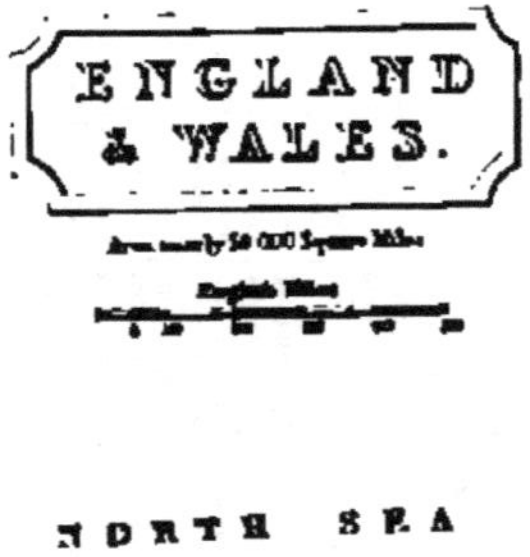
ENGLAND
& WALES.
NORTH SEA

deep dells, narrow ravines, and rumbling torrents; ... in which every variety of nature is a different charm, the vicissitude of black barren moors and wide uninhabited heaths.' (Aikin's England Described, p. 2.)

The distinguishing peculiarity in the aspect of England is, however, the exuberance of its vegetation, and the rich luxuriant appearance of its lower and far most extensive portion. It owes this distinction partly to nature and partly to art. The humidity and mildness of the climate maintain the fields in a constant state of verdure: in winter they are seldom covered with snow, or blighted by long-continued frosts, and in summer they are rarely withered and parched by droughts. In this respect England is as superior to the finest countries of continental Europe — to Italy and Sicily, for example — as she is superior to them and to every other country in the amount of labour that has been expended in beautifying, improving, and fertilising her surface. It is no exaggeration to affirm, that thousands upon thousands of millions have been laid out in making England what she now is. In no other nation has the combination of beauty with utility been so much regarded. Though without any extensive forests, England is extremely well wooded. The country is portioned out into innumerable fields; and these being all, or nearly all, surrounded with hedges and rows of trees, it has, even in the best cultivated districts, a woody appearance, and sometimes almost resembles a vast forest. Since the middle of last century, a great deal has been effected in this way. Most of the extensive, bare, and nearly worthless commons, that were then everywhere met with, have been in the interval subdivided, enclosed, and brought under tillage; making a vast addition to the productive capacities of the kingdom, and materially improving its appearance.

Another peculiar feature in the physiognomy of England is the number and magnificence of the seats of the nobility and gentry. These superb mansions, many of which are venerable from their antiquity, and all of which are surrounded with fine woods and grounds, give to the country an appearance of age, security, and wealth, that we should in vain look for any where else. The farmhouses and cottages have mostly also a substantial, comfortable look; and evince that taste for rural beauty, neatness, and cleanliness, that eminently distinguish their occupiers.

The number, and the prodigious size and splendour of many of the cities and towns of England, justly excite the admiration and astonishment of foreigners, and even of natives. They are the classes seats of opulence, art, science, and civilisation. All the gratifications that wealth can command, or the caprices of taste or fashion require, may there be had in the utmost profusion; at the same time that art and industry are carried in them to the highest perfection to which they have attained, and are aided by every invention and discovery, how remote the country or distant the era of their origin.

Description of the Country. — Though the mountains of England nowhere attain an alpine elevation, they form one of its most interesting, as well as most prominent features. The principal chains, which are found in its N. and W. portions, have received the names of the Pennine, Cumbrian, Cambrian, and Devonian ranges. The first of these ranges extends from the Scottish border, where it is connected with the Cheviots, S., to near Derby: it occupies the W. portion of the cos. of Northumberland, Durham, and York, and the E. portion of Cumberland, Westmorland, Lancaster, Chester, and the middle part of Derbyshire. Its highest summits are Cross Fell, in Cumberland (2,901 ft.); Shunnor Fell, on the confines of Yorkshire (2,329 ft.); Great Whernside (2,385 ft.); Ingleborough (2,361 ft.); and Pen-y-Gant (2,270 ft.), in Yorkshire: at either end, however, the range declines considerably, so that at the part traversed by the old Roman wall, and the modern railway between Newcastle and Carlisle, its height does not exceed 445 ft.; and on the S. side, where the Liverpool and Leeds Canal is conducted across it, the elevation is not more than 500 ft.; still further S., the Derbyshire portion of the chain again becomes more elevated, attaining at Castleton and Great Axehill, 1,751 ft.; and at the Weaver Hill, near Ashbourne (the S. extreme), 1,154 ft. The breadth of the range between Sheffield and Macclesfield is about 22 m., and it comprises, in this portion, some very picturesque scenery; but such is very far from being the character of the N. portion of this mountain system, which may be generally described as with rounded summits, of gradual ascent from either side, having a scanty peat soil, covered mostly with ling, and undulating in dreary succession; the patches of green sward being few and far between, and the aspect of the whole cheerless and monotonous. With the exception of the Thames and Severn, most of the great rivers of England have their sources in this chain; being much nearer the W. than the E. side of the island, the rivers that rise in its E. acclivities have generally the longest course, and are the largest and most important. Of the latter, the Tyne, Tees, the affluents of the Ouse, the Aire, Don, and Trent are the principal; the Eden, Ribble, and Mersey are the principal rivers flowing W. from the Pennine chain. The beautiful vale of the Eden, which separates the Pennine from the Cumbrian range, gradually expands into the Cumbrian plain, which extends N. to the Solway Frith, and occupies the whole tract from Brampton, Caughts, and Renwick, at the base of the Pennine chain W. to the sea, comprising an area of about 300,000 acres. On the E. side of the Pennine chain, from its N. extreme to the Coquet, the district, though hilly, has tolerably good pasture, and comprises a few breadths of well-cultivated land; S. of that stream, a large moorland tract extends through Northumberland, the middle of Durham and Yorkshire, to the Holm Moss in Cheshire, varying in breadth from 10 to 30 m., and in elevation from 600 to 1,000 ft.: its N. is its most sterile portion; but the whole tract consists of a series of monotonous wastes, furrowed, in the two N. cos., by a few narrow glens only: towards the S., these widen and become more frequent, but without much affecting the general aspect, which is preserved, for the most part, through the whole extent of the district. Betwixt it and the sea are the vales of the Tyne and Tees, and the great Yorkshire plain; the latter extending N. and S. between 60 and 70 m., with an average breadth of between 14 and 20 m.: it widens towards the S., and everywhere presents a gently undulating surface of fertile and well-cultivated land. The E. moorlands and wolds, bounding the York plain on that side, have, as their N. limit, the fertile vale of Pickering, extending about 35 m. E. and W., and 10 m. in the opposite direction. It presents the appearance of a drained lake, enclosed between the Hambleton hills and the Yorkshire wolds. The last-named tract, together with the Lincoln wolds, S. of the Humber, occupy about half the space between the German Ocean on the E. and the rivers Derwent and Trent on the W.: generally speaking, they form good pasture lands, interspersed in parts by a few sterile moors, and, in others, by moderate breadths of good arable land. The plain of Holderness, N. of the Humber, and extending from the base of the wolds to the sea, has a strong clayey soil, producing heavy crops of wheat and beans, as well as luxuriant pasture, and ranks amongst the most productive districts in the kingdom. An alluvial tract, of somewhat similar character, also extends along the base of the Lincoln wolds between the Humber and Wash. The low line of coast, forming the E. limits of these tracts, has a submarine forest stretching along it, which is traceable for 1 or 2 m. in breadth between the high and low water-marks.

The Cumbrian group of mountains occupies the central and S. portions of Cumberland, the W. and largest portion of Westmoreland, and the N. and insulated portion of Lancashire. It extends N. and S. about 37 m., and E. and W. about as much. It contains the most elevated summits in the kingdom, and is intersected by deep narrow glens, some of which are occupied by lakes, that radiate in all directions from the central portion of the mass, so as to form several distinct ranges: the whole system declines more rapidly on the N. than the S. side. The highest and most remarkable summits are, Helvellyn (3,055 ft.), Scafell (3,166 ft.), Bowfell (2,911 ft.), Coniston Fell (2,577 ft.), High Pike (2,101 ft.), in the central part of the group; at the N. extreme are Skiddaw and Saddleback (3,022 ft. and 2,787 ft. respectively); and at the SW. end, Blackcombe rises 1,919 ft. above the sea. The Cumbrian mountains are mostly bold, steep, and rugged; their slopes are in general covered with a fine green sward, affording good pasture for sheep, and have little of the tame, monotonous character that belongs to the Pennine range. Except in some of the glens, opening on the N. and W. sides, the cultivable land among these mountains is not very considerable. The lakes embosomed in these mountains rather resemble the reaches of a large river than the expanded figure usually considered as belonging to a lake. Winander Mere, the most extensive of these sheets of water, is between 10 and 11 m. long, and from 1 to 1½ m. broad, with a depth, in some parts, of 85 fathoms. It has 13 or 14 small islets or holms, the largest of which contains about 30 acres; its area, including these, is about 2,574 acres. Ulswater, the next in size, is about 8½ m. in length, by 1 m. at the broadest part, and zigzags in a NE. direction from Patterdale. Derwentwater, Bassenthwaite, Buttermere, Wastwater, Ennerdale, and Coniston Mere are the names of the more considerable amongst the remainder; all of them abound in fish, chiefly trout, perch, pike, and eel; Ulswater and one or two of the smaller tarns have char; and Bassenthwaite salmon, which find their way thither by the Derwent. The scenery of the district occupied by the Cumbrian mountains is perhaps the most interesting and romantic of any in England; and in many parts, as at the head of Ulswater and the Kirkstone Pass, between that lake and Winander Mere, it assumes features of great power and magnificence. The line of road between Ambleside and Keswick, through the vale of St. John, is also interesting for its picturesque and beautiful scenery, well-known through the poems and delineations of Southey and Wordsworth.

The Cambrian mountains extend on the W. side of the kingdom, from the Irish Sea to the Bristol Channel, occupying nearly the whole of Wales. Of these, the Snowdonian range is the chief; its principal chain stretches NE. and SW., the whole length of Carnarvonshire, from Penmanmawr on the N. to the point of the peninsula of Lleyn on the S. Several of its summits exceed 3,000 ft. in height: that of Wyddva (the highest pinnacle of the huge mountain mass bearing the general name of Snowdon) has an elevation of 3,571 ft.; and commands a view of surpassing grandeur, which is only limited by the horizon. Two or three other chains branch from this main one, in a S. direction, many of whose summits reach 2,400 ft., and one (the Arennig Mawr) 2,809 ft. The country included between these ranges has a few picturesque and well-sheltered vales, such as those of Festiniog and Dolgelley; but its general character is that of a partially unreclaimed pasture tract, comprising most magnificent mountain scenery. Anglesea, on its W. side, has several small ridges and detached hills and peaks, but it cannot be called mountainous. On its E. side the beautiful vale of Clwyd extends between the Hiraethog hills and another parallel range stretching between it and the estuary of the Dee; the vales of Mold and Llangollen, also celebrated for their beauty and fertility, extend on the same side, towards the great Cheshire plain.

The Berwyn mountains stretch across the whole principality, S. of the Snowdon ranges, from Llangollen to the middle of Cardigan Bay; the highest summit, Cader-Idris (2,914 ft.), gives its name to the portion of the chain between it and the sea, which narrows to a mere ridge, in parts, not more than 4 or 5 m. across. The general character of the country comprised within the Berwyn range is of the same kind as the former, though with less elevated and abrupt outlines; towards the vale of the upper Severn, and between it and the Plynlimmon chain, a few strips of cultivated land occur. The famous mountain, whence this chain takes its name, is 2,463 ft. in height, and gives birth to the two great rivers, the Severn and Wye, flowing S. to the Bristol Channel, and to the Rheidiol, which has its embouchure at Aberystwith, on Cardigan Bay. From Plynlimmon the chain extends in a curve to the Breiddin hills, W. of the Shropshire plain, whose highest summit reaches 1,350 ft. The whole of the Plynlimmon range is characterised by smooth gradual slopes, and a succession of regularly rounded summits, clothed with a fine green sward, that supports numerous flocks of a small fine-woolled breed of sheep. The hilly tract extending through the S. of Shropshire to Wenlock Edge, may be considered as a continuation of this range, and is characterised by the same general features: its highest summit (Clee Hill) attains 1,805 ft. The mountain region extending N. of the Plynlimmon chain to the Towy, and stretching E. and W. between the Wye and Dyfi, forms the largest waste in the kingdom, and consists of a succession of rounded, barren hills, enclosing vast morasses, amongst which a few spots covered with coarse herbage are sparingly scattered, and afford summer pasturage to a small hardy breed of sheep: Drygarn Hill, near the centre of this cheerless region, is the highest summit, and attains 2,071 ft. The Epynt hills, on its S. border, enclose many strips of good arable land, and are themselves clothed with fine pasture; but the country on the W. side of this great waste, on to Cardigan Bay, is mostly of a rugged, desolate aspect, and comprises a series of table lands, with broken surfaces and scanty vegetation. On the N. side the Yarwith, however, and along the course of that stream and the Rheidiol, especially near Hafod, the scenery is picturesque, and includes many fine cataracts; and along the coast are several large pasture tracts of various degrees of fertility. S. of this, on to St. David's Head and the Bristol Channel, the country consists mostly of unreclaimed table lands of unequal surface, with occasional ridges and detached hills, all of a rugged sterile aspect, with the exception of the district round Milford Haven and the Peninsula of Gower, between the bays of Swansea and Caermarthen in the Bristol Channel, which are fertile and well cultivated.

The Radnor and Black Forest ranges, that stretch S. from the centre of the Plynlimmon chain, on either side the Wye, are mostly covered with verdure, and form good sheep-walks; their offsets stretch into Herefordshire and terminate in that fertile and undulating plain. The districts on either side the range, especially the vales of the Wye and Usk, include much cultivated land. Two other main ranges complete the Cambrian mountain system,—those of the Forest Fawr and Glamorgan; the former stretches through Caermarthenshire and Brecknockshire to Abergavenny, on the Usk; the highest summits are the beacons named from these counties, which are respectively 2,596 ft., and 2,862 ft. high. It comprises excellent and extensive sheep-walks. The Glamorgan range extends N. of the last, in an E. and W. direction, from Pontypool on the Usk to Swansea, about 30 m., and in the widest part (from Merthyr-Tydvil to Llantrisant) about 15 m. The summits are mostly table lands, with steep declivities on either side, intersected by deep narrow ravines, the whole having a rugged, cheerless aspect, but enclosing the most extensive coal and iron deposits in the kingdom. The tract between the two last-named ranges is also of the same sterile character, and wholly unreclaimed; but the plain stretching from the S. declivity of the Glamorgan chain to the Bristol Channel has a rich productive soil, and may, independently of its vast mineral treasures, be considered as the best and most fertile district of the principality. An alluvial tract, 3 or 4 m. in width, extends from the Taff to the Monmouth plain, and is of a similar character. The Welsh lakes are numerous, but for the most part small and uninteresting, rather absorbed by the majestic scenery round them than forming one of its essential features, as is the case with the Cumbrian lakes. The Bala Pool or Llyn Tegid, is the largest of the Welsh sheets of water, and extends 4 m. from SW. to NE., with an average breadth of 1 m. and depth of 40 ft.; its waters cover an uneven rocky bed, and are remarkable for their purity and clearness. In common with most of the others, it abounds in red trout, pike, and eel; but the grayling, or silver skate, is peculiar to it. The Dee issues from its NE. end, flowing by the vale of Llangollen and the Cheshire plain to the Irish Sea; the Clwyd and the Conwy, discharging on the same side; the Nelint, Maw, and Teify, in the St. George's, and the Towy, Wye, and Severn, in the Bristol Channel, are the other chief rivers that originate in this the wildest and most mountainous portion of the kingdom.

The Devonian chain, stretching through the SW. peninsula of England, between the Bristol and the British Channels, is the last that requires any especial notice in this sketch. Dartmoor Forest, forming its wildest and most elevated portion, is an unreclaimed and extensive waste, affording summer pasturage for the store cattle of the lower and more fertile tracts surrounding it: the whole may be considered as a table-land (the average height of which is above 1,600 ft.), with an unequal surface, rising in large rounded swells, with corresponding concavities, and strewed with large boulders and fragments of granite, which also rises through the soil in irregular masses, or tors. Exmoor, at the NE. extreme of the range,

and considerable tracts intermediate between the two, are also unreclaimed, and for the most part of a sterile character. The same description also applies to the central and northern parts of Cornwall, onward to the Land's End; but the less elevated districts on either side the range contain many extensive breadths of fertile land, more especially on the S. One of these, extending from Dartmoor to the sea, between the Dart and Yealm, and known as the South Hams, ranks among the most fertile corn districts in the kingdom. The chain gradually declines from Dartmoor to the Land's End, and also becomes more contracted in that direction. The chief summits are—Dunkerry Beacon, on Exmoor (1,668 ft.), Cawsand Hill (1,792 ft.), Rippon Tor (1,549 ft.), Butterton (1,203 ft.), all on Dartmoor; and in Cornwall, Brown Willy (1,368 ft.), Carnmarth (819 ft.), Carn Brea (697 ft.); and, lastly, the cape itself (about 70 ft.). The Taw and the Torridge, which discharge in the Bristol Channel, and the Pal, Fowey, Tamar, Plym, Dart, Teign, and Exe, descending to the British Channel, are the chief rivers of the district. On the N. coast sand accumulates rapidly in many of the creeks and inlets, forming in some places extensive dunes, beneath which the remains of ancient churches and villages have been discovered. On the beaches of Bude Bay, and a few others, this sand is chiefly composed of comminuted shells, and forms the chief manure of those localities.

The surface features of the central region of England, whence her wealth and importance are mainly derived, though outwardly diversified, are almost wholly devoid of the magnificence and romantic beauty of those previously described. The great plain of Cheshire and Shropshire, on its W. side, extends about 50 m. in a N. and S. direction, and from 25 to 30 m. in the opposite; a few heathy mountains occur within its limits, but by far the greater portion is very fertile; the soil is either rich sand, of a reddish colour, or strong loam. This plain is remarkable for its verdure, and is one of the principal grazing districts, being largely appropriated to the dairy husbandry. Of a similar character are the vales of Severn, Evesham, and Gloucester. The first of these extends about 70 m. on either side the Severn, with a breadth varying from 5 to 12 m., and is alike fertile and beautiful. The district N. of these last has probably the most broken and irregular surface of any part of the kingdom; it is, however, for the most part fertile and well cultivated. Beyond it are the Mendip, Quantock, and Black Down hills, and the fertile and beautiful vales of Taunton and Exe.

The basins of the Trent and Thames occupy the remainder of the central region; the former, in a general point of view, may be considered as forming an extensive plain, with gradual swells and broad intermediate vales, but with very few remarkable elevations. The vale of Belvoir is one of its most fertile portions. In the district forming the basin of the Thames, and drained by that great river and its various tributaries, the surface is, for the most part, greatly undulating, forming wide vales, often extending into plains; the principal elevations are near the valley of the Thames, but none of their summits reach the height of 1,000 ft. The geological character of the tract is greatly diversified, which causes a corresponding variety in the soils. These, however, on the whole, are of a light chalky nature, and moderately fertile, with but few absolute wastes of any extent; the higher constitute the least fertile portions, most of which are obviously indebted to skilful cultivation and the humidity of the climate for a great proportion of their productiveness. The most fertile tract is the vale of Aylesbury, which has a fine loamy soil, not surpassed in fertility by any in the kingdom. The chalk hills, which (with some interruptions) range from the N. side of the Wash to the Thames, between Goring and Henley, to which part the name of the Chiltern Hills applies, form the SE. limits of the basin, sloping gradually in this direction to the Thames, but with many abrupt escarpments on the other; whence extensive views are commanded of the country between the basins of the Trent and Thames, through which the Ouse, Nen, and Welland flow NE. to the Fens, draining Bedfordshire, Hants, Northampton, and Rutland, in their course through a district possessing very few striking inequalities of surface.

The courses of these rivers to their outfalls in the inlet of the German Ocean, called the Wash, are by channels and embankments, artificially formed, through the whole of the extensive flat and marshy district known as the Fens. (See Bedford Level.) Deposits of mud and sand are constantly and rapidly accumulating on this portion of the E. coast, so that it is not without considerable difficulty that the outfalls of the rivers are kept open, and the harbours accessible. Additions are always being made to the surface of the district by encroachments on the sea. Within the 25 years from 1840 to 1865, no less than 170,000 acres of fertile land, extending seaward between the ports of Boston, Wisbech, and Lynn Regis, were reclaimed in this manner.

The great plain NE. of the Fens, comprising Norfolk, Suffolk, and Essex, has an undulating surface throughout; but the inequalities are greater towards the N. extreme, where, in some places, an elevation of 200 ft. above the sea is attained. In this quarter it is not very fertile, but it has been wonderfully improved; and many parts of Norfolk and Suffolk that half a century ago were mere sandy wastes, have, by dint of marling and the introduction of the turnip culture, become among the best and most productive barley lands in the kingdom. The soil of Essex is mostly a strong clayey loam, ranking in the first class of wheat and bean lands. That portion of England extending from Bagshot Heath to Salisbury Plain, and comprising both, may be considered as a sort of elevated table-land, no part of which, probably, is less than 500 ft. above the sea: Thorney Hill is 610 ft., and Westbury Downs 775 ft. Both these eminences are on Salisbury Plain, the highest portion of the tract. This celebrated plain extends about 22 m. from E. to W., and 15 m. in the opposite direction; it is traversed by many considerable depressions, and has a light scanty soil, ill-adapted to cultivation, but affording good sheep-walks. The part of the country of this tract between Chichester and Southampton Water has a fair proportion of tolerably fertile and well-cultivated land; but further W., the Hants and Dorset downs occupy the surface nearly to Dorchester, and form a continuous heathy, dreary, and sterile tract, with but a scanty proportion even of sheep pasture.

To the E. of the Anton river are the chalk ranges of the N. and E. Downs, which extend round the weald district of Sussex, Kent, and Surrey; Beachy Head forming the E. extreme of N. Downs, and the bald chalk cliffs of the Dover Straits that of the N. Downs. The Alton Hills extend between and connect the two. The first are clothed with fine pasture, and form excellent sheep-walks: at their base extends the fertile plain of Chichester. The tract of which the N.

Downs forms the W. portion is, for the most part, well cultivated, and here and there attains considerable fertility, though generally speaking, the soil is meagre and arid. The weald district, enclosed by the last ranges, has in some parts an undulating unequal surface; and there are a few detached hills that attain considerable elevation: taken as a whole, however, it may be considered as forming an extensive plain of about 1,000 sq. m. in extent, the more level portions of which are from 100 to 200 ft. above the sea. The soil is principally clay; in parts very stiff and adhesive, in others mixed with sand in various proportions. The whole is under cultivation, and includes many breadths of luxuriant pasture. At the E. extreme is Romney Marsh, an alluvial tract of about 50,000 acres, which has been reclaimed from the sea, and is defended from its encroachments by embankments. This marsh is, for the most part, remarkably fertile.

Geology.—A brief sketch of the geological structure of England will be best accomplished by commencing with the mountain ranges on its W. side, and thence following the general direction of the successive rock strata: of these, the primary and transition, or (as they are now more correctly designated) Plutonic and metamorphic formations, constitute the mass in the Cumbrian and Cambrian groups, and that of the NW. peninsula, all of which have a general resemblance in their mineral composition, though presenting some points of local and minor difference: thus, granite, which is only traced to a very limited extent in one or two parts of the Cumbrian system, and scarcely at all in Wales, is extensively developed in the SW. peninsula, where it occupies a considerable part of the most elevated portion of the range, in large interrupted masses from Dartmoor to the Land's End; beyond which the Longships Rocks and the Scilly Islands continue the formation in the same general direction, and are supposed (with much probability) to have once formed continuous portions of the range. The veins of tin ore also appear to be limited to this last district. Neither gneiss nor mica slate (so abundant in the Grampians) occur, to any extent, in either of the ranges under consideration; clay and greywacke schists, of very various composition and texture, forming the prevailing rocks in all of them. The whole of these strata are traversed by beds and veins of porphyry, hornblende, and trap, and are for the most part considerably inclined and contorted, everywhere presenting indications of powerful disturbing causes, and of having been upheaved; but there are no traces of volcanic action. In the Carnarvonshire ranges elevated beaches occur at the height of 1,000 ft. and upwards above the sea-level, which are formed of gravel and fragments of recent shells, precisely similar to the present marine beaches. Similar beaches also occur on the N. coast of Cornwall and S. coast of Devon, from 20 to 30 ft. above the present reach of the tides.

The veins of tin and copper which intersect the strata in Devon and Cornwall make the SW. peninsula one of the most important mining districts in the kingdom. These veins, or lodes, have all a general E. and W. direction, and are intersected by others in an opposite (hence called cross-courses), which, by heaving or disturbing the regular course of the lodes, are often the cause of great perplexity and expense in mining operations. A large dyke of this kind traverses Cornwall, from one coast to the other, through its chief mining district, intersecting and disturbing the course of every one of its lodes. Besides these lodes of tin and copper, which furnish the chief mineral riches of this range, lead ore occurs in some of the cross-courses, and has been extensively worked at Beer Alston on the Tamar, and one or two other localities; iron is also found in similar dikes near Lostwithiel in Cornwall, and at the Herrybead on the coast of Devon; from each of which places many thousand tons are annually shipped for the supply of the Welsh furnaces.

Plumbago and manganese occur on the E. side of Dartmoor, both which are worked to some extent, and shipped at Exeter for the manufacturing districts. Porcelain, pipe, and common potters' clay, are also productions occurring in this tract, and are largely shipped for the Staffordshire and other potteries: granite and roofing slate are also quarried in a few localities. This last forms the most important production in the corresponding rock formations of Wales, the quarries of Penrhyn and Llanberris, in Carnarvonshire, being the largest, and furnishing the finest slates in the kingdom. Some copper veins also occur in various parts of this group, though of very minor importance compared with those of Cornwall: in the Parys mountain, however, on the N. side of Anglesea, a very extensive deposit of that ore was discovered in the course of last century, and formed for a considerable period the most productive mine in the kingdom; it is still worked, though at present the produce is very limited. (See ANGLESEA.) On the W. side of the same island, Mona marble, or vert antique, is quarried for various ornamental purposes, at the termination of a large porphyry dike which traverses the district.

In the Cumbrian group, the most remarkable mineral production is the famous graphite, or plumbago, which occurs in an irregular pipe-vein at Borrowdale in Borrowdale. A few lead veins also occur, and are worked to a limited extent, on the N.E. side of the range. At Coniston, copper veins are wrought on a small scale; and, near Ulverston, hematitic ore, which produces iron of a very ductile quality, which is used in the manufacture of ranting-wire; a few quarries of roofing slate are also worked in the same neighbourhood. Beyond the limits of the three main groups we have been describing, similar rock formations occur in a few isolated ridges, of which the most prominent are the Malvern Hills, that traverse the cos. of Worcester and Hereford; the Lickie Hill, S.E. of the last; the Charnwood range in Leicestershire; and a few intermediate rocks along the N. side of Warwickshire. Basaltic rocks also occur in the Wrekin and Caradoc hills, and along the limits of the mountain line, both in Derby and Durham: a large basaltic dike also traverses Yorkshire, from Middleton to the sea-coast N. of Whitby.

The mountain lime and coal formations are the next in order, being limited on the W. by those last described; and on the E. by the line, which formation may be traced, by a waving but continuous line, through the kingdom, from the N.E. coast (between the mouth of the Tees and Whitby), by Charnwood Forest, Evesham, Gloucester, Bath, and Axminster, to the S.W. coast at Lyme Regis. All the mineral riches of the kingdom, as well as the greater part of its manufacturing establishments, are situated on the W. side of this line, by which the three lower of what are usually termed secondary formations are limited. In the mountain line of the Pennine range are the chief lead mines of the kingdom: in that part of it which extends through Allendale and Alston Moor, on the E. side of Cross Fell, the ore occurs in E. and W. veins, that are heaved and disturbed by N. and S. courses, as those in Cornwall. In the Derby portion of the range many lead mines also

occur, that have been wrought from a very remote era; and others in the same formation in Flintshire, near the estuary of the Dee. The coal fields to which England, and, indeed, the empire, is mainly indebted for her manufacturing superiority may be thus briefly enumerated:—Those of Northumberland and Durham extend from the Tweed to the Tees, between the mountain line and the sea-coast: the most northerly has only been partially explored, and is worked, on a limited scale, chiefly for local purposes. The coal field of S. Northumberland and Durham extends about 50 m. N. and S., with an average breadth of from 12 to 15 m. The seams or beds dip N.E., and crop out successively in an opposite direction, so that none of the beds extend through the entire limits of the district. The two thickest and best (high and low main) are 6 ft. thick, and are separated by strata of shale, sandstone, and smaller seams of coal, of the aggregate average thickness of 360 ft. The mines in this district furnish annually a vast quantity of coal, amounting to nearly one-third the produce of the United Kingdom. In the year 1864, there were raised from the mines of Durham and Northumberland not less than 23,218,367 tons of coal. (Hunt, Robert, Annual Report on the Mineral Statistics of the United Kingdom.) Various and very discordant estimates have been framed of the period that will probably be required to exhaust this vast deposit of fuel. But the district has not been sufficiently explored to admit of such estimates being framed on any thing like solid grounds; and, no doubt, were any deficiency in the supply of coal apprehended, methods would be found for materially diminishing the immense quantities now left in the mines, as well as for reducing the waste.

The Whitehaven is a small but valuable field, between the Cumbrian mountains and the Irish Sea, under which the adits of several of its mines are driven: the coal is exported in considerable quantities to Ireland and elsewhere. The Yorkshire and Derby fields extend N. and S. about 70 m., from Leeds onward; their breadth, between Halifax and Aberford, being about 25 m., but it diminishes considerably through the Derbyshire part, to its N. extreme, near Nottingham.

Most of the coal raised in Yorkshire is consumed in its extensive woollen, iron, and hardware manufactories, and in the domestic economy of its numerous population. The Derby field supplies, through the medium of canals, many of the midland cos. The Lancashire field is parted by a range of hills from that of Yorkshire, and extends along their base from Blackinfield to Oldham, thence N. to Rochdale and Colne, and W. to Prescott near Liverpool, having Manchester on its S. border. Coal is excavated in various parts of this extensive field, which affords all but inexhaustible supplies for the various uses of the most important manufacturing district in the kingdom. The produce of the Lancashire district, in 1861, amounted to 11,530,000 tons, S. of the above, occur some smaller fields in Leicestershire and Warwickshire, in the vicinity of Ashby-de-la-Zouch, Tamworth, Atherstone, and Coventry. The Staffordshire field extends N. and S. about 10 m., with a breadth varying from 5 to 7 m. Numerous beds of coal are worked in various parts of this field, which also furnishes potters' clay, and is the site of the potteries. The Wolverhampton and Dudley field, in the same co., extends about 14 m. N. and S., with an average breadth of 4 m., and is the most valuable of any in the central part of the kingdom. Two beds of ironstone, each of considerable thickness, also traverse the field, and supply the innumerable fur-name of the district. The whole rests on transition lime, abounding in beautifully preserved fossils. The total produce of coal of the districts of Staffordshire and Worcestershire amounted to 11,459,850 tons in 1864. A few small fields also occur in Shropshire and Herefordshire, of which the chief is that of Colebrook Dale, 6 m. long by 2 m. in breadth. This formation is also traversed by ironstone, and many furnaces and foundries are established in the locality, though of late years it has not maintained its former importance. The forest of Dean includes a valuable coal basin, which (geologically considered) is the most perfect of any; the different beds of coal and sandstone being regularly arranged in a rounded concavity, and resting on millstone grit and mountain lime: its extent is about 10 by 6 m. The most S. of the English fields extends about 25 m. with a breadth of 5 or 6 m., through the cos. of Gloucester and Somerset, on either side the Avon; but it is overlaid in many parts by more recent formations, and the coal seams are thin, and of inferior quality. The last of these valuable deposits now left to notice, and probably also the most extensive and important, is the great coal field of S. Wales, extending about 100 m. from the Usk, in Monmouthshire, through the cos. of Glamorgan, Carmarthen, and Pembroke, to St. Bride's Bay, and comprising an area of about 1,200 sq. m. There are 23 available seams, of the aggregate average thickness of 95 ft., arranged in a prolonged concavity, and dipping both from the N. and S. towards a central axis; the deepest part of the field being near Neath, where coal occurs 700 fathoms below the highest portions of the seams. The quality of the yield varies considerably; that on the W. side being mostly what is termed dry coal, and the other bituminous coking coal, adapted to the smelting of iron ore. The total produce of coal in the South Wales and Monmouthshire district, in the year 1864, amounted to 10,976,500 tons. Extensive beds of iron also occur in this district, which is now the seat of the greatest iron-works in the empire.

The new red sand occupies nearly all the remaining portion of the surface on to the lias: it consists of beds of clay, marl, gravel, sand, &c., of various texture—the debris of older rock strata. Extensive deposits of gypsum, and vast and all but inexhaustible beds of rock salt occur in this formation, which will be elsewhere noticed. The upper secondary strata occupy the surface from the W. limits of the lias, previously described, to those of the chalk formation on the E.; which last has the same general direction, though forming a much greater curve, which terminates at either extreme of the lias. The strata included within these limits are of very various character, and abound in fossil remains: in some parts they furnish fine freestone for building purposes; in others, lime, fullers' earth, and pipe clay. Their aggregate thickness has been estimated at between 2,000 and 3,000 ft., all formed by deposition in an oceanic basin, as the character and abundance of the fossil remains clearly indicate. The green sand formation rests on those of the oolite, and is succeeded by that of chalk, both abounding in marine testaceous remains; the average thickness of the latter, when fully developed, being about 1,000 ft. It occupies the S. coast, from the Reculver Cliffs to Folkestone, and from Beachy Head to Brighton, stretching inland from the former round the weald district, and from the latter inland towards Salisbury, and thence over the Hampshire and Dorset downs nearly to Dorchester. The Inkpen, in Hants (1,011 ft.), is the highest summit of the chalk. The general dip of all the strata, from the lias to the chalk inclusive, is SE. and very gradual: a line from the N. escarpment of the chalk in Berkshire to the Malvern Hills would intersect the Basset edges of the entire series. The deposits above these are chiefly limited to the SE. cos., and have little of variety or well-marked character to distinguish them; the chief are those of the London and plastic clays, occupying the basin of the Thames. Formations of similar character extend along the sea-coast, from Brighton to Southampton, and occupy a portion of the Isle of Wight. The sandy strata that occur in several of the S. cos., and known by the general name of Bagshot sand; the mixture of ferruginous sand and clay with chalk fragments (crag) that occupy the E. parts of Norfolk and Suffolk; and the still more recent alluvial deposits of Holderness, the Fens, and Romney Marsh; all obviously constituted of the debris of older rock formations, and the latter resulting from atmospheric, oceanic, and other presently existing influences now in active operation. There are other interesting geological phenomena; such as the boulders and fragments of rocks from the Cumbrian and Welsh mountains, that are strewed over some of the midland cos.; the beds of chalk, flint, and gravel that occur at great distances from the main formations, and are found capping summits of others, wholly different; and the remains of mammaliferous animals in the lime caverns of York and Devon; such as those of the extinct species of the elephant, hyæna, bear, &c., which must once have ranged over the districts in which these relics are deposited.

The only medicinal springs of importance are those of the cos. Derby, Gloucester, and Somerset, which will be found fully described under the heads of their respective localities.

Climate.—The British sky is truly said by Tacitus to be *crebris imbribus ac nebulis fœdum*, but also to be without the *asperitas frigorum*. (Vit. Agric., § 12.) The climate of England is chiefly characterised by the absence of extremes in temperature, by humidity, and by almost incessant variations within a limited range, peculiarities ascribable to the geographical position of the country, in contiguity with an extensive continent on the one hand, and a vast ocean on the other: the latter with nearly the same temperature throughout the year, and exerting an equalising influence over the contiguous atmosphere; the other with a varying temperature, above that of the ocean in summer, and lower during the winter months. Hence the origin and direction of the prevailing winds at different periods of the year, according to which ever of these great surfaces exert most rarefying power: those blowing from the continent being comparatively dry, whilst those from the ocean, being charged with its exhalations, bring the chief part of the rain that descends, 2-3ds of the whole of it falling on the W. side of the kingdom. Rains are more prevalent during the summer and autumnal months, when the higher relative temperature of the continent, and greater rarefaction of the atmosphere in contact with it, cause aërial currents from the ocean to set in that direction, in order to supply the comparative vacuum; whilst the E. and NE. winds, that frequently prevail in winter and spring, are attributable to the higher temperature of the sea at those periods. According to a series of observations made under the direction of the Royal Society, the SW. is the most frequent wind in every month of the year, but is more prevalent in July and August than in any other: the NE. prevails most in Jan., and from March to June, inclusive; whilst the NW. is most frequent from Nov. to March, and least so in Sept. and Oct. It also appears, from the same observations, that rain is less prevalent in March than in Nov., in the proportion of 7 to 12; in April than Oct., in the ratio of 1 to 2; and in May than Sept., in the ratio of 8 to 4; hence the summer, autumn, and earlier part of winter, are the most humid portions of the year. The minor differences of climate that exist within the kingdom itself are wholly in accordance with the above views and observations. In Cornwall, the annual average quantity of rain falling is 45 in., and in the W. part of the kingdom, generally, it is found to vary from 30 to 51 in.; in the SE. counties, and also in the metropolis and its vicinity, the quantity is only from 20 to 25 in.; whilst Norfolk has, in all probability, the least humid climate in the kingdom. As yet, however, sufficient data do not exist to make other than an approximate calculation of the average that falls in any of the districts, and of course the general average of the whole can only be stated in the same qualified way. The estimate made by Dr. Dalton appears to be, on the whole, the most precise and satisfactory on this point; and he makes the whole annual quantity falling on the surface of England and Wales, 31 in.; to which he adds a depth of 5 in. supplied from the atmosphere in the form of dew, and calculates that 23 in. of the whole are carried off by evaporation, and the remaining 18 in. through the medium of the various rivers to the ocean. There has been previously noticed the limited range of the thermometer, which at the coldest period (Jan.) seldom falls much below the freezing point, and at the warmest (July and Aug.) as rarely rises higher than 80 Fahr., though occasional instances of greater variation may be cited. In the N. sea, from their contiguity to the sea on either side, the range is still more limited, rarely exceeding 75° or falling more than 8° or 4° below zero; so that their mean annual temperature is within 2° or 3° of those on the S. coast. In a general view, however, the influence of the ocean in tempering the atmosphere (as well as in the humidity it imparts) is greatest on the W. side of the kingdom, and most so within the limits of the SW. peninsula; the temperature of the ocean on that side being, during the coldest season, rarely so low as 50°, whilst that of the German Ocean, on the other, except in the height of summer, seldom exceeds 45 Fahr. On the whole, the most obvious difference that occurs in the local climates of the N. and S. parts of the kingdom is the lateness of spring in the former as compared with the latter; at an average about a fortnight between the sea, N. of the Mersey and Humber, and those of the S. and SW. The local effect of the W. mountain ranges is considerable, and tends to increase, in a greater ratio than would otherwise be the case, the quantity of rain falling in their vicinity; but, as a whole, the elevation of the surface is nowhere so considerable as to have any remarkable influence on the general character of the climate. The fens on the E. coast, and the wolds of Kent and Sussex, are the only tracts of any extent where the superfluous moisture would, but for artificial means, be retained long enough to generate miasma. In almost every other part of the country the surface has sufficient elevation and inequality to facilitate the free percolation of water, and to conduct the superfluity by natural means to the numerous streams that intersect it; so that no where can its physical structure be said to exert an injurious influence on the climate. The more general enclosure and cultivation of the surface within the last century must also have greatly augmented these facilities, and improved

the salubrity of the climate, which, however, as regards its chief characteristics, seems to be much the same as when Cæsar and Tacitus described it. There appears but little foundation for the notion once prevalent that the climate has deteriorated and become colder; an inference from the fact of vineyards having once been cultivated to some extent in various parts of the country. The same accounts also prove that rejuice formed no inconsiderable part, and in some measure constituted the only produce of these vineyards. It is probable that a better result than this might be obtained in the present day, were favourable spots selected, and any probable advantage to be derived from the culture of the vine. The mean daily range of the thermometer on an average of the whole year has been estimated at 11° for the metropolis, 14° for the midland counties generally, and 8° for Cornwall; but the extent of the daily range of course varies with the different seasons, being greatest when the sun has most influence, and the processes of evaporation and radiation are in most active operation. The mean difference between the coldest and the warmest months of the year has been stated at, for London 26°, Cornwall 18½°, and England generally 24½°; but these, and similar calculations, can only be considered as probable approximations to the truth, deduced from such series of observations as exist; which, however, are far too few and limited to make further details or generalizations of any practical utility.

The great drawbacks upon the climate are the prevalence of cold, biting NE. winds in April, May, and June, which frequently render them the most disagreeable season of the year; and the occasional occurrence of wet summers and harvests. The crops in England are very rarely injured by droughts; but they not unfrequently suffer from excess of humidity. In Cornwall, where the climate is most equal, and the winters the mildest, the moisture and coolness of the summers are such that the fruit is inferior in flavour to that raised in the more E. and midland counties at the same time that it arrives later at maturity.

Vegetable Productions.—The Flora of the kingdom comprises between 1,400 and 1,500 indigenous species of phanerogamous plants, of which upwards of 100 belong to the grass family; these, together with the ferns (*Filices europaeae* and *mcana*), the three common heaths (*tetralix, cinerea*, and *vulgaris*), and the different kinds of rushes and sedges, occupy a very large surface, and perhaps characterise better than any other the nature and capabilities of the tracts they occupy. The oak (*Quercus robur*) is the king of native British trees, and supplies the timber of which our finest ships are built. Hence the oak is intimately associated with the maritime glories of England. Take it for all in all, it is probably the best timber of which we have any certain knowledge. Some is harder, some more difficult to rend, and some less capable of being broken across; but none contains all the three qualities in such great and equal proportions; and thus, for at once supporting a weight, resisting a strain, and not splintering by a cannon-shot, it is superior to every other timber. In favourable soils it will flourish at an elevation of 700 ft. The ash, alder, and hawthorn thrive, under similar circumstances, at 810 ft.; the fir (*P. sylvestris*—the only indigenous species) at 1,000 ft.; the mountain ash, and some of the smaller and prostrate varieties of the willow tribe, ascend nearly to the highest summits; whilst the hornbeam, lime, maple, poplar, and elm flourish only in localities much less elevated than any of the preceding. The beech and sycamore reach 750 and 800 ft. respectively; but these, though long perfectly naturalised, are foreign introductions, as are also the larches, pines, chestnut, horse-chestnut, and many others that flourish and attain to considerable size in the extensive parks and plantations of the kingdom. Of indigenous fruits, the list is very scanty; the pear, crab, medlar, wild cherry, bullace, raspberry, blackberry, gooseberry, currant, strawberry, and cranberry, being nearly, if not quite, all that can be so called; and the greater part of these, in their natural state, can scarcely claim to rank as such in the more common acceptation of the word.

Every one is ready to admit that England is indebted to commerce, or to her intercourse with other nations, for a very large proportion of her superior wealth and comforts. But it will be found on examination that her obligations in this respect are really much greater than is generally supposed. England is not only indebted to foreign countries for mere luxuries and superfluities, but for the greater part of those plants and vegetables that supply the largest portion of the food of her inhabitants. England has received from abroad bread corns, potatoes, garden stuffs and fruits, with hops, turnips, and a vast variety of useful and ornamental vegetable products. But it is the good fortune of England that most of these are so admirably suited to the soil, that, unless the contrary were known, they might be supposed to be indigenous. Those species of fruits that require a powerful sun to bring them to maturity do not, indeed, answer in our climate, except artificial means be employed in ripening them. But the more useful kinds attain to perfection. '*Solum praeter oleam vitemque, et cetera calidioribus terris oriri sueta, patiens frugum, fecundum; tarde mitescunt, cito proveniunt; eademque utriusque rei causa, multus humor terrarumque coelique.*' (Tacit. Agric. § 12.)

Of the smaller herbaceous plants it remains only to notice a few most characteristic of the features of an English landscape, or which are otherwise distinguished for their beauty and rarity. Of these, the various grasses that may almost be said to be in a state of constant growth, and to cover so large a portion of the surface, in a literal sense, with perpetual verdure, claim precedence; of flowers, the daisy, primrose, cowslip, violet, and lesser celandine, are the most common and most universal favourites; next which, perhaps, the woodbine, eglantine, hyacinth, harebell, and goldcups, contribute most largely to the adornment of the scenery. Of the less common kinds, the tamarisk, musk, gentian, and a few others, are limited to the SW. peninsula; the hop, briony, and pheasant eye, to the midland cos.; the juniper, parnassia, and a few others, to the more elevated regions of Wales and the N. cos.; and the water lily can only be considered in its native locality when expanding its fine flowers on the surface of the Cumbrian lakes. The foxglove, henbane, hemlock, nightshade (*Atropa*), and the *Cicuta virosa* of the Cambridge fens, are almost the only species that possess active medicinal qualities. The woad, madder, teasel, hop, flax, buckwheat, clovers, tares, and melilots; together with the carrot, parsnip, cabbage, sea kale, and asparagus, comprise the chief indigenous species available for economical purposes; whilst the mistletoe and ivy claim a passing notice from traditional associations. Of cryptogamous plants there exist about 800 distinct species of moss, and upwards of 500 of *Algæ*, that is, lichens and sea weeds: two or three of the lichens are identical with those extensively used as dyes, but they scarcely occur anywhere in sufficient abundance to be worth collecting. Various species of sea weed are thrown abundantly on some parts of the sea-coast, and collected as a manure, but are seldom converted into kelp or barilla.

Zoology.—Without recurring to an older geological period, when the animals, whose bones are found in the limestone caves of Yorkshire and Devon, occupied the country, we find that, within a comparatively recent period, it was in a great degree overspread with vast forests, the abodes of many wild animals, which, as well as their coverts, have now wholly disappeared. The bear may be traced as an inhabitant of these down to A.D. 1057, and may have existed later: the beaver to 1188, when its habits were noticed and described by Giraldus Cambrensis, as witnessed by him on the Teifi: many notices of wolves occur so recently as the reign of Edw. I., notwithstanding the war of extirpation waged against them by Edgar. The New Forest in Hampshire was the latest resort of the wild boar, which must have been extirpated during the period of the last civil war: the wild cat has disappeared from its latest haunts—Cumberland and Westmorland—within living memory, but is still found in Scotland. The wild ox (*Urus*) has only escaped a similar fate, by having been preserved as a curiosity in Chillingworth Park near Berwick, and in one or two other localities, where this fierce and distinct breed (with cream-coloured hide, black muzzle, and downward-bent horns) may still be seen. The stag, fallow deer, and roe have also been saved by similar interference and protection. The indigenous quadrupeds now existent in a wild state are—the fox, badger, polecat, beech and pine martens, otter, weasel, stoat, hedgehog, mole, land and water shrews, squirrel, hare, rabbit, dormouse, field and water lemmings, black rat, common field and harvest mice, and six species of the bat tribe. Of cetaceous mammalia, the razor-back and porpoise are the only species that occur with any frequency, though such lists as comprise stragglers notice thirteen or fourteen others, as of occasional occurrence on the coasts. The Norway rat is an ascertained immigrant, which has warred with the indigenous species, and made it by far the scarcest of the two. The domesticated animals and poultry will be noticed under another head. The great bustard appears to be almost the only species of bird that has been banished from the kingdom by the extension of enclosures and cultivation; although the bittern and two or three others have become scarcer, and have wholly deserted many of their ancient haunts, yet they may still be met with, whilst the former has wholly disappeared from its latest locality, the Wiltshire downs. Pennant, in 1777, notices bustards as occurring in flocks of fifty or more, on most of the open tracts of the S. and E. counties, from the Dorset downs to the Yorkshire wolds. Of those species which are either indigenous or habitual visitors, 20 are birds of prey; of gallinaceous birds (grouse, pigeons, &c.) there are 18 species; of *insessores*, of the tooth-billed tribe (shrikes, ouzels, &c.), 51; of the conical-billed tribe (finches, &c.), 31; the climbers (woodpeckers, &c.) include 9 species; the *fissirostres* (swallows, &c.), 7; the wading tribe, 60; and the swimming water-fowl, 78 species. In all, 274 species have been enumerated; the numbers of land and water-birds being nearly equal—a singular circumstance, attributable to the insular position and extensive coast-line of England. Puffins' eggs on some parts of the Welsh coast, and in a few other localities, are objects of some traffic. The different species of fish frequenting the seas, lakes, and rivers of England, have been estimated at about 170; a large portion of these, as well as of birds, are migratory. The more important species will be subsequently specified. There are between 450 and 500 species of *testacea*; of which the oyster, scallop, cockle, periwinkle, whilk, limpet, and mussel, are the principal edible kinds. The *crustacea* include the crab, lobster, crawfish, prawn, and shrimp; but the former are limited to the more rocky portions of the coast. Upwards of 10,000 insects have been enumerated by Mr. Stephens, whose catalogue does not however include the whole. Of three *diptera* (flies) comprise about 1,700 species, *hymenoptera* (bees, wasps, &c.) 2,000, *coleoptera* (beetles) above 3,000, and *lepidoptera* (butterflies, moths, &c.) about 1,900. None of the latter is of any great size; but several of the British butterflies can boast of considerable beauty and variety of hue. *Annelides* comprise the medicinal and horse-leech, &c. Radiated animals and *zoophytes* are abundant.

In the order of reptilia England is fortunately very deficient. The hawksbill turtle has occasionally been found wandering near our northern coasts, and two instances of the coriaceous turtle having been caught on the western shores are recorded by Borlase. Of the more elegant family of lizards, one species only, the *Lacerta agilis*, is admitted by some authors; others, like Ray, consider that under this name are included four or five distinct species (see Linn. Trans., vol. v. p. 49), a supposition much more probable than the first. The efts are common; two species inhabit our clear pools, where they may be seen swimming about in summer, while the other is strictly terrestrial, and is met with at the roots of rank weeds, growing on the side of walls, or among rubbish. The only serpents are—1. the common snake; 2. the viper; and 3. the blind-worm. The first is harmless, and never exceeds 4 ft. in length. Although habitually inhabiting the land, it is yet known to enter the water, and to swim with facility. Considering its fecundity (it lays from 10 to 20 eggs), it is surprising that the snake is not more frequent; but it has a deadly enemy in the hedgehog, which feeds upon this reptile; thus establishing the fact that as the hedgehog, in its own tribe, represents the real hog in the order of *ungulata*, so there should be some striking point of agreement between them. The viper is the only reptile in England whose bite is poisonous. There are three or four prominent varieties, which some have considered different species; but most modern naturalists regard them but as one. Lastly, the *Anguis fragilis*, or blindworm, is also of rare occurrence, and probably derives its name from the smallness of its eyes: the body is greyish, with two dark brown stripes upon the back; the belly also is brown; and the usual length of the animal is a foot. Of the batrachian reptiles, or true amphibia, the list is equally scanty, comprising only the frog, toad, and natterjack; all these are perfectly harmless, useful in the economy of nature, and serviceable even to man.

Population and Civil Divisions.—Since the days of the great Alfred, England has been divided into counties or shires, and these again generally into hundreds, and always into parishes. Sometimes, however, instead of being divided into hundreds, a co. is divided into wards, as is the case in the N. counties; sometimes it is divided into ridings (a corruption of trithings), as is the case with Yorkshire; and sometimes into lathes and sokes, as in Kent and Lincoln. The subjoined table shows, in alphabetical order, the whole of the counties of England and Wales, their area in statute acres, the number of inhabited houses, and the population (exclusive of army and navy) according to the census taken on April 8, 1861:—

	Area Acres	Inhabited Houses	Population Apr. 8, 1861
England and Wales	37,324,883	3,739,505	20,066,224
England	32,590,397	3,513,431	18,954,444
Wales	4,734,486	226,074	1,111,780
ENGLAND.			
Counties.			
Bedford	295,542	27,177	135,287
Berks	451,210	35,761	176,256
Buckingham	468,822	34,309	167,993
Cambridge	525,182	37,541	176,016
Chester	707,078	97,874	505,428
Cornwall	873,610	78,354	369,390
Cumberland	1,001,273	40,532	205,276
Derby	654,603	69,297	339,327
Devon	1,657,100	101,753	584,373
Dorset	632,025	37,709	188,789
Durham	672,474	64,907	508,666
Essex	1,004,549	81,961	404,851
Gloucester	805,102	97,851	485,770
Hereford	611,873	33,816	123,711
Hertford	391,141	34,893	173,280
Huntingdon	229,544	13,704	64,250
Kent	1,033,419	129,271	733,887
Lancaster	1,219,271	434,503	2,429,440
Leicester	514,184	51,984	237,412
Lincoln	1,714,437	84,829	412,246
Middlesex	180,136	279,145	2,206,485
Monmouth	366,599	32,077	174,644
Norfolk	1,354,301	96,879	434,798
Northampton	631,356	48,511	227,704
Northumberland	1,290,399	64,585	343,025
Nottingham	526,016	62,519	297,067
Oxford	472,717	36,734	170,944
Rutland	85,665	4,641	21,861
Salop	861,935	48,391	240,566
Somerset	1,041,770	87,836	411,976
Southampton	1,070,218	86,476	481,615
Stafford	729,668	117,105	746,943
Suffolk	947,681	71,915	337,070
Surrey	474,792	180,662	831,085
Sussex	810,911	65,578	363,735
Warwick	561,948	116,351	561,555
Westmorland	485,482	11,785	60,817
Wilts	867,402	53,055	249,311
Worcester	472,185	53,126	307,397
York, East Riding	769,615	49,103	240,277
— City	2,130	8,212	40,433
— North Riding	1,330,191	40,178	345,154
— West Riding	1,709,801	315,721	1,807,196
WALES.			
Counties.			
Anglesey	193,453	12,398	54,609
Brecon	469,156	17,913	61,627
Cardigan	443,387	14,736	72,245
Carmarthen	606,351	21,070	111,796
Carnarvon	370,273	20,236	95,694
Denbigh	386,852	18,112	100,778
Flint	184,246	14,112	69,737
Glamorgan	547,484	58,254	317,752
Merioneth	385,271	8,495	38,963
Montgomery	484,523	13,501	66,919
Pembroke	401,651	19,418	96,278
Radnor	272,128	4,604	25,382

All information with respect to the number of people in England antecedent to 1801, when the first census was taken, is extremely vague and unsatisfactory. According to 'Domesday Book,' England, exclusive of Wales and the four N. counties of Northumberland, Cumberland, Durham, and Lancaster, contained immediately after the Conquest, 300,785 families, which, at an average of five persons to each family, will give about 1,500,000 individuals. Adding to this number 650,000 for the pop. of Wales and the excluded English counties, and other omissions, the entire pop. of the kingdom, at that epoch, will be 2,150,000. From the poll-tax returns in 1377, it appears that 1,367,576 persons paid the assessment levied upon every lay person, whether male or female, of 14 years of age, mendicants only excepted. But Wales, Chester, and Durham are not

Included in these returns; and there are doubtless many omissions in the returns that were given in. Little dependence can, therefore, be placed on them; but Mr. Chalmers has thence concluded that the pop. at the period in question amounted to 2,850,000. Perhaps, however, this estimate is rather under the mark; for, in 1377, the country could hardly have recovered from the disastrous influence of the great pestilence of 1349; and it is highly probable that the children and persons under age then exceeded a third part of the pop. at which they are estimated by Mr. Chalmers. Harrison and Sir Walter Raleigh set down the number of fighting men in the kingdom in 1575 and 1583 at 1,172,000. But this was probably little better than a rough guess; and unless it included all the able-bodied individuals between certain specified ages, it would afford but slender means by which to estimate the pop. Perhaps, however, we may conclude, that it was then somewhere about 4½ or 5 millions. There is no reason to suppose that the pop. was materially affected by the civil war under Charles I.; and the period from the Restoration to the Revolution was one of considerable prosperity. Previously to the Revolution, a hearth tax had been imposed; and the celebrated Gregory King, founding on returns obtained under this act, estimated the pop. of England and Wales, in 1690, at 5,500,000; which probably was not far from the mark. A great deal of discussion took place in the course of last century with respect to the progress of pop., Dr. Price and others contending, on the one hand, that it was progressively diminishing; while Mr. Howlett, Mr. Wales, and others, contended, on the other, that there were really no grounds for this conclusion, and that, instead of diminishing, the pop. was steadily increasing. The census of 1801 put an end to these disputes, and showed that, supposing Gregory King's estimate to have been nearly correct, the country had gained an accession of about 3,875,000 inhab. in the course of the 18th century. The subjoined table shows the result of the seven official enumerations held in the present century, giving the date of each census, the number of the population (inclusive of army and navy) and the increase between each decennial period:—

Date of Census	Population	Increase during Decennial Period
March 10th, 1801 . .	9,168,171	
May 27th, 1811 . .	10,464,598	1,296,444
May 28th, 1821 . .	12,172,664	1,718,188
May 29th, 1831 . .	14,081,964	1,879,829
June 7th, 1841 . .	16,035,198	1,963,919
March 31st, 1851 . .	18,064,170	2,016,073
April 6th, 1861 . .	20,226,497	2,174,237
Total Increase 1801 to 1861 . . .		11,073,326

Until the year 1837, when a new system of registration was established under the direction of the registrar general, there were no means by which to form a correct estimate of the numbers of births and deaths. In 1538, the clergy were required to keep registers of them, as well as of marriages, in their respective parishes; and in 1603 the injunction was renewed; but the rite of baptism in the parish church being objected to by numerous sects of Dissenters, the registration of births has been at all periods very defective. The same was the case, though in a less degree, with respect

to the registers of deaths, various classes of Dissenters having their own cemeteries, in which their own forms of burial were adopted; and it happened that in many places a reference to the parish registers merely supplied the means of making an approximate estimate of the number of deaths. The statute of the 26th Geo. II., which made registration indispensable to the validity of a marriage, having come into operation in 1754, the registers of marriages have been since nearly correct. The following table, compiled from the official returns, embraces the fullest information it has been possible to bring together with respect to the proportion of the two sexes in every year from the beginning of the century. The numbers are calculated, from the registration returns, for the middle of the year:—

Years	Total Population	Males	Females
1801	9,000,000	4,604,000	4,656,000
1802	9,178,000	4,641,000	4,660,000
1803	9,251,000	4,696,000	4,760,000
1804	9,358,000	4,830,000	4,867,000
1805	9,413,000	4,631,000	4,861,000
1806	9,656,000	4,742,000	4,912,000
1807	8,784,000	4,760,000	4,976,000
1808	9,976,000	4,901,000	5,092,000
1809	10,044,000	4,901,000	5,141,000
1810	10,145,000	4,957,000	5,270,000
1811	10,357,000	5,045,000	5,792,000
1812	10,479,000	5,104,000	5,375,000
1813	10,649,000	5,191,000	5,456,000
1814	10,828,000	5,301,000	5,479,000
1815	11,094,000	5,375,000	5,609,000
1816	11,198,000	5,476,000	5,721,000
1817	11,327,000	5,508,000	5,809,000
1818	11,858,000	5,850,000	5,831,000
1819	11,718,000	5,341,000	5,970,000
1820	11,883,000	5,441,000	6,000,000
1821	12,103,000	5,358,000	6,155,000
1822	12,339,000	6,060,000	6,306,000
1823	12,599,000	6,158,000	6,376,000
1824	12,700,000	6,266,000	6,414,000
1825	12,900,000	6,334,000	6,469,000
1826	13,074,000	6,417,000	6,617,000
1827	13,247,000	6,500,000	6,716,000
1828	13,426,000	6,581,000	6,846,000
1829	13,675,000	6,801,000	6,844,000
1830	13,935,000	6,887,000	7,057,000
1831	13,994,000	6,958,000	7,186,000
1832	14,164,000	6,843,000	7,203,000
1833	14,324,000	7,023,000	7,300,000
1834	14,599,000	7,116,000	7,444,000
1835	14,795,000	7,313,000	7,510,000
1836	14,979,000	7,310,000	7,418,000
1837	16,103,000	7,552,000	7,711,000
1838	14,291,000	7,678,000	7,868,000
1839	14,511,000	7,906,000	7,977,000
1840	15,730,000	7,049,000	7,941,000
1841	15,929,000	7,784,000	8,144,000
1842	16,180,000	7,867,000	8,242,000
1843	16,382,000	7,900,000	8,341,000
1844	16,643,000	8,193,000	8,442,000
1845	16,703,000	8,195,000	8,542,000
1846	16,841,000	8,298,000	8,745,000
1847	17,150,000	8,400,000	8,745,000
1848	17,346,000	8,542,000	8,853,000
1849	17,544,000	8,643,000	8,950,000
1850	17,772,000	8,707,000	8,866,000
1851	17,967,000	8,103,000	9,174,000
1852	18,103,000	8,949,000	9,241,000
1853	18,504,000	9,110,000	9,341,000
1854	18,616,000	9,111,000	9,504,000
1855	18,929,000	9,221,000	9,617,000
1856	19,417,000	9,311,000	9,751,000
1857	19,736,000	9,403,000	9,446,000
1858	19,411,000	9,500,000	9,962,000
1859	19,666,000	10,279,000	10,279,000
1860	19,997,000	9,794,000	10,199,000
1861	20,119,000	9,901,000	10,218,000

The preceding table is condensed, and, at the same time, somewhat more fully illustrated in the following table, which gives the proportion of males to every 100 females in England and Wales at each of the seven decennial periods when the census was taken:—

Years	Number of Males, including Army, Navy, and Merchant Service abroad, to 1,000 Females	Number of Males, excluding Army, Navy, and Merchant Service abroad, to 1,000 Females
1801	97·426	94·683
1811	97·403	94·361
1821	97·932	96·555
1831	97·304	94·144
1841	97·076	94·444
1851	97·391	95·004
1861	99·465	95·041

The sexual proportion of the population may finally be illustrated by the following table, which shows the number of men in the army, navy, and merchant service abroad; the excess of females over males at home and abroad, and over males at home in England and Wales, 1801–61:—

Years	Men in the Army, Navy, and Merchant Service abroad	Excess of Females over Males at home and abroad	Excess of Females over Males at home
1801	181,817	110,431	251,248
1811	144,186	126,778	271,979
1821	84,648	127,170	211,858
1831	76,771	199,316	275,447
1841	131,650	237,926	338,976
1851	126,441	235,446	365,140
1861	163,373	351,433	519,706

To complete this view of the proportions of the two sexes living at home, their ages must be taken into account. There is an excess of boys over girls living under the ages of 15; and an excess of men is provided all through the middle period of life; but that surplus is overdrawn by emigration, so that the women exceed the men in number to a considerable extent in the early, and middle, and still more in the advanced ages, when their longevity comes into play. The excess of the emigration of males over females accounts for the present difference in the proportions of the sexes. (Census of England and Wales, 1861, vol. iii.; General Report, 1863.)

Agriculture.—Tenures and Estates.—The tenures under which land is held have varied very much at different periods. At present, they may be divided into freehold, copyhold, and leasehold. By the first, an estate is held unconditionally, under the constitutional laws of the kingdom, liable to neither fine nor forfeiture. By the second mode, estates are held of corporate bodies, or of individuals, as portions of some manor or other possession, and subject to certain claims and customs. Leaseholds are either long, as for 1,000 years; life leaseholds, contingent on one or more lives, or subject to certain fines or conditions, but at all times giving a power of alienation or transfer to the lessee. Such leases as do not convey this power do not strictly come under the designation of tenures; they form, however, a large and important class of holdings, usually varying from terms of 7 to 14 years, and the conditions and stipulations in them have a powerful influence over agriculture and the value of property, in the districts in which they prevail. Lands held merely from year to year, at the option of either party, are said to be held at will, and form a large proportion of the lands of the country. The size of estates varies exceedingly; but, despite the great number of very large estates, it is still true that landed property in England is very much divided, by far the largest portion of the kingdom being portioned out into estates under 1,000l. a year. Dr. Beeke, in 1801, estimated the number of proprietors in England and Wales at 200,000; and supposing this estimate to be nearly accurate, and that the total gross rental of the kingdom amounts to 40,000,000l. a year, it will give 200l. as the average annual value of each estate. But as a great number of estates are much above this average, it follows that the majority must be proportionally below it.

According to the census of 1851, the total number of farmers and graziers, in-door farm servants, shepherds, and agricultural labourers, was 1,440,016. There was a considerable decline in the ten years, 1851–1861, for the census of 1861 showed the number to be 1,347,387. The total population connected with agriculture—called in the census reports the 'agricultural order'—was much larger than that above enumerated, embracing not only farmers and their assistants and dependents, but persons engaged in floriculture and horticulture, as well as land surveyors and makers of agricultural implements. The subjoined table shows the total number of persons engaged in these various branches of the 'Agricultural Order,' according to the returns of the registrar general, on the 31st of March, 1851, and on the 8th of April, 1861:—

	1851	1861
Total of Agricultural Order	2,011,447	1,924,110
Land Proprietor	39,815	30,766
Farmer, Grazier	249,431	249,735
Farmer, Grazier's Wife	164,815	162,765
Farmer's Son, Grandson, Brother, Nephew	111,704	97,831
Farmer's Daughter, Granddaughter, Sister, Niece	108,147	89,639
Farm Bailiff	10,561	12,696
Agricultural Labourer (outdoor)	952,997	856,945
Shepherd (out-door)	12,517	23,559
Farm Servant (in-door)	286,772	204,962
Land Surveyor, Land, Estate Agent	5,054	4,792
Agricultural Student	104	420
Hop Grower	80	24
Willow Rod Grower, Dealer	40	34
Teasle Grower, Merchant	33	41
Agricultural Implement Proprietor	55	746
Agricultural Engine and Machine Worker	—	1,505
Land Drainage Service (not in Towns)	11	1,761
Colonial Planter, Farmer	16	91
Others connected with Agriculture	129	117
Woodman, Wood Gatherer	7,772	8,916
Others connected with Arboriculture	228	10
Gardener (not Domestic Servant)	71,804	76,822
Nursery Man—Woman (Horticulturist)	2,883	2,917
Watercress Grower	80	86
Others connected with Horticulture	97	27

The decline which took place in many branches of the agricultural order, notably in that of in-door farm servants, in the ten years 1851–61, is not a little striking as well as suggestive.

Arthur Young, in 1770, estimated the capital employed in agriculture at 4l. per acre; at present it may, perhaps, be taken at about 6l.; which, on 31,000,000 acres, will give 186,000,000l. The rental of the land in England and Wales may be estimated at about one-fourth part of the value of the total produce. It amounted, in 1815, to 54,330,412l.; and it appears from the subsequent returns, that the present rental exceeds 60,000,000l. a year; the fall that has taken place in the interval

In prices having been everywhere partially, and in most parts more than fully countervailed by the spread of improvement, and the opening of new and better markets for all sorts of products. Under the property tax act the profits of the farmers are supposed to amount to half the rent; and though this rate be frequently most unjust in its application to individuals, it may not, as an average, be very wide of the mark; and supposing this to be the case, the aggregate profits of the farmers would exceed 30,000,000l. a year. Farmers holding lands let under 100l. a year, are exempted from the tax. Farms in England are of a medium size, their average being probably about 150 or 160 acres. Wheat, barley and oats, but especially the first, which may be emphatically said to be the bread-corn of England, are the principal crops. The best wheat, as well as the greatest quantity, is raised in Kent, Essex, Suffolk, Rutland, Herts, Berks, Hants, and Hereford. From 2½ to 3 Winch. bushels per acre are required for seed, and the average produce in the above cos. may vary from 26 to 40 bush. per acre. Barley is grown principally in the eastern and some of the midland cos., and chiefly for malting; oats are principally in demand for horses; and the increase of the latter has occasioned a proportional increase in the culture of oats. They are grown more especially in the N. and NE. cos.; in the midland cos. their culture is less extensive, but it is prevalent throughout most parts of Wales. Rye is scarcely at all raised for bread, except in Durham and Northumberland; where, however, it is usually mixed with wheat, and forms what is called maslin, a bread-corn in considerable use in the N. Pease and beans are important crops, and in some parts are pretty largely raised. The potatoe has, unluckily, become pretty general throughout the kingdom, but is most extensively raised in Lancashire and Cheshire, where it also enters to the greatest perfection. The introduction and general extension of the turnip husbandry has effected a revolution in the agriculture of England, second only to that which the inventions of Arkwright have effected in manufactures. They have now all but superseded fallows on the lighter lands. But the giving a valuable crop to the farmer, where there was none, without in any degree diminishing the facilities for clearing the land, is but a part of the advantages resulting from the turnip culture; for, while it enables the farmer to keep and fatten a much larger stock, it also enables him to accumulate a vastly greater supply of manure—of that invigorating power which adds to the productiveness of the best lands, and without which the husbandman's toil. It is not easy to estimate the prodigious additions that have been, in this way, made to the productive capacities of the soil; and the recent introduction of guano and the application of bone manure to turnip husbandry, have already had a wonderful influence, and, no doubt, will continue to become still more and more important. Rape is grown for its oil, or as food for sheep, in all parts except the cos. N. of Yorkshire; and cabbages and carrots are chiefly produced in the E. Flax and hemp are at present but little raised, being found less profitable crops than most of the foregoing. Hops are for the most part confined to Kent, to the vicinity of Farnham in Surrey, and to Herefordshire: their crop is the most uncertain of any, varying in the same localities, in different years, from 1 to 20 cwt. an acre. The apple orchards of Devon, Somerset, Gloucester, and a few other neighbouring cos., are important, on account of the cider they furnish. Perry is made chiefly in Worcestersh. Kent is

famous for its cherries and filberts. The total assessment on real property, for the property and income tax (schedule A.), amounted to 4,177,650l. in the year ending Apr. 5, 1812, and to 4,195,474l. in the year ending Apr. 5, 1803.

The best farmed counties are on the E. coast; and Northumberland, Lincoln, and Norfolk may bear a comparison with Berwickshire or E. Lothian. Such, however, is not the case in very many districts; and we believe it may be safely affirmed that the available produce of the kingdom might be doubled, were it generally cultivated on the principle, and according to the practice, followed in the best farmed districts. Winter wheat sowing usually takes place from Sept. to Nov.; drilling is more in use for barley than wheat, which is mostly sown broadcast. The grain harvest is commonly at its height in Aug. and Sept. Potatoes are taken up and stored for winter use in Oct. and Nov., which are also the chief cider months.

The farm implements in common use in England are decidedly superior to those of most other countries, though a good deal remains to be done in the way of their improvement. Perhaps few classes of people maintain their prejudices with such obstinacy as agriculturists, and especially agricultural labourers; and to this must be mainly attributed the continued use of the old-fashioned clumsy ploughs which are to be seen in some districts; and, what is far less excusable, the employment of 3, 4, 5, 6, and sometimes even 7 horses, to do what might be as well or better done by 2! The use of horses in farm labour is universal, except in Sussex, and some of the W. counties; and machines for thrashing, &c. have become common.

Britain has been celebrated from the era of Cæsar for the extent and excellence of her pastures, and the abundance of her cattle. A full half or more of the arable land of England is applied to grazing husbandry. The best grazing lands are in the vale of Aylesbury, the Fens, Romney Marsh in Kent, and some of the midland and W. counties. Hay is made from natural grasses, and from clover, rye-grass, and in the N. counties sainfoin and lucern: the natural sward yielding from 1 to 1½ tons an acre, and the artificial crops from 1 to 3 tons. The hay-harvest throughout the country takes place pretty generally in June and July.

There are several breeds of horses, the aggregate stock of which, at the present time, probably reaches 1,000,000 head, worth, perhaps, about 10,000,000l. sterling. Of this number it may be estimated that two-thirds are employed in agricultural labour. The old English road-horse is now nearly extinct: the large dray-horse, so admirably adapted for draught, which is believed to have been originally imported from the Low Countries, is bred in considerable numbers in some of the midland counties. Yorkshire is celebrated for its carriage horses, especially the Cleveland bays; and the farm breed of Suffolk is also excellent. The English race-horse, derived from the Arab, Persian, and Barb, is superior to every other breed in speed, and inferior to none in bottom and beauty. Mules and asses are very little used in England; the former are almost unknown, and the latter belong chiefly to the poor.

The stock of cattle may be estimated at little short of 4,500,000, about a fourth part of which are annually slaughtered. They are divided into long-horned, short-horned, and polled: the first division comprising the Lancashire; the second, the Holderness, Northumberland, Durham, N. Devon, Hereford, and Sussex; and the last, the Suffolk duns, &c. Butter and cheese are most important products: Epping Forest, in Essex, Cambridgeshire, and Dorset are the districts most celebrated for the former; and Cheshire, Gloucestershire, Wilts, and other W. counties, and Leicestershire, for the latter. The rich and fine cheese, called Stilton, is made wholly in Leicestershire. Milk is an important marketable article in the vicinity of large towns, and the cows kept for the supply of this article to the metropolis have been estimated to amount to 12,000, yielding milk to the value of 700,000l. sterling a year. Sheep, the total number of which in England and Wales may be about 26,000,000, are divided into long-woolled and short-woolled; the former, including the Romney Marsh, Teeswater, Lincoln, and New Leicester breeds; and the latter (which far excel the former in the quality of the mutton), the South-Down, Dorset, Wilts, Hereford, &c. breeds. The merino breed, introduced from Spain towards the end of the last century, has been chiefly useful in crossing and improving the fleece of other breeds. In some parts of England sheep are kept on fallows, for the benefit of their manure. Great numbers are fed on the open chalk downs of the S. counties. The total annual produce of wool in England is estimated at about 170,000 packs of 240 lbs. each. Hogs are fattened on most farms, and are also kept with advantage by millers, dairymen, brewers, distillers, &c., whose refuse they consume. The Hants, Berks, Gloucestersh. and Herefordsh. are the best of the large breeds, and that of Suffolk is distinguished among the smaller ones. Yorksh. and Westmoreland are famous for their hams; Hants, Wilts, and Berks for their bacon. Poultry are reared on most farms, and by the majority of agricultural cottagers. Large flocks of geese are kept in the Lincoln fens, and plucked once a year for their quills, and four or five times for their feathers. Fowls are largely reared at Oakingham in Berks, and Dorking in Surrey has acquired a name for a fine and large five-clawed variety. Ducks are plentiful in Bucks, and pigeons in almost every co. Since the foundation of our W. India colonies, and the importation of sugars, the demand for honey has declined; this, however, has not affected wax, so that bees still keep their ground as appendages to almost every farm, and to many cottage gardens. Goats are not reared except in the few mountainous parts of England, and deer are now mere articles of luxury, kept in the parks of noblemen and gentlemen. There are still some extensive rabbit-warrens in Norfolk and Cambridgeshire, but they have greatly decreased. About 127,620 acres of land are occupied by the royal forests, 62,020 of which are enclosed for the growth of timber. As already observed, England is very well wooded, especially the N. and W. cos. Oak, the most valuable species, grows in the greatest perfection in the weald of Kent, Sussex, and Surrey. The oak-bark harvest takes place in May.

Agriculture received its first great impulse in England during the reign of Henry VII, from the policy of that monarch; and together with all kinds of commercial enterprise throughout Europe, it derived a stimulus from the great discoveries of the period. But the breeding of sheep was the branch of rural industry the first to extend, and throughout this and the succeeding reigns for a lengthened period wool was extensively exported. The first English treatise on agriculture was written in the reign of Henry VIII, and the hop, as well as several of the common garden vegetables, are introductions of the same period. Mr W. Raleigh has the credit of introducing the potato, which, in the early part of last century, appears to have been a tolerably frequent crop in Lancashire, from which its culture extended to other parts of the kingdom. Turnips seem to have been first cultivated on a large scale in Norfolk, also, in the early part of the same century. Pope speaks of 'All Townsend's turnips.' The old duties and restrictions on the exportation of corn were abolished at the Revolution, and a bounty was then also given on its export. During the latter years of the war with France prices were comparatively high in England; but on the renewal of the intercourse with the Continent, in 1814, vast quantities of corn being imported, prices suddenly gave way. This fall occasioned a great deal of agricultural distress, which, however, was but of temporary duration. In no long time improvements began to be prosecuted with greater vigour than ever; and from 1832 to 1837, a sufficient supply of corn was grown for home consumption. The subsequent recurrence of bad seasons and the failure of the potato, led to the modification and final repeal of the corn laws in the year 1846. The repeal took effect on the 1st of February, 1849, from which date only a nominal duty of one shilling per quarter was levied on corn.

Fisheries.—These are not commensurate, either in extent or importance, with the extent of coast, and have never been a principal source of national wealth. The herring fishery is the principal; but until the middle of last century most of the fish taken on the E. coast (its chief seat) were captured by Dutch smacks. Yarmouth bay is the principal resort of the herring, and about 100 smacks, of from 40 to 50 tons each, belong to the town of Yarmouth, where the fish, smoked for sale, have obtained some celebrity under the name of 'Yarmouth bloaters.' At Sunderland, Whitby, Scarborough, and Harwich, there are also extensive herring fisheries. The cod fishery, including that of haddock, whiting, ling, and hake, ranks next in importance. The pilchard fishery is exclusively confined to the coasts of Devon and Cornwall. A portion of the fish caught are used fresh or salted in those counties; and the rest, to the amount of about 17,000 hhds. a year, are salted and exported chiefly for the Italian markets. The pilchard fishery, by means of seans, employs about 1,500 hands, and that by drift nets employs from 800 to 1,000 men, and 230 boats, exclusive of the women assisting on shore in curing the fish. The total number of fishermen in England and Wales was stated in the census report of 1841 to be 17,737. Mackerel are very abundant, and extensively consumed during the season; sprats, which arrive in immense shoals on the E. and SE. coasts, are taken in great numbers for manure. Oysters, which meet with so rapid and extensive a sale in the markets of the metropolis and other large towns, are found on many parts of the coast; and are largely bred near Milton on the Kentish shore of the estuary of the Thames, at Whitstable and Herne Bay, and in the tideways of the creeks on the Essex shore, particularly in those between the Colne and Blackwater rivers, and in the neighbourhood of Mersea island, famous from the time of the Romans, for this produce. Some very fine oyster-beds also exist at Emsworth, in Hampshire; others of a larger kind come from Poole and Jersey.

Mining Industry.—Coal stands at the head of the mineral products of England; and the country is probably more indebted to its inexhaustible supplies of this valuable mineral than to any thing else, for the extraordinary progress it has made in manufacturing industry. The coal-mines are all in the N. and W. parts of the kingdom; and these, consequently, are the great seats of English manufactures.

The following table shows the quantity of coal raised in the various mining districts of England and Wales in the year 1841:—

Coal Districts	Tons.
Durham and Northumberland	22,345,387
Lancashire	11,329,000
Staffordshire and Warwickshire	11,449,650
South Wales and Monmouth	10,976,500
Yorkshire	8,909,80?
Total	**65,074,147**

Iron ranks next in importance to coal. It was known to exist at a very early period; and the Romans, and perhaps, also, the Britons, had iron-works in the Forest of Dean, and elsewhere in the kingdom. Iron ore is very generally diffused; at present, however, all the great iron-works are situated in the coal districts, an abundant supply of coal being indispensable to the extensive production of iron. But in the infancy of the iron trade, when timber was the only fuel employed in smelting the ores, Kent and Sussex being the best wooded counties, were also those in which most iron was made. In 1740, the total quantity of pig iron made in England and Wales did not exceed the trifling quantity of about 17,000 tons, and we were then, and for a considerable time afterwards, mainly dependent on foreign supplies. But about this period coal began to be successfully substituted for timber in the preparation of iron, and its production was, in consequence, materially augmented. In 1750, the quantity produced did not, however, amount to 20,000 tons; but in 1788 it had increased to 68,000 tons, and in 1796, to 125,000 tons. The progress of the trade has since been rapid beyond all precedent. In 1800, a project was entertained for laying a tax on pig iron; and it was then ascertained that the production amounted to about 250,000 tons a year. In 1820 the produce had increased to about 400,000 tons; and in 1830, it was estimated at about 641,000 tons. But owing to the great demand for iron for railways and other public works, the increase of production continued on an enormous scale, and in the year 1864 amounted to—

	Tons
In England	3,630,472
— Wales	664,779
Total	**3,609,2?1**

Of this immense quantity of pig iron produced in 1864, only 464,351 tons were exported; all the rest was converted into merchant iron. This was effected at 127 iron-works, where 6,762 puddling furnaces were in activity, and 718 rolling mills performing their herculean labours of producing bars and rails. (Report of Mr. Robert Hunt, keeper of Mining Records at the Museum of Practical Geology.)

It may be mentioned as evincing the extraordinary progress of the iron trade, that it could hardly be said to exist in S. Wales previously to 1760. So much, indeed, was this the case, that in 1755, the land and minerals for several miles round Merthyr Tydvil—then an inconsiderable village, but now the seat of the greatest iron works in the kingdom—were let for 99 years for a rent of 200l. a year.

Next to coal and iron, the most important minerals of England are copper, tin, and lead, the latter containing quantities of silver. There were obtained, in 1864, from 192 mines in South Western England, 214,004 tons of copper ore, producing 13,302 tons of metallic copper.

The production of tin is confined to Cornwall and Devonshire; these are also the great copper ores, but copper is likewise produced, though in smaller quantities, in N. Wales, and some other parts. The tin obtained from the mines of Corn-

wall and Devonshire in 1864 was in excess of that ever before procured, although the tin mines and stream works of this district have been diligently worked for more than 7,000 years. 16,211 tons of tin ore were raised by the miners, the largest quantity from very deep mines. This produced of metallic tin 10,108 tons. The price of tin during 1864 was lower than it has been during any year since 1853, and more than 1l. a ton below the price of 1859. The system of mining which prevails renders it imperative on the managers of mines to use every effort to satisfy the shareholders by the regular payment of dividends, or, at all events, to prevent a depreciation in the value of the shares by avoiding 'calls.' To obtain this end tin ore has been raised, 'dressed,' and sold in an already glutted market at whatever price the smelter could offer. Hence the value of the ore sold in 1864 was but 925,909l., or upwards of 58,000l. less than the money value of the block tin sold in 1863.

Lead mines have been wrought in England from a very remote epoch. At present the most productive are in the N. cos. Lead, when first extracted from its ore, always contains a certain proportion of silver, varying from a few grains to 15 oz. or more in the ton. When the silver mixed up with the lead is sufficient to repay the expense, it is usual to separate it, which is effected by the process termed refining. The lead of some of the English mines, especially those of Cornwall, and also of the Isle of Man, contains very considerable quantities of silver. In the year 1864, no less than 94,453 tons of lead ore, principally galena, were dressed, sold, and smelted. This produced 91,283 tons of lead, and gave 641,045 oz. of silver.

Of zinc ores, nearly all being the sulphide of zinc (commonly called black jack), 15,047 tons were mined in 1864, producing 4,040 tons of metal.

Of iron pyrites—ores used for the sulphur they contain in sulphuric acid and soda works—there were procured, in 1864, 94,450 tons. In addition, there were raised small quantities of manganese and wolfram, together with arsenic, ochres, barytes, porcelain and pottery clays, and salt.

Salt, one of the most important of the British minerals, is procured in immense quantities from both fossil beds and brine springs, in Cheshire and Worcestershire. Previously to the discovery of the fossil beds, during the 16th century, and subsequently, a good deal of salt continued to be made by the evaporation of sea-water in salt pans at Lymington, near Portsmouth, and at other places; but the works at these places are now wholly abandoned, while the article in question has become greatly improved in quality; and instead of being imported, as formerly, is very largely exported. The consumption of Great Britain only, exclusive of Ireland, amounts to about 200,000 tons per annum, while the exports are of three times the amount. In 1859, the exports of salt were 665,644 tons; in 1861, they were 703,182 tons, and 624,745 tons in 1863. Before 1823, an oppressive tax of 15s. a bushel, or about thirty times the original cost price of the article, was imposed on salt; but in that year this enormous tax was totally repealed. Alum, fullers' earth, chalk, and lime are amongst the remaining useful minerals; clay for bricks, tiles, and earthenware, is also a product of considerable importance. Freestone is very abundantly diffused; but most English buildings being constructed of brick, its use is limited, except for pavements. Bath or Portland stone is that which has hitherto been mostly used for building. There are granite quarries at Dartmoor, Haytor, and several other places.

Manufactures.—Of these the most useful is that of woollen, the chief seats of which are the W. Riding of Yorkshire, and the cos. of Gloucester, Wilts, Devon, Lancaster, and Somerset. The first impulse towards the improvement of the woollen manufacture was given in the 14th century by Edward III., who invited a number of Flemish manufacturers to settle in England. But the manufacture laboured, down almost to our own day, under a number of vexatious and oppressive restrictions; and it did not begin to make any very rapid progress, or to participate in the wonderful improvements made in the cotton trade, till the introduction of the gig-machine, in 1802, and the repeal of the prohibitory acts of Edward VI. and Mary, in 1837. Leeds, Wakefield, Huddersfield, and Saddleworth, are the great centres of the broad cloth manufacture; Halifax is noted for its flannels and baizes, and Bradford for worsted spinning. Narrow cloths are made at and near Huddersfield; and blankets, flushings, &c., between that town and Leeds. At Dewsbury and Batley there are large establishments, called *shoddy mills*, in which old woollen rags are torn to pieces, rospun, and manufactured, sometimes with and sometimes without an admixture of new wool, into various descriptions of coarse cloth. (See DEWSBURY.) Rochdale in Lancashire is also a great seat of the woollen manufacture.

Gloucestershire has numerous fine broad cloth factories; but Bradford in Wilts is the principal centre of the superfine cloth trade. The cloths of Somerset are of inferior quality. Serges, or long ells, are made in almost every town and village in the co. of Devon, and also to a considerable extent at Wellington, in the co. of Somerset. Carpets are principally made at Axminster, Kidderminster, Ashton, and Wilton. Salisbury is noted for its flannels, and Witney in Oxfordshire for its blankets; though most of what are called Witney blankets are in reality made in Wales. Norwich was long the principal seat of the worsted manufacture; but the command of coal, and the greater facilities for carrying on the business enjoyed in Bradford, and other places in the West Riding of Yorkshire, have given them a decided superiority. The manufacture of woollen and worsted stockings is principally carried on in Leicestershire, about 14,000 stocking-frames being supposed to be at work in that county. Coarse woollens and druggets are made in Cumberland, baizes in Essex and Suffolk, and a few articles are made in North Hants and Surrey; but the woollen manufactures of the S. cos. are comparatively unimportant. The total value of the exports of woollen goods and worsted goods and yarn in 1864 amounted (for the United Kingdom) to no less than 18,566,078*l.* The produce had more than doubled in the course of 15 years, amounting to 8,568,690*l.* in 1850; to 9,500,629*l.* in 1856; and to 16,409,564*l.* in 1863. There are no separate returns of the exports of England and Wales. According to the census reports of 1861, the woollen manufacture at that time employed 130,034 persons, and the worsted manufacture 79,242.

More important still than the woollen manufacture is that of cotton. Vast as this manufacture now is, it may be said to have almost entirely grown up since the accession of George III. In 1760. The first grand stimulus was given to it in 1767, by the invention of the spinning-jenny; and the subsequent and almost miraculous inventions of Arkwright, Watt, Cartwright, Crompton, and others, have carried it to the extraordinary state of improvement to which it has now arrived.

Cotton goods of great beauty and excellent quality have been so much reduced in price, as to be within the command of all but the merest beggars. Hence the astonishing increase in the demand for them; the produce of the British manufacture being now widely diffused over the remotest countries of America and Asia. Lancashire is the grand seat of the English cotton manufacture; and next to it, but at a great distance, are Cheshire, Derbyshire, and Yorkshire.

Various estimates have been given of the value of this great manufacture, and of the number of persons employed in and dependent on it. The census returns of 1861, state the total number of persons engaged in cotton manufacture at 456,646. In fact, according to these returns, the people engaged in producing cotton fabrics are the third most numerous class in England and Wales—the first being agricultural labourers, and the second domestic servants. But, probably, the number actually engaged in, various ways, in cotton manufactures, is considerably larger than that shown in the census. Perhaps it will not be far wrong to estimate the total value of the various descriptions of cotton fabrics and yarn now annually produced at 85,000,000*l.*; and the total number of persons of both sexes, and all ages, employed in all departments of the business, at about 500,000. If right in this latter estimate, it will follow that from 1,000,000 to 1,200,000 individuals may be regarded as depending for support on this great manufacture.

Estimating the entire annual value of the cotton fabrics of Great Britain at 85,000,000*l.*, the value of those annually produced in Scotland may, perhaps, be estimated at nearly 5,000,000*l.*; for, as a large proportion of the fabrics made in Scotland are of a comparatively fine description, their value exceeds what might be inferred from the amount of yarn produced in Scotland as compared with that produced in England.

Subjoined are some statistics of factories for cotton goods, extracted from a return laid before Parliament in 1861:—

Cotton Factories	Number of Factories	Number of Spindles	Number of Operatives
ENGLAND:			
Lancaster .	1,979	21,390,482	310,527
York .	309	2,414,584	27,310
Chester .	212	3,373,113	40,561
Derby .	70	603,000	17,863
Cumberland .	16	130,212	3,393
Middlesex .	10	3,831	375
Stafford .	9	81,118	3,592
Leicester .	9	6,460	719
Nottingham .	24	88,000	2,183
Flint .	1	81,000	190
Suffolk .	1	—	52
Warwick .	7	—	441
Surrey .	2	—	62
Gloucester .	1	64,084	1,614
Norfolk .	2	—	84
Total .	3,315	30,651,925	407,488

The above figures are probably incomplete, the number of operatives being given at nearly 50,000 less than in the census returns—the latter necessarily under the mark, owing to the vast subdivision of labour and the complexity of trades more or less connected with cotton manufactures. Altogether it seems most likely that there are at least half a million individuals directly engaged in the cotton trade.

The *flax manufacture* is seated chiefly in Yorkshire, Lancashire, Salop, Cumberland, Westmoreland, Durham, Dorset, and Somerset. In 1848 the linen factories employed 19,340 hands, and the census returns of 1861 state the number of individuals engaged in the manufacture as 22,050. The silk manufacture is more important. The metropolitan district of Spitalfields, Manchester, and Macclesfield are the chief places in which broad silks and handkerchiefs are made. Coventry is celebrated for its riband factories. Crapes are made mostly in the E. cos.; but this branch of manufacture is declining. A great revolution was effected in the silk manufacture in 1826. Previously to that epoch the legislative enactments with respect to it were the most contradictory and impolitic that can well be imagined. The importation of foreign silks was prohibited under the severest penalties; but the advantage that this prohibition was believed, though most erroneously, to confer on the manufacturer, would, under any circumstances, have been more than neutralised by the imposition of oppressive duties on the raw material. This vicious system was productive of a twofold mischief; for, by teaching the manufacturers to depend on custom-house regulations for protection against foreign competition, it made them indifferent about new discoveries and inventions, while, owing to the exorbitant duties on the raw material, and the want of improvement, the price of silks was maintained at such a price as to restrict the demand for them within comparatively narrow limits. In 1825, however, a new and more reasonable order of things was introduced. The duties on the raw material were greatly lowered; at the same time that foreign silk goods were allowed to be imported on payment of a duty of 30 per cent. *ad valorem*. This new system was vehemently opposed at its outset, and it was confidently predicted that it would occasion the ruin of the manufacture; but the result has shown the soundness of the principles on which it was bottomed. The manufacturers were now, for the first time, compelled to call all the resources of science and ingenuity to their aid; and the result has been that the manufacture has been more improved during the last dozen years than it had been in the whole previous century; and that it has continued progressively to increase. In 1846 the duty was further reduced to 15 per cent. *ad valorem*.

The total quantity of raw silk imported into the U. K. in 1864 amounted to 5,655,401 lbs., being a decrease over imports of the preceding years, which, in 1862, amounted to 10,572,123 lbs., and, in 1863, to 9,271,146 lbs. The total number of persons engaged in the silk manufacture in England and Wales was stated to be 101,678, in the census returns of 1861.

The *hardware manufacture* is one of the most important carried on in England; it comprises all kinds of articles, from the anchor of a man-of-war down to the delicate furniture of a lady's work-box. The more ponderous are wrought in Staffordshire, Coalbrook Dale, and elsewhere; cutlery, and the finer kinds of articles, are made chiefly in Birmingham and Sheffield. In Birmingham steam engines of the largest size are also produced, and the whole tract to the NW. for a considerable distance is one immense field of smithies and forges, and the goods made in it and its immediate neighbourhood are estimated to be worth more than 3,500,000*l*. annually. Firearms, needles, &c., are made in great quantities in London. Watch and clock works are made especially in Lancashire. The total number of hands employed in this great branch of industry is probably upwards of 350,000, and the total value of the goods produced amounts perhaps to 17,000,000*l*. a year. The census returns of 1861 report that there are 125,771 persons engaged in England and Wales in the iron manufacture, as moulders, founders and similar workers, besides which 108,165 individuals are registered as blacksmiths, 60,862 as engine and machine makers, 26,130 as engaged in the nail manufacture, and 20,757 as clock and watchmakers. The value of the hardware exported in 1864 amounted to 13,214,294*l*. The manufacture of leather is nearly equal in importance to that of hardware. Worcester and Yeovil are the principal seats of the glove factories, and furnish together upwards of 800,000 pairs of leather gloves annually. Many more are manufactured in Woodstock, London, Nottingham, and Ludlow. Shoes, harness, and saddlery are made in most large towns, and especially in the metropolis. Boots and shoes are also made in great numbers in Northamptonshire and Staffordshire. The leather manufacture is estimated to employ, in all, nearly 234,000 hands, and to produce goods to the value of 13,000,000*l*. a year, or upwards.

The NW. part of Staffordshire is devoted almost wholly to potteries. British earthenware, so highly improved by the intelligence and ingenuity of Mr. Wedgwood in the last century, now rivals the best produced on the Continent in elegance and excellence, and is much superior in point of cheapness. Hence the earthenware and china of England are very widely diffused; and are the only varieties to be seen in most parts of the Continent and of America. China ware is made at Derby, Worcester, &c. The total value of the articles manufactured in both branches of the trade is estimated at about 2,500,000*l*. The number of persons engaged in the earthenware manufacture in England and Wales, is given at 38,073 in the census returns of 1861. The glass manufacture has grown up since the end of the 16th century, but has especially increased within the last 60 years. Bottle and crown glass are made chiefly at Newcastle and S. Shields; plate-glass exclusively at the former town, and at Ravenhead in Lancashire. There are numerous glass-works in the W. and NW. cos.; and the total produce of glass articles is valued at 2,300,000*l*. a year, their manufacture employing 50,000 hands. The census of 1861, however, only distinguished 15,046 as engaged in glass manufacture. This branch of industry has been materially injured by oppressive duties. But these, after being reduced in 1835, were finally repealed in 1845. Paper is made in Herts, Kent, Lancashire, Surrey, and Yorkshire. It is estimated that there are in all about 312 mills at work in England, employing about 25,000 individuals, and producing goods to the value of about 5,000,000*l*. a year. Paper, like glass, has been subjected to oppressive duties; but these were reduced in 1836, and entirely repealed in 1861. Hats to the value of between 2,500,000*l*. and 3,000,000*l*.; bricks and tiles to an enormous extent; soap, candles, gunpowder, starch, vinegar, dyes, coaches, furniture, and straw plait, are amongst the remaining principal articles produced by the manufacturing industry of England. Extensive sugar-refineries are established in London, Liverpool, Bristol, and other towns.

The conversion of grain into fermented liquors gives rise to three very important and extensive branches of manufacture and trade, those of malting, brewing, and distilling. The breweries throughout England yield upwards of 5,000,000 barrels of beer of all sorts; and from 1,800,000 to 2,200,000 barrels are supposed to be annually supplied by the porter breweries of London, which are the largest and finest establishments of the kind in the empire, and, indeed, in the world. There is also a very large consumption of spirits, particularly gin, in England. But, notwithstanding the numerous allegations to the contrary,

ENGLAND & WALES

English Miles

there can be no doubt, comparing the consumption with the population, that it is decidedly less at present than it was in the reign of George II., and at more recent periods. This is established beyond all question by the statements made in parliament in the debates on the Gin Act in 1743, and by the details given in the tract of the celebrated Henry Fielding on the Increase of Robberies (London, 1752), and other authentic documents. No doubt there is still, in this respect, ample room for improvement. Nothing, however, can be more unfounded than the complaints so often put forth of the increase of drunkenness: that the lower classes are not so temperate as could be wished for, is most true; but they have improved, and are now less given to intoxication than at any former period of our history.

The subjoined statement shows the total quantities of spirits, both home-made and foreign, consumed in England and Wales in 1831 and in 1861, as well as the quantities consumed at the same periods in the United Kingdom.

BRITISH AND FOREIGN SPIRITS.

	1831 Galls.	1861 Galls.	Increase per cent.
British	7,434,047	11,098,352	49·2
Foreign and Colonial	4,697,892	5,715,591	21·65
Total	12,131,779	16,813,341	38·5

UNITED KINGDOM.

British	21,645,531	20,496,102	7· dec.
Foreign and Colonial	4,592,793	6,295,150	24·0
Total	26,158,324	26,791,258	0·14

It will be seen at a glance that the increase in the consumption of spirits has been less than the increase of population.

Internal Communication.—The turnpike roads of England are at present, perhaps, the best in the world. They are placed under the direction of trusts, and kept in repair by tolls levied on passengers and carriages, and rates, which the surveyors of roads are empowered to levy by the act 5 and 6 Will. IV. c. 50. Many of these acts, however, have recently been repealed, and in particular all the turnpikes near the metropolis have been abolished, the maintenance of the roads being left to the parishes, to be defrayed by local assessment. The construction of canals in England originated during the latter half of the last century. Most of them are in the NW., or manufacturing districts. The principal are the Lancaster canal, from Kendal to Wigan; the Liverpool and Leeds; Burnley and Shipton; Aire and Calder Navigation; Duke of Bridgewater's canal, from the head of the Mersey estuary to Manchester; those connecting Bolton and Bury with the latter town; the Rochdale from Manchester; Huddersfield, from Manchester by Ashton-under-Lyne; Peakforest; Trent and Mersey; Ellesmere; Hereford and Gloucester; Thames and Severn; Berks and Wilts; Arundel; Grand Junction from the Thames at Brentford to Northampton; and the Paddington and Regent canal, on the N. side of the metropolis. The total length of the canals traversing England exceeds 2,200 m. All have been constructed by private companies or individuals, and several exhibit splendid triumphs of engineering art; as, for instance, the Ellesmere, which in one place is carried over the Dee at an elevation of 125 ft. above that river, by means of a course of cast-iron plates supported on 19 pairs of stone piers. The Grand Junction has a tunnel, 3,080 yds. long, and the Duke of Bridgewater's canal is excavated subterraneously for a total distance of several m. But the extension of

canals has been nearly suspended since railways came into use. These originated also in the latter half of the last century in the N. mining district. The construction of railways, following upon that of canals, has originated a new era in the industrial and social life of England. The wooden rails at first used gave way to others of iron. The Stockton and Darlington railway, opened in 1825, was the first intended for public use; but it was not till 1830, when the Liverpool and Manchester railway was opened, that the vast importance became manifest. The formation of railways would, however, have been of comparatively little value, but for the invention of locomotive engines, which being successfully introduced on the Liverpool and Manchester railway, made its opening a memorable era in the history of internal communication. By means of these engines long trains of carriages, loaded with passengers and goods, are now impelled along railways at a speed varying from 25 m. to 60 m. or upwards an hour. Hence it is that time and space are nearly annihilated in as far as railway travelling is concerned. This extraordinary speed has also been attained with a great increase of comfort and security; the accidents by railways being very decidedly fewer, as compared with the number of passengers, than those arising out of travelling by common coaches. The latter, in fact, are almost wholly superseded on all the great lines of road.

The total length of railways open in England and Wales, together with the total paid-up capital, in each of the years 1854 to 1863, was as follows:—

Years	Length of Lines open at the End of each Year	Total Capital paid up (Shares, Loans, &c.) at the End of each Year
	Miles	£
1854	6,114	240,235,795
1855	6,210	249,045,378
1856	6,447	253,683,191
1857	6,173	265,104,298
1858	7,001	270,471,643
1859	7,309	277,645,618
1860	7,262	288,629,411
1861	7,570	298,446,103
1862	8,178	319,237,038
1863	8,668	338,514,818

The number of passengers, including the holders of yearly or season tickets, who were conveyed by railway in England and Wales, and the total traffic receipts in each of the years 1854-63, were as follows:—

Years	Total Number of Passengers conveyed (excluding Season-Ticket Holders)	Total of Traffic Receipts
	No.	£
1854	92,368,149	17,348,925
1855	99,174,073	18,383,360
1856	104,365,361	18,729,309
1857	115,361,998	20,487,744
1858	119,536,847	20,344,895
1859	121,541,392	21,749,899
1860	138,989,444	22,479,846
1861	145,831,425	24,071,579
1862	153,491,377	24,529,993
1863	172,844,474	29,313,479

The enormous increase of passengers, far more than that of mileage, within the ten years 1854-63, is very striking, and allows fair conclusions as to the increasing importance of this comparatively new mode of locomotion.

In close connection with railways, and scarcely less important, are the thousands of miles of telegraph wires which have spread over England like a network in the course of little more than a generation. Without tracing the growth of this great auxiliary of modern locomotion, it may suffice to give the length in miles of telegraph wires in England and Wales, in the three years 1861–63.

Telegraph Companies	Length in Miles of Telegraph Lines		
	1861	1862	1863
Electric & International	4,727	7,697	6,730
British & Irish Magnetic	3,963	4,126	4,188
South-Eastern Railway	302½	314	314
London, Brighton, and South-Coast Railway	197	199½	813
London District	92½	103	101
Submarine (Telegraph to Calais, 24 miles ; to Boulogne, 26 m. ; to Dieppe, 74 m. ; to Jersey, 30 m. ; to Ostend, 70 m. ; to Hanover, 60 m. ; and to Denmark, 360 m.)	591	617	607

The number of telegraph stations open to the public in 1863 was 1,707, while the number of messages sent in the same year amounted to nearly three millions.

Constitution and Government.—The legislative power, by the constitution of Great Britain, is vested in the great council of parliament, consisting of the King and the three estates; that is, the Lords Spiritual, Lords Temporal, and Commons.

The early history of the parliament of England is enveloped in great obscurity. This much, however, is certain, that previously to the Norman invasion it was usual to consider and debate matters of public importance in the *Wittenagemote*, or great council of the nation. After the Norman invasion, and the establishment of the feudal system, the king, as lord paramount, was assisted by a great council composed of the principal feudal superiors, or tenants *in capite*, whose concurrence was necessary in matters of general or national importance. In *Magna Charta*, signed by King John on the 15th of June, 1215, it is stipulated that 'no scutage or aid shall be imposed on the kingdom, beyond the ordinary liabilities of the feudal tenure, unless by the common council of the kingdom.' This shows that even at this early period the principle was recognised, that the nation should not be taxed except by its own consent. The great number of tenants *in capite*, or of those who, as they held directly from the crown, were entitled to a seat in the great council or parliament, and the disinclination and inability of many of them to attend, gave rise to the practice of commoning, by name, a few only of the most distinguished, or of those called the greater barons, whence originated baronies by writ; while the others, who were not summoned, adopted, in no very long time, the practice of sending representatives. The latter consisted of two knights for each shire, and of one or more burgesses for the three boroughs, or of those holding of the crown. Different opinions are entertained as to the period when these important innovations took place; but, at all events, there is undoubted evidence to prove that burgesses attended the parliament summoned by Simon de Montfort, earl of Leicester, in 1265. At this time, also, the clergy were summoned to attend by their procurators (proctors); but they struggled successfully to rid themselves of the burden (as it was then considered), and obtained the privilege of meeting in convocation for each of the two provinces, the bishops and mitred abbots only continuing to attend parliament.

Under the reign of Edward I., knights and burgesses were regularly summoned; and in that of Edward II., parliament appears to have been divided into two houses; that is, into the House of Lords, consisting of the great feudal lords who directly attended; and the House of Commons, consisting of the representatives of the smaller tenants and burgesses. In the same reign parliament seems for the first time to have exercised, in a regular manner, the functions of a legislature. In the reign of Henry IV. we first find the right of the Commons to originate all supplies noticed as an existing institution. From this period, the history of parliament is closely interwoven with that of the nation. The number of burgesses was gradually increased by the enfranchisement of fresh boroughs; and the popular influence in the legislature progressively gained strength with the increasing wealth and intelligence of the nation. But for a lengthened period the nature of the government was not well defined, and the rival powers of the crown and of parliament were frequently coming into contact. During the reigns of Henry VIII. and Elizabeth, the regal power attained to a maximum. But the growth of commerce under the latter, combined with the powerful influence of the Reformation, and other causes, not only gave a great accession of strength to the bulk of the people, but made them better acquainted with their rights, and less disposed to submit to their invasion. The princes of the House of Stuart wanted sagacity to appreciate the changes that had thus taken place in their position with respect to the public. Their maxims of government were as arbitrary as those of the Tudors, but they had neither their ability nor their power. Their attempts to govern without a parliament, and in defiance of principles that had been sanctioned from the earliest periods of the monarchy, produced, in the end, a civil war, that happily terminated in favour of the popular party. But it was not till the Revolution of 1688, when the Stuarts were finally expelled from the throne which they had shown themselves unfit and unworthy to fill, that the principles of the constitution were clearly established. The celebrated statute, called the Bill of Rights (1 Will. & Mary, sess. 2, 1689), declared that the suspension of laws, or their execution by regal authority, without the consent of parliament, was illegal; that parliament had the exclusive right to levy money from the subjects; that the debates or proceedings in parliament were not to be questioned in any court or place out of parliament; that it was the right of subjects to petition the king; that jurors were to be duly panelled and returned; and that parliaments should be held *frequently*. By the Triennial Act (1694) the duration of parliaments was limited to three years. In 1715 it was extended to seven, at which period it has continued fixed. The union with Scotland (1707) and Ireland (1800) increased the number of members to 658. We have elsewhere adverted to the circumstances that occasioned the passing of the Reform Act of 1832. (See Vol. I. p. 560.) This important statute made some material changes, by enfranchising some of the greater and disfranchising some of the smaller boroughs; and by modifying the electoral franchise, and creating a new right of voting in all occupiers of premises of the value of 10l. a year in boroughs throughout the three kingdoms.

The king, as a constituent part of the parliament, has the prerogative of giving a final assent or negative to any bill which has passed the two houses. But the royal veto, though conceded by the theory of the constitution, has long ceased to be exercised; and the assent of the sovereign is now nothing more than a formality, necessary to give an act of parliament the force of law.

The descent of the crown of England is limited partly by customary law, partly by statute. By the Act of Settlement (12 & 13 W. III.) it is vested in the descendants of the Princess Sophia, youngest daughter of Elizabeth, queen of Bohemia, and granddaughter of James I., being Protestants; and every person marrying a Papist is rendered incapable of possessing or enjoying it. Subject to these limitations, the crown descends, as of hereditary right, first to the male, then to the female issue in succession. There is no minority in the case of an heir to the crown; and whenever a minor is likely to be called to it, it is usual for parliament to make beforehand a special provision for the emergency.

The House of Lords consists of the lords spiritual and temporal.

The lords spiritual are, the 2 archbishops and 24 bishops of England; with 1 archbishop and 3 bishops of Ireland, who succeed in rotation, and sit for a session only. Before the Reformation, 27 abbots and 2 priors sat in the English parliament. In consequence of the distinction between the two estates (spiritual and temporal), doubts were felt, even so late as the time of Coke, as to the validity of bills which might pass the House of Lords by the votes of one estate only, against or without the voice of all the spiritual or temporal peers. But such scruples are no longer entertained, and no distinction remains between the two estates.

The temporal lords of parliament are, 1. English peers, distinguished in rank as dukes, marquises, earls, viscounts, and barons. Peerages are said to be held by tenure, or created by writ or by patent. The former, which appears to have been the most ancient species of peerage, consisted in the holding of certain baronial estates or "honours," which are supposed to have entitled the owner to be summoned by name as of right to parliament. It has been in effect long obsolete; a few baronies are still asserted to be held by tenure, but it is doubtful whether the claim, if preferred, would be admitted. Creation by writ is a summons to the individual, by the name and style of the peerage conferred, to attend parliament. Creation by patent, at present the ordinary mode, is the grant of a peerage by the crown, with specific limitations as to the descent, usually, in modern times, to the heirs male of the body of the peer, with or without remainder to other branches. The right to a contested or claimed peerage is tried by the House of Lords. 2. Sixteen Scotch peers are elected every parliament by the whole peerage of that country. 3. Twenty-eight Irish peers are elected in like manner for life. Scotch or Irish peers, who have also English peerages, sit and vote in parliament by the title of these peerages. The chancellor, by virtue of his office, is speaker of the House of Lords. This house claims the privilege of originating all bills for the restitution of honours or blood.

The number of members of the House of Commons has been, since the union with Ireland, 658. The number of English representatives was fixed by ancient usages and charters, and that of Scotch and Irish by the respective Acts of Union of those two countries with England; but the distribution of members was materially altered by the changes introduced by the Reform Act of 1832. Aliens and dissenters are disqualified from sitting and voting in the house; as are peers of parliament, and Scotch (but not Irish) peers,

the clergy, and the holders of various offices; while other offices only render it necessary to vacate a seat in parliament, the holder remaining eligible. Bankrupts, persons attainted of treason, and felony, and outlaws (in criminal cases), are also excluded. Formerly the necessary qualification of estate was for counties, the possession of 600l. a year issuing out of land (held for the life of the member, or a greater estate); for boroughs, that of 300l. This property qualification for members was not disturbed by the Reform Bill; but, being of no great importance, and giving rise, moreover, to fraud, it was repealed by 21 & 22 Vict. c. 26, of June 29, 1858. There is, therefore, now no property qualification whatever for members, and a person may sit in Parliament who has not even a vote in the elections.

Of the English boroughs, 50 return 1 member each; London 4, the remainder 2; 6 counties return 2 members each; 7 return 3 members each; 26 return 4 members each, being 2 for each of the districts or divisions into which they are apportioned by the Reform Act. Yorkshire returns 6 members, being 2 for each riding. The Isle of Wight has 1 member. Welsh counties, 1 each; with the exception of Caernarvon, Carmarthen, Glamorgan, 2. Welsh boroughs, 1 each. Scotch counties, 1 each; boroughs, 1 each, with the exception of Edinburgh and Glasgow, which return 2. Irish counties, 2 each; boroughs of Dublin, Cork, Belfast, Limerick, Galway, Waterford, 2; the remainder, 1. The right of voting for county members, in England, is in all freeholders possessing land of the value of 40s. per ann., if of inheritance, or in actual occupation, and not acquired by purchase; the latter conditions being introduced to guard against the creation of fictitious votes. An estate for life of 10l. per annum is sufficient under any circumstances. Copyholders to a certain amount, and leaseholders to a certain amount and duration, are now also in the possession of the franchise; as are all tenants, whether with or without leases, who pay a land fide rent of 50l. a year. In Scotland, besides certain votes on account of ancient rights of a peculiar description, termed superiorities, freeholders of 10l. per annum have the right of voting, and tenants nearly as in England. The right of voting for counties in Ireland is also fixed at 10l. per annum, for freeholders, leaseholders and copyholders nearly as in England.

In English boroughs a uniform franchise, created by the Reform Act of 1832, is possessed by the occupiers of a house or other building, or building with land, of the value of 10l. per annum. In cities that are counties of themselves, freeholders vote as in counties. Besides these, there are in all the boroughs, except such as were enfranchised by the Reform Act, certain ancient rights, reserved to those who were in the possession of the franchise at the passing of that act. These vary according to the usage of particular boroughs. Such are the ancient franchises of pot-wallopers, or pot-boilers, payers of scot and lot, freeholders, burgage tenants, and freemen admitted to the freedom of corporations. But in all these cases provision is made for the gradual extinction of the ancient franchises, no new claimants being registered unless they have acquired the right in certain excepted ways. In Scotland and Ireland, also, the occupiers of houses of the value of 10l. per annum in boroughs possess the franchise, with reservation of certain ancient rights in the latter country. Voters for the universities are such as have attained the degree of master of arts, and have kept their names on the books.

The following is, perhaps, a tolerably fair estimate of the mode in which the present House of Commons is constituted. There may be still from 20 to 25 nomination boroughs: that is, boroughs the members for which are habitually appointed without opposition, by individuals, generally large landed proprietors in the neighbourhood, who possess a paramount influence over the constituency. There are 6 or 7 in which government exercises great or paramount influence. In the remainder, the local influences, as, for instance, of landed property, great manufacturing or mercantile establishments, vary in degree (frequently acting so as to counterbalance each other), according to the comparative 'openness' of the constituency, until in the greater boroughs they can scarcely be said to exist, unless in the case of freemen who are swayed by old corporation politics. Counties are generally, though with differences of more and less, governed by the landed aristocracy of their soil.

The House of Commons is summoned by warrant of the king to the lord high chancellors of Great Britain and Ireland, or to the keepers or commissioners of the great seal, to issue their writs for the election of knights, citizens, and burgesses. When a new parliament is summoned, forty days must elapse between the date of the writ and the return to it. On vacancies occurring during the session, the writ is issued to the clerk of the crown in chancery, on warrant from the speaker; and the speaker also makes out a writ on vacancy occurring during the recess. The writs are delivered to the sheriffs of counties and returning officers of boroughs, who are bound to give the proper notices. If the member or members be not elected by show of hands on the nomination day, a poll is demanded, and takes place on the next day but two in counties, and lasts two days; on the next day in boroughs, lasting one day only. Votes are publicly given and recorded. The name of every elector who is admitted to poll must (since the Reform Act) appear on the register of voters, which is constructed in a different manner, and by different officers, in the three countries. Where votes are equal, it is usual for the returning officer to make a double return. The validity of a return may be questioned by petition to the House of Commons, not only on the ground of irregularity in making it out, but also of riot, of treating or bribery by the member elected, and of the admission of unqualified votes or rejection of good ones. The trial is before a committee of the House of Commons, regulated in its appointment and proceedings by a variety of statutes.

The House of Commons claims the privilege of electing its speaker; and also that most important one, which is a fundamental principle of the constitution, the right to originate all bills imposing any tax or levy on the subject, technically called money bills, or bills of supply. Should the lords, in amending bills sent up from the commons, introduce clauses containing such impositions, they are invariably rejected by the latter, as infringing on their undoubted privilege.

By the Septennial Act (1 George I. 1715), a new parliament must be summoned every seven years; but as the crown has the prerogative of arbitrary dissolution, and as there is a dissolution also on the demise of the crown, the length of a parliament has rarely approached that limit. The sessions of parliament are annual. It does not appear that there is any statutable necessity for its being summoned oftener than once in three years; but as the supplies are annually voted, it is impossible to dispense with a yearly summons. The ordinary session of parliament lasts about six months, from January, or February, to August; but it has of late years been often of longer duration. Parliament is *adjourned* from day to day, or over a short recess (as at *Easter*), by the authority of each house separately. It is *prorogued* by the king's authority, and frequently in his presence, by the lord chancellor, at the close of the session; but, in practice, for two months only, at the end of which it again meets *pro formâ*, and is again prorogued, and so on to the commencement of the session. It is dissolved, as has been said, at the will of the crown, or by efflux of the seven years, or on the demise of the crown; in the latter case, it continues six months after such demise, unless sooner prorogued.

Privilege of Parliament is a comprehensive term, embracing both the personal privileges enjoyed by every member of the legislature, and also the general rights of the body. To the former class belongs the freedom from arrest in civil proceedings enjoyed by every member; to the latter, the freedom of debate, authority to punish for contempts, and the various other safeguards to the liberty and power of the legislative assembly, which have from time to time been acknowledged as principles of our law. The extent of these privileges, and the mode by which they are to be ascertained, are questions to which as yet no authoritative decision has been given. The assertion of those who rate the authority of parliament highest is, that when either house claims a privilege, other tribunals have no authority, either to reject such claim if the privilege be proved, or to decide whether the privilege be proved or not; parliament alone (that is, each house for itself) being the judge of its own privileges.

The great council of parliament possesses exclusive legislative authority. In this character it is said to be legally omnipotent; that is, that there is no recognised power in the constitution to check or overrule it. This legislative authority is commonly exercised, not only in matters of public interest, but also in the passing of laws at the request and for the benefit of private individuals or associations, to give them powers which without such authorisation they could not possess.

The course of legislation in ordinary cases is as follows:—Any member of parliament may bring in a bill, or draught of a law, which (except in certain cases, before enumerated) may commence in either house. In the House of Commons, however, it is necessary first to obtain leave from the house to bring in the bill. The bill is then 'read for the first time,' with or without discussion, which, except on questions of great public interest, does not usually take place on this first stage. It is then printed, and a day fixed for the second reading. The principal debate ordinarily takes place on this occasion. If it pass the second reading, it is referred, if of public importance, to a 'committee of the whole house;' private bills, and others of less consequence, are usually referred to select committees. In committee, the clauses of the bill are considered one by one. On the report of the committee, the third reading of the bill takes place, with the amendments which the committee may have made upon it. If it be not rejected on the third reading, it is sent to the other House, where it passes through similar stages. If the other House amend, the bill is sent back to that in which it originated. If the two Houses disagree as to the amendments, a succession of 'conferences' may take place; and if no agreement be thus effected, the bill drops; otherwise it proceeds to receive the royal assent, and thus becomes an act of parliament, or a statute law.

Bills of supply, or for the providing of the funds required for the carrying on of government, must originate, as has been said, in the Commons. They most always begin in a committee of the whole House, moved for at the commencement of every session by the chancellor of the exchequer. All applications for grants of public money come in the form of messages from the crown. Bills of supply, when they have received the assent of the Lords, return again to the Commons.

Committees are either of the whole House, in which case the principal departure from the usual course of business are, that a private member is voted into the chair, instead of the speaker, and that the same strictness is not observed in the usages of debate, members being allowed, for example, to speak more than once; or permanent, nominated by each House at the commencement of the session, which has now become a mere formality; or consisting of a small number of members selected by the Houses, at their discretion, for the purpose of having bills referred to them. Committees have power to examine witnesses; but those of the House of Lords only examine on oath.

Parliament, and especially the House of Commons, exercises an extensive control over the conduct of the executive, not merely by legislation, but by various established methods of expressing satisfaction or dissatisfaction. Such are motions made by individual members, either founded on petitions (which it is a peculiar part of the business of both Houses to receive and consider), or otherwise; on which resolutions may be adopted by the House, addresses to the crown moved, committees appointed to examine and report, and so forth. The right of parliament to exercise this species of superintendence is unquestionable.

Should the prime minister for the time being happen to be a peer, as is very frequently the case, some member of the cabinet, usually the home secretary or chancellor of the exchequer, acts as 'leader' of the ministerial body, and principal representative of the government in the House of Commons, in which the conflict of parties is chiefly fought. No convenient is this species of leadership found, that any considerable body in opposition usually find it advisable to select a similar head. A certain majority in the Commons, on ordinary occasions, however small, is absolutely necessary for carrying on the government, which may be said to be strong or weak according to the magnitude of this majority. The truth is, whatever may be said in theory of the balance of power in the different branches of the legislature, that the House of Commons has been, since the Revolution of 1688, and still more emphatically since the Reform Act of 1832, the paramount power in the state. Supposing the majority of the H. of Commons to be decisive and firm to its purpose, it may compel either the Crown or the H. of Lords to give way; for, by resorting to the extreme measure of stopping the supplies, it might, were its demands not acceded to, stop the whole machine of government.

Acts of parliament are either public or private. There is no distinction between these two classes as to the binding character of their authority; the only difference being that judicial tribunals are bound to take cognisance of all acts declared 'public,' but not of others, unless specially exhibited and proved before them.

The Executive.—The whole executive and administrative functions of government, as well foreign as domestic, are performed in the name of the sovereign. The sovereign has the sole power of making war and peace; and, as incident to that power, the command and disposal of the army, navy, and other forces of the kingdom. The sovereign is conservator of the public peace, in which character all criminal prosecutions are carried on in his or her name. The sovereign is the head of

the judicial system of the country; and, by fiction of law, is supposed to be present in all his courts when justice is administered. The sovereign has the power of granting pardons for offences, with some exceptions created by statute. The sovereign is commonly called the 'fountain of honour;' to which character all honours, titles, and privileges are conferred by him or her. The sovereign can also erect and dispose of offices, but no remuneration can be attached to them without consent of parliament. The sovereign is also supreme head and governor of the national church. The sovereign has the regulation of internal commerce—establishes fairs and markets, regulates weights and measures, and coins money.

Substantially and in fact, however, the power of the crown is comparatively limited. It is a constitutional principle that 'the king can do no wrong;' but, though he be not, his ministers are held to be responsible for all illegal or unconstitutional acts committed in his name. It is farther indispensable that his ministers should be able to command a majority in ordinary times in the H. of C. Unless they can do this, the countenance and approbation of the sovereign will avail them but little; and the king will be compelled to dismiss them to make room for other ministers, which, though less acceptable to himself, are more agreeable to the majority of the House. The latter has therefore, in effect, a veto on the choice of the king. He appoints ministers; but it belongs to the representatives of the people to confirm these appointments, to inquire into the fitness of ministers for their situations, and to determine whether they shall continue in office or be displaced to make room for others.

Practically, too, the power of the crown to elect ministers is a great deal narrowed by the necessity of choosing those individuals only for the more prominent situations who are members of the House of Lords, or can procure their return to the House of Commons. However well qualified an individual might be to fill the office of secretary of state, for example, he could not be appointed unless he were a peer, or could recommend himself to some constituency; and the chancellor of the exchequer, the attorney-general, lord advocate, and other chief officers, must necessarily be members of the H. of C. Previously to the passing of the Reform Act this was a less serious control over the free choice of the sovereign than it has since become, a much greater number of nominating boroughs being then at the disposal of the crown. Now, however, it frequently happens that few competent individuals have to be appointed in preference to others, merely because they are able to command seats in the H. of C. To obviate this inconvenience it has been proposed to give ministers ex officio seats in the H. of C., which should entitle them to speak but not to vote; and probably, on the whole, this would be an improvement.

Every peer of the realm of England is, according to the theory of the constitution, an hereditary counsellor of the sovereign, and may be called to give his advice, whether parliament be sitting or not; but this principle has no practical consequences.

Privy Council.—To understand the manner in which this body was formed out of the great council of the nation or parliament, it must be remembered that one of the original objects of that institution was the summary redress of grievances which the ordinary legal forms did not avail to meet. The privy council was thus, in its origin, a species of committee of the great council, but nominated by the king, to which such plaints were preferred; and in the course of time its sittings became permanent, to afford relief when parliament was not assembled. From the reign of Richard II. to that of Charles I, we find the privy council (consisting usually of some of the chief officers of state, and some inferior members personally nominated by the king), exercising, in various ways, a very extensive jurisdiction, especially in matters, whether civil or criminal, in which the state was, however remotely, concerned. Under the Tudors and first Stuarts, the privy council was in the habit of granting warrants for the arrest, imprisonment, and even torture of the subject. The court of Star Chamber, and other tribunals of the same description, were offsets of the privy council. Its political functions were also extensive, though not admitting so easily of definition. In the reign of Charles I. (1640) the writ of habeas corpus was granted to persons arrested under warrants from the privy council; and its power in this respect was thus placed on a level with that of ordinary magistrates. The judicial functions of the council were thus effectually annulled; nor have they been revived, except as a court of appeal from the civil law courts, and from the local tribunals subsisting in our colonies and foreign dependencies. The number of privy counsellors, originally inconsiderable, was in the course of time greatly extended; limited by Charles II. to thirty, it has since his time again become indefinite. The political functions of the privy council are now virtually annihilated, and the title of privy counsellor is only one of distinction. The appellate jurisdiction already alluded to is exercised by a body selected from the mass, termed the Judicial committee of the privy council.

The cabinet council is a body which, though without any recognised legal existence, directs, in effect, the government of the country. It consists of a certain number of privy counsellors, usually consisting of the principal ministers of the crown for the time being, summoned to attend at each meeting. The name is said to be derived from the cabinet of Queen Henrietta, in which the ministers of Charles I. were accustomed to meet. The number is usually from 12 to 15. The first lord of the treasury, the chancellor, the chancellor of the exchequer, the president of the council, the three secretaries of state (home, foreign, and colonial), are always, in practice, members of the cabinet; some other officers are usually, but not invariably, accompanied by a seat in it.

The influence which the sovereign exercises over the deliberations of the cabinet, and the degree of executive power that centres in him or her personally, necessarily differ very greatly at different periods, inasmuch as they must materially depend on his character and capacity, and on the state and character of parties. At different periods since the Revolution, Parliament has compelled the crown to dismiss one set of ministers and choose another in opposition to its own predilections; but such ministries have rarely enjoyed much real power or been very lasting. Whichever party in the state was known to have the countenance and to enjoy the confidence of the crown, has generally contrived, in no very long period, to secure a majority in parliament. Hence it is that from the Revolution down to the accession of George III., the Whigs, with the exception of a few short intervals, were constantly in power; and that the Tories held, with similar exceptions, the reins of government from the accession of George III. down to the introduction of the Reform Bill. But it is doubtful whether such will be the case in future. It was comparatively easy for the crown to deal with the proprietors or patrons of nomination bo-

roughs; but the support of such persons is no longer sufficient to secure a majority: the favourable opinion of the constituents must now be also conciliated; and no ministry whose proceedings were disapproved by the bulk of the middle classes could hope to obtain a majority in the event of a dissolution, however high they might stand in court favour. Whether the nation shall be better or worse governed in time to come than it has been since the Revolution, experience only can decide; but there can be no doubt, speaking generally, that the government must now be conducted more in accordance with the opinion of the public. Still, however, the influence of the crown is very considerable; and when parties are nearly balanced in the country and in the H. of C., it may be able to turn the scale in favour of whichever party it espouses. But it is no longer in the power of the crown to make any effectual resistance to a decided majority in parliament, otherwise than by enlisting the public sympathies in its favour. If it cannot do this, there is nothing for it but to submit to be dictated to by the leaders of the dominant party for the time being. And this, in fact, is the decisive criterion of a free government—that the highest authority in the state should be obliged to act in accordance with the public voice as expressed by its representatives.

Not only are the legislative measures proposed by the crown, and the conduct of the internal government of the country and its foreign relations with other states, entrusted to ministers, but they have also the disposal of all or by far the greater part of the patronage belonging to the crown. Officers involving no political responsibility, such as those of the household, have been sometimes excepted from this rule, and left to be filled up by the sovereign according to his personal predilections; but this is not by any means a uniform practice, and ministers have repeatedly required and obtained the disposal of these offices.

Generally speaking, patronage in a country like England is always exercised with a view to the acquiring or preserving parliamentary support. Rulers like the emperors of Austria and Russia might select individuals to fill offices on the sole ground of their superior fitness to discharge their duties. But in a free country suitableness for office is not the only thing to be attended to in deciding as to the comparative claims of candidates for official preferment: if they possess it, so much the better; but the primary consideration is, how is the government to be carried on? Now that, it is plain, will be best effected by securing the active support of the friends of government, and by weakening the party of their opponents; and the distribution of patronage is one of the principal means by which these objects are to be realised. A government that should neglect to avail itself of this power could not long exist. Hence in England nine out of every ten situations are disposed of on the recommendation of persons possessed of parliamentary influence. This, in fact, is here the via regia to preferment and state distinction. In filling up the more conspicuous situations, the talents and acquirements of the candidates, as well as their recommendations, must necessarily be taken into account; but in the great majority of cases parliamentary patronage is the sine quâ non.

Officers of State and King's Ministers.—In England, as in other countries, the sovereigns early found the advantage of surrounding themselves with councillors, or rather with servants, more submissive, and more useful for their purposes, than those great functionaries of state whose dignity nominally entitled them to the chief weight and influence in their several departments. Hence, of the ancient great offices of state, one only can be regarded as now subsisting in the full extent of its power and importance.

Some have become altogether obsolete; others are kept in commission, and their duties then divided among several persons; others confer little more than titular dignity.

The great officers of state were—

1. The lord high steward. This officer is now only nominated on the occasion of a coronation, or an impeachment, in which case he acts as president of the House of Lords.

2. The lord high chancellor. He is entrusted with the care of the king's great seal. If there be no chancellor, the seal is in the hands of an officer styled the lord keeper, or is put in commission. In precedency, he ranks next to the archbishop of Canterbury, and above all other lords temporal and spiritual. He acts as speaker of the House of Lords; he is always a member of the cabinet, and generally has great influence. Besides various other important duties, he exercises the functions of chief judge of the court of chancery, in which capacity he will be afterwards noticed.

3. The lord high treasurer. For a very long period this office has not been filled. It is placed in commission, in the hands of officers styled lords of the treasury. The first lord of the treasury is usually prime minister for the time being. The treasury has the control of all matters connected with the receipt and expenditure of the public money, the appointment and superintendence of the boards and offices of customs and excise, stamps and taxes, post-office department, &c.

4. The lord president of the council (privy council), an office of great antiquity, revived in the reign of Charles II., and which has ever since continued. Its duties are little more than nominal; but it is attended, by custom, with a seat in the cabinet.

5. The lord privy seal. This officer has the custody of the king's privy seal, for the purpose of affixing it to charters, &c., as the lord chancellor has of the great seal. He also usually sits in the cabinet.

6. The lord great chamberlain. This office is hereditary, and has passed in succession to several great families. It is at present vested in females, by whom the deputy chamberlain is appointed. It is now merely a titular office, and not to be confounded with that of the lord chamberlain of the household.

7. The lord high constable was also a hereditary officer, and had extensive military authority. None has been appointed, except on special occasions, such as coronations, &c., since the attainder and execution of Stafford, duke of Buckingham, in 1521.

8. The earl marshal. This dignity is hereditary in the family of Howard, duke of Norfolk. The earl marshal has various ceremonial duties, and a jurisdiction extending for a certain distance round the king's palace at Westminster, which is executed by deputy.

9. The lord high admiral. This office has gradually, although not uniformly, been in commission since the Revolution. The commissioners are styled lords of the admiralty, and the first lord is usually a member of the cabinet. The board of admiralty has the control and direction of all matters relating to the navy of the kingdom, the naval dockyards and all matters relating thereto.

The office of secretary of state appears to have originated, or rather to have first assumed a character of importance, in the reign of Queen

Elizabeth. At that time, however, the secretary of state was not yet elevated to the rank of a member of the privy council, but attended its deliberations in an inferior capacity. The number of secretaries of state has varied at different times; but the office has continued to increase in importance, and at present may be said to discharge most of the higher functions of the executive in these kingdoms and their dependencies. It is divided into four branches—the offices of the secretary of state for the home department, foreign department, colonies, and the secretary to the lord lieutenant of Ireland. Each office has two under-secretaries: one permanent, for the discharge of the regular business of the office; the other a political functionary, depending on the changes in the cabinet. These offices exercise a general superintendence over the police and magistracy of the country, and over the execution of justice. The duties of the foreign and colonial offices extend to all the general business of those departments. The secretary of state for Ireland is the representative, in parliament, of the Irish government, and is usually, in effect, the officer principally charged with its conduct. All four are members of the cabinet. The government of Scotland is, in effect, vested in the lord advocate, as principal law officer for that part of the kingdom.

The secretary at war has a distinct department, being the ordinary channel of communication between the government and the military authorities. By an act passed in 1863, 26th Vict. c. 12, called 'An Act to abolish the office of Secretary at War, and to transfer the duties of that office to one of Her Majesty's principal officers of state,' the appointment was regulated as here expressed. The affairs of India were formerly transacted, according to the provisions of Mr. Pitt's act of 1784, by a board of commissioners, commonly termed the board of control; but an entire change in this respect was made in 1858, by act 21 and 22 Vict. c. 106, called 'An Act for the better Government of India.' This act left to a secretary of state for India all the powers previously exercised by the board of control.

10. The board of trade and plantations is a committee of the privy council; it has cognisance of all matters relating to the commerce and navigation of the country.

11. The post-office is under the control of an officer styled the postmaster-general.

12. Executive officers of the crown, employed in the administration of justice. Of the lord chancellor and the judges more will be said under the head 'Courts of Law.' In each county the sheriff is the principal executive officer. He is annually appointed by certain officers of the crown. His principal duty is to carry into effect the process of the law within his local jurisdiction. He is also judge of the county court; decides the elections of knights of the shire and coroners; and performs various other duties. There appears to be no strict legal qualification for the office of sheriff; but, in practice, it is usual to appoint men possessed of considerable landed property; and, as the exceptions and legitimate excuses are numerous, and the expenses are sometimes heavy, the appointment is felt as a burden by those on whom it falls. The legal duties of the sheriff are executed in practice by his under sheriff, usually a solicitor, appointed by him.

The custos rotulorum has the custody of the rolls and records of the sessions in each county. This office is usually joined with the military dignity of lord lieutenant. His deputy is the clerk of the peace, who performs the ministerial business of the court of quarter sessions in his behalf.

The coroner is chosen by the freeholders in the county court: the office is sometimes filled by an attorney, and sometimes by a medical practitioner, surgeon, or physician. His chief duty consists in holding inquisitions in cases of sudden death, where the body is found; for which purpose he summons a jury of four, five, or six persons.

The justices of the peace are commissioners, appointed under the great seal. Their general duty is to keep the peace, and any two or more of them to inquire of and determine felonies and misdemeanors. New commissions are always made out on the demise of the crown, and on other occasions when deemed advisable. The only legal qualification seems to be property to the amount of £100 per annum; but, in practice, the principal gentry of the counties, and respectable inhabitants of the towns, discharge these important and gratuitous functions. The powers of justices of the peace are extended and defined by a great variety of statutes. They have summary jurisdiction, either singly or in their petty or district sessions, over various minor offences, and in some civil disputes, as between masters and servants respecting wages. They hold, four times a year (in some counties more frequently), courts of general sessions, for the trial of felonies and misdemeanors, and other business. They levy rates, and direct the application of the funds thus raised to purposes of county expenditure.

In towns having municipal corporations, the municipal officers were formerly ex officio magistrates; but since the act of 1835, the crown issues commissions of the peace in such boroughs. Police magistrates (stipendiary) are appointed in the metropolis under various acts of parliament, and may be appointed, on petition, in any borough.

Constables are either high, appointed by the justices of the peace for the several hundreds; or petty, inferior officers charged to keep the peace in each town or parish. They are chosen by the jury at the court leet; or, in default of such court, appointed by two justices of the peace. The police force established in London and the principal English towns was created by Sir Robert Peel, in 1829. It is under the superintendence of commissioners of police, and acts under the direction of the magistrates.

Churchwardens, and overseers of the poor, are officers appointed by the inhabitants of every parish, meeting in vestry, under the authority of various statutes; the first to superintend the preservation of the church, the latter the affairs of the poor. Their duties are much curtailed by recent changes in the poor laws, under which a number of parishes are united, so as to form a district; and every union has its guardians of the poor, partly magistrates—who act ex officio—partly chosen by the vestry for every parish.

13. Municipal corporations are bodies established for the purposes of municipal government in borough towns. The limits of boroughs, to which their jurisdiction extends, are fixed by act of parliament, or by prescription. Municipal franchises began to be granted at an early period of our history, and generally to the whole body of townsmen in every place which obtained them. But, in the course of centuries, their charters became more narrowly interpreted, or were renewed, with different and more oligarchical provisions. Hence, in most towns in the kingdom, exclusive governing bodies were formed, to which the right of admission (freedom of the borough) was vested in the municipality itself. But of these bodies the mayor and aldermen, or other governing magistrates, were chosen according to the usage of each particular place.

The business of these corporations consisted in superintending the administrative government, and preserving the peace of the town; managing the corporate funds, which were often considerable; and exercising (by properly appointed officers) judicial functions, in courts both of criminal and (in some instances) civil jurisdiction. The Municipal Reform Act of 1835 effected a most extensive change, by abolishing the exclusive government of the English boroughs, and extending the municipal franchise to occupiers in general. The common council, or deliberative body, the aldermen, and the mayor, are now chosen by open election; the recorder, who exercises the judicial functions of the corporation, and the magistrates, are appointed by the crown.

Courts of Justice.—The sovereign, as head of the executive, is also the fountain of justice. He or she is, by a fiction of law, supposed to be present in courts of justice by the persons of the judges. No court of justice can be erected, except by the commission of the sovereign. This, however, cannot be issued without the authority of parliament.

In early times it was customary for the sovereign to hear and decide cases in person; but this function has been long delegated to judges, whose jurisdiction is regulated by certain established rules, which cannot be altered except by statute. In England, previously to the Revolution, judges held their situations *durante bene placito*, and might be removed by the sovereign; but when this is the case, as it still is in many countries, it would be too much to expect that the judges should manifest much independence in cases in which the crown is concerned. Subsequently to the Revolution it was enacted, in order to provide in as far as possible for the independence of the judges, by the stat. 13 William III, cap. 2, that the commissions of the judges should be made *quamdiu se bene gesserint*; that their salaries should be ascertained and established; and that they should not be removable except by an address from both houses of parliament. Their commissions, however, continued to be vacated by the demise of the sovereign till the accession of George III., when it was enacted that the demise of the crown should no longer vacate the judges' commissions.

But the great security for English liberties, and for the fair and impartial administration of justice, depends not so much on the laudable precautions taken to secure the independence of the judges, and to prevent their being biassed in favour of the crown, as on the institution of juries. In the common law and criminal courts, juries are the only judges of the facts of any case, and they may also decide as to the law. So long, therefore, as the grand institution of jury trial is preserved, and as juries are fairly and impartially selected, there is little to fear from the weakness or corruption of judges. It is the proud distinction of the English people, that they are self-judged as well as self-governed.

Courts of justice are either general or local. The first of these are—1. The courts of common law; 2. The courts of equity; 3. The court of bankruptcy; 4. The ecclesiastical courts; 5. The court of divorce; 6. The courts maritime. To these may be added the courts of assize and of quarter and general sessions, and county courts, which, although each, strictly speaking, is limited to its own locality, are parts of the general system, and subject to the same general principles of law.

Courts of Common Law.—1. The superior courts of common law are three,—the king's or queen's bench, common pleas, and exchequer. Each consists of a chief justice and five inferior or puisne judges—in the last court termed chief and puisne barons. They must be barristers of the degree of serjeant. Their appointment is nominally in the crown, but substantially, like all other appointments, in the minister for the time being. Criminal jurisdiction, and a general power of superintendence over inferior courts, corporations, and magistrates, throughout the kingdom, are reserved to the court of king's or queen's bench. That of common pleas has the exclusive jurisdiction in real actions, now (through the effects of various statutes) becoming obsolete. Suits in matters relating to the king's revenue are mostly determined in the exchequer. With these exceptions, no difference now exists between the authority of the three courts, in either of which ordinary civil actions may be carried on indiscriminately.

From the decision of any one of the three courts, an appeal (by way of writ of error) lies to what is termed the court of exchequer chamber,—a court of appeal, formed by the judges of the two other courts; thus, decisions of the K. B. are reviewed by the C. P. and exchequer, and so forth. This court derives its name from the apartment in which it commonly sits, an appendage of the court of exchequer. From the exchequer chamber, a writ of error lies to the house of lords, the highest appellate authority of the country.

If the inferior courts of common law, of general jurisdiction, those principally deserving of notice are the courts of sessions, held by the magistrates of the several counties, vested, by various statutes, with a civil jurisdiction in certain matters of public interest (such as questions of the settlement of paupers between parishes), and with a criminal jurisdiction, assisted by juries. The nature of the courts of assize and gaol delivery will be best explained when describing the course of the administration of justice.

2. The courts of equity, originally established, as the name implies, to render substantial justice in cases where an injury would be inflicted by abiding by the strict rules of law, are now divided into two: 1. The court of chancery, consisting of two subordinate courts—one presided over by the vice-chancellor, the other by the master of the rolls; and one superior, presided over by the lord chancellor, which in part adjudicates on matters brought before it on appeal from the other two divisions, and has in part an original jurisdiction. The chancellor is also judge of appeal in the last resort from the court of bankruptcy. 2. What is termed the equity side of the court of exchequer, i. e. a court presided over by a single baron of the exchequer, and subject likewise to appeal to the lord chancellor. From a decree of the chancellor, appeal lies only to the house of lords.

3. The court of bankruptcy, as reorganised under the act of Aug. 6, 1861 (24 & 25 Vict. c. 134), entitled 'An Act to amend the Law relating to Bankruptcy and Insolvency in England,' comprises—1. Of six commissioners, who carry on the ordinary legal proceedings consequent on the state of bankruptcy in a trader in the metropolis; 2. Of a court of review, which reviews their judgments, with further appeal to the chancellor. The court sits in judgment on all bankrupts and insolvent debtors, whether traders or non-traders. Previous to the act of 1861, there existed besides an 'insolvent court,' consisting of three commissioners, who sat in London, and also held circuits in the country, for the discharge of prisoners detained in execution for debt, on delivery of their property to creditors under certain statutes. This court ceased its functions in 1862.

4. The ecclesiastical courts have jurisdiction in some civil causes; some that are termed mixed, of which suits for tithes are the principal; and some termed purely spiritual, viz. in the correction of certain offences, both of the clergy and laity. Justice is administered in them according to the civil and canon law. The principal ecclesiastical courts are—1. The provincial courts of the two archbishoprics, of which the court of arches, in that of Canterbury, is the supreme court of appeal; 2. The diocesan or consistorial courts of each diocese; 3. The courts of the archdeacons; 4. Peculiars (which indeed are local courts), of a small exclusive jurisdiction, which are very numerous.

5. The divorce court, the functions of which are implied in its name, was instituted by the Divorce Act of 1857 (20 & 21 Vict. cap. 85). Subsequent statutes (21 & 21 Vict. cap. 108 and 22 & 23 Vict. cap. 61) defined the jurisdiction of this court.

6. The court of admiralty is held before the lord high admiral or his deputy: it consists of the instance court, which takes cognisance of contracts, and injuries on the high seas; and the prize court, which adjudicates on prizes taken in war.

Local Courts, both of criminal and civil jurisdiction, used to be extremely numerous, and were governed by a variety of usages. At present, most of the inferior and local courts have been superseded by the

County Courts, established under 9 & 10 Vict. c. 95, and subsequent statutes. Under their provisions England and Wales are divided into 491 districts, which are classed into 60 circuits. To each of the latter a judge is appointed, who must hold a sitting in each of his courts, at least once a month, for the trial of causes without the intervention of a jury. The jurisdiction of these courts extends to all actions for debt and damage not involving more than 50*l.*; and actions of more importance may be tried in them by consent of the litigants. Appeals may be made to the superior courts of common law on points of law, and as to the validity of evidence in actions for more than 20*l.*; but an action is not removable by certiorari, except by leave of the judge of the county court, and then the claim must exceed 5*l.* By the Bankruptcy Act of August 6, 1861, before cited, the county court judges exercise in the country all the powers of the former district commissioners of the court of bankruptcy.

The machinery of courts in general will perhaps be best understood by the following sketch of the mode in which justice is administered by their means; which, for the sake of brevity, must be confined to the superior courts.

1. If a party have a complaint of civil injury against another, either in a matter of contract, or tort, i. e. civil wrong, such as trespass and the like, (unless for a debt below a certain amount, for which, by various statutes and customs, the plaintiff may sue, if he please, before various local and inferior tribunals—or for certain small trespasses cognisable by magistrates,) he commences a suit in one of the superior courts of common law. The first step in the action is technically termed a writ of summons. If the suit were for a sum certain, the plaintiff had formerly the right to arrest or hold to bail the defendant; but this right is now extinguished, and the ordinary (or 'non-bailable') process substituted for it, except in certain peculiar cases. The writ of summons is followed by a statement of the cause of action, termed a declaration; which the defendant answers by one or more pleas; and these reciprocal allegations are continued (being drawn

up in a technical form, and shown by the one party to the other) until a direct contradiction (technically an issue) is arrived at, either in point of law or of fact. If the former, the case is argued before the court in which the action is commenced, and judgment given; if the latter, the cause is sent to be tried before a jury.

The three courts of common law hold four terms in the year (each of about three weeks' duration), during which the judges of each sit together. In these sittings they decide on issues of law; hear applications in causes already decided by juries, to have them sent down again for what is termed a new trial; set aside, or maintain, the verdicts of juries on grounds of law; and perform other business, which it is impossible here to particularize. The court of K. B. also exercises at this time its appellate jurisdiction over inferior courts.

To try issues of fact, juries are summoned—1. In London and Middlesex, four times a year, before each of the three courts, for a certain number of days during and after each term. A single judge (usually the chief) of the court in which the action is commenced, presides at its trial by the jury. 2. The remainder of England and Wales is divided into seven circuits; two of these (the Welsh) are travelled by a single judge each, who meet in the county of Chester. In the remaining five, two travel together. These circuits are held twice a year—spring and summer—occupying from seven to four weeks. In the course of them, the judges visit every county town. The selection of circuits is left to the choice of the judges according to seniority. They hold several commissions, of which the principal are those technically termed of assize, nisi prius, oyer and terminer, and general gaol delivery. The first of these is now nearly obsolete. By virtue of the two second (through various fictions originating in ancient usage), they hold courts at which juries are summoned to try causes, in the manner before explained, in each county. It is evident, from the foregoing sketch, that the issues of fact in an action are not necessarily tried before a judge of the court in which the action was commenced; but if it be sought to set aside that verdict, or obtain a new trial, application must be made to that court.

Persons are qualified to serve as jurors by the possession of certain species of property; chiefly freeholders of 10l. per annum, and householders of a certain value. There are numerous causes of exemption, which practically extend to all the higher classes of society. Jurors are summoned by the sheriff, on a system intended to take all qualified persons in the county as nearly as possible in rotation; and twelve are selected by ballot from the list of those in attendance for the trial of each cause,—challenges being allowed under certain legal restrictions, but to such an extent as to exclude all individuals who can be fairly supposed to be biassed in favour of either party, or in a situation to hinder them from bringing in a conscientious verdict. Plaintiffs or defendants may, if so inclined, pray for a special jury; persons qualified to serve on which belong to a higher class of society. Witnesses are examined viva voce, in open court. On verdict given, the court pronounces judgment, with damages and costs, according to the principles of law applicable to each case.

Such is the course of an action at common law; but if the question arising between the parties touch on matters of equitable jurisdiction (which, in technical language, is said to extend to trusts, charities, matters of account, fraud, accident, and mistake,) in some cases the preferable, in others the exclusive, mode of obtaining justice, is by application to a court of equity. That applica-

tion is by a suit commenced by bill or information: questions arising in the progress of the suit are determined on petition or motion. Not only the pleadings, as in courts of common law, but the examination of witnesses, are conducted in writing. The judgment of the court is styled a decree. When a doubtful question of fact arises, the judge will sometimes send the question to be tried by way of issue before a jury in a common law court; but he is not bound by its verdict in making his decree.

It is a general principle in courts of law and equity, that all the proceedings in a cause (with some very trifling exceptions) may be carried on by plaintiff or defendant in person; but this is very rarely done, from obvious causes. If not in person, the party can only carry them on by the authorized officers of the court—viz. 1. Attorneys, or solicitors, who are employed in carrying on all or most of the preliminary proceedings; 2. Barristers, or counsel retained by the former to conduct the proceedings in court. Without entering into technical distinctions, it is sufficient to state that barristers (beginning with the lowest order) are classed as—1. Utter, or within the bar, ranking by seniority; 2. Serjeants, a body formerly possessing the exclusive right to practise in the court of common pleas—now confounded in practice with the next, or third class; 3. Counsel within the bar,—to which rank they are admitted by patent either as king's or queen's counsel or of precedency, enabling them to take rank according to the date of their patent. The attorney and solicitor general rank at the head of the bar. These officers are the counsel employed by the crown in various contingencies, and considered as forming part of the administration—going out of office along with it. There are also other classes of practitioners, not necessarily barristers, viz. pleaders, employed in drawing pleadings at common law; and conveyancers, whose business consists in drawing deeds relating to property.

In the ecclesiastical and admiralty courts, the pleadings are according to forms derived from the civil law; evidence is documentary. The duties of the attorney are executed by officers styled proctors; and the counsel are doctors of civil law, graduates of the universities.

Criminal Process.—Crimes are divided by the ancient customary law of England into treasons, felonies, and misdemeanours; the latter being generally offences of inferior importance (such as breaches of the peace, riots, and attempts to commit certain other offences), are punishable by fine or imprisonment only. Parties suspected of criminal acts may be apprehended on the warrant of a justice, granted only on the sworn testimony of one witness at least, directed to the constable or other peace officer of the district; but any one may lawfully arrest one who has committed felony, or breach of the peace, in his presence. The offender is then carried before a justice of the peace. Unless the case be one of those minor offences for which the justice has power to punish on summary conviction, without the aid of a jury, the party charged is committed to gaol, or admitted to bail, according to the nature of the offence. He is committed to take his trial in most cases at the next ensuing sessions of the peace (either in boroughs or counties), or at the next gaol delivery, by the judges at the assize, whichever may happen first; but capital, and in general the most serious, class of offences are tried at the assize only. In Middlesex and certain adjoining parts, offences are now tried by the Central Criminal Court, which sits twelve times a year at least, and is usually attended by two or more judges of the superior courts, and the judicial authorities of the city of London. The prosecution is then carried on, in the name of the king, by indictment before the grand jury. This body, consisting of from 12 to 23 persons (at the assizes, persons of rank in the county; at the sessions, persons of somewhat inferior station), receives all indictments, and hears the evidence on the part of the prosecution. If the indictment be dismissed, it is returned to the court with the endorsement 'no bill,' and the accused is free. If the evidence appear to them *primâ facie* satisfactory, the bill is said to be found, and the prisoner or defendant is put on his trial. The grand jury is also summoned to find bills against parties not in custody or on bail for offences for which there is no previous arrest, such as perjury; and these are tried at the ensuing gaol delivery. There is also, in certain offences, chiefly of a public nature, a mode of proceeding by *information*, which supersedes the necessity of an indictment.

The accused, when brought into court under this preliminary process, is arraigned before a petty jury, summoned in the same manner as the jury in civil causes just described. If he plead guilty on arraignment, his plea is recorded, and judgment given. If he plead not guilty, the trial proceeds. There are also certain pleas in bar, or defences to the prosecution of a technical nature, rarely resorted to, as the accused by pleading them waves the trial by jury. The witnesses are then heard; and if the jury find the prisoner 'not guilty,' he is released; if 'guilty,' he is convicted, and judgment passes. A judgment may be reversed for error of law by the superior court; and pardon may be granted, either by act of parliament or by letters patent of the sovereign, under the great seal. Pardon, and remission of part of the sentence, is, in point of fact, obtained through the agency of the Home Office. The sheriff is the officer to whom the execution of the sentence of the law is entrusted.

The criminal law of England might formerly, perhaps, have been justly characterised as sanguinary; but in this respect a great change has been effected within these few years, and capital punishments are now never inflicted except for murder. Among the secondary punishments, transportation long occupied a prominent place, but a notion had latterly been gaining ground unfavourable to its efficiency, and it has now been relinquished. It was, indeed, no longer possible to maintain it by sending criminals to Australia; for that, instead of being a punishment, was a favour to the wrong-doers.

Subjoined is a table which shows the total number of criminal offenders committed for trial, convicted and acquitted, in the fifteen years, 1849 to 1863, in England and Wales.

Years	Committed for Trial	Convicted	Acquitted, or otherwise of Persons charged and discharged, or detained
1849	27,816	21,001	5,766
1850	26,813	20,537	4,316
1851	27,960	21,579	4,328
1852	27,510	21,304	6,174
1853	27,057	20,756	6,261
1854	29,359	22,007	4,774
1855	25,972	19,971	5,367
1856	19,437	14,734	4,673
1857	20,269	15,307	4,777
1858	17,855	13,246	4,316
1859	16,674	12,470	4,175
1860	15,999	12,044	3,907
1861	14,726	13,579	4,603
1862	20,001	15,317	4,651
1863	20,318	16,729	4,906

The number of women committed for trial is, on the average, about one-fourth that of men. Among the 20,818 individuals committed for trial in 1863, were 16,461 males and 4,357 females.

Church of England.—The sovereign is head and supreme governor of the national Church of England; has the right to assemble, prorogue, and dissolve all synods and convocations of the clergy; is the ultimate judge of appeal in ecclesiastical causes (an authority exercised by the lord chancellor); and has the nomination to bishoprics and some other ecclesiastical preferments.

The clergy of the Church of England are divided into three degrees or orders—bishops, priests, and deacons. There are two archbishops and 24 bishops within the realm of England. They are nominated to their respective dioceses by the crown; the election being by a writ of *congé d'élire*, or licence to elect, addressed to the dean and chapter of the diocese, accompanied by a letter from the sovereign, directing them to elect a certain specified individual. By the custom of the church, every candidate for holy orders must be examined and approved by a bishop. The bishop has episcopal jurisdiction in his court in ecclesiastical matters, and the general superintendence over the clergy. An archbishop is the chief of the clergy in his province; has the inspection of the bishops and inferior clergy; and exercises an appellate jurisdiction from the episcopal courts.

The archbishop of Canterbury is the primate of all England. He has within his province the bishoprics of Canterbury, Rochester, London, Winchester, Norwich, Lincoln, Ely, Chichester, Salisbury, Exeter, Bath and Wells, Worcester, Lichfield, Hereford, Llandaff, St. David's, Bangor, St. Asaph, Gloucester and Bristol, Peterboro', and Oxford. He has the privilege of crowning the kings of England. He is the usual channel of communication with the crown or the ministers on constitutional questions affecting the interests of the church. The archbishop of York's province consists of the six northern counties, with Cheshire and Nottinghamshire; and includes the bishoprics of York, Chester, Durham, Carlisle, Ripon, Manchester, and the Isle of Man. He has the privilege to crown the queen consort, and to be her perpetual chaplain. The archbishops are the chiefs of the clergy in their provinces, and have within them the inspection of the bishops, as well as of the inferior clergy, for which purpose they make their visitations, which are now, however, practically episcopal, not archiepiscopal, and made only as bishops within their own dioceses. They have, assisted by at least two other bishops, the confirmation and consecration of the bishops. They have also each his own particular diocese, wherein they exercise episcopal, as in their provinces they exercise archiepiscopal, jurisdiction. As superior ecclesiastical judges, all appeals from inferior jurisdictions within their provinces lie to them. They have also each a court of original jurisdiction. They have power, by stat. 25 Hen. VIII. c. 21, but now only exercise in most accustomed occasions, of granting dispensations. This power is the foundation for the grant of special licences to marry, to sanction the holding of two livings, now restricted to the Archbishop of Canterbury. The bishop is the chief of the clergy in his diocese. He has the power of ordaining priests and deacons, of consecrating churches, of confirming the baptized, of granting licences to marry, and of visiting and inspecting the manners of his clergy and people. The bishop is also an ecclesiastical judge; but he appoints a chancellor to hold his court for him, and assist him in matters of ecclesiastical law. In case of complaint against a clerk in holy orders, he is empowered by the Church Discipline Act (3 & 4 Vict. c. 86) to hold a court in his own person, assisted by three assessors. After the archbishops, the bishops of London, Durham, and Winchester have respectively precedence; and then the bishops of both provinces, according to their seniority of consecration, or translation to an English see from that of Sodor and Man, which ranks lowest. Colonial bishops of the established church have been appointed by the crown in forty-two of the principal British colonies. By stat. 59 George III. c. 60, the archbishops of Canterbury and York and the bishop of London are permitted to certain persons specially to reside and officiate in the British colonies. The discrepancy that prevailed in ancient times in the size of bishoprics, though somewhat diminished by the erection of new sees at the Reformation, has continued down to the present time, and the inconveniences thence resulting have been greatly augmented by the wonderful increase that has taken place since 1760 in the population of certain districts compared with others. To remedy this evil to some extent, parliament appointed a committee in 1834, which recommended that two new bishoprics—those of Manchester and Ripon—should be formed in the principal manufacturing districts, chiefly out of territories included in the dioceses of York and Chester. The commissioners also recommended that, saving the rights of the (then) existing incumbents, the bishoprics of Gloucester and Bristol should be united, and the bishoprics of Sodor and Man suppressed. They recommended further that, according as opportunity offered, sundry deductions should be made from the revenues of the sees of Canterbury, York, London, Durham, and Winchester; and that the surplus revenue so arising should be formed into a fund for the endowment of the two new bishoprics, and for raising the income of the poorer class of sees to from 4,000l. to 5,000l. a year. These recommendations were confirmed and carried out in most particulars by the act 6 & 7 Will. IV. c. 77, and by the orders in council issued under its authority. The income of the bishop of Durham was reduced in 1836; and, in the course of the same year, Ripon was formed into a bishopric. The sees of Gloucester and Bristol have also been united. The bishopric of Manchester was formed in 1847.

Every diocese has a chapter, consisting of a dean and a certain number of canons and prebendaries. The chapter is often styled the council of the bishop; but it exercises, in point of fact, no sort of interference with the ecclesiastical jurisdiction, or with the general superintending authority of the bishop. The chief duty of its members consists in maintaining the constant celebration of divine service in the cathedral church. Deaneries are in the gift of the crown; some by the form of election by the chapter (as in the case of bishops), others by the king's letters patent. The canons are variously appointed—by the crown, by the bishop, or by election among themselves. Besides the chapters in cathedral churches, there are also chapters in a few others, which are styled collegiate churches.

Archdeacons are church officers, appointed (in most cases) by the bishops for their assistance in various matters connected with the superintendence of the diocese.

For the management of ecclesiastical affairs, the provinces have each a council, or convocation, consisting of the bishops, archdeacons, and deans in person, and of a certain number of proctors, as the representatives of the inferior clergy; each chapter, in both provinces, sending one; and the

parochial clergy of each diocese in the province of Canterbury, and of each archdeaconry in the province of York, sending two. These councils are summoned by the respective archbishops, in pursuance of the queen's mandate. When assembled they must also have the queen's licence before they can deliberate, as well as the sanction of the crown to their resolutions, before they are binding on the clergy. In the province of Canterbury the convocation forms two houses; the archbishop and bishops sitting together in the upper house, and the inferior clergy in the lower. In the province of York all sit together in one.

Parsons are the incumbents of parish churches. They must be priests; and derive their title by presentation, induction, and institution. They are termed rectors or vicars; the former being such as are entitled to the whole tithes of the parish, the latter only to a certain portion. The number of parochial benefices in England and Wales amounts to about 12,000, besides which there are 200 extra-parochial places. The advowson, or right of presentation, to about one-half the benefices is in the hands of private owners; the remainder belong to the crown (of which the patronage is exercised, as respects livings of inferior value, by the chancellor), to archbishops and bishops, ecclesiastical corporations, and universities. The residence of incumbents in their benefices, and the restriction of the right to hold more than one benefice, have been the objects of a variety of regulations both in canons and statutes. Incumbents may be deprived either by sentence in the ecclesiastical courts for particular offences, or in pursuance of certain penal statutes. Curates are likewise priests, licensed by the bishop of the diocese, and nominated to serve cures. Stipendiary curates are such as are appointed by rectors, either to supply their place in case of non-residence, or to assist them; whose salary is regulated by statute, or episcopal authority. Perpetual curates are appointed to churches in which there is neither rector nor vicar; or to chapels of ease, parochial chapels, and free chapels, that is, district churches in large parishes.

The order of deacon, in the constitution of the English church, serves merely as a necessary preliminary to that of priest. By the canons of the church no bishop can admit any one to holy orders, 'who is not of his own diocese, except he be of either of the universities of this realm, or except he bring letters dimissory from the bishop of whose diocese he is.'

The canons of the Church of England were made by the archbishop and clergy of the province of Canterbury convened in convocation in 1603, and ratified by James I. They have not been established by act of parliament, and consequently are binding on the clergy only.

The revenues of the church are derived partly from land, and partly from tithes. The latter formed the original endowment of every parochial church. But a very large proportion of them fell gradually into the hands of ecclesiastical corporations; and a part of them again, at the dissolution of monasteries, into the hands of private individuals. Out of the 10,500 benefices, more than 8,000 have had their 'great' tithes, or those of corn, wood, &c., appropriated or impropriated: in most of these instances, however, the 'small tithes,' as they are termed, or those of fruit, milk, pigs, and such like articles, are reserved for the maintenance of the church. Nearly a third part of the land of England and Wales is wholly tithe-free, owing to exemptions enjoyed in former times by religious houses. Tithe is now, by an act passed in 1837, under a course of commutation for an invariable corn rent, to be converted into money, at the price of the day.

Although the Church of England be still recognised as the national establishment, the exclusive privileges formerly enjoyed by its members, and, indeed, all legal distinctions between different classes of subjects on account of religious opinions, have, by a series of changes, been nearly abolished. The chief remaining rights, privileges, and liabilities, which respect the church with the state, are nearly as follows:—

1. The headship of the king: as a necessary consequence of which the sovereign himself must be a member of the national church. This headship, all persons taking certain offices are required to recognise, by the oaths of abjuration and supremacy, for which a declaration is substituted in the case of Roman Catholics. Roman Catholics are also specifically excluded from the office of chancellor, and a few other high dignities.

2. The form of public prayer and administration of the rites of the church, its articles of belief, and various points in its discipline, originally settled by convocation, are established by the authority of parliament.

3. The archbishops and bishops sit and vote in the House of Lords.

4. Although the free enjoyment of their different forms of worship is now guaranteed to all Christian dissenters, and that of others (as Jews) tacitly tolerated, there are still some legislative provisions respecting them, by which the superiority of the established church is recognised. Thus, Roman Catholic archbishops and bishops are forbidden to assume the titular dignities of their respective dioceses; and public functionaries are forbidden to attend dissenting places of worship, with the insignia of their office.

5. The clergy of the Church of England have long acted as officers of the civil power, in the character of registrars of births, marriages, and deaths; but the late act, by establishing a new system of registration, has materially altered their position in this respect, and their exclusive authority is now taken away.

A great deal of discussion has taken place at different periods with respect to the right of parliament to interfere with the property and revenues enjoyed by the church. It is now generally admitted, that parliament is entitled to alter the distribution of the church revenue; but it is contended by many that it has no right to take away any portion of such revenue. But a pretension of this sort is totally inadmissible. Whether it would be wise and proper to make any such diversion is a matter dependent on circumstances, and to be judged of at the time; but there is no principle or right of any kind to hinder parliament, should it be so disposed, from dealing with church property as it would deal with anything else. An established church is neither part nor parcel of religion: it is a mere human institution, with functionaries appointed and paid by the state; and should parliament be honestly impressed with the conviction that the great interests of religion and morality will be better promoted by diverting a portion of the church property to other purposes, it is not entitled merely, but it is its bounden duty, so to divert it. The rights of existing incumbents ought, of course, to be protected; but provided this be done, parliament is quite as much entitled to remodel the church, and dispose of its property, as it is to remodel the army or the navy, or to disband a regiment, or pay off a line of battle-ship.

Dissenters from the Church of England are now, after more than a century of struggles, placed entirely on an equal footing with its members in respect of political rights and privileges. The dissenters consist principally of,—

1. The Roman Catholics, who have increased, chiefly through the immigration of Irish labourers, from 60,000 to about 2,000,000 since the accession of George III. 2. The members of what are commonly called the three denominations—Presbyterians, Independents, Baptists: of these, the first, since the period of the civil wars, when for a short time they had political power and the revenues of the church in their hands, have rapidly diminished. Many of their churches have become Unitarian. The Independents or Congregationalists are so termed from asserting, as their fundamental principle, the independence of each separate congregation. They are numerous, and have, for the most part, retained the fundamental doctrines professed by the great majority of Christians. The Baptists are divided into general (or Arminian) and particular (or Calvinistic). 3. Of the Methodists, there are likewise two principal divisions. The Wesleyans, the most powerful and important, whose origin was about a century ago, now number about a million and a half of members. Their dissent from the church is less complete than that of other sects. The Calvinistic Methodists are chiefly established in Wales. 4. The Quakers are more remarkable for the singularity of their tenets and observances (although their strictness in the latter appears to be on the decline), than for their numbers. 5. Jews are not numerous in England; but are supposed to have augmented considerably of late years.

No information regarding the number of persons belonging to the episcopal church and those adhering to other religious creeds in England is given in the last official census. It appears, however, from the returns of the registrar general that, in the year 1851, out of a total number of 163,706 marriages, 100,897 were solemnised according to the rites of the established church. Of the latter number 102,855 were after publication of banns; 29,050 by licence; 4,048 by superintendent registrar's certificate; and 16 by special licence. But this statement does not represent the real numbers with perfect accuracy, as 5,588 marriages were not distinguished in the registers in respect to these particulars. The number of marriages performed otherwise than agreeably to the forms of the established church was 33,809. Roman Catholic marriages were, 7,782; those in the registered chapels of other religious denominations, 13,182; those of Jews, 262; of Quakers, 58; while marriages contracted in superintendent registrars' offices were 11,724. In 1851 the marriages in the established church were about 131,000, and in 1861 they were nearly the same number. In 1851 those not performed in the established church were about 21,000; in 1861 they were 83,000. It appears from these figures that an increase of 9,000, which the total marriages in 1861 exhibited, as compared with those in 1851, was appropriated by persons who married according to other rites than those of the established church.

In 1861 there were in England and Wales 4,561 buildings belonging to Roman Catholics and dissenting denominations, and registered for the solemnisation of marriages. A third part of that number belonged to Independents, 1,000 to Baptists, 895 to Wesleyan Methodists, 551 to Roman Catholics, 193 to Calvinistic Methodists, 175 to Unitarians, 137 to Scottish Presbyterians, and 141 to various other bodies who have not yet acquired a numerical importance.

The number of Roman Catholics in England has greatly increased within the last 50 years. The late Cardinal Wiseman stated at the Congress of Malines, Aug. 25, 1863, that, in the year 1830, there were, in England and Wales, 434 priests; and that in 1863 there were 1,242. In 1830 the churches were 410; in 1863 they were 872. There were 16 convents in 1830; the number has arisen in 1863 to 162. In 1830 there were no houses for religious men, but in 1863 there were eleven. In 1863 the number amounted to 53. Another report—in the 'Catholic Directory,' Lond. 1864—gives the following statistics regarding the number of Roman Catholic priests, churches, and communities in Great Britain:—

	1863	1864	Increase
Roman Catholic Clergy in England	922	1,267	345
Do. in Scotland	131	178	44
Total	1,056	1,445	843
Churches and Stations in England	678	907	229
Do. in Scotland	131	191	67
Total	812	1,098	286
Communities of Men in England	17	46	29
Convents in England	84	172	88
Do. in Scotland	—	12	12
Total	84	184	107
Commissioned Army Chaplains	—	18	18

The present Roman Catholic population of Great Britain is estimated at 2,000,000.

Public Education.—In England no system of public instruction has been established by authority of the legislature. Schools have, however, been established in most parishes, and very large sums have been left by private individuals for the purpose of supplying gratuitous instruction. Almost all the grammar schools in the kingdom owe their origin to this source; and there is, perhaps, no country in which so great an amount of property has been appropriated for the education of youth. Many of these bequests have not been subjected to any controlling authority, so that they have not unfrequently been embezzled and diverted to other purposes than those for which they were originally destined. Still, however, the amount of property applicable to educational purposes in England is very large. It is believed that, at this moment, the incomes of the estates and other property left for educational purposes would amount, if properly managed, to about 400,000l. a year. But it is well known that the management of such property is far from efficient; and the utility of the funds that are realised is greatly impaired by the conditions and restrictions under which they are applied.

The grammar and endowed schools appear to have been principally intended for the use of the upper and middle classes, especially the latter; and it was not till a comparatively recent period that any vigorous effort was made to supply the lower classes with education, or to bring this most important instrument of civilisation and advancement within the command of the children of the poor. But during the present century a great many schools have been founded, having this object in view. These consist principally of what are called National, British, and Foreign, and Sunday schools. The first, under the control of the National Society, are conducted on the system recommended by Dr. Bell of Madras, and use the catechism of the Church of England, with which they are closely connected. The schools of the British and Foreign Society are not connected with any religious sect, but are open to all pupils of whatever creed. Sunday Schools, so called from their being open only on Sundays, belong to all denominations of dissenters, as well as to the Church of England.

Subjoined is a table giving an account of the primary schools in England and Wales, showing the number visited by the government inspectors, the accommodation for the children, and the number of children present at inspection, in the ten years 1854-63:—

Years (ended 31st August)	Number of schools inspected	Number of children that can be accommodated	Number of children present at inspection
1854	3,147	511,174	410,304
1855	3,851	704,193	683,814
1856	4,237	766,137	552,778
1857	4,444	841,213	602,357
1858	5,465	1,061,597	695,311
1859	5,831	1,051,913	757,182
1860	6,032	1,155,777	859,271
1861	6,749	1,215,747	879,884
1862	6,113	1,252,540	904,154
1863	6,777	1,313,588	912,693

It is sometimes said that, speaking generally, education in England is of an inferior description; and that, in point of quality, it is below the standard of Prussia, Holland, and some other countries. It is doubtful whether this be the fact; though at the same time it cannot be denied that it is to the freedom of political institutions, and the scope given to talent and enterprise to elevate their possessors in the scale of wealth and distinction, and not to educational systems, that the progress made by Englishmen, and the triumphs they have achieved in all departments of industry, science, and literature, are to be ascribed.

The superior grammar schools, and the two great universities of Oxford and Cambridge, are especially appropriated to the education of the higher classes. An account of them will be found under the articles CAMBRIDGE, ETON, OXFORD, &c. The London University is, strictly speaking, only a board authorised to examine individuals educated at certain places, and to grant degrees to qualified persons.

Poor Laws.—A compulsory provision for the support of the poor has long existed in England. It grew out of the impotent attempts made in the reigns of Henry VIII., Edward VI., and the earlier part of that of Elizabeth, to suppress mendicancy, and at the same time to provide for the poor by voluntary contributions. At length, the earlier statutes on the subject were consolidated, and the principle of compulsory provision carried to the fullest extent by the famous statute of the 43 Eliz. c. 2, which enacted, that all maimed and impotent persons should be provided for at the expense of their respective parishes, and that employment should be found for the unemployed able-bodied poor. From this remote period, the law of England has regarded every parish in the light of a family, the richer members of which were bound to provide for those who, through inability, misfortune, or want of work, could not provide for themselves. This also is the principle embodied in the law of Scotland with respect to the poor; and provided the means for carrying it into effect be so contrived that indigroce and suffering may be relieved, without at the same time encouraging indolence and vice, the system would seem to be quite unexceptionable. Practically, however, this has been found to be a problem of exceedingly difficult solution, and not a few have concluded that, however administered, all systematic attempts to relieve the poor are necessarily, in the end, productive of increased want and misery.

The poor, no doubt, are naturally anxious that the compulsory provision for their support should be raised to the highest limit, and that their necessities should not only be relieved, but that they should be able, without moderation, to eat the bread of idleness. But wherever the assessment and administration of the provision for their support is left to the care of those on whom the burden of its payment really falls, this tendency to abuse is not long in being effectually provided against, and the sustaining and beneficial influence of the system alone remains. The complicated code of laws respecting settlements, and the establishment of workhouses, owes its origin to this principle—to the wish of the legislature to relieve the poor, and, at the same time, to prevent the abuse of the rates; and there is unquestionable evidence to show, that, from the establishment of the system in 1601 down to about 1780, the devices in question were effectual for their object; and that while poverty was relieved, no encouragement was given to sloth, or to early and improvident unions.

But soon after this period various innovations were made on the old law, which broke down most of the securities against the abuse of the rates; and, in 1795, the principle was adopted, of mixing together wages and poor-rates, and of eking out what was supposed to be a deficiency in the former by payments from the latter. In consequence of this subversion of the principle on which the poor rates had been previously administered, they began rapidly to increase, and threatened to swallow up the whole, or, at least, a very large part of the surplus produce of the land. Various devices were resorted to, with the view of checking the evil; but not one of them had for its object to revert to those practices and mode of administering the law, which the experience of more than 230 years had shown were fully effectual for the prevention of abuse. At length the Poor Law Amendment Act was passed in 1834, which introduced a totally new system for the administration of the poor laws. Under this act the country has been divided into unions of parishes, according to circumstances, the administration of all matters relating to the poor in these unions being entrusted to a board of guardians elected by the rate-payers. These guardians are themselves controlled by, and, in fact, are merely the executive officers of a central board of three commissioners established in London, who have power to issue rules and regulations as to the management of the poor, which all guardians, and other inferior officers, are bound to obey. The central board is assisted by deputy commissioners, who attend at meetings of guardians, explain the law, and adjudicate or report upon extraordinary cases, and see that the rules laid down by the central board are complied with.

To the Poor Law Amendment Act of 1834, some additions, having for principle the enlargement of the respective unions, and the more equal spread of the burthens of taxation, were made in 1861.

Subjoined is a table of the sums expended for the relief and maintenance of the poor of England and Wales at different periods since 1748, with an estimate of the pop. at these periods:—

Years	Annual expended on Poor	Population
Average	£	
1748, 1749, 1750	690,971	6,890,000
1775, 1776	1,530,000	7,000,000
1783, 1784, 1785	2,004,239	8,000,000
1801	4,017,871	9,187,000
1813	6,656,100	10,150,000
1821	6,959,249	11,978,000
1831	6,798,888	13,897,000
1841	4,760,929	14,770,000
1849	5,792,963	17,516,000

The following table shows the number of paupers (exclusive of vagrants) in receipt of relief in the several unions and parishes under boards of guardians, in England and Wales, on the 1st of January in each year:—

(1st Jan.) Years	Number of Unions and Parishes	Number of Paupers		
		Indoor	Out-door	Total
1849	590	119,778	814,014	821,419
1850	606	118,659	641,964	920,583
1851		110,665	761,728	860,893
1852		106,418	729,811	835,874
1853	608	104,196	654,586	759,672
1854	670	112,276	702,661	814,327
1855	674	171,563	729,500	851,063
1856	626	173,597	742,156	871,757
1857	624	159,702	724,476	849,576
1858	629	126,441	743,705	860,146
1859	641	174,885	737,164	900,970
1860	646	110,026	731,904	841,720
1861	644	130,961	759,662	890,173
1862	619	141,191	807,973	848,164
1863	653	146,197	976,427	1,142,671
1864	653	127,363	844,900	981,263

Population in 1851 of 621 unions and parishes, 16,960,441

		479		16,670,739
		617		17,46-,577
		644		17,610,845
	1861	610		19,814,000
		633		19,976,000
		623		19,048,000

The year 1849 is the first year for which the actual number of persons receiving relief on a given day can be returned.

Public Amusements.—There are few things, probably, in which national character and habits are displayed more truthfully than in popular sports and amusements; and though none of these be in any way associated amongst us with civil or religious polity, as in ancient Greece and Rome, and some modern Catholic nations, they are still of sufficient importance to justify and require a short notice. Field sports comprised almost the whole pastimes indulged in during the early period of our history; they were materially modified by the game laws introduced at the Norman period, and which have descended down to our own times. Many generations have passed since the chase was ministry of war; but so far as danger and excitement are concerned, fox-hunting and steeple-chases may be considered as substitutes for the chase of the wolf and the boar. Archery ranked amongst the most popular and important of the old English sports, and constituted the peculiar boast of the ancient yeomanry, as is shown by the ballads, that form so peculiar and valuable a portion of our earlier literature, and which give by far the most faithful and striking illustrations of this and other matters connected with the habits and manners of the commonalty. By their means the fame of the outlaw, Robin Hood, has already outlived that of many a legitimate hero, and bids fair to outlive that of many more. The wild boar was an object of the chase down to the Stuarts, and deer may, in a limited sense, be still considered as such. The fox, hare, and otter are at present the only wild animals that can in a general sense be said to perpetuate the chase. Fox-hunting may, in fact, be said to be in an especial manner the out-door sport of the country gentlemen of England at the present day. Coursing and falconry were sports restricted to those of gentle blood; the latter has long been obsolete, in any popular sense, though, like archery, it is indulged in by a few individuals. Partridge and grouse shooting are universally popular amongst the country gentlemen; and fishing is also extensively practised. Amongst the lower classes wrestling is one of the most popular of the out-door sports: it is followed principally in the northern and western cos., but the modes are essentially distinct, that of the former more nearly resembling the fashion of the ancient athletæ. Quarter staff and boxing must also be considered as peculiarly English in their character; but prize-fighting appears to have got somewhat out of fashion, and the same may be said of bull-baiting and cock-fighting, once so general, with other sports of a like brutalising character. Horse races are of comparatively modern origin, and have attained to the acme of popularity with all classes, from the highest to the lowest. The races at Epsom, Doncaster, Ascot, Goodwood, and other places, attract vast crowds of visitors; and nowhere is there to be seen such a display of magnificence, numbers, good humour, and love of enjoyment, as may be witnessed at these meetings. Cricket grounds and bowling greens maintain their popularity amongst the more respectable classes, as those for skittle playing do amongst the lower. Boat-racing may be noticed as a popular amusement that seems to be gaining ground; and the regattas, or boat-matches, that are now annually got up in several seaport towns, are very well attended. The holyday fair or wake comprises most of the various amusements in vogue, and is usually annual: some of the more ancient of these appear to have had a religious origin, others had business, wholly or partly, in view: as popular festivals, they are all fast degenerating, and bid fair to disappear altogether, at no very distant period. In-doors, billiards have replaced the ancient shuffle-board, and cards, though still pretty general, are rapidly giving way to music and conversation: many minor customs, once universal favourites, might be named, that have wholly disappeared, and though the holly and the mistletoe are still seen in the majority of houses at Christmas, few other of the older customs or sports can be said to survive, in any state of vigour: even blindman's buff and hunt the slipper are now but rarely practised, and never in genteel society. Dancing keeps its ground, changing its character according to the changes of fashion. The mutations that have occurred in dramatic representations from the time when they were associated with the festivals of religion, and chiefly limited to scriptural subjects, down to the allegorical masques of B. Jonson, which Milton's 'Comus' shows lingered on to a later period, form a wide field for investigation. The Christmas mumming (that may still, perhaps, be witnessed in some remote corners, improving the stories of St. George and the Dragon, or Fair Rosamond) seems a lineal descendant of the old mysteries: the mountebank and his merry Andrew, and Punch and Joan, appear to have made their final exit from the scene. It would require far more space than can be afforded, to mark the various changes in what has been called the legitimate drama, whose master-pieces were produced in the earlier part of its career during the reigns of Elizabeth and James I. The drama has, at the present time, lost the greater part of its predominating influence, for which two causes may be assigned,—the theatres have ceased, owing to the greater lateness of the dinner hour, to be the resort of fashion, whose influence in England is all but omnipotent; and they are monopolies which, in the long run, always have a deteriorating influence over literature as well as industry.

In so far as the metropolis may be considered an index to the community at large, music would seem to be rapidly advancing in popular favour; nor can it escape notice, that the means taken to popularise sculpture and painting are fast neutralising the propensities which naturally resulted from want of perception and sympathy with works of art. The results of all this are sufficiently obvious, in the more tranquil and refined tone that pervades most places of public resort, and in the greater taste for country excursions, reading, and quiet amusements.

Language.—The *English Language* is principally of Anglo-Saxon origin. The Gothic or Belgic occupants of the lower and more fertile parts of the country, at the epoch of the Roman invasion, and the Saxons by whom the country was overrun subsequently to the withdrawal of the Romans, were congenerous races; so that the language of the latter easily became that of the people generally. After the conquest of the kingdom by William duke of Normandy, the Norman language became that of the court, the nobility, and the courts of law; but the influx of Norman settlers was too inconsiderable to have any material influence over the language of the bulk of the people, which continued to be essentially Saxon. In the course of time, the prejudices of the English monarchs of the Norman line in favour of their continental dominions and subjects, lost much of their influence, and being bred in England, they began to become familiar with, and use the language of, their English subjects. In 1216, Henry III. issued a writ in the Anglo-Saxon tongue. Not long after, English was substituted instead of French, in viva voce proceedings in the courts of law; and though written proceedings in courts were carried on for a considerable period longer in a barbarous jargon, half French and half English, the latter was ordered by statute, in the reign of Edward III., to be thenceforth exclusively used in the courts. It is curious, however, that, notwithstanding this enactment, the statutes continued to be promulgated in French down to the reign of Richard III.

During all this period the Anglo-Saxon was gradually acquiring the distinctive characters of the English language. A great many French (and consequently Latin) words, through our long-continued connection with France, successively found their way into the language; and it also was enriched by words derived from the Greek and other tongues. Wycliffe and Geoffrey Chaucer, who flourished in the 14th century, may be regarded the one as the father of English prose, and the other of English poetry. During the 15th century, the language advanced but little; but, previously to the close of the 16th century, it had, in a great measure, attained to the form and standard which it now exhibits. During the reigns of Mary and Elizabeth, many additional French, and some Italian and Spanish words were introduced. Still, however, of about 38,000 words, which, excluding the preterites and participles of verbs, the English language is supposed to comprise, about 21,060 are of Anglo-Saxon origin; and the remainder Latin, Greek, and French, in different but uncertain proportions.

It is foreign to our subject to enter into any details as to the works that have been written in English. Suffice it to say, that there is no department of literature, philosophy, or science, in which English writers have not attained to high, and, in not a few instances, to unrivalled excellence. For a lengthened period, our philosophical and political literature has had much more of a practical than of a theoretical or speculative character; and the taste for metaphysical inquiry has almost entirely disappeared. Within the last few years the great object has been to diffuse literature, and to secure the suffrages of a wide, rather than of a select circle of readers. The influence of this change on the character of our literature, and the taste of the public, cannot yet be fairly appreciated; perhaps the former will gain in clearness and lose in depth; and it is not impossible that a lower standard may be formed of philosophical and literary eminence.

Condition of the People.—Some remarks have already been made illustrative of the improved condition of the great bulk of the people in the present times, as compared with their condition at more remote periods. Speaking generally, all classes are now incomparably better fed, better lodged, and better clothed, than at any former epoch in our history. The increase in the consumption of butchers' meat since 1770 has been more than double as compared with the increase of the population; and the increase in the consumption of tea, sugar, coffee, &c., since the same period, has been quite unprecedented. In fact, the poorest individuals are now in the daily enjoyment of many descriptions of luxuries that were, no further back than the 17th century, unattainable even by the richest lords. Tea and sugar are now luxuries necessaries of life; every cottage is well furnished with glass windows, and maids of all work are now quite as well and neatly dressed as the duchesses that figured at the court of Queen Anne. It is not, however, to be denied, that, notwithstanding this signal increase of prosperity, considerable distress exists among certain classes—especially among the agricultural labourers. The depressed condition of this important class of the population appears to be owing to a variety of causes, which, however, may all, or nearly all, be traced to the one great evil of want of education. However, this is an evil resulting from year to year, and, without indulging in too sanguine anticipations, it may be said that, provided tranquillity, good order, and that perfect security essential to all great undertakings, be maintained at home, the fair presumption is, that the prosperity of the country will go on increasing for a very long period, and that England will indefinitely maintain the proud distinction of being the richest, most industrious, and happy of European nations.

Historical Sketch.—After all the discussions which have taken place respecting the history and character of those native tribes which occupied Britain at the period when the Romans first reached its shores, thus much only appears to have been determined with any degree of probability, that they belonged partly to that great family of the human race called the Celtic, and partly to another great family called the Gothic. To the former belonged the Cymry, or inhabitants of Wales, and of the Western, and, perhaps, also, the Northern counties: to the latter, or Goths, belonged the Belgæ, who, having emigrated from the Continent, occupied the eastern, lower, and most fertile portions of the country.

The visit of Julius Cæsar to Britain occurred 55 years before Christ. From that time it remained unmolested by the Romans for nearly 90 years. In A. D. 43, Aulus Plautius, despatched by the emperor Claudius, began its conquest, which, in the space of about 40 years, was completed, with the exception of the northern part of Scotland, into which the Romans scarcely penetrated.

The Romans introduced, to a great extent, their arts and civilisation into this remote province. Thirty-three large towns, and many military stations, were connected together by magnificent roads, constructed by the labour of the Roman soldiers and provincials. It is probable that, between those several centres of civilisation, much of the country remained in that state of forest to which the Romans had found it. Still the population of Roman Britain must have been large, and its progress in refinement considerable, for two centuries after the conquest. After that time, the declining power of Rome yielded to the fierce attacks of the northern tribes of the island, and Britain became, to a certain extent, independent of the empire, but only to suffer the more from these fierce assailants.

Of the history of the long period which elapsed between the retirement of the Roman armies from our island and its conquest by the Saxons, we possess no memorials sufficiently authentic to form a connected narrative. It appears probable that the hereditary chiefs of the ancient British tribes, who had lost their authority during the period of colonial government, resumed it to a certain extent: that in the larger towns, the elected order, together with a council of magistrates and citizens, exercised almost republican authority. The exact era of separation from the empire of Rome cannot be fixed; it seems to have been effected A. D. 409, when the letters of the emperor Honorius commanded the cities of Britain to 'provide for their own defence.' About forty years later, we find no distinct trace of municipal government left; and the country under the government of a number of petty chieftains or kings, and overrun even to the extreme north by the incursions of the Caledonian tribes. At this period (A. D. 449), Hengist and Horsa, Saxon leaders, ranging the coast of the British Channel with three of their piratical vessels, were invited by Gurtheyrn (Vortigern), a British prince of Kent, to serve against three northern invaders. Five thousand auxiliaries soon arrived; quarrels arose between the Britons and their guests; and (A. D. 457) the latter conquered Kent for themselves. Such are the outlines of the ancient story recorded by Gildas. In our critical times some have contended, that the names of the leaders (both signifying a horse) prove that these personages are themselves as fabulous as the well-known tale with which they are connected of the marriage and dowry of the beautiful Rowena, the defeat of the Saxons, their return, and the treacherous release of Vortigern; all of which are mentioned only by later British writers.

The conquest of the greater part of Britain by the Saxons, Jutes, and Angles, occupied a space of about 130 years, from the landing of Hengist. Five British states, Strath-Clyde, Cumbria, North and South Wales, and Cornwall, maintained their existence for a somewhat longer time. Three Saxon kingdoms (Sussex, Wessex, Essex), one Jutish (Kent), four Anglian (Bernicia, Deira, East Anglia, Mercia), were formed in this period.

The Anglo-Saxons were a people divided into various castes. The kings, or ealdormen, reigned by a sort of hereditary right, without any strict adherence to the laws of succession, but all claimed descent from the original race of Odin. The nobility, 'earls,' 'eorlcundmen,' of 'thanehorn,' were a class apart, like those of continental nobles at the present day. The third class was that of the 'ceorls,' or knaldes; placed in a state of dependence on the nobility, yet freemen according to the law. Their rank, as compared with that of the nobles, was estimated by the different value of their compurgatory oaths, in giving evidence, and of their lives and persons, according to the 'were-gild,' or legal compensation for blood; namely, one-sixth. Every freeman was presumptively attached to some 'lord,' and designated as his 'man.' A class of these 'ceorls,' possessed of landed property, yet not 'lords,' seems to have occupied, like the equites of the Romans, a sort of intermediate rank between the patricians and plebeians, under the various titles of sithcundmen, lesser thanes, &c. They seem to have had the privilege, denied to the ceorl, of choosing their own lords. The ceorls, on the other hand, were 'bonde,' attached to the glebe; and might be the subjects of gift or bequest along with it, not as slaves, but as appurtenant to the property. They took no part in the political government of the realm. Lastly, the theowes, or serfs, were slaves in the full import of the word.

The territorial division of England, under the Anglo-Saxons, into counties, hundreds, and tythings, is of very great antiquity, and formed the basis of their civil institutions. The earl, the hundreder, and the tything-man, presided respectively over these divisions. Each of these officers held a court of justice, which was attended by the landed proprietors; and by the well-known custom of 'frank-pledge,' the superior or noble was rendered responsible for the acts of his inferior or man; and the vicinage, collectively, for those of its members. The witenagemote, or assembly of the wise men, seems to have been, in its original nature, rather a high court of justice, for the redress of complaints by or against the great men of the realm, than a legislative assembly. The earls, aldermen, and higher prelates attended it; and it is probable, though uncertain, that the burghs sent deputies to it. Together with the king, it constituted the sovereign power of the empire; the Saxon kings usually promulgated their laws, as regarded by themselves, with the advice of their 'witan;' and the succession to the crown was fixed by their determination. The conversion of the Anglo-Saxons to Christianity was commenced by Augustine and his companions, missionaries despatched by Gregory VII, in the beginning of the seventh century, and proceeded with great rapidity to completion. Although the religion of Christ had been introduced for five centuries at least into the country among the Britons, it had sunk so completely into decay in the revolution which followed the fall of the Roman empire, that, from the reconversion of the island under its Saxon masters, we date our episcopal succession, and the foundation of our religious establishments.

The first appearance of supremacy among the numerous chieftains of the Anglo-Saxons occurs in the instance of Ella, king of Sussex, who having, in consequence of a great victory, obtained a temporary authority over the Britons, assumed the title of Bretwalda (ruler of the Britons), about A.D. 491. During the long period of the Saxon conquest, several independent states were founded, of which the principal and best known are Wessex, Sussex, Kent, Essex, Deira, East Anglia, Northumberland; and hence has arisen the well known term of heptarchy; which, however, is substantially erroneous, inasmuch as

at no particular point of time did these seven states exist independently of each other. Several princes, who by connection or conquest obtained a superior power to the rest, assumed in succession the title of Bretwalda, which eventually (A.D. 830) was held by Egbert, king of Wessex, commonly regarded as the first king of England.

The accession of Egbert to this dignity was contemporary with the first invasion of the Danes. For a century and a half from that time, their inroads were continually repeated. All substantial progress in civilisation was effectually arrested by this terrible evil. The Danes were always at hand; the intervals between their incursions, instead of being employed by the Anglo-Saxon princes and people in forming powerful combinations for defence, were spent in civil wars; and the weaker party habitually called upon this powerful foreign enemy for support. The reign of the great Alfred, the most brilliant in the Saxon annals, took place towards the middle of this period (871-901). Under his successors (Athelstan and Edmund) the Saxon sway was extended, both by the repulse of the Danes, and by the subjugation of the Britons of Cumbria and Devonshire. But the Danes again succeeded in overrunning almost the whole of England; and became, in fact, not only the chieftains, but the progenitors of a large proportion of the population of the country N. of the Humber, and of the coast between that river and the Wash (the ancient Danelage). At length, under Sweyne and Canute (A.D. 1017), the Danes became masters of the kingdom, which, however, they only held for 24 years, or till 1041, when the crown devolved on an Anglo-Saxon prince, Edward, surnamed the Confessor. Six powerful earls, Danes and Englishmen, divided the country between them, under his authority, which, during the greater part of his reign, was little more than nominal. At his death, Harold, one of these chieftains, disregarding both the claims of Edward's natural successors, and those of William, duke of Normandy, his kinsman, to whom he had bequeathed the crown, seized it by force. William, having determined to vindicate his pretensions by force of arms, invaded England with a powerful army; and having defeated and killed Harold in the decisive battle of Hastings, on the 14th October, 1066, succeeded to the throne.

The Norman Conquest is the great æra to which reference is ordinarily made as the beginning of a new order of things in English history. The immediate change, however, consisted chiefly in the division of the lands of the kingdom into 60,000 knights' fees or counties, among the followers of the Conqueror, as feudal lords. Feudality existed among the Saxons as well as the Normans, but the tie which connected the inferior with the superior was more one of personal service, and less strictly territorial in its nature; nor were the peculiar incidents of military tenure, as understood in France and Germany, known among the Anglo-Saxons. Soon after the Conquest, the greater part of the territory of England became in fact, as well as by the gift of the sovereign, the property of the Norman knights. But it seems certain that a large proportion still remained in the hands of Saxon and Danish thanes, who either kept possession of the lands in defiance or evasion of the royal grants, or by composition with the Normans to whom they had been assigned. The class immediately under the nobles, —the freemen or ceorls of the Anglo-Saxon period—if the cifbmi, hordarii, and coturii of Domesday Book be rightly considered as representing that class—appears to have comprised the great bulk of the population; the *servi*, or slaves, mentioned in that record, amount only to about an eighth part of the former class. Eighty-two boroughs are named; and, allowing for those parts of England of which the survey is not preserved, the number was probably about 100. These boroughs seem to have been small, ill-fortified places, inhabited by a population partly governed by municipal customs, and partly under the protection of the king, or of some neighbouring noble or prelate, from whom, in after times, they generally purchased their franchises. The population of England, at the end of the reign of William the Conqueror, has been estimated at about 2,000,000; and considering that the whole northern part lay almost waste, and that many towns, manors, and villages are mentioned as having lost half their inhabitants since the time of Edward the Confessor, through the calamities attending the invasion, it has been supposed that the population under that prince fell little short of 3,000,000; though we incline to think that this is considerably beyond the mark. It may, however, be inferred from other facts, that England, in that early time, was almost wholly reclaimed and cultivated, since nearly all the villages and hamlets with which its surface is so thickly strewn seem to derive their origin from the Saxon age.

From the Norman Conquest to the accession of Edward I. (1066 to 1272), the principal circumstances which fix the attention of the reader of British history are—the disputes between the Norman and Plantagenet kings and their barons, together with the development of the feudal system; the quarrel between the sovereigns and the church; and the foreign relations of England, arising out of the French provinces held by its kings as feudal lords. As, according to the principles of the feudal law, every superior lord had a court, consisting of all those who held land immediately of him, so the king's tenants in chief formed the highest court of common council of the realm. It consisted, consequently, not only of the greater barons, but of such inferior ones as were under no superior lord, but held directly of the king. But the former naturally acquired a preponderating share in it. Backed by the people, they contended with their sovereign for the rights which were finally established by Magna Charta, in 1215. The greater part of this celebrated instrument is directed against the abuses of the king's power as feudal lord; but it established the two great principles, that no one should undergo the judgment of his peers, from which, through a variety of changes, adapted to the necessities of particular times, we derive our modern trial by jury; and that no 'scutage' (originally a pecuniary contribution assessed in lieu of military service) should be levied, except by consent of the great council of the realm. This provision, framed on behalf of the king's tenants in chief only, has become the basis of the popular right of taxation by representatives. Continued disputes respecting the extent of these privileges, and the pride of the nobility, led to the barons' war in the reign of Henry III., in which Simon de Montfort, earl of Leicester, for a time governed the kingdom, and convened the first meeting of the great council, or 'parliament,' to which representatives of the commons distinctly appear to have been summoned. This was in 12**—six centuries ago.

The two great points on which the clergy and the crown were at issue, from the reign of William Rufus to that of Henry III., were those of investitures, and of the jurisdiction over ecclesiastics. The first, in point of fact, involved the

VOL. II.

question, whether the temporalities annexed to the higher offices of the church, such as bishoprics and abbeys, should be in the gift of the crown or the pope; the second, whether clergymen, in criminal proceedings, should be subject to the royal courts or their own. Stephen conceded the point of investiture; but Henry II. strongly resisted the demands of the church, and, by the Constitutions of Clarendon (1164), abrogated many privileges which it had previously enjoyed, under pretence of restoring ancient laws. But the opposition and martyrdom of Becket turned the scale against the royal authority. Henry was forced to recede from his demands. The quarrel of investitures was again renewed in the reign of John; and that prince, pressed by the difficulties of his position, not only yielded the point, but owned the feudal superiority of the see of Rome. But the power of that church seemed suddenly to decay, after attaining the full recognition of her rights: in the long reign of Henry III. the jurisdiction of the royal courts was silently extended over ecclesiastics, and the prize of so protracted a struggle was partially yielded with little resistance.

William I. and his immediate successors possessed no continental dominions except Normandy, for which they owed fealty to the crown of France. But the house of Plantagenet, to which Henry II. belonged, were masters of the provinces of Anjou, Touraine, and Maine; to which that king added Guienne and Poitou by marriage, and Brittany by conquest; so that above a third part of France was under the immediate jurisdiction and sovereignty of the kings of England. Henry was succeeded by his eldest son, Richard, surnamed for his bravery, *Cœur de Lion*. After greatly distinguishing himself, and adding to the glory of the English arms by his exploits in Palestine, he was arrested and imprisoned at Vienna, on his way home, and did not recover his freedom till he had agreed to pay an enormous ransom. He soon after died from the effects of a wound he received in an attack on a castle near Limoges.

Richard, having no issue, was succeeded, in 1199, by his brother John, surnamed Lackland, whose reign is one of the most inglorious in the English annals. During its continuance, Philip Augustus, king of France, an able and politic prince, re-united to the French crown almost all those possessions in France that had been under the feudal sovereignty of the kings of England. But this loss was in some measure counterbalanced by the conquest of Ireland, commenced in 1172, by the Norman chieftains of Henry II. The subjugation of that island was not, however, completely accomplished till about four centuries after.

But the reign of John was chiefly remarkable for the concession of the *Great Charter* (*Magna Charta*), signed at Runnymede in 1215. In the following reign, under Henry III., the commons, as already stated, were expressly summoned as constituent members of parliament. The foundations of the constitution were thus laid; and means prepared for that gradual reduction of the realm under a more regular form of government, which was in great measure effected during the long reign of Edward I. (1272 to 1307), one of the ablest and most successful princes who ever sat on the throne of England. Under him, the great council of the realm assumed a form resembling that of the modern parliament, by the separation of the greater barons, from whom the modern peerage is derived, from the great body of the tenants in chief; the former being personally summoned to parliament, the latter ceasing to be summoned at all, and being present only through their representatives. These, however, continued for a lengthened period to yield a reluctant attendance, and seldom interfered in public affairs, except to vote or refuse the supplies demanded by the sovereign. The commons and lords appear to have sat in separate bodies, at least occasionally, as early as this reign. The power of the great barons, at the expense of the lesser, was materially increased in consequence of the statute termed 'de donis,' which tended to create perpetuities in feudal estates; while, on the other hand, the statute 'quia emptores,' prevented the owners from increasing the number of their vassals by subinfeudation. The combined operation of the two tended to throw the land more extensively into large demesnes, and to diminish the number of the small feudal chieftains, retainers of the higher nobles. With the church Edward was generally at peace, though in his reign considerable steps were made towards the repression of its temporal usurpations, by the subordination of the ecclesiastical to the royal tribunals, and by laws of mortmain. In its foreign relations, the reign of Edward was eminently glorious, unjustifiable as many of his acts must be esteemed. He subdued Wales; interfered with dignity in the affairs of the continent; and, taking part in the dispute respecting the succession to the throne of Scotland, nearly subjugated that country, on the borders of which he died, while engaged in the active prosecution of hostilities against it.

His son Edward II. lost, in a few unfortunate campaigns, the footing which his father had gained in Scotland; the crown of which was triumphantly worn by Robert Bruce, the conqueror of Bannockburn (1312). The rest of Edward's reign was occupied by a lengthened struggle in support of his favourites against the barons and his queen. In the end, he was dethroned, in 1327, by the prelates and nobles, who assumed the power of a parliament, and perished miserably in Berkeley Castle, shortly after his son Edward III. had been raised to the throne, at the age of fourteen.

The reign of this great prince is chiefly celebrated on account of his wars in France, which he conducted with much valour and brilliant, though only temporary, success. The right which he asserted to the crown of France was derived through his mother Isabella, who stood nearest in the line of succession, but was herself excluded by the Salic law from its inheritance. It was maintained in favour of the claim of Edward, that a title derived through a female, though herself incapable of reigning, is valid. The nation, as in the wars of Edward I. against Scotland was carried away by the excitement of foreign conquest, and for a long time aided its sovereign with subsidies, tallages, and loans, prodigally lavished in support of his pretensions. These wars lasted, with few interruptions, from 1337 to 1374; but, notwithstanding the great victories of Cressy (1346) and Poictiers (1356), the capture of a king of France, and the desolation of the greater part of that kingdom, Edward retained at their termination only Bordeaux, Bayonne, Calais, and an insignificant district of Gascony.

It was during these wars, and in the court of Edward, that the spirit of chivalry attained its highest point of exaltation. Although this characteristic of that brilliant era was but of a temporary nature in itself (for the knights of Edward's court left no successors), yet it had very important results in modelling and refining the taste and character of the higher orders. Meanwhile the mass of the people was undergoing a still more

important change, under the influence of different causes. The wars with France, for the first time since the battle of Hastings, thoroughly awakened the spirit of English nationality. The distinction between Norman and Saxon was thenceforth merged in the character of Englishman. The language now contemporaneously with the nation; for though the change of speech from Saxon to English was a very slow process—extending, at least, from the reign of Henry II. to that of Edward III. —the written dialect may be said to have passed at once from barbarism to a high degree of perfection in the poems of Chaucer, whose career began in this reign; a point from which it receded, rather than advanced, for a century afterwards. The royal prerogative declined during the latter part of this reign, owing chiefly to the necessities of the king, whose great expenditure rendered him dependent on his parliaments, which, for the first time, were now directed by statute to be summoned annually. The lowest class, on the other hand, greatly rose in importance.

The great pestilence that raged in England in 1349, is supposed to have cut off a half, or more, of the inhabitants. This is probably exaggerated; however, whether one-half, or a quarter, or a tenth of the inhabitants perished, it seems certain that as the services of those that survived became more valuable, they demanded and received higher wages. This rise was, however, regarded as a grievous hardship; and the king, with the advice of his prelates, nobles, and learned men, issued an edict, by which all labourers were, under severe penalties, ordered to work at their old occupation for the same wages that they received before the pestilence. But 'the servants, having no regard to the said ordinance, but to their ease and singular covetise,' refused to serve unless for higher wages than it allowed. In consequence of this resistance, the famous statute of the 21st Edward III. c. 1, commonly called the statute of labourers, was passed. It enacted, that every able-bodied person under 60 years of age, not having sufficient to live on, being required, shall be bound to serve him that doth require him, or else shall be committed to gaol till he finds surety to serve. If a servant or workman depart from service before the time agreed on, he shall be imprisoned; and if any artificer take more wages than were wont to be paid, he shall be committed to gaol. But the increase of wages having originated in natural causes, could not be checked by such enactments. Their inefficacy did not, however, lead to the adoption of a policy more consistent with justice or common sense. On the contrary, fresh efforts were made to give effect to the statute of labourers; and to prevent its being defeated by the peasantry taking refuge in towns, or emigrating to a distant part of the country, it was enacted by the 34th Edward III. that if any labourer or servant flee to any town, the chief officer shall deliver him up; and if they depart for another country, they shall be burned in the forehead with the letter F. The injustice done to the labourers by these oppressive statutes was the more glaring, as Edward, to obtain funds to prosecute his schemes of conquest in France, had recourse to the disgraceful expedient of enfeebling the standard of the coin. Not only, therefore, did the regulations as to wages, so far at least as they were effectual, deprive the common people of that increased payment to which they were entitled from the diminution of their numbers, but they also hindered them from being compensated for the fraud practised on the coin. It was attempted, indeed, to obviate the effects of the diminution of the latter by fixing the prices of most articles; but this was only to bolster up one absurdity by another, and it is not possible that such limitations could have any material influence. Notwithstanding the degradation and ignorance of the mass of the people, the oppressions to which they were subjected made them at length rise en masse against their oppressors. So long indeed as Edward III. lived, the public tranquillity was preserved, and the villeins and labourers submitted to the injustice of which they were the victims. But the growth of towns and manufactures, during the lengthened reign of this monarch, having materially increased the number of free labourers, a new spirit began to actuate the peasantry, who, contrasting their servile condition with the condition of the citizens, became sensible of their inferiority, and more alive to the oppressions they suffered. An attempt to enforce the provisions of the statute of labourers, in the reign of Richard II., was the ground work of the famous rebellion headed by Wat Tyler. The demands made by the peasantry show the grievances under which they laboured. They required the abolition of slavery, freedom of commerce in market-towns without tolls or imposts, and a fixed rent on lands, instead of the services due by villanage. The rebellion, after having attained to a formidable magnitude, was suppressed with much bloodshed. But though re-established, the servitude of the peasantry was relaxed, and the class of free labourers became gradually more numerous.

How far this national movement was aided by the religious excitement which began at the same time to prevail, has been much debated. About 1360, Wycliffe began his attacks upon the mendicant friars, and upon many abuses of the church as it then existed. He was supported in the royal council by Edward's third son, John of Gaunt, and by many of the principal nobility, through jealousy of the prelates: but his chief reliance for the propagation of his tenets was on the people, among whom he distributed the Scriptures in the vulgar tongue; and despatched his disciples, called his 'poor priests,' who appealed to their homely sense in their own idiom, and by arguments suited to their capacities. He died in 1384; his followers were soon distinguished by the title of heretics; and the increasing prevalence of their opinions was fully testified, in 1400, by the enactment of the statute, 'De Heretico comburendo,' the commencement of a long series of persecutions directed against them.

Richard II. was dethroned in 1399 by Henry of Bolingbroke, his cousin, and murdered shortly afterwards. The usurping monarch, Henry IV., was chiefly occupied, during his reign, with domestic troubles, which were with difficulty overcome by his great abilities; but it is remarkable for two important events in the development of the constitution, though not much noticed at the time—the fixing, by statute, of the parliamentary right of election for counties in all freeholders, afterwards restrained under Henry VI., to those who were worth 40s. per annum; and the recognition of the two houses as bodies possessing distinct privileges, not to be interfered with by each other.

Henry V., son of Henry IV., renewed the claims of his ancestor to the crown of France, and gained the great victory of Agincourt, in 1415, which laid most of that kingdom at his mercy. But this success was productive of no real advantage. France, indeed, was reduced to a state of great distress, but England participated largely in the miseries inflicted on her neighbours. The draughts of men and money required for the reinforcement and maintenance of the armies in France, and the licence given to all sorts of dis-

orders at home, by the absence of the sovereign, could not fail of having a most mischievous influence. A statute of the 9th of Henry V. recites, that 'whereas at the making of the act of the 14th of Edward III. (1340) there were sufficient of proper men in each county to execute every office; but that owing to pestilence and wars, there are not now (1421) a sufficiency of responsible persons to act as sheriffs, coroners, and escheators.' The success of the French arms under the celebrated Joan of Arc and Count Dunois, during the minority of Henry VI., at length put a period to the attempts of the English to conquer France. Unfortunately, however, the tranquillity they enjoyed subsequently to the termination of the French wars, was but of short duration, as England soon after became the theatre of civil war.

Henry IV. was the son of John of Gaunt, third son of Edward III. The title which he set up against Richard II. was derived through his mother, great-granddaughter of Edward, earl of Lancaster, whom a popular tradition represented as the eldest son of Henry III., and excluded from the succession on account of deformity. On the other hand, the Duke of Clarence, second son of Edward III., had also female descendants. Richard, duke of York, through one of them, acquired a title clearly preferable to that of the descendants of Henry IV., if their apocryphal claim through the Earl of Lancaster were rejected. The partisans of the house of Lancaster assumed the red rose for their symbol; those of York, the white. The parties attached to the rival factions were pretty equally balanced, and for nearly forty years, with the exception of a few short intervals, one-half the nation may be said to have turned its arms against the other. Richard, duke of York, fell in the field, leaving his claims to Edward IV., who, after various changes of fortune, dethroned Henry VI. in 1461. His son, Edward V., a minor, is believed to have been murdered in the Tower, after a reign of 13 days, by his uncle, the Duke of Gloucester, afterwards Richard III. This able but sanguinary prince lost his crown and his life in the decisive battle of Bosworth Field, gained in 1485, by Henry Tudor, earl of Richmond. This event put a period to the civil wars, the victor uniting in his person the title of Lancaster through his mother, Margaret Beaufort, and that of York acquired through his marriage with Elizabeth, daughter of Edward IV.

The reign of Henry VII. is one of the most important in the history of the country. This public and able prince completely destroyed the power and influence of the feudal aristocracy. From a very remote period, the great lords had been accustomed to maintain vast numbers of servants and retainers, partly for the purpose of displaying their grandeur, and partly as the means of security and of attack. The retainers generally lived on the estates of their masters, who supplied them with badges and liveries, and with provisions while in service. These persons were not only ready upon all occasions, when called upon, to support the cause of their lords, to execute their orders, and to give evidence for them in courts of law, but, trusting to their influence to screen them from justice, they scrupled not, whenever an opportunity offered, to attack those they considered as their master's enemies. The predatory habits acquired in such a mode of life could not be easily laid aside; and when dismissed from service, or not employed by their masters, they generally supported themselves by theft and robbery. Many statutes had been passed

u 3

for repressing so enormous an abuse, but without any perceptible effect; and during the civil wars the evil attained to a frightful excess. No provision being made for disbanded soldiers or retainers, it was not unusual to expose liveries for sale, and the competition for them amongst idle and disorderly persons was such that they occasionally brought considerable sums. Henry VII. determined to abate this nuisance; and his sagacity and firmness, and the circumstances under which he was placed, enabled him to succeed. Many of the principal nobles had perished in the struggles terminated by the battle of Bosworth; and their power had been impaired by repeated confiscations, and by the extraordinary expenses they had had to sustain. They were, therefore, but ill-fitted to defend their privileges against so able and powerful a prince as Henry, who perceived and made use of his advantage. The laws against giving badges and liveries, and employing retainers, were renewed and enforced with a rigour that none could expect to elude. At the same time, too, that the barons were compelled to lay aside their feudal pomp, and to dismiss their vassals, the improved and more luxurious habits that began to be diffused throughout the nation disposed them to receive money payments, instead of personal services, from their tenants and dependants; and the lower ranks of people being thus, as it were, abandoned by their feudal superiors, were obliged, instead of trusting to them for support and protection, to resort to some species of industry, and to respect those laws they could no longer trample upon with impunity. The change that was thus effected was of the greatest importance, and had the most decisive and beneficial influence on all ranks and orders. Had the practice of maintaining crowds of retainers continued, order and tranquillity could never have been established.

The power of the great lords was undermined by another law, which, though less felt at the time, has been hardly less important, perhaps, in its consequences than any other passed in the reign of Henry VII. This was the legitimation of the practice, introduced in the reign of Edward IV., of breaking entails by a fine and recovery. 'By means of this law,' says Hume, 'joined to the beginning luxury and refinement of the age, the great fortunes of the barons were gradually dissipated, and the property of the commons increased in England. It is probable that Henry foresaw and intended this consequence; because the constant scheme of his policy consisted in depressing the great, and exalting the churchmen, lawyers, and men of mean families, who were more dependent on himself.' But, however this may be, the depression of the higher classes having taken place before the increasing influence of the commons formed a sufficient counterpoise to the power of the crown, Henry VII. was long previously to his death the most powerful of the English monarchs; and left to his son, Henry VIII., a regal influence which was but feebly, if at all, controlled by the popular branch of the legislature.

The great event for which the reign of Henry VIII. is memorable was the Reformation. The event was, to some extent, accelerated by his passions. Unable to procure from the see of Rome a consent to his divorce from his wife, Catherine of Aragon, Henry involved the kingdom in a rupture with the pope, and assumed the title of 'Head of the Church.' During a century of occasional persecution, the Lollards, or sectaries of Wycliffe, had continued to exist, probably in considerable numbers. Thus the Reformation, when first it broke out in Germany, found in England ready adherents among the lower classes; while the more learned part of the clergy, connected by various ties with the continental reformers, readily embraced it; and, finally, the quarrel of the king with Rome drew over the timid and subservient, while the powerful were gratified with the plunder of abbeys and other ecclesiastical establishments. Not that Henry himself embraced the peculiar doctrines of the Reformation; on the contrary, during great part of his reign, belief in them was constantly punished as a crime; but, by assuming the headship of the church, and rejecting all spiritual domination, he made ready the way for the transition of the nation from Popery to Protestantism. At the commencement of his reign Henry was popular; but his passions were violent, and being unrestrained in their indulgence, he came to be the most sanguinary tyrant that ever ruled England.

His only son, Edward VI., succeeded in 1547, at the age of nine years. Being educated in the care of divines strongly attached to Protestantism, the Reformation, in his reign, tended towards the extreme side; and had it been prolonged, it is probable that the Church of England would have approximated much more, both in doctrine and discipline, to that established at a later period by Calvin and his disciples, than it eventually did. He died, however, in 1553; and after an unsuccessful attempt to place Lady Jane Grey, heiress to the house of Suffolk, on the throne, the Princess Mary, eldest daughter of Henry VIII., and a bigoted catholic, was proclaimed without opposition. She immediately restored the Roman Catholic religion, and her reign was distinguished by a fierce persecution of Protestants, of whom more than 200 suffered at the stake. She married Philip II., king of Spain, but happily died without issue in 1558. During her reign, in 1558, Calais, the last possession of the English in France, was taken by the Duke of Guise.

On the death of Mary, her half sister, the Princess Elizabeth, daughter of Henry VIII. by the unfortunate Anne Boleyn, succeeded to the throne. Attached, originally, to the doctrines of Protestantism, this able princess was, nevertheless, disinclined towards the more violent reformers, who had possessed so much influence in the reign of her brother Edward; and, resuming the headship of the church, she succeeded in creating an establishment, dependent rather on the state for support than exercising a control over it. The Roman Catholics, on their part, did not find it necessary to secede from the national church until about 1570; from which period they formed a sect apart. The early part of Elizabeth's reign was prosperous, notwithstanding some discontent on the part of the lower classes. In fact, besides the contest occasioned by the religious changes, the government of that century had social difficulties of a serious nature to struggle with. The suppression of monasteries deprived a number of mendicants, and of the poorest classes, of their ordinary dependence. The decay of the feudal system rendered the great landlords less desirous than heretofore of multiplying their tenantry, and more anxious to enrich themselves; and from this, among other causes, inclosures multiplied, and much arable land was thrown into pasturage, producing complaints of poverty and depopulation. But by far the most important source of change, though at the time the least observed, was the extraordinary diminution of the value of money, occasioned by the discovery of the mines of America. This was followed by a rapid rise of prices: and though wages ultimately rose, their rise was comparatively slow, and much distress accompanied the transition. This great crisis produced numerous enactments for the relief of the poor, which ended in the well-known statute of the 43rd of Elizabeth, enacted in 1602.

The disputes between Elizabeth and the court of Rome grew now more inveterate, and led to two important events in English history—the war with Spain and defeat of the Spanish armada in 1588, and the exertions of Mary queen of Scotland, next in succession to the throne. Being a Catholic, the cause of Mary, who having sought an asylum in England had been imprisoned by Elizabeth, was embraced by most of the Roman Catholics of the country, and produced various unsuccessful plots and conspiracies which ended in the execution of Mary. During this reign, England was joined in alliance, first with the Protestants of France, afterwards, and more closely, with those of the Low Countries; and when Henry IV. ascended the throne of the former country, the combination against the power of Spain and Rome had been completely and successfully organised. Ireland was, also, reduced to a state of greater submission than at any previous period. The taste for naval enterprise was fully awakened, and the commerce and naval power of the country grew rapidly into importance. The last years of the long reign of this illustrious princess were darkened by the intrigues of the court, the rebellion of her favourite Essex, and her unavailing sorrow for his death.

James VI. of Scotland, the son of the unfortunate Mary, being next in succession to the crown, ascended the English throne on the decease of Elizabeth, without opposition, his peaceful but inglorious reign of twenty-four years appears to have been a period of considerable national prosperity; and as it were laid the foundations of that colonial empire in the new world that subsequently attained to so vast a magnitude. But through this whole reign, the struggle was preparing between the rising power of parliament—which in the latter years of Elizabeth had already begun to assume some degree of independence—and that of the crown. The Puritans were the most zealous and steady supporters, at this period, of the authority of parliament. This sect, or rather class (for dissent was not yet recognised by law), originated in the reign of Elizabeth; being composed, in great measure, of the disciples of the more zealous divines of Edward VI.'s reign, and approximating in opinion to the Protestants of Holland and Switzerland. Episcopal government, and the ceremonies of the church, were particularly opposed by them. They continued to increase throughout the reign of James, especially in the larger towns; and in some parts, as the eastern counties, they also became numerous among the country population.

Charles I., who succeeded his father in March, 1625, ascended the throne under the complicated disadvantages of a union with a Roman Catholic princess; the dominion of an unpopular favourite, the Duke of Buckingham; and an exchequer much disordered by the prodigalities of his predecessor. To these adverse circumstances were added a want of sincerity and directness of purpose. But his great defect, and the chief source of the disasters he entailed on himself and the country, consisted in his arbitrary principles of government. He could not brook the growing power and influence of parliament; and was infatuated enough to suppose that a nation so rich, populous and enlightened as England now was, and which had long possessed a representative assembly, would submit to be governed in the same way as in the reigns of Henry VIII. and Elizabeth. An ill-conducted war with France added to his difficulties. Three parliaments were summoned and dissolved during

the first four years of his reign; after which he governed for eleven years (1630 to 1640) without a parliament. During this lengthened period, the discontent of the popular party was continually increasing; especially in consequence of the efforts of the higher clergy, under Archbishop Laud, to suppress the preaching of Puritan ministers, and the spread of their opinions. Devotion to the views of this party involved the king, in 1638, in a war with his Scotch subjects, on whom he had endeavoured to impose episcopacy. The difficulties that grew out of this quarrel compelled Charles, in 1640, to summon that parliament, afterwards so famous in English history by the name of the Long Parliament. The Presbyterians, having gained an ascendancy in this body, forced Charles to retract the unconstitutional acts of his former government; expelled the bishops from the house of lords; and impeached and procured the execution of the Earl of Strafford, his ablest minister. At length the breach became irreconcilable, and both parties prepared for war. This eventful struggle commenced in 1642. It was waged for some time with doubtful advantage on either side, till Cromwell and Fairfax, leaders of the Independent party, obtained the command. With the assistance of the Scotch, they defeated the royal armies at Marston Moor (1644) and Naseby (1645). Charles soon afterwards fell into the hands of the army, and after a variety of intrigues and negotiations between that body, the parliament, and the king, he was condemned and executed by warrant of judges nominated by the parliament, on the 30th Jan. 1649. A republican government was next formed, styled the Commonwealth of England, which ended in the protectorate of Oliver Cromwell (1651). That able and successful general and statesman died in 1658; and a short period of turbulence and intrigue was closed by the restoration of Charles II., son of the executed monarch, in May, 1660.

The restoration was effected amidst the teeming joy of the people; and the first movements of national feeling set strongly in favour of monarchy and the church. Several of the regicides were punished with death; and the ministers of the Presbyterian persuasion who refused to comply with the Act of Uniformity, were universally ejected from their benefices. The test and corporation acts, long considered as the bulwarks of the church, were also enacted in this reign. But after a few years had elapsed, it was evident that the sudden impulse of loyalty which had accompanied the restoration was not congenial with the habitual feelings of the country. Since the accession of Charles I. every thing had been changed: those fundamental notions of rights and duties, both on the part of the sovereign and the people, which now constitute what are termed the principles of the constitution, grew and ripened in this reign into a consistent code, which was ratified at the Revolution. The private life of the king, his vices, and, still more, his follies, and his mean and mercenary dependence on France, were among the causes of his unpopularity. Sanguinary wars with the Dutch served only to exercise the warlike and naval spirit of England, without producing any direct benefit or acquisition. It was about the year 1678, that the houses of Lords and Commons came, for the first time in English history, into a state of permanent collision and opposition; the first containing a majority attached to the court, the latter being governed by its opponents. In 1679, for the first time, the names of Whig and Tory were used to designate the two great parties which then divided the kingdom, and which have ever since found successors in states, if not in spirit. The violent conduct of the Commons, in the matter of the Popish Plot, and their interference with the succession, by entertaining measures for the exclusion of the Duke of York, the king's brother, on account of his religion, produced at last a re-action in favour of the crown. Lord Russell and Algernon Sydney were the victims of this re-action, being executed for participation in a supposed plot; and the king, by proceeding against the corporation of the city of London for the alleged abuse of its franchise, brought all bodies similarly circumstanced throughout the kingdom to a state of submission. The charters were surrendered, and new ones granted on a more oligarchical model. Charles II. died during these temporary successes, in 1685.

He was succeeded by his brother, James II., an avowed Papist, and strongly attached to his religion, to which it was his continual endeavour to obtain proselytes. This circumstance, even more than his steps towards the assumption of absolute power, roused against him a spirit of almost universal discontent. An unsuccessful rebellion, headed by the Duke of Monmouth, a natural son of the late king (who perished on the scaffold), served for a while to strengthen his authority; but the extreme severity with which those who had engaged in it were punished, greatly increased his unpopularity. Having dissolved, in the first year of his reign, that parliament which had proved so favourable to the views of his predecessor, he obtained from the judges an acknowledgment of his right to dispense with acts of parliament, which, in effect, amounted to a recognition of arbitrary power. But the servile, time-serving opinions of the judges were heartily repudiated by the nation at large. The other proceedings of the king were of a still more violent and despotical character, till at last he succeeded in disgusting and alienating all his Protestant subjects. Some of the principal persons in the country retired to Holland, where they found a secure asylum through the protection of William, prince of Orange, son-in-law of James. Had King James succeeded in establishing arbitrary power in England, his subserviency to Louis XIV., then in the zenith of his power, would have been of the most serious consequence to Holland; and to avert this danger, and strengthen the Protestant party, William resolved on the invasion of England. No project was ever more completely successful. James, deserted by his subjects, and by the army on which he had mainly depended, fled to France. The Convention Parliament—so called from its assembling, of necessity, without the royal summons—declared that James had abdicated the crown, and raised William of Orange and his consort Mary to the throne.

A solemn seal was set on the proceedings of the Revolution by the Bill of Rights, which recited and ratified the constitutional liberties of the country, and by the Act of Settlement, which excluded James and the greater part of his family from the succession, and fixed it, eventually, in the Protestant line of Hanover. Such was the end of that fifty years' struggle which commenced with the meeting of the Long Parliament in 1640. The great liberal party which conducted the struggle throughout, in the end successfully vindicated the supremacy of the nation, and the sacred right of resistance to unconstitutional power. Their example has had a powerful influence in all civilised countries, and the form of polity which they established has been introduced into the United States, and, more or less, in the countries of the west of Europe. Its influence in England has been beneficial beyond all that could have been

anticipated; and the country is mainly indebted for by far the larger part of its comfort and wealth, and for the distinguished place it occupies among the nations of the earth, to the triumph of those free principles of government that were consolidated by the Revolution. From this period, English domestic history assumes a new aspect: the conflict of parties succeeds to that of principles. It is true that, for some time after the Revolution, speculative opinions respecting the royal prerogative continued to vary; and the adherence of a considerable body to the cause of the exiled family, although generally passive, placed the state in constant danger; but the fundamental doctrines of the inviolability of the sovereign, the responsibility of ministers, and the supremacy of parliament, were never afterwards practically contested. Force was never abandoned; and government, maintained in ordinary times by influence, was controlled in crises of importance by public opinion.

In the reign of William III., England was involved, in a more serious manner than before, in the politics of the Continent, by becoming a party to the general coalition provoked by the ambition of Louis XIV.; and the feelings of the English people, excited by that prince's persecution of his Protestant subjects, coincided with the continental interests of the king, and made the war be vigorously prosecuted. Louis, on the other hand, gave support and countenance to the exiled family. The peace of Ryswick (1697) put a stop for a short period to the hostilities.

In order to provide for his military expenditure, William III. was forced to have recourse to the system of loans; and by so doing he engaged, to a great extent, the mercantile interest of the country in the support of the revolutionary establishment. That interest, though long influential in England, may be said to have now come prominently forward, for the first time, as a distinct and powerful element in the state. Its increase during the 17th century, relatively to that of the other classes, may be partly judged of by the fact, that London, which in all probability possessed about 250,000 inhabitants at the end of the reign of Elizabeth, had more than half a million in that of William; while there is reason to believe that the number of inhabitants of the whole country—almost five millions and a half in the latter reign—had undergone but a slight augmentation. The Bank of England was founded in 1694.

Hostilities recommenced shortly after the accession of Anne, the surviving daughter of James II., in 1702, and continued until 1713, with some accession of 'glory,' but little else to the British arms, directed by the Duke of Marlborough. The peace of Utrecht ended these hostilities. England obtained by it little except some extension of territory in North America, and Minorca and Gibraltar in Europe. The union with Scotland (1706) was the great domestic event of the reign.

The accession of George I., elector of Hanover, to the throne, according to the limitations contained in the Act of Settlement (1715), again threw power into the hands of the party of the Revolution; and the suppression of a Scotch rebellion strengthened his authority. The Septennial Act, passed in the same year, extended the duration of parliament to seven years, at which term it has since remained fixed. Their power being confirmed by this enactment, the Whigs maintained the ascendancy to the end of the reign; and the tranquillity of the country was undisturbed, except by the excitement produced by the famous South Sea scheme (1721), and the violent though temporary mercantile distress which followed. The peace of Western Europe was guaranteed by the alliance of the new line of English sovereigns with France; first under the regency, and afterwards under the powerful administration of Cardinal Fleury; a short war between Spain and Great Britain, in 1727, alone interrupted it.

In that year George I. died, and his son, George II. ascended the throne. This event made no change in the politics of the government, the new king being equally with his father attached to Sir Robert Walpole, one of the most powerful ministers the country has ever known. For twelve years longer he continued to maintain peace; but public clamour, excited by his political enemies, drove him, in 1739, into hostilities with Spain. This war was wholly of a commercial character, and had its origin in the desire of the British merchants to participate in the trade with those vast American provinces, which the policy of Spain kept closed against foreign commercial enterprise. It proved the ruin of Walpole, who was driven from power, in 1742, by a combination of discordant Whigs, Tories, and Jacobites. About the same time the interests of the sovereign, as elector of Hanover, involved the nation in war with France as well as Spain. In 1745, Charles Edward, grandson of the expelled James II., landed in Scotland, and was immediately joined by the greater number of the Highland clans. At the outset he met with some extraordinary successes, and advanced at the head of a body of Highlanders as far south as Derby. But being joined by but few Englishmen, and having received no support from France, he was obliged to retreat to Scotland, where the battle of Culloden terminated his ill-starred enterprise, and the last civil war that has taken place in Great Britain. The measures that were adopted, in consequence of this outbreak, for abolishing clanship in the Highlands, and putting an end to hereditary jurisdictions in Scotland, were of great advantage to that part of the kingdom.

In 1748, this desultory war was closed by the peace of Aix la Chapelle. The combinations in which England had engaged on the Continent had been in general unsuccessful; nor were the terms of the peace particularly favourable to her interests. But she may be said to have attained in it, what she has never since lost, a decided maritime supremacy over all the other powers of Europe. She entered it as a competitor, and closed it as mistress of the sea. Thenceforward England has fought to preserve, rather than extend, her naval dominion.

At this period France was peculiarly anxious to recover her lost maritime power, in consequence of that desire for extended colonial conquest which then swayed her councils, and seems, indeed, to have been the most active principle of European politics towards the middle of last century. Disputes in the E. Indies and in N. America, together with the continental quarrels of Prussia and Austria, brought about the great contest which commenced in 1756, commonly called the Seven Years' War. Fortunately for England, the management of her affairs soon afterwards fell into the hands of one of those extraordinary men whose influence over their age, from their power of inspiring and directing enthusiasm, is far greater than the highest talents, aided by the most powerful connections, but destitute of this peculiar faculty, have ever acquired. Under the guidance of Pitt (Lord Chatham), her arms triumphed in every quarter of the globe. George II. died in the middle of this war (1760),

and was succeeded by George III. This prince, ill-educated, obstinate, and strongly imbued with anti-popular prejudices, withdrew his confidence from the ministry of his grandfather. Pitt, unable any longer to carry his measures, retired from the cabinet. A new ministry succeeded; and a glorious war was terminated by an inglorious peace, which, however, secured to England the possession of Canada and some other inferior acquisitions.

The foreign dominions for which the seven years' war had been undertaken had now acquired an enormous extension, and were increasing rapidly in population and importance. Founded partly by commercial adventurers, partly by religious and political refugees, the colonies of England on the mainland of America, exclusive of Canada, part of Louisiana, and Nova Scotia, acquired from the French, were divided into thirteen provinces or states, and had 2½ millions of inhabs. In the West Indies, England possessed Jamaica, than the most fertile and best cultivated of the West Indian islands, and a number of smaller colonies. In Hindostan, Lord Clive had laid the foundation of our empire, by the acquisition of the important prov. of Bengal in 1757. Such was the extent of the realms, to the government of which George III. succeeded.

The internal history of England, during the reigns of George I. and II., evinces a gradual and steady increase of national prosperity, without rapid change. Little of the violent political and social emotions which had agitated the preceding age, and were again to agitate the next, was then felt by the community. The Jacobite party were gradually cut, and was in fact, nearly extinct in England before the Scottish outbreak of 1745. The laws against dissenters, which still remained on the statute-book, were so modified by usage, that little political distinction remained in practice between them and members of the church. This period has been regarded by some writers, though probably on no sufficient grounds, as being, on the whole, the most favourable on record as respects the economical condition of the lower classes. Its beneficial influence, in this respect, was probably owing in part to the extraordinary circumstance of a long and steady continuance of productive years. In fifty years, from 1715 to 1765, only five deficient harvests are said to have occurred; and the price of wheat was generally little more than half what it had been in the middle of the 17th century. The population of the country during this period increased only at a moderate rate, or from 5,800,000 in 1700, to 6,400,000 in 1760; and the labouring classes consequently reaped the full benefit of this prosperity in the shape of high wages. The poor-laws, as managed at that time, certainly contributed to prevent a more rapid augmentation. Moral and orderly habits, on the whole, characterised the period; the violence of earlier times had disappeared; and the peculiar vices attending on great wealth and manufacturing industry had scarcely, as yet, begun to prevail.

A novel order of things began with the accession of George III. New moral and social impulses arising at the same time with an extraordinary spread of wealth and industry, materially altered, in a few years, the character of the community. The disputes respecting the expulsion of the demagogue Wilkes from parliament, though unimportant in themselves, were the precursors of great events: they, for the first time (at least since the commonwealth), brought into action a democratic party in the state, hostile to the old aristocratic legislature. This party spread most widely and rapidly to the trans-Atlantic dominions of Great Britain. Exasperated by attempts, on the part of the mother country, to impose on them a system of taxation, and incited by the sympathy of a considerable party in England, the thirteen provinces of N. America revolted in 1776, and openly proclaimed their independence. Notwithstanding a gallant resistance, they might, perhaps, have been subdued, had not France, Spain, and Holland, espoused their quarrel. England was again involved in war with the chief continental nations, and maintained, even against that formidable combination, her maritime supremacy. But she was forced to relinquish her dominion over her revolted colonies, which the peace of 1783 raised to the dignity of an independent federal republic. On the other hand, in India, the arms of Great Britain continued to make a sure and gradual progress.

The close of the American war was followed by ministerial changes of unusual importance. A coalition was formed between Lord North, the unpopular minister, who had conducted the war, and Charles James Fox, who had been its most violent opponent, which embraced most of those great family interests that had, for a series of years, predominated in parliament. The king disliked, however, the coalition ministry; and an attempt to invade his prerogative, by a bill which threatened to transfer the government of India, in some measure, to parliament, afforded a pretext for its dismissal. William Pitt, younger son of the Earl of Chatham, was then called to the direction of affairs, at the early age of twenty-four. He had to contend at the outset with a hostile majority in the H. of C., but the country, in which the coalition was exceedingly unpopular, was decidedly in his favour; and this and the declared support of the court enabled him to dissolve parliament, and to secure a great majority in the new H. of C. Pitt now became the most powerful minister who had swayed the cabinet since Walpole. He called to his assistance new interests, and a new school of politicians; the members of the old oligarchy either came gradually into his views, or continued in permanent opposition. The country continued in the enjoyment of peace, and in a state of great prosperity, during the first ten years of his administration. But, in 1793, it was involved in war with France, then in the crisis of a tremendous revolution.

Between 1750 and 1770, the great system of canals, which now intersects the whole of England, was commenced, and carried a considerable way towards completion. In 1767, the first great step was made in the manufacture of cotton by the invention of Hargreave's spinning jenny. Watt's first patent for improvements in the steam-engine was taken out in 1769; which is also the date of Arkwright's patent. These great industrial inventions, taking place about the same time, may be regarded at once as causes and effects of the sudden spread of commercial activity. At the accession of George III., the exports of England amounted to about 15,000,000; at the breaking out of the revolutionary war, to 25,000,000. During the same period, the national debt had more than doubled, chiefly in consequence of the heavy expense of the American war.

The events of the three and twenty years, between the commencement of the revolutionary war and its final conclusion in 1815, are far too varied and manifold to be more than alluded to in this brief summary. At first the British navy obtained the undisputed sovereignty of the seas; and most of the remaining colonies of France and Holland were conquered. But military operations on the continent, and the combinations which England formed, in conjunction with the great European powers, in opposition to the French, were almost uniformly unfortunate. Pitt, suffering from the ill success of his measures, and determined not to make overtures to France, retired from office; his place was supplied by a ministry which was broken up by the renewal of hostilities in 1803, and he returned once more to power. The last great act of his first ministry was the union with Ireland, a measure long contemplated, but hastened by the unfortunate insurrection that broke out in that country in 1798. The union abolished the separate legislature of Ireland, and introduced 100 new members for Ireland into the Imperial H. of C., and the representative peers of Ireland into the H. of Lords.

The renewed war was but little successful as the outset, except that the fleets of Spain and France were totally destroyed by Nelson, at Trafalgar. Pitt died in 1806, after the last of the great continental confederacies had been dissolved by the battle of Austerlitz. But, shortly afterwards, affairs took a favourable turn. Napoleon, whose ambition was as boundless as his genius was transcendent, having prevailed on the Bourbon princes of Spain to abdicate the crown, resolved to place his brother Joseph on the Spanish throne. But in doing this he provoked a resistance that could hardly have been anticipated. Though the abdicated princes were the merest imbeciles, and their government a theme of censure, the Spaniards took arms in defence of their rights, and of the independence of the nation thus wantonly violated. The English fanned the flame that had thus been excited, and threw supplies of money and ammunition and powerful armies into Spain. At first these had but little success; but no sooner had their command been entrusted to General Wellesley, subsequently duke of Wellington, than the whole aspect of affairs was changed. Possessing in an almost unprecedented degree all those qualities that go to form a consummate commander, the English general successively baffled and defeated all the French troops that were opposed to him; and finally expelled them from the Peninsula.

Meanwhile the colossal power of Napoleon, which had so long triumphed over every combination formed for its overthrow, was irretrievably broken by the frosts and snows of Russia. The invasion of France by the allies in 1814, was followed by Napoleon's abdication; and his short reign after his return from Elba was terminated by the battle of Waterloo, which raised the glory of the English arms and of the English general to the highest pinnacle.

The treaty of Vienna restored, in as far as the altered circumstances of the world would permit, Europe to its state previously to the breaking out of the French Revolution. Except the important advantage of being secured against the danger of attack by a too formidable neighbour, England gained little by the war. She restored Java, and most of the foreign colonial possessions that had fallen into her hands during its progress, retaining only Malta, the protectorate of the Ionian Islands, the Cape of Good Hope, Demerara, Trinidad, and some other places in the West Indies. In India the conquest of Mysore in 1799, and successful wars with the Mahrattas, left her mistress of the whole peninsula of Hindostan, either in direct sovereignty, or as protector of the native princes.

The sacrifices made by the British nation during this protracted struggle were on the most gigantic scale. During its latter years the public revenue amounted to nearly 50,000,000l. a year, and nearly 500,000 men were employed in the national service by sea and land; and, in addition to the sums raised by taxation, above 600,000,000l. were added to the national debt during the course of the contest.

The reign of George III., the longest in English annals, ended in 1820. For several years before his death, the king had laboured under mental alienation, the royal authority being exercised by his son, with the title of prince regent. During the ten years of the reign of George IV., one of the most selfish and sensual of English monarchs, the peace of Europe, in as far as Great Britain was concerned, was interrupted only by the short hostilities of 1827 against the Turks, in behalf of the insurgent Greeks. At home the country was agitated by the unsuccessful effort made by the king to procure a divorce from his wife, Caroline of Brunswick, and by a continued struggle between the two great Whig and Tory parties, taking the terms in their widest acceptation. But the progress of the country—the vast increase of manufactures and commerce, and consequently of the town population, since the commencement of the French war, in 1793—had greatly strengthened the Whig, or popular party. Civil disabilities of all kinds were loudly objected to; the abuses incident to the nomination, or, as it was called, rotten borough system, were denounced; and a demand for a remodelling of the elective system and of the H. of C. was raised, which, being supported by the great bulk of the town pop., and being, also, in itself just and reasonable, could not be long resisted. In 1828, the Test Act, which, though obsolete in fact, still imposed nominal disabilities on Protestant dissenters, was repealed; and, in 1829, the barriers which had so long excluded Roman Catholics from the legislature were removed. These changes, by increasing the popular influence, paved the way for the reform of the constitution of the House of Commons, the most important act of the reign of William IV. The emancipation of slaves throughout the British dominions, and the introduction of the new system for the administration of the poor-laws, were the only other measures of importance in this reign, which terminated on the 20th of June, 1837; when the Princess Victoria, daughter of the Duke of Kent, and grand-daughter of George III., succeeded to the crown. Queen Victoria was married on the 10th of Feb., 1840, to Prince Albert of Saxe-Coburg-Gotha, who died Dec. 14, 1861. Her reign belongs to contemporary history.

TABLE, showing the Commencement, Length, and Termination of the Reigns of the Kings and Queens of England, since the Conquest, with the Date of their respective Birth, and their Ages.

Kings and Queens	Born	Reigns began	Reigned Y. M. D.	Reigns ended	Age
NORMAN MONARCHS.					
Will. Conq.	1027	1066 Dec. 25	20 6 15	1087 Sept. 9	60
Will. Rufus	1057	1087 Sept. 26	12 10 7	1100 Aug. 2	43
Henry I.	1068	1100 Aug. 5	35 3 27	1135 Dec. 1	67
Stephen	1105	1135 Dec. 26	18 10 9	1154 Oct. 25	49
HOUSE OF PLANTAGENET.					
Henry II.	1133	1154 Dec. 19	34 8 16	1189 July 6	56
Richard I.	1156	1189 Sept. 5	9 7 8	1199 April 6	42
John	1165	1199 May 27	17 4 24	1216 Oct. 19	50
Henry III.	1207	1216 Oct. 28	56 0 19	1272 Nov. 16	65
Edward I.	1239	1272 Nov. 20	27 7 17	1307 July 7	67
Edward II.	1284	1307 July 8	19 6 18	1327 Jan. 20	43
Edward III.	1312	1327 Jan. 25	50 4 27	1377 June 21	65
Richard II.	1367	1377 June 22	22 3 7	1399 Sept. 29	33
HOUSE OF LANCASTER.					
Henry IV.	1367	1399 Sept. 30	13 5 20	1413 Mar. 20	46
Henry V.	1388	1413 Mar. 21	9 5 10	1422 Aug. 31	33
Henry VI.	1421	1422 Sept. 1	38 6 4	1461 Mar. 4	49
HOUSE OF YORK.					
Edward IV.	1442	1461 Mar. 4	22 1 5	1483 April 9	41
Edward V.	1471	1483 April 9	0 2 16	1483 June 25	13
Richard III.	1443	1483 June 26	2 1 23	1485 Aug. 22	42
HOUSE OF TUDOR.					
Henry VII.	1456	1485 Aug. 22	23 7 20	1509 Apr. 21	52
Henry VIII.	1491	1509 April 22	37 9 6	1547 Jan. 28	55
Edward VI.	1537	1547 Jan. 28	6 5 8	1553 July 6	15
Queen Mary	1516	1553 July 6	5 4 11	1558 Nov. 17	42
Queen Elizabeth	1533	1558 Nov. 17	44 4 7	1603 Mar. 24	69
HOUSE OF STUART.					
James I.	1566	1603 Mar. 24	22 0 8	1625 Mar. 27	59
Charles I.	1600	1625 Mar. 27	23 10 8	1649 Jan. 30	49
COMMONWEALTH.					
	1649 Jan. 30	11 3 29	1660 May 29	
HOUSE OF STUART RESTORED.					
Charles II.	1630	1660 May 29	24 8 8	1685 Feb. 6	54
James II.	1633	1685 Feb. 6	3 10 5	1689 Dec. 11	67
HOUSE OF ORANGE-STUART.					
William III.	1650	1689 Feb. 13	13 0 20	1702 Mar. 8	52
Mary II.	1662		12 4 24	1714 Aug. 1	33
Queen Anne	1665	1702 Mar. 8		1714 Aug. 1	49
HOUSE OF HANOVER.					
George I.	1660	1714 Aug. 1	12 10 10	1727 June 11	67
George II.	1683	1727 June 11	33 4 14	1760 Oct. 25	77
George III.	1738	1760 Oct. 26	59 3 4	1820 Jan. 29	82
George IV.	1762	1820 Jan. 29	10 4 23	1831 June 26	68
William IV.	1765	1830 June 26	6 11 25	1837 June 20	71
Victoria	1819	1837 June 20			

ENKHUYSEN, or **ENKHUIZEN**, a sea-port town of Holland, prov. N. Holland, cap. cant., on a small peninsula in the Zuyder-Zee, 27 m. NE. Amsterdam. Pop. 5,590 in 1861. The town is fortified on the land side, and has a harbour formerly much frequented by trading vessels, but which is now nearly useless, from having been filled up with sand. It contains several churches, a fine town-hall, and a large cannon foundry; and, by means of a canal, it still commands a considerable trade, particularly in salt fish. During the 16th and 17th centuries, the herring fisheries employed many of the inhabitants, whose number at that time amounted to 17,000. Enkhuysen was founded in 1200; in 1514 it was all but destroyed by an inundation.

ENNIS, an inland town and parl. bor. of Ireland, co. Clare, of which it is the cap.; prov. Munster, on the Fergus, which is here crossed by three bridges; 20 m. NW. Limerick, on the railway from Limerick to Galway. Pop. 7,711 in 1831, and 7,175 in 1861. The town is meanly and irregularly built, and most part of the houses in the suburbs are mere cabins. The public buildings are the parish church, an extensive R. Cath. chapel, used as the cathedral for the diocese of Killaloe, 2 convents, meeting-houses for Independents and Methodists, a school on the foundation of Erasmus Smith, a Catholic college, a national school, the county court-house, gaol, infirmary, fever hospital, a house of industry, and a barrack. The union workhouse, opened in 1841, has accommodation for 1,200 inmates. The constabulary and the revenue police have stations here. Races are held annually in the neighbourhood. Under the charter of James I, of 1612, the corporation consisted of a provost and 12 free burgesses. This body returned 2 mems. for the bor. to the Irish H. of C. down to the Union; and it subsequently returned 1 mem. to the Imp. H. of C. till the Reform Act, when the limits of the bor. were enlarged, and the 10l. freeholders admitted to the franchise. The corporation has now become extinct, and its functions are exercised by commissioners. Registered electors 191 in 1865, all 8l. rated occupiers. The assizes for the co. are held here; as are general sessions in Jan., April, and Oct., petty sessions on Fridays, and a manor court in the suburb of Clonroad occasionally for pleas to the amount of 10l. The co. prison, built on the radiating plan, contains 73 cells and 12 other prison rooms. There are no manufactures, but there is a considerable trade in agricultural produce, part of which is conveyed down the river by lighters to Clare, 2 m. distant, where the river becomes navigable, and is thence shipped for England and other parts. Markets are held on Tuesdays and Saturdays, and fairs on 8th April, 9th May, 1st Aug., 3rd Sept., 11th Oct., and 3rd Dec.

ENNISCORTHY, a town of Ireland, co. Wexford, prov. Leinster, on the Slaney, 13 m. NNW. Wexford, on the railway from Dublin to Wexford. Pop. 5,533 in 1831, and 5,336 in 1861. The town is romantically situated on the declivities of steep hills on each side of the river, here crossed by a bridge; it is navigable by large barges, to facilitate the loading and unloading of which extensive quays have been constructed. The public buildings are the parish church, R. Cath. chapel, convent, meeting-houses for Quakers and Methodists, a large school, almshouses, a fever hospital and dispensary, a market-house, and a court-house: the ancient castle is still standing. The corporation, under a charter of James I., in 1611, consists of a portreeve, 12 burgesses, and a commonalty: it sent 2 mems. to the Irish H. of C. till the

Union, when it was disfranchised. General sessions are held at Easter and Michaelmas, and petty sessions on Thursdays. The town is a constabulary station. There is here an earthenware manufactory, tan-yards, breweries, a ropewalk, flour-mills, and a distillery. A brisk trade is kept up with Wexford by the river. Markets on Thursdays and Saturdays; fairs on 20th Jan., 21st Feb., 21st March, 25th April, 10th May, 7th June, 5th July, 26th Aug., 19th Sept., 11th Oct., 15th Nov., and 21st Dec.

Enniscorthy owes its origin to the castle, still in good preservation, built here by Raymond le Gros, who married a sister of Strongbow. In 1649 it was taken by Cromwell. On the 28th of May, 1798, it fell, after a sanguinary conflict, into the hands of the rebels. The latter afterwards established their head-quarters on Vinegar Hill, which commands the town. Here they were attacked, and driven from their position with great loss by the royal forces under Lord Lake, on the 21st of June, 1798.

ENNISKILLEN, an inland town and parl. bor. of Ireland, co. Fermanagh, of which is is the cap. prov. Ulster, beautifully situated on an island in the river or strait connecting the two principal divisions of Lough Erne; 85 m. NNW. Dublin, on the railway from Dublin to Londonderry. Pop. 6,116 in 1831, and 5,870 in 1861. The town has suburbs on its E. and W. sides on the mainland, with which it communicates by two handsome bridges: it consists principally of a main street, and is pretty well built. Under a charter of James I., in 1613, the corporation consisted of a provost, 14 burgesses, and a commonalty; but it is now extinct, and the town property is vested in commissioners. The bor. sent 2 mems. to the Irish H. of C., and since the Union has sent 1 mem. to the Imperial H. of C. Registered electors 280 in 1865, all 8l. rated occupiers. A bor. court is held on Thursday for the recovery of small sums. The co. assizes are held here, as are the general and petty sessions. The public buildings are the parish church, a R. Cath. chapel, meeting-houses for Presbyterians and Methodists, an infirmary, with a dispensary, a linen-hall, barracks, and the co. court-house and prison. The prison, on the radiating plan, has 56 single cells, and 10 other rooms for prisoners. The Union workhouse, opened in 1845, has accommodation for 1270 inmates. Leather is manufactured to a small extent, and there are 2 distilleries and a brewery. The trade consists in timber, coal, and slate, brought partly by railway, and partly by water from Belleek. Markets are held on Tuesdays and Thursdays, fairs on the 10th of every month, except March, May, and August.

Mr. Inglis speaks in the most favourable terms of the beauty of the country round Enniskillen, and of the town itself. 'I found it one of the most respectable towns I had seen in Ireland; and its population by far the most respectable-looking that I had anywhere yet seen. It abounds in respectable shops; and I never saw shops better filled than they were on the market-day. I understand that many of the tradespeople are wealthy, and that the retail trade is brisk and profitable. The town stands almost wholly on the estate of Lord Enniskillen.' (Inglis's Ireland, ii. 152.) The corporation revenue, derived principally from tolls, amounts to about 600l. a year.

From its position, the possession of Enniskillen has always been of importance in Irish contests. It distinguished itself during the war of 1689, by its attachment to the liberal side, and by its resisting and defeating a superior force sent to re-dare it by James II. Part of the brave defenders of Enniskillen were subsequently formed into a regiment of cavalry, which still retains the name of the Enniskillen dragoons.

ENNS, or ENS, a town of Upper Austria, circ. Traun, on the Enns, near its junction with the Danube, 10 m. SE. Linz, on the railway from Linz to Vienna. Pop. 3,755 in 1858. The town, which is placed upon a steep hill, is well built, and contains a lofty tower erected by the Emperor Maximilian. The expense of building the old walls of Enns was defrayed by a part of the ransom of Richard I. of England. Enns possesses some linen, steel, and hardware manufactories, and breweries. It is supposed to stand upon or near the site of the an. *Lauriacum*, where a persecution of the Christians took place under Galerius in 304. Many Roman antiquities have been found in its vicinity.

ENOS (an. *Ænos*), a marit. town of Turkey in Europe, sanjac of Gallipoli, at the extremity of a long, low, narrow tongue of land forming the S. boundary of the Gulf of Enos, 86 m. NW. Gallipoli; lat. 40° 41' 58" N., long. 25° 54' 44" E. Enos, pop. 8,000. Being situated near the mouth of the Maritza, it is, to some extent, the sea-port of Adrianople, and is very advantageously situated for commerce. However, a sand-bank, which increases every year, has been allowed to form at the entrance to the port. The consequence is, that the town stands in pools and swamps of water, which not only produce pestilential fevers that extend to Adrianople, but are the greatest impediments to trade. Formerly, large vessels used to enter the port; but now even the small craft from Smyrna are obliged to unload outside the bank. The Maritza is navigable up to Adrianople in winter and spring for vessels of considerable burden, but in summer the sea craft only ascend as far as Demotica. (ADRIANOPLE.)

EPERIES, or PRESSOVA, a fortified town of Upper Hungary, co. Saros, of which it is the cap., on the Tarcza, an affluent of the Theiss, and near the Carpathians, 140 m. NE. Pesth; lat. 48° 58' 45" N., long. 21° 13' 49" E. Pop. with its suburbs, 9,610 in 1858. It is one of the best built towns in this part of Hungary, and contains four Catholic churches, a Lutheran church, chapter-house, synagogue, &c. and town halls; a Catholic gymnasium and high-school, Lutheran college, episcopal library, and a place of resort for the religious termed 'Calvary.' It is the residence of a bishop of the United Greek church, and the seat of the board of government for Hungary on this side the Theiss. Its inhab. manufacture linen fabrics, for which this town is noted, woollen and hempen cloths, earthenware, and beer; and have a considerable trade in wine, corn, and cattle. Eperies is surrounded with gardens, and a great deal of flax is raised in its vicinity. The town is ill supplied with water for drinking; near it are some warm chalybeate springs used as baths; and at no great distance is the royal salt mine of Soovár.

EPERNAY (an. *Aquæ Pernanæ*), a town of France, dép. Marne, cap. arrond., near the Marne, which is here crossed by a handsome stone bridge of seven arches, 20 m. WNW. Chalons, on the railway from Paris to Chalons. Pop. 10,521 in 1861. The town was formerly a place of some strength, but its walls are now fallen into a state of decay. Though irregular, it is neat and well built. In one of its open squares is a handsome new church, of the Doric order. Epernay has a theatre, a communal college, and a public library, containing 10,000 vols.; and is the seat of a sub-prefecture and of tribunals of primary jurisdiction and commerce. But the grand distinction of Epernay consists in its being the principal entrepôt for the wines of Champagne, the best of which are produced in its immediate vicinity. Its celebrated wine vaults are excavated in the chalk rock on which the town is built. They are admirably fitted for the stowage and improvement of the wine, are of vast extent, and as solid as if they were supported by arches. The wines are classed *par treilles*, otherwise *par crus*, or growths. Few travellers stop at Epernay without visiting these vaults. This, however, is not always free from danger, especially with newly bottled wines, in the months of June and August, when the vine is in blossom, and when the grape begins to ripen. At such periods the bottles frequently explode with great violence; and fatal accidents have in consequence happened to workmen and visitors, who have neglected to use the precaution of covering themselves with iron masks provided for the purpose.

Epernay has sustained several sieges, especially that in which it was taken by Henry IV. in 1592. Previously to that period it had been burnt by Francis I. to prevent its falling into the hands of Charles V.

EPHESUS, an ancient and now ruined city of Asia Minor, called by Pliny the light of Asia—*lumen Asiæ* (Hist. Nat. lib. v. § 29), and famous alike in sacred and profane history, on the S. side of the Cayster, near its embouchure on the W. coast of Ionia, and near the modern village of Aiasaluck, 88 m. SSE. Smyrna. The epoch of its foundation is very remote, being ascribed by some to the Amazons; but it subsequently received a colony of Ionian Greeks under Androclus, the son of Codrus; and thenceforth occupied a distinguished place among the twelve confederated Ionian cities of Asia Minor. From the remotest period, Ephesus was celebrated for a temple of Diana, hence called the Ephesian goddess, in its immediate vicinity; and on being besieged by Crœsus, the inhab. made an offering of their city to Diana, uniting it to her temple by a rope seven stadia (7,000 m.) in length. (Herod. lib. i. § 26.) Subsequently to this period the original city was gradually abandoned, and a new one grew up round the temple; but its situation was again changed, especially by the interference of Lysimachus, who is said to have compelled a portion of the inhab. to resort to a new town he had built on higher ground. Ephesus, Miletus, and the other Ionian cities, were early distinguished by their commerce, and became among the greatest emporiums of the ancient world. The wealth they had thus accumulated enabled the Ionians to erect, at their joint expense (*factum a tota Asia*, Plin. lib. xxxvi. § 21), a noble temple in honour of Diana, in which was placed her image in ivory, said to have been sent down from heaven by Jupiter, but which was really the work of an artist named Canitia. (Plin. lib. xix. § 4.) This sacred edifice, accounted one of the finest structures of its time, escaped that destruction in which all the other Greek temples of Asia Minor were involved through the impotent fury of Xerxes, after his expulsion from Greece. But it soon after fell a sacrifice to the insane rage for notoriety of an obscure individual of the name of Herostratus, who, to perpetuate his memory, set fire to the temple. (Val. Max., lib. viii. § 14.) The Grand Council of Ionia endeavoured to disappoint the incendiary, by passing a decree that his name should not be mentioned. (Aul. Gell. Noct. Attic., lib. ii. § 6.) But it was divulged by the historian Theopompus. (Val. Max., ubi supra.) This event is said to have occurred on the night in which

Alexander the Great was born. (Cicero de Nat. Deorum, lib. ii. § 27.) At a subsequent period, Alexander offered to rebuild the temple, provided he were allowed to inscribe his name on the front; but this was declined by the Ephesians, who, principally at their own cost, but partly, also, by the voluntary contributions of others, raised a new temple to the goddess far transcending its predecessor, and such as entitled it to be ranked among the seven wonders of the world. To lessen the risk of injury from earthquakes, it was built on the margin of a marsh, its foundations costing an immense expense. It was 425 ft. in length, 220 do. in breadth, and adorned by 127 columns of the Ionic order, each 60 ft. in height. (Plin. Hist. Nat. lib. xxxvi. § 14.) The altar was the work of Praxiteles; the famous sculptor Scopas also contributed to the embellishment of the fane, which, among other *chefs-d'œuvres* of art, could boast of a noble picture of Alexander the Great, by Apelles, a native of the city. An extensive sanctuary was attached to the temple; but this privilege was annulled by Tiberius, on account of the abuses to which it led.

The worship of Diana was entrusted to the care of a number of priests (*Essenes and Essenes*), and a select band of virgin priestesses; and to prevent the chance of any breach of that chastity so dear to the goddess, the former were emasculated. (Strabo, lib. xiv. p. 641.) A great festival in honour of Diana was annually celebrated at Ephesus, under the presidency of *Asiarchs*, or deputies sent by the different Ionian cities, which was resorted to not only by crowds of visitors from all parts of Ionia, but also from all parts of Greece and Magna Græcia, or S. Italy. Games were then celebrated with extraordinary magnificence; and the city was crowded with the votaries of pleasure and traffic, as well as of religion.

Owing to the gradual silting up of the harbour by the deposits brought down by the river, the commerce of the city was laid under considerable difficulties; but every one knows that, though it had undergone many vicissitudes, it had lost nothing of its ancient fame and celebrity when it was visited by St. Paul. Although the city then was, 'threat is Diana of the Ephesians!' (Acts xix. 28, 34,) her worship was doomed speedily to decline. St. Paul resided here for three years, and founded a church that became, as it were, the metropolis of Asia. (Acts xx. 31.) Among his other enormities, Nero is said to have despoiled the temple of Diana of several of its sacred offerings, and of a large amount of treasure. But it recovered, in some degree, from this attack; and continued to attract some portion of its ancient veneration, till it was finally burned by the Goths in the reign of Gallienus. Besides Apelles, his great rival Parrhasius, Heraclitus the philosopher, Hipponax the poet, and Artemidorus the geographer, were natives of Ephesus; but its inhabitants were distinguished more by their voluptuousness, refinement, and traffic, than by their taste for learning or philosophy. They are also said to have been addicted to sorcery, and such like arts. What were called the *Ephesian letters*, appear to have been magical symbols inscribed on the crown, girdle, and feet of the statue of Diana, in the great temple; and it was believed that whoever pronounced them, had forthwith all that he desired! (Gibbon, cap. 10; Dictionnaire de Trevoux, art. *Ephèse*.)

The walls, which may be still traced, embrace, according to Pococke, a circuit of about 4 m. Besides its temple, Ephesus had many noble buildings, among which may yet be seen the ruins of a circus, a theatre, and gymnasium. But the ravages of earthquakes and other convulsions of nature have completed the ruin of this once famous city; and her ancient magnificence is indicated by the extent, rather than the preservation, of her remains. The ancient aqueduct, of which a portion still exists, is ascribed to the Greek emperors. Her 'candlestick has been removed out of his place,' (Rev. ii. 5.) In 1764, when Ephesus was visited by Dr. Chandler, 'its population consisted of a few Greek peasants, living in extreme wretchedness, dependence, and insensibility; the representatives of an illustrious people, and inhabiting the wreck of their greatness; some the substructure of the glorious edifices which they raised; some beneath the vaults of the stadium, once the crowded scene of their diversions; and some in the abrupt precipice, in the sepulchres which received their ashes. . . . Ephesus was a ruinous place when the Emperor Justinian filled Constantinople with its statues, and raised the church of St. Sophia on its columns. Since then it has been almost quite exhausted. Its streets are obscured and overgrown. A herd of goats was driven to it for shelter from the sun at noon; and a noisy flight of crows from the quarries seemed to insult its silence. We heard the partridge call in the area of the theatre and of the stadium. The glorious pomp of its heathen worship is no longer remembered; and Christianity, which was there nursed by apostles, and fostered by general councils, until it increased to fulness of stature, barely lingers on in an existence hardly visible.' (Tour in Asia Minor, p. 150, 4to. ed.; see also the Antiquities of Ionia by the Dilettante Society, where plates and measurements are given of the principal extant ruins; Tournefort, ii. 613–623; Ancient Universal History, vii. 416, 8vo. ed.; Cramer's Asia Minor, i. 363, &c.)

EPINAL, a town of France, dép. Vosges, of which it is the cap., on both banks of the Moselle; 36 m. SSE. Nancy, 65 m. NNE. Besançon, and 290 m. ESE. Paris, on the railway from Nancy to Belfort and Besançon. Pop. 11,857 in 1861. The town was formerly fortified with ramparts, and defended by a castle; but of these, only the ruins of the latter now remain. It is tolerably well built, and, though ill paved, is clean: it has quays and fine promenades along the river. The principal public buildings are the barracks, hotel of the prefecture, 3 hospitals, the church, theatre, public library with 17,000 vols., and a museum of paintings and antiquities. Epinal is the seat of a tribunal of primary jurisdiction, a chamber of manufactures, and a communal college. It has a society of emulation, schools of linear design and music, and a gratuitous course of midwifery; manufactures of embroidery and lace, linens, stockings, paper, paper, and oil; and some trade in corn, cattle, iron, deals, and other timber. It is said to have been founded in the 10th century; in 1466, it came into the possession of John, duke of Lorraine, who granted it many privileges; in 1670 it was taken by the French.

EREKLI. See HERACLEA.

ERFURT, a fortified town in a nearly isolated portion of the Prussian dominions, prov. Saxony, formerly a free imperial city, and now the cap. of a reg. and circ. of same name, on the Gera, a tributary of the Unstrut, about midway between Gotha and Weimar, on the railway from Leipzig to Cassel and Frankfort-on-the-Mayne. Pop. 37,012 in 1861, excl. of a garrison of 4,464. The town is somewhat irregularly laid out, and has no street or square worthy of notice, except the market-place, with a small obelisk, erected in honour of one of the last electors of Mayence, and the Goudroplets, leading to the cathedral. This building, originally a fine Gothic structure, has been seriously injured by the hostile attacks to which the town has been exposed; but considerable sums have recently been expended on its repair. In its tower is a bell 10½ in. thick, 10 ft. high, about 32 ft. in circumference, and weighing 275 cwt. There are 14 other churches; and an Ursuline convent, to which a girl's school is attached. The Augustine convent, in which Luther passed several years of his life, is now converted into an orphan asylum; but the apartment of the Reformer is preserved as nearly as possible in its original condition, and contains his Bible, portrait, and other relics. The town has another orphan asylum, with institutions for the blind; the deaf and dumb; a school for poor children, and a house of correction. Its university, founded in 1392, and suppressed in 1816, has been replaced by gymnasiums for Catholics and Protestants: it has, besides, a teachers' seminary, an academy of sciences, with a library; and a botanic garden with a library of 40,000 vols., which formerly belonged to the university. There are also schools for drawing, mathematics, architecture, commerce, and midwifery; several scientific and literary associations, and cabinets of natural history, medals, and other objects of art. Erfurt is a fortress of the second class, and important from its position on the high road between Frankfort and Leipzic. In addition to its outer ramparts and ditches, it is defended by the fort of Petersberg, built on a hill in its interior, and that of Cyriaksberg without its walls, on a height about 800 ft. in elevation. In the time of Charlemagne, Erfurt was one of the chief commercial cities of Germany, and so late as the end of the 16th century, it is said to have had as many as 58,000 inhab. The business of shoemaking is extensively carried on, and it has manufactures of woollen and cotton cloths, silk ribands and other fabrics, vermicelli, pearl-barley, liqueurs, vinegar, and leather. It is the seat of a local government for its reg. and circle, a board of taxation council and tribunal for the town and circle, but not of a judicial court for its distr. or reg. It first formed part of the Prussian dom. in 1803; from 1807 to 1813 it was occupied by the French, and in 1808 a memorable interview took place in it between Napoleon and Alexander, emperor of Russia. It was restored to Prussia in 1814.

ERIE (LAKE), one of the five great lakes of N. America, between Canada and the U. States, included in the middle portion of the basin of the St. Lawrence. It lies between lat. 41° 22' and 42° 52' N., and long. 79° and 83° W., having N. the fertile peninsula of Upper Canada, and S. and E. the states of Ohio, Pennsylvania, and New York. Its shape is elliptical; length SW. to NE. about 265 m.; breadth varying from 10 m. to about 63 m. in its centre. Its area is estimated in the 'American Cyclopædia' at 12,000, but by Darby at only 9,000 sq. m. It receives near its W. extremity the superabundant waters of the lakes St. Clair, Huron, and the upper lakes by the Detroit river, its own surplus waters being conveyed to Lake Ontario by means of the Niagara, celebrated for its stupendous waterfall. Its mean height above the level of the ocean is estimated at 565 ft., being about 62 ft. below that of Lake Michigan and Huron, and 372 ft. above that of Ontario. Its depth, which is less than that of any of the other great lakes of the St. Lawrence basin, is no where more than 270 ft., and in most parts is considerably under 200 ft. It is also said to be gradually becoming shallower; and in proof of this it is stated in a late Buffalo journal, that Long Point had in three years gained 3 m. on the water;

and that the land is also rapidly gaining along its S. shore. Its bottom appears to be composed of an alluvial deposit of sand and mud, resting on secondary schistose sandstone. (Darby.) Its N. shore is rocky and dangerous; the opposite one has also long lines of rock; and, except at either extremity, none of its shore-harbours afford a safe and steady entrance of 7 ft. water. (Darby.) In addition to other impediments to navigation, a current, not perceptible in the other great lakes of the St. Lawrence system, sets constantly W. and NW. or SW. winds continually prevail; besides which, in consequence of its shallowness, a part of Lake Erie is frozen over every winter, and traffic on it is obstructed by ice for some weeks in the spring after the navigation of the other lakes is open and unimpeded. Towards the W. extremity, there are several groups of small islands, and one —Cunningham Island, belonging to the U. States, —has an excellent harbour called Put-in-Bay, with 12 ft. water. On the N. shore, several promontories stretch into the lake, the principal of which are the N. and S. Forelands, and Point Landguard. Except the Detroit, Lake Erie receives few rivers of any consequence, and all, without exception, have bars at their mouths. The Ouse or Welland, which unites with its E. extremity, is its principal affluent, and has been taken advantage of for the construction of the Welland canal, of which it forms a part, connecting the Lakes Erie and Ontario, and avoiding the Falls of Niagara. (See CANADA.) The Erie canal, 363 m. in length, runs from the town of Buffalo to the Hudson river; the Ohio canal, 334 m. in length, extends from Cleveland at the mouth of the Cuyahoga to the Scioto, a little N. of Columbus. The former of these canals places Lake Erie in communication with the Atlantic; the latter connects it with the Gulf of Florida. (For further particulars respecting these important canals, see CLEVELAND, UNITED STATES, NEW YORK, and OHIO.) Buffalo, Dunkirk, Ashtabula, Erie, Cleveland, Sandusky, Portland, and Detroit are the principal towns on Lake Erie, within the territories of the U. States; and Port Talbot, Dover, and Sherbrooke in those belonging to Great Britain.

ERIVAN, or IRWAN, a town of Asiatic Russia, being the cap. of Russian Armenia, on the Zenghir, an affluent of the Araxes, 31 m. NNE. Mount Ararat, and 106 m. N. by W. Teflis. Pop. 9,510 in 1831. The town contains about 2,000 houses, interspersed with numerous gardens, and ruins of various dates, the whole fortified, and protected by a citadel placed on a steep rock, more than 600 ft. in height, overhanging the river. This fortress, which is about 3,000 yds. in circuit, is encompassed by a double rampart of earth flanked with towers; it contains the ancient palace of the khans, now the residence of the governor; a fine mosque, a cannon foundry, and barracks. The houses in the town are mostly mean, and irregularly built. Erivan has, however, a large and handsome caravanserai, with 780 shops, besides 4 Armenian churches, one Russo-Greek ditto; an Armenian convent, 3 mosques, some aqueducts of a curious construction, and a good stone bridge of several arches, across the river. The town has some manufactures of cotton stuffs, leather, and earthenware. It is a station for caravans from Tiflis and Erzeroum, and has a considerable trade with Russia and Turkey. The epoch of the foundation of Erivan is unknown. It was taken by the Persians from the Turks in 1635. The latter retook it in 1724; but it was again taken by the Persians, under Nadir Shah, in 1748. The Russians were repulsed in an attempt to take it in 1808; but they succeeded in 1827, and were confirmed in its possession by the treaty with Persia of the following year.

ERLANGEN, a town of Bavaria, circ. Central Franconia, on the Regnitz, 23 m. N. Bamberg, on the railway from Bamberg to Nuremberg. Pop. 10,925 in 1861. It is walled and divided into the old and new towns: the latter, which is one of the best-built towns of Germany, was founded by Christian Ernest, margrave of Bayreuth, in 1686. It contains the celebrated Protestant university, the only one in the kingdom, established 1743, and usually attended by about 280 students. This institution occupies the ancient palace of the margraves of Bayreuth, and has connected with it schools of theology, moral philosophy, midwifery, medicine, and the fine arts, a polytechnic school, a gymnasium, general and lying-in hospitals, cabinets of natural history, a botanic garden, and a library of 100,000 vols. The palace gardens are very handsomely laid out, and adorned with statues. Woollen goods, stockings, hats, leather and leathern articles, are made in the town; which has also a large plate-glass manufactory, and a brewery, besides some trade in cattle. Most of the pop. are Protestants. Many French refugees settled in Erlangen after the revocation of the edict of Nantes, and in 1666, the first learned society in Germany was established here.

ERLAU (Hung. *Eger*, Slav. *Jager*), a fortified town of Hungary, co. Hevres, of which it is the cap., on both sides the Erlau, an affluent of the Theiss, 63 m. NE. Pesth. Pop. 18,815 in 1856. The town has 2 suburbs; is entered by 6 gates; and contains a cathedral and 6 other churches, an archbishop's palace, lyceum, with a library and observatory, an archiepiscopal seminary, gymnasium, Catholic high school, conventual hospital, asylum for infirm clergymen, and various other public institutions. The neighbourhood of Erlau is very fertile and highly cultivated; it produces from 180,000 to 200,000 *eimers* of fine red wines annually, and the best tobacco in Hungary. The culture of these articles, together with manufactures of woollen and linen fabrics, leather and leathern goods, employ most of the inhab.

ERNE (LOUGH), a celebrated lake of Ireland, co. Fermanagh, which it divides into two nearly equal portions. It consists of two principal lakes, the Upper and Lower, connected by a broad winding channel. It contains in all an area of about 40,000 acres; and stretches NW. and SE. 20 or 25 m. The lower lake is the largest; and both it and the upper lake are full of islands, some of them large and thickly inhabited, many of them well wooded, and the whole so disposed and accompanied by such a diversity of coast, as to form a vast number of rich and interesting prospects. Enniskillen stands on an island in the channel between the upper and lower lakes; and on another island is the magnificent seat of the Marquis of Ely. The lake is elevated about 140 ft. above the level of the sea. It receives the Erne and several other rivers; and discharges itself at its NW. extremity by a rapid current of about 8 m., which after falling over many ledges of obstructing rocks, precipitates itself down a grand cataract into the sea at Ballyshannon. It has been proposed to open a navigable channel from the sea to the lake by means of a canal, which would certainly be of material service.

ERZEROUM or ERZ-RUM (Arab. *Arzen-el-Room*), an important city of Turkish Armenia, cap. of an extensive pachalic of the same name, and residence of a seraskier pacha; in a plain at the foot of the Tchehlir mountains, near the sources of the N. arm of the Euphrates, from

6,000 to 7,000 ft. above the level of the sea, 134 m. SE. Trebizond, 144 m. NE. by E. Diarbekr, and 150 m. WNW. M. Ararat; lat. 39° 59' 80" N.; long. 41° 46' 15" E. Its pop. has been variously estimated at different periods; but previously to the ravages of the plague in 1821, it amounted to near 100,000, and at the time of the Russian invasion in 1829, it is supposed to have been about 70,000 or 80,000. But having been abandoned by most Armenian families, previously to its being again delivered up to the Turks, it had not, in 1835, according to Mr. Brant, above 15,000 inhab. (Geog. Journal, vL 201.) Probably, however, this estimate was below the mark, and the pop. has since increased.

Only the citadel, which occupies a low eminence within the city, is now fortified. A trench and two wells once surrounded it; but the inner wall only is now entire. It is solidly built of stone, and does not suffer in comparison with Turkish fortresses in general. Besides the bazaars, the principal mosques, and many private dwelling-houses, it formerly enclosed the palace of the pacha; but that extensive building was demolished by the Russians. (Missionary Researches in Armenia, pp. 63, 64.) According to Mr. Kinneir (Asia Minor, p. 808), the citadel is 3 or 4 m. in circ. Capt. Wilbraham entered it by a strong and massive gateway, flanked by two mutilated though still beautiful minarets. Most of the Turkish inhab. reside within the citadel. The streets of the city, which may be regarded as a suburb attached to the citadel, are narrow, crooked, filthy, and infested with troops of hungry dogs. The houses are mostly constructed of mud, wood, or sun-dried bricks, being, in general, only one story high. A green sward has grown over the terraces of dirt, by which, instead of roofs, they are all covered, and gives them, when viewed from an eminence above, almost as much the aspect of a meadow as of a city. The environs are singularly destitute of trees, the dried faeces of the cattle being the only fuel. Water is good and abundant, but wine, according to Tournefort, is execrable. (Lettres du Levant, ii. 259, 4th ed.) Erzeroum has two Armenian churches, a Greek church, and about 40 mosques, the largest of which will accommodate 3,000 people. It has an extensive custom-house and 36 khans or inns, many of which are large and solidly constructed. Its bazaars are poor and small, though its markets appear to be well supplied with provisions; and a great many oxen are killed weekly. The city is well situated for trade, on the high road between Asia Minor, Georgia, and N. Persia; and it was once the thoroughfare for most part of the overland commerce between Europe and the East, which survived the discovery of the passage round the Cape of Good Hope. Recently its commerce has been diminished from a variety of causes; but mainly from the emigration of its Christian inhab., who were its mechanics and tradesmen, to the different possessions of Russia. The manufacture of copper utensils, which once formed the principal branch of industry, is now almost abandoned; but it still continues to have some trade in furs, galls, &c. The amount of goods that passes through Erzeroum, in transitu, is very considerable; and Capt. Wilbraham, who visited the city in 1837, says that it had materially increased since the establishment of steam-boats on the Black Sea. From the E., the shawls of Cashmere and Persia, silk, cotton, tobacco, rice, indigo, madder, rhubarb, and a variety of drugs, are brought to Erzeroum; and, from the W., broadcloths, chintzes, shawls, and cutlery. Little, however, is seen of any of these goods, except at the custom-house and in the khans; so much is this the case, that, according to Tournefort, a person might die for want of a few grains of rhubarb at the very moment that there are bales of it in the town. (Tournefort, p. 202.) The limited extent and meanness of the bazaars evince the small importance of the retail trade.

This city is a principal halting-station for caravans of pilgrims from Tehran and elsewhere, to Mecca. Of its 70,000 inhab. previously to 1829, it was estimated that 23,000 were Armenians, and the rest principally Turks, with about 250 Greeks. The city had no Jewish inhab. Of the Armenians, about 4,000 belonged to the Rom. Cath., and 19,000 to the Armenian church. The diocese of the Armenian bishop includes the whole pachalic of Erzeroum, which, since the late war, has been much extended, and now comprises the former pachalic of Kars. There was in 1829 an Armenian grammar school in the city, with 6 or 7 teachers, and from 500 to 600 scholars, besides a seminary for the instruction of the Armenian clergy; and a comparatively large proportion of the pop. were then reputed to possess the rudiments of education.

Owing to the elevation of the place, the winters are long and severe. In the neighbourhood, however, cattle, sheep, horses, &c., of superior kinds, are reared in great numbers; and in the adjacent plain, corn of a very excellent quality is grown, which forms one of the principal articles of export.

Erzeroum was founded, about 415, by a Byzantine general of Theodosius II., after whom it was named Theodosiopolis. It derives its present name from the an. Arez or Ardaze, a populous city which stood not far to the E., but which, having been destroyed by the Seljukians, the surviving inhab. transferred their residence and the commerce and name of their city to the present site. (Missionary Researches in Armenia, 63.) This was anciently the strongest of the Armenian possessions of the Lower Empire; and it is at present considered the bulwark of those belonging to Turkey.

ESCURIAL, or **ESCORIAL**, a celebrated palace, convent, church, and mausoleum of the sovereigns of Spain, Old Castile, prov. Segovia, 25 m. NW. Madrid, on the railway from Madrid to Burgos. The name of the place, according to Casiri, is of Arabic origin, signifying a place full of rocks; though others derive it from a Spanish word implying the scoria, or scum of melted metal, some iron mines having been formerly wrought in the locality. Its situation certainly bears out the former etymology. It has a most gloomy site, surrounded by the bare crags of the Sierra Guadarrama. The view from it, though extensive, is not pleasing; and the facility of procuring stone for its construction would seem to have been the only inducement to the choice of its site. It was commenced in 1563 by Philip II., and finished in 21 years, under the superintendence of two architects. It is one of the largest and most magnificent edifices in Europe, though far from being externally the most elegant. It is dedicated to St. Lawrence; and as this saint is said to have been broiled alive on a gridiron, in the 3rd century, the founder chose to have the building on the plan of that culinary instrument, the bars of which form several courts, while the handle contains the royal apartments. (Twiss, p. 99.) The handle is about 460 Spanish feet in length; the principal front of the main portion of the building is 657 ft. (740 Span.) broad; the sides 494 ft. (580 Span.) in depth; and the general height of the edifice is about 60 ft.; a square tower, about 200 ft. in height, flank-

ing each angle. It is wholly built of a grey stone, called Berroqueña, resembling a kind of granite, though not so hard. The Doric order prevails in its architecture. The most striking part of the Escurial is the church in its centre. It is built with a cupola and two towers, after the manner of St. Peter's at Rome; its dome is 330 ft. high. Mr. Inglis, who visited it in 1830, observes that its interior exceeded in richness and magnificence any thing that he had previously imagined. 'It is quite impossible,' he says, 'to enter into minute descriptions of all that composes this magnificence: the riches of Spain and her ancient colonies are exhausted in the materials; marbles, porphyries, jaspers, of infinite variety and of the most extraordinary beauty,—gold, silver, and precious stones; and the splendid effect of the whole is not lessened on a nearer inspection; there is no deception, no glitter—all is real. The whole of the altar-piece in the Capilla Mayor, upwards of 90 ft. high and 50 broad, is one mass of jasper, porphyry, marble, and bronze, gilded; the 16 pillars that adorn it, each 18 ft. high, are of deep red and green jasper, and the interrals are of porphyry and marble of the most exquisite polish, and the greatest variety of colour.' (Inglis, i. 265.) The celebrated crucifix of Benvenuto Cellini, formerly in the possession of the Medici family, is, or was, in this church. The ceiling is covered with the admirable frescoes of L. Giordano, comprising a consecutive history of the Christian religion and other subjects, and which are considered to be excelled only by the works of M. Angelo. The sacristy, for its decorations, equals in beauty any part of the Escurial; and contains some of the choicest works of the most illustrious painters. Of the 42 pictures that adorn the sacristy, it may be said, what can rarely be said of any collection, that 'there is not one that is not a chef-d'œuvre.' There are 8 of Raphael, including the celebrated La Perla, and the Madonna della Pesce, 7 of Leonardo da Vinci, 6 of Titian, and many of Tintoretto, Guido, Paul Veronese, &c. The reliquary of the convent contains, of course, an abundance of relics. The library of printed books contains about 24,000 vols., many of which are very scarce. The manuscript library, more valuable than the former, comprises about 4,000 MSS., in Arabic, Latin, Greek, Hebrew, &c., including several of the 9th and 10th centuries. This library suffered greatly from a fire in 1661. The royal mausoleum beneath the church is a most magnificent sepulchre. It is of a circular form; the walls of jasper and black marble; and in rows, one over another, are ranged the coffins of the sovereigns of Spain. Here are the urns of 8 kings and 8 queens, on opposite sides of the mausoleum; the former including the emperor Charles V. and his son Philip II.: numerous other royal personages are buried in a chapel in the Escurial, called the Pantheon of the Infantas. The palace adjoining the monastery would any where else be considered a splendid edifice, but here it is comparatively little worthy of notice, from its inferiority to the rest of the Escurial. The total expense of raising this immense pile of building is said to have amounted to 6,000,000 piastres. The French carried away a great quantity of gold, silver, gems, and other valuables from the Escurial; but, on the whole, they treated the edifice with greater forbearance than might have been anticipated. When Mr. Inglis visited it in 1830 there were about 100 resident monks of St. Jerome living, not as ascetics, but in a state of luxurious indulgence. The revenue of the monastery formerly amounted to 12,000l. a year. A straggling village of 1,000 inhab., called Escurial, or San Lorenzo, adjoins this wonder of Spain. (For an elaborate and excellent account of the curiosities of the Escurial, see Tate's Travels in Spain, 96-154; also Inglis's Spain in 1830, 262-281; and Townsend, ii. 119-172, &c.)

ESKI-SAGRA (an. *Beraea?*), a town of Turkey in Europe, prov. Roumelia, on a tributary of the Tundja, near the N. foot of the Balkhans, on the high road from Constantinople, and Adrianople, to Widin; 68 m. NW. Adrianople, and 76 m. SE. Nhumla. Estimated pop. 15,000. The town is finely situated on the declivity of some well cultivated hills, but is very indifferently built, with narrow dirty streets: it is surrounded by a rampart of earth, has 8 mosques, with manufactures of carpets and coarse cloth. There are numerous orchards in its vicinity, and, at a short distance, are some well-frequented warm mineral baths.

ESNEH (the *Latopolis* of the Greeks), a town of the Thebaid or Upper Egypt, on the W. bank of the Nile; 24 m. N. Thebes; lat. 25° 17' 30" N., long. 32° 29' 55" E. The valley of the Nile is here about 4 m. in width; it is, however, too much elevated to be covered by the inundation; and the canals by which it had been irrigated having been allowed to fill up, it had become in a great degree barren. But Mehemet Ali has succeeded in reopening these canals, so that the ancient fertility of the district has been in part recovered, and it has become the seat of extensive cotton plantations. The town, seated on a mound of débris, 30 ft. in height, is the principal commercial place in Upper Egypt. It is the entrepôt for the Sennaar caravan, while the Ababdie camel breeders of the desert bring their camels, and the Berbers from Nubia their commodities, to sell in its markets. It has also some manufactures, particularly of *melayeh* or cotton shawls, much worn in the country, and pottery. It is the seat of a Coptic bishop, and numbers among its inhab., from 300 to 400 Christian families, who have two churches, and a third further up the country. There is a Coptic monastery to the N. of the town. (Ritter's Africa, iii. § 20; Jowitt's Christian Researches.)

In the centre of the town is a famous temple, built of sandstone, and of colossal magnitude. Having been made a magazine for the warehousing of the cotton of the surrounding district, it has fortunately escaped the destruction that has lately overwhelmed some of the finest Egyptian monuments. The walls of this temple are covered (crept) with the mud of the Nile; and it is so encumbered with mud walls, sand, filth, and cotton, that it is difficult to form a correct idea of its form and vast size. It has a zodiac somewhat resembling that at Denderah; and from the mode of interpreting the figures on it, this temple was long supposed to be the most ancient in Egypt; but so far from this being the case, it is, according to Champollion, '*le plus moderne de ceux qui existent encore en Égypte; car les bas-reliefs qui le décorent, et les hiéroglyphes surtout, sont d'un style tellement grossier et tourmenté, qu'on y aperçoit, au premier coup d'œil, le point extrême de la décadence de l'art.*' (Lettres, 199.) This conclusion is established by the hieroglyphic inscriptions, which show that the oldest part of the temple, a small portion of the pronaos or portico, was built by Ptolemy Epiphanes; but that the portico was principally constructed by the Emperor Claudius; and that the other parts of the structure belong to a still later era, or to that of various Roman emperors, from Claudius to Septimius Severus and Geta. It appears, however, notwithstanding the comparative lateness of the temple, that Esneh had been a place of much importance under the

Pharaohs, fragments of edifices having been discovered bearing hieroglyphical inscriptions that refer to their era. Champollion supposes that these ancient edifices had been destroyed during the Persian invasion. Immediately opposite to Esneh, on the opposite side of the river, at what was called Contra Latu, was a small temple; but this interesting relief no longer exists. It was demolished about a fortnight before Champollion visited the place, and its stones carried off to repair the quays at Esneh. (Lettres, 107.)

ESSECK, or ESSEG (Slav. *Ossick*; an. *Mursia*, or *Mursa*), one of the most strongly fortified towns in the Austrian empire, the cap. of Slavonia, and seat of the government of that prov., on the Drave, 19 m. from its confluence with the Danube, 68 m. WNW. Peterwardein, and 131 m. S. by W. Buda; lat. 45° 34' 13" N., long. 11° 42' 5" E. Pop. 13,000 in 1848. The greater number of the inhabitants are of German descent. The modern fortress was erected upon the site of a previous one, by the Emperor Leopold I., between 1712 and 1719; it is not extensive, but is well constructed, contains an arsenal and barracks capable of accommodating 30,000 men, and is strengthened by a *tête de pont* on the opposite side of the river; the houses and other buildings within it are generally lofty and massive. It is surrounded by a broad glacis, and communicates on the NW., by a long avenue, with the Ober-Varos, or upper town; on its E. side is the *Unter-Varos*, or lower town, on the site of the an. Mursia, and on the W. the *Meinhófe*, or new town, in which suburb most of the trade is conducted. Essech has a due military parade, and contains five Catholic churches, a united Greek church, four chapels, a town council house, county hall, engineers' college (*Ingenieurschau*), military school, Catholic gymnasium, high and other schools, and various other public establishments. In the arsenal numerous banners and other trophies, taken at different times from the Turks, are exhibited. The Drave, and the swampy country on the side opposite the town, are crossed by a long wooden bridge. It has manufactures of silk stuffs and twist; but the chief commercial importance of Essech is derived from its large and well-frequented fairs for corn, horses, cattle, and hides, held four times a year.

Mursia was founded by Hadrian, anno 125, and became the Roman cap. of Lower Pannonia; it was erected into a bishopric by Constantine.

ESSEN, a town of Rhenish Prussia, distr. Düsseldorf, circ. Duisburg, on the Berne, 18 m. NE. Düsseldorf, and 42 m. SE. Cleves, on the railway from Düsseldorf to Hanover. Pop. 20,811 in 1861. The town is walled, and has several Catholic and Lutheran churches, a Capuchin convent, a gymnasium, hospital, workhouse, and orphan asylum. It is the seat of a municipal court of justice, and the mining board for the towns of Essen and Werden; as it was formerly of the diets of the librarish princes and other distinguished assemblies. The inhab. of this industrious and thriving town are employed in a great many different manufactures, including those of woollen and linen goods, leather, vitriol, arms, cast-iron and steel articles, gas apparatus, and steam-engines, as well as in dyeing woollen stuffs, and coal mines in the vicinity. The celebrated cast steel manufactory of Herr Krupp, the largest in the world, turning out annually above 12,000,000 lbs., is near Essen.

ESSEQUIBO. See GUIANA.

ESSEX, a marit. co. of England, having E. and S. the German Ocean and the Thames, N. the cos. of Suffolk and Cambridge, and W. Herts and Middlesex. Length, 47 m.; breadth, 52 m.; area, 1,057 sq. m., or 1,055,549 acres. Surface generally flat, but in parts undulating. Soil mostly loam, and extremely fertile; but in the NW. part of the co. there is some chalk land; the low grounds along the Thames and the ers are in parts marshy and very rich. In parts of the coast the land is indented by arms of the sea, forming a series of islets and peninsulas: some salt marshes along the shore are protected from inundation by embankments. The low grounds are subject to fever and ague, but otherwise the co. is sufficiently healthy. Tillage husbandry in an advanced state. Wheat and barley are the principal corn crops; the ground is in most parts unsuitable for turnips, and fallowing is very extensively practised: beans, however, are frequently substituted for fallows on the heavy loams; and this practice is gaining ground. Potatoes are extensively cultivated. The quality of Essex wheat is very superior. The suckling of calves for the London markets, and the grazing and dairy business, are both carried on to a considerable extent. The district of Epping is celebrated for its butter, which is probably superior to that of any other part of England. The total stock of sheep is estimated at between 500,000 and 550,000 head, and the annual produce of wool at between 8,000 and 9,000 packs. Farms of all sizes, from 5l. to 20,000l. a year. Many small and moderate sized farms occupied by their owners. Some of the hired farms in this co. are amongst the largest of any devoted to tillage in the empire. Leases when granted are usually for 7 and 14 years; but they are not so common now as formerly. Minerals, with the exception of the lime and chalk quarries at Purfleet, unimportant. Manufactures, principally of baize and other woollen stuffs, were formerly carried on at Colchester, Coggeshall, and other places, but they have now nearly disappeared. Principal rivers, Roding, Crouch, Chelmer, Blackwater, Colne, which intersect the co., exclusive of the Thames, Lea, and Stour, which bound it on the SW. and N. Oysters are raised in large quantities in the Essex rivers, especially the Crouch and Blackwater. Principal towns, Colchester, Chelmsford, Maldon, and Harwich. Essex contains 70 hunds. and 408 parishes. It sends 10 mems. to the H. of C., viz. 4 for the co., and 2 each for the bors. of Colchester, Harwich, and Maldon. Registered electors for the co., 12,650 in 1865, of whom 5,434 for the northern, and 7,166 for the southern division. Pop. 404,851 in 1861, living in 81,261 houses. Gross annual value of real property assessed to income tax—Northern division, 830,474l. in 1857, and 959,612l. in 1862; southern division 1,065,077l. in 1857, and 1,286,556l. in 1862.

ESSLING, a village of Lower Austria, on the left bank of the Danube, about 7 m. below Vienna, opposite the island of Lobau. This and the contiguous village of Aspern were the scene of a tremendous engagement of two days' duration (21st and 22d May, 1809), between the great French army, under Napoleon, and the Austrians, under the Archduke Charles. (See ASPERN.)

ESSLINGEN, a town of Würtemberg, circle Neckar, cap. of a distr., on the Neckar, in a fertile plain, 6 m. ESE. Stuttgard, on the railway from Stuttgard to Ulm. Pop. 12,521 in 1861. The town is walled, and has 5 suburbs and 9 churches, one of which, a Gothic edifice built in 1440, has a tower 230 ft. high; a handsome town-hall, a court of justice, a richly endowed hospital, with a high school, and teachers' seminary. The Neckar here divides into 2 arms, and is crossed by 2 bridges: on the island which it encloses, one of the suburbs is placed. An old castle above the town commands a fine view of the surrounding country. There are manufactures of woollen cloth and other stuffs,

cotton and woollen yarn, lacquered tin ware, and glue; there are also some breweries, and a factory for bleaching. Vineyards, orchards, and kitchen gardens are numerous in the vicinity. Esslingen is a very ancient town, and previously to 1803 ranked as one of the five cities of the German empire.

ÉTAMPES, a town of France, dép. Seine-et-Oise, cap. arrond., in a fertile valley, on the banks of two small rivers, 23 m. S. Versailles, on the railway from Paris to Orleans. Pop. 8,720 in 1861. The town is well built, and consists, together with its suburbs, of one street, extending for 2 m. along the road between Paris and Orleans. It has 4 par. churches, a hospital, a theatre, and a tower, the only remains of a ancient castle. It is the seat of a sub-prefecture, a tribunal of primary jurisdiction, and a communal college; has straw-hat, soap, leather, and woollen manufactures, many flour mills, and a large trade in corn. In middle-age Latin this town was called Stampae: anno, 604, Thierry II. defeated his uncle Clotaire near it in a sanguinary battle.

ESTE (an. Ateste), a town of Austrian Italy, prov. Padua, cap. of a distr., at the foot of the Euganean Hills, on the Restara canal; 15 m. SW. Padua, and 42 m. SE. by Verona, on the railway from Padua to Ferrara. Pop. 10,631 in 1858. The town is well built, has a fine market-place, several handsome edifices, numerous churches, a hospital, and a large barrack; with manufactures of silk-twist and hats. The town is chiefly known from its having given its name to the illustrious family of Este, allied with the Guelphs, different branches of which now fill the thrones of Great Britain, Brunswick, and Hanover.

ESTELLA, a city of Spain, prov. Navarre, 25 m. SW. Pampeluna, on the Ega, a little below its confluence with the Amescoa. Pop. 5,593 in 1857. The town is situated in a pleasant valley, surrounded by hills clothed with vines and olives, and producing wheat, barley, oats, maize, and other grain. Streets ill-paved and dirty. It has 8 churches, 7 convents, and a hospital. In former times it had a castle that was deemed impregnable, and was the head-quarters of the military force of the king of Navarre. There were formerly 4 bridges over the river; but one of them was swept away in 1801. In its centre is a handsome promenade, planted with elms, limes, and poplars. It has manufactures of woollen cloths and cassimeres, with oil presses and brandy distilleries. A fair is held here from the 11th to the 30th of November. At a short distance from the town is the university of Larche, which has the same privileges as those of Salamanca and Valladolid.

ESTEPA (an. Astapa), a town of Spain, prov. Seville, cap. of a dep., on a hill surrounded by plains, planted with olive trees, 16 m. W. Osuna, and 50 m. W. city of Seville. Pop. 8,133 in 1857. The town is regularly built, and the houses are in tolerable condition. It has 2 churches, 3 convents, a hospital, a public granary, and a palace of the marquises of the same name. Astapa was a place of importance in the time of the Romans, and was burnt by its inhab. when besieged by Scipio's generals.

ESTEPONA, a sea-port town of Spain, prov. Granada, on the Mediterranean, 24 m. NE. Gibraltar. Pop. 9,316 in 1857. The town is tolerably well built; has a church, a hospital, a public granary, and a castle. The chief support of the place is its coasting trade: it exports raisins, figs, sweet potatoes, oranges, lemons, and wine; for which it receives wheat and other grain. In this ways it employs about 100 vessels. It has also a productive fishery of sardines.

ESTERHAZY (Hung. *Eszterhaz*), a village of Lower Hungary, co. Oedenburg, near the SE. extremity of the Neusiedl lake, 14 m. SE. Oedenburg, and 36 m. S. by W. Presburg. Pop. 405 in 1858. The village is celebrated for a magnificent palace, belonging to Prince Esterhazy, built in 1700, in the florid Italian style. It comprises 168 different apartments, and is surrounded by a gallery adorned with numerous vases, statues, &c. It formerly contained fine collections of paintings, engravings, Chinese porcelain, and a library; but most of them have been removed. It has attached to it an observatory, riding school, stabling for 100 horses, and an opera-house, in which the incident occurred which opened to the composer Haydn his subsequent career of celebrity. The palace is surrounded by a noble park, and has an orangery, numerous fountains, fish-ponds, and a pheasantry; but the gardens are overgrown with weeds; and the numberless pleasure-houses with which the grounds are crowded are fast falling into decay, the family having, for the most part, abandoned this noble seat for that of Eisenstadt. This, which also adjoins the lake, is, like Esterhazy, in the Italian style, of large dimensions, and well fitted for a princely residence. It was rebuilt in 1805, and is distant 24 m. NW. Esterhazy. The grand ball-room is a noble apartment. Its park and gardens are much admired; and the botanical collections in the large hot-houses of the latter are surpassed by few in Europe; they comprise no less than 70,000 exotics, and are particularly rich in Australian species. The Leopoldine temple in the park has a statue of the Princess of Lichtenstein, by Canova.

The estates of Prince Esterhazy are said to equal the kingdom of Wurtemberg in size; and contain 130 villages, 40 towns, and 34 castles. But the annual revenue from these vast possessions is said not to exceed 200,000l. per annum, though it is capable of considerable increase. The family of Esterhazy professes to trace its descent from Attila.

ESTHONIA, or REVEL, a marit. gov. of Russia in Europe, in the NW. part of which it is situated, forming one of the Baltic prov. It lies between lat. 58° 20' and 59° 30' N., and long. 22° 30' and 28° 20' E., having E. the gov. of Petersburg, S. the lake Peipus and the gov. of Riga, W. the Baltic, and N. the gulf of Finland. Area, inclusive of the islands belonging to it, about 6,870 sq. m. Pop. 310,400 in 1846, and 308,474 in 1858. Surface generally flat, but diversified in parts with undulating hills; it contains many small lakes and streams, but has no navigable river: its shores are bold and rocky, climate rigorous, the winters are long, and fogs and violent winds are common throughout the year. Soil in great part sandy, and rather infertile; the cultivable lands are supposed to be in the unproductive, forests, &c. as 1 to 3. Agriculture is the chief employment of the pop., and more corn is produced than is sufficient for home consumption: it is principally rye, barley, and oats; but wheat and buck-wheat, besides flax, hemp, hops, and tobacco, are also raised. Most part of the corn not required for food is set aside for the purpose of distillation. Different species of pulse are extensively cultivated, and form a large proportion of the nourishment of the peasantry. Fruit trees are neglected; but certain wild fruits are very abundant. The pine and fir are the most common forest trees; but the oak, elm, and beech are met with. A good many head of live stock are reared, and some are driven into this prov. from distant ones, to be fattened for the Petersburg markets. The oxen and horses

of Esthonia are very indifferent, as well as the sheep, and goats, though active endeavours have been made to improve the breed of the latter. Poultry is abundant. The bear, wolf, badger, and fox inhabit the forests, and there are a few elks. The lakes do not contain many fish; but the fisheries on the coast are of importance to the inhabit. A few mineral products are obtained, but they are of no great consequence. Nearly all the manufactures are domestic, the peasantry weave their own coarse woollens, and some very tolerable linen stuffs. In the islands, the building of boats is a principal employment; distilleries are common in every part of the country, the free use of mills being one of the most important of their ancient privileges that the Esthonians preserve. The chief exports are corn, spirits, salt-fish, and hides; amongst the chief imports are herrings and salt. Revel (which see) is the centre of the trade of the government. The prov. is under the political superintendence of the governor-general of Riga; but has its own provincial council and judicial court. Nearly all the inhab. are Lutherans; only about 1 in 140 of the pop. are educated. The upper classes, both in the towns and the country, are mostly of German or Danish descent. The Esthonians are of the Finnish stock, and having been in a state of slavery till a recent period, have, it is alleged, contracted most of the vices incident to such a state. This country was sold by the Danes to the Teutonic knights in 1347, conquered by Sweden in 1561, and finally annexed to Russia by Peter the Great in 1710.

ESTREMADURA, an extensive prov. of Spain, lying between 87° 54' and 40° 58' N. lat., and 4° 50' and 7° 24' W. long. It has Salamanca, and part of Avila, on the N.; Toledo, La Mancha, and part of Cordova, on the E.; Seville, on the S.; and Alentejo and Beira, in Portugal, on the W. Its length, from N. to S., is 100 m.; and mean breadth, from W. to E., about 80 m. Area, 14,379 sq. m.; pop. 707,115 in 1857. Estremadura is divided at present into the two provinces of Badajoz and Caceres, the former with a pop. of 404,001, and the latter with 303,184, according to the census of 1857. It consists of immense plains, terminated on the N. by the Sierras de Gredos, de Bejar, and de Gata; and, on the S., by those of Constantina, a continuation of the Sierra Morena. Another branch of the latter chain runs along the boundary N., from the confines of Seville and Cordova to the river Guadiana, from which a branch of the mountains of Guadalupe again extends as far as the Tagus. These two rivers, each of which is here joined by several affluents, cross the prov., from E. to W., and an extension of the Castiling or Toledo mountains, under the names of the Sierras de Guadalupe, San Benito, and San Pedro, lying in the same direction, divides it into two nearly equal parts, the N. (Estremadura Alta) bring in the basin of the Tagus, and the S. (Estremadura Baja) in that of the Guadiana. The summers are hot; there is then but little rain; the nights, however, are cool, and the dew, which is abundant, is sufficient to moisten the ground. Although the high mountains are covered with snow at the end of November, the winter is not severe. In summer, the heat often brings on epidemic fevers, particularly with strangers. The soil is very fertile, and might be rendered highly productive by a proper use of the water of the many rivers that intersect it; but a combination of causes, at the head of which are to be placed bad government, have extinguished all industry. Agriculture is wholly neglected; and the noble plains, that might yield abundance of all sorts of products, are devoted to pasturage only. It is stated that about 4 millions of merino sheep come every year from other parts to winter in the plains, according to the ancient institution of the *Mesta* (see SPAIN), besides those that belong to the country, and immense herds of swine. The produce of corn, wine, oil, hemp, and flax is insufficient for the consumption; but there is an abundance of chestnuts, from which the population of this naturally fine country derives a considerable part of its scanty subsistence. The plains of Plasentia, the vicinities of Coria and La Serena, and the territory between Badajoz and Llerena, are the best peopled and most productive, and shew what the rest might be under any thing like a good system of husbandry. Immense plains are found all over the prov. covered with various species of buckthorn, myrtle, marjoram, and other medicinal and odoriferous plants, which are good for nothing unless it is to feed great numbers of bees. Here and there woods of noble evergreen oaks are met with, whose acorns feed the herds of swine whose flesh is so highly esteemed throughout Spain. It has mines of lead, copper, silver and iron, but they are all, or mostly all, neglected.

The manufactures of Estremadura are hardly worth notice. Hats are made at Badajoz and Zafra, and there are a good many tanneries in the latter place and at El Casar de Caceres. Commerce is also very small; the chief article of export is the flesh of its hogs, its trade in cattle and sheep with Madrid and Andalusia being of slight consideration. The state of the roads and the want of internal navigation would, in fact, be all but insuperable obstacles to traffic.

The prov. is governed by a captain-general, with various subaltern military governors; its ecclesiastical jurisdiction is divided into three bishoprics, those of Badajoz, Plasentia, and Coria. The people are among the most taciturn and grave of the inhabitants of Spain, uneducated, and sunk in indolence. But it is said that, when excited by hope, or any other stimulus, they are persevering and indefatigable. They are robust and vigorous, frank, honourable, and honest; slow to receive an impression, but firm in following it up.

Cortes, the conqueror of Mexico, the two Pizarros, the Almagros, and other adventurers, were natives of Estremadura. It anciently formed part of the kingdom of Leon.

ESTREMADURA, a prov. of Portugal, which see.

ESTREMEZ, a town of Portugal, prov. Alentejo, partly in a plain, and partly on the slope of a hill, and in a well cultivated country, 27 m. W. Elvas, 76 m. NE. Evora. Pop. 6,920 in 1838. The town is ill-built, but has a large open square in the centre, and is strongly fortified with an ancient castle on a commanding eminence, an arsenal, and quarters for a regiment of cavalry. There are also four parish churches, five convents, a hospital with a church attached, and a house of charity. It has manufactures of Delftware, especially of water coolers, and has some trade in hardware.

ETIENNE (ST.), a celebrated manufacturing town of France, dép. Loire, cap. arrond., on the torrent of the Furens, an affluent of the Loire, 30 m. SE. Montbrison, and 51 m. SW. Lyons, on the railway from Lyons to Le Puy. Pop. 92,750 in 1861. The population has more than doubled in the course of thirty years, having amounted to but 41,584 in 1831. The town is, on the whole, well-built; streets wide and straight; houses good, though blackened with the smoke of its numerous coal fires. It has no public edifice worthy much notice; it contains nine churches, one of which dates from the 6th century; a town-hall, court of justice, theatre, public library, cabinet of natural history, and several benevolent institutions. A handsome fountain in the form of an obelisk ornaments the principal square. The railroad 36½ m. in length, from Lyons to St. Etienne, was the first railroad constructed in France, and it was followed by another 54 m. in length, from St. Etienne to Andrezieux and Roanne. The manufactures are various; they include those of arms (in a government manufactory originally established in 1505, besides some private establishments), hardware, cutlery, nails, files, and other tools, and numerous kinds of steel articles. These manufactures, if they do not owe their origin, are, no doubt, mainly indebted for their rapid extension to the supplies of coal and iron-stone found in the vicinity. The waters of the Furens, which are said to be particularly well adapted for the tempering of steel, supply a great many factories. Exclusive of hardware, silk fabrics are largely manufactured; and lace, embroidered muslins, tulles, cotton yarn, eau-de-Cologne, and lamp black are produced. There are, besides, some bleaching and dyeing establishments, with tanneries, and glass and paper factories. The silk, and especially the silk-riband manufacturers who comprise a large proportion of the whole, have, of late years, for the most part, removed from the town of St. Etienne into the adjacent country, where their fabrics are uninjured by the smoky atmosphere, and the weavers live cheaper and better, by avoiding the crowds, or town duties. Nearly one-half the inhabitants are connected with the riband or silk haberdashery trade. The quantity of silk consumed annually in the riband manufacture is estimated at about 500,000 kilogr., principally of the superior qualities. More than three-fourths of the produce are exported. The price of labour at St. Etienne is in general less than at Lyons, and said to be about equal to three-fourths of that at Coventry; but it is very difficult to institute any comparison between them, except by comparing the cost of the work performed in each. The wages of the riband weaver vary from 1s. to 3s. 6d. a day; but the average may be about 1s. 8d. This average is less than that earned in most of the other trades at St. Etienne; the reason assigned being that the riband weavers, not residing in the town itself, mostly divide their time between the manufacture and agriculture. The proprietors of 18,000 single hand-looms in the mountainous distr. round St. Etienne and St. Chamond are, in reality, little farmers. Few cottages are without one or more looms, at which the inmates work when not employed in the business of the small farm. Entirely different from this class are the *passementiers*, or small master weavers, who possess from two to five, and sometimes ten or twelve looms each, and devote themselves wholly to the manufacture. There is, at St. Etienne, an establishment called a *Condition*, in which silks are submitted to a temp. of from 73° to 77° Fahr., to test their quality, and bring them into a certain state of dryness. The average quantity of silk sent to this establishment annually is estimated at 8,070 bales, or 595,000 lbs.; that sold without passing through the *Condition* amounts to about 1,740 bales, or 267,000 lbs. per annum. The latter consist chiefly of foreign silks, which supply the factories of St. Etienne in the proportion of nearly one to two of French silk.

Some authors have supposed that this town occupies the site of the ancient *Furanum*, built by the Romans anno 65 B.C.; but this is very doubtful, and no annals of St. Etienne go further back than the 10th century. In 1441, the town

x 2

consisted of only 200 indifferent houses, which Charles VII., a few years afterwards, suffered the inhabitants to surround with a wall to protect them against the incursions of the English. A few vestiges of this wall still exist; but it did not prevent M. Etroma from suffering greatly in the religious wars of the 16th century. The plague destroyed 7,000 of its inhab. in 1595, and 8,000 in 1628–29. Since the peace of 1815, it has increased rapidly both in pop. and wealth.

ETNA (Lat. *Ætna*, Ital. *Mongibello*), a mountain and volcano of Sicily, by far the most celebrated in both respects, either in ancient or modern times, rising from the E. shore of the island, prov. Catania, between the river Alcantara on the N. and the Giarrita on the S., the crater being in lat. 37° 40′ 31″ N., long. 15° E. It is entirely distinct from, and independent of, any other mountain range. Its base is about 87 m. in circ., but its lavas have extended over a much larger space. It consists of a congeries of mountains rising one above another. Not only is it the highest mountain of Sicily, but it is also one of the highest in Europe, being, according to Sir J. F. Herschel, with whom Captain Smyth's measurement almost exactly coincides, 10,872 ft. above the level of the sea. Its largest diameter runs from E. to W. The ascent is various on its different sides; that from Catania being about 24 m., from Linguagrossa 18, and from Randazzo scarcely 12. The extent of the base gives so easy an inclination to the sides, in most places, as greatly to facilitate the ascent; but at the same time it diminishes the grandeur of its aspect at first sight, and its commanding elevation is scarcely perceived, until the traveller has got nearly half way up, and begins to look down on the rest of Sicily, while the summit still seems as far from him as at first; then, indeed, the mountain assumes an appearance so noble, majestic, and imposing, that, associated with the considerations of its cause and effects, it excites the most intense interest, mixed with a degree of awe that elevates the mind, and inspires sublime feelings. (Smyth's Memoir, p. 146.)

The multitude of minor cones distributed over its flanks, and which are most abundant in the woody region, is, according to Mr. Lyell, 'a grand and original feature in the physiognomy of Etna. These, although they appear but trifling irregularities, when viewed from a distance as subordinate parts of so imposing and colossal a mountain, would, nevertheless, be deemed hills of considerable altitude in almost any other region. There are about eighty of these secondary volcanoes, of considerable dimensions: 52 on the W. and N., and 27 on the E. side of Etna. One of the largest, called Monte Minardo, near Bronte, is upwards of 700 ft. in height; and a double hill near Nicolosi, called Monti Rossi, formed in 1669, is 450 high, and the base 2 m. in circ.; yet it ranks only as a cone of the second magnitude amongst those produced by the lateral eruptions of Etna. On looking down from the lower borders of the desert region, these volcanoes present us with one of the most beautiful and characteristic scenes in Europe. They afford every variety of height and size, and are arranged in beautiful and picturesque groups. However uniform they may appear when seen from the sea or the plains below, nothing can be more diversified than their shape when we look from above into their craters, one side of which is generally broken down. There are, indeed, few objects in nature more picturesque than a wooded volcanic crater. The cones situated in the higher parts of the forest zone are chiefly clothed with lofty pines; while those at a lower elevation are

adorned with chestnuts, oak, beech, and holm.' (Principles of Geology, ii. 112, 3rd ed.)

The mountain is, in general, of a symmetrical form, but is broken on its E. side by a deep and extraordinary valley, called the *Val del Bove*, which, commencing near the summit of the mountain, descends into the woody region, and is thence continued by other and smaller valleys to the confines of the fertile region. The Val del Bove is 4 or 5 m. across, and is surrounded by nearly vertical precipices from 1,000 to 5,000 ft. in height. This gigantic chasm has been repeatedly traversed by torrents of lava; and in 1763 it was swept by a tremendous inundation caused by the melting of the snows near the summit of the mountain. It has a singularly dreary and blasted appearance.

The structure of Etna is chiefly of the tertiary period antecedent to the present epoch; it consists partly of volcanic, partly of sedimentary rocks; but to what extent is not known, they being so much covered by modern lavas, interstratified with layers of tuff and breccia; around its base is a line of hills formed of bluish marl, and clays enclosing marine shells and yellowish sand, from 800 to 1,000 ft. above the level of the sea; about Paterno, and elsewhere, these are capped with basalt, tufa, and volcanic conglomerates.

The mineral products of Etna are chrysolite, zeolite, selenite, copper, mercury, alum, nitre, vitriol, specular iron, amianth, pozzolana, and a fine puller's earth; there are many hot, chalybeate, and sulphurous mineral springs; but no rivers, except what are subterraneous, descend from this region, owing to the rapid absorption of the soil. (See Sicily.)

The mountain is naturally divided into three regions or zones, viz. the Fertile (*La Regione culta* or *Piemontese*), the Woody (*Nemorum* or *Sylvana*), and the Desert (*La Regione Deserta* or *Scoperta*); to which might be added the Fiery region (*Regione di Fuoco*), consisting of the central cone and crater. These regions differ widely from each other in their products and general character. The lower, or fertile, zone varies greatly in width, being 11 m. broad above Catania, but no more than 1½ m. on the N. side. It is composed almost entirely of lava, which, in the course of ages, has been decomposed and converted into a very fertile soil. It is comparatively well cultivated and peopled. All travellers speak in the highest terms of the beauty and fertility of this region. 'No language,' says Mr. Hughes, 'can do justice to the savoury and luxuriant fertility of this tract; whose bosom, heated by subterranean fires, and situated in the most favourable climate, teems with every flower, and plant, and tree, that can delight the eye, and every species of fruit that can gratify the palate; fields covered with golden grain, or the purple vine, villages, and convents embosomed in groves of chestnuts and oriental plains, many fountains, and transparent streams; exhausted craters covered with a canopy of foliage, and nameless other beauties, invite the tourist to these charming scenes. Here, also, the sportsman will meet with every species of game that he can desire; and the botanist or mineralogist find inexhaustible sources of amusement.' (Travels, i. 113, &c. &c.) But here, as in most parts of Italy and Sicily, there is a painful contrast between the richness and beauty of the country and the appearance and condition of the inhab. The latter are squalid, slovenly, and dirty; a consequence, it appears, of the ashes and dust that pervade the air, soil their persons, and injure their eyes; and of the want of water, which is absorbed, as soon as it falls, by the porous soil.

The woody region is 6 or 7 m. in width, and reaches to about 6,400 ft. perpendicular height; it begins and terminates abruptly; in the lower parts the trees are principally oak and chestnut; in the middle they are almost entirely oaks, some of them attaining to an immense size; in the upper part the oaks decrease in size, and are intermixed with pines (*Pinus tæda*); as we ascend the mountain the oaks nearly disappear, the firs become stunted, and at length all vegetation ceases, and we enter on the desert. The ground in the greater part of the woody region is covered with aromatic plants and fern. Tillage soon ceases; there are no corn fields, but here and there a few vineyards, and very rich pasture land on which numerous flocks of sheep are fed.

In this region, near Carpinetto, stands the celebrated chestnut tree, *Castagno di cento cavalli*, so called from its being supposed capable of sheltering 100 horses under its boughs. It consists of five great arms, which, however, are all united in a single stem a little below the surface. The estimates of the size of this enormous tree vary considerably, probably from their not being taken in the same way. Swinburne makes it 196 ft., and Smyth 163 ft. in circular above ground. A house of ample dimensions for the accommodation of travellers has been constructed in the interior of the tree. Several other large chestnut trees grow in the vicinity, the principal of which is 57 ft. round. The products of the woody zone are chiefly tar, honey, cantharides, and charcoal; and its inhabitants are herdsmen and charcoal burners.

The minor volcanic caves abound principally in this region. Caverns are numerous; and one of them, the *Grotta dei Capri*, or grotto of the goats, from its affording shelter to these animals, was formerly resorted to by travellers, as a resting-place in their ascent. In the vicinity are deep reservoirs of snow, whence Catania and other cities derive their supplies of that article, which is there really a necessary of life; being packed in straw, it is carried to a great distance on mules and oxen. (Hughes, I. 117.) Wild boars, wolves, badgers, wild goats, deer, martens, and all kinds of game, eagles, vultures, and falcons, belong to this region.

The desert region, or zone, is a dismal tract, full of gloomy and rocky hollows and immense chasms, formed of black lava, scoriæ, ashes, and volcanic sand; covered, for the greater part of the year, with snow and ice, which are always to be found in the hollows. 'In this lofty region the air is chill and piercing; every sign of life and vegetation ceases; not an insect crawls over the cold surface of the ground, not a lichen adheres to the grey masses of the lava; not even the eagle's wing soars so high, to disturb the awful solitude of nature: here only the thunder and the tempest, or the still more tremendous explosions of the volcano, are heard.' (Hughes, I. 113.) In the midst of this gloomy region the principal cone, forming the summit of the mountain, rises to the height of about 1,100 ft.: it is very precipitous, and as it consists of loose scoriæ and ashes, which frequently yield under foot, the ascent is extremely laborious. At the foot of the cone is a house, with rooms and stabling, erected in 1811, at the expense of the British officers then in Sicily, for the accommodation of travellers, to whom it is a very great convenience. The cone at its base is from 7 to 8 m. in circumference; but at its summit its circ. is reduced to about 4 m. It consists of a horizontal plain, with a vast central crater, or breathhole, 3½ m. round, agreeing in this respect with the dimensions assigned to it by Pliny:

'Crater ejus patet ambitu stadia xx.' (Sat. Hist., lib. iii. § 8.)

The view from the summit of Etna is superb beyond description. Sicily is spread out like a carpet at the spectator's feet, who traces every river through all its windings, from its source to the sea. The strait that separates Italy and Sicily, the Calabrian shores, and the Lipari Islands, are distinguishing features in this magnificent panorama, which, it is said, sometimes extends to Vesuvius on the one hand, and Malta on the other. The wonderful extent of view, and the unequalled sublimity of the scene, is owing partly to the great altitude of the mountain, partly to the highly interesting nature of the objects, but more than all to Etna being 'alone in its glory,' and having no other mountain in its vicinity to detract from its grandeur, or to interrupt the immensity of the prospect.

The enjoyment of the spectacle of sunrise is the grand object of travellers who ascend to the summit of the mountain. Brydone has described it in terms not unworthy of the glorious scene, though doubts have been entertained whether he really saw what he depicted, or trusted to the reports of others. It is probably one of the grandest, if not the grandest, of all the views of natural scenery that it is possible to behold in Europe. Not the least interesting portion of the extraordinary prospect from the summit of Etna is the distinct image of the mountain itself, seen at the extremity of the shadow that it projects across the island. (Brydone, Letter x.; Hughes, I. 170.)

The crater, when Captain Smyth visited the mountain, was of an oval form, directed from NE. to SW., its conjugate diameter being about 493 yards; but its size and form are perpetually varying from the accumulation and falling in of volcanic matter. Its interior is encrusted with extensive efflorescences of ammonia, sulphur, and vitriolic salts, to the depth of 100 yards on the E., but less on the W. side: those of an orange colour are the most common. Its bottom is flat, and tolerably hard; near its centre are two mounds of scoriæ and ashes, surrounded by several fissures, 'whence,' says Captain Smyth, 'at intervals issue volumes of thick smoke, with a rumbling noise, and hissing sound. There is also a light thin vapour occasionally oozing from the bottom and sides of the huge amphitheatre in every direction. 'I endeavoured,' he adds, 'to look into the principal chasm; but the rapid ejection of the cinders, and the strong sulphureous vapours that exuded, prevented me from attaining my object.' (Memoir, p. 151.)

Mr. Hughes, however, has supplied a more minute account of the principal spiraculum or funnel. It has three stages of descent; the first, which extended only a few hundred yards, terminated in a shelf or ridge of cinders; the second, more precipitous than the first, extended to a similar shelf; the third being the perpendicular and unfathomable abyss. Between the two principal spiracula are several smaller conical mounds, constantly smoking. The ground here is so hot round the crater that visitors are obliged constantly to shift their places, and yet even here, in the interior of the crater, snow is seen in immense ridges, 'disputing, as it were, the pre-eminence of fire, in the very centre of its dominions.'

Before eruptions local earthquakes are felt, hollow internal noises heard, irregular clouds of smoke burst forth, and *ferilli*, or volcanic lightnings, are seen darting from the top of the mountain: the agitations increase, till at length, either from the great crater, or from some other part of the mountain, a terrific discharge of red-hot stones, flakes of fire, ashes, sand, or other substances, accompanied with vast volumes of smoke, suddenly takes place with tremendous violence.

> '———— horrifici juxta tonat Ætna ruinis,
> Interdumque atras prorumpit ni æthera nubem,
> Turbine fumantem piceo, et candente favilla;
> Attollitque globos flammarum, et sidera lambit:
> Interdum scopulos avulsaque viscera montis
> Brigit eructans, liquefactaque saxa sub auras
> Cum gemitu glomerat, fundoque exæstuat imo.'
> _Æneid_ iii. line 571.

Some of the matters thrown up during an eruption are occasionally projected to an immense distance. They not unfrequently rise to the height of 4,000 or 6,000 ft. above the summit; stones of 18 oz. weight have fallen 16 m. from the crater; and in the great eruption of 1669 a stone 50 cubic ft. in size was ejected with such prodigious force that it fell a mile from the crater. Ashes are said to have sometimes fallen in Malta, about 180 m. distant. The eruptions are generally followed of accompanied by the outbreak of a torrent of lava. If this current of liquid fire be stopped by inequalities of ground, a portion cools, and the rest topples over it; sometimes it overwhelms whole cities, villages, and tracts of country: the torrent of lava that partly destroyed Catania in 1669, was stopped by the city walls, 60 ft. in height; but the burning flood accumulated till it rose to the top of the rampart, and then fell over it in a fiery cascade. This mass was so enormous that, according to the reports—the truth of which, however, seems somewhat doubtful—it was eight years in cooling.

Generally, however, it soon congeals, and when mixed with scoriæ, cracks, decomposes, and forms an extremely fertile soil. Sometimes inundations of boiling water occur, through the melting of the snow in the upper regions by contact with the lava; and the strange phenomenon has also occurred of a body of snow and ice being covered with a layer of ashes, and then with a torrent of burning lava, and so preserved for an indefinite period. (Lyell, ii. 122.) About one eruption in three takes place from the principal crater, and these are generally the least dangerous, the lava being mostly retained in the immense hollows of the upper region.

Though Homer has made Sicily the scene of some of the most interesting adventures in the travels of Ulysses, and has described the island and the strait of Scylla and Charybdis, he does not so much as allude to Etna. It has thence been inferred that the mountain had not then been an active volcano; for it can hardly be supposed, had it been such, that so careful an observer would have failed to notice it, and to avail himself of the means which it afforded of embellishing his verses by a topic so well suited to the dignity of epic poetry. No doubt it is very difficult to reconcile the silence of Homer, with the fact of the mountain being at the time eruptive, though it would be rash thence to conclude positively that it was not; it had then, perhaps, been long quiescent, and its eruptions forgotten. Pindar is the oldest extant author (about 500 years B.C.) who takes any notice of the eruptions of Etna; and his account is peculiarly interesting, inasmuch as it appears from his representing its summit as supporting the heavens, and being covered with perpetual snows and frost, that it must then have been about as high as at present. According to the ancient poets, Jupiter, after the overthrow of the giants, buried the hundred-headed Typhœus under this mountain; and its earthquakes and eruptions were said to be occasioned by the struggles of the monster. The passage in which Pindar

alludes to Etna has been rendered by West as follows:—

> 'Ever under snowy-topt Cuma's sea-bound coast,
> And vast Sicilia lies his shaggy breast;
> By snowy Etna, nurse of endless frost,
> The pillar'd prop of heav'n, for ever press'd:
> Forth from whose nitrous caverns issuing rise
> Pure liquid fountains of tempestuous fire,
> And veil in ruddy mists the noon-day skies,
> While wrapt in smoke the eddying flames aspire;
> Or, gleaming through the night with hideous roar,
> Far o'er the redd'ning main huge rocky fragments pour.'
> _First Pythian Ode_, dated, &.

Thucydides mentions three eruptions of Mount Etna, but he leaves the date of the first uncertain; the second occurred four or five years previously to the period when Pindar wrote the above ode. Since then there have been a great many eruptions, both in antiquity and in modern times. One of the most tremendous occurred in 1669, when the hill of Monti Rossi was formed; but the most extraordinary phenomenon in this eruption, was the opening of a fissure about 6 ft. wide, and of unknown depth, which stretched from the plain of S. Lio to within a mile of the summit of the mountain, a distance of 12 m.: it emitted an intensely vivid light. Five other parallel fissures also opened, and gave out tremendous noises. The lava that burst forth on this occasion overwhelmed 14 towns and villages, filled up the port of Ulysses, and, as already stated, partly destroyed Catania. About 27,000 persons are supposed to have lost their lives in this convulsion. The last great eruption occurred in 1832, when the town of Bronte narrowly escaped being overwhelmed by a current of lava. An eruption in 1852, though most violent, caused little damage. (Besides the authorities already referred to, numerous works have been written on Etna; one of the best is Ferrara, _Storia Generale dell' Etna_, 8vo. Catania, 1793.)

ETON, a town and par. of England, on Berks, hand. Stoke, on the N. bank of the Thames, immediately opposite to Windsor, with which it is connected by a neat iron bridge; 23 m. SSE. Aylesbury, 21 m. W. London by road, and 23½ by London and South-Western railway. Pop. of town 2,940, and of par. 3,122 in 1861. The town consists principally of a single street, well paved and lighted, and which of late has been much improved, many of the houses having been rebuilt. The establishment to which Eton owes all its importance is its college, founded by Henry VI. in 1440. That monarch, by whom it was liberally endowed, intended it principally for the education of 'poor and indigent boys,' destined for the church. By his second charter, dated Oct. 21, 1441, the foundation consisted of a provost, 10 priests or fellows, 4 clerks, 6 choristers, a master, 25 scholars, and 25 alms or beads-men; but about 1443, the date of the college statutes, he increased the number of scholars from 25 to 70, added an usher, clerk, and two choristers, and reduced the number of beadsmen to 13. Various changes were made in the succeeding reigns, and the establishment suffered considerable spoliation, especially from Edward IV.; but it was particularly excepted in the act of Parliament for the dissolution of colleges and chantries in the reign of Henry VIII. The foundation at present consists of a provost, appointed by the crown; 7 fellows, one of whom acts as vice-provost; 2 chaplains, called conducts; 2 lay-clerks, 10 choristers, 3 masters (each of whom has 4 assistant masters), and 70 scholars, who since the reign of George III. have been called 'king's scholars.' Besides the latter, the different masters have a number of stipendiary pupils, not on the foundation, but who receive instruction in the col-

large. These are called oppidans, and generally consist of members of families, superior in rank or wealth to those of the king's scholars. Their number is variable, but at an average may be estimated at about 350. Under a recent head master, the number of boys at Eton, of both classes, at one time exceeded 600. The buildings of the college surround two quadrangles: the outer quadrangle, or school-yard, is enclosed by the chapel, schools, dormitories of the scholars, and masters' chambers; and has in its centre a bronze statue of the royal founder of the college. The inner or lower quadrangle is bounded by the cloisters, containing the residences of the fellows, the library, hall, and various offices. Between the two is the provost's lodge, appertaining to which is an ancient tower and a gateway in the centre, connecting the two courts. The chapel, on the S. side of the outer court, is a handsome Gothic edifice, 175 ft. in length, including the ante-chapel, and in its style and ornaments greatly resembles the chapel of King's College, Cambridge. The par. church of Eton having fallen to decay, the inhabitants attend public worship in the college chapel, the provost having archidiaconal jurisdiction in the par.; but there is also a chapel of ease in the town, at which one of the conducts officiates. The college library contains a large and valuable collection of books, engravings, drawings from the antique, and medals; it is a fine apartment, and fitted up in a superior style. The dining hall for the scholars on the foundation is spacious, but little ornamented; it contains, however, two large ancient pieces of tapestry. The upper school, on the W. side of the outer court, was designed by Sir C. Wren, and is supported by an arcade with double columns of the Doric order. The school-room is spacious and of fine proportions, but fitted up in a plain manner. The school-room of the lower school is of considerable length, but not of a proportional height, with a range of ancient oak arches on either side, and the seats of the scholars behind them. It is beneath a part of the principal dormitory, called the long chamber. To the E. of the cloisters are the college gardens; to the N. the playing fields, and adjacent to the latter the shooting fields, in which cricket matches and other games are played.

The scholars on the foundation are lodged and boarded by the establishment. They are eligible from the ages of eight to fifteen, and are elected separately by the individuals of a body composed of the provosts of Eton and King's College, Cambridge, the vice-provost and master of Eton, and two prsrs (M.A.s) of King's College. This body meets on the last Monday in July of every year, when usually twenty-four boys are nominated to fill up vacancies as they may occur in Eton Lower School, and twelve of the best boys in the same establishment are nominated in a similar manner to King's College, Cambridge, according to the statutes of the founder. Those who go to King's are, after three years, entitled to fellowships. Eton College also sends two scholars to Merton College, Oxford, where they are called Postmasters, or, by corruption, postmasters. Failing an appointment to either university, Eton collegians are superannuated at eighteen or nineteen, and for scholars so superannuated there are a few exhibitions, and some other means of slightly augmenting their income, in the gift of the college. By statute, the education of King's scholars should be gratuitous; but some innovation has taken place on this head, and the average annual expense to the parents is estimated at 60l.

The oppidans board either in the houses of the lower master or assistants, or, at a somewhat lower charge, in the boarding houses attached to the school; some few, chiefly of noble birth, in private lodgings, under the care of private tutors. The total expenses of a boy educated at an oppidan may perhaps average from 150l. to 200l. a year. Without the boundaries of the college, the oppidans are comparatively little under the control of the college functionaries; but within its walls they are in no respect distinguished from the King's scholars, and mix with them in the same classes. The entire school is divided into Upper and Lower. The latter comprises, together with the junior classes, the third and fourth forms, each consisting of three subdivisions or removes. Each of these is under the control of a separate assistant master; and as boys of various ages come to Eton, they are placed at the bottom of whatever remove in the lower school they may seem fit for by their previous acquirements and age, passing into the superior ones according to their proficiency. The upper school consists of the fifth and sixth forms, and is under the immediate control of the head master. The number of boys in the sixth form is limited to twenty-two; and of these the ten highest are styled monitors, and act in some measure as assistants to the masters. The head of the whole school, who arrives at his post by seniority, is called the 'captain.'

The course of instruction at Eton is almost wholly classical. The only entire works read are those of Homer, Virgil, and Horace, but extracts from those of numerous others are occasionally made use of. The well-known Eton Latin and Greek Grammars, committed to memory, form the basis of grammatical instruction. In the Upper School the boys are engaged in writing Latin and Greek themes and verses, for the best of which rewards are given; and a play of some Greek author is usually in the course of reading. Mathematics form a part, but a very small one, of the school discipline; and though there are masters in French, writing, arithmetic, &c., such studies are wholly unconnected with the general business of the school, and only attended at extra hours. All the boys attend chapel twice on Sundays, and once on saints' days and holidays; and, in addition, the collegers attend prayers every evening, after which they are confined to their several dormitories. The system of fagging, by which the boys of the Lower School are fags, or servants, to those of the Upper, out of school hours, prevails; but its supposed severity and degradation have been much exaggerated.

Eton College has in its gift nearly forty ecclesiastical preferments, besides several presentations. The provost, though as rector he derives no emolument from the par., has very extensive powers within it; for, by an act passed in 25 Henry VI., no inhabitant is allowed to take a lodger without his permission, under penalty of 10l., which fine may also be levied upon the individual engaging lodgings without such permission. In 1461 a charter was granted to Eton for a market on Wednesdays, with considerable privileges, but this has been long discontinued. There were formerly also two fairs, but only one is now kept up—that on Ash Wednesday for horses and cattle.

EU, an inland town of France, dép. Seine Inférieure, cap. cant., on the Bresle, about 2 m. from its mouth in the British Channel, 16 m. NE. Dieppe, and 43 m. NNE. Rouen. Pop. 4,416 in 1861. The town is generally well built, and has a fine square; it has several churches, one of which, a fine Gothic edifice, is remarkable for a subterranean chapel, a college, and a hospital. In its neighbourhood, in a noble park surrounded by large gardens, is the magnificent Château d'Eu, containing a fine collection of historical portraits.

There are several Roman remains in and about Eu. The town is the seat of a tribunal of commerce; has manufactures of lace, serges, linseed oil, and soap; is an entrepôt for the corn of the Somme, and has some trade in hemp, flax, timber, and linens, exported at Treport, at the mouth of the river. A large forest, which takes its name from the town, extends to the E. and N. Eu was burnt by Louis XI. in 1443, to prevent its falling into the hands of the English, who meditated a descent into Normandy: it is said never to have recovered its original prosperity.

EUPATORIA, or KOSLOFF, a sea-port town of Russia in Europe, W. coast of the Crimea, lat. 45° 9' N., long. 33° 9' 20" E. Pop. 6,580 in 1838. The town has a considerable trade; exporting salt, wheat, barley, hides, and lambskins. The houses, with the exception of a very small number built in the European style, are altogether of Asiatic architecture. The roadstead is a sandy circular bay, and affords no shelter with the winds at N. and E.

EUPEN, a town of Rhenish Prussia, immediately within its W. border, cap. circle of same name; on the Weser or Vesder, a tributary of the Meuse, 7 m. S. by W. Aix-la-Chapelle, on the railway to Verviers. Pop. 13,190 in 1861. The town is principally inhabited by the descendants of French Protestants who took refuge here subsequently to the revocation of the edict of Nantes; and is one of the principal manufacturing towns in the Rhenish provinces of Prussia, having some very extensive broad cloth and kerseymere factories, with others of nitric acid, chicory, &c. It is the seat of a council for the circle, and of a court of primary jurisdiction; and has a superior citizens' school.

EUPHRATES and TIGRIS, two famous rivers of Turkey in Asia, which, rising in Armenia, flow generally parallel to each other in a SE. direction, and finally unite in lat. 31° 0' 28" N. and long. 47° 40' E., in the Shat-al-Arab, or 'River of Arabia,' which discharges itself into the bottom of the Persian Gulf.

The Euphrates (Gr. Εὐφράτης), so called from εὐφραίνω, to exhilarate or make glad, because its waters, like those of the Nile, fertilise the adjacent lands, is the most considerable river of W. Asia, and its basin, exclusive of that of the Tigris, is computed to comprise about 109,000 sq. geog. m. After watering on either side the territories belonging to Turkey as far S. as near lat. 36°, it forms, from that point to about lat. 33° 30', the boundary between them and the newly acquired Asiatic dominions of the pasha of Egypt; it next divides Turkey from Arabia; and lastly, from its union with the Tigris to its mouth in the Persian Gulf, about lat. 30° and long. 49° 30', it separates Arabia and Persia.

The ancients seem to have had no correct information respecting the sources either of the Euphrates or the Tigris; and there is the greatest obscurity and discrepancy in the statements they have put forth respecting them. The popular opinion seems to have been that their sources were identical. (Lucan, lib. III. v. 257); and though this notion was rejected by Strabo, Mela, and Pliny, none of them appear to have had any precise information on the subject. (See Cellarii Notit. Orbis Antiqui, ii. 876.)

Both rivers have their sources in the table-land of Armenia. The Euphrates rises in the pachalic of Erzeroum, and is formed by the junction of two great arms—the Frat and the Murad. The former, which is also the most N., has its principal sources about 20 m. NE. from Erzeroum, in the Tchehlir mountains, near the sources of the Araxes.

The Murad has its sources on the N. declivity of the Argish-dagh mountains, 45 m. NE. from the nearest point of Lake Van. Both these rivers pursue a W. course, inclining to the S., till they unite near Kebban, in about the 39th deg. of lat. and 39° 25' E. long. The united stream thence flows SW. to Samisat (Samosata) in lat. 37° 31', long. 38° 25', having received on the right the Kara-su, and forced a passage for itself through the main range of Taurus, and formed a double cataract 15 m. above Samisat. From the latter point the river pursues a nearly S. course to Rakka, about 50 m. E. from Aleppo, its course being thence almost uniformly SE. At its source the Frat, or N. arm of the Euphrates, is only 90 m. from the Black Sea, but a very mountainous country intervenes between them. During its S. course the Euphrates approaches within 122 m. of the Mediterranean, and as the interjacent country is for the most part level or undulating, it would, perhaps, present no very serious obstacles to the formation of canals or carriage roads. From Hillah (Babylon) to its mouth it flows through a perfectly level country, which was anciently intersected by numerous canals. At Hit, 107 m. NE. Antioch, the Euphrates is 678 ft. above the level of the Mediterranean (Ainsworth, p. 109), the rate of inclination from which being estimated to average only about 6¼ inches a mile. The total length of the river, measured from the sources of the Murad, is estimated at about 1,800 m. (Geog. Journal, iii. 243.) Its breadth at Malatia is 100 yds., and at Hit 130 yds. At Ul Der (an. Thapsacus) (Kinneir's Memoir on the Persian Empire, p. 9) the Euphrates is 800 yds. wide; at Hillah its bed is contracted to about 200 yds.; but below the latter it frequently spreads out to a considerable breadth, and the Shat-al-Arab ranks amongst the noblest rivers of the Asiatic continent. The Euphrates is navigable to the cataract above Samisat; at Hillah it has seldom less than 18 ft. water, even in the lowest season, and a vessel drawing 15 ft. water may ascend to Korna, where it is joined by the Tigris. The principal tributary of the Euphrates is the Tigris, which, indeed, is but little inferior to itself; its next greatest tributaries are the Kara-su, Khabur (an. Chaboras), and Kerah, which joins the Shat-al-Arab.

The banks of the Euphrates were in antiquity the seat of many noble cities. The small mean town of Hillah occupies a minute portion of the site of the once mighty Babylon, 'the glory of kingdoms, the beauty of the Chaldees' excellency;' Hit (an. Is or Aeopolis), Anna (an. Anatho), Kerkisiya (Circesium), and Bir are amongst the other towns on its banks; but Bussorah or Basra, on the Shat-al-Arab, is at present the only large city on the Euphrates.

The Tigris is throughout its whole course comprised within the Turkish dom. It rises in the pachalic of Diarbekr, from numerous sources on the S. side of the Taurus chain, by which it is separated from the Murad, in about lat. 38° 40' N., and at an elevation of about 3,624 ft. above the level of the sea. (Ainsworth, p. 110.) Its course, to its junction with the Euphrates, is, with very little deviation, SE. It runs at first through a mountainous country, with great rapidity; at Mosul it is no more than 253 ft. above the level of the Persian Gulf; from Bagdad it flows, with a moderate current, through a nearly level plain. Its distance from the Euphrates varies from 18 to 85 m.; the two rivers enclose the province in antiquity called, from that circumstance, Mesopotamia. The entire length of the Tigris is estimated at 1,146 m. At Mosul it is 150 yds. wide; between Bagdad and Korna its average breadth is 200 yards. It brings down great quantities of mud, which it deposits in shoals and islands in the lower part of its course; and between Mosul and Bagdad it passes over several ledges of rock, which form rapids of more or less difficulty. It is neither so deep nor so suitable for navigation as the Euphrates. It is, however, navigable for vessels, drawing 4 ft. water as far as the ruins of Opis near the mouth of the Adhaym (Lynch in Geogr. Journ.); and, in Dec. 1836, it was ascended considerably above Bagdad by Col. Chesney's steamer 'Euphrates.' Its principal affluents are the Kabour, the Great and Little Zab (an. Zabatus and Zerbus Minor), the Adhaym (an. Physcus?), the Diala (an. Dêlas or Arba). In antiquity its banks were studded with cities of the first rank, as Nineveh, Seleucia, Ctesiphon, Opis, &c. Bagdad may be considered as the modern representative of Seleucia and Ctesiphon, as Mosul is that of Nineveh, opposite the site of which it is placed. Diarbekir is the only other important town on its banks.

The Tigris derives its name from the rapidity of its course, the term Tigris signifying 'an arrow,' in the language of the Medes and Armenians. So late as the age of Alexander the Great, the Tigris did not unite with the Euphrates, and each river preserved a separate course to the sea. But they not long after became united, and have since found their way to the sea in a collective stream. The ground in the lower part of their course being soft and alluvial, and their waters being also diverted into new channels by means of canals, the courses of both rivers must necessarily have differed materially at different periods. (Rennell's Geog. of Herodotus, i. 365.)

The Euphrates and Tigris run through chalky formations of a very friable nature, easily disintegrated by the action of the elements. Both rivers have their regular inundations, rising twice a year—first in Dec., in consequence of the autumnal rains; and next, from March till June, owing to the melting of the mountain snows. (Rich, p. 54.) They bring down immense quantities of alluvium; and the extent of land covered by their deposits is supposed to exceed 32,000 sq. m. The ancient writers have not failed to notice this resemblance between the Euphrates and the Nile. Cicero says, 'Mesopotamian fertilem efficit Euphrates, in quam ... singulis annis novos agros inmittit,' (De Nat. Deorum, lib. ii.) And Lucan—

'—— spargos in arva
Fertilis Euphrates, Phario vice fungitur undae.'
Lib. iii. v. 259.

Mr. Ainsworth found the maximum of sediment mechanically suspended in the waters of the Euphrates, in Dec. and Jan. 1836 (in which months most mud is brought down), to be equal to 1-80th part of the bulk of the fluid. A good deal of this mud is deposited in the marshes of Lemlúm (an. Paludes Babylonia), a swampy tract about 40 m. long by as many broad, commencing 50 m. S.W. of Babylon, and which has existed from the remotest period to the present day. The quantity of mud brought down by the Tigris was found, in Jan. 1837, to be equivalent to 1-108th part of the suspending fluid; but as it is not dispersed in marshes, more is carried down by this than by the Euphrates to the mouth of the Shat-ul-Arab. The rapidity of the Upper Tigris frequently causes it to break down its banks; Mr. Rich says, that when at its height it has a current of near seven knots an hour. In the alluvial plain, however, it averages only 1½ m. an hour throughout, and in many places it is less than 1 m. The Euphrates above Samisat is, perhaps, as rapid as the Tigris; and at Hillah, where its bed is narrowed, its rate

is from 3 to 4 m. an hour; but, in the low plain, this rate is diminished to about 1 or 1½ m.

Lower Mesopotamia, or Babylonia, was, as already stated, anciently intersected by canals in every direction, for the purposes both of navigation and irrigation. Many connected the Tigris with the Euphrates; those which still exist are especially numerous near Bagdad, where the rivers approach within 25 m. of each other; and some, as the Nahr Malcha, might be easily repaired. (Rich's Babylon, p. 57.) In fact, the Euphrates steamer passed from the Euphrates to the Tigris by the Isa canal, which leaves the former a few miles above Feluga, and enters the latter a short way below Bagdad. The Shat-el-Hie, which connects the two rivers, is also navigable in spring by large boats. The most celebrated of the ancient canals, that of Pallacopas, cut by the earliest Assyrian monarchs, partly through solid rock, extended for a very considerable distance parallel to the Euphrates on its S.W. side. Nitochris supposed it had commenced at Hit. It may still be traced, almost continuously, from a little below Babylon to its probable mouth in the Persian Gulf (Khore Abdallah). Remains of aqueducts and towns, and various other ruins, abound in this region; and the ancient Median wall which ran from Macepracta on the Euphrates, to near the site of Opis on the Tigris, is still clearly traceable. (See Messrs. Ross and Lynch, in Geog. Journal, vol. ix.)

The steam navigation of the Euphrates is of considerable importance; it may be navigated, as high as Hit, by steamers drawing 4 ft. water. To establish a new connection between Europe and India by means of the Euphrates route has long been a favourite scheme of merchants and statesmen. The proposed line has recently gained in political importance by the opening of (1861) of the electric telegraph, which skirts the Tigris and Euphrates from Bagdad to the Persian Gulf.

EURE, a dép. of France, in the N. part of the kingdom, being one of the five comprised in the ancient prov. of Normandy; between lat. 48° 39' and 49° 29' N., and long. 0° 15' and 1° 45' E.; having N. the estuary of the Seine and the dép. Seine Inférieure, E. the dép. Oise and Seine-et-Oise, S. and S.W. Eure-et-Loire and Orne, and W. Calvados. Length E. to W. 65 m., breadth varying from 26 to 52 m. Area 599,765 hectares. Pop. 398,661 in 1861. Surface nearly flat. There are a few ranges of low hills, principally in the N., none of them reaching an elevation of more than 830 ft. These ranges divide the dép. into several distinct plateaus, presenting a great variety of aspect. It is well watered; the Seine flows through its E. portion, and along its N.E. border. The Eure, whence it derives its name, rises in Orne, and after running at first E. and then N. falls into the Seine 6 m. N. Louviers. The Iton, Rille, and Charentonne are the other principal streams. Climate mild, but damp and variable; W. winds are the most prevalent. Soil chiefly calcareous or marly; but on the banks of the Seine it is sandy, and rather sterile. Iron ore is abundant, and there are numerous mines. According to official tables, the arable lands comprise about two-thirds of the department. Property is less subdivided in this than in most other déps.; still, however, of 181,517 properties, subject to the contribution foncière, nearly one-half are assessed at less than 5 fr. Previously to the revolution the estates were much larger, but most of them have since been repeatedly subdivided by the operation of the law of equal succession. (See France.) Farms vary in size from 20 to 150 hectares. Agriculture, though more improved than in many other parts of France, is still very backward. The farm-buildings and cottages of the peasantry are in many instances of the very worst description, being frequently ill situated, built of wood, thatched with stubble, and surrounded by dunghills and filth. The fences are not well kept; but, notwithstanding these drawbacks, the country has, on the whole, a considerable resemblance to England. Wheat, oats, maslin, and rye are the principal kinds of grain cultivated. In some parts flax is grown; in others, hemp, pulse, and woad. Little wine is made, but apples and pears are very plentiful, and cider and perry are the ordinary drink of the pop. The stock of sheep is estimated at about 415,000 head, producing annually about 420,000 kilogs. of wool. The mining and manufacturing establishments of this dep. rank amongst the most extensive and important in France. The various works for smelting and working iron, copper, and other metals, employ about 50,000 hands; the copper and zinc works at Romilly are very extensive. The cotton and woollen manufactures are also important. The broad cloths of Louviers are celebrated in foreign countries as well as in France, and, in addition to them, cottons, flannels, druggets, baize, velvets, glass, paper, and leather are largely manufactured. This is one of the very few deps. of which the pop. has been decreasing, in the ten years 1851-1861, having amounted, at the former period, to 415,777. It is divided into five arronds., 36 cantons, and 794 communes. Chief towns, Evreux the cap., Louviers and Bernay. The women of this dep., as in other parts of Normandy, are good-looking and tidy; they wear dresses of remarkably bright colours, and lofty pyramidal caps, called bonnets cauchoises, ornamented with a great quantity of lace. Eure contains some Celtic and many Roman antiquities; but those of the middle ages were mostly destroyed during the Revolution.

EURE-ET-LOIRE, a dép. of France, in the N. part of the country, between lat. 47° 57' and 48° 57' N. and long. 0° 44' and 1° 59' E., having N. the dép. Eure, E. those of Seine-et-Oise and Loiret, S. the last named and Loir-et-Cher, and W. Sarthe and Orne. Length N. to S. 60 m., greatest breadth about 55 m.; area 587,430 hectares; pop. 290,465 in 1861. There are only a few scattered heights in this dep., nearly the whole of which consists of an undulating plain. Principal rivers, the Eure towards the N., and the Loire in the S. Small lakes are numerous. Climate temperate and healthy. As much as 310,000 hectares of the surface consists of rich alluvial soil, and this dep. contains a greater extent of cultivable and less waste land than any other department of France. Of 114,901 properties subject to the contribution foncière, about one-third are assessed at less than 5 fr.; the number of considerable estates is, however, above the average of the dep. This is especially a corn-growing dep., producing principally wheat and oats. Good flax and hemp, pulse, turnips, onions, melons, and woad are grown, but few potatoes. In some cantons the vine is cultivated, and in ordinary years about 200,000 hectolitres of inferior wine are made, as well as about the same quantity of cider. In 1861, about one-tenth of the surface consisted of pasture land, and the dep. contained 86,000 oxen and 700,000 sheep; the latter furnishing about 1,000,000 kilog. a year of wool. There are some iron mines, but they are little wrought. Manufactures of no great importance; the chief are those of ironware, earthenware, paper, cotton and woollen fabrics, beet-root sugar, and leather. This dep. is divided into 4 arrond., 24 cantons, and 437 communes.

Chief towns, Chartres the cap., Chateaudun, Dreux, and Nogent-le-Rotrou.

EUROPE, the most populous, but, with the exception of Australasia, the smallest of the divisions of the globe, being about a fifth part of the size of Asia or America, and a third part of that of Africa. However, though inferior in point of size, Europe is vastly superior to the other continents in the enterprise, intelligence, and civilisation of her inhabitants, and perhaps also in her physical advantages. ' Altera orbaris omnium gentium pugnaÔ, longéque terrarum pulcherrima.' (Plin. Hist. Nat. lib. iii. § 1.) Europe is mostly situated within the temperate zone, and no part of her surface approaches within many degrees of the intertropical regions. The climate is, therefore, rather inclined to cold; but it is comparatively temperate, and is neither so cold in winter nor so hot in summer as the countries in the corresponding latitudes of Asia and America; so that while comfortable lodging and warm clothing are indispensable, the exertions of the inhabitants are not impeded by the too great intensity of cold on the one hand, or of heat on the other. The surface, too, of the country is infinitely varied and picturesque; and it has the advantage of being more intersected than any other continent by great arms of the sea, supplying facilities to internal and foreign commerce, that are all but wholly denied to Asia, Africa, and Australasia, and not enjoyed in an equal degree even by America. The soil of Europe seems also to be of the quality best suited to stimulate and reward the efforts of the husbandman; for though it is nowhere so fertile as to produce crops without laborious diligence, and, consequently, does not foster indolence or a want of attention, it never fails liberally to reward the efforts of the industrious and skilful cultivator. Hence it is that this continent has everything that seems best fitted to call forth and develope human genius and resources. But the advanced civilisation and superior influence of Europe in the affairs of the world seems, after all, to be owing in no small degree to the superior capacity of her inhabitants, as evinced in their enterprise, invention, perseverance, and power of combination. In all these respects they seem to be decidedly in advance of the most improved Asiatic nations; while the difference between them and the most civilised native nations of Africa, America, and Australia, appears almost as great as the difference between man and the least advanced of the lower animals. Europe is the only part of the world in which civilisation and the arts have, generally speaking, been uniformly progressive. Important discoveries have been made, at remote periods, in China, India, and other Asiatic countries, but these would seem to have been the result of accident only, and, at all events, have had comparatively little influence; it is here only that they have been appreciated, improved, and perfected, and made instrumental in the production of further discoveries. It is characteristic of the European that he is never satisfied with what he has achieved; he is always pressing forward with unabated ardour in the career of industry and invention; and is as anxious to advance himself at this moment as his semi-barbarous ancestors 3,000 or 4,000 years ago. How much of this distinctive character and superiority of the European is to be ascribed to different and favourable circumstances, and how much to difference of race, is an inquiry not easily solved. Most probably a good deal is ascribable to both causes; but, at all events, his superiority is alike great and obvious. It would seem, too, that he is destined to extend his dominion over every other

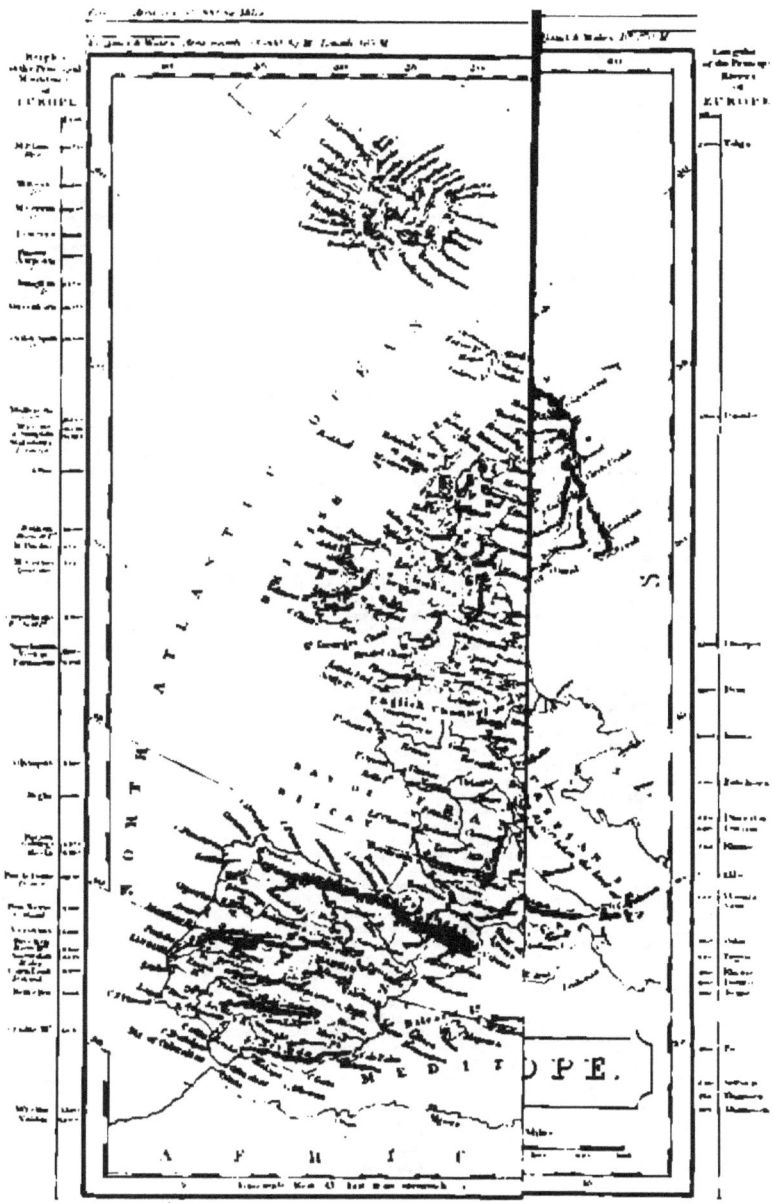

part of the world, with the exception, perhaps, of the bulk of the African continent. The European is already master of by far the largest portion of America; he has also laid the foundations of settlements in Australia that will, no doubt, at no very distant period, spread over every part of that remote and barbarous continent; and some of the oldest, most extensive, and richest countries of Asia are already in his power; and the presumption seems to be that he will in the end extend his conquests over every part of that great continent. Hence the prodigious preponderance of Europe in a moral and political point of view. It is to the world at large what Rome was to Italy, or Athens to Greece—the favoured land *unde homonitas, doctrina, religio, fruges, jura, leges atte atque in omnes terras distribute putantur.*

Situation and Limits of Europe.—Europe forms the NW. portion of the old or E. continent, having Asia on its E. and partly on its S. border; Africa, parted from it by the Mediterranean Sea, on the S.; the Atlantic Ocean, separating it from America on the W.; and the Arctic Ocean on the N. Its limits are extremely well defined upon the S. and W., but in other directions doubts exist as to what is or is not Europe. Had the early Greek geographers, indeed, been aware that for more than 1,500 m. it was joined to Asia, the probability is that no man would have been imposed to distinguish it from that division of the world; but the first observers on the shores of Greece and Asia Minor having adopted terms to designate the countries N. and S. of the narrow seas in that quarter, the subsequent discoverers applied the same as generic appellations to all the lands which gradually became known to them. Believing themselves to be permanently separated by the sea, the European naturally included in his Europe, and the Asiatic in his Asia, the discoveries made by each along the N. and S. shores of the Euxine; till, in their progress, they met on the banks of the Phasis, which thence became the first arbitrarily assumed line of demarcation. (Herodotus, Mel. 37, 38.) Even in the time of Herodotus, however, this division was growing uncertain (Mel. 45), and a line, formed by the Cimmerian Bosphorus, the Palus Mæotis, and the Tanais (Strait of Yenikale, Sea of Asoph and Don), was superseding it. This line was subsequently adopted universally as the E. limit of Europe. (Strabo, ii. 127; Pliny, iii. 1; Ptolemy, iii. 5, 6, 7, 9; Pomponius Mela, i. 2.) Little or nothing was known of this region during the middle ages; and when the arms of Russia laid it open to observation, the winding course of the Don, with which the ancients were but very vaguely acquainted, betrayed the geographers of the last century, in their anxiety to accommodate their systems with those of the Greeks, into an inextricable labyrinth of contradictions and absurdities. At length the academy of St. Petersburgh having, with great judgment, fixed the Oural Mountains as the NE. limit of Europe, proposed to continue the line of demarcation, upon their meridian, by the river Jaik or Oural, as far N. as the commencement of the great salt plains N. of the Caspian; thence the boundary was an imaginary line running NW. to Zarcuin, where the Wolga approaches nearest to the Iset; crossing the former river at that point, and then following the old limit, along the bank of the Catza, to the Sea of Asoph. (Acta Acad. Pet. 1778, p. 6; Pallas's Observations on Mountains, p. 28.) But the latter part of this boundary has two obvious defects; it is not sufficiently marked by natural features, and it divides the sources of three great rivers, the Oural, Wolga, and Don, leaving a part

of each in Europe, and a part in Asia. Malte-Brun (Abrégé de Géographie, p. 174) proposes to follow the Oural to its mouth, and then to take the Caspian for its E. border, as far as the outlet of the Kuma; thence to follow that river and the Manytch across the Caucasian plain to the junction of the latter with the Don, the lower course of which he also leaves in possession of its old destination. He considers this line as preferable to that which would follow the Terek and Kuban, because its depression is somewhat greater; but this line is hardly less arbitrary than that of the Russian academicians, and, like theirs, it is not marked by any grand natural feature. It is, indeed, not a little extraordinary, that neither looked to the gigantic chain of the Caucasus for a boundary; but it is evident that it forms one that is in all respects unexceptionable. It divides, as if by a wall (Strabo, lib. xi. p. 843), the Isthmus between the Euxine and Caspian seas, stretching between Anapé on the former, and Cape Abscharun on the latter, forming a well-defined and indestructible barrier between Europe and Asia. It would not, in fact, be more absurd to extend the boundaries of France to the Ebro, or of Spain to the Garonne, losing sight of the Pyrenees, than it is to fix the limits of Asia and Europe either to the S. or N. of Caucasus. Nature has obviously intended that that great chain should be the limit between the two continents, and by adopting it all difficulties as to their boundaries vanish. The SE. and E. frontiers of Europe are then marked by the shores of the Ægæan Sea, the Hellespont, the Propontis, or Sea of Marmara, the Bosphorus of Thrace, the Euxine, round to the Caucasus, and the ridge of that mountain system to the Caspian, thence along the shore of that sea to the Oural (from its mouth to its source), and the Oural Mountains, which, being continued to the Frozen Ocean and even further, to the high lands of Nova Zembla, complete the outline in this direction. Still it is evident that Europe is so connected with Asia, being in fact nothing but a peninsular prolongation of the larger mass of land, that no division can be quite satisfactory on physical principles; and, were it not for the vast difference in the races by which they are inhabited, we might be disposed to agree with Herodotus, who objects to giving different names to what is substantially one and the same continent. (Mel. pam. 45.)

At the first glance, it may appear that nature had marked the limits of Europe too strongly towards the N. to admit of any doubt regarding them; but Iceland having been discovered and colonised long before the voyage of Columbus, was considered as belonging to Europe; though, as it lies much nearer to the American coast, or rather to that mass of land beginning with Greenland, which appears to be divided from the American main by Baffin's Bay and Barrow's Strait, it is properly an American island. On the other hand, Spitzbergen has been sometimes considered as belonging to America, though lying on the meridian (the 20th), which passes through the very heart of Europe; and Nova Zembla has been, in like manner, included in Asia, notwithstanding the comparatively wide sea of Kara flows between it and that continent, while it is parted from Europe merely by a strait, which is moreover broken by an island (Vaigats) of some size. According to the principle, then, which considers as belonging to a continent those islands which lie nearest to it, Nova Zembla and Spitzbergen should be included in Europe, and Iceland in America; and the same arrangement, perhaps, requires that the Azores, though very distant, should also be included in Europe. According to this distribution, Europe and its islands extend from the rock of Cufonia, S. of Crete, in lat. 34° 49' N., to Little Table Island, the most N. of the Spitzbergen group, in 80° 48' 21" N.; and from Flores, the most W. of the Azores, in long. 31° W., to Jelania Noss or Cape Desire, the most E. point of Nova Zembla, in 77° E. The continental portion lies in much narrower limits, its extremes in lat. being the Tarifa Rock, W. of Gibraltar, in 36° N., and Nordkun in Finmark, 71° N. In long. the European continent extends from Cape Da Roca, near Lisbon, 9° 30' W., to the mouth of the Kara river, 64° E. (Admiralty Charts; Great Russian Map, 1800; Parry's Fourth Voyage, p. 42; Arrowsmith's Atlas, pl. 5, &c.) Its extreme length, ENE. to WSW., from the Ouralian Mountains, near Orsk in Russia, to Cape St. Vincent in Portugal, is nearly 3,400 m.; its greatest breadth, N. to S., from the North Cape to Cape Matapan in Greece, 2,450 m. Its area, pop., subdivisions, &c., will be stated hereafter.

*Physical Geography.—General Aspect.—*Europe, as already stated, is distinguished from all the other continents of the globe by the great irregularities of its shape and surface, and by the great number of its inland seas, gulfs, harbours, peninsulas, promontories, and headlands. This circumstance tends not only to influence very materially the climate and natural products of this continent, but to promote commerce and navigation.

The great indentations on the boundaries of Europe, especially on its NW. and N. sides, being its most important natural feature, the seas, on which these indentations depend, deserve to be first noticed. These seas are not very extensive. The Mediterranean, the noblest of all inland seas, is sometimes reckoned among the strictly European seas; but it would be quite as correct to describe it as belonging to Africa or Asia as to Europe. It is common to them all; and cannot justly be said to belong to one more than another. This also is nearly the case with the Black Sea and the Caspian; though, as they are mostly surrounded by countries belonging to Asia, they must be considered as belonging rather to that continent than to Europe. The great arm of the Mediterranean called the Adriatic, and the Sea of Azoph, being almost wholly encircled by European countries, are most properly said to be European seas. The Baltic, however, is the real Mediterranean of Europe; and has, including its gulfs and bays, an immense extent of coast. The Zuyderzee and the White Sea are also nearly landlocked by European countries, and consequently add to the number of European seas.

The chief of the bays of Europe are the Gulf or Sea of Kara in N. Russia, the Bays of Archangel and Onega, belonging to the White Sea; the Gulfs of Bothnia, Finland, and Riga, belonging to the Baltic; the Bay of Biscay, forming a part of the Atlantic; the Gulf of Lyons, in the S. of France; those of Genoa, Naples, Taranto, Venice (head of the Adriatic), and Trieste, in Italy; of Arta, Lepanto, Egina, Volo, and Saloniki, in Greece.

Having so irregular an outline, Europe necessarily presents numerous peninsulas and headlands. In the S. the principal peninsulas are, Spain, with Portugal; Italy, with its sub-peninsula of Calabria and Otranto; Turkey, with Greece, which includes the sub-peninsulas of the Morea and Salonica, and the Crimea. In the N. of Europe, the great Scandinavian peninsula, and those of Lapland and Jutland are the principal; and in the W. are the much less considerable ones of Brittany and Cotentin in France, and that including

the counties of Devon and Cornwall in England. The principal capes or headlands, proceeding from N. to S., are—Cape Gelania, in Nova Zembla; the North Cape and the Naze, in Norway; Cape Skagen, in Denmark; Cape Wrath, in Scotland; the Land's End, in England; Cape Clear, in Ireland; Capes La Hogue and Finisterre, in France; Rocca, St. Vincent, and the rock of Gibraltar, in Spain and Portugal; Spartivento and Leuca, in Italy; Passaro, in Sicily; and Matapan and Colonna, in Greece. (Malte-Brun, l'Europe, pp. 444-451; Balbi, Abrégé de Géogr., pp. 81-86.)

The principal islands forming part of Europe (Iceland being excluded) are—Great Britain and Ireland, with their dependent groups in the Atlantic and North Sea; Sicily, Sardinia, Corsica, Candia, the Cyclades and Sporades, the Ionian Islands, Dalmatian Archipelago, Malta, Elba, Majorca, Minorca, Ivica, the Lipari Isles, &c., in the Mediterranean and its cognate seas; Zealand, Funen, Laland, Bornholm, Oland, Gottland, Oesel, Dagö, and the Aland Archipelago, in the Baltic; the Loffoden and other islands, on the coast of Norway; Spitzbergen and Nova Zembla, in the Arctic Ocean; Jersey, Guernsey, Alderney, and Sark, in the British Channel; Ushant, Belleisle, and a few others, on the W. coast of France; and perhaps the Azores in the Atlantic, and Lampedusa, Linosa, &c., in the Mediterranean.

Mountains.—The European mountains are divided by Bruguière, in his Orographie de l'Europe, into seven distinct systems—the Hesperic, Alpine, Sardo-Corsican, Tauric, Sarmatian, British and Hibernian, and Scandinavian. The Ouralian and Caucasian chains are omitted in this enumeration, being boundary ridges between Europe and Asia, and consequently belonging as much to the latter as to the former. We have already, however, briefly noticed Caucasus (see ante, pp. 20-24); and both it and the Oural are fully described in separate articles. The Alps compose the great central table-land of Europe, over a sixth part of which their ramifications are estimated to extend. (Malte-Brun, Europe, p. 454.) The summits of the Alpine system yield in elevation only to those of the Caucasus; Mont Blanc, in Savoy, the culminating point is 15,732 ft. in height. (Bruguière.) The Alps divide into nine principal branches, which spread over Switzerland, France, Germany, the Austrian empire, Turkey, Greece, and Italy; the Apennines, Carpathians, and Balkhan, all belong to, or are intimately connected with, this system. The next in order is the Hesperic or Pyrenean system, which extends throughout Spain, Portugal, and a part of France. Its ranges, for the most part, run E. to W., through the Iberian peninsula; its culminating point is the Cerro de Mulhacen in the Sierra Nevada, 11,660 ft. high. (Bruguière: Malte-Brun.) The Sardo-Corsican system is confined, as its name implies, to the islands of Sardinia and Corsica; its highest summit appears to be that of Monte Rotondo, in Corsica, 9,068 ft. above the level of the sea. The Tauric system is comprised within the Crimea; its greatest elevation is 5,052 ft. The British and Irish system has but few summits of any considerable height; the principal are—in England, in Wales, Snowdon, 3,555 ft., and Cader-Idris, 3,550 ft.; in Scotland, Ben Nevis (Inverness-shire, 4,370 ft., and Ben Macdhu and Cairntoul (Aberdeenshire), 4,397 and 4,245; and in Ireland, Carrau Tual (co. Kerry), 3,410 ft. in height. The Scandinavian system is spread over Norway, Sweden, Lapland, and Finland; its principal chains run mostly N. and S.; its highest point, the Sneehaetta, is 8,120 ft. in elevation. The Sarmatian system consists of a few scattered hill chains in Russia, Poland, and the N.E. part of Prussia: its greatest elevation in the plateau of Valdai does not, however, reach more than 1,118 ft. above the level of the sea. (Brugaliere, l'Orographie de l'Europe.)

Plains and Valleys.—The whole of Lower Europe,—by which may be understood the entire extent of country from the Ouralian mountains and Astrakhan W. to the longitudes of Paris and London, including the greater part of Russia in Europe and Poland, Prussia Proper, the N. of Germany, Holland, Belgium, the N. of France, and the E. part of England, consists of an immense plain, interspersed only here and there with a few detached hill ranges of no great magnitude. This plain is very little elevated above the level of the sea; and we have elsewhere shown (see BALTIC), that it may be certainly concluded that at a comparatively recent period in the history of our planet, it formed part of the bed of a vast ocean, of which the Baltic is now the only considerable remaining portion. The innumerable shallow lakes in the N. of Germany, and between the Baltic and the White Sea, are smaller remnants of this great ocean; and independently of this, the morasses, abounding in marine plants, and the sands of N. Germany and Prussia, are incontestable evidences of the former submersion of the land. The more inland and easterly parts of this plain, which seem to have first emerged from the sea, particularly in the Russian governments of Kiev, Poltawa, Kharkov, Koursk, Orel, Kalouga, Toula, Tambof, and Voronije, are covered with a rich vegetable soil, varying from 8 to 5 ft. in depth. This highly fertile region, whose vast capabilities are as yet but little known, has been estimated to comprise an extent of surface equal to that of France and Austria united. Next to this great plain, rank those watered by the Lower Danube (Wallachia and Bulgaria), the Middle Danube (the Greater and Less Hungarian plains), and the Upper Danube (the plain of Bavaria); the plain watered by the Lower Rhine, that of Lombardy, and the Bohemian basin. The valleys of Europe generally are but insignificant, compared with those of Asia; but those of the Rhine, Upper Rhone, and Drave, deserve notice, as well for their extent as their picturesque beauty. Those of Norway and Scotland are commonly long and narrow, and their bottoms are often occupied by lakes, having the appearance of rivers.

Europe has no *deserts* at all similar to those of the other great divisions of the globe. There are, however, some very extensive heaths or wastes. The principal are the *steppes* of Ryn, between the Wolga and Oural, and of the Wolga, between that river and the Don; the *pustas* of Hungary, the wilds of Sweden, Norway and Lapland, the sterile districts of Stade, Hanover, Luneburg, and Zell, in the kingdom of Hanover; and of Pomerania and Brandenburg in Prussia. The greater portion of the deps. Landes and Gironde, in France, are covered with unproductive heaths, as is also a considerable part of the Terra di Bari in Italy.

Rivers.—The great watershed of Europe, or the ridge dividing the waters which flow into the Mediterranean, or Black Sea, from those which flow into the Baltic and North Sea, runs through the continent in the general direction of N.E. and SW. The courses of the principal rivers are, therefore, for the most part SE. or NW.; of the six largest, the Wolga, Danube, Dnieper, Don, Rhine, and Dwina, the four first flow in the former, and the two last in the latter direction. The chief rivers of Europe may be classed according to the seas into which they discharge themselves. The Wolga

(with the Kama) and the Oural, fall into the Caspian; the Don, Dnieper, Dniester, and Danube, into the Black Sea, and Sea of Azoff; the Petchora and Dwina into the Arctic Ocean and White Sea; the Neva, Duna, Niemen, Vistula, and Oder (Russia, Poland, and Prussia), into the Baltic and its gulfs; the Elbe, Weser, Rhine, Meuse, Scheldt (N. Germany), into the North Sea; the Loire, Garonne, Douro, Tagus, and Guadalquivir, into the Atlantic; and the Ebro, Rhone, and Po, into the Mediterranean and its gulfs. Nearly all the great rivers are in the E. and N.E. parts of the continent. Western Europe has but few rivers that have a course of more than 500 or 600 m. Still, however, this part of the continent is extremely well watered; and some of the shortest rivers, as the Thames and Shannon, afford the greatest facilities to internal navigation and commerce. If the length of the Danube be represented by 100 parts, the length of the other principal rivers will be, Wolga 130, Dnieper 72, Don 69, Rhine 49, Elbe 42, Vistula 41, Loire 37, Tagus 32, Rhone 38, Po 21, Tiber 10, and Thames 9, of these parts.

Lakes.—The lakes of Europe are situated chiefly in Russia, Finland, Sweden, Switzerland, Italy, Hungary, Austria, Prussia, Scotland, Ireland, and Greece. Among the principal are the following:—

Lakes	Sq. Miles	Lakes	Sq. Miles
Ladoga (Russia)	6,330	Geneva (Switzerland)	240
Onega (do.)	3,270		
Wener (Sweden)	2,143	Constance (do.)	200
Wetter (do.)	840	Garda (Italy)	180
Meelaren (do.)	760	Maggiore (do.)	110
Saima (Finland)	1,600	Balaton (Hungary)	140
Enara (Lapland)	665		

Lagunes are numerous along the S. coasts of the Baltic, and some parts of the Mediterranean and Adriatic shores; and Holland is full of dykes and pools. The coasts of Norway and a part of Sweden abound with inlets of the sea, which often stretch a long distance inland; these, however, do not consist of stagnant waters. There are some extensive swamps in Europe, as that occupying nearly all the basin of the Pripec in Poland, those along the course of the Danube and the Theiss in Hungary, and at the mouths of the Danube, Po, and other rivers. Many of minor extent are to be found in the great plain of the continent, in the E. part of England, Touraine in France, Italy (in particular the Pontine Marshes), Sicily, Western Greece, and on the shores of the Black Sea. (Malte-Brun, Balbi.)

Climate.—The whole of Europe, with the exception of parts of Lapland, Sweden, Norway, and N. Russia, being situated within the temperate zone, it suffers but little from the extremes either of cold or heat. Its average temperature is higher than that of those parts of Asia or America, situated within the same latitudes. This circumstance is probably owing to various causes—as the fact of its general elevation being less than that of Central Asia; its being surrounded by seas, the waters of which are warmer than those of the oceans which surround the other continents; the agency of the gulf-stream in the Atlantic, which not only brings towards Europe a continual warm current from the torrid zone, but prevents the ice of the Arctic Ocean reaching its shores; and the powerful influence of civilisation and culture exhibited in the drainage of marshes. But within the limits of Europe, there are vast differences of climate, and independent of the changes consequent on difference of latitude, the temperature diminishes so much in proportion as we proceed eastward, that the inhabitants of Turkey, in lat. 42°, often experience a degree of cold unknown in the N. of England in lat. 54°. The hottest part of Europe is its SW. extremity: in Portugal the heat is often very oppressive. The S. of Europe, shut off from the cold N. and E. winds by the great Alpine ranges, has generally a warm climate, and occasionally suffers from the indurance of the sirocco. Humidity is the chief characteristic of the atmosphere in the W. of Europe, as frigidity is of that in the E. With respect to the duration of the different seasons of the year, Europe may be divided into three zones. Southward of lat. 45° the winter is mostly confined to rainy weather from Oct. or Nov. to Jan. or Feb.; snow rarely falls, and vegetation is scarcely impeded; the spring lasts from the latter months till April or May; and the summer, during which the temperature often rises to 107° (Fahr.), and autumn, the remainder of the year. Between lat. 45° and 55° the winter is the longest season, lasting generally from Nov. to March or April; the spring continues from the latter month till June; the summer, the heats of which frequently rise to 98° Fahr., lasts till Sept.; the autumn is the shortest season of all. North of lat. 55° the seasons are for the most part confined to two—winter and summer. In the more northern parts of this zone, the snow lies on the ground, and the rivers are frozen for more than six months of the year. Beyond the arctic circle, mercury freezes in the thermometer in Sept.; and the desolation of winter is broken only by two or three months of intense heat, during which the sun is perpetually above the horizon. The absence of this luminary for the rest of the year is compensated for by the magnificent phenomenon of the aurora borealis, which shines in three regions with the utmost brilliancy. (See Malte-Brun, Géogr. de l'Europe, pp. 455-461; Balbi, Abrégé, p. 94.)

The following table is taken from Humboldt (Annals of Philos., xl. 190); the first division shows the temperature of the year, and of the various seasons in places having the same latitude; the second shows the different distribution of heat through the various seasons in places having the same mean annual temperature.

Places	Mean Temperature					
I. Lat. 60°						
Edinburgh		47·5	39·4	48·2	58·4	40·4	38·3
Copenhagen		45·6	30·9	41·9	62·4	48·4	45·0	47·7
Moscow		40·2	19·9	44·9	67·1	53·3	70·8	8·9
Lat. 48°						
St. Malo		54·4	54·2	57·2	66·0	55·4	67·4	41·2
Vienna		50·4	32·9	51·2	69·7	50·4	70·6	76·4
II. Lat.						
Dublin	49° 31′	48·2	38·2	47·2	50·4	50·0
Prague	50° 5′	48·4	31·4	47·4	69·9	52·2

Geology.—According to the map in Lyell's Principles of Geology (i. 399), the following parts of Europe consist chiefly of primitive or transition formations: the Uralian mountains; Lapland; nearly all Sweden, Finland, and Norway; most part of Scotland; the W. part of Wales; about the half of Ireland; the N.W. cos., and those of Devon and Cornwall in England; Brittany, the W. of Normandy, and a great portion of the centre and NE. parts of France; the high ranges of the Alps; Corsica; most of Sardinia; the western shores of Central Italy, the former grandduchy of Tuscany, Calabria Ultra, and the NE. parts of Sicily; Bohemia; Carinthia; Styria; parts of Hungary and Transylvania; the E. half of Turkey and Greece; and the central chain of the Caucasus. Those parts principally occupied by secondary formations are, the lowlands of Scotland; the central half of Ireland; the NE., central, and most of the S. cos. of England; most part of France, and W. Germany; the loftiest summits of the Pyrenees; the country on either side of the central chain of the Alps; central and S. Italy; the N. of Sicily; Istria; Dalmatia; the W. half of Turkey and Greece; Galicia, and the E. parts of Transylvania; some considerable tracts on the Wolga and Kama; and the N. declivity of the Caucasus. The rest of Europe, comprising nearly the whole of Russia, Poland, and the Prussian dominions; a large extent of country on both sides of the Gulf of Bothnia; all Denmark, NW. Germany, and Holland; a great part of Belgium; the E. and many of the W. cos. of England; the basins of Paris, and of the Rhone, Loire, and Garonne in France; the N. part of Switzerland; the plains of Lombardy, Hungary, Wallachia, and Bulgaria; most of Apulia; and the S. and W. parts of Sicily, is composed chiefly of tertiary, alluvial or diluvial formations; and has been obviously submerged at no very remote geological period. (Lyell, pp. 209-214.) Among the chief primary rocks of the great table-land of Europe are granite, gneiss, and sienite. In the alpine ranges W. of St. Gothard, calcareous rocks abound, often interminated with clay-slate and mica-slate; E. of St. Gothard the central chain is accompanied by lofty calcareous ranges, full of caverns. Granite is abundant in most European countries, where primary formations are met with; gneiss is the rock in which the Saxon, Bohemian, and Austrian metallic mines are principally situated. Transition limestone, which furnishes some of the best ornamental marbles, occurs in the N. and W. of England, S. of France, the Hartz mountains, Alps and Pyrenees; greywacke, in which numerous metallic ores reside, abounds in Germany, Transylvania, and the NW. parts of Italy. Coal exists extensively in the British Islands, Sweden, France, Belgium, Germany, and Bohemia. Chalk is a formation almost peculiar to Europe, extending throughout a great part of England, the N. of France, and parts of Poland, Russia, Sweden, Ireland, and Spain. Tertiary beds, containing a great number of fossils, have been discovered in various parts of Europe; the most noted of them are the London and Paris basins. The volcanic region of Europe (Iceland being excepted) appears to be principally included within the limits of Italy and its islands. There are three active volcanoes, Etna, Vesuvius, and Stromboli; but of these, only one, Vesuvius, is situated on the continent. There are, however, obvious traces of former volcanic activity in France, Greece, Germany, and some other countries; and a considerable part of central Italy is geologically composed chiefly of volcanic products. Mineral springs in great variety abound in Europe.

Natural Products.—Minerals.—If nature has denied to Europe the precious metals in any very great quantity, their absence has been fully counterbalanced by the presence of iron, coal, salt, copper, tin, lead, and mercury, in greater abundance, perhaps, than in any other region of similar extent. Iron and salt are pretty universally diffused; coal, the most important of all the minerals, is most plentiful in W. Europe, and especially in Great Britain. Copper abounds chiefly in the N. and W., in Sweden, and the extreme W. counties of England; and the tin mines of Cornwall are not only the most productive, but probably also the most ancient in the world, since it is nearly certain that they were wrought in the time of the

Phœnicians. Lead is most plentiful in Spain and England: the quicksilver mines of Idria in the Austrian empire, and of Almaden in Spain, are extremely rich. Gold, silver, and platina are found, the first chiefly in Transylvania, Hungary, and Russia; the second in various parts of Central and W. Europe; and the last has been recently discovered in the Caucasian and Ouralian mountains. Zinc, cobalt, arsenic, and nearly all other metals are found within the limits of Europe, with almost every variety of precious stones. North Italy yields the finest statuary marble, and the south part of the same country and Sicily supply immense quantities of sulphur, vitriol, sal-ammoniac, and various other volcanic products. Nitre is found in great quantities in Hungary. Besides these products, Europe furnishes the finest granite and building stone of various kinds, serpentine, slate, porcelain clay, rock crystal, alabaster, amianthus, and most of the minerals that are in the highest degree useful to man. (Balbi: Tableau Minéralogique in Abrégé de Géogr., p. 95.)

Vegetable Products.—The Flora of the extreme southern parts of Europe have a great analogy with that of the contiguous parts of Africa. In Sicily, the date, palm, sugar-cane, and cotton-plant (*Gossypium herbaceum*), several euphorbias, rare in this continent, the prickly pear, American aloe (*Agave americana*), and castor oil plant (*Ricinus africanus*), flourish. The same plants are met with in the S. parts of Spain and Portugal, in which peninsula many common to the Azores, and others, originally natives of America, grow freely without culture. In Greece, Turkey, and the S. of Russia, a large intermixture of Asiatic plants is found. The orange and lemon grow to perfection in the sheltered valleys of W. Europe, as far N. as 43° 30'; the olive ceases at about 44°; but the vine affords excellent wine in the W. as high as 50°, and its fruit comes to tolerable perfection in the open air for several degrees beyond that point in France and England. Where the vine, however, ceases to come to perfection, apple and pear trees begin to flourish, and cider occupies an important place as a beverage in the region in which wine has ceased, and beer is not in general use. The mulberry, pistachia, pomegranate, and melons, abound in the S.; peaches preserve their full flavour in the open air to lat. 50°, and the fig grows a little further N. Rice is cultivated to about 47°, but it requires a peculiar soil and climate; maize has nearly the same range. The limits of the culture of the common cerealia, or bread corn, are not very well defined, as the necessities of man oblige him to raise corn under the most unfavourable circumstances. Generally, however, the parallel of 57° or 58° may be regarded as the N. limit of the cultivation of wheat in Europe; though in some favoured spots of Finland it is raised as far N. as 60° or 61°. The hardier grains, as rye, oats, and barley, are cultivated in some sheltered situations on the coast of Norway as high as the lat. of 69° 30'; but farther E. in Russia their cultivation has not been found practicable beyond 67° or 68°. The introduction of potatoes, which are now widely diffused over almost all parts of Europe, promised until recently, when they degenerated by the attack of a peculiar disease, of great advantage to the N. regions. In ancient times, nearly the whole surface of Europe was covered with dense forests; these, however, have in a great measure disappeared in the better cultivated and more populous countries. Germany, Poland, Russia, Sweden, Norway, and some parts of the Austrian empire, are at present almost the only parts of Europe which contain forests of any considerable extent. The natural orders of *Amentaceæ*

and *Coniferæ*, comprise the greatest number of the noblest trees in the woods of Northern and Central Europa. In these regions, the oak (*Quercus pedunculata and sessiflora*) is the lord of the forest, and often attains to an enormous size. It disappears about lat. 60°; the ash does the same at 61°; the beech and lime are seldom found farther N. than 65°, or few and poor beyond 70°. The tree that grows in the highest lat. is the dwarf birch (*Betula alba*): and the last plant met with towards the pole in Europe is considered to be the 'red snow' (*Palmella nivalis*), a cryptogamic species. The vegetable products of the N. of Europe are, however, by no means confined exclusively to that region. On the Alps, the Pyrenees, and other elevated mountain ranges in Central and S. Europe, similar products are met with at the different degrees of elevation, the temperature of which corresponds with that which the various plants require; and on the declivity of Etna, at different heights, the Flora of the torrid zone, and that of the Arctic circle, are both met with. But a marked difference from that of the rest of Europe takes place in the vegetation S. of about lat. 44°. The mountains there are covered with chestnut woods; evergreens take the place of oaks, and the maritime and stone pines of other conifers; the plane tree flowering ash, carob, laurels, lentisks, oleanders, cistus, and a host of dyeing, medicinal, and aromatic plants, abound, and the surface of the earth is almost continually covered with a carpet of brilliant and odorous flowers. (Balbi; Malte-Brun; Dict. Géogr.)

The superficial extent of Europe may be estimated at about 3,654,000 sq. m. If we draw a curved line from a point in the Ouralian mountains, about the lat. of 60° or 61°, to the W. coast of Norway, in the lat. of 69°, passing through the lake Onega, and a little to the N. of the Gulf of Bothnia, this line will mark the extreme limits of cultivation, and will cut off a space equal to about 550,000 sq. m., or about 1-7th part of the entire surface of Europe. The culture of rye, oats, and barley is confined to the region N. of this line, and includes more than 5-6ths of Europe; but in the N. parts of this zone only a very small proportion of the land will bear corn. The region adapted to the cultivation of wheat comprises about 4-7ths of Europe, and includes all the densely peopled parts. The region of the vine extends over 3-7ths of Europe. (Encyc. Brit., art. ' Europe.')

Animals.—The numbers of the higher classes of animated beings are less numerous and varied in Europe than in either Asia or Africa. Some of these species known to the ancients as inhabiting this continent, as the *urus* and *aurock*, or bison, have become extinct, or nearly so; and the great increase of population and cultivation, and the clearing of forests, which have been going on from an early period, have greatly checked the increase and diminished the numbers of those which at present exist. According to Cuvier, the total number of the species of mammalia inhabiting this portion of the earth is only 150, and of this number only 58 are peculiar to Europe. The most formidable wild animals are the white bear, confined to the Arctic circle; the brown bear, which was once common in England (though long since extirpated), and is so still in the Alps, Pyrenees, and other remote mountainous and wooded regions; the wolf, still inhabiting many parts of Europe, and the wild boar. The largest animals, exclusive of whales, which inhabit the northern seas, are the elk and rein-deer, the latter of which is of the most essential service to the inhabitants of the north: these kinds of deer give place, in Central Europe, to the red-deer and roebuck; and the latter again, in the Alpine regions, to the chamois and ibex. The other principal wild animals are, the lynx, met with chiefly in the N.; the wild cat, fox, martin, otter, beaver, polecat, glutton, porcupine, hedgehog, various kinds of weasels, squirrels, hares, rabbits, rats, and mice.

The domestic animals deserve more notice. The black cattle of Europe have attained to the highest perfection; their size is in general dependent on the goodness of the pasture. The sheep, so universally diffused, is believed by some to have originated from the mouflon, or musmon, a wild animal now confined to the mountainous districts of Sardinia, and a few other Mediterranean islands. According to other writers, it was originally introduced from Asia by way of Africa; but certain ancient authorities bear testimony to the existence at one period of an indigenous breed of sheep in Great Britain. The chief races of sheep at present existing are the Spanish merino, Cretan, Wallachian, and English. The merinos are the most celebrated for their wool; but, taken altogether, the various English breeds are the most valuable, since the whole of the products they furnish bear a high character for excellence. The domestic goat was believed by Cuvier to have been derived from the Capra ægagrus, a wild species inhabiting the Alps and Illyria; the domestic hog is evidently the descendant of the European wild boar. The European horse has been supposed by some naturalists to be of Tartar origin; but no satisfactory reasons have been assigned for this opinion, which is, most probably, entirely unfounded. The English heavy horses are unrivalled for draught, and the race-horses for speed and bottom; the latter, and the hunters, have been crossed with Arab horses, the first of which was imported so late as the reign of James I. The ass degenerates in the colder parts of Europe, but in the S. it is a fine animal, and greatly valued for the breeding of mules, the sure-footedness and hardiness of which render them highly valuable. Dogs are more numerous in Europe than anywhere else; and, by frequent crossings, very numerous varieties have been produced. The domestic cat appears to be the lineal descendant of the wild species. The birds are much more various than the quadrupeds of Europe; as many as 400 different species have been enumerated; more of them, however, are birds of passage than in other continents. Four species of vultures inhabit the Alpine ranges, but are seldom seen in higher latitudes; in the rocky and mountainous parts of the N., their places are supplied by enormous eagles, falcons, large owls, and other birds of prey. Most of the birds in the Arctic regions are aquatic; in the N. there is a great intermixture of the birds of Africa and Asia, as the Balearic crane, pelican, flamingo, &c. The common sorts of game are generally diffused throughout Europe; but the red grouse is confined to Scotland, and is said to be the only species peculiar to Great Britain. Bustards abound in some parts of Turkey and Greece. In general, the European birds cannot boast of very brilliant plumage, but they excel all others in melody. Reptiles are not numerous, and few are either large or venomous. In the Mediterranean a very delicate species of turtle (Testudo raretta) is found; and in some of the Austrian lakes, the Proteus anguinus, a singular link between reptiles and fishes. Of the latter-named class of animals, the principal are the herring, cod, whiting, mackerel, haddock, mullet, anchovy, and tunny, in the ocean and seas; and the salmon, pike, trout, carp, and perch, in fresh waters. The anchovy and tunny are almost confined to the Mediterranean, where their capture forms a valuable branch of industry;

Crustacea are particularly numerous in the N., and Mollusca in the S.; the latter are especially abundant and various in the Gulf of Taranto, anciently so famous for the murex, affording the Tyrian dye. In the same part of Europe, scorpions and tarantulas are sometimes troublesome; mosquitoes infest the N.; and Europe generally is considered by naturalists as the grand region of butterflies. The European Annelides include the medicinal leech, so plentiful in the pools of Germany and Poland. Radiated animals, Zoophytes, &c., are particularly abundant on the N. coasts, where some of them, as Actiniæ, are used for food, and where the coral fisheries employ many hands. (Murray's Encycl. of Geography; Malte-Brun; Balbi, 90–100; Dict. Geographique.)

Races of Men.—To trace and define the original races of mankind, and to describe their generic and specific characters as we do those of the lower animals, is everywhere most difficult in consequence of the nice shades of distinction which prevail among some of those that approach nearest each other. But this difficulty is, perhaps, greatest of all in Europe, where, from the superior enterprise of the people, intermixture of blood, through conquest and emigration, has taken place to a greater extent than in any other part of the world. The great mass of the people of Europe belongs to the race which Blumenbach, and after him Cuvier, have called the Caucasian, under the idea not only that its type is best exhibited in the inhabitants of the Caucasian range, but that this was its original seat, and that the race thence spread itself throughout Europe. But this last supposition appears to be wholly without foundation. The inhabitants of the Caucasus have been, in all ages, unenterprising semi-barbarians, who have never emigrated beyond their own bounds; nor, through the medium of language, can a trace of them be discovered in any part of Europe. Even language, the best guide elsewhere, often fails wholly in this part of the world. Thus, through the greater part of the southern portions of Europe, the foundation of all the modern languages is Latin, originally the language of an inconsiderable nation of Central Italy; but spread by conquest, and the destruction, or absorption of the local idioms, to its present wide extent. In the same manner the German language has spread from the northern confines of France and Italy, through the central part of Europe, comprising its whole north-west portion as far as the North Cape, and including Iceland and the greater portion of the British Islands, to say nothing of the modern diffusion of the same language in America, and elsewhere.

The farther we go back in history, the greater number of distinct families of the European race will be discovered, and consequently the greater number of languages will be found to exist. In Italy and its islands, where but one language is now spoken, there were in ancient times, but after the people had made considerable advances in civilisation, six distinct native tongues, which had each a written character and a literature, besides foreign dialects; and Strabo enumerates, in all, not less than forty Italian nations, each of which, in all probability, had its own peculiar language, or at least dialect. In France, where there are now but two spoken languages, Cæsar describes three as existing in the independent part, exclusive of one, at least, in the Roman province, while Strabo enumerates no fewer than 70 different nations as inhabiting it. Within the Alps the same author gives the names of at least 30 tribes; and in the Spanish peninsula, where there are now but two languages, he enumerates

27 nations. In perusing such statements, we rather fancy ourselves reading of American, Malay, or Hindoo nations, and tribes, than of the people of Europe. The ancients were incurious both in regard to language and physical form, except their own; but comparing the few facts known to us, with the present condition of nations in a rude state of society, the probability is that, 2,000 years ago, the inhabitants of Europe, like the people now referred to, had a great diversity of languages, and might be distinguished by much difference of physical form, which in the intermixture of families can be no longer satisfactorily traced. Even in the early period alluded to, and, indeed, in a far earlier one, the intermixture of families and languages must have already made considerable progress. The Greeks had settled in Italy and its islands. The inhabitants of Gaul had colonised a considerable portion of northern Italy. The Italians, in their turn, had settled and colonised in the south of France; and the Germans, by whole tribes, had formed settlements in Gaul and Britain.

It does not seem likely, however, notwithstanding the extinction of some languages and the substitution of others, that any conquered European nation was ever exterminated; and it seems probable, that the greatest change that took place through conquest was in those cases in which the conquerors being more numerous than the conquered, a mixed race was the result, bearing a nearer resemblance to the first than to the last. Of this the Saxon conquest of England, or, at all events, the German conquest of a portion of it, which preceded the arrival of the Romans, affords the most striking example. In the great revolutions now referred to, the near approach in physical forms of the European families, and their approximation, moreover, in manners and customs, would make amalgamation a matter of little difficulty,— very different, in short, from what would have been the case had there existed a wide discrepancy, as we see in the case of the Turks and Greeks, and still more strikingly in the case of the African and European races in the New World.

The European race is distinguished from the African, Mongolian, Semitic, Tartar, Hindoo, Indo-Chinese, Chinese, Malayan, and American, by traits so obvious and distinct as not to be mistaken. The skin is white, and the colouring matter of the rete mucosum so small in amount, that in the cheeks, and some other parts of the body where the skin is thinnest, it can be seen through, and hence blushing, or, rather, visible blushing, is peculiar to the European. The hair varies in colour in different individuals, and, for the most part, is of a soft texture and undulating; the eyes also vary in colour from a light blue, or light grey, up to a dark blue or dark brown. These three characters of the skin, the hair and the eyes, are peculiar to the European, and never to be found in any other race of mankind. Variety, at least in complexion, if not in features also, is the peculiar physical characteristic of the European race, as distinguished from the other inhabitants of the globe considered by classes. The intellectual powers, as they have been developed in this race in all periods of their history, from their first emanation from the woods down to the highest point of the civilisation of Greece and Rome, or of modern Europe, exhibit a singular superiority over the other races. They display a higher degree of energy, intrepidity, enterprise, and invention, than any other. They are the only race that has as yet exhibited, in the highest degree, the peculiar prerogative of mankind, that of always continuing to accumulate

Vol. II.

knowledge, and who, notwithstanding many oscillations in their history, still continue to advance. Other races have continued stationary, or retrograded; but, as previously stated, it is a distinctive trait of the European race to have constantly moved onwards, and gained in civilisation in periods when it appeared to be retrograding; for even in the dark ages, when the fine arts, and science, and polite literature were nearly lost, the foundations were being laid of a far better constitution of society and of government. The very mixture of races conduced to intellectual advancement, and, most probably, contributed, as it is known to do with the lower animals, to physical improvement. It is in vain, therefore, that naturalists class the Semitic, Tartar, and Hindoo races along with Europeans, merely because the form of their skulls, and the shape of their faces, do not materially differ. There are other, and quite as important characteristics, that show them to be essentially different.

In attempting the following classification, it is taken for granted, that emigration and conquest have not so completely altered the physical form of the different families of men now inhabiting Europe, but that they are still, in some considerable degree, to be distinguished by the form which belonged to each in its original locality:

1. Beginning from the south-west, the first family which occurs is the Spanish or Iberian, including the whole inhabitants of the Peninsula, the Portuguese and Basques, as well as the true Spaniards. Notwithstanding the double admixture in this case of Semitic blood, and of Italian and Gothic, this family is sufficiently distinguished by colour, features, and intellectual character, from its neighbours across the Pyrenees, and those farther up the Mediterranean. They have displayed the peculiar characteristics of the European race in their resistance to and final conquest of the Arabs, in their conquest and settlement of South America, in their progress in the fine arts, and in the production of such a genius as Cervantes.

2. The next race is the Italian: its ancient type has been well preserved, notwithstanding much admixture of Greek and German blood: this is to be found in the numerous, and obviously faithful representations of its men and women of the classical ages, which exist in the statues of the Vatican and Capitol; and which do not appear to differ in any material respect from the well-formed and handsome peasantry of Italy in the present day. We may refer, as examples of the highest order of the Italian form, to the statues of Augustus and of Napoleon, which, by the way, so much resemble each other, that the likeness can hardly escape the most inattentive observer. Of the distinguished men produced by this family it is almost needless to speak: suffice it to mention the names of Cæsar and Cicero, of Dante, Raphael, Columbus, and Napoleon.

3. Proceeding eastward, we come to the Greek family. This comprises the inhabitants of the Grecian continent and islands, including the Illyrians, Albanians, Thessalians, &c. The ideal type of these is to be found in the Apollo, the Venus de Medici, and other fine remains of antiquity: and the reality in the statues of great men in the museums of Italy, and in the modern Greeks. Notwithstanding a subjugation of nearly 4 centuries, the Greeks have gained very little with their conquerors; and have preserved their language and physical form wonderfully distinct, and are now, as of old, remarkable for personal beauty. It would be idle to speak of the genius of the family which produced Homer and Demosthenes, Themistocles and Epaminondas; which

Y

routed and expelled from Europe the hordes of Asia, carried its conquest to the Indus, diffused arts and civilisation over Western Europe, and is the parent of all rational literature and sound science.

4. The next family, proceeding eastward, is the Turkish or Tartar, the only oriental race that ever succeeded in forming by conquest a great permanent establishment in Europe. Though with a considerable mixture of Semitic and European blood they still closely resemble their brethren who inhabit Transoxiana. Invariably dark eyes, and dark hair of a coarse texture, with a squatter form and an intellectual listlessness, distinguished them from all the genuine European families.

5. Turning again to the W., we find N. of Spain, and NW. of the Mediterranean, the Celtic family. Inhabiting France, Belgium, a small part of Western Switzerland, and a part of the British islands. Physically and intellectually, the general character of this people (allowance being made for the influence of civilisation) is probably, in most essential particulars, the same as that of the Gauls of Cæsar, and of the Caledonians and Silures of Tacitus. They are distinguished from the German race by darker complexions, a far greater prevalence of brown hair and dark eyes; and intellectually by superior vivacity, as exemplified in the French and Irish; but at the same time, perhaps, by less constancy and solidity. The statues of Voltaire, and the portraits of Francis I. and Sully, may be taken as examples of this family in modern times; while the dying Gladiator, now commonly considered a Gaul, may be held as representing it in antiquity. Language affords us test in regard to this family; for we know nothing of the ancient dialects of France, while the modern language is formed on that of the Roman conquerors, with the exception of about two millions of people inhabiting Brittany, who still speak a tongue which is, in reality, the same as the Welsh. The Welsh, again, is as remote from the Erse of Scotland or the Irish of Ireland, as the languages of any two American, Oceanic, or Indian tribes, a hundred miles apart from each other; while the Celtic dialects of Scotland and Ireland are, in fact, nearly identical. It must, indeed, be admitted, that there are great, if not insuperable, difficulties, even in a physical point of view, in classing all the natives now enumerated under one head; the Welsh and Scotch Highlanders being short in stature, and the French not tall, while the Irish are remarkable for their stature.

6. We come next to the German family, at present the most powerful and possessing the greatest influence of any in Europe, though two thousand years ago it was almost unknown. This family is characterised by the great prevalence of blue eyes, yellow or flaxen hair, and a very fair skin. It embraces the Swedes, Norwegians, Danes, Dutch, all the inhabitants of Germany, with the exception of a few Bohemians, and the great bulk of the Scotch and English. Along the banks of the Rhine, and in Britain, there has been much admixture of Celtic and probably, also, of Italian blood; and it is only in the northern parts of Europe, as in Sweden, Denmark, and along the coast of the North Sea, that the peculiar characteristics of the German race are still found pure and unmixed. Generally speaking, however, the German family, in its native seat, is less intermixed with foreign blood than any other European family. Its own country has never been conquered; while the Germans have been the most extensive and permanent of all conquerors, as is shown by their conquests of France, England, Italy, and Spain, and by the still more extensive conquests they are now achieving across the Atlantic, and in Australasia. The German family has probably exhibited greater enterprise, perseverance, and genius for invention, than any other family, as evinced by its discoveries in arts and sciences, its military enterprises, and its political institutions. For the last two thousand years, and probably even before it was known to the rest of the world, it has gone on steadily advancing in civilisation, and in the accumulation of knowledge. The portraits of Luther, Milton, Newton, and Goethe are favourable representations of this family, and those of Gustavus Adolphus and Charles XII. exhibit its ruder and more vulgar form.

We may here observe that there are really no grounds whatever for the common and favourite hypothesis of the German or Gothic family having emigrated at some remote period from Asia. This is a purely gratuitous and, apparently, most unfounded supposition. There is not, in fact, so much as the shadow of any kind of evidence to prove that the Germans described by Tacitus were not the original occupants of the country they then occupied; and the fair presumption is, that such was really the case. At all events, if they ever inhabited any part of Asia, their emigration must have been of the most effectual description, as not a trace of any cognate people is now to be found in that continent.

7. The next great family is the Slavonic, embracing the Russians, Poles, Lithuanians, and a portion of the Bohemians, the Wends, Dalmatians, Croatians, Slavonians, Bosnians, Servians, and Bulgarians. Swarthy complexions, as compared with the German family, dark brown hair, with a light reddish beard, a round face, high cheek bones, and eyes somewhat Mongolian, characterise this family, which, as yet, though greatly superior in energy, enterprise, and power of combination, to any Asiatic people, has made no very remarkable progress in civilisation. Peter the Great is, perhaps, the most remarkable man that this family has produced, and his portrait is a favourable specimen of it.

8. The Finnish is another family, comprising chiefly the Finns and Laplanders, with some smaller nations, the whole extending from the Gulf of Finland to the Ouralian mountains. This family is short in stature, of a strong and robust make, with a flat face, high cheek bones, light brown hair, and a thin beard. It is said to be of oriental origin, but apparently with no good foundation. The Finns have made little progress in civilisation, and many of them are to this day in the nomadic state. The whole number of this family is not estimated at above three millions.

9. Hungary, Transylvania, Wallachia, Moldavia, and Bessarabia are inhabited by a variety of races not very easily classified. The ancient inhabitants were the Pannonians and Dacians, whose robust and manly forms are well represented in the statues of their kings and warriors, many of which are still to be found among the ancient remains which exist in Rome and other cities of Italy. The genuine Hungarians of the present day are tall and handsome, with dark complexions and brown or black hair. They are said to be descended from the Magyars, who are themselves represented as emigrants from Central Asia; but if the ancestors of the Hungarians really emigrated from any country E. of the Wolga, it is certain that there is now nothing oriental in their descendants either in mind or body.

10. The NE. portion of Europe is inhabited by a portion of the Mongolian race, either in a nu-

madic or other rude state, such as the Samoyedes, the Kriyanes, Permians, Wiguls, Wotyaks, Kalmaks, and Kirghises. The Jews, a portion of the Semitic family, are found dispersed throughout all Europe, but are most numerous in some of the rudest parts of it, as Poland and Russia. It is probable, indeed, that their numbers at present far exceed what they ever amounted to before their conquest and dispersion, and when they were an independent nation. The dark complexion, black eyes, and black hair, with aquiline nose, show generally to what extent the purity of the original race has been preserved. Still, as we find not unfrequently among them, especially when living among the German family, fair hair and blue eyes, which no Asiatic ever possessed, it admits of little question that a considerable intermixture of blood has taken place. Some have gone so far as to assert that the fair Germanic type is the original one of the Jewish race.

Population.—The progress of population in Europe in modern times has been vastly greater than in any other quarter of the world, except those parts of America and Australasia that are occupied by Europeans. This increase has been at once a consequence and a cause of the progress of industry; and it has, accordingly, been greatest in those countries in which industry has been most developed,—in Great Britain, Germany, Russia, and France; and nearly stationary where industry has been stationary, as in the Peninsula and Turkey. In some countries, however, there has been a considerable increase of population without any corresponding increase of industry. However, it does not appear that the increase of population has been anywhere accompanied by a deterioration in the condition of the inhabitants. On the contrary, it has been in most countries signally improved. In Great Britain, France, Germany, Russia, and, in fact, nearly all other European countries, the great bulk of the inhabitants are now better fed, better clothed, and better lodged than at any former period. The rate of mortality has been also very materially diminished; so that there has been almost everywhere not only a great increase of comfort and enjoyment, but also of health and longevity. It may be further observed that the extensive intercourse that now prevails among different countries has almost wholly nullified the influence of those deficient harvests in particular countries that used now and then to sweep off a large proportion of their inhabitants. Scarcities are never general; and it is always found that when the crops are deficient in one quarter they are unusually productive in quarters having a different soil and climate. And commerce, by setting the surplus of one country against the deficiency of another, produces, as it were, perpetual plenty; and exempts civilised nations from those vicissitudes in respect to the supply of food that are so destructive in rude societies.

Government.—Various forms of government may be found in Europe; but, speaking generally, they may all be distributed into the three great classes of absolute and limited monarchies, and republics. It is necessary, however, to observe, that the term of absolute monarchy is not meant to express a form of government where the sovereign is really absolute, or may act as his judgment or caprice may dictate. There is no such government in any part of Europe, not even in Turkey. All that is meant by an absolute monarchy is a government where the legislative and executive functions are administered by the sovereign, without his being subject to the control of any legally constituted or recognised public body. But every

country in Europe has laws and institutions which the sovereign must respect, and public opinion has everywhere vast influence. The most absolute of the European monarchs are aware of its power, and all of them would hesitate in adopting any line of conduct that they suppose would be likely to be disapproved by any considerable proportion of their subjects. The checks on the power of the sovereign in the different limited monarchies to be found in Europe, are different both in kind and degree. They mostly, however, consist of organised bodies that share, to a greater or less extent, in the legislative authority. In some countries, as in the United Kingdom, one of the bodies that shares in the legislative authority is elected for a specified period by a pretty widely diffused system of suffrage, and has, consequently, very great influence. The distribution of power in republics is, as well known to be, quite as complete as in monarchies.

The subjoined table furnishes a condensed view of the public expenditure of the various governments of the European states in the year 1861-63, the list being arranged in the order of expenditure:—

Names	Total Expenditure	Population	Average contribution of each inhabitant
	£		£ s. d.
France	63,316,355	37,382,225	1 4 3
Great Britain	68,233,000	29,070,932	2 2 1
Russia	60,764,310	73,997,378	0 16 0
Italy	39,743,702	21,777,334	1 14 11
Austria	36,457,715	33,919,000	1 3 10
Spain	76,176,800	16,301,851	1 15 0
Prussia	20,543,455	18,497,158	1 2 2
Turkey	18,605,877	35,350,000	0 7 0
Netherlands	8,162,635	3,372,683	1 8 7
Sweden and Norway	7,797,180	5,231,013	1 9 7
Belgium	6,500,970	4,529,560	1 8 9
Bavaria	3,863,597	4,689,817	0 16 3
Portugal	3,450,370	3,641,027	0 19 0
Hanover	8,111,274	1,888,070	1 13 10
Saxony	1,853,457	2,225,240	0 16 0
Denmark	1,614,804	1,609,201	1 3 7
Baden	1,261,560	1,369,791	0 10 1
Würtemberg	1,763,649	1,720,708	0 14 10
Greece	704,044	1,329,226	0 11 11
Switzerland	771,441	2,534,242	0 6 1
Hesse-Darmstadt	735,605	852,150	0 16 0
Hesse-Cassel	721,270	749,454	0 19 3
Nassau	670,400	452,531	0 19 9
Oldenburg	337,810	295,255	1 4 3
Anhalt	277,310	181,874	1 10 6
Holstein and Lauenburg	266,720	604,448	0 5 10
Brunswick	247,150	282,450	0 17 8
Saxe-Weimar	246,100	271,254	0 14 7
Mecklenburg-Schwerin	240,000	548,449	0 8 11
Mecklenburg-Strelitz	90,000	99,060	1 1 8
Saxe-Meiningen	165,757	172,841	0 17 8
Saxe-Altenburg	124,061	137,093	0 17 9
Schwarzburg-Sondershausen	92,811	64,894	1 8 11
Hesse-Coburg-Gotha	90,976	160,481	0 11 4
Schwarzburg-Rudolstadt	66,429	71,918	0 10 10
Waldeck	66,573	59,984	1 2 9
Reuss-Homburg	45,997	35,617	1 1 8
Reuss-Schleiz	41,896	85,300	0 10 1
Schaumburg-Lippe	31,046	20,374	1 1 9
Lippe-Detmold	83,142	106,813	0 4 8
Reuss-Greiz	29,560	42,130	0 14 11
Liechtenstein	8,850	7,190	0 14 4

Europe is divided into 42 independent states, specified in the preceding table. At the head of

T 2

these states are Great Britain, Russia, France, Austria, and Prussia, called, par excellence, the five great powers. The states of the second rank are Italy, Spain, Sweden, and Turkey; those of the third, the Netherlands, Belgium, Portugal, Bavaria, Denmark, Saxony, Würt-emberg, Hanover, and the Swiss Confederation; and those of the fourth comprise the remainder.

Languages.—The principal languages at present spoken in Europe may be classed as follows:—

1. GRECO-LATIN FAMILY:
  a. Modern Greek.
  b. Italian, Spanish, Portuguese, French, Moldavian and Wallachian.

2. CELTIC FAMILY:
  a. Gaelic of Scotland, Erse or Irish Gaelic.
  b. Welsh, Armorican (Brittany). Cornish (allied to the two latter dialects) extinct.

3. TEUTONIC FAMILY:
  a. High German, Low German (Dutch), Swedish, Danish and Norwegian, Icelandic.
  b. English and Lowland Scotch, very mixed, especially the former, but founded on the old Anglo-Saxon or other Teutonic dialects.

4. SLAVONIC FAMILY:
  Russian, Polish, Bohemian, Servian, Sorab or Wendish, two dialects spoken in E. Saxony, Croatian, Bosnian, and Bulgarian.

5. OURALIAN FAMILY:
  a. Finnish, Lappish, Esthonian, Carelian.
  b. Magyar or Hungarian.
  c. Turkish.

6. PECULIAR LANGUAGES:
  a. Basque, spoken in Pyrenean districts of Spain and France—quite unlike any other European tongue.
  b. Albanian, belonging to neither the Slavonian nor Greek family, but intermixed with both.
  c. Lithuanian, Lettish, Livonian, Samogitian, quite peculiar, though containing many Slavonic words.

Density of Population.—The subjoined table furnishes a comprehensive view of the relative density of population of the chief European states. The statistics of pop. are nearly all of 1861, when a census of the inhabitants was taken in the United Kingdom, Germany, France, Prussia, the Netherlands and most other European countries.

States	Population	Area in Eng. Sq. Miles	Population per Sq. Mile
Belgium	4,529,560	11,313	401
Netherlands	3,317,852	10,905	308
Great Britain and Ireland	29,070,932	119,924	242
Italy	21,777,334	99,764	272
German States	18,971,786	95,347	148
France	37,382,225	211,859	176
Switzerland	2,534,242	14,723	167
Prussia	18,497,278	107,940	156
Austria	35,019,000	238,315	148
Denmark	1,600,551	14,492	110
Portugal	3,584,677	34,510	94
Spain	16,301,850	182,758	89
Turkey in Europe	15,500,000	200,678	76
Greece	1,279,736	19,360	68
Russia in Europe	65,848,221	2,062,399	32
Sweden & Norway	5,551,673	291,363	19
Total	290,000,000	3,701,323	75

It will be seen that the density of population varies enormously over the European continent, its north-eastern portion, comprising Russia, Sweden and Norway, or one-fourth of the surface, having only one-third of the average population—25 to 75. It may be interesting, for the sake of comparison, to show the density of population of the five divisions of the globe, which is as follows:—

Divisions of the Globe	Population	Area of Terr. Town in Eng. Sq. Miles	Population per Sq. Mile
Europe	290,000,000	3,701,323	75
Asia	780,561,000	17,846,146	44
Africa	80,000,000	11,675,000	7
America	74,100,000	15,846,000	5
Australasia	1,500,000	2,565,078	1
Total	1,311,000,000	51,603,432	25

The above statistics of the area and pop. of the five divisions of the globe are, except as far as Europe is concerned, only estimates, which however are on the best authorities.

Civilisation.—Though the least civilised state of Europe is, on the whole, more advanced in all that respects mental cultivation and improvement in the arts, than the most improved native state founded in any other part of the world, there is a wide difference in the degrees of civilisation that obtain among the different European communities. The Italian republics were the first to emerge through the barbarism that involved Europe after the Roman empire had fallen a prey to the attacks of the Germans and other Northern invaders. It was in them that commerce, arts, and literature again rose to such excellence as to rival or excel their state in the most brilliant periods in the annals of Greece and Rome. The invention of printing in the 15th century gave to the moderns a power of diffusing, increasing, and perpetuating information of which the ancients were wholly destitute, and which has contributed incomparably more than any thing else to accelerate the progress of civilisation. It is, perhaps, not going too far to say, that we are indebted to the invention of printing for the Reformation—that great event which restored to mankind the right of thinking and judging for themselves on matters of religious belief; and broke to pieces the shackles which churchmen and bigots had forged to enchain and weigh down the energies of the human mind. But though the invention of printing and the Reformation have everywhere had a powerful influence, it has been much greater in some countries than in others. Only a very short time elapsed after books began to be multiplied, till governments, beginning to be sensible of the importance of this new power, endeavoured to make it subservient to their views, by enacting laws for its regulation, and preventing any work from being published without a licence, or till it had been revised by a censor; and it was not till Holland had emancipated herself from the blind and brutal despotism of old Spain, and the Stuarts had been expelled from England, that the press began to be really free; and that periodical literature, and especially newspapers, began to acquire some portion of the vast importance to which they have since attained. But the jealousy of the doctrines broached by the early Reformers was still greater than that of the freedom of the press. They attacked principles that had been long regarded as sacred, and which, in fact, had been looked upon by most persons as part and parcel of the Christian faith. In addition to this religious feeling, most princes believed that the government derived a strong support from the church; and that, were its foundations unsettled, the whole frame-work of society would, most likely, be shaken to pieces, and their power and authority might fall to the ground. We need not, therefore, be surprised that almost all the great sovereigns of Europe, as the kings of France and Spain, the Kaiser of Germany, &c., were determined enemies of the Reformation. In England, the licentiousness of Henry VIII. effected a separation from the church of Rome, which otherwise it might have been impossible, or, at all events, very difficult to bring about; and in France, the extinction of the line of Valois by the death of Henry III. in 1589, and the elevation of Henry IV. to the throne, secured to the country the advantages of a toleration that could not be obliterated, even by the revocation of the edict of Nantes. But in the Peninsula, Austria, and Italy, the efforts of the enemies of the Reformation prevailed. Philip II., though he failed in his attempt to extirpate the principles of civil and religious liberty in the Low Countries, completely succeeded in Spain and Portugal; where he not only consigned every adherent of the new doctrines to the stake, and established the formidable tribunal of the Inquisition, but also suppressed the free institutions that had previously existed in Aragon and other Spanish kingdoms. The result has been such as might have been anticipated. Spain, deprived of those means of instruction and improvement that she once possessed, and which have been enjoyed by other countries, has not merely been outstripped by her rivals in the career of wealth and improvement, but has positively retrograded; and is infinitely less industrious and civilised at this moment than in the reign of Charles V. She has been, in fact, a prey to every species of misgovernment; and affords a striking and impressive example of the incalculable injury that an enfeebling and degrading superstition and an irresponsible government may entail upon a people. In no other country has the freedom of the press and of religious opinion been so completely rooted out as in Spain; and none, consequently, has fallen into such a deplorable state of weakness and decrepitude. In general, it may be affirmed of the different countries of Europe, that their civilisation is proportioned to the amount of freedom they have practically enjoyed. Other things have, no doubt, had a material influence in advancing and retarding their progress; but it has, notwithstanding, mainly depended on the freedom of the press and of public opinion.

Of the secondary causes that have influenced the progress and diffusion of civilisation, commerce has undoubtedly been by far the most powerful. An extensive commerce is only another name for an extensive intercourse with foreigners; and it is impossible that this should take place without partiality, at least, obliterating local and national prejudices, and expanding the mind. Commerce is also a powerful means of promoting industry and invention. An agricultural people having little communication with their neighbours, may be either stationary or but slowly progressive; but such cannot be the case with a commercial people. They necessarily become acquainted with all the arts and inventions of those with whom they carry on trade, and with the endless variety of their peculiar products and modes of enjoyment. The motives which excite, and the means of rewarding superior industry and ingenuity, are thus prodigiously augmented. The same producers exert themselves to increase their supplies of disposable articles, that they may exchange them for those of other countries and climates. And the merchant, finding a ready demand for such articles, is stimulated to import a greater variety, to find out cheaper markets, and thus constantly to supply new incentives to the vanity and ambition,

and consequently to the industry, of his customers. Every power of the mind and body is thus called into action; and the passion for foreign commodities—a passion which some shallow moralists have ignorantly censured—becomes one of the most efficient causes of industry, wealth, and civilisation.

Commerce, and the manufactures to which it gives rise, and by which again it is indefinitely extended, are always most advantageously carried on in great towns; which, consequently, are uniformly most numerous in commercial countries. These great towns are the grand sources of civilisation. The competition that takes place in them, the excitement that is constantly kept up, the collision of so many minds brought into immediate contact, and all endeavouring to outstrip each other in their respective departments, develops all the resources of the human mind, and renders a great city a perpetually radiating focus of intelligence and invention.

At no former period in the history of the world has commerce been nearly so extensive as at present; and it is all but certain that it will continue to increase, with the increase of intelligence, population, and wealth, all over the world. But the tendency of an extensive commercial intercourse among different nations is to diffuse the advantages of civilisation equally amongst them all; and the fair presumption seems to be, that the differences that now exist in the social condition of the people of the various European states, except in so far as they may depend on differences of soil or climate, or other natural causes, will gradually decrease, and finally unite them into one European family.

EUSTATIUS (ST.), one of the Caribbee or W. India islands in the group called the Leeward Islands, belonging to the Dutch, in lat. 17° 30' N., long. 67° 40' W., between St. Christopher's and Saba, about 9 m. NW. the former, and 15 m. SE. the latter island. Area, 109 sq. m. Pop. 3,270 in 1861. The island is evidently an extinct volcano; it rises out of the ocean in a pyramidal form, and has a depression in its centre, apparently its ancient crater, which now furnishes a plentiful cover for numerous wild animals. The coast is almost wholly inaccessible, except on the SW., where the town of St. Eustatius has been built. Climate generally healthy, but terrific hurricanes and earthquakes are frequent. The island suffers also the great drawback of a deficiency of spring water. Soil very fertile, and the industry of the Dutch has brought almost every portion of it into culture. Tobacco, which is the principal product, is raised on the sides of the pyramid to its very summit. Sugar, cotton, indigo, coffee, maize, yams, potatoes, &c., are also grown; and hogs, kids, rabbits, and all kinds of poultry, being reared in much greater numbers than required for the use of the pop., the island furnishes them to others. But we have no accurate information respecting the amount or value of the annual produce, or of the export or import trade. Formerly it used to be the seat of an extensive contraband traffic with the adjacent islands and the continent of S. America. This island was taken possession of by the Dutch, early in the 17th century; it has, since then, several times changed hands between them, the French, and the English; it was finally given up to Holland in 1814.

EUXINE. (See BLACK SEA.)

EVESHAM, a parl. bor. and market town of England, co. Worcester, hund. Blakenhurst, in the fertile vale of Evesham, on the Avon, 19 m. SE. Worcester, and 106 m. NW. London by Great Western and West Midland railway. Pop. 4,680 in 1861. The bor. extends over three parishes, comprising in all an area of 2,150 acres. The bor. is situated on both sides the river, the communication between its two divisions being kept up by a fine stone bridge. Evesham is very ancient, a monastery having been founded here anno 703. It was a mitred abbey, and at the dissolution its revenues amounted to 1,183l. a year. Few vestiges of the building now remain, with the exception of a magnificent tower, now used as a belfry, built not long before the dissolution. This tower is a square, 22 ft. by 22, and 117 ft. in height: it is reckoned the finest extant specimen of the pointed ecclesiastical style of the 16th century. The town consists principally of a main street in the line of the bridge, and of another nearly at right angles to it. It is paved and lighted under the provisions of a local act, which also provides for the watching of the town and the care of the bridge. There are three churches, with chapels for Baptists, Wesleyans, Unitarians, and Quakers. It has a well endowed free grammar school. Archdeacon Beadle's charity school, with national, infant, and Sunday schools, an apprentice fund, and sundry benefactions to the poor. The stocking manufacture is carried on to some extent; parchment is also made; but gardening is the principal business of the inhabs. Evesham claims to be a bor. by prescription; it sent 2 mems. to the parl. holden in the 21st of Edward I., but it was not again represented till the early part of the reign of James I., who gave a charter to the bor. Since then it has continued to send 2 mems. to the H. of C. Previously to the Reform Act, the right of voting was in the mayor, aldermen, capital and other burgesses, members of the corporation. Registered electors 338 in 1865, of whom 63 freemen. Gross annual value of real property assessed to income tax 21,788l. in 1857, and 20,224l. in 1862. The corporation revenue, amounting to 460l. in 1862, is mostly derived from bor. rates, tolls, and dues. The mayor and four senior aldermen of the old corporation were justices of the peace, and had power to hold sessions of oyer and terminer, and to try and punish all crimes other than high treason. So late as 1740, a woman was burned here for petty treason.

Near Evesham was fought, on the 4th of August, 1265, the battle between Edward, prince of Wales, afterwards Edward I., and the confederated barons under Simon de Montfort, earl of Leicester; the latter were totally defeated, and their leader and his eldest son killed.

EVREUX (an. *Mediolanum*, and subsequently *Eburovices*), a town of France, dép. Eure, of which it is the cap., on the Iton, an affluent of the Seine, 28 m. S. Rouen, and 51 m. WNW. Paris, on the railway from Paris to Cherbourg. Pop. 12,245 in 1861. The town is generally well built; but the streets are rather narrow, and its houses have an antiquated appearance; it is surrounded by fine promenades, and is well supplied with water. The chief public building is the cathedral, one of the most ancient and curious in France; it is in the figure of a cross, its centre surmounted by an octagonal dome and pyramid, the summit of which is 230 ft. above the ground; 16 pillars on either side separate the nave and choir from the lateral part of the building; the left entrance, which is flanked by two octagonal towers, is greatly admired. The other principal structures are the church of St. Taurin, probably as ancient as the cathedral; the great clock-tower, built in 1417; the town-hall, hôtel de préfecture, episcopal palace, prison, theatre, and public library, with 10,000 vols. Evreux is the seat of a court of assize, of tribunals of primary jurisdiction and commerce,

of a bishopric, which had its origin so early as the 3rd century, a chamber of manufactures, a departmental college, and a primary normal school. It has a fine botanic garden, societies of agriculture, science, and arts, and of medicine; and various courses of lectures. Its situation on one of the principal roads in France greatly facilitates its trade, and affords ready outlets for its manufactures of woollen and cotton cloths, leather, tickings, satinettes, &c. Evreux has suffered many vicissitudes. It was frequently in possession of the English; and has been repeatedly sacked by them and by the French. It was assigned by Richelieu to the Duc de Bouillon in exchange for the principality of Sedan.

EXETER, a city, co. of itself, sea-port, and parl. bor. of England, co. Devon, hund. Wonford on the Exe, 9 m. NW. from its embouchure in the English Channel; 167 m. WSW. London, by road, and 194 m. by Great Western railway. Pop. 31,479 in 1821; 31,312 in 1841; and 41,749 in 1861. The city is built on the acclivity and summit of a hill rising from the E. bank of the river, amidst a remarkably broken and irregular, but fertile district. The two principal lines of street cross at right angles near the centre of the city; numerous smaller ones intersect these and each other in various directions; these are for the most part narrow, with many ancient houses. The principal street, leading in a direction from E. to W., is broad, and has lofty modern houses and handsome shops; it is connected with an ancient suburb on the opposite side of the river by a fine bridge of three arches, built in 1776. Bedford Circus, the terraces of Northernhay and Southernhay (forming part of the city), and the suburbs, especially those of Heavitree and St. Leonard's on the E. and S., consist also of elegant modern residences; in these directions, terraces of a similar character, and detached villas, are fast increasing; the beauty of the immediate neighbourhood, the contiguity of several favourite watering-places, and the excellent markets, inducing the residence of many wealthy and respectable families. The city is well paved, lighted by gas, and supplied with water by a company, under an act passed in 1831. Exeter is the seat of a bishopric, founded in 1049. The cathedral, begun in 1280, is one of the finest in the kingdom; it is in the pointed style of different periods, with two massive Norman towers. The W. front has a façade, with numerous statues of saints and kings in niches adorned with a profusion of tracery; over it is a magnificent painted window; a corresponding one at the E. end, and those of the aisles and transepts also display great diversity and beauty. The interior is very striking, from its exquisite proportions and simple grandeur; a richly ornamented screen parts the nave from the choir, and is surmounted by a very large organ; St. Mary's chapel, the chapter-house, the bishop's throne, and several ancient monuments are also worthy of especial notice. There is a valuable ancient library, in which, amongst other muniments, is the *Exeter Domesday Book*, published by the Record Commissioners in a supplementary vol. to the great *Domesday*. The cathedral suffered much during the civil war, but has been carefully renovated. The bishop's palace (of the reign of Edward IV.) adjoins it on the SE. The other buildings in the close are modern, obstructing the view of the cathedral on the W. and S.; in the other sides is an open area, planted with trees. There are 29 other churches and episcopal chapels in the city and suburbs; the only one requiring notice is the modern church of St. Sidwell, in the pointed style, which, with its spire, forms a conspicuous orna-

sumi on the N. side of the city. It has also a Catholic and several dissenting chapels, and a synagogue. The principal charitable institutions are, the Devon and Exeter hospital, established in 1743, and accommodating above 200 patients; a deaf and dumb institution for poor children of any of the four W. counties, who are maintained, educated, and taught various trades; a lunatic asylum, a blind asylum, an eye infirmary, city dispensary, and several sets of endowed almshouses; a female penitentiary, humane society, and numerous others. There is a free grammar-school, founded in 1633, with a revenue of about 700l. a year, and six exhibitions of 36l. each, one of 22l., three of 25l., and six of 8l. a year each to either university; a blue-coat school, founded in 1661, for 32 boys and 4 girls, and 80 day scholars; St. Mary Arches school, founded in 1646, and educating 52 boys on Bell's plan, of whom 30 are partly clothed; the episcopal charity school, established 1709, and clothing and instructing 180 boys and 180 girls; the ladies' school, for 40 girls; the national or Bell's school, for 562 boys and 360 girls; an infant school; a dissenting charity school, for 55 children; and many large Sunday schools. The Devon and Exeter scientific and literary institution has a valuable library and museum; there is also an athenaeum; public subscription rooms for balls and concerts; public baths; a good theatre, usually open in winter, and during the county assizes; and an ancient guildhall, near the centre of the High Street. Annual races take place in August, on Haldon Hill, 6 m. SW. of the city. On the site of the ancient Norman castle (of which the remains of the gateway are still preserved) is a modern county sessions-house, where the assizes are held; and in the large area before it, the election of members for N. Devon, and other public meetings, take place. It is surrounded amphitheatrically by the old ramparts, the slopes of which are planted with trees, N. of the ramparts is a fine public avenue, near which are the county gaol and bridewell, and also those of the city; all of them are well-built modern structures; near the former are large cavalry barracks, and on the N. side of the city still more extensive ones for artillery. The principal market is on Friday for corn, cattle, woollen goods, and general provisions; a smaller one on Tuesday for the last named, of which there is also a considerable daily supply, especially on Saturday; a great cattle market on the second Friday in each month; and annual fairs the third Wednesday in February, May, and July, and the second Wednesday in December, chiefly for cattle. At Alphington, about 1 m. from the city, a large horse fair is held yearly in October. Woollen goods formed the ancient staple of Exeter, and during last century it exported large quantities to the Peninsula, and various parts of the Mediterranean; but this trade has wholly ceased. The cotton and shawl manufacture, introduced more recently, has also been given up; and though the weekly meetings of the woollen manufacturers of Devon are still held at Exeter, the work executed there is limited to serges. There are several large breweries and iron-foundries in the city; and tan-yards and paper-mills, employing many hands in the immediate neighbourhood. Its chief business originates in its being the provincial capital, where the public business of the co. is transacted, as well as the daily concerns of the populous and fertile districts immediately round it. It is also a great thoroughfare; and has derived much advantage from the opening of the several lines of railway, which place it in direct communication with London, Bristol, Plymouth, and the chief towns

of England. The custom-house, quays, bonding and other warehouses connected with the shipping trade, are at the SW. end of the city; where the river, confined by a weir, forms a floating harbour connected with a ship canal excavated in 1673, and originally 3 m. long; this has been deepened and extended 2 m. lower, so that vessels of 300 tons now ascend to the city; a large floating basin has also been formed, in addition to the haven, and is the terminus of the Bristol and Exeter railway. These improvements, effected not many years ago, have cost the city corporation upwards of 100,000l. The sea entrance to the harbour has a shifting bar, and is narrow and intricate, but it is well buoyed, and within the narrow neck of land, between it and the English Channel, is a spacious and safe anchorage called the Bight. There belonged to the port on the 1st of Jan., 1864, 36 sailing vessels under 50, and 110 over 50 tons, of a total burthen of 17,283 tons; there was, besides, 1 steamer of 37 tons. Gross amount of customs duty received 102,151l. in 1859; 99,713l. in 1861; and 118,084l. in 1862.

Exeter is a corporation by prescription. Its earliest charter was granted by Henry I.; its last in the 25th George III. The city is divided into six wards, and is governed by a mayor, 12 aldermen, and 36 councillors. The annual revenue of the corporation amounts to about 10,000l., derived partly from lands and houses, but chiefly from market, town, and canal dues. The charities in the city are divided into 'church' and 'general charities,' and are governed by two distinct bodies of trustees, selected from lists submitted to the lord chancellor. Exeter has returned 2 mems. to the H. of C. since 1296; the right of election, previously to the Reform Act, being in freeholders and in freemen by heirship, servitude, and presentation. The Boundary Act extended the limits of the parl. bor., so as to embrace the suburbs of Heavitree and St. Thomas, and some other districts. Registered electors, 3,305 in 1864. The limits of the municipal bor. do not coincide with those of the parl. bor. The courts of justice, or quarter sessions for the city, have jurisdiction, under the powers given by the Municipal Reform Act; they are held four times a year; the recorder presides, and barristers plead in them. There are law courts of civil jurisdiction, the provost's court having jurisdiction to any amount. A court of requests for debts under 40s., established in 13th George III., is held once a fortnight, and much resorted to. The general sessions and assizes for Devonshire are also held here. The city poor are under a corporation established in the reign of William III. The rates average upwards of 5,000l. a year. The annual value of real property assessed to income tax was 204,972l. in 1857, and 197,056l. in 1862.

Exeter is the Isca Damnoniorum of the Roman period, and is first mentioned in the second century; numerous coins and other relics of that people have been discovered. During the Saxon period it was for some time the capital of Wessex, and was noted for the number of its religious establishments. It has undergone several sieges. Archbishop Baldwin, Sir T. Bodley, founder of the Bodleian library, Lord Chancellor King, Lord Gifford, Sir V. Gibbs, were natives of Exeter; it gives the titles of Marquis and Earl to the Cecil family.

EYE, a bor. town and par. of England, co. Suffolk, rape Pevensey, hund. Hartismere, in a low fertile tract, intersected by several streams, about 2 m. from the main line of road from London to Norwich, 75 m. NE. London by road, and 92 m. by Great Eastern railway, viâ Mellis. Pop. 7,038 in 1861. The town is of some importance to the neighbourhood as a market town, but has no pretensions to be considered as a place of trade. The white-washed houses, thatched roofs, and unpaved streets, give it the appearance of a large handsome agricultural village. The church is a spacious cruciform structure, with a noble tower in the later Gothic style: there are also two dissenting chapels; an almshouse for four poor women; a free grammar-school (with two exhib. to the university of Cambridge); a national school, supported by subscription; a house of industry, adjoining which is a handsome modern guildhall. Market, Tuesday for corn; Saturday for general provisions. The inhab. are chiefly employed in agriculture; formerly, hand-made lace employed a majority of the females, but since the introduction of machinery for the purpose, this has declined. It claims to be a bor. by prescription; the earliest charter was granted in the reign of John, and subsequently eight others were conferred. It returned two mems. to the H. of C. from the earliest period down to the passing of the Reform Act, by which it was deprived of one mem. The right of voting was formerly in the burgesses, bailiffs, and commonalty. The Boundary Act extended the limits of the parl. bor. so as to include ten additional parishes, comprising an area of 15,150 acres. Registered electors, 372 in 1865. According to the Municipal Act, the limits of the bor. for municipal purposes are restricted to about 150 acres; and it is governed by four aldermen and twelve councillors.

Corporation revenue about 500£. a year, chiefly derived from rents.

EYEMOUTH, a market town, and the only sea-port in Berwickshire, Scotland, on the German Ocean, at the mouth of the small river Eye; 7 m. N. Berwick-upon-Tweed, and 42 m. E. by S. Edinburgh. Pop. 1,721 in 1861. The town has generally a thriving and respectable appearance, but the only public building worth notice is the parish church. It carries on some fishery business; but latterly it has greatly fallen off. The harbour of Eyemouth lies at the corner of a bay, into which ships may work in and out at all times of the tide, or lie at anchor, secure from all winds except from the N. or NE. Spacious granaries have been erected, in connection with the corn trade, on the quay; and a large building, once used as barracks for soldiers, is employed as a granary. Owing to its near vicinity to England, being the first harbour on the Scotch side, Eyemouth was formerly famous for smuggling; but illicit traffic has long disappeared.

Eyemouth is a place of considerable antiquity; but the most important fact in its history is that the Duke of Somerset, in his expedition against Scotland in 1547, caused a fort to be erected on a bold promontory to the S. of the town, the remains of which can still be traced. The great Duke of Marlborough, though not otherwise connected with this place, was created Baron Eyemouth by William III.; but the title, being limited to heirs male, is now extinct.

F

FABRIANO, a city of Central Italy, prov. Ancona, at the E. foot of the Apennines, 50 m. WSW. Ancona. Pop. 17,730 in 1861. The town has a cathedral and numerous convents. Felt cloth of good quality, for printers, distillers, and paper makers, is produced here; and it is celebrated for its paper and parchment. It has been supposed that this was one of the first places at which paper from linen rags was manufactured. Glue and some other articles are also produced. It has three annual fairs, and markets twice a week.

FAENZA (an. Forentia), a town of Central Italy, prov. Ravenna, on the Emilian Way, at the junction of the canal of Zanelli with the Lamone; 9 m. NW. Forli, and 30 m. SE. Bologna, on the railway from Bologna to Ancona. Pop. 35,592 in 1861. The town is surrounded with walls, and defended by a citadel. It has four well built streets, leading to a square in its centre, in which are the cathedral, town-hall, new theatre, and many handsome private residences, with a fine marble fountain in the middle. The rest of the town consists of miserable courts and lanes. There are twenty-six churches, fifteen convents, two schools of painting, a lyceum, hospital, and two orphan asylums. The manufacture of a kind of porcelain which has derived its name (fayence) from this town, still continues to be carried on, but to a much less extent than formerly. There are some factories for silk fabrics, and twist and paper mills. Its trade, which is tolerably active, is facilitated by the canal, which leads to the Po-di-Primaro. Faenza was sacked by the Goths in the sixth century; nearly ruined by the emperor Frederick II.; and annexed to the papadom by Julius II. in 1509. It was the residence of Torricelli, the inventor of the barometer.

FALAISE, a town of France, dép. Calvados,

cap. arrond. on the Ante, 21 m. SSE. Caen, on a short branch of the railway from Caen to Mans. Pop. 8,561 in 1861. The town is built on the declivity of a hill, the summit of which is crowned by its castle, now in part a ruin, but anciently the residence of the dukes of Normandy, and the birthplace of William the Conqueror. The town was formerly pretty well fortified, and is still surrounded with walls. It is clean and well built; has 3 long streets, 4 squares adorned with modern fountains, 3 churches, 2 hospitals, a theatre, and a public library with 4,000 vols. Falaise has a tribunal of original jurisdiction, and a communal college. Its manufactures consist of lace, tulles, and cotton fabrics. Its suburb of Guibray is celebrated for a large fair held in it each year, from the 10th to the 25th August. The value of the commodities disposed of at this fair has been estimated at 15,000,000 fr., or 600,000£.

FALKIRK, a market town, parl. bor., and par. of Scotland, co. Stirling, on an eminence, 3 m. NSW. Frith of Forth, at Grangemouth, at the SW. extremity of the fertile tract of land called 'the Carse of Falkirk,' 22 m. W. by N. Edinburgh, and 10 m. S. by E. Stirling, on the railway from Edinburgh to Stirling. Pop. of bor. 5,379, and of district 20,576 in 1861. The district includes Airdrie, Hamilton, Lanark, and Linlithgow, as well as two villages within less than a mile each of the town, namely, Camelon on the W., and Laurieston on the E. The Carron Iron Works (see CARRON) are within 2 m. of the town. Grangemouth, situated at the junction of the Forth and Clyde canal with the river Carron, about ½ m. from the Forth, forms the port of Falkirk. The canal in question runs past the N. extremity of Bainsford village, and is joined by the Union canal from Edinburgh, at Lock 16, within less than a m. of Falkirk. (See GRANGEMOUTH.) The Edinburgh and Glasgow and the Scottish Central railways pass within a short distance of the borough.

Falkirk consists of one well-built street, about ½ m. in length, with various cross lanes, and of Grahamston and Bainsford, which stretch in a continuous line 1 m. to the N. The only public buildings are the parish-church, built in 1811, with a steeple 180 ft. in height; the town-house, and chapels belonging to the Free Church, the Relief, Associate Synod, and the Baptists. There are 32 schools, male and female, in the parish, of which 22 belong to the town. There is, also, a flourishing school of arts, in which courses of lectures on different branches of science are delivered every winter.

Falkirk can hardly be said to possess any manufactures. There are sundry printing presses, tanneries, breweries, and some small manufactories of pyroligneous acid, with a few muslin weavers who work for Glasgow manufacturers, and weave linen or cloth from yarn spun by families in the district. But the neighbourhood of the town teems with manufactures and other sources of employment. In addition to the Carron Works, there is the Falkirk Foundry, at the N. extremity of Bainsford, in which about 500 persons, young and old, are engaged. Near it are various extensive collieries which not only supply the district, but furnish, to a considerable extent, the Edinburgh market. There are also saw-mills, several flour-mills, and a small ship-building yard. Camelon is principally occupied by nailers, their number varying from 240 to 250. Bainsford is almost exclusively inhabited by the workmen belonging to the Carron Works, and to the Falkirk Foundry.

Falkirk is celebrated chiefly for its trysts, which are the greatest fairs or markets for cattle of any in Scotland. There are three trysts annually, beginning respectively on the 2d Tuesday of Aug., Sept., and Oct.: the last being by far the largest. They continue at least two days each time, and sometimes for nearly a week. The cattle are chiefly from the Highlands, and sold for feeding in the N. of Scotland, or in England. The entire value of the stock annually disposed of at these trysts cannot be much, if at all, under 1,000,000l. These trysts were established upwards of 200 years ago.

The town is of considerable antiquity. The old church, on the site of which the new one was built in 1811, was founded by Malcolm Canmore in 1057. In the valley between Falkirk and the Carron, a battle was fought by the Scotch, under Sir William Wallace, against the English, under Edward I., in which the latter prevailed, and Sir John Graham and Sir John Stewart fell. The tomb of Graham, which the gratitude of his countrymen has thrice renewed, is to be seen in the churchyard of Falkirk. On a moor, within ½ m. of the town on the SW., Charles Stuart the Pretender, in 1746, gained a victory over the royal army, under General Hawley. Camelon was once a Roman station; and near this the famous Roman wall began, commonly called 'Graham's Dyke,' which was erected anno 140, in the reign of the emperor Antoninus Pius, and which extended across the island from the Carron to the Clyde. Falkirk was a burgh of barony till the year 1833, when it obtained a constitution from parliament; and it is now governed by a provost, three bailies, a treasurer, and seven councillors. It returns a mem. to the H. of C. in union with the burghs of Linlithgow, Lanark, Hamilton, and Airdrie, and in 1865 had 1,510 registered voters.

FALKLAND, an ancient bur. of royalty of Scotland, co. Fife, at the N. base of the East Lomond Hill, 21 m. N. by W. Edinburgh. The hill in question so far overshadows it, that the rays of the sun cannot reach it for about 10 weeks in the middle of winter. Pop. 1,142 in 1831, and 715 in 1861. The town consists of a single street with some cross lanes; the houses being in many cases thatched, and of an antique primitive description. Falkland is remarkable only for its having been a royal residence, and for the many historical recollections connected with it. The palace, which was originally a stronghold belonging to the Macduffs, thanes of Fife, was attached to the crown in 1424, on the forfeiture of that ancient house, and became a hunting seat of the Scottish monarchs. It stood on the E. of the town; and the present, which is but a fragment of the original building, was erected by James V. This monarch died here in 1542. It was a favourite residence of his grandson, James VI. The last sovereign who visited it was Charles II. in 1650. It was afterwards allowed to fall into decay; but what remained of it has recently been renovated. In 1715, after the battle of Sheriff-Muir, the famous Rob Roy M'Gregor seized on and garrisoned the palace with a party of the M'Gregors, and successfully laid the burgh and country in the vicinity under contribution. Falkland was erected into a royal burgh by James II. in 1458; but it is one of four royal burghs in Scotland (viz. Elie, Earlsferry, Newburgh, and Falkland) that were excused, on their own application, from sending representatives to parliament, owing to their poverty, or inability to afford the necessary expense of an election, and of supporting their members when elected. They still, however, enjoy all the other privileges of royal burghs. 'Falkland Wood,' the royal park, has long disappeared. Falkland gives the title of Viscount to the noble family of Carey, Lord Hunsdon.

FALKLAND ISLANDS (Fr. Malouines, Span. Malvinas), a group in the S. Atlantic belonging to Great Britain, consisting of about 50, or, according to some authorities, as many as 200, large and small islands; between lat. 51° and 52° 45' S., and long. 57° 20' and 61° 16' W.; about 1,000 m. SSW. from the estuary of the La Plata, 240 m. NE. Tierra del Fuego, and about 7,000 m. distant from London. Only two of these islands are of any considerable size,—the E. and W. Falklands. The greatest length of the former, NE. to SW., is nearly 130 m.; greatest breadth, about 60 m. The latter is about 100 m. in length, by 50 m. in its greatest breadth, in the same directions. Their united area is estimated at 13,000 sq. m. Between the two main islands is Falkland Sound, whence the whole archipelago has derived its name: this channel is from 7 to 12 m. in breadth, and navigable for ships of any class; many of the smaller islands are situated in it. Next to E. and W. Falkland, the principal islands are, the Great Swan island on the W., Saunders, Keppel, and Pebble islands on the N., and the Jason isles at the NW. extremity of the group. A small English garrison is stationed at Port Louis, at the head of Berkeley Sound, towards the NE. extremity of E. Falkland; and the islands are further occupied by a few Spanish Ayrean gauchos, Indians brought from the S. American continent, and Europeans; and frequented by numerous American, English, and French whalers and sealers; but most of them are uninhabited, and the pop. of the others is variable and uncertain.

The shores of these islands are for the most part low, except on the W. side of the group, where there are many high precipitous cliffs and ridges of rocky hills about 1,000 ft. in elevation. The average height of the W. is greater than that of the E. island; though the highest hills seem to be in the latter, where they rise to about 1,700 ft. above the sea. All the Falklands are of a very irregular shape, and much indented with bays and inlets. Excellent harbours, easy of access, affording good shelter, with the very best holding ground, abound among them, and, with due care, offer ample protection from the frequent gales. (Fitzroy, p. 248.) The sea around the Falklands is mostly deep, but in general much deeper near the S. and W. shores than on those of the N. The climate is variable, but not so much so as that of England, and it is said to be quite so healthy. The thermometer at port Louis rarely rises in summer above 70° Fah., or sinks in winter below 30°; snow seldom remains on the ground more than 48 hours, except on the mountain tops, and is never freezes so hard as to produce ice capable of sustaining any weight. Excess of wind is the principal evil; a region more subject to its violence, both in summer and winter, it would be difficult to mention. The winds generally freshen as the sun rises, and die away with sunset; the nights are in general calm, and as beautifully clear and starlight as in tropical countries. The prevalent winds are westerly; E. winds are not frequent; gales and squalls come principally from the S. Rain falls more frequently than in England; but the showers are lighter, and the evaporation is quicker. Thunder-storms are unusual. Falkland is the island that has been the most explored. Its most elevated parts are composed of a compact quartz rock. In the lower country, clay-slate and sandstone are intermixed, and are often covered by excellent clay fit for making bricks and earthenware. In many places very solid peat in layers, varying in depth from 2 to 10 feet, has been discovered; and this valuable product appears to be plentiful throughout the whole of the archipelago, where it may for ages supply the deficiency of timber. The soil consists principally of a black mould, from 6 in. to 2 ft. in depth; in many places, and especially near the foot of the hill ranges, there are extensive bogs. Fresh water is good and plentiful; there are plenty of ponds and small lakes, but no rivulets worthy of note. Copper and iron have been discovered.

The aspect of these islands is unprepossessing; but it is said that the barrenness is only apparent; that most of the land is abundantly fertile, and covered with a coarse, long, and brown, but sweet grass; while, in the interior, there are numerous sheltered valleys, feeding large herds of wild cattle. In various parts along the sea-shore, a tall rocky grass called tussock, growing to 6, or sometimes nearly 10 ft. in height, is plentiful; of this the cattle are very fond, and it is also well adapted for thatching buildings, and for the manufacture of mats and baskets. Timber of all kinds is wanting; and though the contrary has been affirmed, we believe that there is but little chance of its succeeding were the attempt made to plant it. Generally, both the soil and climate are unsuitable for corn, though it has been raised in some sheltered spots near Port Louis, where potatoes, onions, turnips, carrots, and other vegetables have also been raised.

Should these islands ever become the seat of a considerable colony, its wealth will probably be derived chiefly from breeding and rearing live stock. For this the country is well adapted. The French, and afterwards the Spanish, colonists turned loose upon the E. Falkland a number of black cattle, horses, pigs, and rabbits, and goats and pigs have been landed upon the smaller islands at different periods. These animals have multiplied exceedingly; and though they have been killed indiscriminately by the crews of vessels, as well as by settlers (who sometimes kill a wild cow merely to get the tongue), there are still many thousand head of all kinds. The wild bulls and horses are very fierce, and apt to attack individuals, who are never secure unless they are well armed, or protected by well-trained dogs. All the wild cattle are very large and fat. The horses are lightly built, and average about 14 hands 2 in. in height. The only formidable wild land animal is the warrah, or wolf-fox. This is as large as an English mastiff, and very fierce; according to Captain Fitzroy, however, it appears to be only a variety of the Patagonian fox. Sea-elephants and seals (both fur and hair seals) abound on the shores in great numbers, and whales are frequent around the coasts. Birds and fish are amazingly numerous.

Amerigo Vespucci has been commonly reputed the discoverer of these islands, but it is most probable that he never saw them. They were in reality discovered by Davis in 1592; Hawkins sailed along their N. shores in 1594; and Strong, in 1690, anchored between the two large islands in the channel, which he called Falkland Sound. In 1600, the Jason or Sebald Islands were discovered by the Dutch. The Falklands were visited during the first half of the 18th century by many French vessels; and in 1763 they were taken possession of by France, who established a colony at Port Louis on the E. island, from which, however, they were, in 1765-67, expelled by the Spaniards. About the same period the English settled at Port Egmont, Saunders' Island, though in 1770 they also were obliged to evacuate the Falklands by the Spaniards. A war with the latter was nearly the consequence of this proceeding; but in 1771 Spain gave up the sovereignty of the islands to Great Britain. Not having been actually colonised by us, the republic of Buenos Ayres assumed in 1820 a right to the Falklands, and a colony from that country settled at Port Louis, which increased rapidly, until, owing to a dispute with the Americans, the settlement was destroyed by the latter in 1831. In 1833 the British flag was again hoisted both at Port Louis and Port Egmont, and a British governor has since been continually resident at the former station, which, however, comprises only a ruined fort, state house, and a few houses. Total pop. 621 in 1862.

The possession of the Falkland Islands offers some advantages. They are situated in a part of the world where there is no other colony intermediate between England and Australia and New Zealand; the harbours are good and easy of approach, and they go far to command the passage round Cape Horn. They are capable of affording a plentiful supply of live stock and good water to ships touching at them. The value of imports amounted to 11,300l. in 1856, and to 23,384l. in 1862. The value of exports was 11,900l. in 1856, and 15,566l. in 1862.

FALMOUTH, a parl. bor. and sea-port town of England, co. Cornwall, NW. division, hund. Kerrier, on the W. side of Falmouth harbour, about 2 m. from Penryn, and 15 m. NNE. the Lizard Point, and 310 m. WSW. London by Great Western railway. Pop. of municipal bor. 5,709, and of parl. bor., incl. Penryn, 14,485 in 1861. The town is, on the whole, well built. There are several churches, the principal one being dedicated to Charles the Martyr, with chapels belonging to the Baptists, Wesleyans, Bryanites, Friends, Unitarians, and Rom. Cath.; a Jews' synagogue, a market-house, town-hall, a good hall built in 1831, good public rooms, a fine hall belonging to the Cornwall Polytechnic Society, a custom-house, a good quay, and numerous schools and charitable institutions. It is lighted with gas, and has with its environs a cheerful and picturesque appearance. The inlet of the sea, called Falmouth Harbour, is one of the finest asylums for shipping in England. Its entrance, between St. Anthony's Head on the E. and Pendennis Castle on the W., is about 1 m. in width, and it thence stretches inland about 6½ m. Falmouth is situated on a creek on its W. and St. Mawes on its E. side, immediately within St. Anthony's Head. It has deep water, and excellent anchorage ground for the largest ships; they may also anchor without the harbour, having it in their power to retreat into it should the wind come to blow from the S., which gives a great facility to ships getting to sea. Ships of large burden unload at the quay at Falmouth. Near the middle of the entrance to the harbour is a large rock covered at high water; but a beacon has been erected upon it to point it out; the usual entrance is between this rock and St. Anthony's Head, on which is a lighthouse. The harbour is defended by Pendennis Castle on its W., and that of St. Mawes on its E. side. The former is constructed on a rock more than 300 ft. above the sea. They were built by Henry VIII.; but have since been much improved and strengthened. The mail-packets for the Mediterranean, Spain, the W. Indies, and N. America, were despatched from Falmouth for about a century and a half, but of late years they have been, for the most part, despatched from Southampton, which has now, in fact, become as it were an out-port of the metropolis. The exports from Falmouth include copper, tin, tin-plates, woollen goods, pilchards, and other fish; a considerable coasting trade is carried on between Falmouth and London, Plymouth, Jersey, Bristol, and other ports. The shipping belonging to Falmouth consisted, on Jan. 1, 1864, of 46 sailing vessels under 50, and of 94 sailing vessels above 50 tons; there were also 4 small steamers, of a total burden of 130 tons. The customs revenue was 9,583l. in 1859; 6,256l. in 1861; and 6,557l. in 1863. In 1850, Falmouth had 115 registered vessels of the aggregate burden of 7,393 tons. Market-day, Thursday, for provisions generally.

Previously to the late Municipal Reform Act, the bar. was limited to the old town, which comprises only about half the modern town; but its limits were then extended so as to embrace the whole town and some adjacent territory, with Pendennis Castle. For parliamentary purposes the Reform Act added Falmouth to the bor. of Penryn. The united bor. sends two members to the H. of C.; registered electors, 703 in 1865.

It is governed by a mayor, four aldermen, and twelve councillors. Corporation revenue, about 400l. a year. In the early part of the seventeenth century Falmouth consisted only of a few fishermen's huts; it owes its subsequent rise to the patronage of the Killigrew family, and the establishment of the packets; which last was a consequence of the excellence of its harbour, and its situation so near the Land's End.

FALSTER, one of the Danish islands in the Baltic, separated by narrow straits from Zeeland on the N., Moen on the NE., and Laland on the W. Length, N. to S., 27 m.; breadth very variable. Area 190 sq. m. Pop. 25,216 in 1860. The surface is almost entirely flat, but it is considerably elevated above the sea, and is comparatively healthy. It is well watered, though it has no stream deserving notice. Its S. portion, a projecting tongue of land, is mostly occupied by the lagoon of Rotstrø. It is the pleasantest of all the Danish islands; is richly wooded, fertile, and well cultivated, and produces so much fruit that it is called the 'orchard of Denmark.' More corn is

grown than is required for home consumption; and flax, hemp, and hops are cultivated. Cattle, hogs, and poultry are plentiful; bee-hives are numerous, honey and wax being important articles of produce. Turf, chalk, and building stone are found. Some vessels are built, but the few manufactures of the island are wholly domestic. Nykiøbing, on its W. side, is the principal town; it has a cathedral, an ancient castle, and 1,400 inhabitants.

FAMAGUSTA, a sea-port town of Cyprus, in what is now a bleak and barren district on the E. shore of the island, a little N. from the mouth of the Pedea, and 40 m. E. Nicosia; lat. 35° 7' 40" N., long. 33° 55' E. It was formerly well fortified; and its works, which are now dismantled, cover a circ. of about 2 m., and consist of a rampart and bastions, defended on the land side by a broad ditch hewn out of the rock. The entrance to the harbour, which appears not to be more than from 80 to 100 yards across, is defended on one side by a beacon, and on the other by a ruined tower. This port once admitted vessels of a considerable draught of water; but since its conquest by the Turks, sand and rubbish have been suffered to accumulate to such an extent that none but small craft now enter it in safety. The town, which is poor and in ruins, has numerous deserted and choked-up streets and decayed churches; indeed, for the number of the latter, Kinneir says it might be compared to Old Goa, though not on so superb a scale. In its centre are the remains of the Venetian palace, near the cathedral of St. Sophia, a respectable Gothic building, in ruins, and in part converted into a mosque. Only a few Turkish families are found in Famagusta, most of its inhabitants being Greeks. During the Venetian régime, it was one of the most populous, commercial, and richest towns in the Levant. Its ruin was completed by an earthquake in 1735. About 5 m. NE. are the ruins of Constantia, occupying the site of the ancient Salamis, now called Eski, or Old Famagusta. These ruins consist of the foundation of the ancient walls, about 3 or 4 m. in circuit; with cisterns, broken columns, the foundations of buildings, &c., which lie scattered along the sea-shore, and over the mouth of the Pedea.

Guy of Lusignan was here crowned king of Cyprus, by order of Richard I., in 1191. It remained in the possession of his family till 1489, and then successively belonged to the house of Savoy and the Venetians. Selim II. took it after a long and memorable siege, in 1571, when its gallant governor, Bragadino, met with treacherous and inhuman treatment.

FANO (an. Fanum Fortunæ, from a temple dedicated to the goddess Fortune), a sea-port town of Central Italy, prov. Urbino, on the Adriatic, at the mouth of the Metauro, and on the Emilian Way, 7 m. NE. Pesaro, and 29 m. NW. by W. Ancona, on the railway from Bologna to Ancona. Pop. 19,672 in 1861. The town presents a lofty bastioned wall towards the sea; and has a large square ornamented with a fountain and a bronze figure emblematic of the town; a cathedral in an enriched style of architecture, which, like some of its other churches, contains paintings by Domenichino; many convents, a college of Jesuits, public school, public library, and a theatre, said to be one of the most elegant in Italy. On the road to Fossombrone is a triumphal arch, erected in honour of the Emperor Augustus, besides some other remains of antiquity. Fano has some fabrics of silk stuffs and twist, and some trade in corn and oil; but its harbour admits only small vessels. It received a colony under Augustus: in its vicinity the Romans gained an important victory over Asdrubal, anno 207 B. C. It had some extensive suburbs destroyed by the Turks in 1487.

FAREHAM, a market town and par. of England, co. Hants, on a creek at the NW. extremity of Portsmouth harbour; 4 m. NNW. Gosport, and 64 m. SE. London by road, and 74½ m. by London and South Western railway. Pop. of town 4,011 in 1861, and of par. 6,197. Area of par. 6,670 acres. The town consists principally of one broad street; and has a church and several dissenting chapels. During the summer months, it is resorted to for sea-bathing, and has every accommodation for the convenience of visitors. It has manufactures of sacking, and ropes for shipping, which are sent to Portsmouth, and vessels of large burden are built. Market, Wednesday. The government is vested in a bailiff, 2 constables, and 2 ale-conners.

FARINGDON (GREAT), a town and par. of England, co. Berks, partly in hund. Faringdon, partly in that of Shrivenham, at the base of Faringdon Hill, in the vale of the White Horse; about 3 m. from the Isis, and 67 m. W. by N. London, by Great Western railway. Pop. of town 2,943, and of par. 3,702 in 1861. Area of par. 6,910 acres. It is a very neat town, paved, lighted, and amply supplied with water from the noted spring of Portwell. The church is an interesting structure; its E. end is of great antiquity; the remainder is in the Gothic style of different periods; its spire was destroyed during the last civil war. There is also a chapel of ease at Coxwell, in the par., and a dissenting chapel in the town; a national school for 300 children, and an infant school. Market, Tuesday, a large one for corn; fairs, February 13, Whit-Tuesday, October 29, for horses, fat cattle, and pigs. Statute fairs are also held the Tuesday before and after Old Michaelmas-day. The chief trade of the town is in bacon, several thousand pigs being annually killed by its butchers. Its position at the junction of 2 main lines of road also occasions a good deal of business and activity. The line of the Great Western railway passes within 3 m. of the town.

FARNHAM, a town and par. of England, co. Surrey, hund. Farnham; 34 m. SW. London by road, and 40½ m. by London and South Western railway. Pop. of town, 3,926, and of par. 9,278 in 1861. Area of par. 10,510 acres. The town, situated near the Wey, on the main line of road from London to Southampton, consists of two principal streets, with a market-place at their intersection, and some smaller streets. It is paved, lighted, and well supplied with water, from springs in the neighbouring hills, conveyed by pipes to a large reservoir in the town. The church, a spacious building in the later Gothic style, was formerly a chapel belonging to Waverley Abbey, in the vicinity. There are also several dissenting chapels; almshouses for eight poor people, founded in 1619, and endowed with lands producing 30l. a year; a free grammar-school, with an endowment producing 30l. a year; and a national school supported by subscription. Market, Thursday; it was formerly one of the largest corn markets in the kingdom, and is still a considerable oat market. Fairs, Holy Thursday, June 24, and November 13, for horses, cattle, sheep, and pigs. The town was anciently noted for its cloth manufacture, but this is quite extinct. It is now celebrated principally for its hops, those produced in the vicinity being of a very superior quality. On the Wey are several large flour mills, whose produce is mostly sent to the metropolis by the Basingstoke canal, which passes within 1 m. of the town.

Farnham, which was a bor. by prescription, returned two mems. to the H. of C. from 4 Edward II. to 34 Henry VI., subsequent to which the privilege has not been exercised. It received two charters from the bishop of Winchester, but virtually lost the distinction of being a bor. from about 1790, or earlier. Petty sessions for the div. are held in Farnham, and there is also a court for recovery of debts under 40s., which sits every third week. Farnham Castle, on a hill N. of the town, is a residence of the bishops of Winchester, and contains a good library and some valuable paintings; it is surrounded by an extensive park, in which is an avenue nearly 1 m. in length, commanding a beautiful prospect, and much resorted to as a public promenade. It stands on the site of a castle built during the reign of king Stephen, by his brother Henry of Blois, and was built subsequently to the Restoration. Some interesting remains also exist in the vicinity of the abbey of Waverley, founded in 1128, for Cistercian monks, and subsisting till the general dissolution under Henry VIII., when its annual revenue was estimated at 174l. 8s. 3d. There is a handsome modern mansion contiguous to the site, amidst fine park scenery.

FANO, a sea-port city of Portugal, on the S. coast of the prov. of Algarve, cap. comarca of same name, on the Valformoso, near its mouth; 45 m. ESE. Lagos, and 20 m. WSW. Tavira; lat. 36° 59' 24" N., long. 12° 31' 18" E. Pop. 9,350 in 1858. The town is surrounded with walls, said to have been constructed by the Moors, and is well built, the streets being wide, and the houses good, and, to appearance, mostly new. It has a cathedral, four convents, a house of charity, seminary, military hospital, custom-house, and arsenal. It is the seat of a corregidor for the comarca, a military governor, of a bishopric, transferred thither from Silves in 1580; and of town and district judicial courts. The harbour is shallow and inconvenient; but it has a good roadstead, formed by three islands, opposite the mouth of the river. It exports figs, raisins, almonds, dates, and other dried fruits, oranges, lemons, wines, cork (the produce of its territory), rummarb, baskets, and anchovies. Many of the inhab. are fishermen. This town received its first pop. from the city of Ossonova, which stood not far distant, destroyed by the Moors on their entrance into the country. It was raised to the rank of a city by John III. in the 16th century.

FAROE, FEROE, or FÆROE ISLANDS, a group of 22 islands belonging to Denmark, in the Northern Ocean; between lat. 61° 15' and 62° 21' N., and long. 6° and 8° W.; about 185 m. NW. the Zetland Isles, and 320 m. SE. Iceland. The principal island, Stromoe, in the centre, is 27 m. long by about 7 broad; the other chief islands are Osteroe, Vaagroe, Bardoe, Sandoe, and Suderoe. Total area, 486 sq. m. Pop. 8,312 in 1860. Only 17 islands of the group are inhabited. The shores are everywhere bold and precipitous; and though there are numerous harbours, most of them are beset with rocks, or exposed to the violence of the winds and waves, so that they afford safe anchorage only in the summer. The whole surface of the land is a succession of hills, the highest of which, Skeelling in Stromoe, is 2,710 ft. in elevation. (Landt.) There are no valleys of any extent, neither are there any streams but such as are generally fordable throughout the year; small fresh-water lakes exist in several of the islands, the largest of which, in Vaagroe, is about 2 m. in circ. Climate very variable; but, notwithstanding the height of the lat., it is said to be milder and more equable throughout the year than in the S. prov. of Denmark, the snow seldom lying for more than eight days at a time. Rain and fogs are very prevalent, and the islands suffer greatly from the violence of the winds and storms. Principal rocks, granitic trap, felspar, and clay-slate; basalt in columns is frequent, peat and coal are abundant, and traces of iron, copper, and some other metals, besides opal, chalcedony, and zeolite, are found. Soil very thin, bring no more than 4 ft. in depth even at the bottoms of the valleys, and, to render it productive, it must generally be manured pretty highly; the proportion of cultivated to uncultivated land is only about 1 to 60. Some barley is grown, but neither oats nor rye will come to much perfection; and what corn is grown has to be dried under cover by means of fires. Most of the supply of corn is therefore brought from Denmark. Turnips and potatoes succeed pretty well, and are important articles of food. As might be expected, agriculture is very backward, and is principally carried on by the spade. Hay is one of the chief vegetable products; there is no timber of any description. The chief wealth of the inhab. is in their flocks of sheep, of which a peasant often possesses from 200 to 300 head; next to their flesh, they are chiefly valuable for their wool and fat; the ewes are never milked. The wool, which is coarse, is principally used in the domestic manufacture of hose and cloth. The cows are small, and no care is taken to improve the breed; every peasant is the owner of at least one. The horses are small, and used only for burdens, the steepness of the country not admitting of their being employed for draught. Hogs are rarely kept. As great numbers of sea-fowl, valuable alike for their flesh and their feathers, build round the coast, fowling is an important pursuit. It is also an extremely hazardous one, and requires great nerve and dexterity. The rocks are in many parts so precipitous that the fowlers have to be let down from the summit by a rope 100 or 200 fathoms in length. In the most inaccessible places the fowl are frequently so tame that they may be taken by the hand; but elsewhere they are taken by a net thrown over them by the fowler. Sealing, whaling, and fishing also employ a good many hands in the season. Manufactures almost wholly domestic; the chief are three of coarse woollen fabrics, woven by a loom of the rudest kind, and knit woollen stockings. Hats, combs, furniture, and other articles of prime necessity are made, and good boats built in many places; dyeing, fulling, and tanning, are also conducted in the country. Principal exports,—beef, tallow, fish, train oil, feathers, skins, and butter: imports,—corn, pulse, bread, malt, spirits, colonial produce, iron, lead, gunpowder, lime, bricks, timber, tar, glass, linen cloth, shoes, and books. About 100,000 pairs of hose are exported annually. Barley bread, dried meat, fish, soup of oatmeal, fat, and water, milk, and turnips, compose the chief articles of food. The people are of Scandinavian origin, and speak a dialect similar to old Danish.

These islands have a civil governor, called amtmann, a judge or landvogt, and a provost with supreme authority in religious matters. The country is divided into 7 parishes and 39 congregations. The only town is Thorshavn, at the SE. end of Stromoe, which is defended by a fort, and has about 1,000 inhab. The land partly belongs to the inhab., and partly to the crown; the public revenue, derived from the royal domains, quit rents, taxes on flocks and fisheries, is paid mostly in kind. There are no schools, except one in Thorshavn; but most of the pop. possess the rudiments of education. The Faroe isles are supposed to have been discovered by the Norwegians

in the 9th century; since the union of Norway with Denmark, in the 14th century, they have belonged to the latter country.

FARS, or FARSISTAN, a prov. of Persia, which, by the change of the s into p, has, in European languages, given its name to the whole country in the S. part of which it is situated; between lat. 27° 40' and 31° N., and long. 49° 30' and 55° E., having N. the prov. Irak, E. that of Kerman, S. Laristan and the Persian Gulf, and W. the latter sea and Khuzistan; length, N. to S., nearly 300 m.; breadth 200 m. Area, perhaps about 55,000 sq. m. Pop. uncertain. A mountain chain, which is a continuation of Mount Zagros, extends from NW. to SE., through this prov., dividing it into the hot and cold regions (Germasser and Sirhad); the former of which, the smaller division, extends with a variable breadth inland along the whole coast; while the latter comprises most of the N., E., and mountainous parts of the prov. The mountain ranges in some places rise from 2,500 to 3,000 ft. above the sea; they are interspersed with numerous plains from 15 to 100 m. in length, though seldom more than from 8 to 10 m. in breadth. These plains are in general fertile, sufficiently well watered, and afford abundance of pasturage and wood; some of them are tolerably well cultivated, but they are, for the most part, and particularly to the N. and W., destitute of inhabitants. In the E. part of the prov. the plains are of greater extent, the soil is more sandy, and water is less plentiful. The central mountain chain divides the rivers into those which flow into the Persian Gulf, and those discharging themselves into Lake Bakhtegan. The principal of the former is the Tab (the Arosis), and of the latter the Benderemeer, or rather Bund-emeer (the Cyrus of Arrian).

Besides the Lake Bakhtegan, which is 70 m. in circ., there are several other lakes, the chief of which is in the neighbourhood of Shiraz. These, as well as some of the rivers, are salt, the soil of Fars being strongly impregnated with that mineral; and the bed of the lake Bakhtegan affords in summer, when it is nearly dry, great quantities of fine salt. The climate of the hot region is unhealthy; fevers, ophthalmia, and other diseases are prevalent; famine for want of rain is not uncommon, and the people are poor, and live wretchedly in mud huts. In the cold region, on the contrary, the climate is temperate and healthy, and agriculture is not in so bad a state as in some other prov. of Persia. The E., though less highly favoured than some other parts of Fars, is that best cultivated; and great quantities of the finest tobacco are raised there. A great deal of corn, and especially rice, dates, raisins, and various other fine fruits; opium, saffron, hemp, cotton, &c., are among the chief agricultural products; silk is produced; the cactus feeding the cochineal is plentiful; and great numbers of roses are cultivated for the manufacture of attar. The wine is of a rather superior quality, and that of Shiraz has attained, perhaps, more celebrity than it deserves. Many cattle and sheep are reared; the horses, oxen, and camels are good; fish, game, and other wild animals, are abundant. There are said to be mines of lead and iron, and quarries of marble and alabaster; borax is obtained, and there are some very productive springs of naphtha. The inhab. are, generally speaking, among the most civilised and industrious in Persia. They manufacture fine woollen, silk, and cotton stuffs, camel skins &c. for exportation. The trade is principally with Hindostan. Chief towns, Shiraz, the cap., Bushire, Firozabad, Darab-jerd, Kazeroun, Bender-righ, &c. In this prov. are also the ruins of Persepolis, Pasargae, and Shahpur. Fars was the ancient patrimony and kingdom of Cyrus the Great, previously to his foundation of the Persian empire.

FAVERSHAM (formerly *Feversham*), a bor., par., and sea-port town of England, co. Kent, lathe of Scray, hund. Faversham; 45 m. SE. by N. London by road, and 48 m. by London, Chatham and Dover railway. Pop. of bor. 5,858, and of par., 6,363 in 1861. Area of par., 2,270 acres. The town, situated near a branch of the Swale, and within ½ m. of the old road from London to Dover, consists chiefly of two irregular streets, crossing at right angles, with a market-place and town-hall at the point of intersection. A suburb called Brent Town consists of cottages built within a recent period; and Ospringe Street, on the above line of road, is another suburb. The village of Preston is also quite contiguous. Faversham is paved and lighted. The church, a spacious structure, with a fine tower and spire, was rebuilt in 1755, on the site of a structure of the reign of Edward III. There are also several dissenting chapels; a free grammar school, founded by Elizabeth, for 8 boys; and 2 other free schools, one for 12 boys, the other for a like number of girls; almshouses for 12 poor people; a theatre, and assembly rooms. Market, Wednesday and Saturday; fairs, Feb. 25, Aug. 12. There are gunpowder mills in the vicinity belonging to private individuals, but the government mills have been discontinued. At present the oyster fishery forms the most important staple of the place, and is conducted by a privileged company, admission to which is obtained by birth, or apprenticeship to a member; but the claimant must be a married man. There belonged to the port on the 1st of Jan. 1851, 2 sailing vessels of under 50, and 193 sailing vessels of above 50 tons, besides two small steamers of 22 tons. The bor., since the Municipal Reform Act, is governed by 4 aldermen and 12 councillors. Average annual corporation revenue, 1,000l. The limits of the old borough (which did not comprise the entire town) have been extended so as to include that and the whole of Ospringe Street. There is a court of requests for debts under 40s., and a union workhouse.

FAYAL, one of the Azores, which see.

FAYOUM, a famous valley and prov. of Central Egypt, anciently the nome of Arsinoë. At about 15 m. WSW. Benisouef; there is a depression in the Libyan or most westerly of the two chains, which accompany the Nile out of Nubia. From this gorge—about 8 m. in length—the hills diverge, making a circular bend to the W. and N., and enclose the valley of Faioum; which is of an oval figure, and forms a low table-land, gradually sloping towards the N. and S.; the N. depression occupied by the Birket-el-Kerûn (the lake Moeris of the ancients), and the S. depression by lake Garah. Thus, unlike other basins, the valley of Faioum has its greatest depressions, not in the middle, but at the sides; its central portion forming a low, slightly convex plateau, extending towards the W. Upon this culminating line runs an arm of the great canal of Egypt, the Bahr Iusf (given out at the narrow pass mentioned above), which at a short distance from Medinet-el-Faioum, the capital of the province, spreads out into various small branches, and gives a fertility to the valley which, though comparatively great, has been much overrated by some travellers. Faioum is about 40 m. in length from E. to W., and 30 m. in breadth from N. to S.

Towns, Villages, and Canals.—At the entrance of the ravine, which affords the only communication between this isolated province and the Nile, stand the village of Illahoun, on the SE. bank of the canal, and the town of Howarah-el-Kebyr, on its SW. bank, connected by a bridge of three arches, and provided with a number of reservoirs to regulate the masses of water during the inundation. Near Illahoun is a dilapidated pyramid 60 ft. high, with a base of 197 ft. square, consisting of calcareous stone, that supports a pile of unbaked bricks. At the other extremity of the gorge, where the valley fairly opens, is Howarah-el-Soghir, near to which two ancient branches of the Bahr Iusf diverge in opposite directions. The waters of the main canal are turned into these branches by means of bridge-dykes, built upon foundations above the ordinary level of the stream, so that at high water the current of the Nile continues its course through the arches; but these canals are so encumbered with mud that their waters never reach the lake except during the inundation. Between El Saguir and Medinet-el-Faioum are strewed the remains of the celebrated Labyrinth, consisting of, first, a brick pyramid, 128 yds. square and 197 ft. high; under which the French discovered a subterranean passage, a sarcophagus, and a salt spring; secondly, the remains of a temple to the E. of the pyramid, presenting the fragments of huge columns of granite, with several sepulchral excavations. A large mass of ruins are buried in earth and rubbish, and have never been explored; the whole forming an oblong parallelogram 804 ft. in length, with nearly as great a breadth. Among another series of ruins, to the N. of Medinet, and occupying an area of about 2½ m., Belzoni found two immense stone pedestals, to which the name of 'Pharaoh's feet' have been given; various granite statues, some wrought iron, and a quantity of half melted glass. At some distance from these stands a syenite obelisk with a circular top, and though 43 ft. high, is covered with a profusion of sculpture. A portion of these remains are believed to have belonged to the Labyrinth, but most of them to the ancient city of Arsinoë, now replaced by Medinet-el-Faioum. This capital is divided by a branch of the Bahr-el-Wady into two parts, connected by five bridges, and much of it is built of the remains of the ancient city. In 1824 Medinet contained 5,000 inhab., partly Copts and partly Moslems. It is the residence of the provincial governor. Some ruins at a short distance from the E. point of Birket-el-Kerûn accord very nearly with the ancient Bacchis or Bacchias, 18 m. WNW. of the village of Nardeh, and 3 m. from the lake, stands a temple, known as Kasr-Kerûn, 94 ft. long, and 63 ft. high, with 14 chambers, having on either side a long passage whose end wall is divided into three narrow cells. (Wilkinson's Topog. of Thebes, pp. 852, 853.) Jomard penetrated one of these avenues, and, finding it skilfully adapted for the conveyance of the voice, inferred that it was designed for the utterance of oracles. This temple is manifestly of Roman origin, as is a smaller one 130 paces to the SE. of it. We pass over the less noticeable villages of Faioum, of which there are altogether not quite 70. (Encycl. Britannica, art. 'Egypt;' Ritter's Africa, vol. iii. p. 35–51, French edition; Letronne's Noar. Annales des Voyages, vi. pp. 153–164; Belzoni's Researches, &c., ii. 145, &c.)

Lake Moeris.—According to the statement of Herodotus, confirmed by that of other historians, this lake occupied in his time a large proportion of the valley, having a circumference of 450 m. (3,600 stadia), and a maximum depth of 150 ft. The basin was filled by the waters of the Nile conducted to it by canals, for it had no springs. The statement as to the size of the lake in antiquity is not inconsistent with its present contracted dimensions; the supply of water has been gradually lessened by the raising of the bed of the Nile, and by the filling up of the lakes and canals, so that very little reaches it at present, even during the inundation; not enough to countervail the copious evaporation which in this hot climate is continually going on. Hence, last century, the lake was 50 m. long and 10 m. broad (Pococke's Travels, i. 62), whereas it is now only 30 m. long and 6 m. broad in the middle or widest part. Herodotus states that the Lake Moeris was artificially excavated by order of the king whose name it bears; but by this he no doubt referred to the excavation of the canals by which the lake was filled, and perhaps also to some excavations made in the lake itself. He says that for six months the waters flowed from the Nile to the lake, and that during the other six months they flowed from the lake to the river; but the level of the lake must always have been too low for the waters to have returned to the Nile; while that of the canals does so to this day. (Herod. lib. ii. § 149; Encyc. Brit., art. 'Egypt;' Wilkinson's Topog., 'Thebes,' p. 351.)

The Labyrinth.—This extraordinary structure is said by Herodotus, by whom it was visited, to have surpassed all the works of the Greeks, including the temples of Ephesus and of Samos, and to have been superior even to the pyramids. (Lib. ii. § 148.) It was divided into 12 courts, corresponding to the 12 nomes or provinces into which Egypt was then distributed, and is said to have contained 3,000 apartments, 1,500 above, and as many below ground. Herodotus visited those above ground, and speaks of them from his own observation, but he was refused admittance to the others, and informed that they were used as sepulchres for the sacred crocodiles, and the kings who had constructed the edifice. (L'bi supra.) The different chambers were connected by an infinite number of winding passages, so artfully contrived as to give the structure its name. The ceilings, walls, and pillars were of the whitest marble, all adorned with sculpture. In fact, one's belief is almost staggered by the accounts of this extraordinary edifice; and nothing less than the authority of the venerable father of history could have made us believe in the existence of such a structure. (For further information as to this extraordinary plan, see the notes to Larcher's Herodotus, tom. ii. 491–505, 2d ed.) There can be little question that the ruins strewed about near Medinet, and between 1 and El Seguir, are those of the Labyrinth, though the position of Kasr Keron was assigned to it by early European travellers.

Faioum is chiefly inhabited by two branches of the Saracenian tribe of Arabs from the W., states of Burkhary, who were able at the end of the last century to supply 2,970 soldiers. (Girard, 'sur les Habits de Faioum,' Desc. de l'Egypte, tome iii. p. 250.) Near the capital large quantities of roses are cultivated, which are converted into rose water of a highly esteemed quality. The land capable of cultivation in Faioum has been estimated at 450 sq. m., of which scarcely the half is at present tilled.

FECAMP, a sea-port town of France, dép. Seine Inférieure, cap. cant., between two ranges of hills, at the mouth of a small river of the same name, 48 m. NW. Rouen, on a short branch line of the Rouen-Havre railway. Pop. 12,211 in 1851. The town consists of little more than a main street, not well built, but upwards of 2 m. in length from the church to the port. The church, a handsome edifice, is the sole remaining part of a celebrated abbey, founded by Richard I.,

duke of Normandy, in 988, and destroyed during the revolution. Fecamp has an exchange, hospital, chamber of commerce, and a gratuitous school of navigation. Its port, though small, is one of the best on the Channel; and it has been very greatly improved by the construction of an inner port, with a fine quay, and a magnificent lighthouse. It has two roadsteads: the *Grand Road*, lying opposite to Cricquebœuf, about 2 m. off shore, with thirteen fathoms, and a good clay bottom, mixed with sand; the *Little Road* lies off the W. side of the harbour, and has from ten to seven fathoms. It manufactures cotton yarn, linen fabrics, seamen's shoes, hardware, rope, cordials, candles, and soda; and has sugar refineries, tanneries, and building docks. It also fits out vessels for the cod, mackerel, and herring fisheries, and is an *entrepôt* for colonial produce, salt, and brandy. The air of this town is celebrated for its purity, its men for their healthy appearance, and its women for their beauty.

FELEGYHAZA, a town of Hungary, between the Danube and Theiss, cap. distr. of Little Cumania, on the road between Pesth and Temeswar, 65 m. SE. the former. Pop. 19,420 in 1857. The town has a Roman Catholic church and gymnasium; and a court of justice, in which the archives of the distr. are preserved. Some Roman antiquities have been discovered in its neighbourhood. The country round produces corn, wine, fruit, &c., and large cattle markets are held in the town.

FELIPE-SAN, formerly JATIVA, or XATIVA (an. *Setabis*), a town of Spain, Valencia, cap. prov. of same name, on the declivity of a hill, near the confluence of the Montesa and Albayda, 44 m. S. by W. Valencia, and 195 m. SE. Madrid, on the railway from Valencia to Alicante. Pop. 15,747 in 1857. The town is well built, and supplied with public fountains. It has a collegiate, 3 par. churches, 10 convents, a hospital, and an asylum for widows. The ancient city stood on the summit of the hill, near the foot of which the modern town is built. It had a strong fortress; and having been a Roman station, contained some Roman edifices, as well as others erected by the Moors, all of which are now in ruins. Inglis, speaking of the latter, says, 'The magnificence and extent of the Moorish remains struck me with astonishment, even after having seen the Alhambra. These crown the hill that rises immediately behind the city; this hill is twice the height of that upon which the Alhambra stands, and the remains at San Felipe are also greatly more extensive. They are not, indeed, like the Alhambra, in preservation, nor do they present the terraces, and arches, and columns, that at once point out its Moorish origin; but they are seen covering the summit of a mountain ridge, 1,000 or 1,200 ft. high, and presenting in fine relief, against the sky, an irregular line of not less than two miles in extent of massive and imposing ruins.' (Spain in 1830, ii. 245.) In 1706, during the war of the succession, Xativa, after it had held out a long time against the French, was taken and burned; it was rebuilt on its present site by Philip V., who gave it his own name. The Moorish style, however, which prevailed in the former city, seems to characterise the edifices and manners of the present one. 'Passing along the streets, I observed many signs of Moorish days, more than either in Seville or Granada: in a court-yard which I entered, mistaking it for that of the posada, I noticed that the walls were arabesque; and looking in at the doors of the shops and houses, I scarcely saw a single person seated upon a chair, or even upon a stool; every one was squatted upon a mat.'

(Inglis, *ubi supra*.) San Felipe has no manufacture; all its inhabitants are said to find employment and subsistence from its contiguous *huerta*, or irrigated valley.

FELIPE (SAN), a town of the repub. of Venezuela, Colombia, dep. Venezuela, on the Yragul, not far from the Gulf of Triste, and 136 m. W. by S. Caracas. Pop. estim. at 6,000. The town is regularly laid out with wide and straight streets, and has a good parish church. Cocoa, cotton, indigo, coffee, &c. grow abundantly in its neighbourhood, and are the chief articles of export. Its climate is, however, oppressive, damp, and unhealthy.

FELTRE (an. *Feltria*), a town of Austrian Italy, prov. Belluno, on a hill at the foot of the Alps, and near the junction of the Colmeda with the Piave, 16 m. SW. Belluno. Pop. 5,450 in 1857. The town is partially fortified, and is tolerably well built; streets broad and well paved. It has a handsome market-place, a cathedral, many other churches, an episcopal gymnasium, a seminary of theology and philosophy, a hospital, and an orphan asylum. It has silk twist and some wax-bleaching factories; and trades in silk, wine, and oil, the produce of the adjacent territory.

FERMANAGH, an inland co. of Ireland, prov. Ulster, having E. Cavan, E. and N. Monaghan, Tyrone, and Donegal, and W. Leitrim. Area, 471,186 acres. Extent of arable land, in sq. m. 452 in 1841; 526 in 1851; and 510 in 1861. Of the total area, above 188,000 acres are unimproved bog and mountain, and 48,797 water, principally consisting of Lough Erne. This, which properly consists of two lakes, joined by a deep and winding channel, is a noble sheet of water. It stretches the whole length of the co., which it divides into two nearly equal portions. See ERNE (LOUGH). Surface varied, and in general better wooded than most Irish cos. Farms of all sizes; but the great majority very small. In the N. part of this co., agriculture is in a forward state; but, elsewhere, it is very backward: a good many cattle are bred on the high grounds. Oats, barley, wheat, flax, and potatoes are the principal crops. Iron ore is found in different places. Manufactures unimportant. Fermanagh contains 6 baronies and 18 parishes, and sends 3 mems. to the imperial parliament, viz. 2 for the co., and 1 for the bor. of Enniskillen, which is the principal. Reg. electors for co. 4,872 in 1862. The pop. amounted to 156,852 in 1841; to 116,441 in 1851; and to 105,768 in 1861. The decrease of pop. was 25·76 per cent. between 1841 and 1851, and 9·17 between 1851 and 1861.

FERMO (an. *Firmum Picenum*), a city of Central Italy, prov. Ascoli, on a hill about 3 m. from the Adriatic, and 32 m. SSE. Ancona. Pop. 18,990 in 1861. The town is surrounded by a wall, of little importance as a means of defence; and has a cathedral, 10 other churches, 15 convents, a palace, built by Jerome Bonaparte, a university founded in 1585, and 2 fine collections of statuary and paintings. The harbour on the Adriatic, called *Porto di Fermo*, is small, and frequented only by a few trading vessels. The exports consist chiefly of corn, silk, and woollen cloth; it has an annual fair, lasting from August 16 to Sept. 6. Fermo is the seat of an archbishopric, and of a court of primary jurisdiction, with appeal to a superior tribunal at Macerata. It was founded by the Sabines, before Rome existed; and colonised by the Romans towards the beginning of the first Punic war, and has been plundered by Alaric, Attila, and other barbarian chiefs. It, however, continued during a blockade of 11 years to hold out against Alboin, and was only obliged, through famine, to yield to his successor, Autharis. Since

the 8th century it has, with few intermissions, belonged to the see of Rome, till it came to form part of the new kingdom of Italy in 1860. Lactantius and Galeazzo Sforza were both natives of Fermo.

FERMOY, an inland town of Ireland, co. Cork, prov. Munster, on the Blackwater, 116 m. SW. Dublin, on a branch of the Great Southern and Western railway. Pop. 6,976 in 1831, and 8,705 in 1861, the Cmb. being in the Protest. In the proportion of about 8 to 1. The town which, till 1791, was but a station for carriers, consists of a square, and several well-built streets on each side the river, which is here crossed by a fine bridge: its rapid improvement is owing to its having been made a military *dépôt* during the last war with France. It has a par. church and a R. Catholic chapel, both spacious and elegant buildings, a convent, a Methodist meeting-house, several large schools, and a court-house; a workhouse, which was formerly turned into barracks for 3,000 men. Races are held annually in the neighbourhood. There are extensive flour-mills; and a considerable trade in flour and agricultural produce, mostly sent to Youghal, whence coal and other produce is received in return. There are also two paper-mills and a brewery; duty is paid, on the average, on 72,000 bushels of malt, and the town is the centre of a considerable retail trade. Markets on Saturdays; fairs on 21st June, 20th August, and 7th November. General sessions are held in January; petty sessions every Monday.

FERNANDEZ. See JUAN FERNANDEZ.

FERNANDO-DE-APURE (SAN), a town of the repub. Venezuela, Colombia, dep. Orinoco, on the Apure, near its junction with the Portuguesa, 164 m. E. by S. Varinas. Estimat. pop. 3,000.

FERNANDO-PO, an island in the Bight of Biafra, 20 m. from the African coast, about 10 m. in length by 20 m. in breadth, now abandoned, but formerly occupied by Great Britain, it having been selected as a military and naval station from its supposed salubrity and from the facilities afforded by its situation for the suppression of the illicit slave trade. 'It is about 120 m. in circ., and, like the adjacent part of the mainland, is exceedingly mountainous; Clarence Peak, the most elevated point, attaining the height of several thousand feet (10,700 ft.). The S. extremity is also intersected by several steep mountains, varying from 1,000 to 3,000 ft., which, with the intervening valleys, are covered with dense forests of large and valuable timber, and watered by numerous rivulets. The wet season commences at the latter end of May, and continues till the end of November: the annual quantity of rain and the temperature are much the same as at the other stations on the coast. The sea breeze is regular, but the land breeze generally deficient, being intercepted by the high range of mountains on the mainland.

'Clarence Town, the principal settlement (on the N. side of the island), lies in lat. 3° 53' N., long. 7° 40' E., and is built close to the sea upon an elevated plain from 100 to 200 ft. in height, embracing two small peninsulas, Point William and Point Adelaide, with a semicircular space extending about a mile in length, and forming a cove well adapted for shipping. All the ground in the immediate vicinity is covered with forest trees and jungle, except to the extent of about 6 sq. m., which was partially cleared on the formation of the settlement. The soil, which is generally argillaceous, resting on a bed of freestone, gives proofs of abundant fertility when cultivated. The water, both of spring and brook, is of the best quality, and there are no marshes in the vicinity, the hilly nature of the ground not admitting of their formation.' At this settlement part of a company of black troops belonging to the Royal African corps was stationed, with more civil officers of government, in 1827–29; and a number of European mechanics went out in these and the succeeding years to aid in the erection of barracks and other buildings. But the climate was soon found to be quite as pestiferous as that of the other settlements on this part of the African coast. Most Europeans were attacked by fever, and the instances of recovery were very rare. In consequence, the detachment of troops was withdrawn in 1834, and from this date Fernando Po ceased to be a military station. (Tulloch's Report on the Sickness of the Troops in Western Africa, p. 121.)

FERNEY, a village of France, dép. Ain. 6 m. NS E. Gex, and 5 m. NW Geneva. Pop. 1,166 in 1861. Ferney is indebted not merely for its celebrity, but even existence, to its having been for a lengthened period the residence of one of the greatest writers of modern times. Voltaire purchased this estate in 1758. The seigniory enjoyed an exemption from all public taxes and burdens; but it would seem that Voltaire wished to establish himself in this retreat, not so much from its enjoying the privilege now mentioned, and its agreeable situation, as from the facility which its vicinity to Geneva afforded of placing himself in a safe asylum in the event of any measures being taken to interfere with his freedom. Voltaire conferred the greatest advantages on Ferney. Out of a paltry village, consisting of a few miserable cottages, he reconstructed a neat little town, in which he established a colony of industrious artizans, principally consisting of watchmakers, from Geneva; he also rebuilt the church; drained and planted the adjoining grounds; defended his vassals in their contests with the revenue officers and the church, and did all that a rich, enlightened, and really benevolent landlord could do to promote the comfort and happiness of those around him. The château, to which a great little theatre was attached, was fitted up in a state of elegant simplicity; and his hospitalities were on the most liberal scale. Voltaire resided here with little interruption for more than 20 years. During the whole of this period, Ferney was to the literary and refined what Mecca is to the Mohammedan world; and the most distinguished personages of the time eagerly resorted to Ferney from all parts of Europe, to pay their respects to its illustrious master. Voltaire quitted Ferney for the last time on the 8th of February, 1778. His château is, or was not long since, preserved nearly in the state in which he left it. He expired at Paris on the 30th May, 1778. (Condorcet, Vie de Voltaire, 203.)

FERRARA, a famous city of Central Italy, cap. prov. of same name, formerly an independent duchy, in a low marshy plain, on the left bank of the Volano, 5 m. N. from the Po, to which it is united by a canal, and 26 m. NNE. Bologna, on the railway from Bologna to Padua. Pop. 67,503 in 1861. The city is well fortified and defended on its W. side by a strong pentagonal citadel. While it was under its native princes of the house of Este, Ferrara was the seat of one of the most polished and refined of the Italian courts, and is said to have had from 80,000 to 100,000 inhab. But it has long been in a state of decay, and numbers of its splendid palaces are uninhabited. In the principal square, or Piazza Nuova, are several statues of two of the dukes of Ferrara. The duomo, or cathedral, was consecrated in 1135: it is a vast but tasteless edifice. The city has an immense number of other churches, mostly in a state

of decay; but several of them, as well as of the palaces, have good pictures. Its university, or rather college, founded in 1390, and revived by pope Leo XII., has two faculties of law and medicine, but it is not well attended. The public library, founded so recently as 1746, has 80,000 volumes and a museum of antiquities; but its most valuable treasures are the manuscripts of the works of Ariosto and Tasso, with other relics of the former. There is here, also, a botanical garden, an anatomical theatre, several charitable establishments, and one of the finest theatres in Italy. The manufactures and trade of the town are inconsiderable.

The celebrity of Ferrara is almost wholly derived from its being intimately, at least, if not honourably, associated with the history of some of the greatest names in the literature of Italy, or indeed of Europe. Ariosto, though born at Reggio, in Modena, resided for a lengthened period in Ferrara: here, in 1516, appeared the first edition of the 'Orlando;' and here, on the 5th of June, 1533, the poet breathed his last. The house in which he lived is still kept up. He was buried in the church of the Benedictines; and it is a curious fact, that the bust on his tomb, being struck by lightning towards the middle of last century, the iron laurels that wreathed the brows of the poet were melted. Lord Byron has alluded to this circumstance as follows:—

  'The lightning rent from Ariosto's bust
  The iron crown of laurel's mimic'd leaves;
  Nor was the ominous electric spared,
  Yet the true laurel wreath which glory weaves
  Is of the tree no bolt of thunder cleaves,
  And the false wreathage but disarmed his brow;
  Yet still, if fondly superstition grieves,
  Know that the lightning sanctifies below
  Whate'er it strikes; —yon head is doubly sacred now.'
               Childe Harold, iv. c. 41.

In 1801, the remains and tomb of Ariosto were conveyed with great pomp to the public library; and here, also, are his manuscripts, arm-chair, and inkstand.

Tasso is another of the glories, but he is also the shame, of Ferrara. A cell in the lunatic hospital of Sta. Anna, about 9 paces by 5 or 6, and 7 ft. high, lighted by a grated window, is shown as that in which the author of the 'Gerusalemme Liberata' was immured from March, 1579, to December, 1586, when he was removed to a contiguous and larger apartment. In 1581 his prison was again enlarged; but it was not till 1586 that he was set at liberty, at the intercession of the Duke of Mantua. It is difficult to ascertain the real cause of this ignominious treatment of, perhaps, the greatest of the Italian poets. The apologists of the house of Este, or rather of the duke Alphonso, by whom, though the pretended patron of Tasso, he was imprisoned, have stated that it was occasioned by his extravagances, and that in shutting him up Alphonso really consulted the safety and honour of the prisoner. (Tiraboschi, vii. 1267, Modena, 1792.) But, though the subject be not quite free from difficulty, there can be very little doubt that the imprisonment of Tasso is ascribable to the vindictive malignity of the duke, who took this method of avenging some unguarded expressions of the poet, provoked by the ungenerous treatment he had received. (See Serassi, Vita di Tasso, p. 207; and the Extracts from Tasso's Letters, p. 283.; see also the learned essay on the imprisonment of Tasso in Mr J. Hobhouse's Illustrations of Childe Harold, pp. 5–32.)

Guarini, author of the Pastor Fido, the cardinal Bentivoglio, and several other distinguished persons, were also natives of Ferrara.

From a small town Ferrara became a walled city, A. D. 670. The family of Este possessed it first as chief magistrates, and afterwards as hereditary sovereigns, from about 1050 to 1597; when, on the death of its last duke, and the extinction of the male line of the family, it was taken possession of by the pope. Under the French régime it was the cap. of the dép. of Basso Po.

FERROL, a sea-port town of Spain, on the NW. coast of Galicia, prov. Betanzos, cap. of a jurisdiction of same name, and of one of the 3 naval departments of the kingdom, on the N. arm of the Bay of Betanzos, or Coronna, 11 m. NE. the latter, and 25 m. SW. Cape Ortegal; lat. 43° 29' 30" N., long. 8° 15' W. Pop. 17,504 in 1857. The harbour of Ferrol is one of the best in Europe in point of depth, capacity and safety. It is approached by a strait about 2 m. in length, and in its narrowest part not quite a quarter of a mile broad; this channel, which has from 8 to 11 fathoms water, will only admit one ship at a time, and is commanded by strong forts on either side. The tides in it run so strong that it is advisable to enter or leave the harbour an hour before high or low water. The town is protected on the land side by strong fortifications. It is well laid out, the streets mostly intersecting each other at right angles; but in some parts they are less regular, the ground enclosed by the fortifications being very uneven. It has 2 hospitals, 3 large churches, a nunnery, consistory, a good prison, academies of navigation and mathematics for pilots, and a school for the naval education of seamen; and contains the residences of the captain and auditor-general, intendant, and superior financial officer of the department, and of the military commandant, who is also the superintendent of police in the jurisdiction, which comprises the adjacent town of La Graña. On the E. side of the town are the royal arsenal and dockyard; the former is the first and largest in Spain, and used to be furnished with all necessary stores for the construction of the navy; the docks rank amongst the finest in Europe. The basin, in which the ships are laid up, is of great extent and solid workmanship, and every ship has its separate storehouse. The naval barracks occupy a large and handsome building, and afford accommodation for 6,000 men. Six hundred galley-slaves are (or were) employed in the most laborious works of the harbour. This port being intended solely for the royal navy, general commerce and all foreign merchant ships are excluded. There are, however, some manufactures of hats, paper, leather, naval stores, and hardware; and corn, wine, brandy, vinegar, pilchards, and herrings, the produce of its own fisheries, are exported; while salted meat, French, English, Irish, Dutch, woollen, linen, and other fabrics are imported; broken inframes from Catalonia, and silks from Valencia. But the trade of the town is principally limited to the supply of the inhab., the navy, and the government officers. Prior to 1752, Ferrol was only a fishing hamlet, frequented by coasting vessels; but, owing to the advantages of its situation, it has since been made the chief naval station of Spain. A railway from Lugo to Ferrol, branching off from the line from Madrid to Coruna, was sanctioned by the Cortes in 1865.

FEVERSHAM. See FAVERSHAM.

FEZ (properly Fas), a city of Morocco, and, next to Morocco and Mequinez, the principal in that empire, cap. of the prov., as it formerly was of the independent kingdom, of the same name, and residence of a kaid or governor. It is singularly and beautifully situated in a funnel-shaped valley, open only to the N. and NE., the sloping sides of which are covered with fields, gardens, orange groves, and orchards, 95 m. from the Atlantic, 225 m. NE. Morocco, and 80 m. SE. Tangier; lat. 34° 6' 3" N., long. 5° 1' 19" W. Its pop. has been very variously estimated; but, according to Count Graberg de Hemso, the resident pop. may be estimated at about 88,000, of whom 65,000 are Moors and Arabs, 10,000 Berbers and cognate tribes, 9,000 Jews, and 4,000 Negroes. More recent estimates state the population at only 50,000. Fez consists of two separate towns, Old and New Fez; the latter standing on a height, and overlooking the former. They are surrounded by decayed walls, which include a large space; and at both its E. and W. extremities are castles, in one of which the governor at present resides. The Wad-el-Jahar (River of Pearls), an affluent of the Sebou, winds through the valley, irrigating a large portion of its surface, and turning a great number of mills, and, after entering Fez, divides into two arms, which furnish water in abundance to the houses and mosques. The Old City is built on sloping ground; its streets are narrow and dark, unpaved, and in wet weather excessively dirty. The houses are lofty, flat-roofed, and built around court-yards; their different stories are surrounded with galleries supported on colonnades. Their cracked, leaning, and bulging walls are propped up by others which stretch at different intervals across the streets. These cross-walls are perforated by arched passages, not over wide; and these being closed at night, the city becomes divided into different quarters, all communication between which is effectually cut off. The New City, called also Medinat-al-bida, or "the White City," founded in the thirteenth century, is somewhat better laid out and built than the old, and is surrounded by fine gardens: it contains several palaces, among which is that of the emperor, some public baths, and several tolerable modern houses. The imperial palace covers a considerable extent of ground: it has a great number of court-yards, some of which are only half finished, while others are half dilapidated. Its interior does not exhibit much splendour. All Bey, early in the present century, reports that the cabinet in which the sultan used to receive visitors was but a poorly furnished room, 15 ft. square; while the office of the minister was a miserable, low, damp apartment, at the bottom of a small staircase, about 5 ft. long by 5 ft. wide, and without any other furniture than an old carpet! The Jews are confined to the New City, where they have a synagogue, and are obliged to keep within their own quarter at night. According to Leo Africanus, Fez is said in the sixteenth century to have contained as many as 700 mosques; but this would appear to be a gross exaggeration: at present the city contains only about 100. All are built on a uniform model: they consist of a courtyard surrounded with arcades, and on the S. side a covered square, in the middle of the wall bounding which there is a niche, where the imam places himself to direct the prayers, and so the left-hand side of the latter a pulpit. The chief mosque, called El Carubin, was erected soon after the foundation of the city. It has a greater number of arches than the large mosque of Tangier, many gates, and upwards of 300 pillars, and in its court there are two handsome fountains. This mosque can boast of the singularity of having a covered place for women who may choose to participate in the public prayers—a circumstance unique in Mohammedan places of worship. Same travellers describe El Carubin as one of the most remarkable edifices of its kind in Africa; but Ali Bey says that it is upon the whole a heavy and mean structure, and far inferior to the great mosque of Cordova. Its minaret contains some clocks, globes, and astronomical instruments, brought from Europe nearly a century and a half ago,—and a library; but, from having been abandoned to dust and damp, most of the instruments and books have become useless. The most frequented mosque is that of Muley Edris, the founder of Fez: it contains the sepulchre of that prince, and the sanctity with which it is thereby invested is so great that it affords perfect security to a criminal guilty of even high treason. Its minaret is the finest and highest in the city: it contains many European articles of mechanism. Public baths are numerous in Fez, and some of them are very good. There are also some tolerably convenient inns, though their outward appearance is not prepossessing. The number of shops, viewed externally, would almost warrant the belief that Fez contained four times its actual pop.; but most of them are mere 'stalls with just room enough for a sedentary Moor, who never moves; and for the packets that are heaped around him, to which he points as purchasers arrive.' (Chenier, i. 77.) Each street is devoted to a separate trade; and it is seldom that more than one species of goods is sold in a single shop. The markets are plentifully supplied; and provisions are both good and cheap. The climate is oppressively hot in summer; in the winter the thermometer often falls to 40° Fah., and the average height of the barometer is 27 in. The atmosphere is almost always damp and misty; and the situation is considered unhealthy (Chenier): the New City is, however, much less so than the Old.

During the struggle with the Moors in Spain, and especially on their expulsion from that kingdom, many Mohammedans sought an asylum at Fez, taking with them new manners, arts, and knowledge. They introduced the Spanish method of dressing and dyeing goat and sheep skins red and yellow (forming the leather then called Cordovan, but now Morocco), as well as the manufacture of milled woollen fabrics. These articles are still manufactured at Fez, and, in addition, gauzes, silks, sashes, gold and silver stuffs, jewellery, slippers, girdles, saddlery, woollen hauks, fine carpets, coarse linen fabrics, arms, copper goods, and earthenware. The trade with the adjacent country is brisk; and twice a year caravans go from this city across the desert to Timbuctoo.

Fez has been always considered one of the principal seats of Mohammedan learning. There are schools attached to many of the mosques; of these, seven are considered superior to the rest; and in these a mixed jargon of religion, morality, legislation, physics, metaphysics, geometry, astronomy, alchemy, and medicine is taught, principally out of the Koran, and the works of Euclid, Ptolemy, and Aristotle. There are several hospitals, the largest of which is appropriated to lunatics. The military government of the city is in the hands of the kaid; the civil and judicial authority is exercised by a cadi; and a minister, entitled el mohtesen, fixes the price of provisions, and decides all points that arise on this branch of the public service.

Old Fez was founded in 793 by Edris II., a descendant of Mohammed, and continued the cap. of an independent kingdom till 1548, when it was, together with its territory, conquered and annexed to Morocco. After a period of decline, it again rose to prosperity on the ruins of the Moorish kingdom of Cordova; and its pop. became afterwards still further augmented, by reason of the edicts of Philip II. against the Mohammedans. It has been always held so sacred by the Arabs and others, that when the pilgrimages

to Mecca were interrupted in the 10th century, the western Moslems journeyed to Fez, as the eastern did to Jerusalem; and even now none but the faithful can enter Fez without express leave from the emperor. (Graberg of Hemso; Specchio dell' Imp. di Marocco, pp. 47-49; Chenier, Morocco, vol. i.; Mod. Trav., vol. xii. &c.)

FEZZAN (an. Phazania Regio, and the country of the Garamantes), a country of Central Africa, immediately S. of Tripoli, to which pachalic it is tributary. It is supposed to reach from about 23½° to 31° N. lat., and from about the 13th to the 16th deg. E. long. But its boundaries are ill defined, and its area and pop. are alike uncertain. The latter, however, has been estimated by Horneman at no more than from 70,000 to 75,000. Fezzan is, as far as we know, the largest oasis, or cultivable tract, in the Great African Desert, by which it is surrounded on all sides; having W. the country of the Tuaricks, and S. and E. that of the Tibboos. A portion of it consists of an extensive valley bounded by an irregular circle of mountains on all sides except the W., where it opens into the desert; but a great part of the mountainous region to the E., as well as of the desert to the W. and S., are nominally included in its territory. The Gibel-el-Assood, or Black Harutech, mountains (an. Mons Ater), the White Harutech, and other ranges, intersect the country generally in the direction of NW. to SE. None of these ranges, however, is of any remarkable height; the first named, in the N. of Fezzan, is no more than about 1,200 or 1,500 ft. in elevation, and the hills elsewhere for the most part appear to be only from 400 to 600 ft. high. Their summits are in general tabular; a few only have conical peaks. Basalt is one of their principal constituents, and especially in the Black Mountains, where, however, the lower stratum of all the hills is invariably limestone, mixed with a reddish clay. Calcareous formations, containing many shells, are generally predominant; the other chief geological rocks are porphyritic clay slate, aluminous schist, and sandstone, frequently intermixed with beds of clay. A large portion of the surface is covered with sand, beneath which, in some places, volcanic substances have been found. Salt and nitre frequently effloresce on the soil, and impregnate many of the small lakes. There is no river or rivulet throughout the country; fresh water is procured by digging to variable depths, but at most to about 8 or 10 ft. under ground, when a plentiful supply is obtained. Rain is very rare, and descends only in small quantities. The heat in summer is oppressive in the highest degree, not only to foreigners but to the natives, rising sometimes to 130° Fahr.; the cold in winter is also sharper than might be expected from the latitude, the thermometer descending occasionally to below 50°, and accompanied with piercing blasts from the N.; added to which, furious tempests frequently occur, overwhelming caravans of travellers with the sands of the desert. The climate of Mourzouk and various other places is decidedly unhealthy. Only a small portion of the surface is under culture, and that only in the valleys, where sufficiently watered. Wheat is raised; but maize and barley are the grains on which the inhabitants chiefly depend for subsistence, and these are not grown in sufficient quantities for their supply. Pot herbs and garden vegetables are plentiful, particularly carrots, cucumbers, onions, and garlic; these, however, as well as most of the corn, are raised only in gardens near the towns, which are watered with great labour from brackish wells. Dates are the staple product, and the tax on the date trees is an important source of the public revenue. Figs, pomegranates, and jujubes are also

E E

grown. The rearing of domestic animals is little attended to: goats are the most numerous; and in the S. there are flocks of hairy broad-tailed sheep, of a light brown colour. Horned cattle are to be found in the most fertile districts, and there only in small numbers: beef is rarely eaten, except by the rich. Horses are few, the most laborious kinds of work being chiefly performed by men. Camels are used for travelling and the conveyance of goods; but these animals are dear, and only kept by large merchants, or other wealthy individuals. Dates form the principal food of all domestic animals. They also compose the chief nourishment of the people, the luxuries of life, even in the cups, being very limited; and, in fact, the necessaries of life, generally speaking, are so scanty, that, to designate a rich man, the common expression is, 'he eats bread and meat every day.' This state of things is mainly owing to the apathy of the inhabitants, many of whom do not, for months together, taste corn: when obtained, they make it into a paste called aseeda. Bread is badly made, and baked in ovens of clay, planted in holes in the earth, and heated by burning embers. Fowls, geese, and ducks are scarce, in consequence of the sovereign having appropriated all he could lay his hands on for his own use. Butter is brought in goats' skins from Tripoli, and is very dear. Tobacco, mixed with trona, is very generally chewed by the women, as well as by the men: smoking is rather confined to the opulent, mild tobacco and pipes being dear; but all the men, though professedly Mohammedans, drink largely of intoxicating liquors, obtained from dates. The principal wild animals met with in the country are the lion, panther, hyena, jackal, tiger cat, immense herds of buffaloes, &c.; and among birds, vultures, falcons, and other rapacious species, ostriches, and bustards. From the products of the animal kingdom, which supply its commerce, are derived a great part of what wealth Fezzan possesses. There are a few manufactures of agricultural implements, coarse woollen fabrics, carpets, and Morocco leather; but Horneman could not find throughout Mourzouk a single artificer skilful in any trade or work. 'The smith fashions without distinction every metal into every form: the same man who forges shoes for the sultan's horses, makes rings for his princesses.' Capt. Lyon, however, remarks that some work in gold and silver is executed with much skill, considering the badness of their tools; and every man is capable of acting as a carpenter or mason. The wood being that of the date tree, and the houses being built of mud, little taste or skill are displayed. Much deference is paid to the artists in leather or metals, who are called puressiffrace, &c., or master, as, iron-master, leather-master, &c. The shuttle is unknown, and woollen cloths are made by the women with the hand only. The chief occupation of the people is commerce and the conveyance of goods. Fezzan derives its chief importance from its situation, which renders it a grand depôt for the commerce carried on between N. and Central Africa. The communication of Egypt as well as Barbary with the vast countries to the E. and S. of the Niger, centres almost entirely in Mourzouk. Thither an annual caravan sets out for did in the time of (Horneman) from Cairo, reaching its destination in about 40 days. From Tripoli to Mourzouk the journey usually occupies about 25 or 27 days. Of the caravans to the S., the principal are those to Bornou, with which country Fezzan maintains a regular and extensive communication, and the caps of which travellers reach in about 50 days. Other caravans go to Cassina, which journey occupies 60 days; and a few proceed still further S., crossing the mountains

to Ashantee. 'The arrival of the great caravans forms a sort of jubilee in the cities of Fezzan; and on reaching Mourzouk, they find the sovereign seated on a chair of state, outside the city, to receive them.' Male and female slaves from Bornou and the adjacent S. countries, gold dust from the banks of the Niger, copper, senna from Agades, civet, three-skins, dyed leather, and some kinds of cotton manufactures are the chief imports from the interior of Africa; which, together with ivory and ostrich feathers, are forwarded to Barbary and Egypt to be exchanged for provisions, and the manufactures of Europe and the East. Many of the latter are re-exported to the S. including firearms, gunpowder, sabres, knives, glass, paper, beads, imitations of coral, toys, and European manufactures of a great variety of kinds, tobacco, snuff, &c. The articles of clothing imported from the N. are principally muslins (partly from India), striped, blue, and white calicoes, woollen cloth, and worsted caps. Salt and dates are, however, the principal articles exported to the S.; the quantity of the former being estimated at 300 or 400 camel loads.

The People are of a mixed race; in the N. many are Arabs, in the S. they are chiefly Negroes. The Fezzanners, who compose the mass of the people, appear intermediate between the two, though more inclining to the latter type. Their colour is black: they are, according to most authorities, tolerably well formed; but neither race has handsome features. They have a very peculiar cast of countenance, which distinguishes them from other blacks; their cheek-bones are higher and more prominent, faces flatter, noses less depressed, and more peaked at the tip than in the negro; eyes generally small; lips protuberant, and somewhat thick; teeth good; hair inclined to be woolly, but not completely frizzled. They are said to be cheerful, and fond of dancing and music, and not prone to sudden anger, nor revengeful; but are at the same time selfish, devoid of hospitality, insincere, and wholly destitute of either physical or mental energy or enterprise. The Arabs, in person and disposition, are much the same as elsewhere; and are greatly superior to the Fezzanners in activity and cleanliness. In Mourzouk there are some white families, descended from the Mamelukes, whose designation they are very proud of preserving. The court and upper classes of Fezzan dress mostly in the costume of Tripoli; the lower orders wear a large shirt of white or blue cotton, with long bare sleeves, trousers of the same, and sandals of camel's hide; and on Fridays they perhaps add a turban, and appear in yellow slippers. The women plait their hair, often mixing it with black wool; they use great quantities of oil and perfumes; and those who can afford it, load themselves profusely with armlets, anklets, and other ponderous ornaments of gold, silver, copper, iron, ivory, glass, and horn, together with cornelians, agates, beads, and coral. Both sexes have a singular custom of stuffing their nostrils with a twisted leaf of onions or cloves. The habits of all classes are said to be debauched and profligate in the extreme.

The Government is in the hands of a chief who exercises unlimited power within his own territory, where he has the title of sultan, though in addressing his superior, the pacha of Tripoli, he assumes only that of sheik. His revenues are derived from taxes on slaves, merchandise, date plantations, gardens, and other cultivated lands; from fines and requisitions, duties on foreign trade, and the crown domains, salt ponds, and natron lakes. For every slave, great or small, he receives, on their entering his dominions, 2 Spanish dollars; and in some years the number of slaves amounts to 4,000. On the sale of every slave, one-fourth of the pur-chose money goes to the sultan, in addition to which he receives a dollar and a half per head, which, at the rate of 4,000, gives about 6,000 dollars annually. The tax on a camel's load of oil or butter entering the country is 7 dollars; on a load of beads, copper, or hardware, 4 dollars; and on one of clothing, 8 dollars. All Arabs who buy dates pay 1 dollar duty on each load; and above 3,000 loads are annually sold to them. Date-trees (with a few exceptions) are taxed at 1 dollar for every 200, and those in the vicinity of the cap, alone yield the sultan an annual profit of 10,000 dollars. The trees, which are his private property, produce about 6,000 camel-loads of dates, each load about 400 lbs. weight, and which may be estimated to fetch 14,000 dollars. He is entitled to one-fifth of all sheep or goats; every garden pays one-tenth of the corn it produces. Each town pays a certain sum, which, altogether, may be averaged at 4,000 dollars. He sends out private parties for slaves; and has alone the privilege to sell horses, which he buys at a cheap rate from the Arabs, and realises a large profit by obtaining slaves for them in exchange. If a man die childless, the sultan inherits a great part of his property. There are various other ways in which he extorts money. The cadi, and other state officers, including the ministers of religion, are supported by lands set apart for the purpose. All the servants of the sultan are maintained by the public; and he has no money to pay, except to the pasha of Tripoli. The tribute was formerly to the amount of about 15,000 dollars a year, till a quarrel between the two sovereigns broke out some years ago; since which it has been much less. It is paid in gold, senna, and slaves, and an embassy is annually sent for it by the pasha. The armed force of Fezzan may usually amount to 5,000 men; but in time of war, all who are able to bear arms are called out, and in this way a tumultuary force of from 15,000 to 20,000 men has sometimes been raised.

The cities and towns of Fezzan are said to exceed 100; but the largest has not more, perhaps, than 3,000 inhab. The principal are Mourzouk the cap., Sockna, Sebha, Hoon, and Wadan. Mourzouk stands in lat. 25° 54' N., long. 15° 52' E. It is surrounded with well-built mud walls, at least 20 ft. high, with round buttresses, loopholes for musketry, and gates wide enough to admit a laden camel. Pop. about 2,500. The street of entrance is about 800 yards long, by 100 broad, and leads to the sultan's castle, an immense, but irregular edifice, built of mud, in the middle of the city. In Mourzouk there are said to be 16 mosques; but most of them are small. Sockna is situated in a plain, on the road between the cap. and Tripoli. It is walled, and may contain 3,000 inhab. Germa has been considered, but without sufficient evidence, to be the an. Garama. No antiquities have been discovered in it; though, in various parts of the country, remains belonging to the Roman and subsequent periods are frequently met with.

The country of the Garamantes was conquered by the Romans under Cornelius Balbus, soon after the Christian era. In the 7th century it fell under the dominion of the Arabs; but in 1800 a portion of it was tributary to the Soudan state of Kanem. Soon afterwards a family of the Sheriffs (descendants of Mohammed) took possession of it, and held it till 1811, when the bey Mukni usurped the throne. (Denham and Clapperton, Trav. in Africa; Oudney, Lyon, Ritchie, Horneman, &c.)

FIESOLE (an. Faesulae), in antiquity a considerable city of Etruria; now a small though celebrated village of Central Italy, prov. Florence, on a precipitously steep hill commanding a fine view of the Val d'Arno, 4 m. NE. Florence. Pop. of dis-

trict, 11,698 in 1861. The face of the hill is cut into a gradation of narrow terraces, enclosed in a trellis of vines, and faced with loose stone walls. It has a cathedral, a seminary, and numerous country houses belonging to the citizens of the Tuscan capital. It is first noticed by Polybius in his account of the early wars between the Gauls and the Romans. It was the head-quarters of Catiline, who retired thither after the discovery of his conspiracy. Near it, in 405, was fought the last great battle gained by the Romans in Italy, in which Stilicho defeated Radagaisus and the Huns. In 1010, the Florentines dismantled and ruined Fiesole, and enlarged their own city with some of its materials; but the ruins of a few of its ancient buildings are still visible, particularly those of its Etruscan walls, and of a vast amphitheatre supposed to be of Roman origin. (Rampoldi, ii. 14; Cramer's An. Italy, i. 177.)

FIFE, a marit. co. of Scotland, consisting of the peninsula lying between the Frith of Forth on the S., the German Ocean on the E., and the Frith of Tay on the N.; having on the W. the cos. of Perth, Kinross, and Clackmannan. Area, 505 sq. m., or 323,361 acres, of which more than two-thirds are cultivated. This is one of the best situated and most beautiful of the Scotch counties, exhibiting every variety of surface and soil, from the mountain to the level plain, and from rams and gravel to the finest loams. The Lomond hills, on its W. border, attain to an elevation of about 1720 ft. above the level of the sea. The E. and SE. parts of the county are comparatively level and fertile; and the district, called the 'How of Fife,' traversed by the Eden, is particularly well cultivated and productive. There is a good deal of moor land in the W. parts of the county along the E. and S. borders of Kinross-shire, and between the latter and Dunfermline; but it is gradually being brought under tillage. Climate dry and good, having been materially improved by drainage and extended cultivation. Generally speaking the soil is superior; and both arable and stock husbandry are well understood and successfully practised. All the new improvements in drainage and in agriculture have been introduced into the county, which has, in consequence, been wonderfully improved.

By the new system of agriculture, and especially by the liberal employment of draining, the land has been brought into the highest state of cultivation; and grounds, which half a century ago would have been thought good for nothing, are now seen waving with the richest harvests. The houses of the peasantry are now equal to what those of the farmers were then; and the mansions of the latter surpass, both in appearance and comfort, such as the smaller proprietors formerly possessed. The Fife breed of cattle is well known, and is one of the most valuable of the Scotch breeds. Property is more subdivided in this than in most Scotch counties. Farms vary in size from 50 to 500 acres: leases for 19 years, and corn rents, general. No county affords finer situations for building, or is better wooded, or has a greater number of gentlemen's seats. Coal and lime are both abundant, and are largely exported. The linen manufacture is carried on very extensively at Dunfermline, Kirkaldy, Dysart, and other towns. A considerable number of people in the smaller towns round the coast derive a subsistence from fishing. Principal rivers, Eden and Leven. Principal towns, Dunfermline, Kirkaldy, and St. Andrew's. Fife contains 13 royal burghs, 61 parishes, and a university, St. Andrew's. It returns 4 mems. to the H. of C., viz. 1 for the county, 1 for the E. district of boroughs, or those of Cupar and St. Andrew's; 1 for the Dysart district, including those of Dysart, Kirkaldy, &c.; and 1 for the W. district of Kinrogins, including Inverkeithing, Dunfermline, Queensferry, Culross and Stirling, of which the last two do not belong to the county. Registered electors for the county, 2,725 in 1863. Pop. 154,770 in 1861, inhabiting 26,079 houses. The old valued rent was 30,200.; the new valuation for 1864–5 was 581,155*l*.

FIGEAC, a town of France, dép. Lot, cap. arrond., on a declivity beside the Cole, 31 m. NE. Cahors, on the railway from Clermont to Montauban. Pop. 8,341 in 1861. The town is surrounded by an amphitheatre of wooded and vine-clad hills, interspersed with numerous habitations and abrupt rocky heights; but the town is generally ill-built, and its streets narrow, crooked and dirty. It was formerly encompassed by ramparts and ditches, but these were demolished in 1672, and only some traces of them exist. It is said to owe its origin to a Benedictine monastery, established here in 755 by Pepin le Bref. The church of this ancient abbey is remarkable for the singularity of its architecture; it has a dome surmounted by a spire, together upwards of 755 ft. in height. At the N. and W. extremities of the town are two obelisks, called *aiguilles*, as to the origin of which several fabulous stories are afloat. These are octagonal, and upwards of 50 ft. in height; and appear to have been intended to support lanterns. Figeac contains numerous ancient buildings, among which is the castle of Balène, an edifice of great extent and solidity, and originally a place of some strength; it is now used as a hall of justice. It has a court of primary jurisdiction, a communal college, and a school of design; and has manufactures of linen and cotton fabrics, dyeing-houses, tanneries, and some trade in wines and cattle. It suffered greatly in the religious wars of the 16th century. It was the birthplace of Champollion, the Egyptian traveller.

FIGUERAS, a town of Spain, near the NE. extremity of the kingdom, Catalonia, prov. Gerona, on the road between Perpignan and Barcelona, 71 m. NNE. the latter. Pop. 10,349 in 1857. Figueras is a long straggling town, situated in the middle of a plain on which an abundance of olive trees are grown. Like almost all Spanish towns, it has its square (plaza); the streets are tolerably wide, but the houses ill-built. It has a parish church, three convents, a hospital, barracks, with a small garrison, and a custom-house. About three furlongs WNW. of the town is the citadel, or castle of San Fernando, constructed at an immense cost, about the middle of the last century, and reckoned one of the finest fortresses in Europe; it stands on a little eminence, commanding the whole plain; all the approaches to it are under-ground, and every building within it is bomb-proof. Its form is an irregular pentagon; the walls are of freestone, and very thick; the moats deep; and wide; its ramparts, magazines, stables, cellars, barracks, and hospital are defended by a casemate; and the firm, bare rock on which it is built has been turned to so great advantage, that trenches can scarcely be opened on any side, the ground being everywhere stony. It will serve as an intrenched camp for from 16,000 to 17,000 men. It has, however, been several times captured: the French took it in 1808; the Spaniards recovered it in 1811; but it was retaken in the same year by the French, who kept possession of it till 1814. They took it again in 1823.

This fortress has a military governor, whose jurisdiction extends over the town. Figueras is the seat of a subdelegation of police; it has some trade with France, manufactures of leather and

paper, mills of various kinds, and a large market every Thursday. Iron and black marble are obtained in its vicinity.

FILIPPO D'ARGIRO (SAN) (an. *Agyrium*), a town of Sicily, not far from the centre of the island, Val di Catania, cap. cant., on a hill near the Trochina, 34 m. W. by N. Catania. Pop. 2,152 in 1861. The town has several churches and convents. The best saffron in Sicily is grown in its environs. Agyrium was of great antiquity, and is celebrated as being the birthplace of Diodorus Siculus.

FINALE, a town of Central Italy, prov. Modena, cap. distr., on an island in the Panaro, 10 m. from its confluence with the Po, 21 m. NE. Modena, and 16 m. W. Ferrara. Pop. 11,693 in 1861. The town derives its name from its having been formerly the last town to the E. in the Modenese doms. It is surrounded by a wall, and has some wide streets, fine bridges, and a college. It has manufactures of silk and woollen fabrics, and some trade in corn, wine, and hemp. In 1872 it suffered much damage from an inundation of the Panaro.

FINDHORN, a village and sea-port of Scotland, co. Moray, on the river of that name, at its mouth, and in the par. of Kinloss; 3 m. N. by E. Forres, and 10 m. W. by N. Elgin. Pop. 891 in 1861. The Findhorn, which falls into the Moray Frith, and which, near its mouth, flows into a loch or arm of the sea, upwards of 1 m. in length by ¼ m. in breadth, is rendered famous by its inundation in the disastrous floods of August, 1829. (Sir Tho. D. Lauder's Morayshire Floods.) The majority of the inhab. are engaged in the herring fishery. Some salmon are also caught here. A considerable quantity of grain is shipped from Findhorn. About 9 m. S. from the village stood the Abbey of Kinloss, belonging to the Cistercian order of monks.

FINDON, or FINNAN, a fishing village of Scotland, co. Kincardine, on the sea-coast, in the par. of Banchory Devenick, 6 m. S. Aberdeen. It is a poor place, but has long been celebrated for its preparation of smoked haddocks, known by the name of 'Finnan haddocks.' This village was at one time unrivalled for the whole process—for gutting, cleaning, splitting, and smoking the fish; but it is admitted that the several white-fishing stations on the coasts of Kincardine and Aberdeen are now about equal to it in this respect. Dunbar and various towns on the Frith of Forth have tried to rival Finnan, but in vain. The most delicate part of the process is the smoking, which should be done by the green branches of fir, particularly spruce, thus communicating to the fish its peculiar odour and bright yellow colour. A somewhat similar result may be effected by the use of pyroligneous acid, but nothing but the fir has ever been used for the purpose at Finnan and the neighbouring coast. The genuine Finnan haddock should never be kept above two or at the farthest three days after it has been caught, should be roasted by a very quick fire, and served up immediately. The inhabitants of Finnan, like those of many other fishing towns on the E. coast of Scotland, are supposed to have had a foreign, most likely a Danish, origin; their physical aspect, dress, manners, language, living peculiar, and remaining unchanged from generation to generation. (The Book of Bon Accord, Aberdeen, 1839, pp. 17, 18, 270; Meg Dods' Cookery, p. 17; Boswell's Life of Johnson, by Croker, ii. 343.)

FINISTERE, or FINISTERRE, the extreme W. dép. of France, formerly a part of the prov. Brittany, between lat. 47° 45′ and 48° 45′ N., and long. 3° 26′ and 4° 50′ W., surrounded on three sides by the ocean and British Channel, and having E. the dep. Cotes-du-Nord and Morbihan. Length, N. to S., 65 m.; breadth about 55 m.; area, 672,112 hectares; pop. 627,304 in 1861. The coasts of this dep. are generally steep, rocky, and indented with many bays and harbours, some of which, as that of Brest, are of the first excellence. Ushant, and many groups of small rocky islands, are situated near the shores. Two hill-chains run through this dep. E. to W., one terminating near Brest, and the other in the opposite peninsula of Crozon. Both chains are granitic, but the summits of neither rise above 990 feet. Rivers numerous; the principal are the Aulne, Landernau, and Oder: there are also a great many small lakes. Climate mild, but humid; fogs are common; W. winds are most prevalent, and violent storms often occur. In the official tables, the extent of rich land in the dep. is set down at 259,890 hectares; arable lands occupy 378,210 hect.; and heath and waste lands 268,578 hect. Agriculture is in a very backward state, and the land is capable of yielding a much larger return if better methods of husbandry were followed; still, however, more corn is produced than is required for home consumption; it consists chiefly of oats, rye, wheat, and barley, in the order now stated. Until very recently, in accordance with a singular superstition, which prevailed from a remote period, one corner of every ploughed field was left fallow, and designated the part du diable. Flax, hemp, and pulse of a good quality are grown: the vine is not raised; but about 70,000 hectol. a year of cider are made. Pasturage is excellent, and three hay harvests are sometimes obtained in a year. Many cattle are reared, principally oxen and cows. Hogs are numerous, and bees are largely reared—honey and wax being important articles in the commerce of the dep. The terres froides, or thin and poor soils, are sown with broom or furze, which furnish at the same time forage, fuel, and manure. The farms in the dep. vary in size, principally between five and forty or forty-five acres. The larger farms are commonly let on leases of nine years, the rent being paid in money at Michaelmas. The rent of poor lands varies from about 6s. to 11s.; and of terres chaudes, or rich lands, from 17s. to about 80s. an acre. Pork, beef, cabbage soup, oatmeal porridge, potatoes, bread, butter, and pudding comprise the chief articles of food. The women spin, and assist in field labour; and the condition of the farmers is said to be prosperous. The pilchard and other fisheries are important; they employ about 840 boats, and 4,400 hands, and are estimated to realise a gross produce of about 2,100,000 fr. a year.

Finistere is rich in metallic products, especially lead. The mines of Poullaouen and Huelgoet are, perhaps, the largest of any in France. The lead is argentiferous; and about 700 kilogr. of silver a year are extracted at an average. Iron, zinc, and bismuth are, amongst the other metals, procured in the dep. There are also numerous granite, porphyry, slate, serpentine, and marble quarries, and beds of coal and potters' earth. The manufactures are principally those of linen and woollen fabrics, sail-cloth, paper, earthenware, cordage, leather, wax-candles, soap, and chemical products. Morlaix has a brisk trade in litharge, butter, &c. The exports generally exceed the imports in value; the latter are chiefly the produce of the more S. deps., as wines, brandy, and oil. Finisterre is divided into 5 arrondissements, 43 cantons, and 291 communes. Chief towns, Quimper, the cap., Brest, and Morlaix.

FINLAND, called by the inhab. Suomen-maa, or Land of Lakes or Marshes, a country of N. Europe, lies, with the exception of part of Lapland, the extreme NW. portion of the Russian empire. It lies between lat. 59° 50' and 69° 25' N., and long. 21° and 32½° E.; having N. Russian Lapland; E. the governments of Archangel and Olonetz; S. the Lake Ladoga, the government of Petersburg, and the Gulf of Finland; and W. Sweden and the Gulf of Bothnia. Length, NNE. and SSW., 800 m.; average breadth about 240 m. Total area, 6,501 geo. sq. m., or about 136,000 Eng. sq. m. Pop. 1,784,193 in 1858, of whom 1,684,191 Lutheran, and 36,061 members of the orthodox Greek church.

Physical Aspect.—Finland consists principally of a table land from 400 to 800 feet above the level of the sea, and interspersed with hills of no great elevation. In the N., however, the Manselka mountains have an average height of between 3,000 and 4,000 ft. The coasts, particularly on the S., are surrounded by a vast number of rocky islets, separated from the main land and from each other by intricate and narrow channels, rendering the shores of Finland easy of defence in case of hostile attack by sea. But the chief natural feature of the country is its myriads of lakes, which spread like a network over a large proportion of its surface; some of them being of very considerable size. The greater number of these are in the S. and E.; they have frequent communications with each other, and generally abound with islands. There are numerous rivers, but none of much importance. Climate rigorous. Even in the S. the winter lasts from six to seven months, and in the N. from eight to nine months. Dense fogs are very frequent; heavy rains take place in autumn, and in May and June the thaws nearly put a stop to all travelling. In the N. the sun is absent during Dec. and Jan.; but during the short summer, while that luminary is almost perpetually above the horizon, the heat is often very great; and near Uleaborg, in about the 65th deg. lat., the corn is sown and reaped within six or seven weeks. Crops, in all parts of the duchy, are exposed to the double danger of being destroyed by sudden frosts, and by the ravages of a variety of caterpillar called turile by the natives. The principal geological formations are granite, which very easily disintegrates, hard limestone, and slate. Soil for the most part stony and poor.

Agricultural and other Produce.—Finland is more productive than the opposite part of the Scandinavian peninsula, and when it belonged to the Swedish crown, it furnished a good deal more corn than was necessary for its own consumption, and was termed the granary of Sweden. Rye and barley are the kinds of grain chiefly cultivated, the rye of Vasa being celebrated for its excellence; wheat is but little grown, but oats are raised in considerable quantities. The peasants are obliged, from the humidity of the atmosphere, to kiln-dry all the grain, after which it will keep for fifteen or eighteen years. Pulse, beans, hemp, flax, and a little tobacco are raised. Potatoes were introduced about 1762, and are now in general use. Only a small proportion of the surface is under culture. The land requires a large quantity of manure, and that in common use is wood ashes, procured by setting fire to the forests and underwood, after which operation heavy crops are sometimes obtained. The natural poverty of the soil is such that, excepting in the S. prov. of Tavastehus, where it is deprived of a continued supply of artificial stimulus, the crops rapidly fall off, and the cleared land is soon abandoned for another portion of soil, the wood on which is purposely destroyed. This plan of manuring the land, though well enough adapted to bring lots covered with brushwood into cultivation, is highly injurious to the forests, and consequently to one of the chief sources of national wealth. The forests, which are very extensive, and stretch to the N. limits of the duchy, consist principally of pine and fir; but contain also beech, elm, poplar, oak, ash and birch. Timber, deals, potash, pitch, tar, and rosin are amongst the most important products of Finland. Cherries and apples ripen at Vasa, and a species of crab-apple grows wild in the W.; but other fruits, except a few kind of berries, are rare. Next to agriculture, cattle-breeding and tillage are the chief occupations of the people. The meadows and pastures, though but little attention is paid to them, are, in general, very good; and furnish, with beans and straw, an abundant supply of food for nearly 900,000 head of cattle. Sheep, with considerable numbers of hogs and goats. Horses, of which there are about 250,000, are small, strong, and hardy. In the N. the peasants possess large herds of rein-deer. Bears, wolves, elks, deer, foxes, beavers, polecats, and various kinds of game abound. Seal and herring fisheries are established on many parts of the coast; and the salmon and streamling (*Clupea harengus*) are caught in great quantities in the lakes, supplying the inhabitants with an important part of their food. Iron mines were formerly wrought, but at present only bog-iron is procured. Lead, sulphur, arsenic, nitre, tin and copper are met with; the last two, but especially copper, being produced in considerable quantities; salt is very scarce, and is one of the chief articles of import. Manufactures, except the products of a few cotton factories (the result of the prohibitive system), iron forges, glass works, salt-cloth, and linen factories, are entirely domestic. The peasant prepares his own tar, potash, and charcoal; constructs his own boat, furniture and wooden utensils; and weaves at home the coarse woollen and other fabrics he uses. He often lives 100 miles from any town, and is, therefore, thrown for the most part upon his own resources and ingenuity for the supply of his wants. In some districts the inhabitants never repair to a town but to obtain salt. The exports of the value of about 3,000,000 silver roubles a year, consist of timber, lumbers' meat, butter, skins, potash, tar, and fish, to Russia and Sweden, with which countries the principal intercourse is maintained. In 1852 the export trade employed 407 vessels of 107,000 tons, and the coasting trade about 900 vessels of 50,000 tons. There are a few good roads made by the Sweden while they were in possession of the country; but they do not extend far into the interior. Post horses are furnished, as in Sweden, by the adjacent farmers. In commercial dealings, the Russian is the currency established by law; but Swedish paper money is also in circulation.

Government.—Since 1831, Finland has been divided into 8 läns, or governments, of very unequal magnitude, the most northerly, Uleaborg, being about as large as all the others; but this is a consequence of the wildness and sterility of the country, the absolute amount of its population and its density, especially the latter, being far below the average of the more southerly governments. The läns are subdivided into *fogderies* or districts, and in *härader* or circles. Chief towns, Helsingfors, the present cap.; Abo, the former cap.; Tavastehus, Vasa, Uleaborg, and Tornea. A Russian military governor resides at Helsingfors. Finland has a diet composed of the orders of the nobility, clergy, citizens, and peasantry, and a

code of laws and judicial system similar to that of Sweden. For more than half a century this diet was inactive, until it was again convoked in Sept. 1863, by the 'Grand-duke Emperor.' The revenue of Finland, which is kept quite distinct from that of the empire, amounted, in 1862, to 8,005,809 roubles, or 429,321L. and the expenditure, during the same year, to 2,811,474 roubles, or 464,197L. Among the privileges of the people is that none but a native Finlander can hold any office of trust in the country. The Finnish troops amounting in time of peace to about 3,000 men, are not intermixed with the ordinary Russian troops. The Finnish fleet, by far the best manned portion of the Russian naval force, forms a distinct squadron under the national flag. Sveaborg, on some small islands in the Gulf of Finland, at a little distance from Helsingfors, is a principal station of the Russian fleet, and is very strongly fortified. Almost all the pop. as before stated are Lutherans, under the bishops of Abo and Borgo; except in the government of Wyborg, where they belong to the Russian church. Public education is very backward. There is, however, a university at Helsingfors, with 3 academies, and 12 superior schools. A society for the encouragement of the Finnish language and literature has been warmly patronised by the Russian government.

People.—On the W. coast, and in the Aland Archipelago (which is included in Finland), the inhab. are mostly of Swedish, and in the SE. of Russian descent; but the great majority of the pop. are Finns. The latter have, by many geographers, been identified with the *Fenni* of Tacitus, and the *Phinni* of Ptolemy. There are, however, circumstances which give rise to considerable doubt respecting such identity. The Finns call themselves *Suomolaiset*, or 'inhabitants of the marshes.' They have no analogy with the Sclavonian or Teutonic races. They are of middle height, robust, flat-faced, with prominent cheek-bones, light, reddish, or yellowish brown hair, grey eyes, little beard, and a dull sallow complexion. They are courageous, hospitable, and honest; but obstinate in the extreme, indolent, dirty, and it is said revengeful. They are grave and rather saturnine. Almost every one is a poet or musician. But they have no taste for dancing, or indeed knowledge of the art, or of games of chance, except in the towns, where they have been introduced by the Swedes. Their amusements consist principally in feats of bodily strength and activity. The customs and habits of the Finns have been handed down time immemorial, and their costume forcibly brought their supposed E. origin to the mind of Mr. Elliot, who observes in his letters from the N. of Europe, 'I could fancy myself in Asia. The peasants wear long loose robes of coarse woollen manufacture, secured by a silken ceinture like the *hummerband* of the Mussulmans. Their dress, except the fan-jam hat, resembles that of the Hospodars of Cstandi. In Russian or Old Finland, the peasants wear a cloak or caftan, sometimes called a *kaimar*, resembling in form, as well as in name, the E. dress.' (pp. 251-252.) The Finns make frequent use of hot vapour baths, and Malte-Brun supposes that they communicated the custom to their Russian conquerors.

History.—The Finns were pagans, living under their own independent kings till the 12th century, about the middle of which Finland was conquered by the Swedes, who introduced Christianity. The province of Wyborg was conquered and annexed to Russia by Peter the Great, in 1721; the remainder of the country became part of the Russian dominions (also by conquest) in

1809. Ever since that period the Russian government has endeavoured, by conciliating the Finnish
party, and promoting objects of national importance, to attach the bulk of the population to its
interests; and in this it is said to have been
eminently successful.

FIORENZOLA, a town of Central Italy, prov.
Piacenza, on the Lardi, 15 m. SE. Piacenza. Pop.
6,132 in 1861. The town was the native place of
Cardinal Alberoni. About 8 m. S., on the right
bank of the Mira, stood the ancient city of Veleia,
buried in the fourth century by the fall of the
mountain at the foot of which it was situated, and
not discovered till 1761. The remains of antiquity that have been dug out of its ruins are more
numerous and perfect than in any other ancient
city of Italy, with the exception of Herculaneum
and Pompeii.

FIUME, a sea-port town of Austria, situated
on the Gulf of Quarnero, at the NE. extremity of
the Adriatic Sea; lat. 45° 19' 35" N., long. 14°
36' 45" E. Pop. 15,119 in 1857. Fiume is the
chief town and seat of government of the dist.
called the Hungarian 'Littorale,' and, with the
minor adjacent harbours of Buccari, Porto Rè, and
Martinschizza, is the point of contact for the rich
and powerful kingdom of Hungary with the Mediterranean. The importance of Fiume was recognised at an early period by the emperor
Charles VI., who constructed a magnificent road
about 75 m. in length, leading to this port from
Carlstadt in Croatia, the spot where the inland
navigation by means of the rivers Save and Culpa
terminates. This road was called, after its founder,
the 'Carolina;' but the difficult task of traversing
the Julian Alps was found to be but imperfectly
accomplished by its means, and the emperor Joseph II. laid down another line of road to the
coast, between Carlstadt and Zeng, in the military
frontier, which was named the 'Josephina.' In
1809, a third line of road, one of the finest undertakings of the kind in Europe, was opened at the
expense of a joint-stock company, the shareholders in which were chiefly magnates of Hungary. This road was named the 'Louisa,' after
the empress Maria Louisa; and, on account of its
comparatively gentle declivity, is the most frequented. Notwithstanding these exertions, and
the outlay of a considerable sum of money in an
endeavour to render the Culpa navigable above
Carlstadt, the trade carried on here is not very
considerable, excepting in years when there is a
large exportation of grain to Great Britain, France,
and other countries.

The branch of the Louisa road which leads to
Buccari is chiefly used for the transport of timber
and staves. The oak timber of Carniola, and the
Littorale, is of the best quality; and the ships
built at Trieste, Fiume, and other ports, being
strong, handsome, and well fitted out, and their
crews expert and temperate, are much sought after
and being high freights. There are some manufactures at Fiume of linen, coarse cloths, leather,
and naughlo; also a wax-bleaching establishment,
and a sugar refinery. The harbour is small, being
only the entrance to a mountain stream of a few
miles in length, which it is very difficult to keep
clear. Large vessels lie in the roadstead, at a few
hundred paces off shore, where the water is deep
enough, and where the high land of the coast
shelters them tolerably well from the efforts of the
bora, or NE. wind.

FLAMBOROUGH HEAD, a bold promontory
of England on the Yorkshire coast, projecting a
considerable distance into the sea; lat. 54° 7' N.,
long. 0° 5' W. This is at once the most striking
and most celebrated headland on the E. coast of

the kingdom. Its high, white, perpendicular,
limestone cliffs render it a most conspicuous object.
Many of the rocks of which it is composed are insulated, of a pyramidal form, and rise to a great
height. Most of them have solid bases, but others
are pierced through and arched. On the N. side
are vast caverns, leading into the body of the
head, the retreat of immense numbers of sea-fowl
and wild pigeons. A light-house, with a revolving
light, having the lantern elevated 214 ft. above
the level of the sea, was erected on this head in
1806.

FLANDERS, the name of a fertile and well-
cultivated district of Belgium, divided into the
provinces of E. and W. Flanders. See BELGIUM.

FLECHE (LA), a town of France, dép. Sarthe,
cap. arrond., on the Loire, 24 m. SW. Le Mans, on
the railway from Paris to Nantes. Pop. 7,077 in
1861. The town is generally well built; streets
broad, clean, and ornamented with fountains supplied by an aqueduct upwards of ½ m. in length.
Its chief public building is a royal military
college, formerly a celebrated Jesuits' college,
founded in 1603, by Henry IV. It is very extensive, and well laid out; contains an elegant
church, a public library with 14,000 vols., a picture-
gallery; and has attached to it a fine park, and
gardens. The church of St. Thomas, town-hall,
hall of justice, and hospital, are the other principal
edifices. La Flèche, though advantageously placed
on a navigable river, is remarkably deficient in
manufactures and trade. It is the seat of a sub-
prefecture, and court of original jurisdiction. Its
environs are exceedingly agreeable. Previously
to the 10th century, it was called Fisus; it owes
its present name to the spire (flèche), placed in
the 12th century on the tower of St. Thomas's
church. One of the greatest of Scotch philosophers,
David Hume, resided at La Flèche in 1735 and
1736, and here composed the greater portion of
his earliest work, the 'The Treatise of Human
Nature.' La Flèche was the birthplace of Descartes.

FLENSBURG, a sea-port town of Germany, on
the E. coast of Schleswig-Holstein, at the bottom
of a deep fiord or bay, 16 m. NNW. Schleswig,
on the railway from Kiel to Fredericia. Pop.
19,682 in 1860. The town is modern, well built,
clean, and thriving. The harbour has water sufficient to float the largest ships. There are sugar-
houses and distilleries, with manufactures of cloth,
cotton, paper, soap, and tobacco; but it is chiefly
celebrated for the tiles made in its immediate
vicinity, of which large quantities are exported.
About 250 vessels belong to, and several are built
at, the port.

FLINT, a marit. co. of N. Wales, consisting of
two separate portions, the largest and most important of which is bounded on the N. by the
Irish Sea, on the E. by the estuary of the Dee
and the Dee itself, and on the S. and W. by Den-
bighshire; the other and smaller portion lies along
the S. bank of the Dee, between Cheshire and
Salop. Area 299 sq. m., or 181,905 acres, being
the smallest of the Welsh counties. The surface
is considerably diversified. The S. part is mostly
flat, and consists in great part of a portion of the
vale of Clwyd. The vale of Mold is also flat and
highly productive, as is the detached portion to
the S. of the Dee. A ridge of hills runs through
the whole extent of the county, mostly parallel
to the Dee and its estuary, which, though externally barren, are valuable from their mines of
lead and other minerals; but, on the whole, there
is a larger proportion of good land in this than in
any other Welsh county. Besides the Dee and the
Clwyd, the county is watered by the Alyn and

other streams. Agriculture, though still rather backward, has been materially improved, and many parts are well cultivated. There has also, within the present century, been a great improvement in the farm buildings and cottages, and in the implements and stock. Manufactures have been introduced into Flintshire, especially that of cotton, which is carried on to some extent at Mold. But the principal branch of industry carried on in this co., next to agriculture, is that of mining: its lead mines are at present the most extensive of any in the empire; those of copper are also of considerable value: and beds of coal exist all along the shore of the Dee, large quantities of which are used in smelting works, in addition to those that are exported. The smelting works in the vicinity of Holywell are very extensive, and employ from 600 to 700 hands. Flint returns 2 mems. to the H. of C., viz. 1 for the co. and 1 for the town of Flint and its contributory bor. Registered electors for the co. 2,895 in 1862. Flintshire is divided into 5 hundreds and 28 parishes; and in 1861 had a pop. of 69,737, living in 15,115 houses. Gross annual value of real property, 255,369£ in 1857, and 331,807£ in 1862.

FLINT, a par. bor. and sea-port town of N. Wales, co. Flint, hund. Coleshill, on the estuary of the Dee, 11 m. NW. Chester, and 195 m. NW. London by London and North Western railway, viâ Chester. Pop. 8,479 in 1861. The borough includes the parish of Flint and the township of Coleshill-Fawr. It is situated within a large quadrangular space, surrounded, on the principle of a Roman encampment, by rampart, and a deep entrenchment, having at the NE. extremity its ancient castle. Two main streets cross at right angles, and are similarly intersected by smaller streets, the frequent gaps and broken walls in which give the town a dilapidated, deserted aspect. It has a new church, dedicated to St. Mary, and five dissenting chapels; a national school for 140 children, several Sunday schools, and a guildhall. The assizes, formerly held here, have been long since removed to Mold, to which the county goal has also been more recently transferred. The coal works and lead mines in the vicinity employ the chief part of the pop. Of late years, in consequence of obstructions in the channel of the Dee, Flint has become, to a considerable extent, the port of Chester; and here the larger vessels (especially those with timber) discharge into lighters, or rafts are formed and floated up to that city. The wharfs, which have been much improved, and extended of late years, are accessible to vessels of 300 tons, at any time of tide. Railways lead from the wharfs to the mines. The exports consist chiefly of coals to Ireland and coastwise; and lead, in pigs, sheets, &c., from the works in the vicinity. During the summer season Flint is a place of some resort for sea-bathing: there are also hot baths for the accommodation of visitors. The castle, now in a state of rapid decay, is a square building, with round towers at three of the angles, and at the fourth is a much larger tower at a little distance from the castle, but originally joined to it by a drawbridge. Formerly the Dee flowed beneath the walls, and rings were fastened in them, to which ships were moored; but it has now receded to some distance. The foundation of this castle is ascribed to Henry II. Flint received its first charter in 1283. Since the 27th Henry VIII. it returned 1 mem. to the H. of C., along with the contributory bors. of Rhuddan, Overton, Caerwis, and Caergwle: to these the Reform Act added St. Asaph, Holywell, and Mold: the right of voting previously to the Reform Act was vested in the inhab. paying poor and church rates. Re-

gistered voters for Flint and its contributory bors. 751 in 1862, among whom 85 'scot-and-lot' voters. The municipal law, is restricted to a small space round the town, and is governed by 4 aldermen and 12 councillors. The numerous relics in the vicinity make it probable that Flint was a Roman station; and the remains of smelting-places and washes on the ancient plan, prove that the lead mines had been worked at a remote period.

FLODDEN, a village of England, co. Northumberland, 5 m. SE. Coldstream, memorable as the scene of one of the most destructive conflicts recorded in British history. James IV., king of Scotland, having invaded England with a large force, was encountered here, on the 9th of Sept., 1513, by an English army under the Earl of Surrey. James, who was destitute of every quality of a general, except bravery, was killed, and his army totally defeated. The loss on the part of the Scotch was extremely great. Besides the king, no fewer than 12 earls, 13 lords, and 5 eldest sons of peers, with a vast number of gentlemen and persons of distinction, and probably about 10,000 common soldiers, were left on the field. The loss on the part of the English was comparatively inconsiderable. This is by far the most calamitous defeat in the Scottish annals, and as there was hardly a family of distinction in the kingdom who did not lose one or more members in it, the whole nation was involved in mourning and despair. (See Tytler, Pinkerton, Histories of Scotland. Sir Walter Scott has given a vivid and generally correct account of this great battle in his 'Marmion.')

FLORENCE (Ital. *Firenze*, an. *Florentia Tuscorum*), a famous city of Central Italy, and, since 1865, capital of the kingdom of Italy, on both sides the Arno, 63 m. S. by W. Bologna, 68 m. ENE. Leghorn, and 187 m. NW. Rome, on the railway from Rome to Milan. Pop. 112,236 in 1862, and estimated at 150,000 in Sept. 1865, after the transfer of the government and court of the king of Italy to the city. Florence stands in a richly wooded, well cultivated, and beautiful valley, encircled by the Apennines, and is well built and agreeable. Its shape is nearly a square, the sides of which almost correspond with the cardinal points: the Arno intersects it from SE. to NW., 5 of the quarters into which it is divided being situated on the right, and the fourth on the left bank of the river. It is enclosed by an old wall about 5 m. in circuit, flanked with towers and pierced by 7 gates, which, besides being useless as a means of defence, is injurious, by preventing the free ingress and egress of the citizens, and checking the circulation of the air. The communication between the opposite sides of the river is maintained by means of 7 bridges. Florence contains a great number of magnificent edifices and squares, generally adorned with statues, columns, or fountains: there are no fewer than 170 churches, 89 convents, 2 royal, and many other palaces, 12 hospitals, and 8 great and small theatres. Altogether Florence bears the aspect of a city filled with nobles and their domestics—a city of bridges, churches, and palaces. Every building has a superb and architectural form. Each angle of a street presents an architectural view, fit to be drawn for a scene in a theatre. Many of the houses are palaces; and a palace in Florence is a magnificent pile, venerable from its antiquity, of a square and bulky form, with a plain front, extending from two to three hundred feet, built of huge dark grey stones, in a massive, gloomy, and imperative style. The roof is flat, with a deep cornice, and bold projected soffits, which gives a grand, square, and magnificent appearance to the edifice. The chimneys are grouped into stacks, the tops of which, increasing in bulk as they rise in height, resemble a crown. Many of these palaces are fitted up with great magnificence, and some of them contain valuable galleries of pictures, that are mostly open to the public. The streets, though in parts narrow, winding, and angular, are mostly wide and straight; and they are admirably paved, after the manner of the old Roman roads, with angular blocks of trap, or sandstone. The houses generally are substantial, more so, apparently, than those of Rome. The Piazza reale is the largest square; it has a fine marble fountain, and an equestrian statue in bronze of Cosmo I. by John of Bologna; the Piazza dell' Annunziata is surrounded by arcades, and has two fine bronze fountains, and an equestrian statue of Ferdinand I. The Piazza del Mercato Vecchio, exactly in the middle of the city, has a marble column from which it is a mile to each extremity. The Arno is decidedly superior to the Tiber at Rome. The bridge, S. Trinità, built of marble in 1557 by Ammanati, is designed in a style of elegance and simplicity unrivalled by the most successful efforts of modern artists. The Ponte Vecchio, built in 1345, has the houses of the street continued over it, so that it is not till they arrive at an open arcade in the centre that passengers become aware of their situation. The bridges and the banks over though not spacious quays by which it is bordered, afford fine views of the river, Florence being in this respect much superior to the 'Eternal City.' The cathedral, or Duomo, a vast edifice, coated with marble, about 500 ft. in length, and 341 ft. in height to the top of the cross, stands in a spacious square. It was begun by Arnolfo di Lapo in 1298, and finished by Brunelleschi in 1426: its cupola is said to have suggested to Michael Angelo the first idea of that of St. Peter's. It is built of brick, and veneered, as it were, with various-coloured marble slabs, arranged in narrow strips or panels. 'There is something,' says a recent traveller, 'imposing in the name of a marble edifice, but not so in the reality; polished marble is worse than rough marble, which, again, is inferior to sandstone or granite; but coloured marble (parti-coloured especially) is worse than all. The Duomo of Florence, built in defiance of all the orders of architecture, is neither Grecian nor Gothic, although of the age of the latter style; and its dimensions alone give it greatness. The interior is very striking, but spoiled by a circular screen of Grecian columns round the altar.' Another traveller says, that this cathedral is to St. Peter's what harlequin is to a Roman senator. The Campanile, or belfry, adjoining the Duomo, but detached from it, is a fine tower 267 ft. in height. Charles V. was so well pleased with it, that he used to say it should be kept in a glass case. With the exception of the Duomo, the other churches have little worth notice in their architecture, and many of them are unfinished and poor. That of Santa Croce, however, called the Pantheon of Florence, is interesting from its containing the remains and tombs of four of the greatest men of modern Italy, or indeed of modern times—Michael Angelo, Galileo, Machiavelli, and Alfieri. The church of San Lorenzo contains the mausoleum of the Medici family, said by Lord Byron to be a 'tawdry, glaring, and unfinished chapel,' and admitted by less severe critics to be meretricious and in bad taste. In a cloister attached to this church is the Laurentian library, containing a peculiarly valuable collection of above 6,000 manuscripts and 170,000 vols.

Among the palaces are the Palazzo Vecchio, or old palace, inhabited by the Medici, when citizens of Florence. It was begun in 1298, and finished in 1560. It is in a massive, severe, and gloomy

style, and has a noble tower 268 ft. in height, which commands a fine view of the surrounding country. This palace is now occupied with the principal public offices. The Palazzo Pitti, erected in 1440, the ordinary residence of the king of Italy, is a vast and heavy structure; it is furnished in the most costly manner, and is enriched with a great number of fine statues, busts, and pictures, and an excellent library. Attached to the Pitti palace are the Boboli gardens, laid out by Cosmo I. in 1550, in the pure classical style; that is, in rectangular walks, flanked with cut trees fashioned into a wall of arrived over head, and furnished with a due quantity of stone steps, stone walls, and stone statues. Connected with these gardens is the botanical garden, a museum of natural history, a splendid anatomical collection modelled in wax by the Abbé Fontana, occupying 15 apartments, and a fine library. Another fine palace, the Riccardi, was built in 1440, after a design by Michelozzo. It has a noble gallery, with a ceiling painted by Luca Giordano, and a select library with 40,000 vols., open to the public. It is now occupied by the Accademia della Crusca and some public departments. But the glory of Florence is its grand gallery: it occupies the upper floor of the Uffizi, a building erected after a design of Vasari by Cosmo I., consisting of two parallel corridors or galleries, each 440 ft. in length, and 72 ft. apart, united at one end by a third corridor, the choicest and most valuable specimens of art being preserved in saloons opening from the corridors on each side. This gallery contains some chefs-d'œuvre of statuary, at the head of which, by universal consent, is placed the Venus de Medici, the goddess who 'lives and loves in stone.' The matchless statue was discovered in the 16th century, in the Villa Hadriana, near Tivoli; and being acquired by the Medici family, was placed in their palace in Rome, whence it was conveyed to Florence by order of Cosmo III. The whole of the left arm, and a part of the right, are modern, having been restored by Bandinelli. An inscription on the base intimates that it is the work of an Athenian artist, called Cleomenes; this, however, is generally discredited. But, whoever may be the sculptor, it is certainly worthy to rank with the famous statue of Venus sold by Praxiteles to the Cnidians, respecting which some rather curious particulars may be seen in Pliny. (Hist. Nat. lib. xxxvi. § 4.) The attitude of the Venus de Medici corresponds with the verses of Ovid, who perhaps had this very statue in his eye:—

'Ipsa Venus pubem, quoties velamina ponit,
Protegitur læva semi-reducta manu.'
De Arte Amandi, ii. v. 614.

Addison says of this famous statue, that 'the softness of the flesh, the delicacy of the shape, air and posture, and the correctness of the design, are inexpressible.' (Travels, art. 'Florence.') And, according to Byron,

'——— the goddess loves in stone, and fills
The air around with beauty.'

Among the other chefs-d'œuvre, the best perhaps are, the Knife-grinder, the Faun, the Wrestlers, and Niobe and her Children. The collection of paintings comprises superb specimens of all the best schools, and is said to surpass even that of the Vatican. Speaking of this gallery, an English traveller observes:—'Persons like myself, with no pretensions to connoisseurship, will feel how poor and vulgar the pictures of the tramontane artists are when placed beside the works of the great Italian masters. Among those who admire the Dutch and Flemish painters for their correct and faithful representation of individual

FLORENCE

nature, and their skill in chiaroscuro, there are few. I venture to think, whose taste, after some weeks spent in perambulating the picture galleries of Rome or Florence, will not undergo a metamorphosis—few who will not feel a strong preference for what is called the "ideal" or "grand style" —for the saints, prophets, Madonnas, holy families, sybils, and goddesses of the Roman, Florentine, and Lombard artists, which are in truth impersonations of the noblest attributes of humanity, —maternal love, hence fortitude, intellectual energy, sublime benevolence, and rapt devotion. The same predilection will probably also create a predilection in sculpture for the naked figure, and induce a belief that the artist's labour is thrown away upon togas and tunics, however gracefully folded,—that it is the kernel, not the husk,—the man, not his drapery, which is the well-spring of beauty and the recipient of character. Such at least was my own experience. The study of the works of the great Italian masters has this fine moral effect, that it ennobles our conceptions of the capabilities and destiny of man. It puts the doctrine of immortality on canvas, and presents it to the eye. I was delighted with Guido's female heads, which seemed to me radiant with grace and sweetness, purity and beauty, even beyond those of Raphael. The Italian schools are less rich in landscape, yet in this department who can surpass Salvator? I had no adequate idea of this great artist's genius till I saw nearly a dozen of his large pieces in the Pitti and Corsini palaces in Florence. They seemed to have all the splendour of Claude's, with the addition of that lofty, bold, mountain scenery which a Scotsman is apt to consider as essential to the highest class of landscape.' The great gallery communicates by a covered passage not only with the Palazzo Vecchio, separated from it by a street, but also with the Pitti palace, though on the other side of the river, being carried over the latter by the Ponte Vecchio, or old bridge.

Besides the Riccardi and Laurentian libraries, the Magliabecchi library, containing a rare, extensive, and valuable collection of books, is open to the public: it is placed below the grand gallery.

Florence is subject to fogs in the winter; but in spring and autumn it is a delightful residence, well provided with everything that can gratify the man of taste and science, or the voluptuary. It has manufactures of silks, straw hats, articles of alabaster, scagliola and pietre dure, perfumery, jewellery, artificial flowers, porcelain, engravings, and other objects of the fine arts. The literary and educational institutions are numerous and important. At the head of these is the academy Della Crusca, established in 1582, to which has been united the ancient university of Florence. The name Crusca (chaff, or husk of corn) has been assumed by this academy, in allusion to the grand object of its institution, the sifting or purifying of the Italian language. This academy, published in 1612, in 1 vol. folio, the first edition of the celebrated lexicon, entitled Vocabolario della Crusca, the fourth and last edition of which appeared in 6 vols in 1729-38; a work which, though perhaps not quite perfect, has been generally admitted to be the standard of the Italian language. (Tiraboschi, Storia della Litteratura Italiana, viii. 511, edit. 1794.) An edition of the Vocabolario della Crusca, including numerous words selected from the authors quoted by the academy, but omitted by them, was published at Naples in 1746, in 6 vols folio. This is preferred by some to the genuine Vocabolario. (Tiraboschi, ubi supra.) Besides this famous academy, there are in Florence

FLORIDA

a Scuola di Belle Arti, or school of the fine arts, a medico-chirurgical school, an athenæum, and a number of other literary societies. A school was opened in 1820 for the instruction of the poorer classes, on the principle of mutual instruction; and another institution was founded in 1823, for the instruction of girls from 7 to 12 years of age: they are educated with great care, and are said to be instructed in all that has a tendency to make them active and provident mothers. There are a great number of other schools and institutions for the instruction of students in the higher branches of education. The charitable institutions are numerous, extensive, and well conducted. Among others is the Monte di Pietà, founded in 1495; a foundling hospital; a workhouse, on a large scale. The Fraternità della Misericordia is an institution in which the higher classes undertake various duties in relation to the poor. The Palazzo del Podestà, the ancient government-house, is now converted into a prison.

The common people of Florence are well clothed, and have a comfortable-like appearance; and there are, as compared with most other Italian towns, few beggars, priests, and monks. The citizens are friendly, cheerful, and hospitable. The encouragement given under the late as well as the present government, to artistic and scientific studies, has conferred advantages on Florence unknown in most other parts of Italy. All sorts of foreign publications are met with here; and the facilities it affords for gratifying a taste for the fine arts, the beauty and security of the town and environs, and its salubrity and cheapness, make it, on the whole, a more desirable residence than Rome.

The origin of Florence is not clearly ascertained; but it owed its first distinction to Sulla, who planted in it a Roman colony. In the reign of Tiberius it was one of the principal cities of Italy, and was distinguished by its writers and orators. In 541 it was almost wholly destroyed by Totila king of the Goths. About 200 years afterwards it was restored by Charlemagne. It then became the chief city of a famous republic; and was for a lengthened period in Italy what Athens had been in Greece in the days of Xenophon and Thucydides. At length, in 1537, the Medici, from being the first of the citizens, became the sovereigns of Florence. The city remained the capital of Tuscany till 1860, when it was annexed to the new kingdom of Italy. On the 2nd of June, 1861, Florence was the scene of a 'first Italian national festival,' in commemoration of the national unity, liberty, and independence. In the spring of 1865, the seat of government of the kingdom was transferred from Turin to Florence.

Florence has produced more celebrated men than any other town of Italy, or perhaps of Europe: among others may be mentioned, Dante (a fine statue of whom was unveiled at the 'Dante Festival' of 1865), Petrarch, Boccaccio, Villani, Cosmo and Lorenzo de' Medici; Galileo, M. Angelo, Leonardo da Vinci, Benvenuto Cellini, Alberti, Lapo Brunelleschi, Giotto, Andrea-del-Sarto, Macchiavelli; Popes Leo X. and XI., Clement VII., VIII., and XII.

FLORIDA, an extensive peninsula of N. America, stretching S. from the 30th to the 25th deg. of lat., forming a state in the extreme S.W. territory of the U. States. The state is comprised between lat. 24° and 31° N., and long. 80° and 87° 35' W.; having N. Alabama and Georgia, E. the Atlantic, S. the channel of Florida, and W. the Gulf of Mexico, and a small portion of Alabama. Length N.W. to N.E. about 640 m.; average breadth about 81 m.; area, 59,268 sq. m. Pop. 140,425 in 1860.

The Gulf Stream, which sets from the Gulf of Mexico round the S. and SE. coasts, has in the course of ages worn away the land, and formed the low sandy islands generally known by the name of the ' Florida Keys,' or Martyrs, separated from the main land by a navigable channel which, however, is both difficult and dangerous. There are a few good harbours, the best of which are those of Pensacola and Tampa on the W., and of St. Augustine and St. Mary's on the E. coast. Florida is naturally divided into two different zones, about the 28 deg. of lat. The surface of the portion N. of this parallel is more elevated, broken, and wooded, than that on its S. side, which is generally level and marshy, and may be termed the true palm-tree section of the U. States. The centre rises into hills of no great elevation, which slope gradually towards the Gulf of Mexico and the Atlantic, and NW. towards the body of the Continent; but as we proceed towards the S., the whole surface becomes a dead flat, and, in great part, inundated plain, terminating at the extreme point of the peninsula in heaps of sharp rocks, partially covered with shrubby pines.

The chief rivers are the St. John's, Appalachicola, Suwanee, St. Mark's, and Conecuh. The St. John's partakes more of the character of an inlet or sound than of a river, from the number of lakes formed by its enlargements. Its chief branch, the Ocklawaha, appears to rise near the centre of the peninsula, and flows in a NW. direction for about 80 m., when it unites with the St. John's proper, which rises within a few miles of the ocean, and the united water, after a tortuous course of 130 m., falls into the Atlantic, near the NE. extremity of the territory. It is a curious fact, that though a fresh-water stream at its mouth, it is often rendered brackish towards its head from the waters of the Gulf of Mexico being driven by the winds into the lagoons and marshes among which it has its sources. Both branches of this river are navigable for some distance above their junction, but have little commercial value. The Appalachicola has its estuary in that portion of the territory W. of the peninsula. It has a course of about 100 m. N. to S. within the territory, but does not possess a depth of water proportionate to its magnitude. This river is considered to form the boundary between E. and W. Florida. There are several lakes, of which the Macaco, near the centre of S. Florida, and Lake St. George, an enlargement of the St. John's river, are the principal.

The whole peninsula appears to rest upon a base of shell-limestone of comparatively recent formation and different degrees of hardness. The soil on the banks of the rivers is often very fertile; but the proportion of good land is, notwithstanding, believed to be but small. In the N. part of E. and in W. Florida there are many finely variegated and fertile tracts, and the country is often richly wooded. The most valuable district of the territory is a tract of about 150 m. in length by 30 m. in breadth in W. Florida, nearly in the centre of which is Tallahassee, the capital. There are some very extensive swamps and savannahs, particularly the swamp of Okefonoco, half in Florida and half in Georgia; and there are also some very extensive marshes. The climate of the N. parts, though hot, has been represented as good, and the air as being always elastic and pure. The winters are so mild that it is never necessary to house cattle. In the N. snow never falls, and frost, although it sometimes occurs, is rare. During July, August, and September, the heat is very oppressive, and fevers are prevalent.

The chief agricultural products are—rice, Indian corn, tobacco, indigo, cotton, and hemp; the olive, vine, lime, chaddock, and other tropical fruits, are successfully cultivated, and in some of the occupied maritime districts the sugar-cane and coffee. Large herds of cattle are reared. Much fine timber, besides pitch, tar, and turpentine, are obtained from the forests; the swamps and rivers produce a great variety of fish and fisheries.

The state is divided into 4 districts and 20 counties. Tallahassee is the capital. Pensacola, St. Augustine, and Jacksonville are the other chief towns: all of these are in the N. From the St. Mary's river, which divides Florida at its NE. angle from Georgia, a canal, 250 m. in length, extends NE. to SW., across the peninsula to Appalachicola Bay. A railroad, 12 m. in length, between Lake Wimico and St. Joseph's, completed in 1836; and another from Jacksonville to St. Mark's, 160 m. in length, were the first railways constructed in the state.

Florida is entitled to send one member to the House of Representatives of the United States, and two to the Senate. The population, at the census of 1860, included 61,745 slaves, being nearly a moiety of the inhabitants. The value of real estate and personal property (including slaves) amounted to 22,862,270 dollars in 1850, and to 73,101,500 dollars in 1860.

Florida derives its name from Pascua Florida, or Palm Sunday, the day on which it was discovered by Juan Ponce de Leon, in 1512. Its name was for some length of time applied by the Spaniards and Italians to the whole W. coast of N. America. It remained a Spanish possession until 1763, when it was ceded to the British, soon after which it was divided into E. and W. Florida. In 1783 the whole territory was restored to Spain. In 1819 negotiations were opened for the transfer of Florida to the United States; and in 1821 a treaty was ratified, by which it became a part of the union as a 'territory,' under the central government. It was admitted into the Union as an independent state March 3, 1845. An Act of Secession from the United States was passed by a convention Jan. 10, 1861; but having been reconquered by the armies of the North, Florida was again incorporated into the Union in 1865.

FLOUR (ST.), a town of France, dép. Cantal, cap. arrond., on a basaltic plateau, 42 m. ENE. Aurillac. Pop. 5,283 in 1861. The town is ill-built; streets narrow and gloomy. Its public edifices are, however, generally handsome. Among these are a cathedral, episcopal palace, diocesan seminary, Jacobin convent, Jesuit college, hospital and sub-prefecture. It is well furnished with water. It is the seat of the departmental court of assize, and of tribunals of original jurisdiction and commerce.

FLUSHING (Dut. Vlissingen), a fortified seaport town of Holland, prov. Zealand, on the W. Scheldt, near the S. extremity of the isl. Walcheren, 4 m. SSW. Middleburgh: lat. 51° 26' 42" N., long. 3° 31' 57" E. Pop. 10,799 in 1861. The town is strongly fortified; besides its own ramparts, it is defended, together with its fine harbour, by several adjacent forts, and provided with sluices, by means of which the surrounding country may be inundated. The town is well built, but presents little worthy of notice, most of its best public buildings having been destroyed during the bombardment by the English in 1809. Its port is extensive, safe, and has deep water. Two canals, communicating with it, enable the largest merchant-vessels to penetrate into the town, and unload on the quays close to the warehouses. A strong wall of masonry protects the town against the sea; the side facing the Scheldt is embanked with great care, and kept in repair at an enormous expense. Flushing has a dockyard, and a naval arsenal; and is the seat of an admiralty board. It has an extensive trade with both the E. and W. Indies, and continual communication, by means of packet-boats, with the other sea-port towns of Holland; it has also a considerable trade with England. Few towns have suffered so severely from war and inundations. It hoisted the standard of revolt against the Spaniards, immediately after the capture of the Briel in 1572. Together with some other towns, it was given to England by the Prince of Orange in 1585, and remained in British possession till 1616. From 1809 to 1814 it belonged to the French. Since 1809 its fortifications have been greatly improved, and, in conjunction with the Fort of Rammekens to the E. and those of Lissekens on the opposite side of the river, it now completely commands the mouth of the W. scheldt, or *Hond*. Flushing was the birthplace of the celebrated Dutch admiral, De Ruyter.

FOCHABERS, a village of Scotland, co. Moray, and par. of Bellie, on a rising ground on the Spey, 4 m. from its embouchure in the Moray Frith, 8 m. E. Elgin, and 60 m. NW. Aberdeen. Pop. 1,146 in 1841. The village consists of two wide streets, crossing each other at right angles, and having a square in the middle. The par. church and a Rom. Cath. chapel are the chief public buildings. An elegant bridge which spans the Spey was partly swept away by the great floods of 1829, but has since been rebuilt. The whole district through which the river flows suffered severely from these inundations. The bridge in question, which was erected in 1801 at a cost of 15,000*l.*, has four arches, of which the two smallest have each a span of 75 ft., and the two in the middle a span each of 95 ft. Gordon Castle, the splendid residence of the ancient house of Gordon, and now the property of the Duke of Richmond, is in the immediate vicinity of Fochabers; a circumstance to which the village owes its origin and any importance that may attach to it. Fochabers is a bungh of barony, governed by a baron-bailie nominated by the noble proprietor of Gordon Castle.

FOGGIA (so-called from its corn magazines, *fosse*), a city of Southern Italy, cap. of province of same name, in the centre of the great Apulian plain, 46 m. E. by S. Campobasso, 21½ m. SW. Manfredonia, and 80 m. NE. by E. Naples on the railway from Ancona to Trani and the Gulf of Taranto. Pop. 25,107 in 1861. The town is well built and paved; the streets are wide and clean; the shops large and well supplied; and the whole has an air of opulence and prosperity. It has a handsome *intendenza*, or palace, where the governor of the province resides; many excellent private houses, a Gothic cathedral, and about twenty other churches; a good custom-house and theatre; and the remains of a palace which, together with a large well, was constructed by the Emperor Frederick II. The corn magazines, for which Foggia is noted, are very extensive; they stretch under all the large streets and open squares, consisting of vaults lined with masonry, and their orifices closed up with boards and earth. Being situated in a fruitful country, and traversed by roads leading to Naples, Bovino, Brindisi, Manfredonia, and Pescara, Foggia has a considerable trade, principally in corn, wool, cheese, cattle, wine, oil, capers, and other agricultural produce. Its consequence always has been, and still is,

owing to its being a staple market for corn and wool. The *dogana*, or register-office, at Foggia has the distribution of a fixed assessment upon the numerous flocks of sheep that descend in autumn from the mountains of Abruzzo into the plains of Puglia, where they winter, and in May return to the high country. (See ABRUZZO.) This duty originated with the ancient Romans, when they obtained possession of the country. It continued uninterruptedly to be collected till the 18th century, after which, for about two centuries, the passage appears to have been open without fee to all shepherds who chose to bring down their flocks. Under Alphonso I., however, the crown resumed its rights; and having purchased a considerable extent of pasture land, formed the *tavoliere*. (See APULIA.) The Abruzzi shepherds, who came down with their flocks into the *tavoliere*, paid a fixed rate per head for their sheep; but had not the power to dispose of their wool, lambs, cheese, or any other commodity produced during their winter residence, in any fair but that of Foggia, where they were to be deposited in the royal magazines, and not touched without a permit. The fair of Foggia, holden from the 8th to the 20th of May, is an important mart, and attended by a great number of commercial and other visitors. Foggia is the seat of the superior criminal court for the prov., and of the tribunal of commerce for Apulia.

This city appears to have been founded in the 9th century, and peopled from *Arpi* or *Argyrippa*, an ancient city 4 m. distant, said to have been founded by Diomed, which surrendered to Hannibal after the battle of Cannae, and of which some faint vestiges are still extant. Foggia was greatly enriched by the Suabian princes of Naples. It was sacked in 1264 by Charles of Anjou, who died there in 1285. It was nearly destroyed by the earthquake of 1731.

FOLDVAR (an. *Lussonium*), a town of Hungary, co. Tolna, on the summit and declivity of a hill, on the right bank of the Danube, 48 m. S. Buda. Pop. 8,400 in 1857. The town has a Rom. Cath. high school and a prison; it belongs, together with its lordship, to the university of Pesth.

FOLIGNO (an. *Fulginium*), a town of Central Italy, prov. Perugia, in the Val Spoletana, and on the Flaminian Way, 20 m. SE. Perugia, and 14 m. N. by W. Spoleto, on the railway from Rome to Ancona. Pop. 12,880 in 1838. The town is walled, but its ramparts and bastions now serve for public promenades. Its streets generally intersect each other at right angles. There are few public buildings worthy of notice. The cathedral, commenced in the last century, is still unfinished: there are 8 other churches, 20 convents, a town-hall, and a cabinet of antiquities. There are numerous paper-mills turned by the Topino; and the town has manufactures of woollen cloth, silks, parchment, and bleached wax, and a considerable trade in cattle. The vicinity abounds with vineyards, and olive and mulberry plantations. This city appears to have been anciently of some importance; it was considerably augmented on the destruction of the adjacent town of *Forum Flaminii*, by the Lombards, in 740. It was given to the see of Rome in 1439.

FOLKESTONE, a bor., sea-port town, and par. of England, co. Kent, lathe Shepway, hund. Folkestone; on the straits of Dover, 62 m. SE. by E. London, and 7 m. W. by S. Dover, on the South Eastern railway, which has here two stations, and a harbour for its steamers to Boulogne. Pop. of munic. bor. 8,507, and of par. 8,674 in 1861. The town is built between two precipitous chalk cliffs, on ground rising gradually from the coast; and consists chiefly of three narrow and irregular streets, principally extending up the acclivities of the W. cliff, on the summit of which is the church, in the early Gothic style, with a tower from the centre. There are also five dissenting chapels, and a free school for twenty poor children, founded in 1674. Market, Thursday, in a commodious market-house, built within a recent period by the Earl of Radnor. The chief employment of the inhabitants is fishing, and occupation in connection with the goods and the passenger traffic of the South Eastern railway. The pier-harbour, formed at an expense of upwards of 50,000l., was originally built by a local company, but subsequently purchased by the railway. It is very safe, but only accessible at high water, so that the steamers going to and coming from Boulogne have to vary their hours of departure and arrival, and run in connection with 'tidal trains.' The journey from London to Paris, by this route, occupies little more than ten hours. There is a strong modern battery on the heights, and the line of coast is defended by three Martello towers. Folkestone has been a member of the cinque port of Dover from a period previous to the reign of Henry I. Average annual corporation revenue 224l. 16s. Under the Municipal Reform Act it is governed by four aldermen and twelve councillors; and its limits, which extended along the coast 2½ m. on the E. side of the town, and to Sandgate on the other, being at the same time considerably contracted in those directions and extended inland, so as to include the hamlet of Ford, on the line of road to Canterbury, and about ¼ m. from Folkestone. The Reform Act associated Folkestone with the bor. of Hythe in the privilege of returning one mem. to the H. of C. The town has suffered much at different periods from encroachments of the sea. William Harvey, the discoverer of the circulation of the blood, was a native of Folkestone, having been born here on the 1st of April, 1578 ; the charity school, endowed by his nephew, was built from a bequest left by him for the purpose.

FONDI (an. *Fundi*), a town of Southern Italy, prov. Caserta, on the high road between Naples and Rome, and on the Appian Way, 11 m. NE. Terracina, and the same NW. Gaeta. Pop. 6,212 in 1861. All travellers agree in speaking in dispraise of Fondi. It is a miserable town, near a pestiferous lake (the an. *Lacus Fundanus*), which renders the air unwholesome ; and its inhab. generally are in a wretched condition, though the neighbourhood is abundantly fertile in every kind of produce. This, in fact, is the *Caecubus ager*, anciently so famous for its wine—

> 'Caecubum et prelo domitam Caleno
> Tu bibes uvam.'   Hor. l. Od. 70.

(See also li. Od. 14 ; and Martial, xiii. Ep. 15.) But, like the town, the wine has really degenerated, and is now quite unworthy the encomiums lavished on its ancient growth.

Fondi is surrounded by the remains of walls of a Cyclopean structure, particularly described by Swinburne (i. 507, 508). It has a Gothic cathedral, a college, and two houses of charity. It obtained the privileges of a Roman city, A. U. C. 417. In 1272, it was burnt by the adherents of the emperor Fred. II. It has several times suffered from invasions by the Turks, especially in 1534, when they made an unsuccessful attempt to carry off Julia Gonzaga, countess of Fondi.

FONTAINEBLEAU, a town of France, dép. Seine-et-Marne, cap. arrond., near the Seine, in the forest of the same name, and on the railway between Paris and Lyons, 37 m. SSE. the former

cir., and 8 m. N. by E. Melun. Pop. 11,930 in 1861. The town is well built ; streets wide, straight, well paved and clean ; but, excepting the principal ones, they are dull. It has several good churches and other public buildings, two excellent cavalry barracks, a hospital founded by Anne of Austria, an asylum for girls established by Mad. de Montespan, a college, public library, with 20,000 vols., public baths, a large reservoir ; and at its S. extremity an obelisk erected in 1786, on occasion of the marriage of Louis XVI. and Marie Antoinette. The town has manufactures of porcelain and other earthenware ; but it owes all its celebrity, and indeed origin, to the palace or château of Fontainebleau, a favourite residence of the former kings of France. The precise date of the foundation of the palace is uncertain. It would appear that Robert-le-Pieux erected a small house of retirement on the spot towards the end of the 10th century ; which edifice, having fallen to decay, was rebuilt in the 12th century by Louis VII. Philip Augustus, Louis IX., and other sovereigns, added to it, and it was in particular enlarged and embellished by Francis I. It grew rapidly under the hands of his successors : Henry IV. expended 2,440,850 liv. on it ; Louis XIII., XIV., and XV. added to and improved it ; Napoleon I. is said to have spent 6,212,000 fr. on it between 1804 and 1813. It is a vast pile, with little harmony among its parts, being, in fact, rather a collection of palaces of different epochs, and in different styles of architecture, than a single edifice. Saracenic, Tuscan, and Greek orders are intermixed and interspersed with the most bizarre and dissimilar ornaments ; yet, upon the whole, the building has a striking air of grandeur and majesty. Its palaces are united by galleries, and enclose six principal courts :—the Cour du Cheval-blanc ; des Fontaines ; Ovale, or du Donjon ; de l'Orangerie ; des Princes ; and des Cuisines. The largest is the Cour du Cheval-blanc, which forms also the principal entrance from the W. and derives its name from an equestrian statue in plaster, erected by Catherine de Medici, but no longer existing. At its upper end there is a remarkably fine flight of stone steps, under which a passage leads to a chapel remarkable for the elegance of its architecture and decoration. It was in this court that Napoleon I. bade adieu to his guard previously to his departure for Elba in 1814. The Cour des Fontaines has on one side the suite of apartments occupied by Charles V. in 1539. The buildings surrounding the Cour Ovale are the most ancient of all ; they comprise the ball-room, adorned with paintings by Primaticcio, Nicolo, &c., the library, the king's and queen's apartments, the throne, and council-halls. In one of these rooms the small round table is still shown on which Napoleon I. signed his act of abdication in 1814. The Cour de l'Orangerie is also called the Cour de Diane, from a fine bronze statue of Diana in its centre ; in the Galerie des Cerfs, one of the buildings surrounding it, Monaldeschi was assassinated by order of Christina of Sweden. The Cour des Princes, the smallest of all, is surrounded by the apartments occupied by Queen Christina. The Cour des Cuisines is large, regular, and enclosed with buildings erected by Henri IV. The palace contains a great number of ancient and modern paintings ; it is surrounded, especially on the S. side, by fine gardens, ornamented with fountains and fish-ponds, and traversed by a canal nearly 3-4ths of a mile in length. The forest of Fontainebleau comprises 82,877 superfs. or about 34,500 acres, a part being on the opposite side of the Seine. Its surface is very varied, and in parts very picturesque. It supplies Paris with a small portion of its wood-fuel, and with a considerable part of its paving stone.

The château of Fontainebleau has been the scene of many historical events: Philip IV., Hen. III., and Louis XIII. were born in it; and the first monarch died there. It was visited by Peter the Great; Louis XV. espoused the daughter of the king of Poland in this palace; Pope Pius VII. was confined within its walls for 18 months; and it is intimately connected with the history of Napoleon. It was comparatively neglected by Louis XVIII. and Charles X.; but King Louis Philippe restored it to somewhat of its ancient grandeur. In 1837 the nuptials of the Duke of Orleans were celebrated here with great pomp. Under Napoleon III., the palace was still more enlarged and embellished, becoming the scene of luxurious autumnal fetes, rivalling those of Louis XIV.

FONTARABIA (properly *Fuenterrabia*), a fortified frontier and sea-port town of Spain, Biscay, prov. Guipuzcoa, on a small peninsula on the left bank of the Bidassoa, at its mouth, 20 m. W. by S. Bayonne. Pop. 3,058 in 1857. The town used to be reckoned one of the keys of Spain; but its walls were levelled by the British troops in 1813. On the side of the sea it is, however, defended by Fort St. Elmo, and on the land side covered by a lofty hill. It has a royal palace, now occupied by the military governor and the civil superintendent, a town-hall, hospital, convent, and a fine par. church dating from the 15th century. On the N.E. side of the town is the harbour, which is shallow, and admits only barks of 40 or 50 tons burden. The principal occupation of the inhab. is fishing. Fontarabia has sustained numerous sieges: its fortifications were greatly augmented by the emperor Charles V.; under Philip IV. it received the rank and title of a city. The auxiliary British legion under General Evans had some severe fighting with the Carlist forces in the vicinity of this town, which they took in 1837.

FONTENAY, a town of France, dep. Vendée, cap. arrond., on the Vendée, at the point where it becomes navigable 42 m. S.E. Napoléon-Vendée. Pop. 7,971 in 1861. With the exception of some modern houses, the town is very ill-built; streets narrow, ill paved, and dirty. The church, with a spire 311 ft. in height, is the object most worthy of notice. The town was originally fortified, and had a castle belonging to the counts of Poitiers, some ruins of which may still be seen. Fontenay is the seat of a sub-prefecture, a court of original jurisdiction, and a communal college; it has linen and cotton cloth factories, tanneries and breweries and some trade in timber, charcoal, Bordeaux and other wines. A regular communication is kept up by steamers between Fontenay and La Rochelle.

FONTENOY, a village of Belgium, prov. Hainault, 4 m. S.E. Tournay. Here, on the 30th of April, 1745, a battle was fought between the allied English, Hanoverian, and Dutch forces, under the Duke of Cumberland, and the French, under Marshal Saxe, Louis XV. and the Dauphin being also with the army. The contest was obstinate and severe. At one time victory seemed to have declared in favour of the allies; and if the English had been properly supported by the Dutch, such would probably have been the case. In the end, however, the French were victorious. ' *Les Anglais*,' says Voltaire, ' *se rallièrent, mais ils cédèrent; ils quittèrent le champ de bataille sans tumulte, sans confusion, et furent toujours avec honneur.*' (Siècle de Louis XV. chap. 15.) The allies lost about 7,000 men killed and wounded, and 2,000 prisoners, on this occasion. The loss of the French amounted to nearly 6,000 men killed and wounded.

FORELANDS (NORTH AND SOUTH), two

headlands on the E. coast of the co. of Kent: the first, or N. Foreland, forms the N.E. angle of the co.; it projects into the sea in the form of a bastion, and consists of chalky cliffs nearly 200 ft. in height. A light house of the first class, having a fixed light, elevated 310 ft. above the level of the sea, was erected on this headland in 1690. This lighthouse is in lat. 51° 22' 45" N., long. 1° 27' W. The S. Foreland, about 16 m. S. from the latter, consists of chalky cliffs. Two lighthouses, with fixed lights, have been erected on this headland, to warn ships coming from the S. of their approach to the Goodwin Sands. The S. Foreland is made by act of parliament the S.E. extremity of the port of London.

FORFAR, or ANGUS, a marit. co. on the E. coast of Scotland, having E. the German Ocean, S. the Frith of Tay and the co. Perth, W. the latter, N. Aberdeen, and N.E. Kincardine. It is of a quadrangular shape, and comprises an area of 889 sq. m., or 568,750 acres. It is naturally divided into four districts, whereof the *first* and most extensive, called the ' Braes of Angus,' comprises all the S. slope of the Grampians, from the summit of the ridge till it loses itself in the valley of Strathmore. The mountains in this division are mostly rounded and tame, but in parts they exhibit bold, terrific precipices. The *second* division consists of that portion of the valley of Strathmore that lies in this co. between the foot of the Grampians and the Sidlaw hills (Low of Angus), and is for the most part a finely diversified, well cultivated country. The *third* division consists of a portion of the range called the Sidlaw hills, parallel to the Grampians, and attaining to a height of 1,200 or 1,400 ft. Some of them are conical, detached, and covered with heath, while others are wholly cultivated. ' Dunsinnan Hill' is found in this group. The *fourth* and last division consists of the rich, low-lying, level land between the Sidlaw hills and the sea and the Frith of Tay. Principal rivers N. and S. Esks and Isla. No where, perhaps, in Great Britain has agriculture and the appearance of the country been more rapidly improved than in this co. The progress made in this respect during the last sixty years has been quite extraordinary. At the beginning of the century the appearance of the country was bare and bleak, and the climate cold and damp, owing to the quantity of water on the land. Most of the houses were at that time of the rudest and meanest kind, built of unhewn stone, and covered with thatch; scarcely one of them warm, or covered with slates. Now the farms are all laid out and enclosed, draining is carried to great perfection, and farm-houses and others are neatly built and covered. Thriving woods and belts of plantations are rising up, and giving a rich and clothed appearance to the co. Along with all this it is gratifying to observe that the habits of the people are improving. There is a greater neatness and cleanliness in their dwellings, and a greater share of the comforts of life amongst them; and though last, not least, there is evidently an increasing desire of information, and, generally speaking, a higher and better tone of moral feeling. The vicious practice of holding land in run-rig (see ANGYLE), that formerly prevailed in all the hill districts of this co., is now comparatively rare; and improvements are beginning to be made even in the cottages among the Grampians. There are some great estates, but property is, notwithstanding, a good deal subdivided. Excepting limestone, minerals are of no importance. This co. has recently become the principal seat of the manufacture of coarse linens, which is carried on to a great extent at Dundee, Arbroath, Forfar, Montrose and other towns. Forfar contains 5 royal burghs, and 56

parishes, and returns 3 mems. to the H. of C.:
viz. 1 for the co., 1 for the town of Dundee, and 1
for Montrose and its contributory boroughs. Registered electors for the co., 2,108 in 1863. Pop.
204,425 in 1861, inhabiting 23,460 houses. The
old valued rent was 14,206l.; the new valuation
for 1864–5 amounted to 458,352l., exclusive of
railways.

FORFAR, a parl. and royal bor., and par. of Scotland, cap. of the above co., in the *How of Angus*,
or valley of Strathmore, 14 m. N. Dundee, on the
Scottish Midland Junction railway. Pop. 9,238
in 1861. The town consists principally of one
long street, and of a shorter one at right angles to
it. Forfar is a bor. of considerable antiquity, and
in the centre of a well-cultivated county, having
excellent communications on all sides. A great
proportion of its inhab. are engaged in weaving,
chiefly in connection with other towns. The chief
trade is the weaving of Osnaburgs and coarse
linens. The town has long been famous for the
manufacture of a particular kind of shoes called
'brogues,' adapted for the use of a Highland district. The streets are well built, and many new
houses are in progress; the tendency of the manufacturer who works at home being to convert his
earnings, as soon as possible, into a new feu, or the
property of a piece of land. There are valuable
quarries here, the products of which are all sent to
a great distance. The means of education are extensive and good; the town having an academy
for languages, a parish school, a mechanics' institute, Sunday schools, and a large infant school:
here also is a subscription news-room, a subscription library, and a mechanics' reading room. Forfar
is governed by a provost, 7 bailies, and 16 councillors. Corporation revenue 2,161l. in 1863–4.
Forfar unites with Montrose, Arbroath, Brechin,
and Bervie, in sending 1 m. to the H. of C. Registered parl. electors, 301 in 1865. Annual value
of real property, 16,955 in 1864–5.

FORLÌ, or FORLÌO, a sea-port town of Southern
Italy, on the W. shore of the island of Ischia, cap.
distr. of same name. Pop. 6,704 in 1861. Streets
very narrow, but the houses are solidly built, and
there are 3 good churches, all very much decorated.
It has a good harbour, and some trade with Naples,
Leghorn, and Genoa. In its vicinity there are hot
mineral springs, used as baths.

FORLÌ (an. *Forum Livii*), a town of Central
Italy, cap. of province of same name, in a fertile
plain between the Montone and Ronco, on the
Emilian Way, 84 m. SE. Bologna, and 15 m. SW.
Ravenna, on the railway from Bologna to Ancona.
Pop. 39,500 in 1861. The town is surrounded by
old walls; is generally well built; has 4 spacious
streets; a square, in which there are several fine
buildings; a cathedral; 9 other churches; and
numerous convents. Many of the private residences are built of marble, and the streets are
ornamented with arcades. The ceiling in the
council-chamber of the town-hall was painted by
Raphael. Forlì is the seat of a provincial governor, and a court of primary jurisdiction dependent
on a superior court at Bologna. It has manufactures of plain silk riband and silk twist, and of oil-
cloth, woollen fabrics, wax, nitre, and refined
sulphur. It also trades in corn, wines, oil, hemp,
and seaweed, which, as well as its manufactures, is
considerably facilitated by the railway, as well as
a canal from Acquaviva. There is a college, a
public library, and some learned societies. Forlì
was founded an. no 205 B.C. It was annexed to the
see of Rome by Pope Julius II. In 1797, the
French made it the cap. of the dep. of the Rubicon.
It was reunited to the Roman dom. in 1815, but
fell to the kingdom of Italy in 1860.

VOL. II.

FORLIMPOPOLI (an. *Forum Popilii*), a town
of Central Italy, prov. Forlì, 5 m. SE. Forlì, on the
railway from Forlì to Rimini. Pop. 4,765 in 1861.
The town has an ancient castle, a cathedral, two
parish churches, and several convents. This and
the other forums in different parts of Italy are
supposed to have been all convent, or swine
towns; but the proximity of those on the Emilian
Way, particularly of Forlì and Forlimpopoli, seems
to contradict that opinion.

FORMOSA (Chin. *Tae-wan*, or 'Terrace Bay.')
an island in the Chinese Sea, belonging partly to
China; between lat. 22° and 25° 30' N., and long.
120° 30' and 122° E.; about 80 m. from the Chinese coast, from which it is separated by the channel of Fo-kien, and 170 m. N. Luzon, the chief of
the Philippine Islands. Length, N. to S., about
250 m.; breadth, in its centre, about 80 m. The
area is estimated at 14,000 sq. m., and the pop. is
probably between 2,000,000 and 3,000,000.

A chain of mountains runs through the island
in its entire length, forming, in general, the barrier between the Chinese on the W. and the independent natives of the unexplored country on the
E. side. On many of its summits snow remains
during most part of the summer, and Humboldt
has supposed that a portion of it reaches an absolute elevation of upwards of 12,000 ft. It exhibits distinct evidence of former volcanic action
in some extinct craters; in other parts fissure,
mephitic gases, &c., burnt out of the earth; and
sulphur, naphtha, and other volcanic products are
abundant. Some parts of the coast present bold
headlands, but all the W. shore is flat, and surrounded with rocks and quicksands. Its harbours,
which were formerly very good, have become
nearly useless, except to junks of very small tonnage, from the rapid increase of the land on the
sea; so that, at present, Formosa has but one good
port, that of Ke-lung, at its N. extremity.

'That portion of Formosa which is possessed by
the Chinese well deserves its name; the air is
wholesome, and the soil very fruitful. The numerous rivulets from the mountains fertilise the extensive plains which spread below; but throughout
the island the water is unwholesome, and, to un-
acclimated strangers, it is often very injurious.
All the large plain of the S. resembles a vast
well-cultivated garden. Almost all grains and
fruits may be produced on one part of the island
or another; but rice, sugar, camphor, tobacco, &c.,
are the chief productions. Formosa has long been
familiarly known as the granary of the Chinese
maritime provinces. If wars intervene, or violent
storms prevent the shipment of rice to the coast,
a scarcity immediately ensues, and extensive distress, with another sure result—multiplied piracies
by the destitute Chinese. The quantity of rice
exported from Formosa to Fuh-keen and Che-
keang is very considerable, and employs more
than 300 junks. Of sugar there annually arrive
at the single port of Teen-tsin (in China) upwards
of 70 laden junks. The exportation of camphor
is likewise by no means small. Much of the camphor in the Canton market is supplied from Formosa.' (Chinese Repository, ii. 419, 420.) Besides
the foregoing products, wheat, maize, millet,
kitchen vegetables of many kinds, truffles, &c.;
rubarum, a kind of arum, the root of which is a
chief article of food in the interior; oranges, bananas, cocoa and areca nuts, peaches, figs, melons,
and numerous other European and Asiatic fruits
are cultivated. Chestnut woods are plentiful; and
in the N. especially, a good deal of timber for
ship-building is obtained. Pepper, aloes, coffee,
a kind of green tea, but different from the Chinese,
cotton, hemp, and silk, are other important articles

of culture. The ox and buffalo are used for tillage and draught; horses, asses, sheep, goats, and hogs are abundant. The leopard, tiger, wolf, &c. inhabit the island, but do not infest its cultivated portions; pheasants, hares, and other kinds of game are very numerous. Gold is supposed to be found in the E. part of Formosa, as it is seen in the hands of the inhabitants; but the chief minerals are salt and sulphur, of which latter a good deal has been sent to China since 1819, for the manufacture of gunpowder.

The Chinese colonists of the island are mostly from the opposite shore, of Fo-kien, and have emigrated principally from poverty. They are a laborious and industrious race, well disposed towards foreigners, but very turbulent in respect to the home authorities, who maintain only a very precarious sway over them,—the Formosans having frequently risen in open rebellion against their mother country. The greater part of them are cultivators of the soil; but many of the Amoy men (from which district a great number of the emigrants have come) are merchants, fishermen, and sailors. The trade with China is very extensive; the chief exports to that country have been mentioned; the principal imports thence are tea, silk, and woollen, and other kinds of manufactured goods. The trade is mostly in the hands of Fo-kien merchants, who have also advanced the chief part of the capital necessary for the cultivation of the soil. As many as 100 junks a month are estimated to leave Fo-kien for the W. coast of Formosa; where, however, they are obliged to lie at a great distance from the shore, while carts with wheels destitute of spokes, drawn by buffaloes, are used to carry the cargoes to them through the water. There are no junks strictly belonging to the island; all the shipping is the property of the Amoy merchants or of foreigners, chiefly English. The largest and export trade is not very large; the principal article of import is opium, of which 222 chests arrived in 1862, and 512 chests in 1861. (Report of Vice-Consul Swinhoe, dated February 1, 1864.)

The native inhabitants of the E. of Formosa bear no resemblance to the Chinese; but they have apparently an alliance with the Malay or Polynesian tribes. They are of a slender shape, olive complexion, wear long hair, are clad with a piece of cloth from the waist to the knees, blacken the teeth, and wear ear-rings and collars. In the S., those who are not civilised live in cottages of bamboo and straw, raised on a kind of terrace 3 or 4 ft. high, built like an inverted funnel; and from 15 to 40 ft. in diameter. In these they have neither chair, table, bed, nor any moveable. They tattoo their skin. In the N. they clothe themselves with deer-skins. . . . They have no books, or written language; neither have they any king or common head, but petty chiefs and councils of elders, and distinguished men, much like the N. American Indians. It does not appear whether they have any separate priesthood, but it is probable that there is none beyond the conjurors and enchanters of all savage tribes, nor any ancient and fixed ceremonies of divine worship, or system of superstition. They are represented by the Chinese as free from theft and deception among themselves, and just towards each other, but excessively revengeful when outraged.' (Chinese Repository, ii. 419.) The Chinese territory in Formosa having, for a lengthened period, been gradually extending, the really independent tribes have receded towards the E. coast; some of the others have become partially civilised, settled in villages, and intermixed with the border Chinese.

Formosa, together with the Pang-hoo islands, comprises a fou, or department, under the prov. Fo-kien, and immediately subject to its governor. It is divided into five dems of districts. The cap., Tae-wan, is ranking among Chinese cities of the first class in the variety and richness of its merchandise, and in pop. It stands on the W. coast, in about lat. 23° N. and long. 120° 37' E., surrounded by a wall and ditch. Its principal streets are from 80 to 40 ft. broad, and for many months of the year are covered with awnings to keep off the sun. On a small island opposite the city, the Dutch, in 1634, built Fort Zealand, which commanded the harbour, the entrance to which is now choked up. The Chinese garrison in Tae-wan amounts to about 10,000 men; the total armed force usually stationed in the island may be estimated at about double that number, all infantry. These troops, however, were incapable of suppressing the insurrection which spread over the whole island in 1861–64, and being an offshoot of the great Taeping rebellion, led to much rapine and bloodshed.

The Chinese appear not to have been acquainted with Formosa till about 1430, after which its coasts became the resort successively of several Chinese pirates. The Japanese had planted colonies in the N., and at one period the greater part of the island belonged to them; but the Dutch, having been allowed to settle on the W. coast, gradually dislodged all their opponents, including the Spanish and Portuguese (both of whom tried to gain a footing), and became sole masters of the island about 1632. After the conquest of China by the Tartars, in 1611, a Chinese chief, with an army of Chinese refugees, determined to conquer Formosa, and finally expelled the Dutch from it in 1662. In 1683, however, the new dynasty was overthrown by the continental Chinese, aided by the Dutch; and the authority of China has been ever since maintained over the island, though assailed by repeated insurrections. A British vice-consul is stationed at Formosa since 1860. (Consular Reports, 1864; Ritter, Asien Erdkunde, iii. &c. &c.; Klaproth; La Perouse; Gutzlaff.)

FORRES, a royal and parl. bor., town, and par. of Scotland, co. Moray, on the E. side of the burn of Forres, about 2½ m. E. from the Findhorn, and 24 m. N. from the loch or inlet of the sea which receives the Findhorn, and 11 m. W. Elgin. Pop. 3,548 in 1861. The town consists of one principal street, with the town-house in its centre, through which the great road to Inverness passes, with several smaller streets branching off from it. It possesses an academy, called Anderson's Institution, which, together with the salubrious climate and cheapness of living, induce many families to reside here. Findhorn is the sea-port of the bor., and of the surrounding district. Besides the academy, there is a good parish school, an elementary school, and a ladies' seminary. On a hill, at the W. end of the town, are the remains of the ancient castle of Forres. About ½ m. N.E. from the town, is a remarkable granite obelisk, called Sueno's Pillar, consisting of a single stone 23 ft. above ground, 3 ft. 10 in. broad, and 1 ft. 3 in. thick. One side is rudely sculptured. It appears to have been erected by the Scotch in memory of some victory over the Danes. A pillar was erected in memory of Lord Nelson, by public subscription, on a hill to the E. of the town. A bridge of four arches over the Findhorn, near this town, was swept away by the great flood in that river in Aug. 1829. Forres unites with Inverness, Fortrose, and Nairn in sending one member to the H. of C. Registered electors in Forres, 176 in 1863. The bor. is governed by a provost, two bailies, and fourteen councillors. Corporation revenue, 1,208*l*. in 1863-4.

FORT AUGUSTUS, a fortress of Scotland, co. Inverness, the centre one of the three forts erected along the great glen of Scotland, now the line of the Caledonian Canal, beautifully situated at the W. extremity of Loch Ness, 31 m. SW. Inverness, and 28½ m. NE. Fort William. Fort Augustus was built in 1730, and was so named in honour of the Prince of Wales, father of George III. It is a regular fortification, with four bastions, and barracks capable of containing 400 soldiers, with proper lodgings for the governor and officers. It was taken by the Highlanders in 1746, but abandoned after having been partially demolished. Here the Duke of Cumberland established his camp after the battle of Culloden; and the ruins of a turf-house which he occupied are still to be seen. The fort was ordered to be demolished in 1818; and is now occupied by three or four veteran artillerymen. In its immediate neighbourhood is a village, originally called Kilcumin, from its having been the burial-place of the ancient and powerful family of Cumin; but now it bears the same name as the fort. The village is meanly and irregularly built, and forms a contrast to the beautiful situation in which it is placed.

FORT GEORGE, a fortress of Scotland, co. Inverness, 11 m. NE. Inverness, on a low sandy peninsula jutting into the Moray Frith, and forming the most E. of the three forts erected along the great glen of Scotland. It is esteemed the most complete fortification in Britain, and not being commanded by any part of the adjacent country, may bid defiance to assault. The work was erected so as to command the entrance to the Moray Frith. The ramparts on three sides rise almost out of the sea, the waters of which may at pleasure be introduced into the fosse, which skirts the fourth side. It has four bastions, mounted with eighty cannon; a bomb-proof magazine, and accommodation for 3,000 men. The buildings are remarkably neat, and disposed in handsome square, with a fine walk round the ramparts. The fort occupies no less than 15 acres. It was begun to be built in 1747, under the direction of General Skinner, and cost upwards of 160,000*l*. It was partially used as a state prison during the late war. Though Fort Augustus and Fort William, the other forts on the line of the Caledonian canal, have been dismantled since the peace, Fort George is kept in good order, and has a governor and a garrison.

FORT WILLIAM, a fortress of Scotland, co. Inverness, at the E. extremity of Loch Linnhe, and the W. end of the Caledonian Canal. This fort, Fort Augustus in the centre, and Fort George at the E. extremity of the great glen in the line of this canal, were built at different times for supporting the authority of the general government, and curbing the turbulence of the Highland clans. It was originally built of turf, by General Monk, in the time of the Commonwealth, being so large as to contain a garrison of 2,000 men. It was called the Garrison of Inverlochy, owing to its situation at the mouth of the Lochy, a stream which falls into Loch Linnhe. In the reign of William and Mary it was rebuilt of stone, but on so small a scale as to afford accommodation to only 800 men. It then received the name, which it has since retained, of Fort William. It is of a triangular form, with two bastions. In the rebellion of 1715, the Highlanders made an unsuccessful attack on it; and in 1746 it stood a siege of five weeks by the adherents of Prince Charles Stuart, who at the end of that time were forced to retreat. The fort was ordered to be dismantled in 1818; and is now tenanted by about a dozen invalids, in order to

keep it from becoming a complete ruin. Within 1½ m. W. of the fort, and on the edge of Loch Linnhe, is the town of Fort William, originally called Maryburgh, and now more generally Gordonsburgh. Pop. 1101 in 1861, of whom 618 females, and but 486 males. The inhabitants are chiefly engaged in the herring and other fishery. Ben Nevis, the highest mountain in Britain, being 4,370 ft. high, is in the immediate vicinity of the fort and the town, both of them being placed at its base.

FORTH, a river of Scotland, which originates in several mountain streams that have their sources on the E. side of Ben Lomond, in Stirlingshire. Its course is E., with many sinuosities, by Aberfoyle, Stirling, and Alloa, till it unites with the arm of the sea, called the Frith of Forth, at Alloa. The Teith, its most important tributary, has its sources a little more to the N., and pursuing a SE. course past Callender and Doune, joins the Forth a little above Stirling, bringing to it a volume of water but little inferior to its own. Its other most important affluents are the Allan, flowing S. from Perthshire; and the Devon, flowing W. from Kinross-shire. During the latter part of its course, the Forth flows with many windings through a low, level, and very rich country; in fact, though the distance from Stirling to Alloa by the road be only about 7 m., it is no fewer than 23 by water. Steam-boats ascend to Stirling, and ships of 300 tons burden come up to Alloa, which may be regarded as its port.

FORTROSE, a sea-port, royal and parl. bor. of Scotland, co. Ross, on a gentle eminence on the N. bank of the Moray Frith, nearly opposite Fort George, from which it is 2½ m. distant, 8 m. NE. Inverness. Pop. 928 in 1861. There is a regular ferry between Fort George and this bor. Fortrose was formerly known by the name of Chanonry, so called from its being the chanonry of Ross, where the bishop resided, and the members of the chapter. About a mile to the W. stands the small town of Rosemarkie; and the two places were united by a charter granted by James II. in 1444, under the common name of Fortrose, now softened into Fortrose, which charter was ratified by James VI. in 1592. Rosemarkie is a meaner place than Fortrose, but is reckoned the parochial capital, inasmuch as it is the site of the parish church. A handsome episcopal chapel, however, has been erected at Fortrose. The academy there is the first seminary of the kind established in the N. of Scotland, and is supported by donations and subscriptions. The late Sir James Mackintosh received his elementary education here. There are two other schools at Fortrose, and two also at Rosemarkie. There are no manufactures in the place. The salmon and white sea fishery gives considerable employment. No mail or stage coach passes through the parish; but the steam vessels plying in the frith call at Fortrose; and it is by them that salmon and other articles are conveyed thence to Aberdeen, Leith, and London.

The bishop of Ross resided at Chanonry, and was termed 'Episcopus Rosmarkensis.' This episcopal see was founded by David I. in the 12th century. Only a small part of the cathedral now remains. Some of the bishops of Ross were men of literary eminence, particularly John Maxwell, author of Sacro-Sancta Regum Majestas, who died in 1646, archbishop of Tuam in Ireland. Fortrose unites with Inverness, Forres, and Nairn, in sending a mem. to the H. of C.; and had 82 registered electors in 1864.

FOSSOMBRONE (an. *Forum Sempronii*), a town of Central Italy, prov. Urbino, on the Metauro, in a fertile district, 7 m. ESE. Urbino. Pop. 7,485 in 1861. The inhabitants are chiefly employed in the manufacture of silk, said to be the finest in Italy. The town has an old fortress; a fine cathedral, containing many good paintings and interesting inscriptions; three other churches, six convents, a handsome one-arched bridge, and the ruins of an ancient theatre. Near the town was fought, anno 191 B.C., the great battle between the Carthaginians under Asdrubal, the brother of Hannibal, and the Romans, in which the former were totally defeated, and their general killed. Tradition has preserved the memory of the event in the name of a hill in the vicinity, called *Monte de Ambrabale*. This victory determined the fate of the long-contested struggle between the Romans and Carthaginians in favour of the former. Fossombrone was destroyed by the Goths, and again by the Lombards, but rebuilt by the Malatesti. That family sold it in 1440 to the Duke of Urbino, with whose territories it was afterwards transferred to the see of Rome.

FOUAH, a town in the Delta of Egypt, on the E. bank of the Rosetta branch of the Nile, prov. Gharbich. Though still a considerable village, it lost its importance in the sixteenth century, when the trade of which it was the seat was transferred to Rosetta. Fouah is most agreeably situated amidst a great number of flourishing villages and productive fields. The river flows past it through one of the widest and most picturesque portions of its channels and banks.

FOUGERES, a town of France, dép. Ille-et-Vilaine, cap. arrond., on a hill near the Nançon, 27 m. NE. Rennes. Pop. 9,844 in 1861. It is well built, has a fine promenade, and is altogether a very agreeable town. A chalybeate spring attracts to it numerous visitors. Fougères was a strong town in the fifteenth century, and was considered one of the keys of Brittany till that prov. was united to the French crown. During the last century it suffered from four destructive fires, on which account few of its ancient buildings exist, excepting the ruins of a Gothic castle, which form a very picturesque object. There are large manufactures of silk cloth and hemp fabrics, known in trade as St. George cloth, flannels of excellent quality, hats, leather, and dye-houses. It is the seat of a sub-prefecture, a court of primary jurisdiction, and a communal college.

FRAMLINGHAM, a town and par. of England, co. Suffolk, hund. Loes, on an eminence, near one of the sources of the Alde, 14 m. NE. Ipswich, and 90½ m. NE. London, by Great Eastern railway. Pop. 2,252 in 1861. Here is an old church, with a tower 96 ft. high; a free school, and several sets of almshouses. Here, also, are the ruins of a magnificent castle, which was a place of importance in the Saxon times, and to which the Princess Mary repaired during the attempt made by the partisans of Lady Jane Grey to place the latter on the throne.

FRANCAVILLA, a town of Southern Italy, prov. Otranto, cap. distr., on a hill, in a fertile but unhealthy territory, 23½ m. WSW. Brindisi, and 17 m. ENE. Taranto. Pop. 15,945 in 1861. The town is large and regularly built: the streets wide and straight; the houses showy, though in a heavy style of architecture. Since the year 1734, when a considerable part of the town was thrown down by an earthquake, the dwellings have not been raised more than one story above the ground floor. The avenues to the gates are well planted, and afford a pleasant chaste. The college is a large edifice, with many handsome halls and galleries. The principal par. church is gay and well lighted; but so overcrowded, fretworked, and flowery, that the whole decoration is a mere chaos. There are two

hospitals, a charitable asylum, and several convents; with manufactures of woollen stuffs, cotton stockings, earthenware, and a hint of snuff similar to that made in Spain. Francavilla was founded in the fourteenth century, and owes its name to an exemption from taxation for ten years, granted to all persons who settled in it.

FRANCE (EMPIRE OF), one of the richest, most important, and powerful of the states of Europe, in the W. part of which it is advantageously situated; between lat. 42° 20′ and 51° 5′ N., and long. 4° 50′ W. and 8° 20′ E.; having NW. and N. the English Channel (*La Manche*), the Straits of Dover (*Pas de Calais*), and the North Sea; NE. Belgium, Dutch Luxemburg, and the Rhenish prova. of Prussia and Bavaria; E. the territories of Baden, Switzerland, and Italy; S. the Mediterranean and Spain; and W. the Bay of Biscay and the Atlantic. Except on its NE. frontier, its actual arr. identical with its natural boundaries; being on the E. the Rhine from the influx of the Lauter to Basle, the Jura mountains, and the Alps to the Mediterranean; the latter and the Pyrenees on the S.; and NW. and W. the English Channel and the ocean. The shape of France is somewhat hexagonal. Its greatest length NW. to SE. (from the extremity of the dép. Finistère to Nice, on the Mediterranean) is about 664 m.; its greatest breadth (a line crossing the former nearly at right angles) is about 620 m. Length, N. to S., Dunkirk to Perpignan, nearly 600 m.; greatest breadth E. to W. (a line passing from near Lauterburg to Brest, through Paris) about the same; least breadth E. to W. about la centre 335 m. Inclusive of Corsica and the three departments ceded to France since the taking of the last census, the total area is estimated, in the official tables published by the French government, at 54,823,897 hectares, or 211,853 English sq. m. The pop. which in 1801 was 27,349,000, had increased in 1821 to 30,461,875; in 1831 to 32,569,223; in 1836 to 33,540,910; in 1846 to 35,400,486; in 1851 to 35,783,059; and in 1861 to 36,713,168. The subsequent addition of the provinces ceded by Italy brought the population to 37,382,225. (Block, Statistique de la France; and official reports in the Moniteur Universel.)

Physical Geography, Position, Frontiers, Coasts, and Islands.—France is indebted not only to her large population, and the active spirit of her people, but in a great measure to her admirable geographical position, for her commanding influence in European affairs. Unlike any of the other states of Central Europe, she has the command of three seas, including those which wash both the N. and the S. shores of that continent. The NW. coast presents the two considerable peninsulas of Brittany and Cotentin, the bay of St. Malo between them, the estuaries of the Seine, and the harbours of Morlaix, Cherbourg, Havre, Boulogne, Calais, and Dunkirk. From Dunkirk to Calais the shore is bordered by sandy downs. From the latter point to the mouth of the Seine, the coast is chiefly characterised by chalk and marl cliffs; further W. granite cliffs alternate with low shelving shores. There is seldom deep water near the shore on this coast: the bay of Cancale near Avranches, for instance, is in a great measure left dry at ebb-tide, and passengers at such times go from the mainland to Mont St. Michel, across the sands, in carriages. The W. part of this coast is beset with rocks; these are especially numerous between the mouths of the Seine and the Vire. Good harbours are few, and navigation is rendered dangerous by violent tides, the force of which is attested by numerous salt marshes along the shore, produced by irruptions of the sea. The W. coast,

FRANCE.

English Miles

formed in part by the peninsula of Brittany, is at first elevated, bold and rocky, but as it proceeds N. it gradually declines; and from the mouth of the Gironde to the foot of the Pyrenees, it presents an unbroken line of sandy downs interspersed with marshes. Besides the Gironde, the Loire disembogues on this coast which is farther indented by numerous bays. The S. coast, except in its E. part, is generally low, sandy, and bordered, where it surrounds the Gulf of Lyons, by numerous lagoons; and its harbours are in general neither well sheltered nor easy of access, though this is by no means the case with Toulon and one or two more. Exclusive of those at the mouth of the Rhone, the islands round France, and belonging to her, are of comparatively little importance: they lie mostly along the W. coast; Oleron, Ré, Yeu, Noirmoutiers, Belle-Ile, and Ouessant (Ushant) being the chief. Those in the Mediterranean are the isles of Hières, Bastuaran, Porquigue, &c., near Marseilles; and the only ones in the Channel are Jershat, and a few rocky groups in the bay of St. Malo, of which that of Chausey is the principal. Guernsey, Jersey, Alderney, &c. belong to England, and are the only remains of the extensive dominions the English sovereigns once possessed in France. (Hugo, France Pittoresque; Dict. Géogr.; Aperçu Statistique.)

Mountains.—According to Bracalière (Orographie de l'Europe), three belong wholly to the Alpine and Pyrenean systems, the line of separation between which is the valley traversed by the canal of Languedoc. The ramifications of the Alpine system in France are therefore far more extensive than those of the Pyrenean; they comprise the mountain ranges throughout the country, except in the SW. The principal mountain chain, or great watershed of France, intersects the country under the names of the Faucilles, plateau of Langres, Côte d'Or, Cevennes, &c., in a general direction NE. to SW.; but running much nearer to the SE. than the NW. extremity of the kingdom, which is thereby divided into two very unequal parts. On the N. and W. sides of this chain several considerable branches are given off, as the Vosges, Moselle, and Argonne ranges, the plateau d'Orléans, and Morvan mountains, which stretch to the extremities of Normandy and Brittany, &c. These ranges separate the principal river basins, those of the Rhine, Moselle, Meuse, Seine and Loire, from each other; the basin of the Rhone is on the SE. side of the Cevennes, enclosed between them and the Alps. Connected by ramifications with the Cevennes, there is a group of mountains of volcanic origin scattered over Puy de Dôme, Cantal, and some adjacent deps. in the centre and S. of France. This group, which Balbi and other geographers regard as a separate system, under the title of *Gallo-Pyrenique*, separates the basin of the Loire from that of the Garonne. The highest points of this group have a somewhat greater elevation than those of the Faucilles and Cevennes chain. The Pic-de-Sancy (M. Dore) is estimated to be 6,218 ft. in height, and the Plomb-de-Cantal, 6,095 ft.; while Mezenc, the loftiest of the Cevennes, is only 5,918 ft. high; Le Ratulet (Jura), 5,638 ft.; and the Ballon de Suls (Vosges), 4,698 ft. The Pyrenees send off numerous lateral branches through the SE. dép.; their loftiest summit within the French territory is M. Perin, 10,884 ft. in height. But the culminating point belongs to the Alps, and is the 'monarch of mountains,' Mont Blanc; the next greatest in height is M. Olen, 4,214 mètres, or 13,825 ft. high; next to which is the Pic-des-Ecrins, 13,468 ft. in elevation.

Rivers.—Leaving out of view the Rhine, which

can scarcely be called a French river, since it merely runs for about 100 m. along a portion of its E. frontier, France possesses no river to rank with the Wolga or the Danube. The principal are the Loire, Rhone, Garonne, Seine, Meuse, and Moselle. Except the Rhone, which has for the most part a southerly course, all the above-named run in a N. or W. direction. The Loire, which is the largest, and traverses the centre of the kingdom, rises in the mountains of the Vivarais (Ardèche), near Mezenc. It runs generally NW. as far as Orleans, and thence mostly WSW., with a somewhat tortuous course to its mouth in the Atlantic. The length of its entire course is about 620 m., of which about 510 m. are navigable. It receives from the N. the Arroux, the Maine (formed by the Mayenne and Sarthe), and the Endre; and from the S. the Allier, Cher, Indre, Vienne, Sèvre-Nantaise, &c. Nevers, Orleans, Blois, Tours, Saumur, and Nantes are situated upon its banks. The Rhone rises in Switzerland, beyond the Simplon, and after traversing the Lake of Geneva, and forcing for itself a passage through the Alps, not far from Chambéry, enters France S. of the Jura range, forming the entire S. and almost all the W. boundary of the dep. of Ain. From Lyons, where it receives the Saône, the direction of the Rhone is nearly due S. to Arles, where its delta commences; and it falls into the Mediterranean by a double set of mouths, after a course of 430 m. within the French dom., more than 310 of which are navigable. Its principal affluent is the Saône, which runs through Franche-Comté and Burgundy, with an entire course of about 213 m., of which about 165 are navigable. Besides the Saône, the Rhone receives from the N. the Ain; it is joined from the E. by the Isère, Drôme, and Durance, famous for its rapidity; and from the W. it receives the Erieux, Ardèche, Gardon, &c. Lyons, Valence, Montélimart, Avignon, Tarascon, and Arles are the chief cities and towns on the Rhone; upon the Saône (which river is augmented by the Doubs), Gray, Chalon-sur-Saône, and Mâcon are situated. The Garonne rises in the Spanish Pyrenees, near M. Maladetta, and runs at first NE. as far as Toulouse, but thence onward its course is generally NW. to its mouth (or rather the mouth of its estuary, which bears the name of the Gironde,) in the Bay of Biscay, about 55 m. NNW. Bordeaux, and 120 m. SSE. the mouth of the Loire. The entire length of its course, including the Gironde, is estimated at about 350 m., nearly 291 of which are navigable. It receives some considerable tributaries; as the Tarn, which is navigable for a distance of 90 m.; the Lot, navigable for 190 m.; and the Dordogne, navigable for 170 m. from the E., and from the S. in the earlier part of its course; the Save, Gimone, Gers, Baïse, &c. Toulouse, Agen, and Bordeaux are situated on the Garonne. The Seine rises in Burgundy, about 18 m. NW. Dijon; its general course is NW., but it is exceedingly tortuous; and though in a direct line its course is no more than about 250 m., from its mouth in the British Channel, the windings of the river make its total length so much as 500 m. It enters the Channel by a wide and capacious mouth, on the N. side of which is the town of Havre; its estuary, and the lower part of its course, is subject to the phenomenon of the bore, which sometimes occasions considerable damage. (See Avranches, Holy-way Firth, &c.) The principal affluent of the Seine is the Marne; besides which, it receives from the E. the Aube and Oise; and from the N. and W. the Yonne, Juine, Eure, Risle, &c. Paris, Chatillon, Troyes, Melun, St. Denis, St. Germain, Andely, Elbeuf, Rouen, Honfleur, and Havre are situated upon its banks. The Marne, which runs chiefly through Champagne, has a navigable course of 215 m.: it receives the Blaise, Ornain, and Ourcq; Chalons-sur-Marne, Epernay, Chateau-Thierry, and Meaux are seated on it. Both the Meuse and the Moselle run N. to join the Rhine beyond the French line; the former has a navigable course of 163 m., and the latter one of about 72 m. within France. These rivers, however, as well as those of the Escaut (Scheldt), Lys, Sambre, and others, belong more properly to Belgium than to France. The Charente, the basin of which lies between those of the Loire and the Dordogne, has a navigable length of about 120 m.; and the Adour, which traverses the deps. of the Pyrenees and Landes, has a great number of tributaries, including the Midouze, l'Arros, Oleron, &c., and a course generally W., which is navigable for 77 m. The other rivers worthy of any notice, as the Somme, Orne, Aisne, Meurthe, Rance, Vilaine, Ariège, Hérault, Var, &c., are referred to under the deps. to which they give their name, or in which their course is chiefly situated. (Hugo; Aperçu Statistique.)

Lakes and Marshes.—Of the former there are remarkably few, and those quite insignificant in point of size. The largest is that of Grand Lieu, in the dep. Loire Inférieure; but it is only two acres. There are a few small lakes amongst the Jura ranges, and others occupy extinct craters in the volcanic district. In Ain and Loire-et-Cher marshes are numerous. The extensive lagunes on the N. and SW. coasts and elsewhere have been already alluded to; they are too shallow to be used otherwise than for fishing and salt-works.

Geology, Soil, and Minerals.—Geologically, the whole of France may be considered as one extensive basin, the circumference and centre of which consist of primitive formations, the intermediate space being filled with those of a secondary and tertiary kind. Primitive rocks abound most in the Alps, the Pyrenees, the peninsula of Brittany, and the mountains of the so-called Gallo-Frangise system in the centre of France. They are, however, met with in a part of Maine and Normandy, in Vendée, in Ardennes, where they are conterminous with a chain of primitive rocks which extend into NW. Germany, in the Vosges, in Dauphiny (Isère), and on the S. coast E. of Marseilles. The most widely diffused primary rocks are granite, gneiss, micaceous and argillaceous schists, and primitive limestone. In Vendée, to the foregoing may be added a great number of others, including porphyry, diorite, eclogite, and serpentines; and in the Dauphiny Alps and the Pyrenees the rocks are said to present a still greater diversity. In the latter mountains calcareous rocks are very abundant; and some of a transition kind contain a great number of organic remains, even at an elevation of 1,600 toises, or 10,230 ft. (Dict. Géogr.) Argillaceous schist, also containing numerous organic remains, is prevalent throughout a part of Brittany; granitic rocks predominate at the extremity of that peninsula. Porphyry of various kinds, some of which exhibit great beauty, is the prevailing rock in the Vosges mountains. In the central group of Limousin and Auvergne, gneiss, granites, and micaceous schists are abundant, but differ greatly in their characters from those of the surrounding mountain chains. The Puy de Dôme and some other adjacent mountains have a base of trachite, and in the Vivarais (Ardèche), especially, groups of gigantic basaltic columns are frequently met with in many places alternating with calcareous strata containing fresh water shells. These rocks, together with the traces of extinct craters, the

existence of lava streams, and other volcanic products, clearly point to a time of volcanic activity in this region, which has probably had place at no very remote period in the history of our planet. It may here be mentioned that traces of volcanic action have also been met with on the banks of the Rhine, in the Vosges, and in the dep. Var.

The interval between the primitive formations of the centre and circumference of France is almost entirely occupied by secondary formations. These are nearly everywhere calcareous or marly, generally compact, and often contain a vast number of shells, madrepores, and other organic remains. They compose many long hill-ranges, of no great height, but frequently steep and bare, or covered only by a thin vegetable soil. All Lorraine, and a great part of Franche-Comté and Burgundy, consist of these formations. It is on this kind of land that the growths yielding the finest Burgundy wines are raised in the Côte d'Or. The secondary formations extend through Dauphiny, and on the left bank of the Rhone as far as the Mediterranean, through Languedoc with the Cevennes quite to the Pyrenees; and surrounding the Paris basin, they reach the sea both on the N. and W. coast.

The tertiary deposits of France are highly interesting; they are mostly calcareous, enclosing great quantities of shells and the remains of fossil mammalia of large size. The most remarkable of the tertiary formations is what is called the 'Paris Basin,' which occupies a somewhat circular area nearly bounded by a line passing through Blois, Orleans, Montargis, Provins, Epernay, Laon, Beauvais, Pontoise, and Chartres. A still larger tertiary district is found at the foot of the Pyrenees, including almost all the valleys of the Adour and Garonne, the deps. Landes and Gironde. There are others in the valleys of the Loire, Rhone and Allier. The most extensive alluvial district is that around the mouth of the Rhone.

The soil of France is, speaking generally, very superior. No doubt she has large tracts of mountainous, heathy, and unproductive land; but her productive soil bears, notwithstanding, a larger proportion to the entire extent of the country than in most other European states. Exclusive of the recently annexed departments of Savoy and Nice, from which as yet no returns have been received, the soil of France is divided as follows:—

Under cultivation:	Per cent.
Grain Crops	28·90
Other	4·08
Artificial Meadows	6·00
Fallow	10·90
Natural Meadows	9·50
Vineyards	4·10
Chestnuts, Olives, Mulberry, &c.	0·20
Pasture and Waste Lands	18·50
Forest, Water, Roads, Houses, and Uncultivated	23·60
	100·00

The greatest extent of mountainous surface is found in the deps. of the Alps and Pyrenees, and those of Ariège, Côte d'Or, Drôme, Doubs, Haute Loire, and Haute Marne; heath land prevails most in Basses Alpes, Landes, Gironde, Finisterre, and Hérault; calcareous chiefly in Oise, Basses Alpes, Dordogne, Marne and Vienne; sandy soils in Cher, Haute Loire, Loiret, and Puy de Dôme; and rich lands in Gers, Aisne, Eure-et-Loire, Eure, Marne, Nord, Tarn, and Yonne. France has considerable mineral wealth. The metal most abundant is iron: in 1857 it was obtained in 64 of the 86 deps. Those in which it is most plentifully produced are Haute Marne, Haute Saône, Nièvre,

Côte d'Or, Dordogne, Orne, Meuse, Moselle, Ardennes, Isère, Cher, Aude, Pyrénées Orientales, Ariège, and Haute Vienne. Two gold mines were formerly wrought, one in the dép. Hte Rhin and the other in Isère, but both have long been abandoned. There are also two silver mines, one in each of the above deps., but only that of Allemonte (Isère) is at present wrought. Silver is, however, frequently found in the lead mines, which are chiefly abundant in Finisterre, and the Ehrenlah, Alpine, and some of the S. deps. Copper, mercury, zinc, tin, antimony, and manganese, both in large quantities, arsenic, bismuth, cobalt, chrome, &c., are met with; and amongst the rarer metals molybdenum and tungsten, titanium in Haute Vienne, and uranium near Autun. Coal is very widely diffused. The principal coal-field is in the dep. du Nord, where it forms part of a coal district 50 leagues in length by 2 broad, extending into Rhenish Prussia. Others exist in the deps. on the Upper Loire, in Aveyron, &c.; coal mines are particularly numerous around St. Etienne. The salt beds, discovered about 20 years since in Lorraine, are supposed to extend beneath a surface of 30 square leagues, and will fully supply France for ages. Turf in the N., asphaltum in the E. and elsewhere, naphtha and sulphur in the S., vitriol, alum, nitre, plaster of Paris, porcelain and other clays, graphite, asbestos, jet, and some gems, lithographic, mill, and building stone, excellent marble, slate, granite, &c., are amongst the valuable mineral products. Mining industry will be treated of hereafter.

There are no fewer than 700 mineral springs of a medicinal character, though only about 60 of these are frequented by visitors. The principal are the warm sulphureous springs of Barèges, Cauterets, Bagnères-de-Bigorre, and de Luchon, in the Pyrenees; the saline springs of Aix, the chalybeates of Bourbon l'Archambault (Vosges), and Plombières, and the cold springs of English.

The climate of France is not excelled by that of any other part of Europe. The air is generally pure, and the winters mild; though the differences of latitude, elevation, soil, and exposure occasion, in this respect, very material differences. Generally, France may be divided into 4 regions. The 1st, or most N.—the region of the olive—is bounded N. and W. by a line passing diagonally from Bagnères-de-Luchon in the Pyrenees to Die in Drôme. The 2d, or region through which the cultivation of maize extends, stretches as far N. as a line drawn from the mouth of the Gironde to the N. extremity of Alsace. The 3d region, which terminates together with the culture of the vine, has for its N. limit, a line extending from the mouth of the Loire to Mézières in Ardennes. The 4th, or N. zone, comprises the rest of the country. The mean annual temperature of different parts of France has been estimated as follows, by Humboldt: at Toulon 62° (Fahr.), at Marseilles 59½°, at Bordeaux 56°, at Nantes 55½, at Paris 51·7°, and at Dunkirk 50·5°. More rain appears to fall during the year on the SE. than on the NW. side of the great watershed, the average being, in Isère 32 inches, in Haut Rhin from 28 to 32 in., at Lyons 29 in., and at Montpelier 30 in.; while at Paris the fall is only 19 in., in Orne 20 in., and in Ille-et-Vilaine 21 in. But notwithstanding this result, the sky is generally bright, and the atmosphere clear in the SE., and there are at least one third fewer rainy days than in the NW., where the atmosphere is almost constantly charged with moisture brought by the W. winds which commonly prevail, and the weather is more or less cold for half the year. The centre of the country enjoys a happy medium of temperature and climate; in the S. the summers are long, dry, and hot. The departments around the Gulf of Lyons are subject to a violent N. wind called, in Provençal, the bise, the mistral of ancient writers. According to Mr. Inglis, 'this wind is the curse of all these provinces, and it is scarcely possible, in travelling through this country, to meet with a greater misfortune than a bise wind, especially if the word "meet" be interpreted literally. . . . Its effect upon the frame is singularly disagreeable: it parches the mouth and throat, creates a feeling of suffocation, and seems to dry up the whole juices of the body.' (Inglis's Switzerland, &c., p. 108; Aperçu Statistique.)

The vegetable products of France are said by Hugo to comprise upwards of 830 genera, and 6,000 species. All these, however, are not indigenous, and many new plants have been introduced within the last two centuries. The most richly wooded parts are the mountainous districts, particularly the Vosges, the plateau of Langres and Orleans, the Cevennes, and the mountains of Auvergne and Limousin. The Alps and Pyrenees, Provence, the S. part of Languedoc, and the W. of France are but indifferently wooded. The principal forest trees are the oak, elm, beech, maple, ash, walnut, chestnut, birch, poplar, larch, pine, fir, box, cornel, &c. In the Vosges and Jura mountains, Brittany, and the Landes, there are extensive forests of fir; the chestnut woods are very fine in Haute Loire. The olive, orange, lemon, pistachio, and carob grow wild in the S., but there only; and the fruits of all are inferior to those of warmer climates. The caper (Capparis spinosa), diffused over Provence, furnishes a well-known article of export. Cherries, apples, and several other fruits grow wild; apples and pears are largely cultivated in the N. departments, and prunes in the centre of France. The culture of these and other fruits will be adverted to in a subsequent section.

Mr. Inglis, who travelled through many countries of Europe on foot, has the following remarks on the scenery of France:—' All panegyric upon the loveliness and laughing fertility of France is rhodomontade. There is more of the beautiful and the picturesque in many a single county of England, or even of Scotland, than in all the scattered beauties of France, were they concentrated within a ring-fence: excepting always the Pyrenees, which I cannot help looking upon as a kind of separate territory—the mere boundary between France and Spain; but at all events the Pyrenees must be excepted. I have travelled through almost every part of France; and truly, I have found its beauties thinly sown. If the banks of some of its rivers be excepted—the Seine, the Loire, the Rhone, and the Garonne—some parts of Normandy, and the departments of the Pyrenees, France is an unromantic, uninteresting, unlovely land. And even in these favoured parts, such as the vaunted Orléannais, where shall we find the green meadows that lie along the banks of our Thames, or Avon, or Severn; or upon which of them shall we pause to admire those romantic views—that charming variety of rock, wood, and mountain—that characterise the banks of the Tamar, the Wye, the Derwent, the Swale, the Wharff, or the Dove? These are nowhere to be found. . . . I pity the man who crosses France in any direction. Thousands know how rampant is the journey from Calais to Paris, but they who never travel farther, suppose that lovely France, panegyrised by so many, lies beyond. No such thing. Let them continue their journey by whichever road they please, and they will find but little improvement. . . . Château also we have in these provinces (those of the S.), but, oh, how different from the châteaux of which we read in the romance writers, and which never existed but in their imagination. The châteaux are for the most part ruins upon a large scale; staring houses with wings, and a parapet wall in front, covered with ruins of flowers. In short, we find the whole a delusion; and our minds revert to the green activities of our own hills, our oak forests, our lakes and rivers, and the beauty and fertility that, along with the picturesque, mingle in an English landscape." But if the indiscriminating panegyrists of France have gone too far on the one hand, Mr. Inglis certainly has as much overshot the mark on the other. Mr. Maclaren, than whom there can be no better authority, says that from Chalons-sur-Marne to Avignon the Rhone flows through one of the most beautiful, picturesque, and delightful regions in the world. And there are many other districts in France the scenery of which will bear a comparison with that of any other country in Europe.

Animals.—The bear, wolf, and wild boar are the only formidable wild animals now inhabiting France, and the numbers of these have been greatly thinned by the increase of pop. and of civilisation. The black bear (*Ursus Pyrenaicus*) is confined to the higher ranges of the Alps and Pyrenees, where the lamb, chamois, and wild goats are also found. Notwithstanding an active war of extermination carried on against the wolves, these animals are still very numerous in some departments, as in Morbihan Sarthe, Vendée, Landes, and the central mountainous departments. In the Cevennes, the lynx is sometimes found, though rarely. The wild boar, roebuck, and fox abound in all well-wooded parts. The red and fallow deer, formerly so plentiful in the royal parks, have become rare; hares and rabbits are extremely abundant. Several kinds of squirrels, the polecat, weasel, otter, marten, hedgehog, and the other small wild animals common throughout Europe, are as numerous in France as elsewhere; in addition to which, the dormouse (*Mus nurculferus*), an aquatic quadruped, inhabits the neighbourhood of Tarbes, as some braver do the islands at the mouth of the Rhone. Seals, dolphins, and sometimes whales, are met with on the coasts.

Birds are very numerous. They include two kinds of eagles and a species of vulture. Several birds not elsewhere met with, are found on the shores of the Mediterranean, as the flamingo, roller, wasp eater, bee-eater, ortolan, &c. Bustards, large and small, inhabit the dkps. of the N.W. and centre. The cock-of-the-wood and red and grey partridges are the principal winged game. Water-fowl are particularly numerous in Vendée and Charente-Inférieure; in cold winters the wild swan visits the country.

Reptiles are few; there are but two venomous serpents. A kind of gecko inhabits the S. shores; the salamander, large green lizard, mud-tortoise, and *Bufo obstetricans*, are the other most remarkable animals of this class. In some dkps. frogs are reared in large numbers as articles of food. The fisheries of turbot, sole, ray, mackerel, herring, pilchard, mullet, &c. in the N. and W. seas, and of the tunny and anchovy in the Mediterranean, furnish employment to numerous families. Oysters are very abundant on the N. and W. coasts, as well as mussels and lobsters. Leeches are exported in large quantities. There are two species of scorpion. Cantharides and the cochineal insect are met with in the S. (Hugo, France Pittoresque; Aperçu Statistique; Dict. Geog.)

Population.—The information with respect to the pop. of France previously to 1784 is exceed-

ingly imperfect. But, according to the most authoritative statements, it amounted, in 1700, to 19,669,000, and in 1762 to 21,769,000, including Corsica. In 1784, it was estimated by Necker at 24,800,000.

The official returns give the following numbers for the undermentioned years:—

1801	.	27,349,003	1828	.	31,540,910
1806	.	29,107,425	1841	.	34,213,970
1771	.	30,461,874	1846	.	35,400,486
1831	.	32,569,272	1861	.	35,783,800

The following table gives the total population of France in the department of the Seine, in towns and in the country, in each year from 1828 to 1861, according to the civil returns of births and deaths:—

	1828	1850	1861
In Department of the Seine	1,358,023	1,836,091	1,953,680
In Towns	8,363,274	9,461,572	8,371,745
In the Country	26,359,055	26,704,741	26,554,800
Total	36,251,047	36,372,404	37,294,812

The subjoined table shows the total number of births, deaths, and marriages, in France, in each year from 1857 to 1861:—

Years	Living Births	Deaths	Marriages
1857	900,709	859,745	295,510
1858	889,543	874,195	307,058
1859	1,107,808	879,838	799,417
1860	938,074	781,535	799,928
1861	1,065,076	946,307	305,703

The following table gives the number of living births, legitimate and illegitimate, in France, in each of the years 1860 and 1861:—

	1860			1861		
	Legitimate	Illegitimate	Total	Legitimate	Illegitimate	Total
Department of the Seine	49,330	16,092	8-04	44,782	16,144	8-78
Town Population	216,461	27,744	1-71	228,799	30,820	1-80
Country Population	623,147	26,881	22-90	636,547	29,828	22-17
Total	887,578	69,397	12-90	929,501	76,887	12-10

The department of the Seine comprises, besides Paris, only a few villages inhabited chiefly by persons from the capital, and has scarcely any rural population. The number of births in the city of Paris during the year 1862 was 52,312, of which 26,805 were boys, and 25,507 girls. Of the whole number 14,591 were illegitimate, and 37,721 born in wedlock. Of the illegitimate children, 6,711 were recognised and allowed to bear their father's name.

France is divided into 89 departments—86 previous to the annexation of Savoy and Nice—with 373 arrondissements, 2,938 cantons, and 37,510 communes. According to the census of 1861—slightly differing from the civil register returns given above—the population of the 89 departments amounted to 37,382,225 souls, inclusive of a number of troops stationed in Syria and at Rome, and counted in the regimental lists. The following table shows the distribution of the population as well as the increase in the ten years 1851-61.

Departments	Old Provinces	English sq. m.	Population in 1831	Population in 1841	Departments	Old Provinces	English sq. m.	Population in 1831	Population in 1841
N E.					**N W.**				
Seine		181	1,472,051	1,253,660	Charente-Inför.	Aunis	2,761	443,201	461,060
Seine-et-Oise		2,721	471,554	413,073	**S W.**				
Seine-et-Marne	De de France	2,293	345,076	332,317	Gironde		4,155	514,347	567,159
Aisne		2,832	556,909	546,581	Dordogne	Guienne	3,459	480,707	601,557
Oise		2,280	401,367	401,417	Tarn-et-Garonne		1,378	237,533	232,551
Somme	Picardie	2,384	550,641	573,446	Aveyron		3,617	394,186	394,025
Pas-de-Calais	Artois	2,671	697,974	716,334	Charente	Angoumois	3,716	363,914	379,941
Nord	Flandre	2,276	1,158,765	1,345,090	Corrèze	Limousin	2,790	220,954	310,114
Ardennes		2,130	331,790	329,111	Lot		1,551	738,724	295,567
Marne	Champagne	3,214	978,309	345,489	Lot-et-Garonne	Gascogne	1,186	341,445	573,083
Aube		2,393	785,347	367,765	Landes		3,590	307,198	400,453
Haute Marne		2,442	205,594	374,413	Hautes Pyrénées		1,200	230,984	240,159
Yonne		2,821	304,133	370,205	Gers		2,416	307,159	441,041
Vosges	Lorraine	3,924	477,809	413,485	Basses Pyrénées	Béarn	2,979	446,207	456,889
Meurthe		2,035	430,125	439,644	Haute Garonne		3,463	441,219	409,291
Moselle		2,509	379,655	391,540	Aude	Languedoc	2,716	709,747	252,611
Bas Rhin	Alsace	1,613	502,434	577,824	Hérault		2,111	309,708	419,291
Haut Rhin		1,392	494,147	413,807	Tarn		1,514	344,072	334,643
Doubs	Franche Comté	2,179	741,879	396,240	Ariège	Foix	2,051	961,118	251,500
Jura		1,910	313,401	728,854	Pyrénées Orientales	Roussillon	1,403	151,958	191,763
Haute Saône		1,792	347,459	317,183	**SE.**				
Saône-et-Loire	Bourgogne	3,721	574,720	543,187	Rhône	Lyonnais	1,347	374,743	642,493
Côte d'Or		3,290	442,797	344,180	Loire		1,091	475,508	417,603
N W.					Puy-de-Dôme	Auvergne	3,359	446,507	616,449
Seine-Inför.		2,300	781,009	769,969	Cantal		2,240	258,579	340,522
Eure		2,014	413,177	394,091	Haute Loire		1,920	306,815	305,371
Orne		2,457	439,044	443,293	Ardèche	Languedoc	2,139	368,558	388,579
Calvados		2,740	441,244	441,671	Gard		2,271	406,184	472,107
Manche	Normandie	2,613	600,847	591,671	Isère	Dauphiné	1,072	144,504	187,347
Ille-et-Vilaine		2,641	471,618	484,930	Ain	Bourgogne	1,761	872,879	879,747
Côtes-du-Nord		2,520	572,913	648,676	Isère		3,254	643,647	577,748
Finistère	Bretagne	2,670	617,710	627,504	Hautes Alpes		2,140	132,415	123,110
Indre-et-Loire	Touraine	3,401	313,641	323,573	Drôme	Venaissin	1,611	324,860	371,611
Eure-et-Loire		2,363	784,902	765,123	Vaucluse		1,340	264,616	268,253
Loire-et-Cher	Orléanois	2,363	761,903	329,029	Bouches-du-Rhône	Provence	2,831	489,860	407,112
Loiret		2,613	341,424	342,713	Basses Alpes		3,470	152,029	146,368
Nièvre	Nivernois	7,691	377,141	379,814	Var		3,705	347,967	315,570
Allier	Bourbonnais	3,009	336,750	354,443	Corsica (Island)		3,705	236,251	138,049
Cher	Berry	2,653	346,761	323,093					
Indre		2,749	271,908	370,954	**Total**		207,177	33,763,038	34,718,164
Morbihan		2,610	474,177	444,404					
Loire-Infér.	Anjou	2,733	532,861	680,201	The newly-annexed Departments				
Maine-et-Loire		2,741	515,552	529,012	Alpes-Maritimes		1,104	..	154,378
Mayenne	Maine et	2,010	514,588	375,143	Savoie		2,271	..	574,608
Sarthe	Perche	2,475	474,071	466,165	Haute-Savoie		1,345	..	700,447
Vienne	Marche	2,244	747,075	770,045					
Haute Vienne	Limousin	2,167	813,517	319,705	**General Total**		211,252	33,763,038	37,381,725
Deux-Sèvres		2,297	374,015	374,817					
Vendée	Poitou	2,614	344,730	395,605					
Vienne		2,862	516,144	177,026					

Distribution of Landed Property.—Agriculture. —In France, previously to the revolution, the property of persons dying intestate was subject, in different parts of the kingdom, to different regulations; but every where estates could be disposed of by will, and settled by entail. At the revolution a nearly total change was made in these respects; the same regulations for the distribution of property were established in all parts of the kingdom; and the power of disposing of property by will was confined within the narrowest limits. It was enacted, 1. That the property of persons dying intestate shall be equally divided among their children, without respect to sex or seniority; and, 2. That when a person possessed of property wishes to make a will, he shall be permitted, provided he have only one child, to dispose of a moiety of his property, the child inheriting the other moiety as matter of right; if the testator have two children, he is allowed to dispose of a third part of his property; and if four children, of a fourth; and so on, the rest being equally divided among the children.

This law was intended to subvert the foundations of that old feudal aristocracy, whose usurpations and oppressive privileges had entailed much misery on the country; and there can be no doubt that it was well fitted to accomplish this object. That the condition of the agriculturists of France has been materially improved since the revolution, is true. But it has not been quite proved that this improvement has been in any respect owing to the law of equal inheritance. Some argue that it has taken place, not in consequence, but in despite, of that law. The abolition of the game laws, and of feudal privileges of the nobility and clergy, and of the gabelle, corvée, and other oppressive and partial burdens, and imposts, was of the greatest service to proprietors and farmers; and, in addition to these advantages, a large extent of common lands was divided, and a great part of the vast possessions belonging to the church and to the emigrants came into their hands at extremely low prices, so that while small properties were generally augmented, farmers were, at the same time, in very many instances, changed into landlords. No wonder, therefore, that fresh energy was given to agricultural pursuits, and that a great improvement has been effected.

Still, however, it is certain that the rapid division of landed property, and the continually increasing excess of the agricultural population, caused by the existing law of succession, have gone far to neutralise the effects of these advantageous circumstances, and form one of the evils in the social condition of the people of France. 'The population of that country,' says Mr. Birkbeck, 'seems to be arranged thus: a town depends for subsistence on the lands immediately around it. The cultivators individually have not much to spare; because, as their husbandry is a sort of gardening, it requires a large country population, and has, in proportion, less superfluity of produce. Thus is formed a numerous but poor country population. The cultivator receives payment for his surplus produce in ease, and he expends only men. The tradesman is on a par with the farmer; as they receive so they expend; and thus 60,000 persons may inhabit a district, with a town of 10,000 inhabitants in the centre of it, bartering the superfluity of the country for the arts and manufactures of the town. Free from generation to generation, and growing continually poorer as they increase in numbers; in the country by the division and subdivision of property, in the town by the division and subdivision of trades and professions; such a people, instead of proceeding from the necessaries to the comforts of life, and then to the luxuries, as is the order of things in England, are rather retrograde than progressive. There is no advancement in French society, no improvement, nor hope of it.' (Tour in France, 4th ed. p. 31.)

In his Tour in France, Mr. J. P. Cobbett makes the following observations with respect to the influence of the law of equal succession in Normandy:—'I bear, on all sides here in Normandy, great lamentations on account of the effects of this revolutionary law. They tell me, that it has dispersed thousands upon thousands of families who had been on the same spot for centuries; that it is daily operating in the same way; that it has, in a great degree, changed the state of the farm buildings; that it has caused the land to be worse cultivated; that it has caused great havoc amongst timber trees; and there are persons who do not scruple to assert, that society in France will become degraded in the extreme, unless the law be changed in this respect.' (p. 169.)

The contribution foncière, though there are great inequalities in its pressure, amounts, at an average, to about a fifth or a sixth part of the rent of the land; and the official returns show that it was assessed in 1835 and 1842 as follows, viz.:—

Properties assessed at less than	1835	1842
5 fr. a year	5,205,111	5,449,580
from 5 fr. to 10 fr.	1,761,934	1,816,174
10 „ 20	1,814,251	1,616,997
20 „ 30	728,246	191,711
30 „ 50	444,165	741,912
50 „ 100	553,230	612,926
100 „ 500	306,714	140,104
500 „ 1000	53,150	86,462
1000 and upwards	13,361	16,746
Total	10,833,500	11,311,541

The first 5,205,111 proprietors belong to about half as many families, who thus derive a mean rental of about 4fr. per annum from their property. There are, besides, about 4,250,000 families (out of the whole 5,449,263 families that are owners of land), who derive an annual income of 6s. only from their portions of the soil. Nothing is more common than for these little

freeholds to become forfeited to the state, from the inability to pay the impôt foncier (which is about 7 per cent. on the rental). The unfortunate defaulter is allowed, on payment however of a registration fee of ten francs, to give up for ever his little plot, in order to save his personal property from the tax-gatherer. There are not quite 11,000,000 separately rated côtes foncières in France. These are divided into 123,360,336 parcels, about eleven to a côte—not enclosures of the same farm in juxtaposition to each other, but more like our own lands lying in common fields in England, perpetually intersected by those of the neighbours. Within ten years, more than half the value of the land of all France, 833,000,000fr. worth of property, has been proved to have changed hands, of which 372,000,000fr. have been by inheritance in the usual course of nature; 33,000,000fr. by donations inter vivos, and with the view of preventing dismemberment. But the residuary 475,000,000fr., i.e. more than a fourth part of the whole fee simple of the country, has passed in these ten years into the hands of complete strangers. At this rate a generation and a half would see the entire kingdom in the hands of another race, totally unconnected with its former owners. There is in France the same intense anxiety to possess a bit of land as in Ireland. It has given rise to the 'bandes noires,' an expressive term for an association of notaries, country bankers, attorneys, land surveyors, and jobbers of all sorts, who combine together when an estate is to be sold, tempt the owner with a good price and an exemption from all trouble to dispose of it to them; they then cut it up into lots to suit the market: a farm of 200 acres is thus parcel off into twenty, fifty, or more allotments, which are paid for partly in money, and partly mortgaged. This accounts for another phenomenon—the enormous extent of debt with which the land is burdened. Considering that almost the whole of it changed hands at the revolution, that there was an entire sweeping away of titles, charges, mortgages, fortunes, &c., and that almost every title in France is now less than fifty years old, it is a remarkable fact that within that time, or, indeed, for less, the owners have charged an income of 60,823,000fr. with a debt bearing an interest of 72,100,831fr. This debt increases, and must continue to do so. The avidity to possess land, the fancied independence that it confers on its owner, acts on the 1½ millions of families and their kindred incessantly. 'These heroic men,' says their friend and admirer, Professor Michelet, in his 'People,' 'fight as it were for their lives, but weary fights against them with a force of 4 to 1; their land brings them in 2 per cent., and they pay 8 per cent. for borrowed money.'

The whole of the area of France may be classed under six grand divisions, each consisting of from thirteen to fifteen departments.

1st Region, the North-West.—This region comprises the departments—

Nord	Seine	Eure
Pas-de-Calais	Seine-et-Oise	Orne
Somme	Seine-et-Marne	Manche
Aisne	Seine-Inférieure	Eure-et-Loire
Oise	Calvados	Loiret

These departments have an area of 21,811,270 acres, with a population of 9,543,047, or about 250 persons to the square mile. This is by far the wealthiest, the most populated, and most commercial section of the empire.

The department du 'Nord' contains 565 persons to a square mile. If all France were equally peopled, the country would have a population of over 100 millions. In this department, the land is divided into large, middling, and small properties, the small predominating; and, notwithstanding the development of industry and commerce, those who live through agriculture amount to about one-half of the whole population.

In the 'arrondissement' of Lille, land is worth 400l. the 'hectare,' and less for 150 francs, or 60l., equivalent to 2l. 8s. the acre.

In 'l'Ile de France,' there are upwards of 3,000 proprietors, each paying a land-tax exceeding 1,000 francs, or 40l., and among them are many proprietors who have an income of from 2,000l. to 4,000l. a year. Estates from 1,250 to 2,500 acres are numerous, particularly in 'Seine-et-Marne.' While the whole of France contains 16,846 proprietors paying a land-tax to the amount of 1,000 francs, or 40l., and above, and 80,862 proprietors paying from 500 to 1,000 francs, the above fifteen departments contain one-half of the first, and very nearly a moiety of the second class. The 'propriétaires cultivateurs' form one-fifth of the rural population, and the farmers with their families another fifth, while the remaining three-fifths are composed of journeymen and servants. In 3d, a day is about the average rate of wages in these departments.

2nd Region, the North-East, comprising the following departments:—

Ardennes	Côte d'Or	Moselle
Aube	Doubs	Meurthe
Marne	Jura	Vosges
Haute Marne	Haute Saône	Haut Rhin
Yonne	Meuse	Bas Rhin

These departments have a total area of 22,453,250 acres, with a population of 3,587,253, or about 160 persons to a square mile. There is a great subdivision of property in these departments. But this subdivision is not of modern origin, and is mainly attributed to the cultivation of the vine. The greater part of the land belongs to those who cultivate it.

In the above fifteen departments agricultural produce has doubled since the year 1789; nevertheless, while the north-west contains 16,000 persons paying a land-tax of from 500 to 1,000 francs, or 20l. to 40l., and 8,000 persons paying 1,000 francs and above, the north-east contains only 4,000 of the first, and less than 2,000 of the second class.

3rd Region, the West, comprising the departments of—

Indre-et-Loire	Côtes-du-Nord	Deux Sèvres
Mayenne	Finistère	Vienne
Sarthe	Morbihan	Charente
Maine-et-Loire	Loire-Inférieure	Charente-Inférieure
Ille-et-Vilaine	Vendée	

which have a total area of 22,761,875 acres, with a population of 6,505,204, or 180 persons to a square mile.

This region contains the valley of the Loire, where the land is of extraordinary fertility, exhibiting a garden culture, but is also much subdivided. The land is commonly sold at 400l. the hectare, and a very small quantity under vine culture suffices for the easy maintenance of a family. About 500,000 persons in this valley live on as many acres. The land is very nearly equally divided between town and country. But if, on the one hand, extreme subdivision characterise the valley of the Loire, on the other, in the department of l'Indre, properties of from 2,500 acres to 5,000 acres are common, while in Anjou the farms have an average of from 75 to 100 acres. There are many smaller, but few very small estates.

In Brittany there are few large properties; the farms average 60 acres. The whole of the province does not contain 800 properties paying a land-tax of 1,000 francs, or 40l. and upwards; but as in Brittany the value of land is only half what it is in Normandy, a land-tax of 500 francs represents the same extent of land, as a land-tax of 1,000 francs in Normandy or Picardy.

In all these departments, the division of the soil has followed the course consequent on the increase of population.

4th Region, the South-East, comprising the following departments:—

Saône-et-Loire	Ardèche	Hérault
Ain	Drôme	Basses Alpes
Rhône	Hautes Alpes	Bouches-du-Rhône
Loire	Vaucluse	Var
Isère	Gard	Corsica

These departments have an area of 22,800,427 acres, and a population of 5,071,209 souls, or 160 to the square mile.

This region agriculturally occupies the fourth place, whereas, commercially, industrially, and through its wealth, it takes the second place. The slopes of this district towards the Saône, between Châlon and Mâcon, contain fine vineyards and fertile plains. As Lyons is approached, fertility increases. In the department of the Ain, at the foot of the Jura mountains, the principal culture is wheat, Indian corn, and the vine. There are also extensive meadow lands. Property is divided, but not in excess, and the number of proprietors is in many circumstances very considerable.

The plain of Nîmes is one of the most productive in France. Land is there worth from 200l. to 240l. the hectare. The departments of the Upper and Lower Alps had, in 1789, about 400,000 inhabitants; they have now only 271,168. This is the poorest and least populated part of France, having only 27 persons to 100 hectares, while Corsica has 22.

With the exception of the four departments of the Bouches-du-Rhône, Gard, Hérault, and Saône-et-Loire, where there are large towns and much manufacturing wealth, the other eleven departments of this region have not among them more than 600 proprietors paying a land-tax of 1,000 francs, or 40l. and upwards. This is less than the number found in one department in the north west region, or in the neighbourhood of the capital.

5th Region, the South-West.—This region comprises the departments—

La Gironde	Gers	Hautes Pyrénées
Lot-et-Garonne	Haute Garonne	Ariège
Lot	Tarn	Aude
Tarn-et-Garonne	Aveyron	Pyrénées Orientales
Landes	Basses Pyrénées	

These departments contain 21,971,125 acres, and a population of 4,751,886 souls, or 140 to the square mile. The rural population, which in the north-west and the south-east attains the half, and in the other regions the two-thirds of the entire population, in this region reaches three-fourths. The whole of the fourteen departments only contain 1,300 proprietors paying a land-tax of 1,000 francs, or 40l., and upwards, so that there are more small proprietors in this than even in the south-eastern region. The average extent of the farms here is 60 acres, and many are low. This great subdivision of property is of old date. It was observed by the 'Parliament de Paris' in 1760, that in Bearn and the neighbouring south-west provinces, every person was a 'proprietor.' This region contains one-third of the vineyards of France. Those of Médoc, extending over 50,000 acres, produce about 40,000 tuns of wine, of which 3,000 are superior, 5,000 are 'vins de bourgeois,' the remaining 30,000 are 'vins ordinaires,'

6th Region, the Centre.—This region comprises the departments—

Loiret-et-Cher	Creuse	Puy-de-Dôme
Cher	Haute Vienne	Cantal
Indre	Corrèze	Lozère
Nièvre	Dordogne	Haute Loire
Allier		

having an area of 21,106,997 acres, and a population of 4,212,997 souls, or 127 to the square mile. The land in these thirteen departments is mostly divided into large properties. The farms are on the average of 600 acres and above. The ancient province of Berri and the neighbouring districts have always been the region of large properties. Over one hundred estates of 2,500 acres are to be found here, while many are still larger; the largest of all, that of 'Valençay,' a property of the Prince de Talleyrand, has not less than 30,000 acres spread over 27 parishes. In the province of Auvergne alone, property is much divided. While the returns from the land have doubled throughout France since 1789, it is doubtful whether in this region they have increased 50 per cent.; and the revolution has exercised but little influence on the state of property in these central departments. In the provinces of Berri, Le Nivernois, and Le Bourbonnais there are large properties; but the department of Cantal has only 18 estates paying a land-tax of 1,000 francs, or 40£, and above; the department of Lozère has only 7 of the same class; the department of Creuse, 5; Corrèze, 4; and Haute Loire, 2; while—to compare these departments with others of similar size—Seine-et-Oise, in the north-west region, has 771, and Seine-Inférieure 748 estates of the first-class. (Lavergne, L'Économie Rurale de la France depuis 1789, Paris, 1861; Block, Statistique de la France, Paris, 1860.)

The best wheat is said to be that of the N. and NE. provs.; but Flanders, Picardy, Normandy, the district of Beauce in Eure-et-Loire, Berri, Touraine, and the vicinity of the Puy-de-Dôme, furnish the greater quantity. Rye, like wheat, is grown in almost every part of the country; but it is principally cultivated in the NE., in Isère, and on the thin soils of Puy-de-Dôme, Creuse, Haute Vienne, Allier, Loire, &c. The culture of maize, though it extends as far N. as the banks of the Loire, is most prevalent in the SW., where the grain is of the best quality. Barley and oats are raised principally in the N. Buckwheat on the worst arable lands of the centre and S. The potato is not yet an article of so much importance as in England or the Low Countries; but within the last twenty years its cultivation has increased very rapidly. It is mostly grown where corn is the least cultivated, as in Lozère, the Vosges, &c., and in the dép. Meurthe and Moselle in the NE., Aube, Côte d'Or, and Eure-et-Loire in the centre, and Ille-et-Vilaine, Vaucluse, and Ariège in the S.

The culture of beet-root for sugar is extensively pursued in the neighbourhood of the capital, and the deps. of the N. and E., and parts of the centre. It is sometimes grown on the same land for several years in succession, though, most commonly, wheat is sown alternately with it every third year, when it yields as much as if the ground had been previously fallowed. The produce of beet varies from 12,000 kilog. per hectare to double that quantity; in the deps. du Nord and Pas-de-Calais, from 25,000 to 30,000 kilog. are reckoned an average crop.

This branch of industry began during the exclusion of colonial products in the reign of Napoleon, and received a severe check at the return of peace, by the admission of West Indian sugars at

a reasonable duty. It is probable, indeed, that it would long since have been entirely extinguished, but for the additions made to the duties on colonial and foreign sugars in 1820 and 1822. After the last-mentioned epoch, however, the production of beet-root sugar began rapidly to increase; and such was its progress, that though, in 1828, its produce did not exceed 4,000,000 kilog., it amounted, in 1838, to 39,199,000 kilog. It rose to 49,781,325 kilog. in 1847; to 81,801,819 kilog. or 1,608,012 cwt. in 1857; to 151,514,435 kilog., or 2,976,177 cwt. in 1860; sunk to 132,640,671 kilog., or 2,645,638 cwt. in 1862; and rose to the unprecedented figure of 152,012,616 kilog., or 2,985,961 cwt. in 1864. The following table shows the number of beet-root factories at work, and the quantity of sugar manufactured by them, in the several departments where this industry is carried on, in the year 1864:—

Departments	Number of Factories	Quantities of Sugar Manufactured
		Kilos.
Aisne	61	74,320,793
Nord	139	43,100,392
Oise	70	8,879,871
Pas-de-Calais	66	30,194,846
Somme	10	11,720,303
Other Departments	28	10,196,492
Total	**834**	**152,012,616**
		Cwt. 2,985,961

Besides the 834 factories here enumerated, there were 21 establishments for the production of beet-root not at work in this year 1864.

Kitchen vegetables are universally grown, and are of excellent quality. In the N. and E. the wild cabbage, rape, and poppy, and other oleaginous products are extensively cultivated; the former especially in the dép. du Nord, where oil is a principal article of trade. Chicory is also raised in this part of France; truffles are cultivated in several parts, but especially in Dordogne, where they enter into the composition of the celebrated patés. Hemp and flax are grown in most deps.; but, along with hops, they are principally raised in the N.

Madder is extensively grown in Vaucluse and some parts of Alsace, and is a prominent article of export. Woad, saffron, and some other dyes, castor oil, &c., are among the other kinds of produce.

Wines.—The growth of these forms a distinctive feature in French agriculture. The vine is cultivated more or less throughout the whole kingdom, to the N. of Brittany, Normandy, Artois, and Flanders; with the exception of the dép. Creuse, in which, owing principally to the poverty of the rural pop., it is little or not at all grown. The cultivation of the vine has been slowly increasing since the revolution. In 1789, there were about 1,600,000 hectares under cultivation; in 1829 the area had amounted to 1,953,397; in 1839, to 2,134,221; and in 1863 to 2,587,870 hectares. These vineyards are distributed over 79 departments, but in 70 of them the culture is merely nominal. The number of growers are estimated at 1,800,000 persons; and the quantity produced at 36,785,000 hectol., or about 800,000,000 imperial gallons, worth 16,760,000£, or about 5d. a gallon. The cost of wine to the people of France is estimated at about 12,000,000£, which supposes it to cost, at an average, each individual of the pop. about 7s. a year. The duties paid on the wine consumed at home amount, at an average, to 2,500,000£ a year. Hence of all the products of France, next to wheat, wine is incomparably the most important. The vineyards occupy, at present,

4·10 per cent. of the entire surface; at the same time that the duties laid on wine amount to nearly one-third part of the land tax, and to one-tenth part of the entire public revenue. In 1861 the official value of the wine exported from France was above 375,000,000 of francs, or 15,000,000*l*. The export of Bordeaux wine in cask increased from 1,158,000 hectolitres in 1862, to 1,214,000 in 1863, and to 1,419,000 in 1864. The other French wines exported rose from 1,156,000 hectolitres in the year 1862, to 1,214,000 in 1863, and to 1,119,000 in 1864. Brandy distilled from wine rose from 162,000 hectolitres in 1862, to 190,000 in 1863, and to 279,000 in 1864. (Official Reports in Moniteur Universel.)

The départs. in which the greatest extent of land is occupied by vineyards are the Gironde, Charente-Inférieure, Hérault, Charente, Dordogne, Gers, Gard, Lot-et-Garonne, and Var; but the départs. of Marne and Aube, forming the ancient prov. of Champagne, and those of Côte d'Or and Saône-et-Loire, comprised in Burgundy, though yielding a less quantity of wine than many others, are highly distinguished for the superior quality of their products. Gironde furnishes the wines known in England by the name of *claret*. A fifth part of the Bordeaux wines is used for the distillation of brandy, exported chiefly to the U. States, England, Russia, Sweden, and Denmark; but they are inferior for this purpose to those of the Charente, which supply the famous Cognac brandy. For more minute details respecting the growth of the French wines, we refer to the arts. devoted to the several départs.

Tobacco.—The cultivation of tobacco is of great importance. In former days, under the old monarchy, the state had reserved to itself the exclusive right of buying, selling, and manufacturing tobacco, and this monopoly was farmed out to a company which paid the amount of 52,000,000 francs to the revenue for the privilege. The revulsion of 1789 abolished the monopoly, and tobacco remained free up to the year 1798, when a duty on its importation and manufacture was imposed. The import duty was 44 and 66 francs per 100 kilogrammes, according as the tobacco was imported in French or foreign bottoms; the duty on the manufacture was 20 francs (per 100 kilogrammes) on tobacco for smoking, and 40 francs on snuff. In 1802 the duty on the manufacture was raised to 40 francs for all kinds of tobacco without distinction. In 1804 the custom-house duties were raised from 44 and 66 francs to 88 and 110 francs. In 1806 the duty on manufacture was doubled, the custom-house duties were raised to 128 and 220 francs, and a duty of 20 francs per 100 kilogrammes was laid on the sale of the article; and lastly, in 1810, the duty on the sale was raised to 180 francs, and the custom-house duties to 256 and 440 francs.

Notwithstanding these great and rapid elevations of the duty, the revenue derived from tobacco hardly reached half the amount yielded under the old system. Smuggling was carried on on a large scale, and great complaints were made by the public of the bad quality of the merchandise. The old system was consequently re-established, and, by a decree of December 29, 1810, the tobacco monopoly was restored to the state. It was, in 1816, confirmed for a period of five years by the law of April 24, 1816, and was afterwards prolonged by successive laws in the years 1819, 1824, 1829, 1835, 1840, and 1852. The last prolonged its existence for a period of ten years, and in 1862 it was extended for another ten years. The tobacco monopoly, therefore, although it has been in existence for more than half a century, is not, like the

salt and powder monopolies, as yet a permanent institution in France. It only exists at present on sufferance, inasmuch as the chambers only vote it for a stated period.

In the report which served as a base for the 'Projet de Loi' of 1862, the government expressed the hope that it would soon be able to proclaim the existence of the tobacco monopoly once for all; but the minister of finance contented himself, for the time, with asking for a further prolongation of the monopoly for a period of ten years, and, after a short debate, this was voted by the corps législatif on the 22nd June, 1862, almost unanimously. It would have been impossible to do otherwise. A sacrifice of such an important source of income could not be contemplated in the face of all the pressing demands on the treasury; nor would it have been easy to find a substitute for the revenue derived from tobacco. During the last few years tobacco has become one of the most lucrative sources of revenue for the French government, in consequence of the great increase in its consumption. An article on this subject in the 'Journal des Débats' by M. Horn, recently published, gives an interesting account of the progress of the consumption of tobacco, in France, of which the following is a summary:—

During the first five years after the re-establishment of the tobacco monopoly (1811 to 1815), the sale of tobacco averaged 14,500,000 of kilogrammes. During the next twenty years, France, now reduced to her ancient limits, consumed but from 10,000,000 to 13,000,000 of kilogrammes annually; and the figure of 13,000,000 was not exceeded before the year 1836. From this date the consumption steadily progressed, and in ten years it increased by 5,000,000. In 1845 it was more than 18,000,000 of kilogrammes. From 1846 to 1850 it remained nearly stationary, but in the ten next years, 1851–60, it made an advance quite unprecedented. The quantities sold at the government establishment, or the 'Régie,' during these ten years, amounted to—

Year	Quantities in Millions of Kilogrammes
1851	19·7
1852	20·1
1853	21·0
1854	22·0
1855	26·7
1856	25·4
1857	27·2
1858	27·9
1859	24·1
1860	29·3

The consumption of tobacco since 1860 has been steadily increasing, in the proportion exhibited in the above table. (Report by Mr. Grey, Her Majesty's Secretary of Embassy, in 'Reports on Foreign Manufactures and Commerce,' presented to Parliament, 1863.)

Fruits.—France is abundantly supplied with fruit. Where the culture of the vine ceases, that of apples and pears becomes of considerable importance; in the N. orchards replace vineyards, and cider and perry are the ordinary beverages of the inhab. The cider of the départs. la Manche and Calvados is the best, and some of it is drunk even in the wine départs. The *Pays de Caux* (Seine-Inférieure) is noted for its numerous and excellent apple-orchards; and the dép. Eure, in which almost all the roads are bordered by a double or treble range of apple-trees, exports large quantities of apples to England and elsewhere. Cider is sometimes used in the distillation of brandy. Several of the central and S. deps. are famous for their dried pears; Aveyron, the Basses Alpes, and Indre-et-Loire for their prunes; and the Limagne (Auvergne) and the valley of Montmorency for their cherries. In the Vosges a small cherry abounds, which is extensively used in the manufacture of *kirschwasser*. Chestnuts are very abundant in some of the central and S. deps., where a portion of the rural pop. live almost entirely on them for half the year. The chestnut crops in Ardèche form a large part of its agricultural wealth. In Haute Vienne chestnut woods occupy more than one-fourteenth part of the whole surface. The olive was formerly more cultivated than at present; the hard winter of 1789 destroyed many of the trees, and the climate even of the most favourably situated parts of France seems not altogether suitable for the plant. The oil of the neighbourhood of Aix, and of the dep. Bouches-du-Rhône, is the best. The culture of the mulberry tree is chiefly carried on in the S.

Pasturage and Cattle.—The mountainous districts of France, especially the Vosges, the mountains of Ardennes, Argonne, the Côte d'Or, and the central mountain system, have extensive natural pastures; the deps. Ardèche and Corrèze, and the entire prov. of Limousin, form together one continuous *prairie*, subdivided by mountain ranges, and interspersed, often sparingly, with fields of buckwheat and rye. Some of the best natural pasture grounds are in the maritime districts of Normandy and Flanders, and in the Isle of Camargue, at the mouth of the Rhône. It has been already remarked that artificial pastures have increased of late years; the plants sown are chiefly lucerne, sainfoin, trefoil, and vetches.

At no very distant period France possessed various superior breeds of horses; but, from want of attention, many of them have deteriorated. Government, however, has latterly been making active efforts to obviate this defect, by the establishment of *haras*, or studs, for the improvement of the breeds, in different parts of the country. Normandy furnishes the best carriage and cavalry horses and hunters. The horses of Brittany are the next in order; the Breton is not so handsome as the Norman horse, but it is stronger and hardier. The deps. du Nord furnishes a good breed for farm labour, and other heavy work; those of Anjou, Maine, and Touraine are also especially suitable for draught. Proceeding southward, the breeds diminish in value, till we come to Limousin, which prov., with those of Auvergne, Périgord, Guienne, and Navarre, produces the best saddle horses. The purity of their blood and their excellence increase, the nearer we approach the Spanish frontier. Alsace, and some of the other E. provs., have a large breed of horses, analogous to those of the N. In Lorraine and Champagne there is a small breed, capable of resisting fatigue for a long period, if well taken care of. Asses and mules are reared chiefly in the deps. of the centre, the W. and S. Those of Deux-Sèvres and Vienne are the best, and many are exported to Spain. The mules used in France are bred chiefly in Auvergne and Provence.

Throughout great part of France, and especially in the mountainous regions of the country, the ox is preferred to the horse for farm labour; and as it furnishes besides the principal supply of animal food, the rearing of horned cattle is everywhere pretty general. Many cows are kept along the banks of the Marne, Yonne, and Seine; in the mountains of Franche Comté (where they yield a great deal of milk, from which cheese similar to that of Gruyère is made); in the Terres mountains; and in Dauphiny, where also the cheese is much esteemed. Flanders, Normandy, Brittany, Alsace, Limousin, Auvergne, and the SW. provs., are

those in which the greatest number of black cattle are reared: many are sent out of Artois and Picardy to be fattened in the two first-named provs. The oxen of Gascony are the largest; their weight varies from 600 to 900 lbs.; the city of Bordeaux and the navy are entirely provisioned from this species. Paris is in great measure supplied from Anjou. The best butter is made in the N. of France, and from Brittany, Normandy, and the vicinity of Boulogne, considerable quantities, both fresh and salted, are exported; the best cheese is made in the N.

According to official returns, the number of each kind of live stock in France, in the years 1851 and 1861, was as follows:—

		1851	1861
Horses:—		No.	No.
Stallions & Geldings	.	1,271,639	634,218
Mares	.	1,194,241	1,107,655
Colts	.	457,636	409,761
Total	.	2,918,406	2,966,034
Mules	.	312,841	312,851
Asses	.	413,419	360,180
Cattle:—			
Bulls	.	599,976	599,997
Oxen	.	1,800,638	1,801,561
Cows	.	5,501,624	5,701,667
Calves	.	2,166,342	2,161,918
Total	.	9,538,331	10,093,127
Sheep:—			
Males	.	575,716	612,578
Wethers	.	9,107,140	9,613,116
Ewes	.	14,544,940	14,466,015
Lambs	.	7,360,309	8,719,446
Total	.	32,151,620	33,391,502
Goats	.	964,300	1,337,940
Swine	.	4,910,571	5,246,643

The consumption of butchers' meat in France generally does not, compared with the population, amount to nearly a third part of its consumption in England.

Next to corn, wine, and silk, wool is the most important article of rural produce. The annual produce in the ten years 1854-64 amounted to 60,000,000 kilogs. It has become of greater value since the native breeds have been crossed with the merino and others; but this improvement has hitherto proceeded to only a small extent, so much so, that it has been alleged that not more, perhaps, than 1-10th part of the entire stock of sheep has experienced its effects. The extreme subdivision of the soil is but little less hostile to sheep, than it is to corn and cattle farming. The imports of wool in 1861 were of the value of 179,170,644 francs, or 7,167,812£. Most of the French wool is coarse and inferior; for the finer sorts the manufacturers are obliged to have recourse to Germany, Spain, and other countries.

Goats are most abundant in the Pyrenean and Alpine deps. In a small district near Lyons a great number are kept in troops of perhaps sixty each, and fed in winter on vine leaves plucked after the vintage, and preserved moist for the purpose. An attempt has been made to acclimate the Thibet goat (*Capra Ægagrus*), for the sake of its wool, so valuable in the shawl manufacture; but it is not possible, owing to the greater moisture of the French climate, that the experiment can succeed. Hogs are largely reared in the N. and E.; in Alsace they furnish almost the only animal food used by the rural pop. They are nutecrous in most parts of France, and in the E. deps. a considerable trade is carried on in them.

Poultry of all kinds is also plentiful, especially in Maine, Normandy, Guienne, and Languedoc. Geese are salted like pork: in the dep, Tarn there is a very large and fine species. Turkeys are almost everywhere plentiful; and the *dindes aux truffes* are important articles of commerce in many towns of Dordogne and Lot. Ducks and fowls are very common: the value of the eggs exported to foreign countries in 1861 amounted to 16,212,555 francs, or 648,502l. The weight of French eggs exported in 1864 amounted to the enormous quantity of fifteen million kilogs., or thirty-three millions of pounds.

Bees are reared, especially in the deps. of Calvados, Basses Alpes, Aude, some of those on the Loire, Sarthe, and Jura. In the deps. on the Loire it is a common practice to move the hives from one district to another, which is supposed both to augment the quantity and improve the quality of the products; this process is effected in the night, and in vehicles built for the purpose. The distance travelled over at a time is often upwards of 30 m., 'and it is not unusual to see in the autumn as many as 8,000 strange bee-hives collected in a little village, where they remain for perhaps two months.' (Aperçu, 69.) The best honey is that of Narbonne; but in several deps. as those of Jura, Basses Alpes, and Calvados, the honey is but little inferior.

Fisheries.—From Dunkirk to St. Valery, the inhabs. of the coast derive a considerable part of their subsistence from the fisheries for sole, ray, turbot, mackerel, herring, &c. The sole and ray fishery lasts from about the beginning of January to that of May; the mackerel fishery then commences, and continues till about the end of July; the herring fishery, the head-quarters of which are at Dieppe, begins early in Oct., and ends towards the 20th of Dec. The pilchard fishery of Brittany employs, during its continuance, a large number of fishermen, besides a number of hands in curing and barrelling the fish. About 8,000 barrels of salted pilchards, the produce of this fishery, are sent into the market annually, and the inhabs. on the coast live in great part on fresh pilchards during the season. The pilchard fishery is also a branch of industry of some consequence along the coast of Charente-Inférieure and La Vendée, as that of the anchovy is on the Mediterranean coast, especially in the dep, du Var. Great numbers of oysters are sent to Paris from Cancale Bay and the mouth of the Seine. Except those already named, the fisheries on the French coast are of comparatively trifling importance, and have only a local interest. The French cod fishery, in 1864, employed 328 vessels, of an aggregate burthen of 72,371 tons; while in the whale fishery there were 8 vessels, of 283 tons burthen. Both fisheries are on the decline since 1858, when they employed 603 vessels of an aggregate burthen of 83,571 tons. (Official Tables.)

Mines and Mineral Products.—These are in France of very considerable value and importance, though inferior to those of the U. Kingdom. The principal products are coal, iron, and salt, with alum, copper, lead, and manganese. Mining industry is placed, in a great degree, under the control of the government. The country is divided into six departments, each under an inspector-general, which six inspectors, together with the minister of public works, compose the council-general of mines. There is a school of mines in Paris, and a practical miners' school at Saint Étienne. The instruction in the latter is wholly gratuitous.

Coal in France is obtained from between 60 and 70 different coal-fields; but of these the greater number are extremely unimportant, and those in the deps. du Nord and Loire are the only ones of any considerable magnitude, or, at all events, they are the only ones that are wrought to any considerable extent. The production of coal has increased very materially of late years. From the report of the committee composed of the owners of French coal mines it appears that in the year 1858 the French coal mines produced 5,000,000 tons, of the value of 59,654,983 francs. In the year 1863 the produce rose to 10,000,000 tons, of the value of 117,500,000 francs, which is only a little more than one-eighth of the produce of the English coal mines. Except in the principal towns, coal is rarely used as fuel in France. The peasantry and occupiers of land, and the inhabs. of the smaller, with many also of those of the larger towns, use little save wood or turf for firing.

The subjoined table exhibits the production of coal in the seven years 1858–64:—

Years	Quantities	Value
	Quint. Met.	Francs
1858	59,329,842	49,857,907
1859	62,779,074	14,877,992
1860	71,354,429	51,957,869
1861	79,857,665	101,921,431
1862	72,017,542	99,587,669
1863	72,745,674	81,745,874
1864	74,955,716	84,950,164

Iron works are carried on in various parts of France. Formerly iron was almost wholly made by means of wood; but of late years coal has been extensively substituted for the former, and at present three-fourths of the iron produced in the country is smelted by its agency. But, despite this improvement and the increase of the manufacture, the iron of France is still comparatively high-priced, and insufficient for the home supply. The production, in the seven years 1858 to 1864, amounted to—

Years	Quantities	Value
	Quint. Met.	Francs
1858	33,145,842	10,857,897
1859	35,466,501	13,601,371
1860	38,765,007	14,854,416
1861	40,481,313	16,455,367
1862	44,937,543	16,152,731
1863	39,531,946	11,191,198
1864	35,482,731	12,146,352

In all, about 70,000 hands are supposed to be employed in the different works connected with the production of iron in France. The other metallic products raised in France, consisting of silver, lead, copper, manganese, are of inconsiderable value and importance. The produce of salt exceeds 4,000,000 m. q., or above 850,000 tons.

Manufactures.—As respects the extent and value of her products, France ranks as a manufacturing country next to Great Britain. But her natural and acquired capabilities for carrying on manufactures are very inferior to those enjoyed by this country. There is a great want of capital in France, so that most establishments are conducted on a comparatively small scale. Coal is found in many parts of France, but, as seen above, the supply is insufficient for the wants of the country, and is comparatively dear; and iron, a cheap and abundant supply of which is so indispensable to manufacturing eminence, is much higher priced than in England.

Arms are principally made at Tulle, St. Étienne,

and Klingenthal. Bronzes of a very superior quality are principally made in Paris. The trade in cutlery, which employs a great many hands, is principally carried on in Paris, Langres, Nogent-le-Rotrou, Chatelherault, Thiers, &c. French cutlery is, speaking generally, very inferior to that of England. The ornamental jewellery trade centres chiefly in Paris; and there, and in other parts of the country, about 50,000 hands are employed in watchmaking.

The silks of France are unrivalled among those of Europe, and are, in many respects, superior to those produced in any other part of the world. At the close of last century, it was ascertained by a series of accurate experiments, that French organzined silk was 25 per cent. superior in elasticity to the best Piedmontese, and its tenacity as 26 to 21 or 20. (Bowring's Second Report, p. 4.) But besides this the French silks are distinguished by superior taste and elegance, and their excellence is sufficiently proved by the fact that 4-5ths of them are exported. According to official returns, the quantity of silk, in cocoons, produced in France, has of late been on the decrease. The five years, 1860–4, showed the following result:—

Years	Quantities	Value
	Kilogrammes	Francs
1860	26,000,000	117,000,000
1861	11,500,000	50,915,000
1862	19,500,000	90,000,000
1863	7,500,000	57,000,000
1864	7,500,000	67,000,000

Lyons, Nimes, Avignon, Tours, Saint Jean-de-Gard, Alais, Le Vigan, Saint Etienne, and Paris are the principal seats of this important manufacture, which will be found more particularly alluded to under those separate heads. The silk manufactures of Paris have increased very considerably of late years.

The woollen manufacture of France is also of great value and importance, and is one of those that seems well adapted to the country. The total value of the woollen fabrics annually produced in France has been estimated by French writers at 420,000,000 fr., or 16,800,000l., which, however, is probably over the mark. The chief seats of the woollen manufacture are Sedan, Louviers, Elbeuf, Rouen, Bernay, Castelsee, Darnetal, Clermont l'Hérault, Lodève, Carcassonne, and Chateauroux; for carpets, Paris, Aubusson, Abbeville, and Amiens.

The progress of the cotton manufacture in France since 1815 has been great. During the later years of the war, the difficulties in the way of importing raw cotton into France were such that its price in Havre was usually twice or thrice more as great as its price in Liverpool. When, therefore, the return of peace enabled the French manufacturers to obtain supplies of cotton at the same rate that was paid for it by others, the manufacture could not fail rapidly to advance; and foreign cotton goods being excluded, it necessarily went on increasing till the home demand was pretty well supplied. But beyond this limit it has not been, and, it is most probable, it will not be, materially advanced. The French excel in the brightness and durability of their dyes; but, with this single exception, they are behind us in all that is indispensable to success in this department. Their machinery is at once more expensive and less improved, as coal, which may be said to be the nerves and sinews of the business, costs double in Rouen what it costs in Manchester or Glasgow. Previously to the revolution little cotton yarn was spun by machinery; but since that epoch, or rather since the peace of 1815, machinery has been imported from England, and cotton spinning has become a principal branch of industry; and, excepting some of the highest numbers for the muslin factories of Tarare and St. Quentin, and the lace manufactures of Calais and Douai, the country now supplies sufficient yarn for its own demand. The total annual value of the cotton manufactures of all kinds has been estimated by some French writers at no less than 600,000,000 fr., or 20,400,000l. The exports of cotton goods of all kinds from France amounted to 172,800,000 fr. in 1856; rose to the enormous sum of 823,200,000 fr. in 1857; but sank again to 111,000,000 fr. in 1858. In the six years 1859 to 1864, the exports fluctuated between 172 and 160 millions of francs. After England, France is the greatest consumer of cotton. Of the 160 millions sterling which represent the average annual produce of the cotton manufacture, in the ten years ending 1864, the share of France amounts to 32 millions sterling, or 800,000,000 francs. According to an official report of Jan. 1865, there are 5,593,765 spindles, with rather more than half a million of hands employed in the cotton manufacture.

Linens are manufactured principally in the N. provinces, and in Brittany, Maine, Dauphiny, and Auvergne. The best cambrics and muslins are made, the former at St. Quentin, Valenciennes, Cambrai, and Solesmes; and the latter at Lyons and Alençon. St. Quentin, Lille, Calais, Tarare, and Douai are particularly noted for their tulles and broderie; but this latter branch of industry has been long in a very depressed state. Valenciennes, Lille, Dieppe, Alençon, St. Lô, and Avranches are noted for their lace; and Caen, Bayeux, Bayeux, Chantilly, and Le Puy for their blondes. Kid gloves are made principally at Paris, Grenoble, Chaumont, Blois, and Vendôme. Other leathern articles, as shoes and saddlery, are, of course, made in large quantities. The French saddlery exported is worth about 4,000,000 fr. a year. The china of Sèvres, and other fine kinds of French porcelain, are much and justly esteemed. There are, in all, twelve manufactories of fine porcelain, at Paris, Sèvres, Limoges, and Toulouse, producing goods to the value of from 8,000,000 to 10,000,000 a year. The value of the exports of fine porcelain, in the year 1865, amounted to 4,921,813 francs, representing 2,050,765 kilogrammes in weight, while of common porcelain the exports, in the same year, amounted to 4,590,215 francs, representing 3,058,514 kilogrammes in weight. The total exports of porcelain and earthenware rose from 8,410,000 francs in 1861, to 10,723,000 francs in 1863, and to 11,045,000 francs in 1865. Glass to the value of above 20,000,000 fr. with bricks, tiles, furniture, mineral acids, and other chemical products, glue, oil-cloth, cordage, soap, musical instruments, liqueurs, paper, paperhangings, and hats, are other important articles of manufacture.

Commerce.—The commercial intercourse of France with other nations has enormously increased since the adoption of the principle of free trade. It was first put into practice in the commercial treaty with Great Britain, signed Jan. 23, 1860, due chiefly to the wise and energetic labours of the late Richard Cobden. This treaty was followed by others, of a similar nature, with the chief continental states. The following table, compiled from the official returns of the French customs, shows the state of commercial relations between France and Great Britain, both before and after the treaty:—

IMPORTS FROM GREAT BRITAIN TO FRANCE.

Before Conclusion of the Treaty.			Since the Conclusion of the Treaty.		
	Value in Millions of Francs			Value in Millions of Francs	
Years			Years		
1857		271·4	1860		334·4
1858		241·6	1861		674·6
1859		372·2	1862		535·7
Total .		881·2	Total .		1,271·7
Average .		297·1	Average .		471·9

EXPORTS FROM FRANCE TO GREAT BRITAIN.

1857		364·7	1860		599·9
1858		474·1	1861		644·4
1859		501·2	1862		619·3
Total .		1,041·1	Total .		1,476·9
Average .		400·0	Average .		458·2

TOTAL OF IMPORTS AND EXPORTS.

1857		706·2	1860		987·2
1858		647·7	1861		864·5
1859		863·9	1862		1,115·7
Total .		2,264·4	Total .		2,947·0
Average .		745·1	Average .		982·2

Taking the three years' average on each side, it will be seen that the imports increased by 157,000,000 francs, or 58 per cent., and the exports by 90,000,000, or 19 per cent.

The total value of the produce and manufactures of the United Kingdom exported to France in the year 1863 was 8,673,392l. against 9,219,367l. in 1862, 8,035,504l. in 1861, 5,249,940l. in 1860, and 4,751,254l. in 1859. The business done with France in 1863 was thus nearly double the corresponding total for 1859. If we carry the comparison back for ten years and compare 1859 with 1863, we shall see that the French demand for British products has more than quadrupled in the last 14 years. Thus in 1858 the value of our exports to France was 1,803,131l.; in 1857, 6,213,552l.; in 1856, 6,432,650l.; in 1855, 6,017,659l.; in 1854, 3,175,294l.; in 1853, 2,636,330l.; in 1852, 2,731,294l.; in 1851, 2,928,462l.; in 1850, 2,401,956l.; and in 1849, 1,951,269l. The chief article of export from Great Britain to France is woollen goods, which were in 1863 delivered to the extent of 3,410,562l. In 1859 the corresponding total was only 213,226l. Of coal, cinders, and coke the value of our exports to France in 1863 was 543,733l. against 615,837l. in 1859; of wrought and unwrought copper, 856,390l. against 493,063l. in 1859; of cotton goods, 556,719l. against 222,262l. in 1859; of wrought and unwrought iron, 825,812l. against 895,133l. in 1859; of steam engines and machinery, 863,812l. against 199,102l. in 1862; of sheep and lambs' wool, 348,771l. against 426,942l. in 1862; and of woollen and worsted yarn, 309,122l. against 176,114l. in 1862.

The foreign trade of France is divided, in the official returns, into the two great divisions of 'general commerce,' including the aggregate of all commercial transactions, and 'special commerce,' embracing only such imports as are consumed in France and such exports as have been manufactured within the country. The transactions are moreover classified according to 'real values' and 'official' values. Official value is fixed according to a basis determined in 1826, which represents the average values with a view to reduce all the merchandise to a common unity, an arrangement which allows of totalizing and comparing, on a uniform and invariable scale, the results obtained at different periods. The real value is, on the contrary, essentially variable, changing with the rise and fall of prices. It is

Vol. II.

fixed by a permanent commission in the Department of Agriculture, Commerce, and Public Works, aided by the Chambers of Commerce; and its object is to determine, as accurately as possible, the average price of each kind of merchandise for the time being.

The official reports of 'general commerce' during the year 1864 show that the exports exceed the imports by more than 453,000,000l., the imports being estimated at 2,640,700,000l., and the exports at 2,939,000,000l. The imports and exports both show a gradual increase as compared with the imports and exports of the three preceding years, 1861, 1862, and 1863. The importation of horses, horned cattle, and sheep has increased from 512,000 in 1862 to 638,000 in 1863, and 775,000 in 1864. The importation of hides nearly doubled since 1862, being almost 80,000,000 kilogrammes in place of 45,500,000. The importation of wool increased from 44,000,000 to 63,000,000. Cotton, of which the importation fell to 50,000,000 kilogrammes in 1862, rose in 1864 to 67,000,000, India and Egypt supplied the greater part. Sugar from the French colonies, of which there were 104,000,000 kilogrammes imported in 1862, and 125,000,000 in 1863, fell in 1864 to 80,000,000, while foreign sugar rose from 108,000,000 in 1862 to 112,000,000 in 1863, and to 133,000,000 in 1864. The importation of coffee rose from 37,000,000 to 44,000,000 kilogrammes.

The exports show that silk goods maintain themselves under the most adverse circumstances. Disease has killed the silkworm, the importation of raw silk has not increased, and still the value of the silks exported has risen from 332,000,000l. in 1861, to 844,000,000l. in 1863, and to 399,000,000l. in 1865. This may be accounted for by the increased value of the article. The prosperity of the woollen manufacturers is more clearly shown. The value of the woollen goods exported increased from 187,000,000l. in 1863 to 855,000,000l. in 1864, and to 876,000,000l. in 1865. The export of Bordeaux wine in cask has increased gradually from 550,000 hectolitres in the year 1862, to 635,000 in 1863, and to 664,000 in 1864. The other French wines exported rose from 1,166,000 hectolitres in 1862 to 1,211,000 in 1863, and to 1,119,000 in 1864. Brandy distilled from wine exported rose from 162,000 hectolitres in 1862 to 190,000 in 1863, and to 223,000 in 1864. Brandy distilled from molasses and rice rose from 209,000 hectolitres in 1862 to 260,000 in 1863, and fell to 204,000 in 1864. The total value of all French wines exported to foreign countries amounted to 195,923,000l. in 1861, to 210,000,000l. in 1862, to 279,730,000l. in 1863, to 234,538,000l. in 1864, and to 260,001,000l. in 1865.

The relative importance of the trade of France with the various foreign countries is shown in the subjoined tables, which give the value of the imports and exports, in millions of francs, for the year 1864:—

Imports from	General Commerce	Special Commerce
	Millions of Francs	Millions of Francs
Great Britain . . .	636	296
Belgium . . .	471	256
Switzerland . . .	270	50
Italy . . .	276	172
Germany—Zollverein	271	188
Turkey . . .	177	130
Russia . . .	95	71
United States . . .	85	84
Brazil . . .	80	62
India—British . . .	76	76
Spain . . .	73	55

U B

Exports to	General Commerce	Special Commerce
	Millions of Francs	Millions of Francs
Great Britain	811	670
Switzerland	345	186
Italy	374	319
Germany—Zollverein	273	219
Belgium	230	206
Spain	204	137
United States	121	106
Brazil	97	83
Turkey	83	40
Russia	43	23

It will be seen that, both in respect to imports and exports, the commercial intercourse with Great Britain is by far the most important for France. In this respect the position of both countries is radically different. For while to France British intercourse is, at the least, thrice as valuable as that of any other country, French commerce is to Great Britain of inferior importance, no less than five other countries, India, the United States, Germany, Australia, and Turkey, taking precedence. (See 'Table of the Chief Markets of Great Britain in their order of importance,' under Great Britain.)

Metrical System, Weights and Measures.—By a law of the French National Assembly in 1795, a uniform system of weights and measures was introduced, all measures being derived by the decimal multiplication or division of the *mètre*, which is equal to the 10-millionth part of the distance between the equator and the pole. According to this data, the measures of length are:—

The Millimètre	=	0·039 inch, Engl.
Centimètre	=	0·394 do.
Decimètre	=	3·937 do.
Mètre	=	3 ft. 3·371 in.
Decamètre	=	32 ft. 9·7 in.
Hectomètre	=	109 yds. 1 ft. 1 in.
Kilomètre	=	1093·633 yds.
Myriamètre	=	10936·331 yds.

In like manner the *are* (100 sq. mètres, or about 1/40th part of an Eng. acre) is multiplied into the *hectare* 2·741 acres), &c.; the *litre* (a cubic *decimètre* and 0·2 of a gallon, or a little more than a quart Eng.) into the *hectolitre* (2·76 bushels), &c.; and the *gramme* (0·0353 of an oz. avoird.) into the *kilogramme* (2·206 lbs. avoird.), and other weights.

But, besides the foregoing, the ancient French measures are still to some extent in use: as the inch (equal to 1·066 Eng. in.); the foot (1 ft. 0·789 in. Eng.); *aune* 1·3 Eng. yds.); *toise* (6 ft. 4·735 in. Eng.); the *league* of 2,000 *toises* (2 miles 743 yds.); the *league* of 25 to the degree (2 m. 1,540 yds.), &c. The *arpent* is equivalent to 1·043 Eng. acre.

The French pound is equal to 1·080 lbs. avoird.; the *muid* = 1·124 hhds.; the *boisseau* = 0·369 bushels; and the *setier* = 4·430 bushels.

Money.—Accounts are kept in France, a silver coin worth 9·69d. Eng. which is divided into 10 *décimes* and 100 *centimes*. The par of exchange with England is very near 25 francs per pound sterling.

Roads.—The aggregate length of roads throughout France is about 65,000 m. The roads are divided into national, departmental, and communal; their expenses being respectively defrayed by the government, and the *dépts.* or communes to which they belong. The national roads have a united extent of about 22,000 m., of which about one-eleventh part is paved, and the rest macadamised, or constructed in the ordinary manner. They are commonly well made, and very direct; their construction and repair, as well as those of

the departmental roads, being under the superintendence of the central board of bridges and public ways, which has a head engineer established in each dép. The communal roads, which are subject to no such control, are mostly in a deplorable state, and are often impracticable for carriages.

Railways.—The first railways in France were coal lines in the valley of the Loire. The earliest 'concession,' or permission on the part of the government to construct a line, was given, under date of Feb. 26, 1823; but it was not till 1830 that railways came to be used for passengers as well as merchandise. Even now there was very little progress in making new lines; yet the subject having attracted public attention, there was a long discussion, extending over twelve years, in the chambers and the press, as to whether railways ought to be constructed by private enterprise, as in England, or by the state, as in Belgium. The discussion ended in a compromise, embodied in the law of June 11, 1842. This law, in principle, gave the construction of railways to private companies, but under a government guarantee, and a condition that the lines thus built should become the property of the state after a certain term of years. This term was originally fixed at 35 years; but subsequently, by a law passed in 1852, enlarged to 99 years. Under the law of 1842, concessions were granted to a number of companies—from Paris to Strasbourg, Tours to Nantes, Bordeaux to Cette, Paris to Cherbourg, Paris to Lyons, and others. Some of these companies were unable to fulfil their engagements, and had to demand assistance from the state. To prevent this as much as possible, the government tried to bring about an amalgamation of existing companies, and the plan succeeded completely. Of 59 companies which had been successively created, there were, at the end of 1851, only 27 in existence, which divided among themselves lines of the length of 8,918 kilomètres, or 145 kilomètres per company. The movement of amalgamation still continuing, on the 31st Dec. 1854, 6 great companies divided among them a connected length of 16,850 kilomètres of lines, or an average of 2,717 kilomètres per company. In 1853, a new arrangement was come to between the government and the various railway companies, by which all the lines were classed under two categories, called distinct categories, under the designation of the old and the new réseau. The following table shows how this division was established:—

Name of Company	Old Réseau	New Réseau	Total
	Kilomètres	Kilomètres	Kilomètres
Orléans .	1,704	1,169	2,873
Lyons and Mediterranean	1,836	2,484	4,320
Great Northern (Nord)	967	618	1,585
Eastern (Est)	845	1,365	2,210
Western (Ouest)	1,192	1,112	2,304
Southern (Midi)	794	825	1,673
Other Companies	234	..	234
Total .	7,774	8,478	16,252

The old réseau, which is entirely constructed and open for traffic, has its own separate accounts, and provides for its expenditure and the interest of its capital from its own resources. On the other hand, the new réseau, of which only a minor portion is completed, is authorised to place the interest of its capital under the head of 'expenses of construction' until the whole réseau is finished. Once completed, the government guarantees to the companies a minimum of 4·65 per cent. on the capital of, in round numbers, about 3,400,000,000 francs, or 120,000,000l., which it is estimated the construction of the 8,678 kilomètres of the new réseau will cost. The old réseau, when its profits exceeded a certain amount fixed beforehand for each company, is bound to contribute towards covering the possible deficit in the revenues of the new réseau, and the expenses of the government on account of its guarantee of interest, will be so much reduced. In return for the state guarantee, the companies have undertaken to complete the great réseau, or 'net-work' of railways, embracing a total length of 2,729 kilomètres of line, of which 1,978 kilomètres are finally decided upon; the rest being contingent. The lines which are finally decided upon are to be constructed within the term of 8 years from 1863; but the state is bound to contribute to them, not only by a guarantee of interest, but by direct subventions of considerable amount; in round numbers about 236,000,000 francs. The following table shows how the concessions and subventions are divided among the five great companies:—

Name of Company	Length of Concessions in Kilomètres			Subventions
	Déjà décidé	Contingent	Total	
				Francs
Paris, Lyons, and Mediterranean	409	431	840	45,700,000
Eastern (Est)	672	84	724	42,800,000
Southern (Midi)	720	9ca	856	70,000,000
Orleans	726	87	873	16,000,000
Western (Ouest)	143	61	204	21,300,000

The Great Northern of France is not included in this list, having been no party to the new arrangements come to in 1862, which make the railways more dependent upon the government. This line holds a rather independent position, the greater number of the shares of the company being held by the house of Rothschild.

The following table shows the progress as well as the income of French railways during the ten years 1852–61:—

Years	Length of Lines Open	Receipts per Kilomètre
	Kilomètres	Francs
1852	3,684	33,117
1853	3,978	41,313
1854	4,358	43,683
1855	4,993	41,417
1856	5,392	48,046
1857	6,084	45,923
1858	8,100	41,330
1859	8,531	44,90
1860	9,271	44,724
1861	9,648	41,343

Whilst the receipts per kilomètre have increased within the ten years, the working expenses have diminished, especially on the lines of the old réseau. Thus, on the Great Northern, the expenses absorbed during 1861 but 37·4 per cent. of the receipts, against 38·4 per cent. in 1860; on the Orleans line, 29·7 per cent. against 30·6 per cent.; on the Southern line, 89·0 per cent. against 42·3 per cent.; on the Western line, 41·6 per cent. against 48·3 per cent.; on the Eastern line, 39·1 per cent. against 40·8 per cent.; on the Mediterranean line, 37·6 per cent. against 89·1 per cent. The six great companies here mentioned include 95 per cent. of the whole French network, or 'réseau' of railways; and in consequence of the diminution of the working expenses, and the increase of the receipts per kilomètre, these companies have been enabled to give their shareholders high returns, as will be seen by the following tabular statement of the dividends given during five years:—

Name of Company	1857		1858		1859		1860		1861	
	Fr.	c.	Fr.	c.	Fr.	c.	Fr.	c.	Fr.	c.
Nord	60	0	61	0	63	80	63	50	66	0
Orléans	50	0	57	0	57	2	104	0	103	0
Midi	70	0	28	0	27	8	33	0	20	0
Ouest	27	50	33	0	37	80	37	80	42	80
Est	40	65	40	45	34	70	40	0	40	0
Mediterranée	53	0	49	50	53	85	63	50	73	0

The railway shares being 500 francs (except the Great Northern railway shares, which were issued at 400 francs), the dividends given in 1861 represent for the original shareholders a rate of interest varying between a minimum of 8 per cent. and a maximum of 20 per cent. in the case of the Orléans line. (Report of Mr. Grey, Her Majesty's Secretary of Embassy, in Reports presented to Parliament, 1864.)

The total receipts on the six principal railways of France in the 52 weeks ending 29th Dec. 1864, amounted to 20,519,121l., and for the same period in 1863, to 19,460,090l., showing an increase of 1,059,031l., or 5·44 per cent. The receipts on the Paris and Mediterranean amounted to 5,871,605l., against 6,791,647l. in 1863, showing an increase of 76,958l.; on the Paris and Orléans the receipts were 3,405,333l., against 3,290,510l., showing an increase of 198,783l. On the Eastern the receipts amounted to 3,317,793l. against 2,997,365l. in 1863, showing an increase of 320,428l. On the Northern the receipts were 2,929,735l., against 2,774,961l. in 1863, showing an increase of 154,774l. On the Western the receipts amounted to 7,483,557l., against 2,248,815l., showing an increase of 234,722l.; and on the Southern the receipts 1,427,190l., against 1,353,770l. in 1864, showing an increase of 73,850l.

Canals.—The entire length of the communications by means of navigable rivers and canals was, in 1864, estimated at 7,668 m., of which extent nearly five-sevenths were contributed by the former. There were then 74 navigable canals completed; 16 more were in progress of construction; and 14 others were projected. The principal existing are as follows:—the Canal du Midi, or the Languedoc Canal, which runs from Cette to Toulouse, where it joins the Garonne, and thus connects the Mediterranean with the Atlantic; the Canal of Charolais, or du Centre, connects the Loire with the Saône; the Canal of the Rhine and Rhone (du Monsieur) forms a communication between those rivers by connecting the Saône with the Doubs, and the latter with the Ille, a tributary of the Rhine; the Canal of Burgundy connects the Saône with the Yonne, and consequently the Seine with the Rhone and Rhine; the Canal of Briare, and that of Orleans, unite the Loire with the Loing, a tributary of the Seine; that of St. Quentin connects the Escaut with the Oise; that of Brittany, the longest of all, being upwards of 230 m. in length, runs between Nantes and Brest. Those of Berri, Ardennes, the Ille et Rance, Nivernais between the Loire and Yonne, d'Ourcq, which supplies Paris with water, and Somme, are the others most worthy of notice. (Official Tables; Encyc. des Gens du Monde.)

Government.—Previously to the revolution of February, 1848, the government of France, as fixed by the charter of 1830, was a limited monarchy, hereditary in the male line only with a representative assembly of 459 members, chosen by the electoral class, and a house of peers. The constitution of 1848 voted by the republican National Assembly, vested the legislative, and part of the executive power in a parliament of 750 members, elected by universal suffrage. This charter had to give way to another, which was decreed 'in virtue of the powers delegated by the French people to Louis Napoleon Bonaparte, by the vote of the 20th and 21st of Dec. 1851.' It bears date of Jan. 14, 1852; was promulgated Jan. 22, 1852, and subsequently modified by the senatus-consulte of Nov. 7, 1852, the imperial decree of Dec. 2, 1852, the 'organic decree' of Dec. 18, 1852, the senatus-consulte of Dec. 25, 1852, of Feb. 2, 1861, and of Dec. 31, 1861. These statutes recognise five powers in the state—namely (as cited in the preamble of the constitution of of Jan. 14, 1852)—

1. The executive power, represented by the emperor.
2. The ministers, nominated solely by the emperor.
3. A council of state, preparing laws under the direction of the ministers.
4. A legislative body, nominated by universal suffrage, 'discussing and voting laws.'
5. A 'second assembly,' formed of eminent men, acting as a moderating power — pouvoir pondérateur—the guardian of the constitution and of the liberties of the nation.

The emperor is irresponsible, and his person is inviolable. He appoints and discharges his ministers, has the right to pardon criminals, and is the fountain of all honours and dignities in the state. He commands in chief the armies and navies; has the right to make peace and to declare war; to enter into commercial, offensive, and defensive alliances with other sovereigns and nations, and to nominate to all charges, appointments, and offices whatsoever in the realm. He has the sole initiative in legislation, and justice is rendered in his name. No law is valid unless sanctioned by the emperor, and no person can hold any employ without taking the oath of fidelity to his majesty.

The ministers are appointed solely by the emperor, and hold office at his pleasure. They are responsible to the nation, but only for their individual acts. There is no community of action between them, each directing the affairs only of his own department. The senate alone can bring a bill of accusation against the ministers.

The council of state is composed of from 40 to 50 members, nominated by the emperor, and liable to be displaced by him. The duty of the council of state consists in preparing, under the direction of the sovereign and his ministers, such projects of law as are to be laid before the legislative body, and 'to solve any difficulties which may arise in administrative matters'—'de résoudre les difficultés qui s'élèvent en matière d'administration.' The council of state has to defend before the senate and the legislative body the laws proposed by the government, a number of members being appointed for this particular purpose every session by the emperor. Each member of the council of state has a salary of 25,000 francs, or 1,000l. per annum. The ministers take part, ex officio, in the deliberations of the council of state.

The members of the legislative body are elected by universal suffrage, at the rate of one member to every 35,000 electors. They are chosen for six years, and receive a salary of 2,500 francs, or 100l. a month, during the period of each session, whether ordinary or extraordinary. It is the duty of the legislative body to discuss and vote any laws sent before it by the council of state, as well as the annual budget of income and expenditure presented by the government. The ordinary session of the legislative body lasts six months, and the sittings are public; but on the

demand of five members, the public may be excluded. The president and vice-president of the legislative body are nominated by the emperor, for the period of a year. The legislative body cannot receive petitions. The emperor summons, prorogues, and dissolves the legislative body; but, in case of dissolution, new elections must take place within six months.

The 'second assembly,' cited in the preamble of the constitution 'formed of eminent men, acting as a moderating power,' is called the senate. The assembly is composed of the cardinals, marshals, and admirals of the realm, and a number of other members, not exceeding 150, nominated by the emperor. Each senator has a salary of 30,000 francs or 1,200l. per annum. The dignity is irrevocable and for life; the members of the senate, however, are allowed to resign their post. No vote of the legislative assembly is effective without the sanction of the senate, and the latter alone has the right to receive petitions. Changes in the fundamental laws of the realm may be proposed by the senate, with the concurrence of the ministers; and, should such modifications be approved of by the emperor, they are called senatus-consulta. The president and vice-president of the senate are nominated by the emperor for the period of one year. It is the special duty of the senate to oppose the promulgation of all laws contrary to the constitution, religion, public morals, freedom of conscience, individual liberty, and equality of all citizens before the law. The senate is summoned, and the duration of its sittings fixed by imperial decree.

There are eleven ministerial departments. According to an imperial decree, promulgated in the 'Moniteur' of Dec. 21, 1860, the ministers take rank according to the length of time during which they have been members of the council, with the exception of the minister of state, who has the precedence of all the rest. (Annuaire Diplomatique; Moniteur Universel.)

The 89 deps. of France are subdivided into 373 arrondissements, and these again into 2,938 cantons, and 37,510 communes. Each dép. is governed by a prefect, with a salary varying from 10,000 to 40,000 fr. a year, except in the dép. Seine, where the salary of the prefect is 100,000 fr. Each arrond. is superintended by a sub-prefect, with a salary of 4,000 fr. a year; and each commune by a mayor and other magistrates, whose services are gratuitous. The prefect is assisted by the council-general of the dép., which consists of a member from each canton, and meets once a year: a great deal of the internal administration of the dép., as the distribution of taxation, is undertaken by this council. The sub-prefects and mayors are also aided by councils elected by the citizens. All the mayors are nominated by the government, or the prefect; but the communes have certain rights and privileges of their own, which cannot be interfered with by the state, though the latter has perfect command over the administration of the deps. and arronds.

Justice.—The administration of justice in France, previously to the revolution, was, in the last degree, partial and corrupt. Justice in fact was, in the vast majority of cases, openly bought and sold; and a poor man without powerful protectors could never hope to succeed in any case. The institution of juries was unknown; and the criminal law was, if possible, in a still more vicious and degraded state than the civil. Happily, however, three things are now matter of history. The revolution swept off every vestige of the old system of jurisprudence, and of the useless and flagrant abuses that had grown up under it. The present civil and criminal law of France has been embodied in codes drawn up, under the auspices of Napoleon I., with singular perspicuity and brevity; and is honestly and impartially administered.

The ordinary judicial tribunals are of six kinds, as follows:—Simple police courts, tribunals of justice of the peace, courts of original or primary jurisdiction (*tribunaux de première instance*), imperial courts, courts of assize, and the court of cassation. The extraordinary tribunals are—citizens' benches called *conseils des prud'hommes*, tribunals of commerce (the *tribunaux de commerce*), courts martial, university and other special courts, and the senate and legislative body. In each commune there is a police court in which the mayor presides; and in every canton there is at least one justice of the peace, appointed by the government, with power to decide in civil causes under the value of 100 francs; his decision in these under 50 francs being without appeal. There is in each arrond. a court of original jurisdiction to decide without appeal in causes not above the value of 1,000 fr., as well as appeals from the simple police courts. These are composed of from three to twelve judges. Superior courts are established in the principal cities and towns, and have jurisdiction throughout a territory including from one to seven dep. They are composed of a president, several vice-presidents, some legal functionaries, and from 20 to 60 counsellors; they are almost exclusively courts of appeal from the last mentioned courts, and the tribunals of commerce. The courts of assize are temporary tribunals which take cognizance of criminal cases; one is holden at certain periods in each dep. In these, tried by jury is adopted; the juries are composed of 12 citizens above 30 years of age, who are either taxed directly to the amount of 200 fr. yearly, or have belonged to certain professions. There are three judges, one of whom is a counsellor belonging to a royal court. The decisions of these courts are commonly without appeal, and can only be annulled by the court of cassation on the plea of informality. The last-named tribunal is a supreme court of appeal in both civil and criminal cases. It is composed of 49 members (including a first president and 3 others), appointed for life by the government. Each member must be at least 40 years of age, and have a legal diploma; but no member may practise in the legal profession, or exercise any public function, but such as may be connected with his duty in the court. The court of cassation is divided into three separate chambers of 15 members and a president each. It may suspend the functions of any subordinate judges, and summon them before the minister of justice to answer for their decisions; and it has the highest and most absolute authority in all judicial matters.

The *cour des comptes* is established to audit and examine all accounts connected with the public revenue and expenditure. It ranks immediately after the court of cassation, and is organized in a similar manner. The *conseils des prud'hommes* and tribunals of commerce are established in the principal manufacturing and commercial towns, being composed chiefly of commercial men. The former tribunals determine disputes between the manufacturers and the workmen employed by them; the latter decide in cases to the value of 1000 fr., but do not themselves see their decisions enforced.

According to the reports published by the minister of justice, the yearly average of heavy crimes committed in France during the 7 years ending with 1864 amounted to 7,690. Female criminals compose about one-fifth of the whole. Crimes against the person are most common in the deps. of the centre and S.; their number being greatest in Corsica. Crimes against property abound most in the N. The prisons are divided into the 4 classes of, 1st, departmental prisons; 2d, *maisons centrales de détention*; 3d, prisons for juvenile offenders; and, 4th, *bagnes*, of which there are 3 at Brest, Rochefort, and Toulon, in which there are about 7,000 criminals. The *dépôts de mendicité*, of which there are 5, are also a species of prisons, or of *maisons de détention*.

Religion.—Religious toleration exists in a widely extended degree. Each citizen professes freely his religion, and receives from the state for the exercise of his worship an equal protection. Hence, when it is said that the Roman Catholic is the dominant religion in France, all that is meant is that it is the religion of the greatest number of the population. The creeds of both Protestants and Jews being recognized by law, their clergy receive public pensions. The population of France, on January 1, 1862, consisted of 35,734,667 Roman Catholics, 1,561,250 Protestants, 116,000 Jews, and 20,815 members of other sects and forms of belief. In Algeria there were, besides, 2,775,291 Mohammedans. In regard to Protestants, this official statement is somewhat at variance with that of the synods and consistories, the heads of which estimate the members of the Reformed Church at 1,300,000, and those of the Lutheran Church at 700,000. All religions are recognized by the state, but only the Roman Catholics, Protestants, and Jews are noticed in the budget; the latter only since 1831. In the budget of 1862 the allowances to the Roman Catholic clergy amount to 42,819,936 francs, or very nearly 2,000,000l. sterling; and those to the Protestant Church, 1,423,436 francs, or 56,737l. The whole income of the Roman Catholic clergy, from public and private sources, is computed to amount to above 100,000,000 francs, or 4,000,000l. sterling; and that of the Protestant ministers at about 150,000l. There are eighty-four prelates of the Roman Catholic Church—namely, seventeen archbishops and sixty-seven bishops. The archbishop of Paris has a salary of 50,000 francs, or 2,000l., and each of the other archbishops of 20,000 francs or 800l.; while the sixty-seven bishops have an income of 16,000 francs, or 600l. each. An extra-allowance of 10,000 francs or 400l. is made to six of these prelates, on account of their being cardinals, and, as all cardinals are *ex-officio* senators, the further sum of 8,000 francs, or 1,200l., is further due to them in this capacity. The other Roman Catholic clergy comprise 178 vicars-general, with salaries of from 1,500 to 2,000 francs, or 60l. to 100l.; 688 canons, with allowances varying from 1,200 to 2,400 francs, or 84l. to 96l.; 3,136 curés, or incumbents with incomes ranging from 1,200 to 1,600 francs, or 48l. to 64l.; and 30,743 desservants, or curates, with stipends of from 900 to 1,200 francs, or 36l. to 48l. The Protestants of the Augsburg Confession, or Lutherans, are, in their religious affairs, governed by a general consistory, established at Strasbourg; while the members of the Reformed Church, also called Calvinists, are under a council of administration, the seat of which is at Paris. The Jewish priesthood consists of ten high rabbis, with salaries of from 3,440l. to 7,000 francs, or 140l. to 240l.; fifty-one rabbis, with incomes ranging from 800 to 1,500 francs, or 32l. to 60l.; and sixty-two precentors, with allowances from 500 to 2,000 francs, or 20l. to 80l. The Lutherans have a seminary and a faculty of theology at Strasbourg, with fifty-three churches; and the Calvinists have consistorial churches in fifty-nine departments, who meet occasionally in synod, and have a faculty of theology at Montauban. The Jews have a central consistory at Paris, and sixty synagogues distributed over the country. (Moniteur; Le Lien, Protestant Paper; Stateman's Year-book.)

Public Instruction.—The proportion of individuals receiving education to the whole pop. was in 1820 estimated at 1 in 37; in 1830 at 1 in 20; in 1840 at 1 in 10; and in 1861 at 1 in 7. Instruction is primary, secondary, or superior. To afford the first, every commune is obliged by law to support at least one primary school, either of its own, or in conjunction with neighbouring communes. Reading, writing, the French language, the first rules of arithmetic, weights and measures, the first lines of geography, and history and drawing, are the principal branches of education in these schools; they are afforded gratuitously. According to official returns, there were, in October, 1863, in France 62,173 establishments of primary instruction, or 16,136 more than in 1818; and the scholastic population, which at this last period was only 3,771,507, had risen in 1862 to 4,731,946, giving an augmentation of nearly a million, or a quarter of the whole. The 36,499 communes provided, in October 1863, with means of instruction, comprised 41,426 public and free schools, special for youths or mixed as to the sexes, of which 37,895, numbering 2,145,420 pupils, were directed by laics, and 3,531, numbering 482,096 pupils, had 'congregationist' masters. Of the 2,627,476 children in these schools, 922,870, or more than one-third, were admitted gratuitously. The number of schools for girls, in October 1863, amounted to 26,321; of which 13,491 were directed by laics provided with diplomas of capacity, and 13,101 by religious sisters, of whom 12,335 had only the 'letter of obedience.' These schools received 1,649,213 pupils, of whom rather more than a third, or 604,247, were in the lay schools, and 1,039,906 in the congregationist establishments. One quarter of these pupils were admitted gratuitously, viz. 130,210 in the lay, and 490,094 in the congregationist schools; total 620,304. The emoluments of the female public teachers amounted to 9,169,000 francs, giving an average annual salary of 655 francs, or 26l. per head. (Exposé de la Situation de l'Empire; Legoyt, Statistique de la France.) There are 26 academies for superior instruction, one in each of the cities and towns in which there is an imperial court of justice, excepting Algeria. Each of these academies is governed by a rector and has 2 inspectors, who visit in turn all the schools, both private and public, within their separate jurisdictions. The faculties of these academies are empowered to grant the degrees of doctor, licentiate, and bachelor. There is in Paris an école normale, or academy for the education of professors for the colleges throughout the country, the institution of which has been of the greatest service. The whole of the foregoing establishments constitute the university of France, which is presided over by the minister of public instruction and a council of 9 members; under whose authority 17 inspectors-general visit all parts of France, to ascertain the state of education. There are some establishments, however, which are beyond the jurisdiction of the university. Such are the College of France, the Museum of Natural History, the École des Chartes, School of Oriental Languages, the Institute, the most celebrated literary association in Europe, and an abundance of societies of all kinds for the advancement of knowledge. The College of France, founded by Francis I. in 1530, enjoys a high celebrity. It is wholly devoted to the pursuit of the highest branches of art and science; most of its courses of instruction

are elementary. It had, in 1845, 78 professors. The public libraries in the dep. contain nearly 4,500,000 vols.; those of Paris (8 in number), in addition to the foregoing, contain 2,100,000 vols. At many of the institutions in the cap. lectures on most branches of science are delivered gratuitously by professors of acknowledged eminence.

Public, Charitable, and other Institutions.—The amount of pauperism varies, as in other countries so in France, in different years with the varying state of the crops, the prosperous or unprosperous condition of commerce, and the facilities for obtaining employment. So much is this the case, that the numbers soliciting relief, and depending in great measure on charity, are sometimes twice or thrice times as great in one year as in another. Mendicancy, notwithstanding the efforts made for its suppression, is still very frequent. The establishments for the relief of pauperism consist of hospitals hospices (asylums), and bureaux de bienfaisance. Their funds are partly derived from the state, and partly from their own landed or other property, endowments and donations of individuals. There were, in 1861, above 2,000 hospitals and asylums, and nearly 10,000 bureaux de bienfaisance, which gave in-door and out-door relief. There are several lunatic asylums, a royal institution, and a royal hospital for the blind in Paris, deaf and dumb establishments at Paris and Bordeaux, maternity societies, others for the assistance of prisoners, the sick, and a vast number of philanthropic societies of all kinds dispersed throughout the country.

It has been estimated that the proportion of foundlings to the total number of births is about 34 to 1,000. The average number of children in the foundling hospitals of France is about 100,000. The annual expense of the foundling hospitals varies from 8 to 10 million fr.; though much reduced, the mortality amongst the foundlings is still very heavy. Among the charitable institutions are about sixty monts-de-piété or government pawnbroking establishments throughout France. They are situated in the chief towns; some, as that at Montpelier, lend money without interest; while that of Paris receives an interest of 12 per cent. on money advanced. In Hautes Alpes there are some similar institutions for the benefit of the agriculturists, in which the pledges received and the interest due are both paid in corn.

According to official documents, the number of savings-banks amounted to 478 at the beginning of 1863, and 485 at the close; connected with these are 392 branch banks, of which 73 were opened during the year. The expenses of management were 1,627,899 fr., being 80,155 fr., or about 6·25 per cent. more than in 1862. After deducting these expenses, the proper funds of the banks amounted to 12,038,225 fr., being 711,694 fr. more than in 1862. The number of depositors has increased from 1,379,374 to 1,471,347, or about 6·64 per cent. The new accounts opened in 1863 were 15,249 more than in 1862, and the average number of depositors for each bank was 3,153, against 3,061 in the preceding year. On the whole population of France, there was 1 depositor in 25 during 1863, while there was only 1 in 27 in 1862. The proportion of depositors to the population is the greatest in the department of the Seine, and the least in the Ariège, being 1 in 8 in the former, and 1 in 224 in the latter. The depositors increased in all the principal banks except five, and in these the falling off was trifling. The total amount due to depositors at the end of 1863 was 447,577,314 fr., showing an increase during the year of 23,767,655 fr., or about 5½ per cent. There were 1,058,192 accounts of 500 fr. and under; 100,221 between 501 fr. and 800 fr.; 110,173 between 801 fr. and 1,000 fr.; and 82,758 above 1,000 fr. With regard to the social position of the depositors, 83,350 are artisans and labourers 89,519 domestic servants, 11,675 clerks and shopmen, 7,663 sailors and soldiers, 55,313 of various professions, 40,583 minors, and 329 mutual benefit societies. During the year 1863 investments were made in the public funds for 13,042 depositors of capital, producing an interest amounting to 857,948 fr.

Co-operative institutions and mutual relief societies have sprung up in all directions in France in the ten years previous to 1864. From an official report addressed to the emperor by the president of the 'mutual relief' societies, for the year ending December 1863, it appears that up to that date there existed in France 4,721 societies either approved or authorised by the government. These societies comprised 676,521 members, of whom 78,511 were honorary and 597,978 participants, and of whom 506,576 were males and 91,612 females. In 1863 there was an increase of 139 over the previous year, composed of 87,478 members, 4,663 of whom were honorary and 82,815 participants, the latter counting 27,521 men and 5,294 women. The fund belonging to these societies, including what is called the retiring fund, consisted of 81,370,772 fr. The receipts for the year amounted to 11,019,519 fr., the expenditure to 8,830,433 fr. The receipts were the subscriptions of honorary members, subsidies, donations, legacies, interest from the funds, subscriptions of participant members, entrance fees, fines, &c. The expenditure consisted in relief to the sick members, doctors' fees, medicines, funeral expenses, relief to widows and orphans, pensions in sickness or old age, expenses of management (this item stands at 484,197 fr.), furniture, extraordinaries, &c. The surplus of receipts over expenditure was 2,189,085 fr. (Moniteur; Exposé de la Situation de l'Empire, 1865.)

Army.—The standing army of France dates from the time of Louis XIV.; but was organised on its present footing during the wars of the revolution and Napoleon I. The army is formed by conscription, to which every man who has reached the age of 21 is liable. An annual decree fixes the number of men to be draughted during the year. Formerly the normal number was 80,000; but during the Oriental war, in the years 1853 to 1855, the amount was raised to 140,000, and in 1857 it was settled to 100,000. At the outbreak of the Italian war it was again raised to 140,000, and remained so till 1861, when 100,000 was once more settled to be the annual number of men to be drawn for the army. The legal time of service is seven years; but the soldiers are kept seldom longer than six years under arms, and are often sent home much earlier to form, together with the young recruits, the army of reserve. Only a portion of the annual contingent of recruits are incorporated with the standing army, and the rest are drilled for six months in the departmental depôts. This period of six months may be, and is usually, extended over three years; so that the annual exercises last but two months on the average. In this manner 30,955 recruits were drilled in 1860, and 83,254 in the year 1861. The method was established by imperial decree in 1820, being a fruit of the personal experiences of Napoleon III. in Switzerland.

Every man drawn for conscription has the right to buy a substitute. Such substitutes were procured formerly through private agencies; but an imperial decree of April 26, 1855, organised a new system, making the right to furnish substitutes a government monopoly. According to this system, the re-enlistment of old soldiers is greatly encouraged, so as to give the army a standing nucleus of experienced troops, who have made the military service their life profession. The government annually fixes the price to be paid for substitutes. It was fixed in 1855 at 2,800 fr., or 112l., was lowered in 1857 to 1800 fr., or 72l., and was subsequently raised again to 2,800 fr., or 112l. In 1863 the pay for a substitute was settled by the minister of war at 2,800 fr., or 72l. This sum, increased by various other items enumerated below, is thrown into an army fund, out of which the substitutes are paid a certain amount at the time of enlistment, besides receiving an increase of pay at the end of seven years, another increase at the end of fourteen, and a premium of one franc or twopence a day after a service of forty-five years. Soldiers are allowed to re-enlist as long as they are fit for service. The number of volunteers for the army—without bounty—is on the decrease. Before the year 1854, there were, on the average, 10,000 volunteers per annum; in 1853 there were 8,000; in 1854 they rose to 10,676; in 1855 they reached the number of 21,055; in 1856 they declined to 19,546; in 1857 to 6,824; in 1858 to 11,845; in 1859 to 2,241; and in 1860 to 2,192. A large number of volunteers engage for the artillery; very few for the cavalry. Advancement to the highest rank of military hierarchy being open to every French soldier, the volunteers, as a rule, make their way rapidly in the army, being distinguished, in the majority of cases, by a superior education.

The subjoined table gives a summary of the French army, as organised in the year 1864:—

Summary of the French army	Peace-footing		War-footing	
	Men	Horses	Men	Horses
Staff	1,771	160	1,841	160
Infantry	732,751	821	515,957	430
Cavalry	67,759	48,143	100,721	65,000
Artillery	39,557	14,646	66,172	49,506
Engineers	5,461	164	13,443	1,100
Gendarmes	24,525	14,769	25,644	15,000
Troops of the Administration	13,686	6,442	89,265	17,400
Total	864,192	98,264	757,737	148,290

The whole of France is divided into six 'arrondissements militaires,' or corps d'armée, each commanded by a field-marshal. These again are separated in military divisions and sub-divisions, the latter of the same circumference as the departments.

France has 119 fortresses, of which 8 are of the first rank—Paris, Lyon, Strasbourg, Metz, Lille, Toulon, Brest, and Cherbourg; 12 of the second rank; 23 of the third; and 76 of the fourth rank. The fortification of Paris is stated to have cost 200,000,000 fr., or 8,000,000l., while 170,000,000 fr., or 6,800,000l., has been spent on Cherbourg. The standing army of France is kept up at a much lower expense than that of Great Britain; for while in the latter country the average cost of each soldier is 101l. 12s., the French soldier cost only 43l. 1s. per annum. The total expenditure for the French army in 1861 amounted to 370,000,000 francs, or 14,800,000l. (Annuaire Militaire de l'Empire, 1865.)

Navy.—The French navy has gone through several remarkable phases in the course of a century. Powerful in the reign of Louis XIV. and his successor, it afterwards declined; but is again rising to a state of high efficiency, since the reforms inaugurated in 1858. In 1760 the fleet of war consisted of 60 first-class ships, 31 second-class, and 103 smaller vessels—altogether 206 ships, with 12,300 guns, and 78,000 sailors. In 1790, the number had sunk to 346 ships, with 51,000 sailors, and less than 10,000 guns; while at the battle of Trafalgar, 1805, in which the greater part of the imperial naval force was engaged, there were only 18 French men-of-war, with 1,752 guns. In 1814 the navy consisted of 226 sailing vessels, and 47 steamers, with 8,639 guns and 24,513 sailors; and this strength was not increased till the year 1855, when a commission was appointed by the emperor Napoleon III. to plan a new organisation of the navy. In conformity with the scheme proposed by this commission and sanctioned by the government, there were constructed—1st, a transition fleet, composed of sailing vessels capable of being transformed; 2nd, a swift fleet of war, composed of 40 ships of the highest type, 20 ordinary frigates for distant expeditions, 90 vessels of inferior rank, in all 150 bottoms; 3rd, a transport fleet, to carry 40,000 men and 12,000 horses—75 bottoms; 4th, a flotilla of small craft—about 125. Lastly, there were built special vessels—about 30—for the defence of the ports. This brought the fleet of war to a total of 380 vessels; and, adding 20 sailing vessels still kept for cheap transports, the number reached the figure of 400. The French navy, at the commencement of 1865, included 31 ironclads, with 776 guns, and of 19,075 horse-power, the largest being the 'Magenta' and the 'Solferino,' of 52 guns and 1,000 horse-power each.

The French navy is manned by conscription, like the army. The marine conscription, however, is of much older date than that of the land forces, having been introduced as early as the year 1683. On the navy lists are inscribed the names of all male individuals of the 'maritime population,' that is, men and youths devoted to a sea-faring life, from the 18th to the 60th year of age.

According to the budget of the minister of marine and the colonies for the year 1863, the French navy was officered by 2 admirals; 12 vice-admirals in active service and 14 on the reserve list; 21 rear-admirals in active service and 20 on the reserve list; 130 captains of first-class men-of-war; 270 captains of frigates; 750 lieutenants; 600 ensigns; 500 midshipmen, or 'aspirants;' 270 under-midshipmen, or 'pupils;' and 75 lieutenants with fixed residence—altogether 2,467 officers. The sailors numbered 32,854, which, together with engineers, navy-surgeons, chaplains, and other personnel, brought the grand total of men engaged in the service of the imperial fleet up to 89,264. The coasts are divided into five marine prefectures, those of Cherbourg, Brest, L'Orient, Rochefort, and Toulon. The principal naval ports, proceeding N. to S., are Dunkirk, Calais, Boulogne, Havre, Cherbourg, St. Malo, Morlaix, Brest, Nantes, L'Orient, Sables d'Olonne, La Rochelle, Rochefort, Bayonne, Port-Vendres, Marseilles, Toulon, and Fréjus. The minister of marine is assisted by an admiralty council and a board of naval works. The principal naval schools are those of Toulon and L'Orient, and that on board a ship in Brest Roads; there are, besides, 44 inferior schools.

Colonies.—These, which are under the superintendence of the minister of marine, comprise the islands of Martinique and Guadaloupe, and some smaller ones, in the Antilles; French Guiana in S. America; the regency of Algiers, Senegal, and the island of Goree in Africa; the isles of Bourbon and St. Marie in the Eastern Ocean; and

Pondicherry, Chandernagore, Karikal, Mahé, and Yanaon in Hindostan. The four principal colonies, Martinique, Guadaloupe, Bourbon, and Guiana, have each a colonial council elected by the resident French above 25 years of age, and having certain property qualifications. In every colony there is a governor appointed by the king as his representative, who controls or dissolves the colonial councils at pleasure, and provisionally accedes to, or suspends, the execution of the decrees passed by them. The French codes of laws are in force, and justice is administered in the colonies, as in France, in tribunals of the peace, of original jurisdiction, royal courts, and courts of assize. (For further particulars, see the separate articles as above.)

Taxes.—The system of taxation that existed in France previously to the revolution of 1789, had every possible defect. It consisted in great part of direct taxes laid on property, from which, however, that of the nobility and clergy, or of the richest classes, was exempted. The indirect taxes were also assessed on the most vicious principles; and the contributions of forced labour, or corvées, fell almost wholly on the peasantry. The collection of the taxes by farmers was also exceedingly unpopular; and, in fact, the whole system was one of partiality, injustice, and oppression. The revolution made an end of these abuses, and established the principle embodied in the charter and the new constitution, that every citizen should contribute, without distinction, to the wants of the state in proportion to his means. To carry out this principle, it was first attempted to raise the greater part of the public revenue by direct taxation; but the practical difficulties were found to be so great that this had to be given up, and indirect taxation was again resorted to, though on an entirely new basis. At the present moment, by far the greatest part of the state income is derived from indirect taxes, one of which, the excise, produces twice as much as all the direct taxes together. The most important direct tax is the land tax, or contribution foncière, assessed on all lands and houses in proportion to their rent. The next important imposition, the contribution personnelle et mobilière, is a mixed tax. The first part being a sort of poll tax, rated at the value of two days' labour, and charged on men of 18 years and upwards; the mobilière is a tax on the occupiers of houses of a certain class, charged according to the rent. The droits des patentes, or licence duties, are charged on all persons following a trade, profession, or business. They are assessed partly according to the rent of the house occupied by the patentee, and partly according to the part of the town in which he carries on business. In every department of France there is an office for the registry of deeds, the fees on which, besides the expenses of the establishment, which is highly useful, yield a considerable revenue to government. The other public taxes are nearly the same in France as in England.

Besides the public taxes, octrois or duties are levied on all articles entering towns of any considerable magnitude, the rate of the duties varying with the part of the towns. These duties are great obstructions to trade and industry; but as their produce is employed to defray indispensable local charges, including the expenses of hospitals and asylums, it has not been possible to repeal them, although attempts to this effect have been made at various times.

The subjoined tabular statement gives a succinct account of the nature and amount of taxes levied in the empire of France for the year 1862.

STATE TAXES:	Francs	£
Direct, viz.		
On Lands . . .	121,600,000	4,864,000
Forests . . .		
Poll tax . . .	46,439,000	1,857,570
Scores . . .	83,456,000	4,702,448
Indirect, viz.		
On Customs . .	166,720,250	7,668,810
Excise . . .	553,940,075	27,277,631
Stamps . . .	65,380,450	2,615,218
Lotteries . . .		
Licences . . .	67,996,785	2,819,829
Sales of Goods .	147,716,050	5,906,522
Sales of Land .		
Inherited Property .	140,556,250	5,622,250
Miscellaneous:		
On Woods and Forests,		
Fisheries, Postage, Uni-	176,952,560	7,828,100
versities, miscellaneous		
PROVINCIAL TAXES:		
Direct, viz.		
On Houses . .	26,690,000	1,035,600
Lands . . .	54,300,000	2,172,370
Assessments .	18,140,000	725,600
Licences . .	10,588,255	479,543
Miscellaneous:		
Casual Provincial Taxes		
Fr. 32,922,000	40,113,600	1,604,536
Primary Instruction,		
Fr. 7,391,000		
MUNICIPAL OR TOWN TAXES:		
Direct, viz.		
On Houses . .	19,750,000	790,000
Lands . . .	37,176,000	1,487,040
Poll Tax . .	9,930,000	397,200
Licences . .	16,690,000	471,901
Indirect, viz.		
On Bread . . .		
Beer . . .	7,606,415	357,979
Meat . . .	42,949,378	1,719,570
Wine . . .	46,600,434	1,867,437
Miscellaneous:		
Combustibles, fr.10,457,175		
Fodder . . 6,497,840		
Materials . 18,728,374	43,488,160	1,779,444
Miscellaneous 3,814,520		

The figures in the above table are official, having been furnished by the French government to Lord Cowley, her majesty's ambassador at the court of the Tuileries. (Report of Lord Cowley, dated Paris, June 24, 1864.)

Public Debt and Budget.—The public debt of France amounted, at the beginning of 1864, to 11,902 millions of francs, or 476 millions sterling, distributed as follows:—

	Millions of Francs	Or
Funded Debt . . .	9,719	£388,760,000
Floating Debt . . .	750	30,000,000
Ancient special Debts and Annuities . . .	50	2,000,000
'Obligations Trentenaires'	150	6,000,000
Life-rents and Pensions representing a Capital of	725	29,076,000
Total . .	11,902	£476,000,000

The funded debt of France increased in the following proportions in the ten years from 1851–61. It amounted, on Jan. 1st,

| 1851 to 6,345,637,200 francs, or £213,875,634 |
| 1852 — 6,516,156,000 — 270,647,144 |
| 1853 — 5,527,501,307 — 222,100,182 |
| 1854 — 5,099,455,013 — 720,704,201 |
| 1855 — 5,907,877,552 — 244,315,114 |
| 1856 — 7,556,069,572 — 302,721,633 |
| 1857 — 8,031,997,464 — 321,279,600 |
| 1858 — 8,452,096,188 — 848,003,421 |
| 1859 — 8,562,360,188 — 342,731,578 |
| 1860 — 9,331,017,000 — 372,300,446 |
| 1861 — 9,719,176,313 — 388,767,070 |

There were, at the last-named period, very nearly a million holders of the funded debt, divided as follows:—

650,221 new 4½ per Cents., representing a Capital of . . 8,830,308,583 frs.
1,744 old do. do. . 19,666,556 „
3,237 Four per Cents. do. . 64,427,573 „
304,274 Three do. do. . . 8,311,218,766 „

945,476 'Inscriptions,' representing a Capital of . . } 9,710,978,913 frs.

The *senatus-consultum* of December 31, 1861, inaugurated the system by which the budgets of the French government are at present regulated. Under this system, the minister of finance distinguishes between three classes of income—namely, ordinary, extraordinary, and special revenue; and he also recognises three sorts of expenditure, viz. ordinary, extraordinary, and supplementary. It is the practice to lay before the legislative body, in the first instance, the budget of ordinary income and expenditure; when this has been voted, after a lapse of time more or less considerable, the extraordinary budget is submitted to the chamber, and, finally, the special budget.

The following are the figures of the budget for the year 1865:—

ESTIMATED REVENUE FOR 1865.

	Francs	£
Ordinary . . .	1,799,061,402	71,997,042
Extraordinary .	106,756,000	4,236,100
Special . . .	229,824,615	9,179,721
Total . . .	2,139,044,007	85,571,761

ESTIMATED EXPENDITURE FOR 1865.

	Francs	£
Ordinary . . .	1,797,265,790	71,910,632
Extraordinary .	106,650,000	4,266,000
Supplementary .	229,484,625	9,179,721
Total . . .	2,133,600,425	85,356,353

The actual revenue and expenditure of the French government, during the nine years from 1855 to 1863, was as follows:—

Revenue	Expenditure	Excess of Revenue over Expenditure	Excess of Expenditure over Revenue
Francs	Francs	Francs	Francs
1855 2,751,278,955	2,989,117,840		196,156,184
1856 1,912,511,167	2,186,261,787		781,851,854
1857 1,799,723,820	1,877,676,311		83,951,276
1858 1,871,301,261	1,854,493,793	12,806,013	
1859 2,179,789,152	2,207,660,695		79,971,793
1860 2,197,061,019	2,489,819,615		44,780,601
1861 2,843,194,761	2,449,511,399	96,412,651	
1862 2,661,861,726	2,621,916,972	63,129,251	
1863 2,583,921,661	2,629,510,989		46,549,128

The constant deficits shown in the table were occasioned entirely by increased expenditure for the army. According to a statement made in the legislative body in 1864, the wars and warlike operations of France, since the accession of Napoleon III., have cost the following sums:—

	Francs	£
Crimean War . .	1,240,000,000	51,970,000
Italian " . .	314,000,000	12,000,000
Chinese " . .	100,000,000	3,000,000
Occupation of Rome .	40,000,000	2,000,000
" Syria .	24,000,000	1,000,000
Supplementary Expenses	80,000,000	3,000,000
Total . .	3,920,000,000	51,000,000

The Mexican war, it will be noticed, is not included in this calculation. The cost of the Mexican expedition, up to the end of 1865, amounted to 270,000,000 francs, or 10,800,000. This brings the total cost of recent French wars to 2,256 millions of francs, or about 92 millions sterling.

The public revenue of France has immensely increased since the time of the Revolution; but the expenditure has kept more than pace with it. The income budget of 1791 amounted to 502 millions of francs; that of 1814 to 800 millions; that of 1831 to 966 millions; and that of 1863, as above given, to 1,799 millions. Consequently, the state income more than trebled in the course of about two generations.

Language and Literature.—It has been estimated that of the total pop., about 33,000,000 speak French, or various patois, having different degrees of analogy with that language; that 1,600,000 use German dialects, 1,250,000 the Breton, and 150,000 the Basque tongue. It is chiefly with reference to these languages that Balbi has divided the inhab. of France into four great families—the Greco-Latin or Gallic, Germanic, Celtic, and Basque; besides the Semitic, including the Jews, and some few individuals of Saracenic origin in the S. dep., and the Hindoo family including the gipsies, or gipsies. The Greco-Latin family, which comprises the great bulk of the pop., speaking dialects derived from the Latin, are probably for the most part descended from the ancient Celtic pop. by whom the country was principally inhabited at the period of the Roman conquest; and who, during the subsequent ages of Roman dominion, gradually adopted the Latin tongue, which forms the basis of the modern French. The Romans, no doubt, intermixed with the native pop., and the latter, in the S., may still have some infusion of Greek blood derived from the Greeks, who founded Marseilles, and other colonies on the Mediterranean coast; the French are also in part the offspring of the Visigoths, Burgundians, Alani, and Franks, who successively became masters of Gaul in the middle ages. But notwithstanding that the modern French are thus descended more or less from all these races, there can be little doubt that the ancient Gallic or Celtic blood predominates, especially in the Central and NW. prov. The intermixture of Roman and Greek blood could not have been very great; the Visigoths, Burgundians, Alani, and other barbarous tribes, swept over the country as conquerors, but maintained themselves in it too short a time to have any material influence on the native pop.; and the Franks, though, like the Normans in England, they established a martial supremacy, gave little to France but its name, and were in too small numbers to impress their own character on the nation, except perhaps in the NE., where the population is less French than elsewhere.

The French have attained to high excellence in almost every branch of literature. Their writers are particularly distinguished by extreme perspicuity, good sense, an attachment to classical models, and perhaps, also, by a deficiency of sentiment. Latterly, however, the public taste has apparently undergone some considerable modifications; and the literature and philosophy of their German neighbours appear to be materially influencing their tastes and pursuits. About 20,000 new books, including pamphlets and new editions, are published annually in France, more than two-thirds of them in Paris. There were, at the end of 1864, 318 political and 652 non-political newspapers, reviews, and magazines published throughout the country.

History.—Before the time of Cæsar, the whole of France was known to the Romans by the name of Transalpine Gaul; but after its conquest, it was divided into the four provinces of *Provincia Romana* (*Provence*), and *Gallia Aquitania*, *Celtica*, and *Belgica*. In the 5th century it was subdivided into 17 provinces, inclusive of all the territory on the E. bank of the Rhine. At the latter epoch the Germanic nations began to pour in an irresistible torrent over Gaul. The Visigoths established themselves in the W. and S. from the Loire to the Pyrenees, where they established a kingdom that lasted till about 500. The Burgundians, in a similar manner, settled in the E. from the Lake of Geneva to the Rhine, and afterwards stretched along the Rhone to the Mediterranean; the independent sovereignty they erected lasted till about 532. The Franks, whose dominion swallowed up those of both the foregoing tribes, had been long settled in the N.; and Pharamond, their chief in 420, is considered the founder of the French monarchy, as he was of the first of Merovingian race of Frankish kings. In 486 Clovis defeated Syagrius, the Roman general, at Soissons, and finally extinguished the Roman power in the W.; and in 507, by his victory over the Visigoths, he rendered himself master of all the country between the Loire and the Garonne. On the death of Clovis, in 511, his dominions were divided into four kingdoms—those of Paris, Metz, Soissons, and Orleans—each governed by one of his four sons; these, however, were reunited in 558. In 752 Charles Martel defeated the Saracens, who had effected the conquest of a great part of the S. of France, in a great battle; and ultimately succeeded in expelling them from the kingdom. In 751 the Carlovingian dynasty commenced in the person of Pepin le Bref, son of Charles Martel, and was carried to the summit of its power by Charlemagne, the son of Pepin.

Under the first race of kings the country was a prey to bloodshed, spoliation, and anarchy; industry and commerce were almost unknown, or extended only to the production and barter of a few indispensable articles. Nor was this condition much ameliorated during the rule of the succeeding race. Charlemagne, indeed, encouraged trade and manufactures in the towns, which before his reign were chiefly confined to the cloister, or practised by isolated individuals; but after his death things returned to their original state of confusion. Under his immediate successor, France was again divided into four parts, and the Normans began to ravage its N. provinces; the power of the nobility also rapidly increased; and the last sovereign of the Carlovingian dynasty, Louis V., in 986-7, possessed only the town of Laon. His successor, Hugh Capet, count of Paris and Orleans, the founder of the third race of kings, governed only the Ile-de-France, Picardy, and the Orleannais. The dukes of Normandy, Brittany, Aquitaine, Gascony, Lorraine, and Burgundy; the counts of Flanders, Champagne, Vermandois, Toulouse; and several minor seigneurs shared among them the rest of the modern kingdom. By degrees, however, all the great fiefs fell in various ways to the crown. Vermandois was united to it by Philip Augustus; Toulouse and Perche by Louis IX.; Champagne in 1274; the Lyonnais, Dauphiny, and Languedoc, in the 14th century; Berri, Normandy, Gascony, Burgundy, Anjou, Maine, and Provence in the 15th; the Bourbonnais, Auvergne, Brittany, Lorraine, and considerable territories in the NW., in the 16th; and Flanders, Artois, Franche-Comté, and Alsace, in the 17th century. The names of the sovereign powers of France, beginning with Hugh Capet, and the dates of their accession, are as follows:—

987	Hugh Capet.	1461	Louis XI.
996	Robert (le Sage).	1483	Charles VIII.
1031	Henri I.	1498	Louis XII. (Père du Peuple).
1060	Philippe I.	1515	Francis I.
1108	Louis VI. (le Gros).	1547	Henri II.
1137	Louis VII. (le Jeune).	1559	Francis II.
1180	Philip Augustus.	1560	Charles IX.
1223	Louis VIII. (Cœur de Lion).	1574	Henri III.
1226	Louis IX. (St. Louis).	1589	Henri IV. (le Grand).
1270	Philippe III. (le Hardi).	1610	Louis XIII.
1285	Philippe IV. (le Bel).	1643	Louis XIV. (le Grand).
1314	Louis X. (le Hutin).	1715	Louis XV.
1316	John I.	1774	Louis XVI.
1316	Philippe V. (le Long).	1792	Republic.
1322	Charles IV. (le Bel).	1799	Consulate.
1328	Philippe VI. (de Valois).	1804	Napoleon I.
1350	John II. (le Bon).	1814	Louis XVIII.
1364	Charles V. (le Sage).	1824	Charles X.
1380	Charles VI.	1830	Louis Philippe.
1422	Charles VII.	1852	Napoleon III.

While the monarchy gained in consistency and extent the royal power was making constant advances. The political rights and privileges which the nobles exercised under the feudal system were the objects of continued attacks on the part of the crown, which, though sometimes defeated, were, in most instances, successful. At length, under the administration of Richelieu, the nobles were stripped of all power; and there being no other body in the state, with the exception of the parliaments, which had degenerated into little else than courts of law, that enjoyed any constitutional privileges, the power of the crown was raised above control. Under the vigorous and, for a lengthened period, prosperous government of Louis XIV., the royal prerogative arrived at a maximum. But the close of this reign was eminently unprosperous; and the wars in which Louis had been long engaged, the burdens they obliged him to impose on his subjects, and the vast debts he had contracted, produced not only great suffering and misery, but also great discontent. During the regency and the subsequent part of the reign of Louis XV., abuses of all sorts multiplied on all hands, and were no longer concealed by the dazzling splendour and magnificence of the preceding period; the most worthless parasites obtained a predominating influence at court; the command of fleets and armies was entrusted to the merest imbeciles, the finances were involved in the greatest disorder; and France and Europe were scandalised and disgusted by the gross immorality and vulgar profligacy of the king and his intimate associates. Louis XVI., who ascended the throne in 1774, was actuated by the best intentions, but he wanted the firmness of purpose and capacity required in so desperate a crisis. The abuses that infected the whole frame of society, though destructive of the public interests, were either really advantageous, or believed to be so, to a vast number of persons, including the nobility and clergy; and it would have required a mind of a very different order from that of Louis to have frustrated the solicitations, intrigues, and cabals of such powerful parties, and to have safely carried through the reforms that had become indispensable. At length, after a variety of futile expedients had been in vain resorted to, it was resolved, in 1789, to hold a meeting of the States-General, which had not been convened since 1614, for effecting the necessary changes, and averting a public bankruptcy. This was the commencement of that tremendous revolution which cost Louis XVI. the crown and his life, and destroyed every vestige of the government and institutions that existed when it broke out.

The atrocities connected with the Revolution were the sad, but not unnatural, excesses of an uninstructed populace, that had suddenly been emancipated from a state of extreme degradation, and which had innumerable grievances to suppress, and wrongs to avenge. It unfortunately happened, that when the nobles were stripped of all political power, and rendered incapable of opposing any effectual resistance to the sovereign, they were, at the same time, left in full possession of their feudal privileges as landlords. These comprised an exemption from those direct taxes that fell with their full severity on every one else; the dispensation of justice in manorial courts; and a host of vexatious privileges connected with the game laws, and the laws respecting mills. The rental of very many estates consisted, previously to the Revolution, of little else than services and feudal tenures, by the hateful influence of which the industry of the occupiers was almost exterminated. The country population was everywhere, in fact, in a situation of predial slavery; and while the nobility and clergy threw the burdens of the taillecarrière, and other oppressive imposts, wholly on the tiers état, they engrossed to themselves every situation of power and emolument; so that down to the Revolution, no individual, how meritorious soever, unless he obtained a patent of nobility, could be made an officer of the army, or be promoted to almost any public employment. Government deprived the nobility and landed aristocracy of all that could have rendered them useful, at the same time that it left them all that could render them little tyrants, and a curse to the country in which they lived. If we add to these grievances the fact, that the peasantry received no efficient protection from the government, and that the administration of justice in the king's courts was, speaking generally, partial, venal, and infamous, we shall be at no loss to understand why the aristocracy was so universally detested in France, and why the Revolution, which was indispensable, was so sweeping, bloody, and destructive.

The proscriptions and anarchy by which the Revolution was accompanied continued till Napoleon attained to the supreme direction of affairs. The talents of this extraordinary man were surpassed only by his ambition, which, by overstepping all bounds, precipitated him into enterprises that ultimately led to his overthrow. In 1814 the family of Bourbon was replaced on the throne; but the elder branch had profited as little as the Stuarts in England, under similar circumstances, by the lessons of adversity, and in 1830 they were re-expelled from the kingdom. The crown was then offered, under certain conditions, to Louis Philippe, duke of Orleans, by whom it was accepted. He has the merit of having contributed, under very difficult circumstances, to maintain, for a lengthened period, the peace of France and of Europe. But he alienated the public by his plans for advancing and enriching his children; and by the corruption which pervaded every department of his government. This led to the revolution of Feb. 21, 1848, and the establishment of the republic, presided over by a Provisional Government. A new constitution having been voted by a 'Constituent Assembly' of 900 members, Prince Louis Napoleon was elected head of the republic, for four years, by 5,562,834 votes, on the 10th of December, 1848. The 'Prince-President' dissolved the National Assembly by a coup d'état, Dec. 2, 1851, and having remodelled the constitution, appealed to universal suffrage, which decreed him president for 10 years, by 7,439,216 votes, on the 21st of December, 1851. Appealed to a third time, Prince Napoleon was chosen emperor of France, by 7,864,189 against 231,143 votes, on the 22d November, 1852. The elect of the people accepted the imperial dignity, and assumed the title, 'Napoleon III., Emperor of the French,' on the 1st of December, 1852.

FRANKFORT, or FRANKFURT-ON-THE-MAYN, a celebrated commercial town and 'Free City' of W. Germany, seat of the diet of the Germanic confederation, on the N. bank of the Mayn, 18 m. NE. by E. from its confluence with the Rhine at Mayence, 49 m. SE. Coblenz, 86 m. SSW. Cassel, and 17 m. N. by W. Darmstadt, on the main line of railway from Hamburg to Basel. Pop. of city 73,591, and of district belonging to the city 87,518 in 1861. Frankfort is oval-shaped, and communicates with Sachsenhausen, on the opposite bank of the river, by a stone bridge, of 14 arches, being about 950 ft. long by 41 broad. Its fortifications were demolished by the French, and their site is now occupied by public walks and gardens. The city is, however, still entered by 9 principal gateways, 3 of which are in the suburb of Sachsenhausen; and some of them are remarkable for their elegant and classic style of architecture.

Frankfort presents many varieties of aspect. The old town, with its narrow streets and quaint wooden buildings, with gables overhanging the basement stories, has an unprepossessing appearance, and the Jews' quarter is filthy. In the new town, however, the Zeil, the new Mayence Street, Allée, and especially the fine quay which stretches along the Mayn nearly the whole length of the city, are beautiful streets and promenades, and not a few of the houses in them are literally palaces. The streets are generally well paved, and lighted with gas. There are some good squares, several, as the Ross-markt (Horse-market), being ornamented with fountains and avenues of trees. Frankfort possesses several interesting public buildings. The Römer, or council-house, is of uncertain origin, but was most probably built by the Frankish emperors. It possesses no architectural beauty, but is deserving of notice, as being the place where the emperors of Germany were elected. The election chamber, on the ground floor, now serves for the sittings of the senate of Frankfort. Above this apartment is the Kaisersaal, or 'Hall of the Emperors,' a large chamber, with a vaulted roof, once the scene of the splendid pageant of the election banquet, at which the emperor was waited on at table by the high dignitaries of the empire. Its walls are surrounded by niches, in which are placed the portraits of the German emperors in the order of their succession, from Conrad I. to Francis II.; the latter, with whom the line of the emperors of Germany crowed, filling up the last vacant space. In this building is preserved the famous 'Golden Bull,' the deed by which Charles IV., in 1356, settled the mode of election of the German emperors, fixed the number of electors at seven, and determined their rights of voting. The present diet of the German confederation assembles in the former palace of the prince of Tours and Taxis, now the residence of the Austrian ambassador; a structure of the last century, containing 140 different apartments, and richly furnished. The cathedral, or church of St. Bartholomew, is an edifice of Gothic architecture, in the form of a cross. 246 German ft. long, by 216 broad. It is said to have been begun in the time of the Carlovingian princes; the greater part of it is, however, the work of the 13th and 14th centuries; the tower, which is 260 ft. in height, is still unfinished. This church has not much beauty, but it contains some curious monuments,

especially that of the emperor Gunther of Schwarzburg, killed by his rival, Charles IV., a fine painting of the Assumption by Holbein, and a Dying Magdalen by Brendel; and the chapel in which the German emperors were crowned. There are 20 other places of worship, including 8 Lutheran, 1 Calvinist, 1 French-Protestant, and 4 R. Cath. churches, besides 3 synagogues, one of which is a very handsome building. In the church of St. Catherine, there is a fine painting, of 'Jesus on the Mount of Olives,' by Ikm. The church of St. Leonard, near the river, occupies the site of a palace built by Charlemagne, but of which no traces exist. The Saalhof, a building of the last century, also near the Mayn, is erected on the site of another palace, built by Louis the son of Charlemagne, and which afterwards became the residence of the Carlovingian emperors of Germany. The modern edifice includes within it the chapel of the original one, which is probably the most ancient structure in Frankfort. The ancient palace of the Knights of the Teutonic Order, in Sachsenhausen, is in a state of decay, and now serves as a barrack for Austrian troops, who, in conjunction with Prussian garrison Frankfort. The Haus am Braunfels, or exchange, is a small neat quadrangle, surrounded by a range of warehouses and shops, thronged during the fair with merchants of all nations. The Staedel Museum and Academy of Painting (so named after its founder, a rich banker and citizen, who, in 1816, bequeathed a million of florins, together with a respectable collection of pictures and engravings for its foundation) occupies a handsome new building in Mayence Street. The gallery, without being first-rate, possesses several good specimens of art, chiefly of the Flemish and Dutch masters. Private collections of pictures are very numerous; and there is scarcely a merchant or banker in Frankfort, of moderate affluence, who has not his little gallery, which, with his snuff, his calèche, and his pipe, forms his favourite recreation from the fatigue of business. The principal work in the fine arts at Frankfort is Dannecker's celebrated statue of 'Ariadne seated on a Tiger,' in the garden of Mr. Bethmann, a banker. Most travellers rank this piece of sculpture among the most distinguished productions of modern art; but it unfortunately happens that the marble in which it is executed is covered with blue veins and spots. Dannecker had this work in hand for 16 years, but only received for it 13,000 florins, or 1,240l. One of the most interesting public monuments is without the Friedberg-gate; it is a colossal mass of granite rocks grouped together, on one of which are inscribed the names of the Prince of Hesse Philipsthal and the Hessians, who fell on the spot defending Frankfort, the whole surmounted by a military device cast from cannon taken from the French, and surrounded by weeping willows. This memorial was erected by the King of Prussia. The Senkenberg Museum of Natural History, and Medical Institute, occupy an imposing building of the 11th century; the museum contains many rare specimens brought by the traveller Rüppell from NE. Africa. The public library, with 80,000 vols.; 5 hospitals, the orphan asylum, lunatic asylum, theatre, with an illuminated clock over the stage, the casino, or principal reading club, and the new cemetery near the city, containing several works by Thorwaldsen, are the remaining objects most worthy of notice. The hotels in Frankfort are amongst its most magnificent edifices, and rank among the first in Germany for elegance and comfort. Many of these are situated in the Zeil.

The chief manufactures are carpets, table-covers, oil cloth, woollen, cotton, and silk stuffs, woollen

yarn, coloured paper, tobacco, playing-cards, gold
and silver articles, and printers black. There are
about twenty printing offices, besides several
stereotype and lithographic establishments. But
the principal source of wealth to the merchants
of Frankfort are commercial transactions, banking, and speculations in the funds. The inhab. of
Sachsenhausen are mostly peasantry of Saxon
descent, and distinguished from the rest of their
fellow-citizens in manners, customs, dress, and
language, as well as occupations. They are gen-
rally employed in garden cultivation, fishing, &c.,
or as porters. Frankfort is one of the great emporiums for the supply of Germany with all kinds
of merchandise, and enjoys therefore a considerable proportion of transit and commission business. Two large and celebrated fairs, at Easter
and Michaelmas, are annually held in this city.
These suffered materially during the occupation
of the country by the French, and since the peace
they have been affected by the improved communications established in all parts of the country,
the greater diffusion of shops and magazines in
all the principal towns, and, in short, by the concurrence of all those causes that tend, as civilisation advances, to lessen the importance of fairs.
However, a large amount of business is still transacted at the Frankfort fairs. Cotton twist and
stuffs, and cutlery, are the British commodities
in greatest demand. The city is included in the
German customs' league. (For an account of the
territory of the city see below — FRANKFORT,
REPUBLIC OF.)

The town and country civil and criminal tribunals, court of appeal, board of taxation, and
most of the administrative establishments of the
republic, are held in the city. There are a great
many educational institutions, including a gymnasium; the medical institute, with a botanic
garden; normal, Jewish, drawing, deaf and dumb,
and trades' schools, and numerous private seminaries; many learned and benevolent associations,
as the Schenenberg society, the society for the encouragement of useful arts, and philosophical,
Bible, and missionary societies. Few towns
abound so largely with public charities. The
Jews are unusually numerous in the city, and
occupy most of the finest mansions here and in
the environs. They were formerly much oppressed — compelled, for centuries, to live in
a dark unwholesome quarter called the Juden-
gasse, or Jews' Lane—but, from taking heirs, they
have now risen to be almost the masters of the
city.

Frankfort is one of the most ancient cities in
Germany. Charlemagne held a council in it in
794, and it was fortified by Louis-le-Debonnaire
in 838. In 813 it became the cap. of the kingdom of Austrasia, and not long afterwards, under
Louis the German, its great fairs originated, and
Frankfort became the commercial cap. of Germany. From this period the increase of its prosperity was rapid, and in 1154 it was made an independent free city. It acquired considerable
privileges during the next two centuries; and in
1356 had obtained nearly its present extent of
territory. From 1806 to 1810 it was the cap. of a
prince-primacy, and from the latter year till the
downfall of Napoleon it was the cap. of the grand
duchy of Frankfort, which comprised a territory
of nearly 2,000 sq. m. It was the native place of
Goethe, born here in 1749, as well as of Amschel
Rothschild, ancestor of the great banking family
of the name, now spread over all Europe.
Amschel Rothschild was born in a wretched
dwelling in the Jews' Lane, in 1772, the son of
very poor parents, and died in 1812, worth above

a million sterling. (Stories of Banks and Bankers,
by F. R. Martin, Lond. 1845.)

FRANKFORT (REPUBLIC OF), a nomi-
nally independent state of W. Germany, and the
smallest in Europe, consisting of the city of
Frankfort-on-the-Mayn, and the country immediately around it, together with some detached
portions of territory, the whole having an area of
43 sq. m., with a pop. of 87,518 inhabitants in
1861. The state includes, besides the city, eight
villages, with a pop. of 11,928 in 1861. The
largest portion of territory belonging to Frankfort
lies on both sides the Mayn, having NW. and N.
the dom. of Nassau and Hesse Cassel; and SE.
and S. those of Hesse Darmstadt. It is quite
level, and very productive and well cultivated,
yielding corn, potatoes, pulse, fruit, and wine, and
feeding many cattle. Much of it is, however,
laid out in gardens; the environs of the city of
Frankfort being completely studded with the
country houses of merchants and others. Previously to the insurrection of 1848—which in
Frankfort was of a very sanguinary nature—the
institutions of the republic were oligarchical; but
they were subsequently changed to others of a
more democratic nature. The present constitution
of Frankfort-on-the-Mayn was proclaimed by the
constituent assembly of the free city, on December 17, 1854, and accepted by a general vote of
the citizens of Feb. 5 and 6, 1855. According to
this charter, the government of the commonwealth
is exercised by two representative bodies—the
senate, consisting of twenty-one life-members, and
the legislative assembly, composed of eighty-
eight deputies, of which fifty-seven are elected by
the burgesses, twenty by the common council of
the city, and eleven by the inhabitants of the
rural districts. Vacancies in the senate are filled
by a ballot-committee of twelve members, six of
whom are appointed by the legislative assembly,
and six by the senate. A president and vice-
president—called elder burgomaster and younger
burgomaster—elected annually, represent the executive authority vested in the senate. The right
of making and altering laws, and that of imposing and distributing financial burdens, belongs
solely to the legislative assembly. The budget is
voted annually.

The budget for the year 1862 comprised an
income of 2,576,483 florins, or £11,707£, and an
expenditure of 2,224,147 florins, or 185,345£.
About one-third of the income is derived from
customs duties, and another third from the excise.
There is a state-lottery, which produces, on the
average, 130,000 florins, or 10,833£, per annum.
The cost of government, including army and
police, forms more than three-fourths of the whole
expenditure; and for educational and ecclesiastical
affairs, 118,482 florins, or 9,874£, are set aside. The
public debt at the commencement of 1862 amounted
to 16,353,000 florins, or 1,362,750£. Very nearly
one-half of this debt—exactly 7,404,060 florins—
was incurred for the establishment of railways.
One million of florins of the capital pays no
interest, it having been advanced, under this condition, by the bank of Frankfort, against a permission to issue notes.

The contribution of Frankfort to the German
confederate army amounts to 1,419 men, nearly
all infantry. The whole of this force is raised by
enlistment, for periods of four years and two
months—formerly six years and two months—
under the offer of a bounty of 300 florins, or 25£.
The men receive 19 kreuzer, or about 6½d. per
diem, with increase of pay at the end of ten years'
service. It is owing to the position of the Free
City, as the seat of the Germanic Diet, that it has

to keep a much larger armed force, in comparison with its population, than any other state of the confederation. The city has also a guard of burgesses, the duties of which, however, are of a strictly civil nature. Frankfort maintains representatives in most of the principal neighbouring states of Germany, a minister at Paris, and consuls in London and some of the American capitals. It has one vote in the full council of the German confederation; and divides one in the lower council, and the 17th place in the diet, with the other Hanse Towns.

FRANKFORT-ON-THE-ODER, a town of the Prussian dominions, prov. Brandenburg, cap. government of same name, circ. Lebus; on the Oder, about 116 ft. above the level of the Baltic, 60 m. E. by S. Berlin, on the railway from Berlin to Breslau. Pop. 86,557 in 1862, excl. of garrison of 2304. Though no longer a fortress of any strength, the town is surrounded by walls, with towers and a ditch. It is well built; the streets are straight and broad; the houses generally good; and many of the public edifices handsome. The town communicates with one of its three suburbs by a wooden bridge across the Oder. It has a good market-place, six Protestant churches, a Roman Catholic chapel, synagogue, government house, council-house, new post-house, gymnasium, high school, school of midwifery, school for neglected children, and various other schools; an orphan asylum, two hospitals, a workhouse, with a house of correction, and a theatre. A university established in it, in 1506, was, in 1810, removed to Breslau. It is the seat of the authorities for its government and circle, of a superior judicial tribunal for the former, and inferior courts for the latter; and the town, a circle council, council of nobility (Ritterschafts-Direction), and boards of taxation, agriculture, and canals. Being situated on the high road from Berlin to Silesia, and on a navigable river communicating, by canals, with the Vistula and the Elbe, it has a considerable trade; though, in commercial activity, it is far inferior to its namesake on the Mayn. It has manufactures of woollen and silk fabrics, stockings, gloves, leather, earthenware, wax, and sugar; with brandy distilleries and mustard-works, for which article it is celebrated. A good deal of wine is grown in its vicinity. Three large fairs are held here annually, in Feb., July, and Nov. They are attended by great numbers of merchants and dealers from foreign countries, as well as from Germany. Besides the woollens, linens, earthenware, silks, and other articles furnished by the town and its vicinity, and the various raw and manufactured products of the Prussian and other German states, very large quantities of British, French, Swiss, and other foreign goods, are disposed of at these fairs, partly for the supply of the surrounding country, but principally, perhaps, for exportation to Poland, Galicia, Russia, and Bohemia.

Many of the inhabitants are employed in navigating the Oder and the communicating streams and canals to Dantzic, Warsaw, Magdeburg, and Hamburg. The village of Kunersdorf, in the vicinity of this town, has been the scene of one of the most sanguinary contests in modern times. On the 12th of August, 1759, Frederick the Great attacked the entrenchments of the Austrians and Russians at that place; but after partially succeeding, and exhausting all the resources of skill and valour, he was compelled to retreat with immense loss; the approach of night having alone saved his army from being completely destroyed. (See Thomas Carlyle's History of Frederick II., called Frederick the Great,

which contains a graphic sketch of the Battle of Kunersdorf.)

FRANKFORT, a town of the U. States of America, Kentucky, of which it is the cap. co. Franklin, on both sides of the Kentucky, which is here crossed by a bridge; 63 m. WSW. Cincinnati. Pop. 3,500 in 1861. The town is buried among steep hills, and the banks of the river are here precipitous, and from 400 to 500 ft. in height. Frankfort is well built, chiefly of stone, but many of the private as well as public buildings are of fine white marble. The principal public edifices are the state-house, with a fine Ionic portico; the penitentiary, having generally about 100 inmates; three churches, an academy, county court-house, and several manufacturing establishments. It is a place of some trade: steam vessels navigate the Kentucky river as far as this town, and at certain seasons three or four are kept in regular employ.

FRANCATI (an. Tusculum), a town of Central Italy, compart. di Roma, 11 m. SE. Rome. Pop. usually about 4,300, but during summer this number is considerably increased by the influx of visitors. It is beautifully situated on the declivity of a hill commanding an extensive view of the surrounding country; but except the piazza in which the cathedral is situated, the town is dirty and inconvenient. Its ruins, and the surrounding villas, constitute its chief attraction; but the latter are now falling into neglect, the present fashion of the Roman nobility being to pass the summer at Albano. The principal villas are those of Conte Aldobrandini, Bracciano, Falconieri, and Ruffinelli, on the grounds of which last are the ruins of the ancient Tusculum. The splendid mansion of the Borghese family, Monte Dragone, is now neglected and in a state of decay. Frascati has a public seminary, and numerous convents, churches, and public fountains. Its bishop is always one of the five members of the highest episcopal council. The ruins of Tusculum (municipium clarissimum, Cicero pro Fonteio, § 14.) comprise an amphitheatre, a theatre, an immense hall, supposed to have been attached to baths, fountains, &c. This was one of the most ancient cities of Italy, its foundation being ascribed to Telegonus the son of Circe. It was strong, as well by its position as by the walls by which it was surrounded, portions of which still exist. It was, also, one of the most faithful of the allies of Rome; and successfully resisted an attack by Hannibal. The top of the hill on which Tusculum was built, 2,079 French ft. above the level of the sea, was surmounted by a citadel, now wholly destroyed. Like Frascati, in modern times, Tusculum was crowded with the villas of distinguished Roman citizens, among which may be mentioned those of Lucullus and Maecenas. But the fame of all the other villas has been wholly eclipsed by that of Cicero, so often mentioned in his works, and from which his beautiful ethical disquisitions, entitled the Disputationes Tusculanae, have received their name. The attempts that have been made to identify the site of this famous villa have had but little success. (Gell's Rome, i. 433, and ii. 293; Cramer's Ancient Italy, ii. 44.)

FRASERBURGH, a town and sea-port, Scotland, co. Aberdeen, on its NE. coast, on a slight eminence N. side of Kinnaird Head, a bold promontory, on which are an old castle and light-house, 120 ft. above the level of the sea at high water; 184 m. E. Banff, and 37 m. N. by E. Aberdeen, on the North of Scotland railway. Pop. 3,101 in 1861. The town is nearly square. Most of the streets cross each other at right angles. A considerable number of new houses have been built within these few years. The chief public buildings are the parish church, the episcopal chapel, and the jail. The cross, which is of a hexagonal form, is reckoned a fine structure: the area of its base is 500 ft. There are no fewer than 10 schools in the parish, of which only one is endowed. The harbour has been much enlarged and improved, partly at the expense of government. It embraces an area of upwards of 6 Scotch acres, nearly a half of which has been excavated along the piers and jetties. It is of easy access; and as it affords excellent anchorage for ships of every size, it has been found to be of great importance to the shipping interest in general on this coast. Dried and pickled cod are exported to the extent of about 7,000l. sterling; grain of various kinds, about 12,000 qrs.; potatoes, 6,000 bolls.

The town and harbour existed above two centuries ago, the former having been erected into a burgh of regality in 1813, called Frederburgh, in honour of Sir Alexander Fraser of Philorth, who obtained the charter. The same Sir Alexander Fraser obtained a charter from the crown, in 1592, for the erection and endowment of a university; and at the west end of the town there is an old quadrangular tower of three stories, which formed part of a building originally intended for this seminary. In 1597, Mr. Charles Ferme, of the University of Edinburgh, was elected principal of this intended college; but from causes not explained, probably from want of funds, the plan was abandoned.

FREDERICKSHALL, a marit. town of Norway, gov. Aggerhuus, at the influx of a small river into the Ide-fiord, near the N.E. angle of the Skagerrack; 67 m. SSE. Christiania. Pop. 5,561 in 1845. Frederickshall is an open town, but immediately above it, on a perpendicular rock, 400 ft. in height, overhanging the sea, is the strong fortress of Frederickstein, at the siege of which Charles XII., king of Sweden, was killed, on the 30th of Nov. 1718. It was doubted for a while whether the king met his death by a ball from the fortress, or had been assassinated; but there seems to be no good grounds for supposing that treachery had anything to do with the matter. Frederickshall spreads irregularly round the rock on which the castle is built; 'it is a strange-looking little town, in which houses, rocks, and water are curiously mingled. One street is terminated by a perpendicular rock; another by a deep creek; and, as there are only three or four little streets in the town, it has at least the praise of being singularly picturesque.' (Inglis's Norway, p. 289.) The streets, though few, are wide and regular, presenting many handsome houses, generally two stories high; all of which appear to have been built since the conflagration, in 1759, by which nearly the whole place was laid in ashes. A considerable trade in timber is carried on, and there are a few manufactures of linens, tobacco, &c.

The castle of Frederickstein is one of the most inaccessible fortresses in Europe. The place suffered greatly by the fire alluded to above, and is now in a state of great neglect. As obvious has been erected on the place where Charles XII. fell.

FREIBERG, a town of the k. of Saxony, and cap. of its mining district, circ. Dresden, near the E. arm of the Mulde; 19 m. SW. Dresden, and 50 m. NE. Leipsic, on the railway from Dresden to Chemnitz. Pop. 17,510 in 1861. Freiberg is an ancient imperial city, and is still surrounded by old walls and a ditch; but the greater part of its fortifications are laid out in gardens and public walks. It is well built, paved, and lighted. It has a cathedral, a handsome Gothic edifice, with a richly ornamented portal, in the Byzantine style, called the Golden Gate; some curiously carved stone pulpits; the tomb of Werner the geologist; a chapel in which the Protestant princes of Saxony, from 1541 to 1694, were buried; and a remarkable monument with an alabaster statue of the Elector Maurice, who died of the wounds he received at the battle of Sievershausen, on the 9th of July, 1553, when he completely defeated the army of the Margrave of Brandenburg. There are 7 other churches, one of which has a spire upwards of 210 ft. high; and an orphan asylum. Without the town is the old castle of Freudenstein, now used as a corn magazine. The rise and fall of Freiberg has been determined by the productiveness of its silver mines, to the discovery of which it owes its origin in the twelfth century. In the height of its prosperity, before the 30 years' war, it is said to have had 52,000 inhabs. Its prop., together with the produce of its mines, has of late fallen off; owing to the richest veins being exhausted, or to the shafts having been driven so deep that it is next to impossible to drain off the water. Still, however, there are in the vicinity numerous mines of silver, copper, lead, and cobalt, employing altogether about 4,500 miners. The principal silver mine is called the Himmelsfürst, that is 'Prince of Heaven,' and is said to be the first in Europe, as well for the quantity of ore it furnishes, as for the excellence of its works. It has been wrought upwards of 440 years, and for 200 yielded silver to the annual amount of 95,000 crowns.

The ore is smelted at the village of Halsbrücke, about 3 m. from Freiberg, where there are numerous furnaces, forges, &c., and where the process of amalgamation is conducted on scientific principles. Freiberg has manufactures of gold and silver lace, employing 700 hands; a woollen cloth and cassimere factory, in which twelve steam engines, 110 hands are employed; manufactures of lace, cotton fabrics, and thread, white lead, litharge, vitriol, leather, copper ware, &c.; some extensive breweries; and a shot foundry, the only one in the kingdom. It is the seat of the high board of mines (Oberbergamt), and that of foundries (Oberhüttenamt), with supreme jurisdiction over all such establishments throughout the kingdom. It has a gymnasium with a good library; but its most celebrated establishment is its mining academy, founded in 1765. It owes its principal celebrity to Werner, appointed professor of mineralogy to it in 1775: his eloquence and the charm of his manner inspired the greatest enthusiasm into his pupils, and besides raising the school of Freiberg to the highest eminence, and attracting to it students from the most distant countries, gave a great stimulus to the science. There are now about 10 professors in the school of Freiberg, who give instruction in the working of mines and of metals, and in chemistry, and all the accessory sciences. A specified number of Saxon pupils receive gratuitous instruction in this school, some of whom work as miners for a certain time each day, receiving higher wages than the ordinary miners. There is also a preparatory school to qualify pupils for the academy. Attached to the latter are many scientific collections, and among others the celebrated collection of precious stones amassed by Werner, and bequeathed by him to the academy.

Freiberg was long the residence of the Saxon princes, who bestowed on it many immunities and privileges. It suffered greatly during both the thirty years' and the seven years' war.

FREIBURG, or FRIBOURG, a canton of Switzerland, the ninth in rank in the confederation, in the W. part of which it is situated, between lat. 46° 27′ and 47° N. and long. 6° 45′ and 7° 27′ E.; having for the most part N. and E. the Bernese territory, and S. and W. that of Vaud. A detached portion to the W. has for its NW. boundary the Lake of Neufchatel, and is everywhere else enclosed by the cant. Vaud. Its greatest length N. and S. is about 40 m., and its breadth varies from 8 to 26 m. Area, 564 sq. m. Pop. 105,970 in 1860, being 177 to the square mile. The northern part of the canton is almost a level plain, or at most only hilly; but proceeding N. the surface becomes more uneven, and the S. half of the canton is covered with mountains, appertaining partly to the Jura system and partly to the Bernese Alps, but none of their summits reach the limit of perpetual snow. The principal are the Dent de Broc, 7,836 ft.; the D. de Folliéran, 7,667 ft.; and M. Moleson, 6,578 ft. high. Nearly the whole canton is included in the basin of the Aar, its centre being traversed by the Saane, or Sarine, an affluent of that river. The Broye is the other principal stream. The chief lake is that of Morat (Murtensee) in the N., through which the last-named river flows; it is 6 m. long by 2 m. broad, and very abundant in fish, particularly the eels. Climate mild in the N., but rigorous in winter in the S. The higher mountains are composed principally of a coarse-grained limestone, containing many flints; those of inferior height of sandstone. A considerable proportion of the land is fertile: it has about 100,000 acres of arable land, 64,800 do. meadow, 20,000 do. of pasture land, 700 do. vineyards; and the forests are supposed to comprise 34,500 acres. Agriculture is the chief pursuit of the inhab. in the N., and cattle-rearing in the S. districts. Enough of corn is grown for home consumption, but the dairy husbandry is the most important branch of industry, and is in a more advanced state than in any other Swiss canton. The annual produce of cheese is estimated at 40,000 cwt., worth 1,700,000 fr. The famous Gruyère cheese, produced in the district of that name, in the valley of the Sarine, stands decidedly at the head of the Swiss cheese, and is highly prized in this and other countries. The average produce is about 25,000 cwt. a year. The breeds of horses and black cattle are considered the best in the confederation, and large markets for the sale of both are held at Romont, Bulle, and Freiburg. Gardens, orchards, vineyards, &c. are most numerous in the N. Tobacco, oleaginous plants, hemp, flax, &c. are grown, but in no great quantities. The produce of timber is important. Turf is procured in many places, coal only in the N., and to an inconsiderable amount. There is a glass factory at Semsales, employing 150 workmen. The other principal manufactures are those of straw hats, leather, and paper, but they are quite insignificant. The chief article of export besides cheese is timber to France, from which about 22,000 or 23,000 cwt. of salt are imported yearly. The people generally are in comfortable circumstances. The public roads, which were formerly very bad, have been of late years greatly improved, and the great line of railway from Berne to Geneva is running right through the canton. Freiburg is divided into 13 circles or distr. Chief towns, Freiburg, the cap., Morat, Gruyères, Estavayer, Bulle, and Romont; but, except the first, none has 1,500 inhab. Seven-eighths of the pop. are Rom. Cath.; the Protestants, about 8,400, reside chiefly in the district of Morat. German is spoken in the NE., and a dialect of Romansche or Italian in the S.; but French is the language most universally employed in the canton, and has been adopted as that of all state proceedings. Since 1831, the government has been wholly democratic. It consists of a great and petty council; the former, which has the sole legislative power, consists of 66 members, or about one for every thousand of the inhab.; all males above 25 years of age, not

servants or subject to foreign powers, have the right to vote in the appointment of the electors of the central body. The petty or executive council is composed of 13 members chosen by the legislative body, who also appoint for life the 13 judges of the supreme court of appeal. The great council is presided over by an Avoyer, who holds office for two years only; while the council itself exists for nine years. Each circle has its own local council, a governor called an Oberamtmann, and a court of justice with appeal to that in the cap. Personal freedom, the privilege of petitioning, and the abolition of feudal rights, have been guaranteed; as also the liberty of the press. Education in this cant. was formerly, and to a great extent is still in a lower state than in many others. The places for superior instruction are chiefly in the town of Freiburg (which see), and the Protestant college at Morat.

Freiburg furnishes a contingent of 1,240 men to the army of the Swiss confederation, and contributes 18,600 francs annually to its expenditure. Besides the above contingent, and an equally numerous corps de reserve, there is a militia of all the male pop. between 16 or 20 and 45 or 50. The total public revenue in 1862 amounted to 1,201,240 francs, and the expenditure, in the same year, to 1,168,789 francs. The canton, at the same period, had a debt of 3,886,400 francs. Before the 11th century this territory formed a part of the kingdom of Burgundy, but afterwards belonged to the dukes of Zæhringen, and other feudal nobles. Its history, after the 15th century, is for the most part that of its cap.

FREIBURG, or FRIBOURG, a town of Switzerland, cap. of the above canton, on both sides the Sarine, 15 m. SW. Bern, and 52 m. NE. Lausanne, on the railway from Bern to Lausanne and Geneva. Pop. 10,454 in 1860. Few towns in Europe are so singularly situated as Freiburg. It is naturally divided into the Upper and Lower town; the former built on the summits of a succession of rocky hills, and the latter in the narrow valley of the Sarine, which is here crossed by two bridges of wood, and one of stone. The upper town is the principal. Many of its houses stand on the very edge of the precipice overhanging the river; and their quaint architecture, the long line of embattled walls, stretching up hill and down dale, varied by the chain of feudal watch-towers and gateways of the ancient fortifications, which still exist in a perfect state, together with the singular and romantic features of the gorge of the Sarine, give the distant view of the town an aspect different from that of any other in Europe, which is at once imposing and highly picturesque. The great glory of the town is its iron suspension bridge, one of the longest and finest in Europe. It is erected across the ravine through which the river flows, and is 905 ft. in length, 22 ft. in breadth, and 174 ft. in elevation; being more than one-third longer, and nearly as much higher, than the Menai Bridge between Anglesea and Carnarvonshire. The materials of which it is composed are almost exclusively Swiss. It was completed in three years, at an expense of about 25,000l., under M. Chaley, an engineer of Lyons; and was thrown open to the public in 1831. Freiburg has 9 convents, and 4 churches, besides numerous chapels. The principal church, that of St. Nicholas, is a rather handsome Gothic edifice, with a spire elevated 376 ft., being the highest in Switzerland. It has some curious bas-reliefs and paintings; and an organ with 7,800 pipes, reckoned one of the finest on the continent. The Jesuits have a monastery at Freiburg, founded in 1581. It was suppressed previously to 1818, when it was restored by a decree of the great council of the canton. It supported, for a time, 60 'fathers,' and had attached to it a college, in which between 200 and 400 pupils were educated, mostly the children of French and German R. Catholic families. The college was suppressed in 1847, after the Sonderbund war, and notwithstanding many efforts to that effect, has not since been re-opened. The extensive buildings belonging to the college occupy the highest site of the town, and tower over all other houses. The remaining objects most worthy of notice are the town-hall, on the site of an ancient castle of the dukes of Zæhringen; the hospital, orphan asylum, workhouse, house of correction, some public baths, several public libraries, and several learned societies. Freiburg is the seat of government, and of the court of appeal for the canton, and the residence of the R. Cath. bishop of Lausanne and Geneva. It has a few manufactories of straw hats, porcelain, tobacco, chicory, paper, hats, and musical instruments, and dyeing houses, tanneries, and breweries. Most of its pop. are Catholics; and it is a singular circumstance that the inhab. of the upper town speak French, while those of the lower speak German; and many understand only one of those languages. The upper town was founded, in 1174, by Duke Berthold of Zæhringen; the lower town had existed previously. In 1277 Freiburg fell into the possession of Rudolph of Hapsburg; but in 1451 it became a free city of the empire. The Duke of Savoy soon afterwards constituted himself its protector; but the Freiburgers having distinguished themselves in the contest against Charles the Bold of Burgundy, the city and its territory were received into the Swiss Confederation in 1481. In 1476 a celebrated battle was set within the walls of Freiburg, and in 1803 another, the latter being that at which the French Act of Mediation was accepted.

FREJUS (an. *Forum Julii*), a town of France, dép. Var, cap. cant., in a spacious plain, 1 m. from the Mediterranean, and 15 m. SE. Draguignan, on the railway from Marseilles to Nice. Pop. 2,047 in 1861. The town offers contrasts painfully with its ancient condition. Formerly it was a league in circ., was surrounded by strong walls flanked with towers, and had 60,000 inhab. Its amphitheatre, the outer circ. of which is 2104 ft. still exists in a ruined state. Its port, which was under its walls, and communicated with the sea by means of a canal 1½ m. in length, was bordered by fine quays, the traces of which still exist; as well as part of a lighthouse, and a large triumphal arch, which formed the entrance from the port into the town. The sites of the port and canal are now occupied by gardens. The town and port were formerly supplied with water from the river Siagne, by means of a fine aqueduct, 16½ m. in length; this noble work is in great part destroyed. Frejus has a church and episcopal palace, both of which are of Gothic architecture, but in part constructed of the materials of Roman edifices. The chapel of the baptistery is an octagonal building, ornamented with eight ancient Corinthian columns. Numerous other remains of antiquity may be seen in the neighbourhood. It has a seminary and a hospital, both modern and handsome buildings. Frejus is the seat of a bishopric, and of a chamber of commerce; it has some bottle-cork factories, and water-works for sawing timber; but its trade is now next to nothing, and its ancient sorts have dwindled down to a few boats.

This town was a place of importance in the time of Julius Cæsar, who gave it his own name.

Vol. II.

Augustus sent thither the 300 galleys taken from Antony at the battle of Actium, made *Forum Julii* a naval station of importance, and planted in it a colony of soldiers of the 8th legion. Agrippa further devoted his endeavours to increase the prosperity of the town. Its strong fortifications protected it for a considerable period against the barbarians; but about the year 940 it was destroyed by the Saracens, nor has it since recovered so much as the shadow of its former prosperity. At St. Raphael, a little fishing village about 1½ m. from Frejus, Napoleon disembarked on his return from Egypt, in 1799, and again embarked for Elba in 1814.

Frejus was the birthplace in antiquity of Julius Agricola, C. Gallus the poet, and Kracius the artist; and in modern times of the Abbé Sieyès.

FREYBURG, or FREIBURG, a city of the grand duchy of Baden, circ. Upper Rhine, of which it is the cap., on the Dreysam, a tributary of the Rhine, within the skirts of the Black Forest, and at the entrance of the Hollenthal, 71½ SSW. Carlsruhe, and 32 m. NNE. Basle, on the railway from Frankfort O. M. to Basle. Pop. 16,883 in 1861. The town was originally fortified by its founder; but its fortifications were levelled by the French in 1744, and their place is now occupied by fine public walks and vineyards, from which excellent wine is obtained. It is generally well-built and lighted, contains several good squares, and has numerous public edifices. The principal of the latter is the minster or cathedral, one of the most perfect Gothic buildings in Germany, and remarkable alike for the delicate symmetry of its proportions, and the good taste of its decorations. It was begun by Conrad of Zæhringen in the 13th, but not completed till towards the end of the ensuing century. The whole edifice is built of red sandstone. The W. front, with a magnificent portal, and the tower and spire, 380 ft. high, which surmount it, were the work of the celebrated Erwin of Steinbach, the architect of Strasburg cathedral. The spire is of the finest openwork tracery, all of stone, and of extreme boldness as well as lightness. The minster contains statues of Berchthold V. and the other dukes of Zæhringen, who were buried in it; several tombs worthy of notice; a remarkable piece of sculpture of the Lord's Supper, by an artist of the 16th century; paintings by H. Grün, a famous artist, also of the 16th century; and some stained glass windows of great beauty. The university, founded in 1454, is in a very flourishing state; it has about 600 students, their number having, for several years past, been on the increase. It is particularly famous as a school of theology, having united with it the high Rom. Cath. seminary of the grand duchy, removed thither from Meersburg. The university possesses a good deal of landed property in Würtemberg, Baden, and Switzerland; besides which it enjoys considerable government grants. It has a library with upwards of 100,000 vols., a cabinet of nat. history, museum, fine collection of philosophical instruments, chemical laboratory, anatomical theatre, school for clinical instruction, and a botanic garden. In the church of the university there are several paintings by Holbein. Freiburg has a grand-ducal and an archiepiscopal palace, 3 hospitals, a custom-house, a venerable old Gothic building, a new archiepiscopal seminary and church, a Lutheran church, new museum, town-hall, theatre, house of correction, foundling and orphan asylums, many other benevolent institutions, a gymnasium, an industrie-schule, or school of forest and garden economy; Herder's institute of arts, for copper-plate engraving, and printing, and lithography; a geographical institute, teachers' seminary, school for girls kept by Ursuline nuns, and a great number of general and primary schools. In the centre of the square called the tob-market, is a fountain surmounted by a statue of the founder of the city, Duke Berchtold III. of Zähringen. Freyburg is the seat of an archbishopric, with jurisdiction over the whole of the grand-duchy, and the bishoprics of Mayence, Fulda, Rothenburg, and Limburg; of an aulic court, and the superior courts of law, and government offices for the circle of the Upper Rhine. Its chief sources of prosperity are its university and other public establishments; but it has also manufactures of chicory, soap, starch, leather, tobacco, bells and other metallic articles, good musical and surgical instruments, earthenware, besides several paper-mills and dyeing-houses. In its vicinity are the fine gardens of Ludwigslust, the ruins of the castle of Zähringen, and many other spots admired for their picturesque beauty. Freyburg was founded in 1118, by Duke Berchtold III.; it was long the cap. of the landgraviate of Breisgau; belonged successively to the house of Austria and the Duke of Modena; and was finally ceded to Baden by the treaty of Presburg.

FRIESLAND, a prov. of Holland; which see.

FRIGENTO, or FRICENTO, a town of Southern Italy, prov. Avellino, 17 m. ENE. Avellino. Pop. 8,638 in 1861. The town has a fine cathedral, containing some excellent paintings. Its inhab. subsist by the sale of sheep, hogs, and corn. Frigento is said by some antiquaries to occupy the site of the ancient *Frequentum*, and by others that of *Æculanum*, besieged by Sulla during the civil wars; but the probability is that it is different from either. Near is a valley, supposed, apparently on good grounds, to be identical with the *Amsancti valles* of Virgil. It is narrow, and is pressed in on both sides by high ridges thickly covered with masses of oak. The bottom of the dell is bare and arid. In the lowest part, and close under one of the hills, is an oval pool, not 50 ft. in diameter, the water in which bolls, and spouts up in jets d'eau, at irregular intervals, to a height of several feet, with a hissing noise, accompanied by strong sulphurous and mephitic exhalations.

It was through this orifice that the fury Alecto descended to Tartarus, and the appearance of the place perfectly corresponds with the admirable description given by Virgil :—

'Est locus, Italiae in medio sub montibus altis,
Nobilis, et famae multis memoratus in oris,
Amsancti valles; densis hunc frondibus atrum
Urget utrimque latus nemoris, medioque fragosum
Dat sonitum saxis et torto vertice torrens.
Hic specus horrendum, et saevi spiracula Ditis
Monstrantur, ruptoque ingens Acheronte vorago
Pestiferas aperit fauces.'
                Æneid vii. line 563.

FROME, or FROME-SELWOOD, a parl. bor., town, and par., of England, co. Somerset, hund. Frome, near the W. border of the co.; 11 m. S. by E. Bath, and 115½ m. W. London by Great Western railway. Pop. of parl. bor. 9,522, and of par. 11,800 in 1861. Frome is situated on an irregular acclivity rising abruptly from the Frome, or stream whence it derives its name, and which is here crossed by a bridge of five arches. The principal street contains many well-built houses, and a good modern market-place; between thirty and forty other streets, mostly very narrow and irregular, being connected with it on either side. It is paved, lighted, and amply supplied with water. The church, dedicated to St. John, is a

spacious structure in the later Gothic style, with a tower and fine octagonal spire 170 ft. in height, has four aumbril chapels, and many interesting monuments; it was restored in 1845. There are three other churches, and six dissenting chapels. There are also asylums for the maintenance, education, and apprenticing forty poor girls; an almshouse for twenty poor men, in a substantial quadrangular building erected in 1790, and endowed with funded property of £600, a year; a free grammar school, founded in the reign of Edward VI.; a charity school, in which thirty-seven boys are clothed and educated for four years, and then apprenticed; an almshouse for thirty-one old women, founded at the same period as the charity school (Edw. IV.), and connected with the same endowment. There is also a national school for 300 boys and 150 girls; and several large Sunday schools. The chief market, Wednesday; a smaller one, Saturday. Fairs, chiefly for cattle and cheese, Feb. 24 and Nov. 25. The woollen manufacture is the ancient staple of the town, and furnishes the chief employment of the pop. The goods consist chiefly of the finer kinds of broadcloth and kerseymeres. Cards for dressing wool are also manufactured, though to a much less extent than formerly, when it supplied them to a great part of the kingdom. There is a canal hence to Stalbridge, with a branch to Wells and Bradford. The Reform Act conferred, for the first time, on Frome the privilege of sending one mem. to the H. of C. The limits of the parl. bor. comprise a nearly square space, extending about 1 m. each way. Registered electors 869 in 1862. The neighbourhood is fertile and picturesque, and contains many old family mansions. Frome has long been celebrated for its excellent ale. Two courtleets are held, one by the Marquis of Bath, the other by the Earl of Cork, lords of the manor. Petty sessions for the division are also held in the town.

FROSINONE (an. *Frusino*), a town of Southern Italy, prov. of same name, at the foot of a high hill near the Cosa, and on the upper road between Rome and Naples; 474 m. ESE. the former city. Pop. 7,800 in 1858. The town is very ill-built, but has many churches and convents; is the seat of a bishopric, and the residence of a card. delegate. It has an annual fair, which begins at Whitsuntide, and lasts twenty days, being near the confines of the former Neapolitan territory, its neighbourhood is infamous for brigandage; to repress which a criminal tribunal, established in it, offers a reward for the heads of brigands.

FUERTE (EL), an inland city of Mexico, state of Sonora, of which it is the cap.; on a river of the same name, 350 m. NW. by W. Durango, and 770 m. NW. Mexico. Pop. estim. at 4,500 in 1862. The town was originally a military station, established by the Spaniards in their progress towards the N. It is now a commercial depôt for goods passing to and from the port of Guaymas; and the seat of the governor, and supreme tribunal of justice. Its importance is wholly due to these circumstances, its local position being far from favourable. It stands on the N. ridge of a vast sandy plain, destitute of vegetation, except in the rainy season, or in spots where the vicinity of the mountains, or the confluence of two large streams, ensure a supply of water; added to which the heat in summer is almost insupportable.

FULDA, a town of W. Germany, cap. prov. Fulda, G. D. Hesse-Cassel, on the river of same name, which is here crossed by a handsome stone bridge, 62 m. SSE. Cassel, and 56 m. NE. Frankfurt-on-the-Mayn, on the railway from Frankfort to Eisenach. Pop. 10,112 in 1861. It is a pretty town in a very agreeable situation; has some good streets, and several squares; of the latter, that in which the cathedral stands is the principal, and is ornamented with two obelisks upwards of 40 ft. high. The cathedral is an elegant edifice, about 336 ft. long, by 213 ft. in breadth; it has a tower 160 ft., and a handsome cupola 100 ft. high, the latter raised upon 16 Ionic columns; a high altar and 15 others, 2 organs, the largest of which is one of the finest in Germany, and the tomb of St. Boniface. There are three other Rom. Cath. churches, a Lutheran church, and some other places of worship, a bishop's palace and garden, a Franciscan monastery, Benedictine convent, Rom. Cath. seminary, public library, gymnasium, lyceum, school of industry, and many other schools; several hospitals, an orphan asylum, various benevolent institutions, an arsenal, house of correction, and workhouse. Fulda is the residence of a R. Cath. bishop, with supreme ecclesiastical jurisdiction throughout Hesse-Cassel; and is the seat of the superior judicial court of the prov. It has factories of stockings, linen and woollen fabrics, tobacco, and leather; dye-houses, and wax bleaching and saltpetre works. About a league S. of the town is Adolphseck, or the 'Phesantry,' a country seat formerly belonging to the prince-bishops of Fulda.

FUNCHAL, a town of Madeira, which see.

FUNEN, or FYEN, an island of the Danish archipelago, the next in size after Zealand, lying between it and continental Denmark; separated from the former by the Great, and from the latter by the Little Belt. It extends between lat. 55° 2' and 55° 38' N., and long. 9° 42' and 10° 55' E. Length, NE. to SW., 50 m. Area, 1,187 sq. m. Pop. 198,811 in 1858. The surface is generally undulating; there are a few hills in the N., but they rise to no considerable height. The shores are very much indented; and in the NE. the Odensee-fiord extends inland for several miles. The chief river is that of Odensee, which runs through the centre of the island; rivulets, lakes, and marshes numerous. Climate humid and variable; soil very productive. 'Funen presents a less agreeable prospect than Zealand, owing not to its more scanty fertility, but to the greater paucity of trees. The crops seem equally abundant, and the flocks equally numerous; and, indeed, Funen is more an exporting country than Zealand, in both corn and cattle.' (Inglis, Norway, 827–329.) Barley, oats, buckwheat, rye, and vegetables are grown in quantities much beyond those required for home consumption; flax and hemp are largely cultivated, and orchards are numerous. The honey is very superior, and an article of considerable export. Turf, clay, and chalk are the only mineral products of value. There are a few manufactures of woollen and linen fabrics, and many domestic ones of stockings, and other articles. Besides corn, cattle, horses, and honey, the chief exports are fruit, lard, butter, leather, salted meat, and some manufactured goods; the trade is brisk, and chiefly with Norway and Sweden. Funen, together with the islands of Langeland, Taasing, &c., forms a prov. of Denmark. Chief towns, Odensee, the cap., Svendborg, and Nyeborg.

FUNFKIRCHEN (Hung. *Pecs*), one of the most an. towns of Hungary, co. Baranya, of which it is the cap., on the declivity of a hill in a rich country, 404 m. SSW. Buda, and 40 m. NW. by N. Essek. Pop. 17,447 in 1857. The town, besides the cathedral, which is the oldest religious edifice in Hungary, and occupies the site of a Roman fortress, has six churches, and several

convents. There are also numerous remains of mosques, baths, and other Turkish edifices, Funfkirchen having been in the possession of the Turks from 1543 to 1686. This town is the residence of a Rom. Cath. bishop, and has a seminary for the R. Cath. clergy, a gymnasium, normal and military schools, a library, and a cabinet of coins. It has also manufactures of woollen cloths, flannels, leather, and tobacco, and a considerable trade, principally in wine, tobacco, and gall-nuts, the produce of the adjacent country. The town boasted for warm mineral baths, and about 7 m. distant from it is a remarkable stalactite cavern.

FURRUCKABAD, a distr. of Hindostan, prov. Agra, presid. Bengal, almost wholly included in the Doab; between lat. 27° and 28° N., and long. 78° 40' and 79° 40' E.; having N. the districts of Moradabad and Bareilly, E. the dooa. of Oude and the distr. of Cawnpore, and S. and W. those of Etawah and Alighur. Area, 1,850 sq. m. The distr. suffered greatly from the anarchy that prevailed in this part of India before the British rule was established.

FURRUCKABAD (Farukhabad, a happy residence), an inland city of Hindostan, prov. Agra, cap. of the above district, near the S. bank of the Ganges, 82 m. ENE. Agra, 156 m. NW. Allahabad, and 80 m. WNW. Lucknow. Pop. estim. at 60,000. The city is considered the chief commercial emporium of the ceded and conquered provinces, and is said to be the common resort of needy and dissolute characters from the rest of Hindostan. It is surrounded by a wall, kept in tolerable repair; streets in parts wide, and many of the open spots and buildings shaded by trees; but, excepting in the principal thoroughfares, most of the houses are of mud. Here, in 1805, Lord Lake surprised and obtained a decisive victory over Holkar's cavalry.

FURTH, a town of Bavaria, circ. Middle-Franconia, on the Regnitz, 20 m. NE. Anspach, on the railway from Nürnberg to Würzburg. Pop. 19,125 in 1861. The town is irregularly built, but contains many good houses; is the seat of a town and district judicial tribunal, and has two Lutheran churches, a Rom. Cath. church, several synagogues, Latin and numerous other schools, besides schools of industry, arts and trades. The Jews, who are interdicted from settling in Nürnberg, enjoy in Furth privileges denied them elsewhere on the Continent; they have here a separate court of justice, a Hebrew college, and two printing presses, exclusively devoted to Hebrew publications. It is principally owing to their exertions that Furth has become, next to Nürnberg, the principal manufacturing town in the Bavarian dominion. It has numerous factories of mirrors, chandeliers, lacquered ware, spectacles, lead pencils, tobacco, gold and silver wire, gold leaf, turned brass, wood, horn, and bone wares, stockings and other woollen and cotton fabrics, leather, liqueurs, coloured paper, buttons, toys, trinkets, and pipes. These articles are exported principally to N. and S. America, the Levant, Holland, Spain, Italy, N. Germany, Denmark, and Sweden. Besides the foregoing, there is a considerable trade in other kinds of produce; and a large fair is annually held here. The first railroad for steam carriages in Germany was completed in 1836–36, between this town and Nürnberg; a distance of 4½ m. About half way between the two towns, the canal which connects the Danube with the Rhine, is carried over the railway. Furth is first mentioned early in the 10th century. Gustavus Adolphus was defeated in 1632, in an attempt to carry the entrenchments of Wallenstein, in the neighbourhood of this city. It was not till 1818 that Furth obtained its municipal rights.

FUTTEGHUR (Fataghur, the fort of victory), an inl. town of Hindostan, prov. Agra, on the W. bank of the Ganges, 3 m. E. Furruckabad; lat. 27° 21' N., long. 79° 30' E. It is a British military station, and the residence of the civil authorities of the Furruckabad collectorate, as well as of several European merchants. Most of the houses are built with mud walls, and a mud fort has been erected for the protection of the arsenal. The cantonments possess an elegant theatre. A government mint has been established here. Tents of a superior kind are manufactured in Futteghur.

FUTTEHPOOR, a large inland town of Hindostan, prov. Allahabad, cap. distr. of same name, on the high road from Bengal to the upper provinces; 60 m. NW. Allahabad, on the railway from Allahabad to Delhi; lat. 25° 56' N., long. 80° 45' E. Some years since it appeared prosperous, and contained, besides several good houses, a recently built and elegant mosque. Like most towns in its

vicinity, it is surrounded with tombs, and on one side of it is a large endowed serai, or hotel for the gratuitous accommodation of travellers.

FUTTIPOOR SIKRA, an inland town of Hindostan, prov. Agra, on the British frontier, 19 m. WSW. Agra; lat. 26° 5' N., long. 77° 34' E. This town was the favourite residence of the emperor Acbar, who built a stone wall of great extent, with battlements and towers round it, the area within which appears never to have been filled up. The town, which is but small, is built of stone. It contains the spacious and tolerably entire remains of Acbar's palace, the tombs of several of his family, and of some Mohammadan saints and statesmen.

FYZABAD (a beautiful residence), an inland town of Hindostan, prov. Oude, of which it was formerly the cap., on the S. bank of the river Kalee; 60 m. E. Lucknow. It is still large and populous; it contains the remains of a fortress, and of the palace of Shuja al Dowlah.

G.

GAETA (an. Caieta), a fortified sea-port town of Southern Italy, prov. Caserta, cap. distr. and cant., at the extremity of a peninsula, on the W. shore of Italy, forming the NW. boundary of the gulf to which it gives name; 4 m. SSW. Mola-di-Gaeta, 41 m. NW. Naples, and 72 m. SE. Rome. Pop. 16,344 in 1861. The town is regarded as one of the keys of Southern Italy, being strong from its position, and defended by walls flanked with bastions and redoubts, and by a square castle situated on a rock. Its suburbs are much more extensive than the town itself.

Gaeta is irregularly built; its streets are narrow and steep; those in the city are, however, greatly inferior to those in the suburbs. It has a cathedral with a fine tower, the construction of which is attributed to the emperor Frederick Barbarossa; nine other churches, several convents, a public seminary, a hospital, and a foundling asylum. On the isthmus connecting the citadel with the mainland stands the Torre d'Orlando, originally the tomb of Plancus; and near the suburb of Castellone is the Tower of Cicero. Its port, which has 7 fath. water, though not the largest, is one of the safest and best in Italy. This city is the seat of a bishopric, under the immediate superintendence of the pope. It is the centre of a considerable trade. Its neighbourhood is extremely beautiful, and covered with villas and country houses.

'Caieta is very ancient. Virgil says it derived its name from the nurse of Æneas buried in it:—

'Tu quoque litoribus nostris, Æneia nutrix,
Æternam moriens famam, Caieta, dedisti.'
Æneid vii. 1.

It became the residence of many opulent patrician families of Rome; and Cicero was put to death, by order of Antony, in its immediate vicinity. After the fall of the Western empire, it had a republican form of government, at the head of which, however, was placed a duke, acknowledging the temporal supremacy of the pope. It coined its own money till 1191; in 1435, it was taken by Alphonso V. of Aragon; and since then has belonged to the crown of Naples. In modern times it has been repeatedly besieged; the last siege of any great note was in 1806, when it fell into the hands of the French. It, however, held out against the Austrians for some time, both in

1815 and 1821, and withstood during a few weeks the troops under General Garibaldi, who besieged it in November, 1860, when the last king of Naples had taken refuge in the fortress.

GAILLAC, a town of France, dép. Tarn, cap. arrond., on the Tarn, 12 m. SSW. Alby. Pop. 7,834 in 1861. The town is ill built, but has been of late considerably improved, and is well lighted. It has extensive suburbs, a tribunal of primary jurisdiction, a communal college, a society of agriculture, two hospitals, and a small theatre. It is the seat of a sub-prefecture; and has manufactures of wine casks, hats, leather, and brandy; besides dyeing houses, and docks for building boats. Its neighbourhood produces some very good, strong-bodied, deep-coloured wines, which are said to bear sea-voyages well. These wines constitute the principal exports of Gaillac.

GAINSBOROUGH, a market-town, river port, and par. of England, co. Lincoln, wap. Corringham in Lindsey, on the Trent, about 21 m. from its embouchure, in the estuary of the Humber; 15 m. NW. Lincoln, 117 m. N. by W. London by road, and 153½ m. by London and North Western railway. Pop. 6,820 in 1861. The town, consisting chiefly of one long street, running parallel with the river, is clean, well-paved, and sufficiently lighted. The church is a modern building, erected by the inhab. in 1748; the living, a vicarage attached to one of the stalls in Lincoln cathedral. There are also several places of worship for dissenters of various denominations. The town-hall, which is used also as an assembly-room, is a convenient brick building; the lower part is used as a gaol, and for shops. There is a small theatre. The bridge over the Trent, at the N. extremity of the town, built in 1791, is of stone, with three elliptical arches. At the NW. end of the town stands a very singular building, known as the Old Hall, and is said to have been a palace of John of Gaunt; but its appearance shows it to be of later date. It is composed of oak timber framing, and forms three sides of a quadrangle, the N. side of which was a chapel; gardens were formerly attached, and a moat surrounded it. About ½ m. N. from the town, on the bank of the river, are the Castle hills, mounds supposed to have been erected during the civil wars under Charles I. The tide ascends the Trent as far as Gainsborough, which being reached by vessels of from 150 to 200 tons, has a considerable coasting and some foreign trade; and it possesses means of communication with the interior by the Chesterfield and Foundyke canals. The shipping belonging to the port consisted, on the 1st of Jan. 1861, of 11 sailing vessels under 50, and of 2 sailing vessels above 50 tons; besides 10 steamers of a total burden of 509 tons. The customs duties received here were to the amount of 11,637*l.* in 1859; of 11,158*l.* in 1860; and of 9,064*l.* in 1858. Vessels of considerable burden have been built here. Markets on Tuesday, and fairs for cattle and toys on Easter Tuesday and October 20. Gainsborough is the birthplace of Bishop Patrick, the well-known commentator on the Bible.

GALACZ, or GALATZ, a town of Moldavia, on the N. bank of the Danube, between the confluence of the Sereth and the Pruth with that river, 80 m. W. its Sculianeh mouth; lat. 45° 21′ N., long. 28° E. Pop. estimated at between 75,000 and 80,000. The town, especially the older parts, is ill-built and filthy. 'Picture to yourself,' says a French traveller, M. St. Marc Girardin, 'upon an eminence sloping rapidly to the waterside, a confused cluster of wooden huts, intersected by irregular streets, paved with trunks of trees, placed from one side to the other; which it is fine weather a tremendous dust,—converted by rain into deep mud. Imagine these cabins, dark and sombre within; and without, filthy with mud; a sorry caravansary by way of inn, with apartments almost without furniture, and as full of dust as the streets; not the least appearance of order, cleanliness, or arrangement; a town constructed like an encampment—such is Old Galacz.' The houses are built of unpainted wood, and roofed with the same material. Most of them are limited to a single floor, with a front open towards the street; and goods exposed for sale are spread out on the ground. But considerable improvements have latterly been effected in all parts of the town; and, within the last few years, a new and superior town has begun to grow up by the side of the former. It is seated on a hill which overlooks the Danube. The houses, two stories high, tiled and white-washed and furnished in good style, are occupied by the richer class of merchants, and by the consuls and other public functionaries resident in the place. About 1,000 of the inhabts. of Galacz are said to be immigrants from the Ionian Islands. It has also, a considerable number of Jews and Armenians. Hitherto, however, the greater part of its trade has been carried on by Greek merchants; but latterly many English and other foreign houses have been established in it.

Galacz has of late years, or since the opening of the trade of the Danube in 1829, become an important emporium. It is the principal port of Moldavia, and along with Ibraila or Ibrahilov (see IBRAHILOV), about 12 m. further up the river, is the chief entrepôt of the vast countries traversed by the Danube from Hungary to the Black Sea. Its great articles of export are wheat and Indian corn, rye, tallow, wool, butter, timber, staves, hides, wax, wine, and skins; the principal imports being olive and other oils, manufactured goods, hardware, and colonial produce. Galacz has also some manufactories, among which a large soap and candle factory, an establishment for preserving meat, and a large steam flour mill. Galacz is the shipping port for almost all the merchandise which enters the principalities by sea. The wheat shipped here is superior to that of Ibraila. Subjoined is a table showing the quantities and value of the principal articles exported from Galacz in the year 1860 :—

Articles		Quantities	Value
			£
Wheat	Qrs.	151,530	247,606
Rye	.	88,770	84,443
Barley	.	6,697	4,118
Indian Corn	.	770,217	593,774
Linseed	.	1,502	4,364
Rapeseed	.	1,814	4,274
Kidney Beans	.	2,190	5,560
Flour	Cwts.	—	—
Tallow	.	—	—
Walnuts	.	—	—
Tobacco	.	8,813	7,636
Preserved Meat	Cases	—	—
Planks and Deals	Pieces	384,847	6,976
Stock Fish	Bales	—	—
Other Articles	Value	—	8,843
Total	.	.	895,866

The shipping, in 1860, consisted of 620 vessels, of a total burden of 169,170 tons, which cleared the port. Among them were 58 British vessels, of 10,355 tons burden.

GALASHIELS, a town of barony and manufacturing town of Scotland, partly in co. Selkirk, and partly in co. Roxburgh, on both sides the Gala, 1 m. from its influx into the Tweed, 27 m. SSE. Edinburgh and 50 m. N. Carlisle, on the Edinburgh and Hawick railway. Pop. 8,189 in 1861. Though a place of considerable antiquity, most of the buildings are new; manufactures, to which it owes its present importance and increased size, have only of late years been carried to any great extent. The town is somewhat irregularly built, but it has a picturesque appearance, being situated in the centre of a fine pastoral district, and flanked in by richly wooded hills, of considerable height. The opposite portions of the town are connected by three bridges. There are no public buildings of importance, except the parish church and a few chapels. The schools are four in number; one of them parochial. There are two subscription libraries, a reading room, and a mechanics' institution. Galashiels is remarkable for its woollen manufacture. Situated in the middle of a pastoral country, which yields abundance of wool, the inhabitants seem to have cultivated this manufacture at an early date, though it was long on a rude and limited scale. Nearly half the raw material is manufactured into stockings and stocking yarn, flannels, blankets, shawls and plaids; the remainder into narrow cloths, of various kinds and colours, and crumb cloths, of grey or mixed colours. To this narrow cloth the general name of tweeds was long given, because it was manufactured on the Tweed, or in its immediate vicinity; but the term is now confined to a particular species, of a mixed indefinite colour. Black and white checks, and tartans of various patterns, are made to a great extent. The tartans made at Haunchfamm are of hard-spun yarn; those made in Galashiels are of soft-spun yarn; the two fabrics being altogether different in their texture and appearance. The cloths manufactured have generally been of a coarse kind, but of late a finer species has been produced; indeed broad cloths of the finest quality have been attempted, and with no inconsiderable success. By the use of foreign wool, the flannels of this place have risen to a degree of fineness surpassing any made in Scotland, and not much inferior to the best produced in the sister kingdom. The shawls, when made of foreign wool, are exceedingly soft and elegant, as also what are called mufflers, or neckcloths, for gentlemen's use. Tanning of leather is also carried on to a considerable extent in Galashiels.

Galashiels was erected into a bor. of barony in

1599, at which date its pop. was 400. But it is mentioned in history nearly three centuries before this date. (Haiko' Annals, apud annum 1337.) Gainshiels was once a royal hunting station, and was used as much when the king came to 'the forest' (Selkirkshire) to enjoy the pleasures of the chase. The tower, called 'the Peel,' a rudely built square edifice, of two stories high, in which he resided, was demolished within the last twenty years. Gala-house, the residence of the feudal superior of the bar... is in its immediate vicinity. Abbotsford, the celebrated residence of Sir Walter Scott, is not above a mile distant, being on the opposite side of the Tweed, in the parish of Melrose. Gala is celebrated in song, 'the braw, braw lads of Gala water;' as are also the Tweed, and its two tributaries in this neighbourhood, the Ettrick and Yarrow.

GALICIA and LODOMERIA (KINGDOM OF), a prov. of the Austrian empire, forming its N.E. portion, between 47° 10' and 51° 50' N. lat., and 18° 50' and 26° 30' E. long. The name Galicia is derived from the Polish 'Halicz,' as Lodomeria is from 'Wladimir,' both being ancient principalities, forming a part of the present province, which also includes the territories of Poland which fell to Austria in the various partitions of that country, and the Bukowina ceded by the Turks in 1774. Galicia lies to the N. of the Carpathian Mountains, by which it is separated from Hungary: on the NW. Galicia is separated from Prussia, the state of Cracow, and a part of the kingdom of Poland, by the Vistula; on the S. and NE. it is open, and has no well defined boundary; the E. frontier towards Volhynia is formed by the little stream Podhorce, which falls into the Dniestr. A range of heights divides the Bukowina from the Turkish part of Moldavia. On the W. the little stream Biala, a tributary of the Vistula, forms the boundary towards Austrian Silesia.

Surface of the Country.—Lying on the N. and E. fall of the Carpathians, from their summits to the great N. plain into which they subside, Galicia is mountainous in the S., hilly in the centre, and in the N., and most extensive portion, a continued plain. (For an account of the Carpathians, see that article.)

Rivers.—Galicia is most advantageously supplied with rivers suited both to the purposes of commerce and irrigation. The Vistula (Weisel), which rises in Silesia, and flows N., to Dantzic, where it falls into the Baltic, enters the kingdom at Dzieditz above Oswiecim, and forms the frontier as far as Zamyrhost, a short distance below its point of junction with the San. Blumenbach states the elevation of its bed above the level of the sea at its entrance into Galicia to be 747 ft., and at Cracow to be 540 ft. Notwithstanding this rapid fall, the Vistula is navigable from Oswiecim for barges, and at Cracow for larger vessels. In this part of its course the Vistula receives the Sola, Skawa, Dunajec, and San, the sources of which are in the northern Carpathian range. The San is the second river of importance in Galicia from the length of its navigable course, which commences at Przemysl. The Bug, whose sources lie in the hills to the N. of Lemberg, leaves the kingdom before it becomes unnavigable. The sources of the Dniestr, which flows NE. till it falls into the Black Sea, are situated in the Carpathians, a little to the W., of those of the San. The course of the Dniestr is at first from SW. to NE., but at Koniaski it changes to a general south-east direction, which it preserves until it leaves the kingdom. The Dniestr is navigable from Koniaski, within 26 m. of the San, where it is

navigable; so that it would not be difficult, by uniting these rivers, to form a channel of communication between the Baltic on the one hand and the Black Sea on the other, enabling the corn and other produce of the prov. to be sent to whichever offered the most profitable outlet. Several other important rivers, such as the Pruth and the Szereth, with the Suinwa and the Moldawa, its tributaries, take their rise in the Bukowina, which, however, they leave before they attain any size.

Lakes.—If all the sheets of standing water which are denominated lakes be numbered, few countries can boast of so many as Galicia. Not only the plain at the foot of the hills, but the valleys that intersect the hilly country, and the shelvy declivities of the granite masses of the Tatra, are full of small lakes. Some of the last mentioned are most picturesquely situated, and furnish water to fine cascades. The most elevated is the Black Lake of about 40 acres in extent, on the north side of the Krivan.

Climate.—The climate of Galicia is, with the exception of the Bukowina, tolerably equal, and in winter is very cold. The greatest heat is + 92° in summer, and the greatest cold is —21° of Fah., according to Blumenbach, who states the mean temperature of Lemberg to be + 45°. In the Bukowina the climate is much milder, notwithstanding the mountainous nature of the country, and the mean temperature is several degrees higher, although not so high as at Vienna, which lies under nearly the same parallel of latitude as Czernowitz, the cap. of the Bukowina. The winds are violent, and thunder-storms, accompanied by hail and torrents of rain, are of frequent occurrence.

Soil.—The most generally fertile portion of the province is the hilly country which occupies its centre; the country rises towards the S., the summits of the mountains presenting little but bleak naked rocks. Towards the N. the fertility of the soil likewise diminishes as the hills subside into the sandy marshy plain. The valleys which intersect the hills are usually filled with swamps, of which much as are drained (and there are now the greater part) have a very fertile soil; but the richest portion of the province is that part of the valley of the Dniestr, which once formed a part of Podolia, including the circles of Stanislawow, Czortkow, Kolomea, and part of Zaleszczyki. Some very fertile tracts are likewise found along the banks of the San.

Products.—The agricultural productions are the most important in point of value, although confined to the common grains and potatoes. Maize is only cultivated in the Bukowina. The forests are chiefly of fir; flax and hemp are grown in great abundance. Of minerals, iron is found all through the range of the Carpathians, although but little mining is carried on; gold and lead, with silver, in small quantities, copper near Dunaporits in the Bukowina, zinc and sulphur; but none of these minerals occur in a quantity proportioned to the riches of the other provinces of the empire. Salt alone is found in extensive, and almost inexhaustible beds, which stretch all along the range of the Carpathians. Coals are found in many places; marble and alabaster of unkilling qualities, and quartz in great abundance, which is used for the manufacture of glass; rock crystal, agate, jasper, and inferior qualities of opal, occur in the mountains. But as the greater part of the secondary formations are covered by the immense bed of sand which forms the Polish plain, it is not easy to ascertain their exact nature, and what minerals they contain.

Political Divisions.—Galicia is composed, as

already stated, partly of Polish and partly of Turkish territories. In the W. parts the duchies of Oswiecsin and Zabor, though belonging, at the time of the partition, to the kingdom of Poland, were claimed as fiefs of the German empire, because anciently the Polish sovereigns occasionally did homage for these possessions. Between them and the San, a Polish race, the Mazurs, inhabit the hilly country, while the mostly level land beyond that river is tenanted by a Russian race, differing in language, manners, and appearance from their Polish, as well as from their Moldavian neighbours in the Bukowina.

The population of the province amounted in 1818, according to official reports, to 3,760,319; in 1837, to 1,299,601; and in 1857, to 4,597,470. The figures show that, although there was an increase of nearly one per annum during the 19 years preceding 1837, there was a slight decrease in the next 20 years. This is explained by the revolutionary movement of 1848–9, which was of extraordinary violence in Galicia, and cost a great number of lives.

According to the census of 1857, there are no less than 24,975 noblemen in Galicia. They claim to be the descendants of the Polish knights who subjugated the original inhabitants, of Russian origin, the Russniaks. All these nobles are very proud, though most of them are wretchedly poor. Their number is on the increase, owing to the unlimited divisibility of real as well as other property, and the fact that their titles no less than estates descend alike to all the children of a family, however numerous.

Occupations of the People.—Agriculture is the principal source of wealth in the province, a great portion of which is very fertile. Of late years improved systems of agriculture have been introduced on nearly all the estates of the larger landed proprietors, and beet-root sugar factories are generally diffused. The principal agricultural products are barley and oats, explained by their immense consumption in distilleries, as whisky and potatoes may be said to be the principal beverage and food of the peasantry. Agriculture is extremely backward. Cattle breeding has been very much improved of late years. Swiss and Tyrolese horned cattle, and merino sheep from Saxony and Silesia, have been introduced by improving proprietors, amongst whom Counts Alfred and Leo Potocki, Sanguszko, Molsarch, and Prince Sanguszko, may be specified. The native breeds of all kinds of cattle are very bad, or have degenerated; the horses are small, but capable of great endurance; and the cows give but little milk. Great quantities of horned cattle are annually imported from Moldavia, but are mostly driven through to the great market at Olmutz, which supplies Vienna. Calves and heifers are, however, also bought of the Moldavians, and fattened either in the fine marsh pastures, which are very numerous, or by stall-feeding, which is in general practice upon large farms, and which is always connected with distilling. Turnips and clover are commonly grown where farming is good. Though large, the estates of the Galician nobles are less extensive than those of Poland. A return, of the year 1860, gives the number of estates belonging to nobles, and possessed of manorial jurisdiction, at 3,172, and those belonging to other proprietors, mostly small in extent, at 780,410. The continued subdivision of the soil seems to produce gradually results similar to those witnessed in France.

Manufactures.—The manufacturing industry of the province is quite inconsiderable. After distilleries and breweries, both of which are united with farming on large estates, mining industry is the most considerable. Salt, which is found in great abundance, is worked only on government account, it being a monopoly of the crown. The salt mines of Wieliczka—connected by railway with Cracow—and Bochnia are celebrated for their immense extent. On the cession of these mines to Austria, a stipulation was made in favour of the kingdom of Poland, to which these mines furnish annually any quantity required, at 2 fl. (4s.) per cwt. for rock salt, and the same price for 140 lbs. of boiled salt. The quantity of salt produced in Galicia amounts, on the average, to 1,500,000 cwt. per annum, of which Wieliczka furnishes nearly 1,000,000 cwt. On the whole, mining may be said rather to decline than to increase. Flints for guns were formerly prepared in large quantities at Nizniow and Podgorze, but this manufacture is now transferred to the territory of Cracow.

The manufacturing industry of Silesia has spread partially into the neighbouring parts of Galicia, and both woollen and cotton spinning mills and factories are established in the circle of Wadowice. Glass is made in several parts of the province, but does not rival that of Bohemia in quality. A great deal of linen is woven by the peasantry, who are not registered as workmen, and it is not unusual both for the peasants to pay a portion of their rent in linen, and for servants to receive them in part payment of their wages. Salaries of bailiffs and superior officers of large households are also in part paid in kind, and this is often the case with the allowances of the clergy, schoolmasters, and even of many civil officers employed by government.

Commerce.—The new roads from Brody to Bialla, and that along the mountains through Dukla to Stanislawow, which unites with the high road from Lemberg to Czernowitz, have all been constructed since Galicia came under the Austrian sceptre, and have conferred the highest possible benefit on the country. More recently, an important line of railway has come to unite Cracow and Lemberg, placing the latter city, together with Jaroslaw, Tarnow, and other places, in direct communication with all the great towns of Europe. It is in contemplation to extend this line farther from Lemberg to Czernowitz, for which purpose an English company was formed in 1861, under a guarantee of a minimum dividend of 7 per cent. per annum. The railway from Lemberg to Czernowitz—built by an English contractor, Mr. Brassey—is to be opened in the spring of 1866. From the fairs of Leipzig, Breslau, and Frankfort on the Oder, manufactured goods from Western Europe, and colonial wares, are transported along the roads and railways of Galicia to the E. part of Europe, and a considerable traffic is kept up by their means with Odessa and the Black Sea. The fine navigable rivers which water Galicia are but little used, except the San and the Vistula, when the exportation from Dantzic are sufficiently extensive to affect this part of the country. The boats on the upper Vistula are small, carrying from 30 to 60 tons. On the San the Ukinow boats carry the lower burdens. The Danajec, Poprad, Wysloka, and Bug are navigable for rafts, as are also the Pruth, Sereth, and other rivers. The little river Stry, which falls into the Dniestr, is navigable for rafts for nearly 50 miles, and its valley offers a good pass across the Carpathians into Hungary, the distance from the Stry to the Theiss in Hungary, not exceeding 70 miles.

The imports into Galicia consist chiefly of cattle from Moldavia, and Turkish wares for inland consumption and for the transit trade from Odessa. Furs, hare-skins, wax, and honey are imported from the Russian provinces; the exports consist of corn, timber, linen, hemp and flax, salt, lime, and other articles. But the trade with the other provinces of the empire is of more importance than the foreign trade, as it includes the colonial wares, wine, metals, and manufactured articles consumed by the inhabitants.

Government.—The government of Galicia is similar to that of the other provinces of the Austrian empire. It is represented in the reichsrath, or council of the empire, by 38 deputies, and has besides, its own provincial diet, elected by the tax-paying inhabitants. (See AUSTRIA.) The agricultural population is known to be very devoted to the government, since the introduction of the constitution of 1860, which, indeed, raised them from a state of serfdom to independent citizenship. In respect to the administration of justice Galicia is placed on a similar footing to the German provinces and Bohemia. The seat of the highest authorities is at Lemberg Polish (*Lwów*), where the courts of justice, both civil and criminal, of last resort are stationed. Criminal courts are held at Lemberg, Wisnicz, Sambor, Stanislawów, Rzeszow, and Czernowitz. In his full title the Emperor of Austria styles himself King of Galicia.

GALICIA, a prov. of Spain, situated at the NW. extremity of that peninsula, lying between lat. 41° 57′ and 43° 47′ N., and between long. 7° 17′ and 9° 14′ W. It is bounded N. and W. by the Atlantic, S. by Portugal, and E. by the Spanish prov. of Leon and Asturias. Area, 15,907 sq. m. Pop. 1,471,382 according to the census of 1857. The country is in general very mountainous, being intersected by the branches of the Asturian mountains, which separate at the Sierra de Peñamarella, and form three ranges running WNW. and NSW. through the prov. In this prov. there are numerous depressions or valleys in every direction, of which those inclining W. and SW. are extensive and fertile, especially those of the Miño, Sil, and Ulla. The rivers, which follow the course of these valleys, and generally give them their names, are neither long nor important, except the Miño, which rises in the Sierra de Mondoñedo, in the NE. part of the prov., and flowing S., with numerous affluents by Lugo, receives the Sil from the mountains of Astorga, and then passing by Orense, Ribadavia, and Tuy, enters the Atlantic in lat. 41° 52′ N., after a course of 166 m. The next in importance are the Tambro, running E. into the Bay of Noya, the Ulla running ENE. into the Bay of Arosa, and the Lima, which enters Portugal near Lindoso. The coast of Galicia, especially on the W. side, is abrupt and much indented, forming numerous capes and bays. Of the former, C. Ortegal and Finisterre are best known; of the latter the Bays of Ferrol, Betanzos, Corunna, Pontevedra, and Vigo, are the most extensive. The temperature varies greatly; in the N. and among the mountains, cold, damp, and rainy; warm and moist on the coast; but warm, dry, and genial in the SW. part of the prov. Although fog and moisture prevail more here than in most other parts of Spain, the climate is not unhealthy, and the people are robust, and capable of heavy and continuous labour. The high lands produce abundance of good forest timber, adapted for shipbuilding. There is good pasturage for cattle, sheep and horses, which are kept in small quantities by even the lowest classes of the rural population, and sold at monthly fairs for removal to other parts of Spain. The produce of the valleys consists of wine, maize, wheat, barley, flax, and potatoes, a part of which are shipped off to Alicante, Malaga, and Barcelona. The sweet chest-

nut grows abundantly, and may be justly called the bread of the Galicians, as it constitutes their common and favourite food. The mineral productions consist of copper, lead, antimony, and tin; white marble and jasper are found in the mountains of the N. part. There are several mineral springs; one is at Orense. Along the coast are anchovy fisheries, chiefly conducted by Catalonians.

The pop. is principally agricultural, and landed property is usually divided into small possessions, so that there are few rich proprietors, but many occupiers tilling their own land and reaping their small stocks of cattle. Manufactures are but little followed, coarse woollens, linens, and sailcloth being the only articles produced. The Galicians, or *Gallegos*, are a quiet, simple, hospitable, and industrious people, grave, sober, and trustworthy; the men are hardy, and patient under fatigue or privation; the women are dark, but handsome, cheerful, and fond of singing their national airs. Like the Swiss, they leave their country in great numbers, sometimes 50,000 in a year, to seek employment in other parts of Spain and Portugal, where labour is better rewarded. The best servants in Madrid and other principal towns come from Galicia, and they are preferred for fidelity and obedience; and the porters and water-carriers of Madrid, Lisbon, and Seville are usually natives of this prov. Indeed, so much more effective are Galicians in getting in the harvest and vintages than the Castilian and Portuguese peasants, that a failure is considered as a necessary consequence of their absence from the work. They make also the best soldiers in the Spanish army. The language spoken in this prov. is the old Castilian (which much resembles Portuguese) mixed with low Latin.

Galicia is divided into the four provinces of Corunna, with 435,670; Lugo, with 357,377; Orense, with 319,636; and Pontevedra, with 360,012 inhabitants—all according to the census of 1857. Principal towns, St. Iago, the cap., Corunna, and Orense. The church discipline is conducted by an archbishop and four bishops.

The Callaici, the ancient inhab. of this district, were first conquered by Decimus Junius Brutus, and wholly subjugated by Augustus, who included the country in the prov. of Tarraconensis. The Visigoths took the country from the Romans, and were in their turn driven from it by the Moors. The princes of Asturias retook it from the Moors, and annexed it to their kingdoms, which was united with Castile in 1039.

GALL (ST.), a canton of Switzerland, in the E. part of which it is situated, occupying the 14th place in the Swiss confederation. It has E. a portion of the Austrian dom. (the Vorarlberg and Lichtenstein), from which it is separated by the Rhine; SE. and S. the Grisons; W. the cantons Glarus, Schwytz, and Zurich, with its lake; and N. Thurgau, and the Lake of Constance. Length N. to S., about 40 m.; breadth varying from 11 to nearly 35 m. Area, 767 sq. m. Pop. 181,091 in 1860. The surface is greatly diversified; in the N. there is an inconsiderable portion of plain country, but the central and S. parts are almost wholly covered with Alpine ranges, the summits of some of which rise above the limit of perpetual snow. Mount Sentis, at the SW. extremity, is estimated to be 10,100 ft. above the level of the sea, and Kamorkberg 7,614 ft.; the Speer, a mountain, near the centre of the canton, is 6,305 ft. in elevation. There are, however, several rapturous and fertile valleys, as that of Toggenburg, watered by the Thur, 56 m. in length, those of the Rhine, and others noted for their wild and picturesque character. Next to the Rhine, the chief rivers are the Thur, Sitter, Sern, and Tamina; all, except the last, have generally a N.W. direction. The principal lake is that of Wallenstadt, mostly comprised within this cant. The plains and valleys are in many parts well cultivated; but the corn produced is insufficient for home consumption. Potatoes are extensively grown. Fruit is largely cultivated, especially in the N. Cider is the ordinary drink of the people; and in the mountainous parts of the country a good deal of Kirschwasser, or cherry-brandy, is made. There are vineyards in many of the districts, in which a red wine is made; and the wine of Buechberg, in the valley of the Rhine, is esteemed the best of German Switzerland. But the principal branch of rural industry is the rearing of cattle. Artificial meadows are well kept in the mountainous parts, but not generally so in the lower parts of the country. The number of horned cattle is very great; and in the S. there are many sheep, goats, and hogs. Every spring considerable flocks of sheep are bought in the Grisons, kept during the summer in St. Gall, and sold in the autumn. Dairy husbandry is not so well attended to in this as in many other cantons; but though the cheese be of an inferior quality, the butter is superior. The forests in the S. are extensive, consisting principally of pine and fir, with some beech trees, and a few oaks. But, at present, the forests are almost useless; since, from the want of roads, and the mountainous nature of the country, it is in most places very difficult to bring the timber to market. There are some iron mines near Sargans, and coal and turf are met with elsewhere. Mineral springs are numerous; amongst them are the celebrated baths of the Pfeffers in the S. (see PFEFFERS). St. Gall is one of the principal Swiss manufacturing cantons; as many as 60,000 of its inhab. being supposed to be employed in its manufactures. These are chiefly of cotton fabrics and thread, especially muslins and linen cloth, which was formerly the principal. Muslins of extreme fineness are woven in large quantities in the town of St. Gall, and are embroidered by the women (a most of the) districts. Cotton thread is spun mostly in the S., but also in the valley of Toggenburg, where many cotton handkerchiefs and other fabrics are made. There are some good cloth bleaching establishments at St. Gall and Herisach, and a few glass and wax-bleaching factories; but the manufactures of the canton have generally diminished since the peace. External commerce is chiefly confined to the import of corn and other provisions, and of raw materials for the manufacturer; and to the export of manufactured goods, raw hides, and cattle. The transit trade is inconsiderable, except on the Wallenstadt lake and Linth canal, which form part of the main channel of communication between Zurich and Italy. St. Gall is divided into eight districts, and has no town, St. Gall the cap., excepted, with 2,000 inhab. The government is one of the most democratic in Switzerland. It is composed of a grand and a petty council: the first consists of 150 members (84 Rom. Cath. and 66 Protestants), chosen in the different circles and communes by the suffrage of those citizens above 21 years of age who are neither bankrupt, receiving aid from public charities, nor against whom a criminal judgment has been pronounced; and who pay taxes on property to the amount of 200 francs. Members of the grand council must be above 30 years of age; they are elected for three years, but are always re-eligible. The petty council which has the executive power, consists of nine members,

chosen from among the grand council, each of whom must pay taxes on property to the amount of 6,000 fr. The grand council passes or rejects laws proposed to it by the petty council; has the superintendence of all the state accounts; appoints all public functionaries, and fixes their salaries; exercises the right of granting pardons; and nominates the president of the petty council, as well as its own, who are called landammans, one being a Catholic and the other a Protestant, and who alternately preside in either assembly for a year. The people at large have, however, the privilege of a veto on any law passed by the councils, if that privilege be exercised within 45 days from the time of its passing. Each commune has a council, composed of from 4 to 12 members, and a syndic, to which the local administration is confided. Members must be 25 years of age, and pay taxes on property of 50 fr. value. There are communal and district judicial courts, and appeal from the latter to a supreme court in the cap., consisting of 13 judges, whose qualifications are similar to those of members of the petty council. The total pop. consists of about two-thirds Catholics and one-third Protestants; the latter reside chiefly in the cap., and the valleys of Toggenburg and the Rhine. They exhibit more activity and intelligence than the Catholics; but the greatest harmony exists between the two persuasions, and in the various public schools teachers belonging to either are indiscriminately employed. Education was till lately very backward, but primary and secondary schools are now established in every district. There are some high schools in the cap. German is the language of the cant. St. Gall furnishes a contingent of 2,630 men to the army, and 89,150 fr. to the treasury of the Swiss confederation. The public revenue in the year 1862 amounted to 1,527,057 francs, and the expenditure to 1,532,112 francs. The canton, at the same time, had a debt of 6,700,000 francs. This canton was first formed in 1798, by the union of the territories of the city and abbey of St. Gall with those of other districts, previously subject to the Swiss confederation, and administered by bailiffs.

GALL (St.), a town of Switzerland, cap. of the above cant., on the Steinach, in a narrow and elevated valley, 6] m. S.W. the Lake of Constance, and 39 m. E. by N. Zurich, on the railway from Zurich to Augsburg. Pop. 14,552 in 1860. The town is surrounded by old walls and a dry ditch, now converted into gardens; and has three suburbs. It is well built, and has broad streets, the remains of a celebrated abbey, 6 churches, an arsenal, hospital, orphan asylum, a Catholic gymnasium with 11 professors, a Protestant college with 14; many learned and benevolent societies, public and private libraries, collections of natural history, and a casino or public reading-room. A magnificent abbey was erected over the tomb of a monk, called Gallus, said to have belonged at one time to Iona, under the auspices of Pepin l'Heristal. This abbey was one of the oldest ecclesiastical establishments in Germany. It became the asylum of learning during the dark ages, and was one of the most celebrated schools in Europe between the 8th and 10th centuries. Here the works of the authors of Rome and Greece were not only read but copied, and we owe to the labour of these obscure monks some of the most valuable classical authors; Quintilian, Petronius Arbiter, Silius Italicus, and Valerius Flaccus having been printed from MSS. found here in 1415. Several of its most valuable MSS. having been lent to the dignitaries attending the Council of Constance, were not returned; but it still contains a collection of letters, in 18 volumes folio, by the most distinguished German and Swiss reform-era. The library, which now belongs to the town, occupies a fine apartment; and, besides its literary treasures, has some busts, portraits, and a cabinet of mineralogy. The abbey church is now the cathedral of the diocese of St. Gall and Appenzell; the ancient palace of the abbots (die Pfalz) at present serves for the public offices of the cantonal government; and the other buildings of the monastery have been appropriated to the Catholic gymnasium. The abbey was secularised after the French revolution, and in 1805 its revenues were sequestrated.

St. Gall is one of the chief manufacturing towns of Switzerland. It has extensive manufactures of muslin; in the centre of the Swiss trade in that article, and of embroidery in gold and silver; and a general depôt for the merchandise of the cantons of St. Gall, Appenzell, and Thurgau. Other cotton fabrics and yarn are also produced, the spinning of the latter employing several factories. In the suburbs there are a great many bleaching establishments. Some pretty extensive banking operations are transacted in the town. A market is held every Saturday, and two fairs of eight days each take place twice a year. The inhab. are generally active and prosperous; about seven-eighths of them are Protestants. About 2 m. SW. St. Gall is the fine bridge over the Sitter, called the Krätzerenbrücke, 550 ft. long, and 85 ft. above the surface of the river.

The abbots of St. Gall about the 10th century began to assume a military character, and surrounded the convent with walls and ditches. From the 13th century they enlarged their dominions at the expense of their neighbours, till they became the most considerable territorial sovereigns in N. Switzerland, and were raised to the rank of princes of the empire. Early in the 15th century, however, Appenzell threw off their yoke, and at the Reformation the town of St. Gall emancipated itself from their control, and acquired a territory of its own. The town was first incorporated in the 10th century: in 1454 it allied itself with the five Swiss cantons, and sent a deputy to the diet; and at the end of the 17th century its civil and political independence was secured.

GALLIPOLI (an. Callipolis), a sea-port town of Southern Italy, prov. Lecce, cap. distr. and cant., on a rocky islet on the E. coast of the Gulf of Taranto, 49 m. SE. Taranto, and 28 m. WSW. Otranto, at the terminus of the railway from Ancona. Pop. 9,208 in 1861. The town is united by a bridge with the mainland, on which is its suburb Lizza. Gallipoli is fortified, and has a castle, bombarded by the English in 1812. It is well built, and has a good cathedral, several churches and convents, a seminary, and some other public schools. About 1 m. W. from the town is the island of Andrea, on which is a lighthouse; and between it and Gallipoli there are from 9 to 10 and 12 fathoms water; but vessels of considerable burden must not come within gunshot of the city. Gallipoli displays an air of great industry, if not of affluence. It is the most frequented of all the sea-ports on the SE. coast of Naples, and the great mart for the oil of Apulia, most of which is shipped here. It being peculiarly well adapted to serve as a depôt for oil. The rock (limestone) on which the town is built is easily excavated; and in caverns thus constructed oil clarifies sooner, and keeps without rancidity much longer than in any other place. Hence numerous oil-houses are established at Gallipoli, and a very considerable portion of the rock is cut into cisterns. A Gallipolitan oil-warehouse generally occupies the ground-floor of a dwelling-house, and has a low arched roof. Some are more extensive; but, on an average, they are about 30

ft. square. In the stone floor you see 4, 6, or more holes, which are circular, about 2 ft. in diameter, and like the mouths of wells. Each of these holes gives access to a separate cistern beneath your feet; and when the oil is poured into them, care is taken not to mix different qualities, or oils at different stages, in the same reservoir. When the oil is to be shipped, it is drawn off the cistern into stern or skins, and so carried on men's shoulders down to the sea-shore. Gallipoli has also manufactures of muslin, cotton stockings, and woollen goods; considerable trade in corn, wine, fruit, and a productive tunny-fishery. It is said to have been originally founded by colonists from Lacedaemon. It suffered greatly at the hands of Charles II. of Naples, the Venetians, and the Turks; but the emperor Charles V. improved its fortifications, and restored to it a considerable share of prosperity.

GALLIPOLI (an. Callipolis), a sea-port town of Turkey in Europe, prov. Roumelia, cap. sanjiack and distr., on a headland called the Braccio di Gallipoli, at the point where the Hellespont unites with the sea of Marmora, 91 m. S. Adrianople, and 128 m. W. by S. Constantinople; lat. 40° 24' 37" N., long. 26° 39' 45" E. Estimat. pop. 11,000. The town was once fortified, but is now destitute of walls; its only defence being, in the works of Tournefort, 'a sorry square castle, with an old tower, doubtless that of Bajazet.' The town consists of miserable houses and dirty streets, intermixed with gardens. The bazaars, however, are extensive and well-furnished. There are two ports, a N. and S., which frequently harbour the Imperial fleets, Gallipoli being the chief station of the capitan-pasha. It is also the see of a Greek bishop; and has manufactures of cottons, silk, earthenware, and the best Morocco leather made in Turkey. A few remains of antiquity are in good preservation, and fragments of sculpture and architecture are seen in every part of the town. The great number of Turkish tombs in its vicinity prove it to have been a favourite place of residence with the Turks. A little corn is grown in its neighbourhood, but not enough for one-fourth part of the pop. Gallipoli was the first European town which fell into the hands of the Turks. They took it in 1357, on which occasion the emperor John Palæologus observed, that he had only lost a jar of wine, and a sty for hogs, alluding to the magazines and cellars built by Justinian. Bajazet I. however, knowing its importance for passing from Prusa to Adrianople, had it repaired and strengthened, and its port improved.

GALLOWAY, a distr. in the S. of Scotland, comprising the cos. of Wigtown and Kirkcudbright. Its dimensions were at one time much more extensive; but for a lengthened period it has been restricted as above.

GALLOWAY (MULL OF), a promontory of Scotland, co. Wigtown, comprising the S. portion of the distr. called the Rhynns. It stretches in a SSE. direction from Portpatrick to the Point of the Mull, about 17 m.: its breadth varies from about 3 to about 5 m. The Point of the Mull, the farthest S. limit of Scotland, in lat. 54° 38' N., long. 4° 52' W., rises about 235 ft. above the level of the sea, and is bold, bleak, and striking. A lighthouse of the first class, with an intermittent light, having the lantern elevated 325 ft. above the level of the sea, has been erected on this headland. The view from the balcony of the lighthouse is very extensive, commanding the whole Isle of Man, the coast of Cumberland, and the Cumberland mountains; a great part of the coast with the mountains of Dumfriesshire and Galloway, the Paps of Jura, and the coast of Ireland, from Fairhead to the Mourne mountains.

GALLOWAY (NEW), a royal and parl. bor. of Scotland, on an acclivity, on the W. bank of the Ken, nearly in the centre of the S. of Kirkcudbright, on the road from Kirkcudbright to Ayrshire by Dalry and Dalmellington, 17 m. NNW. Kirkcudbright. Pop. 462 in 1861. Though finely and romantically situated, it is a poor, mean place, without trade or importance of any kind. Kenmure Castle, the residence of the viscounts Kenmure, within a ¼ m. of the bor., stands on a conical mound at the head of Loch Ken, through which the river of the same name flows. The song 'Kenmure's on and awa.' refers to the viscount Kenmure who was beheaded for rebellion in 1715. In the bor. or neighbourhood were born Thomas Gordon, author of "Cato's Letters," the 'Independent Whig,' &c., and translator of Sallust and Tacitus; John Lowe, author of 'Mary's Dream;' and Robert Heron, author of a 'History of Scotland,' in 6 vols., and various other works.

New Galloway was erected into a royal burgh in 1633. It unites with Stranraer, Wigtown, and Whithorn in sending one mem. to the H. of C. Registered voters 18 in 1861.

GALWAY, a mar. co. on the W. coast of Ireland, prov. Connaught, having S. Galway Bay and the cos. Clare and Tipperary; E. King's County and Roscommon; N. the latter and Mayo; and W. the Atlantic Ocean. Area, 1,510,592 acres, of which 476,492 are mountain and bog, and 77,822 water, principally consisting of Loughs Corrib and Mask. Extent of arable land, in sq. m., 1,161 in 1841; 1,358 in 1851; and 1,507 in 1861. (Census of Ireland for 1861, part v.) The coast of the co. is deeply indented in its W. and SW. portions by numerous bays and arms of the sea, affording fine but neglected asylums for shipping, and good, but also neglected, fishing stations. Climate mild, but humid. The co. presents every variety of surface and soil; the country lying to the W. of Loughs Corrib and Mask, including the districts of Connemara, Jarconnaught, and Joyce's Country, being one of the most rugged and wildest portions of Ireland. The other portion of the co. or that lying to the E. of Galway town and of the above-mentioned lakes, is comparatively flat and fertile. After the Shannon, which bounds the co. on the NE., the most considerable rivers are the Suck and the Black River. Agriculture is very backward. A great extension of tillage has taken place of late years; but it is doubtful whether this be any improvement, and whether it be not wholly ascribable to the improvident breaking up of old pasture land. Principal crops, oats and potatoes; but a good deal of wheat is now also raised. Estates mostly very large. Tillage farms mostly very small, and very generally let on the village or partnership system, which is destructive alike of agriculture and of the interests of the occupiers. A good deal of work is performed by the hoy or spade. Excellent long-horned cattle are met with in this co., which, indeed, is much better fitted for grazing than for tillage. The farm-houses and cottages are, generally speaking, wretched in the extreme; and the cottiers are quite as badly off as in most other parts of Ireland. Manufactures can hardly be said to exist; and, with the exception of limestone and marble, the minerals are of no importance. In many districts the Irish language is in all but universal use. Galway is the only considerable town. The co. is divided into 16 baronies and 116 parishes, and returns four mems. to the H. of C., viz. two for the co., and two for the bor. of Galway. Registered electors for the co., 6,082 in 1862. Pop. 440,048 in 1841; 298,229 in 1851; and 254,511 in

1861. The decrease, it will be seen, amounted to 29.50 per cent. between 1841 and 1851, and to 14.66 per cent. between 1851 and 1860.

GALWAY, a town, sea-port, and parl. bor. of Ireland, on its W. coast, prov. Connaught; it is a co. of itself, but is locally situated in the above co., on both sides the river flowing from Lough Corrib to the sea, at its mouth, and at the NE. extremity of Galway Bay; 113 m. W. Dublin, on the terminus of the Midland Great Western railway. Pop. 33,120 in 1831, and 25,161 in 1861. Galway, from a remote period, has been a place of considerable importance, both as a military station and a commercial mart. It underwent various vicissitudes during the civil war of 1641, when it was taken by the parliamentary army, and in that of 1690, when it surrendered to the forces of King William. The town is situated principally on the E. side of the river: that portion of it which was included within the old walls is built chiefly in the Spanish fashion, the houses being of stone, in a quadrangular form, with an open area in the centre, to which the entrances from the street are through arched gateways. In this part the streets are narrow, ill paved, and dirty. The river is crossed by two bridges, one built in 1342, and still in excellent condition, the other of modern construction. The walls were taken down in the beginning of last century, with the exception of the N. bastion, which has been preserved in its original state. The New Town, E. from the Old Town, built according to the modern fashion, contains a square and several wide streets. The extensive suburb of Claddagh, inhabited exclusively by fishermen, lies on the W. side of the river. The town, with the surrounding district, comprising the parish of St. Nicholas and seven others in the vicinity, constitutes the wardenship of Galway, a separate ecclesiastical jurisdiction exempted from that of the bishop, and subject only to the archbishop's triennial visitation. According to the R. Cath. arrangements, the town is the head of the newly erected see of Galway, comprising 17 parishes. The parish church of St. Nicholas is a large and venerable cruciform structure in the pointed Gothic style, of considerable antiquity, having been founded in 1320. The R. Cath. chapel of the same parish, which is also the bishop's cathedral, is a spacious modern edifice. The Franciscans, Dominicans, and Augustines, have monasteries here, to each of which a chapel is attached, as is one to the nunnery of the order of the Presentation. The Presbyterians have also a meeting-house. The educational establishments comprise one of the new Queen's Colleges opened in 1849; a classical school, on the endowment of Erasmus Smith; a large parochial school for boys and another for girls, under the care of the nuns of the Presentation, and several private schools. The charitable establishments are the house of industry, with a dispensary, the Protestant poor-house, the widows' and orphans' asylum, and the Magdalen asylum.

The town is governed by the high-sheriff, recorder, local magistrates, and a board of twenty-one commissioners, elected triennially. Town revenue above 2,000l. a year. A court of record for pleas to any amount is held on Mondays and Fridays. The assizes, both for the co. and the town, are held here, as are the general sessions of the peace for the co., in April and October, and those for the town four times a year. The bor. sent two members to the Irish H. of C., and one to the imperial H. of C., down to the passing of the Reform Bill, which gave it again two members. The parl. bor. includes the entire co. of the town, and had 1,284 regis. voters in 1865.

The court-houses for the co. and for the town are elegant buildings of modern construction. The workhouse, opened in 1842, has room for 1,800 inmates, which is not more than required.

Galway is not a manufacturing town. The linen manufacture was attempted, but failed. Its trade at present consists almost exclusively in the export of agricultural produce, fish, kelp, and marble, beautiful slabs of large size being sent to England and the U. States. It is sawn and polished in mills in the town. It has several flour-mills, two foundries, two breweries, a paper-mill, and two distilleries. The salmon fishery is valuable; the fishery of cod, hake, and haddock is less valuable than it might be, in consequence of the poverty of those engaged in it, and their pertinacious adherence to rules devised by themselves for the exclusion of strangers from the business. The progress of Galway was long checked by the insufficiency of its harbour, which dries at low water, so that vessels of any considerable burden were obliged to anchor between the town and Mutton Island, where they are exposed to the SW. gales. To obviate these defects, an extensive dock has been constructed, which admits vessels drawing 14 ft. water. A lighthouse has also been erected on Mutton Island; and the bay north of the island now forms an excellent roadstead, used for a time by the large mail steamers which, by contract with the government, ran from Galway to America. The steamers ceased to run in 1864.

The shipping belonging to the port of Galway consisted, on the 1st Jan. 1864, of 12 sailing vessels under 50, and 6 above 50 tons; besides 1 steamer of 36, and another of 67 tons. The gross amount of customs duties received at the port was 29,732l. in 1859; 31,802l. in 1861; and 23,372l. in 1863. The total exports of home produce amounted to 85,141l. in 1859; 77,179l. in 1860; 4,650l. in 1861; 892l. in 1862; and 7,597l. in 1863—altogether a most extraordinary variation, such as is visible at no other maritime town in the U. K.

Until of late years, Galway had but little connection, owing to the want of roads with the extensive country W. from it. This defect is now, however, in a great measure obviated by the carrying of roads into Connemara. Joyce's Country, and other wild districts, affording an easy transit for their produce to Galway.

The inhabs. of the Claddagh suburb constitute a separate community; their number is from 5,000 to 6,000, and they are so exclusively fishermen that their cottages have scarcely even a potato garden attached to them. The community is governed by a mayor, elected by themselves, whose authority is so highly respected that appeals from his decisions to the constituted authorities are almost unknown. Their dress is comfortable and substantial, but of a peculiar make. When at home, the men are wholly unemployed. They leave the entire superintendence of their pecuniary affairs to the women, who receive the cargoes of fish on the arrival of the boats, dispose of the produce, and supply the male part of their families with clothing, food, and spirits. The men indulge in whisky; but riots or tumults originating in excess are notwithstanding infrequent, and when they go to sea, which they do in a body, commanded by a leader to regulate their movements, they strictly prohibit any whisky being brought aboard their boats. Their strong religious feeling is evinced by the erection of a large chapel out of their earnings, and by the liberality of their contributions to the support of its officiating clergymen; as also from the custom, undeviatingly adhered to, of having a prayer offered up by a clergyman, according to a specified form, previously to the sailing of their fleet of fishing craft.

GANDIA, a town of Spain, Valencia, distr. Denia, on the Mediterranean, 34 m. SSE. Valencia. Pop. 6,473 in 1857. Gandia is an agreeable town, and is noted for the industry of its inhabitants. It has a fine collegiate church, convent, college, and cavalry barracks. There is a small harbour, and an active fishery is carried on; besides which, there are some linen manufactures. Gandia is in the centre of one of the best cultivated districts of Spain, where much hemp and flax is grown, and the culture of the silk worm is carried on extensively.

GANGES, the principal river, or, as it has been expressively termed, the Nile, of Hindostan, through the N. and E. parts of which it flows, watering its most fertile region, and extending through 18 degrees of long. and nearly 10 degrees of lat. from the central chain of the Himalaya to the Bay of Bengal. Its course is almost wholly comprised within the British presidencies of Bengal and Agra. It rises by two principal heads, the Bhagirathi and Alcananda, about lat. 31° N., and between long. 78° and 80° E. The Bhagirathi, or W. branch, though neither the longest nor largest, is considered by the Hindoos as the 'true Ganges.' It issues about 14 m. above Gangootri, and 200 m. NNW, Delhi, from under a low arch called the 'Cow's Mouth,' at the base of a mass of frozen snow, about 13,800 ft. above the level of the sea; with a mean breadth of 27 ft., and a medium depth of 12 inches. It forms a junction with the Alcananda at Deoprayag, about 9 m. SW. Srinuggur; lat. 30° 9′ N., long. 78° 33′ E. The resulting stream, with a width of about 80 yards, assumes the name of the Ganges; and at Hurdwar enters the great plain of Hindostan at an elevation of only 1,024 ft. above the level of the sea. It flows thence, with a smooth navigable stream, to the ocean, a distance of about 1,350 m., diffusing abundance on all sides by its waters, its products, and the facilities it affords for internal transit. As far as Hurdwar its course is mostly N. or SW.; thence to its confluence with the Jumna, in lat. 25° 25′, long. 81° 10′, it runs generally SE.; from Allahabad to Rajemahal its course is mostly E.; and it then turns SE., and lastly S., till it enters the Bay of Bengal by numerous mouths, between lat. 22° and 21° 30′ N., and long. 88° and 90° 40′ E. Its entire course may be about 1,500 m. The chief tributaries of the Ganges are the Jumna, Ramgunga, Goomuty, Goggra, Sone, Gunduck, Cusi, Mahanunda, and Teesta. They vary in length from 800 to 600 m.; and except the Sone, flow towards the Ganges from the N.

About 200 m. from the sea, the delta of the Ganges (which is twice as large as that of the Nile) begins to be formed. Of its two principal arms, which form the outermost of the whole series, the E. is the larger, and preserves the original direction of the main stream, together with the name of the Ganges; but the W. arm, or Cossimbazar branch, called afterwards the Hooghly, is considered by the natives the true Bhagirathi, and invested by them with the greatest portion of sanctity. The whole of the delta between the two principal arms is a vast alluvial flat, nearly 200 m. in breadth, intersected by numerous rivers interlacing each other in all directions, and which enter the sea by from 12 to 20 mouths. The region round the mouths of the Ganges, termed the Sunderbunds, is a pestiferous tract, covered with jungle, and swarming with tigers and other beasts of prey.

Between Hardwar and Allahabad the course of the Ganges is tolerably straight, the breadth of its bed generally being from 1 to 1½ m. Thenceforward it winds more, and having received the Gogra, Sone, and Gunduck, attains its greatest magnitude. For the last 600 m, of its course its bed varies from ¼ m. to 8 m. in width, and at the lowest season the mean breadth of its channel is about 8-4ths of a m. Above its confluence with the Jumna it is sometimes fordable; below that confluence it is generally of considerable depth, for the additional streams bring a greater accretion of depth than width. At 500 m. from the sea the Ganges is 30 ft. deep, and is continues of that depth, at the least, till it approaches very near its mouth. The rate of descent from Hardwar to the sea averages about 9 inches a m., but nearly 2-3rds of the entire fall takes place before reaching Cawnpore. The mean rate of the current in the dry months is less than 8 m. an hour, but in the wet season it is often from 5 to 6 m., and in particular situations from 7 to 8 m. The banks of the Ganges are commonly precipitous on the side on which the current impinges, and shelving on the other side. The force of the stream, when the river is at its height, sometimes breaks down the banks, which are composed of a loose and yielding soil, with such rapidity that an acre of land has been seen to disappear in less than half an hour. From the great quantity of mud brought down by the river in the latter season, and other causes, its mouths are encumbered with bars and shoals. The Hooghly is less so than the E. arm, but no ship drawing more than 15 ft. water can navigate the latter with safety; and the E. I. Co's ships, that were usually from 1,000 to 1,200 tons burden, and drew above 27 ft. water, loaded and unloaded at Kedgeree Island. (Crawfurd's Miss. to Siam, i. 8.) The Cossimbazar branch, also, is almost dry from Oct. to May; and the Churnina, which enters the sea by the Hooringotta mouth, is the only branch that is at all times navigable.

The annual inundation of the Ganges is owing chiefly to the tropical rains. These prevail successively throughout all the countries through which the Ganges flows; and in this respect its inundation differs from that of the Nile, whose waters are augmented by rains falling along the upper part of its course only. The Ganges, and other rivers in Bengal, begin to rise in consequence of the rains in the mountains at the end of April, their rate of increase for the first fortnight being about an inch a day; this gradually augments to 2 or 3 inches a day, and the total rise amounts, by the end of June, to between 15 and 16 ft., or half the entire height it attains, before any quantity of rain falls in Bengal. But from the latter period, when the rains there become general, the medium increase of the water is about 5 inches a day; and by the end of July all the lower parts of Bengal, contiguous to the Ganges and Brahmaputra, are under water. The progress of the inundation, in consequence of the flatness of the country, is very slow, being no more than half a mile an hour. Owing to this and other physical causes, the difference in the height of the waters adjacent to, and at a distance from, the sea, is very considerable, but increases in proportion to the distance. In the lower part of the Sunderbunds, the influence of the inundation is at ordinary times little or not at all felt: at Luckipore, about 10 m. inland, it is when highest about 5 ft. in elevation; at Dacca 14 ft.; at Custee 31 ft., and at Jellingley, near the apex of the delta, 32 ft. The total increase at the latter place may however, in medium years, be set down at 31 ft. The rise of the inundation continues till nearly the middle of Aug. For a few days pre-

ceding the 15th of that month, its height is nearly stationary; but it then begins to decrease, notwithstanding that great quantities of rain continue to fall for the next six or seven weeks. During the latter half of Aug. and the whole of Sept. the decrease is from 3 to 4 inches a day; from Sept. till the end of Nov. is gradually lessens from 3 in. to 1½ in. The decrease of the inundation, however, does not uniformly keep pace with that of the river, by reason of the height of the banks; but after the beginning of Oct. when the rains have nearly ceased, the remainder of it goes off quickly by evaporation, leaving the lands highly manured. The Ganges decreases at the average rate of half an inch a day from the end of November to the latter end of April, when it is lowest in Bengal, though the rains in the mountains have already begun to augment it in the upper part of its course. Major Rennell estimated the quantity of water discharged by the Ganges per second in the dry season at 80,000 cubic ft., and in the rainy season at 405,000 cubic ft.; being for the average of the year 180,000 cubic ft. per second. But, according to some observations made at Ghazipore, above Calcutta, by Mr. Everest in 1831, it would appear that in the four months of the flood season (June to Sept.) about 500,000 cubic ft. per second are discharged; while the average for the remainder of the year is only 100,000 cubic ft. per second. The quantity of earth brought down by the river is very great. According to Mr. Everest, the solid matter suspended in the water during the rains weighs about 1-12th part of the water, and occupies about 1-80th part of its bulk; giving a discharge of about 577 cubic ft. of mud per second, or 6,082,041,600 cubic ft. for the discharge in the 122 days of rain. The total annual discharge of mud is estimated at 6,368,077,440 cubic ft.; the weight of which, according to Mr. Lyell, would exceed sixty times that of the great pyramid of Egypt. (Journal of the Asiatic Society, No. 4, p. 238; Lyell's Geology, i. 361—364.)

A very striking effect of the inundation of the Ganges is the change produced by it, year after year, in the bed of the stream. This happens in numerous parts of its course through the lower portion of the great plain of Hindostan; but particularly in Bengal, where the soil is the most loose and yielding. The different branches of the river constantly shifting their places, a number of extensive jheels or marshes are continually being produced; and the geographical face of the country, the condition and extent of private property, &c., change in the same proportion. In consequence also of the looseness of the soil through which it flows, the river is alternately forming and destroying islands in its bed, some of which are 4 or 5 m. in extent, yet formed or removed in the space of a few years. Certain tracts are preserved from the inroads thus by being surrounded by dykes, the collective length of which was estimated in Rennell's time at upwards of 1,000 m. The policy of their erection has been considered very doubtful, for the land has to be irrigated at certain periods, when the dykes must be cut; besides which, they do not always answer their purpose, owing to the want of tenacity in their materials; and they are maintained at a great expense. The country has, however, been brought by them into so artificial a state, that there is now no alternative but to persist in keeping them up.

The Ganges, like the Brahmaputra, the Amazon, several European rivers, the Gulf of Cambay, on the opposite side of Hindostan, is subject to the phenomenon of the bore, or a rapid rush of the tide in a perpendicular face, up the river to a con-

siderable distance. It is especially strong at spring
tides. This occurs in all the mouths of the Ganges,
and particularly in the Hooghly, through which
branch it ascends as far as Culna, or even Nuddea,
200 m. from the sea. The column of water is
sometimes a dozen feet in height near the mouth
of the river, and often 5 ft. high opposite Calcutta.
Its appearance is that of a monstrous billow in a
storm, or the dash of a foaming surf: its sound
resembles that of a steamboat, but is infinitely
louder. Sometimes it takes one side of the river,
sometimes the other: it never extends over the
whole basin. 'The time of its approach being
well known, hundreds of boats may then be seen
rowing, as for life, towards the middle of the river,
the crews urging on each other with wild shouts
or shrieks, though at the moment no danger appears; but soon afterwards the spectator is made
sensible how necessary was the precaution, as the
bore passes by with tremendous noise and velocity.'
(Heber, in Ind. Trav., ix. 108.)

But, in the words of Mr. Crawfurd,—' With all
the difficulties and dangers of the Ganges, the
English, if their Indian conquests be of any advantage to them, owe almost as much gratitude
to the Ganges as the Hindoos themselves, for
unquestionably to it they are indebted for their
Indian empire. It is the great military highway
which enabled us to conquer the richest provinces
of Hindostan,—the acquisition of which enabled
us eventually to conquer and maintain the rest of
our possessions.' (Embassy to Siam, &c. i. 7.)
Its value to the natives of Hindostan is immense.
It is, and always has been, the grand route of communication and traffic in that country, throughout
which the roads adapted for the conveyance of
goods are very few. Not only the main stream,
but all its tributaries from the N. are navigable
for large or small boats, to the very foot of the
mountains, for more than half the year; thus
forming a most extensive system of inland navigation. Sixty years ago, Major Rennell estimated
the number of boatmen employed on the Ganges,
in Bengal, &c., at 30,000, and the value of the
commercial exports and imports conveyed by its
means at 2,000,000l. a year. But ten times the
above number of boatmen would apparently be
nearer the mark in such a region of rivers, where
almost every cultivator and fisherman is also occasionally a navigator. And at present the gross
amount of the imports and exports embarked on its
waters varies between 12,000,000l. and 16,000,000l.
annually, independent of the inland trade, which
has, doubtless, not a little increased with the
increase of the pop., and the greater degree of
security afforded to commerce under the English
rule.

Perhaps no river in the world has on its banks
so many populous cities. On different branches
of the delta are placed Calcutta, Moorshedabad,
and Dacca, the three great cities of Bengal, with a
united pop. of little short of a million; besides
Chinsurah, Chandernagore, Hooghly, Cutwa, Ber-
hampore, Cossimbazar, Kishenagur, and Jessore.
Proceeding up its course, we find on its banks
Rajemahal, Monghir, Patna, Ghazipore, Benares,
Allahabad, Cawnpore, and Furruckabad; with
myriads of villages, temples, and bungalows.

The native craft used in the Ganges vary greatly
in different parts of its course. The flat clinker-
built vessels of the W. districts give way about
Patna to lofty, deep, and heavy boats, which navigate the river thence to Calcutta. In the Sunderbunds, again, the shallowness of the streams
requires that the vessels should be without keels;
and the banks there being impracticable for the
tracking-rope, rowing is the chief method of pro-

pulsion. The boats on the main arm of the Ganges,
and others in the E. part of the delta, are better
than those on the Hooghly, though all are of a
very rude and cheap kind. Within the last 20
years, the river has been extensively navigated
by steamers, some of them, in the upper parts,
being so constructed as to draw but a few feet of
water, in order to safely navigate the shallows.

The Ganges, from Gangoutri to Sauger Island,
is considered holy by Hindoos of all castes, though
in some places much more so than in others.
Hindoo witnesses in British courts of justice are
sworn upon the water of the Ganges, as the Christians and Mussulmans are upon their sacred books.
The Ganges water is believed by the Hindoos to
purify from all sins; many ablutions and suicides
accordingly take place in it; and the feet of the
dying, when they are sufficiently near Hindrants, are
in most instances immersed in it. (Rennell's Memoir on a Map of Hindostan, pp. 335–355; Hamilton's Hindostan and E. I. Gaz.; Ritter's Erdkunde von Asien, iv. part 2, pp. 1108–1248;
Asiat. Researches; Coleridge; Heber; Prinsep,
passim.)

GANJAM, a distr. of British Hindostan. (See
CIRCARS, NORTHERN.)

GANJAM, a town of Hindostan, cap. of the
above distr., near the coast of the Bay of Bengal,
84 m. SW. Cuttack, and 535 m. NE. Madras. It
formerly had a considerable pop., as well as numerous and excellent private houses belonging to
British civil officers, a fort, and cantonments; but it
has now, in great part, been deserted and fallen
into decay.

GANNAT, a town of France, dép. Allier, cap.
arrond., on the Andelot, 33 m. S. Moulins, on the
railway from Moulins to Clermont. Pop. 5,559
in 1861. The town is ill built. It was formerly
fortified, and the remains of its ancient castle still
serve as a prison. It has a hospital, and a tribunal of primary jurisdiction.

GAP (lat. Vapincum), a town of France, dép.
Hautes-Alpes, of which it is the cap., in a wide valley, nearly 2,500 ft. above the sea, surrounded by the
lower Alpine ranges, and on the road from Paris to
Marseilles by way of Grenoble, 44 m. SE. Grenoble.
Pop. 8,319 in 1861. It is a very ill-built and generally disagreeable town. Its principal public edifices
are, the cathedral, prefecture, town-hall, bishop's
palace, hall of justice, and barracks, some of
which are good buildings. The cathedral is in
the Gothic style, and richly ornamented: it contains the tomb and effigy of the celebrated constable de Lesdiguières. The tomb is a sarcophagus
of black marble, surrounded with bas-reliefs in
alabaster, repre-enting the principal actions of
that warrior. It has several other churches, a
communal college, a society of emulation, a
museum of natural history, collections of paintings,
sculpture, antiquities, and physical objects, and a
small theatre.

Gap is the seat of tribunals of primary jurisdiction and commerce; it has manufactures of
woollen cloth, linen fabrics, silks, chamois and
other kinds of leather, and cotton yarn. Its immediate vicinity is very fertile; there are in it
marble quarries known to the ancient Romans,
and many mineral springs. The town is of very
great antiquity; it was the cap. of the Tricorii,
under the name of Vap. It became the seat of a
bishopric in the 4th century, and belonged for a
lengthened period to its own prince or count
bishops. It suffered greatly in the middle ages,
from the devastations of the Lombards and
Saracens, and from repeated sieges, fires, the
plague, and religious wars, but, more than all,
from the revocation of the edict of Nantes. Pre-

viously to 1630, Gap is said to have had 16,000 inhab.

GARD, a dép. of France, in the S. part of the kingdom, formerly comprised in the prov. of Languedoc; between lat. 43° 27′ and 44° 27′ N., and long. 3° 17′ and 4° 50′ E.; having N. the déps. Lozère and Ardèche; E. the Rhone, separating it from Vaucluse and Bouches-du-Rhône; W. Aveyron; and S. Hérault, the Mediterranean, and the Isle de Camargue. Area, 583,538 hectares; pop. 472,107 in 1861. The N. and W. parts are occupied by ramifications of the Cevennes, the general slope of the dép. being from NW. to SE., in which latter part of its surface there is a considerable extent of level country, with numerous and extensive pools and marshes. Most of the rivers have a SE. direction; the principal are the Gard or Gardon (whence the dép. derives its name), the Cèze, and the Vidourle. The Gard, which rises in the Cevennes from several sources, runs through the centre of the dép, and falls into the Rhone not far from Beaucaire, after a course of about 60 m. It at first passes through a succession of deep mountain gorges; and when the snows in the Cevennes begin to melt, it subjects the lower parts of the country through which it passes to extensive and often very destructive inundations. Its bed sometimes increases in width to nearly a mile; and its waters not unfrequently rise from 18 to 20 ft. in a few hours. The Hérault has its source in this dép. Climate variable, but for the most part hot and dry. The N. wind, or bise, blows sometimes with great impetuosity, and the sirocco is by no means rare. The arable lands comprise about 157,500 hectares; vineyards about 71,000 do.; forests, 106,474 do.; and heaths about 150,000 do. Though the arable land is in general pretty fertile, the produce of corn, owing to the extremely backward state of agriculture, is estimated at only 650,000 hectolitres, being about half the quantity required for home consumption. It is principally wheat, oats, and barley. A good many potatoes are also grown, and in the mountain region chestnuts go far to supply the place of corn; the Cevennes being covered with chestnut woods. The plough described by Virgil, drawn by two mules, is in common use. The annual produce of wine is estimated at 1,120,000 hectolitres, a third part of which is consumed in the dép.; the wines of St. Gilles and Favel are those most esteemed. The date, jujube, pistachio, and pomegranate flourish in the open air; oranges and lemons are grown, but a good deal of care is required in their culture. Olive trees are grown on low hills with a S. aspect; but they suffer severely from cold winters, and their number has decreased of late. Gard is the principal dép. in France for the culture of the mulberry; the quantity of cocoons collected amounts to about 3,000,000 kilogrammes per annum.

More than 600,000 sheep belong to this dép., yielding about 800,000 kilog. of wool. Many of the flocks are sent to feed on the Alps in the summer. The greater part of the déps. is parcelled out into very small estates, and the number of considerable properties is greatly below the average of the kingdom. Gard is rich in minerals; and mining, though ill-conducted, constitutes one of the chief sources of its wealth. Iron, argentiferous lead, antimony, zinc, and manganese are found in the mountains; and gold is met with in the sands of some of the rivers. Iron and coal are abundant; the forges of the arrond. of Alais alone employ from 1,000 to 1,200 hands. Great quantities of salt are obtained from the salt marshes on the coast, which altogether occupy a surface of 86 hectares. In the salt pans of Peccais, during June and July, as many as 2,000 hands are employed to wash the produce. About 1,200 workmen are employed in the gypsum, mill-stone, and other quarries. Gard stands at the head of the déps. in the N. of France for manufacturing industry; it is especially distinguished for its manufacture of silk. The principal seat of these is Nîmes, where they employ about 18,000 hands. The other manufactures are those of cotton and woollen fabrics, hats, paper, pasteboard, brandy, leather, glass, and earthenware; there are besides many tawing and dyeing establishments. In the neighbourhood of the coast canals are numerous, and include those of Beaucaire, Sylvereal, and Grand Roubine. There is, however, but one sea-port, Aigues-Mortes, and this is 4 m. from the Mediterranean, with which it communicates by the last-named canal. 162 fairs are annually holden in the dép.; among them is the celebrated one of Beaucaire (which see). Gard is divided into four arrondissements, 38 cantons, and 438 communes. The chief towns are Nîmes, the cap., Alais, Uzès, and Le Vigan.

The antiquities in the dép. belong principally to the Roman period. The principal is the amphitheatre (see Nîmes), and the Pont du Gard. The latter is an aqueduct, and one of the most splendid relics of the Roman power, built over the Gardon, about 10 m. NE. Nîmes. Mr. Inglis thus describes it (Switzerland, &c., ch. xxii.): 'The aqueduct is formed upon three bridges, one above another; the total height, from the level of the river to the top of the aqueduct, being 1.45 ft. The underpart of the bridges consists of six arches, through the largest of which the river passes. The middle bridge has eleven arches; and the uppermost has thirty-five arches (but these are much smaller than those of both the other tiers). Above this is the aqueduct, which is 64 ft. high, and 4 ft. wide. The arches both of the lower and middle bridge are unequal; which, if it does not increase the architectural beauty of the structure, certainly adds to its picturesque effect. The two lower stories of the bridge are formed of hewn stones, placed together without the aid of any cement; but the masonwork underneath the aqueduct is of rough stones cemented, by which all filtration was of course prevented.' After the decline of the Roman power, the Vandals, Visigoths, Saracens, and Franks successively possessed this dép.

GARDA (LAKE OF), an. Lacus Benacus, a famous lake of Austrian Italy, bounded by the provs. of Mantua, Brescia, and Verona, and the circ. of Roveredo in the Tyrol. From Peschiera, at its SE. extremity (15 m. W. Verona), it stretches NNE. to Riva, a distance of about 35 m. Its lower or S. portion is about 12 m. across where broadest; but its upper or N. portion is not more than from 3 to 4 m. across. It is everywhere enclosed by ramifications of the Alps, except on the S., where the luxuriant plain presents a striking contrast to the magnificent mountain scenery that closes round its upper waters.

On the S. shore of this lake, between Peschiera and Rivoltella, the narrow peninsula of Sirmione projects about 4 m. into the lake. It is joined to the mainland by a low slender neck, but behind this it rises into a hill covered with olives, at the extremity of which are some ruins, said to be those of the villa of Catullus. But whether this be so or not, it is, at all events, certain that the poet had a country-house in this singularly beautiful situation; and he has expressed his admiration of and attachment to it in some fine verses:—

'Proinsularum Sirmio, insularumque
Ocelle, quascumque in liquentibus stagnis
Marique vasto fert uterque Neptunus:
Quam te libenter, quamque laetus inviso!'
Catull. 83.

'The soil of this peninsula,' says Eustace, 'is
fertile, and its surface varied; sometimes shelving
in a gentle declivity, at other times breaking in
craggy magnificence; and thus furnishing every
requisite for delightful walks and luxurious baths;
while the views vary at every step, presenting
rich coasts or barren mountains, sometimes confined to the cultivated scenes of the neighbouring
shore, and at other times bewildered and lost in
the windings of the lake and the recesses of the
Alps.' (i. 283, 8vo, ed.)

The surface of this lake is elevated about 820 ft.
above the Mediterranean; it is generally deep; its
waters are remarkably pure and limpid; and it is
well stocked with fish. In the beginning of summer the level of its surface is raised 4 or 5 ft. by
the melting of the snow on the Alps. It receives
the waters of the Sarco at its N. extremity near
Riva; but none of its other feeders are of such importance as to merit any special notice. Its surplus waters are carried off by the Mincio, which
issues from it at Peschiera. A great number of
towns and villages are built upon its banks, of
which the principal, besides Peschiera, are Desen-
zano, Salo, Gargnano, Riva, Garda, whence the
lake has its modern name, &c. The greater number of these towns have safe and commodious
harbours, and a good deal of trade is carried on
upon the lake. Like all Alpine lakes, it is subject
to violent storms and gusts of wind, a peculiarity
to which Virgil has alluded—

'Fluctibus et fremitu assurgens Benace marino.'
Georg. ii. line 160.

GARLIESTOWN, a village and sea-port of
Scotland, co. Wigtown, at the head of a small bay,
W. coast of Wigtown Bay. Pop. 685 in 1861. The
main street is in the form of a semicircle, facing
the sea. The harbour, which is tolerably safe and
commodious, is the centre of a good deal of coasting trade; and it is the only port in Wigtown-
shire at which the steamer that plies between
Galloway and Liverpool touches. Galloway House,
the seat of the noble family of Galloway, is in the
immediate vicinity of the village. Patrick Han-
nay, a poet of the 17th century, was born at Sorbie
Place (of which his father was proprietor), near
Garlieston.

GARMOUTH, a sea-port of Scotland, co. Elgin,
at the mouth of the Spey, 35 m. NW. Aberdeen,
and 64 m. NE. Elgin. Pop. 802 in 1861. The
harbour was injured by Morayshire floods in 1829,
by the deposition of gravel in the bay; but it is
still the principal shipping place in the co. The
chief exports are timber, grain, and salmon. The
value of timber (which is floated down the Spey
from forests in the interior) exported here was, at
one time, estimated at 40,000l. a year; but it now
seldom exceeds 1,000l. About 20,000 qrs. of
grain, chiefly oats and wheat, are annually exported. The Spey Fishing Company's salmon
smacks do not come into the harbour, but load in
the bay. From 8 to 12 such smacks are employed
in conveying salmon, chiefly to the London
market.

GARONNE, a river of France, which see. See
also GIRONDE, Dép.

GARONNE (HAUTE), a dép. of France, region
S., formerly comprised in the prov. of Languedoc,
between lat. 42° 40' and 43° 55' N., and long. 0°
27' and 2° 5' W., having N. the dép. Tarn-et-Garonne, E. those of Tarn and Aude, SE. Ariège, W.
Gers and Hautes Pyrénées, and S. the Pyrenees.

Area, 628,888 hectares; pop. 484,081 in 1861. Its
SW. portion is covered with lofty mountains, the
highest of which, M. Maladetta, is 11,150 ft. above
the level of the sea; and among which there are
numerous glaciers. In the NE. there are some
plains of considerable extent. The Garonne rises
a little beyond the Spanish border; but most of
the upper part of its course is in this dép., which
hence derives its name. The other chief rivers are
the Tarn, Arize, and Salat, all of which have a N.
course, and are tributary to the Garonne. Climate
generally temperate; but none of the Pyrenean
déps. suffers so much from hail-storms. This is
an essentially agricultural dép., and is reckoned
one of the most productive of grain. The arable
land amounts to almost 352,000 hectares; and the
produce of corn is, in general, nearly double what
is required for home consumption. Vineyards occupy 48,000 hectares, and about 470,000 hectol. of
wine are made annually; the best kinds are those
of Frontou, Villaudric, and Montesquieu. Garden
cultivation is well attended to. Near Toulouse,
corn-fields, vineyards, gardens, and country-houses
occupy every mob of land; and the appearance of
the country people bespeaks a healthy and happy
condition. Orange trees are grown for the sake of
their flowers; the culture of the mulberry tree is
very little pursued. The mountains and valleys
afford good pasturage; but there are few artificial
meadows; and the number of cattle is smaller than
in the contiguous déps. Near Toulouse, a fine
breed of horses was formerly raised for the dragoon
service, but it has been suffered to degenerate.
Poultry are plentiful; the pâtés de Toulouse, made
of ducks' livers, enjoy a high reputation. There
are a great many small estates; the number of
considerable properties is, however, above the average of the déps. Mines numerous, especially those
of iron; but there are others of copper, antimony,
bismuth, zinc, and lead, and some important
marble quarries. Mineral springs are abundant;
many of them are visited by invalids; among
which may be specified those of Bagnères de
Luchon. Manufactures various, but not extensive
or flourishing; the chief are of tools and other
metallic articles, coarse woollens, cotton and linen
fabrics, leather, sail-cloth, hats, watches, and mathematical instruments. The trade is greatly augmented by the Canal du Midi, which commences
in this dép. Toulouse is also the entrepôt for supplying the N. of Spain with the products of Central and N. Europe. The dép. is divided into 4
arronds., 39 cantons, and 597 communes. The
chief towns are—Toulouse, the capital, Muret,
St. Gaudens, and Villefranche.

GASCONY, the name of one of the old prov.
of France, which comprised, previously to the revolution, the country now included in the déps. of
the Hautes Pyrénées, Gers, and Landes, and portions
of the country now included in the déps. of Basses
Pyrénées, Haut Garonne, and Lot et Garonne.

GATEHOUSE, a bor. of regality, river-port
and market-town of Scotland, stewartry of Kirkcudbright, on the Fleet, a little above where it
falls into Fleet Bay; and on the high road from
Dumfries to Portpatrick, 28 m. SW. Dumfries,
and 64 m. W. by N. Kirkcudbright. Pop. 1,633
in 1861. The town is beautifully situated, in a
romantic valley opening on the S. to the sea, and
bounded on both sides by finely wooded, picturesque hills. It consists principally of three
parallel streets, and is remarkably neat, clean,
and well built. The par. church, erected in 1817,
adjoins the town on the N.; and it has also a secession meeting-house, and a place of worship for
Independents. There are 4 schools in the parish,
one of which is parochial. The Fleet is navigable

to Gateshouse by vessels of 180 tons burden. It was created a bor. of barony in 1735; and is governed by a provost, 2 bailies, and 4 counsellors. A bor. court for the recovery of debts not exceeding 3£. is held once a fortnight. Market-day, Saturday, and a rural fair 4 times a year.

Callyhouse, the magnificent seat of the Murray family, lies about 1 m. S. from the town. It is wholly of granite, finely polished; it was constructed after a design by Adams.

GATESHEAD, a parl. bor., town, and par. of England, E. div. of Chesterward, co. Durham, on R. bank of the river Tyne, which divides it from Newcastle; 276 m. N. London by Great Northern railway. Pop. 33,587 in 1861. Gateshead is substantially a suburb of Newcastle, with which it is connected by two bridges, one of them known as Stephenson's High Level Bridge. The latter is 130 ft. above the river, and carries the railway, with road beneath for vehicles and foot passengers. The town consists principally of one good and wide street, and which is the high road to the N. Several narrow streets and lanes which compose the remainder of the town branch off on each side of the principal street, and the pop. on the different sides is not very unequal. There is not any appearance of wealth or houses belonging to the richer classes; the town is densely populated with the families of the manufacturers and pitmen; the master manufacturers or proprietors of the coal pits reside in Newcastle, or in the neighbourhood of the two Fells, where they can enjoy a better atmosphere than in the town. The best street runs in a curve by the church to the river, making the steep descent of the High Street. The par. church is a spacious cruciform structure, regularly built, having a handsome and lofty tower; the interior was wholly repaired at great expense, in 1834. The rectory, which was until lately close to the church, is situated in the suburbs. In Gateshead-fell, which was made a separate par. in 1809, is a well-built church, opened in 1825. There are several other churches, besides eight chapels, belonging to various denominations of Dissenters, of whom the Wesleyan Methodists are the prevailing body. St. Edmund's hospital, founded in 1248, was in early times a considerable monastery; it was re-established by James I. in 1611, and now consists of a master, chaplain, and ten brethren, who divide the funds, about 400£. annually. The rector for the time being is the master, who appoints the brethren. There are also almshouses for old women.

Gateshead is a place of considerable importance, not only from its proximity to, and commercial connection with, Newcastle, but more particularly from the numerous glass manufactories and ironworks within the town, and from the coal pits in its immediate neighbourhood. The old bor. of Gateshead is supposed to have been incorporated in very early times; but there is no charter extant. It was originally governed by a bailiff appointed by the Bishop of Durham, and subsequently by two stewards, who managed the borough property, subject to the borough-holders' and freemen's control; but, under the Municipal Reform Act, it is divided into three wards, and is governed by six aldermen and eighteen councillors. The Reform Act conferred on this bor., for the first time, the privilege of sending one mem. to the H. of C. Registered electors, 549 in 1866. The bor. rev. including rates, amounted, in 1862, to 4,545£. The gross annual value of real property assessed to income-tax was 75,749£ in 1857, and 83,411£ in 1862.

GATTON, a bor. and par. of England, co. Surrey, hund. of Reigate, 17 m. N. London. Area of par., 1,110 acres; pop. of do. 191 in 1861. The bor. is

VOL. II.

quite inconsiderable, and was formerly one of the most perfect specimens in England of a nomination or rotten bor. It sent 2 mem. to the H. of C. from 1451 down to the passing of the Reform Act, by which it was disfranchised. The right of voting was nominally in the inhab. paying scot and lot, but really in the lord of the manor.

GAUDENS (ST.), a town of France, dép. Haute Garonne, cap. arrond., on a hill near the Garonne, 48 m. SW. Toulouse. Pop. 5,103 in 1861. The town consists principally of one spacious well-built and well-kept street. It has several churches, one of which is among the oldest in France, several convents, tribunals of primary jurisdiction and commerce, a communal college, and a society of agriculture. It has manufactures of coarse serge and tape, water-mills for sawing, and a brisk trade in the natural produce of its neighbourhood.

GAWELGUR, a fortress of Hindostan, in the N. part of the Nizam's dom., cap. of a distr. of the same name, on a high and rocky hill, 11 m. NW. Ellichpoor. It is very elaborately fortified, and was formerly considered very strong; but it was taken by storm in 1803, by the forces under General Wellesley (duke of Wellington) and Colonel Stevenson, after a siege of only two days.

GAYA, a town or city of British Hindostan, presid. Bengal, prov. and distr. Bahar, of which last it is the modern cap., on the Phalgu, a tributary of the Ganges, 44 m. SW. Bahar, and 56 m. SW. by S. Patna. It is estimated to contain nearly 7,000 houses, most of which are densely inhabited; but the pop. is very variable and uncertain, Gaya being frequented by great numbers of pilgrims and devotees, often amounting to several thousands. It consists of two parts, the old town of Gaya and the modern one of Sahebgunge. The former, which is the residence of numerous Brahmins and others, and considered by the natives as a place of great sanctity, stands on a rocky eminence; the latter, chiefly laid out by the British, and the seat of trade as well as of the European residents, is situated in a plain between the former and the river. The old town of Gaya is a strange looking place, but its buildings are much better than those of the quarter named Sahebgunge, the greater part of the houses being of brick and stone, and many of them two or three stories high. The architecture is very singular, with corners, turrets, and galleries, projecting with every possible irregularity. The streets are narrow, dirty, crooked, uneven, and encumbered with large blocks of stone, or protruding angles of rock.'

Gaya is uncommonly hot, and in spring obscured by perpetual clouds of dust. The streets in Sahebgunge are wide, perfectly straight, and kept in good order, though not paved, with a double row of trees, leaving in the middle an excellent carriage road, with a footpath on each side.

In the immediate vicinity are the ruins of Buddha-Gaya, traditionally supposed to have been the place of the residence and apotheosis of Buddh (the Gautama of the Indo-Chinese nations). These ruins consist mostly of irregular and shapeless heaps of brick and stone. The number of vaulted caverns cut out of immense masses of solid granite is incredible, as is the number of images scattered around to the distance of 15 or 20 m. Among the latter is a gigantic figure of Buddh, in the usual sitting posture. There are now, however, no Buddhists at Gaya, the worship of the Brahminical deities—many groups of which are sculptured on the rocks—having entirely superseded that of the rival divinity. The present town of Gaya contains no ancient monuments whatever, and appears to have derived all its sanctity from its contiguity to the site of the ancient city.

D D

GEFLE, a sea-port town of Sweden, cap. of a län, and at the mouth of a river of the same name, on the Gulf of Bothnia, 50 m. NNW. Stockholm, and 14 m. NE. Talun, with which it is connected by railway. Pop. 11,322 in 1861. The town is divided into four quarters by the river, which separates itself into three branches, and forms two islands, on which, as well as on either bank, the town is built. The houses are well built, some of stone, others of wood, and the streets, though irregular, are wide and well paved. The market-place is remarkable in point of size. The chief buildings are the church, the government-house, the town-hall, and the hospital. There are, besides, a gymnasium of some celebrity, two schools, an orphan asylum, and some unimportant manufactories of linen cloth, leather, and tobacco. The excellence of its harbour, defended by a long jetty, and having a depth of 18 ft. a little from the shore, gives it great advantages for trade. Its exports are fir, timber, pitch, tar, and iron; and its chief imports, wheat and salt.

GENEVA (CANTON OF), the smallest canton of Switzerland, at the SW. extremity of which, and of the lake which bears its name, it is situated; having N. the canton Vaud, E. and S. Savoy, and W. France. Area, 913 sq. m.; pop. 83,310 in 1860. It is the densest populated of any of the Swiss cantons, there being 702 inhabitants to the sq. m., or about twenty-three times as many as in the Grisons, which has but a pop. of 30 per sq. m. The canton, which ranks 22nd in the confederacy, is composed of the territory of the ancient republic of Geneva, together with some communes formerly belonging to Savoy and France, annexed to it in 1815. Its surface is flat, or but slightly uneven. It is enclosed between the Jura mountains on the NW., and some Alpine ranges in the opposite direction. The Rhone and Arve are the principal rivers. The climate is mild, but the land is not very productive. The cultivable soil comprises about 565,850 acres; of which the lands in crop make about a half, grass lands somewhat less than one-fifth, and woods about one-tenth. In average years from 29,000 to 32,000 imperial quarters of corn may be produced; but, as neither this nor any other species of agricultural produce is grown to an extent sufficient for home consumption, considerable quantities are imported.

Geneva is essentially a manufacturing canton; but its manufactures and trade belong to the town. The government is representative. The legislative power is exercised by a body of 278 members, elected by all citizens above 25 years of age. Four syndics preside over this body, and are members of the executive council, or council of state, which is composed of 28 mem., elected from among the council of representatives, usually for life, but subject to a vote of censure, and removable at pleasure. The magistrates of the different judicial courts are appointed for a certain number of years by the former council. The canton is divided into three districts, in each of which there is a court of audience; besides these, there are in the cap. a court of appeal from the foregoing, a tribunal of commerce, and a supreme court of justice, composed of nine judges. All trials are public. The French code of laws is generally operative.

The press is free. Education is in a flourishing state. The canton furnishes a contingent of 800 men to the army of the Swiss confederation, and a contribution of 22,000 Swiss fr. a year to its treasury. The public revenue of the canton, in the year 1862, amounted to 2,742,000 fr. and the expenditure to 2,466,000 fr. The canton had, at the same time, a debt of 16,000,000 fr. Except the city of Geneva, the canton contains no town

of importance. (For further details, see succeeding article.)

GENEVA (Germ. Genf), the most populous city of Switzerland, cap. of the above canton, situated in a picturesque country, abounding in the most enchanting and magnificent prospects, at the SW. extremity of the lake of Geneva, 81 m. SW. Berne, and 70 m. SE. by E. Lyons, on the railway from Berne to Lyons. Pop. 41,115 in 1860. The Rhone divides Geneva into three parts; the city on the right bank, the quarter of St. Gervais on the left, and the island between them, enclosed by two arms of the river. The city, or upper town, is the largest portion, and is in part built on an eminence, rising to nearly 100 ft. above the level of the lake. Its streets are narrow, crooked, and steep; but many of its private edifices are good: it consists almost entirely of the residences of the burgher aristocracy. The lower town, or quarter of St. Gervais, is the chief seat of commercial activity. It has narrow streets and lofty houses. Some of the latter are furnished with a shed or pent-house, called a chine, which projects from the roof over the street, supported by wooden props reaching from the pavement. The island is upwards of a furlong in length, by about 200 ft. broad, and connected with the other quarters by several bridges. The aspect of Geneva from the lake is very beautiful. Of late years, an entirely new quarter has sprung up on the right bank of the Rhone, called the Quartier des Bergues, displaying a handsome front of tall houses, among which is the Hotel des Bergues, lined with a broad and fine quay, towards the lake. The unsightly houses that formerly lined the margin of the lake in the lower town have been repaired and beautified; and a broad belt of land has been gained from the water to form a quay. This is connected with the Quai des Bergues, on the opposite bank, by a handsome suspension bridge, and another bridge communicating with a small island, situated at the point where the Rhone leaves the lake, is ornamented with a bronze statue of Rousseau. Geneva is surrounded on the land-side by ramparts and bastions, constructed about the middle of last century: these are of little use as fortifications, the city being commanded by some adjacent heights; but they serve as public promenades, and three iron suspension bridges have been thrown over them to facilitate the intercourse between the city and the surrounding country. The gates of Geneva are closed from midnight to day-break; and after sunset a toll is levied on all horses or carriages.

Geneva has but few fine public buildings. The principal is the cathedral or church of St. Peter: it is in a conspicuous situation, has three steeples, and is an interesting specimen of the Gothic style of the 11th century; but a Corinthian portico, in imitation of that of the Pantheon at Rome, has been inconsistently enough added to it. This church contains the tombs of Agrippa d'Aubigny, the friend of Henri IV., and of the Count de Rohan, a leader of the French Protestants in the reign of Louis XIII. There are, besides, three Calvinist and two Lutheran churches, a Catholic church, and a synagogue. The town-hall and general hospital are almost the only other edifices worth notice. The last is an extensive and spacious building; in the chapel belonging to it the service of the English Church is performed on Sundays. The Musée Rath, so named after its founder, is a neat building, containing a collection of paintings by native and other artists. The museum of natural history contains the geological collections of Saussure, Brongniart, and Decandolle, the collections of M. Necker, a cabinet of antiquities, and a reading room well supplied with the best

European journals. The academy, founded by Calvin, has faculties of jurisprudence, theology, natural science, and literature, and 39 salaried or honorary professors. It has attached to it a library of 40,000 vols., including many valuable MSS. Geneva has also a college for classical education; a school preparatory for the academy; a school of manufactures, established 1822; schools of watchmaking, drawing, music, &c., and many private schools. It has a public observatory; a society for the advancement of arts; societies of medicine and natural history, and other learned associations; lunatic and deaf and dumb asylums; and various other charitable institutions. The working classes have united in several benefit societies for mutual assistance, and a savings' bank was established in 1816. The ancient palace of the bishops of Geneva has been converted into a prison; but in 1825 a new prison was established on the panoptic system, the first of the kind founded on the Continent. The prisoners on arriving are detained in solitary cells for a longer or shorter period, and afterwards set to regular work, during which, as well as at all other times, they are obliged to observe a profound silence. Each occupies a chamber by himself, and solitary confinement is the usual punishment for refractory behaviour. The prisoners have books distributed to them from the prison-library. A part of the produce of their labour is put aside for their own use; and when they finally leave the prison, a committee furnishes them with employment. Geneva has an arsenal and a theatre; it is well lighted, and is supplied with water by a hydraulic machine situated in the island. There are various public walks within as well as without the walls, which command noble views of the Alps and the lake; amongst them are the Terrace de la Treille, the square of St. Antoine and Maurice, and the fine botanic garden, laid out in 1816. Geneva is a favourite place of resort of the English.

The main source of the prosperity of this city consists in its manufactures; the principal of these are watches, jewellery, musical boxes, and objects of taste in the fine arts. The number of working watchmakers and jewellers is estimated at nearly 6,000. The number of watches annually made is estimated at upwards of 70,000, and of these at least 60,000 are of gold. In watchmaking and jewellery, it is estimated that between 70,000 and 80,000 oz. of gold, and about 50,000 oz. of silver, are used annually. The gems (most of which are pearls) used in jewellery and the embellishment of watches may be worth perhaps 20,000l. a year. The watchmaking business is divided into two branches; that of haute horlogerie, comprising chronometers, stop-watches, and other articles in which the perfection of the machinery is the highest kind; and that of horlogerie du commerce, in which the beauty of the work is its chief recommendation. The articles of the latter class are by far the most numerous. 'The great advantage which the Swiss possess in competition with the watchmakers in England,' says a consular report, 'is the low price at which they can produce the flat cylinder watches, which are at present much in request. The watches of English manufacture do not come into competition with those of Swiss production, which are used for different purposes, and by a different class of persons. Notwithstanding all the risks and charges, the sale of Swiss watches is large, and it has not really injured the English watchmaking trade. The English watches are far more solid in construction, fitter for service, and especially in countries where no good watchmakers are to be found, as the Swiss watches require delicate treatment. English watches, therefore, are sold to the purchaser who can pay a high price; the Swiss watches supply the classes to whom a costly watch is inaccessible.' The works or machinery of the watches are often made in the neighbourhood of Geneva, at Fontainemelon and Travancourt in France. The unfinished work is called an ébauche, and is polished and perfected by the Genevese artisan. Almost everything is done by the piece, and not by daily wages. The other manufactures of Geneva and its canton are principally horn and tortoiseshell combs, carriages, saddlery, agricultural implements, tools of all kinds, cutlery, fire-arms, enamels, musical instruments, printing types, and philosophical instruments of a very superior description. Lithography and engraving medals and vignettes are flourishing branches of art. Some factories of woollen cloth have been recently established; the produce of various spinning establishments finds consumption in Switzerland; and printing would form a very important and very productive branch of industry, were it not for the impediments thrown in the way of exportation to neighbouring countries. Tanning is carried on in but a small extent, though the very superb quality of the leather always insures it a preference in foreign markets, particularly in Italy. At the beginning of the seventeenth century the Genevese also carried on an extensive trade in silk stuffs and lace; and before the French revolution there were many extensive establishments for the manufacture of printed cottons, besides factories of various other kinds. These no longer exist, most of them having been crushed by the system of prohibitions and high duties established by the continental powers during the ensuing period. The watches and jewellery manufactured at Geneva are subjected to a strict system of supervision, to prevent a falling off in the reputation of this important branch of trade of the republic. A committee of master-workmen, with a syndic at their head, called the commission de surveillance, are appointed by the government to inspect every workshop, and the articles made in it, to guard against fraud in the substitution of metals not of the legal standard. By a law of 1815, the manufacture of any gold work of a lower standard than ·750 is forbidden; and the legal standards for silver are fixed at ·800, ·875, and ·950. Geneva is the seat of the council of state; the supreme court of justice for the cant.; a court of appeal from the district courts; and a chamber of commerce. The last has a very extensive jurisdiction; every commercial transaction, of whatever description, may be brought before it; and a private individual, who may have brought more of an article than he requires, and sells the surplus, becomes responsible to it.

Geneva is very ancient. 'Extremum oppidum Allobrogum est, proximumque Helvetiorum finibus, Geneva,' are the words of Cæsar in speaking of this city, (De Bello Gallico, i. § 6.) Many Roman antiquities have been discovered in and near it; and in the island traces may still be discovered of a Roman structure, supposed to be the foundations of one of the towers erected by Cæsar to prevent the Helvetians crossing the river. In 470 Geneva was taken by the Burgundians, and became their cap.; it afterwards belonged successively to the Ostrogoths and the Franks, and formed a part of the kingdom of Arles, and the second kingdom of Burgundy. On the fall of the latter it fell under the sole dominion of its bishops, between whom and the counts of the Genevois, in Savoy, there existed incessant contests for its possession. At the reformation the bishop was expelled, and the town, with its territory, became a republic. Calvin, having sought refuge in Geneva in 1538, was ad-

D D 2

licited to settle there, and was soon afterwards raised to the highest rank in the state, which he in a great measure governed for 23 years, with a severity and strictness that impressed deep and abiding traces on its jurisprudence and manners. In 1553 the famous Michael Servetus, who had been arrested at Geneva, at the instigation of Calvin, was accused of blasphemy in regard to the Trinity, and being tried and convicted, was ordered to be committed to the flames, which barbarous sentence was immediately carried into execution. The conduct of Calvin in this deplorable affair, though in part excused by the spirit and temper of the times, was directly hostile to every principle for which he had been contending against the Church of Rome, and will ever remain a dark blot upon his character, and that of the early reformers. In 1792, in consequence of internal dissensions, Geneva was occupied by the troops of France, Sardinia, and Berne. In 1798 it was taken by the French revolutionary forces, and subsequently became the cap. of the dép. Leman. It was, with its territory, united to Switzerland as an independent canton in 1814. Few cities have produced more eminent individuals; amongst others may be specified J. J. Rousseau; Casaubon, the critic; Lefort, the friend of Peter the Great; Necker, and his daughter, Mad. de Staël; the naturalists Saussure, De Luc, Bonnet, and Jurine; Tremembley and Huber; Dumont, the friend and editor of the best works of Bentham; the philosopher Abauzit; J. B. Say, the political economist; and Simonde de Sismondi, the historian.

GENEVA (LAKE OF), or LAKE LEMAN (Germ. *Genfersee*, an. *Lacus Lemanus*), the largest lake of Switzerland, near the SW. extremity of which it is situated. It has N., E., and SE. the canton of Vaud or Leman; NW. that of Geneva; and S. Savoy. It fills up the lower portion of a somewhat extensive valley enclosed between the Alps and the Jura. It is crescent-shaped, the convexity being directed NNW., and the horns facing SSE. Its greatest length—a curved line passing through its centre from Geneva at its W. extremity, to Villeneuve at its E.—is about 45 m.; but along its N. shore the distance from end to end is almost 56 m., while along its S. it is no more than about 40 m. Its breadth varies from 1 to 8 m.; its area is estimated at about 240 sq. m. Its greatest depth, near Meillerie, towards its E. extremity, is said to be 1,012 (950 Fr.) ft.; its level is about 1,200 ft. above that of the Mediterranean. In Aug. when its waters are the highest, its surface is often 4½ ft. above its level in March, when it is lowest. It is divided, in common parlance, into the Great and Little lake; the latter is more exclusively called the Lake of Geneva, and extends from that city for a distance of 14 m., but with a breadth never more than 3½ m., to Point d'Yvoire; beyond which, Lake Leman widens considerably. The Rhone enters it near its E. extremity, bringing with it so much alluvial soil, that considerable encroachments are continually made on its upper end. Port Vallais, now 1½ m. distant, was formerly on the margin of the lake, the basin of which is said to have originally extended upwards as far as Bex. The Rhone emerges from the Lake of Geneva at its SW. extremity, where its waters like those of the lake itself, are extremely clear, and of a deep blue colour, circumstances which have been often adverted to by Byron. (See Childe Harold, lib. iii. 85, &c.) Lake Leman receives upwards of forty other rivers; the principal of which are, the Veveyse, from the N., and the Drance, on the side of Savoy. It seldom freezes, and has never been known to be entirely frozen over. It is subject to

a curious phenomenon called the *seiches*. This consists in a sudden rise of its waters, generally for 1 or 2 ft., but sometimes as much as 4 to 5 ft., followed by an equally sudden fall; and this ascent and descent goes on alternately, sometimes for several hours. This phenomenon is most common in summer, and in stormy weather; its cause has not been satisfactorily ascertained, but it would seem to depend on the unequal pressure of the atmosphere upon different parts of the lake.

Lake Leman abounds with fine fish. Its banks are greatly celebrated for their picturesque beauty and sublimity. Their scenery is the most imposing at its E. extremity; but the whole of the S. shore exhibits great boldness and grandeur. The N. shore is of a softer character; it is adorned with a succession of low hills covered with vineyards and cultivated fields, and interspersed with numerous towns, villages, and habitations. Nyon, Rolle, Morges, Ouchy (the port of Lausanne), Vevay, Clarens, and the Castle of Chillon, are on the N. bank; on the S. or Savoy side, are Meillerie; Ripaille, the place of retirement of Pope Felix V.; Thonon and the *Champagne Division* in the Genevese territory (the residence of Lord Byron in 1816). The first steam vessel in Switzerland, the William Tell, was launched on the Lake of Geneva in 1823; in 1836, there were four steam boats plying on it; and in 1864, there were above twenty. A line of railway encircles the whole of the lake.

GENOA (Ital. *Genova*, an. *Genua*), a celebrated marit. city of Northern Italy, once the cap. of an indep. repub., and now of a prov. or division of the kingdom of Italy, at the head of the gulf of the same name; 75 m. SE. Turin, and 90 m. NW. Leghorn, on the railway from Turin to Rome. Pop. 119,610 in 1862. Genoa is built round, but principally on the E. side of its port, which is semicircular, the cord being about 1 m. in length. Two gigantic moles (the *Molo vecchio* and *Molo nuovo*) project into the sea from either angle, and enclose and protect the harbour. The land on which the city is built rises amphitheatrewise round the water's edge, to the height of 500 or 600 ft., so that its aspect from the sea is particularly grand and imposing. The white snowy houses form streets at the lower part of the acclivity, while the upper part is thickly studded with detached villas. Behind all, the Apennines are seen towering at the distance of 10 or 12 m., their summits during a part of the year covered with snow. Genoa has a double line of fortifications. The inner one encloses merely the city itself on the N. and E. sides of the port; the outer walls extend from either angle of the port back to the summit of the hills, on the declivity of which the city is built, and are 8 or 10 m. in length. The old or E. portion of the city consists of a labyrinth of excessively narrow, crooked, and dark streets, their breadth being generally no more than from 6 to 12 ft. They run between a succession of lofty houses, 5, 6, and even 7 stories high, each story being from 12 to 16 ft. deep, the cornices under the roof of which sometimes project so far as to meet, and thereby exclude all daylight. 'In these streets you meet with vast numbers of mules and some asses, carrying all sorts of articles, bricks, firewood, &c. on their backs; for wheeled carriages are only used in the broad streets, which are rare, except in the suburbs. The streets are paved with broad flags of lava, which are laid in mortar, and have the smoothness and durability of good masonry. In the middle of this pavement there is a pathway laid with bricks set on edge, about 2 or 3 ft. broad, and a little higher than the lava. This is for the accommodation of the mules, the lava being considered too smooth to afford their feet a sufficient hold.' (Maclaren's Notes on France and Italy, p. 46.) The streets, narrow and steep as they are, are very clean, cool, and quiet. The newer part of the city, which stretches along the N. side of the port, is more regularly laid out, and contains some broad and very handsome streets, in particular that running from the *Piazza delle Fontane* to the *Piazza dell' Acquaverde*, near the W. gate, and including the *Strada Nova* and *Novissima*, the *Piazza del Vastato*, and the *Strada Balbi*. The last of these, says M. Simond, is entirely formed of palaces, more magnificent than those of Rome, and neater in their interior. Each is built round a court, and the best apartments are on the third floor, for the benefit of light and air. The roofs, being flat, are adorned with shrubs and trees, as myrtle, pomegranate, orange, lemon, oleanders, &c., 25 ft. high, growing not in boxes only, but in the open ground several feet deep, brought hither and supported on arches. Fountains play among these artificial groves, and keep up their verdure and shade during the heat of summer. In Italy, Genoa has acquired, and deserves, the title of *la Superba*. It exhibits fewer remains of ancient splendour than Venice, but more actual wealth and comfort. 'Its architecture is grand in its style, and admirable in its materials. Its palaces are numerous, and many of their princely gates 40 ft. high, with marble columns, courts paved with various coloured marbles in mosaic, broad staircases all of marble, rooms 30 ft. high with arched ceilings, adorned with gilded columns, large mirrors, superb crystal lustres, mosaic floors, the roofs panelled, and the panels filled with finely executed frescoes or paintings in oil, and divided by sculptured figures. Behind are orangeries. I visited four or five of these palaces; but there are multitudes.' (Maclaren's Notes, p. 46.) The common houses are of stone plastered with stucco, the floor of marble. Of its palaces, that of Doria, built by and still belonging to the illustrious family of that name, is the largest and finest: it opens into large gardens which extend along the shore; but it is said not to be well kept, and to be falling into decay. It has a noble colonnade supporting a terrace facing the gardens, the whole in white marble: its interior is very richly ornamented. The emperors Charles V. and Napoleon both made this palace their residence during their stay in Genoa. Another *Palazzo Doria* is now a residence of the king of Italy.

There are two palaces originally belonging to the Durazzo family. That on the *Strada Balbi* is now a royal mansion; its front is about 250 ft. in length; it has a court, rich in architectural embellishments, and a famous gallery 100 ft. long, ornamented with frescoes, and containing a curious collection of statues and sculptures ancient and modern, numerous portraits of the Durazzi, historical paintings, and others by Carlo Dolci, Titian, Vandyck, A. Durer, and Holbein. In another room is the *chef-d'œuvre* of Paul Veronese, 'Mary Magdalen at the feet of our Saviour.' The other Durazzo palace is scarcely less rich; its gallery contains some fine works by P. Veronese, L. and A. Caracci, Guercino, Titian, Domenichino, several by Guido, and Rubens. The ancient palace of the Doges was almost wholly destroyed by fire in 1777; but the modern building, on its site, is a fine structure, and contains the city council-hall, 125 ft. by 65, and 64 ft. high. The Serra, Spinola, Balbi, Brignole, Carega, Mari, and Pallavicini palaces are amongst the most remarkable of the others. But if the palaces of Genoa be superior to those of Rome, its churches are generally inferior; though some of them would be beautiful, if less profusely ornamented. That of the *Annunziata*, founded in the 13th century, is the finest, and contains some good paintings. The cathedral or church of St. Lorenzo, built in the 11th century, is of Gothic architecture; its exterior has a strange appearance from being cased with black and white marble in alternate horizontal stripes. The church of St. Ciro, the old cathedral, is very ancient; that of St. Stefano has a famous altar-piece, the joint work of Raphael and Julio Romano. The church of *San Filippo Neri*, and the chapel of the Carmelite nuns, are both greatly admired for their chaste style. The church of *Santa Maria Carignano* is also a structure in the best taste, erected by one of the princely citizens of Genoa, whose son, in the 16th century, united two elevated parts of the town by a bridge, the *Ponte di Carignano*, 100 ft. in height, and which passes, 'with three giant strides, over houses six stories high that do not come up to the spring of the arches.' (Simond, p. 599.) There are said to be, altogether, 37 par. churches, and 69 convents and monasteries. There are 3 large hospitals richly endowed; the principal of which, the *Albergo di Poveri*, is a large quadrangular edifice immediately N. the inner city walls. In this institution 1,500 or 1,800 individuals, orphans and old people, are provided for; the children are brought up to different trades, and some others are educated; at a proper age, they are allowed half the produce of their labour, with which they in part provide for themselves. The establishment is generally well conducted; the building is handsome, spacious, and clean; it contains numerous busts and statues of its benefactors, and a 'Dead Christ,' in *alto relievo*, by Michael Angelo; probably the finest piece of sculpture in Genoa. Among the other chief public buildings, are the exchange, the old bank of St. George, and one of the three theatres,—that of *Carlo Felice*, recently built. The opera in Genoa is said to be indifferent. The university in the *Strada Balbi* (founded in 1812) is a fine edifice, and has a large library and botanic garden; but it is not otherwise remarkable. Around the port is a rampart, affording an excellent promenade. On the N. side of the harbour is the *Darsena*, a double basin enclosed by piers, and destined for a refitting dock; adjoining it is the arsenal.

From the centre of the city several quays and jetties stretch into the port, bounded on the NE. by the old mole, projecting into the sea W. by N. about 260 fathoms; it has a battery near its middle. The new mole, on the N. or opposite side of the port, adjoins the S. extremity of the suburb of St. Pietro d'Arena, and projects from the shore ESE. about 210 fathoms. The mole heads bear from each other NE. by E. and NW. by W., the distance between them, forming the entrance to the harbour, being about 350 fathoms. A conspicuous lighthouse is erected without the port on its W. side, on a high rock at the extremity of a point of land contiguous to the bottom of the new mole. There is no difficulty in entering the harbour; the ground is clean, and there is plenty of water, particularly on the side next the new mole; care, however, must be taken, in coming from the W., to give the light-house point a good offing. Moderate sized merchantmen commonly anchor inside the old mole, contiguous to the *porto franco*, or bonded warehouses. Men-of-war, and the largest class of merchantmen, may anchor inside the new mole, but they must not come too near the shore. Ships sometimes anchor without the harbour, in from 10 to 75 fathoms, the light-house bearing N. ¼ W., distant 2 or 3 m. The

NW. winds occasion a heavy swell, but the bottom is clay, and holds well. Public fountains are few in Genoa, but the city is well supplied with water brought by an aqueduct from the little river Bisagno immediately E. of the outer walls. The atmosphere is pure; and the climate of the city and its neighbourhood is healthy, and appears to be particularly favourable for the rearing of silk-worms.

Genoa is the entrepôt of a large extent of country; and her commerce, though inferior to what it once was, is very considerable, and has latterly been increasing. She is a free port; that is, a port where goods may be warehoused, and exported, free of duty. The exports consist partly of the raw products of the adjacent country, such as olive oil (an article of great value and importance), rice, fruits, cheese, rags, steel, and argol; partly of the products of her manufacturing industry, such as silks, damasks, and velvets (for the production of which she has long been famous); thrown silk, paper, soap, works in marble, alabaster, and coral; the printed cottons of Switzerland, and the other products of that country, and the W. parts of Lombardy, intended for the S. of Europe, and the Levant; and partly of various foreign products brought by sea, and placed in porto-franco. The imports principally consist of cotton and woollen stuffs; cotton wool, mostly from Egypt; corn from the Black Sea, Sicily, and Barbary; sugar, salted fish, spices, coffee, cochineal, indigo, hides, iron, and naval stores from the Baltic; hardware and tin plates from England; wool, tobacco, lead (principally from Spain), and wax. Corn, barilla, Gallipoli oil, cotton, valloues, sponge, galls, and other products of the countries adjoining the Black Sea, Sicily, the Levant, &c., may in general be had here, though not in so great abundance as at Leghorn. The various duties and custom-house fees formerly charged on the transit of goods through Genoa and the Italian territories have recently been abolished.

The bank of St. George, in Genoa, was the oldest bank of circulation in Europe, having been founded in 1407. It was conducted by a company of shareholders; and having gradually advanced immense sums to the government, a large proportion of the public revenue was assigned to it in payment of the interest. On the invasion of Genoa by the Austrians, in 1746, a part of the treasure of the bank was carried off. Finally, on the union of Genoa with France, the bank was suppressed; the government of France becoming responsible for an annual dividend of 3,100,000 Genoese livres payable to its creditors.

Genoa is the residence of a general-commandant and an archbishop, and the seat of the superior judicial court for the prov., an admiralty-council, and a tribunal and chamber of commerce. It has a royal college; a naval school, the first established in Italy; an excellent deaf and dumb establishment; a public library, with 50,000 vols. and 1,000 MSS.; several learned societies, and various schools.

Genoa is of great antiquity. After a variety of vicissitudes she became, in the 11th century, the cap. of an independent republican state; and was early distinguished by the extent of her commerce, and by her settlements and dependencies in various parts of the Mediterranean and of the Black Sea. Their conflicting pretensions and interests involved the Genoese in long-continued contests with the rival republics of Pisa and Venice. The struggle with the latter, from 1376 to 1382, is one of the most memorable in the Italian annals of the middle ages. The Genoese having defeated the Venetians at Pola, penetrated to the lagoons which

surrounded Venice, and took Chiozza. Had they immediately followed up this success, the probability is that they would have taken Venice; but having procrastinated, the Venetians recovered from the consternation into which they had been thrown, and the Genoese were ultimately compelled to retire. The ascendancy of Venice dates from this epoch. (Koch, Tableau des Révolutions, i. 253.)

The government of Genoa was long the most turbulent that can be imagined; and the city was agitated by continual contests between the nobility and the citizens, and between different sections of the nobility. The mischiefs arising from these struggles were such, that to escape from them, the citizens not unfrequently called in the aid of foreigners; and placed themselves, at different periods, under the protection of France, the Marquis of Montferrat, and the dukes of Milan. Indeed, from 1464 down to 1528, Genoa was regarded as a dependency of the latter. In the latter year, however, it recovered its independence; and was, at the same time, subjected to a more aristocratical government. But the republic continued to be agitated by internal dissensions down to 1576. At that period further modifications were made in the constitution, after which it enjoyed a lengthened period of tranquillity. (A very full account of the revolutions of Genoa is given in the Modern Universal History, xxviii. 353-533; see also Sismondi, Républiques Italiennes, passim.)

The conquest of Constantinople, and of the countries round the Black Sea by the Turks, and the discovery of the passage to India by the Cape of Good Hope, proved destructive of a great part of the trade of Genoa. She was, also, successively stripped of all her foreign possessions. Corsica, the last of her dependencies, revolted in 1730, and was ceded to France in 1768. In 1797, Genoa was taken by the French. After the downfal of Napoleon, the congress of Vienna, in 1815, assigned Genoa and the adjacent territory to the king of Sardinia, of whose dominions they formed a part, till incorporated, with the rest, in the new kingdom of Italy.

GEORGIA (Pers. Gurdjistan, Russ. Grusia, an. Iberia), a country of W. Asia, and formerly the centre of a monarchy of some extent, but now a government of the Russian empire. It occupies a considerable portion of the isthmus, between the Black Sea and Caspian; extending from lat. 40° to 42° 30' N., and long. 43° 10' to 46° 50' E.; separated on the N. by the central chain of the Caucasus from Circassia; E. by the Alazan and Kursk, two tributaries of the Kur, from Shekin and Guilistan; S. and SW. by the Kapan mountains from Armenia; and W. from Imeritia, by a transverse Caucasian range. Thus surrounded on three sides by mountain ranges, Georgia is in a great measure shut out from communication with the neighbouring countries, there being but one pass either across the Caucasus into Circassia, or across the W. range into Imeritia. (See Caucasus, p. 20.) The length of Georgia NW. to SE., measured on the best maps, is about 175 m.; its average breadth from 110 to 110 m. The area has been estimated at about 18,000 sq. m., and the pop. at between 300,000 and 400,000.

The surface is mostly mountainous, consisting of table lands and terraces, forming a portion of the S. and more gradual slope of the Caucasus. The country, however, slopes from the E. and W., as well as the N., to the centre and SE., which are occupied by the valley of the Kur, an undulating plain of considerable extent and great fertility. Between the mountain ranges there are also numerous fertile valleys covered with fine forests, dense underwood, and rich pasturages watered by an abundance of rivulets. All the rivers have more or less an E. course. The principal is the Kur, or Mtkwari (an. Cyrus). This river rises in the range of Ararat, a little NW. of Kars. It runs at first N., and afterwards NE. to about lat. 42° N., and long. 44° E.; from which point its course is generally NE. to its mouth, on the W. shore of the Caspian. It is in many places of considerable breadth, and sometimes several fathoms deep; but its great rapidity prevents its being of much, if any, service for navigation; and only rafts are used upon it. Its principal affluents are the Aragwi from the N., which unites with it at Mtskethi, the ancient capital of Georgia, and undoubtedly the 'Armozica of Strabo, about 10 m. NW. Tiflis; and the Aras (an. Araxes) from the S., which joins it not far above its mouth, where its course deflects southward. Tiflis, the cap. of Georgia, is situated on the Kur.

The climate of Georgia of course varies greatly, according to elevation; it is, however, generally healthy and temperate, being much warmer than that of Circassia, or the other countries on the N. slope of the Caucasus. The winter, which commences in Dec., usually ends with Jan. The temperature at Tiflis, during that season, is said not to descend lower than about 40° Fahr.; and in the summer the air is excessively sultry, the average temperature at the end of July, 1830, having been, at 3 P.M., 79°, and at 10 P.M., 71° Fahr. (Miss. Researches, p. 121.) The soil is very fertile; and agriculture and the rearing of cattle are the chief employments of the inhab. Wheat, rice, barley, oats, maize, millet, the holcus sorghum and h. bicolor, lentils, madder, hemp, and flax are the most generally cultivated articles; cotton is found in a wild state, and is also cultivated.

Georgia is noted for the excellence of its melons and pomegranates; and many other kinds of fine fruits grow wild. Vineyards are very widely diffused, and the production of wine is one of the principal sources of employment. It is strong and full-bodied, with more bouquet than Port or Madeira; but from having generally little care bestowed on its manufacture, it keeps badly; and casks and bottles being for the most part unknown, it is kept in buffalo-skins, smeared inside with naphtha, which not only gives it a disagreeable taste, but disposes it to acidity. But notwithstanding these drawbacks, and its extensive consumption in the country, considerable quantities are imported. Mr. Wilbraham says, that 'the Georgians have the reputation of being the greatest drinkers in the world: the daily allowance, without which the labourer will not work, is four bottles; and the higher classes generally exceed this quantity; on grand occasions the consumption is incredible.' (Travels in the Caucasus, Georgia, p. 192.) According to Smith and Dwight, 'the ordinary ration of an inhab. of Tiflis, from the mechanic to the prince, is said to be a tonä, measuring between five and six bottles of Bordeaux. The best wine costs but about four cents the bottle, while the common is less than a cent. The multiplied oppressions to which the inhab. have been long subjected, and the fertility of the soil, have gone far to extinguish all industry. The peasant thinks only of growing corn enough for the support of himself and family, and a small surplus to exchange at the nearest town for other articles of prime necessity. The plough in use is so heavy as to require six or eight buffaloes for its draught, and often double the number are used; the harrow is nothing more than a felled tree; and

a great quantity of the produce is wasted owing to the corn being trodden out by buffaloes. Domestic animals of all kinds are reared: the horses and horned cattle equal the best European breeds in size and beauty; and the long-tailed sheep afford excellent wool. Game, including the stag, antelope, wild boar, hares, wild goats, pheasant, partridge, &c., is very abundant; bears, foxes, badgers, jackals, lynxes, and it is said leopards, are common. The forests consist of oak, beech, elm, ash, linden, hornbeam, chestnut, walnut, and many other trees common in Europe; but they are of little or no use. The mineral products of the country, though scarcely unexplored, are believed to be various; iron is plentiful on the flank of the Caucasus, and coal, naphtha, &c., are met with. The houses of the peasantry, even in the most civilised parts, are nothing more than slight wooden frames, with walls made of bundles of reeds covered over with a mixture of clay and cowdung, and a roof of rush. 'A room 30 ft. long and 20 broad, where the light comes in at the door; a floor upon which they dry madder and cotton; a little hole in the middle of the apartment, where the fire is placed, above which is a copper cauldron attached to a chain, and enveloped with a thick smoke, which escapes either by the ceiling or the door, is a picture of the interior of these dwellings.' (Malte-Brun.) In the houses even of the nobility, the walls are sometimes built only of trunks of trees cemented with mortar, and the furniture consists of a very few articles. The roads, except that across the Caucasus to Tiflis, which has been improved by the Russians, are in a wretched state. The vehicles in use are of the rudest kind, and all commodities, except straw or timber, are transported upon horses, mules, asses, or camels. The inhab. never ride, except on horseback. Coarse woollen, cotton, and silk fabrics, leather, shagreen, and a few other articles, are manufactured; the arms made at Tiflis have some reputation; but most of the other goods are very inferior, and only enter into home consumption.

Georgia composes one of the five Trans-Caucasian governments of Russia. Their government is wholly military; and how little soever it may square with our notions of what a government should be, it is not ill-fitted for the circumstances of the country; and there cannot be a question that its establishment has been most advantageous to the population.

The Georgian ladies have usually oval faces, fair complexions, and black hair; and though not generally reckoned handsome by Europeans, they have long enjoyed the highest reputation for beauty in the East: the men are also, on the whole, well formed and handsome. This superiority in the physical form of the Georgians, and other contiguous Caucasian tribes, and the low state of civilisation that has always prevailed amongst them, explains the apparently unaccountable fact, that these countries have been, from the remotest antiquity down to our times, the seat of an extensive slave-trade. Latterly the harems of rich Mussulmans of Turkey and Persia have been wholly or principally supplied by female slaves brought from Georgia, Circassia, and the adjoining provinces; and they also furnished male slaves to supply the Mameluke corps of Egypt and various other bodies with recruits. In modern times the Georgians have been divided, with the exception of a few free commoners, into the two great classes of the nobles and their vassals or slaves. Previously to the Russian conquest, the latter were the absolute property of their lords, who, besides employing them in all manner of manual and laborious occupations, derived a considerable part of their revenue from the sale of their sons and daughters. Indeed, the daughters of the nobles not unfrequently shared the same fate, being sacrificed to the necessities or ambition of their unnatural parents. (Tournefort, ii. 303; Missionary Researches, p. 151.)

The Russians have put an end to this traffic; and they have also deprived the nobles of the power capitally to punish their vassals, and set limits to their demands upon them for labour and other services. There cannot therefore be, and there is not, a doubt with any individual acquainted with the circumstances, that the Russian conquest has been of signal advantage to the bulk of the Georgian people. We believe, however, that the Russians are quite as much disliked by the nobles of Georgia as by those of Circassia; and those travellers who live with them, and credit their stories, will be amply supplied with tales of Russian barbarity and atrocity.

With a settled state of affairs, Tiflis might again become, as in the days of Justinian, a thoroughfare for the overland commerce between Asia and Europe. The Georgians belong to the Greek church, and since becoming subject to Russia, have been subordinate in ecclesiastical matters to a Russian archbishop at Tiflis, who has three suffragans S. of the Caucasus. The clergy are generally very ignorant. A high school in the capital has been recently erected into a gymnasium; and, in addition to it, there are a few small schools, in which, however, very little is taught. No serf is, or at least used to be, instructed in reading, but all the nobility are more or less educated; the females of this class teach each other, and are commonly better informed than the males. The Georgian language is peculiar, differing widely from the languages spoken by the surrounding nations.

Georgia was annexed to the Roman empire by Pompey the Great, anno 65 B.C. During the 6th and 7th centuries it was long a theatre of contest between the E. empire and the Persians. In the 8th century a prince of the Jewish family of the Bagratides established the last Georgian monarchy, which continued in his line down to the commencement of the present century. The last prince, George XI., before his death in 1799, placed Georgia under the protection of Russia; and, in 1802, it was incorporated with the Russian empire. (Tournefort; Klaproth; Wilbraham; Letters from the Caucasus; Smith and Dwight; Missionary Researches.)

GEORGIA, one of the U. States of N. America, and, with the exception of Florida, the most S. territory in the Union; between lat. 30° 21′ and 35° N., and long. 81° and 85° 30′ W.; having N. Tennessee and a small portion of N. Carolina; NE. and E. N. Carolina and the Atlantic; S. Florida; and W. Alabama. Length N. to S., 300 m.; breadth variable. Area 58,000 sq. m. Pop. 1,057,286 in 1850, of which number there were 691,550 whites, 3,500 free-coloured people, 381 Indians, and 462,198 slaves. Along the coast of Georgia lies a range of low, flat, sandy islands. The mainland for about 30 m. towards the interior is perfectly level; and, for several miles from the shore, consists of a salt marsh of recent alluvion; the whole of the flat country is intersected by swamps, which are estimated to constitute 1-10th part of the whole state. Beyond the swamps which line the coast occurs an extensive range of pine barrens, similar to those of S. Carolina. The Okefinoke swamp, 50 m. long by 30 broad, lies at some distance inland, upon the borders of and partly within, Florida. This swamp is regularly inundated during the rainy season. At the

extremity of the low country there is a barren sandy tract of rather greater elevation, which extends N. as far as the river falls, and is generally regarded as dividing the upper from the lower country. Farther N. the surface becomes gradually more hilly and broken, and the N. extremity of the state comprises some of the most S. ridges of the Appalachian mountain chain, which here rise to about 1,500 ft. above the level of the Atlantic. There are only three harbours on the coast capable of receiving vessels exceeding 100 tons burden, viz. those formed by the mouths of the rivers Savannah, Atalamaha, and St. Mary's. The first of these is navigable by large ships as far as the city of Savannah, 17 m. from its mouth. Three of the principal rivers form the boundaries between Georgia and the adjoining states. The Savannah rises in the S. declivity of the Appalachian mountains, and running along the NE. border of the state, separates it from S. Carolina. The Chattahoochee has its source near that of the Savannah, runs chiefly S., and forms for a considerable distance the boundary between Georgia and Alabama. At the SW. angle of the state it unites with the Flint, and, on its entrance into Florida, is called the Appalachicola. On the E. the St. Mary's, with a tortuous course of 110 m., forms the boundary of the state for about 80 m. The Atalamaha, formed by the junction of several streams which traverse the centre of the state, falls into the Atlantic, after a course of about 280 m.

Soil, for the most part, very productive. In the low country and the islands, it consists of a light grey sand, gradually becoming darker and more gravelly towards the interior. Farther N. it is a black loam mixed with red earth, called the mulatto soil; this is succeeded in the more remote districts by a rich black mould of great fertility. As the elevation of the N. part of the state is estimated at from 1,200 to 1,500 ft. above the level of the islands on the coast, a difference of more than 7 degrees is estimated to exist between the mean temp. of the two extreme points. The N. parts are very healthy, the winters mild; frost and snow frequently occur, but are not severe or of long continuance. In the low country the usual tropical diseases are prevalent. Hurricanes and thunder storms frequently occur in the autumn, at which season the cultivators with their families generally remove either to the islands, or the most N. districts of the state. In the low tracts the thermometer usually ranges during the summer from 76° to 80° (Fahr.); but it has been known to stand as high as 102° (Fahr.).

Principal agricultural products, cotton, wheat, and other European grains, maize, tobacco, the sugar-cane, indigo, and rice. The coast islands were formerly covered with extensive pine barrens; but they now yield large quantities of sea-island cotton, which is not only far superior to that grown on the mainland, but is, in fact, superior to, and fetches a higher price than, any other description of cotton to be found in the market. (See CAROLINA, S.) Wheat and other corn are grown chiefly in the central parts along the bottoms of the rivers, and on the slopes of the hills nearly to their summits. The proportion of productive land is much greater in the hilly country than in the plains. The tops of the hills are mostly crowned with forests, composed chiefly of the pine, palmetto, oak, ash, cypress, hickory, black walnut, mulberry, and cedar trees. Bears and deer inhabit the forests; alligators infest the swamps and mouths of the rivers; honey bees are very numerous in the N.

Gold has been found in considerable quantities

in the N. part of the state; iron and copper exist in different parts; and there are several valuable mineral springs; good millstone is met with in the central districts.

Cotton is the great staple; and it and tobacco, indigo, canes, timber, deer skins, and maize form the chief exports; the sugar-cane has hitherto been cultivated mostly for home consumption only. From the distance between the N. part of Georgia and its ports, and the difficulty of communication by water, the corn and other produce of the interior have a very limited outlet.

The imports consist chiefly of manufactured goods, E. India produce, wines from the S. of Europe; butter, cheese, and fish from the N. states. The value of the real estate and personal property, including slaves, was £15,425,714 dollars in 1850, and 615,855,237 dollars in 1860, being an increase of 98 per cent within the ten years. Subsequently, however, there must have been an equally large decrease, an immense destruction of property having taken place in the civil war 1861-65. The state is divided into 76 counties: Milledgeville, near the centre, is the cap.; Savannah, Augusta, Washington, and St. Mary's are the other chief towns.

The University of Georgia, called Franklin's College, at Athens, was founded in 1785-89. It was intended to embrace the whole system of public education in the state, including the establishment of an academy in each county; but this project has never been accomplished. It was reorganised in 1802, and possesses two edifices, a philosophical and chemical apparatus, a cabinet of minerals, a good library, and a botanic garden. There is a medical college at Augusta. The state has a school fund, and there are numerous and flourishing academies in Savannah, Augusta, and the other chief towns. Several manual-labour schools have been successfully established in different parts. A canal 16 m. in length, from Savannah to the Ogeechee river, was completed in 1829; another, 12 m. in length, between Brunswick and the Atalamaha, is in active progress. The legislature consists of a senate of 93 members, and a house of representatives of 207 members, chosen by all the citizens and inhabitants of the state of full age who have resided in it for the year preceding the election, and paid taxes. The governor is elected by the people, and holds office for two years; the senators and representatives are chosen annually. For the administration of justice the state is divided into ten circuits, each of which has a superior court, and a judge elected by the legislature. There is an inferior court in each county, presided over by five justices, chosen by the people every four years; the justices have no salary. There are courts of oyer and terminer at Savannah and at Augusta.

Georgia was the last settled of the present U. S., founded by the British. It was first colonised by them in 1733, in which year the city of Savannah was commenced by General Oglethorpe. It suffered much during the early period of its settlement from the incursions of the savages, and it was not until 1835 that the Cherokees, the last remnant of the Indian pop., had entirely disappeared. In 1776, it enlisted in the struggle for independence, but continued in the occupation of the British until 1782. Georgia joined the insurrection against the government of the United States on the 19th of January, 1861, when an Act of Secession was passed by a convention called for the purpose. Having been overrun by the armies of the North, the state was compelled to join the Union again in 1865.

GERA, a town of Central Germany, principality

of Reuss (younger branch), cap. of the lordship of the same name, on the Elster, 21 m. NE. Schleitz, and 31 m. SW. by S. Leipzig, with which it is connected by railway. Pop. 13,302 in 1861. The town is well built, is surrounded with walls, and has several suburbs. It possesses six public squares, a fine town-hall, two churches, two hospitals, an orphan asylum, a house of correction, a richly-endowed gymnasium, with a library and cabinet of natural objects, a teachers' seminary, some good citizens' schools, evening and Sunday schools, &c. It has been long noted for its commercial activity; and has manufactures of woollen and cotton fabrics, hats, leather, tobacco, soap, oilcloth, porcelain, and other earthenware, coaches, and other vehicles; and many cotton-printing and dyeing establishments, breweries, and brick-kilns. In its immediate neighbourhood there are some greatly frequented baths. In 1780 Gera was almost wholly destroyed by fire; but it has since been laid out and rebuilt in a much better manner than previously.

GERACE (an. Locri), an inland town of Southern Italy, prov. Reggio, cap. distr. and cant.; on a hill within 4 m. of the Ionian Sea, 46½ m. SSW. Catanzaro, and 29 m. NNE. Cape Spartivento. Pop. 6,430 in 1861. Though rebuilt since the earthquake of 1783, its streets are narrow, mean, and filthy. It has the remains of a castle, a cathedral, nine par. churches, a hospital, and a foundling asylum: its public edifices were greatly injured by the earthquake alluded to. The ruins of its castle, demolished at an anterior period, show it to have been a fortress of great size and strength. It is said to have been built by the Saracens, and to have been capacious enough to contain a garrison of 10,000 men. The cathedral was formerly a handsome Gothic edifice, but it is now so dilapidated that only a portion of its crypt remains available for public worship. Its ruins contain many fine marble columns, which originally belonged to the ancient city. Gerace is generally supposed to stand either upon or near the site of Locri Epizephyrii, so called from its founders being Locrians, and its situation adjacent to Cape Zephyrium. This was one of the oldest, largest, and most prosperous of the Greek cities in S. Italy or Magna Graecia. It was mainly indebted for its prosperity and fame to its great legislator Zeleucus, one of the most illustrious of the Grecian political philosophers. Some ruins still remain to attest its former grandeur, among which are those of an aqueduct, of a celebrated Greek temple of Proserpine (sacked by Pyrrhus), and of a temple of Castor and Pollux.

Locri never recovered from the injuries inflicted on her by Pyrrhus. In the second Punic war she sided with the Carthaginians; and having been conquered by the Romans she continued progressively to decline. The present town is supposed to have been founded in the 8th or 9th century.

GERMAIN-EN-LAYE (ST.), a town of France, dép. Seine-et-Oise, cap. cant.; on a hill adjoining the Seine, 6 m. N. Versailles, and 9 m. W. by N. Paris, on the railway from Paris to Rouen. Pop. 17,708 in 1861. Though laid out without any fixed rule, it is well built, and its streets are wide and well paved. It has several large hotels, a public library with 8,500 vols, a theatre, a new corn-market; with manufactures of horse-hair goods and leather; and an active retail trade. It is, however, chiefly noted for its royal residence, originally built by Charles V. in 1370; reconstructed by Francis I.; and embellished by many succeeding sovereigns, especially Louis XIV., who added to it five extensive pavilions, and constructed the fine terrace which extends from it with a breadth of nearly 96 ft. for a distance of 1½ m. between the forest of St. Germain and the Seine. That sovereign expended in all upon St. Germain's the sum of 6,155,611 livres; but it is said that he afterwards became disgusted with, and abandoned, the palace, because he could see St. Denis, the burial-place of the kings of France, from its windows. Charles IX. and Henri II., as well as Louis XIV., were born in this palace; it was the residence of Madame de la Vallière; and James II. of England, with most of his family, passed their exile, and died in it. It is now used as barracks and a military prison. Henri IV. constructed a palace, call the Château Neuf, about ¼ m. distant from the above: of this there now exist only the ruins. A castle, built here in the 11th century by King Robert, was destroyed by the English in 1346.

The Forest of St. Germain, one of the finest of its kind in France, extends N. of the town, enclosed W., N., and E. by the Seine. It is 9 m. in length by 6 m. in breadth; covers an extent of 8,665 English acres; and is traversed by roads, the aggregate length of which is not less than 1,180 m.

GERMAN'S (ST.), a bor., market town, and par. of England, co. Cornwall, hund. East, on the Till, near Lynher creek, 19 m. ENE. Bodmin, and 196 W. by S. London. Area of par., 10,050 acres (being the largest par. in Cornwall). Pop. of par. 2,812 in 1861. The town is built on a slope, and consists chiefly of one street. The par. church, formerly conventual, and now containing an episcopal chair and prebendal stalls, is a fine old specimen of Saxon architecture, consisting of two aisles and a nave: the W. front has two towers, between which is an ancient arched doorway, the entrance to the church. The living is in the gift of the dean and canons of Windsor. A free grammar school and a parochial library have been founded and endowed by the Eliot family, whose seat, Port Eliot, near the church, occupies the site of the ancient priory, and is surrounded by delightful grounds watered by the Tidi. The inhab. chiefly gain their livelihood by fishing and agriculture. Previously to the Reform Act, by which it was disfranchised, this bor. sent two mems. to the H. of C.; the right of election was vested in the proprietors of burgage tenements; but of these there were very few, so that the mems. were, in fact, nominated by Lord St. Germans. Markets on Friday; fairs, May 28th and August 1st, for cattle.

GERMANY (Germ. Deutschland or Teutschland; Fr. Allemagne; It. Germania, Slavonia). The word Germany is as uncertain in its derivation, as it is often vague and indefinite in its application. The Germans call themselves Deutsche, or Teutsche, and their country Deutschland. The first syllable of this name is derived by those who use this orthography from the verb deuten, signifying to interpret or explain; so that Deutsche means the people who were intelligible to one another, in contradistinction to the Walsche (Welsh), or Celtic nations, whose language they did not understand. Those who write Teutschland derive the name of the country from the God Tuisco or Teut, mentioned by Tacitus. The Latin denomination of the country, which English-speaking people have adopted, is supposed to be derived from the Roman manner of pronouncing the word Wehrmann, which signifies soldier—the character in which the Germans were mostly known to the Romans.

The extent of country comprised under the term Germany has varied in every century since it first

became known to the Romans. At present Germany comprises the chief countries of Central Europe, and is bounded N. by Denmark and the Baltic; E. by Prussian Poland, Galicia, and Hungary; S. by the Tyrol and Switzerland; and W. by France, Belgium, Holland, and the German Ocean.

Physical Aspect.—The surface of Germany is much diversified; its mountain tracts lie chiefly in the SE. and E., while W. and N. the land spreads in spacious sandy plains, intersected by the rivers which run in the same direction from the higher lands towards the sea. The mountains, which may be considered as a N. branch of the great Alpine system of Europe, bear no comparison with the Alps in point of height, for the loftiest summits are only 5,000 ft. high; but they occupy a great space, and diverge in so many various directions through the country that it is difficult to trace them without the aid of a map. The Fichtelgebirge, however, in the N. part of Bavaria, may be considered as the centre and nucleus of the mountains in Central Germany; and from it branch, in four directions, the ranges composing the watershed that divides the rivers of the Black Sea from those of the Baltic and German Ocean. 1. The Erz-gebirge, diverging NE., forms the boundary between Saxony and Bohemia, and has its westward side N. towards the Eger. Its E. continuations, called the Sudeten-gebirge, join the Carpathian ridge near the sources of the Oder and Vistula. 2. The Bohemian Forest range separates Bohemia from Bavaria. It runs SE. about 150 m., and then turning NE. joins the Sudeten-gebirge, near the sources of the March, in long. 16° 40' E. These ranges, by their reunion, enclose an elevated plain, constituting the kingdom of Bohemia, and drained by the Elbe and its branches, the Eger and Moldau. 3. The Suabian Alps are a low range, branching off SW. from the central point, and forming the watershed between the affluents of the Rhine and those of the Danube. 4. they join the Black Forest range, the connection of which with the Alps is effected by a low chain skirting the Lake of Constance, and joining the main ridge at Mount Septimer. 4. The Thuringian range runs NW. from the Fichtel-gebirge, and after a course of 50 m., divides into two chains, one running N. into Hanover, and forming the Harts chain which divides the waters of the Weser from those of the Elbe; the other running W. under various names, nearly as far as the Rhine, and separating its waters from those of the Weser and its affluents. The hills W. of the Rhine are continuations of the Vosges system. (See FRANCE.)

The rivers of Germany are numerous and important. The largest of these is the Danube (1,800 m. long), which rises in the Black Forest, and is navigable from Presth to its mouth in the Black Sea. The chief tributaries of the Upper Danube are the Altmuhl, the Naab, and the March on its N. bank, rising on the S. slopes of the German mountains; and the Iller, the Lech, the Isar, and the Inn on its S. bank, all rising in the Tyrolese Alps. The Rhine, which rises on Mont St. Gothard, flows through the Lake of Constance, and thence W. to Basle: navigable from this place, it turns N. in which general direction it runs as far as Bingen, whence it pursues a course NNE. into the German Ocean. Its chief affluents, with the exception of the Moselle and the Maas, are on the E. bank: of these the Neckar and the Main rise in the Suabian Alps; the Lahn, the Ruhr, and the Lippe in the hills of W. Germany. The Weser is formed by the junction, at Munden, of the Werra and Fulda, which rise in the Habia-gebirge; its course is N. by W. till the junction of the Aller, at which

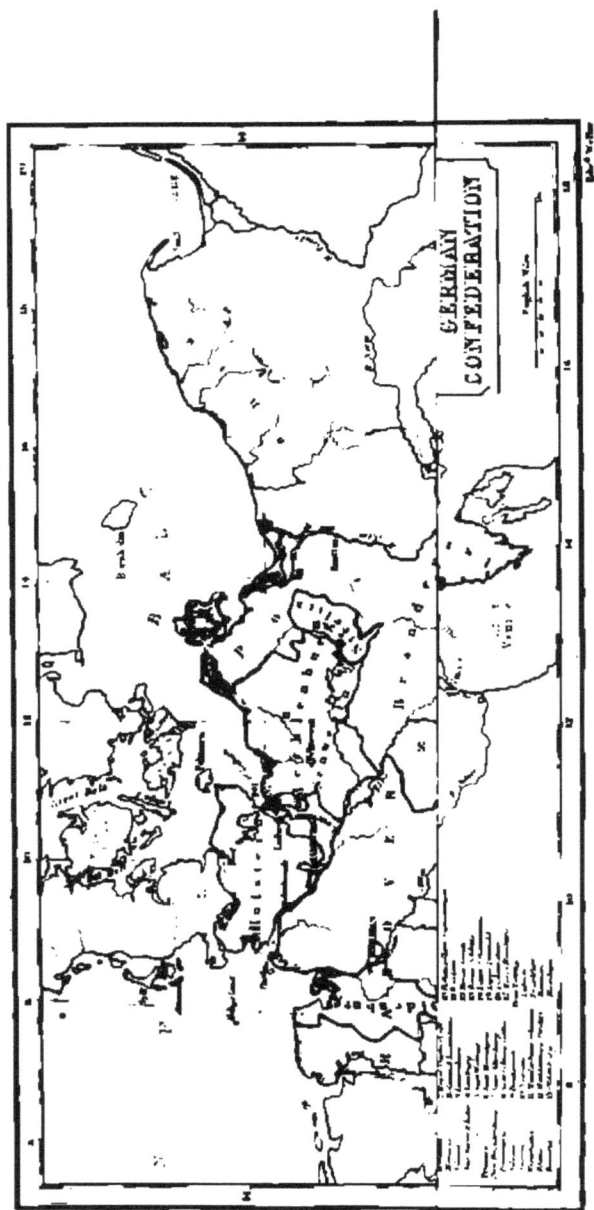

GERMAN CONFEDERATION

point it turns NE., and falls in the German Ocean about 40 m. below Bremen. The Elbe rises on the N. side of the plateau of Bohemia, which, after receiving the Moldau and the Eger, it leaves at Schandau, and enters the great NW. plain of Germany, which it traverses to the German Ocean; its largest affluents from the S. are the Mulda and Saale from the Erzgebirge, and its chief N. tributary is the Havel. The Oder rises on the N. side of the Carpathian range, near its W. termination, and after a general NNW. course, and receiving many affluents, falls through the Great Haffe into the Baltic Sea. Besides these rivers, which of themselves constitute a most extensive watersystem, there are numerous lakes connected with the rivers; such are the lakes of N. Bavaria and Austria, and the many sheets of water lying on the low plain of N. Germany, between the Oder and the Elbe.

Climate.—The climate of Germany is far less variable than the nature of its mountain system, and the range of latitudes in which it lies, would lead us to suppose. If the small strip of Illyria which borders on the Adriatic Sea, near Trieste, be excepted, scarcely any diminution of warmth is observable between the southern and northern parts. There are only two degrees difference between the mean temperature of Vienna and that of Hamburg. The vegetation of Germany resembles, in its general character, that of the N. of France. In the S. river valleys the vine flourishes, and walnuts, chestnuts, and plums grow abundantly; but the severity of the winter injures the growth of garden shrubs and flowering plants. Only hollies and some of the hardier species of junipers thrive, as even the rivers in the warmest parts freeze, and the Rhine, near Mannheim, as well as the Danube, near Vienna, are usually covered with a road of ice, notwithstanding their great breadth and the rapidity of their currents. The extreme cold of the winter, although it only lasts in all its violence, in common winters, for a few days, is rendered often very destructive from the continuance of a low, but still considerable cold, which often lasts uninterruptedly for months. The thermometer usually falls once or twice in the course of the winter as low as —5° Fahr., but seldom continues at that figure during twentyfour hours successively. A few degrees below the freezing point is the temperature which frequently lasts for months together in the winter season.

The fall of rain is stated by Berghaus to be, in the four principal regions of Germany, as follows:—

In the region of the Rhine . . . 24 In. 7"' Paris meas.
Weser . . 73   4   "
Elbe & Oder 27   3   "
Danube . 20   "

The quantity of rain which falls in summer is more than double the fall of the winter, throughout Germany. The number of rainy days averages 150, that of thunder-storms averages 19 for all Germany; but the latter are very unequally divided. The greatest number of thunder-storms is said to take place in Silesia, where the average amounts to 28 in the year. The smallest number is found in Lower Austria, where their annual number does not exceed 8. The prevailing winds are the W. and NW.

Geographical and Political Divisions.—The first Carlovingian sovereigns of Germany were hereditary monarchs; but, as early as 887, the states, or great vassals of the crown, deposed their emperor, Charles le Gros, and elected another sovereign in his stead. And from that remote period the emperors of Germany continued to be elected, down to the beginning of the present century. Several of the great vassals of the empire had thus early

attained to all but unlimited power; and it consisted of a vast aggregation of states of every different grade, from large principalities down to free cities and the estates of earls or counts. The federal tie by which these different states were held together was exceedingly feeble. Their interests and pretensions were often conflicting and contradictory, and they were frequently at war with each other and with the emperor. There was, in consequence, a great want of security; and the wish to repress the numberless disorders incident to such a state of things led, at an early period, to the formation of leagues among the smaller states, and the institution of secret tribunals. The privilege of voting in the election of emperor was restricted to a few of the most powerful vassals, being confined, by the Golden Bull issued by Charles IV., in 1356, to the archbishops of Mayence, Treves, and Cologne, the duke of Saxony, the count palatine of the Rhine, the margrave of Brandenburg, and the king of Bohemia. The sovereigns of Bavaria, Hanover, and Hesse did not acquire a right to vote till a much later period. Most of the great offices in the empire were hereditary; and the public affairs were transacted in diets or assemblies of the great feudatories and of the representatives of the free cities. But as the diet had no independent or peculiar force to carry its decisions into effect, they were very frequently disregarded. At length, in the reign of Maximilian I., an attempt was made to introduce a more regular system of administration and a better police into the empire.

As the political division of Germany at this period was independent of the territorial subdivisions which the changes in families produced, it lasted as long as the empire itself preserved its unity as a political body; and even after the assumption of independence by the king of Prussia, that part of the kingdom of Prussia which previously formed a part of the empire was still included, nominally at least, in the circles to which it belonged. By their refusal to join in this arrangement of internal police, and to become amenable to the decrees of the Aulic Chamber (*Reichs Kammergericht*), the Swiss cantons finally severed the last tie which united them to the empire. The influence of the kings of Poland caused a similar separation between the empire and the lands belonging to the Teutonic order, on the right bank of the Vistula.

At this period of the outbreak of the French revolution, in 1789, the ten circles of Germany were subdivided into the following territories:—

I. The circle of Austria, belonging entirely to the house of Austria, contained—
  1. The duchy of Lower Austria.
  2. Inner Austria, or the duchies of Styria, Carinthia, and Carniola, with Friuli and the district of Trieste.
  3. Upper Austria, or the county of Tyrol, with the bishopric of Trent and Brixen.
  4. Fore-Austria, containing the Austrian Brisgau, the margraviate of Burgau, the landgraviate of Nettenburg, the city of Constance, the manors of Altorf and Ravensburg, the towns of Riedlingen, Mengen, and the lordships of Vorarlberg.

II. The circle of Burgundy, belonging to Austria:—
  1. The duchies of Brabant, Limburg, Luxemburg, and Gueldres.
  2. The Counties of Flanders, Hainault, and Namur.

III. The circle of Westphalia, divided between clerical and lay princes:—
  1. The bishoprics of Munster, Paderborn, Liège and Osnabrück.

2. The abbeys of Corvey, Stablo, and Malmedy, Werden, St. Cornelis-Münster, Essen, Thorn, and Herford.

3. The duchies of Cleves (Prussian), Juliers and Berg (to the elector palatine), Oldenburg (to the bishop of Lübeck).

4. The principalities of Minden (Prussia), Werden (elector of Hanover), Nassau (counties of Dietz, Siegen, Dillenburg, and Hadamar, belonging to the stadtholder of Holland), East Friesia, Mörs, and Oserbize (Prussian).

5. The counties of Mark, Ravensberg, Tecklenburg, and Lingen (Prussian); Schaumburg (Hesse-Cassel and Lippe); Bentheim, Steinfurt, Hoya, and Diepholz (Hanover and Cassel); Hückeburg (Prince Kaunitz); Pyrmont (Count Waldeck); Wied, Sayn, Virnenburg, Reineberg, Orsoldeck, Reckheim, Holzapfel, Blankenheim and Gerolstein, Kerpen, Lommersum, Schleiden, Halkermond.

6. The lordships of Anholt, Wilten, Winneberg and Beilstein, Gehmen, Gimborn and Neustadt, Wickerad, Myllendorck, Reichenstein.

7. The free imperial cities Cologne, Aix-la-Chapelle, and Dortmund.

IV. The circle of the Palatinate, divided between one lay and three clerical princes:—

1. The electorate of Mayence, the archbishop of which ranked as the first elector and primate of the German empire. The electorate consisted of the archbishopric of Mayence, the city of Erfurt, the district of Eichsfeld, and of the town and district of Pristarr.

2. The electorate and archbishopric of Trèves.

3. The archbishopric of Cologne and the duchy of Westphalia, which constituted the electorate of Cologne.

4. The palatinate of the Lower Rhine.

5. The principalities of Arenberg.

6. The bailiwick of Coblenz (Teutonic order).

7. The lordship of Brilstein.

8. The landgraviate of Reineck.

9. The county of Lower Isenburg.

V. The circle of the Upper Rhine, divided amongst a number of territorial lords, the most powerful of whom was the landgrave of Hesse-Cassel:—

1. The bishoprics of Worms, Spires, Strasburg, Basle, and Fulda.

2. The abbey of Weissenburg.

3. The principality of Heitersheim.

4. The abbeys of Prüm and Odenheim.

5. The principalities of Nassau, Lautern, Veldenz, and Deuxponts.

6. Landgraviate of Hesse, in two lines, Cassel and Darmstadt.

7. The principality of Hersfeld (Hesse-Cassel).

8. The counties of ... (Palatinate and Hesse-Baden), Salm and Nassau; Waldeck, Hanau-Münzenberg, Solms, Königstein (elector of Mainz and Count Stolberg); Upper Isenburg; the possessions of the count of the Rhine and the Wildgrave, viz. the county of Salm, the lordship of Grumbach, &c.; the counties of Leiningen, Wittgenstein, Falkenstein (belonging to the emperor), Reipoldskirchen, Kirchingen, Wartenberg.

9. The lordships of Hanau-Lichtenberg, Bretzenheim, Dachstuhl, and Ollbrück.

10. The free imperial cities Worms, Spires, Frankfort-on-the-Main, Friedberg, and Wetzlar.

VI. The Swabian circle. Amongst the many princes of this circle, the duke of Wirtemberg and the margrave of Baden were the most powerful:—

1. The bishops of Constance and Augsburg.

2. The abbeys of Elbwangen, Kempten, Lindau, and Buchau.

3. The duchy of Wirtemberg.

4. The margraviate of Baden.

5. The principalities of Hohenzollern (Hechingen and Sigmaringen).

6. The county of Tengen, the counties of Heiligenstadt and Baar (prince of Fürstenberg), lordships of the princes and counts of Oettingen, the landgraviate of Klettgau (Prince Schwarzenberg), and the principality of Liechtenstein.

7. The domains of 17 abbots and 4 abbesses.

8. The lordships of 18 counts and barons.

9. Thirty-one free imperial cities, viz. Augsburg, Ulm, Esslingen, Reutlingen, Nördlingen, Schwäbisch Hall, Ueberlingen, Rothweil, Heilbronn, Gmünd, Memmingen, Lindau, Dinkelsbühl, Biberach, Ravensburg, Kempten, Kaufbeuren, Weil, Wangen, Leutkirch, Wimpfen; Giengen, Pfullendorf, Buchhorn, Aalen, Bopfingen, Buchau, Offenburg, Gengenbach, and Zell am Hammersbach.

VII. The circle of Bavaria, in which the elector of Bavaria and the bishop of Salzburg took the lead:—

1. The archbishop of Salzburg.

2. The bishoprics of Freising, Ratisbon, and Munich.

3. The abbeys of Berchtesgaden, and Mödre and Ober-Münster.

4. The duchy of Bavaria, with the Upper Palatinate.

5. The principality of Neuburg and Sulzbach.

6. The landgraviate of Leuchtenberg and Sternstein.

7. The counties of Haag and Ortenburg.

8. The lordships of Mörnsfeld, Salzburg, Pyrbaum, Hohenwaldeck, and Breitenek.

9. The free imperial city of Ratisbon (Regensburg).

VIII. The circle of Franconia included:—

1. The bishoprics of Bamberg, Würzburg, and Eichstädt.

2. The master of the Teutonic order's territories at Mergentheim (Deutschmeister).

3. The principalities of Bayreuth and Anspach.

4. The counties of Henneberg and Schwarzenberg.

5. The principality of Hohenlohe.

6. The counties of Castell, Wertheim, Rieneck, and Erbach.

7. The lordships Limburg, Seinsheim, Reichelsberg, Wiesentheid, Welzheim, and Hausen.

8. The five cities Nuremberg, Rothenburg-on-the-Tauber, Windsheim, Schweinfurt, and Weissenburg.

IX. The circle of Lower Saxony:—

1. Duchy of Magdeburg (Prussia).

2. Duchy of Bremen, principalities of Lüneburg, Grubenhagen, and Kalenberg (elector of Hanover).

3. Duchy of Wolfenbüttel, principality of Blankenburg.

4. Principality of Halberstadt (Prussia).

5. Duchy of Mecklenburg-Schwerin and Strelitz.

6. Duchy of Holstein, with the county of Ranzau, and the lordship of Pinneberg (king of Denmark).

7. The bishopric of Hildesheim.

8. The duchy of Saxe-Lauenburg (Brunswick).

9. The bishopric of Lübeck.

10. The principality of Schwerin (duke of Mecklenburg-Schwerin).

11. The principality of Ratzeburg (duke of Mecklenburg-Strelitz).

12. The free imperial cities Lübeck, Hamburg, Bremen, Goslar, Mühlhausen, Nordhausen.

X. The circle of Upper Saxony:—

1. The duchy of Pomerania (of which that part beyond the Peene belonged to Sweden, the remainder to Prussia).

2. The mark of Brandenburg (belonging to the king of Prussia).

3. The principality of Anhalt, divided among four princes:—Dessau, Bernburg, Kothen, and Zerbst.

4. The electorate of Saxony (Saxe Albertine line).

5. Principalities of Weimar, Eisenbach, Coburg-Gotha, Altenburg (dukes of Saxony, of the Ernestine line).

6. The abbey of Quedlinburg.

7. The county of Schwarzburg (Sondershausen and Rudolstadt), Mansfeld (Prussia and Hesse), Stolberg and Wernigerode.

5. The lordships of Reuss and Schönburg and the county of Hohenstein.

The Slavonic countries, which were not included in any circle, were—

The kingdom of Bohemia.

The margraviate of Moravia.

The duchy of Silesia, — so far as it was Austrian, the margraviates of Upper and Lower Lusatia, the duchy of Silesia (Prussian), and the county of Glatz.

The lands held directly of the emperor, and not included in any circle, were—

The counties of Mümpelgard and Homburg.

The lordships Aerb, Wasserburg, Frehberberg, Illeria, Jever, Dyck, Schinan, Wylve, Rickoldi, Fürst, Dreyss, Landskron, Rhede, Saffenberg, Schaumburg, Obermein, Schauen, Kniphausen, and Hörstein.

The abbeys and convents Elten, Kappenberg, and Burtscheid.

The free imperial towns, Aschenhausen in Suabia, Althausen in Franconia, Sulzbach and Baden near Frankfort-on-the-Maine, with the freeholders on the heath of Leutkirch.

Such were the territorial divisions of the German empire at the period of the outbreak of the French revolution, according to Berghaus, from whom the following statements are likewise taken :—

Every circle had its diet, in which the clerical and secular princes, the prelates, the counts and barons, and the free imperial cities, formed five benches or colleges.

Affairs of general importance to the empire at large were treated by the imperial diet, which the emperor had the power of summoning wherever he pleased; but which, since 1663, has been constantly assembled at Ratisbon. In 1789, the members of the diet were as follows :—

1. The college of electors : Mayence, Trèves, Cologne, the Palatinate, Brandenburg, Saxony, Bavaria (since 1623), and Brunswick Lüneburg (since 1692).

2. The college of the clerical and secular princes, bishops, margraves, counts, &c. : the members of the clerical members being 33, and of the secular 100, 82.

3. The colleges of the free imperial cities, then 51 in number.

In 1791 began the memorable contest with revolutionary France, which ended in the overturn of the old Germanic constitution. The treaty of Campo Formio, the first that history records in which the Rhine was acknowledged as the frontier of France, decreed an indemnification to those princes who lost by the cession ; and this indemnification could only be obtained by the spoliation of some others whose rights were equally indefensible, in the heart of the empire itself. On the 25th of January, 1803, a decision was come to by the plenipotentiaries assembled for the arrangement of this matter, the import of which was as follows :—

The Holy Roman Empire, as that of Germany was styled, remained as it was divided into circles, but which, with the total loss of the circle of Burgundy, and of the lands on the left bank of the Rhine, were reduced to nine, whose boundaries it was proposed to regulate anew. This regulation was, however, prevented by the wars which so quickly succeeded each other. The right to sit and vote in the diet remained, as formerly, attached to territories held directly as fiefs of the empire ; and the place of the convocation of the diet remained at Ratisbon. The colleges remained also three in number ; the first being the College of Electors, who were ten in number ; one clerical,—the elector archchancellor ; and nine secular,—Bohemia, Bavaria, Saxony, Brandenburg, Brunswick - Lüneburg, Salzburg, Wirtemberg,

Baden, and Hesse-Cassel. The electorate of Mayence had merged into that of the archchancellor, and the Palatinate into the electorate of Bavaria ; Trèves and Cologne had disappeared, and four new electorates had been created.

The second College—of Princes—counted 131 votes.

The College of Towns was composed of six with votes : Hamburg, Lübeck, Bremen, Frankfort-on-the-Maine, Augsburg, and Nuremberg. The other territories, enumerated above as not being included within the circles, remained as they were, nor did any change take place in the extent or position of the Slavonic countries.

Napoleon who, since 1799, had directed the foreign policy of the French, not satisfied with this reduction of the power of the empire, now conceived the design of effecting its final dissolution. The treaty of Presburg, in 1805, which followed the battle of Austerlitz, gave him the means of carrying this project into effect, by forming a confederation of German princes, called the Confederation of the Rhine, who, uniting into a corporate body, in 1807, placed themselves under the protectorate of the emperor of the French. The wars which followed, with Prussia in 1807, and with Austria in 1809, gave Napoleon the power of altering the territorial distribution of Germany at pleasure. He accordingly created for his brother Jerome the new kingdom of Westphalia, and for his brother-in-law Joachim Murat, the grand duchy of Berg, and raised those members of the Confederation of the Rhine who supported his cause to new dignities and an openly recognised independence as sovereigns. Under these circumstances, the emperor, Francis II., by a solemn act, renounced the style and title of Emperor of Germany, on Aug. 6, 1809. In the following year, Napoleon incorporated the coasts of the German Ocean with the French empire, and divided them into departments ; thus separating from Germany a district peopled by more than 1,100,000 inhabitants.

The termination of the war with Russia, or as called in Germany, 'the war of liberation,' restored Germany to its geographical and political position in Europe, but not as an empire acknowledging one supreme head. A confederation of 35 independent sovereigns and 4 free cities replaced the elective monarchy, that fell under its own decrepitude. In the choice of the smaller princes, who were to become rulers, as well as of those who were obliged to descend to the rank of subjects, more attention was paid to family and political connexion than to the old territorial divisions under the empire. The clerical fiefs, and the greater part of the free imperial cities, were incorporated into the states of the more powerful princes, upon the dissolution of the empire, and were not re-established. Only four cities remained in the enjoyment of their political rights.

The signing and ratification of the Act of Confederation took place, after long discussion, on 8th June, 1815. The following are the principal stipulations of the treaty :—

' 1. The sovereigns and free cities of Germany, including their majesties the emperor of Austria, and the kings of Prussia, Denmark, and the Netherlands ; the emperor and king of Prussia for the whole of their territories, formerly belonging to the German empire ; the king of Denmark, for Holstein ; and the king of the Netherlands, for Luxemburg, agree to unite to form an internal league, to be denominated the German Confederation (der Deutsche Bund).

' 2. The object of this confederation is the maintenance of the security of Germany, internally and externally, and the assertion of independence and integrity of the respective Germanic states.

'3. All members of the confederation have, as such, equal rights. They all bind themselves equally to observe inviolably the act of union.

'4. The affairs of the confederation are managed by the diet, in which every member is represented, either by a separate or by a joint vote, in the order of the annexed list, but without prejudice to the rank of the sovereigns. In the committee the members are represented by 17 plenipotentiaries.

'5. Austria enjoys the right of presiding in the diet. Every member of the diet has the right of making propositions, and of bringing forward measures for discussion; and the president is bound to submit them to consideration within a certain term, to be hereafter fixed.

'6. Propositions relating to the adoption or alteration of the fundamental laws of the confederation, or which concern its organisation, or the adoption of establishments calculated in any way to be generally advantageous to the members, must be submitted to a full assembly of the diet, in which every individual member has one or more votes according to the size of each state.

'7. The committee decides by a majority of votes in how far a subject is adapted for the consideration of the full assembly.

'The propositions to be subjected to the decision of the full diet must be prepared and brought to maturity in the committee. The decision in both assemblies is by a majority of votes, but in the plenum, the majority must amount to two-thirds of the votes.

'When the votes are equally divided in the committee, the president has the casting voice.

'But where the adoption or alteration of fundamental laws is concerned, or the rights of individual members, or in religious matters, no resolution can be adopted by the committee alone, nor can the full assembly decide by a mere majority of votes.

'The committee of the diet is constantly assembled, but may adjourn its sittings when the affairs that have been submitted to its consideration are disposed of.

'8. Respecting the order in which the votes of the members are collected, no discussion shall take place during the organisation of the confederation, nor shall any accidental order which may arise during this period be prejudicial to the rights of the members, or be considered as establishing a precedent.

'When the organisation of the league is concluded, the diet will take up the question of precedence for definitive arrangement, and will then adhere as closely as possible to the usage of the former diet of the empire, but especially to that fixed by the decree of the imperial deputation (of 1803). But this order of voting is to have no influence in fixing the rank of the individual members, nor upon their order of precedence on other occasions than that of voting in the diet.

'9. The place of assembly for the diet is Frankfort-on-the-Maine.

'10. The first subject which shall take up the attention of the diet upon its opening must be the drawing up of the fundamental laws of the confederation, and its organisation in respect to its relations with foreign powers, its military and internal arrangements.

'11. All the members bind themselves for the protection of Germany against the attacks of any foreign power, as well as for the security of each individual state; and guarantee to each other mutually the possessions of each state which are comprehended within the confederation.

'When war has been declared by the confederation, no member of the confederation can enter into separate negotiations with the enemy; nor can separate truces or treaties of peace be concluded by individual members.

'The members of the league reserve to themselves the right of making alliances of every kind, but bind themselves not to enter into any which could be prejudicial to the security of the confederation, or of any of its members.

'12. The members further bind themselves under no pretence to declare war against one another, nor to pursue their mutual differences by force of arms, but engage to submit them to the diet.

'The diet is in such cases competent to attempt a reconciliation, by the appointment of a select committee; and should this not prove successful, to procure a decision from a well-organised court of arbitration, whose sentence is implicitly binding upon the disputing parties.

'13. In all the states of the confederation, a constitution based on representation by estates shall be introduced (Landständische Verfassung).'

A further and more detailed declaration of the objects of the league, as well as of the mode of conducting the affairs of the confederation, was published on the 15th May, 1820. This document, with the original act, as given above, and the resolutions of the diet, principally relating to affairs of internal policy, published in 1832, may be regarded as the fundamental laws of the confederation.

To give the diet a more representative form, a plan has been laid before the committee of confederation for creating a lower house of parliament at the side of the now existing upper chamber. According to this plan, Austria shall send thirty deputies, divided among the assemblies of her German provinces; Prussia thirty, and Bavaria ten, to be chosen among the members of their chambers; Saxony, Hanover, and Wurtemberg, each six; Baden, five; Electoral Hesse and Grand-Ducal Hesse, each four; Holstein, Luxemburg, Brunswick, Mecklenburg, Nassau, and Weimar, each two; Meiningen, Coburg-Gotha, Altenburg, Oldenburg, the two Anhalts, the two Schwarzburgs, Waldeck, Lippe, Lichtenstein, Frankfort, Bremen, and Hamburg, each one—in all 17* popular delegates.

As settled by the treaty of Vienna, Germany was divided into thirty-nine sovereign states, or portions of states; but the number is now reduced to thirty-four. The five missing members are—1st, The Saxon princedom of Gotha, which became extinct in 1825, by the decease of the last Herzog, whose territories were divided by compact among his collateral relatives, the princes of Coburg and Meiningen; 2nd, the duchy of Anhalt-Cöthen, which, in 1847, became annexed to Anhalt-Dessau; 3rd and 4th, the principalities of Hohenzollern-Hechingen and Hohenzollern-Sigmaringen, both which states were united to Prussia in 1849, in consequence of the simultaneous abdication of the two reigning princes in favour of their kinsman the King of Prussia, head of the house of Hohenzollern; and, 5th, the duchy of Anhalt-Bernburg, the reigning house of which became extinct with Duke Alexander, who died Aug. 19, 1863, leaving the succession to the last remaining princes of Anhalt, formerly called of Anhalt-Dessau.

The following are the members of the confederation as now constituted, with their votes in the general assembly, their votes in committee of confederation, and their place or rank in the diet:—

Title of Sovereign	Members of the Confederation	Votes in Council	Votes in Confederation	Place in Diet
Emperor	Archduchy of Austria, Bohemia, Styria, Tyrol, Moravia, and part of Illyria	1	3	I.
King	Prussia, exclusive of the provinces of Posen and Prussia	4	3	II.
,,	Bavaria	4	3	III.
,,	Saxony	4	3	IV.
,,	Hanover	4	3	V.
,,	Würtemberg	4	3	VI.
Grand-duke	Baden	3	3	VII.
Elector	Hesse-Cassel	3	3	VIII.
Grand-duke	Hesse-Darmstadt	3	3	IX.
Duke	Holstein and Lauenburg	3	1	X.
Grand-duke	Luxemburg and Limburg	3	1	
Duke	Brunswick	2	1	XI.
Grand-duke	Mecklenburg-Schwerin	2	1	XIII.
Duke	Nassau	2	1	XIV.
Grand-duke	Saxe-Weimar	1		XIII.
Duke	Saxe-Meiningen	1		XII.
,,	Saxe-Altenburg	1		
,,	Saxe-Coburg-Gotha	1		
Grand-duke	Mecklenburg-Strelitz	1		XIV.
,,	Oldenburg	1		
Duke	Anhalt	1		XV.
Prince	Schwarzburg-Sondershausen	1	1	
,,	Schwarzburg-Rudolstadt	1		
,,	Lichtenstein	1		
,,	Waldeck	1		
,,	Reuss-ältere	1		XVI.
,,	Reuss-Schleiz	1		
,,	Schaumburg-Lippe	1		
,,	Lippe-Detmold	1		
Landgrave	Hesse-Homburg	1		
Free City	Lübeck	1		XVII.
,,	Frankfort	1		
,,	Bremen	1		
,,	Hamburg	1		
	Total—Thirty-four States	63	17	

The committee of confederation, consisting of the ambassadors of the thirty-four states, is sitting in permanence at Frankfort-on-the-Maine; but of late has exercised very little political influence. An attempt to reconstitute the confederation on a more liberal basis, made by the Emperor of Austria, and laid before a congress of German sovereigns which met at Frankfort in August, 1863, led to no result, owing chiefly to the opposition of Prussia.

German Zollverein, or Customs' League.—Until a recent period, each of the states into which Germany is divided had its own custom-houses, and its own tariff and revenue laws; which frequently differed very widely indeed from those of its neighbours. The internal trade of the country was, in consequence, subjected to all those vexatious and ruinous restrictions that are usually laid on the intercourse between distant and independent states. Each state endeavoured either to procure a revenue for itself, or to advance its own industry, by taxing or prohibiting the productions of those by which it was surrounded; and custom officers and lines of custom-houses were spread all over the country. Instead of being reciprocal and dependent, everything was separate, independent, and hostile; the commodities admitted into Hesse were prohibited in Baden, and those prohibited in Würtemberg were admitted into Bavaria. The disadvantages of the old system had long been seen and deplored by

well-informed men; but so many interests had grown up under its protection, and so many deep-rooted prejudices were enlisted in its favour, that its overthrow seemed to be hopeless, or, at all events, exceedingly distant. The address and resolution of the Prussian government, however, triumphed over every obstacle. The first treaties in furtherance of this object were negotiated by Prussia with the principalities of Schwarzburg-Sondershausen and Schwarzburg-Rudolstadt, in 1819 and 1819, on the principle that there should be a perfect freedom of commerce between these countries and Prussia; that the duties on importation, exportation, and transit, in Prussia and the principalities, should be identical; that these should be charged along the frontier of the dominions of the contracting parties, and that each should participate in the produce of such duties in proportion to its population. All the treaties subsequently entered into have been founded on this fair and equitable principle; the only exceptions to the perfect freedom of trade in all the countries comprised within the league or tariff alliance being confined, 1st, to articles constituting state monopolies, as salt and cards, in Prussia; 2nd, to articles of native produce, burdened with a different rate of duty on consumption in one state from what they pay in another; and 3rd, to articles produced under patents, conferring on the patentees certain privileges in the dominions of the states granting the patents. With these exceptions, which are not very important, and are daily decreasing, the most perfect freedom of commerce exists among the allied states.

Since 1818, when the foundations of the alliance were laid, it progressively extended. Ducal Hesse joined the alliance in 1828, and electoral Hesse in 1831; the kingdoms of Bavaria, Saxony, and Würtemberg joined it afterwards. The successive formation of the Zollverein took place in the following order:—

1828, Feb. 14	.	Union of Prussia with Hesse-Darmstadt.
,, July 11	.	Adherence of Anhalt-Dessau.
1829, July 5	.	Saxe-Meiningen and Saxe-Coburg-Gotha.
1831, April 14	.	Waldeck.
,, Aug. 24	.	Hesse-Cassel.
1833, Mar. 22	.	Bavaria and Würtemberg.
,, Mar. 30	.	Saxony.
,, May 11	.	Saxe-Weimar, Saxe-Altenburg, and the two Schwarzburgs.
1835, Feb. 25	.	Hesse-Homburg.
,, May 12	.	Baden.
,, Dec. 10	.	Nassau.
1836, Jan. 20	.	Frankfort.
1841, Oct. 18	.	Lippe-Detmold.
,, Oct. 19	.	Brunswick.
1847, April 2	.	Lauenburg.
1851, Sept. 7	.	Hanover, Oldenburg, and Schaumburg-Lippe.

The treaties which bind all these states into the Zollverein are not of a permanent nature, but open to dissolution at stated terms. The treaties now in force will expire with the end of the year 1877.

The Zollverein includes, at present, the whole of the states of the confederation except Austria, the two duchies of Mecklenburg, Holstein, Lichtenstein, and the free cities of Hamburg, Lübeck, and Bremen. The whole of Prussia forms part of the Zollverein, including that portion not belonging to the Confederation.

An assembly of representatives from the allied states meets annually, to hear complaints, adjust difficulties, and make such new enactments as may seem to be required. The duties are received into a common treasury, and are apportioned according to the population of each of the allied states.

Population.—The last general census of the

states of the confederation took place Dec. 8, 1861. In two states, however, Holstein-Lauenburg and Mecklenburg-Strelitz, the most recent enumeration is of the year 1860. The following table contains the area, in English square miles, and number of inhabitants of the thirty-four states, according to three last official returns, compared with the census of 1852:—

	Area in English Sq. Miles	Population in 1852	Population in line A 1861
Austrian States of the Confederation	73,873	13,210,380	13,072,064
Prussian States of the Confederation	73,080	19,837,325	14,188,404
Bavaria	89,838	4,449,482	4,808,697
Saxony	5,746	1,987,872	2,225,240
Hanover	14,776	1,819,253	1,990,974
Würtemberg	7,510	1,733,263	1,720,708
Baden	5,901	1,356,943	1,369,291
Hesse-Cassel	9,514	755,350	758,454
Hesse-Darmstadt	3,242	854,314	856,907
Holstein and Lauenburg	3,730	466,020	484,560
Luxemburg and Limburg	1,904	394,762	431,900
Brunswick	1,388	267,177	282,460
Mecklenburg-Schwerin	4,934	542,763	548,449
Nassau	1,802	429,020	462,571
Saxe-Weimar	1,421	267,524	273,252
Saxe-Meiningen	953	166,364	172,341
Saxe-Altenburg	540	132,849	137,844
Saxe-Coburg-Gotha	816	150,451	160,551
Mecklenburg-Strelitz	997	99,750	99,060
Oldenburg	2,117	785,278	295,742
Anhalt	849	164,496	161,578
Schwarzburg-Sondershausen	318	74,946	64,888
Schwarzburg-Rudolstadt	340	68,030	71,919
Liechtenstein	64	6,994	7,140
Waldeck	464	59,697	59,604
Reuss-Greiz	116	41,944	45,139
Reuss-Schleiz	387	79,826	83,260
Schaumburg-Lippe	221	29,409	30,774
Lippe-Detmold	445	106,014	108,513
Hesse-Homburg	108	21,971	26,017
Lübeck	117	48,424	49,642
Frankfort	43	73,550	87,216
Bremen	166	99,700	99,571
Hamburg	148	211,350	229,941
Total	342,087	43,778,111	44,013,564

The great majority of the inhabitants of the confederation belong to the Teutonic race. Only in two states—Austria and Prussia—are natives of other races, nearly all Slavonians. They number 6,460,000 in the Austrian provinces of the confederation, for the greater part in Bohemia; and 825,000 in Prussia, the whole of them in the county of Posen, a former integral part of the kingdom of Poland.

Of the forty-five million inhabitants of Germany, about twenty-six millions are Roman Catholics, and the rest Protestants, with 470,000 Jews. In Austria, there are only 300,000 Protestants, and in Bavaria but one-third of the inhabitants belong to the Reformed Church. On the other hand there is not a single Roman Catholic in the little principality of Schaumburg-Lippe, and but thirty members of the same confession in the grand duchy of Mecklenburg-Strelitz. As a rule, the Protestants are more numerous in the northern states of the confederation, and the Roman Catholics in those of the south.

Army.—In the act of the congress of Vienna of June, 1815, the contribution of the various states to the army of the Germanic Confederation was fixed at one per cent. of the population, that is, the population possessed at that particular period,

without taking into account a further increase or decrease of numbers. One-seventh of this army was to consist of cavalry; and ten pieces of artillery, with a proportionate number of men, were to be furnished with every 1,000 soldiers. On this basis, the army of the confederation consisted of 301,637 men, rank and file. Various changes were introduced into this military organisation subsequently; and by a vote of the diet of March 10, 1853, the total strength of the army was increased one-sixth per cent. The actual strength of the army of the confederation now consists of 391,684 infantry, 60,768 cavalry, and 50,620 artillery, divided into ten corps d'armée, and a number of troops of reserve—in total, 503,072 men. The distribution is as follows among the thirty-four states:—

	Corps d'Armée	Infantry	Cavalry	Artillery	Total
Austria	I., II., III.	122,073	19,766	18,216	160,057
Prussia	IV., V., VI.	103,174	16,809	13,846	133,769
Bavaria	VII.	44,938	7,415	6,997	59,814
Saxony	IX. 1	15,767	2,847	1,755	20,900
Hanover	X. 1	14,993	2,730	3,846	21,761
Würtemberg	VIII. 3	17,472	2,304	2,876	20,280
Baden	VIII. 7	10,163	2,003	2,418	14,862
Hesse-Cassel	IX. 3	7,455	1,188	630	9,466
Hesse-Darmstadt	VIII. 3	8,071	1,201	968	10,226
Holstein and Lauenburg	X. 1	4,556	750	691	6,580
Luxemburg & Limburg	IX. 1	1,736	887	—	2,700
Brunswick	X. 1	2,765	487	801	4,623
Mecklenburg-Schwerin	X. 2	4,582	767	307	5,857
Nassau	IX. 1	5,422	14	604	6,100
Saxe-Weimar		3,516	—	44	3,668
Saxe-Meiningen		1,880	—	14	2,010
Saxe-Altenburg		1,631	—	17	1,638
Saxe-Coburg-Gotha		1,841	—	19	1,860
Mecklenburg-Strelitz	X. 2	997	—	200	1,197
Oldenburg	X. 2	3,916	460	370	4,746
Anhalt		3,016	—	—	3,066
Schwarzburg-Sondershausen		744	—	7	731
Schwarzburg-Rudolstadt		890	—	—	890
Liechtenstein		91	—	—	91
Waldeck		847	—	—	848
Reuss-Greiz		1,279	—	15	1,241
Reuss-Schleiz		847	—	—	886
Schaumburg-Lippe		1,100	—	19	1,372
Lippe-Detmold		686	—	—	353
Hesse-Homburg	X. 3	860	44	100	673
Lübeck		1,110	—	—	1,110
Frankfort	X. 3	628	101	—	744
Bremen	X. 2	1,745	466	31	2,162
Hamburg					
		391,684	60,768	50,620	503,072

Under the sole command of the diet, and garrisoned by federal troops, are five German fortresses, namely, Mayence, Luxemburg, Landau, Rastadt, and Ulm. The strongest and most important of these fortresses, Mayence, is garrisoned by Austrian and Prussian, in equal moieties; Luxemburg, by Prussians chiefly; Landau, by Bavarian troops; Rastadt, by troops of Baden, Austria and Prussia; and Ulm, by troops of Würtemberg, Bavaria and Austria.

GERONA (an. *Gerunda*), a fortified city of Spain, Catalonia, cap. corteg. of same name; on the declivity, and at the foot of a steep mountain, on the Ter. 50 m. NE. Barcelona. Pop. 12,905 in 1857. The plan of the city is nearly triangular. At the commencement of the Peninsular War, during which it suffered greatly, Gerona was an important fortress; being surrounded with old walls in good repair, and further defended by the citadel of Monjuic, a square fort 720 ft. in length on each side, with bastions, outworks, &c., placed on an eminence about 60 fathoms distant, commanding the city; besides which there were four forts, with redoubts upon the high ground above it. It is still partly surrounded by walls; but Monjuic is in only outwork remaining. It is now chiefly noted for the number of its religious edifices; there are no fewer than 13 churches, besides the cathedral, and 11 convents. The principal buildings are the cathedral and the collegiate church of San Felice. The former, built on the ridge of the mountain, displays a majestic front at the top of three grand terraces ornamented with granite balustrades. The ascent to it is by a superb flight of 81 steps, of a breadth the whole extent of the church. The front is decorated in bad taste with three orders of architecture—Doric, Corinthian, and Composite, and flanked with two hexagon towers. The interior is large and handsome; it has only a nave in the Gothic style. It contains the monuments of Raymond Berenger, count of Barcelona, and his consort. The treasury of this church afforded a rich spoil to the French, on their gaining possession of Gerona. The collegiate church is of Gothic architecture, consisting of a nave and two aisles divided by pillars; connected with it there is a very lofty and ancient tower. In the Capuchin convent there is an Arabian bath of elegant construction. The streets of Gerona are narrow and gloomy, but clean and well paved; and the houses are tolerably well built. There are three squares, one of which is of considerable size; two hospitals, a seminary, college, with a good public library (formerly belonging to the university, founded 1521 by Philip II., and abolished 1715 by Philip V.), a Beguine seminary for poor girls, and several other schools. The place wears, however, a dull and melancholy look. 'The inhabs. have no theatre, no public amusements, no common rendezvous. Every one seems to live alone. One-fourth of the pop. are priests, monks, nuns, and students. They carry on very little trade. The only manufactories are a few looms for weaving coarse woollen and cotton stuffs and stockings, which have been established in the asylum within the last twenty years.' (Mod. Trav. xviii. 57.)

Gerona is the seat of a bishopric, which is richer than that of Barcelona; of an ecclesiastical tribunal; a sub-deleg. of police; and a military governor. It is of great antiquity, and formerly gave the title of prince to the son of the king of Arragon. It has sustained numerous sieges, and been famous for the brave defence it has always made; but especially for that it sustained under Mariano Alvarez, in 1809, for upwards of seven months, against the French.

GERS, a dep. of France, reg. SW., between lat. 43° 17' and 44° 1' N., and long. 0° 18' W., and 1° 11' E.; having N. the dép. Lot-et-Garonne, E. those of Tarn-et-Garonne and Haute Garonne, S. the latter and the Hautes and Basses Pyrénées, and W. Landes. Length, E. to W. 74 m., by about 54 m. in breadth. Area, 629,053 hectares. Pop. 299,931 in 1861. The last ramifications of the Pyrenees cover most of this dép., the slope of which is mostly from S. to N. Rivers numerous; all of them have more or less a N. direction, and are affluents of the Garonne or Adour. There are many ponds and small lakes. Gers, like the neighbouring deps., is subject to violent storms; its

soil is, however, in general fertile. It has 383,600 hect. of arable land, 60,000 hect. meadows, and 87,800 hect. vineyards. Agriculture is very backward, but it has been much improved of late years. Only about 3-4ths of the corn required for home consumption is raised in the dép.; it is chiefly wheat, maize, and oats. Garlic, onions, various other vegetables, hemp, and flax, are plentifully cultivated. Fruit is good. About 800,000 hectol. of wine are produced in ordinary years, but it is mostly of mediocre quality; about a half is consumed at home, and the other half converted into brandy, which ranks next after that of Cognac. There were estimated to be about 800,000 sheep, and 141,350 oxen, in the dep.; considerable attention has been paid to the improvement of the breed of the former, which yield annually about 540,000 kilog. wool. Poultry are plentiful; and, as well as hogs, fattened, or killed and salted, they form an important article of trade. Manufactures unimportant; there a few fabrics of glass and earthenware, leather, starch, linen, woollen, and cotton cloth, and thread. The trade is, however, chiefly in the products of the soil, with which this dep. supplies the neighbouring ones and Spain. Gers is divided into 5 arronds. 29 cantons, and 684 communes. Chief towns, Auch, the cap., Condom, Lombez, and Mirande. The number of large properties is much below the average of the deps. of France.

GHAZIPOOR, or GHAZEEPORE, a dist. of British Hindostan, presid. Bengal, prov. Allahabad; between lat. 25° 10' and 26° 20' N. and long. 83° 10' and 84° 30' E.; having NW. and N. the dist. Azimghur and Goruckpore, NE. Sarun, SE. Shahabad, and W. Benares and Jumpoor. Area 2,850 sq. m. Estimated pop. 1,600,000. The Ganges runs through its S. part; the Goggra bounds it on the N. It is one of the most fertile divisions of Hindostan, and the sugar-cane, corn, and fruit trees are extensively cultivated. It has long been celebrated for the excellence of its rose-water and attar. 'The rose of Ghazeepore are planted in large fields, occupying many hundred acres of the adjacent country. They bloom sparingly, upon a low shrub, which is kept to a dwarfish size by the gardener's knife, and the full-blown flowers are carefully gathered every morning. The first process which the roses undergo is that of distillation. They are put into the alembic with nearly double their weight of water. The rose-water thus obtained is poured into large shallow vessels, which are exposed, uncovered, to the open air during the night. The jars are skimmed occasionally; the essential oil floating on the surface being the attar. It takes 700,000 flowers to produce the weight of a rupee in attar. This small quantity, when pure and unadulterated with sandal oil, sells upon the spot at 100 rupees (10l.); an enormous price, which, it is said, does not yield very large profits. A civilian, having made the experiment, found that the rent of land producing the above-named quantity of attar, and the purchase of utensils alone, came to 5l.; to this sum the hire of labourers remained still to be added, to say nothing of the risk of an unproductive season. The oil produced by the above-mentioned process is not always of the same colour, being sometimes green, sometimes bright amber, and frequently of a reddish hue. When skimmed, the produce is carefully bottled, each vessel being hermetically sealed with wax; and the bottles are then exposed to the strongest heat of the sun during several days. Rose-water, also, when bottled, is exposed to the sun for a fortnight at least.' (Roberts's Scenes of Hindostan, ii. 115, 115.) Rose-water which has been skimmed is

reckoned inferior to that which retains its essential
oil, and is sold at Ghazeepore at a lower price;
though, according to many, there is scarcely, if
any, perceptible difference in the quality. A ser
(a full quart) of the best may be obtained for 8
annas, or about 1s. It enters into almost every part
of the domestic economy of the natives of India,
being used for ablutions, in medicine and cookery,
as presents, &c. The chief towns in this district
are Ghazipoor, the cap., Azimpoor, and Dacri-
ghaut. (Hamilton's E. I. Gaz.; Roberts's Hindostan; Parl. Rep. on E. I. Affairs, &c.)

GHAZIPOOR, a large town or city of Hindostan,
prov. Allahabad, cap. of the above divl., on the N.
bank of the Ganges, 40 m. ENE. Benares, and
100 m. W., Patna; lat. 25° 35′ N., long. 8.19 43′ E.
From the river it has a very striking appearance,
though, like other Indian cities, its noblest buildings turn out, on approach, to be ruins. The
native city is better built and better kept than
many places of more importance. The bazaars
are neat, well supplied, and famous for their
tailors, whose excellent workmanship is celebrated
in the adjacent districts. A very considerable
number of the inhab. are Mussulmans, though the
neighbouring pop. is chiefly Hindoo; their mosques
are numerous and handsome, and the former
grandeur of Ghazipoor is evinced by a superb
palace, built by the Nawab Cossim Ali Khan,
which occupies a considerable extent of ground
overlooking the Ganges. This noble building is
now in a melancholy state of dilapidation, neglected by the government, who have turned it
into a custom house, and have converted many of
its suites of apartments into warehouses, and the
residences of police peons belonging to the guard.
Though thus rendered useful, it is not thought
worthy of repair; its splendid banqueting-hall
and cool verandahs, replete with architectural
beauty, abutting into the river, are deserted, and
left to the swift devastations of the climate. In a
very short period the whole of this magnificent
fabric will become a heap of ruins. (Roberts's
Scenes, &c. ii. 154.) At the other extremity of
the town are the houses of the civil servants of
the company. These are spacious and well built,
and surrounded by gardens. The military cantonments adjacent are, however, low, ugly bungalows, with sloping roofs of red tile, but derive
some advantage from being intermingled with
trees; very different from the stately but naked
barracks of Dinapoor. On the parade ground, a
little NE. the city, is the mausoleum of the Marquis Cornwallis, who died at Ghazipoor. This
edifice consists of a dome supported upon pillars;
and is entirely constructed of large blocks of Chu-
nar freestone. It cost a lakh of rupees, and fifteen
years were spent upon its erection; but its style
and execution have been found much fault with;
and it is insignificant when compared with the
native sepulchral edifices of Hindostan. The gaol
of Ghazipoor is large, strong, airy, and commodious, and usually crowded with delinquents of all
castes and denominations; this district being noted
for the turbulence of its inhabitants, and their insubordination to the laws. The E. I. Company
have a breeding stud of horses near the city,
Ghazipoor is garrisoned by two or three companies
of a native regiment; it is famous for its salubrity,
and is well supplied with European and native
products. Its environs are planted with fine forest
trees, the haunts of innumerable monkeys and
birds. (Roberts's Scenes in Hindostan, vol. ii.;
Heber; Hamilton's E. I. Gaz, &c.)

GHENT (Germ. Gent, Fr. Gand), a celebrated
city of Belgium, cap. E. Flanders, at the confluence of the Scheldt and Lys, 30 m. NW. Brussels,

30 m. WSW. Antwerp, and 23 m. SE. by E.
Bruges, on the railway from Brussels to Ostend.
Pop. 120,515 in 1846. The city is the seat of a
bishopric, of a court of appeal, a tribunal of first
resort, and a chamber of commerce; is a fortress
of the second class, and the residence of a high
military commandant for the provs. of E. and W.
Flanders. The pop. is not characterised by the
decayed and listless appearance of its neighbour
Bruges. Under Charles V., this city probably
covered more ground than any other in W. Europe,
whence the boast of the emperor, "that he could
put Paris in his glove (gant)." At present the
circ. of its walls is between 7 and 8 m.; but much
of the enclosed space is occupied by fields, gardens, and orchards. The shape of the city is
somewhat triangular; it is entered by seven gates.
The Scheldt and Lys, together with the Lieve and
Moere, after having supplied the ditches surrounding the fortifications, enter the city, and, in conjunction with some artificial canals, divide it into
twenty-six islands, most of which are bordered
by magnificent quays. Of the bridges connecting
these islands, seventy are of considerable size, and
of these forty-two are of stone, and twenty-eight
of wood. In general, the streets are wide, though
a few of the most frequented are so narrow that
two carriages cannot pass each other. There are
numerous fine public edifices; and many of the
private houses are well built and handsome. Their
antique appearance, and the fantastic variety of
the stair-like gable ends, ornamented with scrolls
and carving, arrest the stranger's eye at every turn.
There are thirteen public squares. The principal
is the Marché au Vendredi, or Friday Market, so
called from its weekly linen market held on that
day. In its centre was formerly a column, erected
in 1660 in honour of the emperor Charles V. The
greatest curiosity in the Friday Market is an enormous iron ring, upon which are exposed the pieces
of linen which, having been found defective on
being brought to market, are confiscated by the
authorities, and given to the hospitals. The square
of St. Peter is one of the largest in Belgium; it
serves as a parade ground for the garrison. In the
Place St. Pharailde is an old turreted gateway,
a relic of the castle of the counts of Flanders,
built in 868, and doubtless one of the oldest existing remains in Belgium. This castle was, in
1338-39, the residence of the family of Edward
III., whose son, John of Gaunt or Ghent, duke
of Lancaster, was born in it. Its existing remains
form part of a cotton factory. The corn market
and the Place de Recollets are the other chief
squares.

The finest promenades in Ghent are, one along
the Coupure, a canal cut in 1758, uniting the Lys
with the Bruges canal; and the Kouter, or Place
d'Armes. The boulevards, anciently the ramparts
surrounding the city, and some of the quays, are
also agreeable promenades. Ghent contains many
churches worthy of notice, not only for their architecture, but for the chefs-d'œuvre of the Flemish
school which they contain. The cathedral, or
church of St. Bavon, near the centre of the city,
was originally founded in 941, and the crypt, or
eglise souterraine, of the original building still
exists. The modern edifice was commenced in
1228, and completed early in the sixteenth century. Its style is simple Gothic; it has no very
striking beauty externally, if we except its tower,
remarkable for its elegance and height—about
285 ft. From its summit, which is ascended by
440 steps, the surrounding country may be seen
in clear weather for a distance little short of 40 m.
The interior of this church is of almost unrivalled
magnificence. It is entirely lined with black

marble, with which the pillars of pure white Italian marble form a strong contrast. Adjoining the cathedral is the beffroi, or belfry, a lofty square tower, founded in 1183. Its lower part is used for a prison; its summit is ornamented with a gilt copper dragon, carried off by the Gantois from Bruges in 1415, to which city it had been taken from Constantinople during the first crusade. The church of St. Michael, containing the celebrated 'Crucifixion' by Vandyck (now much injured), the 'Annunciation' by Lens, and the finest organ in Belgium; that of St. Nicholas, the oldest church in Ghent, and those of St. Peter, St. Martin, St. James, St. Sauveur, the Dominicans, and the Augustines, all contain excellent paintings, and are well worth notice. There are in all fifty-five churches, including an English Protestant church and a synagogue. Ghent contains the only large nunnery that survived the dissolution of conventual institutions by the emperor Joseph II. This establishment, called the Grand Beguinage, founded in 1234, is of great extent, forming almost a little town of itself, with streets, squares, and gates, surrounded by a wall and moat. It is inhabited by about 600 nuns. The Beguines are not bound by any vow; they may return into the world whenever they please; but it is said to be their boast that no sister has been known to quit the order after having once entered it. The sisters attend the sick as nurses in the hospitals and elsewhere.

The finest public building in Ghent is the palace of the university, founded by William I., king of Holland, in 1816, and attended by about 350 students. In front it has a fine portico raised upon eight Corinthian columns. It has a noble entrance-hall under a vaulted roof 91 ft. in height, a grand staircase, an amphitheatre capable of accommodating 5,700 persons, in which academic prizes are distributed; a court of classes, or square, surrounded by lecture rooms and cabinets of natural history, comparative anatomy, mineralogy, and natural philosophy. The library, containing 60,000 vols, besides many valuable MSS., the royal college, and the botanic garden, occupy the ancient abbey of Baudeloo and its grounds. Besides these institutions, Ghent has a museum and royal academy of drawing, the expenses of which are defrayed by voluntary contributions; societies of rhetoric, the fine arts and literature, music, botany and horticulture, agriculture and commerce, and various clubs for gymnastic and other pursuits. The museum contains a good many paintings, though none are of first-rate excellence; but there are several private collections of great merit, the principal being that of M. Van Schamp, containing a fine 'Annunciation' by Correggio, several paintings by Rubens, including portraits of himself and family; and others by Rembrandt, Vandyck, Teniers, and other distinguished masters.

The town-hall is a large and, at first sight, an imposing building. It has a double front; one in the Moorish-Gothic, and the other in the classic style. The architecture of the latter is incongruous; its 3 principal stories being ornamented successively with Doric, Ionic, and Corinthian columns. This front has upwards of 70 windows, exclusive of those in the roof. In the interior is the throne-room, in which the treaty called the Pacification of Ghent was signed in 1576. This large and fine apartment is now used for the distribution of prizes given by the town to those who attain excellence in the arts, &c. There are 22 public hospitals. The principal, the Byloke (enclosure), founded 1225, is capable of containing 600 sick persons, and has attached to it asylums for aged individuals of both sexes. There are some hospitals for idiots, 2 lunatic asylums, 2 deaf and dumb institutions, and many other charities. The great prison, remarkable for its size and admirable arrangement, has served as a model for several others in London, Prussia, and the U. States. It was begun under Maria Theresa in 1771, and finished in 1824; it stands on the Coupure. Its form is that of a perfect octagon, in the middle of which is a large court-yard, communicating with the different parts of the prison. Each division has a yard, and in the centre of that appropriated to the women is a basin for washing linen. Each prisoner sleeps alone in a small room, looking into an extensive and well-lighted gallery. These apartments are kept very neatly, and are ventilated when the prisoners go to work. One half of the produce of the prisoners' labour is reserved by the government for the expenses of the establishment, and the remainder is divided into 2 portions, one of which is given to the prisoners for pocket-money, while the other accumulates, and is given to them on leaving the prison. The ruins of the citadel, constructed by Charles V., are still to be seen near the Antwerp gate. The modern citadel, built between 1822 and 1830, the cavalry barracks, Hôtel de l'Octroi, workhouse, meat-market, fish-market, slaughter-house and shambles, theatre, and the celebrated piece of cannon—the largest in Europe—18 ft. long, 10½ in circumference, its bore 2½ ft. in diameter, and weighing 34,000 lbs., are the remaining objects most worthy of remark in Ghent. The climate of the city is healthy and temperate; the average heat of winter is 21°, of summer about 70° Fahr.

Manufactures and Commerce.—At the commencement of the 15th century, Ghent is said to have contained 40,000 weavers; but this, like most statements of the sort, is most probably much exaggerated. At the commencement of the present century, the manufacture of cotton yarn was introduced, and furnished employment for a time to more than 30,000 workmen. In 1804, while united to France, Ghent was ranked by Napoleon as the chief manufacturing town in his dominions after Lyons and Rouen. In 1819, the importation of spinning-jennies and high-pressure steam-engines from England afforded an additional stimulus to the cotton manufacture. The separation of Belgium from the Netherlands gave the first check to this flourishing industry, most of the capital engaged in it being with the Dutch. Thus the manufacture gradually declined, and has continued declining to the present day, entailing immense suffering upon the working classes. Previously to the French Revolution, lace was the staple manufacture of Ghent, great quantities of it being sent to Holland, England, France, Spain, and the colonies. Lace-making has now greatly diminished; but it still ranks, together with the manufacture of silk, linen, and woollen fabrics, amongst the principal branches of industry, after that of cotton. The sugar refineries employ annually from 10,000,000 to 12,000,000 lbs. of the raw material. There are numerous gin-distilleries, soap-manufactories, breweries, tanneries, and salt-works. Sail-cloth, oil-cloth, gold and silver stuffs, masks, gloves, pins, bronze articles, mineral acids, white lead, Prussian blue, and other colours, buttons, cards, paper, tobacco and tobacco-pipes, glue, surgical instruments, machinery, cutlery, articles of various kinds, in wood, stone, ivory, &c., are made in Ghent, and there are some good building docks. The city is admirably situated for commerce, besides being one of the centres of the Belgian railway system, it is connected by a ship-canal with Bruges, and by another, which passes by Bas-Van Ghent, with the Schelde at Termeuse. The latter gives the city all the advantages of a sea-port; vessels drawing 10 ft. water may anload in the basin under its walls. The country in the vicinity produces a great deal of corn, flax, tobacco, and madder; and besides its manufactured produce, Ghent has a large trade in three articles, especially the flax, and a very extensive transit trade.

The origin of Ghent is involved in uncertainty; but it is tolerably well ascertained that it became a place of considerable importance early in our era. In 872-80, the Danes under Hastings, repulsed from England, plundered Ghent, and obtained an immense booty. Ghent belonged successively to the counts of Flanders and the dukes of Burgundy; but the allegiance of its citizens appears to have been little more than nominal, since, whenever the seigneurs attempted to impose an unpopular tax, the great bell sounded the alarm, the citizens flew to arms, and killed or expelled the officers of the sovereign. The city became subsequently the cap. of Austrian Flanders; but having, in 1539, unwarily rebelled against the authority of its sovereign, the emperor Charles V., and even offered to transfer its allegiance to his rival, Francis I., king of France, it brought on itself a punishment, from the effects of which is never fully recovered. In 1678, it was taken by Louis XIV.; in 1706, by Marlborough. In 1708, it was again taken by the French, and was, till 1814, the cap. of the dép. of the Schelde. Ghent has given birth to many distinguished individuals, at the head of whom must be placed the emperor Charles V., born here on the 24th February, 1500; among the others may be mentioned John of Gaunt, son of Edward III.; the popular leader, Jaques Van Artavelde, the "brewer of Ghent," and his son Philip; Hemlinus the critic; and the sculptor Delvaux.

GHILAN, a prov. of Persia, in its NW. part; between lat. 36° 25' and 37° 45' N., and long. 48° 35' and 50° 47' E.; having NW. the Russian distr. of Talish, NW. the Elbourz mountains, separating it from Azerbijan and Irak, SE. Mazanderan, and NE. the Caspian. Length NW. to SE., about 120 m.; area probably about 5,000 sq. m. Pop. estimated at from 400,000 to 600,000. It is one of the most beautiful portions of the Persian empire. Climate mild and healthy, except in certain districts in summer. It is well watered, and abounds with forests of oak, pine, boxwood, &c.; along the Caspian, there are extensive morasses. The soil is very rich, and yields hemp, hops, many kinds of fruit, corn, rice, &c. in great quantities. The vine grows with the greatest luxuriance; but the chief product of the prov. is silk of excellent quality, the culture and manufacture of which employs most of the pop. The only town of any consequence is Reshd, on the Caspian, which has a considerable trade in silk with Astrakhan.

GHIZNI, or GHUZNEE, a fortified town or city of Afghanistan, and formerly the cap. of an empire reaching from the Tigris to the Ganges, and from the Jaxartes to the Persian Gulf: though now containing only about 1,500 houses, exclusive of suburbs without the walls, it is still considered throughout Central Asia as a fortress of the highest importance. It stands on a slight elevation, in a plain nearly 7,000 ft. above the level of the sea, at the foot of a narrow range of hills, 54 m. SW. by S. Cabul, and 156 m. NE. Candahar. From its great height its climate is very cold; for a great part of the year the inhab. seldom quit their houses, and the snow has been known to lie deep on the ground long after the vernal equinox. It is surrounded by stone walls, flanked with numerous

towers, and entered by three gates, outside which it is encompassed by a fausse-braye, and wet ditch. On the W. side the walls are elevated to a height of 240 ft. above the level of the plain; and the rock on which they are built might be warped so as to render it thoroughly inaccessible on that side. (Vigne.) The Ghizni river, a pretty large stream, runs along its W. face; and previously to its capture by the British, a new outwork had been constructed commanding its bed. Ghizni has three bazaars, of no great breadth, with high houses on each side; a covered chowrana, and several dark and narrow streets. A citadel, enclosing a palace, is the only edifice worth notice. About ½ m. N. of the city stands a minaret, and about 600 yards farther, another of the same kind, erected by Sultan Mahmoud of Ghizni. Both are of brickwork, elegantly ornamented, and covered in many parts with Cufic inscriptions. Adjoining them is the site of Old Ghizni, a city which, in the 10th century, was, according to Ferishta, adorned beyond any other in the E. The adjacent plain is covered with ruins. About 3 m. from the modern town, in the midst of a village, is the tomb of Mahmoud, a specimen but not a magnificent building, covered with a cupola. But of all the antiquities of Ghizni, the most useful is an embankment across a stream which was built by Mahmoud, and which, though damaged by the insane fury of the Ghuree kings, still supplies water to the fields and gardens round the town. (Elphinstone, Conolul, i. 141, 142.)

The empire of which Ghizni was the cap. was founded by Sebuctagini in 975, and lasted under 13 successive sovereigns till 1171, when the city was conquered by Mahomed Ghoree, and burned. Recently it has acquired some celebrity from having been taken by storm by the British. 23rd July, 1839, after a siege of less than 48 hours; the town being garrisoned by about 3,500 Afghuns, under the command of a son of Dost Mahomed Khan. Our loss on that occasion amounted to 17 men killed, and 182 wounded. Of the enemy, about 600 were killed, many wounded, and 1,600 taken prisoners, including the governor and his staff.

GIANTS' CAUSEWAY, a basaltic promontory of Ireland, N. coast of the co. Antrim, between Bengore Head on the E., and the embouchure of the Bush river on the W. This extraordinary promontory consists of a vast mole or quay, formed of polygonal basaltic columns, projecting from the base of a steep promontory to a great distance into the sea. It is divided into three distinct portions: the first, which is seen at low water, is about 1,000 ft. in length, and the others not quite so much. The pillars are from 40 to 55 ft. in length, and have from three to eight sides; but those having six sides are by far the most common. The surface formed by the summits of the pillars is so smooth, and the joints so close, that the blade of a knife can hardly be introduced into them. The pillars are divided into segments, admirably fitted to each other, varying from 6 in. to a foot in thickness. At Fair Head and Bengore Head, in the immediate vicinity, the columns are higher; but the angles are not so sharp, and they are altogether of a coarser texture than those of the Giant's Causeway. The same sort of basaltic columns, though of a less perfect form, extend along the coast for several miles, and, being sometimes detached from the shore, have, at a distance, the most grotesque appearance. Rathlin Island contains similar columns, and they extend a good way inland.

GIAVENO, a town of Northern Italy, prov. Susa, cap. mand. on the Sangone, 16 m. W. by S. Turin. Pop. 9,951 in 1861. The town is encircled by an old wall, and has manufactures of silk and linen, with tanneries and iron forges, some transit trade, and a market for linens, which is well attended.

GIBRALTAR, a town and very strong fortress belonging to Great Britain, in the S. part of Spain, adjoining the narrowest part of the strait joining the Atlantic and Mediterranean, to which it gives name; 61 m. SE. Cadiz, 98 m. S. by E. Seville, and 312 m. SSW. Madrid; lat. 36° 6′ 30″ N., long. 5° 21′ 12″ W. Pop., exclusive of the military, 12,182 in 1831, and 15,462 in 1861. Area, 1½ sq. m. The fortress stands on the W. side of a mountainous promontory or rock (the *Mons Calpe* of the ancients), projecting into the sea S. about 3 m., being from ¼ to ⅜ m. in breadth. The S. extremity of the rock, 11¼ m. N. Ceuta, in Africa, is called Europa Point. Its N. side, fronting the low narrow isthmus which connects it with the mainland, is perpendicular, and wholly inaccessible; the E. and S. sides are steep and rugged, and extremely difficult of access, so as to render any attack upon them, even if they were not fortified, next to impossible; so that it is only on the W. side, fronting the bay, where the rock declines to the sea, and the town is built, that it can be attacked with the faintest prospect of success. Here, however, the strength of the fortifications is such that the fortress seems impregnable, even though attacked by an enemy having the command of the sea. The town, which lies on a bed of red sand, at the foot of the rock, on its NW. side, has a principal street, nearly a mile long, well built, paved and lighted; and of late years many of the narrow streets have been widened, the alleys removed, and the general ventilation improved. Still, however, the houses are constructed for the latitude of England, not of Africa; for, instead of patios, fountains, and open galleries, admitting a free circulation of air, closed doors, narrow passages, wooden floors, small rooms, and air-excluding windows, keep out the fresh, and keep in the foul air.' (Inglis, ii. 121.) These circumstances seem, in part at least, to account for the contagious fevers by which the town is sometimes scourged. The principal buildings are the governor's house and garden, the admiralty, the naval hospital, the victualling-office, and the barracks. There is an excellent public library, founded in 1793, and a small theatre. A steam corn-mill has been erected. The Protestant church, situated on the Line-wall, will contain 1,018 persons, and the governor's chapel at the convent, 500 more; the Rom. Cath. church, when full, contains about 1,500. The Wesleyan Methodists and other dissenters have places of worship, and there is a Jews' synagogue. The fortifications are of extraordinary extent and strength. The principal batteries are all casemated, and traverses are constructed to prevent the mischief that might come from the explosion of shells. Vast galleries have been excavated in the solid rock, and mounted with heavy cannon; and communications have been established between the different batteries by passages cut in the rock, to protect the troops from the enemy's fire. In fact, the whole rock is lined with the most formidable batteries, from the waters to the summit, and from the Land-gate to Europa Point; so that, if properly victualled and garrisoned, Gibraltar may be said to be impregnable.

The bay of Gibraltar, formed by the headland of Cabrita and Europa Point, 4 m. distant from each other, is spacious and well adapted for shipping, being protected from all the more dangerous winds: the extreme depth within the bay is 110 fathoms. To increase the security of the harbour, two moles have been constructed which respectively extend 1,100 and 700 ft. into the bay. The Spanish town and port of Algeciras is on the W. side of the bay. As a commercial station, Gibraltar is of considerable consequence. Being made a free port in 1704, subject to no duties and restrictions, it is a convenient entrepôt for the English and other foreign goods destined to supply the neighbouring provinces of Spain and Africa. Gibraltar, however, is fallen and falling as a place of commerce; and there is no prospect of its revival. This decay is owing to a variety of causes, partly and principally to the protectionist policy of the Spanish government. The exports from Gibraltar to the United Kingdom were of the value of 152,511*l*. in 1860; 133,834*l*. in 1861; 97,553*l*. in 1862; and 60,136*l*. in 1863. The imports from the United Kingdom are considerably larger; they amounted to 1,244,233*l*. in 1860; 1,169,142*l*. in 1861; 1,144,094*l*. in 1862; and 1,471,453*l*. in 1863. The military expenditure amounted to 450,695*l*. in 1863. The advantage which the possession of Gibraltar confers on Great Britain, though wholly of a political character, is most important. It is, as it were, the key of the Mediterranean; and while its occupation gives the means of effectually annoying enemies in war, it affords equal facilities for the protection of British commerce and shipping.

Gibraltar, the *Calpe* of the Greeks, formed with Abyla on the African coast 'the pillars of Hercules.' Its name was changed to *Gibel-Tarif*, or mountain of Tarif, in the beginning of the 8th century, when Tarif Ebn Zara landed with a large army to conquer Spain, and erected a strong fortress on the mountain side. During the Moorish occupation of Spain it increased in importance, but was at length taken by Ferdinand, king of Castile, in the 14th century. It was soon recaptured, and did not become the appanage of Spain till 1462. Its farther history till its conquest by the English in 1704 is unimportant. During the war of the Spanish succession the English and Dutch fleets, under Sir George Rooke and the Prince of Hesse-Darmstadt, attacked the fortress, which surrendered after some hours' resistance. The Spaniards, during the nine following years, vainly tried to recover it; and in 1713 its possession was secured to the English by the peace of Utrecht. In 1727 the Spaniards blockaded it, for several months, without success. The most memorable, however, of the sieges of Gibraltar is the last, begun in 1779, and terminated in 1783. The batteries on the rock were known to be most formidable; and yet the bold, not to say extravagant, project was entertained of attempting to silence them by the fire of ten enormous floating batteries ingeniously constructed by the Chevalier d'Arçon. A powerful combined French and Spanish fleet and army was collected to co-operate in the attack, which excited an extraordinary interest in all parts of Europe. The grand effort was made on the 13th of Sept. 1782; and the only thing to be wondered at is, that the floating batteries should have so long resisted, as they actually did, the tremendous fire of red-hot shot to which they were exposed. At length, however, two of them took fire, and their terrible explosion terminated the conflict. The garrison, and their gallant commander, Sir Gilbert Elliot, afterwards Lord Heathfield, were not more distinguished by their brave defence than by their generous efforts to rescue their enemies from the flames and the waves. No farther attempt has been, nor is it likely will be, made to deprive us of this fortress.

GIEN, a town of France, dép. Loiret, cap. arrond., on the Loire, and on the high road between Orleans and Nevers, 37 m. SE. the former. Pop. 6,609 in 1861. The town is irregularly built on a hill, on the summit of which is its church, and an ancient castle now serving for the sub-prefecture, the residence of the mayor, and the seat of a tribunal of original jurisdiction. The Loire is here crossed by a handsome stone bridge, and the town has a spacious quay, and a good bath establishment. Here also are manufactures of earthenware, serge, and leather. It has some trade in agricultural produce.

GIESSEN, a town of W. Germany, Hesse-Darmstadt, cap. prov. Upper Hesse; on the Lahn, which is here crossed by a stone bridge, 47 m. N. by E. Darmstadt, and 49 m. E.N.E. Coblentz on the railway from Cassel to Frankfort-on-the-Maine. Pop. 9,310 in 1861. The town was formerly fortified; but its ramparts have been levelled, and their site is now laid out in public walks. It is irregularly built; but has some good edifices, including the castle, now the seat of the provincial government, the university, arsenal, town-hall, and a new church. Giessen is chiefly noted for its educational establishments, which constitute one of its principal resources. Its university, founded in 1607, is now usually attended by between 300 and 400 students; a few years ago the ordinary number was upwards of 500. The town has, besides, a teachers' seminary, several other seminaries of a superior kind, schools of forest economy and midwifery; a lying-in hospital, philological institute, two public libraries, a cabinet of natural history, and a botanic garden. It is the residence of the governor of Upper Hesse, the seat of a superior judicial court for the prov., a council of mines, board of taxation and committee of public instruction. It has a few manufactures of tobacco and woollen goods.

GILOLO, one of the Molucca islands, which see.

GIOVENAZZO (an. Natiolum), a sea-port town of Southern Italy, prov. Bari, cap. cant., on a high rock which advances sufficiently into the Adriatic to afford shelter on its N. side to a considerable number of fishing boats; 9½ m. NW. Bari, and 4½ m. NE. Molfetta. Pop. 8,075 in 1862. The town is surrounded by strong turreted walls, and farther defended by a castle. Its streets are narrow, dark, and dirty, and crossed by frequent deep archways, which render them still more gloomy. Giovenazzo is the seat of an archbishopric, united to that of Terlizzi: it has a cathedral, three other churches, several convents, two hospitals, and an asylum for foundlings. Without the town, an avenue of immense cypress and pine trees, nearly a mile in length, leads to a very large, but dilapidated palace, formerly belonging to the Cellamare family.

GIRGENTI, a town of Sicily, adjacent to the ruins of the ancient Agrigentum or Agrigentum, cap. intend., in the Val di Mazzara, 54 m. SSE. Palermo, and 95 m. W. by N. Syracuse, lat. 37° 19' 25" N., long. 13° 27' E. Pop. 16,412 in 1861. The modern city stands on the slope of one of the highest hills of S. Sicily, called Monte Camico, about 1,200 ft. above the sea, and nearly 4 m. from the port at the mouth of the small river which divides the present city from the ruins. The mode of building Girgenti, with its streets rising in terraces, and the cathedral crowning the whole, gives it an imposing aspect from the sea; but the interior is irregular and dirty; most of the streets are ill-paved, and difficult of access. Besides the cathedral, there are forty-five churches and fifteen convents, a fact which fully explains the extraordinary number of ecclesiastics met with here, and the consequent poverty of the people. The cathedral, a large, heavy building of the 13th century, is in the Norman style, barbarously mixed with a modern imitation of the Greek orders: its chief curiosity is an echo, or porta voce, by which a whisper is conducted from the entrance to the cornice over the high altar (240 ft.). It has a beautiful font of carved stone, and some pictures, one of which is a Madonna by Guido. Bishop Lucchesi, a great benefactor to Girgenti, among other acts of enlightened policy, founded a seminary for the clergy, and a good public library, to which he bequeathed a valuable collection of antique vases, coins, and medals. The country round is delightful, producing corn, wine, and oil in great abundance, with a great variety of fruits, as oranges, lemons, pomegranates, almonds, &c. The port of Girgenti has a mole built by Charles III. in 1756; a lighthouse has been erected on the mole-head, and another on an adjacent cliff, but they are so badly constructed and lighted as to be nearly useless. There are here very extensive curiatori, or magazines, dug in the rock, for the ware-housing of corn, considerable quantities of which are shipped from this port, and which, under an intelligent government, capable of calling forth the productive energies of the country, might be vastly increased; it is, also, a principal port for the shipment of sulphur. In ordinary years about seventy British ships clear out from Girgenti, mostly loaded with brimstone.

The ancient Agrigentum was not only one of the largest and most famous cities of Sicily, but of the ancient world. According to Polybius, it surpassed most other cities in its advantageous situation, its strength, and the beauty and grandeur of its buildings. Its ruins, so interesting to the historical student for the reminiscences they suggest, and to the antiquary and artist for their instructive lessons on ancient architecture, stand between the Rupe Athenea, a high rock E. of Girgenti, and the two branches of the river anciently called Agragas, in the midst of orchards, gardens, and groves of the most luxuriant foliage. The S. wall stood on a rock, having adjoining to it a triangular plain, in which may still be seen the tomb of Theron, one of the most illustrious of all the princes, or tyrants, who ruled over Agrigentum, (see Pind. Nic. lib. xi.) It is about 38 ft. high, and 15 ft. square at the base, consisting of a square pilaster on a triple plinth, with a cornice, and fluted Ionic columns in the Attic story; but Mr. Smyth describes it as 'neither magnificent nor elegant, a strange mixture of architectural peculiarities.' At the E. angle of the S. wall, on a bold rock, stands the temple of Juno, or rather the Doric columns that formed a part of it. Their situation on a gently swelling eminence, and surrounded by fruit trees, is highly picture-que. On the W. front a grand flight of steps leads up to the vestibule, which was supported by six fluted Doric columns; at the sides are thirteen others not fluted. Within this temple were preserved some of the most valuable pictures of antiquity, among which was one by Zeuxis of the goddess herself. W. of these ruins is the temple of Concord, which presents the most perfect specimen extant of the earliest epoch of Greek architecture. It is composed of a parallelogram, like the last, six columns broad in front, and thirteen columns at the sides. It is peripteral, that is, has a colonnade all round the building. In each of the side walls of the cella are six arched openings without any appearance of doors, and on each side of the transverse wall of the pronaos a flight of steps leads to the summit of the architrave; the whole temple, with the exception of part of the entablature and roof, is so nearly perfect, as to be a favourable specimen of the beauty of uninterrupted lines in architecture. Its dimensions are:—

	Feet	Inches
Length	171	6
Breadth	44	6
Length of Cella	44	6
Width of do.	24	6
Height of Columns	22	1
Diameter of ditto at base	4	7

The last king of Naples repaired the most damaged parts of this structure, and it is now used as a Christian church. His name and work recorded on the front entablature in large bronze characters, on a glaring white ground, ill agree with the ustiness and chastity of the old building. W. of the temple of Concord, and near the sea-gate, stood the temple of Hercules; but the foundations and one single dilapidated column are all that remain. Cicero (in his fourth oration against Verres) speaks in rapturous terms of a statue of the god, the face of which had been worn by the kissing of devotees. Nearly opposite the ruins of the last temple, are the gigantic remains of the temple of the Olympian Jupiter, now known as *il Tempio di Gigenti*, which, although never completed, was the largest religious edifice of Sicily. Diodorus says that it was 360 ft. long, 60 ft. broad, and 120 ft. high to the commencement of the roof; but it appears probable, from a comparison with other temples and their proportions, that 160 was meant instead of 60, for the breadth, an error that might easily have crept into the early manuscripts (Smyth's Sicily, p. 211); and an examination of the ruin justifies the supposition. Enormous blocks of stone testify its former grandeur. One of the columns measured by Russell in the upper part of its length had flutings, the girth of which was 23 in., a circumstance which goes far to confirm the statement of Diodorus, 'that a man might easily place himself in one of them.' The lower half of a human face, apparently part of a statue that ornamented the pediment, measured a foot from the chin to the middle of the mouth, and 2 ft. across from cheek to cheek, dimensions much larger than those of the Egyptian Memnon in the British Museum. (Smyth, p. 212.) The Sicilian government, about the commencement of the last century, when the mode of the harbour was built, gave orders that the stones of this noble ruin should be removed and used in its construction; and this circumstance, to a greater extent than any other, accounts for the paucity of the present remains. The dimensions of the temple, as computed by Mr. Cockerill, who devoted great labour to ascertain the facts, are as follows:—

	Feet	Inches
Length of Basement	363	8
Its width	182	6
Ditto of Cell	58	8
Ditto of Temple Interior	144	8
Height of Basement	9	6
Ditto of Columns	61	
Pedestal	78	4
Tympanum	72	
Total height		**220 ft.**

Near these ruins are those of the temple of Vulcan, and that of Castor and Pollux; the latter is singular, as being the only one of the Ionic order. The celebrated spring of petroleum, and the fishpond excavated by the Carthaginians prisoners, after the disasters of Himera, B.C. 480, still exist; this pond was more than 40 ft. deep (Diodorus says 120), and about 4,500 ft. in circumference, and is stated to have amply supplied the tables of the rich and luxurious Agrigentines, of whom Plato wittily said, that 'they built as if they were going to live for ever, and ate as if directly about to die.' Ασα οἱ Αγραρρεντιvoι οικοδομεωντις μιν ὡς αει ζησομενοι, οικουντες δι ὡς αει τοθνηξομεννο. (Ælian, lib.

xii. 29.) The pond is now dry and used as a garden, as it was also in the time of Diodorus, who, therefore, must greatly have mistaken its dimensions. Besides the ruins thus described in detail, there are fragments dispersed over the entire site of the city, respecting which conjecture has been busy, but which need no particular mention. It is curious, however, that in the whole space within the city walls there are no ruins that can be presumed to have belonged to places of public entertainment. (Swinburne, ii. 291.) On the whole Agrigentum may be truly said to be surpassed by few cities, either in respect to the beautiful and magnificent Grecian temples and other antique monuments still existing, or the wild and romantic scenery with which it is surrounded. (Russell, p. 101.)

Vast as the public revenue of a city must have been capable of erecting such splendid structures, the wealth of its private citizens appears to have been still greater than could have been anticipated from the national magnificence. The accounts of the riches of Gellias, Antisthenes, and other citizens of Agrigentum, are such as almost to stagger belief. The former, who lived in more than regal splendour, is reputed to have had 300 wine cisterns, excavated in the rock on which the city is built, kept constantly full of the choicest wines; and at the marriage of the daughter of the latter, upwards of 800 carriages were in the nuptial procession. The return of Exænetus, a victor in the chariot-race of the 92nd Olympiad, was celebrated with a splendour of which we can form no adequate idea; in proof of which it is enough to mention, that, among myriads more, no fewer than 800 carriages in the triumphal procession were drawn by white horses. (Mitford's Greece, v. 357, 8vo. ed.)

It is much to be regretted that we have no authentic information as to the means by which such vast wealth was acquired. No doubt, however, it must mainly have been the result of extensive commercial and manufacturing industry; for, notwithstanding its great fertility, the territory belonging to the city was far too limited, and probably, also, too much subdivided, to allow of the accumulation of such gigantic private fortunes. It is clear, too, that a city possessed of such extraordinary riches must have had wisely contrived institutions, and been, on the whole, well governed.

This great city was founded, anno 580 B.C. by a colony from Gela, another Sicilian city, which had itself been founded by a colony of Cretans and Rhodians. (Herod. vii. 153; Thuc. vi. 4.) Most probably its government was at first republican; but it early became subject to tyrants, or princes, of which Phalaris is one of the most ancient, and also the most celebrated. The accounts of him are, however, too much mixed up with fable to be relied on. After his death the republican form of government appears to have been restored, and maintained for a considerable period, till Theron, an able and politic citizen, attained to the supreme direction of affairs. This prince, having carried off the prize in the chariot race at the Olympic games, has been the theme of the glowing eulogy of Pindar:—

'Theron, hospitable, just, and great,
Famed Agrigentum's honour'd king,
The prop and bulwark of her tottering state!'
West's Pindar, Ode ii.

And he obtained and deserved the respect and esteem of the nation by his justice and moderation, and his success in defeating, with the aid of his son-in-law Gelon, the Carthaginians in a great battle. The construction of the piscina, and of other great works at Agrigentum, has, as already stated, been ascribed to the captives taken on this occasion.

After the death of Theron, who was succeeded by his son Thrasydæus, a foolish and licentious prince, the Agrigentines once more asserted their independence, and established a republican government. During the invasion of Sicily by the Athenians, Agrigentum remained neuter, but does history again mention it till B.C. 406, when, if we take Diodorus's account, it seems to have been most flourishing, the population being 800,000; but this, most probably, is much beyond the mark. At this time it was attacked and blockaded by 120,000 Carthaginians, headed by Hamilcar, who desired to separate Agrigentum from the cause of Syracuse. After eight months' siege the inhabitants were forced by hunger to evacuate the place during the night, and made for Gela, which they reached in safety. Hamilcar and his troops made Agrigentum their winter quarters, and in the following spring, everything valuable was either taken to Carthage or sold. Timoleon, according to Plutarch (rather a doubtful authority in these matters), rebuilt the city B.C. 340, and, about 30 years after, the Agrigentines attempted to regain their ancient power in Sicily, but were defeated by the Syracusans. Its history during the Punic wars is very imperfectly ascertained. In the first, it was the ally of Carthage; and during the struggle which made Sicily the seat of war, it was alternately in the hands of the Romans and Carthaginians. Its later history must be learnt by a perusal of Cicero's orations against Verres, particularly the fourth of those eloquent invectives. Little more is known of the history of Agrigentum.

GIRONDE, a marit. dép. of France, and the largest in the kingdom, in the SW. part of which it is situated; between lat. 44° 12' and 45° 35' N., and long. 0° 18' and 1° 16' W.; having N. the estuary of the Gironde, and the dép. Charente Inférieure; E. Dordogne, and Lot-et-Garonne; S. Landes; and W. the Atlantic (Bay of Biscay). Length, N. to S. about 100 m.; average breadth, between 80 and 60 m. Area, 974,032 hectares; pop. 607,120 in 1861. There are a few hills in the E.; but the surface generally is level, and all its W. portion is a vast sandy flat, termed the 'Landes,' bounded towards the sea in its whole extent, by a range of sandy downs or dunes, adjacent to which extends a line of extensive lagoons and marshes. The coast has generally a remarkably straight outline, but near the S. extremity of the dép. it presents a considerable inlet, the Bassin d'Arcachon, which communicates with some of the lagoons before-mentioned, and contains numerous islands. The port of La Teste de Buch is situated on its S. side. Chief rivers, Garonne, Dordogne, and the Isle and Dronne, affluents of the latter. The river or estuary of the Gironde, whence the dép. derives its name, is formed by the union of the Garonne and Dordogne, near Bourg. It has a NNW. direction to its embouchure in the ocean, 45 m. distant. Its breadth varies from 2 to 6 m.; at its mouth, however, it is only 3 m. wide. It is navigable throughout, though at some points is led to encumbered with sandbanks. It is stated that 826,110 hectares, or about 1-3rd of the dép. consists of heaths and wastes; 228,555 hect. of arable lands, 138,823 hect. of vineyards, and 166,709 hect. of woods. Only about half the corn necessary for home consumption is grown; it is chiefly wheat. The culture of the vine is by far the most important branch of industry carried on in this dép. The annual produce of the wines of Gironde, the red growths of which are known in Britain by the

general name of claret, amounts to about 2,500,000 hectolitres, or about 55,000,000 imp. gallons. The vineyards are the property of about 12,000 families, and the expenses of their cultivation are estimated to amount to 45 or 46 millions of francs a year. The best growths are from the confines of the 'Landes,' behind Bordeaux; the secondary growths are chiefly the produce of the country between the Garonne and Dordogne, and the palus, a district of a strong and rich soil bordering the banks of those rivers.

The first growths of the red wines are denominated Lafitte, Latour, Château-Margaux, and Haut-Brion. The first three are the produce of the district of Haut Médoc, NW. of Bordeaux, and the last of the district called the Graves. These wines are all of the highest excellence; their produce is very limited, and in favourable years sells at from 3,000 to 3,500 fr. the tun, which contains 210 imp. gallons; but when they have been kept in the cellar for six years the price is doubled, so that even in Bordeaux a bottle of the best wine cannot be had for less than 6 fr. The Lafitte is the most choice and delicate, and is characterised by its silky softness on the palate, and its charming perfume, which partakes of the nature of the violet and the raspberry. The Latour has a fuller body, and, at the same time, a considerable aroma, but wants the softness of the Lafitte. The Château-Margaux, on the other hand, is lighter, and possesses all the delicate qualities of the Lafitte, except that it has not quite so high a flavour. The Haut-Brion, again, has more spirit and body than any of the preceding, but is rough when new, and requires to be kept 6 or 7 years in the wood; while the others become fit for bottling in much less time.

Among the secondary red wines those of Rozan, Gorce, Leoville and Larose, Brane-Mouton, Pichon-Longueville, and Calon are reckoned the best. The third-rate wines comprise those called Paveillac, Margaux, St. Julien, St. Estèphe, and St. Emilion. It is but seldom that any of these growths are exported in a state of purity. The taste of the English, for example, has been so much modified by the long-continued use of port, that the lighter wines of the Gironde would seem to want body. Hence it is usual for the merchants of Bordeaux to mix and prepare wines according to the markets to which they are to be sent. Thus the strong rough growths of the Palus and other districts are frequently bought up for the purpose of strengthening the ordinary wines of Médoc, and there is even a particular manufacture, called travail à l'Anglaise, which consists in adding to each hogshead of Bordeaux wine three or four gallons of Alicant or Benicarlo, half a gallon stum wine, a bottle of alcohol, and sometimes a small quantity of Hermitage. This mixture undergoes a slight degree of fermentation; and when the whole is sufficiently fretted in, it is exported under the name of claret. This mixture chiefly consists of secondary wines, the first-rate growths falling far short of the demand for them. (Henderson on Wines, p. 184; Jullien, Topographie des Vignobles, p. 203.) But even the first-class wines are frequently intermixed with the best secondary growths; and it is customary to employ the wines of a superior to mix with and bring up those of an inferior vintage.

The white wines of the Gironde are of two kinds; those called Graves, which have a dry, flinty taste, and an aroma somewhat resembling cloves; of these, the principal are, Sauterne, Barsac, Preignac, and Langon. These are said by Jullien to be très moelleux, ou, pour mieux dire, semi-liquoreux, et assez spiritueux. The white wines of the Gironde have for several years past been advancing in estimation and value; and may be said, speaking generally, to come to us in a less adulterated state than the red wines.

About half the wines of the Gironde are sent to other parts of France; one-fifth part is consumed in the dép.; one-fifth is exported, the direct growths to England, but the larger quantity to the N. of Europe and Holland; and about one-fifth is supposed to be converted into brandy.

Some excellent fruit and good hemp are grown in this dép. The forests furnish a great deal of timber for deals and masts, together with resin, pitch, and turpentine. The number of considerable properties is about the average of the deps. There are no mines, but several furnaces and forges of considerable size, for the reduction and manufacture of metallic products. There are some stone quarries; a great deal of good turf is found, besides sand and clay suitable for the manufacture of earthenware; salt is obtained in the marshes of Medoc. The total annual value of the mineral products is officially estimated at about 4,500,000 fr. Manufactures various; they include woollen and cotton fabrics, cordage, iron, steel, copper, gold, and silver articles, glass, pottery ware, and liqueurs. Sugar refiners and dyeing establishments are numerous; and many merchant ships are built at Bordeaux and elsewhere. (For farther details respecting the trade, which is extensive, see BORDEAUX.) The dep. is divided into six arrond., 48 cantons, and 580 communes. Chief towns, Bordeaux, the cap., Libourne, and Bazas. This dep. contains several fine Roman antiquities. It was ravaged by the Saracens in the 8th, and the Normans in the 9th century; it belonged to the English from the time of Henry II.'s marriage with Eleanor of Guienne, till it was annexed to the French crown by Charles VII.

GIRVAN, a sea-port, market town, and par. of Scotland, co. Ayr, in the S. bank of the river of the same name, near its influx into the sea, 17½ m. S. Ayr, with which it is connected by railway, and 55 m. N. Stranraer. Pop. 6,921 in 1861. The town commands a beautiful view of the sea, the N. coast of Ireland, the rock of Ailsa, the Mull of Cantyre, and the various islands lying in the Frith of Clyde. Though large, it is meanlooking, consisting mostly of houses of one story. The inhab. are mainly employed in weaving cotton for the Paisley and Glasgow manufacturers. The people are generally poor and ill lodged, so many as two or three families being, in some instances, crowded together in one end of a house, while the other is filled with the looms on which they work. No fewer than two-thirds of the inhab. are Irish or of Irish extraction, attracted by the facility of learning the business of hand-loom weaving, and the miserable remuneration which it affords. It is not uncommon, while the father is working on the loom, for the mother and children to set out as beggars. While weaving is the staple business, both salmon and white herring affords employment to not a few, and is being prosecuted with great energy. The harbour has been improved by the erection of a new quay; and both grain and coal are exported to a considerable extent. There are five schools in the par., exclusive of the parochial school, in which ten boys and ten girls are educated gratuitously, on an endowment left by Mrs. Crawford of Ardmillan. There are also two subscription and two circulating libraries, and twelve friendly societies. Girvan was erected into a bar. of barony in 1668; but, owing to its diminutive size, the charter lay dormant till 1785.

GIULIANO (SAN), a town of Sicily, Val-di-Trapani, occupying the site of the ancient Eryx, on the summit of the mountain of the same name, 5 m. NE. by E. Trapani, and 40 m. W. by S. Palermo. Pop. 11,679 in 1861. The town has 9 convents, 15 churches, a hospital, and a monte di pietà. From its elevated situation it commands a fine prospect, and has a pure atmosphere; the inhab. enjoy excellent health, the women being remarkable for their beauty and clearness of complexion; a circumstance which rendered it an appropriate situation for the temple of Venus, which existed here in antiquity. Mons Eryx, on which the temple was built, rises 2,175 ft. above the level of the sea, and was said by Polybius to be the largest mountain in Sicily. Etna excepted; and he adds that the temple far excelled all the other temples in the island, in splendour, wealth, and magnificence. (lib. i. § 55.) The accounts of the origin of this famous temple are obscure and contradictory. According to Virgil it was founded by Æneas (Æneid. lib. v. lin. 760); and at all events it was extremely ancient, as Dædalus is said to have built the Cyclopean walls that surround part of the mountain, and to have enriched its treasury with some extraordinary works of art. The votaries of the goddess, thence frequently called Venus Erycina, including persons of the highest distinction, resorted thither in crowds, not only from all parts of Sicily, but also from Italy and Greece. It was, in fact, one of the most celebrated seats of superstition, pleasure, and dissipation in the ancient world. According to Diodorus Siculus, 17 cities contributed to the support of the temple. The priestesses of the goddess were slaves, but some of them became rich enough to purchase their freedom. The temple was plundered by Hamilcar, a Carthaginian general, who, being afterwards taken by the Syracusans, explated his sacrilege by the most cruel torments. But this seat of superstition and debauchery having lost its attractions, was in Strabo's time nearly deserted. It was, in some measure, restored by Tiberius; but it never recovered its former splendour, and was in no very lengthened period wholly abandoned. (See the article on this temple in the learned Mémoire sur Vénus, by Larcher, pp. 166-191, and the authorities referred to in it.)

"Eryx is at present an abrupt and sterile mountain, with but few vestiges of its former magnificence: those still existing are principally a few granite pillars, and some remains of a Cyclopean wall; there is also a kind of cistern, now dry and filled with weeds and brambles, in the castle court, called the well of Venus; and coins, vases, amphora and patera, are frequently found, as are also many leaden bullets for slings inscribed with imprecations." (Smyth, Sicily, p. 242.) Wild pigeons still resort to the mountain in great numbers, as in ancient times.

GIURGEVO, a town of Wallachia, on the N. bank of the Danube, opposite Rustchuk, and 38 m. SSW. Bucharest. Estime. pop. 10,000. It is a miserable place, composed of dirty, narrow streets, and houses built of mud, with here and there one a little more pretending in its appearance, commanded by a wooden verandah. It was formerly fortified, but its ramparts were levelled by the Russians in 1829. The coffee-houses are numerous, and apparently afford more comfort than the private residences. Giurgevo carries on a considerable trade with some of the Austrian towns; and a great part of the commerce of Bucharest, of which it may be regarded as the port, flows through it.

GIVET, a town of France, dép. Ardennes, cap.

rant, on both sides the Meuse, close to the Belgian frontier, 25 m. NNE. Mezières on the railway from Mezières to Namur. Pop. 6,404 in 1861. The two divisions of the town are connected by a fine stone bridge of 5 arches; and both are fortified. The town is in general well-built, especially the grand square. Among the public buildings are commodious barracks, in which English prisoners were detained during the last war; a military hospital, and a public library with 5,000 vols. Givet has a tolerable port, a brisk trade, and manufactures of acetate of lead, sealing-wax, glue, earthenware, pipes, and leather. In its canton is the gorge, 4 m. in length, through which the Meuse flows; the overhanging rocks on either side of which are called the *Dames de Meuse*.

GLADOVA (Turk. *Fet-Islam*), a town of Servia on the Danube, immediately below the 'Iron Gate,' and at present one of the chief stations of the Danube Steam Navigation Company. It is destitute of any house capable of affording accommodation to travellers, being a mere collection of wretched huts. Its inhabs. find constant employment in the conveyance of merchandise, &c., by land to and from Orsova, the station above the rapids of the Danube, a journey of nearly 10 m., which most passengers perform by land. About 2½ m. below Gladova are the remains of Trajan's bridge. (See DANUBE.)

GLAMORGAN, a co. of S. Wales, being the most southerly in the principality, having S. the Bristol Channel, E. the co. of Monmouth, from which it is separated by the Remny, N. Brecknock, and W. Caermarthen. It is about 52 m. in its greatest length W. to E., and 26 m. in its greatest breadth. Area 850 sq. m., or 547,194 acres, of which nearly 100,000 acres are supposed to be waste lands. On the N. and NE. the county is mountainous; but in S. portion, consisting of the vale of, more properly speaking, great level of Glamorgan, stretching from the mountains to the sea, is by far the most fertile part of S. Wales. The soil of this level is a reddish clay resting on a limestone bottom, and is most excellently adapted for the growth of wheat. But the agricultural capacities of this co. are surpassed by its all but inexhaustible mineral treasures. In fact, the whole of this co. N. of Llantrisent, is comprised within, and forms the largest portion of, the coal basin of S. Wales—the greatest depôt of coal in the empire, and capable, it is believed, of alone supplying its present rate of consumption for above 2,000 years. This co. has also less than usual supplies of lime and ironstone, and is the seat of the Merthyr-Tydvil, Aberdare, Hirwain, and numerous other iron works, the greatest establishments of their kind in the empire. The energies of the inhabs. being thus principally directed to mining pursuits, agriculture is not in a very advanced state. A great deal of excellent wheat is, however, produced; barley, oats, and potatoes being the other principal crops. Lime is the principal manure. Estates and farms vary very much in size. The latter are most commonly held under leases of 7 or 14 years. The hills afford good pasture for sheep and cattle, and great quantities of cheese and butter are made. The Glamorgan cattle are the largest of the Welsh breed. Recently they have been crossed with the Ayrshire breed; and the mixed breed thence resulting are found to yield a greater quantity of milk than the old Glamorgan; at the same time that they are hardier, and can be kept at a good deal less expense. The cottages in many cases are said to be amongst the best in the empire. The custom of white-washing houses, office-houses, walls, &c., is universal; and it is alleged that, occasionally, even hedges have

been subjected to this favourite operation. Principal rivers, Tawe, Neath, and Taffe. There are several canals and railways in the co. by which an easy communication is kept up between the mining districts in the N. and the ports of Swansea, Neath, Cardiff, &c. Near Swansea and Neath are the greatest smelting works in the empire. Cardiff is the principal port in the principality for the shipment of coal and iron. Principal towns, Merthyr-Tydvil, Cardiff, Swansea, and Neath. This co. returns 5 mems. to the H. of C., viz. 2 for the co., 1 for Merthyr-Tydvil, and 1 each for Cardiff and Swansea and their contributory bors. Registered electors for the co. 6,745 in 1862. Pop. 317,752 in 1861, inhabiting 59,524 houses. Gross annual value of real property assessed to income-tax 563,735£. in 1857, and 655,351£. in 1862. Glamorgan is divided into 10 hundreds, and 127 parishes.

GLARUS, or GLARIS, a canton of Switzerland, in the E. part of which it is situated, and ranking seventh in the confederation; between lat. 46° 47' and 47° 10' N., and long. 8° 51' and 9° 15' E.; having N. and E. the cant. St. Gall, SE. and S. the Grisons, and W. Uri and Schwytz. Length, N. to S., 27 m. Area, 279 sq. m.; pop. 33,159 in 1860. This canton is one of the most singular in Switzerland, not only in its geographical position and natural features, but also in its political constitution, and some of its laws and usages. Its central portion consists of the long narrow valley of Linth, into which there is but one road; and of two small lateral valleys, to neither of which there is any access but by the principal valley. The rest of the surface is mostly covered with mountains belonging to different ranges, which, in general, rise higher than those in the neighbouring cantons. The Dœdiberg, at its S. extremity, the loftiest summit in E. Switzerland, is 11,765 ft. in height; the Glarnish is 9,600 ft.; and the Wiggis, 7,444 ft. high. The Linth, its principal river, rises beneath the Dœdi, and runs in a N. direction through the whole canton, into the lake Wallenstadt, which forms a part of its N. boundary. Besides this and the lake of the Klœnthal, there are many other small lakes in the mountains. Glaciers are also numerous, and the scenery generally is very striking. Not 1-10th part of the land is arable; orchards of plum, pear, cherry, apricot, almond, and other trees are sufficiently plentiful, and in some parts the vine is cultivated; but very little grain, or other agricultural produce, is obtained. The pasturages on the mountain sides are fine, and feed during the summer about 10,000 cows and 5,000 sheep. A great many goats are kept. This canton is the peculiar seat of the manufacture of the *Schabzieger*, or green cheese. This article is made of cows' milk, and not of goats', as its name might seem to imply. The peasants, who feed their cows in the mountains, bring down the curd in sacks, each containing about 200 lbs., for which they get about 50s. The cheese owes its peculiar appearance, smell, and flavour to the blue *melilot* (*Trifolium Melilotus cœrulea*). This herb is grown in small enclosures beside most of the cottages; dried, ground to powder, and in that state thrown into the mill along with the curd, in the proportion of 3 lbs. of herb to 100 lbs. of the latter. After being turned for about 24 hours, the mixture is ready to be put into shapes, where it is kept until it dries sufficiently to be ready for use. When sold wholesale, it fetches about 3½d. per lb. This is considered a very lucrative trade, and the richest people in the canton are cheese manufacturers. A good deal of Schabzieger cheese is exported to America. The possessor of twenty or twenty-five cows in Glarus is considered to be in very easy circumstances, and yet his whole property does not amount to more than 160l., the usual price of a cow being 7l. or 8l. at most. But with a single cow, and a little potato land, or with three or four goats, an individual is above poverty. A person possessing property to the amount of 3,000l. is considered very wealthy, and there is said to be not one in the canton worth 8,000l. The woods, which chiefly consist of fir and larch trees, belong for the most part to the communes. They have, however, been ill managed, and timber has become dear. Several mines of copper, iron, and silver exist, but they are not wrought. Fine black and other marbles, slate, quartz, and gypsum, are found, and there are some sulphureous springs. The inhabitants are very active and industrious: they manufacture cotton and linen goods and print muslins, pretty extensively, and have established silk manufactures. They formerly traded in the more precious European woods and marquetry-work; but the demand for these has greatly diminished. The chief exports of Glarus are about 2,000 head of cattle and 200 or 300 horses annually, Schabzieger and other kinds of cheese, butter, honey, dried fruits, manufactured articles, and slates. The principal imports are corn, wines, salt, metals, wool, colonial produce, glass, earthenware, and straw hats, muslins, silks, and Lyonese goods, which the traders sell in the fairs of Italy, Germany, and the N. of Europe. It is estimated that 1,500th part of the pop. are engaged in business out of the canton; some travel for Zurich merchants, and others on their own account; and natives of Glarus are settled in many of the large commercial cities of Europe. The cant. is divided into fifteen communities; chief towns, Glarus, Mollis, Schwanden, and Ennenda; the last has risen up since 1780 to be a place containing 2,000 inhab., the most thrifty in the canton. The constitution is purely democratic. The government is in the hands of the whole body of the male pop. above sixteen years of age, who meet annually on the first Sunday in May, in a general assembly, to appoint their magistracy, and to accede to or reject the laws proposed to them by the executive body. The latter consists of a council of about 60 members, of whom 3 4ths are Protestants, and the remainder Catholics. The two persuasions enjoy the same rights, and alternately elect the presidents of the general assembly and council. Some very singular laws prevail in Glarus. One is, that only a son or daughter can inherit property, unless such have been purchased by the testator. Property otherwise falls to the government, by which it is let out to the poor at the rate of 15 batzen (2s. 1d.) for 36 fl. sq. A large proportion of the land is held in this way, and generally planted with potatoes or blue pansy. This law gives general satisfaction. The laws respecting marriage are curious. Whatever may be the age of persons desirous of marrying, they cannot do so without the consent of their respective parents. This law is, however, partially neutralised by another. If a young woman is marriageable, the person in fault is obliged to marry her; or, in case of a refusal, he is declared incapable of being elected to a seat in the council; his evidence is inadmissible in a court of justice; and, in short, he is deprived of civil rights. Both the Catholic and Protestant clergy are paid by the government; but the strictest economy prevails in all the public departments; the chief magistrate receives but 20l. a year. Taxation is very low; the state expenditure is defrayed by a poll-tax of 4 batzen (about 6d.) upon every cow above sixteen years of age; a property-tax of 2 batzen upon every

1,000 florins, rent of state property, customs, post-office, excise, and fines. The public revenue, in 1832, amounted to 208,837 francs, or 8,353l., and the expenditure to 176,524 francs, or 7,001l. There is no direct poor-law, but something very like one. On Sundays there are what are called voluntary subscriptions for the poor; but if any one known to have the means of giving be observed not to give, he may be summoned before the council, and compelled to contribute. There are one or more schools in every commune, for the ordinary useful branches of education, the masters of which are paid by government about 35l. a year. Parents are obliged to send their children to school; but all instruction is gratuitous. Glarus furnishes 472 men to the army, and 3,615 francs to the treasury of the Swiss confederation. As early as the 5th century, the territory of Glarus belonged principally to the abbey of Seckingen on the Rhine; but it fell in the 13th century into the possession of the house of Austria. In 1351, it was occupied by the troops of the confederated Swiss cantons, and soon afterwards joined the confederacy; its independence being consolidated by the memorable battle of Naefels, in 1388. After the Reformation, it was the seat of continual religious wars; and, in 1799, was the theatre of a contest between the Austrians and Russians and the French. The historian, Tschudi, was a native of this canton.

GLARUS, a town of Switzerland, cap. of the above cant., in the narrow valley of the Linth, between two Alpine mountain ranges, 33 m. SE. Zurich, and 6½ m. S. the Lake of Wallenstadt. Pop. 4,797 in 1833. The town is well built, and cheerful; the houses, many of which are antiquated, are chiefly of stone, and frequently ornamented on the outside with fresco paintings. The par. church, an old Gothic edifice, is used by both Protestants and Catholics. The Linth is here crossed by two bridges. Glarus has a hospital, town-hall, a free school for 100 children, erected by private subscriptions; public library, and reading-room. Most of its inhab. are engaged in commerce, and it has a brisk trade; besides manufactures of printed cotton goods, muslins, woollen cloth, and Schabzieger cheese.

GLASGOW, a city, river-port, and the most populous and important manufacturing and commercial town of Scotland, co. Lanark, on both sides the Clyde, 42 m. W. by S. Edinburgh, and 14 m. ESE. Greenock, on the terminus of the Edinburgh-Glasgow and the Caledonian railway. Pop. 329,964 in 1841, of whom 209,925 females, and but 101,939 males; inhabited houses 33,068. The greatest extent of the city from E. to W. is nearly 4 m., and from S. to N. nearly 3 m. The site on which Glasgow is built is a dead level on the S. of the river, and also for about ½ m. on the N., after which the ground rises with considerable rapidity, till, at the extremity of the town, in this direction, it is 150 ft. above the level of the Clyde.

The town originally stood on the elevated ground, adjoining the cathedral erected in the 6th century (by Kentigern, or St. Mungo, the tutelar saint of the city), on the banks of the ravine intersected by the Molendinar rivulet (Sacittie burn), which formed for centuries its W. boundary. From this point the buildings gradually extended downwards till they occupied the whole of the intervening space N. of the Clyde, and ultimately in every direction, including the large suburb (the Gorbals) S. of the river. Other extensive suburban villages, such as Calton, Anderston, Bridgeton, Camlachie, &c., are now regarded as forming part of the city, being continuously attached to it. The houses both of the city and suburbs are of stone, covered with slate. The principal street, running E. and W., parallel to the river, bearing the several names of Argyle Street, Trongate, and Gallowgate, is above 1½ m. in length; and, though not of uniform width, is everywhere of ample dimensions. It is lined on either side with well-built houses, from three to five stories in height, having handsome shops on a level with the cause-way; and is, in fact, one of the best streets and most crowded thoroughfares in Europe. Parallel to this are many fine streets, as Ingram Street, St. Vincent Street, George Street, &c.; and these are intersected by other streets running N. and S., of which the principal and most ancient is the High Street and Saltmarket. All that part of the city W. of George's Square and NW. from Argyle Street to the canal, is comparatively modern. Here, within the last forty or fifty years, a city, of noble streets, squares, and palaces, has been raised. Blythswood Square, on rising ground N. from the Stockwell, is splendidly built, and may be regarded as the most fashionable part of the town—the Belgrave Square of Glasgow. The other principal squares are St. Andrew's, St. Enoch's, and St. George's. On the extreme W. of the city, on elevated ground, are Woodside Crescent, Woodside Terrace, Claremont Terrace, and other splendid ranges of buildings, commanding an extensive view of the basin of the Clyde and the adjacent country.

But while the newer and more fashionable parts of Glasgow will bear a comparison with the finest quarters of any of the best built cities of Great Britain, it has other quarters which, till lately, did not rank above, if they were not below, the worst parts of the liberties of Dublin, St. Giles's in London, or the wynds leading from the High Street in Edinburgh. The principal district of this sort lies in the centre of the city, between the Trongate on the N., the Saltmarket on the E., the Clyde on the S., and Stockwell Street on the W. It consists of a labyrinth of narrow lanes or wynds, whence numberless entrances lead off to small square courts or 'closes.' These wynds and courts are formed of old, ill-ventilated, and mostly dilapidated houses, varying from two to four stories in height, without water, and let out in stories or flats; one of the latter often serving for the residence of two or three families. Frequently, however, the flats are let out in lodgings, as many as fifteen or twenty individuals having been occasionally found huddled together in a single room. The whole district is occupied by the poorest, most depraved, and worthless part of the pop. Latterly, however, a great deal has been done to introduce cleanliness into these recesses, and to improve their sanatory condition; and, though still susceptible of much improvement, they are now in a comparatively satisfactory state.

In 1817 gas was introduced into the city. The city was served very insufficiently with water by public and private wells till 1806, when the 'Glasgow Water Company' was formed by act of parliament. But the water of this company, and of another formed in 1808, was drawn from the Clyde, and therefore full of impurities; and to furnish a better supply, a gigantic undertaking was accomplished in 1859, by which an abundant quantity of the purest water was brought from Loch Katrine, thirty-six miles distant. These new waterworks, which furnish above twenty million gallons daily, were opened with much ceremony by Queen Victoria, on her visit to Glasgow in 1859.

Glasgow can boast of many magnificent public buildings, of which the cathedral, or high-church,

is entitled to the first notice. The original edifice, built by St. Mungo, having gone to decay, the present structure was begun by John Achaius, bishop of Glasgow, in 1123, in the reign of David I., but was not completed for upwards of three centuries. As the building stands on an elevation (on the W. bank of the Molendinar rivulet), 104 ft. above the level of the Clyde, it is seen at a great distance in almost all directions. It is a large oblong structure, in what is called the early English style, which, notwithstanding the different eras of the building, is well kept up. The greatest length, from E. to W., is 319 ft., the breadth 63 ft., the height of the choir 90 ft., and of the nave 85 ft. A square tower, which rises from the centre of the building to the height of 30 ft. above the roof, is surmounted by an octangular tapering spire, terminating in a ball and vane 225 ft. above the floor of the choir. It has in all 157 windows, many of which are of exquisite workmanship. The crypt, under the choir and chapter-house, is not to be equalled by any in the kingdom. It was formerly used as a church, but since 1798 has been used as a cemetery only. This venerable and magnificent structure, the most perfect by far of the ancient religious edifices still existing in Scotland, narrowly escaped falling a sacrifice at the era of the Reformation to the destructive zeal of the mob; but was fortunately saved by the timely and vigorous interposition of the trades. It has recently been thoroughly repaired and renovated partly and principally at the expense of government, and partly by subscriptions from the corporation, and other public bodies and private individuals. It formerly contained three churches, one of which, as already stated, was in the crypt; but now it contains only one. The bishop's palace, or castle, as it was called, erected in 1430, stood a little NW. from the cathedral, and was enclosed by a strong wall. The ruins were removed, in 1788, to make way for the infirmary, one of the finest buildings in the city.

Most of the churches, both established and dissenting, are fine buildings, particularly St. Enoch's, St. Andrew's, St. David's, and the Tron; St. Andrew's, episcopal chapel; and the R. Catholic chapel, a magnificent Gothic edifice, in West Clyde Street. The University, including the houses for the accommodation of the professors, situated on the E. side of the High Street, is of considerable extent, having a front of 305 ft. to the High Street, and extending 782 ft. from E. to W. The buildings, occupying four quadrangular courts, are generally three stories high, diversified with turrets and appropriate ornaments. In connection with the college and near it, on the SE., is the Hunterian Museum. The building is one of the most perfect specimens of a pure classical structure to be found in the empire. It was erected in 1804, from funds (£8000) left for the purpose by the celebrated Dr. William Hunter, a native of the parish of Kilbride, near Glasgow, for the reception of the various articles he bequeathed to the university. They comprised a library of from 10,000 to 12,000 vols., embracing many rare and splendid editions of the classics and other standard works; a choice, and not easily matched, cabinet of Greek and Roman coins and medals; about 60 capital pictures; and a magnificent assortment of anatomical preparations, shells, minerals, zoological specimens, and other scientific collections. This noble collection is said to have cost Dr. Hunter £100,000, and since it was placed in its present situation it has received many additions. The adjoining ground on the E. of the college, though called the College (Larden, is a park containing several acres, enclosed by a high wall, and laid out in walks for the use of the professors and students. The Macfarlane Observatory stands near its E. end; but a new observatory has recently been erected in the Gartnavel grounds from 2 to 3 m. W. of the city. The Royal Exchange, in Queen Street, is a splendid fabric, built in the florid Corinthian style, and surmounted by a lantern, one of the most conspicuous objects in the city. The colonnade, one of the boldest and most imposing structures of the kind in the kingdom, consists of a double row of fluted Corinthian pillars of great height. The apartment appropriated to a news-room is 100 ft. in length by 40 in breadth, with a richly ornamented arched roof, supported by fluted pillars. The Royal Exchange is placed in the centre of an area, two sides of which are lined with magnificent ranges of buildings; while behind it is the Royal Bank, a Grecian structure, much admired for the simplicity and chasteness of the design. On each side the bank two superb Doric arches afford access to Buchanan Street, one of the principal streets of the city. Amongst the other public buildings are the gaol and court-houses; the new city and county buildings in Wilson Street; the tontine buildings at the E. end of the Trongate, opposite the statue of William III., constructed in 1781, as its name implies, by a company of subscribers, on the principle of survivorship. The news-room on the lower floor is of very large dimensions, and, previously to the erection of the new exchange, was the grand resort of the mercantile body; the upper part is occupied as an hotel. The old lunatic asylum to the N. of the city, a large and massive structure, has been converted within these few years into a workhouse for the city parish; and a new lunatic asylum, on a still larger scale, has been erected, in a conspicuous situation, in the Gartnavel grounds to the W. of the city. The bridewell, merchants' hall, town hospital, trades' hall, assembly rooms, the Andersonian university, high school, National Bank of Scotland, Union Bank, surgeons' hall, barracks, theatre, Hutcheson's hospital, house of refuge, and lyceum deserve notice.

In connection with public buildings may be mentioned the bridges over the Clyde. The finest of them is the new Victoria Bridge, of granite, on 5 arches, the middle one 70 feet span, and of the width of 60 feet; it stands on the site of Bishop's Bridge, constructed 1345, and taken down 1850. Of the other bridges, the handsomest is Glasgow Bridge, built in 1836, on the site of a former bridge, removed for the purpose. It is of Aberdeen granite, 560 ft. in length, of 7 arches, and 60 ft. in width over the parapets. Hutcheson's Bridge, near the Green, was built in 1833, and there is also a suspension bridge below King's Park.

Public Places and Monuments.—An equestrian statue in bronze of William III., the gift of James Macrae (1735), a citizen of Glasgow, and governor of the presidency of Madras, stands in the Trongate; an obelisk in honour of Lord Nelson, in the public green; an equestrian statue of the Duke of Wellington, by Marochetti, in front of the exchange, and a statue of Queen Victoria, by the same sculptor, near the Western Club; a statue of Sir John Moore (a native of Glasgow), in bronze, on a granite pedestal, by Flaxman; a similar statue of James Watt, by Chantrey, both in George Square. In the centre of the same square is a fluted Doric pillar, about 100 ft. in height, in honour of Sir Walter Scott, with a colossal statue of the great minstrel at the top; in the town-hall is a statue of William Pitt, in marble, by Flaxman.

The Necropolis, formed by the Merchant Company, in 1830, in an elevated park (rising suddenly to the height of 300 ft. on the E. side of the Molendinar rivulet, opposite the cathedral) is tastefully laid out. Of the various monuments which it contains, an obelisk erected on the summit of the eminence, in honour of John Knox, surmounted by a statue of the reformer, is the most striking; like the cathedral, it is visible at a great distance in every direction. The Green is the Hyde Park of Glasgow, lies between the Clyde and the Calton and Bridgeton, and contains about 185 acres, appropriated to the recreation of the citizens. Another fashionable resort is Kelvin Grove Park, at the west end, laid out by the late Sir Joseph Paxton, at a cost of £100,000. There is also the Queen's Park, at the south side of the city, enclosing 120 acres.

Churches and Chapels.—Glasgow contains altogether 180 places of worship, including 72 churches of the establishment; 44 free churches; 35 united Presbyterians; 5 episcopal churches; 12 Roman Catholics, and 11 Baptist chapels. The number of dissenters is very great, comprising not only the members of the United Presbyterian Synod, a very important body, and every denomination of Protestant dissenters, but many R. Catholics, with Unitarians, Jews, 'Hervuns,' 'Universalists,' and other singular sects.

Education.—The university was founded by Bishop Turnbull, under a papal bull, dated 1450; and its privileges have been subsequently confirmed and extended by royal charters and parliamentary statutes. The discipline of the university is administered by the court of the rector (or vice-rector), and by assessors nominated by him, who have for many years been the principal and professors. The public affairs of the university are under the management of the senate, which is composed of the rector, dean of faculties, the principal, and all the professors, the latter being 21 in number. The business of the college as a subordinate corporation, is conducted by the principal and 13 professors, called the Faculty, who, with the rector and dean, dispense the college patronage. The rector, who is generally an erudite literary or political character, who seldom resides, or even appears, except at his inauguration, is chosen annually by the matriculated students. The office, which is now one of distinction only, has been filled by Burke, Adam Smith, Francis Jeffrey, Sir Robert Peel, and other distinguished men. There is also a sinecure officer, named chancellor, nominated for life by the senate, who is generally a nobleman of distinction. The chancellor appoints a vice-chancellor, but neither has any rights or privileges either in the discipline of the institution or in the exercise of its patronage. In addition to the 21 professors, there is a lecturer on the structure, functions, and diseases of the eye. Government also instituted, in 1840, a professorship of mechanics and civil engineering, and endowed it with a salary of 250l. a year. The principal presides as chairman at meetings of the senate, and generally over the institution, and is honorary professor of theology, but teaches no class. The crown is patron of the principality, and of 14 professorships, including that newly instituted; the faculty, rector, and dean being patrons of the remaining 8 professorships. The professors derive their incomes partly from the fees paid by the students (which vary from 2 to 5 guineas), and partly from funds belonging to the college. In addition to these sources of income, government annually gives a grant, varying in amount, to augment the income of several of the chairs. It is required by law, that all the professors be members of the established church; the law, however, is not strictly enforced, except in the case of the principal and theological professors. Religious distinctions are of no consequence in the case of students; those only who belong to the national church and whose parents do not live in town, are required to attend public worship in the college chapel. The curriculum, or course of study, is divided into the four faculties of Arts, Divinity, Medicine, and Law; which last is confined to a single professorship. There is only one session in the year, beginning 10th Oct., and terminating 1st May. There are 30 bursaries, the benefits of which are extended to 65 students. Their average annual income is 1,163l. 16s. 4d.; the highest is 50l.; the lowest 4l. 10s. Mr. Snell, of Warwickshire, about a century ago, left a landed estate in that county for the purpose of founding ten exhibitions in Balliol College, Oxford, in favour of students of the episcopal church, who have attended at least two sessions at the University of Glasgow, or one session there and two at some other Scotch university. Among the distinguished persons who have been educated on Snell's foundation, may be mentioned Dr. Douglas, bishop of Salisbury, Adam Smith, and Dr. Matthew Baillie. Each exhibition is of the yearly value of 120l., and lasts for ten years. As in the other Scotch universities, there are no apartments for the residence of the students within the college. The number of students varies from 1,200 to 1,300. The university library, which was founded in the 15th century, contains about 70,000 volumes, and is open to all the students. The botanic garden attached to the chair of botany in the university is in the Great W. Road, about 2 m. W. from the city. It occupies an elevated situation in the vicinity of the new observatory. Some of the most illustrious names in the literature of Scotland have been professors in the University of Glasgow; amongst others may be specified Hutcheson, Adam Smith, Simson, Millar, and Reid.

Anderson's University, or Andersonian Institution, was founded by Dr. John Anderson, professor of Natural Philosophy in the University of Glasgow, who died in 1796, leaving his effects, including his museum and philosophical apparatus, to the institution. It is under the management of a large body of trustees, elected periodically. It possesses a fine building in George Street, embracing suitable class rooms, a large hall, chemical rooms, and a museum. It consists of three distinct apartments:—1. General branches for youth, consisting of mathematics, logic and ethics, natural philosophy, chemistry, French, German, geography, drawing, and painting. 2. A medical school, embracing all the branches for the various colleges of surgeons, and public boards. 3. Mechanics' classes; comprising 50 lectures on mechanics and chemistry in alternate winters, and drawing. Excepting those in the mechanics' classes, the lecturers pay rents for their rooms. The staff of professors numbered 14 in 1862. There is a good library, to which the students have access. The classes for mechanics in this institution were the first established in the empire.

The Glasgow mechanics' institution was founded in 1823, chiefly by some members of the mechanics' class in Anderson's University, who felt dissatisfied with the management. A ticket, price 4s., admits to the classes of natural philosophy and chemistry, on each of which there are courses of lectures. Mathematics and other branches are also taught. A scientific and literary reading-room is attached to the institution. The Athenæum in Ingram Street was established in 1847. Its object is to place within reach of the public the most

recent information on all subjects of general interest, whether commercial, literary, or scientific. It has a good library, and its large hall is supplied with newspapers and other periodical publications.

An immense number of public and private schools and academies, some supported in whole or in part by subscriptions, and others depending entirely on fees—are scattered over all parts of the city. But the means of elementary instruction are, nevertheless, said to be still rather deficient.

The High School of Glasgow was formerly an exclusively classical seminary, with the exception of a writing class, having 5 teachers for Latin and Greek, with 1 for writing; the time devoted to classical literature being from 5 to 6 hours daily. But in 1834 it was resolved to modify the course of instruction in the school, so as to make it more suitable to the wants of a great manufacturing and commercial city. In consequence, the classical department was limited to 2 teachers, and the time to 4 hours; and teachers of English literature, geography, mathematics, modern languages, and drawing were introduced. In 1836 a chemical class was established; and soon afterwards this department was made to embrace natural philosophy and natural history.

A normal school, or a school for instructing teachers in the art of tuition, was founded by the Glasgow Educational Committee in 1836, and was the first seminary of the kind in Scotland. Its directors must, according to its constitution, belong to the national church; but there is no such exclusion in regard to those who are instructed in it. The fee is 3l. 3s. for the course of training, which may extend over a whole year. The Free Church has, also, a well attended and efficient normal school.

Notwithstanding their devotion to commercial pursuits, the merchants of Glasgow have always been distinguished by their attention to and patronage of literature and science. The Literary and Commercial Society was established nearly a century ago, and can exhibit in the list of its members, at different times, the names of Dr. Francis Hutcheson, Adam Smith, Dr. Joseph Black, Mr. Millar, professor of law, and other distinguished individuals. It has, since its origin, been attended by the leading citizens of Glasgow, both literary and commercial. In the range of its discussions, it includes every subject except theology and party politics. The Glasgow Philosophical Society, instituted in 1802, is also an important association. The Maitland Club, instituted in Glasgow in 1828, is similar to the Bannatyne Club of Edinburgh and the Roxburghe Club of London, printing for the use of its members MSS. and rare works illustrative of the early history, commerce, and literature of Scotland. It was originally limited to 50 members, but has been extended to 100. Glasgow has also two statistical societies, a geological society, and several others. In addition to those belonging to the university, to Anderson's institution, the Athenæum, and the mechanics' institute, there are numerous subscription and circulating libraries.

Letterpress printing was not introduced into Glasgow till 1638, upwards of 100 years after it had been established in Edinburgh; nor did it flourish for nearly a century after its introduction. But about the middle of last century the Messrs. Foulis raised the Glasgow press to the highest eminence, and their editions of some of the principal Greek and Latin classics are valuable alike for the beauty of their typography and their accuracy. Glasgow is not, however, a literary mart; and its authors usually make arrangements with Edinburgh or London houses for printing and publishing their works.

Charitable and Reformatory Institutions.—The charitable institutions of the city are too numerous to be minutely specified. They comprise, amongst others, two lying-in hospitals and dispensaries, a row-pox institution, Magdalen asylum, deaf and dumb institution, blind asylum, eye infirmary, lunatic asylum, house of refuge, humane society, &c. In addition to Hutcheson's hospital for the maintenance of decayed burgesses and their widows, and the education of boys, sons of burgesses, there are numerous free schools for the poor, and similar institutions.

A regular police establishment was first organised in Glasgow, by act of parliament, in 1800. This was followed by a separate act for Gorbals, another for Calton, a fourth for Anderston, and a fifth for the river and harbour, all included within the present parliamentary limits. These five separate and independent establishments were not found to work uniformly or satisfactorily—frequent jarrings occurring amongst the various officers—till the magistrates and town council, in 1846, found it necessary to apply to parliament for a bill to abolish these separate jurisdictions, and unite the whole into one municipality, with one set of magistrates and police officers; which bill, after considerable opposition, passed both houses of parliament in the summer of 1846; and after the election of councillors and magistrates, in November of that year, came into operation.

The city is divided, for police purposes, into five districts or divisions. To each division an assistant superintendent is attached, who is responsible to the chief superintendent for the men under his charge, and for the quiet and order of the district. In each district a police court is held every morning for the trial of offenders; one magistrate presiding in the central police court, while another magistrate holds a court in each of three other districts, at different hours. A court is also held for the trial of offences against the river and harbour regulations.

The prison of Glasgow consists of two branches, the north, formerly denominated the city and county bridewell, situated in Duke Street,—and the south, in connection with the court house, fronting the Green, or public park. The Glasgow bridewell is said to be one of the most perfect establishments of the kind in the empire, and, according to the official report of the inspector of prisons, it leaves, in respect of cleanliness and economy, nothing to desire, and is a pattern for Europe.

Harbour and Shipping.—Glasgow owes its present greatness to its advantageous situation on a fine river, in one of the richest coal and mineral districts in the empire. Originally, however, the Clyde was much encumbered by fords and shallows, and for a lengthened period it served rather to excite and disappoint expectation, than to confer any real commercial advantage on the city. In 1662, after several other schemes had failed, the magistrates of Glasgow purchased the ground on which Port Glasgow (16 miles lower down the river, now stands, where they formed a harbour and a graving dock, the first work of its kind in Scotland. For a considerable period the intercourse between Glasgow and its newly acquired port was principally carried on by land carriage; but from 1665 attempts were every now and then made to deepen the river. In 1668 a quay was formed at the Broomielaw; but even as late as 1775 no vessel drawing 6 ft. water could reach Glasgow, except at spring tides. At length, however, a plan proposed in 1769 by Mr. Golburn, engineer of

Chester, for deepening the river to 7 ft, as near this, was adopted. He proceeded to accomplish his task, partly by the employment of dredging machines, and partly by constructing dams and jetties, so as to confine and strengthen the course of the river. These measures have since been continuously and energetically followed up, particularly of late years; and with such success that there are now usually 15 or 16 ft. water in the river at high water neaps. The total cost of this undertaking has been above two millions, and the work of deepening and straightening the river is still vigorously prosecuted. The river, for 7 m. below the city, is very much contracted, and forms nearly a straight line; the sloping banks, formed of whinstone, being constructed in imitation of ashlar. The accommodation for shipping at the Broomielaw, or harbour, is now, also, very greatly extended. It comprises about 55 acres of water. The quays, on both sides the river, are nearly 8 m. in length, and are amply furnished with sheds for goods, cranes, &c., and have the important advantage of being directly connected, by means of the General Terminus line, with the various railways that centre in the city. The revenue of the Clyde trust, in 1862, was 111,492l.

The influence of these improvements on the shipping and trade of Glasgow has been most striking. Dr. Cleland says that, 'less than 50 years ago, a few gabbards, and these only 80 or 90 tons burden, came up to Glasgow; and I recollect the time when, for weeks together, not a vessel of any description was to be found in the port of Glasgow.' (Former and Present State of Glasgow, 30.) Now, however, a greater number of sailing vessels and of steamers belong to Glasgow than to any other Scotch port; and the harbour is constantly crowded with ships from foreign ports, coasting vessels, and steamers. The steam-packets belonging to the Clyde that ply to Liverpool, Dublin, and Belfast, are amongst the finest vessels of their class in the empire. In all there belonged to Glasgow, on the 1st January, 1864, 167 sailing vessels under 50 and 373 sailing vessels above 50 tons—the former of a total burden of 5,861, and the latter of 191,282 tons. Of steamers there were, at the same date, 37 under 50, and 161 above 50 tons—the former of a total burden of 1,156, and the latter of 63,169 tons. In the course of the year 1863, there cleared at the port 184 British vessels, of a total burden of 65,953 tons, and 4 foreign vessels of a burden of 1,634 tons. The foreign shipping, it will be seen, is but small.

The gross amount of customs duties received at various periods exhibits the growth of the commerce of Glasgow in a striking manner. These customs duties amounted to but 3,124l. in 1812; they had risen to 16,147l. in 1822; to 69,741l. in 1832; to 828,101l. in 1842; to 640,203l. in 1850; to 848,621l. in 1860; and to 979,950l. in 1863. In respect to the value of exported home produce, Glasgow ranks as the fourth port of the United Kingdom. The value of such exports was 5,758,909l. in 1862, and 6,770,366l. in 1863.

Canals and Railroads.—In addition to river navigation, the city enjoys the advantage of several canals and railroads. Of the former, the Forth and Clyde, generally called the Great Canal, begun in 1768, but not completed till 1790, is by far the most important. It unites the two seas on the E. and W. of Scotland, extending from Grangemouth on the Frith of Forth, to Bowling Bay on the Clyde, a distance of 35 m., with a collateral cut of 2½ m. to Port Dundas, at the N. extremity of the city of Glasgow. Its medium width at the surface is 56 ft., at the bottom 27, and the depth of water 10 ft.; thus serving for the transit of vessels of upwards of 100 tons burden. The Union Canal from Edinburgh joins this canal 4 m. E. Grangemouth. The other canals are, the Monkland, length 12 m., which connects Glasgow with the coal and iron mines in the pars. of Old and New Monkland; and the Glasgow, Paisley, and Johnstone Canal. The depth of these canals is 6 ft. With regard to railways, Glasgow is amply furnished, and is, indeed, a principal centre of railway communication. The city is, in fact, either the source or is intimately connected with all the principal Scottish lines; at the same time that she is supplied by numerous smaller lines with the products of the adjacent mineral and other districts. Among the principal lines may be specified the Caledonian, uniting Glasgow with Carlisle, and consequently, with Manchester, Liverpool, and London; the lines to Edinburgh, Greenock, and Ayr; the Scottish Central leading to Stirling, Perth, Dundee, and Aberdeen. There are five termini for different railways. The Caledonian has a joint terminus with the Garnkirk railway, on the north side of the city, as well as a terminus on the south side, near that of the Glasgow and South Western line.

Commerce and Manufactures.—Prior to 1300, Glasgow was a fishing village, that part of it lying on the river, now the Briggate, being called the Fisher Row. The business was long on a small scale, and limited to the home market; but, in 1420, the trade of fishing and curing salmon and herrings for the French market was introduced; a traffic that was followed with varied success for almost two centuries. Indeed, this seems to have been the only important branch of business carried on here till 1656, when a person of the name of Fleyming, and partners, proposed to erect a weaving factory, provided the municipal authorities would grant them encouragement. On considering this offer, the town-council gave them a lease of suitable premises, for 17 years, free of rent; an act of liberality that ran great risk of being defeated by the opposition of the freemen weavers, who protested against the grant, on the ground that the factory would be injurious to their interests. In the end the company, to get rid of the opposition, agreed not to employ any weavers other than freemen. This was the origin of weaving factories in Glasgow. But nearly a century elapsed before the manufacture of lawns, cambrics, and such like fabrics, was introduced. These, however, were extensively produced from about 1740, till the business was superseded by the introduction of the cotton manufacture.

The situation of Glasgow as to trade, in 1651, may be accurately learned from the statement of Tucker, who had been commissioned by Cromwell's government to draw up a report on the revenue of customs and excise in Scotland. 'With,' says he, speaking of Glasgow, 'the exception of the collegiators, all the inhabitants are traders; some to Ireland, with small smiddy coals, in open boats, from four to ten tons, from whence they bring hoops, rungs, barrel staves, meal, oats, and butter; some to France, with plaiding, coals, and herring, from which the return is salt, pepper, raisins, and prunes; some to Norway for timber. There hath likewise been some who ventured as far as Barbadoes, but the loss which they sustained by being obliged to come home late in the year, has made them discontinue going thither any more. The mercantile genius of the people is strong, if they were not checked and kept under by the shallowness of their river every day more and more increas-

ing and filling up, so that no vessel of any burden can come up nearer the town than 14 m., where they must unlade, and send up their timber on rafts, and all other commodities by 3 or 4 tons of goods at a time, in small cobbles, or boats of 3, 4, or 5, and none above 6 ton a boat. There is in this place a collector, a cheque, and four waiters. There are 12 vessels belonging to the merchants of this port, viz. 3 of 150 tons each, 1 of 140, 2 of 100, 1 of 50, 3 of 30, 1 of 15, and 1 of 12, none of which come up to the town. Total 957 tons.'

A company for carrying on the whale fishery and making soap was formed in 1674. They employed five ships, and had extensive premises at Greenock for boiling blubber and curing fish. The whale fishery has long been given up, but the soap manufacture has ever since been extensively carried on. This is evinced by the fact, that the quantity of soap made in Glasgow in 1838 amounted to 9,248,140 lbs. of hard, 4,246,922 lbs. of soft, and 593,110 lbs. silicated soap, being about 2-3rds of the whole quantity of soap made during the same year in Scotland. The manufacture of ropes was commenced in 1696; and two years afterwards an act of parliament was obtained in favour of this business, imposing a duty on all ropes imported from the Sound or E. seas; and, in return, the company were to advance a capital of 40,000l. Scots, and to bring in foreigners to the work. The manufacture of ropes and cordage is now also an extensive branch of industry, in which large capitals are invested. The tanning of leather and the brewing business were introduced previously to the Union (1707), and have ever since, particularly the latter, formed important branches of manufacture. Almost the whole of the Scotch ale imported into our colonies is produced at Glasgow.

But it was not till after the Union, in 1707, when the trade to the American and West Indian colonies was, for the first time, opened to the enterprise and activity of the Scotch, that the commercial energies of Glasgow began to be fully developed. Her merchants immediately embarked in the trade to the W. Indies and America, especially in that to Maryland and Virginia; and such was the success that attended their efforts in this new department, that in a few years Glasgow became the grand entrepot through which the farmers general of France principally received their supplies of tobacco. But for a considerable time they carried on their colonial trade in vessels chartered from English ports; and it was not till 1718, that a ship, built in the Clyde, the property of Glasgow merchants, crossed the Atlantic. To such an extent was this branch of commerce carried on, that, for several years prior to 1770, the annual import of tobacco into the Clyde ranged from 35,000 to 45,000 hogsheads. In 1771, the quantity was 49,016 hogsheads; and in 1775, 57,143. The American war put an end to a traffic from which Glasgow had reaped great advantages. But no sooner had this business been cut off than the merchants directed their energies to other channels; and found in the extension of the W. Indian trade, and still more in the introduction of the cotton manufacture, new and far more productive sources of employment and wealth. The wonderful inventions and discoveries of Hargreaves, Arkwright, and Watt, powerfully attracted the attention of the more enterprising and intelligent citizens of Glasgow; and in a few years the cotton manufacture was introduced and established. The manufacture of linens, lawns, cambrics, &c., having been already extensively carried on, the work people had little difficulty in applying themselves to the new business; at the same time that the favourable situation of the city for trade, and its unlimited command of coal and iron ore, gave it every facility for successfully prosecuting the manufacture. Hence is it that for a lengthened period Glasgow has been second only to Manchester in this great department of industry. Her cotton mills are on the largest scale, her machinery is of the most perfect description, and in the fineness of her muslins and other fabrics she is, perhaps, unrivalled. The following table, compiled from official returns (*Miscellaneous Statistics of the United Kingdom*, Part V. 1864), shows the average wages earned by the workmen in these various branches of industry, in the year 1862 :—

Nature of Employment	Per Day
COTTON SPINNING :	
Carding Masters	5s.
Carders (3rd)	1s. 3d.
(1st)	3d. to 5d.
Spinning Masters	4s. to 10s.
Spinners, 1st Class	5s. 6d.
2nd Class	4s. 6d. to 5s.
3rd Class	3s. to 3s. 6d.
Piecers, 1st Class, (Girls chiefly)	1s. 3d.
2nd Class	1s. 3d.
3rd Class	9d. to 1s. 3d.
Self-actor Overlookers	4s. 6d.
Mechanics	4s.
Millwrights	4s.
POWER-LOOM WEAVING :	
Tenters	4s. 3d.
Searchers or Dressers	4s. 6d.
Weavers, 1st Class (Girls)	1s. 3d.
2nd Class (do.)	1s. 6d.
Mechanics	4s.
Millwrights	4s.
HAND-LOOM WEAVING, WARPING, &c. :	
Warpers	2s. 3d.
Winders (by Machine)	1s.
Weavers, Cotton Fabrics	1s. 6d.
Mixed Fabrics	1s. 3d.
BLEACHING :	
Men employed in Firing, Boiling, Wheel-washing, Mangling, &c.	2s. 3d. to 2s. 9d.
Rubbers and Cloth Lappers	2s. 3d. to 3s.
Women	1s. to 1s. 3d.
Boys	6d. to 1s. 3d.
CALICO PRINTING :	
Block Printers, Journeymen	3s. 6d. by 4s. 3d.
Apprentices	1s. 6d. to 3s. 6d.
Tierers	1s. 6d. to 1s. 10d.
Cylinder Printers, Journeymen	4s. 3d. to 8s. 6d.
Apprentices	1s. 3d. to 6s. 6d.
Flat Press Printers	3s. to 4s.
Lead Plate Dischargers	2s. 6d. by 3s. 6d.
Hand Engravers, Journeymen	3s. 6d. to 6s. 6d.
Apprentices	3d. to 2s. 3d.
Machine Engravers	4s. 3d. to 6s. 6d.
Die Cutters	4s. 3d. to 6s. 6d.
Block Cutters, Journeymen	3s. 6d. to 4s. 6d.
Apprentices	6d. to 2s. 6d.
Pattern Designers	5s. to 10s.
Putters on and Sketch Makers	3s. 6d. to 4s. 10d.
Colour Mixers, Journeymen	4s. 2d. to 5s.
Apprentices	10d. to 3s. 6d.
Bleachers	2s. 6d. to 3s. 6d.
Dyers	2s. 6d. to 3s. 2d.
Wheelmen and Washers	2s. to 3s.
Warehouse Women	10d. to 1s. 3d.
Mechanics	3s. 6d. to 4s. 6d.
Joiners	3s. 6d. to 4s. 6d.
Masons	3s. 6d. to 4s. 6d.
Firemen	2s. 6d. to 3s. 6d.
Labourers	1s. 9d. to 2s. 6d.
Boys and Girls	6d. to 10d.

Glasgow is not only a grand centre of the cotton manufacture, but if a circle with a radius of 15 m.

be drawn around Glasgow, it will embrace the whole cotton district of Scotland except a few miles scattered up and down in more distant localities. There are, on the average, 25,000 persons employed in the cotton factories of Glasgow. Glasgow has likewise become the centre of a most extensive iron trade. The production of iron in the neighbourhood of the city exceeds that of the whole of S. Wales. The banks of the river and the suburbs are marked by many large ship-building yards, print and dye works, cotton factories, chemical works, and by about 130 blast furnaces and iron factories; among which the most conspicuous are Napier's ship-yards, Dixon's iron-works, Napier's Vulcan Foundry, and Higginbotham's cotton factory.

The chemical works at St. Rollox, for the manufacture of sulphuric acid, chloride of lime, soda and soap, are considered the most extensive in Europe. They extend over 11 acres of ground, and contain upwards of 100 furnaces, retorts, or fireplaces. Distillation, the manufacture of earthenware, sugar-refining, and many minor branches of industry, are successfully prosecuted. The relative importance of the various manufactures carried on in Glasgow may be judged, to some extent, from the export tables, which show, for the year 1863, the total value of home produce to have been 6,705,000l., in which sum cotton piece goods figured to the amount of 2,536,769l., linsey and small wares 111,892l., linen piece goods 453,607l., and haberdashery and millinery 149,000l. The importance of the cotton manufacture may be further seen from the fact that the exports of Glasgow represent very nearly the whole of Scotland, the exports from the kingdom, in 1863, having been to the amount of 2,658,405l., and three of Glasgow alone 2,536,769l.

Ship-building, except in respect to iron steamboats, can scarcely be said to exist in Glasgow, being chiefly confined to Paisley, Govan, Dumbarton, Bowling Bay, Port Glasgow, and Greenock. But Glasgow and the Clyde generally are more celebrated for the manufacture of steam machinery and the building of iron steamers than perhaps any other place in the empire. They have, also, supplied machinery to some of the largest and finest vessels belonging to foreign powers, as well as to the navy of Great Britain.

Progress of Population and Pauperism.—According to the best attainable information, the pop. of Glasgow, at different periods down to 1861, has been as follows :—

Years	Inhabitants	Years	Inhabitants
1689	4,500	1801	77,385
1610	1,644	1811	100,749
1660	14,678	1821	147,043
1708	12,766	1831	202,426
1740	17,034	1841	282,134
1763	28,300	1851	329,097
1780	42,842	1861	394,864

It will be seen from the preceding table that during the interval between 1801 and 1861, the increase of the pop. of Glasgow has been no less than 385 per cent.—a progress wholly unexampled in any old settled country.

The increase of pop. has, of course, been mainly occasioned by the still more rapid increase of wealth and employment. It has not, however, depended wholly on this; and there can be no doubt that the increase of pop. has in some degree exceeded the increased demand for labour, and that increase has been. This has been principally a consequence of the prodigious influx of labourers from Ireland. There are, probably, but few in-

manners in which the Irish have been improved by the change; but they have had, partly by the effect of their competition in reducing wages, and partly and principally by their habituating the Scotch, though their example, to become contented with a lower standard of comfort, the most pernicious influence over the condition of the Scotch part of the labouring pop. At the same time, too, that Irish labourers have been pouring into the city, the weavers, who form a large portion of the pop., have had to bear up against the competition of the power-loom. In fact, but for the reduction of wages occasioned by the Irish immigration, it is probable that the race of hand-loom weavers in Glasgow would have been nearly extinct. And considering the fluctuations to which this business is exposed, the facility with which it is learned, and the comparatively low wages which those engaged in it have always earned, no one could regret its annihilation. But the moment a Scotch family has withdrawn from the business, its place has been supplied by an Irish one; and the extension of power-looms has been checked by the extreme lowness of the wages paid to the hand-loom weavers, a clear picture of which is given in a preceding table, drawn up from official documents.

In consequence of this depressed state of the weaver pop., of the fluctuations incident to manufacturing employment, and of the crowded, filthy, and miserable lodgings occupied by the pauper portion of the pop., Glasgow is frequently visited by the most destructive fevers, and the rate of mortality has of late years been very high. It is usual to ascribe much of the want and suffering of the poor of Glasgow, as of other great towns, to the prevalence of drunkenness; but it can be shown from official returns that drinking, instead of increasing, has considerably diminished.

Parliamentary Representation and Municipal Government.—Previously to the Reform Act, the representation of Glasgow was in the worst possible state. This great city had not even a representative of its own, but was united with the insignificant bors. of Rutherglen, Renfrew, and Dumbarton, in sending a town, to the H. of C.; the vote of each of these bors. having equal weight with that of Glasgow. The Reform Act made an end of this preposterous arrangement, and conferred on Glasgow the privilege of sending 2 mems. to the H. of C. The parl. bor. includes Gorbals, Calton, Bridgeton, Anderston, Cambachie, and part of Port Dundas, and had 16,578 registered electors in 1865. The corporation revenue amounted to 15,051l. in 1868-4. The gross annual value of real property assessed to income-tax was 1,972,901l. in 1857, and 2,611,933l. in 1862. The government of the city is vested in a provost and 50 councillors.

With regard to the history of Glasgow, little need be added to what has already been incidentally said. So insignificant at first was this great city, that it was included in the privileged boundaries of Rutherglen, which was made a royal bor. in 1202. Nor was it till 1611 that a similar privilege was conferred on Glasgow, though it had long enjoyed the rank and importance of a bor. of barony, originally bestowed on it by Bishop Jocelin about the year 1172. The see was made archiepiscopal towards the end of the 15th century. From the time of Achaius, the restorer of the bishopric, till the Reformation, Glasgow was governed by 26 bishops and 4 archbishops; and between the Reformation and the final establishment of Presbytery, in 1690, by 14 Protestant archbishops. The town was, in former times, frequently visited by the plague. Leprosy also pre-

vailed; there was a leper hospital in the Gorbals. The famous General Assembly of the Kirk of Scotland, which, in 1638, displaced Episcopacy, deposed and excommunicated the bishops and established Presbytery, was held in Glasgow. On the occasion of the Union, in 1707, the citizens manifested great discontent, and could with difficulty be restrained from outrage; but that event, by opening new sources of trade, eventually proved of the most signal advantage to their city. They raised 2 battalions of 600 men in defence of government, in 1715, but the city was, notwithstanding, taken by the Pretender, and had to submit to heavy exactions. At the commencement of the American war, in 1775, the citizens of Glasgow raised, at their own expense, a regiment of 1,000 men; and during the revolutionary war with France, they kept on foot several regiments of volunteers.

In more recent times the contests between masters and their workmen, resulting, on the part of the latter, in strikes and combinations for an advance of wages, have been pretty frequent. In some instances, these strikes have been supported with great obstinacy; and, on one occasion, they were productive of fatal results, and were found to involve principles of the most destructive tendency. Among recent incidents in the history of Glasgow may be mentioned the visit of Queen Victoria to the city on the 14th of August, 1849.

Glasgow (Port). See Port Glasgow.

GLASTONBURY, a bor. town, and par. of England, co. Somerset, hund. Glaston-twelve-hides, on the Brue, 22 m. S.W. Bath, and 112 m. W. by S. London, on the Dorset central railway. Pop. of par. 3,593, and of munic. bor. 3,496 in 1861. The town is situated in the valley which separates the Poldew and Mendip Hills, and stands chiefly on a low peninsula (once the Isle of Avalon) formed by the turnings of the river; it consists of two streets, the chief of which runs from E. to W., the other from N. to S., forming the road to Bridgewater and Exeter; and in both of these streets the fronts and other parts of many houses are composed of stone from the ruins of the abbey. Of these the most remarkable are the George Inn, a curious building probably of the 15th century, given by Abbot Selwood in 1480 to the chamberlain of the abbey; the Tribunal, having a fine oriel window adorned with the arms of abbots and other benefactors; the abbey-house, built in 1714 from the materials of the abbey's buildings; and the great tithe-house, now one of the inns of the town. The hospital of St. John, on the Bridgewater Road, was founded in 1246. The cross, now a mere ruin, stands at the intersection of the chief streets. Of the two parish churches, which are both old, that of St. John the Baptist is remarkable for a fine lofty tower, which forms the most ornamental feature of the place. The abbey belonged to the Benedictines, situated on the S. side of High Street, was surrounded with a high wall containing about 60 acre. which, however, is now scarcely traceable. The great church joined the W. front, and was 580 ft. long; and in other parts were various lodgings for the abbot, prior, and other inmates of the abbey; the great hall was 111 ft. long by 50 ft. broad. The ruins of the church are extensive, and serve to give an idea of its size. The abbots' kitchen, which is in better preservation than any other part, is octagonal, and in the roof rises an octangular turret crowned with a lantern. This abbey, founded by Augustine of Canterbury in 605, was re-established and chiefly built during the 12th century, the hall and chapter-house being added in the 14th century.

At the dissolution of the monasteries in 1539, the last abbot being unwilling to surrender his abbey, was hanged without trial, and the site was granted by Edward VI. to the Duke of Somerset. At this time the revenues were valued at 8,511l. On a hill a little N.E. of the town, is a curious tower, called the Tor of St. Michael, which, from its elevation and peculiar shape, serves as a landmark in navigating the Bristol Channel. On the W. side is a figure of Michael the Archangel.

The town has but little trade, though it has, besides the railway, a canal to the mouth of the river Brue, near Highbridge, the point where the Brue runs into the Parret; it is for vessels of 70 to 100 tons. Timber, slate, tiles, and coal are the principal articles at present conveyed upon it. The bor. is governed by a mayor, 4 aldermen, and 12 councillors. The mayor was formerly a magistrate within the bor., and provided of quarter sessions; but, in consequence of the removal of the police business to Wells, the commission of peace has been taken from Glastonbury. The local act of 51 Geo. III. is that by which the paving and improvement of the town is regulated. The rates levied under this act amount to almost 2400l. per annum. The poor-rates average 1,800l. a year, and the contribution to county rate about 240l. Market on Tuesday. Fairs Sept. 10 and Oct. 11, the former being for horses and cattle.

The history of the town is intimately connected with that of the abbey, on which its prosperity has mainly depended. It was burnt down in the 12th century, with part of the abbey; and, after having been rebuilt by Henry III., was once more destroyed by (as is said) an earthquake, after which it was gradually restored, chiefly by the help of the abbey. The abbots of Glastonbury lived in great splendour, and possessed great political power; they were always parliamentary barons, and, till 1161, had precedence of all other mitred abbots in England. Sharpham Park, in the vicinage of this town, was formerly a manor-house belonging to the abbots of Glastonbury. Before the Reformation, Glastonbury was a part. bor., and sent 2 mems. to the H. of C.

GLATZ (Slav. Kladsko), a fortified town of Prussian Silesia, gov. Breslau, cap. circ. of same name, on the Neisse, near the Austrian frontier, 52 m. SSW. Breslau, with which it is connected by railway. Pop. 11,415 in 1861, exclusive of a garrison of 2,162. The town is strongly walled, and being situated between two adjacent heights, is further defended by an old castle placed on one, and a new and regular fortress on the other. It has four R. Catholic, and two Lutheran churches, a hospital, Catholic gymnasium, royal citadel, arsenal, large barracks, and other buildings for military service. It is the residence of a military commandant, and the seat of the council and courts of justice for the circ. and town, and commissions for the superintendence of public works and navigation. It has manufactures of woollen cloth, damasks, plush, ribands, muslins, leather, and tanners, and some linen-printing establishments. Glatz surrendered to Frederick the Great in 1742; it was retaken by the Austrians in 1760, but restored to Prussia at the peace of 1763.

GLOGAU (GREAT), a strongly fortified town of the Prussian dominions, prov. Silesia, gov. Liegnitz, cap. circ. of same name; on the Oder, 33 m. N. Liegnitz, and 83 m. NE. Frankfort-on-the-Oder, on a branch of the railway from Frankfort to Breslau. Pop. 17,538 in 1861, exclusive of garrison of 1,190. The town is connected by a wooden bridge with the Dominsel (cathedral-island) in the Oder, which is also fortified. Besides the cathedral it has several other R. Cath.

and Protestant churches, and a synagogue. It has a royal citadel and a large garrison. It is the seat of the superior judicial court for Lower Silesia, of tribunals for the circle and town, a board of taxation, circle council, board of agriculture, &c.; it has a Catholic and a Protestant gymnasium, and a school of midwifery. Except a large beet-root sugar establishment, it has few manufactories; its inhab. among whom there are a very large number of Jews, derive their principal resources from the supply of the garrison, general trade, and the navigation of the Oder. Glogau has a large corn-market. It came into the possession of Prussia in 1741.

GLOUCESTER, a marit. co. of England, on both sides the Severn, having N. the channel of that river, the co. Somerset, from which it is principally separated by the Avon and Wilts; E. a point of Berks and Oxford; S. Warwick and Worcester; and W. Hereford and Monmouth. Area, 1,256 sq. m., or 803,102 acres, of which about 760,000 are arable, meadow, and pasture. It is naturally divided into the Vale, Cotswold, and Forest districts. The vale, which comprises the low lands from Stratford-on-Avon to Bristol, is commonly divided into the vales of Gloucester, Evesham, and Berkeley; the Cotswold district comprises the hilly country parallel to the Severn from Chipping Camden to Bath, dividing the sources of the Isis, Windrush, Coln, Churn, and other remote feeders of the Thames from the Stroud and other streams flowing W. The forest district includes the greater portion of the land on the W. side the Severn, and was formerly for the most part included within the Forest of Dean, whence its name. The Vale of Gloucester, taking the term in its widest sense, is one of the most fertile districts in the kingdom; the soil consists in part of a sandy loam, and in part of a reddish clay; and the climate is remarkable for its mildness. The soil of the other two districts is, for the most part, light and comparatively poor. Agriculture is not in an advanced state; there is a great waste of labour in ploughing, and a great want of an effective system of drainage. There are, however, some exceedingly productive meadows, especially along the banks of the Severn below Gloucester.

This county has been long famous for its dairies, and for the peculiar description of cheese that bears its name. The average yield of a cow in the dairies is estimated at from 3½ to 4½ cwt. of cheese a year. The sheep of the Cotswold hills are large, and yield long combing wool; the total stock of sheep in the co. is estimated at from 550,000 to 600,000 head. This is one of the principal cider cos. Estates and farms of all sizes. Gloucester is not only a great agricultural but also a great manufacturing co. It is especially famous for its manufacture of fine broad cloths. The principal clothing districts are Stroud, Wotton, and Dursley. Iron ore is abundant in the Forest of Dean; but notwithstanding it is also well supplied with coal, the ironworks carried on in it are of comparatively little importance. Principal river the Severn, which intersects the co.; the Wye divides it from Monmouth, and the upper Avon skirts it on the N., and the lower Avon on the S.; the Isis, as already stated, has its sources in the Cotswold Hills. (For an account of the Gloucester canal and railway, see following article.) Principal cities and towns, Bristol, Bath, Gloucester, Cheltenham, and Stroud. Gloucestershire is divided into 28 hunds. and 339 pars.; it returns 15 mems. to the H. of C., viz. four for the co., two each for the cities of Bristol and Gloucester, and the boro. of Cirencester,

G G 2

Stroud, and Tewkesbury, and one for Cheltenham. Registered electors for the co., 16,779 in 1865, of which number 7,374 for the Eastern division, and 9,405 for the Western division. Pop. 485,770 in 1861, living in 92,831 houses. Gross annual value of real property assessed to income-tax—Eastern division 766,202*l.* in 1857, and 800,87*l.* in 1862; Western division 787,080*l.* in 1857, and 826,302*l.* in 1862.

GLOUCESTER, a city, co., parl. bor., and river port of England, on the E. bank of the Severn, locally situated in the above co., hund. of Dudstone and King's Barton, 32 m. N. by E. Bristol, and 83 m. W. by N. London by road, and 114½ by Great Western railway. Pop. 16,512 in 1861. The city is situated on a high eminence, gently falling to the N. and S., and towards the river it consists of four principal streets, crossing each other at right angles. It possesses some good streets, and has a general appearance of wealth and business. The river, which is here divided into two channels by Alney Island, is crossed, at the NW. end of the city, by two fine bridges, one over each channel. There are several handsome public buildings, among which, besides the cathedral, the shire-hall, the tolsey or town-hall, the co. gaol, and market-house, deserve notice. The shire-hall, in which the assizes and county sessions are held, has a fine front of Ionic architecture, and is well constructed for the purposes of business. The county gaol, built in 1791, at an expense of 35,000*l.*, on the site of the old castle, covers about three acres; it was constructed on a plan suggested by Howard. But though it has been much enlarged of late years, and large sums have been expended upon it, it is still objected to as being extremely deficient in the means of accommodating and classifying prisoners. The market-house, which is commodious and of plain exterior, cost 40,000*l.* A spa having been discovered in 1814, a highly ornamental pump-room and other edifices have been built near it. Several of the churches are old and handsome structures. The cathedral of abbey church, a magnificent fabric, occupying one side of the college-green, is 427 ft. in length, by 154 in breadth. On its site was formerly a monastery of Benedictines; the present building was partly erected about 1048, but not completed till the close of the 13th century. Hence it exhibits the various gradations of style during the great era of church architecture, from the Norman conquest downwards. The crypt, the nave, and north aisle being the oldest parts, are in the Anglo-Norman style, with round-arched windows; the windows of the south aisle, built two centuries later, are of the obtuse lancet shape; the W. front, and the continuation of the nave, erected in the 14th century, exhibit a yet later and more elaborate style than the other parts. Under the tower (which is square, flanked with four highly ornamented pinnacles, and 224 ft. in height), at the E. end of the nave, is the approach to the choir; and from this point is one of the best views of the interior, the highly finished choir, with its curiously wrought roof, forming a remarkable contrast with the simpler architecture of the nave and transepts. The arching of the choir, nave, and transepts is so contrived that, while the eye beholds the massive pillars as they branch upwards, the whole structure has an extraordinary lightness and beauty. The high altar is ornamented with angels playing on musical instruments, and behind it is the great E. window, said to be the largest in England, and containing 2,000 square ft. of glass. It was set up in the reign of Edw. III., and is now much mutilated. The floor in front of the altar is of

curiously painted tiles, representing the arms of the Plantagenets, and of the earls of Gloucester. A monument of Edward II., near the altar, is well carved, and in good preservation. The choir is 140 ft. long, and has 81 stalls on either side, of exquisitely wrought tabernacle work. The lady chapel, added to the choir in 1475, and rebuilt in 1498, is a peculiarly elegant structure, and most ingeniously united to the church. The cloisters are remarkable for their rich workmanship and beautiful windows; they were begun in 1351, and finished about 1398. (See Dallaway's Anecd. Arch. pp. 50–55.)

Gloucester was made a bishop's see by Henry VIII. in 1541. In consequence of recent ecclesiastical changes, it is united with Bristol. The churches of St. Mary de Crypt, St. Michael, St. John, and the comparatively new one of Christchurch, are all edifices ornamental to the town. There are altogether twelve churches, including two district churches in the suburbs, one at Barton Terrace, and the other at High Orchard, near the docks. The Wesleyans, Independents, Baptists, R. Catholics and others, have also places of worship, and there is a Jews' synagogue. Here are three foundation schools:—1, the college school, founded by Henry VIII. held in the N. transept of the cathedral; 2, the crypt school, founded by Dame Cook, and sending two exhibitioners to Pembroke Coll. Oxford; 3, the bluecoat school, founded in 1666. Besides these, there are National, British, and other schools, which furnish instruction to great numbers of children. It deserves to be mentioned, that Sunday schools originated in the city in 1781. They were first suggested and set on foot by Mr. Raikes, a printer, a benevolent and intelligent individual, who rendered by this act an essential service to humanity. Here are four hospitals, of ancient monastic foundation, used as almshouses; besides which, there is an infirmary and a lunatic asylum.

Gloucester is situated in a fertile and populous district, and enjoys an extensive command of internal navigation. Latterly, also, its importance as a port has been much increased, owing to the greater facilities given to it by the excavation of the Gloucester and Berkeley canal, by which the intricate and, sometimes, dangerous navigation of the Severn is avoided. This canal, opened in 1826, is 18 m. long; it commences at Sharpness Point, about 2½ m. from Berkeley, and ends in a commodious basin, a little S. of Gloucester; it is 60 ft. wide, and being 18 ft. deep is capable of floating vessels of above 500 tons burden. The shareholders, finding their subscribed capital insufficient, applied to government for a loan, with the interest of which they are still burdened. Gloucester, since the opening of this canal, has had considerable trade with the West Indies and Baltic. On the 1st of January, 1861, there belonged to the port 370 sailing vessels under 50, and 71 above 50 tons, besides 5 steamers under and 2 above 50 tons. The gross amount of customs duties received was 79,560*l.* in 1859; 71,642*l.* in 1861; and 68,956*l.* in 1863. Gloucester is well supplied with railway accommodation, being united on the N. with Cheltenham, Worcester, Birmingham, &c., and on the S. with Bristol and Exeter. It is united with the Great Western railway, and consequently with the metropolis, by a cross line leading by Stroud to Swindon. The manufactures of Gloucester are but inconsiderable.

Gloucester has returned two mem. to the H. of C. since the 2nd of Edward I. Previously to the passing of the Reform Act, the franchise was vested in the freemen of the bor., who became so by birth, purchase, or apprenticeship. Registered electors 1,745 in 1862.

Gloucester possesses numerous charters of early date; but that by which it was formerly governed, and on which its privileges are founded, was granted in 1673 by Charles II., who received from the city 673l. in return. The local acts, by which the lighting and improvement of the city are regulated, are the 4th, 17th, and 21st of Geo. III., and the 1st and 2nd of Geo. IV. By the provisions of the Municipal Corporation Act, the bor. is divided into three wards, and is governed by a mayor, recorder, 6 aldermen, and 18 councillors. Corporation rev. 9,874l. in 1862. Annual value of real property 89,000l. in 1857, and 93,757l. in 1862. The custom of borough-English, whereby estates descend to the youngest son, prevails here. Markets, which are well supplied, are held on Wednesday and Saturday. Fairs for cheese, cattle, horses, &c., are held on April 5, July 5, Sept. 28, and Nov. 28.

The history of Gloucester carries back to the time of the Romans. It is mentioned in Antonine's Itinerary as Glevum Glevum, and was founded by Claudius, A.D. 44, to repel the wild Celts of S. Wales. Roman coins and antiquities are constantly dug up near the supposed site of the old encampment. In Anglo-Saxon times it surrendered to the king of Wessex in 577, being then called Gleaw-ceaster. In the war between Robert and William, the sons of the Norman conqueror, it was nearly destroyed, and was rebuilt, when the present cathedral was commenced. In the wars between Charles I. and his parliament the inhab. sided zealously with the latter; and hence, at the Restoration, the city fortifications were ordered to be destroyed. The gates continued to stand for many years subsequently; but even of these only the name remains.

GLUCKSTADT, a town of the duchy of Holstein, Germany, in a marshy tract on the left bank of the Elbe, about 30 m. from its mouth, and 20½ m. NW. Hamburg, on a branch of the railway from Hamburg to Kiel. Pop. 5,752 in 1860. Glückstadt was formerly a fortress of some strength; but, since 1814, its works have been nearly dismantled. It is regularly built and has a good harbour. It is traversed by several canals, but has a very deficient supply of good drinkable water, on which account the rain has to be carefully preserved in cisterns. It is the seat of the council, and of the superior judicial courts of the prov.; and has a school of navigation, and various other schools. Since 1830, Glückstadt has been a free port. Its inhab. are principally engaged in trade, navigation, and the Greenland whale fishery.

GOA, a city of Hindostan, and the cap. of the Portuguese dominions in the East. prov. Bejapoor, on an isl. of the same name, at the mouth of the Mandova, 250 m. SSE. Bombay; lat. 15° 30' N., long. 74° 2' E. Pop. reduced to about 4,000, it having been nearly supplanted by New Goa or Panjim, built on the sea-shore about 5 m. distant, which has a pop. of about 20,000. The old city, now almost deserted except by priests, is "a city of churches; and the wealth of provinces seems to have been expended in their erection." The ancient specimens of architecture at this place far excel any thing that has been attempted in modern times in any other part of the East, both in grandeur and taste. The chapel of the palace is built after the plan of St. Peter's at Rome, of which it is said to be an accurate copy. The church of St. Dominick is decorated with paintings of Italian masters; and that of the Jesuits contains the tomb of St. Francis Xavier, a sepulchre of black marble, richly sculptured in bas-

relief, representing various passages of his life. The cathedral is worthy of one of the principal cities of Europe; and the Augustine church and convent is also a noble pile of building. Most of the churches are, however, going rapidly to ruin, and the ancient palace of the viceroys has been long unoccupied; the building formerly occupied by the Inquisition, though entire, has been shut up for many years. (Buchanan's Christian Researches, p. 244.)

New Goa, founded early in the 18th century, and now the residence of the viceroy and the principal Portuguese inhab., is a well-built town, the houses being of stone, and roofed with tiles, a circumstance unusual in Hindostan. Thin layers of oyster shell generally supply the place of glass in the windows. A fine causeway, 3 m. in length, connects the town with San Pedro (the present residence of the archbishop of Goa), and serves to shut out the sea from an extensive tract, partly in cultivation, and partly occupied by salt-pits.

New Goa has a harbour, reckoned one of the best in India, but, during the rainy season, so much mud is brought into it by the river, that ships of large burden find it difficult to enter. Like another harbour on the S. side of Goa island, it is defended by several forts and batteries; both the towns are also fortified, but not strongly.

The inhab. of Goa are principally the mixed descendants of the Portuguese and the natives, and African slaves; there are some Jews; native Portuguese are few. The wholesale trade is in the hands of the Christian pop., the retail in those of the Jews and Hindoo natives. Though formerly the centre of eastern commerce, Goa has now only an inconsiderable trade with the mother country and the Portuguese settlements in China and on the coast of Africa. Its imports are chiefly piece-goods, raw silk, ivory, sugar, woollens, glass, and a few other European articles. Its exports are very trifling, and are chiefly betel, betel nut, cowries, and toys, beads, &c., for Africa.

The territories possessed by Portugal in Hindostan, exclusive of Damaun and Diu, are confined to the district around Goa, 40 m. in length by 20 in breadth, below the W. Ghauts, having N. the dom. of Sattarah, E. and S. the British territories, and W. the ocean; with a total pop. of about 417,000 inhab. Goa was taken from the Hindoo sovereigns of Bijanagur by a Mohammedan prince of the Baluneer dynasty in 1489; and in 1510 was besieged and taken by Albuquerque, who made it the cap. of the Portuguese possessions in India. During the 16th century, the Portuguese were masters of a number of places on the sea-coasts of India, but their territories at no period extended far inland. In 1807, Goa fell into the hands of the English, who held it till 1815. During the late civil war in Portugal, this colony declared itself in favour of Donna Maria I.

GODALMING, a bor., town, and par. of England, co. Surrey, hund. of the same name, on the river Wey, 4 m. SSW. Guildford, 31 m. SW. London by road, and 43 m. by London and South Western railway. Pop. of mun. bor. 2,321, and of par. 5,775 in 1861. The town, situated in a valley, is nearly surrounded by high and steep ground. It consists principally of one street, which extends about ¾ m. along the high road from London to Portsmouth, but it is narrow, badly paved, and insufficiently lighted. The village of Crownpits stands about ¼ m. SE., and that of Ferncomb about the same distance NE. of the town; and both are nearly united to it by houses. The church is spacious, with a lofty steeple containing eight bells; the living a vicarage in the patronage of the Dean of Salisbury. There are places of worship for Wesleyan Methodists. Baptists, Independents and Quakers. On the common, about 1 m. from the town, is an almshouse for ten old men, founded in 1672. The bridge, which is of brick and stone, was opened in 1763. Attached to the church is a good charity school, and there are several Sunday schools in the town.

Godalming, anciently a clothing town of some note, at present possesses very little importance beyond being a place of considerable thoroughfare. There are four or five mills on the river for the manufacture of paper, parchment, and leather; and the manufactory of cotton stockings gives employment to a few persons. Timber, bark, and hops are exported. The river Wey is made navigable from Godalming under the 3rd of George II.; and coals are brought up here in considerable quantities.

The old corporation of this town consisted of a warden and eight assistants, and was chiefly governed by a charter granted in the 13th of Charles II. The present government is vested in four aldermen (one being warden) and twenty-one councillors. The local act, regulating the paving, &c. of the town, is the 6th of George IV.; and the rates levied under it average about 310l. a year. Markets on Saturday; fairs for horses and farming stock, Feb. 13 and July 10.

GODAVERY, a considerable river of Hindostan, through the central part of which it flows, extending through nearly 1° of long. Its course lies between those of the Nerbudda and Nahasuddy on the N. and the Krithna on the S., chiefly through the dominions of the rajah of Berar. It rises by numerous streams in the W. Ghauts, about lat. 20° N. and long. 74° E., and runs in a direction generally E., but with a slight inclination southward, to near long. 80° E. From this point, it turns mostly SE. for about 90 m. [passing] the prov. Hyderabad NE., and separates near Rajahmundry (N. Circars) into two arms, which fall into the Bay of Bengal, between lat. 16° 20' and 16° 40', enclosing a fertile delta, with an area of about 500 sq. m. The entire length of the Godavery is estimated at about 800 m., and during the rainy season it is in many parts 1½ m. wide. Its chief affluents are the Wynengunga, with its numerous tributaries, from the N., and the Mungera from the S. Its banks abound with timber, but no very important towns are situated on them.

GOLCONDA, a town and fortress of Hindostan, prov. Hyderabad, on a hill about 3 m. W. of the city of that name, and formerly the cap. of an extensive Hindoo kingdom. It is chiefly noted as a depôt for diamonds, which are brought to it to be polished and prepared for sale from other marts, mostly in the Balaghaut districts. Its immediate vicinity contains no diamond mines.

GOLDBERG, a town of Prussian Silesia, gov. Liegnitz, cap. circ. of Goldberg-Hainau; on the Katzbach, a tributary of the Oder, at the foot of the Hummelberg, 18 m. SW. by W. Liegnitz. Pop. 6,866 in 1861. The town is the seat of the council and judicial courts for the circle and town, and has a high school (bürger schule) at which Wallenstein was educated. The inhabitants are chiefly occupied in weaving woollen cloth, but have also manufactures of flannels, woollen stockings, gloves, and considerable dye-works. The town derived its name from a neighbouring goldmine, now abandoned, but formerly very productive. The hamlet of Wahlstadt, about 6 m. E. of this town, is memorable in Prussian history for the decisive and important victory gained on the 26th of August, 1813, by Marshal Blucher and the landwehr under his command over the French under Macdonald. The latter lost 15,000

men, killed and wounded, and 102 pieces of cannon fell into the hands of the conquerors.

GOLNITZ, a market town of Hungary, co. Zips, 72½ m. SW. Eperies. Pop. 4,237 in 1857. The town is the seat of a mining council and tribunal, and has considerable mines of iron and copper, iron forges, and cutlery and iron-wire factories. Its inhabitants are partly Rom. Catholics and partly of the Reformed Church.

GOMBROON, or HUSSEIN-ABBAS (' Port of Abbas,' an. Harmuza of Harmuzia), a seaport town of Persia, prov. Kerman, but at present belonging to the Imâm of Muscat, on the Persian Gulf, nearly opposite the island of Ormus, and 160 m. S. Kerman. Pop. from 4,000 to 5,000, chiefly Persians, Arabs, and Kurds, with a few Armenians and Bedouins. The town stands on a slope approaching the sea, in a barren and desolate country: it is about three-fourths of a mile in circ., and surrounded by a mud wall. The houses are few and wretchedly constructed, and the people are mostly lodged in huts. Gombroon appears to have been a town of very little importance before 1622, when Shah Abbas, assisted by the English, drove the Portuguese from the island of Ormus, and transferred the commerce to this port. Instead of being carried by sea up to Bussorah and the N. parts of the gulf, many of the imports from India and Africa were now landed at Gombroon, and transported by caravans to the interior, so that it became for a time the emporium of Persia. The English, Dutch, and French, for a long period, had large factories here; but towards the close of the seventeenth century, the route to the interior having become interrupted by wars and commotions, the factories were left to decay or destruction, and the European merchants removed to Bushire, now the centre of the trade. Some remains of the English factory still exist, but the Dutch is the only one in a tolerable state of preservation: it is used by the Imâm as an occasional residence. Gombroon appears to present more natural advantages for a commercial town than Bushire, the route from it leading by natural passes into the heart of Persia; and when, some years ago, Bushire remained in a disturbed state, commerce speedily found its way again into this channel. Even now its trade is considerable, and is said to be increasing. Persian carpets, tobacco, and dried fruits form its exports; its imports are chiefly piece goods, Indian cloths, and China ware. The Imâm collects a revenue of from 8,000 to 10,000 dolls. a year from the town. Immediately without the walls are the cemeteries of the former European inhabitants, and in their neighbourhood are some very extensive tanks excavated by the Portuguese, the length of the largest of which has been estimated at ½ m. (Whitelock; Kempthorne in Geog. Journal, v. and viii.)

GOMERA, one of the Canaries, which see.

GONDAR, a large city, commonly called the cap. of Abyssinia, kingd. Amhara, prov. Dembea, on the Angrab, about 20 m. N. Lake Tsana or Dembea, 270 m. E. by S. Sennar, and 1,260 85E. (Cairo) lat. N. 12° 31' 30'', long. E. 37° 30' 15''. Estimated pop. 6,000. The city stands on a lofty eminence, surrounded on all sides by low lands, and, when seen from a distance, resembles more a forest than a city, on account of the quantity of trees that surround its churches. The city is built in a straggling manner, occupying a space about 11 m. in circ.; the houses, which are mean and wretched, are either of plaster or stone, having one story and a high thatched roof. The only structure worth notice is the royal palace, a square Gothic stone building, flanked with towers, and once consisting of four stories; it was built under the direction of Jesuit missionaries, in the latter part of the sixteenth century. A great part is now in ruins; but the lower floors still contain ample accommodation. One room, used as an audience chamber, is 120 ft. long. The churches, of which there are above forty, have no pretensions either to beauty or convenience. There are very few shops, and all goods for sale are exposed in the great square. The people of Gondar have for some years been subject to the ravages of the wild tribes by which it is surrounded. The city is now in the hands of the Gallas, who, for a long period, have been the scourge of the Abyssinians. (Ritter's Africa, i. 230; Gobat's Abyssinia, 79, 168, 176.)

GOOD HOPE (CAPE OF). See CAPE OF GOOD HOPE.

GOODWIN SANDS, famous and formerly dangerous sand-banks, off the E. coast of the co. Kent, about 4 m. E. Deal, and stretching NE. and SW., about 10 m. These sands are supposed by some to have once made part of the Kentish land, and to have been submerged about the end of the reign of William Rufus, or the beginning of that of Henry I. Formerly the sands were held to be very dangerous; vessels riding in the Downs being sometimes driven upon them, and generally wrecked; occasionally through the ignorance and carelessness of pilots, but more frequently from the violence of the SE. and NE. winds. They are divided into two principal parts by a narrow channel: in many places they are dry at low water, and some spots appear even wetter. The N. division is of a triangular form, lying N. and S., being about 3½ m. long, and 2½ m. broad; the N. end, called the North Sand Head, is about 7 m. from the coast. Its position being marked by a light-vessel. The Hand Head, on the W. side, is very dangerous. The largest spot that dries on this sand has got from seamen the name of Jamaica Island. The S part of the Goodwin Sands is about 7 m. in length; at its N. end it is about 2½ m. in breadth, gradually diminishing towards the SW. till it terminates in the narrow point called South Sand Head, marked by a light-vessel, moored about 3 m. from shore. But the position of these sands varies more or less every year, through the joint influence of storms and tides.

GOOLE, a town and river port of England, W. Riding, co. York, on the Ouse, 22 m. W. Hull, and 175 m. N. London, by Great Northern railway. Pop. 5,950 in 1861. Less than fifty years ago Goole was an obscure hamlet; and is indebted for its rapid rise to its situation on the Ouse, at the point where it is joined by the canal, belonging to the Aire and Calder Navigation Company, from Ferrybridge; and to its also being contiguous to the junction of the Don with the Ouse. To accommodate the shipping engaged in three great lines of internal navigation, two extensive docks, and a harbour communicating with them and with the river, have been constructed. Warehouses of sufficient security having also been built, Goole was made a bonding port in 1828; and it has since continued to increase in size and importance. In 1859 the gross customs' duties collected at Goole amounted to 30,747l.; in 1861 to 56,753l.; and in 1863 to 66,226l. The distance inland, and the difficulty of navigating the Ouse, are the principal drawbacks on Goole; but vessels drawing 15 and 17 ft. water have, by taking advantage of the tide, reached it in safety. In January, 1861, there belonged to Goole 209 sailing vessels under 50, and 813 above 50 tons, besides 6 steamers under, and 10 above 50 tons burthen.

GOREE, an isl. and town adjacent to the W. coast of Africa, in lat. 14° 39' 55" N., long. 17° 26'

25" W., on the S. side of Cape de Verd, belonging to the French, and forming a part of their colony of Senegal. The island is merely a barren rock, about 3 m. in circuit, very steep on its W., S., and E. sides, and having in its centre a small elevated plateau, on which is fort St. Michael, commanding the town. On the NE. side of the island is a small harbour, affording good anchorage for eight months of the year. The town of Goree occupies more than 2-3rds of the island. Pop. 4,100 in 1861, of whom but 102 were Europeans. Its streets are rather narrow, but straight and clean; its houses, built of basalt cemented with mortar, are terraced in the Italian style. It has a civil and commercial tribunal, and is an entrepôt for gum Senegal, ivory, gold-dust, and other productions of the coast. The island is deficient in water, which has to be brought from the mainland; but it is said to be healthy. It was taken possession of by the French in 1677.

GORITZ (Germ. Görz, Ital. Gorizia), a town of Austria, prov. of Illyria, gov. Trieste, cap. circ. of same name; on the Isonzo, 12 m. from the Adriatic, and 21 m. NNW. Trieste on the railway from Trieste to Venice. Pop. 13,299 in 1857. Goritz is composed of an upper and a lower town. The first, situated on a hill, is the more ancient; it is surrounded with walls, and has a partly ruined castle, formerly belonging to the counts of Goritz, now used as a prison; the second, situated beneath the former, is a well-built town, its houses being mostly modern, and its streets clean and furnished with foot-paths. Goritz has a fine cathedral, 4 other churches, a handsome bishop's palace, and other noble residences, some barracks, occupying what was formerly a Jesuit's college, a circle-hall, town-house, alms-houses, and an elegant new theatre. It is the seat of the superior tribunal of the circle, and of a non-suffragan bishop; and has an episcopal seminary for the whole gov. of Trieste, a philosophical academy, gymnasium, superior female school, belonging to Ursuline nuns, a Piarist college, Jews' school, teachers' academy, and a society of agriculture and arts. It has three sugar-refineries, silks, rosoglio, leather, and various other factories, dye-houses, and a brisk general trade. The exiled king of France, Charles X., died at Goritz in 1836.

GORLITZ, a town of Prussian Silesia, gov. Liegnitz, cap. circ. of same name, on the Neisse, 32 m. W. by S. Liegnitz, on the railway from Liegnitz to Dresden. Pop. 27,963 in 1861, exclusive of a garrison of 1,449. The town is walled, and is entered by six gates, and has three suburbs. The town is in general well-built, and in a flourishing state, with wide streets and spacious squares. It has several fine public edifices, including the church of Sts. Peter and Paul, an edifice of the 15th century, and the town-hall. There are four hospitals, a prison, orphan asylum, gymnasium, and three public libraries. It is the seat of the council for the circle of the courts of justice for the town and the principality of Goritz, a board of taxation, and the Oberlausitz association of arts and sciences. A good deal of linen and woollen cloth is made here; there is also an active trade in the linen fabrics and wool of the surrounding districts. The manufacture of steel and iron wares, bell-casting, tanning, lithographic and other printing, and linen bleaching, are the other chief branches of industry.

GORUCKPORE, a distr. of British Hindostan, presid. Bengal, prov. Oude, between lat. 25° 40' and 27° 40' N., and long. 81° 60' and 84° 30' E., having N. Nepaul, E. the distr. Sarun, S. those of Ghazipoor and Jaunpore, and W. the dom. of the nabob of Oude. Area, 9,589 sq. miles. Pop.

8,087,874 in 1852. The Gogora divides the district into two portions, Asinghar and Gorackpoor Proper. The former division some years ago contained about 850,190 begas of land in cultivation, assessed at 951,133 rupees; and the latter, 863,872 begas in cultivation, assessed at 792,205 rupees. A great extent of the surface consists of jungle-forest, inhabited by elephants and other formidable wild animals; and at the foot of the hill ranges there is a very extensive, low, marshy, and unhealthy tract of country called the terraical. Chief towns, Gorackpore the cap, and Asinghar. This territory came into the possession of the British by cession from the nabob of Onde, in 1801.

GOSLAR, a town of the k. of Hanover, distr. Hildesheim, on the Gose, a tributary of the Ocher, at the N.E. foot of the Harz, 44 m. S.E. Hanover, near the railway from Hanover to Hardurg. Pop. 7,619 in 1861. Goslar is one of the most ancient towns of Germany, and was, till 1801, a free town of the empire; often the residence of the emperor, and formerly the seat of the diet. It is walled, and has a very antique appearance. Like most old towns its interior is gloomy; and the streets narrow, crooked, and dirty. Its greatest curiosity, a cathedral finished in 1050, was almost wholly pulled down in 1820; little now remaining of it except a small chapel, containing an ancient Saxon altar, and some other curiosities. Part of a palace, built in the 9th or 10th century, is now used as a corn-warehouse. Goslar is the seat of the mining council for the Harz, and of the corn magazines for the same district. It has several churches, an hospital, gymnasium, several breweries, the brew of which enjoys great celebrity; manufactures of virriol, sheet lead, shot, copper, and iron wares. Most of the inhab. are Lutherans, and employed in the mines of the Rammelsberg, about 1 m. from the town.

GOSPORT, a sea-port and market town of England, co. Hants, hund. Tichfield, par. Alverstoke, opposite to and separated from Portsmouth by the mouth of Portsmouth harbour, 14 m. S.E. Southampton, and 89 m. S.W. London by London and South Western railway. Pop. 7,789 in 1861. The town is surrounded by fortifications, which appear to be a segment of those of Portsmouth. These fortifications include, not only the town of Gosport, but the government establishment of Weovil, separated from the former by enclosed fields. Gosport and Weovil together occupy the E. extremity of a point of land between two inlets of Portsmouth harbour; the northern of which is called Forton Lake, and the southern, Haslar and Alverstoke Lake. The town consists chiefly of one broad street, containing many good houses, running W. from the shore through its whole extent; one or two other streets running parallel with the former; and several more crossing them mostly at right angles. It is in general pretty well built and paved, clean, well lighted with gas, and well supplied with water. Towards its N. side, it has a tolerably good square, termed Cold Harbour; it has few public buildings worthy of remark. The church, a neat and spacious edifice, is a curacy of Alverstoke; there are Independent, Rom. Catholic, Baptist, and Methodist chapels, an academy for ministers of the first-mentioned sect, several charity schools, some almshouses for poor widows, an extensive bridewell, and an assembly-room at the principal hotel. A large building was, in 1811, erected by shares, in a conspicuous situation on the shore, for a market-house; but it proved a losing speculation, and is no longer devoted to that purpose. Its lower part has long been shut up; its upper part is at present used for the meetings of the Ferry Committee and the

Philosophical Society. The town is quite open on the side of the harbour; there is a floating bridge, propelled by steam, and of large dimensions, for the conveyance of goods and passengers to Portsmouth.

On the land side, beyond the gates, is the populous suburb of Hingham-Town, in which is the terminus of the London and South-Western railway. The inhabitants of Gosport are of the same description as those of Portsmouth, follow the same pursuits, and partake equally of the benefits which result from the public establishments. In time of war, Gosport shares in the commercial activity that prevails on the other side of the harbour. Some vessels and boats are built, but there are no other manufactures of consequence. There are 2 fairs annually, but they are of no importance.

The establishment of Weovil comprises the royal brewery and cooperage; storehouses for provisions of all kinds for the navy; an extensive ship-biscuit manufactory, wrought by machinery; and the general victualling department, removed thither from Portsmouth in 1827-8. It communicates with the sea by a large basin and canal, where ships of large burden take in stores. Near Weovil are some extensive military barracks. N. of Forton Lake is Priddy's Hard, where is a large powder magazine. At Forton there was formerly a brick edifice of considerable size, in which many French prisoners were detained during the late war; but it has been pulled down. On the N. side of Haslar Lake stands Haslar Royal Hospital, a magnificent asylum for sick and wounded seamen. It was commenced in 1746 and finished in 1762. It is built of brick, and consists of a central portion 570 ft. broad, with two wings, each about 550 ft. in length, the whole surrounded by a high wall, enclosing an area of nearly a mile in circuit. It is capable of at once accommodating 2,000 patients; and has, besides, apartments for the numerous officers connected with it, a neat chapel, and a fine museum of natural objects. The annual expenses of Haslar Hospital are estimated at about 40,000l. At Stoke Bay, about 2½ m. S.W. Gosport, a little watering-place has grown up since 1825, and is rapidly rising into importance. Gosport is a polling-place for the S. division of Hants.

GOTHA (PRINCIPALITY OF). See Coburg-Saxe-Gotha.

GOTHA, a town of Central Germany, cap. of the above principality, and, conjointly with Coburg, the residence of the sovereign prince; on the declivity of a hill, the summit of which is crowned by the palace of Friedenstein, 46 m. N. by W. Coburg, and 17½ m. W. by S. Erfurt, on the railway from Leipzig to Frankfort-on-the-Main. Pop. 15,105 in 1861. This is one of the best laid out and best built towns of Germany, and is surrounded by handsome boulevards, which replace its ancient fortifications. Being situated from 900 to 1,050 ft. above the level of the sea, its climate is cold, the mean temperature of the year not exceeding 49° Fahr. The palace, called Friedenstein, is an imposing building, conspicuous at a distance, not unlike Windsor Castle in its situation, and surrounded by similar terraces, commanding fine views. It contains a picture-gallery, in which there are some good paintings by Italian masters, though the works of the old German and Dutch schools predominate; a collection of copperplate engravings; a library of 150,000 vols.; a cabinet of coins; a museum of natural history and the fine arts; and a Japanese and Chinese museum, containing Chinese and Japanese books, articles of furniture and weapons, including a part of the collection of the eastern traveller, Neviern. The cabinet of coins and medals is both extensive and complete, and considered one of the finest collections of the kind in Europe; it comprises nearly 20,000 ancient and 52,000 modern coins, 13,000 impressions in sulphur, a numismatic library of 6,000 vols., and 9,000 drawings of medals. The town of Gotha has seven churches, an arsenal, a gymnasium, with an excellent library, a new ducal gymnasium, orphan and lunatic asylums, a house of correction, an institution for the improvement of neglected children, the Caroline establishment for poor girls, a teachers' seminary, school of trades, society for the encouragement of arts and trades, and a fire and life insurance office, from which policies may be obtained for any part of Germany. Gotha has a large manufactory of porcelain; and produces cotton, woollen, and linen fabrics and yarn, sail-cloth, leather, tin and lacquered wares of all kinds, fire engines and buckets, coloured paper and furniture, and has numerous dyeing-houses. It has an active and extensive trade, and amongst other articles, Gotha sausages are sent to all parts of Germany. A little to the N.E. is the observatory of Seeberg; and not far from the palace is a pleasure-house, with a fine garden and orangery, and a ducal park ornamented with statues. The foundation of Gotha is attributed to William, archbishop of Mayence, in 961.

GOTTENBURG, or GOTHENBURG (Swed. *Göteborg*), a sea-port city of Sweden, and the second in that kingdom, in the W. part of which it is situated; at the head of a fiord, near the Cattegat, which receives the Gotta, about 260 m. S.W. by W. Stockholm, with which it is connected by railway, and 157 m. N.N.E. Christiania. Pop. 34,218 in 1860. The town stands principally in a marshy plain, surrounded by precipitous ridges of naked rocks, from 100 to 340 ft. high; but partly on the heights to the W.; being thus divided into the Lower and Upper town. The former is intersected by numerous canals, and has an appearance very similar to that of the towns in Holland. The entrance to Gottenburg from the S. is extremely fine: the slope of the hill, along which the road winds, is covered with houses whose shaded gardens spread beautifully up the height behind, while in front are long terraces, and neatly-clipped harbour walks, all mingling richly among large trees of southern foliage. The city is entered by a good bridge, and the lofty flat-roofed houses, all built of stone, or of well-stuccoed brick,—the wide streets, regularly paved, with foot-walks,—the deep canals, with which the place abounds, displaying rows of trees on either bank—all help to keep up the illusion that Gottenburg is a southern city. Since 1854 the town has rapidly increased; most of the empty spaces inside have been built upon, and the rent of houses has risen 35 and 40 per cent. Many of the houses in the Upper town are erected upon the steepest ridges of the rock, rising one above another in situations apparently the most perilous and insecure; these, however, together with the bold scenery round the city, and the harbour thronged with vessels and boats in front of it, give Gottenburg a very picturesque appearance. It is defended by three forts. The suburbs are larger than the town itself, and stretch for a considerable distance along the fiord. The city has several large squares and market-places, and some tolerable hotels; there are, however, few public edifices or other objects worth notice. The principal are the exchange, the extensive buildings belonging to the E. India Company, an hospital, and a magnificent church, built since 1812, with stone from Scotland. The exchange is handsome, large, and splendid enough for a commercial city of the first class. The city

has five churches, one being a cathedral, a Moravian chapel, two orphan asylums, a gymnasium, Prince Oscar's school, in which 100 soldiers' children are educated, a free school for the education of 200 poor children, and the board of 200 do.; with Sunday-schools and many benevolent institutions. It has also an arsenal, custom-house, 2 banks, a theatre, barracks, and docks for ship-building, and is a place of considerable manufacturing activity. Within the last thirty years 3 large cotton mills and 1 large sail-cloth and linen manufactory have been built, and are in full operation; the machinery was brought principally from England and Belgium. There are, also, several factories for weaving common printed cotton goods. In addition to these, there are manufactures of tobacco, refined sugar, glass, and paper; but most of these are upon a limited scale. There is a considerable porter brewery, the produce of which is famous throughout the N. of Europe.

The harbour is the most conveniently situated for foreign trade in Sweden. It is formed by two long chains of rocks, and protected at its mouth by the fort of Nya-Elfsborg, built at the extreme projection of a long rocky island, running into the Cattegat. Immediately within this fort, where the road is not half a mile wide, the larger vessels trading to the port usually remain, while those of smaller burden proceed some distance further, to Klippen, an extensive suburb of Gottenburg, from whence the inner harbour commences. Vessels do not come close to the city, but lie in the river or harbour at a short distance from the shore, goods being conveyed from and to them by lighters that navigate the canals of the Lower town. The depth of water in the port is 17 ft.; and there is no tide, bar, or shallow. A vessel entering the Gotha (fiord) must take a pilot on board, whose duty it is to meet her half a league W. of Wingo Beacon. After Stockholm, Gottenburg has the most extensive commerce of any town in Sweden. According to an official report (from Mr. Gregstrom, British Consul at Gottenburg, in Consular Reports, No. XI.), the total value of the exports during the five years, from 1855 to 1859, was estimated as follows:—

		£
1855	.	1,101,000
1856	.	850,000
1857	.	765,000
1858	.	875,000
1859	.	1,100,000

The same report stated the value of imports as follows:—

		£
1855	.	1,461,000
1856	.	1,750,000
1857	.	1,720,000
1858	.	1,120,000
1859	.	1,542,000

Iron and steel, the former excellent, the latter inferior to that made in England, form the principal articles of export. They are brought from the rich mines of Wermeland, distant about 200 m., being conveyed by the lake Wener, the Trollhetta canal, and the Gotha. The next great article of export is timber, particularly deals, which are also furnished by Wermeland. The other articles of export are linen, sailcloth, tar, copper, alum, glass, cobalt, manganese, litharol, oak bark, bones, juniper berries, cranberries, and pork meats for dyeing. The principal articles of import are sugar, coffee, tobacco, cotton yarn and twist, salt, indigo, dye-woods, South Sea oil, rice, wine, spirits, and herrings. Gottenburg was, at no distant period, to be one of the principal seats of the herring fishery; but at present this branch of industry has become extinct. It has always been very capricious, the fish alternately swarming on, or altogether deserting the coast. Since 1842 they have entirely disappeared; so that Gottenburg, instead of exporting, at present imports considerable supplies of herrings.

There belonged to the port, in 1850, exclusive of river craft, 145 ships, measuring 13,281 tons (Consular Report). The opening of the Götha canal, by which Gottenburg communicates with a large part of the interior of Sweden by means of an extensive system of inland navigation (respecting which, see SWEDEN), has exercised a material and beneficial influence upon its commercial destinies. Still more important has been the construction of a railway to Stockholm, undertaken at the cost of the government, and opened in 1862. The trade with England is extensive, and English is generally understood in Gottenburg. Steamers run once a week between Gottenburg and Hull for eight months of the year; but in winter intercourse takes place only by the tedious route of Lubeck and Hamburg. Goods may be landed for any length of time in the warehouses of the city, on payment of ½ per cent. ad valorem.

Gottenburg is the see of a bishop, the residence of a military governor, and the seat of various courts of justice, and a chamber of manufactures. It has an academy of sciences and literature, incorporated 1778. It was built on its present site by Gustaphus Adolphus, in 1611.

GOTTINGEN, a town of W. Germany, k. Hanover, cap. prine. of same name, distr. Hildesheim, on the railway from Hanover to Cassel; 64 m. S. Hanover, and 24 m. NE. Cassel. Pop. 12,516 in 1861. The town is pleasantly situated on the banks of the Leine, in a beautiful and fertile valley, 512 ft. above the sea, at the foot of the mountain of Hainberg. It is divided into three parts, the old and new town, and March is walled round, and has four gates. The ramparts are planted with trees, and form a pleasant walk for the inhab. Streets broad and well paved; but the houses, though old, appear neither venerable nor picturesque. There are three squares, the largest being the market-place, with a handsome esplanade and fountain in the centre, three Lutheran churches, a Reformed church, and a Rom. Cath. chapel. The church of St. John's has two steeples, each 300 ft. high; and St. James's is 300 ft. high. The University church was opened in 1822. The other chief buildings are, the university-hall, finished in 1837; the court of justice; the lying-in hospital; the observatory, in the SE. suburb of the town; and the theatre of anatomy; of these, the first and last two are chaste and elegant structures. A school of industry was founded in 1765. The trade of the place, independently of the university, is quite insignificant; the sale of books, and the manufacture of tobacco-pipes, are the only thriving branches.

The university, founded by George II. in 1734, and chartered in 1736, as the Academia Georgia Augusta, with an endowment out of the revenues of some secularised monastic property, was, down to 1801, fully entitled to its appellation, 'the queen of German universities,' both on account of the celebrity of its professors, and the number of students flocking thither from all parts of Europe. It is chiefly indebted for its early prosperity to the fostering care of its first curator, Baron Munchausen, the king's home minister; and its subsequent success has been owing to the judicious liberality of its sovereigns, who, while cautiously watching its progressive efficiency, have not changed the direction of their bounty, or

doled out its supplies with a niggard hand. The first course of lectures was begun by Gebauer the civilian, in 1734; and in the century since elapsed no less than 730 professors have given instruction, most of them in every branch, possessing a higher degree of talent than those attached to any other university in the country; among them were Heumann, Mosheim, Schleusner, Michaelis, Eichhorn and Ewald, in theology; Gebauer, Spangenberg, Waldi, Hugo, and Bergmann, in law; Gesner, Heyne, Schlozer, Müller, Grimm, and Heeren, in philology and history; Haller, Blumenbach, Langenbeck, Schröder, Baldinger, Conradi, and Osiander, in medical science; Gmelin and Stromeyer, in chemistry; Zeun, Hoffman, and Schroeder, in botany. The entire number of matriculated students during the first century of the university's existence was 39,756; the greatest attendance being between 1822 and 1826, when the average was 1,491 annually. Since 1831, however, in consequence of the political disturbances at Gottingen, in which the professors and students were implicated, the university has fallen into disrepute, and the number of students has greatly declined. The oppressive measures of king Ernest in 1837, which drove Grimm, Ewald, Dahlmann, and other professors to other universities, still further injured it. The gross annual expenditure of the university is about 160,000 thalers (about 24,000l.), nearly half of which goes to enrich the library and museum, the rest being divided among about 52 ordinary professors, whose salaries vary from 80l. to 350l. a year. The professors altogether, including private tutors (privatim docentes), are reckoned at ninety. The students in Gottingen are not compelled to reside within college, nor tied to stated hours of discipline, nor forced to oaths of orthodoxy; each student may live in any part of the town he likes, take his meals how, when, and where he pleases, and even pursue his own course in the choice of his academical studies. Their age at entrance varies from seventeen to twenty, and they usually continue here for four years, the periods of study occupying ten months in each year; the winter semester lasts from Oct. to March, that in the summer from April till the end of Aug. There is a preliminary examination for the Hanoverian students, called Maturitätsprüfung, which all must pass who wish to serve the state in the learned professions. This probation, however, is not required of foreigners. The matriculation fee is one louis-d'or, or 17s., and this admits to the use of the library and to attend lectures. Of the lectures, some are public, and may be attended without any additional fee; but the greater number are private, the fee being a louis-d'or for each semestral course of daily lectures. The medical fees are higher. Many of these lectures are delivered in public auditories, especially those of the medical faculty; some professors have private class-rooms. The medical and public lectures are very numerously attended; the attendance of the rest varies from fifty to twelve. Not less than 140 courses are delivered by the whole body of teachers during each semester, and several have two or three courses on different subjects proceeding contemporaneously. The expenses of students greatly depend on their habits. Saalfeld, in his edition of 'Putter's History of the University,' mentions 360 thalers a year as sufficient for respectable maintenance; but this is too low a calculation, 400 or 450 thalers (about 90l.) being, it is alleged, the lowest sum that can be spent consistently with comfort and convenience for study. For the poorer scholars there are 204 Freitische-stellen, or stipendships (sums paid for board), and a number of scholarships (*Stipendien*). With reference to degrees, the university is composed of four faculties—divinity, law, medicine, and philosophy, each of which confers its own degrees. The faculty of divinity confers the degree of licentiate in theology and doctor of divinity; that of law, the degree of doctor of laws; that of medicine creates doctors of medicine; while the philosophical faculty confers the degree of doctor of philosophy and master of arts. All these degrees are conferred on disputations and examinations approved by the deans of the respective faculties. These degrees, however, though generally pre-requisites, confer of themselves no right of practising the learned professions in Hanover. This is gained by a subsequent government examination. The members of these faculties consist altogether of twenty professors, from whom ten are chosen to form the Senatus Academicus. The judicial government of the university, which acknowledges no control beyond that of the king of Hanover, its *rector magnificentissimus*, and his two curators, who appoint the salaried professors, is conducted by the pro-rector, or principal, an officer elected each semester by the professors from among themselves, who is assisted in his duties by two judges, a secretary and recorder, all of whom, likewise, are professors.

The chief academic establishments of Göttingen are,—1. The library, consisting of 320,000 printed books and 5,000 MSS., admirably selected and arranged, to which the students have full access, with the additional privilege of taking the books home; 2. The academical museum (founded in 1773, and removed to its present depository in 1783), consisting of fourteen rooms, filled with several thousand specimens of zoology, mineralogy, and geology, besides others explanatory of the manners and customs of different nations, and a curious collection of models; 3. The observatory, first erected in 1751, and removed to its present site in 1816, containing an apparatus of excellent modern instruments, and every accommodation for astronomical observers; 4. The botanic garden, first laid out under Haller's superintendence in 1739, but now more than quadrupled in extent, and provided since its renewal with beautiful green-houses, adapted to plants of all temperatures, and ponds for aquatic plants; 5. The chemical laboratory, constructed by Gmelin, and perfected by Stromeyer, who provided it with an apparatus for experimental students; 6. The school of anatomy, first established by Haller in 1736, and since 1828 held in a fine building containing a spacious theatre and dissecting rooms; 7. Two infirmaries for medical and surgical cases, and a lying-in hospital, accommodating about 120 pregnant women a year; 8. The *Spruch-Kollegium*, or court of equity, composed of a president and several subordinate members appointed by government, which serves the double purpose of a court of judicial advisers in legal questions sent from all parts of Germany, and of a school for the legal students; 9. The Homiletic seminary, for the instruction of divinity students in preaching and pastoral duties; 10. The philological seminary, founded by Gesner in 1737, and under the direction of three professors, which gives minute philological instruction to eleven stipendiary students (paid fifty thalers each every year), and as many more as the director pleases to admit, after the requisite examination. The last three establishments have been eminently successful in raising up useful and able men in the professions to which their instruction tends. Nearly connected with the university is the Royal Society of Sciences, established by George

II. in 1751, on a plan suggested by Haller, and well known to the savans of Europe. Its transactions are published in Latin, and may be considered a repertory of all the original views in literature and science started in Göttingen by the professors of the medical and philosophical faculties. An annual prize of fifty ducats (21l.) is open to persons of every country for the best essay on mathematics, physics, and history alternately. This society is the patron and superintendent of the Göttingen Literary Review (*Göttingische gelehrte Anzeigen*), which, having risen to eminence under the editorship of Haller, has since been conducted by Heyne, Eichhorn, and Heeren. Two large 8vo. vols. are published yearly, and the work has throughout been distinguished not only for exalted talent, but for a tone of moderation and strict adherence to truth. These qualities alone have enabled it to outlive the various and important political changes of the country.

GOTTLAND, an island of the Baltic, belonging to Sweden, in the Län of the same name, lying between lat. 56° 57′ and 57° 56′ N., and between 18° 5′ and 19° 8′ E., dist. 60 m. from the continent of Sweden. Greatest length 75 m.; ditto breadth, 26 m. Area, 1,334 sq. m. Pop. of the Län, sometimes called Wisby Län, which includes the small surrounding islands, 42,571 in 1830. This island presents the appearance of a large plateau, varying from 150 ft. to 200 ft. above the sea. Its sides, which in some places gently slope towards the sea, are in others steep and precipitous in others, as to look like artificial walls. The coasts are indented by several bays, the largest of which are Kapelhamn on the N., and Slitehamn on the E. The high lands, except the barren summits of Thorsburg and Hoburg, are generally well wooded. There are several small lakes. The rivers are few and inconsiderable. In some parts swamps occur, but of no great extent. The geological features of the island, though generally calcareous, vary extremely, especially in the N., where occur large masses of hard grey sandstone containing mica, and susceptible of a high polish. The soil is either calcareous or sandy, and would be very productive if better cultivated. The chief products are wheat, barley, oats, turnips, potatoes, and hops, which are grown only for home consumption. The forest trees are large and handsome, and they furnish timber for exportation. The only other exports are marble, sandstone, and lime, which are sent to Stockholm. The rearing of cattle occupies a considerable share of the people's attention. Horses, goats, and sheep are reared in large numbers; and the breed of sheep has been improved by the introduction of Merinos. Game is very plentiful. There are no manufactures on the island. The Län, of which Gottland forms a part, is divided into 20 districts; and the sea-port town of Wisby, on the W. side of the island, is the capital.

The epoch of the foundation of Wisby is uncertain; but during the 14th and 15th centuries it was a principal fortress of the Hanseatic League, and attained to considerable wealth and importance. It is famous in the history of maritime jurisprudence, for the Code of Sea Laws which bears its name. The date of this compilation is uncertain, and some of the northern jurists contend that the Laws of Wisby are older than the Rules of Oleron; but it has been repeatedly shown that there is no foundation for this statement. Grotius has spoken of the Laws of Wisby in the most laudatory manner. '*Quæ de maritimis negotiis*,' says he, '*inesse Gothlandiæ habitatoribus jura*

placuerunt, tentum re se habent, tum equitatis, tam
prudentiæ, ut omnes servani arcula eo, nam temporum
proprio, ord eclui gratiam jure sinantur.' (Prolego-
mena ad Procopium, p. 64.) The text of these
laws, with a translation and an elaborate introduction and notes, is given in the excellent Collection des Loix Maritimes of M. Pardessus (t. pp.
425–562).

In 1361, Vladimir III., king of Denmark, took
Gottland from the Swedes. By the treaty of
1644, it again became their property; and since
then has continued in their possession, with the
exception of a short period in 1807, when it was
occupied by the Russians.

GOUDA, or TERGOUW, a town of S. Holland,
cap. cant., on the Yssel, at the influx of the
Gouw, 10½ m. N.E. Rotterdam, on the railway
from Rotterdam to Utrecht. Pop. 15,205 in
1861. Gouda is a neat town, with beautifully
wooded environs. It is known only in England
by its cheeses and tobacco pipes; but in Holland
it is famed for its painted windows, chiefly the
work of the two brothers Krabeth, and reckoned
the finest specimens of their kind in Europe.
They are the windows of the old church of St.
John, a large gothic structure, kept in excellent
repair, and particularly clean. The windows are
31 in number, each measuring about 30 ft. in
height, with the exception of those of the transepts, which are nearly double that altitude, and
all illuminated with pictorial representations, in
colours of the most brilliant hues. The subjects
are either scriptural or allegorical, and are full of
figures, whose robes in blue, purple, and red, shine
with extraordinary lustre. The faces are the
best part of the execution, the remainder of the
figures being painted in a stiff and formal style,
though nevertheless interesting from their antiquity. Besides the large windows, there are
several of a smaller size, chiefly blazoned with
the coats of arms of the old Netherlandish nobility.
These paintings were mostly executed in the 15th
and 16th centuries; and amongst others are introduced portraits of Philip II., and the Duke of
Alva. Besides St. John's (the cathedral) there
are 6 other churches in Gouda; and it has, also,
a handsome town-hall, an hospital for men, an
orphan asylum, and a foundling hospital. A
Latin school, and a library containing several
curious MSS., belong to the town. There are
upwards of 170 tobacco manufactories in Gouda,
some employing 80 workmen; and numerous
brick kilns in its neighbourhood. It has manufactures of woollen cloth, sailcloth, and cordage,
and large markets for cheese, flax, hemp, corn,
timber, and other produce.

GOUR (probably the Gangia Regia of Ptolemy),
a ruined city of Hindostan, and the ancient cap.
of Bengal, distr. Dinagepoor, on the E. side of
the Ganges, about 50 m. S. by W. Moorshedabad;
lat. 24° 52' N., long. 88° 14' E. Its ruins extend
in a direction NNW. to SSE., coincident with
the ancient bed of the Ganges, the main stream
of which formerly washed its ramparts; at present, however, from a change in the course of the
river upwards of 200 years ago, no part of the
ruins is less than 4 m., while other parts are as
much as 12 m. from the Ganges. The city
appears, from the extent of the old embankments,
which enclosed it on every side, to have been
10 m. long, and from 1 to 1½ m. broad. Beyond
those boundaries, however, a smaller embankment
has been carried forward for 7 m. farther S., in
which space are found mosques, tanks, and the
remains of habitations; and the same indications
are evident for 2 m. to the N. The city and its
suburbs thus extended in length about 19 m.,

with an average breadth of about 1½ m.; and,
according to the estimates of both Major Rennell
and Mr. Creighton, would appear to have anciently occupied an area of 30 sq. m.! The embankments surrounding the city, some of which
are faced with bricks, were sufficient to guard it
from floods during the inundation, and a good
defence against hostile attacks: they are mounds
of earth from 30 to 40 ft. high, and 150 to 200 ft.
in breadth at their base, with broad ditches on
their outside. Additional embankments were
made on the E. side, probably for greater security
against a large lake in that quarter, which in
stormy weather dashes with great violence against
them. Two high brick gateways, in an imposing
style of architecture, at the N. and S. ends of
the city, and several others, are still standing,
and the remains of some that have been destroyed are still traceable. Two grand roads,
raised with earth, and paved with brick, led
through the city in its whole length, crossing in
their course various canals and drains, by means
of bridges of brick, the ruins of several of which
remain in some degree of perfection. The whole
area of the city is furnished with a multitude of
tanks, of various sizes, and intersected with drains
and ditches in every direction. On the earth
thrown up in forming these, which raised the
ground considerably above its previous level, the
houses, &c. were built as in the cities and villages
of Egypt; the excavations supplying good water,
sufficient for every purpose. One of these reservoirs is a mile in length by half a mile broad,
and there are several others of considerable size.
All of them are, however, overgrown with reeds
and swarm with alligators and other reptiles.
Towards the centre of the city is the fort, an in-
closure rather less than a mile in length by about ½
a m. in breadth, surrounded with an earth rampart,
10 ft. high, with bastions, and a deep ditch encircling it. The handsome gate, flanked by two
towers, forming its N. entrance, is still standing.
Within this enclosure is part of a brick wall,
42 ft. high, which surrounded a space 700 yards
long by 300 wide, supposed to have been occupied
by the palace. Few other remains of that edifice
exist, and the whole site is so covered with trees
and brambles, as to render it not only difficult
but dangerous to explore, from the number of
tigers and other wild beasts that infest it.

There are scarcely any antiquities of a remote
date extant at Gour; most of the buildings that
remain are of Mohammedan origin, erected, indeed,
with the materials of the ancient Hindoo edifices.
Toiling through bush and long grass, now crows-
ing a field that some great has farmed, now wading
through pools of water, or ferrying across them,
you make your way from point to point, and find
only the ruins of seven or eight mosques, the half-
broken down walls of a large Moorish fortress, and
two strikingly grand and lofty gates of a citadel
evidently built by Mohammedans.' (Sketches of
India, p. 165.) Of the religious edifices, the finest
and largest is the 'Great Golden Mosque.' This
building, situated N. of the fort, is 170 ft. long,
by 76 ft. broad, and 20 ft. high, exclusive of the
domes, of which there are 44, rising 10 ft. above
the roof. (Creighton, Pl. V.) Its walls are 8 ft.
thick; it is built of brick, and has been wholly
cased with hornblende, little of which is now remaining. Eleven painted arches open into an
area divided by another similar row of arches, and
20 stone pillars arranged in 2 rows, into 4 aisles,
each surmounted by 11 domes. This beautiful
edifice is now going rapidly to decay, not only from
the effects of wanton dilapidations, but also from
banian and other trees insinuating their roots in-

tween the bricks of which it is composed. The small 'Golden Mosque' is built and cased in a somewhat similar style, but has only 5 arches in front, and 3 aisles instead of 4. Many of the inferior mosques are in higher preservation than the first mentioned; their domes are still perfect, and lined within by tiles painted of the most vivid colours, and highly glazed; and one of the smallest has a tessellated pavement of great beauty. The *Nutti Musjeed*, or 'Painted Mosque,' is an elegant edifice, having its walls cased both inside and out with glazed bricks about 3 or 4 in. square, of different colours, wrought in different patterns. Its interior is a handsome apartment, about 36 ft. square, the four walls closing above, and forming a majestic dome from 40 to 50 ft. above the ground, and unsupported by pillar, beam, or rafter. Within the fort is the tomb of Hossnin Shah, one of the kings of Gour in the 16th century, a fine mausoleum, now much dilapidated; and at a short distance without the citadel is the obelisk or tower erected by Firose Shah at the latter end of the 15th century. This structure is 21 ft. in diameter at its base, and as much as 17 ft. at the floor of its fourth story, 71 ft. high. Its entire original height was probably about 100 ft.: it was surmounted by a cupola, of which Mr. Creighton gives a representation, but since his time the dome has completely disappeared.

This city, called *Lakshmanvati* (by the Mohammudans, *Lucknorty*), from its last Hindoo sovereign Lakshman, was first taken by the Mohammudans in 1204. In 1575 it was repaired and beautified by the emperor Akbar, by whom it was called Jeonutabad (the abode of paradise); but in 1564, the seat of government of Bengal was removed to Tanda, a little higher up the river, owing to which event, and the desertion of it by the Ganges, Gour speedily declined. It, however, appears to have suffered less from the want of time than from active demolition. For centuries the materials of its structures were extensively removed to construct other towns; Maursheedabad, Makhah, Rajmahal, Dacca, &c., are in a great part built of them; and many portions of its fine buildings have been taken away to erect the cathedral of Calcutta, and to supply tombstones and monuments for the cemeteries of that city. A few straggling villagers are scattered here and there over the site of Gour; but it is now for the most part only an uninhabited waste, which strongly reminds the spectator of the desolation of Babylon. (See Creighton's Ruins of Gour; Rennell's Memoir; Mod. Trav., &c.)

GOZZO, a small island of the Mediterranean, contiguous to and dependent on Malta (which see).

GRAMMONT (Flemish *Geerardsbergen*), a town of Belgium, prov. E. Flanders, arrond. Audenaerde, cap. cant., on the Dender, which divides it into the upper and lower town, 21½ m. SSE. Ghent, on the railway from Ghent to Tournay. Pop. 8,795 in 1856. The town is walled, and has two churches, several chapels, a town-hall, convent, prison, hospital, orphan asylum, college, several schools, and manufactures of cotton yarn, lace, linen and woollen fabrics, paper, tobacco, some bleaching, dyeing, and tanning establishments, with distilleries, breweries, and mills for various purposes. It was founded and fortified by Count Baldwin de Mons in 1068.

GRAMPIANS (THE), a celebrated mountain chain forming the line of demarcation between the Lowlands and Highlands of Scotland. Its limits are not very well defined; but it may be regarded as commencing on the E. side of Loch Etive in Argyleshire, and as stretching across the island, till it terminates between Stonehaven and the mouth of the Dee on the E. coast. It forms, as

it were, a natural rampart, bounding the entire frontier of the Highlands. Its N. acclivity rises from the great valley of Strathmore. The summit of the ridge marks the line that separates the waters that flow into the Forth, the Tay, and its numerous tributaries, and the South Esk, from those that flow into the Spean, the Spey, and the Dee. With the exception of Ben Nevis, the highest mountains of Scotland are comprised in the Grampian range. The principal summits, beginning at the W. and proceeding E. are Cruachan Ben, at the head of Loch Awe, 3,390 ft. above the level of the sea; Ben Lomond, on the E. side of Loch Lomond, 3,195 ditto; Ben More, at the head of Glen Lochart, 3,870 (?) ditto; Ben Lawers, on the N. side of Loch Tay, 3,945 ditto; Schiehallion, at the E. end of Loch Rannoch, 3,550 ditto. But the most elevated part of the Grampian chain lies at the head of the Dee, between Ben Gloe, in Perthshire, and Cairngorm, on the confines of Aberdeenshire and Inverness-shire. Ben Macdhui, the most elevated of the mountains in this vicinity, is 4,327 ft. high, being only 48 ft. lower than Ben Nevis; and the adjoining mountains of Cairngorm, Cairntoul, and Ben Avon, are respectively 4,095, 4,245, and 3,867 ft. high. From this central point, the principal branch of the Grampians runs along the S. side of the Dee, gradually declining in height till it reaches Garlock Hill, near Stonehaven, 1,890 ft. high. The coast from Stonehaven to the Dee is high and precipitous, and may be considered as the extreme limit of the Grampians on the E. The branch of the Grampians to the N. of the Dee is of comparatively small extent, terminating at the Buck, above Glenbucket, on the N., and near Tarland, on the S.

The Grampians are, in general, remarkable for their sterility, and the desolate aspect which they present. Their sides are in some places extremely precipitous, exhibiting vast perpendicular ledges of rock. Their summits are frequently rounded, sometimes nearly flat, entirely covered by disintegrating blocks and stones, together with grit and sand, except where the granite rocks present the singular appearance of large tabular protruding pinnacles, having their blocks seemingly arranged in regular strata.

Of the Grampian passes, the principal are those of Aberfoyle, Leni, Glenalvie, and Killicrankie. The latter, which is the most celebrated, is about 15 m. from Dunkeld. It is about half a m. in length. The road is cut out of the side of one of the contiguous mountains; and below it, at the foot of a high precipice, in the bottom of the ravine, the river Garry dashes along over rugged rocks, but so shaded with trees as hardly to be seen. At the N. extremity of this pass, the revolutionary army, under Mackay, was defeated in 1689 by the troops of James II., under the famous Graham of Claverhouse, Viscount Dundee, who fell in the moment of victory.

GRAMPOUND, a bor. and market town of England, co. Cornwall, W. div. hund. of Powder, par. of Creed, on the Fal, 12 m. NE. Falmouth. Pop. 751 in 1831, and 573 in 1861. This inconsiderable place sent 2 mems. to the H. of C. from the reign of Edward VI. down to 1821, when it was disfranchised for gross bribery and corruption.

GRAN (Hungar. *Esztergom*, anc. *Strigonium*), a city of Hungary, cap. co. same name, on the Danube, nearly opposite the mouth of the river Gran, 80 m. ESE. Presburg, and 26 m. NE. by N. Pesth, on the railway from Pesth to Vienna. Pop. 11,216 in 1857. Gran consists of the royal free town, the archiepiscopal town occupying the site of the former citadel, the adjacent market-towns of St. George and St. Thomas, and several suburbs.

Gran was once the finest city in Hungary, and the residence of its kings, some of whose tombs are still to be seen. It is now the seat of the Prince-primate of Hungary, who ranks next to the palatine, and had formerly the privilege of crowning the king and of granting letters of nobility. The superb new cathedral, the palace of the archbishop, and the houses of the chapter, occupy a commanding position, overlooking the town and river, on the summit of a high and precipitous rock, on which an old fortress once stood. The cathedral, the most splendid modern building in Hungary, was commenced, in 1821, by the late archbishop Rudnay at his own expense; but, by his death, was left unfinished. It is in the Italian style, surmounted by a dome, and having a handsome portico of 24 pillars. The interior is lined with polished red marble, and supported by 54 columns. The dome is 82 ft. in diameter. The altar-piece, by Hesz, a Hungarian artist, represents the baptism of St. Stephen, the first Christian king of Hungary, a native of Gran, who founded the archbishopric in 1001. Under the church is the primate's burial vault. The see of Gran is perhaps the richest in Europe: its actual revenue is unknown, but common rumour generally estimates it at 100,000l. per annum; though some reduce it to 80,000l., or even 60,000l. The Danube is here of great breadth, but is crossed by a flying bridge, which communicates with the opposite market-town of Parkany. Besides the cathedral it has 2 Rom. Cath. churches, a Greek church, and 4 chapels, town-hall, house of assembly, hospital for poor citizens, a Rom. Catholic gymnasium, female school, and a good printing establishment. It is the seat of the assembly and judicial courts of the county. Its inhab. are partly Magyars and partly Germans. Their chief resources are derived from trading in wine; but they also manufacture and dye woollen stuffs. At the bottom of the rock on which the cathedral stands are some warm mineral baths.

Gran was several times taken by the Turks, who destroyed most of its ancient edifices. It was for a long period the advanced posts of their armies in Europe; but was finally taken from them, in 1683, by Sobieski and Prince Charles of Lorraine.

GRANADA, a prov. and part of an ancient roy. kingdom of Spain, consisting of the S.E. part of Andalusia, between lat. 36° 17′ and 38° 21′ N., and between long. 1° 61′ and 3° 57′ W.; and bounded E. by Murcia; N. and W. by Seville, Cordova, and Jaen; and S. by the Mediterranean. Its general shape is that of an acute-angled triangle, whose base faces the E. Its length is about 210 m., and its breadth varying from 25 to 80 m. Area 9,623 sq. m.; pop. 820,155 in 1846, and 1,298,987 in 1857. Granada is at present divided into three sub-provinces, namely, Almeria, with a pop. of 315,664; Malaga, with 444,408, and Granada, with 411,517, according to the census of 1857. The prov. consists chiefly of high land; but three chains may be distinguished—one forming the N. boundary of the prov. and connecting itself eastward with the Sierra Morena; a second and principal one, traversing the centre of the prov., (called the Sierra Nevada in the highest part, and the Sierras de Loxa, de Antequera, and de Ronda, E. and W. of the culminating point); and a third, nearer the shore, called the Alpujarras. The line of perpetual snow here is at 9,915 ft., and in the principal chain are several summits rising above it, the highest of which are the Cerro de Mulahacen, 11,660 ft., and the Picacho de Veleta, 11,387 ft.; from the last the Sierra Morena, distant 55 m., and the coast of Africa, distant 112 m., may be discerned in clear weather. The slope in the principal chain is more gradual northwards,

while on the Alpujarras the S. side is scarped and the gentle descent is towards the sea. The Sierra de Gador, in the latter chain, is 6,570 ft. high. From the N. side of the principal chain flows the Xenil, measuring 120 m. to its junction with the Guadalquivir; and farther E. are the smaller streams, the Guadix and the Barbata, both affluents of the same river. The rivers on the S., with the exception of the Guadalfeo and Almeria, are little better than torrents. In this mountainous district are several valleys of considerable extent, the largest of which is the Vega of Granada, a plain 30 m. long, and 16 m. broad, elevated about 2,000 ft. above the sea, surrounded by mountains, and watered by numerous affluents of the Xenil, which traverses it in its whole length, and essentially contributes to its extraordinary fertility. In the E. of the prov. is another valley—the Hoya de Baza—which, though smaller, is extensive, well watered, and fertile. There are others of more confined extent. The temperature, on account of the varying altitude of the country, is much diversified, but the climate is generally healthy, except occasionally on the coast, where the miasma produces fever among the inhabitants. The geology of the Granadian mountains is imperfectly known: the Sierra Nevada is of mica slate, gneiss, and clay slate, the whole overlaid on the N. side by black transition limestone containing sulphuret of lead, which here, as well as elsewhere in the prov., is worked to advantage. The mountains generally are rich in jasper and marbles, especially about the city of Granada, where they eclipse most countries in the beauty, transparency, and polish of the slabs. Precious stones are often found in the quarries. The chief mineral springs of the prov. are at Alhama and Almeria. The soil on the hills is calcareous, that on the plains light and easily tilled, while that on the coast is sandy. The forests produce oaks, cork-trees, chestnuts, and firs; and the plains bear the vine, the fig-tree, the strawberry-tree, the olive and mulberry trees, and others. Tillage, where possible, is pursued according to the Moorish plan of irrigation, and occupies great attention. The fruits of the S. of Europe—oranges, citrons, pomegranates, melons—grow here in great abundance, mingled with the productions of the N., Wheat, barley, maize, rice, hemp, flax, and the sweet potatoe, are raised in large crops; and on the coast of the Mediterranean indigo, cotton, coffee, and the sugar-cane are cultivated. Mr. Inglis, on the authority of General O'Lawley (manager of a large estate in the Vega of Granada called Soto de Roma, given to the Duke of Wellington by the Cortes in 1813), speaks as follows of the usual rotation of crops in the Vega (ii. p. 188):—'After the land has been fully manured, hemp is put in: and two, or sometimes three, crops of wheat, according to the nature of the land are taken in the same year; a crop of flax, and a crop of Indian corn, follow the next year, and beans and Indian corn are taken the third year. For the last crop the land is half manured and then it is fully manured for the hemp, to begin the next rotation. The hemp is considered necessary to prepare the land for wheat, which otherwise would come up too strong after the manure. This is the rotation on land subjected to the process of irrigation.' As to the value of land, he says:—'Ten years ago, land in the Vega of Granada was worth from 50 to 100 dollars per acre; at present, it does not average above 16. Wheat sold, ten years ago, at three dollars the fanega; now it does not average, year by year, more than one dollar and a half. Rents are, of course, fallen in proportion; and, low as rents are, they are difficult to be recovered. Upon the lands not capable of being irrigated, the crops are extremely precarious; and where a money rent is required, it is next to impossible to find a cultivator for the land. As a remedy for this, proprietors of high lands are contented to receive a certain proportion of the crop, generally a fifth; and upon land subject to irrigation, a tenant is willing to pay one-fourth of the produce. Land generally, in the Vega of Granada, returns 4 per cent. taxes paid; but a considerable quantity returns as much as 6 per cent. The return from land under tillage is greater than from meadow land. The estates belonging to the Duke of Wellington lie in the lower part of the Vega, about two leagues from Granada, and all the land is capable of irrigation. His grace's estates return about 15,000 dollars a year; his rents are paid in grain; a fixed quantity, not a proportion of the crop, a plan beginning to be universally followed by other landholders. The duke has 300 tenants; from which it appears that very small farms are held in the Vega; for if the whole rental be divided by 300, the average rent of the possessions will be but 50 dollars each. The tenants upon the duke's estate are thriving; they pay no taxes; and these estates are exempt from many of the heavy burdens thrown upon land. A composition of 6 per cent. is accepted from the Duke of Wellington in lieu of all demands.' The mountain regions afford good pasture; but grazing is less understood here than in most other parts of Spain. The horses of Granada are inferior to those of Cordova; and sheep, though plentiful, have very coarse wool. The asses are superior to most others, both in height and strength. Goats are very numerous, and thrive well. Pigs of a black breed are reared in vast numbers in the woods near Alhama. The anchovy and the tunny fisheries give full employment to the inhabs. of the sea-shore.

Except in the articles of wine and oil, the produce of this once fertile prov. does not equal the local consumption. Coarse linen and woollen cloths, silk, paper, leather, and gunpowder are made in small quantities; but no branch of industry is thriving. Its exports, through Malaga and Almeria, chiefly consist of wines, oil, dried fruits, wax, anchovies, and lead; its imports, of hardware and cutlery from England, lace from France, and cloths from England and Holland.

Granada formed a part of the ancient Boetica; and on the destruction of the Ibero-African empire, it became a new state, founded by Mohammed Alhamar, in 1238. It remained in the possession of the Moors for 254 years, which comprise the season of its prosperity. In 1492, it surrendered to Ferdinand the Catholic, being the last province that opposed his arms. The Moors were, by the treaty of peace, to enjoy freedom of religious worship; but this condition was soon broken, and ultimately they were expelled the prov.

GRANADA (an. Illiberis), a famous city of Spain, cap. of the above prov. and kingdom, on the N. side of the Sierra Nevada, and at the junction of the rivers Darro and Xenil, in a mountainous region, not less than 2,210 ft. above the sea, 116 m. E. by S. Seville, with which it is connected by railway, and 217 m. S. by W. Madrid. Pop. 65,993 in 1857. The city stands on the edge of a fertile and extensive expe of plain, which three rivers traverse, on two hills, one of which, between the rivers is crowned by the palace of the Alhambra and the Torres Bermejos; the other, N. of the Darro, by the Albaycin and the Alcazaba. It still covers a considerable extent of ground, though certainly far less than it must have occupied when swarming with half a million Mohammedans. The approach to it on the Malaga side

is particularly fine; a handsome stone bridge, built by the French during the war of independence, spans the Xenil, and immediately beyond rise orchard walls, and terraced gardens, domes, minarets, and shining steeples, reaching to the base of the rock which bears the Alhambra. Every thing within the precincts of the city bears the marks of Moslem hands: the narrow, crooked, and badly-paved streets, and gushing fountains, the lofty flat-roofed houses and heavy projecting balconies, are all quite Oriental; whilst here and there the entrance of some old mosque or ruined bath bears in its horse-shoe arch the peculiar stamp of the nation. The city contains a cathedral, a chapel of the Catholic kings, and twenty-three parish churches, of which those of San Geronimo and San Juan de Dios are best worth seeing. In all of them are to be seen specimens of variegated marble, not equalled elsewhere, perhaps, except in Italy. The cathedral is a clumsy-looking building, 425 ft. long, and 250 ft. broad; the interior is heavy, excessively gaudy, and fitted up in the worst possible taste. The high altar, flanked by its gilded pillars, is insulated after the Roman fashion, under a dome 170 ft. high, and the area round its base is conspicuous by reason of its light iron railing, and marble pavement. In this church is an exquisite Holy Family by Murillo. The chapel of the kings, which adjoins the cathedral, is of Gothic architecture, is noted for a flat arch of remarkable boldness, which supports its roof. Ferdinand and Isabella, and their successors Philip and Joanna, are buried in front of the altar, and their tombs are superbly sculptured. (Swinburne's Spain, i. 301; Scott, i. 261.) The Carthusian convent, about a mile from town, which had till lately great wealth and immense revenues, has a fine marble altar, and some excellent paintings by Murillo and Cano. The palace of the Alhambra (al-hamra, the red) is, however, the building by which the travellers' attention is chiefly arrested. This irregular mass of houses and towers, perched on a very high hill, which projects into the plain, and overlooks the city, is said to have been erected about 1224. The walls of the fortress follow the various sinuosities of the cliffs, which bound the plateau on which it stands. The chief entrance, which is approached through a long avenue of elms and myrtles, in one of the towers on the S. front, is called the Gate of Judgment; and over it is embossed a key, the memorial ensign of the Andalusian Moors. The first object seen on entering, in the centre of the plateau, is the palace of the emperor Charles V., built by Verreguete. It is a complete square of 185 ft., having two orders of pilasters, Doric and Ionic, upon a rustic base, the whole measuring 62 ft. from the higher entablature to the base. An oblong vestibule leads into the circular court, forming the centre of the palace; a colonnade of two stories each supported by thirty-two columns, runs round its circumference. This building, remarkable for magnificence, elegance, and unity of design, was never completed; the pillars are much damaged, and the whole will soon fall to the ground. N. of this building, and strangely contrasted in appearance, stands the palace of the Moorish kings, externally a huge heap of as ugly buildings as can well be seen. A plain unornamented door admits to the interior. The first place entered is an oblong square, having a deep reservoir for water in the middle, and baths at the sides also, with parterres and rows of orange-trees ranged around; the ceilings and walls bring ornamented with intricate stucco and fretwork painted, gilt, and lettered, as in other parts of the building, in the most delicate manner. Beyond this is the

Court of Lions, an oblong enclosure, 100 ft. by 50 ft., once paved with white marble, but now converted into a garden, and surrounded by a colonnade of about 130 slender white marble pillars, irregularly placed, and supporting low arches that run round the place. In the centre is a fountain, supported by thirteen lions, or rather panthers, who disgorge water into a basin of black marble. The arabesque work here is most elaborate. N. of the last-mentioned court is the tower of the two sisters, a range of apartments having a beautiful ceiling stuccoed in stalactites, and beautifully gilded, and a large window opening to the country; and on the opposite side is the Hall of the Abencerrages, where the chiefs of that noble line are said to have been massacred. The Hall of Ambassadors, however, may be truly called the pride of the Alhambra: it is a square of 36 ft., and is 60 ft. high to the top of the cupula, having a ceiling vaulted in a singularly graceful manner, and inlaid with mosaic of mother of pearl; its walls, also, being adorned with groups of flowers, and these intermingled with arabesques of curious workmanship. Highly finished inside, it has also the advantage of extensive views over the city, the dark valley of the Darro, and some other parts of the palace. The gardens, which abound with orange and lemon trees, pomegranates, and myrtles, lead by a low postern gate to the summer palace of the generalife, situated on the steep declivity of the opposite hill. In the building itself there is nothing particularly worthy of observation; but the myrtle groves and terraces are agreeable, and from the latter there is a charming view over the Alhambra and its gardens. Above the palace, near the summit of the rock, is a seat cut in the rock, which the Moorish kings are said to have used as a point of observation during the siege of Granada. In the city are several hospitals, the largest being that of San Juan de Dios. The university, founded in 1531, has, on the average from 900 to 1,000 students. There are six colleges and two academies; one for mathematics, the other for design. The walks about the city are most beautiful; especially two alamedas, one on the Xenil, above which fine orange groves, cypress alleys, and clusters of laurels grouped together; the other on the Darro, flowing through a deep romantic ravine, whose scenery equals that of Switzerland.

Granada, many years ago, had extensive factories for velvets, silks, and ribands, employing 2,000 hands, and working up the produce of the neighbourhood (not less than 2,600,000 lbs. of silk), with large paper-mills, and a flourishing oil trade. But at present its industry is in a very low state. This decline in the manufactures and trade of Granada has been ascribed to the emancipation of S. America; and this, probably, may have had some effect. But they had long previously been in a state of paralysis and decay, occasioned by the vicious regulations and the oppressive and injurious imposts to which they were subject. The principal existing business is carried on in the market-place, surrounded with small houses inhabited by the poorer orders, and in a narrow crooked street called El Zacatin, the little market, which in better times was the great silk mart. Towards the centre of the city is a bazaar in the Eastern fashion, each stall being boarded off from the rest; but in none of these is there much apparent activity.

The Granadians (called the Gascons of Spain) are proud of their city, and boast not a little of its antiquities and faded grandeur, reckoning themselves at the same time most constitutional citizens. The women are handsome and elegant,

like the rest of the Andalusians, but are spoiled by adopting French costumes. Like the rest of their countrywomen, they are fond of theatres, masked balls, and the indispensable tertulia. Granada is the see of an archbishop, who formerly possessed a revenue of above 25,000l. a year, and the residence of a captain-general, and is governed by a corregidor and two alcaldes. The Alhambra has its separate governor.

The early history of Granada is hidden in obscurity. Under the Romans, Illiberis was a place of some importance, being made by them a municipal colony entitled Municipium Florentinum Illiberitanum. The Goths changed the Roman name into Eliberi, and allowed the place to fall into decay. The present city was founded by the Moors in the 10th century, and became a part of the kingdom of Cordova. In 1236 it was strengthened and augmented, in consequence of being selected by Mohammed Alhamar as the capital of his new kingdom. The throne continued in the family of that prince till 1492, when, after a year's siege, it surrendered to Ferdinand the Catholic. Many Moorish families continued to reside here for a century and a half after its conquest, and contributed to its prosperity and importance. Various attempts to convert them to Christianity were made subsequently to the conquest of Granada; but these having proved, as is alleged, totally unsuccessful, the imbecile, priest-ridden government of Philip III, resolved, at the instigation of a few bigoted ecclesiastics, to expel the Moors from all parts of Spain. This insane resolution, by which the kingdom was deprived of a large number of its most industrious and valuable citizens, was carried into effect in 1609 and 1610, under circumstances of the greatest barbarity. This act may be said to have consummated the degradation of Spain; and her vicious institutions have prevented her recovering, down even to the present hour, from the wounds inflicted by the bigotry and stupidity of her rulers.

GRANADA (NEW). See COLUMBIA.

GRANARD, an inland town of Ireland, co. Longford, prov. Leinster, 13 m. W. by N. Longford. Pop. 2,058 in 1831, and 1,671 in 1861. The town consists of one street, and has in it the par. church, a Rom. Cath. chapel, a market-house, and dispensary. Adjoining the town is a remarkable rath or mount, called the Moat of Granard, which commands extensive views of the surrounding country. Markets, well supplied with agricultural produce, are held on Mondays, and fairs on May 3 and Oct. 1. Petty sessions on Thursdays. It is a constabulary station.

GRANGEMOUTH, a sea-port town of Scotland, co. Stirling, par. Falkirk, at the E. extremity of the Forth and Clyde Canal, at a point where this line of communication unites with the small river Carron, ½ m. from the Frith of Forth, 11 m. SE. Stirling, and 18 W. by N. Edinburgh, on the Scottish Central railway. Pop. 1,789 in 1861. The town is substantially built; public buildings, the custom-house, and a large Presbyterian church, in connection with the Kirk of Scotland. Grangemouth has spacious warehouses, commodious quays for shipping, and a dry dock. The Carron Iron Company, distant 2 m. inland, has a wharf here for its vessels, varying from 15 to 20 in number. The place may, indeed, be regarded as the emporium of the trade, not only of Carron, Falkirk, and other places in its vicinity, but of Stirlingshire, as it possesses the best harbour in the county, though no vessels drawing above 12 ft. water can with ease or safety approach it. The chief exports are iron goods, grain, and wool; but the manufacturers of Stirling and St. Ninians, also, send their goods by land carriage to be exported at Grangemouth. The chief article of foreign import in timber and ship-building is carried on to a considerable extent. Timber imported for Stirling, and, even sometimes for Leith, is landed here, and conveyed to its final destination by means of rafts. The custom-house of Grangemouth, established in 1810, includes the subsidiary port of Alloa, on the opposite side of the Forth. Gross customs' duties received, 27,469l. in 1859; 10,512l. in 1861; and 15,341l. in 1864.

Grangemouth was founded in 1771, in connection with the Forth and Clyde canal, and has long superseded Airth, which had previously been the chief sea-port of Stirlingshire. The inhabitants are all employed in connection with the trade of the place or the canal, except a few who engage in fishing. Kinnaird House, the seat of the late Mr. Bruce, the celebrated Abyssinian traveller, is in the neighbourhood, and Kerse House, an elegant seat of the Earl of Zetland, is within ½ m. of Grangemouth.

GRANTHAM, a parl. bor., market town, and par. of England, co. Lincoln, soke Grantham, on the Witham, 96 m. N. by W. London by road, and 105½ m. by Great Northern railway. Pop. of munic. bor. 4,953 and of parl. bor. 11,121 in 1861. The town, consisting chiefly of four streets, is neat, clean, and well lighted, but not remarkable for its buildings, and is wholly situated on the W. bank of the river. An increase of buildings has taken place and is still going on, principally in the Spittlegate end of the town. The church, a fine specimen of the Gothic style of the 13th century, has an elegant spire 270 ft. high, and in the interior an elaborately carved font, and some splendid monuments; in the vestry is a public library, left by Dr. Newcombe, master of St. John's Coll., Cambridge. The living, a vicarage, is divided, and is in the gift of two prebends of Salisbury Cathedral. The guildhall was rebuilt in 1787, with the addition of a spacious assembly-room. The grammar-school, at which Sir Isaac Newton was partly educated, was founded and endowed by Henry VIII. and his son Edward VI., out of the spoils of a monastery of grey friars in the town. Grantham is not a manufacturing town; but it is said to be flourishing, and its trade to be increasing. The principal trade is that of malting, which is carried on to a great extent. There is a canal, uniting the town with the Trent, by means of which an extensive export of corn and other agricultural produce takes place, and an import, principally of coal, with which the neighbouring towns to a considerable distance are supplied.

The bor., which was formerly ruled by 2 aldermen, 13 com. burgesses, and 12 second burgesses, according to a charter granted in the 7th of Charles I., is now under 4 aldermen and 12 burgesses. Grantham has returned 2 mem. to the H. of C. since the 7th of Edward IV. Previously to the Reform Act, the parl. bor. was identical with the old bor.; the right of voting was vested in freemen not receiving alms, and the average number of electors for 30 years before 1831, was 864. The Boundary Act extended the limits of the parl. bor., so as to make it include the whole par. Registered electors, 789 in 1865, of whom 175 freemen. Markets on Saturday, and fairs for sheep and cattle, 5th Monday in Lent, Ascension Day, July 10, Oct. 26, and Dec. 17.

Grantham is situated on the old Roman road called Ermine Street, and was a strong Roman station. At the time of the Norman survey it was a royal demesne. It was first incorporated by Edward IV. in 1463, and received, in addition, 12 charters of later date.

GRANVILLE (an. *Grannonum*), a fortified sea-port town of France, dep. Manche, cap. cant., built on and adjoining to a steep rocky promontory projecting into the English Channel 30 m. SW., St. Lo, and 46 WNW. Caen, with which it is connected by railway. Pop. 17,160 in 1861. Granville is the only fortified town on the coast between Cherbourg and St. Malo; it is encircled by strong walls, which shut the citadel off from a suburb on the E. and SE.; and though irregularly laid out with precipitous and narrow streets, contains many venerable edifices, among which is a Gothic par. church. It has no hospital, and some good baths. The port, on the S. side of the town, is spacious and secure, being defended W. and SW. by a large and handsome granite pier, which cost 2,500,000 francs. The harbour is partially dry at low water. There is regular steam communication between Granville and St. Helier, Jersey, 30 m. distant. Granville is the seat of a tribunal of commerce, and of a school of navigation; and the residence of a commissary of marine. Its chief trade is in the cod and oyster fisheries. The latter of these employ about 800 hands, in 50 boats, of about 12 tons each. In the cod-fisheries of Newfoundland about 70 vessels, of 100 to 350 tons each, are employed, with about 8,000 men; besides which, about 15 vessels are engaged in supplying the French colonies with salt fish. Thirteen vessels are employed in trading with the E. and W. Indies, of the burden of 4,100 tons. About 33 smaller vessels are employed in the coasting and channel island trade. The total burden of the shipping of this port amounts to 23,000 tons. Eggs are largely exported from Granville to London. Granville was bombarded and burned by the English in 1695; and was partly destroyed by the Vendean troops in 1793.

GRASSE, a town of France, dép. Var, cap. arrond., on the S. declivity of a hill facing the Mediterranean, from which it is about 7 m. distant, and 23 m. NE. Draguignan. Pop. 12,613 in 1861. The situation of Grasse is highly picturesque; from the S. it rises in successive terraces of white houses, having at its summit the principal church, and a large Gothic tower, the only remnant of the walls by which it was surrounded in the middle ages. It commands extensive and beautiful prospects, and enjoys a healthy climate; though the heat in summer is oppressive. The buildings of the town are generally good; but the streets are steep, narrow, crooked, and dirty; it has, however, a large open market-place, clean, and surrounded by good shops; and at its W. extremity is a fine public promenade. The town is extremely well-furnished with water by a rivulet which rises above it; and which supplies not only the public fountains, and two considerable reservoirs, but turns many mills, and supplies various factories. The principal church is a large, but low-heavy Gothic building; it has a curious crypt cut out of the rock, a marble altar, and some good paintings. There are 3 hospitals, in the chapel of one of which are 3 paintings by Rubens; a town hall, exchange, theatre, communal college, public library with 5,000 vols., and gallery of paintings. Some Roman antiquities exist here; particularly a small edifice about 30 ft. in diameter, formerly used as a chapel, but supposed to have been originally a temple of Jupiter. Grasse is the seat of a sub-prefecture, and of tribunals of original jurisdiction and commerce. It is noted for its manufactures of perfumery, and has a large trade in that article, which dates from about the middle of the last century. Great quantities of orange-flower water and essences of various kind are distilled; and extensive purchases of Italian perfumery are made by the inhabitants, who also buy up the flowers of the principality of Monaco, and the dep. of Nice, and the oil of their own arrondissement. In the latter article, as well as fruits, Grasse has an active trade; it has also manufactures of coarse woollen stuffs, organzined silk, linen, thread, leather, soap, liqueurs, and brandy. Fine marble and alabaster are found in its neighbourhood. The present town is said to have originated in 543, from a colony of Sardinian Jews, who had embraced Christianity. In the succeeding ages, the adjacent coasts being frequently ravaged by the Saracens, Grasse received great accessions to its population in emigrants from Frejus and Antibes.

GRATZ (Slav, *Nemetzki Gradetz*, 'the mountain fortress of Niemezki'), a city of the Austrian empire, cap. Styria, near the centre of which it is situated, on both sides the Mur, a tributary of the Drave, 82 m. NE. Laybach, 88 m. NNW. Agram, and 89 m. SW. Vienna, on the railway from Vienna to Trieste. Pop. 63,176 in 1861. Gratz is, next to Vienna, Prague, and Trieste, the largest, most populous, and most important city of the German portion of the Austrian empire. It stands in the N. part of an oblong plain, and consists of the city proper on the E. bank of the Mur; and four extensive suburbs, the Murvadt on the W. bank of the river, connected with the opposite side by two bridges, and three others. The ancient fortifications were finally levelled by the French in 1809. A great bluff lump of rock, which rises to the height of 300 ft. at the N. extremity of, or rather within, the city itself, and whereon once stood the citadel, serves now only as an occasional promenade for the inhabitants, thence to survey the singular beauty of the surrounding scenery. After Salzburg and Innsbruck, Gratz boasts of a more picturesque situation than any other city in the Austrian dominions. All around its plain, through which the Mur, a large and rapid river, flows amidst fields of corn and rural hamlets, then an amphitheatre of hills, none very high, but finely diversified in form, green, and wooded; and beyond these again are beheld, towards the N. and W., the lofty mountain masses of Upper Styria and Carinthia, rising in rugged grandeur, and for the greater portion of the year covered with snow.

Gratz, with its suburbs, is about 1½ German, or nearly 7 English m. in circ.; but the city itself forms but a very small part of the whole, being only 220 fathoms in length by 420 in breadth, and containing about 30 streets and open spaces, with little more than 400 houses. The interior is like that of most ancient towns. The streets are generally narrow and dark, opening occasionally into large irregular 'Places.' The shops are tolerable; the houses of the higher classes, all of stone, are spacious and gloomy; and such is the character also of the churches, many of which are highly decorated within. The inner city, like that of Vienna, is surrounded by high ramparts, now of no use as fortifications, and is entered by six gates. The ramparts, together with the glacis or esplanade beyond them, form the favourite walks of the inhabitants. The esplanade is planted with chestnut trees, and is well kept. The city and its suburbs generally are tolerably well built, and contain many good private, as well as some fine public edifices; but the thoroughfares, especially in the inner town, are mostly ill-paved and ill-drained.

Gratz has twenty-three churches and chapels, besides seven monasteries. The cathedral, or church of St. Ægidii, a Gothic edifice built in 1356, contains many handsome marble monuments. Near it is a chapel in the Italian style,

containing the mausoleum of Ferdinand II., a native of Gratz. Opposite this edifice is the Convent, the largest building in Gratz, formerly a Jesuits' college, now a public school belonging to the university. The latter institution, founded by Charles Francis, duke of Styria, in 1586, was closed by Joseph II. and reopened by the emperor Francis in 1827. It is one of the second order, having faculties of theology, law, and philosophy. In medicine lectures are given, but no degrees are conferred. The library, according to Turnbull, comprises about 60,000 vols., 2,000 MSS., and several literary curiosities. It is kept partly in some smaller rooms, but principally in a lofty, spacious, and elegant saloon, which, at the period when the university was under the direction of the Jesuits, was not unfrequently used as a theatre, for the performance of 'Mysteries.' The ordinary students attending the university exceed 300. The Burg, or ancient palace of the Styrian dukes, now the residence of the governor; the par. church, with the highest tower in the town, and an altarpiece by Tintoretto; the *Landhaus*, a very ancient edifice, in which the estates or parl. of Styria meet, and in which the ducal hat of Styria is preserved; the new council-house, built in 1807; the theatre, and the palaces of various noblemen, are the other principal buildings. One wing of the Landhaus is called the 'arsenal,' and is filled with many thousand suits of rusty armour.

But the pride of Gratz and of Styria is the *Johanneum*, one of the most valuable establishments of the kind in Europe. It owes its origin to the late archduke John, whence its name; by whom it was founded in 1811, and who presented to it the whole of his extensive collections in art and science. Its object is the encouragement of the arts and manufactures of Styria, by means of collections, lectures, and a public library. The museum of natural history occupies thirteen rooms, some very spacious. The departments of mineralogy and zoology have very complete collections of the minerals and animals of Styria, and the botanical department contains a *hortus siccus* of more than 15,000 plants. There are collections of the manufactured articles of Styria, and of the agricultural and mechanical implements used in the duchy; besides which, are specimens or models of the principal instruments and machines of all kinds adopted for similar purposes in foreign countries. One room is devoted to antiquities, comprising many Roman, Styrian, and other coins, and Persian, Babylonian, and other antiquities. Near this room is a fire-proof apartment for the custody of records, containing, among other documents, several charters of the ninth and tenth centuries, especially one of 878 by the emperor Carloman. An extensive botanic garden is now attached to the building. The salaries of the eminent professors, who give lectures on mineralogy, geology, botany, chemistry, agriculture, and the useful arts, are defrayed by the *Stande*, or provincial parliament, the students attending *gratis*. The library, which is open to the public at large, comprises the best standard works of all countries. There is another reading-room and library attached to the Johanneum, to which strangers are admitted gratuitously, and natives on payment of about 2s. 6d. a month. It receives newspapers and periodical publications from all parts of Germany, Italy, France, and Great Britain; in all, more than a hundred journals.

Besides the foregoing educational establishments, Gratz has a gymnasium, episcopal academy, military school, a school for teachers, female seminaries, a school kept by Ursuline nuns, schools of music, dancing, oratory, the fine arts, and many

Sunday schools, and others for the instruction of the poor. There are five convents and two monasteries. The splendid abbey, built by Ferdinand II. for the Capuchin monks, and intended to commemorate the fact of his burning 20,000 Protestant bibles by the hands of the common hangman, was converted by Joseph II. to the purpose of a madhouse. Gratz has six hospitals, besides others belonging to some of the monastic establishments, a foundling hospital, orphan and deaf and dumb asylums, and various other benevolent institutions; a provincial gaol, workhouse, some military magazines, a society for the furtherance of agriculture, other learned associations, and several collections of paintings. It is the seat of the highest civil authorities for the duchy of Styria; of the military commandant for Styria, Illyria, and the Tyrol; the prov. part. of the duchy; the council for the circle of Gratz; and the residence of the prince-bishop of Seckau. Its principal manufactures are cotton, silk, and woollen fabrics, leather, iron wire, nails, and other metallic goods; it has, however, others of starch, hats, manglin, paper, and earthenware. Its trade in timber, iron, clover-seed, and the other products of Styria, with Hungary, Croatia, Transylvania, and Turkey, is considerable; and it has a large share of the transit trade between Vienna and Trieste. It has two large fairs yearly. The Mur, though it often greatly injures the city and its vicinity by its inundations, renders the latter very fertile. Gratz is well supplied with all kinds of provisions, and is one of the cheapest towns in the Austrian dominions: many of its inhabitants are retired officers of the army, and persons of rank but with limited means. As early as the ninth century, Gratz was a town of some consideration; in 1127 it became the residence of the dukes of Styria. It was taken by the French in 1809, after a siege of seven days. After the revolution of 1830, it was for a while the residence of Charles X., and the exiled royal family of France.

GRAUDENZ (Slav. *Grudziuadz*), a town of the kingd. and prov. Prussia, gov. Marienwerder, cap. circ. same name, on the Vistula, which is here crossed by a bridge of boats; 60 m. N. by E. Dantzic, near the railway from Dantzic to Berlin. Pop. 12,701 in 1861, exclusive of a garrison of 2,669. The town is walled, and farther defended by a strong fortress erected on the Vistula in 1776. It has three suburbs, five Rom. Cath. churches, a Lutheran church, two superior schools, a teachers' seminary, house of correction for W. Prussia, with which an establishment for the treatment of juvenile felons is connected, circle council, board of taxation, judicial court of the first class for the district and town, and manufactures of tobacco and carriages, with extensive breweries, and some trade in corn and woollen cloth.

GRAVESEND, a bor., market-town, sea-port, and par. of England, co. Kent, hund. Toltingtrough, on the S. bank of the Thames, 20 m. E. by N. London by road, and 24 m. by North Kent railway. Pop. of par. 7,885, and of munic. bor. 18,782 in 1861. That part of the town which adjourns the river has steep, narrow, inconvenient, dirty-looking streets; but the upper and more recent part is built in better taste, with wide streets, neat and cheerful residences, and pretty gardens. The principal edifices are the old church, built of brick, in 1730; a town-hall and market-place, handsomely built, but pent up amid mean and dirty houses; a custom-house, and a small theatre. A battery lies to the E. of the town, nearly facing Tilbury Fort, on the Essex shore. Two or three hotels are amongst the handsomest buildings in the place. W. of the town, on the river bank, are some baths, beautifully as well as commodiously constructed, and forming a highly ornamental feature from the water. The pier, which is of iron, is a modern erection, built by the corporation, and bringing in a large income by the tolls levied on the visitors and others landing there. Another pier, or jetty of wood, has been erected 800 yards E. of the former, by parties opposed to interest to the corporation: both are extensive proprietors of steam-boats plying between London and this place. Nearly ½ m. S. of the river is a suburb, called Windmill Hill, with tea-gardens and archery grounds: from the summit is a fine view of the river and surrounding parts of Kent. The village of Milton is chiefly known by its picturesque church, nearly 1 m. E. from the town. Northfleet, lying 1½ m. W. is a favourite place of resort for those who dislike the bustle of Gravesend. The Sard pop. consists principally of ship-carpenters, bargemen, watermen, and people employed in the chalk-works.

Gravesend some years ago placed its main dependence on the trade brought to it by ships wanting supplies of various kinds, and by captains and passengers passing through and staying in the town: since the establishment of steam-boats, however, and the erection of the pier, it has been rapidly increasing in size and importance, the cheap and speedy communication having rendered it a place much resorted to in summer by the middle classes, many of which have houses here, to which they come daily or weekly at the close of business. The crowds of visitors on Sunday, in fine weather, are very great. Much of the land about the town is occupied by market-gardeners, who raise vegetables for the London market.

Gravesend, which was incorporated with Milton in the reign of Elizabeth, was, before the Mun. Reform Act, under the local jurisdiction of a mayor, 12 jurats, and 24 common councilmen, with a recorder, and other officers. By that act the bor. was enlarged, by the addition of a part of Northfleet parish, and divided into two wards, governed by six aldermen (one of whom is mayor) and 18 councillors. It is one of the polling places for W. Kent. Markets, Wednesdays and Saturdays. Fairs, May 4 and Oct. 24.

The town is called *Gravesham* in Domesday Book, and its later name was Graves-end, supposed to be derived from the Saxon *gerefa*, or German *greve*, ruler, and *ende*, boundary, because the town was the limit of the ancient portreeve's authority. The high bailiff was called the portreve in the 14th century. In the time of Richard II. the town was burnt by the French, and many of the inhab. carried into captivity. In the same reign the watermen of Gravesend obtained the exclusive right of conveying passengers to London, which right is still acknowledged, by a yearly compensation from the steam-packet companies. The town was first defended towards the river in the reign of Henry VIII. when Tilbury Fort was erected.

GRAVINA, a town of Southern Italy, prov. Bari, on a river of the same name, 334 m. SW. Bari. Pop. 10,460 in 1862. The town is a bishop's see, has a cathedral, and eight other churches, several convents, and a college. Two large fairs are held annually. It was formerly a place of some strength, having been unsuccessfully besieged by the Saracens in 975.

GRAY, a town of France, dép. Haute-Saône, cap. arrond., on the declivity of a hill on the Saône, 28 m. SW. Vesoul on the railway from Vesoul to Dôle. Pop. 7,031, in 1861. The town has a fine quay, and a handsome bridge across the Saône; but its streets are narrow, crooked, and steep. It is well furnished with public fountains, has an ancient residence of the dukes of Burgundy, cavalry barracks, a town-hall, built in 1505, an exchange, par. church, communal college, public library, with 4,000 vols., and a remarkable water-mill serving various purposes. Gray has an extensive trade, being an entrepôt for the produce of the S. destined for the E. of France. It has 4 large annual fairs.

GREECE, a modern kingdom of Europe, and the most celebrated state of antiquity. In its flourishing period Greece comprised the S. portion of the great E. peninsula of Europe, and extended N. to about lat. 42°, including Thessaly, and a part of modern Albania, with the Ionian Islands, Crete, and the islands of the Archipelago. ' Hæc contra Græcia, quæ fama, quæ gloria, quæ doctrina, quæ plurimis artibus, quæ etiam imperio et bellica laude floruit, parvum quemdam locum Europæ tenet, semperque tenuit.' (Cicero pro Flacco, § 27.) This famous region was originally called Hellas ('Ελλάς), and received the name of Greece from Græcus, a Thessalian prince. (Plin. Hist. Nat., lib. iv. § 7.) The modern kingdom of Greece, though less extensive than the country anciently so called, comprises the territories of all the most celebrated and interesting of the Grecian states. It includes that portion of the continent S. of the gulfs of Arta and Volo, and an imaginary line drawn between them nearly due E. and W., with the islands of Eubœa, the Cyclades, and the N. and W. Sporades. These dominions lie between lat. 36° 16' and 39° 54' N., and long. 20° 12' 30' and 26° 28' E.; the continental portion having N. the Turkish pachalics of Trikhala (Thessaly), and Albania (Epirus), and being surrounded every where else by the Mediterranean, denominated on the W. the Ionian Sea; and on the E. the Ægean or Levant. Total area of the kingdom, including the Ionian Islands, 19,340 sq. m.; area of the kingdom, 1,096,810 in 1861, and of the Ionian Islands—annexed in 1864—232,426 in 1864.

Population.—Continental Greece is naturally divided into two principal portions: the northern, or Hellas, comprising what has been called E. and W. Greece; and the southern, comprising the Morea, or Peloponnesus. The political division:—Greece is divided into ten provinces, or monarchies—subdivided into eparchies—with the following population, according to the census of 1861 and that of 1838:—

Monarchies	Population	
	1838	1861
Attica and Bœotia	87,231	116,071
Euboea	65,721	77,368
Phthiotis	81,850	107,751
Acarnania and Ætolia	80,619	108,592
Argolis and Corinth	115,714	135,749
Achaia and Elis	114,281	173,719
Arcadia	130,872	86,514
Messenia	96,805	117,184
Laconia	87,101	112,910
Cyclades	108,631	118,120
Total	904,122	1,096,810

The decennial increase of population in some of the provinces, and the decrease in others, point to a continual migration of the inhabitants, caused chiefly by the unsettled state of landed property. Thus the mountainous province of Arcadia is gradually getting depopulated, by migration seaward into Laconia and Argolis. The same movement is taking place in many districts of the Ionian Islands. Of the 1,096,810 inhabitants of continental Greece, registered at the census of 1861, there were 567,834 males and 528,976 females. It

is a curious fact that the number of male inhabitants should exceed in Greece the females, while the reverse occurs in all other European countries.

Physical Geography.—Greece possesses, in a high degree, those geographical features which distinguish Europe at large. No country is more remarkable for the irregularity of its shape, its shore, and its surface. Its N. portion, Hellas, stretches WNW. to ESE. for about 200 m., gradually decreasing in breadth from Acarnania to Cape Colonna in Attica. Its S. portion, the Morea, is a peninsula, said to derive its modern name from its supposed resemblance to a mulberry leaf. Its actual shape, however, is more like that of a vine leaf; it is united NE. to Hellas by the Isthmus of Corinth. The greatest length of the Morea, N. to S., is about 140 m.; its breadth varies from 80 to 135 m.; it comprises about half the area of the newly erected kingdom.

The surface of Greece is throughout mountainous, and scarcely any room is left for plains. Such of the latter as exist are principally along the sea-shore, or near the mouths of rivers, or else are mere basins, once forming the beds of mountain lakes, enclosed on all sides by mountains, or communicating with each other only by deep and narrow gorges. Such are the plains of Mantineia, Orchomenus, Stymphalus, Tripolis, and Copais. The most extensive tracts of plain country are in W. Hellas, and on the NW. and N. shores of the Morea. These are also the most productive parts of the country; but other very fertile, though small, plains are scattered through the E. of Greece, as those of Bœotia, E. Phocis, Marathon, and many others, which are still, as anciently, the granaries of the country. The most flourishing cities of antiquity, as Athens, Eleusis, Megara, Corinth, Argos, Sparta, and Thebes, were situated in the midst of or on the borders of these plains; and others, as Tripolitza, Leondari, Mistra, Gastouni, Patras, Mesolonghi, Zeitoun, and Livadia, which, in modern times, have ranked amongst the principal towns in Greece, have been similarly located.

The Mountains belong to the Alpine system, being a continuation of the Julian Alps, so remarkable in their whole extent for their numerous grottoes and caverns. The principal chain—that of Pindus—runs NW. to SE. through the centre of Hellas, as far as the Isthmus of Corinth. On entering Greece, the Pindus chain is supposed to be nearly 7,500 ft. in height. It sends off on its W. side some ranges through Acarnania and Ætolia, and the range of Mount Zagora or Behcua in Bœotia; but its offsets on this side are of very inferior height. The mountains of Acarnania in general are estimated at only about 1,500 ft. in height; and Mount Paleo Vouna, the summit of Helicon, has only 5,730 ft. of elevation. On the E. side the branches of Pindus are more lofty; Mount Guiona, the highest point in Greece, and near its N. boundary, is 8,239 ft. high; and Katahothra (Œta), 7,061 ft. The celebrated Mount Parnassus is a part of the central mountain chain: its principal summit, Liakoura, is 8,068 ft. in height. Mount Elaten (Cithæron) is 4,629 ft.; and in Attica, Parnes 4,636, Pentelicus 3,642, and Hymettus (Trelo-vouni) 3,370 ft. high. A mountain chain runs through Eubœa in its whole length nearly parallel to that of Pindus; its highest point, Mount Delphi (Dirphosuno), near its centre, reaches the elevation of 5,725 ft. A chain passes through the isthmus, and nearly through the Morea E. to W., giving off lateral branches, which reach quite to the extremities of the four S. promontories of the peninsula. The culminating point in this part of Greece is Mount St. Elias (Taygetus), in Maina.

7,500 ft. high. No mountain in Greece reaches the limit of perpetual snow. (Bruguière, Orographie; Peytier, in Geogr. Journal, viii. part 3; Expédition Scientifique de Morée et Atlas.)

Rivers and Lakes.—Greece has no navigable river, nor would any be worth notice, but for the classical recollections which attach to every portion of the soil and waters of this celebrated country. The Aspro-Potamos (*Achelous*) between Etolia and Acarnania, is the largest; the principal remaining ones are the Gurrha Mavro-Potamos (*Cephissus* of Bœotia), which runs into the lake Topolias, the Hellada (*Spercheus*) Asopo, the Athenian Cephissus and Ilissus,—in the Morea, the Rouphia (*Alpheus*), Vasilico (*Eurotas*), Iliaco (*Pencus*), Planizza (*Inachus*), Mavro-Nero (the ancient *Styx*), &c. The principal lake is that of Topolias (*Copais*), in W. Bœotia, said by Thiersch to be 1,000 ft. above the sea. It is of a very irregular shape, and in winter is sometimes 15 m. long, by 10 m. broad; but its size varies considerably at different periods of the year. In summer, it is reduced to a mere swamp, partly cultivated, and partly covered with reeds, and omitting pestiferous exhalations. It contains several small islands, and has a subterranean outlet for its waters under Mount Ptoon into the Channel of Talanti. There are a few insignificant pools in the Morea, including the Lernean and Stymphalian lakes, so famous in classic fable. The former of these is formed by several clear and copious springs (the veritable heads of the *Hydra*), which rush out of a rock at the foot of a hill. The lake is, however, so diminutive, and so much concealed by reeds and other aquatic plants, that it might easily be passed without attracting the attention of the traveller.' (Dodwell.) Marshes are numerous. Nearly the whole N. shore of the Morea, from Corinth to Patras, is low and marshy; and the inlets of both these towns, as well as of Nauplia, Argos and Zeitoun, the plains of Marathon, and a portion of that of Athens, suffer, at certain seasons of the year, from malaria generated by stagnant pools.

The want of navigable rivers in Greece is obviated by the numerous gulfs and inlets of the sea, which indent its coasts on every side, and afford unusual facilities to commerce, while they add to the variety and beauty of the scenery. The principal gulfs or bays are those of Volo, Zeitoun, Egina, or Athens (*Sin. Saronicus*), and Argos or Nauplia on the E.; Kolokythia and Koron on the S.; Arkhadia, Patras, and Arta, on the W.; and the extensive and beautiful Gulf of Corinth, between Hellas and the Morea. Between Lutraca and the main land are the Channels of Talanti and Egripo, united by the ancient *Euripus*. The shores of Greece are mostly abrupt. The chief headlands are, Capes Mantelo in Eubœa, Colonna (*Sunium*), and Skyllo (*Scyllœum*) on the E.; St. Angelo (*Malea*), Matapan (*Tœnarum*), and Gallo (*Acritas Pr.*), on the S.; and Klarenza and Khrophia on the W. coasts. (Leake, Col., Travels in N. Greece and the Morea; Hoffmann's Europa und seine Bewohner.)

Geology and Minerals.—The central chain of Pindus is composed in great part of primitive rocks, as serpentine, covered with a yellowish-green steatite, granite, gneiss, mica, and other schists. Rocks of this kind are also met with in E. Hellas; and they are plentiful in the higher mountain ranges of the Morea and the islands, particularly Mycone and Delos. Slate occurs in the ridge of Œta and several of the mountain-masses of Messenia and Arcadia. By far the greater portion of the country, however, consists of secondary formations. Greece, generally speaking, is a region of compact grey limestone. This material ascends to a considerable height above the level of the sea, and the chain of Œta, as well as Mounts Parnassus and Helicon, is almost entirely composed of it. The calcareous formations are similar in appearance to those of the S. of Ireland; and contain in many places great quantities of silex. The shores of the Morea are bordered by tertiary formations, containing an abundance of small shells. Volcanic action is clearly traceable, particularly in some of the islands. The whole of Greece abounds with caverns and fissures, whence sulphureous and other mephitic vapours arise, which were taken advantage of in antiquity, at Delphi and elsewhere, for practising religious deceptions. There are numerous hot and cold mineral springs, both saline and sulphureous; but few have yet been analysed. In some parts the soil is impregnated with nitre; this is especially the case near Corinth and Kalavrita. Marble of various colours, red and green in the Morea, and white at Pentelicus in Attica, porphyry, slate, gypsum, zinc, lead, iron, gold, and silver, in small quantities, cobalt, copper, manganese, zinc, sulphur, and asphaltum, are amongst the principal mineral products; but the quantities of any of them at present obtained are quite insignificant. It is the opinion of the most competent authorities that the gold, silver, copper, and lead mines of Attica and the islands of Siphnos and Seriphos are far from being exhausted. Iron abounds in Scyros, at Tænarum, and in Eubœa, where, also, as well as in Elis, there are abundant seams of coal.

The climate is temperate, and for the most part healthy, except in the low and marshy tracts round the shores and lakes, some of which are very unhealthy. The mean temperature, in a country the surface of which is so uneven, must, of course, vary considerably; but the medium temperature of the year in the plains of N. Greece may be about 60°, and in those of the S. about 64° 5' Fahr. At Athens the thermometer not unfrequently rises in July above 100° Fahr. Snow falls in the mountains by the middle of Oct., and even in the plains it is occasionally six inches deep; but it never lies long in the latter. The winters at Athens are confined to the two first months of the year. Both spring and autumn are rainy seasons; and in Dec. the rains are generally so heavy that many parts of the whole summer, which may be said to comprise half the year, a shower, or a cloud in the sky, is rare in several parts of the country. The harvest usually takes place in June, but it is nearly a month earlier in Attica than in other parts of Greece. The latter province enjoys the driest atmosphere of any, to which circumstance the better preservation of its splendid specimens of ancient art is mainly owing. Its climate is much more agreeable in every respect than that of some of the other provs., as Bœotia, Arcadia, &c. Violent tempests often occur in autumn, and storms of thunder and lightning in spring; earthquakes are not uncommon. Intermittent fevers, *elephantiasis*, and lepra, are amongst the most prevalent diseases; Greece has been occasionally visited by the plague. (Peytier in Journ. de Travaux; Leake, Hughes, Lord Byron, Cochrane.)

The *vegetable products* are for the most part similar to those of S. Italy. The country may, in this respect, be considered as divided into four distinct zones or regions, according to its elevation. The first zone, reaching to 1,500 ft. above the level of the sea, is adapted to the culture of the different kinds of grain, vines, figs, olives, dates, oranges, citrons, melons, pomegranates, and other fruits, cotton, indigo, tobacco, and almonds besides in evergreens, as the cypress, bay, myrtle, arbutus, oleanders, tentisks, with the oriental plane, manna, ash, several kinds of oaks and pines, and a multitude of aromatic herbs. The second zone is the *region of oak and chesnut*; it extends from 1,500 to 3,500 feet perpendicular, and produces, besides the trees above named, the white fir, several kinds of pine, and the manna-ash. The third zone is the *region of beech and pine*; it reaches to the height of 5,500 ft., and contains numerous woods consisting of these trees, interspersed with a few cornfields. The fourth zone, including all the surface above 5,500 ft. in height, is the *sub-alpine region*, and yields only a few wild plants. Among the extracts from Dr. Sibthorp's papers, given in Mr. Walpole's Memoirs, is a very complete list of Grecian plants, with an account of their medicinal and economic uses. A great deal of the surface abounds with aromatic plants peculiarly adapted for the honey-bee; and the *pirnari* (the *pirnus* of the ancient Greeks), which feeds the cochineal insect, is found of every size, from a low shrub to a large forest tree, both in the plains and on the mountains. Acarnania, Elis, Messenia, and the W. parts of Greece generally, are the most richly wooded; the islands are mostly destitute of wood. (Hoffmann, Europa and seine Bewohner, lii. 61; Leake, N. Greece and Morea.)

Animals.—The wolf, jackal, lynx, badger, fox, wild boar, wild goat, red deer, roebuck, moufflon (?) &c., inhabit the wilder and more inaccessible and densely wooded parts of Greece; and bears are sometimes met with on the N. frontier, and in the lofty regions of Arcadia and Maina. Hares are very numerous, and their skins are a considerable article of export from the Morea. The otter inhabits the rivers and marshes of Livadia; and plover and porpoises are seen around the coasts, and sometimes in the Corinthian Gulf. The large vulture frequents the cliffs of Delphi, and the woods and precipices of Parnassus. There are several species of the falcon tribe. The little owl (*Strix micarrina*), anciently the bird of Minerva, is still as common round Athens as in antiquity. The red-legged partridge, quails, woodcocks, snipes, wood-pigeons, &c., are plentiful; pheasants are to be found in the W. and N.; and large flocks of bustards are often seen in Thrace. The coasts and lakes abound with wild fowl; storks and many other birds of passage sojourn in Greece. Sturgeons, salmon, mullet, tunny, mackerel, anchovies, and abundance of shell-fish, are caught around the coasts. Large and delicate white eels (often weighing 12 lbs.) are still found, as anciently, in the lake Copais. They are salted, and sent in large quantities to Constantinople, and into the marts of Greece. The coast-fisheries afford employment and subsistence to no inconsiderable number of the population; but their produce is notwithstanding insufficient to supply the demand during the long fasts prescribed by the Greek church, and a good deal of salted fish is imported. Poisonous vipers, and other serpents, infest certain localities; leeches are very plentiful in some of the brooks, which are therefore farmed out by the government as a means of revenue. The insect tribes of Greece include several Asiatic and African as well as European species; especially of the order *Orthoptera*. Wild bees are abundant; clouds of locusts occasionally do great damage to the crops. (Pouqueville; Hughes; Leake; Cockmore.)

Scenery.—Travellers in Greece generally speak in high terms of its scenery. It has everywhere the finest views, and is interesting not less from its natural beauties, than its classical associations,

and the ruins of ancient art and splendour scattered over it.

'Yet see thy skies so blue, thy crags so wild;
Sweet are thy groves, and verdant are thy fields,
Thine olive ripe as when Minerva smiled,
And still his honied wealth Hymettus yields;
There the blithe bee his fragrant fortress builds,
The freeborn wanderer of thy mountain air;
Apollo still thy long, long summer gilds,
Still in his beam Mendeli's marbles glare;
Art, Glory, Freedom fail, but Nature still is fair.

'Where'er we tread, 'tis haunted, holy ground;
No earth of thine is lost in vulgar mould,
But one vast realm of wonder spreads around,
And all the Muse's tales seem truly told,
Till the sense aches with gazing to behold
The scenes our earliest dreams have dwelt upon;
Each hill and dale, each deepening glen and wold,
Defies the power which crush'd thy temples gone;
Age shakes Athena's tower, but spares gray Marathon.'
　　　　　　　　　　Childe Harold, canto ii.

The richly wooded and well-watered provinces of Acarnania and Etolia are succeeded towards the E. by the lofty, rugged, and forest-clad chains of Parnassus and Œta, alternating with the fertile valleys of the Cephissus and Hellada. Bœotia, consisting of two elevated basins, has been uniformly celebrated for its fertility, and was considered the granary of ancient Greece. Athens has been said to surpass all the other capitals of Europe; not only in ancient celebrity, but also in the beauty and variety of the surrounding country. It is much to be regretted that the fine forests which once clothed the hills of Greece have been so extensively ravaged, partly by the wanton rapacity of the inhabitants, partly by the Turkish troops, who carried fire and sword into the remote fastnesses of the mountains. Still, however, on Parnassus, Helicon, Taygetus, in Megaris and Arcadia, oak-forests and pines are found of great extent. (Thiersch's Athens and Attica.)

Distribution of Land and Agriculture.—Mr. Urquhart (Turkey and its Resources, 1833) estimated Hellas (E. and W.) to contain 3,548,200 stremmata of arable land, 199,710 str. vineyards, 4,430 str. garden ground, and 854,000 olive trees. About 2½ stremmata are equal to an English acre. This estimate, often quoted, was, however, probably under the mark. According to an official statement of the year 1862—given by Mr. Humboldt, British secretary of legation, in a report dated Athens, July 10, 1863 (Reports, no. viii.)—the total area of the kingdom is reckoned at 46,429,000 stremmata, or 13,129 square kilometres. No general cadastral survey of the country has as yet been attempted. The vineyards, olive grounds, currant plantations, &c., have alone, to some extent, been measured and valued. 'It is thus impossible,' says Mr. Humboldt, 'to know with any precision the cultivated area of Greece. The clumsy machinery of the dîme tax, or tax of the tenth of the agricultural produce levied in kind, alone affords some means of arriving at a conclusion on the subject. Nothing can be more vague than the delimitation of property in Greece. Landed proprietors themselves are often at a loss to determine the limits or even the site of their property. A case recently came under my notice where the owner of a piece of waste land could not with any certainty ascertain its position. All he knew was that he was the owner of some ground situated between the Pnyx and the hill of Philopappus; and an offer of purchase made to him by a friend of archæological taste who wished to make excavations, fell to the ground in consequence of the inability of the proprietor to point out the exact spot. About one-third of the country consists of mountains and rocks. One-fifth is covered with forests, in which great havoc is yearly made by the wandering shepherds, who ruthlessly set fire to the woods in order to obtain more pasture-land for their flocks. Probably one-half of the entire superficies of the soil is available for cultivation, and of this barely half has been turned to account. Yet with the climate and the richness of the soil the agriculture of Greece ought to thrive, whilst at present it is in the rudest, most hopeless condition. In order to reclaim it from its present state it will, above all, be necessary to alter the system of taxation, and to construct roads. As one of the many instances of the bad effect of the want of communications, it may be mentioned here, that though the country is in many parts rich in forests, one of the chief articles of importation is timber for ship-building and other purposes.

There is no regular succession of crops; and two years' fallows are common. Hellas is a better corn country than the Morea; and corn is extensively grown in Acarnania, Ætolia, and Bœotia: in the last-named prov. there is always a good crop, the soil being continually moist, even though drought prevail throughout the rest of Greece. As many as six different species of wheat are grown; returning, it is said, after a dry spring, from 3 to 5, or in a very favourable season, as many as from 10 to 15 for 1. The wheat of the Morea has long been highly prized in the adjacent islands; the lands on either side the Gulf of Corinth, and in a part of Attica, are favourable to the growth of barley, as well as celebrated for their olives. The culture of oats and rye is unimportant. Maize is grown in Bœotia, and the Morea. Rice is cultivated in the plains of Marathon, Argos, and other marshy tracts along the coasts; and the rice of Argolis is said to be esteemed next after that of Damietta in the markets of Constantinople, to which it is exported from Nauplia. Marathon, though forgotten in almost every other respect, is still celebrated, as before the era of its glory, for being the granary of Athens. The demand for the currant-grape in Great Britain and other N. countries of Europe, has brought it into extensive culture in the Morea; and the S. shore of the Corinthian Gulf from Corinth to Patras is in great part covered with currant-vineyards. The hills of Greece are admirably adapted for the vine (*Vitis vinifera*), yet few vines are grown, except in low situations. The wines of Metaxa and Corinth, Elis and Arcadia, the valley of Helicon, the islands of Naxos, Santorin, &c., have a rich and delicate flavour; but they have comparatively little body, and are almost universally ruined (for other European palates), by the addition of resin or turpentine, a practice handed down from the ancients. Most part of the wine used in continental Greece is brought from the islands of the Archipelago, which are rich also in fruits of various kinds. The olive-oil of Greece would be good if well-prepared; the best is said to be furnished by Attica, Ægina, and Maina. Cotton of good quality is grown in Messenia, Laconia, and other parts of the Morea, but especially in the plain of Argos. Madder and tobacco in Bœotia, flax and hemp, figs in Attica (so famous in antiquity), and elsewhere, pomegranates, oranges, lemons, peaches, almonds, and a great variety of shell-fruit, apricots and other pulses; tomatas, cucumbers, artichokes, potatoes, and the pot-herbs common in the rest of Europe, are among the remaining articles of culture. The collecting of gall-nuts and valonea bark, which formerly received a considerable share of attention, has been latterly much neglected; and but little pains are bestowed on mulberry plantations, though the annual export of silk be estimated at 60,000 okes. Large quantities of wax are exported

from Nauplia. Honey is a highly important product; that of Attica, and especially of Mount Hymettus, is now, as of old, the best in Europe. It is transparent, and has a delicious perfume.

The fertility of the soil of Greece appears to be as great now as it was in ancient times. Mr. Mumbold, British secretary of Legation, in his report before quoted (of July 10, 1863), says in this respect:—' Notwithstanding the excessive dryness of the climate and the torrid heat of summer, the soil, when turned up and only superficially raked as by the rude plough of the time of Hesiod, to this day used by the Greek husbandman, is generally found to be most fertile. A proprietor in Euboea bought some land which had been under cultivation, but had been left fallow for some time previous to his purchase. Although contiguous to his former property, and the soil being to all appearance similar, the crops on his new acquisition were much heavier and yielded superior grain. When the causes of this difference were inquired into, it was found that the former owner had cultivated madder or 'garance,' a plant largely used for the dye of the scarlet habiliments of the far-famed 'fantassin' of France, and which requires a far more searching investigation of the soil than the superficial scratches which constitute the furrows of Greek husbandry. As in Euboea, so in other parts of Greece. Even beneath the desolate stony wastes of Attica in many places lie all the wealth of a virgin soil. Remove but the hard sun-dried surface, and a rich brown loam will turn up, at sight of which the hearts of our English farmers would be gladdened. But nothing is done: no water is brought from the neighbouring ranges of Parnes and Pentelicus to refresh it; no hand is raised to weed out the stones and cut down the rank overgrowth of evergreens and brushwood; and all the year round the cold blasts from the north sweep over the dreary plain, and the pitiless sun pours down its scorching rays on a parched stony desert. The old myth of Deucalion is forgotten indeed.

Manufactures.—Manufactures are almost wholly domestic, every peasant's family producing, with few exceptions, the articles required for their consumption. A few silk, cotton, and woollen stuffs, household pottery, some cutlery, leather, and soap are made in the larger towns, carpets in the Isle of Andros, and sail-cloth and straw hats in that of Siphnos. Goat skins are prepared for holding wine, oil, and honey; brandy, liqueurs, vinegar, meerschaum pipes, and arms may also be mentioned. Saddlery and house-furniture have deteriorated since the departure of the Turks; and these, as well as most articles of luxury, are now imported from other parts of Europe. The art of dyeing in bright colours, for which the ancient Greeks were so celebrated, has, however, been perpetuated to the present day; and the Greek women excel in embroidery. Salt sufficient for the consumption of the country is produced in the lagoons near Missolonghi and elsewhere. Ship-building is extensively carried on in many places.

Commerce.—The Greeks have particularly distinguished themselves by the spirit and success with which they have engaged in naval and mercantile enterprises. Their commerce, next to their freedom, was the grand source of the prosperity of Athens, Corinth, and other Greek cities of antiquity. And in this respect the modern Greeks have been no unsuccessful imitators of their illustrious progenitors. The great articles of export from Greece consist of currants, silk, figs, wool, olive-oil, valonea, wine, sponge, wax, and tobacco; the principal imports being manufactured cotton and woollen goods, corn, with a great variety of subordinate articles, principally from England, but partly also from France and Germany. The exports and imports of Greece amounted in the eight years from 1851 to 1858 to 12,571,851l., or 1,571,457l. per annum, being about 30s. per head. The following tables represent the value of the imports and exports for the years 1858, 1859, and 1860:—

Imports

Years	Tons	
	Drachmas	£
1858	44,301,311	1,572,893
1859	49,382,317	1,741,704
1860	47,650,727	7,354,954

Exports

Years	Tons	
	Drachmas	£
1858	28,843,165	1,026,999
1859	27,392,247	988,081
1860	30,447,479	1,088,123

The following table shows the total number and tonnage of vessels entered at ports in Greece, in the foreign and coasting trades, during the year 1860:—

Nationality of Vessels	1860	
	Vessels	Tons
British	851	127,433
American	7	699
Austrian	612	195,721
Egyptian	7	1,713
French	207	133,441
Ionian	8,547	488,436
Italian	864	87,671
Dutch	17	2,473
Wallachian and Moldavian	36	6,632
Russian	708	37,654
Turkish	8,926	45,251
Other Countries	11	655
Greek	69,157	1,849,311
Total	71,955	2,795,153

The Greek mercantile marine, in 1858, consisted of 3,920 vessels, measuring 250,600 tons, and manned by 23,128 seamen. Of these, two were small steamers of 850 tons; 2,660 vessels of the first class, of only 26,567 tons; and 1,258 of the second class, measuring 241,897 tons (this class includes all vessels above 60 tons). In 1857, the tonnage was 826,000, with 25,000 sailors; but 96 vessels, measuring 19,000 tons, were sold to foreigners. The number and tonnage of vessels in 1860 are shown in the following table:—

Description	1860	
Sailing Vessels, 1st Class (under 60 tons)	Number	2,657
	Tonnage	79,193
Sailing Vessels, 2nd Class (of 60 tons and upwards)	Number	1,318
	Tonnage	271,802
Total		4,070
		763,072
		23,842

The commerce of the Ionian Islands, not included in the above tables, amounted, in the year 1861, to imports valued at 1,317,605l., and exports at 735,981l.

Weights and Measures.—The weights in use are—

The Oke = 43·5 oz. avoirdupois.
Kilo = 71 okes.
Cantar or quintale 44 okes.
Stremma (of land) nearly 1·3 acres.
Arpent nearly 1½ acre.

The Greeks ordinarily reckon distance by the hour; thus they say 'an hour distant,' meaning about 3 m. They calculate time by the old style, i.e., twelve days later than we do.

Money.

Gold pieces of 10, 20, 40, and 80 drachmas.					
Silver	Obol, or 5 drachma piece	= 3s. 6½d.			
	Drachma	= 0	8½		
	Half and quarter dr.	= 0	3		
Copper	Piastre	= 0	3	(Urquhart)
	Para, 40 to the piastre, 100 to the drachma.	= 0	0 1·2		
	Aspri	= 0	0 1·3 of a para.		

Government.—Since the establishment of its independence, in 1830, Greece has undergone many vicissitudes of government. The rule of King Otho, which lasted a whole generation, from 1833 to 1862, was, in theory, meant to be strictly constitutional, and was, perhaps, so in reality for a few years, but ended as a sort of feeble despotism. One of the first acts of Otho's successor, King George, was to get a new charter framed, in substitution of the old one which had worked so ill. Accordingly a constituent assembly, elected in December, 1862, was occupied, during the whole of the year 1863, in elaborating a new constitution for the kingdom, on the basis of universal suffrage. The assembly decided, on September 19, 1864—by 211 votes against 62—that the whole legislative power of the realm should be vested in a single chamber of deputies, to the exclusion of a senate or upper house. The constituent assembly of Greece consisted, in October 1864, of 282 members, including 84 deputies from the Ionian Islands, elected, by universal suffrage, in June, 1864.

The executive of the kingdom is in seven departments—those of the royal household and foreign affairs, the interior, religion, and public instruction, justice, finance, war, and maritime affairs. The council of state appointed to assist the king in his duties, consists of 8 vice-presidents, 17 ordinary, and 14 extra-ordinary councillors. The synod of the clergy, elected annually, consists of a president and 3 members, with 2 secretaries, the government being represented by a state officer called the Procurator. There are 33 bishops of the Greek church in the kingdom; and they elect from themselves 8 synties, composing the above synod. The 4 Ilian, Catholic bishops of Naxos, Tinos, Syra, and Santorin have no political existence. The towns of Greece, from the earliest periods, have enjoyed municipal rights and privileges under different modifications; as did their foreign rulers interfere much with the patriarchal system by which their society is governed. Even during the Turkish rule, the heads of families in every town, village, and commune, throughout the Morea, chose a demogeront or mayor, who took cognisance of all civil judicial matters. No tax can be levied without the concurrence of these demogeronts; and they were sometimes called in to assist in council with the primate; and the coirode appointed by the pacha, who jointly superintended the provinces. Maina was at the same period ruled by its own capitani, the chief of whom had the title of Bey. N. Greece was governed, with little difference, in the same mode as the Morea, till Ali Pacha destroyed its liberties. In the islands the demogeronts were entitled archontes, and were criminal as well as civil judges. Count Capo d'Istrias suspended altogether the municipal rights of the towns, and placed over each eparchy a creature of his government; but on his fall, those individuals were expelled, and the towns and communes everywhere resumed their privileges, which were confirmed by the crown in 1834. The administration of each demos or borough is consequently still exercised by one or more demogeronts, assisted by a municipal council. The demogeronts are elected annually from amongst the heads of families,—one in each commune or rural district, and three in each town. They next assemble in the chief town of their several eparchies, when three or more are elected to form, in conjunction with the demogeronts of that town, the eparchial or provincial council for the ensuing year. The government of each eparchy is administered by an officer named an eparch, subordinate to the monarch, whose authority, in the same manner, extends over a monarchy.

The mayors, aided by the communal tribunals, composed of respectable inhabitants of the commune, have authority in cases of petty misdemeanors, and arbitrate, without appeal, in civil transactions to the amount of 20 drachmas. There are eparchial courts presided over by a judge, appointed by the government; and a court of original jurisdiction is established in the chief town of each monarchy, as before the subdivision of the kingdom into 30 governments, an event which appears to have had but little practical influence as to the internal arrangements. Formerly there were 3 courts of appeal—at Nauplia, Missolonghi, and Chalcis; but since 1833 their number has been reduced to 2—those of Athens, for Hellas and Euboea, and Tripolitza for the Morea, &c. The decisions of these are subordinate to the authority of the Court of Cassation and criminal court, established in the cap., composed of judges, a state-attorney and a registrar. Besides these, there are 10 primary tribunals, and 3 commercial courts. There is no regularly organised court of laws, but the decision of the judges are mostly guided by the Code Napoleon and established customs. Trial by jury has been introduced, and is said to be generally understood, and to work well.

Religion and Education.—The great mass of the pop. belong to the Greek church; but since 1833, Greece has been independent of the authority of the Patriarch of Constantinople. The king is titular head of the church, the affairs of which are conducted by a synod. The Greek priesthood are, speaking generally, poor and illiterate. Their habits are, however, said to be simple and exemplary. Monasteries are by no means so numerous as formerly. The national congress, held at Argos in 1829, wisely abolished 394, which contained, at an average, nearly 5 monks each; there are now 82 in all, with a total of 1,500 or 2,000 inmates, besides about 30 convents. There are about 15,000 Roman Catholics in Greece; some Protestants, and about 4,000 Jews. Full religious toleration is guaranteed by the constitution.

An edict was issued in the early part of King Otho's reign for the establishment of elementary schools in each commune, to which the inhabitants should be obliged to send their children from 5 to 12 years of age. This edict has not been fully carried into effect; nevertheless, education has made great progress within the last thirty years. According to a report of Mr Lytton, British secretary of Legation (dated Athens, Jan. 26, 1865), there were at that time 'three principal public schools or gymnasiums in the Morea; one at Patras, one at Nauplia, and one at Tripolitza, in which were taught Greek, Latin, mathematics, geography, natural history, physics, and French. Each school had a head master and five assistants.

There were also two similar schools at Athens, one at Syria and one at Lamia, the whole maintained at an expense to the state of 200,000 drachmas per annum. In addition to these, there were in the several towns in Greece seventy-nine minor schools for boys, having 5,112 scholars, in which were taught ancient Greek, Latin, catechism, the Scriptures, geography, and history, first principles of physics, natural history, and drawing. Each school had a head master and two assistants, and the whole cost the government 267,512 drachmas, or 10,625l. annually. There were also in Greece 443 communal schools for the education of boys, maintained at an expense to the state of 115,251 drachmas, and to the communes of 224,829 drachmas. There were also forty private schools, thirty-one public schools for girls, having 4,300 scholars, where nearly the same lessons were taught as to the boys. In addition to which there were 300 schools where only reading was taught, having about 11,000 scholars, besides seventeen private schools. There were also two schools for forming schoolmasters and schoolmistresses for the primary schools, as well as an ecclesiastical school, and several schools for orphans, founded by private individuals; also an agricultural school at Tyrens, in Argolis.' At present, Mr. Lytton says, ' the chief impediment to the diffusion of knowledge in Greece exists rather in the poverty of the communes than in the apathy of parents, who, however illiterate they may be themselves, value and desire instruction for their offspring.' This, too, is asserted by Mr. Aubrey de Vere, who gives the following account of a school at Athens which he visited in 1849:—

'I visited, with equal surprise and satisfaction, an Athenian school which contains 700 pupils, taken from every class of society. The poorer classes were gratuitously instructed in reading, writing, and arithmetic, and the girls in needle-work likewise. The progress which the children had made was very remarkable; but what particularly pleased me was, that air of bright alertness and good-humoured energy which belonged to them, and which made every task appear a pleasure, not a toil. The greatest punishment which can be inflicted on an Athenian child is exclusion from school, though but for a day. About 70 of the children belonged to the higher classes, and were instructed in music, drawing, the modern languages, the ancient Greek, and geography. Most of them were at the moment reading Herodotus and Homer. I have never seen children approaching them in beauty; and was much struck by their oriental cast of countenance, their dark complexions, their flashing eyes, and that expression at once apprehensive and meditative, which is so much more remarkable in children than in those of a more mature age.'

Armed Force.—Previously to 1850 the army amounted to nearly 10,000 men; but by the new law of conscription the regular army consists of 8,000 men, levied by a conscription of 2,000 in each year. The duration of service is fixed at four years, and all individuals are liable to serve, from the age of 18 to 30, unless those claiming exemption as married men, university students, ecclesiastics, civil servants of the state, only sons, or the guardians of minors. Service by substitute is allowed. The troops consist of 3 battalions of infantry of the line, 3 of light infantry, 4 squadrons of cavalry, a corps of artillery, and another of pioneers. They are chiefly garrisoned at Athens, Argos, Corinth, and Nauplia: at the last-mentioned place is a military school.

The prefecture of the Marine at Poros has 10

members. There being the fleet, about 2,400 officers, sailors, and marines. The government dock-yards are at Poros and Nauplia. At the commencement of 1862, the navy consisted of one frigate, of 50 guns; two corvettes, of 22 and 26 guns; one paddle steamer, of 170 horse-power, with 6 guns; six screw steamers, of 85 horse-power each, with a total of 10 guns; and 22 smaller vessels, of various sizes, including gun-boats.

Revenue and Expenditure. — The finances of Greece are and have long been in the greatest disorder. The revenue may be estimated at about 26,000,000, and the expenditure (including interest of debt) at 28,000,000 drachms. The revenue is principally derived from direct taxes, including the rent of the public lands. Previously to the revolution these belonged to the Turkish inhab., and on their expulsion, they became the property of the public; and, notwithstanding their continued illegal appropriation, they are still supposed to amount to 2-3rds of the cultivated, and to 4-5ths of the uncultivated lands. This immense national property, were it well administered, would furnish a large amount of revenue; but it is a prey to all sorts of abuse. The rent of the public lands is raised at from 10 to 15 per cent. of the gross produce; but owing to the venality and corruption of the officers, it is frequently reduced to a mere nominal sum; and does not, perhaps, on the whole, amount to 1-4th part of what it should do. The other items of revenue consist of customs' duties, a tax on cattle, a tax on salt, stamp duties, &c. The following was, according to official statements, the amount of the several branches of revenue and expenditure of Greece in the years 1861 and 1862:—

Amount of Revenue	1861
	Drachms
Direct Taxes	9,473,750
House Taxes, Licences, &c.	7,780,000
Customs	4,700,000
Stamps	2,100,000
Sundries	440,000
Post and Printing	852,750
Mines	676,750
Salt, &c.	730,000
Fisheries	143,475
Wood and Timber	567,746
Olive Trees	273,165
Grapes and Currants	718,575
Gardens, &c.	842,957
Shops and Manufactories	84,566
National Domains	667,691
Revenue from Courts of Justice, Dividends, &c.	473,899
Ecclesiastical Income	763,300
Miscellaneous	1,560,000
Total { Drachms	38,998,763
{ £	892,741

The public debt of Greece amounted, in July, 1864, to 6,892,361l., chiefly due to Eng. creditors.

Manners and Customs. — The following statements embody the valuable testimony of Thiersch as to the habits and state of the people when he visited Greece in 1831–32: 'There is a pretty marked distinction among the inhabs. of the three great divisions of Greece — Greece N. of the Isthmus, the Peloponnesus, and the Islands. The inhabitants of N. Greece have retained a chivalrous and warlike spirit, with a simplicity of manners and mode of life, which strongly remind us of the pictures of the heroic age. The soil here is generally cultivated by Bulgarians, Albanians, and Wallachians. In E. Greece, Parnassus, with its natural bulwarks, is the only place where the Hellenic race has maintained itself; in the mountainous parts of W. Greece there are also some remnants of Hellenic stock. In these parts the language is spoken with more purity than elsewhere. The pop. of the Peloponnesus retains nearly of the same races as that of N. Greece, but the Peloponnesians are more ignorant and less honest than the inhabitants of Hellas. The Albanians occupy Argolis and a part of the ancient Triphylia. Among the rest of the inhab., who all speak Greek, there are considerable racial differences. The pop. of the town is of a mixed character, as in N. Greece; where there is an active and intelligent body of proprietors, merchants, and artisans in the towns, and among them some of Greek stock. The Mainotes form a separate class of the pop.; they are generally called Maïnotes from the name of one of their districts; but their true name, which they have never lost, is Spartans. They occupy the lofty and sterile mountains between the Gulfs of Laconia and Messenia, the representatives of a race driven from the sunny valley of the Eurotas to the bleak and inhospitable tracts of Taygetos, though the plains which are spread out below them are no longer held by a conqueror, and the fertile lands lie uncultivated for want of labourers. In the islands there is a singular mixture of Albanians and Greeks. The Albanians of Hydra and Spezzia have long been known as active traders and excellent mariners. The Hydriotes made great sacrifices for the cause of independence in the late war; the Spezziotes, more prudent and calculating, increased their wealth and their merchant navy. The island of Syra, which has long been the centre of an active commerce, now contains the remnant of the pop. of Ipsara and Chios. The Ipsariots are an active and handsome race, and skilful seamen; the Chiots, following the habits of their ancestors, are fond of staying at home and attending to their shops and mercantile speculations; they amass wealth, but they employ it in founding establishments of public utility, and in the education of their children. In Tinos, the peasants, who are also the proprietors, cultivate the vine and the fig even amidst the most barren rocks; in Syria, Santorin, and at Naxos, they are the tenants of a miserable race of nobility, whose origin is traced to the time of the crusades, and who still retain the Latin creed of their ancestors. Besides these, there are various bodies of colonists, of people from the heights of Olympus, Candiotes, many Greek families from Asia Minor, Fanariotes, and others, who have emigrated, or been driven by circumstances within the limits of the new kingdom. The Fanariots are those who are supposed to have the least intermixture of foreign blood. They have the fine and characteristic Greek physiognomy, as preserved in the marbles of Phidias and other ancient sculptors; they are 'ingenious, loquacious, lively to excess, active, enterprising, vainglorious, and disputatious.' The modern Greeks are generally rather above the middle height, and well shaped; they have the face oval, features regular and expressive, eyes large, dark and animated, eyebrows arched, hair long and dark, and complexions olive-coloured.

The islanders are commonly darker, and of a stronger make than the rest; but the Greeks are all active, hardy, brave, and capable of enduring long privations. Generally speaking, the women of the islands and of Hellas are much handsomer than those of the Morea. The character of the Greeks, while under the Turks, was thus summed up by Mr. Hope. (Anastasius, i. 78–80.) 'The complexion of the modern Greek may receive a different cast from different surrounding objects; the core is still the same as in the days of Pericles. Credulity, versatility, and the thirst of distinctions, from the earliest periods formed, still form, and ever will form, the basis of the Greek character. ... When patriotism, public spirit, and pre-eminence in arts, science, literature, and warfare, were the road to distinction, the Greeks shone the first of patriots, of heroes, of painters, of poets, and of philosophers. Now that craft and subtlety, adulation and intrigue, are the only paths to greatness, the same Greeks are—what you see them!'

The Albanians are of a much more serious and passive disposition than the Greeks; and it has been remarked that they may be considered to bear the same relation to the latter that the Doric did to the Ionic population in ancient times. The language of the modern Greeks (for the Albanian is of Illyrian origin) is called Romaic. It has a greater similarity to the ancient Greek than the Italian to the Latin; but many of the alterations from the ancient tongues which distinguish both the modern languages are analogous. Many of the popular customs of the Greeks bear the impress of antiquity; various superstitious observances are kept up, and even the ordinary amusements of the people are the same which were popular in ancient times. The far-famed Romaica, for instance, the theme of so many travellers, is obviously the same as the Cretan or Pyrrhian dance; and another modern dance, the Albanatica, is supposed to resemble the Pyrrhic dance of the ancients.

History.—The Greek nation boasts of the highest antiquity, and in the mythic period of their history it is often impossible to separate fable from fact. We infer, however, that the Hellenes were not the earliest inhabit. of Hellas, which was previously the abode of the Pelasgi, who migrated not only into Greece, but Italy, and the islands of S. Europe, and there practised tillage and other simple arts of early industry; the remains of Cyclopean walls, scattered in different parts, denote them to have had some knowledge even of architecture. Over these people the Hellenes gradually gained the superiority, and drove them from the continent to the islands, while they peopled it with their own nation, divided into the 4 tribes of Æolians, Achæans, Ionians, and Dorians, and spreading in different directions over the country, were joined soon afterwards by colonists from Egypt and Phœnicia. The first constitution of Greek cities is beyond the reach of exact history; but it seems that monarchy was the earliest form, and Sicyon is said to have been founded B.C. 2000, Argos, Thebes, Athens, Sparta, and Corinth, claiming an origin not much later. The expedition of Cadmus to Colchis, the siege of Thebes, and the Trojan war (B.C. 1200), are the principal events of the mythic or heroic period. The confusion arising from the last event deprived many kingdoms of their princes, and encouraged the ambition of the Dorian Heraclidæ to get possession of the Peloponnesus, and expel its inhabit. A fresh impulse was thus given to emigration; large bodies of the people crossed the Ægean, and colonised the shores of Asia Minor; governments changed with their rulers, and the states now partook more of that republican form which was afterwards their characteristic feature.

The civil jealousy of Sparta and Athens, whose growing power now began to lessen the influence of the other states, was most successful in calling forth the public energies, and in making small means produce great results. The progress of military knowledge and of the more refined arts was contemporaneous with that of politics; most departments of science and of the fine arts, pursued with impatient zeal by the highly sensitive Greeks, were carried by them to a higher pitch of perfection than elsewhere in ancient, and in some respects even than in modern times; and their commerce, conducted by means of their colonies on the Illyric Sea and on the coasts of Italy, Sicily, and Gaul, was extensive and important. Their pride, activity, and enterprise, and, above all, their love of liberty, bore them triumphant through all the difficulties of the Persian war (closed B.C. 469); and the same features of character, differently developed, involved them in intestine feuds. The Peloponnesian war, which lasted nearly thirty years (B.C. 431-404), by destroying their union, and exhausting their strength, paved the way for their subjugation by Philip of Macedon, who won the decisive battle of Chæronea, B.C. 338. The brilliant conquests of Alexander engaged them for a few years; but their courage was now enervated, and their love of liberty all but extinguished. The Achæan league proved a vain defence against the power of Macedon, and, when this kingdom fell, Greece was wholly unable to cope with the arms of Rome. The contest was brief, and ended with the capture of Corinth, anno 146 B.C., from which time, during 1550 years, it continued to be either really or nominally a portion of the Roman empire. Literature and the arts, long on the decline, were at last destroyed by Justinian, who closed the schools of Athens. Alaric the Goth invaded the country in the year 400, followed by Genseric and Zaber-khan in the 6th and 7th, and by the Normans in the 11th century. After the Latin conquest of Constantinople, in 1204, Greece was parted into feudal principalities, and governed by a variety of Norman, Venetian, and Frankish nobles; but in 1261, with the exception of Athens and Nauplia, it was re-united to the Greek empire by Michael Palæologus. In 1438 it was invaded by the Turks, who finally conquered it in 1461. The Venetians, however, were not disposed to allow its new masters quiet possession, and the country during the 16th and 17th centuries was the theatre of obstinate wars, which continued till the treaty of Passarovitz, in 1718, confirmed the Turks in their conquest. With the exception of Maina, the whole country remained under their despotic sway till 1821; when the Greeks once more awoke from their protracted lethargy, and asserted their claim to a national existence, and to the dominion of the land preserved and ennobled by their ancestors. The heads of the nobler families and others interested in the regeneration of their country, formed an hetæria for concerting patriotic measures; and, in 1821, Ypsilanti proclaimed that Greece had thrown off the yoke of Turkey. The revolution broke out simultaneously in Greece and Wallachia; and war continued with various success and much bloodshed till the great European powers interfered, and the battle of Navarino (Oct. 20, 1827) insured the independence of Greece, which was reluctantly acknowledged by the Porte in the treaty of Adrianople, in 1829. The provisional government, which had been set on foot during the revolutionary struggle, was agitated by discontents and jealousies, and the president, Count Capo d'Istrias, was assassinated in 1831. The allied powers having previously determined on erecting Greece into a monarchy, offered the crown to Prince Leopold of Saxe-Coburg, who declined it; finally, it was conferred on Otho, a younger son of the king of Bavaria, who was proclaimed at Nauplia, Aug. 30, 1832. The long and inglorious reign of King Otho lasted till the commencement of 1863, when a revolutionary movement, which broke out while he was engaged in a tour through the islands, forced him to leave the country. He was solemnly deposed by decree of the Greek National Assembly, of Feb. 4, 1863, and, after protracted negotiations, a successor for him was found in the person of Prince George, born 1845, second son of King Christian IX. of Denmark. He landed in Greece Nov. 2, 1863, and nominally assumed the reins of government, having been declared of age by a resolution of the National Assembly of May 15, 1863.

GREENLAND, an extensive territory forming part of N. America, and partly occupied by Danish colonies, extending N. from Cape Farewell, in lat. 59° 49' N., between long. 20° and 75° W., having W. Baffin's Bay and Davis' Straits, N. and E. the N. Atlantic Ocean, and N. the unexplored Arctic regions. Pop. estimated at only 6,000 or 7,000, all Esquimaux, except about 150 Europeans. Greenland was long supposed to be united on the NW. to the continent of America; but the discoveries of recent navigators render it more probable that it is an island. Shape, somewhat triangular with the apex towards the S. It is high and rocky, its surface presenting a chaotic assemblage of sterile mountains, bare or covered with ice, which also occupies a great portion of the intervening valleys. The centre is said to be traversed by a range of lofty mountains, by which the country is divided into E. and W. Greenland. Of the former, from lat. 53° to 68°, little or nothing is known, the shore being constantly beset by vast accumulations of ice. All this coast appears to be colder, more barren and inhospitable than the W. coast. It may be said to consist of one uninterrupted glacier, exhibiting only a few patches of vegetation, generally on the banks of the rivers; and often advancing far into the sea and forming promontories of ice, large masses of which frequently fall in avalanches. The W. shore is high, rugged and barren, and rises close to the water's edge into precipitous cliffs and mountains, seen from the sea at a distance of 80 m. The whole coast is indented with a series of bays or fiords, interspersed with a number of islands of various form and size. The principal of these is the island of Disco, in the bay of the same name, on the W. coast, between lat. 69° and 70°. Only the coasts and islands are yet ascertained to be inhabited, no other part having been explored by Europeans. The air is pure, light, and healthy; but the cold during the long winter is often very intense. More snow falls, and the climate is more severe on the E. than the W. coast. In S. Greenland the cold seldom exceeds 16° or 18° Réaumur, but in the N. the thermometer sometimes stands at 30° Réaum. The sun has considerable power during the summer, but fine weather is never of long continuance. Lightning sometimes occurs, and hail, but the latter seldom. Violent storms are frequent in autumn. The rare occurrence of rain, and the intense degree of cold produced by the NE. wind, has given reason to believe that the most E. parts of Greenland form a great archipelago, encumbered with perpetual ice. The aurora borealis has at some seasons a light equal to that of the full moon. The rocks are principally granite, gneiss, clay-slate, porphyry, sandstone, &c., arranged in vertical beds. They have been found to contain a rich copper ore, black lead, marble, asbestos, serpentine, garnets, crystals, and some other valuable stones. There are no volcanoes; but three hot springs have been found in an island on the W. coast. Coal is found in the island of Disco. Vegetation, even in the S., is limited to a few stunted birch, elder, and willow

trees, moss, lichens, grasses, fungi, &c. Proceeding N. the surface becomes more sterile, and at last nothing is met with except bare rocks. Several kinds of wild berries attain tolerable perfection, and the soil on the W. coast towards the S. has been found fit for the cultivation of various culinary vegetables: the growth of the potato has latterly been attempted with some success. Among the animals are the reindeer in the N., the polar bear in the N., white hares, foxes of various colours, and dogs; seals abound in the N., where the walrus also is met with; whales of various kinds inhabit the seas, chiefly towards the N.; and the sea, fiords, and rivers abound in fish, especially turbot, herrings, salmon-trout, halibuts, rays, &c., with a great variety of shell-fish. Fishing and seal-hunting are the principal occupations of the native inhab.

In 1837 there were in W. Greenland 13 colonies, 15 minor commercial, and 10 missionary, establishments. The most N. station is Uppernavic, in lat 72° 30', Good Hope, the most ancient of the settlements, in lat. 61° 10', has an excellent harbour. The trade gives employment to about five or six vessels. The exports consist chiefly of whale-oil, seal, bear, and reindeer skins, eider down, &c. The Greenlanders are believed to be of the same race as the inhab. of the coasts of Hudson's Bay, Labrador, the NW. coasts, Kamtchatka, &c., from whom they differ little in person, manner, and language. On the W. coast they do not much exceed 5 ft. in height. They have long black hair, small eyes, and a yellow or brown skin. The inhab. of the E. coast differ from the former in being taller, fairer, and more active and robust; but they do not exceed a few hundreds in number. There is no European colony on the E. coast, and little or no intercourse is maintained between it and the W. coast. The inhab. display considerable skill in the structure of their fishing boats and hunting implements, which are made of the drift wood brought in vast quantities to the coasts. Many have embraced a species of Christianity; and their superstitious belief in sorcery, &c., is now giving way to a rude kind of civilisation. Their kayaks or fishing boats are from 12 to 14 ft. long, and only about 1½ ft. broad, sharp at both ends, and covered with skins, except a small round opening in the middle, where the Greenlander, having a single oar, takes his seat. Their houses are from 6 to 8 ft. high, and vary in size according to the number of families they are intended to accommodate, which sometimes amount to seven or eight. The interior is divided by skins into different compartments; the walls are lined with brown and hung with skins, and the floor paved with flat stones. Their domestic arrangements are simple, and more remarkable for a want of cleanliness than any thing else. The food of the natives is principally the dried flesh of the seal, with a little game and fish; coffee, tobacco, snuff, and brandy are esteemed the greatest luxuries.

Greenland is said to have been discovered by an Icelander, near the commencement of the 10th century; and the first colonisation of the country, according to the old chronicles, dates from the year 983, when it was settled by the Norwegian Icelanders. It has long been a subject of discussion whether colonies were established on both coasts; but from the accounts of recent adventurers it is pretty certain that no European colony was ever founded to the E. of Cape Farewell; at all events, no ruins indicative of any ancient settlements have been discovered on that coast, though numerous traces of them remain on the W. coast. Under the Norwegian colonists, the country was governed by Icelandic laws, and had its own bishops. An intercourse was maintained between Norway and these settlements till the end of the 11th or the beginning of the 15th century, when the trade with Greenland was interdicted. Of the subsequent history of the country, and the fate of the colonies, we have no certain accounts. Several expeditions have from time to time been undertaken for the discovery of the lost colonies, but without success. The first of the modern settlements was established in 1721, under the auspices of the Danish crown, by Hans Egede, a Norwegian, who has written an interesting work on Greenland. (For further particulars, see Egede's work; Malte-Brun's Geography; Crantz's History of Greenland; Graah's Voyage to Greenland, 1837; and Journal of R. Geog. Society.)

GREENOCK, a parl. bor. and sea-port town of Scotland, co. Renfrew, on the S. bank of the Frith of Clyde, 18 m. WNW. Glasgow, with which it is connected by railway. Pop. 42,098 in 1861. The situation of Greenock is interesting and picturesque. Immediately behind it the land rises rapidly to a height of 800 ft.; and though the town is built mainly on a strip of level ground stretching upwards of 2 m. along the shore, it ascends at one place about 500 yards up the ridge. In its front the Clyde is about 4 m. in width; and its magnificent estuary, which seems land-locked on every side, with the picturesque mountain scenery of Argyle and Dumbarton on the opposite coast, form a noble view. Crawfurdsdyke, or Cartsdyke, on the E., once a rival bor., is now incorporated with Greenock. The progress of the town has been very rapid, it having nearly doubled in the forty years 1821 to 1861.

The town is upwards of 2 m. in length. The width is inconsiderable, except near its centre, where, as already stated, it stretches up the hill. It is pretty regularly built, particularly in the more modern parts. The leading streets run E. and W. The houses are of stone, covered with slate. The streets, which are causewayed, have foot pavements of convenient breadth on both sides. The town is rapidly stretching towards the W., where the best streets have been erected. A number of elegant villas are scattered in this direction, and along the heights behind the town. Greenock, however, is not remarkable for cleanliness. From its situation on the W. coast, and its vicinity to the mountains, the climate is moist, the average fall of rain being about 55.84 inches annually. It is lighted with gas.

Of the public buildings, the most distinguished is the custom-house, erected in 1818 at a cost of 30,000l. It is advantageously situated in the centre of the quay, about 40 yards from its edge, and being unconnected with any other building, is seen in all directions. It is in the Grecian style, and its portico fronting the quay is particularly handsome. The other more prominent public buildings are, the town-hall, erected in 1766; the gaol, built in 1810; the infirmary, erected at a cost of nearly 5,000l., with accommodation for 150 patients; the Tontine hotel, built in 1801, at an expense of 10,000l.; the exchange buildings and assembly rooms, the sheriff-court-hall; the Watt monument; the mechanics' institute; the Highlanders' Academy; and the workhouse. The mansion-house of Greenock, once the residence of the ancient family of Shaw, the superiors of the place, is situated on an eminence overhanging the town. Part of the building is old, but additions at different times have been made to it. It is now used as chambers for conducting the business of the superior and harbour bailie. Greenock contains three parishes; the Old

or West parish, originally taken from the neighbouring parish of Inverkip, and the Middle and East parishes, both taken from the West, and erected into parishes quoad sacra by the Court of Teinds. Of the parish churches, two are not without architectural pretensions; the Middle church, built in 1741, and the new West church, which cost about 9,000l. Besides seven churches and a Gaelic chapel belonging to the establishment, there are twenty-one other places of worship, including Episcopalian, Free Church, United Presbyterian, Reformed Presbyterian, Congregational, Wesleyan, Baptist, Evangelical Union, Roman Catholic, and Catholic Apostolic chapels.

There are thirty-five common schools in Greenock; two of them, the grammar school and the mathematical school, under the management of the town council; and one, the Highlanders' Academy, under the management of an educational society. The others are either congregational or adventure schools. There are also two charity schools and a ragged school. The whole number of pupils at school in Greenock is not supposed to exceed one-tenth of the population. There are three libraries in the town; the Cartsdyke mechanics' library, with about 2,100 vols.; the library of the mechanics' institution, with nearly 4,000 vols.; and the Greenock library, founded in 1783, and the property of a body of shareholders. This last contains about 10,000 vols. of miscellaneous literature, besides a foreign library, and a scientific library (one of the first in the kingdom), composed of the collection of Spence, the celebrated mathematician, who purchased his books to his native town, and of more recent works purchased with funds left for the purpose by James Watt. The Greenock library occupies the principal apartment in the Watt monument, a beautiful edifice dedicated to the memory of the most distinguished native of Greenock. Besides the library, this building contains a marble statue of Watt by Chantrey, a museum, and a lecture-room. Though the inhabitants are eminently distinguished for education, intelligence, and commercial enterprise, literature, in the strict sense of the term, is not much cherished by them, and the town can boast of few great names besides those of Watt and Spence. In 1707, when Wilson, the author of Clyde, a poem, was appointed master in the grammar-school of Greenock, the magistrates stipulated that he should renounce what they called 'the profane and unprofitable art of poem-making.'

An extraordinary work has been constructed in the vicinity of Greenock, by which not only the town is abundantly supplied with water, but machinery to a great extent may be impelled. To accomplish this an artificial lake, covering 291½ imp. acres, has been excavated in the bosom of the neighbouring alpine district, behind the town, by turning the courses of several streams and collecting the rain into a basin prepared for their reception. From this, as from a common source, an aqueduct or canal is conducted along the mountain range for several miles, at an elevation of 520 ft. above the level of the Clyde; and when within less than a mile of the town, it pours down a torrent in successive falls, the whole length of the aqueduct being 5½ m. In addition to the principal basin, there is a compensation reservoir occupying 40 acres, besides several of smaller dimensions, to secure a plentiful supply of water in seasons of the greatest drought. A series of self-acting sluices has been constructed in a most ingenious manner, by which all risk of overflow is obviated, at the same time that every drop of rain, even during the greatest floods, is preserved. This magnificent public work was planned by the late Mr. Robert Thom, of Rothesay, and executed at the cost of the Shaw's Water Company. It has more than realised the expectations of the projectors, though the cost from first to last has not fallen short of 80,000l. There are two lines of falls, each with a descent of 512 ft. The water sent down amounts to 1,300 cubic ft. per minute, being equal to 1,843 horse power.

The docks of Greenock were first projected in 1696, and the first part, forming a small harbour, was finished in 1710, at an expense of ———. Greenock being, in the same year, made a custom-house port, and a branch of the neighbouring and then more flourishing bur. of Port Glasgow. A new dock was built in 1785 at a cost of 4,000l. In 1821, two markets and docks were added, which cost 119,000l.; and in 1850 a new tidal harbour was constructed, with 14 ft. water at low ebb in ordinary spring tides. The latter cost about 150,000l. The foundation of another dock, called the Albert Harbour, was laid in August, 1862. The harbour is managed by trustees, including the provost, magistrates, and town council.

The Clyde is navigable to Greenock for vessels of any burden, at any time of the tide; but a submarine bank extends from a spot opposite Greenock 9 m. up the river to Dumbarton; and the channel for navigation, though deep, is only 300 ft. wide. The system, often pursued, of towing by steamboats, obviates, in great measure, this inconvenience. Government has recently completed a survey of the river.

The trade of Greenock has kept pace with the improvements made on its harbour. The union of the kingdoms (1707) opened the colonies to the enterprising inhabitants of this town, and generally of the W. of Scotland; but it was not till 1719 that the first vessel, belonging to Greenock, crossed the Atlantic. The tobacco trade with Virginia and Maryland was prosecuted with great vigour and success for fully half a century after this date; but it was to a considerable extent carried on upon account of, and in connection with, Glasgow merchants. The war with the American colonies depressed, for a lengthened period, the trade of Greenock, but other sources of commerce were gradually taken advantage of; and, at present, ships from this town may be found in almost every considerable port to which British enterprise has extended. The first application to government to open the East India trade went from Greenock; and its merchants were also among the first to take advantage of the opening. The trade of Greenock is at present chiefly with Newfoundland, North America, and the West and East Indies. The gradual increase of trade may be seen from the following account of the gross receipts of customs' duties at the port of Greenock in various years:—

Years	Duty	Years	Duty
1775	15,321	1833	450,425
1779	87,234	1844	407,964
1805	215,347	1859	516,454
1822	362,164	1865	1,221,279

There belonged to Greenock, on the 1st of January, 1864, 174 sailing vessels under 50, and 201 above 50 tons; there were also 18 steamers under 50, and 9 above 50 tons, the latter of a total burthen of 1,544 tons.

The herring fishery, the trade in which the inhab. of the town first engaged, is still prosecuted to a considerable extent. The Greenland whale-fishery was begun in 1752, but has been long since discontinued.

The principal trade of Greenock is sugar refining, which is carried on more extensively here than in any part of the kingdom out of London. Shipbuilding is also extensively pursued. There are six building yards, of which two, those of John Scott and Sons, and Robert Steele and Co., are among the largest in the empire. At the former, the first iron steam frigate, Greenock, was built; at the latter many of the Cunard steamers, which ply between Liverpool and New York, were constructed. Among the other branches of business may be mentioned foundries for the manufacture of steam engines, chain-cables, anchors, and other iron work; several extensive roperies and sail-cloth factories, breweries, soap and candle-works; the manufacture of hats of felt, silk, and straw; pottery, boat-building, block-making, brass-founding, cork-cutting, copper-work, and many others common to the other large towns throughout the country.

Greenock originally consisted of a few thatched houses stretching along the bay; and Cartsdyke, now incorporated with it, was long a place of greater consideration. It was created a bur. of barony in 1635, and Cartsdyke in 1669. Sir John Shaw, the feudal superior, gave power by charter to the feuars, subfeuars, and bongauss to be afterwards admitted, to meet yearly for the purpose of choosing nine managers of the public funds of the town, viz. two bailies, a treasurer, and six councillors. The united bur. is now governed under the Scotch municipal reform act, by a provost, four bailies, and sixteen councillors, of whom one fills the office of treasurer. The gross revenue of the corporation amounted to 17,220l. in the year 1863-4. The annual value of real property in 1863-4 was 163,070l. The Reform Act raised Greenock to the dignity of a parl. bur., by conferring on it, for the first time, the privilege of sending 1 mem. to the H. of C. Registered voters, 1,763 in 1865. Greenock and the three adjoining parishes of Inverkip, Port Glasgow, and Kilmacolm, were in 1815 constituted the Lower Ward of Renfrewshire, and placed under the jurisdiction of a sheriff-substitute, who resides and holds courts in the town.

GREENWICH, a parl. bor., town, and par. of England, on the S. bank of the Thames, co. Kent, lathe Sutton-at-hone, hund. Blackheath, 4½ m. ESE. London by South Eastern railway. Pop. of par. 40,002, and of parl. bor. 139,436 in 1861. Greenwich, which, in fact, is now a mere suburb of the metropolis, is a thriving town, but without any particular trade or manufacture; the business of the place being derived from its public establishments, from families of fortune residing in or near it, and from the shipping and craft on the river. The streets are in some places narrow and irregular; but within the last few years many handsome houses have been erected, and the town has been greatly improved. It is lighted with gas, and supplied with water from the Kent water-works at Deptford. The par. church is a handsome stone fabric, with a noble portico, and an interior richly ornamented in the Corinthian order. It appears from Willis's Notitia Parl. (vol. iii. p. 83) that the bor. of Greenwich sent two burgesses to parl. in the reign of Philip and Mary; but neither the extent of the bor., nor the nature of the franchise, nor the reason why it ceased to be exercised, has been specified. The Reform Act again conferred on Greenwich the right to send 2 mems. to the H. of C.; but the parishes of Deptford and Woolwich, and about two-thirds of that of Charlton, are included with it in the modern park, bor., which had 8,062 registered electors in 1861.

Greenwich Hospital, the noblest establishment of its kind in Europe, occupies the site of a palace, erected by Humphry, duke of Gloucester, in 1433, and was long a favourite residence of the Tudor family. The present building, originally intended for a palace, was commenced by Charles II., who erected one wing at an expense of 36,000l. In the reign of William III. the case of the disabled seamen of the navy engaged the attention of the king and queen, and, in consequence, this palace was granted as an asylum for their relief. Commissioners were appointed to carry out the royal intentions; Sir Christopher Wren undertook to superintend the completion of the building without charge, and voluntary contributions were requested in aid of the public grant, which last amounted to 58,200l. In 1715, the confiscated estates of the Earl of Derwentwater, amounting to 6,000l. a year, were given to it by parl., and their value has immensely increased within the last half century. The hospital was partly also supported by the several contributions (by act passed 7 and 8 William III.) of 6d. a month from the wages of all seamen in the king's and merchants' service. But since 1835, merchant seamen have been exempted from this contribution, in lieu of which the sum of 20,000l. a year is advanced from the consolidated fund to the hospital. The entire building consists of four magnificent detached quadrangular piles, of Portland stone, called King Charles's, Queen Anne's, King William's, and Queen Mary's: the interval between the two former is the grand square, 273 ft. wide, in the centre of which is a statue of George II. by Rysbrach; the space between the two latter is filled up by two colonnades supported by 300 double columns and pilasters. The principal front, on the N. side towards the river, comprises the sides of King Charles's and Queen Anne's buildings; and before it, extending 865 ft. in length, is a spacious terrace, with a double flight of steps in the middle, commanding a fine view of the building, and forming a handsome landing place to the hospital. King Charles's building, in the SW. angle, was erected after Inigo Jones's designs; in it are the council-chambers and residences for the governor and lieutenant-governor. Queen Anne's building contains 21 wards for the pensioners, and some officers' apartments. King William's building, designed and directed by Sir C. Wren, contains the great hall, with its vestibule surmounted by a fine cupola and 11 wards. The hall is 106 ft. long by 56 broad, and 50 high; the roof and walls were painted by Sir James Thornhill, at a cost of 6,685l. Several pictures of great naval actions, with portraits and statues of distinguished officers, give interest to this noble apartment. Opposite the hall in Queen Mary's building is the chapel, with a vestibule and cupola corresponding with those of the hall. The roof and inside having been destroyed by fire, were ably restored by 'Athenian Stuart,' in 1780. A flight of fourteen steps leads to the interior, which is 111 ft. long by 52 broad, and accommodates 1,000 persons. The carving of the pulpit and other parts is exquisitely finished. The altar-piece, by West, represents the Shipwreck of St. Paul. This hospital supports about 1,700 old or disabled seamen in the house at an average cost of 22l. per annum, and gives pensions varying in amount, but which average about 17l. a year, to a much more numerous body of out-pensioners. The nurses are all seamen's widows. The revenues of the hospital being required for the support of the in-pensioners, the expense of the out-pensioners is

defrayed by an annual parliamentary grant. Connected with the hospital, in a building contiguous to the park, part of which was intended for a ranger's lodge, is the Naval Asylum, for the education of 800 boys, sons of commissioned and warrant officers, private seamen and marines. The management of the hospital revenues is vested in incorporated commissioners; and the interior regulations are under the superintendence of a governor, lieutenant-governor, chaplain, and numerous other officers.

Greenwich Park, which was attached to the old palace, and is now in the hands of the crown, contains nearly 200 acres; it is well stocked with timber and deer, and furnishes from its higher part magnificent views of the metropolis and its vicinity. On an eminence 160 ft. above the river, about ¼ m. from the park-gates, is the royal observatory, erected by Charles II. for the celebrated Flamsteed, and fitted up with telescopes and other astronomical instruments, which have been successively improved and increased by Graham, Bradley, Hooke, Herschell, Dollond, and others. The upper part of the building consists of rooms well adapted for observations: the lower part being used as the residence of the astronomer royal. This important and honourable situation has been held by some highly distinguished astronomers, as Flamsteed, Halley, Bradley, Bliss, Maskelyne, Pond, and Airy, who at present (1866) enjoys that honour. The longitudes of all English charts and maps are reckoned from this observatory; and the captains of ships take their time as given here at 1 P.M. daily. It is 2° 20′ 15″ W., from Paris, and 18° 9′ 45″ E. from Ferro or Hierro, the most W. of the Canary Islands.

Greenwich has for many years been a favourite resort of holyday-seekers from the metropolis, and the means of access have been greatly facilitated by steam-boats, and by a branch of the South-Eastern railway terminating at Charing Cross. Greenwich markets, on Wednesday and Saturday, are well supplied. The fairs, held at Easter and Whitsuntide, are well known, for the various amusements furnished to the crowds that resort thither from all parts of London and its neighbourhood.

GREIFSWALD, a town of the k. of Prussia, prov. Pomerania, cap. circ. of same name, on the Ryck, about 3 m. from the Baltic, and 18 m. SE. Stralsund, on the railway from Berlin to Stralsund. Pop. 15,714 in 1861. The town is the seat of a superior court of appeal, the high judicial tribunal for the territory, formerly Swedish Pomerania (New-Vor-Pommern), others for the circle and town, a circle-council, a high board of customs, consistory, orphan-tribunal, and board of agriculture. It has a harbour at the mouth of the Ryck, which is navigable for small vessels; manufactures of salt and tobacco, oil-mills, distilleries, and a brisk trade both by land and sea. A university was founded here in 1456, and some new buildings were erected for it in 1750, but the number of students is inconsiderable. Greifswald was taken by the Elector of Brandenburg in 1768.

GREIZ, a town of Central Germany, cap. princ. of Reuss (elder branch), on the White-Elster, 40 m. S. Leipzig, on the railway from Leipzig to Nürnberg. Pop. 10,509 in 1861. Greiz is a walled town, and is tolerably well built. It is the residence of the sovereign prince, who has a summer palace here, built on an eminence, and surrounded with fine gardens. The church is the only other public building. There are Latin and normal schools; and it has manufactures of coarse woollen cloths and leather, with distilleries. It is the seat of the government, and of a judicial consistory.

GRENADA, one of the W. Indian islands belonging to Great Britain, and the most southerly of the windward group, Tobago and Trinidad excepted, between lat. 11° 58' and 12° 14' N., and long. 61° 20' and 61° 35' W., about 90 m. N. Trinidad, and 68 m. SSW. St. Vincent. Greatest length, 20 m.; greatest breadth, 10 m. Area, 133 sq. m. Pop. 28,927 in 1851, and 31,900 in 1861.

A chain of rather lofty hills run through the island, in which many small rivers have their sources. There are some small lakes, which appear to occupy the craters of extinct volcanoes. The soil is, on the whole, very fertile, and adapted to every kind of tropical product; but the climate is decidedly unhealthy. About five-eighths of the surface is cultivated. Indigo, tobacco, sugar, coffee, cocoa, and cotton thrive well. Game, and birds of numerous species, are very abundant. The exports from Grenada to the United Kingdom, comprising chiefly coffee, cocoa, rum, sugar, and molasses, were of the value of 64,214l. in 1850; of 110,682l. in 1860; and of 102,702l. in 1863. The imports from the United Kingdom amounted to but 55,331l. in 1850; 53,238l. in 1860; and 36,944l. in 1863.

Grenada, like most other W. Indian islands, has its governor, council, and assembly, by whom it is governed. Its cap. St. George, on a spacious bay on the S. side of the island, is a well-built town, and has one of the safest and most commodious harbours in the British W. Indies. The sum awarded by government, in 1835, for the manumission of slaves in Grenada amounted to 616,444l. 17s., being about 26l. 4s. per head. This island was discovered by Columbus in 1498, and colonised by the French about 1650, at first as a private speculation, but after 1674 it belonged to the French crown, till taken by the British in 1762. In 1779 it was retaken by the French, but restored to Great Britain at the peace of 1783.

GRENOBLE (an. *Gratianopolis*), a fortified city of France, dép. Isère, of which it is the cap.; on both sides the Isère, 64 m. SE. Lyons, and 230 m. SE. Paris, on a branch of the Paris-Mediterranean railway. Pop. 34,726 in 1861. The portion on the left bank of the river (the city, properly so called) is the larger and more ancient: it is surrounded by bastioned ramparts, and has a citadel; but these defences are at present very much out of repair. The portion on the right bank, originally built by the emperor Gratian, called the Faubourg St. Laurent, is confined between the river and the foot of an abrupt mountain, and consists of little more than one spacious street. It is, however, comparatively the more populous division, and the chief seat of commercial activity. St. Laurent is enclosed by only an indifferent wall, but is defended by the new fortress of Bastille on the mount above it. The two parts of the city are connected by two bridges; one of wood, the other of stone. Grenoble is ill laid out, and ill paved; but is generally well built and clean; many improvements have taken place in it of late years. It contains numerous squares and handsome public fountains; and near its centre is a spacious garden laid out in public walks, planted with trees, and having a quay on the river. Many other agreeable promenades surround the city. The chief public buildings are the cathedral, the episcopal palace, hotel of the prefecture, formerly the residence of the celebrated Constable de Lesdiguières, the general hospital, hall of justice, royal college, theatre, and a public library with 40,000 printed vols. and 600 MSS. Here are 4 par. churches, a Protestant church, several convents and seminaries, a foundling and another hospital, a university academy, schools of

medicine, drawing, &c., cabinets of natural history and antiquities, and a fine collection of paintings. In the Place St. André is a colossal bronze statue of the Chevalier Bayard, the knight '*sans peur et sans reproche*,' who is interred in a contiguous church. Grenoble is the seat of a prefecture, a royal court, and of tribunals of original jurisdiction and commerce. It is the see of a bishop, the rendezvous of the 7th military division of France; and has a chamber of manufactures, arts, and commerce, faculties of law and sciences, and a Society of Arts &c. It is noted for its manufacture of kid gloves; and has others of liqueurs, linen fabrics, &c.; and some trade in hemp, iron, marble, and timber. It originally bore the name of Cularo, till Gratian enlarged it and gave it his own name. It was long the cap. of Dauphiny. Its inhab. warmly espoused the popular cause against the court of Louis XVI.; and were afterwards devoted partisans of Napoleon, in whose favour they made a very vigorous stand against the allies in 1815.

GRETNA GREEN, a small village of Scotland, parish of Graitney, co. Dumfries, famous until recent times for the celebration of irregular marriages, on the border of England, near the Sark, 9 m. NW. Carlisle, and 22 m. E. by S. Dumfries. The old marriage ceremony merely amounted to an admission before witnesses that certain persons were man and wife; such acknowledgment being sufficient, provided it be followed or preceded by cohabitation, according to the law of Scotland, to constitute a valid marriage. A certificate to this effect having been signed by the officiating priest (who was seldom above the rank of a tradesman), and by two witnesses, the union, under the above condition, became indissoluble. The marriages of this sort celebrated at Gretna Green, when the place was most flourishing, were estimated at between 300 and 400 a year. The people were generally from England, and of the lowest ranks; though there were a few instances of persons of the higher ranks, and even of a lord chancellor having had recourse to the services of the *so-called* parsons of Gretna Green. A trip to Gretna, or the presence of a self-dubbed parson, was not, however, at all necessary. Parties crossing the Scottish border, and declaring before witnesses that they were man and wife, were, under the old law of Scotland, held to be duly married. This law, however, was altered a few years ago, in so far that a short residence in the country became necessary for the validity of the contract, and this, of course, was sufficient to destroy the objectionable custom of Gretna Green marriages. The practice began at Gretna Green about 100 years ago by a person named Paisley, a tobacconist, who died in 1811.

GRIMSBY (GREAT), a bor. and sea-port, in the co. of Lincoln, on the S. side of the estuary of the Humber, which at this point is about 7 m. across, 7 m. W. from the lighthouse on Spurn Head, 15½ m. N. London by road, and 155 m. by Great Northern railway. Pop. of munic. bor. 11,067, and of parl. bor. 15,060 in 1861. The town stands on the flat shores of the Humber, opposite Spurn Head. The long, low, narrow, hooked tongue of land, which terminates in the Head, protects a capacious roadstead, with good holding ground, extending to within a mile of the new works at Grimsby, and well known as a harbour of refuge to those who navigate the North Sea. The entrance to the river is marked by the lighthouse on Spurn Head, and by two light-ships in the Channel. Grimsby has, in consequence, the double advantage of a secure roadstead and of proximity to the open sea. The utility of this harbour is evident from the scarcity of ports along this portion of the coast of England; for, except the landing port of Hull, which also lies on the Humber, but 15 m. further inland, there is no other port with docks but that of Grimsby, between Hartlepool in Durham and King's Lynn in Norfolk, a distance of fully 150 m.

Grimsby is a borough of considerable antiquity, and was formerly a port of such importance that in the reign of Edward III. it sent 11 ships to the siege of Calais. Owing, however, to the gradual filling up of its harbour, it latterly sank into comparative insignificance. In 180? a harbour was constructed; but being accessible only at high water, it was not productive of all the advantages that were expected. But, in 1846, a new harbour, on a large scale, accessible at all times of the tide, was commenced; and in anticipation of its being finished, Grimsby was made the terminus of two important railways.

The old dock or floating basin, constructed in 1802, measures about 17 acres; but being placed at the high water margin of a flat shore, and being consequently accessible only towards high water, it is of very limited utility. To secure a proper depth of water at the entrance of the new works was an object of the first importance; and to attain it they were projected ¾ of a mile into the estuary in advance of the old dock, reclaiming at the same time and enclosing 130 acres of land. The new works comprise a wet dock of upwards of 25 acres in extent, with two entrance locks, having in front a tidal basin of 15 acres. The latter, formed by two timber piers, which are together about 2,000 ft. in length, is provided with landing slips. It has a depth of 9 ft. at low water springs, and of 12½ ft. at low water neaps; the rise of tide at the former being about 18, and at the latter about 12 ft. The facility of ingress and egress afforded by this basin is especially useful to steamers, which, as they usually convey passengers or light merchandise, do not require to enter a dock. Here they lie afloat alongside the piers at all times of the tide.

The new dock, opened in May, 1852, is entered from the basin by two locks, furnished with double sets of gates for ebb and flood tides, the larger of which, constructed (by special agreement with the government) to admit the largest class of war steamers, is of the following dimensions, viz. length between the gates 300 ft.; breadth from wall to wall 70 ft.; depth of water on cill, at low water spring tide, 7 ft.; depth of water on cill, at low water neap tides, 10½ ft.; depth of water on cill, at high water spring tide, 25½ ft.; depth of water on cill, at high water neap tides, 27½ ft. At half tide the average depth of water on the cill of this lock is 16 to 17 ft., and at three-quarters tide 20 to 22 ft. The Royal Docks, opened in 1852, occupy 140 acres, near the railway terminus, and, including the wet dock, afford ample accommodation for more than 1,200 sail. There is also a graving dock, 400 ft. long; the wharfs and quays extend 1,200 yards. A tower, 300 ft. high, serves as a lighthouse, and also as a hydraulic press for opening the floodgates.

There belonged to the port on the 1st of Jan., 1854, 131 sailing vessels under 50, and 78 above 50 tons, besides 4 steamers under and 6 steamers above 50 tons. The gross amount of customs duties received was 47,800l. in 1859; 20,000l. in 1861; and 83,847l. in 1863. The principal foreign trade is with the Baltic. There are mills for grinding bones and tanneries. Connected with the harbour are large warehouses and timber-yards, and on the shore N. of the harbour is an extensive nursery. The other manufactures are local and unimportant.

The old town of Grimsby, which was co-exten-

sive with the township, sent 2 mem. to the H. of C. from the reign of Edward III. down to 1837, the right of voting being vested in resident freemen paying scot and lot, of whom, in 1831, there were 400. The Reform Act deprived the bor. of one of its mem.; and, at the same time, enlarged its boundaries by the addition of eight other pars. Registered electors, 1,092 in 1852.

The bor. is governed by four aldermen (one of whom is mayor) and twelve councillors. Petty sessions are held on Thursdays, and quarter sessions by the recorder. A court of requests, for the recovery of debts under 5l., was established in 40 of George III. Markets on Wednesday, fairs 17th June for sheep, 15th September for horses.

GRINSTEAD (EAST), a market-town and par. of England, co. Sussex, rape Pevensey, on the high road between London and Brighton, 20 m. N. the former, and 22 m. N. the latter. Area of par. 13,390 acres. Pop. 4,266 in 1861. The town is pleasantly situated close to the N. border of the co. on an eminence commanding fine views of the country to the S. The streets, which are narrow and irregular, contain many good modern houses. The church, on the E. side of the main street, is a large, handsome building, of modern date, the old edifice having been destroyed by the fall of the tower in 1785. The present tower is lofty and well proportioned, having pinnacles at the corners. The living is a vicarage in the gift of the Duke of Dorset, the lord of the manor. There are also places of worship for Wesleyan Methodists and Baptists. The town-hall, which is large and commodious, was used as an assize court, till the Lent assizes were removed to Horsham. At the E. end of the town is Sackville College, endowed by Robert earl of Dorset with an income of 330l. a year, and erected in 1616, for the support of twenty-four unmarried persons of both sexes, each of whom has a comfortable room and 3l. a year in money. A free grammar-school was founded in 1709, and endowed with a freehold farm in the parish, the rent of which is taken to pay the master's salary. Markets, chiefly for corn, on Thursday. Fairs, April 21, July 13, and Dec. 11, for horned cattle and poultry.

East Grinstead, before the passing of the Reform Act, by which it was disfranchised, sent 2 mem. to the H. of C., a privilege which it had enjoyed since the first of Edward II.; the electors were 35 burgage-holders, nominated by the Duke of Dorset, whose bailiff was the returning officer.

GRISONS (Germ. Graubündten or Bünden, an. a part of Rhætia), a canton of Switzerland, and, excepting that of Bern, the most extensive in the union, of which it occupies the N.E. portion. It ranks fifteenth in the confederation, and lies between lat. 46° 15' and 47° 4' N., and long. N° 40' and 10° 29' E.; having N. the cants. Glarus and St. Gall, the principality of Lichtenstein, and the Vorarlberg; E. the Tyrol; S. the Val-Tellina, Lombardy, and the cant. Ticino; and W. the last-named cant. and that of Uri. In the greater part of its extent, it is enclosed by the Austrian territories; but is cut off from them, as well as from the rest of Switzerland, at nearly every point, by lofty mountain ranges. Length E. to S., 86 m.; greatest breadth about its centre 58 m. Area. 2,968 sq. m. Pop. 91,177 in 1860. It has the thinnest population of any of the cantons of Switzerland, there being but 30 inhabitants to the square mile. The whole canton is one mass of mountains and valleys; there is not a single plain worthy of notice. The main chain of the Rhætian Alps crosses the canton from W. to E., at first separating it from Ticino and Italy, and afterwards dividing it into two unequal parts, the valley of the Rhine, being the larger, on the NW., and that of the Inn, on the NE. A great portion of this chain is above the limit of perpetual snow. The *Maschelhorn*, 10,640 ft., the *Piz Val Nhin*, 10,290 ft., M. Maloya, 11,480 ft. high, form parts of it, and it is crossed by the passes of the Splügen, St. Bernard, Albula, and Braletta. From the E. extremity of the canton, a chain, little inferior in height, passes off NE., separating the Grisons from Tyrol, Glarus, and St. Gall. Another chain bounds the Engadine on the SE., to which belong the Monte dell' Oro, 10,554 ft., and M. Bernina, 7,654 ft. high; and which is crossed by the Pass of Bernina, about 6,460 ft. above the level of the sea. A fourth chain, called the *Rhætikon*, also including many elevated peaks, forms the boundary between the Grisons and the Vorarlberg. Both the Rhine and the Inn rise in the Grisons, as do several tributaries of the Upper Adige, Po, and Adda: the Rhine receives most of the minor Rhætian rivers. Climate and soil very various; but where the Rhine, Inn, and other rivers leave the canton, the general temperature is sufficiently high to admit of the cultivation of the vine. The scenery is peculiarly grand and magnificent; the canton contains upwards of 240 glaciers, comprising the largest in Switzerland. The nature of the country generally unfits it for agriculture; but in the Engadine, where the inhab. are very industrious, every patch of land is cultivated that is worth the pains. The corn raised is chiefly rye, barley, oats, and Turkish wheat; but not half the quantity required for home consumption is produced, and it is consequently imported to the annual value of about 300,000 florins. Hemp and flax, also, though generally grown, are not produced in sufficient quantities for home demand. Potatoes have been cultivated only of late years. Fruit and wine are among the articles of export. The chief wealth of the canton consists in its cattle. Its pasture lands are estimated to feed, in the summer, 100,000 head of cows and oxen, besides from 60,000 to 70,000 goats, and perhaps 100,000 sheep, many of which are driven from Italy to feed in the Alpine pastures for about three months, under the care of Bergamasque shepherds. The best breed of cattle is that of the Prottigau (or valley of the Lanquart); but the best cheese is made in the Engadine. A great many hogs are kept, most of them for home consumption. Rural economy, and the condition of the peasantry, vary greatly in different parts. Throughout the Engadine, the land belongs to the peasantry, and each individual usually supplies his family with provisions and clothing entirely from the produce of the territory belonging to him. Poverty is here rare, and beggary unknown. Indeed, many of the inhab. of the Engadine are possessed of considerable property, which they have amassed in some of the commercial cities of Europe, chiefly as confectioners. Schools are numerous, and few of the children in the valley of the Inn are uneducated. In the valley of the Rhine, the peasants are also the proprietors of the soil, living upon the produce of their own lands; but, as in most other parts of the Grisons, they are not industrious. Their land is badly tilled; garden cultivation is ill-conducted; and the forests are neglected. In the *Terciera,* that there is a good deal of squalid misery. Wages are, notwithstanding, high throughout the Grisons. There are some rich veins of metal, especially iron; but they are not wrought. Manufactures few, and mostly domestic; the principal are those of cotton fabrics, some of which are exported. The most profitable branch of commerce is the transit trade between Zurich and Italy, the

route of which passes through the Grisons and over the Splügen, and is a source of wealth to Chur, the cap. The chief exports from the Grisons are timber, of the value of about 180,000l., and cattle, mostly to Italy, to the amount of 70,000l. a year: the principal imports are corn, salt, oil, sugar, coffee, tobacco, foreign manufactured goods, and iron.

This canton comprises a confederation of little republics in itself. It consists of a number of communes, exercising within themselves rights almost independent. These are united into 26 *Hochgerichte,* or high-jurisdictions, each of which is, in many important respects, independent, not only of the rest, but even of the supreme council. These high-jurisdictions are united into the 3 leagues of the *Grau Bünden* (Grey League), containing 8; the *Gottes-haus Bund* (League of the House of God), 11; and the *Zehngerichte* (League of Justice), 7 high-jurisdictions. The whole unite in electing a supreme federal legislative council of 65 members, chosen in the different jurisdictions and communities, by the universal suffrage of the male pop. above 18 years of age. The supreme council or diet of the leagues meets at Chur every year, in June, and appoints a commission of 9 members to prepare matters for its own consideration; and a minor council of 3 members, one from each league, to whom the executive duties are entrusted. It also elects the public officers of the canton generally, concludes treaties, and appoints 9 judges to form a central court of appeal; though, for the most part, the communities and petty municipalities themselves exercise full judicial powers, and in each of the high-jurisdictions there is a power of life and death in criminal cases, which is sovereign and without appeal. The common law is different in each jurisdiction: every one has its own peculiar laws and usages, and by them the questions within their boundaries must be determined. The decisions of the supreme council have also to be submitted for approval to the jurisdictions and communities at large. The inhab. of the Grisons are fond of boasting of the liberties they enjoy; but, in point of fact, they are destitute of some of the most important rights of the citizens of really free states. A free press, and trial by jury, are unknown; and both the supreme council and the courts of law deliberate and determine with closed doors. There is, however, no direct taxation of any kind; the state revenues are derived from customs and duties on the transit trade, a monopoly of salt, and some other sources. The total public revenue of the canton, in the year 1862, amounted to but 731,000l., or 29,240l. The annual surplus is devoted to the payment of a small cantonal debt. About two-fifths of the pop. are of German, and one-tenth of Italian origin. The different communities elect and support their own clergy. The canton furnishes a contingent of 1,600 men to the army, and 12,000 ft. annually to the treasury of the Swiss confederation. It has a militia of all its male inhab. from the ages of 17 to 60. Chur, Mayenfeld, and Ilanz are the only places worthy the name of towns. Few countries abound so much with ruined castles and other feudal remains. These belonged, in the middle ages, to the nobles, who for a long period were possessors of the soil. In 1396, a number of communities revolted against the feudal nobles, and, headed by the bishop of Chur, formed the *Gottes-haus* Bund; in 1424, the *Grauband* was formed in a similar manner in the W. part of the Grisons; and in 1428, the *Zehngerichte* in the E. In 1471, the three leagues entered into a common union; and, in 1497-8, formed an alliance with the Swiss confederacy, though it was not till

1798 that the Grisons became a canton of Switzerland.

GRODNO, a government of Russia, formerly included in the old k. of Poland; between lat. 51° 30′ and 54° 20′ N. and long. 23° 7′ and 26° 43′ E., having N. the gov. Wilna, E. Minsk, S. Volhynia, and W. Bialystok and the k. of Poland. Greatest length NE. to SW., about 200 m.; average breadth nearly 76 m. Area, 696 geo. sq. m., or about 14,700 Eng. sq. m. Pop. 891,891 in 1858. The surface is an alluvial or sandy plain, broken only by a few undulating chalk hills. The Niemen, Bug, Narew, and Pripec are the principal rivers; in the N. there are some large marshes. The climate is damp, and the atmosphere cloudy and foggy. The principal agricultural product is rye, about 6,975,000 hectolitres of which are said to be produced annually, a third part of which is exported. Few other kinds of grain or vegetables are grown for food, but flax, hemp, and hops are raised in considerable quantities. There is a large extent of pasture land; cattle-breeding is pretty well understood; and the native breed of sheep, which has been much improved by crossings with the breeds of Silesia and Germany, yields good wool, which is a principal article of export. The forests are extensive. Many belong to the crown, and that of Bialovieja, a royal domain, occupies nearly 86,200 hectares. Iron, lime, nitre, and building-stone are found. Manufactures are hardly worth notice; the principal are those of woollen cloth, leather, and felt. The exports consist of corn, flax, cattle, and wool; much of the produce is sent to Memel, Königsberg, Windau, and Riga, by the canal of the Niemen, and by land. The greater part of the inhab. are Russniaks, except in the N., where Lithuanians prevail. The nobles comprise about 1-20th part of the whole pop., and are principally Poles. Jews are very numerous. There are some Tartars and colonies of German artisans. The dominant religions are the Rom. Catholic and the United Greek church. Education is at a low ebb. Chief towns, Grodno the cap., Novogrudek, Slonim, and Brzesc (Brest Litofski).

GRODNO, a town of Russian Poland, and cap. of the above gov. in the NW. part of which it is situated, on a hill on the Niemen, 85 m. SW. Wilna, and 154 m. NE. Warsaw on the railway from St. Petersburg to Warsaw. Pop. 16,970 in 1858. Grodno was formerly considered the second town of Lithuania, and even disputed the superiority with Wilna. Its houses are partly of stone and partly of wood; and the greater number of its streets are extremely filthy. It has a fine castle, built by Augustus III. of Poland, the ruins of a more ancient fortress, 9 Roman Catholic and 2 Greek churches, a synagogue, and some handsome residences of the nobility, a gymnasium, an academy of medicine founded by Stanislaus Augustus, many other schools, a good public library, cabinets of mineralogy and physical objects, and a botanic garden. There are some inconsiderable manufactures in the town and its vicinity; and it has some well frequented fairs.

GRONINGEN, a fortified city of Holland, cap. prov. of same name, and the most important town in the N. Dutch provs., on the Hunse, at the influx of the Aa, 40 m. E. by N. Harlingen, and 80 m. NE. Amsterdam, on the railway from Leeuwarden to Emden. Pop. 36,192 in 1861. The town is well built, and clean; its market-place (Bree-Markt) is one of the largest and handsomest squares in Holland; and there is a fine public promenade, called the Plantage. It has a strong citadel, built in 1607, and is surrounded by ramparts and ditches, kept in good condition. Many of the public buildings are handsome, especially the great church of St. Martin, a Gothic structure, the spire of which is the loftiest in Holland; and the town-hall, erected in 1793. The university, founded in 1615, is usually attended by about 180 students, a much greater number than formerly; it possesses an excellent museum of natural history, a library, and a botanic garden. Groningen has an academy of painting, sculpture, and architecture, a seminary for deaf and dumb, another for the instruction of the blind, societies of natural history and chemistry, poetry, literature and jurisprudence, and a branch of the society of "public good." It has a large paper manufactory, besides some factories of woollen and silk stuffs, cotton stockings, &c., and yards where merchant-vessels are sometimes built. It has also an active trade in cattle and butter; and by means of a canal large vessels come from the estuary of the Ems, quite up to the town.

This town is not mentioned previously to the ninth century, and it was not fortified for several ages afterwards. It was first attached to the United Provinces in 1576: it afterwards fell into the hands of the Spaniards, but was finally retaken by Prince Maurice in 1594.

GRUYÈRE (Germ. Greyers), a town of Switzerland, cant. Freiburg, 16 m. S. Freiburg. Pop. 952 in 1860. The town is situated on a hill, the summit of which is crowned by the ancient castle of the counts of Gruyère, a fortress said to have been founded in the 8th century, and which is one of the most extensive and best-preserved feudal monuments in Switzerland. The town is walled, and contains a handsome parish church, a rich hospital, and a public library. The district around Gruyère is famous for its cheese, of which it produces about 25,000 cwt. a year. It is made on a chain of mountains about 10 leagues in length and 4 in breadth: all the cheese, though made in the same manner, is not of the same quality; the lower pastures not being in such estimation as those in the more elevated situations. The very finest qualities are said to be too delicate for exportation. The whole district is divided into greater or lesser farms, which the proprietors let out on leases of 3 or 6 years, at rents varying according to the nature and elevation of the ground; the lower pastures, though not of the best quality, bring the dearest, because, being sooner freed from the snow, and later covered with it, they afford food to the cattle for a longer time. The farmers, who rent pastures, hire from the different peasants in the canton from 40 to 60 cows, from the 15th of May to the 8th of Oct., paying for them certain rates per head. Each cow, at an average, yields daily from 20 to 24 quarts of milk, and supplies 300 Swiss pounds of cheese during the five months. On the 14th of October the farmer restores the cows to the different proprietors. The cattle are then pastured in the meadows, which have been twice mowed, until the 10th or 11th of November, when, on account of the snow, they are usually removed to the stables, and fed during winter on hay and after-grass. Throughout the commune of Gruyère the inhab. are above poverty. Having a part of the year there are not so many hands in the cheese country as are required, and these are borrowed from other and poorer communes. Wages are very high, in comparison with most other parts of Switzerland, being about 2s. 6d. a day, exclusive of living. (Inglis's Switzerland, p. 163; Coxe's Switzerland, ii. 270.)

GUADALAXARA, or GUADALAJARA, an inland city of Mexico, cap. of the state of same name (otherwise called Xalisco), in a rich and extensive plain, on the Rio Grande de Santiago.

H H 2

130 miles from the Pacific and 275 miles WNW. Mexico; lat. 21° 9′ N., long 103° 7′ 15″ W. A superficial enumeration of the year 1864 showed the number of inhabitants to be 62,350, so that it is, in point of pop., the second city in the republic. It covers a great extent of ground, and at a distance has a very picturesque appearance. Its interior is also handsome; its streets are airy and well laid out, and many of the houses extremely good, though mostly of only one story. There are 14 squares, the principal of which, the *Plaza de Armas*, has in it the government-house, in which the congress assembles; the cathedral, a fine edifice, though much injured by the earthquake of 1818; and the *Portales de Comercio*, consisting of piazzas or arcades built around three large square blocks of houses. Within the town the Portales are the principal rendezvous, as, besides a number of handsome shops, well provided with European and Chinese manufactures, they contain a variety of stalls covered with domestic productions, fruits of all kinds, earthenware from Tonala, shoes in quantities, mangas, saddlery, birds in cages, "dolces" of Calabazate, and a thousand other trifles, for which there seems to be an increasing demand. As each of these stalls pays a small ground rent, the contents to which the Portales belong derive from them a considerable revenue. They are the counterpart of the Parian in Mexico, but infinitely more ornamental, being both with equal solidity and good taste.' (Ward's Mexico, ii. 562.) Besides this public promenade there is the *Paseo*, an extensive avenue shaded by double rows of fine trees, having a stream flowing through it, and leading to the *Alameda*, a public walk 'very prettily laid out, for the trees, instead of being drawn up in battle array, in lines, intersecting each other at right angles, like the streets, are made to cover a large tract of ground in irregular alleys, while in summer the intervening spaces are filled with flowers, particularly roses, which give both life and variety to the scene. There is a fountain too in the centre, and a stream of water all round.' (Ward, ii. 561, 562.) Many of the public places are adorned with fountains. Besides the cathedral there are several churches, with numerous monasteries and convents, a college maintained at the public expense on the most liberal footing, and for which a magnificent building has been erected, two ecclesiastical establishments for the education of young women, three for young men, five boys' schools, a public hospital, bishop's palace, mint (a fine building), and a neat theatre. A large pile of building, erected during the Spanish rule, for a workhouse, now serves as a barrack for about 500 men. The coffee-houses are tolerable, and the shops and market place are well supplied with provision, but the last, which is large, is very ill kept. The city is supplied with water from the Cerro de Col, three leagues distant; it is lighted at night, except at the time of the full moon, and watched by a patrol. Many of the streets look melancholy and deserted, most of the lower orders being occupied in their own houses, where they exercise various trades in a small way, as in San Juan. They are good blacksmiths, carpenters, silversmiths, and hatters, and are famous for their skill in working leather, as well as in manufacturing a sort of porous earthenware, with which they supply not only all Mexico, but the neighbouring states upon the Pacific. Shawls of striped calico, much used by the lower orders, are made in considerable quantities, as were formerly blankets; but this branch of trade, after suffering much in 1812, when the port of San Blas was opened by General Cruz, has been destroyed entirely by importations from the United

States.' (Ward, ii. 557.) There is at present little or no foreign trade, San Blas having been nearly abandoned for the ports of Mazatlan and Guaymas; and foreign goods are brought overland, chiefly from San Luis or Mexico. The city was founded in 1551, and in 1570 was erected into a bishopric. Under the Spaniards it was the cap. of an intendency of the same name, and the seat of a royal *audiencia*, as well as of some flourishing manufactures.

GUADALAXARA, a town of Spain, and cap. prov. of same name, on the E. bank of the Henares, 35 m. NW. Madrid, on the railway from Madrid to Sevilla. Pop. 6,538 in 1857. The town was once walled, and fragments of its walls still remain. It is wretchedly built; the only buildings of any consideration being the palace of the Duke del Infantado, a large edifice, constructed with very little taste; and the church of the Franciscans, which contains a superb mausoleum of the duke's family, said to be second only in splendour to that of the Escurial. Here is a bridge over the Henares, originally built by the Romans, and restored in 1758. A woollen cloth factory, established here by Philip V., is said to have employed, in 1786, 4,000 hands, besides giving employment in spinning to no fewer than 40,000 in the adjacent villages. But the whole trade is now nearly extinct. The town is the seat of a corregidor, and is governed by an alcalde of the first class.

GUADALQUIVIR, a river of Spain, having its sources in Murcia and La Mancha, and flowing SW. through Andalusia. The source called the Guadalquivir is in the Sierra de Cazorla, lat. 37° 51′ N., and long. 2° 56′ W.; but the true source, and that most distant from the mouth, the Guadarmena, rises in the Sierra de Alcaraz, not far from the town so called; lat. 38° 48′ N. long. 2° 30′ W. The length of the river from this point is 140 m. direct distance, and 320 m. along the channel. The general direction is SW., by W., as far as Sevilla, where it takes a turn nearly S., and, after forming two islands, Isla Mayor and Isla Menor, flows through a marshy and most unhealthy flat into the Atlantic, at San Lucar. It is navigable for vessels of 100 tons as far as Sevilla, and for boats as high as Cordova, 774 ft. above the sea. The chief affluents are, the Jandula, Guadiato, Bembezar, and Biar, on the r. bank; and the Guadalimar, Guadiana Menor, and Xenil, on the L. Of these the Xenil, flowing through Granada, is the longest, being 170 m. long. The ancient name was Betis: the present appellation is Arabic, *Wady-al-kebir*, the great river.

GUADELOUPE, one of the Windward Islands, in the W. Indies, and one of the most valuable colonies belonging to France, lying (inclusive of Grande-Terre) between lat. 16° 30′ and 16° 13′ N., and long. 61° 15′ and 61° 56′ W., 40 m. SE. Antigua, and 30 m. N. Dominica. The area of Guadeloupe, together with its dependencies, the adjacent islands of Marie-Galante, La Desirade, and Les Saintes, and two-thirds of the island of St. Martin (Leeward Islands), is 635 Eng. sq. m., and the population amounted, in 1861, to 130,000, including 85,000 negroes. Guadeloupe is divided into two unequal parts by the *Rivière salée*, or Salt River, an arm of the sea about 5 m. in length, and varying in width from 80 to 120 yards. The division SW. of this inlet is Guadeloupe Proper; that on the NE. is called Grande-Terre; the former is of an oblong shape; length, N. to S., about 25 m.; average breadth, about half as much; area, 82,280 hectares. A chain of volcanic mountains, covered with woods, runs through the centre of the island, nearly in its entire length. The medium height of its summits is somewhat more than 3,000 ft.

but, near its S. extremity, the Soufrière, a volcano still exhibiting a smouldering activity, rises to 5,108 ft. above the level of the ocean. A multitude of rivulets, by which every part of the island is well watered, run down the flanks of this mountain chain; two of them, the Goyave and Lezarde, are navigable for small craft, and highly useful for the conveyance, upwards, of sea-meat, to manure the lands, and downwards, of the produce of the land. Guadeloupe contains many mineral springs.

The island of Grande-Terre is of a triangular shape, and has an area of about 55,928 hectares. It is little raised above the level of the sea, and differs remarkably in its features from Guadeloupe. It is almost a level plain, with only a few scattered hills. It is destitute of woods, and its rivers are insignificant, in consequence of which the rain, which is much less frequent than in Guadeloupe, is obliged to be carefully preserved in cisterns. Marie-Galante, a circular-shaped island about 17 m. to the SE., is traversed, E. to W., by a chain of hills, which, like those of Guadeloupe, abound in timber.

The mean temperature of the year at Basse-Terre is about 81° Fah.; its annual range is between 70° and 90°. In the sun, the thermometer sometimes rises to 130° Fah.; the heat is, however, tempered by land or sea-breezes. The atmosphere is remarkable for humidity. About 85 inches of rain fall annually, on an average, chiefly between the middle of July and the middle of October. Like the other Antilles, Guadeloupe is very subject to hurricanes, and shocks of earthquakes are frequent. The soil is light and easy of tillage, but its productiveness is owing more to the heat of the climate and the abundance of water than to its richness. The soil of Grande-Terre is, on the other hand, very rich. Almost every part of that island is capable of cultivation, and, notwithstanding the deficiency of water, it is very productive. The greater part of the island is laid out in sugar plantations, mostly belonging to great proprietors. It appears, however, from official returns (given in 'Statistical Tables relating to Foreign Countries,' Part IX. p. 252, Lond. 1864), that the extent of land under this cultivation is on the decrease. The sugar plantations of the island of Guadeloupe embraced 18,641 hectares in 1858; 17,898 hectares in 1860; and 17,868 in 1861. The produce of 1861 consisted of 81,219,220 kilogs. of sugar; 1,784,717 litres of syrup and molasses; 3,664,809 litres of rum, or tafia. Next to sugar, coffee is the most important produce, there being devoted to it 2,009 hectares of land in 1859; 1,501 hectares in 1860; and 1,676 hectares in 1861. The produce, in 1861, consisted of 992,932 kilogs. of coffee. The sugar-cane, at present grown, is of the Otaheitian variety, and was introduced in 1790, after the other kinds were found to have degenerated. Most of the kitchen vegetables of Europe are raised in the gardens at Basse-Terre; but they degenerate rapidly; tropical fruits, and others of the S. of Europe, attain considerable perfection. Agriculture has been much improved of late years by the introduction of the plough and the use of manure, including lime, salt, and phosphates. The sugar manufacture has been also greatly improved by the introduction of steam-mills. The live stock consists principally of black cattle, sheep, and mules. Guinea grass is the only forage grown.

The manufacturing establishments are limited to a few tanneries, potteries, and limekilns. The various trades and handicrafts in the colony are exercised chiefly by whites. There is no fishery on any extended scale; but about 30,000 kilogs. of fish are annually taken.

Nearly all the exported articles are sent to

France, whence 9-10ths of the imports are derived. The imports are chiefly salted meat and fish, wheat flour, maize, pease and beans, olive oil, cotton, linen, and silk fabrics, wine, timber, candles, perfumery, hats, and wrought metals. The total imports into Guadeloupe were of the value of 26,920,651 francs, or 1,076,825l. in 1861, and the total exports in the same year amounted to 18,403,997 francs, or 736,359l. The trade is carried on almost entirely by French shipping. The principal roadsteads and ports are those of Basse-Terre, and Mahault, in Guadeloupe; Pointe-à-Pitre, and Moule, in Grande-Terre; the roadstead of Saintes, and a few others.

Guadeloupe and its dependencies are divided into 3 arrondissements, 6 cantons, and 24 communes. The legislature consists of a governor and a colonial council of 30 members, elected for 5 years, by natives of France resident in the island, above 25 years of age, paying taxes of 500 fr. a year, or having a capital of the value of 80,000 fr. To be eligible for a member of council, an individual must be 30 years of age, and pay taxes to the amount of 600 fr., or possess property of the value of 60,000 fr. There is a royal court at Basse-Terre; the other tribunals are 2 courts of assize, 3 of original jurisdiction, and 6 tribunals of justices of the peace. The colony has a military commandant, and an armed force of 2,198 men, including 190 officers. There are about 80 ecclesiastics, upwards of 60 public schools and hospitals in the chief towns. A bishopric of Guadeloupe was formed in 1850. Slavery was abolished throughout the colony by decree of the Republican government of France in the year 1848. The town of Basse-Terre, the cap. of Guadeloupe, and the seat of government, on its NW. shore, is clean, well built, and contains 5,600 inhab. It has two parish churches, a government house, hall of justice, a large hospital, an arsenal, some good public fountains and promenades, and a fine colonial garden. It is defended by several batteries on the side of the sea. Gayetterre, on the E. side of the island, is its other chief town. Point-à-Pitre, a town of 12,000 inhab., is situated at the W. end of Grande-Terre. It owes its prosperity to its excellent port. It is regularly built, has a handsome church, and many good private edifices. Several forts protect its harbour. The other towns are insignificant; but three of them, besides the foregoing, have their own municipal councils.

These islands were discovered by Columbus in 1493: the French took possession of them in 1635. Guadeloupe has, on several occasions, been taken by the English, and was occupied by British troops from 1810 to 1815, when it was restored to France.

GUADIANA (an. *Anas*, Arab. *Wady-Ana*), a river of Spain, rising in the mountains of La Mancha, about 15 m. NW. of Villaharrosas, lat. 3°0 55′ N., long. 2° 48′ W., and flowing through New Castile, Estremadura, and a part of Portugal. It has several sources, which form small connected lakes, called the Lagunas de Ruidera. Its direction at first is NNW. for about 80 m.: it then disappears among the marshes, and is not traceable for 14 m. It rises again NE. of Daymiel, at a place called Los Ojos de Guadiana, with a general E. direction past Merida, as far as Badajoz, where it turns S., and after a very tortuous course of 424 m., enters the Atlantic by two mouths. It is navigable about 45 m. as far as Mertola, to the falls called El Salto del Lobo. The chief affluents are the Giguela, the Guadarrampe, and the Oxirra, on the r., and the Jabalon, the Guadalema, the Ardilla, and the

Chanza, on the l. bank. With the exception of
the Giguela, the affluents on the l. bank are by
far the largest.

GUADIX (an. *Acri*), a town of Spain, prov.
Granada, on the river of same name, 32 m. W.
by S. Granada, and 216 m. S. Madrid. Pop.
11,056 in 1857. Guadix is an old walled town,
with steep, narrow, and badly-paved streets. It
has a cathedral, built in the Corinthian and Composite orders, with a handsome portico, 5 par.
churches, 7 convents, and a hospital. The approach to the town is through a fine avenue of
trees, and the surrounding land is rich, and subjected to irrigation. The chief branch of industry
is the manufacture of large cheap knives.

GUAMANGA, or HUAMANGA, called also
San Juan de la Victoria, or *de la Frontera*, a city
of Peru, cap. prov., on the river of same name, in
an extensive and beautiful plain, 210 m. ESE.
Lima, and 185 m. WNW. Cusco. Estim. pop.
28,000. The town is well built, has good squares
and streets, and the houses, which are of stone,
have gardens and orchards attached to them. It
has a cathedral, with several other churches and
convents; and a university with faculties of philosophy, divinity, and law. Guamanga is the seat
of an intendant, and the see of a bishop. It was
founded by Pizarro, on the site of an Indian village of the same name, for the convenience of the
trade between Cusco and Lima.

GUANARE, a town of the repub. Venezuela,
dep. Orinoco, prov. Varinas, on a river of the
same name, 15 m. SE. Truxillo, and 65 m. NNE.
Varinas. Pop. estim. at 18,000. The town has
wide and straight streets, and neatly built houses.
A handsome church, the interior of which is splendidly adorned, contains a shrine of our Lady of
Consolation, much resorted to by pilgrims. The
chief wealth of the inhab. is derived from their
trade in cattle, of which they possess large herds;
and which, together with mules, they export by
way of Coro and Puerto Cabello.

GUANAXUATO, or GUANAJUATO, an in-
land and mining city of Mexico, cap. of the state
of same name, in the Sierra de Santa Rosa, 6,836
ft. above the level of the sea, and in the very
centre of the richest mining district in the whole
country, 156 m. NW. Mexico; lat. 21° 0′ 15″ N.,
long. 79° 23′ 53″ W. Pop., including its suburbs,
according to Humboldt, in 1803, 70,600, which
number had, however, diminished to 33,000 in
1851. The town is very irregularly built; the
streets are full of ascents and descents, many of
which are so steep as to render the use of four
mules in the carriages of the more wealthy inhabitants almost universal. The open spaces cannot be called squares, for they are of irregular
forms; the whole city, in short, is distributed
here and there, wherever vacancies at all adapted
for building have been left by the mountains.
One part is so hidden from another, that, viewed
from the streets, it appears to be a small town.
'It is only by ascending the heights on the opposite side that a view is gained of the whole valley,
broken into ravines along the sides of which the
town is built. Surveyed from this point, the no-
velty of its situation strikes the stranger with
astonishment. In some places it is seen spreading out into the form of an amphitheatre; in
others, stretching along a narrow ridge; while
the ranges of the habitations, accommodated to
the broken ground, present the most fantastic
groups.' (Mod. Trav. xxvi. 2.) The houses also
have a singular appearance: they are large and
well built of hewn stone, but disfigured by their
fronts being painted of the gayest colours. Some
of the residences belonging to the principal families

are, however, really magnificent, as are the
churches, and the Alhondiga, or public granary.
But the civil war, and the decay of the mines,
has inflicted great and, perhaps, irreparable injury
on the city. The town and its suburbs have
numerous amalgamation works, one of which
sometimes occupies a whole ravine, the space
above, on either side, being crowded with miners'
huts. Guanaxuato suffers two serious inconveniences; one is, a scarcity of water, there being
within the city only a few cisterns belonging to
wealthy individuals; so that most part of this
important necessary has to be brought a distance
of 2 m. upon the backs of asses; the other is,
that during a portion of the year it is liable to
inundation from the torrents which descend from
the mountains, and, though works to prevent this
have been constructed at a great expense, few
years pass without some accidents occurring,
some of the public highways have been strangely
neglected. On approaching Guanaxuato from the
S., there is, indeed, a raised path for foot-pas-
sengers, but coaches and animals of all kinds have
to proceed up the bed of a river, which during
the rainy season rushes along with dangerous impetuosity.

The town has been entirely created by the
mines which surround it. In the vicinity of some
of them, little pueblos, as Valenciana, Rayas, and
Serena, have been formed, which may be considered as its suburbs. The first mine—that of
St. Barnabe—was opened in 1548; but it is
only within the last 70 or 80 years that the mines
of Guanaxuato have become so famous. In 38
years, viz. from 1766 to 1803, they produced gold
and silver of the value of 165,000,000 piastres, or
12,720,000 lbs. tr.; the annual average produce
being 656,000 marcs of silver, or 364,911 lbs. tr.,
and from 1,500 to 1,600 marcs of gold. The *Veta-
Madre*, or great 'mother-vein,' is composed of
several parallel veins running NW. and SE. for
rather more than 6 leagues, within which distance
there have been upwards of 100 shafts opened.
According to Humboldt, the mother-vein has
yielded more than a fourth part of the silver of
Mexico, and a sixth part of the produce of all
America. The principal mines situated on this
vein are those of Valenciana, San Juan de Rayas,
Mellado, Necho, Cata, Tepeyac, and Serena.
When Humboldt visited these works in 1803,
they employed 5,000 workmen, 1,896 grinding-
mills, and 14,618 mules; and before the revolution
of 1810, they yielded, in all, 10,000 mule-loads
of ore, of 11 arrobas (275 lbs.) each, weekly,
making 62,562 parcels of 82 quintals of ore yearly,
worth 7,727,500 dollars. Of this quantity, the
mine of Valenciana alone produced from 5,000 to
6,000 loads, Rayas 1,500, and the other mines the
remainder. 'The mine of Valenciana,' say Humboldt, 'is the sole example of a mine which, for
forty years, has never yielded less to its proprietors than from 2,000,000 to 3,000,000 fr. (80,000l.
to 100,000l.) annual profit.' (Polit. Essay, iii.
188.) It is at the NW. extremity of the mother-
vein. After having been abandoned for a long
period as unpromising, it began again to be
wrought about 1762, by M. Obregon, a young
Spaniard without capital, but with good credit
and great perseverance. In 1768 considerable
quantities of silver began to be extracted from it;
and from 1771 till 1804, it constantly yielded an
annual produce of 600,000l., the net profit to the
proprietors being in some years as much as
250,000l. At that period, 1,800 men were employed in the interior of the mine, besides 900
men, women, and children employed without in
different ways; and Valenciana (a town which

afterwards contained 22,000 inhab.) at an early part of these proceedings sprang up, and had between 7,000 and 8,000 inhab. on the very spot where gold had been brought 10 years before. The machinery of this celebrated mine was much injured by Hidalgo in 1810, and destroyed by Mina after his unsuccessful attack on Guanajuato in 1817. When the Anglo-Mexican Mining Association undertook to drain and work the mine, it was nearly 2-4ths filled with water, and the town of Valenciana had become a ruined place, with only about 4,000 inhab.; and notwithstanding the expenditure of vast sums by the association it has not hitherto recovered its former productiveness.

Much of the landed property in this and the neighbouring states belongs to the great mining families resident in Guanajuato. The vicinity of this city abounds with tillage-land, yielding rich crops of wheat, as well as splendid gardens. Agriculture, however, has been much depressed through the injury done to the mines, and the suspension of mining labours. Guanajuato was founded in 1545, constituted a town in 1619, and a city in 1741.

GUATEMALA, one of the republics of Central America, formerly a part of the Mexican Confederation, but erected into a separate state in 1847. Guatemala extends between lat. 14° and 17° N., and long. 89° and 94° W., having N. Yucatan and Mexico; E. Honduras and San Salvador, and S. the Pacific Ocean. Area 40,777 sq. m.; pop. 850,000, according to a superficial enumeration of the year 1856. The physical features of the country are mountainous throughout, and although no very distinct mountain chain traverses Guatemala, an elevated plateau occupies the central parts of the country, forming a kind of chain of communication between the Cordilleras of N. America and the mountain chains of Mexico. This plateau rises much more precipitously from the side of the Pacific than the Atlantic, the general slope of the country being to the N.E. The table-land averages perhaps 5,000 ft. in height above the ocean; the loftiest summits, which are either active or extinct volcanoes, bring in that part of the confederation. The Water Volcano, near Guatemala, so called from its frequently emitting torrents of hot water and stones, but never fire, is 12,620 ft. above the Pacific. There are two large plains—those of Nicaragua and Comayagua, besides many of less size on the banks of the larger rivers and along the shores; these principally consist of extensive savannahs with rich pasturage interspersed with clumps of trees. All the larger rivers flow N.E. or E., the proximity of the high mountain range to the Pacific permitting but a short course to those flowing W. The Motagua is of considerable size, and useful for the conveyance of European and other goods into the interior of Guatemala. The principal lakes are the Golfo-Dolce, and those of Leon or Managua, Peten, Atitan, and Amatitan. The Golfo-Dolce, 24 m. long by 10 broad, receives several rivers, and discharges itself by the Rio Dolce into the Bay of Honduras.

The coast plains are subject to violent tropical heats, and are very unhealthy, especially those on the east coast, on the Caribbean Sea, where fevers incessantly prevail. These are chiefly inhabited by the Indian pop., whose constitutions are better able to resist the pestiferous nature of the atmosphere than those of Europeans. The climate of the table-land varies according to its elevation, but an equable, moderate, and agreeable temperature may be obtained there all the year round, with a perfectly healthy climate. The dry season lasts from October to the end of May, during which N. winds prevail; and in the table-land, in November and December, water exposed to the open air at night is sometimes, though rarely, covered with a thin pellicle of ice. The rest of the year is enabled the wet season; but the rains, though heavy, last only during the night, and the days are fair and cloudless. Earthquakes are very frequent.

The forests yield many valuable kinds of timber, including mahogany, cedar, palo di maria, a species of wood well adapted for ship-building, &c. But the log wood tree (Hæmatoxylon Campechianum, Linn.) is by far the most valuable of the products of the forests. It is found here and in the adjoining peninsula of Yucatan in the greatest perfection, and is a most important article of export; a species of Brazil wood is also exported. Among the other vegetable products may be enumerated the dragon's blood, mastic, palma Christi, and other balsamic, aromatic, and medicinal plants; with the sugar-cane, cocoa, indigo, coffee, tobacco, and cotton, which are extensively cultivated. The crops vary according to the elevation of the surface. Below the level of 3,000 ft., indigo, cotton, sugar, and cocoa are the principal. The last is chiefly grown along the shores of the Pacific. The district is also distinguished for the growth of indigo, to which the agriculturists devote their attention so exclusively, as almost wholly to neglect the cultivation of articles of prime necessity. The culture of indigo is, however, very general throughout Central America, and, according to Humboldt, it was formerly produced to the value of 12 millions of livres a year. Between the heights of 3,000 and 5,000 ft., the Nopal, or cochineal plant, is a favourite object of cultivation, particularly in the neighbourhood of Guatemala. Maize is generally grown, but wheat only in the high table-land in the N. Flax and hemp, though they grow luxuriantly, receive little attention, owing to the superior facilities for growing and manufacturing cotton; and vanilla is suffered to run to waste for want of hands to gather and prepare it. Among the remaining kinds of produce are tamarinds, cassia, long pepper, ginger, and others, which, though highly useful, are little known in commerce. The subjoined table exhibits the principal articles of home produce, exported from the republic in each of the years 1859 and 1860:—

Principal Articles		1859	1860
Indigo	lbs.	807,080	772,490
Sugar	cwts.	98,504	17,419
Cochineal	lbs.	1,764,544	1,777,290
Hides	no.	165,498	88,741
Deerskins		8,083	6,085
Wool & Mahogany	logs	7,817	1,356
Sarsaparilla	lbs.	Bales 540	Bales 774

Agriculture, and cattle and sheep breeding, are the chief occupations of the people; but the manufactures are not quite unimportant. While it belonged to Spain, Guatemala produced most of the cotton and woollen fabrics required for its own consumption; at present the former are chiefly imported from Great Britain, but coarse woollens are still manufactured, together with some cotton cloths, caps, and hats. A good many hands are also employed in making earthenware, furniture, wooden articles in cabinet work, &c., and an inland trade is carried on in mats, woven of different colours by the Indians, and used at Guatemala as carpets.

The commerce of Guatemala, comprising, is regard to exports, the produce already enumerated, and in imports the ordinary manufactured articles is chiefly with Great Britain and the United States. The subjoined two tables exhibit the value of the imports as well as the exports in each of the five years 1856 to 1860, distinguishing between the total imports and exports and those from and to Great Britain:—

Years	Imports			
	Total		From Great Britain	
	Tons.	£	Tons.	£
1856	1,604,139	913,624	570,102	146,971
1857	1,134,866	739,502	836,824	161,361
1858	1,542,816	314,564	742,895	114,566
1859	1,570,404	364,921	1,075,172	245,544
1860	1,634,671	796,934	809,353	181,851

Years	Exports			
	Total		To Great Britain	
	Cwt.	£	Cwt.	£
1856	1,742,255	844,458	1,474,542	795,844
1857	1,814,352	571,824	1,310,578	216,914
1858	1,823,926	840,784	1,304,799	379,701
1859	1,755,824	351,105	1,014,854	397,971
1860	1,916,826	863,363	1,361,056	374,319

It will be seen that while the exports of the Republic are taken almost wholly by Great Britain, the imports are furnished to not the same extent. The latter come in part from the United States.

The government of Guatemala is in the hands of a president elected for life; a council of state, composed of 13 members, and of house of representatives of 54 members, elected for six years. Five ministers of state, appointed by the president, superintend the departments of foreign and home affairs, finance, justice, and war. In Guatemala the Spanish laws have been entirely abolished, and the code compiled by Mr. Livingstone, of the U. States, substituted in their stead.

The Roman Catholic is the established religion, but complete religious toleration exists. The monastic orders have been wholly suppressed; and the few nunneries that exist are not permitted to enforce the residence of their inmates against their will. Each of these establishments has attached to it a free school for the education of the poor in reading, writing arithmetic, and religious principles. Slavery is entirely abolished.

The Indians of Guatemala preserve to a great degree their aboriginal languages and customs. The chief occupation of the settled tribes is agriculture; some are engaged as workmen in various manufactures. They live in great harmony with the whites, but entertain a dislike to the ladinos. The latter are a mixed breed between the whites and Indian tribes; their complexions are much fairer than those of the W. Indian mulattoes, and many are little distinguishable in appearance from the whites. The latter are mostly of Spanish descent.

History.—The N.E. coast of this region was discovered by Columbus in 1502. Most part of it was conquered by the Spaniards about 1524, and erected into a captain-generalship by the emperor Charles V., in 1527. The policy adopted by Spain towards Guatemala was attended with unintentional benefits to the latter. Being only a captain-generalship, the scale of its public expenditure was kept down in deference to the higher pretensions of the Spanish viceroyalties, and as its financial wants were few, taxation pressed lightly on the people. It was not, however, permitted to export more of its native produce than were sufficient to pay for the articles which the merchants of Cadiz thought necessary to send for its consumption. Guatemala, together with the other states of Central America, became independent in 1821, and was subsequently incorporated with Mexico. The Mexican Confederation was again broken up in 1823, and the Central American states formed a league by themselves in 1842. From this union Guatemala seceded March 21, 1847, and has since continued a separate state.

GUATEMALA (SANTIAGO DE), or NEW GUATEMALA, a city of Central America, capital of same name, in the spacious plain of La Virgen, in the valley of Mexico, 4,990 ft. above the level of the sea, 105 m. WNW. San Salvador, and 655 m. ESE. Mexico: lat. 14° 37' N., long. 90° 35' W. Pop. 60,000, according to a rough enumeration of the year 1853. Viewed at a distance from the surrounding mountains, few cities present a more beautiful aspect. It lies in the midst of sloping meadow lands and rich plantations; its walls, domes, and steeples being covered with a white and glittering cement. It forms a square divided into 4 quarters, each of which is again divided into two *barrios*, or wards, superintended by their own alcaldes. The streets, which are 12 yards broad, are mostly paved, and in their centre is usually a streamlet of water. To obviate the danger of earthquakes, the houses are only one story high; but they occupy a considerable space, being built in squares, round one or more open courts. The roofs are flat. The *Plaza*, or Great Square, is a rectangle, 150 yards each way, surrounded on three sides with colonnades, and having in it the cathedral, with the archbishop's palace, the *College de Infantes*, the old royal palace, and various government offices, including the supreme court of justice, treasury and mint; the town-hall, prisons, markets, public granary, and custom-house. In the middle is a large stone fountain, of very superior workmanship, supplied with water brought by pipes from the mountains upwards of 2 leagues distant; the same source supplying 12 public reservoirs. In different parts of the city, besides many belonging to convents and private houses. Besides the cathedral, there are a great number of highly ornamented churches. There is a university, but it is on a limited scale. Girls' schools are attached to the nunneries, and there are some endowed schools for boys. On the N.E., adjoining the city, is an extensive suburb, divided into two quarters and four *barrios*. Guatemala has manufactures of fine muslins, gauzes, calicoes, and common cotton goods, earthenware, and china of very good quality. Among the females are excellent embroiderers, dress-makers, and florists; many also are employed in the manufacture of cigars, and spinning cotton yarn of all degrees of fineness. The inhabitants possess an aptitude for the arts, and are particularly noted as workers in silver, sculpture, and musicians. Their chief entertainments are picnic parties to the surrounding country; bull fights, a circus for which stands about half a mile from the city; and the theatre, an edifice partially open to the sky, the performances in which take place during daytime. Religious festivals have always been celebrated in this city with great magnificence. On Sundays, from sunrise till 11 o'clock, the churches are devoted to public worship, and filled with successive congregations; but at the latter hour a new scene commences. The church doors are shut; the plaza, which till then had been filled with crowds hurrying to and from their devotions, is suddenly converted into a fair: stalls

and booths are erected in all parts of it, and the
remainder of the day is devoted to business or
pleasure.

Guatemala is the seat of the government of the
repub. of the same name, and also the see of the
primate. It was founded in 1776, after the destruction by an earthquake of old Guatemala,
25 m. W. by N. from the new city. But the latter
has been again rebuilt, and is a favourite place of
resort, having seldom fewer than from 12,000 to
15,000 inhab.

GUAYAQUIL, a city, and the chief sea-port of
the republic of Ecuador, South America, on the
river of the same name, 157 m. SSW. Quito, and
43 m. N. by E. the isl. Puna, in the Gulf of
Guayaquil; lat. 2° 10′ 21″ S.; long. 79° 45′ W.
Pop. estimated at 25,000. It is built principally on
the N. bank of the river, and is divided into the
old and new town, the former being occupied by
the poorer classes. The city is tolerably well laid
out; and as its houses are of wood, and it has frequently suffered from fires, much of it is comparatively modern, and has a good appearance. Its
private residences are mostly tiled and furnished
with arcades. It contains several good edifices,
including the custom-house, three convents, a college, and hospital; but from being situated on a
dead level, and intersected by many creeks, the
drainage is bad, and the streets are so swampy as
to be sometimes unpassable. Many of the inhab.
live on the river, on balsas, or floating rafts, from
50 to 80 ft. long. The river opposite the city is
about 3 m. wide, and has on its N. bank a dry
dock, where several ships of a superior construction have been built. The city is unhealthy, and
is ill supplied with water, which has to be brought
from a considerable distance on balsas, which,
indeed, are used for the conveyance of all kinds of
goods. The port of Guayaquil is one of the best
on the Pacific, ships of large size coming up close
to the town. It is defended by three forts, one
being on the opposite side of the river. Ships
bound for Guayaquil usually call at Puna for
pilots. The principal articles of export are cocoa,
timber, hides, cattle, tobacco, cviba wool, and the
other produce of the country.

According to an official statement of Mr.
Mocatta, British vice-consul at Guayaquil (Consular Reports, No. IX., Lond. 1862), the imports
and exports of Guayaquil, in the five years 1856 to
1860, were as follows:—

Year	Imports	Exports
	£	£
1856	604,782	380,644
1857	651,000	741,102
1858	509,456	474,694
1859	599,318	662,642
1860	419,373	427,598

The exports of 1860 were shipped in the following manner:—

Nationality	Vessels	Tons	Value of Cargoes
British	54	44,572	£189,469
Equatorian	55	1,676	4,000
Peruvian	102	8,045	57,670
Chilian	8	1,411	8,680
Spanish	18	4,642	217,206
French	2	976	12,080
United States	6	7,681	14,500
Sardinian	1	211	1,000
Danish	1	963	1,000
Prussian	1	577	10,000
Dutch	1	330	8,000
Total	**250**	**63,513**	**£523,405**

The merchandise imported during the year 1860
consisted of the following articles:—

	Estimated Value
	£
Cotton Manufactures	144,465
Linen ditto	14,723
Woollen ditto	77,741
Silk ditto, and Raw	11,341
Haberdashery and Hosiery	4,150
Thread and Tape	3,614
Wearing Apparel	451
Hardware	20,046
Metals—Iron, Copper, Lead, &c.	8,261
Earthenware, Porcelain	6,741
Glassware	2,771
Naval Stores	1,861
Oil, Paint, &c.	4,716
Soap, Candles, Wax	9,779
Grocery	41,810
Flour	21,780
Wine	19,647
Spirits and their Compounds	23,763
Ale and Porter	1,601
Drugs, Spices, &c.	1,271
Dye Stuffs	1,913
Stationery and Books	6,153
Furniture	8,449
Arms and Ammunition for private use	3,173
Miscellaneous Articles	32,750
Total	**£432,676**

It may be noticed that this total is not the same
as that given in the preceding table—a fact not
otherwise accounted for in the report of the British
vice-consul as that of the latter being estimated
value.

GUAYMAS, a sea-port town of Mexico, state
Sonora, at the mouth of a considerable river, on
the E. shore of the Gulf of California, 230 m.
WNW. El Fuerte. Lat. 27° 50′ N., long. 112° W.
Pop. estimat. at 5,000. The town has grown up
since the revolution, and owes its origin and rise
to its magnificent harbour, the best in Mexico.
This inlet is capable of accommodating 200 vessels, and is sheltered from all winds by the lofty
hills which surround it, and the island of Pasaron,
which forms a natural breakwater before its entrance. Close to the pier there are 6 fathoms
water, and deeper soundings, with good anchorage, are found a short distance further off shore.
The more modern houses are large and well built;
the rest are chiefly of mud, and flat-roofed. The
climate is healthy, though hot. Water, with provisions, have to be conveyed to the town from a
distance of about 5 m., the immediate neighbourhood being arid and sterile. But the great commercial advantages of the place countervail these
drawbacks, and will probably render it the
principal commercial depôt on the W. coast of
Mexico; it being much superior as a port to either
Mazatlan or San Blas, and easier of access than
Acapulco to vessels from China to Calcutta,
which from the prevalence of particular winds in
the Pacific, seldom make the Mexican coast S.
of Guayman. At this port and Mazatlan, indeed,
all the trade between Mexico and E. Asia is now
transacted.

GUAYRA (LA), the principal sea-port town of
the repub. Venezuela, South America, gov. Caraccas,
on the Caribbean Sea, 11 m. NNW. Caraccas; lat.
10° 36′ 19″ N., long. 67° 6′ 45″ W. Pop. estimat.
at 8,000. Humboldt observes:—"The situation of
La Guayra is very singular, and can only be compared to that of Santa Cruz, in Teneriffe. The
chain of mountains that separates the port from
the high valley of Caraccas, descends almost directly into the sea; and the houses of the town
are backed by a wall of steep rocks. There scarcely
remains 100 or 140 fathoms' breadth of flat ground
between this wall and the ocean. The town is
commanded by the battery of Cerro Colorado, and
its fortifications along the sea-side are well dis-

paved and kept in repair. The aspect of this place
has something solitary and gloomy. . . . The heat
is stifling during the day, and most frequently
during the night." (Tere, Narrat. Trans. vol. iii.
343, 344.) In 1812 the town was nearly destroyed
by an earthquake, from the effects of which it has
not yet wholly recovered. Its port is a mere roadstead, open to the N. and E., and slightly sheltered to the W. by Cape Blanco. Vessels anchor
in from 6 and 7 to 25 and 30 fathoms, according
to their distance off shore; but though the anchorage be open, and there is a considerable surf,
the holding-ground is good, and vessels properly
found in anchors and cables are seldom driven
from their moorings. The trade of La Guayra is
extensive. The principal articles of export are
coffee, cocoa, indigo, and hides. The imports consist principally of manufactured goods from England, and provisions from the United States.
According to a report of Mr. Mathison, British
vice-consul (Report dated March 22, 1864, in
'Commercial Reports received at the Foreign Office,'
p. 542), the trade of La Guayra for the year 1863
comprised exports to the value of 62,925l., and
imports of 31,215l. The town is unhealthy in
summer, especially to strangers; and it is extremely hot, the mean temp. of the town being
nearly 83° Fah. La Guayra was founded by Osorio
in 1588.

GUBEN, a town of the Prussian dom., prov.
Brandenburg, gov. Frankfort, cap. circ. of same
name on the Neisse, 27 m. SSE. Frankfort-on-the-
Oder, on the railway from Frankfort to Breslau.
Pop. 15,929 in 1861. The town is the seat of the
courts of justice for the circ., town, and district, a
board of forest economy, and a gymnasium. It is
one of the most populous and flourishing towns in
the gov.; and, besides producing woollen and linen
stuffs, yarn, and stockings, has tanneries, water-
mills, and a copper foundry, with building docks,
and a considerable trade in cattle, wool, and agricultural produce: it has also a brisk transit trade.

GUERNSEY, an island in the English Channel,
belonging to Great Britain, 75 m. S. the Isle of
Portland, 32 m. E. the coast of Normandy in France,
27 m. SW. Alderney, and 15 m. WNW. Jersey.
Shape triangular; greatest length, 9 m.; breadth,
5 m.; area, 16,000 acres. Pop. 29,804 in 1851, and
29,846 in 1861. The surface of its N. part is level
with a low irregular line of coast; but the S. part
is more lofty, varied with deep gullies; the coast
is bold and precipitous, presenting fine marine
scenery. The geological formation is almost entirely granitic, and quarries of gneiss and granite,
at Grande Rocque, are extensively worked; on
the W. side of the island, trap-rocks and micaceous
schist occur. There are no metals of any kind.
The climate, though inconstant and uncertain, is
very moist, is not unhealthy. The winters are
mild, snow seldom lying on the ground more than
two or three days, and the summer heats are less
oppressive than on the neighbouring coast of France,
or even in the SW. of England. The thermometer
ranges from about 30° to 87°; prevailing winds
are E. in spring, and W. the rest of the year. The
water is excellent, and the lands are well watered
by streams running in every direction towards the
sea. Guernsey, in point of fertility, does not equal
Jersey, neither is it so well covered with timber;
and it contains, especially in the N., considerable
portions of waste, or imperfectly reclaimed land.
There is a great division of properties, which vary
from 5 to 12 acres, 30 acres being considered a
large farm. This division is owing to the law, or
custom, which gives to each son an equal share of
his father's landed property. The annual growth
of wheat is estimated at 4,000 quarters; the growth

of barley amounts to 3,500 quarters. These quantities supply only about a fourth part of the home
consumption, the deficiency being made up by importations from France and the Baltic. Barley is
chiefly employed in malting. Oats and rye are
little grown; but parsnips, beet-root, and potatoes
are extensively grown. The principal manure is
vraic, a kind of sea-weed, gathered by the people
twice a year. With the exception of draining
marsh-lands, several hundred acres of which have
been brought into cultivation, the art of tillage is
to a great extent stationary. Garden produce forms
a main part of the cottager's subsistence. Melons,
figs, peaches, and even oranges are abundant. The
breeding of cattle is the most profitable branch of
farming; the price of Guernsey cows varies from
10l. to 16l., according to their excellence, and they
yield about 7 lbs. of butter weekly. The cows,
which are milked three times a day, are universally tethered; about 1½ acres being reckoned sufficient for the support of each. The law forbidding
the importation of foreign breeds is strictly enforced, and thus the purity of the native race is
maintained. Hogs are numerous, and of great size;
sometimes attaining from 50 to 80 stone weight.

The trade of Guernsey is very inferior to that of
Jersey, and has greatly decreased since the French
war. Before the introduction of the bonding system,
Guernsey was used by merchants as a depôt for
foreign wines and other goods; besides which it
had a most extensive smuggling trade, which,
however, has now wholly ceased. The shipping
is at present chiefly employed in exchanging the
wines of Spain and the Mediterranean for the sugar,
coffee, spices, &c. of S. America, which they take
to Hamburg or Rotterdam, and again exchange
for corn. The exports consist chiefly of cider,
apples, potatoes, building-stone, and wine; the
imports are wheat and flour, British manufactures,
wines, sugar, and coffee. (For particulars, see
Jersey.) There are some manufactures in Guernsey of cement, bricks, cordage, paper, and soap;
but all on a small scale.

The military government of the island is vested
in a lieutenant-governor, who represents the sovereign in the assembly of the states. The legislative body, called the states, is composed of the
bailiff, the procureur or attorney of the royal court,
12 jurats, the rectors and constables of parishes,
total 32; and of these the first two are appointed
by the crown, and the rectors by the governor;
while the jurats and constables are chosen by the
islanders. The states vote money for ordinary
public expenses; but new taxes must be sanctioned
by the crown: indeed all new laws and constitutional changes can be effected only by application
to the privy council. The 'royal court,' the
supreme tribunal, consists of a bailiff appointed by
the crown, and 12 jurats elected by the people.
The language spoken in court is French. Juries
are not known; and the powers of the court are
extensive, undefined, and sometimes oppressively
used. Guernsey is a deanery, in the diocese of
Winchester, and comprises eight livings; but as
the great tithes belong to the government, the
clergy are wretchedly paid, and have little personal influence.

The natives of Guernsey, like those of Jersey
(both of whom, in the lower ranks, speak a Norman
patois), are thrifty, parsimonious, clean and neat
in person and dress, simple in their manners, and
generally honest. They are credulous, many still
believing in witchcraft. The estab. of schools,
however, in every parish has greatly raised the
moral feelings of the lower orders; these schools
have been repaired, and are partly supported by
public money. Queen Elizabeth's college, founded

in 1563, and greatly enlarged in 1821 at an expense of 16,000l., is now in a flourishing state, and furnishes a first-rate classical and scientific education to about 200 students, at an expense of about 17l. a year each. The improvement of this establishment is conducing materially to the prosperity of the island, both by its direct influence on the natives, and by bringing new residents from England.

The only considerable town of Guernsey is Peter-le-port, its cap. situated on the E. side of the island. Being built on the slope of a hill, it looks well from the sea; but the streets, except in Hauteville, the modern and best built quarter, are narrow, steep, and crooked, lined with old and very lofty houses. The chief buildings are, the government-house, Queen Elizabeth's college, the court-house, the town hospital, and a handsome fish market. The par. church was built in 1312. The harbour, formed by two piers, is considered sufficient for the trade of the place, and there is great anchorage in the roadstead. Fort George, a strong fortress, stands ½ m. S. of the town.

Guernsey, as well as the other Channel Islands, was included in the duchy of Normandy, which once belonged to Great Britain. The French have made several attempts to conquer it, but without success. The last was in 1780.

GUIANA, GUYANA, or GUAYANA, an extensive region of S. America, embracing, in its widest acceptation, all the territory between the Amazon and Orinoco, and extending between lat. 4° N. and 8° 40′ N., and long. 50° and 68° W. By far the greater portion of this region (formerly called Spanish and Portuguese Guayana) belongs to the Venezuelan and Brazilian territories; and the term Guiana is now generally understood to refer only to the country between lat. 0° 40′ and 8° 40′ N., and long. 57° 30′ and 60° W., divided among the English, Dutch, and French.

GUIANA (BRITISH) is the most westerly portion of the above territory, and the largest, if we include within its limits the entire territory claimed by the British. The latter extends between lat. 0° 40′ and 8° 40′ N., and between the 57th and 61st deg. of W. long., having E. Dutch Guiana, from which it is separated by the Corentyn; S. Brazil; W. Venezuela; and N. and N.E. the Atlantic. Area, 76,000 sq. m.; pop. 127,695 in 1851, and 148,026 in 1861. The latter census stated a considerable preponderance of males, of which there were 79,644, against 68,382 females.

Physical Geography.—An alluvial flat extends from the coast inland, with a breadth varying from about 10 to 40 m., terminating at the foot of a range of sand hills, from 50 to 120 ft. high. Parallel with this range run several detached groups of hills, seldom more than 700 ft. high, which cross the Essequibo in lat. 6° 15′, being continuous with the Sierra Imataca in Venezuela. About lat. 6° a mountain chain, composed of granite, gneiss, and other primitive rocks, an offset of the Orinoco mountains, runs W. to E. through Guiana, forming large cataracts where it is crossed by the bed of the rivers, and rising frequently to the height of 1,000 ft. above the ocean. About a degree farther N. are the Pacaraima mountains, which in a similar manner run W. and E. and are of primitive formation. This chain forms many rapids and cataracts in the larger rivers, and contains the sources of several rivers, of secondary importance, including the Berbice and Mazaruni. Its highest point, M. Roraima, lat. 5° 9′ 31″ N., long. 60° 47′ W., near the W. extremity of the territory claimed by the British, is 7,500 ft. high. The Canuco or Canuco chain, running NE., connects the Pacaraima with the Sierra Acarai. The

latter is a densely wooded chain of mountains, forming the S. boundary of Guiana, and the watershed between the basins of the Amazon and Essequibo. Mr. Schomburgh estimated the elevation of the highest summits of this chain at 4,000 ft. The Essequibo and Corentyn rise in it.

'The whole surface of the coast lands of British Guiana is on a level with the high water of the sea. When these lands are drained, banked, and cultivated, they consolidate, and become fully a foot below it. It requires, therefore, unremitting attention to the dams and sluices to keep out the sea, one inundation of which destroys a sugar estate for 18 months, and a coffee one for 6 years. The original cost of damming and cultivating is fully paid by the first crop, and the duration of the crops is from 30 to 50 years; so that, though great capital is required for the first outlay, the comparative expense of cultivation is a mere trifle compared with that of the (W. India) islands, notwithstanding that the expense of works, buildings, and machinery may be treble or quadruple, being built on an adequate scale for half a century of certain production.' (Hillhouse on the Warow Land, Geog. Journ., iv. 332.)

Between the first and second chains are some extensive savannahs, which approach the sea-shore E. of the river Berbice. S. of the Pacaraima chain and the Rupununy are others still more extensive, but not so well watered. In the latter region are situated the small lake of Amucu and the frontier settlement of Pirara. With the exception of these savannahs, and the swamps on the Berbice, the interior is mostly covered with hill-ranges and dense forests.

The greatest slope of the country is towards the N., in which direction run the principal rivers. The chief of these is the Essequibo, which rises in the Sierra Acarai, about 40 m. N. the equator, and discharges itself into the ocean by an estuary nearly 20 m. wide, after a course of at least 620 m. Its entrance is much impeded by shoals, and is navigable for sailing vessels for only about 50 m. from its mouth. According to the volume of water, its current is more or less strong, but it is seldom more than 4 knots an hour, even during the rainy season. The Corentyn rises about lat. 1° 30′, and long. 57°, and discharges itself also by an estuary 20 m. wide. Between these two rivers run the Berbice and the Demerara; the former may be ascended for 165 m. by vessels drawing 7 ft. water; the latter is navigable for 85 m. above Georgetown, which is situated near its mouth. The Mazaruni, Cuyuni, &c., affluents of the Essequibo, are the other principal streams. All the large rivers bring down great quantities of detritus, which being deposited around their mouths and estuaries renders the whole coast shoal. For 12 or 15 m. seaward the mud bottom is covered by only 3 or 4 ft. water.

Geology and Minerals.—These deposits extend the coast rest upon deep strata of strong clay of different kinds, alternating with others of sand, and beds of small shells; and these again upon a granitic formation, which begins to appear on the surface in the second chain of mountains. The granite rocks in the interior often assume the most imposing and singular forms; mural precipices, with cascades 1,400 or 1,500 ft. high descending over them; granite boulders of huge size, spread over extensive tracts, &c.; and in lat. 3° 53′ is a natural pyramid, called the Ataraipu, wooded to the height of 850 ft., and rising from that limit in naked grandeur to an elevation of about 800 ft. Mr. Schomburgh gives a sketch of this pyramid in the 'Geog. Journ.,' i. 163. The other chief rocks are porphyry, and various kinds of trap, gneiss, clayslate, sandstone, coloured ochres, &c.: there is a total absence of limestone and its modifications. Traces of iron are frequent, but none of the precious metals has been discovered. Next to granite, excellent pipe and other clays are the most valuable mineral products.

Climate.—The mean temperature of the year at Georgetown is 81° 2' Fahr., the maximum 89°, the minimum 74° on the coast. Two wet and two dry seasons constitute the changes of the year. The great dry season begins towards the end of August, and continues to the end of Nov., after which showers of rain follow to the end of Jan.; the short dry season then commences, terminating about the middle of April, when the rains begin to descend in torrents, and the rivers to inundate their banks. The winds during the rains are generally westerly; in the dry season they blow mostly from the ocean, particularly in the day-time. Hurricanes are unknown, gales unfrequent; thunder-storms occur at the changes of the seasons, but, like a few occasional shocks of earthquakes, are not attended with danger. The low and swampy coast-lands are unhealthy, but the interior is quite otherwise; and the insalubrity of Georgetown, and other sea-port towns, has been greatly aggravated by the quantity of refuse suffered to collect and decompose on the shore.

Vegetable Products.—The forests abound with trees of immense size, including the mora cornata, siparl or green-heart, and many others, yielding the most valuable timber, and an abundance of medicinal plants, dye-woods, and others of excellent quality for cabinet-making. Arnotto, so extensively used in the colouring of cheese, grows wild in profusion on the banks of the Upper Corentyn. That magnificent specimen of the American Flora, the Victoria regia, was discovered by Mr. Schomburgk, on the banks of the Berbice. (Geog. Journ.) Another indigenous plant deserving of mention, is the Ani-arry, a papilionaceous vine, the root of which contains a powerful narcotic, and is commonly used by the Indians in poisoning waters to take the fish. The Indians beat the roots with heavy sticks, till it is in shreds, like coarse hemp; they then infuse it, and throw the infusion over the area of the river or pool selected. In about 30 minutes, every fish within its influence rises to the surface, and is either taken by the hand or shot with arrows. A solid cubic foot of the root will poison an acre of water, and the fish are not thereby deteriorated. (See Hillhouse, in Geog. Journ, iv.)

Wild Animals.—The jaguar, puma, peccari, and wild hog, tapir, and many kinds of deer, abound in Guiana: the sea-cow is met with in the larger rivers, which are also inhabited by the cayman, alligator, and guana. There are several kinds of formidable serpents, but they are fortunately of a sluggish and inactive nature. The birds have the most magnificent plumage. Turtles are plentiful. The rivers teem with fish; the low-low, a species of silurus, often weighs from 200 to 300 lbs. The insect tribes are but excessively annoying.

Trade and Commerce.—The staples of the colony are at present sugar, coffee, and cotton; the two latter were formerly almost exclusively grown, but their culture is now in a great measure superseded by that of the sugar-cane. The coast regions are the only parts cultivated for sugar; but many tracts in the interior seem to be equally well fitted for that purpose; coffee, also, is grown only on the coast, but, according to Mr. Schomburgk, no tract appears better suited for it than the central ridge of the mountains. The Indians have generally some indigenous cotton growing round their huts, and among the Macusis (on the

Rupununi) it is raised to a considerable extent. It comes to perfection in most parts of the colony; but is cultivated by the colonists chiefly on the coast. There are numerous other products, which as yet neither form articles of export, nor of internal consumption, for which both the soil and climate are suitable, and which might be raised with advantage, were it not for the want of labour. Among these are rice, maize, Indian millet, Victoria wheat, cocoa, vanilla (a native of Guiana), tobacco, and cinnamon. Between the Berbice and the Essequibo there is a tract of many thousand acres, possessing the means of constant irrigation, on a small portion of which three crops a year have been repeatedly raised; but at present it is nearly all a complete wilderness, and will so continue till labour becomes more abundant and cheaper. The coast region, which is covered by a deep layer of vegetable mould, forming what is called a pegass soil, is so extremely fertile that 6,000 and even 8,000 lbs. of sugar, and from 20,000 to 30,000 lbs. of plantains, are sometimes produced on an acre; but in order to cultivate this soil, dams and embankments as before stated are necessary, and agriculture is conducted at a great outlay, and on large estates.

Large herds of horses and cattle wander wild on the wide but ill-watered savannahs beyond the Pacaraima; and, with little exception, have hitherto afforded food only for beasts of prey. The savannahs between the Berbice and the Demerara occupy upwards of 3,000 sq. m.; they are clothed with nutritious grasses, plentifully irrigated, and interspersed with shady woods. Were these stocked with cattle from the interior, beef might be obtained as cheaply as in the U. States. From 1,800 to 7,000 individuals, 7-10ths Indians, are employed in cutting timber, which is in great demand within the colony, though its export has hitherto been very trifling.

Since 1837, there has been a rapid decrease in the quantities of the staples grown and exported. Different circumstances have probably conspired to bring about this result; but there can be no manner of doubt that it is mainly ascribable to the nature of the climate, and the aversion of the emancipated negroes to severe labour. The total value of the exports which in 1836, amounted to 2,135,379l., had sunk in 1860 to 1,513,452l.; in 1861, to 1,583,644l.; and in 1862, to 1,365,275l. The imports amounted to 1,145,959l. in 1860, to 1,319,713l. in 1861, and to 1,107,181l. in 1862. Very nearly the whole of the exports are sent to Great Britain or to British America and the W. Indies. There are about 250 m. of public roads. Dutch and English measures, and Spanish, Dutch, and English money are in use.

Government.—The government is vested in a governor, and a court of policy, consisting, besides the governor, of the chief justice, attorney-general, collector of the customs, and government secretary, and an equal number of unofficial persons elected from the colonists by the college of electors. This college is a body of seven members, appointed by the inhabit. for life, whose qualification is the payment of taxes to the amount of 2l. sterling a year. The unofficial members of the court of policy serve for three years, and go out by rotation. There is a college of financial representatives of six members, with the same qualifications as the members of the college of electors, chosen by the inhab. for two years. The court of policy decides on all financial regulations; but when they have prepared an estimate of the expenses for the year, and the mode of taxation, and the different items have been discussed and acceded to by a majority, the estimates are handed over to the financial representatives, who, in concert with the court of policy, examine the charges. In this assembly, which is called the Combined Court, every member, whether of the court of policy or financial representatives, has an equal vote. The court of policy, combined with the financial representatives, having approved of and sanctioned the ways and means, they are passed into a law. The governor not only has a casting vote, as president of the court of policy, but an absolute veto on all laws passed by a majority. The supreme civil court consists of a chief judge, two puisne judges, a secretary, registrar, and accountant. It is a court of appeal from the rolls court in each, in which one of the judges of the supreme court presides. The supreme criminal court is composed of three civil judges and three assessors, chosen by ballot. Its judgments are decided upon by a majority of votes, and are delivered in open court. Inferior criminal courts are holden by the sheriffs of each county, with whom three magistrates are associated. Special magistrates, appointed from England, decide between the masters and labourers in the different districts; three superintendents of rivers, and six post-holders are appointed for the protection of the Indians in the interior. The criminal law is the same as that of Great Britain, but civil cases are ruled by the Roman-Dutch law, in so far as it has not been modified by orders in council and local ordinances. The military force consists of one regiment of the line, and a detachment of another. The colonial militia has been disbanded.

The public revenue is derived from taxes on produce; on incomes of 500 dollars and upwards; on imports not of the origin or manufacture of Great Britain; and from assessed taxes on horses, carriages, wine and spirit licences. The total revenue amounted to £61,263, in 1863, and the expenditure, in the same year, to £51,164. The portion of the 20 millions sterling falling to this colony, as compensation for the freedom of slaves, amounted to 4,500,000£.

The only towns worthy of mention are Georgetown and New Amsterdam. Georgetown, formerly Stabroek, the cap. and seat of government, is on the E. bank of the Demerara, near its mouth; lat. 6° 49′ 20″ N., long. 58° 11′ 30″ W. Except Water Street, which is built close to the river, the streets are wide and traversed by canals; the houses are of wood, seldom above two stories high, shaded by projecting roofs, having verandahs and porticos, and surrounded by gardens separated by trenches. An edifice facing the river, built of brick and stuccoed, which cost the colony upwards of 5,000£, comprises all the government offices near it are the Scotch church, market-house, and town guard-house. Within a mile of the town, near the mouth of the river, is Fort William Frederick, a small mud fort. A handsome Gothic church, which cost 13,000£, has been erected at Georgetown; another episcopal church stands on the jumale ground, besides which it has a Roman Catholic cathedral, Wesleyan chapel, 3 public, an infant, and 8 private schools, a colonial hospital, an excellent seaman's hospital, a savings' bank, two commercial banks, and an amateur theatre. Shops and stores are numerous, and European goods of all kinds plentiful; no duty being laid on English merchandise. The markets are good, and a new market-house is being erected. New Amsterdam, on the Berbice, in lat. 6° 15′ N., long. 57° 27′ W., extending about 1½ m. along the river, is intersected by canals, and has about 8000 inhabitants. It has English, Scotch, and Dutch churches, Rom. Catholic and Wesleyan chapels, a free school, court-house, barracks, fort, many commodious wharfs and ware-

houses, and two commercial banks. It is less unhealthy than Georgetown.

History.—According to some, Columbus discovered Guiana in 1498; others gave that honour to Vasco Nunez in 1504. The Dutch, who were its first European settlers, established some settlements near the Pomeroon and elsewhere in its neighbourhood, in 1580, and several further to the E. a few years afterwards. The English began to form settlements about 1630. Most of Guiana, however, remained in the hands of the Dutch till 1796, when Demerara and Essequibo surrendered to the English. They were restored to the Batavian republic in 1802; and re-taken by the British in 1803. The territory called British Guiana has belonged to us ever since that period; that called Dutch Guiana was given up to Holland at the conclusion of the late war.

GUIANA (DUTCH). This territory is intermediate, both in size and position, between British and French Guiana. It extends between the 2nd and 6th deg. of N. lat., and the 53rd and 57th deg. W. long., having E. French Guiana, from which it is separated by the Maroni, S. Brazil, W. the Corentyn, which divides it from British Guiana, and N. the Atlantic. Length, N. to S. 250 m.; average breadth, about 155 m. Area about 38,500 sq. m. Pop., exclusive of Indians and Maroons, estimated at 65,000, of whom 6,000 are whites or free coloured people, chiefly Dutch, French, and Jews, and the remainder negro slaves. The maroons of the interior are the descendants of runaway negroes, and were very troublesome during the past century; they have now, however, adopted much more settled habits than formerly, and receive annual presents of weapons and arms from the Dutch, the territory they occupy forming a kind of military frontier to the colony. The physical geography, climate, and productions of Dutch Guiana are pretty much the same as those of British Guiana. All the rivers have a N. direction; the chief is the Surinam, which runs through the centre of the country, and falls into the Atlantic, after a course of nearly 300 m. It gives its name to the N. portion of the territory, and is navigable for large ships for about 4 leagues from the coast. Paramaribo is situated near its mouth. About 60 ships are employed in the transport of the produce of the colony to Europe. Sugar is the chief staple, and about 25,000,000 lbs. are produced annually; the export of coffee may be estimated at about 3,000,000 lbs. a year; cocoa, cotton, rice, cassava, and yams are also grown in considerable quantities; and plentiful supplies of various descriptions of timber, and of woods for cabinet work, with gums, balsams, and other drugs, are procured from the interior. Provisions, arms, and manufactured goods are imported from Holland; provisions are also imported from the U. States, to which the exports are syrup and rum; there is some commerce with the W. Indies, and a smuggling trade is carried on with Colombia. The government is vested in a governor-general and a high council. The cap. and seat of government is Paramaribo, a town of 20,000 inhabitants, three-fourths of whom are blacks, or of mixed descent. It is neatly laid out in the Dutch style, and has R. Catholic, English, and Lutheran churches, a German, and a Portuguese Jewish synagogue, and an exchange, and is the centre of the trade of the colony. The fort of Zeelandia, a little N. of the town, is the residence of the governor, and the seat of most of the government establishments.

GUIANA (FRENCH). This, which is the most E. and smallest division of Guiana, lies between the 2nd and 6th deg. N. lat., and 51½ and

51½ deg. W. long., having E. and S. Brazil, W. Dutch Guiana, and N. and NE. the Atlantic. Length, N. to S., 240 m.; breadth varying from 100 to 190 m. Area 27,560 sq. m. Pop. 19,559 in 1841.

The coast plain (*basses terres*) is an alluvial tract of extreme fertility, interspersed with a few isolated hills, apparently of volcanic origin, and some ranges of low hillocks. The uplands (*terres hautes*) are also very fertile, their soil being generally argillaceous, more or less intermixed with granite, sand, and tufa, and in some parts highly ferruginous. The mountain chains run E. and W.; they are almost wholly granite, but no where reach any great elevation; in the centre of the colony they rise from 1,600 to 2,400 ft. above the level of the sea. Few countries are more abundantly watered. There are upwards of 20 rivers of tolerable size, all of which have a N. course. Their mouths are obstructed by sandbanks, and do not admit of the entrance of vessels drawing more than 12 or 15 ft. water; they cease to be navigable, except for canoes, at a distance of from 45 to 50 m. inland. In the rainy season they inundate the low country to a great extent, but are then innavigable from their rapidity. The coasts are low, and, except at the river mouths, ships cannot approach the shore. There is only one roadstead, that of Cayenne, where vessels can ride in security. Several small rocky or wooded islands lie off the coast, among which is Cayenne, at the mouth of the Oyapok, on which the cap. is built. The climate is similar to that of British Guiana; but the coast lands appear to be less unhealthy. About 50 or 60 m. from the coast the country begins to be covered with vast forests. The lowlands are in a great part uncleared, and covered with underwood. The settled and occupied lands were dispersed in 1836 over a surface of 250 sq. leagues, or about 1-80th part only of the whole surface of the colony, the rest of which is tenanted by wild beasts and roving Indians. The cultivated lands are chiefly given up to the growth of sugar cane, coffee, cocoa, and spices. The sugar plantations, in 1841, covered 452 hectares, and produced 509,061 kilogs. of sugar; coffee was grown, in the same year, on 498 hectares, and produced 74,700 kilogs.; and cloves, weighing 31,311 kilogs., came from 350 hectares of land. (Statistical Tables relating to Foreign Countries, Part IX, p. 252.)

The sugar-cane was introduced by the earliest colonists, and its culture has been greatly extended since 1822; it is grown only on the low lands. Coffee is very inferior to that of the W. Indies, and its culture has rather diminished of late years. Cotton, cocoa, annatto, and vanilla are indigenous. The clove succeeds pretty well, especially on the uplands; other spices have met with only doubtful success. Cocoa is unfit for the French markets, and most of what is grown is exported to the U. States; indigo and tobacco are of very inferior quality. Manioc, rice, maize, and bananas are grown, but the quantities produced fluctuate greatly, and are often insufficient for home consumption. Nearly the whole of the exports, except cocoa and a small quantity of sugar, are sent to France. Of sugar, the total exports in 1841 were of the value of 168,709 fr., of which the amount of 135,653 fr. went to France. The entire exports of French Guiana, in 1841, was of the value of 1,239,115 fr., or 51,976l., while the imports in the same year amounted to 6,413,950 fr., or 256,000l.

French Guiana is divided into two districts, those of Cayenne and Sinnamary; and fourteen communes, comprising six electoral arrondisse-

ments, and sending sixteen deputies to the colonial council, Cayenne, the seat of government (which see), is the only town worth notice. The government is vested in a governor, assisted by a privy council of seven of the highest official functionaries; and the colonial council, composed of sixteen members, elected for five years, by inhab. of French descent, twenty-five years of age, born, or having resided in Guiana for two years, and paying direct taxes to the amount of 200 fr. a year, or the possession of property to the value of 20,000 fr. Slavery was abolished in French Guiana by decree of the Republican government of France in the year 1848.

Guiana was colonized early in the 17th century. Some French adventurers first settled at Cayenne in 1604; and with only a few short interruptions from the Dutch and English, the French held that station and the rest of the colony till 1809; it was then taken possession of by the English and the Portuguese, and held by the latter till 1816, when, in pursuance of the Treaty of Paris, it was restored to France.

GUIENNE, one of the provs. into which France was divided previously to the Revolution. It was situated in the SW. part of the kingdom, on both sides the Gironde; and is now distributed among the deps. of the Gironde, Lot-et-Garonne, Dordogne, Lot, and Aveyron.

GUILDFORD, a parl. bor. and market town of England, co. Surrey, of which it is the cap., hund. Woking, on the Wey, 27 m. SW. London by road, and 30¼ m. by London and South Western railway. Pop. 8,020 in 1861. Guildford, as seen from the W., has an imposing appearance, being principally situated on the declivity of a chalk down, at the foot of which runs the Wey, crossed by a bridge of five arches. It consists chiefly of one long, broad, and well-built, but inconveniently steep, street, which is crossed by several other streets of inferior dimensions. It is well paved, lighted with gas, and supplied with water forced up from the river. It has three par. churches, all ancient structures; a handsome co. hall, townhall, council-chamber, a gaol, rebuilt in 1763; chapels belonging to Baptists, Presbyterians, Quakers, Rom. Catholics, &c.; a large free grammar school, founded by Edward VI., with an endowment for a scholar at Cambridge and at Oxford; a charity-school, at which twenty-five boys are educated and clothed; and a theatre. Guildford was a residence of the Anglo-Saxon kings, and the ruined keep of a castle, consisting of a quadrangular tower, 70 ft. high, and built of flint, ragstone, and Roman bricks, forms a picturesque object at the S. extremity of the town. The traces of an ancient palace are also clearly discoverable. Since the passing of the Municipal Corporation Reform Act, Guildford has been governed by four aldermen, one of whom is mayor, and twelve councillors. Petty sessions are held here, and the assizes in the summer circuits for E. and at Croydon alternately. Guildford has sent two mems. to the H. of C. since the time of Edward I. Previously to the Reform Act, the right of voting was in the freeholders and freemen resident in the town, paying scot and lot. The Reunting Act considerably extended the limits of the parl. bor. Registered electors, 721 in 1865. Corporation revenue, 2,100l. Guildford has a considerable trade with the metropolis in corn, timber, malt, &c., sent to London by the railway. Market-day, Saturday, for corn, and other necessaries. Fairs, May 4 and Nov. 22, for horses and cattle.

GUILSBOROUGH, or GUISBOROUGH, a market town and par. of England, co. York, N.

Riding, E. div., Langbaurgh lib., 89 m. N. York, and 21 m. E. Darlington. The par. comprises five townships. Area of township of Guisborough, 6,120 acres. Pop. of do. 4,081 in 1861. The town stands in a small but beautiful and very productive valley near the river Tees, and at the foot of the Cleveland hills. It consists of a single wide and handsome street, lined with old but substantial houses. The church is a modern edifice, supposed to occupy the site of one attached to the Austin Priory, established here in 1129, some ruins of which still remain in the meadows S. of the town. In the church-yard are the grammar-school and hospital, founded by the last prior, and chartered by Queen Elizabeth, in 1561. The hospital lodges and clothes six old men and six old women, and gives them a money allowance for food and coals. Guisborough is a quiet country town, with little trade, except on Monday, the market-day, and its six fair days (last Tuesday in April and May, third Tuesday in May, Aug. and Sept., and second Tuesday in Nov.). It used, however, to have a considerable trade in alum, and the first alum-works in England were begun here about 1600. This mineral is worked in some of the neighbouring parishes, especially Lofthouse; but it has for many years ceased to be a branch of industry at Guisborough.

GUIMARAENS, a town of Portugal, prov. Entre Douro-e-Minho, cap. of a comarca of same name, 28 m. NNE. Oporto, and 196 m. N. by E. Lisbon. Pop. 8,612 in 1858. The town is built on a slight elevation in the midst of a beautiful and productive plain between two small rivers, the Ave and Vizella, and is surrounded with fortifications. The streets, which are wide and straight, are lined with well-built houses, and there are several handsome piazzas, or squares. Among the public buildings are four churches, one of which is collegiate, and remarkable for its fine architecture; there are also five convents and four hospitals. It has some small manufactures of cutlery, hardware, and linen. There are thermal springs in the neighbourhood, which were known to the Romans. The ancient town is said to have been founded, anno 600 B.C., under the name of Araduca; the modern one was the first capital of the Portuguese monarchy.

GUINEA, a name applied by European geographers to designate a portion of the W. coast of Africa. The origin of the word is not correctly ascertained, nor are writers agreed respecting the limits of coast to which the name should extend. D'Anville, and the older geographers, apply it to the line of coast from the mouth of the Gambia to that of the Quorra; whereas Ritter, and the more modern authors, extend its confines from C. Verga, lat. 10° 30′ N., to the mouth of Nourse's river, lat. 17° S., and call the district S. of C. Lopez, lat. 1° S., comprising Congo, Angola, and Benguela, by the name of S. Guinea; while under N. Guinea, or Guinea Proper, are comprehended Sierra Leone, Liberia, the Grain and Ivory Coast, Ashantee, Dahomey, Benin, and Biafra. The description of this extensive line of coast will be found under the heads of the countries above mentioned.

GUINGAMP, a town of France, dép. Côtes-du-Nord, cap. arrond. on the Trieux, in an extensive plain, 17 m. WSW. St. Brieuc. Pop. 7,350 in 1861. The town was formerly surrounded with walls, parts of which still exist; a spacious street intersects it from end to end, about the middle of which is a singular par. church, with a square tower, surmounted by a dome. The town contains several good edifices, and is surrounded by agreeable walks. It has manufactures of the fabrics named from the town ginghams, linen cloth, thread,

&c., and twelve fairs yearly, at which large quantities of corn, cattle, flax, hemp, and manufactured goods, are sold.

GUIPUZCOA. See Biscay.

GUJERAT, GUJRAT, or GUZERAT (Gorjura Mandira), an extensive prov. of W. Hindostan, chiefly between lat. 21° and 24° N., and long. 69° and 74° E.; having N. Rajpootana, E. Malwah and Candeish, S. Aurungabad and the Gulf of Cambay, and W. the Indian Ocean, the Gulf of Cutch, and the Runn. It comprises the N. districts of the British presidency of Bombay, part of the Guicowar's dom., and the territories of many smaller chieftains. Its length, E. to W. may be estimated at 340 m., by an average breadth of about 180. Total area 41,536 sq. m.; pop. estimat. at 3,500,000. Gujerat is bounded on the N. and NE. by steep and craggy mountains of difficult access, sending out many ramifications, the intervals between which are filled with jungle, into this part of the prov. the Mahrattas were never able to penetrate; but they conquered the S. part, consisting of an open fertile plain, apparently level, but in reality intersected by numerous ravines and chasms, and watered by numerous rivers. The W. part consists of the Peninsula of Gujerat, stretching into the ocean between the Gulfs of Cambay and Cutch, about 190 m. in length, by 100 broad, and which forms the great theatre of the Guicowar's territories.

The NW. part of the prov. is in part a swampy plain where it adjoins the Runn, and an arid desert continuous with that of NW. India. The climate is oppressively hot in summer, but, in winter, temperate and agreeable. Though in parts there is a great deal of barren land, it is, upon the whole, one of the richest parts of Hindostan, both as respects its productiveness and the condition of its pop. Of 1,452,000 acres in tillage in British Gujerat, it was estimated, some years since, that 157,770 were under cotton culture, 4,256 under sugar-cane, 1,928 under indigo, 10,765 under tobacco, and the rest appropriated to the growth of grains and garden produce. All the foregoing articles of growth are of excellent quality; indigo was, however, grown formerly to a much greater extent than now. Oil, hemp, flax, and pulse, are the other principal kinds of produce. In the British districts, nearly all the land is cultivated that is capable of yielding an adequate return; in some parts of the prov. there are fine pasture lands, on which many good horses and draught cattle are reared. The land is assessed on the village system, the tax being collected through the medium of putails, or head-men. (See Bombay Presid.) In the British territories, most of the land is occupied by permanent tenants; leasehold lands are few. There are also few landholders of any extent; and in Kattywar, property is very much subdivided. The inhab. are mostly Hindoos, amongst whom the Jain sect are more numerous than in any other prov. of India. The pop. is, however, extremely mixed, and includes numerous tribes of Grassias, Katties, Coolies, Bheels, Mewassies, Bhatts, and other lawless races, who acted an important part during the wars of the Mahrattas and other dynasties that long troubled this part of India. Many of these tribes still lead a roving life; but most of them have now adopted peaceful occupations. Besides the native tribes, Gujerat (with Bombay) is the chief seat of the Parsees, a people who emigrated from Persia in the 7th century, after the overthrow of the Sassanide dynasty by the Muhammedans. (For some details respecting them, see Bombay.) The Mohammedans in Gujerat make about 10 per cent. of the pop. Almost all the castes of this prov.

work at the loom occasionally, and cotton fabrics, went in considerable quantities to Bombay, form, in fact, the chief export of the prov., after corn and raw cotton. The Sugat manufactures, of various kinds, have long been famous for their cheapness and good quality. The principal imports of the prov. are sugar, raw silk, pepper, cocoa-nuts, cochineal, and woollen goods. During the period of its independence in the 15th and 16th centuries, Gujerat enjoyed a much more flourishing trade than at present; but there are still many rich native merchants in the towns, the chief of which are Surat, Ahmedabad, Baroach, Baroda, Cambay, Gogo, Bhownuggur, Champaneer, and Junaghur. Gujerat was subjected by the Mahommedans under Mahmoud, of Ghizni, about 1025; from 1390 to 1572, it belonged to a native Rajpoot dynasty, which had revolted from the Moguls; but at the latter date it fell into the hands of the Emperor Acbar. After the death of Aurungzebe, in 1707, it was conquered by the Mahrattas, and remained a part of their empire till the destruction of their power by the British.

GUMBINNEN, a town of Prussia, prov. Prussia, cap. gov. of the same name, on the Pissa, 70 m. E. Königsberg, on the railway from Königsberg to Wilna. Pop. 8,010 in 1861. The town is regularly built, and has several churches, 2 hospitals, a public library, a gymnasium, and schools of midwifery and architecture. It is the seat of the superior courts, and central for its gov., and has manufactures of woollen cloths and stockings, distilleries, breweries, and some trade in corn and linseed.

GUNDWANA, a large prov. of the Deccan, Hindostan, extending between lat. 18° and 23° N., and long. 77° 30' and 86° E.; having N. the prova. Malwah and Allahabad, E. those of Bahar and Orissa, S. the Northern Circars and Hyderabad, and W. Berdar, Berar, and Candeish. It comprises the N.E. portion of the table land of Central India, and is chiefly included in the dominions of the rajah of Berar (the Nagpoor rajah) and the ceded and almost unexplored territories in the SW. parts of the British presidency of Bengal. A large proportion of its surface is mountainous, and some of the largest secondary rivers of Hindostan rise within its limits: as the Nerbudda, Sone, Mahanuddy, &c. while the Wurda and Godavery bound it W.; but in general it is ill-watered, unhealthy, covered with jungle, and thinly inhabited. The prov. consists chiefly of Goands, apparently an aboriginal people, at a rude period partly conquered and converted by the Hindoos, and the remainder driven to the hills and jungles, where they live nearly in a state of nature, the country continuing to be for the most part a sort of primeval wilderness. Their broad flat noses, thick lips, and often curly hair, distinguish them from the other native tribes of Hindostan. Some are domesticated in the plains, where they make good agricultural labourers; those who live wild, on the contrary, have no agriculture, and subsist on roots, vegetables, bamboo shoots, and whatever animal food they can obtain. Their own idols are of the rudest description, but they have also borrowed many objects of worship from the Hindoos, to which they offer up animal, and even human, sacrifices; in many parts they divide themselves into castes, like the Hindoos, and have adopted various institutions and practices from the Mahommedans. Their language contains, among its elementary words, many of Telinga and Tamul origin. The chief

towns in Gundwana are Nagpoor, Rambhalpoor, Deoghur, Mundlah, &c. Deoghut was formerly the seat of an extensive Hindoo empire; but the S. part of the prov. was included in the kingdom of Telingana, which, with Deoghur, afterwards constituted a portion of the Bahmunee empire of the Deccan; while the N. parts of the country were tributary to the Mogul emperors. There are, however, no remains in the prov. to indicate that it ever flourished as a highly civilised or cultivated country.

GÜSTROW, a town of N. Germany, G. D. Mecklenburg Schwerin, cap. prov. of same name, on the Nebel, 31 m. ENE. Schwerin, with which it is connected by railway. Pop. 9,212 in 1846. The town is walled, has an anc. castle, now converted into a workhouse and house of correction, and several handsome public edifices, among which are, the cathedral, 2 other churches, and the government house. It is the seat of a court of chancery, and boards of taxation and police, and is a town of considerable commercial importance. It has between 50 and 60 manufactories of different kinds, including many breweries and distilleries. Two large fairs for cattle and wool are held yearly.

GWALIOR, a strong fortress and town of Hindostan, and the modern cap. of Scindia's dom. prov. Agra, 61 m. SE. Agra, and 260 NE. by N. Oujein; lat. 26° 15' N. long. 78° 1' E. It stands on a precipitous, isolated hill, close around the base of which its defences of stone are carried. This hill is rather more than 1½ m. in length; but its greatest breadth does not exceed 300 yards: the height at its N. end is 342 ft. At this end is a palace; and about the middle of the fort are two remarkable pyramidal buildings of red stone, in the most ancient style of Hindoo architecture. The only gate is towards the N. extremity of the E. side; from which, by several flights of steps, you ascend to the top of the rock. Within the citadel there are large natural excavations, which furnish a supply of excellent water. The town, which runs along the E. side of the hill, is large, well inhabited, and contains many good houses of stone, which is furnished in abundance by the neighbouring hills. E. of the town runs the river Sasmera, beyond which is a large Mohammedan tomb, a handsome stone building, with a cupola covered with blue enamel. There are numerous caves adjacent to the fort, said to contain many Buddhist sculptures. Gwalior, from its position, must always have been a military post of great importance, but by no means impregnable: for it has frequently changed masters. It was taken by escalade in 1780 by the British; but finally ceded in 1805, to Scindia, and has since been the permanent residence of his court.

GYONGYOS, a market-town of Hungary, co. Heves, at the foot of the Matra mountains; 28 m. SW. by W. Erlau, and 42 m. NE. Pesth. Pop. 15,450 in 1857. The town has several churches, a Franciscan gymnasium, and a Roman Catholic high school; manufactures of woollen cloth, leather, hats, brandy, &c. and a large trade in agricultural produce and cattle. Good wine is made in its vicinity.

GYULA, a market-town of Hungary, cap. co. Bekes, on the White Körös, 85 m. NNW. Arad. Pop. 16,637 in 1857. The town consists of two parts, Hungarian and German Gyula, separated by the river; it has a fortress, a county-hall, several churches, some oil-mills, and a large trade in cattle.

H.

HAARLEM, or HARLEM, one of the principal cities of the Netherlands, prov. N. Holland, cap. arrond. and cant., on the Spaarn, 10 m. W. Amsterdam, on the railway from Rotterdam to Amsterdam. Pop. 19,426 in 1841. The city is now in great part destitute of defences, but was formerly a place of some strength, having been fortified in the 16th century with brick walls, parts of which, with an old gateway, still remain. It has an ancient and somewhat dingy aspect. The architecture of some of the houses is remarkably picturesque, with sharp-pointed gables; and the roofs show several rows of small attic windows, like what one is accustomed to see in old Flemish pictures. The streets are arranged in an irregular manner, with cross alleys and back courts, and few of them have havens in the centre, which is quite a singularity in a Dutch town. Its pop. at present is greatly below what it formerly contained. It has a large paved market-place surrounded by several of the principal edifices of the city, as the church of St. Bavon, a vast Gothic structure, with a high square tower; the flesh-market, and the Stadthaus, a fine building. Opposite the church is a statue of Laurence Coster, the reputed inventor of moveable types, a citizen of Haarlem. St. Bavon's has somewhat of a naked appearance inside; but its organ has long been considered one of the finest and largest in Europe. It is supported on ponderous pillars, and fills up the whole of one end of the church, reaching up to the roof. It has nearly 5,000 pipes; its tones are remarkably fine, and its power very great; but in the diameter of some of its pipes, it has been surpassed by organs built at York, Birmingham, and other English towns. Immediately under the organ, and between two masses of pillars, is a group of figures the size of life, in white marble, representing Faith, Hope, and Charity. The remaining chief public buildings and institutions in Haarlem are several churches, public charities and schools, the Teylerian Museum, with a good collection of philosophical instruments, and others of fossils, at which lectures on different scientific subjects are delivered; the academy of sciences, and many other schools. There are several good private collections of paintings. Haarlem is the residence of a civil governor and a military commandant; is a bishop's see, and the seat of tribunals of original jurisdiction and commerce. It has manufactures of silk, linen, and cotton fabrics, velvets, rugs, carpets, lace, ribands, soap, and oil. Many of these have greatly declined; but several cotton factories, which have been established in its neighbourhood, appear to be flourishing, and the manufacture of cotton goods has increased materially since the separation of Holland and Belgium. In one of these factories the king is a shareholder; steam-engines are employed to turn the machinery. There are 8 factories on a similar scale at Haarlem, employing in all 2,000 individuals, men, women, and children. In the environs of Haarlem are extensive bleaching grounds for linen, and here were at one time prepared those fine fabrics long known in England as Holland cloths. An important branch of trade in Haarlem is the sale of flowers and roots, of which traffic it is the chief seat. Near the city, on the S., are the 'Bloemen-Tuin,'

Vol. II.

or gardens for rearing these products. Each garden is secluded from the public road by a high wall, or a brick house thickly painted, containing the offices or warehouses devoted to the business of drying and packing the roots. Each garden stretches out to the length of perhaps a quarter of a mile by a breadth of 100 yds., and is separated from other gardens, as well as frequently divided across by partitions of wood 6 ft. high. In the many square spots thus sectioned off, are all the varieties of tulips, dahlias, hyacinths, ranunculuses, and various other flowers. The drying-houses are filled with shelves, on stands, on which are spread myriads of roots, and in adjacent apartments men are kept constantly busy packing for exportation. In packing, each root is first twisted into a small piece of paper, and then a hundred are put together in a paper bag, according to sorts. The bags are afterwards packed in cases, and are thus sent to all parts of the world. The Dutch are very fond of flowers, and during the time of the 'tulip mania,' the most extravagant prices were given for these roots; but 100 florins, or about 8l., is now considered a very large sum for one, and the greater part of the tulips cultivated and sold by the florists of Haarlem, are valued at from 1d. to 10d. each. The city was once celebrated for its printing; but at present this branch of industry is not more active than in an English country town. It has still, however, a type-foundry, chiefly for Greek and Hebrew characters, from which the Jews principally supply themselves with the latter.

The neighbourhood round Haarlem is carefully laid out in plantations and public walks, and for several miles on the road to Leyden the country is sprinkled with numerous neat villas. Immediately on the S. of the city is a wood of considerable extent, in which is a large and elegant mansion in the Grecian style, called the Pavilion. It formerly belonged to Mr. Hope, the banker, who sold it for 500,000 guilders (about 44,000l.) to Napoleon for his brother Louis. At the peace it was sequestrated by the government.

The epoch at which Haarlem was founded is uncertain. In 1572 it was besieged by a Spanish force under Toledo, a worthy son of the Duke of Alva. The city held out for seven months, when it being known that the garrison intended to make a desperate sortie as a forlorn hope, terms of capitulation were offered and accepted; but no sooner had the Spaniards obtained possession of the town, than they commenced a massacre of the inhabs., and upwards of 1,000 individuals were either put to the sword, or tied in pairs and thrown into the lake. In 1577 the town was retaken by the Dutch. Haarlem was the birthplace of Ostade, Wouverman, Berghem, Van der Helst, and Schrevelius.

HACKNEY, a town and par. of England, co. Middlesex, hund. Ossulstone, forming a suburb of the metropolis. Area of par., including the hamlets of Clapton, Homerton, Dalston, Shacklewell, and Kingsland, 3,327 acres. Pop. of par., 76,687 in 1841. Hackney consists chiefly of two wide streets, running nearly at right angles to each other, from which other streets diverge. There are many large and substantial residences, both detached and connected with the lines of street; but the houses generally are of inferior size. The par, has 7 churches, including the mother-church, St. John's, and 2 chapels of ease. All are commodious; but none are remarkable for architectural elegance. The dissenters have several places of worship, among which is one rendered illustrious by the ministerial labours of Bates, Matthew Henry, Priestley, and Price. At Homerton is an academy for Independent ministers. There are 3 charity schools, educating in the whole about 600 children; a school of industry for 60 children; and 3 hospitals or almshouses for aged people. At Clapton is the London Orphan Asylum, where 500 children, the orphans of respectable parents, are boarded, clothed, and educated; and at Hackney-wick is an establishment supported by the Society for the Suppression of Juvenile Vagrancy. The land about Hackney is chiefly occupied by nurserymen and market-gardeners; the rest is employed in cow-pastures and brick-fields. The part of Hackney and Stoke Newington form a union under the Poor-Law Amend. Act. (See Lonpons.)

HADDINGTON (CO. OF), see Lothlan.

HADDINGTON, a parl. and royal bor. and market town of Scotland, cap. co. Haddington, 16 m. E. by N. Edinburgh, and 10½ W. by S. Dunbar on a branch of the Edinburgh-Berwick railway. Pop. 5,897 in 1861. The town lies at the foot of the Garleton hills, bounded by the Tyne on the E., which stream divides it from the suburb of Nungate, to which it is joined by a bridge of 4 arches. It consists principally of two parallel streets, running E. and W., and a long cross street which bounds one of these, and intersects the other nearly at right angles. The main parallel street, which is a continuation of the road from Edinburgh, is spacious; the general character of the town, as to buildings and appearance, is superior to that of most others of its size. The streets are paved, and lighted with gas. The principal buildings are the town-hall, with a lofty spire 150 ft. in height; the county buildings, which contain accommodation for the sheriff's court, the meetings of the county, and apartments for the preservation of the public records; and a Gothic parish church, supposed to have been erected in the 13th or 14th century. It is 210 ft. in length; the choir and transept are in a somewhat dilapidated state; it has square towers, and is 90 ft. high. The western part of the cross is used as the parish church. Fordun styles it Ferrum Landoniæ, the lamp of Lothian. The parish church of Haddington is one of the few churches in Scotland, not in Edinburgh, that are collegiate. There are chapels belonging to the Scottish Episcopalians, to the United Associate Synod, to the Old Light Burghers, the Independents, and Methodists. Haddington can boast of one of the earliest schools established in Scotland, and it possesses an excellent classical seminary under the direction of the magistrates, and 6 other schools. A mechanics' institution was established here in 1823. The number of benevolent, friendly, and religious societies is great. There are no manufactures in the town, but there is a considerable trade in wool, in tanning, and currying leather, in preparing bones and rape-cake for manure, and various minor branches of industry. Haddington is celebrated for its weekly grain market, which is the second in point of importance in Scotland. Dalkeith being the first. The agricultural and horticultural societies of the county hold their meetings in the town.

Haddington is very ancient. A castle on its W. boundary was used as a royal residence in the 12th and 13th centuries, and here Alexander II. was born in 1198. A convent of Cistercian, or Bernardine, nuns was founded here in 1178;

and a monastery of Franciscan, or Grey Friars, in the subsequent century. (Keith's Cat. of Scot. Bishops, 449 and 462.) The suburb of the Nungate obtains its name from the former of these institutions. It was in this nunnery that the Scottish Parliament was convened (1548), when its assent was given to the marriage of Queen Mary with the Dauphin of France, and to her education at the French court. Haddington has often suffered severely from the overflowing of the Tyne. The last inundation was in 1775, when the river rose 17 ft. above its usual level, and flooded more than half the town. In 1244, the town, then composed of wooden buildings, was totally consumed from fire. It was again nearly consumed from the same cause in 1598. Haddington unites with N. Berwick, Dunbar, Lauder, and Jedburgh, in sending a member to the H. of C. In 1865, its registered voters were 725. The municipal income was 1894. in 1848–4.

Various eminent men have been connected with Haddington. John Knox, the famous reformer, is generally believed to have been born in the suburb of Giffordgate in 1505, and received his education at the burgh school; but some writers regard the village of Gifford, five miles distant, as his birthplace. The Maitlands of Lethington, a place within a mile of the town, are known both in literary and general history. Sir Richard Maitland, lord privy seal of Scotland, and a lord of session, was himself a poet, and a collector of ancient Scottish poetry. His eldest son William is well known in history as secretary of state during the reign of Queen Mary; his second son John was lord high chancellor of Scotland; and Thomas, his youngest son, is celebrated both for his Latin poems (Delitiæ Poet. Scot.), and for being one of the interlocutors in Buchanan's dialogue De jure regni apud Scotos. The Duke of Lauderdale, the capricious and tyrannical secretary of state for Scotland in the time of Charles II., was a descendant of Sir Richard; also John, earl of Lauderdale, author of 'the Works of Virgil translated into English Verse.' The only eminent man of more modern times connected with Haddington was the Rev. John Brown, author of the 'Self-Interpreting Bible,' and other theological works, who died in 1787.

HADLEIGH, a market town and par. of England, co. Suffolk, hund. Cosford, on the Bret, a tributary of the Stour, 8 m. W. Ipswich, and 58 m. NE. London by road, and 69½ by Great Eastern railway. Pop. of town 2,779, and of par. 3,608 in 1861. Area of par. 8,640 acres. It is an ancient-looking town, exhibiting, both in brick and wood, many curious specimens of old house architecture. The church, a handsome structure with a fine steeple, forms the principal ornament of the town. There are also 12 almshouses, and a curious brick gate-house, with hexagonal turrets, erected at the end of the 15th century. This town had formerly a flourishing clothing trade; but the chief manufacture at present carried on is the spinning of yarn for the Norwich weavers. Hadleigh was formerly a corporate town, but lost its character by a quo warranto in the reign of James II. Markets on Monday; fairs on Whit-Monday, and Oct. 4.

HAGUE (THE), (Dutch, Gravenhaag, 'the count's meadow;' Fr. La Haye), a town of the Netherlands, of which it is the cap. and usual residence of the king and court, prov. S. Holland, on a branch of the canal and on the railway between Leyden and Rotterdam, 10 m. SW. the former, and 18 m. NW. the latter city, Pop. 87,820 in 1861. The Hague is an open town, being surrounded only by a moat crossed by drawbridges. It has the usual features of a Dutch town; its houses and pavements are of brick, and several of its streets are intersected with canals, and planted with rows of trees; its general appearance, however, is much superior to that of the commercial cities of Holland. The N. end of the town is the fashionable quarter, and in it is the Vyverberg, a fine open space, ornamented with a lake and wooded island in its centre. Around and adjacent to this square are all the chief public edifices. The first of these is the National Museum, occupying the former palace of Prince Maurice, an elegant building of the 17th century. Its extensive picture gallery is reached by a noble staircase; the paintings here are mostly confined to works of the Dutch school, but in that department the collection is almost unrivalled. The grand object of attraction is Paul Potter's Bull, a picture which occupies nearly the whole end of one of the rooms. 'The representation is that of a young bull with brown and white spots, a cow reclining on the green sward before it, two or three sheep, and an aged cowherd leaning over a fence—all as large as life; the background being a distant landscape. The chief animal in the group appears to stand out in bold relief, with a liveliness in its air that is perfectly startling; each hair is the minuteness of the touching, in order to make every hair on the hide and forehead of the creature tell, that the picture will endure the closest inspection. This highly-prized work of art was carried off to Paris by order of Napoleon, and hung in the Louvre.' (Chambers, Holland, p. 22.) The Royal Museum of curiosities, occupying the lower part of the building, consists principally of a large and unique collection of Chinese and Japanese articles. One apartment is devoted to objects of interest connected with Dutch history, containing, among other similar articles, the armour and weapons of De Ruyter. The king's palace, in an adjacent street, presents little than is remarkable either without or within: it is an edifice in the Grecian style, its centre and two wings forming three sides of a square. There is in it a good suite of state rooms, in which the king gives audience, every Wednesday, to his subjects indiscriminately. The palace of the Prince of Orange is a large but plain edifice; it contains, however, a good collection of Dutch paintings, and the valuable assemblage of chalk drawings by the old masters, formerly the property of Sir Thomas Lawrence. On one side of the Vyverberg is the Binnenhof, an irregular pile of buildings of various dates, comprising a handsome Gothic hall, the only existing remnant of the ancient palaces of the counts of Holland. It is occupied by various government offices, and the chambers in which the states-general and states of Holland meet. The Binnenhof served for the prison of Grotius and Barneveldt; the latter of whom was executed in front of it in 1619. There are 14 churches, several chapels, 7 synagogues, an orphan asylum, state prison, house of correction, 3 poor schools, several intermediate and superior private schools, a royal library with 100,000 vols., a museum of medals, gems, cameos, &c., many private galleries of paintings, and learned and benevolent associations, and a theatre for Dutch, German, and French plays. The favourite promenade is the Voorhout, a fine wide road, lined with elegant mansions, planted with rows of trees, furnished with benches, &c., which leads from the N. quarter of the town to the Bosch. The latter is a finely wooded park, belonging to the king of Holland, and immediately adjacent to the Hague. In the centre of the grounds, which are embellished with artificial

sheets of water, and winding walks amongst the trees, stands the Huys in den Bosch (house in the wood), the summer palace of the royal family. It is an edifice of an unpretending character externally, but within are many excellent pictures, and it has a ceiling partly painted by Rubens. Almost 3 m. W. of the Hague is Scheveningen, a fashionable but dreary Dutch watering-place; and about 1½ m. NE. the town is the castle of Ryswick, which gave its name to the treaty of 1697.

The Hague has never been a place of much commercial importance. The inhab. derive their resources chiefly from supplying or being employed by the court and government establishments; and they suffered very considerably from the transfer of the seat of government to Amsterdam on the erection of Holland into a kingdom by Napoleon. The manufacture of porcelain, and the printing of books, especially those in the French language, are almost the only branches of industry. There is, however, a cannon foundry, established in 1668.

The Hague became the residence of the feudal lords of Holland in 1250, from which period it continued the seat of government till 1808; it again assumed the rank of a capital on the restoration of the Orange family. It was the native place of the astronomer Huygens, the naturalist Huysch, and William III., king of England.

HAGUENAU, a town of France, dép. Bas-Rhin, cap. cant., on the Moder, 15 m. N. Strasburg, on the railway from Strasburg to Mannheim. Pop. 9,679 in 1861. The Moder here divides into two arms, one of which intersects the town, while the other encircles it on the N. Haguenau is surrounded by old and ill-constructed walls, and a wide ditch: it was originally fortified by the emperor Frederick Barbarossa in the 12th century. A fine Gothic church erected about the same period, and ornamented with some elegant sculptures, is its chief public edifice; it has several other churches, a synagogue, civil and military hospital, some good cavalry barracks, many oil, madder, and other mills, and manufactures of cotton fabrics and yarn, woollens, soap, &c. The forest of Haguenau is one of the largest in France; it extends over an area of 17,000 hectares.

HAINAN, or HAI-LAM (Chinese, 'South of the Sea'), a large island of the Chinese Sea, between lat. 18° and 20° N., and long. 108° 20' and 111° E., belonging to the Chinese empire, and forming a dep. of the prov. of Canton, but separated from the continent by a strait from 15 to 30 m. wide, probably identical with what was called 'the Gates of China,' by the Mohammedan authors of the 8th and 9th centuries. (Chinese Repository, i. 37.) The island is of a somewhat oval shape; greatest length, NE. to SW., about 180 m.; average breadth, nearly 70 m. Area, perhaps about 12,000 sq. m. Pop. estimated in 1831 at little short of a million, independent of unconquered tribes in the interior. A mountain chain runs through Hainan in the direction of its length, and near its centre rises above the limit of perpetual snow. In this part of the island the principal rivers take their origin, some of which are of considerable size. The E. coast is bold and rocky; the W. low; the S. has some good harbours; but Hainan generally, like Formosa, is surrounded with many rocks and shoals dangerous to shipping. The climate is very hot; the heat is, however, tempered by sea-breezes, frequent fogs, and abundant dews. The soil is mostly sandy; the W. side of the island is more productive than the E., but the country is, upon the whole, barren; and, except timber, rice, and sugar (the latter principally sent to the N. of China), its articles of export are very few. Its chief wealth consists in its timber: the forests which cover the mountains abound with sandal, cocoa, rose, and other cabinet woods, braziletto, ebony, &c. Tobacco, cotton, and indigo are raised, but in no great quantities. Various fruits are grown, and the sweet potato forms an important article of culture and food. Hens are very plentiful, and wax is a valuable item of produce. Pearl oysters and coral abound around the shores, on many parts of which extensive salt works are established. Small quantities of gold and silver are obtained in the interior. The natives carry on some trade with Anam, Siam, and Singapore. On their voyages to Siam, they cut timber along the coasts of Tsiampa and Cambaja, with which they build junks at Bankok. These junks are then laden with cargoes saleable at Canton or Hainan, and both cargoes and junks bring sold, the profits are divided among the builders. Most part of the pop. are Chinese, who are similar to the inhabts. of the opposite coast; but the interior is inhabited by a different race, supposed to be aboriginal, some of whom have submitted to the Chinese government, while others still hold a savage independence. The island is subdivided into 13 districts. The cap., Kiong-tchou, a populous city, and the residence of the Chinese governor, is on the N. coast. Several other towns have a pop. of some thousand inhabs. Hainan appears to have been discovered by the Chinese about some 108 B.C., and conquered by them soon afterwards. It was annexed to the prov., of which it now forms a part, in 1381. (Ritter, Asien Erdkunde, iii. 881-895; Parsloy, in Asiat. Researches, vol. xii.)

HAINAULT, a prov. of Belgium, which see.

HALBERSTADT, a town of Prussia, prov. Saxony, gov. Magdeburg, cap. circ. and principality of same name, on the Holtemme, a tributary of the Bode, 32 m. SW. Magdeburg, with which it is connected by railway. Pop. 22,810 in 1861, exclusive of garrison of 1,186. The town is very ancient; is built chiefly in the Gothic style, and is surrounded with walls, outside which are three suburbs. It has a cathedral, an edifice of the 15th century, remarkable for its paintings and stained glass windows, ten other Protestant, and two Rom. Catholic churches, a synagogue, a handsome mansion house (formerly a royal palace), gymnasium, superior town and girls' schools, a teachers' seminary, two large public libraries, a school of midwifery, an orphan asylum, house of correction, theatre, and several fine private collections of paintings, medals, and antiques. It is the seat of the superior courts of the gov., of town and distr. courts, and a board of tolls and taxation, and has numerous factories for woollen stuffs of secondary quality, carpets, linen fabrics, leather gloves, straw hats, starch, tobacco, and soap, with extensive oil refineries, numerous breweries, lithographic printing establishments, and a considerable trade in corn and wool. Its commercial importance appears to have increased of late years. The epoch of its foundation is uncertain. It was made a bishop's see in 804. A great part of it was destroyed in 1179, by Henry the Lion. It was ceded to Prussia, together with its principality, at the peace of Westphalia, and has ever since belonged to that power, except during the existence of the short-lived kingdom of Westphalia, of which it formed a part.

HALES-OWEN, a par. and market-town of England, partly in an insulated portion of co. Salop, hund. Brimstrey, and partly in co. Worcester, lower div. hund. Halfshire, 104 m. NW. London, 7 m. WSW. Birmingham, and 24 m. NE. Worcester. Pop. of town 2,911, and of par. 70,283 in 1861. Area of par., 11,790 acres. The town,

which consists of a handsome main street, crossed by several others of inferior character, stands on the Stour, in a beautiful and well-wooded valley, and bears the appearance of a busy and thriving place. The church is of Norman architecture and has a light spire curiously supported on four arches. St. Kenelm's chapel, situated outside the town, was originally erected in the time of the Saxons, and a part yet remains apparently of that early date. The far larger part, however, was built in the reign of Henry III., and the tower, with its ornamental pinnacles, is an elegant specimen of the Gothic style. Few buildings so small present such striking architectural contrasts. There are three places of worship for dissenters. A free grammar-school was established here during the Commonwealth by a chancery commission, which provided it with an endowment, the present yearly value of which is about 100l. Shenstone, the poet, who was also the proprietor of 'the Leasowes,' a beautiful villa in the neighbourhood, was educated at this school; his monument is in the church. The manufacture of nails and the coarser kinds of hardware and tools constitutes the chief employment of the working classes. Steel is extensively made in the hamlet of Congreaves; and coal mines are worked within the parish.

Hales-owen is under the jurisdiction of the co. magistrates, who hold petty sessions here. A high bailiff, headborough, and constable are annually elected at the court leet of the lord of the manor, and these officers govern the internal economy of the town. A court of requests is held every third week for the recovery of debts under 5l., the power of which extends to five other pars. Markets on Monday; fairs on Easter and Whit-Monday for horses, cattle, and cheese.

An abbey of Premonstratensian monks was founded here in the reign of King John out of funds provided by that monarch. Its revenues, at the dissolution of the religious houses, amounted, according to Speed, to 33l. The ruins are extensive, and have partially been converted into farming premises. A few very fine lancet windows at the gable end of the chapter-house indicate the style of building to have been early English.

HALIFAX, a market-town, par., and parl. bor. of England, co. York, W. Riding, wap. Morley, on the Hebble, a branch of the Calder, 36 m. WSW. York, 43 m. WSW. Leeds, 170 m. NNW. London by road, and 193½ m. by Great Northern railway. Pop. of bor., 37,014, and of par. 147,394 in 1861. The entire parish is one of the most extensive in the kingdom, and nearly equals in size the county of Rutland. It includes 21 townships, and 73,749 acres. For rating, it is divided into three parts: the parish district of Halifax, the chapelry of Heptonstall, and the chapelry of Elland. The parl. bor. includes the township of Halifax, with small contiguous portions of the townships of N. and S. Owram, lying along the E. side of the Hebble brook. The town is built on a gentle slope, in a valley surrounded by hills. In many parts the streets are narrow and irregular; but some, as Broad Street and Waterhouse Street, are handsome and spacious. It is well paved, and lighted with gas. The houses are almost exclusively built of stone from the quarries of N. and S. Owram; but a few still remain, built in the reign of Henry VIII., of plaster, with carved oak framework. Within the entire parish there are above 20 episcopal, and 90 dissenting places of worship; but some of these are in the rural districts. Within the town are seven churches, the largest of which, St. John's, the parish church, built in the fifteenth century, is of pointed Gothic architecture. It has a lofty nave, side aisles, and chancel; and 2 side chapels were added in the sixteenth century. There is a handsome painted window, similar to the Marygold window in York cathedral. The tower, which is highly ornamented, contains a peal of ten bells, and is 117 ft. high. Among the other churches in the new Gothic edifice of All Souls, built at the cost of Mr. E. Akroyd, with a spire 236 ft. high. In Sowerby Church is a monumental statue of Archbishop Tillotson, a native of that township. Trinity Church, built in 1798, is a Grecian edifice, with Ionic pilasters, surmounted by a tower and cupola at the W. end. St. James's, opened in 1832, is a pseudo-Gothic structure, with square turrets at the W. end. Besides the episcopal places of worship, there are chapels for Independents, Wesleyan Methodists, Methodist New Connexion, Roman Catholics, Primitive Methodists, Unitarians, and the Society of Friends, among which one, belonging to the Independents, is remarkable for classical elegance and good taste. Connected with the churches and chapels are many Sunday schools; and the Halifax S. S. Union comprises numerous schools, attended by upwards of 5,000 children. The National School, built in 1815, near Trinity Church, is attended by about 800 boys; and the Lancastrian School, opened in 1818, has more than 300 of both sexes. The parish has 7 free or endowed schools; but of these only one, Smith's charity school, founded in 1726, is situated in the town. Queen Elizabeth's grammar school, in the township of Skircoat, was chartered in 1585, and is under the direction of 12 governors, chosen from among the inhab. The rental of the school property is considerable. The school is free to the sons of all parishioners; but the number of scholars was recently only about 60. The grammar schools at Hipperholme, N. Owram, and Heptonstall are attended not only by the free boys, but others, who pay for their schooling. Wheelwright's school at Heshworth is a noble establishment, supported at an expense of more than 2,000l. a year, and providing a liberal education for 80 boys, with 2 exhibitions of 150l. a year at the universities; it is superintended by 2 masters and a matron. There are numerous charities for the relief of the poor and aged, none of which need any particular mention, except Waterhouse's almshouse and blue-coat school, established in 1627 for 12 aged persons, and 20 orphan children. The largest public building is the Piece Hall, a very extensive quadrangular stone structure, occupying more than 2 acres of ground; it has a rustic basement story, above which are two other stories fronted by colonnades having walks within them leading to the various storerooms, of which there are 315. In these rooms the manufacturers keep their cloths for sale. This building, erected in 1779, cost 12,000l. The infirmary, built in very elegant style, furnishes excellent accommodation for the many sick who resort thither. The baths on the Huddersfield road are well adapted for their purpose, and have a bowling-green attached. The building in Harrison Lane, called the Public Rooms, has elegant assembly rooms, and other accommodations, both for pleasure and business. There are two subscription libraries, one of which has apartments in the Public Rooms. The town possesses, among other public establishments, a Literary and Philosophical Society, established in 1830, and a Mechanics' Institution, opened in 1856. The Odd Fellows' Hall, in St. James's Road, erected in 1848, has a large room adapted for lectures. The theatre, though small, is quite large enough for a pop. which seems to feel little interest in such amusements. Outside the town, on the W., is Gibbet Hill, where formerly, in consequence of a

local law designed principally for the protection of the clothiers, felons convicted of depredating upon their property were executed by a machine like the French guillotine. The gas works are in S. Owram, and in Ovenden are the springs and reservoirs which supply the town with excellent water. A public cemetery has been laid out, with a path of 15 acres; to which baths are attached. The park is the gift of Mr. Frank Crossley, owner of a carpet factory employing 3,000 to 4,000 persons.

The magistrates of Halifax are also county magistrates. Petty sessions are held every Saturday, and there is a court for the recovery of debts under 10l.; a county court is established in the town. During the Commonwealth, Halifax sent 2 members to the H. of C.; but the franchise was withdrawn at the Restoration; and, notwithstanding its growing and universally acknowledged importance, it had no voice in the legislature till the Reform Act again conferred on it the privilege of sending 2 representatives to the H. of C. The parl. bor. includes small portions of N. and S. Owram, as well as the township of Halifax: registered electors, 1,639 in 1865. Market on Saturday. Fairs, June 24, and the first Saturday in Nov. for cattle and horses.

For the administration of the poor laws, the par. is formed into 2 unions. Halifax Union comprising 20 townships, and the Todmorden Union, including the Heptonstall district and the chapelry of Todmorden.

The rise of Halifax is attributable wholly to its manufacturing industry, which is itself mainly a consequence of its unlimited command of coal and of the means of internal navigation. The cloth-weavers first settled here in the beginning of the 15th century, since which time it slowly, but gradually, increased till the American and French wars, when extraordinary activity prevailed, and the pop. was proportionally enlarged. The introduction of steam-engines and power-looms has also, of late years, contributed in no little degree to increase its importance as a place of trade. The town is united by a canal with the Rochdale canal and the Calder and Hebble navigation; and has, consequently, a navigable communication with Hull on the one hand, and Liverpool on the other. The establishment of railways has much contributed to the rise of Halifax, situated as it is in the very centre of this new network of roads. Its constant increase of wealth is shown in its income-tax returns, which show an enormous rise in the annual value of real property. It amounted to 129,786l. in 1857, and to 160,906l. in 1862.

The staple manufactures of the town and neighbourhood are shalloons, tammies, and draw-boys, best known under the title of figured lastings and amens, superfine quilted everlastings, double russets and serges, all which are made of combing wool. They are brought in an unfinished state to the Piece-hall, where the merchants attend every Saturday to make their purchases. There is, besides, a very considerable manufactory of kerseys and half-thicks, also of burkings and baise, chiefly carried on in the vale of Hipponden, whence comes a large portion of the cloth used for clothing the British navy. Large quantities are also sent to Holland and all parts of America. The most promising branch of manufacture, however, is that of cloth and coatings, which was also introduced at the end of the last century by persons of enterprise, who, at vast expense, erected mills on the Calder and its tributaries. The success of these factories was such as to excite the jealousy of the Leeds merchants, who had been previously used to buy the same articles from the lower manufacturers at their cloth-hall, and parliament was petitioned, in 1794 and 1806, to prevent any merchant from becoming a manufacturer. The legislature very properly refused to cramp the energies of Halifax, to serve private interests in Leeds. Bombazine also and crapes, together with other fabrics of silk and worsted mixed, are manufactured here; and the manufacture of cottons is becoming a rapidly increasing and most important branch of industry. A great number of hands are employed in making machinery.

HALIFAX, a marit. city of British N. America, on a small peninsula on the N.E. coast of Nova Scotia, of which it is the cap. Pop. 21,860 in 1861. The town stands on the declivity of a hill about 250 ft. in height, rising from the W. side of one of the finest harbours in the American continent. The streets are generally broad; the principal, which runs next the harbour, is well paved, and most of the others are macadamised.

The front of the town is lined by wharfs. Warehouses rise over the wharfs, as well as in different parts of the town; and dwelling-houses and public buildings rear their heads over each other as they stretch along and up the sides of the hill. Among the public edifices is Province Building, a handsome stone edifice 140 ft. long, by 70 ft. broad, and ornamented with a colonnade of the Ionic order. It comprises chambers for the council and legislative assembly, the supreme court, various government offices, and the Halifax public library. In the S. part of the town is the Government House, a sombre, but solid-looking building, over which is the residence of the military commandant. On the N. side of the town is the admiral's residence, a plain stone building. The dockyard, at the end of a straggling suburb, covers 14 acres, and forms the chief depot of naval stores in the British N. American colonies. It is peculiarly fitted for the shelter, repair, and outfit of the fleets cruising on the American coast and in the W. Indies. The N. and S. barracks may accommodate three regiments; and attached to them is a good library. The other government buildings are the ordnance and commissariat stores, and the military hospital, erected by the late Duke of Kent. Dalhousie College is a handsome edifice of freestone, but not yet efficient as a seat of education. There are 3 churches, a large R. Catholic chapel, 3 Presbyterian, and 4 other chapels belonging to different sects, a poorhouse, house of correction, an exchange, some assembly rooms, and a small theatre. The markets are well supplied with provisions, but the inns and boarding-houses are reported to be very indifferent.

The harbour opposite the town, where ships usually anchor, and where, at medium tides, there are 12 fathoms water, is rather more than a mile wide. After narrowing to ½ m., about 1 m. above the upper end of the town, it expands into Bedford Basin. This sheet of water, which is completely landlocked, occupies a surface of 10 sq. m., and is capable of containing the whole British navy. Halifax harbour is accessible at all seasons, and its navigation is scarcely ever interrupted by ice. The best mark in sailing for it is Sambro lighthouse, on a small island off Sambro Head, about 13 m. S. by E. Halifax, with a fixed light 210 ft. high. Another lighthouse stands on Magher's Beach, a spit extending from M'Nab's Island, a wooded and cultivated island, at the very entrance of the port. When the latter light is seen, ships may run in without fear. The passage on the W. side of M'Nab's Island is for large ships, the other on the E. has only water for schooners. There are several other small islands further in, on one of which, nearly opposite the town, some strong

batteries are mounted. Some other forts defend the harbour. North West Arm, which bounds Halifax peninsula on the W., is 4 m. long, nearly ½ m. wide, and has from 10 to 20 fathoms depth of water, with safe anchorage. Near its head lies Melville Island, some buildings on which were formerly used for the detention of prisoners of war. A joint-stock company's canal, in aid of which the legislature contributed 12,000l., connects the harbour of Halifax with Cobequid Bay and the Bay of Fundy.

Since its first settlement, in 1749, Halifax has continued to be the seat of a profitable fishery and trade. The latter, especially, is in as prosperous a condition as that of any town in British America; and this city may be said to engross the whole foreign trade of Nova Scotia. The chief trade is with the W. Indies, and other British colonies, the U. States, and Great Britain. The vessels belonging to this and the other ports of Nova Scotia are principally engaged in the fisheries and in the timber and lumber trade.

In 1817, Halifax was declared a free port to a certain extent, and has since acquired the privilege of warehousing. Some ships of large size are employed in the South Sea fishery; but, generally speaking, the inhab. are less enterprising and successful traders than the New Englanders. Halifax has some manufactures, but they are of no great importance, and confined to articles of immediate consumption; as soap, candles, leather, paper, snuff, rum, gin, whisky, porter, ale, and refined sugar. Packets sail between Halifax and Falmouth, and others regularly to Liverpool, Boston, New York, and the W. Indies; steam ferry-boats also ply constantly to and from Dartmouth, on the opposite side of the harbour.

HALL, a town of Würtemberg, circ. Jaxt, on both sides the Kocher, which is here crossed by a stone bridge, 34 m. N.E. Stuttgart, on the railway from Stuttgart to Anspach and Nuremberg. Pop. 6,862 in 1861. Hall—sometimes called Suabian Hall, to distinguish it from other towns of the name—is ancient, and was formerly a free imperial city. It has seven churches, a fine town-hall, a richly endowed gymnasium, an ancient mint, a hospital, and public libraries. Next to Ulm, it has the greatest number of sugar refineries in the kingdom; it has also some soap and other factories, and a large trade in corn and hops; but its chief article of commerce is salt, procured from the saline springs in its vicinity.

HALLE, a town of Prussian Saxony, distr. Merseburg, cap. circ. same name, on the Saale, 9¼ m. S.W. Berlin, 65 m. N.E. Gotha, and 18 m. N.W. Leipsic, on the railway from Leipsic to Magdeburg. Pop. 42,976 in 1861, exclusive of a garrison of 1,469. The shape of the town is an irregular parallelogram, and contains three quarters, viz. Halle, Glaucha, and Neumarkt, each of which has its own magistrates. It has few remarkable edifices. The Gothic church of St. Mary was built in the sixteenth century, and that of St. Maurice as early as the twelfth. In the market place is a singular structure, 250 ft. high, called the Red Tower. The other principal buildings are Franke's institute, the university hall, and the hospitals. Outside the walls, E. of the town, is an elegant monument in honour of the Germans who fell in the battle of Leipsic. The old castle of Moritzburg, where the archbishops of Magdeburg used formerly to reside, was mostly destroyed in the thirty years' war; the military remaining wing is used as a Calvinistic church. Halle is not remarkable as a place of trade; but hardware and starch-making are more followed than any other branch of industry. In a valley near the river are two

large salt springs, which formerly were extremely
productive; at present, however, they yield only
about 16,000 quintals a year.

The university was founded by Frederick I, in
1694, and soon after its establishment became
known as the seat of the great Pietist divines of
Germany, who have exercised in subsequent times
a most powerful and beneficial influence over the
morals of the people; and since that time it has
always been known as a great theological university, though the sentiments of its professors
have verged more and more towards Rational-
ism. At the beginning of the present century, the
university of Halle had reached the height of its
prosperity; but Napoleon's victory at Jena led to
its dissolution, nor can it be said to have regained
a positive existence till after his overthrow in 1815,
when it was united with that of Wittenberg, and
called the *United Frederick-University of Halle-
Wittenberg*. In 1828 there were 1,461 students,
944 of whom belonged to the theological faculty.
Subsequently to this period, however, the university of Berlin attracted many of its students,
whose numbers have fallen to about 800. Francke,
Wolff, Vater, Semler, Wegscheider, Gesenius, Thiluk,
and Tholuck are a few among its theologians; besides whom, Meckel and other medical professors
have contributed to raise its character as a school
of medicine. The library contains about 50,000
volumes; and there are, besides, museums of various kinds, an anatomical theatre, chemical laboratory, botanical garden and observatory. Three
hospitals connected with the medical school furnish
the students with ample opportunities of acquiring
practice. Besides the university, there are several
institutions for education, the chief among which
is the institute founded by Francke in 1698. It
consists—1, of an orphan school, educating about
160 children, three-fourths of whom are boys; 2,
of a royal pædagogium, for educating children of
the upper classes, and which has trained since its
establishment upwards of 3,000 children; 3, of a
Latin school, intended chiefly to impart sound
grammatical instruction to the sons of the citizens;
4, of a bible press, which has sent forth some
millions of copies of the Scriptures at a cheap
rate, and at which also certain classical works are
printed for the use of the students. The profits
are continually applied to increase the usefulness
of the establishment. The building has been
recently enriched with an excellent bronze statue
of the founder, by Rauch. Its cost was defrayed
by a subscription, headed by the King of Prussia.
Halle has a society of natural history and an Oriental society.

HALSTEAD, a market town and par. of England, co. Essex, hund. Hinckford, on rising ground,
near the Colne, 46 m. NE. London, and 25 m. W.
by S. Ipswich, on the Great Eastern railway.
Pop. of town 5,707, and of par. 5,917 in 1861.
Area of par. 6,230 acres. The town has wide and
clean streets, and a good market place in its centre.
The church is a fine old building, in the Gothic
style, having a tower and wooden steeple. Besides
the church, there are three places of worship for
dissenters. A grammar school was founded here
in 1594, for the education of forty poor children
within this or the adjoining parishes, the governors of Christ's Hospital, in London, being the
trustees. The town has six other schools. A
baize manufacture, formerly flourishing, has almost wholly decayed; but there are many hand
looms employed on figured and plain silk velvets.
Winding silk employs numerous females; many
of the poor people are engaged in straw plaiting.
Hops are abundantly raised in the neighbourhood.
Halstead, under the Poor Law Amendment Act,

is the chief town of a union comprising sixteen
parishes. Markets on Friday, chiefly for corn;
fairs on May 6 and Oct. 29, for cattle, &c.

HAM, a town of France, dep. Somme, cap.
cant., in a marshy plain near the Somme, and on
the canal d'Angoulême, 35 m. ESE. Amiens.
Pop. 2,875 in 1861. Ham is celebrated for its
castle, a strong fortress used as a state prison, in
which Prince Polignac and other ministers of
Charles X. were confined for six years. Subsequently, Ham became the prison of Prince Louis
Napoleon, from Oct. 10, 1840, to May 25, 1846; and
the prince having become ruler of France, he himself sent there, after the *coup d'état* of Dec. 2,
1851, some of his political adversaries. The castle
of Ham is visible from a great distance; it has a
large round tower, built in 1470, 100 ft. in height,
and as many in diameter, with walls of extraordinary thickness. The lordship of Ham was
annexed to the possessions of the crown by Henri
IV.; Louis XIV. demolished the fortifications of
the town, but preserved the castle.

HAMADAN (an. *Ecbatana*), a town of Persia,
prov. Irak, and cap. beglerbeglik same name, 190
m. WSW. Teheran, and 260 m. NW. Ispahan;
lat. 34° 53' N., long. 48° E. It stands on a slope
near the small river Hamadan-tchai, and at the
foot of Mount Elwend (the Orontes of antiquity).
Its pop. is variously stated at from 25,000 to 40,000,
the smaller number being perhaps nearest the
mark. It is meanly built, and occupies a considerable space, the houses being profusely interspersed with trees. The ruins of walls and houses
show that it must formerly have been an immense
city, filled with splendid edifices; but it now contains only a single good street, the rest being inferior to those seen in other eastern towns. The
largest public building is the *Mesjid-Jumah*, in a
large square, used as a market place; there are
also several other mosques, an Armenian church,
a Jews' synagogue, some public baths, bazaars and
caravanserais, all of which indicate, by their ruinous state, the fallen prosperity of the place. Near
the great mosque, in a Jews' grave-yard filled
with tombs, stands a building which claims, by
its Hebrew inscription, to be the sepulchre of
Esther and Mordecai; but Morier is of opinion that
the structure is Mohammedan; and it was, perhaps, raised or rebuilt after the sack of Hamadan
by Timour. Within the town also are the tombs
of the celebrated physician Avicenna, of the Persian poet Attar, and of the Arabic poet Abul-
Hasif; and on this account it is much resorted to
by pilgrims from all parts of Turkey and Persia.
On a height commanding a complete view of the
town are the ruins of a castle destroyed by Aga
Mahomad Khan; and a little below are some remains, considered by Morier to have belonged to
the ancient palace of the kings of Media. The
same writer observes, that 'Hamadan presents
more objects of research to the antiquary than
any other city that he had visited in Persia.' The
modern town is famed for its manufacture of leather,
in which it has a large trade, and carpet and silk
weaving is also pursued to some extent; but its
chief wealth is derived from its situation on the
great commercial road between Bagdad, Teheran,
and Ispahan. The environs are highly productive;
but the absence of forest timber deprives the
scenery of a picturesque character, and causes
wood to be so expensive that dried cow-dung is
usually substituted for it as fuel.

There is every reason to believe that Hamadan
stands on or near the site of ancient Ecbatana,
Agbatana, or Apobatana; though Sir W. Jones
fixed it at Tabriz, and Dr. Williams, of Edinburgh,
at Ispahan. No position, however, accepts Hama-

dan, will suit the descriptions of Isidore Carax and Diodorus Siculus, as has been clearly proved by the reviewer of 'Williams's Geog. of Asia Minor,' in the 'Journal of Education.' (ii. p. 805.) Ecbatana of Media was founded, or rather enlarged, by Dejoces, circa B.C. 680. The Medes, says Herodotus, 'obedient to the command of their king, erected that great and strong city now known under the name of Agbatana, where the walls are built circle within circle, and are so constructed, that each inner circle overtops its outer neighbour by the height of the battlements alone. This was effected partly by the nature of the ground, a conical hill, and partly by the building itself. The number of the circles was seven, and within the innermost were built the palace and the treasury. The circ. of the outermost wall was almost equal to that of Athens. The Median nation were ordered to construct their houses in a circle round the outer wall,' (Herod. l. 95-130.) We are told in the Apocrypha, that in the reign of Arphaxad (Phraortes) it was besieged and taken by Nebuchadnezzar, who 'spoiled the streets thereof, and turned the beauty thereof into shame.' (Judith, i. 14.) From the days of Darius to those of Jenghiz Khan it was, on account of the coolness of its climate, the favourite residence of the kings of Persia during those months of summer in which the heat of Susa and Ispahan is almost insupportable. It was reduced by the caliph Othman, nearly destroyed by Jenghiz Khan, and again taken and ravaged by Timour at the end of the 14th century. It was rebuilt, however, and appears to have been a city of considerable importance under the Sophi dynasty. In 1722 it suffered greatly during the wars that took place after the dethronement of Shah-Hussein, and more recently from the pillage of the Turks under Ahmed, pacha of Bagdad. It remained subject to the Turks till Nadir Shah drove them beyond the Tigris, and again annexed it to the kingdom of Persia. Its present ruinous appearance is attributable to the fact of its having been so often the theatre of war, and the object of plunder. This, the great Median Ecbatana, must not be confounded with the Atropatenian Ecbatana, the site of which has been fixed by Major Rawlinson at Takhti-Suleiman, 180 m. NNE. Tabriz. (Geog. Journal, x.; Kinneir's Persia; Ker Porter's Travels, ii.; Morier's Travels, ii.)

HAMAH (an. Epiphania), a city of Syria, and cap. of a sanjiack, on the Orontes, 76 m. NE. Tripoli, and 81 m. S. Aleppo; lat. 34° 55' N., long. 37° 6' 15" E. Pop. estimated at 45,000. The city is pleasantly situated on both banks of the Orontes, or Aasy, which is here crossed by four bridges. It is walled and otherwise well defended, and some agreeable suburbs give it externally a prepossessing appearance. But the streets, as in most cities of Syria, are narrow, irregular, and dirty; and the houses, though handsome inside, present to the street only unattractive mud brick walls. The principal buildings are the palace of the Mutsellim and the mosques, one of which is remarkable for a fine old minaret. There are several bazaars, three public baths, and some handsome residences with spacious gardens. Some curious hydraulic works for supplying the town with water have been constructed on the river, one of the wheels of which is 70 ft. in diameter. The industry of the town comprises silk and cotton fabrics; it trades largely with Aleppo in European and colonial merchandise, and being on a great caravan route has considerable commerce with the interior of Asia and Africa. The place suffered much from an earthquake in 1157, in common with other Syrian towns; and hence there are few antiquities, a

square mound of earth in the middle of the city being the only vestige of the older buildings. There is no doubt, however, that Hamah stands on the site of the Hamath mentioned in Scripture, and reputed to have been founded by Hamath, son of Canaan. It was known in the time of Moses; and at a later period it was relieved from the oppression of a neighbouring prince by the virtuous David, to whom, in testimony of his gratitude, 'the king sent Joram, his son, to salute him and to bless him.' (2 Sam. viii. 9, 10.) The prophet Amos (vi. 2) styles it 'Hamath the Great.' Its name was changed by the Macedonians, in honour of Antiochus Epiphanes; and during the expedition of Pompey into Apamea and Cœle-Syria, it became subject to the Romans, anno 63 B.C.

HAMBURG (REPUBLIC OF), an indep. state of NW. Germany, the territories of which comprise the city of Hamburg and the country immediately surrounding it; the town of Bergdorf, with the district called the Vierlande (the sovereignty over which is, however, shared with Lübeck), Ritzebuttel, Cuxhaven, and the island of Neuwerk, at the mouth of the Elbe, some islands in that river opposite the cap., and several small detached territories, chiefly situated N. of the rest, and enclosed by the duchy of Holstein. Total area, 148 sq. m.; pop. 210,973 in 1850, and 229,941 in 1861. Nearly the whole of the inhab. are Lutherans except some 2,000 Calvinists, 4,000 Roman Catholics, and above 7,000 Jews. The little state is bounded on all sides by the duchy of Holstein except on the N. and SW., where the Elbe separates it from Hanover. Besides the Elbe, it is watered by the Alster and Bille. It is generally a level plain; not particularly fertile, excepting the Vierlander, to the SE. The islands in the Elbe called the marsh-lands are very productive. A good deal of land is devoted to fruit, flower, and vegetable gardens; and the entire country round Hamburg is dotted over with flourishing villages and plantations. The rural pop. is in a good comfortable condition.

The government of the republic was, until the year 1848, of an oligarchical character, but owing to civil commotions then breaking out, it had to be changed in a democratic sense. Several draughts of charters having been discussed, the present constitution of the state was published on the 28th September, 1860, and came in force on the 1st of January, 1861. According to the terms of this fundamental law, the government—Staatsgewalt—is intrusted, in common, to two chambers of representatives, the senate, and the Bürgerschaft, or house of burgesses. The senate, which exercises chiefly, but not entirely, the executive power, is composed of 18 members, one-half of which number must have studied jurisprudence, while seven out of the remaining nine must belong to the class of merchants. The members of the senate are elected for life by the house of burgesses; but a senator is at liberty to retire at the end of six years. A first and second burgomaster, chosen annually in secret ballot, preside over the meetings of the senate. No burgomaster can be in office longer than two years; and no member of the senate is allowed to hold any public office whatever. The house of burgesses consists of 192 members, 84 of which are elected in secret ballot by the votes of all tax-paying citizens. Of the remaining 108 members, 48 are chosen, also by ballot, by the owners of house property in the city valued at 2,000 marks, or 187 l., over and above the amount for which they are taxed; while the other 60 members are deputed by various guilds, corporations, and courts of justice. All the members of the house of burgesses are chosen for six years, in such a manner that every three years new elections take place for one-half the number. The house of burgesses is represented, in permanence, by a burger-convention, or committee of the house, consisting of 20 deputies, of which no more than five are allowed to be members of the legal profession. It is the special duty of the committee to watch the proceedings of the senate, and the general execution of the articles of the constitution including the laws voted by the house of burgesses. In all matters of legislation, except taxation, the senate has a veto; and in case of a constitutional conflict, recourse is had to an assembly of arbitrators, chosen in equal parts from the senate and the house of burgesses. There are in the cap. an upper court of justice, which takes cognisance of all suits above 2,000 marks; appeals from which can, however, be made to the superior court of the Hanse Towns at Lübeck; a lower court of justice, which tries criminal cases, and derives in civil causes under 2,000 marks; and a commercial tribunal, a final appeal from the decision of which lies to the upper court of justice. The inferior towns have their own magistracy, and police courts subordinate to 2 directors of police in Hamburg. The armed force consists of about 2,000 regular troops, enrolled by enlistment, after the English fashion, and rather well paid, and besides of a burgher militia, not salaried, including all the citizens between the ages of 18 and 45. The contingent furnished to the army of the German Confederation is 1,298 men. The public revenue for the year 1842 amounted to 10,751,267 marks, or 613,610l. and the expenditure to the same sum. There was a public debt, on the 1st of Jan. 1862, of 50,135,326 marks, or 3,568,181l. Hamburg has as many as 60 consuls in different parts of the world; it enjoys a separate vote in the full German Diet, and together with Lübeck, Bremen, and Frankfort, has one in the Lower Council of the Confederation.

HAMBURG, the principal commercial city and seaport of Germany, cap. of the above republic, and one of the three existing Hanse Towns, and four free imperial cities, of Germany; on the N. bank of the Elbe, at the point where it receives the Alster, 60 m. SE. from its mouth, 60 m. NE. Bremen, and 86 m. SW. Lübeck, on the railway from Berlin to Kiel. Pop. 179,841 in 1861. The city is oval shaped; is about 4 m. in circ. and was formerly fortified; but having suffered severely during its occupation by the French in the last war, its ramparts have been levelled since the peace, and converted into public walks. The principal ornament of Hamburg is the Alster. This river rises in Holstein, some miles above the city, and spreads out into a wide lake, which flows through deep broad ditches, some of which encircle the ramparts, and communicate with the Elbe by sluices, while others intersect the city in all directions, forming numerous canals navigable for barges of considerable size. This lake is called the Outer Alster. The Inner Alster is a large square sheet of water connected with the former by a narrow channel, spanned by a single arch. On three sides of the Inner Alster there are broad walks, with rows of trees, the favourite resort of the Hamburgers of all classes and all ages. The best houses in the city are to be found in its immediate neighbourhood. The Jungfernstieg occupies its N. and W. sides.

The whole of the city has been very nearly rebuilt since May, 1842, when it was visited by a tremendous fire, which raged for three days, destroyed the buildings on two sides of the Alster basin, the Hathaus Bank, and other public edifices, and 1,748 private houses. This visitation elicited the deepest sympathy in all European countries; and as much as 400,000l. was subscribed by foreigners (about 41,000l. in England), and remitted to Hamburg in alleviation of the distresses of the sufferers.

But, however severe at the time, this conflagration, like the great fire of London, proved in the end for the advantage of the city. The system of mutual insurance having been generally adopted, the proprietors of houses and other property were subjected to a tax to defray the interest of a loan of 32,180,000 marks banco, raised to indemnify the sufferers, and to enable them to rebuild their houses. The work thus vigorously commenced has since been successfully carried on. The ground that had been cleared by the flames has been laid out on an improved plan, with wider and straighter streets, and other essential ameliorations. The finest of the new buildings are near the Alster. Many of them are of vast extent, and have been constructed at an enormous cost. The foundations are mostly of granite, the superstructure of brick and stucco. The arcade opening out of the Jungfernstieg deserves attention for its extent and beauty. Other improvements consist in conducting the drains to the Elbe without allowing them to enter the canals, and in the conversion into a new quarter of the town of a low marshy tract on the right bank of the Elbe. Its surface was raised 4 ft. by covering it with the rubbish of the fire.

The city proper is divided into five parishes, those of Saints Peter, Nicholas, Catherine, James, and Michael, the churches of which are amongst the principal edifices. The church of St. Peter, originally built in the 12th century, was burnt down in the great fire, but it has since been rebuilt, and is a fine holy edifice. St. Nicholas, also, was burnt down; but was re-erected, in the Gothic style, and far more magnificent and on a grander scale than before. The church of St. Michael is the most interesting in the city. It is 245 ft. long, by 180 ft. broad; and has a tower 456 ft. in height, ascended by a stair of nearly 600 steps. Its interior is capable of accommodating 6,000 persons; it has a fine altar-piece, an organ with 5,600 pipes, and a large crypt supported by 69 granite columns. There are about twenty other places of worship, including the chapels of the German, French, and English Calvinists, and the English Episcopal, Calvinist, and Roman Catholic churches. The new exchange opened in 1841, escaped the ravages of the fire. It contains a magnificent hall for the assemblage of the merchants; a hall for the meetings of the merchant company; rooms for the use of the commission, or board of trade, and for the extensive commercial library belonging to the latter institution. Hamburg has a great many charitable institutions, some of which are on a splendid scale. The general infirmary, erected in 1823 in the suburb of St. George, on the Lübeck road, cost about 85,000l. Its yearly expenditure is about 16,500l. the greater part of which is supplied from the city funds. It contains 140 sick wards, the majority about 40 ft. long, 25 ft. broad, and 14 ft. high, and various apartments for different offices, with apartments for officers. It may accommodate from 4,000 to 5,000 patients; invalids of the middle ranks are attended to in it on their paying a proportionate subscription. In the new orphan asylum, 600 orphans are received into the establishment, and 500 more are provided for elsewhere. There are, also, asylums for aged persons, deaf and dumb, the blind, sailors and their widows; and a private hospital, in which, besides medical attendance, a superior education is also given to deformed children and cripples, of whom Hamburg contains a large number. The old Rathhaus, or senate-house, was burned down in 1842; but a new Rathhaus, on an improved plan, was erected adjacent to the exchange. The bank, also, was destroyed in 1842, but its treasure, which was in its vaults, escaped untouched. The bank was founded in 1619: it is a bank of deposit only, and is extremely well-managed. The Eimbeck-house, workhouse, prison, town-hall, arsenal, and two theatres, are amongst the remaining chief buildings. The new theatre is one of the largest in Germany, and the performances and music are generally good. A commodious new building, the Schulgebäude, opened in 1840, contains the Gymnasium, or college for instruction in philosophy, philology, history, physics, and natural history; the Johanneum, or high school, an excellent and well-directed institution, founded in 1529; and the city library, containing 180,000 vols., open to every burgher and literary man. Hamburg has also an observatory and a botanic garden, academies of design, commerce, navigation, anatomy; museums of physical objects and works of art, and several learned societies, especially one for the promotion of the fine and useful arts.

In 1813, while the town was occupied by the French, a series of wooden bridges, and a chaussee connected by ferries with the N. and S. shores, were thrown across the swamps and islands of the Elbe, separating Hamburg from Hanover. Having been fitted only for temporary purposes, they were removed in 1816, and the communication is now maintained by steamboats. The arm of the Elbe opposite the city is not very wide, but it is deep enough for vessels of considerable burden. The maintenance of floating lights, buoys, &c., for the safe navigation of the river, costs the city a large sum every year. The city harbour presents an animated scene: a forest of ships of all nations, and from every quarter of the globe, while the face of the stream is covered with boats moving about in every direction. The tide rises at the quays from 5 to 12 ft., and flows about 70 English miles above the city. Between Hamburg and Altona, an adjoining town belonging to Holstein, is the suburb of St. Pauli, a narrow strip of about ½ m., called Hamburgerberg, which is in fact a kind of 'Wapping.' The environs of Hamburg abound with the villas of merchants, public cemeteries, pleasantly laid out, hotels, tea-gardens, and places of public entertainment.

The manufactures of the town are in some respects not so flourishing as formerly. Sugar refining is the chief branch of industry, but is not carried on to such an extent as in the first quarter of the present century. Besides sugar refineries, there are breweries, distilleries, calico printing, dyeing, lime-kilns, rope-walks, anchor and other iron forges. Glue, cork, sailcloth, leather, whalebone, feathers, hats, tobacco, soap, cotton-yarn, woollen, linen, cotton, and silk fabrics, tin ware, gold, silver, and copper articles, needles, waxlights, surgical and musical instruments, dice, &c., are amongst the remaining articles of manufacture. The shipping belonging to Hamburg is small as compared with its trade. The English shipowners engross most part of the direct trade with England. The Hamburg ships are almost entirely employed in transatlantic commerce, and in the coasting trade with continental Europe.

Commerce.—Hamburg is the greatest commercial city of Germany, and perhaps of the continent. She owes this distinction principally to her situation. The Elbe, which may be navigated by lighters as far as Melnick in Bohemia, renders her the entrepôt of a vast extent of country. Advantage, too, has been taken of natural facilities, that extend still further her internal navigation; a water communication having been established, by means of the Spree, and of artificial cuts and sluices, between the Elbe and the Oder, and between the latter and the Vistula; so that a considerable part of the produce of Silesia destined for foreign markets, and some even of that of Poland, is conveyed to Hamburg. There is, also, a communication by means of a canal with the Trave, and consequently with Lubeck and the Baltic, by which the necessity of resorting to the difficult and dangerous navigation of the Sound is obviated. Vessels drawing 14 ft. water may safely come up to the town at all times, and vessels drawing 18 ft. may come safely up with the spring tides. There are no docks nor quays at Hamburg; and it is singular, considering the great trade of the port, that none have been constructed. Vessels moor in the river outside of piles driven into the ground a short distance from shore; and in this situation they are not exposed to any danger unless the piles give way, which, though rarely, sometimes happens. Hamburg is joined by railways with the principal towns of the Continent. There is a sort of an inner harbour formed by an arm of the Elbe which runs into the city, where small craft lie and discharge their cargoes. The largest vessels sometimes load and unload by means of lighters at Cuxhaven. The trade of Hamburg embraces every article that Germany either sells to, or buys from, foreigners. The exports principally consist of linens, grain of all sorts, wool and woollen cloths, leather, flax, glass, iron, copper, smalts, rags, staves, wooden clocks and toys, Rhenish wines, spelter, &c. Most sorts of Baltic articles, such as grain, flax, iron, pitch and tar, wax, &c., may generally be bought as cheap at Hamburg, allowing for difference of freight, as in the ports whence they were originally brought. The imports consist principally of sugar, coffee, which is the favourite article for speculative purposes; raw cotton; woollen and cotton stuffs and yarn; tobacco, hides, indigo, wine, brandy, rum, dye-woods, tea, pepper, &c. The following table (from Report of Mr. Ward, British Consul-General) shows the imports into Hamburg from Great Britain and Ireland in 1862:—

Article		Weight	Value
			Sterl. Rms.
Cotton Yarn and } Twist }	centners	144,017	18,798,190
Woollen & Mixed } Woollen Yarn }	"	173,367	24,984,210
Cotton Wool	"	903,451	18,414,810
Furs and Peltry	"		4,674,770
Linen Yarn and } Thread }	centners	57,794	5,689,040
Coals and Cinders	lasts	779,500	4,949,880
Indigo	centners	5,090	2,736,110
Silk	"	1,448	7,475,580
Sheep's Wool	"	20,744	1,749,106
Leather	"		1,500,010
Kip and Cow Hides	centners	47,583	1,964,080
Dry American Hides	"	27,540	1,193,290
Pig and Smelting Iron	"	813,926	1,207,870
Linseed Oil	"	64,379	1,193,240
Forged Iron	"	191,941	1,093,130
Tea	"	18,368	1,279,930
Raw Sugar	"	84,574	849,060
Cotton Manufactures			12,684,380
Woollen and Mixed Woollen } Manufactures }			17,888,790
Various Manufactured Stuffs			3,990,190
Linen and Linen Manufactures			5,367,710
Silk and Mixed Silk			5,764,080
Machinery and Parts of Machinery			4,344,400
Fine Iron Manufactures			1,486,170
Hard and Small Wares			164,690
Earthenware and Pottery			468,630
Coarse Iron Manufactures		78,038	810,560
Bullion and Coin			7,845,890

Subjoined is a statement of the weight and value of the British goods imported into Hamburg in the years referred to:—

Years	Weight	Value
	Commons	Marks Banco
1857	12,244,416	170,730,340
1858	17,259,325	145,717,340
1859	11,914,473	144,141,340
1860	12,549,376	161,497,060
1861	12,949,051	157,547,020
Total	62,259,641	779,107,810
Average of the 5 years	12,417,326	154,481,562
Year 1862	17,640,640	166,154,000

The exports from Hamburg can no longer be ascertained, inasmuch as no official accounts of them have been kept since the year 1858, when the export duty was abolished. There is, however, every reason to believe that the exports have kept pace with the imports at the usual rate. (Report of Mr. Consul-General Ward.)

The number and burthen of the British ships which arrived at and sailed from Hamburg in each of the five years 1858 to 1862 inclusive, were as follows:—

BRITISH SHIPS INWARDS.

Year	Ships Arrived	Burthen in Commercial Lasts
1858	1,677	189,179
1859	1,579	187,080
1860	1,780	210,311
1861	1,979	771,656
1862	1,813	214,069

BRITISH SHIPS OUTWARDS.

Year	Ships Sailed	Burthen in Commercial Lasts
1858	1,880	189,764
1859	1,441	182,494
1860	1,860	210,437
1861	1,910	772,543
1862	1,818	217,443

The British flag has long participated much more largely than any other in the shipping and navigation of this great commercial mart. The average of the five years 1858 to 1862 inclusive was, ships arrived, average of the whole, 4,249; ditto of British ships, 1,776; their burthen in commercial lasts, average of the whole, 408,021; ditto of British ships, 204,904. The number of the ships crews was upon an average of the four years 1859 to 1862, total men, 49,660; men in British ships, 22,605.

The number of sea-going ships belonging to the port of Hamburg amounted, at the close of the year 1862, to 505, and their total burthen to 69,571 commercial lasts; comprising 417 square-rigged ships, barques, and brigs, 69 schooners, sloops, galliots, &c., and 20 steamers. In the same year the number of sea-going ships registered at Altona was 48, and their burthen 7,875 old lasts, equal to 5,250 commercial lasts.

The flourishing state of the commerce of Hamburg is owing, to a great extent, to the absence of almost all fiscal impositions on the liberty of intercourse. The only tax existing is an import duty of one-half per cent. ad valorem. Exports, as well as transit goods, are totally exempted from duty. The liberty of transit is limited to the term of three months from the time of receiving the transit ticket; but, upon application being made for a prolongation of the term previously to the expiration of the first three months, it is granted on payment of ¼ per cent. on the value of the goods; but under no circumstances is the term extended beyond six months. If the goods be not then exported, they become liable to the ordinary duties. No warehousing system has been introduced at Hamburg; nor, from the smallness of the duties, is it necessary.

Accounts are kept in marks divided into 16 schillings, and these into 12 pfennings each; or else in pounds, shillings, and pence. The money in circulation is from 25 to 75 per cent. under the value of bank money (banco). There is no coin representing the latter in circulation, all payments made in it being effected by transfer in the books of the bank. The rate of exchange is continually varying; but at an average the rix dollar banco is worth 4s. 0½d.; the rix dollar current, 3s. 8½d. nearly; the mare banco, 1s. 5½d.; and the mark current, 1s. 2½d. The Hamburg gold ducat is worth about 9s. 4d. 100 Hamburg lbs. = 106·8 lbs. avoird. The ohm is equivalent to 38½, and the fuder to 234½ English gallons. The Hamburg foot = 11·7 English inches.

Hamburg is well supplied with provisions, and the traveller is little inconvenienced by those vexatious custom-house regulations so common throughout most parts of the Continent. The activity that constantly prevails, and the gaiety and cheerfulness of the inhabit. render this city an agreeable residence to a visitor. Mr. Hodgskin says, "it resembles Paris on a Sunday; and on week days, when the quays, the streets, and the 'change, are crowded with people of all countries, it resembles London.' (Tour in the N. of Germany, i. 158.) Certain customs prevail that arrest the attention of most visitors. Among others, funerals are attended by bodies of hired mourners, some of whom are attired in a black Spanish habit, a large wig, a ruff about their neck, and a sword by their side. These individuals also attend weddings and other festive meetings. The Vierland flower girls, who wear a peculiar costume, market women, and female servants, all carry in the streets an oblong wicker basket, covered with a printed cotton shawl of the brightest colours. The public baths, and the dancing saloons, are among the principal features of the city; especially the latter, which are fitted up in most elegant style, and are the most popular places of public resort. Some of them are of questionable reputation; but others are frequented by the families of highly respectable citizens.

The climate of Hamburg is rather damp, but otherwise healthy. The drainage of the city was formerly as bad as possible; but extensive improvements have, in these respects, been recently commenced. The police is good, and beggars are not suffered to infest the streets. The city gates are shut at dusk, but are opened afterwards on payment of a toll, which increases in amount with the lateness of the hour. The water gate is, however, absolutely closed at dark.

This city was founded by Charlemagne towards the close of the 8th century. After the extinction of his dynasty, it became successively subject to the dukes of Saxony and the counts of Holstein. Early in the 13th century it joined with Lubeck in the formation of the Hanseatic league; in 1258 it obtained a portion of territory; and acquired the right to legislate for itself in 1292. In 1528 it adopted Lutheranism. It was long subject to attacks from the Danes, but in 1768 it purchased a resignation of all claims upon it from Denmark, and a security against future attacks. In 1806 it was occupied by the French, and in 1810 made the cap. of the dép. Bouches de l'Elbe. It suffered considerably from the exactions of the French troops under Marshal Davoust; but at the peace it was partially indemnified for its losses, and has since gradually retrieved its former flourishing condition.

HAMELN, a fortified town of N. Germany, k. Hanover, distr. Hanover, on the Weser, at its confluence with the Hamel, 25 m. SW. Hanover, on the railway from Hanover to Cologne. Pop. 6,620 in 1861. The Weser here forms an island, and on it a large sluice was constructed by Geo. II. in 1731, for the convenience of shipping: the town, by its position, commands the navigation of the Upper Weser, and has extensive communications with different parts of Germany. It is defended by Fort Georges, a strong fortress on a hill on the opposite side of the river. Its inhab. many of whom are wealthy, and have a considerable trade, carry on various branches of manufacture.

HAMILTON, a parl. bor., market, and manufacturing town of Scotland, co. Lanark, being the cap. of the Middle Ward, on the Clyde, on a rising ground gently sloping towards the E., 10 m. SE. Glasgow, and 12 m. NW. Lanark, on the Caledonian railway. Pop. 10,666 in 1861. The town stands about 1 m. W. of the conflux of the Avon with the Clyde, is intersected by the Cadzow burn, and is about 80 ft. above the level of the highwater mark at Glasgow. The town is not regularly but substantially built, and has an appearance of respectability, wealth, and comfort. It is paved and lighted with gas. The most important of its public buildings are the two parish churches, both elegant structures, particularly the older, in an elevated situation near the centre of the town; and the trades' ball and guel. The last edifice, which stands on high ground W. of the town, and was built in 1836, has in connection with it suitable apartments for all the public offices, municipal and civil. The court-room, common to the sheriff of the district and magistrates of the burgh, is 87 ft. long by 32 broad. In the vicinity are extensive cavalry barracks.

But the great object of attraction connected with this place is Hamilton Palace, the magnificent seat of the Dukes of Hamilton, separated from the town on the E. by a wall and plantation. The pleasure-grounds round the mansion, lying between the town and the Clyde, comprise 1,460 acres, and are the most extensive in Scotland. The oldest portion of the palace was erected about 1591, but the greater part of the building is comparatively modern, some very extensive additions having recently been made to it. The front, which faces the N., is 264 ft. 8 in. in length, adorned by a noble portion, consisting of a double row of Corinthian pillars, each of a single stone 25 ft. high, surmounted by a lofty pediment. The interior decorations are not less splendid than the exterior; and altogether it forms one of the largest and most superb structures of its kind in Britain. The collection of paintings, in particular, has long been considered as unrivalled in Scotland. It contains above 2,000 pieces. There is, also, a vast number of antique vases, antique cabinets, slabs of porphyry, and other similar relics. Within a mile of the town are Chatelherault, a venerable building, and still an occasional residence of the Dukes of Hamilton, and the ruins of Cadzow Castle, the original seat of this noble family, on the summit of a precipitous rock 200 ft. in height, the base of which is washed by the Avon.

Besides the par. churches, there are several meeting-houses belonging to the Relief, to the Associated Synod, and to the Independents. The Cameronians and Rom. Catholics have each places of public worship. The old par. church was uncollegiated in 1836; and a new church built for one of the ministers. About two-thirds of the pop. are dissenters.

The grammar or classical school of Hamilton is of ancient date, and has uniformly been an efficient seminary. There are in the parish about twenty other schools, including several for young ladies. There are also several subscription libraries and a mechanics' institution. The charitable institutions, and other provisions made for the poor, are considerable. There are two hospitals, and a good deal of property has been left in mortmain for behoof of the poor.

Hamilton has been the principal seat of imitation cambric weaving since the introduction of the cotton trade into Scotland. The reeds run from 1,200 to 3,000, which are the finest setts that cotton has been wrought into. But the trade has for years been on the decline. The average wages of a hand-loom weaver are never above 1s. 6d. per day; out of which must be deducted 1s. per week for expenses, and 10s. per annum for loom-rent. A house with a room and kitchen, and a four loom shop, lets at from 5d. to 6d. The females are employed in winding weft, and in tambouring, sometimes in weaving. The work is executed for the Glasgow manufacturers. The lace manufactory was introduced here many years ago, but it had become almost extinct, when a manufactory of the same kind was introduced, which has continued to prosper. About twenty houses are now engaged in this branch of trade; and it employs upwards of 3,000 females in this and the neighbouring parishes. Vast quantities of black silk veils of peculiar patterns are also manufactured here. A weaver's wife makes higher wages in these trades than her husband. Many thousand check shirts have of late been manufactured, chiefly for the Australian market. The other branches of trade are of minor importance.

In the park attached to Cadzow Castle are still preserved genuine specimens of the old Scotch breed of wild cattle; they are milk white, with black muzzles, horns, and hoofs, and are ferocious and untameable. They are not taken and killed like other cattle, but shot in the field. Similar cattle are to be found in Chillingham Park and in Chartley Park.

Cadzow was a royal residence for at least two centuries previously to the battle of Bannockburn in 1314; immediately after which it was conferred on the chief of the Hamilton family, in whose possession it has since continued. In 1414, James, first lord Hamilton, married the Princess Mary, eldest daughter of James II.; by which connexion his descendants came to be declared in parliament, on the demise of James V., in the event of the death of his only child Mary, next heirs to the crown. In consequence of the marriage of Anne, duchess of Hamilton, to Lord W. Douglas, eldest son of the Marquis of Douglas, the Hamilton family now represent the male line of the Douglases. On the death of the last Duke of Douglas, in 1761, the house of Hamilton, as male representatives of the Douglases, laid claim to the estates, under the plea that Mr. Douglas, the alleged son and heir of the only sister of the Duke of Douglas, was a supposititious child, taken at Paris from the real parents. A long lawsuit, well known by the name of the 'Douglas cause,' was the result. It was decided in Paris, and in the court of session in Scotland, in favour of the Hamiltons; but, on an appeal to the House of Lords, it was ultimately decided in favour of Mr. Douglas, afterwards created Lord Douglas. Cadzow Castle has been made the scene of one of Scott's finest ballads. It turns on the assassination of the Regent Murray by Hamilton of Bothwellhaugh.

Hamilton was created a royal burgh in 1548; but the magistrates, having consented to resign that privilege, in 1676, accepted of a charter from Anne duchess of Hamilton, by which it was constituted the chief burgh of the regality and dukedom of Hamilton. An attempt was made by the magistrates, in 1723, to get the original privilege restored, but in vain. Since the passing of the Reform Act it has been a parliamentary burgh, and unites with Airdrie, Linlithgow, Falkirk, and Lanark, in returning 1 mem. to the H. of C. In 1864 it had 403 registered voters. Municipal revenue, 1,176l. in 1863-4.

Among historical events connected with Hamilton, the battle of Bothwell Bridge, fought between the Covenanters and the royal forces, under the Duke of Monmouth, in 1679, deserves mention. The result of the engagement was unfavourable to the former, about 400 of whom were killed on the spot, while 1,200 were taken prisoners. (Laing's Hist. of Scotland, iv. 104.)

In addition to various distinguished characters that the noble house of Hamilton has produced, this burgh has given birth to several eminent persons: Dr. Cullen, the celebrated physician, born here in 1714; Professor Millar, of Glasgow, author of an 'Historical View of the English Government,' and other works; the late Dr. Matthew Baillie, of London, and his sister, Miss Joanna Baillie, authoress of 'Plays on the Passions.'

HAMME, a town of Belgium, prov. E. Flanders, arrond. Dendermonde, cap. cant., on the Durme, 10 m. ENE. Ghent, on the railway from Ghent to Antwerp. Pop. 9,812 in 1856. The town has manufactories of linen, soap, starch, and cordage, with numerous breweries and oil-mills, and a brisk trade with the surrounding country. Some antiquities have been discovered in its neighbourhood.

HAMMERSMITH, a village and chapelry of England, par. Fulham, co. Middlesex, hund. Ossulston, near the N. bank of the Thames, and on the great W. road out of London, from which it is distant 4 m. W. by S. Area, 2,140 acres. Pop. of par. 24,519 in 1861. The village is well paved and lighted with gas; but the streets are irregular, and the majority of the houses inferior. Many handsome mansions, however, lie scattered in different parts, and more especially by the side of the river, and along the great road which forms its main street. The church, erected in 1631, is a plain brick building with a low tower; and the interior is old fashioned and inconvenient. The living is a perpetual curacy, in the gift of the Bishop of London. A district church was erected in 1839. The dissenters also have several places of worship, and there is a Jews' synagogue. Close to the R. Cath. chapel is a small Benedictine nunnery, originally a boarding-school, established in 1669; and the monastic rules are strictly observed. Among the charity schools, one founded by Bishop Latimer has revenues amounting to 800l. a year. Other day and Sunday-schools are supported both by subscents to the church and by dissenters. The most striking feature in Hammersmith is the suspension bridge over the Thames, completed in 1827 at an expense of 80,000l. It consists of a horizontal roadway, suspended from iron chains carried over stone piers and archways, and secured by substantial abutments. The roadway is 822 ft. long, and 20 ft. wide, exclusive of a foot-path 5 ft. wide. The West Middlesex Water Company has its engines and reservoirs a little above this bridge. The grounds in the neighbourhood are chiefly occupied by nurserymen and market-gardeners, who

supply London with some of the choicest flowers and vegetables. (See LONDON.)

HAMPSHIRE, HANTS, or SOUTHAMPTON, a marit. co. on the S. coast of England; it includes the Isle of Wight, and has Berkshire on the N., Surrey and Sussex on the E., Wilts and Dorset on the W., and the English Channel on the S. Area, 1,672 sq. m. or 1,070,216 acres, of which 1,040,000 are arable, meadow, and pasture, and 30,000 forest. Hants is one of the most agreeable cos. in England, the surface being finely varied with gently rising hills and fruitful vales, and its climate being at the same time peculiarly mild and genial. Soil various; in the N. districts on the borders of Berks, it is hilly and poor; but between Basingstoke and Silchester is some fine wheat and bean land; a broad mass of chalky downs, intersected by numerous valleys, extends across the co. In the S. and middle parts of the co., and particularly in the vales watered by the Anton, Itchen, and other rivers, are large tracts of fine land, and some of the best water meadows in England. The SW. district, or that lying between Southampton Water and Dorsetshire, is principally occupied by the New Forest, and by extensive heaths. Principal crops, wheat, barley, oats, and beans; turnips are extensively cultivated, especially on the light soils. Farms till lately have been mostly let on leases, but the practice of holding them at will is gaining ground. Tenants are prohibited from taking two wheat crops in succession; but two white crops in succession have not been usually objected to, and it is common to take a crop of oats after wheat. This erroneous practice is, however, beginning to be corrected, and agriculture in this co. is generally good, and the condition of the land such as to reflect credit on the occupiers. Cattle of various breeds; the dairy is not an object of much attention. Stock of sheep large. Weyhill, near Andover, in this co., has the greatest sheep fair in England. Hants is famous for its bacon; and excellent honey is produced in different parts of the co. Estates mostly large; farms of all sizes, from 25 to 800 acres. The co. is everywhere particularly well wooded. The New Forest comprises about 92,000 acres, but only about 67,000 are now the property of the crown, the rest having been assigned to individuals. About 6,000 acres have been inclosed and set apart for the growth of timber. There are the remains of other extensive forests; and brushwoods are met with on most of the chalk lands. Minerals of little importance. If we except the building of ships at Portsmouth, and the various works subordinate to their outfit, the other manufactures are but of trivial importance; there are, however, silk mills at Overton, and straw hats are made in different parts of the co. Principal rivers, Avon, Anton, and Itchen. Portsmouth harbour and the road of Spithead lie in the Sound between the mainland and the Isle of Wight. Principal towns, Portsmouth, Southampton, Winchester, and Lymington. Hampshire, including the Isle of Wight, has 44 hundreds and 317 parishes. It sends 17 mems. to the H. of C.; viz. 2 for each division of the co.; 2 each for the bors. of Portsmouth, Winchester, Lymington, Southampton, and Andover; 1 for the Isle of Wight; and 1 each for the bors. of Petersfield and Christchurch. Registered electors for the co. 11,575 in 1865, namely, 3,650 for the Northern division; 5,696 for the Southern division; and 2,229 for the Isle of Wight. Pop. 481,815 in 1861. Annual value of real property assessed to income tax in 1862—Northern division 669,770l.; Southern division, 842,020l.; and Isle of Wight, 288,097l.

HAMPSHIRE (NEW), one of the U. S. of America, in the N.E. part of the Union (New England), and between lat. 42° 40′ and 45° 10′ N., and long. 70° 40′ and 72° 28′ W.; having N. Lower Canada, E. Maine, W. Vermont, S. Massachusetts, and SE. the Atlantic, on which, however, it has a coast of only 18 m. Length, N. to S., about 170 m.; breadth very variable. Area, 9,280 sq. miles. Pop. 326,073 in 1860. The coast is indented by small inlets, but has only one harbour of value, that of Portsmouth. It is skirted by a narrow sandy plain, which, at no great distance inland, rises rapidly into a hilly country. In the interior, the state is covered with mountains of granitic formation. The White Mountains, towards the N., which attain a height of more than 7,000 ft., are the highest in the Appalachian system, and, consequently, in the U. S. But between the mountains are many green and sheltered valleys, and the state contains a considerable proportion of fertile land, as well as a great deal of beautiful and picturesque scenery. Several of the principal rivers of New England rise in this state; among which are the Connecticut, Merrimac, Piscataqua, Androscoggin, and Saco, which have a general N. direction. The Connecticut forms the W. boundary of the state. There are several considerable lakes, the largest of which, the Winnipiseogee, 28 m. in length, is situated near the centre of the state. With the exception of the alluvial lands bordering the rivers, the soil is, perhaps, more adapted for pasture than cultivation. The country was originally densely wooded, and such is still the character of the interior. Climate very healthy, but cold. The lakes and rivers are generally frozen for four months in the year, and winter lasts from November to April. Wheat, rye, maize, barley, oats, pulse, and flax are grown; cattle-breeding is pursued to a considerable extent. Manufactures have greatly augmented of late years; they include cotton and woollen fabrics, nails and other hardware, paper, glass, &c. The exports consist principally of cattle, pork, flax seed, lime, timber, fish, beef, granite, and manufactured goods. The foreign trade is inconsiderable.

New Hampshire is divided into eight counties; Concord, on Merrimac, being its political cap. Portsmouth is the largest town, and the only seaport. Dover, Exeter, Hanover, New Ipswich, Keene, and Haverhill are increasing places, already of some size. Dartmouth College, at Hanover, established in 1770, ranks third among the literary institutions of New England. It has attached to it a medical school, library, and philosophical apparatus; and had, in 1862, upwards of 500 students. There is a theological seminary at New Hampden, besides upwards of 30 incorporated academies. The state has a literary fund, the income arising from which, with the produce of a tax on banks, is devoted to the support of free schools. There are established on the same system as in the other Atlantic states. A lunatic asylum is at Portsmouth. Several canals have been constructed connected with the Merrimac, which, by its communication with the Middlesex Canal, affords a navigable route between many parts of the state and Boston.

The legislature consists of a senate of 12 mems. and a house of representatives which had 333 members in the session of 1862–63. The latter, as well as the governor, are chosen annually by the electors of each district, consisting of every white male citizen above the age of 21 years who pays taxes and has resided in the state for three months. Together, they are styled the General Court of New Hampshire, and assemble annually on the 1st Wednesday of June, at Concord. The governor is assisted in his executive duties by a council of 5 mems., elected for a similar period with himself. The poor in this, as in other N.E. states, are supported by a direct tax on the towns to which they belong. The militia, comprising 10 brigades, consisted, in 1863, of an aggregate body of 29,503 men. Justice is administered in a superior court, and county courts of common pleas, presided over by the judges of the superior court, and two justices selected from each county. The judges hold their offices during good behaviour, until 70 years of age; but may be removed by impeachment, or by address of the two houses of the legislature.

New Hampshire was first colonised by the British in 1622. It was twice united to Massachusetts; and the final separation between them did not take place till 1741. New Hampshire was one of the first states to take a decided part in the war of independence. A temporary constitution was formed in 1784, which, in 1792, was altered and amended nearly to that now in force. The state sends three representatives to Congress.

HAMPSTEAD, a par. and village of England, co. Middlesex, hund. Ossulston, 4 m. NNW. London, of which it forms a kind of suburb. Area of par. (which includes part of Kilburn), 2,070 acres; pop. of do., 19,106 in 1841. Hampstead lies on the brow and S. slope of an irregularly formed hill, on the summit of which (460 ft. above high water mark) is an extensive heath, covering about 240 acres, which commands fine views of the metropolis, Kent, and Surrey southward, and of the highly cultivated lands of Bucks and Herts on the NW. The streets are mostly crooked and irregular, lined with houses of every size and quality, from the spacious mansion to the mere cottage; and the subordinate streets, connecting High Street with the other parts are narrow, inconvenient, and in some places even dangerous. The church, which has been parochial since 1598 (when Hampstead was separated from Hendon), was rebuilt by subscription in 1747; it is a plain brick building, having at its E. end a low tower and spire. The living is a vicarage, and there is a lectureship founded for the benefit of the curates. A chapel of ease, in Well Walk, occupies what was a century back the most fashionable assembly-room in the town, and a favourite place of resort for all who came to drink the chalybeate waters. There are places of worship for Independents, Wesleyan Methodists, Unitarians, and Roman Catholics. Besides churches and chapels, there are no public buildings; but numerous large private mansions, in different parts within and round the town, attest its importance as a fashionable suburban retreat. A large square house, on an eminence to the left of the London road, with a row of elms in front, once belonged to Sir Harry Vane, one of the regicides, who, at the Restoration, was here seized, and soon after executed: it was subsequently occupied by Bishop Butler. In the upper part of the town, near the Terrace, is Branch-hill Lodge, once the residence of the Earl of Macclesfield and Lord Loughborough; but its fine collection of painted glass windows, procured from various convents at the period of the French revolution, has been removed by Sir Thomas Neaves, to his home at Dagenham, in Essex. The Upper Flask Inn, in High Street, formerly the resort of the celebrated Kit-cat Club, and subsequently inhabited by G. Steevens, the editor of Shakspeare, is now a private residence. The inns receive hundreds of visitors on Sundays and holidays during summer.

The manor of Hampstead was given by King Ethelred to the Abbey church of Westminster, by whom it was retained till 1550, when Edward VI. took possession of it and presented it to a layman, from whom the present lord of the manor is descended. In the reign of Henry VIII. Hampstead was an obscure hamlet, 'chiefly inhabited by washerwomen;' and being well covered with wood, and abounding with game, it was often visited by hunting parties from court. James II. is said to have had a hunting-seat here, still known as Thicken House, and now let out to several poor people. About 1640, Hampstead became a fashionable watering-place, and concerts, balls, and races were established for the amusement of the visitors. The wells (the water of which is a simple carbonate chalybeate) were in high repute during the 17th century, but they have long since ceased to attract attention. The election of mayor for the co. was held on the heath from 1640 to 1701, when it was removed to Brentford.

HAMPTON, a village and par. of England, co. Middlesex, hund. Spelthorne, on the N. bank of the Thames, opposite the point where it receives the Mole, 12 m. WSW. London, and 3 m. W. by N. Kingston on the London and South Western railway. Area of par., 3,190 acres. Pop. 5,355 in 1861, and including the hamlet of Hamptonwick, close to Kingston, 7,340. The town, which is a favourite resort for anglers, is not remarkable either for the width of streets or regularity of the buildings; but many beautiful villas ornament the neighbourhood, among which is one formerly the property of the celebrated David Garrick. A wooden bridge, built across the Thames in 1753, joins the town to E. Moulsey. The church is a very handsome structure, having a square tower at the W. end. A free grammar school was founded here in 1556, and the original endowment has been subsequently so much enlarged, as to furnish the master with a salary of 230l., and a sum of 36l. yearly for six poor men.

HAMPTON COURT. About 1 m. from the village of Hampton, close to the Thames, is Hampton Court, respecting which Grotius has not scrupled to say,—

'Si quis opes cernit (nec quis tamen ille ?) Britannum,
Hampton Court, or Cardinal Wolsey's, hunc consulat ille larem;
Contulerit toto cum quæro Palatia mundo,
Dicet ibi Reges, hic habitare Jove!'

The palace was begun by Cardinal Wolsey, who, in 1526, presented it to Henry VIII. The original edifice consisted of five quadrangles, of which two only remain. The W. quadrangle, little altered since Wolsey's time, presents a good specimen of Tudor architecture: the middle or clock-court is of mixed style, Sir C. Wren's Ionic colonnade strangely contrasting with the massive construction of the old building: the third quadrangle was erected by William III. The king's entrance in the clock-court leads to the grand staircase and state apartments. The ceiling and walls of the former were painted by Verrio, in his usual glaring style: the rooms, which open from each other, and are partially furnished, consist of the guard-chamber, presence and audience chambers, public dining-room, state drawing-room and bed-rooms. Notwithstanding the removal of some of the best specimens to Windsor, there is still at Hampton Court an extensive and excellent collection of pictures. It comprises many by the principal Italian and Flemish masters; and an extensive collection of portraits connected with English history by Holbein, Lely, Kneller, and West. The great glory of Hampton Court, the Cartoons or drawings executed by Raphael, by order of Pope Leo X., for patterns for tapestry intended to decorate the Vatican, were carried away in the spring of 1865, and are now in the Kensington Museum. It is promised, however, that they shall be returned to Hampton Court Palace. They are called cartoons from being painted on sheets of large paper, cartoon. These noble drawings, of which there were originally 25, being left neglected at Brussels, the greater number of them appear to have been lost or destroyed. Fortunately, however, seven were purchased by Rubens for Charles I.; but even since their arrival in this country they have been exposed to numerous vicissitudes, and would seem to owe their preservation as much to accident as to anything else. The gallery in which they were for many years placed at Hampton Court, was built for their reception by William III.; but George III. removed them first to Buckingham Palace and thence to Windsor, whence they were at length brought to Hampton Court, which, indeed, appears to be the most fitting place for these splendid works of art. They represent some of the most striking incidents recorded in the New Testament, and are unrivalled for sublimity of conception and purity of design. They have been well engraved by Holloway.

Among the parts of the palace not usually shown to the public are the chapel and hall, the former of which was refitted after the ravages of the fanatics during the Commonwealth, and handsomely paved with oak by Q. Anne. The latter, built by Wolsey, and still retaining his name, is a finely proportioned room 160 ft. long, and 40 ft. broad, having two large gabled windows, and an elaborately carved wooden roof, similar to that of Westminster Hall and that of Christ Church Hall, Oxford. This room was thoroughly restored to the old model in 1841. Close to the Hall is the Board of Green Cloth, a small, though very beautiful Gothic chamber, which furnished Sir Walter Scott with the pattern for one of the finest rooms at Abbotsford. 'The garden front of the palace, though disfigured by modern windows, is still very magnificent. The gardens comprise about 44 acres: the pleasure-grounds were laid out by William III., in the Dutch taste: the terrace is ½ m. long, and the first view of it is very striking. The home-park, immediately adjoining the gardens, is 5 m. in circuit, and its soil produces very fine herbage. The canal, which is ¾ m. long, and 40 yds. broad, is lined with an avenue of lime trees, and other avenues intersect the park in every direction, through one of which is a good view of the tower of Kingston church.' (Jesse's Gleanings, 3d. ser.) The green-houses contain, among many valuable exotics, a vine said to be the largest and most productive in Europe; and a maze or labyrinth furnishes much amusement to young visitors. The palace, in which Wolsey maintained a more than regal state, was afterwards the favourite residence of Henry VIII. and his children, and of James I. and his son Charles I., who escaped from his imprisonment here in 1647. The Protector Cromwell resided here during the Commonwealth; and it afterwards became the usual abode of William III. and his queen, and of the princess, afterwards Queen Anne. George II. was the last monarch by whom it was inhabited. Of late years it has been mostly divided into private dwellings given to court-pensioners, and the state rooms have been fully opened to the public. All individuals are now freely admitted to view the public apartments and grounds, without any demand being made upon them; and without, as formerly, being hurried from one apartment to another, at the caprice of some mercenary cicerone. In consequence, Hampton Court is resorted to in summer by crowds of visitors, and is deservedly one of the principal points of attraction in the vicinity of the metropolis. Near Hampton Court palace is Bushy

Park, comprising 1,100 acres, with a central avenue 1 m. long. 'The numerous chestnut-trees, though of great age, are still healthy and vigorous; and, when they are in blossom, they appear at a short distance as if covered with snow.' (Jesse.) The house on the right of the grand avenue was during many years the favourite retreat of William IV. when duke of Clarence, and is still at times inhabited by members of the royal family. (Lysons's Environs of London; Jesse's Gleanings.)

HANAU, a town of W. Germany, electorate of Hesse, cap. prov. of same name, and seat of its superior courts, on the Kinzig, near its junction with the Main, 11 m. E. by N. Frankfort, and 72 m. SSW. Cassel, on the railway from Frankfort to Nuremberg. Pop. 17,108 in 1861. Hanau is no longer fortified; and its ancient castle is now used for the purposes of the Wetteravian Society of Natural History. It is divided into the old and new towns; the former is ill-built, but the latter has broad and regular streets, modern-built houses, and, near its centre, a good market place. There are four Calvinistic parish churches, a Rom. Cath. church, a large hospital, handsome theatre, gymnasium, free school, drawing academy, many scientific and benevolent associations, and a school of trades. Hanau is the most industrious town, and the place of the greatest commercial activity, in the electorate. Its manufactures are numerous and extensive, including silk stuffs, cambric, leather, gloves, stockings, hats, excellent carpets, cotton fabrics, tobacco, playing-cards, gold and silver wares, brass musical instruments, and carriages. It has a large trade in timber, barrels, and wine. Many of its inhab. are descendants of Dutch and Flemish emigrants, who fled thither from the persecutions in the low countries, under Philip II., early in the 16th century. Very near it are the mineral springs of Wilhelmsbad. Here, on the 30th October, 1813, Napoleon, on his retreat from Leipsic, gained a decisive victory over a very superior force of Bavarians, and other allied troops, under Marshal Wrede. The combined army lost about 10,000, while the loss of the French did not exceed 3,000 or 4,000 men; but the opening of the route to France was the most important advantage gained by the latter. The principality of which Hanau was the cap. was, after the extinction of its princes in 1736, divided between Hesse-Cassel and Hesse-Darmstadt.

HANG-TCHEOU, one of the largest and richest cities of China, cap. prov. Tche-kiang; on the Tsien-tang-kiang, 30 m. from its mouth in the Eastern Sea, and 100 m. SE. Nankin; lat. 30° 20' 20" N., long. 119° 48' E. Its pop. was estimated by Du Halde at upwards of a million, without, apparently, including the pop. of the suburbs; but this estimate is most probably much beyond the mark, especially as the houses are but one story high, and there are gardens of large size interspersed among them. The city is surrounded with high and thick walls, said to be as much as four leagues in circuit. The W. part of this enclosure is taken up by a fort or citadel, in which the officers of the government reside, and a garrison of 10,000 men is maintained. The Grand Canal has its S. terminus here, in a large commodious basin. This city has, in consequence, a direct communication with Pekin, and a vast command of internal navigation, which it has turned to good account. On its W. side is a lake highly celebrated for its natural and artificial beauties. Barrow, by whom this city was visited, says, 'the city of Hang-tcheou-foo being particularly famed for its silk trade, we were not surprised to meet with extensive shops and warehouses; in point of size, and the stock contained

within them, they might be said to vie with the best in London. In some of these were not fewer than ten or twelve persons serving behind the counter; but in passing through the whole city not a single woman was visible, either within-doors or without. The crowd of people, composed of the other sex, appeared to be little inferior to that in the great streets of Pekin.' (Travels, p. 527.) The streets are not so wide as Cranbourn Alley, but as well paved. They are ornamented in many places with triumphal arches, and monuments to eminent individuals, and are kept remarkably neat and clean. Barrow says: 'In every shop were exposed to view silks of different manufactures, dyed cottons and nankins, a great variety of English broad-cloths, chiefly however blue and scarlet, used for winter cloaks, for chair covers, and for carpets; and also a quantity of peltry, intended for the northern markets. The rest of the houses, in the public streets through which we passed, consisted of butchers' and bakers' shops, fishmongers, dealers in rice and other grain, ivory cutters, dealers in lacquered ware, tea-houses, cook-shops, and coffin-makers; the last of which is a trade of no small note in China. The number of inhab. in the suburbs, with those that constantly resided upon the water, were, perhaps, nearly equal to those within the walls.' (Barrow; Du Halde, vol. i.; Dict. Geographique.)

HANLEY, a town and market. tw. of England, belonging to the par. of Stoke-upon-Trent, co. Stafford, hund. Pirehill-north; 2½ m. SE. Stoke-upon-Trent, 16 m. N. Stafford, and 150½ m. NNW. London, by London and North Western railway. Pop. of town 14,578, and of market. bor. 31,953 in 1861. The town consists of one main street, intersected by various others; and many good houses have recently been built, though the pop. is chiefly confined to the working classes. The church is handsome, and has a fine tower 100 ft. high. Good schools are connected both with the church and the three dissenting places of worship. The inhab. are chiefly employed in the potteries, which alone have raised this district to its present importance. (For further particulars, see STOKE-UPON-TRENT and POTTERIES.)

HANOVER, a kingdom of NW. Germany, situated between lat. 51° 16' and 53° 52' N., and long. 6° 43' and 11° 45' E., bounded N. by the German Ocean and the Elbe, E. by Prussia and Brunswick, S. by Prussia and Hesse-Cassel, and W. by Holland. Its boundary line is very irregular, and a portion on the W. is almost divided from the rest of the kingdom by the grand duchy of Oldenburg. Length, from the mouth of the Elbe S., 172 m.; breadth, E. and W., 180 m.

Hanover is divided into seven landdrosteien, or administrative divisions superintended by a Landdrost, or high-bailiff. The seventh of these districts, however, the mining district of the Harz, is not under a landdrost, but a berghauptmann, or captain of the mountain. The area of the provinces and population, according to the census of 1852 and of Dec. 1861, is as follows:—

Landdrosteien	Area in Engl. sq. m.	Population 1852	Population 1861
Hanover	2,852	319,974	368,973
Hildesheim	1,776	341,603	364,766
Lüneburg	4,344	338,764	347,689
Stade	2,879	279,541	276,629
Osnabrück	2,474	261,905	262,314
Aurich	1,154	165,179	179,279
Mining District	711	33,370	31,391
Total	16,848	1,819,335	1,890,070

Of the population in 1861, the last census, 943,581 were males and 944,089 females, living in 875,851 hamshaltungers, or families, and 273,362 separate dwellings. In twenty-one towns—the largest, Hanover, with 71,170 inhabitants; the smallest, Münden, with 4,432—there lived 507,156 persons, while the country was inhabited by 1,603,124. Through emigration, particularly to the United States, the country loses, on the average, between four and five thousand souls per annum. The number of emigrants amounted to 4,562 in 1859, to 4,917 in 1860, and to 4,296 in 1861.

Surface.—Hanover, physically considered, is an inclined plain, gently sloping from SE. to NW., and nowhere, except on a few of its eminences, more elevated than 800 ft. above the sea. The districts of Stade, Lüneburg, Hanover, and part of Osnaburg belong to the N. plain of Germany, which stretches from the North Sea E. into Russia. No hill in the central provinces reaches 1,400 ft. In the S. part of Hildesheim are the Harz mountains, the highest summit of which, Königsberg, is 5,800 ft. high. The well-known Brocken (3,660 ft.) is within the Prussian dominions. This mountain mass forms the watershed between the Elbe and the Weser. Its geological formation is chiefly granite overlaid by grauwacke, grauwacke slate, and clay slate; and in these latter formations the mineral riches, hereafter described, are mostly found. Above these strata lie the Stone and tertiary formations. The great plain of the N., with the exception of a few limestone hills in Lüneburg and Stade, is of diluvial formation, and consists either of extensive tracts of sand covered with furze, or of vast moors and marsh-lands. The heath of Lüneburg, in its whole extent, comprises about 1-6th of the kingdom; granite boulders are found in different parts of it, some of very extraordinary size. Of the peat-moors the largest are the Bourtanger moor, on the Ems, and the Hoch moor, in E. Friesland. The lowlands on the sea-coast are below the sea-level, and hence are kept dry by means of dykes similar to those of Holland and the Bedford Level; the maintenance of which occasions an expenditure of several thousand dollars yearly. These lands, however, are by far the most productive of the kingdom.

Rivers and Lakes.—Hanover is traversed by three large rivers, all of which fall into the German Ocean:—1, the Elbe, which, rising in the plateau of Bohemia, enters the kingdom at Schnackenburg, and forms, with a slight exception, its whole N. boundary, as far as its mouth; the chief affluents within Hanover are the Jetze, Ilmenau, Este, and Oste, all on the S. bank: 2, the Weser, formed by the junction of the Werra and Fulda at Münden, flowing NW. as far as the junction of the Aller, and its tributary the Leine, and thence N., past Bremen into the German Ocean: 3, the Ems, rising in Westphalia, and flowing N. through the moorlands of Meppen, and E. Friesland to Emden, at its mouth. Throughout the flats of N. Germany there are numerous lakes and stagnant pools, in which the water subsides after the floods, which extensively cover the country in winter and spring; the chief of these in Hanover are the Steinhuder-meer, 5 m. long by 2½ broad, the Dümmer-see, and, the Neuburger-see. In E. Friesland the subterranean lake Jonian is so thickly coated with vegetation, that waggons can pass over it. The mountain lake Odertich, in the Harz, is 2,200 ft. above the sea.

Soil and Climate.—The nature of the soil of Hanover will be best understood from the distribution of the land, as stated by Marvard, which, though a number of years ago, is still quite correct, owing to the stationary character of the po-

pulation. The Hanoverian morgen is equal to 64 English acres.

	Morgen
Arable Meadow, and Garden Land .	5,811,000
Forests .	2,247,000
Waste Land, Lakes, and Rivers .	6,516,000
Total of the kingdom .	14,580,000

The waste lands, which form so large a proportion of the whole country, consist principally of vast sandy tracts wholly unavailable for tillage. They extend in a broad belt across the kingdom, of which they occupy about 1-6th part. This band of sand is aptly termed 'the Arabia of Germany.' The sandy districts are covered with heath, on which a very small and hardy breed of sheep, known by the name of *Haidschnucken*, find a scanty subsistence. They yield wool of the coarsest description, but their flesh is well-flavoured.

The proportion of land under cultivation to the whole extent in each province, except the mining district, is as follows:—

Hanover .	.63	Stade .	.49
Hildesheim .	.56	Osnabrück .	.31
Lüneburg .	.47	Aurich .	.10

The richest land of the kingdom is the alluvial soil and weald-clay of the Hadeln-land at the mouth of the Elbe, and of E. Friesland at the mouth of the Weser. It is taxed as belonging to the highest class. The soils of the secondary classes are found in the limestone districts of Hildesheim, Göttingen and Grubenhagen, Bremen and Werden. The least productive of all, belonging to the lowest class, is that of the duchy of Arenberg-Meppen. Much of this land, however, is laid out in meadow, especially the rich soil of E. Friesland, as the following table, giving the proportion of meadow to the whole cultivable soil, will show:—

	Per cent.		Per cent.
E. Friesland .	52	Lüneburg, Dannenberg, & Lauenburg	25
Bremen and Werden .	36		
Osnabrück .	31½	Grubenhagen .	21
Hoya .	27	Kalenberg .	18½
Diepholz .	26	Göttingen .	16
		Hildesheim .	11

In E. Friesland 46 cwt. of hay are reckoned as the produce of a morgen of meadow land, and 7·45 morgen of summer pasture are reckoned in that province to one cow. In Hildesheim, the morgen yields half a cwt. of hay, and 6-10 morgen are deemed enough to pasture one cow.

The climate is damp and unwholesome in the low country about the coast; but the winters are not so severe as in the interior, where, especially near the Harz, they begin in September and last till May. The spring is the most gloomy and disagreeable part of the year, owing to the long prevalence of NE. and E. winds. NW. winds prevail in the summer months. The temperature of the kingdom is thus stated by Von Reden, in his 'Statistical Description of Hanover,' i. 24:—

Place	Mean Temperature (Réaum.)				
	Year	Spring	Summer	Autumn	Winter
Cuxhaven . .	6·4°	5·8°	13·4°	7·7°	0·6°
Lüneburg . .	7·2	7·04	13·8	7·4	0·7
Göttingen . .	6·4	6·1	14·1	7·4	0·7
Harz District .	4·9				
Average of Kingdom	7·4	6·4	14·5	7·7	0·6

The fall of rain during the year averages 22·5 in.; but it is very unequal in different parts of the kingdom. Fogs prevail in the dyke-lands; and in the winter violent storms frequently occur, causing great damage to the embankments and drainage.

K K

Agriculture and Grazing.—The soil, on account of its general mediocre quality, requires effective cultivation to make it profitable to the proprietor; this is evident to be met with, owing to the smallness of the estates into which the land is divided.

The following table exhibits an estimate of the proprietorship of the soil of Hanover:—

Proprietors	Arable and Garden Grounds	Grazing Lands	Forests	Prop. Total
	Morgen	Morgen	Morgen	Morgen
Royal Domains	204,483	1,417	1,209,516	17·5
Monasteries	47,235	191	20,653	·9
Nobles	221,360	346	171,230	6·1
State Officers & Corporations	75,680	374	735,731	5·7
Clergy & Schools	141,461	14	16,175	1·2
Small Proprietors	3,744,917	180	84,825	61·6
Total of Land (except Turf-moors)	3,822,608	2,461	2,717,576	100·

The number of small proprietors in 1861 amounted to about 266,000, and it appears, therefore, that three-fifths of the land is in the hands of owners the average property of whom is only 20 morgen, or 12 acres. These small landowners, called *Bauern*, are a race of hard-working men, and reported to be, on the whole, very happy and comfortable, poverty being unknown amongst them.

The best cultivated lands belong to the crown and the nobility, and on these estates as much attention is given to improved systems of tillage as in Pomerania and Prussia. In the land held by small proprietors, the best farms are in the marsh-lands, and they both yield abundant crops and support numerous cattle. The freeholds in the principalities of Hildesheim, Göttingen, Grubenhagen, part of Kalenberg, and near the large towns, are next in order as respects tillage. Among these the system prevails, called *Koppel-schlag-wirthschaft*, which consists in parcelling the land out into a number of fields for a rotation of crops proportioned to the numbers of the owner's cattle, and his consequent power of keeping the land properly dressed. The small proprietors in the sandy districts, and the *Meier* (stewards), who farm small parts of the crown lands and of the nobles' estates, abide by the old fashion of three courses—fallow, winter corn (chiefly rye), and summer corn (barley or oats), with clover on the fallow, where the land will bear it. Potatoes are universally grown, and constitute the chief food of the poor. Rye is generally grown for bread, the raising of wheat being confined to the rich weald soils, and the quantity is insufficient for the demand. Barley and oats are largely cultivated, and, when in demand, are exported to England in considerable quantities. Clover and lucern are much grown on good farms, and even by the peasants, on dry soils. Turnips are a favourite article of production, and flax, hemp, tobacco, and hops are more or less cultivated in different parts. The cranberries, abounding on the heath-lands, are gathered for exportation. The forest-land, which amounts to 2,242,574 morgen (equal to about 1,400,000 acres), yields about 51,873,000 cubic ft. of timber yearly, not including inferior wood. The timber in the Harz district consists of fir; large beech and oak forests are found in Kalenberg, the duchy of Bremen, and the Upper Weser. These forests are under special control, and even when forming a part of private property, are confided to foresters scientifically educated and licensed for the purpose.

The breeding of horses is a very important occupation, large numbers being annually sold to the French and Italian armies. The following table gives a return of the total number of stallions used for breeding purposes in the kingdom of Hanover in the year 1862, as compared with the four preceding years, specifying the number of those belonging to the government breeding studs, and of those belonging to private individuals, and likewise the number of live foals bred:—

Years	Number of Stallions belonging to the Government Studs kept for Breeding Purposes	Number of Stallions belonging to private Owners allowed for Breeding Purposes	Total Number of Stallions used for Breeding Purposes	Number of live Foals bred
1862	214	270	484	19,100
1861	216	372	488	19,370
1860	219	373	492	17,846
1859	217	360	671	19,861
1858	217	361	678	22,490

The number of foals bred in the kingdom of Hanover in 1862, viz. 19,100, was slightly below the average of the last ten years, but about 1,000 more than were bred in 1861 and 1860. The above table is from an official report of Mr. Petre, British secretary of Legation, dated Hanover, January 1864. (Reports of Her Majesty's Secretaries of Legation, No. VII, p. 219.)

The rearing of cattle and sheep, though not of the same importance as horse-breeding, is attended to extensively. Bees are a favourite addition to a farm throughout the kingdom, and thrive well on account of the quantity of flowering heath and buck-wheat in the sandy districts. The annual produce of honey is valued at 40,000l. Large flocks of geese are kept in moist situations; their flesh is salted for domestic use, and the feathers are preserved. Leeches, which formerly abounded in the marsh-lands, have become nearly extinct, from being too eagerly fished. Fish are caught in all the ponds and rivers, and contribute to the support of no small number of the poorer orders. The herring and eel fisheries at Emden used to employ about 1,500 hands, taking 13,000 tons annually; but the produce at present is not one-third part of the former amount.

Mining.—This is the most extensive branch of Hanoverian industry. Mr. Petre, British secretary of Legation, in a report dated Hanover, January 1862 (Reports, No. V, p. 233), gives an interesting account of the state of this industry. He says:—'The mineral wealth of this country, which is considerable, lies, with the exception of coal, for the most part in the mountainous districts, thickly clothed with forests, which constitute the Hanoverian portion of the Harz, and in that part of what is called the Lower Harz which is held in joint proprietorship by Hanover and Brunswick. The mines, foundries, and salt-works, &c. in these districts, with the exception of a few proprietary claims vested in individuals, are the property of the state, and are exclusively worked under its immediate direction, represented by the ministry of finance. The revenue derived by the state from these sources, as it figured in the budget, amounted for the year 1859–60 to 2,889,580 dollars.

'The entire population of the Harz, about 82,000, is connected, directly or indirectly, with the mining industry, and depends wholly upon government employment for even bare subsistence. Corn is supplied to the population from the government magazines established for that purpose, and sold, somewhat on the principle of the "Caisses des Boulangers" at Paris, at a moderate and uniform price. Until the year 1848, the inhabitants of the Harz were exempted from contributing any share of the public burdens, either in the shape of taxes or military service, and even their subsequent liability to taxation has been no real gain to the public treasury, as it was necessary to make good the loss to them by increase of wages—wages paid by the state.

'This anomalous condition of a large labouring population engaged in the most important branch of Hanoverian industry has long since given rise to strictures upon the policy of the government, which perpetuates a system no longer suited to the times, a system intrinsically wrong on economical grounds, and one which, in a country so richly endowed with mineral wealth, stifles all incentive to individual enterprise.

'The opponents of the present system argue, apart from the general objections which exist to a government entangling itself in industrial undertakings, that the departure in the present instance from sound economical laws involves an evident loss to the general wealth of the country, inasmuch as there is little doubt but that the mines, foundries, and forests of the Harz would be infinitely more productive if, instead of being worked as they are now by the state, they were leased to companies or individuals. Such a course, moreover, would develope private enterprise, by which the state must eventually profit. On the other hand, the advocates of the present tutelary system say in its defence, that the state has inherited a large mining population, living under anomalous conditions, and that an application of the ordinary economical laws would in this instance be both unwise and cruel.

'The Harz possesses no agriculture, and produces no food for its inhabitants; they are dependent, and have been for centuries, on their labour in the mine, the foundry, and the forest, for their bare sustenance. Any disturbance, therefore, in the regular and constant demand for labour which is secured to them by the state would result in misery and starvation. They must either find constant work, starve, or emigrate. Whilst many of the mines are worked at a large profit, others yield but little or none, and some are even worked at a loss. Were the mining industry of the Harz to be given over to private speculation, the two latter classes of mines would be closed altogether, and all the hands connected with these thrown permanently out of employ, whereas the profitable mines would be worked at a rate which in the course of a few years might leave them exhausted and valueless. Whenever that contingency should arrive, the whole population of the Harz would be thrown for subsistence upon the state. Whatever may be reasonably urged against the policy of the usurpation by the state of what legitimately belongs to private enterprise, it must be allowed that any change to a better system than the one which has been so long interwoven with the social and economical condition of the Harz, ought to be a gradual one. Looking at the question from a financial point of view, it is very doubtful whether the revenue which figures in the budget as derived from the mines, &c. of the Harz is any but a nominal one. The salaries of mining officials, and all the miscellaneous outlay connected with the mining administration, added to the value of the fuel consumed in the smelting houses and foundries, which is supplied by the government forests, must be subtracted from the nominal gain.

Trade and Manufactures.—Hanover, though furnished, by its mineral wealth and navigable rivers, with means for carrying on a considerable commerce, holds a very low station among the trading countries of Europe. Its inhab. have little enterprise or ardour for business, and even that which they might exert is effectually checked by restrictions. The manufacture of linen is, perhaps, more extensive than any other. Spinning and weaving form the great in-door employment of the rural pop., and large quantities both of yarn and thread are the work of private hands. In Osnabruck, for instance, where large crops of flax are raised, no less than 1,375,000 skeins (456,750 lbs.) are spun annually by the farmers' and peasants' families. The number of professional weavers is nearly 5,000, using 7,200 looms, and the linen cloths produced by them are known in the markets by the name of Osnaburgs and white rolls, there being different qualities of each. The hempen cloths are known as Tecklenburgs, being put up in pieces of 100 double ells (125 yds.). These cloths, when bleached and ready for sale, are taken to the various *Lege-Anstalten*, or cloth-marts of Hanover (chiefly in the district of Osnabruck), where, after being measured, stamped, and valued, they are bought, chiefly by Bremen and Hamburg merchants, who export them to England, Spain, and Portugal, N. America, and the W. Indies. Prior to her connection with the Zollverein, Hanover was destitute of any manufacturing industry but that here enumerated. However, since 1851, with the Zollverein for a home market, and under the influence of its protective tariff, a manufacturing industry has sprung up and prospers. Some new iron works, engine and machine manufactories, cotton-spinning factories, chemical works, india-rubber, gutta percha, and cigar manufactories, testify to the rising industry and trade of the country; whilst the large sums of money which have been expended by Government at Harburg and upon the new part of Geestemünde, at the mouth of the Weser, show the importance attached to their development. The subjoined table gives the number and tonnage (in lasts) of vessels entered and cleared from ports in Hanover, from various countries, distinguishing tonnage with cargoes, in the year 1862:—

		1862		
Countries	Vessels	Tonnage, in Lasts of 2 Tons	
		Total	With Cargoes
ENTERED.			
Russia	48	7,658	2,559
Norway	853	11,978	11,461
Denmark	73	1,995	1,726
Hamburg	169	8,546	2,221
Bremen	812	8,190	4,178
Holland	493	12,618	11,223
Great Britain	996	66,774	64,991
France	10	1,160	1,108
Naples	7	765	765
Prussia	123	11,988	11,165
Oldenburg	574	3,493	3,485
Hanover	2,764	81,428	21,160
Other Countries	41	3,146	7,941
Total	6,885	168,684	141,671
CLEARED.			
Russia	48	8,951	347
Norway	376	11,360	103
Denmark	67	2,802	765
Hamburg	365	9,728	3,413
Bremen	744	8,911	7,541
Holland	443	14,752	12,845
Great Britain	1,178	73,383	87,609
France	69	4,771	3,162
Prussia	844	4,784	8,444
Oldenburg	2,323	89,614	16,965
Other Countries	8	381	716
Total	6,865	181,810	98,511

The chief imports of the kingdom are English manufactures (such as cotton and woollen, hardware and cutlery), colonial produce, wine and spirits. The returns of trade are included in those of the Zollverein. (See GERMANY.)

Coins, Weights, and Measures.—By the new mint regulations of 1834, the coinage has been fixed as follows:—

GOLD.

1 George-pistole	=	16s. 4d.	Eng.
1 William-pistole	=	8s. 2d.	„
1 Ducat	=	4s. 1d.	„

SILVER.

1 Thaler	= 30 Groschen	= 2s. 11¼d.	Eng.
½ Thaler	= 4 „	= ½d.	„

WEIGHT.

1 Zentner	= 46⅔ Kilog.	= 103 lbs.	Avoird.
1 Pfund	= 46·7 Grms.	= 1·03 lb.	„
1 Loth	= 14·6 „	= 1 lb. 4oz.	„

MEASURES—LENGTH.

1 Foot	= 12 Zoll	= 11½ Eng.	Inches
1 Ell	= 2 „	= ·639 „	Yard
1 Ruti	= 16 Feet	= 5·1 „	Yards
1 Mile	= 25,000 Feet	= 4·6 „	Miles

SURFACE.

1 Sq. Foot	= 91 Eng. sq.	Foot
1 Morgen	= 64 Eng.	Acre

Condition of the People.—Although the soil and climate of Hanover is unfavourable to agriculture, the condition of the peasantry in the hereditary provs. of the house of Brunswick has, until very lately, been such as to confine them almost exclusively to the cultivation of the soil: indeed, the trading resources offered by the rivers of the kingdom are only beginning to be appreciated by the people. In the sandy districts the prop. is necessarily scanty and indigent; in the better soils of Hildesheim, Göttingen, and Grubenhagen, the peasants are in a comfortable condition. The most prosperous districts are E. Friesland, and the rich lands along the Elbe, where good agriculture, mixed with activity and enterprise in trade, serves to enrich the prop. The people are everywhere industrious and temperate, labouring, without illfeeling, for the smallest possible remuneration. They are mostly descendants of the ancient Saxons, and, as such, speak the Low German dialect, excepting the inhab. of the Harz, who came from Upper Germany. The nobility possess large privileges as regards the right of holding property and civil and criminal jurisdiction. Hanover was one of the last states of Germany—the last, except Mecklenburg—in which serfdom and legal torture were abolished. The jurisdiction of the nobility on their own estates was done away with, to a great extent, in 1831 and 1848, but remnants of it still exist. The feudal service of the agricultural population was abrogated in 1831, on the condition that the value of such service be paid to the owner of the land at the rate of 25 years' income. It being impossible, in many cases, for the labouring people to raise the necessary capital, the redemption has not been accomplished to more than onehalf in the course of thirty years. According to an official return, 72,863 allotments of land were redeemed from 1819 to 1860, at a price of 11,178,909 thalers, or 1,676,836l.

Government.—Before Prussia ceded Hanover to France, in 1801, the form of government was monarchical, and the various territories were subject to feudal lords. The peasants of the marsh-lands had more freedom, and in E. Friesland the constitution of the country was almost republican. In the territories of the princes of the empire, the representation of the people by estates, composed of the nobles, prelates, and deputies from the towns, served to check the power of the sovereign, as in other parts of Germany. In 1808, when Napoleon

erected the kingdom of Westphalia, the territories of Hanover, with the districts of Hildesheim and Osnabrück, formed a part of it, and the Code Napoleon took the place of the ancient laws, and a sham representative government was established. On the return of the legitimate sovereign to Hanover, in 1813, the French institutions were summarily abolished, and the old forms re-established; and in 1815 the estates, summoned upon the ancient footing, drew up the form of a new constitution, modelled on that of England and France, and substituting a uniform system of representation for the various representative forms which prevailed under the empire. The chief change that excited disapprobation arose from the arbitrary decision of the sovereign (George IV.), advised by Count Munster, that there should be two chambers instead of one, contrary to the proposal of the estates, and the universal custom of Germany. The respective rights of the sovereign and of the country to the crown land revenues were not clearly defined by this fundamental law; but the interests of the people were supposed to be sufficiently consulted by the institution of a national treasury, the commissioners of which, named for life, were *ex officio* members either of the upper or of the lower chamber.

This constitution, however, contained no properly defined statements respecting either the rights of the people, or the prerogatives of the crown; and as the new system of representation was not sufficiently consolidated to resist the encroachments of a monarch supported by powerful foreign influence the necessity of a more definite fundamental law, in which the rights of the citizens should at least be declared, was felt on all sides. This feeling led to the drawing up of the constitution of 1833, which differed in but few, though most essential, points from that of 1819. The principal points of difference were a fuller acknowledgment of the right of the chambers to control the budget, and to call the ministers to account for their conduct; the restriction of the king's expenditure, by a regulated civil list: and the reservation, for the use of the nation, of the surplus revenue of the crown demesnes. These modifications rendered the treasury, whose functions thus devolved upon the chambers, wholly unnecessary, and it was dissolved. The new fundamental law, after being discussed by both chambers, received the assent of William IV. in 1833, who, however, by the same act, modified 14 articles of the bill. New elections followed, and the new chambers were exhibiting their activity in reforming abuses, and introducing economy into the state disbursements, when the death of William IV. interrupted their proceedings. As the Salic law, excluding females from the succession to the throne, prevails in Hanover, William IV. was succeeded by his eldest surviving brother, Ernest, duke of Cumberland, in England. Immediately on taking the government, the new king declared the chambers dissolved; and, previously to their re-assembling, he abolished, by proclamation, the fundamental law which had been adopted under the reign of his predecessor, and in the most arbitrary manner, insulting alike his brother's memory and the whole country, declared the fundamental law of 1819 to be alone valid. Under the last-named law, he summoned a fresh parliament; but he found the spirit of the nation aroused and indignant; for not only the courts of law, but the highest legal authority in Germany, and several faculties of universities, declared his proceedings illegal; many towns refused to send representatives to the parliament, and those which met signed a memorable protest, declaring their opinion that the fundamental law of 1833 was as still the law of the land. As the chambers could not be convened, for decency's sake, they were declared dissolved.

The present Constitution of Hanover is embodied in the 'Landesverfassung's Gesetz' of July 31, 1840, with modifications introduced April 10, Sept. 5, and Oct. 28, 1848; Aug. 1, 1855; Sept. 7, 1856; and March 24, 1857. According to these fundamental laws, the crown is hereditary in the male line of the house of Brunswick-Lüneburg, the sovereign coming of age at eighteen. The whole legislative and executive power is vested in the king, and the representatives of the people can only give advice to the crown, or afford co-operation—'Mitwirkung.' These consultative and co-operative functions are vested in two bodies, an Upper and a Lower Chamber. The former consists of the princes of the royal house; the heads of five families of the upper nobility of the kingdom; the hereditary court marshal; four members nominated by the king; thirty-three deputies of the largest landed proprietors; ten deputies of chapters and colleges; ten deputies of commercial bodies, and four deputies of inns of court. One-half of the elected members of the Upper House quit their seats every three years, to be replaced by deputies nominated in new elections. The Lower House consists of two members nominated by the king, who must be ministers; of thirty-eight deputies of towns, and of forty-four of country districts. The members are elected only for the term of one session, which, however, by prorogation, may extend over several years. Consultative functions are assigned, besides, to seven provincial diets, representing—1, the principality of Kalenberg and Göttingen; 2, the principality of Lüneburg; 3, the counties of Hoya and Diepholz; 4, the duchy of Bremen and Verden; 5, the principality of Osnabrück; 6, the principality of Hildesheim; and 7, the principality of East Friesland. Three provincial diets meet every three years, and in them the ultra-conservative element is very largely represented.

The executive power is entirely and unreservedly in the hands of the sovereign, acting through irresponsible ministers. For facilities of administration, the ministry is divided into six departments, the limits between which, however, are not kept up with strictness. The supreme court of justice is at Celle, and under it are nine chanceries or district courts, besides the magistracies of the towns, and the manorial and minor royal courts, as primary tribunals.

Religious matters are directed by Calvinist consistories at Hanover, Stade, Aurich, and Northorn, with the subordinate consistories of Hadeln and Neustadt; the Lutheran consistory at Osnabrück, and the Roman Catholic consistory of the same sev., which is alternately filled by a Roman Catholic and by a secularised Protestant bishop; lastly, the bishop and consistory of Hildesheim, for the Roman Catholic inhab. of that district. Education has been much attended to in Hanover. Public education is placed under the direction of a superior council for 'Unterricht's Angelegenheiten.' In the year 1861 there were 4,781 primary schools, besides numerous secondary, elementary, and industrial schools. The highest seat of learning is the university of Göttingen, established in 1734 by King George II., and re-chartered in 1836 as the 'Academia Georgia Augusta.'

Finances.—The budget period embraces a term of two years. In the revenue account beginning July 1, 1862, and ending June 30, 1864, the public income amounted to 39,583,000 thalers, or 5,517,450l., and the expenditure to 40,023,000 thalers, or 5,553,450l., leaving a deficit of 440,000

thalers, or about 86,000l. The expenditure for the financial year 1864–65 was calculated at 21,486,040 thalers, or 3,150,906l., and the expenditure for the year 1865–66 at 24,745,150 thalers, or 3,111,774l., giving a total for the two years of 41,752,190 thalers, or 6,262,774l., or an increase of nearly a million sterling over the preceding financial period.

The revenue and expenditure for the financial year 1863–64 were made up of the following items:—

INCOME FOR THE YEAR 1863–64.

	Thalers
Produce of Public Domains	1,991,498
Taxes and Custom Duties	7,440,940
Mines and Forests in the Upper Harz	2,797,089
Mines in the Lower Harz	397,910
Coal Mines	841,376
Salim and other Works	87,691
Shipping Dues	891,000
Post-office	1,079,000
State Railways and Telegraphs	4,950,000
Tolls on Roads and Bridges	190,000
Profit on Public Lotteries	60,000
Miscellaneous Items	851,412
Total	19,837,566
Or	£2,945,176

EXPENDITURE FOR THE YEAR 1863–64.

	Thalers
Ministry of State	725,497
Chamber of Representatives and Provincial Diets	64,189
Ministry of Foreign Affairs	121,700
of War	2,676,500
of Justice	1,544,320
of Education and Ecclesiastical Affairs	935,790
of the Interior	5,757,907
of Commerce	37,470
of Finance	5,941,807
Salaries and Pensions	5,301,314
Miscellaneous and Extraordinary Expenses	1,177,078
Total	19,656,330
Or	£2,974,343

The changes of the government of Hanover have necessarily, and in the most important degree, affected its finances. The re-establishment of the ancient order of things, in 1815, brought upon the country the whole mass of abuses belonging to a past age, which had been abolished by the French. Amongst the most obnoxious was the claim of the nobles to exemption from the land-tax; and this, as well as many other points, had to be arranged by the estates assembled under the constitution of 1819. Between 1821–26, a measurement and valuation of the country and its soil was made; and the amount of annual produce, after deducting expenses, being taxed at 10½ per cent., was calculated to yield 1,810,000 dolls.; but in this loose estimate, the values undoubtedly fell much below the reality. An indemnity was, at the same time, granted to the nobles, in lieu of exemption, to the amount of 1 per cent. on the revenue taxed. This charge appeared in the budget of 1826–27, and amounted to 65,000 dolls. The revenues claiming exemption amounted, consequently, to 6,500,000 dolls, nearly equalling the amount of taxable property belonging to peasants and burghers, and which, in 1819, was found (exclusive of E. Friesland) to amount to 6,899,717 dolls. Thus, half the nation was obliged to purchase justice from the other half, after the re-establishment of the so-called constitution of 1819 had been granted. The other direct taxes are the house-tax, which is 4 per cent. on the appraised rent; the personal tax, rated in 6 classes; an income-tax, which likewise includes all salaries, and the rate of which is ½ per cent. below 100 dolls., rising to 2 per cent. above 2,000 dolls. annual income; and lastly, the industry-tax, which is paid by all tradesmen, in 7 classes, the lowest paying ½ doll., the highest 80 dolls. The indirect taxes include the customs, the tax on spirits, beer, &c., the monopoly of the sale of salt, the stamp and legacy duties, besides duties levied on the grinding of corn, and unslaughtered beasts.

The published budget does not include the civil list of the king; nor are the other expenses of the court and royal family accounted for to the chambers. The whole of this expenditure is drawn from vast domains claimed to be the private property of the royal house, but not admitted to be such by the decisions of former parliaments. Numerous debates of the national representatives at 1848 and at subsequent periods, have not been able to settle the so-called question of 'Ausscheidung des Kronguts.'

The public debt of Hanover has been increasing for many years, chiefly through the establishment of a network of state railways. On January 1, 1862, the debt amounted to—

Old Debt	18,721,760 thalers,	or £2,856,264
Railway Debt	37,672,075 —	or 4,605,640
Total	56,244,835 thalers,	or £8,451,714

The gross produce of the railways in the financial period 1860–61 amounted to 5,115,492 thalers, and the expenses to 3,111,868 thalers, leaving a net income of 1,973,724 thalers, equivalent to 5·96 per cent. This, however, as will be seen, varies considerably from the figures of the official budget above given, in which the gross income, including state telegraphs, is set down at a considerably lower rate. The expenditure connected with the railway traffic is enumerated under the department of the minister of the interior. The telegraph lines of the state, erected at a cost of 725,849 thalers, according to a return made July 1, 1862, give a net income of 6·78 per cent, exclusive of the free despatches of the government authorities and other public bodies.

Previous to the separation of the crowns of Great Britain and Hanover, one-half of the public income was derived from the state domains, and the contributions of the tax-payers amounted to scarcely one-fifth of the present sum. The following was the budget for the year 1834, three years before the accession of the Duke of Cumberland to the throne of Hanover:—

Income for 1834:—
From Domains   3,170,632 thalers
Taxation, &c.   3,446,262 —

Total   6,516,894 thalers, or £866,283
Expenditure for 1834:—6,373,564 — or £846,284

Comparing the income from the domains in the period 1861–2, with the produce of 1834, the sums drawn at present for the civil list and similar expenses may be closely estimated. Exclusive of these sums, the public expenditure is seen to have risen from six and a half million thalers to very nearly twenty millions, or from 966,533l. to 2,978,753l.

The question of the income of the kings of Hanover has never been satisfactorily settled. Since the death of King William IV. of Great Britain, and the accession of Ernest Augustus, duke of Cumberland, to the throne of Hanover, the states and the sovereign have been in conflict on this subject. The constitution of 1833 settled a civil list of 500,000 thalers, or 75,000l., upon the king; but Ernest Augustus declared this sum to be wholly insufficient, and his demands for the possession of the state domains not being acceded to, he overthrew the constitution, chiefly on this account. From 1841 to 1848 the royal family

enjoyed the whole province of the crown property; but in the last-named year the king was compelled to give up this source of income, and to accept the grant of the civil list of 500,000 thalers as sole income. In 1848, however, the constitution was once more overthrown, and by a royal decree part of the state property was assigned to the king's use; besides the interest of a sum of 500,000l. invested by the Hanoverian government in English stocks, in the years 1764 and 1790, and that of a so-called 'Schatullenkapital' of 2,400,000 thalers, formed of the accumulated excess of state income over expenditure during a period of 40 years. At present the income of the sovereign of Hanover, as far as it is known, amounts to about 830,000 thalers, or 125,000l.

Armed Force.—The army of the kingdom is formed partly by conscription and partly by enlistment, the former supplying any insufficiency of the latter mode of raising soldiers. All citizens above 20 are liable to be drawn for conscription; but, as a rule, only a very small percentage are called up for active service in the infantry; a great portion of these troops, and nearly the whole of the cavalry and artillery, bring forward of volunteers. The cavalry, especially, is a branch of service much sought after by the sons of peasants and small farmers, on account of the advantages connected therewith. The privates in these regiments, as soon as the short term of drill and first practice is passed, are sent home on furlough, being allowed to take their uniforms and their horses. They must keep their horses partly at their own expense; but they may use them in agricultural and other labour, taking due care of the health of the animals under their charge. The term of service is seven years in the infantry, and ten years in the cavalry; but about three-fourths of this period may be spent on furlough, interrupted only by a short annual practice of arms.

On July 1, 1862, the army consisted of:—

	Men	Troops of Reserve
8 Regiments of Infantry, numbering	17,304 with 7,640	
3 Brigades of Cavalry	2,749	500
3 Battalions of Artillery	2,871	..
3 Companies of Engineers	655	..
Staff	40	..
Total	23,614	8,144

There are 10 garrison towns, a cannon foundry at Hanover, and a manufactory for small arms at Herzberg.

History.—The kingdom of Hanover is formed out of the duchies formerly possessed by several families of the junior branch of the house of Brunswick. The reigning family derives its origin from the union of the Marquis d'Este, in the eleventh century, with a wealthy princess of Bavaria, the heir of which received the surname Guelph, from his maternal ancestors, and inherited the dukedom of Bavaria. Henry the Proud, third in descent from him last mentioned, married Gertrude, the ruling princess of Brunswick: their son, well known in the history of the crusades as Henry the Lion (born 1129), was the first Guelph duke of Brunswick. He married a daughter of Henry II., king of England; and from this marriage both the houses of Brunswick and Lüneburg are descended. The history of Hanover for the two centuries preceding the Lutheran reformation presents little interest, except in the connection of its princes with the wars of the Guelphs and Ghibellines, in the latter end of the fourteenth century; little or nothing is known of its internal history. The Reformation numbered the princes of Brunswick among its most zealous supporters, and their subjects, during the thirty years' war, warmly seconded their anti-papal efforts. Ernest of Zell, the reigning duke, was one of the most eloquent defenders of Luther at the diet of Worms. His endeavours to improve the people by establishing clerical and general schools, when learning was esteemed only by the few, show him to have been a man of enlightened views. His grandson, Ernest Augustus, married Sophia, a grand-daughter of James I. of England (by his daughter Elizabeth, the wife of the elector-palatine); and on this marriage was founded the claim of the elder branch of the house of Brunswick to the English crown, acknowledged by parliament in 1701. George Louis was the issue of this marriage, and became king of England in 1714; from which time till 1837, year of the death of William IV., England and Hanover had the same sovereigns. The Salic law in 1837 conferred the Hanoverian crown on Ernest, duke of Cumberland, fifth, but eldest surviving son of George III. During the reigns of George I. and II. the territory of the electors of Hanover was increased by the conquest and purchase of many adjoining districts; Bremervörden and Wildeshausen in 1719 and the Hadeln-land in 1731. George III. added Hohnstein and the bishopric of Osnabrück, which, by the treaty of Westphalia, was held by his house as a secularized bishopric alternately with a Rom. Catholic prelate. In 1801 Prussia took possession of Hanover, but ceded it in the same year to the French, who constituted it a part of the kingdom of Westphalia, established in 1806. At the peace of 1815 the king of Great Britain reclaimed his dominions, which were much enlarged by the stipulations of the treaty of Vienna, and formed into a kingdom. On the definitive settlement of the kingdom, the district of Lauenburg was ceded by Hanover, which obtained in return the bishopric of Hildesheim, the principality of East Friesland, the districts of Lingen and Harlingen. In consequence of a family treaty dating back to the seventeenth century, ratified by the German diet, and renewed between the houses of Hanover and Brunswick on March 5, 1862, it is settled that the crown of Hanover, in the event of the extinction of the male line, shall fall to the ducal house of Brunswick, and vice versa. The present Duke of Brunswick having no male heirs, it is probable that this treaty will have to be executed before long in favour of the royal family of Hanover.

HANOVER, a city of W. Germany, cap. of the above kingdom, on the Leine, a branch of the Weser, 44 m. S. Hamburg, 62 m. SE. Bremen, 85 m. W. Brunswick, on the railway from Brunswick to Bremen. Pop. 71,170 in 1861. The city is built in an extensive sandy plain, and is divided by the river (over which are several bridges) into an old and new town, each of which is governed by a separate magistrate. The old town, on the right bank, has crooked and narrow streets, and is ill-built and dirty: the streets of the new town are more regular, and are lined with handsome houses, particularly George Street and Frederick Street, opening on a fine esplanade; the latter is adorned with the handsome monumental rotunda of Leibnitz, and the column, 156 ft. high, sacred to the memory of the Hanoverians who fell in the battle of Waterloo. The chief public buildings are the royal palace, of good exterior architecture, and splendidly fitted up within, especially the Ritter-saal, or knights' hall; the opera-house attached to the palace; the viceroy's palace; the house of assembly of the states (Landständehaus); the mint; the arsenal; the Gewerb-schule (trade school); the royal stables, where the well known breed of black and cream-coloured Hanoverian horses is kept; and the town-hall and record-office, containing a library of 80,000 printed books, besides about 2,000 valuable MSS., chiefly given by Leibnitz, who was a great benefactor to this town. Besides this, there are seven other public libraries, attached to various national establishments. There are 7 churches, 4 Lutheran, 2 Calvinist, and 1 Roman Catholic: of these the handsomest are the court and city church in the new town, and the Schloss-kirche, which contains the remains of the electress Sophia and her son George I., king of England. Outside the town are two suburbs, Linden and Gartengemeinde, in the latter of which are upwards of 500 houses with gardens. About ½ m. distant is Mount Brillant, the king's country residence, and formerly the seat of Count Walmoden, who enriched it with a gallery of fine pictures. About 1 m. distant is the old palace of Herrnhausen, once the favourite residence of George I. and George II.: it is heavy and tasteless, and appears to be going to decay. The gardens, which are laid out in the old French style, formerly contained a fine collection of rare plants; but they were dispersed during the late war. Hanover has several establishments for education, among which are the Georgianum, founded in 1776, for educating 40 sons of the nobility free of expense, the lyceum, the normal school (the earliest of its kind, founded in 1751), several elementary schools, and a girls' school of industry. Among the charitable institutions are a large almshouse, an orphan asylum, and several hospitals, one of which has been only lately erected. There are also a Bible Society, founded in 1804, a Society of Natural History, an Hist. Society, an Art Union, which annually exhibits specimens of Hanoverian artists, and a trade union. The manufactures are of trifling importance. The transit trade with Bremen and the interior of Germany is very considerable: there is an exchange, a chamber of commerce, and a Berghauptbury, or market for mining produce. Commercial activity, however, prevails more among the Dutch and foreign German merchants settled in the town, than amongst the Hanoverians. Some of the bankers are considerable capitalists. The town is not considered healthy: N. and E. winds are prevalent, and much rain falls. Longevity is said to be rare.

The foundation of Hanover, though attributed to the eleventh century, is most probably of still earlier date. In 1803 it is mentioned as having some trade in cloth, skins, and salt. Little more of it is recorded till 1566, when its inhabitants distinguished themselves by their zeal for the Reformation. It escaped the devastations of the thirty years' war, and even refused admission to the victorious troops of Tilly in 1625. The old royal palace was built early in the 17th century, and in 1641 it became the residence of Duke Christian Louis, since which it has always been the capital of the electorate and kingdom, and has made great advances in size and splendour. The ramparts being found useless as a means of defence, were in 1760 converted into a handsome esplanade, and planted with trees.

HARBOROUGH (MARKET), a market town and chapelry of England, par. Gt. Bowden, co. Leicester, hund. Gartree, on the S. bank of the Welland, which divides it from Northamptonshire 14 m. SE. Leicester, and 81½ m. N. London by Midland railway. Pop. 2,303 in 1861. The town consists of a well-built street, crossed by several others of inferior character; and near the middle is a handsome town-hall, with shops below, and a justice-room above, in which the county magistrates transact their business. The church is fine and spacious, and its octangular spire is one of the most elegant in England. The dissenters have 5 places of worship, attached to which, as well as to the church, are Sunday schools, giving instruction altogether to about 500 children. Considerable trade takes place on the market-days and at the October fairs; which, not less now than in the time of Camden, are famous for the show of beasts. Silk and shalloon weaving and the manufacture of carpets are carried on here, but not extensively. Market-Harborough is one of the polling-places for the S. division of the co. and is the chief town of a poor law union, comprising 41 pars. or townships. Markets on Tuesday; fairs Jan. 6, Feb. 16, April 29, and July 31, Oct. 19 and 8 following days, for cattle, leather, cheese, &c. Other fairs are held on the Tuesdays after March 2, after Midlent Sunday, and before Nov. 22 and Dec. 8.

HARBURG, a town of Germany, kingdom of Hanover, landt, and 23 m. NW. Lüneburg, on the Elbe, at the influx of the Seeve, and on the Hanover and Brunswick railway, 4½ m. S. Hamburg. Pop. 14,079 in 1861. The town has a citadel with drawbridges, and a custom-house, gunpowder mills, sugar refinery, manufactures of woollens, linens, hosiery, and a flourishing transit trade.

HARLINGEN, a sea-port town of Holland, prov. Friesland, on the Vliestroom, or entrance to the Zuyder Zee, opposite the Texel, and at the mouth of the canal of Leewarden, 15 m. W. by N. that town, at the terminus of the Northern railway of Holland. Pop. 9,772 in 1861. The town is fortified, and is strong by its position, the surrounding country being readily laid under water. Streets regular, well built, clean, and intersected with canals bordered with trees. Chief edifices the Admiralty, a large par, church, and the town-hall. It has a good harbour; but the entrance to it is blocked up with sand-banks, so as not to admit large vessels. It has manufactures of sailcloth, salt, bollands, paper, bricks, and lime, with building docks, and a brisk trade in corn, butter, cheese, flax, hemp, glue, pitch, and tar. It is the seat of the naval office for the prov., and suffered severely from a violent storm in 1825.

HARROW-ON-THE-HILL, a village and par. of England, co. Middlesex, hund. Gore, 10 m. NW. by W. London by road, and 11½ m. by London and North Western railway. Pop. of par. 5,525 in 1861. Area of par. 9,070 acres. The hill on which the village stands rises singly out of an extensive and fertile vale; it is considerably depressed in the centre, but has two very conspicuous eminences at the extremes. On the more N. of these stands the church, with its tower and lofty steeple, a prominent feature throughout Middlesex, and some of the adjoining counties. Part of this building is Norman, belonging to the 11th century; but the main fabric, with the tower, belongs to the 14th century. Immediately below the church lies the village, chiefly consisting of one street running down the slope of the hill. The best houses are occupied either by assistant-masters, or other teachers, who accommodate the scholars attending the free school, to which Harrow is wholly indebted for its celebrity. This school was founded in 1571, by Mr. John Lyon, a wealthy yeoman of the neighbouring hamlet of Preston, and received a royal charter, by the terms of which the management of the property and the appointment of the master were committed to six trustees as a body corporate. The school-buildings are of brick, and have no claim to particular mention. The head master's house has a Gothic porch, and is a fine old mansion. The primary object of this establishment was the gratuitous instruction of the poor children of Harrow, without limitation of

number; but the founder expressly directs 'that the master may receive, over and above the youth belonging to the par., as many foreigners as can be well taught and accommodated, for such stipends and wages as be can get, so that be take pains with all indifferently, as well of the par. as foreigners, as well of poor as of rich.' This liberality of the founder, and the judicious choice by the trustees of able and learned men as its masters, have chiefly conduced to its present very high reputation as a school for the English aristocracy; but, at the same time, there can be no doubt that the founder's intentions, as respects the poor of the par. itself, have been wholly frustrated. A classical education is quite unsuitable to the pop. of a village, and hence the school has been little need of late years by the parishioners. A petition of the latter to the Court of Chancery, in 1810, for the reformation of these abuses was unsuccessful, (See Vesey's Chancery Reports, xvii. 498.) The revenue strictly applicable to the school amount to nearly 2000l. a year, in the hands of trustees, usually noblemen or gentlemen living in or near the par. The education furnished was exclusively classical till within the last 30 years, when Drs. Butler and Longley ventured to introduce a little modern history and arithmetic, neither of which, however, is considered at all important: beyond these trifling attempts at reform no deviation has been made from the beaten path of the old grammar-schools. The routine of grammar, classes, and school hours, very much resembles that pursued at Eton, owing, no doubt, to the appointment of several head-masters from that school; the Eton grammar is used; verse-making supersedes the more useful study of prose composition; learning-by-heart is a favourite employment; and the private-tuition system, the chief object of which seems to be to save the master's labour, and fill the tutor's pocket, prevails at Harrow no less than at Eton and Westminster. The masters originally were two only, the master and the usher or under-master, both of whom were permitted to take 'foreigners' as boarders; but as the school increased, further assistance became from time to time necessary, and there are now six assistant masters, paid either by the high or lower master, according to the school in which they teach; and besides these there is a mathematical teacher. All the masters receive boarders; but the head-master does not furnish tuition, and hence arises the difference in the terms; for at a tutor's house they amount to 130l., whereas at the head-master's they are little more than 100l. All, however, are compelled to procure tuition, which is a part of the system. At least 60l. a year must be added to complete the necessary annual expenses of boys educated at this school. The governors have given prizes for verses, and the late Sir R. Peel established a prize for Latin prose composition. The speech-days, on which these papers are read or recited, are the first Wednesdays in June and July. The University scholarships attached to Harrow-school are four, established by the founder, of 60 guineas each, either to Oxford or Cambridge, and two of the same value, founded by the late Mr. Sayer, to Caius College, Cambridge—all tenable for four years; they are gained by an impartial examination. The number of boys attending the school fluctuates at present between 350 and 450. Among the many public characters educated in this school may be mentioned Sir William Jones, Spencer Perceval, Dr. Parr, Lord Byron, Marquis of Hastings, and Sir Robert Peel. Harrow had formerly a weekly market, which is now decayed; but a pleasure fair is still held on the first Monday in Aug. Bentley

Priory, a fine seat belonging to the Marquis of Abercorn, is within this par.; it occupies the site of a monastery, dissolved at the Reformation.

HARROWGATE, a town of England, celebrated for its mineral waters, co. York, W. riding, wap. Claro, forming with Bilton a chapelry of the par. of Knaresborough, 178 m. N. London, 11 m. N. Leeds, and 20 m. W. by N. York, on the Midland railway. Pop. 4,737 in 1861. The town is divided into High and Low Harrowgate. High Harrowgate is built on an elevated plain, which 100 years ago was properly described by Smollett as 'a wild common, bare and bleak, without tree or shrub, or the least signs of cultivation.' At the close of last century, however, Lord Loughborough made large plantations; houses have since been built in different directions; and the situation is now extremely pleasant, commanding a most extensive view of the distant country, finely varied by towns, villages, fields, and woods. The cathedral of York is distinctly seen at the distance of 20 m., and the view W. is terminated by the mountains of Craven, and E. by the Hamilton Hills and Yorkshire wolds. The air is pure and bracing, and the climate dry and salubrious. Low Harrowgate is situated in a valley, and has many handsome stone buildings, erected either for hotels or private lodging-houses for visitors. An almost continuous series of these houses unites the upper and lower parts of the town. The church of High Harrowgate is a well-built structure, erected in 1749 by subscription; that in the lower village was built in 1824. There are besides two chapels for Independents, and one for Wesleyan Methodists. A bath hospital was erected in 1826, which has been subsequently enlarged; it accommodates about 40 patients, who have the benefit of the waters free of charge.

The springs of Harrowgate are both chalybeate and sulphureous. The chalybeate springs rise in both villages, the sulphur springs only in Low Harrowgate. The chalybeate waters are principally tonic and alterative, the sulphureous waters strongly purgative. The latter are also used externally in rheumatism and scorbutic cases. The wells are covered with elegant cupolas, and surrounded by promenades, for the accommodation of those who come to drink the waters. Races are held in summer on the high ground to the W., where also is a high tower or observatory, from the top of which is a very extensive prospect of the surrounding country.

HARTFORD, a town of the U. S., Connecticut, of which it is joint cap. with Newhaven, co. Hartford, on the W. bank of the Connecticut river, 50 m. from its mouth, and 82 m. NNE. Newhaven; lat. 41° 46′ N., long. 72° 50′ W. Pop. 29,150 in 1860. The town is advantageously situated, the river being navigable for sloops up to this point. It is generally well built, particularly the main street, and is connected with E. Hartford, on the other side of the river, by a bridge of six arches, 974 ft. long. It has a handsome state-house, three banks, including a branch of the U. S. bank, an arsenal, academy, museum, college, nine places of worship, and an asylum for deaf and dumb. The last named, the first institution of the kind established in America, was founded in 1817, and in 1819 was presented with a grant of 23,000 acres of land by congress; besides which it is possessed of other donations and sources of revenue. It is open to patients from the whole union, at a charge of only 115 dollars a year, and many are provided for and educated gratuitously. It occupies a large and commodious brick building, on an eminence about ½ m. W. of the city; is surrounded by grounds between seven

and eight acres in extent, and has attached to it some workshops, in which the male pupils are taught mechanical trades. A little N. of the town is an asylum for the insane, a spacious stone edifice, with extensive grounds. Washington Episcopal College, established 1826, is another of the public institutions at Hartford. It has a president, eight professors, generally from 80 to 100 students, and a library of 6,200 vols. Hartford is the seat of the state assembly for Connecticut, alternately with Newhaven. It has manufactures of leather, shoes, woollen and cotton goods, saddlery, brass-work, and carriages; many printing houses, a large inland trade, and daily communication with New York by steam-boats and stage-coaches. A railroad connects Hartford and Newhaven.

HARTLAND, a market town and par. of England, co. Devon, hund. same name, 44 m. WNW. Exeter, and 190 m. W. London. Area of par. 11,030 acres; pop. of do. 1,916 in 1861. The town is situated in a bleak district close to the borders of Cornwall, and 2 m. from the Bristol Channel, with which it is connected by a steep road that leads down to a quay lying under the cliffs, and much frequented by fishermen. The church, which stands on the cliffs, about a mile from the town, is a large building, and serves as a landmark to mariners. The inhabs. are employed in fishing and agriculture; the herring fishery on the coast is of some consequence, and the market is well attended. The town became a mar.-port by an act made in the reign of Elizabeth, and is governed by a portreeve. In a fine valley near it is Hartland Abbey, formerly a monastery of Black Canons, but now converted into a modern mansion. NW. of the town is Hartland Point, a very high cliff, forming the W. boundary of Bideford Bay, and near it is a ridge of rocks, on which the sea breaks very heavily. Markets on Sat.; fairs, Easter Wed. and Sept. 23, for cattle.

HARTLEPOOL, a munic. bor., par. and sea-port of England, co. Durham, ward Stockton, near the mouth of the Tees, 17 m. SE. Durham, 16 m. N. by E. Sunderland, and 2564 m. N. London by Great Northern railway. Pop. of the bor. of Hartlepool 12,245 in 1861, and of West Hartlepool 12,503. The town stands on a peninsula, connected with the mainland by a narrow neck at the N. end, which at high water assumes a crescent shape, stretching N. and SW., forming a natural harbour, secure from the E. wind. The cliffs towards the sea N. are bold and abrupt, and their summits command a magnificent view of the sea, and the coasts both of Durham and Yorkshire. The town, which occupies the SW. portion of the peninsula, has latterly been very much enlarged and improved. It has, in fact, increased with extraordinary rapidity, for the pop. in 1831 was only 1,250. This has been partly and principally a consequence of the facility afforded by the situation of Hartlepool for the formation of a harbour, and partly of its having been made a terminus of railways connecting it with Durham and the adjacent coal fields. A wet dock, about 20 acres in extent, has been formed within the harbour, and another wet dock has been constructed by a rival company about ½ m. SW. of the old dock on the W. side of the bay. In consequence of the accommodation thus afforded, Hartlepool has become a leading port for the shipment of coal. The total value of the exports amounted to 4,018,521l. in 1859; to 4,355,894l. in 1861; and to 1,543,715l. in 1863. There belonged to the port, on the 1st of Jan., 1864, six sailing vessels under 50, and 130 above 50 tons. To the port of West Hartlepool there belonged, besides nine sailing vessels under 50, and 55 above 50 tons, in addition to 17 steamers, the latter of a total burden of 6,916 tons. Hartlepool was formerly fortified, as the old Durham gate and the ruins of walls abundantly testify. The church stands on a rising ground at the E. end of Southgate, and appears to have been built at different periods. A free school was founded by John Crooken, in 1712, for the education of 30 poor children. The school-house was built in 1780. At no great distance from the town are two strongly fortified batteries, N. of which is a chalybeate spring. Fishing was formerly the chief occupation of the people, who were described as free, bourn, industrious, and much attached to their town. Hartlepool was governed by a mayor, aldermen, and common council, under two charters, granted by King John in 1709, and by Queen Elizabeth in 1593; but the power of the corporation was destroyed by the Municipal Reform Act in 1834. The local act by which the town is regulated is 63 Geo. III., c. 35. Markets on Sat.; fairs, May 14, Aug. 21, Oct. 9, and Nov. 27.

Hartlepool is a very old town, and, during the 13th and 14th centuries, was a place of considerable importance. In the reign of Edward III. it furnished five ships to the royal navy, and was the second town of the county palatine of Durham.

HARWICH, a market town, parl. bor., and sea-port of England, co. Essex, hund. Tendring, on a point of land at the N.E. extremity of the estuary of the Stour, 66 m. E.N.E. London, 94 m. N.E. Ipswich, on the Great Eastern railway. Pop. 5,070 in 1861. The bor. includes the parishes of St. Nicholas and Dover Court. Area, 2,060 acres. There are three principal streets, and several smaller; the houses are of brick, and the town is well paved, and lighted with gas. The church, a large brick structure, with stone buttresses and steeple, was erected in 1821, on the site of an older building. The grammar-school was founded in 1780 for 32 boys. The principal public buildings are the town-hall, gaol, and custom-house. The old gates and fortifications were demolished during the civil war, and there are very few traces of them. The harbour of Harwich is the best on the E. coast of England; the access to it is, however, a good deal encumbered with rocks, but ships properly navigated need apprehend no danger; there is water to float the largest men-of-war, and the harbour is at once capacious, safe, and commodious. It is said that 100 ships of war, and above 300 colliers, have been anchored here at the same moment. The excellence of the harbour, and its convenient situation, made Harwich be selected as the station for the old sailing packets carrying the mails for Hamburg and Helvoetsluys. The town is defended by a battery and by Landguard Fort, on the opposite side of the estuary. The entrance to the harbour is indicated by two lighthouses with fixed lights, and is well buoyed. The sea has made great encroachments on the peninsula on which Harwich is built; and the battery, which, when constructed about half a century ago, had a considerable space of ground between it and the sea, is now partially undermined.

On the 1st of January, 1864, there belonged to the port of Harwich 67 vessels under 50, and 54 above 50 tons. There were no steamers belonging to the port at this date; but steam communication between Harwich and Rotterdam has since been established by the Great Eastern railway company, and greatly contributed to the commercial prosperity of the town.

Under the Municipal Reform Act the bor. is go-

verned by a mayor, four aldermen, and twelve councillors. Harwich returned two mems. to the H. of C. in the reign of Edward III.; but the privilege was very soon withdrawn and not restored till the 12th of James I. The franchise was vested in the resident members of the corporation, and it was, in fact, a nomination bor., in the patronage of the existing government. Under the Reform Act it still returns two mems., and its limits continue unaltered. Registered electors, 356 in 1865. The boundaries of the municipal and parl. bor. are co-extensive, and include the parish.

The town is said to be of Roman origin, and in the time of the Saxons was used as a fortress. The earls of Norfolk were the lords of the manor, and through their agency its chief mun. and parl. privileges were originally obtained.

HARZ (*Silva Hercynia*, Tac.), a mountain-chain of Germany, on the S.W. frontier of Hanover, connected by low hills with the Thuringer-wald, a W. offset from the Fichtelgebirge, the great centre of the German mountain-system. (See GERMANY.) It extends farther N. than any other chain, and immediately at its foot commences the great plain which stretches N. to the Baltic and from the N. Sea to the Wolga. It is a mass of mountain-land rather than a succession of ridges, and has no summits so high as Snowdon in N. Wales; its length is about 60 m., and average breadth 24 m.; area, 8,150 sq. m. Mansfield and Nevers are considered as the limits of the Harz; and it is divided into two sections by the watershed of the Weser and Elbe, which takes a direction from S.S.W. to N.N.E., and cuts the range at the Brocken (3,489 ft.). The higher summits are N.W. of the Brocken, and this section is, therefore, called the Upper Harz. It contains the chief mineral wealth of the range, and its forests consist of pines and other resinous trees. Its chief summits are the Heinrichshöhe, 3,409 ft., and the Königsberg, 3,307 ft. The Lower Harz, which lies E. of the Brocken, is much less elevated, and its sides, covered with oaks, beeches, and other deciduous trees, are remarkable for beautiful scenery. The hills flanking its range, and beyond its strict limits, are called the Vor-harz. The geological composition of the Harz is granitic, overlaid by graywacke and clay-slate, in which the mineral wealth is wholly found. The Vor-harz is composed of the 50ths, or old red-sandstone formation. The mineral products of the Harz are considerable. (See HANOVER, pp. 450–3.)

HASLEMERE, a bor., market town, and par. of England, par. Chiddingfold, in the S.W. angle of co. Surrey, hund. Godalming, 40 m. S.W. London, and 17 m. N. Chichester, on the London and South Western railway. Pop. of par. 952 in 1861. The town, only partly paved, stands on the side of a steep hill, and consists of a wide main street, crossed by two others, at the intersection of which is an ancient-looking town-hall. The houses are generally old and ill built, interspersed here and there with handsome residences. The church is ancient, with a low square tower; the independents have a chapel; and there is a good national school. This place once possessed rather extensive manufactures of silk and crape; but these have disappeared; but it has still some large paper-mills about 1 m. distant. Its importance has greatly diminished since the alteration of the London and Portsmouth road, which withdrew from it the traffic incidental to a great thoroughfare. Markets (ill provided and thinly attended) on Tuesdays; fairs for cattle, May 13 and Sept. 26. This small and unimportant town sent two mems. to the H. of C. from the 27th of Elizabeth down to the passing of the Reform Act, by which it was disfranchised. The electors were the burgage-holders; but it was, in fact, a mere nomination bor. of the Earl of Lonsdale, the chief proprietor.

HASLINGDEN, a market town and chapelry of England, par. Whalley, co. Lancaster, hund. Blackburn, 160 m. NNW. London, and 7 m. SE. Blackburn, on the East Lancashire railway. Pop. 6,929 in 1861. The town is pleasantly situated on the slope and at the foot of a hill. Most part of the houses are of stone; and it has the appearance of industry and prosperity. The church is modern, with an old tower. The dissenters have several places of worship, and in the Sunday schools are taught about 1,700 children. A free school, having a scanty endowment for ten children, furnishes instruction to about fifty. The increase of the town (which in 1831 had doubled itself since 1801) is attributable to the introduction of the cotton manufacture, which now employs the bulk of the working classes almost to the exclusion of the woollen manufacture, which formerly was the staple of the town. Haslingden is the chief town of a poor law union, comprising eleven parishes. The surrounding country abounds in good building stone, and slate is quarried about 1 m. S. of the town.

HASSELT, a town of Belgium, prov. Limborg, cap. arrond., on the Demer, 14½ m. WNW. Maestricht, on the railway from Maestricht to Antwerp. Pop. 10,918 in 1846. The town is well built, and was surrounded with walls in 1282. It is the residence of the chief courts and civil authorities for the Belgian div. of the prov., and has several churches and hospitals, a college, prison, numerous distilleries, a large salt refinery, with other manufacturing establishments, and a considerable trade in spirits, tobacco, and madder, and two weekly markets.

HASTINGS, a cinque port, parl. bor., and town of England, co. Sussex, rape same name, 54 m. SSE. London, and 32 m. E. Brighton, on the South Eastern railway. Pop. of munic. bor. 22,857, and of parl. bor. 22,910 in 1861. Hastings is pleasantly situated in a vale, surrounded on every side, except towards the sea, by hills and cliffs, the latter of which shut E. of the town, close on the shore, those on the W. sloping more towards the interior; and it owes chiefly to its mild climate, consequent on this sheltered position, its high rank among the watering-places of the N. coast of England. Less than a century ago it consisted of two chief streets, lined with ancient-looking houses; but within the present century many handsome streets and squares have been built, for the accommodation of visitors, and the appearance of the beach has been much improved by the removal of some old tenements which obstructed the sea-view. The two par. churches are ancient structures; but there are three modern churches, among them an edifice in Pelham Crescent, erected at the expense of the Earl of Chichester. There are also places of worship for Wesleyan Methodists, Independents, and other dissenters. There is a handsome town-hall. A grammar-school, founded in 1619, is attended by upwards of 100 boys; and there is a free school for 70 boys and 80 girls, with an endowment for apprenticing them. The chief public buildings are the town-hall and custom-house; there are also extensive baths, well-assorted libraries, a handsome assembly-room, a theatre, a literary institution, and a savings' bank. Races were established in 1827. The suburbs are very beautiful, furnishing delightful drives and walks. Connected with Hastings in one continuous row of houses,

and forming its western suburb, is the village of St. Leonard's, built according to the plans of Mr. D. Burton, and comprising a fine church, a large market-place, and many handsome houses and villas, occupied during the season by people of property and fashion. There is a Rom. Catholic training college at St. Leonard's; also a nunnery. The trade of Hastings arose, from the charters, to have been once very extensive; and its port or road was anciently protected by a pier destroyed by a storm in the reign of Elizabeth, and not rebuilt. Considerable quantities of fish are taken, and sent to the London market; a good deal of boat-building is also carried on, and lime is extensively produced in the neighbourhood. The mun. gov. of the town, which was vested in a mayor and twelve other jurats, and regulated by the charter of the cinque ports (20 Charles II.), and by one peculiar to itself (80 Eliz.), is now, under the Mun. Reform Act, committed to a mayor, five other aldermen, and eighteen councillors, the town being divided into three wards. Petty and quarter sessions are held here, at the latter of which the recorder presides. Hastings has sent two mems. to the H. of C. since the 23rd of Edward III., the franchise till the Reform Act having been vested in all resident freemen (made so by birth or election) not receiving alms; the number of electors being small, it had for many years been a mere nomination bor., in the patronage of the gov. for the time being. The present parl. bor. comprises the town and port, the liberty of the Sluice, and a detached part of the par. of St. Leonard's. Reg. electors, 1,432 in 1865.

Hastings is a place of high antiquity, having already, in the time of Athelstan, attained such importance as to be made the residence of a mint-master. On the edge of the W. cliff are the walls of an ancient castle, apparently of great strength, and the traces of walls indicate the town to have been fortified. On a hill E. are banks and trenches, supposed to have been constructed by William the Norman during his contest with Harold II., which terminated the Saxon dynasty. Its subsequent history is closely connected with that of the cinque ports, among which it ranked first. The cinque ports, or trading towns, which were selected from their proximity to France, and early superiority in navigation, to assist in protecting the realm against invasion, were vested with chartered privileges from a very early period. The ports are, Hastings, Romney, Hythe, Dover, Sandwich, Winchelsea, and Rye. Seal was afterwards incorporated, and made subject in some particulars to Sandwich. In early times they furnished among them all the navy required by the state, and even after the formation of a national navy, were compelled to assist it with their vessels. In return for these services, which have long ceased to be rendered, these corporate towns, together with twenty-three others subordinate to them, enjoyed the privilege of exemption from service on county juries and in the militia, and the power of criminal and civil jurisdiction, even in capital cases, in courts peculiar, held under the authority of the lord warden. These exclusive privileges were suffered to continue, much to the injury of the community at large, and even of the towns themselves, till the Parl. and Mun. Reform Acts reduced them, with the reservation of the sessions-court and the exemption from serving on county juries, to the level of other towns.

HATFIELD, a town and par. of England, co. Hertford, hund. Broadwater, near the Lea, 18 m. NNW. London, and 7 m. E. St. Albans, on the Great Northern railway. Pop. of par. 3,871 in 1861. This place was granted in the 10th cen tury to the Abbey of Ely; and on the conversion of the latter into a bishopric the manor-house became a palace of the bishops, whence it has been called Bishops Hatfield, Queen Elizabeth, who had resided in the bishop's palace for some time previously to her accession to the throne, and was very much attached to the place, prevailed on the bishop of Ely to alienate it to the crown, in exchange for other property. In the succeeding reign, James I., exchanged the manor of Hatfield with his minister, Robert Cecil, earl of Salisbury, for the manor and park of Theobalds. Its new master erected the present magnificent quadrangular mansion, one of the finest specimens of the baronial buildings of that age. A few years since it was materially injured by fire; but it has been restored, with great taste, quite in the old style. The town is small, and unimportant; it has a handsome church, with an embattled tower.

HAVANNAH, or HAVANA (Span. *Habana*, 'the harbour'), a large and flourishing marit. and commercial city, the cap. of the isl. of Cuba, and, perhaps, next to New York, the greatest emporium in the W. hemisphere. It stands on the NW. coast of the island, and on the W. side of one of the finest harbours in the world; lat. 23° N. long. 82° 21' 48" W. The pop. of the city and suburbs amounted in 1791 to 44,337, in 1810 to 96,304, and in 1827 to 94,023, of whom 40,621 were whites, and 23,562 free mulattoes and blacks, the residue being slaves. According to a rough enumeration of the year 1861, the pop. of the city, including all its suburbs, amounted to 201,000.

From its position, which commands both inlets to the Gulf of Mexico, its great strength, and excellent harbour, the Havannah is, in a political point of view, by far the most important marit. station in the W. Indies. For a long period it engrossed almost the whole foreign trade of Cuba; but since the relaxation of the old colonial system, various ports (such, for instance, as that of Matanzas), that were hardly known 50 years ago, have become places of great commercial importance. The rapid extension of the commerce of the Havannah is, therefore, entirely to be ascribed to the freedom it now enjoys, and to the great increase of wealth and pop. in the city, and generally throughout the island. The port of Havannah is the finest in the W. Indies, and one of the best anywhere to be met with. The entrance is narrow, but the water is deep, without bar or obstruction of any sort, and within, it expands into a magnificent bay, capable of accommodating 1,000 large ships; vessels of the greatest draught of water coming close to the quays. The city lies along the entrance to and on the W. side of the bay; the suburb Regla is on the opposite side. The Morro and Punta castles, the former on the E., and the latter on the W. side of the entrance of the harbour, are strongly fortified, as is the entire city; the citadel is also a fortress of great strength; and fortifications have been erected on each of the neighbouring heights as command the city or port. The city-proper, which stands upon level ground, is about 2,100 yds. in length by 1,200 broad, and contains but a small portion of the total pop. It is separated on the W. by a ditch and glacis from its suburbs of Salud, Guadalupe, Jesu-Maria, Cerro, and Horcon. Within the walls, the streets are narrow, crooked, and mostly unpaved; but in the suburbs, particularly Salud, they are wider and better laid out. The Havannah was formerly very much exposed, in the autumn, to the ravages of the yellow fever, owing partly to

the filth of the city, the want of common sewers, and the contiguity of marshes; but of late years, the cleanliness and police of all parts of the town have been very materially improved, and fever is much less prevalent and fatal. The houses, within the walls, are all of stone; without, they are of various materials. The public edifices, such as the cathedral, government house, admiralty, arsenal, general post-office, and royal tobacco-factory, are less remarkable for beauty than solidity of construction. Besides the cathedral, which contains the ashes of Columbus, removed thither from St. Domingo in 1796, there are 9 par. churches, 6 others connected with hospitals and military orders, 3 chapels or hermitages, 11 convents, a university, 3 colleges, a botanical garden, anatomical museum and lecture-rooms, an academy of painting, a school of navigation, and above 70 ordinary schools for both sexes. The charitable institutions consist of the *Casa Real de Beneficencia*, a penitentiary or magdalen asylum, a foundling asylum, and 7 hospitals, one of which comprises a lunatic asylum. The *Casa Real* also has within its walls two other lunatic asylums, with about 180 patients, an hospital for the aged and infirm, and boys' and girls' schools. The revenues of this institution, derived from landed and household property, donations, subscriptions, government grants, taxes on the flour imported at the Havannah and Matanzas, on public billiard-tables, landing-places, a poll tax, and various other sources, amount to from 58,000 to 60,000 dollars a year, the whole of which sum is annually expended on objects of the charity. There are 3 theatres, an amphitheatre for bull-fights, and several handsome public promenades. The arsenal and dock yard are at the S. extremity of the city. In the latter, ships of the line, frigates, and war brigs and schooners have been built. The saw-mills there are turned by water from an aqueduct, which also supplies the shipping in the port.

At the village of Casa Blanca, on the opposite side of the harbour, there are also some wharfs and shipyards, at which vessels of all classes may be laid up, fitted out, or repaired. This village is notorious as the resort of the slavers frequenting the Havannah, at which port a considerable number of the slaves brought into Cuba are landed. (For accounts of the articles of import and export at the Havannah, the duties levied on Spanish and foreign trading vessels, &c., see CUBA.)

The Havannah is an episcopal see, the seat of the provincial government, and the residence of all the colonial authorities, except the judges of the supreme court of justice, which sits at Puerto Principe. The principal nations of Europe and America have consuls resident at this city. It has an extensive manufacture of cigars, for which it is widely celebrated; its other manufactures, of coarse woollens, straw hats, &c., are comparatively unimportant. This city was founded in 1511, by Diego Velasquez; it was taken by a French pirate in 1563; afterwards by the English, French, and buccaneers; and again by the English in 1762, by whom it was restored to Spain at the peace of 1763.

HAVERFORD-WEST (called by the Welsh *Hulfordd*, a parl. bor., market town, river-port, and co. of itself in S. Wales, locally in the co. Pembroke, of which it is the cap., on the Cleddy, near where it falls into a creek stretching from the N. side of Milford Haven, 255 m. W. by N. London, and 87½ m. by Great Western railway. Pop. 7,019 in 1861. The town lies, in a very picturesque manner, on the sides and at the bottom of very steep hills: the river Claddy passes through its E. part, terminating in the creek. It is paved and lighted with gas; but High Street and Market Street, however, notwithstanding the improvements in paving, are still dangerously steep. The handsomest of the churches is St. Mary's, a cathedral-like structure of pointed architecture, surmounted by a large square tower. St. Martin's is an extensive and lofty structure, apparently an appendage to the castle, and has a tower and spire. Outside the town, at the top of the hill, is St. Thomas's, said to have been built in 1725; and there is a low turreted church at Prendergast. There are several chapels for Methodists, Presbyterians, Baptists, and the Society of Friends. A charity school, for clothing and educating 74 boys and 12 girls, was founded in 1691; and a free grammar school was established in 1614, and endowed with lands for the gratuitous education of the sons of poor burgesses. The town-hall is a respectable building, but placed so as to obstruct the view of St. Mary's church. A market-house, built by the corporation, was opened in 1825. A modern gaol stands on the green, near St. Thomas's church. Overhanging the town is the ruined keep of an old castle; and within the precincts of an old priory of Black Canons, some ruins of which are yet standing. A dockyard and quays have been constructed for the convenience of the shipping. Vessels of 100 tons can come up to the town at spring tides; but at neaps, vessels much exceeding 30 tons cannot come up. Hard coal, for malting, is exported to the S. coast of England, and to London; also goods are brought by water; and about half a dozen timber ships unlade here in the year. Butter and oats are exported; but the most important native commodity is the cattle, a great quantity of which is sold for the English market.

Haverford-west was first chartered in the reign of Richard II.; but its governing charter, down to the passing of the Municipal Reform Act in 1845, was that granted in 7 James I. The bor. is now governed by a mayor, 8 other aldermen, and 12 councillors: corporation revenue in 1847, 815l. Haverford-west has sent 1 mem. to the H. of C. since the 17th of Henry VIII. Previously to the Reform Act, the right of voting was vested in the inhab. of the town and co. paying scot and lot, and in the burgesses, who became so by birth, servitude, or election. The Boundary Act enlarged the limits of the parl. bor., by adding to the old bor., or town and co. of Haverford-west, portions of the par. of Prendergast and Uzmaston: the towns of Fishguard and Narberth were thus also made contributory boroughs. Registered electors in the three boroughs, 832 in 1848. The assizes and quarter and petty sessions are held here. Markets on Tuesday and Saturday; fairs for horses and live stock, May 12, June 12, July 18, Sept. 23, Oct. 18. This town was anciently the cap. of the Flemish possessions in Pembrokeshire. Its castle was erected by Gilbert de Clare, first earl of Pembroke, in the 14th century.

HAVRE (LE) (formerly Havre-de-Grace), a fortified town, and the principal commercial port on the W. coast of France, dép. Seine Inférieure, cap. arrond., on the N. bank of the estuary of the Seine, at its mouth in the English Channel, 42 m. W. Rouen, and 109 WNW. Paris, on the terminus of the Paris-Rouen-Havre railway. Pop. 74,336 in 1861. The town is built on a low alluvial tract of ground formerly covered by the sea, and is divided in two unequal parts by its outer port and basin, which stretch into the town and insulate the quarter of St. Francis. A fine main street, the Rue de Paris, wide, clean, and lined with good

houses and numerous shops, completely traverses the town N. to S., from the Place de la Bourse, one of the quays, to the Ingouville gates: this is the chief seat of commercial activity; the other streets present nothing remarkable. There are nine quays, which, with the High Street, form the favourite promenades. The fortifications, begun by Louis XII., continued by many succeeding sovereigns, and perfected by Napoleon, are about 3½ m. in circuit, and consist of bastioned ramparts surrounded by trenches. The tower of Francis I., a heavy round edifice of freestone, built by that monarch, nearly 70 ft. in height, and 85 in diameter, guards the entrance to the harbour on one side, and a small battery, mounting six pieces of cannon, on the other. The citadel, constructed by Richelieu in 1564, comprises the barracks, military arsenal, and residence of the governor. Havre has few other public buildings worth notice; the chief are—the church of Notre Dame, a singular edifice of the 16th century, the marine arsenal, new theatre, commenced 1817, exchange, custom-house, entrepôt-général, royal tobacco-manufactory, and a public library with 15,000 vols. It has numerous public fountains, and is well supplied with water, conveyed by pipes from the vicinity.

The port, which is the best and most accessible on the coast, consists of 5 basins separated from each other, and from the outer port, by 4 locks, and capable of accommodating about 450 ships. A large body of water being retained by a sluice, and discharged at ebb tide, clears the entrance of the harbour, and prevents accumulations of filth. Two lighthouses, 60 feet high, 825 feet apart, and exhibiting powerful fixed lights, stand on Cape de la Hève, a promontory about 2 m. NNW. Havre, and 390 feet above the level of the sea; and there is also a brilliant harbour light at the entrance of the port, on the extremity of the western jetty. Havre has 2 roadsteads; the great, or outer, is about a league from the port, and the little, or inner roadstead, about half a league. They are separated by the sand bank called l'Éclat, between which and the bank called Les Hauts de la Rade, is the W. passage to the port. In the great road there are from 6 to 7½ fathoms water at ebb; and in the little, from 8 to 3½. Large ships always lie in the former. The rise of the tide is from 21 to 27 feet, and by taking advantage of it the largest class of merchantmen enter the port. The water in the harbour does not begin perceptibly to subside till about 3 hours after high water—a peculiarity ascribed to the current down the Seine, across the entrance to the harbour, being sufficiently powerful to dam up for a while the water in the latter. Large fleets, taking advantage of this circumstance, are able to leave the port in a single tide, and get to sea, even though the wind should be unfavourable. Havre being the sea-port of Paris, most of the colonial and other foreign products destined for its consumption are imported thither. The chief imports are cotton, sugar, coffee, rice, indigo, tobacco, hides, dyewoods, spices, drugs, timber, iron, tin, dried fish, grain, and flour. The chief exports are silk, woollen and cotton stuffs, lace, gloves, trinkets, perfumery, Burgundy, Champagne, and other wines, brandy, glass, furniture, books, and articles de Paris. Havre receives seven-tenths of the cotton imported into France; more than half the tobacco, and wood for cabinet work, half the potash and indigo, more than two-fifths of the rice and dye-woods, and more than a third part of the sugar and coffee. As respects cotton, Havre is to France what Liverpool is to England.

Most of the goods imported at Havre are destined for the internal consumption of France. The coasting trade has increased very largely of late years, as is proved by the great increase of French wines, soaps, and other produce imported at Paris from Havre, instead of being sent to the cap, by land. The coasting vessels in many cases transfer their cargoes to large barges, called chalands, which are towed by steam as far as Rouen, and by horses for the rest of the way to Paris.

The number of British vessels that arrived in the port in 1863 with cargoes, including 212 passenger steamers from London and Southampton, amounted to 1,111, against 1,026 in the year 1862. Of this number, 437 were laden with coal, against 400 similarly laden in 1862. Of vessels bearing the French flag, 4,326 (including the coasting trade) arrived in the port in 1863, against 4,941 in the year 1862. Of vessels bearing the flag of other nations, 254 arrived in 1863, against 316 in 1862.

Havre has manufactures of chemical products, furniture for the colonies, earthenware, starch, oil, and tobacco, besides good building docks, rope-walks, breweries, &c.; and many females are occupied with making lace.

On a bright immediately N. of Havre is its well built and pleasant suburb of Ingouville. In that village is the Hospice d'Havre, founded by Henry II., 1554, and removed to Ingouville in 1649, at which establishment it is estimated that about 120 sick persons, and upwards of 500 aged, orphan, or infirm, are annually provided for.

HAWICK, a bur. of barony, and eminent manufacturing town of Scotland, co. Roxburg, on level ground, on the banks of the Teviot, 45 m. SE. Edinburgh, and 43 m. N. by E. Carlisle, on the Edinburgh-Carlisle railway. Pop. 8,191 in 1861. A small mountain stream, called the Slitering, falls into the Teviot, towards the extremity of the town. The country round is mountainous and pastoral, except the narrow valley through which the two rivers flow. The town was originally confined to the bank of the Teviot, and to the parish of its own name, but its boundaries now extend to the opposite side of the river, in the parish of Wilton.

Hawick consists chiefly of a single street, ½ m. in length, which forms the line of the public road; but there are several suburban streets, of which the largest and the most elegant is the Crescent, built on the right bank of the river. The town, the houses of which are of stone and slated, has a substantial thriving appearance; and the transparent waters of the Teviot and Slitering flowing over a pebbly bed, with the mountains which so closely environ it, give it a high degree of picturesque beauty. The streets are paved, and lighted with gas. Being a border town, and consequently of old exposed to attacks from the English, the houses were anciently built with stone walls and vaulted below, without any door to the street, but having an archway, giving access to a court-yard behind, from which alone entrance to the house was obtained. Of these structures a few specimens yet remain. There are two bridges over the Teviot, and two over the Slitering, one of the latter being supposed to be of Roman origin. The only public buildings are the subscription rooms, the town-house, the parish church, with a small square spire, and several dissenting meeting-houses.

Hawick has establishments for the manufacture of thongs, gloves, candles, machinery for tanning of leather, and other branches; but the woollen manufacture is that for which the town is chiefly distinguished, a department of industry which owes its origin to the command of water-power which the Teviot and Slitering afford, and to the wool-growing district in the middle of which Hawick is situated. The manufacture of carpets was established in 1752; the inkle (a species of tape) manufacture in 1783, and that of cloth in 1787. But these have very generally given way to the manufacture of stockings and under-clothing, introduced in 1771. But comparatively trifling progress was made in the manufacture till the introduction of machinery, which took place about the beginning of this century, since which the business has been steadily advancing.

Hawick has been a bur. of barony from an early date. But its present charter was granted by William Douglas, of Drumlanrig, in 1537, and confirmed by Queen Mary, in 1545. The feudal superiority of the bur. descended to the barons of Buccleugh till 1747, when, all hereditary jurisdictions being abolished by act of parliament, the Duke of Buccleugh received 400l. in compensation for the regality. From its situation near the English border, Hawick was exposed to that continual hostility and commotion which for centuries distinguished that portion of the empire. It was burnt down in 1418. It suffered severely in 1544, when the whole district of Teviotdale was laid waste by the English. To prevent its occupation by the troops of the Earl of Surrey, in 1570, the inhabitants themselves tore off the thatch from the roofs of the houses, and set fire to it on the streets, by which, with the exception of the Black Tower, the whole town was completely consumed. The inhabs. of Hawick mustered strong in the battle of Flodden, and were there nearly extirpated; but the survivors succeeded in rescuing their standard, which is still carefully preserved.

There is an artificial mound of earth situated at the W. extremity of the town called 'the Mote,' used, in ancient times, for meetings both judicial and deliberative. Branxholm Castle, the ancient seat of the Scotts of Buccleuch, and celebrated in The Lay of the Last Minstrel, is situated within 2 m. of the town. Several eminent persons have been born in or connected with Hawick. Gavin Douglas, afterwards bishop of Dunkeld, and the translator of Virgil's Æneid, was rector of Hawick in 1496; Dr. Thomas Somerville, minister of Jedburgh, and author of a History of Queen Anne, and other works, was born in the burgh; the Rev. Mr. Young, author of Essays on Government, was a dissenting clergyman here; and Mr. Robert Wilson, author of the History of Hawick, a native of the burgh, died here in 1837.

HAYE (LA), a small town of France, dép. Indre-et-Loire, cap. cant., on the Creuse, 30 m. S. Tours. Pop. 1,620 in 1861. The town is worthy of notice as the native place of Descartes, born here on the 31st March, 1596. The house in which he first saw the light has been carefully preserved, and is the subject of an almost religious care and veneration. To distinguish it from other small places of the name, in the departments of Vosges and Eure, this town is often called La Haye Descartes.

HAYTI, or HAITI (Carib. the mountainous country), the original and now revived name of one of the W. India Islands, being, next to Cuba, the largest of the Greater Antilles. Columbus gave it the name of Hispaniola, and it was frequently also called St. Domingo, from the city of that name on its SE. coast. The French bestowed on it the deserved epithet of la Reine des Antilles. It lies between lat. 17° 40′ and 19° 54′ N., and long. 68° 20′ and 74° 35′ W.; having N. the Atlantic, E. the Mona Passage, separating it from Porto Rico, from which it is 76 m. distant, S. the Caribbean Sea, and W. the Windward Pasage, which lies between it and Cuba and Jamaica, its NW. point being 44 m., E. of the former, and its SW, 112 m., E. of the latter. Its shape is somewhat triangular, the apex directed eastward; but it has several considerable peninsulas and promontories, which render its outline very irregular. Greatest length, W. to E., about 400 m.; its breadth varies from 40 m., near its E. extremity, to 165 m., about its centre. The island is divided into two states, the first, the republic of Hayti, having an area of 658 geographical sq. m., with an estim. pop. of 700,000; and the latter, known as San Domingo, and, since 1861, a dependency of Spain, with an area of 810 geogr. sq. m., and an estim. pop. of 700,000.

Physical Geography.—The surface of Hayti is, as its name implies, generally mountainous; but there are some extensive plains, especially in the E. The mountain system is complicated, and it is difficult to give a clear idea of it without the aid of a map. A great mountain knot, the Cibao, occupies the centre of the country, from which two parallel chains, running E. and W., extend through the island in its entire length. The loftiest summits of the Cibao are considerably more than 6,000 ft. in height. In the SW. is an additional mountain chain, which stretches W. to the extremity of the long and narrow peninsula terminating in Cape Tiburon. Between this peninsula and the NW. promontory of the island is the spacious bay of Gonaive, including the island of the same name, and having at its head Port Republicain (or Port-au-Prince). Tortuga is opposite the NW. promontory. The shores of Hayti are in general bold, except on the E., where low and swampy lands prevail. They are almost every where surrounded by small uninhabited islands and dangerous reefs, but they have, notwithstanding, many excellent harbours, especially along the N. and W. coasts. The largest plain, called by the Spaniards Los Llanos, in the SE. extends along the coast for 80 m., with a breadth varying from 20 to 25 m. It is said to be well adapted to the culture of most tropical products, but has always remained chiefly of wide savannahs, used for pasture lands. N. of it, enclosed between two mountain ranges, is the more productive plain of Vega Reale, little inferior in size to the foregoing. In the W. half of the island are the large plains of Artibonite and the Cul-de-Sac. The last named, E. of Port-au-Prince, is from 30 to 40 m. long, by about 9 broad, and was formerly one entire magazine-garden, though now almost wholly waste. There are several plains of less extent. Hayti is in most parts profusely watered; its has numerous rivers, the largest being the Yaque, Yuna, Neive, and Artibonite, which disembogue on the N., E., S. and W. coasts. These are navigable for a great part of their course; they are generally deep, and two or three of them are, near their mouths, as wide as the Thames at Vauxhall. Three lakes of considerable size exist at no great distance from the N. coast of Henriquillo; the largest is about 60 m. in circuit, and has salt water, while the adjacent lake of Azuey is fresh.

The climate of the low lands is very unhealthy to Europeans; and Mackenzie says that 'the yellow fever would effectually secure the island, in case of external attack, if the policy of abandoning the coasts and destroying the towns were acted on.' (Notes on Haiti, vol. ii.) The excessive heats of the plains are, however, tempered by fresh sea breezes at night. The temperature, of course, decreases with the elevation, and in the mountains the cold is often piercing. The year, as elsewhere between the tropics, is divided between the wet and dry seasons. The change of

the seasons is accompanied by stormy weather; but hurricanes are not so frequent as in most of the other Antilles, nor are earthquakes common, though in 1770 a convulsion of that kind destroyed Port-au-Prince.

Little is known of the geology; a limestone somewhat analogous to that of Cuba, containing vestiges of marine shells, is a prevalent formation. The soil is almost universally a deep vegetable mould, the fertility of which is scarcely equalled. The mountains, even to their summits, are, according to Mackenzie, capable of cultivation. The greater part of the island is covered with dense forests of mahogany, iron-wood, logwood, cedars, and other large and useful trees, or an impenetrable underwood. The plantain, potato, vanilla, manioc, &c. are indigenous; as is the palmetto, or cabbage-tree. The latter is 'truly the prop of the E. Haytian, who eats the upper portion of it, builds and covers his house with its various parts, and fashions his furniture out of its trunk.' Of several kinds of quadrupeds found by the first European settlers, the agouti is the only one remaining. Parrots, and other birds of brilliant plumage, and waterfowl, are very abundant; the alligator, cayman, iguana, turtles, &c. abound in the larger rivers; several kinds of serpents are met with; and the crustacea and testacea afford a plentiful supply of food to the inhabitants of the coasts. Hayti produces gold, silver, copper, tin, iron of good quality, and rock-salt. The principal copper-mine yields an ore containing a considerable admixture of gold, and the sands of many of the rivers contain a good deal of gold-dust, small quantities of which are collected; the working of gold mines has, however, entirely ceased. The mines of Cibao, which have long been unproductive, are said by Robertson to have yielded for many years a revenue of 460,000 pesos (nearly 100,000l.) annually; but it deserves to be remarked, that notwithstanding the extensive destruction of the original inhabitants in the working of these and other mines, the Spaniards derived so little advantage from them, that when Sir Francis Drake made a descent on the island in 1586, the inhabitants were so wretchedly poor as to be compelled to use pieces of leather as a substitute for money. (Edwards, i. 110, ed. 1819.)

History and Resources.—The island was discovered by Columbus, on the 5th of Dec. 1492, at which time it is said to have been divided into five states. Having taken possession of it in the name of Spain, Columbus founded the town of La Isabella on the N. coast, and established in it, under his brother Diego, the first colony planted by Europeans in the new world. The city of St. Domingo, which subsequently gave its name to the entire island, was founded in 1494. The island is believed to have contained, at the epoch of its discovery by the Spaniards, above 1,000,000 inhabitants of the Carrib tribe of Indians. But in consequence of their wholesale butchery by the Spaniards, and of the severe drudgery they were compelled to undergo in the mines, the natives were reduced to about 60,000 in the short space of fifteen years. (Robertson's America, i. 185, r-l. 1777.) The aboriginal inhabitants were soon, in fact, wholly destroyed; and their place was at first very inadequately supplied by Indians forcibly carried off from the Bahama islands, and adventurers from Spain and other European countries, and in the following century by the importation of vast numbers of negroes from Africa. The Spaniards retained possession of the whole island till 1665, when the French obtained a footing on its W. coasts, and laid the foundations of that colony that afterwards became so flourishing. In

1691, Spain ceded to France half the island; and in 1776 the possessions of the latter were still farther augmented. It was not, however, till 1722, when the monopoly of trading companies was put an end to, that the French part of the island began rapidly to advance in pop. and wealth. From 1776 to 1789 the colony had attained the acme of its prosperity; and its produce and commerce were then equal or superior to those of all the W. India islands. Unhappily, however, this prosperity was as brief as it was signal; and the ruin that has overwhelmed the colony may be said to be complete.

To attempt to give any intelligible sketch, how might soever, of the events by which this destruction was brought about, and by which the blacks of Hayti have emancipated themselves from the dominion of the whites, and founded an independent state, would far exceed our limits. At the epoch of the French revolution, the negroes in the French part of St. Domingo were estimated at from 480,000 to 500,000. That a good deal of dissatisfaction existed amongst them is certain; but there was no disposition to revolt, and the rash and injudicious proceedings of the mother country, the debates and proceedings of the colonial assembly, and the deep-rooted animosities of the whites and mulattoes, were the prominent causes of the revolution. The proscriptions, ruin, bloodshed, and atrocities by which it was accompanied and brought about, are, perhaps, hardly to be paralleled. In 1800, Hayti was proclaimed independent; and its independence was consolidated by the final expulsion of the French in 1803. This was effected by Dessalines, who erected the French or W. part of the island into an empire, of which he became emperor, with the title of James I. His despotism and cruelty having rendered him universally detested, Dessalines was slain in an insurrection in 1806, and Hayti was divided among several chieftains, the principal of whom were Christophe in the N.W. and Petion in the S.W. In 1811, the former made himself be proclaimed king, under the title of Henry I.; Petion continued to act as president of a republic till his decease in 1818, when he was succeeded by Boyer. The latter, after the suicide of Christophe, in 1820, took possession of his dominions, and the Spanish portion of the island having, in 1821, voluntarily placed itself under his government, he became master of the whole of Hayti.

The whole extent of land under cultivation in the three provinces was 763,923 carreaux, equal to 2,289,400 English acres, about two-thirds of which were situated in the mountains. The French, who justly considered this their most valuable colony, cultivated its territory with the greatest care. Every plantation was laid out with the utmost neatness, and so arranged as to bring every portion of the soil into use in its proper order of succession. Artificial irrigation was effected on a large scale, and the remains of the aqueducts in the plain of Cayes are really magnificent. The growth of sugar engaged the largest share of attention; the immense fertility of the soil making the average produce about 2,712 lbs. an acre, or nearly two-thirds more than the general yield of the land in cases in Jamaica. (Edwards, p. 135.) The coffee plantations were also exceedingly productive, and those of cotton, indigo, and cocoa had begun to be prolific sources of wealth to individuals, and of revenue to the state. Besides these staples, large quantities of Indian corn, rice, pulse, and almost every description of vegetables required for the consumption of the inhabitants were grown. The live stock in the French colony consisted of about 40,000 horses, 50,000 mules, and 250,000

cattle and sheep. The Spaniards never paid much attention to the culture of their portion of the island. The example of the French, indeed, stimulated them to grow tobacco, sugar, cocoa, and some of the other staple products of the Antilles; but their chief source of wealth consisted in the herds of cattle they reared on their extensive savannahs. With these they supplied their French neighbours, whose demands were large; besides which, they exported a good many to Jamaica and Cuba. Hides were also one of their chief articles of export, and, according to Edwards, many cattle were slaughtered for their hides only. The occasional cutting of mahogany, cedar, and other kinds of timber, made up nearly all the rest of their resources. It is stated that the French purchased annually upwards of 25,000 head of horned cattle, and about 2,500 mules and horses; and that the Spaniards also transmitted upwards of half a million of dollars in specie, during the year, for the purchase of goods, agricultural implements, and negroes. Large shipments of mahogany and dye-woods found their way to Spain and different parts of Europe, the U. States, and Jamaica, and a considerable intercourse was kept up with Porto Rico and the Spanish main. Most of the trade of the Spanish colonists was, however, illicit, the facilities for smuggling being quite as great as the advantages derived from evading the heavy duties imposed on commerce.

The following is an estimate of the average exports from the French part of St. Domingo during each of the three years ending 1789:—

Articles	Quantities	Value in Livres
Clayed sugar - lbs.	46,647,314	41,049,549
Muscovado do.	93,549,573	34,419,331
Coffee	71,663,187	71,663,187
Cotton	6,886,634	19,397,716
Indigo - lbds.	651,807	9,644,443
Molasses	23,061	1,767,296
Rum	2,600	312,000
Raw Hides - No.	6,500	57,000
Tanned ditto	7,800	114,000
Total Value at Ports of Shipping		171,544,666 = £1,765,129

One of the first effects of the revolution which abolished the slavery of the blacks was an enormous decrease in the amount of agricultural produce. From 1794, the year in which the slaves were declared free by the National Convention of France, to 1796, the value of the exported produce had sunk to 8,606,720 livres, being only about 5 per cent. of what it had been in 1789; and seven years afterwards, the country had become almost a desert, not only from the waste of civil war, but also from the indolence of the black pop. The famous Toussaint l'Ouverture adopted coercive measures to restore agriculture; and it is, we believe, idle to suppose that any other will ever be effectual in such a country to impel the negro to labour. By an edict issued in 1800, Toussaint obliged every Haytian not a proprietor of land (with a few exceptions) to hire himself as an agricultural labourer to some proprietor, without the power subsequently to withdraw himself from his service. The labouring classes were thus again rendered slaves in fact, though not in appearance. The use of the whip was abolished; but, on the other hand, the sabre, musket, and bayonet, in the hands of a military police, were employed to keep the peasantry at work. This object was enforced with the most rigid severity; the hours of labour were to continue from sunrise to sunset, with a few intervals; and both the cultivator and pro-prior he were visited with heavy pains and penalties; the former if he refused to work, and the latter if he did not oblige the former to do so. By such means, with a labouring pop. not exceeding 750,000, according to Humboldt, the exports in the most productive year during the short sway of Toussaint were raised to the following amount:—

Sugar .	51,640,000 lbs.	Cocoa .	752,000 lbs.
Coffee .	34,370,000 „	Indigo .	37,000 „
Cotton .	4,950,000 „	Molasses .	9,128 hhds.

This compulsory system was followed both by Dessalines, who at one period raised the value of the exports to 69,181,000 livres, or to a third part what it was in 1789; and by Christophe, an able, though a brutal and sanguinary tyrant. Petion, on the contrary, abandoned the coercive plan; and, in consequence, while the NW. part of the island had the appearance of industry and cultivation, the SW. displayed little more than occasional spots of culture. Boyer, during the first few years of his rule, continued the lax system of his predecessor, and the total value of the exports of the entire island amounted, in 1825, to no more than 4,783,758 dollars (1s. 2d. each). The state of agriculture at that period was most deplorable; every branch requiring systematic industry had fallen into decay; the sugar plantations had become almost annihilated; the plain of Cul-de-Sac, formerly an immense sugar-garden, had on it only four plantations of any extent; little or no sugar was made, the juice being either used as syrup for domestic purposes, or distilled into tafia, the favourite liquor of the natives; coffee, in the W. part of the island, was grown only around Cayes, and in some small patches in the mountains; and in the former locality at least two thirds of what was raised was lost for want of hands to gather the produce; all other products were obtained in small quantities only; maize, the only species of corn grown, was frequently scarce, and sometimes imported from the U. States. In the course of the next generation, and under manifold changes of government, Hayti made some progress, though it never recovered the industrial activity of the reign of Toussaint L'Ouverture. President Boyer was deposed in 1843, when the state of San Domingo separated from Hayti, and formed itself into a separate republic, electing General Santana president in 1844. The next change was into that of an empire, President Soulouque assuming the title of Emperor Faustin I. in 1849. Forced to abdicate in 1859, Hayti became once more a republic; while the state of San Domingo gave itself up to Spain in 1861. There are not wanting efforts to re-unite the whole island to the colonial possessions of Spain.

Commerce.—The foreign trade is entirely in the hands of European or American merchants, towards whom, however, the most restrictive policy is adopted. The coasting trade, on the other hand, wholly belongs to Haytian citizens. The interior is supplied with imported goods by means of hucksters (usually females), the agents of the foreign merchants, with whom they balance accounts weekly. Beasts of burden are commonly used for the conveyance of goods, the roads, except in the NW., being generally bad, and carriages few. The principal foreign trade is with the United States, Great Britain, France, Holland, and Germany; besides which there is a considerable smuggling trade between Cayes and Cuba, and Jamaica. The chief British imports are printed cottons, muslins, ginghams, coffee bagging, woollens, cutlery, tin, and hardware, earthen and glass wares, cordage, army accoutrements, and ammunition. France supplies wines, liqueurs, silks,

shawls, gloves, brandy, porcelain, perfumery, and other manufactured goods. The small imports from Holland and Germany include linen fabrics, bagging, inferior woollens, Rhenish wines, Spa and Seltzer waters. The U. States supply lumber, provisions, hides, and colonial produce. The total value of the imports into the republic of Hayti amounted, in the year 1862, to 80,668,936 francs, or 1,547,586l. Very nearly one-half of these imports—19,204,917 francs in value—came from the United States. The total exports, in the same year, amounted to 43,856,163 francs, or 1,785,846l. The exports to the United Kingdom amounted to 123,067l. in 1860; to 187,471l. in 1861; and to 161,719l. in 1862.

The government of the republic of Hayti is vested in a president, senate, and chamber of representatives. The president, who must be 35 years of age at the time of his election, holds his office for life; is charged with all the executive duties; commands the army and navy; makes war, peace, and treaties, subject to the sanction of the senate; appoints all public functionaries; proposes to the commons all laws except those connected with taxation; and directs the receipt and issue of taxes: but in case of malversation, may be denounced by the senate, and tried by the High Court of Justice. The ministry consists of a secretary-general, and a financial and a judicial secretary. The senate consists of 36 mems. above 30 years of age, each chosen by the chamber of representatives, from lists furnished by the president. The senate sits 9 years; and its previous mems. are re-eligible after a lapse of three years. The chamber of representatives consists of 58 mems. chosen every five years by the electoral colleges of the respective communes. Its mems. must be 25 years of age, and each receives 200 dollars a month, besides a dollar a league for travelling expenses. The session of the chambers is limited to three months annually.

The High Court of Justice, composed of 15 judges, has jurisdiction in all charges preferred by the legislative bodies against their own mems., or against the high state functionaries. There is no appeal from its decision, but the accused has the privilege of rejecting two thirds of his judges. There are 8 provincial, civil, and criminal courts—at Cape Haytien, Cayes, St. Domingo, Gonaives, Jeremie, Jacquiel, Port-au-Prince, and St. Jago, composed of a president, 8 judges, and a government commissary, appeal from which lies to a court of cassation in the capital. Ordinary legal cases are decided by justices of the peace, who decide without appeal. The legal code is a modification of the old colonial laws of France.

The Roman Catholic is the established religion; but all other sects are tolerated. The church is under the archbishop of St. Domingo, four vicars general, and 31 parish priests. The government has appropriated to its own use all the property formerly belonging to the church; the monasteries have been suppressed; the chapter of St. Domingo has now only six canons; and the clergy, who are said to be in the last degree ignorant and corrupt, rely for support on voluntary contributions and fees, two thirds of which they must pay into the treasury.

The armed force consists of about 28,000 men, exclusive of staff officers. There is, besides, the national guard, composed, with few exceptions, of all the males from 15 to 60 years of age. These form a body of perhaps 40,000 men, the superior officers of which are chosen by the president or emperor, and the inferior ones by the privates. The navy, in 1862, consisted of 3 steamers and 3 sailing brigs.

The *public revenue* is derived from import and
export duties, territorial imposts, wharfage dues,
taxes on democrats farmed out, the land-tax,
stamps, patents, registry taxes, sale of demesnes,
and various other sources. It amounted, in 1862,
to 291,382,; while the expenditure, in the same
year, was 289,909,.

Hayti is divided into 6 departments and 52
arrondissements. Next to Cape Haytien and Port-
au-Prince, which have been alternately the capitals, the chief towns are St. Domingo and Cayes.

St. Domingo, a sea-port, on the SE. coast of the
island, at the mouth of the Ozama, which forms
its harbour, lat 18° 28' 40" N., long. 69° 59' 37"
W., was the first permanent settlement made by
Europeans in America, and, though greatly
diminished in importance, has still above 12,000
inhabitants. It is surrounded by old ramparts
strengthened by bastions and outworks. Its
interior is regularly laid out; the streets, which
intersect each other at right angles, are spacious,
but not all paved. The houses are in the Spanish
style, and many of them are fine substantial
buildings. Besides the cathedral, a Gothic edifice,
finished in 1540, and reported to have formerly
contained the remains of Columbus, there are
3 other churches, 2 convents, 2 hospitals, some
large barracks, an arsenal, lighthouse, and old and
new national palace. The handsome Jesuits' college has been converted into a military storehouse.
No monks are to be seen, but in other respects the
town has very much the air and character of a
Spanish city. The whites and coloured inhabitants far outnumber the blacks. The climate is agreeable, the air being continually cooled by sea-
breezes. The harbour is both capacious and secure;
it has from 10 to 12 ft. of water; but, owing to a
bar at the mouth of the Ozama, large ships are
obliged to anchor in the roadstead outside, exposed
to the S. winds. St. Domingo has a considerable
trade with the interior, but its external commerce
is now very limited. Cayes, one of the most
flourishing towns in the island, is built close to
its SW. shore, lat. 18° 11' 10" N., long. 73° 50' 15"
W. Its harbour admits ships drawing 13 ft. water;
those of larger size lie in the roadstead of Cha-
tandin, half a league W. Several British houses
are established at this port.

HAZEBROUCK, a town of France, dép. du
Nord, cap. arrond., in a fertile tract, 23 m. WNW.
Lille, on the Northern of France railway. Pop.
8,273 in 1861. The greater part of the town is not
well laid out; but there are several handsome
public buildings, including the par. church, with
a lofty and elegant spire, the town-hall, finished
in 1820, a fine specimen of classic style, the sub-
prefecture, and Augustine convent now occupied
by a college, primary school, house of charity, and
depôt of tobacco. It has manufactures of linen
fabrics, thread, starch, soap, leather, salt, beer, oil,
and lime, and a large market for these and other
kinds of goods.

HEBRIDES (THE), or WESTERN ISLES
OF SCOTLAND (the *Hebudes* or *Ebudes* of the
ancients), a series of islands and islets lying along
the W. coast of Scotland, partly and principally
in the Atlantic Ocean, but partly also in the Frith
of Clyde, between 55° 35' and 58° 31' N. lat., and
between 5° and 7° 57' W. long. The islands
(seven) in the Frith of Clyde constitute a county
(Buteshire), the others belong respectively to the
counties of Argyle, Inverness, and Ross. The
Hebrides consist of about 200 islands, great and
small, and are usually divided into the Inner and
Outer Hebrides; the former embracing all those
islands which lie nearest to the mainland, including those in the Frith of Forth; the latter con-

sisting of a long continuous range of islands,
stretching NNE. and SSW., from Barra Head, in
lat. 56° 48' N., to the Butt of the Lewis, in lat. 58°
31' N. The strait which divides the Outer Hebrides from the Inner, and from the mainland of
Scotland, is called the Minch, and in some
narrowest, from 15 to 16 m. across. The Outer
Hebrides are commonly called the Long Island,
and appear, in fact, as if they had originally consisted of one long-horned island, divided at a remote
era into its present portions by some convulsion
of nature. Lewis and Harris (which are more
extensive than all the rest put together), though
considered as separate, form, in fact, only one
island; and the sounds, or arms of the sea, which
intervene between the larger islands of the group,
are so interspersed with islets, that the range is
still nearly continuous. The following table contains a list of the principal islands of which the
Inner and Outer Hebrides are respectively composed, with their estimated extent in sq. m. :—

Inner Hebrides	Sq. m.	Outer Hebrides	Sq. m.
Bute, Arran, and the other islands constituting Buteshire	16½	Barra, including the islets Vater-say, &c.	
Coll	7¾	Pabbay, Minga-lay, and others dependent on it	34½
Colfonsay and Oransay	12	Benbecula, with its subsidiary islets	45
Gigha and Cara	8	Harris, with do.	184
Iona or Icolmkill	10	Lewis, with do.	457
Islay	245	North Uist, with do.	118
Jura	81	South Uist, with do.	127
Lismore	16	St. Kilda	3
Luing islands, or Scarba, Luing, Lung, Seil, Shuna, Eisdale, Kerrera, &c.	30	Add, for several islets, or rocks not included in the foregoing	20
Mull	301		
Raasay	31½		
Skye	535		
Ulva, Kerrera, and other islets depending on Raasay or Skye	70		
Small islands, or Canna, Rum, Eig, and Muck	65	Inner Hebrides	1,604
Staffa	¼		
Tyree	42	Total extent of Hebrides	2,750
Ulva	20		
Total	1,604		

Of the total extent of the Hebrides, estimated,
as above stated, at about 2,750 sq. m. or 1,760,000
acres, 54,000 are lakes. The island group is
divided into 30 parishes, of which 5 are in the
islands in the Frith of Clyde, 17 in the Inner
Hebrides, and 8 in the Outer Hebrides.

In the census of Scotland for 1861, there is no
distinct classification made between the Hebrides,
spread as they are over several counties, and the
other islands belonging to the kingdom. The
total population of all the islands, 196 in number,
was found at the census to be 164,215, exclusive
of the shipping, and 164,924 inclusive of the
dwellers on board vessels. Buteshire, in 1861,
had a population of 16,331, against 11,791 in
1841, and 14,151 in 1851. The total pop. has
considerably increased since the census of 1841.
Of the 200 islands of which the Hebrides consist,
more than half are so small, or so sterile, as not
to be inhabited. In 1861, only 79 were regularly
inhabited during the whole year; while 8 were
tenanted during the summer, and abandoned on
the approach of winter. The greater portion of
the people reside within a mile of the sea-shore;
in fact, except in the islands of Bute and Islay,

scarcely an inhabited house can be seen 1,000 yards from the sea-shore, or 500 feet above the level of the sea.

From the thinness of the pop., it is not to be expected that schools should be very common, or be easily accessible to the inhab. of every district; but each par. has at least one parochial school.

Gaelic or Celtic is the language spoken throughout the whole extent of the Hebrides; and in some of the more remote or thinly inhabited islands, it is still the only language used or known. But both English and Gaelic are now taught in almost every one of the schools, and the former is becoming common, and, in some instances, has almost superseded the use of the Gaelic. A few families, chiefly farmers from the lowlands of Scotland, have, of late years, settled in different parts of the Hebrides; and this, combined with the increased facilities of communication with the low country and with England which steam navigation affords, has had the effect of diffusing a more general knowledge of the English tongue than would otherwise have been the case. In the more populous portions of the Hebrides, there are few persons, if any, under 30 years of age, who do not understand English, though, with slight exceptions, Gaelic continues the language of common conversation. Gaelic was not, till about the beginning of last century, a written language; but the Bible, and a great variety of religious as well as miscellaneous books, have since been translated into it; and Gaelic grammars and dictionaries have also been published. These things have been done, not with the view of perpetuating the knowledge of a rude language, but of diffusing information among the inhabitants.

The 30 parishes of which the Hebrides consist have each a parish church, and a resident clergyman. There are besides, 14 quoad sacra chapels belonging to the established church, 6 chapels belonging to the R. Catholics, 3 to Presbyterian dissenters, 2 to the Episcopalians, and 1 to Independents; the total number of places of worship being 56. In some of the islands, particularly Barra, Eig, and S. Uist, Catholicism abounds, to the entire exclusion of almost every other creed. The Catholic priests do not confine their labours to the islands in which they have their head quarters, but periodically visit all those in their neighbourhood where a single member of their church is to be found. Missionaries, belonging both to the established church and to the dissenters, are common throughout the Hebrides.

Though a poor law has existed in Scotland since 1579, it is practically unknown in the Hebrides. Limited as are the means of the inhabs., the poor are supported exclusively by the collections made at the church doors on Sunday, by other voluntary contributions, and by occasional funds; a legal assessment for their behoof having never been adopted. It appears, from official returns, that the poor receiving relief are only as 1 to 51 of the inhab.; that the average annual amount given to each individual is 31s. 6d.; and that the cost averages rather less than 2½d. to each head of pop. This insignificant degree of assistance is scarcely appreciable, and shows how extremely destitute the people are, and how low their estimate of physical comfort.

The climate of the Hebrides is more humid, variable, and inhospitable, than that of any other part of the British dominions. 'The temperature of the atmosphere is variable, the climate very rainy, and the air extremely moist; insomuch that when a person walks by the sea-side, in a hazy atmosphere and under a cloudy sky, the saline particles rests like dew on the pile of his coat. The dampness of the air is such, that in rooms wherein fires are not constantly kept, the walls emit a hoary down of a brinish taste, resembling pounded saltpetre, when brushed off. The climate is an enemy to polished iron and to books. Frequent and heavy rains fall at all seasons, especially after the Lammas term, whereby the hopes of the husbandman are often blasted, and the fruit of his toil and industry in a great measure lost.' (New Stat. Account of Scotland, No. 12, p. 116.) In the Outer Hebrides winter lasts for six months, from the end of Oct. to the end of March; spring, summer, and autumn occupy the other half of the year. 'During the spring, E. winds prevail, at first interrupted by blasts and gales from other quarters, accompanied by rain or sleet, but ultimately becoming more steady, and accompanied with a comparative dryness of the atmosphere, occasioning the drifting of the sands to a great extent. Summer is sometimes fine, but as frequently wet and boisterous, with S. and W. winds. Frequently the wet weather continues, with intervals, until Sept., from which period to the middle of Oct. there is generally a continuance of dry weather. After this W. gales commence, becoming more boisterous as the season advances. Dreadful tempests sometimes happen through the winter, which often unroof the huts of the natives, destroy their boats, and cover the shores with immense heaps of sea-weeds, shells, and drift timber.' (Macgillivray's Acc. of the Outer Hebrides; Edinburgh Quarterly Journ. of Agric. No. 11, p. 274.) These remarks are applicable, with very slight modification, to the whole range of the Hebrides, the islands in the Frith of Clyde excepted; in which latter, the climate, though damp and variable, is comparatively genial and mild.

In addition to the unfavourable climate, the Hebrides are remarkable for their rugged and sterile soil, more than six sevenths of their superficial content consisting of irreclaimable mountains, morasses, &c.; while the extent of arable and meadow land under grass, corn, and potatoes, is little more than a ninth part. Assuming the whole extent of the islands to be equal to 1,592,000 Scotch acres, or about 2,000,000 English (an estimate somewhat different from that given in this article), Mr. M'Donald, in his Agricultural Survey of the Hebrides, supposes it may be distributed as follows:—

	Acres
Mountains, morasses, and unreclaimed lakes, scarcely yielding any specified rent to the proprietors	610,000
Hill pasture, appropriated to particular farms, and sometimes restored, or at least limited by acknowledged marches, as lakes, rivulets, &c., and paying rent	700,000
Arable and meadow land, under grass, hay, corn, and potatoes	180,000
Kelp shores, dry at ebb-tide, regularly divided among the tenantry, and producing 5,000 tons of kelp, besides manure, annually	50,000
Ground occupied by villages, farm-houses, gardens, gentlemen's parks, &c.	20,000
Ground occupied by peat-mosses annually; and by roads, ferry-houses, and boats	37,000
Barren sands, tossed about by the winds, and pernicious to their vicinity	12,000
Ground occupied as glebes, or, in lieu of glebes by established clergymen, missionaries, churches, and churchyards	8,000
Ground occupied by schoolmasters	2,000
Ground under natural woods, coppices, and new plantations, chiefly in Bute, Islay, Mull, and Skye	500
Total	1,447,000

L L 2

But while the arable and meadow land is so limited, it is, at the same time, light, sandy, and poor, with some exceptions. In Islay and a few other islands, and unsusceptible of much improvement. The ordinary produce is black oats, barley or bigg, and potatoes. Mr. M'Donald distributes the arable land as follows:—

INNER HEBRIDES.

	Acres
Bute, &c.; Arran, 10,597—	10,244
Gigha	1,244
Islay	22,000
Jura	8,000
Colonsay and Oronsay	3,500
Kerrera, &c.; and the other Lorn Islands, &c.	6,000
Mull and Dependent Isles	10,442
Lismore	4,000
Coll and Tyree	8,500
Skye and Dependent Isles	91,000
Small Islands, or Canna, Rum, Eig, and Muck	3,500
Barra and Benn	8,000
OUTER HEBRIDES.	
North and South Uist, and Barra, with the Islets N. of the Sound of Harris	40,000
Lewis and Harris	76,000
St. Kilda	500
Total	300,000

Not only are the soil and climate unpropitious, but the tenure on which lands are held is, with some exceptions, as objectionable as possible. A very great majority of the farmers are tenants at will or from year to year, in other words, having no lease, they are liable to be turned out at the end of any year. This wretched system prevails almost universally in the Outer Hebrides. In the islands in the Frith of Clyde, it was laid aside in 1815, and superseded by leases; but in the remaining Inner Hebrides it still holds about three fourths of the land under its fetters, and nine tenths of the farmers. Besides, where leases are given, they generally range from 5 to 7 years, seldom extending to 9 or 12. Wherever this system extends, there is of necessity a total apathy to agricultural improvement.

Hence, with the exception of the islands in the Firth of Clyde, and of Islay, Colonsay, and some portions of Skye and Mull, in all which large farms and other improvements have been more or less introduced, agriculture is in as backward a state as can be imagined. Generally there is nothing like a rotation of crops. The grains usually cultivated are bear or bigg, and the old Scotch grey oat. In the outfield, which means that portion of a farm nearest the hills, and farthest from the farm-house and offices, one miserable crop follows another, till the ground be thoroughly exhausted. It is then allowed to rest, yielding for several years nothing but weeds; and as soon as these begin to disappear, by the return of grass and heath, it is again broken up, to undergo the same exhausting process. In the cultivation of the infield, the system pursued is nearly as injudicious. No regular rotation is followed; but the general rule is,—1. oats; 2. oats; 3. potatoes and peas; 4. barley or bigg, with manure; 5. peas; 6. oats; 7. two years of pasture choked with weeds, unaided by sown grasses, and therefore deficient both in quality and quantity. In a few places only has draining been practised; and without a very extensive system of drainage, no material alteration can be made for the better. In places not drained or levelled, the implements of husbandry are of the same rude and barbarous description that they were nearly a century ago. In the Outer Hebrides, small tenants and cotters generally till the ground with the Chinese plough, of one stilt or handle, and the caschrom, a clumsy instrument, like a large club, shod with iron at the point, and a pin at the ankle for the labourer's foot. This antediluvian implement will soon be superseded by the spade, which has now come into almost general use. But the plough is never seen, except in case of large farms. The common mode of turning the ground is by what is called dressing, forming a kind of lazy beds, such as are made in Ireland for the planting of potatoes. At this work two persons are employed, one on each side the ridge, which is seldom in a straight line, collecting the earth; and the earth, harrowed in this way, makes a proper bed for the seed. The ground being prepared, the seed is sprinkled from the hand in small quantities; the plots of ground being so small, narrow, and crooked, should the seed be cast as in large long fields, much of it would be lost. After sowing the seed, a harrow, with a brush brush at the tail of it, is used, which men and women drag after them, by means of a rope across their breasts and shoulders. The women are miserable slaves: they do the work of brutes, carry the manure in creels on their backs from the byre to the field, and use their fingers as a five-pronged grip, to fill them. In harvest, when the crop is ripe, no sickle is used for the barley among the small tenants. The stalk is plucked; the ground is left bare; and consequently the soil is injured. When the sheaves are dry, and conveyed to the barn-yard, the sickle is then used to cut off the heads or ears. After this operation, all the heads are formed into a little stack covered with the roots of the straw, which had been cut off.' (*New Stat. Acc.* § Lewis, pp. 131–133.)

Pennant's account of the initials, of Islay, through no longer applicable to them, Islay having been most materially improved in the interval, is still strictly applicable to those of most of the other islands. 'A set of people worn down by poverty, their habitations scenes of misery, made of loose stones, without chimnies, without doors, excepting the faggot opposed to the wind at one or other of the apertures, permitting the smoke to escape through the other, in order to prevent the pains of suffocation. The furniture perfectly corresponds: a pot-hook hangs from the middle of the roof, with a pot pendant over a grateless fire, filled with fare that may rather be called a permission to exist, than a support of vigorous life; the inmates, as may be expected, lean, withered, dusky, and smoke-dried.' (*Tour in Scotland.* ii. 253.)

Those who compare this striking paragraph with the description given in the New Statistical Account of Scotland of the houses in the Lewis and other islands, will find that it is, if any thing, really too favourable. There the dwellings of the people are, speaking generally, wretched huts, that afford shelter not only to the cotters and their families, but also to their cattle and pigs:—

'— — Ignavum, fucos, pecus, a praesepibus arcent.'

These huts, which are only half thatched, and without windows or chimnies, are indescribably filthy, and are, in fact, inferior even to the wigwams of the American Indians. The dung and other filth collected is and round the hut, is only removed once a year, when it is carried to the potato or barley field; and where also it is not unusual to strip the thatch off the hut, and to apply it to the same purpose. (*New Statistical Account,* art. 'Ross and Cromarty,' pp. 129, 147, &c.)

It is right, however, to state, that these miserable huts have nearly disappeared from the estates of Mr. Campbell of Islay, of Lord Macdonald in the Isle of Skye, of the Duke of Hamilton in Arran, &c.; and the probability is, that they would in no very long period wholly disappear, were it not for the embarrassed circumstances of many of the landlords, and their inability to undertake any improvement that require any considerable outlay.

The dress of the people corresponds with their food and houses. The kilt and trews, the characteristic Highland dress, are rapidly disappearing, and are no longer to be found in Skye and some other islands. Home-made woollen stuffs, checked or blue, are the universal dress both of men and women. Cotton and linen shirts are not generally in use, except on Sundays; but the dress, as well as the manners of the more civilised parts of the empire, is beginning to make its way into those acquainted recesses. Whenever a steamer is near, Manchester or Glasgow cottons will be found not long after.

The manufacture of kelp and the fishery, once the principal employments in the Hebrides, have declined very much of late years. Kelp is formed by burning sea-weed, previously dried in the sun; the alkaline substance thus formed being used in the manufacture of glass, soap, and alum. The annual produce of kelp, towards the close of the late war has been estimated at about 6,000 tons. Its price was sometimes as high as 20£ a ton; but its average price, during the 21 years ending with 1832, was 10l. 18s. 7d. (Encyc. Brit. art. Scotland.) And such was the influence of the manufacture, that the kelp shores of the island of N. Uist let at one time for 7,000l. a year! But the foundations on which this manufacture rested were altogether unsound. The repeal of the exorbitant duties laid on barilla and salt, especially the latter, virtually annihilated the manufacture of kelp. Its price, instead of averaging upwards of 10l. per ton, has been so low as 4l. 10s., but ranges generally between 3l. and 4l. The manufacture is still carried on in some of the islands, though in some instances to a considerable loss, instead of a profit. (Fullarton and Baird.—App. table iv.) The loss to the Hebrides, however, has been only apparent. The manufacture withdrew the attention of the islanders from what would have been more profitable pursuits. Being engaged during summer and harvest at the kelp shores, their crofts and crops were both neglected; and the sea-weed, which, had it been laid on the land, would have been the best possible manure, was carefully collected and carried off. Although, therefore, the ruin of the kelp trade was injurious to several proprietors, and was extensively felt at the time, it was productive of no real injury to the islands; but, on the contrary, will, in the end, conduce materially to their advantage.

The rearing of black cattle and sheep is the most extensive and profitable business in the Hebrides. The introduction of large farms into some of the islands has given a powerful stimulus to grazing, and black cattle are, in fact, the staple product of the Western Islands. The Kyloes, or West Highlanders, are the general breed, of which the best specimens are to be found in Skye; they are hardy, easily fed, not injured by travel, and, when fattened, their beef is finely grained, and is, perhaps, superior to any brought to table. The stock is estimated at not less than 120,000 head, exclusive of the islands in the Clyde, of which about a fifth part are annually exported lean to the mainland for fattening. When sold lean, their weight ranges from 13 stones to 30; but when fattened, it often rises to 50; but the average is from 24 to 36. The native breed of sheep is small, weighing only from 15 to 20 lbs.; weight of fleece (which is of various colours, even in the same fleece), from ½ to 1 lb. Both the black-

faced, or mountain breed of sheep, and Cheviots have been latterly introduced with success; the former to the greatest extent. The Hebridean horses are small and hardy; but they are not so handsome as those of the Shetland Isles. They are, however, extensively exported.

In the Outer Hebrides there are no trees; and, except in a very few spots, none can be raised. Turf or peat is the common fuel in all the islands; in some islands, as Tyree, Iona, and Canna, coals being deficient, the greater part (in Tyree, the whole) of the fuel has to be imported, chiefly from Mull, a third part of the industry of the inhabits being required to supply themselves with this indispensable article. Limestone is found in several of the islands, particularly Islay, whence it is exported in considerable quantities. Lead mines have also been long wrought in Islay, but not with any spirit. Marble is found in Tyree and other places, and slate in Easdale and the adjacent islands; both are pretty largely exported.

Manufactures, in the usual meaning of the word, are entirely unknown in the Hebrides, except a few cotton mills at Rothesay, and some distilleries in Islay. The people manufacture their own clothing from wool and flax of their own raising; and each head of a family makes the greater part of the utensils, implements, and furniture they require. Boat-building is carried on to a small extent at Tobermory, Stornoway, and several other places. With the exception of one or two common trades, such as those of a tailor, shoemaker, and joiner, the division of employments is nearly unknown; every person carrying on different kinds of business at different seasons of the year, and even at different hours of the day. In some of the smaller islands there are no day-labourers, the small farmer and his family doing all kinds of work.

The introduction of steam navigation has contributed largely to the improvement of the Hebrides, particularly the islands in the Clyde, with which there is a regular steam communication every day, and the Inner Hebrides generally; but the Outer range is scarcely ever visited by steamers. Not only are the former resorted to by numbers of strangers, from whose superior intelligence the inhabitants derive much advantage, but the steam-boats create a taste, and open a market for various articles for which there was previously no demand, and afford a ready means of conveying articles of native produce to Glasgow, Greenock, and other places. These facilities of intercourse and exchange are continually extended, and have a most beneficial effect on the character and circumstances of the inhabitants.

The Hebrides have few remains of antiquities, excepting those of the cathedral and other religious buildings of Iona, a small but famous island (3½ m. long by 1 m. broad), situated 8 m. SE. Staffa, and 1 m. from the SW. point of Mull. Three ecclesiastical ruins are most interesting. St. Columba, who introduced Christianity here from Ireland in 563, and whose successors, and those who adopted his creed, are known under the name of Culdees, is said to have built the cathedral; but it is abundantly evident that it was erected at a considerably later period. Of the buildings, some belong to the Roman, some to the Gothic, and others to the Norman style. The successors of Columba were expelled from the island by the Danes in 807; but two orders of monks, the Benedictines and the Augustines (nuns), took possession of the place in the 13th century, and flourished there till the general abolition of monasteries at the Reformation, when the island became the property of the family of Argyle, to which it still belongs. The remains of three various establishments, which still cover several acres of ground, consist of the cathedral, St. Oran's chapel, the chapel of the nunnery, five smaller chapels, and other dependent buildings. The cathedral is cruciform, with a tower 70 ft. high; the length from E. to W. is 160 ft. the breadth 24 ft.; the length of the transept 70 ft. Within the precincts of the cathedral are two crosses, the one called St. Martin's, the other St. John's. A large space around these buildings was used as a cemetery, in which were interred the remains not only of their religious inmates, and of several Highland chieftains and families of distinction, but (it is said, though the statement is probably much exaggerated) of 48 Scottish and 16 Norwegian kings and 4 French and 4 Irish sovereigns. Of 360 native crosses erected on the island, only 4 remain. (Keith's Cat. of Scot. Bishops, ed. 1824, pp. 414, 436; Pennant's Scotland, ii. 285.) There were five other monasteries in the Hebrides, viz., in Oronsay, Colonsay, Oransay, Lewis, and Harris; but of their history nothing is known, and few remains can be traced of their existence. (Keith, pp. 383-393.)

Iona was visited by Dr. Johnson in his tour to the Western Islands. He has described his sensations on visiting it in the following passage:— 'We were now treading that illustrious island which was once the luminary of the Caledonian regions, whence savage clans and roving barbarians derived the benefit of knowledge, and the blessings of religion. To abstract the mind from all local emotion would be impossible if it were endeavoured, and would be foolish if it were possible. Whatever withdraws us from the power of our senses; whatever makes the past, the distant, or the future, predominate over the present, advances us in the dignity of thinking beings. Far from me, and from my friends, be such frigid philosophy as may conduct us indifferent or unmoved over any ground which has been dignified by wisdom, bravery, or virtue. That man is little to be envied whose patriotism would not gain force upon the plains of Marathon, or whose piety would not grow warmer among the ruins of Iona.'

Of the early history of the Hebrides nothing certain is known. They recognised for a length ened period the sovereignty of the Norwegian kings, but were, in 1264, annexed to the crown of Scotland. Owing, however, to their remote and inaccessible situation, their chieftains were for centuries afterwards lawless and turbulent, and assumed and exercised almost regal authority. Indeed, it was not till the abolition of hereditary jurisdictions, in 1748, that a final blow was given to the influence of the independent chieftains of the Western Islands. The Hebrideans, in 1715 and 1745, were almost to a man in favour of the exiled family of Stuart. Charles landed on the small island of Gruier, to the E. of S. Uist; and after the battle of Culloden, he took refuge, first in the Outer Hebrides, and afterwards in Skye, previously to his escape to France.

HECLA, or HEKLA (MOUNT), a famous volcano of Iceland, near the SW. coast of the island. Its height was estimated by Sir G. Mackenzie at about 4,000 ft., or probably less; but, according to later authorities, it has an actual elevation of 5,210 ft. 'On approaching,' says Sir G. Mackenzie, 'Hecla from the W., it does not appear remarkable; and has nothing to distinguish it among the surrounding mountains, some of which are much higher, and more picturesque. It has 3 distinct summits, but they are not much elevated above the body of the mountain.' The

crater, of which the highest (or N.) peak forms a part, does not much exceed 100 ft. in depth. The bottom is filled by a large mass of snow, in which various caverns have been formed by its partial melting. The middle and lower peaks form the sides of similar hollows, and on the ascent are numerous other craters, whence flames and other matter have at different times been ejected. Hecla, like the Snæfell Jökul, near the W. extremity of the island, terminates in a long group of comparatively low hills. These, and others surrounding, are almost wholly composed of tufa, closely resembling that of Italy and Sicily; but the mountain itself consists chiefly of columnar basalt and lava, which latter forms a rugged and vitrified wall around its base. All the upper part of the mountain is covered with a layer of loose volcanic matter, slag-sand, and ashes, which increases greatly in depth towards the summit. In this part, indeed, few traces of any other substances are to be seen. Mackenzie says, 'We could not distinguish more than four streams of lava, three of which have descended on the S., and one on the N. side; but there may be some streams on the E. side, which we did not see.' (Travels, p. 249.) The view from the summit is one extended scene of frightful desolation. Towards the N. the country is low, except where a jökul here and there towers into the regions of perpetual snow. Several large lakes appear in different places, and among them the Fiske Vatn is the most conspicuous. In this direction the prospect reaches nearly two-thirds across the island. The Blæfell and the Lange Jökula stretch themselves in the distance to a great extent, presenting the appearance of enormous masses of snow heaped up on the plains. The Skaptar Jökul, whence the great eruption in 1783 broke forth, bounds the view towards the N.E.; this is a large, extensive, and lofty mountain, and appears covered with snow to its very base. The Tisa, Tindalla, and Eyafialla Jökuls limit the view to the E. To the S. is an extensive plain covered with lava, ragged with sharp stones and other volcanic substances, imbedded in the soil, and bounded by the sea.

There is, perhaps, no country where volcanic eruptions have been spread over so large a continuous surface as in Iceland, no part of the island being wholly free from the marks of their agency. But the distribution of the volcanic energy over so wide a space is doubtless the reason that the eruptions of Hecla are far behind those of Etna and Vesuvius, both in frequency and magnitude. Since 1004, only 22 eruptions from Hecla have been recorded, but some of these lasted for a considerable length of time; 8 or 9 eruptions have also taken place within the same period from the Kattlagiau, Eyafialla, and Skaptar Jökuls in the immediate vicinity of Hecla; and it is a curious fact, that out of 42 eruptions mentioned by native authors as having occurred in different parts of Iceland since the year 900, 5 were simultaneous, or nearly so, with eruptions of Vesuvius, 4 with those of Etna, and 1 (in 1766) with eruptions of both Etna and Vesuvius. (Sir G. Mackenzie's Travels in Iceland, pp. 236-254; Henderson's Encyc. des Gens du Monde; Lyell's Principles of Geology.)

HEDON, or HEYDON, a bor., market-town, and par. of England, co. York. E. riding, middle div. of wap. Holderness, on the Bevanside, 9 m. S. Hull. Area of par., with which the bor. is co-extensive, 1,410 acres; pop. 1,080 in 1831, and 975 in 1861. The town is small and mean-looking, with little business or trade. It was formerly of greater importance, and its decay is owing to the closing up of its harbour, and the greater advantages enjoyed by the neighbouring part of Hull. A church, dissenting chapel, and charity school are its only public buildings. This inconsiderable place returned 2 towns, to the H. of C., from the 1st of Edward VI. down to the Reform Act, by which it was disfranchised. The franchise was vested in the freemen, who became such by descent, apprenticeship, or gift; the seats were usually sold to the highest bidder.

HEIDELBERG, a city of S. Germany, duch. Baden, and the seat of a town and district bailiwick, at the foot of the Kaiserstuhl, on the Neckar, about 11 m. above its confluence with the Rhine at Mannheim, 30 m. N. Carlsruhe, and 48 m. S. Frankfort-on-Main, on the railway from Frankfort to Basel. Pop. 16,389 in 1861. The town is picturesquely situated at the entrance of the beautiful winding valley of the Neckar, and overlooked by well-wooded hills at the back, while rich vineyards cover the rising ground as far as the Heiligenberg on the opposite side of the river. The town lies close to the bank, and the principal street (Hauptstrasse), into which most of the others run, is nearly a mile long. The streets are narrow and gloomy, and the public buildings have no pretensions to grandeur. The church of the Holy Ghost, a large structure with a very lofty steeple, is divided so as to furnish accommodation both for Protestant and Rom. Catholic worship. St. Peter's church is the oldest in the town, and on its doors Jerome of Prague nailed his celebrated theses expounding the doctrine of the Reformers. There are two other churches and a Jews' synagogue. The University-house is a plain building, in a small square near the centre of the town, and contiguous to it is the library. In the same square is the Museum Club, where the members of the University dine, and meet for various purposes. The Anatomical and Zoological Museum, in the suburbs, was formerly a Dominican convent. Connected with the medical school are 3 hospitals, small and ill-ventilated, and not accommodating, in the whole, more than about 50 patients. The river, only navigable here for barges and rafts, is crossed by a stone bridge of 9 arches, 750 ft. long, and 34 ft. broad; and at its foot, within the town, is a heavy-looking building with towers, formerly used as a prison for riotous students. The Schloss, or electoral palace, stands on the side of the Giersberg, S. of the town, from which its ruins have a most imposing aspect. This castle was sacked and partly burnt by the French in 1689, and afterwards struck by lightning in 1764; since which time it has been wholly uninhabited; it is now roofless, and presents a mass of red-sandstone walls perforated with windows. The styles of architecture partake of all the successive varieties belonging to the 14th, 15th, and 16th centuries. The most ancient part is the E. front, part of which was built in the 14th century by the Elector Otho Henry: it is a mild square building, with towers at each end, one low and round, the other higher and of octagonal shape. A more modern part, less injured than the rest, is remarkable for its tall gables, curious pinnacles, and richly ornamented windows, showing it to belong to the 17th century. The front towards the Giesberg is a mere mass of mouldering buttresses and crumbling walls. Within the ruined hall it has long been the custom to hold a sacred concert once in three years; it is got up in the most splendid style, and is attended by all the people of the surrounding country. The cellars of the castle are very extensive, and are even said to communicate with the town below; in one of them is the famous Heidelberg tun, now empty, but said to be capable of holding 800 tbls. The terrace and gardens furnish the most magnificent views of the Neckar and its windings, and of the Rhine glittering here and there in the distance; spires and towers of numerous cities and villages dot the landscape which is bounded N. by the dusky outline of the Vosges. Heidelberg has no trade of any importance. The most curious objects in the neighbourhood of Heidelberg are the Wolfsbrunnen, the Heiligenberg and its ruined castle, and the Kaiserstuhl. From the top of the tower on this last hill the spire of Strasburg Cathedral, 90 m. distant, may be seen.

The university, called Ruperto-carolina, is, except Prague, the oldest in Germany. It was founded by the elector Rupert II. in 1346, and after the ravages of the thirty years' war, and that of the Palatinate, was restored by the elector Charles Louis, under whom it reckoned Spanheim, Freinsheimius, and Puffendorf among its professors. In 1802, when Heidelberg was ceded to the grand duke of Baden, he accepted the office of rector; through his munificence the university funds were greatly increased, and a fresh spur was given to the exertions of its professors. Its present income from the government is 10,000 florins (about 4,000£.), which, together with the income arising from fees, &c., is applied to the payment of professors' salaries, and the enlargement of the library. There are four faculties (divinity, law, medicine, and philosophy); and to these are attached forty ordinary and extraordinary professors, and twenty-one private tutors. The faculties of law and medicine are those most attended. The fees commonly paid for daily lectures during one semester are from twelve to twenty florins; and the necessary expenses of a student during a university session may be estimated at about 45£. Many of the Germans, however, live at a still lower rate. The library, which in the unhappy period of Heidelberg's history, was pillaged of its most valuable treasures to enrich the papal library, a part only of which were returned by Pius VII. in 1815, now contains 120,000 vols., besides a large number of rare and very valuable MSS. Connected with the university is an homiletic seminary, a philological seminary, and a spruch-collegium, or practical school for law students. There is a good gymnasium for junior students, and seventeen elementary schools are supported by the government.

The date of the foundation of Heidelberg is not known; but it ranked only as a small town in 1225. The court-palatine, however, enlarged it in 1362, and the period reaching thence to the thirty years' war appears to have been the era of its prosperity; for it then displayed, in its handsome buildings, all the splendour arising from a flourishing trade, and the residence of the court of the electors palatine of the Rhine. In 1622, during the thirty years' war, the town was taken by count Tilly, after a month's siege, and given up to be sacked for three days; the library was sent to the duke of Bavaria, and the imperial troops retained possession of the place during eleven years, at the end of which it was retaken by the Swedes under Gustavus Adolphus, and kept by them till the peace of Westphalia, in 1648. In 1674, in consequence of disagreements between Louis XIV. and the elector Charles Louis, French army under Turenne invaded the Palatinate, sacking and setting fire to its towns and villages. The sufferings of Heidelberg at this time, however, bore no comparison to the severe treatment which it met with in 1689 and 1693, when Melac and Chamilly ravaged and burnt the place. (See

Voltaire, Siècle de Louis XIV., ch. 10.) These repeated calamities, and the removal of the elector's residence and court to Mannheim, in 1719, contributed to diminish its importance among the towns of Germany; and it has never since recovered either its trade or population. In 1802, at the peace of Amiens, Heidelberg was attached to the grand-duchy of Baden.

HEILBRONN, a town of S. Germany, k. of Würtemberg, circ. of the Neckar, and near that river, 25 m. N. Stuttgard, on the railway from Carlsruhe to Anspach. Pop. 11,653 in 1861. The most interesting public edifice of the town is the church of St. Kilian, remarkable for the pure Gothic architecture of its choir, and its beautiful tower, built in 1529, 230 ft. high. The town-hall is an antique edifice, in which many imperial charters, bulls, and other ancient records are deposited. In the outskirts of the town is a tall square tower, in which Götz of Berlichingen, celebrated in one of Göthe's dramas, was confined in 1525. The house of the Teutonic Knights is now a barrack; on the other hand, the orphan asylum has been converted into a royal residence. There are three Rom. Cath. and two Protestant churches, a richly endowed hospital, a house of correction, and a gymnasium with a library of 12,000 vols. Heilbronn retained the privileges of a free city of the empire, originally conferred upon it by the emperor Fred. Barbarossa, down to the beginning of the present century. It was formerly a place of importance, from its position near the frontiers of the circles of Swabia, Franconia, and the Lower Rhine, and it still has an active trade, being an entrepôt for the merchandise sent from Frankfort for the supply of S. Germany. It has manufactures of woollen cloth, white lead, tobacco, hats, brandy, paper, oil, gold-and silver articles; and some trade in woollen and cotton goods. The Wilhelms canal, carried into the town, facilitates the traffic between it and the Neckar. Great quantities of wine, some of very tolerable quality, are grown in the neighbourhood, and coal is said to abound in the vicinity.

HELDER (THE), a marit. town of N. Holland, on a projecting point of land at the N. extremity of that prov., opposite the Texel, 40 m. N. by W. Amsterdam; lat. 52° 57′ 47″ N., long. 4° 41′ 55″ E. Pop. 2,950 in 1861. Being important from its position, commanding the Mars-Diep, or channel to the Zuyder Zee, and having almost the only deep water harbour on the coast of Holland, it is strongly fortified. It has a few manufactures, and some trade with Amsterdam, with which city it communicates by the Helder canal, the noblest work of the kind in Holland. (See Amsterdam.) The famous Van Tromp was killed in an engagement off the Helder in 1653. It was taken by the British under Sir R. Abercrombie in 1799.

HELENA (ST.). See St. Helena.

HELIER'S (ST.), the cap. of the Island of Jersey on the S. coast, 90 m. S. Portland Bill, 35 m. NW. Granville, and 39 m. N. St. Malo; lat. 49° 18′ N., long. 2° 13′ 45″ W. Pop. of town and par. 29,528 in 1861. The town stands on the E. side of St. Aubin's Bay, on a slope facing the shore between two rocky heights, on one of which is the citadel, Fort Regent, overlooking the harbour. It is not well built, and in the old and central parts the streets are irregular and narrow; but in the outskirts they are regular and well built, with ornamented garden ground in front. The Royal Square, the chief open space within the town, contains the par. church, built in 1341, the court-house, reading-rooms, and a large hotel.

The principal public buildings besides these, are the theatre, gaol, and two chapels, one being of Gothic architecture. This chapel and the theatre are the only edifices that have any claim to architectural beauty. The market-place is an enclosure within a wall and iron palisades, and the market on Saturday presents a magnificent display of vegetables, fruit and flowers, besides poultry and game from France. Fort Regent, which cost £637,000, was erected in 1806, and possesses all the usual defences of a regular fortress; but it has little accommodation for troops, and is said to have been injudiciously planned. Another fortress, Elizabeth Castle (so called because it was first built in queen Elizabeth's reign) stands on a rocky island ¼ m. from the shore, which at low water may be reached on foot by means of a long natural causeway; it contains extensive barracks, and appears to be a strong position. Lord Clarendon resided here two years while writing his history of the Rebellion. The harbour of St. Helier's is formed by two piers jutting out into the bay at the S. end of the town.

HELIGOLAND or HELGOLAND (an. Heligoland), an island belonging to Great Britain, in the North Sea, 26 m. from the mouths of the Elbe and Weser. Area 54 sq. m. Pop. 2,172 in 1861, of whom 1,034 males and 1,138 females. The island is divided into two parts, a high cliff and a low plain communicating with each other by a flight of rocks, on which is cut a flight of 190 steps. The elevated part is about 4,000 paces in circ., a precipitous rock of red conglomerate, varying from 90 to 170 ft. in height, and covered on the top with thin herbage, but without tree or shrub; the lower part is much smaller, and the entire circ. of the island is less than 4 m. The dimensions are continually lessening, owing to the encroachments of the sea, which, in 1770, separated a part of the island, now an uninhabited sandbank. Lyell (Geol., b. i., ch. 7.) attributes its destruction to the contest between the waters of the Elbe and Weser, and the strong ocean-tides of the North Sea. On the summit of the cliff stands the lighthouse, lat. 54° 11′ 54″ N., and long. 7° 53′ 19″ E. maintained from dues paid by British vessels entering the port of Hamburg. The church also, and the batteries, are conspicuous objects from the sea. Since 1821, when the military establishment was broken up, the batteries have been dismantled, and are falling to decay. The church is a plain structure, erected in 1682, the duties of which are performed by a Lutheran clergyman salaried by government, who is likewise the head master of the free school, which is attended by 320 children. The little town on the cliff consists of about 350 houses, chiefly inhabited by small traders and fishermen. On the lower part of the island are about seventy fishermen's huts, the only remains of the numerous storehouses standing here during the war, when this island was the centre of an extensive contraband trade. Heligoland has two good natural harbours, one on the N., the other on its S. side; and the E. of it is a roadstead, where vessels may anchor in 48 fathoms. The people, who are of Frisian extraction, and speak a dialect of that language, are chiefly employed in the haddock and lobster fisheries, the produce of which is taken to Hamburg, and exchanged for those necessaries which this island does not supply: some thousands of the lobsters come, also, to the London market. Many of the people are excellent pilots, and, being licensed by the island authorities, procure lucrative employment from vessels of all nations entering the Elbe. The females of the pop. are chiefly engaged in raising a little barley and oats on spots where vegetation will thrive, and in tending the few sheep that graze on the downs. The island is under a governor appointed by the crown; he is assisted by an executive and a legislative council, established by an order in council, in the year 1864. The total civil and military expenditure of the colony amounted to 908l. in 1861. The dependency, though useless in time of peace, serves in war for a point of observation, and a depôt for produce.

Heligoland, in ancient times, was the residence of a chief of the Sicambri or N. Frieslanders, and was the seat of worship of the Saxon goddess Phoseta, from which circumstance its name (holyland) was derived. It was in the possession of Denmark till 1807, when it was taken by the British government.

HELLESPONT. See DARDANELLES.

HELMSTADT, a town of NW. Germany, duchy Brunswick, distr. Schöningen, and cap. circle same name, 22 m. E. by N. Brunswick, and 30 m. W. Magdeburg, on a short branch of the railway from Magdeburg to Hanover. Pop. 8,820 in 1861. Helmstadt is an old-fashioned walled town, with four gates; the fortifications are turned into public walks, lined with lime-trees. Its two suburbs are called Ostendorf and Neumark. The places most worthy of note are the principal square, the Lutheran church of St. Stephen, the town-hall, and the circle-tribunal, once the university building. Besides these, there are three other churches, three hospitals, and an orphan asylum. Near the town, in the forest of Marienburg, are some medicinal springs; and on the Comelisaberg are the Lubbensteine, four enormous altars of Thor and Odin, surrounded with a circle of stones somewhat similar to that area at Aubury, in Wiltshire. It was once the seat of a university, founded by Julius, duke of Brunswick, in 1575, which was in a most flourishing state till the establishment of Göttingen university thinned its members. It was suppressed in 1809 by Jerome Bonaparte, and a portion of its library removed to Göttingen. A gymnasium and a normal school are the only existing establishments for education. Helmstadt is a place of considerable trade for its size. Flannels, hats, tobacco-pipes, soap, spirits, and liqueurs are its chief manufactures. It has four markets in the year. It is believed to have been originally built by the emperor Charlemagne, in 782.

HELSTONE, a parl. bor., market town, and par. of England, co. Cornwall, hund. Kerrier, 242 m. W. by S. London, and 10 m. SW. Truro. Area of par. 130 acres. Pop. of municipal bor. 3,843, and of parl. bor. 8,497 in 1861. The town stands on the side of a hill sloping to the river Low or Cober, which is here crossed by a bridge. The houses are chiefly ranged along four streets, which cross each other at right angles; it is well paved, lighted with gas, and abundantly supplied with water by streams running through the streets. Near the centre of the town is an ancient town-hall, and there is a coinage hall, now disused and let for private dwellings. The church is a modern structure, on high ground, having a fine pinnacled tower 90 ft. high. The dissenters also have several places of worship, and the Sunday schools are attended by 500 children. The grammar school has a high character; and there is a good national school. Helstone is the market for an extensive farming district, and also participates in the advantages derived from the mining speculations in the immediate neighbourhood; the mechanics are numerous, especially shoemakers, and the town is, on the whole, in a thriving state. Looe Pool, about

1 m. below the town, dries at low water; but facilities have been afforded to the trade by sea by the improvement of the harbour of Porthleven, about 3 m. distant. Iron, coal, and timber are imported in large quantities, for the use of the neighbouring mines. A singular custom prevails here, called the Furrey-dance, a kind of joyous procession, celebrated May 8, which is always observed as a holiday. The town received its first charter from king John; and Edward I. made it a coinage town, with the privilege of sending two members to the H. of C. The governing charter of the corporation, previously to the Municipal Reform Act, was granted in 1774. The last-mentioned act vested the government in four aldermen and twelve councillors. Corp. revenue 1,120l. in 1862. Previously to the Reform Act, the elective franchise was vested in the freemen, elected by the mayor and aldermen; but it had been for many years a mere nomination bor. belonging to the duke of Leeds. The Boundary Act added to the old bor. the entire par. of Sithney, and a large portion of the par. of Wendron. Registered electors, 353 in 1865. Markets on Wednesdays and Saturdays; fairs on the Saturdays before Mid-lent Sunday and Palm Sunday, and on Whit Monday, July 20, Sept. 9, Oct. 28, and the first three Saturdays in December.

HELVOETSLUYS, or HELLEVOETSLUIS, a fortified town and port of Holland, prov. S. Holland, on the Haring-vliet, the largest mouth of the Rhine, 16 m. SW. by W. Rotterdam. Pop. 4,213 in 1861. An excellent harbour, capable of accommodating the whole Dutch navy, runs through the centre of the town, and, being bounded by a pier on either side, extends a considerable way into the river. It has also a large arsenal, and docks for the construction and repair of ships of war, and a naval school. It used to be the regular station for the English and Dutch packet boats, which sailed to and from Harwich twice a week, till the adoption of steam-packets for the conveyance of the English mail to Rotterdam, in 1823. William III. embarked at Helvoetsluys for England in 1688.

HEMEL-HEMPSTEAD, a market town and par. of England, co. Hertford, hund. Dacorum, 22 m. NW. London, and 16 m. W. Hertford, near the London and North Western railway. Pop. of par. 7,946 in 1861. Area of par. 7,310 acres. The town stands on the slope of a hill, close to the small river Gade, and consists of a main street, lined with tolerably good houses. The church, in a spacious churchyard, is cruciform, with an embattled tower surmounted by a high octagonal steeple; the architecture was originally Norman, and the W. door is considered by Dallaway one of the finest specimens in England. Many alterations and enlargements have, however, been made at subsequent periods, which greatly diminish the beauty of the edifice. The town-hall, the only other public edifice, is a long narrow building, with an open space underneath for the accommodation of the farmers, who bring thither large quantities of corn for sale on Thursday, the market day. Within the par. are two endowed free schools, one for boys, the other for girls; besides which there are two infant schools, two national schools, and two schools of industry. The chief employment of the female part of the pop. is straw-plaiting, and this art is taught to children in dame-schools. In the neighbourhood are some large paper-mills; and within 4 m. of the town there are numerous flour-mills. The Grand Junction canal and North Western railway are 1½ m. SW., and greatly contribute to increase the traffic of the place, by the facility they afford for the tranmit of corn and other agricultural produce. Henel-Hempstead was incorporated by Henry VIII., and the inhabitants are empowered to have a bailiff and to hold courts of pie-powder during fairs and markets. This corporation, however, is mentioned neither in the commissioners' report, nor in the schedules of the Municipal Reform Act. Markets on Thursday; fair for sheep, Holy Thursday; statute fair, third Monday in September.

HENLEY-ON-THAMES, a market town, municipal bor., and par. of England, co. Oxford, hund. Binfield, on the W. bank of the Thames, 22 m. SE. Oxford, 35 m. W. London by road, and 35¼ m. by Great Western railway. Pop. of town 3,419, and of par. 3,676 in 1861. Area of par. 1,970 acres. The town is beautifully situated at the foot of the Chiltern range, which is here well covered with beech and other forest timber. The E. entrance is by a handsome stone bridge of five arches, built in 1786; and the first object presenting itself to the view, on entering from London, is the church, a handsome though irregular Gothic structure, built at different times, and having a lofty tower, ornamented at the angles with taper octagonal turrets, rising to a considerable height above the basement. It contains some curious monuments, and a library bequeathed by Dean Aldrich in 1737. The High Street, which runs W. from the bridge, is wide, well paved, and lighted, and lined with good houses; at its further end, on the rise of a hill, stands the town-hall, a neat building, on pillars, having on the upper story a hall, council chamber, and other rooms; its lower part, which is open, being used as a market house. Crossing the High Street at right angles are two other streets, much narrower, and lined with inferior houses. There are places of worship for Independents and Wesleyan Methodists, some almshouses endowed by Longland, bishop of Lincoln, and several schools. The principal of the latter are the United Charity Schools, founded in 1644, and endowed with land. The chief industry of Henley is malting, but the trade has much declined of late years; and the town can scarcely be said to possess any peculiar manufacture at the present time. It is a corp. town, its governing charter being granted in 1722, having a recorder, ten aldermen (one of whom is mayor), and sixteen burgesses. Quarter sessions and a court for the recovery of small debts are held here. Markets on Thursday, for corn and other grain; fairs, March 7, Holy Thursday, Thursday in Trinity week, and the Thursday after Sept. 21, chiefly for horses, cattle, and sheep.

HERACLEA PONTICA, also called PERINTHUS, a famous marit. city of antiquity, now called Erekli, on the S. coast of Asia Minor, on the Euxine Sea: lat. 41° 16' N., long. 31° 30' E. 'Heraclea,' says Major Rennell, 'has filled the page of history with its grandeur and misfortunes; and its remains testify its former importance.' Dionysius Siculus describes it as situated on an elevated neck of land about one stadium in length, the houses thickly set, and conspicuous for their height, out-topping one another, so as to give it the appearance of an amphitheatre. This is exactly the appearance that it exhibits at the present day; and the harbour, though neglected, is magnificent, forming a semicircle like a horse-shoe. The walls are now in a ruinous condition, and constructed chiefly of the remains of a former rampart. In the part fronting the sea, where are the remains both of an inner and an outer wall, huge blocks of basalt and limestone are piled one on another and intermingled with columns and fragments of Byzantine cornices with Christian inscriptions. The castle upon the height is in ruins.

Only a part of the ancient city was contained within the wall, the outer portion extending, in the form of a triangle, to a small river-valley, in which was formerly a harbour defended by two towers. The modern town comprises five mosques, two khans, two public baths, and about 800 houses, 50 of which belong to Greek Christians and the rest to Mahommedans. According to the Dict. Géog. it manufactures linen cloth, and exports flax, silk, wax, and timber; importing coffee, sugar, rice, tobacco, and iron.

The ancient Heraclea, founded by the Megareans, early attained to considerable wealth and importance as a place of trade. The inhab. maintained their independence for several years, subject only to a tribute paid to the Persian monarch. The Heracleots supplied the 10,000 Greeks, under Xenophon, on their memorable retreat, with vessels to carry them back to Cyzicus. The republican government was overthrown, about 360 B.C., by Clearchus, one of the chief citizens, in whose family the government continued nearly a century. Heraclea furnished succours to Ptolemy against Antigonus; and afterwards, notwithstanding the aid furnished to Rome by its marine, and a treaty of alliance, both offensive and defensive, with that powerful state, it was pillaged by Cotta, under the pretext that it had resisted the exactions of the publicans (or tax farmers) of Rome. Its splendid library, temple, and public baths were plundered and set on fire, and many of the inhab. put to death by the conqueror. The city, however, continued to flourish under the Roman emperors, and coins of Trajan and Severus are extant, in which it is styled metropolis and augusta. The fleet of the Goths waited here for the return of the second expedition that, in the time of Gallienus, ravaged Bythynia and Mysia; and it is mentioned as still prosperous even so recently as the reign of Manuel Comnenus. Athenæus informs us that it was celebrated for its wine, almonds, and nuts. (Tournefort, ii.; Walsh's Constant. 101.; Geog. Journ. ix.)

HERAT, or HERAUT, formerly HAREE (an. Aria or Artacoana), a city of W. Cabúl, in antiquity the cap. of Ariana, and one of the most renowned cities of the E., and still the largest and most populous town of the modern prov. of Khorasan, and the cap. of an independent chiefship. It stands on the Herirúd (an. Arius), in a fertile plain, 360 m. W. by N. Cabúl, 370 m. NW. Candahar, 410 m. NE. Yezd, and 410 m. SSW. Bokhara: lat. 34° 40' N., long. 62° 27' E. Pop. estimated some years since by Christie at 100,000; but at present it does not probably exceed 45,000, of whom 2-3rds are native inhab.; about 1-10th part Dooraunee Afghans, and the rest Moguls, Eimauks, Hindoo merchants, Jews, and other strangers. Previously to 1824, when the city was besieged by the Candahar troops, it covered a large extent of ground, having had some considerable suburbs outside the walls. It now consists of only the fortified town, 3-4ths of a m. square, surrounded with lofty walls of unburnt brick, erected upon a solid mound formed by the earth of a broad wet ditch, which goes entirely round the city, and is filled by springs within itself. There are five gates, each defended by a small outwork; and on the N. side of the fortress is the citadel, a square castle of burnt brick, flanked with towers at the angles, and, like the town itself, built on a mound enclosed by a wet ditch. The interior of Herat is divided into quarters by four long bazaars, covered with arched brick, which run from four of the gates, and meet in a small domed quadrangle in the centre of the city. (Conolly.) It is said to have about 4,000 dwelling-houses, 1,200 shops, 17 caravanserais, and 20 baths, besides many mosques, and fine public reservoirs. But, notwithstanding a plentiful supply of water, and abundant means for insuring cleanliness, Herat is one of the dirtiest places in the E. Many of the small streets which branch from the main ones are built over, and form low dark tunnels, containing every offensive thing. No drains having been contrived to carry off the rain which falls within the walls, it collects and stagnates in ponds which are dug in different parts of the city. The residents cast out the refuse of their houses into the streets, and dead cats and dogs are commonly seen lying upon heaps of the vilest filth. Huma ast—"it is the custom"—was the only apology I heard from those even who admitted the evil. (Conolly, ii. 3, 4.) The residence of the prince is a mean building, standing before an open square, in the centre of which is the gallows and the great mosque. The latter, a lofty and spacious edifice, supposed to date from the twelfth century, surmounted with elegant domes and minarets, and ornamented with shining painted tiles, is going to decay. 'But though the city of Herat,' says Conolly, 'be as I have described it, without the walls all is beauty. The town is 4 m. distant from hills on the N., and 1½ from those which run S. of it. The space between the hills is one beautiful extent of little fortified villages, gardens, vineyards, and corn-fields. A canal is thrown across the Herirood; and its waters, being turned into many canals, are so conducted over the vale of Herat, that every part of it is watered. The most delicious fruits are grown in the valley; the necessaries of life are plentiful and cheap; and the bread and water of Herat are proverbial for their excellence.' (Ib. ii. 4, 5.) Herat, from its extensive trade, has acquired the appellation of bunder, or emporium, it being a grand centre of the commerce between Caubul, Cashmere, Bukhara, Hindostan, and Persia. From the N., E., and S., the chief goods received are shawls, indigo, sugar, chintz, muslins, leather, and Tartary skins, which are exported to Meshed, Yezd, Kerman, Ispahan, and Tehran; whence dollars, tea, china-ware, broad cloth, copper, pepper, and sugar candy, dates and shawls from Kerman, and carpets from Ghaen, are imported. The staple commodities of Herat are saffron and assafoetida; silk is obtainable in the neighbourhood, but not in sufficient quantity for commerce. Many lamb and sheep skins are made up into caps and cloaks; and when Conolly visited the city, there were in it more than 150 shoemakers' shops. The latter were, however, inadequate to supply the demand of the poor, and many camel loads of slippers were brought from Candahar. The carpets of Herat are in great repute for their softness, and brilliancy of colour: but the trade in them has declined of late years. The greatest capitalists here are the Hindoo merchants. A mile N. of the town are the remains of what anciently was the wall of Heri, not far from which are the magnificent ruins of a place of worship, built by a descendant of Timour. The princes of his house constructed several palaces, gardens, and cemeteries on the hill range N. of Herat, traces of which still exist. Herat is capable of being made a place of great strength. An army might be garrisoned in it for years with every necessary immediately within its reach; and the influence of any W. power in possession of this fortress would be felt over all the country E., as far at least as Candahar. It long formed the cap. of an extensive empire transmitted by Timour or Tamerlane to his sons. It thence passed under the rule of

Persia; was taken in 1715 by the Dourannee Afghans: in 1731, by Nadir Shah; and retaken by the Afghans, under Ahmed Shah, in 1749. Since then, the Persians have often attacked it unsuccessfully.

HERAULT, a marit. dép. of France, in the S. part of the kingdom, formerly a part of the prov. of Languedoc, between lat. 43° 18' and 43° 57' N., and long. 2° 33' and 4° 13' E.; having NW. the déps. Tarn and Aveyron, NW. Aude, NE. Gard, and SE. and S. the Mediterranean. Length, NE. to SW., 73 m.; average breadth, about 50 m. Area, 619,729 hectares; pop. 409,391 in 1861. The slope of this dép. is from NW. to SE., and most of its rivers run in that direction; but the Herault, from which it derives its name, has mostly a SW. course from the dép. Gard, in which it rises, to its mouth in the Mediterranean, 15 m. SW. Cette. Its total length is 81 leagues, 3½ of which are navigable. A long succession of lagoons, occupying an area of more than 40,000 hectares, lines the coast, on which there are several good ports, including those of Agde and Cette. The climate, though hot and dry, is generally healthy; the soil is mostly calcareous. It appears from official returns that 158,666 hectares are arable, and 8,537 in pasture; that there are of vineyards 103,683 hect., woods, 77,644 hect., and heaths and wastes upwards of 214,000 hect. The growth of wine is the principal branch of industry. About 2,000,000 hectol. are made annually, 600,000 hectol. of which are exported, and a similar quantity used for home consumption; the rest is converted into brandy. The best kinds are the red wines of St. George and Viragues, and the white wines of Frontignan and Lunel. Corn, which is chiefly wheat, with some oats and rye, is not grown in sufficient quantity for home consumption; the annual produce is about 1,000,000 hectol. Oil, olives, figs, and dried fruits form important articles of commerce. There are some 237,000 mulberry-trees in the dép., from which 500,000 kilog. of silk cocoons were obtained. Bees are largely reared; and wax to the value of nearly a million of francs is annually exported. The number of large properties is greatly above the average of the dép. The pilchard and other fisheries in the Mediterranean and the lagoons, employ a great many hands; and it is estimated that 75,000 quintals of fish are annually taken, worth 545,000 fr. Herault is rich in mineral products; iron, copper, and coal mines, and quarries of marble, alabaster, gypsum, and granite are wrought. The principal manufactures are those of woollen cloths, silk and cotton fabrics, of which Montpellier is the chief seat; there are others of paper, chemical products, perfumery, and liqueurs; many distilleries and dyeing establishments, and a good deal of salt is made in the marshes. Herault is, however, much more an agricultural and commercial, than a manufacturing dép. Montpellier, Cette, and Agde have extensive trade, and their intercourse with the interior is promoted by several navigable canals, of which the Canal du Midi is the chief. Herault is divided into 4 arronds., 36 cantons, and 338 communes; chief towns, Montpellier, Beziers, Lodève, and St. Pons. This dép. anciently formed a part of Narbonese Gaul, and contains many Celtic and Roman antiquities.

HERCULANEUM, or HERCULANUM (Class. Art. vii. 8), an anc. and now buried city of Campania, in Italy, close to the Bay of Naples, and 5 m. SE. that city. The date of its foundation is unknown, and its early history fabulous; but there is little doubt that it was held by Osci, Pelasgi, and Samnites, before it came into the possession of the Romans. Velleius Paterculus tells us that its inhabs. took an active part in the social and civil wars, and that the city suffered considerably in consequence. Little more is known about it except its destruction with Pompeii and Stabiæ, by an eruption of Mount Vesuvius. The volcano had for some centuries been inactive, and even covered with verdure; but in the first year of the reign of Titus, A.D. 79, it burst forth with great violence, and caused those terrible disasters so well described by the younger Pliny, in two entire epistles (vI. 16, and 20), and more briefly by Tacitus:—' *Lserum effubit atrar et continuus tremor terrer, quem errsts est horrenda Vesurii montis conflagratum. Pulcherrima Campaniæ ora misrri fatains: obruter quæ urbes Herculanium et Pompeii; ruste hominum strages, quam inter perdire Agrippam ejusque mater Drusilla. At studiorum famin more C. Plinii fuit insignitior.'* (App. Chron.) Martial alludes also to the fate of Herculaneum:—

' Hic locus Hercvleo nomine clarus erat ;
Cuncta jacent flamsis et tristi merse favilla.'
Epigr. iv. 44.

The city appears to have been completely buried under showers of ashes, over which a stream of lava flowed, and afterwards hardened. The figure of the coast itself was altered by the burning torrent; and thus, when the local features were so wholly changed, all knowledge of the city, beyond its name, was soon lost. After a concealment of more than sixteen centuries, accident led to the discovery of its ruins. In 1713 the Prince d'Elbœuf, a French nobleman, who was building a palace at Portici, having need of materials for stucco, procured large quantities of marble and terra cotta from the sinking of a well on his estate. As the sinking proceeded, the workmen, when about 70 ft. below the surface, came to fragments of statues; and the prince then ordered excavations to be made, with the view of ascertaining the extent of the remains. A vault, a marble door-way, and several statues of vestals, were disclosed with little labour; but the works were soon afterwards stopped by the jealousy of the court of Naples. Twenty-five years after, on the accession of Don Carlos, the Infanta of Spain, to the throne of Naples, the works were resumed on a grander scale, and a theatre, chalcidicum, two temples, and a villa, were successively discovered and excavated. Owing, however, to the clumsy manner in which the mining was conducted, discreditable alike to the engineer and the government employing him, the statues and columns were needlessly injured and demolished, and, strange to say, the earth, instead of being brought to the surface, was used to fill up one part as another was searched. In consequence of this procedure, a small portion of the theatre is all that is now accessible; and the works, together with the interest excited by them among the Neapolitans, have long been discontinued. The whole extent of the ground explored was about 600 yards from NW. to SE., by 300 yards in breadth. The largest street was the NE. limit, beyond which it was supposed the mining could not be carried without endangering the town of Resina. Parallel with it was another street, and three others cut them at right angles. These streets appear to have been paved with lava, like those of modern Naples, a fact which proves that there must have been an eruption of Vesuvius prior to that which overwhelmed the city. The theatre was situated at the N. end of the town, which is supposed by Winkelmann to have extended nearly 2 m. along the shore, but without any great breadth. The theatre appears, from an inscription on its architraves, to have been built by Memmius, and its dimensions are as follow:—External circumference, 290 ft.; internal ditto as

far as stage, 230 ft.; internal diameter, 150 ft.; width of stage, 70 ft.; height, not known.

There were 18 rows of benches, besides 3 above the portico; and the entrance to them was by corridors or passages leading from the three tiers of arched corridors which ran round the building, and communicated by steps with the exterior. Its walls were cased with polished marble; both inside and outside beautiful statues and highly wrought columns were found. The floor was composed of thick squares of yellow marble, many of which still remained when Winkelmann examined the place. The theatre is supposed to have been capable of accommodating 3,000 spectators, and was therefore very much smaller than many others, the ruins of which are still extant. In the chief street, which is 86 ft. wide, having a raised foot-way on either side, with portions of columns shewing the existence of an old colonnade, are the remains of a forum, or chalcidicum, and of two temples. The forum is an oblong building, 228 ft. long and 132 ft. broad, with a colonnade of 42 pillars running round its exterior; and it had 5 entrances, 3 in front, formed by 4 great pilasters decorated with equestrian statues, and 2 smaller entrances at the sides. The buildings are all cased with marble except under the colonnade, where the walls are covered with frescoes. One of the equestrian statues formerly at the front entrance has been restored, and is reckoned quite a chef-d'œuvre of ancient art. The two temples are united under a single roof, and the entire length of both is 192 ft., and the breadth 60 ft. They are very unequal in size; but are highly ornamented internally with columns, frescoes, and inscriptions. Among the private buildings excavated, all of which were small, with only one story, was a suburban villa most profusely decorated with statues and frescoes paintings. It seems to have been extensive, having rooms extending along the side of the garden; but they are all on the same story. Here were found the celebrated papyri, upwards of 640 in number, the unrolling of which has given so much trouble to the learned, and which would appear to be little better than thrown away, if the value of the 40 already unrolled and partly published may be taken as any criterion of the value of the others. The subjects are various; but the works and their authors are alike uninteresting. (Phil. Transac. for 1753; Mr H. Davy's Report in the Journal of the Royal Institution for April, 1819.) Close to this villa a large tank, or piscina, was discovered, 254 ft. long and 27 ft. broad, with semi-circular ends, and enclosed by a balustrade on which were ranged many exquisitely wrought bronze figures, now in the museum of the royal palace at Naples. The ornamental beds and arrangements of the garden were still discoverable, and at its extremity towards the sea was a pavilion floored with African marble and jaune antique. The precious relics of antiquity, as far as they were capable of removal, were taken to Naples, and are now deposited, with the other relics from Pompeii, in a large museum in a wing of the king's palace. The collection is most extensive, and comprises not only frescoes, statues, and works of art, but also articles of household furniture, such as tripods, chandeliers, lamps, basins, paterae, mirrors, articles of the toilet, medical and surgical instruments, and even cooking utensils. Engravings and descriptions of them will be found in David and Maréchal's Antiquités d'Herculaneum, 12 vols. 4to, and also in that instructive little work, Pompeii, in the Library of Entert. Knowledge. The paintings which have been cut from the walls on which they were originally executed have, since their restoration to the light, lost somewhat of their brightness; but the colours are still wonderfully fresh. Their merit of coarse varies extremely, and many are incorrect in drawing; but the vigour of the touches by which some of the figures are expressed, and the graceful elegance of the attitudes selected by the painter are truly astonishing. The most beautiful of these were taken from the walls of the theatre at Herculaneum, and the subjects may be understood at a glance, by those acquainted with Grecian history and mythology. Among the statues, the palm is generally given to a Mercury and a drunken Faun; but there are many, of bronze as well as marble, of most exquisite beauty; both the statues and lamps are very numerous. In the collection of medals, a gold medallion of Sicily, struck in the 15th year of the reign of Augustus, is considered by virtue to be the most rare and curious. On the whole, the remains of Herculaneum, so varied and perfect, throw a light on the arts and domestic customs of the Romans, which no mere description by a classic author could give. Antiquity here seems to revive, and we are carried back to the days when Rome was the mistress of the world. (Encyc. Metrop. art. Herculaneum, by Rev. G. C. Renouard; Winkelmann's Letters on Herculaneum, passim; Gell's Pompeii; Moore's Italy, &c.)

HEREFORD, an inland co. of England, on the borders of Wales, having N. the co. Salop, E. Worcester, and Gloucester, S. the latter and Monmouth, and W. Brecknock and Radnor. Area, 836 sq. m., or 534,823 acres, of which about 500,000 are arable, meadow, and pasture. The aspect of this co. is every where rich and beautiful; the surface is finely diversified with gentle eminences and valleys, magnificent woods, orchards, and meadows, enclosed with hedges and rows of trees. It is usually represented as being every where remarkable for fertility; but it has probably been in this respect overrated, and though the soil in many districts be not surpassed by any in the kingdom, it has, notwithstanding, a considerable extent of inferior land. It produces excellent crops of wheat and barley, and is one of the principal cyder cos. Its wool is also esteemed equal, if not superior, to any produced elsewhere in England. The Hereford breed of cattle are deservedly held in high estimation: they are of a dark red colour, with white faces, throats, and bellies, and fatten easily; are excellent workers, and are remarkably quiet and docile; but as respects the dairy, they are good for nothing. Nearly half the field labour of the co. is performed by the cattle. The wool of the Ryland sheep, formerly so celebrated for its fineness, has been injured by crossing by the Leicesters; but the carcass of the animal has been, in consequence, materially improved, and the weight of the fleece increased. Agriculture is in a pretty advanced state in this co., but there is a great want of drainage. Turnips are pretty extensively cultivated; and a vast improvement has been effected in many districts by means of irrigation. Hops are largely grown, particularly on the banks of Worcestershire, from 12,000 to 12,500 acres being under this crop. Property is variously divided; there are a few large estates, with many of a medium, and some of a small size. The tenures of gavelkind and but. English exist in some districts, but are usually nullified by will. The farms, which are mostly large, are usually held from year to year. All the more modern farm buildings are of brick and slated, those of older date being principally thatched. Oak bark is an important product. Iron ore has been discovered, but it is not wrought; and the other minerals seem to be of no importance. If we except cyder, which is produced to a greater extent here than in any other county, manufactures are inconsiderable;

gloves, however, are made at Hereford and Leominster, and some coarse woollens in a few places. Principal rivers Wye, Lug, and Munnow. The Wye is navigable to Hereford for barges carrying from 18 to 20 tons, but the navigation is difficult, and but little to be depended on. Hereford is divided into 11 hunds., and 219 pars.; it sends 7 mems. to the H. of C., viz., 3 for the co., and 2 each for the burs. of Hereford and Leominster. Registered electors for the co. 7,525 in 1863. The census of 1861 showed a pop. of 123,712, living in 25,314 houses. The gross annual value of real property assessed to income-tax was 781,444l. in 1857, and 811,680l. in 1862.

HEREFORD, a city and parl. bor. of England, co. same name, of which it is the cap., hund. Grimsworth, on the N. bank of the Wye, 118 m. WNW. London, and 56 m. SW. Birmingham, on the Great Western railway. Pop. 15,585 in 1861. The city stands on a gravelly soil, in a valley, near the centre of the co. The parl. bor., which is co-extensive with the old mun. bor., comprises the entire pars. of All Saints, St. Peter's, St. Owen's, St. Nicholas, with parts of St. Martin's, and St. John the Baptist, exclusive of out-townships; and extends about 8¼ m. from N. to S., and nearly 4 m. from E. to W., enclosing an area of about 2,840 acres. The new municipal borough excludes about 2-5ths (chiefly rural tracts) of the above district. The streets are wide, straight, macadamised, flagged, and well lighted with gas. The private dwellings, almost entirely of brick, are generally old-fashioned, some few only being of modern construction. Among the many public edifices the largest is the cathedral, founded in 825, rebuilt in 1072, and thoroughly restored in 1862-3. It is a cruciform structure of the Saxon and early Norman style, and at the points of intersection rises a fine square tower 160 ft. high. The fall of the tower and a part of the nave in 1786 led to the erection of a very plain W. end. The extreme length of the cathedral is 860 ft., length of the great transept 180 ft., breadth of nave and side aisles 71 ft., height of nave 68 feet, height of entire building 91 ft. The nave is divided from the aisles by two rows of massive columns, sustaining semicircular arches, over which are rows of arcades with pointed arches. At the E. end are the Ladye Chapel, an octangular chapter house, and a well-stocked and valuable library. The N. porch is generally admired as a specimen of the ornamental Gothic style. Within the church are many fine monuments, among which that of Bishop Cantelupe (who died in 1287) is beautifully ornamented with the most delicate sculpture. Adjoining the cathedral are the college and bishop's palace, in the former of which are apartments for the vicars and other officers of the establishment. The cloisters connecting the palace with the church are considered curious and handsome. A triennial musical festival takes place within the cathedral, the profits of which are given to charitable institutions within the county. A side chapel is used as the parish church of St. John Baptist, the living of which is held under the dean and chapter. Of the other parish churches, that of All Saints, which is united with St. Martin's, has a tall and well-proportioned steeple, but is otherwise uninteresting. St. Peter's, which is united with St. Owen's, is a plain building with a spire. The church of St. Nicholas is old-fashioned and uninteresting; the rectory is in the gift of the crown. The dissenting places of worship belong to Wesleyan Methodists, Independents, Quakers, and Roman Catholics. Sunday schools are connected both with the churches and chapels, and there is a good charity school for clothing and educating

50 boys and 30 girls. The free grammar school, locally known as the College School, was either founded or enlarged by Queen Elizabeth: but it appears to have fallen into disrepute, and to be now almost useless, notwithstanding the 20 exhibitions which it offers to the universities of Oxford and Cambridge; connected with this school is Dean Langford's charity, which clothes and educates four children, and sends them to Brazennose College, Oxford, with scholarships of 85l. per annum for four years. Among the numerous and richly endowed charities of Hereford, the principal are:—1. St. Ethelbert's Hospital for 10 aged persons, having an income of 100l. yearly. 2. Coningsby's Hospital, founded in 1625, on the site of an ancient monastery, and providing lodging, clothing, and 15l. a year each to 11 old soldiers, and a salary of 20l. for a chaplain. 3. Lazarus's Hospital, once used for lepers and others afflicted with contagious diseases, but now an almshouse for six poor women, who divide 19l. yearly. 4. St. Giles's Hospital, established in 1290, as a monastery of Grey Friars, and given by Richard II. to the corporation, by which it was formed into an almshouse for five poor men, who are clothed, and share 20l. yearly. 5. William's Hospital, providing six decayed tradesmen with good lodgings, and 3l. 10s. each per month, and a chaplain, at a salary of 30l., who also officiates in the last-mentioned hospital. 6. Price's Hospital, for 12 men, who are lodged and paid 2l. a month each. 7. Trinity Hospital, a handsome brick building, in which 16 poor people are lodged, clothed, and pensioned, at 5s. each per week. The last five of these charities are in the patronage of the corporation, who, according to the statements of the municipal commissioners, formerly used their influence for the most corrupt purposes. Many other minor endowments belong both to the corporation and the parishes; indeed few cities in England possess so many charitable trusts as Hereford. (Charity Comm., 32nd Rep.) A large infirmary, supported by subscriptions and benefactions, and containing accommodation for 70 patients, stands NE. of the city, near the Castle Green. The union workhouse, completed in 1838, stands on the NE. side, outside the city. The chief public buildings not yet noticed are the shire-hall, designed by Sir R. Smirke, having a fine Doric portico. The ancient town-hall, an old-fashioned wood and plaster building, supported on pillars forming an arcade, was pulled down in 1861. Of other public edifices there are the guild-hall, built of brick; the theatre; the co. gaol, a well-arranged prison, in which the silent system and hard labour are rigorously enforced; and the town gaol, which is very small. Though the principal streets contain many good dwelling-houses and shops, there are no evidences of any very active or thriving establishments. The Wye is navigable by barges up to the city, except in dry summers or during heavy floods.

Hereford received its first charter of incorporation in 1189, from Richard I., but the governing charter, previously to the Municipal Reform Act, was granted by William III., in 1697. The corporation now comprises a mayor, six aldermen, and eighteen councillors; the city is divided into three wards. Hereford has sent two mems. to the H. of C. since the 23rd Edward I., the franchise, previously to the Reform Act, being vested in freemen, resident or non-resident, who became so by birth, marriage, apprenticeship, gift, or purchase. Reg. electors 959 in 1852. Gross annual value of real property assessed to income tax 84,058l. in 1857; and 61,012l. in 1862. Quarter and petty sessions, and a mayor's court, for the recovery of debts, are held within the city. The local acts are 14 Geo. III. c. 84, and 56 Geo. III. c. 23. Market-days on Wed. and Sat., the Wed. after St. Andrew's day being the 'great market.' Fairs, first Tuesday after Feb. 2, and Oct. 3 for cattle, cheese, and farming produce, being among the largest in England. Cattle fairs are also held on Wed. in Easter week, and July 1. The May fair, called the bishop's fair, lasts nine days.

During the disputes between Henry III. and his barons, and in the wars of York and Lancaster, Hereford was repeatedly the seat of hostilities; and its fine castle and strong walls, according to Leland, were so much injured, that in the time of Henry VIII. they were going fast to ruin. During the parliamentary wars it was garrisoned by Charles I., and twice besieged: in 1643 it surrendered to the parl. troops under Sir W. Waller, and being retaken by the royalists, was nearly the last that opened its gates to the parliament. The ancient fortifications and castle are wholly destroyed, and their site is now occupied by a public promenade, maintained by subscription and forming the favourite resort of the inhabitants.

HERFORD or HERVORDEN, a town of the Prussian dom., prov. Westphalia, gov. Minden, cap. circ. of the same name, on the Werra, 15 m. SW. Minden, on the railway from Minden to Düsseldorf. Pop. 10,714 in 1861. The town has courts of justice for the circle and district, a large prison, a gymnasium, and Rom. Cath. high school, and manufactures of cotton cloth and yarn, leather, tobacco, and linen goods. The central museum of arts, antiquities, and manufactures for Westphalia is established at Herford.

HERISAU, a town of Switzerland, cant. Appenzell, div. Outer Rhodes, 5 m. WNW. Appenzell. Pop. 9,510 in 1860. Herisau is cap. of the canton, jointly with Trogen, these towns being alternately the seat of the legislature. It stands on a height, at the junction of two small streams, which turn the machinery of numerous factories. The principal manufactures are those of cottons and silks, the last of recent introduction. It has an ancient church, in which the archives of the Outer Rhodes are kept, a pretty large public library, orphan asylum, court of justice, and arsenal. Near it is the Heinrichsbad, one of the most frequented watering-places in E. Switzerland.

HERMANSTADT (Hung. Nagy-Szeben), a town of Transylvania, cap. of the Saxon land, in an extensive and fertile plain, on the Tibin, a branch of the Aluta, 71 m. SSE. Clausenburg, and 70 m. W. Cronstadt. Pop. 18,500 in 1857. The town partly stands on an eminence, and is thence divided into an upper and a lower town. It is pretty well built, mostly in the Gothic style, and has a square ornamented with a statue and fountain; but still it has a dull and stagnant appearance. It has three suburbs, and is surrounded by a double wall, having a fosse and five gates. The most remarkable public buildings are the palace of Baron Bruckenthal, the favourite minister of the Empress Maria Theresa, containing an extensive library and fine museum; the churches, eleven in all (among them five Lutheran, two Calvinist, three Rom. Cath., and one Greek); the barracks, the military hospital, and the orphan asylum. The Lutherans have a gymnasium, in which the study of divinity, law, and philosophy is pursued, and a free school; besides which, there is a Rom. Cath. gymnasium, and a normal school. Hermanstadt is the head quarters of the commander-in-chief of the troops in Transylvania, and several departments of the government, as the customs and post-superintendence, are located here. It is a place of considerable trade, having three markets in the year, and it has manufactures of linen and woollen cloths, and hats. There is a brisk overland trade through Wallachia into Turkey. The Hermanstadters are said to be of Flemish origin. There are not less than seven distinct dialects among these Saxons, supposed to have been derived from the different parts of Germany from which they originally came.

The town, which takes its name from Hermann, the Saxon chief who conquered Transylvania, is said to have been founded in 1160, and to have early possessed many valuable rights and privileges under the Hungarian government; the greater part of the town, however, was built in the 16th century. It was once the capital of Transylvania, and was then in its most flourishing condition.

HERTFORD, an inland co. of England, having N. Middlesex, E. Essex, N. Cambridge, and W. Buckingham and Bedford. It has a very irregular outline, and a detached portion at Coleshill is wholly surrounded by Buckingham. Area, 611 sq. m., or 391,141 acres, of which about 353,000 are arable, meadow, and pasture. A ridge of chalk hills, from 800 to 900 ft. high, runs along the S. frontier of the co., and the rest of its surface is beautifully diversified with uplands and valleys; it has many thriving plantations, and a more than ordinary proportion of fine seats, among which Ashridge and Hatfield occupy the first rank. The sub-soil is generally chalk. It has every variety of soil, and may, on the whole, be said to be of about an average degree of fertility. By far the greater portion of the land is in tillage; and the wheat and barley of this co. are reckoned equal to those of any other district in England. Agriculture is not, however, in a very advanced state. Two white crops not unfrequently follow each other; and the land is mostly ploughed very shallow. Drill husbandry is but little introduced. Meadow land is in general much better managed than the arable, the quantity of hay produced being large, and the quality superior. Few cattle are raised or fed in this county; but the stock of sheep is considerable. Few large estates. Farms of various sizes, but not generally large. Leases, where granted, are usually for seven or fourteen years. With the exception of chalk, the minerals are of no importance. Manufactures not very important. Paper, however, is made on a large scale, of the best quality, and by the most improved machinery, near Watford and Rickmansworth. Malting is extensively carried on at Ware, Hitchin, and other towns; and a good deal of straw plait is made in different parts of the county; silk and cotton are also spun, and ribands made, at Tring, Watford, and St. Albans. In 1861, the numbers of persons engaged in the leading manufactures and occupations were as follows:—Paper, 626; straw plait, 8,753; malting, 437; silk, 968; farmers, 1,839; shepherds, 383. Principal rivers, Lea, Rib, Beane, Colne, Gad, &c. The Grand Junction Canal passes through its W. parts, and it is also traversed by the London and North Western and Great Eastern railways. There are castle ruins at Berkhampstead and Hertford; and a fine abbey church at St. Albans, the Roman Verulamium, from which Bacon took both his titles. Hertfordshire was in Flavia Caesariensis, and on the borders of the Mercian and the E. Saxon kingdoms. Hertfordshire has 8 hunds. and 135 pars.; it sends 7 mems. to the H. of C., viz. 3 for the co., and 2 each for the boro. of Hertford and St. Albans. Registered co. electors, 4,779 in 1865. The census of 1861 showed a pop. of 173,280 inhab. living in 31,869 houses. The gross annual value of real property assessed to income-tax was £856,878 in 1857, and 974,866l. in 1862.

HERTFORD, a parl. bor. and market-town of England, cap. of the above co., hund. Hertford, on the Lea, 10 m. N. London by road, and 26 m. by Great Eastern railway. Pop. 6,769 in 1861. The parl. bor. includes, besides the old bor. and liberties, portions of the parishes of Brickendon and Bengeo. The town, which stands in a valley, though irregularly laid out, is respectable in appearance, well paved and flagged, abundantly supplied with water, and lighted with gas. There are 2 churches, which serve for all the parishes, the others having been demolished. All Saints, the corp. church, is a spacious cruciform structure in the later English style, with a square tower and spire; and St. Andrew's, at the S. end of the town, though smaller in extent, is handsome, and has a low embattled tower and spire, and a large gallery within, for the accommodation of the children belonging to Christ's Hospital. There are 5 chapels: the Independents, the Wesleyan Methodists, and the Society of Friends, have commodious places of worship. Among the public charities in Hertford, the chief are,—1. A well endowed free grammar-school, founded in the reign of James I., having seven scholarships at Peter-house, Cambridge. 2. The Green-coat School, founded and endowed in 1700, in which about 50 boys are educated. 3. The branch school of Christ's Hospital, occupying a large brick building with wings, and accommodating 500 of the younger pupils of that great establishment. 4. A girl's charity school, attended by about 60 children. 5. An Infant school. 6. An almshouse for aged people, built and endowed with 50l. a year. The principal public buildings are—the castle, originally built in 905, afterwards enlarged, and now the property of the Marquis of Salisbury; the shire-hall, erected in 1771, under which is the corn-market; the sessions-house, in which the assizes are held; and the gaol, on the E. side of the town.

Hertford is a busy town, and there are sever. l mills on the Lea, the principal trade being meal-ing and malting, the produce of which it exchanges with London for coals and other commodities. There are also some large breweries, and an extensive distillery. The markets, held on Saturday, are among the largest in the S. of England for corn; fairs for cattle are held on the Saturday fortnight before Easter, and on May 12, July 5, and Nov. 8. This bor. received its earliest corporate privileges from William the Conqueror; its markets were granted by Edward III. The corporation now consists of a mayor, 3 other aldermen, and 12 councillors, and holds a commission of the peace; corp. rev., 1,464l. in 1862. Hertford sent 2 mems. to the H. of C. from the reign of Edward I. to the 50th of Edward III., when it was relieved from the burden, on the plea of poverty, and did not regain the privilege till the 23d of James I., since which time it has exercised the franchise. Down to the passing of the Reform Act, the electors were the householders and freemen resident, when they received their freedom. Registered electors 590 in 1862. Gross annual value of real property assessed to income-tax 27,132l. in 1857, and 27,561l. in 1862.

The date of the foundation of Hertford is uncertain. At the time of the Domesday survey, the town and lands were divided between the Conqueror and eight of his followers. In the wars between John and his revolted barons, the castle, originally built by Edward the Elder, was taken from the king, after a month's siege. It was restored in the following reign, and in 1345 was granted, with the earldom of Hertford, to John of

Gaunt, who made it his usual residence. The castle was afterwards inhabited by the queens of Henry IV., V., and VI.; and here, also, 150 years later, Queen Elizabeth occasionally resided and held her courts.

HESSE-CASSEL or ELECTORAL HESSE (Germ. *Kurhessen*), a state of W. Germany, consisting of a central territory (having NW. Prussian Westphalia and Waldeck, NE. Hanover and Prussian Saxony, E. Weimar, SE. and S. Bavaria, and W. Frankfort, Nassau, and Hesse-Darmstadt), and several small detached portions, the chief of which are the co. of Schaumburg to the N., and the lordship of Schmalkalden to the E. The whole territory lies between lat. 50° 5′ and 51° 25′ N., and long. 8° 30′ and 10° 40′ 30″ E.

The electorate of Hesse-Cassel is divided, for administrative purposes, into four provinces, of the following area and population, according to the census of 1855 and of 1861:—

Provinces	Area in Eng. sq. Miles	Population	
		1855	1861
Lower Hesse	1,743	259,548	259,508
Upper Hesse	878	119,950	119,483
Fulda	887	135,188	136,572
Hanau	883	121,588	137,383
Total	4,430	724,886	786,466

The population of the country was 567,868 in the year 1818, and kept on slowly increasing till 1849, when came a period of decline. The census of 1819 showed a population of 759,751, which had sunk, in 1852, to 755,850. The next census of 1855 showed a further diminution to 736,892, or a loss of 18,958 souls. The census of 1858, shown above, registered the disappearance of another 10,000 inhabitants. Thus, in nine years, the country lost nearly 3 per cent of its population, mostly by emigration to North America.

The greater part of Hesse-Cassel belongs to the table-land of central Germany, of which it forms the N. extremity, sometimes called the 'Hessian terrace.' Its N. part is traversed by the Werra mountains; its central portion is occupied by the plateau of Fulda; and its territory towards the SE. and S. covered by the Rhön, Spessart, and other mountain ranges, which enter Hesse from Bavaria. No summit, however, rises higher than the Meissner, belonging to the Werra range, which is 2,837 ft. above the level of the sea. The detached district of Schmalkalden, between the Prussian, Saxe-Meiningen, and Saxe-Gotha territories, is covered by the Thuringian forest mountains, and Schaumburg, between Hanover, Lippe Detmold, and Prussia, by ramifications of the Harz. Electoral-Hesse belongs principally to the basin of the Weser, which bounds it on the N., and receives the Fulda, Werra, Eder, Schwalm, Diemel, and Lahn; the Main bounds it on the S., and receives the Kinzig and Nidda. There are many large ponds, especially in the N., though none is large enough to be called a lake. The climate is healthy, but in winter the cold is severe, except in the prov. Hanau, S. of the elevated plateau of Fulda, and in the vale of the Werra, where some wine of an inferior sort is grown. The medium temp. of the year throughout the Electorate is about 50° Fahr. The soil is stony, sandy, and no where particularly fertile, except in Hanau. It is there very productive, and rye is reported to yield 16 or 20 fold, and wheat and barley in good situations as much as 71 fold, but such statements are uniformly almost greatly exaggerated. The whole country,

however, is capable of being rendered much more productive than at present; only the narrow valleys and the lower portions of the hill slopes are cultivated, and the valleys, which, from their confined extent are exposed to excessive moisture, are very imperfectly drained. A degree of indolence pervades the people in the rural districts; the villages have more of the Bavarian than the Saxon character, being often composed of mere ruinous wooden hovels; and the inhab. are commonly dirty, squalid, and slovenly. Agriculture is their chief occupation; it is in the most forward state in the valleys of the larger rivers. More corn is grown than is required for home consumption; it is principally rye, barley, and oats. These are every where cultivated; wheat is grown chiefly in Lower Hesse; the yearly produce of these four species of grain is estimated at 4,000,000 *scheffel*. Buckwheat is grown only in Schaumburg, and some parts of Fulda; and maize is confined to Hanau. About 350,000 *scheffel* of pulse of various kinds are annually grown, and from 700,000 to 800,000 *sch*. of potatoes; these products comprise the chief articles of food in the higher districts, besides which, potatoes are used to some extent in distilleries. Tobacco, esteemed the best in Germany, is grown in Hanau, and on the banks of the Werra in Schmalkalden; its annual produce averages from 17,000 to 20,000 cwt. Flax, also, of good quality, is largely cultivated in the S.N. provs., and about 160,000 *stris* are obtained yearly. Wine, which is almost exclusively produced in Hanau, does not amount to above 1,000 *eimers* a-year. Orchards are every where numerous; hemp, hops, chicory, poppy-seed, and culinary vegetables, are the remaining articles of culture. Hesse-Cassel is one of the most richly-wooded countries of Europe; nearly 1-3d of its surface, particularly in Fulda, Hanau, and Schmalkalden, is covered with forests. In the Thuringian forest, and in Hanau, firs are the principal trees; in the more level country oak, elm, and beech, predominate; the oaks are in some parts very fine. Juniper berries form an article of considerable export from Lower Hesse. The pastures are good, but cattle are not numerous. There were, in 1861, 91,167 oxen, 123,465 cows, and 500,317 sheep. Hogs and poultry are plentiful, not so hens. Game is not very abundant, and fisheries contribute but little to the support of the inhab. The peasantry, like their neighbours throughout Westphalia, are principally hereditary tenants; and there are men among them who boast of being able to prove that they still cultivate the same farms on which their ancestors lived before Charlemagne conquered the descendants of Herrman (*Arminius*), or, for any thing they know, before Herrman himself, drawing his hordes from these very valleys, annihilated the legions of Varus.

Mining is pursued, more or less, in all the provs. About 35,000 cwt. of iron, 5,140 cwt. of cobalt, and 1,000 cwt. of copper are obtained annually. There were formerly some tolerably productive silver mines near Frankenberg, in Upper Hesse, but they had long ceased to be wrought; a small quantity of silver still, however, is obtained near Bieber, in Hanau. About 235,000 cwt. of rock-salt, 300,000 cwt. of coal, 400,000 cwt. of brown coal, and turf in large quantities are annually produced. Coal of a good quality is abundant throughout the country. Manufactures have not reached any high degree of importance, but they are rapidly increasing. Linen weaving and spinning are the most widely diffused, and form throughout the country the common auxiliary employments of the small farmers and their families. The fabrics are of every quality, from the coarsest household cloths to the finest damask. The town and prov. of Fulda are the chief seats of this branch of industry, and it is estimated that from them alone 200,000 pieces of linen are exported, a large proportion of which are sold under the denomination of Osnaburgs. Schmalkalden is, however, the only district in which there is any approach to manufacturing establishments on a large scale; it is the seat of extensive iron works, and manufacture of fire-arms, cutlery, hardware, &c. Iron and steel wares are also made in the valley of the Werer. Coarse woollens, stockings, camlets, carpets in Hanau; leather, tobacco, glass, crucibles, porcelain and earthenware, paper, hats, gunpowder, tar, wooden wares, and musical instruments are among the other chief articles of manufacture. There are many bleaching and dyeing establishments, breweries, and distilleries. Cassel and Hanau are the principal manufacturing as well as commercial towns.

The great article of export is linen cloth, sent by way of Bremen and Frankfurt, chiefly to Holland, Denmark, and America. The other principal exports are linen yarn, woollen cloths, hats, jewellery, hides, sheep-skins, paper, iron and steel wares of all kinds, crucibles, timber, corn, dried fruits, and spirits. The chief imports are colonial goods, drugs, wine, flax and hemp seed, silk, fine wool, and woollen fabrics, mirrors and other glass wares, herrings, stock fish, horses, cattle, tin, gold, silver, and tobacco. The imports and exports nearly balance each other; but the most profitable branch of commerce to the Electorate is the transit of trade; the grand routes of communication between Frankfurt and Hamburg, Berlin and Dresden, passing through the territories of Hesse-Cassel. The dollar current is that of Prussia = 8s. The Hessian ell = ·623 English yards, the foot = ·943 English. The viertel of corn = ·55 English qr.; the cwt. is nearly equivalent to the English.

The Government is a limited monarchy, hereditary in the male line only. The different orders in the state are represented in one parliamentary chamber, composed of 52 members, consisting of the heads of the collateral branches of the electoral family, the mediatised nobles, the family of Riedesel (hereditary lords-marshal) and the hereditary marshals of Kaufungen and Wetter, six deputies from the nobles and knights of Hanau, Fulda, and Hersfeld; 16 from the towns, and 18 deputies sent by the peasantry. The present constitution was proclaimed Jan. 5, 1831; it was abrogated for some time, a new and less democratic charter being substituted by the Elector in 1852, but in consequence of general dissatisfaction, threatening insurrection, the government was forced, in 1862, to re-establish the fundamental law of 1831. The inhab. of Electoral Hesse in the last century suffered much from the oppression and rapacity of their rulers, who were accustomed, amongst other acts of tyranny, to traffic largely in the blood of their subjects, by hiring out their troops in the service of other European powers. The supply of Hessian troops to England during the American war brought to the electoral treasury the sum of 21,276,780 crowns between 1776 and 1784. The conquest of the country by the French put an end to this slave trade. Though popular at first, the obstinate attachment of the late elector to abuses, and the growing demand of the people for reforms, produced a revolt in 1830, which brought in its train at least the promise of better government—promise, however, but inadequately fulfilled. The constitution of 1831 guarantees equality under the laws, the free exercise of religion, free right of appeal, and eligibility to every

VOL. II.

office under government. For civil and criminal justice there is a high court of appeal in Cassel, and a superior provincial court of the cap. of each of the provinces. With each of these a forest court is connected, and subordinate to them are the district judicial and rural police courts. The town police is under a separate commission; and each of the provincial caps. has a head police court, as well as medical, manufacturing, and commercial tribunals, subordinate to head tribunals of the same kind in the cap. About four-fifths of the pop. are Protestants, one-sixth part Rom. Catholics, and the remainder chiefly Jews. Except the latter, and between 1,000 and 2,000 individuals, the descendants of emigrants from France, at the revocation of the edict of Nantes, all the pop. are of the German stock. The reigning family is Lutheran, but three-fourths of the Protestant inhab. are Calvinists. Since 1818, both Calvinists and Lutherans have been united for ecclesiastical government under 3 consistories, at Cassel, Marburg, and Hanau; the Rom. Catholics are under the bishop of Fulda. The principal establishment for education is the university of Marburg, founded in 1527, which has 57 professors, and is usually attended by from 350 to 400 students. There are lyceums, or colleges of arts, at Cassel and Fulda, teachers' seminaries in Cassel, Marburg, and Hanau; gymnasia, or high grammar schools, in the 5 principal towns; several schools of drawing, forest economy, and numerous primary schools. Education was formerly more backward in the Electorate than in any other state in Germany, but such is no longer the case. The armed force is raised by conscription, and every male under 30 years of age capable of bearing arms is liable to be called on to serve. The contingent furnished to the army of the Germ. Confederation is 9,406 men, of which 7,455 infantry.

Financial System.—The budget period embraces a term of three years. Divided into annual periods, the budget for the years 1861 to 1863 was made up of the following items:—

INCOME FOR THE YEAR.

	Thalers
Direct Taxes	801,500
Indirect Taxes	1,103,250
Public Lands	344,570
Mines and Salt Works	351,450
Forests and Fisheries	867,550
Post-office	43,500
Tax on Realty and other waters	93,250
Interest on State Property	843,020
Revenue of State Railways	814,940
Miscellaneous Income	185,250
Total	5,117,340
Or	£767,601

EXPENDITURE FOR THE YEAR.

	Thalers
Electoral Court	845,570
Allowance to Members of the Reigning Family	25,870
Ministry of Foreign Affairs	51,680
" of Finances	1,840,772
" of Justice	381,170
" of the Interior	1,047,979
" of War	903,550
Pensions and Annuities	810,720
Total	4,800,701
Or	£721,828

The budget granted by the chamber for the whole of the three years, 1861 to 1863, amounted to 16,357,020 thalers, or 2,302,052 revenue, and to 15,403,098 thalers, or 2,310,464 expenditure. According to the convention of 1831, half the revenues of the electoral property belongs to the public treasury; the other half is at the free disposal of the elector; but fresh disputes have since

M M

arisen between the electoral house and the nation, respecting the claim to the property of the landgrave of Hesse Hatenburg.

History.—The house of Hesse-Cassel was founded by William the Sage, in 1567. The landgrave was raised to the dignity of elector by the treaty of Luneville, in 1801, which title he retained when restored to his dominions in 1815, though there was no longer an emperor to elect. From 1808 to 1813 Hesse-Cassel formed a part of the kingdom of Westphalia, of which Cassel was the cap. The electorate holds the 8th rank in the German confederation, having three votes in the full council, and one in the committee.

HESSE-DARMSTADT, or the GRAND DUCHY OF HESSE, a state of W. Germany, consisting of two principal and not very unequal tracts of country, separated from each other by the territories of Hesse-Cassel, and Frankfort on the Main, and of some smaller detached portions chiefly inclosed within the territory of Waldeck, the whole lying between lat. 49° 12′ and 51° 19′, and long. 7° 53′ and 9° 40′ E. Upper Hesse, the most N. of the two principal tracts, is bounded W. by Prussian Westphalia and Nassau, and encircled on all other sides by Hesse, Cassel; the other principal tract has N. Nassau, Frankfort, and Hesse Cassel; E. Bavaria; S. Baden; and W. Rhenish Bavaria and Prussia; and is separated by the Rhine into the provs. of Starkenberg and Rhenish Hesse.

The grand duchy is divided into three provinces; the area and population, according to the census of 1858 and of December 3, 1861, are as follows:—

Province	Area in Eng. sq. miles	Population	
		1858	1861
Upper Hesse	1,510	310,781	314,714
Starkenberg	1,145	319,472	327,904
Rhenish Hesse	575	226,808	234,642
Total	3,240	845,571	852,250

For the three years previous to 1855, the population decreased to the number of 17,810; since then there has been a gradual increase.

The surface is very diversified. Rhenish Hesse and the W. part of Starkenberg consist mostly of a level plain of great fertility; the E. part of Starkenberg is occupied by the richly wooded Odenwald, a hilly tract, along the foot of which runs the picturesque and celebrated *Bergstrasse*, a very ancient line of road, extending in nearly a straight direction from Frankfurt to Heidelberg. Upper Hesse is hilly or uneven throughout, being intersected by the Taunus, Westerwald, Vogelgebirge, and other mountain ranges, the last named of which separates the basin of the Weser from that of the Rhine. The loftiest summits of the Vogelgebirge are about 2,500 ft. in elevation. Next to the Rhine, the chief rivers are its tributaries, the Main, Weschnitz, Nidt, and Nahe, in Starkenberg and Rhenish Hesse; and in Upper Hesse the Wetterau, Nida, Lahn, Eder, Fulda, &c. There are many large ponds, but none worthy of the name of a lake. The climate is generally healthy, but varies very much in different parts. The mean temp. of the year in the plain of the Rhine is about 53° Fah.; in Upper Hesse it is little more than 51°, and snow lies on the Vogelgebirge for 8 or 9 months of the year.

Hesse-Darmstadt is especially an agricultural country. The plains of Rhenish Hesse and Starkenberg, with the adjacent parts of Baden and Nassau, are amongst the best cultivated, as well as most fertile tracts of Germany; a circumstance which accounts for their supporting a pop. nearly as dense as that of Ireland in comparative comfort, without manufactures, and with but little trade. Rhenish Hesse, in particular, is covered with corn fields, vineyards, orchards, and villages; and besides supplying the demand for home consumption, exports corn in considerable quantities. Wheat is the principal produce of the low lands, buckwheat of the Odenwald, and rye of Upper Hesse; but in the higher parts of the latter province little else than barley and oats are grown. In Rhenish Hesse the rotations of crops are various, and studied with constant reference both to the soil and seasons, and the land is never fallow. Poppy seed, rape, tobacco of good quality, and fruit are extensively cultivated in this province; and its vineyards yield some of the finest growths on the Rhine. The total produce of wine in Hesse Darmstadt, is estimated at 180,000 ohm (6,342,500 imp. galls.), two thirds of which are exported. Flax, hemp, hops, and garden vegetables are the other chief objects of culture. Cattle-breeding is practised most extensively in Upper Hesse, where there is an active trade in live stock, including sheep, and hogs; but many cattle, &c. are also fattened in the Odenwald, chiefly for the supply of Frankfurt. The principal forest trees are beech, oak, hornbeam, pine, and fir; and in the Vogelgebirge, maple, elm, and larch. Large quantities of timber and wooden wares are sent from Upper Hesse and Starkenberg, down the Main and the Neckar. In Rhenish Hesse, however, timber is exceedingly scarce and dear, owing to the great destruction of the woods during the French dominion; and nearly all the material required for fuel has to be brought from the Black Forest or Spessart mountains. The forests are mostly either communal or grand ducal property; they belong to the communes, especially in Rhenish Hesse, where, from their scarcity, they are highly valued. In the latter province, and in Starkenberg, property is very much sub-divided.

The condition of the lower classes of agriculturists, who are here, as all over Germany, a kind of copyhold possessors of the land, has been very much improved since the year 1815. Personal services of all kinds have been redeemed on easy terms, by the interference of the government, which began by giving up those due to crown lands at a moderate valuation. The tithes on new enclosures were voluntarily resigned both by the crown and by land-owners, and the existing tithes were converted into fixed redeemable rent-charges, for the purchase of which the state advances capital at the rate of 5 per cent. interest to the land-owner. A charge to cover this outlay appears annually in the budget.

Mining is the occupation next in importance. Salt mines are wrought at Wimpfen, in a detached portion of territory to the N. enclosed between Baden and Wirtemburg, where this mineral is found in great abundance; and for the supply of Rhenish Hesse, two mines near Kreutznach on the Nahe have been rented from Prussia. Berghaus estimates the produce of salt at 180,000 cwt. annually. Copper is obtained at Thalitter in Upper Hesse, where a vein is profitably wrought, though the ore yields only from 1·6 to 2 per cent. of metal. At Bieberkopf, and on the estates of Prince Solms, in the mountainous parts of Upper Hesse, and in the Odenwald, extensive iron mines are wrought. Coal of inferior quality is abundant in Upper Hesse, and in scattered beds through the other provs.; but the total yearly produce is not more than 280,000 cwt. Turf, building stone, slates, marble, gypsum, and potter's clay, are the other chief mineral products, and there are traces of lead and mercury.

Manufactures on any extended scale cannot be said to exist in the grand duchy. Spinning and weaving linen and hemp are, as above mentioned, an auxiliary occupation of the agricultural classes, particularly in the N. and NW., parts of Upper Hesse, at Lauterbach, Schlitz, and Herbstein. Among these are damasks and other fine fabrics; but the linens of Hesse-Darmstadt cannot compete with those of Westphalia or Silesia. Some silk-weaving is carried on at Offenbach, and stockings are woven there and at Baben-Hausen. Coarse woollens are manufactured in several places, principally in the N. Tobacco is prepared for use at Offenbach, the principal manufacturing town in the grand duchy. Few metallic articles are made, except needles and pins. Paper, glazed pasteboard for export to Russia, brandy, vinegar, dyes, leather (not enough for home consumption), earthenware, and chemical products, comprise most of the remaining manufactures. The chief articles of import are colonial goods, burrs, cattle, hides, leather, leaf-tobacco, and wine. But the transit trade is the most considerable trade of commercial industry. It was very profitable to Mayence so long as obstacles existed to the free navigation of the Rhine, and all wares were forced to be shifted into boats owned in that city. This barbarous privilege has been given up of late years, but a toll is still raised upon boats passing up and down the river. Mayence is the emporium of the fruitful districts of the Upper Rhine, as well as of those on the Maine and Neckar. Hesse-Darmstadt was a mem. of the German Customs' Union for many years before it was joined by Frankfort; and a successful attempt was made, while that city held out against the proposals of the Union, to establish a rival fair at Offenbach. The government of the Grand Duchy raised the tolls on the Maine, and the mart of Offenbach was making a considerable progress towards prosperity, when the adhesion of Frankfort to the Union occasioned the abandonment of the experiment.

The florin in circulation, equivalent to 1s. 8d., is divided into 60 kreutzers. The chief weights and measures are the pound—1·1 lb. Eng., the ahm—37·5 galls, the malter—4·4 Eng. qrs., the fuoi—·82 ft. Eng., and the morgen—·62 Eng. acre.

The Government is a limited monarchy, hereditary in the male line. The States, according to the constitution of Dec. 17, 1820, slightly modified in 1848 and in 1856, consist of two chambers. The first is composed of members of the Grand Ducal house, the mediatised nobility, the R. Catholic bishop, the head Protestant ecclesiastic, the chancellor of the university of Giessen, and ten citizens nominated for life by the grand duke. The second chamber consists of six deputies from the knights or inferior nobility, who pay direct taxes to the amount of 300 florins annually, ten deputies from the towns, and thirty-four from the freehold land-owners, contributing each direct taxes of 100 florins a year. The deputies are elected every six years, and the chambers meet at least once in three years. No changes in the laws can take place without their sanction, but they never assume the initiative in legislation; they have only the right of petitioning for new laws, which are then submitted to them by the minister. By the constitution of 1820, every subject enjoys freedom of person and property, and the free exercise of religion; all are equal under the law; and all, except the members of the mediatised noble houses, are liable to military service from

20 to 25 years of age. This service may, however, be performed by substitute, and there is a government officer, through the agency of which substitutes are obtained on moderate terms. The contingent furnished to the army of the confederation is 10,825 men, made up of 8,071 infantry, 1,791 cavalry, and 963 artillery. Mayence, the most important fortress in Germany, is garrisoned by equal numbers of Austrian and Prussian troops. The press is free, and the abuse of its freedom is cognisable only by the civil law. The executive powers are in the hands of a prime minister and three others. Justice is administered in municipal and cantonal tribunals; high courts in the capitals of the provinces; a military tribunal at Mayence, and a superior court and court of appeal in Darmstadt. In Rhenish Hesse the courts of justice are modelled upon the French system, and trial by jury is in force, on which privilege a high value is placed.

Almost five-sevenths of the pop. are Protestants, one-fourth R. Catholics, and 72,000 Jews, besides whom there are a few Mennonites and other sects. The Catholics reside principally in the S., and are subordinate to the bishop of Mayence. The two Protestant confessions have been organised into one, and have assumed the ritual and discipline of the Prussian evangelical church. The reigning family is Protestant. Public instruction has advanced rapidly within the last fifty years, especially in Rhenish Hesse, where formerly the inhabitants generally were greatly ignorant.

In Mayence, which was the seat of a university, there was, in 1816, not a single bookseller, and mass-books and catechisms were the only works printed. The institutions for education are now excellent. One elementary school at least exists in every parish, besides which there are four citizens' schools, seven gymnasia, three seminaries for schoolmasters, four colleges, a military academy, a university at Giessen, attended usually by from 300 to 400 students, and many special academies for the arts and sciences. The communes elect their own headboroughs, and the usual restrictions with respect to marriage and settlement are enforced, as in the neighbouring German states. Commissions for the support of the poor are appointed in the towns, and, in Mayence especially, the charitable establishments are very well organised. A house of correction for secondary punishment has been established on an improved principle at Marienschloss, in which 350 convicts are confined, who both contribute by their labour to the support of the establishment, and earn a sum which is paid to them on their discharge.

The budget is granted for the term of three years: the items for the year 1862 were—

INCOME FOR THE YEAR 1857.

				Florins
State Property	.	.	.	2,601,625
Direct Taxes	.	.	.	2,756,034
Indirect Taxes	.	.	.	3,603,737
Miscellaneous Revenue	.	.	.	374,784
Total	.	.	Or	9,098,046
				£758,065

EXPENDITURE FOR THE YEAR 1862.

				Florins
Civil List and Grand-ducal Court	.	.	751,000
Ministry of Foreign Affairs	.	.	121,142
" the Interior	.	.	1,810,594
" Justice	.	.	850,848
" Finance	.	.	2,316,694
" War	.	.	1,660,649
Interest on Public Debt	.	.	505,576
Pensions and Annuities	.	.	450,161
Miscellaneous Expenses	.	.	507,320
Total	.	.	Or	9,046,706
				£753,568

The public debt amounted, at the commencement of 1863, to 15,245,000 florins, or 1,270,000l., the greater part of which was incurred for the establishment of a network of state railways.

The grand duke is descended from Philip the Magnanimous, between whose four sons the dominions of Hesse became separated towards the end of the sixteenth century. The grand duchy of Hesse-Darmstadt holds the ninth rank in the German Confederation, having three votes in the full diet, and one in the committee.

HESSE - HOMBURG (LANDGRAVIATE OF), a state of W. Germany, and one of the smallest in the Confederation, consisting of two detached portions, Homburg and Meisenheim, about 45 m. apart; the former enclosed between Hesse-Darmstadt and Nassau, and the latter surrounded by the territories of Prussia, Oldenburg, and Rhenish Bavaria. United area, 105 sq. m. Pop. 26,817 in 1861. The Homburg division is on the S. declivity of the Taunus mountains, the highest point of which, the Feldberg, is within its limits. The soil is not in general rich, but it has been rendered sufficiently productive by the industry of the inhabitants to furnish more corn than is required for home consumption, besides fruit, garden vegetables, flax, and timber. There are manufactures of woollen stuffs, linen fabrics, and stockings, which, after supplying the home demand, find a ready sale at Frankfurt. Meisenheim, W. of the Rhine, is partially covered with ranges from the Hundsrück mountains. Its N. part is high, and its climate cold; but the surface of its S. portion is much less elevated, its temperature mild, and it yields a good deal of wine. Corn and cattle are plentiful, as are timber, coal, iron, and building stone. A little linen cloth, some linen and woollen yarn, and glass, are made; and there are a few iron-forges. There is a superior court of justice in Homburg, with appeal to the high court of appeals in Darmstadt. The pop. is mostly Calvinist; there are, however, about 6,000 Lutherans, 3,000 Rom. Catholics, and 1,020 Jews. The public revenue in 1862 amounted to 528,507 florins, or 44,856l., and the expenditure to 519,687 florins, or 43,307l. The contingent furnished to the army of the Confederation is 533 men. Hesse-Homburg is united, in the slender tie of 'personal union,' to Hesse-Darmstadt, the grand duke of the latter country being also landgrave. The last independent landgrave died early in 1866, without leaving any direct heirs; and by a treaty made previous to his death between him and his collateral heirs, the rulers of Hesse-Darmstadt, it was settled that the landgraviate should remain a separate state for 25 years longer, or till 1891.

HETTON-LE-HOLE, a village and township of England, par. Houghton-le-Spring, co. Durham, N.E. div. of Easington ward, 6 m. N.E. Durham. Area of township, 1,590 acres. Pop. 6,419 in 1861, having increased from 919 in 1821. This astonishing increase is wholly attributable to the establishment of a large colliery, connected by a railway with the port of Sunderland. This populous village, chiefly inhabited by pitmen, consists, like most other pit-villages in Durham, of numerous cottages fronted by little gardens and interspersed here and there with houses of a better character. A church, dependent on that of Houghton-le-Spring, several places of worship for dissenters, and some good and well-attended schools, have been established since the place has risen to its present importance. (See HOUGHTON-LE-SPRING.)

HEXHAM, a market town and par. of England, co. Northumberland, S. div., Tyndale ward,

20 m. W. Newcastle, and 23 m. E. Carlisle, on the Newcastle-Carlisle railway. Pop. of town 4,655, and of par, 6,479 in 1861. Area of par., 29,370 acres. The town stands on a high bank S. of the Tyne, a little below the confluence of its N. and S. branches, and in the midst of a rich and well cultivated country. A handsome stone bridge of nine arches connects it with the N. bank of the river. The streets, though narrow and irregular, contain several good houses; and the market-place, with the conduit in the centre, is a handsome quadrangle, on the S. side of which is an old market-house, supported by pillars, and beneath it are stalls for butchers and country dealers; on the E. side, surmounted by a stone tower, formerly used as the town gaol, is the ancient town-hall, where the manor court and petty sessions are held; and on the W. side is the Abbey church, partly in ruins, and now consisting only of a transept and choir of mixed Norman and Gothic architecture, with a square tower, 90 ft. high, rising from the centre of the building. The living is peculiar to the prov. of York, and the great tithes are appropriated to one of the stalls in York cathedral. The R. Catholics have a handsome chapel, besides which there are places of worship for Wesleyan Methodists, Independents, and others. A free grammar school, founded by queen Elizabeth in 1598, was subsequently endowed with property for the education of the youth of this and of the adjoining towns and parishes. The foundation boys, whose number is not limited, pay a stipend of 7s. 6d. a quarter, and about forty more are educated with them, the instruction not being exclusively classical. A mechanics' institute, a savings' bank, and a dispensary have been established of late years.

Hexham has long been famous for a peculiar description of gloves, called 'tan-gloves;' they were formerly much worn, but of late years have fallen into comparative disuse. Hats and coarse worsted goods are also made in considerable quantities; and about half the pop. are employed in these branches of industry. Markets on Tuesday and Saturday, but chiefly on the former; and cattle markets on every alternate Tuesday. Fairs, Aug. 5, and Nov. 8, for live stock and woollen goods. The annual sales in the Hexham market average 4,000 qrs. of wheat, 2,000 qrs. of oats, and 1,500 qrs. of rye.

The site of the town close to Hadrian's wall, and the discovery of many Roman inscriptions, altars, and other monuments, have led to the supposition that it occupies the site of the Roman station Axelodunum. St. Wilfrid, archbishop of York, introduced into Hexham the arts of France and Italy. This prelate made it a bishop's see and a co. palatine; but in 883 it was united with Lindisfarne, and finally, in 1112, was annexed to one of the prebends in York cathedral. David, king of Scotland, shortly before the battle of Neville's Cross, halted here for three days. The church, which had been ruined, was rebuilt by Thomas, Archbishop of York, who also founded a priory of Augustine canons, the annual revenues of which amounted, at the dissolution of the monasteries, to 138l.

HIERES or HYERES, a town of France, dép. Var, cap. cant. on the S. declivity of a conical hill, 3 m. from the Mediterranean, and 34 m. SW. Draguignan. Pop. 10,360 in 1861. The town commands beautiful and extensive views, but its internal appearance is far from corresponding with its situation, its streets being steep, narrow, crooked, dark, and very badly paved. Its highest point is crowned by the ruins of an ancient fortress, from which descend on either side the traces of a line of thick walls, that formerly surrounded the whole town. In the *Place Royale*, a large but gloomy-looking square, is a column, surmounted with a fine marble bust of the most illustrious of its citizens, Massillon, born here on the 24th of June, 1663. The suburb at the foot of the hill is much pleasanter, and more frequented by visitors, than the town itself; it has some excellent hotels. It is said that Hieres was formerly a sea-port; at present, a plain of great fertility intervenes between it and the sea, covered with orange plantations, the best in France, vineyards, and olive grounds. The town has manufactures of orange-flower water and other perfumes; brandy, oil, and silk twist; and trades in these articles, olives and other fruits, and wine. Under the name of *Arrea*, this was one of the colonies anciently established by the Greeks on the shores of the Mediterranean; the Romans called it *Hierea*, but the monuments with which they embellished the city have entirely disappeared.

HIERES, ISLES OF (an. *Stoechades*), a group of four small islands in the Mediterranean, about 10 m. SE. Hyeres, and 14 m. ESE. Toulon. Porquerolles, the largest, is 5 m. long by 2 m. broad; it is fortified, and has about 100 inhab. Port-Cros has also a garrison, and about 50 inhab. The other islands are surrounded by several rocky islets. None of them is fertile.

HIGHAM-FERRERS, a bor., market town, and par., of England, co. Northampton, hund. of same name, near the Nen, 14 m. ENE. Northampton, and 83 m. N. London by London and North Western railway. Pop. of par. 1,152 in 1861; area of par. 1,871 acres. The town stands on a rocky height, commanding a fine view over the valley of the Nen. The church has a finely ornamented W. front, and a tower and spire 160 ft. high. A monastic college founded here in 1472 was surrendered in 1543, and a portion of its revenues was devoted to the endowment of the present free school, recently rebuilt in a handsome style. Higham-ferrers, which, many years ago, had a respectable lace-trade, is now quite insignificant as a place of industry. The place, which is a bor. by prescription, sent two mems. to the H. of C., from the reign of Philip and Mary down to the passing of the Reform Act, by which it was disfranchised. The franchise, though nominally vested in the freemen, was really exercised by earl Fitzwilliam, the proprietor of the greater part of the borough.

HIGHGATE, a village and chapelry of England, partly in Hornsey, and partly in St. Pancras par., co. Middlesex, hund. Ossulston, 5 m. N. London. Pop. of eccles. distr. 4,547 in 1861. The village stands on the top and sides of a hill about 450 ft. high; and many of the houses are well built, being occupied by opulent merchants and others belonging to London. On the top of the hill, on the road towards Barnet, is the Gatehouse, formerly a toll-gate at the boundary of the bishop of London's estates. For many years a tavern existed here, in which strangers were 'sworn at Highgate;' that is, in which an old custom was kept up of swearing them not to drink small beer when they can get strong 'unless they like it better.' The old chapel, built in 1565 as a chapel-of-ease to Hornsey, was replaced in 1832 by a church in the pointed style, contiguous to which is a spacious cemetery. The dissenters have three places of worship, to all of which are attached large Sunday schools. The grammar school, founded in 1562, was for many years almost useless; last, in consequence of the representations of the charity commissioners, a reform was

effected in its management, and it has lately become an efficient well-attended classical school. Many good boarding-schools for boys and girls are established in and about the village. There are almshouses for twelve poor persons, and two well-supported charity schools. E. of Highgate runs the old great north turnpike-road in an excavated hollow, about 60 ft. deep at one spot, where it is crossed by a bridge or archway, forming the thoroughfare to Hornsey. Close to the opening of the archway-road is the Foresters' hospital, a handsome Elizabethan structure, with two wings, and a chapel in the centre. Caen-wood, the beautiful seat of the earl of Mansfield, lies between Highgate and Hampstead.

HIGHLANDS. See SCOTLAND.

HILDESHEIM, a town of Hanover, cap. of prince. and landdrostei, on the Innerste, a tributary of the Leine, 19 m. SSE. Hanover, and 41 m. N. Göttingen, on a branch of the railway from Hanover to Göttingen. Pop. 17,134 in 1861. Hildesheim is an old town, surrounded with ramparts, now used as public promenades, irregularly built, and having extremely narrow streets. Among its churches, the cathedral, erected by Louis the Pious, in 818, is remarkable for its fine bronze gates of the 11th century, its paintings on glass, and for a hollow pillar of greenish stone, supposed to have been a heathen idol, and now surmounted by an image of the Virgin Mary. This, and three other churches, belong to the Roman Caths., who have also a consistory and a divinity college, attended by forty-two students. The other educational establishments are a Lutheran gymnasium with a good library, nine schools, and a large and admirably regulated poor-school connected with a house of industry. Among public buildings and institutions are the episcopal palace, council-hall, treasury, lunatic asylum, three orphan houses, and an establishment for the deaf and dumb. The trade of Hildesheim is inconsiderable, except in coarse linen cloths and yarn; its other products are leather, soap, starch, snuff, bleached wax, and earthenware: but cattle-fairs are held here said to be the largest in the kingdom.

HILLAH. See BABYLON.

HIMALAYA MOUNTAINS (THE), (San. *Himalaya*, abode of snow; ar. *Imaus* or *Emodus*,) an extensive mountain range of Asia, and the loftiest of which we have any knowledge, bounding the low and level plain of Hindostan on the N., and separating it from the table-land of Thibet, which stands 10,000 ft. above the sea. This chain is continuous westward with the Hindoo-koosh and Belut-tagh, and E. with the table-land of Yun-nan: but the term Himalaya is usually restricted by geographers to that portion of the range lying between the passages of the Indus and Brahmaputra, or Sanpoo; the former being in lat. 35° N., and long. 75° E., and the latter in lat. 29° 15' N., and long. 96° E. The direction of the range, as thus defined, is SE. from the Indus to the Gundak, and thence E. to its termination. Its entire length is 1,500 m. its average breadth 90 m., and the surface which it covers is estimated at 160,000 sq. m. The NW. extremity of the chain, called the Gomele mountains, extends in a NE. direction along the sources of all the Punjab rivers, except the Sutledje, and separates the hilly part of Lahore from Little Thibet. E. of the Sutledje, which cuts a passage through the mountains, in lat. 31° 30' N., and long. 77° 40' E., the range, still running NE., crosses the heads of the Jumna and Ganges; it then, in its course E., gives rise successively to the Gogra, Gundak, Cosi, Mahanuddu, and Teesta, and is bounded on both sides at its E. extremity by the circuitous channel of the San-poo, to which, however, it contributes few affluents of importance. The average height of the Himalaya chain has been estimated at 15,700 ft.; but numerous peaks far exceed in altitude the Chimborazo of the Andes, so long supposed to be the highest point on the globe. The principal of these are as follows, with their situation and height from the sea.

Name	W. Lat.		E. Long.		Height
					Feet
Jumnotri, in Gurwhal	31°	7'	78°	56'	21,154
Budrinath, do.	31	42	79	20	71,441
Dhawahir, in Kumaon	30	72	73	67	71,749
Dhawalagiri, in Nepaul	78	31	63	30	76,492
Kunchinganga, E. Peak, in Sikkim	774	9	66	0	98,178
Do. W. Peak, in do.	—		—		77,876
Chamalari, in Bootan	28	4	60	52	74,3(...)

The passes over the main ridge, so far as we know at present, amount to about twenty, a few only of which are practicable for horses, sheep being chiefly used as beasts of burden over the steeper passes. Their height above the sea varies from 10,000 to 18,000 ft.; the principal are, the Kandrihall pass, between Cashmere and Ladak; the Paralaha (16,500 ft. high) leading from the Upper Chenab valley to Ladak; the Shatool, Boorendo, and Pirting passes, all much frequented, on the road N. up the valley of the Sutledje; the Ghang-tang-ghaut (10,150 ft.), practicable for horses, and leading up the bed of the Bhagirathi to Chaprung, a Chinese post on the Upper Sutledje; the Neetee-ghaut (16,814 ft.), used by the great caravans passing between Thibet and N. Hindostan; the Ibraghaut (17,790 ft.), also a much frequented route, connecting the valley of the Kalee with Dumpa, in Thibet; and the Mantang pass, near the source of the Gunduk: the passes to the E. of this river are little known. The glens, through which these mountain-tracks run, are usually at right angles with the main range, and the NW. face is invariably rugged, and inclined at an angle of 50°, while the SE. slope is more smooth, and has an inclination of only 27° or 30°. (Lloyd and Gerard, ii. 29, 61.) The limits of perpetual congelation in the Himalaya chain, which, according to Leslie's theory, would be 11,400 ft. above the sea, have been ascertained, by the observations of Webb, Gerard, &c., to be generally higher; and they have likewise proved that, while the snow-line on the S. slope is at an elevation of 12,000 ft., the mountains on the side of Thibet are free from snow in summer as high as 16,600 ft. This unexpected circumstance is attributed by some to the difference between the serene climate of Thibet and the foggy atmosphere of Hindostan; but by Lyell and others, with more probability, to the influence of the heat radiated by a great continent in ameliorating the cold. (Lyell's Geol., i. 181.)

Geology.—The only rock sufficiently extensive to characterise the geological formation of the great chain is gneiss, which constitutes the substance of the highest ridges and crests. Granite veins occur on the surface only in some directions, intersecting the gneiss; but Captain Johnson and other travellers are of opinion, that granite forms the base of the mountains, and that gneiss is superimposed on the general bed. On leaving the centre of the range, schistus and clay-slate, primitive and secondary limestone, and red sandstone are successively met with on either side. Even in the centre of the chain, however, masses of limestone and sandstone have been found at an elevation of 16,000 and 18,000 ft., locked here and there in unraised crystalline rocks, a phenomenon observable also in the Alps and Pyrenees. (Geog. Journal, iv. 64.) The fossil remains found in the Himalaya mountains consist of bones of many different species of ruminating animals (some of which were found by Captain Webb at an elevation of 16,000 ft.), of ammonites, belemnites, and various kinds of land and fresh-water shells. The chief minerals hitherto found are sulphur, alum, rock-salt, gold dust, copper, lead, iron, antimony, and manganese; and the mines of Nepaul are reported by Buchanan Hamilton to produce large quantities of lead, copper and sulphur. (Hamilton's Nepaul, Introd.) There are no direct traces of volcanoes in the districts explored by the English; but the numerous thermal springs (that of Jumnotri having a temperature of 194° Fahr.), and many shocks of earthquakes felt by travellers in different parts of the range, indicate it to be the focus of subterraneous movements and derangements of the earth's crust. Among the physical phenomena observed on this great chain may be mentioned the falls of the Pabar, the highest known, and exceeding 1,500 ft., and the dripping rock of Sanodarmah, near Deyra Doahl, in Gurhwal, resembling, though on a larger scale, those of Knaresborough in Yorkshire, and Roslyn, near Edinburgh. This rock, situated in a glen surrounded by mountains rising almost perpendicularly to the height of 6,000 ft. and clothed to the very top with the most beautiful wood, overhangs a small basin of water like the roof of an open piazza, extending about 50 yards in length; and above it is a small stream, which being absorbed by the marshy nature of the soil, is filtered through it, and falls into the basin in a continual shower. The roof of the rock, and also of a neighbouring cave, are covered with stalactitic incrustations, which in some cases have descended to the floor, having the appearance of sparkling pillars. (Capt. Johnson, in Geog. Journ. iv. 43.; and Hamilton's Gaz.)

Vegetation.—The height at which plants and trees flourish on the Himalaya range varies on the N. and S. slopes, nearly proportionally to the difference in the altitude of the snow-line. On the N. slope grain cultivation is not attempted higher than 10,000 ft.; the highest habitation is at an elevation of 9,500 ft.; pines (which form by far the largest proportion of forest in every place) show their best growth at a height of 10,500 ft.; but beyond 11,000 ft. they grow in smaller quantities, and are of less girth and growth. The rhododendron grows up to 12,000 ft., and birches are found as high as 13,000 ft. above the sea. (Gerard and Lloyd, i. 345., ii. 9.) On the S. side, villages are found between 11,000 and 13,000 ft. high, and grain cultivation advances to a height of 13,500 ft.; birch-trees rise to 14,000 ft.; and vegetation is found up to an elevation of 17,500 ft. that is, upwards of 3,000 ft. higher than on the S. slope. The grains found on these heights are wheat and barley, bhatoo (Amaranthus anardhana), cheenah (Panicum miliaceum), khula (Paspalum scrobiculatum), oat (Hordeum cadrate), and phapar (Polygonum tartaricum). Strawberries and currants thrive on the N. side at a height of 11,600 ft., and 1,000 ft. higher on the opposite side.

Zoology.—The mammalia of the Himalaya range are chiefly confined to ruminating animals, a few varieties only of the horse and cat tribe being found in these regions. The wild horse is seen on the N. side of the mountains; but the principal tenants of the hilly pastures are the yak (Bos grophagus), much used as a beast of burden by the Tartars, the ghoral (Capra aegagrus), of which the Cashmere and Thibet goats are varie-

tra, the musk-deer, the Nepaul stag, the black deer, the *Cervus Caprœlus*, the chru or one-horned antelope, the goral, and the nylghau. Among the birds of the Himalaya may be mentioned the lammer-geyer (*Gypaetus barbatus*), the chorcotee (*Perdix rufa*), the common cuckoo, the Impeyan pheasant (*Lophophorus refulgens*), the red-legged crow, and the sand-pigeon. (Ritter's Asia, ii., iii.; Geog. Journ., iv.; Lloyd and Gerard's Tour in the Himalaya; and Bergham's Asien, with Maps.)

HINCKLEY, a market town and par. of England, co. Leicester, hund. Sparkenhoe, 12 m, SW, Leicester, and 102 m. NWN, London, by London and North Western railway. Pop. 6,544 in 1861. The town stands on a commanding eminence close to Warwickshire, from which it is divided by the old Roman Watling Street. It is well built, though old, and near the centre stand an ancient town-hall and school-house. The church is a fine old Gothic building, with a tower and steeple 130 ft. high. The dissenters have several places of worship, connected with which and the church are Sunday schools. There are also national and infant schools. The staple manufacture of the place is hosiery, introduced about 1640, and now employing in the town and neighbourhood upwards of 2,500 hands. Coarse substantial stockings are said to be made here in larger quantities than in any other part of England. Markets (well attended) on Monday; Shire lot, 2nd, and 3rd Monday after Epiphany; Easter Monday, Monday before Whit-Sunday, and Whit-Monday, for horses and live stock; Aug. 26, and Monday after Oct. 26.

Near the Ashby-de-la-Zouch canal, which passes close to the town, are the remains of a Roman fortification, and the remains of a wall and ditch, traceable all round, indicate Hinckley to have been formerly a place of some importance.

HINDOSTAN, or INDIA ON THIS SIDE THE GANGES or BRAHMAPUTRA. *Name and Limits.*—The ancient inhabitants of India had no common name for themselves or their country; but their Persian neighbours called the people Hindoos, and the country, as far as they knew it, Hindostan; words which, in old English, would have been accurately as well as literally rendered, 'Negro,' and 'Negroland.' The comprehensive sense in which the term Hindostan is now employed, as distinctive of the entire territory S. of the Himalaya mountains over which the institution of castes prevails, is of European origin; the people of the country confining the term to the territory lying N. of the Nerbuddah, and calling all to the N. of that river the Deccan, a word derived from the Sanscrit, and meaning 'the right hand,' and also 'the south.' In the European sense, Hindostan comprises the whole of that vast triangular country extending from the borders of Little Thibet, in about the 35th deg. of N. lat., to Cape Comorin, in about the 8th deg. It is bounded on the N. by the highest range of mountains in the world, the Himalaya; and by the two great rivers, the Brahmaputra and Indus, on the NE. and NW.; and in every other direction by the ocean. It comprises in all an area of between 1,200,000 and 1,300,000 sq. m., or about a third part of the estimated area of Europe; but from the absence of gulfs, inland seas, and lakes, the proportion of solid land is greater.

Surface and Geology.—The surface of Hindostan, taking this word in its widest acceptation, is of a very marked character. On the N., constituting the base of the triangle, we have three great ranges of mountains, with elevated valleys between. These chains rise, the one higher than the other as we proceed northward, the last

constituting the highest mountains hitherto discovered. For 1,000 m., from China to Cashmere, a plain might be extended, resting on peaks 21,000 ft. high, while some are even 4,000 ft. above this elevation. The valleys themselves are from 3,000 to 4,000 ft. above the level of the sea. Primitive rocks alone compose the higher ranges. Gneiss predominates; but with it is found granite, mica slate, hornblende schist, chlorite slate, crystalline limestone, and marble. On these repose clay slate and flinty slate. In the lower or southern range, sandstone composes that portion which terminates in the plain of the Ganges. Crossing this plain, and proceeding southward, we come to another chain of mountains, the Vindhyan range, running nearly E. and W., across the centre of Hindostan, in about the 23d deg. of lat. This is the basis of a triangle of mountain ranges which supports the vast table-land of Central India. The formation here is primitive, consisting chiefly of gneiss; but where it terminates in the plain of the Ganges, and forms the N. barrier of the latter, the formation is sandstone, as on the N. side of the same plain. The great W. range of mountains commonly called Ghauts, commences on the NW., where the Vindhyan range terminates, and runs in a direction nearly N. and S., to between the 10th and 11th deg. of latitude, until at Coimbatore they meet the E. range, or Ghauts. The formation of this chain is primitive; but to the N. there is a great extent of overlying trap, columnar, prismatic, tabular, and globular. To the S., again, the overlying rock to a great extent is laterite, or clay iron-ore. The W. is much more elevated and continuous than the E. Ghauts, and some of its highest granitic peaks rise to the height of from 6,000 to 8,700 ft. It is remarkable for the absence of valleys of denudation, and of rivers running W., but is covered with extensive rivers flowing E. In fact, the sea, in some situations, comes up to the very foot of the mountains, and nowhere leaves anything more than a narrow belt of low land, much broken by deep and narrow inlets. This is the coast of Malabar, exposed to all the violence of the SW. monsoon, blowing without interruption for six months from the coasts of Africa and Arabia. Where the E. and W. Ghauts meet, commences the remarkable valley or gap of Coimbatore, which leaves a clear breach in the mountain chains, extending from the E. to the W. sea. A single chain of the same formation as the E. Ghauts then runs all the way to Cape Comorin, leaving the plain of Travancore to the W., and the more extensive plain of Madras and Tinnevelly to the E. The E. chain, or Ghauts, may be said to commence at the Neilgherry hills, which are among the highest mountains of S. India. From this point they diverge in an E. direction, and soon break into a succession of parallel ranges less elevated and more broken than the W. Ghauts. In their further progress to the N., the E. Ghauts break into subordinate ranges and valleys, which give passage to the great rivers that drain nearly all the waters of the peninsula into the Bay of Bengal. This range terminates nearly in the same parallel of latitude to the W. Granitic rocks, especially sienite, form the basis not only of the E. chain, but of the range which runs from the gap of Coimbatore to Cape Comorin. The sienite discovers itself at all the accessible summits, from Cape Comorin to Hyderabad, from the 8th up to the 17th deg. of latitude. Resting on the granite, gneiss, and talc-slate, that form the sides and bases of the E. chain, are sometimes seen clay, hornblende, flinty and chloride slate, with primitive marble of various colours. At the Pennar river, in the 14th and 15th deg. of latitude, clay, iron-ore, or laterite, expands over a large surface, and sandstone begins to appear. At Vizagapatam, Ganjam, and Cuttack the same formation continues, and the laterite extends through Midnapore up to Beerbhoom, sometimes reposing upon sandstone. A cellular carbonate of lime, called kankar, peculiar to the geology of India, is found over all the district now named, as well as in many other parts of Hindostan. The great coal-field runs for 65 m. in length, and 12 in breadth, on both sides the river Damuda. It is supposed to cross the Ganges, and to extend all the way to Sylhet and Cachar, from which places abundant specimens of surface coal have been brought. The rock formation here consists of sandstone, clay-slate, and shale, the latter, as usual, lying immediately over the coal. Mr. Jones, an English miner, opened the first colliery in India, in the year 1815, at this place. The pits are to the depth of 90 ft.; seven seams of the mineral have been met with, one of them of the thickness of 9 ft.: coal is now largely consumed in Calcutta, chiefly for forges and steam navigation. From the Damuda river to Benares granitic rocks prevail. On approaching the river Soane, however, sandstone becomes the surface rock, and, one interval excepted, extends to the N. of Agra, as far as the 28th deg. of latitude. The exception occurs in the lower portion of the province of Bundelcund, where granite again prevails, while the upper consists of sandstone. The great surface formations of the table-land itself are granitic, including always gneiss and sienite, with sandstone and the overlying rocks. Basaltic trap extends over the provinces of Malwa and Nagpur, proceeds by Nagpore, sweeps the W. portion of the Hydrabad territory down to the 15th deg. of lat., where it bends to the NW., and running all the way to the coast of Malabar, forms the shores of the Concan. In all, it seems to cover an area of about 200,000 sq. m. We may observe here that the geological formation of India is extremely simple, compared with that of European countries, consisting only of four classes of rocks, viz. the granitic, the sandstone and clay-slate, the trap, and the alluvial. Of the latter an example on a great scale is in the plains of the Ganges and Indus, which meet between the 28th and 31st deg. N. lat., and the 76th and 77th deg. E. long.; as well as in the plain lying between the E. Ghauts and Bengal from Cape Comorin to Cuttack.

The natural geographical divisions of Hindostan are as follows:—1. The ranges of the Himalaya with their valleys. 2. The Gangetic plain, comprising only the tract of inundation, and which rises very little above the level of the sea. 3. The upper plain of the Ganges, from the province of Bahar inclusive, up to the foot of the range of the Himalayas, where the Ganges and Jumna issue from the hills to the N., bounded to the S. by the Vindhyan range, and to the W. by the great desert. The height of the E. portion of this division may be about 600 ft. above the level of the sea, and the land rises gradually as we proceed N., until, where the great rivers emerge into the plain, it has an elevation of 1,000 ft. 4. The N. portion of the great central table-land, as far S. as the valley of the Nerbudda, which generally interpects the table-land in question from E. to W. The height of this portion of the table-land ranges from 1,700 to 2,000 ft., as at the towns of Oojein, Indore, and Mhow. 5. The portion of the table-land which lies S. of the valley of the Nerbudda, down to the junction of the E. and W. Ghauts, and the valley of Coimbatore. The height of the table-land ranges here from 2,000 ft. to 3,100 and

8,000, as at Poonah, Seringapatam, and Bangalore. 6. From the gap of Coimbatore inclusive to Cape Comorin. 7. The narrow strip of low land lying between the W. Ghauts and the sea, or coast of Malabar, including the W. acclivities of the mountains themselves. 8. The alluvial plain, of unequal breadth, which lies between the E. Ghauts and the Bay of Bengal, generally called the Carnatic, rising gradually from the shore to the foot of the mountains: at the town of Arcot, 60 m. inland, it is 490 ft. above the level of the sea;—and 9. The peninsula of Gujrat, with the adjacent country, containing much mountain-land and a few plains. All these differ so materially in their physical aspect, climate, geological formation, animal and vegetable productions, as well as in the character of the nations and tribes which inhabit them, as fully to warrant this distribution.

Rivers.—The rivers of India have their sources either in the Himalaya mountains, or within the great central table-land. The first class are by far the largest and most important. Beginning from the E., the first great river which occurs is the Brahmaputra. The source of this stream is not exactly ascertained; but its course has been estimated at about 800 m., and it is believed to discharge a larger volume of water than even the Ganges. Its course in the plain of Bengal, from Goyalpara, to the bottom of the Bay of Bengal, where it debouches, is but 350 m.; and having a rapid current, and passing generally through a wild and uninhabitable country, it is of comparatively little service to commerce or navigation. The Ganges, called Gonga by all the Indians, has its origin in two principal branches, about 31° N., lat., and between 79° and 80° E. long. Its whole course is reckoned at about 1,350 m.; but from its entrance into the plain at Hurdwar, its course to the sea, into which it falls within a few m. of the Brahmaputra, is about 1,200 m. Within the plain all its branches are navigable for boats, and the Bhagheruttee, its most W. branch, usually called by Europeans the Hooghly, is navigable for ships of 400 tons burden, as far as Calcutta, 100 m. from the sea. According to Major Rennel, the principal branch discharges 80,000 cubic ft. of water per second. The greatest of the affluents of the Ganges is the Jumna. It also has its origin in two branches within the highest masses of the Himalaya, to the W. of the sources of the Ganges. Its course within the mountains is about 120 m.: it issues into the plain about 30 m. W. of the Ganges, and here its bed is about 1,200 ft. above the level of the sea. In the course of a few miles, however, passing over some falls, it takes a lower level. After a course of 450 m., passing by the Mohammedan capitals of Delhi and Agra, and being navigable for a great part of its course, it joins the Ganges at Allahabad. The other principal affluents of the Ganges which take their source from the Himalaya, are the Ram Ganga, which joins the Ganges above Canoge; the Goomtee, which passes by Lucknow, and after a winding course, whence it derives its name, joins the Ganges between Benares and Ghazeepoor; the Gogra, with a course of 600 m., and the largest of the affluents of the Ganges on this side the Himalaya, after passing through Fyzabad and Oude, joins the Ganges above the town of Chupra; the Gunduck, which has a course of 450 m.; the Bagmutty, which passes close to Catmandoo, the capital of Nepaul; and the Cosy, originating in the table land of Tibet, and which enters the Ganges at Boglipoor. The great delta of the Ganges may be said to commence at Sicligully. The first bifurcation of the Ganges itself commences at Soory, 70 m. below Rajemahal, at which last place the river is pressed in by some low hills of that name. The Ganges receives, after this, from the Himalaya, the Mahanada and Teesta, which have their sources in the mountains of Nepaul and Bootan, with courses of from 230 to 300 m. After the junction of these, the Ganges communicates with the Brahmaputra by a variety of branches. The rivers which fall into the Ganges, or its affluent the Jumna, from the N., acclivity of the central table-land, are the Soane, the Betwah, and the Chumbul; the latter has a course of 400 m. Both it and the Betwah fall into the Jumna. The Soane is an affluent of the Ganges, and falls into that river a little above Patna. (See Ganges.)

Lakes.—India is remarkably deficient in lakes, and in fact contains no large collections of water, fresh or salt, such as the lakes of N. America, N. Asia, Switzerland, or even Scotland. In the N. parts of Bengal there are a few freshwater lakes of some extent, but the greater number of this description found throughout the country are supposed to be nothing more than the old channels of rivers which have taken a new course. Of the same character, in some respects, are the Chilka lake in Cuttack, and the Colair lake in the Circars; the first of which communicates with the Mahanuddy, and the last with the Godavery and Kistna. The Chilka lake is 35 m. long and 8 broad, and contains several islands, and abounds in fish; it is separated from the sea by a sand-bank not above ½ m. broad. The Colair lake is 24 m. by 12 in the dry season, but during the periodical rains, expands from 40 to 50 m. in length. During the latter period, the whole flooded country, including the islands of the lake, are fertilised by the deposit of mud brought down by the two rivers; and hence Major Rennel, with some propriety, compares the neighbouring country to the delta of the Nile, or sandy desert to the W. of the plain of the Ganges several salt lakes occur, the largest of which, however, does not exceed 20 m. in length. Collections of salt water, more or less connected with the sea, are of more frequent occurrence. Several considerable ones of this nature are to be found on the lower E. coast of the continent; but the greatest and most remarkable is the Runn, lying between the Gulf of Cutch and the mouths of the Indus, which is believed to occupy a space of 6,000 sq. miles.

Coast Outline.—The outline of the coast of Hindostan is comparatively little broken by any considerable inlet of the sea. From the mouths of the Indus to those of the Ganges there are but three great gulfs, those of Cutch, Cambay, and Bengal; if the latter, indeed, which, though it breaks the coast of Asia, does not break the coast of Hindostan, can be reckoned in this class. Harbours are even less frequent. Along the W. coast, over 14° of lat., there is but a single good one, Bombay; and from Cape Comorin to the W. mouths of the Ganges, a distance of 1,500 m., there is not one. In this unfavourable feature of its geography India resembles more the W. coast of America, or the E. and W. coast of Africa, than the E. coast of America, or the shores of the S. countries of Europe. The Indian coasts are also in a great measure destitute of islands. Unless Ceylon be admitted as belonging to Hindostan, which can hardly be done, there is not one on the E. coast; and on the W. there are very few, and these of inconsiderable size. In this respect, Hindostan is remarkably distinguished from the two great corresponding Asiatic promontories of Malacca and Cambodia, the coasts of which are thickly studded with islands, many of them of considerable magnitude.

Climate.—In a country which embraces 27° of latitude, which contains extensive plateaus, elevated from 2,000 to 3,000 ft. above the level of the sea—some of the most extensive plains in the world, almost on a level with, or but a few hundred ft. above, the sea—the highest range of mountains in the world—tracts of bare rock—deserts of mere sand, and deep primeval forests,—it is needless to say that there must exist a very great diversity of climate. But besides the diversity arising from these causes, the distribution of rain is another source. The whole continent of India, up to the 35th deg. of lat., is subject to the influence of the monsoons, which blow from the N.E. during the severe temperate months of winter, and from the S.W. during the tempestuous and hot or rainy months of summer and autumn. This is the general rule; but in India, as in other countries of Asia under the influence of the monsoons, and where there are ranges of mountains running N. and S. of sufficient elevation to intercept the clouds, the time of the periodical fall of rains is reversed. To the W. of the great chain of the W. Ghauts, on the one hand, over 11° of lat., the periodical fall of rains corresponds with that of other parts of India, or takes place during the W. monsoon. E. of the Ghauts, on the other hand, over 8° of lat., the fall of rain takes place during the E. monsoon; while the table-land which lies beyond the two ranges partakes, to a moderate degree, in both falls. As a general rule, the year is divided in India into three well-defined seasons; a hot, corresponding with part of spring and summer; a wet, corresponding with part of summer and autumn; and a cold, corresponding generally with our winter months. With respect to temperature, much of India being within the tropics, and the remaining portion within 13° of the tropic, the whole is entitled to the designation of a hot country. On the low plains within the tropic, and up to about the 18th deg. of lat., winter is scarcely perceptible, and the year may be said to be divided into wet and dry. From that parallel N., winter becomes more and more distinct, and beyond the 27th deg. lasts for six months, during which the climate is not inferior in point of agreeableness or salubrity to that of Italy. This is, however, counterbalanced by the severity of the hot and dry season, which lasts for three months, and is so intense as nearly to destroy all appearance of vegetation. On the elevated central plateau, the temperature is generally from 8° to 10° Fahr. lower than in the same latitudes on the low lands, and the fall of rain being more equally distributed, the necessary effect is a climate in general temperate and agreeable, though not always salubrious. In the valleys between the two great chains of the Himalaya, the same order of seasons generally prevails as in the plains, and here the thermometer is rarely less than 18° or 20° lower than in the plains under the same parallels. A few examples may be given of temperature, as indicated by the thermometer. The mean temperature of Bombay is 82° Fahr., and in the table-land in the same latitude, at an elevation of 1,700 ft., it is 77°. At Madras the mean annual temperature is 84°, and at Barwar on the table-land it is 75°. At Ustarunnd, in the Neilgherry mountains, 7,000 ft. above the level of the sea, the mean temperature is 56°, or 28° lower than that of Madras. Here the thermometer sometimes rises as high as 69°, and rarely falls as low as 20°. In the peninsula of Gujrat, and on the level of the sea, the thermometer occasionally rises to 100° in summer, and falls to 45° in winter. The mean annual temperature of Calcutta is 79° Fahr. In May, the hottest month, it is 84°, and in Jan., the coldest, 67°. In summer, however, the thermometer frequently rises above 100°, and in winter falls so near the freezing point that, with a trifling assistance from evaporation, ice is easily obtained. Within the upper portion of the plain of the Ganges, both the latitude and elevation contribute to reduce the temperature. From the middle of Dec. to the middle of Feb. the thermometer sinks every day below the freezing point, and small pools of water are covered with ice, and the average temperature of Jan. is 57°. From April till the middle of June, when the rain falls, the thermometer gradually rises to 94°, and even to 110°; and at Delhi, Agra, and other places on the W. bank of the Jomna, in the whole period from March to June, scorching SW. winds, proceeding from the desert, prevail. It is in these same countries that, during the whole period from the beginning of Nov. to that of March, the climate equals that of S. Italy.

Nations and Tribes.—Besides foreigners, who, as peaceful emigrants, or conquerors, have settled in India during the last twelve centuries, but chiefly during the last eight, the number of aboriginal races distinguished by differences of language, manners, states of society, and great variation, if not difference of religious belief, is still very great; and undoubtedly was much greater before the blending which must have been more or less the result of the extensive conquests of the N. invaders. These have been in active operation for nearly seven centuries, and, in all likelihood, have been materially promoted by the conquests of the more powerful Hindoo states over the smaller. There are at present spoken in India, by the most civilised races, not less than 25 distinct languages or dialects, indicating the existence of as many distinct nations; but, including tribes more or less savage or barbarous, at least 50 languages, indicating the presence of at least as many distinct tribes. Of the more civilised nations, eight may be said to be distinguished from the rest by some superiority of civilisation, as implied in the possession of a national literature, a national alphabet, superior population, superior industry, a greater progress in the useful arts, with the richer and more extensive territory which they are found to occupy. These are the Bengalee, Ooriya, Mahratta, Gujratee, Telinga, Tamul, Karnata, and Hindi or Hindostanee nations. The Bengalee nation occupies above 80,000 sq. m. of fertile land, chiefly within the delta of the Ganges, and amounts in number to above 25,000,000. The Tamul nation occupies 56,000 sq. m. at the S. extremity of the peninsula, and numbers between 6,000,000 and 7,000,000 people. The Telinga nation occupies 100,000 sq. m. of the NE. portion of the peninsula, and numbers probably between 7,000,000 and 8,000,000 people; and the Ooriya nation occupies at least 17,000 sq. m. of the low land which connects the delta of the Ganges with the S. peninsula, and numbers about 4,000,000. The Mahratta nation extends probably over 200,000 sq. m. of territory, laying between the 72nd and 23rd degrees of N. lat. and its numbers may be roughly computed at 12,000,000. The Karnata or Canara nation, occupying a central portion of the table-land S. of the 16th degree of lat., may occupy about 75,000 sq. m. of territory, and their numbers may be taken at about 5,000,000. The nation speaking the Hindostanee or Hindu language occupies at least 100,000 sq. m. of the upper portion of the valley of the Ganges, and cannot amount to less than 20,000,000, physically and intellectually the most vigorous of all the Indian races. The most enterprising of these nations, it is to be observed, have occasionally passed, either

as conquerors or colonists, into the territories of each other or of their neighbours. Thus we find colonies of the Tamuls settled in Malayalim; of Telingas in Karnata and the Tamul country; of Mahrattas in the Tamul, Telinga, and Karnata countries; of Karnatas colonised in the countries below the E. Ghauts; and colonies from the upper plain of the Ganges settled as far as Gujrat, Bengal, Nepaul, and even Malabar. These colonies, of whatever nation, not unfrequently preserve their national language, their original manners, and even the purity of their descent, in their adopted countries. The barbarous and savage tribes of India are universally to be found in the recesses of mountainous and hilly regions, never within the fertile plains or extensive table-lands; and there is scarcely any considerable range throughout India in which some of them are not to be found. They are, however, most numerous on the E. frontier of Bengal, in the fastnesses of the mountains and sterile region of Gundwana, and generally in the ranges of hills which lie between the Gangetic plain and the great central plateau. These barbarous tribes have been supposed by some observers to be the aboriginal natives of the country driven from the plains to the hills by strangers and invaders; but this hypothesis seems little better than a gratuitous assumption; the mountaineers are no doubt aboriginal, in common with the inhab. of the plains, and their barbarous condition seems naturally enough accounted for by the unfavourable circumstances of their situation, and their remaining in that condition to the hostility of the powerful occupants of the lower and more fertile lands.

Foreign Settlers.—Besides the original and peculiar inhab. of Hindostan, a crowd of foreign colonists or settlers of different nations, either scattered indiscriminately over the country or confined to particular spots, from the accident of their arrival or other chance, forms a considerable proportion of the general population of the country. These, following generally the order of their arrival, or supposed arrival, are as follows:—Jews, Syrian Christians, Arabs, Armenians, Persians, Afghans, Tartars, Turks, Abyssinians, Portuguese, English, Dutch, French, Danes, and Chinese.

Hindoo Religion.—The forms of religious worship which prevail are the Brahminical, Buddhist, Jain, Seik, Mohammedan, Jewish, and Christian. These, and especially the most prevalent of them, are again divided into many sects. But besides national, colonial, and religious distinctions, there are other nearly innumerable divisions of the great mass of the people. Many are distinguished by the profession which they have immemorially followed; many by their condition as slaves; and many as outcasts, without being slaves: some are in the hunter, and a few in the pastoral state: some are freebooters, others pirates; and there are whole tribes who have, time immemorial, been illustrious as thieves, robbers, highwaymen, and professional assassins. These distinctions into tribes and families are all hereditary; each section and even subsection is isolated by nearly impassable limits from the rest of the society. In the prov. of Malabar, for example, which contains but 6,000 sq. m. and about 900,000 inhab., there are almost 800 different tribes, few of which are founded on distinctions strictly religious or national. In Canara, with an area of 7,700 sq. m. and 657,000 inhab., there are, exclusive of strangers, and foreign settlers, 104 native castes; and in the rural district of Burdwan, in Bengal, it was found that in 36 villages, containing a pop. of about 40,000, there existed, independent of strictly religious distinctions, no fewer than 44 castes, chiefly discriminated by the trades or professions which they followed, each caste being known by a distinct name, each being hereditary, and each—at least theoretically—incapable of eating, drinking, intermarrying, or in any other manner intimately associating with the others. The circumstances on which this almost infinite distinction is founded are often trivial, and sometimes even ludicrous; and yet the practical separation is not therefore the less real. For example: one tribe of oilmakers in Telingana, who use two oxen in the mill, will hold no intercourse with another following the same profession, but who use one only; they will neither follow the same gods nor the same leaders. The great division of the right and left hand, which prevails throughout the S. parts of India, but which is not known in the N., does not appear to be of a religious character. One of these tribes ranges itself on one side, and another on the opposite; and serious disturbances of the public peace are not unfrequently the result of quarrels which concern neither religion nor politics.

Under the general name of the Hindoo religion are comprised many different doctrines, and an infinity of sects and castes, which it would be almost impossible to describe, or even to enumerate. This religion, perhaps beyond any other, pervades the entire frame of civil society, and mixes itself up with every concern of life, public, private, and domestic. A Hindoo can neither be born, die, eat, drink, or perform any of the most ordinary or even vulgar functions of the animal economy, unembarrassed by its trivial and unmeaning ceremonies: military enterprises, the details of commerce, and the operations of agriculture, are more or less under its guidance; it is part and parcel of the code of laws, or, to speak more correctly, it is itself the law. Almost every act of a Hindoo may, in fact, be said to be more or less a religious act. The most civilised and instructed of the Hindoos, but these only, believe in the immortality of the soul, and in a future state of rewards and punishments. The belief in the transmigration of souls is somewhat more general, but far from universal. There are reckoned to be four orthodox sects, whose principles are determined by the preference they give to their worship to some one of the greater gods of the Hindoo pantheon; for there are gods, great and small, some almost omnipotent, particularly for mischief, and others so feeble as to be all but contemptible, and no match even for an ordinary Brahmin. According to the best authorities, the Hindoo pantheon is peopled by precisely 333,000,000 deities; but as no one has attempted to name them, it can only be concluded that the Hindoo deities are in reality innumerable. They consist of three principal gods, who are supposed to represent (but their powers and functions are frequently interchangeable at the caprice of their votaries) the powers of creation, destruction, and preservation; and of the families of these, with deifications of the elements and powers of nature, of heroes, and especially of saints and abstract ideas. Among the lower orders of the people, and especially among the ruder tribes, a sort of fetichism prevails; and trees, rocks, and rude masses of stone are worshipped or abandoned, according to the fears, hopes, or caprices of their votaries. The present race of Hindoos are tolerant in all matters of religion, or, to speak more correctly, they are indifferent: in fact, they go even beyond indifference, and in cases of emergency are ready to invoke any strange god, or strange saint, by whose aid they may hope to profit. The Mahratta chiefs are in the frequent practice of invoking Mohammedan

saints; and Madajee Scindia, the chief of the Mahratta state, a shrewd and politic prince and a great conqueror, was in the habit of making frequent offerings at the tomb of a celebrated saint in Ajmeer, the same to whom shrine Akbar, the most illustrious of the Mogul emperors, walked 230 m. barefooted. The Mohammedans of the lower orders, who in some parts of the country are indeed little better than Hindoos, return the compliment, and in their need propitiate the gods of the Hindoos; and each will join in the religious festivals and processions of the other. In the N. of India the Hindoos, in their distress, will not unfrequently propitiate even the Catholic Christian saints, and the Christian Hindoos reciprocate. It is not, as already stated, to matters of doctrine or morality, that the Hindoos attach importance. In the same tribe, or even family, will be found sectarians of the Destroying Power, of his consort, of the Preserver in several of his incarnations (the Creator among the Hindoos has no worshippers), all intermarrying with each other, and the wife adopting the opinions of the husband without any difficulty. Some of the Christians of S. India intermarry with the Hindoos of their own tribe, without any forfeiture of caste on either side, provided external observances be attended to. Persecution in recent times is the exception; but the sectaries of Nanak or the Sriks, have been considerable persecutors in their way; they have destroyed most of the mosques within their territory, and will seldom allow Mohammedans to assemble in the few that remain: they forbid them from eating beef or praying aloud, according to law. What, however, the Hindoos really attach importance to are not doctrinal matters, but distinctions of caste, ceremonies connected with marriage and funeral rites, and the whimsical observances respecting supposed purity and impurity in regard to food and other matters connected with ordinary domestic life. The distinctions of caste are the most remarkable of these, and form indeed the characteristic mark of Hindoo society. Every one has heard that the Hindoos are divided into four great classes or castes, founded upon the great distinctions which prevail amongst all people in their first advance towards civilisation; that is, into priests, soldiers, traders, and labourers. As such a distinction into tribes is natural, and indeed known to have existed among other people, it is highly probable that it prevailed with the first rude tribe or nation with which the Brahminical form of worship originated, and that it constituted the foundation of the present superstructure of the castes.

The first in rank among the four great classes, of course, is the Brahmin or priest; and next to him comes, very naturally, the soldier; at a great distance follows the industrious capitalist or trader; and far removed from all is the labourer. These divisions are hereditary, impassable, and indefeasible. Such is the theory of the distinctions of Hindoo society; but the practical and real distinctions are very different indeed. The attributes of the different classes, as they are described in the ancient books of the Hindoos, we may be sure never could have been practically in operation. These books, it must be recollected, were written by Brahmins who claimed an exclusive right to expound them, and all but the monopoly of reading them; and it was their interest to dwell on the immeasurable superiority of their own order; but it is hardly credible that any society should be able to hold together for a moment, in which laws such as we find in the Hindoo sacred books were bona fide enforced. For example, it is enacted among myriads of the same sort, that if a labouring man sit upon the carpet of a priest, he shall be punished, either by having a hot iron thrust into his buttock, or by being branded, or banished the kingdom, or having the offending buttock cut off. Many, in fact, of the Hindoo laws appear to have been framed by the Brahmins more for the purpose of deterring, through the terrors of superstition and punishment, the other classes from interfering with their privileges, than for any other object. Whatever may have been the original attributes and privileges of the great classes, at their first institution, it is certain that these classes themselves can hardly be said at present practically to exist. In the advance of society, the increase of population, the extension of commerce and conversion, and the operation of the human passions, they have given way to a different order of things. In the most ancient Hindoo work extant, the *Institutes of Menu*, which has been computed to be about 2,700 years old, there already existed nearly a hundred castes; and it must be supposed that the enumeration was confined to that part of the country in which the work was written. The number of the castes not coming within the pale of the four great divisions, suggested the notion of the mixed castes, supposed to originate from an illegitimate intercourse between the four great orders, with the crosses which again sprung from these. This was clearly an afterthought—a new theory made for the occasion, and wholly inadequate to explain the actual state of society as we find it. This may be made sufficiently plain by a few examples. Among the 26,000,000 of people who speak the language of Bengal, there are none who even pretend to be of the second or third order, that is, of the military or mercantile classes: all who are of these two orders are comparatively recent immigrants from the north, and identified as such. In so far as the four great orders are concerned, the native inhabitants of Bengal consist, in fact, of Brahmins, and those who are not Brahmins. The Brahmins themselves consist here of two classes; viz. those who can trace their pedigree to the N. of India, and who are held in the highest repute, and of those who cannot—far less esteemed. The Brahmins of Bengal, including all of both these classes, consist of no less than 169 subdivisions, claiming various degrees of purity, and not one of which will eat, drink, or intermarry with another. The next most important caste in Bengal is denominated Chyeta, and is chiefly engaged in mercantile pursuits: this is the tribe whose name has been supposed by many to have furnished the English language with the word *caste*; but this is a mistake, for the term is simply the Spanish and Portuguese word *casta*, meaning race or lineage. They are reckoned pure Sudras, or persons of the original servile class, and amount to 88 subdivisions, equally unsocial among themselves with the Brahmins. Among the people speaking the Orissa language, the military order is altogether wanting, and there are but a few families generally reputed of the third class. In Malabar and Canara the second and third orders are wanting, the first of these being supplied by the military aristocracy of the Nairs, who are considered to be pure Sudras, or of the servile class, that is, of the class represented by the Hindoos as being in the last degree of degradation: they are, notwithstanding, the lords of the soil, and, before very recent conquest, the real sovereigns of the country. The celebrated Rajpoots, the most distinguished military order among the Hindoos, have, according to the Sanscrit writings, a vulgar origin; they are sprung from the mercantile classes on the paternal side, and from one of the mixed classes on the maternal. All the warlike and conquering nation of the Mahrattas, who are not Brahmins, are deemed to be of the fourth, or servile order. In every part of India there is a considerable portion of the inhab. who are utter outcasts, or, at least, beyond the pale of the Brahminical religion, condemned to this exclusion by their servile condition, their poverty, or the meanness of the employment in which they are engaged.

In the district of Dinagepore, in Bengal, out of a Hindoo population of 800,000, it was found that 8¼ per cent. only were considered pure tribes, 46½ impure, 18¾ very low, and 26½ abominable. In Malabar, out of a population of 720,000, 100,000 are in a state of slavery, and treated by the Brahmins and Nairs as if they were hardly human. Even a great proportion of the free and industrious classes must not approach, owing to their alleged impurity, the person of a Nair nearer than a prescribed number of paces, this Nair himself being, as already mentioned, the lowest of the original classes; a being, according to the ancient Hindoo writings, expressly created for the purpose of performing servile offices to the Brahmins and other superior classes. What is still more remarkable, and the same thing obtains with respect to many other impure classes in the S. of India, the Brahmins refuse to afford them instruction or spiritual comfort: in fact, they are not of the Brahminical religion at all; never enter the Brahminical temples, or offer worship to the gods of the Brahmins; but have their own peculiar deities, priests of their own caste, and, contrary to the creed of the Brahmins, usually have no knowledge of a future state. The Brahmins, although they are to be found throughout India, and have a vast influence everywhere, are divided into more numerous families and varieties than any of the other classes; and while each is revered by its own immediate followers among the laity, they almost all hold each other in contempt as pretenders. They go the length of reckoning no less than 2,000 separate distinct families of their order. The order of the Brahmins composes the very essence of Hindooism: the Brahmin who lives by charity, or the voluntary contributions of the laity, and who performs no part of the common ritual of the Hindoo worship, is held in the highest repute. Next to him comes the Brahmin who lives by his industry and temporal employments, provided they be such as become the dignity of the order; but which commonly exclude holding the plough, and performing any of the manual employments of agriculture. The lowest rank of all is assigned to those Brahmins who perform the common ritual of the Hindoo worship; and among these last, the meanest office of all is that of performing the service of the gods in the temples. To exercise even the office of astrologer or village priest, is far more respectable. The service of the temples, indeed, has fallen into such disrepute, that the Brahmins in some cases have abandoned it to the inferior classes. Fastidiousness in respect to food is a characteristic mark of purity of caste, and no people ever carried this matter to so absurd and extravagant a length as the Hindoos. On this point the most essential thing of all is to abstain from eating the flesh of the cow. He who eats beef is no Hindoo, but an utter outcast. He who kills an ox by accident ought to be excommunicated; and he who kills one designedly ought to suffer death. This is perhaps the only religious precept which is of universal acceptance among all Hindoos; pretty much in the same way as an abstinence from the flesh of the hog is imperative upon all Jews and Mohammedans. The higher classes commonly abstain from eating the flesh of all domestic animals except that of the goat or sheep. Hindoos generally par-

Dissenting Forms of Religion.—These are the Jain, Buddhist, Seik or Singh, the Mohammedan, and Christian...

Population.—Of the whole territory of Hindostan, supposed to contain about 1,300,000 sq. m., the population may be estimated at about 150,000,000, or more than half the population of Europe...

to 71. In the narrow plain between the W. Ghauts and the sea, and from the 10th deg. of lat. up to the 20th, it is estimated at about 100. Of the whole table-land, extending from the 18th deg. of lat. up to the Vindhyan range, and N. border of the Gangetic plain, probably the pop. does not exceed 50 to the sq. m. The pop. of the great peninsula of Gujrat rises to about 170. More than half the whole pop. of Hindostan is contained in the great plain of the Ganges; comprising the area of this tract at 250,000 sq. m., and the pop. at 80,000,000, the average rate per sq. m. exceeds 700, which is a higher ratio than that of our own island. Within this wide range, however, there is a great difference in the rates of population. From the bottom of the Bay of Bengal up to the W. confines of Bahar, which comprises, of course, the tract of inundation, a territory of upwards of 40,000 sq. m. contains a pop. of more than 300 to the sq. m. The tract of inundation itself far exceeds this. Thus the district of Burdwan has a density of 593; that of Hooghly, 514; the districts of which Calcutta is the centre, 549; and that of Moorshedabad, above 400. As the country becomes mountainous to the E., the population diminishes. Thus Backergunge has but 450; Chittagong, 235; and Tipperah, 200 to the sq. m. In the low lands to the S. of Bengal, including Midnapore and Cuttack, the ratio is but 225. From the W. confines of Bengal to the confluence of the Jumna with the Ganges, the country is far beyond the reach of inundation, and although very fertile, the pop. is only at the rate of 220 to the sq. m.; but in this is included the large, hilly, and wide district of Rhamgur, which has no higher ratio than 100. The whole of the plain to the W., from the confluence of the Jumna till it terminates in the Great Desert, may be computed to have a density of population not exceeding 180 to the sq. m., and the proportion generally diminishes as we proceed westwards. The Punjab, or plain watered by the five affluents of the Indus, probably does not contain a pop. of more than 100 to the sq. m., and 50 would be a large estimate for the delta of the Indus. The extensive desert lying between the western limit of the Gangetic plain most probably does not contain 10 inhabitants to the sq. m.

History.—The Hindoos, it is generally admitted, have no history; they do not even possess any rational, connected, and authentic narrative of their own affairs for a single century. The oldest inscription found in Hindostan, and it is of doubtful authenticity, dates but 23 years before Christ: one of the next authentic era dates last 57 years before that of Christ; and another of extensive currency dates 78 years after Christ, the origin of both being buried in fable. The first of these dates is but three centuries after the invasion of Alexander, and almost five centuries more recent than the commencement of authentic history in Europe. The temple of Juggernaut is but 640, and a ruin connected with it 1,142 years old, the latter being, however, a date which rests on tradition only. In so far, then, as history is concerned, had it not been for the companions and successors of Alexander, who describe the Hindoos as in many respects resembling what they are at the present day, we might, for all that their own history teaches, be led to believe that they were not an ancient, but a comparatively recent people. Independent of history, however, there remains abundant evidence to show that the Hindoos had been very early civilised. The most remarkable, perhaps, is the existence amongst them of the literature of at least three languages, which have long ceased to be spoken by any living people.

These are the Sanscrit, a language of complex grammatical structure, like the Greek, Latin, or Arabic; the Sariswati, or Pracrit, a language derived from the Sanscrit, but of simpler structure, and bearing something like the relation to it which the Italian does to the Latin; and the Pali, a language also of a simpler structure, derived from the Sanscrit, but formed in a different part of the valley of the Ganges. The first of these is at the present day the sacred language of all who follow the Brahminical religion, as the last is that of those who follow the Buddhist worship, whether in India or beyond it. All these languages appear to have been dialects of people who lived in the upper portion of the valley of the Ganges. The Hindoos and their ancient writings point very distinctly to the territory lying W. of Delhi, on the right bank of the Jumna, the principal affluent of the Ganges, as the seat of the people who spoke the Sanscrit. There are certainly many arguments in favour of the belief that the Brahminical worship originated in this quarter, and that the nation that propagated it, and spread civilisation over India, inhabited this country. Thus, the upper and elevated portion of the plain of the Ganges is as much the principal scene of all the great events of Hindoo mythology as Greece was of those of the Greek mythology. Here are the scenes of the wars of the Mahabarat, of the kingdom of Rama, of the localities of the adventures of Krishna. Hastinapura, Ayuta and Mathura. The principal holy places are also here; as Oya, Allahabad, Benares, Hurdwar; not to mention the great Ganges itself, the Jumna, and their sacred tributaries. The evidence afforded by language and religion tends to corroborate this supposition. Thus, the Sanscrit most abounds, and exists in greatest purity in the dialects of the upper portion of the valley of the Ganges, and gradually diminishes both in amount and purity in proportion as we recede from it to the E., and particularly to the S. The distinction of castes is also most strongly marked in this part of Hindostan, and diminishes away from it. The country itself, from its fertility, salubrity, and freedom from rank vegetation and forest, must at all times have been more favourable to the development and progress of an early civilisation than any other portion of India. Although the incursion of Alexander (B.C. 325) made India known to the European world, its effect upon the people of India was scarcely greater than that of any one of the thirteen expeditions of Mahmud of Ghizni. It is highly probable, however, that the influence of the kingdom which his successors established in Bactria, and which lasted for 130 years, was much greater. The Greek princes of Bactria appear to have conquered several of the NW. provinces of India; and from this source, in all likelihood, the Hindoos derived their knowledge of astronomy. The real history of India commences with the first Mohammedan invasion, in the year 1,000, between thirteen and fourteen centuries after the invasion of Alexander. The hero of these invasions, for there were thirteen of them, was Mahmoud, sovereign of Ghizne, in Afghanistan, the son of a man who had been a Turkish slave, but who had raised himself to sovereign power. Mahmoud pushed his conquests, or rather incursions, as far as Canoge, Bundlecund, and Gujrat. India was at this time divided amongst many sovereigns, most of them petty ones; and the resistance made to the conqueror was hardly more formidable than that which the Americans offered to the Spaniards. Towards the close of the twelfth century, the Afghans made their first appearance on the theatre of Indian history. A chief of this nation, of the district of Gaur, raised himself to independent sovereignty, and while the Turkmans seized upon the provinces of the Ghaznian empire, he and his successors seized upon the capital and its eastern provinces, while the second prince of the race, Mahomed Gauri, invaded Hindostan. His favourite general, Cootub, originally a Turkish slave, pushed the Afghan conquests as far as Gujrat; and Mahomed dying without children, Cootub seized upon the Indian conquests of his master, and fixed the seat of his government at Delhi in the year 1183. This may be considered as the date of the first effectual conquest of Hindostan. From this period down to 1525, or in 342 years, twenty-six Afghan princes reigned in Delhi. But it is not to be supposed that the Delhi sovereigns of this race ever ruled over all Hindostan; for in the Deccan, Gujrat, Malwah, Junapoor, and Bengal, there were independent Mohammedan princes, who conquered, and ruled for themselves, and many Hindoo sovereigns continued unsubdued. During the reign of the Afghan princes of Delhi, in 1398, Timour invaded India, but his expedition was a mere plundering incursion. In 1525, India was invaded by Baber, the fifth in descent from Timour, and the sovereign of the little principality of Firghana, a territory lying between the Imaus mountains and river Jaxartes to the N., and Kashgar and Samarcand to the E. and W. He had first conquered Caubul and Candahar, and from the first of these entered Hindostan, defeated and killed the last Afghan sovereign, and seated himself on the throne of Delhi. With him began the race of princes improperly called Mogul by Europeans and Indians, for neither Baber nor his ancestor Timour were Moguls, but Turks. All the conquerors of Hindostan, in fact, who were not Afghans, were Turks, or natives of the great province or kingdom of Transoxiana, whose native tongue was Turkish. Neither were any of them Persians, though the language of the latter people, being a more cultivated tongue than their own, was adopted by both the Turkish and Afghan races of princes. It will be observed that the last Mohammedan conquest of India took place 37 years after Vasco de Gama found his way to that country. The Mogul empire was consolidated under Aurungzebe, who died in 1707, and it began to decline immediately on the death of his son and successor. In 1712. The Mohammedan power acquired its greatest extent under Aurungzebe; but even under him was much inferior, not only in resources but in extent, to the empire now held by Britain in the same country. The passage by the Cape of Good Hope opened the way to a new and more formidable race of conquerors. The Portuguese, by whom it was effected, never acquired more than a petty territory on the W. coast; and the continental acquisitions of the Dutch were limited to a few commercial factories. The French, at one time, seemed to be on the high road to the establishment of a great Indian sovereignty; but, in the end, they were completely worsted by the greater resources and superior maritime strength of the English, and by the extraordinary talents, courage, and enterprise of Clive. The first territorial acquisition of Great Britain consisted of a patch of 5 sq. m. of land on the Coromandel coast, where Madras now stands. The real foundations of the British Indian empire were laid in the interval between 1750 and 1765, when Clive defeated the lieutenants of the Mogul and the Mogul himself, and acquired Bengal, the richest of all the Indian provinces, the most easily defended, and that which has afforded, throughout, those resources which have enabled Great Britain to conquer and to preserve all our subsequent acquisitions.

The total area and population of British India according to official returns of 1862 are as follows:—

Presidencies	Area in Eng. Square Miles	Population
Governor-General's district	170,530	14,145,161
Bengal	200,300	41,486,606
Madras	137,505	73,197,855
Bombay	127,741	11,257,613
Punjaub	100,498	14,784,411
North-west Provinces	116,483	30,110,197
Total	853,772	185,634,044

The above numbers of the population are but the result of estimates, as an accurate enumeration has never been made—and, probably, cannot be made—owing to religious prejudices, and the peculiar mode of life of the natives of India. Some authorities estimate the population of the British Indian empire at close upon 200 millions.

The English population in India amounted, according to the returns made by the several governments, to only 125,945 persons in 1861. Of these 125,915 people, 84,083 went to compose the British officers and men of the Indian army; while 27,556 consisted of men and boys in civil life, including the civilians in the public service; the remaining 19,306 being females, of whom 9,773 were over 20 years of age. When the census was taken, the number of females of English origin in India above the age of 15 was 11,636, including 8,556 wives and 1,146 widows. Of the officers and men of the royal army 93 per cent. of all ages were unmarried, while the proportion of civilians above the age of 20 unmarried amounted to 50 per cent.

According to returns published in April 1862, the whole Indian army numbered nearly 200,000, of which number 3,062 were European officers, and 76,489 European non-commissioned officers and men; the native officers and men amounting to 108,502, exclusive of 11,852 men in the Punjaub local force. The distribution of these troops was as follows: 88,000, in round numbers, in Bengal, the north-west provinces and the Punjaub; 42,000 in the Bombay Presidency, and 54,000 in Madras. (See INDIA.)

Languages.—It has been stated, that there are no fewer than 25 native languages spoken throughout Hindostan, independent of the dialects of tribes in a very rude state of society. The Hindoos of the N. portion of Hindostan are acquainted with three dead languages, viz. the Sanscrit, the Saruswaty, or Prucrit, and the Pali. Of these three the Sanscrit contains internal evidence of being the oldest. It was the language of a people who, according to a very probable Hindoo tradition already referred to, occupied the right bank of the Jumna, a little way to the NW. of the city of Delhi, and with it probably originated the Brahminical religion, and the first dawn of Hindoo civilisation. The Saruswaty or Prucrit was the language that succeeded it in the same country, and it seems to bear the same sort of relation to it that the Italian does to Latin. The Pali is a language which sprung up in the province of Bahar. Of this, also, the Sanscrit forms the groundwork, and the relation between them may be supposed to bear a similar relation to that which subsists between the Spanish, or French, and the Latin tongues. With the people speaking the Pali language sprung up the religion of Buddh; and Pali is, to the present day, the sacred language of all the Asiatic nations who have Buddhism for their national worship. The existence of these three languages, that have successively ceased to be spoken, affords, as before observed, satisfactory evidence of the great antiquity of Hindoo civilization. One or other of the languages in question is more or less mixed up, not only with every language of Hindostan, but also with the languages of most of the neighbouring countries. To the N. they form the groundwork of these languages, as Latin does of Italian; to the S., on the contrary, they are engrafted on the language in something like the manner in which the French is engrafted on our own Saxon tongue. The literary Hindoo reckon that there are ten cultivated languages, having a written character and a literature, viz. five to the N., called the five Gaura, and five to the S., called the five Dravira. The enumeration, however, is not very clear and distinct, at least as applicable to present times. The Gaura are the Saraswatty, Canoj, Gaura or Bengalee, Maithila or Tirutiya, and the Outima. The first of these is the dead language already mentioned. The Maithila is confined to a small portion of the district of Tirhoot, the Gaura is the language of the numerous people of Bengal, already mentioned, and the Oorissa or Urya, of the people of Cuttack. The Canoj, as such, is an extinct language, but is considered, on good grounds, to be the parent of the modern Hindee, the most cultivated and generally spoken of all the native languages of Hindostan. Upon the language of Canoj has been grafted the Persian, the court and literary language of the Mohammedan conquerors of India. This language, in fact, is found to exist in the Hindee, very much as the French is found in our own Saxon tongue, its introduction having been effected exactly in the same manner. Besides the local language of each district, the Hindee is commonly spoken by all persons of education throughout all parts of India, and almost universally by all persons of the Mohammedan persuasion. Its prevalence, it may be observed, is probably owing as much to the parent language having been, previously to the conquest, the language of a numerous and powerful nation, as to the subsequent influence of the conquerors. Without this supposition, it is difficult to believe that, in the comparatively short period which elapsed from the first permanent conquest of the Afghans, at the end of the 12th century, until it acquired its existing form, it should have acquired so wide an extension as it is found to possess.

The five Dravira are the Tamul, called by Europeans, very improperly, the Malabar; the Maharashtra or Mahratta; the Karnata or Canara; the Telinga or Telugu, improperly called by Europeans, the Gentoo; and the Gujrati. The groundwork of all these languages is peculiar; but upon all of them is engrafted more or less of the Sanscrit language, or its derivative, the Pracrit; the amount of words decreasing, as we proceed S., until, in the ancient Tamul, it disappears altogether. The Tamul, the Telinga, and the Canara are divided into two dialects, an ancient and a modern; the first containing the national literature, and being nearly unintelligible to the people at large.

Besides these more cultivated tongues, there are at least 70 languages spoken by nations tolerably civilised, and of considerable numbers, as the Assami, spoken in Assam; the Nepali, Bwali, and Jhagari, three languages spoken in Nepaul; the Cashmeri, spoken in the celebrated valley of Cashmere; the Punjabi, spoken in the country of the five affluents of the Indus; the Multani, the dialect of the prov. of Multan; the Sindhi, spoken by the Sindhians, at the mouth of the Indus; the Hindawi; the Marwari; the Jaya-

pari; the Oriyari; four languages spoken in Rajpootana; the Haroti; and the Braja, spoken in the higher portions of the valleys of the Ganges and Jumna, and derivatives of the Saraswatty or Pracrit; the Magadhi, spoken in the S. portion of the prov. of Bahar; the Malwa, spoken in the prov. of the same name; and the Bundela, spoken in the prov. of Bundlecund. Many of these languages are in course of gradual extinction and absorption by the Hindee, as the Celtic dialects of our own country are in progress of extinction by the English; the Armorican by the French, and the Basque by the Spanish. To the S. we have the Konkani, the language of the Concan; the Tulawa, or language of the country which Europeans call Canara; and the Malayalim, spoken by the inhab. of the S. portion of coast lying below the W. Ghauts, as far as Cape Comorin.

Of the languages of rude or savage tribes, such as the Garrows, Coolies, Cattias, Gonds, and Coles, not less than 50 may be enumerated. Besides the three dead languages, one of them, the Sanscrit, as much studied as Latin is in Europe, there are in India eight languages, each spoken by a numerous pop.; 20 spoken by people less numerous, but still civilised; and at least 30 spoken by rude tribes; making in all 58 living languages. This simple fact goes far to prove the generally admitted fact that all India never was subject to one government, or never even thoroughly united in large masses. To the native languages now enumerated must be added the Persian, still as much studied, and much more generally written, than Latin is in Europe; the Arabic, often studied, from religious motives, although not spoken; the Portuguese is a good deal spoken on some parts of the maritime coast, especially by the converts to Christianity; and the English, which has begun to make considerable progress.

Literature. — The best and largest portion of Hindoo literature is contained in the dead Sanscrit; that which is contained in the seven living languages already enumerated bring for the most part little else than translations, or rather paraphrases, from it. To Hindoo literature in any language, prose composition is hardly known. Every thing is in verse, from works of imagination to history, to treatises on theology, astronomy, medicine, grammar, and even dictionaries. These facts are at once evidence of antiquity and of rudeness, while they show that, for 2,000 or 3,000 years at least, native literature has made little progress. The Hindoos have been said to be, at the present moment, in the condition, in reference to literature, of the Europeans of the middle ages; who had no books but such as they inherited from the Greeks and Romans. But it is obvious that they are in a much worse condition, inasmuch as their models are incomparably inferior. The two most celebrated works of Hindoo literature are the Mahabharat and the Ramayana; the one giving an account of the wars of the sons of Bharat, and the other the adventures of Rama, king of Ayodhya or Oude, a supposed incarnation of Vishnu, the 'Preserver' of the Hindoo Triad. The scene of both is laid in the upper portion of the valley of the Ganges. Mr. Mill's description of these poems, some of the best specimens of which have been translated into English, is not unjustly deprecatory:—'These fictions,' says he, 'are more extravagant and more unnatural, not only less correspondent with the physical and moral laws of this globe, but, in reality, less ingenious, more monstrous, with less of any thing that engage the affection, awaken sympathy, or excite admiration, reverence, or terror, than the poems of any other, even the rudest, people with whom our knowledge of the globe has yet brought us acquainted. They are excessively prolix and tedious. They are often, through long passages, trifling and childish to a degree which those acquainted with only European poetry can hardly conceive.' (History of British India, i. 362, 4to, edition.)

Science.—The sciences in which the Hindoos have made some progress are arithmetic, algebra, geometry, and astronomy. The first and second are probably the only ones in which, perhaps, they are entitled to lay any claim to originality. They are probably the inventors of the system of notation, which the Arabs borrowed from them, and we from the Arabs. It is not necessary, however, to add that the Hindoos are clumsy arithmeticians; and that, as in the case of gunpowder, certainly invented in China, it is in Europe only that the art has been perfected.

In geography, medicine, botany, and the physical sciences generally, the Hindoos, like other Asiatic nations, may be considered as profoundly ignorant. In metaphysical and ethical speculations, more consonant to the genius of such a people, they have indulged to a much greater degree; and their speculations in grammar especially, if not distinguished for utility, are remarkable for ingenuity. The Sanscrit language, distinguished for the complexity and variety of its structure, has afforded an ample field for such discussions. It may be remarked that it is the only one of their languages that is subjected to rules, and that they have never composed a grammar of any of the living languages. Geometry is another science, the invention of which is ascribed to the Hindoos; but their earliest treatises are of the seventh century, 1,000 years after they had been in contact with the Greeks of Bactria, and at least 15 centuries after the first knowledge of the science in Greece itself. In astronomy, the Hindoos make large claims to antiquity, reckoning their tables from the commencement of the Caliyuga, or iron age of the Hindoo mythology, 3,102 years before Christ. Of such an antiquity, however, there are great doubts; and the more general opinion seems now to be, that the astronomy of the Hindoos was either derived from the Bactrian Greeks, or intermediately from the Arabs of the middle ages. The coincidence between it and the Greek astronomy, is at all events, both remarkable and suspicious. Thus, the days of the week are seven in number, and named after the seven planets; while they follow in the same order as they do in the Greek. The ecliptic is divided, as among the Greeks, into 12 signs, with the same names, emblems, and arrangement; and the signs are also divided into 30 degrees. As these matters are purely arbitrary, they cannot but have had the same source. Two things seem to be agreed upon by all parties; viz., that the Hindoo astronomy is empirical, and not founded on general principles; and that, among the Hindoos, astronomy has only been used as an auxiliary to astrology, and never applied to any useful practical purpose; with the exception, and this in a very rude manner, of reckoning time.

Arts.—The arts in which the Hindoos have made the greatest progress are, agriculture, weaving, dyeing, and architecture. The ox, buffalo, horse, ass, elephant, hog, dog, sheep, and goat have been domesticated, and used by the Hindoos from the earliest antiquity. The camel, probably, has been equally long known in Upper Hindostan. The common poultry is also of great antiquity among the Hindoos; and is supposed, and most likely with good reason, to have spread from them to the W. world. The buffalo and ox

only are used for agricultural purposes; the horse generally for war or pleasure, now and then for burthen; the elephant for pleasure or burthen; the camel and ass, with few exceptions, for burthen only. With the exception of the horse, camel, sheep, and goat, every one of the animals above enumerated are still found in many parts of India in the wild state. The agricultural implements used by the Hindoos are simple and rude, such as might naturally be expected among poor occupants, cultivating each a small patch of land upon an uncertain tenure; and the process is equally rude. But neither the one nor the other are so much inferior to those of the S. part of Europe as a native of this country, accustomed to the more perfect implements and processes of English husbandry, would expect to find them. The greatest exercise of the skill and labour of the Hindoos in agriculture is displayed in works of irrigation; and the reader will not be surprised at this, when he understands that through means of irrigation the produce of the land is, according to circumstances, always multiplied never less than five fold, and often as much as ten. The works for this purpose consist of immense embankments, reservoirs or tanks, and wells. The delta of the Ganges, and the celebrated mound of the Cavery in S. India afford examples of the first description of works: reservoirs or tanks are sometimes of vast extent, and capable of converting 4,000 or 5,000 acres of what is often a dreary desert of sand into productive corn-fields: these are most frequent in S. India. Wells, which are often sunk to the depth of between 200 and 300 ft., afford the principal means of irrigation in the upper portion of the valley of the Ganges. In a few cases there exist canals for irrigation resembling those of Lombardy, but these are of Mohammedan, not Hindoo, origin.

The articles cultivated by the Hindoos from very early times, are wheat, barley, rice, millet, several pulses, the sugar-cane, sesame, mustard, the cocoa, areca, and other palms; cardamoms, ginger, black pepper, cotton, the mulberry, indigo, madder, the mango, and the banana. From the Mohammedans they received the rice, the fig, the apple, peach, and pear; the pomegranate, limes, and oranges; the carrot, onion, and melon, with the opium poppy. From Europeans they have received maize, oats, common potatoes, the batata or sweet potato, the ground pulse of arachis, the capsicum, guava, and pine-apple, by way of America; the shaddock, from Java; the litchi, from China; and most of the common pot-herbs, direct from Europe. The sugar-cane is most probably a native of Hindostan, and the art of manufacturing coarse sugar from it is traced by the etymology of the word *gur*, to Bengal. The art of granulating sugar, and separating it from the molasses, was probably introduced into India from China, as the name of the commodity *Chíni*, would seem to imply. The art of candying or crystallising sugar, the only mode of refining practised in the East, was taught the Hindoos by the Mohammedans, who themselves appear to have first practised the art in Egypt, as the name of the article *Misri* (that is, Egyptian), would seem to import.

The Hindoos had made a far greater progress in the art of weaving, than in any other. It was confined to materials which their country either produced in great abundance, or of great excellence; or of which, in fact, in ancient times, they may be considered to have possessed nearly a monopoly, viz., cotton, silk, and the hair of the Tibetian goat. With the exception of silk, which they had in common with China, India may be considered as the native country both of the material and manufacture of the others. The cotton plant is grown almost every where, from the S. extremity of India up to the valleys of the most N. range of the Himalaya, and it may be traced from India to every warm country by its original Sanscrit name. The quality and nature of the fabric varies every where with the quality of the plant; and hence a vast variety of fabrics known by the names of the districts producing the raw material; thus the fine textures known in Europe as Dacca muslins, were produced only in that district in which it is cultivated, within narrow limits, a variety of the plant, with a staple remarkable for fineness and beauty, not found any where else.

Silk weaving, like that of cotton, is an art which has been practised from remote antiquity in India. In the Sanscrit language there is a peculiar name for the class of persons exclusively employed in the feeding of silk worms. The variety of the latter bred in India differs from that of China and Europe; and the species of mulberry grown for the food of the worm is a distinct one from that used either in Europe or China. But as the Hindoos are much inferior in skill and ingenuity to the Chinese, the silk fabrics of Hindostan have never equalled those of China; nor is the raw material, even now, equal to that of the Chinese, though under the superior care and skill of Europeans. The Cashmerians, the manufacturers of the well-known shawls which bear their names, are descended from genuine Hindoos; and though the shawl goat be not a native of their country, they were the nearest civilized people to the rude nomadic tribes, to whom it belonged. They naturally, therefore, became the manufacturers; and the invention of the shawl manufacture may, therefore, be fairly ascribed to the Hindoos. From these statements, it will appear that the discoveries now described, and the progress in manufacturing industry which they imply, are rather owing to the accident of position than to any superiority of skill and ingenuity. This is at once apparent, by the little skill which the Hindoos evince in arts, where they possess no superiority in the raw material, as in woollen textures, iron fabrics, and earthenware, in respect to which there are few nations ruder and more unsuccessful. Orme, who is followed by Mill, ascribes the superiority of the Hindoos in the manufacture of cotton fabrics to the peculiar softness and delicacy of the Hindoo hand; but this is a fancy for which there seems to be no ground whatever. The Hindoos, comparing them with other nations in the same state of society, and to Europeans until comparatively recent periods, had attained considerable skill in the art of dyeing, producing colours that are both fast and brilliant. Here also, however, they had several advantages of the same nature as those already described, such as the possession of indigo, lac, and madder, three of the finest and most durable of all known colouring materials. Inferior dyes, such as the carthamus, morinda, turmeric, and saffron, are also natives of the country. Their dyeing processes, however, have always been, and are, tedious, operose, and empirical.

Nearly the whole architecture of the Hindoos which deserves notice is dedicated to religion. The people have always lived in huts, and even their chiefs and princes were satisfied with very mean accommodation; and the only palaces have been those of the gods. But even their temples are more distinguished for magnitude, the substantial nature of the materials, and the elaborate character of the ornaments, than for beauty, grandeur, or propriety. Many of the most remarkable consist of caves, or subterranean

grottoes; and the rest here, for the most part, a pyramidal form. One class of religious monuments which makes so conspicuous a figure in the architecture of Christians and Mohammedans, is wholly wanting among the Hindoos, — those erected in honour of the dead; a circumstance no doubt arising from the universal practice of burning the corpse, and the belief in the doctrine of the metempsychosis.

Of a far higher order is the architecture introduced into India by the Mohammedans, particularly since the time of the Turkish dynasty, the descendants of Timour. These consist of mosques and mausoleums. In the style of architecture introduced by the Arabs into Spain; and are so remarkable for beauty and chasteness of design, grace of proportion, and excellence of material and workmanship, as to be entitled to be compared with the finest remains of Grecian or Roman art. In these Mohammedan buildings, white and coloured marbles are largely employed, a material never seen in any Hindoo building, though very abundant in many parts of the country. The most remarkable of the Mohammedan monuments, well known to Europeans by the name of the Tajemahal, is situated near the city of Agra, on the right bank of the Jumna. It is a mausoleum occupying, with its gardens, a quadrangle of forty acres: the principal building, with its dome and minarets, being almost wholly of white marble. This was built by the Emperor Shah-Jehan, about two centuries ago. Even the palaces of the Mohammedan princes, and the houses of the rich were built in a very superior style to those of the Hindoos of the same rank. In fact, the Mohammedan architecture exhibits unquestionable evidence of superior science, taste and civilisation.

In useful architecture, such as the construction of roads, bridges, and public accommodation for travellers, the Hindoos have made very little progress, as may be seen by an examination of the most N. portion of India, which Mohammedan influence hardly reached. The ancient Hindoos were unacquainted with the arch, and hardly ever built a bridge of any sort. Down to the present day the principal rivers of the Deccan are crossed on wooden floats, or in baskets covered with leather. Now and then a few miles of good road lead to some celebrated place of pilgrimage, and on the ways leading to such places inns for the accommodation of travellers, called chaultries, are not unfrequently met with. These consist of bare walls and a roof, without food, furniture or attendance. Both these roads and inns have been constructed from religious motives only. In this department of architecture, also, the Mohammedans have made considerable improvements; the only bridges existing in India are of their construction; and the same thing may be said of public roads.

Effects of British Rule. — The great body of the Indian people had, for six centuries before the commencement of British government, been under the dominion of foreigners; but of foreigners more energetic than themselves, and a good deal more civilised. Upon a fair retrospect of what they have lost and gained by the Mohammedan dominion, they must, upon the whole, be considered as having been considerable gainers. The conquerors being Asiatics, and approaching to the native inhabitants in complexion, manners, customs, and state of civilisation, assimilated with the latter, and, to a certain extent, adopted their language and customs. Even in matters of religion, where the difference was widest, a considerable share of toleration was established; and Hindoos converts to Moham-

mechanism, and mixed races were in time admissible to the highest offices of the state, and not unfrequently permitted to them. This condition of things was superseded by the British rule, which may now be considered as having been practically consolidated for a period of about a century. The British government, as established in India, and as it is now in operation, may be considered an enlightened despotism, a good deal controlled by the public opinion of Englishmen on the spot, and to a smaller extent by parliament and public opinion in England, and possessing some advantages over, but also many disadvantages which did not belong to, the Mohammedan government, which it superseded. It may be divided into four periods. The first was that which intervened between the victory of Plassy in 1767, and the effectual interference of parliament in 1781, but not practically enforced till 1793, an interval of 86 years. This was a period of pretty general anarchy, accompanied by constant, or at least frequent, wars. The government was carried on upon the principles of the Mohammedan system, and did not pretend to be based upon any other. The taxes were levied with more than Mohammedan rapacity; and the administration of justice followed the Mohammedan law with less than Mohammedan intelligence. The only modification in any of these particulars depended wholly on the moral and intellectual character of a few public functionaries. At the same time the industry of the country was subjected to a commercial monopoly, exercised by the government itself, and the aim of which, as of all similar institutions, was to obtain possession of as much as possible of the produce of the country at less than it cost, and to sell it for more than it was worth. It cannot be supposed that the British government during the period in question could possibly be productive of beneficial results to the native inhabitants of the country; and it certainly produced none to the parent country, whose resources were wasted, and whose commerce was not augmented by the possession of India.

The next period of British administration embraces the twenty years from 1793 to 1813. During this time the land tax, the greatest burden of the Indian people, was established in perpetuity throughout the greater part of the Indian territory. Regular courts of justice were instituted, and the judicial and fiscal administrations were carefully and completely separated, after the example of European nations. The commercial monopoly continued as in the previous period, but it was exercised with greater leniency and forbearance, except in so far as concerned the settlement and resort of British subjects to India, the laws against which were most rigorously carried into effect than ever. Parliament never effectually interfered in the affairs of India during this period; every thing was presumed to be going on prosperously. The wars that were carried on in India in the meantime nearly doubled the extent of the British dominions, and raised the territorial debt to 30,000,000l. sterling. But instead of reaping any direct advantage from these acquisitions, parliament was obliged, on the lapse of the charter, to exonerate the E. I. Company from a long arrear of a tribute of about half a million sterling a year, which it was wholly unable to pay. The entire advantage conferred upon the people of India, during this period resolves itself into the permanency of the land-tax, with some ameliorations in the administration of justice, and freedom from foreign aggression and invasion. The English nation derived no benefit whatever from India; the commerce with it, which was

but of trifling importance, continued stationary; Great Britain paid a monopoly price for every Indian commodity and even was obliged to forego the whole of the paltry tribute bargained for.

The third period commenced in 1814, and continued for 41 years, till 1858. In 1814 the Indian trade was, in a great measure, thrown open; and in 1834 the last vestige of monopoly, and even the company's commercial character, was finally put an end to,—a measure which, with some drawbacks, had been productive of much advantage both to the people of India and of England, though in a greater degree to the latter. The exports of India to this country more than doubled; and the people of India and of England respectively received each other's productions for about from a half to a third part of what they cost them under the monopoly. The influx of Europeans into India was followed by a great influx of British capital; and something like a public and independent opinion sprung up at the principal seats of commerce, to control the despotism of a virtually absolute government. A system of effectual native education may also be said to have begun in 1814; and the native inhabitants of the principal towns, who before considered all education to be comprised in the study of the Persian, a foreign language, or of the Sanscrit, a dead one, betook themselves with great ardour to the study of the language of the conquerors. Finally, the fourth period, following in the wake of a great military mutiny, scarcely, if at all, shared in by the bulk of the population, drew India closer to Great Britain than ever it had been before. The commencement of the fourth period dates from the Act 21 and 22 Vict. cap. 106, called 'An Act for the better Government of India,' sanctioned Aug. 2, 1858. By the terms of this Act, which cannot be but the herald of a happier future and real 'better government' of the immense British empire in the East, all the territories hitherto under the rule of the East India Company were vested in the sovereign of the United Kingdom. (See INDIA, BRITISH.)

From the third period of history, here sketched, may be dated the abandonment, on the part of many of the most wealthy and enlightened inhabitants of the towns, of the gross superstitions of their forefathers, and the adoption of rational opinions in matters of religion ; and it may be remarked as extraordinary, that this species of conversion has been most frequent with the Brahminical order. Commerce, the great engine by which civilisation, as well as improved morals, have been produced in Europe, has begun to do its work in Hindostan also. The value of knowledge and of character has begun to be felt, and already there may be counted among the merchants of Calcutta, Bombay, and other places where commerce is carried on upon a large scale, Hindoo, Mohammedan, and Parsee merchants, as faithful to their engagements, and of as strict probity, as any community can boast of.

The disadvantages of Great Britain for carrying on the administration of India are sufficiently obvious. The British, in the first place, is not a national government, nor is it as yet a government carried on by conquerors who have made the slightest progress towards naturalisation or amalgamation with the people governed. The rulers are aliens in blood, in manners, in language, and in religion, carrying on the administration of 100,000,000 of people, and exercising a control over 50,000,000 more, at a distance of 12,000 m. The local government is purely vice-rial, and the essential administration rests with men residing at a vast distance, who never saw the country, and who have no accurate knowledge of its manners and institutions. These men themselves are perpetually changing, and look upon Indian affairs as matters of very secondary importance to domestic and European politics. The local governments, instead of being responsible to the people whose administration they conduct, are only amenable for their acts to their political friends in Europe, while the affairs of India are too complex, too extensive, and too remote, to be understood by, or, for the most part, to excite any interest in, the people and parliament of England. In India, generally, the acts of the local governments are secretly prepared without consulting or attempting to conciliate the subjects for whom the laws are made.

One of the great disadvantages of the British government in India is the vast expense at which it is conducted, and the consequent weight of taxation to which the people are necessarily subjected. In India there are six local governments, and in England another central administration, all paid for out of the Indian revenue, on a scale of expense of which the rest of the world affords no example. Thus the salary of the governor-general is equal to five times that of the first lord of the treasury, while an Indian secretary is more highly paid than an English secretary of state. There are about 1,000 civil officers engaged in the judicial, magisterial, and fiscal administration of India, every one of whom costs the Indian people, including his pension on retirement, more than a puisne judge of the Court of King's Bench costs the people of England. As British rule is still maintained, not through the affections and goodwill of the people, but partly through their docility, and partly by the sword, a vast army becomes necessary. An important part of it must be carried over the Atlantic and Indian oceans, and, mortality included, is maintained at double the expense of the same force in Europe. The officers of the whole Indian army amount to about 8,000, and these, retiring pensions included, cost about three times what the same number would cost in Europe.

It is not, however, to be supposed, that the large salaries allowed to those engaged in the administration of the Indian government originate in extravagance merely. It may, in fact, be doubted whether it be possible, on any reasonable ground, to make any sensible diminution in their amount; and whether the excess that might be deducted from some departments should not go to balance a deficiency in others. The salaries of Europeans in India must be high; first, because of the expensive style of living in the country, and the immense number of servants and retainers that a person in any prominent situation must keep; and second, because of the many expenses attending the training and fitting out of a young man for the Indian service. Till one or both of these sources of expenditure be diminished, of which there is but little prospect, it is idle to talk of materially reducing the cost of European functionaries in India.

The greatest revenue which a colonial empire ever yielded, and, in fact, the largest public revenue in the world, that of Britain and France excepted, is unequal to meet so enormous an expenditure; and one of the worst forms in which bad government can present itself, oppressive and grinding taxation, is the necessary consequence. Nor is it, perhaps, in the power of the best disposed administration much to ameliorate this state of things, so long as government is conducted on the principles hitherto persevered in. The Indian revenue approaches to 46,000,000l., and con-

sidering the poverty of the people, as indicated by the low rate of wages, and the comparatively small amount of capital and industry in the country, this is said to be equivalent to an annual public revenue in England of twice the amount. Moreover, the Indian revenue never diminishes, but, on the contrary, may be considered a perpetual war taxation, from which there is no relief or abatement.

One advantage the people of India certainly derive from British rule, which they never enjoyed, at least to the same extent, before—freedom from civil war, and from foreign aggression and invasion. But it must, at the same time, be acknowledged that these benefits have been purchased at no inconsiderable price—the suppression of all competition and emulation between different parts of the country; and the entire sacrifice of national independence, accompanied with an utter hopelessness of those successful insurrections by which other Asiatic people rid themselves of tyranny, and procure, at least, a momentary melioration of their condition. What probability, it may be asked, is there of the stability and permanence of British dominion? This is a question more easily put than answered. No people under the same circumstances ever possessed such an empire before, or anything resembling it; and there exists, therefore, no precedent to give a reply. However, it may be fairly said that India appears to be unassailable, except by a nation that has the command of the sea. Her land frontier is fenced by impassable mountains, and by deserts and rivers that could not be traversed by an invading army without great difficulty and loss. No doubt, however, if British troops cross the natural barriers that protect India, and advance into Central Asia, they may meet Russian troops on ground congenial to them. But so long as Great Britain confines herself within the proper limits of India, there is little to fear from foreign aggression. An attack by Asiatic powers is not of the question; and the danger of French and Russian invasion is far more chimerical than real.

Bodily and intellectual endowments.—The Hindoos, as already stated, constitute sixth-sevenths of the population of Hindostan; but the remaining inhabitants, though the stocks were in many cases originally different, are now so much assimilated with them through a mixture of blood, and the adoption of Indian manners and customs, that for our present purpose the whole population may be considered under one head. In point of race, the Hindoos have been regarded by naturalists as belonging to what they call the Caucasian, and even to the same family of that race as the white man of Europe. But this is a fantastical notion, for which there is hardly even so much as the shadow of a foundation. The only three points in which any analogy has been discovered between the Hindoo and European are the oval form of the face, the shape of the head, and traces of a certain community of language. In every other respect the points of contrast are incomparably more decisive than those of resemblance. The European is white, the Hindoo dark. The European (and his is the only race that is so distinguished) has an infinite variety in the colour of the hair, from flaxen to black, and great variety in the colour of the eye, from light blue or grey up to dark brown; with the Hindoo the colour of the hair is ever black, and the colour of the eye ever dark brown. The European is taller than the Hindoo, more robust, and more persevering. Even in the rudest states of civilisation, the European has exhibited a firmness, perseverance, and enterprise which strikingly contrast with the feeble, slow, and irresolute character of the Hindoo. In the performance of ordinary labour in those employments where there are means for drawing a just comparison, the labour of one Englishman is equal to that of three ordinary Indians. Three Indian seamen will hardly perform the work of one English seaman, and three battalions of sepoys would not in any case, supply the place of a single battalion of Europeans. There is little doubt but that an equal inferiority would have been the result of a trial of strength with a Roman legion or a Greek phalanx. When the skill required in any particular employment rises in amount, and the European is enabled to avail himself of improved tools, which the Hindoo either cannot or will not use, the disparity becomes still greater. In physical force and continuity of labour the Hindoo is unquestionably not only below the European, but below the Arab, the Persian, and, above all, the Chinese.

In one physical quality there is a striking distinction between the Hindoo and European. The European is born with an inflexible and comparatively rigid fibre; the Hindoo with a finer, more pliant and soft than that of European women. This distinction, however, is a mere matter of climate, for the quality supposed in this instance to be peculiar to the Hindoo frame is common to that of natives of every warm climate; even Creole Europeans, in the very first generation, are distinguished by it. This flexibility in the animal fibre has been supposed by some observers to be accompanied with great sensibility and acuteness in the organs of sense, conferring upon the Hindoo a remarkable advantage in some of the nicest of the manual arts. But there seems no truth in this hypothesis any more than there would be in imagining, contrary to all experience, that the delicate and more pliant fingers of a woman confer upon her an advantage in skilled labour over man. In the finer processes of mechanic art, habit now gives to the rigid hand of the European artisan a nicety of touch and a dexterity of execution which no Hindoo has ever yet attained; in general, the Hindoos possess more agility than the Europeans, and their nimbleness is assisted by the lightness of their persons. They are, to a remarkable degree, the best runners, the best wrestlers, and the best climbers of Asia. In these respects the Persians, Arabs, and Chinese, are not to be compared with them. Hence it follows that, as ordinary seamen, they are far more dexterous and useful than any of these nations, yet a certain want of firmness and presence of mind incapacitate them for officers, or even for steersmen, and, in this latter capacity, the natives of the Philippine Islands are so preferable to them, that, whenever they can be obtained, they are always employed, to the total exclusion of the Hindoos. A Hindoo cannot be urged to any personal exertion for a great length of time without producing failure or exhaustion. Even in their own country and climate the sepoys have been beaten by European troops, in a long succession of forced marches.

Among the Hindoo nations, though the common features of their physical and intellectual character are generally well preserved, much variety exists—more, probably, than among the nations of Europe. This variety has been ascribed to difference of latitude and climate, and to diversity of aliment; it has been affirmed that the inhabitants of the south, whose chief aliment is rice, are smaller and feebler than those of the north, whose chief bread corn is wheat and millet. Experience shows that this opinion is without any foundation. The smallest and the feeblest family of Hindoos are the natives of Bengal, whose locality is between the 21st and 24th deg. N. lat.; those living a dozen degrees farther south, and upon the same vegetable aliment, are taller, more robust, energetic, and hardy. The natives of the table-land, whose vegetable aliment is neither rice nor wheat, are equal but not superior to the inhabitants of the Carnatic, or of the low damp coast of Malabar. The tallest and most robust, but not the most active or agile, are the inhabitants of the upper portion of the valley of the Ganges, where a few of these in easy circumstances live only on wheat; the majority of the people on barley or millet.

It is the quantity and not the quality of the vegetable aliment which has the most material influence in India; it may be said, that in Hindostan generally there is a wider distinction in physical development between the classes in easy circumstances and the poor, than in any other country. The Hindoos of the upper and more distinguished classes, are almost invariably larger, stouter, and handsomer than the poor and degraded classes. The most inattentive observer cannot fail to notice the superiority of the military, mercantile, and above all the sacerdotal classes over the common labouring poor. The sepoys of the army of Bengal, who are a selection from the numerous yeomanry of the northern and central provinces, though very inferior in strength and energy, are equal, if not superior, in stature and personal appearance to the common run of European troops; and even in the streets of Calcutta, a stranger cannot fail to be struck with the disparity in the appearance of the well-fed merchant, or broker, and the squalid half-starved labourer or artisan. The mountaineers, and generally all the semi-barbarous tribes, are short, emaciated, and ill-looking, particularly those who gain their livelihood by the chase, or by collecting the natural objects of the forests, such as honey, wax, and drugs. Where slaves are few in number, and this is the case in all the populous parts of the country, they are in personal appearance nearly on a level with the rest of the peasantry, and not to be distinguished from them. Where, however, they are numerous, and whole tribes are in a servile state, they may be easily distinguished from the rest of the community by their ugliness, small stature, and feeble frame. As a general rule it may be laid down, whatever be the climate, and whatever the general aliment, that wherever the price of labour is low, and the people consequently compelled by necessity to live upon the lowest description of food, or upon the smallest possible quantity of a better description that will support life, the great mass of the inhabitants are the most degraded in body, as well as in mind.

It is a popular but erroneous notion that the Hindoos live almost entirely on a vegetable diet; such a fact would be inconsistent with the physical nature of man, who, in reality, is omnivorous. The most fastidious of the Hindoos in point of diet are great eaters of milk and butter; fish is also extensively used near all the sea-coasts, and on the shores of the principal rivers; and none of the people of India hold this description of food as abominable, except the inhab. of the remote interior, who have no means of procuring it. Even flesh, however capricious in the selection, is occasionally eaten by the greater portion of the Hindoo people, and it is the want of means, rather than religious scruples, that makes them refrain from it. In cases of urgent necessity, even religion authorizes any kind of food, and in the event of famine, a Brahmin may eat the limb of a dog.

Upon the intellectual and moral qualities of the Hindoos, a very few words will suffice. The more educated classes, and it is from a consideration of the character of these only that any fair conclusion can be drawn, may be pronounced without hesitation to be a shrewd, wary, and acute people. Subtlety, perhaps, more than strength, is the prominent character of their intellect. Good imitators, they have hitherto discovered no original powers of invention. They have little imagination, for the poor distempered dreams of their theology and literature are not entitled to this name. In practical good sense they are decidedly below the Chinese. In vigour and manliness of mind, they are below the Arabs, the Persians, and those Mohammedan nations of Tartary who sent forth the men that invaded and conquered them. No comparison with European nations can be made, because the contrast is too great to admit of any parallel. The departments of industry, in which their intellectual faculties appear to most advantage, and for which they seem best fitted, are the administration of justice and finances, and such branches of trade as do not imply the possession of comprehensive knowledge and bold enterprise.

The moral character of the Hindoos is the growth of probably many thousand years of anarchy and oppression. Such a condition of society produces no demand for candour, integrity, or ingenuousness; and among the Hindoos these qualities can hardly be said to exist. Rapacity, violence, fraud, and injustice characterised the native rulers; and the usual weapons of defence, viz., falsehood, artifice, chicane, and deceit, have, consequently, sprung up in abundance among the people. In reality, for generations, integrity may be said to have been at a discount in India, and dissimulation at a high premium. Probity and candour are virtues which, in fact, could not be practised with any regard to personal freedom, life, or property; in such a state of things, such a simpleton as an honest man would have become the inevitable prey of a host of knaves, and would have been laughed at and despised. Generally it may be said that the Hindoos seldom speak the whole truth without some mental reservation. Judicial perjury is practised in Hindostan perhaps on a wider scale than in any other country in the world. The British courts of justice have been blamed for encouraging the crime, and probably, to a certain extent, they do so; but, upon the whole, they can only be looked upon simply as an arena for the exhibition of this vice upon a great scale. Falsehood and equivocation are inseparable from such a condition of society as that of Hindostan, and have characterised the manners of the Hindoos from the era when Europeans first acquired any authentic information respecting them. The description which Bernier, one of the most accurate of travellers, has given of the Hindoos under Aurungzebe, is strictly applicable to the present times. Sir William Jones, often their indiscriminate eulogist, declared from the bench his conviction, that falsehood of every imaginable sort might as easily be procured in the streets and markets of Calcutta as any other article of traffic; adding, on the subject of oaths, that even if a form the most binding on the consciences of men were established, there would be found few Hindoo consciences to be bound by it.

Among the better qualities of the Hindoos may be reckoned frugality, patience, docility, and even industry. But the first of these virtues makes, in many cases, too near an approach to avarice. This is a quality of the Hindoo character which it is not very easy to explain. The usual effect of bad government, by rendering property insecure, is to make the people prodigal, and if not indifferent to

possession, at all events careless of accumulating. Undoubtedly opposite effects have been the result among the Hindoos. A thoughtful writer, endeavouring to account for it, says, ' Slavery has sharpened the natural fineness of all the spirits of Asia. From the difficulty of obtaining, and the greater difficulty of preserving, the Germans are indefatigable in business, and masters of the most exquisite dissimulation in all affairs of interest.' This states the fact very correctly, but leaves the cause wholly unaccounted for; for undoubtedly slavery has produced no such effect on the Arabs, the Turks, the Persians, the Chinese, or even the Mohammedans of India. The docility, too, of the Hindoos is very much akin to passiveness; they are almost as easily trained to submit to oppression and rapacity, as to endeavour to improve and amend their condition. (For further details, see INDIA (BRITISH), and the arts. BENGAL, BOMBAY, and MADRAS.)

HIRSCHBERG, a town of Prussian Silesia, and a considerable emporium for the linen manufactures of that prov., cap. circ., on the Bober, near the Riesengebirge, 25 m. NW. Liegnitz. Pop. 8,940 in 1861. The town stands in a very high situation, 1,050 ft. above the level of the sea. It is handsomely and well built; has 4 suburbs, 5 churches, one of which is Protestant; a gymnasium, deaf and dumb and orphan asylums; and is the seat of the council, and superior courts for the circle. Fine linen is woven in the neighbourhood, in which there are also many sugar refineries, bleaching establishments, and paper-mills. Its manufactures, however, have fallen off very much since the middle of last century. Warmbrunn, the most celebrated watering-place of Silesia, is at no great distance from this town.

HIRSCHFELD, or HERSFELD, a town of Hesse-Cassel, prov. Fulda, cap. distr. and principality of the same name, on the Fulda, which is here crossed by a stone bridge, 32 m. NNE. Cassel. Pop. 7,410 in 1861. The town is walled, and has 2 churches, an hospital, an orphan asylum, numerous other charities, and the best-conducted Calvinist college in the electorate. It has also some woollen cloth factories and tanneries.

HITCHIN, a market town and par. of England, co. Hertford, hund. Hitchin and Pirton, on the Great Northern railway, 31 m. N. by W. London, and 13¼ m. NW. Hertford. Area of par. 6,150 acres; pop. of do. 7,577, and of town, 6,330 in 1861. The town stands at the foot of a steep hill belonging to the Chiltern range, and consists of several streets, irregularly laid out, and lined with old but well-built houses. The church, in the ornamental Gothic style, has a low embattled tower, surmounted by a spire, and a N. porch, a fine specimen of Tudor architecture: the interior, which is richly ornamented, contains a curious font, and many splendid monuments. There are several places of worship for dissenters, 2 endowed schools, 2 Lancastrian schools, an infant school, and some almshouses. The trade of Hitchin, which in the 14th, 15th, and 16th centuries, was a large wool-staple, is now chiefly confined to mealing and malting, its markets being well attended and abundantly supplied with grain. Straw-plaiting employs many hands; and there is a silk-mill. The town is divided into 3 wards, each governed by 2 constables and 2 headboroughs, appointed by the lord of the manor. Petty sessions are held by the county magistrates every Tuesday, the market day. Fairs, Easter and Whit Tuesday, for sheep and poultry.

HOANG-HO, or YELLOW RIVER. See CHINA.

HOCHSTADT, a small village of Bavaria, circ. of the Upper Danube, on the N. side of the Danube, 23 m. NW. Augsburg, and 2 m. W. Blenheim.

Pop. 687 in 1861. The great victory gained here on the 13th Aug. 1704, by the English and Imperialists under the Duke of Marlborough and Prince Eugene, over the French and Bavarians, is called by the French and Germans the battle of Hochstädt; we call it the battle of Blenheim. (See BLENHEIM.)

HOF, a town of Bavaria, circ. Upper Franconia, cap. of a distr. on the Saale, 37 m. NE. Baireuth, on the railway from Leipsic to Nuremberg. Pop. 12,018 in 1861. The town was formerly walled, and has two suburbs, a gymnasium, with an extensive library, and several charitable institutions. Its manufactures consist of muslins and other cotton fabrics, on an extensive scale; and of cotton yarn, woollen stuffs, leather, paper, and colours. It has 2 annual fairs. Iron mines and marble quarries are wrought in its vicinity.

HOGUE, or HAGUE (CAP DE LA), a bold prominent headland of France, on the English Channel, at the NW. extremity of the dep. la Manche, 16 m. W. by N. Cherbourg, lat. 49° 43′ 43″ N., long. 1° 45′ 15″ W. This cape is famous in naval history, from the great battle fought in the adjacent area on the 19th, 20th, and 22nd of May, 1692, between the combined English and Dutch fleets under Admiral Russell, and the French under Tourville. The allies, who were superior in force, gained a decisive victory; about 20 of the French ships, including that of the admiral, were taken or destroyed. This engagement may be considered as the era of the naval preponderance of England over France.

HOHENLINDEN, a village of Bavaria, circ. Isar, 19 m. E. Munich. Pop. 260 in 1861. Near this village took place, on the 3rd of Dec., 1800, one of the greatest conflicts of the revolutionary war, between a French and Bavarian army, under Moreau, and the Austrians, under the archduke John. The former gained a complete victory. Besides killed and wounded, the Austrians lost 10,000 prisoners and 100 pieces of cannon. Campbell's tale, entitled Hohenlinden, has rendered the name of this battle familiar to most Englishmen.

HOLBEACH, a market town and par. of England, co. Lincoln, wap. Elloe, parts of Holland, 37 m. SSE. Lincoln, 89 m. N. London by road, and 105 m. by Great Northern railway. Pop. of town, 2,000, and of par. 4,956 in 1861. Area of par., 20,240 acres. The town, situated on the Holland Level, between the Glen and the Nen, and about 6 m. from the sea, is old and badly built. The church is large and handsome, having a tower surmounted by a light octagonal spire, which is visible from a great distance across the fens. Among the public buildings is a chapel for Wesleyan Methodists, a well-endowed free school, and an hospital for 11 poor old men. Holbeach is one of the polling places for the N. division of the co. Markets on Thursday; butter-fairs, well attended, May 17, Sept. 11, and Oct. 11.

HOLLAND, or THE NETHERLANDS, comprising the territories formerly included within the RIVER UNITED PROVINCES, now a secondary European kingdom, but which, in the 17th and 18th centuries, was an independent republic, raised by the industry, economy, and enterprise of its inhabitants to the first rank as a commercial and maritime power. The kingdom of Holland (exclusive of Dutch Limburg and Luxemburg) lies in NW. Europe, between lat. 51° 12′ and 53° 30′ N., and long. 3° 22′ and 7° 12′ E.; having E. Hanover and Rhenish Prussia, S. Belgium, and W. and N. the North Sea. Length, NE. to NW., about 200 m.; average breadth about 65 m. The W. half of Limburg, which belongs to Holland, joins

the above territory on the SE., and is enclosed by Belgium W. and S., and Rhenish Prussia E. That part of the grand duchy of Luxemburg which belongs to Holland is situated between lat. 49° 29′ and 50° 13′ N., and long. 5° 45′ and 6° 30′ E.; it is detached from the rest of the Dutch dominions, and surrounded by those of Prussia, Belgium, and France. The kingdom is divided into the following ten provinces:—

Provinces	Area sq. miles	Population in 1840	Population Dec. 31, 1861
North Holland .	978	474,873	454,819
South Holland .	1,106	549,664	635,181
Zealand . . .	688	145,582	176,131
Utrecht . . .	542	140,574	162,315
Guelderland . .	2,018	336,504	414,401
Overyssel . . .	1,394	191,863	240,207
Drenthe . . .	1,044	76,371	96,507
Groningen . . .	778	172,437	211,466
Friesland . . .	1,251	257,615	276,439
North Brabant .	1,653	368,102	411,866
Total . . .	**10,905**	**2,860,899**	**3,372,657**

The pure Dutch, or Netherlanders, numbering about 2½ millions, inhabit the provinces of North and South Holland, Zealand, Utrecht, and Guelderland; the Frisians, speaking a dialect of the Dutch language, are dispersed, to the number of half a million, through Overyssel, Drenthe, Groningen, and Friesland; while North Brabant is almost entirely inhabited by a Flemish population.

Physical Geography.—With the exception of some insignificant hill-ranges in Guelderland and Utrecht, and a few scattered heights in Overyssel, the whole k. of Holland is a continuous flat, partly formed by the deposits brought down by the rivers intersecting it, and partly conquered by human labour from the sea, which is above the level of a considerable portion of the country. Holland is consequently at all times liable to dangerous inundations. The W. coast, however, from the Helder to the Hook of Holland, is partially protected by a natural barrier composed of a continuous range of sand-banks, or *dunes*, thrown up by the sea, of great breadth, and frequently 40 or 50 ft. in height. As the sand, which is very fine, is easily blown about by the winds, the dunes are carefully planted with the *Arundo arenaria*, or bent, which binds them firmly together, obviating the injury that would otherwise be caused by their spreading over the country, and rendering them an effectual barrier against the encroachments of the sea. But, in other parts of the country, particularly in the prov. of Zealand, Friesland, and Guelderland, the sea is shut out by enormous artificial mounds or dykes, any failure in which would expose extensive districts to the risk of being submerged. In nothing, indeed, is the industry and perseverance of the people so conspicuous as in the construction and maintenance of those dykes. It being necessary to shut out not only the sea, but the rivers, the channels of which are in parts elevated considerably above the level of the land, the extent of dykes is immense, and the expense and labour required to keep them in repair is very great. They are constructed principally of earth and clay, sloping very gradually from the sea or the river, and usually protected in the more exposed parts by a facing of wicker-work formed of interwoven willows; sometimes their bases are faced with masonry; and they are in parts defended by a breastwork of piles, intended to break the force of the waves. The most stupendous of these dykes are those of W. Capelle, in the island of Walcheren, and that of the Helder;

but there are many others of hardly inferior dimensions.

The *rivers* of Holland have mostly a W., or N. direction. The principal is the Rhine, which, for the most part, separates N. Brabant from Guelderland and N. Holland, and after receiving the Meuse, divides into two principal arms, called the Lower and Hollands-Diep. Before reaching Nimeguen, it has given off a branch to the N., which, though of less size, preserves the name of the Rhine, instead of the main stream, and itself gives off the Yssel; these two branches discharge themselves into the Zuyder Zee. The main stream from the above point, near Nimeguen, takes the name of the Waal, and after its junction with the Meuse is called the Merwe. A branch called the Leck unites the lower Rhine with the Merwe E. of Rotterdam. (See RHINE.) The Meuse traverses the SE. part of Holland; the Scheldt, its SW. extremity. The Meuse, true Rhine, and Scheldt discharge themselves into the North Sea. The estuary of the Ems forms the NW. boundary of Holland. Lakes are extremely numerous, especially in the N. prov.; and there are some extensive marshes, as the *Bourtang* on the NE. frontier, the Peel in N. Brabant and Limburg, &c. The islands may be classed in two groups: the S. group, comprising a great part of the prov. Zealand and a portion of S. Holland, is formed at the mouths of the principal rivers, and comprises Cadsand, N. and S. Beveland, Walcheren, Schouwen, Tholen, Over-Flakkee, Voorn, Beyerland, Yssermond, &c.; the N. group follows the line of coast stretching from the Helder to near the mouth of the Ems, and includes the Texel, Vlieland, Ter Schelling, Ameland, &c. There are several small islands in the Zuyder Zee. (Balbi, Abregé, pp. 252, 353; Dirt. Geog.; De Clos, Descr. Géog., des Pays Bas.)

Climate.—Holland is colder than any part of England in the same lat., and all passage for ships on the great canal between Amsterdam and the Helder is annually stopped by ice for three months. The mean temperature of the year throughout the country is stated in the 'Journal de Travaux' of the French Statistical Society to be 47° Fahr. The climate generally is variable, and the atmosphere much loaded with moisture, especially in the W. prov., where intermittent fevers, dropsies, pleurisies, rheumatisms, and scurvy are frequent diseases. Guelderland is the healthiest prov., but all the E. parts of the country are warmer and more salubrious than the others. Holland is continually subject to strong winds, without which, indeed, to remove the exhalations from the stagnant marshes and numerous canals, the country would be very unhealthy. This circumstance is also taken advantage of for turning innumerable windmills, by the help of which the drainage of the land is chiefly effected. In winter the winds sometimes rise to violent tempests, and in spring are often very high. They are particularly liable to cause inundations by raising the tides on the coast higher than usual, when they blow strongly from the W., or NW. In winter N. or NE. winds are the most common; snow falls abundantly, and even the Zuyder Zee is sometimes frozen over. In summer cold nights often succeed to days of intense heat. (De Clos, p. 85; Letters sur la Hollande, LVI, &c.)

Natural Products.—The soil is almost everywhere alluvial clay and sand. Holland possesses little, if any, mineral wealth. It has no mines of any description. Some bog-iron is met with, but no other metal. No coal deposits are found, but, extensive beds of marine peat, of a most excellent quality, abound, especially in Friesland and Hol-

land. Potters' clay, fullers' earth, and some calcareous products, are met with, but scarcely any stone is found from one end of the kingdom to the other. Holland, however, is abundantly supplied with granite and limestone, conveyed from Limburg by the Meuse; but the greater part of the lime used in the marit. provs. is obtained by burning sea shells. The country contains very little wood. There is some timber in the E. provs., and at the Hague, Utrecht, and Haarlem, there are woods of oak, elm, and beech, but, speaking generally, most of the trees have been planted. The principal canals, especially in and near the towns, are lined with rows of willows and poplars; and in various places along the sandy shore firs are produced. In other respects the vegetation is very similar to that of England. The fringed buckbean (*Menyanthus nymphoides*), however, which is rare in the latter country, here floats in the greatest profusion on the surface of the canals, and the more rare *Senecio paludosus* is not unfrequently met with. The zoology, also, is in most respects like that of the southern and central parts of Great Britain. The larger kinds of wild animals are not met with. Hares and rabbits are plentiful, but not winged game. The preservation of game is an object of great interest to most proprietors; and notices to that effect are fixed up, and great vigilance exercised to prevent the trespassing of sportsmen and others. In dry seasons, in some districts, field mice multiply to such an immoderate degree as to produce serious loss to the farmers, by destroying the roots of the grass in the meadows, where they harbour by millions. The pools and marshy grounds abound with frogs and other reptiles, which are a favourite food of storks. These birds are particularly numerous in Holland, where they remain from the middle of February to the middle of August. They are great favourites, and severe penalties are enforced upon their wilful destroyers. In the towns they build their nests on the houses; and in those parts of the country that are destitute of trees, buildings, or other means of protection, an old cart-wheel is very often raised upon a high pole, to afford them facilities for the same purpose. Water-fowl are very abundant. The principal fish that frequent the Dutch coasts are cod, turbot, soles, and other flat fish. The herring fishery is a most important source of wealth.

Public Works, Dykes, and Canals.—There is perhaps no country for which nature has done so little, and man so much, as the Netherlands. The first and greatest of the works of art are the stupendous *dykes*. The construction and repair of these prodigious bulwarks is placed under the control of a particular department of the government (*Waterstaat*), and a corps of engineers especially appointed for this important service. The expenditure of this department amounts to a large sum annually. The cost of each dyke is defrayed by a tax laid on the surrounding lands, assessed according to long-established usage, and levied by commissioners appointed for the purpose. The expenditure in labour, though great, is generally much exceeded by that in willows and timber. The former are raised in extensive plantations near the places where they are wanted.

If there be any danger of an inundation, the inhab., on a signal being given, repair en masse to the spot. There is never any backwardness on these occasions, every one being fully aware, not only that the public interests are at stake, but that his own existence perhaps, and that of his family and friends, would be involved in extreme hazard should the waters break through the dykes. Hence, the most strenuous efforts are made to ward off the impending danger, and every possible device is adopted by which the dykes may be strengthened, and the threatened inroad prevented, or its violence mitigated. In despite, however, of these precautions and efforts, Holland has on numerous occasions sustained extreme injury from inundations. That extensive arm of the sea called the Zuyder Zee, between the provs. of Holland, Guelderland and Friesland, occupying an area of about 1,300 sq. m., was formed by successive inundations in the course of the 13th century. The so-called Haarlem Meer or Lake, which in recent years has been artificially dried, and, therefore, ceased to be a lake, owed its origin to an inundation in the 16th century, which proved fatal to great numbers of the inhab.; and many inundations have taken place within a comparatively modern period. Owing, however, to the improved construction of the dykes, and the greater skill in engineering, these calamities are now neither so frequent nor so destructive as formerly.

Some of the interior parts of the country traversed by the great rivers are even more exposed to the dangers of inundation than those contiguous to the shore; and when the débacle, or breaking up of the ice, takes place in the upper part of the river, before it has begun to reach the sea, as is sometimes the case, the risk of inundation is extreme. On such occasions every effort is made, not excepting even the employment of artillery, to break the ice and facilitate the exit of the water, but sometimes without the desired effect. The following is an instance of this sort of calamity. 'One of the richest tracts of country, in the vicinity of Arnheim, has been often exposed to tremendous inundations. These are frequently felt at the breaking up of a long frost; but in no instance so calamitously as in the winter 1808–9. A violent tempest from the NW. had raised the waters of the Zuyder Zee some feet above the highest mark of the spring tides, and the waves beat with unusual violence against the dykes constructed to break their fury. The thaw on the Upper Rhine had increased the quantity, and the force of its waters, which brought down masses of ice 14 ft. in height, and more than half a mile in length; to which the embankments, softened by the thaw, and somewhat injured, presented an insufficient barrier. A breach made in one part soon extended itself, and the torrent quickly covered the country, bearing before it by its force the villages, the inhab., and the cattle. The height of the Zuyder Zee prevented the water from finding an outlet; and it consequently remained on the ground for a long period, in spite of the exertions of the surviving inhab. By this event, more than 70 houses were totally destroyed, a far greater number irretrievably damaged; and of 900 families, more than 500 were rendered utterly destitute; more than 400 dead bodies were left on the borders of the current; and at the city of Arnheim 500 persons, mostly women and children, with many hundred head of cattle, were rescued from a watery grave, by the hazardous heroism of the inhab. who ventured in boats to their rescue.' (Jacob's View of the Agric. of Holland, pp. 57, 58.)

The general aspect of Holland is different from that of any other country in Europe. Its surface presents one immense network of canals, which are there as numerous as roads in England, the purposes of which, indeed, they for the most part answer. The greater number are appropriated to the drainage of the land; many, however, are navigable by large vessels. The principal is the Grand Ship Canal of N. Holland, between Amsterdam and Nieuwdiep, near the Helder. This noble

work, the greatest of its kind in Europe, is about 50¼ m. long, 125 ft. broad at its surface, and 36 at bottom, with a depth of 20 ft. 9 in.; it extends from Amsterdam to the Helder, and was completed between 1819 and 1825, at an expense of 950,0000l. It has a towing path on each side, and admits of two frigates or merchant vessels of the largest size passing each other. By means of this canal, ships avoid the delay and danger they were formerly subject to in navigating the Zuyder Zee, and reach the Texel from Amsterdam in 18 hours. As a commercial speculation, it has been but indifferently successful; but it is of incalculable benefit to Amsterdam, to which it has given all the advantages of a deep-water harbour on the most accessible part of the Dutch coast. The other chief canals are—the Zederik, in S. Holland, from Vianen to Gorcum; that from Bois-le-Duc to Maestricht, available for vessels of 800 tons; and that between the Ems and Harlingen, in Friesland. As they run through an entirely level country, locks are generally unnecessary, except at their mouths. One of the finest monuments of scientific skill in Holland, is a succession of locks or sluices of enormous size and strength, constructed in 1809, at the mouth of that branch of the Rhine on which Leyden is situated. This mouth was for a long time choked up with sand, but it is now kept quite clear, the locks being closed with the flow and thrown open by the ebb of the tide. The larger canals are commonly about 60 ft. broad, by six deep; and though often below the level of the sea, not only their surface, but their bottom, is frequently higher than the adjoining country. The smaller canals, by which the country is drained, traverse and surround sections of land protected from inundation by means of dykes.

Such sections are termed polders. A tract of land on being rescued from the sea or a river is in the state of a morass or marsh; and the 'next process is to dry it, so as to render it suitable for tillage or pasture. To effect this, the marsh is intersected by water-courses, and windmills are employed, as in the Fens in England, to lift up the water. These mills are erected on the dyke or rampart, excluding the sea or river, and raise the water to a ditch or canal on the other side. Pumps are seldom employed for this purpose, wheels being by far the most generally used. Sometimes the marsh is too extensive to be drained simultaneously, in which case it is divided into compartments by subordinate ramparts and water-courses; and mills being erected on them, each portion is separately divested of water. In many cases, however, the depth of the marsh below the level of the sea or river is too great to allow of the drainage being effected by one series of ramparts and ditches; and in these cases, two or more series of ramparts, ditches, and mills are constructed at different elevations, the water being lifted up successively from one to another, till it be finally brought to the desired level and conveyed away. The labour and patience required in an undertaking of this kind is shown by the fact that the surface of some of these polders is as much as 14 ft. below high-water mark, and 30 ft. below the level of the highest tides. The soil of the polders is of various sorts. Where it is clayey, and the drainage perfect, they are extremely fertile, and are not unfrequently cultivated; but where the soil is mossy, or the drainage incomplete, they are employed as meadows.

In sailing along the arms of the sea, the rivers or canals of this singular country are seen at a considerable elevation above the surrounding fields, reminding the traveller of Goldsmith's verses:—

' To turn of other minds my fancy flies.
Enthroned in the deep where Holland lies ;
Methinks her patriot sons before me stand,
Where the broad ocean beats against the land,
And, solicitous to stop the rushing tide,
Lift the tall rampire's artificial pride,
Onward, methinks, and diligently slow,
The firm compacted bulwark seems to grow ;
Spreads its long arms around the watery roar,
Scoops out an empire, and usurps the shore ;
While the pent ocean rising o'er the pile
Sees an amphibious world beneath him smile ;
The slow canal, the yellow-blossom'd vale,
The willow-tufted bank, the gliding sail,
The crowded mart, the cultivated plain,
A new creation rescued from his reign.'

The facility with which the country may be laid under water, contributes materially to its strength in a military point of view. This, indeed, is not a resource to be resorted to, except on extreme occasions; but it was repeatedly made use of in the war of liberation, and also in 1672, when Louis XIV. invaded Holland. It is said that in 1830-32 every thing was prepared for an inundation, had the threatened inroad of the French taken place.

The roads and private estates are commonly fenced by canals or ditches about; hedges are rare. The highways in the central provs. are among the best in Europe. They run for miles in a straight line along the summits of the dykes, and are thus at once dry and elevated, so as to command extensive views. Between the large cities they are level, and usually paved with a kind of small hard bricks called clinkers, mostly made of sand mixed with the clayey mud obtained in cleaning the canals. They are fitted so exactly to each other, when laid down, that scarcely a crevice is to be seen; and being well covered with sea sand, they sustain little injury from carriages. Elsewhere, the roads are made of sea-shells and the common soil, well compounded together; which mixture though soft, is not much cut by the wheels. Where water conveyance is so abundant, it may be easily supposed that few carriages will travel on roads burdened with tolls so high as to amount to nearly as much expense as the post-horses. In fact, the transport of the greater part of farm-produce and other bulky goods is carried on by means of water; and persons travelling, unless they use the railways, which intersect the country nearly so much as canals, commonly make use of the canal barges, or trekschuits, towed by horses. This is especially the case in the NE. provs.

Distribution of Land.—Of about 7,800,000 acres, which the total surface of Holland comprises, there are estimated to be 5,310,000 acres of cultivated land; 2,000,000 ditto uncultivated; 270,000 ditto occupied by canals and ponds; and the residue by roads, buildings, and public walks. The richest lands are in the S. and central provs.; the poorest, for the most part, in the NE.; in Over-Yssel and Drenthe, especially, heath and waste lands prevail to a great extent. A good deal of waste land, originally of a very unpromising quality, has, of late years, been brought into cultivation by the pauper population settled upon it. For the purposes of the land-tax, a cadastral valuation of landed property has been made, and continued from time to time according to the changes which have taken place by bringing waste lands into cultivation, and by the increase of buildings. Newly reclaimed lands, however, and new buildings have the benefit of an exemption from the tax for ten years. The amount of the valuation in 1859 of all income from land subject to the tax in the Netherlands was

71,541,171 florins, or about 6,080,000l. sterling; and the amount of revenue raised in the same year from this source was 10,536,766 florins, about 875,000l., or at the rate of 2s. 11d. in the pound. Landed property is divided for this purpose into such as is occupied with buildings ('gebouwde'), and such as is not so occupied ('ongebouwde'). The respective extent of the latter in the several provinces, with the number of buildings, distinguishing those which are liable to the tax from the others, was, for the year 1859, as follows :—

Provinces	Land not occupied with Buildings			
	Number of Buildings		Extent in Acres	
	Liable to Tax	Not liable to Tax	Liable to Tax	Not liable to Tax
North Brabant	854,101	3,730	1,165,768	89,877
Gelderland	826,512	4,348	1,704,710	56,365
South Holland	597,600	7,—	704,747	34,650
North Holland	365,600	3,305	636,776	34,118
Zeeland	184,256	1,958	414,186	19,470
Utrecht	178,561	1,240	354,548	5,440
Friesland	266,210	8,101	779,652	46,673
Overyssel	220,435	1,662	612,750	6,563
Groningen	277,654	1,575	534,169	12,—
Drenthe	196,364	675	522,—	9,—
Limburg	685,—	8,407	421,761	93,671
Total	5,850,347	37,245	7,792,454	369,700

The above table is from a report by Mr. Ward, British Secretary of Legation, dated The Hague, June 29, 1861. (Reports of Secretaries, N. V. 1862.)

In N. Holland the proportion of pasture to arable land is about 2 to 1. The average size of farms is from 40 to 50 bunders (the same as the French hectare, nearly 2½ acres each); large farms run from 70 to 100 bunders. The principal proprietors usually let their land on lease to the peasantry; the proprietor paying the property-tax, and the dues on dykes, polders, and water-mills; and the farmer a personal tax and the tax on servants. In some instances the landlords furnish or pay for seed and manure, and go halves in the crops with the tenants on the metayer principle; but even when this is not the case, the rent is always paid in kind. The leases are commonly for 6 years. In N. Holland, farms average no more than 70 bunders, or 50 acres, each; on which from 16 to 18 cows, 4 calves, a horse, and 15 or 20 sheep, besides a few hogs, may be kept if the soil be good. The rent of pasture land varies from about 18 to 30 florins, of arable land from 35 to 50 florins the bunder; garden grounds near the towns let somewhat higher. In Friesland, the quantity of pasture is more than 8 times greater than that of arable land. The common size of a farm is from 75 to 100 acres; but some are nearly twice as large. They are generally let on leases of 5 or 7 years, the proprietor paying the land-tax, and the cultivator the other assessments; though in some parts the proprietor contributes his quota to the maintenance of dykes and dams. Few proprietors cultivate their own land. The best clay pasture in that province fetches a rent of from 3l. to 4l. the bunder; but a considerable proportion of the soil is sandy and inferior, and lets for only from 2s. to 3s. the bunder; there are also about 300 bunders marshy and unproductive, some yielding a rent of no more than 10d. a bunder. In Gelderland there is some good land, but a great deal more is very indifferent; and in the SW. vast tracts have been planted with Scotch firs, and Weymouth pines; many hundred acres have also been sown with acorns, without any hope of the oaks ever reaching the size of timber, but merely for the sake of the underwood.

Crops and Mode of Agriculture.—The principal grains cultivated are rye and buckwheat; next to these come oats and barley. About 1,800,000 lasts of wheat are produced yearly, 10 per cent. of which is estimated to be consumed in breweries, distilleries, and starch and other manufactories. Wheat is a good deal grown round Utrecht, the country there being more elevated and suitable for it, than most other parts of Holland; the wheat of Friesland, however, is extremely good; and the prov. of Zealand yields more than is required for its own consumption. In both the last-named provs. garden vegetables are abundantly grown, besides woad and madder in the former, and millet and horse-radish in the latter. Flax is raised in large quantities in the S., and especially round Dort, which is the centre of a considerable trade in that article. There is an abundance of fruit in Guelderland and Holland; but in the N. provs. only apples and pears come to any perfection. The vine is cultivated only in Luxemburg. Utrecht and Guelderland are noted for their tobacco; 30,000 quintals yearly were formerly sent into the market, from these provs. Potatoes, rapeseed, hemp, chicory, mustard, hops, lac-root, and some medicinal plants, are the other principal articles of produce. The ancient passion of the Dutch for tulips and other bulbous plants still exists, though now confined within reasonable limits; there are some large flower-gardens, in the neighbourhood of Haarlem especially, from which great numbers of bulbs are annually exported. (See Haarlem.)

In S. Holland wheat is the grain most cultivated, the quantity of it raised being double that of barley, which comes next to it in importance. Wheat is said to produce from 12 to 15 fold, and other grains in proportion; but such statements are seldom worthy of much confidence. The rotation in this prov. is usually as follows:—rapeseed, winter barley, or rye, succeeded by rapeseed, barley, or wheat; flax, beans, or oats, succeeded by summer grains; and these by potatoes; rye, oats, beans and clover; and the last year the remainder of the clover—after which the ground is fallowed. In N. Holland, rape and mustard seeds, barley, oats, peas, and beans and pigeon beans are generally grown in the rotation, though no fixed rule is observed. There are no fallows in this prov. In Friesland, the better sorts of land are appropriated to wheat, barley, rye, and rapeseed, and the inferior to summer grains, as buckwheat and oats. Rapeseed, after fallows, is succeeded next year by wheat or barley; on wheat lands the alternate crops are barley or beans, flax or potatoes; on rye lands, buckwheat and oats. Near the W. border of Guelderland, the land when cleared is manured and sown with buckwheat; after that, a second dressing of dung is administered; and after a single ploughing, rye is sown. The rye is usually harvested in July, when turnips are sown after a single ploughing. There are thus regularly 3 crops in every 2 years. The average produce of buckwheat is from 20 to 22 bushels per acre, and rye 2 bushels more. Probably 7 or 8 cart-loads of manure are applied to an acre of land before buckwheat or rye. Further E. the land improves considerably. Near Doesburg the usual rotation is—first beans; then wheat, in which clover is sown; and after the clover, oats. Some of these lands are of a stiff texture, and on these it is usual to make a year's clean fallow; after which the same rotation is pursued. Madder is very extensively grown in S. Holland, and usually produces 4,000 lbs. to the acre, but it tends to exhaust the most fertile soils. It is frequently followed by rapeseed, sometimes by turnips; to these succeed wheat or oats; after which the land is laid down to grass, the growth of which in a short period becomes very luxuriant. The land destined to the culture of tobacco in Guelderland is laid out in very small patches of not more than a quarter of a rood each, slightly fenced by a few dry sticks, around which scarlet runners are trained, to protect the plants against the wind.

Pasture-farms, Cattle, and Dairy-husbandry.—The rearing of live-stock and dairy-husbandry is a much more important source of national wealth than tillage. Between the capital and Utrecht the land is almost wholly rich pasture, on which numerous cows are kept. The farms there seldom comprise more than from 50 to 100 acres. Their price, including buildings, averages 60l. an acre, though the rent they yield is scarcely more than 2½ per cent. interest on the capital. On these farms numerous cows are kept. The lean cattle, brought from Denmark and Germany, fatten with great rapidity in the Dutch polders, and an important branch of the trade of Friesland is the supply of the capital with fatted cattle. Artificial grasses are but little cultivated, and cattle are seldom stall-fed; indeed, it is too common to suffer the cows to remain in the open damp fields, both day and night, except in winter. The horned cattle of Holland are remarkable for their beauty; in N. Holland they resemble the Devonshire breed, but are rather larger, not, however, equalling the size of the Lincolnshire or Sussex cattle. The Dutch horses are good, and well adapted for draught; the best are those of Friesland; but many are reared in Groningen to be sent to Amsterdam. The breeds of sheep are bad or indifferent; they are mostly long-woolled, with white faces, pulled, and long heads and legs. They yield a great deal of coarse wool.

In the neighbourhood of large towns it is found to be most profitable to retail the milk produced on the farms; but at a distance from such markets, it is nearly all appropriated to the making of butter and cheese. In some of the dairy farms near the Hague, the average stock is about 60 cows; and a good cow may be estimated to produce 80 lbs. of butter, and 180 lbs. of cheese, during the six summer months. Throughout the greater part of Holland, butter is made of the cream only, and cheese of the skimmed milk; but in some districts the whole produce of the cow is devoted to making cheese. A good deal of butter is sent to England. The yearly export of cheese is estimated at 350,000 cwts. The dairy, the cows, and the cow-keeper's family occupy the same building, and in many instances the same apartment; but the cleanliness of the Dutch dissipates any feeling of repugnance that the idea of such an arrangement might produce in a stranger.

A farm of 62 bunders in S. Holland requires, at an average 5 servants, the family of the farmer included. The wages of servants vary from 50 to 150 florins a year; those of a maid-servant understanding the making of butter and cheese average 100 florins. The women are employed in the dairy business, in weeding, hay-making, and binding sheaves in harvest-time. The severer labour required in the making of common-sized cheese is generally performed by men, to whom also milking is often left. All regular servants board and lodge with the farmer, and eat at the same table with the family. Their food chiefly consists of wheat and rye bread, potatoes, turnips, French beans, bacon, fresh and salt beef, and pancakes of buckwheat flour and bacon. Fewer servants are generally required on the farms in N.

Holland. On one on which 20 cows are milked, a man and a woman, exclusive of the farmer and his wife, are sufficient. The wages of regular servants in N. Holland vary from 80 to 100 florins a year; they board and lodge with the farmer, but their food is hardly so substantial as in the last-named year. The clothing of the labouring classes generally is much the same as in England—fustians, velveteens, and stout woollens for the men, and cottons and linsey woolsey stuffs for the women. Wooden shoes are, however, in general use.

Fisheries.—The herring fishery formerly carried on by the Dutch was a considerable source of wealth and employment. It is now, however, confined within comparatively narrow limits, not employing more than about 80 busses of 50 or 60 tons burden, manned by 12 or 14 men each. The herrings cured by the Dutch are decidedly superior to those of the English or any other people. 'The whole process,' says an observant traveller, 'is conducted on shipboard. Immediately on being caught, the herrings are bled, gutted, cleansed, salted, and barrelled. The bleeding is effected by cutting them across the back of the neck, and then hanging them up for a few seconds by the tail. By being thus relieved of the blood, the fish retains a certain sweetness of flavour and delicacy of flesh which unbled herrings cannot possibly possess. The rapidity of the process of curing must likewise aid in preserving the native delicacy of the animal; for the herring is salted and in the barrel in a very few minutes after it has been swimming in the water. The first herrings caught and cured, to the extent of two or three barrels, are instantly despatched by a fast-sailing vessel, for Holland where their arrival is anxiously expected. On their landing at Maas-sluis, one barrel, decorated with flowers, and with flags flying, is despatched to the Hague, as an offering to his majesty, who on this occasion presents the fortunate takers with 1,000 guilders. The other barrels are sold by public auction, and generally fetch from 900 to 1,100 guilders. These precious barrels are then subdivided among the dealers, who retail them at a high price. A single herring of this first importation brings 1½ to 2 guilders,—that is 2s. 6d. to 3s. 4d. each. So highly are they esteemed, that a single herring is considered a handsome present; and it is a custom to make such gifts to friends and acquaintances on this auspicious occasion. Livery servants may be seen passing through the streets with a plate, on which lie one or two herrings, covered with a fine white cloth and a neat card of presentation.' (Chambers, p. 43.)

Manufactures.—The government of Holland is anxious to encourage manufactures; and coal, on which a heavy duty is ordinarily levied (in order to promote the use of peat, and the collateral formation of polders), may be imported duty free, if for their use. The principal manufactures are those of cotton and woollen cloths, particularly the former. The total exports of cotton in the year 1841 amounted to 13,226,021 guilders, or 1,102,251l. About 97 per cent. of the cotton exports go to the Dutch East India possessions, this being a protected trade, as goods furnished with a certificate of Netherland origin pay a duty in those possessions at one-half of the rates paid by foreign goods. Monthly returns are published by the colonial department of the goods for which certificates are passed with that object. The present rates of duty paid upon linens of cotton, linen, and wool, and other manufactures of the same articles, are 12 per cent. on such as are furnished with a certificate of Netherland origin, and

25 per cent. on all others. By a newly projected tariff which has been laid before the chambers, it is proposed to reduce these rates gradually to 6 per cent. on Netherland goods, maintaining, however, the differential duty on foreign goods at the rate prescribed by the treaty of 1824 with Great Britain, that is to say, double that paid by their own manufacturers, or 12 per cent. Silk goods which now pay 6 per cent. ad valorem when imported from a Netherland port, and 12 per cent. when imported from elsewhere, will, according to this proposal, be charged at the rate of 6 per cent. without any differential duty in favour of the port of shipment.

Besides cotton and woollen manufactures, there are others of silks and velvets, in Utrecht, Haarlem, and Amsterdam; of paper, leather, contains, hats, ribands, needles, white lead (the best made in any country), borax, glue, vermillion, saltpetre, tobacco, and liqueurs. There are numerous distilleries, and the town of Schiedam in S. Holland is particularly celebrated for its Geneva or Hollands. In Amsterdam, and other places, there are many sugar refineries. Haarlem has extensive bleaching factories, for which its water is supposed to be especially adapted. At Utrecht and Leyden, large quantities of tiles and bricks are made. Amsterdam is famed for its lapidaries and diamond cutters. Steam engines are employed to turn the machinery in some new and extensive factories; but in general windmills are used to perform offices to which steam engines are applied in Great Britain. Though most of the windmills are for the purpose of draining the land, a great many saw timber, crush rapeseed, grind snuff, &c. They are of larger dimensions than in England; the length of their sails varying from 80 to 120 ft.; they are always in sight in a Dutch landscape, and even in the suburbs of the larger cities there are vast numbers. They have all moveable roofs, so as to present their front to the wind at every change. The Dutch have attained to the highest excellence as millwrights, and some of their draining mills are of sufficient power to raise 200 tons of water to the height of 4 ft. in a minute. At an average, they discharge 250 tons a minute. The ships constructed by the Dutch are built mostly at Rotterdam and Amsterdam. They are neat without being clumsy or heavy; and round sterns, and the other modern improvements in naval architecture followed in our own dock-yards, are also practised in those of Holland. The Dutch E. Indiamen are handsome ships, well rigged, manned, and armed; and are not surpassed either in speed or durability by any similar class of merchantmen in Europe.

Commerce.—The commerce of Holland was formerly the most extensive carried on by any European state; and the wealth which it brought into the country furnished her with the means of supporting the vast expense of her lengthened struggle with Spain, and of her subsequent contests with France and England. The circumstances under which the Hollanders have been placed, the natural poverty of their country, and the necessity of unremitting vigilance to prevent its being submerged, made industry and economy a condition of their existence. Holland being destitute of iron, coal, timber, and many other indispensable articles, the prosecution of commerce is there not a matter of choice but of necessity; and hence it is that, in the earliest periods, we find the Batavians distinguished for their fisheries, their shipping, and their commercial enterprise. For a lengthened period they engrossed nearly the whole sea-fishery of Europe; and they were long the carriers and factors of the principal European states. In 1504, the Dutch appeared, for the first time, in India; and, in the course of a few years, they wrested Amboyna and the Moluccas from the Portuguese; and having obtained with them the monopoly of the spice trade, laid the foundations of an empire in the East, second only in magnitude and importance to that established at a later period by the English. Holland had long, also, a preponderating influence in the trade with the Baltic, from which she has, at all times, drawn a large supply of some of the principal necessaries. It may be stated, as illustrative of the former extent of the trade of Holland, that, in 1650, when it had attained to a maximum, Sir William Petty estimated the whole shipping of Europe at 2,000,000 tons, of which he supposed the Dutch to possess 900,000 tons; and it is believed, that this estimate was rather within than beyond the mark.

The decline of commerce in Holland was occasioned partly and principally by the natural growth of trade and navigation in other countries, and partly by the increase of taxation occasioned by the numerous contests in which the republic was engaged. During the occupation of Holland by the French, first as a dependent state, and subsequently as an integral part of the French empire, her foreign trade was almost entirely destroyed. Her colonies were successively conquered by England; and, in addition to the loss of her trade, she was burdened with fresh taxes. But such was the vast accumulated wealth of the Dutch, their prudence and energy, that the influence of these adverse circumstances was far less injurious than could have been imagined; and, notwithstanding all the losses she had sustained, and the long interruption of her commercial pursuits, Holland was still, at her emancipation from the yoke of the French, in 1814, the richest country in Europe. Java, the Moluccas, and most of her other colonies, were then restored, and she entered again upon a large foreign trade.

The connection of Holland with Belgium, settled at the Congress of Vienna, was, however, an unfortunate one for both countries. The union was not agreeable to either party, and was injurious to the former. Belgium was an agricultural and manufacturing country; and was inclined, in imitation of the French, to lay restrictions on the importation of most sorts of raw and manufactured produce. This protectionist policy was directly opposed to the interests and the ancient practice of the Dutch. But though their deputies prevented the restrictive system from being carried to the extent proposed by the Belgians, they were unable to prevent it from being carried to an extent that materially affected the trade of Holland. On the whole, there can be little doubt that the separation between the two divisions of the kingdom of the Netherlands will eventually redound to the advantage of Holland.

The imports principally consist of sugar, coffee, spices, tobacco, cotton, tea, cochineal, indigo, wine and brandy, wood, grain of all sorts, timber, pitch and tar, hemp and flax, iron, hides, linen, cotton and woollen stuffs, hardware, rock salt, tin plates, coal, and dried fish. The exports consist partly of the produce of Holland, partly of the produce of her possessions in the East and West Indies, and other tropical countries, and partly of commodities brought to her ports, as to convenient entrepôts, from different parts of Europe. Of the first class, are cheese and butter, madder, clover, rape, hemp, and linseed, rape and linseed oils, and linen. Geneva is principally exported from Schiedam and Rotterdam; oak bark principally from the latter. Of the second class are spices, Mocha

and Java coffee; sugar of Java, Brazil, and Cuba; cochineal, indigo, cotton, tea, tobacco, and all sorts of Eastern and colonial produce. And of the third class, all kinds of grain, linens from Germany, timber, and all sorts of Baltic produce; Spanish, German, and English wools; French, Rhenish, and Hungarian wines. The trade of Holland may, indeed, be said to comprise every article that enters into the commerce of Europe.

The subjoined table shows the aggregate value of the trade of the Netherlands with the chief countries of Europe during the year 1863:—

	Imports from	Exports to
	£	£
Great Britain	6,017,542	8,800,278
Germany	6,793,163	10,738,104
Belgium	8,045,216	4,909,221
France	1,730,511	818,212
Russia	1,467,189	641,928
Sweden and Norway	725,381	272,971
Italy	276,763	1,015,888

The imports from Great Britain, in each of the two years 1862 and 1863, consisted of the following articles:—

	1862	1863
	£	£
Alkali—Soda	47,899	67,797
Coals	104,932	92,733
Cottons	643,298	436,341
Cotton Yarn	1,263,716	1,118,571
Hardware and Cutlery	100,532	179,116
Linens	81,744	37,328
Lawn Yarn	107,412	231,390
Machinery	192,063	144,615
Iron	456,740	504,453
Copper, unwrought	30,367	89,384
" wrought	66,540	91,811
Oil seed	136,371	107,116
Salt, thrown	514,172	246,617
Silk Twist and Yarn	84,169	54,843
Woollens	412,620	872,516
Woollen Yarn	799,177	1,345,994
Other Articles	1,188,108	1,277,562
Totals	6,946,242	6,811,462

The mercantile marine of Holland consisted, on the 1st of Jan., 1863, of 2,299 vessels, of a total burden of 504,244 tons. Among the number were 54 steamers, of 12,036 tons.

Money—Weights and Measures.—The most common coin, and that by which accounts are generally reckoned, is the guilder, or Dutch florin, equivalent to 1s. 8d., and divided into 20 stivers (1d.) and 100 cents (cent=1-5th of a penny). The dollar is worth 2s. 6d., and the rix-dollar 4s. 4d. Eng. The William, a gold coin, is valued at 17s. The Dutch schippound is 8 quintals, the quintal 100 lbs., and the livepound 15 lbs.; 100 lbs. Dutch are equivalent to 108 lbs. English. The Dutch quart is equal to 6 8-10ths gall. Eng. The Dutch foot=11·7 in. Eng.; the ell=27·1 in. Eng. The Dutch mile, or league=3½ Eng. ta nearly.

Government.—Previously to its occupation by the French in 1795, and its subsequent erection into a kingdom by Napoleon, Holland was a republic, governed by the states-general, with the executive power lodged in the hands of a stadtholder. There can be no question that the great commerce of the Dutch in the 16th, 17th, and 18th centuries, their wealth and industry, were materially promoted by their free institutions and the nature of their government. At a time when England, France, and most other European states, were a prey to civil wars, caused by religious and political differences, the Dutch had the wisdom to establish and maintain a system of universal toleration, and to make their country an asylum for all persecuted and oppressed strangers. Though complex and not very popular, in practice the constitution gave free scope to all deserving individuals to attain to the highest dignities, at the same time that it effectually secured them against violence and oppression. The utmost latitude was given to every one to dispose of property by will as he thought best; justice was speedily and impartially administered; and though taxation was heavy, the revenue was faithfully and economically expended. Hence the political conquest with the physical circumstances under which the Hollanders were placed to call forth their talents and enterprise, and to render them industrious and economical. That the difficulties incident to their situation, the *duris urgens in rebus egestas*, have done much to make them what they have been and what they are, cannot be disputed; but it is easy to see that they are, at the same time, largely indebted to the freedom of their civil and religious institutions. By decree of the Congress of Vienna, dated May 31, 1815, the ancient form of government of the Netherlands was changed into a constitutional monarchy, and the royal dignity was made hereditary in the family of the Princes of Orange. The king is also grand duke of Luxemburg, in which capacity he belongs to the German confederation. His person is inviolable, his ministers alone being responsible; he nominates to all civil and military offices, prepares and promulgates the laws, declares war or makes peace. The states-general consist of 2 chambers: the first is composed of 59 members, nominated by the provincial diets from among the most highly assessed inhabitants of the various provinces. The second chamber of the states-general numbers 72 members, elected by ballot. All citizens, natives of the Netherlands, paying taxes to the amount of 120 guilders, or 10£, are voters. Clergymen, military officers in active service, and judges, are debarred from being elected. The members of the second chamber receive an annual allowance of 2,000 guilders, or 166£, besides travelling expenses. Every two years one-half of the members of the second chamber, and every three years one-third of the members of the upper house retire by rotation. The sovereign has the right to dissolve either of the chambers separately, or both together, at any time. The constitutional advisers of the king, having a seat in the cabinet, must attend at the meetings of both houses, and have a deliberative voice; but, unless they are also members, cannot take an active part in the debate. All financial measures must originate in the second chamber; the assent of both the sovereign and the upper house is required before any bill which has passed the house of representatives becomes law. The royal veto is seldom, if ever, brought into practice.

The executive authority is in the hands of the sovereign, and exercised by him through a responsible council of ministers. There are seven departments in the ministerial council. Each of the ministers has a salary of 12,000 guilders, or 1,000£ per annum. Whenever the sovereign presides over the deliberations of the ministry, the meeting is called a 'Cabinet Council,' and the privilege to be present at it is given to all princes of the royal family who are of age. There is also a privy council of 14 members, all nominated by the government, which the sovereign may consult.

The different provinces have their own local magistracy and laws established for their own states; the judges are nominated by the king for life, on the recommendation of the provincial states, or the states-general. The provinces are divided into arrondissements, cantons, and communes, similar to those of the French *départ.*, and superintended in like manner. The local courts are also similar to those of France; in each canton there is a court of justices of the peace, and in each arrond. one of original jurisdiction; there are tribunals of commerce in the principal commercial districts. The supreme judicial court, and high board of taxation (*cour des finances*), sit at the Hague, which is also the usual residence of the court. The police is under the control of a central director, a subdirector in each prov., and commissaries in the arrondissements. No medicants or disorderly persons are suffered to offend the public eye, and education is carefully administered to juvenile offenders. There is no imprisonment for debt.

Church Administration.—According to the terms of the constitution, entire liberty of conscience and complete social equality is granted to the members of all religious confessions. The royal family, and a majority of the inhabitants, belong to the Reformed Church; but the Roman Catholics are not far inferior in numbers. In the census of 1849—more recent enumerations do not show the religious creeds—the number of Calvinists, or members of the Reformed Church, is given as 1,906,614; of Lutherans, 66,176; of Roman Catholics, 1,220,667; of Greek Catholics, 41; of diverse other Christian denominations, 41,161; and of Jews, 64,070. The government of the Reformed Church is Presbyterian; while the Roman Catholics are under an archbishop, of Utrecht, and four bishops, of Harlem, Breda, Roermond, and Herzogenbusch. The salaries of several British Presbyterian ministers, settled in the Netherlands, and whose churches are incorporated with the Dutch Reformed Church, are paid out of the public funds. The ministers of the Dutch Reformed Church are allotted to certain districts in proportion to the pop.; there being 1 pastor generally to about every 2,000 or 3,000 people. Their maximum salary is 200£; their minimum 50£.

The Military Force amounted, in 1863, to 57,520 men, under 1,435 officers, excl. of colonial troops. The army is formed partly by conscription and partly by enlistment, in such a manner that the volunteers form the stock, as well as the majority of the troops. The men drawn by conscription at the age of twenty have to serve, nominally, five years; but practically, all that is required of them is to drill for a few months, and, returning home on furlough, meet for a fortnight annually for practice, during a period of four years. Besides the regular army, there exists a militia—'schutters'—divided into two classes. To the first, the 'active militia,' belong all men from the twenty-fifth to the thirty-fifth year of age; and to the second, the 'resting (rustende) militia,' all persons from thirty-five to fifty-five. The principal fortresses, next to Luxemburg, are Maestricht, Breda, Bergen-op-Zoom, Bois-le-Duc, Flushing, and the Helder. Luxemburg is garrisoned by Prussian troops.

The Navy consisted, at the commencement of 1864, of 59 steamships, with a total of 785 guns, and 81 sailing men-of-war with 856 guns. The navy was manned, at the same period, by 6,187 sailors, recruited by voluntary enlistment.

The Public Revenue is derived from a land-tax, or *contribution foncière*, from numerous personal and assessed taxes, excise duties, which, among other articles, are imposed on turf, coal, &c., and from taxes on stamps, registrations, tolls, harbour dues, customs, the post-office, lotteries, &c. The budget for the year 1864-65 estimated the income and expenditure of the kingdom as follows:—

Estimated Revenue for 1864-65	Guilders
Direct Taxes:	
Land Tax	10,435,170
Personal Taxes	7,122,000
Tax on Trades and Professions	3,347,600
Total	21,749,000
	£1,727,168
Excise:	
Sugar	9,000,000
Wine	1,700,000
Spirits	6,200,000
Salt	2,280,000
Soap	1,100,000
Beer and Vinegar	810,000
Butcher's Meat	1,300,000
Fowls	1,280,000
Turf	1,680,000
Total	[illegible]
	£1,466,610
Indirect Taxes:	
Stamps	1,500,000
Registration	4,600,000
Mortgages	[illegible]
Succession and Inheritance	1,800,000
3½ per cent. on these Duties	3,450,000
Total	12,450,000
	£1,040,726
Import and Export Duties:	
Duties on Imports, &c.	3,600,000
Lights and Buoys	591,000
Stamps on Instruments	7,710
Per Centage	1,735
Total	3,979,950
	£176,578
Assay and Tax on Articles of Gold and Silver	756,500
	£31,375
Public Domain:	
From the ordinary Domain, Tithes, &c.	485,000
From the Domain in Possession of the War Department	84,000
Roads and Canals	681,000
Total	1,250,000
	£104,160
Post-office	3,100,000
Telegraphs	504,700
Lotteries	410,000
Game Licenses	100,000
Pilotage	700,000
Mines	200
Miscellaneous	1,201,829
Contribution from Belgium, payment to Treaty of Nov. 5, 1842	400,700
Second Instalment of Purchase-money of Michelin Dues	3,882,228
Interest on Balance of Purchase-money of Belgium Dues not yet paid up	412,672
Colonial Surplus	19,064,000
	£1,621,815
Contribution from the East Indian Revenue towards Payment of the Interest on Debt charged on the East Indian Possessions	2,400,000
Contribution from the East Indian Revenue to pay Debt on the Budgets of the other Colonies	9,975,000
From Balances of former Years	4,000,000
Total Revenue	105,737,945
	£8,644,417

The following was the expenditure sanctioned by the States-General.

Estimated Expenditure for 1864-65	Guilders	£
Civil List of the King	600,000	50,000
Allowance of the Queen Dowager	150,000	12,500
" " Prince of Orange	100,000	8,333
Subsidy for the maintenance of the royal palaces	50,000	4,170
Superior Departments of State	500,000	41,666
Foreign Affairs	600,000	45,000
Roman Catholic Worship	500,000	25,000
Department of Justice	2,549,744	212,478
Protestant Worship	1,340,000	140,000
Home Department	14,376,000	1,172,000
Marine	8,740,000	728,000
National Debt	39,976,000	3,331,415
Finance Department	4,719,000	402,000
War Department	12,735,000	1,061,000
Colonial Department	7,925,000	313,758
Total Expenditure	74,188,918	6,152,625
Surplus	5,514,931	467,027

Of this surplus, the sum of 5,250,000 guilders, or 437,500l., was to be added to the sinking fund for the extinction of the national debt.

According to a statement of the Minister of Finance, made when laying the budget of 1864 before the States-General, the reduction of the national debt, from 1819 to 1864, amounted to 196,000,000 guilders, or 16,416,667l.

Provision for the Poor.—Though pauperism is discouraged, and mendicancy punished, the Dutch are very charitable and liberal in their support of the poor. The institutions for the relief of the indigent consist of hospices for the aged and infirm, orphan-houses, workhouses for towns and districts, the poor colonies, and private charitable institutions. The funds for their support are mostly derived from endowments and voluntary contributions. Boxes, inviting the donation of by-passers for their relief, are stationed in many public ways; the establishment of any new public work excites a fresh call on behalf of the poor; and a tax of about a penny in a shilling, to the same end, is levied on tickets to all places of public amusement. The hospitals, asylums, and other charitable foundations, are very numerous in the towns.

An institution worthy of particular mention is the 'Society for the Promotion of the Public Good,' an association which originated in 1784 with a few benevolent individuals, but which has now 300 branches throughout Holland, and is supported by 20,000 members, each of whom pays a small sum (about 10s.) yearly. Under the direction of this society, savings-banks, libraries, schools of various kinds, including those for the higher branches of knowledge, are established; prizes and rewards are given for superior essays, works of art, or acts of humanity; and in the winter season, public lectures on literary, scientific, or moral subjects are delivered. The establishments of this society formerly extended into Belgium; but since the revolution of 1830, they have mostly ceased to exist in that country.

Among the classes able to labour, a state of even temporary dependence is considered disgraceful, and great exertions are made by the labouring population to avoid it. No sense of degradation attaches to asylum establishments. There are 3 great workhouses for the whole of Holland—one at Amsterdam, another at Middleburg, and a third at Steuve-Pekel-A, in Groningen. In these the inmates work at looms, &c.; the sexes are kept strictly separated; the food is very inferior and somewhat scanty, the clothing coarse; and the inmates are not suffered to go abroad. All beggars are apprehended by the police; if aged or nfirm, they are sent to the workhouses—if able to work, to the penal colonies. In the latter establishments, the paupers labour with the spade, in brick-making, or in manufactures. Guards on horseback, who patrol the boundaries of the colony, reward given to those who bring back any colonist that has attempted to escape; and a uniform dress, are the means adopted to prevent desertion from these settlements.

Public Education.—Holland has been much and deservedly celebrated for its system of public education. There is scarcely a child 10 years old, of sound intellect, who cannot both read and write; almost every one receives instruction at some period, the expense of which is for the most part, and in some instances entirely, defrayed by the state, without the inculcation of any particular religious creed; the interference of the government being exerted only to exclude improper and incompetent teachers, and to regulate the mode of instruction by a system of inspection.

The department of education is under the superintendence of the minister of the interior, assisted by the inspector-general of instruction, from whom all changes and new regulations emanate. The inspection of schools is devolved chiefly upon local inspectors, of whom there are 70, or one for each school district into which the kingdom is divided. These inspectors are assisted by local boards, and each inspector is responsible to the provincial board for the efficiency of the schools within his district; the provincial board being itself responsible for its proceedings to the minister of the interior and the inspector-general. In Holland, no person can open a public school, or even receive private pupils, without first having received a certificate of his ability to teach, granted after inquiry and examination by a board of examiners consisting of district surveyors, who meet for this important purpose. This board grants four sorts of certificates, but one only is granted at a time; and to obtain the highest certificate, four successive examinations must be undergone at different intervals. Having obtained his certificate, the candidate must next apply for leave to open a school to the school committee of the town or district in which he proposes to establish it, who do not grant his request unless when they think such additional school is really required. Very grave doubts have been and may be entertained as to the policy of this last regulation, but there can be none as to the justice of subjecting all persons intending to open schools to the necessity of undergoing an examination as to their fitness. The district inspectors assemble three times a year in the chief town of their respective provinces, where they hold a conference, each inspector making a report, in the presence of the provincial governor, on the state of education in his district. Sometimes the government assembles a council at the Hague, consisting of deputies from each provincial board of education, when everything pertaining to the system is discussed and reviewed in presence of the minister of the interior and the inspector-general. In 1861, there were in the kingdom 2,555 primary schools, attended by 338,000 pupils of both sexes. Besides these, there were 950 higher educational establishments, with 41,932 male, and 40,632 female pupils. A fuller education than these imparted by additional Latin schools, with 1,872 pupils. Above them are the three universities of Leyden, Groningen and Utrecht, with 1,327 students in 1861. The ecclesiastical training schools comprise six Roman Catholic and two Protestant seminaries. There are also three military, one naval, and one veterinary school. The proportion of attendance in the public schools is one in eight of the entire population. There are

two normal schools for the education of teachers in Holland; one at Groningen for the N. prov., and the other at Haarlem for the centre and S. of Holland. The primary schools are divided into *Armen*, or poor, and *Tusschen*, or intermediate, schools. In both much the same kind of instruction is afforded, including reading, writing, arithmetic, geography, the history of Holland, and vocal music; but the latter are attended by the children of parents above the condition of the poor, and the fee, though still very trifling, is somewhat higher. In the poor-schools, as in all the rest, a small sum is generally paid, and in many instances daily, by the parents of the children educated. This circumstance does not retard the progress of education amongst the poor, but has perhaps rather a contrary effect, inasmuch as it removes that sense of degradation which frequently associates itself with the notion of receiving eleemosynary instruction. No law, as in Prussia, exists in Holland directly compelling parents to send their children to school; but the poor are not allowed relief from the public funds unless they comply with this regulation. There is, however, little need of such a provision, since a just sense of the great value of education is found to exist amongst all classes. In the superior private schools, German, French, English, and other modern languages are taught, in addition to the ordinary elementary branches of knowledge. In the Latin schools, which are analogous to the *gymnasia* of Germany and the colleges of France, pupils are instructed in Latin and Greek, the modern languages, mathematics, physics, geography, history, and the other higher branches of education, for the most part as preparatory to their studies at the athenaeums or universities.

In these seminaries, people of all religious persuasions are received indiscriminately, and at stated times attend their respective clergymen for religious instruction. The monitorial system of teaching is scarcely at all introduced. The public schools, like the public charities, make little or no outward display, and are conducted on the most rigid system of economy. The efficiency of the elementary instruction supplied by the schools in Holland is universally admitted; but, with all its excellence, the course of education comprises only the more elementary divisions of mental culture; the study of philosophy, of the principles of politics and political economy, of the higher branches of literature—of all those pursuits, in short, that tend to expand and elevate the mind, is comparatively neglected.

The Dutch school of painting has attained to great celebrity. Its masters excel chiefly in delineations of common life and animated objects: in accuracy and excellence of colouring, and the management of lights and shade, they are surpassed by none. But the subjects of their pictures are, not unfrequently, as very coarse, vulgar, and low, as to be, in many respects, the antithesis of those of the Italian school. The Dutch school can boast of Rembrandt, Teniers, Jan Steen, Ostade, Gerard Dow, Mieris, &c.; besides whom, Wouvermans, Paul Potter, Berghem, and Huysmael excel in landscapes and cattle; Vandervelde and Backhuysen in sea-views; and Weenix, Houbraken, Vandenheyden, Heemskirk, Breughel, &c., in other departments. Many of the best works belonging to this school are to be found in Holland, and especially in the galleries of the Hague, Amsterdam, and other chief towns.

Manners and Customs.—In stature, the Dutch are much the same as the English: the women are comparatively taller than the men; they are decidedly handsome, and, when young, have pa-

turally great complexions, which they might preserve to a later period, did they take more exercise
in the open air, and abandon some injurious customs, such as the incessant use of the *chauffepied*,
a box of burning peat, which accompanies them
everywhere. 'Nothing,' says Mr. Nicholls, 'can
exceed the cleanliness, the personal propriety,
and the apparent comfort of the people of Holland.
I did not see a house or fence out of repair, or a
garden that was not carefully cultivated. We met
no ragged or dirty persons, nor any drunken man;
neither did I see any indication that drunkenness
is the vice of any portion of the people. I was
assured that insanity was almost unknown; and
although we were, during all hours of the day,
much in the public thoroughfares, we saw only
two beggars, and they in manners and appearance
scarcely came within the designation. The Dutch
people appear to be strongly attached to their
government, and few countries possess a population to which the domestic and social duties are
discharged with such constancy. A scrupulous
economy, and cautious foresight, seem to be the
characteristic virtues of every class. To spend
their full annual income is accounted a species of
crime. The same systematic prudence pervades
every part of the community, agricultural and
commercial; and thus the Dutch people are enabled to bear up against the most formidable physical difficulties, and to secure a larger amount of
individual comfort than probably exists in any
other country.' (Report on the Poor of Holland,
in 1838.)

The women are very domestic in their habits,
and carry cleanliness in their houses to the greatest
possible extent; though personal cleanliness does
not always receive the same attention. The ancient
national costume, the wide breeches, full petticoats,
and broad hat, are now mostly confined to the
fishers and peasantry; in the towns, the people
dress like the French and English. The most remarkable element of costume is now in the head-
dress of the Friesland women. The latter, who
are the descendants of the ancient *Frisii*, so often
referred to by Tacitus, and whose blue eyes, flaxen
hair, and fresh ruddy complexions declare them
to be of the Gothic race in perhaps its greatest
purity, wear on both sides of the head large plates
of gold or silver, connected together by a band of
the same metal passing behind, and ornamented
with two singular appendages, of a ram's horn
shape, to which are attached pendants of various
kinds. The whole is covered by a rich cap of
lace; it not unfrequently costs 10l. or 20l. and
often compares the whole dowry of a Friesland
girl. The Dutch, though in general frugal, live
well and substantially. Coffee, tea, beer, and native gin, but especially the first, are the favourite
drinks: the tobacco-pipe is in universal use
amongst all classes. The houses in the towns do
not aim at any external grandeur, and are in
general plainly furnished; but those who can
afford it are extremely fond of collecting china
and other kinds of curiosities. The *laints*, or pleasure houses forming the residences of retired merchants, are mostly built on the same plan. These
edifices are usually of brick, plastered and painted
to look as trim and tidy as if just taken out of a
box; and, with their close-shaven bit of lawn in
front, their narrow wet ditch separating the domain from the public thoroughfare, their little
bridge, dashing wooden gateway, clusters of dahlias, and fresh painted summer-house, form the
beau-idéal of a Dutchman's wishes. On the gateway there is invariably some motto, indicative of
the taste or temper of the owner.

The Dutch are very regular in their habits;
precision, decorum, and a fixed routine govern
every thing. Intoxication is, generally speaking,
rare; but in September an annual festival takes
place, which lasts for ten days, during which great
excesses are committed. As seen, however, as
this festival terminates, the people return at once
to their former habits of sobriety till the next
yearly occasion. Their amusements are not very
intellectual, nor do they include many sports out
of doors. They are mostly similar to the entertainments afforded by the tea-gardens and secondary theatrical establishments in England.

History.—In the time of the Romans, Holland
was inhabited chiefly by the *Batavi* and *Frisii*,
the former of whom, after the conquest of Belgium
by Julius Cæsar, concluded an alliance with the
Romans. This was afterwards silently changed into
subjection to Rome, and it is said that Claudius
Drusus, a Roman governor, about the beginning of
the Christian era, erected the first dyke to ward off
the encroachments of the sea. In the reign of Vitellius, the Batavians endeavoured unsuccessfully
to throw off the Roman yoke; in the second century their country was overrun by the Saxons; in
the eighth it was conquered by Charles Martel; and
it subsequently formed a part of the dominions of
Charlemagne. From the tenth to the fourteenth
century, the Netherlands were divided into many
petty sovereignties, under the dukes of Brabant,
the counts of Holland and Flanders, &c. In 1345,
however, by marriages and otherwise, the whole
passed into the hands of the dukes of Burgundy;
thence to the house of Austria; and lastly, in
1548, under the rule of the emperor Charles V.
The union with Spain was a most unfortunate
event for Holland. The Dutch had long been in
the enjoyment of many political rights and privileges; they had extensive fisheries and trade, and
they had for the most part embraced the doctrines
of the early reformers. Philip II., who regarded
the privileges enjoyed by the Dutch as usurpations
on his own prerogative, and who detested the reformed faith, resolved to recover the former, and
to suppress or extirpate the latter. To accomplish
this purpose, he sent, in 1567, Ferdinand de Toledo,
duke of Alva, with a powerful army into the Low
Countries. But the proscriptions and massacres
with which this sanguinary though able soldier
filled the country, failed of their object. The
Dutch, instead of being subdued, were at length
driven into open rebellion. The malcontents captured the Brill in 1572; and after a struggle unequalled for duration, for the sacrifices it imposed
on the weaker party, and for the importance of its
results, the independence of the republic was acknowledged by Spain in 1609. Except that it was
occasionally darkened by internal feuds, the half
century that succeeded this event is the brightest
in the Dutch annals. The commerce of Holland
attained to an unrivalled magnitude; and while
she extended her colonies and conquests over some
of the most valuable provinces in the E. and W.
Indies, she successfully resisted the attacks of
Louis XIV., contended with England for the empire of the sea, and was justly regarded as one of
the bulwarks of the Protestant faith.

From the death of Louis XIV. down to the
French revolution, the influence of Holland gradually declined, not so much from any decay of
her own resources as from the growth of commerce
and manufactures in other states, especially in
England. The policy of Holland had long been
peaceful; but that could not protect her from
being overrun by revolutionary France. In 1806,
she was erected into a kingdom for Louis, a brother
of Napoleon; and, on the downfall of the latter,
she was united with Belgium, and formed into a

kingdom under the family of Orange, the founders of her liberties. But this union was never cordial. The Dutch and Belgians are, in fact, totally dissimilar in their religion, character, and pursuits; and the connection between them was dissolved by the revolt of the Belgians soon after the French revolution of 1830. Holland, therefore, has now nearly the same limits as before her occupation by the French in 1795.

HOLLAND (NEW). See AUSTRALIA.

HOLSTEIN, a duchy at the NW. extremity of Germany, forming part of the German confederation, bounded W. by the N. Sea, N. by the Elbe, E. by the Baltic, and S. by Schleswig. It is of a compact form, comprising an area of 8,255 sq. m., with a pop. of 544,419 in 1860. Surface and soil considerably diversified; the E. part is somewhat hilly, and, besides fertile plains, has woods, lakes, and picturesque scenery; the middle part is comparatively barren, and is in many parts covered with heath; the W. district, along the Elbe and the German Ocean, consists principally of flat, low-lying, rich marsh land, secured by dykes and sluices against the overflowings of the sea. Principal rivers, Elbe and Stor; the only lake worth notice is that of Plön. The canal of Kiel separates Holstein from Schleswig, and is of great importance, as well for inland as for foreign navigation. (See KIEL.) The lat. of Holstein being the same as that of the N. of England, its productions are also similar, consisting of wheat, barley, and oats; potatoes, hemp and flax, with hops and fruit; but it is chiefly celebrated for its excellent cattle and horses, raised in large numbers in the luxuriant pastures of the marsh-land, and which are an important article of export. The half-dried beef, so abundant in Hamburg, and which is decidedly superior to anything of the sort met with in England, is principally derived from Holstein. Agriculture has been much improved; and the country being in many parts enclosed and well cultivated, is little inferior in appearance to the best districts of England. Minerals not very important. Lime is, however, met with; and there is a brine spring at Oldesloe. Fishing is prosecuted to some extent along the coasts. The duchy has two very good sea-ports, namely, Kiel and Altona, near Hamburg. Glückstadt, a much smaller seaport, situated lower down the Elbe, is the cap. of the duchy. The other principal towns are Rendsburg and Itzehoe. Exclusive of cattle and horses, wheat, oats, and barley, with butter and cheese, are exported. Having been wrested from the crown of Denmark in the war of 1863-4, the duchy was placed under the protection of Austria by the convention of Gastein, concluded Aug. 20, 1865, between the king of Prussia and the emperor of Austria. By the terms of this convention, Rendsburg was made a federal fortress, and Kiel a federal port, the latter under the command of Prussia.

HOLYHEAD (in Welsh Caer-Gybi, 'the castle of Gybi'), a sea-port, park bor., market town, and par. of N. Wales, on a peninsula at the W. extremity of the isle and co. Anglesey, 24 m. W. Bangor, 67 m. W. Liverpool, 224 m. NW. London, and 264 m. by North Western railway. Pop. of park bor., 8,183 in 1861. The peninsula, on the N. side of which the town stands, and which is insulated at high water, rises towards the sea, in an immense precipice of serpentine rock, hollowed out here and there into many magnificent caves, the haunts of innumerable sea-fowl. The town is clean and well paved, comprising two main and several cross streets; it has a fine open market-place, public baths, and government establishments, and contains many superior residences.

The church, formerly collegiate, and now in the patronage of Jesus College, Oxford, is an embattled cruciform structure, in the decorated English style, with a square tower and low steeple; and the churchyard is enclosed by a low wall, said to have formed part of a Roman fortification. There are also four places of worship for dissenters, a free school, established in 1745, and several other day and Sunday schools, furnishing instruction to a great many children. Holyhead has no particular branch of commerce or manufacture: its importance principally depends upon its being one of the most important stations in the great mail route between England and Ireland. The erection of the Menai Bridge, the improvement of the Holyhead road, and the establishment of steam-packets to Dublin, caused a great increase of the intercourse by Holyhead, in the years 1826-38, but it was not until the opening of the great tubular bridge across the Menai Straits, in March, 1850, which carried the railway trains direct to the steamers, that the importance of Holyhead as one of the main stations on the road from London to Dublin became established. Formerly, the harbour, which forms a basin in the shape of a horseshoe used to dry at low water; but great efforts have been made to improve it, and a pier has been projected about 300 fathoms into the sea, having 12 ft. water at its head at low springs. This pier, formed on the rocky island of St. Gybi, is joined to the town by a bridge, and at its other extremity is a lighthouse. The peninsula of Holyhead is terminated by a high rocky promontory called the N. Stack, surmounted by a lighthouse with a revolving light, 211 ft. above low-water mark. The Skerries, a small island 7 m. N. of Holyhead, is also marked by a lighthouse. The town of Holyhead, with a small surrounding suburb, is a park bor. contributory to Beaumaris, which returns 1 mem. to the H. of C. Markets on Saturday.

HOLY ISLAND (an. Lindisfarne), a peninsula, wholly insulated at high water, on the NE. coast of England, co. Durham, ward Islandshire, 11 m. SE. Berwick-on-Tweed. Area, 3,320 acres; pop. 835 in 1861. The form of the peninsula is that of an irregular four-sided figure, more than half of it towards the N. being covered with sand, and abounding with rabbit-burrows; the remainder, however, has been very productive since its enclosure in 1790. The prospect from the island is extremely beautiful, commanding views, northward, of Berwick, and of Hamborough Castle, at nearly the same distance, southward. At the SW. angle of the island is a small fishing village, formerly more extensive, near which are a small harbour and an old castle, situated on a high conical rock, of primitive formation. The inhabitants are chiefly engaged during winter in catching lobsters for the London market, and in other times in getting coal, ling, and haddock. Limestone, coal, and iron ore are abundant; but the index of the tide makes the working of them exceedingly laborious. The great glory of the island, highly esteemed by Anglo-Saxon scholars, is the abbey (with its connected church), formerly the residence of many literary monks. It was founded by St. Aidan in 635, under the patronage of Oswald, king of Northumbria, who erected Lindisfarne into a bishopric. The monastery was all but demolished by the Danes in 867, and was then removed (with the bishop's see) to Durham, a few monks only remaining at the establishment after the partial rebuilding of the church and abbey. The ruins of the abbey, which had been constructed of red freestone, and aptly termed by Sir W. Scott, 'a solemn, large, and dark red pile,' show that it was built at different periods. It cannot be better described than in the words of the great minstrel in the 3nd canto of Marmion :—

> 'In Saxon strength that abbey frown'd,
> With massive arches broad and round,
> That rose alternate, row and row,
> On ponderous pillars short and low,
> Built ere the art was known,
> By pointed aisle and shafted stalk,
> The arcades of an alley'd walk,
> To emulate in stone   •   •
>   •   •   •   •   •
> Not but that portions of the pile,
> Rebuilded to a later style,
> Show'd where the spoiler's hand had been.'

Various fragments of the monastery are extant, and traces of walls are scattered over a space of nearly 4 acres. The main walls on the N. and S. sides of the church still remain, the measurement of the building being 138 ft. in length, and 86 ft. in breadth. N. of Holy Island are 17 small islands, called the Farne Islands, on the largest of which is a lighthouse. (Hutchinson's and Surtees's Durham ; Views of Coast and Harbours of England.)

HOLYWELL, a market town, parl. bor., and par. of N. Wales, co. Flint, hund. Mold, 14½ m. W. by N. Chester, 36 m. NNW. Shrewsbury, and 176 m. NW. London, on the Chester and Holyhead railway. Pop. of parl. bor. 5,335 in 1861. The town is pleasantly situated on the slope of a mountain extending towards the estuary of the Dee, and is large, well paved, and lighted with gas. The streets are irregular ; but there are many good and substantial houses. The church, a plain structure, with a strong embattled tower, stands quite at the bottom of the hill ; there are also 2 Roman Catholic chapels, and several places of worship for dissenters. A beautiful Gothic chapel, dedicated to the legendary saint, Winifred, who lived in the 7th century, and now used as a schoolhouse, is erected over a well, from which water issues so copiously as to turn a large portion of the mill machinery in the town. The lower part of the building is open, and the sanatory virtues of its holy water are even at the present day not wholly discredited by the inhab. The town, which was inconsiderable till the commencement of the present century, is now the largest in the co., and remarkable for its activity in mining and manufactures. Lead, zinc, copper, and coal are extensively worked in several very productive mines close to the town. These mines and the smelting-houses, foundries, &c., in the vicinity of the town, employ from 600 to 700 hands. The chief metallic products are copper wire and copper bolts, nails, and sheathing, which are sent to Liverpool, and shipped in large quantities for the W. Indies and S. America. There are also several extensive cotton mills. A small trade is carried on in the manufacture of galoons and doubles. A short distance from the town is the Mark, a kind of quay, on the Dee, unapproachable by ships at low water, and at all times inconvenient. Holywell was made by the Reform Act a parl. bor., contributory to Flint, which sends 1 mem. to the H. of C., and its boundaries comprise parts of the townships of Holywell and Greenfield.

HONDURAS (BRITISH), a colony belonging to Great Britain, on the E. coast of Central America, chiefly between lat. 16° and 18° N., and long. 86° and 89° W., having N. Yucatan, W. and S. Guatemala, and E. the Bay of Honduras. It is very extensive, but the pop., by the census of April 7, 1861, amounted to but 24,635. The coast is flat, and surrounded with an abundance of reefs and low verdant islands, called keys. The approach to the shore is very dangerous, especially during

N. winds, and the different keys resemble each other so much as to make the navigation of the channels between them extremely difficult, except to experienced pilots. Proceeding inland, the surface rises gradually from the coast into an elevated region, covered with primeval forests, interspersed with marshes. Rivers numerous, and some of them large ; the principal, the Balize, is navigable for 200 m. The climate is moist, but is reported to be more healthy than that of the West India Islands, especially in the wet season. The heat during most part of the year is moderated by sea breezes ; the average annual temp. is about 80° F. The rains are so heavy that the Nbaso river sometimes rises 50 ft. in a few hours ; they are frequently accompanied with violent thunderstorms. Volcanic products, and marble or other limestone formations, are found in various parts ; the shores and banks of the rivers are covered with a deep and rich alluvial soil, capable of growing most European as well as tropical products. The forests abound with some of the finest timber trees, including mahogany, logwood, and many other valuable trees. The two now specified are the staple product of the settlement, and their cutting forms the chief occupation of the settlers. The mahogany (Swietenia mahogani) is one of the most majestic of trees, and is probably 200 years in arriving at maturity. It is seldom found in clusters or groups, but single, and often much dispersed; so that what is termed a mahogany work extends over several sq. miles. There are two seasons in which the trees are cut down; one beginning shortly after Christmas, or at the end of the wet season, and the other about the middle of the year. At such periods all is activity, the pop. being mostly employed in felling or removing the trees. The gangs of negroes employed in the work consist of from 10 to 50 each, at the head of whom is the huntsman, whose chief occupation is to search the woods, and find labour for the whole. An expert negro of this description was formerly often valued at 500l.

'About the beginning of August the huntsman is despatched on his errand. He cuts his way through the thickest of the woods to the highest spots, and climbs the highest tree he finds, from which he minutely surveys the surrounding country. At this season the leaves of the mahogany tree are invariably of a yellow-reddish hue ; and an eye accustomed to this kind of exercise can discover, at a great distance, the places where the wood is most abundant. He now descends, and to such places his steps are now directed; and without compass or other guide than what observation has imprinted on his recollection, he never fails to reach the exact point to which he aims.' The mahogany tree is customarily cut about 12 ft. from the ground. The body of the tree, from the dimensions of the wood it furnishes, is deemed the most valuable; but for purposes of an ornamental kind, the branches or limbs are generally preferred, the grain of these being much closer, and the veins more rich and variegated. Part of the wood is rough-squared on the spot ; but this work is generally postponed till the logs are rafted to the entrance of the different rivers. The rafts often consist of more than 200 logs, and are floated as many miles. 'When the floods are unusually rapid, it sometimes happens that the labour of a season, or perhaps of many, is at once destroyed by the breaking asunder of a raft, the whole of the mahogany being hurried precipitately to the sea.' (Henderson.) The logwood and mahogany do not grow adjacent to each other ; the former inhabits a swampy soil, while the latter flourishes most in high and exposed situations. Every settlement at Honduras has its plantain walk, and many of these comprise an extent of at least 100 acres. Cassava, yams, arrow root, and maize are grown, but only for home consumption; the sugar-cane, coffee, and cotton succeed well, but are little cultivated; cocoa, and an inferior kind of indigo, are indigenous. European cattle, and other domestic animals, thrive greatly. The American tiger, the tapir, armadillo, racoon, grey fox, deer of various kinds, and a vast number of monkeys, inhabit the settlement; birds and fish are in great variety, and found particularly plentiful. Many turtles are taken by the inhabitants living upon the keys, or islands of the coast, a few of which find their way to London.

The value of the exports of Honduras was 292,176 in 1861; 356,389 in 1862, and 390,648 in 1863. The imports amounted to 221,744l. in 1861; 231,857l. in 1862, and to 265,751 in 1863.

Honduras is governed by a Lieutenant-Governor, nominated by the crown, and a legislature assembly of 18 elected and 8 nominated members. Trial by jury is in force. From decisions of the central court, an appeal lies to the sovereign in council. Total public rev. 85,849l. in 1863, total expenditure 28,641l. Amount of compensation received by the proprietors of slaves at their emancipation, 101,958l. The average value of a slave, from 1807 to 1830, was 12l. 4s. 7d., being a larger sum than in any other colony.

The only town in the settlement is Belize, at the mouth of the river of the same name, in lat. about 17° 29' N., and long. 88° 8' W. It consists of about 700 houses, chiefly of wood; the streets are regular, and the whole town is shaded by groves of cocoa-nut and tamarind trees. Its chief edifices are the government house, a church, and several chapels.

This coast was discovered by Columbus, in 1502; the date of its first settlement by Europeans is uncertain. It was transferred from Spain to England by treaty, in 1670, but its occupation was contested at different times by the Spaniards, down to 1798, since which it has remained quietly in the possession of Great Britain. Honduras, formerly a settlement, was erected into a colony on the 12th May, 1862.

HONFLEUR, a sea-port town of France, dép. Calvados, cap. cant.; on the estuary of the Seine, nearly opposite Havre, from which it is 6 m. SE., and 80 m. NE. Caen, on a branch line of the railway from Paris to Cherbourg. Pop. 9,553 in 1861. The town is ill-built, its streets mostly narrow, crooked, and ill-ventilated, and its public edifices more remarkable for antiquity and oddity than elegance. Its port, enclosed between two jetties, is difficult of entrance, and encumbered with mud, so as to be inaccessible, except at high water. It has two basins connected with it, which serve as harbours for numerous fishing boats and coasting vessels. Many of the inhab. are engaged in the herring, mackerel, and whiting fisheries, and numerous vessels sail annually from Honfleur for the cod, whale, and seal fisheries. It is more a commercial than a manufacturing town; it has, however, some building docks, rope walks, and manufactures of coppers, nails, ship biscuit, and lace. Its export and import trade is considerable; butter, fruit, and eggs, in large quantities, are sent to England from Honfleur. A good deal of corn, and melons of very fine quality, are grown in its vicinity. Honfleur was taken from the English by Charles VII. in 1440.

HONITON, a parl. bor., market town, and par. of England, co. Devon, hund. Axminster, near the Otter, 142 m. W. by S. London, and 16 m. ENE. Exeter, on the London and South-Western railway. Pop. 3,301 in 1861, against 3,895 in 1841. Area of par. and parl. bor., which are co-extensive, 2,580 acres. The town, which stands in an extensive vale celebrated for fertility and beauty, consists chiefly of a single well-paved and lighted street, nearly a mile long, lined with neat and respectable houses, built in the middle of the last century, after a destructive fire which laid nearly the whole place in ruins. The inhab. are supplied with water from a brook that runs along the whole length of the street. The church, a quarter of a mile distant, is a small but neat structure, enlarged in 1482, and remarkable for a curiously carved screen separating the nave and chancel. All-hallows Chapel, built of flint in 1765, is a compact building with a square embattled tower. There are 4 chapels for dissenters, a free grammar-school, scantily endowed, a boys' national school, and a girls' working school, and a hospital.

The industry of Honiton consists of serge-weaving and lace-making; but both branches are on the decline. Some years ago, more serge was woven here than in any other town of Devon, and at the beginning of the present century the lace manufacture had arrived at that perfection, was so tasteful in the design, and so delicate and beautiful in the workmanship, as not to be excelled even by the best specimens of Brussels lace. At the beginning of the century, veils of Honiton lace were sold in London at from 20 to 100 guineas, whereas they may now be obtained for 8 or 10 guineas. The competition of the bobbin-lace machinery, which became active in 1820, greatly impaired the trade of Honiton, though not to the extent that it impaired the lace trade of Bedfordshire and Buckinghamshire. Shoemaking and coarse pottery employ several hands, and there is a large trade in butter, the chief portion of which is sent to the London market. Markets on Tuesday, Thursday, and Saturday: the largest on the latter day; an annual fair, the first Wednesday after July 19th, for cattle.

Honiton was granted by Henry I. to Richard de Rivers, from whom it descended to the Courtenays, earls of Devon, who for many years have been the patrons and lords of the manor. It is a bor. by prescription. A portreeve and bailiff are annually elected at the manor court, the civil jurisdiction, however, is vested in the county magistrates. This bor. first sent mems. to the H. of C. in the 28th of Edward I.; but it was only twice represented prior to the reign of Charles I., since which time it has continued to send 2 mems. Previously to the passing of the Reform Act, the franchise was vested in the inhabitant householders. The Boundary Act extended the limits of the parl. bor., so as to make it include the whole par. of Honiton. Registered electors 270 in 1865.

HOOGLY, a town of the Deccan, Hindustan, prov. Beejapoor, presid. Bombay, 13 m. S. Darwar; lat. 15° 20' N., long. 75° 15' E. Pop. estimated at 15,000. It has long been a place of great trade, its merchants and bankers frequently transacting business at Surat, Hyderabad, and Seringapatam. It has two forts, but neither is very strong, and there are no public buildings worthy of notice. It was taken by Sevajee in 1673, and by a son of Aurungzebe in 1685.

HOOGHLY, a distr. of Hindustan, presid. and prov. Bengal, between lat. 22° 15' and 23° 10' N., and long. 87° 30' and 88° 45' E.; having N. the districts Burdwan and the jungle Mehals, E. Nuddea, Calcutta, and the 24 pergunnahs, W. Midnapore, and S. the Bay of Bengal. Area

2,260 sq. m. Pop. 1,580,840 in 1861. The district is a low, level tract of great fertility, but much of it is waste; and the sea-coast, which is very unhealthy, is densely covered with jungle. Besides the Hooghly river, a great many other branches and tributaries of the Ganges intersect it; it has therefore an extensive inland navigation. On the banks of the rivers, near the sea, a good deal of salt of excellent quality is made. About 3-4ths of the pop. are Hindoos, and 1-4th Mohammedans. Education is more extended in this than in most districts in Bengal.

HOOGHLY, a considerable town of Hindostan, presid. and prov. Bengal, cap. of above distr., on the river of the same name. 23 m. N. by W. Calcutta. 'It occupies an elevated and commanding site, and is picturesque in its broken and irregular disposition; the buildings being in one place clustered together in thick groups, in other places wide and straggling, and divided by trees and patches of lawn, &c. A handsome Christian church rises with bold and imposing effect, conspicuous above the temples of the Hindoos and the ghats upon the bank, to the style and architecture of which it forms a striking contrast.' (Heber, i. 241.) The town was once of much greater importance, having been, under the Moguls, the station for collecting the custom and river duties; it is still large, prosperous, well inhabited, and a government civil station. It has a madrassa or college, in which English, Persian, and Arabic, are taught. The Dutch established a factory here in 1625, and the English founded another in 1640; the Portuguese and Moors had also settlements at Hooghly. It was at Hooghly that the first serious quarrel occurred between the Moguls and Europeans, in 1632, when a large Portuguese fleet was destroyed by the Mohammedans; it was here also that the first engagement took place between the British and the Moguls, in 1686; on which occasion the English fleet cannonaded the town, and burned 600 houses.

HOOGHLY RIVER. (See Ganges.)

HOORN, a sea-port town of N. Holland, cap. distr., on the Zuyder-Zee, 20 m. N. by E. Amsterdam, on the railway from Utrecht to Kampen. Pop. 9,259 in 1861. The town is surrounded with old ramparts, is tolerably well built, and has 10 churches, and various other public buildings. Its port is the best along the coast on which it is situated, and large quantities of butter and cheese, cattle, herrings, and other kinds of provisions are exported from it. Hoorn has manufactures of woollen cloths and carpets, and ship building is carried on in it to a considerable extent. It was the birthplace both of the navigator Schouten, who in 1616 discovered Cape Horn, and of Tasman, the discoverer of Van Diemen's Land and New Zealand.

HORNCASTLE, a market town and par. of England, co. Lincoln, sub same name, parts of Lindsey, on the navigable river Bain, 18 m. E. Lincoln, 126 m. N. London by road, and 130½ by Great Northern railway. Pop. of town 4,846, and of par. 4,944 in 1861; area of par. 2,610 acres. The town, which stands in a valley, and is almost surrounded by streams connected with the Witham navigation, comprises a well-built principal street, crossed by others of inferior character, and has a church, three places of worship for dissenters, a grammar school, founded in 1571, two charity schools, a large dispensary, and a union workhouse. Tanning is extensively carried on, and the Horncastle navigation gives rise to a considerable traffic with the surrounding districts. Petty sessions are held here, and it is one of the polling places for the N. division of the co. Horncastle is

the chief town of a poor-law union comprising 88 parishes. Markets on Saturday: large horse-fairs, June 22, Aug. 21, and Oct. 29.

HORNSEY, a par. and village of England, co. Middlesex, hund. Ossulstone, 5 m. N. London, on the Great Northern Railway. Pop. of par. 11,082 in 1861. The par. comprises the hamlets of Muswell-hill, Crouch-end, the chief part of Highgate, and a part of Finchley. The village is long and straggling, containing many handsome and picturesque residences, inhabited chiefly by residents from London; and the New River, which meanders through it, adds greatly to the beauty of the scenery. The church, a building of the 16th century, and 'restored,' comprises a nave, S. aisle, and chancel, with a handsome 'ivy-mantled' tower at the W. end. The living is a rectory, in the gift of the bishop of London, and several bequests have been made at different times for the relief of the church poor. A good charity school is attached to the church. Dissenters have several places of worship within the village.

HORSHAM, a town, parl. bor., and par. of England, co. Sussex, rape Bramber, hund. Singlecross, on the Adur, a tributary of the Arun, in the centre of a fertile and richly-wooded tract, 181 m. NW. Brighton, 31½ m. SSW. London by road, and 37 m. by London and South Coast railway. Pop. 6,747 in 1861. The town consists of two streets, crossing each other at right angles, with an open space on the S., in which stands the courthouse, and a green on the N. The mixture of trees among the houses gives it a more sylvan aspect than most other country towns have. The houses are generally timber-built, but new faced with brick, and in the street leading to the church rows of trees afford to the dwellings an agreeable shade. The town is well paved with stone, obtained from the excellent quarries in the neighbourhood, and is as well supplied with water. The par. church, at the S. extremity of the town, is a spacious and venerable structure, of early English architecture, with a tower surmounted by a lofty spire: it contains some interesting monuments. The town-hall and court-house, a castellated building, with a stone front, was enlarged and improved by the Duke of Norfolk, in 1805, but since that period has been greatly neglected. The county gaol, near the E. extremity of the town, is a commodious prison, built partly with brick, and partly with stone from the neighbourhood, comprising 56 wards, besides dayrooms, and has accommodation for about 180 prisoners. It is under the jurisdiction of the high sheriff of the co., who appoints the governor. Adjacent to the gaol were formerly some barracks, and a magazine, but these have been long removed. Horsham has chapels belonging to the General and Particular Baptists, Independents, Wesleyans, Friends, and Rom. Catholics; and many charitable endowments for the poor, the chief of which is Collier's school, founded in 1532, for 60 scholars. There are also a Lancasterian and some other free schools, an infant school, and several superior private seminaries. Horsham was formerly the seat of the spring assizes for the co., and the midsummer quarter sessions for the W. div. of Sussex are still holden in it. Until the passing of the Mun. Corp. Act the town was governed by a steward and two bailiffs, chosen annually at the court-leet of the lord of the manor. Horsham is a bor. by prescription, and sent 2 mems. to the H. of C. from the time of Edward I. till the passing of the Reform Act, which deprived it of one mem. Previously to that act the right of voting was vested in the holders of burgage tenures; but it was, in fact, a mere nomination, bor. at the disposal of the Duke of Norfolk. The limits of the parl. bor. are now made identical with those of the par. Registered electors, 398 in 1865. Horsham is a polling-place for the W. div. of the co. The town has neither manufactures nor wholesale trade of any consequence; the inhab. deriving their chief support from the retail of goods to the surrounding district. There are two tolerably large weekly markets; one on Saturday for corn, and on Monday for poultry, a good many of which are reared for the London market.

HOUNSLOW, a market town of England, situated partly in Heston and partly in Isleworth par., co. Middlesex, hund. Isleworth, 11 m. WSW. London by road, and 13½ by London and South Western railway. Pop. 5,760 in 1861. The town stands on the W. edge of an extensive heath, bearing the same name, but now to a great extent enclosed: it consists of a single street, in which are numerous inns and posting-houses, once busy and prosperous, but comparatively deserted since the opening of the railway. The church is a modern erection at the W. end of the town, built on the site of an old priory; and connected with it is a charity school attended by 200 children of both sexes. There are several places of worship for dissenters. On the heath are cavalry barracks erected in 1793, for the accommodation of 600 men; and in another part of the heath are two extensive powder-mills. Market-day, Thursday.

HOWDEN, a market town and par. of England, a dependency of the co. of Durham, but situated in the E. rid. co. York, wap. and lib. same name; 17 m. SSE. York, 155 m. S. London by road, and 183 by Great Northern railway via Milford junction. Pop. of par. 5,209 in 1861. The entire par., which contains 14 townships, has an area of 14,510 acres: the township of Howden contains 2,820 acres, and had 2,567 inhabitants in 1861. The town stands in a low but richly cultivated plain, about a mile N. of the Ouse, where there is a small harbour for boats, and a ferry. Streets narrow, badly paved, and only partially lighted: houses mean, and the supply of water insufficient. The church, formerly collegiate, is a spacious cruciform structure, in the decorated English style, with an elegant square embattled tower, 135 ft. high, rising from the centre upon pointed arches, supported by clustered pillars. The chapter-house, built in the middle of the 14th century, is of octagonal shape, resembling the chapter-house at York, but of much less extent. The delicacy, richness, and symmetry of its architecture are equalled by few specimens of the kind in the country, except Melrose Abbey, in Scotland. (Hutchinson's Hist. of Durham, iii. 466.) On the N. side of the church are the remains of an ancient palace, formerly used as a summer residence by the bishops of Durham, especially the celebrated Hugh de Pudsey, who died here in 1195. The ruins consist of a centre, front, and W. wing, with some detached parts, used as granaries. The site of this palace is held on lease from the see of Durham, and the venerable ruins, patched up with modern building, are now converted into a farmhouse. Besides the church there are several places of worship for dissenters. There is an endowed grammar-school, and a national school supported by subscription. Numerous other charities and benefactions exist for the relief of the poor of the par. and township. Market on Saturday. A great horse-fair, the largest in the E. riding, is held here on Sept. 25, and six following days; besides this, there are fairs on every alternate Tuesday for horses and cattle. Howden is

one of the polling places appointed in the Reform Act for the election of members for the E. riding.

HUDDERSFIELD, an important manufacturing town, parl. bor. and par. of England, W. riding, co. York, wap. Agbrigg, on the Colne, a tributary of the Calder, 152 m. N. by W. London, and 15 m. SW. Leeds, on the Great Northern railway. Pop. of par. 52,254, and of parl. bor. 34,877 in 1861. The par. which lies chiefly in the river-valley, extends nearly 12 m. N. of the town, and includes 7 townships, with an area of 15,060 acres; while the township of Huddersfield, which is co-extensive with the parl. bor., extends over 3,960 acres. The present town has little appearance of antiquity, and appears to be wholly the result of manufacturing industry. It is situated on the slope and summit of an eminence rising from the Colne, and is surrounded by other hills of greater height: the streets are regular, well paved, and lighted with gas; and the best houses, which are numerous, built of a light-coloured stone. The market-place is spacious, and surrounded by handsome buildings. The town is well supplied with water from reservoirs about 4 m. W., in the township of Golcar. The chief edifices of Huddersfield are its churches, cloth-hall, and other public buildings. The par. church, built in the reign of Henry VIII., was taken down in 1834, and rebuilt by public subscription, at the cost of 8,552l. Trinity Church, built and endowed at private expense, and opened in 1819, is in the pointed Gothic style, and has an embattled tower at the W. end; it holds conveniently 1,500. Its situation, on an eminence, NW. of the town, renders it a striking object from any point overlooking Huddersfield. St. Paul's Church, erected in 1831, and fitted to accommodate 1,350 persons, is a good modern imitation of the early English style: it may be distinguished by its lower surmounted by a light spire. This, and another church at the Paddock, have been built by funds provided by the parl. commissioners. There are 10 places of worship for dissenters; the most capacious is one belonging to the Wesleyan Methodists: it will hold 2,400 persons. Sunday-schools are attached to all the churches and chapels. Among the secular buildings the chief is the cloth-hall, erected in 1765 by Sir John Ramsden, and enlarged by his son in 1780. It is a circular edifice 2 stories high, bisected, as respects its lower story, by an arcade, on one side of which are separate compartments or warehouses, let out to the larger manufacturers; on the other, an open space taken up by stalls held by the country weavers, and subdivided by passages between the rows of stalls. The attendance on a market-day (Tuesday) averages 640 traders, and the rules of the market make all the business be completed half an hour after noon. The removal of goods is allowed after 3 P.M. The light of the building is wholly admitted from within, a contrivance intended to secure it the better both from fire and depredation. Among the other public buildings may be mentioned the Philosophical Hall, a Grecian structure, erected in 1837 by a thriving mechanics' institute founded in 1825. The Huddersfield and Agbrigg infirmary is an elegant stone edifice with wings, having a portico supported by 4 fluted Doric columns. A dispensary enables the infirmary in giving medical relief to the poor of the town. About ½ m. from the town, on the Sheffield road, is a sulphureous spa, over which have been built spacious and beautiful rooms fitted up with every convenience for bathers. Among the educational and religious institutions of Huddersfield are a church-collegiate school, intended to supply the want of a regular grammar-school; a proprietary college furnishing a good general education, open to all sorts; with technical and other schools, furnishing instruction for a great many children.

Huddersfield is one of the principal seats of the woollen manufacture. It owes its importance to this respect partly to nature and partly to art. It stands in the midst of a rich coal-field, and there is an ample supply of water for mills from the neighbouring rivers. The means of cheap and convenient transit for its products, and the raw materials of its industry, have also been provided.

Sir J. Ramsden, on whose estate the town is built, obtained, in 1774, an act for making a canal to connect this town with the Calder. It commences at King's Mill, close to Huddersfield, and running N.E. for 34 m., joins the Calder navigation at Cooper's bridge, from which point there is a communication with the Humber estuary. The connection with the towns and ports of Lancashire is effected by means of the Huddersfield canal, completed in 1806; it takes a S.W. direction past Slaithwaite to Marsden, where, at a summit level of 656 ft. above the sea (the highest canal level in England), it enters a tunnel 5,450 yards long, cut through Standedge Hill, and thence runs down the vale of Diggle, in Saddleworth, and past Staley bridge to its junction with the Ashton and Oldham canal. Its entire length is 194 m., and it cost £290,000.

This facility of intercourse has since been vastly increased by the completion of railways between the town, Manchester, and Leeds, and which, consequently, connect it with all parts of the kingdom. The goods manufactured in the par. are narrow, and broad cloths of superfine and inferior qualities, kerseymeres, doeskins, and carded cloths of all descriptions. Cloths of wool and cotton mixed, especially fancy articles, are an increasing object of industry, and large quantities are now sent to the foreign markets. Valencias and twills for waistcoats, of stuff and silk, are also much made, and highly prized for superior texture and elegance of pattern. In recent years shawl-making and merino-weaving from British wool have been introduced with advantage.

The Reform Act conferred on Huddersfield, for the first time, the privilege of sending 1 mem. to the H. of C. Registered voters, 1,911 in 1865. Petty sessions are held here every week; and there are two courts, for the recovery of debts under 15l., one for the honour of Pontefract, and the other by a recent local act for the parish, along with certain adjoining parishes. Gross annual value of real property assessed to income tax, 129,897l. in 1857, and 149,714l. in 1862. The cloth-market is held on Tuesday, which is always a day of great bustle. Fairs for cattle, March 31, May 4, Oct. 1.

Huddersfield is said by Dr. Whitaker (Hist. of Leeds, p. 347), to be identical with the Oderfelt of Domesday Book, and to have been at that time 'a mere waste.' The parish, according to the same authority, was, like Halifax, 'separated from Dewsbury, and erected into an independent parish, by the influence of one of the earlier Laceys, to whose piety and munificence this neighbourhood has been greatly indebted, as the founders of its parish churches.' The manor of Huddersfield, which originally belonged to the earls of Halifax, came into the possession of the Barton family, who sold it in the 16th of Eliz. to Sir Gilbert Gerrard. How soon the Ramsden family, its present possessors, acquired it, is uncertain; but one of them applied, as lord of the manor, during the reign of Charles II., for the privilege of holding a market in the small town of Huddersfield: from this time forward it has been a market town. It is indeed indebted to the Ramsden family for many privileges, which

have greatly contributed to raise it to its present importance.

HUDSON, a town and port of entry of the U. States, New York, co. Columbia, of which it is the cap., built chiefly on a rocky promontory on the Hudson river, 90 m. N. by E. New York. Pop. 7,560 in 1860. The town is regularly laid out; the streets are spacious, and cross each other at right angles; Warren Street, the principal, is upwards of a mile in length. Opposite the river is a handsome promenade, and on either side the promontory forming the site of the town is a spacious bay, with depth enough for vessels of any burden, and on which some quays and docks have been constructed. Here is a new and handsome court-house, comprising also a gaol and other offices. Hudson has several places for public worship. Lancastrian and other schools, a private lunatic asylum, many good hotels, several printing establishments, and stores of various kinds. It is a place of considerable trade, but is a port of delivery only, dependent upon the port of New York. Many of the vessels belonging to the port are engaged in the whale fishery. There are manufactures of cotton and woollen fabrics, with establishments for calico printing and bleaching. The town was founded in 1784, and incorporated under a mayor, recorder, and aldermen, in the succeeding year.

HUDSON'S BAY, a large bay or inland sea of N. America, extending between 51° and 61° N. lat., and 78° and 95° W. long., and surrounded on all sides by the partially explored British territories N. of Canada. Its length, N. to S., is about 800 m.; greatest breadth, estimated at 600 m.; area, probably near 300,000 sq. m. Its S. extremity is called James's Bay. It communicates with the Atlantic by Hudson's Straits, a sea about 500 m. in length, and generally upwards of 100 m. in breadth. Hudson's Bay is navigable for only a few months in the year, being at other times frozen over or obstructed by drift ice. It is full of sand-banks, reefs, and islands, and inhabited by few fish. Its shores are rocky and barren. On its W. coast are several settlements of the Hudson's Bay Company, which monopolises nearly all the fur trade of British N. America. This company was incorporated by a charter from Charles II., in 1669.

HUDSON RIVER, the principal river of the state of New York, U. States, through the E. part of which it flows, generally in a S. direction, from near lat. 44° N. to its mouth in the Atlantic, below New York city, about lat. 40° 40′ N. Throughout the greater part of its course (that is, from where it passes over a ledge of primitive rock, and forms what are called Glenn's Falls, in lat. about 43° 16′) it runs through a very remarkable depression or valley. This valley extends from the Atlantic to the St. Lawrence, having in its N. part the Lake Champlain with its outlet the Richelieu river, and though enclosed by lofty mountain ranges on either side, the highest level of its surface is only 147 ft. above the level of the tides in the Hudson. The total length of Hudson River is about 280 m., 120 of which, or up to 5 m. beyond the town of Hudson, are navigable for the largest ships. Sloops pass as far up as Troy, 150 m. from the sea, to which distance the influence of the tide is felt, and thence through a lock to Waterford, a few miles further. Near the head of the tide the mean breadth of the Hudson does not reach a mile; but in the lower part of its course it is much wider, and below New York it expands into a spacious basin 4 m. broad, which forms the harbour of that city. Its only tributary worthy of notice in the Mohawk, which joins it from the W. Owing to its small rate of descent, the current of the Hudson below tide is slow; and, except in the season of floods, it appears rather like an inland bay. At Albany, about the middle of its course, its navigation is at an average closed by frost for about 90 days annually.

The banks of this river are almost everywhere abrupt and lofty. The chief towns on it are New York, Albany, Newbury, Hudson, and Catskill. It is connected with the basin of the St. Lawrence by the Champlain and the Erie canals.

HUÉ, or HUÉ-FO, the capital city of the empire of Anam, on the river of same name, about 10 m. from the Chinese Sea; lat. 16° 19′ N., long. 107° 12′ E. Pop. estimated at from 80,000 to 100,000. This remarkable city, which has probably no parallel in the East, was fortified early in the present century, in the European style, and, it is said, upon the model of Strasburg. The work was undertaken by the king of Cochin China, and was carried on under the instructions of some French officers previously in his service. The new city is completely insulated, having the river on two sides of it, and a spacious canal of from 50 to 40 yards broad on the other two. The circumference of the walls is upwards of 5 m. The form of the fortification is nearly an equilateral quadrangle, each face measuring 1,180 toises. The fortress has a regular and beautiful glacis, extending from the river or canal to the ditch, a covert way all round, and a ditch which is 30 yards broad, with from 4 to 5 feet water in it all through. The rampart is built of hard earth, cased on the outside with bricks. Each angle is flanked by 4 bastions, intended to mount 36 guns apiece. To each face there are also 4 arched gateways of solid masonry, to which the approach across the ditch is by handsome arched stone bridges. The area inside is laid out into regular and spacious streets, at right angles to each other. A handsome and broad canal forms a communication between the river and the fortress, and within is distributed by various branches, so as to communicate with the palace, arsenal, granaries, and other public edifices. By this channel the taxes and tributes are brought from the provinces, and conducted at once to the very doors of the palace or magazines. In the whole of this extensive fortification there is scarcely anything slovenly, barbarous, or incomplete in design. The banks of the river and canal, forming the base of the glacis, are not only regularly sloped down everywhere, but wherever the work is completed, they are cased from the foundation with a face of solid masonry. The canal within the walls is executed in the same perfect manner; and the bridges which are thrown over it have not only neat stone balustrades, but are paved all over with marble brought from Tonquin. (Crawfurd's Embassy to Siam, i. 884–886.) The palace is situated within a strong inner citadel, consisting of two distinct walls or ramparts. The barracks surround the whole of the outer part of the citadel. The arsenal contains a vast number of cannon, shot and shells, &c., all manufactured in the country. The public granaries are also of enormous extent, and kept full of corn. The fortress of Hué, from its immense size, which is its greatest fault, would require at least 50,000 troops to garrison it, in case of an attack from Europeans; against Asiatic enemies it is impregnable. There are some building-docks on the river, and a large fleet of galleys is usually stationed at Hué. The river is not above 400 yards wide at its entrance, but within is little inferior in breadth to the rivers of Saigon or Bankok; owing to a bar at its

mouth, however, it is fitted only for ships of small draught. Its entrance is completely commanded by a stone quadrangular fort, built in the European style. Its banks are well raised, and in some places extremely picturesque. The neighbourhood of the capital is everywhere in a high state of cultivation, with rice, mulberry trees, and cotton, and thickly interspersed with villages. Hué is the only city in India, in the vicinity of which there are numerous good roads, bridges, and canals. About 10 leagues N. is the royal mausoleum, surrounded by magnificent grounds, laid out by a late king of Cochin China. (Crawfurd's Embassy, i. 368–400; White's Voyage; Finlayson; Ritter, Asien Erdkunde, iii. 1008–1012.)

HUESCA (an. Osca), a town of Spain, prov. Aragon, cap. partido same name, and a bishop's see, 35 m. N.E. Saragossa, and 135 m. W. by N. Barcelona, on a branch line of the railway from Saragossa to the Pyrenees. Pop. 10,059 in 1857. The town stands on a slope close to the Isuela, a tributary of the Cinca, is surrounded by walls now falling into decay, and contains many respectable houses. The chief public buildings are a cathedral, 4 par. churches, 15 convents, a foundling hospital, cavalry barracks, 2 schools, and a university. The latter, entitled Sertorianos, comprising 4 colleges, was founded, in 1354 by Peter IV. of Aragon, and further endowed by subsequent monarchs; but the endowment, as in most Spanish universities, is small, and the education is of a very inferior description. The industry of the town is confined to tanning and the weaving of coarse linens; but the neighbourhood abounds in grain, wine, and other fruits, and large flocks of sheep graze on the surrounding hills. An annual fair is held there, and much frequented. The town was originally founded by Quintus Sertorius, anno 77 B.C., and was known in the time of Augustus as urbs victrix Osca. It subsequently fell into the hands of the Moors, from whom it was taken by Peter I. of Aragon, after the battle of Alcoraz, in 1096.

HULL (KINGSTON ON), a large and important commercial town, river-port, mun. and parl. bor. of England, and co. of itself, locally situated in co. York, E. riding, Harthill wap., on the N. bank of the Humber estuary, 22 m. from the Spurn-head, 84 m. S.E. York, 155 m. N. London by road, and 173 by Great Northern railway. Pop. of parl. bor. 97,661 in 1861. The parl. bor. includes, besides the town para., those of Sculcoates and Drypool, and a portion of the par. of Sutton. The town, which stands close to the confluence of the navigable river Hull with the Humber, has been greatly enlarged and improved during the last half century. It is well paved and lighted with gas; the principal streets extend more than 2 m. along the Humber, and about the same distance along the W. bank of the Hull; and from these others branch off, crossing each other in different directions, and covering an extensive area. Almost the whole town is built with brick; the older streets are inconveniently narrow; but many recently laid out are wide and regular, containing handsome residences. The public buildings are numerous, but, generally speaking, not remarkable for beauty; the principal, besides the churches, are the Mansion-house (in which is the court-house and court of requests), the guildhall, exchange, corn-exchange, custom and excise offices, the Trinity-house, the gaol, the theatre, and the citadel, a regularly-garrisoned fort on the E. side of the river Hull, which is have crossed by a stone drawbridge of 3 arches. There is a good market-house, and in the market-place stands an equestrian statue of William III. The town has also a handsome Doric column, surmounted by a colossal statue of Wilberforce, the great advocate for the abolition of slavery. Within the parl. bor. are 14 churches, among which that of the Holy Trinity, in the market-place, begun in the 14th century, is remarkable as one of the best specimens in England of the Gothic style, at different periods. It is a cruciform cathedral-like building, from the centre of which rises a highly ornamented embattled tower with pinnacles, 140 ft. in height. The interior is 280 ft. long, and 72 ft. broad. St. Mary's, in Lowgate, was originally built at nearly the same time as that last mentioned; but having been partly destroyed by Henry VIII., it was afterwards restored at different periods, and with little taste in the architecture. There are also 20 places of worship for dissenters, a Jews' synagogue, and a floating chapel for the use of dissenters; to all of three large Sunday schools are attached, which furnish instruction to upwards of 7,000 children. The principal schools are, the Grammar School, founded by Bishop Alcock, in 1486, and chartered by Queen Elizabeth, in which the instruction is general as well as classical, the Vicar's School, established in 1754 for 60 boys; Cogan's charity school, endowed with 400l. a year for the maintenance and instruction of 40 girls; the nautical school for 36 boys, attached to the Trinity House; with National, Lancastrian, and other schools, attended by a great many children. The means of procuring a sound education have been greatly increased of late years, by the establishment of several proprietary colleges, which furnish instruction in classics, history, and natural science, on a plan similar to that pursued at the University and King's College, London. Among the numerous endowed charities of the town, the oldest is the Trinity House, founded in 1369, for the support of decayed seamen and their widows, and chartered by Henry VIII. The building, erected in 1753, consists of 4 sides enclosing a square; the E. front is an elevation of the Tuscan order, and the interior comprises 2 large and well-proportioned council-chambers, besides offices and apartments for 82 pensioners. A school within the building gives a useful nautical education to the sons of seamen intended for the merchant service. The Charterhouse hospital (originally endowed in 1580 for poor monks) was re-established in 1640, and devoted to the maintenance of poor pensioners. Six other endowed hospitals or almshouses give relief to about 70 persons. The charity-hall is a kind of poor-house, established by an act obtained in 9 and 10 William III.: it was built by subscription, and is now maintained by the poor-rates raised within the bor. The infirmary, a brick building ornamented with stone, was erected in 1782; it accommodates 70 in-patients, and furnishes advice and medicine to an unlimited number of out-patients; the expenses are defrayed by voluntary subscription.

The port of Hull, which ranks fourth amongst those of the British empire as regards tonnage, and third as regards value of exports—the order being, in 1865, as regards tonnage, 1. London, 2. Liverpool, 3. Newcastle, and 4. Hull; and, as regards value of exports of British produce, 1. Liverpool, 2. London, and 3. Hull—has extensive accommodations for shipping, which have been greatly enlarged during the present century. The old dock formed in 1775 occupies the place of the old wall and ramparts: it is 1,700 ft. long, 250 ft. broad, and 24 ft. deep. Its wharfs and quays occupy an area of 12 acres, and the entrance is on the E. side from the Hull, about 800 yards above its mouth. In

1807, the accommodation was further increased by the construction of a dock opening directly into the Humber: its dimensions are 920 ft. in length, 350 in breadth, and 30 ft. in depth, the wharfs, &c., covering an area of 9 acres. A third dock, connecting those above mentioned, was completed in 1829, at an expense of 200,000l.; its water-surface exceeds 6 acres, and affords accommodation for about 70 square-rigged vessels. Besides these there are the Humber Dock, 814 ft. long; the Junction Dock, 645 ft. long, and 2 basins, and the Victoria Dock. All three, however, are found to be insufficient to accommodate the ever-growing commerce of the town, and new docks and basins are projected. There is anchorage in the Humber in 4 to 6 fathoms.

The commerce of Hull, which is very large, depends principally on her advantageous situation. The town is the chief emporium of the extensive and fertile countries situated on the Humber estuary, and those traversed by the numerous and important rivers that have their embouchures in it, including the Trent, Don, and Ouse. The natural facilities for internal communication thus enjoyed by Hull, have been greatly extended by artificial means. Hull is now united by rivers, canals, and railways, with Sheffield, Leeds, Manchester, and Liverpool; so that it has become not merely the principal port for the W. Riding of Yorkshire, but also for a considerable portion of the trade carried on between Lancashire and the N. parts of the Continent. The great articles of export are cotton stuffs and twist, woollen goods, hardware, and earthenware. Of imports, the leading articles are wool, bones, timber, hemp and flax, corn and seeds, madder, bark, turpentine and skins. The value of the exports of Hull amounted to 12,890,587l. in 1859; to 13,909,125l. in 1861; and to 13,565,264l. in 1863. The gross amount of customs duties received was 297,597l. in 1859; 239,990l. in 1861; and 240,134l. in 1863. In the year 1863, there cleared at the port 247 British sailing vessels, of 60,547 tons, and 796 foreign sailing vessels of 141,076 tons. Of steamers, there cleared, in the same year, 894 British vessels, of 814,879 tons, and 269 foreign vessels, of 84,429 tons. On the 1st of January, 1864, there belonged to Hull 371 sailing vessels under 50, and 123 sailing vessels above 50, tons; there were, besides, 15 steamers under 50, and 64 steamers above 50 tons, the latter of a total burthen of 23,366 tons.

The mun. bor., which received its first charter in the 27th of Edward I., was enlarged by the Mun. Reform Act, so as to be co-extensive with the parl. bor., and was divided into seven wards, the government being vested in 14 aldermen (one of whom is mayor) and 42 councillors. Corp. revenue, 29,870l. in 1861. Quarter and petty sessions are held under a recorder. Hull has sent 2 mems. to the H. of C. since the 23rd of Edward I., and the franchise, previously to the passing of the Reform Act, was vested in freemen, by birth, servitude, purchase, or gift. The limits of the present parl. bor. include (besides the old bor.) the entire par. of Sculcoates and Drypool, a small portion of the par. of Sutton, and the extra parochial district called Garrisonside. Reg. electors, 6,610 in 1865, including 1,549 freemen. The gross annual value of real property assessed to income tax was 843,153l. in 1857, and 872,972l. in 1862. The name of Kingston-on-Hull was given to the town by Edward I., who, seeing its eligibility for becoming an important station, erected a fortress, and constituted it a chartered town and port. When Edward III. invaded

France, in 1859, Hull contributed 16 ships and 470 mariners. The fortifications, commenced early in the 14th century, were completed by Sir Michael de la Pole, a great benefactor to this town during the reign of Richard II. The plague made great ravages here during the 16th, 16th, and 17th centuries.

In the reign of Charles I., Hull was the first to close its gates against the king, who shortly after besieged it, and would have taken it by stratagem, if the treachery of Sir John Hotham, its governor, had not been discovered in time to prevent its surrender to the royalists. The town was afterwards besieged by the Marquis of Newcastle, and successfully defended by Lord Fairfax. The fortifications were greatly improved by Charles II., and the citadel was occupied by a large body of troops in order to keep in awe the inhabitants, who were considered to be disaffected to the Stuart dynasty. At the close of the reign of James II., the town, fort, and garrison being in the hands of the Jacobite party, the place was surprised, and the Prince of Orange proclaimed king; the anniversary of which event is still kept as a holiday.

HULME, a chapelry and township of England, co. Lancaster, par., and 1½ m. SW. Manchester, at the termination of the Duke of Bridgewater's canal. Pop. of township 68,423 in 1861. The increase of pop. has been extraordinary; the census of 1801 showed but 1,677, and that of 1831, but 9,624 inhabitants. Hulme is within the boundaries of the parl. bor. of Manchester, with the exception of a small piece of land near the village of Cornbrook, where the Cornbrook after passing the Bridgewater canal, runs on to the river Irwell. The Manchester Botanic Garden, opened in June, 1831, is here, and occupies 17 acres, beautifully laid out. There are also cavalry barracks. The township is divided into seven ecclesiastical districts, the largest, St. George, with a pop. of 27,795 in 1861. (For public buildings, manufactures, and other particulars, see MANCHESTER.)

Hulme Hall, on a bank above the Irwell, is an ancient half-timbered house, with an inner court. It was the seat of the Prestwiches, baronets, and of the ancient family of Prestwich, in the time of the Conqueror. This family, by embarking in the royal cause, during the civil wars of Charles I., lost most of their property; and the last baronet, Sir John Prestwich, a profound antiquary, died in absolute poverty about the year 1800. Hulme Hall, after passing from the original proprietors, came into the hands of the Duke of Bridgewater, whose heirs still possess the estate.

HUMBER, a great river, or rather estuary, on the E. side of England, between Yorkshire and Lincolnshire. It extends from Goole E. to Hull; and thence SE. to its embouchure between the Spurn Point on the N. and the opposite coast of Lincoln on the S. This estuary receives the waters of some of the most important of the English rivers. At its W. extremity it is joined by the Ouse (after the latter has been augmented by the Derwent, the Aire, &c.), and by the Don; and a little lower down it is joined by the Trent, and still lower down by the Hull river. Hull is the principal part of the Humber, and boat to it are Goole and Great Grimsby. At Hull spring tides rise about 17, and neaps about 13 ft.; and as there is at all times a considerable depth of water in the fair-way of the channel, Hull is accessible by very large vessels. Goole, which is about 72 m. more inland, may be reached by vessels drawing 15 and 17 ft. water, provided they take advantage of the tide. The basin of the Humber, or the country drained by the Ouse, Trent and other rivers falling into this great estuary, embraces an extent of about 10,000 sq. m., comprising some of the most populous and fertile districts in the kingdom.

HUNGARY (Hung. Magyar Orszag, Germ. Ungarn), a kingdom of Central or SE. Europe, which, taken in its widest acceptation, includes—besides Hungary Proper, Croatia, Slavonia, the military frontier provinces, and Transylvania. In a more limited sense, it denotes Hungary Proper, with Croatia and Slavonia, to the exclusion of the other provs. Hungary, thus considered, is situated between 44° 5′ 5″ and 49° 30′ N. lat., and between 14° 29′ and 26° 50′ E. long. The chain of the Carpathians forms the boundary of Hungary on the NW., N., and NE. They stretch from the Danube, near Presburg, in the form of a circle, towards Moravia, Galicia, and Transylvania, until they meet the Danube a second time at the ravine called the Iron Gates. On the S., the Danube and the Save separate the kingdom from the Turkish provs. of Servia and Bosnia, to the junction of the latter river with the Una; which thence continues to mark the boundary. Hungary may be considered generally as a large plain sloping to the S., and surrounded on every side by heights of different elevation, but most considerable in the N. sections of the kingdom.

Mountains.—The first group of hills which run N. from the Danube, near Presburg, is named the Little Carpathians, and is of small extent and inconsiderable elevation. Granite and gneiss, overlaid by grauwacke, form a large portion of this group. The adjoining group, named the Savorian, is also composed of grauwacke. A third group called the Jablunka range, terminates with the Paw of Jablanka, through which the high road from the valley of the Waag passes into Silesia. The formations in the last-named group are grauwacke on primitive limestone, which reaches a height of 1,500 to 2,000 ft. On the E. side of the Jablunka Pass a chain of mountains commences, which stretches E. to the banks of the Dunajec. The formations of this chain are, as far as Neumarkt, the same with the Jablunka; the summit being all of limestone, with grauwacke superimposed. At Neumarkt the great sandstone formation commences, and, for an extent of more than 400 m., constitutes the leading feature of the E. Carpathians. Between the Dunajec and the Poprad, a branch of the Magura chain, situated altogether in Galicia, stretches to the SW., and connects with the chain now described an isolated group of lofty mountains, the naked summits of which rise, like so many gigantic sugar-loaves, from the vale of the Waag and the plain of Zips. This is the Tatra group, in which some of the highest summits of the Carpathians are found. The summits of the Tatra are of granite and gneiss, bare of vegetation, and varying annually in elevation, from the effects of thunderstorms and the melting of the snow which covers them for a great portion of the year. The large mountain group, of which the Krivaan Horn forms the highest summit, covers a large portion of NW. Hungary. In some parts, the hills sink low upon the plain, allowing easy passage to the railroad from the capital of Hungary to Debreczin and the frontier of Russia. On the E., the Tatra chain is bounded by the valley of the Gran, on the W. by the Waag. The principal portion of the Tatra group is likewise formed of trachyte, mingled occasionally with granite.

Branching from the N. Carpathians, in the beginning only as a succession of heights, traversing the level country of Zips, another trachyte mountain chain of considerable elevation runs S. between the rivers Hernad and Bodrog, and joins the Theiss near Tokay. This mountain chain, named the Hegyalla, is famous for the opals found within it, as well as for the wine grown upon its N. slope. On the E. bank of the Poprad, a long unbroken chain of the Carpathians stretches E. as far as the sources of the Save, and thence SE. to the sources of the Theiss.

On the W., Transylvania is divided from Hungary by a chain of low mountains, lying between the Szamos and the Maros, two rivers which flow W. to join the Theiss. Though the summits of this chain nowhere exceed 3,600 ft., it is yet extremely rugged and precipitous. In the S. part, limestone rises above the sandstone; and in the N. summits, gneiss and granite break through the upper strata.

These mountains are composed of Jura limestone, resting on transition limestone and mica slate, with occasional interruption of syenite, porphyry, and other volcanic matters, rich in veins of metal of various kinds. They stretch between the Maros, Czerna, and Danube. The frontier of the Banat towards Wallachia and Transylvania is formed by the last offsets of the Carpathians towards the Danube, in the valley of which river the mica slate of the Banat gives place to limestone. The rocks that close in the river as it leaves Hungary, and which are named the Clissura, are composed of limestone, traversed by broad veins of quartz. This passage, between the E. Carpathians and the N. offsets of the Balkan, which meet them on the Servian side, is more than 70 m. in length, and ends with the dangerous rapid named the Iron Gate. (See DANUBE.)

On the N. side of the Danube, near Presburg, are the Leitha mountains, which form the boundary towards Austria, and are offsets from the Alps, as they subside from Styria towards the Danube. Granite and gneiss appear in the highest summits, on which sandstone and limestone formations lie superimposed. The Bakony Forest hills stretch from the Danube towards the N., dividing the lesser from the great plain of Lower Hungary. Near the mouth of the Drave, this chain dividing that river from the Save, subsides to the plain, but rises soon after on the right bank of the Danube, which turns E. as soon as it reaches these heights. The summits of the greater part of these offsets from the Alps are limestone, overlaid by tertiary formations, except on the banks of the Danube, where serpentine and schist rise in bold masses above the secondary rocks. This chain of heights, called the Fruska Gora, terminates at Szenhamien, opposite the mouth of the Theiss.

The Julian Alps and their offsets cover Croatia and the Hungarian coast districts, the Capella and Vistebich being the last branches of this range towards the N.

Valm.—In the N. of Hungary, the valleys are very numerous, and highly picturesque. The glens in the Tatra mountains are wildly romantic, offering every variety of rocky scenery, and being interspersed with numerous lakes and waterfalls. The valley of the Waag is most extensive, being more than 700 m. long. The rocks of Sulyo, where the Waag crosses the ridge of the Tatra, are amongst the most picturesque in Europe. The valley of Kohlbach, that of the Jablunka Pass, and of the five lakes in the high Carpathian groups, the vale of the Cserna, in the hills of the Banat, near the baths of Meludia, are all highly beautiful, and, in mountain chains of less extent, would be deemed grand. The valleys of the Save (the Syrmia) and the Drave contain some of the finest land and scenery of Europe. The climate is like that of the N. of Italy, and the fertility of the soil is unparalleled.

Plains.—The plains of Hungary are very remarkable, the greater part of the kingdom consisting of two extensive levels. The plain of Upper Hungary, by far the smaller of the two, is bounded N. by the Lesser Carpathians and the mountainous districts of the NW. counties; W. by the Leitha mountains, and the offsets of the Styrian Alps, which, as well as the Croatian Hills, confine it also on the S.; the Bakony Forest forming its E. boundary on the E. as far as the Danube. This plain is traversed by the Danube from W. to E., and is watered besides by the Raab, Waag, and Neitra. The Lake of Neusiedler-See, at the foot of the Leitha hills, issues from great marshes lying between it and the Danube. The soil of this plain is more fertile on the N. than on the S. side of the Danube, but it everywhere produces good and abundant crops of corn.

Near Buda, the Danube, breaking through the mountains of the Bakony Forest and the Matra chain, enters the large plain of Hungary, which it traverses N. to S. from Waitzen to Dalya, whence its course is E. The great plain is bounded W. by the Bakony Forest hills; N. by the Hegyalla and offsets of the Carpathians; the frontier hills of Transylvania bound it E.; and the high lands of Servia and Slavonia on the S. The extent of this plain is estimated at 1,700 sq. German miles, or 86,000 sq. English miles, and is consequently about 4,000 sq. m. larger than Ireland. In the whole plain scarcely a single point is more than 100 ft. above the level of the Danube, which, in this part of its course, is 800 ft. above the Black Sea. This plain is watered by the Danube and its tributaries, the Drave and Save, the Theiss, with its affluent the Szamos, Maros, Körös, &c. The fall is everywhere very trifling, and the greater part of these streams have a winding course, through a country flooded by the slightest increase of their waters. Many, such as the Körös and Theiss, form a succession of swamps, and the whole marshy land of the plain is estimated to cover a surface of 2,425 sq. m., which is wholly reclaimable. The Balaton Lake lies at the SW. extremity, at the fall of the Bakony Forest hills. With the exception of some extensive sandy tracts near Debreczin, and in the re. of Pest, the whole of this plain contains some of the richest soil of Europe.

Rivers.—The numerous rivers which water Hungary fall, with one sole exception, into the Danube, which traverses the kingdom in a general SE. direction. The distance along the stream from Presburg, where it enters, to Orsova, where it leaves, Hungary, is 580 m. Its direction from Presburg to Waitzen is E.; but here it makes a sudden turn E., and runs N. to the juncture of the Drave, from which point its general course to Orsova is E. by S. Of the 30 navigable rivers which are its tributaries, several of the largest belong to this country. The largest and most important is the Theiss, 420 m. long, rising in Transylvania, and flowing NW. to lat. 48° 50′ N., and long. 22° 10′ E., whence it runs S. by W., in a very irregular channel, which, for about 100 m. is parallel to that of the Danube. Its chief tributary is the Maros. (See THEISS.) The other affluents on the N. side are the Waag and Neutra, the Gran and the Eipel. Of the S. affluents, the most important is the Drave, which rises in the Pusterthal of the Tyrol, and has an E. course of 330 m. through a plain country; it is navigable from Villach, in Carinthia. (See DRAVE.) The second in size is the Save, which rises in the Julian Alps, and runs E. by S., joining the main stream near

Belgrade. Length about 840 m. The Raab is of considerable size; but the rest are unimportant. Since 1831, regular lines of steamers are running on all the navigable rivers which fall into the Danube. (For further particulars, see DANUBE.)

The only river which rises in Hungary and does not belong to the region of the Danube, is the Poprad, the source of which is in the Krivan, very near that of the White Waag. The Poprad traverses the level country of Zips, passes through the mountains near Murayna, into Galicia, and unites with the Dunajec, which falls into the Vistula. At Lehlo, in Zips, the Poprad is navigable for rafts.

Canals.—No country is better adapted for, or more needs, canals than Hungary. The greater number of those hitherto made have been cut to regulate the course of winding rivers. Such are the Leitha canal, in the co. of Wieselburg; the Albert-Karwicza canal, in the co. of Raruay, and the cuts for the regulation of the Kőrös, in Heves co., and of the Bernava, in the Banat. Other cuts, on a large scale, regulate the course of the Latorcza in the co. of Beregh, and of the Servia, in the cos. of Wesprim, Schulwei-zuburg, Tolna, and Kalimegh. The most remarkable canal in Hungary, however, is the Francis or Bacs canal, between the Theiss and the Danube. It is nearly 70 m. long, and at the level of the water is 8 ft. deep and 60 ft. broad. The difference between the levels of the Danube and the Theiss is 27 ft., which is carried off by locks. The entire cost of this undertaking was 300,000l. A similar canal between the Theiss, near Szegedin, and the Danube, near Pest, is projected.

The Bega canal, between the Temes, near Temesvar, and the Theiss, near Tittel, is on a smaller scale, but a very useful undertaking, and a source of great prosperity to the Banat.

Lakes.—Hungary possesses two of the largest lakes of Europe: the Neusiedler-See (Hung. Fertő-Tava), in Upper Hungary, lying S. of the Danube, in the cos. of Oedenburg and Eisenburg, is 25 m. long, 12 m. broad, and from 9 to 13 ft. deep. Its waters rise and fall without apparent cause, often receding from the banks, and then again filling and overflowing them. Lake Balaton, situated in the great plain, at no great distance from the Neusiedler-See, is nearly 50 m. long by 10 m. broad, and receives the river Szala on the W. side. The water is very slightly tainted with salt. Besides large lakes, Hungary possesses an almost innumerable number of stagnant sheets of water. Some in the Carpathian mountains, though small, are especially worthy of notice; these are the White, the Green, and the Red lakes. The Green Lake is 4,765, the White Lake 5,276 ft. above the sea, and both are enclosed by high and precipitous granite rocks. There are many mineral springs in Hungary, the principal of which are at Mehadia, in the Banat, at Trentchin on the Waag, and at Banfeld, in the N. chain of the Carpathians.

Climate.—The climate of Hungary may be divided into three kinds, or degrees, varying according to the surface of the country. The climate of the Carpathians, including the high lands of NW. Hungary, is coldest, and that of the great plain in the warmest; the climate of the high lands S. of the Danube being a mean between both. The mean temperature of Buda, which represents the mean climate of Hungary, is stated to be 10° Reaumur, or 54° 30′ Fahr., corresponding nearly with the mean temp. of Nantes. At Nantes, however, the difference between the winter and summer averages 13° Reaumur, and the range is 17°; whereas, at Buda, the average difference is 21°,

and the range 23°. In the great plain the mean temp. is 13° 67′ Reaumur, or the same as at Milan. (Bergham.) The mean fall of rain at Buda is 16 inches, the number of rainy days being about 112; the average of all Germany being 150 days. In the high Carpathians, the yearly average is doubtless very much greater; whereas the summer and autumn, in the low lands, are usually seasons of drought, unfavourable alike to agriculture and river navigation.

Vegetable Productions.—The products of Hungary embrace all the plants indigenous to Europe, from the Iceland moss, gathered on the Carpathians, to the rice and cotton plant, so successfully cultivated in the Banat, and the olive, which thrives in the coast district. In the hills, especially in the Carpathian district, fir forests abound; but along the plains and valleys of the Save and the Drave, extensive oak and beech forests are found. The oak forests yield large quantities of gall apples, and large herds of swine are fattened on the acorns and beech mast. The increase of property everywhere introduces improved fruit plantations, and the S. slope of every elevation is found covered with vines and orchards. The well-known liqueur Sliwowitza (Sliwa plum) is made from the plums grown in the S. parts. The grapes are of various kinds, and one species, the furmint grape, of which the Tokay wine is made, is peculiar to Hungary. The extent of the wine country, including the fall of the hills, to the two plains and the valleys of the Save and Drave, is more than 7,000 English miles long, measured in a straight line. Many districts, such as the Franka Gora hills in Slavonia, and the hills near Buda, yield a heavy red wine, which, with care, might readily be fitted for exportation. The water melon in the great plain has obtained a kind of national celebrity; it often attains a weight of 30 lbs. and upwards. Tobacco is particularly fine, the plants of all kinds, madder, woad, and safflower, succeed wherever they are cultivated; but what is of far more consequence, the soil is particularly adapted to the cultivation of wheat, which is largely exported. Of other cereal plants, little more is grown than is required for local consumption, excepting maize, much of which is sent to Italy. Rapeseed and hemp, also the produce of the marshes, are objects of trade; and poppies, for oil, are much cultivated. The laurel, the laurus tinus, arbutus, cedar, and other evergreens, are too tender to bear the winter cold.

Animals.—Among the animals, the bear of the Carpathians is the most remarkable; and in autumn he often visits the oak and beech forests of the low countries; wolves are more numerous. The small lynx, wild cat, and wild boar are found in all parts. There are many varieties of the dog; one of the finest is the wolf-dog, found in every shepherd's cottage. The chamois and marmot are inhabit of the Carpathians; and stags, roebucks, foxes, and hares are common, though seldom preserved for game. Among birds, the golden eagle, as a stray visitor, and the stone eagle, more frequently, various kinds of kites, hawks, bustards, and woodcocks, partridges, and black game; and all kinds of domestic fowls thrive remarkably in the S. parts, and have beautiful plumage. Herons' plumes are taken as rent in some parts of Transylvania. Fish abound in the rivers of Hungary, especially in the Theiss, which is said to be the richest fish-river in Europe; amongst these, the sturgeon and the fogosch of Lake Balaton (Perca lucioperca) are much esteemed. The entomology of Hungary is richer than in any other part of Europe, owing to the extensive forests and large swampy tracts of the warmer districts. In the forests along the Save, cantharides are gathered. Wasps and hornets build enormous nests in the sandy plains, which are not exterminated without difficulty and danger. Swarms of gnats of peculiar kinds occur in the Banat. One kind, which is harmless, is peculiar to the river Theiss, and increases so rapidly at the breeding time, as to cover the stream like a thick coat of moss, and even to impede the navigation. In this state, the masses of insects are collected by the peasantry, and given as food to the cattle. Another more formidable insect, the *Columbacz* gnat, issues from the caverns of limestone rocks on the banks of the Danube, and spreads in swarms over the adjacent plains, to the great annoyance of the cattle. Locusts are often met with; and the destruction of their eggs, which they lay deep in the earth, is a work of great labour. The leeches of N. Hungary, especially those from the Neusiedler-See, form a considerable article of trade. (Paget's Hung. i. 59.)

Minerals.—The minerals are very important. Nearly all the metals are met with in the kingdom. They are mostly found in the central trachyte groups of NW. Hungary. Gold is found at Schemnitz, in a whitish compact limestone, alternating with syenite and porphyry. At Königsberg, Telke Banya, and in the still richer mines of Nagy Banya, on the frontier of Transylvania, the ore is found in small conglomerations, or thin veins, in soft sandlike masses of decayed pumice-stone, lying on and in excavations of the trachyte, or on the porphyry, exactly under the same circumstances as the ores described by Humboldt, in the Mexican mines of Villalpando. Silver, copper, and lead are found mingled with gold at Kremnitz, Schemnitz, Nagy Banya, Telke Banya, in the trachite group of the Hegyalla, near Tokay, and in the Banat. A solution of copper, locally known as cement-water, is found in many parts; and from this copper is easily obtained. Sulphur and arsenic are found at all the above-named places; the former in masses at Radobol, in Croatia. Another mineral peculiar to the trachyte and porphyry rocks is the alum-stone, found in the vicinity of Beregh, near Tokay, and Parad, in the N. part of the Matra mountains, under similar circumstances of position and quality with the alum-stone of the Apennines. Cobalt is a valuable mineral, which occurs in many parts, but especially at Dobschau, in the N. of Hungary. In the extensive sandstone hill stretching from the Danube to the Transylvanian frontier, coal-beds occur, containing large quantities of the carbonate of iron, some of which yield 31 per cent. of metal. Mineral salt is found extensively in the same sandstone in the N. of Hungary and Croatia. The richest mines are those of the county of Marmaros. Indeed, the remarkable fertility of the great plain of Hungary is by some attributed to the abundance of the various salts, muriates, and others, that mingle with the soil, and which serve to explain the appearance of the numerous ponds which yield soda, and from their colour are termed white lakes. These pale-lakes are scattered over the great plain, from the county of Szathmar to that of Bacs; and on the W. side of the Danube, in the counties of Stuhlweissenburg and Oedenburg. Nitre is found in these counties in sufficient quantities to supply the whole empire. The last mineral production to be mentioned is opal, found in clumps of a siliceous stone, met with in pearl-stone rocks. (Beudant.) The pearl stone presents itself in connection with trachyte and porphyry, in several parts of Hungary, over a range of 600 sq. m.; and rising 900, and even 1,200 ft. above the adjacent plains. The clumps above mentioned

are hollow, the inside surface coloured, and consisting of delicate siliceous substances—sometimes chalcedony, sometimes the stone called half-opal. The opal is found within it, lying in the hollows, like a kernel in a nutshell, exactly as Humboldt, in similar geological strata, found the fire opal, at Zimapan, in Mexico. The hyalite partakes both of the nature of the opal and of the chalcedony; and, as well as the garnet, is found in the clefts of the pearl-stone rocks. The greatest extent of pearl-stone rocks occurs in the Hegyalla, or Tokay group, where the celebrated opal mines of Czerwenitza are situated, not far from Eperies, which annually yield a considerable quantity; but, being farmed by a private speculator, nothing is suffered to transpire respecting their product. The most beautiful are the Iris opals, which are seldom found larger than a finger piece, and whose beauty seems to depend on the water with which they are saturated, as they lose their brilliancy on being heated, but regain it when laid in water. The largest opal of which we have any account (weight 17 oz.) is preserved in the mineralogical cabinet of Vienna. The fire opal is next in price; then come the half-opals, the jasper opal, and wood opals, which are very abundant, and which, as was before observed, are found in many other spots; not being, like the Iris opal, confined to the hills of Czerwenitza.

Area and Population.—The official population returns of Hungary, as given by the Austrian government, are founded on a survey of the country made in the reign of Joseph II., to which additions have been annually made. The area of Hungary is estimated to comprise 78,872 Eng. sq. m. According to the first enumeration, made in 1787, the country had a population of 7,120,794, which was found to have increased in 1805, to 7,501,414, or 12 per cent. in 18 years. A rough enumeration or rather estimate of the year 1820, gave a pop. of 8,904,717, which was stated to have increased, in 1837, to 10,975,830. But the census of the Austrian empire, of Oct. 31, 1857, only found a population, exclusive of military, of 9,900,785, showing that either the number of inhabitants had decreased, or, what seems more probable, that the previous estimate had been too high. Hungary contains several large cities. Pest, at the census of 1857, had a pop., excl. military, of 136,556; Buda, on the bank of the Danube, immediately opposite, 53,240; and Debreczin, 57,480. Several towns count between 20,000 and 30,000 inhabs.; and even many villages are equally populous. In winter, the rural pop. is usually collected in the villages; but in summer they are scattered according to their occupations and pursuits, living either in small hamlets on the Pusztas, where the cattle graze, or in detached farming establishments, which are often at a considerable distance from the villages. During the grazing season, the peasants, in large numbers, spend their time with the flocks and herds intrusted to them, in the extensive pastures. The increasing subdivision of property has a tendency to diminish this nomadic system. The herdsmen are distinguished by different names, such as the horse-herd, the cow-herd, and the swine-herd.

The people of Hungary consist of six distinct races, namely, the Magyars, the Germans, the Slovacks, or Slavonians, the Croats, the Wallachians, and the Russniaks. It is impossible to give the exact numbers of each, as there is a great blending of races through all parts of the country, particularly in the west, where the German and Magyar elements fuse into each other. It is commonly asserted that one-half the inhabitants are true Magyars.

Condition of the People.—In the provs. on this

side the Theiss, the Magyars come into contact with the Russniaks. In the prov. beyond the Theiss, with the Wallachians and Illyrian or Servian Slavonians; in the prov. on this side the Danube, with the Croatians, and in that beyond the Danube with the Slowacks, or Slavonians. The Magyars thus occupy the heart of a country bounded on every side by other nations, which, separately taken, are inferior to them in point of numbers, and are, besides, disunited by religious differences. Of the 4,000,000 or 5,000,000 of Magyars, more than one-half are Protestants, the Calvinistic confession being that most spread amongst them. They are a manly and active race, possessing frankness of character, and many other estimable qualities. Their general manner is serious; but in the hours of gaiety and feasting they indulge in tumultuous joy. The advantages possessed by the Magyar over his neighbours of other races is altogether one of character, for, in learning, the peasantry, as well as the middle classes, are behind the Germans. The hussar jacket, with light pantaloons, and the cravates, or light boots, and a huge brimmed hat, form the costume of the lower orders. The Hungarian costume, as worn in full dress by the higher classes is well known, and has been adopted in part for the uniform of hussar regiments in almost every country. The attila, or frock, and the mente, or long surtout, trimmed with fur, are often substituted for the dolman, or short hussar jacket. The kalpak, or fur cap, with the costly heron's feather, forms the national head-gear; and, on official occasions, the sabre is an indispensable addition to a gentleman's attire.

The Slowack, or Slavonian inhabitant of the NW, parts of Hungary, belongs to the same family with the Moravians, whom he resembles in appearance, and whose customs and language he preserves. The Croatian peasant is not so fortunate in the tenure of his land as the Slowack, and feels more acutely the pressure both of his temporal and spiritual lords. Still the Wallachn in E. Hungary, and the Russniak Slavonians of the N., are far behind both the Slowacks and Croatians in point of education, and have a language that has no literature. The Wallachians almost universally profess the Schismatic, and the Russniaks the United Greek confession. The Illyrians, or Servian emigrants of the Banat, use a Slavonian dialect, similar to that of the Croatians, and the majority of the books printed in Servia are written in this province. In fact the written characters constitute the only difference, the Servians using the Russian, while the Croatians adhere to the Roman character. The external appearance of the Wallachians at once declares them to be strangers amongst the Slavonian and Hungarian inhabitants. Their light active figures, dark complexion, and the resemblance to Italian in their dialect, proclaim their Romanic descent. They name themselves Roumani, are poor, light-hearted, but mostly ignorant peasants, fond of brilliant colours in their dress, when their means allow of it, and submissive under oppression.

The nobles and landed proprietors, with the exception of the few foreigners who have purchased property in Hungary, are of Magyar origin in the Hungarian provinces, and mostly Slavonians in Croatia and Slavonia. Their privileges are more extensive than those enjoyed by the nobles of the continent generally, and the rank is held by great numbers, whose property does not exceed that of a peasant. Their numbers can only be learned approximatively, as they refuse to submit to any continued registration. Of late years, the higher classes have been laudably active in endeavouring to ameliorate the condition of the lower orders by the foundation of schools and the distribution of useful works, and their private beneficence has been effectually aided by the legislative measure of 1836, which so much extended the civil rights of the peasants. By the act of the diet of that year, called the Urbarium, the nobles gave up in principle two of the most obnoxious privileges of their order—freedom from taxation, and the right of being judges in their own causes in manorial courts; and agreed that disputes between peasants and their lords should be referred to a court formed of indifferent proprietors of magisterial rank, headed by the Vice-Ashdrichter, or deputy-lieutenant of the county. The former heavy penalties for slight offences were modified, and appeals were admitted from these to the higher courts of the kingdom. The exemption from taxation was waived, not by a voluntary acceptance of burdens, which would have occasioned a vast revolution in property, and endangered one of the most valuable advantages of the Hungarian constitution, but by the enactment, that if a noble purchased a peasant's holding liable to taxation, the noble should continue to pay the impost. In some respects the lords were placed in a disadvantageous position by the new law, as the peasants may leave, sell, or transfer their holdings at will, whereas the lord has no power over them, except that of execution for rent. The amount of rent payable for peasants' holdings was then, also, fixed by the custom of each county. The extent of a session, or full peasant's holding, varies in different parts; 18 jochs of arable land, with 6 jochs of pasture (together 80 acres), being the smallest, and the largest (in the county of Arva) being 40 jochs. The right of drawing wood from the seignorial forests, of fattening pigs on the acorns, and other privileges, still remain to attest the patrimonial tie which once existed between the lord and his dependents. The peasant gives for his holding one day's labour in the week, with a waggon and two horses, or two days' hand labour in all counties excepting the Banat and Slavonia. These last-named districts have peculiar customs respecting tenures. A small sum of money and a part (1-7th to 1-8th) of the produce are likewise paid to the lord, which may be redeemed, or converted into a rent-charge. The small tithes and the tithe of reclaimed land were abandoned by the landlords. To this decree of the diet, which, as a voluntary act of self-renunciation by the nobles, has no parallel in the annals of any other nation, other measures have since been added of scarcely less importance. A decree of the diet of 1839 secures to the peasant the right of disposing by will of all kinds of property. In 1840 the diet passed a bill, declaring Catholics and Protestants to stand upon an equal footing in contracts of marriage, neither confession being suffered to impose restraints upon the other, and admitting Jews to equal rights with other commoners throughout the kingdom. It cannot be matter of wonder, if the Hungarian nation set a high value upon a constitution which has procured them so many advantages, without exposing the country to the trials and disturbances to which states under a strictly monarchical government are constantly subject. To the Magyars as a nation, rather than to the Slavonians, is the merit due of firmly upholding their national institutions.

The Germans, as settlers, are most numerous in the county of Zips, in the Banat, and in the mining districts; they are chiefly found in the towns, where the greater part of the trading population is German. In the country parts the innkeepers are mostly Germans.

Mr. Paget, in his work, 'Hungary and Transylvania,' gives the following description of the various customs of the peasantry :—' The cottage of the Hungarian peasant (Magyar), for the most part a long one-storied building, presenting to the street only a gable end, which is generally pierced with two small windows,—or rather peep-holes, for they are very rarely more than a foot square, —below which is a rustic seat, overshadowed by a eave. The yard is separated from the street, sometimes by a handsome double gateway and stately wall; sometimes by a neat fence formed of reeds, or of the straw of maize: and sometimes by a broken hedge, presenting that dilapidated state of half freedom, half restraint, in which pigs and children so much delight, where they can at once enjoy liberty and set at nought control. Passing through the gateway of one of these cottages, we entered the first door which led into the kitchen, on either side of which was a good-sized dwelling-room. The kitchen, whitewashed like the rest of the house, was itself small, and almost entirely occupied by a hearth 4 ft. high, on which was blazing a wood fire, with preparations for the evening meal. The room to the left, with the two little peep-holes to the street, was evidently the best, for it was that into which they were most anxious to show us. In one corner was a wooden seat, fixed to the wall, and before it an oaken table, so solid that it seemed fixed there too; on the opposite side stood the large cumbrous stove; while a third corner was occupied by a curious phenomenon—a low bedstead, heaped up to the ceiling with feather-beds. The use of this piece of furniture completely puzzled us—to sleep on it was impossible; and we were obliged to refer to the count for an explanation, who assured us it was an article of luxury, on which the Hungarian peasant prided himself highly. For sleeping he prefers to lay his hard mattress on the wooden bench, or even on the floor, but, like other people who think themselves wiser, an exhibition of profuse expenditure in articles of luxury—feather beds are his fancy—flatters his vanity. These beds are generally a part of his wife's dowry. In the favourite corner we commonly observed—for the peasants of Zinkendorf are Catholics—a gilded crucifix, or a rudely coloured Mater dolorosa, the Penates of the family, while all round hung a goodly array of pots and pans, a modest mirror, perhaps even a painted set of coffee cups, and sometimes a drinking cup of no ordinary dimensions. A Protestant peasant supplies the place of saints and virgins with heads of Kaiser Franzel and Prince Schwartzenberg, and not unfrequently Napoleon and Wellington look terrible things at each other across the room.

'The corresponding apartment on the other side of the kitchen was furnished with more ordinary benches and tables, and served for the common eating and sleeping room of the family. Beyond this, but still under the same roof, was a store-room and dairy, and below it a cellar. The store-room well deserved its name; for such quantities of corn (kind of cheese), lard, fruits, dry herbs, and pickles laid up for winter use, I never saw; and in some houses the cellar was not less plentifully supplied, and that too with very tolerable wine. The cow-house was rarely without one or two tenants; the stable boasted a pair, or sometimes four horses; the pig-stye, it is true, were empty, but only because the pigs had not yet returned from the stubble-fields; and to these most of the houses added sheepfolds and poultry-pens—presenting altogether perhaps as good a picture of a rich and prosperous peasantry as one could find in any part of the world.' (i. 247.)

'It would be easy,' adds the same writer, 'to find a contrast to this:—Take G——, a small village of the N. of Hungary, difficult of access from the bad roads in the neighbourhood, and not favoured by nature with the richest of soils. The peasants love the brandy-bottle and hate their landlord. The Baron B—— lives in Vienna, and lets his village to a greedy Jew, who grinds out of the people every particle of possible profit, no matter how injurious ultimately such conduct may prove to them or to their master. The dingy cottages are built of unhewn fir, carelessly put together, and plastered with mud on the inside; they rarely consist of two, and generally only of one chamber, where the whole family must live. Attached to the house is a shed for the oxen and pigs; horses and sheep they have none. I confess I cannot speak so minutely of the interior of the cottages here as at Z——; for in going towards them I stopped up to the knees in a mass of putrefying hemp; which, with the filthy appearance of the children crowding the threshold, effectually cooled my curiosity. Such are the varieties to be found among the Hungarian peasantry; nor have I in Z—— or B—— chosen exaggerated instances of either class.' (i. 291.)

Of the Slowack peasantry Mr. Paget does not give so favourable a picture. 'The peasant's house is almost always built of the unhewn stems of the pine, covered with straw thatch, carelessly and ill made; its interior is not over clean, and the pig, oxen, and goats are on far too familiar terms with the rest of the family. It is rare amongst them to see those neatly fenced farm-yards, large barns and stables, and well-made corn stacks, which are so often met with among the Magyars. How far this may depend on the poverty of the soil, it is difficult to say; that it does not depend on any greater severity of the landlord in one case than in the other, as I have heard insinuated, my own observations convinced me. The men are in general about the middle size, strongly formed, of a light complexion, with broad and coarse features half-shaded by their long flaxen hair; in some particular districts, however, there are found among them singularly fine and handsome men—as a military friend of mine observed, ready-made grenadiers. The peasant women when young sometimes are pretty, but hard labour and exposure to the sun soon deprive them of all pretensions to comeliness.' (i. 86.)

The Wallachians, according to the same authority, stand still lower in the scale of civilisation. 'The Magyar peasant holds the Wallack in the most sovereign contempt. He calls them a people who let their shirts hang out, from the manner in which they wear that article of clothing over the lower part of their dress; and classes them with the Jews and Gipsies. Even when living in the same village, the Magyar never intermarries with the Wallack.

'That the Wallack is idle and drunken, it would be very difficult to deny. Even in the midst of harvest, you will see him lying in the sun, sleeping all the more comfortably because he knows he ought to be working. His corn is always the last cut, and it is very often left to shell on the ground for want of timely gathering, yet scarcely a winter passes that he is not starving with hunger. If he have a waggon to drive, he is generally found asleep at the bottom of it; if he have a message to carry, ten to one but he gets drunk on the way, and sleeps over the time in which it should be executed. But if it be difficult to deny these faults, it is easy to find a palliation for them. The half-forced labour with which the Hungarian peasants pay their rent, has a natural tendency to produce, not only a disposition, but a determination, to do as little as possible in any given time. Add to this, that at least a third part of the year is occupied by feasts and fasts, when, by their religion, labour is forbidden them; that the double tithes of the church and landlord check improvement; that the injustice with which they have been treated has destroyed all confidence in justice, and every sentiment of security; and it will not then be difficult to guess why they are idle. The weakness of body induced by bad nourishment, and still more by the facts of the Greek Church, which are maintained with an austerity of which Catholicism has no idea, and which often reduces them to the last degree of debility, and sometimes even causes death, is another very efficient cause. Like the Turks, the Wallacks ornament their burial-places by planting a tree at the head, and another at the foot of every grave; but instead of the funeral cypress, they plant the Nectarken, or plum, from which they make their brandy.—A very literal illustration of seeking consolation from the tomb. For the death of near relations they mourn by going bareheaded for a certain time,—a severe test of sincerity in a country where the excesses of heat and cold are so great as here. (ii. 213.)

The dress of the Wallachian women consists of a long white linen shirt, embroidered with red or blue wool at the collar and cuffs; two aprons, bound before and behind, serving in place of petticoat and gown; and these aprons are not unfrequently formed of coloured laces, hanging down like a fringe to the ankles. The colours are sometimes very brilliant, and the stripes run both horizontally and perpendicularly, forming the pattern of a Scotch plaid. The Wallachians of Transylvania dress more showily than those of Hungary; and their costume is often ornamental, and even rich. A small sheep-skin jacket, trimmed and richly embroidered, at times, is occasionally worn by the women in both countries.

Of late years, the exertions of writers in the Magyar language have furnished elementary works fitted for schools, as well as newspapers and other periodicals. The foundation of the National Casino at Pesth, which originated with Count Stephen Szechenyi, furnished the lituals and visitors of the capital for the first time with a place of meeting; and the example has been imitated by nearly every town in the kingdom. The national prints, with German and French newspapers and reviews, are now to be found in these clubs. In remote corners of the country; and small provincial theatres are, perhaps, more numerous in Hungary than in any other country. Scientific societies have also sprung up of late years; that for the Magyar language and literature was endowed by Count Szechenyi with the sum of 60,000 fl.; and this noble example was followed, on a smaller scale, by other magnates.

Agriculture.—Agriculture, owing to the richness of the soil, is the most important branch of national industry; and there can be no doubt, that if a market could be found for the produce, the resources of the land would be fully developed, which is not the case at present, one-fourth part of the best land lying wholly uncultivated. The soil, indeed, constitutes a source of wealth in Hungary, which bad laws alone prevent from being adequately worked. In the NW. counties, among the hills, the Moravian systems of farming are met with,—a natural consequence of the Slowack's general resemblance to the Moravian in customs and language. This part of Hungary does not produce corn enough, in ordinary years, to supply its own consumption, and imports corn, &c. from the adjoining level districts. The lower plain of

Upper Hungary contains many fertile tracts, especially N. of the Danube, as well as the islands 'Gross and Kleine Schütt.'

The following is an estimate of the distribution of the soil of Hungary:—

Arable Land	4,387,770 jochs
Gardens	649,000
Vineyards	911,200
Meadows and Pasture	7,715,730
Ponds	850,000
Forests	8,842,000

The rich soil of Lower Hungary is productive and generally well tilled. The black vegetable mould of the Banat, or the district between the Maros, Theiss, and the Danube, extending also over the counties of Bacs, Arad, Bekes, and Csongrad, is peculiarly well adapted to the growth of wheat, which consequently is grown as often as possible, that is, according to the present system, once in three years; a crop of summer corn follows, after which the land either lies fallow or is sown with maize. The immense tracts sown with grain in the great plain present a singular spectacle at harvest, owing to the great number of hands requisite to get in the crop. A square piece of ground is usually well beaten at one end of the field; and if horses and oxen can be got to tread out the corn immediately, it is carried thither at once, and trodden out by their unshod hoofs. This practice of treading out the grain is, however, most wasteful: not only is the work badly done, but, being performed in the open field, it exposes the crop to the chance of plunder, and to all the vicissitudes of the weather; sudden thunder-storms often destroy the greater part of a crop.

Notwithstanding the abundance of the crops in many parts of the plain, and the difficulty of finding a market for produce, but little money is invested in farm buildings to preserve the grain. Holes dug in the earth, and shaped something like a bottle, with a narrow entrance or neck, are dried by burning straw in them, and after being lined with fresh straw, are filled up with wheat dried in the sun. These rude granaries are common in the plain N. of the Theiss, but have given way to the Banat to regular granaries, as the foreign trade in that fertile district has gradually become regular.

The average of a number of years well ascertained gives a produce of 16 metzen per joch, which valued at 1 florin in silver (about 11s. 4d. per quarter) would leave 4 fl. per joch (or about 6s. per acre) profit to the landholder. Small tracts of land, let on short terms to peasants in the best-cultivated parts and in the neighbourhood of towns, are sometimes paid for at the rate of 4, and even 6 or 7 fl. per joch; but large estates are farmed out by the government at 1 fl. per joch, or about 1s. 6d. per acre. From the end of June the ground lies idle till the following April, sheep being turned into the stubble after the September rains. The grazing of the second year is more valuable, and may be let at about 1 fl. per acre, where wool-growing does not form a regular part of the farming system.

The great drawback on the landowner's profit in these productive countries is the difficulty and expense of forwarding the produce to market. The soil of the great plain is so singularly free from stones, that road-making is extremely difficult, and demands a large outlay. The navigable rivers and canals are by no means in a state to allow of their being used at all seasons; and these difficulties, added to the wasteful manner of getting in the corn crops, make it wonderful that as much as sixteen-fold can be returned from the ground. Manure is in these parts scarcely used, as it makes the plant too rank, and forces it up into straw.

In the greater part of the great plain cattle-dung is cut into bricks like turf, and used for firing. Wheat-straw is likewise used for fuel, as wood is scarce.

Tobacco is successfully cultivated, especially in the counties of Heves, Neugedin, and Czongrad, and it has a high character in Germany. The annual produce is reckoned at 250,000 centners, of which only 50,000 are kept for home consumption. On being exported, it is subjected to heavy imposts, levied by government. The annual produce of the Hungarian vineyards is said to be 24,000,000 eimers, or 96,000,000 gallons. These wines, which are strong and fiery, requiring to be kept before they reach perfection, are of two sorts; the sweet wines (*Ausbruch*), and the red and white table wines. Of the former, the Tokay (grown about Tokay, on the Theiss) is unequalled for delicacy and flavour. It is a sweet, rich, but not cloying wine, strong, full bodied, but mild, bright, and clear, seldom to be procured of the finest quality, and then only at the private tables of the nobility. There are three distinct kinds of Tokay. The annual produce of the Tokay vineyards is 250,000 eimers, of which only 1-6th part is of the best quality. Good old Tokay costs, even in Hungary, from 4s. to 8s. a bottle. Next to Tokay comes the Menes wine and the Ruszt, Carlowitz, and St. George. Of the red wines, that of Buda (*Offner-Wein*) is considered equal to the best Burgundy; and next to it are the Pösing, Szekó, Mirkolez, Neustadt, and other wines. The best white wines are those of Somlyo and Neszmely, which, it is alleged, equal any of the white wines of France, except champagne. The cultivation of the mulberry-tree for silkworms was introduced by the Empress Maria Theresa; and in the military frontier a large quantity of silk is produced; but neither this article, nor the cultivation of dye-plants, such as woad, madder, and saffron, is well attended to. The farmers, discouraged by the various difficulties in the way of a sale for their crops, have of late years devoted themselves to sheep-grazing, and the breed has been greatly improved by the introduction of the Merinos. The number of sheep grazed in Hungary is said by Czaplovics to be 20,000,000; and the quantity of wool exported averages 200,000 centners of 123 lbs., which, at the ordinary medium price of 170 florins the centner, would amount to 2,000,000l. The horned cattle bred on the Hungarian plains are among the largest and handsomest in Europe; they are a race peculiar to the country, grey-white in colour, with wide-spreading horns. The horses generally are small and weak, and of an inferior breed; but in some parts considerable attention is given to breeding, especially in the county of Czanad, where nearly 10,000 horses are kept, and stallions of all the best breeds in Europe. Hundreds of thousands of swine are bred in the forests, and on the great heath of Debreczin there are some millions of geese. Poultry and game of every kind abound throughout Hungary.

Manufactures and Trade.—Hungary has, except mining industry, but few manufactures of any importance, and there is but little prospect of her importance in this respect being speedily increased. A small amount of linen manufacture is carried on in the N. and mountainous districts; but it little more than supplies the home consumption of the district. Wool is every where manufactured into coarse cloth, for country consumption. Tobacco, leather, paper, soda, alum, and saltpetre manufactories, with numerous iron-works, are the principal other branches of manufacturing industry.

Mines.—Mining industry ranks next to agriculture. The greater number of the old mines are

worked by the government, but in recent years many new mines have come to be exploited by private owners. In 1852 there were 19,850 individuals employed on government account in the mines of Hungary. The chief produce of the mines are coal and iron, the former averaging 15,000 tons, and the latter 400,000 cwt. annually. Next to coal and iron in importance is copper, producing about 50,000 cwt. annually. Lead and alum are also produced to the amount of 10,000 cwt. a year; besides which zinc, antimony, cobalt, and various other minerals are found. Auriferous and argentiferous ores are met with in considerable quantities, and were formerly much worked, the yield of both amounting in value to about 80,000 marcs annually. But the vast influx of gold from Australia and California has led to a just neglect of the auriferous metals of Hungary, in favour of the infinitely more valuable subterranean stores of coal and iron. The mines are divided, from their position, into four districts: the Schemnitzer, Schmöllnitzer, Nagy-Banyaer, and Bonater, of which the first is by far the most considerable; and hence Schemnitz is considered as the mining capital of Hungary. This town possesses an excellent school for miners. Each of the districts has its government and separate establishment of smelting-houses; but all send their produce to be assayed to Kremnitz, in the Schemnitz district.

River Communications.—The trade by way of the Black Sea was not commenced till the establishment of the steam navigation of the Danube. The Hungarian peasant, with corn, wool, and flax about him in abundance, lives in poverty, for want of a market. The articles imported from Wallachia and Moldavia are wax, honey, wool, bristles, and some metals. The agricultural produce of the great plain along the Save is conveyed to Sziszek, in Croatia, whence the more expensive articles are forwarded along the river to Agram, and thence, by land carriage, to Laibach and Trieste; those more bulky are sent up the Culpa to Carlstadt, and thence conveyed to Fiume by the Louisa Road. (See FIUME.)

The navigation on the rivers is as well managed as permit on the peculiar circumstances of the country will allow. The barges are of great size, usually from 100 to 150 ft. long, by 17 to 24 ft. broad, and drawing 5 to 5½ ft. They are built of Croatian oak, either at Sziszek, on the Save, or at Szegdin, on the Theiss, and cost between 600l. and 700l. They have a high pointed roof, like a house, and serve the purpose of granaries in the interior of the country. The peasants bring their corn for sale to the river's bank, and it is at once laden in sacks into the barge, and sorted into various partitions. These barges hold from 1,500 to 2,000 qrs. of wheat. They are drawn up the Danube or the Save by 20 to 25 small horses, and are often months on their way from the mouth of the Theiss to Raab, or Wieselburg, or Sziszek, when the water is low; but, under favourable circumstances, the trip from the Theiss to Sziszek may be made in 14 or 15 days; and the freight is commonly 15 kreutzers, or 6d. per cwt. All navigation up the stream is, of course, interrupted during floods or hard frosts, and thus there are many months in the year when no navigation is possible. The improvement of the beds of the Save and the Culpa, the use of steam tow-boats instead of horses, and the adoption of smaller craft, has of late much increased the traffic along the rivers of Hungary.

Trade.—A considerable trade is carried on along the course of the Danube, the grand highway of Hungary. Commerce received a considerable imprise from the employment of steamers on the Danube and tributaries, and a still greater from the establishment of railways in many parts of the country. The main line of railway, which crosses the whole of Hungary from north-west to south-east, following the course of the Danube, enters the country near Presburg, and runs, by way of Pesth and Szegedin, into Servia, touching the lower Danube near Belgrade. The chief branch of this great Hungarian railway runs northward from Czegled, near Pesth, to Debreczin, Tokay, and the Russian frontier, while another branch goes southward, from Buda to Stuhlweissenburg and lake Balaton, falling into the great railway from Vienna to Trieste. The construction of these important iron high roads has been of vast advantage to Hungary; still, however, the central situation of the country, and its great distance from the ports accessible to foreign ships, lay its commerce under many disadvantages. The exports consist almost wholly of raw produce, inc. corn, wool, wine, tobacco, cattle and sheep. The imports comprise most species of manufactured goods, with colonial products, dye-stuffs, spices, and hardware. The principal trade is carried on with the Austrian dominions, inc. Galicia, Fiume, the nearest port on the Adriatic to Hungary, is an open roadstead, in which ships cannot lie when either the Bora or Scirocco winds are violent.

Hungary has no commercial town to compare with Cracow or Vienna for bill and banking business; but the transmission of money, &c. is much facilitated by branches of the National Bank of Vienna established at Pesth and other places.

Coins, Weights, and Measures.—The Hungarians use the same standards as the Austrians, in most respects. The florin of 60 kreutzers is equal to 2s. English. The gold ducat of Kremnitz, consisting of 4½ florins, is worth 6s. 6d. English. The ort contains 12 kreutzers, and the poltorack 1½ kr. The Hungarian yard, used in measuring cloth, is 4-5ths of the Austrian yard, or about 34 Engl. inches. The joch, or Austrian acre, contains 1,600 sq. klafter, and is equal to 1·46 Engl. acre. The metz of Presburg, commonly used for measuring dry substances, is 1·75 imp. bushel. The eimer (for liquids) varies; for the wine eimer is equal to 19½ Engl. gallons, while that used in Lower Hungary is equivalent only to 15 gallons. The antal, used in the Tokay district, is equal to 13·8 Engl. gallons.

Constitution and Form of Government.—The ancient constitution of Hungary, which, though abrogated by the decree of the "King-Emperor" in 1849 and 1851, is still held by the leading men of the Magyar race the valid fundamental charter of the kingdom, is strictly monarchical. The prerogatives of the monarch, however, are greatly limited by the power of the aristocracy. All that concerns the security of the country against foreign attacks, in other words, the defence of the nation, is monarchical in principle. The armed force is consequently altogether dependent upon the king. The internal government of the nation is a mixed monarchy and aristocracy. Laws can only be enacted by the joint consent of the king and the diet; and, although the executive power be said to lie with the king, yet the sovereign has only the nomination of *Obergespann* (Obergespanner) of counties and administrators; since every other public officer is either elected by the county itself, or named by its lord-lieutenant—a nomination, however, which is often successfully disputed. Justice is administered on the principle, *rex est fons et origo jurisdictionis*, in the name of the king, who has, however, no further

influence than the power of appointing the president and councillors of the curia regia, that is, of the septemviral and royal courts; but to these courts the crown dignitaries likewise depute their representatives. The king nominates the presidents and councillors of the district courts, watches over the course of justice in all courts, and enjoys in civil suits the exercise of certain prerogatives, and the power of issuing mandates founded upon them; such are the power of ordering a suit to be recommenced (*mandatum novi cum gratia*), and of issuing moratoria: in criminal cases the king has the power of pardoning.

The royal dignity is hereditary in the house of Austria since the year 1526, and confirmed in the female as well as the male line. Since 1723, the succession to the throne of Hungary is placed upon the same footing with that of the other hereditary states of the empire. The chief prerogatives of the crown are,—1st, The power of making laws, after consulting the estates assembled in the diet, and in common with them. The king assembles the diet, and dissolves it at pleasure. 2nd, The highest executive authority in every thing which is in accordance with the laws, or which involves no violation of them. 3rd, The right of patronage, or the nomination to all bishoprics and other clerical dignitaries. 4th, The highest judiciary authority, which the crown, however, only mediately exercises through its officers. 5th, The full power of declaring peace and war. 6th, The right of levying troops, of erecting fortresses, and of demanding warlike subsidies. 7th, The right of calling out the general insurrection of the country for its defence, in the prescribed legal manner. 8th, The right of pardoning. 9th, The right of coining money. 10th, The right of granting patents. 11th, The right of nominating to all offices, except those of palatine of Hungary, of the two guardians of the crown, and of the county officers above mentioned. 12th, Of legitimising bastards. 13th, The *jus praefectionis*, or the power of transferring the right of succession to a daughter, on the extinction of male heirs in a family. 14th, The *jus successionis*, or the inheritance of all noblemen's estates when there are no male heirs. 15th, The right of abrogating decrees of infamy pronounced by the courts of justice. 16th, Of granting letters of prosecution. 17th, The supreme guardianship of orphans. 18th, The post. 19th, The right of sending special commissions to inquire into the faulty administration of the counties.

The prelates, magnates, nobles, and free cities are comprised under the name of estates. The free cities are regarded as nobles in their municipal capacity. Under the kings of the reigning house of Hapsburg a great portion of Hungary and of the annexed districts was conquered from the Turks by great exertion on the part of the other imperial states; and many important alterations, in the relations of the king and the estates, took place at different times. What are called the cardinal privileges of the nobles and of the clergy, who are looked upon as equal to the nobility, have been preserved to the present day to an extent unparalleled in any other country in Europe.

By the decree of the king-emperor Francis Joseph I., published the 20th Oct. 1860, and the 26th Feb. 1861, and which established one constitution for the whole empire, the ancient constitution of Hungary was virtually overthrown. The people refused to acquiesce in this change, and the steadfast opposition which ensued had the consequence that by another decree, issued on the 20th Sept. 1865, the constitution of the whole empire was suspended, for the express purpose of coming to an arrangement with Hungary.

Religion.—The pop., considered in relation to its religious belief, is divided into four grand classes. The religion of the state is Rom. Cath., to which faith 5-10ths of the pop. are attached. The Protestants, according to the census of 1857, number 2,916,822; the adherents of the Greek church about 2,680,000; and the Jews about 253,000. By the decree of Joseph II., who dissolved 600 monasteries, and endowed with their funds various universities and schools, religious toleration, if not absolute equality, was granted to the professors of all Christian creeds; and this liberal policy has been maintained by his successors. The Rom. Catholics are spiritually governed by 3 archbishops and 14 bishops, who are all members of the diet: there are well provided for; but the inferior clergy are poor, and are said to be not remarkable for their liberality of feeling, or exemplary morals. (Paget I. 111.) The archbishop of Gran, who has a very large revenue, is primate of all Hungary. The United Greeks have 4 bishops, and the Orthodox-Greek churchmen, 1 archbishop (abp. of Carlowitz) and 6 bishops, all of whom have had seats in the diet since 1791. The Protestants are not under episcopal jurisdiction, but have 4 superintendents or presidents of synods. They are divided into two classes; the Lutherans, who adhere to the confession of Augsburg, and the Reformed, who follow the doctrines of Calvin. The former are principally found in the N., and among the Slowacks; the latter are almost entirely Magyars, and chiefly inhabit the towns and villages of the Puszta. There are upwards of 700 Jewish synagogues in Hungary. By the law of 1840, they are admitted to all civil rights and privileges.

Education, Crime, and Courts of Law.—No minute a survey cannot be given of the educational institutions of Hungary as of those in other provinces of the empire, owing to the more local character of the Hungarian municipal and parish jurisdictions. The University of Pesth is one of the most richly endowed of Europe, but its services are by no means in proportion to the magnitude of its revenues. The family of Marschany has the credit of founding 100 bursarships for poor students. There are Rom. Catholic lyceums or colleges at Agram, Kaschau, Grosswardein, Presburg, Raab, and Erlau; and Protestant colleges at Presburg, Oedenburg, Karsmark, Eperies, Raab, Debreczin, Neues Pataki, and Papa; they have faculties of law and arts. The largest of these is at Debreczin, founded in 1792. The colleges of Szegedin and Stein am Anger have faculties of arts only. At Schemnitz is a mining college, similar to that at Freiburg, supported by government, with 7 professors and 64 scholarships. The Ludovici academy at Waitzen, and 67 Catholic and 13 Protestant gymnasia or grammar-schools, complete the list of higher schools. The academy at Carlowitz, intended to educate the priests of the Greek Schismatic confession, belongs strictly to the military frontier. There are, moreover, 2 schools for sons of nobles, 1 at Agram, and 1 at Kaschau; 24 Catholic and Greek united clerical seminaries; 14 regimental schools; a nunnery for education at Pesth; and an excellently conducted school for the deaf and dumb at Waitzen. Besides this, every village has its elementary school, or *Trivial schule*, and the larger villages more than one, where instruction is given in the language of the inhabitants. In Pesth and other places subscription infant schools have been established.

All estimates of the state of crime in Hungary are extremely loose, owing to the want of proper returns. The poverty and ignorance of the lower orders are great inducements to offences against property. Cattle-stealing is a common offence, and the insecurity of gardens and field-crops is much complained of by the industrious peasant. Murder, however, is of rare occurrence except in cases of popular tumult.

The prisons, formerly wretched, have of late years attracted attention in Hungary, not less than in other countries. The landowners of the country raised the sum of 30,000 florins by subscription, for the erection of a penitentiary on the American system. A society of ladies likewise raised the sum of 16,000 fl., to erect a workhouse for mendicants. At Gyarmat, Arad, Szexard, Miskolcz, and Jaszbery, the old system of imprisonment has been changed, at the expense of the nobles, into the better one of prison labour. This laudable spirit is spreading rapidly in all parts of the kingdom.

The court of lowest jurisdiction for the peasant is the manorial court of his lord; but in disputes between the peasant and the manor, a special court is formed from members of neighbouring manorial courts, with the 'Vicestuhlrichter,' or police magistrate of the district, and from their decision an appeal lies to the 'Stuhlrichter's' court. This court is the tribunal of first instance for the nobles; but causes involving more than 3,000 fl. come before the court of the 'Vice-gespann,' or sheriff of the county, whence an appeal lies, as well as from the 'Stuhlrichter' to the 'sedes judiciaria,' or 'sedria,' the proper county court of session, and thence to the royal table, or court of king's bench. The 'Septemviraltafel' is so called from its having formerly consisted of 7 judges; it is now composed of 4 prelates, 10 magnates, and 4 nobles, or their representatives, of whom 11 must be present to form a court. This is the highest tribunal of the kingdom.

Local Government.—The local taxation is very slight in most of the towns of Hungary. Many of them, such as Pesth, Debreczin, and Szegedin, have extensive town-lands; which, if properly managed, ought to produce large revenues, but which, under defective municipal systems, do not defray the charges of lighting and paving. The town-lands of Szegedin exceed 10 German sq. m. in extent, and yet 4 strong horses are scarcely able to pull a carriage through the streets of the town, so much are they neglected. The excise and octroi, or consumption dues, levied on the larger Austrian towns, are either unknown in Hungary, or are raised by the municipalities as town-dues, for local purposes. Salt and playing-cards are royal monopolies. Tobacco is free. Even the tithe of all minerals claimed by the crown in the other states of the empire is disputed by the owners of iron and coal mines. A peculiar feature of Hungarian financial economy is the pride which the nobility feel in not being compelled to pay road and bridge tolls. The principal of this absurd exemption has, however, been abandoned in the new chain-bridge between Pesth and Buda, where all classes are to pay toll indiscriminately.

The county meetings, which are the nursery of patriotism in Hungary, are of two kinds, *restaurations* and *congregations*. In the former, the county officers are elected; in the latter, accounts are passed, and the county business discussed. The number of nobles or electors is between 700,000 and 800,000; and as their qualifications are limited neither to property nor instruction, the tumultuous scenes which present themselves at elections, and on other occasions, bear a good deal of resemblance to the occurrences in England at such meetings which took place in a bygone age. The magistrates have an ingenious way of manufacturing votes for their friends. On the candidate's demand to vote, the claim, if opposed on the ground of non-qualification, is referred to the county court as a disputed point, where the magistrates generally have influence enough to settle the matter as they wish. In this way the number of the nobles is annually increased. Whoever purchases land of the crown becomes, by so doing, a noble. At the county congregations a large amount of business consists in the making out of instructions for their representatives during the session of the diet; these are, in fact, delegates without any will of their own, being bound to adhere to that of their constituents, to whom they apply for directions on all doubtful and difficult questions. The county meeting may also recall a refractory member, and send another in his stead. The rights of the nobles are based on the 'Aurea Bulla,' granted to the armed barons by King Andrews, in 1272, in a manner similar to the Magna Charta of England.

The internal management of the cities is wholly dependent on the government, which has power to appoint and remove their officers: they are on this account a constant object of jealousy to the nobles, who consider this dependence as opposed to the principle of constitutional liberty. They reproach the citizens for their financial economy, and for allowing the majority of the inhab. to be excluded from a voice in all public business. The distribution of Hungary into counties is attributed to King Stephen, about the year 1000. The Fö Ispan, or lord-lieutenant, is the only officer named by the crown. The Al Ispan, or deputy-lieutenant, of whom there are usually two, is, however, the common president of the county meetings: he holds the supreme direction of the county police, and presides as chief judge in the county courts; being, in fact, a kind of sheriff. The small salary attached to all county offices seems rather intended to defray extra expenses, than as a remuneration.

The most important national institution, next to the county meetings, is the diet, at which the prelates and magnates formerly assembled with the deputies from the counties and towns. Since 1562, the chambers have been divided. The chamber of magnates is composed of the prelates, with the archbishop of Gran, as primate, at their head; the 'barones et comites regni,' or peers of the realm, in two classes; the great officers of the crown, with the lords-lieutenant of the fifty-two counties; and the barons, summoned by royal letters, including every prime count and baron of twenty-five years of age. The palatine is the president of the chamber of magnates. Magnates who are absent depute representatives, as do also the widows of magnates; but these deputies sit in the second chamber, where they can speak, but have no vote. The business transacted in the lower chamber is previously discussed in a kind of committee of the whole house, called a 'circular session,' in which strict forms are not observed, and each member speaks as often as he can get a hearing. The speeches in both chambers are usually made in Hungarian. Among the magnates some few speak Latin; but this language has almost entirely fallen into disuse. The 'personal,' or president of the lower chamber, who is at the same time chief judge of the 'royal table,' is appointed by the crown. When the diet assembles, the propositions of the crown are first presented to it for consideration, and these form the great busi-

ness of each session; but proposals also originate in the lower chamber, which, when agreed to by the magnates, are also sent to the king, who, if he approve them, communicates his assent by a royal 'resolution.' Many propositions rejected by the crown are voted anew in every diet, under the title of Gravamina; and their number has accumulated to such an extent as to make it expedient to make a selection of the most pressing, which are denominated preferentialia.

The Hungarians attach great importance to their country's being recognised an independent kingdom. The sovereign is styled 'king' in all public acts, and the regalia of the crown are guarded by a special corps appointed for the purpose in the palace at Buda, whence they are only removed, and that with great ceremony, for the sovereign's use on state occasions. The grand officers of the court and household are numerous, and are termed 'aulae ministeriales.' These are the grand justiciary (index curiae), the ban of Croatia, the arch-treasurer (thesaurarius regalium scopiales), the great cup-bearer (pincernarum reg. mag.), the grand carver (dapiferorum reg. mag.), the master of the household (agazonum reg. mag.), the lord chamberlain (cubiculariorum reg. mag.), the grand porter (janitorum reg. mag.), the master of the ceremonies (curie reg. mag.), and the captain of the body-guard (regiminis nobilis turmae praetorianae). The king is represented by his viceroy the palatine, who resides at Buda, but the grand chancery of the kingdom has its seat at Vienna, where the government business is transacted. The exchequer is managed by the 'Hof-kammer,' which has its seat at Buda, and under which are the collectors of taxes, the mining boards, and the direction of the crown domains.

History.—The oldest inhabitants of Hungary, mentioned in history, were known to the Greeks and Romans by the name of Pannonians. Of its history during the time of the Western and Eastern empires, and the various wars and invasions which are said to have taken place between the third and tenth centuries, there is no certain information. Hungary, however, had assumed the form of an independent kingdom in the eleventh century, the sovereign power being vested in the house of Arpad, a chief of the Magyar race. This family having become extinct in 1301, the Hungarians, through the influence of Pope Boniface VIII, elected Charles of Anjou, brother of Louis IX. of France. One of his sons became king of Poland in 1370, and thus his dominions extended from the Baltic to the Adriatic. A few reigns subsequently, under Matthias I., Hungary comprised about 256,000 sq. m., the extent of the present Austrian empire. The Turks, soon after their establishment in Europe, began to assail Hungary. They were, for a lengthened period, vigorously resisted, particularly by the famous John Hunniades. In 1526, however, Louis II. king of Hungary, was totally defeated and slain by the Turks, in the battle of Mohacz, and a large part of his dominions fell into their hands. On his death, Ferdinand I. of Austria, his brother-in-law, succeeded to the throne, and was crowned king of Hungary in 1527, since which time the monarch has always been emperor of Austria: but the Turks continued for many years to hold the greater part of the kingdom. The despotic conduct of the Austrian princes was most distasteful to the Hungarian nobles; and so great was their antipathy to the Austrian yoke, that, in 1683, they rose, with Tekeli at their head, and called upon the Turks to relieve them from servitude. Austria, however, succeeded, by the help of John Sobieski and Prince Eugene, in expelling the

Turks from these countries, and they were finally arrested in it by the treaties of Carlowitz and Passarowitz, in 1718. Hungary remained loyal to the house of Austria until the year 1848, when the long-standing effort of assimilating all the territories of the empire into one homogeneous mass brought about a crisis. An insurrection broke out at Pesth in July, 1848, and, on the 28th Sept. following, the Hungarian diet proclaimed a provisional government, under Batthyani and Kossuth. Austrian troops thereupon marched into the country, and the Hungarians were defeated in several encounters, Dec. 28 and 29. The tide of victory turned at the beginning of the following year, when the insurgents gained the battle of Giran, April 17, 1849. This led the Austrian government to call in the aid of Russia, and the latter power having sent a large army into the field, the Hungarian commander-in-chief was forced to surrender his forces on the 13th of Aug., 1849. The last stronghold of the insurrection, the fortress of Komorn, opened its gates on the 28th of August, and Kossuth having fled into Turkey, and Batthyani been led to the scaffold, October 6th, the Austrian government found itself absolute master of the kingdom. The attempt to unite all the territories of the empire into a uniform state was now again undertaken, and ended in the constitution of 1860-61. But once more the Hungarian nation protested—not in arms, however, but by quiet steadfast opposition, and constant refusal to send deputies to the central parliament, or reichsrath, at Vienna. The result, as already mentioned, was that the king-emperor, on the 20th Sept., 1865, suspended the constitution of Austria, with the sole object of conciliating Hungary.

HUNGERFORD, a market town and par. of England, partly in co. Berks, hund. Kintbury Eagle, and partly in co. Wilts, hund. Kinwardstone, on the Kennet, 36 m. E. Bath, and 61 m. W. London by Great Western railway. Pop. of town 2,931, and of par. 3,081 in 1861. Area of par. 6,940 acres. The town consists chiefly of one long street, in the centre of which is the market house, open below, and having a room above for the transaction of the town business. The church, which stands at the end of a shady avenue on the W. side of the town, is a handsome structure, erected in 1816, and near it is the grammar school. There are also places of worship for Wesleyan Methodists and Independents. Hungerford has no manufactures; but there are some extensive breweries, and a considerable traffic arises from the Great Western railway and the Kennet and Avon canal which passes close to the town. It is a bor. by prescription, and is governed by a constable elected annually by the inhabitants, who are called together by a brass horn, known as the 'Hungerford Horn,' and given by John of Gaunt with the charter. Hungerford Park, at the E. end of the town, is a finely wooded domain, with a mansion in the Italian style, erected on the site of a house built by Queen Elizabeth or the Earl of Essex. Markets on Wednesday; fairs, last Wednesday in April, Aug. 10, and Monday before Michaelmas.

HUNTINGDON, an inland co. of England, partly included within the great level of the Fens; being surrounded by the cos. of Northampton, Cambridge, and Bedford; the latter bounding it only on the NW. Area 229,551 acres, of which 70,000 belong to the Bedford Level fens. Surface in the W. and S. parts gently varied, but the N. and NE. portion, included in the fens, is quite flat. This latter portion of the co. is mostly in grass, the other parts being about equally divided between tillage and pasturage. Chief crops, wheat, oats, and beans. Agriculture, though much improved, is not very advanced. The land is ploughed in immense ridges, by which a great deal is lost; and it is frequently also foul and out of order; turnips little cultivated. A great deal of fine cheese and butter is made. The sheep, the stock of which is estimated at about 200,000 head, produce long combing wool. Estates generally extensive; there are many large farms; but small ones predominate. Pigeon houses are extremely abundant. There are neither minerals nor manufactures of any importance. Principal rivers, Ouse and Nene. There are in the fens two shallow lakes, Whittlesea Mere, and Ramsey Mere; the former containing above 1,550 acres, and the latter about half as much; measures, however, are now in progress for draining the former. Huntingdonshire has four hundreds and 103 pars.; it sends four members to the H. of C., two for the co. and two for the bor. of Huntingdon, the principal town in the co. Registered electors for the co. 3,123 in 1865. In 1861, Huntingdon had 13,704 inhabited houses, and a pop. of 61,250. The pop. in 1851 was 61,183. Annual value of real property, 875,167l. in 1857, and 467,841l. in 1863.

HUNTINGDON, a parl. and mun. bor. and market town of England, co. Huntingdon, of which it is the cap., hund. Hurstingstone, on the Ouse, 57 m. N. London, and 17 m. NW. Cambridge, on the Great Northern railway. Pop. of munic. bor. 3,816, and of parl. bor. 8,254, in 1861. Area of parl. bor., which includes the old bor. and the adjoining par. of Godmanchester, 6,820 acres. The town, which stands on a gentle slope N. of the Ouse, crossed here by a causeway and bridge of six arches, consists principally of a long range of brick houses, running from N. to S., which commences immediately from the bridge, and line each side of the N. road from London. A few streets and lanes branch off on each side; but these are mostly composed of inferior houses. Of fifteen churches once standing, only two remain, to which a new one has been added in recent years. The principal church is All Saints, built in the perpendicular style, and containing some interesting memorials of the ancestors of Oliver Cromwell, who, as is well known, was born at Huntingdon on the 24th of April, 1599. Huntingdon, besides its three churches, has several places of worship for dissenters, an old grammar school, with two exhibitions at Cambridge, attended by 50 boys; a green-coat school, for 30 boys and 12 girls; a national school, with 150 children; and 5 Sunday schools. The town-hall, behind which are the shambles, is a stuccoed building, comprising two court rooms and an assembly room; and close to it is the county gaol, a very large building. There is also a small theatre and a race-course.

Godmanchester, on the opposite side of the river, which seems to have been once an important bor., is now a mere suburb of Huntingdon, chiefly inhabited by farmers and farm labourers. 'Huntingdon, as it were, looks over into the fens; Godmanchester, just across the river, already stands on the black bog.' (Carlyle, Th., 'Cromwell's Letters and Speeches, i. 34.) Both Huntingdon and Godmanchester were chartered in the reign of John; the present officers in each are four aldermen and twelve councillors; but neither of the bors. as now constituted, has a commission of the peace. The county magistrates hold petty and quarter sessions in the town-hall, the chief local act of the town being that of 25 George III. Huntingdon has sent two members to the H. of C. since the reign of Edward I., the franchise, till the passing of the Reform Act, being vested in freemen by birth, grant, or purchase. The boundaries of the present parl. bor. include the entire parish of Godmanchester, as well as the old borough. Registered electors, 417 in 1865. Markets at Huntingdon on Saturday; Godmanchester cattle-fair on Easter Tuesday.

HUNTLY, a bnt. of barony, market town, and par. of Scotland, co. Aberdeen, on the peninsula formed by the confluence of the Deveron and Bogie, 35 m. NW. Aberdeen, on the Great North of Scotland railway. Pop. 3,444 in 1861. The town is neatly built, consisting of two principal streets crossing each other at right angles, having a handsome square or market-place in the middle. The Deveron is crossed by an ancient bridge of a single arch. On occasion of the great floods of 1829, when the waters of the river rose 22 ft. above their usual level, only 6 ft. of the arch remained unoccupied; but it received no injury, and stands apparently as firm as ever. A modern bridge of three arches spans the Bogie. In addition to the par. church, the Episcopalians, Catholics, and Independents have each chapels. In the immediate vicinity of the town are Huntly Lodge and Huntly Castle; the former a seat of the late duke of Gordon; the latter, which is in ruins, an ancient seat of the Gordon family; both are now the property of the duke of Richmond, feudal superior of the town. Huntly was once celebrated for its manufacture of linen; but it has nearly disappeared. There is a thriving bleach-field on the banks of the Bogie. The business of brewing and distillation is carried on to a considerable extent in the town and neighbourhood.

HURDWAR, HARI-DWAR, or GANGA-DWARA ('the gate of the Ganges'), a town of Hindostan, presid. Bengal, prov. Delhi, in lat. 29° 57' N., and long. 78° 2' E.; 105 m. NE. Delhi, and famous from its being one of the principal places of Hindoo pilgrimage, and the seat of the greatest fair in India. The town, which is but inconsiderable, is situated on the Ganges, at the point where that sacred stream issues from the mountains. The pilgrimage and the fair are held together, at the vernal equinox; and Europeans, no less addicted to exaggeration, who have been repeatedly present on these occasions, estimate that from 200,000 to 300,000 strangers are then assembled in the town and its vicinity. But every twelfth year is reckoned peculiarly holy; and then it is supposed that from 1,000,000 to 1,500,000, and even 2,000,000 pilgrims and dealers are congregated together from all parts of India and the countries to the N. In 1819, which happened to be a twelfth year, when the auspicious moment for bathing in the Ganges was announced to the impatient devotees, the rush was so tremendous that no fewer than four hundred and thirty persons were either trampled to death under foot, or drowned in the river.

The foreigners resorting to Hurdwar fair, for commercial purposes only, consist principally of the natives of Nepaul, the Punjab, and Peshawur, with Afghans, Uzbeck Tartars, &c. They import vast numbers of horses, cattle, and camels, Persian dried fruits, shawls, and drugs: the returns are made in cotton, piece goods, indigo, sugar, spices, and other tropical productions. The merchants never mention the price of their goods, but conduct the bargain by touching the different joints of their fingers, to hinder the bystanders gaining any information. During the Mahratta sway, a kind of poll-tax and duties on cattle were levied; but all is now free, without impost or molestation of any sort. Owing to the precautions adopted by the British government, the most perfect order is preserved. Antecedent to the British occupa-

tion, the fairs usually ended in disorder and bloodshed.

HURON (LAKE), one of the five great lakes of N. America, belonging to the basin of the St. Lawrence, second in size only to Lake Superior, and intermediate in position between that lake and Michigan, on the NW. and W., and lakes Erie and Ontario, on the S. and SE. It is of a somewhat triangular shape, extending between lat. 43° and 46° 15' N., and long. 79° 40' and 85° W., surrounded, W. and SW., by the Michigan territory, and on all other sides by the territory of Upper Canada; and divided into two unequal parts by a long peninsula and the Manitoulin chain of islands, the parts to the N. and E., of which are called North Channel and Georgian Bay. The total length of Lake Huron, N. to S., is rather more than 200 m., and its greatest breadth about the same. Area estimated at 19,000 sq. m. Elevation above the surface of the ocean 606 ft., or less by 45 than that of Lake Superior, and by 4 than that of Lake Michigan. Greatest depth towards its W. shore at least 1,000 ft., and its mean depth is estimated at 900 ft., or about 300 ft. below the level of the Atlantic. In various parts it abounds with islands, their total number being said to exceed 32,000, the largest, the Great Manitoulin (Evil Spirit) island, is nearly 90 m. long, and in one part almost 30 m. wide. Lake Huron receives the superabundant waters of Lake Superior, by the river St. Mary, at its NW. angle, and those of Michigan at Michilimackinac; and discharges its own towards Lake Erie, by the St. Clair, at its S. extremity. Lakes Nipissing and Simcoe communicate with it by the Francis and Severn rivers, except which, however, Lake Huron receives no river worthy of mention. The banks of this lake are mostly low, especially along its S. and W. sides. Few towns of consequence exist on its shores, and its navigation is rendered dangerous by sudden and violent tempests.

HYDERABAD, a town and fortress of Hindostan, prov. Sinde, of which it is the cap., though not the largest city. It stands upon a rocky precipice upon an island formed by the Indus and the Fullelee, one of its tributaries, 48 m. NE. Tatta. Pop. estimated at 40,000 in 1843. Hyderabad has a station on the Sinde railway, and this, and a well-organised system of steam navigation on the Indus, extending over a length of 670 m., gives considerable importance to the town, and has led to a great increase of inhabitants. Hyderabad is famous for its fortress, which has an imposing appearance, and is considered very strong by the Sundians; but it could not oppose any effectual resistance to European troops. Its shape is an irregular pentagon; its walls, which are of brick, are about 25 ft. high, very thick at the bottom, but tapering to the top, and flanked with round towers from 800 to 400 paces apart. On one side it is enclosed by a ditch about 10 ft. wide and 8 deep. In its centre is a massy tower unconnected with the works, in which a great portion of the treasure of Sinde are deposited. Formerly, there were 70 pieces of cannon mounted on the ramparts, and 2,500 houses and several handsome mosques within the citadel; at present the fortress is a mere shell, and its walls are going rapidly to decay. N. of it is the suburb or unfortified town, in which most of the inhabs. reside in mud huts; there are, however, some well supplied shops. Hyderabad has manufactories of arms of different kinds, employing many of its inhabs.; and estates of embroidered cloths and leather.

HYDERABAD, a city of the Deccan, Hindostan, former cap. of the Nizam's dom.; on the Mussi, a tributary of the Krishna, 197 m. WNW. Masoliputam, 270 m. SE. Aurungabad, and 190 m. WSW. Bejapoor. Pop., including its suburbs, estimated at 200,000. The town is about 4 m. in length by 3 in breadth, and surrounded by a stone wall, capable of resisting the attacks of predatory cavalry, but no adequate defence against artillery. Streets narrow, crooked, and badly paved; houses mostly of one story only, and built of mud and other combustible materials. A large arched bridge, wide enough for two carriages abreast, here crosses the Moosh. The chief public buildings are the palace and numerous mosques, Hyderabad having long been the stronghold of Mohammedanism in the Deccan. Within the city are also some large magazines belonging to the Nizam, filled with European manufactures. Hyderabad (then called Bhaanuggur) was founded by Cuttub Shah, about 1585. It was taken and plundered in 1687, by the troops of Aurungzebe.

The territory of which Hyderabad once was the cap., known as the Nizam's dom., extends between the 16th and 21st degs. of N. lat. and the 75th and 82nd degs. of E. long.; embracing, together with the prov. Hyderabad and Berder, part of Bejapoor, Aurungabad, and Berar; having an area of 100,000 sq. m., with a pop. of at least 8,000,000. By all accounts, this territory was very badly governed by its native princes, and insurrections were frequent. But, according to a treaty made in 1800, it was provided that the military power of Great Britain should be employed not only in the suppression of rebellion, but also in the collection of the revenue. Since then, the country has been virtually British territory.

HYDRA, an island of the Grecian Archipelago, off the coast of Argolis, from which it is 6 m. distant; lat. 37° 20' N., long. 23° 30' E. Area, 50 sq. m. Pop. 25,000 in 1861. Hydra is a mere rock, so utterly barren as to contribute nothing whatever to the maintenance of its inhabitants, nor, in all probability, would it ever have been peopled, unless its insular situation and the excellence of its harbour had pointed it out as a safe place of refuge from the oppressions of the Turks, and a favourable situation for commercial pursuits. The town of Hydra, which, with the exception of two adjacent villages on the coast, is the only inhabited part of the island, is situated on the NW. side, and rises in successive tiers, like an amphitheatre, over the harbour, presenting from the sea an extremely beautiful prospect. The streets are precipitous and uneven; but the houses are most substantially built of stone, with spacious and well-furnished interiors, and are extremely neat and clean. The harbour, defended by a battery, is crescent-shaped, and, though small, is deep and safe; it is lined, through its entire sweep, with storehouses and shops, some of which, however, are now empty, only showing, by their number, the former consequence of the port. Several Greek churches (two of which have fine marble steeples) and a hall of commerce are the chief public buildings; and the educational establishments, instituted in the days of Hydra's prosperity, comprise a well-regulated college, for instruction in the classical Greek and the modern languages, several elementary schools, and a mathematical seminary. The commerce of Hydra before the war of independence was very considerable, employing, in 1816, according to Pouqueville, 120 vessels, and more at a later period, trading in wheat with Spain and Portugal, and in oil, wine, and other goods, with different ports of the Mediterranean; but it has now greatly fallen off, and in all probability

will never recover its former prosperity, having been chiefly transferred to the more advantageously situated ports of Nauplia and the Piræus. The Hydriots, most of whom are Albanians and not true Greeks, were, during their prosperity, which commenced in the beginning of the French war, the boldest seamen of all Greece, and acquired large sums by privateering. During the war of independence they earned for themselves the character of being the most efficient and intrepid sailors in the Greek navy, and their bravery contributed in no small degree to the successful issue of that contest.

HYMETTUS (MOUNT), a mountain of Greece, gov. Attica, 4½ m. ESE. Athens. Height 3,500 feet. The honey collected here has been in high repute in ancient as well as modern times.

HYTHE, a cinque port, parl. bor., market town, and par. of England, co. Kent, lathe Shepway, hund. same name, 15 m. E. Canterbury, and 59 m. SE. London, near the South-Eastern railway. Pop. of munic. bor. 2,001, and of parl. bor. 21,367 in 1861. The parl. bor. includes Sandgate, Folkestone, and four other small parishes. The town stands near the E. extremity of Romney Marsh, and consists chiefly of one long street, parallel to the sea-coast, which is about ¼ m. distant, the beach lying between being considerably higher than the town. The church, a cruciform structure, built in the early English style, and having two towers, is remarkable for its elegant architecture. There are also places of worship for Wesleyan Methodists and Independents, and national schools supported by subscription. The chief buildings are the court-house, gaol, and theatre. Hythe prospered during the great war with France, in consequence of the large military force quartered in the neighbourhood, and of the expenditure in the formation of the military canal, and of the forts and martello towers which this part of the coast is studded; but its prosperity has declined since it has ceased to be a military station. It has no manufactures; and the beach being open and exposed, the colliers, which are the only vessels trading to the town, are obliged to land their cargoes during the summer months. Fishing employs a few of the inhabs. The corporation, which received its constitution from the general charters granted to the Cinque Ports, especially that in 20th Charles II., has consisted, since the passing of the Municipal Reform Act, of 4 aldermen, one of whom is mayor, and 12 councillors. Corporation revenue 61l. in 1862. The bor. sent 2 mems. to the H. of C. from the 42nd of Edward III. down to the passing of the Reform Act, which deprived it of 1 mem.; previously to that act the franchise was vested in the freemen, made so by birth, marriage, or gift. The Boundary Act enlarged the limits of the parl. bor., by adding to it the liberties of Folkestone, and the parishes of West Hythe, Saltwood, Cheriton, and Newington. Registered electors 1,176 in 1865. Markets on Saturday; fairs July 30 and Dec. 1.

Hythe was formerly, and up to the reign of Henry VIII., one of the principal ports on the English Channel. The quota furnished by it towards the general armament of the Cinque Ports was five ships, with twenty-one men and a boy to each. Leland speaks of Hythe as 'a very great towne, two good miles in length all along the shore.' The name, in Saxon, signifies a port or haven.

I.

IBARRA, a town of Ecuador, Colombia, in a delightful plain, on the Tacuendo, at the foot of the volcano Imbarra, 50 m. NE. Quito, and on the high road between that city and Popayan. Pop. estimated at 12,000. It was founded in 1597, is well built, and has a large and well built church, several convents, a college, formerly belonging to the Jesuits, a hospital, and many good private residences. Without the city are some suburbs, inhabited by the Indian pop. It manufactures fine cotton and other fabrics. The district of which it is the cap. produces sugar and wheat of the finest quality, and a good deal of cotton, the weaving of which into stockings and caps, employs many of its inhab.

IBIALLA. See Braillow.

ICELAND, a large island under the dominion of Denmark, in the N. Atlantic Ocean, on the confines of the polar circle, generally considered as belonging to Europe, but which should, perhaps, be reckoned in America, between lat. 63° 20' and 66° 40' N., and long. 16° and 23° W. It is of a very irregular triangular shape, and is estimated to contain about 30,000 sq. miles. Pop. 54,803 in 1850. The population is spread over about two-thirds of the island, the central portion being totally uninhabited, and imperfectly explored. Iceland appears to owe its existence to submarine volcanic agency, and to have been upheaved at intervals from the bottom of the sea. It is traversed in every direction by vast ranges of mountains; the principal ridges run chiefly E. and W., and, from these, inferior mountains branch off towards the coasts, often terminating in rocky and bold headlands. All the coasts, but more especially the N. and W., are deeply indented with fiords, similar to those of Norway. The most extensive tract of level country is in the SE. It is estimated that about a third part of the surface is covered with vegetation of some kind, while the other two-thirds are occupied by snowy mountains or fields of lava. The general aspect of the country is the most desolate and dreary imaginable. The height of very few of the mountains has been correctly ascertained, and those said to attain an elevation of 7,000 feet are not the most lofty. The Vatkula, or enormous ice-mountains, are among the greatest elevations; the most extensive of these is the Klofa Yökul in the E. It lies behind the heights which line the SE. coast, and forms, with little or no interruption, a vast chain of ice and snow mountains covering a surface of perhaps 8,000 sq. m. The W. quarter contains, among other lofty heights, the Snafel Yökul, 4,580 ft. high. In the N. the mountains are not very high; but in the E. the Orvefa Yökul, 6,240 ft. in elevation, is the most lofty of which any accurate measurement has been obtained. The celebrated volcano Hecla is in the SW. quarter, and about 50 m. inland. It is more remarkable for the frequency and violence of its eruptions than for its elevation, which is only about 5,200 ft. (See Hecla.)

The bays and harbours along the coast are numerous and secure, but little known or frequented; the most so are those of Eyafiorde on the N., Fyrarbacka on the S., and Reikiavik on the W. coast. The rivers, which are numerous and comparatively large, have mostly a N. or S.

course. Although sufficiently wide, they are generally obstructed by rocks and shallows, and are too rapid to admit of navigation. There are several large lakes, of which Mivatn Lake, in the NE., is the most considerable; it is estimated at about 40 m. in circ., and has upwards of 30 islands composed of lava. In no country have volcanic eruptions been so numerous as in Iceland, or spread over a larger surface. Besides more than 30 volcanic mountains, there exists an immense number of small cones and craters, from which streams of melted substances have been poured forth over the surrounding regions; 9 volcanoes were active during the last century, 4 in the N., and the rest lying nearly in a direct line along the E. coast. Twenty-three eruptions of Hecla are recorded since the occupation of the island by Europeans; the first of these occurred in 1004. The most extensive and devastating eruption ever experienced in the island happened in 1783; it proceeded from the Skaptar Yökul, a volcano (or rather volcanic tract having several cones) near the centre of the country. This eruption did not entirely cease for about two years. It destroyed no fewer than 20 villages and 9,000 human beings, or more than one-fifth part of the then pop. of the island! On the S. and W. coasts numerous islands have been from time to time thrown up; some of which still remain, while others have received beneath the surface of the ocean, forming dangerous rocks and shoals. The Vestmanna Islands, which lie about 15 m. from the E. coast, are a group consisting almost entirely of barren vitrified rocks; only one of them is inhabited.

Tracts of lava traverse the island in almost every direction. This substance chiefly occurs in isolated streams, having apparently flowed from the mountains; but in some parts there are continuous tracts, and along the S. coast, for 100 m. inland, the lavas that spread over the country have been ejected from small cones rising immediately from the surface. The ground in this part is frequently broken by fissures and chasms, some of which are more than 2 m. in length, and upwards of 100 ft. in width. Besides the common lavas, Iceland abounds in other mineral masses indicative of an igneous origin; of these the most prevalent are tufa and submarine lava, obsidian, and sulphur. Whole mountains of tufa exist in every part. Sir G. Mackenzie observes, that the instance of tufa excepted, he saw no marks of stratification in any rock in the island, all the substances appearing to have been subjected to a degree of heat sufficient to reduce them to fusion; and thus some, if not all, the Icelandic masses, which are not the produce of external eruptions, are really submarine lavas. The rocks not bearing external marks of heat are mostly of trap, and contain all the varieties of zeolite, chalcedony, greenstone, porphyry, slate, &c.; the celebrated double refracting calcareous spar is found chiefly on the E. coast. Basaltic columns occur in many parts, especially on the W. coast, where they form several grottos; and that of Stappen bears a great resemblance to the cave of Fingal, in the island of Staffa.

Few metals are met with: iron and copper have

been found; but the mines are not wrought. The supply of sulphur is inexhaustible: large mountains are incrusted with this substance, which, when removed, is again formed in crystals by the agency of the hot steam from below. Large quantities were formerly shipped; but latterly the supplies sent to the foreign market have been comparatively small.

By far the most remarkable phenomena of Iceland are the intermittent hot springs met with in several parts, and of all degrees of temperature. The water in some of these springs is at intervals violently thrown into the air to a great height. They have therefore received the name of geysers, from the Icelandic verb geysa, to rage. The most celebrated of these springs are situated in a plain, about 16 m. N. from the village of Skalholt. The great geyser, or principal fountain of this kind, rises from a tube or funnel, 78 ft. in perpendicular depth, and from 8 to 10 ft. in diameter at the bottom, but gradually widening till it terminates in a capacious basin. After an explosion the basin and funnel are empty. The jets take place at intervals of about 6 hours; and when the water, in a violent state of ebullition, begins to rise in the pipe or funnel, and to fill the basin, subterranean noises are heard like the distant roar of cannon, the earth is slightly shaken, and the agitation increases till at length a column of water is suddenly thrown up, with vast force and loud explosions, to the height of 100 or 200 ft. After playing for a time like an artificial fountain, and giving off great clouds of vapour, the funnel is emptied, and a column of steam rushing up with great violence and a thundering noise, terminates the eruption. Such is the explosive force, that large stones thrown into the funnel are instantly ejected, and sometimes shivered into small fragments. (For an explanation of this phenomenon, see Lyell's Geology, ii. 304, 3d ed.) Some of the hot springs, near the inhabited parts of the island, are used for economical purposes; food is dressed over them; and in some places huts are built over small fountains, to form steam baths. In other parts of the island vast cauldrons of boiling mud are seen in a constant state of activity, sending up immense columns of dense vapour, which obscure the atmosphere a great way round.

That Iceland had formerly some extensive forests is apparent from authentic records, but they no longer exist; in fact, the climate seems to be now unsuitable for the growth of trees, those that are found at present being stunted and diminutive, and little better than underwood. Vast quantities of surturbrand, or fossil wood, are frequently found buried at a great depth beneath the surface.

Of the wild animals, foxes are the most numerous. Reindeer, which were introduced from Norway in 1770, with the intention of being domesticated, have increased very rapidly; but they are entirely wild, and are very difficult to kill. Bears are frequently brought down from the arctic regions on masses of floating ice; they sometimes commit great devastations, but are generally destroyed almost immediately after making the land. Nearly all kinds of waterfowl inhabit the coasts and islands; and plovers, curlews, snipes, and a variety of game, are found in the interior. The eider duck is very plentiful; and the down taken from the nest is an important article of export. The birds are so familiar as to build their nests all round the roofs, and even inside the huts. A severe penalty is inflicted on those who kill them. The peasantry entertain a superstitious reverence, mingled with aversion, for the seal. The coasts, rivers, and lakes pro-

duce an abundance of fine fish; and it is from the sea that the Icelanders derive great part of their subsistence. Their fisheries are prosecuted with great activity; and at Niardvik, one of the fishing stations on the E. coast of the island, there are said to be 800 boats. Cod and haddock are plentiful on the coasts; of these, as well as of the other seafish, part is salted for exportation, but by far the greater part is dried for winter provision. The herring fishery is much neglected, as well as the inland fishery on the lakes and rivers.

The climate is more variable than that of the same latitudes on the continent. Great and sudden changes of temperature often occur, and it has frequently happened that, after a night of frost, the thermometer during the day has risen to 70° Fah. The intensity of the cold is much increased by the immense quantities of floating ice, which, being drifted from the polar regions, accumulate upon the coast. Fogs are frequent; but the air, on the whole, is reckoned wholesome. Thunder is seldom heard, but storms of wind and rain are frequent; and the aurora borealis and other meteors are much more common and brilliant here than in countries further to the N. The sun is visible at midnight, at the summer solstice, from the hills in the N. parts of the island. There is a prevalent opinion in Iceland, that the seasons in former ages were less unfavourable; but there is probably no good foundation for this belief. The summers are necessarily short; but Dr. Henderson states that the cold is rarely more intense than in the S. of Scandinavia, and the winter he passed in the island was as mild as any he had experienced in Denmark or Sweden.

No grain is now cultivated, though traces exist of its having been formerly raised. Agriculture is limited to the rearing of various grasses for cattle, and haymaking is consequently the most important branch of rural industry. Potatoes have been introduced with some success; and several kinds of culinary vegetables are raised, but, with the exception of red cabbage, few attain perfection. The grasses are of the sorts common in other N. climates, and keep horses and other cattle in good condition during the summer. Many of the low mountains are covered with a coarse grass, which yields pretty good summer pasturage; and the meadows and valleys through which the rivers flow produce grass in tolerable abundance, which, when the weather allows of its being harvested, is made into hay. Seaweed and moss are eagerly devoured by the cattle in winter, when other food fails, which is often the case. It is estimated that there are about 800,000 head of sheep; from 36,000 to 40,000 head of black cattle; and from 30,000 to 60,000 horses in Iceland; goats are kept only in the N. The number of sheep appears to be increasing; they have remarkably fine fleeces, which are not shorn, but cast off entirely in the spring. The horses are hardy and small, seldom standing more than 14 hands high. There being no carriages of any description, they are practically used for carrying burdens; and the poorest peasant has generally 4 or 5 of these animals. Rents are paid mostly in produce; on the coasts in fish, in the interior in butter, sheep, and other agricultural produce. Tenants who are in easy circumstances generally employ one or more labourers, who, besides board and lodging, have from 10 to 17 specie dollars a year as wages. The whole pop. is employed either in fishing or feeding cattle, or both; those who breed cattle being, as compared with those who live by fishing, nearly as 3 to 1.

No manufactures, of any kind, are carried on for the purposes of trade. Every branch of industry is domestic, and confined chiefly to articles of clothing, such as coarse cloth, gloves, mittens, and stockings. The peasantry supply themselves with such furniture as their cottages require, and some manufacture silver trinkets and small wares, and forge implements of iron. Every man can shoe his own horse; and, in this land of primitive simplicity, even the bishop and chief justice are sometimes employed in this necessary occupation. The greater part of the trade is carried on by means of barter; the quantity of money in circulation is very small, few of the peasants possessing any. The merchants receive the articles for exportation at regulated prices, according to the state of the market, and pay for them in such foreign commodities as the inhabitants may require. The peasantry of the neighbourhood assemble annually at Reikiavik and the other principal settlements, and bring down with them wool, woollen manufactured goods, butter, skims, tallow, Iceland moss (*Lichen Islandicus*), and sometimes a few cattle. In return for these they take back coffee, sugar, tobacco, snuff, a little brandy, rye, rye bread, wheaten flour, salt, and soap. The better class purchase linens and cotton goods, which have latterly come more into use. Those who live near the coasts bring to market dried cod and stock fish, dried salmon, whale, shark, and seal oils, and seal skins. The domestic produce has, of late years, been considerable, and the export of wool amounts to from 3,000 to 4,000 shippounds annually.

The Icelanders are of Norwegian origin; they are tall, have a frank open countenance, a florid complexion, and flaxen hair. They seldom attain to an advanced age, but the females generally live longer than the men. They are hospitable, devotedly attached to their native land; remarkably grave and serious; and, indeed, apparently phlegmatic, but extremely animated on subjects which interest them. They have retained, with few innovations, the ancient modes of life and the costume of their race. Their principal articles of food are fish, fresh and dried, bread, made of imported corn, great quantities of rancid butter, game, and, in some parts, a porridge made of the Icelandic moss. They sometimes use the flesh of the shark or sea-fish, when it has become tender from putrescence. Their huts, though larger, are not unlike those of the Irish: their dampness, with the darkness, filth, and stench of the fish, render them uninhabitable by strangers. The Icelandic, or original Scandinavian tongue, has been here preserved in all its ancient purity. The Icelanders are extremely attentive to their religious and domestic duties, and display in their dealings a scrupulous integrity. Perhaps there is no country in which the lower orders are so well informed. Domestic education is universal; and there are very few among them who cannot read and write, and many among the better class would be distinguished by their taste and learning in the most cultivated society of Europe. Even many of the peasantry are well versed in the classics; and it is reported that the traveller is met unfrequently attended by guides who converse with him in Latin. In winter nights it is customary for a whole family to take their places in the principal apartment, where they proceed to their respective tasks, while one, selected for the purpose, reads aloud some of their sagas (ancient tales), or such other historical narrative as can be found. Their stock of books is not large, but they lend to each other, and frequently copy what they borrow.

The island was formerly divided into four amts, or provinces, answering to the four cardinal points. The N. and E. are now merged into one, and the W. is presided over by the governor in person. This officer has the title of *stiftsamtmand*: he is sometimes a native, but more frequently a Dane. Under him are the amtmen, or provincial governors, who possess a similar jurisdiction over their quarters. Each province is divided into *syssels* or *shires*, presided over by *syssel-men*, with authority similar to that of sheriffs; these collect taxes, hold petty courts, and regulate assessments. Under the *syssel-men* are *repp-men*, who are overseers of the poor, and constables. The *landvoed*, or chief justice, resides at Reikiavik, but very few cases are tried in the island, and all capital punishments are inflicted at Copenhagen. Crimes are rare, petty theft and drunkenness are the most common; the latter has been introduced chiefly by the crews of the Danish vessels that visit the coasts.

The island constitutes one bishopric; the bishop's salary does not exceed 500l. per annum. There are about 191 pars.; but the clergy amount to upwards of 300; their incomes are very small, and they are frequently among the poorest of the community. The only charitable institutions are four hospitals, for the reception of those afflicted with leprosy, which, in the form of elephantiasis, was formerly very prevalent. Small-pox was formerly also very destructive. There are no workhouses, the sick and poor being almost universally supported by their own families. The principal school at Bessestad, near the W. coast, has three masters, who teach classics, theology, and the Danish language; and several young men, after attending this school, go to Copenhagen to finish their studies. Reikiavik, the capital on the SW. coast, has little more than 500 resident inhabitants, chiefly Danes. Most of the villages are situated on the coasts, at convenient spots for the receipt and transport of merchandise.

The early and successful application of the Icelanders to the cultivation of literature is an anomaly in the history of learning. When most parts of continental Europe were in a state of rude ignorance, the inhabitants of this remote island were well acquainted with poetry and history. The most flourishing period of Icelandic literature appears to have been from the 12th to the end of the 14th century. During the last three centuries, however, Iceland has produced many learned men, some of whom have risen to great eminence. The literature of the island in the present day may perhaps be said rather to have changed its character than declined from its ancient fame; the inhabitants now attend more to solid branches of learning than to the poetical and historical romances of the ancient Icelandic sagas. Domestic education is carefully attended to: there is no want of modern books in Icelandic; and a printing press is actively employed in the island of Videy.

The discovery of Iceland by Europeans is attributed to a Norwegian pirate, about the year 861; but the earliest permanent settlement was effected by the Norwegians in 874. In little more than half a century, all the coasts were occupied by settlers; and about the year 928 the inhabitants formed themselves into a republic, and established the *Althing*, or General Assembly of the Nation, which was held annually at Thingvalla, in the SW., and not abolished till 1800. The Icelanders maintained their independence for nearly 400 years; but during the 13th century became subject to Norway, and on the annexation of that kingdom were transferred with it to Denmark.

IDRIA, a town of the Austrian empire, in Illyria, duchy Carniola, circle Adelsberg, in a valley of the Carnic Alps; 28 m. W. by N. Laybach.

Pop. 4,500 in 1857. The inhabs. are principally engaged in mining; the quicksilver mines of Idria belonging to the Austrian government being, after those of Almaden in Spain, the richest and most celebrated in Europe. They yield annually from 2,500 to 3,000 cwt. of metal, about a sixth part of which is converted on the spot into vermilion, corrosive sublimate, and other preparations of mercury. The mine is rather more than 1,800 ft. in depth. The formation in which it is situated is transition limestone, alternating with clay-slate, in which latter rock the quicksilver is found. It exists partly pure, in globules among the slate; but it is mostly found in combination with sulphur, forming veins of cinnabar, which vary greatly in thickness. The cinnabar ore is considered too poor to be wrought when it contains only from 14 to 18 per cent. of quicksilver, and is then usually abandoned in search of a better vein. The richest ore yields from 50 to 70 per cent. of metal. From 600 to 700 workmen are employed, of whom about 600 are miners. These are enrolled in a corps, and have a regular uniform. They are divided into three sections, which relieve each other, each working below for 8 hours in the 24, the work incessantly going on. Within his 8 hours, the labourer is required to perform a certain measurement of work, for which he receives 17 kreutzers (nearly 7d.). If he performs less or more than his measured extent, his pay is proportionally reduced or increased; but the number of those who gain less than the fixed sum is greater than of those who gain more. Besides their money pay, the miners get an allowance of corn sufficient for themselves and their families; and in illness, gratuitous medical aid. No lodging is found them; but they may purchase at a government store a number of articles of prime necessity, at fixed charges, generally below the ordinary market prices. The miners usually enter the service at 15 years of age. After 40 years' service, or earlier, if ill health overtake them, they are allowed to retire on full pay, and enjoy various privileges. The widows and orphans of miners are entitled to a pension, and about 35,000 florins are thus expended annually. The process of mining is very unhealthy; the heat of the mine, varying from 80° up to 86° Fah., impregnates the atmosphere with volatilised mercury, which soon exerts all its characteristic effects on the constitutions of the miners. In some parts, the heat is so great, and the atmosphere so vitiated, that the workmen are obliged to relieve each other every two hours. The mine is very clean, and in its lower parts remarkably dry. In 1803, a violent conflagration broke out in the mine, destroying the whole of the works, with several of the workmen.

Of the mercury produced at Idria a small part goes to Trieste, whence it is exported chiefly to America; but by far the largest portion is sent to Vienna, partly for the plating of mirrors, but principally for the use of the gold and silver mines of Hungary and Transylvania.

At the beginning of the present century, Idria was a place of banishment for state prisoners and criminals, who were condemned to work in the mines. It is so no longer; no coercion is used, and no convicts are sent thither; the supply of labourers petitioning to be admitted is considerably greater than can be received into the service. The town and district of Idria is a mining intendency, with its own government, consisting of a director-general, an imperial comptroller of accounts, a secretary-general, and four comptrollers, who superintend all the departments of the public service, under the council of mines in Vienna. Idria has some

German, primary, and other schools, and a small theatre. It had a school for instruction in mining, but it was abolished on the restoration of the Illyrian provs. to Austria. The aspect of the place is thus described by a traveller who visited it not many years ago. 'We perceived the white church with its little steeple, perched on a small green knoll, and not far from it another insulated height, crowned with an antique-looking castle, erected by the Venetians during the time that they possessed Illyria, and which now serve as a residence for the Bergrath, or director of the mines, and for the government officers connected therewith. Between these two heights, the town straggles along on very unequal ground, with a stream rushing through it, a second church in a sort of open market place, some large buildings connected with the public administration, but scarcely any good shops or private houses.' The mine was discovered by accident in 1497; it was afterwards wrought by a company of Venetian merchants, and purchased by the house of Austria, who accorded the miners considerable privileges in 1575, since which the prosperity of Idria has been generally on the increase.

IGUALADA (an. *Aqua latæ*), a town of Spain. prov. Barcelona, 37 m. NW. Barcelona, and 250 m. ENE. Madrid. Pop. 12,439 in 1857. The town stands on the Noya, a trib. of the Jorel, in a rich plain, abounding with corn-fields and olive-grounds. It has some well-built streets, and a handsome suburb, the chief buildings being a par. church, two convents, a clerical college, hospital, and cavalry barracks. The inhab. are among the wealthiest and most industrious in Spain; and their manufactures, by which they are almost wholly supported, comprise cotton and woollen yarns and cloths, hats, and fire-arms, the last of which are highly esteemed. In the neighbourhood are several considerable paper-mills. Fairs, well attended, for manufactured produce, are held here in the beginning of January and at the end of August.

ILCHESTER, a bor., market town, and par. of England, co. Somerset, hund. Tintinhull, on the Yeo or Ivil (whence its name is derived), 16 m. E. Taunton, and 116 m. WSW. London. Pop. of par. 781 in 1861, against 1,095 in 1831. The town comprises 4 indifferently built streets, and has but few public buildings. The church is remarkable for its octangular tower. A national school and almshouses for 16 women are the only public charities. The co. court-house is handsome, and conveniently arranged. The gaol, built on Howard's plan, is large and well regulated, and capable of accommodating upwards of 200 prisoners, and was often quite full, when employed, as formerly, for a state prison and house of correction: it is now chiefly used for untried prisoners and debtors. The town, which has no manufactures and little trade, derives its chief importance from the fact, that a large portion of the county business is transacted here, the assizes being held at Ilchester alternately with Taunton, Wells, and Bridgewater. It is altogether, however, in a low, declining state. Ilchester is a bor. by prescription, and sent 2 mems. to the H. of C. from the 26th of Edw. I. down to the passing of the Reform Act, when it was disfranchised: it was a mere nomination bor., in the patronage of the Duke of Cleveland.

Distinct traces of a Roman station, and the discovery of numerous Roman coins and antiquities, have led to the belief that this town occupies the site of the *Ischalis* of Ptolemy, the principal military station of the Romans in the West of England. It had 108 burgesses at the time of the Norman Conquest. Still later, it was a place of considerable consequence, and was made, by patent of Edw. III., the assize town of Somerset.

ILDEFONSO (ST.), or **LA GRANJA**, a celebrated palace of the sovereigns of Spain, Old Castile, prov. Segovia, 42 m. NNW. Madrid, and 6 m. SE. Segovia, on the N. declivity of the Sierra Guadarrama, built by Philip V., as a place of retirement during the hottest months of summer. It is placed in a spot where the mountains fall back, leaving a recess sheltered from the hot air of the S. and from much of its sun, but exposed to whatever breeze may be wafted from the N.; the immediate activity towards the S. being occupied by the garden, which, though somewhat formal, is full of shade and coolness. The palace, which is of brick, plastered and painted, occupies three sides of a square, in the centre of which is the royal chapel. The principal front, looking towards the garden is 530 ft. long, having 2 stories, with 19 rooms in a suite; the great entry, with its iron palisade, very much resembling that of Versailles. The interior is, in every thing, regal: the ceilings of the apartments are painted in fresco, the walls decorated with noble mirrors, and the floors chequered with black and white marble, while the furniture, though somewhat antiquated, is highly enriched with jasper, verd-antique, and rare marbles. The upper rooms are adorned with the works of the first masters, chiefly of the Italian school, the lower apartments being used as a repository for sculpture. Many, however, of the best specimens once belonging to this palace, both in painting and sculpture, have been removed to the royal gallery of Madrid. The gardens are laid out in the French style, with formal hedges and walks; and the trees, notwithstanding the labour with which the formation of these grounds was attended, are poor and starved; the chief feature, indeed, in these gardens is the quantity of fine water, disposed in a variety of ways, and especially in the formation of fountains and works. The expense of constructing the garden alone, a large part of which was made by blasting out of the solid rock, must have been very great; and the entire expenditure on the palace gardens and water-works is stated to have exceeded 6,000,000l. In the town of St. Ildefonso (pop. 1,815 in 1857), which lies a little distance below the palace, is a manufactory of mirrors, supported by the government.

ILFRACOMBE, a sea-port, market town, and par. of England, co. Devon, hund. Braunton, on the Bristol Channel, 9 m. N. Barnstaple, 41 m. NW. Exeter, and 172 m. W. by S. London. Pop. of town 8,031, and of par. 3,851 in 1861. Area of par. 3,620 acres. The town, consisting of one long street and a noble terrace facing the sea, extends W. from the harbour along the shore. The church, which stands at its upper end, is a large plain building containing some fine monuments. There are places of worship for Independents and Wesleyan Methodists, a large national school, and a girls' school of industry. The harbour is a natural basin formed by the curve of a very rocky shore, and a bold mass of rocks stretching nearly half way across the entrance of the recess shelters it from the northern storms. A battery and lighthouse stand on the top of this rocky mass, and the harbour is further defended by a pier 850 ft. in length, which is kept in excellent repair. There is safe anchorage for vessels of 250 tons, and ships can easily enter here when they cannot get up the Taw to Barnstaple; the consequence of which is, that Ilfracombe has taken away a great part of its coasting trade. The trade with Bristol,

Swansea, and other ports in the Bristol Channel, is considerable, and many vessels are employed in the herring fishery. The town, however, depends, in a great measure, for its support on the numerous wealthy families that resort thither in summer, since it has attained celebrity as a watering-place. The bathing is excellent, and the neighbourhood abounds with romantic scenery. Steam-packets run daily to and from Swansea, Tenby, and Milford. The town is governed by a portreeve appointed by the lord of the manor. Markets, well-supplied with fish, on Saturdays; fairs April 14, and the first Saturday after Aug. 22.

ILLE-ET-VILAINE, a marit. dep. of France, in the NW. part of the kingdom, formerly included in the prov. of Brittany; having W. Côtes-du-Nord and Morbihan, S. Loire Inférieure, E. Mayenne, and N. La Manche and the English Channel. Length, N. to S., about 70 m. Area, 672,583 hectares. Pop. 584,930 in 1861. The Menez mountains run through this dep. from E. to W.; but they rise to no great height, and the surface elsewhere is not hilly. The chief river is the Vilaine, which has mostly a SW. course, and falls into the Atlantic in the dep. Morbihan: the Ille is one of its affluents. The Rance, which has its mouth in this dep., is connected with the Ille by a canal, extending from Dinan to Rennes, 52 m. in length, and wide and deep enough for vessels of 70 tons. Climate temperate, but very damp; fogs are frequent, and from 36 to 38 in. rain fall annually. Soil thin, and not generally fertile. About 597,496 hectares of land are arable, and 73,340 in pasture; forests, heaths, and waste lands occupying 146,070. Agriculture is in a backward state. Throughout the greater part of the dep. the land is parcelled out into small farms, one of 30 hectares being considered large. Principal crops, rye, oats, and barley; the dep. is not so suitable for wheat; and but little maize is grown: the annual quantity of grain produced is about 3,436,000 hectolitres, which is scarcely sufficient for home consumption; and the peasantry add to their corn chesnut flour, potatoes not being in general use; 18,200 hectares are in gardens and orchards; fruit is plentiful, and some very good cider is made; but the agricultural products of the greatest importance are flax and hemp, and the linen thread of the dep. is very highly valued. Both cattle and horses are of good breeds; many oxen from this dep. are fattened in Normandy for the Paris market. Dairy husbandry occupies a good deal of attention, and the beurre de Prévalaye, made in the neighbourhood of Rennes, is highly esteemed throughout France. The sheep are of an inferior kind. The sole, cod, mackerel, and other fisheries on the coast are extensive; and Cancale Bay is celebrated for its oysters, with which Paris is in great part supplied. From 50 to 60 boats go annually from this dep. to the cod fishery of Newfoundland. Some copper, iron, argentiferous lead, and coal mines, and quarries of marble, granite, slate, and limestone are wrought, but not to any great extent. The manufactures consist chiefly of hemp and linen thread, packing and sail-cloth, cordage, flannels at Fougères, and leather. In the arrond. of Fougères there is a large government glass factory, some of the products of which are equal to any made in Lyons. The dep. is divided into six arronds.; chief towns, Rennes, the cap., St. Malo, Fougères, Redon, Montfort, and Vitré. This part of Brittany has produced many celebrated men, including M. de la Bourdonnaye, Maupertuis, Savary, Vauban, Chateaubriand, and Broussais.

ILLINOIS, one of the U. States of America, the fourth in the Union in point of extent; between lat. 37° and 42° 30' N., and long. 87° 30' and 91° 30' W., having N. the Wisconsin territory, E. Lake Michigan and Indiana, S. Kentucky, from which it is separated by the Ohio river, and W. Missouri and the Sioux territory, the Mississippi forming the whole of its boundary on that side. Length, N. to S., 380 m.; average breadth about 155 m. Area 55,409 sq. m., 54,000 of which are supposed to be susceptible of cultivation. Pop. 1,711,951 in 1860. In the N., its surface is uneven and broken, and in parts of the S. also it is hilly; but, on the whole, next to Louisiana and Delaware, Illinois is the most level state in the Union. It consists mostly of vast undulating prairies, or rich plains, called by the settlers 'barrens,' producing stunted oak, hickory, pine, and other trees. Many tracts in the S. are densely wooded, especially those lying along the rivers; and the prairies are sometimes interspersed with copses, though much more frequently studded with isolated trees at short distances. The state is well watered; next to the Mississippi and Ohio, the chief rivers are the Illinois, its tributary, the Sangamon, the Kaskaskia, Great Wabash, and Rock River. The Illinois rises in the NE. part of the state, and intersecting it in a SW. direction, falls into the Mississippi 25 m. above its junction with the Missouri, after a course of 450 m., most part of which is navigable for steam-boats. The Sangamon has a course of about 100 m., with a boat navigation of 120 or 130 m. The Kaskaskia rises in the centre of the state; runs with a SW. course for nearly 300 m., and falls into the Mississippi 150 m. below Vandalia, to which city it is navigable. The Great Wabash belongs more properly to Indiana, but it forms the lower 2-3ths of the E. boundary of Illinois, and falls at its NE. angle into the Ohio. The Rock River runs through the NW. portion of the state. It has a SW. course, like the Illinois, Kaskaskia, and other tributaries of the Mississippi, which river it enters about lat. 41° 30', after a course of nearly 400 m., for about 200 of which it is navigable. The total length of the navigable rivers is estimated at 4,000 m. Small lakes are numerous, and in the N. is Winnebago Swamp, a considerable extent of marsh-land.

In the W., and probably throughout most of the central and N. parts, the geological strata succeed each other in the following order: — a vegetable mould from 9 to 30 in. in depth, clay, limestone, shale, bituminous coal, generally from 4 to 5 ft. thick, sandstone, and sandstone. Limestone appears to be a universal formation; and coal and sandstone are found almost everywhere. In the NW. a mineral district, very rich in lead, &c., extends for 100 m. N. and S., by a breadth of half that distance, communicating with a tract of a similar character across the Mississippi. The smelting of lead ore on the banks of the Rock River began only in 1822; but it rapidly increased, and the produce of that metal is at present estimated at 20,000,000 lbs. a year. After lead, iron, copper, coal, salt, and lime are the chief mineral products. Copper and iron are found in various parts. The salt springs near Shawneetown yield 60 lbs. of table salt from 160 galls. of water. Other salt springs, and sulphurous and chalybeate mineral waters, are found in many places. The climate is healthy, except in the marshy tracts along the rivers or elsewhere. The winter is, in most parts, short and mild; and the summer heat not oppressive. Probably no portion of the territory has a mean annual temperature of more than 51° Fahr.; and the mean of the state at large is not above 51°.

This state is supposed to possess a larger proportion of first-rate cultivable land than any other in the Union. All the grains, fruits, and roots of temperate regions grow luxuriantly; and in none of the W. states is corn raised with greater facility and in more abundance. Wheat yields a good and sure crop, especially on the banks of the Illinois and in the N. It weighs upwards of 60 lbs. a bushel, and is preferred in the markets of New Orleans to the wheat of Ohio and Kentucky. Indian corn is a great staple, and hundreds of farmers grow nothing else. Its average yield is 50 bushels an acre, and sometimes the produce amounts to 75 or even 100 bushels. Oats, barley, buckwheat, common and sweet potatoes, turnips, rye for horsefeed and distilleries, tobacco, cotton, hemp, flax, the castor bean, and all other crops common in the middle states are raised. Hemp is indigenous in the S., and cotton is grown both for exportation and home use. Fruits of various kinds are very abundant, and the climate of the S. is favourable to the growth of the vine. Great numbers of cattle are reared in the prairies, and hogs in the woods. Sheep generally thrive well; but little has been done to improve the breed by crossing. Poultry are abundant, as are also bees, and the silkworm succeeds well. Deer roam the prairies in large herds. In the Military Bounty tract, in the NW., large tracts of land of the best quality may be had at the government price of 1¼ dollar an acre. This tract was, at a former period, mostly appropriated, by the general gov., in grants to the soldiers who served in the war against Great Britain; but a great part of it has again come into the possession of the gov., having been resumed for arrears of taxes, or disposed of by those to whom it had been granted. All lands in this state purchased of the general gov. are exempted from taxation for five years after purchase.

Many large and flourishing settlements have been formed in its W. part since the introduction of steam navigation on the Mississippi; these, however, are almost exclusively agricultural. Manufactures are not very numerous, and principally domestic. In every town and county artisans for all the trades of prime necessity are to be met with; and boat-building is carried on to some extent on the Mississippi. Grain, cattle, butter, cheese, and other agricultural products form the chief articles of export; and sugar, tea, coffee, wines, woollen cloths, and other manufactured goods are the chief imports. The external trade is carried on principally through New Orleans, to which emporium the articles of export are forwarded by the Mississippi, the imports being also received by the same channel. Illinois presents great facilities for a most extensive system of inland navigation, and much has already been accomplished to forward this object. In 1828, the legislature granted 300,000 acres of land for the construction of a canal to unite Lake Michigan with the head of the steam navigation on the Illinois. This canal, which was begun in 1836, runs from Chicago to the town of Peru, a distance of 95 m. Several sums of money have been also appropriated by the government for the improvement of the river navigation. The state is crossed, in all directions, by lines of railway, the total length of which is nearly 2,000 m.

Illinois is divided into 70 cos., in 60 of which courts are held. Vandalia, on the Kaskaskia, was the cap. till, in 1837, the seat of government was removed to Springfield, near the centre of the state — birthplace of the great President Abraham Lincoln. Jacksonville, Chicago, Kaskaskia, and Albion are the other chief towns. The legislative part of the government is vested in a senate, composed, in 1865, of 25 members, chosen for 4 years; and a house of representatives, having in the same year 85 members. All white male inhabitants above the age of 21, having resided in the state for 6 months, are privileged to become electors. Elections for representatives and the sessions of the legislature are held biennially. The executive duties are discharged by a governor and a lieutenant-governor, chosen by universal suffrage every 4 years. The high judicial functions are exercised by a supreme court composed of a chief justice and three inferior judges. The governor and judges of the supreme court constitute a council of revision, to which all bills that have passed the assembly must be submitted. If objected to by the council of revision, the same may, notwithstanding, become law by the vote of the majority of all the members elected to both houses. Slavery does not exist, having been prohibited by the constitution of 1818. A 36th part of every township of land, and a tax on some reserved lands belonging to the U.S. government, have been appropriated for public instruction, the funds of which amounted, in 1865, to 4,973,842 dollars. A college, founded at Jacksonville, occupies two extensive buildings, and many other lyceums and seminaries are established in different parts of the state.

During most part of the 18th century the name of Illinois was applied to all the country N. and W. of the Ohio. The territory comprised in the present state was discovered in 1670 by a party of French colonists, who made their first permanent settlements at Kaskaskia and Cahokia in 1673. This tract of country was ceded by the French to the English at the same time with Canada, in 1763, and by Virginia to the U. States in 1787. It was admitted, as a state into the Union, on the 3rd of December, 1818; and sends 14 members to congress, under the census of 1860.

ILLYRIA (KINGDOM OF), a territory forming part of the Austrian empire, comprising the provs. of Carinthia, Carniola, and Istria, the islands of the Gulf of Quarnero, and the Illyrian Littorale. It lies between lat. 41° 25' and 47° 7' N., and long. 13° 14' and 16° E., having N. Austria and Styria; E. the latter prov., and Croatia; W. the Tyrol and Italy; and S. the Adriatic Sea. It is divided into the govts. of Laybach and Trieste.

The northern part of Illyria is covered by the central chain of the Alps, and likewise by various offsets, constituting the southern limestone girdle of the Alpine system. The N. portion of the kingdom, comprising the gov. of Trieste, occupies the S. slope of this mountain-range towards the Adriatic. The main chain at the Gross Glockner (14,000 ft. high) takes the name of the Noric Alps, stretching its lofty peaks, here called Tauern, as far as the Ankogel, 10,431 ft. high. All this region contains extensive ice fields and glaciers. At the Ankogel the Noric Alps, taking a N.E. course, enter Styria; but a branch bounds the vale of the Drave on the N., and that of the Lavant on the E., separating their waters from those of the Mur. The Carnic Alps form the S. boundary of the valley of the Drave, dividing it from that of the Save. Various summits in this chain are from 6,000 to 8,000 ft. high; and over one of them, the Loibel, the emperor Charles VI. constructed the road connecting the Drave and the Save valleys; its summit-level is 5,477 ft. above the sea. At Mount Terglou, the Julian Alps break off, running N.E. towards the Adriatic and Dalmatia; E. of Idria they decline in height, forming an elevated plateau, remarkable for drought and sterility, owing to the porous nature of its constituent limestone. Besides the pass over the Loibel, various others

connect the fruitful valleys of this romantic country, the most remarkable being the Katscher, 5,280 ft. high, between the Drave and the Langau; the Wurzen, 3,100 ft., and the Pass of Tarvis, 2,800 ft., leading from the valley of the Drave to that of the Tagliamento. The valleys of the Gail (an. Vallis Julia), the Lavant, and Saan (Vallis Jaunia), in Carinthia, and of the Save and Wocheim in Carniola, offer all the varieties of Alpine beauty, while in the N. those of the Isonzo and Wippach, especially the former, present a picture of the richest Italian cultivation. The only level tracts of any considerable extent lie N. of the Julian Alps towards the Adriatic, and in the Istrian peninsula.

The Carnic and Julian Alps are perforated by very numerous subterranean cavities, which, by draining the surface of water, condemn whole districts to a melancholy sterility. Several of these caverns are celebrated for their great size and curious natural phenomena, as the cave of Adelsberg in Carniola and the neighbouring Magdalen Cavern, in which the *Proteus Anguinus* is found. Through several of these the mountain torrents find subterranean channels, to the great detriment of agricultural prosperity. (See ADELSBERG.)

The N. portion of Illyria is well watered. The Dran or Drave, rising in Tyrol, traverses Carinthia in all its length, and receives tributaries from both the N. and S. mountain barriers of that province. It is navigable from near Klagenfurt to its mouth in the Danube. The river second in importance is the Sau, or Save, which traverse Carniola with an E. course parallel to that of the Drave. The banks of the Upper Save are mostly level; but the mountains close in on the river near Reichenberg. It is navigable from near Laybach; and receives various affluents, both in Carniola and Croatia. The rivers falling on the S. side of the Alps to the Adriatic, are the Isonzo, Ausa, and Timavo. The Isonzo, traversing the beautiful vale of Friuli, and taking near its mouth the name of Sdobba, falls into the sea near Monfalcone. The Ausa falls into the sea near Buso; and the Timavo (Timavus), with a course of scarcely more than 1,000 yards, is navigable up to its source. Istria is very scantily watered; the Quieto, its principal stream, falls into the sea near Cittanova, and, as well as the Arsa, on the E. side of the peninsula, is navigable for some miles of its course.

There are several lakes in the N., but none of any great extent. The lake of Klagenfurt, 1½ m. long, is united with the neighbouring city by a canal. At a short distance from it is the Ossiach lake, 7 m. long, and connected with the Drave by the Laybach. Further NW. lies the Millstadt lake, 10 m. in length, and 1 m. broad, with very picturesque banks. The Weissensee, the Feldersee (an. Lacus Auricius), and, lastly, the remarkable Zirknitzer-see, are of smaller extent. The lake of Zirknitz has 2 islands, and receives its waters through subterranean channels. During the spring and the autumn rains, it presents a sheet of water 4 m. long, and 1 m. broad; but in summer the waters recede, and leave a dry fertile surface, either used for hay, meadows, or raising summer corn. The openings by which the water rises and retires are then visible, and various names have been given them by the peasantry; such as Kottse (the kettle), Reteka (the rack), Rezin (the corn sieve), Rewheto (the great sieve), Nittasze (the hair sieve), &c. When the lake is full it has an abundance of fish, which disappear and return with the water. In Istria there is only one lake, that of Zypristch, near Chersano. The climate of Carinthia is most inclement. The mean temp. of the year at Klagenfurth is estimated by Blumenbach at 7° Réaum.; while, at Obervillach, the mean is 6°. The snow lies in the lower parts of the valley of the Drave till the middle or end of April; but in the valley of the Save the climate is much milder. At Laybach the temperature of the year is 8·7 Réaum. The temperature of the government of Trieste presents a great contrast to that of the mountain districts. In the valley of the Isonzo, as well as in Istria, the olive, vines, and other productions of a southern climate, are largely cultivated.

Occupations of the People.—Agriculture.—Illyria has two distinct agricultural systems; that of the N. government, which is Alpine, and that of the S. districts, which are cultivated in the Italian fashion. The mountainous districts of Carinthia, situated in a cold and damp climate, and having a short summer, are tilled with difficulty. Rye and summer corn are the most usual crops; and the three-course system, according to which 1-3rd part of the land is in fallow, is generally prevalent. The corn, in order to dry thoroughly, requires to be hung up on poles or railings, of a peculiar construction; and these erections (called *Harfen*, Germ., and *Nag* or *Kasar*, Slav.) are often covered with a roof like that of a house. The most productive corn region is the valley of the Lavant, and the district of Krappfeld. In the higher parts of the valley of the Drave, near Gottschee, the climate is so severe as not to allow of winter crops. Cambiola, on the other hand, especially the valley of the Save, and the circle of Istria, has a warm climate, and is highly cultivated. Excellent wheat and maize, especially the 'corn-quantino,' are grown to a great extent; and there is a judicious rotation of crops.

Good flax is grown in all the valleys, and hemp chiefly in Friaul. Fruits of all kinds, especially chesnuts (*maroni*) and figs, are abundant in the coast district. The best wines are those of Montfalcone and Prosecco, grown near Trieste; but very little wine is exported. The oil of Istria is considered equal to that of Provence. The stones and refuse of the olive are used for fuel, and are even exported to Austria. The olive is also extensively cultivated in the Quarnero islands, especially Veglia and Cherso.

The chief wild animals of the northern districts are the chamois, red deer, and roebuck, and less frequently the wolf, bear, and small lynx. In the N. parts the ortolan and the common partridge, quails, water-fowls, and birds of passage are common. The fishery in the Gulf of Quarnero, and in the channels between the islands, furnishes an abundance of fish peculiar to those waters.

Mines.—The chief wealth of Illyria consists in the rich metallic veins found in its mountains. The N. mountain chain separating Carinthia from Styria consists of transition formations, overlying mica slate, which composes the great spine of the Norie Alps, and contains vast quantities of a very superior iron ore. This chain opens S. into several valleys, sending tributaries to the Drave; and in these secluded districts the various mining operations are carried on, favoured by the water-power afforded by the mountain torrents. In the valleys in the Lieser, Gurk, Glan, Metnitz, and Lavant, iron is the chief product. The mountains near Huttenberg are rivalled in productiveness only by the most prolific of the Swedish veins. The ore is chiefly the carbonate of iron. The average annual produce of iron amounts to 500,000 cwt., and of coal to 100,000 cwt. Lead is found to the extent of 70,000 cwt. per annum. But the quantity of metals and minerals annually produced corresponds neither with the wealth of the mines nor with the wants of the empire. There are rich mines of lead at Bleiberg, and of quicksilver at Idria. The latter are situated in the E. portion of the Julian Alps, on the right bank of the Isonzo. The ore is found in a schistose rock, breaking through the predominant limestone of that chain; and as the veins get deeper they are said to become richer. Blasting is the usual method employed for obtaining the ore; and the workmen, on account of the depth and consequent heat of the mines, work by relays of eight hours each gang. The lowest point in the mine is 900 ft. below the bed of the adjacent Idrizza.

Trade.—Istria abounds with ports, many large enough to shelter whole fleets, the principal of which are Capo d' Istria, Pirano (Porto Rose), Quinto, Pola, Parenzo, and Rovigno; but these are only frequented by the barks conveying salt, wine, oil, gall nuts, charcoal, bark, and other productions of the peninsula to Trieste and Venice. There are likewise some tolerable harbours in the Quarnero islands, among which the port of Lussin Piccolo is, perhaps, the most capacious.

The roads of Illyria are as good as in most parts of the Austrian empire. The valleys of the Drave and Save are used for communications between Tyrol and Salzburg, and Carinthia and Carniola. Two main lines of common road lead from the capital to Trieste, one by Klagenfurth and Görtz, the other by Laybach. From Görtz the former has a branch to Venice and other parts of Italy, while the latter is connected by roads following the vales of the Save and Drave, with Hungary and the military frontier provs. But the most important means of intercommunication is the great line of railway from Vienna to Trieste, which runs right through the heart of Illyria, and branches of which extend to Agram, in Croatia, on the one side, and to Venice on the other. The ascent of the main line over the Alps is effected by an unusually steep granite, and powerful locomotives of extraordinary size. The Vienna-Trieste railway is the chief outlet of the exports of Illyria. The internal navigation is limited to rafts on the Save and Drave, by means of which rivers and their tributaries, much timber is floated down from the forests to the Danube.

Population.—The pop. of Illyria, in the course of 20 years, has increased in Carinthia and Carniola at the rate of 17·1 per cent., and in the Littorale at 30·3 per cent.

The inhab. (with the exception of the German settlers and of the Italians who have immigrated into the southern circles) are of Slavonian origin, and the vernacular language of Carniola, which is used as a written dialect, is one of the purest of the Slavonic idioms. Carniola is divided into Upper and Lower, the seats of the *Gorenzi Krainzi* and the *Dolenzi Krainzi*; the former of which are the mountaineers of the Julian Alps, the latter the inhab. of the valley of the Save. The *Uyparci*, in the valley of the Wippach; the *Kranbauzi*, on the Karst; the *Pinzchene*, in the Poik valley; and the *Znizzka*, are perhaps only local names. The general denomination for the Illyrian-Slavonians is 'Wündi' or 'Wenden' (*Venedi*). The inhab. of Friaul call themselves 'Furlani;' the peninsula is occupied by the 'Istriani,' and the Quarnero islands by 'Liburnzi.' Nearly one million of the inhab. are Slavonians.

The condition of the Illyrian pop., though certainly improving, is by no means prosperous. Like so many of the Slavonian inhab. of the empire, they speak a language which has not for centuries been the vehicle of intellectual improvement, and from an early period they were governed by tyrants, who availed themselves of their feudal rights, to the injury of the people, without conferring on them any of the advantages incidental to that system. In fact the Illyrians had no national existence till the time of Napoleon. The ephemeral kingdom of Illyria which he established infused a spirit into all classes, which awakened them from the lethargy of ages. Much still remains to be done towards ameliorating the condition of the peasant, yet the change in his condition for the better within the present century is very great. The mountaineers of Carinthia and Upper Carniola are the poorest and worst fed of the inhab. Amongst them 'cretins,' or idiots are of frequent occurrence, and are recommended to their neighbours' charity by the superstitious notion that their presence in a family indicates good fortune. Goitre is common amongst the mountaineers, and the mortality is so great as scarcely to admit of any increase in the pop. The inhab. of the valleys, especially those living near the Save, are in a better condition, and in the district of Görtz enjoy a considerable degree of prosperity. Istria, with all its natural advantages, is worse cultivated, and less civilised, than the rest of Illyria. The dress of the mountaineers resembles that of the peasant of Tyrol and Salzburg. The women wear peaked, broad-brimmed hats; and in Carniola, instead of stays, they wear a red girdle, sewn to the linen tunic or shift, which is seen between the upper part and skirts of the gown worn over it. Formerly the men of the Gail valley wore a gay dress of motley colours, from which the costume of Harlequin in the Italian comedy is said to be derived; indeed, many of the figures in pantomimes are believed to have been originally caricatures of the Illyrian peasantry.

ILMINSTER, a market town and par. of England, co. Somerset, hund. Abdick and Bulstone, on the Ivel, 10 m. SE. Taunton, 4 m. S. by W. Ilch, and 127 m. W. by S. London, on the Great Western railway. Pop. of town 2,194, and of par. 3,241 in 1861. Area of par. 4,580 acres. The town comprises two streets, intersecting each other at right angles, one of which is nearly a mile long: the houses are irregularly built, some being of stone or brick, and the greater part merely thatched. The church, formerly conventual, is cruciform, in the decorated Gothic style, and has a square embattled and pinnacled tower. There are also places of worship for Wesleyan Methodists and Independents, to which, as well as to the church, are attached well-frequented Sunday schools. A free grammar-school was founded in 1550, and endowed with considerable estates; there is also a hospital for the maintenance of clergymen's widows. Ilminster was formerly an important woollen clothing town; but its industry is now confined to the weaving of narrow cloths, and is of little importance. Lace-net mills have been recently established, and give employment to several hands. Petty sessions are held in the market-house. Markets on Saturday; fairs for horses, live-stock, and cheese, the last Wednesday in August.

IMOLA (an. Forum Cornelii), a town of North Italy, prov. Bologna; on the Santerno and the Emilian Way, 18 m. NW. Furli, and 20 m. SE. Bologna, on the railway from Bologna to Ancona. Pop. 25,919 in 1862. It is a town of some consideration, being a bishop's see, surrounded by ancient walls and ditches, and further defended by an old castle. It is tolerably well built, and has a cathedral and 15 other churches, numerous convents, a hospital, theatre, college, and a literary academy, of some celebrity, termed de' Industriosi, which has included among its members several distinguished individuals. It has manufactures

of cream of tartar, called tartaro di Bologna, and some trade in agricultural produce.

INDIA (BRITISH), a very extensive empire, situated in S. Asia, comprising the province under the Governor-General of India, Bengal, Oude, the Central Provinces, British Burmah, the North-Western Provinces, Madras, Bombay, and the Punjaub, besides a number of native states as well as of foreign states under British protection. These vast dominions lie between lat. 8° 20' and 35° 15' N., and long. 65° 43' and 140° E.; their principal boundaries being, NW. the Indian Desert; N. the Himalaya, which, in the upper prov. of Agra and in Assam, separates them from the Chinese empire, Nepaul, and Bootan; E. the Birman empire and Siam, and S. and W. the Indian Ocean, the Bay of Bengal, and the Arabian Sea. The area and pop. of the principal political divisions of British India are stated as follows, in official returns of the year 1864:—

Political Divisions	Area Sq. Miles	Population
Under Governor-General	16,670	3,592,194
Bengal	261,290	40,464,859
Oude	27,490	5,911,073
Central Provinces	108,560	7,941,480
British Burmah	90,070	1,867,467
North-west Provinces	86,360	29,674,887
Madras	140,917	23,100,273
Bombay	142,063	12,072,344
Punjaub	106,406	14,784,611
Total	1,004,616	143,171,210
Native States	801,030	47,649,193
Foreign States	1,244	517,149

To the foregoing territories, under the immediate rule of the British, there may be added the tributary states of Berar, Oude, Mysore, Travancore, Cochin, Suttarah, the dom. of the Nizam, and of the Rajpoot and Bundlecund chiefs, which are substantially administered by British rulers, and are either entirely or in part surrounded by British territories.

The physical geography, products, inhabs. industry, &c., of the several divisions, provinces and districts of British India, will be found treated of under the head HINDOSTAN, and on separate articles appropriated to each. The present article will, therefore, be principally occupied with topics, such as the general government, the revenue system, army, and commerce of British India, that could not be conveniently introduced under any other head.

Government.—Previously to 1773, the government of that part of India which then belonged to the British was vested in the E. India Company. The body of proprietors of E. India stock, assembled in general court, elected 24 directors, to whom the executive power was entrusted, the body of proprietors reserving exclusively to themselves all legislative authority. A vote in the court of proprietors was acquired by the holders of £500 of the company's stock; but, to be a director, it was necessary to hold 2,000l. stock. The directors, with their chairman and deputy chairman, were chosen annually, and subsequently subdivided themselves, for despatch of business, into ten separate committees. As early as 1707, the three principal presidencies into which British India was then divided—those of Bombay, Madras, and Bengal, were in existence. Each was governed by a president or governor, and a council of from 9 to 12 members, appointed by commission of the company. All power was lodged in the president and council jointly, every question that came before them being decided by a majority of votes.

In 1726, a charter was granted, by which the company were permitted to establish a mayor's court at each of the presidencies, consisting of a mayor and nine aldermen, empowered to decide in civil cases of all descriptions, with an appeal from their jurisdiction to the president and council. The latter were also vested with the power of holding courts of quarter sessions, for the exercise of penal judicature, in all cases excepting those of high treason, as well as a court of requests, for the decision, by summary procedure, of pecuniary questions of inconsiderable amount. Added to this, the powers of justices of the peace were granted to the members of the council, and to them only, the president being, at the same time, commander-in-chief of all the military force stationed within his presidency. The officers of the company were thus recognised as judges in their own cause in all cases; and, notwithstanding the establishment of the mayors' courts, they still held all the judicial as well as the executive functions, both civil and military, in their own hands.

In 1773, the great increase in the territorial possessions of the company attracted the attention of the government at home; while the financial embarrassments of the company, and the abuses which had crept into the government of India, furnished ample grounds for interference. In consequence, the ministry introduced two bills into parliament, distinctly asserting the claim of the crown to the territorial acquisitions of the company, raising the qualification to vote in the court of proprietors from the possession of 500l. to that of 1,000l. stock; giving to every proprietor possessed of 3,000l. 2 votes, of 6,000l. 3 votes, and of 10,000l. 4 votes; limiting the annual election of the whole 24 directors to that of 6 only; vesting the government of Bengal, Bahar, and Orissa, in a governor-general, with a salary of 25,000l. a year, and 4 councillors, of 8,000l. each; rendering the other presidencies subordinate to that of Bengal; and establishing at Calcutta a supreme court of judicature, consisting of a chief justice, with 8,000l. a year, and three puisne judges, with 6,000l. a year each, appointed by the crown. As subsidiary articles it was proposed, that the first governor-general and councillors should be nominated by parliament in the act, and hold their office for five years, after which the patronage of those great offices should revert to the directors, but still subject to the approbation of the crown; that every thing in the company's correspondence from India which related to civil or military affairs, to the government of the country, or the administration of the revenues, should be laid before ministers; that no person in the service either of the king or of the company should be allowed to receive presents; and that the governor-general, councillors, and judges should be excluded from all commercial speculations and pursuits.

Mr. Pitt's India bill of 1784 established the board of control, consisting of six members of the privy council, appointed by the king, two of the principal secretaries of state being always members. The president of the board was, in fact, secretary of state for India, and is the officer responsible for its government, and for the proceedings of the board. The superintendence of the latter extended over the whole civil and military transactions carried on in India.

Mr. Pitt's bill was followed by the act of 1833 (3 & 4 William IV. cap. 85), under which the company held, by the superintendence of the board of control, the political government and patronage of British India. The supreme authority was vested in the governor-general. He was nominated

by the court of directors, the nomination being subject to the approval of the sovereign, and was assisted by a council of five members, three of whom were appointed by the court of directors, from amongst persons who were or had been servants of the company; the fourth was also chosen in a similar manner, but from amongst persons unconnected with the company; and the fifth was the commander-in-chief, taking rank and precedence immediately after the governor-general. The other presidencies had also their governors and councils, subordinate to the governor and council of the Bengal presidency; the presidency of Agra, however, comprising the upper provinces of Bengal, was administered by a lieut.-governor only. The governor-general in council was competent to make laws for the whole of British India, which were binding upon all the courts of justice, unless annulled by higher authority. Parliament reserved to itself the right to supersede or suspend all proceedings and acts of the governor-general; and the court of directors had also power to disallow them.

This constitution remained in force till the year 1858, when the present form of government of the Indian empire was established by the Act 21 and 22 Victoria, cap. 106, called 'An Act for the better government of India,' sanctioned August 2, 1858. By the terms of this act, all the territories heretofore under the government of the East India Company are vested in her majesty, and all its powers are exercised in her name; all territorial and other revenues and all tribute and other payments are likewise received in her name, and disposed of for the purposes of the government of India alone, subject to the provisions of this act. One of her majesty's principal secretaries of state, called the secretary of state for India, is invested with all the powers hitherto exercised by the company or by the board of control, and all warrants and orders under her majesty's sign-manual must be countersigned by the same. The executive authority in India is vested in a governor-general or viceroy, appointed by the crown, and acting under the orders of the secretary of state for India.

The administration of the Indian empire is entrusted by the charter of August 2, 1858, to a council of state for India. The council consists of fifteen members, of whom seven are elected by the court of directors from their own body, and eight are nominated by the crown. Vacancies in the council, if among those nominated, are filled up by the government, and if among the elected, by an election by the other members of the council; but the major part of the council must be of persons who have served or resided ten years in India, and not have left India more than ten years previous to the date of their appointment; and no person not so qualified can be elected or appointed, unless nine of the continuing members be so qualified. The office is held during good behaviour, but a member may be removed upon an address from both houses of Parliament. No member is to sit or vote in Parliament. The salary of each is fixed at 1,200l. a-year, payable, together with that of the secretary of state, out of the revenues of India.

The duties of the council of state are, under the direction of the secretary of state, to conduct the business transacted in the United Kingdom in relation to the government of and the correspondence with India; but every order sent to India must be signed by the secretary, and all despatches from governments and presidencies in India must be addressed to the secretary. The secretary has to divide the council into committees, to direct what departments shall be under each committee respectively, and to regulate the transaction of business. The secretary acts as president of the council, and has to appoint from time to time a vice-president. The meetings of the council are held at times fixed by order of the secretary; but at least one meeting must be held every week, at which not less than five members must be present.

The government in India is exercised by a 'supreme council,' sitting at Calcutta, and consisting of five ordinary and from six to ten extraordinary members, presided over by the governor-general. The ministry, divided in the departments of foreign affairs, finance, the interior, military administration, and public works, forms part of the supreme council. The appointment of the ministers, the members of the council, and the executive governors and lieutenant-governors of the various territories and provinces of the empire rests with the governor-general.

Revenue System.—The land tax constitutes the principal source of the revenue of British India, as it has always done of all eastern states. The governments of such countries may, in fact, be said to be the real proprietors of the land; but in India, as elsewhere, the cultivators have a perpetual, hereditary, and transferable right of occupancy, so long as they continue to pay the share of the produce of the land demanded by the government. The value of this right of occupancy to the rural population depends on the degree of resistance which they have been able to oppose to the exactions of arbitrary governments. In Bengal and the adjacent provinces of India, from the peculiarly timid character of the inhabitants, and the open and exposed nature of the country, this resistance has been trifling indeed, and, consequently, the value of the right of occupancy in the peasant, or ryot (an Arabic word, meaning subject), has been proportionally reduced. This, also, may be considered, though with some modifications, as being nearly the condition, in this respect, of the inhabitants of every part of the great plain of the Ganges, comprising more than half the population of Hindustan. But where the country is naturally difficult, the people have been able more effectually to resist the encroachments of the head landlord, or state, and to retain a valuable share in the property of the soil. This has been particularly the case along the ghauts, as in Bednore, Canara, Malabar, &c.; the inhabitants of which territories not only lay claim to a right of private property in the soil, but have been generally ready to support their claim by force of arms. There can be no question, indeed, that the same modified right of property formerly existed every where; and it is indeed impossible that otherwise the land should ever have been reclaimed from the wilderness. But, in those parts of India which could be readily overrun by a military force, the right of property in the soil has long been little else than the right to cultivate one's paternal acres for behoof of others, the cultivators reserving only a bare subsistence for themselves.

Under the Mogul emperors, the practice in Bengal was to divide the gross produce of the soil, on the *metayer* principle, into equal shares, whereof one was retained by the cultivator, the other going to government as rent of tax. The officers employed to collect this revenue were called *zemindars*; and in the course of time their office seems to have become hereditary. It may be remarked that, in Persian, zemindar and landholder are synonymous; and this etymology, coupled with the hereditary nature of their office, which brought them exclusively into contact with the ryot, or

occupier, as well as with the government, led many to believe that the zemindars were in reality the owners of the land, and that the ryots were their tenants. This, however, it is now admitted on all hands, was an incorrect opinion. The zemindars in reality were tax-gatherers, and were, in fact, obliged to pay to the government *nine-tenths* of the produce collected from the ryots, retaining only one-tenth as a compensation for their trouble; and, so long as the ryots paid their fixed contribution, they could not be ousted from their possessions, nor be in anywise interfered with.

But notwithstanding what has now been stated, the perpetual or zemindary settlement, established by Lord Cornwallis in Bengal, in 1793, was made on the assumption that the zemindars were the proprietors of the soil. His lordship, indeed, was far from being personally satisfied that such was really the case; but he was anxious to create a class of large proprietors, and to give them an interest in the improvement and prosperity of the country. It is clear, however, that this wish could not be realised without destroying the permanent rights of the ryots, for, unless this were accomplished, the zemindars could not interfere in the management of their estates. The interest of the zemindars, and the rights of the ryots, were plainly irreconcilable; and it was obvious that the former would endeavour to reduce the latter to the condition of tenants at will. But this necessary consequence was either overlooked or ineffectually provided against. The zemindars became, under condition of their paying the assessment, or quit-rent, due to government, proprietors or owners of the land. The amount of the assessment was fixed at the average of what it had been for a few years previously, and it was declared to be *perpetual* and *invariable* at that amount. When a zemindar fell into arrear with government, his estate might be either sold or resumed.

That the assessment was at the outset too high cannot well be doubted; and it must ever be matter of regret that the settlement was not made with the ryots, or cultivators, rather than with the zemindars; but, notwithstanding these and other defects, the measure was, on the whole, a great boon to India. Until the introduction of the perpetual system into Bengal, the revenue was raised by a *variable* as well as a most oppressive land-tax. In France, Italy, and other parts of Europe, where the metayer system is introduced, the landlord seldom or never gets half the produce, unless he also furnish the stock and farming capital, and, in most cases, the seed. But in India, neither the government nor the zemindars do any thing of the sort; they merely supply the land, which is usually divided into very small portions, mostly about 6, and rarely amounting to 24 acres. A demand on the occupiers of such patches for half the produce is quite extravagant, and hence the excessive poverty of the people, which is such as to stagger belief. Still, however, the perpetual system was vastly preferable in principle, and also in its practical influence, to any other revenue system hitherto established in India. It set limits to fiscal rapacity, and established, as it were, a rampart beyond which no tax-gatherer dared to intrude. The enormous amount of the assessment and the rigour with which payment was at first enforced, ruined an immense number of zemindars. But their lands having come into new and more efficient hands, a better system of management was introduced, and the limitation of the government demand gave a stimulus to improvement in Hindustan.

The land revenue in most parts of British India is assessed under the system now described; but

In some parts of the Bengal provinces, in the ceded districts on the Nerbudda, and in the greater number of the native states, a different plan is adopted, which has received the name of the *village system*. This system, though defective in many respects, is superior to the ryotwar system, and in some points, is even preferable to the perpetual system. It is a settlement made between the government and the cultivators, through the medium of the native village officers, who apportion the assessment without any direct interference on the part of the government functionaries. It is difficult to state the proportion of the produce of a village paid to government. The authorities know little of the precise property of any of the proprietors; it is not the interest or the wish of the village that they should; and if any member of the community fail to pay his share, that is a matter for the village at large to settle, and they usually come forward and pay it for him. These, however, are private arrangements; and the mocuddim, or headman, through whom the government settles with the cultivators, has no power from government to enforce the assessment on the particular defaulter. The tax to be paid by each village is settled by the villagers amongst themselves; the total assessment being calculated after inquiry into the property of the village—what it has paid and what it can pay—regular surveys of the village boundaries, and of its lands, having been previously made by government. The mocuddim or potail (headman) is elected by the villagers; and, if the latter become dissatisfied with him, they turn him out of office. This system may have, and doubtless has, its disadvantages: the potail may, from various motives, unequally assess the villagers; and the tendency to cultivate waste lands will not be so strong as under the perpetual settlement; but the latter effect is much more likely to be brought about under this than under the ryotwar system; nor does the village system involve the same inquisitorial acts on the part of government.

Besides the lands subject to the foregoing systems of assessment, a considerable extent of land in India is held rent-free. Throughout Hindostan, and indeed throughout Asia, China perhaps excepted, a considerable portion of the land-tax is assigned to a great variety of parties, and for various purposes. Lands have been given to public officers as the reward of their services; to men of learning; to the favourites of sovereigns; for the maintenance of civil and military public establishments; and for the endowment of charitable, educational, and religious institutions. The grants, especially those for the use of temples, mosques, and shrines, were in perpetuity; and others became so through the usage of India. Inscriptions on stone and brass, found in most parts of India, attest the antiquity of these grants. One of them is supposed to be nearly coeval with the invasion of Britain by Julius Cæsar, and hundreds are of dates antecedent to the Norman invasion. (Asiat. Researches, i.; Trans. of the Royal Asiat. Soc., passim.) The extent of these free tenure lands throughout India is very great. In the ceded territory under the Madras presidency, they are estimated to amount to one-fifth part of the entire surface. In the N.W. provinces of the Bengal presidency, the free tenure lands were ascertained by the British commissioners to amount to 44,951,770 begahs; the land-tax of which, if assessed in the usual manner, would have amounted to 1,226,000l. From an inquiry made in 1775, it appeared that the rent-free lands, in Bengal Proper, amounted to 8,673,542 begahs, or 2,164,551 acres, which would have yielded a tax of 1,226,350l. a

year. It is deserving of notice, that the rent-free lands under the Agra presidency were at the very threshold, as it were, of the Mohammedan power; and the territory in which they are included was in the possession of the Mohammedans *for six centuries*. But, notwithstanding their bigotry and despotism, they respected the free tenure. They also, much to their honour, respected them in a singular degree in Bengal, where most of them had originally consisted of tracts of waste or wild land, reclaimed by the labour and capital of the grantees, or their heirs and successors. Lord Cornwallis, and the Indian council of his day, confirmed the possession of the rent-free lands to their holders, on the same perpetual tenure as the taxed lands; and it was enacted that those that held under a free tenure prior to 1765 should remain untaxed 'for ever.'

The following table gives the total receipts of the government of India from the land-tax, including *sayer* and *moturpha*—the first comprising variable imposts, such as tolls and town duties, and the latter taxes on houses and shops—in the various territories in each of the years 1860, 1861, and 1862. The last column gives the entire net receipts, from all sources, during the same three years:—

Years ended 30th April	Territories or Provinces	Land Tax, Sayer, and Moturpha	Total Net Revenue from all sources
		£	£
1860	Territories and Departments under the immediate control of the Government of India	1,814,884	
	Bengal	3,070,000	33,349,853
	North-west Provinces	4,150,333	
	Madras	1,830,944	
	Bombay	2,472,746	
	Punjaub	1,741,795	
1861	Territories and Departments under the immediate control of the Government of India	1,988,318	
	Bengal	3,070,200	
	North-west Provinces	3,854,557	33,078,765
	Madras	3,430,494	
	Bombay	2,970,944	
	Punjaub	1,861,717	
1862	Territories and Departments under the immediate control of the Government of India	2,240,884	
	Bengal	4,849,109	
	North-west Provinces	4,411,651	34,681,712
	Madras	4,001,672	
	Bombay	3,067,313	
	Punjaub	1,729,915	

The total land revenue of British India, exclusive of *sayer* and *moturpha*, amounted to 18,757,400l. in 1860; to 18,509,991l. in 1861; and to 19,644,666l. in 1862. It will be seen that, in each of these three years, the land-tax produced more than one half of the total net receipts. Next to the land-tax, the most important sources of revenue of the Indian government are the opium and salt monopolies. The net receipts from opium were 5,169,770l. in 1860; 5,759,292l. in 1861; and 4,919,802l. in 1862; and those from salt amounted to 2,313,218l. in 1860; to 3,064,982l. in 1861; and to 3,915,151l. in 1862. The net receipts from customs, the last of the notable sources of Indian revenue, amounted to 3,701,210l. in 1860; to 3,996,433l. in 1861; and to 7,632,591l. in 1862.

According to the Act of 1858, the revenue and expenditure of the Indian empire are subject to the control of the secretary in council, and no grant or appropriation of any part of such revenue can be made without the concurrence of a majority of the council.

Such parts of the revenues of India as may be remitted to England, and moneys arising in Great Britain, must be paid into the Bank of England; and paid out on drafts or orders signed by three members of the council, and countersigned by the secretary or one of his under-secretaries. The sovereign of Great Britain is empowered to appoint from time to time an auditor of the accounts, with power to inspect all books and examine all officers, and his report must be laid before parliament. The accounts of the whole revenue and expenditure of the Indian empire must be laid annually before parliament.

The subjoined table gives the total gross amount of the actual revenue and expenditure of India, in each of the years ending April 30, from 1858 to 1862:—

Years ended April 30	In India		Home Charges
	Revenue	Expenditure	
	£	£	£
1858	31,705,776	35,074,528	6,163,043
1859	36,000,760	43,580,734	7,406,100
1860	39,205,972	44,672,209	7,328,654
1861	42,501,344	42,579,975	8,311,602
1862	43,449,472	37,745,754	6,244,344
1863	43,143,752	36,830,406	6,515,609

Adding together the Indian expenditure and the home charges, the financial accounts of India for the year 1862–63 stood as follows:—

The Total Gross Revenue of 1862–63 was . £45,141,752
The Total Expenditure 43,216,407

Surplus . . . £1,927,345

The cost of the army, of the civil and political establishment, and the interest of the public debt, form the chief items of expenditure in India. They amounted to the following sums, in each of the years 1860, 1861, and 1862:—

Years ended 30th April	Military Charges	Civil and Political Establishments	Interest of Debt
	£	£	£
1860	20,340,568	3,114,872	3,124,378
1861	14,730,932	3,721,078	2,272,104
1862	13,641,248	3,642,284	3,043,294

According to returns published in April, 1862, the whole Indian army numbered nearly 200,000, of which number 3,502 were European officers, and 70,009 European non-commissioned officers and men; the native officers and men amounting to 108,502, exclusive of 11,613 men in the Punjaub local force. The distribution of these troops was as follows: 80,000, in round numbers, in Bengal, the North-west Provinces, and the Punjaub; 42,000 in the Bombay Presidency, and 50,000 in Madras.

The interest on the registered debt of India amounted to 3,134,897£ on April 30, 1863. A return issued by the secretary of state for India, in Sept. 1864, stated the debt of India at 116,721,152£; but this included 2,031,976£ capital of railway companies remaining in the home treasury. On the other hand, it did not include the charge for the dividend on the 6,000,000£ capital stock of the East India Company, which is subject to redemption by parliament under the act of 1833.

Roads and Railways.—Throughout the whole of the immense basin of the Ganges there is an extensive inland navigation; and this, also, is the case in the valleys of the larger rivers in the S.; but elsewhere the inland trade, where railways have not been established, is greatly impeded by the want of roads, and the imperfect means of conveyance. With the exception of various military roads, but very few fit for carriages have been constructed in any part of the country. The internal commerce of India, however, has been greatly developed of late years by the construction of several great lines of railways, made under the guarantee of the government. On June 30, 1863, the system of guaranteed railways comprised a length of 4,917 m., of which 3,186 were open for traffic. The net profits in the year ending June 30, 1863, on 2,151 m. of railway, amounted to 610,834£; and to 915,077£ in the year ending June 30, 1864, on 2,489 m. The number of passengers conveyed in the latter year was 11,741,683, compared with 8,242,540 in the former. The total expenditure of capital on the lines which were open, or in course of construction, amounted on May 1, 1865, to 59,912,924£. The expenditure in 1865 amounted to rather more than 5,000,000£,—about 1,500,000£ expended in England, and 3,500,000£ in India. The total amount estimated to be required for the undertakings will reach 77,500,000£. The number of shareholders at the end of the year 1864 was 29,303 in England, and 777 in India; the latter number consisting of 304 Europeans and 323 natives. There were also 6,153 debenture holders. Up to the end of 1864, the government had advanced 13,160,582£ to the railway companies for guaranteed interest, but about 3,500,000£ had been paid back out of the earnings of the railways, leaving nearly 10,000,000£ still due to the government. The charge upon the government was 2,567,743£ in the year 1863; but the receipts from traffic which went in diminution of this charge amounted to about 1,000,000£, and in 1865 reached 1,500,000£.

Trade and Commerce.—Corn, cotton, oleaginous plants, and sugar are the most important objects of inland commerce. The chief trade in raw takes place within the tract of the inundation of the Ganges N. of lat. 25°, it is superseded by that of wheat and barley. Cotton is grown in every latitude in India. It is, speaking generally, coarse, dirty, and short in the staple, and inferior to most other kinds brought to the markets of Europe. But this is not owing so much to any natural incapacity on the part of India to produce good cotton, as to the want of care in selecting the seed, and the culture of the plant. In these respects, too, some very material improvements have been effected of late years; and a good deal of the vast amount of cotton brought from India during the American civil war, was greatly superior to the old produce. But it is still susceptible of much improvement.

Next to cotton, the most important articles of export are spices, opium, rice, dyes, and seeds. Sugar is a principal article of internal culture and trade, but is not exported in large quantity. It is principally raised in the great plain of the Ganges. The average annual consumption of sugar in Hindostan has been estimated at between 11 lbs. and 12 lbs. a head. The average consumption of salt is estimated at 15 lbs. per head. The other staples of the inland trade are indigo, opium, silk, tobacco, nitre, oil-skins, drugs, hides, lime, and timber. The commercial progress of British India within recent times is shown in the subjoined two tables, which give the quantities and values of the principal articles imported, as well as of those exported in each of the two years 1851 and 1863.

IMPORTS INTO INDIA, 1851 AND 1862.

Principal Articles		1850	1862
Apparel	£	192,878	499,984
Books and Stationery	£	132,270	380,646
Cotton Twist and Yarn	lbs.	20,961,144	16,275,361
	£	1,131,696	1,179,843
Cotton Piece Goods	£	3,571,818	4,243,553
Fruits and Nuts	£	168,294	300,506
Jewellery	£	64,178	865,374
Malt Liquors	galls.	100,192	8,414,111
	£		618,876
Machinery	£	8,079	494,117
Metals, Manufactured	£	166,180	418,283
Raw:—			
Copper	cwts.	154,961	
	£	858,909	1,166,683
Iron	cwts.	647,658	
	£	312,143	664,643
Spelter	cwts.	74,750	61,698
	£	104,053	94,851
Steel	cwts.	19,707	72,060
	£	17,108	79,624
Tin	£	45,340	99,077
Military Stores	£	25,907	437,268
Naval	cwts.	83,185	104,231
Salt	£	944,770	3,703,321
	£	461,801	550,471
Silk Goods	£	112,801	944,784
Spices	lbs.	5,257,178	12,977,063
	£	98,258	177,645
Spirits	galls.	205,994	443,493
	£	129,960	443,907
Tea	lbs.		7,279,064
	£	53,810	171,613
Woollen Goods	£	166,144	251,016
Wines	galls.	256,976	361,460
	£	211,574	829,178
Bullion and Specie (Treasure)	£	8,356,089	20,475,080
Total Value of principal and other Articles	£	18,688,904	47,858,284

EXPORTS FROM INDIA, 1850 AND 1862.

Principal Articles		1850	1862
Coffee	lbs.	5,382,344	21,045,755
	£	78,100	512,257
Cotton, Raw	lbs.	181,661,720	479,845,993
	£	2,301,178	19,757,389
Cotton Goods, incl. Twist and Yarn	£	791,062	765,104
Dyes	lbs.	14,046,455	34,731,439
	£	1,507,081	3,217,103
Gunnies & Gunny Bags	£	111,448	179,350
Hides and Skins	£	319,266	879,531
Jewellery and Precious Stones	£	81,628	77,831
Jute	cwts.	991,789	1,266,994
	£	99,809	750,438
Oils	galls.	—	8,871,665
	£	109,947	362,673
Opium	chests	41,947	73,318
	£	5,973,396	13,494,178
Rice	qrs.	818,892	2,201,182
	£	648,973	2,264,061
Saltpetre	cwts.	634,501	644,320
	£	404,294	897,723
Seeds	qrs.	130,543	644,000
	£	216,610	1,861,461
Shawls, Cashmere	pieces	—	18,819
	£	147,091	333,157
Silk, Raw	lbs.	1,655,116	1,377,644
	£	866,194	822,897
Silk Goods	£	441,749	164,266
Spices	lbs.	18,777,683	10,557,851
	£	181,704	176,638
Sugar and Sugar Candy	cwts.	1,871,576	704,568
	£	1,935,092	837,042
Timber and Woods	£	94,515	66,865
Wool, Raw	lbs.	8,153,854	
	£	44,924	841,223
Bullion and Specie (Treasure)	£	971,214	1,159,414
Total Value of principal and other Articles	£	18,383,543	67,882,863

The increase in the exports of India in the short period 1851–62 has been truly extraordinary, and almost unparalleled in the commercial history of any other country. The augmentation of the exports of raw cotton alone, nearly fourfold in quantity, and more than eightfold in value, is quite without precedent.

The chief commercial intercourse of India is, as may be expected, with the United Kingdom. The total value of the imports from, and the exports to, the United Kingdom at the four annual periods, 1850, 1855, 1860, and 1862, is given in the subjoined tabular statement:—

Years	Imports from United Kingdom	Exports to United Kingdom
	£	£
1850	7,645,671	7,141,350
1855	8,852,446	7,556,911
1860	26,563,499	11,281,375
1862	19,119,726	26,676,673

The above figures tell, more eloquently than words, the material results of British rule in India.

INDIA-BEYOND-THE-GANGES, sometimes called INDO-CHINA, an extensive region of Asia, forming the eastern of its three great peninsulas, extending between the 7th and 79th degs. of N. lat., and the 92nd and 109th of E. long., comprising Birmah, Siam, and Anam, the Malay peninsula, Laos, the Tenasserim provs., Aracan, Cachay, Cachar, Assam, and the Bengal districts of Sylhet, Tipperah, and Chittagong.

INDIANA, one of the United States of America, in the NW. part of the Union, having N. the lake and state of Michigan, E. Ohio, W. Illinois, and S. Kentucky, from which it is separated by the Ohio. Length, N. to S., 270 m.; average breadth, 130 m.; Area, 33,809 sq. m.; pop. 1,350,428 in 1860. Surface generally level or undulating; there are, however, some extensive hilly tracts in different parts. The chief elevations in the state are the bluffs which skirt the Ohio; and these, and the country immediately N. of them, are densely wooded. The central and N. parts consist chiefly of level prairies, interspersed with small lakes and swamps. Next to the Ohio, the principal river is the Wabash. It rises in the NE., and, flowing first W. and afterwards S., in the lower part of its course divides this state from Illinois, and falls into the Ohio after a course of 480 m., the greater part of which is navigable. It has several tributaries, including the White and the E. Fork, which also are navigable for a considerable distance. The other principal rivers are the St. Joseph, which falls into Lake Michigan, and the Kankakee, an affluent of the Illinois. The climate differs little from that of Ohio and Illinois; but Indiana is somewhat less subject to the extremes of heat and cold than the latter state. The winters seldom last longer than six weeks; the Wabash, however, is at that season frozen over so as to be crossed with safety. In the valleys of the Ohio and Wabash, bilious fevers, agues, &c. are very prevalent during summer.

Soil in most parts very fertile. The agricultural products are the same as in the adjoining states on the E. and W. Little is known of the metallic resources of the state. Large quantities of sulphate of magnesia are met with in the S. along the banks of the Ohio. The state possesses an extensive system of internal navigation, including the Wabash and Erie Canal, extending from the W. end of Lake Erie to La Fayette, on the Wabash, a distance of 187 m., with a prolongation down the Wabash to Evansville. The railway

system is also very complete. The Atlantic and
Great Western railway, with its prolongation, the
Ohio and Mississippi line, runs through the southern part of the state, while the north and centre
are intersected by seven different lines, centering
at Indianapolis.

Indiana is divided into 91 cos. Indianapolis, on
White River, near the centre of the state, is the
cap. and seat of government; the other chief towns
are New Albany, Madison, and Vincennes. The
government consists of the governor, lieutenant-
governor, secretary of state, treasurer, auditor,
attorney-general, and superintendent of public instruction, the whole of whom are chosen by the
people at the general elections held on the second
Tuesday in October. They hold their offices for
two years. Senators, 50 in number, and representatives, 98 in number, in the year 1865, constitute the legislature, the style of which is the
general assembly of Indiana. The legislature is
required to hold a regular session biennially, commencing in January in the odd years, such as 1863
and 1865. The general assembly meets at Indianapolis. Judges are elected for a term of seven
years. By an act of March 1, 1859, the state was
divided by counties into 21 districts, in each of
which, in October, 1860, a judge and a prosecuting
attorney were elected. The judges are elected for
four years, and the salary of each is 1,000 dollars.
Three terms of each court of common pleas are
held each year, beginning on the first Monday in
January, and on the first Monday of every fourth
month thereafter, unless the circuit court be in
session, and then on the Monday succeeding the
term of the circuit court. The governor is chosen
for three years, and is only twice eligible.

The earliest permanent occupation of Indiana
was made by the French, about 1702, when Vincennes and several other small settlements were
established by them along the Wabash. Previously to 1800, it was included in the NW. territory, and from that year until 1809 was governed
with Illinois, under the title of the Indiana territory. It was admitted into the Union on the 11th
of December, 1816. The state sends eleven representatives to congress.

INDIANAPOLIS, a city of the United States,
and cap. of the state of Indiana. Pop. 2,692 in
1840, and 18,600 in 1860. The town stands on
the E. side of White river, and is the centre of the
most important roads and railways of the state.
It is regularly laid out, more than a mile square,
within a circular area, with the governor's house
in the centre.

INDIES (WEST). Under this term were formerly included not only the Caribbee and other
islands in the Atlantic near the coast of America,
but also all the countries included under the name
of the Spanish Main. But at present the term is
restricted so as to signify only the islands between
lat. 10° and 27° N., and long. 60° and 85° W.,
comprising the larger and smaller Antilles; the
former consisting of Cuba, Hayti, Jamaica, and
Porto Rico; and the latter of the Virgin, Leeward
and Windward groups, with the Bahamas, Trinidad, Tobago, and a few other islands. Of these,
Hayti alone is independent. Cuba and Porto Rico
belong to Spain; Jamaica, the Bahamas, Trinidad,
Barbadoes, Antigua, Dominica, Grenada, St. Lucia,
&c., to Great Britain; Guadaloupe, Martinique,
Marie Galante, &c., to France; St. Eustatius, Saba,
and Curaçoa, to the Dutch; St. Croix, St. Thomas,
and St. John, to the Danes; and St. Bartholomew
to the Swedes. For further details, see the several
islands above named.

INDORE, a city of Hindostan, prov. Malwah,
former cap. of Holkar's dom., a little N. of the

Vindhyan mountains, and 30 m. S. by E. Oujein;
lat. 22° 42' N., long. 75° 50' E. Pop. estimated
at 15,000. Indore is a place of small importance.
It stands at nearly 2,000 ft. above the level of the
sea, in a well wooded, pleasant, and healthy tract,
and has been wholly built within the present century. Some of its streets are tolerably spacious,
paved with granite slabs, and its houses often of
two stories, and constructed partly of brick; but,
speaking generally, it is mean and ill built, and
contains no public edifice worthy of remark, except the palace, a massive quadrangular granite
building, with decorations of carved wood.

INDRE, an inland dep. of France, reg. centre,
formerly included in the prov. Berri, between lat.
46° 22' 30" and 47° 15' N., and long. 0° 51' and
2° 13' E.; having N. Loire-et-Cher, E. Cher, S.
Creuse, and W. Vienne and Indre-et-Loire. Average length and breadth, 60 m. each. Area 679,530
hectares. Pop. 270,054 in 1861. Its surface is
generally level, with a slope towards the NW., in
which direction nearly all its rivers run to join
the Loire or the Cher. The Creuse bounds its W.;
the other chief river is the Indre, whence it derives its name. The latter rises in the dep. Creuse,
and has a course of about 91 m. through the centre
of this and the succeeding deps. to its mouth in the
Loire, below Tours. Châteauroux and Loches
stand on its banks; but, like the other streams of
this dep., it is innavigable. A tract of ponds and
marshes, called the Brenne, extends throughout
the centre and W. part of the dep., occupying
about one-tenth part of the whole surface, and a
more extensive tract towards the E. end, called
the Pays de Champagne, is quite bare of wood, and
infertile; but the remainder is mostly either under
culture, or covered with forests. The arable land
comprises 101,251 hectares, meadows 85,380 h.,
and forests and heaths 182,332 h. Agriculture is
very backward; but more corn is grown than is
required for home consumption, a result owing to
the thinness of the pop. The produce of wine
amounts to about 450,000 hectol. a year, which
also is more than is consumed by the inhabitants.
Fruits are good, and excellent hemp is raised.
There are about 950,000 sheep in the dep., large
flocks being fed on the Pays de Champagne. A
good many oxen are fattened for the supply of
Paris, and hogs for the markets of Auvergne and
Limousin. Geese and other poultry are reared in
large numbers, particularly in the Pays de Brenne.
Fish are abundant; and leeches form an article of
trade. Iron of good quality is found, and forges
are numerous. Good gun-flints are obtained at
Châteauroux. Next to iron goods and woollen
cloths, the principal manufactures are those of
cottons, woollen yarn, leather, tiles, earthenware,
beer, paper, and parchment. The dep. exports
corn, wine, cattle, wool, woollen cloths, and iron
and iron goods, to double the value of its imports.
The number of considerable properties is somewhat
below the average of the deps. The peasantry are
strongly attached to routine practices, and therefore little likely to better their condition. Education is little diffused. Indre is divided into four
arronds.; chief towns Châteauroux, the cap., Le
Blanc, Issoudun, and La Châtre.

INDRE-ET-LOIRE, a dep. of France, reg. of
the W., formerly included in the prov. Touraine,
comprising a tract on both sides the Loire, between
lat. 46° 46' and 47° 43' N., and long. 0° 2' and
1° 21' E., having N. Sarthe and Loire-et-Cher, E.
the latter dep. and Indre, S. Indre and Vienne,
and W. Maine-et-Loire. Area 611,679 hectares;
pop. 323,572 in 1861. Surface almost an entire
plain, with a slope from both the N. and S. to the
Loire, which runs through it, near its centre, from

E. to W. The part of the dep. watered by the Loire is as productive and beautiful that it has been termed the garden of France; but the soil elsewhere is generally dry, thin, and poor, and in the NW. there are some extensive pools and marshes. Heaths and wastes occupy nearly one-sixth part of the surface, and forests more than one-tenth. There are 831,910 hectares arable, 84,463 pasture land, 36,044 vineyard, and 23,673 otherwise cultivated. Agriculture is tolerably well conducted, having been much improved of late years. The corn now produced is more than adequate to the supply of the dep. Beans, pease, &c. are of excellent quality. Wine is annually made of the value of 9 or 10 millions of francs, or about double what is required for home consumption; but it is generally inferior. About 110,000 quintals of hemp, worth 4,600,000 fr., are raised yearly; and liquorice, aniseed, coriander, angelica, and truffles are cultivated. The culture of the mulberry-tree is increasing rapidly. The chief exports of the dep. are its agricultural products; cattle are not reared in any great number, and most kinds of live stock are inferior. Manufacturing industry is in a rather active state. The woollen, leather, and silk manufactures of Tours have materially increased within the last ten years. There is a large file and rasp factory at Amboise. The manufactures of cut steel and iron goods are important; and near Montlouis is the gunpowder factory and saltpetre refinery of Ripault, at which 500,000 kilog. of gunpowder are made annually. Indre-et-Loire is divided into three arronds., the chief towns of which are Tours, the cap. Chinon, and Loches. This is the native country of Descartes, who was born in La Haye on the 31st of March, 1596. Indre-et-Loire has also produced Rabelais and Balzac, Agnes Sorel, Gabrielle d'Estrées, and the Duchess de la Vallière.

INDUS (Sindhu, Sansc.; Aub Sind, Pers.), a large river of S. Asia, forming during great part of its course the proper NW. boundary of Hindostan, and lying between the 23rd and 35th parallels of N. lat., and between the 67th and 81st degrees of E. long. The source of the river is on the N. declivity of the Caïlas branch of the Himalaya range, near the Chinese frontier town of Gertos, and not far from the lake Mansarowra, and the sources of the Sutledje. The stream, called by the Chinese Singh-tcho, takes a general WNW. course past Ladak, and receives the larger river Shyook, NW. of Ladak, whence the united stream run through the country of Little Thibet, and after cutting a passage through the great Himalaya range, in lat. 35° 30' N., and long. 74° 20' E., are joined, about 170 m. S. of the mountains, by the Abce Sern, and lower down at Attock, where it is 260 yards wide, and both deep and rapid, by the river of Cabul. The river is crossed here by a bridge of boats, constructed like that used by Alexander, and described by Arrian (lib. v. cap. 7). The bridge is only allowed to remain between November and April, when the river is low; and the construction of it is completed in the course of six days. S. of Attock the Indus enters a plain, but soon afterwards winds amongst a group of mountains as far as Harrabah, whence it pursues a southward course to the sea, uninterrupted by hills, and expanding over the plain into various channels, which meet and separate again, but are rarely united into one body.

The breadth of the river at Kaharee Ghât, in lat. 31° 27' N., was found to be about 1,000 yards, the deep part of the channel being only 100 yards across, and 13 ft. deep. The banks in this vicinity are very low, and in summer are so much overflowed that the stream expands in many places to a breadth of 15 m. (Elphinstone, vol. ii. p. 416.) In lat. 28° 55', the Indus receives the Punjab rivers, and rolls past Mittun with a width of 2,000 yards, and a depth near the left bank of 4 fathoms. From this point to Bukkur the main stream takes a SW. course, with a direct channel, but frequently divided by sandbanks. Various narrow crooked branches also diverge from the parent stream, retaining a depth from 8 to 15 ft. of water; and these are navigated by boats accrediting the Indus in preference to the great river itself. The country on both sides of the richest nature, but particularly on the E. bank, where it is studded from innumerable channels, cut for the purpose of throwing the water SE. into the interior. (Burnes' Bokhara, vol. i. p. 260-261.) About 17 m. S. of Bukkur, in lat. 27° 10', the Indus sends off a branch to the W. called the Lurkhaun river, which, after making a circuit, and expanding in one place into a large lake 12 m. broad, rejoins the main stream 50 m. below the point of separation. The insulated territory, called Chandukar, is one of the most fertile in the Sinde dominions. About 100 m. below Bukkur is Seburun, in lat. 26° 22'; and between these points the river flows in a zig-zag course nearly SW., the intervening country being richly watered and divided by its ramifications into numerous inlets of the finest pasture. The distance between Seburun and Hyderabad is 105 m.; the banks seldom exceed 8 ft. in height, and the neighbouring grounds are covered with tamarisk. The river throws off no branches to this part of its course, except the Fulailee (generally an unimportant stream), which leaves the Indus 12 m. above Hyderabad, and crossing the W. extremity of the Runn of Cutch, enters the Indian Ocean by the Khorre mouth. The main river opposite Hyderabad is 800 yards broad, and 5 fathoms deep; but the channel becomes narrower and deeper as it approaches Tatta, 45 m. below Hyderabad. Shifting sandbanks also occur in many parts between these towns, to such an extent as to perplex the navigator.

The course of the stream from Hyderabad is NW. by S., with one decided turn below Jarruk, where it throws off the Pinyaree leading to Mughribee, and entering the sea by the Neer mouth. The country N. of Tatta, which might be rendered one of the richest and most productive in the world, is devoted to sterility, presenting to the eye only dense thickets of tamarisk, saline shrubs, and other underwood. About 5 m. S. below Tatta is the commencement of the Delta of the Indus. The river here divides into two branches, that to the right being called Buggaur, while that to the left is known as the Sata. The latter is by far the larger of the two, and a little below the point of division has a breadth of 1,000 yards; it divides and subdivides itself into many channels, and precipitates its water into the sea by 7 mouths, within the space of 35 m.; yet such is the violence of the stream, that it throws up sandbanks or bars; and only one mouth of this many-mouthed arm is ever entered by vessels of 50 tons.' (Burnes' Bokhara, vol. i. p. 207.) The Buggaur, on the other hand, flows in one stream as far as Darajee, within 6 m. of the sea, at which point it bifurcates, forming two arms, which fall into the ocean about 25 m. apart. A sandbank, however, which crosses its upper part, close to the apex of the Delta, renders it unfit for navigation. The land embraced by the Buggaur and Sata extends at the junction of these rivers with the sea to about 70 m.; and so much, correctly speaking, is the existing Delta; but the river covers with its waters a much wider space, and has two other mouths still farther E., viz. the Seer and Khoree, from which, however, the waters have been diverted by the rulers of Sinde into canals for the purposes of irrigation. If, therefore, these forsaken branches be included, the base of the Delta, measured in a straight line from the W. to the E. embouchure, extends 110 m. in a SSE. direction. Arrian estimates its extent at the time of Alexander's expedition at 1,800 stadia, or nearly double that now assigned to it; but it seems doubtful whether we are to attribute this difference to any great changes in the bed of the river, or to the miscalculation of the Macedonian admiral, Nearchus.

The incessancy of the stream through the Delta makes the navigation both difficult and dangerous. The water is cast with such impetuosity from one bank to the other, that the soil is constantly falling in upon the river, and huge masses of clay hourly tumble into the stream, often with a tremendous crash. In some places the water, when resisted by a firm bank, forms eddies and gulphs of great depth, in which the current is really terrific; and, in a high wind, the waves dash as in the ocean. It appears, indeed, from the *Report of the State and Navigation of the Indus,* by Lieuts. Carless, Wood, and Pottinger, that banks and bars offer such great obstructions, as effectually to prevent the river from ever becoming extensively available for the purposes of commerce. Vessels drawing 8 ft. water find themselves aground at the very entrance of the Heeta mouth; the employment of ships is out of the question, and the navigation of the dhoonies, or small native boats, is so tedious, that no communication of any importance can be kept up between Hyderabad and the sea, except by steamers. The introduction of steamers has accordingly been attempted and with great success. By Act of Parliament 20 and 21 Vic. cap. 160 (25th August, 1857) a mercantile association, called the Indus steam flotilla company, was authorized to run steamers on the Indus for a length of 570 miles. The steamers are flat-bottomed, and perform the service exceedingly well. The extension of commerce in recent years has also led to plans of railways along the Indus, and in the summer of 1858 a survey was completed of an 'Indus Valley railway,' which is to connect the Sinde and the Punjab lines, by a line running along the left bank of the Indus.

The tides rise in the mouths of the Indus about 9 ft. at full moon, and both flow and ebb with great violence, particularly near the sea, where they flood and abandon the banks with equal and incredible velocity. This phenomenon was an object of great surprise to Alexander's fleet, and Arrian remarks (lib. vi. cap. 19) that 'the ebbing and flowing of the waters was so in the great ocean, inasmuch that the ships were left upon the dry ground, but what still more astonished Alexander and his friends was, that the tide, soon after returning, began to leave the ships, so that some were swept away by the fury of the tide and dashed to pieces, while others were driven on the banks and totally wrecked.'

The tides are not perceptible more than 75 m. from the sea, or about 25 m. below Tatta. The quantity of water discharged by the Indus is stated to amount to 846,043 cubic ft. per second, nearly as much as is discharged by the Mississippi, and *four times* as much as is discharged by the Ganges, the other great river of Hindostan. This discharge must be attributed chiefly to the greater length of its course in high and snowy regions, to its numerous and large tributaries,

and to the barren arid nature of the soil through which it passes; while the Ganges, on the other hand, expends its waters in irrigation, and blesses the inhabitants of its banks with rich and exuberant crops.

The Indus has numerous affluents, none of which, however, deserve any particular mention except the Sutledje, and the other rivers of the Punjab. Of these rivers, the Sutledje (the *Zaradrus* of Ptolemy), which is the most easterly of all, takes its rise near Garoo, on the great plain N. of the Himalaya mountains, enters the chain at Shipkee (where it is 10,484 ft. above the sea), runs in a narrow mountain valley for upwards of 100 m., and enters the S. plain at Ropur, whence its course is south-westward to its junction with the Indus. The other rivers of the Punjab, besides the Beas (the *Hyphasis* of Arrian), which is an affluent of the Sutledje, are, proceeding westward, the Ravee (the *Hydraotes* of Arrian), the Chenáb (*Acesines*), and the Jylum or *Hydaspes*. The last three, all of which rise on the S. slope of the great mountain range of N. India, join their waters with those of the Sutledje in lat. 29° 10' N., and long. 71° 12' E. The rivers of the Punjab are in general navigable up to the place where they issue from the mountains.

INGOLSTADT, a town of Bavaria, circ. Ratisbon, on the Danube, 33½ m. SW. Ratisbon, on the railway from Augsburg to Ratisbon. Pop. 15,712 in 1861. The town has recently been restored to the condition of a fortress, by the construction of very strong works on an improved plan. Its old fortifications had withstood sieges from the troops of the League of Schmalkald, from Gustavus Adolphus, and Duke Bernard of Saxe Weimar, and resisted Moreau for three months; but he, succeeding at length, caused them to be demolished. Ingolstadt had its university, at which the celebrated Dr. Faustus studied in 1490; it is now transferred to Munich. It still possesses, however, a royal residence, nine churches, in one of which the Bavarian general, Tilly, was buried, and several hospitals and charitable institutions. It had formerly a considerable manufacture of woollen cloths; but this and its other branches of industry and trade has fallen into decay.

INNSBRUCK (Fr. *Inspruck*), a city of the Tyrol, of which it is the cap., on the Inn, 80 m. N. by E. Trent, and 240 m. W. by S. Vienna, on the railway from Munich over the Brenner to Verona. Pop. 14,374 in 1858. The situation of the town is highly picturesque. It stands in the middle of a valley, the sides of which are hemmed by mountains from 6,000 to 8,000 ft. high, and the Inn is crossed by a bridge (whence the name of the city) from which a magnificent prospect is obtained. On and round this bridge one of the severest actions took place during the war of the Tyrolese, under Hofer, against the French. Innsbruck is divided into the old and new towns, and has five suburbs. The latter are larger and better built than the city itself, though badly paved. The houses of Innsbruck are mostly four or five stories high, built in the Italian style, with flat roofs, and are frequently ornamented with frescoes. Many have arcades below, occupied with shops. The object most attractive to strangers is the Franciscan, or Court church, an edifice containing numerous fine works of art. Among others, is the tomb dedicated to the emperor Maximilian. It is ornamented with 24 bas-reliefs, representing the principal actions of his life, and is surrounded by 21 colossal bronze statues of persons celebrated in history, including Clovis, Theodoric, Arthur, Charles the Bold, Duke of

Burgundy, Godfrey of Bouillon, Rodolph of Hapsburg, and many of the emperors of Austria, his descendants. Here, also, is the mausoleum of the archduke Ferdinand of the Tyrol and his wife, also adorned with bas-reliefs; and the grave of Hofer and his statue in white marble. There are numerous other churches, several of which are worth notice. The palace, an extensive building, has gardens extending along the Inn, which form a public promenade. In front of the Old Palace, the former residence of the archdukes of the Tyrol, and of some of the German emperors, is the 'Golden Roof,' a kind of oriel window, covered with a roof of gilt copper, and one of the curiosities of the place; this edifice is now used for the chancery-chamber (Kanzleigebäude). Innsbruck has a university of the 2nd order, in which instruction is entirely gratuitous. It occupies an extensive and fine edifice, and has 25 professors, and exhibitions to the amount of 12,000 fl. yearly. It has attached to it a valuable library, botanic gardens, and normal school. The Ferdinandeum, founded in 1823 upon the model of the Johanneum of Grätz, is a museum devoted to the productions of the Tyrol in both art and natural history, and contains some interesting collections, particularly in the dep. of mineralogy. The seminary for noble ladies, founded by Maria Theresa in 1771, the gymnasium, ancient Jesuits' college, and various convents, provincial house of correction, council chamber, town-hall, theatre, and a handsome ball-room, are the other chief public buildings; a statue of Joseph II., and a triumphal arch raised by Maria Theresa, are among the most conspicuous ornaments of the city. Innsbruck is the seat of the state assembly, high judicial court, and other superior departments of the public service for the Tyrol and Vorarlberg. It has manufactures of silk, woollen and cotton fabrics, leather, glass, and steel goods, and sealing-wax; and is the seat of a considerable trade between Italy and the countries N. of the Alps.

INVERARY, a royal and parl. bor. and sea-port of Scotland, co. Argyle, of which it is the cap., on a bay on the W. shore, and near the bottom of the arm of the sea called Loch Fyne, 40 m. NW. Glasgow. Pop. 1,075 in 1861, and 1,233 in 1841. Inverary consists principally of two rows of houses, one of them fronting the bay, the other at right angles with it, running inward, and having a northern exposure. The houses, built on a uniform plan, are large and commodious; and the town is one of the neatest and cleanest, and its situation the most picturesque in Scotland. The public buildings are the par. church, and a handsome edifice by the water side, containing the court-house and other offices. In the immediate vicinity of the town, on the N., is Inverary castle, the chief residence of the ducal family of Argyle. It was built after a design by Adams in 1749; but it is hardly worthy of the situation. It is an embattled structure, of two stories and a sunk floor, flanked with round overtopping towers, and surmounted with a square-winged pavilion. There is in the saloon a curious collection of old Highland arms, including some of those used by the Campbells in the battle of Culloden.

The staple commodity of Inverary is herrings, those of Loch Fyne being celebrated for their superior excellence; but the fishing in the Loch has latterly declined, and with it the population of the town.

Inverary was erected into a bor. of barony in 1648. In a garden beside the church is a small obelisk, commemorative of the execution in this place, in 1685, of several gentlemen of the name of Campbell, on account of their adherence to

Presbyterianism. This bor. unites with Campbelltown, Oban, and Irvine, in sending a mem. to the H. of C.; and in 1864 had 36 reg. voters. Edmund Stone, a self-taught mathematician, editor of 'Euclid's Elements,' and author of a 'Treatise on Fluxions,' and other works, was a native of Inverary.

INVERKEITHING, a royal and parl. bor., par., and sea-port of Scotland, co. Fife, beautifully situated on rising ground on a bay on the N. bank of the Frith of Forth, 10 m. WNW. Edinburgh, on the railway from Edinburgh to St. Andrew's. Pop. 1,417 in 1861, and 1,827 in 1841. The town consists of a main street, and a smaller one branching off it, besides several wynds or lanes. Many of the houses are extremely old, and an air of antiquity generally marks the place. The only public buildings are the par. church, a dissenting chapel, the borough school, and the town-house. About 10 in every 100 of the inhab. are, at an average, at school; a larger proportion than generally obtains elsewhere. There are three libraries in the bor. The par. abounds with coal, most of which is exported from St. David's, on Inverkeithing Bay. A number of English and foreign vessels resort to Inverkeithing for coal, bringing in exchange bark, timber, and bones for manure. There are, in the immediate vicinity of the town, a distillery, tan-work, ship-building yard, a magnesia manufactory, and a brick work.

Inverkeithing was created a royal borough by William the Lion in the 12th century. Its privileges included right of customs over a considerable district of country lying on the Frith of Forth; but these have fallen into desuetude, with the exception of the duties at the markets held at Kinross and Tullibole, and the customs at North Queensferry. Even Edinburgh, at one time, paid an acknowledgment of superiority for some parts of the Calton Hill, but it was bought up, or relinquished. In the ridings of the Scottish parliament, the provost of Inverkeithing was entitled to precedence next to the provost of Edinburgh. Before the convention of royal burghs was appointed to be held at Edinburgh, Inverkeithing was the place of its meeting. This bor. unites with S. Queensferry, Dunfermline, Culross, and Stirling, in sending a mem. to the H. of C., and in 1864 had 57 registered voters.

INVERLEITHEN, a par. and village of Scotland, famous for its mineral well, co. Peebles, 22 m. S. by E. Edinburgh, and 6 m. E. by S. Peebles, on the Edinburgh and Hawick railway. Pop. 1,150 in 1861. The village is situated in a romantic pastoral country, within 1 m. of the N. bank of the Tweed, and on both sides the Leithen, a tributary of that river. It has long been known as a 'watering place,' and its celebrity was greatly enhanced by the publication (in 1824) of Scott's novel, entitled 'St. Ronan's Well,' of which it was supposed to be the prototype. A yearly festival has been since instituted at Inverleithen, for the celebration of 'the St. Ronan's Border Games;' and the name of almost every street, or separate edifice, in the village, such as 'Abbotsford Place,' 'Waverley Row,' 'Marmion Hotel,' &c., refers to the illustrious novelist. Traquair-house, the seat of the noble family of that name, is in the immediate vicinity of Inverleithen. The first earl of Traquair, lord treasurer of Scotland in the time of Charles I., was one of the most eminent statesmen of his day. Dr. Russell, author of the 'History of Modern and Ancient Europe,' was born near the village, and was educated in it. The woollen manufacture has been introduced into Inverleithen.

INVERNESS, a marit. co. of Scotland, and the most extensive in that part of the U. Kingdom;

it stretches quite across the island from the E. to the W. sea, having N. the Moray Frith and Ross-shire, W. the Atlantic Ocean, S. Argyle and Perth, and E. Aberdeen, Banff, Moray, and Nairne. But it includes, exclusive of the mainland, the large island of Skye, with the smaller islands of Harris, N. and S. Uist, Benbecula, &c. Area, 4,256 sq. m., or 2,723,501 acres, of which 1,913,920 belong to the mainland, and 773,700 to the islands; the former having 81,400, and the latter 87,700 acres of water. Inverness-shire is, speaking generally, wild, mountainous, and rugged. It is supposed that there is not more than 2½ per cent. of its surface not naturally covered with heath. Ben Nevis, which, next to Ben Macdhu, is the highest mountain in Great Britain, being 4,370 ft. above the level of the sea, is situated near Fort William, in this co. Mealfourvony, on the N. side of Loch Ness, is 2,730 ft. above the sea. The arable land, which is of very limited extent, is principally comprised in the low districts contiguous to the town of Inverness, in Strathspey (the low country on both sides the Spey), and in narrow glens along the other rivers and lakes. Climate various; but generally it may be said to be wet and stormy on the W. coast, severe in the interior, and comparatively mild and dry on the shore of the Moray Frith. Principal rivers, Spey, Ness, and Beauly; all which, but especially the first, have valuable salmon fisheries. The arable land of this co. was formerly divided into small patches, having usually a greater or less extent of hill pasture attached to them, and occupied by tenants at will. The latter lived in miserable huts; and were at once excessively poor, idle, and disorderly. But the abolition of hereditary jurisdictions and clanship in 1748, and the carrying of good military and other roads into districts that were formerly quite impervious, by enabling the law to be everywhere brought into full operation, have completely repressed the feuds and disorders that formerly disgraced this and other Highland cos. The small holdings have also been very generally consolidated into sheep-farms, some of which are very extensive, and which are mostly stocked with cheviots. Arable farming has, also, been very much improved; and, in consequence, there has been a very great increase in the quantity of disposable produce, and in the rent and value of the land. Good wheat is raised round the Moray Frith; but oats is the principal crop. The stock of black cattle is very large; and cattle, sheep, and wool constitute the principal articles of export. In some districts there are extensive forests. There are no manufactures of any importance, nor any considerable town, except Inverness. Illicit distillation, that was once very prevalent, is now all but suppressed. Limestone, slate, and marble abound in most places; but the want of coal renders the limestone of little value. Gaelic is the common language; and in the W. parts of the co., and some of the islands, it is the only one that is generally understood. Owing to the thinness of the pop., the co. is but ill supplied with schools, though in this respect, as in others, it is very much improved.

This co. is divided into two nearly equal portions, by a remarkable glen or valley, stretching NE. and SW. from the town of Inverness to Loch Linnhe, opposite the island of Mull, on the W. coast. This glen, which is very narrow, consists principally of a chain of lakes comprising Loch Ness, Loch Oich, and Loch Lochy. Its surface being nowhere more than 94 ft. above the level of the sea, advantage was taken of this circumstance, and of the continuous chain of lakes, to open a navigable communication between the E. and W. seas, avoiding, consequently, the lengthened and dangerous navigation by the Pentland Frith. The entire length of this navigation, or of the Caledonian Canal, inclusive of the lakes, is rather more than 60 m.; but the excavated part is little more than 23 m. It cost about 1,000,000l., and is on a larger scale than any work of a similar class in any other part of the empire.

Inverness co. has 35 parishes; it sends 1 mem. to the H. of C. for the co.; and the bor. of Inverness joins with Fortrose, Nairne, and Forres in sending a mem. Registered electors for the co., 879 in 1865. Inverness-shire had, in 1861, a pop. of 88,888, living in 16,615 houses. The old valued rent was 6,099l.; the new valuation, for 1864-5, was 237,349l.

INVERNESS, the cap. of the above co., and of the Northern Highlands, a royal and parl. bor. and sea-port of Scotland, on both sides the Ness, within a mile of its influx into the Moray Frith, and at the NE. extremity of the Great Glen of Scotland, forming the line of the Caledonian Canal, 119 m. NW. by N. Edinburgh, and 81 m. N. by W. Aberdeen, on the Great North of Scotland railway. Pop. 12,509 in 1861, and 11,580 in 1841. The situation of Inverness is peculiarly striking and picturesque, in the middle of a beautiful plain, of unequal extent in different directions, with the Moray Frith on one side, and the back ground, consisting of variously shaped hills, some of which are richly wooded, while others are bleak and rugged. The Ness, on whose banks the borough stands, is the shortest river in Scotland, flowing between Loch Ness and the Moray Frith, a distance of only 8 m. The most important portion of the town is on the right bank of the Ness. A handsome stone bridge of seven arches was erected across the river in 1685; there is also a wooden bridge, built in 1808. The principal streets lie E. or N., and consist generally of substantial buildings. The streets, which are lighted with gas, are causewayed and flagged. The public buildings are the exchange and town-house, near the centre of the town; the gaol, surmounted by a tower 150 ft. high; the assembly-rooms of the Northern Meeting; infirmary, academy, the united charity institutions; which last edifice occupies an elevated situation in the vicinity of the town. With regard to ecclesiastical buildings, the High Church is the most conspicuous; the square tower attached to it was built by Oliver Cromwell, and the bell brought from the cathedral of Fortrose, on the N. banks of the Moray Frith. There are three churches, in one of which Gaelic alone is used, and two chapels of ease. There are, also, five churches, dissenting chapels, belonging respectively to Episcopalians, the United Associate Synod, the Independents, Baptists, Methodists, and Rom. Catholics.

Gaelic was formerly the only language spoken in Inverness and its neighbourhood; and it is still to some extent the speech of the lower orders, all of whom, however, understand and can speak English. It is admitted that the English language is spoken in greater purity by the middle and upper ranks in Inverness than in any other place in Scotland; a distinction which is said, whether correctly or not, to have originated in the circumstance of Cromwell having stationed and long maintained an English garrison in the town. 'The soldiers seem to have incorporated afterwards with the inhab., and to have peopled the place with an English race; for the language of the town has been long considered as peculiarly elegant.' (Johnson's Tour to the Hebrides.) The Highland character, however, still predominates in the borough. In addition to the Gaelic language, the speech of the common people, their dress is more or less of Celtic fashion, and of home manufacture, such as the short coat, blue bonnet, plaid rig and fur stockings, all of the coarsest materials. The married women usually walk the streets and go to church without a bonnet; the maidens without either cap or bonnet; while the other parts of their dress are of the most simple and homely description.

Inverness has some manufactures of linen, plaidings, and woollen stuffs, and a small hemp manufactory, on the site of Cromwell fort. Shipbuilding is carried on to some extent. There are breweries, distilleries, and tan-works. The shipping is considerable. There belonged to the port, on the 1st of January, 1864, 148 sailing vessels under 50, and 83 above 50 tons, besides one steamer of 70 tons. The gross amount of customs' revenue was 8,000l. in 1859; 8,572 in 1861; and 6,744l. in 1863. The town has regular traders, both steamers and sailing smacks, to Aberdeen, Leith, and London, on the E. coast; she has a similar communication, by means of the Caledonian Canal, with Glasgow, and Liverpool, on the W. coast; and also with Ireland. The canal passes within less than a mile of the bor.; and Clachnaharry, where it joins the Moray Frith, is not more than a mile distant. There are three harbours, one of them for small craft, near the town, the others at the mouth of the river; while the canal wharfs at Clachnaharry are also used for the loading and unloading of goods. Grain used to be imported to Inverness; but oats are now largely exported. Coal, almost the only kind of fuel used, is imported both from England and the Frith of Forth. Inverness has several fairs; but the wool fair, in the month of July, attended by all the principal Highland sheep farmers, as well as by wool staplers and agents from England and the S. of Scotland, is the most eminent. Fully 100,000 stones of wool are annually sold at this market; while above the same number of sheep are also disposed of. The prices paid at this fair generally regulate those of all the other markets in the country.

Inverness is very ancient. In the 6th century it was the capital of the Pictish kingdom, when St. Columba of Iona went thither, according to Neue, with the view of converting the Pictish king to Christianity. An ancient castle stood on a rising ground E. of the town; but it was destroyed in the 11th century by Malcolm III., who built another on a commanding eminence near the

river, which continued to be a royal fortress, till blown up, in 1746, by the troops of the Pretender. Inverness was erected into a royal bor. by David I.; and various royal charters, confirming or extending its privileges, were subsequently conferred on it. The town was often an object of plunder to the lords of the Isles and other Highland chiefs. A monastery, belonging to the Black Friars, existed in this place; but all traces of it have long since disappeared. The citadel referred to above, as constructed by Cromwell, was built in 1652–57, N. of the town, near the mouth of the river. Part of its ruins are still standing. Culloden Moor, the scene of the battle that decided the fate of the Pretender, Charles Stuart, is within 5 m. of the town. Since 1745, great improvements have been effected here. Previously to 1745, the post from Edinburgh to Inverness was conveyed by a man on foot. In 1740, the magistrates advertised for a saddler to settle in the bor.; and in 1778 a cart, purchased by subscription, was first seen in the bor. No plan of regularly cleaning the streets was adopted till about the beginning of the present century. Inverness is now, however, superior perhaps to any town of its size in Scotland as to all the necessaries, comforts, and luxuries of life. Corp. revenue, 2,269l. in 1863–4. This bor. unites with Forres, Fortrose, and Nairn in sending a mem. to the H. of C. Registered voters, 567 in 1864.

INVERURY, a royal and parl. bor. and par. of Scotland, co. Aberdeen, in the angle formed by the confluence of the Don and Ury, 16 m. NW. Aberdeen on the Great North of Scotland railway. Pop. 3,520 in 1861, and 1,679 in 1841. The inhabitants are chiefly agriculturists. The Aberdeenshire canal, begun in 1796, and completed in 1807, commences in the tide-way of the harbour of Aberdeen, and terminates at Port Elphinstone near Inverury. The entire length is 18½ m.; the surface width is 23 ft.; the depth 5½ ft.; it has 17 locks; and its highest level is 168 ft. above low water-mark. Keith Hall, the seat of the Earl of Kintore, who also holds the title of Lord Inverury, is in the immediate vicinity of the bor. Arthur Johnston, editor of the 'Deliciæ Poetarum Scotorum,' and who holds the next place to Buchanan among the Latin poets of Scotland, was born in the neighbourhood of Inverury in 1587. This bor. unites with Elgin, Banff, Cullen, Kintore, and Peterhead, in sending a mem. to the H. of C. Registered voters 130 in 1865.

IONA. (See HEBRIDES.)

00568355